MW01631180

PRENTICE HALL

Chemistry

ANTONY C. WILBRAHAM • DENNIS D. STALEY • MICHAEL S. MATTA • EDWARD L. WATERMAN

Boston, Massachusetts
Upper Saddle River, New Jersey

About the Authors

Antony C. Wilbraham

Antony Wilbraham is an Emeritus Professor of Chemistry. He taught college-level chemistry for more than 25 years at Southern Illinois University at Edwardsville, Illinois, where he was also Director of Hazardous Waste Management. In 1978, he received the University Teaching Excellence Award. Dr. Wilbraham has been writing high school and college-level chemistry textbooks and related ancillaries for over 25 years and has published extensively in science journals for over 30 years. He is a member of the American Chemical Society, a member of the National Science Teachers Association, and a Fellow of the Royal Society of Chemistry.

Dennis D. Staley

Dennis Staley was an Instructor in the Department of Chemistry and the Office of Science and Math Education at Southern Illinois University at Edwardsville, Illinois. He has been teaching high school and college-level chemistry for more than 25 years. In 1981 he received the University Teaching Excellence Award. Mr. Staley has been writing high school and college-level chemistry textbooks and related ancillaries for over 25 years. He is a member of the American Chemical Society and a member of the National Science Teachers Association.

Michael S. Matta

Michael Matta is an Emeritus Professor of Chemistry. He was a professor of chemistry at Southern Illinois University at Edwardsville, Illinois, from 1969 to 1996, and served as Department Chair from 1980 to 1983. In 1973, Dr. Matta received the University Teaching Excellence Award. He has been developing and writing high school and college-level chemistry textbooks and related ancillaries for over 25 years and has published extensively in scientific journals. He is a member of the American Chemical Society, the National Science Teachers Association, and the American Association for the Advancement of Science.

Edward L. Waterman

Edward Waterman has taught chemistry, advanced placement chemistry, and organic and biochemistry since 1976 at Rocky Mountain High School in Fort Collins, Colorado. Mr. Waterman conducts workshops for teachers on such topics as small-scale chemistry labs, advanced placement chemistry, block scheduling, Internet for the chemistry classroom, and virtual chemistry on CD-ROM. Mr. Waterman holds a Bachelor of Science degree in chemistry from Montana State University and a Master of Science degree in chemistry from Colorado State University.

ISBN 0-13-251211-4

4 5 6 7 8 9 10 11 10 09

BRIEF CONTENTS

Teacher's Edition

Student Edition

A Science Program Backed by Research

In developing ***Prentice Hall Chemistry***, research studies were used as a central, guiding element. Research indicated key elements of a textbook program that ensure students' success: support for reading and mathematics in science, consistent opportunities for inquiry, and an ongoing assessment strand. This research was conducted in phases and continues today.

Exploratory: Needs Assessment

Along with periodic surveys concerning state and national standards as well as curriculum issues and challenges, Pearson Prentice Hall conducted specific product development research, which included discussions with teachers and advisory panels, focus groups, and quantitative surveys. Prentice Hall explored the specific needs of teachers, students, and other educators regarding each component of the ***Prentice Hall Chemistry*** program.

Formative: Prototype Development and Field-Testing

During this phase of research, Prentice Hall worked to develop prototype materials, then field-tested the materials with students and teachers and performed qualitative and quantitative surveys. Early prototype testing resulted in feedback about the lesson structure. Results were channeled back into the development of the program.

Summative: Validation Research

Finally, Pearson Prentice Hall continues to conduct long-term research based on scientific, experimental designs under actual classroom conditions. This research identifies what works and what can be improved upon in the next revision of ***Prentice Hall Chemistry***. Prentice Hall also continues to monitor the program in the market through ongoing dialogue with users. This allows us to perpetually refine our products and provide the best support to today's and tomorrow's teachers and students.

In preparation for the NCLB mandates for science anticipated by 2007, Pearson Prentice Hall is committed to providing scientific research to support the efficacy of our science programs in the classroom.

Our study designs will closely follow the criteria of No Child Left Behind. Since NCLB specifies a minimum level of improvement that students must achieve each year, actual Adequate Yearly Progress (AYP) will be reflected in the research used to validate the program.

Foundational Research: Inquiry in the Science Classroom

"Inquiry is the heart and soul of science education, but what is it?"

Defining Science Inquiry

Simply put, inquiry is the intellectual side of science. It's thinking like a scientist—being inquisitive, asking why, and searching for answers. The National Science Education Content Standards define inquiry as the process in which students begin with a question, design an investigation, gather evidence, formulate an answer to the original question, and communicate the investigative process and results. Since it's often difficult to accomplish all this in one class period, the standards also acknowledge that at times students need to practice only one or two inquiry components.

Understanding Inquiry

The National Research Council in Inquiry and the National Science Education Standards (2000) identified several "essential features" of classroom inquiry. We have modified these essential features into questions to guide you in your quest for enhanced and more thoughtful student inquiry.

- **Who asks the question?** In most curricula, these focusing questions are an element given in the materials. As a teacher you can look for labs that, at least on a periodic basis, allow students to pursue their own questions.
- **Who designs the procedures?** To gain experience with the logic underlying experimentation, students need continuous practice with designing procedures. Some labs, in which the primary target is content acquisition, designate procedures. Others should ask students to do so.
- **Who decides what data to collect?** Students need practice in determining the data to collect.
- **Who formulates explanations based upon the data?** Students should be challenged to think—to analyze and draw conclusions based on their data, not just copy answers from the text materials.
- **Who communicates and justifies the results?** Activities should push students to not only communicate but justify their answers. Activities also should be thoughtfully designed and interesting so that students want to share their results and argue about conclusions.

Inquiry in Prentice Hall Chemistry

Prentice Hall Chemistry offers the most opportunities to get students to think like scientists and enables students to enhance their understanding by participating in the discovery.

Prentice Hall Chemistry encourages students to develop inquiry skills across the spectrum from teacher-guided to open-ended. Flexible lab options are included in every chapter, structured from directed to open-ended—providing the flexibility to address all types of learners and accommodate your class time and equipment requirements. Some focus on content acquisition, and thus the question and most of the procedures are specified.

READING

Reading Comprehension in the Science Classroom

Q: Why are science texts often difficult for students to read and comprehend?

A: In general, science texts make complex literacy and knowledge demands on learners. They have a more technical vocabulary, a more demanding syntax, and place a greater emphasis on inferential reasoning.

Q: What does research say about facilitating comprehension?

A: Studies comparing novices and experts show that the conceptual organization of experts' knowledge is very different from that of novices. For example, experts emphasize core concepts when organizing knowledge, while novices focus on superficial details. To facilitate comprehension, effective teaching strategies should support and scaffold students as they build an understanding of the key concepts and concept relationships within a text unit.

Q: What strategies can teachers use to facilitate comprehension?

A: Three complementary strategies are very important in facilitating student comprehension of science texts. First, guide student interaction with the text using the built-in strategies. Second, organize the curriculum in terms of core concepts (e.g., the Key Concepts in each section). Third, develop visual representations of the relationships among the key concepts and vocabulary that can be referred to during instruction.

Nancy Romance, Ph.D.
Professor of Science Education
Florida Atlantic University
Fort Lauderdale, Florida

"Effective teaching strategies should support and scaffold students as they build an understanding of the key concepts and concept relationships within a text unit."

Evaluator's Checklist

Does your science program promote reading comprehension with—

- ✓ Text structured in an outline format?
- ✓ Key Concepts highlighted in boldface type?
- ✓ Real-world applications to activate prior knowledge?
- ✓ Key Concepts, critical vocabulary, and a reading skill for every section?
- ✓ Relevant photos and carefully constructed graphics with questions?
- ✓ Checkpoint questions that appear in each section?
- ✓ Section Assessments that support Key Concepts?

Reading Support in *Prentice Hall Chemistry*

Before students read—

The Guide for Reading introduces students to the key concepts and key terms they'll find in each section. The Reading Strategy is identified.

During the section—

- Boldface sentences identify each key concept and encourage students to focus on the big ideas, or Key Concepts, of science.
- Checkpoint questions reinforce students' understanding by slowing them down to review after every concept is discussed.
- Caption questions draw students into the art and photos, helping them connect the content to the images.

After students read—

Section Assessment questions allow students to review Key Concepts.

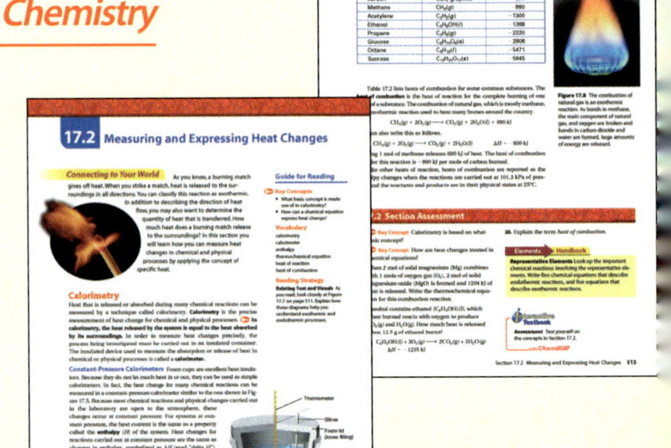

17.2 Measuring and Expressing Heat Changes

William Tate, Ph.D.
Professor of Education and American Culture Studies
Washington University
St. Louis, Missouri

"Mathematics is the language used by scientists to model change in the world."

Math in the Science Classroom

Why should students concern themselves with mathematics in your chemistry class? Good science requires good data from which to draw conclusions. Technology enhances the ability to measure in a variety of ways. Often scientists must measure large amounts of data, and thus an aim of analysis is to reduce the data to a summary that makes sense and is consistent with established norms of communication—i.e., mathematics. Calculating measures of central tendency (e.g., mean, median, or mode), variability (e.g., range), and shape (graphic representations) can effectively reduce 500 data points to 3 without losing the essential characteristics of the data. Scientists understand that a trade-off exists between precision and richness as data are folded into categories, so margins of error can be quantified in mathematical terms and factored into all scientific findings.

Evaluator's Checklist

Does your science program promote math skills with—

- ✓ Giving students opportunities to collect data?
- ✓ Providing students opportunities to analyze data?
- ✓ Enabling students to practice math skills?
- ✓ Using sample problems to apply science concepts?

Math Support in *Prentice Hall Chemistry*

Recognizing the need for math support, ***Prentice Hall Chemistry*** integrates mathematics instruction throughout the program and gives students ample opportunities to practice their math skills. The success of math instruction and practice in ***Prentice Hall Chemistry*** can be traced to nearly 20 years of interacting with chemistry teachers and students. Over the life of this chemistry program, Prentice Hall has developed strategies that allow students to see math as a tool, not an impediment, in learning and understanding chemistry concepts.

In the Student Edition

Sample Problems, CHEMath, Interpreting Graphs, and a *Math Handbook* all provide the practice and instruction which encourages students to read and understand, plan and solve, and then evaluate their answers.

In the Teacher's Edition

Math support in the Teacher's Edition is geared toward the math objectives on high stakes tests. Teaching notes provide the chemistry teacher with strategies for alternative instruction and additional practice. Sample Problems in the Student Edition are supported by additional Practice Problems in the Teacher's Edition.

In the Guided Reading and Study Workbook

These unique worksheets include Guided Practice Problems to help students master reading and enhance their study and math skills. Students can create a record of their work for study and review.

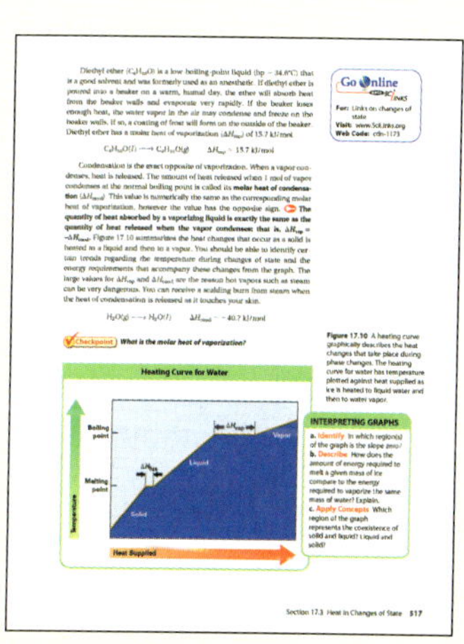

Problem Solving in the Chemistry Curriculum

In chemistry problem solving is an important set of skills that students must learn and then develop through practice. In ***Prentice Hall Chemistry***, students are taught an approach to problem solving that applies to both numeric (quantitative) and conceptual (qualitative) problems. These same problem-solving skills can be used to solve relatively simple problems or complex problems whose solutions require multiple steps. As students practice and gain confidence in their abilities, they will be able to apply their problem-solving skills in new situations and trust that their answers make sense.

When tackling a numeric problem, students must first analyze and come to understand exactly what it is that they are trying to calculate. This planning stage includes identifying what is known and unknown and making decisions regarding the best approach to solve the problem—typically using an equation or dimensional analysis.

Having analyzed the problem and planned a solution, students then solve for the unknown. The plan may require solving an equation for the unknown, substituting the known quantities, and calculating the answer. If the plan is to use dimensional analysis, the answer is calculated by multiplying the known by one or more conversion factors.

An important step that is often overlooked is checking that the planned solution is valid—it allows the calculation of the unknown from the known—and that the calculations have been correctly carried out according to the plan. Many times, going back and checking the work identifies simple mistakes, such as orders of magnitude, that are easily corrected.

Dennis D. Staley

"By having the skills of understanding and utilizing problem solving at their disposal, students can approach brand new situations and trust that their results make sense."

Evaluator's Checklist

Does your chemistry program include problem solving that helps students—

- ✓ Identify what is known and unknown?
- ✓ Calculate an answer?
- ✓ Check validity of their solution?

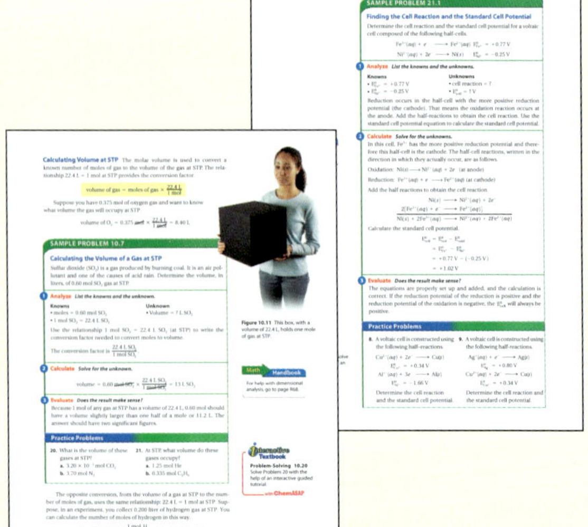

Problem Solving in the *Prentice Hall Chemistry* Student Edition

Prentice Hall Chemistry has been utilizing a consistent problem solving strategy since the program's inception in 1987. And for the first time, in this new edition, ***Prentice Hall Chemistry*** has identified some problems as Conceptual Problems. In these types of qualitative problems, solutions are easier to derive if students can rely on a consistent, systematic process. Because of the basic differences between solving math-based and Conceptual Problems, we've decided to modify the way in which they're solved.

First of all, where possible, we've tied Conceptual Problems to real-world examples to which students can relate. Students are much more interested in topics if they can truly understand how they relate to their lives! Students read about something they already know, which helps them identify the problem. They then pull from the same sorts of tools available to them for solving math-based problems, but they look for a qualitative answer, not a quantitative one. Because their solution is qualitative, it is substantiated with drawings or descriptions, which provides them with an opportunity to look back and evaluate their answer.

By utilizing these proven strategies for solving all types of chemistry problems, students gain the experience and confidence for understanding and solving problems in chemistry, science, and beyond. They can apply these skills to new situations and have confidence in the process and the results.

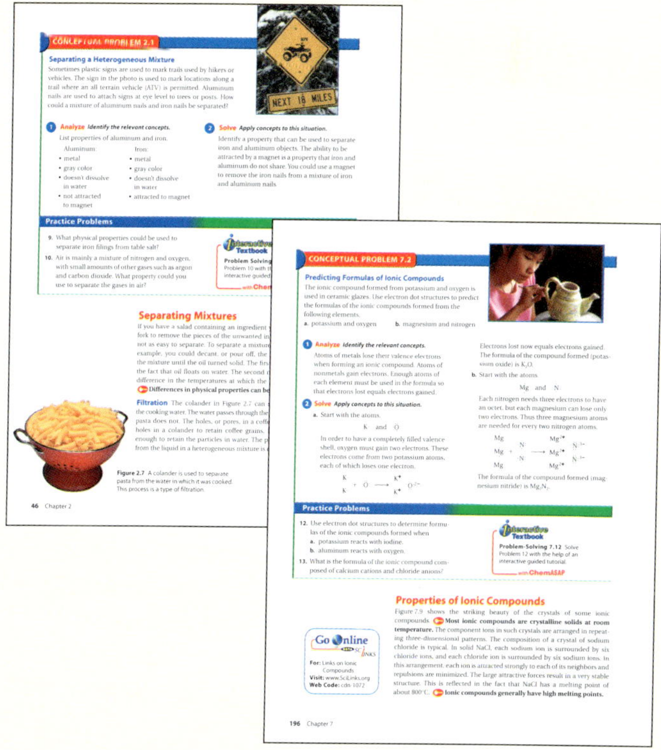

CONCEPTUAL PROBLEM 2.1

Separating a Heterogeneous Mixture

Sometimes plastic signs are used to mark trails used by hikers or vehicles. The sign in the photo is used to mark locations along a trail where an all terrain vehicle (ATV) is permitted. Aluminum nails are used to attach signs at eye level to trees or posts. How could a mixture of aluminum nails and iron nails be separated?

1 Analyze *Identify the relevant concepts.*

List properties of aluminum and iron.

Aluminum:	Iron:
• metal	• metal
• gray color	• gray color
• doesn't dissolve in water	• doesn't dissolve in water
• not attracted to magnet	• attracted to magnet

2 Solve *Apply concepts to this situation.*

Identify a property that can be used to separate iron and aluminum objects. The ability to be attracted by a magnet is a property that iron and aluminum do not share. You could use a magnet to remove the iron nails from a mixture of iron and aluminum nails.

Practice Problems

9. What physical properties could be used to separate iron filings from table salt?
10. Air is mainly a mixture of nitrogen and oxygen, with small amounts of other gases such as argon and carbon dioxide. What property could you use to separate the gases in air?

Separating Mixtures

Figure 2.7 A colander is used to separate pasta from the water in which it was cooked. This process is a type of filtration.

46 Chapter 2

CONCEPTUAL PROBLEM 7.2

Predicting Formulas of Ionic Compounds

The ionic compound formed from potassium and oxygen is used in ceramic glazes. Use electron dot structures to predict the formulas of the ionic compounds formed from the following elements.

a. potassium and oxygen b. magnesium and nitrogen

1 Analyze *Identify the relevant concepts.*

Atoms of metals lose their valence electrons when forming an ionic compound. Atoms of nonmetals gain electrons. Enough atoms of each element must be used in the formula so that electrons lost equals electrons gained.

2 Solve *Apply concepts to this situation.*

a. Start with the atoms.

K and O

In order to have a completely filled valence shell, oxygen must gain two electrons. These electrons come from two potassium atoms, each of which loses one electron.

$$K + K + O \longrightarrow K^+ + K^+ + O^{2-}$$

Electrons lost now equals electrons gained. The formula of the compound formed (potassium oxide) is K_2O.

b. Start with the atoms.

Mg and N

Each nitrogen needs three electrons to have an octet, but each magnesium can lose only two electrons. Thus three magnesium atoms are needed for every two nitrogen atoms.

$$Mg + Mg + Mg + N + N \longrightarrow Mg^{2+} + Mg^{2+} + Mg^{2+} + N^{3-} + N^{3-}$$

The formula of the compound formed (magnesium nitride) is Mg_3N_2.

Practice Problems

12. Use electron dot structures to determine formulas of the ionic compounds formed when
    a. potassium reacts with iodine.
    b. aluminum reacts with oxygen.
13. What is the formula of the ionic compound composed of calcium cations and chloride anions?

Interactive Textbook — **Problem-Solving 7.12** Solve Problem 12 with the help of an interactive guided tutorial. with ChemASAP

Properties of Ionic Compounds

Figure 7.9 shows the striking beauty of the crystals of some ionic compounds. **Most ionic compounds are crystalline solids at room temperature.** The component ions in such crystals are arranged in repeating three-dimensional patterns. The composition of a crystal of sodium chloride is typical. In solid NaCl, each sodium ion is surrounded by six chloride ions, and each chloride ion is surrounded by six sodium ions. In this arrangement, each ion is attracted strongly to each of its neighbors and repulsions are minimized. The large attractive forces result in a very stable structure. This is reflected in the fact that NaCl has a melting point of about 800°C. **Ionic compounds generally have high melting points.**

Go Online SciLinks — For: Links on Ionic Compounds. Visit: www.SciLinks.org Web Code: cdn-1072

196 Chapter 7

HANDS-ON ACTIVITIES

Active Learning in the Lab

Active learning puts students in situations that encourage them to read, speak, listen, think critically, and write. While placing the responsibility of organizing what is to be learned in the hands of the learners themselves, active learning has the ability to reach students of all ability levels. Engaging, relevant labs serve to support active learning through interaction with content.

The ***Prentice Hall Chemistry*** lab program provides students with numerous opportunities to engage in the content and encourage them to develop skills across the spectrum from teacher-guided to open-ended experiences. The Inquiry Activities, Quick Labs and Small-Scale Labs provide visual and tactile cues, offer opportunities for peer communication, and encourage skills that are necessary for critical thinking. Interaction with the content before, during, and at the end of each chapter involves students in the lesson while ensuring understanding along the way.

The goals of a laboratory program should include

- Making chemistry more interesting, so motivation to learn increases.
- Facilitating understanding of chemistry concepts .
- Underscoring the lack of absolute certainty in scientific results.
- Promoting critical-thinking skills.
- Promoting psychomotor and organizational skills.

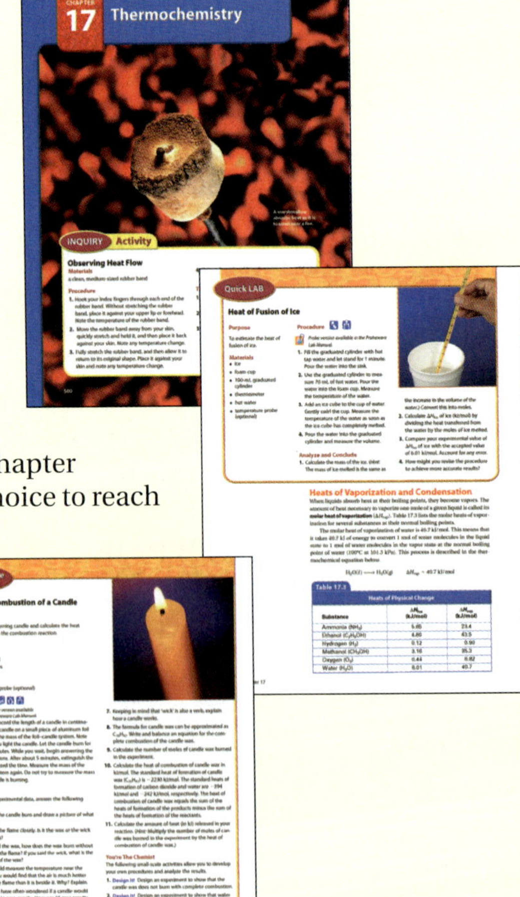

In the Student Edition

All labs from the Student Edition are labeled by level of difficulty in the Chapter Planning Guide in the Teacher's Edition, offering teachers the variety of choice to reach all learners.

An Inquiry Activity at the start of every chapter enables students to explore and discover concepts before reading about them in the chapter.

One Quick Lab per chapter helps students apply chemistry concepts and skills. Each lab can be performed in less than one class period.

Small-Scale Labs, developed with safety and ease of use in mind, encourage students to be inventive with chemistry. Immediate results engage students in the learning process.

Active Learning In the *Prentice Hall Chemistry* Program Resources

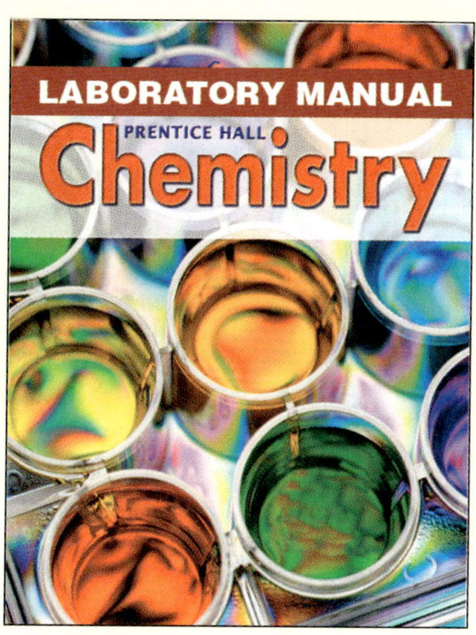

The ***Laboratory Manual*** provides full-scale labs that enhance lab experiences through a variety of learning styles.

Small-scale chemistry labs in the ***Small-Scale Chemistry Lab Manual*** enable students to perform labs quickly, easily, and inexpensively.

The ***Probeware Laboratory Manual*** provides easy-to-use labs with step-by-step instructions for Vernier, Texas Instruments, and PASCO probes and sensors.

The ***Virtual Chem Lab CD-ROM*** helps chemistry concepts come to life in a hands-on lab environment simulated on the computer screen, rendering safety and equipment needs obsolete. ***The Virtual Chem Lab Workbook*** provides an additional 30 labs that are correlated to existing labs in the Student Edition and ***Laboratory Manual***.

Assessment in the Science Curriculum

No Child Left Behind clearly challenges school districts across the nation to raise expectations for all students with testing of student achievement in science to begin in 2007–2008.

A primary goal of NCLB is to provide classroom teachers with better data from scientifically valid assessments in order to inform instructional planning and identify at-risk students who require intervention. It has been a common practice to teach a science lesson, administer a test, grade it, and move on. This practice is a thing of the past. With the spotlight now on improving student performance, it is essential to use assessment results as a way to identify student strengths and challenges. Providing student feedback and obtaining student input is a valuable, essential part of the assessment process.

Assessment is a never-ending cycle, as shown in the diagram below. Although you may begin at any point in the assessment cycle, the basic process is the same.

An important assessment strategy is to ensure that students have ample opportunities to check their understanding of skills and concepts before moving on to the next topic. Checking for understanding also includes asking appropriate, probing questions with each example presented. This enables students and teachers to know whether the skills or concepts being introduced are actually understood.

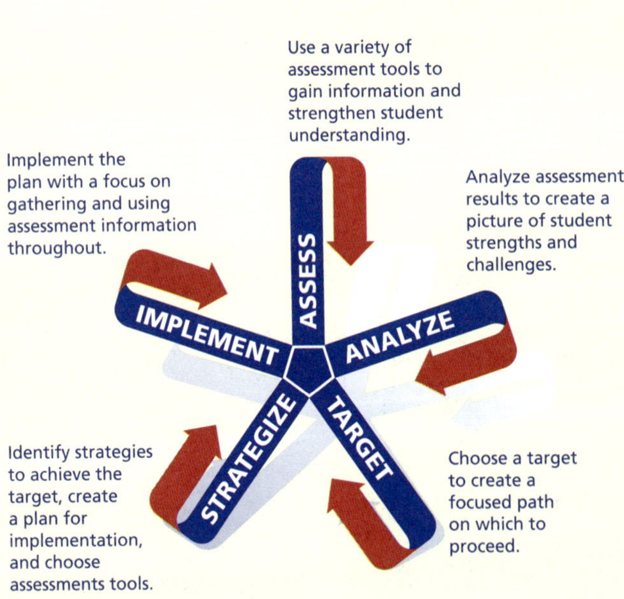

Eileen Depka
Supervisor of Standards and Assessment
Waukesha, Wisconsin

"Meeting the NCLB challenge will necessitate an integrated approach to assessment with a variety of assessment tools."

Evaluator's Checklist

Does your science program include assessments that—

- ✓ Are embedded before, during, and after lesson instruction?
- ✓ Align to standards and to the instructional program?
- ✓ Assess both skill acquisition and understanding?
- ✓ Include meaningful rubrics to guide students?
- ✓ Mirror the various formats of standardized tests?

Assessment in *Prentice Hall Chemistry*

Pearson Prentice Hall has designed a program that gives all students many opportunities to monitor their own progress every step of the way—as they read and after they read. The assessment strategies in ***Prentice Hall Chemistry*** ensure student success in content mastery as well as high-stakes test performance.

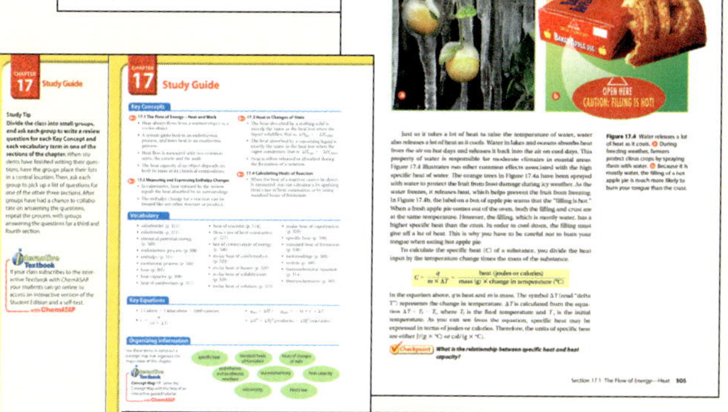

In the Student Edition

Caption questions enhance critical-thinking skills and maximize the effectiveness of art, graphics, and narrative.

Checkpoints reinforce students' understanding of the material just covered.

Section Assessment questions model the way students think and review and assess the understanding of the key concepts.

Comprehensive Study Guides and Chapter Assessments opportunities let students check their own understanding and practice valuable high-stakes test-taking skills.

In the Program Resources

ExamView® Computer Test Bank CD-ROM provides teachers access to thousands of modifiable test questions, hundreds of which contain algorithmically-generated variable data.

ExamView®

Interactive Textbook with ChemASAP

In addition to electronic animations and interactivities, the *Interactive Textbook* provides a wealth of assessment tools. Students can monitor their progress at point of use with ongoing assessment, help tutorials, and instant feedback.

Standardized Test Preparation Workbook includes topic-specific test prep items and practice tests that help teachers focus on improving test scores.

Progress Monitoring Assessments allow you to perform diagnostic tests to identify the strengths and weaknesses of students' content and investigation skills. Benchmark tests monitor the mastery of standards and guide remediation.

Technology in the Chemistry Classroom

The interactivity available through technology environments makes it easy for students to revisit parts of the environments, explore them more fully, test ideas, and receive feedback, thus allowing students to reexamine problems individually and collaboratively. In addition, students are offered more opportunities for practice, feedback, and self-assessment, all linked to their specific needs.

Cathy Dickerson
Humbolt County Office of Education
Eureka, California

"Teachers who make wise use of Web resources can extend the breadth and depth of classroom learning and prepare students for academic and technical futures."

Using Web Resources

While there is a wealth of information on the Internet, scaffolding Web experiences for students can reduce the chances for misdirection and wasted time and increase the educational value of the activity. Successful scaffolding strategies used by teachers include:

- Pointing students to educational Web sites or portals rather than general search engines as starting points.
- Providing links to pre-selected Web resources for assignments or embedding pre-selected links in student assignments.
- Selecting or designing projects and research questions that require synthesis and analysis of gathered information.

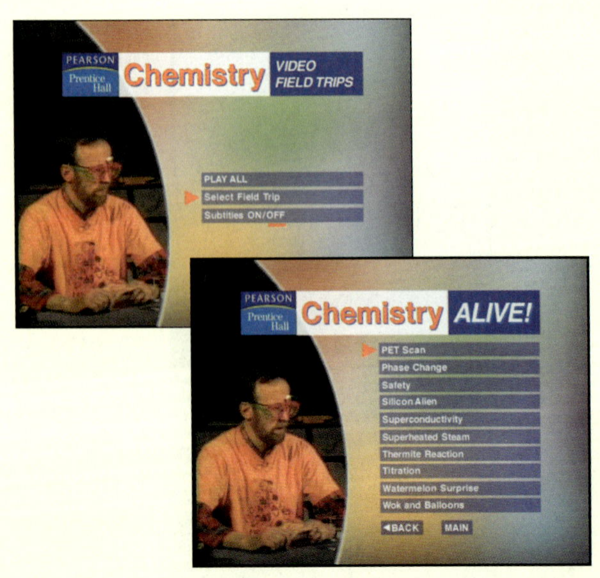

Prentice Hall Chemistry integrates technology into the program

Pearson Prentice Hall develops technology products that tie directly to our Student Editions. Whether it's using online resources, videos, or CD-ROMs, we give students more ways to grasp chemistry concepts.

sciLinks at point of use in the Student Edition

Through sciLINKS, the National Science Teachers Association's topic-based Web resource service, students can access Web resources that have been reviewed by science teachers for their grade-appropriateness and relationship to the topics of study. Students can download a worksheet to guide them through a process of reviewing the Web resources.

Interactive Textbook with Chem ASAP— The Student Edition online and on CD-ROM

- Features built-in videos, activities, and animations to help students visualize tough-to-understand concepts.
- Contains narrated animations and simulations allow students to manipulate variables, observe results, and draw conclusions.
- Includes interactive assessment questions and interactive tutorials with Guided Practice Problems engage learners to navigate chemistry concepts in a variety of ways.

Virtual Chem Lab CD-ROM

Hands-on virtual chemistry laboratory provides students with opportunities to utilize safe, proper lab practices on a limitless combination of chemistry experiments without the set-up and expensive materials and supplies.

Chem Alive! Labs and Field Trips DVD

- 27 chemistry demonstrations performed by Lee Marek and Robert Lewis bring solid chemistry instruction to the classroom in a fun and unusual way.
- Chemistry Field Trip segments feature scientists who explain how chemistry plays a role in students' everyday lives.

Research on Differentiated Instruction

What's different about each of the students in your classroom? Just about everything. A typical classroom is comprised of students whose differences include culture, language, interest, motivation and knowledge.

What is differentiated instruction?

When you begin your school year, you quickly observe the spectrum of your students—limited science background to nearly science experts, basic English skills to English proficiency, independent workers to "fidgets and roamers," and a host of others. How can the needs of all of these students be met? The simple answer is: Differentiated Instruction—focusing on students' individual needs and providing varied materials and grouping patterns that accommodate individual learning differences. Differentiated instruction emphasizes these principles:

1. All students are unique and their strengths and needs change as their literacy develops and as they grow in their knowledge of science.
2. Instruction that attempts to meet the needs of all students must be flexible and adaptable for each individual.
3. Assessment of students' strengths and needs must be continuous to ensure that each student learns all they can.
4. Instruction must be multi-sensory with learning opportunities that rely on all of the senses, and it must be scaffolded for each student's learning level.
5. Instruction with multi-media enhances the probability of each student's learning.

What techniques help differentiate instruction?

Whenever possible, enhance learning with one-to-one teaching episodes, conferring individually with one student while other students are working independently. In addition, small group instruction of students with similar instructional needs where you are in close proximity with students so that quick, supportive responses can be immediate and exact is very effective.

A wide array of materials written at different levels makes information accessible to students. This range of materials should include:

- Print materials that are written at many different reading levels and exhibit many different genres and formats (textbooks, magazines, newspaper articles, computer messages, reference materials).
- Media including videos, computer programs, multi-media productions, and virtual reality explorations.
- Artifacts or realia that make information accessible to students, e.g. pictures and videos, music and objects that can be touched and studied.

Students require different levels of pacing. Differentiated instruction requires that lessons are accelerated for some students and decelerated for others, and often requires the repetition of a lesson.

James Flood
School of Teacher Education San Diego State California

Diane Lapp
School of Teacher Education San Diego State California

"Differentiated instruction focuses on students' individual needs related to culture, language, interest, motivation and knowledge."

Evaluator's Checklist

Does your science program promote differentiated instruction by—

- ✓ Reinforcing reading skills throughout the Student Edition?
- ✓ Integrating varied learning experiences into the program resources?
- ✓ Providing teaching suggestions for differentiating instruction?
- ✓ Labeling resources according to student ability levels?
- ✓ Guiding lesson pacing for varied ability levels?

How *Prentice Hall Chemistry* Meets Individual Needs

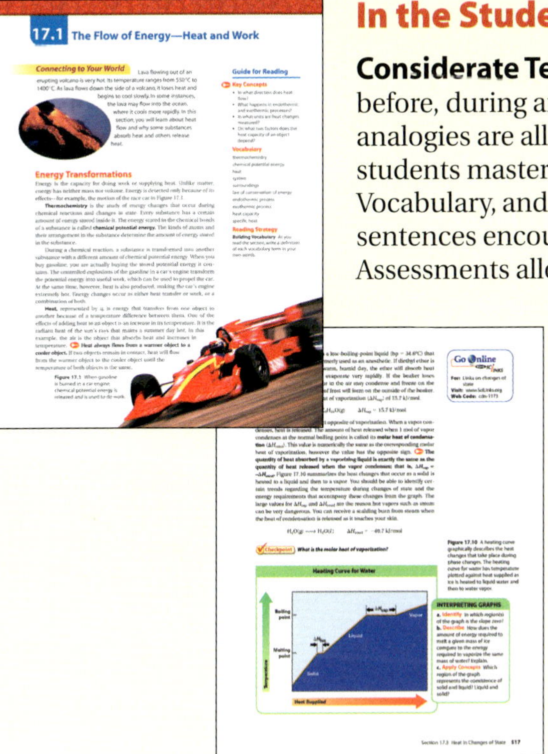
17.1 The Flow of Energy—Heat and Work

Connecting to Your World

Guide for Reading

Energy Transformations

Heating Curve for Water

INTERPRETING GRAPHS

In the Student Edition

Considerate Text Structure Ample, built-in reading support with elements before, during and after each section of each lesson, and frequent use of analogies are all part of a considerate text. The Guide for Reading helps students master concepts by introducing them to the Key Concepts, Vocabulary, and a Reading Strategy. The Key Concepts highlighted as boldfaced sentences encourage students to focus on the big ideas of chemistry. Section Assessments allow students to review Key Concepts at the end of every lesson.

Visual Learning

Graphs, charts, illustrations and photos work hand-in-hand with the text to clarify complex topics for students who think visually.

Active Learning

- The Inquiry Activities, Quick Labs and Small-Scale Labs in the Student Edition provide visual and tactile cues, offer opportunities for peer communication, and encourage skills that are necessary for critical thinking. Interaction with the content before, during, and at the end of each chapter involves students in the lesson while ensuring understanding along the way.
- Internet connections through the phschool.com web site and a partnership with NSTA brings timely, relevant, and appropriate chemistry topics to the classroom.

In the Teacher's Edition

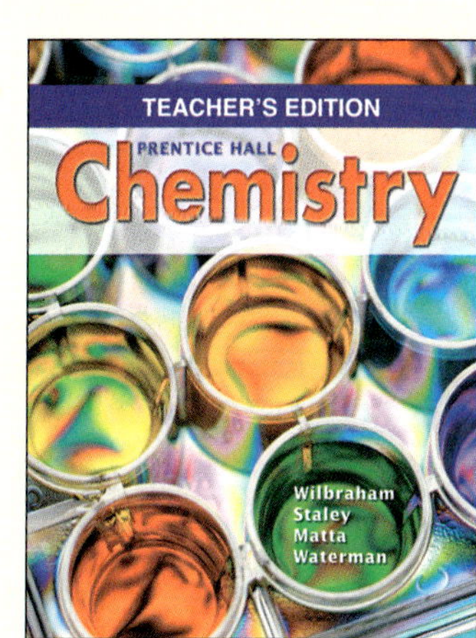

Chapter Planning Guides label activities by levels of difficulty, meeting the needs of all students. All program resources are identified by levels for Conceptual, for Standard, and for Honors.

The Teacher's Edition includes Differentiated Instructional Support Ideas. In order to facilitate use, these practical ideas for visual learning, English language learners, less proficient readers, advanced learners, and inclusion/special needs are placed at point-of-use areas throughout.

In the Program Resources

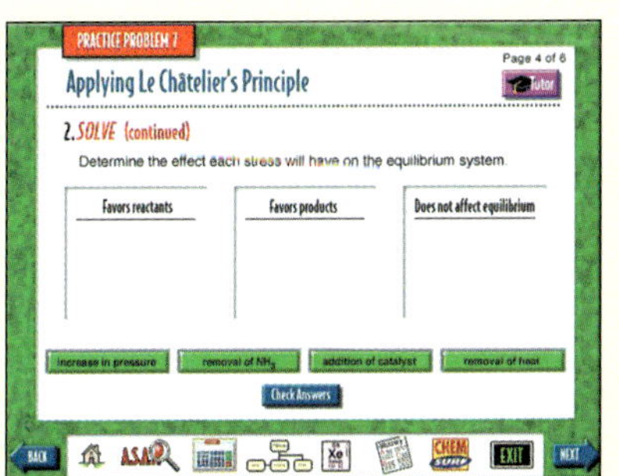

Guided Reading and Study Workbook This ancillary supports every section of the text, helping students with reading and math skills. Instructional graphics provide visual reinforcement of complex topics.

Laboratory Manual and Small-Scale Chemistry Laboratory Manual Lab options help you find just the right lab for every learning style.

Interactive Textbook with ChemASAP The Student Edition Online and on CD-ROM allows students to interact with content, including reading aids, visual and interactive learning tools, and instant feedback assessments.

PACING GUIDE

The Pacing Guide below suggests one way to schedule your instructional time for Basic (students who need additional help), Average (all students), and Enriched (students who need to be challenged) students.

Week	Basic Students	Average Students	Enriched Students
	CHAPTERS/SECTIONS		
1	1	1	1
2	2	2	2
3	3.1, 3.2	3.1, 3.2	3.1, 3.2
4	3.3, 3.4	3.3, 3.4	3.3, 3.4
5	4.1, 4.2	4	4
6	4.3, 5.1	5	5
7	5.2, 6.1, 6.2	6	6
8	6.3, 7.1	7.1, 7.2	7.1, 7.2
9	7.2, 7.3	7.3	7.3
10	8.1, 8.2	8.1, 8.2	8.1, 8.2
11	8.3, 8.4	8.3, 8.4	8.3, 8.4
12	9.1, 9.2	9.1, 9.2	9.1, 9.2
13	9.3, 9.4	9.3, 9.4	9.3, 9.4
14	10.1	9.5, 10.1	10
15	10.2, 10.3	10.2, 10.3	11
16	11.1, 11.2	11.1, 11.2	12
17	11.3	11.3, 12.1	13.1, 13.2
18	12.1, 12.2	12.2, 12.3	13.3, 13.4
19	12.3, 13.1	13	14
20	13.2, 13.3	14	15
21	13.4, 14.1	15	16.1, 16.2
22	14.2, 14.3	16.1, 16.2, 16.3	16.3, 16.4
23	15	16.4, 17.1, 17.2	17
24	16.1, 16.2	17.3, 17.4	18
25	16.3, 17.1	18.1, 18.2, 18.3	19
26	17.2, 17.3	18.4, 18.5, 19.1	20
27	18.1, 18.2	19.2, 19.3, 19.4	21
28	19.1, 19.2, 19.3	19.5, 20	22.1, 22.2, 22.3
29	19.4, 19.5	21	22.4, 22.5
30	20	22.1, 22.2	23.1, 23.2
31	21	22.3, 22.4, 22.5	23.3, 23.4
32	22	23	24.1, 24.2, 24.3
33	24.1, 24.2	24.1, 24.2, 24.3	24.4, 24.5, 24.6
34	24.3, 24.4	24.4, 24.5, 24.6	25
35	24.5, 24.6	25	Project
36	25	Project	Project
Total	**36 Weeks**	**36 Weeks**	**36 Weeks**

1 Week = 2 1/2 blocks or 5 periods

COURSE PLANNING OPTIONS

Most chemistry teachers are faced with the challenge of educating students of varied academic abilities. It takes careful planning to map out a method by which students can best be exposed to the major topics in the high school chemistry curriculum.

Prentice Hall Chemistry is organized in a way that can meet the needs of a variety of students and teachers. The textbook is comprehensive enough to satisfy the most comprehensive high school chemistry curriculum, while remaining readable, interesting, and accessible to less-academically motivated students. In addition, the division of chapters into manageable numbered sections provides today's teachers with the flexibility needed to teach classes comprised of students with diverse learning levels.

Pages T20 and T21 contain section planning options to assist teachers in designing lessons that meet the needs of all students. Symbols are used as follows.

The Conceptual Curriculum

In recent years, more students have been required to take chemistry, either in a mainstream setting or in a conceptual chemistry course. Thus, classroom populations have more students with below-average academic abilities in reading comprehension and math skills. The conceptual curriculum allows for a more concrete approach to ensure these students achieve their maximum potential. Course options are designated with symbols as follows.

● **Concrete concepts**—These sections typically describe fundamental (concrete) chemical concepts without relying on highly theoretical (abstract) or mathematical principles. These topics will help students focus on qualitative aspects of chemistry.

○ **More abstract concepts or math/problem-solving**—These sections offer a combination of concrete topics plus theoretically abstract topics and/or mathematically challenging problem solving. In these sections, the emphasis on theoretical aspects or the degree of problem solving can be determined by the teacher.

Some sections have no symbol because they are beyond the scope of a typical conceptual curriculum and do not include concepts that are prerequisites for later chapters.

The Standard Curriculum

The standard curriculum is designed for students considered average in ability and achievement levels. These students should have little difficulty reading and understanding most of the material presented, and they will have sufficient math skills for tackling the problems in the textbook. Course options are designated with symbols as follows.

■ **Core content**—These sections are suitable for all students in a standard curriculum. A course that includes most of these sections will prepare students for other science courses and for becoming scientifically literate citizens.

□ **Extension topics**—These sections may be covered as time allows or according to the interest of the teacher and the needs and/or problem-solving skills of the students.

Some sections have no symbol because a teacher would not have sufficient time to cover the topic well, or because the material is too challenging or abstract, or is not usually required in state or national curricula. Assume if there is no symbol that the section is not a prerequisite for later chapters.

The Honors Curriculum

The standard curriculum is designed to be used for students considered above average in their reading comprehension and general scholastic abilities. These students should be able to master most of the content in the textbook with less time and effort. Course options are designated with symbols as follows.

◆ **Core honors content**—These sections are suitable for all students in an honors curriculum. A course including these sections will prepare students for advanced science courses and for college coursework.

◇ **Options to accelerate**—These sections may be covered more rapidly using homework assignments. Students may only need to review this material because they have studied it in a previous course or a previous chapter.

1 Introduction to Chemistry

		C	S	H
1.1	Chemistry	●	■	◇
1.2	Chemistry Far and Wide	●	■	◇
1.3	Thinking Like a Scientist	●	■	◇
1.4	Problem Solving in Chemistry	●	■	◇

2 Matter and Change

		C	S	H
2.1	Properties of Matter	●	■	◇
2.2	Mixtures	●	■	◆
2.3	Elements and Compounds	●	■	◆
2.4	Chemical Reactions	●	■	◆

3 Scientific Measurement

		C	S	H
3.1	Measurements and Their Uncertainty	○	■	◆
3.2	The International System of Units	●	■	◇
3.3	Conversion Problems	○	■	◆
3.4	Density	●	■	◆

4 Atomic Structure

		C	S	H
4.1	Defining the Atom	●	■	◇
4.2	Structure of the Nuclear Atom	●	■	◆
4.3	Distinguishing Among Atoms	○	■	◆

5 Electrons in Atoms

		C	S	H
5.1	Models of the Atom	●	■	◆
5.2	Electron Arrangement in Atoms	○	■	◆
5.3	Physics and the Quantum Mechanical Model	○	□	◆

6 The Periodic Table

		C	S	H
6.1	Organizing the Elements	●	■	◆
6.2	Classifying the Elements	●	■	◆
6.3	Periodic Trends	●	■	◆

7 Ionic and Metallic Bonding

		C	S	H
7.1	Ions	●	■	◆
7.2	Ionic Bonds and Ionic Compounds	●	■	◆
7.3	Bonding in Metals	●	■	◆

8 Covalent Bonding

		C	S	H
8.1	Molecular Compounds	●	■	◆
8.2	The Nature of Covalent Bonding	●	■	◆
8.3	Bonding Theories		□	◆
8.4	Polar Bonds and Molecules	●	■	◆

9 Chemical Names and Formulas

		C	S	H
9.1	Naming Ions	●	■	◆
9.2	Naming and Writing Formulas for Ionic Compounds	●	■	◆
9.3	Naming and Writing Formulas for Molecular Compounds	●	■	◆
9.4	Naming and Writing Formulas for Acids and Bases	●	■	◆
9.5	The Laws Governing Formulas and Names	●	■	◆

10 Chemical Quantities

		C	S	H
10.1	The Mole: A Measurement of Matter	○	■	◆
10.2	Mole–Mass and Mole–Volume Relationships	○	■	◆
10.3	Percent Composition and Chemical Formulas		■	◆

11 Chemical Reactions

		C	S	H
11.1	Describing Chemical Reactions	●	■	◆
11.2	Types of Chemical Reactions	●	■	◆
11.3	Reactions in Aqueous Solution	○	■	◆

12 Stoichiometry

		C	S	H
12.1	The Arithmetic of Equations	●	■	◇
12.2	Chemical Calculations		■	◆
12.3	Limiting Reagent and Percent Yield		□	◆

13 States of Matter

		C	S	H
13.1	The Nature of Gases	○	■	◆
13.2	The Nature of Liquids	●	■	◆
13.3	The Nature of Solids	○	■	◆
13.4	Changes of State		□	◆

Conceptual Curriculum	Standard Curriculum	Honors Curriculum
● Concrete concepts	■ Core content	◆ Core honors content
○ More abstract concepts or math/problem-solving	□ Extension topics	◇ Options to accelerate

14 The Behavior of Gases

		C	S	H
14.1	Properties of Gases	●	■	◇
14.2	The Gas Laws	○	■	◆
14.3	Ideal Gases		■	◆
14.4	Gases: Mixtures and Movements	○	■	◆

15 Water and Aqueous Systems

		C	S	H
15.1	Water and Its Properties	●	■	◆
15.2	Homogeneous Aqueous Systems	●	■	◆
15.3	Heterogenous Aqueous Systems		□	◇

16 Solutions

		C	S	H
16.1	Properties of Solutions	●	■	◆
16.2	Concentrations of Solutions	○	■	◆
16.3	Colligative Properties of Solutions		■	◆
16.4	Calculations Involving Colligative Properties		□	◆

17 Thermochemistry

		C	S	H
17.1	The Flow of Energy—Heat and Work	○	■	◆
17.2	Measuring and Expressing Enthalpy Changes	○	■	◆
17.3	Heat in Changes of State	○	□	◆
17.4	Calculating Heats of Reaction			◆

18 Reaction Rates and Equilibrium

		C	S	H
18.1	Rates of Reaction	●	■	◆
18.2	Reversible Reactions and Equilibrium	○	□	◆
18.3	Solubility Equilibrium		□	◆
18.4	Entropy and Free Energy			◆
18.5	The Progress of Chemical Reactions			◆

19 Acids, Bases, and Salts

		C	S	H
19.1	Acid-Base Theories	●	■	◇
19.2	Hydrogen Ions and Acidity	○	■	◆
19.3	Strengths of Acids and Bases		□	◆
19.4	Neutralization Reactions		□	◆
19.5	Salts in Solution		□	◆

20 Oxidation-Reduction Reactions

		C	S	H
20.1	The Meaning of Oxidation and Reduction	●	■	◆
20.2	Oxidation Numbers		■	◆
20.3	Balancing Redox Equations		□	◆

21 Electrochemistry

		C	S	H
21.1	Electrochemical Cells	●	■	◆
21.2	Half-Cells and Cell Potentials	○	□	◆
21.3	Electrolytic Cells	○	■	◆

22 Hydrocarbon Compounds

		C	S	H
22.1	Hydrocarbons	●	■	◆
22.2	Unsaturated Hydrocarbons	○	■	◆
22.3	Isomers		□	◆
22.4	Hydrocarbon Rings	○	□	◆
22.5	Hydrocarbons From Earth's Crust	●	■	◇

23 Functional Groups

		C	S	H
23.1	Introduction to Functional Groups	○	■	◆
23.2	Alcohols and Ethers		□	◆
23.3	Carbonyl Compounds		□	◆
23.4	Polymerization		□	◆

24 The Chemistry of Life

		C	S	H
24.1	A Strategy for Life	●	■	◇
24.2	Carbohydrates		□	◆
24.3	Amino Acids and Their Polymers		□	◆
24.4	Lipids		□	◆
24.5	Nucleic Acids		□	◆
24.6	Metabolism		□	◆

25 Nuclear Chemistry

		C	S	H
25.1	Nuclear Radiation	●	■	◆
25.2	Nuclear Transformations	○	■	◆
25.3	Fission and Fusion of Atomic Nuclei	●	■	◆
25.4	Radiation in Your Life	●	■	◇

Support for the Chemistry Teacher

Pearson Prentice Hall develops teaching support materials that offer unique solutions for the classroom. From presentation tools to lesson planners, our easy-to-use technology for teachers helps you save time and enhance learning. Planning and managing lessons has never been easier with a variety of timesaving resources. Prentice Hall's easy-to-use technology for teachers helps you save time and enhance learning.

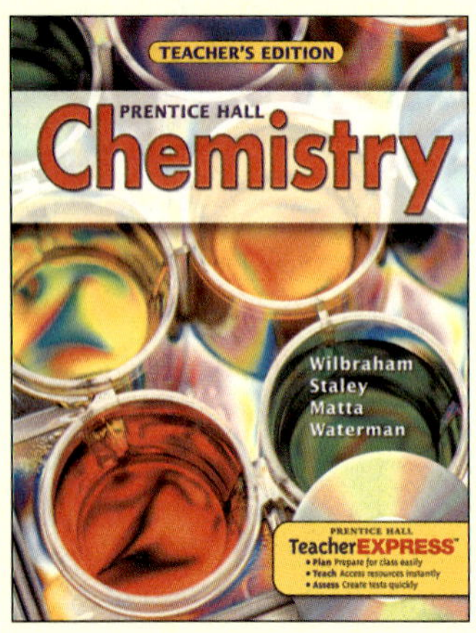

Teacher's Edition

Provides a wealth of additional resources and lesson planning support Chapter Planning Guides and wraparound notes meet the needs of all learners by

- Allowing you to plan for the type of course you are teaching.
- Identifying available print and technology resources for each lesson.
- Offering numerous activities and teacher demonstrations for hands-on learning.

Teacher Express CD-ROM

Provides the latest in planning and resource management software Powered by LessonView, this easy-to-use, time-saving technology lets you

- Create custom lesson plans for any length class in seconds.
- Edit lessons and incorporate personal resources to customize plans.
- Access the complete interactive Teacher's Edition to choose and print resources, manage appointments, organize addresses, and generate reports.

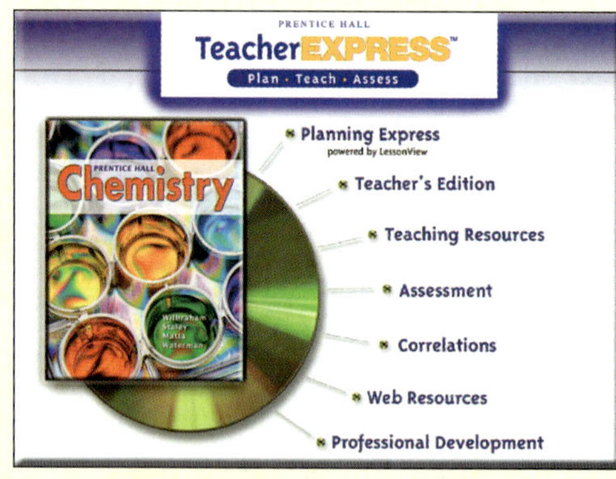

The ExamView® Test Bank CD-ROM

Powered by the ExamView® test generator, this easy to use and teacher-friendly technology lets you

- Choose from over four thousand test items.
- Create tests, worksheets, and study guides in minutes.
- Easily customize the layout of any test in seconds.
- Create paper, LAN-based, and Internet tests.

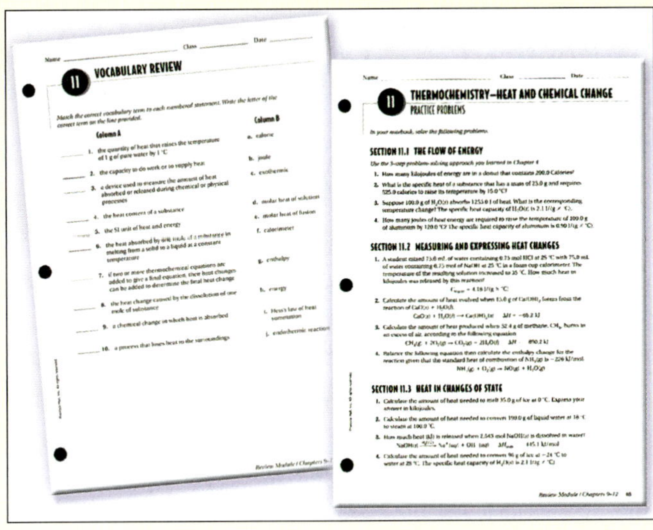

VOCABULARY REVIEW

THERMOCHEMISTRY—HEAT AND CHEMICAL CHANGE
PRACTICE PROBLEMS

SECTION 11.1 THE FLOW OF ENERGY

SECTION 11.2 MEASURING AND EXPRESSING HEAT CHANGES

SECTION 11.3 HEAT IN CHANGES OF STATE

Core Teaching Resources

Complete blackline masters support including:

- Section Reviews
- Interpreting Graphics
- Vocabulary Review
- Chapter Quizzes
- Chapter Tests A and B
- Practice Problems
- Solutions for Chapter Assessments
- Lab Record Sheets

Presentation Express

An easy-to use instructional tool that helps you deliver visually exciting content to your students, meet your course objectives, and save you time. It includes:

Chem**ASAP**

- **A**nimations illustrate important concepts.
- **S**imulations that allow you to manipulate variables, observe results, and draw conclusions.
- **A**ssessments provide opportunities to check student understanding.
- **P**roblem-Solving tutorials provide detailed, step-by-step solutions to over 100 practice problems.

Process and Animated Art
Important processes from the text are animated to reveal step-by-step concepts—helping students better understand key concepts.

QuickTake
Section Quizzes assess students' understanding of important concepts and are available in both QuickTake and PowerPoint format.

Save to Disc
Save presentations and all associated resources to your own disc in one easy step.

LESSON CYCLE

Learning is an ongoing process that occurs in cycles. Similarly, each lesson should follow a cycle that encourages and enhances the learning process. In the Teacher's Edition, the lesson cycle of engage, teach, and assess helps make learning fun and meaningful.

1 Focus

In the first step of the lesson cycle, the teacher engages the student's interest and inspires curiosity. Students are encouraged to activate prior knowledge and ask what they can learn about the new topic. During this stage of the lesson, misconceptions are explored and the teacher can identify target areas on which to focus. The ***Prentice Hall Chemistry Teacher's Edition*** employs the following techniques to engage the students in the first step of the lesson cycle.

Use the Visual provides suggestions for launching a discussion using the photograph and text that introduces the section.

Check Prior Knowledge, which appears in many sections, helps teachers determine students' previous knowledge of the subject to be studied. This feature allows the teacher to assess the students' background and unveil misconceptions.

Motivate and Relate provides suggestions for connecting content and visuals to real-world situations and applications, with a focus on students' relevant experiences.

Teacher Demo or Class Activity serves to pique students' interest with a relevant, active experience. The activity inspires the students to seek ways to answer "how" and "why" questions, rather than to simply acquire information.

2 Instruct

During the second step of the lesson cycle, the teacher serves as a guide while the students explore phenomena, attempt to answer questions raised during the first step of the cycle, and begin to construct an understanding of a concept through hands-on participation. The teacher helps students to develop understanding through communication, construction of models, and critical thinking. Connections are made to other disciplines as well as to social and community issues.

Discuss presents topics that relate to or reinforce the concepts in the student text and provides suggested topics for discussion among students.

Use the Visual provides suggestions for incorporating a photograph, diagram, or illustration into the lesson, which helps to engage the visual, spatial, or nonverbal learner in concept development.

Teacher Demo and Class Activity provide an opportunity for students to witness or experience the concepts described in the text. All students should participate in class activities, which may include the use of manipulatives or role playing.

Build Writing Skills helps teachers to encourage writing as an integral part of learning. Writing activities may include research, letters to the media, and creative writing projects.

Think Critically encourages students to think beyond what is written on the page. This feature may present a problem or provide a series of questions to help guide students.

Cooperative Learning gives practical suggestions for group work in situations such as problem-solving and hands-on activities.

Connections to other sciences and disciplines, which appear in many sections, extend the learning cycle. These strategies give teachers an opportunity to integrate a variety of sciences and other disciplines into their curriculum.

3 Assess

During the assessment step of the lesson cycle, students demonstrate their understanding and mastery of concepts and skills by answering questions based on their observations and experiences during the lesson. This activity allows teachers to identify areas needing further clarification, and allows the students to evaluate their own success in achieving learning goals.

Evaluate Understanding uses questioning, student participation, written explanations, and student generated materials to demonstrate learning.

Reteach presents the teacher with an alternative strategy to reinforce section content and concepts.

NATIONAL SCIENCE EDUCATION STANDARDS

Content Standard A Science as Inquiry	
A-1 Abilities Necessary to Do Scientific Inquiry	Sections 1.3 and 1.4
A-2 Understandings About Scientific Inquiry	Chapter 1

Content Standard B Physical Science	
B-1 Structure of Atoms	Chapters 4, 5, and 25
B-2 Structure and Properties of Matter	Section 2.1 Chapters 4, 5, 6, 7, 8, 13, 20, and 23
B-3 Chemical Reactions	Sections 5.3 and 25.4 Chapters 20 and 21
B-4 Motions and Forces	Sections 7.2, 8.4, 13.1, and 18.4
B-5 Conservation of Energy and Increase in Disorder	Chapters 13, 14, and 17
B-6 Interactions of Energy and Matter	Sections 5.3 and 25.4 Chapters 20 and 21

Content Standard C Life Science	
C-1 The Cell	Chapter 24
C-2 Molecular Basis of Heredity	Section 24.5
C-3 Biological Evolution	Section 24.1
C-4 Interdependence of Organisms	
C-5 Matter, Energy, and Organization in Living Systems	Chapter 24
C-6 Behavior of Organisms	

Continues

Content Standard D Earth and Space Science	
D-1 Energy in the Earth System	Section 17.1, pages 518 and 519
D-2 Geochemical Cycles	pages 764, 765, 789, and 790
D-3 Origin and Evolution of the Earth System	Section 22.5
D-4 Origin and Evolution of the Universe	Section 1.2

Content Standard E Science and Technology	
E-1 Abilities of Technological Design	pages 109, 147, 168, 204, 242, 259, 313, 340, 376, 405, 430, 478, 518, 623, 644, 685, 716, 791, and 841
E-2 Understandings About Science and Technology	pages 18, 43, 88, 329, 404, 452, 463, 548, and 754

Content Standard F Science in Personal and Social Perspectives	
F-1 Personal and Community Health	pages 10, 15, 16, 18, 24, 256, and 452
F-2 Population Growth	
F-3 Natural Resources	Chapter 15, Section 22.5, pages 612 and 754
F-4 Environmental Quality	pages 329, 452, 463, and 611
F-5 Natural and Human-Induced Hazards	page 340
F-6 Science and Technology in Local, National, and Global Challenges	pages 18 and 518

Content Standard G History and Nature of Science	
G-1 Science as a Human Endeavor	pages 87, 120, 145, 256, 329, 366, 404, 452, 579, 611, and 701

Small-scale chemistry is an innovative, low cost, safe, time efficient, conservation-based teaching model for chemistry. It is not a curriculum, but a highly flexible approach to teaching chemistry at all levels. Small-scale science is actually a misnomer—it describes an attitude more than it does a size. (It is **not** microscale!) The goal of small-scale science is to have every student involved in hands-on learning become more efficient in process and more knowledgeable in content while using this highly effective discovery approach to learning.

Small-scale chemistry is more than just a methodology. It is a systemic medium that creates a learning environment that fosters creativity, invention, and scientific problem-solving. As you and your students perform the Small-Scale Labs included in each chapter of the Student Edition, you will see that small-scale chemistry does not use traditional equipment like Bunsen burners, beakers, and test tubes. Instead, small-scale chemistry employs non-traditional, inexpensive, student-designed and built laboratory equipment made from readily available materials. The ***Small-Scale Chemistry Laboratory Manual*** has instructions for building balances, spectroscopes, pH meters, electrolysis devices, and digital burets.

Small-scale chemistry is not designed to simply do better what traditional laboratory methodologies try to do. (It does not, for example, miniaturize traditional experiments.) Small-scale science creates an environment where students and teachers participate together in the interplay of logic and experiment. It promotes the interchange of information among students and between students and teachers.

This approach is different from other student-centered approaches in an important way: it emphasizes that asking good questions is more important than giving right answers. It is a unique and effective blend of process and content. Each time a teacher supervises a lesson, the lesson might be different. In each class, students often contribute new insights, useful innovations, and fresh ideas.

Why Use Small-Scale Chemistry?

Small-scale chemistry is relatively safe. Hazardous reagents and procedures were eliminated before experiments were designed. Small-scale builds in safety at the design stage of an experiment rather than consigning teachers to try to control hazards at the execution stage. For example, solution concentrations are usually about $0.2M$ or less, and concentrations never exceed $1.0M$. Plastic thin-stemmed pipets, originally designed for biomedical research, serve both as liquid-storage vessels and transfer devices. Each pipet delivers only about 20 microliters of dilute solution. The potential for major accidents in traditional labs is replaced by minor inconveniences in small-scale labs.

Small-scale chemistry is time-efficient. Setup is made easy for the teacher and the student because the experimental format does not rely on lengthy written instructions. Instead, the directions are easy to read and follow. Experimental results are immediate and obvious, and the time students take to set up, manipulate, and clean equipment is minimal. Teachers are available to spend their valuable time and energy interacting with students during the laboratory. The simplicity of the experiments also enhances learning outcomes. The simple organization under the reaction surface helps students accurately record, compare, and correlate their data. Students focus on what's happening rather than on what to do. Best of all, the teacher decides and chooses which learning outcomes are most practical for each class.

Small-scale chemistry labs have less environmental impact than do conventional chemistry labs. Clean-up with a paper towel takes seconds, and the minute quantities of chemicals can usually be disposed of safely, usually in the waste basket. As many school districts trim budgets and tighten regulations on the use and disposal of school-laboratory chemicals, small-scale chemistry offers immediate solutions. The small amounts of common, relatively safe chemicals employed never become a waste-management problem.

Small-scale chemistry is a lab-based program that can be used as the central focus of a high school chemistry course. Teachers can build the conventional content around lab where it seems to fit. For example, students write and balance only those chemical equations for reactions that they have carried out in the lab. Quantitative concepts, such as stoichiometry, are well suited for lab-centered teaching. Thus teachers can bring the content of chemistry into the laboratory. Through small-scale chemistry, students can and do learn much more chemistry than they can through conventional labs. In addition, they learn to be creative and inventive with chemistry; to be good problem solvers; and to communicate effectively with the teacher, with other students, and with themselves.

How to Set Up and Run a Small-Scale Chemistry Laboratory

Efficient organization on the part of the teacher is a key to running a successful small-scale chemistry lab program. Here are some suggestions for the setup and day-to-day operations of a small-scale laboratory.

1. Assemble the following list of basic equipment. The quantities here assume you plan to convert your laboratory to small-scale chemistry entirely and further, that you will teach five to six sections of chemistry per day, with 24 students per section. You will need to adjust the numbers if you plan to use fewer small-scale labs your first year to augment your traditional program.
   - 4 boxes of thin-stemmed plastic disposable pipets (500 per box), to store and deliver liquids
   - 50 plastic reagent bottles (250 mL), to store prepared solutions
   - 10 plastic reagent bottles (1.0 L), to store prepared solutions
   - 60 plastic screw-capped specimen jars (60 mL), to fill pipets
   - 60 plastic cups (12-ounce), to store sets of pipets
   - 120 plastic pill vials (30–60 mL), to store and dispense solid reagents
   - 1 plastic "egg crate" light panel (2 feet × 4 feet, cut) to hold pipets for easy access
   - 48 small-scale reaction surfaces (file protectors or overhead projection transparencies), to mix chemicals qualitatively
   - 24 well plates (1 × 8), to calibrate pipets
   - 12 pairs of scissors, to construct instruments
   - 12 office punches (1/4-inch) to construct instruments
   - 6 standard staplers, to construct instruments
   - 150 plastic cups (1-ounce), for titrations and balance pans
   - 150 plastic cups (3 1/2-ounce), for titrations and chromatography
2. Prepare the reagent solutions described in the Master Solutions List on pages T32–33. Store each reagent solution in an appropriately sized plastic reagent bottle. Label each bottle with the concentration, the chemical formula of the contents, the number of grams of solute per volume of solution, and the date of preparation (for example:1*M NaCl, 14.6 g per 250 mL, 01/20/04).* When a reagent bottle is empty, simply make a new solution from the labeled directions and change the date. You do not have to make all the solutions at once; prepare the necessary solutions only for the first lab. Then as the year progresses, you can add solutions as you need them. By the end of your first year of small-scale chemistry, you will have a complete set that will need only occasional maintenance. Once a small-scale lab is set up, a teacher in charge of 5 sections of 24 students each should count on replenishing about 10 to 15 solutions a year.
3. Store each reagent bottle in a designated place. The storage area should be secure from students but easily accessible by a teacher during a laboratory period. The prepared solutions and solids in this book require approximately 36 feet of shelf space.
4. For a classroom of 24 students, prepare and label 15 small-scale pipets and one 60-mL screw-capped plastic specimen jar (reagent-filling container) for each of the solutions listed in the Master Solutions List on pages T32–33. Fill each set of pipets and store them stem-up in a labeled 12-ounce plastic cup on the shelf along with the reagent bottle and plastic specimen jar. To save time, you may want to spend a few minutes in class having students pull pipets and tape labels on them. Use a computer to make labels or print the number of labels you need before you cut them out and tape them to the pipets.
5. Cut the egg-crate plastic light panel into small sections to use as pipet racks. Wear your safety glasses and use a hacksaw. This is a job you have to do only once. It's convenient to cut sections that are 3 squares × 10 squares, which will hold as many as 30 pipets.
6. When it is time to set up a lab, set out a 12-ounce storage cup containing 15 pipets of each solution needed in the lab. Students can pick up one pipet from each cup and quickly assemble a complete set of solutions needed for the lab. You may wish to have two students share a set of reagents, even if

they are doing the lab individually. At the end of the lab, students can return the pipets to their proper cups, ensuring that everything is ready for the next class.

7. Fill empty pipets as students call your attention to them. Because students often need only one or two drops of each reagent, many pipets will not need filling until after several sections of students have done the lab. For labs that require large quantities of reagents and, thus, frequent refills of pipets, you may want to teach students how to prepare pipets for some reagents, as described in "Safe and Efficient Techniques for Using Small-Scale Equipment" in the ***Small-Scale Chemistry Laboratory Manual***.
8. When the lab is finished, return the pipet storage cups to the storage compartment. You may want to leave one or two sets out for another day or two to accommodate students who were absent during the lab.
9. For a classroom of 24 students, prepare and label plastic pill vials (30–60 mL) for each solid on the Master Solids list on page T34. Punch small holes in the lids and cover them with tape. To use the vials, students can remove the tape and carefully shake out the desired quantity. Store the solids in the plastic pill vials with the openings taped. You can store each set in a large plastic cup or in a plastic ice-cube tray.

Evaluation of Student Learning Through Small-Scale Laboratory

The following list provides some suggestions for how you could evaluate what your students learn from doing small-scale chemistry laboratories.

1. **Lab Notebook** As an alternative to using the Laboratory Recordsheets, you may want to have students keep a laboratory notebook. The lab notebook is a real-time, permanent, hand-written record of a student's entire laboratory experiences, including dates, titles, experimental procedures, data, conclusions, and further research and investigations. Pick up the lab notebooks periodically, and grade them for completeness rather than for specific details. With only a quick glance you can verify that a student has kept his or her work current. You may be less interested in format than in how well students can use lab records to solve problems. In this case, look for signs of creativity, invention, problem-solving, and good thought processes. Be sure to tell students your expectations. You may want to give extra credit for evidence of creativity or invention. For example, students might invent interesting new questions or novel ways to solve problems. Or a student might choose to write the lab procedure in cartoon format.
2. **Open-lab Notebook Tests** To check for details and understanding, you may want to give students regular lab tests and allow them to use their own hand- written notes to answer the questions. Try to balance the range of questions from knowledge-level questions to those that evaluate comprehension, analysis, synthesis, evaluation, and creativity. Students soon learn that to score well on these exams, they must have a well-written lab book and show that they can use it to interpret their results.
3. **Closed-lab Notebook Tests** Some skills learned in lab need to be internalized by any student of chemistry. Use closed-book tests to evaluate how well students have learned these skills. For example, test students' abilities to write and balance equations.
4. **Performance-based Assessment** Small-scale chemistry lends itself well to the direct hands-on evaluation of problem-solving using both qualitative and quantitative analysis. Identification of unknowns are always a good source of performance-based assessment as are quantitative titrations of unknowns. The Student Edition small-scale labs have activities in the "You're the Chemist" sections that you can use to assess students' reasoning skills. The 30 practicals in ***Laboratory Practicals*** can be used to test both laboratory skills and comprehension.
5. **Writing Essays** Consider having students write short essays on their conclusions for various labs. Try to assign essays that encourage creativity and original thought—ones that prompt students to write something that would be interesting for you to read.

LAB SAFETY

Chemistry teachers are no longer exempt from successful prosecution for accidents that occur in their classrooms. That threat, coupled with our increased awareness of the insidious effects of chemical exposure, explains the demand for safety education, particularly in science classrooms.

When it comes to laboratory chemistry, the safety of students and teachers is the primary concern of Prentice Hall. To that end, both the traditional ***Laboratory Manual*** and the ***Small-Scale Chemistry Laboratory Manual*** give detailed safety information for the student and the teacher at the beginning of each lab and in the front of each laboratory manual. In the Student Edition, the labs display appropriate safety icons, which are defined for the students in Appendix D.

Safety in the Chemistry Laboratory

The following information about safety issues in the high school chemistry laboratory is intended to be used by classroom chemistry teachers. This reference section is intended to be accessible to the teacher for planning demonstrations and preparing for lab. The material included here is by no means comprehensive. Prentice Hall takes no responsibility for the completeness of this information or for the implementation of these suggestions.

The safety of the students in a high school laboratory is traditionally assigned to the teacher. But the teacher teaches safety awareness by holding each student accountable for the safety of all other participants in the lab, including the teacher. Have students sign a safety contract at the beginning of the year. There is a sample contract in each laboratory manual. Of course, the teacher will most effectively teach safety awareness by modeling safe practices and by making a discussion of safety conerns the first priority for all lab activities. The following suggestions are intended to help you establish a program of safety awareness for staff and students.

Education

- Put together a safety committee for your school. Include faculty, students, and administrators.
- Schedule regular meetings of the safety committee to discuss issues, accident reports, and the results of lab inspections.
- Implement a safety education program for staff and students.
- Require students to know the lab safety rules, assess this by giving a safety quiz, and have the students sign a safety contract, saying that they know the rules and will comply with them. Send the contract home with a set of rules and require that one parent or guardian sign the contract, having read the rules.
- Give a safety grade with each lab, based on each student's behavior during the lab. Penalize unsafe behavior by removing students from the lab.
- Encourage school staff to become certified in CPR and first aid.
- Require that science staff read any of several publications on safety in the high school science classroom.

Management

- Report all accidents to the safety committee.
- Periodically inspect laboratories and report findings to the safety committee.
- Establish an emergency plan and review it periodically with students and staff.
- Keep phone numbers for the fire department, police department, local ambulance, and poison control center next to every phone in a lab or stockroom.
- Display warning signs and safety posters.
- Prohibit student access to stockrooms.
- Require students to keep aisles clear and counters clean. There should be two exits in a laboratory or stockroom, and they should be kept free of obstacles.

- Keep a current chemical inventory. Update the chemical inventory often. Include a Material Safety Data Sheet (MSDS) for each chemical.
- Label all chemicals with the date and possible hazards. Establish a policy for disposing of and buying chemicals.
- In prelab discussions, stress safety issues relevant to each specific lab.

Lab Techniques

- Do not allow any student or staff member to work alone in the laboratory.
- Do not leave any experiment unattended.
- Do not eat, smoke, or drink in the lab.
- Do not store food in chemical refrigerators. Store chemicals only in chemical refrigerators.
- Never taste a chemical in the laboratory.
- Always add acids to water, never water to acids.
- When pipetting, use a suction bulb. Never use your mouth.
- When heating the contents of a test tube, direct the mouth of the test tube away from yourself and away from others.

Personal Safety

- Always wear goggles in lab. Require students and anyone entering the lab to wear goggles.
- Provide lab coats and aprons for all participants in the lab.
- Take into account student health problems such as allergies or asthma.
- Do not allow untied long hair, loose clothing, bare feet, open-toed or -heeled shoes, or dangling jewelry to be worn in the lab.
- Strongly discourage the wearing of contact lenses in lab. In the event of a chemical splash into the eye, the lenses may hold chemicals in contact with the eye, potentially causing greater injury.

Equipment

- Make available the following items in all labs and demonstration areas: an ABC fire extinguisher, a smoke alarm, a safety shower, an eye-wash faucet, a fume hood.
- Maintain and check all safety equipment regularly.
- Keep a first-aid kit stocked and available.
- Keep flammable chemicals in an approved fireproof cabinet.
- Attach compressed gas cylinders to counters or walls. Move them on dollies on which they are secured.
- Use safety shields when there is a possibility of explosion, implosion, chemical splashing, or fire.
- Use hot plates in lieu of open flames as often as possible.
- Use dispensing devices such as syringes, spigots, or droppers, rather than having students pour chemicals from their containers.
- Be prepared for spills, especially acid spills. Spill-control devices are available from suppliers. To avoid mercury spills, use alcohol thermometers.

Storing Chemicals

- Store chemicals in stockrooms, not in the lab.
- Do not store chemicals alphabetically. Use an approved system of chemical storage, and post legends in the stockroom.
- Store only the chemicals you will use during a school year.

Disposal

- Dispose of broken glass in a dedicated container kept out of the way and clearly labeled.
- Dispose of chemicals in the following ways:

  Neutralize acids and bases and flush down the drain.

  Flush light metal salts and their solutions down the drain with excess water.

  Precipitate and filter heavy metals. Evaporate heavy metal solutions down to salt. Keep heavy metals salts in labeled containers and dispose of them by following local regulations.

  When in doubt, check with local authorities about local regulations.

- Have a disposal plan for each lab, and include directions for disposal in your prelab discussion.

Master Solutions List

The following list gives directions for preparing all the chemical solutions necessary for doing all the small-scale labs in ***Prentice Hall Chemistry.*** The most efficient way to organize your small-scale lab is to make up the appropriate aqueous solutions, then use and replace them as needed. Store each solution in a plastic reagent bottle. Depending on the labs you do, some solutions will have to be replaced several times during the year; others will last the entire year. Most solutions are 0.1*M* or 0.2*M*. In no case is any solution of greater concentration than 1.0*M*. Except for sodium chloride, the labs do not require you to prepare multiple solutions for the same chemical. For example, all the experiments that use sodium hydroxide use 0.5*M* NaOH.

Label each reagent bottle with the chemical formula, the concentration, the number of grams of solute per volume of solution, and the date. When the bottle is empty, simply replenish it by following the directions on the label, and change the date. When making dilute acid solutions from concentrated acids, always add acid to water slowly and carefully. The Teacher's Edition of the ***Small-Scale Chemistry Laboratory Manual*** has a combined list of solutions for the labs in the ***Student Edition*** and the ***Small-Scale Chemistry Laboratory Manual.***

Solution	Preparation	Labs
alizarin yellow R, 0.02%	50 mg AYR in 250 mL	617
ammonia, 1.0*M*	67 mL of 15*M* NH_3 in 1.0 L **Caution! Using proper ventilation, add ammonia to water carefully and slowly!**	400, 458
borax, 4%	10 g $Na_2B_4O_7 \cdot 10H_2O$ in 250 mL	753
bromcresol green*, 0.04%	100 mg BCG in 14.3 mL 0.01*M* NaOH; dilute to 250 mL	617
bromphenol blue*, 0.04%	100 mg BPB in 14.9 mL 0.01*M* NaOH; dilute to 250 mL	617
bromthymol blue*, 0.04%	100 mg BTB in 16.0 mL 0.01*M* NaOH; dilute to 250 mL	400, 437, 617, 684
calcium chloride, 0.5*M*	13.9 g $CaCl_2$ in 250 mL	200, 267, 345
copper sulfate, 0.2*M*	12.5 g $CuSO_4 \cdot 5H_2O$ in 250 mL	56, 267, 684, 774
ethanoic acid, 0.8*M* CH_3COOH	undiluted white vinegar	245, 458, 655, 400
hydrochloric acid, 1.0*M*	82 mL of 12*M* HCl in 1.0 L **Caution! Always add acid to water carefully and slowly!**	200, 367, 458, 655, 774
hydrogen peroxide, 3% H_2O_2	undiluted household supply	56
iron(III) chloride, 0.1*M*	6.8 g $FeCl_3 \cdot 6H_2O$ in 25 mL of 1.0*M* NaCl; dilute to 250 mL	200, 267
lead(II) nitrate, 0.2*M*	16.6 g $Pb(NO_3)_2$ in 250 mL	200, 267, 345
magnesium sulfate, 0.2*M*	6.0 g $MgSO_4$ in 250 mL	267
metacresol purple*, 0.04%	100 mg MCP in 26.2 mL 0.01*M* NaOH; dilute to 250 mL	617
methyl orange, 0.05%	125 mg MO in 250 mL	617

Master Solutions List *(continued)*

Solution	Preparation	Labs
nitric acid, 1.0*M*	63 mL of 15.8*M* HNO_3 in 1.0 L **Caution! Always add acid to water carefully and slowly!**	200, 458, 655
pH buffer, Solution A	14.9 g boric acid and 12.6 g citric acid in 1.2 L	617
pH buffer, Solution B	45.6 g $Na_3PO_4 \cdot 12H_2O$ in 1.2 L	617
phenolphthalein, 0.1%	250 mg phenolphthalein in 250 mL of 70% 2-propanol	617
potassium bromide, 0.2*M*	6.0 g KBr in 250 mL	684
potassium iodide, 0.1*M*	4.2 g KI in 250 mL	56, 200, 684
potassium thiocyanate, 0.1*M*	2.4 g KSCN in 250 mL	200
2-propanol, 70%	undiluted rubbing alcohol	245, 458
silver nitrate, 0.05*M*	2.1 g $AgNO_3$ in 250 mL	200, 267, 345
sodium carbonate, 1.0*M*	26.5 g Na_2CO_3 in 250 mL	267, 345
sodium chloride, 0.1%	1.0 g NaCl in 1.0 L	245
sodium chloride, 1.0*M*	14.6 g NaCl in 250 mL	345, 684
sodium hydrogen sulfite, 1.0*M*	4.2 g $NaHSO_3$ in 250 mL	437
sodium hydroxide, 0.5*M*	20.0 g NaOH in 1.0 L	200, 267, 345, 367, 458, 774
sodium hypochlorite, 1%	50 mL liquid bleach (NaClO) in 200 mL	56
sodium phosphate, 0.1*M*	9.5 g $Na_3PO_4 \cdot 12H_2O$ in 250 mL	200, 267, 345
sodium sulfate, 0.2*M*	7.1 g Na_2SO_4 in 250 mL	200, 267, 345, 684
starch, 20%	50 mL liquid starch in 200 mL	56, 684
sulfuric acid, 1.0*M*	56 mL of 18*M* H_2SO_4 in 1.0 L **Caution! Always add acid to water carefully and slowly!**	458, 655
thymol blue*, 0.04%	100 mg TB in 21.5 mL 0.01*M* NaOH; dilute to 250 mL	367, 617

*If the indicator is available as the sodium salt, dissolve directly in 250 mL of deionized or distilled water.

Master Solids List

For each solid that can be poured, fill six 30- to 60-mL plastic pill vials. Use a 1/4-inch office punch to make a hole in each lid. Cover the holes with plastic tape. Label and store each set in a plastic ice cube tray.

Name	Labs
ammonium chloride, NH_4Cl	574
calcium carbonate, $CaCO_3$	304
calcium chloride, $CaCl_2$	400, 574
guar gum	753
iron, Fe (staples)	200, 655
magnesium, Mg	655
magnesium sulfate, $MgSO_4$	458
potassium chloride, KCl	458, 574
potassium iodide, KI	458
sodium carbonate, Na_2CO_3	458
sodium chloride, NaCl (table salt)	56, 304, 458, 574
sodium chloride, NaCl (lab grade)	497
sodium hydrogen carbonate, $NaHCO_3$ (baking soda)	367, 458, 574
sodium phosphate, Na_3PO_4	574
sucrose (table sugar)	458, 497
zinc, Zn (galvanized nails)	655

Master Equipment List

Name	Labs
balance and weighing paper	94, 120, 304, 367, 497, 533, 774
beaker, small	304
chromatography paper	245
colored filters	137
conductivity tester	458
diffraction grating	137
electrolysis device	684
Petri dish, plastic	400, 437
pipet	56, 94, 200, 367, 497, 753
spot plate, 96-well	179
volumetric flask, 50-mL plastic	497
well strip, 8-well	94

Master Household Products List

Store liquids in their original containers.

Name	Labs
alcohol thermometer	574
aluminum cans	94
aluminum foil	533
antacid tablets	56
baking powder	367
beverages, assorted	458
candles, votive	533
candy (colored)	120, 245
cereal	56
cereal box	137
construction paper, colored	56
cornstarch	458, 753
cotton swabs	437
cups, plastic	94, 120, 245, 367, 574, 753
dice	94
eggs	774
food coloring	245
glue, white	753
graph paper	400, 809
ice, cubes or crushed	400, 574
marking pens, fine line	56, 179
matches, safety	553
medicine droppers	56, 200, 458
milk, powdered	774
modeling clay	708
pennies (pre- and post-1982)	94, 655, 809
pickle juice	458
powdered drinks	245
reaction surface	56, 200, 267, 345, 437, 458, 617, 655, 684
scissors	137, 179, 245
spoons, plastic	304, 574, 753
starch, liquid laundry	753
straws	179, 367, 753
toothpicks	245, 708

Quick Lab and Inquiry Activity Materials List

Material	Quick Lab	Inquiry Activity
air freshener, solid	402	
aluminum foil	682	326
ammonia, household	604	
ammonia, 6*M* 100mL 15*M* NH_3 in 250 mL **Caution! Using proper ventilation, add ammonia to water carefully and slowly!**	746	
antacid tablets, effervescent, varied brands	428	540
baking powder		38
baking soda ($NaHCO_3$)	199, 604	38, 186, 586
balance, centigram	308, 372	286, 724
ball-and-stick molecular model kit	707	
balloons, round, different colors	372, 428	212, 412
batteries, D-cell	199	
9-volt battery	682	
beakers, 100-mL	326	62
bell wire	199	
bleach, powder	653	
borax	604	724
bowl		6
box	108	
Bunsen burner	142, 308 544	
calcium chloride ($CaCl_2 \cdot 2H_2O$)	308	
calcium chloride, 0.5*M* 13.9 g $CaCl_2$ in 250 mL	142	
calculator, scientific		798
can, 6-oz and 16-oz	226	
candle		692
carbonated beverage	604	
cardboard, thick, cut in strips	402	
cardboard tube from paper towel roll	780	
clock or watch	428	38, 540
cloth, clean white	604	
coat hanger	226	
colored fabric	653	
copper(II) chloride, 0.1*M* 3.4 g $CuCl_2$ in 250 mL	142	
copper strip		662
copper(II) sulfate ($CuSO_4 \cdot 5H_2O$)	308	
copper wire		630
cornstarch	489	6
cotton swabs	142	
cups, 8-oz plastic, clear	23, 45, 199, 402, 489, 574, 604	38, 186, 444, 540,
cup, plastic foam	522	724
dish or other shallow container	402, 448	630
dishwashing liquid	23, 448, 604	444
distilled water	199, 489	186
dropper bottles, plastic	279	
duct tape	818	
Erlenmeyer flasks, 250-mL	372	
ethanol	746	
filter paper	45, 682	
flashlight	489, 818	
flour		38
flower petals, different colors	653	
food coloring, red		62
freezer		428
fruits and vegetables		586
glass plates, square		692
glasses, drinking		384, 724
glucose solution	746	
glue, white		724
graduated cylinders, 10-mL, 100-mL, and 250-mL	372, 522	62, 384, 724
grape juice	653	
graph paper	175, 226, 428, 818	540, 798

Continues

Quick Lab and Inquiry Activity Materials List *(continued)*

Material	Quick Lab	Inquiry Activity
grass stain on piece of white fabric	653	
grocery bag, plastic	226	
heat-resistant surface	544	
hydrochloric acid, 1.0*M* 20.5 mL of 12*M* HCl in 250 mL **Caution! Always add acid to water carefully and slowly!**	372	
hydrogen peroxide, 3% H_2O_2	653	762
ice	402, 522	470, 540
index cards, 3 inch × 5 inch	72, 87	384
ingredients labels from several products		252
iodine solution, 1% I_2 in 2% KI 2.5 g I_2 plus 5.0 g KI in 250 mL	653	
iron(III) chloride, 0.1*M* 6.8 g $FeCl_3 \cdot 6H_2O$ in 250 mL	279	
jar, cylindrical with lid	489, 604	62, 798
knife	604	586
laundry detergent	604	
lead(II) nitrate, 0.2*M* 16.6 g $Pb(NO_3)_2$ in 250 mL	279	
lemon		662
lemon juice	604	
lithium chloride, 0.1*M* 1.1 g LiCl in 250 mL	142	
magnesium ribbon	372	
magnifying glass		186
marble	108	
masking tape	199, 489	38, 540
measuring cups	23, 604	38
measuring spoons	326	6, 38
meat, red (cooked and raw)		762
magnesium sulfate		186
medicine droppers	428, 604, 653,746	62, 762
metric ruler, meter stick	45, 72, 226, 604, 780, 818	100, 186, 444
milk	604	
milk of magnesia	604	
mineral oil		692
mirror		186
mouthwash	604	
nails, iron finishing	682	630
nickel coin	682	
oxalic acid solution, 1% 2.5 g in 250 mL	653	
paper clips, metal	226, 448	62, 286, 352
paper clips, vinyl coated, different colors		320, 352
paper towels		586, 630
pen	87	38, 540
pencil	45	
pennies	682, 818	798
pens, colored marking, green included	45, 780	412
petroleum jelly		630, 692
plastic foam shapes	108	
plastic wrap	45, 604	630
plate, large paper		586, 762
plate, plastic		470
pliers		126, 630
poster board, white	818	
potassium permanganate, 0.05*M* 0.40 g $KMnO_4$ in 250 mL	653	
propanone	746	

Quick Lab and Inquiry Activity Materials List *(continued)*

Material	Quick Lab	Inquiry Activity
reaction surface	682	
red cabbage leaves	604	
rubber band, No. 25 (about 2 inches long)	226, 448	504
rubbing alcohol	45	62, 692
rusty water	653	
sandpaper, fine		630, 662
sandwich bag, clear plastic		352
scissors	818	630
shampoo	604	62
silver nitrate, 5% 12.5 g $AgNO_3$ in 250 mL	746	
silver nitrate, 0.05*M* 2.1 g $AgNO_3$ in 250 mL	279	
sodium carbonate, 1.0*M* 26.5 g Na_2CO_3 in 250 mL	279	
sodium chloride (NaCl)	23, 199, 604	186, 470, 630
sodium chloride, 0.1*M* 0.6 g NaCl in 100 mL	142	
sodium hydrogen carbonate ($NaHCO_3$)	326, 489	
sodium hydroxide, 1*M* 10 g NaOH in 250 mL	746	
sodium hydroxide, 0.5*M* 5 g NaOH in 250 mL	279	
sodium hypochlorite (NaClO), household liquid bleach	653	
sodium phosphate, 0.1*M* 9.5 g $Na_3PO_4 \cdot 12H_2O$ in 250 mL	279	
sodium sulfate ($Na_2SO_4 \cdot 10H_2O$)	308, 682	
sodium thiosulfate solution, 0.2*M* 12.4 g $Na_2S_2O_3 \cdot 5H_2O$ in 250 mL	653	
spatula	308	
spoon	604	62, 186, 724
spot plate	653	
steel wool (0000 weight)	544	
stirring rod	489	
straw, drinking	23	
string		212, 470
sucrose (table sugar)	23, 199	38, 186
tape measure, metric	428	412
tape, transparent	45, 604	100, 126
tarnished silver item	326	
teaspoon	23, 489, 604	444
test tube holder	308	
test tube rack	142, 279, 308, 746	
test tubes, medium	308	
test tubes, small	142, 279, 746	
thermometer	522	540
thumbtack	780	
tissue paper	544	
tongs	142, 544	
toothpaste	604	
toothpicks, wooden	708, 780	692
vegetable oil	448	692
vinegar, white (CH_3COOH)	199, 604	
voltmeter		662
water, distilled	199	186
waxed paper		444
wintergreen mints, 3 different brands		126
wire for flame tests (10-cm length of nichrome wire)	142	
zinc strip		630, 662

Audiovisual and Technology Suppliers

Audiovisual Distributors

AIMS Multimedia
9710 DeSoto Avenue
Chatsworth, CA 91311-4409
800-328-6700

Charles W. Clark Co., Inc.
4540 Preslyn Drive
Raleigh, NC 27616-3177
800-247-7009

Chem Study
Ward's Natural Sciences Est., Inc.
5100 West Henrietta Road
West Henrietta, NY 14586-9729
800-962-2660

Churchill Films/SVE
6677 North Northwest Highway
Chicago, IL 60631-1304
800-624-1678

Coronet/MTI Films & Video
108 Wilmot Road
Deerfield, IL 60015
800-777-8100

Encyclopedia Brittanica Edu. Corp.
310 S. Michigan Ave.
Chicago, IL 60604-9839
800-621-3900

Films for the Humanities
& Science, Inc.
Box 2053
Princeton, NJ 08543-2053
800-257-5126

The Media Guild
11722 Sorrento Valley Rd., Suite E
San Diego, CA 92121
858-755-9191

National Film Board of Canada
350 Fifth Avenue, Suite 4820
New York, NY 10118
212-629-8890

National Geographic Society
Educational Services
1145 17th St. NW
Washington, DC 20036
800-627-5162

Prentice Hall
One Lake Street
Upper Saddle River, NJ 07458
800-848-9500

University of California Extension
Media Center
2000 Center St., Suite 400
Berkeley, CA 94704
510-642-4124

Equipment Distributors

Flinn Scientific
P.O. Box 219
Batavia, IL 60510
800-452-1261
E-mail: flinn@flinnsci.com
www.flinnsci.com

PASCO® scientific
P.O. Box 619011
10101 Foothills Boulevard
Roseville, CA 95661-9011
800-772-8700
www.pasco.com

Science Kit & Boreal Laboratories
P.O. Box 5003
Tonawanda, NY 14151-5003
800-828-7777
www.sciencekit.com

Technology Distributors

Cambridge Development Lab, Inc.
214 Third Avenue
Waltham, MA 02154
800-637-0047

JCE Software
Department of Chemistry
University of Wisconsin, Madison
1101 University Avenue
Madison, WI 53706-1396
800-991-5534

Falcon Software, Inc.
One Hollis St.
Wellesley, MA 02482
781-235-1767

Project Seraphim
Department of Chemistry
University of Wisconsin, Madison
1101 University Avenue
Madison, WI 53706-1396
608-263-2837

Sunburst Technology
101 Castleton Street
Pleasantville, NY 10570
800-321-7511

Texas Instruments
800-TI-CARES (800-842-2737)

Vernier Software & Technology
13979 SW Millikan Way
Beaverton, OR 97005-2886
503-277-2299
E-mail: info@vernier.com
www.vernier.com

In ordering software, remember to check your system type and system requirements, memory, drives (number and type), and color or monochrome monitor. Check with the distributor about the availability of lab packs, network versions, site licensing, individual program disks versus multiple-disk packages, and the availability of preview disks.

TEACHER'S EDITION

PRENTICE HALL

Chemistry

Wilbraham
Staley
Matta
Waterman

PRENTICE HALL
TeacherEXPRESS™
- **Plan** Prepare for class easily
- **Teach** Access resources instantly
- **Assess** Create tests quickly

ANTONY C. WILBRAHAM • DENNIS D. STALEY • MICHAEL S. MATTA • EDWARD L. WATERMAN

Boston, Massachusetts
Upper Saddle River, New Jersey

PRENTICE HALL Chemistry

PRINT COMPONENTS

Student Edition
Teacher's Edition
Progress Monitoring Assessments
Lesson Plan Book
Core Teaching Resources
Guided Reading and Study Workbook
Guided Reading and Study Workbook, ATE
Laboratory Manual
Laboratory Manual, ATE
Laboratory Practicals
Small-Scale Chemistry Laboratory Manual
Small-Scale Chemistry Laboratory Manual, ATE
Graphing Calculator Problems
Test-Taking Tips With Transparencies
Test Preparation Blackline Masters
Standardized Test Preparation Workbook
Transparencies

TECHNOLOGY COMPONENTS

Interactive Textbook w/Chem ASAP
StudentExpress w/Chem ASAP CD-ROM
TeacherEXPRESS CD-ROM
ExamView® Computer Test Bank
Virtual Chem Labs
PresentationExpress CD-ROM
Chemistry Alive Labs and Chemistry Field Trips DVD
Probeware Lab Manual
Small-Scale Chemistry Lab Video
PHSchool.com Web site

About the cover: The sample-well plate filled with colorful solutions illustrates the beauty and excitement of chemistry. Image from Getty Images, Inc.

Acknowledgments appear on page R145, which constitutes an extension of this copyright page.

ISBN 0-13-251210-6

6 7 8 9 10 12 11 10 09 08

About the Authors

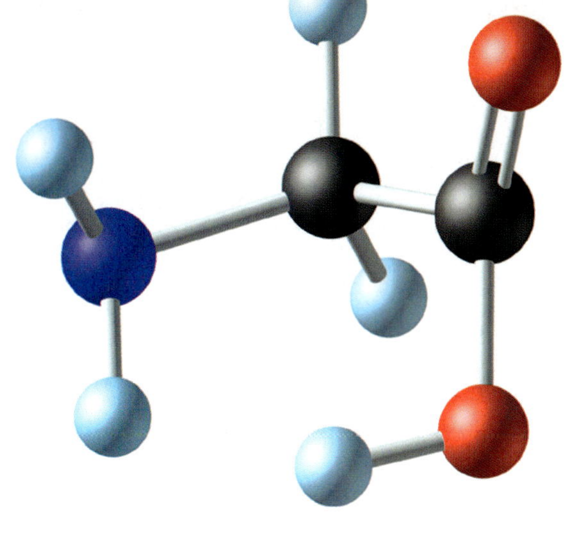

ANTONY C. WILBRAHAM

Antony Wilbraham is an Emeritus Professor of Chemistry. He taught college-level chemistry for more than 25 years at Southern Illinois University at Edwardsville, Illinois, where he was also Director of Hazardous Waste Management. In 1978, he received the University Teaching Excellence Award. Dr. Wilbraham has been writing high school and college-level chemistry textbooks and related ancillaries for over 25 years and has published extensively in science journals for over 30 years. He is a member of the American Chemical Society, a member of the National Science Teachers Association, and a Fellow of the Royal Society of Chemistry.

DENNIS D. STALEY

Dennis Staley was an Instructor in the Department of Chemistry and the Office of Science and Math Education at Southern Illinois University at Edwardsville, Illinois. He has been teaching high school and college-level chemistry for more than 25 years. In 1981 he received the University Teaching Excellence Award. Mr. Staley has been writing high school and college-level chemistry textbooks and related ancillaries for over 25 years. He is a member of the American Chemical Society and a member of the National Science Teachers Association.

MICHAEL S. MATTA

Michael Matta is an Emeritus Professor of Chemistry. He was a professor of chemistry at Southern Illinois University at Edwardsville, Illinois, from 1969 to 1996, and served as Department Chair from 1980 to 1983. In 1973, Dr. Matta received the University Teaching Excellence Award. He has been developing and writing high school and college-level chemistry textbooks and related ancillaries for over 25 years and has published extensively in scientific journals. He is a member of the American Chemical Society, the National Science Teachers Association, and the American Association for the Advancement of Science.

EDWARD L. WATERMAN

Edward Waterman has taught chemistry, advanced placement chemistry, and organic and biochemistry since 1976 at Rocky Mountain High School in Fort Collins, Colorado. Mr. Waterman conducts workshops for teachers on such topics as small-scale chemistry labs, advanced placement chemistry, block scheduling, Internet for the chemistry classroom, and virtual chemistry on CD-ROM. Mr. Waterman holds a Bachelor of Science degree in chemistry from Montana State University and a Master of Science degree in chemistry from Colorado State University.

Reviewers

Safety Consultant

Dr. Kenneth R. Roy
Director of Science and Safety
Glastonbury Public Schools
Glastonbury, Connecticut

Content Reviewers

J. Phillip Bowen
Department of Chemistry and Biochemistry
University of North Carolina
Greensboro, North Carolina

Martha D. Bruch
Department of Chemistry
SUNY – Oswego
Oswego, New York

Pamela J. Delaney
Department of Chemistry and Biochemistry
Montclair State University
Upper Montclair, New Jersey

Alison J. Frontier
Department of Chemistry
University of Rochester
Rochester, New York

Patricia L. Lang
Department of Chemistry
Ball State University
Muncie, Indiana

Chi H. Mak
Department of Chemistry
University of Southern California
Los Angeles, California

M. Uljana Mayer, Ph.D.
Pacific Northwest National Lab
Richland, Washington

David J. Merkler
Department of Chemistry
University of South Florida
Tampa, Florida

Jeffery W. Seyler
Department of Chemistry
University of Southern Indiana
Evansville, Indiana

Harry A. Stern
Cornell University
Ithaca, New York

Carey E. Stronach, Ph.D.
Department of Physics
Virginia State University
Petersburg, Virginia

Aaron W. Todd, Ph.D.
Professor of Chemistry, Emeritus
Middle Tennessee State University
Murfreesboro, Tennessee

Mark E. Welker
Department of Chemistry
Wake Forest University
Winston-Salem, North Carolina

Teacher Reviewers

Paulette Berger
Zionsville Community High School
Zionsville, Indiana

Tonia Bonner
Village Christian Academy
Fayetteville, North Carolina

Malcolm S. Cheney
Supervisor, Mathematics and Science
Wethersfield Public Schools
Wethersfield, Connecticut

William G. Clarke
Asst. Principal, Science Supervision
Franklin D. Roosevelt High School
Brooklyn, New York

Jeanne M. Dyer
Sickles High School
Tampa, Florida

Amy A. Fisher
R. Nelson Snider High School
Fort Wayne, Indiana

Stephen E. Gerofsky
Lake Weir High School
Ocala, Florida

Diane W. Johnson
Hillside High School
Durham, North Carolina

Rod Larsen
West Orange High School
Winter Garden, Florida

Scott McCord
Astronaut High School
Titusville, Florida

Walter McDonald
Churchland High School
Portsmouth, Virginia

Dr. Mary L. Moller
Science Dept. Chair Emerita
The Chapin School
New York, New York

Dr. Kenneth A. Nealy
Science Curriculum Leader
Windsor Public Schools
Windsor, Connecticut

Aimee Pulido
La Joya Senior High School
La Joya, Texas

Anne Purvis
Arapahoe High School
Littleton, Colorado

Michael Richardson
Seymour High School
Seymour, Indiana

Robert W. Rollings
Floyd Central High School
Floyds Knobs, Indiana

Martha Gwen Sibert
Roanoke Valley Governor's School for Science and Technology
Roanoke, Virginia

Rhonda W. Smith
Enloe High School
Raleigh, North Carolina

Nora C. Ugalde
Jupiter High School
Jupiter, Florida

Jan Carroll Weir
Science Dept. Chairman
Lawrence Central High School
Indianapolis, Indiana

Yvonne Whatley
Science Department Head
Hudson High
Hudson, Florida

Thomas C. Zebovitz
Walla Walla Community College
Walla Walla, Washington

Contents

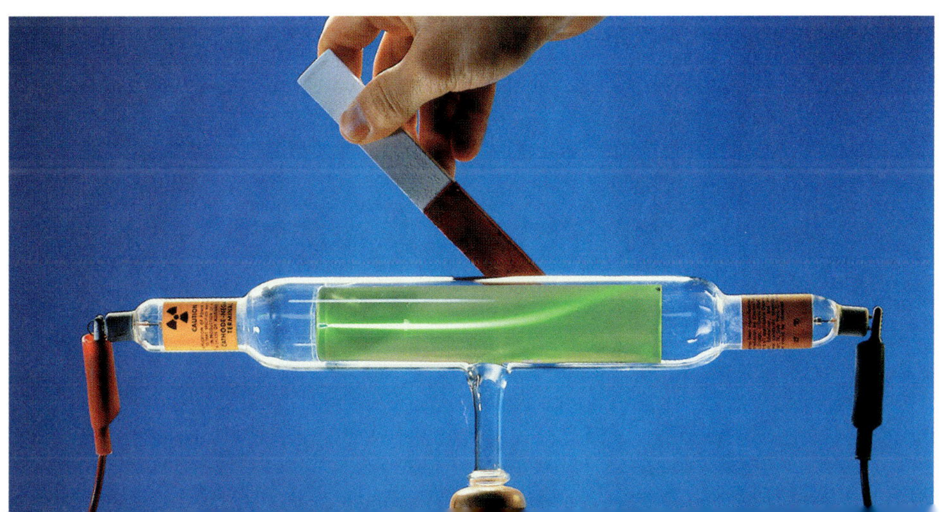

CHAPTER 9

CHEMICAL NAMES AND FORMULAS 252

CHAPTER 10

CHEMICAL QUANTITIES 286

CHAPTER 11

CHEMICAL REACTIONS 320

CHAPTER 12

STOICHIOMETRY 352

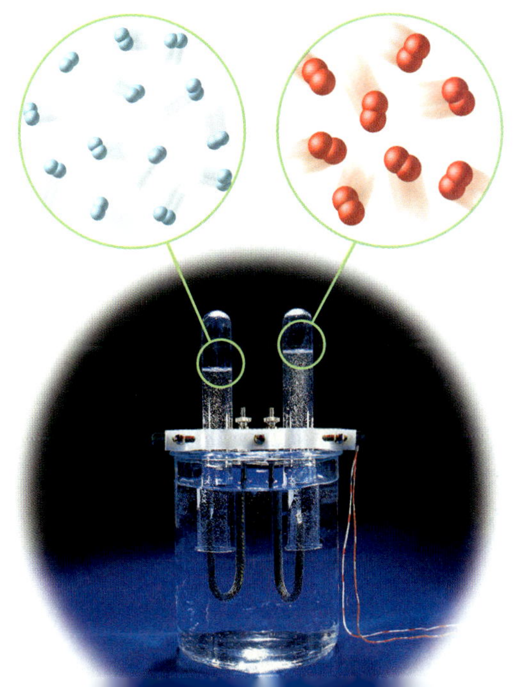

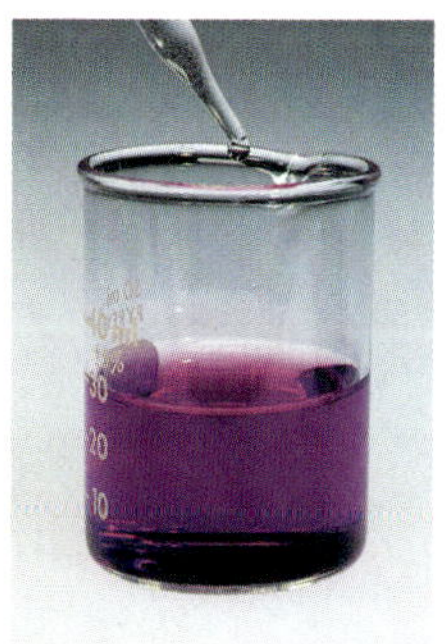

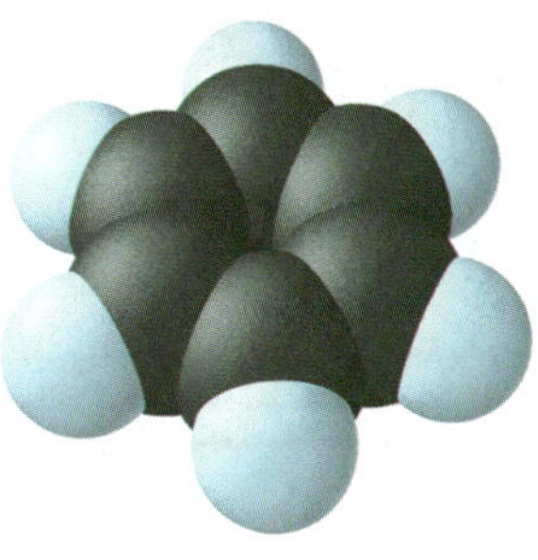

CHAPTER 21

ELECTROCHEMISTRY 662

CHAPTER 22

HYDROCARBON COMPOUNDS 692

CHAPTER 23

FUNCTIONAL GROUPS 724

CHAPTER 24

THE CHEMISTRY OF LIFE 762

CHAPTER 25

NUCLEAR CHEMISTRY 798

Reference Section

Integrated Technology

- Complete textbook and key art and photos online and on CD-ROM
- Online textbook powered by tried-and-true teacher resource and student tool, ChemASAP
- **Animations** illustrate important concepts
- **Simulations** allow students to manipulate variables, observe results, and draw conclusions
- **Assessment** provides over 500 unique test and quiz items
- **Problem-solving** tutorials provide detailed, step-by-step solutions to over 100 practice problems

- Referenced at point-of-use in the Student Edition, these Web links are maintained and updated by the National Science Teachers Association—with accompanying worksheets

- Goes hand-in-hand with the Student Edition for additional support—student activities, chapter assessment, teacher support, and many more resources—with every chapter

- Inquiry-based lab simulation environment allows students and teachers to perform any chemistry lab accurately and efficiently, without costly equipment and cleanup
- 30 pre-set labs correspond to Small-Scale Labs in the Student Edition and labs from the Laboratory Manual
- Virtual ChemLab environment models proper procedures and safety and allows students to better observe and understand events at the atomic and molecular level

Small-Scale LABS

Readily available materials with easy setup to produce immediate, obvious lab results.

 Probe or sensor versions available in the Probeware Lab Manual.

Quick LABS

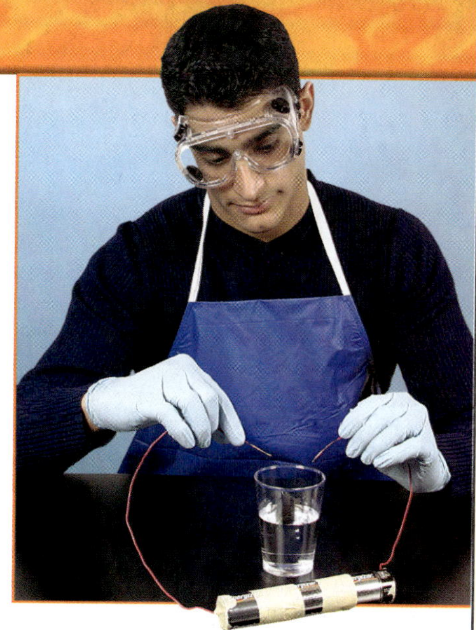

Apply chemistry concepts and skills with these quick, effective hands-on opportunities.

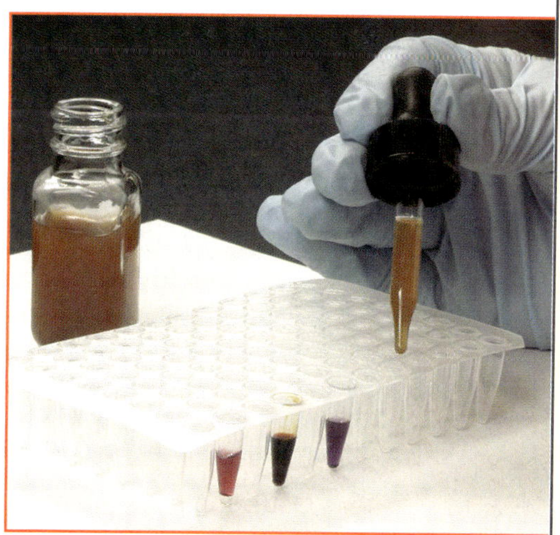

Probe or sensor versions available in the Probeware Lab Manual.

INQUIRY Activities

Begin each chapter with an activity that sets the purpose for reading.

INTERPRETING GRAPHS

Organize and interpret data while building critical thinking skills.

Atomic Radius Versus Atomic Number

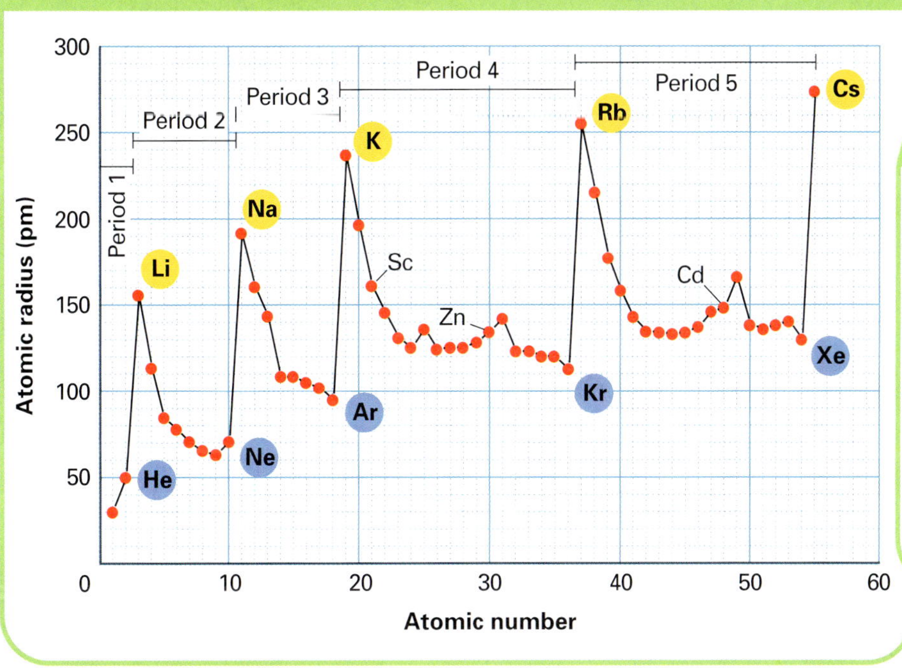

INTERPRETING GRAPHS

a. Analyzing Data Which alkali metal has an atomic radius of 238 pm?

b. Drawing Conclusions Based on the data for alkali metals and noble gases, how does atomic size change within a group?

c. Predicting Is an atom of barium, atomic number 56, smaller or larger than an atom of cesium (Cs)?

CHEMath

Reinforces math concepts needed for problem solving with these point-of-use refreshers.

CHEMath

Logarithms

The common logarithm (log) of a number (N) is the exponent (*x*) to which the base 10 must be raised to yield the number. If $N = 10^x$, then $\log N = x$.

The use of logarithms allows a large range of values to be conveniently expressed as small, nonexponential numbers. For example, $\log(10^3) = 3$ and $\log(10^6) = 6$. Although there is a range of three orders of magnitude ($10 \times 10 \times 10$ or 1000) between the numbers 10^3 and 10^6, the range of the log values is only 3.

The concentration of hydrogen ions in most aqueous solutions, although small, can vary over many orders of magnitude. Scientists use pH, a logarithmic scale that ranges between 0 and 14, to more conveniently express the hydrogen ion concentration of a solution.

Math Handbook

For help with logarithms, go to page R78.

CHEMath

Conversion Problems

A conversion factor is a ratio of two quantities equal to one another. When working conversion problems, the conversion factors should be written so that the unit of a given measurement cancels, leaving the correct unit for your answer. Note that the equalities needed to write a particular conversion factor may be given in the problem; in other cases, you will need to know or look up the necessary equalities.

Math Handbook

For help with conversion problems, go to page R66.

CHEMath

Significant Figures

The significant figures in a measurement are all the digits known with certainty plus one estimated digit. The number of significant figures in the measurements used in a calculation determines how you round the answer.

When multiplying and dividing measurements, the rounded answer can have no more significant figures than the least number of significant figures in any measurement in the calculation.

The product of $3.6\ m \times 2.48\ m = 8.928\ m^2$ is rounded to $8.9\ m^2$ (2 significant figures).

When you add and subtract measurements, the answer can have no more decimal places than the least number of decimal places in any measurement in the problem. The difference of $8.68\ cm - 2.1\ cm = 6.58\ cm$ is rounded to 6.6 cm (one decimal place).

Math Handbook

For help with significant figures, go to page R59.

CONCEPTUAL PROBLEMS

Problem-solving through applications that relate to students' experiences.

Technology & Society

Relates content to current technological advances.

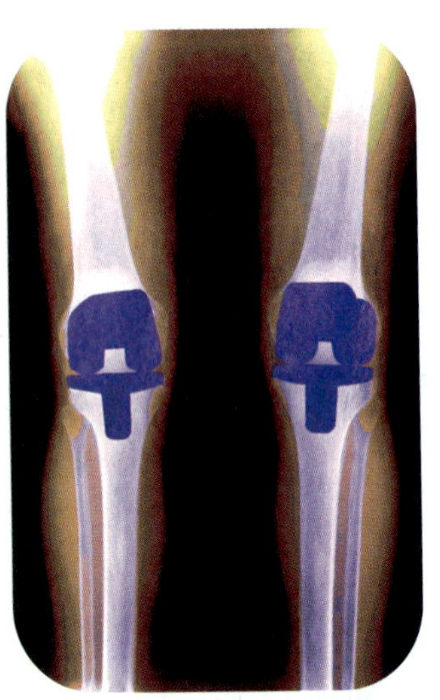

Careers *in* Chemistry

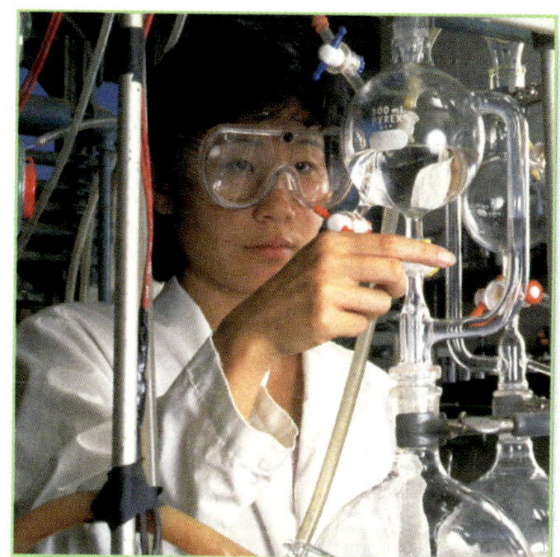

Career opportunities that require an understanding of chemistry.

Use This Book for Success

Read for Meaning

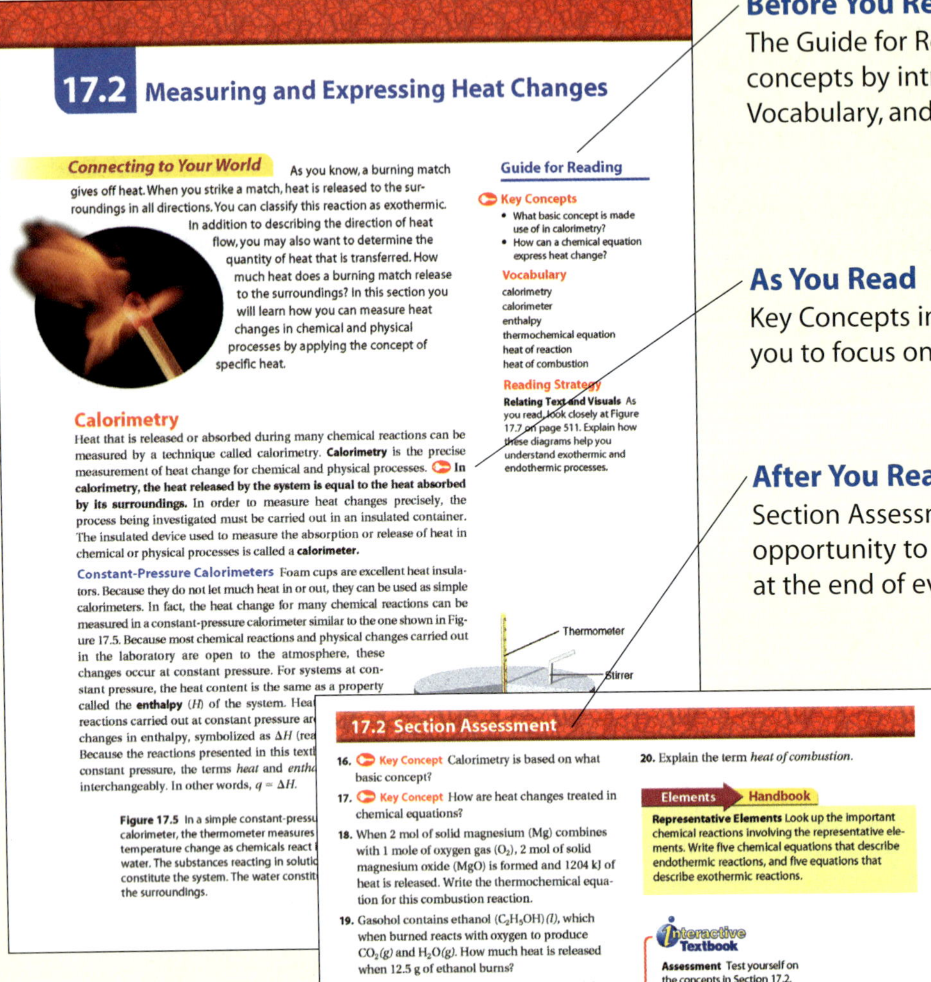

17.2 Measuring and Expressing Heat Changes

Connecting to Your World As you know, a burning match gives off heat. When you strike a match, heat is released to the surroundings in all directions. You can classify this reaction as exothermic. In addition to describing the direction of heat flow, you may also want to determine the quantity of heat that is transferred. How much heat does a burning match release to the surroundings? In this section you will learn how you can measure heat changes in chemical and physical processes by applying the concept of specific heat.

Guide for Reading

Key Concepts
- What basic concept is made use of in calorimetry?
- How can a chemical equation express heat change?

Vocabulary
calorimetry
calorimeter
enthalpy
thermochemical equation
heat of reaction
heat of combustion

Reading Strategy
Relating Text and Visuals As you read, look closely at Figure 17.7 on page 511. Explain how these diagrams help you understand exothermic and endothermic processes.

Calorimetry

Heat that is released or absorbed during many chemical reactions can be measured by a technique called calorimetry. **Calorimetry** is the precise measurement of heat change for chemical and physical processes. **In calorimetry, the heat released by the system is equal to the heat absorbed by its surroundings.** In order to measure heat changes precisely, the process being investigated must be carried out in an insulated container. The insulated device used to measure the absorption or release of heat in chemical or physical processes is called a **calorimeter.**

Constant-Pressure Calorimeters Foam cups are excellent heat insulators. Because they do not let much heat in or out, they can be used as simple calorimeters. In fact, the heat change for many chemical reactions can be measured in a constant-pressure calorimeter similar to the one shown in Figure 17.5. Because most chemical reactions and physical changes carried out in the laboratory are open to the atmosphere, these changes occur at constant pressure. For systems at constant pressure, the heat content is the same as a property called the **enthalpy** (H) of the system. Hea[...] reactions carried out at constant pressure ar[...] changes in enthalpy, symbolized as ΔH (rea[...] Because the reactions presented in this text[...] constant pressure, the terms *heat* and *enth*[...] interchangeably. In other words, $q = \Delta H$.

Thermometer
Stirrer

Figure 17.5 In a simple constant-pressu[...] calorimeter, the thermometer measures [...] temperature change as chemicals react [...] water. The substances reacting in solutio[...] constitute the system. The water constit[...] the surroundings.

17.2 Section Assessment

16. Key Concept Calorimetry is based on what basic concept?
17. Key Concept How are heat changes treated in chemical equations?
18. When 2 mol of solid magnesium (Mg) combines with 1 mole of oxygen gas (O_2), 2 mol of solid magnesium oxide (MgO) is formed and 1204 kJ of heat is released. Write the thermochemical equation for this combustion reaction.
19. Gasohol contains ethanol (C_2H_5OH)(l), which when burned reacts with oxygen to produce $CO_2(g)$ and $H_2O(g)$. How much heat is released when 12.5 g of ethanol burns?
$$C_2H_5OH(l) + 3O_2(g) \longrightarrow 2CO_2(g) + 3H_2O(g)$$
$$\Delta H = -1235 \text{ kJ}$$
20. Explain the term *heat of combustion.*

Elements Handbook

Representative Elements Look up the important chemical reactions involving the representative elements. Write five chemical equations that describe endothermic reactions, and five equations that describe exothermic reactions.

Interactive Textbook

Assessment Test yourself on the concepts in Section 17.2.

with ChemASAP

Before You Read

The Guide for Reading helps you master concepts by introducing Key Concepts, Vocabulary, and a Reading Strategy.

As You Read

Key Concepts in boldface sentences allow you to focus on the big ideas of chemistry.

After You Read

Section Assessment gives you the opportunity to review Key Concepts at the end of every lesson.

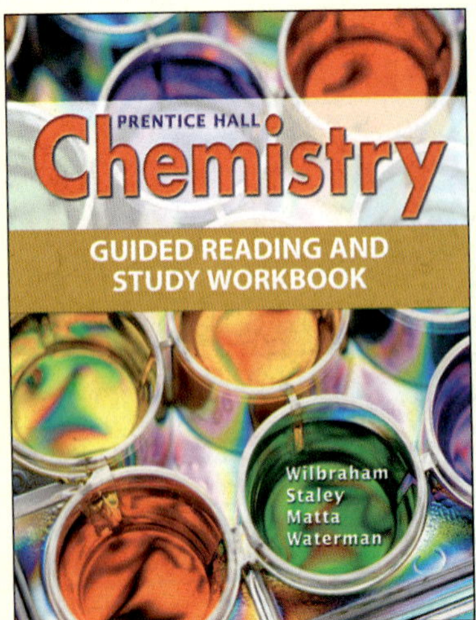

Check Your Understanding

The **Guided Reading and Study Workbook** gives you extra support with vocabulary, reading and math skills, and graphics to enhance your understanding.

Practice Your Math Skills

SAMPLE PROBLEM 17.4

Using the Heat of Reaction to Calculate Heat Change

Using the thermochemical equation in Figure 17.7b on page 511, calculate the amount of heat (in kJ) required to decompose 2.24 mol $NaHCO_3(s)$.

1 Analyze ***List the knowns and the unknown.***

Knowns
- 2.24 mol $NaHCO_3(s)$ decomposes
- $\Delta H = 129$ kJ

Unknown
- $\Delta H = ?$ kJ

Use the thermochemical equation,

$$2NaHCO_3(s) + 129\text{ kJ} \longrightarrow Na_2CO_3(s) + H_2O(g) + CO_2(g),$$

to write a conversion factor relating kilojoules of heat and moles of $NaHCO_3$. Then use the conversion factor to determine ΔH for 2.24 mol $NaHCO_3$.

2 Calculate ***Solve for the unknown.***

The thermochemical equation indicates that 129 kJ are needed to decompose 2 mol $NaHCO_3(s)$. Use this relationship to write the following conversion factor.

$$\frac{129\text{ kJ}}{2\text{ mol } NaHCO_3(s)}$$

Using dimensional analysis, calculate ΔH.

$$\Delta H = 2.24\ \cancel{\text{mol } NaHCO_3(s)} \times \frac{129\text{ kJ}}{2\ \cancel{\text{mol } NaHCO_3(s)}} = 144\text{ kJ}$$

3 Evaluate ***Does the result make sense?***

Because $\Delta H = 129$ kJ refers to the decomposition of 2 mol $NaHCO_3(s)$, the decomposition of 2.24 mol should absorb about 10% more heat than 129 kJ, or slightly more than 142 kJ. The answer of 144 kJ is consistent with this estimate.

Practice Problems

14. When carbon disulfide is formed from its elements, heat is absorbed. Calculate the amount of heat (in kJ) absorbed when 5.66 g of carbon disulfide is formed.

$$C(s) + 2S(s) \longrightarrow CS_2(l) \quad \Delta H = 89.3\text{ kJ}$$

15. The production of iron and carbon dioxide from iron(III) oxide and carbon monoxide is an exothermic reaction. How many kilojoules of heat are produced when 3.40 mol Fe_2O_3 reacts with an excess of CO?

$$Fe_2O_3(s) + 3CO(g) \longrightarrow 2Fe(s) + 3CO_2(g) + 26.3\text{ kJ}$$

CHEMath

Conversion Factors

A conversion factor is a ratio of two quantities that are equal to one another. When doing conversions, write the conversion factors so that the unit of a given measurement cancels, leaving the correct unit for your answer. Note that the equalities needed to write a particular conversion factor may be given in the problem. In other cases, you will need to know or look up the necessary equalities.

Math Handbook

For help with conversion factors go to page 905.

Problem Solving
Solve Problem 15 with the help of an interactive guided tutorial.
with **ChemASAP**

Practice for Test-Taking Success

A variety of problems, utilizing a three-step problem-solving approach, helps you understand chemistry concepts while practicing vital math skills.

Support Where You Need It

You'll find additional help nearby with the **ChemMath** refreshers in the margins and the **Math Handbook** in the Appendix.

Math Support Online

The **Interactive Textbook with ChemASAP** contains practice and problem-solving tutorials to give you extra math reinforcement.

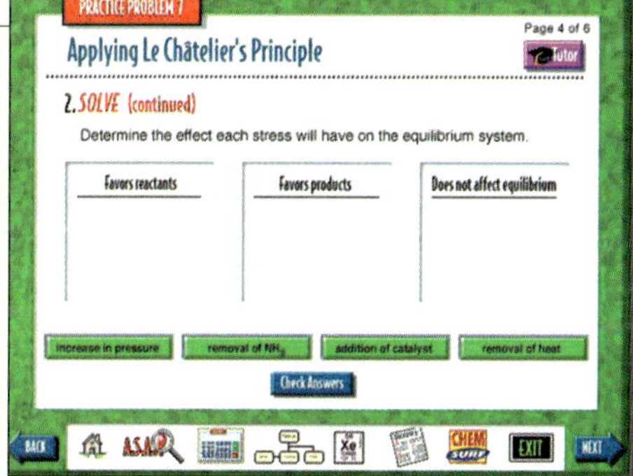

CONCEPTUAL PROBLEM 17.1

Recognizing Exothermic and Endothermic Processes

On a sunny winter day, the snow on a rooftop begins to melt. As the meltwater drips from the roof, it refreezes into icicles. Describe the direction of heat flow as the water freezes. Is this process endothermic or exothermic?

1 Analyze ***Identify the relevant concepts.***

Heat always flows from a warmer object to a cooler object. An endothermic process absorbs heat from the surroundings. An exothermic process releases heat to the surroundings.

2 Solve ***Apply concepts to this situation.***

Define the liquid water as the system and the air as the surroundings. Heat flows from the (warmer) liquid water to the (cooler) air. Because heat is released from the system to the surroundings, this process is exothermic.

Practice Problems

1. A container of melted paraffin wax is allowed to stand at room temperature until the wax solidifies. What is the direction of heat flow as the wax solidifies? Is the process exothermic or endothermic?

2. When solid barium hydroxide octahydrate ($Ba(OH)2{\bullet}8H_2O$) is mixed in a beaker with solid ammonium thiocyanate (NH_4SCN), a reaction occurs. The beaker quickly becomes very cold. Is the reaction exothermic or endothermic?

Two-Step Problem Solving

Conceptual Problems encourage you to solve problems using critical thinking and analytical skills.

Prepare for Tests

CHAPTER 17 **Assessment**

Reviewing Content

17.1 The Flow of Energy—Heat

38. Explain in your own words the law of conservation of energy.

39. What always happens when two objects of different temperatures come in contact? Give an example from your own experience.

40. Define potential energy in terms of chemistry.

41. What factors determine heat capacity?

42. What is the relationship between a calorie and a Calorie?

43. Make the following conversions.
a. 8.50×10^2 cal to Calories
b. 444 cal to joules
c. 1.8 kJ to joules
d. 4.5×10^{-1} kJ to calories

44. Why do you think it is important to define the system and the surroundings?

45. Describe the sign convention that is used in thermochemical calculations.

46. Two substances in a glass beaker chemically react, and the beaker becomes too hot to touch.
a. Is the reaction exothermic or endothermic?
b. If the two substances are defined as the system, what constitutes the surroundings?

47. Classify these processes as exothermic or endothermic.
a. condensing steam **b.** evaporating alcohol
c. burning alcohol **d.** baking a potato

17.2 Measuring and Expressing Heat Changes

48. What name is given to a heat change at constant pressure?

49. What is the function of a calorimeter?

50. There are some obvious sources of error in experiments that use foam cups as calorimeters. Name at least three.

51. What device would you use to measure the heat released at constant volume?

52. Give the standard conditions for heat of combustion.

53. What information is given in a thermochemical equation?

17.3 Heat in Changes of State

54. Explain why ice melts at 0°C without an increase of temperature, even though heat flows from the surroundings to the system (the ice).

55. Calculate the quantity of heat gained or lost in the following changes.
a. 3.50 mol of water freezes at 0°C.
b. 0.44 mol of steam condenses at 100°C.
c. 1.25 mol NaOH(*s*) dissolves in water.
d. 0.15 mol $C_2H_5OH(l)$ vaporizes at 78.3°C.

56. Sodium acetate dissolves readily in water according to the following equation.

$$NaC_2H_3O_2(s) \longrightarrow NaC_2H_3O_2(aq)$$
$$\Delta H = -17.3 \text{ kJ/mol}$$

Would this process increase or decrease the temperature of the water?

17.4 Calculating Heat Reaction

57. Explain Hess's law of heat summation.

58. A considerable amount of heat is required for the decompo...

$2Al_2O$...

a. What is the ... 1 mol of al...
b. Is the react...

59. Calculate the ... lead(IV) chlo... chloride with...

$PbCl_2$(...

Use the follo...

Pb(*s*)...

Pb(*s*...

60. What is the s... element in its...

61. Consider the ... value of ΔH_f... Is this statem...

Check Understanding at Every Step

You'll have many opportunities to assess your understanding with Checkpoint questions and at the end of every section and chapter.

temperature. As you can see from the equation, specific heat may be expressed in terms of joules or calories. Therefore, the units of specific heat are either J/(g × °C) or cal/(g × °C).

What is the relationship between specific heat and heat capacity?

Standardized Test Prep

Test-Taking Tip

Eliminate Wrong Answers If you don't know which response is correct, start by eliminating those you know are wrong. If you can rule out some choices, you'll have fewer left to consider and you'll increase your chances of choosing the correct answer.

Select the choice that best answers each question or completes each statement.

1. The ΔH_{fus} of ethanol (C_2H_5OH) is 4.60 kJ/mol. How many kilojoules are required to melt 24.5 g of ethanol at its freezing point?
a. 2.45 kJ **b.** 5.33 kJ
c. 245 kJ **d.** 18.8 kJ

2. How much heat, in kilojoules, must be added to 178 g of water to increase the temperature of the water by 5.0°C?
a. 890 kJ **b.** 36 kJ
c. 3.7 kJ **d.** 0.093 kJ

...ormation of a free ...d state is always
b. positive.
d. higher for solids than gases.

...$HgO(s) \longrightarrow 2Hg(l) +$ ...
...ΔH for the reaction ...$O(s)$ is
b. −90.83 kJ.
d. −181.66 kJ.

...city of grain alcohol is ten ...pecific heat capacity of ...er with a mass of 55 g is ... mass of cool alcohol. If ... silver bar drops 45 °C, ... alcohol
b. decreases 4.5°C.
d. decreases 45°C.

...rine gas react to form ... Calculate the heat ...onversion of 15.0 g of ...tant pressure.

$H_2(g) + F_2(g) \rightarrow 2HF(g)$

The lettered choices below refer to Questions 7–10. A lettered choice may be used once, more than once, or not at all.

(A) kJ/mol
(B) (J × °C)/g
(C) J/(g × °C)
(D) kJ
(E) kJ/°C

Which unit is appropriate for each of the following measurements?

7. heat of reaction
8. heat capacity
9. molar heat of fusion
10. specific heat capacity

Use the graph and table to answer Questions 11–14. Assume 1.00 mol of substance in each container.

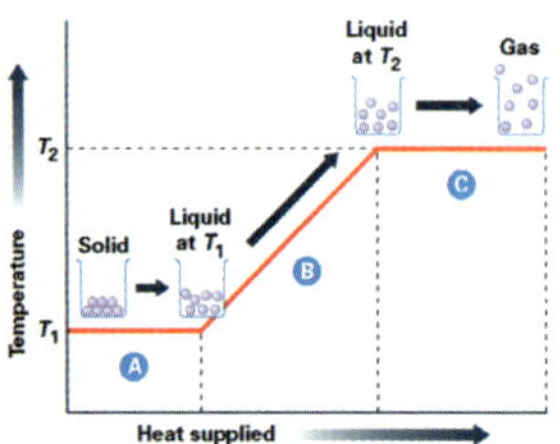

Substance	Freezing point (K)	ΔH_{fus} (kJ/mol)	Boiling point (K)	ΔH_{vap} (kJ/mol)
Ammonia	195.3	5.65	239.7	23.4
Benzene	278.7	9.87	353.3	30.8
Methanol	175.5	3.16	337.2	35.3
Neon	24.5	0.33	27.1	1.76

11. Calculate heat absorbed in region A for neon.
12. Calculate heat absorbed in region C for benzene.
13. Calculate heat absorbed in regions B and C for methanol. [specific heat = 81.6 J/(g × °C)]
14. Calculate heat absorbed in regions A, B, and C ...

17.2 Section Assessment

16. Key Concept Calorimetry is based on what basic concept?

17. Key Concept How are heat changes treated in chemical equations?

18. When 2 mol of solid magnesium (Mg) combines with 1 mole of oxygen gas (O_2), 2 mol of solid magnesium oxide (MgO) is formed and 1204 kJ of heat is released. Write the thermochemical equation for this combustion reaction.

19. Gasohol contains ethanol (C_2H_5OH)(*l*), which when burned reacts with oxygen to produce $CO_2(g)$ and $H_2O(g)$. How much heat is released when 12.5 g of ethanol burns?

$$C_2H_5OH(l) + 3O_2(g) \longrightarrow 2CO_2(g) + 3H_2O(g)$$
$$\Delta H = -1235 \text{ kJ}$$

20. Explain the term *heat of combustion.*

Elements Handbook

Representative Elements Look up the important chemical reactions involving the representative elements. Write five chemical equations that describe endothermic reactions, and five equations that describe exothermic reactions.

Interactive Textbook

Assessment Test yourself on the concepts in Section 17.2. with ChemASAP

Interpreting Graphs

Practice analyzing data in charts and graphs—important skills you'll need to succeed on tests.

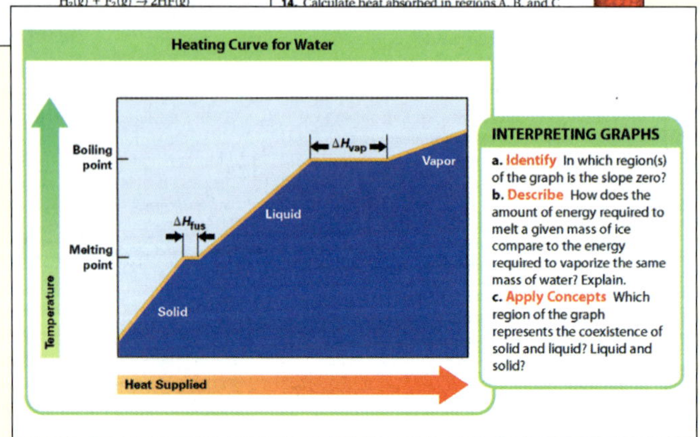

INTERPRETING GRAPHS

a. Identify In which region(s) of the graph is the slope zero?
b. Describe How does the amount of energy required to melt a given mass of ice compare to the energy required to vaporize the same mass of water? Explain.
c. Apply Concepts Which region of the graph represents the coexistence of solid and liquid? Liquid and solid?

Use the Illustrations and Photos

Figure 17.2 Heat flow is defined from the point of view of the system. **a** In an endothermic process, heat flows into the system from the surroundings. **b** In an exothermic process, heat flows from the system to the surroundings. In both cases, energy is conserved. **Interpreting Diagrams** ***In which process does q have a negative value?***

Art and Text Work Together

As you read the text, you'll find helpful photos and illustrations that make concepts more easily understood. Diagrams simplify complex ideas and demonstrate abstract processes.

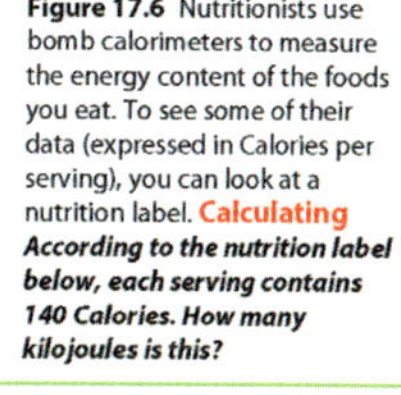

Figure 17.6 Nutritionists use bomb calorimeters to measure the energy content of the foods you eat. To see some of their data (expressed in Calories per serving), you can look at a nutrition label. **Calculating** ***According to the nutrition label below, each serving contains 140 Calories. How many kilojoules is this?***

Nutrition Facts
Serving Size 1 Cookie (26g / 0.9oz.)
Servings Per Container 8

Amount Per Serving	
Calories 140	Calories from Fat 70
	% Daily Value*
Total Fat 8 g	**12 %**
Saturated Fat 2.5 g	**13 %**
Polyunsaturated Fat 0 g	
Monounsaturated Fat 3 g	
Cholesterol 10 mg	**3 %**
Sodium 80 mg	**3 %**
Total Carbohydrate 16 g	**5 %**
Dietary Fiber 0 g	0 %

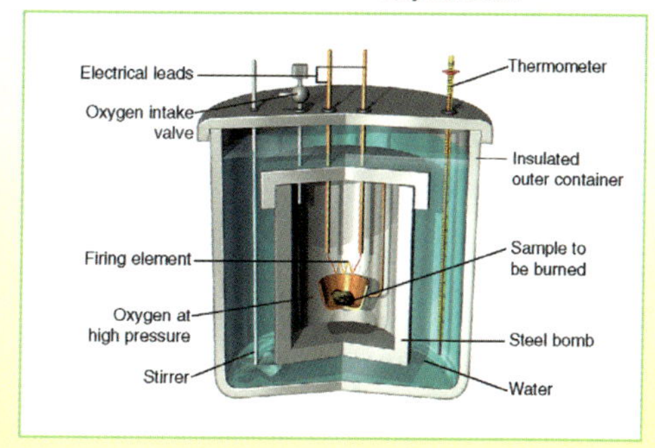

Explore the Technology at Your Fingertips

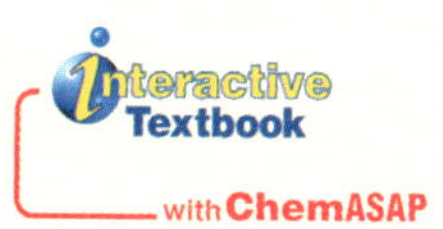

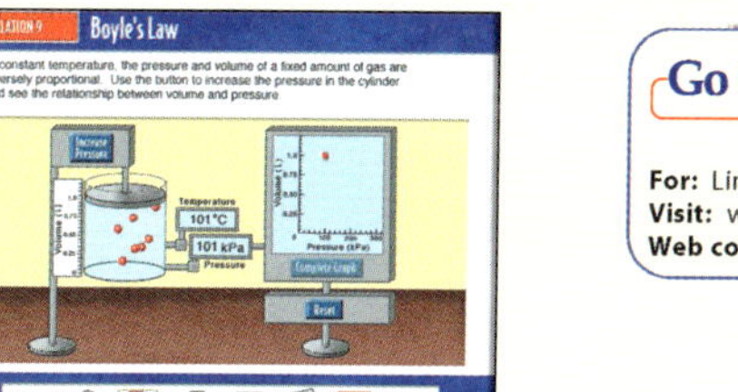

Go Interactive!

Explore chemistry online or on CD-ROM with this truly interactive textbook. **ChemASAP** brings you Animations, Simulations, Assessment, and Problem Solving Tutorials to enrich your studies.

Go Online
NSTA SCiLINKS
For: Links on specific heat
Visit: www.SciLinks.org
Web code: cdn-1171

A World of Resources

Go Online Web links bring you quick access to a variety of chemistry topics. Continuously, updated with current information, these sites show you how chemistry impacts your world.

A Simulated Lab Environment

These hands-on lab activities let you perform exciting experiments without the worry of equipment or clean-up.

CHAPTER 1

Introduction to Chemistry Planning Guide

SECTION OBJECTIVES	STANDARDS NATIONAL	STANDARDS STATE	ACTIVITIES and LABS
1.1 Chemistry, pp. 7–11 1 block or 2 periods **1.1.1 Identify** five traditional areas of study in chemistry. **1.1.2 Relate** pure chemistry to applied chemistry. **1.1.3 Identify** reasons to study chemistry.	A-2, B-3, E-2, F-2, F-4, F-5, F-6, G-1, G-2, G-3		**SE** Inquiry Activity: Solid or Liquid? p. 6 L2 **TE** Teacher Demo: Explaining the Natural World, p. 10 L1 **TE** Class Activity: Chemistry in the News, p. 11 L2
1.2 Chemistry Far and Wide, pp. 12–19 1 block or 2 periods **1.2.1 Identify** some areas affected by chemistry research. **1.2.2 Describe** some examples of research in chemistry. **1.2.3 Distinguish** between macroscopic and microscopic views.	A-2, C-1, C-2, E-2, F-1, F-2, F-3, F-4, F-6, G-3		**TE** Class Activity: Biodiesel, p. 13 L2 **TE** Class Activity: Studying Gasoline Additives, p. 16 L2 **TE** Class Activity: Composition of the Sun, p. 17 L2
1.3 Thinking Like a Scientist, pp. 20–27 1 block or 2 periods **1.3.1 Describe** how Lavoisier transformed chemistry. **1.3.2 Identify** three steps in the scientific method. **1.3.3 Explain** why collaboration and communication are important in science.	A-1, A-2, E-2, G-1, G-2, G-3		**SE** Quick Lab: Bubbles! p. 23 L2 **SE** Small-Scale Lab: Laboratory Safety, pp. 26–27 L2 **TE** Class Activity: Invisible Ink, p. 21 L2 **TE** Class Activity: Lorenzo's Oil, p. 22 L2 **TE** Class Activity: Researching Collaborative Science Projects, p. 24 L2 **LM** Lab 1: Observing and Inferring L2 **SSLM** Lab 1: Making Observations of Matter L2
1.4 Problem Solving in Chemistry, pp. 28–32 1 block or 2 periods **1.4.1 Identify** two general steps in problem solving. **1.4.2 Describe** three steps for solving numeric problems. **1.4.3 Describe** two steps for solving conceptual problems.	A-1, A-2		**TE** Teacher Demo: Fit an Ice Cube in a Soda Bottle, p. 31 L2

Ability Levels

- L1 For students who need additional help
- L2 For all students
- L3 For students who need to be challenged

Components

SE	Student Edition	**LM**	Laboratory Manual	**GCP**	Graphing Calculator Problems	**T**	Transparencies
TE	Teacher's Edition	**SSLM**	Small-Scale Chemistry Lab Manual	**TTT**	Test-Taking Tips with Transparencies, Grades 9–12 Science	**CA**	Chem Alive and Chemistry Field Trips DVD
GRSW	Guided Reading and Study Workbook	**LP**	Laboratory Practicals	**STP**	Standardized Test Preparation Workbook	**iT**	Interactive Textbook with ChemASAP
CTR	Core Teaching Resources	**PLM**	Probeware Lab Manual			**GO**	Go Online Internet Resources
SSG	Spanish Study Guide	**VCL**	Virtual Chemistry Labs				
		CTB	Computer Test Bank				

RESOURCES PRINT and TECHNOLOGY / PROBLEMS and ASSESSMENT

GRSW Chapter 1, Section 1 L1
T T1–T2 L2
GO Chapter 1, Section 1  L2

SE Section 1.1 Assessment, p. 11
iT Section 1.1

GRSW Chapter 1, Section 2 L1
 T T3–T4  L2

SE Section 1.2 Assessment, p. 17
iT Section 1.2

GRSW Chapter 1, Section 3  L1
 T T5–T6  L2
 GO Chapter 1, Section 3  L2

SE Section 1.3 Assessment, p. 25
iT Section 1.3

GRSW Chapter 1, Section 4  L1
T T7–T9  L2

CP 1.1, Running Errands
SP 1.1, Estimating Walking Time
SE Section 1.4 Assessment, p. 32
 iT Section 1.4

Go Online

Web Code: cdn-1011
Web Code: cdn-1012

MATERIALS List

Quantities for each group

STUDENT EDITION

Inquiry Activity, p. 6
tablespoon, 3 tablespoons cornstarch, small bowl, water

Quick Lab, p. 23
3 plastic drinking cups, liquid dish detergent, measuring cup and spoons, water, table sugar, table salt, drinking straw

TEACHER'S EDITION

Teacher Demo, p. 10
slides that show a wide variety of events or structures, projector

Class Activity, p. 11
current national and local newspapers, scissors

Class Activity, p. 21
250 mg of phenolphthalein, 250 mL of 70% 2–propanol, small artist's paintbrush, sheets of copy paper or white butcher paper, household glass cleaner containing ammonia

Class Activity, p. 22
copy of the movie *Lorenzo's Oil*

Teacher Demo, p. 31
ice cubes, bowl, empty soda bottle

Chapter Assessment

CHAPTER ASSESSMENT
SE Chapter Assessment, pp. 34–36
CTB Chapter 1
iT Chapter 1

STANDARDIZED TEST PREP
SE Chapter 1, p. 37
iT Interactive Textbook with Assessment at PHSchool.com

CHAPTER 1

Introduction to Chemistry

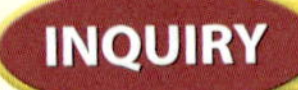

Solid or Liquid?

L2

Purpose

Students make systematic observations of a mixture.

Prep Time 5 minutes
Class Time 10 minutes

Safety Remind students to always wear safety goggles when working in the lab and to wash their hands thoroughly with soap or detergent before leaving the lab.

Think About It

1. It is likely that both the finger and fist reached the bottom of the bowl, but the fist met more resistance.
2. When the mixture is jabbed, it feels solid. Neither the finger nor the fist reached the bottom of the bowl.
3. When a ball of the mixture is squeezed, the ball behaves like a solid. When the pressure is released, the ball behaves more like a liquid.
4. Pressure is the factor that appears to determine whether the mixture behaves more like a solid or more like a liquid.

After students study Section 1.3, return to the Inquiry Activity. Ask students to rewrite their answers to Question 4 as hypotheses. Discuss which variables were manipulated in the Inquiry Activity and which variables were controlled.

CHAPTER 1

Introduction to Chemistry

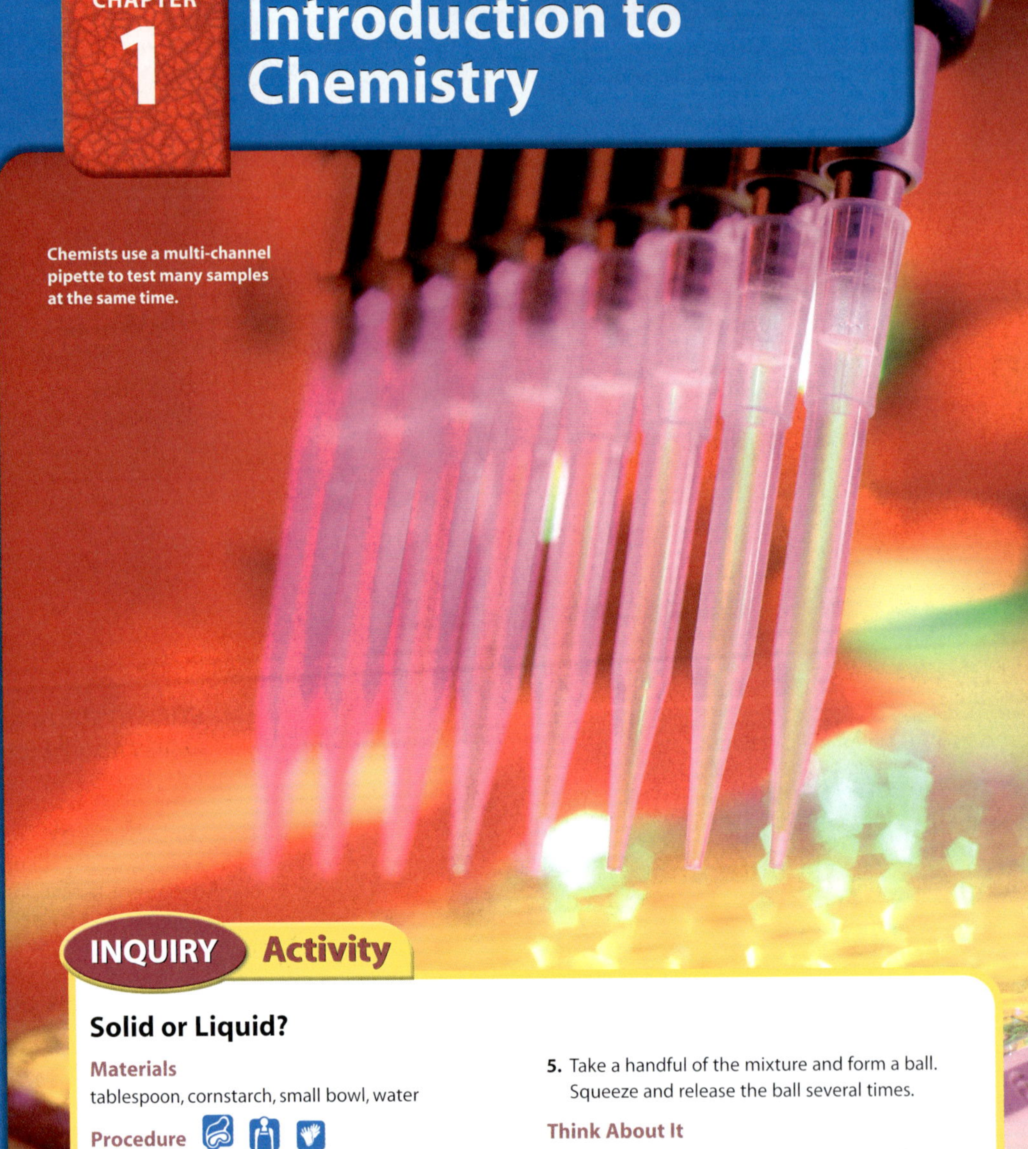

Chemists use a multi-channel pipette to test many samples at the same time.

INQUIRY Activity

Solid or Liquid?

Materials

tablespoon, cornstarch, small bowl, water

Procedure

1. Add 3 heaping tablespoons of cornstarch to the bowl, and then add 3 tablespoons of water.
2. Use the spoon to thoroughly stir the contents of the bowl. Then let the mixture stand for 5 minutes.
3. Slowly push your finger into the mixture. Repeat with your fist.
4. Quickly jab your finger into the mixture. Repeat with your fist.
5. Take a handful of the mixture and form a ball. Squeeze and release the ball several times.

Think About It

1. What happened when you slowly pushed your finger or fist into the mixture?
2. What happened when you jabbed your finger or fist into the mixture?
3. What happened when you squeezed the ball? When you released the ball?
4. What factor seems to determine whether the mixture behaves like a solid or like a liquid?

1.1 Chemistry

Connecting to Your World

The *Galileo* spacecraft was placed in orbit around Jupiter to collect data about the planet and its moons. Instruments aboard *Galileo* analyzed the atmosphere of the moon Io. They found large amounts of sulfur and sulfur dioxide. These chemicals are usually released when volcanoes erupt on Earth. So the presence of these chemicals verified that the volcanoes on Io's surface are active. Chemistry helped scientists to study the geology of a distant object in the solar system. In this section, you will learn about chemistry in general and ways you can use your knowledge of chemistry.

Guide for Reading

Key Concepts

- Why is the scope of chemistry so vast?
- What are five traditional areas of study in chemistry?
- How are pure and applied chemistry related?
- What are three general reasons to study chemistry?

Vocabulary

matter
chemistry
organic chemistry
inorganic chemistry
biochemistry
analytical chemistry
physical chemistry
pure chemistry
applied chemistry
technology

Reading Strategy

Relating Text and Visuals As you read, look closely at Figure 1.2. Explain how this illustration helps you to understand the traditional areas of study in chemistry.

What Is Chemistry?

In autumn thousands of visitors travel to New England to view vivid colors like those in Figure 1.1. These colors appear as the trees approach the winter months when growth no longer takes place. The bright pigments are produced by a complex chemical process, which depends on changes in temperature and hours of daylight. The color pigments in leaves are an example of matter. Matter is the general term for all the things that can be described as materials, or "stuff." **Matter** is anything that has mass and occupies space. You don't have to be able to see something for it to qualify as matter. The air you breathe is an example of "invisible" matter.

Chemistry is the study of the composition of matter and the changes that matter undergoes. **Because living and nonliving things are made of matter, chemistry affects all aspects of life and most natural events.** Chemistry can explain how some creatures survive deep in the ocean where there is no light, or why some foods taste sweet and some taste bitter. It can even explain why there are different shampoos for dry or oily hair.

Figure 1.1 Chemical changes that occur in leaves can cause brilliant displays of color.

Section Resources

Print

- **Guided Reading and Study Workbook,** Section 1.1
- **Core Teaching Resources,** Section 1.1 Review, Interpreting Graphics
- **Transparencies,** T1–T2

Technology

- **Interactive Textbook with ChemASAP,** Assessment 1.1
- **Go Online,** Section 1.1

1.1

1 FOCUS

Objectives

1.1.1 Identify five traditional areas of study in chemistry.

1.1.2 Relate pure chemistry to applied chemistry.

1.1.3 Identify reasons to study chemistry.

Guide for Reading

Build Vocabulary 

Paraphrase Help students translate the definition of *technology* into their own words. For example, *means* might be "a way that something gets done" as in "means to an end."

Reading Strategy 

Summarize Have students write a summary of this section and then check the summaries to see if the summaries answer the key concepts questions on p. 7.

2 INSTRUCT

Connecting to Your World

Have students study the photograph and read the text that opens the section. Ask, **Why would large amounts of sulfur and sulfur dioxide be a sign that the volcanoes on Io's surface are active?** *(These chemicals are usually released when volcanoes erupt on Earth.)* On September 21, 2003 *Galileo's* mission ended when its supply of propellant was almost depleted. Without propellent, *Galileo* couldn't point its antenna toward Earth or avoid a collision with Jupiter's moon Europa. Scientists programmed *Galileo* to collide with and disintegrate in Jupiter's atmosphere.

What Is Chemistry?

Discuss 

To find out if students understand the scope of the category *matter*, ask them to classify items as matter or nonmatter.

Section 1.1 (continued)

Areas of Study

Discuss 

Ask students to consider their activities during a single day. Ask them to give examples of things they do that involve chemical processes or contact with chemicals. Then, ask if they can think of four items or activities that do not in some way involve chemistry. *(It will be hard to find any that do not! Students could consider what they eat, what they wear, the products they use for personal hygiene, their normal mode of transportation, and their residence.*

FYI

The area that is most difficult to describe at the start of a chemistry course is physical chemistry, which is highly theoretical. Physical chemists investigate the underlying scientific principles behind the changes that occur in matter. A physical chemistry course includes topics such as quantum mechanics, thermodynamics, kinetic molecular theory, and reaction mechanisms.

Address Misconceptions

Based on how the term *chemical* is used in the media (and how it is defined in most dictionaries), students may assume that the term *chemical* refers exclusively to harmful materials. Explain that *chemical* describes all types of matter, including life-sustaining substances, such as water and oxygen.

Areas of Study

Because the scope of chemistry is vast, chemists tend to focus on one area. **Five traditional areas of study are organic chemistry, inorganic chemistry, biochemistry, analytical chemistry, and physical chemistry.**

Most of the chemicals found in organisms contain carbon. Organic chemistry was originally defined as the study of these carbon-based chemicals. Today, with a few exceptions, **organic chemistry** is defined as the study of all chemicals containing carbon. By contrast, **inorganic chemistry** is the study of chemicals that, in general, do not contain carbon. Inorganic chemicals are found mainly in non-living things, such as rocks. The study of processes that take place in organisms is **biochemistry.** These processes include muscle contraction and digestion. **Analytical chemistry** is the area of study that focuses on the composition of matter. A task that would fall into this area of chemistry is measuring the level of lead in drinking water. **Physical chemistry** is the area that deals with the mechanism, the rate, and the energy transfer that occurs when matter undergoes a change.

The boundaries between the five areas are not firm. A chemist is likely to be working in more than one area of chemistry at any given time. For example, an organic chemist uses analytical chemistry to determine the composition of an organic chemical. Figure 1.2 shows how research in these areas of study can be used to keep humans healthy.

Figure 1.2 Chemists study structures and processes in the human body. **Inferring** ***Does a bone contain mainly organic or inorganic chemicals?***

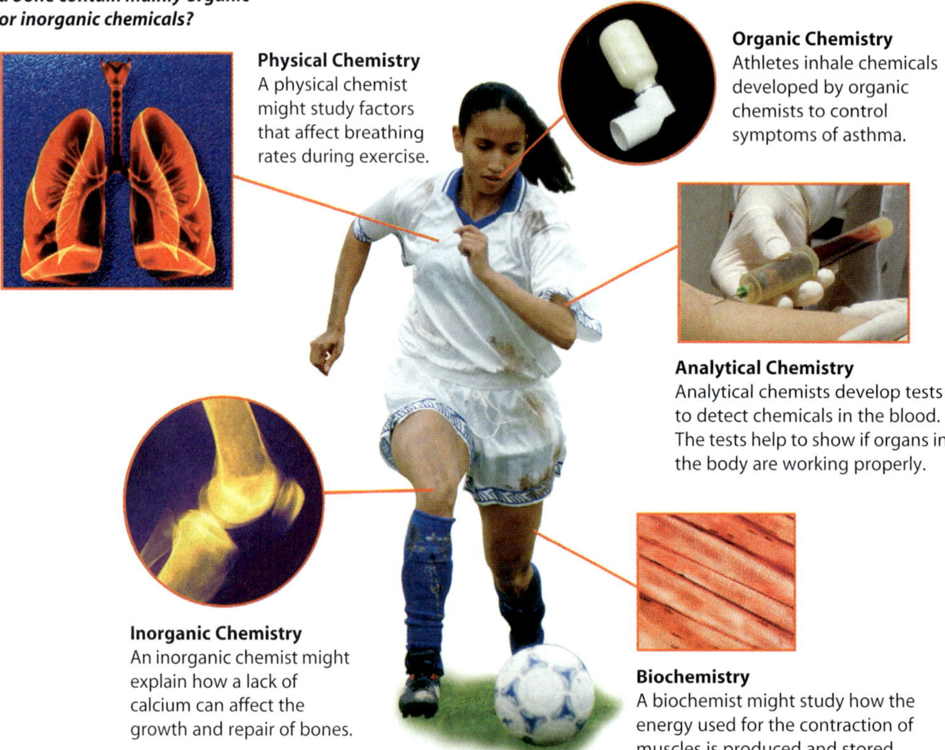

Differentiated Instruction

Gifted and Talented 

Challenge students to conduct research on the traditional areas of study in chemistry. Encourage students to set up an interview with a chemist at a nearby college or local research laboratory. They could also search for information online. Ask students to share their research with the class.

Pure and Applied Chemistry

Some chemists enjoy doing research on fundamental aspects of chemistry. This type of research is sometimes called pure chemistry. **Pure chemistry** is the pursuit of chemical knowledge for its own sake. The chemist doesn't expect that there will be any immediate practical use for the knowledge. Most chemists do research that is designed to answer a specific question. **Applied chemistry** is research that is directed toward a practical goal or application. In practice, pure chemistry and applied chemistry are often linked. **Pure research can lead directly to an application, but an application can exist before research is done to explain how it works.** Nylon and aspirin provide examples of these two approaches.

Nylon For years, chemists didn't fully understand the structure of materials such as cotton and silk. Hermann Staudinger, a German chemist, proposed that these materials contained small units joined together like links in a chain. In the early 1930s, Wallace Carothers did experiments to test Staudinger's proposal. His results supported the proposal. During his research Carothers produced some materials that don't exist in nature. One of these materials, nylon, can be drawn into long, thin, silk-like fibers, as shown in Figure 1.3. Because the supply of natural silk was limited, a team of scientists and engineers were eager to apply Carothers's research to the commercial production of nylon. By 1939, they had perfected a large-scale method for making nylon fibers.

Figure 1.3 Long, thin nylon fibers are woven into the fabric used in this backpack. Other objects that can be made from nylon are jackets, fishing lines, toothbrush bristles, and ropes.

Aspirin Long before researchers figured out how aspirin works, people used it to relieve pain. By 1950, some doctors began to recommend a low daily dose of aspirin for patients who were at risk for a heart attack. Many heart attacks occur when blood clots block the flow of blood through arteries in the heart. Some researchers suspected that aspirin could keep blood clots from forming. In 1971, it was discovered that aspirin can block the production of a group of chemicals that cause pain. These same chemicals are also involved in the formation of blood clots.

Technology The development of nylon and the use of aspirin to prevent heart attacks belong to a system of applied science called technology. **Technology** is the means by which a society provides its members with those things needed and desired. Technology allows humans to do some things more quickly or with less effort. It allows people to do things that would be impossible without technology, such as traveling to the moon. In any technology, scientific knowledge is used in ways that can benefit or harm people and the environment. Debates about how to use scientific knowledge are usually debates about the risks and benefits of technology.

Checkpoint *Which material found in nature does nylon resemble?*

Go Online SCILINKS

For: Links on Applied Chemistry
Visit: www.SciLinks.org
Web Code: cdn-1011

Pure and Applied Chemistry

Use Visuals L1

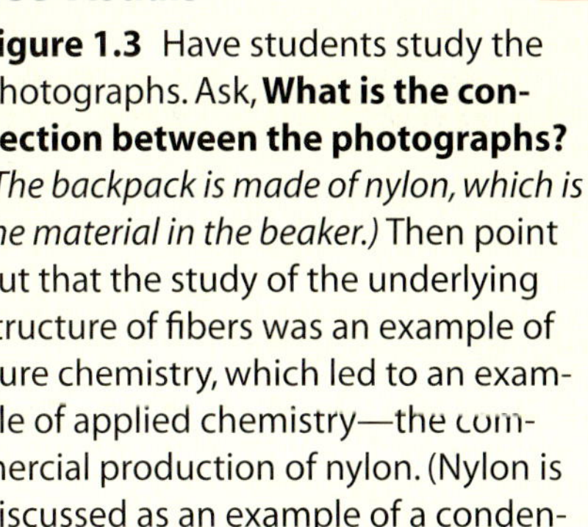

Figure 1.3 Have students study the photographs. Ask, **What is the connection between the photographs?** *(The backpack is made of nylon, which is the material in the beaker.)* Then point out that the study of the underlying structure of fibers was an example of pure chemistry, which led to an example of applied chemistry—the commercial production of nylon. (Nylon is discussed as an example of a condensation polymer in Section 23.4. There is a demo for making nylon on p. 751.)

Relate L2

The public's perception and expectations of science have changed dramatically over the last two hundred years. Advances in technology can lead to increased public concern about the role that science plays in shaping the future. These societal concerns are reflected in the work of 19th- and 20th-century novelists such as H. G. Wells, Edward Bellamy, Jules Verne, George Orwell, and Mary Shelley. Have students select a novel by one of these novelists and critically examine it to determine how the author addressed the role of science, technology, and society in his or her era.

Go Online NSTA SCILINKS

Download a worksheet on **Applied Chemistry** for students to complete, and find additional teacher support from NSTA SciLinks.

Facts and Figures

Aspirin

In the late 1940s, Dr. Lawrence Craven observed that gum containing aspirin, which was used to relieve pain after the removal of tonsils, caused excessive bleeding in children. He hypothesized that aspirin prevented the blood from clotting. He then began to prescribe aspirin to prevent heart attacks.

In 1985, the FDA approved aspirin for patients who had suffered a heart attack. In 1996, the FDA proposed using aspirin during a suspected attack. In 1997, an advisory committee recommended daily low doses of aspirin for people at high risk of a heart attack.

Answers to...

Figure 1.2 Students should infer that a bone contains mainly inorganic materials.

Checkpoint Silk

Why Study Chemistry?

TEACHER Demo

Explaining the Natural World L1

Purpose Students see that chemistry is involved in many natural processes.

Materials slides showing events or structures such as a volcanic eruption, a forest fire, a limestone cave, a flood, lightning, a plowed field, a fireworks display, someone skateboarding, pizza being removed from an oven, someone pumping gasoline

Advance Prep Collect a set of slides that show a wide variety of events or structures. Pick slides for which you can make a connection to chemistry without having to supply details that students will study later. Public libraries and visitor-information bureaus often have slides that can be borrowed.

Procedure Run through the set of slides once without discussion. Discuss the common thread among the images. Then go through the slides a second time and explain the connection to chemistry.

Relate L2

Ask students to volunteer to talk about careers they are considering. Have each volunteer briefly describe what kinds of activities are involved in each career. Challenge the class to think of ways that knowledge of chemistry would be helpful for each career.

Why Study Chemistry?

Should you use hot water or cold water to remove sunblock from a shirt? How could studying chemistry help you to be a better nurse, firefighter, reporter, or chef? If your local government wanted to build a solid waste incinerator in your town, what questions would you ask about the project? Chemistry can have an impact on all aspects of your life. **Chemistry can be useful in explaining the natural world, preparing people for career opportunities, and producing informed citizens.**

Explaining the Natural World You were born with a curiosity about your world. Chemistry can help you satisfy your natural desire to understand how things work. For example, chemistry can be seen in all aspects of food preparation. Chemistry can explain why peeled apples turn brown upon exposure to air. It can explain why the texture of eggs changes from runny to firm as eggs are boiled or scrambled. Chemistry can explain why water expands as it freezes, sugar dissolves faster in hot water, and adding yeast to bread dough makes the dough rise. After you study this textbook, you will know the answers to these questions and many more.

Preparing For a Career Being a chemist can be rewarding. Section 1.2 will present some examples of how chemists contribute to society. In this book, you will find features on careers that require knowledge of chemistry. Some of the choices may surprise you. You do not need to have the word *chemist* in your job title to benefit from knowing chemistry. For example, a firefighter must know which chemicals to use to fight different types of fires. A reporter may be asked to interview a chemist to gather background for a story. Turf managers are admired for the patterns they produce on a ball field while mowing grass, but their more important task is keeping the grass healthy, which requires an understanding of soil chemistry. A photographer, like the one in Figure 1.4, uses chemical processes to control the development of photographs in a darkroom.

Figure 1.4 Even after the invention of the digital camera, many photographers still work with film. They use chemical processes to develop film and produce prints in a darkroom. **Inferring** ***Why isn't film developed under natural light conditions?***

Being an Informed Citizen Industry, private foundations, and the federal government all provide funds for scientific research. The availability of funding can influence the direction of research. Those who distribute funds have to balance the importance of a goal against the cost. Because there is a limit to the money available, areas of research often compete for funds.

For example, space exploration research could not take place without federal funding. Critics argue that the money spent on space exploration would be better spent on programs such as cancer research. Those who support space exploration point out that NASA research has led to the development of many items used on Earth. These include smoke detectors, scratch-resistant plastic lenses, heart monitors, and flat-screen televisions. What if all the money spent on space exploration was used to find a cure for cancer? Are there enough valid avenues of research to take advantage of the extra funding? Would there be qualified scientists to do the research?

Like the citizens shown in Figure 1.5, you will need to make choices that will influence the development of technology. You may vote directly on some issues through ballot initiatives or indirectly through the officials you elect. You may speak at a public hearing or write a letter to the editor or sign a petition. When it comes to technology, there is no one correct answer. But knowledge of chemistry and other sciences can help you evaluate the data presented, arrive at an informed opinion, and take appropriate action.

Figure 1.5 By registering to vote, these citizens in Chicago, Illinois, can have a say in the decisions made by their government. Those decisions include how much money to provide for scientific research.

1.1 Section Assessment

1. **Key Concept** Explain why chemistry affects all aspects of life and most natural events.
2. **Key Concept** Name the five traditional areas into which chemistry can be divided.
3. **Key Concept** Describe the relationship between pure chemistry and applied chemistry.
4. **Key Concept** List three reasons for studying chemistry.
5. Workers digging a tunnel through a city find some ancient pots decorated with geometric designs. Which of the following tasks might they ask a chemist to do? Explain your answer.
   a. Determine the materials used to make the pots.
   b. Explain what the designs on the pots represent.
   c. Recommend how to store the pots to prevent further damage.
6. Would a geologist ask a biochemist to help identify the minerals in a rock? Explain your answer.
7. Explain how knowledge of chemistry can help you be a more informed citizen.

Writing Activity

Describing Technology Pick one activity that you can do faster or with less effort because of technology. Write a paragraph in which you describe the activity, identify the technology, and explain how the technology affects the activity.

Assessment 1.1 Test yourself on the concepts in Section 1.1.

with ChemASAP

Section 1.1 Assessment

1. Living and nonliving things are made of matter, and chemistry is the study of matter.
2. organic chemistry, analytical chemistry, biochemistry, physical chemistry, and inorganic chemistry
3. Pure research can lead directly to an application; an application can exist before research is done to explain how it works.
4. explaining the natural world, preparing people for career opportunities, and producing informed citizens
5. a and c
6. No, a biochemist studies processes that take place in organisms.
7. A possible answer is that knowledge of chemistry helps a citizen evaluate data and arrive at an informed opinion about a public issue that involves technology.

CLASS Activity

Chemistry in the News L2

Have students scan local and national newspapers for current news stories related to chemistry. Ask students to cut out one article and prepare a short presentation on the content of the article and how it relates to chemistry.

3 ASSESS

Evaluate Understanding 

To determine students' understanding about the field of chemistry, ask, **In your own words, how would you define chemistry?** *(Chemistry is the study of the composition of matter and the changes it undergoes.)* **What do chemists do?** *(Chemists study the composition and behavior of matter.)*

Reteach 

Point out that learning about the principles of chemistry will enable students to better understand the modern world. Ask students what is essential in their lives, and make a short list on the board. Have students think of one or two examples of how chemistry plays a part in each essential product or activity.

Writing Activity

Students' answers will vary, but should reveal an understanding of the definition of technology.

If your class subscribes to the Interactive Textbook, use it to review key concepts in Section 1.1.

with ChemASAP

Answers To . . .

Figure 1.4 Students may infer that natural light causes unwanted changes to the film.

1.2

1 FOCUS

Objectives

1.2.1 Identify some areas of research affected by chemistry.

1.2.2 Describe some examples of research in chemistry.

1.2.3 Distinguish between macroscopic and microscopic views.

Guide for Reading

Build Vocabulary L2

Word Parts *Macroscopic* and *microscopic* share the root *scop*, which means "to look at" in Greek. They differ in their prefixes: *macro* is from the Greek *makr*, large; *micro* is from *mikr*, small.

Reading Strategy L2

Visualize Point out that each page in this section is devoted to a different broad area of research. Ask students to visualize topics they would like to investigate as they read about each area.

2 INSTRUCT

Connecting to Your World

Point out that curiosity and a tendency to "tinker" with materials are not traits restricted to professional scientists. Ask, **Why would the carved mussel shells attract fish?** *(The carved shells resemble small fish that larger fish might eat.)* **Why was woven cloth more desirable than animal skins?** *(It was softer and dried more quickly.)*

Materials

Use Visuals L1

Figure 1.6 Have students study the photograph. Ask, **What material is shown in the photograph?** *(hook-and-loop tape)* **What does the microscopic view show that wouldn't be visible in a macroscopic view?** *(how the hooks and loops fit together)* Use this example to discuss how chemists use a microscopic view to explain what is observed at the macroscopic level.

1.2 Chemistry Far and Wide

Guide for Reading

Key Concepts

- What impact do chemists have on materials, energy, medicine, agriculture, the environment, and the study of the universe?

Vocabulary

macroscopic
microscopic
biotechnology
pollutant

Reading Strategy

Monitoring Your Understanding After you read this section, identify something you learned that is important to your life and explain why it is important to you.

Connecting to Your World

The first tools were objects such as a stone with a sharp edge. In time, people learned to reshape these objects to produce better tools. Native Americans in what is now Illinois drilled holes in mussel shells and carved lines onto the shells to make them look like small fish. The shells were likely used as lures for ice fishing. People also began to produce materials that did not exist in nature. By weaving plant fibers together, they made cloth, which is softer and dries more quickly than animal skins. Chemistry plays a key role in the production of new materials. In this section, you will learn how chemistry affects many aspects of modern life.

Materials

The search for new materials continues. **Chemists design materials to fit specific needs.** Often they find inspiration in nature. In 1948, while hiking through the woods of his native Switzerland, George de Mestral took a close look at the pesky burrs that stuck to his clothing. When he looked at the burrs under magnification, he saw that each burr was covered with many tiny hooks that could latch on to tiny loops in the woven cloth of his clothing. George had a weaver make two cloth tapes. On the surface of one tape were hooks, and on the surface of the other tape were loops that the hooks could fit into, as shown in Figure 1.6. In 1955 George patented the design for his hook-and-loop tapes. These tapes are used as fasteners for items such as shoes and gloves.

This story illustrates two different ways of looking at the world—the macroscopic view and the microscopic view. The burrs that George de Mestral used as a model for his tapes are small compared to many objects in nature. However, they were large enough for George to see. Burrs belong to the **macroscopic** world, the world of objects that are large enough to see with the unaided eye. George needed more than his own vision to see the hooks on a burr. The hooks belong to the **microscopic** world, or the world of objects that can be seen only under magnification.

Figure 1.6 This is a magnified view of hook-and-loop tape. Color was added to the photo to highlight the structures. **Classifying** ***Does the photograph show a macroscopic or a microscopic view of the tape? Explain.***

Section Resources

Print

- **Guided Reading and Study Workbook,** Section 1.2
- **Core Teaching Resources,** Section 1.2 Review
- **Transparencies,** T3–T4

Technology

- **Interactive Textbook with ChemASAP,** Assessment 1.2

Energy

Energy is necessary to meet the needs of a modern society. It is used to heat buildings, manufacture goods, and process foods. It is used to transport people and goods between locations. With population growth and more industrialization around the globe, the demands for energy continue to increase. There are two ways to meet the demand for energy—conserve energy resources and produce more energy. **Chemists play an essential role in finding ways to conserve energy, produce energy, and store energy.**

Conservation One of the easiest ways to conserve energy is through the use of insulation. Much of the energy consumed is used to keep houses warm and freezers cold. Insulation acts as a barrier to heat flow from the inside to the outside of a house or from the outside to the inside of a freezer. The foam used in drink cups provides excellent insulation because it contains pockets of trapped air. One of the most exciting modern insulation materials devised by chemists is SEAgel, which is a foam made from seaweed. SEAgel is very lightweight. In fact, SEAgel is so light that it can float on soap bubbles, as shown in Figure 1.7.

Figure 1.7 This insulation is light enough to float on soap bubbles yet is very effective at preventing heat transfer.

Production The burning of coal, petroleum, and natural gas is a major source of energy. These materials are called fossil fuels because they formed from the remains of ancient plants and animals. Scientists are always looking for new sources of energy because the supply of fossil fuels is limited. One intriguing possibility is fuels obtained from plants. Oil from the soybeans shown in Figure 1.8 is used to make biodiesel. Regular diesel fuel is a petroleum product that produces an irritating black exhaust when it burns. When biodiesel burns, the exhaust smells like French fries!

Storage Batteries are devices that use chemicals to store energy that will be released as electric current when the batteries are used. Batteries vary in size, power, and hours of useful operation. For some applications, it is important to have batteries that can be recharged rather than thrown away. One application that benefits from rechargeable batteries is cordless tools. These tools were first developed for NASA. Astronauts in the Apollo program needed a way to drill beneath the Moon's surface to collect samples. Other devices that use rechargeable batteries are digital cameras, wireless phones, and laptop computers.

Checkpoint ***What plant is a source of biodiesel?***

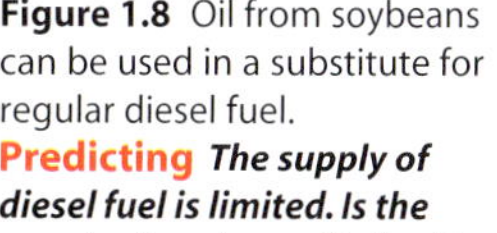

Figure 1.8 Oil from soybeans can be used in a substitute for regular diesel fuel. **Predicting** ***The supply of diesel fuel is limited. Is the supply of soybeans limited?***

Facts and Figures

Insulation

Air is a good barrier to the movement of thermal energy as long as the air is not moving. In fiberglass, air is trapped by a network of thin fibers that are woven together. In foam insulation, the air pockets are completely enveloped by the solid that makes up the framework of the foam. SEAgel, is made of agar, which is derived from seaweed. A gelatin-like mixture of agar and water is freeze-dried to remove the water. What is left is a honeycomb of dried agar filled with air. SEAgel has a density approximately equal to that of air. It is also biodegradable.

Energy

Biodiesel L2

Soybeans are not the only source of biodiesel. Have students research what materials in general are used to make biodiesel *(vegetable oils and animal fats)*. Have them find examples to share with the class of cases where biodiesel is being used as an alternative fuel.

FYI

Fossil fuels are addressed in depth in Section 22.5; batteries are discussed in Section 21.1.

Answers to...

Figure 1.6 microscopic; It shows what cannot be seen by the unaided eye.

Figure 1.8 The supply of soybeans is limited only by the amount of land devoted to growing soybeans.

Checkpoint soybean plant

Medicine and Biotechnology

Relate L2

Use the Technology & Society feature, *Nature's Pharmacy,* on pp. 18–19 to illustrate the contributions that chemists have made to medicine. Explain that the active ingredient in a prescription drug or over-the-counter remedy often is different from the natural chemical on which it is based. For example, the active ingredient that Johann Buchner isolated from willow bark in 1828 was salicin, which chemists used to produce salicylic acid. Later, chemists developed acetylsalicylic acid (aspirin), which had beneficial properties that were similar to those of salicylic acid but with less irritation. (When students study functional groups in Chapter 23, you could return to this topic and compare the structural formulas for salicin, salicylic acid, and acetylsalicylic acid.)

FYI

The detailed discussion of DNA is in Section 24.5. All students need to know at this point is what is stated in the text.

Medicine and Biotechnology

No field has benefited more from advances in chemistry than medicine. **Chemistry supplies the medicines, materials, and technology that doctors use to treat their patients.** Work in the field of medicine is often done by biochemists. Their overall goal is to understand the structure of matter found in the human body and the chemical changes that occur in cells. To accomplish their goal, they work with biologists and doctors.

Medicines There are over 2000 prescription drugs. They are designed to treat various conditions including infections, high blood pressure, and depression. Other drugs, such as aspirin and antacids, can be sold without a prescription. Many drugs are effective because they interact in a specific way with chemicals in cells. Knowledge of the structure and function of these target chemicals helps a chemist design safe and effective drugs.

Materials Chemistry can supply materials to repair or replace body parts. Diseased arteries can be replaced with plastic tubes. Artificial hips and knees made from metals and plastics can replace worn-out joints and allow people to walk again without pain. Burn patients may benefit from a plastic "skin" that can heal itself when the plastic cracks. Chemicals that repair the damage are released from tiny capsules in the plastic.

Biotechnology Figure 1.9a shows a model of a small piece of DNA. Segments of DNA called genes store the information that controls changes that take place in cells. From 1990 to 2003, scientists worldwide worked on the Human Genome Project. They identified the genes that comprise human DNA—about 30,000. They determined the sequence of the genes in DNA. Some tools these scientists developed are used in biotechnology.

Biotechnology applies science to the production of biological products or processes. It uses techniques that can alter the DNA in living organisms. It may depend on the transfer of genes from one organism to another. When genes from humans are inserted into bacteria, the bacteria act as factories. They produce chemicals of importance to humans, such as insulin, which is used to treat some types of diabetes. Production takes place in large versions of the bioreactors in Figure 1.9b. In the future, scientists expect to use gene therapy to treat some diseases. A gene that is not working properly would be replaced with one that will work properly.

Figure 1.9 The discovery of the structure of DNA led to the development of biotechnology. **a** This computer graphics model shows a small segment of DNA. **b** The conditions in a bioreactor are controlled so that the bacteria produce as much of the product as possible.

14 Chapter 1

Facts and Figures

Human Genome Project

Identifying all the genes in human DNA and determining the sequences of base pairs were not the only goals of the Human Genome Project. Other aims were to develop analytical tools that could be transferred to the private sector and address any ethical, legal, and social issues (ESLI).

Agriculture

The world's population is increasing, but the amount of land available to grow food is decreasing. Land that was once used for agriculture is now used for homes and industries. So it is important to ensure that land used for agriculture is as productive as possible. **Chemists help to develop more productive crops and safer, more effective ways to protect crops.**

Productivity One way to track productivity is to measure the amount of edible food that is grown on a given unit of land. Some factors that decrease productivity are poor soil quality, lack of water, weeds, plant diseases, and pests that eat crops. Chemists can help with many of these problems. They test soil to see if it contains the right chemicals to grow a particular crop and recommend ways to improve the soil. They use biotechnology to develop plants that are more likely to survive a drought or insect attack.

Chemists can also help to conserve water. In many regions, water is not an abundant resource. Finding reliable ways to determine when a crop needs water is important. The jellyfish in Figure 1.10 has a gene that causes it to glow. If that gene is inserted into a potato plant, the plant glows when it needs to be watered. These altered plants would be removed from the field before the rest of the crop was harvested.

Crop Protection For years, farmers have used chemicals to attack insect pests. In the past, these chemicals were nonspecific; that is, a chemical designed to kill a pest could also kill useful insects. Today, the trend is toward chemicals that are designed to treat specific problems. These chemicals are often similar to the chemicals that plants produce for protection.

Chemists sometimes use chemicals produced by insects to fight insect pests. Female insects may produce chemicals that attract male insects. This type of chemical has proved effective in combating pinworms. The worms leave holes and black blotches when they tunnel into tomatoes. Pinworms mate when they are in the moth stage of development. The plastic tube wrapped around the stem of the tomato plant in Figure 1.11 contains the chemical that a female pinworm moth emits to attract male moths. When the chemical is released from these tubes, it interferes with the mating process so that fewer pinworms are produced.

Checkpoint ***What jellyfish gene did scientists transfer to a potato?***

Figure 1.10 If genes from this jellyfish *(Aequaria victoria)* are transferred to a potato plant, the plant glows when it needs to be watered. **Predicting** ***How does the modified plant help a farmer to conserve water?***

Figure 1.11 In the plastic tube wrapped around the tomato stem, there is a chemical that attracts male pinworm moths. This process reduces the rate of mating between female and male moths, and the number of pinworms produced.

Agriculture

Use Visuals

Figure 1.11 Point out the tube that is wrapped around the stem of the tomato. Ask, **What does the plastic tube contain?** *(a chemical that attracts male pinworms)* **How does this chemical help reduce the damage to tomato plants from pinworms?** *(With fewer males available to mate with female pinworms, there are fewer offspring to feed on the tomatoes.)*

FYI

A dehydrated altered potato plant glows under a black light. Although potatoes tend to grow larger when the plants are over-watered, it may be necessary to conserve water. Also, excess water can cause nutrients to be leached from soil.

Answers to...

Figure 1.10 Because the modified plant glows when it needs to be watered, the farmer can supply just enough water to keep the plants healthy.

 the gene that causes the jellyfish to glow

Section 1.2 (continued)

The Environment

CLASS Activity

Studying Gasoline Additives L2

Have students find out what chemicals are used as additives in gasoline today. Students might start by looking at labels on gas pumps, by contacting the companies that manufacture these products, and by looking up literature published by the Environmental Protection Agency. Additional questions students should answer include, Why are these alternatives to lead safer? Are there any disadvantages to using these alternative chemicals?

Interpreting Graphs L2

a. 88.2%

b. 440,000

c. After lead was banned in gasoline and in public water supply systems, less lead entered the environment.

Enrichment Question L3

Ask students, **Why has the reduction in lead levels slowed down since the first dramatic drop between 1980 and 1988?** *(The major remaining source of lead is the existing paint in old houses, not a new product whose composition can be controlled.)*

Figure 1.12 This poster was used to warn people about the danger to children from lead-based paint.

The Environment

One unintended consequence of new technologies is the production of pollutants. A **pollutant** is a material found in air, water, or soil that is harmful to humans or other organisms. **Chemists help to identify pollutants and prevent pollution.**

Identify Pollutants Lead is a pollutant with a long history. The Romans used lead pipes for plumbing and stored their wine in lead-glazed vessels. Brain damage from lead poisoning may have caused Roman rulers to make bad decisions, which led to the fall of the Roman Empire. Until the mid-1900s, lead was used in many products, including paints and gasoline. A study done in 1971 showed that the level of lead that is harmful to humans is much lower than had been thought, especially for children. Low levels of lead in the blood can permanently damage the nervous system of a growing child. This damage causes many problems, including a reduced ability to learn.

Prevent Pollution The use of lead paint in houses was banned in 1978. Using lead in gasoline and in public water supply systems was banned in 1986. Today, the major source of lead in children is lead-based paint in about 39 million homes built before 1978. When children play with flakes of peeling paint or touch surfaces covered with paint dust, they can transfer the paint to their mouths with their fingers. The strategies used to prevent lead poisoning include testing children's blood for lead, regulation of home sales to families with young children, and public awareness campaigns with posters like the one in Figure 1.12. The graph in Figure 1.13 shows the results of these efforts.

Checkpoint ***When was the use of lead paint in houses banned?***

Figure 1.13 This graph shows data on children in the United States with higher than acceptable levels of lead in their blood.

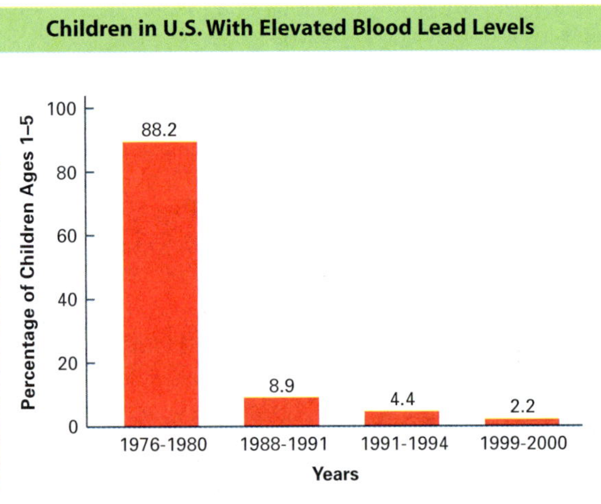

INTERPRETING GRAPHS

a. Analyzing Data What percent of children had elevated lead levels in the 1970s?

b. Calculating If a percentage point equals 200,000 children, how many children had elevated lead levels in 2000?

c. Drawing Conclusions Explain the dramatic drop in the percentage of children affected by lead poisoning between 1980 and 1988.

Figure 1.14 With help from NASA, chemists study matter from other bodies in the solar system. **a** Apollo astronauts brought rocks from the moon back to Earth. **b** This artist's drawing shows the robotic vehicle *Opportunity* on the surface of Mars.

The Universe

Scientists assume that the methods used to study Earth can be applied to other objects in the universe. **To study the universe, chemists gather data from afar and analyze matter that is brought back to Earth.**

In the early 1800s, scientists began to study the composition of stars by analyzing the light they transmitted to Earth. In 1868, Pierre Janssen discovered a gas on the sun's surface that was not known on Earth. Norman Lockyer named the gas helium from the Greek word *helios*, meaning "sun." In 1895, William Ramsay discovered helium on Earth.

Because the moon and the planets do not emit light, scientists must use other methods to gather data about these objects. They depend on matter brought back to Earth by astronauts or on probes that can analyze matter in space. Chemists have analyzed more than 850 pounds of moon rocks that were brought back to Earth. The large rock in Figure 1.14a is similar to rocks formed by volcanoes on Earth, suggesting that vast oceans of molten lava once covered the moon's surface. Figure 1.14b is a drawing of the robotic vehicle *Opportunity*. The vehicle was designed to determine the chemical composition of rocks and soil on Mars. Data collected at the vehicle's landing site indicated that the site was once drenched with water.

1.2 Section Assessment

8. **Key Concept** When chemists develop new materials, what is their general goal?
9. **Key Concept** Name three ways chemists help meet the demand for energy.
10. **Key Concept** How do chemists help doctors treat patients?
11. **Key Concept** What role do chemists play in agriculture?
12. **Key Concept** How do chemists help protect the environment?
13. **Key Concept** Describe two ways that chemists study the universe.
14. Use lead as an example to explain the meaning of the term *pollutant*.
15. Use an example to compare and contrast the terms *macroscopic* and *microscopic*.

Cleaning Up Pollutants Read about phytoremediation on page R44. Write a paragraph summarizing this method for dealing with pollutants. Explain how it could be used to clean up soil containing high levels of lead.

Interactive Textbook

Assessment 1.2 Test yourself on the concepts in Section 1.2.

with ChemASAP

Section 1.2 Assessment

8. design materials to meet specific needs
9. They find ways to conserve, produce, and store energy.
10. They supply technology, medicines, and materials that doctors use to treat patients.
11. They help to develop more productive crops and safer, more effective ways to protect crops.
12. They identify pollutants and help to prevent pollution.
13. They gather data from afar and analyze matter brought back to Earth.
14. A pollutant is a material found in air, soil, or water that is harmful to organisms. Lead can cause permanent damage to a growing child's nervous system.
15. Answers will vary, but should show knowledge of the difference between macroscopic and microscopic views.

The Universe

Composition of the Sun L2

Have students look at page R4 of the Elements Handbook for data on the composition of the sun. Have them explain how scientists can know the composition of the sun without collecting matter from the sun. Ask students to explain why it would be difficult to present the data as a pie chart or bar graph.

3 ASSESS

Evaluate Understanding L2

Ask students to name some recent technological advances that were made possible through the study of chemistry.

Reteach L1

Have students suggest other ways to classify areas impacted by chemistry other than the system used in this section. One alternative might be to focus on the human need for food, clothing, shelter, and transportation.

Elements Handbook

Phytoremediation involves the use of plants to remove pollutants from soil or water. A plant can absorb lead through its roots and the lead can accumulate in the leaves and stems of the plant. After harvesting, plants can be burned and the lead residue buried in an approved landfill or recovered.

Interactive Textbook

If your class subscribes to the Interactive Textbook, use it to review key concepts in Section 1.2.

with ChemASAP

Answers to...

Checkpoint 1978

Nature's Pharmacy

People used ingredients extracted from plants as folk remedies for centuries before chemists were able to isolate the active ingredients from these extracts. The first active ingredient to be isolated was morphine in 1804. Salicylates occur naturally in plants of the genus *Salix,* which includes willow, poplar, and beech trees. Salicin was isolated from willow bark in 1828 by Johann Buchner, a pharmacy professor at the University of Munich.

Discuss L2

Point out the two names given for each organism, a common name and a scientific name. Pronounce the scientific names for the students.

Cinchona succirubra: chin KOH nuh soo chee ROO bruh

Digitalis purpurea: dih gih TAH lis pur PUR ee uh

Salix babylonica: SAY lix ba bih LOH nih kuh

Leiurus quinquestriatus: lay OOR us kwin kwuh stree AH tus

Desmodus rotundus: dez MOH dus roh TUN dus

Conus striatus: KOH nuhs stree AH tuhs

Dendrobates reticulatus: den droh BAY teez reh tik yoo LAH tus

CLASS Activity

Approval of New Drugs L2

Point out that it is one thing to produce new effective medicines and quite a different challenge to test them, get them approved, and produce them in quantity. Have the students research the steps involved in getting approval for a new medicine and how much time is usually required.

Technology & Society

Nature's Pharmacy

About 40 percent of all modern medicines come from chemicals produced by plants or animals. With any source, chemists must first identify the effective, or active, ingredient. Then they must purify the chemical and show that it is safe for human use. Chemists often modify a chemical to make it more effective or less toxic.

Applying Concepts ***Are the chemicals derived from plants and animals organic or inorganic?***

Cinchona tree
Cinchona succirubra
Bark from this tree was used for centuries to treat malaria, a disease with recurring bouts of fever and chills. The active ingredient in the bark is quinine—the chemical that gives tonic water its bitter taste.

Foxglove
Digitalis purpurea
The poison produced by this plant (digitalis) is used in small doses to treat congestive heart failure. It causes heart muscle cells to contract with more power, which increases the ability of the heart to pump blood.

Willow bark
Salix babylonica
For centuries, people made a tea from willow bark to treat headaches and other ailments. By 1828, chemists had isolated the active ingredient in willow bark. For the next 70 years, chemists worked to produce the most effective drug based on this chemical. What they produced is aspirin (acetylsalicylic acid).

Death stalker scorpion
Leiurus quinquestriatus
This scorpion's narrow pincers are weak, but the venom in its tail is very powerful. Scientists have isolated a chemical from this venom, which is used to treat an incurable form of brain cancer.

Vampire bat
Desmodus rotundus
After this bat bites an animal with its sharp teeth, it laps up blood from the wound. A chemical in the bat's saliva keeps the blood from clotting. Doctors use this chemical to dissolve blood clots that cause strokes.

Cone snail
Conus striatus
Cone snails produce toxins that paralyze their prey. Chemists are studying these toxins as possible treatments for chronic pain and nervous system disorders, such as Parkinson's disease.

Poison dart frogs
Dendrobates reticulatus
Poisons from the food these tiny (1/2-inch-long) frogs eat collect in their skin. Chemists use these poisons to study the human nervous system. A chemical found in a frog from Ecuador served as a model for a drug that is as powerful a painkiller as morphine but is not addictive.

Nature's Products Can Be Harmful
Chemicals from natural sources are not always effective or harmless. In 2004, the FDA banned weight-loss products based on the herb ephedra. This herb contains the chemical ephedrine, which is associated with increased blood pressure, abnormal heart rates, a higher risk of stroke, and even death.

Discuss L2

Recently, scientists have turned their attention to animal venom as a source of drugs. The venom can contain numerous fast-acting toxins that target muscles and nerves. Ask, **Why do you think that the death stalker scorpion produces venom?** *(Venom is used to kill or immobilize prey.)* **Why do you think that a poison dart frog produces venom?** *(Venom is also used to protect an animal from predators.)*

FYI

Prior to the ban on products containing ephedra, the federal government had not blocked the sale of an OTC dietary supplement.

Facts and Figures

Toxins

The 5-inch long death stalker scorpion produces a toxin that can block the chloride channels in glioma cancer cells. This blockage keeps the cells from shrinking and migrating to other locations in the brain. Vampire bats are found in Central and South America. The medicine derived from their saliva is called Draculin™. The venom of the Malayan pit viper contains a similar anticoagulant. There are more than 500 species of cone snails. Each produces venom containing 50–200 toxins. Areas of research based on these toxins include treatments for epilepsy, schizophrenia, and stroke. Poison dart frogs tend to be highly colored as a warning to predators. Nearly 600 alkaloid toxins have been isolated from glands in the skin of these frogs. The frogs collect the toxins from ants, mites, millipedes, and other arthropods.

Answers To . . .

Applying Concepts The chemicals derived from plants and animals are organic.

1.3

1 FOCUS

Objectives

1.3.1 Describe how Lavoisier transformed chemistry.

1.3.2 Identify three steps in the scientific method.

1.3.3 Explain why collaboration and communication are important in science.

Guide for Reading

Build Vocabulary L2

Word Forms When discussing the two types of variables, explain that one meaning of *manipulate* is "to manage or control" and one meaning of *response* is "to answer or act in turn."

Reading Strategy L2

Use Prior Knowledge Before students read the section, ask them what they know about the scientific method.

2 INSTRUCT

Connecting to Your World

Have students read the opening paragraph and look at the photo. Ask, **What made Fleming different from other scientists who had seen this mold?** *(Fleming recognized the importance of this discovery. He assumed that the mold had released a chemical that prevented the growth of bacteria.)*

Alchemy

Relate L2

Students who have read the Harry Potter books will be familiar with the name of one alchemist, Nicolas Flamel (1330–1414), who wrote a book about the philosopher's stone. Alchemists believed that the philosopher's stone could change base metals into gold.

1.3 Thinking Like a Scientist

Guide for Reading

Key Concepts

- How did alchemy lay the groundwork for chemistry?
- How did Lavoisier help to transform chemistry?
- What are the steps in the scientific method?
- What role do collaboration and communication play in science?

Vocabulary

scientific method
observation
hypothesis
experiment
manipulated variable
responding variable
theory
scientific law

Reading Strategy

Building Vocabulary After you read this section, explain the difference between a theory and a scientific law.

Connecting to Your World

In 1928, Alexander Fleming, a Scottish scientist, noticed that a bacteria he was studying did not grow in the presence of a yellow-green mold. Other scientists had made the same observation, but Fleming was the first to recognize its importance. He assumed that the mold had released a chemical that prevented the growth of the bacteria. That chemical was penicillin, which can kill a wide range of harmful bacteria. In 1945, Fleming shared a Nobel Prize for Medicine with Howard Florey and Ernst Chain, who led the team that isolated penicillin. In this section you will study the methods scientists use to solve problems.

Alchemy

The word *chemistry* comes from *alchemy*. Long before there were chemists, alchemists were studying matter. Alchemy arose independently in many regions of the world. It was practiced in China and India as early as 400 B.C. In the eighth century, Arabs brought alchemy to Spain, from where it spread quickly to other parts of Europe.

Alchemy had a practical side and a mystical side. Practical alchemy focused on developing techniques for working with metals, glass, and dyes. Mystical alchemy focused on concepts like perfection. Because gold was seen as the perfect metal, alchemists were searching for a way to change other metals, such as lead, into gold. Although alchemists did not succeed in this quest, the work they did spurred the development of chemistry.

Alchemists developed the tools and techniques for working with chemicals. Alchemists developed processes for separating mixtures and purifying chemicals. They designed equipment that is still used today, including beakers, flasks, tongs, funnels, and the mortar and pestle in Figure 1.15. What they did not do was provide a logical set of explanations for the changes in matter that they observed. That task was left for chemists to accomplish.

Figure 1.15 A bowl-shaped mortar and a club-shaped pestle are used to grind or crush materials such as herbs, spices, and paint pigments. The mortar and pestle in the photograph are made of porcelain, which is a hard material.

Section Resources

Print

- **Guided Reading and Study Workbook,** Section 1.3
- **Core Teaching Resources,** Section 1.3 Review
- **Transparencies,** T5–T6
- **Laboratory Manual,** Lab 1
- **Small-Scall Chemistry Laboratory Manual,** Lab 1

Technology

- **Interactive Textbook with ChemASAP,** Assessment 1.3
- **Go Online,** Section 1.3

20 Chapter 1

An Experimental Approach to Science

By the 1500s in Europe, there was a shift from alchemy to science. Science flourished in Britain in the 1600s, partly because King Charles II was a supporter of the sciences. With his permission, some scientists formed the Royal Society of London for the Promotion of Natural Knowledge. The scientists met to discuss scientific topics and conduct experiments. The society's aim was to encourage scientists to base their conclusions about the natural world on experimental evidence, not on philosophical debates.

In France, Antoine-Laurent Lavoisier did work in the late 1700s that would revolutionize the science of chemistry. **Lavoisier helped to transform chemistry from a science of observation to the science of measurement that it is today.** To make careful measurements, Lavoisier designed a balance that could measure mass to the nearest 0.0005 gram.

One of the many things Lavoisier accomplished was to settle a long-standing debate about how materials burn. The accepted explanation was that materials burn because they contain phlogiston, which is released into the air as a material burns. To support this explanation, scientists had to ignore the evidence that metals can gain mass as they burn. By the time Lavoisier did his experiments, he knew that there were two main gases in air—oxygen and nitrogen. Lavoisier was able to show that oxygen is required for a material to burn. Lavoisier's wife Marie Anne, shown in Figure 1.16, helped with his scientific work. She made drawings of his experiments and translated scientific papers from English. Figure 1.17 shows a reconstruction of Lavoisier's laboratory in a museum in Paris, France.

At the time of the French Revolution, Lavoisier was a member of the despised royal taxation commission. He took the position to finance his scientific work. Although he was dedicated to improving the lives of the common people, his association with taxation made him a target of the revolution. In 1794 he was arrested, tried, and beheaded.

Checkpoint ***What long-standing debate did Lavoisier help settle?***

Figure 1.16 This portrait of Antoine Lavoisier and his wife Marie Anne was painted by Jacques Louis David in 1788. The painting includes some equipment that Lavoisier used in his experiments.

Figure 1.17 This reconstruction of Lavoisier's laboratory is in a museum in Paris, France.
Interpreting Photographs ***What objects do you recognize that are similar to objects that you use in the laboratory?***

An Experimental Approach to Science

CLASS Activity

Invisible Ink L2

Purpose Students make hypotheses based on a set of observations.

Materials phenolphthalein, 70% solution of 2-propanol, small artist's paintbrush, sheets of copy paper or white butcher paper, household glass cleaner containing ammonia

Advanced Prep Prepare a solution of phenolphthalein in alcohol (250 mg of phenolphthalein in 250 mL of 70% 2-propanol). Use the brush to letter messages on sheets of paper. Try messages such as "CHEM-IS-TRY" and "This is a LABOR-atory, not a lab-ORATORY." Allow the paper to dry until the messages are invisible. Then post the messages around the room in well-ventilated areas.

Procedure At intervals, spray each sheet with the glass cleaner. Ask students to hypothesize why the pink messages appear, disappear after a few minutes, and then reappear when the paper is sprayed again.

Expected Outcome Students are likely to infer that something in the cleaner caused a reversible change to something on the paper. They may infer that the material in the cleaner is volatile.

FYI

The Royal Society is the oldest scientific society in continuous existence. Robert Boyle (1627-1691) was a founding member. In *The Sceptical Chymist*, Boyle emphasized the necessity of doing experiments to test ideas that were obtained by reason. The motto of the Royal Society is *nullius in verba,* which literally means "nothing in words." The motto stresses the need for evidence.

Facts and Figures

How Oxygen Got Its Name

The ancient Greeks thought that flammable objects contained the element fire, which George Stahl (1660–1734) named *phlogiston.* During burning, phlogiston transferred to the air. Phlogiston-rich air (now called nitrogen) did not support burning; objects burned brightly in phlogiston-poor air. Lavoisier measured the mass of metals before and after heating in a closed container. He showed that the mass gained by the metal was lost by the air. Thus, the process of burning involved a gain of matter, not a loss of phlogiston. Lavoisier named the portion of air that supported combustion *oxygen.*

Answers to...

Figure 1.17 Students are most likely to recognize the two-pan balance

how materials burn

CLASS Activity

Lorenzo's Oil 

The story of how Augusto and Michaela Odone, on their own, developed an oil that relieved the symptoms of adrenoleukodystrophy (ALD), an inherited neurological disease afflicting their son, is an excellent example of "ordinary people" applying the scientific method. This story is told in the movie *Lorenzo's Oil*. Rent or borrow a copy of the movie. Ask students to take notes as they watch the movie on how the Odones use observations, hypotheses, and experiments to help their son.

Word *Origins* 

An expert is a person who has considerable knowledge about a particular field. This knowledge is built on experience. Scientists gain experience by doing experiments or reading about experiments done by other scientists.

Discuss L2

Students often think that an experiment is a failure if they do not get the "right" (expected) results. As students do experiments, help them analyze results that do not fit a hypothesis or vary widely from those of other students. Often, you can identify experimental errors that explain deviation. Also point out that scientists can gain important insights from "failed" experiments.

Using Visuals 

Figure 1.19 If students are not clear on the difference between compare and contrast, tell students that *compare* comes from a Latin word *comparare*, meaning "to make equal with." When you compare two items, you focus on how they are similar. Then say that *contra* means "against" in Latin. Ask, **When you contrast two items, what do you focus on?** *(how the objects differ)*

Figure 1.18 Observation is an essential step in the scientific method.

The Scientific Method

A Nobel Prize winner in science once said that science is about "ordinary people doing ordinary things." Scientists have a powerful tool that they can use to produce valuable, sometimes spectacular, results. Like all scientists, the biochemist shown in Figure 1.18 is using the scientific method to solve difficult problems. The **scientific method** is a logical, systematic approach to the solution of a scientific problem. **Steps in the scientific method include making observations, testing hypotheses, and developing theories.** Figure 1.19 shows how these steps fit together.

Making Observations The scientific method is useful for solving many kinds of problems because it is closely related to ordinary common sense. Suppose you try to turn on a flashlight and you notice that it does not light. When you use your senses to obtain information, you make an **observation.** An observation can lead to a question: What's wrong with the flashlight?

Word *Origins*

Experiment contains the Latin root *peri*, meaning "to try or test." The words *expert* and *experience* contain the same root. **How could experiments provide the experience for someone to become an expert?**

Testing Hypotheses If you guess that the batteries are dead, you are making a hypothesis. A **hypothesis** is a proposed explanation for an observation. You can test your hypothesis by putting new batteries in the flashlight. If the flashlight lights, you can be fairly certain that your hypothesis is true. What if the flashlight does not work after you replace the batteries? A hypothesis is useful only if it accounts for what is actually observed. When experimental data does not fit a hypothesis, the hypothesis must be changed. A new hypothesis might be that the light bulb is burnt out. You can replace the bulb to test this hypothesis.

Replacing the bulb is an **experiment,** a procedure that is used to test a hypothesis. When you design experiments, you deal with variables, or factors that can change. The variable that you change during an experiment is the **manipulated variable,** or independent variable. The variable that is observed during the experiment is the **responding variable,** or dependent variable. If you keep other factors that can affect the experiment from changing during the experiment, you can relate any change in the responding variable to changes in the manipulated variable.

For the results of an experiment to be accepted, the experiment must produce the same result no matter how many times it is repeated, or by whom. This is why scientists are expected to publish a description of their procedures along with their results.

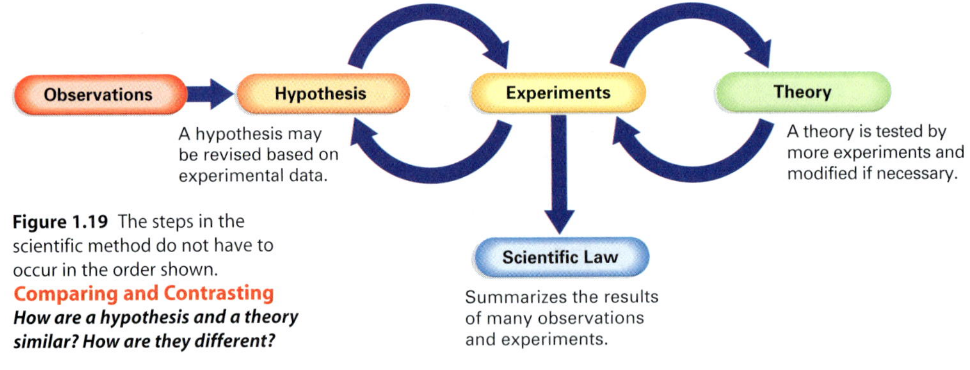

Figure 1.19 The steps in the scientific method do not have to occur in the order shown.
Comparing and Contrasting ***How are a hypothesis and a theory similar? How are they different?***

22 Chapter 1

Differentiated Instruction

English Learners 

Have English learners work with a partner to describe, act out, or illustrate in a drawing the different steps that make up the scientific method. Challenge students to identify how they used each step to solve problems in the past.

Quick LAB

Bubbles!

Purpose
To test the hypothesis that bubble making can be affected by adding sugar or salt to a bubble-blowing mixture.

Materials
- 3 plastic drinking cups
- liquid dish detergent
- measuring cup and spoons
- water
- table sugar
- table salt
- drinking straw

Procedure  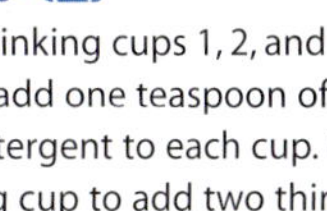

1. Label three drinking cups 1, 2, and 3. Measure and add one teaspoon of liquid dish detergent to each cup. Use the measuring cup to add two thirds of a cup of water to each drinking cup. Then swirl the cups to form a clear mixture. **CAUTION** *Wipe up any spills immediately so that no one will slip and fall.*
2. Add a half teaspoon of table sugar to cup 2 and a half teaspoon of table salt to cup 3. Swirl each cup for one minute.
3. Dip the drinking straw into cup 1, remove it, and blow gently into the straw to make the largest bubble you can. Practice making bubbles until you feel you have reasonable control over your bubble production.
4. Repeat Step 3 with the mixtures in cups 2 and 3.

Analyze and Conclude

1. Did you observe any differences in your ability to produce bubbles using the mixtures in cup 1 and cup 2?
2. Did you observe any differences in your ability to produce bubbles using the mixtures in cup 1 and cup 3?
3. What can you conclude about the effects of table sugar and table salt on your ability to produce bubbles?
4. Propose another hypothesis related to bubble making and design an experiment to test your hypothesis.

Developing Theories Once a hypothesis meets the test of repeated experimentation, it may be raised to a higher level of ideas. It may become a theory. A **theory** is a well-tested explanation for a broad set of observations. In chemistry, one theory addresses the fundamental structure of matter. This theory is very useful because it helps you form mental pictures of objects that you cannot see. Other theories allow you to predict the behavior of matter.

When scientists say that a theory can never be proved, they are not saying that a theory is unreliable. They are simply leaving open the possibility that a theory may need to be changed at some point in the future to explain new observations or experimental results.

Scientific Laws Figure 1.19 shows how scientific experiments can lead to laws as well as theories. A **scientific law** is a concise statement that summarizes the results of many observations and experiments. In Chapter 14, you will study laws that describe how gases behave. One law describes the relationship between the volume of a gas in a container and its temperature. If all other variables are kept constant, the volume of the gas increases as the temperature increases. The law doesn't try to explain the relationship it describes. That explanation requires a theory.

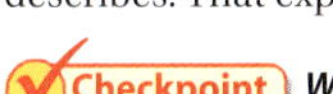

When can a hypothesis become a theory?

For: Links on Scientific Methods
Visit: www.SciLinks.org
Web Code: cdn-1012

Quick LAB

Bubbles! 

Objective After completing this activity, students will be able to:
- test the hypothesis that bubble making can be affected by adding sugar or salt to the bubble-blowing mixture.

Skills Focus Observing, formulating hypotheses, drawing conclusions

Prep Time 5 minutes
Class Time 15 minutes

Safety Remind students to be careful not to draw any liquid into their mouth through the straw.

Expected Outcome Students conclude that sugar has no effect on bubble production and salt prevents bubble production.

Analyze and Conclude

1. no
2. Yes; no bubbles formed from the liquid in cup 3.
3. Sugar has no effect on bubble production, but salt stops it completely.
4. Answers might include examining the effect of temperature or dilution of the bubble-making mixture. For example, diluting the mixture can reverse the salt effect.

For Enrichment 

Facilitate a class discussion about the experiments that students proposed for Question 4. Discuss whether or not the experiments will work. Perform two of the experiments and discuss the results. Emphasize the need to learn from "failed" experiments.

Download a worksheet on **Scientific Methods** for students to complete, and find additional teacher support from NSTA SciLinks.

Answers to...

Figure 1.19 A hypothesis and theory are both explanations. A hypothesis is only a proposed explanation; a theory is a well-tested explanation for a broad set of observations.

When a hypothesis meets the test of repeated experimentation, it may be become a theory.

Collaboration and Communication

Researching Collaborative Science Projects L2

Have students do research to find examples of successful collaborative projects that involved a multidisciplinary approach to solving a scientific problem. One example is the series of NASA missions to the moon in the 1970s.

Discuss L2

Stress that students should not avoid reading about science, but they should look for reliable sources and approach the news with a certain amount of healthy skepticism. Ask, **How can the Internet help people learn about advances in science?** *(Anyone can access information on the Internet.)* **What is one disadvantage of getting information from the Internet?** *(The information is not always reliable.)*

Figure 1.20 For a volleyball team to win, the players must collaborate, or work together.

Collaboration and Communication

No matter how talented the players on a team, one player cannot ensure victory for the team. Individuals must collaborate, or work together, for the good of the team. Think about the volleyball players in Figure 1.20. In volleyball, the person who spikes the ball depends on the person who sets the ball. Unless the ball is set properly, the spiker will have limited success. Many sports recognize the importance of collaboration by keeping track of assists. During a volleyball game, the players also communicate with one another so it is clear who is going to do which task. Strategies that are successful in sports can work in other fields, such as science. **When scientists collaborate and communicate, they increase the likelihood of a successful outcome.**

Collaboration Scientists choose to collaborate for different reasons. For example, some research problems are so complex that no one person could have all the knowledge, skills, and resources to solve the problem. It is often necessary to bring together individuals from different disciplines. Each scientist will typically bring different knowledge and, perhaps, a different approach to bear on a problem. Just talking with a scientist from another discipline may provide insights that are helpful.

There may be a practical reason for collaboration. For example, an industry may give a university funding for pure research in an area of interest to the industry. Scientists at the university get the equipment and the time required to do research. In exchange, the scientists provide ideas and expertise. The industry may profit from its investment by marketing applications based on the research.

Collaboration isn't always a smooth process. Conflicts can arise about use of resources, amount of work, who is to receive credit, and when and what to publish. Like the students in Figure 1.21, you will likely work on a team in the laboratory. If so, you may face some challenges. But you can also experience the benefits of a successful collaboration.

Figure 1.21 Working in a group can be challenging, but it can also be rewarding. **Applying Concepts** ***What steps in the scientific method are these students using?***

24 Chapter 1

Communication The way that scientists communicate with each other and with the public has changed over the centuries. In earlier centuries, scientists exchanged ideas through letters. They also formed societies to discuss the latest work of their members. When societies began to publish journals, scientists could use the journals to keep up with new discoveries.

Today, many scientists, like those in Figure 1.22, work as a team. They can communicate face to face. They also can exchange ideas with other scientists by e-mail, by phone, and at international conferences. Scientists still publish their results in scientific journals, which are the most reliable source of information about new discoveries. Articles are published only after being reviewed by experts in the author's field. Reviewers may find errors in experimental design or challenge the author's conclusions. This review process is good for science because work that is not well founded is usually not published.

The Internet is a major source of information. One advantage of the Internet is that anyone can get access to its information. One disadvantage is that anyone can post information on the Internet without first having that information reviewed. To judge the reliability of information you find on the Internet, you have to consider the source. This same advice applies to articles in newspapers and magazines or the news you receive from television. If a media outlet has a reporter who specializes in science, chances are better that a report will be accurate.

Figure 1.22 Communication between scientists can occur face to face. These chemists are using the model projected on the screen to discuss the merits of a new medicine.

1.3 Section Assessment

16. **Key Concept** What did alchemists contribute to the development of chemistry?
17. **Key Concept** How did Lavoisier revolutionize the science of chemistry?
18. **Key Concept** Name three steps in the scientific method.
19. **Key Concept** Explain why collaboration and communication are important in science.
20. How did Lavoisier's wife help him to communicate the results of his experiments?
21. Why should a hypothesis be developed before experiments take place?
22. Why is it important for scientists to publish a description of their procedures along with the results of their experiments?
23. What is the difference between a theory and a hypothesis?
24. What process takes place before an article is published in a scientific journal?
25. In Chapter 2, you will learn that matter is neither created nor destroyed in any chemical change. Is this statement a theory or a law? Explain your answer.

Connecting Concepts

Being an Informed Citizen Write a paragraph explaining how you can learn about the research that is done by scientists. Then explain how this information could help you be an informed citizen.

Assessment 1.3 Test yourself on the concepts in Section 1.3.

with ChemASAP

FYI

The peer review of journal articles can be painful for a scientist, but it is good for science as a whole because work that is not well founded is usually not published. Some journals publish preliminary results on the Internet because the review process can take from six months to a year, or even longer. Preliminary results should be treated with caution because such results are subject to change.

3 ASSESS

Evaluate Understanding 

Ask students to choose an everyday type of problem and explain how they would use the scientific method to solve the problem.

Reteach L1

Students may mistakenly infer that a theory "grows" into a law by constant testing and refinement. Or they may confuse a theory with a hypothesis. To clarify these concepts, ask students to describe *theory* and *law* in their own words.

Connecting Concepts

Students can learn about research from news reports, from specialized journals, and from the Internet. The more reliable information people have, the better able they are to effectively address public issues related to technology

Interactive Textbook

If your class subscribes to the Interactive Textbook, use it to review key concepts in Section 1.3.

with ChemASAP

Section 1.3 Assessment

16. They developed the tools and techniques for working with chemicals.
17. He helped transform chemistry from a science of observation to a science of measurement.
18. making observations, testing hypotheses, and developing theories
19. They help increase the likelihood of a successful outcome.
20. She made drawings of his experiments and translated scientific papers.
21. It guides the design of the experiments.
22. so that other scientists can repeat the experiments and confirm the results
23. A theory is a well-tested explanation of a broad set of observations; a hypothesis is a proposed explanation for an observation.
24. Articles are reviewed by experts in the author's field of research.
25. a law. It is not an explanation.

Answers to...

Figure 1.21 testing a hypothesis and making observations

Laboratory Safety 

Objective After completing this activity, students will be able to

- demonstrate their knowledge of safe laboratory practices.

Class Time 40 minutes

Skills Focus
Observing, communicating results

Teaching Tips

- Students need to read Appendix D to answer the questions in the lab.
- Use this activity as part of an orientation in which you present your rules for working in the laboratory.
- Provide a floor plan of the room and have students record the location of safety equipment such as an eyewash fountain or a fire extinguisher.
- After you discuss the safety rules, have students sign a safety contract. For an example, see the safety contract in the *Small-Scale Chemistry Laboratory Manual*, p. 15.

Answers to Questions

 Wear safety goggles at all times when working in the lab.

 Tell your teacher and nearby classmates. Dispose of the glass as instructed by your teacher.

 Stand back, notify your teacher, and warn other students.

 Tie back long hair and loose clothing. Never reach across a lit burner. Keep flammable materials away from the flame.

 Wash your hands thoroughly with soap and water.

 It isn't always appropriate to dispose of chemicals by flushing them down the drain. Follow your teacher's instructions for disposal.

Small-Scale LAB

Laboratory Safety

Purpose
To demonstrate your knowledge of safe laboratory practices.

Procedure
While doing the chemistry experiments in this textbook, you will work with equipment similar to the equipment shown in the photographs. Your success, and your safety, will depend on following instructions and using safe laboratory practices. To test your knowledge of these practices, answer the question after each safety symbol. Refer to the safety rules in Appendix D and any instructions provided by your teacher.

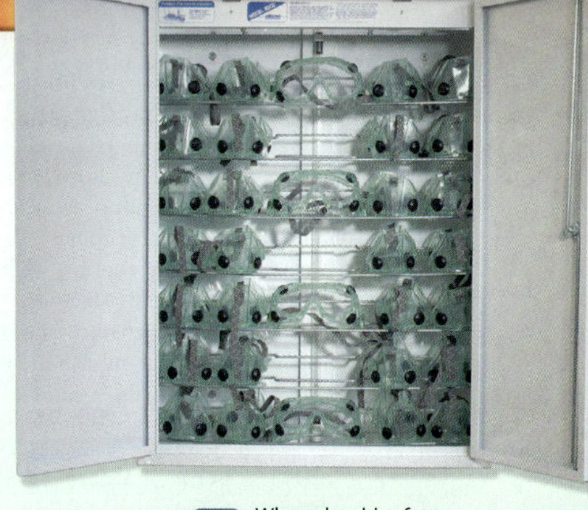

 When should safety goggles be worn?

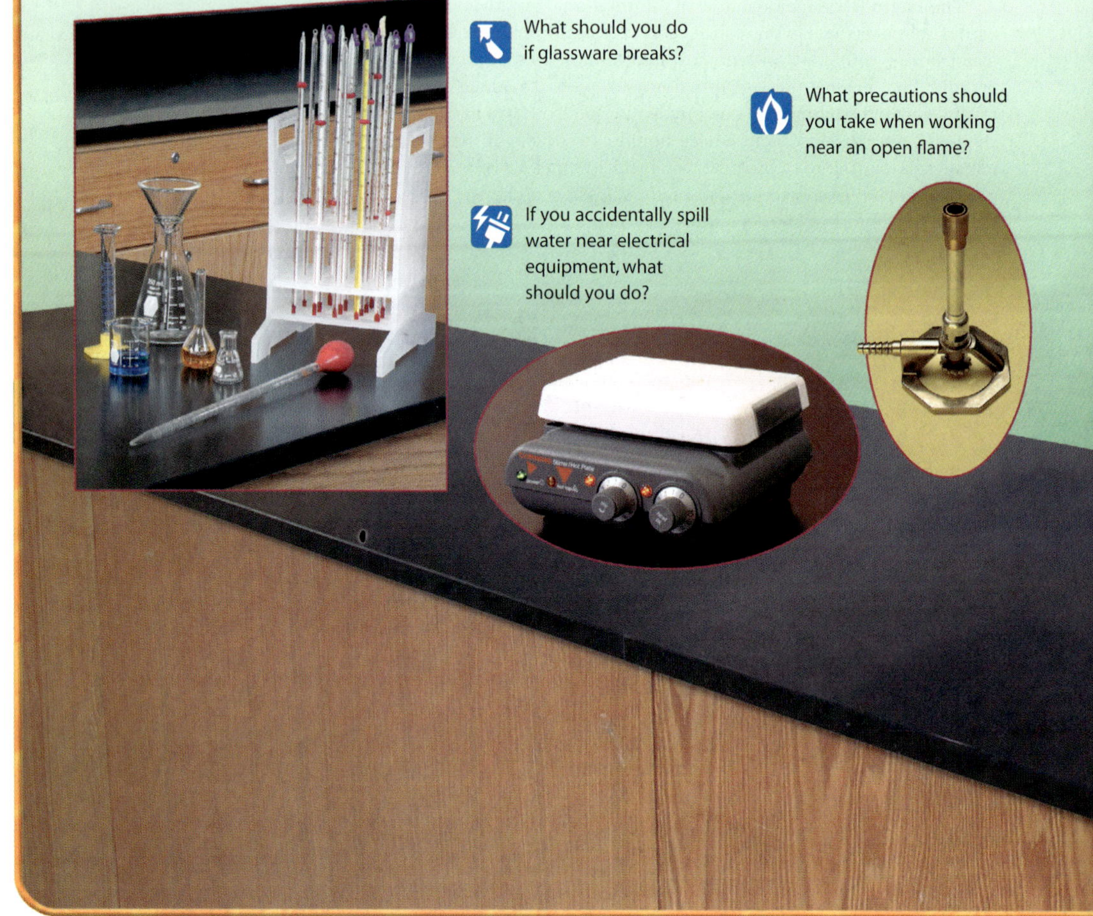

What should you do if glassware breaks?

What precautions should you take when working near an open flame?

If you accidentally spill water near electrical equipment, what should you do?

After you clean up your work area, what should you do before leaving the laboratory?

Is it always appropriate to dispose of chemicals by flushing them down the sink? Explain.

Background

The small-scale chemistry experiments in this book are designed to help you teach students important chemical principles, not just process. For most experiments, the procedure is short and simple. In many cases students are asked to construct a grid, place a small-scale reaction surface over the grid, do the experiment, and record their results in a similar grid. Sometimes the students are told to mark the grid with black Xs. These Xs provide black and white backgrounds against which students can observe reaction mixtures.

The YOU'RE THE CHEMIST activities ask students to apply what they learned in the basic experiment. Some of these activities could be used for performance-based assessment.

For Enrichment L3

Have students look at the safety rules in Appendix D. Have each student choose five of the rules in Appendix D and make a list of what things could happen if the rule is not followed. Divide students into groups of 3 and have them share their answers with each other.

1.4

1 FOCUS

Objectives

1.4.1 **Identify** two general steps in problem solving.

1.4.2 **Describe** three steps for solving numeric problems.

1.4.3 **Describe** two steps for solving conceptual problems.

Guide for Reading

Build Vocabulary L2

Word Forms Students may benefit from a reminder that certain key words and phrases in each word problem indicate the unknown quantity and its units. These include *how much, determine, what is, find,* and *how long.*

Reading Strategy L2

Think Aloud Some students are convinced they cannot do word problems because they are too hard. Ask students to describe the process verbally as they attempt to solve a problem.

2 INSTRUCT

Connecting to Your World

Have students study the photograph and read the text that opens the section. Ask students how they would approach the situation pictured. Ask, What is the most effective way to solve problems? *(Approaches vary depending on the given problem; effective approaches are generally organized and well-planned.)*

Skills Used in Solving Problems

Discuss L2

Explain that the memorization of facts is a relatively small part of learning chemistry. A person who succeeds in chemistry has become a good problem solver. Encourage students to share problem-solving methods and techniques. Discuss the "supermarket" problems. Stress that for these problems, unlike most problems in a chemistry course, answers can vary and still be correct.

1.4 Problem Solving in Chemistry

Guide for Reading

Key Concepts

- What is a general approach to solving a problem?
- What are the three steps for solving numeric problems?
- What are the two steps for solving conceptual problems?

Reading Strategy

Identifying Main Idea/Details Under the heading Solving Numeric Problems, there are three main ideas presented as subheads. As you read, list two details that support each main idea.

Connecting to Your World

Shape-sorter toys fascinate young children. Typically, the children try placing a shape in different holes until they find the right one. They may try to place an incorrect shape in the same hole over and over again. An older child has enough experience to place the correct shape in each hole on the first try. The trial-and-error approach used by young children is one method of problem solving, but it is usually not the best one. In this section, you will learn effective ways to solve problems in chemistry.

Skills Used in Solving Problems

Problem solving is a skill you use all the time. You are in a supermarket. Do you buy a name brand or the store brand of peanut butter? Do you buy the 1-liter bottle or the 2-liter bottle of a carbonated beverage? Do you choose the express line if there are five customers ahead of you or the non-express line with a single shopper who has lots of items?

When you solve a problem you may have a data table, a graph, or another type of visual to refer to. The shopper in Figure 1.23 is reading the label on a can while trying to decide whether to buy the item. She may need to avoid certain ingredients because of a food allergy. Or she may want to know the amount of Calories per serving.

The skills you use to solve a word problem in chemistry are not that different from those you use while shopping or cooking or planning a party. **Effective problem solving always involves developing a plan and then implementing that plan.**

Figure 1.23 A shopper must make many decisions. Some of those decisions are based on data, like the information on a food label.

Section Resources

Print

- **Guided Reading and Study Workbook,** Section 1.4
- **Core Teaching Resources,** Section 1.4 Review
- **Transparencies,** T7–T9

Technology

- **Interactive Textbook with ChemASAP,** Problem-Solving 1.27, 1.29; Assessment 1.4

Solving Numeric Problems

Because measurement is such an important part of chemistry, most word problems in chemistry require some math. The techniques used in this book to solve numeric problems are conveniently organized into a three-step, problem-solving approach. This approach has been shown to be very helpful and effective. So we recommend that you follow this approach when working on numeric problems in this textbook. **The steps for solving a numeric word problem are analyze, calculate, and evaluate.** Figure 1.24 summarizes the three-step process and Sample Problem 1.1 shows how the steps work in a problem.

1 Analyze To solve a word problem, you must first determine where you are starting from (identify what is known) and where you are going (identify the unknown). What is known may be a measurement. Or it may be an equation that shows a relationship between measurements. If you expect the answer (the unknown) to be a number, you need to determine what units the answer should have before you do any calculations.

After you identify the known and the unknown, you need to make a plan for getting from the known to the unknown. Planning is at the heart of successful problem solving. As part of planning, you might draw a diagram that helps you visualize a relationship between the known and the unknown. You might need to use a table or graph to identify data or to identify a relationship between a known quantity and the unknown. You may need to select an equation that you can use to calculate the unknown.

2 Calculate If you make an effective plan, doing the calculations is usually the easiest part of the process. For some problems, you will have to convert a measurement from one unit to another. Or you may need to rearrange an equation before you can solve for an unknown. However, you will be taught these math skills as needed. There will also be reminders throughout the textbook to use the Math Handbook in Appendix C.

3 Evaluate After you calculate an answer, you should evaluate it. Is the answer reasonable? Does it make sense? If not, reread the word problem. Did you copy the data correctly? Did you choose the right equations? It helps to round off the numbers and make an estimate of the answer. If the answer is much larger or much smaller than your estimate, check your calculations.

Check that your answer has the correct unit and the correct number of significant figures. You may need to use scientific notation in your answer. You will study significant figures and scientific notation in Chapter 3.

Checkpoint ***How can making an estimate help you evaluate an answer?***

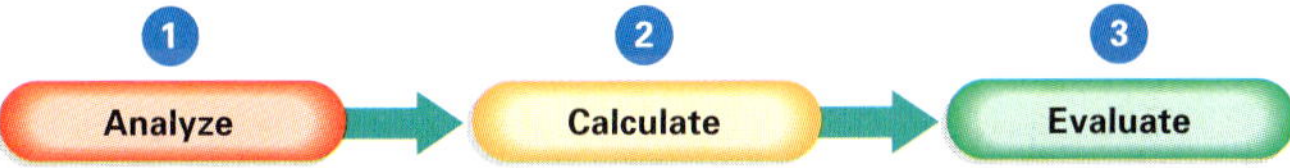

Figure 1.24 This flowchart summarizes the steps for solving a numeric problem. **Predicting** ***In which step do you make a plan for getting from what is known to what is unknown?***

Facts and Figures

How Teachers Solve Math Problems

A Professor of Mathematics at Stanford University, G.Polya (1887–1985), wrote a classic book about problem solving in 1945. The second edition of *How to Solve It* is still in print. Polya used common sense and humor to present problem-solving tactics.

The general four-step approach that Polya outlined in his introduction (understand, plan, try it, and look back) has much in common with the three-step approach for solving numeric problems (analyze, calculate, and evaluate).

Solving Numeric Problems

Discuss 

Point out that in the laboratory, as well as during students' normal routines, they are often presented with more data than is needed to solve a problem. Explain that they will need to sort essential data from extraneous data. In cases with more than the required data, sorting the data into knowns and unknowns helps students to determine the answer to "What data do I need to solve this problem?"

Answers to...

Figure 1.24 Step 1, analyze

If the calculated answer is much larger or smaller than the estimate, check the calculations.

Sample Problem 1.1

Answers

26. 24 short blocks

27. 24 minutes

Discuss L2

Write the sample problem on the board and carry out the three-step problem-solving approach. Stress the discipline of writing down the steps that make the solution possible. You may wish to use chalk or markers of different colors for the knowns and unknowns to make the process clearer. Stress the importance of evaluating whether the result makes sense. Students should make an initial mental estimate to compare with the final result displayed by a calculator.

FYI

Conversion problems are the most common type of numeric problem in an introductory chemistry course. They will be discussed in detail in Chapter 3. In that chapter, students will learn how to combine the two calculate steps in Sample Problem 1.1 into a single step.

This view of Indianapolis, Indiana, shows part of the historic central canal in White River State Park.

SAMPLE PROBLEM 1.1

Estimating Walking Time

You are visiting Indianapolis for the first time. Because it is a nice day, you decide to walk from the Indiana State Capital to the Murat Centre for an afternoon performance. According to the map in Figure 1.25, the shortest route from the capital to the theater is 8 blocks. How many minutes will the trip take if you can walk one mile in 20 minutes? Assume that 10 short city blocks equals one mile.

1 Analyze ***List the knowns and the unknown.***

Knowns

- distance to be traveled = 8 blocks
- walking speed = 1 mile/20 minutes
- 1 mile = 10 blocks

Unknown

- time of trip = ? minutes

This problem is an example of what is typically called a conversion problem. In a conversion problem, one unit of measure (in this case, blocks) must be expressed in a different unit (in this case, minutes).

Divide the distance to be traveled (in blocks) by the number of blocks in one mile to get the distance of the trip in miles. Then multiply the number of miles by the time it takes to walk one mile.

2 Calculate ***Solve for the unknown***

$$8\ \cancel{\text{blocks}} \times \frac{1\ \text{mile}}{10\ \cancel{\text{blocks}}} = 0.8\ \text{mile}$$

$$0.8\ \cancel{\text{mile}} \times \frac{20\ \text{minutes}}{1\ \cancel{\text{mile}}} = 16\ \text{minutes}$$

3 Evaluate ***Does the result make sense?***

The answer seems reasonable, 16 minutes to walk 8 short blocks. The answer has the correct unit. The relationships used are correct.

Practice Problems

26. Using the information in the sample problem, how many short blocks can be walked in 48 minutes?

27. There is an ice cream shop 6 blocks north of your hotel. How many minutes will it take to walk there and back?

Interactive Textbook

Problem-Solving 1.27
Solve Problem 27 with the help of an interactive guided tutorial.
with ChemASAP

30 Chapter 1

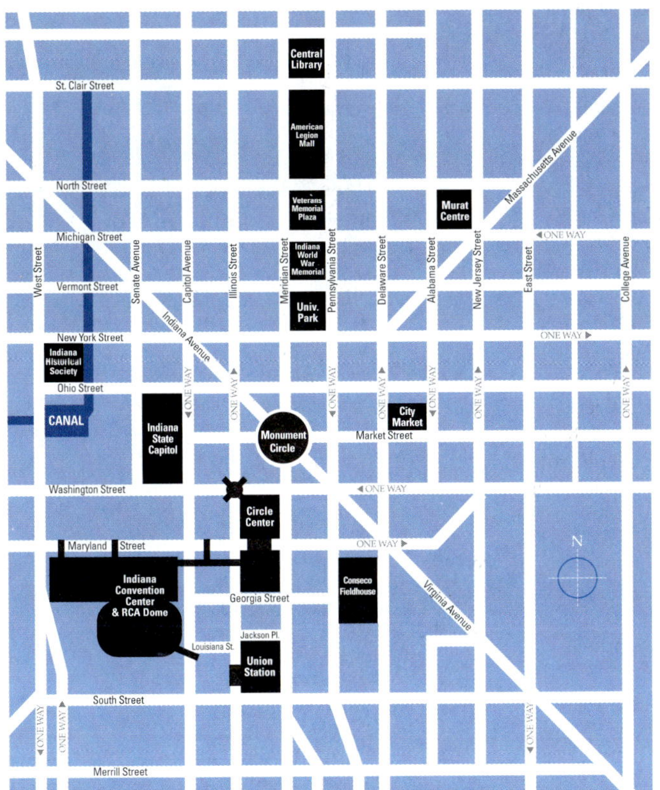

Figure 1.25 Refer to this map of Indianapolis, Indiana, while you do Sample Problem 1.1.
Interpreting Diagrams
In the section of downtown bounded by north, east, south, and west streets, the main streets and avenues are named for states. What are the five exceptions to this pattern?

Solving Conceptual Problems

Not every word problem in chemistry requires calculations. Some problems ask you to apply the concepts you are studying to a new situation. In this text, these nonnumeric problems are labeled conceptual problems. To solve a conceptual problem, you still need to identify what is known and what is unknown. Most importantly, you still need to make a plan for getting from the known to the unknown. But if your answer is not a number, you do not need to check the units, make an estimate, or check your calculations.

The three-step problem-solving approach is modified for conceptual problems. **The steps for solving a conceptual problem are analyze and solve.** Figure 1.26 summarizes the process, and Conceptual Problem 1.1 on the next page shows how the steps work in an actual problem.

Figure 1.26 This flowchart shows the two steps used for solving a conceptual problem.
Comparing and Contrasting
With a conceptual problem, why is the second step called Solve rather than Calculate?

Solving Conceptual Problems

Fit an Ice Cube in a Bottle L2

Purpose Students suggest different approaches for solving a problem.

Materials bowl of ice cubes, empty narrow-neck bottle

Procedure Place a bowl of ice cubes and an empty soda bottle in front of the class. Explain that the problem is to transfer the ice to the inside of the bottle. Have students analyze the problem and suggest different approaches.

Expected Outcome Students may suggest crushing the ice so that pieces are small enough to fit through the mouth of the bottle; or melting the ice, pouring the water into the bottle, and placing the bottle in a freezer.

Answers to...

Figure 1.25 Meridian Street, Senate Avenue, Market Street, Jackson Place, and Capitol Avenue

Figure 1.26 Solving a conceptual problem usually does not require any calculations.

CONCEPTUAL PROBLEM 1.1

Answers

28. The order of the morning errands cannot vary. The order of the afternoon errands can vary, as long as the haircut takes place before 3 pm.

29. Possible answers: Do an errand during lunch hour or extend the hours during which he does errands.

Discuss L2

Discuss using classification schemes as tools for organizing information. Ask students to suggest classification schemes they encounter in their daily lives (the ways that goods are organized in stores or books in libraries).

3 ASSESS

Evaluate Understanding L2

To determine student understanding of the three-step problem solving approach, ask students to suggest ways to evaluate an answer. *(Reread the problem to be sure the answer supplies the requested unknown; round off the numbers and do a quick estimate.)*

Reteach L1

Have students try working in pairs to solve problems. One student thinks aloud while trying to solve a problem. The other keeps a careful record of the process. Then they reverse roles and work on another problem.

Writing Activity

In both cases, the solver analyzes the problem, makes a plan, and carries out the plan. Problems with numeric answers require that the answer be evaluated to see if it is reasonable.

If your class subscribes to the Interactive Textbook, use it to review key concepts in Section 1.4.

with ChemASAP

CONCEPTUAL PROBLEM 1.1

Running Errands

Manny has to run 6 errands between 10 and 5 on Saturday. He must get a haircut, wash his car, buy stamps, rent a video, return a library book, and buy some groceries. Assume that each errand will take 30 minutes and that Manny will do only one errand per hour. Manny will stop for a lunch break between 12 and 1. Use the information in the drawing to figure out a way for Manny to accomplish all 6 tasks.

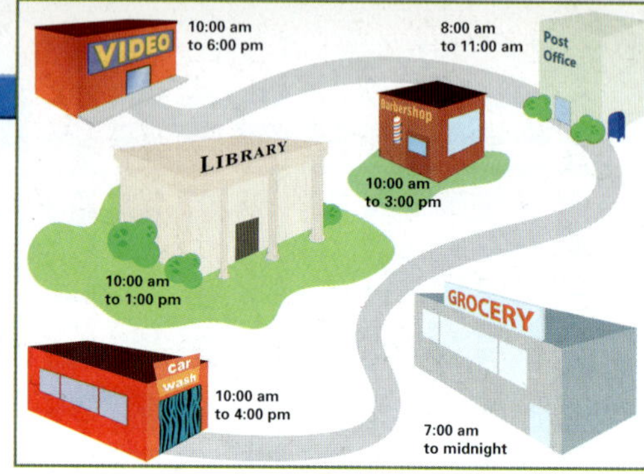

1 Analyze ***Identify the relevant concepts.***

Each place that Manny needs to visit is open for a limited number of hours on Saturday. Manny must do his errands between 10 and 12, and between 1 and 5. At a rate of one errand per hour, Manny must do 2 errands before lunch and 4 errands after lunch.

2 Solve ***Apply concepts to this situation.***

The post office and library are open only in the morning. The barbershop and the car wash close earlier than the video store. The supermarket is open late. One possible order for the errands is post office, library, barbershop, car wash, video store, and supermarket.

Practice Problems

28. Describe two alternative orders in which Manny could complete his errands.

29. What if Manny had 7 errands instead of 6? What would he need to do to adjust for the extra errand?

Problem-Solving 1.29
Solve Problem 29 with the help of an interactive guided tutorial.

with ChemASAP

1.4 Section Assessment

30. **Key Concept** What are the two general steps in successful problem solving?

31. **Key Concept** List the three steps for solving numeric problems.

32. **Key Concept** List the two steps for solving conceptual problems.

33. Read the conversion problem and then answer the questions. "There are 3600 seconds in an hour. How many seconds are there in one day?"
- **a.** Identify the known and the unknown.
- **b.** What relationship between the known and unknown do you need to solve the problem?
- **c.** Calculate the answer to the problem.
- **d.** Evaluate your answer and explain why your answer makes sense.

Writing Activity

Compare and Contrast Paragraph Write a paragraph comparing the processes for solving numeric problems and conceptual problems. How are the processes similar? In what way are they different?

Assessment 1.4 Test yourself on the concepts in Section 1.4.

with ChemASAP

Section 1.4 Assessment

30. Develop a plan and implement the plan.
31. analyze, calculate, and evaluate
32. analyze and solve
33. a. known: 3600 s = 1 h; unknown: ? s = 1 day
b. 24 h = 1 day
c. 3600 s/h × 24 h/day = 86,400 s/day
d. 86,400 seconds in one day seems reasonable in relationship to 3600 seconds in one hour. The answer has the correct units and the relationship used is correct.

CHAPTER 1

Study Guide

Key Concepts

1.1 Chemistry

- Because living and nonliving things are made of matter, chemistry affects all aspects of life and most natural events.
- Chemistry can be divided into five traditional areas of study: organic chemistry, inorganic chemistry, biochemistry, analytical chemistry, and physical chemistry.
- Pure research can lead directly to an application, but an application can exist before research is done to explain how it works.
- Chemistry can be useful in explaining the natural world, preparing people for career opportunities, and producing informed citizens.

1.2 Chemistry Far and Wide

- Chemists design materials to fit specific needs. Chemists play an essential role in finding ways to conserve energy, produce energy, and store energy.
- Chemists supply the medicines, materials, and technology that doctors use to treat patients. Chemists help to develop more productive crops and safer, more effective ways to protect crops.
- Chemists help to identify pollutants and prevent pollution.
- To study the universe, chemists gather data from afar and analyze matter that is brought back to Earth.

1.3 Thinking Like a Scientist

- Alchemists developed tools and techniques for working with chemicals.
- Lavoisier helped to transform chemistry from a science of observation to the science of measurement that it is today.
- Steps in the scientific method include making observations, testing hypotheses, and developing theories.
- When scientists collaborate and communicate, they increase the likelihood of a successful outcome.

1.4 Problem Solving in Chemistry

- Effective problem solving always involves developing a plan and then implementing the plan.
- The steps for solving a numeric word problem are analyze, calculate, and evaluate. The steps for solving a conceptual problem are analyze and solve.

Vocabulary

- analytical chemistry (p. 8)
- applied chemistry (p. 9)
- biochemistry (p. 8)
- biotechnology (p. 14)
- chemistry (p. 7)
- experiment (p. 22)
- hypothesis (p. 22)
- inorganic chemistry (p. 8)
- macroscopic (p. 12)
- matter (p. 7)
- microscopic (p. 12)
- manipulated variable (p. 22)
- observation (p. 22)
- organic chemistry (p. 8)
- physical chemistry (p. 8)
- pollutant (p. 16)
- pure chemistry (p. 9)
- responding variable (p. 22)
- scientific law (p. 23)
- scientific method (p. 22)
- technology (p. 9)
- theory (p. 23)

Organizing Information

Use these terms to construct a concept map that organizes the major ideas of this chapter.

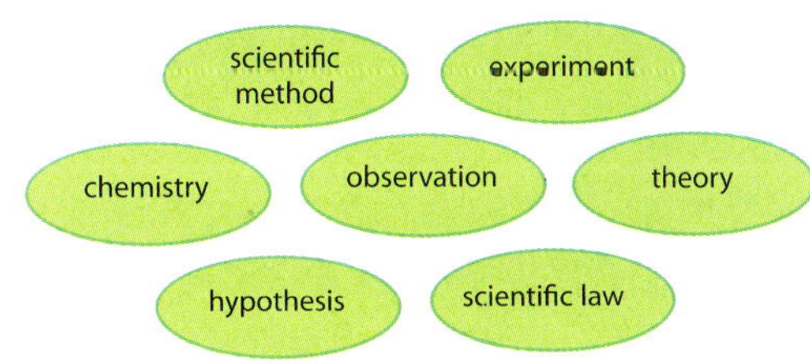

Concept Map 1 Solve the Concept Map with the help of an interactive guided tutorial.

with ChemASAP

CHAPTER 1

Study Guide

Study Tip

Study with a friend Encourage students to find a study partner. Tell them to review each section and quiz each other on the key concepts. Suggest that they separately create a summary for each section, then compare notes.

If your class subscribes to the Interactive Textbook, your students can go online to access an interactive version of the Student Edition and a self-test.

with ChemASAP

Chapter Resources

Print

Core Teaching Resources, Chapter 1
Practice Problems, Vocabulary Review, Quiz, Chapter Test A, Chapter Test B

Technology

Computer Test Bank, Chapter 1 Test
Interactive Textbook with ChemASAP, Chapter 1

CHAPTER 1

ASSESSMENT

Reviewing Content

34. Although air is "invisible," it has mass and occupies space.
35. Chemistry is concerned with the changes that matter undergoes.
36. Organic chemistry is the study of chemicals that contain carbon; inorganic chemistry is the study of chemicals that do not contain carbon.
37. pure chemistry; His experiments tested a hypothesis of another chemist. His results led to applied chemistry—the large-scale production of nylon.
38. The group of chemicals in the body that produce pain are also involved in the formation of blood clots. Aspirin blocks the production of these chemicals.
39. A firefighter needs to know which chemicals to use to fight different types of fires; knowledge of chemistry will help a reporter gather information during an interview with a chemist.
40. In a macroscopic view you can see an object like a burr with the unaided eye; to observe the tiny hooks on a burr's surface requires magnification (a microscopic view).
41. Insulation acts as a barrier to heat flow. If heat flow is reduced, energy is conserved.
42. to understand the structure of matter found in cells and chemical changes that occur in cells
43. gene therapy and production of chemicals such as insulin
44. Testing reveals if soil contains the right chemicals to grow a crop and suggests way to improve soil.
45. A pollutant is a material found in air, soil, or water that is harmful to living organisms.
46. Lead poisoning can cause nervous system damage, especially in growing children.
47. by analyzing the light they transmit to Earth

CHAPTER 1

Assessment

Reviewing Content

1.1 Chemistry

34. Explain why air is classified as matter.

35. The Japanese characters for chemistry literally mean "change study." Why are these appropriate characters to represent chemistry?

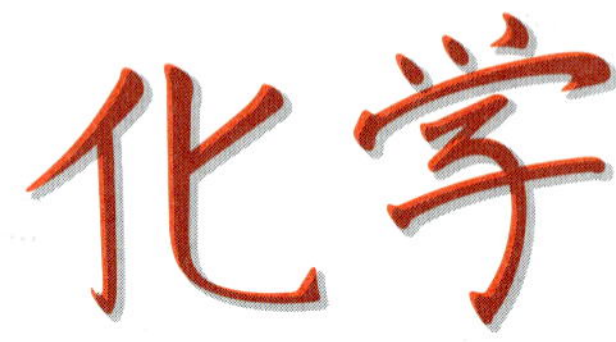

36. Describe the main difference between inorganic chemistry and organic chemistry.

37. Was Wallace Carothers doing pure chemistry or applied chemistry when he discovered nylon? Explain your answer.

38. Explain how chemists were able to connect the ability of aspirin to relieve pain to its ability to reduce the risk of a heart attack.

39. Why would a firefighter or a reporter need to understand chemistry?

1.2 Chemistry Far and Wide

40. George de Mestral used burrs as a model for his hook-and-loop tapes. Using burrs as an example, explain the difference between a macroscopic and a microscopic view of matter.

41. How does the use of insulation help to conserve energy?

42. What is the overall goal of scientists who work with biotechnology?

43. Describe two ways that biotechnology can be used to treat diseases.

44. How can testing soil help to increase the productivity of farmland?

45. What is a pollutant?

46. Why is it important that young children not be exposed to lead-based paint?

47. How can scientists study the composition of distant stars?

1.3 Thinking Like a Scientist

48. What did the scientists who founded the Royal Society of London share with Lavoisier?

49. What is the most powerful tool that any scientist can have?

50. What is the purpose of an experiment?

51. Which of the following is not a part of the scientific method?

a. hypothesis **b.** experiment
c. guess **d.** theory

52. How do a manipulated variable and a responding variable differ?

53. You perform an experiment and get unexpected results. According to the scientific method, what should you do next?

54. Explain how the results of many experiments can lead to both a scientific law and a theory.

55. List two general reasons why scientists are likely to collaborate.

1.4 Problem Solving in Chemistry

56. Identify the statements that correctly describe good problem solvers.

a. Read a problem only once.
b. Check their work.
c. Look up missing facts.
d. Look for relationships among the data.

57. What do effective problem-solving strategies have in common?

58. In which step of the three-step problem-solving approach for numeric problems is a problem-solving strategy developed?

59. On the average, a baseball team wins two out of every three games it plays. How many games will this team lose in a 162-game season?

60. If your heart beats at an average rate of 72 times per minute, how many times will your heart beat in an hour? In a day?

61. How many days would it take you to count a million pennies if you could count one penny each second?

48. They based their conclusions on experimental evidence.
49. the scientific method
50. to test a hypothesis
51. c
52. The manipulated variable is changed during an experiment; the responding variable is observed during an experiment.
53. Repeat the experiment. If you get the same result, you must propose a new hypothesis.
54. A scientific law summarizes the results of many experiments; a theory explains them.
55. to share knowledge across disciplines and resources between industry and academia
56. b, c, and d
57. developing a plan and implementing that plan
58. Step 1, analyze
59. 54 games (one third of 163)
60. 4320 times an hour, 103,680 times a day
61. 12 days

Understanding Concepts

62. Match each area of chemistry with a numbered statement.

a. physical chemistry **b.** organic chemistry
c. analytical chemistry **d.** inorganic chemistry
e. biochemistry

(1) Measure the level of lead in blood.
(2) Study non-carbon-based chemicals in rocks.
(3) Investigate changes that occur as food is digested in the stomach.
(4) Study carbon-based chemicals in coal.
(5) Explain the energy transfer that occurs when ice melts.

63. Explain why chemistry might be useful in a career you are thinking of pursuing.

Use this photograph of a javelin thrower to answer Questions 64 and 65.

64. Explain how chemistry has affected the ability of this athlete to compete.

65. What type of chemist might study how an athlete uses energy during a competition? Give a reason for your answer.

66. A doctor examines a patient's sore throat and suggests that the patient has strep throat. She takes a sample to test for the bacteria that cause strep throat. What parts of the scientific method is the doctor applying?

67. You perform an experiment and find that the results do not agree with an accepted theory. Should you conclude that you made an error in your procedure? Explain.

68. A student is planning a science fair project called Does Temperature Affect How High a Basketball Can Bounce?

a. Based on the project title, identify the manipulated variable and the responding variable.

b. Name at least two factors that would need to be kept constant during the experiment.

69. Describe a situation in which you used at least two steps in the scientific method to solve a problem.

70. Pure water freezes at 0°C. A student wanted to test the effect of adding salt to the water. The table shows the data that was collected.

Effect of Salt on Freezing Point of Water

Salt Added	Freezing Point
5 g	−4.8°C
10 g	−9.7°C
15 g	−15.1°C
20 g	−15.0°C

a. What was the manipulated variable?
b. What was the responding variable?
c. Why must the volume of water be the same for each test?
d. Based on the data, the student hypothesized, "As more salt is added to water, the temperature of the water decreases." Is this hypothesis supported by the data? Explain.

71. In the time a person on a bicycle travels 4 miles, a person in a car travels 30 miles. Assuming a constant speed, how far will the car travel while the bicycle travels 40 miles?

Understanding Concepts

62. Answers will vary, but possible answers are 1c, 2d, 3e, 4b, and 5a.

63. Answers will vary but should demonstrate an understanding that chemistry is the study of matter and the changes it undergoes.

64. One possible answer is development of the materials to produce an artificial limb.

65. Students are likely to choose biochemist because biochemistry is the study of processes that take place in organisms. However, a student might choose physical chemist because physical chemistry includes the study of energy transfer as matter undergoes a change.

66. The doctor's hypothesis is that the sore throat is the result of bacteria that cause strep throat. She tests the hypothesis by testing a sample for the presence of the bacteria.

67. Your experiment may be correct, but your hypothesis may be wrong. You should reexamine your hypothesis and repeat the experiment.

68. **a.** The manipulated variable is the temperature of the basketball. The responding variable is the height of the basketball's bounce.
b. The method for dropping the basketball before it bounces; the method for measuring the height of the bounce

69. Answers will vary but should reflect knowledge of the steps in a scientific method including making observations and testing hypotheses.

70. **a.** amount of salt added **b.** freezing point **c.** Changing the volume of the water would also change the concentration of the salt solution and affect the results. **d.** Yes, to a point. The freezing point appears to be leveling off at about 15°C.

71. 300 miles

CHAPTER 1 ASSESSMENT *continued*

Critical Thinking

72. One possible answer is that with both areas of study, students slowly build up a vocabulary and a set of concepts that relate the new terms. Both areas rely on a systematic approach—grammar versus the scientific method. Chemistry deals with matter and words do not qualify as matter. Science is independent of a specific culture; language varies from culture to culture.

73. A possible answer is that scientists accept hypotheses that are supported by experimental results and reject hypotheses that are not supported by experimental results.

74. **a.** the brand of paper towel
b. Possible answers include strongest, most absorbent, strongest when wet, largest area per sheet, and easiest to tear from roll.
c. Sample answer: The best towel will absorb the most liquid.
d. Possible answers include size of paper towel sample, type of liquid, amount of liquid, method for measuring absorbency, method for measuring strength.

75. A person who is educated in the theories and practice of chemistry would be more likely to recognize the significance of an accidental discovery and have the means and motivation to develop that accidental discovery into an important scientific contribution.

76. Divide the weight of 4 beakers by 2 or multiply the weight of 1 beaker by 2 (2.0 lb/2 or 0.5 lb x 2 = 1 lb).

77. A theory can never be proven. It is a well-tested explanation of a broad set of observations. A theory may need to be changed in the future to explain new observations.

78. b

CHAPTER 1 Assessment *continued*

Critical Thinking

72. Compare and contrast the study of chemistry with the study of a language.

73. Comment on the idea that science accepts what works and rejects what does not work.

74. You are asked to design an experiment to answer the question: "Which paper towel is the best?"
a. What is the manipulated variable in your experiment?
b. List three possible responding variables that could be used to define "best"?
c. Pick one of the responding variables and rewrite the question as a hypothesis.
d. List at least five factors that must be kept constant when you test the hypothesis.

75. Important discoveries in science are sometimes the result of an accident. Louis Pasteur said, "Chance favors the prepared mind." Explain how both of these statements can be true.

76. Four beakers have a total weight of 2.0 lb. Each beaker when full holds 0.5 lb of water. Describe two different methods you could use to calculate the weight of two full beakers of water. Then try both methods and compare the answers.

77. Explain what is wrong with the statement, "Theories are proven by experiments."

78. The air you breathe is composed of about 20% oxygen and 80% nitrogen. Use your problem solving skills to decide which drawing best represents a sample of air. Explain your choice.

a.
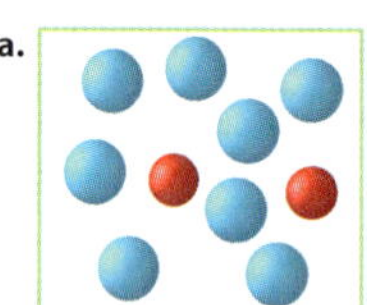

b.
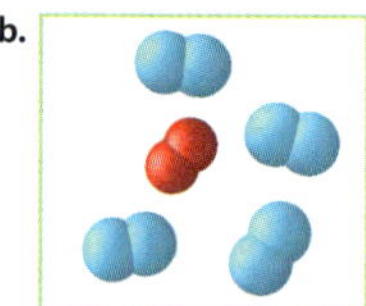

c.

d.
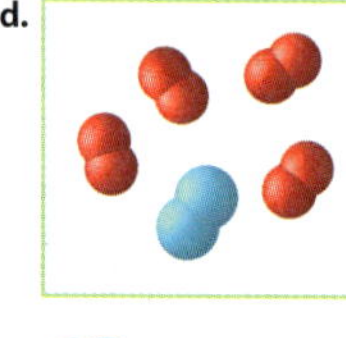

Oxygen

Nitrogen

Concept Challenge

79. You find a sealed box with strings protruding from three holes, as shown in the diagram. When you tug string A, it becomes longer and string C becomes shorter. When you tug string B, it becomes longer, but strings A and C are not affected. Make a diagram showing the arrangement of the strings inside the box.

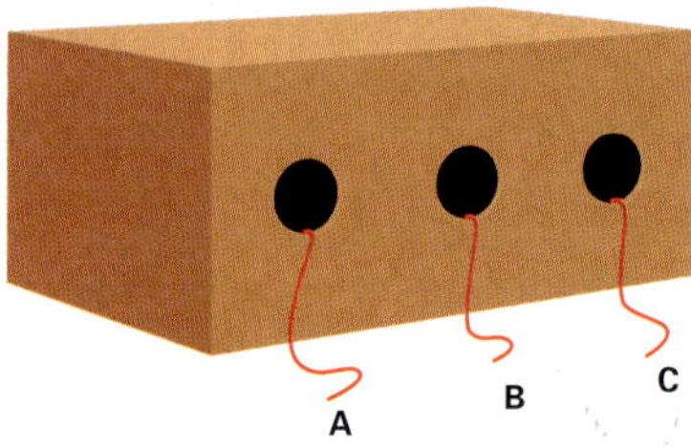

80. A certain ball when dropped from any height, bounces one-half the original height. If the ball was dropped from a height of 60 in. and allowed to bounce freely, what is the total distance the ball has traveled when it hits the ground for the third time? Assume the ball bounces straight up and down.

81. Eggs are shipped from the farm to market by truck. They are packed 12 eggs to a carton and 20 cartons to a box. Four boxes are placed in each crate. Crates are stacked on a truck 5 crates wide, 6 crates deep, and 5 crates high. How many eggs are on a truck?

82. An oil tanker containing 4,000,000 barrels is emptied at the rate of 5000 gallons per minute. What information do you need to figure out how long it would take to empty the tanker?

83. A crate of envelopes sells for $576.00. A package of envelopes contains 250 envelopes. Six packages are packed inside a carton. Twelve cartons are packed in a box. Eight boxes are packed in a crate.
a. What does a package of envelopes cost?
b. What fact given in the problem was not needed to calculate the answer?

Concept Challenge

79. Students' diagrams should show one string that is threaded through both holes A and C. The string at hole B is a separate thread from the string passing through holes A and C.

80. 150 in.

81. 144,000 eggs

82. the number of gallons in a barrel

83. **a.** $1.00 per package **b.** number of envelopes in a package

Standardized Test Prep

Test-Taking Tip

True-False Questions When the word *because* is placed between two statements that may or may not be true, you need to decide two things. First, are the statements both true, are they both false, or is one statement true and the other false. If both statements are true, you must then decide if the second statement is a correct explanation for the first statement.

Select the choice that best answers each question or completes each statement.

1. The branch of chemistry that studies chemicals containing carbon is ____ chemistry.
   a. physical
   b. inorganic
   c. analytical
   d. organic
2. An analytical chemist is most likely to
   a. explain why paint is stirred before it is used.
   b. explain what keeps paint attached to the steel frame of an automobile.
   c. identify the type of paint chips found at the scene of a hit-and-run accident.
   d. investigate the effect of leaded paint on the development of a young child.
3. Chemists who work in the biotechnology field are most likely to work with
   a. X-ray technicians.
   b. geologists.
   c. physicians.
   d. physicists.

Respond to each statement in Questions 4–6.

4. Someone who wears contact lenses does not have to wear safety goggles in the lab.
5. Eating food that is left over from an experiment is an alternative to discarding the food.
6. For a student who has read the procedure, the teacher's pre-lab instructions are unnecessary.

Use the flowchart to answer Question 7

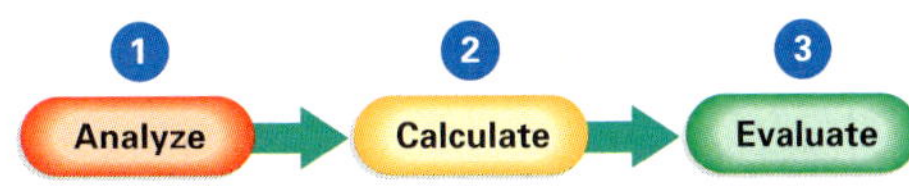

7. What should you do before you calculate an answer to a numeric problem and what should you do after you calculate the answer?

Use this paragraph to answer Questions 8–10.

(1) On a cold morning, your car does not start. (2) You say, "Oh no! The battery is dead!". (3) Your friend who works on cars uses a battery tester and finds that the battery has a full charge. (4) Your friend notices a lot of corrosion on the battery terminals. (5) Your friend says, "Maybe corrosion is causing a bad connection in the electrical circuit, preventing the car from starting." (6) Your friend cleans the terminals and the car starts.

8. Which statements are observations?
9. Which statements are hypotheses?
10. Which statement describes an experiment?

For each question there are two statements. Decide whether each statement is true or false. Then decide whether Statement II is a correct explanation for Statement I.

Statement I		Statement II
11. A hypothesis may be rejected after an experiment.	BECAUSE	Experiments are used to test hypotheses.
12. The supply of fossil fuels is limited.	BECAUSE	Scientists are always looking for new sources of energy.
13. Theories help you make mental models of objects that cannot be seen.	BECAUSE	Theories summarize the results of many observations and experiments.
14. Ideally, chemicals used to attack insect pests should be nonspecific.	BECAUSE	Scientists are looking for safer, more effective ways to protect crops.
15. All Internet sites that provide scientific information are equally reliable.	BECAUSE	All information on these sites is reviewed by qualified scientists.

Standardized Test Prep

1. d
2. c
3. c
4. Contact lenses increase the possibility of eye damage because chemicals can become trapped behind a lens. Safety goggles also protect the eyes from flying debris.
5. No food should be eaten in the laboratory because food may become contaminated with a toxic substance.
6. The teacher may modify the instructions to suit available equipment or chemicals. The teacher will stress any safety precautions. Accidents are more likely to happen when instructions or warnings are ignored.
7. Analyze the question and the given information; evaluate the answer against a reasonable estimate and check the units.
8. 1 and 4
9. 2 and 5
10. 3
11. true, true, correct explanation
12. true, true
13. true, false
14. false, true
15. false, false

CHAPTER 2

Matter and Change Planning Guide

SECTION OBJECTIVES	STANDARDS NATIONAL	STANDARDS STATE	ACTIVITIES and LABS
2.1 Properties of Matter, pp. 39–43 1 block or 2 periods **2.1.1 Identify** properties of matter as extensive or intensive. **2.1.2 Define** physical property and **list** several common physical properties of substances. **2.1.3 Differentiate** among three states of matter. **2.1.4 Describe** a physical change.	A-2, B-2, E-2		**SE** Inquiry Activity: Classifying Matter, p. 38 L2 **TE** Teacher Demo: Volume and Mass, p. 40 L2 **TE** Teacher Demo: States of Chocolate, p. 41 L2 **TE** Class Activity: Comparing Solids and Liquids, p. 41 L2
2.2 Mixtures, pp. 44–47 1 block or 2 periods **2.2.1 Categorize** a sample of matter as a substance or a mixture. **2.2.2 Distinguish** between homogeneous and heterogeneous samples of matter. **2.2.3 Describe** two ways that components of mixtures can be separated.	A-1		**SE** Quick Lab: Separating Mixtures, p. 45 L2 **TE** Teacher Demo: Metallic Breakfast, p. 46 L2
2.3 Elements and Compounds, pp. 48–52 1 block or 2 periods **2.3.1 Explain** the difference between an element and a compound. **2.3.2 Distinguish** between a substance and a mixture. **2.3.3 Identify** the chemical symbols of elements, and **name** elements, given their symbols.	A-1, B-2, B-3, G-3		**TE** Teacher Demo: Decomposition of Sugar, p. 49 L2 **TE** Class Activity: Substances, p. 50 L2
2.4 Chemical Reactions, pp. 53–56 1 block or 2 periods **2.4.1 Describe** what happens during a chemical change. **2.4.2 Identify** four possible clues that a chemical change has taken place. **2.4.3 Apply** the law of conservation of mass to chemical reactions.	A-1, B-2, B-3, G-3		**SE** Small–Scale Lab: 1 + 2 + 3 = Black!, p. 56 L2 **TE** Teacher Demo: Identifying a Chemical Change, p. 54 L2 **LM** Lab 2: Physical and Chemical Change L2 **LM** Lab 3: Observing a Chemical Reaction L2 **SSLM** Lab: 2 A Study of Chemical Changes L2 **LP** Lab Practical 2-1: Chemical and Physical Properties L2

Ability Levels	
L1	For students who need additional help
L2	For all students
L3	For students who need to be challenged

Components

SE	Student Edition	LM	Laboratory Manual	GCP	Graphing Calculator Problems	T	Transparencies
TE	Teacher's Edition	SSLM	Small-Scale Chemistry Lab Manual	TTT	Test-Taking Tips with Transparencies, Grades 9–12 Science	CA	Chem Alive and Chemistry Field Trips DVD
GRSW	Guided Reading and Study Workbook	LP	Laboratory Practicals	STP	Standardized Test Preparation Workbook	iT	Interactive Textbook with ChemASAP
CTR	Core Teaching Resources	PLM	Probeware Lab Manual			GO	Go Online Internet Resources
SSG	Spanish Study Guide	VCL	Virtual Chemistry Labs				
		CTB	Computer Test Bank				

RESOURCES PRINT and TECHNOLOGY		PROBLEMS and ASSESSMENT
GRSW Chapter 2, Section 1	L1	SE Section 2.1 Assessment, p. 42
T T10–T11	L2	iT Section 2.1
GO Chapter 2, Section 1	L2	
GRSW Chapter 2, Section 2	L1	CP 2.1, Separating a Heterogeneous Mixture
T T12–T14	L2	SE Section 2.2 Assessment, p. 47
		iT Section 2.2
GRSW Chapter 2, Section 3	L1	CP 2.2, Classifying Materials
T T15–T17	L2	SE Section 2.3 Assessment, p. 52
GO Chapter 2, Section 3	L2	iT Section 2.3
GRSW Chapter 2, Section 4	L1	SE Section 2.4 Assessment, p. 55
T T18–T19	L2	iT Section 2.4
GO Chapter 2, Section 4	L2	

Go Online

Web Code: cdn-1021
Web Code: cdn-1023
Web Code: cdn-1024

MATERIALS List

Quantities for each group

STUDENT EDITION

Inquiry Activity, p. 38
4 clear containers, 4 small cups, tape or self-adhesive labels, pen, half-cup measure, water, half-teaspoon measures, sugar, baking powder, flour, baking soda, stirring rod, clock or watch

Quick Lab, p. 45
green marking pen, filter paper strip, metric ruler, clear plastic tape, pencil, rubbing alcohol, clear plastic drinking cup, clear plastic wrap

Small-Scale Lab, p. 56
Paper, metric ruler, reaction surface, 0.1*M* KI, 0.2*M* $CuSO_4$, 3% H_2O_2, 1% NaClO, liquid starch, cereal; pipette, medicine droppers, spatulas, iodized and non-iodized salt, antacid tablets, colored marking pens, foods

TEACHER'S EDITION

Teacher Demo, p. 40
plastic syringe (50 mL or larger), balance that reads to 0.01 g, Erlenmeyer flask with tightly fitting 1-hole rubber stopper, funnel, water

Class Activity, p. 41
foam balls or miniature marshmallows

Teacher Demo, p. 41
piece of solid chocolate, small beaker, large beaker, boiling water

Teacher Demo, p. 46
iron-fortified breakfast cereal, 400-mL beaker, distilled water, magnetic stirrer with stirring bar

Teacher Demo, p. 49
powdered sugar, 100-mL beaker, 18*M* sulfuric acid, glass stirring rod

Class Activity, p. 50
20 index cards, masking tape

Teacher Demo, p. 54
Bunsen burner, match, tongs, magnesium ribbon, 4 test tubes, solutions of 1*M* $AgNO_3$, 0.1*M* NaCl, 0.1*M* K_2CrO_4, and 3*M* H_2SO_4, mossy zinc, marble chip, spatulas, cobalt blue glass filters

Chapter Assessment

CHAPTER ASSESSMENT

SE	Chapter Assessment, pp. 58–60
CTB	Chapter 2
iT	Chapter 2

STANDARDIZED TEST PREP

SE	Chapter 2, p. 61

iT Interactive Textbook with Assessment at PHSchool.com

CHAPTER 2

Matter and Change

Classifying Matter 

Purpose

Students will observe some properties of substances and distinguish solids by their properties.

 **Prep Time** 5 minutes
Class Time 10 minutes

Safety

Remind students never to taste anything in the lab, even those materials they think are harmless. Materials that are harmless can be contaminated by other chemicals in the laboratory.

Teaching Tips

- Place containers of the solids with assigned half-teaspoon measures at several locations around the room.
- You may want to substitute powdered sugar for table sugar so that all the solids will have a similar texture.

Think About It

1. The four white solids cannot be distinguished by color or state. (Students may note that table sugar has a different texture.)
2. baking powder; It produces bubbles when the solids are added without stirring.
3. No; The flour floats after the mixtures are stirred, but the baking soda and sugar both dissolve.
4. Students may not use appropriate scientific terms to describe the properties they have observed. Accept any reasonable substitutes.

FYI

This chapter introduces many topics that students will study in greater depth in later chapters, including states of matter, solutions, and chemical reactions. Students need to learn just enough about these topics to prepare them for the content in subsequent chapters.

CHAPTER 2

Matter and Change

As this Indonesian potter shapes a pot, she causes a physical change in the clay.

INQUIRY Activity

Classifying Matter

Materials

4 clear containers, 4 small cups, tape or self-adhesive labels, pen, half-cup measure, water, half-teaspoon measures, sugar, baking powder, flour, baking soda, stirring rod, clock or watch

Procedure

1. Label the clear containers and the small cups Sugar, Baking powder, Flour, and Baking soda.
2. Use the small cups to collect half a teaspoon of each solid. Be sure to use the half-teaspoon measure that is assigned to each solid.
3. Add half a cup of water to each clear container.
4. Without stirring, add each solid to its corresponding container. Observe and record any change that occurs after the solids are added.
5. Stir the contents of each container for at least 45 seconds. Rinse the rod with water between stirrings. Record what you observe.

Think About It

1. Can you use the appearance of the four solids to distinguish them? Explain your answer.
2. Which solid could you distinguish from the other three after Step 4 of the procedure? How could you distinguish this solid?
3. Can you distinguish completely among the remaining three solids based on the results of Step 5? Explain your answer.
4. Make a list of properties, such as color, that can be used to describe materials.

2.1 Properties of Matter

Connecting to Your World

The more than 1200 species of bamboo belong to a family of grasses that includes wheat and corn. In tropical regions, bamboo plants grow rapidly to great heights. The tender shoots of some bamboo plants are a favorite food of pandas. People use the woody stems of mature plants to make furniture, fishing rods, and flooring. Because bamboo is inexpensive and abundant, disposable chopsticks are usually made from bamboo. Bamboo has properties that make it a good choice for use in chopsticks. It has no noticeable odor or taste. It is hard, yet easy to split, and it is heat resistant. In this section, you will learn how properties can be used to classify and identify matter.

Guide for Reading

Key Concepts

- How can properties used to describe matter be classified?
- Why do all samples of a substance have the same intensive properties?
- What are three states of matter?
- How can physical changes be classified?

Vocabulary

mass
volume
extensive property
intensive property
substance
physical property
solid
liquid
gas
vapor
physical change

Reading Strategy

Using Prior Knowledge Before you read, write a definition for the term *liquid*. After you read this section, compare and contrast the definition of *liquid* in the text with your original definition.

Describing Matter

Understanding matter begins with observation and what you observe when you look at a particular sample of matter is its properties. Is a solid shiny or dull? Does a liquid flow quickly or slowly? Is a gas odorless or does it have a smell? **Properties used to describe matter can be classified as extensive or intensive.**

Extensive Properties Recall that matter is anything that has mass and takes up space. The **mass** of an object is a measure of the amount of matter the object contains. The mass of a bowling ball with finger holes is five or six times greater than the mass of the bowling ball shown in Figure 2.1, which is used to play a game called candlepins. There is also a difference in the volume of the balls. The **volume** of an object is a measure of the space occupied by the object. Mass and volume are examples of extensive properties. An **extensive property** is a property that depends on the amount of matter in a sample.

Intensive Properties There are properties to consider when selecting a bowling ball other than mass. Beginning bowlers want a bowling ball that is likely to maintain a straight path. They use bowling balls with a hard surface made from polyester. Experienced bowlers want a bowling ball they can curve, or hook, toward the pins. Often, they use a polyurethane ball, which has a softer surface. Hardness is an example of an intensive property. An **intensive property** is a property that depends on the type of matter in a sample, not the amount of matter.

Figure 2.1 This bowling ball and candlepin are used in a game played mainly in New England.

Section Resources

Print

- **Guided Reading and Study Workbook,** Section 2.1
- **Core Teaching Resources,** Section 2.1 Review
- **Transparencies,** T10–T11

Technology

- **Interactive Textbook with ChemASAP,** Animation 1, Assessment 2.1
- **Go Online,** Section 2.1

2.1

1 FOCUS

Objectives

2.1.1 **Identify** properties of matter as extensive or intensive.

2.1.2 **Define** physical property and **list** several common physical properties of substances.

2.1.3 **Differentiate** among three states of matter.

2.1.4 **Describe** a physical change.

Guide for Reading

Build Vocabulary 

Imagine a Picture Before students read this section, have them visualize the structure of solids, liquids, and gases at the microscopic level. Then have them compare their mental pictures with the drawings in Figure 2.3.

Reading Strategy 

Active Comprehension Read the first paragraph of *Describing Matter.* Ask students what they would like to know about extensive and intensive properties. Students can look for answers as they read the rest of *Describing Matter.*

2 INSTRUCT

Connecting to Your World

Tell students that the characters on the chopsticks translate as good fortune, prosperity, longevity, and good health. **Why are bamboo chopsticks usually disposable?** *(Bamboo is inexpensive and abundant.)* **What characteristic makes bamboo useful for cooking tools?** *(It is heat resistant.)*

Describing Matter

Discuss 

To help students distinguish *extensive* from *intensive,* you may want to note that the prefix *in-* can mean "within." Also, the related word *extend* often refers to an increase in a dimension, such as length, area, or time.

TEACHER Demo

Volume and Mass

L2

Purpose Students observe that air takes up space and has mass.

Materials plastic syringe (50 mL or larger), balance that reads to 0.01 g, Erlenmeyer flask with tightly fitting 1-hole rubber stopper, funnel, water

Procedure Ask students whether or not air takes up space and has mass. Determine the mass of the empty syringe. Then determine the mass when the syringe is full of air. Subtract the two masses to find the mass of air in the syringe. Fit a stopper and funnel into an Erlenmeyer flask. The equipment must be airtight. Quickly pour water through the funnel.

Expected Outcome The mass of the air will be about 0.06 g. Some water will run into the flask (air can be compressed), but not very much, because air takes up the space in the flask.

Identifying Substances

Use Visuals

L1

Table 2.1 Ask, **What is the relationship between the state of a substance at room temperature and its boiling point?** *(Substances that are gases at room temperature have boiling points below 20°C; liquids and solids have boiling points above 20°C.)* **Are the properties listed in the table intensive or extensive?** *(intensive)*

FYI

Some texts use *substance* to describe all materials and *pure substance* to describe elements and compounds. In this text, *substance* and *pure substance* are synonyms. A sample of matter can be described as pure if the amount of impurities in the sample is negligible.

Download a worksheet on **Physical Properties of Matter** for students to complete, and find additional teacher support from NSTA SciLinks.

Figure 2.2 This gold falcon standard from Egypt is about 3000 years old. The copper kettles are about 150 years old. **Analyzing Data** ***Which of the properties listed in Table 2.1 could not be used to distinguish copper from gold?***

Identifying Substances

Each object in Figure 2.2 has a different chemical makeup, or composition. The sculpture of a falcon is mainly gold. The kettles are mainly copper. Matter that has a uniform and definite composition is called a **substance.** Gold and copper are examples of substances, which are also referred to as pure substances. **Every sample of a given substance has identical intensive properties because every sample has the same composition.**

Gold and copper have some properties in common, but there are differences besides their distinctive colors. Pure copper can scratch the surface of pure gold because copper is harder than gold. Copper is better than gold as a conductor of heat or electric current. Both gold and copper are malleable, which means they can be hammered into sheets without breaking. But gold is more malleable than copper. Hardness, color, conductivity, and malleability are examples of physical properties. A **physical property** is a quality or condition of a substance that can be observed or measured without changing the substance's composition.

Table 2.1 lists physical properties for some substances. The states of the substances are given at room temperature. (Although scientists use room temperature to refer to a range of temperatures, in this book it will be used to refer to a specific temperature, 20°C.) Physical properties can help chemists identify substances. For example, a colorless substance that was found to boil at 100°C and melt at 0°C would likely be water. A colorless substance that boiled at 78°C and melted at −117°C would most certainly not be water. Based on Table 2.1, it would likely be ethanol.

Checkpoint ***Which is a better conductor of electric current—gold or copper?***

Table 2.1

Physical Properties of Some Substances

Substance	State	Color	Melting point (°C)	Boiling point (°C)
Neon	gas	colorless	−249	−246
Oxygen	gas	colorless	−218	−183
Chlorine	gas	greenish-yellow	−101	−34
Ethanol	liquid	colorless	−117	78
Mercury	liquid	silvery-white	−39	357
Bromine	liquid	reddish-brown	−7	59
Water	liquid	colorless	0	100
Sulfur	solid	yellow	115	445
Sodium chloride	solid	white	801	1413
Gold	solid	yellow	1064	2856
Copper	solid	reddish-yellow	1084	2562

For: Links on Physical Properties of Matter
Visit: www.SciLinks.org
Web Code: cdn-1021

Differentiated Instruction

Gifted and Talented

Ask students to identify two states of matter other than solid, liquid, and gas. Have students prepare an oral or written report describing the general characteristics of these states and the conditions under which matter is likely to exist in these states. (Plasmas are likely at extremely high temperatures; Bose-Einstein condensates are possible at extremely low temperatures.) Students should explain why these states of matter are not common on Earth.

States of Matter

Depending on the circumstances, you use three different words to refer to water—water, ice, and steam. Water, which is a common substance, exists in three different physical states. So can most other substances. **Three states of matter are solid, liquid, and gas.** Certain characteristics that can distinguish these three states of matter are summarized in Figure 2.3.

Solids A **solid** is a form of matter that has a definite shape and volume. The shape of a solid doesn't depend on the shape of its container. The particles in a solid are packed tightly together, often in an orderly arrangement, as shown in Figure 2.3a. As a result, solids are almost incompressible; that is, it is difficult to squeeze a solid into a smaller volume. In addition, solids expand only slightly when heated.

Liquids Look at Figure 2.3b. The particles in a liquid are in close contact with one another, but the arrangement of particles in a liquid is not rigid or orderly. Because the particles in a liquid are free to flow from one location to another, a liquid takes the shape of the container in which it is placed. However, the volume of the liquid doesn't change as its shape changes. The volume of a liquid is fixed or constant. Thus, a **liquid** is a form of matter that has an indefinite shape, flows, yet has a fixed volume. Liquids are almost incompressible, but they tend to expand slightly when heated.

Interactive Textbook

Animation 1 Relate the states of matter to the arrangements of their particles.

with ChemASAP

Solid
Definite shape
Definite volume
Not easily compressed

Liquid
Indefinite shape
Definite volume
Not easily compressed

Gas
Indefinite shape
Indefinite volume
Easily compressed

Figure 2.3 The arrangement of particles is different in solids, liquids, and gases. (a) In a solid, the particles are packed closely together in a rigid arrangement. (b) In a liquid, the particles are close together, but they are free to flow past one another. (c) In a gas, the particles are relatively far apart and can move freely.
Relating Cause and Effect ***Use the arrangements of their particles to explain the general shape and volume of solids and gases.***

States of Matter

CLASS Activity

Comparing Solids and Liquids L1

Purpose Students construct models of a solid and liquid.

Materials foam balls or miniature marshmallows

Procedure Use Figure 2.3 as a guide for the three-dimensional models.

Expected Outcome The particles in each model should be closely packed, but the arrangement in the solid should be orderly.

TEACHER Demo

States of Chocolate

Purpose Students observe three states of matter when chocolate is heated.

Materials Piece of solid chocolate, small beaker, large beaker, boiling water

Safety Use a thermal mitt while handling the boiling water.

Procedure Show students pieces of solid chocolate. Place them in a small beaker surrounded by boiling water in a larger beaker. After a few minutes, ask, **How many physical states can you discern in the mixture?** *(three)*

Expected Outcome Three states can be discerned as the chocolate melts: solid, liquid, and vapor. Remind students that their ability to smell the chocolate means that some components of the mixture must have vaporized.

FYI

There is a formal definition of *compressibility* in Section 14.1.

Facts and Figures

Thixotropic Materials

Changes of state are typically associated with changes in temperature. However, the state of matter can also be affected by other variables. Thixotropic materials are solidlike materials that liquefy when subjected to shearing forces. For example, many paints are thixotropic; they thin out when brushed on a surface and thicken when the brush strokes stop, thus keeping the paint from sliding off the wall! A shearing force has an opposite effect on quicksand. Quick movements "thicken" the quicksand and make it much more difficult for a person or animal trapped in it to move. (A shearing force causes two layers to slide in opposite directions along a plane of contact.)

Answers to...

Figure 2.2 State; both are solids.
Figure 2.3 Solids have a definite shape and volume because their particles are in a rigid, orderly arrangement; gases have no definite shape or volume because their particles are relatively far apart and can move freely.

copper

Physical Changes

Use Visuals 

Figure 2.4 Have students study the photograph. Refer them to page R14 of the Elements Handbook for data on the properties of gallium. Ask, **Is the melting of gallium a reversible or irreversible change?** *(reversible)* **What is the physical state of gallium at room temperature?** *(solid)*

3 ASSESS

Evaluate Understanding 

To assess students' knowledge of states of matter, ask, **How are three states of matter involved when a candle burns?** Tell students to include the definitions for each state. *(Solid wax has a definite shape and volume; the melted liquid wax is shapeless and runs down the side of the candle; the vapor that forms above the wick has no shape.)*

Reteach L1

If possible, bring samples of some of the substances listed in Table 2.1 to class. Describe the physical properties of selected samples and have students practice identifying them by referring to Table 2.1.

Elements Handbook

The melting point of indium is 157°C. Both indium and gallium (30°C) have relatively low melting points compared with gold (1064°C).

If your class subscribes to the Interactive Textbook, use it to review key concepts in Section 2.1.

with ChemASAP

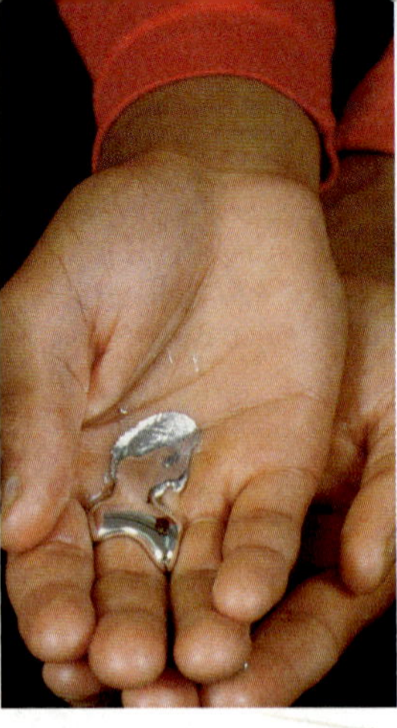

Figure 2.4 The silvery substance in the photograph is gallium, which has a melting point of 30°C. **Inferring** ***What can you infer about the temperature of the hand holding the gallium?***

Gases Like a liquid, a gas takes the shape of its container. But unlike a liquid, a gas can expand to fill any volume. A **gas** is a form of matter that takes both the shape and volume of its container. Look back at Figure 2.3c. As shown in the model, the particles in a gas are usually much farther apart than the particles in a liquid. Because of the space between particles, gases are easily compressed into a smaller volume.

The words *vapor* and *gas* are sometimes used interchangeably. But there is a difference. The term *gas* is used for substances, like oxygen, that exist in the gaseous state at room temperature. (*Gaseous* is the adjective form of *gas*.) **Vapor** describes the gaseous state of a substance that is generally a liquid or solid at room temperature, as in water vapor.

Checkpoint ***When should the term vapor be used instead of gas?***

Physical Changes

The melting point of gallium metal is 30°C. Figure 2.4 shows how heat from a person's hand can melt a sample of gallium. The shape of the sample changes during melting as the liquid begins to flow, but the composition of the sample does not change. Melting is an example of a physical change. During a **physical change,** some properties of a material change, but the composition of the material does not change.

Words such as *boil, freeze, melt,* and *condense* are used to describe physical changes. So are words such as *break, split, grind, cut,* and *crush.* However, there is a difference between these two sets of words. Each set describes a different type of physical change. **Physical changes can be classified as reversible or irreversible.** Melting is an example of a reversible physical change. If a sample of liquid gallium is cooled below its melting point, the liquid will become a solid. All physical changes that involve a change from one state to another are reversible. Cutting hair, filing nails, and cracking an egg are examples of irreversible physical changes.

2.1 Section Assessment

1. **Key Concept** Name two categories used to classify properties of matter.
2. **Key Concept** Explain why all samples of a given substance have the same intensive properties.
3. **Key Concept** Name three states of matter.
4. **Key Concept** Describe the two categories used to classify physical changes.
5. Which property in Table 2.1 can most easily distinguish sodium chloride from the other solids?
6. In what way are liquids and gases alike? In what way are liquids and solids different?
7. Is the freezing of mercury a reversible or irreversible physical change? Explain your answer.
8. Explain why samples of gold and copper can have the same extensive properties, but not the same intensive properties.

Elements Handbook

Read about the metal indium on page R16. What is the melting point of indium? Which other metal has a similar melting point—gallium or gold? Provide data to support your answer.

Assessment 2.1 Test yourself on the concepts in Section 2.1.

with ChemASAP

Section 2.1 Assessment

1. intensive and extensive properties
2. Every sample of a given substance has the same chemical composition.
3. solid, liquid, gas
4. Physical changes are either reversible or irreversible. Reversible changes can be "undone," or reversed. Irreversible changes cannot be undone.
5. Color; sodium chloride is the only white solid listed.
6. Liquids and gases have an indefinite shape. The shape of a solid is definite; the shape of a liquid is indefinite.
7. The freezing of mercury is reversible because solid mercury can be melted.
8. Samples of gold and copper can have the same mass and volume (extensive properties). They cannot have the same set of intensive properties because they have different chemical compositions.

Hanging by a Thread

Strands in a spider web are about one tenth the diameter of a human hair. Yet a golden orb spider web can withstand the impact of a insect, or even a small bird, flying at high speed because the silk in the web's frame and spokes is stronger than steel, more elastic than nylon, and tougher than rubber. Scientists are always looking for lightweight materials with these properties, but they cannot set up farms to harvest spider silk because a spider will fight to defend its territory. Instead, scientists use biotechnology to produce spider silk. **Interpreting Diagrams** ***Where are the silk glands located in a spider?***

Spider anatomy A spider releases its silk through spinnerets at the tip of its abdomen. Inside each spinneret are tens or hundreds of spigots. Silk travels from a silk gland through a duct to a spigot. As the silk is released from a spigot, it changes from a liquid to a solid.

Female Golden Orb Spider
Nephila clavipes
Body length: 24 mm to 40 mm

Spider silk from goat's milk Scientists have identified the spider genes that contain the instructions for producing silk. When these genes are transferred to goats, the goats produce milk containing spider silk. Scientists separate the silk from the milk, purify it, and spin it into fibers.

Technology & Society

Hanging By a Thread

Purpose

This Technology & Society feature builds on the discussion of technology and biotechnology in Chapter 1, where an operational definition of a gene was given, and on the discussion of properties in Section 2.1.

Background

- Only about one-third of spider species spin webs, but all spiders produce silk. Different types of silk glands produce silk used for wrapping prey, silk used for wrapping eggs, non-sticky silk used for draglines, and sticky strands or droplets used for trapping prey.
- Silk is a composite of proteins, which explains why the silk can be strong, tough, and elastic.
- Semi-permanent webs of golden orb spiders are large (40–80 cm diameter). A cable of golden orb dragline silk that is slightly thicker than a garden hose would be strong enough to support two jet planes full of passengers and cargo without breaking.
- Historically, people have collected spider silk to catch fish or birds and to use as bandages to stop blood flow from wounds. Possible commercial applications of recombinant spider silk are for sutures, biodegradable fishing lines, soft body armor, and in composites.

Answers to...

Interpreting Diagrams The silk glands are in the spider's abdomen.

Figure 2.4 Because gallium is shown in the liquid state, the temperature of the hand must be greater than 30°C.

Checkpoint The term *vapor* is used to refer to the gaseous state of a substance that is usually a liquid or solid at room temperature.

2.2

1 FOCUS

Objectives

2.2.1 **Categorize** a sample of matter as a substance or a mixture.

2.2.2 **Distinguish** between homogeneous and heterogeneous samples of matter.

2.2.3 **Describe** two ways that components of mixtures can be separated.

Guide for Reading

Build Vocabulary L2

Word Forms Have students look up the term *homogenize* and explain how this process could be used to turn a heterogeneous mixture into one that is more homogeneous.

Reading Strategy L2

Using Context Clues Tie the Reading Strategy to the Build Vocabulary strategy. Have students use the definitions of homogeneous and heterogeneous mixtures to answer the question posed in the Reading Strategy.

2 INSTRUCT

Connecting to Your World

After students read the introduction, ask **What is the result of panning?** *(Gold is separated from a mixture of materials.)* **What property is used to separate the mixture?** *(heaviness of the materials) Heavier* and *lighter* are used to compare the materials because density isn't discussed until Chapter 3.

Classifying Mixtures

Discuss L2

Bring orange juice or liquid salad dressing to class and compare the properties of these mixtures to those of pure water or pure NaCl. Explain that mixtures are variable in composition.

2.2 Mixtures

Guide for Reading

Key Concepts

- How can mixtures be classified?
- How can mixtures be separated?

Vocabulary

mixture
heterogeneous mixture
homogeneous mixture
solution
phase
filtration
distillation

Reading Strategy

Building Vocabulary After you read this section, explain the difference between homogeneous and heterogeneous mixtures.

Connecting to Your World

In 1848, gold was discovered in California. This discovery led to a massive migration, or rush, of people to California. Panning is one way to separate gold from a mixture of gold and materials such as sand or gravel. A pan containing the mixture is placed underwater and shaken vigorously from left to right. This motion causes heavier materials, such as gold, to move to the bottom of the pan and lighter materials, such as sand, to move to the top where they can be swept away. In this section, you will learn how to classify and separate mixtures.

Classifying Mixtures

A salad bar, like the one in Figure 2.5, provides a range of items, such as cucumbers and hot peppers. Customers choose which items to use in their salads and how much of each item to use. So each salad has a different composition. A **mixture** is a physical blend of two or more components.

Most samples of matter are mixtures. Some mixtures are easier to recognize than others. You can easily recognize chicken noodle soup as a mixture of chicken, noodles, and broth. Recognizing air as a mixture of gases is more difficult. But the fact that air can be drier or more humid shows that the amount of one component of air—water vapor—can vary. Chicken noodle soup and air represent two different types of mixtures. **Based on the distribution of their components, mixtures can be classified as heterogeneous mixtures or as homogeneous mixtures.**

Figure 2.5 You can choose the amount of each item you select from a salad bar. So your salad is unlikely to have the same composition as other salads containing the same items.

Section Resources

Print

- **Guided Reading and Study Workbook,** Section 2.2
- **Core Teaching Resources,** Section 2.2 Review
- **Transparencies,** T12–T14

Technology

- **Interactive Textbook with ChemASAP,** Problem-Solving 2.10, Assessment 2.2

Quick LAB

Separating Mixtures

Purpose
To separate a mixture using paper chromatography.

Materials
- green marking pen
- filter paper strip
- metric ruler
- clear plastic tape
- pencil
- rubbing alcohol
- clear plastic drinking cup
- clear plastic wrap

Procedure    

1. Use the marking pen to draw a line across a strip of filter paper, as shown in the drawing. The line should be 2 cm from one end of the strip.
2. Tape the unmarked end of the filter paper to the center of a pencil so that the strip hangs down when the pencil is held horizontally.
3. Working in a well-ventilated room, pour rubbing alcohol into a plastic cup to a depth of 1 cm.
4. Rest the pencil on the rim of the cup so that the ink end of the strip touches the rubbing alcohol, but does not extend below its surface. Use plastic wrap to cover the top of the cup.
5. Observe the setup for 15 minutes.

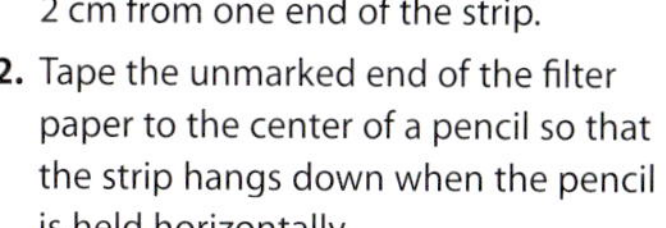

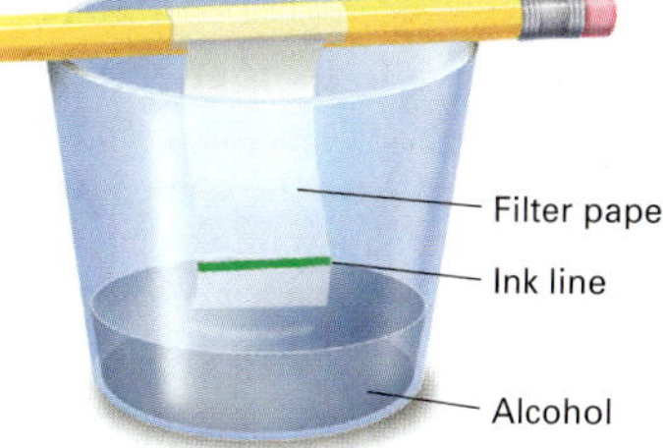

Analyze and Conclude

1. How did the appearance of the filter paper change during the procedure?
2. What evidence is there that green ink is a mixture?
3. How could you use this procedure to identify an unknown type of green ink?

Heterogeneous Mixtures In chicken noodle soup, the ingredients are not evenly distributed throughout the mixture. There is likely to be more chicken in one spoonful than in another spoonful. A mixture in which the composition is not uniform throughout is a **heterogeneous mixture.**

Homogeneous Mixtures The substances in the olive oil and vinegar in Figure 2.6 are evenly distributed throughout these mixtures. So olive oil doesn't look like a mixture. The same is true for vinegar. Vinegar is a mixture of water and acetic acid, which dissolves in the water. Olive oil and vinegar are homogeneous mixtures. A **homogeneous mixture** is a mixture in which the composition is uniform throughout. Another name for a homogeneous mixture is a **solution.** Many solutions are liquids. But some are gases, like air, and some are solids, like stainless steel, which is a mixture of iron, chromium, and nickel.

The term **phase** is used to describe any part of a sample with uniform composition and properties. By definition, a homogeneous mixture consists of a single phase. A heterogeneous mixture consists of two or more phases. When oil and vinegar are mixed, they form layers, or phases, as shown in Figure 2.6. The oil phase floats on the water phase.

 Checkpoint ***How many phases are there in a homogeneous mixture?***

Figure 2.6 Olive oil and vinegar are homogeneous mixtures. The substances in these mixtures are evenly distributed. When olive oil and vinegar are mixed, they form a heterogeneous mixture with two distinct phases.

Quick LAB

Separating Mixtures  L2

Objective

After completing this activity, students will be able to:
- separate the components of a mixture using paper chromatography.

Skills Focus Observing, inferring, drawing conclusions

 **Prep Time** 15 minutes

Advance Prep

- Cut the paper strips in advance to save time. Strips of paper toweling can be used in place of filter paper.

Class Time 25 minutes

Safety Rubbing alcohol is poisonous and flammable. It is also an irritant when inhaled. Keep containers covered and away from heat. If the room is not well ventilated, use a fume hood.

Expected Outcome Bands of color will separate on the filter paper.

Extension Students could repeat the lab using different brands of markers, different colors, or water in place of rubbing alcohol.

Analyze and Conclude

1. Bands of colors appear as the alcohol moves up the paper.
2. The bands of colors indicate that green ink is a mixture.
3. The color pattern of the unknown ink can be compared with color patterns from known types of green ink.

For Enrichment L3

Students can research R_f values of different dyes and how they are calculated (ratio of distance traveled by dye to distance traveled by solvent). They can then quantify this lab by calculating the R_f values of the dyes in the ink.

FYI

In Section 15.3, solutions, suspensions, and colloids are compared.

Differentiated Instruction

English Learners L1

Encourage English-learner students to compile a glossary in which they define each key term in English and in their native language. Suggest that students also include illustrations when appropriate.

Facts and Figures

Chromatography

The basis of chromatography is the partitioning of components between a stationary phase and a moving phase based on differences in solubility. The components travel with different speeds in the moving solvent.

Answers to...

Checkpoint one

CONCEPTUAL PROBLEM 2.1

Aluminum nails are used on trail markers so that if a tree is harvested, the nail can be sliced through at the lumber mill without causing injury to a worker.

Answers

9. Iron is magnetic; table salt is not. Table salt will dissolve in water; iron will not.
10. By lowering the temperature to below the boiling point of each gas, you could condense each substance and separate the gases.

Practice Problems Plus 

Classify the following mixtures as homogeneous or heterogeneous.

a. granite rock *(heterogeneous)*
b. salt water *(homogeneous)*
c. paint *(heterogeneous)*
d. a silver ring *(homogeneous)*

Separating Mixtures

Metallic Breakfast 

Purpose Students will observe the separation of iron filings from iron-fortified breakfast cereal.

Materials iron-fortified breakfast cereal, 400-mL beaker, distilled water, magnetic stirrer with stirring bar

Safety Remind students not to eat the cereal.

Procedure Place a stirring bar in a 400-mL beaker. Add about 30 g of cereal to the beaker and add distilled water until the beaker is about half full. Using a magnetic stirrer, mix gently for about 20 minutes. Retrieve the stirring bar and observe the black iron filings attached to it.

Expected Outcome Iron filings will cover the stirring bar. They are added to cereal as an iron supplement. Explain that stomach acid changes the iron into a form the body can use.

CONCEPTUAL PROBLEM 2.1

Separating a Heterogeneous Mixture

Sometimes plastic signs are used to mark trails used by hikers or vehicles. The sign in the photo is used to mark locations along a trail where an all terrain vehicle (ATV) is permitted. Aluminum nails are used to attach signs at eye level to trees or posts. How could a mixture of aluminum nails and iron nails be separated?

1 Analyze ***Identify the relevant concepts.***

List properties of aluminum and iron.

Aluminum:	Iron:
• metal	• metal
• gray color	• gray color
• doesn't dissolve in water	• doesn't dissolve in water
• not attracted to magnet	• attracted to magnet

2 Solve ***Apply concepts to this situation.***

Identify a property that can be used to separate iron and aluminum objects. The ability to be attracted by a magnet is a property that iron and aluminum do not share. You could use a magnet to remove the iron nails from a mixture of iron and aluminum nails.

Practice Problems

9. What physical properties could be used to separate iron filings from table salt?
10. Air is mainly a mixture of nitrogen and oxygen, with small amounts of other gases such as argon and carbon dioxide. What property could you use to separate the gases in air?

Problem Solving 2.10 Solve Problem 10 with the help of an interactive guided tutorial.

with ChemASAP

Separating Mixtures

If you have a salad containing an ingredient you don't like, you can use a fork to remove the pieces of the unwanted ingredient. Many mixtures are not as easy to separate. To separate a mixture of olive oil and vinegar, for example, you could decant, or pour off, the oil layer. Or you might cool the mixture until the oil turned solid. The first method takes advantage of the fact that oil floats on water. The second method takes advantage of a difference in the temperatures at which the olive oil and vinegar freeze.

Differences in physical properties can be used to separate mixtures.

Filtration The colander in Figure 2.7 can separate cooked pasta from the cooking water. The water passes through the holes in the colander, but the pasta does not. The holes, or pores, in a coffee filter are smaller than the holes in a colander to retain coffee grains. But the holes are not small enough to retain the particles in water. The process that separates a solid from the liquid in a heterogeneous mixture is called **filtration.**

Figure 2.7 A colander is used to separate pasta from the water in which it was cooked. This process is a type of filtration.

Facts and Figures

Distillation

A working distillation apparatus was described in the writings of Maria of Alexandria, an alchemist who lived and worked nearly two thousand years ago. The city of Alexandria, located on the Nile River in North Africa, was a world center of science and culture at that time. Maria of Alexandria is also credited with inventing other chemical apparatus, such as the water bath, which to this day bears her name: the *bain marie.*

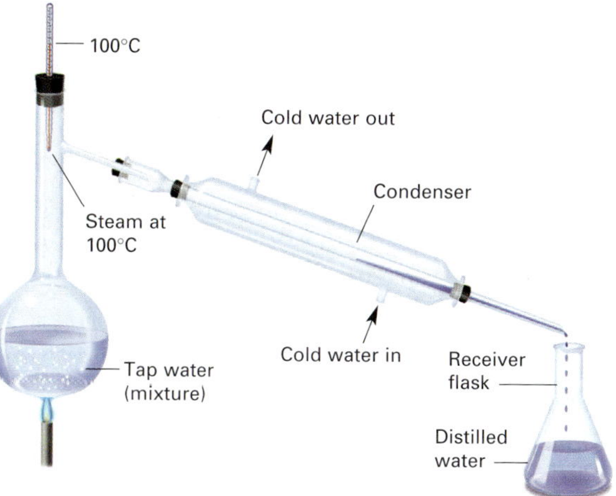

Figure 2.8 A distillation can be used to remove impurities from water. As liquid water changes into water vapor, substances dissolved in the water are left behind in the distillation flask. **Inferring** ***What can you infer about the boiling points of substances dissolved in the impure water?***

Distillation Tap water is a homogeneous mixture of water and substances that dissolved in the water. One way to separate water from the other components in tap water is through a process called distillation. During a **distillation,** a liquid is boiled to produce a vapor that is then condensed into a liquid. Figure 2.8 shows an apparatus that can be used to perform a small-scale distillation.

As water in the distillation flask is heated, water vapor forms, rises in the flask, and passes into a glass tube in the condenser. The tube is surrounded by cold water, which cools the vapor to a temperature at which it turns back into a liquid. The liquid water is collected in a second flask. The solid substances that were dissolved in the water remain in the distillation flask because their boiling points are much higher than the boiling point of water.

2.2 Section Assessment

11. **Key Concept** How are mixtures classified?
12. **Key Concept** What type of properties can be used to separate mixtures?
13. Explain the term *phase* as it relates to homogeneous and heterogeneous mixtures.
14. Classify each of the following as a homogeneous or heterogeneous mixture.
    a. food coloring
    b. ice cubes in liquid water
    c. mouthwash
    d. mashed, unpeeled potatoes
15. How are a substance and a solution similar? How are they different?
16. In general, when would you use filtration to separate a mixture? When would you use distillation to separate a mixture?
17. Describe a procedure that could be used to separate a mixture of sand and table salt.

Writing Activity

Writing to Persuade Write a paragraph in support of this statement: Dry tea is a mixture, not a substance. Include at least two pieces of evidence to support your argument.

Interactive Textbook

Assessment 2.2 Test yourself on the concepts in Section 2.2.

with **ChemASAP**

Section 2.2 Assessment

11. as heterogeneous or homogeneous
12. differences in physical properties
13. A phase is any part of a sample with uniform composition. There is one phase in a homogeneous mixture and two or more phases in a heterogeneous mixture.
14. **a.** homogeneous **b.** heterogeneous **c.** homogeneous **d.** heterogeneous
15. Both have a uniform composition throughout. A substance has a definite composition; a solution has a variable composition.
16. Filtration separates solids from a liquid in a heterogeneous mixture. Distillation can separate a liquid from substances dissolved in the liquid.
17. Add water to dissolve the salt. Pour the resulting mixture onto a piece of closely woven cloth. The sand will remain on the cloth, and the salt solution will pass through. Use evaporation to remove the water from the salt solution, leaving solid salt behind.

Use Visuals 

Figure 2.8 Review the distillation process, and describe the components of the apparatus. Ask, **What might be an advantage to having a long condenser in a distillation apparatus?** *(There would be more surface area on which the vapor can condense.)* **Is distilled water that contains dissolved gases a pure substance?** *(No, distilled water is not pure unless the dissolved gases are removed.)*

3 ASSESS

Evaluate Understanding 

Have students identify five items that fit each of the following categories.

a. substance
b. homogeneous mixture
c. heterogeneous mixture
d. solution

Have students select three items (one each from categories b, c, and d). For each item selected, students should outline a method for separating the components.

Reteach L1

Explain that distillation can be used to separate a mixture of gases. Show students the drawing of the fractional distillation of liquid air on R24 of the Elements Handbook. Ask, **What physical property is used to separate the gases?** *(boiling point)*

Writing Activity

Possible evidence includes the fact that only some substances in the tea leaves dissolve in water, and tea is sold both with and without caffeine.

Interactive Textbook

If your class subscribes to the Interactive Textbook, use it to review key concepts in Section 2.2.

with **ChemASAP**

Answers to...

Figure 2.8 *They are much higher than the boiling point of water.*

2.3

1 FOCUS

Objectives

2.3.1 **Explain** the difference between an element and a compound.

2.3.2 **Distinguish** between a substance and a mixture.

2.3.3 **Identify** the chemical symbols of elements, and **name** elements, given their symbols.

Guide for Reading

Build Vocabulary L2

Word Forms Compare the common meanings of *element, elemental,* and *elementary* in relation to the specific use of *element* in chemistry. In their broadest sense, the terms refer to fundamentals, first principles, or basics.

Reading Strategy L2

Preview Before students read the section, have them preview the headings to get an overall sense or the content.

2 INSTRUCT

Connecting to Your World

Ask students if they have ever eaten cotton candy. Ask those who have to describe its properties. Explain that its sweet taste is due to a sugar called sucrose, which is a compound.

Distinguishing Elements and Compounds

Use Visuals L1

Compare the chemical change of sugar shown on R29 with the one shown in Figure 2.9. Ask, **What is similar about the changes?** *(Sugar breaks down into solid carbon and water vapor.)* **What is different?** *(In one case, heating causes the change; in the other, the addition of an acid.)* You can also compare the electrolysis of water to the distillation of water (Section 2.2).

2.3 Elements and Compounds

Guide for Reading

Key Concepts

- How are elements and compounds different?
- How can substances and mixtures be distinguished?
- What do chemists use to represent elements and compounds?

Vocabulary

element
compound
chemical change
chemical symbol

Reading Strategy

Relating Text and Visuals As you read, look at Figure 2.10. Explain how this illustration helps you understand the relationship between different kinds of matter.

Connecting to Your World Take two pounds of sugar, two cups of boiling water, and one-quarter teaspoon of cream of tartar. You have the ingredients to make spun sugar. Add food coloring and you have the sticky, sweet concoction sold at baseball games and amusement parks as cotton candy. Sugar is a substance that contains three other substances—carbon, hydrogen, and oxygen. In this section, you will learn how substances are classified as elements or compounds.

Distinguishing Elements and Compounds

Substances can be classified as elements or compounds. An **element** is the simplest form of matter that has a unique set of properties. Oxygen and hydrogen are two of the more than 100 known elements. A **compound** is a substance that contains two or more elements chemically combined in a fixed proportion. For example, carbon, oxygen, and hydrogen are chemically combined in the compound sucrose, the sugar in spun sugar. (Sometimes sucrose is referred to as table sugar to distinguish it from other sugar compounds.) In every sample of sucrose there are twice as many hydrogen particles as oxygen particles. The proportion of hydrogen particles to oxygen particles in sucrose is fixed. There is a key difference between elements and compounds. **Compounds can be broken down into simpler substances by chemical means, but elements cannot.**

Breaking Down Compounds Physical methods that are used to separate mixtures cannot be used to break a compound into simpler substances. Boil liquid water and you get water vapor, not the oxygen and hydrogen that water contains. Dissolve a sugar cube in water and you still have sucrose, not oxygen, carbon, and hydrogen. This result does not mean that sucrose or water cannot be broken down into simpler substances. But the methods must involve a chemical change. A **chemical change** is a change that produces matter with a different composition than the original matter. Heating is one of the processes used to break down compounds into simpler substances. The layer of sugar in Figure 2.9 is heated in a skillet until it breaks down into solid carbon and water vapor. Can the substances that are produced also be broken down?

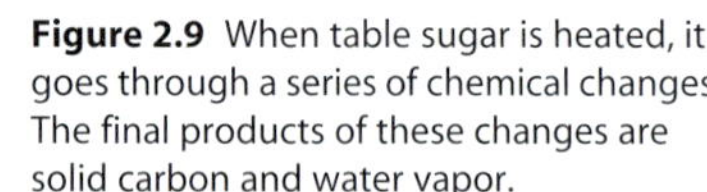

Figure 2.9 When table sugar is heated, it goes through a series of chemical changes. The final products of these changes are solid carbon and water vapor.

Section Resources

Print

- **Guided Reading and Study Workbook,** Section 2.3
- **Core Teaching Resources,** Section 2.3 Review, Interpreting Graphics
- **Transparencies,** T15–T17

Technology

- **Interactive Textbook with ChemASAP,** Problem-Solving 2.19, Assessment 2.3
- **Go Online,** Section 2.3

There is no chemical process that will break down carbon into simpler substances because carbon is an element. Heating will not cause water to break down, but electricity will. When an electric current passes through water, oxygen gas and hydrogen gas are produced. The following diagram summarizes the overall process.

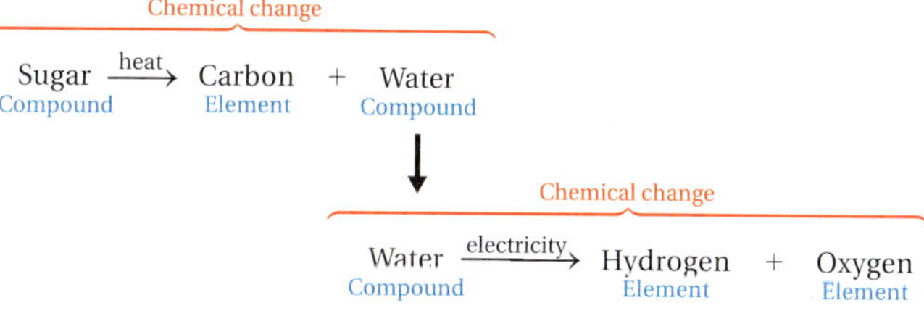

Chemical change

Water —electricity→ Hydrogen + Oxygen
Compound Element Element

Properties of Compounds In general, the properties of compounds are quite different from those of their component elements. Sugar is a sweet-tasting, white solid, but carbon is a black, tasteless solid. Hydrogen is a gas that burns in the presence of oxygen—a colorless gas that supports burning. The product of this chemical change is water, a liquid that can stop materials from burning. Figure 2.10 shows samples of table salt (sodium chloride), sodium, and chlorine. When the elements sodium and chlorine combine chemically to form sodium chloride, there is a change in composition and a change in properties. Sodium is a soft, gray metal. Chlorine is a pale yellow-green poisonous gas. Sodium chloride is a white solid.

Checkpoint ***What process can be used to break down water?***

Word *Origins*

Compound comes from a Latin word *componere,* meaning "to put together." Elements are put together, or chemically combined, in compounds. **What items are put together in a compound sentence?**

Figure 2.10 Compounds and the elements from which they form have different properties. **Observing** ***Based on the photographs, describe two physical properties of sodium and two of chlorine.***

Chlorine is used to kill harmful organisms in swimming pools.

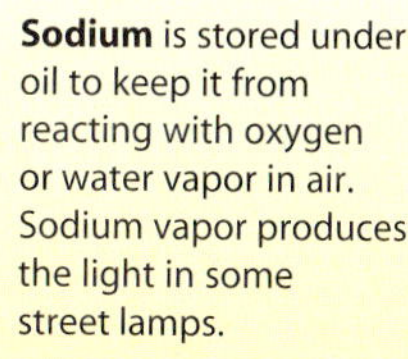

Sodium is stored under oil to keep it from reacting with oxygen or water vapor in air. Sodium vapor produces the light in some street lamps.

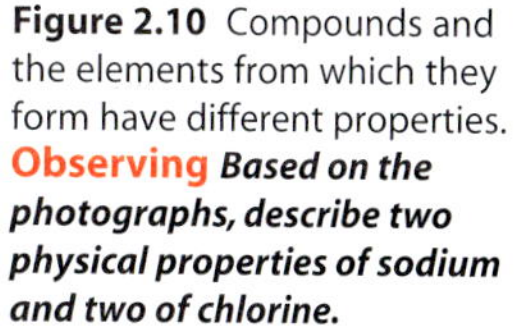

Sodium chloride (commonly known as table salt) is used to season or preserve food.

Differentiated Instruction

Gifted and Talented L3

Other forms of energy besides electricity can be used to bring about the decomposition of water into hydrogen and oxygen. Ask students to consider whether heat could be used for this purpose. Tell them that early chemists considered water to be an element because it would not decompose when heated. Have students do research on the conditions necessary to decompose water using heat. What hazards, if any, might preclude the routine use of this method?

TEACHER Demo

Decomposition of Sugar L2

Purpose Students will observe the decomposition of sugar by acid.

Materials 100-mL beaker, powdered sugar, 18*M* sulfuric acid, glass stirring rod

Safety Perform the demo in a fume hood or in a well-ventilated area and be sure to wear safety goggles. **CAUTION** *This is an extremely exothermic reaction.*

Procedure Place about 25 g of powdered sugar in a 100-mL beaker. Carefully add about 10–15 mL of 18*M* sulfuric acid, and stir rapidly with a glass rod.

Disposal Place the cooled beaker and its contents in a large beaker and add water. Neutralize with NaOH and pour the liquid down the drain, flushing with excess water. Throw the solid residue in the trash.

Expected Outcome The mixture turns dark and a column of carbon rises out of the beaker.

Word *Origins* L2

A compound sentence contains two independent clauses, each with a subject and a verb. The two clauses are joined by a conjunction.

Relate L2

Refer students to *Salt of the Earth* on page R9 of the Elements Handbook for a discussion of the historic importance of table salt.

FYI

In the Small-Scale lab in Chapter 21, students will learn that the electrolysis of water doesn't work with pure water.

Answers to...

Figure 2.10 sodium: gray color, shiny, solid at room temperature; chlorine: yellow-green color, gas at room temperature.

Checkpoint passing an electric current through water

Distinguishing Substances and Mixtures

Use Visuals L1

Figure 2.11 Have students study the figure. Then ask, **Can compounds be separated into their component elements by physical processes?** *(No, they must be separated by chemical processes.)*

CLASS Activity

Substances L2

Purpose To establish the definition of *substance*, using an inductive approach.

Materials 20 index cards, masking tape

Procedure Make signs with the names of each of these items: 1 oxygen, 2 neon, 3 apple, 4 sand, 5 iron, 6 water, 7 air, 8 paint, 9 sodium chloride, 10 sucrose, 11 carbon dioxide, 12 granite, 13 laundry detergent, 14 citric acid, 15 cereal, 16 salad, 17 salad dressing, 18 copper, 19 ocean water, 20 gold. Using masking tape and the board or wall, create two columns. Place items 1 and 2 under Column A and items 3 and 4 under Column B. Ask students to think about the criteria for each column. Then place item 5 in Column A. Students may have to change their criteria. Place item 6 in Column A. Ask students what they think the criteria are now. Continue placing one item up at a time in a column until students understand the criteria. Everything in Column A has a uniform and definite composition.

Expected Outcome Column A is for substances. Column A: 1, 2, 5, 6, 9, 10, 11, 14, 18, 20. Column B is for mixtures. Column B: 3, 4, 7, 8, 12, 13, 15, 16, 17, 19. Make sure students understand the difference between the uniform composition of a substance and of a solution. The composition of the substance is fixed. The composition of the mixture can vary.

Distinguishing Substances and Mixtures

Deciding whether a sample of matter is a substance or a mixture based solely on appearance can be difficult. After all, homogeneous mixtures and substances will both appear to contain only one kind of matter. Sometimes you can decide by considering whether there is more than one version of the material in question. For example, you can buy whole milk, low-fat milk, no-fat milk, light cream, and heavy cream. From this information, you can conclude that milk and cream are mixtures. You might infer that these mixtures differ in the amount of fat they contain. Most gas stations offer at least two blends of gasoline. The blends have different octane ratings and different costs per gallon, with premium blends costing more than regular blends. So gasoline must be a mixture.

You can use their general characteristics to distinguish substances from mixtures. **If the composition of a material is fixed, the material is a substance. If the composition of a material may vary, the material is a mixture.** Figure 2.11 summarizes the general characteristics of elements, compounds, and mixtures.

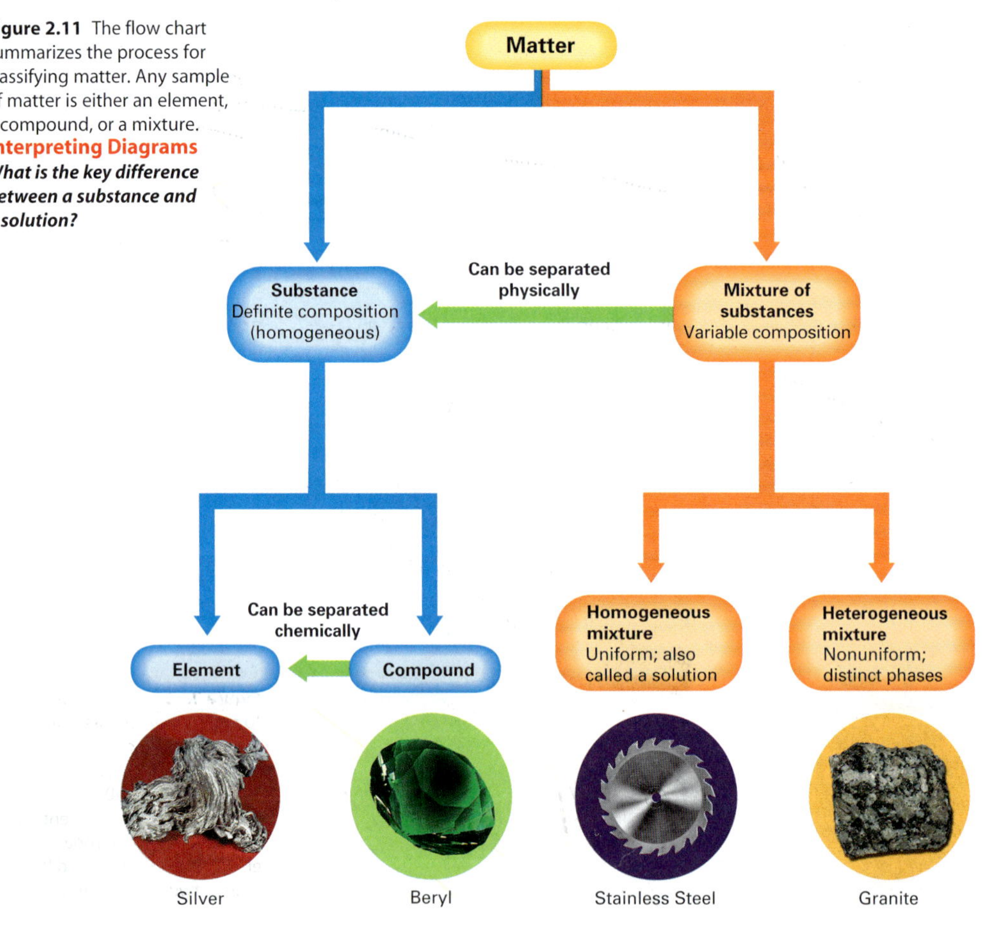

Figure 2.11 The flow chart summarizes the process for classifying matter. Any sample of matter is either an element, a compound, or a mixture.
Interpreting Diagrams ***What is the key difference between a substance and a solution?***

Facts and Figures

Origins of Element Names

Many elements were named for the people who discovered them or the places where they were discovered. Some elements were given descriptive names taken from classical Latin or Greek. Others were named for figures in mythology. Polonium is named for Poland, the native land of Marie Curie, who discovered radium. Californium was discovered at the University of California. The name chlorine comes from the Greek *chloros*, meaning greenish-yellow. Chlorine is a greenish-yellow gas. The name calcium is derived from the Latin *calx*, meaning lime. Calcium is a major component of limestone. Students can find the origins of element names by consulting a dictionary or encyclopedia.

CONCEPTUAL PROBLEM 2.2

Classifying Materials

When the blue-green solid in the photograph is heated, a colorless gas and a black solid form. All three materials are substances. Is it possible to classify these substances as elements or compounds?

1 Analyze ***Identify the relevant concepts.***

List the known facts and relevant concepts.
- A blue-green solid is heated.
- A colorless gas and a black solid appear.
- A compound can be broken down into simpler substances by a chemical change, but an element cannot.
- Heating can cause a chemical change.

2 Solve ***Apply concepts to this situation.***

Determine if the substances are elements or compounds. Before heating, there was one substance. After heating there were two substances. The blue-green solid must be a compound. Based on the information given, it isn't possible to know if the colorless gas or black solid are elements or compounds.

Practice Problems

18. Liquid A and Liquid B are clear liquids. They are placed in open containers and allowed to evaporate. When evaporation is complete, there is a white solid in container B, but no solid in container A. From these results, what can you infer about the two liquids?

19. A clear liquid in an open container is allowed to evaporate. After three days, a solid is left in the container. Was the clear liquid an element, a compound, or a mixture? How do you know?

Interactive Textbook

Problem Solving 2.19 Solve Problem 19 with the help of an interactive guided tutorial.

with ChemASAP

Symbols and Formulas

The common names water and table salt do not provide information about the chemical composition of these substances. Also, words are not ideal for showing what happens to the composition of matter during a chemical change.  **Chemists use chemical symbols to represent elements, and chemical formulas to represent compounds.**

Using symbols to represent different kinds of matter is not a new idea. Figure 2.12 shows some symbols that were used in earlier centuries. The symbols used today for elements are based on a system developed by a Swedish chemist, Jöns Jacob Berzelius (1779–1848). He based his symbols on the Latin names of elements. Each element is represented by a one- or two-letter **chemical symbol.** The first letter of a chemical symbol is always capitalized. When a second letter is used, it is lowercase.

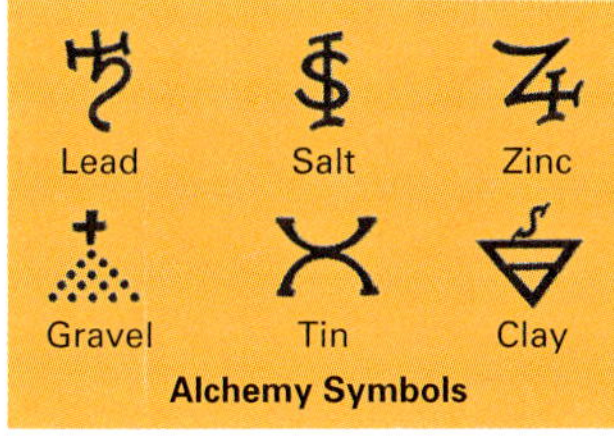

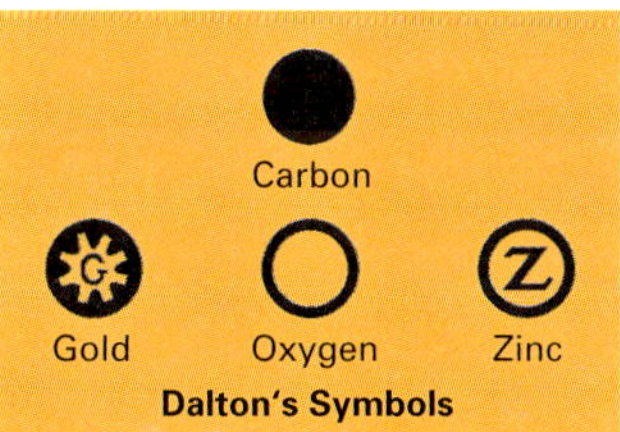

Figure 2.12 The symbols used to represent elements have changed over time. Alchemists and the English chemist John Dalton (1766–1844) both used drawings to represent chemical elements. Today, elements are represented by one- or two-letter symbols.

CONCEPTUAL PROBLEM 2.2

Answers

18. Liquid A is probably a substance. Liquid B is a mixture.

19. The liquid was not an element because a solid was left when the liquid evaporated. A physical process, such as evaporation, cannot be used to break down a compound. Therefore, the liquid was a mixture.

Practice Problems Plus L2

Classify the following materials as an element, compound, or mixture. Give reasons for your answers.

a. table salt (NaCl) *(compound; its formula shows that it is composed of at least two elements)*

b. salt water *(mixture; its name suggests that it is composed of at least two substances)*

c. sodium (Na) *(element; its symbol shows that it is composed of only one kind of matter)*

Symbols and Formulas

Discuss L2

Point out that the use of chemical symbols is an example of the "international language of chemistry." Discuss why people all over the world use the same set of chemical symbols. Show how symbols are used in chemical formulas. Emphasize that a compound is always made up of the same elements in the same proportions.

CLASS Activity

Timeline of Discovery L2

To help students explore the history of science, have the class work as a group to make a timeline for the discovery of elements. Refer them to the data on dates of discovery in the Elements Handbook. Students could use the timeline to detect shifts in naming conventions.

Differentiated Instruction

Less Proficient Readers/ English Learners L1

Have students make flash cards to help learn the names and symbols of at least 30 elements. If an element has a different common name in the student's native country, have the student include that information on the flash card.

Answers to...

Figure 2.11 *The composition of a substance is fixed; the composition of a solution may vary.*

Download a worksheet on **Element Names** for students to complete, and find additional teacher support from NSTA SciLinks.

3 ASSESS

Evaluate Understanding L2

Ask students to explain in their own words the difference between an element and a compound.

Reteach L1

Set up a display of elements in the classroom to help students relate abstract chemical symbols to actual samples. Have students present oral reports about the physical properties of elements and refer to the display.

Writing Activity

Elements and compounds are alike in that they both are substances of fixed composition. They are different in that elements cannot be separated into simpler substances through chemical changes.

If your class subscribes to the Interactive Textbook, use it to review key concepts in Section 2.3.

with **ChemASAP**

Table 2.2

Symbols and Latin Names for Some Elements

Name	Symbol	Latin name
Sodium	Na	*natrium*
Potassium	K	*kalium*
Antimony	Sb	*stibium*
Copper	Cu	*cuprum*
Gold	Au	*aurum*
Silver	Ag	*argentum*
Iron	Fe	*ferrum*
Lead	Pb	*plumbum*
Tin	Sn	*stannum*

For: Links on Element Names
Visit: www.SciLinks.org
Web Code: cdn-1023

If the English name and the Latin name of an element are similar, the symbol will appear to have been derived from the English name. Examples include Ca for calcium, N for nitrogen, and S for sulfur. Table 2.2 shows examples of elements where the symbols do not match the English names.

Chemical symbols provide a shorthand way to write the chemical formulas of compounds. The symbols for hydrogen, oxygen, and carbon are H, O, and C. The formula for water is H_2O. The formula for sucrose, or table sugar, is $C_{12}H_{22}O_{11}$. Subscripts in chemical formulas are used to indicate the relative proportions of the elements in the compound. For example, the subscript 2 in H_2O indicates that there are always two parts of hydrogen for each part of oxygen in water. Because a compound has a fixed composition, the formula for a compound is always the same.

2.3 Section Assessment

20. **Key Concept** How is a compound different from an element?

21. **Key Concept** How can you distinguish a substance from a mixture?

22. **Key Concept** What are chemical symbols and chemical formulas used for?

23. Name two methods that can be used to break down compounds into simpler substances.

24. Classify each of these samples of matter as an element, a compound, or a mixture.
a. table sugar **b.** tap water
c. cough syrup **d.** nitrogen

25. Write the chemical symbol for each element.
a. lead **b.** oxygen
c. silver **d.** sodium
e. hydrogen **f.** aluminum

26. Name the chemical elements represented by the following symbols.
a. C **b.** Ca **c.** K **d.** Au **e.** Fe **f.** Cu

27. What elements make up the pain reliever acetaminophen, chemical formula $C_8H_9O_2N$? Which element is present in the greatest proportion by number of particles?

Writing Activity

Compare and Contrast Paragraph Compare and contrast elements and compounds. Compare them by saying how they are alike. Contrast them by describing how they are different.

Assessment 2.3 Test yourself on the concepts in Section 2.3.

with **ChemASAP**

Section 2.3 Assessment

20. Compounds can be broken down into simpler substances by chemical means, but elements cannot.
21. A substance has a fixed composition. The composition of a mixture may vary.
22. Chemical symbols are used to represent elements. Chemical formulas are used to represent compounds.
23. heating or an electric current
24. a. compound **b.** mixture **c.** mixture **d.** element
25. a. Pb **b.** O **c.** Ag **d.** Na **e.** H **f.** Al
26. a. carbon **b.** calcium **c.** potassium **d.** gold **e.** iron **f.** copper
27. Carbon, hydrogen, oxygen, and nitrogen; hydrogen is present in the greatest proportion by number of atoms.

2.4 Chemical Reactions

Connecting to Your World

Iron is an element with many desirable properties. It is abundant, easy to shape when heated, and relatively strong, especially when mixed with carbon in steel. Iron has one main disadvantage. Over time, objects made of iron will rust if they are left exposed to air. The brittle layer of rust that forms on the surface of the object flakes off, exposing more iron to the air. In this section, you will learn to recognize chemical changes and to distinguish them from physical changes.

Guide for Reading

Key Concepts

- What always happens during a chemical change?
- What are four possible clues that a chemical change has taken place?
- How are the mass of the reactants and the mass of the products of a chemical reaction related?

Vocabulary

chemical property
chemical reaction
reactant
product
precipitate
law of conservation of mass

Reading Strategy

Predicting Before you read, predict what will happen to the mass of a sample of matter that burns. After you read, check the accuracy of your prediction and correct any misconceptions.

Chemical Changes

The compound formed when iron rusts is iron oxide (Fe_2O_3). Words such as *burn, rot, rust, decompose, ferment, explode,* and *corrode* usually signify a chemical change. The ability of a substance to undergo a specific chemical change is called a **chemical property.** Iron is able to combine with oxygen to form rust. So the ability to rust is a chemical property of iron. Chemical properties can be used to identify a substance. But chemical properties can be observed only when a substance undergoes a chemical change.

Figure 2.13 compares a physical change and a chemical change that can occur in a mixture of iron and sulfur. When a magnet is used to separate iron from sulfur, the change is a physical change. The substances present before the change are the same substances present after the change, although they are no longer physically blended. Recall that during a physical change, the composition of matter never changes. **During a chemical change, the composition of matter always changes.** When the mixture of iron and sulfur is heated, a chemical change occurs. The sulfur and iron react and form iron sulfide (FeS).

A chemical change is also called a chemical reaction. One or more substances change into one or more new substances during a **chemical reaction.** A substance present at the start of the reaction is a **reactant.** A substance produced in the reaction is a **product.** In the reaction of iron and sulfur, iron and sulfur are reactants and iron sulfide is a product.

Figure 2.13 A mixture of iron filings and sulfur can be changed. **a** A magnet separates the iron from the sulfur. **b** Heat combines iron and sulfur in a compound. **Classifying** ***Which change is a chemical change? Explain.***

Section Resources

Print

- **Guided Reading and Study Workbook,** Section 2.4
- **Core Teaching Resources,** Section 2.4 Review
- **Transparencies,** T18–T19
- **Laboratory Manual,** Labs 2–3
- **Laboratory Practicals,** 2-1
- **Small-Scale Chemistry Laboratory Manual,** Lab 2

Technology

- **Interactive Textbook with ChemASAP,** Assessment 2.4
- **Go Online,** Section 2.4

2.4

1 FOCUS

Objectives

2.4.1 Describe what happens during a chemical change.

2.4.2 Identify four possible clues that a chemical change has taken place.

2.4.3 Apply the law of conservation of mass to chemical reactions.

Guide for Reading

Build Vocabulary L2

Compare and Contrast Table Have students make a table to compare physical and chemical changes.

Reading Strategy 

Predict Before students read the section, have them predict the meanings of *reactant* and *product.* Ask students what they are basing their predictions on.

2 INSTRUCT

Connecting to Your World

What evidence do you see in the photo that a chemical reaction has occurred? *(The metal has changed color.)*

Chemical Changes

Use Visuals 

Figure 2.13 Discuss the difference between physical and chemical changes. Ask, **What substances are present before and after the change in photo a?** *(iron and sulfur)* **What substances are present before and after the change in photo b?** *(iron and sulfur before, iron sulfide after)*

FYI

The Roman numerals were omitted from iron(III) oxide and iron(II) sulfide to avoid having to explain the Stock system, which is explained on p. 254.

Answers to...

Figure 2.13 *b; A new substance is formed.*

Recognizing Chemical Changes

TEACHER Demo

Identifying a Chemical Change L2

Purpose Students will practice identifying chemical changes.

Materials Bunsen burner, match, tongs, magnesium ribbon, 4 test tubes, solutions of 0.1*M* $AgNO_3$, 0.1*M* NaCl, 0.1*M* K_2CrO_4, and 3*M* H_2SO_4, mossy zinc, marble chip, spatulas, cobalt blue glass filters

Safety For Step 3, wear gloves to avoid stains on your skin from silver nitrate. **CAUTION** *Students should not look at burning magnesium without the cobalt blue glass filters.*

Procedure As you do each step, have students identify the clue for chemical change. Emphasize that the clues point to a chemical change but do not confirm that a change has taken place.

a. Light a Bunsen burner. *(heat and light produced)*

b. Using tongs, hold a piece of magnesium ribbon in a burner flame until it ignites. Remove from heat and observe. *(The product is a white powder; heat and light are produced.)*

c. Put 5 mL of silver nitrate solution in each of two test tubes. Add a small amount of sodium chloride solution to one tube and potassium chromate solution to the other. *(color change, formation of a precipitate)*

d. Put 5 mL of the sulfuric acid in each of two test tubes. Add a piece of mossy zinc to one tube and a marble chip ($CaCO_3$) to the other. *(formation of a gas)*

Download a worksheet on **Chemical and Physical Changes** for students to complete, and find additional teacher support from NSTA SciLinks.

For: Links on Chemical and Physical Changes
Visit: www.SciLinks.org
Web Code: cdn-1024

Recognizing Chemical Changes

How can you tell whether a chemical change has taken place? There are four clues that can serve as a guide. **Possible clues to chemical change include a transfer of energy, a change in color, the production of a gas, or the formation of a precipitate.**

Every chemical change involves a transfer of energy. For example, energy stored in natural gas is used to cook food. When the methane in natural gas combines with oxygen in the air, energy is given off in the form of heat and light. Some of this energy is transferred to and absorbed by food that is cooking over a lit gas burner. The energy causes chemical changes to take place in the food. The food may brown as it cooks, which is another clue that chemical changes are occurring.

You can observe two other clues to chemical change while cleaning a bathtub. The ring of soap scum that can form in a bathtub is an example of a precipitate. A **precipitate** is a solid that forms and settles out of a liquid mixture. Some bathroom cleaners that you can use to remove soap scum start to bubble when you spray them on the scum. The bubbles are a product of a chemical change that is taking place in the cleaner.

If you observe a clue to chemical change, you cannot be certain that a chemical change has taken place. The clue may be the result of a physical change. For example, energy is always transferred when matter changes from one state to another. Bubbles form when you boil water or open a carbonated drink. The only way to be sure that a chemical change has occurred is to test the composition of a sample before and after the change. Figure 2.14 shows examples of practical situations in which different clues to chemical change are visible.

 ***What energy transfer takes place when food cooks?***

Figure 2.14 Clues to chemical change often have practical applications. **a** Bubbles of carbon dioxide gas form when a geologist puts acid on a rock that contains compounds called carbonates. **b** When a test strip is dipped in urine, the color change is used to estimate the level of the sugar glucose in urine. **c** One step in the production of cheese is a reaction that causes milk to separate into solid curds and liquid whey.

Facts and Figures

Urinalysis Test Strips

A test-strip urinalysis is semi-quantitative, meaning that the test provides information about the level of a material present in the urine, but not a specific quantity. The strips can also be used to test for ketones, protein, blood, nitrite, bilirubin, urobilinogen, pH, white blood cells, and specific gravity. All the tests involve a color change.

In one glucose test, the hydrogen peroxide produced when glucose is oxidized reacts with potassium iodide. The oxygen produced in the second reaction binds with a dye to produce colors that range from blue to dark brown. You may want to relate what happens with test strips to what happens in the Small-Scale lab on p. 56.

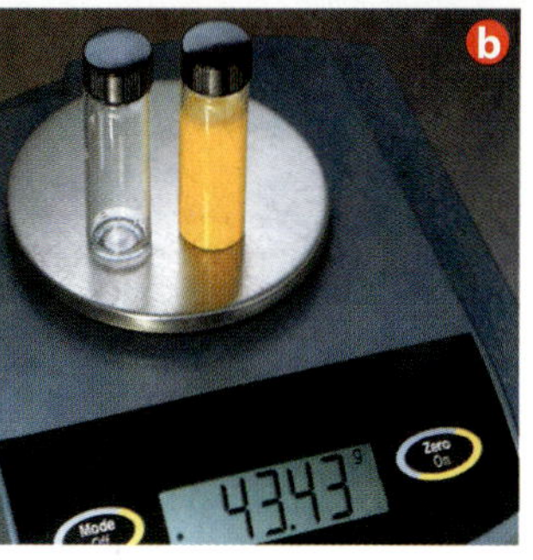

Figure 2.15 When the liquids in photograph A are mixed, they react. None of the products are gases. **Analyzing Data** ***How do you know that a reaction took place and that mass was conserved during the reaction?***

Conservation of Mass

When wood burns, substances in the wood combine with oxygen from the air. As the wood burns, a sizable amount of matter is reduced to a small pile of ashes. The reaction seems to involve a reduction in the amount of matter. But appearances can be deceiving. **During any chemical reaction, the mass of the products is always equal to the mass of the reactants.** Two of the products of burning wood—carbon dioxide gas and water vapor—are released into the air. When the mass of these gases is considered, the amount of matter is unchanged. Careful measurements show that the total mass of the reactants (wood and the oxygen consumed) equals the total mass of the products (carbon dioxide, water vapor, and ash).

Mass also holds constant during physical changes. For example, when 10 grams of ice melt, 10 grams of liquid water are produced. Similar observations have been recorded for all chemical and physical changes studied. The scientific law that reflects these observations is the law of conservation of mass. The **law of conservation of mass** states that in any physical change or chemical reaction, mass is conserved. Mass is neither created nor destroyed. The conservation of mass is more easily observed when a change occurs in a closed container, as in Figure 2.15.

2.4 Section Assessment

28. **Key Concept** How does a chemical change affect the composition of matter?
29. **Key Concept** Name four possible clues that a chemical change has taken place.
30. **Key Concept** In a chemical reaction, how does the mass of the reactants compare with the mass of the products?
31. What is the main difference between physical changes and chemical changes?
32. Classify the following changes as physical or chemical changes.
    - **a.** Water boils.
    - **b.** Salt dissolves in water.
    - **c.** Milk turns sour.
    - **d.** A metal rusts.
33. According to the law of conservation of mass, when is mass conserved?
34. Hydrogen and oxygen react chemically to form water. How much water would form if 4.8 grams of hydrogen reacted with 38.4 grams of oxygen?

Connecting Concepts

The Scientific Method Lavoisier proposed the law of conservation of mass in 1789. Write a paragraph describing, in general, what Lavoisier must have done before he proposed this law. Use what you learned about the scientific method in Section 1.3.

Interactive Textbook

Assessment 2.4 Test yourself on the concepts in Section 2.4.

with **ChemASAP**

Conservation of Mass

Use Visuals 

Figure 2.15 Explain that the photos are before and after views of a reaction. Ask, **Why are the containers covered?** *(The covers prevent evaporation of the solutions and loss of mass.)*

3 ASSESS

Evaluate Understanding 

Write this word equation on the board.

methane + oxygen →
carbon dioxide + water + energy

Have students identify the reactants and products and one possible clue for a chemical change.

Reteach 

Write a word equation on the board for the reaction of magnesium and oxygen to produce magnesium oxide. Ask students to explain how mass is conserved in this reaction. *(The difference between the mass of the magnesium and the mass of the magnesium oxide is the mass of the oxygen with which magnesium combines.)*

Connecting Concepts

Lavoisier would have done a series of experiments in which he measured the masses of reactants and products before he proposed the law of conservation of mass.

Interactive Textbook

If your class subscribes to the interactive Textbook, use it to review key concepts in Section 2.4.

with **ChemASAP**

Section 2.4 Assessment

28. The composition of matter always changes during a chemical change.
29. a transfer of energy, a change in color, the production of a gas, the formation of a precipitate
30. The mass of the products is always equal to the mass of the reactants.
31. In a physical change, the chemical composition of a substance does not change. In a chemical change, the chemical composition of the reactants changes as one or more products form.
32. **a.** physical **b.** physical **c.** chemical **d.** chemical
33. Mass is conserved in every physical change or chemical reaction.
34. 43.2 g

Answers to...

Figure 2.15 change of color and formation of a precipitate; The balance reading did not change.

 **Checkpoint** The burning of methane produces heat that is transferred to the food.

1 + 2 + 3 = Black! 

Objective In this activity, students will begin to:
- familiarize themselves with small-scale equipment and methodology.
- observe and analyze some chemical reactions.

Skills Focus Observing, Classifying

 **Prep Time** 1 hour

Materials Paper, metric ruler, reaction surface, pipette, dropper, spatulas, KI, liquid starch, cereal, NaClO, H_2O_2, $CuSO_4$, foods, iodized and non-iodized salt, antacid tablets, color marker pens

Advance Prep

Solution	Preparation
0.1*M* KI	4.2 g KI in 250 mL water
0.2*M* $CuSO_4$	12.5 g $CuSO_4 \cdot 5H_2O$ in 250 mL water
1% NaClO	50 mL household bleach in 200 mL water
3% H_2O_2	Use undiluted household hydrogen peroxide
20% starch	50 mL liquid starch in 200 mL water

Class Time 40 minutes

Teaching Tips
- Demonstrate how to use a pipette as a stirrer. This information, as well as a discussion of reaction surfaces, can be found in the *Small-Scale Chemistry Laboratory Manual.*
- Explain that using a mixture of KI and NaClO to test for starch is an example of qualitative analysis.

Expected Outcome

	NaClO	H_2O_2	$CuSO_4$
KI	yellow	yellow	brown ppt
KI + starch	black	black	black
KI + paper	black	black	black
KI + cereal	black	black	black

For Enrichment 

Have students design and carry out an experiment to quantify the amount of KI in iodized salt.

Small-Scale LAB

1 + 2 + 3 = Black!

Purpose
To make macroscopic observations of chemical reactions and use them to solve problems.

Materials
- paper
- metric ruler
- reaction surface
- materials shown in grid
- pipette, medicine droppers, and spatulas

Procedure    

1. Draw two copies of the grid on separate sheets of paper. Make each square in the grid 2 cm on each side.
2. Place a reaction surface over one of the grids. Use the second grid as a data table to record your observations.
3. Use the column and row labels to determine which materials belong in each square. Depending on the material, add one drop, one piece, or a few grains.
4. Stir each mixture by forcing air from an empty pipette as directed by your teacher.

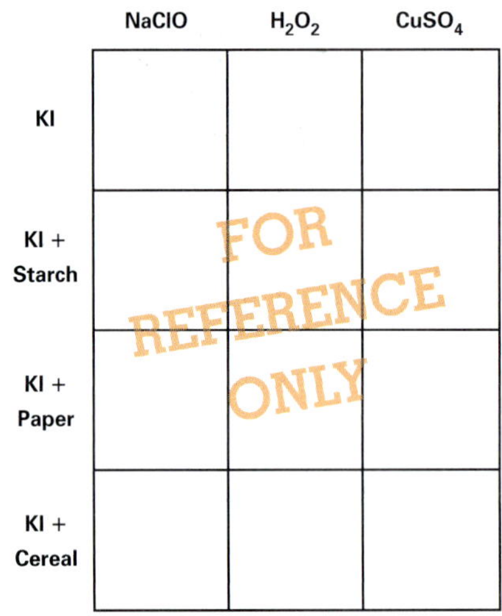

Analyze
Using your experimental data, record the answers to the following questions below your data table.
1. What color is a mixture of sodium hypochlorite (NaClO) and potassium iodide (KI)?
2. What happens when you mix NaClO, KI, and starch?
3. What do NaClO, H_2O_2, and $CuSO_4$ have in common?
4. What substance is found in both paper and cereal? How do you know?
5. If you used NaClO instead of $CuSO_4$ in reactions other than the reaction with KI and starch, would you expect the results to always be identical? Explain your answer.

You're The Chemist
The following small-scale activities allow you to develop your own procedures and analyze the results.
1. **Design It!** Design and carry out an experiment to see which foods contain starch.
2. **Design It!** Read the label on a package of iodized salt. How much KI does iodized salt contain? Design an experiment to demonstrate the presence of KI in iodized salt and its absence in salt that is not iodized.
3. **Design It!** Antacid tablets often contain starch as a binder to hold the ingredients in the tablet together. Design and carry out an experiment to explore various antacid tablets to see if they contain starch.
4. **Analyze It!** NaClO is a bleaching agent. Such agents are used to whiten clothes and remove stains. Use different color marker pens to draw several lines on a piece of white paper. Add one drop of NaClO to each line. What happens? Try inventing a technique that you can use to make "bleach art."

Analyze and Conclude
1. yellow
2. The mixture turns a blue-black color.
3. They all turn a mixture of KI and starch black.
4. Starch; both turn blue-black, which suggests the presence of starch.
5. The results may be the same in reactions that are similar to the one with KI and starch, but different in other reactions.

You're The Chemist
1. Add KI + NaClO to various foods. A black color indicates the presence of starch.
2. Most table salt contains 0.01% KI. Wet only a portion of a small pile of salt with starch. Add $CuSO_4$ or H_2O_2. A black color indicates the presence of KI.
3. If an antacid tablet contains starch, it will turn black when treated with KI + NaClO.
4. The color fades. A picture can be drawn with colored ink. Areas can be treated with NaClO to bleach parts of the picture.

CHAPTER 2 Study Guide

Key Concepts

2.1 Properties of Matter

- Properties used to describe matter can be classified as extensive or intensive.
- Every sample of a given substance has identical intensive properties because every sample has the same composition.
- Three states of matter are solid, liquid, and gas.
- Physical changes can be classified as reversible or irreversible.

2.2 Mixtures

- Mixtures can be classified as heterogeneous mixtures or as homogeneous mixtures, based on the distribution of their components.
- Differences in physical properties can be used to separate mixtures.

2.3 Elements and Compounds

- Compounds can be broken down into simpler substances by chemical means, but elements cannot.
- If the composition of a material is fixed, the material is a substance. If the composition may vary, the material is a mixture.
- Chemists use chemical symbols to represent elements, and chemical formulas to represent compounds.

2.4 Chemical Reactions

- During a chemical change, the composition of matter always changes.
- Four possible clues to chemical change include a transfer of energy, a change in color, the production of a gas, or the formation of a precipitate.
- During any chemical reaction, the mass of the products is always equal to the mass of the reactants.

Vocabulary

- chemical change (p. 48)
- chemical property (p. 53)
- chemical reaction (p. 53)
- chemical symbol (p. 51)
- compound (p. 48)
- distillation (p. 47)
- element (p. 48)
- extensive property (p. 39)
- filtration (p. 46)
- gas (p. 42)
- heterogeneous mixture (p. 45)
- homogeneous mixture (p. 45)
- intensive property (p. 39)
- law of conservation of mass (p. 55)
- liquid (p. 41)
- mass (p. 39)
- mixture (p. 44)
- phase (p. 45)
- physical change (p. 42)
- physical property (p. 40)
- precipitate (p. 54)
- product (p. 53)
- reactant (p. 53)
- solid (p. 41)
- solution (p. 45)
- substance (p. 40)
- vapor (p. 42)
- volume (p. 39)

Organizing Information

Use these terms to construct a concept map that organizes the major ideas of this chapter.

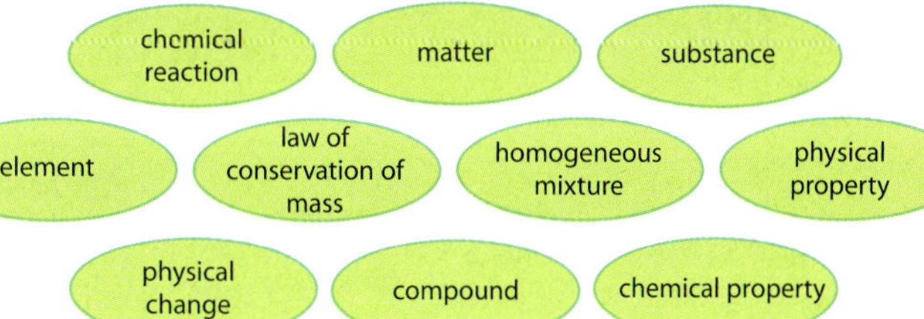

Concept Map 2 Solve the Concept Map with the help of an interactive guided tutorial.

with **ChemASAP**

Chapter Resources

Print

- **Core Teaching Resources,** Chapter 2 *Practice Problems, Vocabulary Review, Quiz, Chapter Test A, Chapter Test B*

Technology

- **Computer Test Bank,** Chapter 2
- **Interactive Textbook with ChemASAP,** Chapter 2

CHAPTER 2 Study Guide

Study Tip

Organizing New Information Have students learn new vocabulary terms by making vocabulary flashcards. Students should write a vocabulary word on one side of a card and its definition on the other side. The cards can be used by students to quiz each other.

If your class subscribes to the Interactive Textbook with ChemASAP, your students can go online to access an interactive version of the Student Edition and a self-test.

with **ChemASAP**

CHAPTER 2

Assessment *continued*

Reviewing Content

35. An extensive property depends on the amount of matter; an intensive property depends on the type of matter. Mass and volume are extensive properties. Color and hardness are intensive properties.
36. Answers could include reddish-yellow color, conductor of heat and electricity, malleable, melting point of 1084°C, and boiling point of 2562°C.
37. melting point and boiling point
38. State; both are gases.
39. **a.** solid **b.** liquid **c.** gas **d.** liquid **e.** gas **f.** liquid
40. A vapor; the term *vapor* is used to refer to the gaseous state of a substance which normally exists as a liquid or solid at room temperature.
41. The particles in a solid are packed tightly together in an orderly arrangement. Particles in a liquid are in close contact, but not in a orderly arrangement. The particles in a gas are relatively far apart.
42. Chlorine condenses. Mercury, bromine and water freeze.
43. Sharpening a pencil is an irreversible change. Making ice cubes is a reversible change.
44. Heterogeneous mixtures have a non-uniform composition with two or more phases. Homogeneous mixtures have a uniform composition.
45. one; Solutions are homogeneous mixtures with uniform composition throughout.
46. **a.** heterogeneous **b.** homogeneous **c.** depends on how well the batter is mixed **d.** homogeneous
47. The goal is to separate the components of a solution. The solution is boiled to produce a vapor, which is condensed to a liquid. Dissolved solids are left behind.
48. Compounds can be separated by chemical means into simpler substances. Elements cannot.
49. **a.** Hydrogen and oxygen are the elements that make up the compound water. **b.** Nitrogen and oxygen are elements present in the mixture air. **c.** Sodium and chlorine are elements in the compound sodium chloride (table salt). **d.** The element carbon and the compound water are the final products of heating table sugar.
50. **a.** nitrogen, hydrogen **b.** potassium, oxygen **c.** carbon, hydrogen, oxygen **d.** calcium, sulfur
51. In W, the single letter is capitalized. In Hg, the first letter is a capital and the second letter is lowercase.
52. The compound water contains two parts hydrogen to one part oxygen.
53. When heated, sulfur and iron react and form iron sulfide. The composition of the reactants in a chemical change is different from the composition of the products. In a physical change, the chemical composition of a sample doesn't change.
54. **a.** physical **b.** chemical (color change) **c.** chemical (production of a gas) **d.** physical
55. chemical property
56. 18 g of water

CHAPTER 2

Assessment

Reviewing Content

2.1 Properties of Matter

35. Describe the difference between an extensive property and an intensive property and give an example of each.

36. List three physical properties of copper.

37. Name two physical properties that could be used to distinguish between water and ethanol.

38. Name one physical property that could not be used to distinguish chlorine from oxygen.

39. What is the physical state of each of these materials at room temperature?
a. gold **b.** gasoline
c. neon **d.** olive oil
e. oxygen **f.** mercury

40. Fingernail-polish remover (mostly acetone) is a liquid at room temperature. Would you describe acetone in the gaseous state as a vapor or a gas? Explain your answer.

41. Compare the arrangements of individual particles in solids, liquids, and gases.

42. Use Table 2.1 to identify four substances that undergo a physical change if the temperature is reduced from 50°C to −50°C. What is the physical change that takes place in each case?

43. Explain why sharpening a pencil is a different type of physical change than freezing water to make ice cubes.

2.2 Mixtures

44. What is the difference between homogeneous mixtures and heterogeneous mixtures?

45. How many phases does a solution have? Explain your answer.

46. Classify each of the following as homogeneous or heterogeneous mixtures.
a. chocolate-chip ice cream
b. green ink
c. cake batter
d. cooking oil

47. What is the goal of a distillation? Describe briefly how this goal is accomplished.

2.3 Elements and Compounds

48. How could you distinguish an element from a compound?

49. Describe the relationship between the three items in each of the following groups. Identify each item as an element, compound, or mixture.
a. hydrogen, oxygen, and water
b. nitrogen, oxygen, and air
c. sodium, chlorine, and table salt
d. carbon, water, and table sugar

50. Name the elements found in each of the following compounds.
a. ammonia (NH_3)
b. potassium oxide (K_2O)
c. sucrose ($C_{12}H_{22}O_{11}$)
d. calcium sulfide (CaS)

51. Not all element names come from English or Latin words. The symbol for tungsten is W from the German word *wolfram*. The symbol for mercury is Hg from the Greek word *hydragyrum*. Use the symbols W and Hg to explain the system of symbols for elements.

52. What does the formula H_2O tell you about the composition of water?

2.4 Chemical Reactions

53. Use the word equation to explain how a chemical change differs from a physical change.

$$\text{iron} + \text{sulfur} \xrightarrow{\text{heat}} \text{iron sulfide}$$

54. Classify each of the following as a physical or chemical change. For any chemical change, list at least one clue to support your answer.
a. A copper wire is bent.
b. Charcoal burns in a grill.
c. Bread dough rises when yeast is added.
d. Sugar dissolves in water.

55. Which type of property cannot be observed without changing the composition of a substance?

56. When ammonium nitrate (NH_4NO_3) explodes, the products are nitrogen, oxygen, and water. When 40 grams of ammonium nitrate explode, 14 grams of nitrogen and 8 grams of oxygen form. How many grams of water form?

Understanding Concepts

57. Explain why mass cannot be used as a property to identify a sample of matter.

58. Is malleability an extensive property or an intensive property? Explain.

59. The state of a substance can change when the substance is heated or cooled. So what does it mean to say that a certain substance is a solid, liquid, or gas?

Use the data table to answer Questions 60–63.

Substance	Color	Melting point (°C)	Boiling point (°C)
Bromine	red-brown	−7	59
Chlorine	green-yellow	−101	−34
Ethanol	colorless	−117	78
Mercury	silvery-white	−39	357
Neon	colorless	−249	−246
Sulfur	yellow	115	445
Water	colorless	0	100

60. Which colorless substance is a liquid at −30°C?

61. Which colorless substance is a gas at 60°C?

62. Which substance is a solid at 7°C?

63. As the temperature rises, which solid will melt before mercury boils?

64. Use the arrangement of particles in solids and gases to explain why solids are not as easy to compress as gases.

65. Imagine you are standing in a kitchen and then in the middle of a park. When you view your surroundings in each location do you see mostly elements, compounds, or mixtures?

66. Identify each of the following items as a mixture or compound. Classify the mixtures as homogeneous or heterogeneous.
a. raw egg **b.** ice
c. gasoline **d.** blood

67. Classify the following properties of the element silicon as chemical or physical properties.
a. blue-gray color
b. brittle
c. doesn't dissolve in water
d. melts at 1410°C
e. reacts vigorously with fluorine

68. How are the items in each of the following pairs similar ? How are they different?
a. copper and silver
b. distilled water and saltwater
c. table sugar and table salt

69. In photograph A, a coil of zinc metal is in a solution of sulfuric acid. In photograph B, a yellow solution of sodium chromate is being added to a colorless solution of silver nitrate. What clues in the photographs indicate that a chemical change is probably occurring?

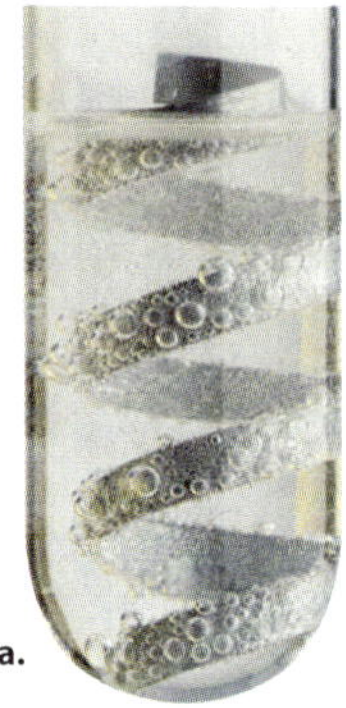
a.

b.

70. Describe clues you might observe during the following events that could support the conclusion that a chemical change is occurring.
a. An antacid tablet is dropped into water.
b. A ring of scum forms around a bathtub.
c. Iron rusts.
d. A firecracker explodes.

71. Explain why the production of a gas does not always mean that a chemical reaction has occurred.

72. The wax appears to disappear as a candle burns. How can the law of conservation of mass apply to this reaction?

Understanding Concepts

57. Mass is an extensive property, which depends only on the amount of matter in the sample, not on the composition of the sample.

58. Malleability is an intensive property, which depends on the type of matter in a sample.

59. Substances are classified as solids, liquids, or gases according to their state at room temperature, which in this book is 20°C.

60. ethanol

61. neon

62. sulfur

63. sulfur

64. The particles in solids are packed tightly together. The particles in a gas are spaced relatively far apart.

65. kitchen, mixtures; park, mixtures

66. **a.** heterogeneous mixture
b. compound
c. homogeneous mixture
d. heterogeneous mixture

67. **a.** physical **b.** physical
c. physical **d.** physical
e. chemical

68. **a.** Both are elements and solids at room temperature; they are a different color and have different melting points and boiling points.
b. Both are clear liquids at room temperature; distilled water is an element and saltwater is a mixture.
c. Both are white, solid compounds. Sugar is composed of carbon, hydrogen, and oxygen. Salt is composed of sodium and chlorine.

69. In photograph A, bubbles indicate the production of a gas. In photograph B, there is a color change and a precipitate.

70. **a.** Gas is produced.
b. formation of a precipitate
c. color and texture change
d. energy change, odor change

71. A gas may be released during a physical change. For example, bubbles form when water boils.

72. The wax appears to disappear because the products of the reaction—carbon dioxide and water vapor—are colorless gases.

CHAPTER 2

Assessment continued

Critical Thinking

73. A gas expands to fill any space; a gas has no shape or volume without a container. A solid has a definite shape and volume; a solid doesn't need a container to maintain its shape and volume.

74. The appearance of a substance will change during a change of state, which is a physical change.

75. Gallium (c) will freeze first; mercury (b) will freeze last.

76. Add sufficient water to dissolve all of the sugar. Separate the charcoal and sand from the sugar water by filtration. Large pieces of charcoal could be separated on the basis of color. Small pieces of charcoal could be burned.

77. Iron rusts when it reacts with oxygen in the air to form an oxide (Fe_2O_3). The mass of the rust is the sum of the mass of the iron and the mass of the oxygen that combined with the iron.

78. Smelling something burning is one possible answer.

79. a. Yes; because the graph is a straight line, the proportion of iron to oxygen is a constant, which is true for a compound.
b. No; a point for the values given wouldn't fall on the line. The mass ratio of iron to oxygen is different.

Concept Challenge

80. a. oxygen and calcium
b. silicon, aluminum, and iron
c. Different; the second most abundant element in Earth's crust, silicon, is not present in the human body, and the second most abundant element in the human body, carbon, is not among the most abundant elements in Earth's crust. If the elements are different, then the compounds must also be different.

CHAPTER 2

Assessment *continued*

Critical Thinking

73. Discuss the statement "A gas requires a container but a solid is its own container."

74. Explain why this statement is false. "Because there is no change in composition during a physical change, the appearance of the substance will not change."

75. Assume that water, mercury, and gallium are all at 40°C. As the temperature drops, which substance will freeze first? Which will be the last to freeze?

76. Devise a way to separate sand from a mixture of charcoal, sand, sugar, and water.

77. When powdered iron is left exposed to the air, it rusts. Explain why the mass of the rust is greater than the mass of the powdered iron.

78. A change in odor can also be a clue that a chemical change has occurred. Describe at least one situation in which you might be likely to detect such a change in odor in a kitchen.

79. The mass of the elements iron and oxygen in four samples of a rust-colored substance was measured in grams (g). The amount of iron and oxygen in each sample is shown on the graph.

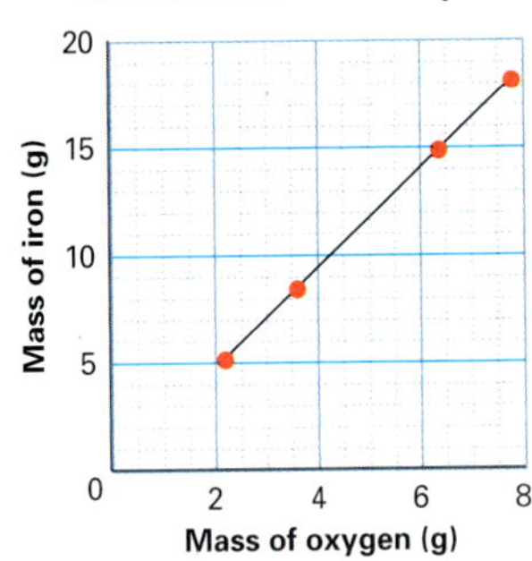

a. Do you think all four samples are the same compound? Explain.
b. Another sample of similar material was found to contain 9.9 grams of iron and 3.4 grams of oxygen. Is this sample the same substance as the other four samples? Explain.

Concept Challenge

80. Five elements make up 98% of the mass of the human body. These elements are oxygen (61%), carbon (23%), hydrogen (10.0%), nitrogen (2.6%), and calcium (1.4%). Compare these data with those in the pie graph below, which shows the five most abundant elements by mass in Earth's crust, oceans, and atmosphere.

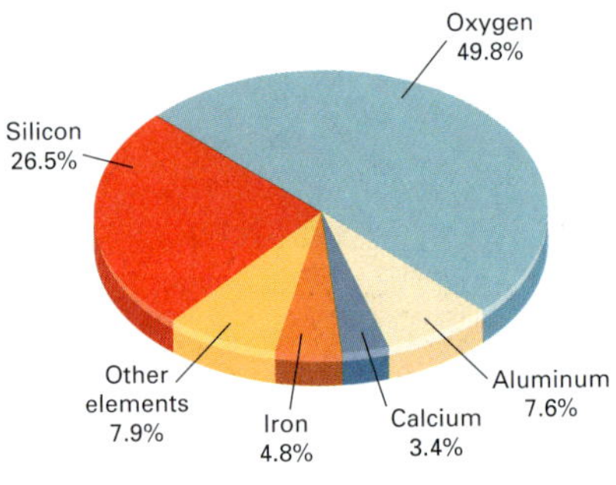

a. Which elements are abundant both in the human body and Earth's crust, oceans, and atmosphere?
b. Which elements are abundant in Earth's crust, oceans, and atmosphere, but not in the human body?
c. Would you expect the compounds found in the human body to be the same as or different from those found in rocks, seawater, and air? Use the data to explain your answer.

81. Use Table 2.1 on p. 40 to answer this question.
a. Which substances in the table are in the liquid state at 125°C?
b. Use the physical properties of one of these substances to explain how you figured out the answer to Question 81a.
c. The substances in the table are listed in order of increasing melting point. Propose another way that these data could be arranged.

82. Each day you encounter some chemical changes that are helpful and some that are harmful to humans or the environment. Cite three examples of each type. For each example, list the clues that identified the change as a chemical change.

81. a. mercury and sulfur
b. Sulfur melts at 113°C and boils at 445°C. Between 113°C and 445°C, it exists as a liquid. Mercury melts at −39°C and boils at 357°C. Between these temperatures, it exists as a liquid.
c. Possibilities include: by color, by boiling point, or in alphabetical order.

82. Many answers are possible. Cooking food (color change, odor change) is an example of a helpful change; the formation of soap scum (precipitate) is an example of a harmful change.

Standardized Test Prep

Test-Taking Tip

Using Models To answer some test questions, you will be asked to use visual models. At first the models may look very similar. Decide which information will help you answer the question. The number of particles, their colors, or their shapes may or may not be important.

Select the choice that best answers each question or completes each statement.

1. Which of the following is *not* a chemical change?
- **a.** paper being shredded
- **b.** steel rusting
- **c.** charcoal burning
- **d.** a newspaper yellowing in the sun

2. Which phrase best describes an apple?
- **a.** heterogeneous mixture
- **b.** homogeneous compound
- **c.** heterogeneous substance
- **d.** homogeneous mixture

3. Which element is paired with the wrong symbol?
- **a.** sulfur, S
- **b.** potassium, P
- **c.** nitrogen, N
- **d.** calcium, Ca

4. Which of these properties could *not* be used to distinguish between table salt and table sugar?
- **a.** boiling point
- **b.** melting point
- **c.** density
- **d.** color

5. The state of matter characterized by a definite volume and an indefinite shape is a
- **a.** solid.
- **b.** liquid.
- **c.** mixture.
- **d.** gas.

The lettered choices below refer to Questions 6–9. A lettered choice may be used once, more than once, or not at all.

(A) compound
(B) heterogeneous mixture
(C) element
(D) homogeneous mixture

Which description correctly identifies each of the following materials?

6. air
7. carbon monoxide
8. zinc
9. mushroom pizza

Use the atomic windows to answer Question 10.

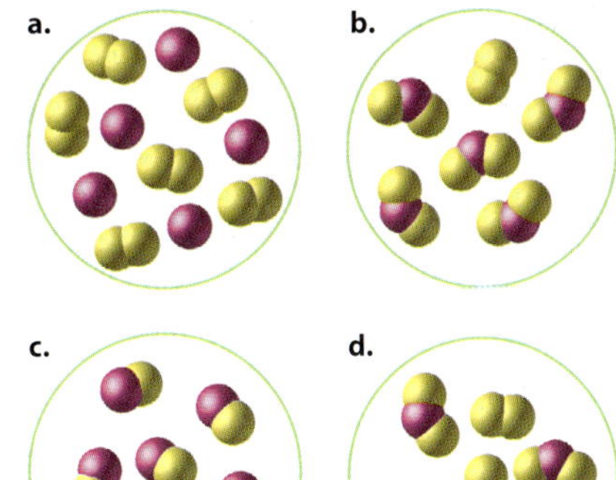

10. The species in window A react. Use the law of conservation of mass to determine which window best represents the reaction products.

Use the data table to answer Questions 11–14.

Mass of magnesium (g)	Mass of oxygen (g)	Mass of magnesium oxide (g)
5.0	3.3	8.3
6.5	(a)	10.8
13.6	9.0	(b)
(c)	12.5	31.5

11. Magnesium metal burns vigorously in oxygen to produce the compound magnesium oxide. Use the law of conservation of mass to identify the masses labeled (a), (b), and (c) in the table.

12. Use the data in the completed table to construct a graph with mass of magnesium on the x-axis and mass of magnesium oxide on the y-axis.

13. How many grams of magnesium oxide form when 8.0 g of magnesium are burned?

14. How many grams of magnesium and oxygen react to form 20.0 g of magnesium oxide?

Standardized Test Prep

1. a
2. a
3. b
4. d
5. b
6. D
7. A
8. C
9. B
10. b
11. (a) is 4.3 g; (b) is 22.6 g; (c) is 19.0 g
12.

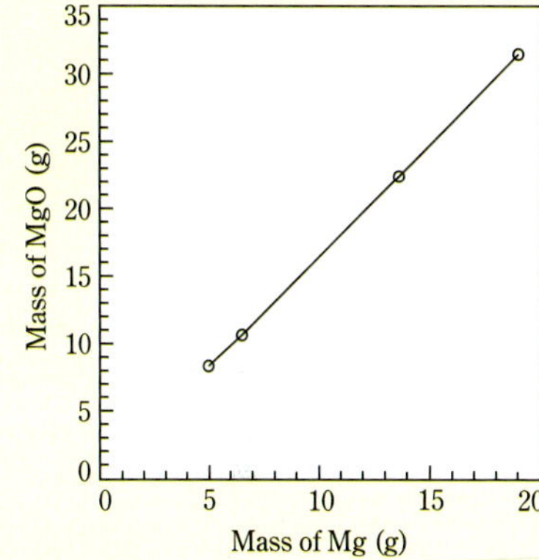

13. 13.3 g magnesium oxide
14. 12.1 g magnesium and 7.9 g oxygen

CHAPTER 3

Scientific Measurement Planning Guide

SECTION OBJECTIVES	STANDARDS NATIONAL	STANDARDS STATE	ACTIVITIES and LABS	
3.1 Measurements and Their Uncertainty, pp. 63–72 1 block or 2 periods **3.1.1 Convert** measurements to scientific notation. **3.1.2 Distinguish** among accuracy, precision, and error of a measurement. **3.1.3 Determine** the number of significant figures in a measurement and in a calculated answer.	A-2		**SE** Inquiry Activity: Exploring Density, p. 62 **SE** Quick Lab: Accuracy and Precision, p. 72 **TE** Class Activity: Precision and Accuracy, p. 64 **TE** Class Activity: Olympic Times, p. 66 **TE** Class Activity: Significant Zeros, p. 68	L2 L2 L2 L2 L2
3.2 The International System of Units, pp. 73–79 1 block or 2 periods **3.2.1 List** SI units of measurement and common SI prefixes. **3.2.2 Distinguish** between the mass and weight of an object. **3.2.3 Convert** between the Celsius and Kelvin temperature scales.	A-2		**TE** Teacher Demo: Volume Measurements, p. 75 **TE** Class Activity: Mass of a Penny, p. 76 **SSLM** Lab 3: Design and Construction of a Small-Scale Balance **SSLM** Lab 4: Design and Construction of a Set of Standardized Weights	L2 L2 L2 L2
3.3 Conversion Problems, pp. 80–88 1 block or 2 periods **3.3.1 Construct** conversion factors from equivalent measurements. **3.3.2 Apply** the technique of dimensional analysis to a variety of conversion problems. **3.3.3 Solve Problems** by breaking the solution into steps. **3.3.4 Convert** complex units, using dimensional analysis.	A-2		**SE** Quick Lab: Dimensional Analysis, p. 86 **TE** Class Activity: Expanding a Recipe, p. 81 **TE** Class Activity: Sports Stats, p. 84	L2 L2 L2
3.4 Density, pp. 89–94 1/2 block or 1 period **3.4.1 Calculate** the density of a material from experimental data. **3.4.2 Describe** how density varies with temperature.	A-2, E-2		**SE** Small-Scale Lab: Now What Do I Do?, p. 94 **TE** Teacher Demo: Density Calculations, p. 90 **TE** Teacher Demo: The Hydrometer, p. 92 **LM** Lab 4: Mass, Volume, and Density **SSLM** Lab 5: Massing Activities for a Small-Scale Lab **LP** Lab 3-1: Basic Measurement **LP** Lab 3-2: Density	L2 L2 L2 L2 L2 L2 L2

Ability Levels
L1 For students who need additional help
L2 For all students
L3 For students who need to be challenged

Components

SE	Student Edition	LM	Laboratory Manual	GCP	Graphing Calculator Problems	T	Transparencies
TE	Teacher's Edition	SSLM	Small-Scale Chemistry Lab Manual	TTT	Test-Taking Tips with Transparencies, Grades 9–12 Science	CA	Chem Alive and Chemistry Field Trips DVD
GRSW	Guided Reading and Study Workbook	LP	Laboratory Practicals			iT	Interactive Textbook with ChemASAP
CTR	Core Teaching Resources	PLM	Probeware Lab Manual	STP	Standardized Test Preparation Workbook	GO	Go Online Internet Resources
SSG	Spanish Study Guide	VCL	Virtual Chemistry Labs				
		CTB	Computer Test Bank				

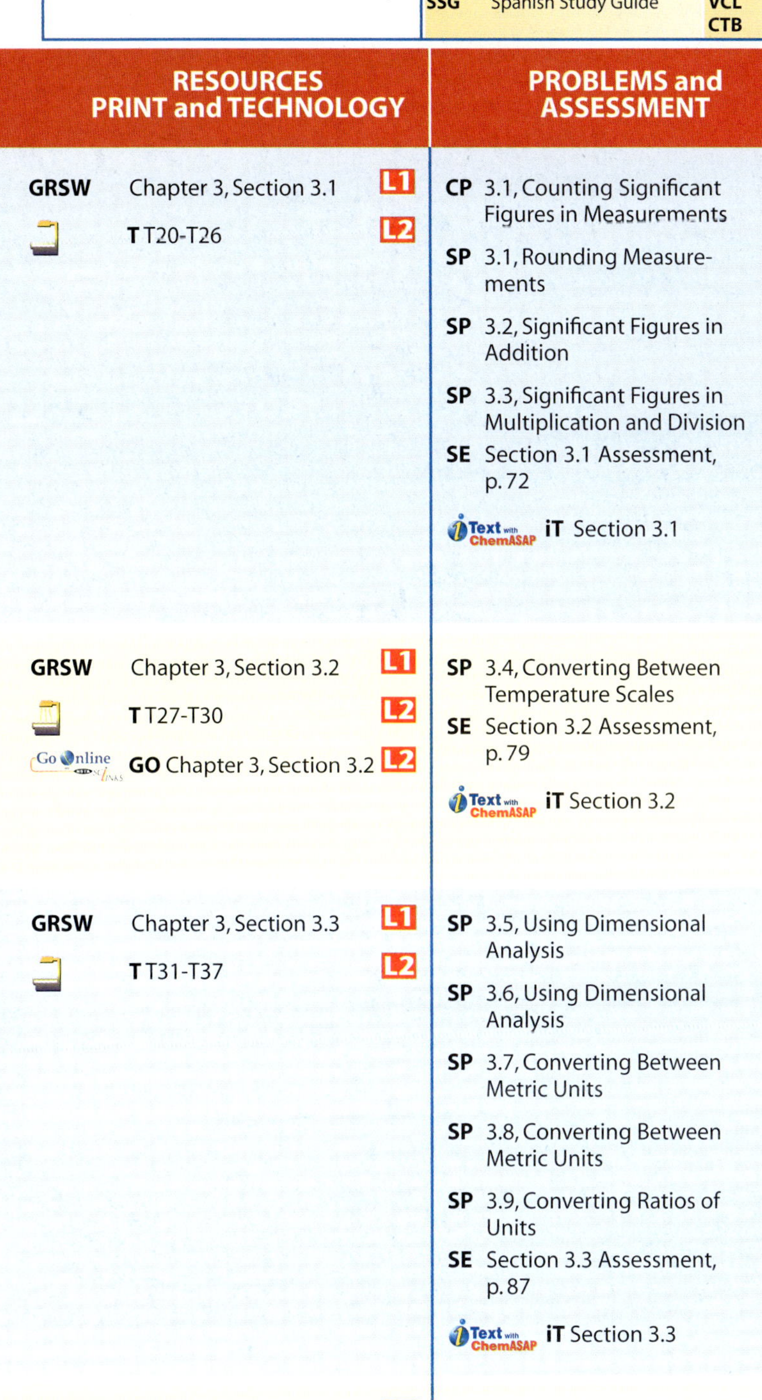
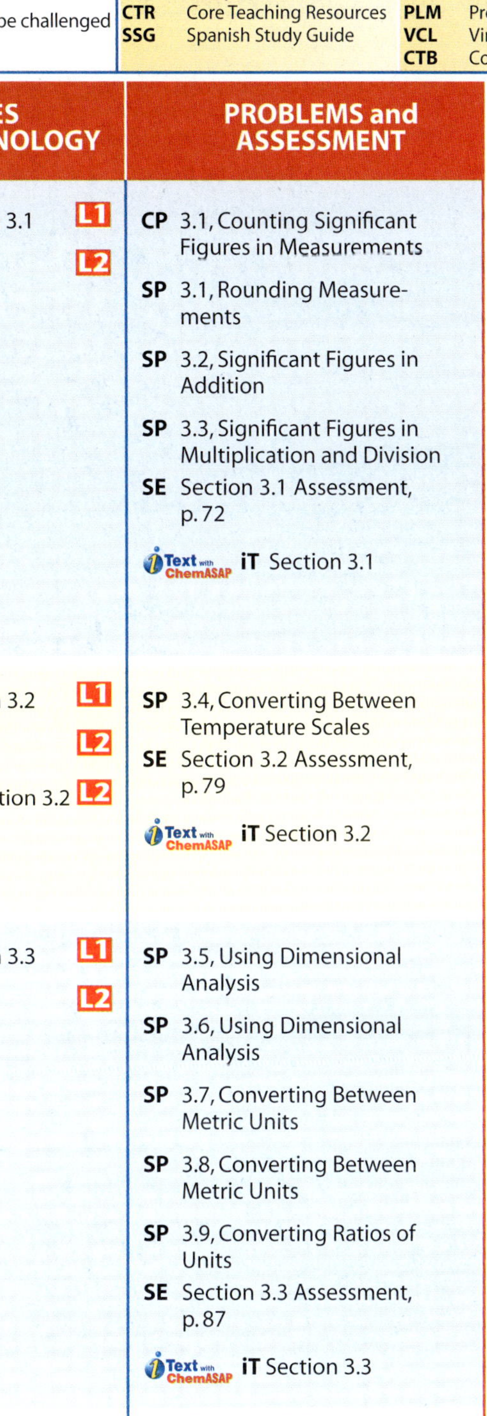

RESOURCES PRINT and TECHNOLOGY		PROBLEMS and ASSESSMENT
GRSW Chapter 3, Section 3.1	L1	CP 3.1, Counting Significant Figures in Measurements SP 3.1, Rounding Measurements SP 3.2, Significant Figures in Addition SP 3.3, Significant Figures in Multiplication and Division SE Section 3.1 Assessment, p. 72 iT Section 3.1
T T20-T26	L2	
GRSW Chapter 3, Section 3.2	L1	SP 3.4, Converting Between Temperature Scales SE Section 3.2 Assessment, p. 79 iT Section 3.2
T T27-T30	L2	
GO Chapter 3, Section 3.2	L2	
GRSW Chapter 3, Section 3.3	L1	SP 3.5, Using Dimensional Analysis SP 3.6, Using Dimensional Analysis SP 3.7, Converting Between Metric Units SP 3.8, Converting Between Metric Units SP 3.9, Converting Ratios of Units SE Section 3.3 Assessment, p. 87 iT Section 3.3
T T31-T37	L2	
GRSW Chapter 3, Section 3.4	L1	SP 3.10, Calculating Density SP 3.11, Using Density to Calculate Volume SE Section 3.4 Assessment, p. 93 iT Section 3.4
T T38-T42	L2	
GO Chapter 3, Section 3.4	L2	
GO Chapter 3, Section 3.4	L2	

Go Online

Go Online SciLinks

Web Code: cdn-1032
Web Code: cdn-1034

Go Online PHSchool.com

Web Code: cdb-1034

MATERIALS List

Quantities for each group

STUDENT EDITION

Inquiry Activity, p. 62
three beakers, a 250-mL graduated cylinder, a colorless and clear cylindrical container, shampoo, water, rubbing alcohol, red food coloring, a medicine dropper, a spoon or stirring rod, and small solid objects (such as a paper clip, bottle cap, or piece of cork.)

Quick Lab, p. 72
3 × 5 in. index cards, metric ruler

Quick Lab, p. 87
3 × 5 in. index cards or paper cut into approximately the same size

Small-Scale Lab, p. 94
meter stick, balance, pair of dice, aluminum can, calculator, small-scale pipet, a pre- and post-1982 penny, 8-well strip, plastic cup

TEACHER'S EDITION

Class Activity, p. 64
a small object, such as a lead fishing weight, triple beam balance

Class Activity, p. 66
almanacs

Class Activity, p. 68
scientific literature, index cards

Teacher Demo, p. 75
sets of Erlenmeyer flask, buret, graduated cylinder, beaker, and volumetric flask, water, food coloring

Class Activity, p. 76
balance with a precision of at least 0.01 g, sets of 10 pennies each separated according to minting dates between 1970 and the present. (Be sure that sets include pre- and post-1982 minting dates)

Class Activity, p. 81
recipe, lists of equivalents and conversions among the following measurements: teaspoon, tablespoon, 1/4 cup, 1/2 cup, and 1 cup

Class Activity, p. 84
media guides containing vital statistics, such as heights and weights, of players

Teacher Demo, p. 90
3 or 4 different-sized cubes of a material such as wood, metal, or marble, metric ruler, balance

Teacher Demo, p. 92
hydrometer, 2 graduated cylinders, water, concentrated solution of sodium chloride

Chapter Assessment

CHAPTER ASSESSMENT

SE	Chapter Assessment, pp. 96–98
CTB	Chapter 3
iT	Chapter 3

STANDARDIZED TEST PREP

SE	Chapter 3, p. 99

iT Interactive Textbook with Assessment at PHSchool.com

CHAPTER 3

Scientific Measurement

 Activity

Exploring Density 

Purpose

To identify difference in densities as a means of separating immiscible liquids

Prep Time 5 minutes
Class Time 15 minutes

Think About It

1. The shampoo, which makes up the bottom layer, is denser than the water, which is denser than the rubbing alcohol.
2. The order in which the liquids are poured does not make a difference as long as the pouring is done carefully, so that mixing does not occur.

CHAPTER 3

Scientific Measurement

Milliliters (mL) are a common metric unit of volume.

INQUIRY Activity

Exploring Density

Materials

three beakers, a 250 mL graduated cylinder, a colorless and clear cylindrical container, shampoo, water, rubbing alcohol, red food coloring, a medicine dropper, a spoon or stirring rod, and small solid objects (such as a paper clip, plastic soda-bottle cap, aluminum nail, or piece of cork).

Procedure

1. Pour about 250 mL each of shampoo, water, and rubbing alcohol into separate beakers.
2. Stir a drop of food coloring into the rubbing alcohol and the shampoo.
3. Carefully pour each liquid (in the following order: shampoo, water, rubbing alcohol) down the inside of the colorless cylindrical container to a depth of 2.5 cm (about 1 in.).
4. Record the positions of the three layers of liquids.
5. Gently add each small solid object to the layered liquids and record its resting location.

Think About It

1. Propose an explanation for the order of layering of the liquids.
2. Do you think it would make any difference if you added the liquids in a different order? Why or why not?

3.1 Measurements and Their Uncertainty

Connecting to Your World

On January 4, 2004, the Mars Exploration Rover Spirit landed on Mars. Equipped with five scientific instruments and a rock abrasion tool (shown at left), Spirit was sent to examine the Martian surface around Gusev Crater, a wide basin that may have once held a lake. Each day of its mission, Spirit recorded measurements for analysis. This data helped scientists learn about the geology and climate on Mars. All measurements have some uncertainty. In the chemistry laboratory, you must strive for accuracy and precision in your measurements.

Using and Expressing Measurements

Your height (67 inches), your weight (134 pounds), and the speed you drive on the highway (65 miles/hour) are some familiar examples of measurements. A **measurement** is a quantity that has both a number and a unit. Everyone makes and uses measurements. For instance, you decide how to dress in the morning based on the temperature outside. If you were baking cookies, you would measure the volumes of the ingredients as indicated in the recipe.

Such everyday situations are similar to those faced by scientists. **Measurements are fundamental to the experimental sciences. For that reason, it is important to be able to make measurements and to decide whether a measurement is correct.** The units typically used in the sciences are those of the International System of Measurements (SI).

In chemistry, you will often encounter very large or very small numbers. A single gram of hydrogen, for example, contains approximately 602,000,000,000,000,000,000,000 hydrogen atoms. The mass of an atom of gold is 0.000 000 000 000 000 000 000 327 gram. Writing and using such large and small numbers is very cumbersome. You can work more easily with these numbers by writing them in scientific, or exponential, notation.

In **scientific notation**, a given number is written as the product of two numbers: a coefficient and 10 raised to a power. For example, the number 602,000,000,000,000,000,000,000 written in scientific notation is 6.02×10^{23}. The coefficient in this number is 6.02. In scientific notation, the coefficient is always a number equal to or greater than one and less than ten. The power of 10, or exponent, in this example is 23. Figure 3.1 illustrates how to express the number of stars in a galaxy by using scientific notation. For more practice on writing numbers in scientific notation, refer to page R56 of Appendix C.

Guide for Reading

Key Concepts

- How do measurements relate to science?
- How do you evaluate accuracy and precision?
- Why must measurements be reported to the correct number of significant figures?
- How does the precision of a calculated answer compare to the precision of the measurements used to obtain it?

Vocabulary

measurement
scientific notation
accuracy
precision
accepted value
experimental value
error
percent error
significant figures

Reading Strategy

Building Vocabulary As you read, write a definition of each vocabulary term in your own words.

$200{,}000{,}000{,}000. = 2 \times 10^{11}$

Decimal moves 11 places to the left. Exponent is 11

Figure 3.1 Expressing very large numbers, such as the estimated number of stars in a galaxy, is easier if scientific notation is used.

Section Resources

Print

- **Guided Reading and Study Workbook,** Section 3.1
- **Core Teaching Resources,** Section 3.1 Review
- **Transparencies,** T20–T26

Technology

- **Interactive Textbook with ChemASAP,** Animation 2, Problem-Solving 3.2, 3.3, 3.6, 3.8, Assessment 3.1

3.1

1 FOCUS

Objectives

3.1.1 Convert measurements to scientific notation.

3.1.2 Distinguish among accuracy, precision, and error of a measurement.

3.1.3 Determine the number of significant figures in a measurement and in a calculated answer.

Guide for Reading

Build Vocabulary L2

Paraphrase Have students write definitions of the words *accurate* and *precise* in their own words. As they read the text, have students compare the definitions with those of *accuracy* and *precision* given in the text.

Reading Strategy L2

Use Prior Knowledge Ask, **What everyday activities involve measuring?** *(Examples include buying consumer products, doing sports activities, and cooking.)* Ask students to recall which units of measure are related to each of the examples they give. Have some students estimate their height in inches. Have other students measure the same students with a yardstick or tape measure. Compare the estimates with measured values. Ask, **What do you think your height is in centimeters?** *(Acceptable answers range from 130 to 200 centimeters.)*

2 INSTRUCT

Connecting to Your World

Have students study the photograph and read the text that opens the section. Ask, **How do you think scientists ensure measurements are accurate and precise?** *(Acceptable answers include that scientists make multiple measurements by using the most precise equipment available. They use samples with known values to check the reliability of the equipment.)*

Using and Expressing Measurements

Use Visuals L1

Figure 3.1 Have students study the photograph and read the text that opens the section. Ask, **How is the exponent of a number expressed in scientific notation related to the number of places the decimal point is moved to the left in a number larger than 1?** *(They are equal.)*

Accuracy, Precision, and Error

Precision and Accuracy L2

Purpose To illustrate the concepts of precision and accuracy

Materials a small object (such as a lead fishing weight), triple-beam balance

Procedure Place the object and a triple-beam balance in a designated area. Set a deadline by which each student will have measured the mass of the object. After everyone has had an opportunity, have students compile a summary of all the measurements. Illustrate precision by having the students find the average and compare their measurement to it.

Expected Outcome Measured values should be similar, but not necessarily identical for all students.

FYI

Other analogies that may be useful in explaining precision vs. accuracy:

- casting a fishing line
- pitching horseshoes
- a precision marching band

Figure 3.2 The distribution of darts illustrates the difference between accuracy and precision. **a** Good accuracy and good precision: The darts are close to the bull's-eye and to one another. **b** Poor accuracy and good precision: The darts are far from the bull's-eye but close to one another. **c** Poor accuracy and poor precision: The darts are far from the bull's-eye and from one another.

Accuracy, Precision, and Error

Your success in the chemistry lab and in many of your daily activities depends on your ability to make reliable measurements. Ideally, measurements should be both correct and reproducible.

Accuracy and Precision Correctness and reproducibility relate to the concepts of accuracy and precision, two words that mean the same thing to many people. In chemistry, however, their meanings are quite different. **Accuracy** is a measure of how close a measurement comes to the actual or true value of whatever is measured. **Precision** is a measure of how close a series of measurements are to one another. **To evaluate the accuracy of a measurement, the measured value must be compared to the correct value. To evaluate the precision of a measurement, you must compare the values of two or more repeated measurements.**

Darts on a dartboard illustrate accuracy and precision in measurement. Let the bull's-eye of the dartboard represent the true, or correct, value of what you are measuring. The closeness of a dart to the bull's-eye corresponds to the degree of accuracy. The closer it comes to the bull's-eye, the more accurately the dart was thrown. The closeness of several darts to one another corresponds to the degree of precision. The closer together the darts are, the greater the precision and the reproducibility.

Look at Figure 3.2 as you consider the following outcomes.

a. All of the darts land close to the bull's-eye and to one another. Closeness to the bull's-eye means that the degree of accuracy is great. Each dart in the bull's-eye corresponds to an accurate measurement of a value. Closeness of the darts to one another indicates high precision.

b. All of the darts land close to one another but far from the bull's-eye. The precision is high because of the closeness of grouping and thus the high level of reproducibility. The results are inaccurate, however, because of the distance of the darts from the bull's-eye.

c. The darts land far from one another and from the bull's-eye. The results are both inaccurate and imprecise.

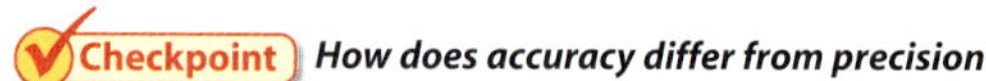
Checkpoint *How does accuracy differ from precision?*

Facts and Figures

Striving for Scientific Accuracy

The French chemist, Antoine Lavoisier, worked hard to establish the importance of accurate measurement in scientific inquiry. Lavoisier devised an experiment to test the Greek scientists' idea that when water was heated, it could turn into earth. For 100 days, Lavoisier boiled water in a glass flask constructed to allow steam to condense without escaping. He weighed the water and the flask separately before and after boiling. He found that the mass of the water had not changed. The flask, however, lost a small mass equal to the sediment he found in the bottom of it. Lavoisier proved that the sediment was not earth, but part of the flask etched away by the boiling water.

Determining Error Note that an individual measurement may be accurate or inaccurate. Suppose you use a thermometer to measure the boiling point of pure water at standard pressure. The thermometer reads 99.1°C. You probably know that the true or accepted value of the boiling point of pure water under these conditions is actually 100.0°C. There is a difference between the **accepted value,** which is the correct value based on reliable references, and the **experimental value,** the value measured in the lab. The difference between the experimental value and the accepted value is called the **error.**

$$\text{Error} = \text{experimental value} - \text{accepted value}$$

Error can be positive or negative depending on whether the experimental value is greater than or less than the accepted value.

For the boiling-point measurement, the error is 99.1°C − 100.0°C, or −0.9°C. The magnitude of the error shows the amount by which the experimental value differs from the accepted value. Often, it is useful to calculate the relative error, or percent error. The **percent error** is the absolute value of the error divided by the accepted value, multiplied by 100%.

$$\text{Percent error} = \frac{|\text{error}|}{\text{accepted value}} \times 100\%$$

Using the absolute value of the error means that the percent error will always be a positive value. For the boiling-point measurement, the percent error is calculated as follows.

$$\begin{aligned}\text{Percent error} &= \frac{|99.1°\text{C} - 100.0°\text{C}|}{100.0°\text{C}} \times 100\% \\ &= \frac{0.9\cancel{°\text{C}}}{100.0\cancel{°\text{C}}} \times 100\% \\ &= 0.009 \times 100\% \\ &= 0.9\%\end{aligned}$$

Just because a measuring device works doesn't necessarily mean that it is accurate. As Figure 3.3 shows, a weighing scale that does not read zero when nothing is on it is bound to yield error. In order to weigh yourself accurately, you must first make sure that the scale is zeroed.

Figure 3.3 The scale below has not been properly zeroed, so the reading obtained for the person's weight is inaccurate. There is a difference between the person's correct weight and the measured value. **Calculating** ***What is the percent error of a measured value of 114 lb if the person's actual weight is 107 lb?***

Word *Origins*

Percent comes from the Latin words *per,* meaning "by" or "through," and *centum,* meaning "100." **What do you think the phrase *per annum* means?**

Word *Origins* 

The phrase *per annum* means "by the year."

Use Visuals L1

Figure 3.2 Have students inspect Figure 3.2. Ask, **If one dart in Figure 3.2c were closer to the bull's-eye, what would happen to the accuracy?** *(The accuracy would increase.)* **What would happen to the precision?** *(The precision would increase.)* **What is the operational definition of error implied by this figure?** *(The error is the distance between the dart and the bull's-eye.)*

Discuss 

Review the concept of absolute value. Ask, **What is the meaning of a *positive* error?** *(Measured value is greater than accepted value.)* **What is the meaning of a *negative* error?** *(Measured value is less than accepted value.)* Explain that the absolute value of the error is a positive value that describes the difference between the measured value and the accepted value, but not which is greater.

Answers to...

Figure 3.3 7%

 Accuracy is determined by comparing a measured value to the correct value. Precision is determined by comparing the values of two or more repeated measurements.

Significant Figures in Measurements

Discuss

L2

Point out that the concept of significant figures applies only to measured quantities. If students ask why an estimated digit is considered significant, tell them a significant figure is one that is known to be reasonably reliable. A careful estimate fits this definition.

FYI

When calibration marks on an instrument are spaced very close together (e.g., on certain thermometers and graduated cylinders), it is sometimes more practical to estimate a measurement to the nearest half of the smallest calibrated increment, rather than to the nearest tenth.

CLASS **Activity**

Olympic Times

L2

Purpose To illustrate how similar measurements fom different eras may vary in precision

Materials Almanacs or Internet access

Procedure Have students look up the winning times for the men's and women's 100-meter dashes at the 1948 and 2000 Olympic Games. Then have them answer the following question. **Why do the more recently recorded race times contain more digits to the right of the decimal?** *(Because the technology used for timekeeping improved to allow for more precise measurements.)*

Expected Outcome Students should find that the race times from 1948 were recorded to the nearest tenth of a second. The race times from 2000 were recorded to the nearest hundredth of a second.

Figure 3.4 The precision of a weighing scale depends on how finely it is calibrated.

Significant Figures in Measurements

Supermarkets often provide scales like the one in Figure 3.4. Customers use these scales to measure the weight of produce that is priced per pound. If you use a scale that is calibrated in 0.1-lb intervals, you can easily read the scale to the nearest tenth of a pound. With such a scale, however, you can also estimate the weight to the nearest hundredth of a pound by noting the position of the pointer between calibration marks.

Suppose you estimate a weight that lies between 2.4 lb and 2.5 lb to be 2.46 lb. The number in this estimated measurement has three digits. The first two digits in the measurement (2 and 4) are known with certainty. But the rightmost digit (6) has been estimated and involves some uncertainty. These three reported digits all convey useful information, however, and are called significant figures. The **significant figures** in a measurement include all of the digits that are known, plus a last digit that is estimated.  **Measurements must always be reported to the correct number of significant figures because calculated answers often depend on the number of significant figures in the values used in the calculation.**

Instruments differ in the number of significant figures that can be obtained from their use and thus in the precision of measurements. The three meter sticks in Figure 3.5 can be used to make successively more precise measurements of the board.

Interactive Textbook

Animation 2 See how the precision of a calculated result depends on the sensitivity of the measuring instruments.

with ChemASAP

Rules for determining whether a digit in a measured value is significant:

1 Every nonzero digit in a reported measurement is assumed to be significant. The measurements 24.7 meters, 0.743 meter, and 714 meters each express a measure of length to three significant figures.

2 Zeros appearing between nonzero digits are significant. The measurements 7003 meters, 40.79 meters, and 1.503 meters each have four significant figures.

3 Leftmost zeros appearing in front of nonzero digits are not significant. They act as placeholders. The measurements 0.0071 meter, 0.42 meter, and 0.000 099 meter each have only two significant figures. The zeros to the left are not significant. By writing the measurements in scientific notation, you can eliminate such placeholding zeros: in this case, 7.1×10^{-3} meter, 4.2×10^{-1} meter, and 9.9×10^{-5} meter.

4 Zeros at the end of a number and to the right of a decimal point are always significant. The measurements 43.00 meters, 1.010 meters, and 9.000 meters each have four significant figures.

a Measured length = 0.6 m

b Measured length = 0.61 m

c Measured length = 0.607 m

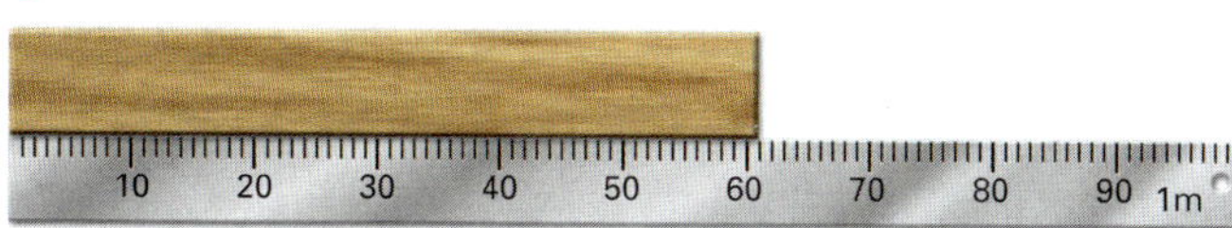

Figure 3.5 Three differently calibrated meter sticks are used to measure the length of a board. **a** A meter stick calibrated in a 1-m interval. **b** A meter stick calibrated in 0.1-m intervals. **c** A meter stick calibrated in 0.01-m intervals. **Measuring** ***How many significant figures are reported in each measurement?***

5 Zeros at the rightmost end of a measurement that lie to the left of an understood decimal point are not significant if they serve as placeholders to show the magnitude of the number. The zeros in the measurements 300 meters, 7000 meters, and 27,210 meters are not significant. The numbers of significant figures in these values are one, one, and four, respectively. If such zeros were known measured values, however, then they would be significant. For example, if all of the zeros in the measurement 300 meters were significant, writing the value in scientific notation as 3.00×10^2 meters makes it clear that these zeros are significant.

6 There are two situations in which numbers have an unlimited number of significant figures. The first involves counting. If you count 23 people in your classroom, then there are exactly 23 people, and this value has an unlimited number of significant figures. The second situation involves exactly defined quantities such as those found within a system of measurement. When, for example, you write 60 min = 1 hr, or 100 cm = 1 m, each of these numbers has an unlimited number of significant figures. As you shall soon see, exact quantities do not affect the process of rounding an answer to the correct number of significant figures.

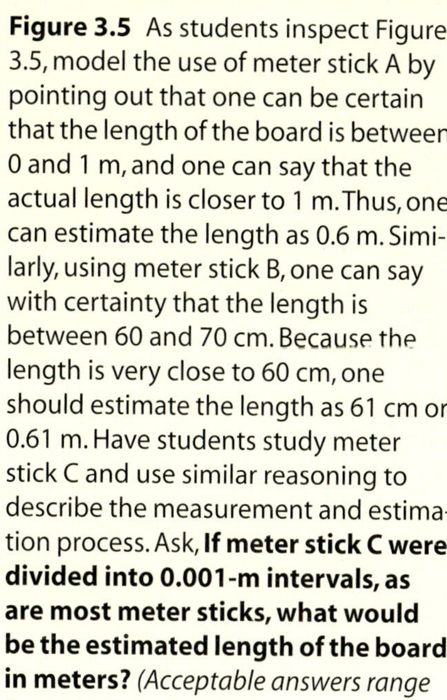

Use Visuals L1

Figure 3.5 As students inspect Figure 3.5, model the use of meter stick A by pointing out that one can be certain that the length of the board is between 0 and 1 m, and one can say that the actual length is closer to 1 m. Thus, one can estimate the length as 0.6 m. Similarly, using meter stick B, one can say with certainty that the length is between 60 and 70 cm. Because the length is very close to 60 cm, one should estimate the length as 61 cm or 0.61 m. Have students study meter stick C and use similar reasoning to describe the measurement and estimation process. Ask, **If meter stick C were divided into 0.001-m intervals, as are most meter sticks, what would be the estimated length of the board in meters?** *(Acceptable answers range from 0.6065 to 0.6074 m)* **In millimeters?** *(606.5 to 607.4 mm)*

Discuss L2

Be sure to review the role of zeros in determining the number of significant figures. When adding or subtracting numbers expressed in scientific notation, remind students that the numbers must all have the same exponent.

Differentiated Instruction

Less Proficient Readers L1

Have students write in their own words the rules for determining the number of significant digits. Help them if necessary. Have them measure several lengths, masses, and volumes; then have them use their rules to determine the correct number of significant figures for each measurement.

Answers to...

Figure 3.5 The measurements are: **a.** 0.6 m (1 significant figure), **b.** 0.61 m (2 significant figures), **c.** 0.607 m (3 significant figures)

CLASS Activity

Significant Zeros L2

Purpose To provide practice in applying the rules governing the significance of zeros in measurements

Materials textbook and scientific literature, index cards

Procedure Have students search their textbooks and other sources for length, mass, volume, or temperature measurements that contain zeros. Have them include some examples written in scientific notation. Ask them to write each measurement on the front of an index card; on the back of each card, have them write (1) all the rules governing the significance of zeros that apply to the measurement, and (2) the number of significant figures in the measurement. Have pairs of students exchange index cards and agree on the appropriateness of the rules and the answers.

Expected Outcome Students should be able to apply correctly rules 2–5 listed on pages 66 and 67.

CONCEPTUAL PROBLEM 3.1

Answers

1. a. 4 **b.** 4 **c.** 2 **d.** 5
2. a. 3 **b.** 2 **c.** 4 **d.** 4

Practice Problems Plus 

Chapter 3 Assessment problem 58 is related to Sample Problem 3.1.

FYI

Due to rounding, there will often be discrepancies between actual values and calculated values derived from measurements. Some students may conclude (correctly) that following the rules for significant figures introduces error in calculated values. Such rounding errors are generally small, but should nonetheless be acknowledged when performing calculations.

CONCEPTUAL PROBLEM 3.1

Counting Significant Figures in Measurements

How many significant figures are in each measurement?

a. 123 m **b.** 40,506 mm
c. 9.8000×10^4 m **d.** 22 meter sticks
e. 0.070 80 m **f.** 98,000 m

1 Analyze ***Identify the relevant concepts.***

The location of each zero in the measurement and the location of the decimal point determine which of the rules apply for determining significant figures.

 **Solve** ***Apply the concepts to this problem.***

All nonzero digits are significant (rule 1). Use rules 2 through 6 to determine if the zeros are significant.

a. three (rule 1) **b.** five (rule 2)
c. five (rule 4) **d.** unlimited (rule 6)
e. four (rules 2, 3, 4) **f.** two (rule 5)

Practice Problems

1. Count the significant figures in each length.
a. 0.057 30 meters **b.** 8765 meters
c. 0.000 73 meters **d.** 40.007 meters

2. How many significant figures are in each measurement?
a. 143 grams **b.** 0.074 meter
c. 8.750×10^{-2} gram **d.** 1.072 meter

Problem-Solving 3.2 Solve Problem 2 with the help of an interactive guided tutorial.
with ChemASAP

Significant Figures in Calculations

Suppose you use a calculator to find the area of a floor that measures 7.7 meters by 5.4 meters. The calculator would give an answer of 41.58 square meters. The calculated area is expressed to four significant figures. However, each of the measurements used in the calculation is expressed to only two significant figures. So the answer must also be reported to two significant figures (42 m^2). **In general, a calculated answer cannot be more precise than the least precise measurement from which it was calculated.** The calculated value must be rounded to make it consistent with the measurements from which it was calculated.

Rounding To round a number, you must first decide how many significant figures the answer should have. This decision depends on the given measurements and on the mathematical process used to arrive at the answer. Once you know the number of significant figures your answer should have, round to that many digits, counting from the left. If the digit immediately to the right of the last significant digit is less than 5, it is simply dropped and the value of the last significant digit stays the same. If the digit in question is 5 or greater, the value of the digit in the last significant place is increased by 1.

 ***Why must a calculated answer generally be rounded?***

SAMPLE PROBLEM 3.1

Rounding Measurements

Round off each measurement to the number of significant figures shown in parentheses. Write the answers in scientific notation.

a. 314.721 meters (four)
b. 0.001 775 meter (two)
c. 8792 meters (two)

1 Analyze ***Identify the relevant concepts.***

Round off each measurement to the number of significant figures indicated. Then apply the rules for expressing numbers in scientific notation.

2 Solve ***Apply the concepts to this problem.***

Count from the left and apply the rule to the digit immediately to the right of the digit to which you are rounding. The arrow points to the digit immediately following the last significant digit.

a. 314.721 meters
↑
2 is less than 5, so you do not round up.
314.7 meters = 3.147×10^2 meters

b. 0.001 775 meter
↑
7 is greater than 5, so round up.
0.0018 meter = 1.8×10^{-3} meter

c. 8792 meters
↑
9 is greater than 5, so round up
8800 meters = 8.8×10^3 meters

3 Evaluate ***Do the results make sense?***

The rules for rounding and for writing numbers in scientific notation have been correctly applied.

Practice Problems

3. Round each measurement to three significant figures. Write your answers in scientific notation.

a. 87.073 meters
b. 4.3621×10^8 meters
c. 0.01552 meter
d. 9009 meters
e. 1.7777×10^{-3} meter
f. 629.55 meters

4. Round each measurement in Practice Problem 3 to one significant figure. Write each of your answers in scientific notation.

Math Handbook

For help with scientific notation, go to page R56.

Problem-Solving 3.3 Solve Problem 3 with the help of an interactive guided tutorial.

with ChemASAP

Significant Figures in Calculations

Sample Problem 3.1

Answers

3. **a.** 8.71×10^1 m
b. 4.36×10^8 m
c. 1.55×10^{-2} m
d. 9.01×10^3 m
e. 1.78×10^{-3} m
f. 6.30×10^2 m

4. **a.** 9×10^1 m
b. 4×10^8 m
c. 2×10^{-2} m
d. 9×10^3 m
e. 2×10^{-3} m
f. 6×10^2 m

Practice Problems Plus 

Round each measurement to two significant figures. Write your answers in scientific notation.

a. 94.592 grams *(9.5×10^1 g)*
b. 2.4232×10^3 grams *(2.4×10^3 g)*
c. 0.007 438 grams *(7.4×10^{-3} g)*
d. 54,752 grams *(5.5×10^4 g)*
e. 6.0289×10^{-3} grams *(6.0×10^{-3} g)*
f. 405.11 grams *(4.1×10^2 g)*

Math Handbook

For a math refresher and practice, direct students to scientific notation, page R56.

Answers to...

 **Checkpoint** A calculated answer must be rounded in order to make it consistent with the measurements from which it was calculated. The calculated answer cannot be more precise than the least precise measurement used in the calculation.

Section 3.1 (continued)

Discuss L2

The rules for rounding calculated numbers can be compared with the old adage, "A chain is only as strong as its weakest link." Explain that an answer cannot be more precise than the least precise value used to calculate the answer. Ask, **In addition and subtraction, what is the least precise value?** (The measurement with the fewest digits to the right of the decimal point.) **In multiplication and division, what is the least precise value?** (The measurement with the fewest significant figures.) If students wonder why addition and subtraction rules differ from multiplication and division rules, point out that in addition and subtraction of measurements, the measurements are of the same property, such as length or volume. However, in the multiplication and division of measurements, new quantities or properties are being described, such as speed (length ÷ time), area (length × length), and density (mass ÷ volume).

Sample Problem 3.2

Answers

5. a. 79.2 m **b.** 7.33 m **c.** 11.53 m **d.** 17.3 m

6. 23.8 g

Practice Problems Plus 

Find the total mass of four stones with the following masses: 10.32 grams, 11.81 grams, 124.678 grams, and 0.9129 gram. *(147.72 g)*

Math Handbook

For a math refresher and review, direct students to scientific notation, page R56.

Addition and Subtraction The answer to an addition or subtraction calculation should be rounded to the same number of decimal places (not digits) as the measurement with the least number of decimal places. Work through Sample Problem 3.2 below which provides an example of rounding in an addition calculation.

Math Handbook

For help with significant figures, go to page R59.

SAMPLE PROBLEM 3.2

Significant Figures in Addition

Calculate the sum of the three measurements. Give the answer to the correct number of significant figures.

12.52 meters + 349.0 meters + 8.24 meters

1 Analyze ***Identify the relevant concepts.***

Calculate the sum and then analyze each measurement to determine the number of decimal places required in the answer.

2 Solve ***Apply the concepts to this problem.***

Align the decimal points and add the numbers. Round the answer to match the measurement with the least number of decimal places.

	12.52	meters
	349.0	meters
+	8.24	meters
	369.76	meters

The second measurement (349.0 meters) has the least number of digits (one) to the right of the decimal point. Thus the answer must be rounded to one digit after the decimal point. The answer is rounded to 369.8 meters, or 3.698×10^2 meters.

3 Evaluate ***Does the result make sense?***

The mathematical operation has been correctly carried out and the resulting answer is reported to the correct number of decimal places.

Practice Problems

5. Perform each operation. Express your answers to the correct number of significant figures.

a. 61.2 meters + 9.35 meters + 8.6 meters

b. 9.44 meters − 2.11 meters

c. 1.36 meters + 10.17 meters

d. 34.61 meters − 17.3 meters

6. Find the total mass of three diamonds that have masses of 14.2 grams, 8.73 grams, and 0.912 gram.

Problem-Solving 3.6 Solve Problem 6 with the help of an interactive guided tutorial.

with ChemASAP

Sample Problem 3.3

Answers

7. a. 1.8×10^1 m^2 **b.** 6.75×10^2 m **c.** 5.87×10^{-1} min

8. 1.3×10^3 m^3

Practice Problems Plus 

Calculate the volume of a house that has dimensions of 12.52 meters by 36.86 meters by 2.46 meters. *(1.14×10^3 m^3)*

Math Handbook

For a math refresher and practice, direct students to using a calculator, page R62.

Multiplication and Division In calculations involving multiplication and division, you need to round the answer to the same number of significant figures as the measurement with the least number of significant figures. The position of the decimal point has nothing to do with the rounding process when multiplying and dividing measurements. The position of the decimal point is important only in rounding the answers of addition or subtraction problems.

 ***How many significant figures must you round an answer to when performing multiplication or division?***

SAMPLE PROBLEM 3.3

Significant Figures in Multiplication and Division

Perform the following operations. Give the answers to the correct number of significant figures.

a. 7.55 meters × 0.34 meter
b. 2.10 meters × 0.70 meter
c. 2.4526 meters ÷ 8.4

1 Analyze ***Identify the relevant concepts.***
Perform the required math operation and then analyze each of the original numbers to determine the correct number of significant figures required in the answer.

2 Solve ***Apply the concepts to this problem.***
Round the answers to match the measurement with the least number of significant figures.

a. 7.55 meters × 0.34 meter = 2.567 $(\text{meter})^2$ = 2.6 meters^2
(0.34 meter has two significant figures)
b. 2.10 meters × 0.70 meter = 1.47 $(\text{meter})^2$ = 1.5 meters^2
(0.70 meter has two significant figures)
c. 2.4526 meters ÷ 8.4 = 0.291 976 meter = 0.29 meter
(8.4 has two significant figures)

3 Evaluate ***Do the results make sense?***
The mathematical operations have been performed correctly, and the resulting answers are reported to the correct number of places.

Practice Problems

7. Solve each problem. Give your answers to the correct number of significant figures and in scientific notation.
a. 8.3 meters × 2.22 meters
b. 8432 meters ÷ 12.5
c. 35.2 seconds × $\frac{1 \text{ minute}}{60 \text{ seconds}}$

8. Calculate the volume of a warehouse that has inside dimensions of 22.4 meters by 11.3 meters by 5.2 meters. (Volume = $l \times w \times h$)

Math Handbook
For help with using a calculator, go to page R62.

Interactive Textbook
Problem-Solving 3.8 Solve Problem 8 with the help of an interactive guided tutorial.
with ChemASAP

Quick LAB

Accuracy and Precision 

Objectives After completing this activity, students will be able to
- measure length with accuracy and precision.
- apply rules for rounding answers calculated from measurements.
- determine experimental error and express it as percent error.

Address Misconceptions

Students may contend that making one measurement of some property, such as length, is satisfactory. Ask, **What possible errors may occur when making only one length measurement?** *(Acceptable answers include misreading the ruler or not holding the ruler parallel to the length of the object.)*

Skills Focus Measuring, calculating

Prep Time 5 minutes
Materials 3 inch × 5 inch index cards, metric rulers
Class Time 15 minutes

Teaching Tips

Emphasize that students should use an interior, marked line, such as 10.0 cm, as the initial point, instead of the end of the ruler, which may be damaged.

Expected Outcome Measured values should be similar, but not necessarily identical for all students.

Analyze and Conclude

1. Four for length; three for width
2. See Expected Outcome.
3. Significant digits for rounded-off answers are area, 3, and perimeter, 4. Some students may not round to the proper number of digits.
4. Errors of ±0.03 cm are acceptable. Such errors yield percent errors of 0.2% for length and 0.4% for width.

For Enrichment L3

Have students devise methods of calculating the volume of one card. Point out that measuring the thickness of one card with a ruler would be very inaccurate. Ask, **How might the measurement of the thickness of the card be improved?** *(Use a more precise instrument, such as a micrometer, or measure the thickness of a number of cards and divide by the number of cards.)* Have students determine the thickness of one card and calculate its volume. Using the class average of the calculated volumes, have each student determine the percent error using the average as the accepted value.

Answers to...

 The same number of significant figures as the measurement with the least number of significant figures.

Section 3.1 (continued)

3 ASSESS

Evaluate Understanding L2

Write the following sets of measurements on the board.

(1) 78°C, 76°C, 75°C
(2) 77°C, 78°C, 78°C
(3) 80°C, 81°C, 82°C

Ask, **If these sets of measurements were made of the boiling point of a liquid under similar conditions, explain which set is the most precise?** *(Set 2 is the most precise because the three measurements are closest together.)* **What would have to be known to determine which set is the most accurate?** *(the accepted value of the liquid's boiling point)*

Reteach L1

Use Figure 3.5 to reteach the method of correctly recording the number of significant figures in a measurement. Then have students convert each measurement into scientific notation. *(6×10^{-1} m, 6.1×10^{-1} m, 6.07×10^{-1} m)*

Writing Activity

Acceptable answers will include the following information: Accuracy compares a measured value to an accepted value of the measurement, precision compares a measured value to a set of measurements made under similar conditions, and error is the difference between the measured and accepzzted values.

If your class subscribes to the Interactive Textbook, use it to review key concepts in Section 3.1.

with ChemASAP

Quick LAB

Accuracy and Precision

Purpose

To measure the dimensions of an object as accurately and precisely as possible and to apply rules for rounding answers calculated from the measurements.

Materials

- 3 inch × 5 inch index card
- metric ruler

Procedure

1. Use a metric ruler to measure in centimeters the length and width of an index card as accurately and precisely as you can. The hundredths place in your measurement should be estimated.
2. Calculate the perimeter [2 × (length + width)] and the area (length × width) of the index card. Write both your unrounded answers and your correctly rounded answers on the chalkboard.

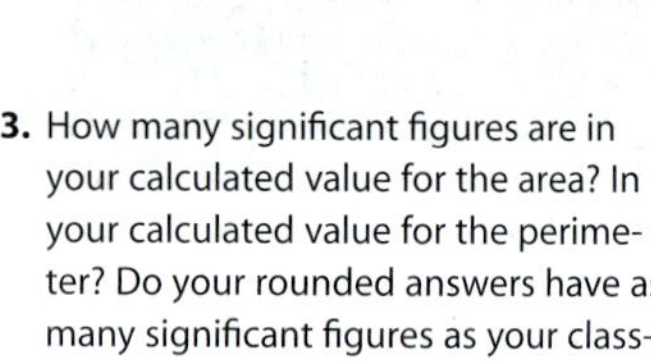

Analyze and Conclude

1. How many significant figures are in your measurements of length and of width?
2. How do your measurements compare with those of your classmates?
3. How many significant figures are in your calculated value for the area? In your calculated value for the perimeter? Do your rounded answers have as many significant figures as your classmates' measurements?
4. Assume that the correct (accurate) length and width of the card are 12.70 cm and 7.62 cm, respectively. Calculate the percent error for each of your two measurements.

3.1 Section Assessment

9. **Key Concept** How do measurements relate to experimental science?
10. **Key Concept** How are accuracy and precision evaluated?
11. **Key Concept** Why must a given measurement always be reported to the correct number of significant figures?
12. **Key Concept** How does the precision of a calculated answer compare to the precision of the measurements used to obtain it?
13. A technician experimentally determined the boiling point of octane to be 124.1°C. The actual boiling point of octane is 125.7°C. Calculate the error and the percent error.
14. Determine the number of significant figures in each of the following.
    - **a.** 11 soccer players
    - **b.** 0.070 020 meter
    - **c.** 10,800 meters
    - **d.** 5.00 cubic meters
15. Solve the following and express each answer in scientific notation and to the correct number of significant figures.
    - **a.** $(5.3 \times 10^4) + (1.3 \times 10^4)$
    - **b.** $(7.2 \times 10^{-4}) \div (1.8 \times 10^3)$
    - **c.** $10^4 \times 10^{-3} \times 10^6$
    - **d.** $(9.12 \times 10^{-1}) - (4.7 \times 10^{-2})$
    - **e.** $(5.4 \times 10^4) \times (3.5 \times 10^9)$

Writing Activity

Explanatory Paragraph Explain the differences between the accuracy, precision, and error of a measurement.

Assessment 3.1 Test yourself on the concepts in Section 3.1.

with ChemASAP

Section 3.1 Assessment

9. Making correct measurements is fundamental to the experimental sciences.
10. Accuracy is the measured value compared to the correct values. Precision is comparing more than one measurement.
11. The significant figures in a calculated answer often depend on the number of significant figures of the measurements used in the calculation.
12. A calculated answer cannot be more precise than the least precise measurement used in the calculation.
13. error = −1.6°C; percent error = 1.3%
14. **a.** unlimited **b.** 5 **c.** 3 **d.** 3
15. **a.** 6.6×10^4 **b.** 4.0×10^{-7} **c.** 10^7 **d.** 8.65×10^{-1} **e.** 1.9×10^{14}

3.2 The International System of Units

Connecting to Your World

"Are we there yet?" You may have asked this question during a long road trip with family or friends. To find out how much farther you have to go, you can read the roadside signs that list destinations and their distances. In the signs shown here, however, the distances are listed as numbers with no units attached. Is Carrieton 44 kilometers or 44 miles away? Without the units, you can't be sure. When you make a measurement, you must assign the correct units to the numerical value. Without the units, it is impossible to communicate the measurement clearly to others.

Guide for Reading

Key Concepts

- Which five SI base units do chemists commonly use?
- What metric units are commonly used to measure length, volume, mass, temperature, and energy?

Vocabulary

International System of Units (SI)
meter (m)
liter (L)
kilogram (kg)
gram (g)
weight
temperature
Celsius scale
Kelvin scale
absolute zero
energy
joule (J)
calorie (cal)

Reading Strategy

Summarizing As you read about SI units, summarize the main ideas in the text that follows the red and blue headings.

Measuring with SI Units

All measurements depend on units that serve as reference standards. The standards of measurement used in science are those of the metric system. The metric system is important because of its simplicity and ease of use. All metric units are based on multiples of 10. As a result, you can convert between units easily. The metric system was originally established in France in 1795. The **International System of Units** (abbreviated **SI,** after the French name, *Le Système International d'Unités*) is a revised version of the metric system. The SI was adopted by international agreement in 1960. There are seven SI base units, which are listed in Table 3.1. From these base units, all other SI units of measurement can be derived. **The five SI base units commonly used by chemists are the meter, the kilogram, the kelvin, the second, and the mole.**

All measured quantities can be reported in SI units. Sometimes, however, non-SI units are preferred for convenience or for practical reasons. In this textbook you will learn about both SI and non-SI units.

Table 3.1

SI Base Units

Quantity	SI base unit	Symbol
Length	meter	m
Mass	kilogram	kg
Temperature	kelvin	K
Time	second	s
Amount of substance	mole	mol
Luminous intensity	candela	cd
Electric current	ampere	A

3.2

1 FOCUS

Objectives

3.2.1 **List** SI units of measurement and common SI prefixes.

3.2.2 **Distinguish** between the mass and weight of an object.

3.2.3 **Convert** between the Celsius and Kelvin temperature scales.

Guide for Reading

Build Vocabulary 

Word Parts Have students predict the meanings of the SI prefixes *deci, centi,* and *milli* from their knowledge of the words *decimal, century,* and *millennium.*

Reading Strategy 

Preview Have students look at Tables 3.1, 3.3., 3.4, and 3.5. Ask students to list any metric units with which they are already familiar.

2 INSTRUCT

Connecting to Your World

Have students study the photograph and read the text that opens the section. Ask, **Do any of the signs list a measurement?** *(No. The signs show only numbers. A measurement contains both a number and a unit.)* **How can you determine what units are implied by the signs?** *(A library or Internet search will show that the names on the signs are towns in Australia, which adopted SI in 1970. The numbers on the signs represent distances in kilometers.)*

Section Resources

Print

- **Guided Reading and Study Workbook,** Section 3.2
- **Core Teaching Resources,** Section 3.2 Review, Interpreting Graphics
- **Transparencies,** T27–T30
- **Small-Scale Chemistry Lab Manual,** Labs 3, 4, 5

Technology

- **Interactive Textbook with ChemASAP,** Problem-Solving 3.17, Assessment 3.2
- **Go Online,** Section 3.2

Measuring with SI Units

Discuss L2

Explain that with the exception of those working in the sciences, the United States is the only country not using the metric system as a routine method of measurement. Ask, **What are some examples of applications in which the metric system is used in the United States.** *(Acceptable answers include 35-mm film; meter distances in track and field; the dimensions of tools (wrenches) and hardware (bolts); power rating in watts of appliances, and road signs.)*

Units and Quantities

Use Visuals L1

Table 3.2 Discuss with students how SI prefixes are always in increments of ten and can be expressed using scientific notation. Ask, **What does the prefix *kilo* mean?** *(The prefix* kilo *means 1000 times larger than the unit it precedes.)* **How is this prefix represented in scientific notation?** *(10^3)* **How much larger is the prefix *centi* compared to the prefix *micro?*** Show students how to make this comparison by writing the following on the board:

$$\text{centi/micro} = 10^{-2}/10^{-6} = 10^{-2-(-6)} = 10^{(-2+6)} = 10^4$$

Remind students that in dividing numbers in scientific notation, the exponent of the denominator is subtracted from the exponent of the numerator. Explain that the prefix *centi* is 10^4, or 10,000, times larger than the prefix *micro*.

Table 3.2

Commonly Used Metric Prefixes

Prefix	Meaning	Factor
mega (M)	1 million times larger than the unit it precedes	10^6
kilo (k)	1000 times larger than the unit it precedes	10^3
deci (d)	10 times smaller than the unit it precedes	10^{-1}
centi (c)	100 times smaller than the unit it precedes	10^{-2}
milli (m)	1000 times smaller than the unit it precedes	10^{-3}
micro (μ)	1 million times smaller than the unit it precedes	10^{-6}
nano (n)	1000 million times smaller than the unit it precedes	10^{-9}
pico (p)	1 trillion times smaller than the unit it precedes	10^{-12}

Units and Quantities

As you already know, you don't measure length in kilograms or mass in centimeters. Different quantities require different units. Before you make a measurement, you must be familiar with the units corresponding to the quantity that you are trying to measure.

Units of Length Size is an important property of matter. In SI, the basic unit of length, or linear measure, is the **meter (m)**. All measurements of length can be expressed in meters. (The length of a page in this book is about one-fourth of a meter.) For very large and very small lengths, however, it may be more convenient to use a unit of length that has a prefix. Table 3.2 lists the prefixes in common use. For example, the prefix *milli-* means 1/1000 (one-thousandth), so a millimeter (mm) is 1/1000 of a meter, or 0.001 m. A hyphen (-) measures about 1 mm.

For large distances, it is usually most appropriate to express measurements in kilometers (km). The prefix *kilo-* means 1000, so 1 km equals 1000 m. A standard marathon distance race of about 42,000 m is more conveniently expressed as 42 km (42 × 1000 m). **Common metric units of length include the centimeter, meter, and kilometer.** Table 3.3 summarizes the relationships among metric units of length.

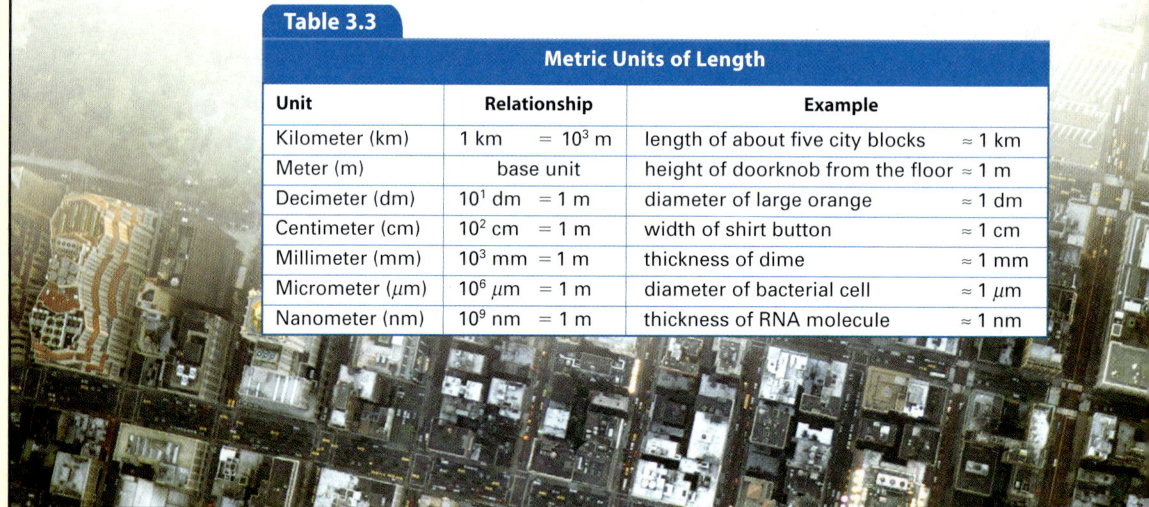

Length of 5 city blocks ≈ 1 km

Table 3.3

Metric Units of Length

Unit	Relationship	Example
Kilometer (km)	1 km = 10^3 m	length of about five city blocks ≈ 1 km
Meter (m)	base unit	height of doorknob from the floor ≈ 1 m
Decimeter (dm)	10^1 dm = 1 m	diameter of large orange ≈ 1 dm
Centimeter (cm)	10^2 cm = 1 m	width of shirt button ≈ 1 cm
Millimeter (mm)	10^3 mm = 1 m	thickness of dime ≈ 1 mm
Micrometer (μm)	10^6 μm = 1 m	diameter of bacterial cell ≈ 1 μm
Nanometer (nm)	10^9 nm = 1 m	thickness of RNA molecule ≈ 1 nm

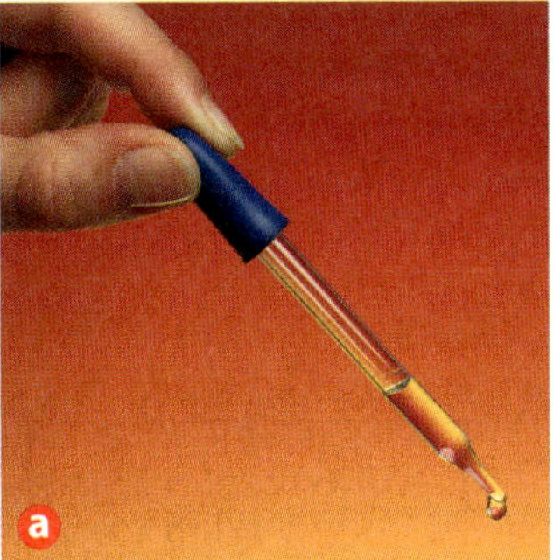

Figure 3.6 These photographs above give you some idea of the relative sizes of some different units of volume. **a** The volume of 20 drops of liquid from a medicine dropper is approximately 1 mL. **b** A sugar cube is 1 cm on each edge and has a volume of 1 cm^3. Note that 1 mL is the same as 1 cm^3. **c** A gallon of milk has about twice the volume of a 2-L bottle of soda.
Calculating ***How many cubic centimeters are in 2 liters?***

Units of Volume The space occupied by any sample of matter is called its volume. You calculate the volume of any cubic or rectangular solid by multiplying its length by its width by its height. The unit for volume is thus derived from units of length. The SI unit of volume is the amount of space occupied by a cube that is 1 m along each edge. This volume is a cubic meter (m^3). An automatic dishwasher has a volume of about 1 m^3.

A more convenient unit of volume for everyday use is the liter, a non-SI unit. A **liter (L)** is the volume of a cube that is 10 centimeters (10 cm) along each edge ($10\ cm \times 10\ cm \times 10\ cm = 1000\ cm^3 = 1\ L$). A decimeter (dm) is equal to 10 cm, so 1 L is also equal to 1 cubic decimeter (dm^3). A smaller non-SI unit of volume is the milliliter (mL); 1 mL is 1/1000 of a liter. Thus there are 1000 mL in 1 L. Because 1 L is defined as 1000 cm^3, 1 mL and 1 cm^3 are the same volume. The units milliliter and cubic centimeter are thus used interchangeably. **Common metric units of volume include the liter, milliliter, cubic centimeter, and microliter.** Table 3.4 summarizes the relationships among these units of volume.

There are many devices for measuring liquid volumes, including graduated cylinders, pipets, burets, volumetric flasks, and syringes. Note that the volume of any solid, liquid, or gas will change with temperature (although the change is much more dramatic for gases). Consequently, accurate volume-measuring devices are calibrated at a given temperature—usually 20 degrees Celsius (20°C), which is about normal room temperature.

What is the SI unit of volume?

Table 3.4

Metric Units of Volume

Unit	Relationship	Example
Liter (L)	base unit	quart of milk ≈ 1 L
Milliliter (mL)	10^3 mL = 1 L	20 drops of water ≈ 1 mL
Cubic centimeter (cm^3)	1 cm^3 = 1 mL	cube of sugar ≈ 1 cm^3
Microliter (μL)	10^6 μL = 1 L	crystal of table salt ≈ 1 μL

Facts and Figures

SI Units of Distance and Time

In 1790, with the establishment of the metric system, the French became the first to adopt measurement standards that were close to being precise. The meter was defined as one ten-millionth of the distance from the equator to the North Pole along the meridian that passes through Paris. The second was defined as 1/86,400 of the average day.

The definitions of these and other base units have since become more precise. The meter is now defined as thedistance traveled by light in a vacuum in1/299,792,458 of a second. The second is now defined in terms of the number of cycles of radiation given off by a specific isotope of the element cesium.

Discuss L2

Point out that data can be reported accurately in various related units (mm, cm, m), but it is best to choose the unit closest in scale to the object being measured.

- The unit of mass most often used by chemists is the gram—not the kilogram. Kilograms of chemicals are too large for laboratory experiments.
- The density of most gases is measured in grams per liter; the densities of most solids and liquids are measured in grams per cubic centimeter.
- The milliliter is the most common unit of volume used by chemists. The cubic meter is much too large a volume for most laboratory experiments.

TEACHER Demo

Volume Measurements L2

Purpose To introduce students to laboratory equipment used to measure volume

Materials sets of Erlenmeyer flasks, buret, graduated cylinder, beaker, and volumetric flask; water; food coloring

Safety Have students wear goggles to emphasize safety in the laboratory.

Procedure Before the demonstration, fill the glassware with water and add a few drops of food coloring. Place the sets at various stations around the classroom. Assign groups of students to each station. As students inspect the glassware, explain that beakers, Erlenmeyer flasks, and graduated cylinders are used to measure approximate volumes. Volumetric flasks and burets allow measurements that are more precise. Ask, **What is the appropriate instrument to use to measure a large volume, such as a 1 L, accurately?** *(Students should choose the volumetric flask.)*

Expected Outcome Students should be able to select the appropriate glassware when presented with the volumetric parameters for an experiment.

Answers to...

Figure 3.6 2000 cm^3

 cubic meter (m^3)

CLASS Activity

Mass of a Penny L2

Purpose To provide practice in weighing groups of objects

Materials balance with a precision of at least 0.01 g, sets of 10 pennies each separated according to minting dates between 1970 and the present. (Be sure that sets include pre- and post-1982 minting dates.)

Safety Have student wear goggles when weighing the pennies.

Procedure Have students weigh three five-penny samples from each set of pennies and record their data. Have students calculate the average mass of a penny and plot the average mass versus the year it was minted.

Expected Outcome Students should observe slight variations in mass from year to year due to chance error, and a dramatic decrease in mass between 1982 and 1983. (Pennies minted before 1982 are mostly copper. Pennies minted after 1982 are mostly zinc, which is less dense than copper.) Ask students to propose a hypothesis for their findings. *(Acceptable answers include a change in the composition of the penny.)* Have students maintain their graph, results, and hypothesis for future reference in Section 3.4.

Use Visuals L1

Figure 3.8 Have students inspect the three thermometers shown. Ask, **Which thermometer is the most precise?** *(liquid-in-glass thermometer)* **Which is the least precise?** *(bimetallic strip thermometer)* **Which probably measures the greatest range of temperatures?** *(bimetallic strip thermometer)*

Table 3.5

Metric Units of Mass

Unit	Relationship	Example
Kilogram (kg) (base unit)	1 kg = 10^3 g	small textbook ≈ 1 kg
Gram (g)	1 g = 10^{-3} kg	dollar bill ≈ 1 g
Milligram (mg)	10^3 mg = 1 g	ten grains of salt ≈ 1 mg
Microgram (μg)	10^6 μg = 1 g	particle of baking powder ≈ 1 μg

Units of Mass The mass of an object is measured in comparison to a standard mass of 1 **kilogram (kg),** which is the basic SI unit of mass. A kilogram was originally defined as the mass of 1 L of liquid water at 4°C. A cube of water at 4°C measuring 10 cm on each edge would have a volume of 1 L and a mass of 1000 grams (g), or 1 kg. A **gram (g)** is 1/1000 of a kilogram; the mass of 1 cm^3 of water at 4°C is 1 g. **Common metric units of mass include the kilogram, gram, milligram, and microgram.** The relationships among units of mass are shown in Table 3.5.

You can use a platform balance to measure the mass of an object. The object is placed on one side of the balance, and standard masses are added to the other side until the balance beam is level. The unknown mass is equal to the sum of the standard masses. Laboratory balances range from very sensitive instruments with a maximum capacity of only a few milligrams to devices for measuring quantities in kilograms. An analytical balance is used to measure objects of less than 100 g and can determine mass to the nearest 0.0001 g (0.1 mg).

The astronaut shown on the surface of the moon in Figure 3.7 weighs one sixth of what he weighs on Earth. The reason for this difference is that the force of gravity on Earth is about six times what it is on the moon. **Weight** is a force that measures the pull on a given mass by gravity. Weight, a measure of force, is different from mass, which is a measure of the quantity of matter. Although the weight of an object can change with its location, its mass remains constant regardless of its location. Objects can thus become weightless, but they can never become massless.

Checkpoint ***How does weight differ from mass?***

Figure 3.7 An astronaut's weight on the moon is one sixth as much as it is on Earth. Earth exerts six times the force of gravity as the moon. **Inferring** ***How does the astronaut's mass on the moon compare to his mass on Earth?***

Differentiated Instruction

Less Proficient Readers L1

Have students search magazines and newspapers for examples of SI units. Have them report seasonal day/night temperature ranges for their countries of origin.

Gifted and Talented L3

Have students research the efforts to make the metric system standard in the United States. Who has promoted the change? What progress has been made? What obstacles remain?

Figure 3.8 Thermometers are used to measure temperature. (a) A liquid-in-glass thermometer contains alcohol or mineral spirits. (b) A dial thermometer contains a coiled bimetallic strip. (c) A Galileo thermometer contains several glass bulbs that are calibrated to sink or float depending on the temperature. The Galileo thermometer shown uses the Fahrenheit scale, which sets the freezing point of water at 32°F and the boiling point of water at 212°F.

Units of Temperature When you hold a glass of hot water, the glass feels hot because heat transfers from the glass to your hand. When you hold an ice cube, it feels cold because heat transfers from your hand to the ice cube. **Temperature** is a measure of how hot or cold an object is. An object's temperature determines the direction of heat transfer. When two objects at different temperatures are in contact, heat moves from the object at the higher temperature to the object at the lower temperature.

Almost all substances expand with an increase in temperature and contract as the temperature decreases. (A very important exception is water.) These properties are the basis for the common liquid-in-glass thermometer. The liquid in the thermometer expands and contracts more than the volume of the glass, producing changes in the column height of liquid. Figure 3.8 shows a few different types of thermometers.

Several temperature scales with different units have been devised. **Scientists commonly use two equivalent units of temperature, the degree Celsius and the kelvin.** The Celsius scale of the metric system is named after the Swedish astronomer Anders Celsius (1701–1744). It uses two readily determined temperatures as reference temperature values: the freezing point and the boiling point of water. The **Celsius scale** sets the freezing point of water at 0°C and the boiling point of water at 100°C. The distance between these two fixed points is divided into 100 equal intervals, or degrees Celsius (°C).

Another temperature scale used in the physical sciences is the Kelvin, or absolute, scale. This scale is named for Lord Kelvin (1824–1907), a Scottish physicist and mathematician. On the **Kelvin scale,** the freezing point of water is 273.15 kelvins (K), and the boiling point is 373.15 K. Notice that with the Kelvin scale, the degree sign is not used. Figure 3.9 on the next page compares the Celsius and Kelvin scales. A change of one degree on the Celsius scale is equivalent to one kelvin on the Kelvin scale. The zero point on the Kelvin scale, 0 K, or **absolute zero,** is equal to −273.15°C. For problems in this text, you can round −273.15°C to −273°C. Because one degree on the Celsius scale is equivalent to one kelvin on the Kelvin scale, converting from one temperature to another is easy. You simply add or subtract 273, as shown in the following equations.

$$K = °C + 273$$
$$°C = K - 273$$

For: Links on Temperature Scales
Visit: www.SciLinks.org
Web Code: cdn-1032

Discuss L2

Have students compare the freezing point of water on the Celsius and Kelvin temperature scales *(0°C and 273.15 K)* Ask, **What temperature scale has its zero point at absolute zero?** *(the Kelvin scale)* **Why do you think scientists use more than one temperature scale?** *(Acceptable answers include that one scale may be more convenient than the other depending on what is being measured; for example, extremely low temperatures are all negative numbers when expressed in degrees Celsius but positive when expressed in kelvins.)*

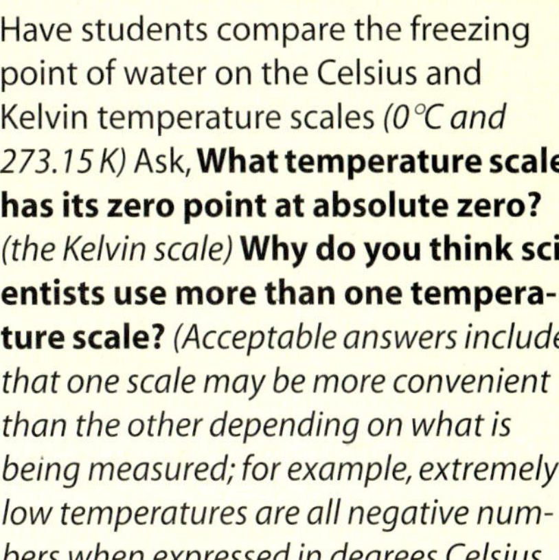

Download a worksheet on **Temperature Scales** for students to complete, and find additional support for NSTA SciLinks.

Facts and Figures

Near Absolute-Zero Temperatures

The coldest temperature achieved so far in a laboratory is one-half nanokelvin, that is 0.000 000 000 5 K above absolute zero. Compare this temperature to the "cold temperature of outer space," which is about 2.7 K.

Answers to...

Figure 3.7 The astronaut's mass is the same both on Earth and in space.

Checkpoint Weight is a force that measures the pull on a given mass by gravity. The weight of an object can change with its location. Its mass remains constant regardless of its location.

Use Visuals 

Figure 3.9 Ask, **Can the 10-K temperature change from 30 K to 40 K be read on either of these thermometers?** *(no)*

Sample Problem 3.4 

Answers

16. −196 °C

17. melting point: 1234 K; boiling point: 2485 K

Practice Problems Plus L2

The boiling point of water on top of Mount Everest is 343 K, while at the bottom of Death Valley, California, water boils at 373.3 K. **Express these temperatures in degrees Celsius.** *(Mount Everest: 70 °C; Death Valley: 100 °C)*

Math Handbook

For a math refresher and practice, direct students to algebraic equations, page R69.

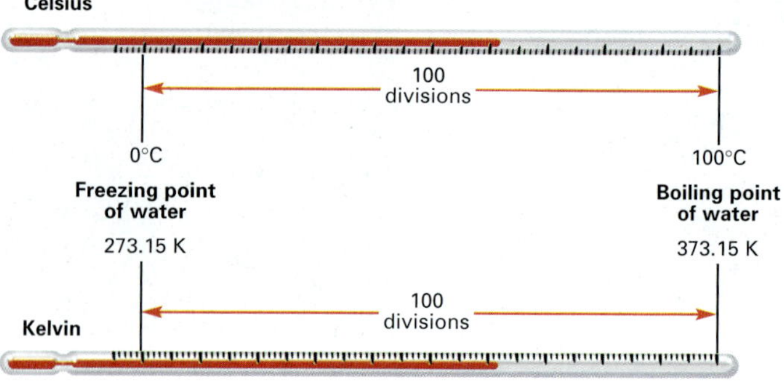

Figure 3.9 These thermometers show a comparison of the Celsius and Kelvin temperature scales. Note that a 1°C change on the Celsius scale is equal to a 1 K change on the Kelvin scale. **Interpreting Diagrams** ***What is a change of 10 K equivalent to on the Celsius scale?***

Math Handbook

For help with algebraic equations, go to page R69.

SAMPLE PROBLEM 3.4

Converting Between Temperature Scales

Normal human body temperature is 37°C. What is that temperature in kelvins?

1 Analyze ***List the known and the unknown.***

Known

- Temperature in °C = 37°C

Unknown

- Temperature in K = ? K

Use the known value and the equation K = °C + 273 to calculate the temperature in kelvins.

2 Calculate ***Solve for the unknown.***

Substitute the known value for the Celsius temperature into the equation and solve.

$$K = {}^\circ C + 273$$
$$= 37 + 273 = 310 \text{ K}$$

3 Evaluate ***Does the result make sense?***

You should expect a temperature in this range, since the freezing point of water is 273 K and the boiling point of water is 373 K; normal body temperature is between these two values.

Practice Problems

16. Liquid nitrogen boils at 77.2 K. What is this temperature in degrees Celsius?

17. The element silver melts at 960.8°C and boils at 2212°C. Express these temperatures in kelvins.

Interactive Textbook

Problem-Solving 3.17
Solve Problem 17 with the help of an interactive guided tutorial.

with ChemASAP

Figure 3.10 Photoelectric panels convert solar energy into electricity.

Units of Energy Figure 3.10 shows a house equipped with solar panels. The solar panels convert the radiant energy from the sun into electrical energy that can be used to heat water and power appliances. **Energy** is the capacity to do work or to produce heat.

Like any other quantity, energy can be measured. **The joule and the calorie are common units of energy.** The **joule (J)** is the SI unit of energy. It is named after the English physicist James Prescott Joule (1818–1889). One **calorie (cal)** is the quantity of heat that raises the temperature of 1 g of pure water by 1°C. Conversions between joules and calories can be carried out using the following relationships.

$$1\ \text{J} = 0.2390\ \text{cal} \qquad 1\ \text{cal} = 4.184\ \text{J}$$

3.2 Section Assessment

18. **Key Concept** Which five SI base units are commonly used in chemistry?
19. **Key Concept** Which metric units are commonly used to measure length, volume, mass, temperature, and energy?
20. Name the quantity measured by each of the seven SI base units and give the SI symbol of the unit.
21. What is the symbol and meaning of each prefix?
    - **a.** *milli-*
    - **b.** *nano-*
    - **c.** *deci-*
    - **d.** *centi-*
22. List the following units in order from largest to smallest: m^3, mL, cL, μL, L, dL.
23. What is the volume of a paperback book 21 cm tall, 12 cm wide, and 3.5 cm thick?
24. State the difference between mass and weight.
25. State the relationship between degrees Celsius and kelvins.
26. Surgical instruments may be sterilized by heating at 170°C for 1.5 hr. Convert 170°C to kelvins.
27. State the relationship between joules and calories.

Elements Handbook

Boiling Points Look up the boiling points of the first four elements in Group 7A on page R32. Convert these temperatures into kelvins.

Interactive Textbook

Assessment 3.2 Test yourself on the concepts in Section 3.2.

with ChemASAP

Section 3.2 Assessment

18. m, kg, K, s, mol

19. *length:* cm, m, km; *volume:* L, mL, cm^3, μL; *mass:* kg, g, mg, μm; *temperature:* °C, K; *energy:* J, cal

20. length (m), mass (kg), temperature (K), time (s), amount of substance (mol), luminous intensity (cd), electric current (A)

21. **a.** m; 10^{-3} of the unit **b.** n; 10^{-9} of the unit **c.** d; 10^{-1} of the unit **d.** c; 10^{-2} of the unit

22. m^3, L, dL, cL, mL, μL

23. $8.8 \times 10^2\ cm^3$

24. Mass is a measure of the amount of matter in an object. Weight is a measure of the force of gravity on an object.

25. °C = K – 273

26. 443 K

27. 1 J = 0.2390 cal

3 ASSESS

Evaluate Understanding L2

Write the following chart on the board.

Student's	Magnitude	SI Unit
age	1.6	K
body temp.	6.2×10^1	kg
height	3.1×10^2	m
mass	4.8×10^8	s

Have students match the appropriate magnitude and SI unit in columns II and III, with the item in column I. *(4.8×10^8 s, 3.1×10^2 K, 1.6 m, 6.2×10^1 kg)*

Reteach L1

Review Tables 3.3, 3.4, and 3.5 with students. Display each on an overhead projector and review the SI base unit listed in each table, pronunciations of the derived units, and the use of the prefixes to determine the numerical relationships between the base unit and the derived units.

Elements Handbook

Group 7A Element	Boiling Point (K)
Fluorine (F)	85 K
Chlorine (Cl)	238 K
Bromine (Br)	332 K
Iodine (I)	457 K

Interactive Textbook

If your class subscribes to the Interactive Textbook, use it to review key concepts in Section 3.2.

with ChemASAP

Answers to...

Figure 3.9 10°C

3.3

1 FOCUS

Objectives

3.3.1 **Construct** conversion factors from equivalent measurements.

3.3.2 **Apply** the technique of dimensional analysis to a variety of conversion problems.

3.3.3 **Solve problems** by breaking the solution into steps.

3.3.4 **Convert** complex units, using dimensional analysis.

Guide for Reading

Build Vocabulary L2

Paraphrase Have students write definitions of the vocabulary terms in their own words. *(Acceptable answers include* **conversion factor:** *a ratio of equivalent measurements used to convert a quantity from one unit to another, and* **dimensional analysis:** *a technique of problem-solving that uses the units that are part of a measurement to solve the problem.)*

Reading Strategy L2

Sequence As the students read the Analyze and Calculate sections of Sample Problems 3.5–3.9, have them write word sequences using the appropriate conversion factors for each problem.

2 INSTRUCT

Connecting to Your World

Have students examine the photograph that opens the section. Ask if any of them has ever noticed a chart or table in a bank or in the newspaper relating the values of foreign currency to the U.S. dollar. Explain that these are conversion tables that allow people to relate one currency to another. Ask, **How would you decide which amount of money would be worth more—75 euros or 75 British pounds?** *(Convert these values to a familiar currency—U.S. dollars.)*

3.3 Conversion Problems

Guide for Reading

Key Concepts

- What happens when a measurement is multiplied by a conversion factor?
- Why is dimensional analysis useful?
- What types of problems are easily solved by using dimensional analysis?

Vocabulary

conversion factor

dimensional analysis

Reading Strategy

Monitoring Your Understanding Preview the Key Concepts, the section heads, and boldfaced terms. List three things you expect to learn. After reading, state what you learned about each item listed.

Connecting to Your World

Perhaps you have traveled abroad or are planning to do so. If so, you know—or will soon discover—that different countries have different currencies. As a tourist, exchanging money is essential to the enjoyment of your trip. After all, you must pay for your meals, hotel, transportation, gift purchases, and tickets to exhibits and events. Because each country's currency compares differently with the U.S. dollar, knowing how to convert currency units correctly is very important. Conversion problems are readily solved by a problem-solving approach called dimensional analysis.

Conversion Factors

If you think about any number of everyday situations, you will realize that a quantity can usually be expressed in several different ways. For example, consider the monetary amount \$1.

$$1 \text{ dollar} = 4 \text{ quarters} = 10 \text{ dimes} = 20 \text{ nickels} = 100 \text{ pennies}$$

These are all expressions, or measurements, of the same amount of money. The same thing is true of scientific quantities. For example, consider a distance that measures exactly 1 meter.

$$1 \text{ meter} = 10 \text{ decimeters} = 100 \text{ centimeters} = 1000 \text{ millimeters}$$

These are different ways to express the same length.

Whenever two measurements are equivalent, a ratio of the two measurements will equal 1, or unity. For example, you can divide both sides of the equation 1 m = 100 cm by 1 m or by 100 cm.

$$\frac{1\text{ m}}{1\text{ m}} = \frac{100\text{ cm}}{1\text{ m}} = 1 \quad \text{or} \quad \frac{1\text{ m}}{100\text{ cm}} = \frac{100\text{ cm}}{100\text{ cm}} = 1$$

conversion factors

A **conversion factor** is a ratio of equivalent measurements. The ratios 100 cm/1 m and 1 m/100 cm are examples of conversion factors. In a conversion factor, the measurement in the numerator (on the top) is equivalent to the measurement in the denominator (on the bottom). The conversion factors above are read "one hundred centimeters per meter" and "one meter per hundred centimeters." Figure 3.11 illustrates another way to look at the relationships in a conversion factor. Notice that the smaller number is part of the measurement with the larger unit. That is, a meter is physically larger than a centimeter. The larger number is part of the measurement with the smaller unit.

Interactive Textbook

Animation 3 Learn how to select the proper conversion factor and how to use it.

with ChemASAP

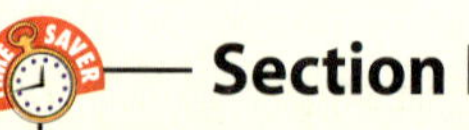

Section Resources

Print

- **Guided Reading and Study Workbook,** Section 3.3
- **Core Teaching Resources,** Section 3.3 Review
- **Transparencies,** T31–T37

Technology

- **Interactive Textbook with ChemASAP,** Animation 3, Problem-Solving 3.28, 3.30, 3.33, 3.35, 3.37, Assessment 3.3

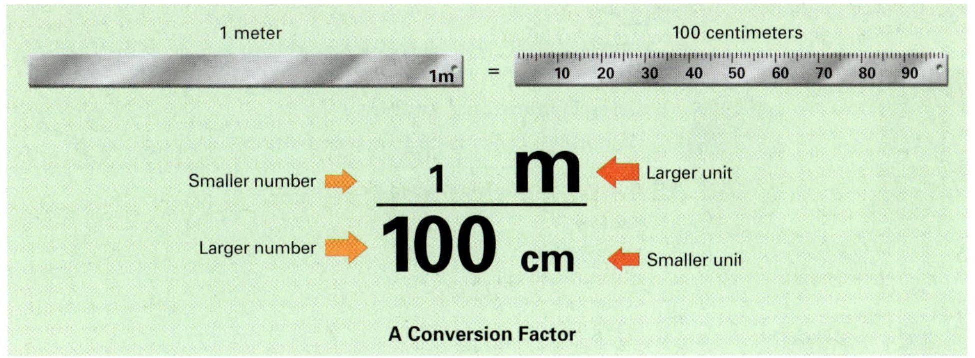

Figure 3.11 The two parts of a conversion factor, the numerator and the denominator, are equal.

Conversion factors are useful in solving problems in which a given measurement must be expressed in some other unit of measure. **When a measurement is multiplied by a conversion factor, the numerical value is generally changed, but the actual size of the quantity measured remains the same.** For example, even though the numbers in the measurements 1 g and 10 dg (decigrams) differ, both measurements represent the same mass. In addition, conversion factors within a system of measurement are defined quantities or exact quantities. Therefore, they have an unlimited number of significant figures, and do not affect the rounding of a calculated answer.

Here are some additional examples of pairs of conversion factors written from equivalent measurements. The relationship between grams and kilograms is 1000 g = 1 kg. The conversion factors are:

$$\frac{1000\text{ g}}{1\text{ kg}} \quad \text{and} \quad \frac{1\text{ kg}}{1000\text{ g}}$$

The scale of the micrograph in Figure 3.12 is in nanometers. Using the relationship 10^9 nm = 1 m, you can write the following conversion factors.

$$\frac{10^9\text{ nm}}{1\text{ m}} \quad \text{and} \quad \frac{1\text{ m}}{10^9\text{ nm}}$$

Common volumetric units used in chemistry include the liter and the microliter. The relationship 1 L = 10^6 μL yields the following conversion factors.

$$\frac{1\text{ L}}{10^6\ \mu\text{L}} \quad \text{and} \quad \frac{10^6\ \mu\text{L}}{1\text{ L}}$$

Based on what you know about metric prefixes, you should be able to easily write conversion factors that relate equivalent metric quantities.

 ***How many significant figures does a conversion factor within a system of measurement have?***

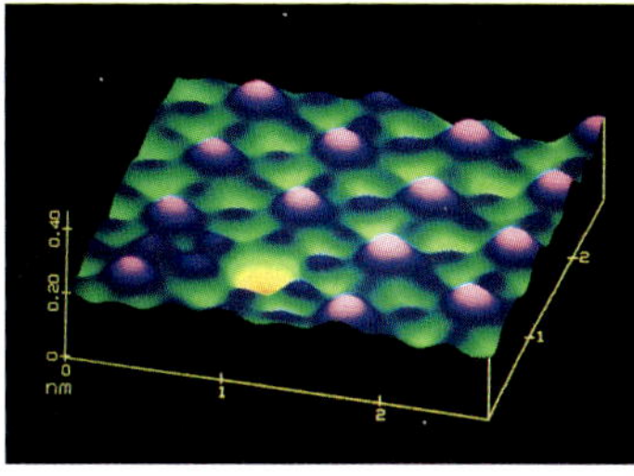

Figure 3.12 In this computer image of atoms, distance is marked off in nanometers (nm). **Inferring** ***What conversion factor would you use to convert nanometers to meters?***

Dimensional Analysis

No single method is best for solving every type of problem. Several good approaches are available, and generally one of the best is dimensional analysis. **Dimensional analysis** is a way to analyze and solve problems using the units, or dimensions, of the measurements. The best way to explain this problem-solving technique is to use it to solve an everyday situation.

Conversion Factors

Use Visuals 

Figure 3.11 Have students inspect the figure. Emphasize that a conversion factor relates two equivalent measurements. Ask, **What two parts does every measurement have?** *(a number and a unit)* Point out that if this is so, every conversion factor must contain two numbers and two units so that one number and its unit equal another number and its unit.

Dimensional Analysis

Expanding a Recipe 

Purpose To use dimensional analysis to convert common units

Materials copies of a recipe, lists of equivalents and conversions among the following measurements: teaspoon, tablespoon, $^1/_4$ cup, $^1/_2$ cup, and 1 cup (These lists are found in most cookbooks.)

Procedure Distribute the recipe and the conversion list to pairs of students. Explain that the students must rewrite the recipe so that it can feed six times the number of serving sizes suggested by the recipe. Point out that it would be tedious to have to measure out a particular ingredient (pick out one) in teaspoons or tablespoons six times, so students must rewrite the recipe in appropriately larger units. After students have rewritten the recipe, have student pairs exchange and compare recipes.

Expected Outcome Students should use the conversion lists to write simple conversion factors, such as 3 teaspoons/1 tablespoon, and then rewrite the recipe using larger measurements.

Facts and Figures

Monetary Exchange Rates

The conversion of chemical units is similar to the exchange of currency. Americans who travel outside the United States must exchange U.S. dollars for foreign currency at a given rate of exchange. These exchange rates vary from day to day. The daily exchange rates affect all international monetary transactions. Each time one type of money is exchanged for another, the current exchange rate serves as a conversion factor. International currency traders keep track of exchange rates 24 hours a day through a linked computer network.

Answers to...

Figure 3.12 1 m/10^9 nm

Checkpoint unlimited

Discuss

L2

Explain that measurements are often made using one unit and then converted into a related unit before being used in calculations. For example, students might measure volume in liters or milliliters in the laboratory, but express it as cubic centimeters in a calculation. Explain to the students that conversions are done using conversion factors. Emphasize that these conversion factors are ratios of *equivalent physical quantities,* such as 1 mL/1 cm^3.

Sample Problem 3.5

Answers

28. 1.0080×10^4 min

29. 1.44000×10^5 s

Practice Problems Plus

At Earth's farthest point from the sun, sunlight takes 8.5 minutes to reach Earth. **How many weeks is this?** *(8.4×10^{-4} weeks)*

Math Handbook

For a math refresher and practice, direct students to dimensional analysis, page R66.

Math Handbook

For help with dimensional analysis, go to page R66.

Problem-Solving 3.29 Solve Problem 29 with the help of an interactive guided tutorial.

with ChemASAP

SAMPLE PROBLEM 3.5

Using Dimensional Analysis

How many seconds are in a workday that lasts exactly eight hours?

1 Analyze ***List the knowns and the unknown.***

Knowns

- time worked = 8 h
- 1 hour = 60 min
- 1 minute = 60 s

Unknown

- seconds worked = ? s

The first conversion factor must be written with the unit hours in the denominator. The second conversion factor must be written with the unit minutes in the denominator. This will provide the desired unit (seconds) in the answer.

2 Calculate ***Solve for the unknown.***

Start with the known, 8 hours. Use the first relationship (1 hour = 60 minutes) to write a conversion factor that expresses 8 hours as minutes. The unit hours must be in the denominator so that the known unit will cancel. Then use the second conversion factor to change the unit minutes into the unit seconds. This conversion factor must have the unit minutes in the denominator. The two conversion factors can be used together in a simple overall calculation.

$$8\ \cancel{\text{h}} \times \frac{60\ \cancel{\text{min}}}{1\ \cancel{\text{h}}} \times \frac{60\ \text{s}}{1\ \cancel{\text{min}}} = 28{,}800\ \text{s}$$

$$= 2.8800 \times 10^4\ \text{s}$$

3 Evaluate ***Does the result make sense?***

The answer has the desired unit (seconds). Since the second is a small unit of time, you should expect a large number of seconds in 8 hours. Before you do the actual arithmetic, it is a good idea to make sure that the units cancel and that the numerator and denominator of each conversion factor are equal to each other. The answer is exact since the given measurement and each of the conversion factors is exact.

Practice Problems

28. How many minutes are there in exactly one week?

29. How many seconds are in exactly a 40-hour work week?

There is usually more than one way to solve a problem. When you first read Sample Problem 3.5, you may have thought about different and equally correct ways to approach and solve the problem. Some problems are easily worked with simple algebra. **Dimensional analysis provides you with an alternative approach to problem solving.** In either case, you should choose the problem-solving method that works best.

SAMPLE PROBLEM 3.6

Using Dimensional Analysis

The directions for an experiment ask each student to measure 1.84 g of copper (Cu) wire. The only copper wire available is a spool with a mass of 50.0 g. How many students can do the experiment before the copper runs out?

1 Analyze ***List the knowns and the unknown.***

Knowns

- mass of copper available = 50.0 g Cu
- each student needs 1.84 grams of copper, or $\frac{1.84 \text{ g Cu}}{\text{student}}$.

Unknown

- number of students = ?

From the known mass of copper, calculate the number of students that can do the experiment by using the appropriate conversion factor. The desired conversion is mass of copper $\longrightarrow$ number of students.

2 Calculate ***Solve for the unknown.***

Because students is the desired unit for the answer, the conversion factor should be written with students in the numerator. Multiply the mass of copper by the conversion factor.

$$50.0 \text{ g } \cancel{\text{Cu}} \times \frac{1 \text{ student}}{1.84 \text{ g } \cancel{\text{Cu}}} = 27.174 \text{ students} = 27 \text{ students}$$

Note that because students cannot be fractional, the result is shown rounded down to a whole number.

3 Evaluate ***Does the result make sense?***

The unit of the answer (students) is the one desired. The number of students (27) seems to be a reasonable answer. You can make an approximate calculation using the following conversion factor.

$$\frac{1 \text{ student}}{2 \text{ g Cu}}$$

Multiplying the above conversion factor by 50 g Cu gives the approximate answer of 25 students, which is close to the calculated answer.

Practice Problems

30. An experiment requires that each student use an 8.5-cm length of magnesium ribbon. How many students can do the experiment if there is a 570-cm length of magnesium ribbon available?

31. A 1.00-degree increase on the Celsius scale is equivalent to a 1.80-degree increase on the Fahrenheit scale. If a temperature increases by 48.0°C, what is the corresponding temperature increase on the Fahrenheit scale?

CHEMath

Conversion Problems

A conversion factor is a ratio of two quantities that are equal to one another. When doing conversions, write the conversion factors so that the unit of a given measurement cancels, leaving the correct unit for your answer. Note that the equalities needed to write a particular conversion may be given in the problem. In other cases, you will need to know or look up the necessary equalities.

Math Handbook

For help with conversion problems, go to page R66.

Problem-Solving 3.30 Solve Problem 30 with the help of an interactive guided tutorial.

with ChemASAP

Differentiated Instruction

Less Proficient Readers L1

Provide as much class time as possible for students to work on problem assignments in cooperative learning groups. Have students explore their own problem-solving styles. Encourage students to draw a diagram or picture of the problem to be solved whenever possible. Some students may want to read the problem aloud or have a partner read it to them. Some may want to work with symbols and equations.

Sample Problem 3.6

Answers

30. 67 students

31. 86.4°F

Practice Problems Plus L2

Chapter 3 Assessment problem 84 is related to Sample Problem 3.6.

CHEMath

Conversion Problems

Remind students that they often convert from one unit to another, both inside and outside of chemistry class. They convert money from cents to dollars and time from minutes to hours. Start out by giving them practice with everyday examples. Ask, **A chicken needs to be cooked 20 minutes for each pound it weighs. How long should the chicken be cooked if it weighs 4.5 pounds?** *(4.5 lb × 20 min/lb = 90 min; 90 min × 1 h/60 min = 1.5 h. Most students will automatically relate 90 minutes to 1.5 hours. This may help them become comfortable with the process.)* If students are having difficulty with conversion factors, you may wish to have them list several conversion factors on the chalkboard. Divide the class in half and have each group challenge the other to write the conversion factor given two related units. Remind them that each conversion factor can appear in two forms depending on which value they put in the numerator.

Math Handbook

For a math refresher and practice, direct students to conversion problems, page R66.

Converting Between Units

CLASS Activity

Sports Stats L2

Purpose To use dimensional analysis to convert between English and metric units

Materials copies of media guides containing vital statistics, such as heights and weights, of players on a sports team (These guides are available from local sports franchises.)

Procedure Group students. Distribute the media guides and assign each group a set of players. Ask the group to convert heights and weights into heights and masses expressed in meters and kilograms, respectively. Have students document their approach, including dimensional analysis expressions, conversion factors, and calculations.

Expected Outcome Students should use conversion factors, such as 2.54 cm/1 inch and 454 g/1 lb to convert their measurements.

Sample Problem 3.7

Answers

32. a. 44 m **b.** 4.6×10^{-3} g **c.** 10.7 cg

33. a. 1.5×10^{-2} L **b.** 7.38×10^{-3} kg **c.** 6.7×10^{3} ms **d.** 9.45×10^{7} µg

Practice Problems Plus 

Make the following conversions.

a. 0.045 L to cubic centimeters *(4.5×10^{1} cm^3)*

b. 14.3 mg to grams *(1.43×10^{-2} g)*

c. 0.0056 m to micrometers *(5.6×10^{3} µm)*

d. 0.035 cm to millimeters *(3.5×10^{-1} mm)*

Converting Between Units

In chemistry, as in many other subjects, you often need to express a measurement in a unit different from the one given or measured initially. **Problems in which a measurement with one unit is converted to an equivalent measurement with another unit are easily solved using dimensional analysis.**

Suppose that a laboratory experiment requires 7.5 dg of magnesium metal, and 100 students will do the experiment. How many grams of magnesium should your teacher have on hand? Multiplying 100 students by 7.5 dg/student gives you 750 dg. But then you must convert dg to grams. Sample Problem 3.7 shows you how to do the conversion.

SAMPLE PROBLEM 3.7

Converting Between Metric Units

Express 750 dg in grams.

1 Analyze ***List the knowns and the unknown.***

Knowns	Unknown
• mass = 750 dg	• mass = ? g
• 1 g = 10 dg	

The desired conversion is decigrams ⟶ grams. Using the expression relating the units, 10 dg = 1 g, multiply the given mass by the proper conversion factor.

2 Calculate ***Solve for the unknown.***

The correct conversion factor is shown below.

$$\frac{1\ \text{g}}{10\ \text{dg}}$$

Note that the known unit is in the denominator and the unknown unit is in the numerator.

$$750\ \text{dg} \times \frac{1\ \text{g}}{10\ \text{dg}} = 75\ \text{g}$$

3 Evaluate ***Does the result make sense?***

Because the unit gram represents a larger mass than the unit decigram, it makes sense that the number of grams is less than the given number of decigrams. The unit of the known (dg) cancels, and the answer has the correct unit (g). The answer also has the correct number of significant figures.

Practice Problems

32. Using tables from this chapter, convert the following.
- **a.** 0.044 km to meters
- **b.** 4.6 mg to grams
- **c.** 0.107 g to centigrams

33. Convert the following.
- **a.** 15 cm^3 to liters
- **b.** 7.38 g to kilograms
- **c.** 6.7 s to milliseconds
- **d.** 94.5 g to micrograms

Problem-Solving 3.33 Solve Problem 33 with the help of an interactive guided tutorial.

with ChemASAP

84 Chapter 3

Multistep Problems Many complex tasks in your everyday life are best handled by breaking them down into manageable parts. For example, if you were cleaning a car, you might first vacuum the inside, then wash the exterior, then dry the exterior, and finally put on a fresh coat of wax. Similarly, many complex word problems are more easily solved by breaking the solution down into steps.

When converting between units, it is often necessary to use more than one conversion factor. Sample Problem 3.8 illustrates the use of multiple conversion factors.

What problem-solving methods can help you solve complex word problems?

SAMPLE PROBLEM 3.8

Converting Between Metric Units

What is 0.073 cm in micrometers?

1 Analyze ***List the knowns and the unknown.***

Knowns
- length = 0.073 cm = 7.3×10^{-2} cm
- 10^2 cm = 1 m
- 1 m = 10^6 μm

Unknown
- length = ? μm

The desired conversion is from centimeters to micrometers. The problem can be solved in a two-step conversion.

2 Calculate ***Solve for the unknown.***

First change centimeters to meters; then change meters to micrometers: centimeters ⟶ meters ⟶ micrometers. Each conversion factor is written so that the unit in the denominator cancels the unit in the numerator of the previous factor.

$$7.3 \times 10^{-2}\ \text{cm} \times \frac{1\ \text{m}}{10^2\ \text{cm}} \times \frac{10^6\ \mu\text{m}}{1\ \text{m}} = 7.3 \times 10^2\ \mu\text{m}$$

3 Evaluate ***Does the result make sense?***

Because a micrometer is a much smaller unit than a centimeter, the answer should be numerically larger than the given measurement. The units have canceled correctly, and the answer has the correct number of significant figures.

Practice Problems

34. The radius of a potassium atom is 0.227 nm. Express this radius in the unit centimeters.

35. The diameter of Earth is 1.3×10^4 km. What is the diameter expressed in decimeters?

CHEMath

Scientific Notation

It is often convenient to express very large or very small numbers in scientific notation. The distance between the sun and Earth is 150,000,000 km, which can be written as 1.5×10^8 km. The diameter of a gold atom is 0.000 000 000 274 m, or 2.74×10^{-10} m.

When multiplying numbers written in scientific notation, add the exponents. When dividing numbers written in scientific notation, subtract the exponent in the denominator from the exponent in the numerator.

Math Handbook

For help with scientific notation, go to page R56.

Problem-Solving 3.35 Solve Problem 35 with the help of an interactive guided tutorial.

ChemASAP

Discuss **L2**

Explain that dimensional analysis is an extremely powerful problem-solving tool. Learning this method requires extra effort on the part of students. They must often use multiple conversion factors. The extra effort can be justified because the proper manipulation of the units assures accurate manipulation of the numbers.

Emphasize that students should use dimensional analysis as a tool for solving all of the problems they encounter in chemistry. Their first question about any quantity should be "What are the units of this quantity?" By comparing the units of various quantities in a problem, students can discover whether they need to perform any unit conversions before proceeding.

Sample Problem 3.8

Answers

34. 2.27×10^{-8} cm

35. 1.3×10^8 dm

Practice Problems Plus **L2**

Chapter 3 Assessment problem 70 is similar to Sample Problem 3.8.

CHEMath

Remind students that writing a number in scientific notation does not change the actual size of the number. In Sample Problem 3.8, the given measurement 0.073 cm can be written as 7.3×10^{-2} cm.

Math Handbook

For a math refresher and practice, direct students to scientific notation, page R56.

Differentiated Instruction

Less Proficient Readers **L1**

Students may benefit from a reminder that certain key words and phrases in each word problem indicate the unknown quantity and its units. Some of these phrases are:
- How much
- What is
- How long
- Determine
- Find

Answers to...

Break the solution down into steps. Use more than one conversion factor if necessary.

Sample Problem 3.9

Answers

36. 1.93×10^4 kg/m^3

37. 7.0×10^{12} RBC/L

Practice Problems Plus L2

1.00 L of neon gas contains 2.69×10^{22} neon atoms. **How many neon atoms are in 1.00 mm^3 of neon gas under the same conditions?** *(2.69×10^{16} atoms)*

Math Handbook

For a math refresher and practice, direct students to dimensional analysis, page R66.

Quick LAB

Dimensional Analysis 

Objective After completing this activity, students will be able to

- develop conversion factors using dimensional analysis.

Address Misconceptions

Students may think that because a conversion factor equals unity, it doesn't matter whether the conversion factor or its reciprocal is used in a calculation. Remind students that a given pair of equivalent measurements yields two different conversion factors, only one of which can be used to obtain the correct conversion.

Skills Focus Calculating

Prep Time 5 minutes

Materials 3 inch $\times$ 5 inch index cards (or paper cut to approximately the same size)

Class Time 15 minutes

Expected Outcome Students should derive appropriate conversion factors among metric units using standard metric quantities and prefixes.

Converting Complex Units Many common measurements are expressed as a ratio of two units. For example, the results of international car races often give average lap speeds in kilometers per hour. You measure the densities of solids and liquids in grams per cubic centimeter. You measure the gas mileage in a car in miles per gallon of gasoline. If you use dimensional analysis, converting these complex units is just as easy as converting single units. It will just take multiple steps to arrive at an answer.

SAMPLE PROBLEM 3.9

Converting Ratios of Units

The mass per unit volume of a substance is a property called density. The density of manganese, a metallic element, is 7.21 g/cm^3. What is the density of manganese expressed in units kg/m^3?

1 Analyze ***List the knowns and the unknown.***

Knowns

- density of manganese = 7.21 g/cm^3
- 10^3 g = 1 kg
- 10^6 cm^3 = 1 m^3

Unknown

- density manganese = ? kg/m^3

The desired conversion is g/cm$^3 \longrightarrow$ kg/m^3. The mass unit in the numerator must be changed from grams to kilograms: g $\longrightarrow$ kg. In the denominator, the volume unit must be changed from cubic centimeters to cubic meters: cm$^3 \longrightarrow$ m^3. Note that the relationship between cm^3 and m^3 was determined from the relationship between cm and m. Cubing the relationship 10^2 cm = 1 m yields $(10^2 \text{ cm})^3 = (1 \text{ m})^3$, or 10^6 cm^3 = 1 m^3.

2 Calculate ***Solve for the unknown.***

$$\frac{7.21 \text{ g}}{1 \text{ cm}^3} \times \frac{1 \text{ kg}}{10^3 \text{ g}} \times \frac{10^6 \text{ cm}^3}{1 \text{ m}^3} = 7.21 \times 10^3 \text{ kg/m}^3$$

Math Handbook

For help with dimensional analysis, go to page R66.

3 Evaluate ***Does the result make sense?***

Because the physical size of the volume unit m^3 is so much larger than cm^3 (10^6 times), the calculated value of the density should be larger than the given value even though the mass unit is also larger (10^3 times). The units cancel, the conversion factors are correct, and the answer has the correct ratio of units.

Interactive Textbook

Problem-Solving 3.37 Solve Problem 37 with the help of an interactive guided tutorial.

with **ChemASAP**

Practice Problems

36. Gold has a density of 19.3 g/cm^3. What is the density in kilograms per cubic meter?

37. There are 7.0×10^6 red blood cells (RBC) in 1.0 mm^3 of blood. How many red blood cells are in 1.0 L of blood?

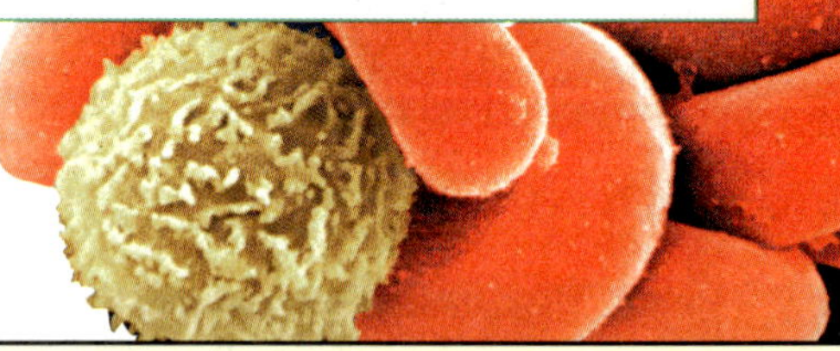

Analyze and Conclude

1. If a measurement of a quantity is being converted, the unit changes. If the equivalence of a quantity is being found, the conversion factor changes the quantity itself.
2. **a.** 0.785 m **b.** 56 cm^3 **c.** 7.7×10^7 mg **d.** 9.8×10^{-10} dm **e.** 9.6×10^3 μm **f.** 6.7×10^3 nm

For Enrichment 

Have students express the ratio in scientific notation between the largest and smallest units listed in each table. *(Table 3.3: 1×10^{12}, Table 3.4: 1×10^6; Table 3.5: 1×10^9)*

Quick LAB

Dimensional Analysis

Purpose

To apply the problem-solving technique of dimensional analysis to conversion problems.

Materials

- 3 inch × 5 inch index cards or paper cut to approximately the same size
- pen

Procedure

A conversion factor is a ratio of equivalent measurements. For any relationship, you can write two ratios. On a conversion factor card you can write one ratio on each side of the card.

1. Make a conversion factor card for each metric relationship shown in Tables 3.3, 3.4, and 3.5. Show the inverse of the conversion factor on the back of each card.
2. Use the appropriate conversion factor cards to set up solutions to Sample Problems 3.7 and 3.8. Notice that in each solution, the unit in the denominator of the conversion factor cancels the unit in the numerator of the previous conversion factor.

Analyze and Conclude

1. What is the effect of multiplying a given measurement by one or more conversion factors?
2. Use your conversion factor cards to set up solutions to these problems.
   a. 78.5 cm = ? m
   b. 0.056 L = ? cm^3
   c. 77 kg = ? mg
   d. 0.098 nm = ? dm
   e. 0.96 cm = ? µm
   f. 0.0067 mm = ? nm

3.3 Section Assessment

38. Key Concept What happens to the numerical value of a measurement that is multiplied by a conversion factor? What happens to the actual size of the quantity?
39. Key Concept Why is dimensional analysis useful?
40. Key Concept What types of problems can be solved using dimensional analysis?
41. What conversion factor would you use to convert between these pairs of units?
    a. minutes to hours
    b. grams to milligrams
    c. cubic decimeters to milliliters
42. Make the following conversions. Express your answers in standard exponential form.
    a. 14.8 g to micrograms
    b. 3.72×10^{-3} kg to grams
    c. 66.3 L to cubic centimeters
43. An atom of gold has a mass of 3.271×10^{-22} g. How many atoms of gold are in 5.00 g of gold?
44. Convert the following. Express your answers in scientific notation.
    a. 7.5×10^4 J to kilojoules
    b. 3.9×10^5 mg to decigrams
    c. 2.21×10^{-4} dL to microliters
45. Light travels at a speed of 3.00×10^{10} cm/s. What is the speed of light in kilometers/hour?

Connecting Concepts

Problem-Solving Skills Reread the passage on solving numeric problems in Section 1.4. Explain how the three-step process might apply to conversion problems that involve dimensional analysis.

Assessment 3.3 Test yourself on the concepts in Section 3.3.

with ChemASAP

Section 3.3 Assessment

38. The numerical value (and the unit) changes; the actual size does not change.
39. Dimensional analysis provides an alternative approach to problem solving.
40. conversion problems
41. a. 1 hour / 60 min b. 10^3 mg / 1 g c. 10^3 mL / 1 dm^3
42. a. 1.48×10^7 µg b. 3.72 g c. 6.63×10^4 cm^3
43. 1.53×10^{22} atoms of gold
44. a. 7.5×10^1 kJ b. 3.9×10^3 dg c. 2.21×10^1 µL
45. 1.08×10^9 km/h

3 ASSESS

Evaluate Understanding L2

To determine students' grasp of conversion factors, ask, **What is the relationship between the numerator and the denominator of any measurement conversion factor?** *(They are equivalent so that the ratio of numerator to denominator equals 1.)*

Reteach L1

Model the conversion of 2 L to 2000 mL. Suggest that students check the answer by explaining that when using a conversion factor, such as 1 L = 1000 mL, the measurement expressed with the smaller unit (mL) should have a larger number associated with it (2000) than the measurement expressed with the larger unit (L).

Connecting Concepts

Acceptable answers include (a) analyze the known and unknowns, devise a set of conversions steps that yields the desired final units, and write a conversion factor for each step; (b) calculate by multiplying (making sure that the appropriate units cancel); and (c) evaluate the magnitude and units of the calculated answer.

If your class subscribes to the Interactive Textbook, use it to review key concepts in Section 3.3.

with ChemASAP

Technology & Society

Scale Models

A scale factor is a ratio that equals the ratio of the model's dimensions to the structure's dimension. Ask, **What is the height of a building if it is represented by a 1.30 m model using a scale factor of 1:48?** *(62.4 m)* Computerized design (CAD) programs have replaced many tedious aspects of technical design. A CAD program usually permits a designer to begin with a few important elements and measurements of the object being designed. As the design progresses, the program computes the dimensions of the added elements and produces drawings of the object from any perspective. The introduction of CAD programs has reduced the time needed to design complex items as well as the number of errors that plague such projects.

Technology & Society

Scale Models

A scale model is a physical or conceptual representation of an object that is proportional in size to the object it represents. Examples include model trains, model airplanes, and dollhouses. Most model trains are built to a scale of 1:87. This ratio means that the model is $\frac{1}{87}$ the size of an actual train. This fraction can be used as a conversion factor. On the model, 1 cm represents 87 cm on the train.

Scale models aren't just for hobbyists—scientists and engineers use them, too. A simple scientific model in the classroom is a globe, which is a small-scale model of Earth. (A globe with a diameter of 30 cm has a scale of 1:42,500,000.) **Applying Concepts** ***How do you use the scale of a model as a conversion factor?***

Scale model
4.5 ft tall, scale 1:192

Architectural drawing

Condé Nast Building New York City
866 ft (264 m) tall, 48 floors

Computer modeling
By testing a model, engineers can make a product better before it is built. Engineers often design scale models on computers. These automotive engineers are using a computer-aided design (CAD) program to view a digital scale model of a car. Physical models of the car's wheels are on the desk.

Model building
Architects use both two-dimensional and three-dimensional scale models to design buildings. A common scale for floor plans is 1: 48.

Answers to...

Applying Concepts The scale of a model can be used as a conversion factor by expressing it as a fraction of equivalent measurements. For example, if a globe has a scale of 1:42,500,000, you can write the conversion factor 42,500,000 cm (actual)/1 cm (model). If the globe has a 30-cm diameter, you can calculate the actual diameter of Earth as follows:

$$3.0 \times 10^1\ \text{cm} \times \frac{42{,}500{,}000\ \text{cm}}{1\ \text{cm}} = 1.3 \times 10^9\ \text{cm}$$

Figure 3.13 lead

Differentiated Instruction

Gifted and Talented 

Have students prepare a display on current CAD applications. Local colleges or technical schools that teach CAD are good sources of information. Other possibilities include library research or articles in magazines or newspapers.

3.4 Density

Connecting to Your World Have you ever wondered why some objects float in water, while others sink? If you think that these lily pads float because they are lightweight, you are only partially correct. The ratio of the mass of an object to its volume can be used to determine whether an object floats or sinks in water. For pure water at 4°C, this ratio is 1.000 g/cm^3. If an object has a mass-to-volume ratio less than 1.000 g/cm^3, it will float in water. If an object has a mass-to-volume ratio greater than this value, it will sink in water.

Guide for Reading

Key Concepts

- What determines the density of a substance?
- How does a change in temperature affect density?

Vocabulary

density

Reading Strategy

Identifying Main Ideas As you read, write the main idea of the text that follows each heading.

Determining Density

Perhaps someone has tricked you with this question: "Which is heavier, a pound of lead or a pound of feathers?" Most people would not give the question much thought and would incorrectly answer "lead." Of course, a pound of lead has the same mass as a pound of feathers. What concept, instead of mass, are people really thinking of when they answer this question?

Most people are incorrectly applying a perfectly correct idea: namely, that if a piece of lead and a feather of the same volume are weighed, the lead would have a greater mass than the feather. It would take a much larger volume of feathers to equal the mass of a given volume of lead.

Simulation 1 Rank materials according to their densities.

with ChemASAP

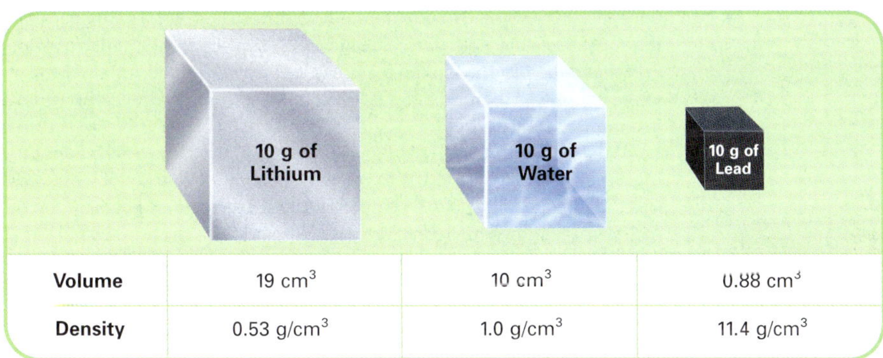

Volume	19 cm^3	10 cm^3	0.88 cm^3
Density	0.53 g/cm^3	1.0 g/cm^3	11.4 g/cm^3

Figure 3.13 A 10-g sample of pure water has less volume than 10 g of lithium, but more volume than 10 g of lead. The faces of the cubes are shown actual size. **Inferring** ***Which substance has the highest ratio of mass to volume?***

Section Resources

Print

- **Guided Reading and Study Workbook,** Section 3.4
- **Core Teaching Resources,** Section 3.4 Review
- **Transparencies,** T38–T42
- **Laboratory Manual,** Lab 4

Technology

- **Interactive Textbook with ChemASAP,** Simulation 1, Problem-Solving 3.47, 3.48, Assessment 3.4
- **Go Online,** Section 3.4

3.4

1 FOCUS

Objectives

3.4.1 **Calculate** the density of a material from experimental data.

3.4.2 **Describe** how density varies with temperature.

Guide for Reading

Build Vocabulary 

Paraphrase After students have read the key concept on page 90, point out that mass, unlike density, is not an intensive property. On the contrary, it is an extensive property. Have students infer the meaning of the term *extensive property. (a property that depends on size)*

Reading Strategy 

Using Context Clues Have students read the key concept statement on page 90. Ask, **What is the meaning of the term intensive property.** *(a property that does not depend on the size of the sample)* **What other properties may be intensive?** *(Acceptable answers include temperature and color.)*

2 INSTRUCT

Connecting to Your World

Have students study the photograph and read the text that opens the section. Ask, **Why do lily pads float in water?** *(The pads are less dense than water.)* **What measurements would you need to make to determine whether an object would float in water?** *(Measure the object's volume and mass; then compute its density and compare the result to the density of water.)*

Determining Density

Use Visuals L1

Figure 3.13 Ask, **What would be the volume of a 10-g sphere of each of the substances shown?** *(Li, 19 cm^3; H_2O, 10 cm^3; and Pb, 0.88 cm^3)*

Relate 

If students performed the *Mass of a Penny* activity on page 76, have them refer to the explanation of the results. Ask, **How can you use density to test the hypothesis that the composition of the penny changed?** *(Determine the volume and mass of a pre- and post-1982 penny. Calculate the density of each and compare them. If the densities changed, the composition of the penny must have changed.)*

TEACHER Demo

Density Calculations 

Purpose To calculate density from a mass-volume graph

Materials 3 or 4 different-sized cubes of a material such as wood, metal, or marble; metric ruler; balance

Procedure Have student volunteers measure the masses of the cubes, measure the lengths of the cubes and calculate their volumes. On the board or an overhead projector, record the masses and volumes. Construct a mass-volume graph (mass on the *y*-axis) and plot the data. Show students how to calculate the slope of the line. Explain that the slope of the line is the ratio of mass/volume and has the unit, g/cm^3, which is the unit of density.

Expected Outcome The mass and volume data should lie along a straight line. Ask, **What is the ratio of mass to volume of a substance?** *(density)*

Table 3.6

Densities of Some Common Materials

Solids and Liquids		Gases	
Material	**Density at 20°C (g/cm^3)**	**Material**	**Density at 20°C (g/L)**
Gold	19.3	Chlorine	2.95
Mercury	13.6	Carbon dioxide	1.83
Lead	11.4	Argon	1.66
Aluminum	2.70	Oxygen	1.33
Table sugar	1.59	Air	1.20
Corn syrup	1.35–1.38	Nitrogen	1.17
Water (4°C)	1.000	Neon	0.84
Corn oil	0.922	Ammonia	0.718
Ice (0°C)	0.917	Methane	0.665
Ethanol	0.789	Helium	0.166
Gasoline	0.66–0.69	Hydrogen	0.084

The important relationship in this case is between the object's mass and its volume. This relationship is called density. **Density** is the ratio of the mass of an object to its volume.

$$\text{Density} = \frac{\text{mass}}{\text{volume}}$$

A 10.0-cm^3 piece of lead, for example, has a mass of 114 g. What, then, is the density of lead? You can calculate it by substituting the mass and volume into the equation above.

$$\frac{114\ \text{g}}{10.0\ \text{cm}^3} = 11.4\ \text{g/cm}^3$$

Note that when mass is measured in grams, and volume in cubic centimeters, density has units of grams per cubic centimeter (g/cm^3).

Figure 3.13 on page 89 compares the density of three substances. Why does each 10-g sample have a different volume? The volumes vary because the substances have different densities. **Density is an intensive property that depends only on the composition of a substance, not on the size of the sample.** With a mixture, density can vary because the composition of a mixture can vary.

What do you think will happen if corn oil is poured into a glass containing corn syrup? Using Table 3.6, you can see that the density of corn oil is less than the density of corn syrup. For that reason, the oil floats on top of the syrup, as shown in Figure 3.14.

You have probably seen a helium-filled balloon rapidly rise to the ceiling when it is released. Whether a gas-filled balloon will sink or rise when released depends on how the density of the gas compares with the density of air. Helium is less dense than air, so a helium-filled balloon rises. The densities of various gases are listed in Table 3.6.

Figure 3.14 Because of differences in density, corn oil floats on top of corn syrup.

 ***What quantities do you need to measure in order to calculate the density of an object?***

90 Chapter 3

Differentiated Instruction

Less Proficient Readers 

Have students write definitions of density in English and, if appropriate, in a native language. Make sure they include the expression density = mass/volume. Have them find photos of objects containing substances from Table 3.6. They could label each photo with the substance's name in multiple languages and the density, whose units are part of a universal language.

Density and Temperature

Experiments show that the volume of most substances increases as the temperature increases. Meanwhile, the mass remains the same despite the temperature and volume changes. Remember that density is the ratio of an object's mass to its volume. So if the volume changes with temperature (while the mass remains constant), then the density must also change with temperature. **The density of a substance generally decreases as its temperature increases.** As you will learn in Chapter 15, water is an important exception. Over a certain range of temperatures, the volume of water increases as its temperature decreases. Ice, or solid water, floats because it is less dense than liquid water.

For: Links on Density
Visit: www.SciLinks.org
Web Code: cdn-1034

SAMPLE PROBLEM 3.10

Calculating Density

A copper penny has a mass of 3.1 g and a volume of 0.35 cm^3. What is the density of copper?

1 Analyze *List the knowns and the unknown.*

Knowns
- mass = 3.1 g
- volume = 0.35 cm^3

Unknown
- density = ? g/cm^3

Use the known values and the following definition of density.

$$\text{Density} = \frac{\text{mass}}{\text{volume}}$$

2 Calculate *Solve for the unknown.*

The equation is already set up to solve for the unknown. Substitute the known values for mass and volume, and calculate the density.

$$\text{density} = \frac{\text{mass}}{\text{volume}} = \frac{3.1\text{ g}}{0.35\text{ cm}^3} = 8.8571\text{ g/cm}^3$$

$$= 8.9\text{ g/cm}^3 \text{ (rounded to two significant figures)}$$

3 Evaluate *Does the result make sense?*

A piece of copper with a volume of about 0.3 cm^3 of copper has a mass of about 3 grams. Thus, about three times that volume of copper, 1 cm^3, should have a mass three times larger, about 9 grams. This estimate agrees with the calculated result.

Practice Problems

46. A student finds a shiny piece of metal that she thinks is aluminum. In the lab, she determines that the metal has a volume of 245 cm^3 and a mass of 612 g. Calculate the density. Is the metal aluminum?

47. A bar of silver has a mass of 68.0 g and a volume of 6.48 cm^3. What is the density of silver?

Math Handbook

For help with algebraic equations, go to page R69.

Problem-Solving 3.47 Solve Problem 47 with the help of an interactive guided tutorial.

with ChemASAP

Discuss — L2

Show how the density equation can be used to determine the mass or volume of an object. For example,

density = mass/volume

(volume)(density) = (mass/volume) × volume

volume × density = mass

Density and Temperature

Relate — L2

Have students refer to the Galileo thermometer shown in Figure 3.8 on page 77. Explain that each partially filled, sealed glass bubble and its attached metal tag are weighted so that it will float at the temperature indicated on the tag. If the air temperature rises, it warms the water in the thermometer and eventually the water reaches the temperature of the air. As liquid water warms, it expands and so its density decreases. Therefore, the glass bubbles that are calibrated to float at lower temperature water will remain at the bottom of the container. Those that are made to float in water at or above the present water temperature will float. Thus, the temperature of the water and the surrounding air is the lowest value of the tag on a floating bubble. Ask, **In the Galileo thermometer shown in Figure 3.8, in what unit is the thermometer most likely calibrated?** *(degree Fahrenheit)*

Sample Problem 3.10

Answers

46. density = 2.50 g/cm^3; no
47. 10.5 g/cm^3

Practice Problems Plus — L2

The density of gold is 19.3 g/cm^3 at 20°C. **What is the mass of a 0.18-cm^3 gold ring?** *(3.5 g)*

Math Handbook

For a math refresher and practice, direct students to algebraic equations, page R69.

Download a worksheet on **Density** for students to complete, and find additional support for NSTA SciLinks.

Answers to...

mass and volume

TEACHER Demo

The Hydrometer L2

Purpose To show the use of a hydrometer

Materials hydrometer, 2 graduated cylinders, water, concentrated solution of sodium chloride

Safety Have students wear goggles to emphasize safety in the laboratory.

Procedure Explain to students that a hydrometer measures the specific gravity of a liquid. Point out that the specific gravity is the ratio of the density of a material to that of water. Float the hydrometer in the water and have students note the submerged height. Read the value of the hydrometer. Now place the hydrometer in the solution and have students note the submerged height. Ask students which liquid is denser.

Expected Outcome Students should predict that the second solution is denser because the hydrometer floated higher in the liquid. Confirm the answer by reading the value of the hydrometer.

Sample Problem 3.11

Answers

48. a. 6.32 cm^3 **b.** 0.342 cm^3

49. See answers for problem 48.

Practice Problems Plus 

Make the following conversions.

a. 2.53 cm^3 of gold to grams (density of gold = 19.3 g/cm^3) *(48.8 g)*

b. 1.6 g of oxygen to liters (density of oxygen = 1.33 g/L) *(1.2 L)*

c. 25.0 g of ice to cubic centimeters (density of ice = 0.917 g/cm3) *(27.3 cm^3)*

Math Handbook

For a math refresher and practice, direct students to scientific notation, page R56.

SAMPLE PROBLEM 3.11

Using Density to Calculate Volume

What is the volume of a pure silver coin that has a mass of 14 g? The density of silver (Ag) is 10.5 g/cm^3.

1 Analyze ***List the knowns and the unknown.***

Knowns	Unknown
• mass of coin = 14 g	• volume of coin = ? cm^3
• density of silver = 10.5 g/cm^3	

You can solve this problem by using density as a conversion factor. You need to convert the mass of the coin into a corresponding volume. The density gives the following relationship between volume and mass.

$$1\ cm^3\ Ag = 10.5\ g\ Ag$$

Based on this relationship, you can write the following conversion factor.

$$\frac{1\ cm^3\ Ag}{10.5\ g\ Ag}$$

Notice that the known unit is in the denominator and the unknown unit is in the numerator.

2 Calculate ***Solve for the unknown.***

Multiply the mass of the coin by the conversion factor to yield an answer in cm^3.

$$14\ \cancel{g\ Ag} \times \frac{1\ cm^3\ Ag}{10.5\ \cancel{g\ Ag}} = 1.3\ cm^3\ Ag$$

3 Evaluate ***Does the result make sense?***

Because a mass of 10.5 g of silver has a volume of 1 cm^3, it makes sense that 14.0 g of silver should have a volume slightly larger than 1 cm^3. The answer has two significant figures because the given mass has two significant figures.

Practice Problems

48. Use dimensional analysis and the given densities to make the following conversions.

a. 14.8 g of boron to cm^3 of boron. The density of boron is 2.34 g/cm^3.

b. 4.62 g of mercury to cm^3 of mercury. The density of mercury is 13.5 g/cm^3.

49. Rework the preceding problems by applying the following equation.

$$\text{Density} = \frac{\text{mass}}{\text{volume}}$$

Math Handbook

For help with significant figures, go to page R59.

Interactive Textbook

Problem-Solving 3.48 Solve Problem 48 with the help of an interactive guided tutorial.

with ChemASAP

Mercury

Facts and Figures

Specific Gravity

Specific gravity is the ratio of the density of a substance to the density of a reference substance, usually at the same temperature. Water at 4°C is often the reference substance. Physicians use the measured specific gravity of a patient's urine to help diagnose certain diseases, such as diabetes. Auto mechanics measure the specific gravity of antifreeze in radiators during tune-ups.

Have students research chemistry-related careers on the Internet. Students can then construct a table that describes the nature of the job, educational requirements, and other necessary information.

Careers *in* Chemistry

Analytical Chemist

Analytical chemists focus on making quantitative measurements. They must be familiar with many analytical techniques to work successfully on a wide variety of tasks. As an analytical chemist, you would spend your time making measurements and calculations to solve laboratory and math-based research problems. You could, for example, be involved in analyzing the composition of biomolecules. Pharmaceutical companies need people to analyze the composition of medicines and research new combinations of compounds to use as drugs. As an analytical chemist, you must be able to think creatively and develop new means for finding solutions.

Many exciting new fields, such as biomedicine and biochemistry, are now hiring analytical chemists. More traditional areas, including industrial manufacturers, also employ analytical chemists. The educational background you need to enter this field is quite extensive. You would need advanced chemical training, including organic chemistry and quantitative chemistry, as well as some training in molecular biology and computer operation. A master's degree in chemistry may be required, and certain positions require a Ph.D.

For: Careers in Chemistry
Visit: PHSchool.com
Web Code: cdb-1034

3.4 Section Assessment

50. **Key Concept** What determines the density of an object?
51. **Key Concept** How does density vary with temperature?
52. A weather balloon is inflated to a volume of 2.2×10^3 L with 37.4 g of helium. What is the density of helium in grams per liter?
53. A 68-g bar of gold is cut into 3 equal pieces. How does the density of each piece compare to the density of the original gold bar?
54. A plastic ball with a volume of 19.7 cm^3 has a mass of 15.8 g. Would this ball sink or float in a container of gasoline?
55. What is the volume, in cubic centimeters, of a sample of cough syrup that has a mass of 50.0 g? The density of cough syrup is 0.950 g/cm^3.
56. What is the mass, in kilograms, of 14.0 L of gasoline? (Assume that the density of gasoline is 0.680 g/cm^3.)

Elements Handbook

Density Look up the densities of the elements in Group 1A on page R6. Which Group 1A elements are less dense than pure water at 4°C?

Assessment 3.4 Test yourself on the concepts in Section 3.4.

with ChemASAP

Careers *in* Chemistry

Analytical Chemist

Analytical chemists ask the important question "How much?" They must make very careful measurements because they deal with small quantities of material. Analytical chemists are employed in a wide range of industrial and research occupations. Encourage students to connect to the Internet address shown on this page to find out more about careers in analytical chemistry.

3 ASSESS

Evaluate Understanding

Ask, **What are the most common units for expressing the density of a substance?** *(g/cm^3)* **How can density be used to determine whether a metal washer is aluminum or zinc?** *(Measure and compare the density of the washer to the accepted density values of aluminum and zinc.)*

Reteach

Remind students that the density equation can be rewritten as *mass = density × volume*, or *volume = mass/density*. Have students examine the application of these alternate equations by providing students with aluminum squares and asking them to measure the area and mass of each. Have students use the density of aluminum in Table 3.6 to calculate the thickness of the foil.

Elements Handbook

Li, Na, and K are less dense than water.

Interactive Textbook

If your class subscribes to the Interactive Textbook, use it to review key concepts in Section 3.4.

with ChemASAP

Section 3.4 Assessment

50. Density is an intensive property that depends only on the composition of a substance, not on the size of the sample. density = mass/volume
51. Density generally decreases when temperature increases.
52. 1.7×10^{-2} g/L
53. All the densities are equal.
54. It would sink because the density of the ball, 0.802 g/cm^3, is greater than the density of gasoline.
55. 52.6 cm^3
56. 9.52 kg

Now What Do I Do? 

Objective After completing this activity, students will be able to
- solve problems in divergent ways.

Skills Focus Measuring, calculating, evaluating, designing experiments

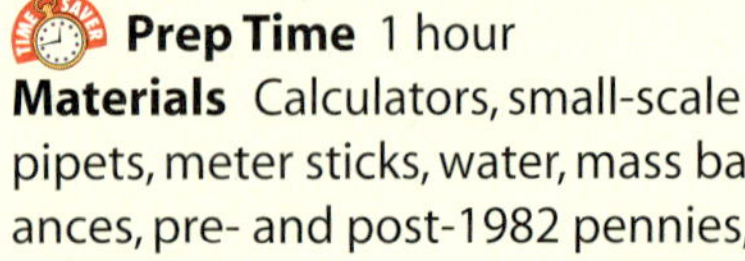

Prep Time 1 hour

Materials Calculators, small-scale pipets, meter sticks, water, mass balances, pre- and post-1982 pennies, dice, 8-well strips, aluminum cans, plastic cups

Advance Prep A day before doing the lab, obtain soda cans from the cafeteria. Wash and let air-dry overnight.

Class Time 40 minutes

Expected Outcome

Students should find that the mass of the pre-1982 penny is 3.11 g and the mass of the post-1982 penny is 2.50 g.

Analyze

1. 19 mg

2. 0.019 cm^3; 0.019 mL; 19 μL

3. 1000 mg/cm^3; 1000 mg/mL

4. 2.95 g Cu; 0.16 g Zn

5. 0.060 g Cu; 2.44 g Zn

6. The new penny is mostly zinc, which has a lower density than copper.

You're the Chemist

1. at 90°, mass of 1 drop: 0.019 g
at 45°, mass of 1 drop: 0.0218 g
at 0°, mass of 1 drop: 0.0242 g
Pipets give different results.

2. The pipet is easiest to control at 90°. Expel the air bubble so that the first drop will be the same size as the rest.

3. Find mass of can and divide by density of aluminum. Sample answer: mass of one can: 14.77 g; $density_{Al}$: 2.70 g/cm^3; V = 5.47 cm^3

4. (1) Measure the mass before and after you fill the can with water. Use the mass and density of water to find the volume.
(2) Measure the height and radius and calculate volume. $V = \pi r^2 h$ (Can is not a perfect cylinder.)
(3) Read label: 12 oz = 355 mL

5. Sample answer:
V = 16.5 m × 3.0 m × 12.8 m
= 630 m^3 × 1000 L/m^3 = 630,000 L
Assume 30 people with an average weight of 130 lb (1 kg = 2.2 lb) and a density of about 1.0 kg/L.
Volume of 30 people = 30 × 130 lb × 1 kg/2.2 lb × 1 L/1.0 kg = 1800 L
The volume of 30 chairs, 15 tables, and 2 desks is about that of 30 people or 1800 L.
The volume of people and furniture is 3600 L.
% error = (3600 L/630,000 L)(100%) = 0.57%.

6. If die measures 1.55 cm on a side:
$V = (1.55\ cm)^3 = 3.72\ cm^3$
A die has 21 holes that are hemispheres with a radius of 0.20 cm.
V of hemisphere = $2/3\pi r^3$ = 0.017 cm^3
V of 21 hemispheres = 0.36 cm^3
V of die = 3.72 cm^3 – 0.36 cm^3 = 3.36 cm^3
Error = 0.36 cm^3
% error = (0.36 cm^3/3.36 cm^3)(100%) = 11%

7. Find weight in pounds and convert to kg. Assume density = 1.00 kg/L.
V = *weight* × 1 kg/2.2 lb × 1L/1.00 kg

Small-Scale LAB

Now What Do I Do?

Purpose

To solve problems by making accurate measurements and applying mathematics.

Materials

- pencil
- paper
- meter stick
- balance
- pair of dice
- aluminum can
- calculator
- small-scale pipet
- water
- a pre- and post-1982 penny
- 8-well strip
- plastic cup

Procedure

1. Determine the mass, in grams, of one drop of water. To do this, measure the mass of an empty cup. Add 50 drops of water from a small-scale pipet to the cup and measure its mass again. Subtract the mass of the empty cup from the mass of the cup with water in it. To determine the average mass in grams of a single drop, divide the mass of the water by the number of drops (50). Repeat this experiment until your results are consistent.
2. Determine the mass of a pre-1982 penny and a post-1982 penny.

Analyze

Using your experimental data, record the answers to the following questions.

1. What is the average mass of a single drop of water in milligrams? (1 g = 1000 mg)
2. The density of water is 1.00 g/cm^3. Calculate the volume of a single drop in cm^3 and mL. (1 mL = 1 cm^3) What is the volume of a drop in microliters (μL)? (1000 μL = 1 mL)
3. What is the density of water in units of mg/cm^3 and mg/mL? (1 g = 1000 mg)
4. Pennies made before 1982 consist of 95.0% copper and 5.0% zinc. Calculate the mass of copper and the mass of zinc in the pre-1982 penny.
5. Pennies made after 1982 are made of zinc with a thin copper coating. They are 97.6% zinc and 2.4% copper. Calculate the mass of copper and the mass of zinc in the newer penny.
6. Why does one penny have less mass than the other?

You're the Chemist

The following small-scale activities allow you to develop your own procedures and analyze the results.

1. **Design It!** Design an experiment to determine if the size of drops varies with the angle at which they are delivered from the pipet. Try vertical (90°), horizontal (0°), and halfway between (45°). Repeat until your results are consistent.
2. **Analyze It!** What is the best angle to hold a pipet for ease of use and consistency of measurement? Explain. Why is it important to expel the air bubbles before you begin the experiment?
3. **Design It!** Make the necessary measurements to determine the volume of aluminum used to make an aluminum soda can. *Hint:* Look up the density of aluminum in your textbook.
4. **Design It!** Design and carry out some experiments to determine the volume of liquid that an aluminum soda can will hold.
5. **Design It!** Measure a room and calculate the volume of air it contains. Estimate the percent error associated with not taking into account the furniture in the room.
6. **Design It!** Make the necessary measurements and do the necessary calculations to determine the volume of a pair of dice. First, ignore the volume of the dots on each face, and then account for the volume of the dots. What is your error and percent error when you ignore the holes?
7. **Design It!** Design an experiment to determine the volume of your body. Write down what measurements you would need to make and what calculations you would do. What additional information might be helpful?

CHAPTER 3 Study Guide

Key Concepts

3.1 Measurements and Their Uncertainty

- Measurements are fundamental to the experimental sciences.
- To evaluate accuracy, the measured value must be compared to the correct value. To evaluate precision, you must compare the values of repeated measurements.
- Calculated answers often depend on the number of significant figures in the values used in the calculation.
- In general, a calculated answer cannot be more precise than the least precise measurement from which it was calculated.

3.2 The International System of Units

- Five commonly used SI base units are the meter, kilogram, kelvin, second, and mole.
- Common metric units of length: cm, m, km. Common metric units of volume: μL, mL, L, cm^3. Common metric units of mass: mg, g, kg. Common units of temperature: °C and K. Common units of energy: J and cal.

3.3 Conversion Problems

- Multiplying by a conversion factor does not change the actual size of a measurement.
- Dimensional analysis provides an alternative approach to problem solving.
- Conversion problems are easily solved using dimensional analysis.

3.4 Density

- Density is an intensive property that depends only on the composition of a substance.
- The density of a substance generally decreases as its temperature increases.

Vocabulary

- absolute zero (p. 77)
- accepted value (p. 65)
- accuracy (p. 64)
- calorie (cal) (p. 79)
- Celsius scale (p. 77)
- conversion factor (p. 80)
- density (p. 90)
- dimensional analysis (p. 81)
- energy (p. 79)
- error (p. 65)
- experimental value (p. 65)
- gram (g) (p. 76)
- International System of Units (SI) (p. 73)
- joule (J) (p. 79)
- Kelvin scale (p. 77)
- kilogram (kg) (p. 76)
- liter (L) (p. 75)
- measurement (p. 63)
- meter (m) (p. 74)
- percent error (p. 65)
- precision (p. 64)
- scientific notation (p. 63)
- significant figures (p. 66)
- temperature (p. 77)
- weight (p. 76)

Key Equations

- Error = experimental value − accepted value
- $\text{Percent error} = \dfrac{|\text{error}|}{\text{accepted value}} \times 100\%$
- $K = °C + 273$ and $°C = K - 273$
- $1\ J = 0.2390$ cal and 1 cal $= 4.184\ J$
- $\text{Density} = \dfrac{\text{mass}}{\text{volume}}$

Organizing Information

Use these terms to construct a concept map that organizes the major ideas of this chapter.

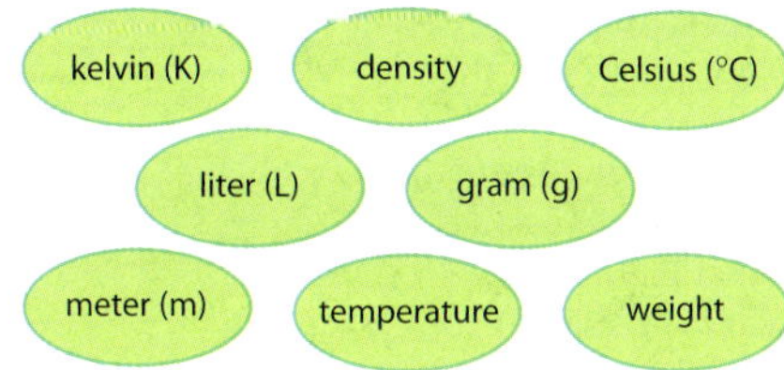

Concept Map 3 Solve the concept map with the help of an interactive guided tutorial.

with ChemASAP

CHAPTER 3 Study Guide

Study Tip

Choose a Quiet Place Tell students that being able to concentrate while studying will increase their comprehension. Therefore, a quiet place where they will not be distracted can help them achieve a better grasp of the subject material and will make their study time more efficient.

If your class subscribes to the Interactive Textbook with ChemASAP, your students can go online to access an interactive version of the Student Edition and a self-test.

with ChemASAP

Chapter Resources

Print

- **Core Teaching Resources,** Chapter 3, *Practice Problems, Vocabulary Review, Quiz, Chapter Test A, Chapter Test B*

Technology

- **Computer Test Bank,** Chapter 3 Test
- **Interactive Textbook with ChemASAP,** Chapter 3

CHAPTER 3 Assessment

Reviewing Content

57. Lissa: inaccurate and imprecise
Lamont: accurate and precise
Leigh Anne: inaccurate and precise

58. **a.** infinite **b.** 4 **c.** infinite **d.** infinite

59. **a.** 98.5 L **b.** 0.000 763 cg **c.** 57.0 m **d.** 12.2°C

60. **a.** 43 g **b.** 225.8 L **c.** 92.0 kg **d.** 32.4 m^3

61. (59) **a.** 9.85×10^1 L **b.** 7.63×10^{-4} cg **c.** 5.70×10^1 m **d.** 1.22×10^1°C
(60) **a.** 4.3×10^1 g **b.** 2.258×10^2 L **c.** 9.20×10^1 kg **d.** 3.24×10^1 m^3

62. The error is the difference between the experimental value and the accepted value. The percent error is the error divided by the accepted value multiplied by 100.

63. **a.** second **b.** meter **c.** kelvin **d.** kilogram

64. pm, nm, μm, mm, cm, dm, m, km; 1 km = 10^3 m, 1 dm = 10^{-1} m, 1 cm = 10^{-2} m, 1 mm = 10^{-3} m, 1 μm = 10^{-6} m, 1 nm = 10^{-9} m, 1 pm = 10^{-12} m

65. **a.** 2.4 mm **b.** 14.33 cm **c.** 27.50 cm

66. 1235 K

67. conversion factor

68. They must equal one another.

69. The unit of the conversion factor in the denominator must be identical to the unit in the given measurement or the previous conversion factor.

70. **a.** 1.57 s **b.** 4.27×10^4 mL **c.** 2.61×10^{-4} mm **d.** 6.5×10^2 dm **e.** 6.42×10^{-3} kg **f.** 8.25×10^9 ng

71. **a.** 7.3 μL/s **b.** 78.6 mg/mm^2 **c.** 1.54 g/cm^3

72. 10^6 mL

73. **a.** 2.83×10^2 mg **b.** 0.283 g **c.** 2.83×10^{-4} kg **d.** 6.6 g **e.** 6.6×10^2 cg **f.** 6.6×10^{-3} kg **g.** 2.8×10^{-1} mg **h.** 2.8×10^{-2} cg **i.** 2.8×10^{-7} kg

74. density = mass/volume

75. Yes, neither mass nor volume of a solid or liquid changes appreciably with location.

CHAPTER 3 Assessment

Reviewing Content

3.1 Measurements and Their Uncertainty

57. Three students made multiple weighings of a copper cylinder, each using a different balance. Describe the accuracy and precision of each student's measurements if the correct mass of the cylinder is 47.32 g.

	Mass of Cylinder (g)		
	Lissa	Lamont	Leigh Anne
Weighing 1	47.13	47.45	47.95
Weighing 2	47.94	47.39	47.91
Weighing 3	46.83	47.42	47.89
Weighing 4	47.47	47.41	47.93

58. How many significant figures are in each underlined measurement?
a. <u>60 s</u> = 1 min **b.** <u>47.70 g</u> of copper
c. 1 km = <u>1000 m</u> **d.** <u>25 computers</u>

59. Round off each of these measurements to three significant figures.
a. 98.473 L **b.** 0.000 763 21 cg
c. 57.048 m **d.** 12.17°C

60. Round off each of the answers correctly.
a. 8.7 g + 15.43 g + 19 g = 43.13 g
b. 853.2 L − 627.443 L = 225.757 L
c. 38.742 kg ÷ 0.421 = 92.023 75 kg
d. 5.40 m × 3.21 m × 1.871 m = 32.431 914 m^3

61. Express each of the rounded-off answers in Questions 59 and 60 in scientific notation.

62. How are the *error* and the *percent error* of a measurement calculated?

3.2 The International System of Units

63. List the SI base unit of measurement for each of these quantities.
a. time **b.** length
c. temperature **d.** mass

64. Order these units from smallest to largest: cm, μm, km, mm, m, nm, dm, pm. Then give each measurement in terms of meters.

65. Measure each of the following dimensions using a unit with the appropriate prefix.
a. the height of this letter I
b. the width of Table 3.3
c. the height of this page

66. The melting point of silver is 962°C. Express this temperature in kelvins.

3.3 Conversion Problems

67. What is the name given to a ratio of two equivalent measurements?

68. What must be true for a ratio of two measurements to be a conversion factor?

69. How do you know which unit of a conversion factor must be in the denominator?

70. Make the following conversions.
a. 157 cs to seconds
b. 42.7 L to milliliters
c. 261 nm to millimeters
d. 0.065 km to decimeters
e. 642 cg to kilograms
f. 8.25×10^2 cg to nanograms

71. Make the following conversions.
a. 0.44 mL/min to microliters per second
b. 7.86 g/cm^2 to milligrams per square millimeter
c. 1.54 kg/L to grams per cubic centimeter

72. How many milliliters are contained in 1 m^3?

73. Complete this table so that all the measurements in each row have the same value.

mg	g	cg	kg
(a)	(b)	28.3	(c)
6.6×10^3	(d)	(e)	(f)
(g)	2.8×10^{-4}	(h)	(i)

3.4 Density

74. What equation is used to determine the density of an object?

75. Would the density of a person be the same on the surface of Earth and on the surface of the moon? Explain.

76. A shiny, gold-colored bar of metal weighing 57.3 g has a volume of 4.7 cm^3. Is the bar of metal pure gold?

77. Three balloons filled with neon, carbon dioxide, and hydrogen are released into the atmosphere. Using the data in Table 3.6 on page 90, describe the movement of each balloon.

76. No, the density of the metal bar is 12 g/cm^3, but the density of gold is 19 g/cm^3.

77. The carbon dioxide–filled balloon would sink. The neon and hydrogen-filled balloons would rise, the hydrogen at a much faster rate.

Understanding Concepts

78. List two possible reasons for reporting precise, but inaccurate, measurements.

79. Rank these numbers from smallest to largest.
    a. 5.3×10^4
    b. 57×10^3
    c. 4.9×10^{-2}
    d. 0.0057
    e. 5.1×10^{-3}
    f. 0.0072×10^2

80. Comment on the accuracy and precision of these basketball free-throw shooters.
    a. 99 of 100 shots are made.
    b. 99 of 100 shots hit the front of the rim and bounce off.
    c. 33 of 100 shots are made; the rest miss.

81. Fahrenheit is a third temperature scale. Plot the data in the table and use the graph to derive an equation for the relationship between the Fahrenheit and Celsius temperature scales.

Example	°C	°F
Melting point of selenium	221	430
Boiling point of water	100	212
Normal body temperature	37	98.6
Freezing point of water	0	32
Boiling point of chlorine	−34.6	−30.2

82. Which would melt first, germanium with a melting point of 1210 K or gold with a melting point of 1064°C?

83. Write six conversion factors involving these units of measure: $1 \text{ g} = 10^2 \text{ cg} = 10^3 \text{ mg}$.

84. A 2.00-kg sample of bituminous coal is composed of 1.30 kg of carbon, 0.20 kg of ash, 0.15 kg of water, and 0.35 kg of volatile (gas-forming) material. Using this information, determine how many kilograms of carbon are in 125 kg of this coal.

85. A piece of wood sinks in ethanol but floats in gasoline. Give a range of possible densities for the wood.

86. The density of dry air measured at 25°C is 1.19×10^{-3} g/cm^3. What is the volume of 50.0 g of air?

87. A flask that can hold 158 g of water at 4°C can hold only 127 g of ethanol at the same temperature. What is the density of ethanol?

88. A watch loses 0.15 s every minute. How many minutes will the watch lose in 1 day?

89. A tank measuring 28.6 cm by 73.0 mm by 0.72 m is filled with olive oil. The oil in the tank has a mass of 1.38×10^4 g. What is the density of olive oil in kilograms per liter?

90. Alkanes are a class of molecules that have the general formula C_nH_{2n+2}, where n is an integer (whole number). The table below gives the boiling points for the first five alkanes with an odd number of carbon atoms. Using the table, construct a graph with number of carbon atoms on the x-axis.

Boiling point (°C)	Number of carbon atoms
−162.0	1
−42.0	3
36.0	5
98.0	7
151.0	9

    a. What are the approximate boiling points for the C_2, C_4, C_6, and C_8 alkanes?
    b. Which of these nine alkanes are gases at room temperature (20°C)?
    c. How many of these nine alkanes are liquids at 350 K?
    d. What is the approximate increase in boiling point per additional carbon atom in these alkanes?

91. Earth is approximately 1.5×10^8 km from the sun. How many minutes does it take light to travel from the sun to Earth? The speed of light is 3.0×10^8 m/s.

92. What is the mass of a cube of aluminum that is 3.0 cm on each edge? The density of aluminum is 2.7 g/cm^3.

93. The average density of Earth is 5.52 g/cm^3. Express this density in units of kg/dm^3.

94. How many kilograms of water (at 4°C) are needed to fill an aquarium that measures 40.0 cm by 20.0 cm by 30.0 cm?

Understanding Concepts

78. Improper calibration or improper use of the measuring device.
79. e, d, c, f, a, b
80. a. accurate and precise
    b. inaccurate and precise
    c. inaccurate and imprecise
81. °F = 1.8°C + 32
82. germanium
83. 1 g/10^2 cg, 10^2 cg/1 g, 1 g/10^3 mg, 10^3 mg/1 g, 10^2 cg/10^3 mg, 10^3 mg/10^2 cg
84. 81.3 kg
85. 0.69 – 0.789 g/cm^3
86. 4.20 x 10^4 cm^3
87. 0.804 g/cm^3
88. 3.6 min lost
89. 0.92 kg/L
90. a. C_2 = −90°C, C_4 = 0°C, C_6 = 70°C, C_8 = 125°C
    b. C_1 through C_4
    c. three
    d. From C_1 through C_9, the increase is approximately 39°C/additional carbon. Over the range C_3 through C_9, the increase is approximately 32°C/additional carbon.
91. 8.3 min
92. 73 g
93. 5.52 kg/dm^3
94. 24.0 kg of water

CHAPTER 3

Assessment continued

Critical Thinking

95. Yes, the mass of an object is constant. The weight of an object varies with location.

96. 16.3 g

97. 31.1 m/s

98. You do not change your estimate. Counting the extra 15 ducks would suggest greater precision than was used in the estimate.

99. Answers will vary. Lakes would freeze solid from the bottom up; aquatic life would be destroyed; possible climate changes.

100. Gasoline is a mixtures and has a variable composition.

101.

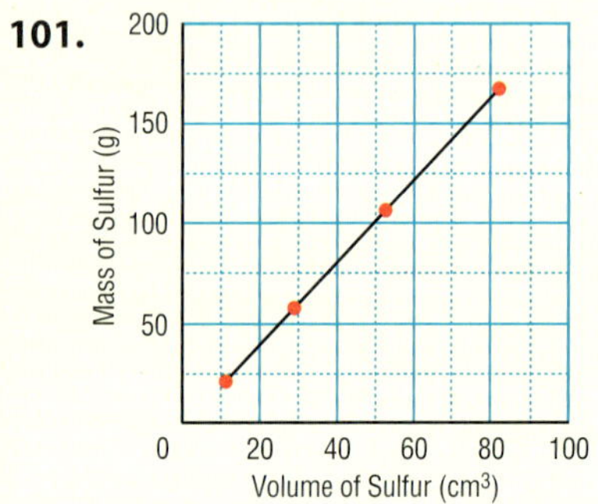

density of sulfur = 2.1 g/cm^3

102. a. The oxygen-filled balloon will sink; the nitrogen-filled balloon will rise.

b. nitrogen; The "weighted" average density of air is closer to the density of nitrogen.

Concept Challenge

103. Volume of iron = 45.1 cm^3
Mass of lead = 514 g Pb

104. 8.0 g Sr

105. 1.8×10^3 kg

106. a. 85 g **b.** 1.3 g/mL

107. 1.79 mL

108. a. corn oil on top of water on top of mercury

b. The density of sugar is greater than the density of water and less than the density of mercury; it floats between the layers of mercury and water.

c. The sugar cube will dissolve in the water over time.

CHAPTER 3

Assessment continued

Critical Thinking

95. Is it possible for an object to lose weight but at the same time not lose mass? Explain.

96. One of the first mixtures of metals, called amalgams, used by dentists for tooth fillings consisted of 26.0 g of silver, 10.8 g of tin, 2.4 g of copper, and 0.8 g of zinc. How much silver is in a 25.0 g sample of this amalgam?

97. A cheetah can run 112 km/h over a 100-m distance. What is this speed in meters per second?

98. You are hired to count the number of ducks on three northern lakes during the summer. In the first lake, you estimate 500,000 ducks, in the second 250,000 ducks, and in the third 100,000 ducks. You write down that you have counted 850,000 ducks. As you drive away, you see 15 ducks fly in from the south and land on the third lake. Do you change the number of ducks that you report? Justify your answer.

99. What if ice were more dense than water? It would certainly be easier to pour water from a pitcher of ice cubes and water. Can you imagine situations of more consequence?

100. Why is there a range of values given for the density of gasoline on Table 3.6 on page 90?

101. Plot these data that show how the mass of sulfur increases with an increase in volume. Determine the density of sulfur from the slope of the line.

Volume of sulfur (cm^3)	Mass of sulfur (g)
11.4	23.5
29.2	60.8
55.5	115
81.1	168

102. At 20°C, the density of air is 1.20 g/L. Nitrogen's density is 1.17 g/L. Oxygen's density is 1.33 g/L.

a. Will balloons filled with oxygen and balloons filled with nitrogen rise or sink in air?

b. Air is mainly a mixture of nitrogen and oxygen. Which gas is the main component? Explain.

Concept Challenge

103. The mass of a cube of iron is 355 g. Iron has a density of 7.87 g/cm^3. What is the mass of a cube of lead that has the same dimensions?

104. Sea water contains 8.0×10^{-1} cg of the element strontium per kilogram of sea water. Assuming that all the strontium could be recovered, how many grams of strontium could be obtained from one cubic meter of sea water? Assume the density of sea water is 1.0 g/mL.

105. The density of dry air at 20°C is 1.20 g/L. What is the mass of air, in kilograms, of a room that measures 25.0 m by 15.0 m by 4.0 m?

106. Different volumes of the same liquid were added to a flask on a balance. After each addition of liquid, the mass of the flask with the liquid was measured. Graph the data using mass as the dependent variable. Use the graph to answer these questions.

Volume (mL)	Mass (g)
14	103.0
27	120.4
41	139.1
55	157.9
82	194.1

a. What is the mass of the flask?

b. What is the density of the liquid?

107. A 34.5-g gold nugget is dropped into a graduated cylinder containing water. By how many milliliters does the measured volume increase? The density of water is 1.0 g/mL. The density of gold is 19.3 g/cm^3.

108. Equal amounts of mercury, water, and corn oil are added to a beaker.

a. Describe the arrangement of the layers of liquids in the beaker.

b. A small sugar cube is added to the beaker. Describe its location.

c. What change will occur to the sugar cube over time?

Standardized Test Prep

Test-Taking Tip

Interpreting Diagrams Diagrams present information in a visual format. Before you answer questions about a diagram, be sure to study the diagram carefully. Ask yourself some questions: *What is the diagram showing? What does the diagram tell me?*

Select the choice that best answers each question or completes each statement.

1. Which of these series of units is ordered from smallest to largest?
a. μg, cg, mg, kg
b. mm, dm, m, km
c. μs, ns, cs, s
d. nL, mL, dL, cL

2. Which answer represents the measurement 0.00428 g rounded to two significant figures?
a. 4.28×10^3 g **b.** 4.3×10^{-3} g
c. 4.3×10^3 g **d.** 4.0×10^{-3} g

3. An over-the-counter medicine has 325 mg of its active ingredient per tablet. How many grams does this mass represent?
a. 325,000 g **b.** 32.5 g
c. 3.25 g **d.** 0.325 g

4. If 10^4 μm = 1 cm, how many μm^3 = 1 cm^3?
a. 10^4 **b.** 10^6
c. 10^8 **d.** 10^{12}

5. How many meters does a car moving at 95 km/h travel in 1.0 s?
a. 1.6 m **b.** 340 m
c. 1600 m **d.** 26 m

6. If a substance contracts when it freezes, its
a. density will remain the same.
b. density will increase.
c. density will decrease.
d. change in density cannot be predicted.

For Questions 7–9, identify the known and the unknown. Include units in your answers.

7. The density of water is 1.0 g/mL. How many deciliters of water will fill a 0.5-L bottle?

8. A clock loses 4 minutes every day. How many seconds does the clock lose in 1 minute?

9. A graduated cylinder contains 44.2 mL of water. A 48.6-g piece of metal is carefully dropped into the cylinder. When the metal is completely covered with water, the water rises to the 51.3-mL mark. What is the density of the metal?

Use the atomic windows below to answer Questions 10 and 11.

a.
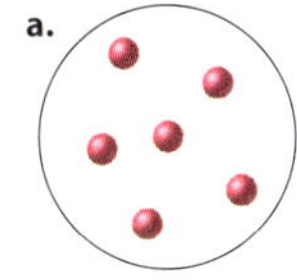
b.
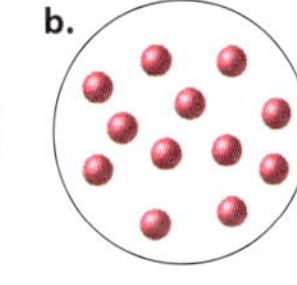
c.
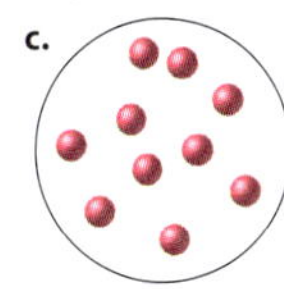

The atomic windows represent particles of the same gas occupying the same volume at the same temperature. The systems differ only in the number of gas particles per unit volume.

10. List the windows in order of decreasing density.

11. Compare the density of the gas in window (a) to the density of the gas in window (b).

For each question there are two statements. Decide whether each statement is true or false. Then decide whether Statement II is a correct explanation for Statement I.

Statement I		Statement II
12. There are five significant figures in the measurement 0.00450 m.	BECAUSE	All zeros to the right of a decimal point in a measurement are significant.
13. Precise measurements will always be accurate measurements.	BECAUSE	A value that is measured 10 times in a row must be accurate.
14. A temperature in kelvins is always numerically larger than the same temperature in degrees Celsius.	BECAUSE	A temperature in kelvins equals a temperature in degrees Celsius plus 273.

Standardized Test Prep

1. b
2. b
3. d
4. d
5. d
6. b
7. Known: volume in liters; unknown: volume in deciliters
8. Known: minutes lost per day; unknown: seconds lost per minute
9. Knowns: mass of metal, initial and final readings of the volume; unknown: density of metal
10. b, c, a
11. The density in a is one-half the density in b.
12. False, False
13. False, False
14. True, True, correct explanation

Atomic Structure Planning Guide

SECTION OBJECTIVES	STANDARDS NATIONAL	STANDARDS STATE	ACTIVITIES and LABS
4.1 Defining the Atom, pp. 101–103 1 block or 2 periods **4.1.1 Describe** Democritus's ideas about atoms. **4.1.2 Explain** Dalton's atomic theory. **4.1.3 Identify** what instrument is used to observe individual atoms.	A-2, B-1, B-2, B-3, E-2, G-2, G-3		**SE** Inquiry Activity: Electric Charge, p. 100 L2
4.2 Structure of the Nuclear Atom, pp. 104–109 1 block or 2 periods **4.2.1 Identify** three types of subatomic particles. **4.2.2 Describe** the structure of atoms according to the Rutherford atomic model.	A-1, A-2, B-1, B-4, E-2, G-2, G-3		**SE** Quick Lab: Using Inference: The Black Box, p. 107 L2 **TE** Teacher Demo: Obsesrving Cathode Rays, p. 105 L2 **TE** Class Activity: Atomic Model Timeline, p. 106 L2 **LM** Lab 5: Atomic Structure: Rutherford's Experiment L2 **LP** Lab Practical 5-1: Modeling Atomic Structure L2 **VCL** Lab 4: Thomson Cathode Ray Tube Experiment L2 **VCL** Lab 5: Millikan Oil Drop Experiment L2 **VCL** Lab 6: Atomic Structure: Rutherford's Experiment L2
4.3 Distinguishing Among Atoms, pp. 110–120 1 block or 2 periods **4.3.1 Explain** what makes elements and isotopes different from each other. **4.3.2 Calculate** the number of neutrons in an atom. **4.3.3 Calculate** the atomic mass of an element. **4.3.4 Explain** why chemists use the periodic table.	A-1, A-2, B-1, B-2, G-1, G-2		**SE** Small-Scale Lab: The Atomic Mass of Candium, p. 120 L2 **TE** Class Activity: Applications of Isotopes, p. 112 L2

Ability Levels	
L1	For students who need additional help
L2	For all students
L3	For students who need to be challenged

Components

SE	Student Edition	**LM**	Laboratory Manual	**GCP**	Graphing Calculator Problems	**T**	Transparencies
TE	Teacher's Edition	**SSLM**	Small-Scale Chemistry Lab Manual	**TTT**	Test-Taking Tips with Transparencies, Grades 9–12 Science	**CA**	Chem Alive and Chemistry Field Trips DVD
GRSW	Guided Reading and Study Workbook	**LP**	Laboratory Practicals	**STP**	Standardized Test Preparation Workbook	**iT**	Interactive Textbook with ChemASAP
CTR	Core Teaching Resources	**PLM**	Probeware Lab Manual			**GO**	Go Online Internet Resources
SSG	Spanish Study Guide	**VCL**	Virtual Chemistry Labs				
		CTB	Computer Test Bank				

RESOURCES PRINT and TECHNOLOGY		PROBLEMS and ASSESSMENT
GRSW Chapter 4, Section 1	L1	**SE** Section 4.1 Assessment, p. 103
T T43–T44	L2	**iT** Section 4.1
GRSW Chapter 4, Section 2	L1	**SE** Section 4.2 Assessment, p. 108
T T45–T47	L2	**iT** Section 4.2
GO Chapter 4, Section 2	L2	
GRSW Chapter 4, Section 3	L1	**CP** 4.1, Understanding Atomic Number
T T48–T56	L2	**CP** 4.2, Writing Chemical Symbols of Isotopes
GO Chapter 4, Section 3	L2	**CP** 4.3, Using Atomic Mass to Determine the Relative Abundance of Isotopes
GO Chapter 4, Section 3	L2	**SP** 4.1, Determining the Composition of an Atom
		SP 4.2, Calculating Atomic Mass
		SE Section 4.3 Assessment, p. 119
		iT Section 4.3

Web Code: cdn-1042
Web Code: cdn-1043

Web Code: cdb-1043

MATERIALS List

Quantities for each group

STUDENT EDITION

Inquiry Activity, p. 100
clear plastic tape, metric ruler

Quick Lab, p. 108
box containing a regularly shaped object fixed in place and a loose marble

Small-Scale Lab, p. 120
several pieces of three kinds of candy, 3.5-ounce plastic cups, balance

TEACHER'S EDITION

Teacher Demo, p. 105
cathode ray tube, magnet

Chapter Assessment

CHAPTER ASSESSMENT

SE Chapter Assessment, pp. 122–124
CTB Chapter 4
iT Chapter 4

STANDARDIZED TEST PREP

SE Chapter 4, p. 125
STP Topic 1
iT Interactive Textbook with Assessment at PHSchool.com

CHAPTER 4

Atomic Structure

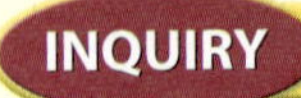

Electric Charge 

Purpose

Students observe the repulsion of like charges and the attraction of unlike charges.

Prep Time 5 minutes
Class Time 10 minutes

Think About It

1. The pieces of tape are drawn together because they have opposite charges.
2. The pieces of tape from Steps 1 and 2 do not have the same charge. When tape is pulled from the desk, it receives a negative charge. When tape is pulled through the fingers, it receives a positive charge.

CHAPTER 4

Atomic Structure

A scanning tunneling microscope was used to produce this color-enhanced image of iron atoms.

INQUIRY Activity

Electric Charge

Materials

four 25-cm lengths of clear plastic tape and a metric ruler

Procedure

1. Firmly stick two of the 25-cm pieces of tape side-by-side, about 10 cm apart, on your desktop. Leave 2 to 3 cm of tape sticking over the edge of the desk. Grasp the free ends of the tapes and pull sharply upward to peel the tape pieces off of the desk. Slowly bring the pieces, which have similar charges, toward one another. Record your observations.
2. Pull the third and fourth pieces of tape between your thumb and forefinger several times, as if trying to clean each one. Slowly bring these two pieces of tape, which now have similar charges, toward one another. Record your observations.

Think About It

1. Predict what might happen if you brought a piece of tape pulled from your desktop close to a piece of tape pulled between your fingers. Try it, and see what happens. Explain your observations.
2. Do you think the pieces of tape used in Step 1 have the same charge as those used in Step 2? Explain.

4.1 Defining the Atom

Connecting to Your World

It often helps to take a closer look. For example, you might walk up to a sign or a poster in order to make out the details. Or you might bring a set of binoculars to a sports stadium so that you can zoom in on the action. The lab technician shown here is using a magnifying lens to examine a bacterial culture in a petri dish. Scientists use many different devices that enhance their ability to see. However, scientists can't always see the details of what they study. In such cases, scientists try to obtain experimental data that helps fill in the picture.

Guide for Reading

Key Concepts

- How did Democritus describe atoms?
- How did John Dalton further Democritus's ideas on atoms?
- What instruments are used to observe individual atoms?

Vocabulary

atom

Dalton's atomic theory

Reading Strategy

Summarizing As you read about early atomic models, summarize the main ideas in the text that follow each red and blue heading.

Early Models of the Atom

Have you ever been asked to believe in something you couldn't see? Using your unaided eyes, you cannot see the tiny fundamental particles that make up matter. Yet all matter is composed of such particles, which are called atoms. An **atom** is the smallest particle of an element that retains its identity in a chemical reaction.

The concept of the atom intrigued a number of early scholars. Although these philosophers and scientists could not observe individual atoms, they still were able to propose ideas on the structure of atoms.

Democritus's Atomic Philosophy The Greek philosopher Democritus (460 B.C.–370 B.C.) was among the first to suggest the existence of atoms. **Democritus believed that atoms were indivisible and indestructible.** Although Democritus's ideas agreed with later scientific theory, they did not explain chemical behavior. They also lacked experimental support because Democritus's approach was not based on the scientific method.

Figure 4.1 Democritus believed that matter consisted of tiny, indivisible, unchangeable particles called atoms. His ideas were later challenged by the Greek philosophers Plato and Aristotle.

Section Resources

Print

- **Guided Reading and Study Workbook,** Section 4.1
- **Core Teaching Resources,** Section 4.1 Review
- **Transparencies,** T43–T44

Technology

- **Interactive Textbook with ChemASAP,** Assessment 4.1

4.1

1 FOCUS

Objectives

4.1.1 Describe Democritus's ideas about atoms.

4.1.2 Explain Dalton's atomic theory

4.1.3 Identify what instrument is used to observe individual atoms.

Guide for Reading

Build Vocabulary 

List the Parts You Know Discuss what a scientific theory is. *(an explanation of the way the world works, based on observation)*

Reading Strategy 

Visualize Ask students to think about these questions as they read this chapter.

- **What could you do if someone asked you to describe something that is too small to see?**
- **How would you find out what it looked like?**

2 INSTRUCT

Connecting to Your World

Ask students to think of objects that require experimental data in order to "picture" them, either because they are small or inaccessible. *(Acceptable answers include: objects in deep space or deep underground)*

Early Models of the Atom

Discuss L2

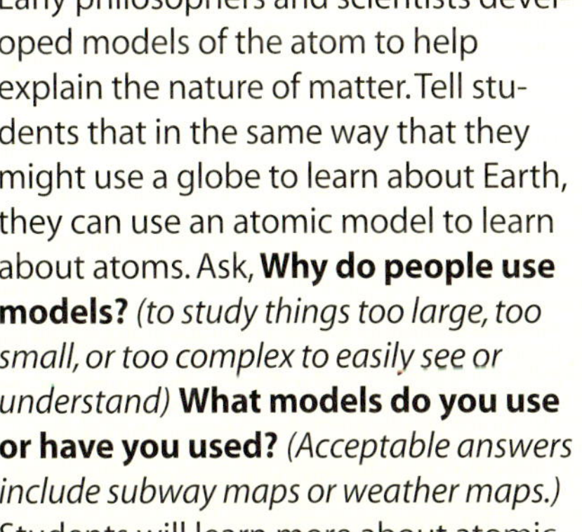

Early philosophers and scientists developed models of the atom to help explain the nature of matter. Tell students that in the same way that they might use a globe to learn about Earth, they can use an atomic model to learn about atoms. Ask, **Why do people use models?** *(to study things too large, too small, or too complex to easily see or understand)* **What models do you use or have you used?** *(Acceptable answers include subway maps or weather maps.)* Students will learn more about atomic models in Sections 4.2 and 5.1.

Section 4.1 (continued)

Discuss 

Explain that John Dalton's work, published in 1808, became the basis for modern atomic theory. This model represented the atom as a simple sphere with no internal structure. Point out how experimental data have been used to test and refine atomic theory over time.

Use Visuals 

Figure 4.2 Point out that the particles in Figure 4.2d represent a compound such as water. Ask, **How many atoms of A and of B form one particle in Figure 4.2d?** *(2 A and 1 B)* **Do you think that a mixture, shown in Figure 4.2c, always leads to compounds, shown in Figure 4.2d? Why?** *(No; some atoms may not readily combine.)*

Word *Origins* 

Atomize means to make something as small as an atom, or at least very small. Ask students to think of other words containing the suffix *-ize* and what they mean (e.g., customize, randomize, itemize).

Sizing Up the Atom

Use Visuals 

Figure 4.3 Point out that the image in Figure 4.3 is greatly magnified. Ask, **What is the radius of the circle of atoms in nanometers?** *(7.13 nm)* **In picometers?** *(7130 pm)*

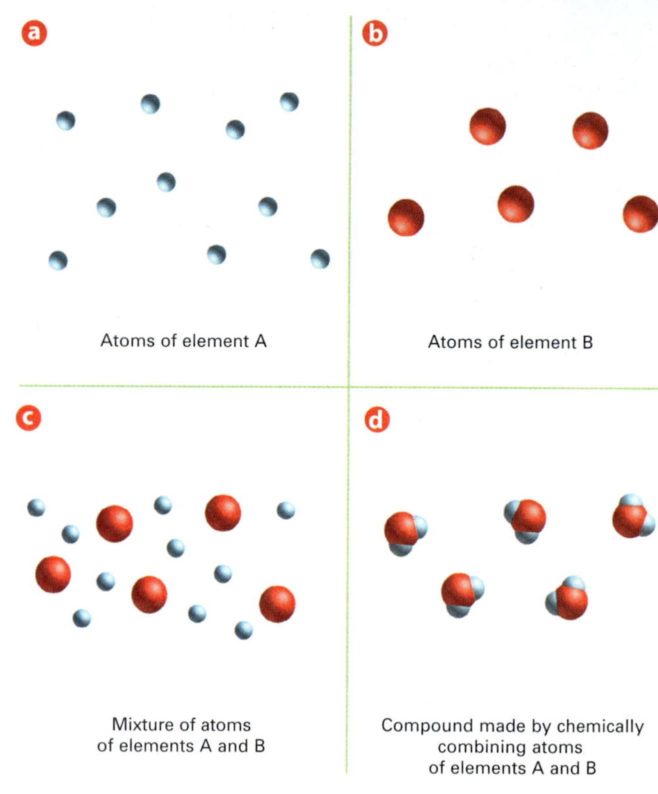

Figure 4.2 According to Dalton's atomic theory, an element is composed of only one kind of atom, and a compound is composed of particles that are chemical combinations of different kinds of atoms. (a) Atoms of element A are identical. (b) Atoms of element B are identical, but differ from those of element A. (c) Atoms of elements A and B can physically mix together. (d) Atoms of elements A and B can chemically combine to form a compound. **Interpreting Diagrams** ***How does a mixture of atoms of different elements differ from a compound?***

Word *Origins*

Atom comes from the Greek word *atomos,* meaning "indivisible." **If the suffix *-ize* means "to become like," what do you think the word *atomize* means?**

Dalton's Atomic Theory The real nature of atoms and the connection between observable changes and events at the atomic level were not established for more than 2000 years after Democritus. The modern process of discovery regarding atoms began with John Dalton (1766–1844), an English chemist and schoolteacher. **By using experimental methods, Dalton transformed Democritus's ideas on atoms into a scientific theory.** Dalton studied the ratios in which elements combine in chemical reactions. Based on the results of his experiments, Dalton formulated hypotheses and theories to explain his observations. The result was **Dalton's atomic theory,** which includes the ideas illustrated in Figure 4.2 and listed below.

1. All elements are composed of tiny indivisible particles called atoms.
2. Atoms of the same element are identical. The atoms of any one element are different from those of any other element.
3. Atoms of different elements can physically mix together or can chemically combine in simple whole-number ratios to form compounds.
4. Chemical reactions occur when atoms are separated, joined, or rearranged. Atoms of one element, however, are never changed into atoms of another element as a result of a chemical reaction.

What happens to atoms in a chemical reaction according to Dalton's atomic theory?

Facts and Figures

Scanning Tunneling Microscope

A scanning tunneling microscope (STM) holds a one-atom-sized tip over a sample that conducts electricity. Electric current flows between the tip and the sample. The current is stronger when the tip and the sample are close and weaker when they are farther apart. This change in voltage is interpreted to show surface topography.

Sizing up the Atom

A coin the size of a penny and composed of pure copper (Cu) illustrates Dalton's concept of the atom. Imagine grinding the copper coin into a fine dust. Each speck in the small pile of shiny red dust would still have the properties of copper. If by some means you could continue to make the copper dust particles smaller, you would eventually come upon a particle of copper that could no longer be divided and still have the chemical properties of copper. This final particle is an atom.

Copper atoms are very small. A pure copper coin the size of a penny contains about 2.4×10^{22} atoms. By comparison, Earth's population is only about 6×10^{9} people. There are about 4×10^{12} times as many atoms in the coin as there are people on Earth. If you could line up 100,000,000 copper atoms side by side, they would produce a line only 1 cm long!

The radii of most atoms fall within the range of 5×10^{-11} m to 2×10^{-10} m. Does seeing individual atoms seem impossible? **Despite their small size, individual atoms are observable with instruments such as scanning tunneling microscopes.** Figure 4.3 shows an image of iron atoms generated by a scanning tunneling microscope. Individual atoms can even be moved around and arranged in patterns. The ability to move individual atoms holds future promise for the creation of atomic-sized electronic devices, such as circuits and computer chips. This atomic-scale, or "nanoscale," technology could become essential to future applications in medicine, communications, solar energy, and space exploration.

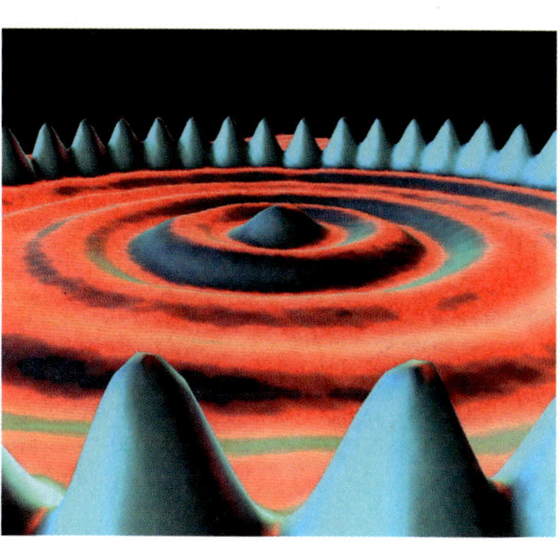

Figure 4.3 Scientists used a scanning tunneling microscope to generate this image of iron atoms, shown in blue. The radius of this circle of atoms is just 7.13×10^{-9} m.

4.1 Section Assessment

1. **Key Concept** How did Democritus characterize atoms?
2. **Key Concept** How did Dalton advance the atomic philosophy proposed by Democritus?
3. **Key Concept** What instrument can be used to observe individual atoms?
4. In your own words, state the main ideas of Dalton's atomic theory.
5. According to Dalton's theory, is it possible to convert atoms of one element into atoms of another? Explain.
6. Describe the range of the radii of most atoms in nanometers (nm).
7. A sample of copper with a mass of 63.5 g contains 6.02×10^{23} atoms. Calculate the mass of a single copper atom.

Connecting Concepts

Scientific Methods Reread the description of scientific methods in Section 1.3. Explain why the ideas on atoms proposed by Dalton constitute a theory, while the ideas proposed by Democritus do not.

Assessment 4.1 Test yourself on the concepts in Section 4.1.

with ChemASAP

3 ASSESS

Evaluate Understanding L2

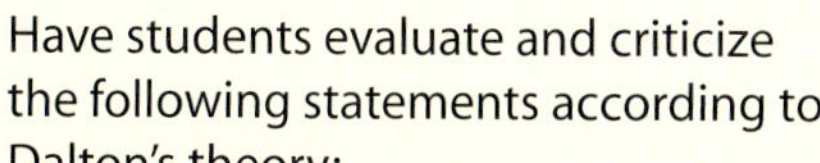

Have students evaluate and criticize the following statements according to Dalton's theory:

- **"All atoms are identical."** *(Dalton said: "All atoms of a given element are identical.")*
- **"Chemical reactions occur when atoms of one element change into atoms of another element."** *(Chemical reactions occur when atoms are separated, joined, or rearranged. The elemental identity of atoms does not change during chemical reactions.)*

Reteach L1

Review Dalton's model of the atom. Discuss how STM could be used to support or contest Dalton's theory. *(No matter how small the sample, you can see that there are many similar atoms in each sample. You could also see where atoms are and whether they change after a reaction.)*

Connecting Concepts

Democritus's ideas were not based on experimental results and did not explain chemical behavior. Dalton's ideas were empirically based and did explain chemical behavior; his experiments showed that the ratios in which elements combined were whole numbers.

If your class subscribes to the Interactive Textbook, use it to review key concepts in Section 4.1.

with ChemASAP

Section 4.1 Assessment

1. as indivisible and indestructible.
2. by using experimental methods
3. a scanning tunneling microscope
4. Answers should include the ideas that all matter is composed of atoms; atoms of different elements differ; and chemical change involves a rearrangement of atoms.
5. Atoms of one element are never changed into atoms of another element as a result of a chemical reaction.
6. 5×10^{-2} nm to 2×10^{-1} nm
7. 1.05×10^{-22} g

Answers to...

Figure 4.2 In a mixture of atoms of different elements, each element in the mixture retains its chemical properties. In a compound, different elements have been combined to form a new substance with chemical properties different from its component elements.

Atoms are separated, joined, or rearranged.

4.2

1 FOCUS

Objectives

4.2.1 **Identify** three types of subatomic particles.

4.2.2 **Describe** the structure of atoms according to the Rutherford atomic model.

Guide for Reading

Build Vocabulary L2

Word Parts Have students think of words that start with the same word parts as the three subatomic particles described in this section: neutron, electron, and proton. *(Sample answer: neutral, electric, protein)* Point out that using words that students already know may help them remember the meaning of new terms.

Reading Strategy L2

Outline Suggest that students use the headings in this section to write an outline. They should include the most important points under each heading.

2 INSTRUCT

Connecting to Your World

Ask students to describe characteristics of the modern cathode-ray tube (CRT)—a television picture tube. *(It has a source of rays at one end. A receiving area at the other end creates a display. Students may know it's a vacuum tube.)*

Subatomic Particles

Relate L2

Mass spectrometers operate like cathode ray tubes. A vaporized sample is bombarded by high-energy electrons. A magnetic field separates the positively charged products based on their mass/charge ratios.

4.2 Structure of the Nuclear Atom

Guide for Reading

Key Concepts

- What are three kinds of subatomic particles?
- How can you describe the structure of the nuclear atom?

Vocabulary

electrons
cathode ray
protons
neutrons
nucleus

Reading Strategy

Comparing and Contrasting When you compare and contrast things, you examine how they are alike and different. As you read, compare different subatomic particles by listing similarities and differences.

Connecting to Your World A simple but important device first used by scientists in the late nineteenth century, the cathode-ray tube would achieve its greatest fame as the picture tube of the common television set. Today, cathode-ray tubes are found in TVs, computer monitors, and many other devices with electronic displays. But more than 100 years ago, scientists were the only ones staring into the glow of a cathode-ray tube. Their observations provided important evidence about the structure of atoms.

Subatomic Particles

Much of Dalton's atomic theory is accepted today. One important change, however, is that atoms are now known to be divisible. They can be broken down into even smaller, more fundamental particles, called subatomic particles. **Three kinds of subatomic particles are electrons, protons, and neutrons.**

Electrons In 1897, the English physicist J. J. Thomson (1856–1940) discovered the electron. **Electrons** are negatively charged subatomic particles. Thomson performed experiments that involved passing electric current through gases at low pressure. He sealed the gases in glass tubes fitted at both ends with metal disks called electrodes. The electrodes were connected to a source of electricity, as shown in Figure 4.4. One electrode, the anode, became positively charged. The other electrode, the cathode, became negatively charged. The result was a glowing beam, or **cathode ray,** that traveled from the cathode to the anode.

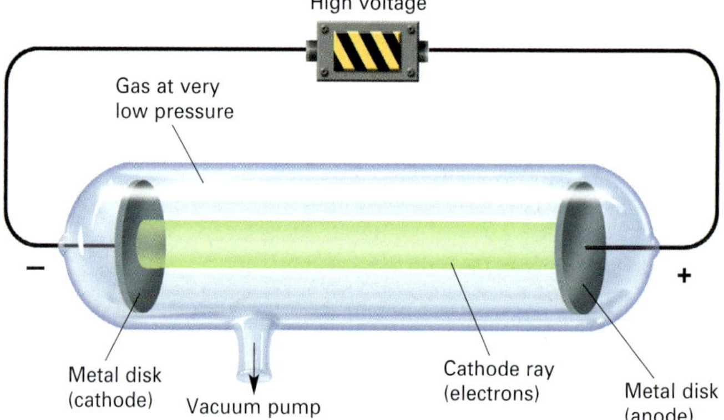

Figure 4.4 In a cathode-ray tube, electrons travel as a ray from the cathode (−) to the anode (+). A television tube is a specialized type of cathode-ray tube.

Section Resources

Print

- **Guided Reading and Study Workbook,** Section 4.2
- **Core Teaching Resources,** Section 4.2 Review
- **Transparencies,** T45–T47
- **Laboratory Manual,** Lab 5

Technology

- **Interactive Textbook with ChemASAP,** Animation 4, Assessment 4.2
- **Go Online,** Section 4.2
- **Virtual Chemistry Labs** 4–6

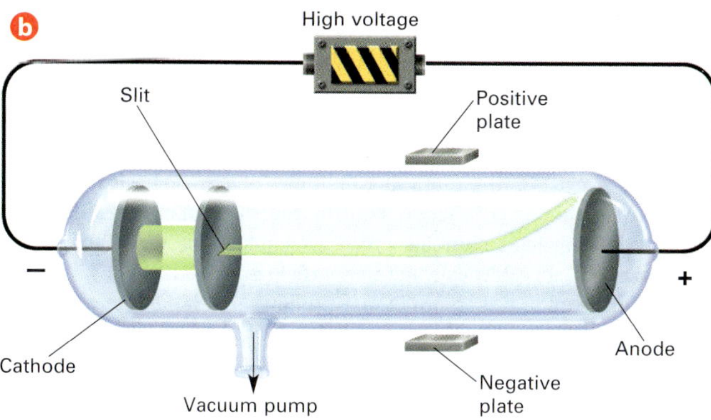

Figure 4.5 Thomson examined two ways that a cathode ray can be deflected: ⓐ by using a magnet, and ⓑ by using electrically charged plates. **Inferring** ***If a cathode ray is attracted to a positively charged plate, what can you infer about the charge of the particles that make up the cathode ray?***

Figure 4.5a shows how a cathode ray is deflected by a magnet. A cathode ray is also deflected by electrically charged metal plates, as shown in Figure 4.5b. A positively charged plate attracts the cathode ray, while a negatively charged plate repels it. Thomson knew that opposite charges attract and like charges repel, so he hypothesized that a cathode ray is a stream of tiny negatively charged particles moving at high speed. Thomson called these particles corpuscles; later they were named electrons.

To test his hypothesis, Thomson set up an experiment to measure the ratio of the charge of an electron to its mass. He found this ratio to be constant. In addition, the charge-to-mass ratio of electrons did not depend on the kind of gas in the cathode-ray tube or the type of metal used for the electrodes. Thomson concluded that electrons must be parts of the atoms of all elements.

The U.S. physicist Robert A. Millikan (1868–1953) carried out experiments to find the quantity of charge carried by an electron. Using this value and the charge-to-mass ratio of an electron measured by Thomson, Millikan calculated the mass of the electron. Millikan's values for electron charge and mass, reported in 1916, are very similar to those accepted today. An electron carries exactly one unit of negative charge, and its mass is 1/1840 the mass of a hydrogen atom.

For: Links on J.J. Thomson
Visit: www.SciLinks.org
Web Code: cdn-1042

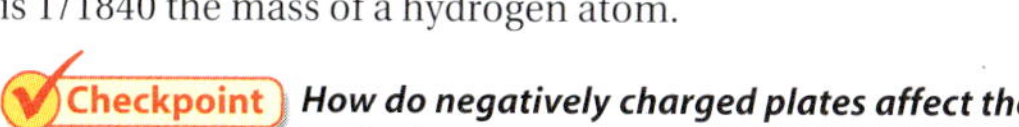

How do negatively charged plates affect the path of cathode rays?

Facts and Figures

Piecing Together Atomic Structure

It took another eighty years after the Dalton model for scientists to begin gathering clues about the structure of atoms. By 1887, the British scientist William Crookes knew that metal atoms contained negatively charged particles. In a cathode ray tube containing hydrogen gas at low pressure, he found that hydrogen contains positive charges. Scientists knew that the mass of the atom was greater than its proton content. They inferred that this extra mass was due to neutral particles, which were called neutrons. James Chadwick provided evidence for neutrons in 1932.

TEACHER Demo

Observing Cathode Rays L2

Purpose To demonstrate a cathode-ray tube and observe properties of cathode rays

Materials cathode-ray tube, magnet

Procedure Demonstrate a cathode-ray tube in class. Use a magnet to deflect the beam of particles. Review the components of a cathode-ray tube, and discuss the connection to television picture tubes and computer monitors.

Expected Outcome Students should be able to explain how the CRT works and see how the cathode ray is deflected by a magnetic field.

FYI

Based on his experiments with cathode rays, J.J. Thomson calculated the charge-to-mass ratio of the electron (e/m_e) to be 1.8×10^8 C/g. (The modern accepted value of e/m_e is 1.758820×10^8 C/g.) This ratio is constant regardless of the gas or the cathode material used in the cathode-ray tube. In 1909, Robert Millikan and his co-workers began a series of experiments in which they studied the motion of charged oil droplets in an electric field. The experiments yielded values for the charge of an electron (e) that were very close to the modern accepted value of 1.602177×10^{-19} C. Based on the values of e/m_e and e, Millikan was able to determine m_e, the mass of an electron. The modern accepted value of m_e is 9.109381×10^{-28} g

Download a worksheet on **J. J. Thomson** for students to complete, and find additional teacher support from NSTA SciLinks.

Answers to...

Figure 4.5 The particles are negative.

A negatively charged plate repels a cathode ray.

Section 4.2 (continued)

The Atomic Nucleus

Use Visuals L1

Table 4.1 Have students compare the masses and charges of the three elementary particles shown in Table 4.1. Ask, **What particles make up most of the mass of an atom?** *(protons and neutrons)* **Why are relative charges and mass useful in talking about subatomic particles?** *(The actual values are unwieldy numbers.)*

CLASS Activity

Atomic Model Timeline L2

Purpose To trace the history of atomic models, and examine the role of the scientific method in the development of such models

Materials Library or Internet access

Procedure Have students create a timeline that traces the development of the atomic model. Have them note the data that led to an existing model being changed.

Expected Outcome Students' timelines should list at least some of the atomic models shown in Figure 5.2 on pages 110 and 111. Students should be able to explain how certain scientific discoveries (e.g., Rutherford's gold foil experiment) resulted in the revision of the prevailing atomic model at the time.

Table 4.1

Properties of Subatomic Particles

Particle	Symbol	Relative charge	Relative mass (mass of proton = 1)	Actual mass (g)
Electron	e^-	1−	1/1840	9.11×10^{-28}
Proton	p^+	1+	1	1.67×10^{-24}
Neutron	n^0	0	1	1.67×10^{-24}

Protons and Neutrons If cathode rays are electrons given off by atoms, what remains of the atoms that have lost the electrons? For example, after a hydrogen atom (the lightest kind of atom) loses an electron, what is left? You can think through this problem using four simple ideas about matter and electric charges. First, atoms have no net electric charge; they are electrically neutral. (One important piece of evidence for electrical neutrality is that you do not receive an electric shock every time you touch something!) Second, electric charges are carried by particles of matter. Third, electric charges always exist in whole-number multiples of a single basic unit; that is, there are no fractions of charges. Fourth, when a given number of negatively charged particles combines with an equal number of positively charged particles, an electrically neutral particle is formed.

Considering all of this information, it follows that a particle with one unit of positive charge should remain when a typical hydrogen atom loses an electron. Evidence for such a positively charged particle was found in 1886, when Eugen Goldstein (1850–1930) observed a cathode-ray tube and found rays traveling in the direction opposite to that of the cathode rays. He called these rays canal rays and concluded that they were composed of positive particles. Such positively charged subatomic particles are called **protons.** Each proton has a mass about 1840 times that of an electron.

In 1932, the English physicist James Chadwick (1891–1974) confirmed the existence of yet another subatomic particle: the neutron. **Neutrons** are subatomic particles with no charge but with a mass nearly equal to that of a proton. Table 4.1 summarizes the properties of these subatomic particles. Although protons and neutrons are exceedingly small, theoretical physicists believe that they are composed of yet smaller subnuclear particles called *quarks*.

Figure 4.6 Born in New Zealand, Ernest Rutherford was awarded the Nobel Prize for Chemistry in 1908. His portrait appears on the New Zealand $100 bill.

The Atomic Nucleus

When subatomic particles were discovered, scientists wondered how these particles were put together in an atom. This was a difficult question to answer, given how tiny atoms are. Most scientists—including J.J. Thomson, the discoverer of the electron—thought it likely that the electrons were evenly distributed throughout an atom filled uniformly with positively charged material. In Thomson's atomic model, known as the "plum-pudding model," electrons were stuck into a lump of positive charge, similar to raisins stuck in dough. This model of the atom turned out to be short-lived, however, due to the groundbreaking work of Ernest Rutherford (1871–1937), a former student of Thomson.

Differentiated Instruction

Gifted and Talented L3

Have students use the Internet or library to find the original papers for the discoveries described in this chapter and write a report on what they have learned.

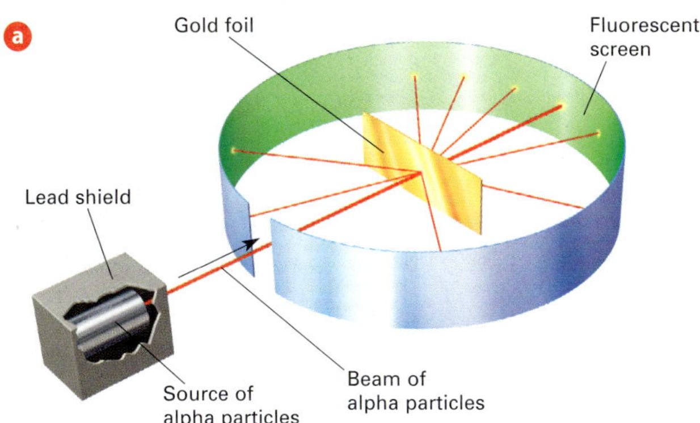

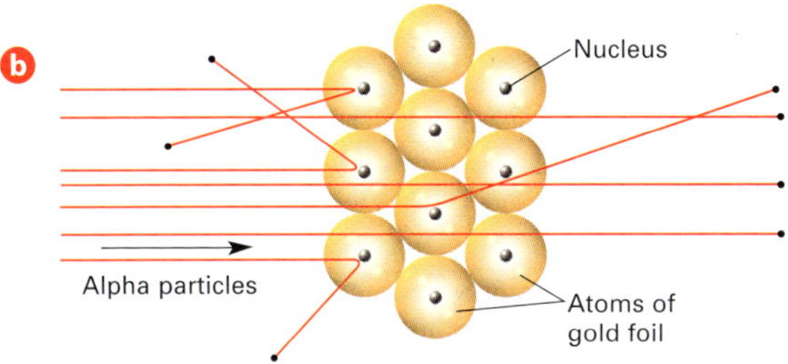

Figure 4.7 Rutherford's gold-foil experiment yielded evidence of the atomic nucleus. **a** Rutherford and his coworkers aimed a beam of alpha particles at a sheet of gold foil surrounded by a fluorescent screen. Most of the particles passed through the foil with no deflection at all. A few particles were greatly deflected. **b** Rutherford concluded that most of the alpha particles pass through the gold foil because the atom is mostly empty space. The mass and positive charge are concentrated in a small region of the atom. Rutherford called this region the nucleus. Particles that approach the nucleus closely are greatly deflected.

Rutherford's Gold-Foil Experiment In 1911, Rutherford and his coworkers at the University of Manchester, England, decided to test what was then the current theory of atomic structure. Their test used relatively massive alpha particles, which are helium atoms that have lost their two electrons and have a double positive charge because of the two remaining protons. In the experiment, illustrated in Figure 4.7, a narrow beam of alpha particles was directed at a very thin sheet of gold foil. According to the prevailing theory, the alpha particles should have passed easily through the gold, with only a slight deflection due to the positive charge thought to be spread out in the gold atoms.

To everyone's surprise, the great majority of alpha particles passed straight through the gold atoms, without deflection. Even more surprisingly, a small fraction of the alpha particles bounced off the gold foil at very large angles. Some even bounced straight back toward the source. Rutherford later recollected, "This is almost as incredible as if you fired a 15-inch shell at a piece of tissue paper and it came back and hit you."

The Rutherford Atomic Model Based on his experimental results, Rutherford suggested a new theory of the atom. He proposed that the atom is mostly empty space, thus explaining the lack of deflection of most of the alpha particles. He concluded that all the positive charge and almost all the mass are concentrated in a small region that has enough positive charge to account for the great deflection of some of the alpha particles. He called this region the nucleus. The **nucleus** is the tiny central core of an atom and is composed of protons and neutrons.

Animation 4 Take a look at Rutherford's gold-foil experiment, its results, and its conclusions.

with ChemASAP

Discuss 

The ratio of the size of the nucleus to the size of an atom is about 10^{-5}. Discuss how small the nucleus is compared to the entire atom. Explain that if a housefly sitting on second base in a baseball stadium represented the nucleus of an atom, the rest of the atom would be the size of the stadium.

Quick LAB

Using Inference: The Black Box L2

Objective After completing this activity, students will be able to:

- determine the shape of the hidden object by analyzing the rebound paths of a marble rolled at the object.

Skills Focus Observing, inferring

 **Prep Time** 5 minutes

Materials box containing a regularly shaped object fixed in place and a loose marble

Advance Prep Cut geometric shapes-such as a triangle, circle, or L, from a sheet of 1-inch plastic foam.

Class Time 10 minutes

Expected Outcome Students' inferences may or may not be different for the same object.

Analyze and Conclude

1. See Expected Outcome.
2. The activity simulates the strategy that Rutherford used to probe the structure of metal atoms. Like the students, Rutherford and his coworkers were also faced with the problem of identifying properties of an object not visible to the eye.

For Enrichment 

Make a more challenging black box for students who have an easy time with the simple boxes. Put two single objects in one box, or a single object with a complex shape.

3 ASSESS

Evaluate Understanding L2

Have students describe the discoveries of Thomson, Millikan, and Rutherford, and relate how those discoveries led to the current understanding of atomic structure.

Reteach L1

Have students review Table 4.1. Ask them to compare electrons, protons, and neutrons, using a table or a diagram.

Writing Activity

In Rutherford's gold-foil experiment, a narrow beam of alpha particles was directed at a very thin sheet of gold foil. Most of the alpha particles passed straight through the gold atoms without being deflected, but a small fraction bounced off the foil at very large angles. To explain his results, Rutherford proposed that an atom is mostly empty space, and that the positive charge and most of the mass of the atom are concentrated in a very small region. Rutherford's experiment yielded evidence of the nucleus, and led to an improved atomic model known as the nuclear atom.

If your class subscribes to the Interactive Textbook, use it to review key concepts in Section 4.2.

with ChemASAP

Quick LAB

Using Inference: The Black Box

Purpose

To determine the shape of a fixed object inside a sealed box without opening the box.

Materials

- box containing a regularly shaped object fixed in place and a loose marble

Procedure

1. Do not open the box.
2. Manipulate the box so that the marble moves around the fixed object.
3. Gather data (clues) that describe the movement of the marble.
4. Sketch a picture of the object in the box, showing its shape, size, and location within the box.
5. Repeat this activity with a different box containing a different object.

Analysis and Conclusions

1. Find a classmate who had the same lettered box that you had, and compare your findings.
2. What experiment that contributed to a better understanding of the atom does this activity remind you of?

The Rutherford atomic model is known as the nuclear atom. **In the nuclear atom, the protons and neutrons are located in the nucleus. The electrons are distributed around the nucleus and occupy almost all the volume of the atom.** According to this model, the nucleus is tiny compared with the atom as a whole. If an atom were the size of a football stadium, the nucleus would be about the size of a marble.

Although it was an improvement over Thomson's model of the atom, Rutherford's model turned out to be incomplete. In Chapter 5, you will learn how the Rutherford atomic model had to be revised in order to explain the chemical properties of elements.

4.2 Section Assessment

8. **Key Concept** What are three types of subatomic particles?
9. **Key Concept** How does the Rutherford model describe the structure of atoms?
10. What are the charges and relative masses of the three main subatomic particles?
11. Describe Thomson's and Millikan's contributions to atomic theory.
12. Compare Rutherford's expected outcome of the gold-foil experiment with the actual outcome.
13. What experimental evidence led Rutherford to conclude that an atom is mostly empty space?
14. How did Rutherford's model of the atom differ from Thomson's?

Writing Activity

Explanatory Paragraph Write a paragraph explaining how Rutherford's gold-foil experiment yielded new evidence about atomic structure. *Hint:* First describe the setup of the experiment. Then explain how Rutherford interpreted his experimental data.

Assessment 4.2 Test yourself on the concepts in Section 4.2.

with ChemASAP

Section 4.2 Assessment

8. protons, neutrons, and electrons
9. A positively charged nucleus surrounded by electrons, which occupy most of the volume.
10. proton, positive charge, relative mass = 1; electron, negative charge, relative mass = 1/1840; neutron, no charge, relative mass = 1
11. Thomson passed an electric current through sealed glass tubes filled with gases. The resulting glowing beam consisted of tiny negatively charged particles moving at high speed. Thomson concluded that electrons must be parts of the atoms of all elements. Millikan determined the charge and mass of the electron.

Technology & Society

Electron Microscopy

Within 30 years of J.J. Thomson's discovery of the electron, scientists were studying how to produce images of objects by using an electron beam. In 1931, German scientists Ernst Ruska and Max Knoll built the first electron microscope. While an ordinary light microscope uses a beam of light and lenses to magnify objects, an electron microscope uses an electron beam and "lenses" consisting of magnetic or electric fields. A typical light microscope is capable of magnifying an object 1000 times. An electron microscope can magnify an object over 100,000 times. **Interpreting Photographs** ***What characteristics of the images below provide the viewer with a sense of scale?***

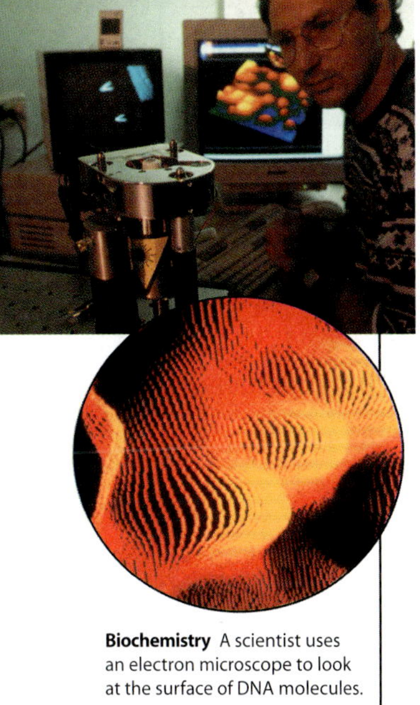

Biochemistry A scientist uses an electron microscope to look at the surface of DNA molecules.

Microelectronics In the colorized electron micrograph below, a wood ant (*Formica fusca*), about 5 mm long, holds a microchip in its jaws. Microelectronics engineers use electron microscopes to measure and analyze the characteristics of microcircuits.

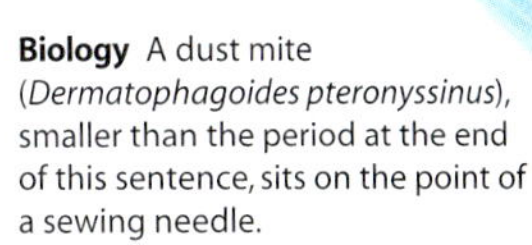

Biology A dust mite (*Dermatophagoides pteronyssinus*), smaller than the period at the end of this sentence, sits on the point of a sewing needle.

12. Rutherford expected all the alpha particles to pass straight through with little deflection. He found that most alpha particles passed straight through, but some particles were deflected at very large angles—and some even bounced straight back.

13. The great majority of the alpha particles passed straight through the gold foil.

14. Rutherford's atomic model described the atom as having a positively charged, dense nucleus that is tiny compared to the atom as a whole. In Thomson's plum-pudding model, electrons were stuck in a chunk of positive charge.

Technology & Society

Electron Microscopy

Point out how a person sees an image made by the electron microscope; the scientist in the photo is not looking into a tube, but sees a projection. Have students research the different types of electron microscopy, what their limitations are, and how specimens are prepared for viewing.

Answers to...

Interpreting Photographs In both images, a small, familiar object is shown so that the viewer can make a size comparison. In the image of the dust mite, showing the tip of the sewing needle provides a sense of scale; in the image of the microchip, showing the ant provides a sense of scale.

4.3

1 FOCUS

Objectives

4.3.1 **Explain** what makes elements and isotopes different from each other.

4.3.2 **Calculate** the number of neutrons in an atom.

4.3.3 **Calculate** the atomic mass of an element.

4.3.4 **Explain** why chemists use the periodic table.

Guide for Reading

Build Vocabulary 

Graphic Organizer Have students use the vocabulary for this section to build a concept map that links and relates the terms.

Reading Strategy 

Visualize Have students revisit Figure 4.10 when they are tackling Sample Problem 4.2. Ask how the visual relates to the problem.

2 INSTRUCT

Connecting to Your World

Have students study the photograph and read the text that opens the section. Ask, **What characteristics can you use to classify different apples?** *(Sample answers: color, size, taste, texture, cooking qualities)* **Which of these involve the apple's chemistry?** *(taste, cooking qualities)*

Atomic Number

Use Visuals 

Table 4.2 Point out that the atomic number is equal to the number of protons for each element. Ask, **Why must the number of electrons equal the number of protons for each element?** *(Atoms are electrically neutral.)* **What is the relationship between the number of protons and the number of neutrons?** *(The number of neutrons tends to rise with the number of protons.)*

4.3 Distinguishing Among Atoms

Guide for Reading

Key Concepts
- What makes one element different from another?
- How do you find the number of neutrons in an atom?
- How do isotopes of an element differ?
- How do you calculate the atomic mass of an element?
- Why is a periodic table useful?

Vocabulary

atomic number
mass number
isotopes
atomic mass unit (amu)
atomic mass
periodic table
period
group

Reading Strategy

Building Vocabulary As you read the section, write a definition of each vocabulary term in your own words.

Connecting to Your World Fruits and vegetables come in different varieties. For example, a grocery store might sell three varieties of apples: Granny Smith, Red Delicious, and Golden Delicious. Apple varieties can differ in color, size, texture, aroma, and taste. Just as apples come in different varieties, a chemical element can come in different "varieties" called isotopes. In this section, you will learn how one isotope of an element differs from another.

Atomic Number

Atoms are composed of protons, neutrons, and electrons. Protons and neutrons make up the nucleus. Electrons surround the nucleus. How, then, are atoms of hydrogen, for example, different from atoms of oxygen? Look at Table 4.2. Notice that a hydrogen atom has one proton, but an oxygen atom has eight protons. **Elements are different because they contain different numbers of protons.**

The **atomic number** of an element is the number of protons in the nucleus of an atom of that element. Because all hydrogen atoms have one proton, the atomic number of hydrogen is 1. Similarly, because all oxygen atoms have eight protons, the atomic number of oxygen is 8. The atomic number identifies an element. For each element listed in Table 4.2, the number of protons equals the number of electrons. Remember that atoms are electrically neutral. Thus, the number of electrons (negatively charged particles) must equal the number of protons (positively charged particles).

Table 4.2

Atoms of the First Ten Elements

Name	Symbol	Atomic number	Protons	Neutrons*	Mass number	Number of electrons
Hydrogen	H	1	1	0	1	1
Helium	He	2	2	2	4	2
Lithium	Li	3	3	4	7	3
Beryllium	Be	4	4	5	9	4
Boron	B	5	5	6	11	5
Carbon	C	6	6	6	12	6
Nitrogen	N	7	7	7	14	7
Oxygen	O	8	8	8	16	8
Fluorine	F	9	9	10	19	9
Neon	Ne	10	10	10	20	10

*Number of neutrons in the most abundant isotope. Isotopes are introduced later in Section 4.3.

Section Resources

Print
- **Guided Reading and Study Workbook,** Section 4.3
- **Core Teaching Resources,** Section 4.3 Review
- **Transparencies,** T48–T56
- **Small-Scale Chemistry Laboratory Manual,** Lab 6

Technology
- **Interactive Textbook with ChemASAP,** Problem-Solving 4.15, 4.17, 4.20, 4.21, 4.24, Assessment 4.3
- **Go Online,** Section 4.3

CONCEPTUAL PROBLEM 4.1

Understanding Atomic Number

The element nitrogen (N), shown here in liquid form, has an atomic number of 7. How many protons and electrons are in a neutral nitrogen atom?

 **Analyze** ***Identify the relevant concepts.***

The atomic number gives the number of protons, which in a neutral atom equals the number of electrons.

2 Solve ***Apply the concepts to this problem.***

The atomic number of nitrogen is 7, which means that a neutral nitrogen atom has 7 protons and 7 electrons.

Practice Problems

15. Complete the table.

Element	Atomic number	Protons	Electrons
K	19	(a)	19
(b)	(c)	(d)	5
S	16	(e)	(f)
V	(g)	23	(h)

16. How many protons and electrons are in each atom?

a. fluorine (atomic number = 9)

b. calcium (atomic number = 20)

c. aluminum (atomic number = 13)

Problem-Solving 4.15 Solve Problem 15 with the help of an interactive guided tutorial.

with ChemASAP

Mass Number

You know that most of the mass of an atom is concentrated in its nucleus and depends on the number of protons and neutrons. The total number of protons and neutrons in an atom is called the **mass number.** Look again at Table 4.2 and note the mass numbers of helium and carbon. A helium atom has two protons and two neutrons, so its mass number is 4. A carbon atom, which has six protons and six neutrons, has a mass number of 12.

If you know the atomic number and mass number of an atom of any element, you can determine the atom's composition. Table 4.2 shows that an oxygen atom has an atomic number of 8 and a mass number of 16. Because the atomic number equals the number of protons, which equals the number of electrons, an oxygen atom has eight protons and eight electrons. The mass number of oxygen is equal to the number of protons plus the number of neutrons. The oxygen atom, then, has eight neutrons, which is the difference between the mass number and the atomic number (16 − 8 = 8). **The number of neutrons in an atom is the difference between the mass number and atomic number.**

Number of neutrons = mass number − atomic number

The composition of any atom can be represented in shorthand notation using atomic number and mass number. Figure 4.8 shows how an atom of gold is represented using this notation. The chemical symbol Au appears with two numbers written to its left. The atomic number is the subscript. The mass number is the superscript.

You can also refer to atoms by using the mass number and the name of the element. For example, $^{197}_{79}$Au may be written as gold-197.

 ***How do you calculate mass number?***

Figure 4.8 Au is the chemical symbol for gold. **Applying Concepts** ***How many electrons does a gold atom have?***

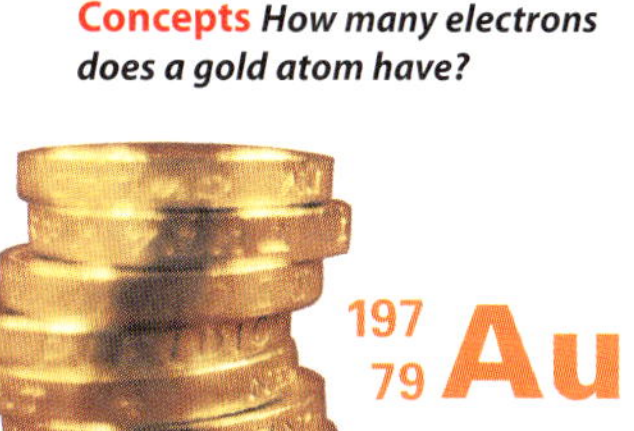

111

CONCEPTUAL PROBLEM 4.1

Answers

15. a. 19 **b.** B **c.** 5 **d.** 5 **e.** 16 **f.** 16 **g.** 23 **h.** 23

16. a. 9 protons and 9 electrons
b. 20 protons and 20 electrons
c. 13 protons and 13 electrons

Practice Problems Plus L2

Chapter 4 Assessment problem 47 is related to Conceptual Problem 4.1.

Mass Number

Discuss 

Discuss why the mass number of an element is defined as the number of protons and neutrons. Explain that chemists have arbitrarily assigned a value of one atomic mass unit to represent the mass of one twelfth of a carbon-12 atom. Ask, **How do you find the number of neutrons in an atom?** *(Subtract the atomic number from the mass number.)*

Differentiated Instruction

Less Proficient Readers L1

Have students make a list of familiar elements and describe at least one use for each element. Have students combine their lists into a master list on the chalkboard.

Answers to...

Figure 4.8 79 electrons

 Mass number = number of neutrons + atomic number

Sample Problem 4.1

Answers

17. a. 8 **b.** 16 **c.** 61 **d.** 45 **e.** 125

18. a. $^{12}_{6}C$ **b.** $^{19}_{9}F$ **c.** $^{9}_{4}Be$

Practice Problems Plus L2

The element strontium has four isotopes, strontium-84, strontium-86, strontium-87, and strontium-88. **Given that strontium has an atomic number of 38, how many neutrons are in each of these isotopes?** *(46, 48, 49, 50)*

Isotopes

CLASS Activity

Applications of Isotopes L2

Purpose To learn about practical applications of isotopes

Materials Library or Internet access

Procedure Have students use the library or Internet to find an isotope of an element that has a practical or everyday use.

Expected Outcome Students' research will most likely focus on applications of radioisotopes such as carbon-14 (used in archaeological dating), americium-241 (used in smoke alarms), iodine-131 (used in the treatment of thyroid disorders), and cobalt-60 (used in the treatment of some cancers). Point out that the instability of these isotopes is what makes them useful. Radioisotopes and radioactivity are discussed in Chapter 25.

SAMPLE PROBLEM 4.1

Determining the Composition of an Atom

How many protons, electrons, and neutrons are in each atom?

	Atomic number	Mass number
a. Beryllium (Be)	4	9
b. Neon (Ne)	10	20
c. Sodium (Na)	11	23

1 Analyze ***List the knowns and the unknowns.***

Knowns
- atomic number
- mass number

Unknowns
- number of protons = ?
- number of electrons = ?
- number of neutrons = ?

Use the definitions of atomic number and mass number to calculate the numbers of protons, electrons, and neutrons.

2 Calculate ***Solve for the unknowns.***

number of electrons = atomic number

a. 4 **b.** 10 **c.** 11

number of protons = atomic number

a. 4 **b.** 10 **c.** 11

number of neutrons = mass number-atomic number

a. $9 - 4 = 5$ **b.** $20 - 10 = 10$ **c.** $23 - 11 = 12$

3 Evaluate ***Do the results make sense?***

For each atom, the mass number equals the number of protons plus the number of neutrons. The results make sense.

Practice Problems

17. How many neutrons are in each atom?

a. $^{16}_{8}O$ **b.** $^{32}_{16}S$ **c.** $^{108}_{47}Ag$

d. $^{80}_{35}Br$ **e.** $^{207}_{82}Pb$

18. Use Table 4.2 to express the composition of each atom in shorthand form.

a. carbon-12 **b.** fluorine-19 **c.** beryllium-9

Problem-Solving 4.17
Solve Problem 17 with the help of an interactive guided tutorial.
with ChemASAP

Isotopes

Figure 4.9 shows that there are three different kinds of neon atoms. How do these atoms differ? All have the same number of protons (10) and electrons (10), but they each have different numbers of neutrons. **Isotopes** are atoms that have the same number of protons but different numbers of neutrons. **Because isotopes of an element have different numbers of neutrons, they also have different mass numbers.** Despite these differences, isotopes are chemically alike because they have identical numbers of protons and electrons, which are the subatomic particles responsible for chemical behavior.

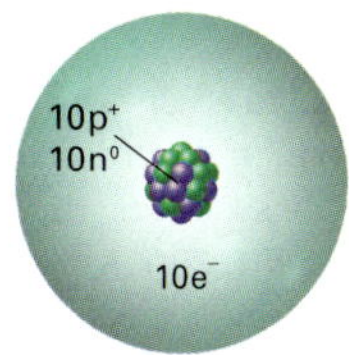

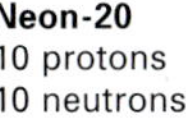

Neon-20
10 protons
10 neutrons
10 electrons

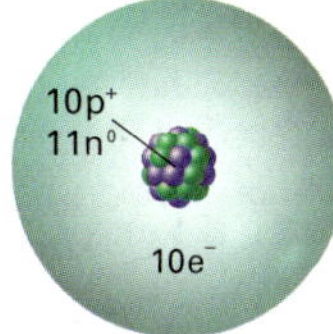

Neon-21
10 protons
11 neutrons
10 electrons

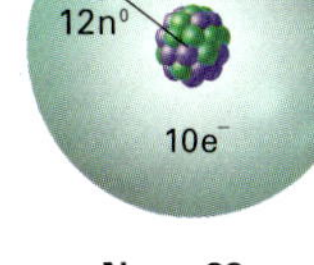

Neon-22
10 protons
12 neutrons
10 electrons

Figure 4.9 Neon-20, neon-21, and neon-22 are three isotopes of neon, a gaseous element used in lighted signs. **Comparing and Contrasting** ***How are these isotopes different? How are they similar?***

There are three known isotopes of hydrogen. Each isotope of hydrogen has one proton in its nucleus. The most common hydrogen isotope has no neutrons. It has a mass number of 1 and is called hydrogen-1 ($^{1}_{1}H$) or simply hydrogen. The second isotope has one neutron and a mass number of 2. It is called either hydrogen-2 ($^{2}_{1}H$) or deuterium. The third isotope has two neutrons and a mass number of 3. This isotope is called hydrogen-3 ($^{3}_{1}H$) or tritium.

 ***What are three known isotopes of hydrogen?***

CONCEPTUAL PROBLEM 4.2

Writing Chemical Symbols of Isotopes

Diamonds are a naturally occurring form of elemental carbon. Two stable isotopes of carbon are carbon-12 and carbon-13. Write the symbol for each isotope using superscripts and subscripts to represent the mass number and the atomic number.

1 Analyze ***Identify the relevant concepts.***

Isotopes are atoms that have the same number of protons but different numbers of neutrons. The composition of an atom can be expressed by writing the chemical symbol, with the atomic number as a subscript and the mass number as a superscript.

 **Solve** ***Apply the concepts to this problem.***

Based on Table 4.2, the symbol for carbon is C and the atomic number is 6. The mass number for each isotope is given by its name. For carbon-12, the symbol is $^{12}_{6}C$. For carbon-13, the symbol is $^{13}_{6}C$.

Practice Problems

19. Three isotopes of oxygen are oxygen-16, oxygen-17, and oxygen-18. Write the symbol for each, including the atomic number and mass number.

20. Three isotopes of chromium are chromium-50, chromium-52, and chromium-53. How many neutrons are in each isotope, given that chromium has an atomic number of 24?

Problem-Solving 4.20
Solve Problem 20 with the help of an interactive guided tutorial.
with ChemASAP

CONCEPTUAL PROBLEM 4.2

Answers

19. $^{16}_{8}O$, $^{17}_{8}O$, $^{18}_{8}O$

20. Chromium-50 has 26 neutrons; chromium-52 has 28 neutrons; chromium-53 has 29 neutrons.

Practice Problems Plus L2

Calculate the number of neutrons in each of the following radioactive isotopes.

a. $^{14}_{6}C$ *(8)*

b. $^{40}_{19}K$ *(21)*

c. $^{238}_{92}U$ *(146)*

d. $^{99}_{42}Mo$ *(57)*

Discuss L2

Point out that isotopes of an element are chemically the same. Ask students why that is the case. *(because the electrons, not the neutrons, determine the atom's chemical properties)*

Answers to...

Figure 4.9 They have different numbers of neutrons but the same number of protons.

 hydrogen-1, hydrogen-2, hydrogen-3

Atomic Mass

Use Visuals

L1

Table 4.3 Point out that the average atomic masses listed in Table 4.3 are based on the masses of stable isotopes and their percent abundance in Earth's crust. Have students study the average atomic masses in the table. Ask, **Which elements exist predominantly as one natural isotope?** *(those with atomic masses closest to a whole number)* **Which element has substantial amounts of each of its natural isotopes?** *(chlorine)*

FYI

Students may ask why the masses (in amu) of most of the isotopes in Table 4.3 are not whole-number values like mass numbers. Although the mass number of carbon-12 exactly equals its mass in amu, this is generally not the case for other isotopes due to the mass defect. The mass defect is the difference between the mass of a nucleus and the sum of the masses of its component protons and neutrons. (This difference can also be expressed in terms of binding energy, based on the relationship $E = mc^2$, in which E = energy, m = mass, and c = the speed of light.) Because mass defect varies with different elements, the masses of isotopes other than carbon-12 in amu (a unit based on the mass of a carbon-12 atom) will generally not be whole numbers. For example, the mass of hydrogen-2 is 2.0141 amu; the mass of oxygen-16 is 15.995 amu. Chapter Assessment problem 82 asks students to calculate the mass defect of an atom in grams by comparing the actual mass of the atom to the sum of the masses of the atom's component protons, neutrons, and electrons.

Atomic Mass

A glance back at Table 4.1 on page 106 shows that the actual mass of a proton or a neutron is very small (1.67×10^{-24} g). The mass of an electron is 9.11×10^{-28} g, which is negligible in comparison. Given these values, the mass of even the largest atom is incredibly small. Since the 1920s, it has been possible to determine these tiny masses by using a mass spectrometer. With this instrument, the mass of a fluorine atom was found to be 3.155×10^{-23} g, and the mass of an arsenic atom was found to be 1.244×10^{-22} g. Such data about the actual masses of individual atoms can provide useful information, but, in general, these values are inconveniently small and impractical to work with. Instead, it is more useful to compare the relative masses of atoms using a reference isotope as a standard. The isotope chosen is carbon-12. This isotope of carbon was assigned a mass of exactly 12 atomic mass units. An **atomic mass unit (amu)** is defined as one twelfth of the mass of a carbon-12 atom. Using these units, a helium-4 atom, with a mass of 4.0026 amu, has about one-third the mass of a carbon-12 atom. On the other hand, a nickel-60 atom has about five times the mass of a carbon-12 atom.

A carbon-12 atom has six protons and six neutrons in its nucleus, and its mass is set as 12 amu. The six protons and six neutrons account for nearly all of this mass. Therefore the mass of a single proton or a single

Crystalline sulfur

Table 4.3

Natural Percent Abundance of Stable Isotopes of Some Elements

Name	Symbol	Natural percent abundance	Mass (amu)	Average atomic mass
Hydrogen	$^{1}_{1}H$	99.985	1.0078	
	$^{2}_{1}H$	0.015	2.0141	1.0079
	$^{3}_{1}H$	negligible	3.0160	
Helium	$^{3}_{2}He$	0.0001	3.0160	4.0026
	$^{4}_{2}He$	99.9999	4.0026	
Carbon	$^{12}_{6}C$	98.89	12.000	12.011
	$^{13}_{6}C$	1.11	13.003	
Nitrogen	$^{14}_{7}N$	99.63	14.003	14.007
	$^{15}_{7}N$	0.37	15.000	
Oxygen	$^{16}_{8}O$	99.759	15.995	
	$^{17}_{8}O$	0.037	16.995	15.999
	$^{18}_{8}O$	0.204	17.999	
Sulfur	$^{32}_{16}S$	95.002	31.972	
	$^{33}_{16}S$	0.76	32.971	32.06
	$^{34}_{16}S$	4.22	33.967	
	$^{36}_{16}S$	0.014	35.967	
Chlorine	$^{35}_{17}Cl$	75.77	34.969	35.453
	$^{37}_{17}Cl$	24.23	36.966	

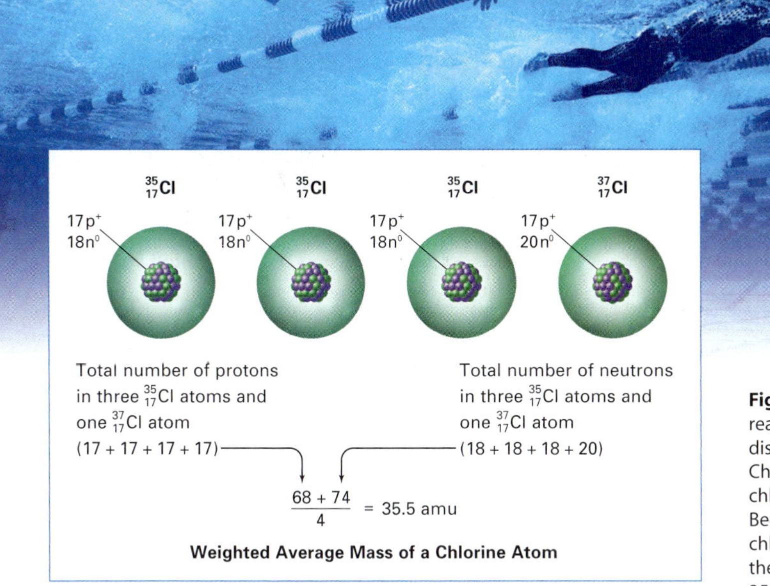

Figure 4.10 Chlorine is a reactive element used to disinfect swimming pools. Chlorine occurs as two isotopes: chlorine-35 and chlorine-37. Because there is more chlorine-35 than chlorine-37, the atomic mass of chlorine, 35.453 amu, is closer to 35 than to 37. **Evaluating** ***How does a weighted average differ from an arithmetic mean?***

neutron is about one-twelfth of 12 amu, or about 1 amu. Because the mass of any single atom depends mainly on the number of protons and neutrons in the nucleus of the atom, you might predict that the atomic mass of an element should be a whole number. However, that is not usually the case.

In nature, most elements occur as a mixture of two or more isotopes. Each isotope of an element has a fixed mass and a natural percent abundance. Consider the three isotopes of hydrogen discussed earlier in this section. According to Table 4.3, almost all naturally occurring hydrogen (99.985%) is hydrogen-1. The other two isotopes are present in trace amounts. Notice that the atomic mass of hydrogen listed in Table 4.3 (1.0079 amu) is very close to the mass of hydrogen-1 (1.0078 amu). The slight difference takes into account the larger masses, but smaller amounts, of the other two isotopes of hydrogen.

Now consider the two stable isotopes of chlorine listed in Table 4.3: chlorine-35 and chlorine-37. If you calculate the arithmetic mean of these two masses ((34.969 amu + 36.966 amu)/2), you get an average atomic mass of 35.968 amu. However, this value is higher than the actual value of 35.453. To explain this difference, you need to know the natural percent abundance of the isotopes of chlorine. Chlorine-35 accounts for 75% of the naturally occurring chlorine atoms; chlorine-37 accounts for only 25%. See Figure 4.10. The **atomic mass** of an element is a weighted average mass of the atoms in a naturally occurring sample of the element. A weighted average mass reflects both the mass and the relative abundance of the isotopes as they occur in nature.

For: Links on Isotopes
Visit: www.SciLinks.org
Web Code: cdn-1043

What is the atomic mass of an element?

Go Online NSTA SCILINKS

Download a worksheet on **Isotopes** for students to complete, and find additional teacher support from NSTA SciLinks.

Facts and Figures

Determining Relative Abundance of an Element's Isotopes

How do you determine relative abundance for an element's isotopes? Look at a sample of the element in a mass spectrometer. The display shows a peak at the mass of each isotope. The height of the peak shows the relative abundance of each isotope.

Answers to...

Figure 4.10 In an arithmetic mean, all the numbers in the calculation are weighted equally. A weighted average takes into account the varying weights of the numbers in the data set.

Atomic mass is the weighted average mass of the atoms in a naturally occurring sample of the element.

CONCEPTUAL PROBLEM 4.3

Answers

21. boron-11

22. Silicon-28 must be by far the most abundant. The other two isotopes must be present in very small amounts.

Practice Problems Plus 

Argon has three isotopes with mass numbers 36, 38, and 40, respectively. **Which of these isotopes is the most abundant?** *(argon-40)*

Relate L2

Teachers sometimes evaluate a student's performance based on the weighted average for different work, for example, a term paper worth 20%, a midterm worth 30%, and a final exam worth 50%. If an A = 4.0, a B = 3.0, and a C = 2.0, have students calculate a grade for a student who receives a B on the term paper, a C on the midterm, and a B on the final exam. *(The student would receive a score of 2.7, a C+.)*

CONCEPTUAL PROBLEM 4.3

Using Atomic Mass to Determine the Relative Abundance of Isotopes

The atomic mass of copper is 63.546 amu. Which of copper's two isotopes is more abundant: copper-63 or copper-65?

1 Analyze ***Identify the relevant concepts.***

The atomic mass of an element is the weighted average mass of the atoms in a naturally occurring sample of the element. A weighted average mass reflects both the mass and the relative abundance of the isotopes as they occur in nature.

2 Solve ***Apply the concepts to this problem.***

The atomic mass of 63.546 amu is closer to 63 than to 65. Because the atomic mass is a weighted average of the isotopes, copper-63 must be more abundant than copper-65.

Practice Problems

21. Boron has two isotopes: boron-10 and boron-11. Which is more abundant, given that the atomic mass of boron is 10.81 amu?

22. There are three isotopes of silicon; they have mass numbers of 28, 29, and 30. The atomic mass of silicon is 28.086 amu. Comment on the relative abundance of these three isotopes.

Problem-Solving 4.21 Solve Problem 21 with the help of an interactive guided tutorial.

with ChemASAP

Now that you know that the atomic mass of an element is a weighted average of the masses of its isotopes, you can determine atomic mass based on relative abundance. To do this, you must know three values: the number of stable isotopes of the element, the mass of each isotope, and the natural percent abundance of each isotope. **To calculate the atomic mass of an element, multiply the mass of each isotope by its natural abundance, expressed as a decimal, and then add the products.** The resulting sum is the weighted average mass of the atoms of the element as they occur in nature.

You can calculate the atomic masses listed in Table 4.3 based on the given masses and natural abundances of the isotopes for each element. For example, carbon has two stable isotopes: carbon-12, which has a natural abundance of 98.89%, and carbon-13, which has natural abundance of 1.11%. The mass of carbon-12 is 12.000 amu; the mass of carbon-13 is 13.003 amu. The atomic mass is calculated as follows.

$$\text{Atomic mass of carbon} = (12.000\ \text{amu} \times 0.9889) + (13.003\ \text{amu} \times 0.0111) = 12.011\ \text{amu}$$

What three values must be known in order to calculate the atomic mass of an element?

Facts and Figures

Carbon-14 Dating

All living organisms contain carbon-12 and carbon-14 in a fixed ratio. After an organism dies, this ratio changes as the carbon-14 decays. Paleontologists and archaeologists use this fact to establish the age of fossils and ancient artifacts.

SAMPLE PROBLEM 4.2

Calculating Atomic Mass

Element X has two natural isotopes. The isotope with a mass of 10.012 amu (^{10}X) has a relative abundance of 19.91%. The isotope with a mass of 11.009 amu (^{11}X) has a relative abundance of 80.09%. Calculate the atomic mass of this element.

1 Analyze ***List the knowns and the unknown.***

Knowns
- isotope ^{10}X:
  mass = 10.012 amu
  relative abundance = 19.91% = 0.1991
- isotope ^{11}X:
  mass = 11.009 amu
  relative abundance = 80.09% = 0.8009

Unknown
- atomic mass of element X = ?

The mass each isotope contributes to the element's atomic mass can be calculated by multiplying the isotope's mass by its relative abundance. The atomic mass of the element is the sum of these products.

2 Calculate ***Solve for the unknown.***

for ^{10}X:	10.012 amu × 0.1991 =	1.993 amu
for ^{11}X:	11.009 amu × 0.8009 =	8.817 amu
for element X:	atomic mass =	10.810 amu

3 Evaluate ***Does the result make sense?***

The calculated value is closer to the mass of the more abundant isotope, as would be expected.

Practice Problems

23. The element copper has naturally occurring isotopes with mass numbers of 63 and 65. The relative abundance and atomic masses are 69.2% for mass = 62.93 amu, and 30.8% for mass = 64.93 amu. Calculate the average atomic mass of copper.

24. Calculate the atomic mass of bromine. The two isotopes of bromine have atomic masses and relative abundance of 78.92 amu (50.69%) and 80.92 amu (49.31%).

CHEMath

Percents

A percent is a shorthand way of expressing a fraction whose denominator is 100. For example, 85% is equivalent to 85/100 or 0.85.

When working with percents, it is usually necessary to convert percents to fractions or decimals before using them in a calculation. For instance, if the natural percent abundance of an isotope is 35.6%, then there are 35.6 g of that isotope in 100 g of the element.

Math Handbook

For help with percents, go to page R72.

Problem-Solving 4.24 Solve Problem 24 with the help of an interactive guided tutorial.

with **ChemASAP**

Bromine

Sample Problem 4.2

Answers

23. 63.6 amu
24. 79.91 amu

Practice Problems Plus 

Chlorine has two isotopes, chlorine-35 (atomic mass 5 34.97 amu, relative abundance 5 75.77%) and chlorine-37 (atomic mass 5 36.97 amu, relative abundance 5 24.23%). **Calculate the atomic mass of chlorine.** *(35.45 amu)*

CHEMath

Percents

Students will need to have a basic understanding of percents in order to calculate atomic mass. Atomic mass is a weighted average that takes into account natural percent abundance. Show students that the relative percent abundances of the isotopes add up to 100%.

Math Handbook

For a math refresher and practice, refer students to percents, page R72.

Answers to...

Checkpoint The number of stable isotopes, the mass of each isotope, and the natural abundance (expressed as a percent) of each isotope.

The Periodic Table—A Preview

Use Visuals L1

Figure 4.11 Have students study the periodic table. Explain that just as items in a supermarket are arranged by food type, such as breads, dairy products, and produce, the chemical elements in the periodic table are arranged in groups that have similar chemical properties. Ask, **Where do you find elements with similar chemical behaviors?** *(in a column)* **What elements are similar chemically to oxygen, O?** *(S, Se, Te, Po)*

FYI

The modern periodic table is discussed in more detail in Chapter 6 and also in Appendix A: Elements Handbook

Careers *in* Chemistry

Archaeologist

Encourage students to find more information on archaeological careers on the Internet. An interesting possibility is nautical archaeology, which sends archaeologists into lakes and oceans. Here's a site for nautical archaeology with links to other web pages. http://www.schoolship.org/careers/archaeologist.html.

Have students research chemistry-related careers. They can then construct a table that describes the nature of the job, requirements, employment outlook, working conditions, and other necessary information.

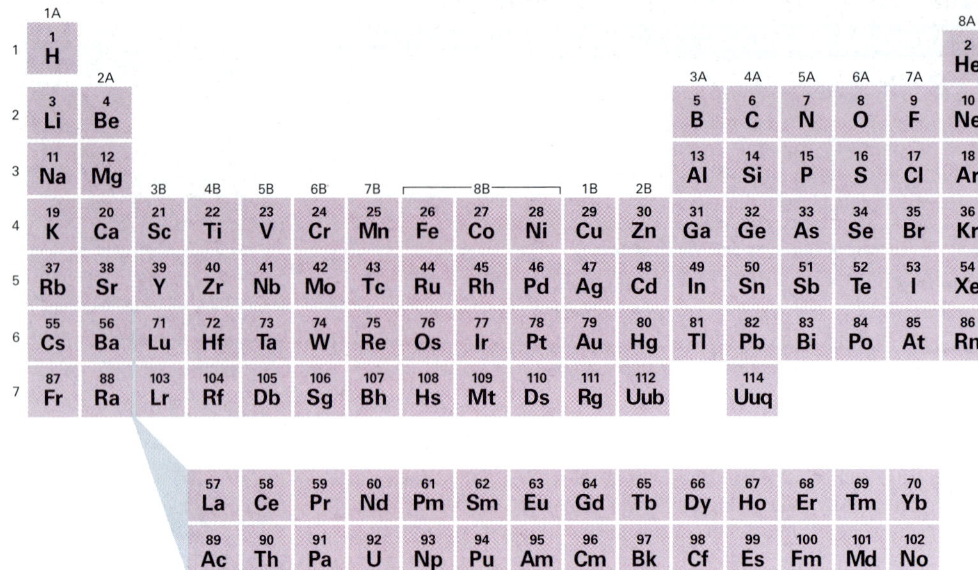

Figure 4.11 Elements are arranged in the modern periodic table in order of atomic number. **Interpreting Diagrams** ***How many elements are in Period 2? In Group 2A?***

The Periodic Table—A Preview

Now that you can differentiate between atoms of different elements and also between isotopes of the same element, you need to understand how the elements are organized with respect to each other. A **periodic table** is an arrangement of elements in which the elements are separated into groups based on a set of repeating properties. **A periodic table allows you to easily compare the properties of one element (or a group of elements) to another element (or group of elements).**

Figure 4.11 shows the most commonly used form of the modern periodic table, sometimes called the long form. Each element is identified by its symbol placed in a square. The atomic number of the element is shown centered above the symbol. Notice that the elements are listed in order of increasing atomic number, from left to right and top to bottom. Hydrogen (H), the lightest element, is in the top left corner. Helium (He), atomic number 2, is at the top right. Lithium (Li), atomic number 3, is at the left end of the second row.

Each horizontal row of the periodic table is called a **period.** There are seven periods in the modern periodic table. The number of elements per period ranges from 2 (hydrogen and helium) in Period 1, to 32 in Period 6. Within a given period, the properties of the elements vary as you move across it from element to element. This pattern of properties then repeats as you move to the next period.

Each vertical column of the periodic table is called a **group,** or family. Elements within a group have similar chemical and physical properties. Note that each group is identified by a number and the letter A or B. For example, Group 2A contains the elements beryllium (Be), magnesium (Mg), calcium (Ca), strontium (Sr), barium (Ba), and radium (Ra). You will learn more about specific trends in the periodic table in Chapter 6.

Careers *in* Chemistry

Archaeologist

Archaeologists are detectives of the past, sifting for clues that uncover secrets of past civilizations. Archaeologists excavate ancient cities and dwellings looking for artifacts that indicate what kinds of foods ancient people ate, what types of tools they used, and how they interacted with one another as a society. Often, archaeologists must draw conclusions based on indirect evidence.

Knowing when an event occurred or when an artifact was made can provide important information. Archaeologists use techniques such as radiometric-dating, in which a sample is dated by measuring the concentration of certain isotopes, such as carbon-14. This method tells them the age of a sample within a certain range, and is used with the greatest accuracy for samples no more than 10,000 years old.

Archaeologists also perform chemical tests on artifacts to determine their composition. For example, archaeologists might analyze the glazes used on pottery, or the dyes used in clothing. Archaeologists may also use chemicals to preserve artifacts that have been unearthed, so that the artifacts can be examined without being damaged.

Archaeology requires a background in both history and science. Archaeologists often spend as much time in the laboratory studying their finds as they do out in the field excavating sites. Archaeologists take courses in archaeological techniques, biology, anatomy, chemistry, math, and history.

For: Careers in Chemistry
Visit: PHSchool.com
Web Code: cdb-1043

4.3 Section Assessment

25. **Key Concept** What distinguishes the atoms of one element from the atoms of another?
26. **Key Concept** What equation tells you how to calculate the number of neutrons in an atom?
27. **Key Concept** How do the isotopes of a given element differ from one another?
28. **Key Concept** How is atomic mass calculated?
29. **Key Concept** What makes the periodic table such a useful tool?
30. What does the number represent in the isotope platinum-194? Write the symbol for this atom using superscripts and subscripts.
31. The atomic masses of elements are generally not whole numbers. Explain why.
32. List the number of protons, neutrons, and electrons in each pair of isotopes.
    a. $^{6}_{3}Li$, $^{7}_{3}Li$ b. $^{42}_{20}Ca$, $^{44}_{20}Ca$ c. $^{78}_{34}Se$, $^{80}_{34}Se$
33. Name two elements that have properties similar to those of the element calcium (Ca).

Elements Handbook

Elements Within You Read page R5 of the Elements Handbook. Identify the five most abundant elements in the human body, and locate them on the periodic table.

Assessment 4.3 Test yourself on the concepts in Section 4.3.

with **ChemASAP**

FYI

Radiometric dating and its applications are discussed in more detail in Chapter 25.

3 ASSESS

Evaluate Understanding 

Write the symbols for isotopes of an element not described in the chapter. Ask students what the superscripts and subscripts refer to and the differences between the atoms shown. Ask, **How would changing the value of the subscript change the chemical properties of the atom?** *(The subscript designates the number of protons in the atoms of that isotope. Changing the number of protons would change the chemical identity of the isotope to that of another element.)*

Reteach 

Review the concept of weighted averages. Work through the following calculations. If 75% of chlorine atoms are ^{35}Cl species and 25% are ^{37}Cl species, this implies that for a sample of 100 atoms, 75 atoms are ^{35}Cl and 25 atoms are ^{37}Cl species. The combined masses of these atoms would be (75 × 35 amu) + (25 × 37 amu) = 3550 amu for 100 atoms, or 35.5 amu for one atom.

Elements Handbook

Oxygen (Period 2, Group 6A), carbon (Period 2, Group 4A), hydrogen (Period 1, Group 1A), nitrogen (Period 2, Group 5A), and calcium (Period 4, Group 2A).

If your class subscribes to the Interactive Textbook, use it to review key concepts in Section 4.3.

with **ChemASAP**

Section 4.3 Assessment

25. Atoms of different elements contain different numbers of protons.
26. mass number − atomic number = number of neutrons
27. They have different mass numbers and different numbers of neutrons.
28. For each isotope, multiply its atomic mass by its % abundance, then add the products.
29. It allows you to compare the properties of the elements.
30. Mass number, $^{194}_{78}Pt$
31. The atomic mass is the weighted average of the masses of its isotopes.
32. **a.** lithium-6: 3 p^+, 3 e^-, 3 n^0; lithium-7: 3 p^+, 3 e^-, 4 n^0 **b.** calcium-42: 20 p^+, 20 e^-, 22 n^0; calcium-44: 20 p^+, 20 e^-, 24 n^0 **c.** selenium-78: 34 p^+, 34 e^-, 44 n^0; selenium-80: 34 p^+, 34 e^-, 46 n^0
33. any two: beryllium (Be), magnesium (Mg), strontium (Sr), barium (Ba), radium (Ra)

Answers to...

Figure 4.11 There are eight elements in Period 2. There are 6 elements in Group 2A.

Small-Scale LAB

The Atomic Mass of Candium 

Objective After completing this activity, students will be able to:

- make measurements to calculate the relative abundances of three types of candy in a mixture.
- use their data to calculate the atomic mass of a candium particle.

Skills Focus Measuring, calculating

 **Prep Time** 15 minutes

Materials mass balance, coated candies (3 different brands), small plastic cups or containers

Advance Prep Prepare in advance a large mixture of the three candies and half fill a clean 3.5-ounce plastic cup for each student. Each sample will contain about 50 total pieces.

Class Time 30 minutes

Safety Discourage students from eating the candies after the experiment. Contamination can easily occur in a lab even if you have taken every precaution to keep the candy free of contamination.

Teaching Tips This lab is similar to the longer Small-Scale Lab "Isotopes and Atomic Mass" found in the *Small-Scale Chemistry Laboratory Manual*.

Expected Outcome Sample data are listed below.

Total Mass: 13.16 g; 13.83 g; 15.40 g; 42.39 g (total).
Number: 15; 13; 20; 48 (total).
Average mass (grams): 0.8773 g; 1.064 g; 0.7700 g; 0.8831 g (total).
Relative abundance: 0.3125, 0.2708, 0.4167, 1.000 (total).
Percent abundance: 31.25%; 27.08%; 41.67%; 100% (total).
Relative mass: 0.2742 g; 0.2883 g; 0.3208 g; 0.8833 g (total).

Analyze

1–5. See Expected Outcome

6. Percent abundance is parts per hundred. Relative abundance is parts per one, or the decimal form of percent. The individual percent abundances add up to 100. The individual relative abundances add up to 1.

7. Relative abundance tells you the decimal fraction of particles.

8. The total in row 3 is an average that ignores the relative abundances of particles. The total in row 6 is a weighted average that best represents atomic mass because it considers differences in mass and abundance among the particles.

9. Another student might not have had the same relative abundance of each candy.

You're the Chemist

1. Any differences are probably due to small variations in the numbers of each kind of candy in the samples, which affects the relative abundances.

2. The larger the samples, the better the results with any of the methods. Mass is likely to provide better results than volume.

Small-Scale LAB

The Atomic Mass of Candium

Purpose

To analyze the isotopes of "candium" and to calculate its atomic mass.

Materials

- sample of candium
- balance
- pencil
- paper

Procedure

Obtain a sample of "candium" that contains three different brands of round, coated candy. Treat each brand of candy as an isotope of candium. Separate the three isotopes into groups labeled A, B, and C, and measure the mass of each isotope. Count the number of atoms in each sample. Make a table similar to the one below to record your measured and calculated data.

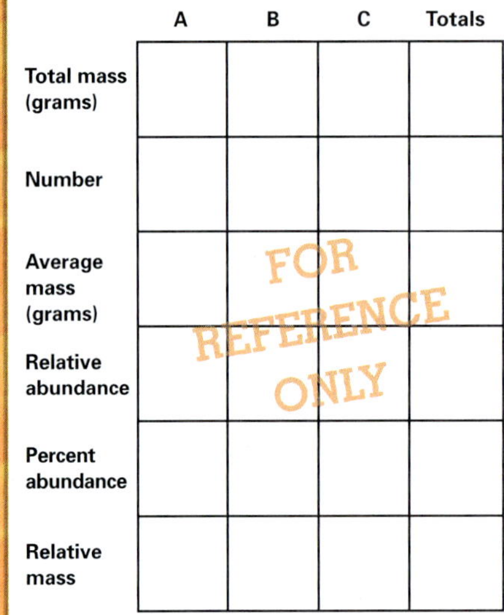

	A	B	C	Totals
Total mass (grams)				
Number				
Average mass (grams)				
Relative abundance				
Percent abundance				
Relative mass				

Analyze

Using the experimental data, record the answers to the following questions below your data table.

1. Calculate the average mass of each isotope by dividing its total mass by the number of particles of that isotope.
2. Calculate the relative abundance of each isotope by dividing its number of particles by the total number of particles.

3. Calculate the percent abundance of each isotope by multiplying the relative abundance from Step 2 by 100.
4. Calculate the relative mass of each isotope by multiplying its relative abundance from Step 2 by its average mass.
5. Calculate the weighted average mass of all candium particles by adding the relative masses. This weighted average mass is the atomic mass of candium.
6. Explain the difference between percent abundance and relative abundance. What is the result when you total the individual relative abundances? The individual percent abundances?
7. The percent abundance of each kind of candy tells you how many of each kind of candy there are in every 100 particles. What does relative abundance tell you?
8. Compare the total values for rows 3 and 6 in the table. Explain why the totals differ and why the value in row 6 best represents atomic mass.
9. Explain any differences between the atomic mass of your candium sample and that of your neighbor. Explain why the difference would be smaller if larger samples were used.

You're the Chemist

The following small-scale activities allow you to develop your own procedures and analyze the results.

1. **Analyze It!** Determine the atomic mass of a second sample of candium. How does it compare with the first? Suggest reasons for any differences between the samples.
2. **Design It!** Design and test methods to produce identical samples of candium. Try measuring mass or volume as a means of counting. Test these methods by counting each kind of candy in each sample you produce. Which method of sampling gives the most consistent results?

CHAPTER 4

Study Guide

Key Concepts

4.1 Defining the Atom

- Democritus believed that atoms were indivisible and indestructible.
- By using experimental methods, Dalton transformed Democritus's ideas on atoms into a scientific theory.
- Scientists can observe individual atoms by using instruments such as scanning tunneling microscopes.

4.2 Structure of the Nuclear Atom

- Three types of subatomic particles are electrons, protons, and neutrons.
- In the nuclear atom, the protons and neutrons are located in the nucleus. The electrons are distributed around the nucleus and occupy almost all the volume of the atom.

4.3 Distinguishing Among Atoms

- Elements are different because they have different numbers of protons.
- The number of neutrons in an atom is the difference between the mass number and atomic number.
- Because isotopes of an element have different numbers of neutrons, they also have different mass numbers.
- To calculate the atomic mass of an element, multiply the mass of each isotope by its natural percent abundance (expressed as a decimal), and then add the products.
- The periodic table lets you easily compare the properties of one element (or a group of elements) to another element (or group of elements).

Vocabulary

- atom (p. 101)
- atomic mass (p. 115)
- atomic mass unit (amu) (p. 114)
- atomic number (p. 110)
- cathode ray (p. 104)
- Dalton's atomic theory (p. 102)
- electron (p. 104)
- group (p. 118)
- isotopes (p. 112)
- mass number (p. 111)
- neutron (p. 106)
- nucleus (p. 107)
- period (p. 118)
- periodic table (p. 118)
- proton (p. 106)

Key Equation

- number of neutrons = mass number − atomic number

Organizing Information

Use these terms to construct a concept map that organizes the main ideas of this chapter.

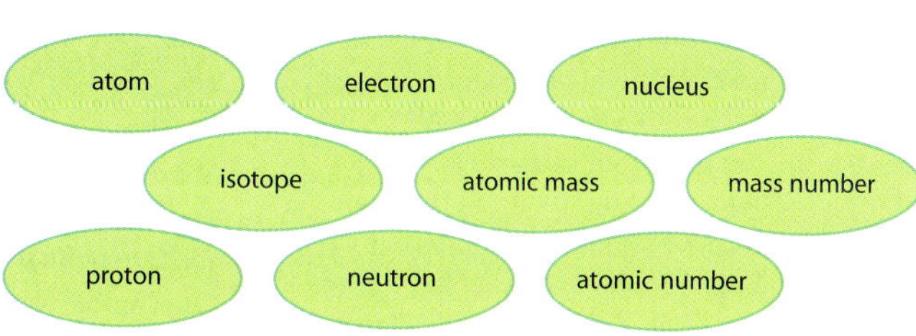

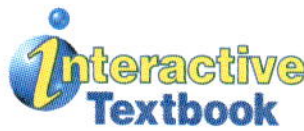

Concept Map 4 Solve the Concept map with the help of an interactive guided tutorial.

with ChemASAP

Chapter Resources

Print

- **Core Teaching Resources,** Chapter 4, *Practice Problems, Vocabulary Review, Quiz, Chapter Test A, Chapter Test B*

Technology

- **Computer Test Bank,** Chapter 4 Test
- **Interactive Textbook with ChemASAP,** Chapter 4

CHAPTER 4 Study Guide

Study Tip

Organize New Information Tell students: Use a reading strategy that you practiced in this chapter to organize the subject matter in a way that helps you study. For example, if you have trouble remembering all the new vocabulary, try listing the new terms and defining them in your own words.

If your class subscribes to the Interactive Textbook with ChemASAP, your students can go online to address an interactive version of the Student Edition and a self-test.

with ChemASAP

CHAPTER 4

Assessment

Reviewing Content

4.1 Defining the Atom

34. What is an atom?

35. What were the limitations of Democritus's ideas about atoms?

36. With which of these statements would John Dalton have agreed in the early 1800s? For each, explain why or why not.

a. Atoms are the smallest particles of matter.

b. The mass of an iron atom is different from the mass of a copper atom.

c. Every atom of silver is identical to every other atom of silver.

d. A compound is composed of atoms of two or more different elements.

37. Use Dalton's atomic theory to describe how atoms interact during a chemical reaction.

4.2 Structure of the Nuclear Atom

38. What experimental evidence did Thomson have for each statement?

a. Electrons have a negative charge.

b. Atoms of all elements contain electrons.

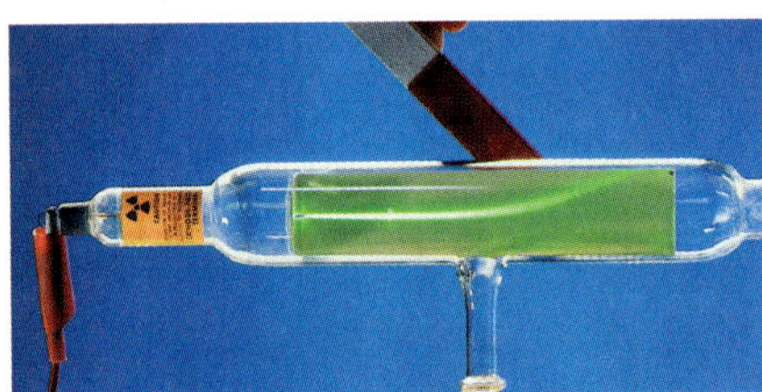

39. Would you expect two electrons to attract or repel each other?

40. How do the charge and mass of a neutron compare to the charge and mass of a proton?

41. Why does it make sense that if an atom loses electrons, it is left with a positive charge?

42. Describe the location of the electrons in Thomson's "plum pudding" model of the atom.

43. How did the results of Rutherford's gold-foil experiment differ from his expectations?

44. What is the charge, positive or negative, of the nucleus of every atom?

45. In the Rutherford atomic model, which subatomic particles are located in the nucleus?

4.3 Distinguishing Among Atoms

46. Why is an atom electrically neutral?

47. What does the atomic number of each atom represent?

48. How many protons are in the nuclei of the following atoms?

a. phosphorus

b. molybdenum

c. aluminum

d. cadmium

e. chromium

f. lead

49. What is the difference between the mass number and the atomic number of an atom?

50. Complete the following table by referring to Figure 4.11 on page 118.

Atomic number	Mass number	Number of protons	Number of neutrons	Symbol of element
9	(a)	(b)	10	(c)
(d)	(e)	14	15	(f)
(g)	47	(h)	25	(i)
(j)	55	25	(k)	(l)

51. Name two ways that isotopes of an element differ.

52. How can there be more than 1000 different atoms when there are only about 100 different elements?

53. What data must you know about the isotopes of an element to calculate the atomic mass of the element?

54. How is an average mass different from a weighted average mass?

55. What is the atomic mass of an element?

56. How are the elements arranged in the modern periodic table?

57. Look up the word *periodic* in the dictionary. Propose a reason for the naming of the periodic table.

CHAPTER 4

Assessment

Reviewing Content

34. The smallest particle of an element that still has the properties of that element.

35. Democritus's ideas were not helpful in explaining chemical behavior because they lacked experimental support.

36. Dalton would agree with all four statements because they all fit his atomic theory.

37. The atoms are separated, joined, and rearranged.

38. a. A beam of electrons (cathode rays) is deflected by an electric field toward the positively charged plate. **b.** The cathode rays were always composed of electrons regardless of the metal used in the electrodes or the gas used in the cathode-ray tube.

39. repel

40. The mass of the proton and neutron are equal; protons are positively charged and neutrons are neutral.

41. Atoms are neutral: number of protons = number of electrons. Loss of an electron means that the number of p^+ is greater than the number of e^-, so the remaining particle is positively charged.

42. The electrons were stuck in a lump of positive charge.

43. He did not expect alpha particles to be deflected at a large angle.

44. positive

45. protons and neutrons (Rutherford suspected there was something in the nucleus in addition to protons—but didn't know them as neutrons.)

46. It has equal numbers of protons and electrons.

47. the number of protons in the nucleus

48. a. 15 **b.** 42 **c.** 13 **d.** 48 **e.** 24 **f.** 82

49. The atomic number is the number of protons. The mass number is the sum of the protons and neutrons.

50. a. 19 **b.** 9 **c.** F **d.** 14 **e.** 29 **f.** Si **g.** 22 **h.** 22 **i.** Ti **j.** 25 **k.** 30 **l.** Mn

51. mass numbers, atomic masses, number of neutrons, relative abundance

52. because of the existence of isotopes

53. which isotopes exist, their masses, and their natural percent abundance

54. Average atomic mass is the arithmetic mean of the isotopes. Weighted average atomic mass considers both the mass and the relative abundance of the isotopes.

55. The atomic mass is the weighted average of the masses of all the isotopes.

56. according to their atomic numbers

57. Sample answer: The table is set up so that chemical properties of elements recur at regular intervals.

Understanding Concepts

58. Characterize the size of an atom.
59. Compare the size and density of an atom with its nucleus.
60. Imagine you are standing on the top of a boron-11 nucleus. Describe the numbers and kinds of subatomic particles you would see looking down into the nucleus, and those you would see looking out from the nucleus.
61. What parts of Dalton's atomic theory no longer agree with the current picture of the atom?
62. Millikan measured the quantity of charge carried by an electron. How did he then calculate the mass of an electron?
63. How is the number of electrons in an atom of a given element related to the atomic number of that element?
64. How is the atomic mass of an element calculated from isotope data?
65. The four isotopes of lead are shown below, each with its percent by mass abundance and the composition of its nucleus. Using these data, calculate the approximate atomic mass of lead.

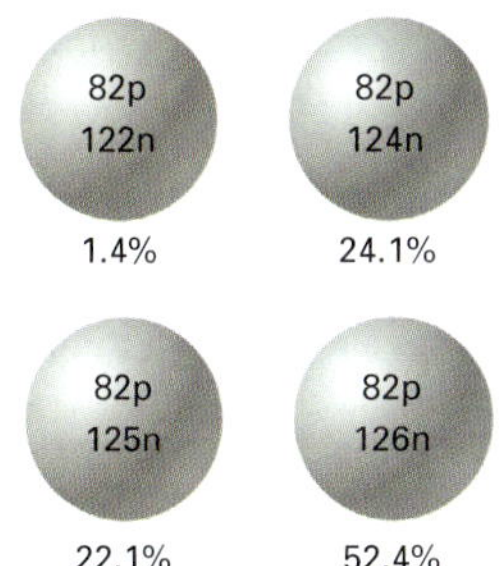

66. Dalton's atomic theory was not correct in every detail. Should this be taken as a criticism of Dalton as a scientist? Explain.
67. Why are atoms considered the basic building blocks of matter even though smaller particles, such as protons and electrons, exist?
68. The following table shows some of the data collected by Rutherford and his colleagues during their gold-foil experiment.

Angle of deflection (degrees)	Number of deflections
5	8,289,000
10	502,570
15	120,570
30	7800
45	1435
60	477
75	211
>105	198

a. What percentage of the alpha particle deflections were 5° or less?
b. What percentage of the deflections were 15° or less?
c. What percentage of the deflections were 60° or greater?

69. Using the data for nitrogen listed in Table 4.3, calculate the weighted average atomic mass of nitrogen. Show your work.
70. What characteristics of cathode rays led Thomson to conclude that the rays consisted of negatively charged particles?
71. If you know the atomic number and mass number of an atom of an element, how can you determine the number of protons, neutrons, and electrons in that atom?
72. What makes isotopes of the same element chemically alike?
73. In the periodic table, what happens to the pattern of properties within a period when you move from one period to the next?

Understanding Concepts

58. very, very tiny—but larger than protons and electrons
59. The nucleus is very small and very dense compared with the atom.
60. 5 protons and 6 neutrons in the nucleus; 5 electrons outside the nucleus
61. All atoms of the same element are not identical (isotopes). The atom is not the smallest particle of matter.
62. He used the quantity of charge value and the charge-to-mass ratio measured by Thomson.
63. They are the same value.
64. The masses of isotopes in a sample of the element are averaged, based on relative abundance. The result is the element's atomic mass.
65. 207 amu
66. No; in general he proposed a valid theory in line with the experimental evidence available to him.
67. Atoms are the smallest particle of an element that retains the properties of that element.
68. **a.** 92.90% **b.** 99.89% **c.** 0.00993%
69. $^{14}_{7}N$: 14.003 amu; 99.63% $^{15}_{7}N$: 15.000 amu; 0.37%
average atomic mass = 14.01 amu
70. They were attracted to a positively charged plate.
71. Atomic number is the same as the number of protons and electrons; mass number minus atomic number equals number of neutrons.
72. Because they have identical numbers of protons, they also have identical numbers of electrons; electrons are the subatomic particles that are responsible for chemical behavior.
73. The pattern repeats.

CHAPTER 4

Assessment continued

Critical Thinking

74. **a.** the nucleus of an atom; **b.** very small volume; almost all the mass of the atom; high density; positive charge; **c.** electron

75. Change the metal used as a target and account for differences in deflection patterns.

76. The following are reasonable hypotheses: The space in an individual atom is large relative to the volume of the atom, but very small relative to an object the size of a hand. There are many layers of atoms in a wall or a desk. The space that exists is distributed evenly throughout the solid, similar to the distribution of air pockets in foam insulation.

77. The theory must be modified and then retested.

78. Yes—but answers will vary.

79. In a chemical change, atoms are not created or destroyed; they are rearranged.

Concept Challenge

80. Because diamond is more dense than graphite, pressure could be used to "squeeze" the carbon atoms closer together.

81. 92.5%

82. 4×10^{-25} g

Cumulative Review

83. Pure chemistry involves the accumulation of scientific knowledge for its own sake. Applied chemistry is accumulating knowledge to attain a specific goal.

84. Scientific theory attempts to explain why experiments give certain results. Scientific law describes a natural phenomenon but does not explain it.

85. **a.** element **b.** mixture
c. mixture **d.** mixture

86. 48 g

87. 6.38×10^{7} cm^3

88. 99.5 g

CHAPTER 4

Assessment continued

Critical Thinking

74. The diagram below shows gold atoms being bombarded with fast-moving alpha particles.

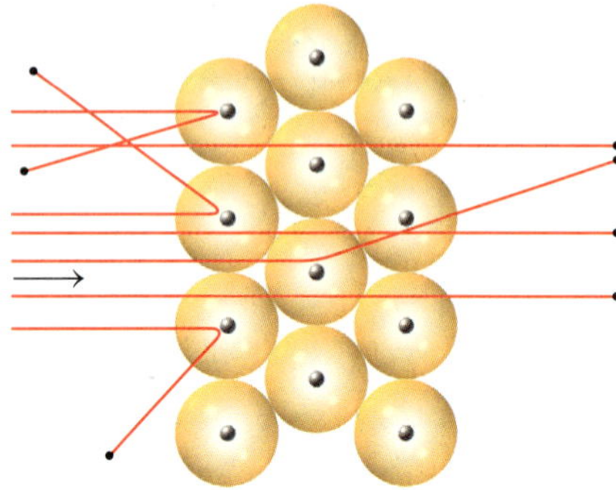

a. The large yellow spheres represent gold atoms. What do the small gray spheres represent?

b. List at least two characteristics of the small gray spheres.

c. Which subatomic particle cannot be found in the area represented by the gray spheres?

75. How could you modify Rutherford's experimental procedure to determine the relative sizes of different nuclei?

76. Rutherford's atomic theory proposed a dense nucleus surrounded by very small electrons. This implies that atoms are composed mainly of empty space. If all matter is mainly empty space, why is it impossible to walk through walls or pass your hand through your desk?

77. This chapter illustrates the scientific method in action. What happens when new experimental results cannot be explained by the existing theory?

78. Do you think there are more elements left to be discovered? Explain your answer.

79. The law of conservation of mass was introduced in Chapter 2. Use Dalton's atomic theory to explain this law.

Concept Challenge

80. Diamond and graphite are both composed of carbon atoms. The density of diamond is 3.52 g/cm^3. The density of graphite is 2.25 g/cm^3. In 1955, scientists successfully made diamond from graphite. Using the relative densities, imagine what happens at the atomic level when this change occurs. Then suggest how this synthesis may have been accomplished.

81. Lithium has two naturally occurring isotopes. Lithium-6 has an atomic mass of 6.015 amu; lithium-7 has an atomic mass of 7.016 amu. The atomic mass of lithium is 6.941 amu. What is the percentage of naturally occurring lithium-7?

82. When the masses of the particles that make up an atom are added together, the sum is always larger than the actual mass of the atom. The missing mass, called the mass defect, represents the matter converted into energy when the nucleus was formed from its component protons and neutrons. Calculate the mass defect of a chlorine-35 atom by using the data in Table 4.1. The actual mass of a chlorine-35 atom is 5.81×10^{-23} g.

Cumulative Review

83. How does the goal of pure chemistry compare with that of applied chemistry? (*Chapter 1*)

84. How does a scientific law differ from a scientific theory? (*Chapter 1*)

85. Classify each as an element, a compound, or a mixture. (*Chapter 2*)
a. sulfur **b.** salad oil
c. newspaper **d.** orange

86. Oxygen and hydrogen react explosively to form water. In one reaction, 6 g of hydrogen combines with oxygen to form 54 g of water. How much oxygen was used? (*Chapter 2*)

87. An aquarium measures 54.0 cm × 31.10 m × 380.0 cm. How many cubic centimeters of water will this aquarium hold? (*Chapter 3*)

88. What is the mass of 4.42 cm^3 of platinum? The density of platinum is 22.5 g/cm^3. (*Chapter 3*)

Standardized Test Prep

Test-Taking Tip

Connectors Sometimes two phrases in a true/false question are connected by a word such as *because* or *therefore*. These words imply a relationship between one part of the sentence and another. Statements that include such words can be false even if both parts of the statement are true by themselves.

Select the choice that best answers each question or completes each statement.

1. An atom composed of 16 protons, 16 electrons, and 16 neutrons is
   a. $^{48}_{16}S$ **b.** $^{16}_{32}Ge$ **c.** $^{32}_{16}S$ **d.** $^{16}_{32}S$
2. Which list contains elements that fall within the same group on the periodic table?
   a. He, Ar, Xe **b.** O, F, Ne
   c. K, Rb, Ba **d.** H, He, Li
3. Which of these descriptions is *incorrect?*
   a. proton: positive charge, in nucleus, mass of ≈ 1 amu
   b. electron: negative charge, mass of ≈ 0 amu, in nucleus
   c. neutron: mass of ≈ 1 amu, no charge
4. Thallium has two isotopes, thallium-203 and thallium-205. Thallium's atomic number is 81 and its atomic mass is 204.38 amu. Which statement about the thallium isotopes is true?
   a. There is more thallium-203 in nature.
   b. Atoms of both isotopes have 81 protons.
   c. Thallium-205 atoms have fewer neutrons.
   d. The most common atom of thallium has a mass of 204.38 amu.

Use the art to answer Question 5.

5. How many nitrogen-14 atoms (^{14}N) would you need to place on the right pan to balance the three calcium-42 atoms (^{42}Ca) on the left pan of the "atomic balance" below? Describe the method you used to determine your answer, including any calculations.

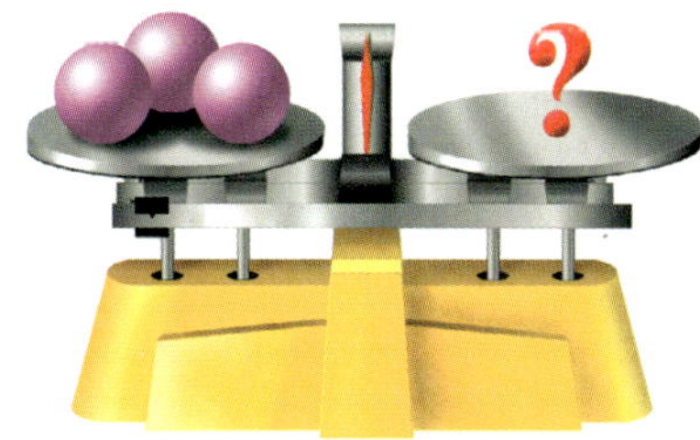

6. Which of the following statements about the periodic table are correct?
   I. Elements are arranged in order of increasing atomic mass.
   II. A period is a horizontal row.
   III. The properties of the elements within a period vary from element to element.
   a. I only
   b. I and II only
   c. I, II, and III
   d. I and III only
   e. II and III only

For each question below there are two statements. Decide whether each statement is true or false. Then decide whether Statement II is a correct explanation for Statement I.

	Statement I		Statement II
7.	Every aluminum-27 atom has 27 protons and 27 electrons.	BECAUSE	The mass number of aluminum-27 is 27.
8.	Isotopes of an element have different atomic masses.	BECAUSE	The nuclei of an element's isotopes contain different numbers of protons.
9.	An electron is repelled by a negatively charged particle.	BECAUSE	An electron has a negative charge.

Standardized Test Prep

1. c
2. a
3. b
4. b
5. 9; three ^{42}Ca atoms have an approximate mass of $3 \times 42 =$ 126 amu; one ^{14}N atom has an approximate mass of 14 amu; $126/14 = 9$ ^{14}N atoms with an approximate mass of 126 amu.
6. e
7. false, true
8. true, false
9. true, true, correct explanation

CHAPTER 5

Electrons in Atoms
Planning Guide

SECTION OBJECTIVES	STANDARDS NATIONAL	STANDARDS STATE	ACTIVITIES and LABS
5.1 Models of the Atom, pp. 127–132 2 blocks or 4 periods **5.1.1 Identify** the inadequacies in the Rutherford atomic model. **5.1.2 Identify** the new proposal in the Bohr model of the atom. **5.1.3 Describe** the energies and positions of electrons according to the quantum mechanical model. **5.1.4 Describe** how the shapes of orbitals related to different sub-levels differ.	A-2, B-1, B-2, B-3, E-2, G-3		**SE** Inquiry Activity: Observing Light Emission from Wintergreen Mints, p. 126 L1 **TE** Teacher Demo: Quantized Energy, p. 128 L2 **TE** Teacher Demo: Energy and Energy Levels, p. 129 L1 **TE** Class Activity: The Shapes of Orbitals, p. 130 L2
5.2 Electron Arrangement in Atoms, pp. 133–137 1 block or 2 periods **5.2.1 Describe** how to write the electron configuration for an atom. **5.2.2 Explain** why the actual electron configurations for some elements differ from those predicted by the aufbau principle.	A-1, B-1		**SE** Small-Scale Lab: Atomic Emission Spectra, p. 137 L2 **TE** Class Activity: Writing Electron Configurations, p. 134 L2 **SSLM** Lab 7: Design and Construction of a Quantitative Spectroscope L2
5.3 Physics and the Quantum Mechanical Model, pp. 138–147 2 blocks or 4 periods **5.3.1 Describe** the relationship between the wavelength and frequency of light. **5.3.2 Identify** the source of atomic emission spectra. **5.3.3 Explain** how the frequencies of emitted light are related to changes in electron energies. **5.3.4 Distinguish** between quantum mechanics and classical mechanics.	A-1, A-2, B-2, B-6, E-2, G-1, G-2, G-3		**SE** Quick Lab: Flame Tests, p. 142 L2 **TE** Class Activity: Black Box Discovery, p. 139 L2 **TE** Class Activity: Properties of Waves, p. 140 L2 **TE** Class Activity: Atomic Scientists, p. 145 L2 **LM** Lab 6: Flame Tests for Metals L2 **LM** Lab 7: Energies of Electrons L2 **LM** Lab 8: Introduction to the Spectrophotometer L2 **SSLM** Lab 8: Visible Spectra and the Nature of Light and Color L2 **LP** Lab Practical 5-1: Flame Tests L2 **LP** Lab Practical 5-2: Spectroscopy with a Spectrophotometer L2 **VCL** Lab 1: Flame Tests for Metals L2 **VCL** Lab 7: Atomic Emission Spectra L2 **VCL** Lab 8: Photoelectric Effect L2 **VCL** Lab 9: Diffraction Experiments L2 **VCL** Lab 10: Electronic State Energy Levels L2

Ability Levels
L1 For students who need additional help
L2 For all students
L3 For students who need to be challenged

Components

SE	Student Edition	**LM**	Laboratory Manual	**GCP**	Graphing Calculator Problems	**T**	Transparencies
TE	Teacher's Edition	**SSLM**	Small-Scale Chemistry Lab Manual	**TTT**	Test-Taking Tips with Transparencies, Grades 9–12 Science	**CA**	Chem Alive and Chemistry Field Trips DVD
GRSW	Guided Reading and Study Workbook	**LP**	Laboratory Practicals	**STP**	Standardized Test Preparation Workbook	**iT**	Interactive Textbook with ChemASAP
CTR	Core Teaching Resources	**PLM**	Probeware Lab Manual			**GO**	Go Online Internet Resources
SSG	Spanish Study Guide	**VCL**	Virtual Chemistry Labs				
		CTB	Computer Test Bank				

RESOURCES PRINT and TECHNOLOGY	PROBLEMS and ASSESSMENT
GRSW Chapter 5, Section 1 L1 T T57 L2	**SE** Section 5.1 Assessment, p. 132 **iT** Section 5.1
GRSW Chapter 5, Section 2 L1 T T58–T60 L2 **GO** Chapter 5, Section 2 L2	**CP** 5.1, Writing Electron Configurations **SE** Section 5.2 Assessment, p. 136 **iT** Section 5.2
GRSW Chapter 5, Section 3 L1 T T61–T64 L2 **GO** Chapter 5, Section 3 L2 **GO** Chapter 5, Section 3 L2	**SP** 5.1, Calculating the Wavelength of Light **SE** Section 5.3 Assessment, p. 146 **iT** Section 5.3

Web Code: cdn-1052
Web Code: cdn-1053

Web Code: cdb-1053

MATERIALS List

Quantities for each group

STUDENT EDITION

Inquiry Activity, p 126
3 different brands of wintergreen mints, a pair of pliers, transparent tape, and a dark room

Small-Scale Lab, p. 137
cereal box, diffraction grating, tape, ruler, pencil, scissors, black construction paper, white notebook paper

Quick Lab, p. 142
bunsen burner, 6 small test tubes, test tube rack, tongs, 6 cotton swabs, 0.1*M* NaCl, 0.5*M* $CaCl_2$, 0.1*M* LiCl, 0.1*M* $CuCl_2$, unknown solution

TEACHER'S EDITION

Teacher Demo, p. 128
a trumpet or trombone, player

Teacher Demo, p. 129
three balls, red, blue, and yellow; stepstool with three steps

Class Activity, p. 130
three metal or wooden skewers, three foam balls (3 or 4 inches in diameter)

Class Activity, p. 139
several sealed boxes containing common items such as rubber bands, golf, tennis, or table-tennis balls, cans of different sizes, and different shapes cut from blocks of wood or foam

Class Activity, p. 140
paper, ruler, pencil

Chapter Assessment

CHAPTER ASSESSMENT

SE	Chapter Assessment, pp. 149–152
CTB	Chapter 5
iT	Chapter 5

STANDARDIZED TEST PREP

SE	Chapter 5, p. 153
STP	Topic 1

iT Interactive Textbook with Assessment at PHSchool.com

CHAPTER 5

Electrons in Atoms

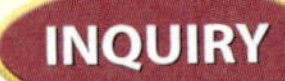 Activity

Observing Light Emission from Wintergreen Mints 

Purpose

Students observe light associated with the movement of electrons.

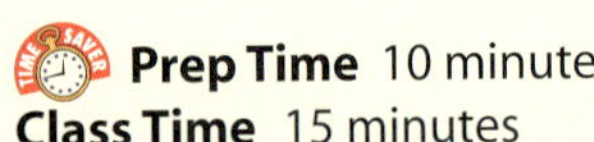

Prep Time 10 minutes
Class Time 15 minutes

Think About It

1. Flashes of blue-green light were seen.
2. All the wintergreen mints emitted light. The wintergreen flavor must be energized and release blue-green light.

Teaching Tips

After students study Section 5.3, have them evaluate and revise their proposed explanations for the behavior of the mint.

CHAPTER 5

Electrons In Atoms

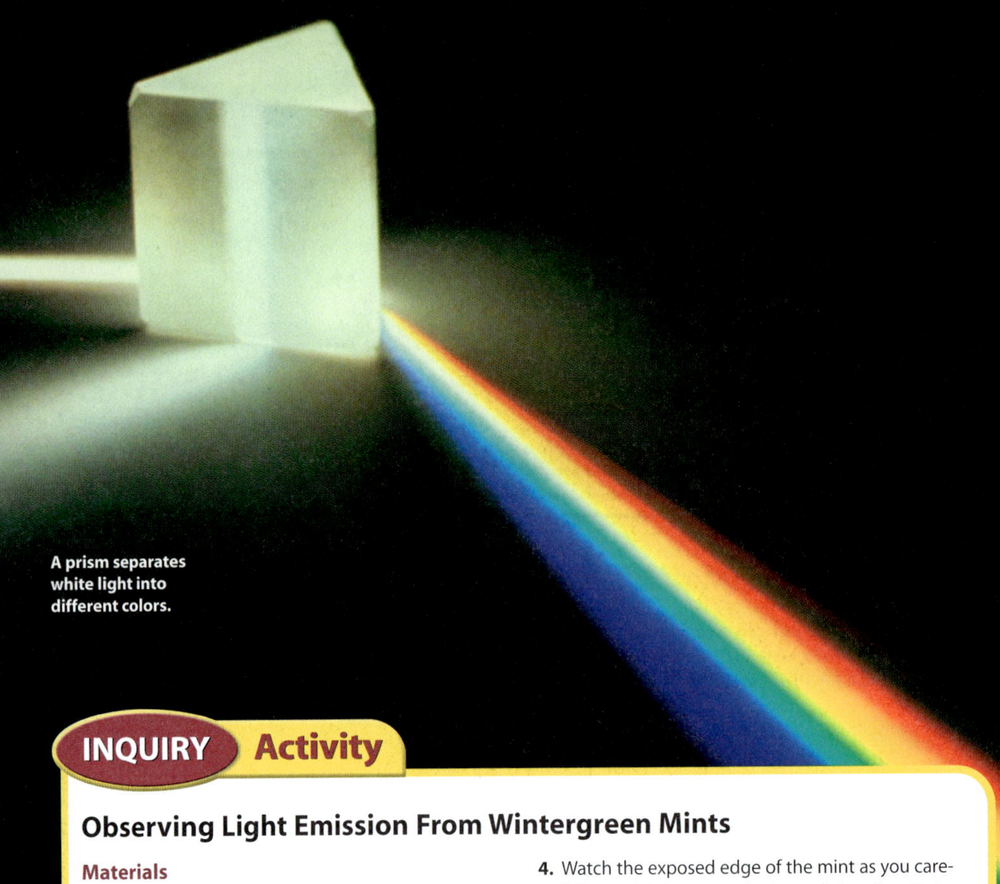

A prism separates white light into different colors.

INQUIRY Activity

Observing Light Emission From Wintergreen Mints

Materials

3 different brands of wintergreen mints, a pair of pliers, transparent tape, and a dark room

Procedure

1. Break each mint in half.
2. Wrap the jaws of the pliers with transparent tape. **Caution** *Handle pliers carefully to avoid pinching your skin.*
3. Turn off the lights and allow your eyes to adjust to the darkness. **Caution** *Never taste any material used in the lab, including candy.*
4. Watch the exposed edge of the mint as you carefully crush it between the jaws of the pliers. Note the color and brightness of any emitted light.
5. Repeat Step 4 for the other two mints.

Think About It

1. What did you observe when you crushed the mints?
2. Did all the mints emit light? Propose an explanation for your observations.

126

5.1 Models of the Atom

Connecting to Your World Aeronautical engineers use wind tunnels and scale models to simulate and test the forces from the moving air on each proposed design. The scale model shown is a physical model. However, not all models are physical. In fact, several theoretical models of the atom have been developed over the last few hundred years. In this section, you will learn about the currently accepted model of how electrons behave in atoms.

Guide for Reading

 **Key Concepts**

- What was inadequate about Rutherford's atomic model?
- What was the new proposal in the Bohr model of the atom?
- What does the quantum mechanical model determine about the electrons in an atom?
- How do sublevels of principal energy levels differ?

Vocabulary

energy levels
quantum
quantum mechanical model
atomic orbital

Reading Strategy

Using Prior Knowledge Before you read, jot down three things you already know about atoms. As you read the section, explain how what you already knew helped you learn something new.

The Development of Atomic Models

So far in this textbook, the model for the atom consisted of protons and neutrons making up a nucleus surrounded by electrons. After discovering the atomic nucleus, Rutherford used existing ideas about the atom and proposed an atomic model in which the electrons move around the nucleus, like the planets move around the sun. Rutherford's model explained only a few simple properties of atoms. It could not explain, for example, why metals or compounds of metals give off characteristic colors when heated in a flame, or why objects—when heated to higher and higher temperatures—first glow dull red, then yellow, then white, as shown in Figure 5.1. **Rutherford's atomic model could not explain the chemical properties of elements.** Explaining what leads to the chemical properties of elements requires a model that better describes the behavior of electrons within atoms.

Figure 5.1 Rutherford's model fails to explain why objects change color when heated. As the temperature of this horseshoe is increased, it first appears black, then red, then yellow, and then white. The observed behavior could be explained only if the atoms in the iron gave off light in specific amounts of energy. A better atomic model was needed to explain this observation.

Section Resources

Print

- **Guided Reading and Study Workbook,** Section 5.1
- **Core Teaching Resources,** Section 5.1 Review
- **Transparencies,** T57

Technology

- **Interactive Textbook with ChemASAP,** Animation 5, Assessment 5.1

5.1

1 FOCUS

Objectives

5.1.1 Identify the inadequacies in the Rutherford atomic model.

5.1.2 Identify the new proposal in the Bohr model of the atom.

5.1.3 Describe the energies and positions of electrons according to the quantum mechanical model.

5.1.4 Describe how the shapes of orbitals related to different sublevels differ.

Guide for Reading

Build Vocabulary 

Word Parts The word *orbital* was coined by scientists in 1932 to describe the space in which an atom has a high probability of being found. It is derived from the word *orbit*, the path around the nucleus that electrons had been postulated to follow.

Reading Strategy 

Relate Text and Visuals Have students follow the Time Line in Figure 5.2 and explain why each model was inadequate or incomplete.

2 INSTRUCT

Connecting to Your World

Have students study the photograph and read the text. Ask, **What are two types of models?** *(physical and theoretical)* **Why are models important?** *(You can test a model to see if your ideas or design are correct.)*

The Development of Atomic Models

Relate 

Compare the use of atomic models to develop the atomic theory to the use of physical or computer-generated models to design, develop, and test products, such as automobiles.

The Bohr Model

Discuss L2

Prepare students for the expanded view of the atom by asking, **What are the three major subatomic particles that comprise an atom?** *(electron, proton, and neutron)* **What are the electrical charges associated with each of these particles?** *(electron, 1–; proton, 1+; neutron, 0)* **Describe the structure of the nuclear atom in terms of the locations of each of the subatomic particles.** *(Every atom, except hydrogen, consists of a small dense nucleus composed of protons and neutrons that accounts for most of the mass of the atom. Hydrogen is an exception because it has only one proton in its nucleus. Negatively charged electrons surround the nucleus and occupy most of its volume. Electrons contribute little to the mass of an atom.)*

Word *Origins* 

The word *quantity* has the same root as *quantum* and means "an indefinite amount or number."

TEACHER Demo

Quantized Energy 

Purpose Students witness an example of quantized energy.

Materials trumpet or trombone and a player

Procedure Have a student blow into a trumpet or trombone. Challenge the student to produce as many different notes as he or she can without depressing any valves.

Expected Outcomes It is not possible to play the entire scale without changing the valve positions. In the open position, no matter how much energy is put in, the instrument will only accept certain specific amounts of energy and produce certain notes. The instrument accepts quantized packages of energy.

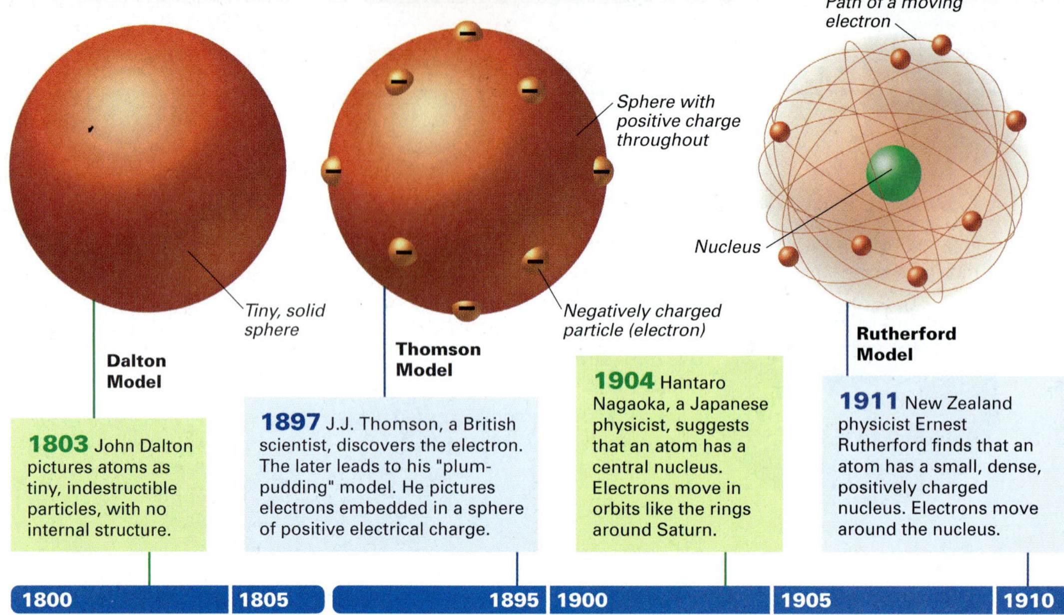

Figure 5.2 These illustrations show how the atomic model has changed as scientists learned more about the atom's structure.

The Bohr Model

Niels Bohr (1885–1962), a young Danish physicist and a student of Rutherford, believed Rutherford's model needed improvement. In 1913 Bohr changed Rutherford's model to include newer discoveries about how the energy of an atom changes when it absorbs or emits light. He considered the simplest atom, hydrogen, which has one electron. **Bohr proposed that an electron is found only in specific circular paths, or orbits, around the nucleus.** The timeline in Figure 5.2 shows the development of atomic models from 1800 to 1935.

Each possible electron orbit in Bohr's model has a fixed energy. The fixed energies an electron can have are called **energy levels.** The fixed energy levels of electrons are somewhat like the rungs of the ladder in Figure 5.3a. The lowest rung of the ladder corresponds to the lowest energy level. A person can climb up or down a ladder by going from rung to rung. Similarly, an electron can jump from one energy level to another. A person on a ladder cannot stand between the rungs. Similarly, the electrons in an atom cannot be between energy levels. To move from one rung to another, a person climbing a ladder must move just the right distance. To move from one energy level to another, an electron must gain or lose just the right amount of energy. In general, the higher an electron is on the energy ladder, the farther it is from the nucleus.

A **quantum** of energy is the amount of energy required to move an electron from one energy level to another energy level. The energy of an electron is said to be quantized. You have probably heard the term *quantum leap* used to describe an abrupt change. The term originates from the ideas found in the Bohr model of the atom.

Word *Origins*

Quantum comes from the Latin word *quantus*, meaning "how much." **What other commonly used English word comes from this root?**

Differentiated Instruction

English Learners 

Have students make a poster to show the evolution of the model of the atom. Have them start with Dalton's model and continue through all the modifications to the current quantum mechanical model. Have them include enough information to make the model clear and understandable.

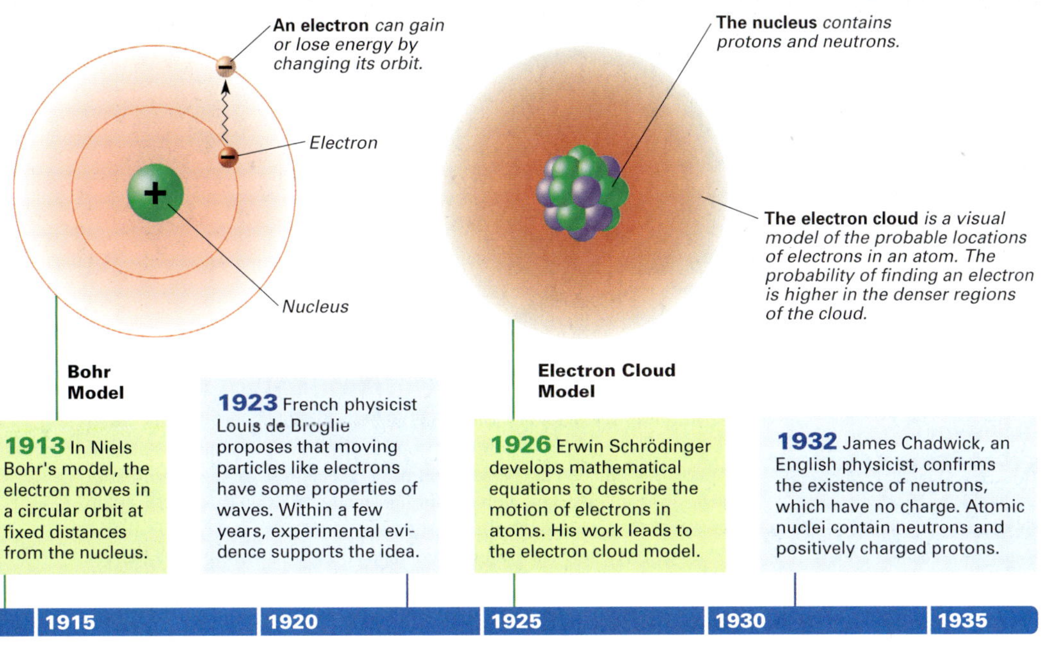

The amount of energy an electron gains or loses in an atom is not always the same. Like the rungs of the strange ladder in Figure 5.3b, the energy levels in an atom are not equally spaced. The higher energy levels are closer together. It takes less energy to climb from one rung to another near the top of the ladder in Figure 5.3b, where the rungs are closer. Similarly, the higher the energy level occupied by an electron, the less energy it takes to move from that energy level to the next higher energy level.

The Bohr model gave results in agreement with experiment for the hydrogen atom. However, it still failed in many ways to explain the energies absorbed and emitted by atoms with more than one electron.

Figure 5.3 These ladder steps are somewhat like energy levels. **a** In an ordinary ladder, the rungs are equally spaced. **b** The energy levels in atoms are unequally spaced, like the rungs in this ladder. The higher energy levels are closer together.

Use Visuals 

Figure 5.3 Ask, **What are the rungs in the ladders compared to?** *(the energy levels in atoms)* **How is the ladder in b different from the ladder in a?** *(The rungs are not evenly spaced. They get closer together nearer to the top.)*

Relate 

When light shines on a fluorescent material, its electrons absorb the energy and move to a higher energy level. Almost immediately, the material begins to emit light as the electrons drop back down to their usual energy level. Have students bring objects painted with fluorescent paint or glow-in-the-dark objects to class and observe them in a darkened room.

TEACHER Demo

Energy and Energy Levels 

Purpose Students observe an analogy for energy transitions in atoms.

Materials one red, one yellow, and one blue ball, step stool with three steps

Procedure Stand on the first step of the step stool. Jump down to the floor while throwing the red ball to a student catcher. Stand on the second step and repeat the process throwing the yellow ball. Repeat jumping from the third step while throwing the blue ball. Ask students to state the analogy. *(The different colored balls are different amounts of energy that correspond to the energy emitted when electrons drop from excited energy levels to a lower level. Blue must be the largest amount of energy and red the smallest.)*

The Quantum Mechanical Model

Discuss L2

Have students imagine a used dartboard that has concentric circles of decreasing value. (The analogy will not work for professional dartboards.) Ask students to describe the distribution of holes. *(More holes are close to the bull's-eye than near the edge, which is analogous to the distribution of electrons in an atom. Probability diminishes with distance from the nucleus, approaching zero as distance approaches infinity.)* Ask, **If the dartboard is a model for an atom, what are its major failures?** *(It is a 2-dimensional model for a 3-dimensional atom; the distribution of holes may not decrease uniformly)* Have students assume that the holes do decrease uniformly. Ask them to compare the probability for two holes on a circle at a specified distance from the center. *(The probabilities are equal.)* This rule holds true for electrons that are equidistant from the nucleus.

The Shapes of Orbitals L2

Purpose Students observe the shapes of *s* and *p* orbitals and their orientation in a three-dimensional coordinate system.

Materials Three metal or wooden skewers, three foam balls (3 or 4 inches in diameter)

Procedure Push three skewers through a ball at right angles to each other and as close to the ball's center as possible. Ask students what type of orbital it represents. (*s*) Ask them to describe its shape. (*Students may say it is circular. Make the point that it is spherical.*) Identify the *x*-, *y*-, and *z*-axes. Push a skewer through two balls to represent a *p* orbital. Tell students that the skewer can represent any one of the three axes. Ask, **Where along the axis is the atomic nucleus?** *(at the point between the two balls)*

Figure 5.4 The electron cloud of an atom is compared here to photographs of a spinning airplane propeller. ⓐ The airplane propeller is somewhere in the blurry region it produces in this picture, but the picture does not tell you its exact position at any instant. ⓑ Similarly, the electron cloud of an atom represents the locations where an electron is likely to be found.

The Quantum Mechanical Model

The Rutherford planetary model and the Bohr model of the atom are based on describing paths of moving electrons as you would describe the path of a large moving object. New theoretical calculations and experimental results were inconsistent with describing electron motion this way. In 1926, the Austrian physicist Erwin Schrödinger (1887–1961) used these new results to devise and solve a mathematical equation describing the behavior of the electron in a hydrogen atom. The modern description of the electrons in atoms, the **quantum mechanical model,** comes from the mathematical solutions to the Schrödinger equation.

Like the Bohr model, the quantum mechanical model of the atom restricts the energy of electrons to certain values. Unlike the Bohr model, however, the quantum mechanical model does not involve an exact path the electron takes around the nucleus. **The quantum mechanical model determines the allowed energies an electron can have and how likely it is to find the electron in various locations around the nucleus.**

How likely it is to find the electron in a particular location is described by probability. If you place three red marbles and one green marble into a box and then pick a marble without looking, the probability of picking the green marble is one in four, or 25%. This means that if you put the four marbles in a box and picked one, and repeated this a great many times, you would pick a green marble in 25% of your tries.

The quantum mechanical model description of how the electron moving around the nucleus is similar to the motion of a rotating propeller blade. Figure 5.4a shows that the propeller blade has the same probability of being anywhere in the blurry region it produces in the picture, but you cannot tell its precise location at any instant. Similarly, in the quantum mechanical model of the atom, the probability of finding an electron within a certain volume of space surrounding the nucleus can be represented as a fuzzy cloud. The cloud is more dense where the probability of finding the electron is high. The cloud is less dense where the probability of finding the electron is low. Though it is unclear where the cloud ends, there is at least a slight chance of finding the electron at a considerable distance from the nucleus. Therefore, attempts to show probabilities as a fuzzy cloud are usually limited to the volume in which the electron is found 90% of the time. To visualize an electron probability cloud, imagine that you could mold a sack around the cloud so that the electron was inside the sack 90% of the time. The shape of the sack would then give you a useful picture of the shape of the cloud.

Facts and Figures

Development of Wave Mechanics

Schrödinger's equation accounted mathematically for de Broglie's discovery of the wave-like properties of the electron. These developments led to a major break away from the Bohr model of the atom. Schrödinger extended de Broglie's work by considering the movement of a particle in an electromagnetic field, which established an area of physics known as wave mechanics.

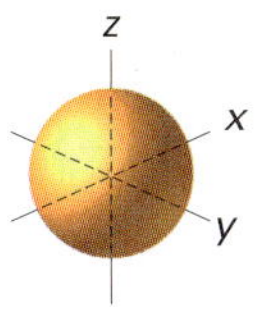

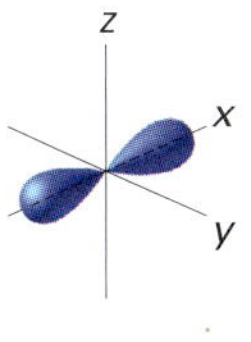

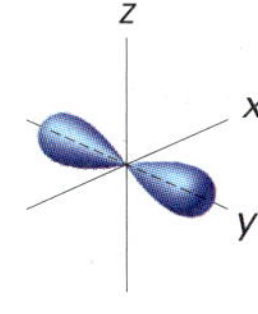

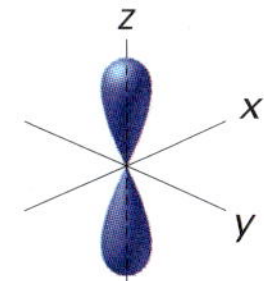

Figure 5.5 The electron clouds for the *s* orbital and the *p* orbitals are shown here.

Atomic Orbitals

Solving the Schrödinger equation gives the energies an electron can have. These are its energy levels. For each energy level, the Schrödinger equation also leads to a mathematical expression, called an atomic orbital, describing the probability of finding an electron at various locations around the nucleus. An **atomic orbital** is often thought of as a region of space in which there is a high probability of finding an electron.

The energy levels of electrons in the quantum mechanical model are labeled by principal quantum numbers (n). These are assigned the values $n = 1, 2, 3, 4$, and so forth. For each principal energy level, there may be several orbitals with different shapes and at different energy levels. These energy levels within a principal energy level constitute energy sublevels.  **Each energy sublevel corresponds to an orbital of a different shape, which describes where the electron is likely to be found.**

Different atomic orbitals are denoted by letters. As shown in Figure 5.5, *s* orbitals are spherical, and *p* orbitals are dumbbell-shaped. Because of the spherical shape of an *s* orbital, the probability of finding an electron at a given distance from the nucleus in an *s* orbital does not depend on direction. The three kinds of *p* orbitals have different orientations in space.

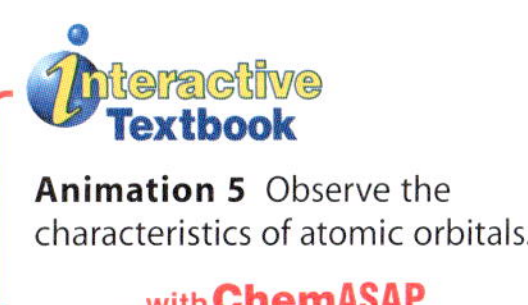

Animation 5 Observe the characteristics of atomic orbitals.

with ChemASAP

Checkpoint *How do s and p orbitals differ?*

Table 5.1

Summary of Principal Energy Levels, Sublevels, and Orbitals

Principal energy level	Number of sublevels	Type of sublevel
$n = 1$	1	1*s* (1 orbital)
$n = 2$	2	2*s* (1 orbital), 2*p* (3 orbitals)
$n = 3$	3	3*s* (1 orbital), 3*p* (3 orbitals), 3*d* (5 orbitals)
$n = 4$	4	4*s* (1 orbital), 4*p* (3 orbitals), 4*d* (5 orbitals), 4*f* (7 orbitals)

Atomic Orbitals

Discuss 

Present the orbital concept as an outgrowth of the quantum mechanical model and describe the shapes of the *s* and *p* orbitals. Stress that the quantum mechanical model predicts the shapes of the various orbitals. The fact that experiments affirm the model's predictions is validation that the model is correct. Remind students that an orbital is only a region of mathematical probability, not a concrete item that can be seen or felt. An orbital does not have the sharp boundaries that models and diagrams seem to indicate.

Use Visuals 

Table 5.1 Explain that every atom has an infinite number of energy levels, but no element, when it is in its ground state, requires more than seven energy levels to describe the probability regions (orbitals) where its electrons can be found. Ask, **How many sublevels are in level 5?** (*5*) **in level 6?** (*6*) **in level 7?** (*7*). **Review the types of sublevels that are in level 5.** [*5s; 1 orbital, 5p; 3 orbitals, 5d; 5 orbitals, 5f; 7 orbitals,* and *5g; 9 orbitals*] Ask, **What is the mathematical relationship between type (*s, p, d, f, g* and so on) and number of orbitals?** (*1, 3, 5, 7, 9 and so on*). Ask, **How many orbitals relate to level 1?** (*1*) **level 2?** (*4*) **level 3?** (*9*) **level 4?** (*16*) **Ask students if they see a pattern emerging.** (*The number of orbitals is the square of the energy level,* n^2.) **How many orbitals would you expect to relate to level 7?** (*49*) Stress that orbitals are mathematical probability regions where electrons may be located. The equation for determining how many electrons can relate to each energy level is $2n^2$ where n represents the energy level.

Differentiated Instruction

Less Proficient Readers 

Have students expand Table 5.1 to include a column with the heading *Number of orbitals* *(n^2)* and another headed *Number of electrons* *($2n^2$)*. Have students fill in the rows. Make sure students do not confuse an orbital with any predictable physical location. Concentrate primarily on *s* and *p* orbitals because they are key to understanding the behavior of representative elements.

Answers to...

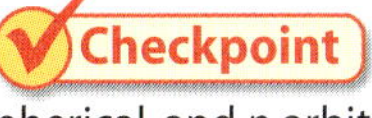 *S* orbitals are spherical, and *p* orbitals are dumbbell-shaped

3 ASSESS

Evaluate Understanding L2

Ask students to discuss the parts of the Bohr model that are valid and the parts that are no longer considered accurate. Ask students to explain what a quantum of energy is and how an electron in an atom moves from one energy level to another.

Reteach L1

Review the following information. Energy levels are designated with the letter *n*. Each energy level has as many sublevels as the number of the energy level. When $n = 2$, there are 2 sublevels. Each energy level relates to n^2 orbitals. When $n = 3$, there are 3 sublevels and 9 orbitals ($1s + 3p + 5d = 9$). Each orbital contains up to 2 electrons. The maximum number of electrons at each energy level is $2n^2$. Thus, energy level 3 has up to 18 electrons.

Connecting Concepts

The Dalton model was a hard sphere; the Thomson model was a sphere of positive charge with embedded negative charges; the Rutherford model was a tiny nucleus containing positive charge with circulating electrons. The quantum mechanical model is of a tiny nucleus with electrons dispersed in a cloud surrounding it. The electron positions are dictated by probability.

If your class subscribes to the Interactive Textbook, use it to review key concepts in Section 5.1

with ChemASAP

Answers to...

Figure 5.6 Both lie in the xy plane. The lobes of the d_{xy} orbital lie between the *x* and *y* axes. Those of the $d_{x^2-y^2}$ orbital lie along the *x* and *y* axes.

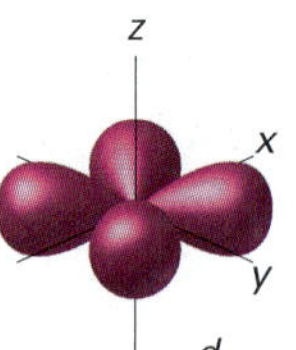

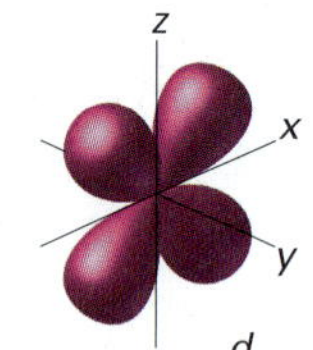

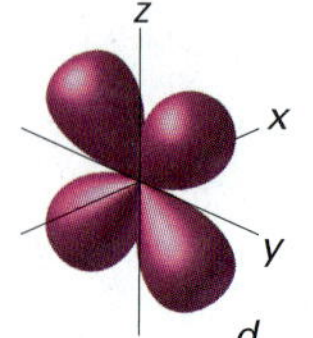

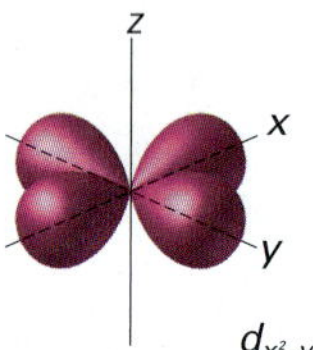

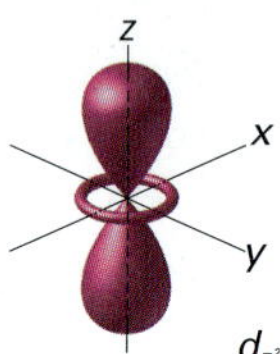

Figure 5.6 The *d* orbitals are illustrated here. Four of the five *d* orbitals have the same shape but different orientations in space. **Interpreting Diagrams** ***How are the orientations of the d_{xy} and $d_{x^2-y^2}$ orbitals similar? How are they different?***

Figure 5.6 shows the shapes of *d* orbitals. Four of the five kinds of *d* orbitals have clover leaf shapes. The shapes of *f* orbitals are more complicated than for *d* orbitals.

The numbers and kinds of atomic orbitals depend on the energy sublevel. The lowest principal energy level ($n = 1$) has only one sublevel, called 1*s*.

The second principal energy level ($n = 2$) has two sublevels, 2*s* and 2*p*. The 2*p* sublevel is of higher energy than the 2*s* and consists of three *p* orbitals of equal energy. The long axis of each dumbbell-shaped *p* orbital is perpendicular to the other two. It is convenient to label these orbitals $2p_x$, $2p_y$, and $2p_z$. Thus the second principal energy level has four orbitals: 2*s*, $2p_x$, $2p_y$, and $2p_z$.

The third principal energy level ($n = 3$) has three sublevels. These are called 3*s*, 3*p*, and 3*d*. As shown in Figure 5.6, the 3*d* sublevel consists of five *d* orbitals of equal energy. Thus the third principal energy level has nine orbitals (one 3*s*, three 3*p*, and five 3*d* orbitals).

The fourth principal energy level ($n = 4$) has four sublevels, called 4*s*, 4*p*, 4*d*, and 4*f*. The 4*f* sublevel consists of seven *f* orbitals of equal energy. The fourth principal energy level, then, has 16 orbitals (one 4*s*, three 4*p*, five 4*d*, and seven 4*f* orbitals).

As mentioned, the principal quantum number always equals the number of sublevels within that principal energy level. The maximum number of electrons that can occupy a principal energy level is given by the formula $2n^2$, where *n* is the principal quantum number. The number of electrons allowed in each of the first four energy levels is shown in Table 5.2.

Table 5.2
Maximum Numbers of Electrons

Energy level *n*	Maximum number of electrons
1	2
2	8
3	18
4	32

5.1 Section Assessment

1. **Key Concept** Why did Rutherford's atomic model need to be replaced?
2. **Key Concept** What was the basic new proposal in the Bohr model of the atom?
3. **Key Concept** What does the quantum mechanical model determine about electrons in atoms?
4. **Key Concept** How do two sublevels of the same principal energy level differ from each other?
5. How can electrons in an atom move from one energy level to another?
6. The energies of electrons are said to be quantized. Explain what this means.
7. How many orbitals are in the following sublevels?
   a. 3*p* sublevel **b.** 2*s* sublevel **c.** 4*p* sublevel
   d. 3*d* sublevel **e.** 4*f* sublevel

Connecting Concepts

Reread the materials on the quantum mechanical model of the atom. Describe how the quantum mechanical model differs from Dalton's model, from Thomson's model, and from Rutherford's model.

Assessment 5.1 Test yourself on the concepts in Section 5.1.

with ChemASAP

Section 5.1 Assessment

1. It could not explain why metals or compounds of metals give off characteristic colors when heated nor the chemical properties of the elements.
2. An electron is found only in specific circular paths or orbits around the nucleus.
3. It determines the allowed energy levels an electron can have and the likelihood of finding an electron in various locations around the nucleus.
4. The sublevels have different shapes.
5. by losing or gaining just the right amount of energy – a quantum
6. In an atom, the electrons can have certain fixed energy levels. To move from one energy level to another requires the emission or absorption of an exact amount of energy, or quantum. Thus the energy of the electron is said to be quantized.
7. **a.** 3 **b.** 1 **c.** 3 **d.** 5 **e.** 7

5.2 Electron Arrangement in Atoms

Connecting to Your World Does this scene look natural to you? Surprisingly, it is. Arrangements like this are rare in nature because they are unstable. Unstable arrangements, whether the grains of sand in a sandcastle or the rock formation shown here, tend to become more stable by losing energy. If this rock were to tumble over, it would end up at a lower height. It would have less energy than before, but its position would be more stable. In this section, you will learn that energy and stability play an important role in determining how electrons are configured in an atom.

Guide for Reading

Key Concepts

- What are the three rules for writing the electron configurations of elements?
- Why do actual electron configurations for some elements differ from those assigned using the aufbau principle?

Vocabulary

electron configurations
aufbau principle
Pauli exclusion principle
Hund's rule

Reading Strategy

Building Vocabulary As you read the section, write a definition of each vocabulary term in your own words.

Electron Configurations

Try to balance a pencil on its point. Each time you try, the pencil falls over. At the end of its fall, its energy has decreased. In most natural phenomena, change proceeds toward the lowest possible energy. In an atom, electrons and the nucleus interact to make the most stable arrangement possible. The ways in which electrons are arranged in various orbitals around the nuclei of atoms are called **electron configurations.**

Three rules—the aufbau principle, the Pauli exclusion principle, and Hund's rule—tell you how to find the electron configurations of atoms. The three rules are as follows.

Aufbau Principle According to the **aufbau principle,** electrons occupy the orbitals of lowest energy first. Look at the aufbau diagram in Figure 5.7. Each box represents an atomic orbital.

Figure 5.7 This aufbau diagram shows the energy levels of the various atomic orbitals. Orbitals of greater energy are higher on the diagram. **Using Tables** ***Which is of higher energy, a 4d or a 5s orbital?***

Section Resources

Print

- **Guided Reading and Study Workbook,** Section 5.2
- **Core Teaching Resources,** Section 5.2 Review
- **Small-Scale Chemistry Laboratory Manual,** Lab 7
- **Transparencies,** T58–T60

Technology

- **Interactive Textbook with ChemASAP,** Simulation 2, Problem-Solving 5.9, Assessment 5.2
- **Go Online,** Section 5.2

5.2

1 FOCUS

Objectives

5.2.1 **Describe** how to write the electron configuration for an atom.

5.2.2 **Explain** why the actual electron configurations for some elements differ from those predicted by the aufbau principle.

Guide for Reading

Build Vocabulary 

Word Forms Explain that a configuration is an arrangement of parts. An electron configuration is the arrangement of electrons in orbitals around the nucleus of an atom

Reading Strategy 

SQ3R Survey, Question, Read, Recite, Review Encourage students to apply SQ3R as they read this section. Model how to employ it for the first two pages.

2 INSTRUCT

Connecting to Your World

Because of gravity, rocks, such as the one shown, are less stable than rocks on level ground. Explain that electrons have a comparable ground level. Ask, **What role do energy and stability play in the way that electrons are configured in an atom?** *(In the most stable atoms, electrons occupy the lowest possible energy orbitals.)*

Electron Configurations

Use Visuals 

Figure 5.7 Ask, **Do all the orbitals at energy level $n = 4$ have the same energy?** *(No, 4p orbitals have higher energy than 4s, and 4d have higher energy than 4p.)* **Which has higher energy, 4s or 3d ?** *(3d)*

Answers to...
Figure 5.7 4d

Download a worksheet on **Electron Configuration** for students to complete, and find additional teacher support from NSTA SciLinks.

Discuss

L2

Develop the electron configurations for several of the simpler elements. Introduce each rule governing the process as needed. Begin with hydrogen. Use the aufbau diagram to explain that electrons enter orbitals of lowest energy first. Show how the orbital notation ($1s^1$) describes the energy level, the orbital, and the number of electrons. Repeat the process for helium. Then continue with lithium, beryllium, and boron. Apply the Pauli exclusion principle to explain why additional orbitals must be used. When you reach carbon, explain and apply Hund's rule. Complete the configurations for the second period elements.

CLASS Activity

Writing Electron Configurations

L2

Purpose Students gain practice in determining the correct order for filling orbitals.

Procedure Have students work with a partner to develop the electron configurations for the third period elements. When the exercise is complete, ask students to compare the configurations of second and third period elements in preparation for future discussions of the periodic table.

The orbitals for any sublevel of a principal energy level are always of equal energy. Further, within a principal energy level the *s* sublevel is always the lowest-energy sublevel. Yet the range of energy levels within a principal energy level can overlap the energy levels of another principal level. Notice again in Figure 5.7 that the filling of atomic orbitals does not follow a simple pattern beyond the second energy level. For example, the 4*s* orbital is lower in energy than a 3*d* orbital.

Pauli Exclusion Principle According to the **Pauli exclusion principle,** an atomic orbital may describe at most two electrons. For example, either one or two electrons can occupy an *s* orbital or a *p* orbital. To occupy the same orbital, two electrons must have opposite spins; that is, the electron spins must be paired. Spin is a quantum mechanical property of electrons and may be thought of as clockwise or counterclockwise. A vertical arrow indicates an electron and its direction of spin (↑ or ↓). An orbital containing paired electrons is written as [↑↓].

Hund's Rule When you use the aufbau diagram to decide how electrons occupy orbitals of equal energy, one electron enters each orbital until all the orbitals contain one electron with the same spin direction. **Hund's rule** states that electrons occupy orbitals of the same energy in a way that makes the number of electrons with the same spin direction as large as possible. For example, three electrons would occupy three orbitals of equal energy as follows: [↑][↑][↑]. Second electrons then occupy each orbital so that their spins are paired with the first electron in the orbital. Thus each orbital can eventually have two electrons with paired spins.

Table 5.3

Electron Configurations for Some Selected Elements

Element	Orbital filling						Electron configuration
	1s	2s	$2p_x$	$2p_y$	$2p_z$	3s	
H	↑						$1s^1$
He	↑↓						$1s^2$
Li	↑↓	↑					$1s^22s^1$
C	↑↓	↑↓	↑	↑			$1s^22s^22p^2$
N	↑↓	↑↓	↑	↑	↑		$1s^22s^22p^3$
O	↑↓	↑↓	↑↓	↑	↑		$1s^22s^22p^4$
F	↑↓	↑↓	↑↓	↑↓	↑		$1s^22s^22p^5$
Ne	↑↓	↑↓	↑↓	↑↓	↑↓		$1s^22s^22p^6$
Na	↑↓	↑↓	↑↓	↑↓	↑↓	↑	$1s^22s^22p^63s^1$

Simulation 2 Fill atomic orbitals to build the ground state of several atoms.

with ChemASAP

Differentiated Instruction

Gifted and Talented

Students may be interested in learning that an electron in an orbital is described precisely by four quantum numbers, the first being the principle quantum number, *n*. Direct them to a college text and ask them to name and explain the three other quantum numbers. *(The azimuthal quantum number, l, defines the shape of the orbital; the magnetic quantum number, m, describes the orientation of the orbital in space; and the spin quantum number,* m_s*, denotes the direction of electron spin, + or –.)*

Look at the orbital filling diagrams of the atoms listed in Table 5.3. An oxygen atom contains eight electrons. The orbital of lowest energy, 1*s*, has one electron, then a second electron of opposite spin. The next orbital to fill is 2*s*. It also has one electron, then a second electron of opposite spin. One electron then occupies each of the three 2*p* orbitals of equal energy. The remaining electron now pairs with an electron occupying one of the 2*p* orbitals. The other two 2*p* orbitals remain only half filled, with one electron each.

A convenient shorthand method for showing the electron configuration of an atom involves writing the energy level and the symbol for every sublevel occupied by an electron. You indicate the number of electrons occupying that sublevel with a superscript. For hydrogen, with one electron in a 1*s* orbital, the electron configuration is written $1s^1$. For helium, with two electrons in a 1*s* orbital, the configuration is $1s^2$. For oxygen, with two electrons in a 1*s* orbital, two electrons in a 2*s* orbital, and four electrons in 2*p* orbitals, it is $1s^22s^22p^4$. Note that the sum of the superscripts equals the number of electrons in the atom.

When the configurations are written, the sublevels within the same principal energy level are generally written together. This is not always the same order as given on the aufbau diagram. The 3*d* sublevel, for example, is written before the 4*s* sublevel, even though the aufbau diagram shows the 4*s* sublevel to have lower energy.

For: Links on Electron Configuration
Visit: www.SciLinks.org
Web Code: cdn-1052

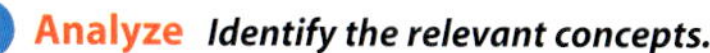

CONCEPTUAL PROBLEM 5.1

Writing Electron Configurations

Phosphorus, an element used in matches, has an atomic number of 15. Write the electron configuration of a phosphorus atom.

1 Analyze ***Identify the relevant concepts.***
Phosphorus has 15 electrons. There is a maximum of two electrons per orbital. Electrons do not pair up within an energy sublevel (orbitals of equal energy) until each orbital already has one electron.

2 Solve ***Apply concepts to this situation.***
Using Figure 5.7 on page 133, place electrons in the orbital with the lowest energy (1*s*) first, then continue placing electrons in each orbital with the next higher energy.

The electron configuration of phosphorus is $1s^22s^22p^63s^23p^3$.
The superscripts add up to the number of electrons. When the configurations are written, the sublevels within the same principal energy level are written together. This is not always the same order as given on the aufbau diagram.

Practice Problems

8. Write the electron configuration for each atom.
a. carbon **b.** argon **c.** nickel

9. Write the electron configuration for each atom. How many unpaired electrons does each atom have?
a. boron **b.** silicon

Problem Solving 5.9
Solve Problem 9 with the help of an interactive guided tutorial.
with ChemASAP

Discuss L2

Students may ask how two negative electrons can occupy the same orbital, or why an orbital never contains more than two electrons. The answer lies in the quantum property known as spin. A spinning electron acts as a tiny magnet, with a north pole at one end and a south pole at the other. If electrons with opposite spins enter one orbital, the attraction between their opposite magnetic poles counteracts some of the electrical repulsive force. If a third electron tries to enter the orbital, its spin will always be the same as one of the existing electrons, and it will be repelled.

CONCEPTUAL PROBLEM 5.1

Answers

8. a. $1s^22s^22p^2$
b. $1s^22s^22p^63s^23p^6$
c. $1s^22s^22p^63s^23p^63d^84s^2$

9. a. $1s^22s^22p^1$; one unpaired electron
b. $1s^22s^22p^63s^23p^2$; two unpaired electrons

Practice Problems Plus L2

What are the electron configurations for atoms of the following elements? How many unpaired electrons does each atom have?
a. neon (*$1s^22s^22p^6$; no unpaired electrons*) **b.** sulfur (*$1s^22s^22p^63s^23p^4$; two unpaired electrons*)

Students may find it helpful to use a diagonal diagram to write electron configurations until they become familiar with the order in which sublevels fill.

Differentiated Instruction

Special Needs L1

Encourage students to draw the aufbau diagram (Figure 5.7) on a 5 x 7 card and keep it handy when writing electron configurations. Also encourage them to make flash cards, each with the name and atomic number of an element on one side and its electron configuration on the other side. Pairs of students can use these to quiz each other.

Exceptional Electron Configurations

Discuss L2

Among the transition elements there are some exceptions to the filling rules. Exceptions can be explained by the atom's tendency to keep its energy as low as possible. These exceptions help explain the unexpected chemical behavior of transition elements.

3 ASSESS

Evaluate Understanding L2

Make a set of small cards, each with the symbol and atomic number of an element. Choose elements from throughout the periodic table. Have students choose cards and write the electron configurations for the elements.

Reteach L1

Have students work in small groups, using the set of element cards to practice writing more electron configurations. Have them refer to Figure 5.7, if necessary.

Writing Activity

Students are likely to write that the two magnets would push each other apart. In the same way, electrons with the same spin would push apart and be unable to occupy the same orbital.

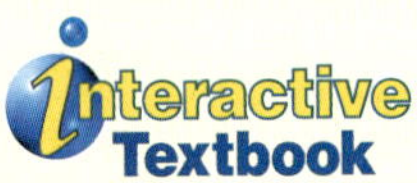

If your class subscribes to the Interactive Textbook, use it to review key concepts in Section 5.2.

with ChemASAP

Answers to...

Checkpoint Exceptions to the aufbau principle are due to subtle electronic interactions in orbitals with similar energies.

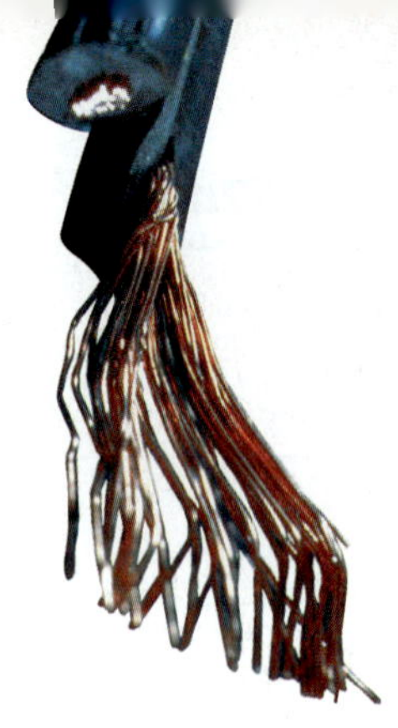

Figure 5.8 Copper is a good conductor of electricity and is commonly used in electrical wiring.

Exceptional Electron Configurations

Copper, shown in Figure 5.8, has an electron configuration that is an exception to the aufbau principle. You can obtain correct electron configurations for the elements up to vanadium (atomic number 23) by following the aufbau diagram for orbital filling. If you were to continue in that fashion, however, you would assign chromium and copper the following incorrect configurations:

$$\text{Cr } 1s^2\,2s^2\,2p^6\,3s^2\,3p^6\,3d^4\,4s^2$$
$$\text{Cu } 1s^2\,2s^2\,2p^6\,3s^2\,3p^6\,3d^9\,4s^2$$

The correct electron configurations are as follows:

$$\text{Cr } 1s^2\,2s^2\,2p^6\,3s^2\,3p^6\,3d^5\,4s^1$$
$$\text{Cu } 1s^2\,2s^2\,2p^6\,3s^2\,3p^6\,3d^{10}\,4s^1$$

These arrangements give chromium a half-filled *d* sublevel and copper a filled *d* sublevel. Filled energy sublevels are more stable than partially filled sublevels. **Some actual electron configurations differ from those assigned using the aufbau principle because half-filled sublevels are not as stable as filled sublevels, but they are more stable than other configurations.** This tendency overcomes the small difference between the energies of the 3*d* and 4*s* sublevels in copper and chromium.

Exceptions to the aufbau principle are due to subtle electron-electron interactions in orbitals with very similar energies. At higher principal quantum numbers, energy differences between some sublevels (such as 5*f* and 6*d*, for example) are even smaller than in the chromium and copper examples. As a result, there are other exceptions to the aufbau principle. Although it is worth knowing that exceptions to the aufbau principle occur, it is more important to understand the general rules for determining electron configurations in the many cases where the aufbau rule applies.

How are energy differences and exceptions to the aufbau principle related?

5.2 Section Assessment

10. **Key Concept** What are the three rules for writing the electron configuration of elements?
11. **Key Concept** Explain why the actual electron configurations for some elements differ from those assigned using the aufbau principle.
12. Use Figure 5.7 to arrange the following sublevels in order of decreasing energy: 2*p*, 4*s*, 3*s*, 3*d*, and 3*p*.
13. Why does one electron in a potassium atom go into the fourth energy level instead of squeezing into the third energy level along with the eight already there?

Writing Activity

Modeling the Pauli Exclusion Principle Write a brief description of how trying to place two bar magnets pointing in the same direction alongside each other is like trying to place two electrons into the same orbital.

Assessment 5.2 Test yourself on the concepts in Section 5.2.

with ChemASAP

Section 5.2 Assessment

10. aufbau principle, Pauli exclusion principle, Hund's rule
11. Half-filled sublevels and filled sublevels are more stable than other configurations.
12. 3*d*, 4*s*, 3*p*, 3*s*, 2*p*
13. The 3*s* and 3*p* orbitals are already filled, so the last electron must go to the next higher energy orbital, which is 4*s*.

Small-Scale LAB

Atomic Emission Spectra

Purpose

To build a spectroscope and use it to measure the wavelengths, frequencies, and energies of atomic emission lines

Materials

- cereal box
- tape
- pencil
- black construction paper
- diffraction grating
- ruler
- scissors
- white notebook paper

Procedure  

Tape together two 2.0 cm × 10 cm strips of black construction paper so that they are parallel and form a narrow slit about 2 mm wide. Remove the top of a cereal box and tape the construction paper slit as shown. Cover the rest of the opening with white notebook paper. Cut a square hole (approximately 2 cm per side) and tape a diffraction grating over the hole as shown. Point the spectroscope toward a fluorescent light. Tape up any light leaks. Your lab partner should mark the exact positions of all the colored emission lines you see on the notebook paper. Measure the distances between the violet line and the other lines you have marked.

Analyze

1. List the number of distinct lines that you see as well as their colors.
2. Each line you see has a property called its wavelength. The prominent violet line has a wavelength of 436 nm and the prominent green line is 546 nm. How many mm apart are these lines on the paper? By how many nm do their wavelengths differ? How many nanometers of wavelength are represented by each millimeter you measured?

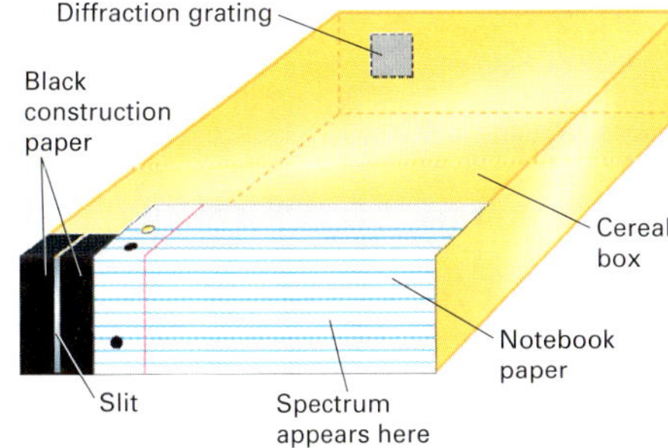

3. Using the nm/mm value you calculated in Step 2 and the mm distance you measured for each line from the violet reference line, calculate the wavelengths of all the other lines you see.
4. Each wavelength corresponds to another property of light called its frequency. Use the wavelength value of each line to calculate its frequency given that $\nu = c/\lambda$ where $c = 2.998 \times 10^{17}$ nm/s (2.998×10^{8} m/s).
5. The energy (E) of a quantum of light an atom emits is related to its frequency (ν) by $E = h \times \nu$. Use the frequency value for each line and $h = 6.63 \times 10^{-34}$ J·s to calculate its corresponding energy.

You're The Chemist

1. **Design It!** Design and carry out an experiment to measure the longest and shortest wavelengths you can see in daylight. Use your spectroscope to observe light from daylight reflected off a white piece of paper. **Caution:** *Do not look directly at the sun!* Describe the differences in daylight and fluorescent light.
2. **Design It!** Design and carry out an experiment to determine the effect of colored filters on the spectrum of fluorescent light or daylight. For each filter tell which colors are transmitted and which are absorbed.
3. **Analyze It!** Use your spectroscope to observe various atomic emission discharge tubes provided by your teacher. Note and record the lines you see and measure their wavelengths.

Small-Scale LAB

Atomic Emission Spectra L2

Objective After completing this activity, students will be able to:
- build a simple spectroscope and use it to determine wavelength, frequency, and energy of emission lines

Prep Time 10 minutes
Class Time 20 minutes

Teaching Tips Have a completed spectroscope for students to examine. Have students bring in their own cereal boxes. Use holographic diffraction gratings of about 700 lines per millimeter. The diffraction grating needs to be positioned so that the lines on the grating are parallel to the slit. Very narrow slits do not let in enough light to see a bright spectrum.

Expected Outcome Students calculate the wavelength, frequency, and energy of emissions from a fluorescent light.

Analyze

1. Most fluorescent lights display 5 lines: violet, blue, green, yellow and red. All fluorescent lights display violet, green and yellow. Older lights will display only two distinct violet and blue lines. A diffuse yellow line is also evident.
2. A typical box will yield a measurement of 30 mm between the violet and green lines; 546 436 = 110 nm; 110 nm/30 mm = 3.67 nm/mm
3. Blue: 436 nm + (16 mm × 3.67 nm/mm) = 495 nm; Green: 436 nm + (30 mm × 3.67 nm/mm) = 546 nm; Yellow: 436 nm + (42 × 3.67 nm/mm) = 590 nm; Red: 436 nm + (50 × 3.67 nm/mm) = 620 nm
4. Violet: $\nu = 3.00 \times 10^{17}$ nm/s /436 nm = 6.88×10^{14} s^{-1}; Blue: $\nu = 3.00 \times 10^{17}$ nm/s /495 nm = 6.06×10^{14} s^{-1}; Green: $\nu = 3.00 \times 10^{17}$ nm/s/546 nm = 5.49×10^{14} s^{-1}; Yellow: $\nu = 3.00 \times 10^{17}$ nm/s/590 nm = 5.08×10^{14} s^{-1}; Red: $\nu = 3.00 \times 10^{17}$ nm/s/620 nm = 4.84×10^{14} s^{-1}
5. Violet: 6.63×10^{-34} J·s × 6.88×10^{14} s^{-1} = 4.56×10^{-19} J; Blue: 6.63×10^{-34} J·s × 6.06×10^{14} s^{-1} = 4.02×10^{-19} J; Green: 6.63×10^{-34} J·s × 5.49×10^{14} s^{-1} = 3.64×10^{-19} J; Yellow: 6.63×10^{-34} J·s × 5.08×10^{14} s^{-1} = 3.37×10^{-19} J; Red: 6.63×10^{-34} J·s × 4.84×10^{14} s^{-1} = 3.21×10^{-19} J

You're The Chemist

1. Sunlight has more red and more violet and lacks the distinct mercury emission lines of fluorescent light. Typical measurements are: Violet line to edge of red = 59 mm; 436 nm + (59 mm × 3.67 nm/mm) = 653 nm
   Violet line to edge of violet = – 4 mm. 436 nm – (4 × 3.67) nm/mm = 421 nm
2. Translucent colored file tabs from an office supply store work well as filters. A yellow filter transmits green, yellow, orange and red and absorbs violet and blue. A green filter transmits blue, green and yellow and absorbs violet, orange and red. A blue filter transmits violet, blue and green and absorbs yellow, orange and red. A red filter transmits yellow, orange and red and absorbs violet, blue and green.
3. Answers will vary. Provide students with gas discharge tubes containing hydrogen and noble gases.

5.3

1 FOCUS

Objectives

5.3.1 **Describe** the relationship between the wavelength and frequency of light.

5.3.2 **Identify** the source of atomic emission spectra.

5.3.3 **Explain** how the frequencies of emitted light are related to changes in electron energies.

5.3.4 **Distinguish** between quantum mechanics and classical mechanics.

Guide for Reading

Build Vocabulary L2

Word Parts The word *emission* comes from the Latin *emittere* meaning "to send out." The word *spectrum* is the Latin word meaning "appearance." An atomic emission spectrum allows the light sent out from an atom to appear or be seen. Ask, **What is an emissary?** (*a person sent out to represent another person*)

Reading Strategy L2

Relate Text and Visuals Refer students to Figure 5.9 and the text on this page. Then, have them write, in their own words, definitions of wavelength, frequency, and amplitude.

2 INSTRUCT

Connecting to Your World

Ask, **What causes an element such as neon to emit light when heated by an electric current?** (*Energy absorbed by electrons is emitted as light.*)

Light

Use Visuals L1

Figure 5.9 Ask students to assume that both waves are traveling at the same speed. Ask, **Which wave would have more complete wave cycles passing a point in a given time interval?** (*The wave on the right.*) Ask, **Which wave has the higher frequency?** (*The wave on the right.*)

5.3 Physics and the Quantum Mechanical Model

Guide for Reading

Key Concepts

- How are the wavelength and frequency of light related?
- What causes atomic emission spectra?
- How are the frequencies of light an atom emits related to changes of electron energies?
- How does quantum mechanics differ from classical mechanics?

Vocabulary

amplitude
wavelength
frequency
hertz
electromagnetic radiation
spectrum
atomic emission spectrum
ground state
photons
Heisenberg uncertainty principle

Reading Strategy

Monitoring Your Understanding Before you read, preview the Key Concepts, the section heads, the vocabulary terms, and the visuals. List three things you expect to learn. After reading, state what you learned about each item you listed.

Connecting to Your World

If you walk in the evening along a busy street lined with shops and theaters, you are likely to see neon advertising signs. The signs are formed from glass tubes bent in various shapes. An electric current passing through the gas in each glass tube makes the gas glow with its own characteristic color. In this section you will learn why each gas glows with a specific color of light.

Light

The previous sections in this chapter introduced you to some ideas about how electrons in atoms are arranged in orbitals, each with a particular energy level. You also learned how to write electron configurations for atoms. In the remainder of this chapter, you will get a closer look into what led to the development of Schrödinger's equation and the quantum mechanical model of the atom.

Rather curiously, the quantum mechanical model grew out of the study of light. Isaac Newton (1642–1727) tried to explain what was known about the behavior of light by assuming that light consists of particles. By the year 1900, however, there was enough experimental evidence to convince scientists that light consists of waves. Figure 5.9 illustrates some of the properties of waves. As shown, each complete wave cycle starts at zero, increases to its highest value, passes through zero to reach its lowest value, and returns to zero again. The **amplitude** of a wave is the wave's height from zero to the crest, as shown in Figure 5.9. The **wavelength,** represented by λ (the Greek letter lambda), is the distance between the crests. The **frequency,** represented by ν (the Greek letter nu), is the number of wave cycles to pass a given point per unit of time. The units of frequency are usually cycles per second. The SI unit of cycles per second is called a **hertz** (Hz). A hertz can also be expressed as a reciprocal second (s^{-1}).

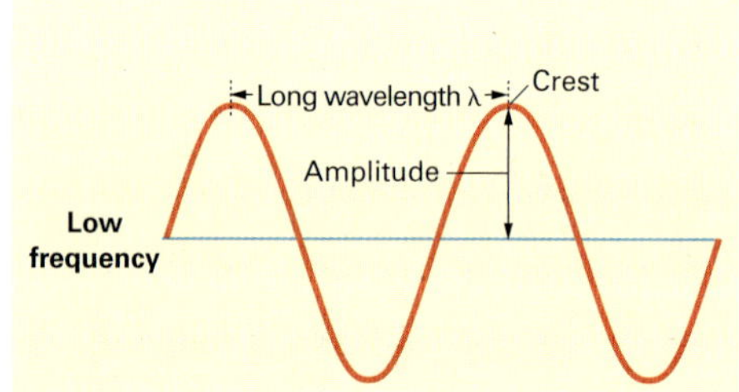

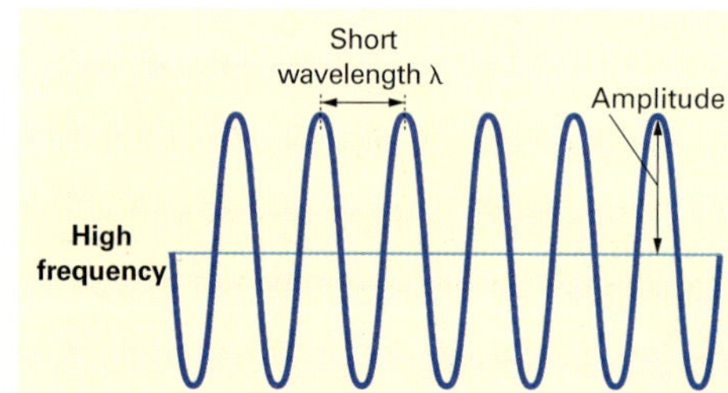

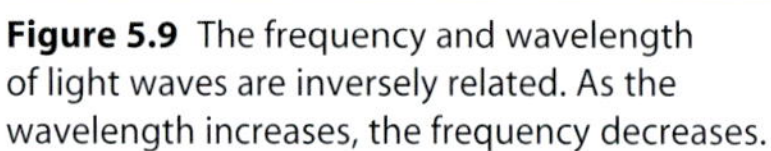
Figure 5.9 The frequency and wavelength of light waves are inversely related. As the wavelength increases, the frequency decreases.

138 Chapter 5

Section Resources

Print

- **Guided Reading and Study Workbook,** Section 5.3
- **Core Teaching Resources,** Section 5.3 Review, Interpreting Graphics
- **Laboratory Manual,** Labs 6–8
- **Small-Scale Chemistry Laboratory Manual,** Lab 8
- **Transparencies,** T61–T64

Technology

- **Interactive Textbook with ChemASAP,** Simulations 3, 4, Problem-Solving 5.15, Animation 6, Assessment 5.3
- **Go Online,** Section 5.3
- **Virtual Chemistry Labs** 1, 7–10

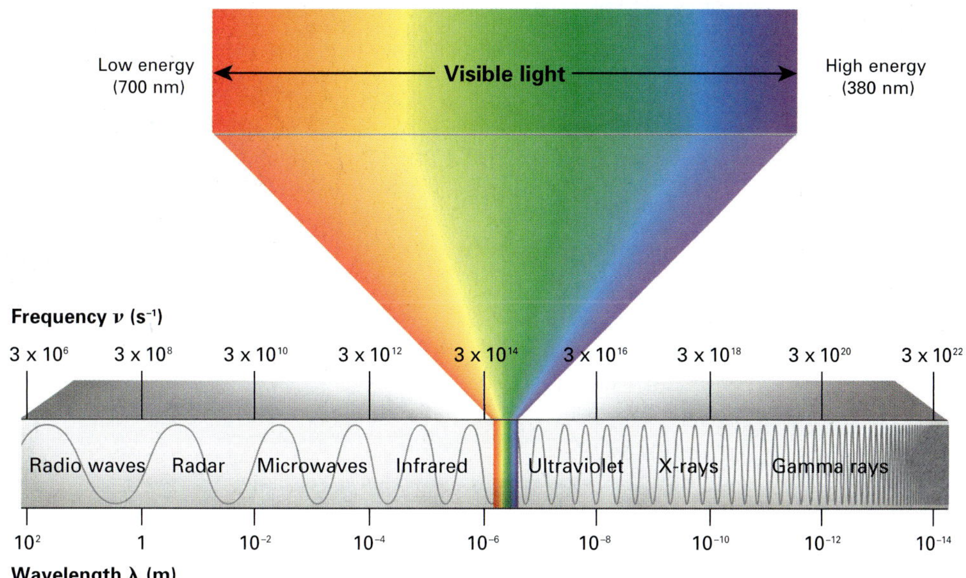

The product of frequency and wavelength always equals a constant (c), the speed of light:

$$c = \lambda\nu$$

The wavelength and frequency of light are inversely proportional to each other. As the wavelength of light increases, for example, the frequency decreases.

According to the wave model, light consists of electromagnetic waves. **Electromagnetic radiation** includes radio waves, microwaves, infrared waves, visible light, ultraviolet waves, X-rays, and gamma rays. All electromagnetic waves travel in a vacuum at a speed of 2.998×10^8 m/s.

Sunlight consists of light with a continuous range of wavelengths and frequencies. As you can see from Figure 5.10, the color of light for each frequency found in sunlight depends on its frequency. When sunlight passes through a prism, the different frequencies separate into a **spectrum** of colors. A rainbow is an example of this phenomenon. Each tiny droplet of water acts as a prism to produce a spectrum. Each color blends into the next in the order red, orange, yellow, green, blue, and violet. In the visible spectrum, as shown in Figure 5.10, red light has the longest wavelength and the lowest frequency.

 ***What color in the visible spectrum has the longest wavelength?***

Figure 5.10 The electromagnetic spectrum consists of radiation over a broad band of wavelengths. The visible light portion is very small. It is in the 10^{-7}m wavelength range and 10^{15} Hz (s^{-1}) frequency range. **Interpreting Diagrams** ***What types of nonvisible radiation have wavelengths close to those of red light? To those of blue light?***

Simulation 3 Explore the properties of electromagnetic radiation.

with **ChemASAP**

CLASS Activity

Black Box Discovery L2

Purpose Students discover information about objects without seeing them.

Materials Several sealed boxes containing different common items such as rubber bands, golf, tennis, or table-tennis balls, cans of different sizes, and different shapes cut from blocks of wood or foam

Procedure Pass the boxes around the room and record student observations and guesses about what is inside each "black box." Allow students to move the boxes and listen to what happens. Point out that they are adding kinetic energy and observing how the hidden object responds. Point out that this method is analogous to the way scientists add energy to atoms and observe how they respond.

Expected Outcome If the items are carefully selected, students will be surprised at how much they can determine without opening the boxes. This activity could also be done as a group activity with groups reporting what they find.

Use Visuals L1

Figure 5.10 Ask students to name some types of radiation that are not visible. (*infrared, ultraviolet, X-rays, gamma rays*) Ask, **Are any of these invisible rays more energetic than visible light, and if so, what evidence do you have?** (*UV causes suntan or sunburns, X-rays penetrate flesh, and gamma rays can penetrate bones.*) Ask, **What do all the rays in the electromagnetic spectrum have in common?** (*All travel at the speed of light.*)

Facts and Figures

Light as a Particle

In 1666, Sir Isaac Newton (1642–1727) carried out important experiments with light that led to the proposal that light consists of tiny particles called corpuscles. Twelve years later, a Dutch scientist, Christian Huygens, suggested a wave theory to explain the properties of light. For more than 200 years, scientists argued about these seemingly contradictory theories. By the year 1900, most scientists had finally accepted that light could be described as a wave. Paradoxically, they soon had to accept that light could, at the same time, be described as a particle.

Answers to...

Figure 5.10 infrared, ultraviolet

 red

 Activity

Properties of Waves 

Purpose Students gain familiarity with wavelength, frequency, and amplitude.

Procedure On a sheet of paper, have students draw a horizontal line approximately 5 inches long. Then, using the line as a midpoint, ask them to draw two waves, superimposed, that have the same amplitudes but different frequencies. Have them draw another horizontal line and two waves with the same wavelength but different amplitudes.

CHEMath

Algebraic Equations

Solving problems in chemistry often involves using algebraic equations such as the one in Sample Problem 5.1. Write the equation $c = \lambda\nu$ on the board. Stress that the equation must be rearranged in order to solve for λ (lambda), that is, λ must be on one side of the equation by itself. Show students that they can do this by dividing both sides of the equation by ν (nu).

Math Handbook

For a math refresher and practice, direct students to algebraic equations, page R69.

Sample Problem 5.2

Answers

14. 2.00×10^{-5} m; longer wavelength than red light

15. $6.00 \times 10^{15}\ s^{-1}$; ultraviolet

Practice Problems Plus L2

What is the frequency of radiation which has a wavelength of 7.00×10^{-5} cm? *($4.29 \times 10^{14}\ s^{-1}$)* **In what region of the electromagnetic spectrum is this radiation?** *(infrared)*

Figure 5.11 Sodium vapor lamps produce a yellow glow.

CHEMath

Algebraic Equations

An algebraic equation shows the relationship between two or more variables. Often, an equation must be solved for the unknown variable before substituting the known values into the equation and doing the arithmetic.

Most equations can be solved if you remember that you can carry out any mathematical operation, such as addition (+), subtraction (−), multiplication (×), or division (x/y or $x \div y$), without destroying the equality, as long as you do it to both sides of the equation.

Math Handbook

For help and practice solving algebraic equations, go to page R69.

Problem-Solving 5.15 Solve Problem 15 with the help of an interactive guided tutorial.

with ChemASAP

SAMPLE PROBLEM 5.1

Calculating the Wavelength of Light

Calculate the wavelength of the yellow light emitted by the sodium lamp shown above if the frequency of the radiation is 5.10×10^{14} Hz (5.10×10^{14}/s).

1 Analyze ***List the knowns and the unknown.***

Knowns
- frequency (ν) = 5.10×10^{14}/s
- $c = 2.998 \times 10^{8}$ m/s

Unknown
- wavelength (λ) = ? m

2 Calculate ***Solve for the unknown.***

Solve the equation $c = \lambda\nu$ for λ.

$$\lambda = \frac{c}{\nu}$$

Substitute the known values and solve.

$$\lambda = \frac{c}{\nu} = \frac{2.998 \times 10^{8}\ \text{m/s}}{5.10 \times 10^{14}/\text{s}} = 5.88 \times 10^{-7}\ \text{m}$$

3 Evaluate ***Does the result make sense?***

The magnitude of the frequency is much larger than the numerical value of the speed of light, so the answer should be much less than 1. The answer should have three significant figures, because the original known value had three significant figures.

Practice Problems

14. What is the wavelength of radiation with a frequency of 1.50×10^{13} Hz? Does this radiation have a longer or shorter wavelength than red light?

15. What is the frequency of radiation with a wavelength of 5.00×10^{-8} m? In what region of the electromagnetic spectrum is this radiation?

Differentiated Instruction

Special Needs 

When introducing the formulas and units in this section, be sure to review mathematical operations for multiplication and division using scientific notation. Also be sure to show that $1/10 = 0.10 = 10^{-1}$. This will help students understand the units used for frequency: cycles per second = cycles/s = s^{-1}, or hertz (Hz).

140 Chapter 5

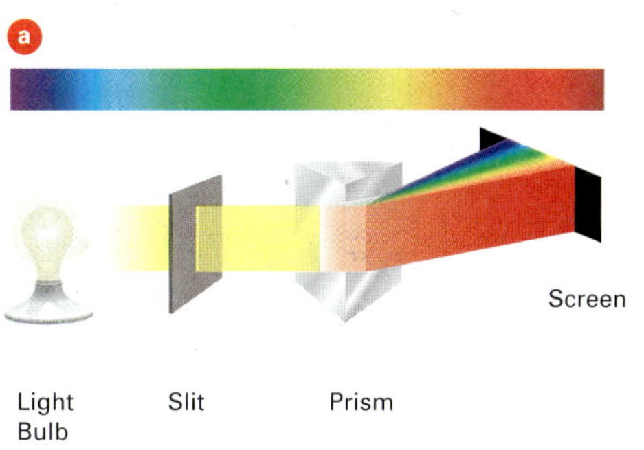

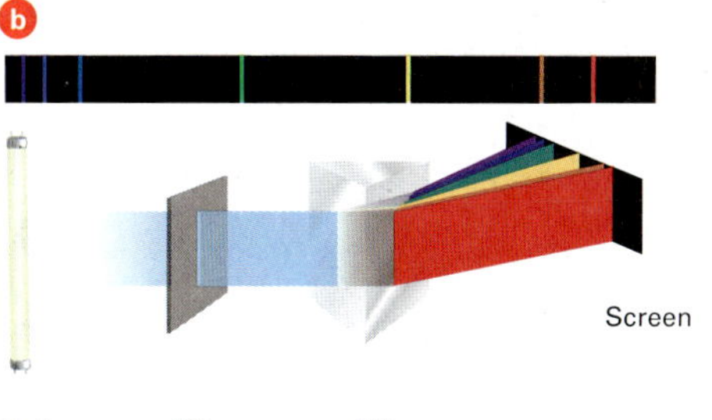

Figure 5.12 A prism separates light into the colors it contains. **a** For white light this produces a rainbow of colors. **b** Light from a helium lamp produces discrete lines. **Identifying** ***Which color has the highest frequency?***

Atomic Spectra

Passing an electric current through a gas in a neon tube energizes the electrons of the atoms of the gas, and causes them to emit light. **When atoms absorb energy, electrons move into higher energy levels. These electrons then lose energy by emitting light when they return to lower energy levels.** Figure 5.12a shows how ordinary light is made up of a mixture of all the wavelengths of light. However, the light emitted by atoms consists of a mixture of only specific frequencies. Each specific frequency of visible light emitted corresponds to a particular color. Therefore, when the light passes through the prism shown in Figure 5.12b, the frequencies of light emitted by an element separate into discrete lines to give the **atomic emission spectrum** of the element.

Each discrete line in an emission spectrum corresponds to one exact frequency of light emitted by the atom. Figure 5.12b shows the visible portion of the emission spectrum of helium.

The emission spectrum of each element is like a person's fingerprint. Just as no two people have the same fingerprints, no two elements have the same emission spectrum. In the same way that fingerprints identify people, atomic emission spectra are useful for identifying elements. Figure 5.13 shows the characteristic colors emitted by mercury and by nitrogen. Much of the knowledge about the composition of the universe comes from studying the atomic spectra of the stars, which are hot glowing bodies of gases.

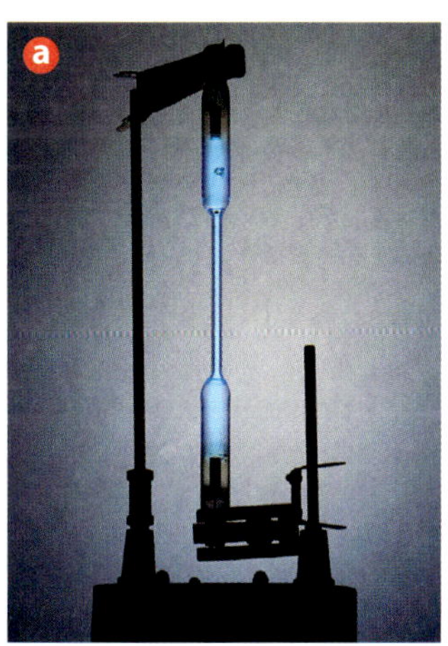

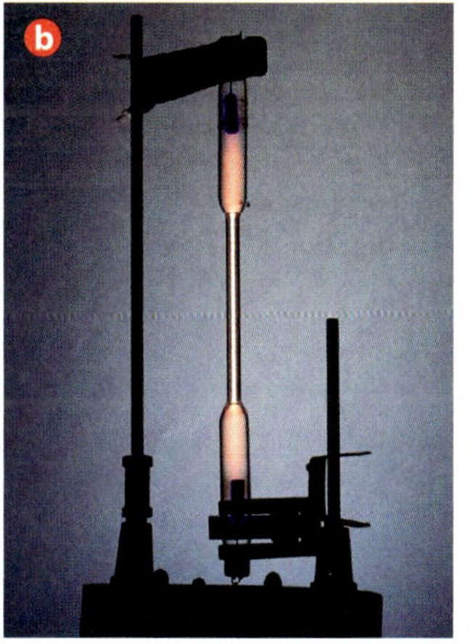

Figure 5.13 No two elements have the same emission spectrum. **a** Mercury vapor lamps produce a blue glow. **b** Nitrogen gas gives off a yellowish-orange light.

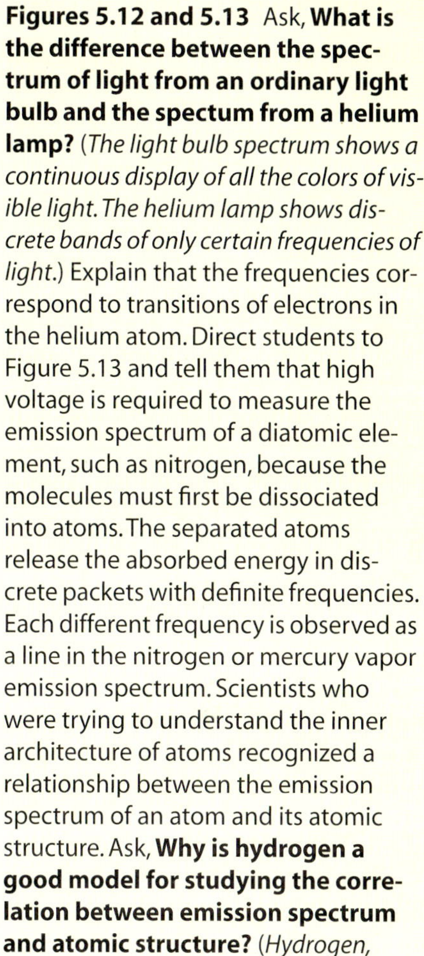

Atomic Spectra

Use Visuals L1

Figures 5.12 and 5.13 Ask, **What is the difference between the spectrum of light from an ordinary light bulb and the spectum from a helium lamp?** (*The light bulb spectrum shows a continuous display of all the colors of visible light. The helium lamp shows discrete bands of only certain frequencies of light.*) Explain that the frequencies correspond to transitions of electrons in the helium atom. Direct students to Figure 5.13 and tell them that high voltage is required to measure the emission spectrum of a diatomic element, such as nitrogen, because the molecules must first be dissociated into atoms. The separated atoms release the absorbed energy in discrete packets with definite frequencies. Each different frequency is observed as a line in the nitrogen or mercury vapor emission spectrum. Scientists who were trying to understand the inner architecture of atoms recognized a relationship between the emission spectrum of an atom and its atomic structure. Ask, **Why is hydrogen a good model for studying the correlation between emission spectrum and atomic structure?** (*Hydrogen, with only one electron, has the simplest electron configuration and thus the simplest emission spectrum.*)

Discuss  L2

Remind students of the Inquiry Activity on page 126. Crushing the mint creates an unequal division of electrons: positive charge accumulates on one piece, negative charge on the other. When an electric spark jumps between the pieces, nitrogen atoms in the air absorb the energy, emitting it as ultraviolet light. The wintergreen flavor molecules absorb part of this energy and release it as visible blue-green light. All wintergreen mints should emit light. To see the light, students must allow enough time for their eyes to adapt to the dark room.

Differentiated Instruction

Gifted and Talented L3

Students may welcome the opportunity to design a system for refracting light and projecting a rainbow on a wall or screen. Depending on their setup, they may need a prism and a bright light source.

Answers to...

Figure 5.12 violet

Quick LAB

Flame Tests

L2

Objective Students use flame tests to determine the identity of the cation in an unknown solution.

Skills Focus Observe, interpret, identify

 **Prep Time** 30 minutes

Materials Bunsen burner, 6 small test tubes, test tube rack, tongs, 6 cotton swabs, 0.1*M* NaCl, 0.1*M* $CaCl_2$, 0.1*M* LiCl, 0.1*M* $CuCl_2$, 0.1*M* KCl, unknown solution

Advance Prep

Solution	Preparation
0.1*M* NaCl	0.6 g NaCl in 100 mL water
0.5*M* $CaCl_2$	5.5 g $CaCl_2$ in 100 mL water
0.1*M* LiCl	0.5 g LiCl in 100 mL water
0.1*M* $CuCl_2$	1.3 g $CuCl_2$ in 100 mL water

Class Time 20 minutes

Safety Students should wear safety goggles and a lab apron and observe all safety procedures.

Expected Outcome

Na^+	bright yellow
Ca^{2+}	orange-red
Li^+	bright crimson red
Cu^{2+}	blue-green

Quick LAB

Flame Tests

Purpose
Use the flame test to determine the identity of the cation in an unknown solution based on its characteristic color in a flame.

Materials
- Bunsen burner
- 6 small test tubes
- test tube rack
- tongs
- 6 cotton swabs
- 0.1*M* NaCl
- 0.1*M* $CaCl_2$
- 0.1*M* LiCl
- 0.1*M* $CuCl_2$
- 0.1*M* KCl
- unknown solution

Procedure    

1. Make a two-column data table. Label the columns Cation and Flame Color. Enter the cation's name for each salt solution in the first column.
2. Label each of 5 test tubes with the name of a salt solution; label the sixth tube Unknown. Add 1 mL of each salt solution to the appropriately labeled test tube.
3. Dip one of the cotton ends of a cotton swab into the sodium chloride solution and then hold it briefly in the burner flame. Record the color of the flame. Do not leave the swab in the flame too long or the plastic will melt.
4. Repeat Step 3 for each of the remaining salt solutions using a new cotton swab each time.
5. Perform a flame test with the unknown solution. Note the color of the flame.

Analyze and Conclude

1. Identify the cation in the unknown.
2. Each salt solution produces a unique color. Would you expect this based on the modern view of the atom? Explain.
3. Some commercially available fireplace logs burn with a red and/or green flame. On the basis of your data, what elements could be responsible for these colored flames?
4. Aerial fireworks contain gunpowder and chemicals that produce colors. What element would you include to produce crimson red? Yellow?

An Explanation of Atomic Spectra

Atomic line spectra were known before Bohr proposed his model of the hydrogen atom. However, Bohr's model not only explained why the emission spectrum of hydrogen consists of specific frequencies of light. It also predicted specific values of these frequencies that agreed with experiment.

In the Bohr model, the lone electron in the hydrogen atom can have only certain specific energies. When the electron has its lowest possible energy, the atom is in its **ground state.** In the ground state, the principal quantum number (n) is 1. Excitation of the electron by absorbing energy raises the atom from the ground state to an excited state with $n = 2, 3, 4, 5,$ or 6, and so forth. A quantum of energy in the form of light is emitted when the electron drops back to a lower energy level. The emission occurs in a single abrupt step, called an electronic transition. Bohr already knew from earlier work that this quantum of energy E is related to the frequency ν of the emitted light by the equation $E = h \times \nu$, where h is equal to 6.626×10^{-34} J·s.  **The light emitted by an electron moving from a higher to a lower energy level has a frequency directly proportional to the energy change of the electron.** Therefore each transition produces a line of a specific frequency in the spectrum.

Analyze and Conclude

1. Answer depends on the unknown.
2. Yes; because the composition of every atom is different, the amount of energy needed for transitions between energy sublevels is different. These energy differences are reflected in the different line spectra.
3. copper (green), lithium (red), or calcium (orange-red)
4. red, lithium; yellow, sodium

For Enrichment

L3

Mix two of the unknown solutions. Ask students to identify the two metal ions.

Figure 5.14 shows the explanation for the three groups of lines observed in the emission spectrum of hydrogen atoms. The lines at the ultraviolet end of the hydrogen spectrum are the Lyman series. These match expected values for the emission due to the transition of electrons from higher energy levels to the lowest energy level, $n = 1$. The lines in the visible spectrum are the Balmer series. These lines result from transitions from higher energy levels to $n = 2$. This generally involves a smaller change in electron energy than transitions to $n = 1$. Transitions to $n = 3$ from higher energy levels produce the Paschen series. The energy changes of the electron, and therefore the frequencies of emitted light, are generally smaller still. The lines are in the infrared range. Spectral lines for the transitions from higher energy levels to $n = 4$ and $n = 5$ also exist. Note that the spectral lines in each group become more closely spaced at increased values of n because the energy levels become closer together. There is an upper limit to the frequency of emitted light for each set of lines. The upper limit exists because an electron with enough energy completely escapes the atom.

Bohr's theory of the atom was only partially satisfactory. It explained the emission spectrum of hydrogen but not the emission spectra of atoms with more than one electron. Moreover, it was of no help in understanding how atoms bond to form molecules. Eventually a new and better model, the quantum mechanical model, displaced the Bohr model of the atom. The quantum mechanical model is based on the description of the motion of material objects as waves.

What is the name of the series of visible lines in the hydrogen spectrum?

Interactive Textbook

Animation 6 Learn about atomic emission spectra and how neon lights work.

with ChemASAP

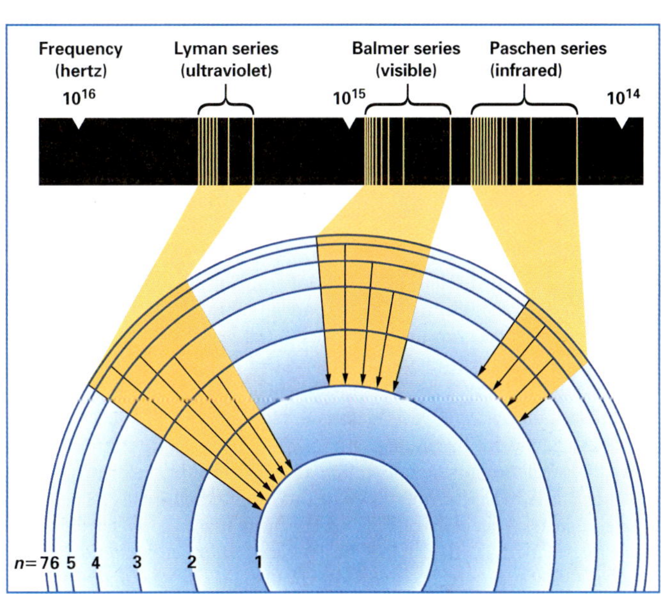

Figure 5.14 The three groups of lines in the hydrogen spectrum correspond to the transitions of electrons from higher energy levels to lower energy levels. The Lyman series corresponds to the transitions to the $n = 1$ energy level. The Balmer series corresponds to the transitions to the $n = 2$ energy level. The Paschen series corresponds to the transitions to the $n = 3$ energy level.

An Explanation of Atomic Spectra

Use Visuals L1

Figure 5.14 Pose this question to students: **Suppose an electron, in its ground state at energy level one, absorbs enough energy to jump to level two. What type of radiation will it emit when it returns to the ground state?** (*UV radiation*) Ask, **If you observed a hydrogen gas discharge tube through a diffraction grating, could you see the line corresponding to this emission?** (*No; the human eye cannot detect radiation in the UV range.*) **Which series of lines could you detect?** (*the Balmer series, which has frequencies in the visible region of the spectrum*) **Compare the energy of the Paschen and Balmer series.** (*The Paschen series has lower energy.*) Ask, **What do you notice about the spacing of the energy levels from $n = 1$ to $n = 7$?** (*The levels are not evenly spaced. The lines get closer as the distance from the nucleus increases.*) Note that each set of lines is a converging series that approaches the ionization energy of hydrogen.

Facts and Figures

The Discovery of Helium

Sometimes discoveries in one area of science, such as chemistry, become important in solving problems in another, such as astronomy. In 1868, Pierre Janssen and Joseph Norman Lockyer discovered an emission spectrum from gases on the surface of the sun that did not match any known element on Earth. In 1895, William Ramsay discovered the existence of helium on Earth. The emission spectrum of helium was found to be identical to that of the unknown gas observed by Janssen and Lockyer almost thirty years earlier. Thus, by combining two discoveries from two different fields of science, a new discovery was made: Earth and the stars have some elements in common.

Answers to...

the Balmer series

Quantum Mechanics

Discuss

L2

Tell students that, according to quantum mechanics, scientists cannot observe an object without changing the object in the process. The change produced by making an observation is the basis for the uncertainty principle. Present the following analogy: Suppose you want to find out if a car is in a long tunnel. In quantum mechanics, the only sort of experiment you could do would be to send another car into the tunnel and listen for a crash. Although it is possible to detect the presence of a car with this method, it is obvious that the car will be changed by the crash.

Relate

L2

The wave–particle behavior of light was difficult for scientists trained in classical physics to accept. However, it became difficult to dispute because it provided an explanation for a previously mysterious phenomenon called the photoelectric effect. In the photoelectric effect, it was shown that metals eject electrons called photoelectrons when light shines on them. But not just any frequency of light will cause the photoelectric effect. Red light ($\nu = 4.3 \times 10^{14}\ s^{-1}$ to $4.6 \times 10^{14}\ s^{-1}$), for example, will not cause the ejection of photoelectrons from potassium, no matter how intense the light or for how long the light radiates the metal surface. But even a weak yellow light ($\nu = 5.1 \times 10^{14}\ s^{-1}$ to $5.2 \times 10^{14}\ s^{-1}$) shining on potassium causes the emission of photoelectrons. Thus, photons of red light do not have sufficient energy to cause the ejection of electrons, but photons of yellow light do.

Download a worksheet on the **Photoelectric Effect** for students to complete, and find additional teacher support from NSTA SciLinks.

Figure 5.15 An electron microscope can produce sharp images of a very small object, such as this mite, because of the small wavelength of a moving electron compared with that of light.

Quantum Mechanics

In 1905, Albert Einstein, then a patent examiner in Bern, Switzerland, returned to Newton's idea of particles of light. Einstein successfully explained experimental data by proposing that light could be described as quanta of energy. The quanta behave as if they were particles. Light quanta are called **photons.** Although the wave nature of light was well known, the dual wave-particle behavior of light was difficult for scientists trained in classical physics to accept.

In 1924, Louis de Broglie (1892–1987), a French graduate student, asked an important question: Given that light behaves as waves and particles, can particles of matter behave as waves? De Broglie referred to the wavelike behavior of particles as matter waves. His reasoning led him to a mathematical expression for the wavelength of a moving particle. The proposal that matter moves in a wavelike way would not have been accepted unless experiments confirmed its validity. Only three years later, experiments by Clinton Davisson and Lester Germer at Bell Labs in New Jersey did just that. The two scientists had been studying the bombardment of metals with beams of electrons. They noticed that the electrons reflected from the metal surface produced curious patterns. The patterns were like those obtained when X-rays (which are electromagnetic waves) reflect from metal surfaces. The electrons—believed to be particles—were reflected as if they were waves! De Broglie was awarded the Nobel Prize for his work on the wave nature of matter. Davisson also received the Nobel Prize for his experiments demonstrating the wave nature of electrons.

Today, the wavelike properties of beams of electrons are useful in magnifying objects. The electrons in an electron microscope have much smaller wavelengths than visible light. This allows a much clearer enlarged image of a very small object, such as the mite in Figure 5.15, than is possible with an ordinary microscope.

For: Links on the Photoelectric Effect
Visit: www.SciLinks.org
Web Code: cdn-1053

Facts and Figures

Planck's Constant

Max Planck (1858–1947) showed that energy could be absorbed or emitted by a body only in quanta whose energy is given by the equation $E = h \times \nu$, where h is a proportionality constant called Planck's constant. Planck's proposal was revolutionary. Everyday experience had led people to believe that no limitation existed for the smallness of permissible energy changes in a system. For example, appearances would lead you to believe that thermal energy may be continuously supplied to heat liquid water to any temperature between 0°C and 100°C. Actually, the water temperature increases by infinitesimally small steps that occur as individual molecules absorb quanta of energy.

De Broglie's equation predicts that all moving objects have wavelike behavior. Why are you unable to observe the effects of this wavelike motion for ordinary objects like baseballs and trains? The answer is that the mass of the object must be very small in order for its wavelength to be large enough to observe. For example, a 50-gram golf ball traveling at 40 m/s (about 90 mi/h) has a wavelength of only 3×10^{-34} m, which is much too small to detect experimentally. On the other hand, an electron has a mass of only 9.11×10^{-28} g. If it were moving at a velocity of 40 m/s, it would have a wavelength of 2×10^{-5} m, which is comparable to infrared radiation and is readily measured. The newer theory is called quantum mechanics; the older theory is called classical mechanics. **Classical mechanics adequately describes the motions of bodies much larger than atoms, while quantum mechanics describes the motions of subatomic particles and atoms as waves.**

German physicist Werner Heisenberg examined another feature of quantum mechanics that is absent in classical mechanics. The **Heisenberg uncertainty principle** states that it is impossible to know exactly both the velocity and the position of a particle at the same time. This limitation is critical in dealing with small particles such as electrons. The Heisenberg uncertainty principle does not matter, however, for ordinary-sized objects such as cars or airplanes.

To understand this principle, consider how you determine the location of an object. To locate a set of keys in a dark room, for example, you can use a flashlight. You see the keys when the light bounces off them and strikes your eyes. Likewise, to locate an electron, you might strike it with a photon of light as shown in Figure 5.16. In contrast to the keys, the electron has such a small mass that striking it with a photon affects its motion in a way that cannot be predicted precisely. So the very act of measuring the position of the electron changes its velocity, and makes its velocity uncertain.

The discovery of matter waves paved the way for Schrödinger's quantum mechanical description of electrons in atoms. Schrödinger's theory leads to the concept of electron orbitals and configurations, and it includes the wavelike motion of matter and the uncertainty principle.

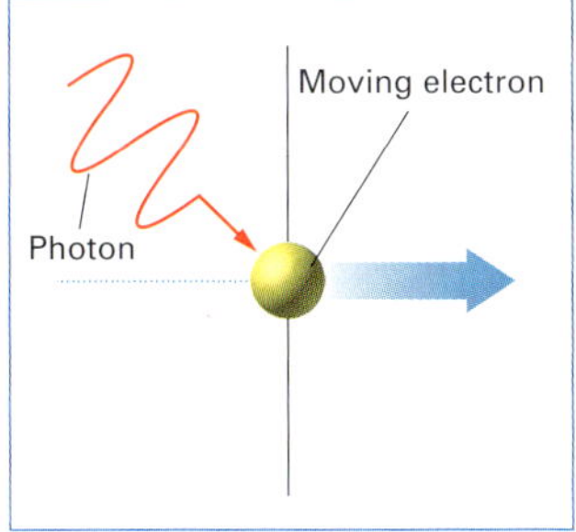

Before collision
A photon strikes an electron during an attempt to observe the electron's position.

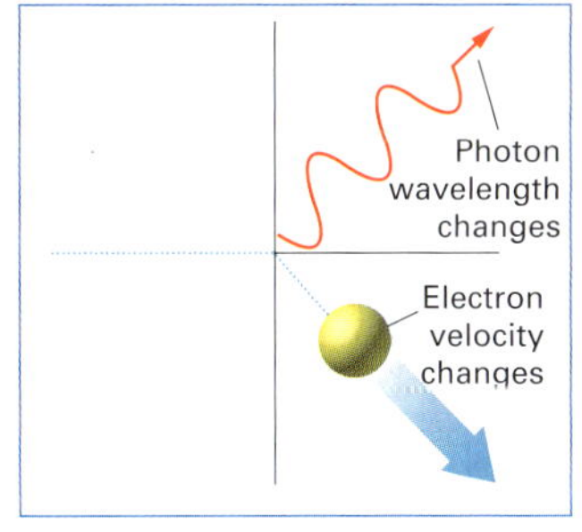

After collision
The impact changes the electron's velocity, making it uncertain.

Figure 5.16 The Heisenberg uncertainty principle states that it is impossible to know exactly both the velocity and the position of a particle at the same time.

Simulation 4 Simulate the photoelectric effect. Observe the results as a function of radiation frequency and intensity.

with ChemASAP

Atomic Scientists L2

Numerous physicists contributed to modern theories of light and atoms. Dalton, Becquerel, Roentgen, Rutherford, Thomson, Millikan, Planck, Einstein, de Broglie, Heisenberg, Schrödinger, Bohr, Chadwick, Balmer, Huygens, Newton, and Foucault all made contributions in one of the most productive periods of scientific discovery.

Purpose Students acquire a deeper understanding and appreciation for the achievements of these scientists.

Procedure Have students work in teams of two or three. Assign scientists to each group and ask students to prepare a brief oral discussion highlighting how the work of the scientists contributed to the understanding of light or the atom. Encourage students to consider the work of other scientists they encounter in their research. Contributions at the periphery can help students understand the interactive and collaborative nature of scientific research.

Expected Outcome Class presentations allow students to share information.

3 ASSESS

Evaluate Understanding L2

Have students draw and label wave diagrams that illustrate the relationship between wavelength and frequency. Ask students to explain the origin of the lines in the Lyman, Balmer, and Paschen series in the atomic emission spectrum of hydrogen. (*Each line arises from light emitted by an atom as electrons move from a higher energy level to a lower one. The Lyman series is in the ultraviolet region of the spectrum, the Balmer is in the visible, and the Paschen is in the infrared.*)

Facts and Figures

A Scientific Controversy

Albert Einstein and Niels Bohr had different views of quantum mechanics. Although Einstein accepted quantum mechanical theory as the best explanation available at the time, he was convinced the theory was incomplete. He believed in the existence of "hidden variables" which, if known, would eliminate problems such as wave/particle duality and uncertainty. Bohr, on the other hand, urged physicists to accept quantum mechanics without qualification.

Careers *in* Chemistry

Spectroscopist

Astronomers use spectroscopy to analyze light from the stars. Atomic absorption spectroscopists study how atoms absorb radiation. X-ray spectroscopists use X-rays to determine the crystal structure of substances. Have students explore these and other fields of study that use spectroscopy for gathering data.

Have students research chemistry-related careers in the library or on the Internet. Students can then construct a table that describes the nature of the work, educational and training requirements, employment outlook, working conditions, and other necessary information.

Reteach L1

Use the illustrations in this section as a basis for review. Starting with Figure 5.9 and continuing to Figure 5.16, question students about what is being illustrated and what is significant about it.

Elements Handbook

Students will find that strontium produces red light and barium produces yellow-green light. Students should recognize that these colors, when separated by a spectroscope, are the colors of the identifying lines for strontium and barium and arise from electrons moving from higher energy levels to lower energy levels.

Interactive Textbook

If your class subscribes to the Interactive Textbook, use it to review key concepts in Section 5.3.

with ChemASAP

Careers *in* Chemistry

Spectroscopist

If you like the idea of finding the chemical content of unknown materials in chemical research, police investigations, and studies of distant stars, you might consider a career as a spectroscopist. Spectroscopy is the recording and analysis of the wavelengths of electromagnetic radiation emitted by samples of materials. Optical emission spectroscopy uses emission lines from atomic transitions in a heated sample of material. Spectroscopists observe emission lines from the sample by using an electronic detector and recording its output. The recorded data gives the wavelength and the intensity of each emission line. The characteristic pattern of wavelengths and intensities is the emission spectrum of the sample.

Spectroscopists use spectrometers, densitometers, and other measuring instruments to collect data. They analyze the densitometer or spectrometer readings to find the ratio of various elements in the sample. They calculate the relative concentrations of substances in the sample by comparing with data for known concentrations. They also use their mathematical skills in statistics to calculate a numerical value indicating the reliability of each analysis.

Spectroscopists usually have an advanced degree in chemistry, along with skills in mathematics and in using scientific equipment.

Spectroscopist

For: Careers in Chemistry
Visit: PHSchool.com
Web Code: cdb-1053

5.3 Section Assessment

16. **Key Concept** How are wavelength and frequency of light related?
17. **Key Concept** Describe the cause of atomic emission spectrum of an element.
18. **Key Concept** How is the change in electron energy related to the frequency of light emitted in atomic transitions?
19. **Key Concept** How does quantum mechanics differ from classical mechanics?
20. The lines at the ultraviolet end of the hydrogen spectrum are known as the Lyman series. Which electron transitions within an atom are responsible for these lines?
21. Arrange the following in order of decreasing wavelength.
    a. infrared radiation from a heat lamp
    b. dental X-rays
    c. signal from a shortwave radio station

Elements Handbook

Color and Transitions Look at the photographs of flame tests on page R11 of the Elements Handbook. List the colors emitted from strontium compounds and from barium compounds when heated in a flame, and explain how electron transitions account for the specific colors being emitted.

Assessment 5.3 Test yourself on the concepts in Section 5.3.

with ChemASAP

Section 5.3 Assessment

16. Frequency and wavelength of light are inversely proportional to each other.
17. Electrons in atoms absorb energy as they move to higher energy levels, then lose the energy by emitting it as light as they drop back.
18. The light emitted in an electronic transition from a higher to a lower energy level has a frequency that is directly proportional to the energy change of the electron.
19. Quantum mechanics describes the motions of atoms and subatomic particles; classical mechanics describes the motions of larger bodies.
20. electron transitions from higher levels to $n = 1$
21. c, a, b

Technology & Society

Lasers at Work

A laser produces an intense beam of light with the unusual property that the waves contributing to it have crests that coincide. The properties of laser light make it useful for many different purposes, including reading CD-ROMs and DVDs in electronic equipment and scanning bar codes at cash registers. Some industries use high-intensity lasers to cut metal. Other lasers are used in surgery. **Inferring** ***Explain how the properties of a laser beam make it useful for the kind of eye surgery shown.***

Flashlight beam vs. laser beam Light from a flashlight travels in different directions, covering a wide area. Light rays from a laser travel parallel to one another, resulting in a narrow beam of intense light in a specific direction.

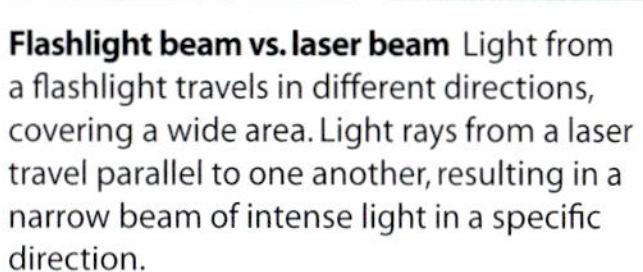

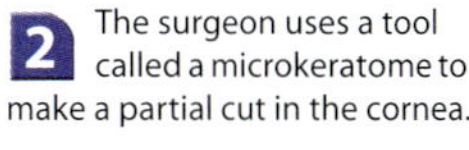

2 The surgeon uses a tool called a microkeratome to make a partial cut in the cornea.

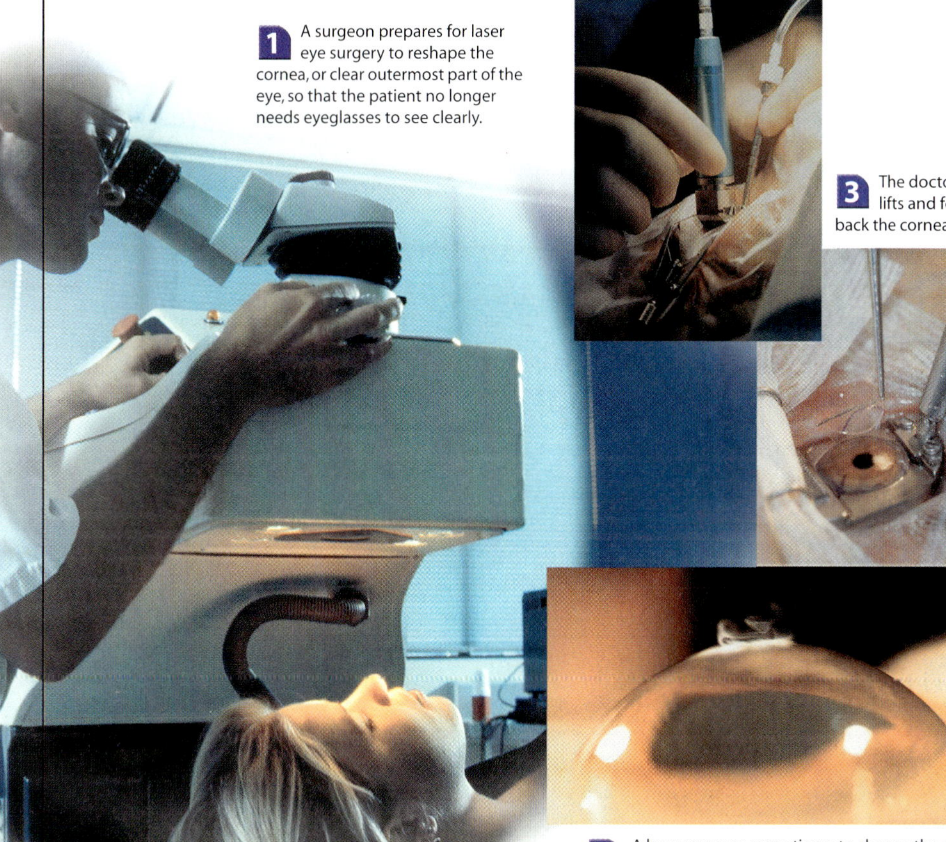

1 A surgeon prepares for laser eye surgery to reshape the cornea, or clear outermost part of the eye, so that the patient no longer needs eyeglasses to see clearly.

3 The doctor lifts and folds back the cornea.

4 A laser removes some tissue to change the shape of the cornea and correct the patient's vision.

Technology & Society

Lasers at Work

Have students note the narrow beam of the laser in the photo compared with the flashlight beam. Explain that the intensity and narrowness of the laser beam allows the surgeon to make a precise surgical cut and forestall bleeding. Laser is an acronym *for light amplification by the stimulated emission of radiation.*

Encourage students to find out how coherent light—light with one wavelength—is produced. (*When a photon of a particular frequency strikes an atom, the photon stimulates the atom to emit a photon of the same frequency. The original photon is unaffected. Now two photons of the same frequency can collide with other atoms and stimulate them to produce identical photons. The process goes on building up an aggregate of photons all with the same frequency.*) Students can investigate what substances are stimulated to create laser light and what wavelengths correspond to each substance. (*Some typical lasers are nitrogen, $\lambda = 337$ nm; helium-cadmium, $\lambda = 441.6$ nm; argon, $\lambda = 476.5$ and 488.0 nm; krypton, $\lambda = 534.5$ nm; ruby, $\lambda = 694.3$ nm; neon, $\lambda = 632{,}8$ nm; gallium arsenide, $\lambda = 840$-930 nm, depending on temperature; neodynium, $\lambda = 1040$ nm; carbon dioxide, $\lambda = 10{,}600$ nm.*)

Some students could research the technology of reading CD-ROMs and DVDs with lasers and the use of lasers in fiber optics and holograms. Others may prefer to find out more about how nearsightedness is cured by laser surgery. Students could research the history of laser development, a "hot" topic of research in the 1950s, and find out what is the magic behind holograms.

Answers to...

The laser beam is a narrow beam of intense light that can be aimed in a specific direction.

CHAPTER 5 Study Guide

Study Tip

Summarizing Summarizing requires students to identify key ideas and state them briefly in their own words. Tell students that they will remember the content of an entire section better even if they summarize only a portion of the section.

If your class subscribes to the Interactive Textbook with ChemASAP, your students can go online to access an interactive version of the Student Edition and a self-test.

with ChemASAP

CHAPTER 5 Study Guide

Key Concepts

5.1 Models of the Atom

- Rutherford's planetary model could not explain the chemical properties of elements.
- Bohr proposed that electrons move only in specific circular paths, or orbits, around the nucleus.
- The quantum mechanical model determines the allowed energies an electron can have and how likely it is to be found in various locations around the nucleus.
- Each sublevel of a principal energy level corresponds to an orbital shape describing where the electron is likely to be found.

5.2 Electron Arrangement in Atoms

- Three rules—the aufbau principle, the Pauli exclusion principle, and Hund's rule—tell you how to find the electron configurations of atoms.
- Some actual electron configurations differ from those assigned using the aufbau principle because half-filled levels are not as stable as filled levels, but they are more stable than other configurations.

5.3 Physics and the Quantum Mechanical Model

- The wavelength and frequency of light are inversely proportional to each other.
- When atoms absorb energy, electrons move into higher energy levels. These electrons then lose energy by emitting light when the electrons drop back to lower energy levels.
- The light emitted by an electron moving from a higher to a lower energy level has a frequency directly proportional to the energy change of the electron.
- Classical mechanics adequately describes the motions of bodies much larger than atoms, while quantum mechanics describes the motions of subatomic particles and atoms as waves.

Vocabulary

- amplitude (p. 138)
- atomic emission spectrum (p. 141)
- atomic orbital (p. 131)
- aufbau principle (p. 133)
- electromagnetic radiation (p. 139)
- electron configurations (p. 133)
- energy levels (p. 128)
- frequency (p. 138)
- ground state (p. 142)
- Heisenberg uncertainty principle (p. 145)
- hertz (p. 138)
- Hund's rule (p. 134)
- Pauli exclusion principle (p. 134)
- photons (p. 144)
- quantum (p. 128)
- quantum mechanical model (p. 130)
- spectrum (p. 139)
- wavelength (p. 138)

Key Equation

- $c = \lambda\nu$

Organizing Information

Use these terms to construct a concept map that organizes the major ideas of this chapter.

energy level · aufbau principle · quantum mechanical model · electron configuration · Pauli exclusion principle · Hund's rule · quantum

Concept Map 5 Solve the concept map with the help of an interactive guided tutorial.

with ChemASAP

Chapter Resources

Print

- **Core Teaching Resources,** Chapter 5, *Practice Problems, Vocabulary Review, Quiz, Chapter Test A, Chapter Test B*

Technology

- **Computer Test Bank,** Chapter 5 Test
- **Interactive Textbook with ChemASAP,** Chapter 5

CHAPTER 5 Assessment

Reviewing Content

5.1 Models of the Atom

22. What was inadequate about Rutherford's model of the atom? Which subatomic particles did Thomson include in the plum-pudding model of the atom?

23. What did Bohr assume about the motion of electrons?

24. Describe Rutherford's model of the atom and compare it with the model proposed by his student Niels Bohr.

25. What is the significance of the boundary of an electron cloud?

26. What is an atomic orbital?

27. How many orbitals are in the $2p$ sublevel?

28. Sketch $1s$, $2s$, and $2p$ orbitals using the same scale for each.

29. How many sublevels are contained in each of these principal energy levels?
a. $n = 1$ **b.** $n = 2$ **c.** $n = 3$ **d.** $n = 4$

5.2 Electron Arrangement in Atoms

30. How many electrons are in the highest occupied energy level of these atoms?
a. barium **b.** sodium
c. aluminum **d.** oxygen

31. What are the three rules that govern the filling of atomic orbitals by electrons?

32. Write electron configurations for the elements that are identified only by these atomic numbers.
a. 15 **b.** 12 **c.** 9 **d.** 18

33. What is meant by $3p^3$?

34. Give electron configurations for atoms of these elements:
a. Na **b.** S **c.** Mg **d.** Ne **e.** K

35. Which of these orbital designations are invalid?
a. $4s$ **b.** $3f$ **c.** $2d$ **d.** $3d$

36. What is the maximum number of electrons that can go into each of the following sublevels?
a. $2s$ **b.** $3p$ **c.** $4s$ **d.** $3d$
e. $4p$ **f.** $5s$ **g.** $4f$ **h.** $5p$

37. Arrange the following sublevels in order of increasing energy:
$3d$, $2s$, $4s$, $3p$.

38. How many electrons are in the second energy level of an atom of each element?
a. chlorine **b.** phosphorus **c.** potassium

39. Write electron configurations for atoms of these elements.
a. selenium **b.** vanadium
c. nickel **d.** calcium

5.3 Physics and the Quantum Mechanical Model

40. List the colors of the visible spectrum in order of increasing wavelength.

41. What is meant by the frequency of a wave? What are the units of frequency? Describe the relationship between frequency and wavelength.

42. Use a diagram to illustrate each term for a wave.
a. wavelength
b. amplitude
c. cycle

43. Explain the difference between the energy lost or gained by an atom according to the laws of classical physics and according to the quantum model of an atom.

44. How are ultraviolet radiation and microwave radiation the same? How are they different?

45. Consider the following regions of the electromagnetic spectrum: (i) ultraviolet, (ii) X-ray, (iii) visible, (iv) infrared, (v) radio wave, (vi) microwave.
a. Use Figure 5.10 to arrange them in order of decreasing wavelength.
b. How does this order differ from that of decreasing frequency?

46. List one way in which each of the radiations listed in Question 45 is used.

47. What happens when a hydrogen atom absorbs a quantum of energy?

48. When white light is viewed through sodium vapor in a spectroscope, the spectrum is continuous except for a dark line at 589 nm. How can you explain this observation?

49. The transition of electrons from higher energy levels to the $n = 2$ energy level results in the emission of light from hydrogen atoms. In what part of the spectrum is the emitted light, and what is the name given to this transition series?

CHAPTER 5 ASSESSMENT

Reviewing Content

22. could not explain why metals and metal compounds give off characteristic colors when heated, nor could it explain the chemical properties of the elements; electrons

23. that electrons traveled in circular paths around the nucleus

24. In Rutherford's model, negatively charged electrons surround a dense, positively charged nucleus. In Bohr's model, the electrons are assigned to concentric circular orbits of fixed energy.

25. An electron is found 90% of the time inside this boundary.

26. a region in space around the nucleus in which there is a high probability of finding an electron

27. 3

28. The $1s$ orbital is spherical. The $2s$ orbital is spherical with a diameter larger than that of the $1s$ orbital. The three $2p$ orbitals are dumbbell shaped and oriented at right angles to each other.

29. a. 1 **b.** 2 **c.** 3 **d.** 4

30. a. 2 **b.** 1 **c.** 3 **d.** 6

31. Electrons occupy the lowest possible energy levels. An atomic orbital can hold at most two electrons. One electron occupies each of a set of orbitals with equal energies before any pairing of electrons occurs.

32. a. $1s^22s^22p^63s^23p^3$ **b.** $1s^22s^22p^63s^2$ **c.** $1s^22s^22p^5$ **d.** $1s^22s^22p^63s^23p^6$

33. The p orbitals in the third quantum level have three electrons.

34. a. $1s^22s^22p^63s^1$ **b.** $1s^22s^22p^63s^23p^4$ **c.** $1s^22s^22p^63s^2$ **d.** $1s^22s^22p^6$ **e.** $1s^22s^22p^63s^23p^64s^1$

35. b and **c**

36. a. 2 **b.** 6 **c.** 2 **d.** 10 **e.** 6 **f.** 2 **g.** 14 **h.** 6

37. $2s$, $3p$, $4s$, $3d$

38. a. 8 **b.** 8 **c.** 8

39. a. $1s^22s^22p^63s^23p^63d^{10}4s^24p^4$ **b.** $1s^22s^22p^63s^23p^63d^34s^2$ **c.** $1s^22s^22p^63s^23p^63d^84s^2$ **d.** $1s^22s^22p^63s^23p^64s^2$

40. Violet, indigo, blue, green, yellow, orange, red

41. Frequency is the number of wave cycles that pass a given point per unit time. Frequency units are cycles/s or s^{-1} or Hertz. Wavelength and frequency are inversely related.

42. diagrams similar to those in Figure 5.9

43. Classical physics views energy changes as continuous. In the quantum concept, energy changes occur in tiny discrete units called quanta.

44. Both travel at the same speed. Ultraviolet is short wavelength and high frequency; microwave is long wavelength and low frequency.

45. a. v, vi, iv, iii, i, ii **b.** It is the reverse.

46. Students may say that ultraviolet is used for tanning the skin and growing plants, X-rays for taking pictures of the interior of the body, visible for seeing, infrared for warmth, radio waves for communication, and microwaves for cooking.

47. The electron of the hydrogen atom is raised (excited) to a higher energy level.

48. The outermost electron of sodium absorbs photons of wavelength 589 nm as it jumps to a higher energy level.

49. visible spectrum, Balmer series

CHAPTER 5

ASSESSMENT continued

Understanding Concepts

50. **a.** Ar **b.** Ru **c.** Gd

51. $1s^22s^22p^63s^23p^63d^{10}4s^24p^3$; level 1, 2; level 2, 8; level 3, 18; level 4, 5; The fourth energy level is not filled.

52. **a.** 2 **b.** 4 **c.** 10 **d.** 6

53. $1s^22s^22p^3$ nitrogen; 3

54. 2.61×10^4 cm

55. **a.** 4.36×10^{-5} cm **b.** visible
c. $6.88 \times 10^{14} s^{-1}$

56. **a.** 5.890×10^{-5} cm and 5.896×10^{-5} cm
b. 5.090×10^{14} s^{-1} (Hz) and 5.085×10^{14} s^{-1} (Hz)
c. yellow

57. **a.** Na, sodium
b. N, nitrogen
c. Si, silicon
d. O, oxygen
e. K, potassium
f. Ti, titanium

58. The frequency is inversely proportional to the wavelength, so if the frequency increases by a factor of 1.5, the wavelength will decrease by a factor of 1.5.

59. It is not possible to know both the position and the velocity of a particle at the same time.

60. 2

61. c

62. c

63. a

64. a

CHAPTER 5

Assessment continued

Understanding Concepts

50. Give the symbol for the atom that corresponds to each electron configuration.
a. $1s^22s^22p^63s^23p^6$
b. $1s^22s^22p^63s^23p^63d^{10}4s^24p^64d^75s^1$
c. $1s^22s^22p^63s^23p^63d^{10}4s^24p^64d^{10}4f^75s^25p^65d^16s^2$

51. Write the electron configuration for an arsenic atom. Calculate the total number of electrons in each energy level and state which energy levels are not full.

52. How many paired electrons are there in an atom of each element?
a. helium **b.** boron
c. sodium **d.** oxygen

53. An atom of an element has two electrons in the first energy level and five electrons in the second energy level. Write the electron configuration for this atom and name the element. How many unpaired electrons does an atom of this element have?

54. Suppose your favorite AM radio station broadcasts at a frequency of 1150 kHz. What is the wavelength, in centimeters, of the radiation from the station?

55. A mercury lamp, such as the one below, emits radiation with a wavelength of 4.36×10^{-7} m.

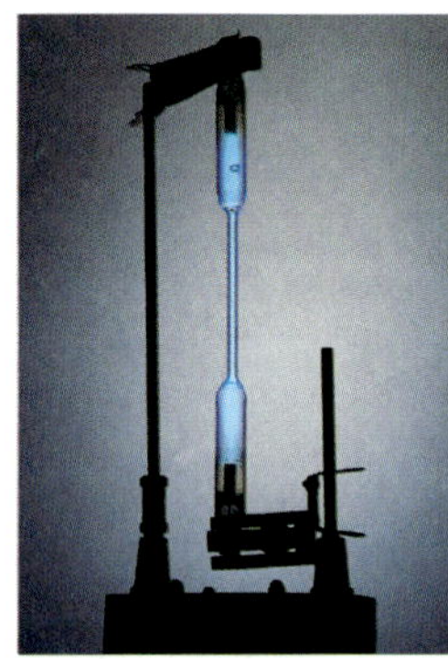

a. What is the wavelength of this radiation in centimeters?
b. In what region of the electromagnetic spectrum is this radiation?
c. Calculate the frequency of this radiation.

56. Sodium vapor lamps are used to illuminate streets and highways. The very bright light emitted by these lamps is actually due to two closely spaced emission lines in the visible region of the electromagnetic spectrum. One of these lines has a wavelength of 5.890×10^{-7} m, and the other line has a wavelength of 5.896×10^{-7} m.
a. What are the wavelengths of these radiations in centimeters?
b. Calculate the frequencies of these radiations.
c. In what region of the visible spectrum do these lines appear?

57. Give the symbol and name of the elements that correspond to these configurations of an atom.
a. $1s^22s^22p^63s^1$
b. $1s^22s^22p^3$
c. $1s^22s^22p^63s^23p^2$
d. $1s^22s^22p^4$
e. $1s^22s^22p^63s^23p^64s^1$
f. $1s^22s^22p^63s^23p^63d^24s^2$

58. Describe how the wavelength changes if the frequency of a wave is multiplied by 1.5.

59. State the Heisenberg uncertainty principle.

60. What is the maximum number of electrons that can be found in any orbital of an atom?

61. Pieces of energy are known as
a. isotopes **b.** particles
c. quanta **d.** line spectra

62. The lowest sublevel in each principal energy level is represented by the symbol
a. f **b.** p **c.** s **d.** d

63. Which electron transition results in the emission of energy?
a. $3p$ to $3s$
b. $3p$ to $4p$
c. $2s$ to $2p$
d. $1s$ to $2s$

64. Which is the ground state configuration of a magnesium atom?
a. $1s^22s^22p^63s^2$
b. $1s^22s^22p^63s^1$
c. $1s^22s^23s^22p^6$
d. $1s^22s^22p^43s^2$

Critical Thinking

65. Explain the difference between an orbit in the Bohr model and an orbital in the quantum mechanical model of the atom.

66. Traditional cooking methods make use of infrared radiation (heat). Microwave radiation cooks food faster. Could radio waves be used for cooking? Explain.

67. Think about the currently accepted models of the atom and of light. In what ways do these models seem strange to you? Why are these models not exact or definite?

68. Orbital diagrams for the ground states of two elements are shown below. Each diagram shows something that is incorrect. Identify the error in each diagram and then draw the correct diagram.

a. Nitrogen $1s$ [↑↓] $2s$ [↑↓] $2p$ [↑↓][↑][ ]

b. Magnesium $1s$ [↑↓] $2s$ [↑↓] $2p$ [↑↓][↑↓][↑↓] $3s$ [ ]

69. Picture two hydrogen atoms. The electron in the first hydrogen atom is in the $n = 1$ level. The electron in the second atom is in the $n = 4$ level.

a. Which atom has the ground state electron configuration?

b. Which atom can emit electromagnetic radiation?

c. In which atom is the electron in a larger orbital?

d. Which atom has the lower energy?

70. Identify the elements whose electrically neutral atoms have the following electron configurations.

a. $1s^2 2s^2 2p^5$

b. $1s^2 2s^2 2p^6 3s^2 3p^6 4s^2 3d^{10} 4p^2$

c. $1s^2 2s^2 2p^6 3s^2 3p^6 3d^3 4s^2$

71. Which of the following is the ground state of an atom? Which is its excited state? Which is an impossible electron configuration? Identify the element and briefly explain your choices.

a. $1s^2 2s^2 2p^6 3s^2 3p^6 5p^1$

b. $1s^2 2s^2 2p^6 3s^2 3p^6 4s^1$

c. $1s^2 2s^2 2p^6 3s^2 3p^7$

72. Why do electrons occupy equal energy orbitals singly before beginning to pair up?

Concept Challenge

73. The energy of a photon is related to its frequency and its wavelength.

Energy of photon (J)	Frequency (s^{-1})	Wavelength (cm)
3.45×10^{-21}	a. ______	5.77×10^{-3}
2.92×10^{-20}	b. ______	6.82×10^{-4}
6.29×10^{-20}	c. ______	3.16×10^{-4}
1.13×10^{-19}	d. ______	1.76×10^{-4}
1.46×10^{-19}	e. ______	1.36×10^{-4}
3.11×10^{-19}	f. ______	6.38×10^{-5}

a. Complete the table above.

b. Plot the energy of the photon (y-axis) versus the frequency (x-axis).

c. Determine the slope of the line.

d. What is the significance of this slope?

74. The average distance between Earth and Mars is about 2.08×10^8 km. How long does it take to transmit television pictures from the Mariner spacecraft to Earth from Mars?

75. Bohr's atomic theory can be used to calculate the energy required to remove an electron from an orbit of a hydrogen atom or an ion containing only one electron. This is the ionization energy for that atom or ion. The formula for determining the ionization energy (E) is

$$E = Z^2 \times \frac{k}{n^2}$$

where Z is the atomic number, k is 2.18×10^{-18} J, and n is the energy level. What is the energy required to eject an electron from a hydrogen atom when the electron is in the ground state ($n = 1$)? In the second energy level? How much energy is required to eject a ground state electron from the species Li^{+}?

76. The energy (E) of a photon absorbed or emitted by a body is proportional to its frequency (ν).

$$E = h \times \nu$$

The constant h equals 6.63×10^{-34} J·s. What is the energy of a photon of microwave radiation with a frequency of $3.20 \times 10^{11}\ s^{-1}$?

Critical Thinking

65. An orbit confines the electron to a fixed circular path around the nucleus; an orbital is a region around the nucleus in which electrons are likely to be found.

66. Answers will vary. Some students may note that radio waves have the lowest energy in the electromagnetic spectrum, and thus would not be energetic enough to cook food. Others may reason that if microwaves cook food faster than infrared radiation, then radio waves would cook food even faster.

67. Answers will vary. The model of the atom is based on the abstract idea of probability. Light is considered a particle and a wave at the same time. Atoms and light cannot be compared to familiar objects or observations because humans cannot experience atoms or photons directly and because matter and energy behave differently at the atomic level than at the level humans can observe directly.

68. a. Electrons in $2p$ boxes should not be paired—there should be one electron in each.
b. Magnesium has 12 electrons. Two more electrons need to be added to $3s$.

69. a. $n = 1$ level
b. $n = 4$ level
c. $n = 4$ level
d. $n = 1$ level

70. a. fluorine **b.** germanium
c. vanadium

71. a. potassium, excited state, valence electron has been promoted from $4s$ to $5p$ **b.** potassium, ground state, correct electron configuration **c.** impossible configuration, $3p$ orbitals can hold a maximum of 6 electrons, not 7

72. The electrons obey Hund's rule.

Concept Challenge

73. a. Frequency(s^{-1}): 5.20×10^{12}; 4.40×10^{13}; 9.50×10^{13}; 1.70×10^{14}; 2.20×10^{14}; 4.70×10^{14}

b.

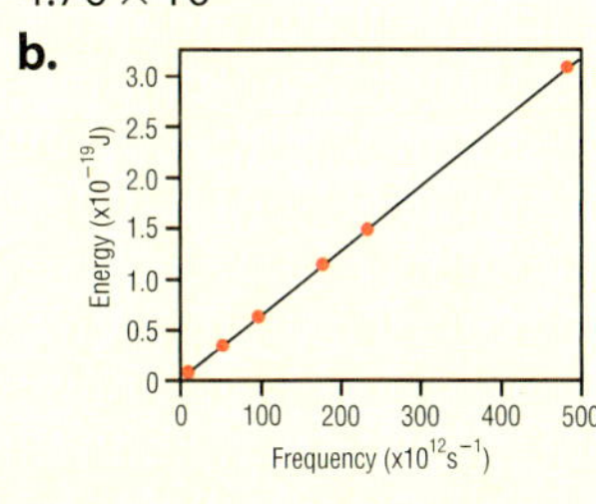

c. 6.63×10^{-34} joule·second.
d. The slope is Planck's constant.

74. 6.93×10^2 s

75. H: ($n = 1$), 1312kJ/mol); ($n = 2$), 328 kJ
Li^{+}: ($n = 1$), 1.18×10^4 kJ

76. 2.12×10^{-22} J

CHAPTER 5 ASSESSMENT continued

Cumulative Review

77. a and b are heterogeneous; c is homogeneous
78. Answers will vary but could include, water is lost as steam and burned meat gives off carbon dioxide.
79. A compound has constant composition; the composition of a mixture can vary.
80. a heterogeneous mixture
81. 7.7×10^{-5} μm
82. 18.9 cm^3
83. the piece of lead
84. a. 3.9×10^{-5} kg
    b. 7.84×10^{2} L
    c. 8.30×10^{-2} g
    d. 9.7×10^{6} ng
85. a and b are exact
86. Mass remains the same; weight decreases because gravity on moon is less than gravity on earth.
87. 8.92 g/cm^3
88. 154 g, 1.54×10^{-1} kg
89. Helium gas is less dense than the nitrogen gas and oxygen gas in the air.
90. a. 55 protons, 55 electrons
    b. 47 protons, 47 electrons
    c. 48 protons, 48 electrons
    d. 34 protons, 34 electrons
91. Accuracy is a measure of how close the value is to the true value; precision is a measure of how close a series of measurements are to one another.
92. a
93. Neon-20 has 10 neutrons; neon-21 has 11 neutrons.
94. The value 35.453 amu is a weighted average. Its calculation is based on the percentage natural abundance of two isotopes, chlorine-35 and chlorine-37.

CHAPTER 5 Assessment continued

Cumulative Review

77. Classify each of the following as homogeneous or heterogeneous (*Chapter 2*)
  a. a page of this textbook
  b. a banana split
  c. the water in bottled water

78. Hamburger undergoes a chemical change when cooked on a grill. All chemical changes are subject to the law of conservation of mass. Yet a cooked hamburger will invariably weigh less than the uncooked meat patty. Explain. (*Chapter 2*)

79. Homogeneous mixtures and compounds are both composed of two or more elements. How do you distinguish between a homogeneous mixture and a compound? (*Chapter 2*)

80. The photo shows a magnified view of a small piece of granite. Is granite a substance or a mixture? (*Chapter 2*)

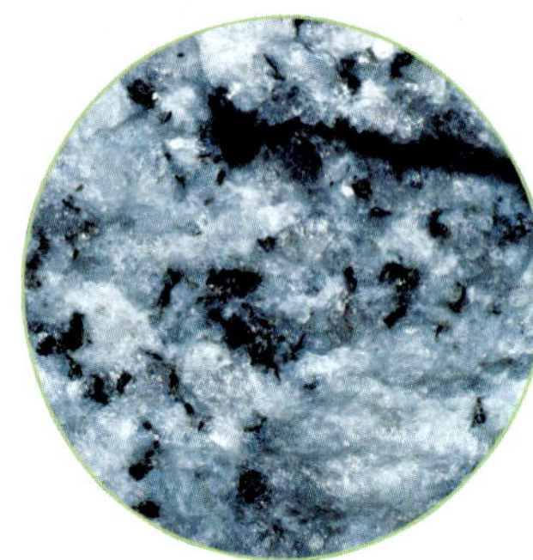

81. The diameter of a carbon atom is 77 pm. Express this measurement in μm. (*Chapter 3*)

82. A silver bar has a mass of 368 g. What is the volume, in cm^3, of the bar? The density of silver is 19.5 g/cm^3. (*Chapter 3*)

83. Which has more mass, a 28.0-cm^3 piece of lead or a 16.0–cm^3 piece of gold? The density of lead is 11.4 g/cm^3; the density of gold is 19.3 g/cm^3. (*Chapter 3*)

84. Express the following measurements in scientific notation. (*Chapter 3*)
  a. 0.000 039 kg **b.** 784 L
  c. 0.0830 g **d.** 9 700 000 ng

85. Which of these quantities or relationships are exact? (*Chapter 3*)
  a. 10 cm = 1 dm
  b. 9 baseball players on the field
  c. a diamond has a mass of 12.4 g
  d. the temperature is 21 °C

86. A one-kilogram steel bar is brought to the moon. How are its mass and its weight each affected by this change in location? Explain. (*Chapter 3*)

87. When a piece of copper with a mass of 36.4 g is placed into a graduated cylinder containing 20.00 mL of water, the water level rises to 24.08 mL. What is the density of copper? (*Chapter 3*)

88. The density of gold is 19.3 g/cm^3. What is the mass, in grams, of a cube of gold 2.00 cm on each edge? In kilograms? (*Chapter 3*)

89. A balloon filled with helium will rise upward when released. What does this show about the relative density of helium and air? Explain. (*Chapter 3*)

90. Give the number of protons and electrons in each of the following. (*Chapter 4*)
  a. Cs **b.** Ag
  c. Cd **d.** Se

91. Explain the difference between the accuracy of a measurement and the precision of a measurement.

92. Which of these was an essential part of Dalton's atomic model? (*Chapter 4*)
  a. indivisible atoms
  b. electrons
  c. atomic nuclei
  d. neutrons

93. How do neon-20 and neon-21 differ from each other? (*Chapter 4*)

94. The mass of an atom should be very nearly the sum of the masses of its protons and neutrons. The mass of a proton and the mass of a neutron are each very close to 1 amu. Why is the atomic mass of chlorine, 35.453 amu, so far from a whole number? (*Chapter 4*)

Standardized Test Prep

Test-Taking Tip

Eliminate wrong answers If you don't know which response is correct, start by eliminating those you know are wrong. If you can rule out some choices, you'll have fewer left to consider and you'll increase your chances of choosing the correct answer.

Select the choice that best answers each question or completes each statement.

1. Select the correct electron configuration for silicon, atomic number 14.
- **a.** $1s^22s^22p^23s^23p^23d^24s^2$
- **b.** $1s^22s^22p^43s^23p^4$
- **c.** $1s^22s^62p^6$
- **d.** $1s^22s^22p^63s^23p^2$

2. Which pair of orbitals has the same shape?
- **a.** $2s$ and $2p$
- **b.** $2s$ and $3s$
- **c.** $3p$ and $3d$
- **d.** More than one is correct.

3. Which of these statements characterize the nucleus of every atom?
- I. It has a positive charge.
- II. It is very dense.
- III. It is composed of protons, electrons, and neutrons.
- **a.** I and II only
- **b.** II and III only
- **c.** I and III only
- **d.** I, II, and III

4. As the wavelength of light increases,
- **a.** the frequency increases.
- **b.** the speed of light increases.
- **c.** the energy decreases.
- **d.** the intensity increases.

5. In the third energy level,
- **a.** there are two energy sublevels.
- **b.** the *f* sublevel has 7 orbitals.
- **c.** there are three *s* orbitals.
- **d.** a maximum of 18 electrons are allowed.

The lettered choices below refer to Questions 6–10. A lettered choice may be used once, more than once, or not at all.

(A) s^2p^6 (B) s^2p^2 (C) s^2 (D) s^4p^1 (E) s^2p^4

Which configuration is the configuration of the highest occupied energy level for each of these elements?

6. sulfur

7. germanium

8. beryllium

9. krypton

10. strontium

Use the atomic models to answer Questions 11–13.

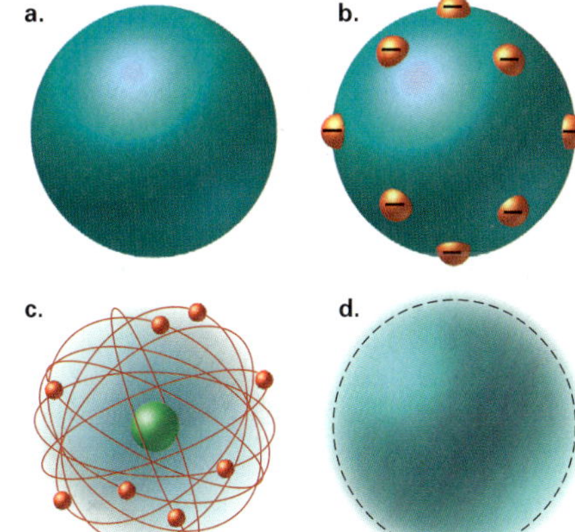

11. Which model best represents J. J. Thomson's model of an atom?

12. Which model was proposed based on data from Rutherford's gold foil experiment?

13. Which model best represents Dalton's model of the atom?

Use the drawings to answer Questions 14–17. Each drawing represents an electromagnetic wave.

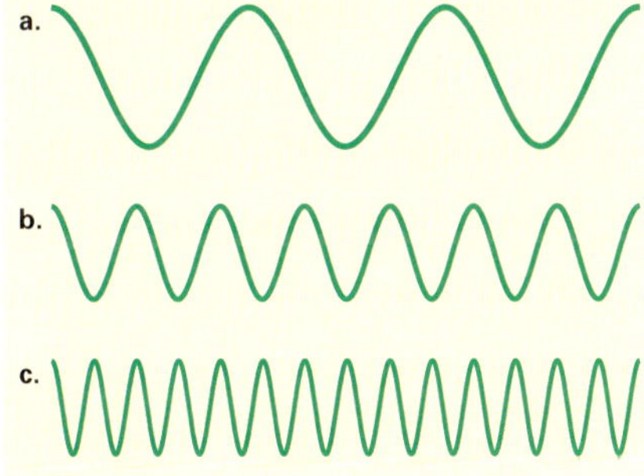

14. The wave has the longest wavelength.

15. The wave has the highest energy,

16. The wave has the lowest frequency.

17. The wave has the highest amplitude.

Write a short essay to answer Question 18.

18. Explain the rules that determine how electrons are arranged around the nuclei of atoms.

Standardized Test Prep

1. d
2. b
3. a
4. c
5. d
6. (E)
7. (B)
8. (C)
9. (A)
10. (C)
11. b
12. c
13. a
14. a
15. c
16. a
17. a
18. According to the aufbau principle, electrons enter orbitals of lowest energy first. According to the Pauli exclusion principle, an orbital may contain at most two electrons. According to Hund's rule, one electron will enter each orbital of equal energy before electrons begin to pair up.

CHAPTER 6

The Periodic Table Planning Guide

SECTION OBJECTIVES	STANDARDS NATIONAL	STANDARDS STATE	ACTIVITIES and LABS	
6.1 Organizing the Elements, pp. 155–160 1 block or 2 periods **6.1.1 Explain** how elements are organized in a periodic table. **6.1.2 Compare** early and modern periodic tables. **6.1.3 Identify** three broad classes of elements.	A-2, B-2, G-1, G-2, G-3		**SE** Inquiry Activity: Trends in Physical Properties, p. 154	L2
			TE Teacher Demo: Organizing Elements, p. 155	L2
			TE Class Activity: Name the Element, p. 159	L1
6.2 Classifying the Elements, pp. 161–169 1 block or 2 periods **6.2.1 Describe** the information in a periodic table. **6.2.2 Classify** elements based on electron configuration. **6.2.3 Distinguish** representative elements and transition metals.	A-1, B-2, E-2, G-2		**TE** Teacher Demo: Observing Differences in Metals, p. 165	L2
			TE Class Activity: Lanthanides in Consumer Products, p. 166	L2
			LM Lab 9: Periodic Properties	L2
6.3 Periodic Trends, pp. 170–179 1 block or 2 periods **6.3.1 Describe** trends among the elements for atomic size. **6.3.2 Explain** how ions form. **6.3.3 Describe** periodic trends for first ionization energy, ionic size, and electronegativity.	A-1, A-2, B-1, B-2, B-3		**SE** Quick Lab: Periodic Trends in Ionic Radii, p. 175	L2
			SE Small-Scale Lab: Periodicity in Three Dimensions, p. 179	L2
			TE Class Activity: Listing Elements, p. 172	L2
			TE Class Activity: Effective Nuclear Charge and Electron Shielding, p. 173	L2
			TE Teacher Demo: Predicting Reactivity, p. 174	L2
			TE Teacher Demo: Trends in Ionic Size, p. 176	L1
			SSLM Lab 9: A Periodic Table Logic Problem	L2
			LP Lab Practical 6-1: Density and the Periodic Table	L2
			LP Lab Practical 6-2: Solubility and the Periodic Table	L2

Ability Levels	
L1	For students who need additional help
L2	For all students
L3	For students who need to be challenged

Components

SE	Student Edition	LM	Laboratory Manual	GCP	Graphing Calculator Problems	T	Transparencies
TE	Teacher's Edition	SSLM	Small-Scale Chemistry Lab Manual	TTT	Test-Taking Tips with Transparencies, Grades 9–12 Science	CA	Chem Alive and Chemistry Field Trips DVD
GRSW	Guided Reading and Study Workbook	LP	Laboratory Practicals	STP	Standardized Test Preparation Workbook	iT	Interactive Textbook with ChemASAP
CTR	Core Teaching Resources	PLM	Probeware Lab Manual			GO	Go Online Internet Resources
SSG	Spanish Study Guide	VCL	Virtual Chemistry Labs				
		CTB	Computer Test Bank				

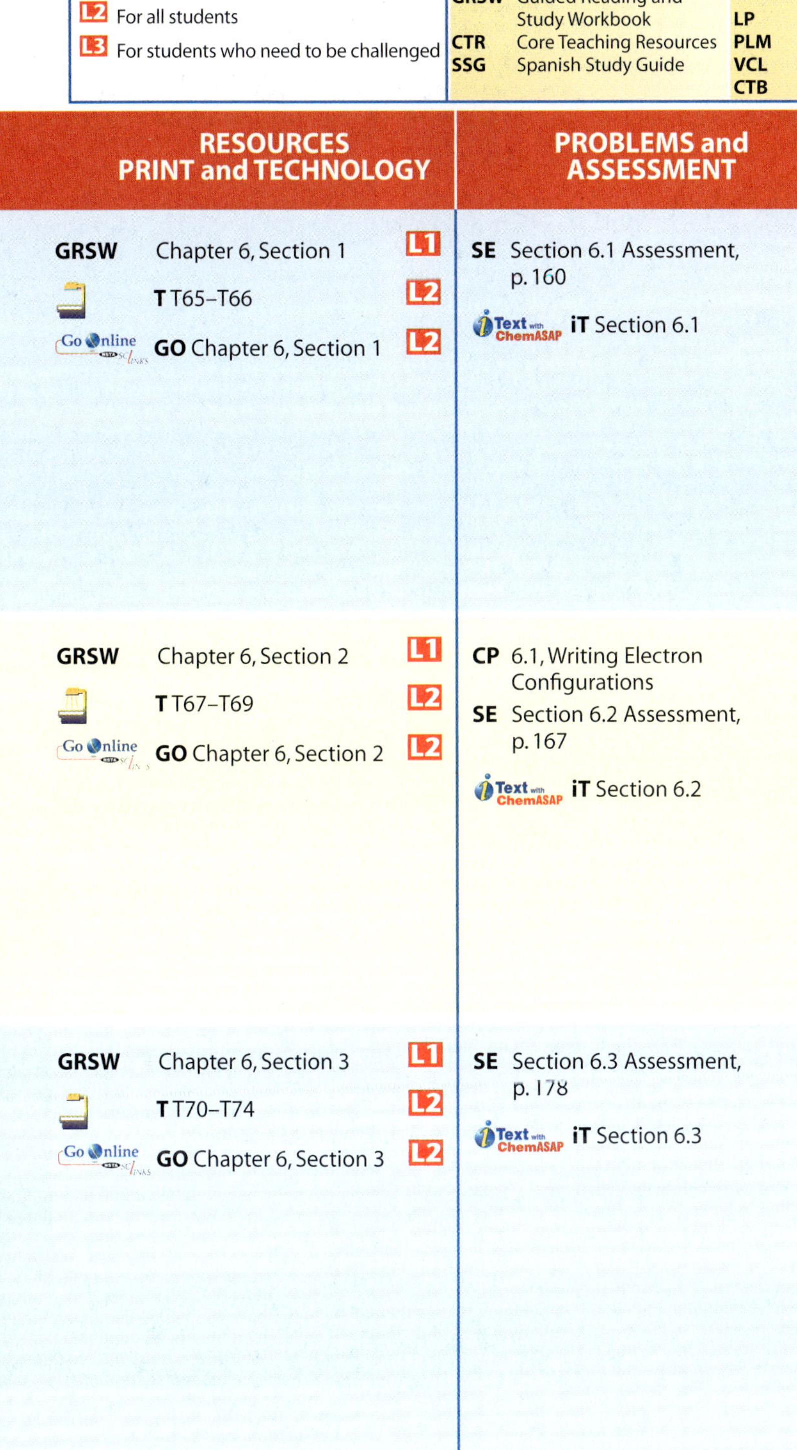

RESOURCES PRINT and TECHNOLOGY		PROBLEMS and ASSESSMENT
GRSW Chapter 6, Section 1	L1	SE Section 6.1 Assessment, p. 160
T T65–T66	L2	iT Section 6.1
GO Chapter 6, Section 1	L2	
GRSW Chapter 6, Section 2	L1	CP 6.1, Writing Electron Configurations
T T67–T69	L2	SE Section 6.2 Assessment, p. 167
GO Chapter 6, Section 2	L2	iT Section 6.2
GRSW Chapter 6, Section 3	L1	SE Section 6.3 Assessment, p. 178
T T70–T74	L2	iT Section 6.3
GO Chapter 6, Section 3	L2	

Web Code: cdn-1061
Web Code: cdn-1062
Web Code: cdn-1063

MATERIALS List

Quantities for each group

STUDENT EDITION

Quick Lab, p. 175
graph paper

Small-Scale Lab, p. 179
96-well spot plate, 20 straws, scissors, metric ruler, permanent fine-line marker

TEACHER'S EDITION

Teacher Demo, p. 155
as many samples of elements as possible, photographs of elements

Teacher Demo, p. 165
0.2*M* HCl; 6 large test tubes; test tube rack; small pieces of clean magnesium, tin, and copper; matches; wood splint

Teacher Demo, p. 174
20 mL of 1*M* HCl, two 50-mL beakers, overhead projector, 20 cm of magnesium ribbon, 1 g of calcium

Teacher Demo, p. 176
washers or other small circular items, smaller item to represent nucleus, overhead projector

Chapter Assessment

CHAPTER ASSESSMENT

SE	Chapter Assessment, pp. 181–184
CTB	Chapter 6
iT	Chapter 6

STANDARDIZED TEST PREP

SE	Chapter 6, p. 185
TP	Topic 2

iT Interactive Textbook with Assessment at PHSchool.com

CHAPTER 6

The Periodic Table

Trends in Physical Properties 

Purpose

Students find trends in the physical properties of elements.

Prep Time none
Class Time 10 minutes

Think About It

1. They both decrease.
2. No, they increase with atomic number.
3. They increase.
4. Two (Cl and F) are gases at room temperature, one (Br) is a liquid, and one (I) is a solid. Their physical states reflect very different densities.

The lab includes the terms *metal* and *nonmetal* to make it easier to refer to the two groups of elements. *Metal* is a familiar term. *Nonmetal* also may be familiar from previous science courses.

After alkali metals and halogens are described in Section 6.2, you could revisit this activity, using the density and melting point graphs on R6 and R32 of the Elements Handbook.

FYI

Chapter 6 expands on the introduction to the periodic table in Section 4.3. It uses atomic structure to explain trends within groups and across periods in the table. It lays the groundwork for the subsequent chapters on bonding.

CHAPTER 6

The Periodic Table

These jeans are organized by color and size. In a periodic table, elements with similar properties are in the same group.

INQUIRY Activity

Trends in Physical Properties

Procedure

1. Make a table with five columns. In the first column, list the metal elements lithium (Li), sodium (Na), potassium (K), rubidium (Rb), and cesium (Cs) in order. Title the other four columns Atomic number, Melting point, Boiling point, and Density.
2. Make a similar table for the nonmetal elements fluorine (F), chlorine (Cl), bromine (Br), and iodine (I).
3. Use Table B.2 in Appendix B to complete each table. Include the appropriate units for each property.

Think About It

1. For the metals, what are the trends for the melting points and boiling points as the atomic number increases?
2. Are the trends for melting and boiling points the same for the nonmetals? Explain.
3. What is the general trend in densities of the metals with increasing atomic number?
4. Why is the range of densities much greater among the nonmetals than among the metals? (*Hint:* The densities listed are for elements at 0°C.)

154

6.1 Organizing the Elements

Connecting to Your World In 1916, a self-service grocery store opened in Memphis, Tennessee. Shoppers could select items from shelves instead of waiting for a clerk to gather the items for them. In a self-service store, the customers must know how to find the products. From your experience, you know that products are grouped according to similar characteristics. You don't expect to find fresh fruit with canned fruit, or bottled juice with frozen juice. With a logical classification system, finding and comparing products is easy. In this section, you will learn how elements are arranged in the periodic table and what that arrangement reveals about the elements.

Guide for Reading

 **Key Concepts**

- How did chemists begin to organize the known elements?
- How did Mendeleev organize his periodic table?
- How is the modern periodic table organized?
- What are three broad classes of elements?

Vocabulary

periodic law
metals
nonmetals
metalloid

Reading Strategy

Comparing and Contrasting As you read, compare and contrast Figures 6.4 and 6.5. How are these two versions of the periodic table similar? How are they different?

Searching For an Organizing Principle

A few elements have been known for thousands of years, including copper, silver, and gold. Yet only 13 elements had been identified by the year 1700. Chemists suspected that other elements existed. They had even assigned names to some of these elements, but they were unable to isolate the elements from their compounds. As chemists began to use scientific methods to search for elements, the rate of discovery increased. In one decade (1765–1775), chemists identified five new elements, including three colorless gases—hydrogen, nitrogen, and oxygen. Was there a limit to the number of elements? How would chemists know when they had discovered all the elements? To begin to answer these questions, chemists needed to find a logical way to organize the elements.

Chemists used the properties of elements to sort them into groups. In 1829, a German chemist, J. W. Dobereiner (1780–1849), published a classification system. In his system, elements were grouped into triads. A triad is a set of three elements with similar properties. The elements in Figure 6.1 formed one triad. Chlorine, bromine, and iodine may look different. But they have very similar chemical properties. For example, they react easily with metals. Dobereiner noted a pattern in his triads. One element in each triad tended to have properties with values that fell midway between those of the other two elements. For example, the average of the atomic masses of chlorine and iodine is [(35.453 + 126.90)/2] or 81.177 amu. This value is close to the atomic mass of bromine, which is 79.904 amu. Unfortunately, all the known elements could not be grouped into triads.

Figure 6.1 Chlorine, bromine, and iodine have very similar chemical properties. The numbers shown are the average atomic masses for these elements.

Chlorine 35.453 amu | Bromine 79.904 amu | Iodine 126.90 amu

Section Resources

Print

- **Guided Reading and Study Workbook,** Section 6.1
- **Core Teaching Resources,** Section 6.1, Review, Interpreting Graphics
- **Transparencies,** T65–T66

Technology

- **Interactive Textbook with ChemASAP,** Assessment 6.1
- **Go Online,** Section 6.1

6.1

1 FOCUS

Objectives

6.1.1 Explain how elements are organized in a periodic table.

6.1.2 Compare early and modern periodic tables.

6.1.3 Identify three broad classes of elements.

Guide for Reading

Build Vocabulary 

Word Parts Point out that *metal, nonmetal,* and *metalloid* all have the same root. Discuss the meanings of the suffix *-oid* (resembling) and prefix *non-* (not).

Reading Strategy 

Use Prior Knowledge Ask students to write down three things they know about the development of the periodic table. After they read the section, have them revise their statements and record three new things that they learned.

2 INSTRUCT

Connecting to Your World

Have students look at the opening photograph and describe the types of characteristics that might be used to group produce in a store. Explain that an initial classification may be broad, as in produce, dairy, and canned goods. Then these categories can be subdivided.

Searching For an Organizing Principle

TEACHER 

Organizing Elements 

Display as many samples of elements as possible. If elements are not available, use photographs. Ask students to use properties of the elements to organize them into groups, and to provide an explanation for their groupings.

Mendeleev's Periodic Table

Use Visuals L1

Figure 6.3 Have students study Mendeleev's version of the periodic table. Show students how the elements with similar chemical properties were arranged in horizontal rows. Point out the gaps in the table where Mendeleev left room for elements that had not yet been discovered. Emphasize that Mendeleev did not make the first periodic table, but he did make the first table that gained wide acceptance.

Figure 6.2 Dimitri Mendeleev proposed a periodic table that could be used to predict the properties of undiscovered elements.

Mendeleev's Periodic Table

From 1829 to 1869, different systems were proposed, but none of them gained wide acceptance. In 1869, a Russian chemist and teacher, Dmitri Mendeleev, published a table of the elements. Later that year, a German chemist, Lothar Meyer, published a nearly identical table. Mendeleev was given more credit than Meyer because he published his table first and because he was better able to explain its usefulness. The stamp in Figure 6.2 is one of many ways that Mendeleev's work has been honored.

Mendeleev developed his table while working on a textbook for his students. He needed a way to show the relationships among more than 60 elements. He wrote the properties of each element on a separate note card. This approach allowed him to move the cards around until he found an organization that worked. The organization he chose was a periodic table. Elements in a periodic table are arranged into groups based on a set of repeating properties. **Mendeleev arranged the elements in his periodic table in order of increasing atomic mass.**

Figure 6.3 is an early version of Mendeleev's periodic table. Look at the column that starts with Ti = 50. Notice the two question marks between the entries for zinc (Zn) and arsenic (As). Mendeleev left these spaces in his table because he knew that bromine belonged with chlorine and iodine. He predicted that elements would be discovered to fill those spaces, and he predicted what their properties would be based on their locations in the table. The elements between zinc and arsenic were gallium and germanium, which were discovered in 1875 and 1886, respectively. There was a close match between the predicted properties and the actual properties of these elements. This match helped convince scientists that Mendeleev's periodic table was a powerful tool.

но въ ней, мнѣ кажется, уже ясно выражается примѣнимость выставляемаго мною начала ко всей совокупности элементовъ, пай которыхъ извѣстенъ съ достовѣрностiю. На этотъ разъ я и желалъ преимущественно найдти общую систему элементовъ. Вотъ этотъ опытъ:

			Ti=50	Zr=90	?=180.
			V=51	Nb=94	Ta=182.
			Cr=52	Mo=96	W=186.
			Mn=55	Rh=104,4	Pt=197,4
			Fe=56	Ru=104,4	Ir=198.
			Ni=Co=59	Pl=106,6	Os=199.
H=1			Cu=63,4	Ag=108	Hg=200.
	Be=9,4	Mg=24	Zn=65,2	Cd=112	
	B=11	Al=27,4	?=68	Ur=116	Au=197?
	C=12	Si=28	?=70	Sn=118	
	N=14	P=31	As=75	Sb=122	Bi=210
	O=16	S=32	Se=79,4	Te=128?	
	F=19	Cl=35,5	Br=80	I=127	
Li=7	Na=23	K=39	Rb=85,4	Cs=133	Tl=204
		Ca=40	Sr=87,6	Ba=137	Pb=207.
		?=45	Ce=92		
		?Er=56	La=94		
		?Yt=60	Di=95		
		?In=75,6	Th=118?		

Figure 6.3 In this early version of Mendeleev's periodic table, the rows contain elements with similar properties. **Observing** ***A fourth element is grouped with chlorine (Cl), bromine (Br), and (I) iodine. What is this element's symbol?***

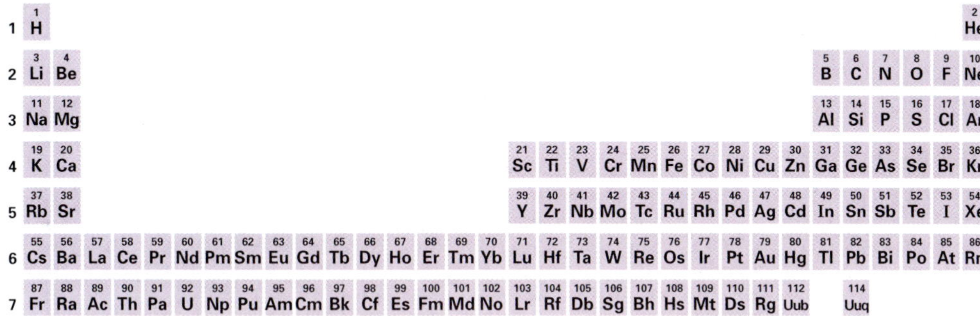

Figure 6.4 In the modern periodic table, the elements are arranged in order of increasing atomic number. **Interpreting Diagrams** ***How many elements are there in the second period?***

The Periodic Law

The atomic mass of iodine (I) is 126.90. The atomic mass of tellurium (Te) is 127.60. Based on its chemical properties, iodine belongs in a group with bromine and chlorine. So Mendeleev broke his rule and placed tellurium before iodine in his periodic table. He assumed that the atomic masses for iodine and tellurium were incorrect, but they were not. Iodine has a smaller atomic mass than tellurium does. A similar problem occurred with other pairs of elements. The problem wasn't with the atomic masses but with using atomic mass to organize the periodic table.

Mendeleev developed his table before scientists knew about the structure of atoms. He didn't know that the atoms of each element contain a unique number of protons. Remember that the number of protons is the atomic number. In 1913, a British physicist, Henry Moseley, determined an atomic number for each known element. Tellurium's atomic number is 52 and iodine's is 53. So it makes sense for iodine to come after tellurium in the periodic table. **In the modern periodic table, elements are arranged in order of increasing atomic number.**

The elements in Figure 6.4 are arranged in order of atomic number, starting with hydrogen, which has atomic number 1. There are seven rows, or periods, in the table. Period 1 has 2 elements, Period 2 has 8 elements, Period 4 has 18 elements, and Period 6 has 32 elements. Each period corresponds to a principal energy level. There are more elements in higher numbered periods because there are more orbitals in higher energy levels. (Recall the rules you studied in Chapter 5 for how electrons fill orbitals.)

The elements within a column, or group, in the periodic table have similar properties. The properties of the elements within a period change as you move across a period from left to right. However, the pattern of properties within a period repeats as you move from one period to the next. This pattern gives rise to the **periodic law:** When elements are arranged in order of increasing atomic number, there is a periodic repetition of their physical and chemical properties.

 ***How many periods are there in a periodic table?***

Word *Origins*

Periodic comes from the Greek roots *peri* meaning "around" and *hodos*, meaning "path." In a periodic table, properties repeat from left to right across each period. The Greek word *metron* means "measure." **What does *perimeter* mean?**

The Periodic Law

Use Visuals 

Figure 6.4 This is the first of three periodic tables in the chapter. The information in this first table is limited so that students can see how all the elements fit in the table, and can focus on atomic number as an organizing principle. Have students compare Figures 6.3 and 6.4. Ask, **How is the modern periodic table shown here similar to the early periodic table?** *(In both tables, elements are arranged in rows and columns. Early periodic tables ordered the elements according to atomic mass. Modern periodic tables order elements according to atomic number.)* **What does an element's position in the table indicate about its properties?** *(The properties of an element are similar to those of other elements in the same group.)*

Discuss 

To emphasize the concept that elements in the same group of the periodic table have similar properties, point out one group and describe some properties of the elements in that group. You can point to the group on the far right, and explain that all of these elements are gases that do not normally form compounds. Using a large poster-size display of the periodic table, trace one period across the table. Describe some physical and chemical properties of each element in the period. Then have students compare these elements to those below them in the next period.

Word *Origins* 

One meaning of *perimeter* is "a line or boundary around an area." It comes from *peri*, "around," and *meter*, "measure."

Answers to...

Figure 6.3 F
Figure 6.4 8
 7

Metals, Nonmetals, and Metalloids

Use Visuals **L1**

Figure 6.5 The focus in Figure 6.5 is on the classification of elements as metals, nonmetals, and metalloids. Point out the sections of the table that correspond to metals, nonmetals, and metalloids. Emphasize the periodic trend across each period from metals to nonmetals. Ask, **Which class of elements do most of the elements belong to?** *(metals)* **What do the three numbers at the top of each column represent?** *(different systems for numbering the groups)*

Download a worksheet on **Metals and Nonmetals** for students to complete, and find additional teacher support from NSTA SciLinks.

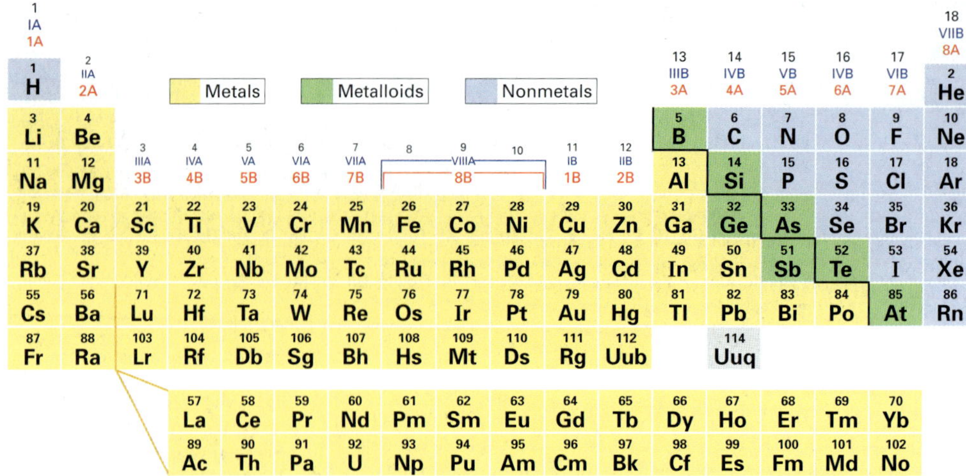

Figure 6.5 One way to classify elements in the periodic table is as metals, nonmetals, and metalloids. **Inferring** ***What is the purpose for the black stair-step line?***

Metals, Nonmetals, and Metalloids

Most periodic tables are laid out like the one in Figure 6.5. Some elements from Periods 6 and 7 are placed beneath the table. This arrangement makes the periodic table more compact. It also reflects an underlying structure of the periodic table, which you will study in Section 6.2. Each group in the table in Figure 6.5 has three labels. Scientists in the United States used the labels shown in red. Scientists in Europe used the labels shown in blue. There is some overlap between the systems, but in many cases two different groups have the same letter and number combination.

For scientists to communicate clearly, they need to agree on the standards they will use. The International Union of Pure and Applied Chemistry (IUPAC) is an organization that sets standards for chemistry. In 1985, IUPAC proposed a new system for labeling groups in the periodic table. They numbered the groups from left to right 1 through 18 (the black labels in Figure 6.5). The large periodic table in Figure 6.9 includes the IUPAC system and the system used in the United States. The latter system will be most useful when you study how compounds form in Chapters 7 and 8.

Dividing the elements into groups is not the only way to classify them based on their properties. The elements can be grouped into three broad classes based on their general properties. **Three classes of elements are metals, nonmetals, and metalloids.** Across a period, the properties of elements become less metallic and more nonmetallic.

Metals The number of yellow squares in Figure 6.5 shows that most elements are metals—about 80 percent. **Metals** are good conductors of heat and electric current. A freshly cleaned or cut surface of a metal will have a high luster, or sheen. The sheen is caused by the metal's ability to reflect light. All metals are solids at room temperature, except for mercury (Hg). Many metals are ductile, meaning that they can be drawn into wires. Most metals are malleable, meaning that they can be hammered into thin sheets without breaking. Figure 6.6 shows how the properties of metals can determine how metals are used.

For: Links on Metals and Nonmetals
Visit: www.SciLinks.org
Web Code: cdn-1061

Differentiated Instruction

Less Proficient Readers **L1**

Make a chart on the board of the three types of elements. Ask students to fill in details about the general characteristics of the three classes as they read about them.

Nonmetals In Figure 6.5, blue is used to identify the nonmetals. These elements are in the upper-right corner of the periodic table. There is a greater variation in physical properties among nonmetals than among metals. Most nonmetals are gases at room temperature, including the main components of air—nitrogen and oxygen. A few are solids, such as sulfur and phosphorus. One nonmetal, bromine, is a dark-red liquid.

The variation among nonmetals makes it difficult to describe one set of general properties that will apply to all nonmetals. However, nonmetals are not metals, as their name implies. So they tend to have properties that are opposite to those of metals. In general, **nonmetals** are poor conductors of heat and electric current. Carbon is an exception to this rule. Solid nonmetals tend to be brittle, meaning that they will shatter if hit with a hammer.

Which type of elements tend to be good conductors of heat and electric current?

Figure 6.6 The metals iron, copper, and aluminum have many important uses. How each metal is used is determined by its properties.

Iron (Fe)
The Gateway Arch in St. Louis, Missouri, is covered in stainless steel containing iron and two other metals, chromium (Cr) and nickel (Ni). The steel is shiny, malleable, and strong. It also resists rusting.

Copper (Cu)
Copper is ductile and second to only silver as a conductor of electric current. The copper used in electrical cables must be 99.99% pure.

Aluminum (Al)
Aluminum is one of the metals that can be shaped into a thin sheet, or foil. To qualify as a foil, a metal must be no thicker than about 0.15 mm.

Name The Element L1

Have each student choose an element without revealing the choice to other students. Each student should write a short description of the chosen element to read to the class. Ask the other students to identify the element from its description. Encourage students to be as specific as possible in their descriptions. Example: bromine is a reddish-brown liquid.

FYI

At a height of 630 feet, The Gateway Arch is the tallest monument in the United States. It commemorates the westward expansion of the U. S. in the nineteenth century.

Answers to...

Figure 6.5 Students are likely to say the line separates the metals from the nonmetals.

metals

Section 6.1 (continued)

Discuss L2

Point out silicon's position in the periodic table and explain that silicon is a metalloid, an element having both metallic and nonmetallic properties. Ask students about the properties that distinguish metals from nonmetals. Have them name other elements that are classified as metalloids. (Chemists do not always agree on which elements to classify as metalloids. Some include polonium (Po) and astatine (At) as metalloids. Some classify polonium as a metal and astatine as a nonmetal.)

3 ASSESS

Evaluate Understanding L2

Have students draw a concept map relating the following terms: *groups, periods, periodic law, periodic table, repeating properties, metals, nonmetals,* and *metalloids.*

Reteach 

Compare and contrast the periodic table and a monthly calendar. Similarity: The same progression of days occurs each week. Difference: Although the length of a day is based on a natural event (Earth's rotation), what happens on a Monday or Saturday depends on human decisions, not on an underlying natural principle.

Connecting Concepts

The atomic number of an element tells you how many protons are in its nucleus. It is a good way to organize elements because atomic number is unique for each element, while other properties, such as atomic mass, can vary for atoms of an element.

If your class subscribes to the Interactive Textbook, use it to review key concepts in Section 6.1.

with ChemASAP

Figure 6.7 Pancake-sized circular slices of silicon, called wafers, are used to make computer chips. Because a tiny speck of dust can ruin a wafer, the people who handle the wafers must wear "bunny" suits. The suits prevent skin, hair, or lint from entering the room's atmosphere.

Metalloids There is a heavy stair-step line in Figure 6.5 that separates the metals from the nonmetals. Most of the elements that border this line are shaded green. These elements are metalloids. A **metalloid** generally has properties that are similar to those of metals and nonmetals. Under some conditions, a metalloid may behave like a metal. Under other conditions, it may behave like a nonmetal. The behavior often can be controlled by changing the conditions. For example, pure silicon is a poor conductor of electric current, like most nonmetals. But if a small amount of boron is mixed with silicon, the mixture is a good conductor of electric current, like most metals. Silicon can be cut into wafers, like those being inspected in Figure 6.7, and used to make computer chips.

6.1 Section Assessment

1. **Key Concept** How did chemists begin the process of organizing elements?
2. **Key Concept** What property did Mendeleev use to organize his periodic table?
3. **Key Concept** How are elements arranged in the modern periodic table?
4. **Key Concept** Name the three broad classes of elements.
5. Which of these sets of elements have similar physical and chemical properties?
   a. oxygen, nitrogen, carbon, boron
   b. strontium, magnesium, calcium, beryllium
   c. nitrogen, neon, nickel, niobium
6. Identify each element as a metal, metalloid, or nonmetal.
   a. gold
   b. silicon
   c. sulfur
   d. barium
7. Name two elements that have properties similar to those of the element sodium.

Connecting Concepts

Atomic Number What does an atomic number tell you about the atoms of an element? Why is atomic number better than atomic mass for organizing the elements in a periodic table? Use what you learned in Section 4.2 to answer this question.

Assessment 6.1 Test yourself on the concepts in Section 6.1.

with ChemASAP

Section 6.1 Assessment

1. Chemists used the properties of elements to sort them into groups.
2. in order of increasing atomic mass
3. in order of increasing atomic number
4. metal, metalloid, nonmetal
5. b
6. **a.** metal **b.** metalloid **c.** nonmetal **d.** metal
7. lithium, potassium, rubidium, cesium, or francium

6.2 Classifying the Elements

Connecting to Your World

The sculptor Augustus Saint-Gaudens designed this gold coin at the request of Theodore Roosevelt. President Roosevelt wanted coins minted in the United States to be as beautiful as ancient Greek coins, which he admired. The coin is an example of a double eagle. The name derives from the fact that the coin was worth twice as much as $10 coins called eagles. A coin may contain a lot of information in a small space—its value, the year it was minted, and its country of origin. Each square in a periodic table also contains a lot of information. In this section, you will learn what types of information are usually listed in a periodic table.

Guide for Reading

 **Key Concepts**

- What type of information can be displayed in a periodic table?
- How can elements be classified based on their electron configurations?

Vocabulary

alkali metals
alkaline earth metals
halogens
noble gases
representative elements
transition metal
inner transition metal

Reading Strategy

Relating Text and Visuals As you read, look carefully at Figure 6.9. After you read the section, explain what you can tell about an element from the color assigned to its square and the color assigned to its symbol.

Squares in the Periodic Table

The periodic table displays the symbols and names of the elements, along with information about the structure of their atoms. Figure 6.8 shows one square from the detailed periodic table of the elements in Figure 6.9 on page 162. In the center of the square is the symbol for sodium (Na). The atomic number for sodium (11) is above the symbol. The element name and average atomic mass are below the symbol. There is also a vertical column with the numbers 2, 8, and 1, which are the number of electrons in each occupied energy level of a sodium atom.

The symbol for sodium is printed in black because sodium is a solid at room temperature. In Figure 6.9, the symbols for gases are in red. The symbols for the two elements that are liquids at room temperature, mercury and bromine, are in blue. The symbols for some elements in Figure 6.9 are printed in green. These elements are not found in nature. In Chapter 25, you will learn how scientists produce these elements.

The background colors in the squares are used to distinguish groups of elements. For example, two shades of gold are used for the metals in Groups 1A and 2A. The Group 1A elements are called **alkali metals,** and the Group 2A elements are called **alkaline earth metals.** The name alkali comes from the Arabic *al aqali,* meaning "the ashes." Wood ashes are rich in compounds of the alkali metals sodium and potassium. Some groups of nonmetals also have special names. The nonmetals of Group 7A are called **halogens.** The name halogen comes from the combination of the Greek word *hals,* meaning salt, and the Latin word *genesis,* meaning "to be born." There is a general class of compounds called salts, which include the compound called table salt. Chlorine, bromine and iodine, the most common halogens, can be prepared from their salts.

Figure 6.8 This is the element square for sodium from the periodic table in Figure 6.9. **Interpreting Diagrams** ***What does the data in the square tell you about the structure of sodium atoms?***

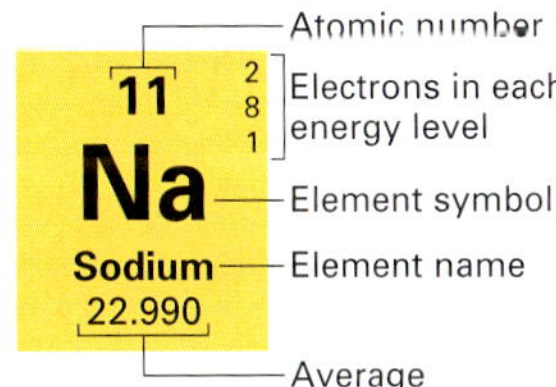

6.2

1 FOCUS

Objectives

6.2.1 **Describe** the information in a periodic table.

6.2.2 **Classify** elements based on electron configuration.

6.2.3 **Distinguish** representative elements and transition metals.

Guide for Reading

Build Vocabulary 

LINCS Have students use the LINCS strategy. In LINCS exercises, students **L**ist what they know about each term, **I**magine a picture that describes the term, **N**ote a reminding "sound-alike" word, **C**onnect the terms to the sound-alike word by making up a short story, and then perform a brief **S**elf-test.

Reading Strategy 

Predict After students preview the section, have them predict the relationship between an element's electron configuration and its position in the periodic table. Students can revise their predictions after they read the section.

2 INSTRUCT

Connecting to Your World

After students read the opening paragraph, have them volunteer other examples of small items that contain lots of information (e.g., train schedules, digital watch face, nutrient labels).

Squares in the Periodic Table

Use Visuals 

Figure 6.8 Remind students what average atomic mass represents. Make sure students understand the vertical column of numbers.

Answers to...

Figure 6.8 There are 11 protons in the nucleus and 11 electrons in the three occupied energy levels.

Section Resources

Print

- **Guided Reading and Study Workbook,** Section 6.2
- **Core Teaching Resources,** Section 6.2 Review
- **Transparencies,** T67–T69
- **Laboratory Manual,** Lab 9

Technology

- **Interactive Textbook with ChemASAP,** Assessment 6.2
- **Go Online,** Section 6.2

Use Visuals

L1

Figure 6.9 Have students examine the periodic table and find the types of information that the table provides. Summarize suggestions on the board. Examples include element name, atomic number, average atomic mass, and physical state at atmospheric pressure and room temperature.

Discuss

Have students examine the periodic table and determine the accuracy of this statement: Atomic mass always increases as atomic number increases. *(The trend is generally true, but there are exceptions. For example, the atomic numbers of Co, Ni, and Cu increase by one unit (27, 28, 29), but Ni has the lowest average atomic mass of these three elements.)*

Periodic Table of the Elements

Representative Elements: Alkali Metals; Alkaline Earth Metals; Other Metals; Metalloids; Nonmetals; Noble Gases

Transition Elements: Transition Metals; Inner transition metals

C Solid; Br Liquid; He Gas; Tc Not found in nature

1 1A	2 2A	3 3B	4 4B	5 5B	6 6B	7 7B	8 8B	9 8B
1 H Hydrogen 1.0079 (1)								
3 Li Lithium 6.941 (2, 1)	4 Be Beryllium 9.0122 (2, 2)							
11 Na Sodium 22.990 (2, 8, 1)	12 Mg Magnesium 24.305 (2, 8, 2)							
19 K Potassium 39.098 (2, 8, 8, 1)	20 Ca Calcium 40.08 (2, 8, 8, 2)	21 Sc Scandium 44.956 (2, 8, 9, 2)	22 Ti Titanium 47.90 (2, 8, 10, 2)	23 V Vanadium 50.941 (2, 8, 11, 2)	24 Cr Chromium 51.996 (2, 8, 13, 1)	25 Mn Manganese 54.938 (2, 8, 13, 2)	26 Fe Iron 55.847 (2, 8, 14, 2)	27 Co Cobalt 58.933 (2, 8, 15, 2)
37 Rb Rubidium 85.468 (2, 8, 18, 8, 1)	38 Sr Strontium 87.62 (2, 8, 18, 8, 2)	39 Y Yttrium 88.906 (2, 8, 18, 9, 2)	40 Zr Zirconium 91.22 (2, 8, 18, 10, 2)	41 Nb Niobium 92.906 (2, 8, 18, 12, 1)	42 Mo Molybdenum 95.94 (2, 8, 18, 13, 1)	43 Tc Technetium (98) (2, 8, 18, 14, 1)	44 Ru Ruthenium 101.07 (2, 8, 18, 15, 1)	45 Rh Rhodium 102.91 (2, 8, 18, 16, 1)
55 Cs Cesium 132.91 (2, 8, 18, 18, 8, 1)	56 Ba Barium 137.33 (2, 8, 18, 18, 8, 2)	71 Lu Lutetium 174.97 (2, 8, 18, 32, 9, 2)	72 Hf Hafnium 178.49 (2, 8, 18, 32, 10, 2)	73 Ta Tantalum 180.95 (2, 8, 18, 32, 11, 2)	74 W Tungsten 183.85 (2, 8, 18, 32, 12, 2)	75 Re Rhenium 186.21 (2, 8, 18, 32, 13, 2)	76 Os Osmium 190.2 (2, 8, 18, 32, 14, 2)	77 Ir Iridium 192.22 (2, 8, 18, 32, 15, 2)
87 Fr Francium (223) (2, 8, 18, 32, 18, 8, 1)	88 Ra Radium (226) (2, 8, 18, 32, 18, 8, 2)	103 Lr Lawrencium (262 (2, 8, 18, 32, 32, 9, 2)	104 Rf Rutherfordium (261) (2, 8, 18, 32, 32, 10, 2)	105 Db Dubnium (262) (2, 8, 18, 32, 32, 11, 2)	106 Sg Seaborgium (263) (2, 8, 18, 32, 32, 12, 2)	107 Bh Bohrium (264) (2, 8, 18, 32, 32, 13, 2)	108 Hs Hassium (265) (2, 8, 18, 32, 32, 14, 2)	109 Mt Meitnerium (268) (2, 8, 18, 32, 32, 15, 2)

Elements 104–114 are the transactinide elements.

Lanthanide Series

57 La Lanthanum 138.91 (2, 8, 18, 18, 9, 2)	58 Ce Cerium 140.12 (2, 8, 18, 20, 8, 2)	59 Pr Praseodymium 140.91 (2, 8, 18, 21, 8, 2)	60 Nd Neodymium 144.24 (2, 8, 18, 22, 8, 2)	61 Pm Promethium (145) (2, 8, 18, 23, 8, 2)	62 Sm Samarium 150.4 (2, 8, 18, 24, 8, 2)

Actinide Series

89 Ac Actinium (227) (2, 8, 18, 32, 18, 9, 2)	90 Th Thorium 232.04 (2, 8, 18, 32, 18, 10, 2)	91 Pa Protactinium 231.04 (2, 8, 18, 32, 20, 9, 2)	92 U Uranium 238.03 (2, 8, 18, 32, 21, 9, 2)	93 Np Neptunium (237) (2, 8, 18, 32, 22, 9, 2)	94 Pu Plutonium (244) (2, 8, 18, 32, 24, 8, 2)

Figure 6.9 In this periodic table, the colors of the boxes are used to classify representative elements and transition elements.

Facts and Figures

Element Nomenclature

Names suggested by those who create new elements must be approved by the International Union of Pure and Applied Chemistry (IUPAC). The process can be lengthy. For example, the names chosen for elements 104–108 by the nomenclature committee of IUPAC in 1994 were not those endorsed by the American Chemical Society (ACS). After years of negotiation, a compromise resulted in the names that appear in the periodic table. Elements beyond 111 have been discovered but are not yet named. Element 110 was discovered in 1994 by a team of scientists in Darmstadt, Germany. The IUPAC Council voted formal approval of the name darmstadtium (Ds) on August 16, 2003.

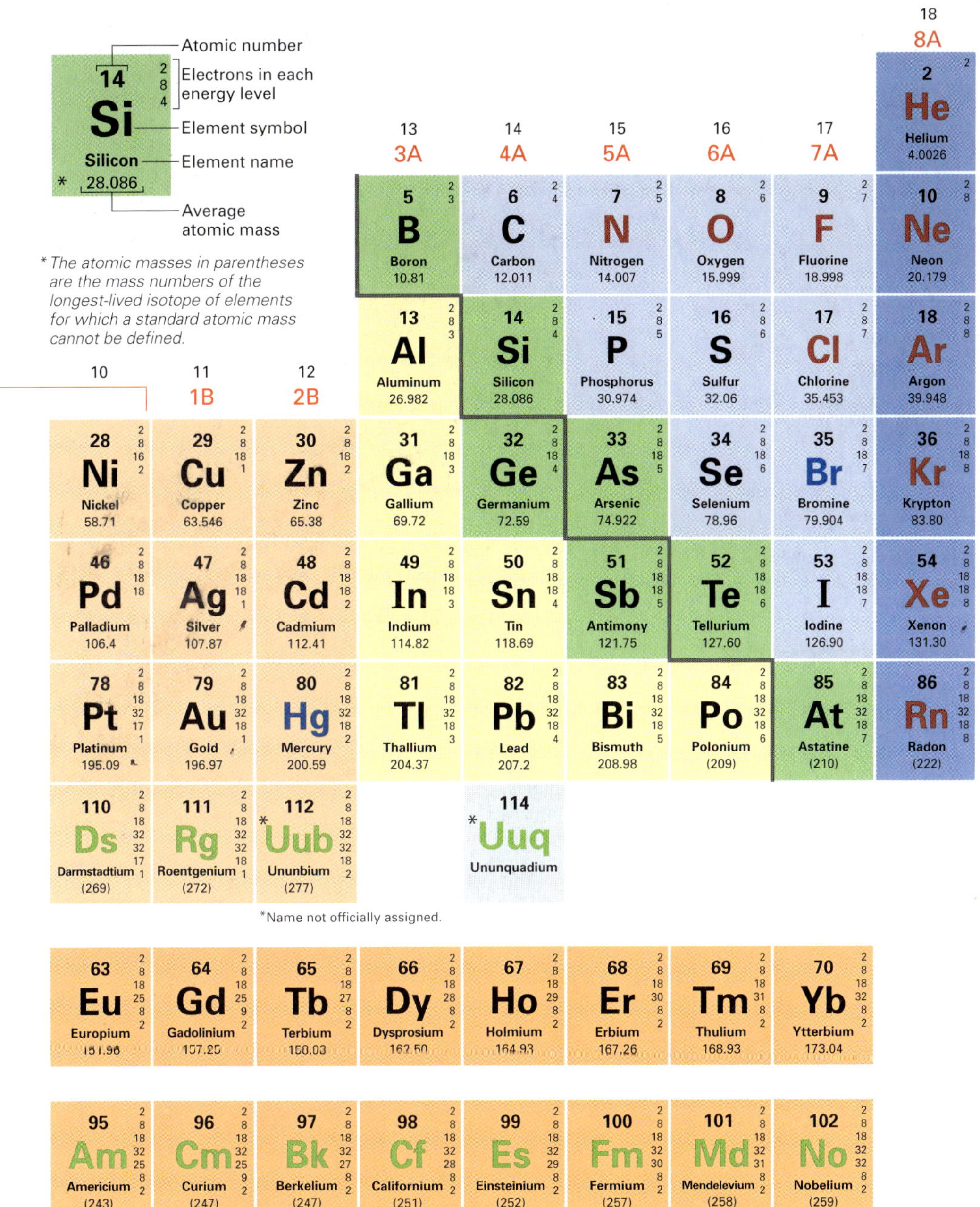

Relate L2

Have students start a list of elements found in the body that are essential for metabolism. Explain that many elements are present in trace amounts but are, nevertheless, essential for survival. Have students do research on such elements as copper, chromium, iodine, and manganese to find out more about their biological roles. They can begin with the discussion of micronutrients on page R45 of the Elements Handbook.

Electron Configurations in Groups

Discuss

L2

Lead a class discussion on electron configurations of noble gases and representative elements. Select some elements and have students write out the electron configurations for those elements. Have students compare the electron configurations for all the elements in a single group. Ask students to identify similarities. (Noble gases are sometimes classified as representative elements because they are in the *p* block of elements.)

FYI

The term *valence electron* will be introduced in Chapter 7.

Download a worksheet on **Chemical Families** for students to complete, and find additional teacher support from NSTA SciLinks.

Figure 6.10 This blimp contains helium, one of the noble gases. **Applying Concepts** ***What does the ability of a helium-filled blimp to rise in air tell you about the density of helium?***

Electron Configurations in Groups

Electrons play a key role in determining the properties of elements. So there should be a connection between an element's electron configuration and its location in the periodic table. **Elements can be sorted into noble gases, representative elements, transition metals, or inner transition metals based on their electron configurations.** You may want to refer to Figure 6.9 as you read about these classes of elements.

The Noble Gases The blimp in Figure 6.10 is filled with helium. Helium is an example of a noble gas. The **noble gases** are the elements in Group 8A of the periodic table. These nonmetals are sometimes called the inert gases because they rarely take part in a reaction. The electron configurations for the first four noble gases in Group 8A are listed below.

Element	Configuration
Helium (He)	$1s^2$
Neon (Ne)	$1s^2 2s^2 2p^6$
Argon (Ar)	$1s^2 2s^2 2p^6 3s^2 3p^6$
Krypton (Kr)	$1s^2 2s^2 2p^6 3s^2 3p^6 3d^{10} 4s^2 4p^6$

Look at the description of the highest occupied energy level for each element, which is highlighted in yellow. The *s* and *p* sublevels are completely filled with electrons. Chapter 7 will explain how this arrangement of electrons is related to the relative inactivity of the noble gases.

The Representative Elements Figure 6.11 shows the portion of the periodic table containing Groups 1A through 7A. Elements in these groups are often referred to as **representative elements** because they display a wide range of physical and chemical properties. Some are metals, some are nonmetals, and some are metalloids. Most of them are solids, but a few are gases at room temperature, and one, bromine, is a liquid.

In atoms of representative elements, the *s* and *p* sublevels of the highest occupied energy level are not filled. Look at the electron configurations for lithium, sodium, and potassium. In atoms of these Group 1A elements, there is only one electron in the highest occupied energy level. The electron is in an *s* sublevel.

Element	Configuration
Lithium (Li)	$1s^2 2s^1$
Sodium (Na)	$1s^2 2s^2 2p^6 3s^1$
Potassium (K)	$1s^2 2s^2 2p^6 3s^2 3p^6 4s^1$

In atoms of carbon, silicon, and germanium, in Group 4A, there are four electrons in the highest occupied energy level.

Element	Configuration
Carbon (C)	$1s^2 2s^2 2p^2$
Silicon (Si)	$1s^2 2s^2 2p^6 3s^2 3p^2$
Germanium (Ge)	$1s^2 2s^2 2p^6 3s^2 3p^6 3d^{10} 4s^2 4p^2$

For any representative element, its group number equals the number of electrons in the highest occupied energy level.

Why are noble gases sometimes referred to as inert gases?

For: Links on Chemical Families
Visit: www.SciLinks.org
Web Code: cdn-1062

Facts and Figures

Using Shorthand Notation

You can use the electron configurations of noble gases to produce a shorthand notation for other electron configurations. The symbol of a noble gas in brackets can stand for its electron configuration. The configuration of sodium can be written as [Ne] $3s^1$; the configuration of fluorine as [He] $2s^2 2p^5$; and the configuration of chromium as [Ar] $3d^5 4s^1$.

Magnesium This magnified view of a leaf shows the green structures where light energy is changed into chemical energy. The compound chlorophyll, which contains magnesium, absorbs the light.

Sodium When salt lakes evaporate, they form salt pans like this one in Death Valley, California. The main salt in a salt pan is sodium chloride.

1A	2A	3A	4A	5A	6A	7A
1 H Hydrogen 1.0079						
3 Li Lithium 6.941	4 Be Beryllium 9.0122	5 B Boron 10.81	6 C Carbon 12.011	7 N Nitrogen 14.007	8 O Oxygen 15.999	9 F Fluorine 18.998
11 Na Sodium 22.990	12 Mg Magnesium 24.305	13 Al Aluminum 26.982	14 Si Silicon 28.086	15 P Phosphorus 30.974	16 S Sulfur 32.06	17 Cl Chlorine 35.453
19 K Potassium 39.098	20 Ca Calcium 40.08	31 Ga Gallium 69.72	32 Ge Germanium 72.59	33 As Arsenic 74.922	34 Se Selenium 78.96	35 Br Bromine 79.904
37 Rb Rubidium 85.468	38 Sr Strontium 87.62	49 In Indium 114.82	50 Sn Tin 118.69	51 Sb Antimony 121.75	52 Te Tellurium 127.60	53 I Iodine 126.90
55 Cs Cesium 132.91	56 Ba Barium 137.33	81 Tl Thallium 204.37	82 Pb Lead 207.2	83 Bi Bismuth 208.98	84 Po Polonium (209)	85 At Astatine (210)
87 Fr Francium (223)	88 Ra Radium (226)					

Figure 6.11 Some of the representative elements exist in nature as elements. Others are found only in compounds.

Sulfur These scientists are sampling gases being released from a volcano through a vent called a fumarole. The yellow substance is sulfur.

Arsenic This bright red ore is a major source of arsenic in Earth's crust. It contains a compound of arsenic and sulfur.

Differences in Reactivity of Metals L2

Purpose Students observe differences in the reactivity of magnesium, tin, and copper.

Materials 0.2*M* HCl; 6 large test tubes; test tube rack; small pieces of clean magnesium, tin, and copper; matches; wood splint

Safety Wear goggles for this demo.

Procedure Place three large test tubes in a test tube rack. To each test tube, add a small, clean piece of a different metal—magnesium, tin, and copper. Carefully add some of the 0.2*M* HCl to each test tube and invert a test tube over it. Point out the appearance of bubbles. After a while, carefully ignite the hydrogen in the inverted test tube from the magnesium or tin reaction.

Expected Outcome The magnesium and tin react with the HCl, producing hydrogen. The hydrogen ignites with an explosive pop. The copper shows no sign of reaction.

Answers to...

Figure 6.10 Helium is less dense than air.

They rarely take part in a reaction.

Transition Elements

Lanthanides in Consumer Products L2

Have students use the library or Internet to compile a list of consumer products that contain lanthanides or require lanthanides for processing.

Discuss L2

Students may think that elements whose names are familiar are always more abundant than less familiar elements. You can use the following data to show why calling inner transition metals "rare-earth elements" is misleading. In Earth's crust, there is 1.7 ppb of osmium, 37 ppb of platinum, 67 ppb of mercury, 1700 ppb of tantalum, 6000 ppb of thorium, 10,000 ppb of lead, 60,000 ppb of cerium, and 34,000 ppb of lanthanum.

Use Visuals L1

Figure 6.12 Emphasize the relationship between the position of an element in the periodic table and the element's atomic structure. Ask, **What do all the elements in Group 1A have in common?** *(They all have one electron in the s orbital in the highest occupied energy level.)* **What do all the elements in Group 3A have in common?** *(They all have one electron in a p orbital in the highest occupied energy level.)* To help students understand the instructions for writing electron configurations for elements in the *d* and *f* blocks, review the aufbau diagram in Figure 5.7 on p. 133.

Transition Elements

In the periodic table, the B groups separate the A groups on the left side of the table from the A groups on the right side. Elements in the B groups, which provide a connection between the two sets of representative elements, are referred to as transition elements. There are two types of transition elements—transition metals and inner transition metals. They are classified based on their electron configurations.

The transition metals are the Group B elements that are usually displayed in the main body of a periodic table. Copper, silver, gold, and iron are transition metals. In atoms of a **transition metal,** the highest occupied *s* sublevel and a nearby *d* sublevel contain electrons. These elements are characterized by the presence of electrons in *d* orbitals.

The inner transition metals appear below the main body of the periodic table. In atoms of an **inner transition metal,** the highest occupied *s* sublevel and a nearby *f* sublevel generally contain electrons. The inner transition metals are characterized by *f* orbitals that contain electrons. Before scientists knew much about inner transition metals, people began to refer to them as rare-earth elements. This name is misleading because some inner transition metals are more abundant than other elements.

Blocks of Elements If you consider both the electron configurations and the positions of the elements in the periodic table, another pattern emerges. In Figure 6.12, the periodic table is divided into sections, or blocks, that correspond to the highest occupied sublevels. The *s* block contains the elements in Groups 1A and 2A and the noble gas helium. The *p* block contains the elements in Groups 3A, 4A, 5A, 6A, 7A, and 8A, with the exception of helium. The transition metals belong to the *d* block, and the inner transition metals belong to the *f* block.

You can use Figure 6.12 to help determine electron configurations of elements. Each period on the periodic table corresponds to a principal energy level. Say an element is located in period 3. You know that the *s* and *p* sublevels in energy levels 1 and 2 are filled with electrons. You read across period 3 from left to right to complete the configuration. For transition elements, electrons are added to a *d* sublevel with a principal energy level that is one less than the period number. For the inner transition metals, the principal energy level of the *f* sublevel is two less than the period number. This procedure gives the correct electron configurations for most atoms.

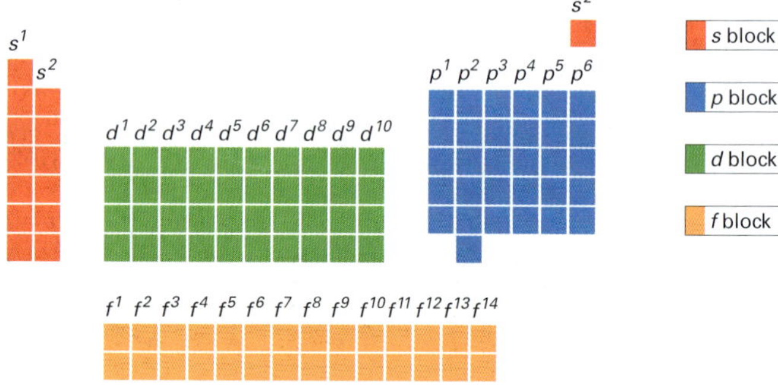

Figure 6.12 This diagram classifies elements into blocks according to sublevels that are filled or filling with electrons. **Interpreting Diagrams** ***In the highest occupied energy level of a halogen atom, how many electrons are in the p sublevel?***

Differentiated Instruction

Less Proficient Readers 

Divide the class into small groups. Have each group write electron configurations for an *s*-block element, a *p*-block element, a *d*-block element, and an *f*-block element. Don't assign any of the *d*-block elements with upredictable configurations. Ask students to explain the method they used to determine specific electron configurations.

CONCEPTUAL PROBLEM 6.1

Using Energy Sublevels to Write Electron Configurations

Nitrogen is an element that organisms need to remain healthy. However, most organisms cannot obtain nitrogen directly from air. A few organisms can convert elemental nitrogen into a form that can be absorbed by plant and animal cells. These include bacteria that live in lumps called *nodules* on the roots of legumes. The photograph shows the nodules on a bean plant. Use Figure 6.12 to write the electron configuration for nitrogen (N), which has atomic number 7.

 **Analyze** ***Identify the relevant concepts.***

For all elements, the atomic number is equal to the total number of electrons. For a representative element, the highest occupied energy level is the same as the number of the period in which the element is located. From the group in which the element is located, you can tell how many electrons are in this energy level.

 **Solve** ***Apply concepts to this situation.***

Nitrogen is located in the second period of the periodic table and in the third group of the *p* block. Nitrogen has seven electrons. Based on Figure 6.12, the configuration for the two electrons in the first energy level is $1s^2$. The configuration for the five electrons in the second energy level is $2s^22p^3$.

Practice Problems

8. Use Figure 6.9 and Figure 6.12 to write the electron configurations of the following elements.
a. carbon **b.** strontium **c.** vanadium
(*Hint:* Remember that the principal energy level number for elements in the *d* block is always one less than the period number.)

9. List the symbols for all the elements whose electron configurations end as follows. Each *n* represents an energy level.
a. ns^2np^1
b. ns^2np^5
c. $ns^2np^6nd^2(n+1)s^2$

6.2 Section Assessment

10. Key Concept What information can be included in a periodic table?

11. Key Concept Into what four classes can elements be sorted based on their electron configurations?

12. Why do the elements potassium and sodium have similar chemical properties?

13. Classify each element as a representative element, transition metal, or noble gas.
a. $1s^22s^22p^63s^23p^63d^{10}4s^24p^6$
b. $1s^22s^22p^63s^23p^63d^64s^2$
c. $1s^22s^22p^63s^23p^2$

14. Which of the following elements are transition metals: Cu, Sr, Cd, Au, Al, Ge, Co?

15. How many electrons are in the highest occupied energy level of a Group 5A element?

Noble Gases Look at the atomic properties of noble gases on page R36. Use what you know about the structure of atoms to explain why the color produced in a gas discharge tube is different for each gas.

Interactive Textbook
Assessment 6.2 Test yourself on the concepts in Section 6.2.
with ChemASAP

CONCEPTUAL PROBLEM 6.1

Answers

8. a. $1s^22s^22p^2$
b. $1s^22s^22p^63s^23p^63d^{10}4s^24p^65s^2$
c. $1s^22s^22p^63s^23p^63d^34s^2$

9. a. B, Al, Ga, In, Tl
b. F, Cl, Br, I, At **c.** Ti, Zr, Hf, Rf

Practice Problems Plus L2

Chapter Review problems 34 and 35 are related to Conceptual Problem 6.1.

3 ASSESS

Evaluate Understanding 

Call out pairs of elements in the same group and have students write their electron configurations. This activity can be made into a game where groups compete with other groups to come up with the answer first. Students should eventually be able to write the electron configurations quickly.

Reteach 

To reinforce the relationship between configurations and position in the periodic table, provide configurations and ask students to identify and locate the elements. Ask students to explain which parts of a configuration proved most useful for determining the identity.

Elements Handbook

When atoms of a noble gas are energized, electrons move into higher energy levels. When the electrons return to lower energy levels, they emit light of specific frequencies. Each noble gas has a unique emission spectrum based on its unique electron configuration.

If your class subscribes to the Interactive Textbook, use it to review key concepts in Section 6.2.

with ChemASAP

Section 6.2 Assessment

10. symbols and names of the elements; atomic number and average atomic mass; information about electron configuration

11. representative elements, noble gases, transition metals, and inner transition metals

12. They are in the same group and have the same number of electrons in the highest occupied energy level.

13. a. noble gas
b. transition metal
c. representative element

14. Cu, Cd, Au, Co

15. 5

Answers to...
Figure 6.12 five

Technology & Society

True Colors

Purpose

The connection of this Technology & Society to Chapter 6 content may not be immediately obvious. It presents an important application of transition metals. Many pigments depend on the tendency of most transition metals to form colored compounds.

Background

Paint is used to cover or hide a surface to which it is applied, decorate a surface, or protect a surface. Egg yolk (in egg tempera paint), gum Arabic (in water colors and gouache), and linseed oil (oil paint) are examples of binders. Many pigments depend on the tendency of transition metals to form colored compounds.

The bison painting is from the Altamira Caves in Spain (about 12,000 B.C.) Although the oldest known paintings are about 35,000 years old, archaeologists in Zambia found pigments and paint grinding equipment that were between 350,000 and 400,000 years old. Some prehistoric artists chemically altered pigments before applying them to cave walls. About one quarter of the samples from Troubat Cave in the Pyrenees were heated in an open fire.

Technology & Society

True Colors

Paint consists essentially of a pigment, a binder, and a liquid in which the other components are dissolved or dispersed. The liquid keeps the mixture thin enough to flow. The binder attaches the paint to the surface being painted, and the pigment determines the color. Pigments may be natural or manufactured. They may be inorganic or organic. The same pigment can be used in a water-based or oil-based paint. **Comparing and Contrasting** ***Describe at least three differences between the cave painting and the painting by Jacob Lawrence.***

Yellow ochre

Red ochre

Natural pigments A prehistoric artist had a limited choice of colors— black from charcoal and red, brown, and yellow from oxides of iron in Earth's crust. These oxides (or ochre) pigments are often referred to as earth tones.

Charred wood is a source of charcoal.

Prehistoric art Around 14,000 years ago, an artist painted this bison on the ceiling of a cave in Spain. It is about two meters long.

Manufactured pigments
Alchemists (and then chemists) made pigments that don't exist in nature. They also made purer versions of natural pigments. Many of these pigments contain transition metals.

From pigments to paint
Artists mixed manufactured pigments with binders and solvents to make paint. Although premixed paints became available around 1800, some artists, including Jacob Lawrence, continued to mix their own paints.

The Builders, 1974, by Jacob Lawrence

Coding Scheme for Paints

Because pigments with the same common name can contain different substances, the Society of Dyers and Colourists in London and the American Association of Textile Chemists and Colorists established a standard coding scheme for pigments. Manufacturers agreed to list both common pigment names and color index codes on labels. See the table below for the composition and codes for the pigments shown on p. 169.

Red iron oxide	iron(III) oxide, Fe_2O_3	PR 101
Cadmium orange	solid solution of cadmium selenide, CdSe, and cadmium sulfate, $CdSO_4$	PO 20
Cobalt yellow	potassium cobalt nitrite, $K_3[Co(NO_2)_6]\cdot H_2O$	PY 40
Zinc white	zinc oxide, ZnO	PW 4
Chromium oxide green	chromium(III) oxide, Cr_2O_3	PG 17
Cobalt blue	cobalt(II) aluminate, $Co(AlO_2)_2$	PB 28
Manganese violet	manganese ammonium pyrophosphate, $Mn(III)NH_4P_2O_7$	PV 16

Answers to...

Comparing and Contrasting
Possible differences include the subject matter, the range of colors, and the anonymity of the artist who painted in the cave.

6.3

1 FOCUS

Objectives

6.3.1 Describe trends among the elements for atomic size.

6.3.2 Explain how ions form.

6.3.3 Describe periodic trends for first ionization energy, ionic size, and electronegativity.

Guide for Reading

Build Vocabulary L2

Graphic Organizer Have students make a compare/contrast table for *cation* and *anion*, and use the table to decide which type of ion an element is likely to form.

Reading Strategy L2

Relate Text and Visuals Tell students that they will learn about trends related to the location of elements in the periodic table. They should look for the visuals throughout the section that summarize the trends.

2 INSTRUCT

Connecting to Your World

Have students read the opening paragraph. Ask, **What does the geometric pattern indicate about the arrangement of particles in salt?** *(It is orderly.)*

Trends in Atomic Size

Use Visuals L1

Figure 6.13 Ask, **Why can't a scientist measure the diameter of a single atom?** *(because an atom does not have a sharply defined border)* Discuss how measuring the distance between nuclei solves this problem.

FYI

In Chapter 8 there are formal definitions of molecule and diatomic molecule. The operational definition of a molecule should be sufficient for a discussion of atomic radii.

6.3 Periodic Trends

Guide for Reading

Key Concepts

- What are the trends among the elements for atomic size?
- How do ions form?
- What are the trends among the elements for first ionization energy, ionic size, and electronegativity?
- What is the underlying cause of periodic trends?

Vocabulary

atomic radius
ion
cation
anion
ionization energy
electronegativity

Reading Strategy

Building Vocabulary After you read this section, explain the difference between a cation and an anion.

Connecting to Your World

An atom doesn't have a sharply defined boundary. So the radius of an atom cannot be measured directly. There are ways to estimate the sizes of atoms. In one method, a solid is bombarded with X rays, and the paths of the X rays are recorded on film. Sodium chloride (table salt) produced the geometric pattern in the photograph. Such a pattern can be used to calculate the position of nuclei in a solid. The distances between nuclei in a solid are an indication of the size of the particles in the solid. In this section, you will learn how properties such as atomic size are related to the location of elements in the periodic table.

Trends in Atomic Size

Another way to think about atomic size is to look at the units that form when atoms of the same element are joined to one another. These units are called molecules. Figure 6.13 shows models of molecules (molecular models) for seven nonmetals. Because the atoms in each molecule are identical, the distance between the nuclei of these atoms can be used to estimate the size of the atoms. This size is expressed as an atomic radius. The **atomic radius** is one half of the distance between the nuclei of two atoms of the same element when the atoms are joined.

The distances between atoms in a molecule are extremely small. So the atomic radius is often measured in picometers. Recall that there are one trillion, or 10^{12}, picometers in a meter. The molecular model of iodine in Figure 6.13 is the largest. The distance between the nuclei in an iodine molecule is 280 pm. Because the atomic radius is one half the distance between the nuclei, a value of 140 pm (280/2) is assigned as the radius of the iodine atom. **In general, atomic size increases from top to bottom within a group and decreases from left to right across a period.**

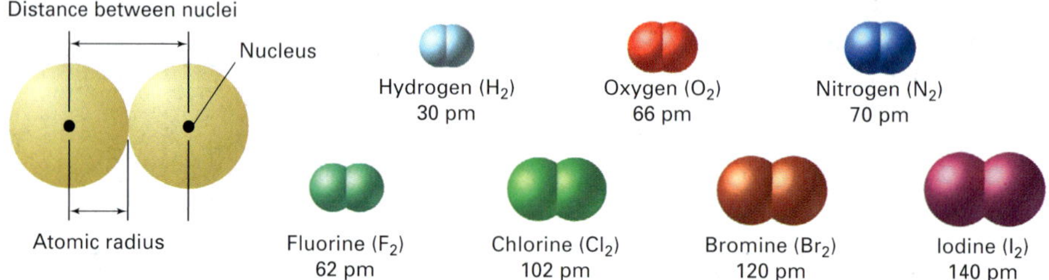

Figure 6.13 This diagram lists the atomic radii of seven nonmetals. An atomic radius is half the distance between the nuclei of two atoms of the same element when the atoms are joined.

Section Resources

Print

- **Guided Reading and Study Workbook,** Section 6.3
- **Core Teaching Resources,** Section 6.3 Review
- **Transparencies,** T70–T74
- **Small-Scale Chemistry Laboratory Manual,** Lab 9
- **Laboratory Practicals,** 6-1, 6-2

Technology

- **Interactive Textbook with ChemASAP,** Animation 7, Assessment 6.3
- **Go Online,** Section 6.3

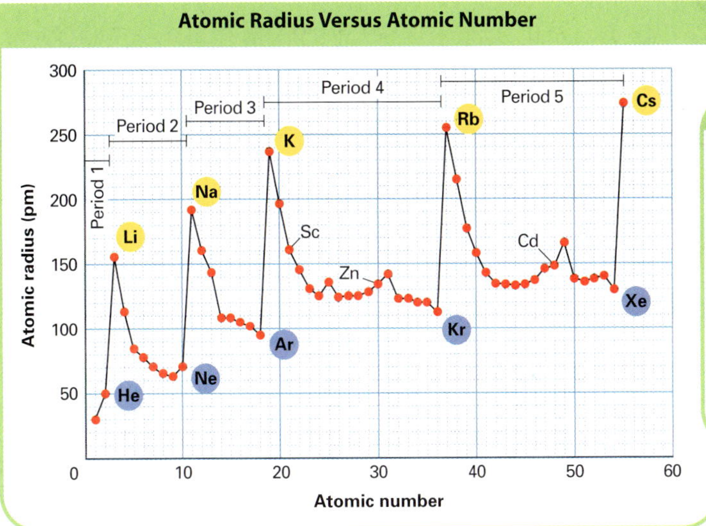

Figure 6.14 This graph plots atomic radius versus atomic number for 55 elements.

INTERPRETING GRAPHS

a. Analyzing Data Which alkali metal has an atomic radius of 238 pm?

b. Drawing Conclusions Based on the data for alkali metals and noble gases, how does atomic size change within a group?

c. Predicting Is an atom of barium, atomic number 56, smaller or larger than an atom of cesium (Cs)?

Group Trends in Atomic Size In the Figure 6.14 graph, atomic radius is plotted versus atomic number. Look at the data for the alkali metals and noble gases. The atomic radius within these groups increases as the atomic number increases. This increase is an example of a trend.

As the atomic number increases within a group, the charge on the nucleus increases and the number of occupied energy levels increases. These variables affect atomic size in opposite ways. The increase in positive charge draws electrons closer to the nucleus. The increase in the number of occupied orbitals shields electrons in the highest occupied energy level from the attraction of protons in the nucleus. The shielding effect is greater than the effect of the increase in nuclear charge. So the atomic size increases.

Periodic Trends in Atomic Size Look again at Figure 6.14. In general, atomic size decreases across a period from left to right. Each element has one more proton and one more electron than the preceding element. Across a period, the electrons are added to the same principal energy level. The shielding effect is constant for all the elements in a period. The increasing nuclear charge pulls the electrons in the highest occupied energy level closer to the nucleus and the atomic size decreases. Figure 6.15 summarizes the group and period trends in atomic size.

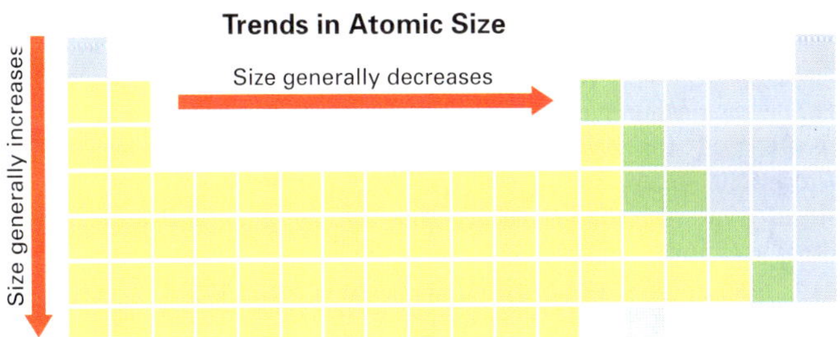

Figure 6.15 The size of atoms tends to decrease from left to right across a period and increase from top to bottom within a group. **Predicting** ***If a halogen and an alkali metal are in the same period, which one will have the larger radius?***

Interpreting Graphs 

a. potassium
b. It increases.
c. smaller

Because of the amount of data in this graph, you may need to help students get oriented before they begin to interpret the graph.

Enrichment Question 

Emphasize the key roles electrical attraction and repulsion play within atoms and ions. Review the effects of increasing nuclear charge and changes in the shielding effect of electrons on the size of an atom. (Nuclear charge increases within groups and across periods; the shielding effect increases within groups, but is constant across periods.) Have students use these effects to describe the trends for atomic size within a period and within groups.

Discuss 

As an analogy to positions and trends in properties of elements in the periodic table, use seating charts and pricing data from local theaters or sports venues to discover trends. Ask students to determine patterns that relate the position of a seat to its price. Students should discover that variables such as distance from the stage or field, location relative to the center of the action, and whether the view will be obstructed, all affect price.

Facts and Figures

Elements and the Big Bang

At the time of the Big Bang, the temperature was many billions of degrees. Neutrons, protons, and electrons may have formed within 10^{-4} second after the Big Bang, and the lightest nuclei formed within 3 minutes. Matter was in the form of plasma, a sea of positive nuclei and negative electrons. It took an estimated 500,000 years for electrons and nuclei to cool enough to form atoms. According to the Big Bang theory, Earth, with its wealth of chemical elements, formed from the debris of supernova explosions.

Answers to...

Figure 6.15 the alkali metal

Ions

Listing Elements L2

Give students a list of elements. Ask them to locate each element in the periodic table, and decide whether its atoms are likely to form positive or negative ions. Have students make a list of elements that are likely to form positive ions and another list of elements that are likely to form negative ions.

FYI

The major discussion of ions and ionic bonding is in Chapter 7. This section presents the limited information that you will need to discuss the trends in ionization energy.

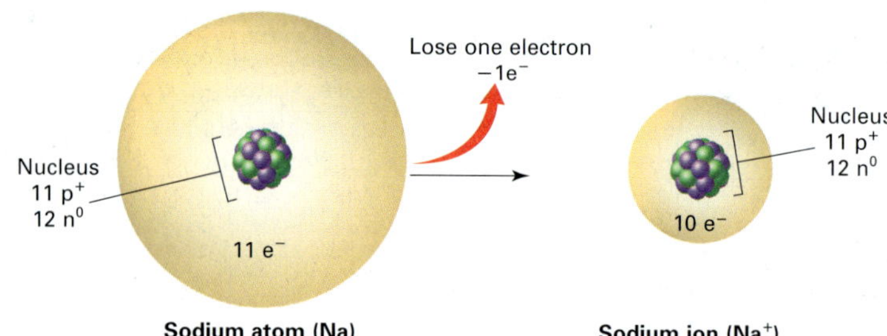

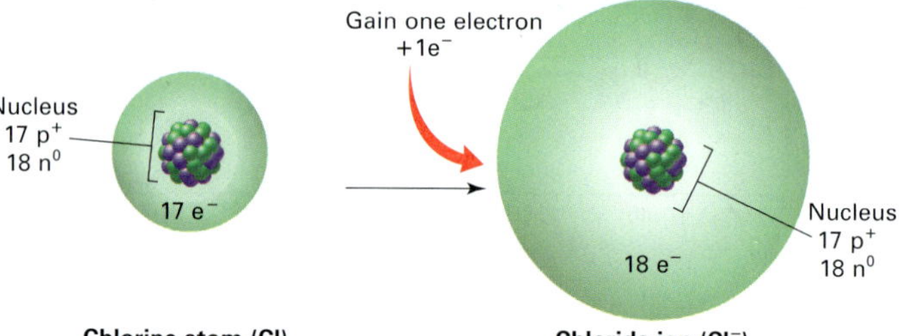

Figure 6.16 When a sodium atom loses an electron, it becomes a positively charged ion. When a chlorine atom gains an electron, it becomes a negatively charged ion. **Interpreting Diagrams** ***What happens to the protons and neutrons during these changes?***

Ions

Some compounds are composed of particles called ions. An **ion** is an atom or group of atoms that has a positive or negative charge. An atom is electrically neutral because it has equal numbers of protons and electrons. For example, an atom of sodium (Na) has 11 positively charged protons and 11 negatively charged electrons. The net charge on a sodium atom is zero $[(11+) + (-11) = 0]$.

Positive and negative ions form when electrons are transferred between atoms. Atoms of metals, such as sodium, tend to form ions by losing one or more electrons from their highest occupied energy levels. A sodium atom tends to lose one electron. Figure 6.16 compares the atomic structure of a sodium atom and a sodium ion. In the sodium ion, the number of electrons (10) is no longer equal to the number of protons (11). Because there are more positively charged protons than negatively charged electrons, the sodium ion has a net positive charge. An ion with a positive charge is called a **cation.** The charge for a cation is written as a number followed by a plus sign. If the charge is 1+, the number 1 is usually omitted from the symbol for the ion. So Na^+ is equivalent to Na^{1+}.

Atoms of nonmetals, such as chlorine, tend to form ions by gaining one or more electrons. A chlorine atom tends to gain one electron. Figure 6.16 compares the atomic structure of a chlorine atom and a chloride ion. In a chloride ion, the number of electrons (18) is no longer equal to the number of protons (17). Because there are more negatively charged electrons than positively charged protons, the chloride ion has a net negative charge. An ion with a negative charge is called an **anion.** The charge for an anion is written as a number followed by a minus sign.

Animation 7 Discover the ways that atoms of elements combine to form compounds. with **ChemASAP**

Checkpoint ***What is the difference between a cation and an anion?***

Trends in Ionization Energy

Recall that electrons can move to higher energy levels when atoms absorb energy. Sometimes there is enough energy to overcome the attraction of the protons in the nucleus. The energy required to remove an electron from an atom is called **ionization energy.** This energy is measured when an element is in its gaseous state. The energy required to remove the first electron from an atom is called the first ionization energy. The cation produced has a 1+ charge. **First ionization energy tends to decrease from top to bottom within a group and increase from left to right across a period.**

Table 6.1 lists the first, second, and third ionization energies for the first 20 elements. The second ionization energy is the energy required to remove an electron from an ion with a 1+ charge. The ion produced has a 2+ charge. The third ionization energy is the energy required to remove an electron from an ion with a 2+ charge. The ion produced has a 3+ charge.

Ionization energy can help you predict what ions elements will form. Look at the data in Table 6.1 for lithium (Li), sodium (Na), and potassium (K). The increase in energy between the first and second ionization energies is large. It is relatively easy to remove one electron from a Group 1A metal atom, but it is difficult to remove a second electron. So Group 1A metals tend to form ions with a 1+ charge.

Table 6.1

Ionization Energies of First 20 Elements (kJ/mol*)

Symbol	First	Second	Third
H	1312		
He (noble gas)	2372	5247	
Li	520	7297	11,810
Be	899	1757	14,840
B	801	2430	3659
C	1086	2352	4619
N	1402	2857	4577
O	1314	3391	5301
F	1681	3375	6045
Ne (noble gas)	2080	3963	6276
Na	496	4565	6912
Mg	738	1450	7732
Al	578	1816	2744
Si	786	1577	3229
P	1012	1896	2910
S	999	2260	3380
Cl	1256	2297	3850
Ar (noble gas)	1520	2665	3947
K	419	3069	4600
Ca	590	1146	4941

*An amount of matter equal to the atomic mass in grams.

Trends in Ionization Energy

Discuss L2

Explain that ionization energy is a measure of the difficulty in removing an electron from the highest occupied energy level. Ask, **Why is the first ionization energy of a nonmetal much higher than that of an alkali metal?** *(Because the nuclear charge increases from left to right across a period and the shielding effect stays the same, it is more difficult to remove an electron.)* There are ionization energy graphs in the Elements Handbook on R7, R11, R15, R19, R25, R29, and R33.

TEACHER Demo

Effective Nuclear Charge and Electron Shielding L2

To help students understand the concepts of effective nuclear charge and electron shielding, choose four students to be "protons" and four to be "electrons." Construct a lithium "nucleus" by having three protons stand together at the front of the room. Note that for purposes of this demo you are ignoring the neutrons. Place two electrons together at a short distance from the nucleus to represent the 1*s* electrons. Place the third electron a bit farther away to represent the 2*s* electron. You should be able to draw a line from the nucleus through the 1*s* electrons to the 2*s* electron. Point out that there are no other electrons between the 1*s* electrons and the nucleus. Thus, these electrons experience the full impact of the 3+ charge. Because the third electron's "view" of the nucleus is partially blocked, it is shielded somewhat from the full force of the 3+ charge.

FYI

Because the unit for ionization energy is kJ/mol, a footnote in Table 6.1 supplies an operational definition of mole, which is introduced in Section 10.1.

Answers to...

Figure 6.16 nothing

A cation has a positive charge. An anion has a negative charge.

Interpreting Graphs 

a. lithium; sodium
b. First ionization energy decreases as atomic number increases.
c. hydrogen; it has only one electron.

Because of the amount of data in this graph, you may need to help students get oriented before they begin to interpret the graph.

Enrichment Question 

Challenge students to explain why the portion of the graph for periods 4 and 5 is different from the portion of the graph for periods 2 and 3. *(Periods 4 and 5 include transition metals, whose atoms have electrons in d orbitals.)*

TEACHER Demo

Predicting Reactivity L2

Purpose Students observe the relative reactivities of magnesium and calcium and predict relative reactivities for other pairs of elements.

Materials 20 mL 1*M* HCl, two 50-mL beakers, overhead projector, 20 cm magnesium ribbon, 1 g calcium

Safety Wear goggles for this demo.

Procedure Pour 20 mL HCl into each beaker. Set the beakers on an overhead projector. Coil the magnesium ribbon and drop it into one beaker. Drop 1 g calcium into the other beaker. Compare the reaction rates in the two beakers. Point out the positions of the two elements in the periodic table, and relate the difference in reactivity to their first and second ionization energies. Ask students to predict the relative reactivities of other pairs of elements in Groups 1A and 2A.

Expected Outcome The calcium fizzes in the HCl. The magnesium reacts more slowly with the HCl.

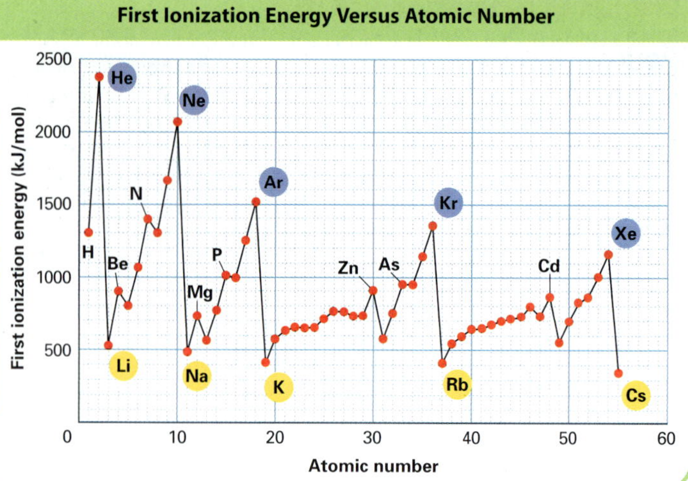

Figure 6.17 This graph reveals group and period trends for ionization energy.

INTERPRETING GRAPHS

a. Analyzing Data Which element in period 2 has the lowest first ionization energy? In period 3?

b. Drawing Conclusions What is the group trend for first ionization energy for noble gases and alkali metals?

c. Predicting If you drew a graph for second ionization energy, which element would you have to omit? Explain.

Group Trends in Ionization Energy Figure 6.17 is a graph of first ionization energy versus atomic number. Each red dot represents the data for one element. Look at the data for the noble gases and the alkali metals. In general, first ionization energy decreases from top to bottom within a group. Recall that the atomic size increases as the atomic number increases within a group. As the size of the atom increases, nuclear charge has a smaller effect on the electrons in the highest occupied energy level. So less energy is required to remove an electron from this energy level and the first ionization energy is lower.

Periodic Trends in Ionization Energy In general, the first ionization energy of representative elements tends to increase from left to right across a period. This trend can be explained by the nuclear charge, which increases, and the shielding effect, which remains constant. The nuclear charge increases across the period, but the shielding effect remains constant. So there is an increase in the attraction of the nucleus for an electron. Thus, it takes more energy to remove an electron from an atom. Figure 6.18 summarizes the group and period trends for first ionization energy.

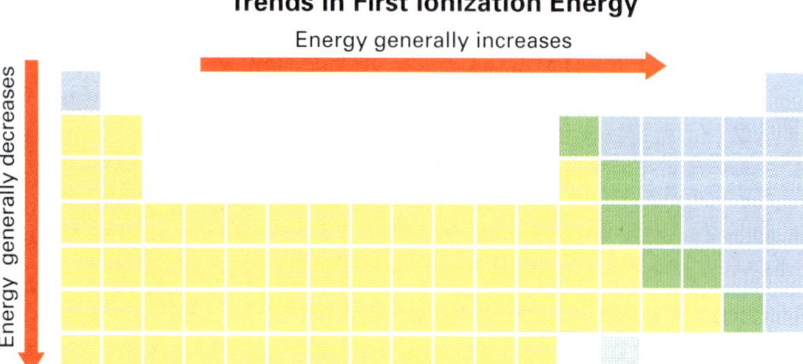

Figure 6.18 First ionization energy tends to increase from left to right across a period and decrease from top to bottom within a group.
Predicting ***Which element would have the larger first ionization energy—an alkali metal in period 2 or an alkali metal in period 4?***

Quick LAB

Periodic Trends in Ionic Radii

Purpose
Make a graph of ionic radius versus atomic number and use the graph to identify periodic and group trends.

Materials
- graph paper

Procedure
Use the data presented in Figure 6.19 to plot ionic radius versus atomic number.

Analyze and Conclude
1. Describe how the size changes when an atom forms a cation and when an atom forms an anion.
2. How do the ionic radii vary within a group of metals? How do they vary within a group of nonmetals?
3. Describe the shape of a portion of the graph that corresponds to one period.

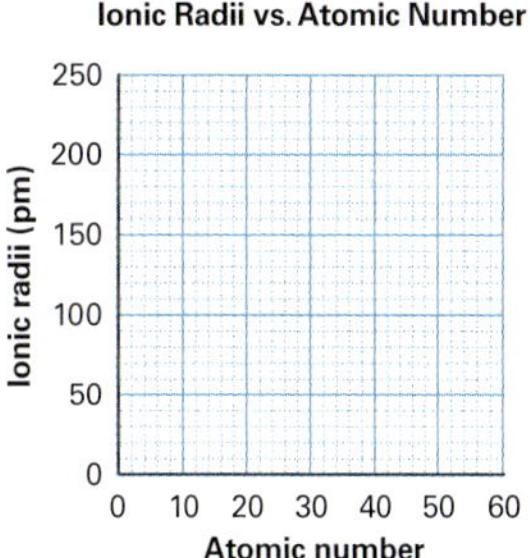

4. Is the trend across a period similar or different for periods 2, 3, 4, and 5?
5. Propose explanations for the trends you have described for ionic radii within groups and across periods.

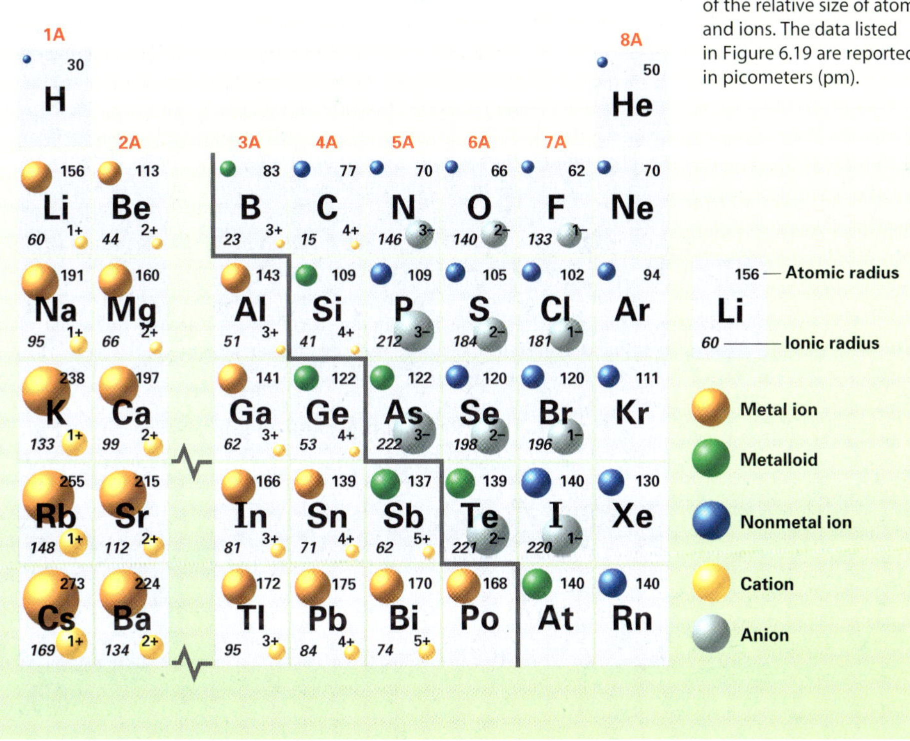

Figure 6.19 Atomic and ionic radii are an indication of the relative size of atoms and ions. The data listed in Figure 6.19 are reported in picometers (pm).

Quick LAB

Periodic Trends in Ionic Radii 

Objective After completing this activity, students will be able to
- identify periodic trends in ionic size.

Skills Focus Using tables and graphs, predicting, drawing conclusions

Prep Time none
Class Time 40 minutes

Teaching Tips
- If time is too limited for students to make the graph, use Figure 6.19 to answer Questions 1, 2, 4, and 5.
- You may want to reference the radii diagrams in the Elements Handbook on R7, R11, R15, R19, R25, R29, and R33.

Expected Outcome
Ionic radii increase from top to bottom within a group. The radii of cations and anions decrease from left to right across a period.

Analyze and Conclude
1. Cations are smaller than their atoms; anions are larger than their atoms.
2. Ionic radii increase from top to bottom within a group of metals or within a group of nonmetals.
3. Two portions of the curve slope down from left to right.
4. The trend is similar for the periods.
5. The radii increase within a group because the number of occupied energy levels increases. The radii of cations decrease across a period because the nuclear charge increases, the shielding effect is constant, and the number of electrons decreases. (The effect is smaller with anions because the number of electrons increases.)

For Enrichment 
Have students use the graph on page R41 to describe the periodic trend in atomic size for transition metals. Ask how the trend for transition metals compares to the trend for representative elements.

Answers to...

Figure 6.18 an alkali metal in period 2

Trends in Ionic Size

Discuss L2

Relate the periodic trends in ionic size to those discussed earlier for atomic size. Explain that the effective nuclear charge experienced by an electron in the highest occupied orbital of an atom or ion is equal to the total nuclear charge (the number of protons) minus the shielding effect due to electrons in lower energy levels. The effective nuclear charge determines the atomic and ionic radii. As you proceed from left to right in any given period, the principal quantum number, *n*, of the highest occupied energy level remains constant, but the effective nuclear charge increases. Therefore, atomic and ionic radii decrease as you move to the right in a period. In contrast, within any group, as you proceed from top to bottom, the effective nuclear charge remains nearly constant, but the principal quantum number, *n*, increases. Consequently, atomic and ionic radii increase from top to bottom within a group.

TEACHER Demo

Trends in Ionic Size 

Purpose Students observe an analogy for the effect of adding or removing electrons from an atom.

Materials washers or other small circular items, smaller item to represent the nucleus, overhead projector

Procedure On the overhead projector, make a circle of washers to represent an electron cloud in a neutral atom. The washers should be touching. Place the "nucleus" in the center of the circle. Add or subtract washers to mimic ion formation. With each change, adjust the circle so that the washers are still touching. Explain that the change in the diameter of the circle is analogous to the change in the effective attraction of the nuclear charge for electrons.

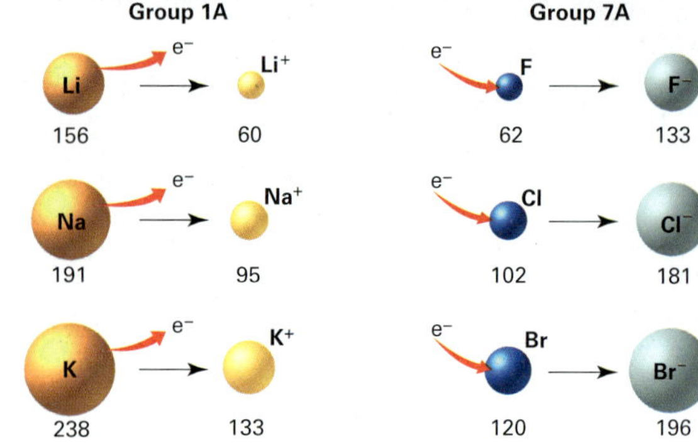

Figure 6.20 This diagram compares the relative sizes of atoms and ions for selected alkali metals and halogens. The data are given in picometers.
Comparing and Contrasting ***What happens to the radius when an atom forms a cation? When an atom forms an anion?***

Trends in Ionic Size

During reactions between metals and nonmetals, metal atoms tend to lose electrons and nonmetal atoms tend to gain electrons. The transfer has a predictable affect on the size of the ions that form. **Cations are always smaller than the atoms from which they form. Anions are always larger than the atoms from which they form.**

Figure 6.20 compares the relative sizes of the atoms and ions for three metals in Group 1A. For each of these elements, the ion is much smaller than the atom. For example, the radius of a sodium ion (95 pm) is about half the radius of a sodium atom (191 pm). When a sodium atom loses an electron, the attraction between the remaining electrons and the nucleus is increased. The electrons are drawn closer to the nucleus. Also, metals that are representative elements tend to lose all their outermost electrons during ionization. So the ion has one fewer occupied energy level.

The trend is the opposite for nonmetals like the halogens in Group 7A. For each of these elements, the ion is much larger than the atom. For example, the radius of a fluoride ion (133 pm) is more than twice the radius of a fluorine atom (62 pm). As the number of electrons increases, the attraction of the nucleus for any one electron decreases.

Look back at Figure 6.19. From left to right across a period, two trends are visible—a gradual decrease in the size of the positive ions followed by a gradual decrease in the size of the negative ions. Figure 6.21 summarizes the group and periodic trends in ionic size.

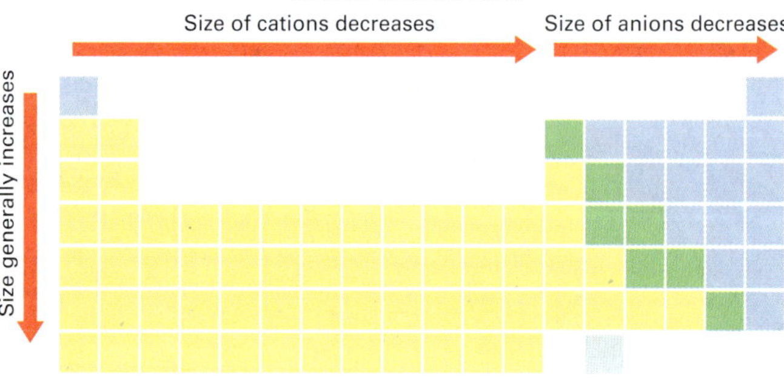

Figure 6.21 The ionic radii for cations and anions decrease from left to right across periods and increase from top to bottom within groups.

Differentiated Instruction

Gifted and Talented 

Have students research and describe the phenomenon of the lanthanide contraction. Ask them to discuss how the lanthanide contraction accounts for the fact that zirconium and hafnium have virtually the same atomic radius even though hafnium is below zirconium in Group 4B of the periodic table.

Trends in Electronegativity

In Chapters 7 and 8, you will study two types of bonds that can exist in compounds. Electrons are involved in both types of bonds. There is a property that can be used to predict the type of bond that will form during a reaction. This property is called electronegativity. **Electronegativity** is the ability of an atom of an element to attract electrons when the atom is in a compound. Scientists use factors such as ionization energy to calculate values for electronegativity.

Table 6.2 lists electronegativity values for representative elements in Groups 1A through 7A. The elements are arranged in the same order as in a periodic table. The noble gases are omitted because they do not form many compounds. The data in Table 6.2 is expressed in units called Paulings. Linus Pauling won a Nobel Prize in Chemistry for his work on chemical bonds. He was the first to define electronegativity.

In general, electronegativity values decrease from top to bottom within a group. For representative elements, the values tend to increase from left to right across a period. Metals at the far left of the periodic table have low values. By contrast, nonmetals at the far right (excluding noble gases) have high values. The electronegativity values among the transition metals are not as regular.

The least electronegative element is cesium, with an electronegativity value of 0.7. It has the least tendency to attract electrons. When it reacts, it tends to lose electrons and form positive ions. The most electronegative element is fluorine, with a value of 4.0. Because fluorine has such a strong tendency to attract electrons, when it is bonded to any other element it either attracts the shared electrons or forms a negative ion.

Checkpoint *Why are values for noble gases omitted from Table 6.2?*

For: Links on Electronegativity
Visit: www.SciLinks.org
Web Code: cdn-1063

Table 6.2

Electronegativity Values for Selected Elements

H 2.1						
Li 1.0	Be 1.5	B 2.0	C 2.5	N 3.0	O 3.5	F 4.0
Na 0.9	Mg 1.2	Al 1.5	Si 1.8	P 2.1	S 2.5	Cl 3.0
K 0.8	Ca 1.0	Ga 1.6	Ge 1.8	As 2.0	Se 2.4	Br 2.8
Rb 0.8	Sr 1.0	In 1.7	Sn 1.8	Sb 1.9	Te 2.1	I 2.5
Cs 0.7	Ba 0.9	Tl 1.8	Pb 1.9	Bi 1.9		

Trends in Electronegativity

Discuss

L2

Lead a class discussion on periodic and group trends in electronegativities. Point out that electronegativity values help chemists predict the type of bonding that exists between atoms in compounds. Ask, **Why aren't the noble gases included in a discussion on electronegativity?** *(because they form very few compounds)* **Which element is the most electronegative and which is the least electronegative?** *(fluorine; cesium)* You may want to reference the electronegativity graphs in the Elements Handbook on R7, R11, R15, R19, R25, R29, and R33.

FYI

The values for electronegativity are often based on values for ionization energy and electron affinity. Ionization energy is a measure of an atom's ability to lose electrons. Electron affinity is a measure of an atom's ability to gain electrons.

Download a worksheet on **Electro–negativity** for students to complete, and find additional teacher support from NSTA SciLinks.

Answers to...

Figure 6.20 The radius decreases for cations and increases for anions.

They rarely form compounds.

Summary of Trends

Use Visuals L1

Figure 6.22 Point out that this diagram incorporates information from several diagrams earlier in the chapter. Ask students whether they find this or earlier diagrams more helpful, and why.

3 ASSESS

Evaluate Understanding L2

Have students compare two elements in the same group in terms of atomic radius, ionic radius, ionization energy, and electronegativity. Repeat the exercise with a metal and nonmetal from the same period. Have students write general statements to summarize the trends revealed by these comparisons.

Reteach L1

Review the terms used in Figure 6.22. Then, use the periodic table and the terms to play a version of "I'm Thinking of . . . ." For example, choose fluorine and say you are thinking of an element that has a very small atomic size and a very high electronegativity. Let students guess, then discuss the correct answer. Have students continue the game in small groups.

Writing Activity

From top to bottom within a group, the effective nuclear charge remains nearly constant and the principal quantum number increases. From left to right in a period, the principal quantum number remains constant, but the effective nuclear charge increases.

If your class subscribes to the Interactive Textbook, use it to review key concepts in Section 6.3.

with ChemASAP

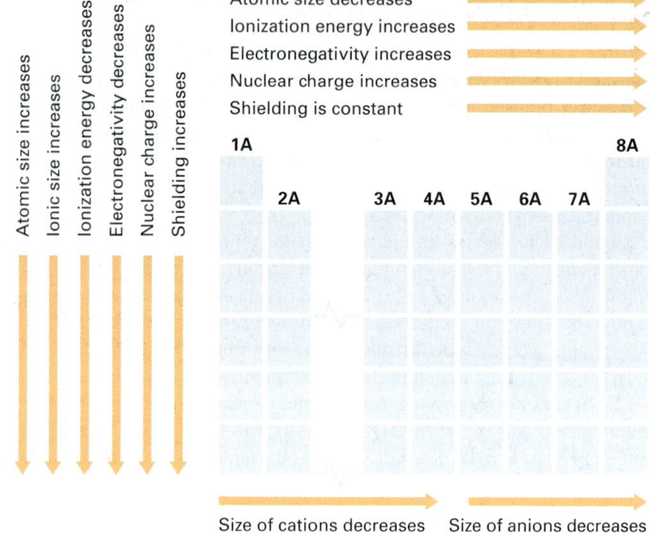

Figure 6.22 Properties that vary within groups and across periods include atomic size, ionic size, ionization energy, electronegativity, nuclear charge, and shielding effect.
Interpreting Diagrams ***Which properties tend to decrease across a period?***

Summary of Trends

Figure 6.22 shows the trends for atomic size, ionization energy, ionic size, and electronegativity in Groups 1A through 8A. These properties vary within groups and across periods. **The trends that exist among these properties can be explained by variations in atomic structure.** The increase in nuclear charge within groups and across periods explains many trends. Within groups an increase in shielding has a significant effect.

6.3 Section Assessment

16. **Key Concept** How does atomic size change within groups and across periods?
17. **Key Concept** When do ions form?
18. **Key Concept** What happens to first ionization energy within groups and across periods?
19. **Key Concept** Compare the size of ions to the size of the atoms from which they form.
20. **Key Concept** How does electronegativity vary within groups and across periods?
21. **Key Concept** In general, how can the periodic trends displayed by elements be explained?
22. Arrange these elements in order of decreasing atomic size: sulfur, chlorine, aluminum, and sodium. Does your arrangement demonstrate a periodic trend or a group trend?
23. Which element in each pair has the larger first ionization energy?
    - a. sodium, potassium
    - b. magnesium, phosphorus

Writing Activity

Explaining Trends in Atomic Size Explain why the size of an atom tends to increase from top to bottom within a group. Explain why the size of an atom tends to decrease from left to right across a period.

Assessment 6.3 Test yourself on the concepts in Section 6.3.

with ChemASAP

Section 6.3 Assessment

16. Atomic size generally increases within a group and decreases from left to right across a period.
17. Ions form when electrons are transferred between atoms.
18. decreases within a group and increases from left to right across a period
19. Anions are larger and cations are smaller than the atoms from which they form.
20. decreases from top to bottom within a group and increases from left to right across a period
21. The trends can be explained by variations in atomic structure.
22. sodium, aluminum, sulfur, chlorine; periodic trend
23. a. sodium **b.** phosphorus

Small-Scale LAB

Periodicity in Three Dimensions

Purpose

To build three-dimensional models for periodic trends.

Materials

- 96-well spot plate
- straws
- scissors
- metric ruler
- permanent fine-line marker

Procedure  

1. Measure the depth of a well in the spot plate by inserting a straw into a well and holding the straw upright as shown in the photograph. Make a mark on the straw at the point where the straw meets the surface of the plate. Measure the distance from the end of the straw to the mark in centimeters. Record this distance as well depth.
2. Cut the straw to a length that is 4.0 cm plus well depth. The straw will extend exactly 4.0 cm above the surface of the plate.
3. Fluorine has an electronegativity value of 4.0. On a scale of one cm equals one unit of electronegativity, the portion of the straw that extends above the surface of the plate represents the electronegativity value for fluorine. Using the same scale, cut straws to represent the electronegativity values for all the elements listed in Table 6.2. Remember to add the well depth to the electronegativity value before cutting a straw. As you cut the straws, mark each straw with the chemical symbol of the element that the straw represents.
4. Arrange the straws in the spot plate in rows and columns to match the locations of the elements in the periodic table.
5. Make a rough sketch of your completed model.

Analyze

Observe your model and record the answers to the following questions below your sketch.

1. Which element represented in your model is the most electronegative?
2. Based on your model, what is the general trend in electronegativity from left to right across a period?
3. Relate the trend in electronegativity across a period to the location of metals and nonmetals in the periodic table.
4. What is the general trend in electronegativity within a group? Are there any notable exceptions?
5. Why do you think that the electronegativity value for hydrogen is so high given its location in the table?

You're the Chemist

The following small-scale activities allow you to develop your own procedures and analyze the results.

1. **Design It!** Construct a similar 3-D model for first ionization energies. Use the data in Table 6.1. Use a scale of one cm equals 300 kJ/mol.
2. **Design It!** Design and construct a 3-D model that shows trends in atomic and ionic radii for the elements in Groups 1A and 7A. Devise a way to display both ionic and atomic radii in the same model.
3. **Analyze it!** Xenon has an electronegativity value of 2.6. Cut and place a straw in your first model to represent xenon. Does xenon support the trend for electronegativity across a period? Is xenon likely to form compounds? Explain your answers.

Periodicity in Three Dimensions 

Objective

After completing this activity, students should be able to:

- build concrete models to reinforce periodic trends.
- apply a procedure to a new variable and design a model on their own.

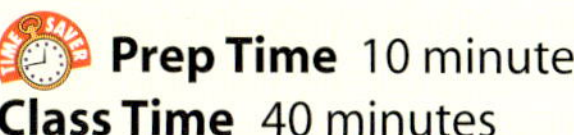

Prep Time 10 minutes
Class Time 40 minutes

Materials 96-well spot plates, straws, scissors, metric rulers, permanent fine-line markers

Advance Prep Straws with a 1/4-inch diameter fit snugly in the wells.

Teaching Tips

- Students can use colored straws to color code groups or periods.
- If you don't have spot plates, press a lump of clay the size of a golf ball flat on a table with a block of wood. Students mark out a 1-cm square grid and insert the straws in the clay.

Expected Outcome Students produce 3-D models for periodic trends.

Analyze

1. fluorine
2. Electronegativity generally increases from left to right along a period.
3. Metals, which are on the left side of the table, have lower electronegativity values than nonmetals, which are on the right.
4. Electronegativity generally increases from bottom to top within a group. Except for boron, the rest of Group 3A shows a reverse in this trend.
5. Hydrogen is placed in Group 1A based on its electron configuration, but is classified as a nonmetal.

You're the Chemist

1. Students divide the values of first ionization energies by 300 and measure the appropriate length of straws.
2. Students must determine their own scale before they begin. Students often use two wells to represent both ionic and atomic radii. Other students cut a straw to a length that represents the larger radius of an atom and mark the straw to show the smaller radius of the corresponding cation.
3. The value for xenon is similar to iodine, which is consistent with the general trend. Based on this value, xenon appears to have the ability to attract electrons and form compounds.

For Enrichment 

Have students use the data on page R41 to make a 3-D model of trends in atomic size for transition metals.

Answers to...

Figure 6.22 size of atoms and ions

CHAPTER 6 Study Guide

Study Tip

Rewrite Class Notes Remind students to take notes during class of the key points that are discussed, and to rewrite the notes so that they are organized and legible. Students can use the textbook to fill in any important points that they missed. They can use the organized notes to review for quizzes and tests.

If your class subscribes to the Interactive Textbook, your students can go online to access an interactive version of the Student Edition and a self-test.

with ChemASAP

CHAPTER 6 Study Guide

Key Concepts

6.1 Organizing the Elements

- Chemists used the properties of elements to sort them into groups.
- Mendeleev arranged the elements in his periodic table in order of increasing atomic mass.
- In the modern periodic table, elements are arranged in order of increasing atomic number. The elements within a group in the table have similar properties.
- Three classes of elements are metals, nonmetals, and metalloids.

6.2 Classifying the Elements

- The periodic table displays the symbols and names of elements, along with information on the structure of their atoms.
- Elements can be sorted into noble gases, representative elements, transition metals, or inner transition metals based on their electron configurations.
- The periodic table can be divided into *s*, *p*, *d*, and *f* blocks that correspond to the highest occupied sublevels in atoms of elements.

6.3 Periodic Trends

- In general, atomic size increases from top to bottom within a group and decreases from left to right across a period.
- Positive and negative ions form when electrons are transferred between atoms.
- First ionization energy tends to decrease from top to bottom within a group and increase from left to right across a period.
- Cations are always smaller than the atoms from which they form. Anions are always larger than the atoms from which they form.
- In general, electronegativity values decrease from top to bottom within a group. For representative elements, the values tend to increase from left to right across a period.
- Trends in atomic size, ionization energy, ionic size, and electronegativity can be explained by variations in atomic structure. The increase in nuclear charge within groups and across periods explains many trends. Within groups an increase in shielding has a significant effect.

Vocabulary

- alkali metals (p. 161)
- alkaline earth metals (p. 161)
- anion (p. 172)
- atomic radius (p. 170)
- cation (p. 172)
- electronegativity (p. 177)
- halogens (p. 161)
- inner transition metal (p. 166)
- ion (p. 172)
- ionization energy (p. 173)
- metalloids (p. 160)
- metals (p. 158)
- noble gases (p. 164)
- nonmetals (p. 159)
- periodic law (p. 157)
- representative elements (p. 164)
- transition metal (p. 166)

Organizing Information

Use these terms to construct a concept map that organizes the major ideas of this chapter.

ionic size
periodic trends
electronegativity
electron configuration
atomic size
ionization energy
periodic table

Concept Map 6 Solve the Concept Map with the help of an interactive guided tutorial.

with ChemASAP

Chapter Resources

Print

- **Core Teaching Resources,** Chapter 6 *Practice Problems, Vocabulary Review, Quiz, Chapter Test A, Chapter Test B*

Technology

- **Computer Test Bank,** Chapter 6 Test
- **Interactive Textbook with ChemASAP,** Chapter 6

CHAPTER 6

Assessment

Reviewing Content

6.1 Organizing the Elements

24. Why did Mendeleev leave spaces in his periodic table?

25. What effect did the discovery of gallium have on the acceptance of Mendeleev's table?

26. What pattern is revealed when the elements are arranged in a periodic table in order of increasing atomic number?

27. Based on their locations in the periodic table, would you expect carbon and silicon to have similar properties? Explain your answer.

28. Identify each property below as more characteristic of a metal or a nonmetal.

a. a gas at room temperature
b. brittle
c. malleable
d. poor conductor of electric current
e. shiny

29. In general, how are metalloids different from metals and nonmetals?

6.2 Classifying the Elements

30. Where are the alkali metals, the alkaline earth metals, the halogens, and the noble gases located in the periodic table?

31. Which of the following are symbols for representative elements: Na, Mg, Fe, Ni, Cl?

32. Which noble gas does not have eight electrons in its highest occupied energy level?

33. Which of these metals isn't a transition metal?

a. aluminum **b.** silver
c. iron **d.** zirconium

34. Use Figure 6.12 to write the electron configurations of these elements.

a. boron
b. arsenic
c. fluorine
d. zinc
e. aluminum

35. Write the electron configuration of these elements.

a. the noble gas in period 3
b. the metalloid in period 3
c. the alkali earth metal in period 3

6.3 Periodic Trends

36. Which element in each pair has atoms with a larger atomic radius?

a. sodium, lithium
b. strontium, magnesium
c. carbon, germanium
d. selenium, oxygen

37. Explain the difference between the first and second ionization energy of an element.

38. Which element in each pair has a greater first ionization energy?

a. lithium, boron
b. magnesium, strontium
c. cesium, aluminum

39. Arrange the following groups of elements in order of increasing ionization energy.

a. Be, Mg, Sr
b. Bi, Cs, Ba
c. Na, Al, S

40. Why is there a large increase between the first and second ionization energies of the alkali metals?

41. How does the ionic radius of a typical metal compare with its atomic radius?

42. Which particle has the larger radius in each atom/ion pair?

a. Na, Na^+ **b.** S, S^{2-}
c. I, I^- **d.** Al, Al^{3+}

43. Which element in each pair has a higher electronegativity value?

a. Cl, F **b.** C, N **c.** Mg, Ne **d.** As, Ca

44. Why are noble gases not included in Table 6.2?

45. When the elements in each pair are chemically combined, which element in each pair has a greater attraction for electrons?

a. Ca or O **b.** O or F
c. H or O **d.** K or S

46. For which of these properties does lithium have a larger value than potassium?

a. first ionization energy
b. atomic radius
c. electronegativity
d. ionic radius

CHAPTER 6

ASSESSMENT

Reviewing Content

24. so that undiscovered elements with similar properties could be placed in the same group

25. The close match between the predicted properties and actual properties of gallium helped gain wider acceptance for Mendeleev's periodic table.

26. a periodic repetition of their physical and chemical properties

27. Yes; both carbon and silicon are in Group 4A, and each has four electrons in the highest occupied energy level.

28. **a.** nonmetal **b.** nonmetal **c.** metal **d.** nonmetal **e.** metal

29. Metalloids have properties that are similar to those of metals and nonmetals. How a metalloid behaves depends on the conditions.

30. Group 1A, Group 2A, Group 7A, and Group 8A, respectively

31. Na, Mg, Cl

32. helium

33. aluminum

34. **a.** $1s^22s^22p^1$
b. $1s^22s^22p^63s^23p^63d^{10}4s^24p^3$
c. $1s^22s^22p^5$
d. $1s^22s^22p^63s^23p^63d^{10}4s^2$
e. $1s^22s^22p^63s^23p^1$

35. **a.** Ar: $1s^22s^22p^63s^23p^6$
b. Si: $1s^22s^22p^63s^23p^2$
c. Mg: $1s^22s^22p^63s^2$

36. **a.** sodium **b.** strontium **c.** germanium **d.** selenium

37. The first ionization energy is the energy needed to remove a first electron from an atom. The second ionization energy is the energy needed to remove a second electron.

38. **a.** boron **b.** magnesium **c.** aluminum

39. **a.** Sr, Mg, Be **b.** Cs, Ba, Bi **c.** Na, Al, S

40. It is relatively easy to remove the first electron from an alkali metal atom; it is much more difficult to remove the second.

41. The ionic radius of a cation is smaller than the atomic radius of the metal atom.

42. **a.** Na **b.** S^{2-} **c.** I^- **d.** Al

43. **a.** F **b.** N **c.** Mg **d.** As

44. Noble gases, with rare exception, do not form compounds.

45. **a.** O **b.** F **c.** O **d.** S

46. a and c

CHAPTER 6

ASSESSMENT *continued*

Understanding Concepts

47. a. 1801–1850 **b.** Mendeleev's periodic table helped scientists predict the existence of undiscovered elements. **c.** 75%

48. a. C **b.** La **c.** Ne, P, Br **d.** Sb, Bi

49. b; Nitrogen and phosphorus are in the same group (Group 5A).

50. Fluorine has a smaller atomic radius than oxygen because fluorine has a larger nuclear charge. Fluorine has a smaller radius than chlorine because it has eight fewer electrons.

51. Nonmetals; The trend is for ionization energy to increase from left to right across a period.

52. a. Ca^{2+} **b.** P^{3-} **c.** Cu^{+}

53. a. H, Li, Na, K, Rb, Cs, Fr **b.** O, S, Se, Te, Po **c.** Zn, Cd, Hg, Uub

54. An element's electron configuration determines its location (group) in the periodic table.

55. c

56. 170 pm

57. First ionization energy increases across a period.

58. It is relatively easy to remove two electrons from magnesium; it is much more difficult to remove a third electron. It is relatively easy to remove three electrons from aluminum; it is much more difficult to remove a fourth.

59. a. The atomic radius increases from top to bottom within the group. **b.** Cations are smaller than their corresponding atoms. The attraction between the nucleus and any remaining electron is greater. There is one fewer occupied energy level.

CHAPTER 6

Assessment *continued*

Understanding Concepts

47. The bar graph shows how many elements were discovered before 1750 and in each 50-year period between 1750 and 2000.

a. In which 50-year period were the most elements discovered?

b. How did Mendeleev's work contribute to the discovery of elements?

c. What percent of the elements were discovered by 1900?

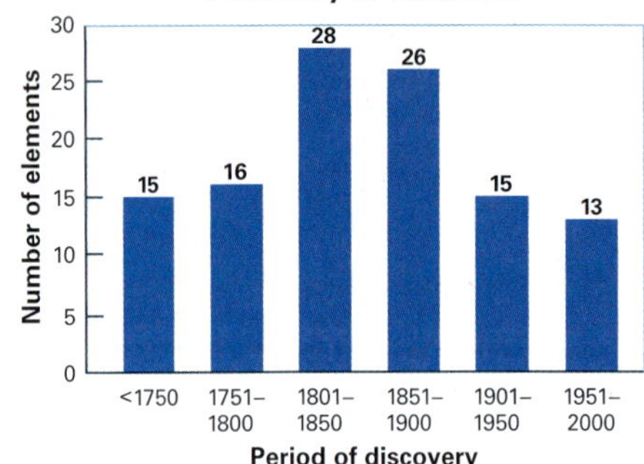

48. Write the symbol of the element or elements that fit each description.

a. a nonmetal in Group 4A

b. the inner transition metal with the lowest atomic number

c. all of the nonmetals for which the atomic number is a multiple of five

d. a metal in Group 5A.

49. In which pair of elements are the chemical properties of the elements most similar? Explain your reasoning.

a. sodium and chlorine

b. nitrogen and phosphorus

c. boron and oxygen

50. Explain why fluorine has a smaller atomic radius than both oxygen and chlorine.

51. Would you expect metals or nonmetals in the same period to have higher ionization energies? Give a reason for your answer.

52. In each pair, which ion is larger?

a. Ca^{2+}, Mg^{2+} **b.** Cl^{-}, P^{3-} **c.** Cu^{+}, Cu^{2+}

53. List the symbols for all the elements with electron configurations that end as follows. Each n represents an energy level.

a. ns^1 **b.** ns^2np^4 **c.** ns^2nd^{10}

54. Explain why there should be a connection between an element's electron configuration and its location on the periodic table?

55. Which equation represents the first ionization of an alkali metal atom?

a. $Cl \longrightarrow Cl^{+} + e^{-}$

b. $Ca \longrightarrow Ca^{+} + e^{-}$

c. $K \longrightarrow K^{+} + e^{-}$

d. $H \longrightarrow H^{+} + e^{-}$

56. Use the graph in Figure 6.14 to estimate the atomic radius of the indium atom.

57. What trend is demonstrated by the following series of equations?

$Li + 520\ kJ/mol \longrightarrow Li^{+} + e^{-}$

$O + 1314\ kJ/mol \longrightarrow O^{+} + e^{-}$

$F + 1681\ kJ/mol \longrightarrow F^{+} + e^{-}$

$Ne + 2080\ kJ/mol \longrightarrow Ne^{+} + e^{-}$

58. There is a large jump between the second and third ionization energies of magnesium. There is a large jump between the third and fourth ionization energies of aluminum. Explain these observations.

59. The bar graph shows the relationship between atomic and ionic radii for Group 1A elements.

a. Describe and explain the trend in atomic radius within the group.

b. Explain the difference between the size of the atoms and the size of the ions.

Comparing Radii of Alkali Metals

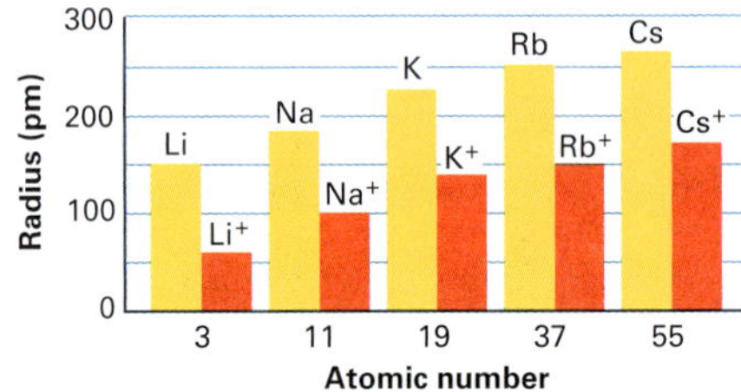

Critical Thinking

60. Do you think there are more elements left to discover? If so, what is the lowest atomic number a new element could have? Explain your answers.

61. The graphs show the relationship between the electronegativities and first ionization energies for period 2 and period 3 elements.

a. Based on data for these two periods, what is the general trend between these two values?

b. Use nuclear charge and shielding effect to explain this trend.

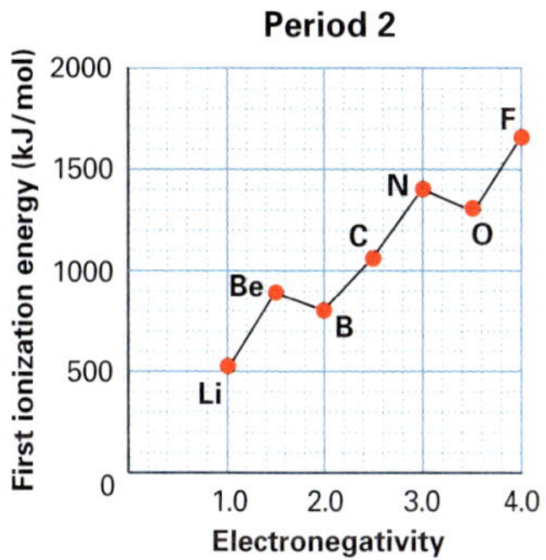

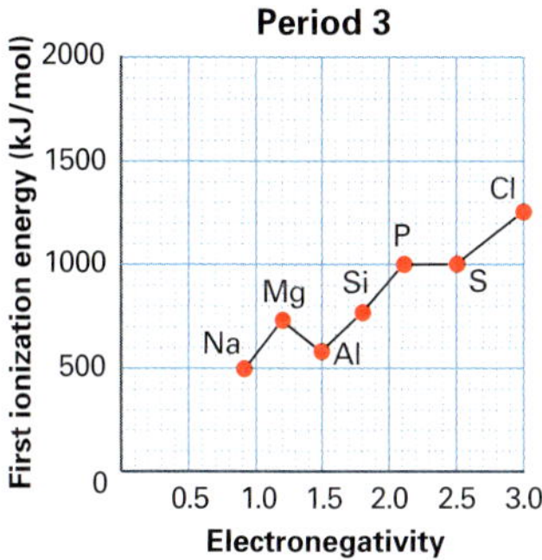

62. Explain why it takes more energy to remove a 4*s* electron from zinc than from calcium.

63. Explain each of the following comparisons.

a. Calcium has a smaller second ionization energy than does potassium.

b. Lithium has a larger first ionization energy than does cesium.

c. Magnesium has a larger third ionization energy than does aluminum.

Concept Challenge

64. The Mg^{2+} and Na^{+} ions each have ten electrons. Which ion would you expect to have the smaller radius? Explain your choice.

65. The ions S^{2-}, Cl^{-}, K^{+}, Ca^{2+}, and Sc^{3+} have the same total number of electrons as the noble gas argon. How would you expect the radii of these ions to vary? Would you expect to see the same variation in the series O^{2-}, F^{-}, Na^{+}, Mg^{2+}, and Al^{3+}, in which each ion has the same total number of electrons as the noble gas neon? Explain your answer.

66. Make a graph of average atomic mass versus atomic number. Choose 11 points (atomic numbers 1, 10, 20, and so forth up to atomic number 100) to make your graph. Use the graph to describe the relationship between average atomic mass and atomic number. Is there a 1:1 correspondence between average atomic mass and atomic number? Explain how you reached your conclusion.

67. The ionization energies for the removal of the first six electrons in carbon are, starting with the first electron, 1086 kJ/mol, 2352 kJ/mol, 4619 kJ/mol, 6220 kJ/mol, 37,820 kJ/mol, and 47,260 kJ/mol.

a. Make a graph of ionization energy versus ionization number. The ionization number indicates which electron is lost.

b. Between which two ionization numbers does the ionization energy have the largest increase? Explain why this behavior is predictable.

68. Atoms and ions with the same number of electrons are called isoelectronic.

a. Write the symbol for a cation and an anion that are isoelectronic with krypton.

b. Is it possible for a cation to be isoelectronic with an anion from the same period? Explain.

69. Electron affinity is a measure of an atom's ability to gain electrons. Predict the trend for electron affinity across a period. Explain your answer.

Critical Thinking

60. Yes, 113; 1 through 112 are known.

61. a. Electronegativity increases as first ionization energy increases.
b. Both properties depend on the attraction between the nucleus and electrons in the highest occupied energy level. Across a period, the nuclear charge increases, but the shielding effect is constant.

62. Zinc has a greater nuclear charge (more protons) than calcium.

63. a. The electrons in calcium are removed from the same energy level. In potassium, the second electron is removed from a lower energy level.
b. Because cesium has a larger atomic radius than lithium, the nuclear charge in a cesium atom has a smaller effect on the electrons in the highest occupied energy level.
c. The third electron removed from a magnesium atom is in a lower energy level.

Concept Challenge

64. Mg^{2+} has more protons; its attraction for electrons is greater.

65. The ionic radii decrease from S^{2-} to Sc^{3+}. The number of electrons and the shielding effect do not change, but the number of protons increases from left to right in this series. So the ionic size decreases. The same is true for the series O^{2-} to Mg^{2+}.

66. There is not a 1:1 correspondence. Explanations will vary.

67. a.

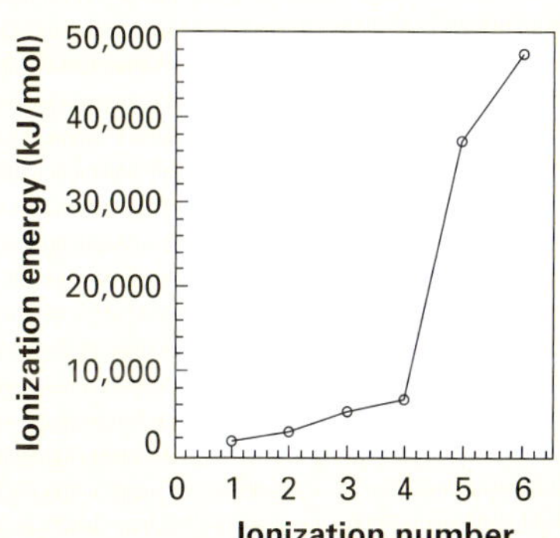

67. b. The largest increase is between numbers 4 and 5 because carbon easily loses the first four electrons from the second energy level. The fifth electron is removed from the first energy level.

68. a. Possible cations are Rb^{+} and Sr^{2+}; possible anions are Br^{-}, Se^{2-}, and As^{3-}.
b. No; a cation is isoelectronic with the noble gas in the preceding period, and an anion is isoelectronic with the noble gas in the same period.

69. Electron affinity increases (becomes more negative) from left to right across a period because the nuclear charge increases and the shielding effect is constant.

CHAPTER 6 ASSESSMENT *continued*

Cumulative Review

70. Answers will vary but are likely to mention the scientific method.

71. a. physical change
b. chemical change
c. physical change
d. chemical change

72. Use a magnet; iron beads are attracted to magnet, copper beads are not. Use tweezers and a magnifying glass; iron beads are silvery black, copper beads are dull red.

73. 4

74. a. 3
b. 8, the tenths place

75. The density of the cube is $0.984\ g/cm^3$. The cube will float.

76. a. 2.24×10^{-9} m
b. 8.13×10^{-2} m
c. 7.4×10^{-12} m
d. 9.37×10^{-3} m

77. 5.2%

78. 5.2×10^3 g

79. The density of the olive is $1.05\ g/cm^3$. The olive will sink.

80. 173

81. The density of sulfur does not vary with mass. The density is constant.

82. 2.57×10^2 mL

83. $4.54\ g/cm^3$

84. a. 48
b. 44
c. 114
d. 110

85. a. silver, 62 neutrons
b. tin, 50 protons
c. molybdenum, 42 electrons
d. scandium, 21 electrons

86. a. none
b. one, 2*p*
c. none
d. none

CHAPTER 6 Assessment *continued*

Cumulative Review

70. Explain why science today depends less on chance discoveries than it did in the past. *(Chapter 1)*

71. Identify each process as a chemical or physical change. *(Chapter 2)*
a. melting of iron **b.** lighting a match
c. grinding corn **d.** souring of milk

72. Describe at least two methods to separate a mixture of small copper and iron beads. *(Chapter 2)*

73. In the United States a typical can of "cola" holds 355 mL. How many 2.00-L bottles could be filled from a 24-can case of cola? *(Chapter 3)*

74. The volume of the liquid in the graduated cylinder is reported as 31.8 mL. *(Chapter 3)*
a. How many significant figures are there in the measurement?
b. In which digit is there uncertainty?

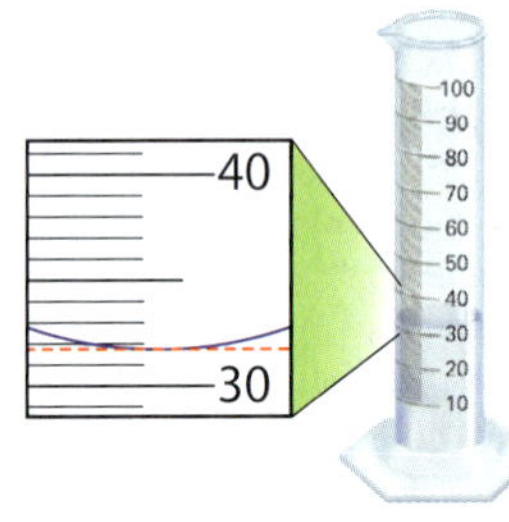

75. A cube of plastic 1.20×10^{-5} km on a side has a mass of 1.70 g. Show by calculation whether this plastic cube will sink or float in pure water. *(Chapter 3)*

76. Convert the measurements to meters. Express your answers in scientific notation. *(Chapter 3)*
a. 2.24 nm **b.** 8.13 cm
c. 7.4 pm **d.** 9.37 mm

77. An apprentice jeweler determines the density of a sample of pure gold to be $20.3\ g/cm^3$. The accepted value is $19.3\ g/cm^3$. What is the percent error of the jeweler's density measurement? *(Chapter 3)*

78. What is the mass of 7.7 L of gasoline at 20°C? Assume the density of gasoline to be $0.68\ g/cm^3$. *(Chapter 3)*

79. A black olive containing its seed has a mass of 4.5 g and a volume of $4.3\ cm^3$. Will the olive sink or float on the water? *(Chapter 3)*

80. The distance is 1.50×10^8 km from the sun to Earth. The speed of light is 3.00×10^8 m/s. How many round trips between Earth and the sun could a beam of light make in one day? *(Chapter 3)*

81. The table shows how the volume of sulfur varies with mass. How does the density of sulfur vary with mass?*(Chapter 3)*

Mass of Sulfur Versus Volume of Sulfur

Mass of sulfur (g)	Volume of sulfur (cm^3)
23.5	11.4
60.8	29.2
115	55.5
168	81.1

82. Calculate the volume of acetone with the same mass as 15.0 mL of mercury. The density of mercury is 13.59 g/mL. The density of acetone is 0.792 g/mL. *(Chapter 3)*

83. A rectangular container has inside dimensions of 15.2 cm by 22.9 cm and is about 1 meter tall. Water is poured into the container to a height of 55.0 cm. When a jagged rock with a mass of 5.21 kg is placed in the container, it sinks to the bottom. The water level rises to 58.3 cm. What is the density of the rock? *(Chapter 3)*

84. How many neutrons does an atom of each isotope contain? *(Chapter 4)*
a. ${}^{84}_{36}Kr$ **b.** ${}^{79}_{35}Br$ **c.** ${}^{190}_{76}Os$ **d.** ${}^{185}_{75}Re$

85. Name the element and calculate the number of requested subatomic particles in each isotope. *(Chapter 4)*
a. neutrons in ${}^{109}_{47}Ag$ **b.** protons in ${}^{118}_{50}Sn$
c. electrons in ${}^{96}_{42}Mo$ **d.** electrons in ${}^{45}_{21}Sc$

86. How many filled *p* orbitals do atoms of these elements contain? *(Chapter 5)*
a. carbon **b.** phosphorus
c. oxygen **d.** nitrogen

Standardized Test Prep

Test-Taking Tip

Interpreting Data Tables Tables present a large amount of data in a small space. Before you try to answer questions based on a table, look at the table. Read the title, if there is one, and the column headings. Then read the questions. As you read each question, decide which data you will need to use to answer the question. You may need to focus on a single entry or column. You may need to find a relationship between data from multiple columns.

Select the choice that best answers each question or completes each statement.

1. Which of the following properties increases as you move across a period from left to right?
I. electronegativity
II. ionization energy
III. atomic radius
a. I and II only
b. I and III only
c. II and III only
d. I, II, and III

2. List the symbols for sodium, sulfur, and cesium in order of increasing atomic radii.
a. Na, S, Cs
b. Cs, Na, S
c. S, Na, Cs
d. Cs, S, Na

3. The electron configuration for an element in the halogen group should always end with
a. ns^2np^6.
b. ns^2np^5.
c. ns^2np^4.
d. ns^2np^2.

Use the spheres to answer Questions 4–6.

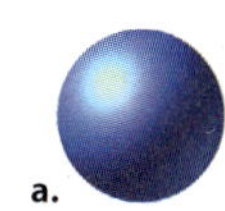
a.

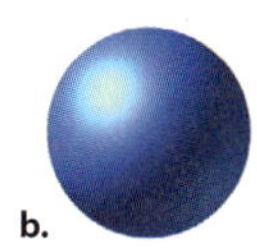
b.

4. Which sphere would most likely represent a potassium atom, K?

5. Which sphere would most likely represent a potassium ion, K^+?

6. If the spheres represent an atom and an anion of the same element, which sphere represents the atom and which represents the anion?

Use the data table to answer Questions 7–9.

Alkali metal	Atomic radius (pm)	First ionization energy (kJ/mol)	Electronegativity value
Li	152	520	1.0
Na	186	495.8	0.9
K	227	418.8	0.8
Rb	244	250	0.8
Cs	262	210	0.7

7. If you plotted atomic radius versus first ionization energy, would the graph reveal a direct or inverse relationship?

8. If you plotted atomic radius versus electronegativity, would the graph reveal a direct or inverse relationship?

9. If you plotted first ionization energy versus electronegativity, would the graph reveal a direct or inverse relationship?

For each question there are two statements. Decide whether each statement is true or false. Then decide whether Statement II is a correct explanation for Statement I.

	Statement I		Statement II
10.	Electronegativity values are higher for metals than for nonmetals.	BECAUSE	Atoms of nonmetals are among the largest atoms.
11.	A calcium atom is larger than a calcium ion.	BECAUSE	Ions are always larger than the atoms from which they are formed.
12.	The element hydrogen is a metal.	BECAUSE	Hydrogen is on the left in the periodic table.
13.	Among all the elements in a period, the noble gas always has the smallest ionization energy.	BECAUSE	Within any period, atomic radii tend to decrease moving from right to left.

Standardized Test Prep

1. a
2. c
3. b
4. The larger sphere represents the potassium atom.
5. The smaller sphere represents the potassium ion.
6. The smaller sphere represents the atom; the larger sphere represents the anion.
7. inverse
8. inverse
9. direct
10. false, false
11. true, false
12. false, true
13. false, false

CHAPTER 7

Ionic and Metallic Bonding Planning Guide

SECTION OBJECTIVES	STANDARDS NATIONAL	STANDARDS STATE	ACTIVITIES and LABS	
7.1 Ions, pp. 187–193 1 block or 2 periods **7.1.1 Determine** the number of valence electrons in an atom of a representative element. **7.1.2 Explain** how the octet rule applies to atoms of metallic and nonmetallic elements. **7.1.3 Describe** how cations form. **7.1.4 Explain** how anions form.	A-1, B-1, B-2		**SE** Inquiry Activity: Shapes of Crystalline Materials, p. 186 **TE** Teacher Demo: Valence Electrons, p. 188 **SE** Class Activity: Forming Cations, p. 190 **SSLM** Lab 10: Electronic Configurations of Atoms and Ions	L2 L2 L2 L2
7.2 Ionic Bonds and Ionic Compounds, pp. 194–200 1 block or 2 periods **7.2.1 Explain** the electrical charge of an ionic compound. **7.2.2 Describe** three properties of ionic compounds.	A-1, B-2, B-3		**SE** Quick Lab: Solutions Containing Ions, p. 199 **SE** Small-Scale Lab: Analysis of Anions and Cations, p. 200 **TE** Class Activity: "Hardness" of Water, p. 196 **TE** Class Activity: Types of Ionic Compounds, p. 197 **TE** Teacher Demo: Form and Structure of Crystals, p. 197 **PLM** Probeware Lab: Solutions Containing Ions **LP** Lab Practical 7-1: Crystal Structures **VCL** Lab 2: Names and Formulas of Ionic Compounds	L2 L2 L2 L2 L2 L2 L2 L2
7.3 Bonding in Metals, pp. 201–205 1 block or 2 periods **7.3.1 Model** the valence electrons of metal atoms. **7.3.2 Describe** the arrangement of atoms in a metal. **7.3.3 Explain** the importance of alloys.	B-2, B-6, E-2		**TE** Teacher Demo: Metals vs. Ionic Compounds, p. 202 **TE** Teacher Demo: Types of Alloys, p. 203 **TE** Teacher Demo: Making an Alloy, p. 205 **LM** Lab 10: Crystal Structures **LP** Lab Practical 7-1: Crystal Structures	L2 L2 L2 L2 L2

Ability Levels	
L1	For students who need additional help
L2	For all students
L3	For students who need to be challenged

Components

SE	Student Edition	LM	Laboratory Manual	GCP	Graphing Calculator Problems	T	Transparencies
TE	Teacher's Edition	SSLM	Small-Scale Chemistry Lab Manual	TTT	Test-Taking Tips with Transparencies, Grades 9–12 Science	CA	Chem Alive and Chemistry Field Trips DVD
GRSW	Guided Reading and Study Workbook	LP	Laboratory Practicals			iT	Interactive Textbook with ChemASAP
CTR	Core Teaching Resources	PLM	Probeware Lab Manual	STP	Standardized Test Preparation Workbook	GO	Go Online Internet Resources
SSG	Spanish Study Guide	VCL	Virtual Chemistry Labs				
		CTB	Computer Test Bank				

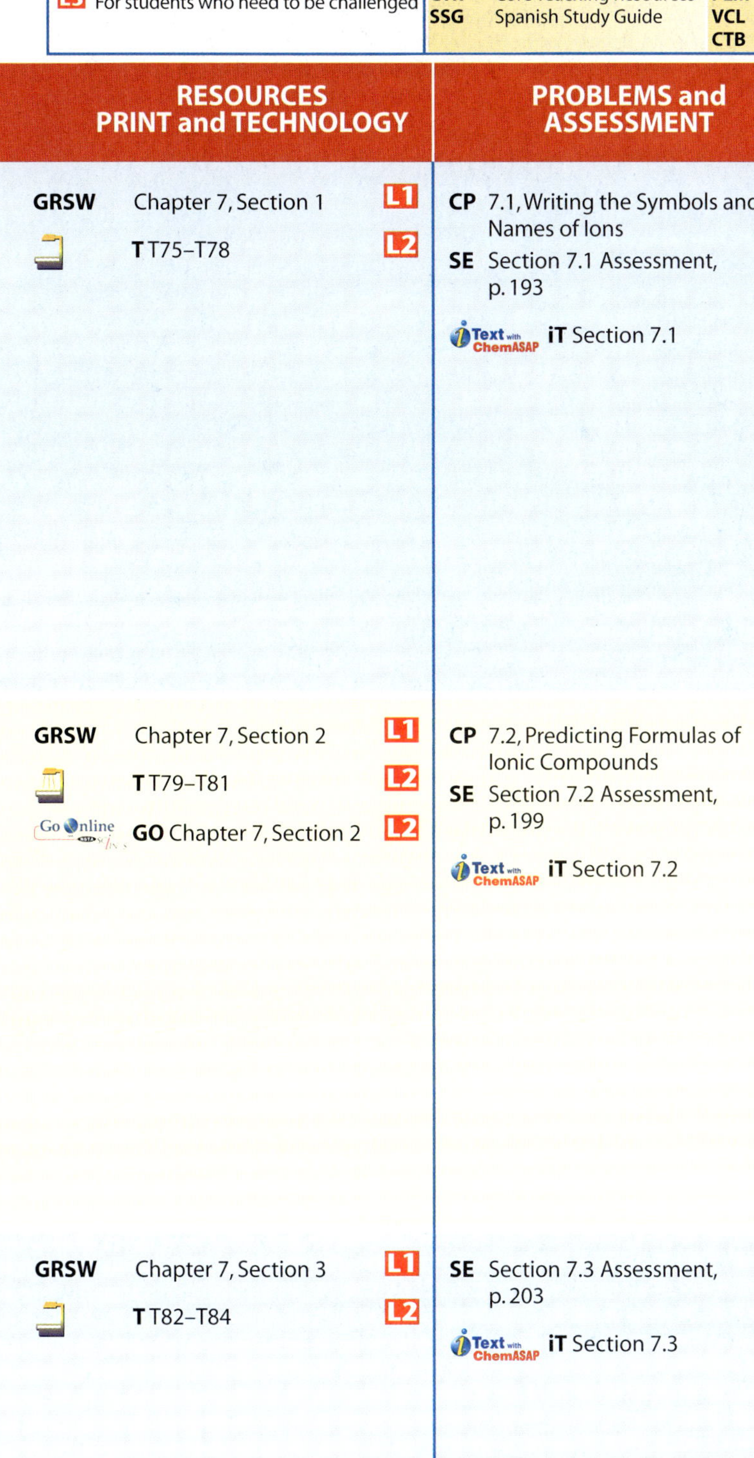

RESOURCES PRINT and TECHNOLOGY		PROBLEMS and ASSESSMENT
GRSW Chapter 7, Section 1	L1	**CP** 7.1, Writing the Symbols and Names of Ions
T T75–T78	L2	**SE** Section 7.1 Assessment, p. 193
		iT Section 7.1
GRSW Chapter 7, Section 2	L1	**CP** 7.2, Predicting Formulas of Ionic Compounds
T T79–T81	L2	**SE** Section 7.2 Assessment, p. 199
GO Chapter 7, Section 2	L2	iT Section 7.2
GRSW Chapter 7, Section 3	L1	**SE** Section 7.3 Assessment, p. 203
T T82–T84	L2	iT Section 7.3

Go Online

Web Code: cdn-1072

MATERIALS List

Quantities for each group

STUDENT EDITION

Inquiry Activity, p. 186
4 small disposable cups, distilled water, a ruler, a spoon, sodium chloride (table salt), sucrose (table sugar), sodium hydrogen carbonate (baking soda), magnesium sulfate, a small clean mirror, a magnifying glass

Quick Lab, p. 199
3 D-cell batteries, masking tape, 2 30–cm lengths of bell wire with ends scraped bare, clear plastic cup, distilled water, tap water, vinegar, sucrose, sodium chloride, baking soda, conductivity probe (optional)

Small-Scale Lab, p. 200
medicine droppers, reaction surface, pipet, 0.2*M* Na_2SO_4, 1.0*M* HNO_3, 0.1*M* Na_3PO_4, 0.5*M* $AgNO_3$, 1.0*M* HCl, 0.2*M* $Pb(NO_3)_2$, 0.1*M* KI, 0.5*M* $CaCl_2$, 0.1*M* $FeCl_3$, 0.5*M* NaOH, 0.1*M* KSCN, solid iron

TEACHER'S EDITION

Teacher Demo, p. 188
plastic egg, 11 marbles

Class Activity, p. 196
tap water samples, 10-mL graduated cylinder, potassium thiocyanate solution, dilute ethanoic acid, sodium oxalate solution, dropper

Class Activity, p. 197
crystals of ionic compounds, watch glasses, magnifying glasses

Teacher Demo, p. 202
copper metal or alloy, copper-containing ionic compound

Teacher Demo, p. 203
foam balls of different sizes and colors, toothpicks

Teacher Demo, p. 205
penny, fine sandpaper, granulated zinc, dilute NaOH solution, evaporating dish, tongs, hot plate, teaspoon

Chapter Assessment

CHAPTER ASSESSMENT

SE Chapter Assessment, pp. 207–210

CTB Chapter 7

iT Chapter 7

STANDARDIZED TEST PREP

SE Chapter 7, p. 211

iT Interactive Textbook with Assessment at PHSchool.com

CHAPTER 7

Ionic and Metallic Bonding

Shapes of Crystalline Materials

L2

Purpose

To observe that crystals form at different rates and have distinctive appearances

Prep Time 20 minutes

Class Time 30 minutes

Think About It

1. Yes, all the solids crystallized.
2. No, the rates differed.
3. The crystals will have different appearances.

CHAPTER 7

Ionic and Metallic Bonding

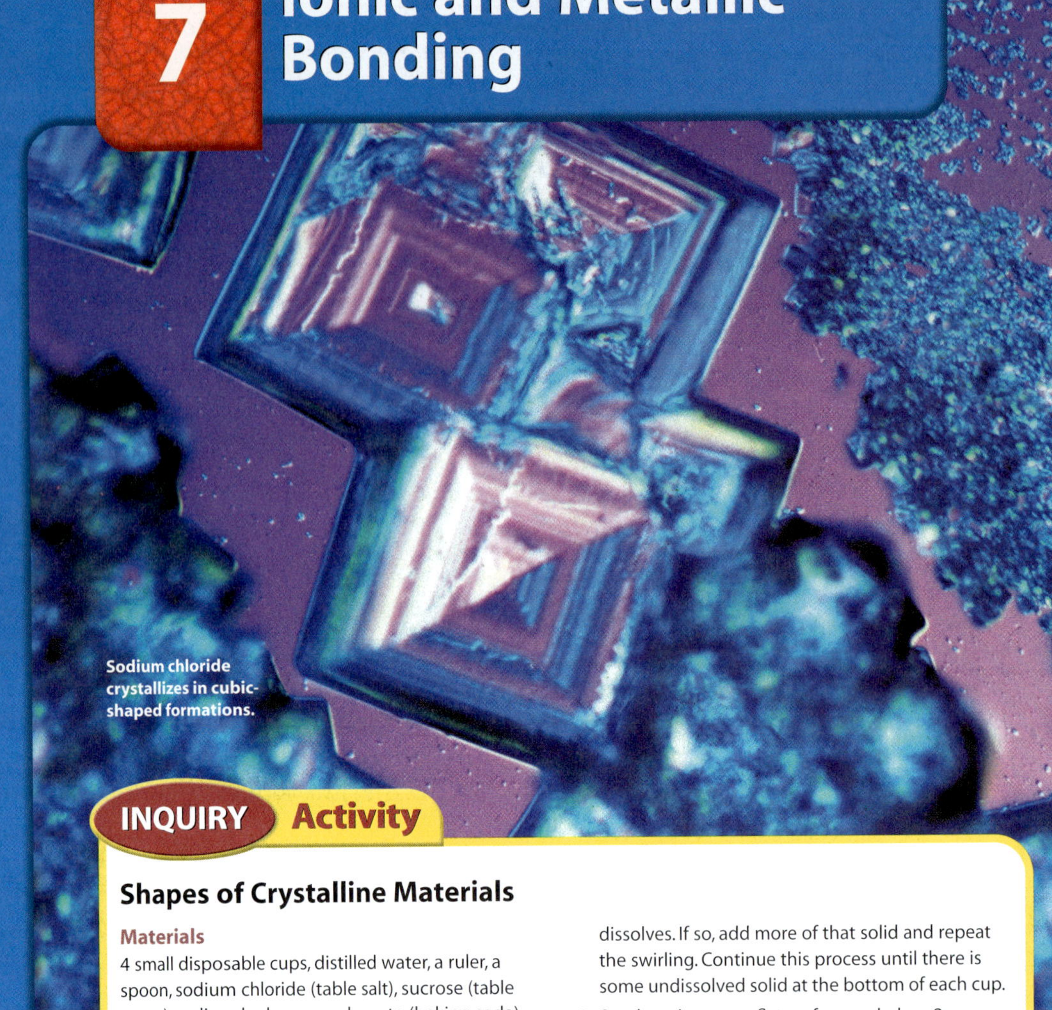

Sodium chloride crystallizes in cubic-shaped formations.

INQUIRY Activity

Shapes of Crystalline Materials

Materials

4 small disposable cups, distilled water, a ruler, a spoon, sodium chloride (table salt), sucrose (table sugar), sodium hydrogen carbonate (baking soda), magnesium sulfate, a small clean mirror, a magnifying glass

Procedure    

1. Label each cup with the name of one of the solids. Pour water into each cup to a depth of 1 cm.
2. Add a spoonful of each solid to its corresponding cup. Swirl each cup for 30 seconds and then let it stand for a few minutes.
3. Swirl the cups at least two more times. Each time, note whether any of the solids completely dissolves. If so, add more of that solid and repeat the swirling. Continue this process until there is some undissolved solid at the bottom of each cup.
4. Set the mirror on a flat surface and place 2 or 3 drops of each liquid onto separate areas of the mirror.
5. Use a magnifying glass to examine each drop of liquid after 15 minutes, and then again after 24 hours.

Think About It

1. For each material, did the solids crystallize?
2. Did the crystals form at the same time?
3. Describe the crystals that formed.

7.1 Ions

Connecting to Your World

Pyrite (FeS_2), a common mineral that emits sparks when struck against steel, is often mistaken for gold—hence its nickname, "fool's gold." Although certainly not worth its weight in gold, pyrite can be used as a source of sulfur in the production of sulfuric acid, a common industrial chemical. Pyrite is an example of a crystalline solid. In crystalline solids, the component particles of the substance are arranged in an orderly, repeating fashion. In this chapter, you will learn about crystalline solids composed of ions that are bonded together. But first you need to understand how ions form from neutral atoms.

Guide for Reading

Key Concepts

- How do you find the number of valence electrons in an atom of a representative element?
- Atoms of which elements tend to gain electrons? Atoms of which elements tend to lose electrons?
- How are cations formed?
- How are anions formed?

Vocabulary

valence electrons
electron dot structures
octet rule
halide ions

Reading Strategy

Summarizing Write a one-paragraph summary of how the octet rule applies to the formation of ions.

Valence Electrons

Mendeleev used similarities in the properties of elements to organize his periodic table. Scientists later learned that all of the elements within each group of the periodic table behave similarly because they have the same number of valence electrons. **Valence electrons** are the electrons in the highest occupied energy level of an element's atoms. The number of valence electrons largely determines the chemical properties of an element.

The number of valence electrons is related to the group numbers in the periodic table. **To find the number of valence electrons in an atom of a representative element, simply look at its group number.** For example, the elements of Group 1A (hydrogen, lithium, sodium, potassium, and so forth) all have one valence electron, corresponding to the 1 in 1A. Carbon and silicon, in Group 4A, have four valence electrons. Nitrogen and phosphorus, in Group 5A, have five valence electrons; and oxygen and sulfur, in Group 6A, have six. The noble gases (Group 8A) are the only exceptions to the group-number rule: Helium has two valence electrons, and all of the other noble gases have eight. Figure 7.1 shows some applications of Group 4A elements.

Figure 7.1 Group 4A elements include carbon, silicon, and germanium. (a) This saw blade contains carbon in the form of diamond. (b) Silicon is used in the manufacture of microchips. (c) Germanium is one of the materials used to make thermoscanning goggles.

187

Section Resources

Print

- **Guided Reading and Study Workbook,** Section 7.1
- **Core Teaching Resources,** Section 7.1 Review
- **Transparencies,** T75–T78
- **Small-Scale Chemistry Laboratory Manual,** Lab 10

Technology

- **Interactive Textbook with ChemASAP,** Problem-Solving 7.1, Assessment 7.1
- **Go Online,** Section 7.1

7.1

1 FOCUS

Objectives

7.1.1 Determine the number of valence electrons in an atom of a representative element.

7.1.2 Explain how the octet rule applies to atoms of metallic and nonmetallic elements.

7.1.3 Describe how cation form.

7.1.4 Explain how anions form.

Guide for Reading

Build Vocabulary L2

Graphic Organizer Have each student choose a halide ion. Then have students sketch a graphic organizer that shows how a halide ion relates to the other three vocabulary terms.

Reading Strategy L2

Preview Have students skim the headings, look at the visuals, and read the boldfaced text to preview the section.

2 INSTRUCT

Connecting to Your World

Have students study the photograph and read the text that opens the section. Discuss the ways in which iron and sulfur ions can interact to form the regular pyrite crystal structure.

Valence Electrons

Use Visuals L1

Table 7.1 Have students identify the total number of electrons and the number of valence electrons in selected elements from Table 7.1. Reemphasize that the group number equals the number of valence electrons in an atom of a representative element.

TEACHER Demo

Valence Electrons 

Purpose To model the valence electrons of an atom.

Materials plastic egg, 11 marbles

Procedure Hold up a plastic egg containing 10 marbles. State that the egg represents a sodium atom and that the marbles represent the 10 electrons making up the $n = 1$ and $n = 2$ energy levels. Explain that these electrons cannot be "removed" without "breaking" the egg. Now hold up one additional marble next to the egg. State that this marble represents the eleventh electron, and occupies the $n = 3$ energy level. This is the valence electron. Explain that if this electron is lost, the resulting atom has an overall 1+ charge.

Expected Outcome Students should be able to distinguish valence electrons from nonvalence electrons.

The Octet Rule

Discuss 

Have students determine the accuracy of this statement: All stable ions of elements result in electronic configurations that are isoelectronic with noble gases. *(Most of the time this statement is true, but there are exceptions. Use Cu(I) as an example. Explain that a noble gas configuration is not generally possible with elements that would have to gain or lose many electrons to become stable.)*

Word *Origins* 

In music, an octet refers to a group of eight performers, or a composition written for eight musicians. In poetry, an octet refers to a group of eight lines of verse.

Table 7.1

Electron Dot Structures of Some Group A Elements

Period	Group 1A	2A	3A	4A	5A	6A	7A	8A
1	H·							He:
2	Li·	·Be·	·Ḃ·	·Ċ·	·N̈·	:Ȯ·	:F̈·	:N̈e:
3	Na·	·Mg·	·Al·	·Si·	·P̈·	:S̈·	:C̈l·	:Är:
4	K·	·Ca·	·Ga·	·Ge·	·Äs·	:S̈e·	:B̈r·	:K̈r:

Valence electrons are usually the only electrons used in chemical bonds. Therefore, as a general rule, only the valence electrons are shown in electron dot structures. **Electron dot structures** are diagrams that show valence electrons as dots. Table 7.1 shows electron dot structures for atoms of some Group A elements. Notice that all of the elements within a given group (with the exception of helium) have the same number of electron dots in their structures.

Checkpoint ***What is an electron dot structure?***

Word *Origins*

Octet comes from the Greek word *okto*, meaning "eight." There are eight electrons in the highest occupied energy level of the noble gases, except for helium. **How do you think the term *octet* might also be applied to music or poetry?**

The Octet Rule

You learned in Chapter 6 that noble gases, such as neon and argon, are unreactive in chemical reactions. That is, they are stable. In 1916, chemist Gilbert Lewis used this fact to explain why atoms form certain kinds of ions and molecules. He called his explanation the **octet rule:** In forming compounds, atoms tend to achieve the electron configuration of a noble gas. An octet is a set of eight. Recall that each noble gas (except helium) has eight electrons in its highest occupied energy level and a general electron configuration of ns^2np^6. The octet rule takes its name from this fact about noble gases. **Atoms of metals tend to lose their valence electrons, leaving a complete octet in the next-lowest energy level. Atoms of some nonmetals tend to gain electrons or to share electrons with another nonmetal to achieve a complete octet.** Although there are exceptions, the octet rule applies to atoms in most compounds.

Formation of Cations

An atom is electrically neutral because it has equal numbers of protons and electrons; an ion forms when an atom or group of atoms loses or gains electrons. **An atom's loss of valence electrons produces a cation, or a positively charged ion.** Note that for metals, the name of a cation is the same as the name of the element. For example, a sodium atom (Na) forms a sodium cation (Na^+). Likewise, a calcium atom (Ca) forms a calcium cation (Ca^{2+}). Although their names are the same, there are many important chemical differences between metals and their cations. Sodium metal, for example, reacts explosively with water. By contrast, sodium cations are quite unreactive. As you may know, they are a component of table salt, a compound that is very stable in water.

Differentiated Instruction

Less Proficient Readers 

Lead a class discussion on writing electron dot structures. Ask, **What is the significance of the electrons that are represented by dots?** *(They are valence electrons.)* **How are they placed?** *(symmetrically around the atom to show placement in orbitals according to pairing rules)* **Why are the non-valence electrons not shown?** *(Generally, they are not available for chemical bonding.)*

The most common cations are those produced by the loss of valence electrons from metal atoms. Most of these atoms have one to three valence electrons, which are easily removed. Sodium, in Group 1A of the periodic table, is typical. Sodium atoms have a total of eleven electrons, including one valence electron. A sodium atom can lose an electron to become a positively charged sodium ion. The sodium ion has an electron configuration that is identical to the noble gas neon. When forming a compound, a sodium atom loses its one valence electron and is left with an octet (eight electrons) in what is now its highest occupied energy level. Because the number of protons in the sodium nucleus is still eleven, the loss of one unit of negative charge produces a cation with a charge of 1+. You can represent the electron loss, or ionization, of the sodium atom by drawing the complete electron configuration of the atom and of the ion formed.

$$\text{Na } 1s^2 2s^2 2p^6 3s^1 \xrightarrow{-e^-} \text{Na}^+ \ 1s^2 \underbrace{2s^2 2p^6}_{\text{octet}}$$

Notice that the electron configuration of the sodium ion ($1s^22s^22p^6$) is the same as that of a neon atom. The diagrams below help illustrate this point.

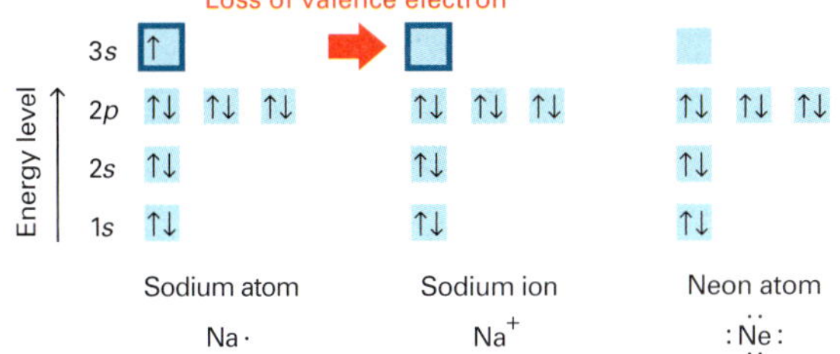

Both the sodium ion and the neon atom have eight electrons in their valance shells (highest occupied energy levels). Using electron dot structures, you can show the ionization more simply.

$$\text{Na}\cdot \xrightarrow[\text{ionization}]{\text{loss of valence electron}} \text{Na}^+ + e^-$$

Sodium atom (electrically neutral, charge = 0)	Sodium ion (plus sign indicates one unit of positive charge)	electron (minus sign indicates one unit of negative charge)

Figure 7.2 The sodium atoms in a sodium-vapor lamp ionize to form sodium cations (Na^+). **Applying Concepts** ***How many electrons are in the highest occupied energy level of Na^+?***

Go Online NSTA SCILINKS
For: Links on Sodium
Visit: www.SciLinks.org
Web Code: cdn-1071

Formation of Cations

Use Visuals L1

Figure 7.2 Display the equation that goes with Figure 7.2 on an overhead projector. Point out that the interaction between atoms that produces bonding involves only the outermost electrons of the atoms. The inner electrons are locked tightly in filled energy levels and do not participate in bonding. Use a colored pen to circle the outermost electron in the sodium atom in this equation. Remind students that the outermost electrons are called valence electrons. Use a different colored pen to circle the octet of electrons in the sodium ion's highest energy level. Circle the corresponding octet of electrons in neon to show the similarity in electron configurations. Have students draw a similar diagram for calcium.

Differentiated Instruction

Gifted and Talented L3

Have students show how the atoms of transition elements become stable with pseudo-noble-gas configurations. These atoms would have to gain or lose too many electrons to achieve a noble-gas electron configuration. For example, show how Ag, Zn, and Ga lose 1, 2, and 3 electrons, respectively to form pseudo-noble-gas configurations.

Answers to...

Figure 7.2 8 electrons

A diagram that shows valence electrons as dots.

Discuss 

One way to determine the number of valence electrons in an atom is to look at the electron configuration of the atom. Explain that any electron in an atom outside the noble-gas core is called a valence electron. Using diagrams such as those on pages 189 and 190, show students several examples of how various atoms of the representative elements form ions and gain a noble-gas electron configuration. Indicate the noble-gas core and valence electrons in your diagrams. Lay pieces of magnesium, zinc, copper, and aluminum on a dry surface in the lab to show that metals do not spontaneously form metal cations.

CLASS Activity

Forming Cations L2

Purpose Students model the formation of cations using equations.

Materials paper, pencil

Procedure Have students write equations similar to that for Mg on student page 190, showing the formation of metal cations from metal atoms. Students should show the electron dot structures for the metal atoms and metal cations that are formed. In addition, you may want students to write out the electron configurations for the metal atoms and cations. Check students' equations to be sure the correct metal ion is formed.

Expected Outcome Students should be able to use electron dot structures to correctly write equations describing the ionization of metal atoms.

Figure 7.3 Walnuts are a good dietary source of magnesium. Magnesium ions (Mg^{2+}) aid in digestive processes.

Magnesium (atomic number 12) belongs to Group 2A of the periodic table, so it has two valence electrons. A magnesium atom attains the electron configuration of neon by losing both valence electrons. The loss of the valence electrons produces a magnesium cation with a charge of 2+.

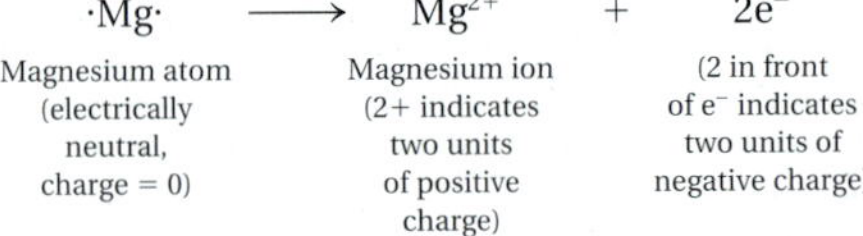

Figure 7.4 lists the symbols of cations formed by metals in Groups 1A and 2A. Cations of Group 1A elements always have a charge of 1+. Similarly, the cations of Group 2A elements always have a charge of 2+. This consistency can be explained in terms of the loss of valence electrons by metal atoms: The atoms lose enough electrons to attain the electron configuration of a noble gas. For example, all Group 2A elements have two valence electrons. In losing these two electrons, they form 2+ cations.

For transition metals, the charges of cations may vary. An atom of iron, for example, may lose two or three electrons. In the first case, it forms the Fe^{2+} ion. In the second case, it forms the Fe^{3+} ion.

Some ions formed by transition metals do not have noble-gas electron configurations (ns^2np^6) and are therefore exceptions to the octet rule. Silver, with the electron configuration of $1s^22s^22p^63s^23p^63d^{10}4s^24p^64d^{10}5s^1$, is an example. To achieve the structure of krypton, which is the preceding noble gas, a silver atom would have to lose eleven electrons. To acquire the electron configuration of xenon, which is the following noble gas, silver would have to gain seven electrons. Ions with charges of three or greater are uncommon, and these possibilities are extremely unlikely. Thus silver does not achieve a noble-gas configuration. But if it loses its $5s^1$ electron, the configuration that results ($4s^24p^64d^{10}$), with 18 electrons in the highest occupied energy level and all of the orbitals filled, is relatively favorable. Such a configuration is known as a pseudo noble-gas electron configuration. Silver forms a positive ion (Ag^+) in this way. Other elements that behave similarly to silver are found at the right of the transition metal block. For example, a copper atom can ionize to form a 1+ cation (Cu^+), as illustrated below.

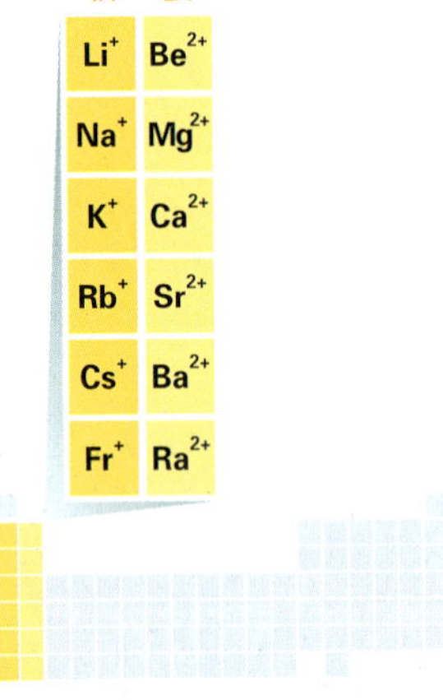

Figure 7.4 Cations of Group 1A elements have a charge of 1+. Cations of Group 2A elements have a charge of 2+.

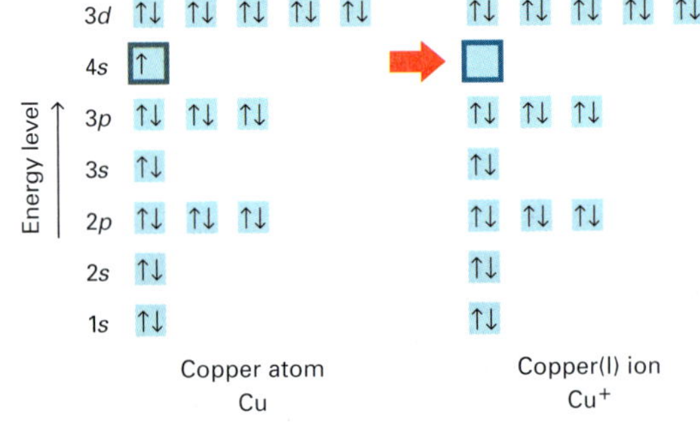

By losing its lone 4*s* electron, copper attains a pseudo noble-gas electron configuration. Likewise, cations of gold (Au^+), cadmium (Cd^{2+}), and mercury (Hg^{2+}) also have pseudo noble-gas configurations.

Differentiated Instruction

Special Needs 

Ask students to work with a partner to practice drawing electron dot structures for Group A elements. Have one student randomly choose an element and write the symbol for that element on a piece of paper. Have another student fill in the electron dots for this element. Students can check their work by referring to Table 7.1. Make sure students understand that the electron dots represent only valence electrons, not the total number of electrons, and that valence electrons are the only electrons involved in chemical reactivity. Also, reinforce the stable octet dot structures for the noble gases and explain how ions of other elements will react to obtain this configuration.

Formation of Anions

An anion is an atom or a group of atoms with a negative charge. **The gain of negatively charged electrons by a neutral atom produces an anion.** Note that the name of an anion of a nonmetallic element is not the same as the element name. The name of the anion typically ends in *-ide*. Thus a chlorine atom (Cl) forms a chloride ion (Cl^-), and an oxygen atom (O) forms an oxide ion (O^{2-}). Figure 7.5 shows the symbols of anions formed by some elements in Groups 5A, 6A, and 7A.

Because they have relatively full valence shells, atoms of nonmetallic elements attain noble-gas electron configurations more easily by gaining electrons than by losing them. For example, chlorine belongs to Group 7A (the halogen family) and has seven valence electrons. A gain of one electron gives chlorine an octet and converts a chlorine atom into a chloride ion.

$$\text{Cl}\quad 1s^2 2s^2 2p^6 3s^2 3p^5 \xrightarrow{+e^-} \text{Cl}^-\quad 1s^2 2s^2 2p^6 \underbrace{3s^2 3p^6}_{\text{octet}}$$

The chloride ion is an anion with a single negative charge. Notice that it has the same electron configuration as the noble gas argon.

$$\text{Ar}\quad 1s^2 2s^2 2p^6 \underbrace{3s^2 3p^6}_{\text{octet}}$$

Chlorine atoms, therefore, need one more valence electron to achieve the electron configuration of the nearest noble gas. The diagrams below illustrate how both the chloride ion and the argon atom have an octet of electrons in their highest occupied energy levels.

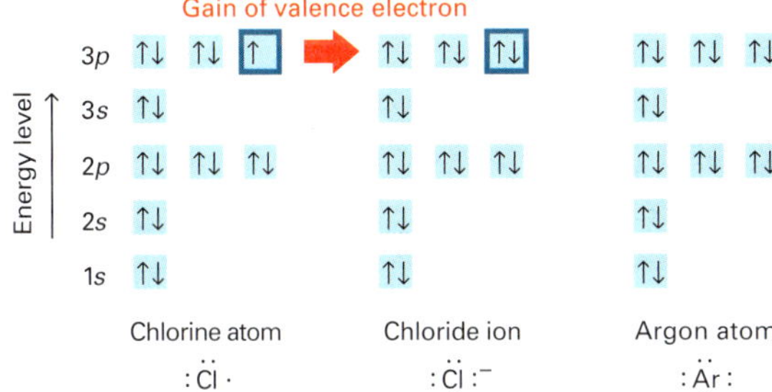

Based on the diagrams above, you use electron dot structures to write an equation showing the formation of a chloride ion from a chlorine atom.

$$:\ddot{\underset{..}{\text{Cl}}}\cdot \;+\; e^- \xrightarrow[\text{valence electron}]{\text{gain of one}} :\ddot{\underset{..}{\text{Cl}}}:^-$$

Chlorine atom — Chloride ion (Cl^-)

In this equation, each dot in the electron dot structure represents an electron in the valence shell in the electron configuration diagram.

Checkpoint ***Which noble gas has the same electron configuration as a chloride ion?***

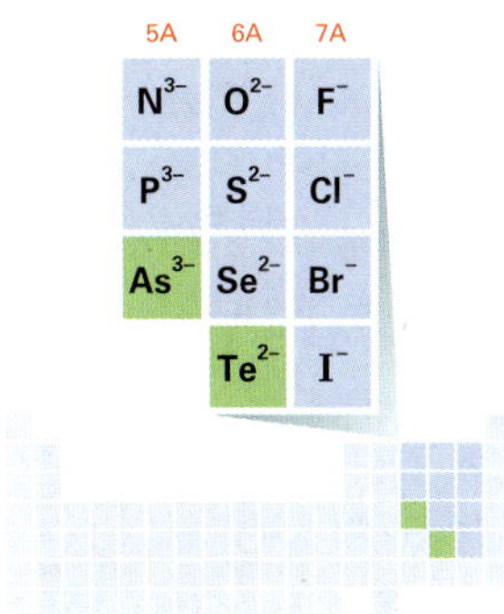

Figure 7.5 Atoms of nonmetals and metalloids form anions by gaining enough valence electrons to attain the electron configuration of the nearest noble gas.
Interpreting Diagrams ***In which group of the periodic table do the elements bromine and iodine belong?***

Formation of Anions

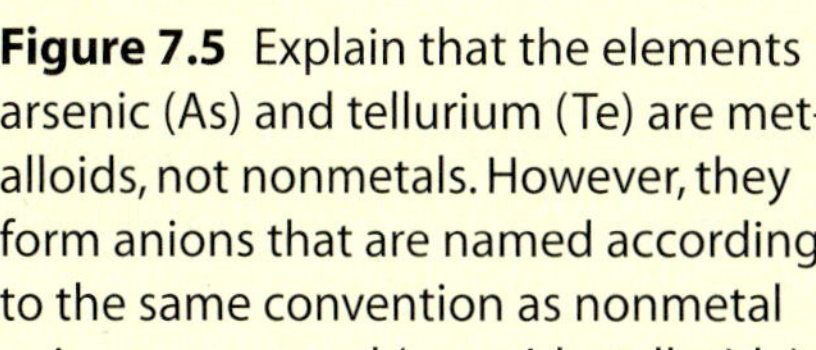

Use Visuals L1

Figure 7.5 Explain that the elements arsenic (As) and tellurium (Te) are metalloids, not nonmetals. However, they form anions that are named according to the same convention as nonmetal anions are named (arsenide, telluride).

Discuss L2

Explain to students that when one atom forms an ion, that ion is called a monatomic ion. Explain that the prefix *mon-* or *mono-* means "one." Ask students to predict what ions might be called if they contained more than one atom. Tell students that they will study such ions in Chapter 8.

Facts and Figures

Water Purification

Chlorine gas is often used to purify drinking water; it kills a variety of microorganisms, including those that carry diseases. But chlorine also reacts with organic substances in the water to produce chlorinated compounds such as chloroacetonitrile. Chloroacetonitrile has been shown to cause inflammation of the digestive tract in laboratory animals. As an alternative to chlorine, some countries have begun purifying water with ozone. Ozone kills microorganisms even more effectively than chlorine. About one percent of the water supply in the United States is now purified with ozone. It is estimated that it would cost $6 billion to switch completely to ozone for treating all the drinking water supplies.

Answers to...

Figure 7.5 Group 7A

Checkpoint argon

Discuss

Discuss with students how certain elements can form either anions or cations. Have students write the electron configuration of nitrogen ($1s^2 2s^2 2p^3$). Ask, **How can a nitrogen atom form a cation that has the electron configuration of a noble gas?** *(It can lose five electrons.)* **How can a nitrogen atom form an anion that has the electron configuration of a noble gas?** *(It can gain three electrons.)* Have them repeat this process for carbon. Explain that atoms that have few electrons to gain or lose to achieve an octet are more likely to form ions. Other atoms are not likely to form ions, and the process they undergo will be studied in Chapter 8.

CONCEPTUAL PROBLEM 7.1

Answers

1. **a.** sulfide ion, S^{2-}
   b. aluminum ion, Al^{3+}
2. **a.** 2 electrons lost
   b. 3 electrons gained
   c. 2 electrons lost

Practice Problems Plus

Give the name and symbol of the ion formed when:

a. a nitrogen atom gains three electrons *(nitride ion, N^{3-})*
b. a calcium atom loses two electrons *(calcium ion, Ca^{2+})*
c. a fluorine atom gains one electron *(fluoride ion, F^-)*

Figure 7.6 The six most abundant ions in seawater are chloride (Cl^-), sulfate (SO_4^{2-}), sodium (Na^+), magnesium (Mg^{2+}), calcium (Ca^{2+}), and potassium (K^+).

The ions that are produced when atoms of chlorine and other halogens gain electrons are called **halide ions.** All halogen atoms have seven valence electrons and need to gain only one electron to achieve the electron configuration of a noble gas. Thus all halide ions (F^-, Cl^-, Br^-, and I^-) have a charge of 1−. The seawater in Figure 7.6 contains many different ions, but the negatively charged ions—the anions—are mostly chloride ions.

Look at another example. Oxygen is in Group 6A, and oxygen atoms each have six valence electrons. Oxygen atoms attain the electron configuration of neon by gaining two electrons, as shown in the diagrams below.

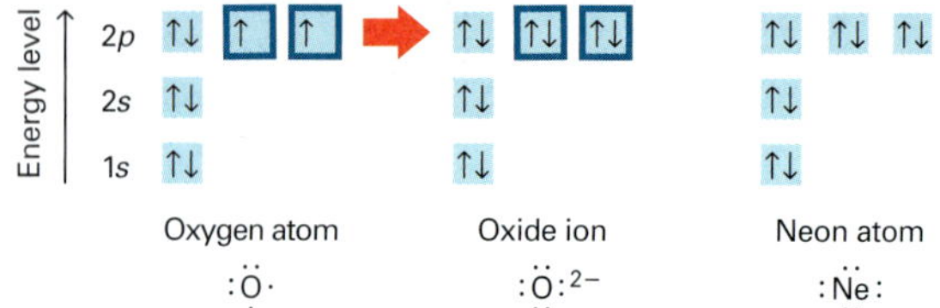

The resulting oxide ions have charges of 2− and are written as O^{2-}. Using electron dot structures, you can write the equation for the formation of oxide ions as follows.

$$:\ddot{O}\cdot + 2e^- \longrightarrow :\ddot{O}:^{2-}$$

Table 7.2 lists some common anions that you will be learning about in this book. Note that not all of the anions listed end with the suffix *-ide*.

Checkpoint ***How many electrons do halogen atoms need to gain in order to achieve the electron configuration of a noble gas?***

Table 7.2 Some Common Anions

1−		2−		3−	
F^-	fluoride	O^{2-}	oxide	N^{3-}	nitride
Cl^-	chloride	S^{2-}	sulfide	P^{3-}	phosphide
Br^-	bromide	SO_4^{2-}	sulfate	PO_4^{3-}	phosphate
I^-	iodide	CO_3^{2-}	carbonate		
OH^-	hydroxide				
ClO^-	hypochlorite				
NO_3^-	nitrate				
$C_2H_3O_2^-$	acetate				
HCO_3^-	hydrogen carbonate				

Differentiated Instruction

Gifted and Talented

Many common anions are polyatomic ions. Have students infer what a polyatomic ion is, then confirm their definitions. Have them list some common polyatomic anions, such as sulfate (SO_4^{2-}), nitrate (NO_3^-), and phosphate (PO_4^{3-}). Tell them that polyatomic ions form ionic compounds and that polyatomic ions will be studied in Chapter 8.

CONCEPTUAL PROBLEM 7.1

Writing the Symbols and Names of Ions

The beaker shown on the right contains iodine vapor. Write the symbol and name of the ion formed when

a. an iodine atom gains one electron.

b. a strontium atom loses two electrons.

1 Analyze ***Identify the relevant concepts.***

a. An atom that gains electrons forms a negatively charged ion (anion). The name of an anion of a nonmetallic element ends in *-ide*.

b. An atom that loses electrons forms a positively charged ion (cation). The name of a cation of a metallic element is the same as the name of the element.

2 Solve ***Apply concepts to this situation.***

a. I^-, iodide ion (an anion)

b. Sr^{2+}, strontium ion (a cation)

Practice Problems

1. Write the name and symbol of the ion formed when
   a. a sulfur atom gains two electrons.
   b. an aluminum atom loses three electrons.
2. How many electrons are lost or gained in forming each ion?
   a. Ba^{2+} **b.** As^{3-} **c.** Cu^{2+}

Problem-Solving 7.1 Solve Problem 1 with the help of an interactive guided tutorial.

with **ChemASAP**

7.1 Section Assessment

3. **Key Concept** How can you determine the number of valence electrons in an atom of a representative element?
4. **Key Concept** Atoms of which elements tend to gain electrons? Atoms of which elements tend to lose electrons?
5. **Key Concept** How do cations form?
6. **Key Concept** How do anions form?
7. How many valence electrons are in each atom?
   a. potassium **b.** carbon
   c. magnesium **d.** oxygen
8. Draw the electron dot structure for each element in Question 7.
9. How many electrons will each element gain or lose in forming an ion?
   a. calcium (Ca) **b.** fluorine (F)
   c. aluminum (Al) **d.** oxygen (O)
10. Write the name and symbol of the ion formed when
    a. a potassium atom loses one electron.
    b. a zinc atom loses two electrons.
    c. a fluorine atom gains one electron.
11. Write the electron configuration of Cd^{2+}.

Connecting Concepts

Ionization Energy Reread page 173 in Section 6.3. How does the octet rule explain the large increase in energy between the first and second ionization energies of Group 1A metals?

Assessment 7.1 Test yourself on the concepts in Section 7.1.

with **ChemASAP**

Section 7.1 Assessment

3. look up the group number of that element
4. Atoms of nonmetallic elements tend to gain electrons; atoms of metallic elements tend to lose electrons.
5. when an atom loses valence electrons
6. when an atom gains valence electrons
7. **a.** 1 **b.** 4 **c.** 2 **d.** 6
8. **a.** K· **b.** ·C̈· **c.** ·Mg· **d.** :Ö:
9. **a.** lose 2 **b.** gain 1 **c.** lose 3 **d.** gain 2
10. **a.** potassium cation, K^+
    b. zinc cation, Zn^{2+}
    c. fluoride anion, F^-
11. Cd^{2+}: $1s^22s^22p^63s^23p^63d^{10}4s^24p^64d^{10}$

3 ASSESS

Evaluate Understanding 

Have students refer to the periodic table on page162. To determine the students' knowledge about the formation of elemental anions and cations, ask the students to determine whether the following ions are likely to exist and why: H^- *(yes, isoelectronic with He);* H^+ *(yes, but without electrons, there is no comparable noble-gas configuration);* Sr^{2+} *(yes, isoelectronic with Kr);* Al^{3+} *(yes, isoelectronic with Ne);* Xe^- *(no, cannot form ions easily due to stable electron configuration);* Zn^{6-} *(no, isoelectronic with Kr but formation would require a gain of too many electrons);* Zn^{2+} *(yes, not isoelectronic with a noble gas but has pseudo-noble-gas electron configuration with eighteen electrons filling up the outer energy level:* $3s^23p^63d^{10}$*)*

Reteach 

Select groups from the periodic table in random order and ask students to predict the common ions that could be formed from elements of each group. Note that predicting is fairly easy for groups at the far left or far right of the table, but more difficult for groups in the center of the table, which have partially filled d and f orbitals.

Connecting Concepts

The amount of energy needed to remove the one valence electron of a Group 1A metal atom (first ionization energy) is low. After this electron is lost, the outer energy level contains an octet and is stable. If it loses another electron, it will be less stable, so the second ionization energy is high.

If your class subscribes to the Interactive Textbook, use it to review key concepts in Section 7.1.

with **ChemASAP**

Answers to...

Checkpoint 1 electron

7.2

1 FOCUS

Objectives

7.2.1 Explain the electrical charge of an ionic compound.

7.2.2 Describe three properties of ionic compounds.

Guide for Reading

Build Vocabulary L2

Paraphase Have students skim through the section to locate the meanings of the vocabulary terms. Then have them paraphrase each definition.

Reading Strategy 

Identify Main Ideas/Details Have students identify the main idea of each paragraph of this section. Have them include these ideas in an outline of the section.

2 INSTRUCT

Connecting to Your World

Have students examine the section-opening photograph and Figure 7.8. Discuss how the reactive (and poisonous) elements sodium metal and chlorine gas can combine to form harmless table salt. Ask, **What characteristics of sodium and chlorine atoms allow them to form the stable compound sodium chloride, also known as table salt?** *(Sodium atoms can lose an electron easily, and chlorine atoms can accept an electron easily. The resulting ions can combine with the other oppositely charged ions.)* Remind students that NaCl is an example of an ionic compound.

Formation of Ionic Compounds

Discuss 

Explain that the formation of positive ions and of negative ions are simultaneous and interdependent processes. An ionic compound is the result of the transfer of electrons from one set of atoms to another set of atoms. An ionic compound consists entirely of ions.

7.2 Ionic Bonds and Ionic Compounds

Guide for Reading

 **Key Concepts**

- What is the electrical charge of an ionic compound?
- What are three properties of ionic compounds?

Vocabulary

ionic compounds
ionic bonds
chemical formula
formula unit
coordination number

Reading Strategy

Previewing Before you read this section, rewrite the headings as *how*, *why*, and *what* questions about ionic compounds. As you read, write answers to the questions.

Connecting to Your World

You have heard of harvesting crops such as wheat or rice—but salt? In many coastal countries that have warm, relatively dry climates, salt is produced by the evaporation of seawater. The salty water is channeled into a series of shallow ponds, where it becomes more concentrated as the water evaporates by exposure to the sun. When the saltwater is concentrated enough, it is diverted into a pan, on which the sodium chloride crystals deposit. Salt farmers then drain the pans and collect the salt into piles to dry. In this section, you will learn how cations and anions combine to form stable compounds such as sodium chloride.

Formation of Ionic Compounds

Compounds composed of cations and anions are called **ionic compounds.** Ionic compounds are usually composed of metal cations and nonmetal anions. For example, sodium chloride, or table salt, is composed of sodium cations and chloride anions. **Although they are composed of ions, ionic compounds are electrically neutral.** The total positive charge of the cations equals the total negative charge of the anions.

Ionic Bonds Anions and cations have opposite charges and attract one another by means of electrostatic forces. The electrostatic forces that hold ions together in ionic compounds are called **ionic bonds.**

Sodium chloride provides a simple example of how ionic bonds are formed. Consider the reaction between a sodium atom and a chlorine atom. Sodium has a single valence electron that it can easily lose. (If the sodium atom loses its valence electron, it achieves the stable electron configuration of neon.) Chlorine has seven valence electrons and can easily gain one. (If the chlorine atom gains a valence electron, it achieves the stable electron configuration of argon.) When sodium and chlorine react to form a compound, the sodium atom gives its one valence electron to a chlorine atom. Thus sodium and chlorine atoms combine in a one-to-one ratio and both ions have stable octets.

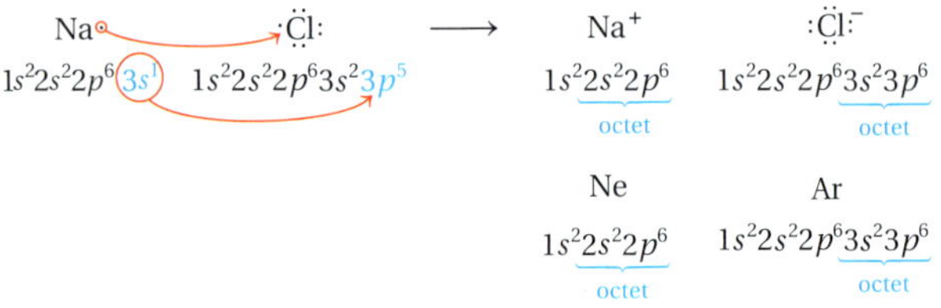

Animation 8 Take an atomic-level look at the formation of KCl.

with ChemASAP

Section Resources

Print

- **Guided Reading and Study Workbook,** Section 7.2
- **Core Teaching Resources,** Section 7.2 Review, Interpreting Graphics
- **Transparencies,** T79–T81
- **Probeware Laboratory Manual,** Section 7.2

Technology

- **Interactive Textbook with ChemASAP,** Animation 8, Simulation 5, Problem-Solving 7.12, Assessment 7.2
- **Go Online,** Section 7.2
- **Virtual Chemistry Lab,** Lab 2

Figure 7.7 shows aluminum and bromine reacting to form the compound aluminum bromide. Each aluminum atom has three valence electrons to lose. Each bromine atom has seven valence electrons and readily gains one additional electron. Therefore, when aluminum and bromine react, three bromine atoms combine with each aluminum atom.

Formula Units The ionic compound sodium chloride is composed of equal numbers of sodium cations (Na^+) and chloride anions (Cl^-). As you can see in Figure 7.8, the ions in solid sodium chloride are arranged in an orderly pattern. There are no single discrete units, only a continuous array of ions.

Chemists represent the composition of substances by writing chemical formulas. A **chemical formula** shows the kinds and numbers of atoms in the smallest representative unit of a substance. NaCl, for example, is the chemical formula for sodium chloride. Note, however, that the formula NaCl does not represent a single discrete unit. Because an ionic compound exists as a collection of positively and negatively charged ions arranged in repeating patterns, its chemical formula refers to a ratio known as a formula unit. A **formula unit** is the lowest whole-number ratio of ions in an ionic compound. For sodium chloride, the lowest whole-number ratio of the ions is 1:1 (one Na^+ to each Cl^-). Thus the formula unit for sodium chloride is NaCl. Although ionic charges are used to derive the correct formula, they are not shown when you write the formula unit of the compound.

The ionic compound magnesium chloride contains magnesium cations (Mg^{2+}) and chloride anions (Cl^-). In magnesium chloride, the ratio of magnesium cations to chloride anions is 1:2 (one Mg^{2+} to two Cl^-). So its formula unit is $MgCl_2$. Because there are twice as many chloride anions (each with a 1− charge) as magnesium cations (each with a 2+ charge), the compound is electrically neutral. In aluminum bromide, described earlier, the ratio of aluminum cations to bromide ions is 1:3 (one Al^{3+} to three Br^- ions), so the formula unit is $AlBr_3$.

Checkpoint ***What is the formula unit for magnesium chloride?***

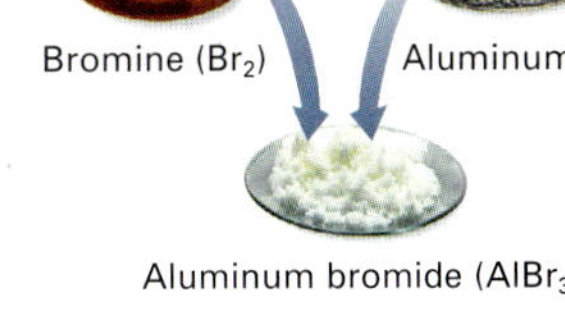

Figure 7.7 Aluminum metal and the nonmetal bromine react to form an ionic solid, aluminum bromide.

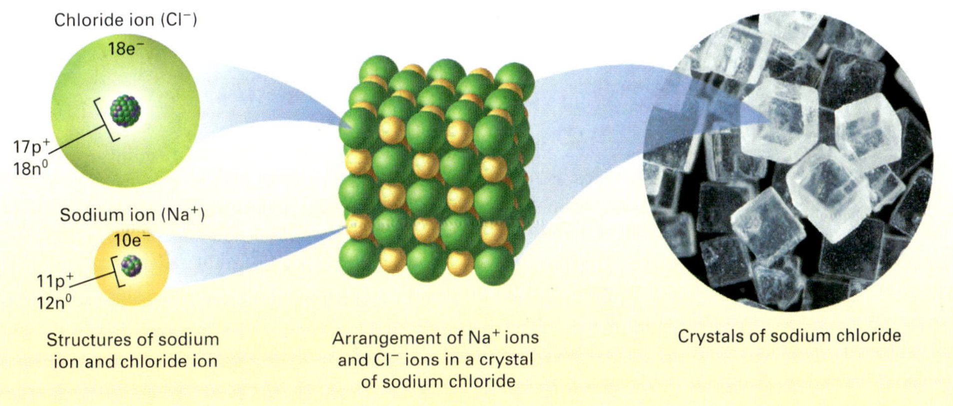

Figure 7.8 Sodium cations and chloride anions form a repeating three-dimensional array in sodium chloride (NaCl). **Inferring** ***How does the arrangement of ions in a sodium chloride crystal help explain why the compound is so stable?***

Discuss L2

Ask, **Why are crystalline ionic compounds generally so rigid and brittle?** *(The crystal is rigid because it is held together by a specific three-dimensional array of relatively strong attractive forces between anions and cations, which is accompanied by minimal charge repulsion of like ions. The crystal is brittle because the attractive interactions are specifically arranged within the crystal structure. If this arrangement is disturbed, as it would be if the crystal were hit with a hammer, charge repulsion between ions of the same charge can force the crystal to fragment.)*

Discuss L2

To assess students' prior knowledge about ionic bonds and crystals, ask, **What is an ionic bond?** *(an electical attraction between ions of opposite charge)* **How do a polyatomic ion and a monatomic ion differ?** *(A monatomic ion is an ion formed from a single atom; a polyatomic ion is a stable unit of two or more tightly bound atoms that carries a charge.)* **Why are crystals of different ionic compounds different shapes?** *(The shapes reflect different geometric arrangements of anions and cations with different sizes and charges.)*

Differentiated Instruction

English Learners L1

Encourage students to look up and define terms used to describe ionic compounds. Students should define terms such as *crystal* and *formula unit* in English and their native language. Emphasize how their strong bonding arrangement accounts for crystals having unique properties.

Answers to...

Figure 7.8 Each ion in the arrangement is strongly attracted to its neighbors and repulsions are minimized.

$MgCl_2$

Section 7.2 (continued)

CONCEPTUAL PROBLEM 7.2

Answers

12. a. KI

b. Al_2O_3

13. $CaCl_2$

Practice Problems Plus L2

Use electron dot structures to determine chemical formulas of the ionic compounds formed when the following elements combine:

a. magnesium and chlorine *($MgCl_2$)*

b. aluminum and sulfur *(Al_2S_3)*

Properties of Ionic Compounds

CLASS Activity

"Hardness" of Water L2

Purpose Students detect the presence of ions in water samples.

Materials tap water samples, 10-mL graduated cylinder, potassium thiocyanate solution, dilute ethanoic acid, sodium oxalate solution, dropper

Procedure Water "hardness" is based on ions present in the water. Have students bring water samples from home to test for hardness. Test 2-mL samples as follows. Add three drops of potassium thiocyanate (KSCN) to the first sample. Add three drops of dilute ethanoic acid, CH_3COOH, and three drops of sodium oxalate, $Na_2C_2O_4$, to the second sample. Mix well.

Expected Outcome A red color from the iron(III) thiocyanate ion, $Fe(SCN)^{2+}$, indicates the presence of Fe^{3+} ions. A white precipitate of calcium oxalate, CaC_2O_4, indicates the presence of Ca^{2+} ions.

Download a worksheet on **Ionic Compounds** to complete, and find additional teacher support on NSTA SciLinks.

CONCEPTUAL PROBLEM 7.2

Predicting Formulas of Ionic Compounds

The ionic compound formed from potassium and oxygen is used in ceramic glazes. Use electron dot structures to predict the formulas of the ionic compounds formed from the following elements.

a. potassium and oxygen **b.** magnesium and nitrogen

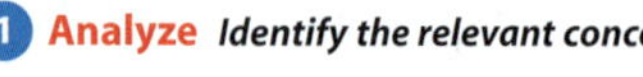

1 Analyze *Identify the relevant concepts.*

Atoms of metals lose their valence electrons when forming an ionic compound. Atoms of nonmetals gain electrons. Enough atoms of each element must be used in the formula so that electrons lost equals electrons gained.

2 Solve *Apply concepts to this situation.*

a. Start with the atoms.

K· and ·Ö:

In order to have a completely filled valence shell, oxygen must gain two electrons. These electrons come from two potassium atoms, each of which loses one electron.

$$2\,\mathrm{K}\cdot + \cdot\ddot{\mathrm{O}}: \longrightarrow 2\,\mathrm{K}^+ \; :\ddot{\mathrm{O}}:^{2-}$$

Electrons lost now equals electrons gained. The formula of the compound formed (potassium oxide) is K_2O.

b. Start with the atoms.

Mg and ·N:

Each nitrogen needs three electrons to have an octet, but each magnesium can lose only two electrons. Thus three magnesium atoms are needed for every two nitrogen atoms.

$$3\,\mathrm{Mg} + 2\,\cdot\dot{\mathrm{N}}: \longrightarrow 3\,\mathrm{Mg}^{2+} \; 2\,:\ddot{\mathrm{N}}:^{3-}$$

The formula of the compound formed (magnesium nitride) is Mg_3N_2.

Practice Problems

12. Use electron dot structures to determine formulas of the ionic compounds formed when

a. potassium reacts with iodine.

b. aluminum reacts with oxygen.

13. What is the formula of the ionic compound composed of calcium cations and chloride anions?

Problem-Solving 7.12 Solve Problem 12 with the help of an interactive guided tutorial.

with ChemASAP

Properties of Ionic Compounds

Figure 7.9 shows the striking beauty of the crystals of some ionic compounds. **Most ionic compounds are crystalline solids at room temperature.** The component ions in such crystals are arranged in repeating three-dimensional patterns. The composition of a crystal of sodium chloride is typical. In solid NaCl, each sodium ion is surrounded by six chloride ions, and each chloride ion is surrounded by six sodium ions. In this arrangement, each ion is attracted strongly to each of its neighbors and repulsions are minimized. The large attractive forces result in a very stable structure. This is reflected in the fact that NaCl has a melting point of about 800°C. **Ionic compounds generally have high melting points.**

For: Links on Ionic Compounds
Visit: www.SciLinks.org
Web Code: cdn-1072

Facts and Figures

Do You Have "Soft" Water?

Explain that water hardness varies with location and source. Generally, water from groundwater sources is harder than water from surface sources. In the United States, most northeastern, southern, and north western states have predominantly soft water. Generally, hard water of varying degrees is found in the southwestern and midwestern states.

Fluorite (CaF_2)

Grossularite ($Ca_3Al_2(SiO_4)_3$)

Aragonite ($CaCO_3$)

Barite ($BaSO_4$) and calcite ($CaCO_3$)

Wulfenite ($PbMoO_4$)

Beryl ($BeAl_2(SiO_3)_6$)

Franklinite ($(Zn, Mn^{2+}, Fe^{2+})(Fe^{3+}, Mn^{3+})$)

Pyrite (FeS_2)

Hematite (Fe_2O_3)

Rutile (TiO_2)

Cinnabar (HgS)

Figure 7.9 The beauty of crystalline solids, such as these, comes from the orderly arrangement of their component ions.

Interactive Textbook

Simulation 5 Simulate the formation of ionic compounds at the atomic level.

with ChemASAP

CLASS Activity

Types of Ionic Compounds L2

Purpose Students investigate properties of different classes of ionic compounds.

Materials library or Internet access

Procedure Divide the class into groups. Have each group choose a different class of ionic compounds to research and write about. For example, one group could work with oxides while another group worked with sulfides. Initially, each student should work alone to discover information such as where the compounds occur in nature, how they are produced, their physical and chemical properties, and any important uses. Finally, students in each group can pool their information to prepare a class display or report.

Expected Outcome Students will discover that different classes of ionic compounds share some properties with other ionic compounds and have some unique properties.

CLASS Activity

Crystal Structures 

Purpose Students observe different types of crystals.

Materials crystals of ionic compounds, watch glasses, magnifying glasses

Safety Use only nontoxic crystals. Remind students to not touch the crystals.

Procedure Pass around crystals of ionic compounds of various types in watch glasses. Have the students examine the crystals with magnifying glasses and write down their observations. Make a list of these observations and then discuss them in terms of the underlying ionic lattice structures.

Expected Outcome Students should observe the different geometries of different crystal structures.

Differentiated Instruction

Gifted and Talented L3

Have students research how minerals are categorized according to their ionic nature. Suggest that their written report include information concerning the physical properties of minerals and how these properties are used in mineral identification. Encourage students to include drawings, photos, or examples of minerals from each category.

Quick LAB

Solutions Containing Ions 

Objective After completing this activity, students will be able to:

- show that ions in solution conduct an electric current.

Address Misconceptions

Students may think that the solutions contain only one ion. Clarify that each solution contains both cations and anions.

Skills Focus Observing, experimenting, concluding

Prep Time 20 minutes

Materials D-cell batteries, masking tape, 30-cm lengths of bell wire with ends scraped bare, clear plastic cups, distilled water, tap water, vinegar, sucrose, sodium chloride, baking soda, conductivity probe (optional)

Class Time 30 minutes

Safety Students should handle wires with caution. The wires may become hot during the activity.

Expected Outcome When ions are present in solution, the solution conducts an electric current.

Analyze and Conclude

1. Solutions of vinegar, sodium chloride, and baking soda (and maybe tap water) contain ions and therefore conduct electric current and produce bubbles.

2. Distilled water and sugar solution (and maybe tap water) do not contain ions and therefore do not conduct an electric current or produce bubbles.

3. Answers will vary but should indicate that a larger number of batteries will increase the current, which will, in turn, cause the rate at which the bubbles appear to increase.

For Enrichment 

Ask, **What gases form the bubbles you observe?** *(hydrogen and oxygen)* **What is the source of these gases?** *(water)* Have students collect the gases produced and check them for the presence of hydrogen and oxygen. Hydrogen will burn when a lit splint is placed in it, and a glowing split will begin to flame when placed in oxygen.

Differentiated Instruction

Gifted and Talented 

Have students write the formulas for ionic compounds formed from selected pairs of cations and anions. Include various polyatomic cations and anions as well.

Examples:

1) K^+ and S^{2-} *(K_2S)*

2) Ca^{2+} and HCO^{3-} *($Ca(HCO_3)_2$)*

---

a Sodium chloride (NaCl) — Cl^-, Na^+

b Cesium chloride (CsCl) — Cl^-, Cs^+

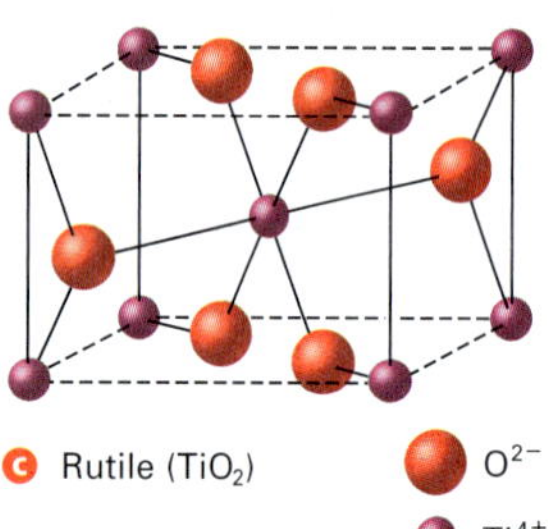

Figure 7.10 Sodium chloride and cesium chloride form cubic crystals. **a** In NaCl, each ion has a coordination number of 6. **b** In CsCl, each ion has a coordination number of 8. **c** Titanium dioxide forms tetragonal crystals. In TiO_2, each Ti^{4+} ion has a coordination number of 6, while each O^{2-} ion has a coordination number of 3.

The **coordination number** of an ion is the number of ions of opposite charge that surround the ion in a crystal. Figure 7.10a shows the three-dimensional arrangement of ions in NaCl. Because each Na^+ ion is surrounded by six Cl^- ions, Na^+ has a coordination number of 6. Each Cl^- ion is surrounded by six Na^+ ions and also has a coordination number of 6. Cesium chloride (CsCl) has a formula unit that is similar to that of NaCl. As Figure 7.10b illustrates, both compounds have cubic crystals, but their internal crystal structures are different. Each Cs^+ ion is surrounded by eight Cl^- ions, and each Cl^- ion is surrounded by eight Cs^+ ions. The anion and cation in cesium chloride each have a coordination number of 8.

Figure 7.10c shows the crystalline form of titanium dioxide (TiO_2), also known as rutile. In this compound, the coordination number for the cation (Ti^{4+}) is 6. Each Ti^{4+} ion is surrounded by six O^{2-} ions. The coordination number of the anion (O^{2-}) is 3. Each O^{2-} ion is surrounded by three Ti^{4+} ions.

Another characteristic property of ionic compounds has to do with conductivity. **Ionic compounds can conduct an electric current when melted or dissolved in water.** As Figure 7.11 shows, when sodium chloride is melted, the orderly crystal structure breaks down. If a voltage is applied across this molten mass, cations migrate freely to one electrode and anions migrate to the other. This ion movement allows electricity to flow between the electrodes through an external wire. For a similar reason, ionic compounds also conduct electricity if they are dissolved in water. When dissolved, the ions are free to move about in the aqueous solution.

Checkpoint ***What is the coordination number of Ti^{4+} in TiO_2?***

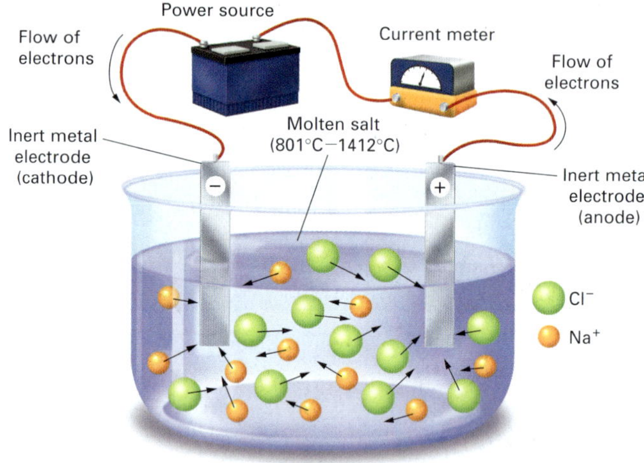

Figure 7.11 When sodium chloride melts, the sodium and chloride ions are free to move throughout the molten salt. If a voltage is applied, positive sodium ions move to the negative electrode (the cathode), and negative chloride ions move to the positive electrode (the anode). **Predicting** ***What would happen if the voltage was applied across a solution of NaCl dissolved in water?***

Quick LAB

Solutions Containing Ions

Purpose

To show that ions in solution conduct an electric current.

Materials

- 3 D-cell batteries
- masking tape
- 2 30-cm lengths of bell wire with ends scraped bare
- clear plastic cup
- distilled water
- tap water
- vinegar
- sucrose
- sodium chloride
- baking soda
- conductivity probe (optional)

Procedure   

Probe version available in the Probeware Lab Manual.

1. Tape the batteries together so the positive end of one touches the negative end of another. Tape the bare end of one wire to the positive terminal of the battery assembly and the bare end of the other wire to the negative terminal. **CAUTION** *Bare wire ends can be sharp and scratch skin. Handle with care.*
2. Half fill the cup with distilled water. Hold the bare ends of the wires close together in the water. Look for the production of bubbles. They are a sign that the solution conducts electric current.
3. Repeat Step 2 with tap water, vinegar, and concentrated solutions of sucrose, sodium chloride, and baking soda (sodium hydrogen carbonate).

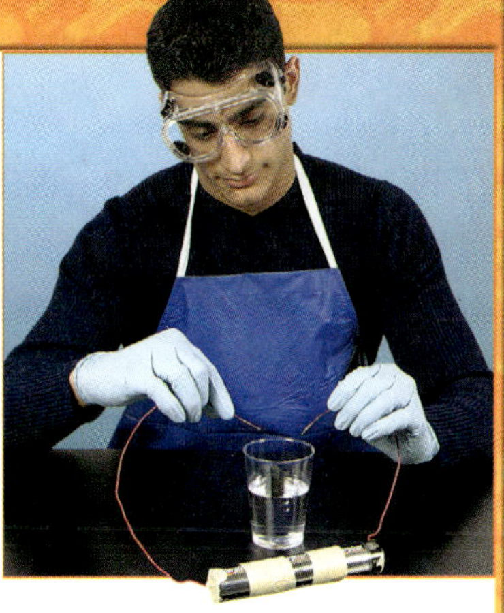

Analyze and Conclude

1. Which solutions produced bubbles of gas? Explain.
2. Which samples did not produce bubbles of gas? Explain.
3. Would you expect the same results if you used only one battery? If you used six batteries? Explain your answer.

7.2 Section Assessment

14. **Key Concept** How can you describe the electrical charge of an ionic compound?
15. **Key Concept** What properties characterize ionic compounds?
16. Define an ionic bond.
17. How can you represent the composition of an ionic compound?
18. Write the correct chemical formula for the compounds formed from each pair of ions.
    a. K^+, S^{2-} b. Ca^{2+}, O^{2-}
    c. Na^+, O^{2-} d. Al^{3+}, N^{3-}
19. Write formulas for each compound.
    a. barium chloride b. magnesium oxide
    c. lithium oxide d. calcium fluoride
20. Which pairs of elements are likely to form ionic compounds?
    a. Cl, Br b. Li, Cl
    c. K, He d. I, Na
21. Describe the arrangement of sodium ions and chloride ions in a crystal of sodium chloride.
22. Why do ionic compounds conduct electric current when they are melted or dissolved in water?

Elements Handbook

Restoring Electrolytes Read about restoring electrolytes on page R8. Write electron configurations for the two principal ions found in body fluids.

Assessment 7.2 Test yourself on the concepts in Section 7.2.

with ChemASAP

Discuss L2

Point out that the rusting of iron is the production of iron oxide from iron metal and oxygen gas. Point out that Fe^{3+} is a stable cation of Fe. (Students should know that O^{2-} is the stable anion of O.) The compound formed from these ions is iron(III) oxide, or Fe_2O_3.

3 ASSESS

Evaluate Understanding 

Name ionic compounds and ask students to identify the cation, anion, and ratio of cations to anions in each compound.

Reteach 

Emphasize that an ionic solid is a collection of independent ions. There is no joining of individual particles to form molecules. Each ion "belongs" as much to one of its nearest neighbors as it belongs to any other. The arrangement in an ionic crystal is such that each ion is surrounded by ions of opposite charge, which produces a strong bonding force.

Elements Handbook

Na^+ $1s^22s^22p^6$
K^+ $1s^22s^22p^63s^23p^6$

If your class subscribes to the Interactive Textbook, use it to review key concepts in Section 7.2.

with ChemASAP

Section 7.2 Assessment

14. electrically neutral
15. usually solids at room temperature; have high melting points; conduct an electric current when melted or dissolved in water
16. electrostatic forces that hold ions together in an ionic compound
17. by writing its chemical formula
18. **a.** K_2S **b.** CaO **c.** Na_2O **d.** AlN
19. **a.** $BaCl_2$ **b.** MgO **c.** Li_2O **d.** CaF_2
20. b and d
21. Acceptable answers should describe a solid containing positive sodium ions and negative chloride ions in an alternating, regular, and repeating three-dimensional pattern.
22. The ions are free to move.

Answers to...

Figure 7.11 The solution would conduct an electric current; the Na^+ ions in solution would migrate to the negative electrode, and the Cl^- ions in solution would migrate to the positive electrode.

Ti^{4+} has a coordination number of 6 in TiO_2.

Analysis of Anions and Cations L2

Objective

After completing this activity, students will be able to:

- develop tests for various ions.
- use the tests to analyze unknown solutions.

Skills Focus Observing, drawing conclusions, designing experiments

Prep Time 40 minutes

Materials pencil, paper, ruler, reaction surface, medicine droppers, pipet, staples or solid Fe, solutions from Figures A & B.

Solution	Preparation
0.05*M* $AgNO_3$	2.1 g in 250 mL
0.2*M* $Pb(NO_3)_2$	16.6 g in 250 mL
0.2*M* Na_2SO_4	7.1 g in 250 mL
0.1*M* Na_3PO_4	9.5 g $Na_3PO_4{\cdot}12H_2O$ in 250 mL
0.5*M* NaOH	20.0 g in 1.0 L
0.1*M* KSCN	2.4 g in 250 mL
0.1*M* KI	4.2 g in 250 mL
0.5*M* $CaCl_2$	13.9 g in 250 mL
0.1*M* $FeCl_3$	6.8 g $FeCl_3{\cdot}6H_2O$ in 25 mL of 1.0*M* NaCl; dilute to 250 mL
1.0*M* HCl	82 mL of 12*M* in 1.0 L
1.0*M* HNO_3	63 mL of 15.8*M* in 1.0 L

CAUTION! Always add acid to water carefully and slowly.

Class Time 40 minutes

Expected Outcome Ag^+ and PO_4^{3-} form a yellow precipitate; Fe in HCl turns yellow with NO_3^-; Pb^{2+} forms white precipitates with SO_4^{2-} and PO_4^{3-}; OH^- and Ca^{2+} form a white product; KSCN and Fe^{3+} form a red one.

Analyze

1. Nitrate ion is the only ion that produces a yellow color with iron in the presence of acid.
2. $AgNO_3$ for PO_4^{3-}; Fe and HCl for NO_3^-; $Pb(NO_3)_2$ for SO_4^{2-}; NaOH for Ca^{2+}; and KSCN for Fe^{3+}.

Small-Scale LAB

Analysis of Anions and Cations

Purpose

To develop tests for various ions and use the tests to analyze unknown substances.

Materials

- pencil
- ruler
- medicine droppers
- chemicals shown in Figures A and B
- paper
- reaction surface
- pipet

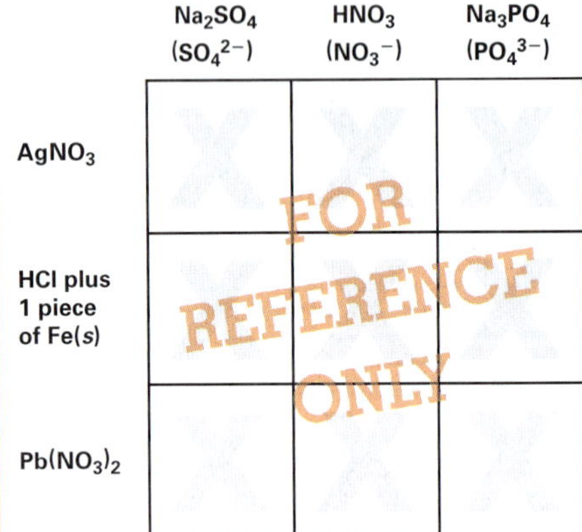

Figure A
Anion Analysis

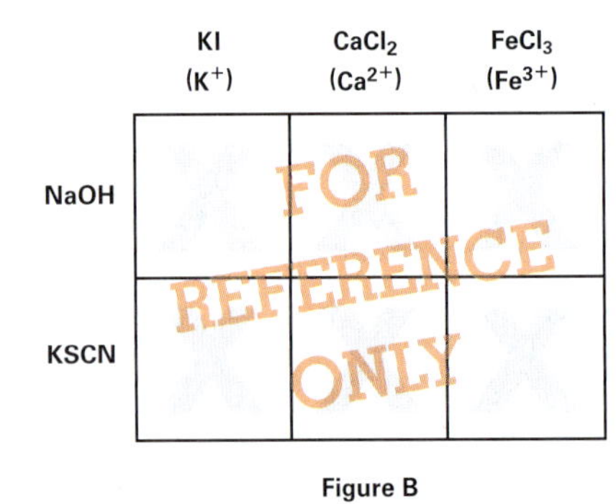

Figure B
Cation Analysis

Procedure     

On one sheet of paper, draw grids similar to Figure A and Figure B. Draw similar grids on a second sheet of paper. Make each square 2 cm on each side. Place a reaction surface over the grids on one of the sheets of paper and add one drop of each solution or one piece of each solid as shown in Figures A and B. Stir each solution by blowing air through an empty pipet. Use the grids on the second sheet of paper as a data table to record your observations for each solution.

Analyze

Using your experimental data, record the answers to the following questions below your data table.

1. Carefully examine the reaction of Fe(*s*) and HCl in the presence of HNO_3. What is unique about this reaction? How can you use it to identify nitrate ion?
2. Which solutions from Figure A are the best for identifying each anion? Which solutions from Figure B are the best for identifying each cation? Explain.
3. Can your experiments identify K^+ ions? Explain.

You're the Chemist

The following small-scale activities allow you to develop your own procedures and analyze the results.

1. **Design It!** Obtain a set of unknown anion solutions from your teacher and design and carry out a series of tests that will identify each anion.
2. **Design It!** Obtain a set of unknown cation solutions from your teacher and design and carry out a series of tests that will identify each cation.
3. **Design It!** Obtain a set of unknown solid ionic compounds from your teacher. Design and carry out a series of tests that will identify each ion present.

3. No, neither of the solutions produced a visible product.

You're the Chemist

1. Mix a drop of each unknown with one drop of each solution from figure A. Compare the results to those with known solutions.
2. Mix a drop of each unknown with one drop of both solutions from Figure B. Compare the results to those with known solutions.
3. Mix an unknown with one drop of each solution from Figures A and B and compare results.

For Enrichment 

Have students hypothesize about why none of the tests in the lab identified the potassium ion. Ask, **Why do you think potassium ions don't form precipitates with any of the anions used in the lab?** *(All potassium compounds are soluble in water.)* Have students read through the Quick Lab on page 142. Have them use this information to design an experiment that will provide a positive test for the potassium ion. The flame test for potassium produces a violet color.

7.3 Bonding in Metals

Connecting to Your World

You have probably seen decorative fences, railings, or weathervanes made of a metal called wrought iron. Wrought iron is a very pure form of iron that contains trace amounts of carbon. It is a tough, malleable, ductile, and corrosion-resistant material that melts at a very high temperature. As you already know, metals often have distinctive, useful properties. In this section, you will learn how metallic properties derive from the way that metal ions form bonds with one another.

Guide for Reading

Key Concepts

- How can you model the valence electrons of metal atoms?
- How are metal atoms arranged?
- Why are alloys important?

Vocabulary

metallic bonds

alloys

Reading Strategy

Using Prior Knowledge Before you read, jot down three things you know about metals. When you have read the section, explain how what you already knew helped you learn something new.

Metallic Bonds and Metallic Properties

Metals are made up of closely packed cations rather than neutral atoms. **The valence electrons of metal atoms can be modeled as a sea of electrons.** That is, the valence electrons are mobile and can drift freely from one part of the metal to another. **Metallic bonds** consist of the attraction of the free-floating valence electrons for the positively charged metal ions. These bonds are the forces of attraction that hold metals together.

The sea-of-electrons model explains many physical properties of metals. For example, metals are good conductors of electrical current because electrons can flow freely in them. As electrons enter one end of a bar of metal, an equal number leave the other end. Metals are ductile—that is, they can be drawn into wires, as shown in Figure 7.12. Metals are also malleable, which means that they can be hammered or forced into shapes.

Animation 9 See how metallic bonding explains some physical properties of metals.

with ChemASAP

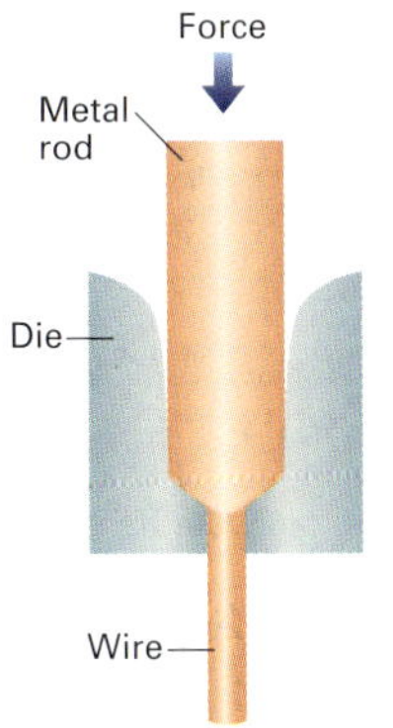

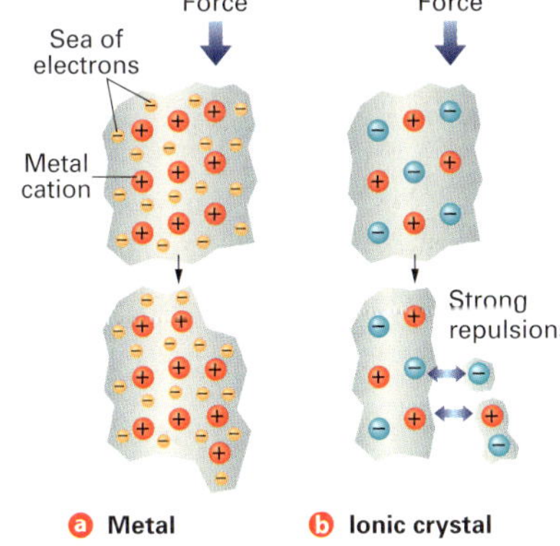

Figure 7.12 A metal rod can be forced through a narrow opening in a die to produce wire. **a** As this occurs, the metal changes shape but remains in one piece. **b** If an ionic crystal were forced through the die, it would shatter. **Interpreting Diagrams** ***What causes the ionic crystal to break apart?***

Section Resources

Print

- **Guided Reading and Study Workbook,** Section 7.3
- **Core Teaching Resources,** Section 7.3 Review
- **Transparencies,** T82–T84
- **Laboratory Manual,** Lab 10

Technology

- **Interactive Textbook with ChemASAP,** Animation 9, Assessment 7.3

7.3

1 FOCUS

Objectives

7.3.1 **Model** the valence electrons of metal atoms.

7.3.2 **Describe** the arrangement of atoms in a metal.

7.3.3 **Explain** the importance of alloys.

Guide for Reading

Build Vocabulary 

Word Forms From what they know about the terms *metallic* and *bond,* have students infer the meaning of *metallic bond.*

Reading Strategy 

Relating Cause and Effect Have students describe the cause-and-effect relationship between metallic bonding and the properties of metals.

2 INSTRUCT

Connecting to Your World

Have students read the text that opens the section. Ask, **What property makes metals good electrical conductors?** *(Metal bonding involves highly mobile electrons, which are shared by all of the nuclei in a metallic solid. Electron mobility accounts for the high electrical conductivity of metals.)*

Metallic Bonds and Metallic Properties

Relate 

To assess students' prior knowledge about metals, ask, **What are the properties of malleable and ductile metals?** *(They can be hammered into different shapes and drawn into wires.)* **What is an alloy? Give an example.** *(An alloy is a mixture of two or more elements, at least one of which is a metal. Examples include steel and brass.)*

Answers to...

Figure 7.12 repulsions between ions of like charge

TEACHER Demo

Metals vs. Ionic Compounds L2

Purpose Students compare copper metal and a copper compound.

Materials copper metal or alloy, copper-containing ionic compound

Procedure Show the class a small sample of elemental copper or a copper alloy and a sample of a copper-containing crystalline ionic mineral such as chalcocite (Cu_2S). Wearing safety glasses and standing far from the students, smash both samples with a hammer. Discuss why the two substances respond differently to the stress of the hammer blow.

Expected Outcomes The elemental copper will flatten but not break because the cations and electrons are mobile; the crystal will shatter.

Crystalline Structure of Metals

Use Visuals 

Figure 7.14 Lead a class discussion on the concept of "closest packing" of metal cations in pure metals. Use the three different closest packing arrangements shown in Figure 7.14 as a reference. Emphasize that the concept of closest packing also relates to more than just metal atoms. Have the students describe other examples of closest packing.

Figure 7.13 These tomatoes illustrate a pattern called a hexagonal close-packed arrangement.

Both the ductility and malleability of metals can be explained in terms of the mobility of valence electrons. A sea of drifting valence electrons insulates the metal cations from one another. When a metal is subjected to pressure, the metal cations easily slide past one another like ball bearings immersed in oil. In contrast, if an ionic crystal is struck with a hammer, the blow tends to push the positive ions close together. They repel, and the crystal shatters.

Crystalline Structure of Metals

The next time you visit a grocery store, take a look at how the apples or tomatoes are stacked. More than likely, they will have a close-packed arrangement, as shown in Figure 7.13. This arrangement helps save space while allowing as many tomatoes to be stacked as possible.

Similar arrangements can be found in the crystalline structures of metals. You may be surprised to learn that metals are crystalline. In fact, metals that contain just one kind of atom are among the simplest forms of all crystalline solids. **Metal atoms are arranged in very compact and orderly patterns.** For spheres of identical size, such as metal atoms, there are several closely packed arrangements that are possible. Figure 7.14 shows three such arrangements: body-centered cubic, face-centered cubic, and hexagonal close-packed arrangements.

In a body-centered cubic structure, every atom (except those on the surface) has eight neighbors. The elements sodium, potassium, iron, chromium, and tungsten crystallize in a body-centered cubic pattern. In a face-centered cubic arrangement, every atom has twelve neighbors. Among the metals that form a face-centered cubic lattice are copper, silver, gold, aluminum, and lead. In a hexagonal close-packed arrangement, every atom also has twelve neighbors. Because of its hexagonal shape, however, the pattern is different from the face-centered cubic arrangement. Metals that have the hexagonal close-packed crystal structure include magnesium, zinc, and cadmium.

 ***What metals crystallize in a face-centered cubic pattern?***

Figure 7.14 Metal atoms crystallize in characteristic patterns. **a** Chromium atoms have a body-centered cubic arrangement. **b** Gold atoms have a face-centered cubic arrangement. **c** Zinc atoms have a hexagonal close-packed arrangement. **Inferring** ***Which of these arrangements is the most closely packed?***

Chromium

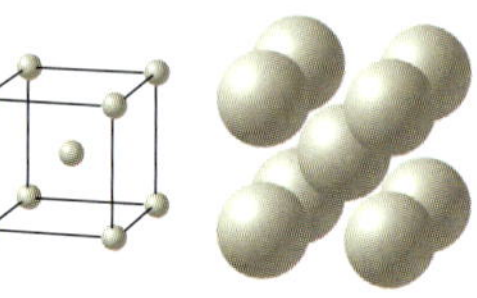

Body-centered cubic

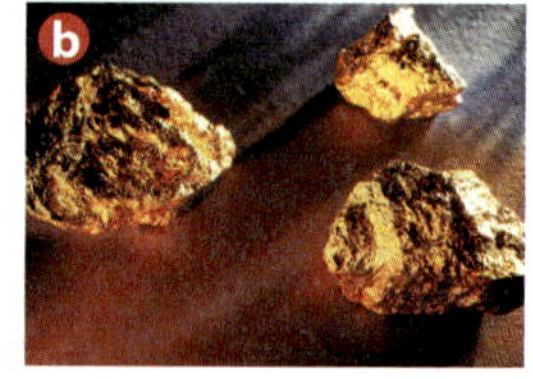

Gold

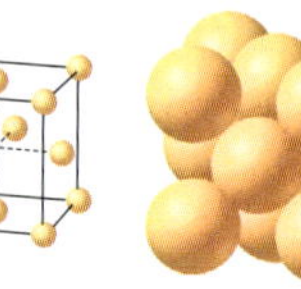

Face-centered cubic

Zinc

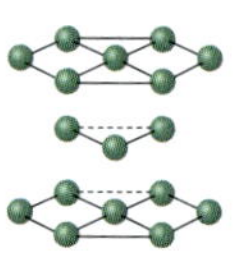

Hexagonal close-packed

Differentiated Instruction

Gifted and Talented 

Have students develop formulas for calculating the density of a metal given the atomic radius of the metal and the cubic unit cell packing arrangement of the metal atoms. Use sodium, potassium, iron, chromium, and tungsten for body-centered cubic metals. Use copper, silver, gold, aluminum, and lead for face-centered cubic metals. The students may need to make models to see geometric relationships more clearly.

Figure 7.15 Bicycle frames are often made of titanium alloys that contain aluminum and vanadium.

Alloys

Although every day you use metallic items, such as spoons, very few of these objects are pure metals. Instead, most of the metals you encounter are alloys. **Alloys** are mixtures composed of two or more elements, at least one of which is a metal. Brass, for example, is an alloy of copper and zinc. **Alloys are important because their properties are often superior to those of their component elements.** Sterling silver (92.5% silver and 7.5% copper) is harder and more durable than pure silver but still soft enough to be made into jewelry and tableware. Bronze is an alloy generally containing seven parts of copper to one part of tin. Bronze is harder than copper and more easily cast. Nonferrous (non-iron) alloys, such as bronze, copper-nickel, and aluminum alloys, are commonly used to make coins.

The most important alloys today are steels. The principal elements in most steel, in addition to iron and carbon, are boron, chromium, manganese, molybdenum, nickel, tungsten, and vanadium. Steels have a wide range of useful properties, such as corrosion resistance, ductility, hardness, and toughness. Table 7.3 lists the composition of some common alloys.

Alloys can form from their component atoms in different ways. If the atoms of the components in an alloy are about the same size, they can replace each other in the crystal. This type of alloy is called a substitutional alloy. If the atomic sizes are quite different, the smaller atoms can fit into the interstices (spaces) between the larger atoms. Such an alloy is called an interstitial alloy. In the various types of steel, for example, carbon atoms occupy the spaces between the iron atoms. Thus, steels are interstitial alloys.

Table 7.3
Composition of Some Common Alloys

Name	Composition (by mass)
Sterling silver	Ag 92.5% Cu 7.5%
Cast iron	Fe 96% C 4%
Stainless steel	Fe 80.6% Cr 18.0% C 0.4% Ni 1.0%
Spring steel	Fe 98.6% Cr 1.0% C 0.4%
Surgical steel	Fe 67% Cr 18% Ni 12% Mo 3%

7.3 Section Assessment

23. **Key Concept** How do chemists model the valence electrons in metal atoms?
24. **Key Concept** How can you describe the arrangement of atoms in metals?
25. **Key Concept** Why are alloys more useful than pure metals?
26. Describe what is meant by ductile and malleable.
27. Why is it possible to bend metals but not ionic crystals?
28. What are three different packing arrangements found in metallic crystals?
29. Describe two widely used alloys.

Writing Activity

Explanatory Paragraph Write a paragraph describing how the sea-of-electrons model is used to explain the physical properties of metals. *Hint:* First write a sentence that summarizes the model. Then discuss how the model applies to specific properties of metals.

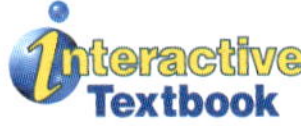

Assessment 7.3 Test yourself on the concepts in Section 7.3.

with ChemASAP

Alloys

TEACHER Demo

Types of Alloys 

Purpose Students compare models of interstitial and substitutional alloys.

Materials styrofoam balls of different sizes and colors, toothpicks

Procedure Use styrofoam balls to illustrate the crystal structures of interstitial and substitutional alloys. Use toothpicks to hold the "atoms" together. Point out that brass is a substitutional alloy in which copper atoms are replaced by similarly sized zinc atoms. Steel is an interstitial alloy in which relatively small carbon atoms occupy the interstices between closely packed iron atoms.

Expected Outcome Students should be able to distinguish between the two types of alloys.

3 ASSESS

Evaluate Understanding 

Ask, **How can the conductivity of metals be explained?** *(Electrons can flow in and out of a metal because the valence electrons are not fixed.)*

Reteach 

Compare chemical bonding in ionic compounds and pure metals.

Writing Activity

Metal cations are surrounded by free-floating electrons. When metals are hammered, the cations move past each other. Conductivity results from mobile electrons.

If your class subscribes to the Interactive Textbook, use it to review key concepts in Section 7.3.

with ChemASAP

Section 7.3 Assessment

23. metal cations surrounded by a sea of mobile valence electrons
24. Atoms in metals are arranged in a compact and orderly manner.
25. The properties of alloys are often superior to their component elements.
26. ductile: can be drawn into wires; malleable: can be hammered into different shapes
27. Under pressure, the cations in a metal slide past each other. The ions in ionic crystals are forced into each other by the rigid structure.
28. body-centered cubic; face-centered cubic; hexagonal close-packed
29. Sterling silver used in jewelry is 92.5% silver and 7.5% copper; bronze used in casting is 7 parts copper and 1 part tin.

Answers to...

Figure 7.14 face-centered cubic

 copper, silver, gold, aluminum, and lead

Technology & Society

Building With Alloys

Explain that materials for a building must be able to withstand the stresses that a building undergoes. Discuss with students several different types of stress. Tensile stress can be observed when a building beam sags. The bottom of the beam undergoes tensile stress as it is slightly stretched. The top of the beam undergoes compression stress, which results when two forces push toward each other through a solid. The top of the beam is compressed as it is slightly shortened. A building undergoes shear stress in a strong wind. In shear stress, forces are applied from different directions, and the building might twist or break. Ask, **Which of these buildings are most likely to undergo shear stress?** *(The Atomium, the Chrysler Building)*

Metals are used as building materials because of strength and durability, but other properties might determine the use of a metal in a building. Ask, **What property of a metal might make it useful on the outside of a building?** *(Answers might include luster when a shiny appearance is desired, malleability when the metal covers another material, or the ability of a certain metal to form a compound that protects the rest of the metal or other materials under it.)* **What metal was chosen for each these buildings? Why do you think that alloy was used?** *(possible answers: the Atomium—aluminum alloy; shiny, corrosion resistant, light, malleable; the Chrysler Building—steel; shiny, malleable, corrosion-resistant; the Jewish Museum Berlin—zinc-titanium alloy: corrosion resistant, light)*

Technology & Society

Building with Alloys

Modern architecture would be a lot shorter if it weren't for steel. Since the late 1800s, using steel columns and girders in construction has allowed architects to design taller, stronger, and lighter buildings. Unlike buildings made of wood, brick, or stone, steel structures are strong enough to accommodate large, open interior spaces that do not require supporting walls. Usually, you can't see the steel used to construct a building; either it's hidden by the floors and walls, or—in the case of a building made of reinforced concrete—it's actually embedded in the floors and walls. The exteriors of buildings often feature lighter alloys, such as alloys of aluminum or titanium that resist corrosion.

Comparing and Contrasting ***How does steel-frame construction differ from reinforced-concrete construction?***

Jewish Museum Berlin
Berlin, Germany
This angular building is covered in thin sheets of zinc-titanium alloy. The untreated alloy will slowly oxidize and change color from exposure to the air and weather.

The Atomium
Brussels, Belgium
Designed to resemble a crystal of iron magnified 165 billion times, the Atomium consists of nine spheres made of aluminum-alloy panels and connected by steel tubes. The top sphere contains an observation deck 92 meters above ground.

Chrysler Building
New York City
Completed in 1930, this steel-frame high-rise stands 319 m tall and features a distinctive spire sheathed in shiny stainless steel.

Facts and Figures

A Good Foundation

Concrete itself is not strong enough to be a good framing material. Reinforcing the concrete by pouring it over steel rods that have been laid out in a grid adds a considerable amount of strength to the concrete. When exceptionally strong concrete is needed, it is poured over steel cables that are stretched. After the concrete dries, the rods are released, and the concrete is compressed as the rods return to their original length. This exceptionally strong concrete is called prestressed concrete.

Steel-frame construction The most common method to build a high-rise is to use a steel framework. Steel frames can be assembled quickly and allow for flexible interior spaces.

Reinforced-concrete construction Another method for building high-rises is to use reinforced concrete. Concrete is more rigid than steel and holds up better structurally in fires. In addition, a concrete frame takes up less vertical space than a steel frame, meaning that a concrete-frame building can contain more floors than a steel-frame building of the same height.

TEACHER Demo

Making an Alloy L2

Purpose Students observe how to make an alloy from copper and zinc.

Materials penny, fine sandpaper, granulated zinc, dilute NaOH solution, evaporating dish, tongs, hot plate, teaspoon

Safety Be sure to use adequate ventilation and have no skin contact with the NaOH solution. Do not touch hot objects.

Procedure Use the sandpaper to clean any tarnish from the penny. Add a teaspoon of zinc to the evaporating dish, and cover the zinc with NaOH solution. Place the penny on the zinc, being sure the penny is also covered by the solution. Heat the dish until the penny changes to a silvery color. Using the tongs, remove and rinse the penny. Place the penny on the hot plate, which should be set to medium heat. When the penny turns a gold color, use tongs to remove it from the hot plate.

Expected Outcome A gold-colored alloy forms from copper and zinc. Ask, **Why did the penny turn a silver color?** *(Zinc was deposited on the penny.)* **At what point was an alloy formed?** *(An alloy was formed when the copper and zinc on the penny were heated to a golden color.)* **Why was the penny heated to form the alloy?** *(Heat increases the kinetic energy of the atoms, allowing them to mix more freely.)*

Differentiated Instruction

Less Proficient Readers L1

Have students create a list of the building materials they find in their school building. If the material is a metal, have them try to determine whether the metal is an alloy or not. Point out to students that most of the metals will be alloys, such as steel, because alloys can be made to have the properties desired in a building material.

Answers to...

Comparing and Contrasting Steel frames allow for more flexible interior spaces. Reinforced-concrete construction takes up less vertical space, so more floors can be contained in a building of a certain height if reinforced concrete is used instead of steel.

CHAPTER 7 Study Guide

Study Tip

Write and answer questions about the content of the chapter. For Chapter 7, you might ask the following questions, which require you to summarize the information contained in the chapter. **What information does the electron dot structure of an element provide?** *(the number of valence electrons around the atom)* **In most cases, what is the stable electron configuration of an elemental ion?** *(In most cases, a stable configuration is characterized by an octet of valence electrons; the configuration is isoelectronic with that of a noble gas.)* **Why does NaCl dissolved in water conduct electricity?** *(In water, the ionic compound breaks apart into Na^+ and Cl^- ions that are free to move in the solution.)* **What is the theory of bonding in pure metals?** *(metal cations held together by a pool of mobile valence electrons)*

If your class subscribes to the Interactive Textbook with ChemASAP, your students can go online to access an interactive version of the Student Edition and a self-test.

with ChemASAP

CHAPTER 7 Study Guide

Key Concepts

7.1 Ions

- To find the number of valence electrons in an atom of a representative element, simply look at its group number.
- Atoms of the metallic elements tend to lose their valence electrons, leaving a complete octet in the next-lowest energy level. Atoms of some nonmetallic elements tend to gain electrons to achieve a complete octet.
- An atom's loss of valence electrons produces a positively charged cation.
- The gain of electrons by a neutral atom produces negatively charged anion.

7.2 Ionic Bonds and Ionic Compounds

- Although they are composed of ions, ionic compounds are electrically neutral.
- Most ionic compounds are crystalline solids at room temperature, and they generally have high melting points. Ionic compounds can conduct an electric current when melted or dissolved in water.

7.3 Bonding in Metals

- The valence electrons of metal atoms can be modeled as a sea of electrons.
- Metal atoms are arranged in very compact and orderly patterns.
- Alloys are important because their properties are often superior to those of their component elements.

Vocabulary

- alloys (p. 203)
- chemical formula (p. 195)
- coordination number (p. 198)
- electron dot structure (p. 188)
- formula unit (p. 195)
- halide ion (p. 192)
- ionic bonds (p. 194)
- ionic compounds (p. 194)
- metallic bonds (p. 201)
- octet rule (p. 188)
- valence electron (p. 187)

Organizing Information

Use these terms to construct a concept map that organizes the major ideas of this chapter.

ionic bond
valence electron
stable electron configuration
properties of ionic compounds
octet rule
metallic bond
ionic compound

Concept Map 7 Solve the Concept Map with the help of an interactive guided tutorial.

with ChemASAP

Chapter Resources

Print

- **Core Teaching Resources,** Chapter 7, *Practice Problems, Vocabulary Review, Quiz, Chapter Test A, Chapter Test B*

Technology

- **Computer Test Bank,** Chapter 7 Test
- **Interactive Textbook with ChemASAP,** Chapter 7

CHAPTER 7

Assessment

Reviewing Content

7.1 Ions

30. Describe two ways that an ion forms from an atom.

31. State the number of electrons either lost or gained in forming each ion.

a. Br^- **b.** Na^+
c. As^{3-} **d.** Ca^{2+}
e. Cu^+ **f.** H^-

32. Name each ion in Problem 31. Identify each as an anion or a cation.

33. Define valence electrons.

34. How many electrons does each atom have? What group is each in?

a. nitrogen **b.** lithium
c. phosphorus **d.** barium
e. bromine **f.** carbon

35. Write electron dot structures for each of the following elements.

a. Cl **b.** S
c. Al **d.** Li

36. How many electrons must each atom lose to attain a noble-gas electron configuration?

a. Ca **b.** Al
c. Li **d.** Ba

37. Write the formula for the ion formed when each of the following elements loses its valence electrons.

a. aluminum **b.** lithium
c. barium **d.** potassium
e. calcium **f.** strontium

38. Why do nonmetals tend to form anions when they react to form compounds?

39. What is the formula of the ion formed when the following elements gain or lose valence electrons and attain noble-gas configurations?

a. sulfur **b.** sodium
c. fluorine **d.** phosphorus

40. How many electrons must be gained by each of the following atoms to achieve a stable electron configuration?

a. N **b.** S
c. Cl **d.** P

7.2 Ionic Bonds and Ionic Compounds

41. Which of the following pairs of atoms would you expect to combine chemically to form an ionic compound?

a. Li and S **b.** O and S
c. Al and O **d.** F and Cl
e. I and K **f.** H and N

42. Identify the kinds of ions that form each ionic compound.

a. calcium fluoride, CaF_2
b. aluminum bromide, $AlBr_3$
c. lithium oxide, Li_2O
d. aluminum sulfide, Al_2S_3
e. potassium nitride, K_3N

43. Explain why ionic compounds are electrically neutral.

44. Which of the following pairs of elements will not form ionic compounds?

a. sulfur and oxygen
b. sodium and calcium
c. sodium and sulfur
d. oxygen and chlorine

45. Write the formula for the ions in the following compounds.

a. KCl **b.** $BaSO_4$
c. $MgBr_2$ **d.** Li_2CO_3

46. Most ionic substances are brittle. Why?

47. Explain why molten $MgCl_2$ does conduct an electric current although crystalline $MgCl_2$ does not.

7.3 Bonding in Metals

48. Explain briefly why metals are good conductors of electricity.

49. Name the three crystal arrangements of closely packed metal atoms. Give an example of a metal that crystallizes in each arrangement.

50. Name some alloys that you have used or seen today.

51. Explain why the properties of all steels are not identical.

CHAPTER 7

Assessment

Reviewing Content

30. by gaining or losing electrons

31. **a.** gain of 1 electron
b. loss of one electron
c. gain of 3 electrons
d. loss of 2 electrons
e. loss of 1 electron
f. gain of 1 electron

32. **a.** bromide, anion
b. sodium, cation
c. arsenide, anion
d. calcium, cation
e. copper, cation
f. hydride, anion

33. electrons in the highest occupied energy level

34. **a.** 7, 5A **b.** 3, 1A **c.** 15, 5A
d. 56, 2A **e.** 35, 7A **f.** 6, 4A

35. **a.** :C̤̈l· **b.** ·S̤̈· **c.** ·Ȧl· **d.** Li·

36. **a.** 2 **b.** 3 **c.** 1 **d.** 2

37. **a.** Al^{3+} **b.** Li^+ **c.** Ba^{2+} **d.** K^+
e. Ca^{2+} **f.** Sr^{2+}

38. Most nonmetals gain 1, 2, or 3 electrons to achieve a noble-gas electron configuration.

39. **a.** S^{2-} **b.** Na^+ **c.** F^- **d.** P^{3-}

40. **a.** 3 **b.** 2 **c.** 1 **d.** 3

41. a, c, e

42. **a.** Ca^{2+}, F^- **b.** Al^{3+}, Br^- **c.** Li^+, O^{2-}
d. Al^{3+}, S^{2-} **e.** K^+, N^{3-}

43. The positive charges balance the negative charges.

44. a, b, d

45. **a.** K^+, Cl^- **b.** Ba^{2+}, $SO_4{}^{2-}$
c. Mg^{2+}, Br^- **d.** Li^+, $CO_3{}^{2-}$

46. Their network of electrostatic attractions and repulsions forms a rigid structure.

47. Ions are free to move in molten $MgCl_2$.

48. They have many mobile valence electrons. Electrons in the current replace the electrons leaving the metal.

49. body-centered cubic: Na, K, Fe, Cr, or W; face-centered cubic: Cu, Ag, Au, Al, or Pb; hexagonal close-packed: Mg, Zn, or Cd

50. Answers will vary and could include tableware, steel in cars and buses, high-speed dental drill bits, solder in stereos and televisions, and structural steel in buildings.

51. The properties of the steel will vary according to its composition. In addition to iron, steel can contain varying amounts of carbon and such metals as chromium, nickel, and molybdenum.

CHAPTER 7

Assessment *continued*

Understanding Concepts

52. Construct a table that shows the relationship among the group number, valence electrons lost or gained, and the formula of the cation or anion produced for the following metallic and nonmetallic elements: Na, Ca, Al, N, S, Br.

53. Write electron dot formulas for the following atoms.

a. C **b.** Be
c. O **d.** F
e. Na **f.** P

54. Show the relationship between the electron dot structure of an element and the location of the element in the periodic table.

55. In terms of electrons, why does a cation have a positive charge?

56. Why does an anion have a negative charge?

57. The spheres below represent the relative diameters of atoms or ions. Rearrange the sequences in a. and b. so the relative sizes of the particles correspond to the increasing size of the particles as shown in the illustration.

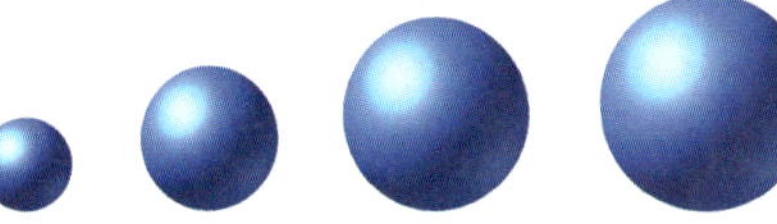

a. oxygen atom, oxide ion, sulfur atom, sulfide ion
b. sodium atom, sodium ion, potassium atom, potassium ion

58. Write electron configurations for the 2+ cations of these elements.

a. Fe **b.** Co
c. Ni

59. Write electron configurations for the 3+ cations of these elements.

a. chromium **b.** manganese
c. iron

60. The atoms of the noble gas elements are stable. Explain.

61. Write the formula for the ion formed when each element gains electrons and attains a noble-gas configuration.

a. Br **b.** H
c. As **d.** Se

62. Write electron configurations for these atoms and ions, and comment on the result.

a. Ar **b.** Cl^-
c. S^{2-} **d.** P^{3-}

63. Write electron configurations for the following and comment on the result.

a. N^{3-} **b.** O^{2-}
c. F^- **d.** Ne

64. Name the first four halogens. What group are they in, and how many valence electrons does each have?

65. Write complete electron configurations for the following atoms and ions. For each group, comment on the results.

a. Ar, K^+, Ca^{2+}
b. Ne, Na^+, Mg^{2+}, Al^{3+}

66. If ionic compounds are composed of charged particles (ions), why isn't every ionic compound either positively or negatively charged?

67. Which of the following substances are most likely not ionic?

a. H_2O **b.** Na_2O
c. CO_2 **d.** CaS
e. SO_2 **f.** NH_3

68. Can you predict the coordination number of an ion from the formula of an ionic compound? Explain.

69. Metallic cobalt crystallizes in a hexagonal close-packed structure. How many neighbors will a cobalt atom have?

70. Explain how hexagonal close-packed, face-centered cubic, and body-centered cubic unit cells are different from one another.

71. The properties of all samples of brass are not identical. Explain.

CHAPTER 7

Assessment continued

Understanding Concepts

52.

Group number	Valence electrons lost or gained	Formula of ion
1A	1	Na^+
2A	2	Ca^{2+}
3A	3	Al^{3+}
5A	3	N^{3-}
6A	2	S^{2-}
7A	1	Br^-

53. a. ·Ċ· **b.** ·Be· **c.** ·Ö:
d. :F̤· **e.** Na· **f.** :Ṗ·

54. For the representative elements the number of electrons in the electron dot structure is the group number.

55. It has lost valence electrons.

56. It has gained valence electrons.

57. a. oxygen atom, sulfur atom, oxide ion, sulfide ion
b. sodium ion, potassium ion, sodium atom, potassium atom

58. a. $1s^22s^22p^63s^23p^63d^6$
b. $1s^22s^22p^63s^23p^63d^7$
c. $1s^22s^22p^63s^23p^63d^8$

59. a. $1s^22s^22p^63s^23p^63d^3$
b. $1s^22s^22p^63s^23p^63d^4$
c. $1s^22s^22p^63s^23p^63d^5$

60. They have little chemical reactivity; their outermost occupied energy levels are filled.

61. a. Br^- **b.** H^- **c.** As^{3-} **d.** Se^{2-}

62. All have the noble-gas configuration of $1s^22s^22p^63s^23p^6$.

63. All are $1s^22s^22p^6$. All have the same configuration as neon.

64. fluorine, chlorine, bromine, and iodine; Group 7A, 7 valence electrons

65. a. $1s^22s^22p^63s^23p^6$
b. $1s^22s^22p^6$;
Each has a noble-gas electron configuration

66. The positively charged cations exactly balance the negatively charged anions.

67. a, c, e, f

68. No, the packing of ions in a crystalline structure depends on a number of factors including the relative sizes of the ions. The coordination number of an element can vary from compound to compound.

69. 12

70. Hexagonal close-packed units cells have twelve neighbors for every atom or ion. Face-centered cubic unit cells also have twelve neighbors for every atom or ion, with an atom or ion in the center of each face. Body-centered cubic units cells have eight neighbors for every atom or ion, with an atom or ion at the center of each cube.

71. Brass is a mixture of copper and zinc. The properties of a particular sample of brass will vary with the relative proportions of the two metals.

Critical Thinking

72. What is the relationship between the number of electrons in the valence shells in an electron configuration diagram for an atom and the number of dots in the corresponding electron dot structure?

73. Why are many elements more stable as ions than they are as atoms?

74. Describe the formation of a cation that is an exception to the octet rule. In your description, compare the electron configuration of the cation to the electron configurations of the nearest noble gases.

75. Is it accurate to describe sodium chloride (NaCl) as consisting of individual particles, each made up of one Na^+ cation and one Cl^- anion? Explain your answer.

76. How do the motions of sodium ions and chloride ions in molten sodium chloride differ from the motions of these ions in sodium chloride crystals?

77. How atoms and ions are arranged in crystals is not just dependent on size. The spheres in each atomic window below are identical in size. The windows have exactly the same area. In which window are the spheres more closely packed? Explain your reasoning.

a.

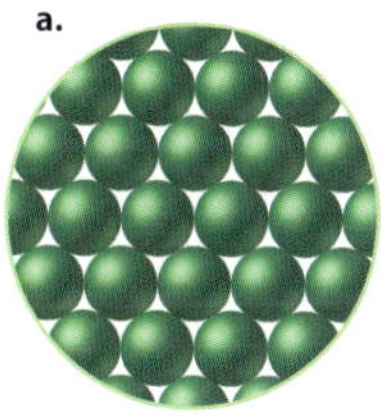

b.

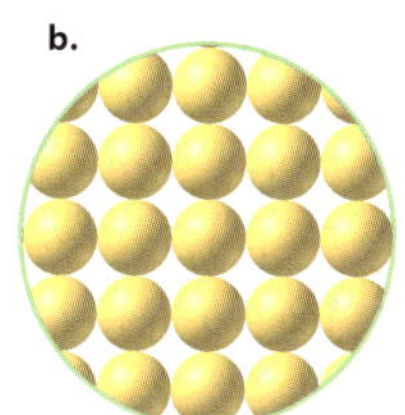

78. Two physical properties of metals are ductility and malleability. Explain these properties based on what you know about the valence electrons of metal atoms.

79. Compare and contrast the physical and chemical characteristics of metals and ionic compounds.

Concept Challenge

80. Classify each atom in the following list. Will each atom form a cation or an anion? Or is it chemically unreactive? For the atoms that do form ions during a chemical reaction, write the number of electrons the atom will gain or lose.

a. lithium
b. sodium
c. neon
d. chlorine
e. magnesium

81. The chemically similar alkali metal chlorides NaCl and CsCl have different crystal structures, whereas the chemically different NaCl and MnS have the same crystal structures. Why?

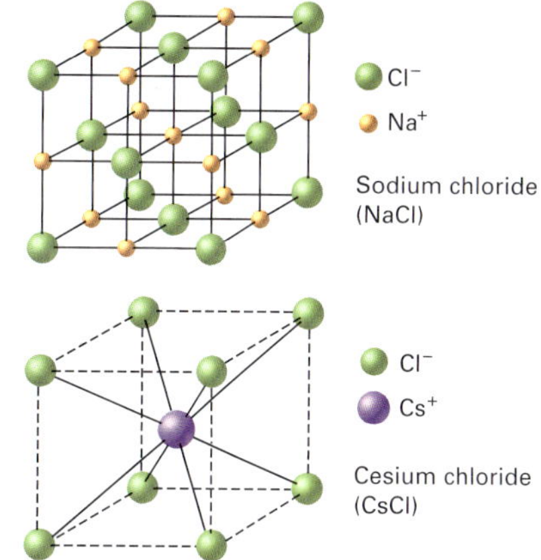

82. Silver crystallizes in a face-centered cubic arrangement. A silver atom is at the edge of each lattice point. The length of the edge of the unit cell is 0.4086 nm. What is the atomic radius of silver?

83. List the elements that are used to make each alloy.

a. brass
b. sterling silver
c. bronze
d. stainless steel
e. surgical steel
f. spring steel

79. Both metals and ionic compounds are composed of ions. Both are held together by electrostatic bonds. Metals always conduct electricity, and ionic compounds conduct only when melted or in water solution. Ionic compounds are composed of cations and anions, but metals are composed of cations and free-floating valence electrons. Metals are ductile, but ionic compounds are brittle.

Concept Challenge

80. a. cation: lose $1e^-$ **b.** cation: lose $1e^-$ **c.** unreactive **d.** anion: gain $1e^-$ **e.** cation: lose $2e^-$

81. Na^+ and Cs^+ differ greatly in size. Na^+ and Cl^- are similar in size to Mn^{2+} and S^{2-}.

82. 0.1445 nm

83. a. Cu, Zn **b.** Ag, Cu **c.** Cu, Sn **d.** Fe, Cr, Ni, C **e.** Fe, Cr, Ni, Mo **f.** Fe, Cr, C

Critical Thinking

72. Each dot in the electron dot structure represents a valence electron in the electron configuration diagram.

73. By gaining or losing electrons the atoms of elements achieve a noble-gas electron configuration.

74. An atom of silver has the electron configuration $1s^22s^22p^63s^23p^63d^{10}4s^24p^64d^{10}5s^1$. To achieve the configuration of the preceding noble gas, krypton, a silver atom would have to lose eleven electrons and form Ag^{11+}. To achieve the configuration of the following noble gas xenon, a silver atom must gain seven electrons and form Ag^{7-}. Because ions with such high charges are unlikely, silver does not achieve a noble-gas configuration. However, if a silver atom loses its $5s^1$ electron, the result is an outer electron configuration of eighteen electrons, written as $4s^24p^64d^{10}$. This configuration is favored and stable. It is known as a pseudo-noble-gas configuration.

75. No, sodium chloride is composed of equal numbers of sodium ions and chloride ions; the ions are in a 1:1 ratio. Each sodium ion is surrounded by chloride ions, and each chloride is surrounded by sodium ions.

76. In sodium chloride crystals the sodium and chloride ions vibrate about fixed points; in the molten state, the ions are free to move.

77. The spheres are more closely packed in (a); there is less empty space in (a), and a rough count shows 25 spheres in (a) compared with 22 spheres in (b).

78. Metals are ductile (can be drawn into wires) and malleable (can be hammered into shapes). These changes are possible because a metal consists of metal cations in a sea of valence electrons. When subjected to pressure, the cations easily slide past one another.

Cumulative Review

84. Organic chemistry is the study of chemicals containing carbon; inorganic chemistry is the study of chemicals that do not contain carbon.
85. an analytical chemist
86. use insulation
87. a, b, and d are chemical changes; c is a physical change
88. b and e; c is not a mixture, it is a pure substance
89. a. liquid, vapor
b. vapor
c. liquid, vapor
d. liquid, vapor
90. a. 56.6 g
b. 0.0048 m
c. 1.81 L
d. 4.0×10^3 mg
91. b
92. –269°C
93. 27.0 cm^3
94. a. $^{64}_{30}Zn$ **b.** $^{37}_{17}Cl$ **c.** $^{3}_{1}H$
d. $^{40}_{20}Ca$
95. 14 amu
96. Each of the isotopes has 8 protons and 8 electrons; oxygen-16 also has 8 neutrons, oxygen-17 has 9 neutrons, and oxygen-18 has 10 neutrons.
97. a. 1
b. 3
c. 1
d. 5
98. a. N, $1s^22s^22p^3$
b. Be, $1s^22s^2$
c. P, $1s^22s^22p^63s^23p^2$
d. K, $1s^22s^22p^63s^23p^64s^1$
99. chlorine, Cl, $1s^22s^22p^63s^23p^5$
100. a. 5×10^{-7} m
b. the visible region, green
101. a. K, $1s^22s^22p^63s^23p^64s^1$
b. Al, $1s^22s^22p^63s^23p^1$
c. S, $1s^22s^22p^63s^23p^4$
d. Ba, $1s^22s^22p^63s^23p^63d^{10}4s^24p^64d^{10}5s^25p^66s^2$
102. the electron
103. sodium (Na), cesium (Cs), rubidium (Rb), lithium (Li)

CHAPTER 7 Assessment continued

Cumulative Review

84. How is organic chemistry distinguished from inorganic chemistry? *(Chapter 1)*

85. What is the name given to a chemist who studies the composition of matter? *(Chapter 1)*

86. Explain an easy way to conserve energy. *(Chapter 1)*

87. Classify the following as chemical or physical changes. *(Chapter 2)*
a. Cookies are baked.
b. A firefly emits light.
c. A figure is carved from wood.
d. Caramel is made from sugar.

88. Which of the following is not a homogeneous mixture? *(Chapter 2)*
a. gold ring
b. spaghetti sauce
c. cane sugar
d. window glass
e. river water
f. bottled water

89. What physical state(s) can each of the following substances become as you raise its temperature? *(Chapter 2)*
a. silver **b.** gasoline
c. ice **d.** wax

90. Round each measurement to the number of significant figures indicated in parentheses. *(Chapter 3)*
a. 56.55 g (3) **b.** 0.004 849 m (2)
c. 1.8072 L (3) **d.** 4.007×10^3 mg (2)

91. Which of the following linear measurements is the longest? *(Chapter 3)*
a. 6×10^4 cm **b.** 6×10^6 mm
c. 0.06 km **d.** 6×10^9 nm

92. Helium has a boiling point of 4 K. This is the lowest boiling point of any liquid. Express this temperature in degrees Celsius. *(Chapter 3)*

93. The density of silicon is 2.33 g/cm^3. What is the volume of a piece of silicon that has a mass of 62.9 g? *(Chapter 3)*

94. Express the composition of each atom in shorthand form. *(Chapter 4)*
a. zinc-64 **b.** chlorine-37
c. hydrogen-3 **d.** calcium-40

95. An atom of carbon and an atom of element Z together have a mass of 6 amu less than double the mass of an atom of oxygen. If an atom of oxygen has a mass of 16 amu and the mass of an atom of carbon is 12 amu, what is the mass of an atom of element Z? *(Chapter 4)*

96. Determine the number of protons, electrons, and neutrons in each of the three isotopes of oxygen. *(Chapter 4)*

97. How many orbitals are in the following sublevels? *(Chapter 5)*
a. 4*s* sublevel **b.** 2*p* sublevel
c. 3*s* sublevel **d.** 4*d* sublevel

98. Give the symbol for each element and write the electron configuration for each atom. *(Chapter 5)*
a. nitrogen **b.** beryllium
c. phosphorus **d.** potassium

99. An atom of an element has 17 electrons. Give the name and symbol of the element and write the complete electron configuration. *(Chapter 5)*

100. A beam of electromagnetic radiation has a wavelength of 500 nm. *(Chapter 5)*
a. What is this wavelength in meters?
b. In what region of the spectrum is this?

101. Give the symbol of the element and the complete electron configuration of the element found at each location in the periodic table. *(Chapter 6)*
a. Group 1A, period 4
b. Group 3A, period 3
c. Group 6A, period 3
d. Group 2A, period 6

102. Which subatomic particle plays the most important role in chemistry? *(Chapter 6)*

103. Give the name and symbol of two elements that have properties similar to those of potassium. *(Chapter 6)*

210 Chapter 7

Standardized Test Prep

Test-Taking Tip

Read through all answers before making your choice Even if you find an answer that looks correct, continue reading until you have looked at *every* answer. There may be more than once correct response, or one may be better than another. Also, "all of the above" may be a possible answer. If you stop reading too soon as you find an answer that is correct, you won't notice this option.

Select the choice that best answers each question or completes each statement.

1. Which of these is not an ionic compound?
   - **a.** KF
   - **b.** Na_2SO_4
   - **c.** SiO_2
   - **d.** Na_2O
2. Which statements are correct when barium and oxygen react to form an ionic compound?
   - I. Barium atoms lose 2 electrons and form a cation.
   - II. Oxygen atoms form oxide anions (O^{2-}).
   - III. In the compound the ions are present in a one-to-one ratio.
   - **a.** I and II only
   - **b.** II and III only
   - **c.** I and III only
   - **d.** I, II, and III
3. How many valence electrons does arsenic have?
   - **a.** 5
   - **b.** 4
   - **c.** 3
   - **d.** 2
4. For which compound name is the incorrect formula given?
   - **a.** magnesium iodide, MgI_2
   - **b.** potassium selenide, K_2Se
   - **c.** calcium oxide, Ca_2O_2
   - **d.** aluminum sulfide, Al_2S_3
5. Which electron configuration represents a nitride ion?
   - **a.** $1s^22s^23s^24s^2$
   - **b.** $1s^22s^22p^3$
   - **c.** $1s^22s^22p^6$
   - **d.** $1s^2$
6. When a bromine atom gains an electron
   - **a.** a bromide ion is formed.
   - **b.** the ion formed has a 1^- charge.
   - **c.** the ion formed is an anion.
   - **d.** all the above are correct.

The lettered choices below refer to Questions 7–10. A lettered choice may be used once, more than once, or not at all.

(A) gains two electrons
(B) loses two electrons
(C) gains three electrons
(D) loses one electron
(E) gains one electron

Which choice describes what happens as each of the following elements forms its ion?

7. iodine
8. magnesium
9. cesium
10. phosphorus

Use the description and the graph to answer Questions 11–13.

Lattice energy is the energy required to change one mole of a crystalline, ionic solid to gaseous ions. The graph below shows the lattice energy for ionic compounds formed between selected alkali metals and halogens.

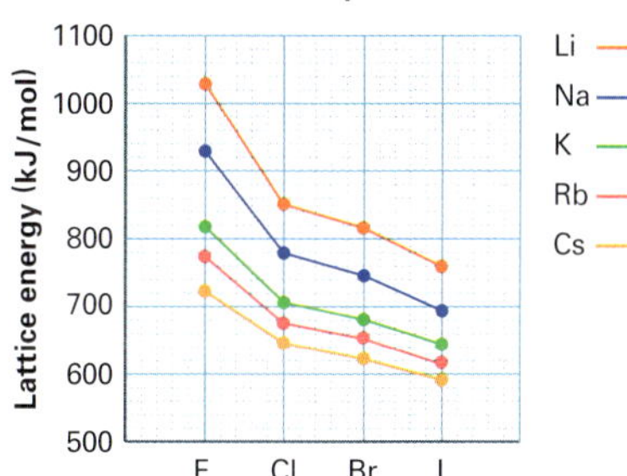

11. For a given alkali metal, what is the trend in lattice energy as the atomic radius of the halogens increases?
12. For a given halogen, what is the trend in lattice energy as the atomic radius of the alkali metals increases?
13. Complete this sentence. "As the atomic radius of either the halogen or the alkali metal increases, the lattice energy _____."

Standardized Test Prep

1. c
2. d
3. a
4. c
5. c
6. d
7. (E)
8. (B)
9. (D)
10. (C)
11. Because the atomic radius increases moving down the halogen group, the lattice energy decreases.
12. Lattice energy decreases as the atomic radius of the alkali metal increases.
13. decreases

CHAPTER 8

Covalent Bonding Planning Guide

SECTION OBJECTIVES	STANDARDS NATIONAL	STANDARDS STATE	ACTIVITIES and LABS
8.1 Molecular Compounds, pp. 213–216 1 block or 2 periods **8.1.1 Distinguish** between the melting points and boiling points of molecular compounds and ionic compounds. **8.1.2 Describe** the information a molecular formula provides.	B-2		**SE** Inquiry Activity: Shapes of Molecules, p. 212 L2 **TE** Teacher Demo: Molecular Structures and Formulas, p. 215 L1
8.2 The Nature of Covalent Bonding, pp. 217–229 3 blocks or 6 periods **8.2.1 Describe** how electrons are shared to form covalent bonds and identify exceptions to the octet rule. **8.2.2 Demonstrate** how electron dot structures represent shared electrons. **8.2.3 Describe** how atoms form double or triple covalent bonds. **8.2.4 Distinguish** between a covalent bond and a coordinate covalent bond and describe how the strength of a covalent bond is related to its bond dissociation energy. **8.2.5 Describe** how oxygen atoms are bonded in ozone.	A-1, A-2, B-2, G-2		**SE** Quick Lab: Strengths of Covalent Bonds, p. 226 L2 **TE** Class Activity: Representing Molecules, p. 218 L2 **TE** Class Activity: Bonding for Second Row Elements, p. 220 L2 **TE** Teacher Demo: Bond Energies, p. 225 L2 **TE** Teacher Demo: A Resonance Hybrid, p. 228 L2 **PLM** Probeware Lab: Strengths of Covalent Bonds L2
8.3 Bonding Theories, pp. 230–236 1 block or 2 periods **8.3.1 Describe** the relationship between atomic and molecular orbitals. **8.3.2 Describe** how VSEPR theory helps predict the shapes of molecules. **8.3.3 Identify** ways in which orbital hybridization is useful in describing molecules.	B-2, E-2		**TE** Class Activity: Making Molecular Models, p. 232 L2 **LM** Lab 11: Molecular Models L2 **SSLM** Lab 11: Paper Chromatography L2 **LP** Lab Practical 8-1: Model Building L2 **LP** Lab Practical 8-2: Empirical Formula of a Hydrated Compound L2
8.4 Polar Bonds and Molecules, pp. 237–245 2 blocks or 4 periods **8.4.1 Describe** how electronegativity values determine the distribution of charge in a polar molecule. **8.4.2 Describe** what happens to polar molecules when they are placed between oppositely charged metal plates. **8.4.3 Evaluate** the strength of intermolecular attractions compared with the strength of ionic and covalent bonds. **8.4.4 Identify** the reason why network solids have high melting points.	A-1, B-2, E-2		**SE** Small-Scale Lab: Paper Chromatography of Food Dyes, p. 245 L2 **TE** Class Activity: A Magnetic Analogy, p. 238 L1 **TE** Teacher Demo: Observing Evidence of Polarity, p. 240 L2 **TE** Teacher Demo: Evidence of Hydrogen Bonding, p. 243 L2

Ability Levels

- **L1** For students who need additional help
- **L2** For all students
- **L3** For students who need to be challenged

Components

SE	Student Edition	LM	Laboratory Manual	GCP	Graphing Calculator Problems	T	Transparencies
TE	Teacher's Edition	SSLM	Small-Scale Chemistry Lab Manual	TTT	Test-Taking Tips with Transparencies, Grades 9–12 Science	CA	Chem Alive and Chemistry Field Trips DVD
GRSW	Guided Reading and Study Workbook	LP	Laboratory Practicals	STP	Standardized Test Preparation Workbook	iT	Interactive Textbook with ChemASAP
CTR	Core Teaching Resources	PLM	Probeware Lab Manual			GO	Go Online Internet Resources
SSG	Spanish Study Guide	VCL	Virtual Chemistry Labs				
		CTB	Computer Test Bank				

RESOURCES PRINT and TECHNOLOGY	PROBLEMS and ASSESSMENT
GRSW Chapter 8, Section 1 L1 **T** T85–T86 L2 **GO** Chapter 8, Section 1 L2	**SE** Section 8.1 Assessment, p. 216 **iT** Section 8.1
GRSW Chapter 8, Section 2 L1 **T** T87–T89 L2 **GO** Chapter 8, Section 2 L2	**CP** 8.1, Drawing an Electron Dot Structure **CP** 8.2, Drawing the Electron Dot Structure for a Polyatomic Ion **SE** Section 8.2 Assessment, p. 229 **iT** Section 8.2
GRSW Chapter 8, Section 3 L1 **T** T90–T92 L2 **GO** Chapter 8, Section 3 L2	**SE** Section 8.3 Assessment, p. 236 **iT** Section 8.3
GRSW Chapter 8, Section 4 L1 **T** T93 L2 **GO** Chapter 8, Section 4 L2	**CP** 8.3, Identifying Bond Type **SE** Section 8.4 Assessment, p. 244 **iT** Section 8.4

Web Code: cdn-1081
Web Code: cdn-1082
Web Code: cdn-1083
Web Code: cdn-1084

MATERIALS List

Quantities for each group

STUDENT EDITION

Inquiry Activity, p. 212
nine spherical balloons, several short pieces of string.

Quick Lab, p. 226
170-g (6 oz) can of food, 2 454-g (16 oz) cans of food, 3 No. 25 rubber bands, metric ruler, coat hanger, plastic grocery bag, paper clip, graph paper, motion detector (optional)

Small-Scale Lab, p. 245
pencil, ruler, scissors, toothpicks, 4 different colors of food coloring, plastic cup, 0.1% NaCl solution, chromatography paper, various colored candies, markers, and powdered drinks, 70% 2-propanol, .8M CH_3COOH

TEACHER'S EDITION

Teacher Demo, p. 215
molecular model kit or foam balls and toothpicks

Teacher Demo, p. 225
10- to 15-cm piece of clean magnesium ribbon, a small piece of charcoal, tongs, Bunsen burner, filters for viewing

Teacher Demo, p. 228
small piece of copper metal, concentrated nitric acid, evaporating dish, fume hood

Class Activity, p. 232
ball-and-stick molecular model kits or foam balls and toothpicks

Class Activity, p. 238
magnets, iron filings, cardboard

Teacher Demo, p. 240
2 burets, 2 beakers, ring stand, buret clamps, turpentine, water, rubber rod, glass rod, fur cloth

Teacher Demo, p. 243
four 300–mL stoppered flasks, each containing 100 mL of one of these liquids: glycerol, ethylene glycol, water, ethanol.

Chapter Assessment

CHAPTER ASSESSMENT

SE	Chapter Assessment pp. 247-250
CTB	Chapter 8
iT	Chapter 8

STANDARDIZED TEST PREP

SE	Chapter 8, p. 251
STP	Topic 1

iT Interactive Textbook with Assessment at PHSchool.com

CHAPTER 8

Covalent Bonding

Shapes of Molecules

L2

Purpose

Students learn about geometric shapes that characterize molecules.

Prep Time 5 minutes
Class Time 20 minutes

Think About It

1. Students may use terms such as linear, triangular, or pyramidal.
2. Two balloons produce a linear arrangement, three balloons produces a trigonal planar arrangement, and four balloons produce a tetrahedral arrangement.

CHAPTER 8

Covalent Bonding

Water droplets result from attractions between water molecules.

INQUIRY Activity

Shapes of Molecules

Materials

nine spherical balloons, several short pieces of string

Procedure

1. Inflate the nine balloons to approximately the same size and tie them off.
2. Use short pieces of string to tie the ends of two balloons together to form a cluster. Tie the ends of three other balloons together to form a second cluster, and then tie the remaining four balloons to form a third cluster.
3. Hold the four balloons from the tied ends and observe how they tend to arrange themselves.
4. Sketch the three-dimensional shape of each balloon cluster.

Think About It

1. What are some geometric terms that you can use to describe the shapes?
2. Some of the terms used by chemists to describe these shapes are trigonal planar, tetrahedral, and linear. Match each term with one of the shapes you made.

8.1 Molecular Compounds

Connecting to Your World

These toy models are made from circular pieces joined together in units by sticks. Although the types of pieces are limited, there can be many different models depending on how many pieces are used and how they are arranged. As with the circular pieces, there are a limited number of different types of atoms. But, atoms can also be arranged in different ways to make a variety of products. In this section, you will learn how atoms can share electrons to form a bond, called a covalent bond, and how the atoms join to form units called molecules.

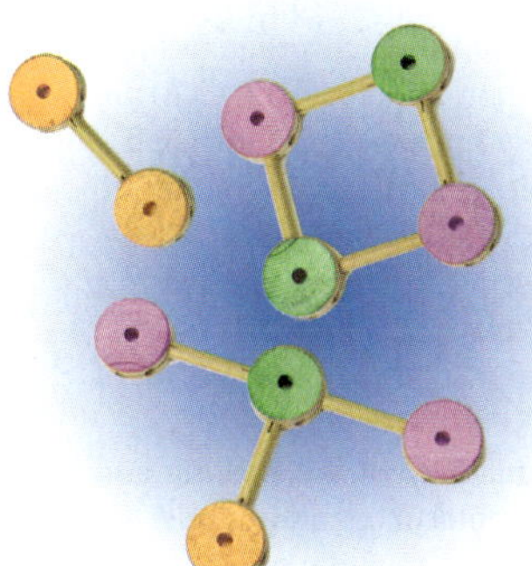

Guide for Reading

Key Concepts

- How are the melting points and boiling points of molecular compounds different from those of ionic compounds?
- What information does a molecular formula provide?

Vocabulary

covalent bond
molecule
diatomic molecule
molecular compound
molecular formula

Reading Strategy

Building Vocabulary As you read the section, write a definition of each vocabulary term in your own words.

Molecules and Molecular Compounds

In nature, matter takes many forms. The noble gas elements, such as helium and neon shown in Figure 8.1, exist as atoms. They are monatomic; that is, they consist of single atoms. You learned in the previous chapter that atoms of some elements combine to form ionic compounds. These compounds are crystalline solids with high melting points.

Other compounds, however, can have very different properties. Hydrogen chloride (HCl), for example, is a gas at room temperature. Water (H_2O) is a liquid at room temperature. These two compounds are so different from ionic compounds that you might correctly suspect that attractions between ions fail to explain their bonding. Such compounds are not ionic. Their combining atoms do not give up electrons or accept electrons.

Instead, a "tug of war" for the electrons takes place between the atoms, bonding the atoms together. The atoms held together by sharing electrons are joined by a **covalent bond.**

Figure 8.1 The noble gases, including helium and neon, are monatomic. That means they exist as single atoms. Helium, being less dense than air, is often used to inflate balloons. The colors produced in what we commonly call neon lights are a result of passing an electric current through one or more noble gases.

Section Resources

Print

- **Guided Reading and Study Workbook,** Section 8.1
- **Core Teaching Resources,** Section 8.1 Review
- **Transparencies,** T85–T86

Technology

- **Interactive Textbook with ChemASAP,** Assessment 8.1
- **Go Online,** Section 8.1

8.1

1 FOCUS

Objectives

8.1.1 Distinguish between the melting points and boiling points of molecular compounds and ionic compounds.

8.1.2 Describe the information provided by a molecular formula.

Guide for Reading

Build Vocabulary

Word Parts The word *covalent* is a combination of the prefix *co-* from the Latin *com* meaning "with" or "together" and the verb *valere* meaning "to be strong." Two electrons together have the strength to hold two atoms together in a bond. Ask students what other words they know beginning with co-. *(e. g. cooperate, coexist, coagulate)*

Reading Strategy

Relate Text and Visuals Encourage students to examine the illustrations carefully as they read and compare them with representations of ionic bonding from Chapter 7. Have them note the differences.

2 INSTRUCT

Connecting to Your World

Ask, **Could you take these models apart and make different models?** *(yes)* **How might the models you make be different from these?** *(The new models could be different shapes and sizes.)*

Molecules and Molecular Compounds

Use Visuals

Figure 8.1 Point out that some other elements are most stable when two atoms are joined together. Ask, **Do you think H_2 could be an ionic compound?** *(No, ionic bonds do not form between two atoms of the same element.)* Tell students that H_2 is a molecule held together not by the transfer of an electron but by the sharing of a pair of electrons.

Use Visuals 

Figure 8.3 Ask students, **How do the submicroscopic illustrations of NaCl and H_2O differ?** *(The ions of sodium and chlorine are separate and arrayed in a regular pattern; two hydrogen atoms and one oxygen atom are bonded together in one molecule.)* **Compare the formulas for NaCl and H_2O.** *(They look similar, but NaCl represents just one unit in the array of sodium and chloride ions, whereas H_2O represents one separate molecule of water.)*

Discuss 

Use the hydrogen molecule to introduce the shared nature of covalent bonding. Remind students that atoms bond to reach a more stable state, one in which the orbitals related to the highest energy levels of atoms are filled, as in the highly unreactive noble gases. A transfer of electrons between hydrogen atoms would not work; what factor could determine which hydrogen atom donated and which received an electron? However, if the hydrogen atoms share their electrons, they can each achieve the stable arrangement of a helium atom. Show the various ways that the bonding in a hydrogen molecule can be represented: molecular formula, structural formula, electron-dot structure, and orbital diagram showing the overlap of the 1*s* orbitals. Then introduce the octet rule and repeat the exercise for a chlorine molecule.

Download a worksheet on **Covalent Bonds** for students to complete, and find additional teacher support from NSTA SciLinks.

Figure 8.2 Water is a molecular compound with molecules composed of two hydrogen atoms and one oxygen atom. Carbon monoxide is a molecular compound with molecules composed of one carbon atom and one oxygen atom.

Some elements found in nature are in the form of molecules. A **molecule** is a neutral group of atoms joined together by covalent bonds. For example, air contains oxygen molecules. Each oxygen molecule consists of two oxygen atoms joined by covalent bonds. A **diatomic molecule** is a molecule consisting of two atoms. An oxygen molecule is a diatomic molecule.

Atoms of different elements can combine chemically to form compounds. In many compounds, atoms are bonded to each other to form molecules. Examples include water and carbon monoxide, which are described in Figure 8.2. A compound composed of molecules is called a **molecular compound.** The molecules of a given molecular compound are all the same. Remember, there is no such thing as a molecule of sodium chloride or magnesium chloride. Instead, these ionic compounds exist as collections of positively and negatively charged ions arranged in repeating three-dimensional patterns. Recall from Chapter 7 that the composition of ionic compounds is expressed as formula units.

Go Online NSTA SciLINKS
For: Links on Covalent Bonds
Web Code: cdn-1081

Molecular compounds tend to have relatively lower melting and boiling points than ionic compounds. Many molecular compounds are gases or liquids at room temperature. In contrast to ionic compounds, which are formed from a metal combined with a nonmetal, most molecular compounds are composed of atoms of two or more nonmetals. For example, one atom of carbon can combine with one atom of oxygen to produce one molecule of a compound known as carbon monoxide. Carbon monoxide is a poisonous gas produced by burning gasoline in internal combustion engines. Figure 8.3 illustrates some differences between ionic and molecular compounds, using sodium chloride and water as examples.

Ionic compound – Table Salt

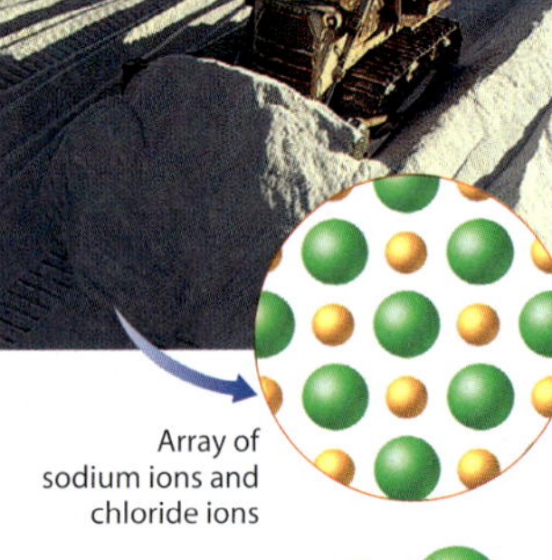

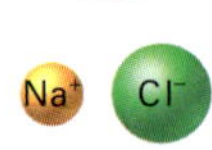

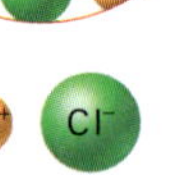

Molecular compound – Water

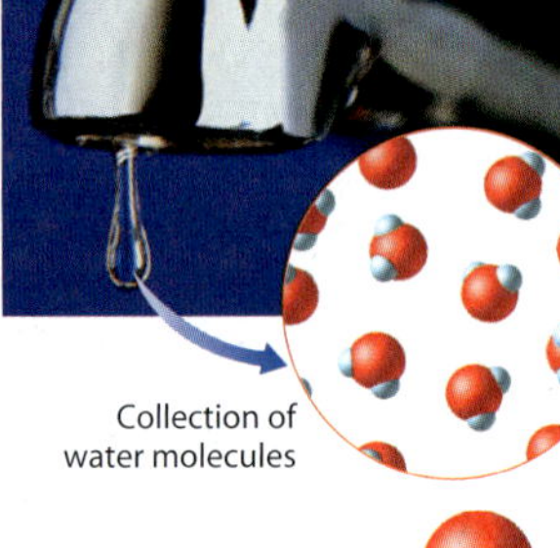

Figure 8.3 Sodium chloride, which is an ionic compound, and water, which is a molecular compound, are compared here. **Interpreting Diagrams** ***How do molecular compounds differ from ionic compounds?***

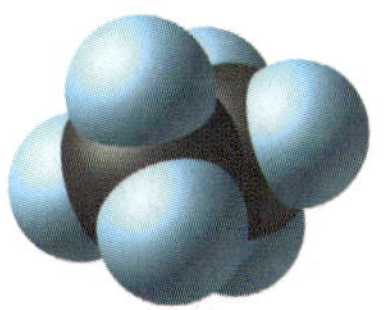

Ethane (C_2H_6)

Figure 8.4 Ethane is a component of natural gas. **Inferring** ***What information about the ethane molecule given by its molecular formula C_2H_6 is also given by the drawing shown here?***

Molecular Formulas

A **molecular formula** is the chemical formula of a molecular compound. **A molecular formula shows how many atoms of each element a molecule contains.** A water molecule consists of two hydrogen atoms and one oxygen atom. The molecular formula of water is H_2O. Notice that a subscript written after the symbol indicates the number of atoms of each element in the molecule. If there is only one atom, the subscript 1 is omitted. The molecular formula of carbon dioxide is CO_2. This formula represents a molecule containing one carbon atom and two oxygen atoms. As shown in Figure 8.4, ethane, a component of natural gas, is also a molecular compound. The molecular formula for ethane is C_2H_6. According to this formula, one molecule of ethane contains two carbon atoms and six hydrogen atoms. A molecular formula reflects the actual number of atoms in each molecule. The subscripts are not necessarily lowest whole-number ratios.

Molecular formulas also describe molecules consisting of one element. Because the oxygen molecule consists of two oxygen atoms bonded together, its molecular formula is O_2.

A molecular formula does not tell you about a molecule's structure. In other words, it does not show either the arrangement of the various atoms in space or which atoms are covalently bonded to one another. A variety of diagrams and molecular models, some of them illustrated in Figure 8.5, can be used to show the arrangement of atoms in a molecule. Diagrams and models like these will be used throughout the textbook.

 ***What is the molecular formula for ethane?***

Ammonia

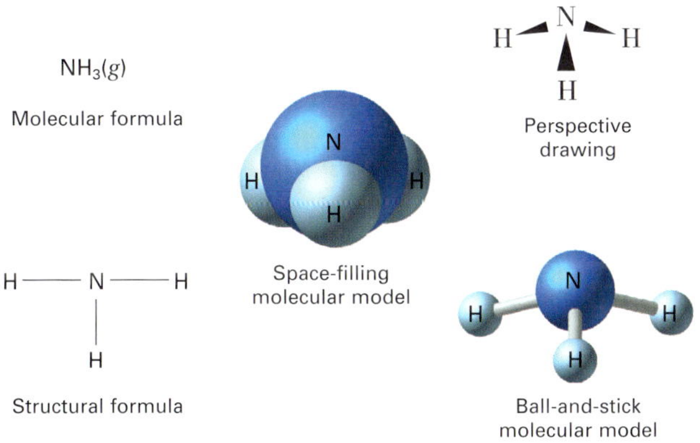

Figure 8.5 Ammonia (NH_3) is used in solution as a cleaning agent. You can represent the ammonia molecule by its molecular formula, its structural formula, a space-filling molecular model, a perspective drawing, or by a ball-and stick molecular model.

Molecular Formulas

TEACHER Demo

Molecular Structures and Formulas 

Purpose Students see the structure associated with molecular formulas.

Materials molecular model kit or foam balls and toothpicks

Procedure Construct models of these pairs of molecules: O_2 and O_3; CO and CO_2; H_2O and H_2O_2. Show students each pair and ask them how many of each type of atom is in each molecule. Write the corresponding chemical formulas on the chalkboard.

Expected Outcome Students will see that the addition of one more oxygen atom to O_2, CO, and H_2O results in new molecules that have very different structures.

Discuss 

Point out that the recipe for a cake provides information about what type and how much of each ingredient is needed. Varying the types and amounts of ingredients leads to cakes that have different tastes and textures. Ask, **How does a molecular formula provide similar information about a compound?** *(The subscripts in a molecular formula tell the number of atoms of each element in a compound.)* Explain that the subscripts are not always in the smallest whole-number ratio of the atoms. For example, ethane is correctly represented by the formula C_2H_6 because the molecule actually contains two carbon atoms and six hydrogen atoms. It would be incorrect to represent ethane as CH_3. Ask, **How many atoms of nitrogen are in N_2H_4?** *(two)* **How many atoms of hydrogen?** *(four)*

Differentiated Instruction

English Learners 

Prepare large, three-dimensional models of ammonia similar to the two in Figure 8.5. Also, write the molecular and structural formulas for ammonia on large cards. Hold the models and cards up one by one and point out the information that each model or formula provides. Make a summary list on the chalkboard.

Answers to...

Figure 8.3 Molecular compounds are made up of molecules and usually have lower melting and boiling points than ionic compounds, which are made up of ions.

Figure 8.4 It contains two carbon atoms and six hydrogen atoms.

 C_2H_6

3 ASSESS

Evaluate Understanding L2

Have students label the following statements true or false:

1. All molecular compounds are composed of atoms of two or more different elements. *(True)*
2. All compounds contain molecules. *(False)*
3. No elements exist as molecules. *(False)*
4. Most molecular compounds are composed of two or more nonmetallic elements. *(True)*
5. Atoms in molecular compounds share electrons. *(True)*
6. Molecular compounds tend to have higher melting and boiling points than ionic compounds. *(False)*

Reteach 

Write a series of formulas on the chalkboard: Br_2, KBr, CH_4, SO_3, N_2H_2, Ne_2. Ask students to identify any substance that does not exist. *(Ne_2)* Ask them to identify any ionic compounds. *(KBr)* Ask, **Which formulas represent molecules?** *(Br_2, CH_4, SO_3, N_2H_2)* Ask, **Which formulas represent compounds?** *(KBr, CH_4, SO_3, N_2H_2)* Ask, **How many atoms are in SO_3?** *(4)*

Writing Activity

Students should write that a molecular compound is a compound composed of molecules. A molecular formula shows how many atoms of each element a molecule contains.

If your class subscribes to the Interactive Textbook, use it to review concepts in Section 8.1.

with ChemASAP

Answers to...

Figure 8.6 carbon dioxide

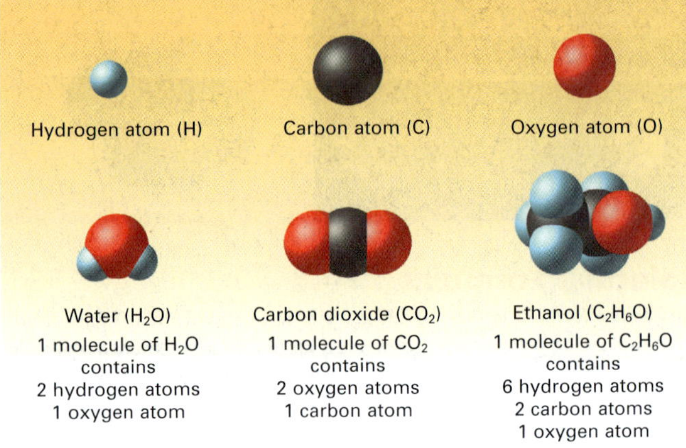

Figure 8.6 The formula of a molecular compound indicates the numbers and kinds of atoms. The arrangement of the atoms within a molecule is called its molecular structure. **Using Models** ***Which of these molecules has the greatest number of oxygen atoms?***

Figure 8.6 shows the chemical formulas and structures of some other molecular compounds. Carbon dioxide, for example, is a gas produced by the complete burning of carbon. It is found in Earth's atmosphere and dissolved in seawater. It is also used to carbonate many beverages. The molecular structure of carbon dioxide in Figure 8.6 shows how the carbon atom in each molecule has two oxygen atoms on opposite sides of it. It shows how the three atoms are arranged in a row. The molecular structure of water shows how the hydrogen atoms, in contrast, are mainly on one side of the water molecule. The molecular structure of ethanol is more complicated, but the molecular structure illustrated in Figure 8.6 also shows how many of each kind of atom are in each molecule, and how the atoms are arranged with respect to one another.

8.1 Section Assessment

1. **Key Concept** How are the melting points and boiling points of molecular compounds usually different from those of ionic compounds?
2. **Key Concept** What information does a molecular formula provide?
3. What are the only elements that exist in nature as uncombined atoms? What term is used to describe such elements?
4. Describe how the molecule whose formula is NO is different from the molecule whose formula is N_2O.
5. Give an example of a diatomic molecule found in Earth's atmosphere.
6. What information does a molecule's molecular structure give?

Writing Activity

Molecular Compounds and Formulas Describe what a molecular compound is. Explain how a molecular formula is the chemical formula of a molecular compound.

Assessment 8.1 Test yourself on the important concepts of Section 8.1.

with ChemASAP

Section 8.1 Assessment

1. Molecular compounds tend to have relatively lower melting points and boiling points.
2. Molecular formulas show how many atoms of each element one molecule of a compound contains.
3. noble gases; monatomic
4. NO has 1 N atom and 1 O atom; N_2O has 2 N atoms and 1 O atom.
5. nitrogen (N_2) or oxygen (O_2)
6. Molecular structures show the arrangement of atoms in a molecule.

8.2 The Nature of Covalent Bonding

Connecting to Your World

You know that without oxygen to breathe, you could not live. But did you know that oxygen plays another important role in your life? High in the atmosphere, a different form of oxygen, called ozone, forms a layer that filters out harmful radiation from the sun. The colors in this map indicate the concentrations of ozone in various parts of Earth's atmosphere. In this section, you will learn how oxygen atoms can join in pairs to form the oxygen you breathe and can also join in groups of three oxygen atoms to form ozone.

Guide for Reading

Key Concepts

- What is the result of electron sharing in covalent bonds?
- How do electron dot structures represent shared electrons?
- How do atoms form double or triple covalent bonds?
- How are coordinate covalent bonds different from other covalent bonds?
- How is the strength of a covalent bond related to its bond dissociation energy?
- How are oxygen atoms bonded in ozone?
- What are some exceptions to the octet rule?

Vocabulary

single covalent bond
structural formula
unshared pair
double covalent bond
triple covalent bond
coordinate covalent bond
polyatomic ion
bond dissociation energy
resonance structure

Reading Strategy

Identifying Main Idea/Details List the main idea in the paragraph following the heading *The Octet Rule in Covalent Bonding.* As you read, list examples of how this rule is followed when a single covalent bond, a double covalent bond, and a triple covalent bond form.

The Octet Rule in Covalent Bonding

Recall that when ionic compounds form, electrons tend to be transferred so that each ion acquires a noble gas configuration. A similar rule applies for covalent bonds. **In covalent bonds, electron sharing usually occurs so that atoms attain the electron configurations of noble gases.** For example, each hydrogen atom has one electron. But a pair of hydrogen atoms share these two electrons when they form a covalent bond in a hydrogen molecule. Each hydrogen atom thus attains the electron configuration of helium, a noble gas with two electrons. Combinations of atoms of the nonmetals and metalloids in Groups 4A, 5A, 6A, and 7A of the periodic table are likely to form covalent bonds. In this case the atoms usually acquire a total of eight electrons, or an octet, by sharing electrons, so that the octet rule applies.

Single Covalent Bonds

The hydrogen atoms in a hydrogen molecule are held together mainly by the attraction of the shared electrons to the positive nuclei. Two atoms held together by sharing a pair of electrons are joined by a **single covalent bond.** Hydrogen gas consists of diatomic molecules whose atoms share only one pair of electrons, forming a single covalent bond.

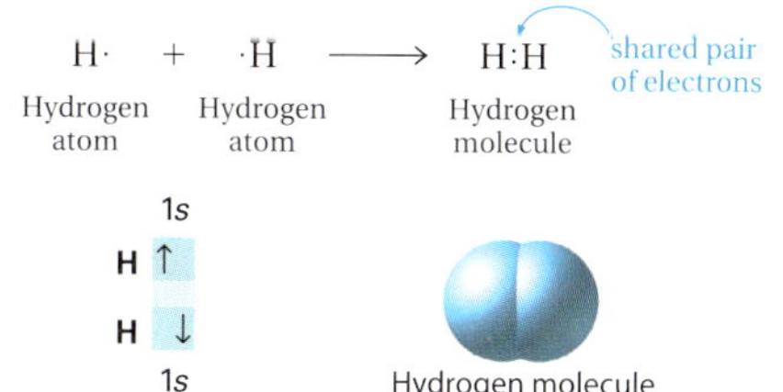

Hydrogen molecule

Section Resources

Print

- **Guided Reading and Study Workbook,** Section 8.2
- **Core Teaching Resources,** Section 8.2 Review
- **Probeware Laboratory Manual,** Quick Lab, p. 226
- **Transparencies,** T87–T89

Technology

- **Interactive Textbook with ChemASAP,** Problem-Solving 8.8, 8.10; Simulation 6, Assessment 8.2
- **Go Online,** Section 8.2

8.2

1 FOCUS

Objectives

8.2.1 Describe how electrons are shared to form covalent bonds and identify exceptions to the octet rule.

8.2.2 Demonstrate how electron dot structures represent shared electrons.

8.2.3 Describe how atoms form double or triple covalent bonds.

8.2.4 Distinguish between a covalent bond and a coordinate covalent bond and describe how the strength of a covalent bond is related to its bond dissociation energy.

8.2.5 Describe how oxygen atoms are bonded in ozone.

Guide for Reading

Build Vocabulary  L2

Word Parts The word *structure* comes from the Latin verb *struere,* which means "to build." A structural formula is one that shows how a molecule is built, that is, how the atoms are joined together by chemical bonds. Ask students to define structural steel. *(Steel is used to form the skeleton of buildings.)*

Reading Strategy  L2

Directed Reading/Thinking Activity Have students read the red headings throughout the section first and ask themselves questions about what they might learn when they read the text.

2 INSTRUCT

Connecting to Your World

Explain that scientists attribute the thinning of the ozone layer, in part, to the action of compounds called chlorofluorcarbons (CFCs), which have been released into the atmosphere. Ask, **Why is ozone important in the atmosphere?** *(It filters out radiation that could harm living beings on Earth.)*

The Octet Rule in Covalent Bonding

Discuss 

Write electron configurations for carbon, nitrogen, oxygen, fluorine, and neon on the chalkboard. Ask, **How many electrons would carbon, nitrogen, oxygen, and fluorine need to share in order to achieve the same electron configuration as neon?** *(4, 3, 2, and 1 respectively)*

Representing Molecules 

Purpose Students practice different ways to represent molecules.

Materials paper and pencil

Procedure Divide students into groups of three or four. Have them practice drawing molecular diagrams, structural formulas, electron-dot structures, and orbital diagrams for molecules such as OF_2, SCl_2, N_2H_4, CCl_4, $CHCl_3$, and C_2H_6.

Single Covalent Bonds

Download a worksheet on **Valence Electrons** for students to complete, and find additional teacher support from NSTA SciLinks.

For: Links on Valence Electrons
Visit: www.SciLinks.org
Web Code: cdn-1082

An electron dot structure such as H:H represents the shared pair of electrons of the covalent bond by two dots. The pair of shared electrons forming the covalent bond is also often represented as a dash, as in H—H for hydrogen. A **structural formula** represents the covalent bonds by dashes and shows the arrangement of covalently bonded atoms. In contrast, the molecular formula of hydrogen, H_2, indicates only the number of hydrogen atoms in each molecule.

The halogens also form single covalent bonds in their diatomic molecules. Fluorine is one example. Because a fluorine atom has seven valence electrons, it needs one more to attain the electron configuration of a noble gas. By sharing electrons and forming a single covalent bond, two fluorine atoms each achieve the electron configuration of neon.

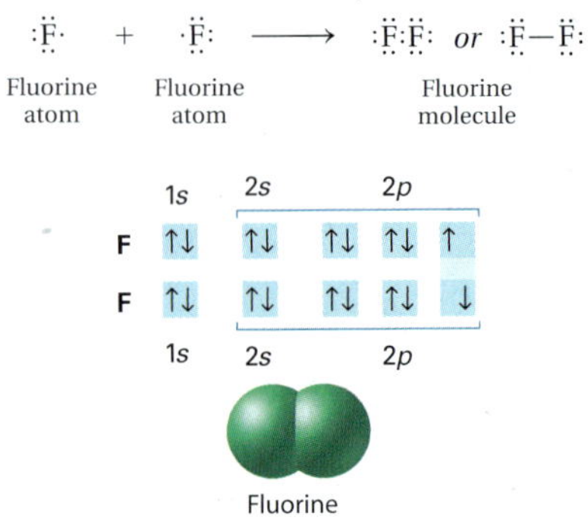

In the F_2 molecule, each fluorine atom contributes one electron to complete the octet. Notice that the two fluorine atoms share only one pair of valence electrons. A pair of valence electrons that is not shared between atoms is called an **unshared pair,** also known as a lone pair or a nonbonding pair.

You can draw electron dot structures for molecules of compounds in much the same way that you draw them for molecules of diatomic elements. Water (H_2O) is a molecule containing three atoms with two single covalent bonds. Two hydrogen atoms share electrons with one oxygen atom. The hydrogen and oxygen atoms attain noble-gas configurations by sharing electrons. As you can see in the electron dot structures below, the oxygen atom in water has two unshared pairs of valence electrons.

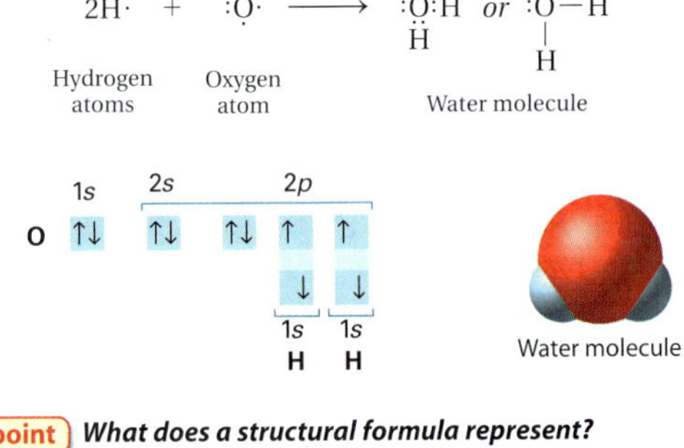

 Checkpoint ***What does a structural formula represent?***

Differentiated Instruction

Special Needs L1

Pair each student with a study partner. Have them use the periodic table and quiz each other on writing electron dot structures for single atoms and bonded atoms. Make sure they understand that the Group number for any atom, 1A to 7A, indicates the number of valence electrons that atom has, and that it is valence electrons that appear in the electron dot structures.

You can draw the electron dot structure for ammonia (NH_3), a suffocating gas, in a similar way. The ammonia molecule has one unshared pair of electrons.

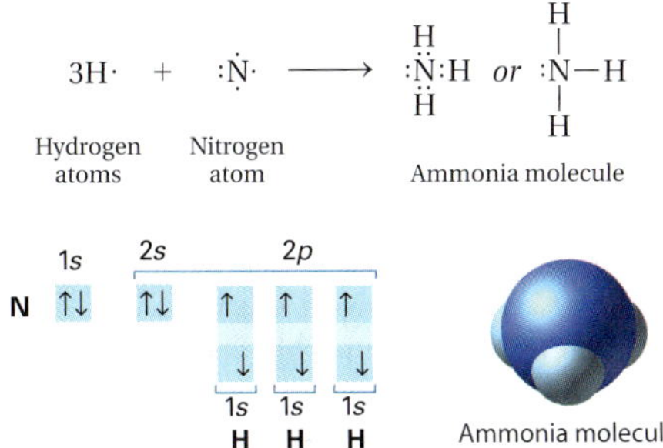

Methane (CH_4) contains four single covalent bonds. The carbon atom has four valence electrons and needs four more valence electrons to attain a noble-gas configuration. Each of the four hydrogen atoms contributes one electron to share with the carbon atom, forming four identical carbon-hydrogen bonds. As you can see in the electron dot structure below, methane has no unshared pairs of electrons.

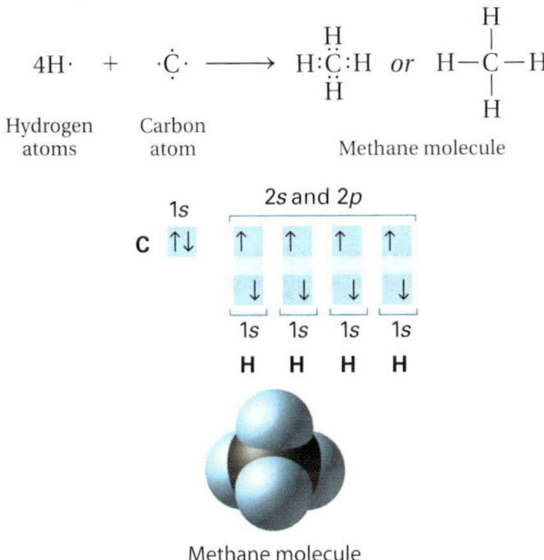

Methane molecule

When carbon forms bonds with other atoms, it usually forms four bonds. You would not predict this based on carbon's electron configuration, shown below.

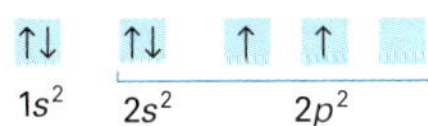

 **Checkpoint** ***What is the electron dot structure of a methane molecule?***

Discuss

L2

First review the molecular and structural formulas, electron dot structures, and orbital diagrams for fluorine, water, and ammonia molecules. If possible, display physical models. Call attention to the fact that fluorine has one half-filled orbital and forms one bond, oxygen has two half-filled orbitals and forms two bonds, and nitrogen has three and forms three bonds. Tell students carbon has two. Ask, **How many covalent bonds do you think carbon forms?** *(Students may logically say two.)* Tell students that CH_2 does not represent a stable molecule, but CH_4 (methane) is a stable molecule. Explain the concept of electron promotion, which allows carbon to form four single covalent bonds. Point out that elements in groups 3A and 4A promote *s* electrons to *p* orbitals, increasing their bonding capacity. For example, boron's electron configuration is $1s^22s^22p^1$. Based on this configuration, students might infer that boron can form only one covalent bond. However, the chloride of boron is BCl_3 rather than BCl. The promotion of one 2*s* electron to the 2*p* orbital allows for the formation of three bonds. Boron does not achieve a noble-gas configuration, but it does achieve added stability by forming three bonds rather than one.

Facts and Figures

Expanding the Octet

Nonmetals in the third row and beyond such as phosphorus, sulfur, and iodine can form more than four bonds because they have empty *d* orbitals. Phosphorus, for example, can unpair a 3*s* electron and promote it to an empty 3*d* orbital. The promotion allows phosphorus to form five bonds. The amount of energy needed to promote an electron is less than the energy released with the formation of an extra bond. Sulfur can promote one 3*s* electron and one 3*p* electron and form two extra bonds.

Answers to...

 **Checkpoint** It shows the covalent bonds as dashes and shows the arrangement of covalently bonded atoms.

CLASS Activity

Bonding for Second Row Elements

L2

Purpose Students gain understanding of covalent bonding and distinguish between covalent and ionic bonding.

Procedure Have students draw electron dot structures for each element in the second row of the periodic table: Li, Be, B, C, N, O, and F. Then have them answer the following:

- **Predict how many bonds each atom must form to attain a noble-gas configuration.** *(1, 2, 3, 4, 3, 2, 1)*
- **Can lithium form a covalent bond and reach stability?** *(no)*
- **Which elements can reach stability by forming covalent bonds?** *(C, N, O, F)*
- **Can fluorine form an ionic bond?** *(yes)*
- **Are the bonds in nitrogen molecules (N_2) ionic or covalent?** *(covalent)*

CONCEPTUAL PROBLEM 8.1

Answers

7. a. :C̤̈l:C̤̈l: **b.** :B̤̈r:B̤̈r: **c.** :Ï̤:Ï̤:

8. a. H:Ö̤:Ö̤:H **b.** :C̤̈l:P̈:C̤̈l:
:C̤̈l:

Practice Problems Plus

The following covalent molecules have only single covalent bonds. Draw an electron dot structure for each.

a. NF_3 :F̤̈:N̈:F̤̈:
:F̤̈:

b. SBr_2 :B̤̈r:S̤̈:B̤̈r:

If you tried to form covalent C—H bonds for methane by combining the two 2*p* electrons of the carbon with two 1*s* electrons of hydrogen atoms, you would incorrectly predict a molecule with the formula CH_2 (instead of CH_4). The formation of four bonds by carbon can be simply explained. One of carbon's 2*s* electrons is promoted to the vacant 2*p* orbital to form the following electron configuration.

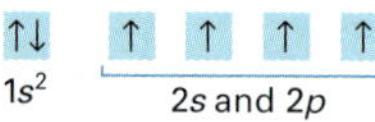

This electron promotion requires only a small amount of energy. The promotion provides four electrons of carbon that are capable of forming covalent bonds with four hydrogen atoms. Methane, the carbon compound formed by electron sharing of carbon with four hydrogen atoms, is much more stable than CH_2. The stability of the resulting methane more than compensates for the small energy cost of the electron promotion. Therefore, formation of methane (CH_4) is more energetically favored than the formation of CH_2.

CONCEPTUAL PROBLEM 8.1

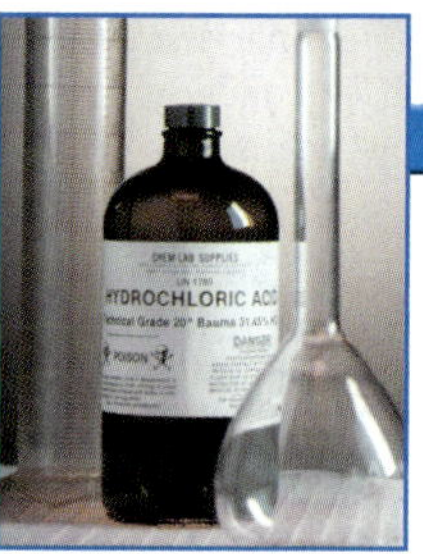

Drawing an Electron Dot Structure

Hydrochloric acid (HCl (*aq*)) is prepared by dissolving gaseous hydrogen chloride (HCl (*g*)) in water. Hydrogen chloride is a diatomic molecule with a single covalent bond. Draw the electron dot structure for HCl.

1 Analyze ***Identify the relevant concepts.***

In a single covalent bond, a hydrogen and a chlorine atom must share a pair of electrons. Each must contribute one electron to the bond. First, draw the electron dot structures for the two atoms. Then show the electron sharing in the compound they produce.

2 Solve ***Apply concepts to the situation.***

H· + ·C̤̈l: ⟶ H:C̤̈l:

Hydrogen atom | Chlorine atom | Hydrogen chloride molecule

In the electron dot structures, the hydrogen atom and the chlorine atom are each correctly shown to have an unpaired electron. Through electron sharing, the hydrogen and chlorine atoms are shown to attain the electron configurations of the noble gases helium and argon, respectively.

Practice Problems

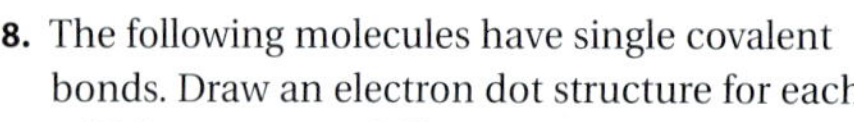

7. Draw electron dot structures for each molecule.
a. chlorine **b.** bromine **c.** iodine

8. The following molecules have single covalent bonds. Draw an electron dot structure for each.
a. H_2O_2 **b.** PCl_3

Interactive Textbook

Problem Solving 8.8 Solve Problem 8 with the help of an interactive guided tutorial.

with **ChemASAP**

Facts and Figures

Inventing Electron Dot Structures

Gilbert Newton Lewis (1875–1946) was an American chemist who invented electron dot structures, which are often called Lewis structures or diagrams in his honor. These structures supported Lewis's theory of the electron pair in chemical bonding. As a professor of physical chemistry, he expanded the theory of acids and bases by defining an acid as an electron pair acceptor and a base as an electron pair donor. The definitions encompass all Brønsted-Lowry acid-base reactions and include many others not previously categorized as acid-base reactions.

Double and Triple Covalent Bonds

Sometimes atoms bond by sharing more than one pair of electrons. **Atoms form double or triple covalent bonds if they can attain a noble gas structure by sharing two pairs or three pairs of electrons.** A bond that involves two shared pairs of electrons is a **double covalent bond.** A bond formed by sharing three pairs of electrons is a **triple covalent bond.**

You might think that an oxygen atom, with six valence electrons, would form a double bond by sharing two of its electrons with another oxygen atom.

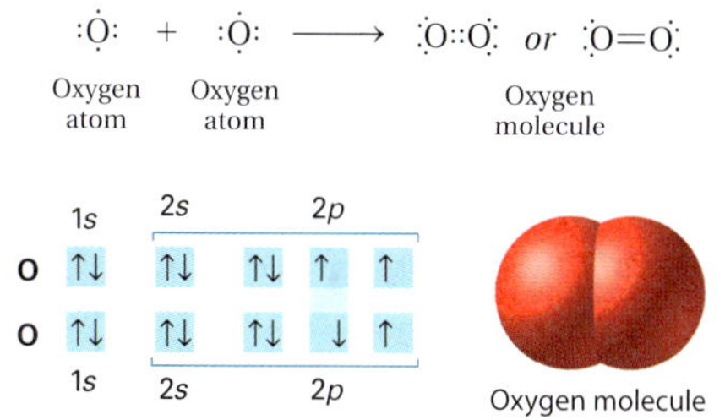

In such an arrangement, all the electrons within the molecule would be paired. Experimental evidence, however, indicates that two of the electrons in O_2 are still unpaired. Thus, the bonding in the oxygen molecule (O_2) does not obey the octet rule. You cannot draw an electron dot structure that adequately describes the bonding in the oxygen molecule.

An element whose molecules contain triple bonds is nitrogen (N_2), a major component of Earth's atmosphere illustrated in Figure 8.7. In the nitrogen molecule, each nitrogen atom has one unshared pair of electrons. A single nitrogen atom has five valence electrons. Each nitrogen atom in the nitrogen molecule must gain three electrons to have the electron configuration of neon.

Figure 8.7 Oxygen and nitrogen are the main components of Earth's atmosphere. The oxygen molecule is an exception to the octet rule. It has two unpaired electrons. Three pairs of electrons are shared in a nitrogen molecule.

Interactive Textbook

Simulation 6 Simulate the covalent bonding between molecules.

with ChemASAP

Double and Triple Covalent Bonds

Discuss L2

To introduce the discussion of multiple covalent bonds, use the electron dot structure for the nitrogen molecule. Ask students to show how the structure of diatomic nitrogen can satisfy the octet rule. *(The nitrogen atoms can share six electrons.)* Have them compare the bonding in ammonia with the bonding in nitrogen gas. Then introduce the oxygen molecule. Ask students to draw a structure that obeys the octet rule. Ask, **Why doesn't oxygen form a triple bond?** *(Each oxygen needs to share only two electrons to achieve a stable electron configuration.)* Explain that although a double bond in the oxygen molecule fulfills the octet rule, it does not fit with experimental evidence that shows that the oxygen molecule contains two unpaired electrons. Thus the structure of O_2 is an exception to the octet rule. Help students draw the electron dot structure and orbital diagram for carbon dioxide. Ask, **What type of bonds does carbon form with the two oxygen atoms in CO_2?** *(double covalent bonds)* Note that carbon can form single, double, and triple bonds, but a quadruple bond is impossible because of geometric restrictions. Have students draw diagrams for hydrogen cyanide (HCN) and formaldehyde (H_2CO). Ask, **What kind of bonds does carbon form in each of these molecules?** *(HCN: one single carbon-to-hydrogen bond and one triple carbon-to-nitrogen bond; H_2CO: two single carbon-to-hydrogen bonds and one double carbon-to-oxygen bond)* As the discussion proceeds, provide models for the molecules discussed.

Differentiated Instruction

Less Proficient Reader L1

Divide students into groups of three or four. Have them draw electron dot formulas for oxygen and fluorine, then combine the drawings into the electron dot formula for OF_2. Have them do the same for SCl_2 and CCl_4. When students have mastered these, continue the exercise with $CHCl_3$, N_2H_4, and C_2H_6.

Relate L1

Earth's atmosphere is approximately 80 percent nitrogen gas, but surprisingly few nitrogen compounds exist compared with the numerous compounds of oxygen, which constitutes only 20 percent of the atmosphere. Students may correctly surmise that the triple bond in N_2 is harder to break than a double bond and considerably harder to break than a single bond. Thus, N_2 is a stable molecule. Plant and animal life depends on nitrogen, but in order to be usable to living systems, the element must be converted to a compound, a process called nitrogen fixing. Nitrogen fixing occurs naturally when lightning provides the energy for atmospheric nitrogen to react with oxygen to form nitrogen oxides. The oxides dissolve in rain and fall to the ground where they can be utilized by plants. Nitrogen fixing bacteria in the soil are also able to convert atmospheric nitrogen to usable compounds.

Table 8.1

The Diatomic Elements

Name	Chemical formula	Electron dot structure	Properties and uses
Fluorine	F_2	$:\ddot{F}-\ddot{F}:$	Greenish-yellow reactive toxic gas. Compounds of fluorine, a halogen, are added to drinking water and toothpaste to promote healthy teeth.
Chlorine	Cl_2	$:\ddot{Cl}-\ddot{Cl}:$	Greenish-yellow reactive toxic gas. Chlorine is a halogen used in household bleaching agents.
Bromine	Br_2	$:\ddot{Br}-\ddot{Br}:$	Dense red-brown liquid with pungent odor. Compounds of bromine, a halogen, are used in the preparation of photographic emulsions.
Iodine	I_2	$:\ddot{I}-\ddot{I}:$	Dense gray-black solid that produces purple vapors; a halogen. A solution of iodine in alcohol (tincture of iodine) is used as an antiseptic.
Hydrogen	H_2	H—H	Colorless, odorless, tasteless gas. Hydrogen is the lightest known element.
Nitrogen	N_2	$:N\equiv N:$	Colorless, odorless, tasteless gas. Air is almost 80% nitrogen by volume.
Oxygen	O_2	Inadequate	Colorless, odorless, tasteless gas that is vital for life. Air is about 20% oxygen by volume.

Figure 8.8 Carbon dioxide gas is soluble in water and is used to carbonate many beverages. A carbon dioxide molecule has two carbon–oxygen double bonds.

Up to this point in your textbook, the examples of single and triple covalent bonds have involved diatomic molecules. Table 8.1 lists the properties and uses of the elements that exist as diatomic molecules. Single, double, and triple covalent bonds can also exist between unlike atoms. For example, consider carbon dioxide (CO_2), which is present in the atmosphere and is used to carbonate many soft drinks as shown in Figure 8.8. The carbon dioxide molecule contains two oxygens, each of which shares two electrons with carbon to form a total of two carbon–oxygen double bonds.

$$:\ddot{O}\cdot + \cdot\dot{C}\cdot + \cdot\ddot{O}: \longrightarrow :\ddot{O}::C::\ddot{O}: \text{ or } :\ddot{O}=C=\ddot{O}:$$

Oxygen atom + Carbon atom + Oxygen atom → Carbon dioxide molecule

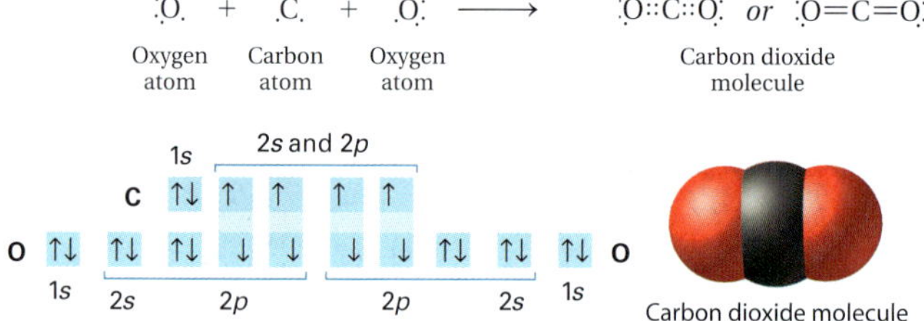

The two double bonds in the carbon dioxide molecule are identical to each other. Carbon dioxide is an example of a triatomic molecule, which is a molecule consisting of three atoms.

222 Chapter 8

Coordinate Covalent Bonds

Carbon monoxide (CO) is an example of a type of covalent bonding different from that seen in water, ammonia, methane, and carbon dioxide. A carbon atom needs to gain four electrons to attain the electron configuration of neon. An oxygen atom needs two electrons. Yet it is possible for both atoms to achieve noble-gas electron configurations by a type of bonding called coordinate covalent bonding. To see how, begin by looking at the double covalent bond between carbon and oxygen.

·C· + ·O: ⟶ :C::O:

Carbon atom | Oxygen atom | Carbon monoxide

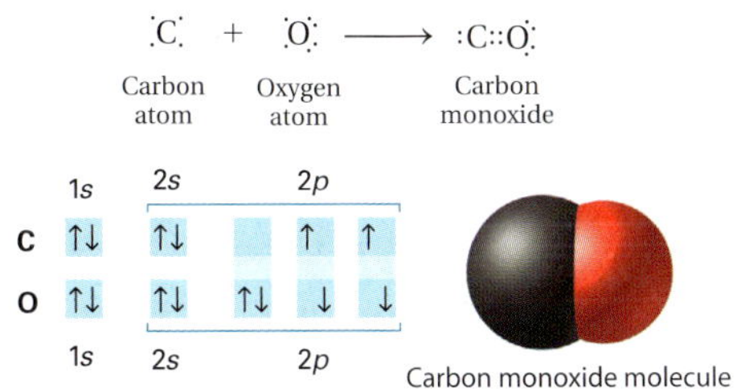

With the double bond in place, the oxygen has a stable configuration but the carbon does not. As shown below, the dilemma is solved if the oxygen also donates one of its unshared pairs of electrons for bonding.

:C::O: ⟶ :C:::O:

Carbon monoxide molecule

A **coordinate covalent bond** is a covalent bond in which one atom contributes both bonding electrons. In a structural formula, you can show coordinate covalent bonds as arrows that point from the atom donating the pair of electrons to the atom receiving them. The structural formula of carbon monoxide, with two covalent bonds and one coordinate covalent bond, is C≡O. **In a coordinate covalent bond, the shared electron pair comes from one of the bonding atoms.** Once formed, a coordinate covalent bond is like any other covalent bond.

The ammonium ion (NH_4^+) consists of atoms joined by covalent bonds, including a coordinate covalent bond. A **polyatomic ion,** such as NH_4^+, is a tightly bound group of atoms that has a positive or negative charge and behaves as a unit. The ammonium ion forms when a positively charged hydrogen ion (H^+) attaches to the unshared electron pair of an ammonia molecule (NH_3). Most plants need nitrogen that is already combined in a compound rather than molecular nitrogen (N_2) to grow. The fertilizer shown in Figure 8.9 contains the nitrogen compound ammonium sulfate.

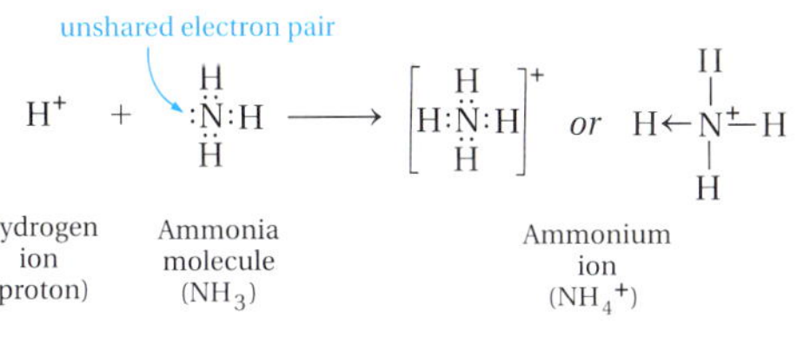

What is a polyatomic ion?

Figure 8.9 The polyatomic ammonium ion (NH_4^+), present in ammonium sulfate, is an important component of fertilizer for field crops, home gardens, and potted plants.

Coordinate Covalent Bonds

Discuss L2

A coordinate covalent bond is an exception to the rule that covalent bonding occurs between two half-empty orbitals of two atoms. In a coordinate covalent bond, one atom has an empty orbital and the other an orbital filled with an electron pair that is not yet involved in a chemical bond. Note that the bonding still involves only one pair of electrons and one pair of orbitals but one atom provides both of the shared electrons.

Discuss L2

Ask students to compare the electron dot structure for ammonia (NH_3) and for a hydrogen ion (H^+). After forming the three nitrogen-to-hydrogen bonds, nitrogen has an unshared pair of electrons. The hydrogen ion has no electrons available for bonding. Ask, **How do you explain the existence of the ammonium ion, NH_4^+?** *(The empty orbital of the H^+ must overlap the filled orbital in ammonia. The electrons from nitrogen are attracted to both the nitrogen nucleus and the hydrogen nucleus. A bond forms when electrons are simultaneously attracted to two nuclei. This type of bond is called a coordinate covalent bond.)* Ask students to explain the bonding in CO_2 and then explain the bonding in CO using coordinate covalent bonding.

Answers to...

a group of atoms that has a positive or negative charge and behaves as a unit

Discuss

L2

Have students write the electron dot structure for SO_2. Emphasize that the structure should satisfy the bonding requirements of all three atoms. Students should find that, to satisfy the octet rule for all the atoms, they must write a structure in which one oxygen atom is double bonded to sulfur. The other oxygen is single bonded by a coordinate covalent bond in which the electrons are donated by sulfur. Point out that experimental evidence indicates that both sulfur-oxygen bonds are identical. Explain that this evidence indicates that the bonding in SO_2 must be some intermediate between a single and double bond. Ask, **How does the formation of a coordinate covalent bond differ from that of a covalent bond?** *(In a covalent bond, each atom provides one electron, in a coordinate covalent bond, both electrons are provided by the same atom.)*

Relate

Have students think of everyday examples in which the shape of an object is as important as its composition. For example, several keys might be made of the same metal, but only one will fit into a particular lock.

Most polyatomic cations and anions contain both covalent and coordinate covalent bonds. Therefore, compounds containing polyatomic ions include both ionic and covalent bonding.

As another example, draw the electron dot structure of the polyatomic ion SO_3^{2-}. Start by drawing the electron dot structures for the oxygen and sulfur atoms, and the two extra electrons indicated by the charge. Then join two of the oxygens to sulfur by single covalent bonds.

:Ö· ·S̤: ·Ö: + ·· ⟶ :Ö:S̤: ·Ö:
·Ö: ·Ö:

Then join the remaining oxygen by a coordinate covalent bond, with sulfur donating one of its unshared pairs to oxygen, and add the two extra electrons. Put brackets about the structure and indicate the 2− charge, giving the result shown.

:Ö:S̤:Ö: + ·· ⟶ [:Ö:S̤:Ö: :Ö:]$^{2-}$
·Ö:

Each of the atoms of the completed structure has eight valence electrons, satisfying the octet rule. Without the two extra electrons, two of the oxygens would be electron-deficient.

Table 8.2 lists electron dot structures of some common compounds with covalent bonds.

Table 8.2

Some Common Molecular Compounds

Name	Chemical formula	Structure	Properties and uses
Hydrogen peroxide	H_2O_2	H–Ö–Ö–H	Colorless, unstable liquid when pure. It is used as rocket fuel. A 3% solution is used as a bleach and antiseptic.
Sulfur dioxide	SO_2	:Ö=S̈→Ö:	Oxides of sulfur are produced in combustion of petroleum products and coal. They are major air pollutants in industrial areas. Oxides of sulfur can lead to respiratory problems.
Sulfur trioxide	SO_3	:Ö=S(→Ö:)→Ö:	
Nitric oxide	NO	:Ö=N̈·	Oxides of nitrogen are major air pollutants produced by the combustion of fossil fuels in automobile engines. They irritate the eyes, throat, and lungs.
Nitrogen dioxide	NO_2	:Ö=Ṅ→Ö:	
Nitrous oxide	N_2O	:Ö←N≡N:	Colorless, sweet-smelling gas. It is used as an anesthetic commonly called laughing gas.
Hydrogen cyanide	HCN	H–C≡N:	Colorless, toxic gas with the smell of almonds.
Hydrogen fluoride	HF	H–F̤̈:	Two hydrogen halides, all extremely soluble in water. Hydrogen chloride, a colorless gas with pungent odor, readily dissolves in water to give a solution called hydrochloric acid.
Hydrogen chloride	HCl	H–C̤̈l:	

Remember, the electron dot structure for a neutral molecule contains the same number of electrons as the total number of valence electrons in the combining atoms. The negative charge of a polyatomic ion shows the number of electrons in addition to the valence electrons of the atoms present. Because a negatively charged polyatomic ion is part of an ionic compound, the positive charge of the cation of the compound balances these additional electrons.

CONCEPTUAL PROBLEM 8.2

Drawing the Electron Dot Structure of a Polyatomic Ion

The polyatomic hydronium ion (H_3O^+), which is found in acidic mixtures such as lemon juice, contains a coordinate covalent bond. The H_3O^+ ion forms when a hydrogen ion is attracted to an unshared electron pair in a water molecule. Draw the electron dot structure of the hydronium ion.

1 Analyze ***Identify the relevant concepts.***

H_3O^+ forms by the addition of a hydrogen ion to a water molecule. Draw the electron dot structure of the water molecule. Then, add the hydrogen ion. Oxygen must share a pair of electrons with the added hydrogen ion to form a coordinate covalent bond.

Solve ***Apply the concepts to this situation.***

$$H^+ + :\ddot{\underset{\cdot\cdot}{O}}:H \text{ (with H above O)} \longrightarrow \left[H:\ddot{\underset{\cdot\cdot}{O}}:H\right]^+ \text{ (with H above O)} \quad or \quad \left[H\leftarrow\underset{\cdot\cdot}{O}-H\right]^+ \text{ (with H bonded above O)}$$

Hydrogen ion (proton) — Water molecule (H_2O) — Hydronium ion (H_3O^+)

The oxygen in the hydronium ion has eight valence electrons, and each hydrogen shares two valence electrons. This satisfies the needs of both hydrogen and oxygen for valence electrons. The water molecule is electrically neutral, and the hydrogen ion has a positive charge. The combination of these two species must have a charge of 1+, as is found in the hydronium ion.

Practice Problems

9. Draw the electron dot structure of the hydroxide ion (OH^-).

10. Draw the electron dot structure of the polyatomic boron tetrafluoride anion (BF_4^-).

11. Draw the electron dot structures for sulfate (SO_4^{2-}) and carbonate (CO_3^{2-}). Sulfur and carbon are the central atoms, respectively.

12. Draw the electron dot structure for the hydrogen carbonate ion (HCO_3^-). Carbon is the central atom, and hydrogen is attached to oxygen in this polyatomic anion.

Problem-Solving 8.10 Solve Problem 10 with the help of an interactive guided tutorial.

with ChemASAP

CONCEPTUAL PROBLEM 8.2

Answers

9. $[H:\ddot{\underset{\cdot\cdot}{O}}:]^-$

10. $\left[:F:B:F: \text{ with } :F: \text{ above and below B}\right]^-$

11. a. $\left[:O:S:O: \text{ with } :O: \text{ above and below S}\right]^{2-}$ **b.** $\left[:O:C \text{ with } :O: \text{ above and } O \text{ below-right of C}\right]^{2-}$

12. $\left[H:O:C \text{ with } O \text{ above and } :O: \text{ below C}\right]^-$

Practice Problems Plus

Draw the electron dot structure of the polyatomic chlorate anion, ClO_3^-.

$\left[:O:Cl:O: \text{ with } :O: \text{ below Cl}\right]^-$

TEACHER Demo

Bond Energies

L2

Purpose Students observe two reactions in which bonds are formed and compare ionic and covalent bond energies.

Materials 10- to 15- cm piece of clean magnesium ribbon, a small piece of charcoal, tongs, Bunsen burner, filters for viewing

Safety Wear safety goggles and lab apron.

Procedure Using tongs, hold the piece of magnesium in a Bunsen burner. **CAUTION!** *Tie back long hair and loose clothing and wear safety goggles. Do not look directly at the flame. Observe through filters.* Discuss with students the large amount of heat and light given off in the formation of MgO. Write the balanced equation on the board.

$$2Mg + O_2 \rightarrow 2MgO + \text{energy}$$

Using tongs, place a small piece of charcoal in the Bunsen burner flame and try to ignite it. Write the balanced chemical equation.

$$C + O_2 \rightarrow CO_2 + \text{energy}$$

Expected Outcomes Students note that much less energy is given off in forming CO_2 than in forming MgO. Ask, **What kind of bonds are in MgO?** *(ionic)* **What kind of bonds are in CO_2?** *(covalent)* Ionic bond energies are, in general, greater than covalent bond energies.

Bond Dissociation Energies

Strengths of Covalent Bonds L2

Objective After completing this activity, students will understand that the dissociation energy of a covalent bond increases in order from single bond to double bond to triple bond.

Skills Focus Solving, interpreting

Prep Time 10 minutes

Advance Prep Obtain the necessary materials in advance. You may wish to ask students to help with this task by bringing cans of food, coat hangers, and plastic grocery bags to class.

Class Time 30 minutes

Teaching Tips

- Use tape to secure the hanger in place if necessary.
- Have groups of students record the data and then create a graph as a class using an overhead projector.

Expected Outcome As the mass of the load increases, the stretch of the rubber band or bands increases. For a given mass, a single rubber band stretches farther than a double and a double rubber band stretches farther than a triple.

Analyze and Conclude

1. Triple covalent bonds are stronger than double covalent bonds, which are stronger than single covalent bonds.
2. The change in bond dissociation energies in going from a carbon-carbon single bond to a carbon-carbon double bond to a carbon-carbon triple bond is nearly constant. The change in length of one, two, and three rubber bands, as given by the slopes of the lines, is not constant. It is large going from one to two rubber bands and small going from two to three rubber bands.

Quick LAB

Strengths of Covalent Bonds

Purpose

To compare and contrast the stretching of rubber bands and the dissociation energy of covalent bonds.

Materials

- 1 170-g (6-oz) can of food
- 2 454-g (16-oz) cans of food
- 3 No. 25 rubber bands
- metric ruler
- coat hanger
- plastic grocery bag
- paper clip
- graph paper
- pencil
- motion detector (optional)

Procedure

Sensor version available in the Probeware Lab Manual.

1. Bend the coat hanger to fit over the top of a door. The hook should hang down on one side of the door. Measure the length of the rubber bands (in cm). Hang a rubber band on the hook created by the coat hanger.
2. Place the 170-g can in the plastic bag. Use the paper clip to fasten the bag to the end of the rubber band. Lower the bag gently until it is suspended from the end of the rubber band. Measure and record the length of the stretched rubber band. Using different combinations of food cans, repeat this process three times with the following masses: 454 g, 624 g, and 908 g.
3. Repeat Step 2, first using two rubber bands to connect the hanger and the paper clip, and then using three.
4. Graph the length difference: (stretched rubber band) − (unstretched rubber band) on the *y*-axis versus mass (kg) on the *x*-axis for one, two, and three rubber bands. Draw the straight line that you estimate best fits the points for each set of data. (Your graph should have three separate lines.) The *x*-axis and *y*-axis intercepts of the lines should pass through zero, and the lines should extend past 1 kg on the *x*-axis. Determine the slope of each line in cm/kg.

Analyze and Conclude

1. Assuming the rubber bands are models for covalent bonds, what can you conclude about the relative strengths of single, double, and triple bonds?
2. How does the behavior of the rubber bands differ from that of covalent bonds?

Bond Dissociation Energies

A large quantity of heat is released when hydrogen atoms combine to form hydrogen molecules. This suggests that the product is more stable than the reactants. The covalent bond in the hydrogen molecule (H_2) is so strong that it would take 435 kJ of energy to break apart all of the bonds in 1 mole (about 2 grams) of H_2. (You will study the mole, abbreviated mol, in Chapter 12.) The energy required to break the bond between two covalently bonded atoms is known as the **bond dissociation energy.** This is usually expressed as the energy needed to break one mole of bonds, or 6.02×10^{23} bonds. The bond dissociation energy for the H_2 molecule is 435 kJ/mol.

A large bond dissociation energy corresponds to a strong covalent bond. A typical carbon–carbon single bond has a bond dissociation energy of 347 kJ/mol. Typical carbon–carbon double and triple bonds have bond dissociation energies of 657 kJ/mol and 908 kJ/mol, respectively. Strong carbon–carbon bonds help explain the stability of carbon compounds. Compounds with only C—C and C—H single covalent bonds, such as methane, tend to be quite unreactive. They are unreactive partly because the dissociation energy for each of these bonds is high.

For Enrichment 

Have students draw a graph of actual bond dissociation energies versus number of bonds for carbon-carbon single, double, and triple bonds. Have students write a statement that relates this graph to their conclusions in the activity.

Resonance

Ozone in the upper atmosphere blocks harmful ultraviolet radiation from the sun. At the lower elevations shown in Figure 8.10, it contributes to smog. The ozone molecule has two possible electron dot structures.

$$:\ddot{O}:\ddot{O}::\ddot{O}: \longleftrightarrow :\ddot{O}::\ddot{O}:\ddot{O}:$$

Notice that the structure on the left can be converted to the one on the right by shifting electron pairs without changing the positions of the oxygen atoms.

As drawn, these electron dot structures suggest that the bonding in ozone consists of one single coordinate covalent bond and one double covalent bond. Because earlier chemists imagined that the electron pairs rapidly flip back and forth, or resonate, between the different electron dot structures, they used double-headed arrows to indicate that two or more structures are in resonance.

Double covalent bonds are usually shorter than single bonds, so it was believed that the bond lengths in ozone were unequal. Experimental measurements show, however, that this is not the case. The two bonds in ozone are the same length. This result can be explained if you assume that the actual bonding in the ozone molecule is the average of the two electron dot structures. The electron pairs do not actually resonate back and forth.  **The actual bonding of oxygen atoms in ozone is a hybrid, or mixture, of the extremes represented by the resonance forms.**

The two electron dot structures for ozone are examples of what are still referred to as resonance structures. A **resonance structure** is a structure that occurs when it is possible to draw two or more valid electron dot structures that have the same number of electron pairs for a molecule or ion. Resonance structures are simply a way to envision the bonding in certain molecules. Although no back-and-forth changes occur, double-headed arrows are used to connect resonance structures.

Checkpoint ***What notation is used to show that the two covalent bonds in O_3 are the same?***

Figure 8.10 Although ozone high above the ground forms a protective layer that absorbs ultraviolet radiation from the sun, at lower elevations ozone is a pollutant that contributes to smog.

Resonance

Discuss 

Although the electron dot structures of O_2 on page 221 show that all electrons are paired, experimental evidence reveals that two electrons are unpaired. Ask, **Would you consider oxygen a reactive gas?** *(yes)* **What could account for oxygen's reactivity?** *(Unpaired electrons are reactive.)* After a brief discussion of the other exceptions to the octet rule that are presented in this section, remind students that many bonds do follow the octet rule.

Relate 

Students may be interested in knowing that substances containing unpaired electrons can be identified through a phenomenon called *paramagnetism*. When molecules with unpaired electrons are placed in a magnetic field, they tend to be drawn into the field. These substances are paramagnetic. In contrast, molecules in which all electrons are paired tend to be pushed from a magnetic field. These substances are *diamagnetic*.

Paramagnetism should not be confused with ferromagnetism, which is the familiar attraction of metals such as iron, cobalt and nickel to a magnetic field. The property of paramagnetism is the evidence that shows that the oxygen molecule has unpaired and thus cannot be described exactly by application of the octet rule. However, the oxygen bond does have double bond character; its bond length and bond energy are similiar to those of double bonds in other molecules that conform to the octet rule.

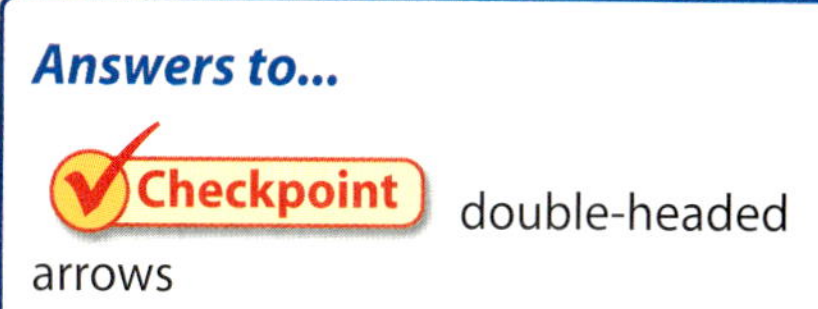

Answers to...

Checkpoint double-headed arrows

Exceptions to the Octet Rule

The octet rule provides guidance for drawing electron dot structures. For some molecules or ions, however, it is impossible to draw structures that satisfy the octet rule. **The octet rule cannot be satisfied in molecules whose total number of valence electrons is an odd number. There are also molecules in which an atom has fewer, or more, than a complete octet of valence electrons.** The nitrogen dioxide (NO_2) molecule, for example, contains a total of seventeen, an odd number, of valence electrons. Each oxygen contributes six electrons and the nitrogen contributes five. Two plausible electron dot structures can be drawn for the NO_2 molecule.

$$:\ddot{O}=\dot{N}-\ddot{\underset{\cdot\cdot}{O}}:$$
$$:\ddot{\underset{\cdot\cdot}{O}}-\dot{N}=\ddot{O}:$$

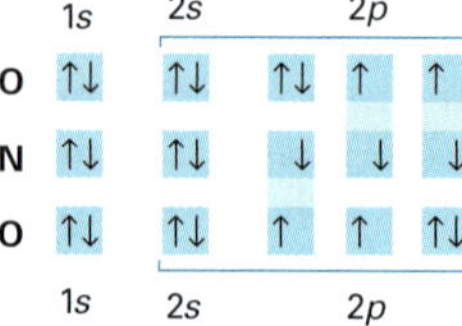

Nitrogen dioxide molecule

An unpaired electron is present in each of these structures, both of which fail to follow the octet rule. It is impossible to draw an electron dot structure for NO_2 that satisfies the octet rule for all atoms. Yet, NO_2 does exist as a stable molecule. In fact, it is produced naturally by lightning strikes of the sort shown in Figure 8.11.

A number of other molecules also have an odd number of electrons. In these molecules, as in NO_2, complete pairing of electrons is not possible. It is not possible to draw an electron dot structure that satisfies the octet rule. Examples of such molecules include chlorine dioxide (ClO_2) and nitric oxide (NO).

Figure 8.11 Lightning is one means by which nitrogen and oxygen in the atmosphere produce nitrogen dioxide.

Several molecules with an even number of valence electrons, such as some compounds of boron, also fail to follow the octet rule. This may happen because an atom acquires less than an octet of eight electrons. The boron atom in boron trifluoride (BF_3), for example, is deficient by two electrons, and therefore is an exception to the octet rule. Boron trifluoride readily reacts with ammonia to make the compound $BF_3{\cdot}NH_3$. In doing so, the boron atom accepts the unshared electron pair from ammonia and completes the octet.

$$:\ddot{F}-B(:\ddot{F}:)_2 + :N(-H)_3 \longrightarrow :\ddot{F}-B(:\ddot{F}:)_2\leftarrow N(-H)_3$$

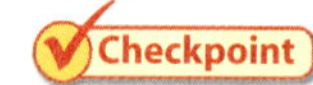

Checkpoint ***Give two examples of exceptions to the octet rule.***

Section 8.2 (continued)

Exceptions to the Octet Rule

TEACHER Demo

A Resonance Hybrid L2

Purpose Students observe the formation of nitrogen dioxide and write its resonance structures.

Materials small piece of copper metal, concentrated nitric acid, evaporating dish, fume hood

Safety Wear safety goggles, gloves, and lab apron. Perform the experiment under an efficient hood.

Procedure Place a small piece of copper in an evaporating dish and cover it with concentrated HNO_3. While students observe the reaction, write the balanced equation for the reaction on the board.

$$Cu + 4HNO_3 \rightarrow Cu(NO_3)_2 + 2H_2O + 2NO_2 + \text{energy}$$

Tell students that NO_2 is one of the pollutants in automobile exhaust. It gives smog its reddish-brown color and is very reactive and poisonous. Have students try to write the electron dot structure for NO_2. They will be unable to find a way to arrange the 17 electrons around the central nitrogen and the two oxygen atoms so that the octet rule is obeyed. Refer students to the two resonance structures on this page, each with an unpaired electron on the nitrogen atom. Ask them to draw another plausible resonance structure. *(They could draw a structure with the unpaired electron on an oxygen atom.)*

Expected Outcome A reddish-brown gas is produced.

Facts and Figures

Free Radicals

Nitrogen dioxide has an odd number of valence electrons (17), so one electron must be unpaired. Molecules with unpaired electrons are called free radicals and tend to be reactive. Two resonance structures for NO_2 appear on this page. These have the unpaired electron on the nitrogen atom. Two other possible structures place the unpaired electron on oxygen atoms. When free radicals interact, they share their unpaired electrons and create a dimer. Thus, dinitrogen tetroxide, N_2O_4 consists of two nitrogen dioxide molecules joined by a N-N covalent bond. The bond is temperature dependent; at higher temperatures the bond breaks as shown in an equilibrium diagram on p. 557.

228 Chapter 8

Figure 8.12 Phosphorus pentachloride, used as a chlorinating and dehydrating agent, and sulfur hexafluoride, used as an insulator for electrical equipment, are exceptions to the octet rule. **Interpreting Diagrams** ***How many valence electrons does the sulfur in sulfur hexafluoride (SF_6) have for the structure shown in the figure?***

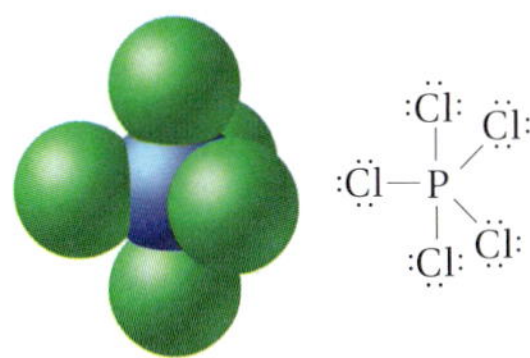

Phosphorus pentachloride

Sulfur hexafluoride

A few atoms, especially phosphorus and sulfur, sometimes expand the octet to include ten or twelve electrons. Consider phosphorus trichloride (PCl_3) and phosphorus pentachloride (PCl_5). Both are stable compounds in which all of the chlorines are bonded to a single phosphorus atom. Covalent bonding in PCl_3 follows the octet rule because all the atoms have eight valence electrons. However, as shown in Figure 8.12, the electron dot structure for PCl_5 can be written so that phosphorus has ten valence electrons.

8.2 Section Assessment

13. **Key Concept** What electron configurations do atoms usually achieve by sharing electrons to form covalent bonds?
14. **Key Concept** How is an electron dot structure used to represent a covalent bond?
15. **Key Concept** When are two atoms likely to form a double bond between them? A triple bond?
16. **Key Concept** How is a coordinate covalent bond different from other covalent bonds?
17. **Key Concept** How is the strength of a covalent bond related to its bond dissociation energy?
18. **Key Concept** Draw the electron dot resonance structures for ozone and explain how they describe its bonding.
19. **Key Concept** List three ways in which the octet rule can sometimes fail to be obeyed.
20. What kinds of information does a structural formula reveal about the compound it represents?
21. Draw electron dot structures for the following molecules, which have only single covalent bonds.
    a. H_2S **b.** PH_3 **c.** ClF
22. Use the bond dissociation energies of H_2 (435 kJ/mol) and of a typical carbon–carbon bond (347 kJ/mol) to decide which bond is stronger. Explain your reasoning.

Elements Handbook

Ozone Read the feature on ozone on page R31 of the Elements Handbook. Describe the effect of CFCs on the ozone layer. Explain why the United States has banned the use of CFCs in aerosols.

Assessment 8.2 Test yourself on the important concepts of Section 8.2.

with **ChemASAP**

3 ASSESS

Evaluate Understanding L2

Ask student to write electron dot structures for SBr_2 and C_3H_8. Have pairs of students decide whether their structures are correct and work together to correct any errors. Have students define a coordinate covalent bond in their own words. *(A coordinate covalent bond is formed when one atom contributes both bonding electrons.)* Ask students to explain resonance structures. *(Resonance structures occur when two or more valid electron dot structures can be written for a molecule.)*

Elements Handbook

CFCs react with ozone, thinning the ozone layer. Banning the use of CFCs should help decrease the amount of CFCs reaching Earth's ozone layer.

If your class subscribes to the Interactive Textbook, use it to review key concepts in Section 8.2.

with **ChemASAP**

Answers to...

NO_2, BF_3

Figure 8.12 In sulfur hexafluoride, the sulfur atom must have twelve valence electrons.

Section 8.2 Assessment

13. the configurations of noble gases
14. Two dots represent each covalent bond.
15. when they can attain a noble-gas structure by sharing two pairs or three pairs of electrons
16. The shared electron pair comes from one of the bonding atoms. In other covalent bonds each bonding atom provides an electron.
17. A large bond dissociation energy corresponds to a strong covalent bond.
18. :Ö:Ö::Ö: ⟷ :Ö::Ö:Ö:
    The actual bonding of oxygen atoms in ozone is a hybrid, or mixture, of the extremes represented by the resonance forms.
19. The octet rule cannot be satisfied in molecules whose total number of valence electrons is an odd number. There are also molecules in which an atom has fewer, or more, than a complete octet of valence electrons.
20. the arrangement of atoms in a molecule
21. **a.** :S̤:H (with H below S) **b.** H:P̤:H (with H below P) **c.** :C̤l:F̤:
22. The H-H bond is stronger because it has a greater dissociation energy.

8.3

1 FOCUS

Objectives

8.3.1 Describe the relationship between atomic and molecular orbitals.

8.3.2 Describe how VSEPR theory helps predict the shapes of molecules.

8.3.3 Identify ways in which orbital hybridization is useful in describing molecules.

Guide for Reading

Build Vocabulary L2

Paraphrase The hybridization of atomic orbitals is the combining of two or more orbitals to form a single orbital that is different from those that were combined.

Reading Strategy 

Compare and Contrast As students read about atomic and molecular orbitals encourage them to make a table in which to list similarities and differences. Have them extend the list to include hybrid orbitals.

2 INSTRUCT

Connecting to Your World

Have students read the opening paragraph. Ask, **Why is very little paint wasted when a car is electrostatically spray painted?** *(The droplets of paint are attracted to the auto body and don't fall to the floor.)* **Can you infer the charge on the auto body?** *(positive)*

Molecular Orbitals

Use Visuals 

Figure 8.13 Have students note the positions of the two nuclei in the two hydrogen atomic orbitals. Ask, **Where is the electron most likely to be in the hydrogen atomic orbital?** *(In the sphere, close to the nucleus.)* **Where are the two shared electrons likely to be in the H_2 molecular orbital?** *(In the region between the two nuclei.)*

8.3 Bonding Theories

Guide for Reading

 **Key Concepts**

- How are atomic and molecular orbitals related?
- How does VSEPR theory help predict the shapes of molecules?
- In what ways is orbital hybridization useful in describing molecules?

Vocabulary

molecular orbitals
bonding orbital
sigma bond
pi bond
tetrahedral angle
VSEPR theory
hybridization

Reading Strategy

Summarizing When you summarize, you review and state, in the correct order, the main points you have read. As you read about bonding theories, write a brief summary of the text following each heading. Your summary should include only the most important information.

Connecting to Your World This car is being painted by a process called electrostatic spray painting. A custom-designed spray nozzle wired up to an electric power supply imparts a negative charge to the paint droplets as they exit the spray gun. The negatively charged droplets are attracted to the auto body. Painting with attractive forces is very efficient, because almost all the paint is applied to the car body and very little is wasted. In this section, you will learn how attractive and repulsive forces influence the shapes of molecules.

Molecular Orbitals

The model for covalent bonding you have been using assumes that the orbitals are those of the individual atoms. There is a quantum mechanical model of bonding, however, that describes the electrons in molecules using orbitals that exist only for groupings of atoms. When two atoms combine, this model assumes that their atomic orbitals overlap to produce **molecular orbitals,** or orbitals that apply to the entire molecule.

In some ways, atomic orbitals and molecular orbitals are similar. **Just as an atomic orbital belongs to a particular atom, a molecular orbital belongs to a molecule as a whole.** Each atomic orbital is filled if it contains two electrons. Similarly, two electrons are required to fill a molecular orbital. A molecular orbital that can be occupied by two electrons of a covalent bond is called a **bonding orbital.**

Sigma Bonds When two atomic orbitals combine to form a molecular orbital that is symmetrical around the axis connecting two atomic nuclei, a **sigma bond** is formed, as illustrated in Figure 8.13. The symbol for this bond is the Greek letter sigma (σ).

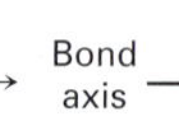
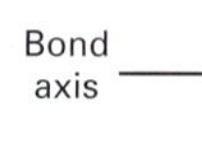

Figure 8.13 Two *s* atomic orbitals can combine to form a molecular orbital, as in the case of hydrogen (H_2). In a bonding molecular orbital, the electron density between the nuclei is high.

Section Resources

Print

- **Guided Reading and Study Workbook,** Section 8.3
- **Core Teaching Resources,** Section 8.3 Review, Interpreting Graphics
- **Laboratory Manual,** Lab 11
- **Small-Scale Chemistry Laboratory Manual,** Lab 11
- **Transparencies,** T90–T92

Technology

- **Interactive Textbook with ChemASAP,** Simulation 7, Assessment 8.3
- **Go Online,** Section 8.3

⊕ represents the nucleus.

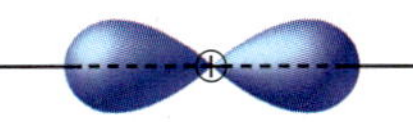

+

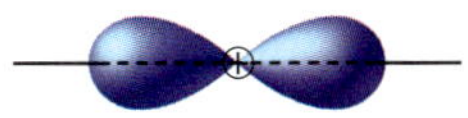

→

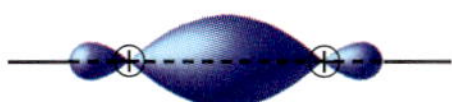

Figure 8.14 Two *p* atomic orbitals can combine to form a sigma-bonding molecular orbital, as in the case of fluorine (F_2). Notice that the sigma bond is symmetrical around the bond axis connecting the nuclei.

In general, covalent bonding results from an imbalance between the attractions and repulsions of the nuclei and electrons involved. Because their charges have opposite signs, the nuclei and electrons attract each other. Because their charges have the same sign, nuclei repel other nuclei and electrons repel other electrons. In a hydrogen molecule, the nuclei repel each other, as do the electrons. In a bonding molecular orbital of hydrogen, however, the attractions between the hydrogen nuclei and the electrons are stronger than the repulsions. The balance of all the interactions between the hydrogen atoms is thus tipped in favor of holding the atoms together. The result is a stable diatomic molecule of H_2.

Atomic *p* orbitals can also overlap to form molecular orbitals. A fluorine atom, for example, has a half-filled 2*p* orbital. When two fluorine atoms combine, as shown in Figure 8.14, the *p* orbitals overlap to produce a bonding molecular orbital. There is a high probability of finding a pair of electrons between the positively charged nuclei of the two fluorines. The fluorine nuclei are attracted to this region of high electron density. This attraction holds the atoms together in the fluorine molecule (F_2). The overlap of the 2*p* orbitals produces a bonding molecular orbital that is symmetrical when viewed around the F—F bond axis connecting the nuclei. Therefore, the F—F bond is a sigma bond.

Pi Bonds In the sigma bond of the fluorine molecule, the *p* atomic orbitals overlap end-to-end. In some molecules, however, orbitals can overlap side-by-side. As shown in Figure 8.15, the side-by-side overlap of atomic *p* orbitals produces what are called pi molecular orbitals. When a pi molecular orbital is filled with two electrons, a pi bond results. In a **pi bond** (symbolized by the Greek letter π), the bonding electrons are most likely to be found in sausage-shaped regions above and below the bond axis of the bonded atoms. It is not symmetrical around the F—F bond axis. Atomic orbitals in pi bonding overlap less than in sigma bonding. Therefore, pi bonds tend to be weaker than sigma bonds.

⊕ represents the nucleus.

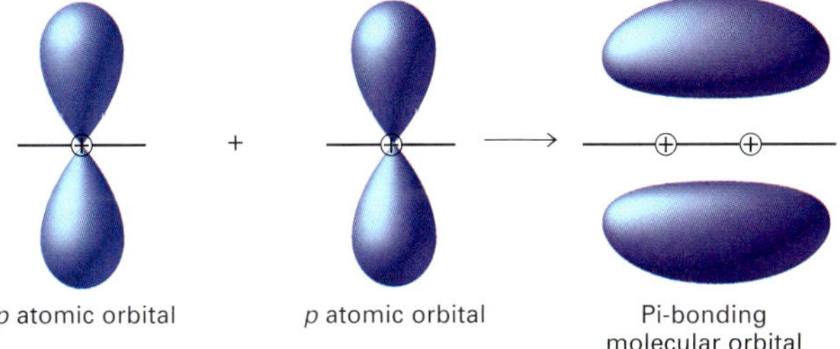

Figure 8.15 The side-by-side overlap of two *p* atomic orbitals produces a pi-bonding molecular orbital. Together, the two sausage-shaped regions in which the bonding electron pair is most likely to be found constitute one pi-bonding molecular orbital.

Discuss 

Ask, **How can you visualize the formation of the covalent bond in H_2 as two separated hydrogen atoms move closer together?** *(As the distance between them decreases, each nucleus is attracted by both electrons, and the potential energy of the system decreases.)* Explain that repulsive forces also come into play as the nuclei and the electrons of both atoms begin to repel each other. An interatomic distance is reached at which the potential energy is at a minimum. If the distance decreases beyond this point, the potential energy increases because of increased repulsion. Thus, a chemical bond forms with a bond length equal to the interatomic distance at which the potential energy is a minimum.

Use Visuals 

Figures 8.14 and 8.15 Ask, **What is different about the ways that the two *p* orbitals overlap in the two illustrations?** *(In Figure 8.14, the* p *orbitals overlap end-to-end and form an orbital that is symmetrical around the axis joining the nuclei. In Figure 8.15, the* p *orbitals overlap side-by-side and form two areas of electron density, one above and another below the axis joining the nuclei.)* Ask, **What are the two bonds called?** *(Figure 8.14: sigma bond; Figure 8.15: pi bond)*

Relate 

Topographic maps are usually drawn as a series of lines tracing the elevations of a terrain. The more closely spaced the lines are, the steeper is the terrain. Scientists who wish to know the positions of atoms in crystals of molecules may use electron densities that are plotted in much the same way that a terrain is shown on a topographic map. From X-ray diffraction data, scientists can map the electron density at any position in a crystal of interest. Where the lines are closest in the map, the electron density of the molecule is greatest.

Differentiated Instruction

Gifted and Talented 

Molecular orbital theory is a third way of treating chemical bonding in addition to the valence bond approach and VESPR theory, both of which are covered in this text. Each approach to bonding contributes different important ideas. Molecular orbital theory is a quantum mechanical approach involving not just localized bonds but bonds that bind entire molecules. Students may want to access a college text and learn more about this theory and how it explains the paradox of bonding in the oxygen molecule.

VSEPR Theory

Discuss

L2

Ask students to write out the electron configuration for carbon when it is forming bonds. ($1s^2$, $2s^1$, $2p^3$) **How many valence electrons does carbon have?** *(4)* **How many bonds does carbon form?** *(4)* Have students describe the three p orbitals. *(dumbbell-shaped orbitals containing two lobes; the three orbitals are perpendicular to each other, oriented on the* x-, y-, *and* z-*axes)* Point out that if carbon forms three of its four bonds using *p* orbitals, it must form molecules that are three-dimensional.

CLASS Activity

Making Molecular Models

L2

Purpose Students achieve an understanding of simple molecular shapes and relate actual three-dimensional geometry to two-dimensional illustrations.

Materials ball-and-stick molecular model kits or foam balls and toothpicks

Procedure Have students make the tetrahedral model of methane. Have them examine the model and compare it with the illustrations in Figure 8.16. Ask them to draw the model and label it tetrahedral. Have students remove one ball leaving the stick in place. The stick can represent an unshared pair of electrons. This is a model for the ammonia molecule. Again have students compare their models with Figure 8.16 and make their own illustration, labeling it pyramidal. Have students remove a second ball. Now the model simulates the water molecule with two sticks (pairs of unshared electrons). Have students compare their models with Figure 8.17 and make their own illustration labeled bent triatomic.

Word Origins

L2

The word *tetrapod* means having four feet. Dogs, cats, and elephants are all tetrapods because they all have four feet.

VSEPR Theory

A photograph or sketch may fail to do justice to your appearance. Similarly, electron dot structures fail to reflect the three-dimensional shapes of the molecules illustrated in Figure 8.16. The electron dot structure and structural formula of methane (CH_4), for example, show the molecule as if it were flat and merely two-dimensional.

```
   H                 H
H:C:H                |
   H             H — C — H
                     |
                     H
```

Methane (electron dot structure) — Methane (structural formula)

In reality, methane molecules are three-dimensional. As Figure 8.16a shows, the hydrogens in a methane molecule are at the four corners of a geometric solid called a regular tetrahedron. In this arrangement, all of the H—C—H angles are 109.5°, the **tetrahedral angle.**

The valence-shell electron-pair repulsion theory, or **VSEPR theory,** explains the three-dimensional shape of methane. **According to VSEPR theory, the repulsion between electron pairs causes molecular shapes to adjust so that the valence-electron pairs stay as far apart as possible.** The methane molecule has four bonding electron pairs and no unshared pairs. The bonding pairs are farthest apart when the angle between the central carbon and its attached hydrogens is 109.5°. This is the H—C—H bond angle found experimentally.

Unshared pairs of electrons are also important in predicting the shapes of molecules. The nitrogen in ammonia (NH_3) is surrounded by four pairs of valence electrons, so you might predict the tetrahedral angle of 109.5° for the H—N—H bond angle. However, one of the valence-electron pairs shown in Figure 8.16b is an unshared pair. No bonding atom is vying for these unshared electrons. Thus they are held closer to the nitrogen than are the bonding pairs. The unshared pair strongly repels the bonding pairs, pushing them together. The measured H—N—H bond angle is only 107°.

Word Origins

Tetrahedral comes from the Greek *tetra-*, meaning "four," and *hedra*, meaning "face." The Greek *pod* means "foot." **What do you think *tetrapod* means?**

a Methane (CH_4)

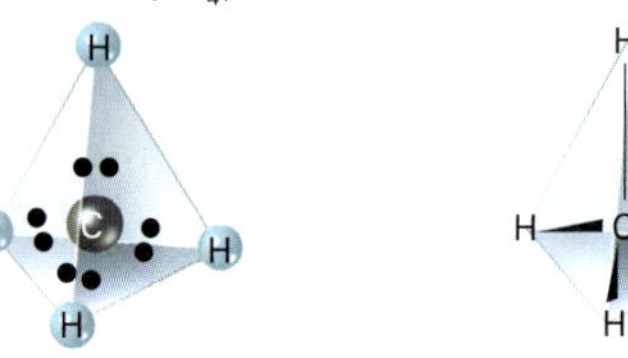

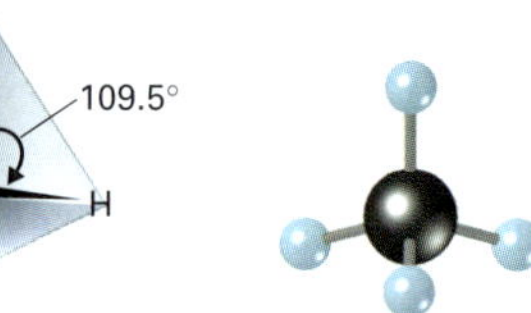

b Ammonia (NH_3)

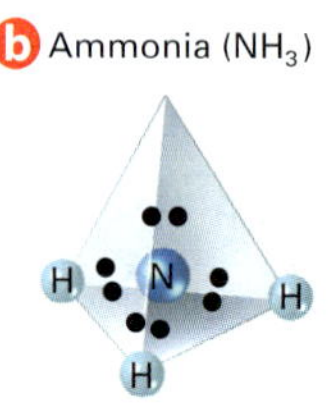

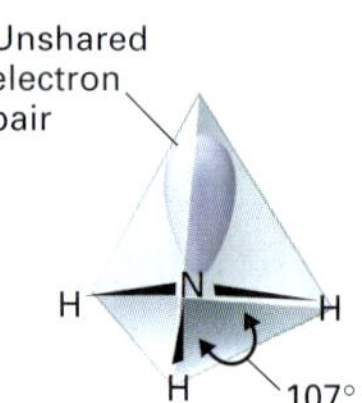

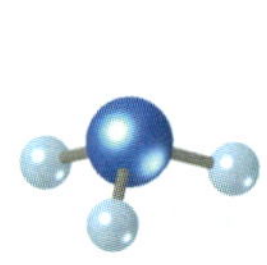

Figure 8.16 Methane and ammonia, represented here, are three-dimensional molecules. **a** Methane is a tetrahedral molecule. The hydrogens in methane are at the four corners of a regular tetrahedron, and the bond angles are all 109.5°. **b** An ammonia molecule is pyramidal. The unshared pair of electrons repels the bonding pairs. **Interpreting Diagrams** ***How do the resulting H—N—H bond angles compare to the tetrahedral angle?***

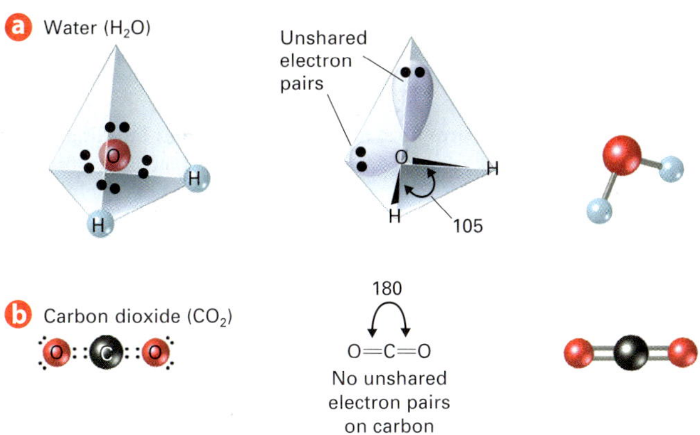

Figure 8.17 This comparison of water and carbon dioxide illustrates how unshared pairs of electrons can affect the shape of a molecule made of three atoms. **a** The water molecule is bent because the two unshared pairs of electrons on oxygen repel the bonding electrons. **b** In contrast, the carbon dioxide molecule is linear. The carbon atom has no unshared electron pairs.

In a water molecule, oxygen forms single covalent bonds with two hydrogen atoms. The two bonding pairs and the two unshared pairs of electrons form a tetrahedral arrangement around the central oxygen. Thus the water molecule is planar (flat) but bent. With two unshared pairs repelling the bonding pairs, the H—O—H bond angle is compressed in comparison with the H—C—H bond angle in methane. The experimentally measured bond angle in water is about 105°, as shown in Figure 8.17a.

In contrast, the carbon in a carbon dioxide molecule has no unshared electron pairs. The double bonds joining the oxygens to the carbon are farthest apart when the O=C=O bond angle is 180° as illustrated in Figure 8.17b. Thus CO_2 is a linear molecule. Nine of the possible molecular shapes are shown in Figure 8.18.

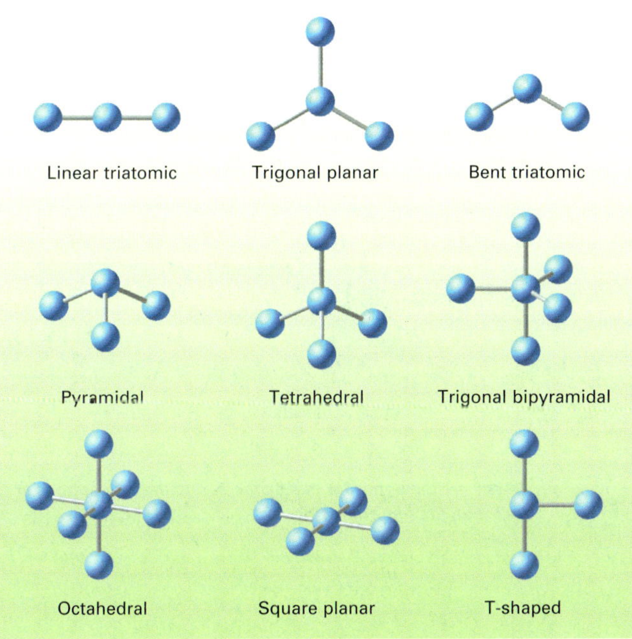

Figure 8.18 Shown here are common molecular shapes.

For: Links on VSEPR Theory
Visit: www.SciLinks.org
Web Code: cdn-1083

Discuss 

In predicting molecular shapes, students may find it helpful to start with an electron dot structure. The electron dot structure shows both the bonding and nonbonding pairs of electrons around the central atom. Point out that when using VSEPR theory to predict molecular shape, double and triple bonds are viewed as single bonds. Each group of electrons in the electron dot structure must be accommodated around the central atom. Because electrons repel each other, they will stay as far apart as possible. When four pairs must be accommodated, the pairs have maximum space when they are arranged tetrahedrally with an angle of 109.5° between them. When three pairs or groups must be accommodated, a trigonal planar arrangement provides the most space with 120° separating them. Finally, when only two groups are present, they will adopt a linear arrangement in which they are separated by 180°. Have students predict the shapes of H_2S and BF_3. *(bent and trigonal planar respectively)*

Go Online NSTA SCiLINKS

Download a worksheet on **VSEPR Theory** for students to complete, and find additional teacher support from NSTA SciLinks.

Differentiated Instruction

Less Proficient Readers 

Give students as many opportunities as possible to see three-dimensional models of molecules. When you write formulas on the chalkboard, try to have a three-dimensional model to hold up or circulate around the class. This will help students interpret two-dimensional drawings.

Answers to...

Figure 8.16 They are smaller.

Hybrid Orbitals

Relate

L2

Students are aware that third row nonmetals such as phosphorus and sulfur, in Groups 5A and 6A respectively, can form more bonds than the number predicted by the octet rule. For example, phosphorus can form PCl_5 as well as PCl_3, and sulfur can form SF_6 as well as SF_2. Second row elements in the same groups, nitrogen and oxygen, do not form compounds that violate the octet rule. The difference lies in the fact that third row elements can promote electrons to empty 3*d* orbitals. Second row elements have no 2*d* orbitals. Bonding in phosphorus and sulfur in compounds such as PCl_5 and SF_6 involves hybrid orbitals. For phosphorus, one *s*, three *p*, and one *d* orbitals combine to form five dsp^3 hybrid orbitals. The geometry of a set of five dsp^3 orbitals is trigonal byprimidal. For sulfur, one *s*, three *p*, and two *d* orbitals combine to form six d^2sp^3 hybrid orbitals. The geometry of six d^2sp^3 orbitals is octahedral.

Hybrid Orbitals

The VSEPR theory works well when accounting for molecular shapes, but it does not help much in describing the types of bonds formed. **Orbital hybridization provides information about both molecular bonding and molecular shape.** In **hybridization,** several atomic orbitals mix to form the same total number of equivalent hybrid orbitals.

Hybridization Involving Single Bonds Recall that the carbon atom's outer electron configuration is $2s^2 2p^2$, but one of the 2*s* electrons is promoted to a 2*p* orbital to give one 2*s* electron and three 2*p* electrons, allowing it to bond to four hydrogen atoms in methane. You might suspect that one bond would be different from the other three. In fact, all the bonds are identical. This is explained by orbital hybridization.

The one 2*s* orbital and three 2*p* orbitals of a carbon atom mix to form four sp^3 hybrid orbitals. These are at the tetrahedral angle of 109.5°. As you can see in Figure 8.19, the four sp^3 orbitals of carbon overlap with the 1*s* orbitals of the four hydrogen atoms. The sp^3 orbitals extend farther into space than either *s* or *p* orbitals, allowing a great deal of overlap with the hydrogen 1*s* orbitals. The eight available valence electrons fill the molecular orbitals to form four C—H sigma bonds. The extent of overlap results in unusually strong covalent bonds.

Figure 8.19 In methane, each of the four sp^3 hybrid orbitals of carbon overlaps with a 1*s* orbital of hydrogen.

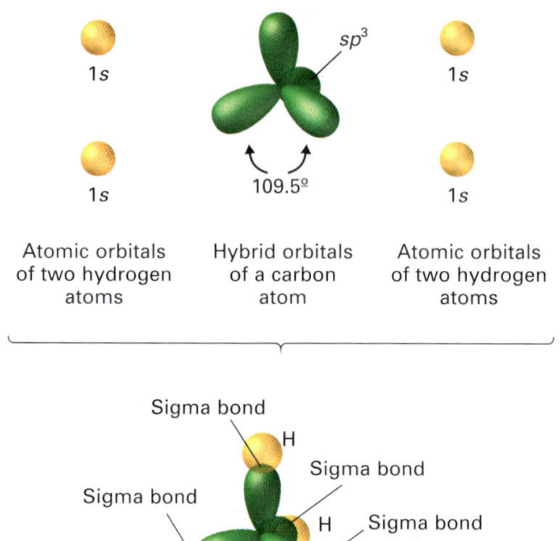

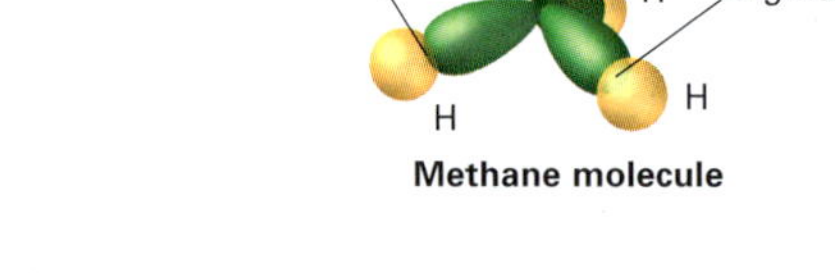

Checkpoint ***Why are the bonds formed by the 2s and the 2p electrons of carbon the same in methane?***

Facts and Figures

Towering 20th Century Chemist

An esteemed chemist and dedicated humanitarian, Linus Pauling (1916–1994) is the only person to have won two Nobel prizes that were unshared. In 1954 he won the prize for chemistry; in 1962 he won the prize for peace. The peace prize recognized his efforts to bring nuclear weapons under international control and halt nuclear testing. Much of his scientific work involved the structure of simple and complex molecules. His book *The Nature of the Chemical Bond and the Structure of Molecules and Crystals* is one of the most important texts of the century. Pauling introduced the idea of hybrid orbitals to explain why the four bonds around a carbon atom are equivalent. He invented the concept of electronegativity and proposed that most covalent bonds have some ionic character. He also taught that the structures of some compounds must be described by two or more resonance hybrids.

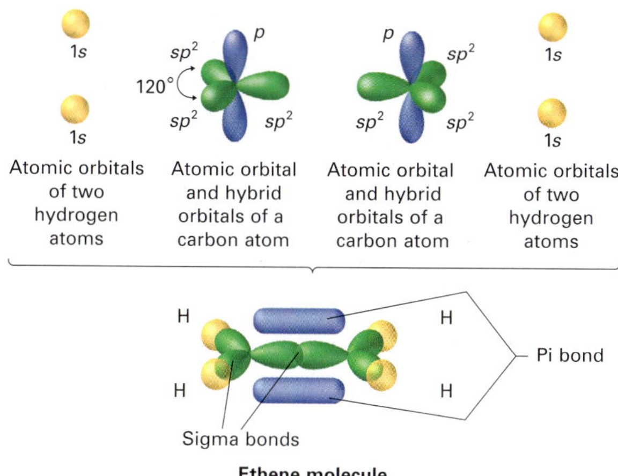

Figure 8.20 In an ethene molecule, two sp^2 hybrid orbitals from each carbon overlap with a $1s$ orbital of hydrogen to form a sigma bond. The other sp^2 orbitals overlap to form a carbon–carbon sigma bond. The p atomic orbitals overlap to form a pi bond. **Inferring** ***What region of space does the pi bond occupy relative to the carbon atoms?***

Hybridization Involving Double Bonds Hybridization is also useful in describing double covalent bonds. Ethene is a relatively simple molecule that has one carbon–carbon double bond and four carbon–hydrogen single bonds.

$$\mathrm{H_2C{=}CH_2}$$

Ethene

Experimental evidence indicates that the H—C—H bond angles in ethene are about 120°. In ethene, sp^2 hybrid orbitals form from the combination of one $2s$ and two $2p$ atomic orbitals of carbon. As you can see in Figure 8.20, each hybrid orbital is separated from the other two by 120°. Two sp^2 hybrid orbitals of each carbon form sigma-bonding molecular orbitals with the four available hydrogen $1s$ orbitals. The third sp^2 orbitals of each of the two carbons overlap to form a carbon–carbon sigma-bonding orbital. The nonhybridized $2p$ carbon orbitals overlap side-by-side to form a pi-bonding orbital. A total of twelve electrons fill six bonding molecular orbitals. Thus five sigma bonds and one pi bond hold the ethene molecule together. The sigma bonds and the pi bond are two-electron covalent bonds. Although they are drawn alike in structural formulas, pi bonds are weaker than sigma bonds. In chemical reactions that involve breaking one bond of a carbon–carbon double bond, the pi bond is more likely to break than the sigma bond.

Hybridization Involving Triple Bonds A third type of covalent bond is a triple bond, which is found in ethyne (C_2H_2), also called acetylene.

$$\mathrm{H{-}C{\equiv}C{-}H}$$

As with other molecules, the hybrid orbital description of ethyne is guided by an understanding of the properties of the molecule. Ethyne is a linear molecule. The best hybrid orbital description is obtained if a $2s$ atomic orbital of carbon mixes with only one of the three $2p$ atomic orbitals. The result is two sp hybrid orbitals for each carbon.

Facts and Figures

Ethene Ripens Tomatoes

Ethene, or ethylene ($H_2C{=}CH_2$), is a gaseous hormone found in most plants. Tomatoes release ethene as they ripen, so a tomato placed in a container with unripe fruit can hasten the ripening process. Because it's often convenient to pick and ship fruits and vegetables before they are ripe, the produce industry uses ethene to control how fast produce ripens.

Use Visuals L1

Figure 8.20 Direct student attention to the sigma bonds formed by the two carbon atoms in ethene. Each carbon forms three bonds using three sp^2 orbitals. Thus, the entire molecule is planar with the unhybridized p orbital forming lobes of electron density above and below the plane.

Discuss 

Have students compare the shapes of the carbon hybrid orbitals in methane (sp^3), ethene (sp^2), and ethyne molecules (sp). The key concept in hybridization is that the atomic orbitals that combine to form the hybrid orbitals lose their identities. Entirely new orbitals form that have different characteristics. Emphasize that the shape and orientation of the hybrid orbitals cannot be predicted from the shapes of the combining atomic orbitals. Hybrid orbitals of a particular type (sp^3, sp^2, or sp) are identical to one another in terms of shape and bonding properties.

3 ASSESS

Evaluate Understanding 

Have students compare and contrast the characteristics of sigma and pi bonds. *(A sigma bond is formed when two atomic orbitals combine to form a molecular orbital that is symmetrical along the axis connecting the two atomic nuclei. The electrons tend to be found between the two nuclei. A pi bond is formed from the overlap of two p atomic orbitals. The electrons in a pi bond are found in sausage-shaped regions above and below the bond axis.)* Ask how unshared pairs of electrons influence the shape of a molecule. *(The unshared pairs take up space in the molecule. They repel the bonding pairs, pushing them closer together and decreasing the bond angle.)*

Answers to...

 because of orbital hybridization

Figure 8.20 the regions above and below the carbons

Section 8.3 (continued)

Reteach 

Divide students into pairs to determine the shapes of molecules by means of the VSEPR theory. Give them the following information about the central atoms of five molecules and ask them to determine the geometry of each. Have them begin by writing electron dot structures.

• four bonding pairs + no nonbonding pairs *(tetrahedral);* • three bonding pairs + one nonbonding pair *(pyramidal);* • two bonding pairs + two nonbonding pairs *(bent);* • two bonding pairs + no nonbonding pair *(linear);* • one bonding pair + any number of nonbonding pairs *(linear)*

Writing Activity

Oxygen is paramagnetic, which means that it tends to be attracted to a magnetic field. This behavior is characteristic of substances that have unpaired electrons.

If your class subscribes to the Interactive Textbook, use it to review key concepts in Section 8.3

with ChemASAP

Answers to . . .

Figure 8.21 Two *pi* bonds are formed.

Atomic orbital of a hydrogen atom

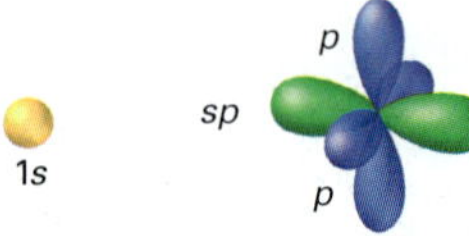

Atomic orbitals and hybrid orbitals of a carbon atom

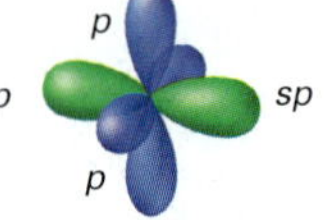
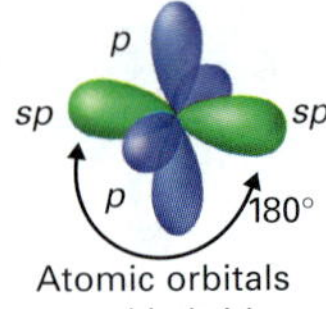

Atomic orbitals and hybrid orbitals of a carbon atom

1s

Atomic orbital of a hydrogen atom

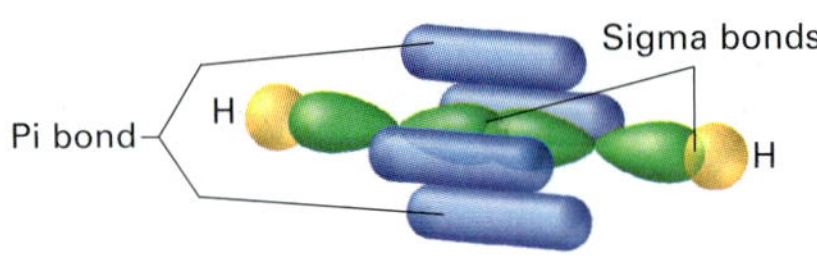

Ethyne molecule

Figure 8.21 In an ethyne molecule, one *sp* hybrid orbital from each carbon overlaps with a 1*s* orbital of hydrogen to form a sigma bond. The other *sp* hybrid orbital of each carbon overlaps to form a carbon–carbon sigma bond. The two *p* atomic orbitals from each carbon also overlap. **Interpreting Diagrams** ***How many pi bonds are formed in an ethyne molecule?***

The carbon–carbon sigma-bonding molecular orbital of the ethyne molecule in Figure 8.21 forms from the overlap of one *sp* orbital from each carbon. The other *sp* orbital of each carbon overlaps with the 1*s* orbital of each hydrogen, also forming sigma-bonding molecular orbitals. The remaining pair of *p* atomic orbitals on each carbon overlap side-by-side. They form two pi-bonding molecular orbitals that surround the central carbons. The ten available electrons completely fill five bonding molecular orbitals. The bonding of ethyne consists of three sigma bonds and two pi bonds.

Interactive Textbook

Simulation 7 Compare *sp*, sp^2, and sp^3 hybrid orbitals.

with ChemASAP

8.3 Section Assessment

23. **Key Concept** How are atomic and molecular orbitals related?

24. **Key Concept** Explain how the VSEPR theory can be used to predict the shapes of molecules.

25. **Key Concept** How is orbital hybridization useful in describing molecules?

26. What shape would you expect a simple carbon-containing compound to have if the carbon atom has the following hybridizations?

a. sp^2 **b.** sp^3 **c.** *sp*

27. What is a sigma bond? Describe, with the aid of a diagram, how the overlap of two half-filled 1*s* orbitals produces a sigma bond?

28. How many sigma and how many pi bonds are in an ethyne molecule (C_2H_2)?

29. The BF_3 molecule is planar. The attachment of a fluoride ion to the boron in BF_3, through a coordinate covalent bond, creates the BF_4^- ion. What is the geometric shape of this ion?

Writing Activity

Molecular Bonding in Oxygen Research how chemists know that an oxygen molecule has unpaired electrons. Write a brief report on what you find.

Assessment 8.3 Test yourself on the important concepts of Section 8.3.

with ChemASAP

Section 8.3 Assessment

23. When two atoms combine, their atomic orbitals overlap to produce molecular orbitals. An atomic orbital belongs to a particular atom, and a molecular orbital belongs to a molecule as a whole.

24. Each molecule assumes the shape that places valence-electron pairs as far apart as possible.

25. Orbital hybridization provides information about both molecular bonding and molecular shape.

26. **a.** trigonal planar **b.** tetrahedral **c.** linear

27. A sigma bond is formed by the overlap of two orbitals along the axis between two nuclei. See Figure 8.13.

28. 3 sigma bonds and 2 pi bonds

29. tetrahedral

8.4 Polar Bonds and Molecules

Connecting to Your World Snow covers approximately 23 percent of Earth's surface. Each individual snowflake is formed from as many as 100 snow crystals. The size and shape of each crystal depends mainly on the air temperature and amount of water vapor in the air at the time the snow crystal forms. In this section, you will see that the polar bonds in water molecules influence the distinctive geometry of snowflakes.

Guide for Reading

Key Concepts

- How do electronegativity values determine the charge distribution in a polar bond?
- What happens to polar molecules between a pair of oppositely charged metal plates?
- How do intermolecular attractions compare with ionic and covalent bonds?
- Why do network solids have high melting points?

Vocabulary

nonpolar covalent bond
polar covalent bond
polar bond
polar molecule
dipole
van der Waals forces
dipole interactions
dispersion forces
hydrogen bonds
network solids

Reading Strategy

Relating Text and Visuals Asyou read, look closely at Figures 8.22 and 8.23. Explain how these illustrations help you understand how molecules attract each other.

Bond Polarity

Covalent bonds involve electron sharing between atoms. However, covalent bonds differ in terms of how the bonded atoms share the electrons. The character of the bonds in a given molecule depends on the kind and number of atoms joined together. These features, in turn, determine the molecular properties.

The bonding pairs of electrons in covalent bonds are pulled, as in the tug-of-war in Figure 8.22, between the nuclei of the atoms sharing the electrons. When the atoms in the bond pull equally (as occurs when identical atoms are bonded), the bonding electrons are shared equally, and the bond is a **nonpolar covalent bond.** Molecules of hydrogen (H_2), oxygen (O_2), and nitrogen (N_2) have nonpolar covalent bonds. Diatomic halogen molecules, such as Cl_2, are also nonpolar.

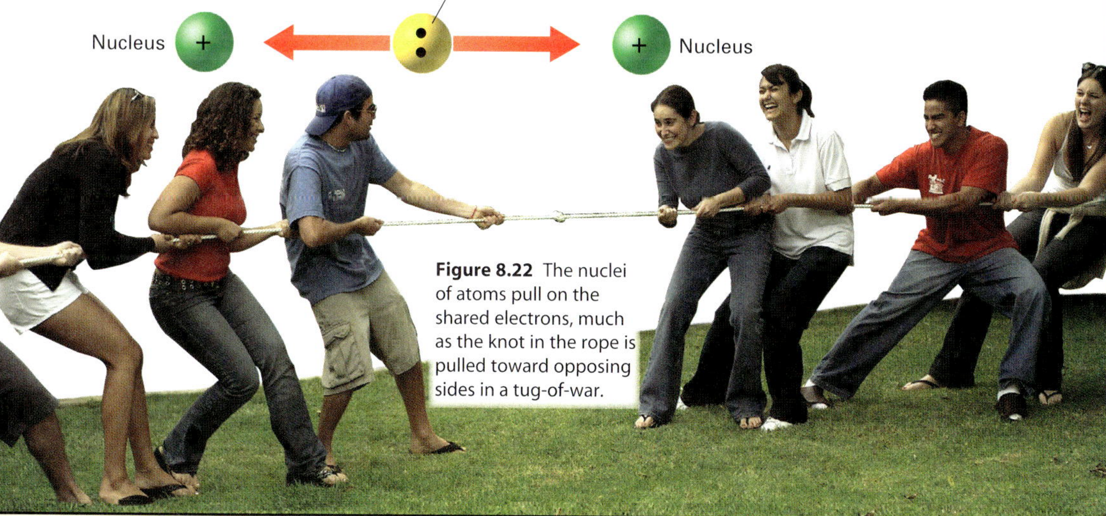

Figure 8.22 The nuclei of atoms pull on the shared electrons, much as the knot in the rope is pulled toward opposing sides in a tug-of-war.

Section Resources

Print

- **Guided Reading and Study Workbook,** Section 8.4
- **Core Teaching Resources,** Section 8.4 Review
- **Transparency,** T93

Technology

- **Interactive Textbook with ChemASAP,** Animation 10, Simulation 8, Assessment 8.4
- **Go Online,** Section 8.4

8.4

1 FOCUS

Objectives

8.4.1 Describe how electronegativity values determine the distribution of charge in a polar molecule.

8.4.2 Describe what happens to polar molecules when they are placed between oppositely charged metal plates.

8.4.3 Evaluate the strength of intermolecular attractions compared with the strength of ionic and covalent bonds.

8.4.4 Identify the reason why network solids have high melting points.

Guide for Reading

Build Vocabulary  L2

Word Parts Polar, nonpolar, and dipole all have the same Latin stem *polus* meaning pole. A pole is part of a system (large or small) that has opposite electric or magnetic positions. Thus, Earth has magnetic north and south poles. Batteries have positive and negative poles or terminals.

Reading Strategy  L2

Predict Having reviewed the root of the words polar, nonpolar, and dipole, ask students to predict what is discussed under the three red headings in this section. Have them check their predictions after reading.

2 INSTRUCT

Connecting to Your World

Have students read the opening paragraph. Ask, **What kind of bonds are in the water molecule?** *(polar)* **What is the shape of the snowflake?** *(pentagon)*

Bond Polarity

Discuss 

Ask, **What does the term *electronegativity* mean?** *(It describes the attraction an atom has for electrons when the atom is in a compound.)* Remind students that electronegativities are calculated values from 0.0 to 4.0 based on the properties of the elements. Ask, **Which element is the most electronegative? Which is the least?** *(Fluorine has an electronegativity value of 4.0; cesium, the element that has the lowest attraction for electrons, has a value of 0.7.)*

Discuss 

When students use Table 8.3 to solve Conceptual Problem 8.3, they may be surprised to discover that bonds previously labeled ionic are now categorized as polar covalent. The reality is that most bonds are neither totally ionic nor totally covalent. Most fall on a continuum from ionic (complete transfer of electrons) to covalent (equal sharing of electrons). In between are polar-covalent bonds in which electrons are displaced more or less toward one atom or the other. Make sure that students understand the two uses of the Greek letter delta. A lowercase delta (δ) is used to denote partial charges; an uppercase delta (Δ) is used to denote a change in a variable.

CLASS Activity

A Magnetic Analogy 

Purpose Students experience the attraction of opposite poles of a magnet as an experience similar to the attraction of opposite electric charges.

Materials magnets, iron filings, cardboard

Procedure Place some iron filings on a cardboard. Hold a magnet under the cardboard and move it to create patterns. Allow students to handle magnets and feel the pull when north and south poles are close. Have them write a sentence relating the experience to polar molecules.

Figure 8.23 This electron-cloud picture of hydrogen chloride shows that the chlorine atom attracts the electron cloud more than the hydrogen atom does. **Inferring** ***Which atom is more electronegative, a chlorine atom or a hydrogen atom?***

A **polar covalent bond,** known also as a **polar bond,** is a covalent bond between atoms in which the electrons are shared unequally. **The more electronegative atom attracts electrons more strongly and gains a slightly negative charge. The less electronegative atom has a slightly positive charge.** Refer back to Table 6.2 in Chapter 6 to see the electronegativities of some common elements. The higher the electronegativity value, the greater the ability of an atom to attract electrons to itself.

Consider the hydrogen chloride molecule (HCl) shown in Figure 8.23. Hydrogen has an electronegativity of 2.1, and chlorine has an electronegativity of 3.0. These values are significantly different, so the covalent bond in hydrogen chloride is polar. The chlorine atom acquires a slightly negative charge. The hydrogen atom acquires a slightly positive charge. The lowercase Greek letter delta (δ) denotes that atoms involved in the covalent bond acquire only partial charges, less than 1+ or 1−.

$$\overset{\delta+}{\mathrm{H}}-\overset{\delta-}{\mathrm{Cl}}$$

The minus sign in this notation shows that chlorine has acquired a slightly negative charge. The plus sign shows that hydrogen has acquired a slightly positive charge. These partial charges are shown as clouds of electron density. The polar nature of the bond may also be represented by an arrow pointing to the more electronegative atom, as shown here.

$$\overset{+\!\!\longrightarrow}{\mathrm{H-Cl}}$$

The O—H bonds in the water molecule are also polar. The highly electronegative oxygen partially pulls the bonding electrons away from hydrogen. The oxygen acquires a slightly negative charge. The hydrogen is left with a slightly positive charge.

$$\overset{\delta-\,\delta-}{\mathrm{O}}\text{ bonded to } {}_{\delta+}\mathrm{H} \text{ and } \mathrm{H}_{\delta+} \quad or \quad \mathrm{H} \nearrow \mathrm{O} \nwarrow \mathrm{H}$$

As shown in Table 8.3, the electronegativity difference between two atoms tells you what kind of bond is likely to form. Remember when you use the table that there is no sharp boundary between ionic and covalent bonds. As the electronegativity difference between two atoms increases, the polarity of the bond increases. If the electronegativity difference is greater than 2.0, it is very likely that electrons will be pulled away completely by one of the atoms. In that case, an ionic bond will form.

Table 8.3

Electronegativity Differences and Bond Types

Electronegativity difference range	Most probable type of bond	Example
0.0–0.4	Nonpolar covalent	H—H (0.0)
0.4–1.0	Moderately polar covalent	$\overset{\delta+}{\mathrm{H}}-\overset{\delta-}{\mathrm{Cl}}$ (0.9)
1.0–2.0	Very polar covalent	$\overset{\delta+}{\mathrm{H}}-\overset{\delta-}{\mathrm{F}}$ (1.9)
≥ 2.0	Ionic	Na^+Cl^- (2.1)

CONCEPTUAL PROBLEM 8.3

Identifying Bond Type

Which type of bond (nonpolar covalent, moderately-polar covalent, or ionic) will form between each of the following pairs of atoms?

a. N and H **b.** F and F **c.** Ca and Cl **d.** Al and Cl

1 Analyze ***Identify the relevant concepts.***

In each case, the pairs of atoms involved in the bonding pair are given. The types of bonds depend on the electronegativity differences between the bonding elements. Use Table 6.2 to find the electronegativity difference, then use Table 8.3 to determine the bond type.

2 Solve ***Apply concepts to this situation.***

From Tables 6.2 and 8.3, the electronegativities, their differences, and the corresponding bond types are as follows.

a. N (3.0), H (2.1); 0.9 moderately polar covalent
b. F (4.0), F (4.0); 0.0; nonpolar covalent
c. Ca (1.0), Cl (3.0); 2.0; ionic
d. Al (1.5), Cl (3.0); 1.5; very polar covalent

Practice Problems

30. Identify the bonds between atoms of each pair of elements as nonpolar covalent, moderately polar covalent, very covalent, or ionic.
a. H and Br **b.** K and Cl **c.** C and O
d. Cl and F **e.** Li and O **f.** Br and Br

31. Place the following covalent bonds in order from least to most polar.
a. H—Cl **b.** H—Br
c. H—S **d.** H—C

Polar Molecules

The presence of a polar bond in a molecule often makes the entire molecule polar. In a **polar molecule,** one end of the molecule is slightly negative and the other end is slightly positive. In the hydrogen chloride molecule, for example, the partial charges on the hydrogen and chlorine atoms are electrically charged regions or poles. A molecule that has two poles is called a dipolar molecule, or **dipole.** The hydrogen chloride molecule is a dipole. Look at Figure 8.24. **When polar molecules are placed between oppositely charged plates, they tend to become oriented with respect to the positive and negative plates.**

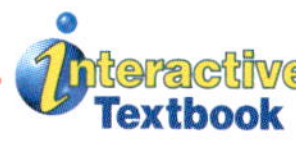

Animation 10 Learn to distinguish between polar and nonpolar molecules.
with **ChemASAP**

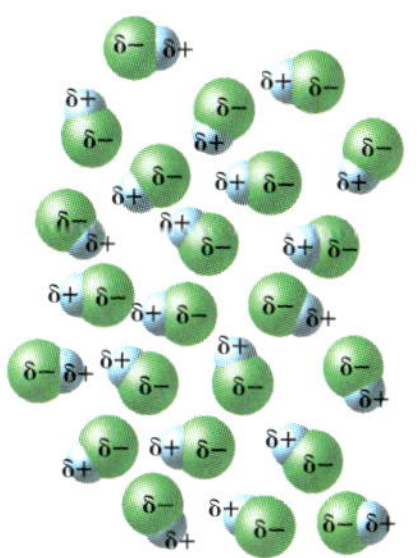

Electric field absent.
Polar molecules orient randomly.

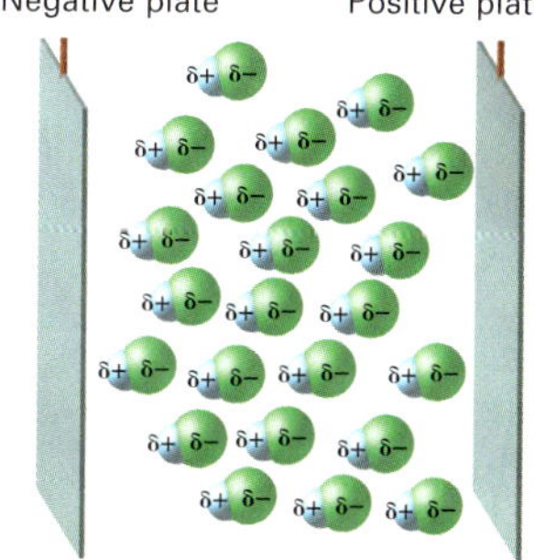

Electric field on.
Polar molecules line up.

Figure 8.24 When polar molecules, such as HCl, are placed in an electric field, the slightly negative ends of the molecules become oriented toward the positively charged plate and the slightly positive ends of the molecules become oriented toward the negatively charged plate. **Predicting** ***What would happen if, instead, carbon dioxide molecules were placed between the plates? Why?***

CONCEPTUAL PROBLEM 8.3

Answers

30. a. moderately polar covalent **b.** ionic **c.** moderately to very polar covalent **d.** moderately to very polar covalent **e.** ionic **f.** nonpolar covalent

31. c and d (tie), b, a

Practice Problems Plus 

Which type of bond (nonpolar, covalent, moderately-covalent, or ionic) will form between each of the following pairs of ions?

a. C and H *(nonpolar covalent)*
b. Be and F *(ionic)*
c. Si and Cl *(very polar covalent)*

Polar Molecules

Discuss 

Draw diagrams of molecules on the board or overhead projector. Use electronegativity values to determine which element has the partial positive charge and which has the partial negative charge. Explain how the attraction of electrons can be described with a vector. The shape of the molecule determines the direction of the vector. If the sum of the vectors acting on the central atom in one direction is cancelled by the sum of the vectors acting in the other direction, the molecule has no dipole. If they don't cancel, the molecule is polar.

Answers to...

Figure 8.23 chlorine
Figure 8.24 Carbon dioxide molecules would not change orientation because the CO_2 molecule is nonpolar.

TEACHER Demo

Observing Evidence of Polarity 

Purpose Students observe what happens to polar molecules when they are placed near positively charged materials

Materials 2 burets, 2 beakers, ring stand, buret clamps, turpentine, water, rubber rod, glass rod, fur cloth

Safety Wear safety goggles and gloves.

Procedure Attach two burets to a ring stand. Fill one with water and the other with turpentine. Place a beaker under each. **CAUTION!** *Turpentine is an irritant.* Rub a rubber rod with a fur cloth to charge it negatively. Bring the rod near the open end of the water-containing buret as you slowly open the stopcock. Have students observe the behavior of the water. Repeat the procedure using turpentine. Then repeat using a charged glass rod for both liquids.

Expected Outcomes The polar water molecules are attracted to both negatively and positively charged materials, but the nonpolar turpentine molecules are not attracted to either.

Attractions Between Molecules

Discuss 

The sharing of electrons binds atoms into molecules. If no other forces were present, all covalently bonded molecules would be gases at any temperature. No attractive forces would tend to pull them together to form a liquid or solid. However, forces do exist between covalently bonded molecules. These include van der Waals forces and hydrogen bonding.

The effect of polar bonds on the polarity of an entire molecule depends on the shape of the molecule and the orientation of the polar bonds. A carbon dioxide molecule, for example, has two polar bonds and is linear.

$$\overset{\longleftarrow\!+}{\mathrm{O{=}C}}\overset{+\!\longrightarrow}{\mathrm{{=}O}}$$

Note that the carbon and oxygens lie along the same axis. Therefore, the bond polarities cancel because they are in opposite directions. Carbon dioxide is thus a nonpolar molecule, despite the presence of two polar bonds.

The water molecule also has two polar bonds. However, the water molecule is bent rather than linear. Therefore, the bond polarities do not cancel and a water molecule is polar.

Attractions Between Molecules

Molecules can attract each other by a variety of forces. **Intermolecular attractions are weaker than either ionic or covalent bonds.** Nevertheless, you should not underestimate the importance of these forces. Among other things, these attractions are responsible for determining whether a molecular compound is a gas, a liquid, or a solid at a given temperature.

Van der Waals Forces The two weakest attractions between molecules are collectively called **van der Waals forces,** named after the Dutch chemist Johannes van der Waals (1837–1923). Van der Waals forces consist of dipole interactions and dispersion forces.

Dipole interactions occur when polar molecules are attracted to one another. The electrical attraction involved occurs between the oppositely charged regions of polar molecules, as shown in Figure 8.25. The slightly negative region of a polar molecule is weakly attracted to the slightly positive region of another polar molecule. Dipole interactions are similar to but much weaker than ionic bonds.

Figure 8.25 Polar molecules are attracted to one another by dipole interactions, a type of van der Waals force.

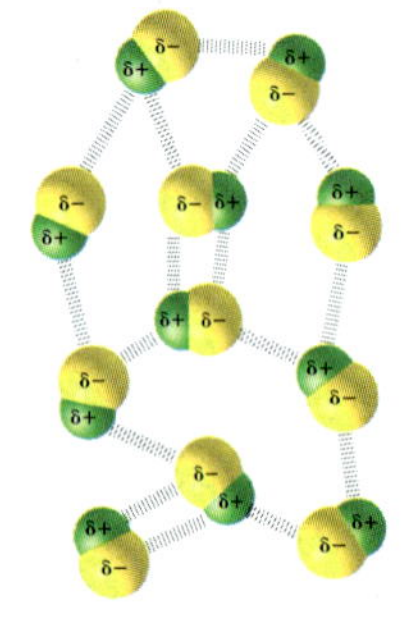

Dispersion forces, the weakest of all molecular interactions, are caused by the motion of electrons. They occur even between non-polar molecules. When the moving electrons happen to be momentarily more on the side of a molecule closest to a neighboring molecule, their electric force influences the neighboring molecule's electrons to be momentarily more on the opposite side. This causes an attraction between the two molecules similar to, but much weaker than, the force between permanently polar molecules. The strength of dispersion forces generally increases as the number of electrons in a molecule increases. The halogen diatomic molecules, for example, attract each other mainly by means of dispersion forces. Fluorine and chlorine have relatively few electrons and are gases at ordinary room temperature and pressure because of their especially weak dispersion forces. The larger number of electrons in bromine generates larger dispersion forces. Bromine molecules therefore attract each other sufficiently to make bromine a liquid at ordinary room temperature and pressure. Iodine, with a still larger number of electrons, is a solid at ordinary room temperature and pressure.

Facts and Figures

Molecular Modeling

The widespread availability of fast computers has greatly benefited chemists interested in the shapes of molecules. Molecular modeling computer programs are becoming an important tool in the pharmaceutical industry for designing drugs. These programs create models of molecules composed of thousands of atoms. Using a special viewing apparatus, chemists can see three-dimensional representations of complex molecules. Models can be rotated, shrunk, and cut in pieces. Through computer technology called virtual reality, a chemist can virtually get inside a molecular structure to examine it from every possible angle.

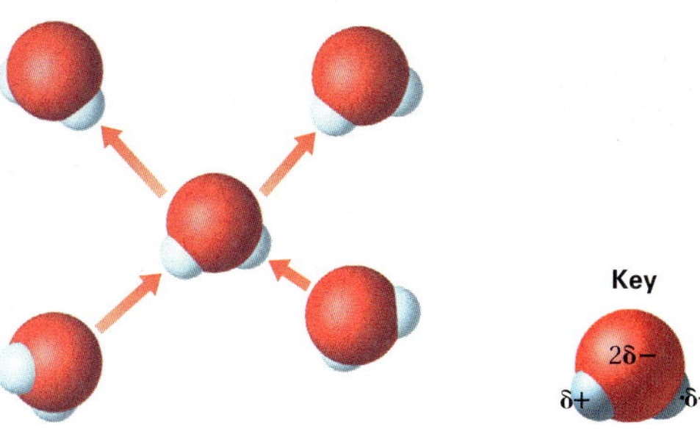

Figure 8.26 The strong hydrogen bonding between water molecules accounts for many properties of water, such as the fact that water is a liquid rather than a gas at room temperature.

Hydrogen Bonds The dipole interactions in water produce an attraction between water molecules. Each O—H bond in the water molecule is highly polar, and the oxygen acquires a slightly negative charge because of its greater electronegativity. The hydrogens in water molecules acquire a slightly positive charge. The positive region of one water molecule attracts the negative region of another water molecule, as illustrated in Figure 8.26. This attraction between the hydrogen of one water molecule and the oxygen of another water molecule is strong compared to other dipole interactions. This relatively strong attraction, which is also found in hydrogen-containing molecules other than water, is called a hydrogen bond. Figure 8.26 illustrates hydrogen bonding in water.

Hydrogen bonds are attractive forces in which a hydrogen covalently bonded to a very electronegative atom is also weakly bonded to an unshared electron pair of another electronegative atom. This other atom may be in the same molecule or in a nearby molecule. Hydrogen bonding always involves hydrogen. It is the only chemically reactive element with valence electrons that are not shielded from the nucleus by other electrons.

Remember that for a hydrogen bond to form, a covalent bond must already exist between a hydrogen atom and a highly electronegative atom, such as oxygen, nitrogen, or fluorine. The combination of this strongly polar bond and the lack of shielding effect in a hydrogen atom is responsible for the relative strength of hydrogen bonds. A hydrogen bond has about 5% of the strength of an average covalent bond. Hydrogen bonds are the strongest of the intermolecular forces. They are extremely important in determining the properties of water and biological molecules such as proteins. Figure 8.27 shows how the relatively strong attractive forces between water molecules cause the water to form small drops on a waxy surface.

Checkpoint *What are hydrogen bonds?*

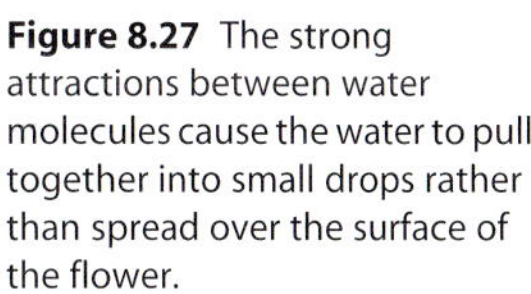

Figure 8.27 The strong attractions between water molecules cause the water to pull together into small drops rather than spread over the surface of the flower.

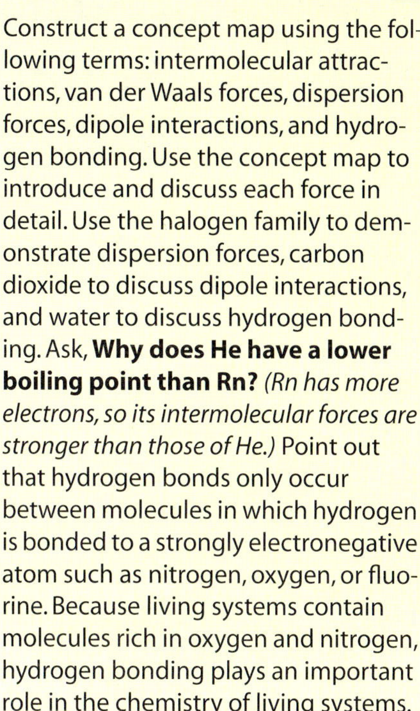

Discuss

L2

Construct a concept map using the following terms: intermolecular attractions, van der Waals forces, dispersion forces, dipole interactions, and hydrogen bonding. Use the concept map to introduce and discuss each force in detail. Use the halogen family to demonstrate dispersion forces, carbon dioxide to discuss dipole interactions, and water to discuss hydrogen bonding. Ask, **Why does He have a lower boiling point than Rn?** *(Rn has more electrons, so its intermolecular forces are stronger than those of He.)* Point out that hydrogen bonds only occur between molecules in which hydrogen is bonded to a strongly electronegative atom such as nitrogen, oxygen, or fluorine. Because living systems contain molecules rich in oxygen and nitrogen, hydrogen bonding plays an important role in the chemistry of living systems. Hydrogen bonding explains many of the physical properties of water.

Download a worksheet on **Intermolecular Forces** for students to complete, and find additional teacher support from NSTA SciLinks.

Facts and Figures

Molecular Modeling

The Dutch physicist Johannes van der Waals (1837–1923) surmised that real gas behavior could not be described exactly by the ideal gas law because the law assumes that gas molecules have no volume and that no attractive forces exist between gas molecules. Van der Waals adjusted the ideal gas equation by introducing two factors: one to take cognizance of the real volume of gas molecules; the other to account for intermolecular forces between molecules. For his work in developing an understanding of intermolecular attractions, he was awarded the Nobel Prize in Physics in 1910, and van der Waals forces, the weak attractive forces he had been studying, were named in his honor.

Answers to...

Checkpoint a hydrogen bond is the attraction between a hydrogen atom bonded to a very electronegative atom and an unshared electron pair of another electronegative atom.

Technology & Society

Make sure that students are familiar with polymerization. Ask, **What is a monomer?** *(a single small molecule)* **What is a polymer?** *(a large molecule made up of many repeated monomer units)* **In an adhesive, what causes the monomer to form a polymer?** *(moisture in the air or on the surfaces to be joined together)* **Why does the adhesive not polymerize in the tube?** *(A stabilizer is added to the monomer that prevents polymerization until the monomer comes in contact with water.)*

Students can use library and Internet resources to find out more about any of the following topics:

- the adhesives used in ancient times, as long ago as 3300 years ago,
- natural adhesives from plants and animals including simple flour and water paste, animal glue, casein glue, and blood albumin glue,
- thermoplastic adhesives that respond to heating and cooling,
- the structure of epoxy resins,
- remoistenable adhesives used on stamps and envelopes.

Students may also want to investigate the adsorption theory of adhesion. What is the mechanism by which substances stick to each other? In theory, all substances can be categorized as adhesives if they undergo intermolecular attractions or bonding with other contacting substances.

Answers to...

Interpreting Diagrams A stabilizer stops the monomers from forming a solid polymer.

Technology & Society

The Chemistry of Adhesives

Most common adhesives are sticky substances that you can use to bind two surfaces together. The diagram shows how some adhesives work. The adhesive is a liquid until the surfaces are in position. Then the adhesive sets into a solid. Adhesion can work in three ways: Molecules of the polymer may fill crevices in the surfaces being connected, the molecules may also become attracted by intermolecular forces, or they may react by forming covalent bonds. **Interpreting Diagrams** ***Explain the purpose of a stabilizer in an adhesive.***

Permanent adhesion An epoxy resin attached this full-size automobile to the billboard. Epoxy resins are often stored in two parts that are mixed just before the epoxy is used. Strong binding forces in these adhesives make them heat- and water-resistant.

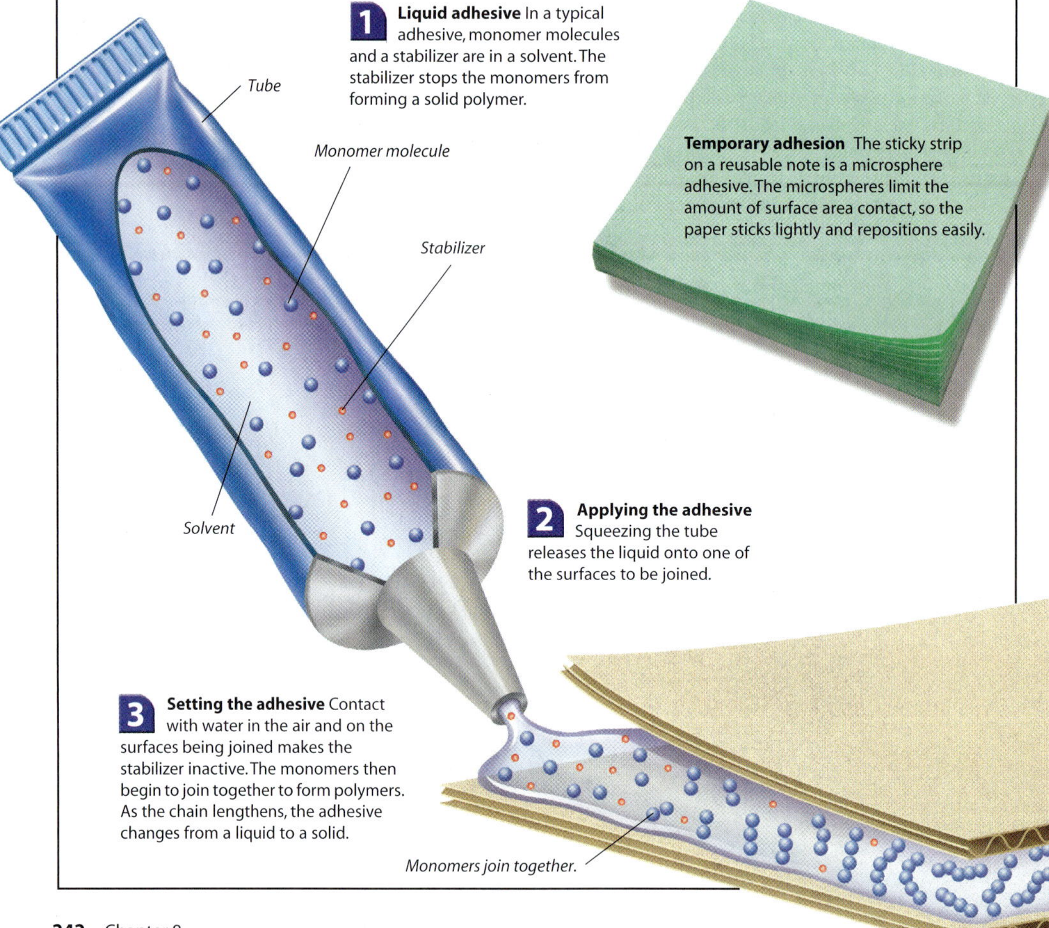

1 Liquid adhesive In a typical adhesive, monomer molecules and a stabilizer are in a solvent. The stabilizer stops the monomers from forming a solid polymer.

2 Applying the adhesive Squeezing the tube releases the liquid onto one of the surfaces to be joined.

3 Setting the adhesive Contact with water in the air and on the surfaces being joined makes the stabilizer inactive. The monomers then begin to join together to form polymers. As the chain lengthens, the adhesive changes from a liquid to a solid.

Temporary adhesion The sticky strip on a reusable note is a microsphere adhesive. The microspheres limit the amount of surface area contact, so the paper sticks lightly and repositions easily.

242 Chapter 8

Intermolecular Attractions and Molecular Properties

At room temperature, some compounds are gases, some are liquids, and some are solids. The physical properties of a compound depend on the type of bonding it displays—in particular, on whether it is ionic or covalent. A great range of physical properties occurs among covalent compounds. This is mainly because of widely varying intermolecular attractions.

The melting and boiling points of most compounds composed of molecules are low compared with those of ionic compounds. In most solids formed from molecules, only the weak attractions between molecules need to be broken. However, a few solids that consist of molecules do not melt until the temperature reaches 1000°C or higher, or they decompose without melting at all. Most of these very stable substances are **network solids** (or network crystals), solids in which all of the atoms are covalently bonded to each other. **Melting a network solid would require breaking covalent bonds throughout the solid.**

Diamond is an example of a network solid. As shown in Figure 8.28, each carbon atom in a diamond is covalently bonded to four other carbons, interconnecting carbon atoms throughout the diamond. Cutting a diamond requires breaking a multitude of these bonds. Diamond does not melt; rather, it vaporizes to a gas at 3500°C and above.

Silicon carbide, with the formula SiC and a melting point of about 2700°C, is also a network solid. Silicon carbide is so hard that it is used in grindstones and as an abrasive as illustrated in Figure 8.29. The molecular structures of silicon carbide and diamond are similar to each other. You can think of samples of diamond, silicon carbide, and other network solids as single molecules.

What substance is an example of a network solid?

Figure 8.28 Diamond is a network-solid form of carbon. Diamond has a three-dimensional structure, with each carbon at the center of a tetrahedron.

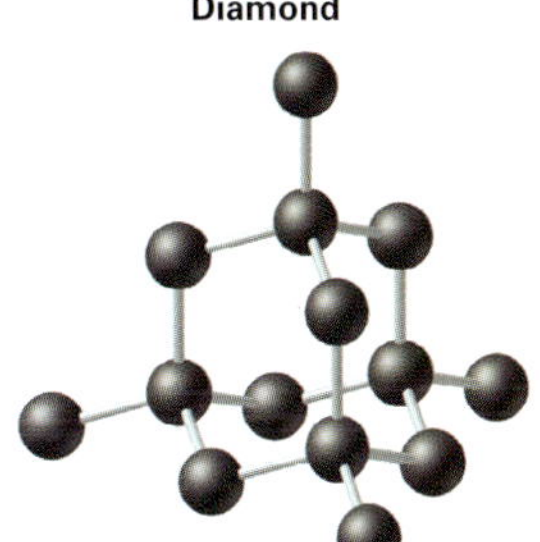

For: Links on Diamond
Visit: www.SciLinks.org
Web Code: cdn-1085

Figure 8.29 Silicon carbide, a network solid, is so hard that it is used in this grindstone to wear down the end of a hardened steel cutting tool to form a sharp edge.

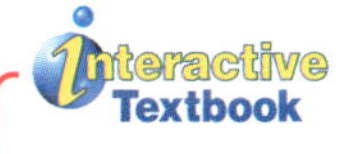
Simulation 8 Relate melting and boiling points to the strength of intermolecular forces.
with ChemASAP

Intermolecular Attractions and Molecular Properties

Download a worksheet on **Diamond** for students to complete, and find additional teacher support from NSTA SciLinks.

TEACHER Demo

Evidence of Hydrogen Bonding 

Purpose Students see how hydrogen bonding affects the surface tension and viscosity of fluids.

Materials Four 300-mL stoppered flasks, each containing 100 mL of one of these liquids: glycerol, ethylene glycol, water, ethanol

Procedure Gently rotate one of the flasks so that its contents begin to swirl. Have students measure the time interval between your ceasing to rotate the bottle and the disappearance of the swirling vortex. Repeat the procedure for each fluid. Be sure to rotate each flask the same number of times.

Expected Outcome From longest time to shortest: ethanol, water, ethylene glycol, and glycerol. Point out that hydrogen bonding increases a liquid's surface tension and viscosity. Thus, liquids with a relatively high degree of hydrogen bonding generally stop swirling more quickly. Have students order the fluids according to increasing hydrogen bonding. *(ethanol, water, ethylene glycol, glycerol)* Show the structural formulas of the compounds to demonstrate the origins of the hydrogen bonding.

Answers to...
Checkpoint diamond

Section 8.4 (continued)

3 ASSESS

Evaluate Understanding L2

Have students account for the generally low melting points of covalent compounds in terms of bonding. *(Intermolecular attractions between covalent compounds are generally not as strong as ionic bonds.)*

Reteach L1

Help students list the types of intermolecular attractions that operate between molecules. Give examples of compounds that would fit into each category. Ask students to name other examples.

Connecting Concepts

In dipole interactions, the slightly positive end of a polar molecule attracts the slightly negative end of another polar molecule. Dispersion forces arise from the movement of electrons in any molecule, polar or nonpolar. If the electrons in one molecule happen to be momentarily more on one side, the electrons in a neighboring molecule can be repulsed. Weak attractions result between the temporary dipoles.

If your class subscribes to the Interactive Textbook, use it to review key concepts in Section 8.4.

with ChemASAP

Table 8.4 summarizes some of the characteristic differences between ionic and covalent (molecular) substances. Note that ionic compounds have higher melting points than molecular compounds. Ionic compounds also tend to be soluble in water.

Table 8.4

Characteristics of Ionic and Covalent Compounds

Characteristic	Ionic compound	Covalent compound
Representative unit	Formula unit	Molecule
Bond formation	Transfer of one or more electrons between atoms	Sharing of electron pairs between atoms
Type of elements	Metallic and nonmetallic	Nonmetallic
Physical state	Solid	Solid, liquid, or gas
Melting point	High (usually above 300°C)	Low (usually below 300°C)
Solubility in water	Usually high	High to low
Electrical conductivity of aqueous solution	Good conductor	Poor to nonconducting

8.4 Section Assessment

32. **Key Concept** How do electronegativity values determine the charge distribution in a polar covalent bond?
33. **Key Concept** What happens when polar molecules are between oppositely charged metal plates?
34. **Key Concept** Compare the strengths of intermolecular attractions to the strengths of ionic bonds and covalent bonds.
35. **Key Concept** Explain why network solids have high melting points.
36. Not every molecule with polar bonds is polar. Explain this statement. Use CCl_4 as an example.
37. Draw the electron dot structure for each molecule. Identify polar covalent bonds by assigning slightly positive ($\delta+$) and slightly negative ($\delta-$) symbols to the appropriate atoms.
    a. HOOH **b.** BrCl **c.** HBr **d.** H_2O
38. How does a network solid differ from most other covalent compounds?

Connecting Concepts

Dipole Interactions and Dispersion Forces
Explain how dipole interactions and dispersion forces are related. First, explain what produces the attractions between polar molecules. Then explain what produces dispersion forces between molecules. Identify what is similar and what is different in the two mechanisms of intermolecular attraction.

Assessment 8.4 Test yourself on the important concepts of Section 8.4.

with ChemASAP

Section 8.4 Assessment

32. The more electronegative atom attracts electrons more strongly and gains a partial negative charge. The less electronegative atom has a partial positive charge.
33. Polar molecules tend to become oriented with respect to the positive and negative plates.
34. Intermolecular attractions are weaker than either ionic or covalent bonds.
35. Melting a network solid requires breaking covalent bonds throughout the solid.
36. The atoms in CCl_4 are oriented so that the bond polarities cancel.
37. **a.** $\delta+\ \delta-\ \delta-\ \delta+$ H:Ö:Ö:H **b.** $\delta+\ \delta-$:Br:Cl: **c.** $\delta+\ \delta-$ H:Br:
    d. $\delta+\ \delta-\ \delta+$ H:Ö:H
38. The atoms in a network solid are covalently bonded in a large array (or crystal), which can be thought of as a single molecule.

Small-Scale LAB

Paper Chromatography of Food Dyes

Purpose

To use paper chromatography to separate and identify food dyes in various samples.

Materials

- pencil
- paper
- ruler
- scissors
- toothpicks
- 4 different colors of food coloring
- plastic cup
- 0.1% NaCl solution
- chromatography paper

Procedure

Cut a 5 cm × 10 cm strip of chromatography paper and label it with a pencil, as shown below. Use a different toothpick to place a spot of each of the four food colors on the Xs on your chromatography paper. Allow the spots to dry for a few minutes. Fill the plastic cup so its bottom is just covered with the solvent (0.1% NaCl solution). Wrap the chromatography paper around a pencil. Remove the pencil and place the chromatography paper, color-spot side down, in the solvent. When the solvent reaches the top of the chromatography paper, remove the paper and allow it to dry.

Analysis

Using your experimental data, record the answers to the following questions below your data table.

1. If a food color sample yields a single streak or spot, it is usually a pure compound. Which food colors consist of pure compounds?
2. Which food colors are mixtures of compounds?
3. Food colors often consist of a mixture of three colored dyes: Red No. 40, Yellow No. 5, and Blue No. 1. Read the label on the food color package. Which dyes do your food color samples contain?
4. Identify each spot or streak on your chromatogram as Red No. 40, Yellow No. 5, or Blue No. 1.
5. Paper chromatography separates polar covalent compounds on the basis of their relative polarities. The most polar dyes migrate the fastest and appear at the top of the paper. Which dye is the most polar? The least polar?

You're The Chemist

The following small-scale activities allow you to develop your own procedures and analyze the results.

1. **Design It!** Design and carry out an experiment to identify the dyes in various colored candies.
2. **Design It!** Design and carry out an experiment to identify the dyes in various colored markers using the paper chromatography method.
3. **Design It!** Design and carry out an experiment to identify the dyes in various colored powdered drinks using the paper chromatography method.
4. **Analyze It!** Use different solvents, such as 2-propanol (rubbing alcohol), vinegar, and ammonia, to separate food colors. Does the choice of solvent affect the results?
5. **Analyze It!** Explore the effect of different papers on your results. Try paper towels, notebook paper, and coffee filters. Report your results. Examine the relative positions of Blue No. 1 and Yellow No. 5. What do you observe?

For Enrichment 

Students could investigate how different carrier fluids affect the movement of the four dyes. For example, they might try pure water or ethanol.

Analyze

1. Red, yellow, and blue are pure.
2. Green is Yellow No. 5 plus Blue No. 1.
3. Blue No. 1; Red No. 40 (No. 3 if supplies haven't been used up); Yellow No. 5 or Yellow No. 6 if the dye appears orange.
4. See the drawing.

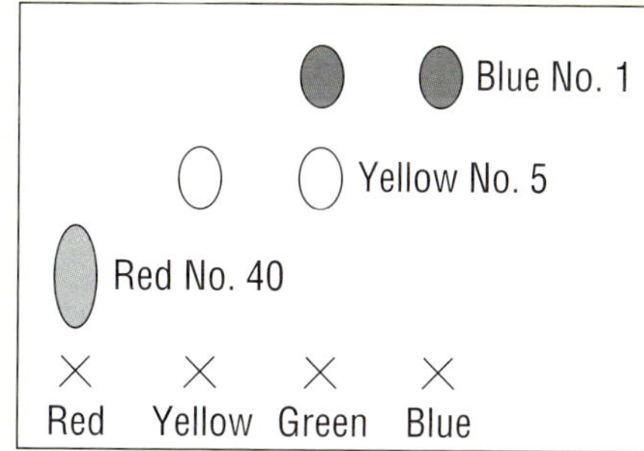

5. Blue is the most polar; red, the least polar.

Paper Chromatography of Food Dyes 

Objective

After completing this activity, students will be able to infer differences in polarity by observing the chromatographic separation of food dyes.

 **Prep Time** 15 minutes

Advance Prep

- To prepare 0.1% NaCl, dissolve 1.0 g NaCl in 1.0 L water.
- Jelly beans and small candy-coated chocolates are good sources of food dyes. Dissolve a package of unsweetened powdered soft drink in a small amount of water and have students use toothpicks to spot this solution on chromatography paper.
- Coffee filters can substitute for chromatography paper, but caution students not to overload their samples.

Class Time 30 minutes

Teaching Tips Tell students to position the chromatography paper so that the lower end of the paper is in the liquid but the spots of dye are above the surface. A similar, more complete lab on chromatography is found in the Small-Scale Laboratory Manual.

Expected Outcome In order of increasing speed of travel: red, yellow, blue. Green separated into yellow and blue.

You're the Chemist

1. Wet candy, blot to remove excess water, press onto chromatography paper, and develop in 0.1% NaCl.
2. Place spots on chromatography paper. Develop in NaCl solution.
3. Use a toothpick to place spots of powdered drink solutions.
4. Rubbing alcohol moves more slowly and gives slightly better separation.
5. With some papers, the positions of Blue No. 1 and Yellow No. 5 are reversed.

CHAPTER 8 Study Guide

Study Tip

Switch Tasks When you feel yourself losing focus, switch the type of task you are working on, the subject that you are studying, or the environment you are in. Take a break and walk around a bit. Stop studying when you are no longer productive.

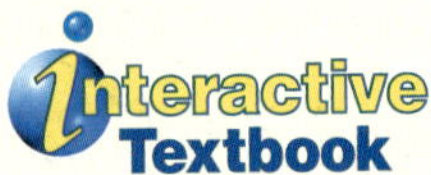

If your class subscribes to the Interactive Textbook with ChemASAP, your students can go online to access an interactive version of the Student Edition and a self-test.

with **ChemASAP**

CHAPTER 8 Study Guide

Key Concepts

8.1 Molecular Compounds

- Molecular compounds tend to have relatively low melting and boiling points.
- A molecular formula shows how many atoms of each element a molecule contains.

8.2 The Nature of Covalent Bonding

- Electron sharing occurs so that atoms attain the configurations of noble gases.
- An electron dot structure shows the shared electrons of a covalent bond by a pair of dots.
- Atoms form double or triple bonds by sharing two or three pairs of electrons.
- In a coordinate covalent bond, the shared electron pair comes from a single atom.
- A large bond dissociation energy corresponds to a strong covalent bond.
- In ozone, the bonding of oxygen atoms is a hybrid of the extremes represented by the resonance forms.
- The octet rule is not satisfied in molecules with an odd number of electrons, and in molecules where an atom has less, or more, than a complete octet of valence electrons.

8.3 Bonding Theories

- Just as an atomic orbital belongs to a particular atom, a molecular orbital belongs to a molecule as a whole.
- According to VSEPR theory, the repulsion between electron pairs causes molecular shapes to adjust so that the valence-electron pairs stay as far apart as possible.
- Orbital hybridization provides information about both molecular bonding and molecular shape.

8.4 Polar Bonds and Molecules

- When different atoms bond, the more electronegative atom attracts electrons more strongly and acquires a slight negative charge.
- Polar molecules between oppositely charged metal plates tend to become oriented with respect to the positive and negative plates.
- Intermolecular attractions are weaker than either an ionic or covalent bond.
- Melting a network solid requires breaking covalent bonds throughout the solid.

Vocabulary

- bond dissociation energy (p. 226)
- bonding orbital (p. 230)
- covalent bond (p. 213)
- coordinate covalent bond (p. 223)
- diatomic molecule (p. 214)
- dipole (p. 239)
- dipole interactions (p. 240)
- dispersion forces (p. 240)
- double covalent bond (p. 221)
- hybridization (p. 234)
- hydrogen bonds (p. 241)
- molecular compound (p. 214)
- molecular formula (p. 215)
- molecular orbital (p. 230)
- molecule (p. 214)
- network solids (p. 243)
- nonpolar covalent bond (p. 237)
- pi bond (p. 231)
- polar bond (p. 238)
- polar covalent bond (p. 238)
- polar molecule (p. 239)
- polyatomic ion (p. 223)
- resonance structure (p. 227)
- sigma bond (p. 230)
- single covalent bond (p. 217)
- structural formula (p. 218)
- tetrahedral angle (p. 232)
- triple covalent bond (p. 221)
- unshared pair (p. 218)
- van der Waals forces (p. 240)
- VSEPR theory (p. 232)

Organizing Information

Construct a concept map that organizes the major ideas of this chapter.

Concept Map 8 Solve the Concept Map with the help of an interactive guided tutorial.

with **ChemASAP**

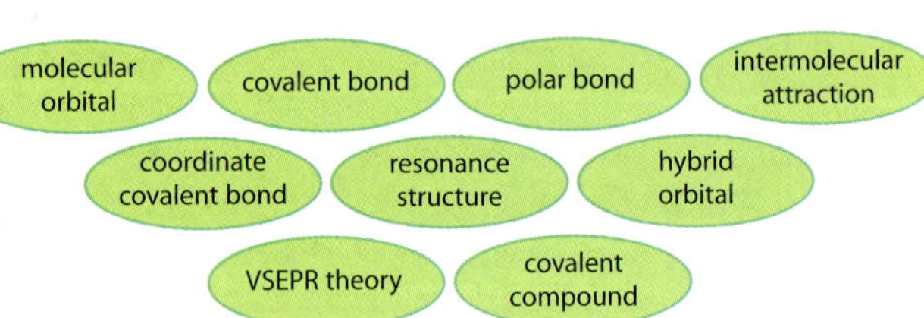

Chapter Resources

Print

- **Core Teaching Resources,** Chapter 8, Practice Problems, Vocabulary Review, Quiz, Chapter Test A, Chapter Test B

Technology

- **Computer Test Bank,** Chapter 8 Test
- **Interactive Textbook with ChemASAP,** Chapter 8

CHAPTER 8

Assessment

Reviewing Content

8.1 Molecular Compounds

39. The melting point of a compound is 1240°C. Is this compound most likely an ionic or a molecular compound?

40. Identify the number and kinds of atoms present in a molecule of each compound.
a. ascorbic acid (vitamin C), $C_6H_8O_6$
b. sucrose (table sugar), $C_{12}H_{22}O_{11}$
c. trinitrotoluene (TNT), $C_7H_5N_3O_6$

41. Which of the following gases in Earth's atmosphere would you expect to find as molecules and which as individual atoms? Explain.
a. nitrogen **b.** oxygen **c.** argon

8.2 The Nature of Covalent Bonding

42. Explain why neon is monatomic but chlorine is diatomic.

43. Classify the following compounds as ionic or covalent.
a. $MgCl_2$ **b.** Na_2S **c.** H_2O **d.** H_2S

44. Describe the difference between an ionic and a covalent bond.

45. How many electrons do two atoms in a double covalent bond share? How many in a triple covalent bond?

46. Draw plausible electron dot structures for the following substances. Each substance contains only single covalent bonds.
a. I_2 **b.** OF_2 **c.** H_2S **d.** NI_3

47. Characterize a coordinate covalent bond and give an example.

48. Explain why compounds containing C—N and C—O single bonds can form coordinate covalent bonds with H^+ but compounds containing only C—H and C—C single bonds cannot.

49. Using electron dot structures, draw at least two resonance structures for the nitrite ion (NO_2^-). The oxygens in NO_2^- are attached to the nitrogen.

50. Which of these compounds contain elements that do not follow the octet rule? Explain.
a. NF_3 **b.** PCl_2F_3 **c.** SF_4 **d.** SCl_2

51. Explain what is meant by bond dissociation energy.

52. What is the relationship between the magnitude of a molecule's bond dissociation energy and its expected chemical reactivity?

8.3 Bonding Theories

53. What is a pi bond? Describe, with the aid of a diagram, how the overlap of two half-filled *p* atomic orbitals produces a pi bond.

54. Use VSEPR theory to predict the shapes of the following species.
a. CO_2 **b.** $SiCl_4$ **c.** SO_3
d. SCl_2 **e.** CO **f.** H_2Se

55. The molecule CO_2 has two carbon–oxygen double bonds. Describe the bonding in the CO_2 molecule, which involves hybridized orbitals for carbon and oxygen.

56. What types of hybrid orbitals are involved in the bonding of the carbon atoms in the following molecules?
a. CH_4 **b.** $H_2C{=}CH_2$
c. $HC{\equiv}CH$ **d.** $N{\equiv}C{-}C{\equiv}N$

8.4 Polar Bonds and Molecules

57. How must the electronegativies of two atoms compare if a covalent bond between them is to be polar?

58. The bonds between the following pairs of elements are covalent. Arrange them according to polarity, listing the most polar bond first.
a. H—Cl **b.** H—C **c.** H—F
d. H—O **e.** H—H **f.** S—Cl

59. What is a hydrogen bond?

60. Depict the hydrogen bonding between two ammonia molecules and between one ammonia molecule and one water molecule.

61. Why do compounds with strong intermolecular attractive forces have higher boiling points than compounds with weak intermolecular attractive forces?

CHAPTER 8

Assessment

Reviewing Content

39. ionic

40. **a.** 6 C, 8 H, 6 O **b.** 12 C, 22 H, 11 O **c.** 7 C, 5 H, 3 N, 6 O

41. Nitrogen and oxygen achieve stability as diatomic molecules. Argon exists as individual atoms because it has a stable noble-gas electron configuration.

42. Neon has an octet of electrons. A chlorine atom achieves an octet by sharing an electron with another chlorine atom.

43. **a.** ionic **b.** ionic **c.** covalent **d.** covalent

44. Ionic bonds depend on electrostatic attraction between ions. Covalent bonds depend on electrostatic attraction between shared electrons and nuclei of combining atoms.

45. A double covalent bond has four shared electrons (two bonding pairs); a triple covalent bond has six shared electrons (three bonding pairs).

46. **a.** :I:I: **b.** :F:O:F: **c.** H:S:H **d.** :I:N:I: with :I: below N

47. One atom contributes both electrons to a coordinate covalent bond, as in CO.

48. An unshared pair of electrons is needed for a coordinate covalent bond. There are no unshared pairs in compounds with only C-H and C-C bonds.

49. $[:\ddot{O}:\ddot{N}::\ddot{O}:]^- \leftrightarrow [:\ddot{O}::\ddot{N}:\ddot{O}:]^-$

50. b and c; assuming only single bonds, the P and S atoms each have 10 valence electrons.

51. Bond dissociation energy is defined as the energy needed to break one covalent bond.

52. Increasing bond dissociation energy is linked to lower chemical reactivity.

53. A pi bond is formed by the side-by side overlap of two half-filled *p* atomic orbitals to produce a pi molecular orbital. In a pi bond, the bonding electrons are most likely to be found in sausage-shaped regions above and below the bond. See Figure 8.15.

54. **a.** linear **b.** tetrahedral **c.** trigonal planar **d.** bent **e.** linear **f.** bent

55. The 2*s* and the 2*p* orbitals form two *sp* hybrid orbitals on the carbon atom. One *sp* hybrid orbital forms a sigma bond between the carbon atom and each oxygen atom. Pi bonds between each oxygen atom and the carbon are formed by the unhybridized 2*p* orbitals.

56. **a.** sp^3 **b.** sp^2 **c.** *sp* **d.** *sp*

57. The electronegativities of the two atoms will differ by about 0.4 to 2.0.

58. c, d, a, f, b, e

59. A hydrogen bond is formed by an electrostatic interaction between a hydrogen atom that is covalently bonded to an electronegative atom, and an unshared electron pair of a nearby atom.

60.
H₃N: ····H—NH₂ (H—N(H)(H): ····H—N:(H)(H))

H—O:(H) ····H—N:(H)(H)

61. More energy is required to separate the molecules.

Assessment continued

Understanding Concepts

62. The 3*s* and three 3*p* orbitals of phosphorus hybridize to form four sp^3 atomic orbitals. The resulting shape is pyramidal with a bond angle of 107° between the sigma bonds.

63. :C̤l̤:S̤:C̤l̤: with :Ö: below S

64. a. C does not have an octet.
[:C::N:]⁻

b. One F has more than an octet.
:F̤:P̤:F̤: with :F̤: below P

65. a. tetrahedral, 109.5°
b. trigonal planar, 120°
c. tetrahedral, 109.5°
d. bent, 105°

66. a. The percent ionic character increases as the difference in electronegativities increases.
b. 1.6
c. (1) 85% (2) 10% (3) 62% (4) 23%

67. a. 109.5°
b. 120°
c. 180°

68. a. trigonal planar
b. pyramidal
c. linear
d. tetrahedral

69. a. Phosphorus in PBr_5 has 10 valence electrons.

70. a. H:C̤:C:Ö:H (H above and below the first C; :Ö: double-bonded above the second C)

b. No, the molecule contains one carbon-oxygen double bond and one carbon-oxygen single bond.
c. polar bond
d. Yes, it has polar oxygen atoms at one end of the molecule and a nonpolar CH_3^- group at the opposite end.

Assessment *continued*

Understanding Concepts

62. Devise a hybridization scheme for PCl_3 and predict the molecular shape based on this scheme.

63. The chlorine and oxygen atoms in thionyl chloride ($SOCl_2$) are bonded directly to the sulfur. Draw an acceptable electron dot structure for thionyl chloride.

64. Explain why each electron dot structure is incorrect. Replace each structure with one that is more acceptable.

a. [:C::N̤:]⁻ **b.** :F̤:P::F̤: with :F̤: below P

65. Use VSEPR theory to predict the geometry of each of the following.
a. $SiCl_4$ **b.** CO_3^{2-} **c.** CCl_4 **d.** SCl_2

66. The following graph shows how the percent ionic character of a single bond varies according to the difference in electronegativity between the two elements forming the bond. Answer the following questions, using this graph and Table 6.2.

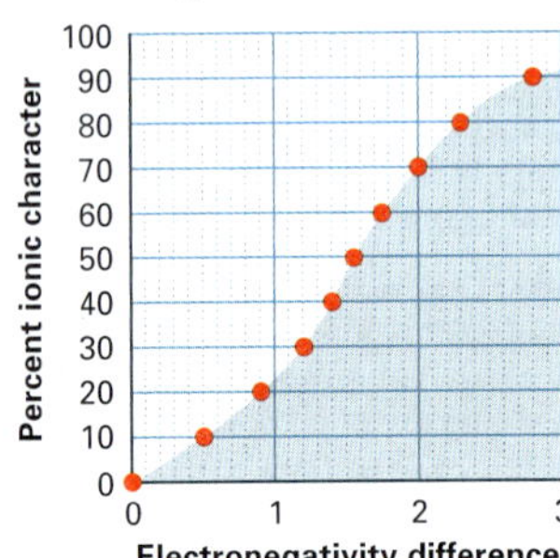

a. What is the relationship between the percent ionic character of single bonds and the electronegativity difference?
b. What electronegativity difference will result in a bond with a 50% ionic character?
c. Estimate the percent ionic character of the bonds formed between (1) lithium and oxygen, (2) nitrogen and oxygen, (3) magnesium and chlorine, and (4) nitrogen and fluorine.

67. Give the angles between the orbitals of each hybrid.
a. sp^3 hybrids
b. sp^2 hybrids
c. sp hybrids

68. What is the geometry around the central atom in each of these simple molecules?

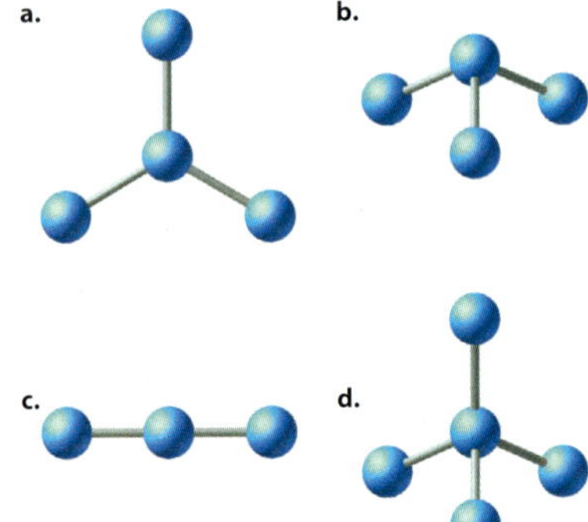

69. Which of the following molecules contains a central atom that does not obey the octet rule?
a. PBr_5
b. AlI_3
c. PF_3
d. $SiCl_4$

70. Vinegar contains the compound ethanoic acid, whose molecular formula is CH_3COOH.
a. Draw the electron dot structure of ethanoic acid.
b. Is the bonding between each of the oxygen atoms and the carbon the same?
c. Is the bonding between the carbon atom and each oxygen atom a polar or nonpolar bond?
d. Is ethanoic acid a polar molecule?

Critical Thinking

71. Make a list of the elements in the compounds found in Table 8.2 on page 224. What do the elements that form covalent bonds have in common?

72. Is there a clear difference between a very polar covalent bond and an ionic bond? Explain.

73. Although the relative positions of the atoms are correct in each of these molecules there are one or more incorrect bonds in each of the electron dot structures. Identify the incorrect bonds. Draw the correct electron dot structure for each molecule.

a. H═C═C═H

b. :F—O—H

c. : I :::Cl :

d. H—N ::: N—H

74. Ethyl alcohol and dimethyl ether each have the same molecular formula, C_2H_6O. Ethyl alcohol has a much higher boiling point (78°C) than dimethyl ether (−25°C). Propose an explanation for this difference.

75. What shape do you expect for a molecule with a central atom and the following?

a. two bonding pairs of electrons and two nonbonding pairs of electrons

b. four bonding pairs and zero nonbonding pairs

c. three bonding pairs and one nonbonding pair

76. Is this statement true or false? "As the electronegativity difference between covalently bonded atoms increases, the strength of the bond increases." Use the table below to justify your answer.

Bond	Electronegativity Difference	Bond Dissociation Energy(kJ/mol)
C—C	2.5 − 2.5 = 0.0	347
C—H	2.5 − 2.1 = 0.4	393
C—N	3.0 − 2.5 = 0.5	305
C—O	3.5 − 2.5 = 1.0	356

Concept Challenge

77. The electron structure and geometry of the methane molecule (CH_4) can be described by a variety of models, including electron dot structure, simple overlap of atomic orbitals, and orbital hybridization of carbon. Draw the electron dot structure of CH_4. Sketch two molecular orbital pictures of the CH_4 molecule. For your first sketch, assume that one of the paired $2s^2$ electrons of carbon has been promoted to the empty $2p$ orbital. Overlap each half-filled atomic orbital of carbon to a half-filled $2s$ orbital of hydrogen. What is the predicted geometry of the CH_4 molecule, using this simple overlap method? In your second sketch, assume hybridization of the $2s$ and $2p$ orbitals of carbon. Now what geometry would you predict for CH_4? Which picture is preferable based on the facts that all H—C—H bond angles in CH_4 are 109.5° and all C—H bond distances are identical?

78. There are some compounds in which one atom has more electrons than the corresponding noble gas. Examples are PCl_5, SF_6, and IF_7. Draw the electron dot structures of P, S, and I atoms and of these compounds. Considering the outer shell configuration of P, S, and I, develop an orbital hybridization scheme to explain the existence of these compounds.

79. Draw the electron dot structure of formic acid, H_2CO_2. The carbon is the central atom, and all the atoms are attached to the carbon except for a hydrogen bonded to an oxygen.

80. Oxalic acid, $C_2H_2O_4$, is used in polishes and rust removers. Draw the electron dot structure for oxalic acid given that the two carbons are bonded together but neither of the hydrogen atoms is bonded to a carbon atom.

81. Draw as many resonance structures as you can for HN_3. (*Hint:* the three nitrogen atoms are bonded in a row and the hydrogen atom is bonded to a nitrogen atom at the end of the row of nitrogens.)

82. Draw an electron dot structure for each molecule and explain why it fails to obey the octet rule.

a. BeF_2 **b.** SiF_6 **c.** ClO_2 **d.** BF_3 **e.** XeF_2

Critical Thinking

71. C, O, H, S, N, F, Cl: These elements are all nonmetals.

72. Answers will vary. Table 8.3 suggests there is no clear difference. The student's argument could be based on chemical properties, such as conductivity of the compound in the liquid state.

73. a. two covalent bonds to both hydrogens; double bond between carbons H:C::C:H

b. Fluorine and oxygen have only four electrons. :F:O:H

c. Halogens form one covalent bond, not three. :I:Cl:

d. Nitrogen forms three covalent bonds, not four. H:N::N:H

74. Ethyl alcohol can form intermolecular hydrogen bonds between its polar –OH groups, but dimethyl ether can not form hydrogen bonds.

75. a. bent

b. tetrahedral

c. pyramidal

76. False. The bond dissociation energies exhibit no particular trend and, in fact, are fairly constant.

Concept Challenge

77. H:C:H (with H above and below); 90°, 90°, ?; 109.5°, 109.5°, 109.5°, 109.5°

The first sketch shows carbon's three p orbitals oriented at 90° angles, resulting in a pyramidal structure for the carbon atom with three hydrogen atoms. The 4th C-H bond, formed with carbon's 2s orbital and a hydrogen atom's 1s orbital, is at unspecified angles to the other three C-H bonds. The second sketch is tetrahedral. The bond angles in the first sketch are not all the same; some are 90°. The bond angles in the second sketch are all 109.5° The second sketch is correct. (Note: The wedge-shaped lines come out of the page; the dotted lines recede into the page.)

78.

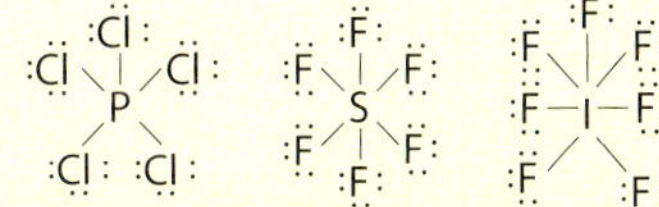

P forms 5 hybrid orbitals (dsp^3), S forms 6 hybrid orbitals (d^2sp^3), and I forms 7 hybrid orbitals (d^3sp^3).

79. H:C:O:H (with :O: above C)

80. H:O:C:C:O:H (with :O: :O: above the carbons)

81. H:N::N::N: ⟷ H:N:N:::N:

82. a. Be has only 4 valence electrons. :F–Be–F:

b. S has 12 valence electrons.

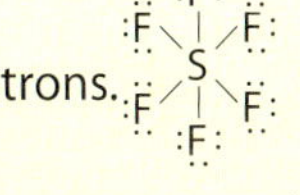

c. Cl has only 7 valence electrons.

d. B has only 6 valence electrons.

e. Xe has 10 valence electrons. :F–Xe–F:

CHAPTER 8

Assessment continued

Cumulative Review

83. formation of a gas, a change in color or odor

84. **a.** 6.65×10^4 micrometers
b. 4 centigrams
c. 5.62×10^{-1} decigram per liter
d. 2.4×10^1 meters per second

85. **a.** 2
b. 2
c. 4
d. 4

86. **a.** 16
b. 12
c. 8
d. 26

87. Isotopes have the same number of protons and electrons, but different numbers of neutrons.

88. Protons and electrons must be equal.

89. **a.** 6
b. 2
c. 5
d. 0

90. The wavelength decreases as the frequency increases.

91. The *d* orbitals related to the third principal energy level contain 5 electrons.

92. **a.** $1s^22s^22p^63s^1$
b. $1s^22s^22p^63s^23p^4$
c. $1s^22s^22p^63s^23p^3$
d. $1s^22s^22p^3$

93. The anion is larger than the corresponding neutral atom.

94. Mendeleev arranged the elements by increasing atomic mass in vertical rows and by similarities in chemical and physical properties. Mosely arranged the elements by increasing atomic number in vertical rows and by similarities in chemical and physical properties.

95. **a.** K, $1s^22s^22p^63s^23p^64s^1$
b. Al, $1s^22s^22p^63s^23p^1$
c. S, $1s^22s^22p^63s^23p^4$
d. Ba, $1s^22s^22p^63s^23p^63d^{10}4s^2 4p^64d^{10}5s^25p^66s^2$

96. **a.** barium
b. silicon
c. sodium

CHAPTER 8

Assessment continued

Cumulative Review

83. Name three indicators of chemical change. *(Chapter 2)*

84. Make the following conversions. *(Chapter 3)*
a. 66.5 mm to micrometers
b. 4×10^{-2} g to centigrams
c. 5.62 mg/mL to decigrams per liter
d. 85 km/h to meters per second

85. How many significant figures are in each measurement? *(Chapter 3)*
a. 0.00052 m
b. 9.8×10^4 g
c. 5.050 mg
d. 8.700 mL

86. How many neutrons are in each atom? *(Chapter 4)*
a. silicon-30
b. magnesium-24
c. nitrogen-15
d. chromium-50

87. How do isotopes of an atom differ? *(Chapter 4)*

88. In a neutral atom, the number of which two subatomic particles must always be equal? *(Chapter 4)*

89. How many electrons are in the 2*p* sublevel of an atom of each element? *(Chapter 5)*
a. aluminum
b. carbon
c. fluorine
d. lithium

90. What happens to the wavelength of light as the frequency increases? *(Chapter 5)*

91. What does the *5* in $3d^5$ represent? *(Chapter 5)*

92. Write correct electron configurations for atoms of the following elements. *(Chapter 5)*
a. sodium
b. sulfur
c. phosphorus
d. nitrogen

93. How does the ionic radius of a typical anion compare with the radius for the corresponding neutral atom? *(Chapter 6)*

94. What criteria did Mendeleev and Moseley use to arrange the elements on the periodic table? *(Chapter 6)*

95. Give the electron configuration of the element found at each location in the periodic table. *(Chapter 6)*
a. Group 1A, period 4
b. Group 3A, period 3
c. Group 6A, period 3
d. Group 2A, period 6

96. Identify the larger atom of each pair. *(Chapter 6)*
a. calcium and barium
b. silicon and sulfur
c. sodium and nitrogen

97. Which of these statements about the periodic table is correct? *(Chapter 6)*
I. Elements are arranged in order of increasing atomic mass.
II. A period is a horizontal row.
III. Nonmetals are located on the right side of the table.
a. I only
b. I and II only
c. I, II, and III
d. I and III only
e. II and III only

98. Which of the following ions has the same number of electrons as a noble gas? *(Chapter 7)*
a. Al^{3+}
b. O^{2-}
c. Br^-
d. N^{3-}

99. What element is likely to form an ionic compound with chlorine? *(Chapter 7)*
a. iodine
b. cesium
c. helium

100. How many valence electrons does each atom have? *(Chapter 7)*
a. argon
b. aluminum
c. selenium
d. beryllium

101. Write the electron configuration of each ion. *(Chapter 7)*
a. oxide ion
b. magnesium ion
c. nitride ion
d. potassium ion

102. An alloy is composed of two or more elements. Is an alloy a compound? Explain your answer. *(Chapter 7)*

97. **e.** II and III only

98. All have the same number of electrons as a noble gas.

99. **b.** cesium

100. **a.** 8
b. 3
c. 6
d. 2

101. **a.** $1s^22s^22p^6$
b. $1s^22s^22p^6$
c. $1s^22s^22p^6$
d. $1s^22s^22p^63s^23p^6$

102. No, an alloy is a homogeneous mixture.

Standardized Test Prep

Test-Taking Tip

Connectors Sometimes two phrases in a true/false question are connected by a word such as *because*. The word implies that one thing caused another thing to happen. Statements that include such words can be false even if both parts of the statement are true by themselves.

Select the choice that best answers each question or completes each statement.

1. A bond in which two atoms share a pair of electrons is not
   a. a coordinate covalent bond.
   b. a polar covalent bond.
   c. an ionic bond.
   d. a nonpolar covalent bond.
2. How many valence electrons are in a molecule of phosphoric acid, H_3PO_4?
   a. 7 **b.** 16 **c.** 24 **d.** 32
3. Which of these molecules can form a hydrogen bond with a water molecule?
   a. N_2 **b.** NH_3 **c.** O_2 **d.** CH_4
4. Which substance contains both covalent and ionic bonds?
   a. NH_4NO_3 **b.** CH_3OCH_3 **c.** LiF **d.** $CaCl_2$
5. Which of these bonds is most polar?
   a. H—Cl **b.** H—Br **c.** H—F **d.** H—I

Use the description and data table below to answer Questions 6–9.

The table relates molecular shape to the number of bonding and nonbonding electron pairs in molecules.

Bonding pairs	Non-bonding pairs	Arrangement of electron pairs	Molecular shape	Example
4	0	tetrahedral	tetrahedral	CH_4
3	1	tetrahedral	pyramidal	NCl_3
2	2	tetrahedral	bent	H_2S
1	3	tetrahedral	linear	HF

6. Draw the electron dot structure for each example molecule.
7. Explain why the arrangement of electron pairs is tetrahedral in each molecule.
8. H_2S has two hydrogen atoms bonded to a sulfur atom. Why isn't the molecule linear?
9. What is the arrangement of electron pairs in PBr_3? Predict the molecular shape of a PBr_3 molecule.

For Questions 10–12, identify the type of intermolecular bonding represented by the dotted lines in the drawings.

10. H_2O

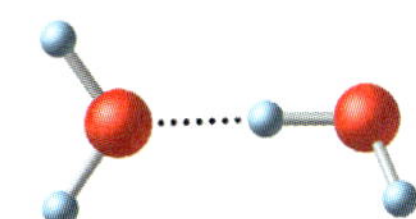

11. BrCl (bromine chloride)

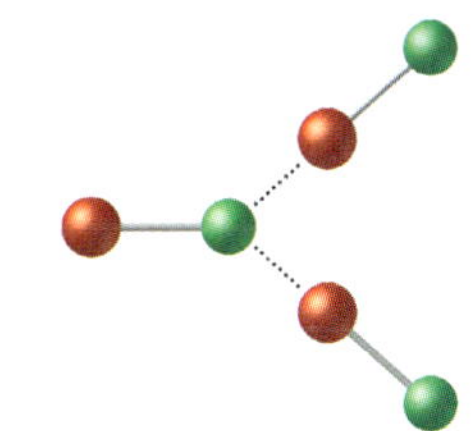

12. CH_3OH (methanol)

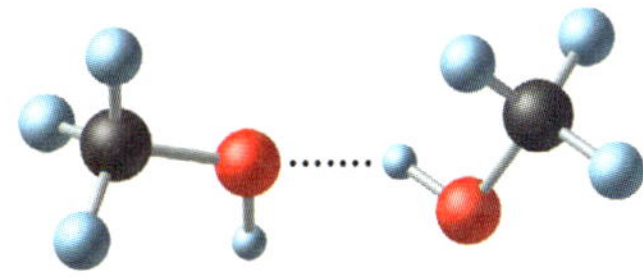

In Questions 13–15, a statement is followed by an explanation. Decide if each statement is true and then decide if the explanation given is correct.

13. A carbon monoxide molecule has a triple covalent bond because carbon and oxygen atoms have an unequal number of valence electrons.
14. Xenon has a lower boiling point than neon because dispersion forces between xenon atoms are stronger than those between neon atoms.
15. The nitrate ion has three resonance structures because the nitrate ion has three single bonds.

Standardized Test Prep

1. c
2. d
3. b
4. a
5. c
6. H:C̤̈:H (with H above and below C)   Cl:N̤̈:Cl (with Cl above N)   H:S̤̈:H   H:F̤̈:
7. Each central atom has four pairs of electrons that, according to VSEPR theory, assume a tetrahedral shape.
8. The two nonbonding pairs repel the bonding pairs; there are still four pairs of electrons around the sulfur atom.
9. The arrangement of electron pairs is tetrahedral. The electron dot structure shows three bonding electron pairs and one non-bonding electron pair; thus, the predicted molecular shape is pyramidal.
10. hydrogen bonding
11. primarily dispersion forces
12. hydrogen bonding
13. True, True
14. False, True
15. True, False

Chemical Names and Formulas Planning Guide

SECTION OBJECTIVES	STANDARDS NATIONAL	STANDARDS STATE	ACTIVITIES and LABS
9.1 Naming Ions, pp. 253–259 1 block or 2 periods **9.1.1 Identify** the charges of monatomic ions by using the periodic table, and **name** the ions. **9.1.2 Define** a polyatomic ion and write the names and formulas of the most common polyatomic ions. **9.1.3 Identify** the two common endings for the names of most polyatomic ions.	A-1, A-2, B-2, E-2, G-1		**SE** Inquiry Activity: Element Name Search, p. 252 L2 **TE** Class Activity: Determining Ionic Charge, p. 254 L2 **TE** Teacher Demo: Colorful Ions, p. 254 L1 **TE** Class Activity: Symbols for Monatomic Ions, p. 255 L2 **TE** Class Activity: Shapes of Polyatomic Ions, p. 258 L2
9.2 Naming and Writing Formulas for Ionic Compounds, pp. 260–267 1 block or 2 periods **9.2.1 Apply the rules** for naming and writing formulas for binary ionic compounds. **9.2.2 Apply the rules** for naming and writing formulas for compounds with polyatomic ions.	A-1, B-2, G-2, G-3		**SE** Small-Scale Lab: Names and Formulas for Ionic Compounds, p. 267 L2 **TE** Class Activity: Naming Ionic Binary Compounds, p. 261 L2 **TE** Teacher Demo: Making and Naming an Ionic Compound, p. 262 L2 **TE** Class Activity: Formulas for Binary Ionic Compounds, p. 263 L2 **TE** Teacher Demo: Making and Naming Lead Carbonate, p. 264 L2 **TE** Class Activity: Naming and Writing Formulas, p. 265 L2 **SSLM** Lab 12: Chemical Names and Formulas L2
9.3 Naming and Writing Formulas for Molecular Compounds, pp. 268–270 1 block or 2 periods **9.3.1 Interpret** the prefixes in the names of molecular compounds in terms of their chemical formulas. **9.3.2 Apply the rules** for naming and writing formulas for binary molecular compounds.	B-2		**TE** Class Activity: Naming Binary Molecular Compounds, p. 269 L2
9.4 Naming and Writing Formulas for Acids and Bases, pp. 271–273 $^1/_2$ block or 1 period **9.4.1 Apply three rules** for naming acids. **9.4.2 Apply the rules** in reverse to write formulas of acids. **9.4.3 Apply the rules** for naming bases.	A-1, B-2		
9.5 The Laws Governing Formulas and Names, pp. 274–279 1 block or 2 periods **9.5.1 Define** the laws of definite proportions and multiple proportions. **9.5.2 Apply the rules** for naming chemical compunds by using a flowchart. **9.5.3 Apply the rules** for writing the formulas of chemical compounds by using a flowchart.	A-1, B-2		**SE** Quick Lab: Making Ionic Compounds, p. 279 L2 **LP** Lab Practical 9-1: Nomenclature L3

Ability Levels

L1 For students who need additional help
L2 For all students
L3 For students who need to be challenged

Components

SE	Student Edition	**LM**	Laboratory Manual	**GCP**	Graphing Calculator Problems	**T**	Transparencies
TE	Teacher's Edition	**SSLM**	Small-Scale Chemistry Lab Manual	**TTT**	Test-Taking Tips with Transparencies, Grades 9–12 Science	**CA**	Chem Alive and Chemistry Field Trips DVD
GRSW	Guided Reading and Study Workbook	**LP**	Laboratory Practicals	**STP**	Standardized Test Preparation Workbook	**iT**	Interactive Textbook with ChemASAP
CTR	Core Teaching Resources	**PLM**	Probeware Lab Manual			**GO**	Go Online Internet Resources
SSG	Spanish Study Guide	**VCL**	Virtual Chemistry Labs				
		CTB	Computer Test Bank				

RESOURCES PRINT and TECHNOLOGY			PROBLEMS and ASSESSMENT
GRSW	Chapter 9, Section 1	L1	**CP** 9.1, Classifying and Naming Cations and Anions **SE** Section 9.1 Assessment, p. 258 **iT** Section 9.1
T	T94–T95	L2	
	GO Chapter 9, Section 1	L2	
GRSW	Chapter 9, Section 2	L1	**CP** 9.2, Writing Formulas for Binary Ionic Compounds **CP** 9.3, Writing Formulas for Compounds with Polyatomic Ions **SE** Section 9.2 Assessment, p. 266 **iT** Section 9.2
T	T96–T97	L2	
	GO Chapter 9, Section 2	L2	
GRSW	Chapter 9, Section 3	L1	**SE** Section 9.3 Assessment, p. 270 **iT** Section 9.3
T	T98	L2	
	GO Chapter 9, Section 3	L2	
GRSW	Chapter 9, Section 4	L1	**SE** Section 9.4 Assessment, p. 273 **iT** Section 9.4
T	T99	L2	
	GO Chapter 9, Section 4	L2	
GRSW	Chapter 9, Section 5	L1	**SP** 9.1, Calculating Mass Ratios **SE** Section 9.5 Assessment, p. 279 **iT** Section 9.5
T	T100–T103	L2	

Go Online

Web Code: cdn-1092
Web Code: cdn-1093
Web Code: cdn-1094

Web Code: cdb-1091

MATERIALS List

Quantities for each group

STUDENT EDITION

Inquiry Activity, p. 252
ingredient labels from items such as cake mixes, baking soda, cereal, soap, shampoo, conditioner, toothpaste, and antacids

Small-Scale Lab, p. 267
reaction surface, 0.05*M* $AgNO_3$, 0.2*M* $Pb(NO_3)_2$, 0.5*M* $CaCl_2$, 1.0*M* Na_2CO_3, 0.1*M* Na_3PO_4, 0.5*M* NaOH, 0.2*M* Na_2SO_4, 0.2*M* $CuSO_4$, 0.2*M* $MgSO_4$, 0.1*M* $FeCl_3$

Quick Lab, p. 279
nine small test tubes, test tube rack, six solutions in plastic dropper bottles containing 0.05*M* $AgNO_3$, 0.2*M* $Pb(NO_3)_2$, 1.0*M* $NaCO_3$, 0.5*M* NaOH, 0.1*M* $Na_3(PO_4)_2$, and 0.1*M* $FeCl_3$

TEACHER'S EDITION

Class Activity, p. 254
a display-size periodic table

Teacher Demo, p. 254
beakers containing solutions of soluble metal salts: $MnCl_2$, $FeCl_2$, $CoCl_2$, $NiCl_2$, and $CuCl_2$.

Class Activity, p. 255
blank photocopy of the periodic table for each pair of students

Class Activity, p. 258
foam balls, wooden sticks

Teacher Demo, p. 262
1 g powdered zinc, 4 g iodine (I_2), watch glass, eyedropper, 8 mL water

Class Activity, p. 263
the list of names of ionic compounds generated in the activity on p. 261

Teacher Demo, p. 264
50 mL 0.1*M* lead(II) nitrate, 50 mL 0.1*M* sodium carbonate, 100-mL beaker

Chapter Assessment

CHAPTER ASSESSMENT

SE Chapter Assessment, pp. 281–284
CTB Chapter 9
iT Chapter 9

STANDARDIZED TEST PREP

SE Chapter 9, p. 285
STP Topic 2
Interactive Textbook with Assessment at PHSchool.com

CHAPTER 9

Chemical Names and Formulas

Element Name Search L2

Purpose

Students become aware of the substances in ordinary products and how they are named.

Prep Time 15 minutes
Class Time 20 minutes

Think About It

1. Salts of sodium and potassium are often found in processed foods and toiletries.
2. Students' lists will vary; sodium, potassium, calcium, phosphorus, and carbon are likely, depending on the products.
3. Some are complex with no names of elements. These are often organic compounds. Others are simpler and contain elements. These are often inorganic chemicals such as the ionic compounds students will be learning about.
4. Usually the element-containing names are simpler.

CHAPTER 9

Chemical Names and Formulas

A pharmacist's knowledge of the names and formulas of chemical compounds is crucial to the safety of her clients.

INQUIRY Activity

Element Name Search

Materials

Ingredient labels from items such as bread, cake mixes, baking soda, cereal, soap, shampoo, conditioner, toothpaste, and antacids

Procedure

1. Select a variety of ingredient labels from the suggested items or from similar ones.
2. List all the ingredients that you recognize. Don't be discouraged by the many unfamiliar names. Thousands upon thousands of substances are used in everyday products.
3. With the help of a periodic table, make another list of all the ingredient names that contain or are derived from the name of an element.

Think About It

1. How often did you find the same element name in more than one product?
2. What element names are on both lists?
3. Did all the ingredients have complex chemical-sounding names?
4. How do the names of element-containing compounds compare with the other names on your list?

9.1 Naming Ions

Connecting to Your World

In the play *Romeo and Juliet*, William Shakespeare wrote, "What's in a name? That which we call a rose/By any other name would smell as sweet." A rose is *rosa* in Spanish, *warda* in Arabic, and *julab* in Hindi. To truly understand another culture, you must first learn the language used in that culture. Similarly, to understand chemistry, you must learn its language. Part of learning the language of chemistry involves understanding how to name ionic compounds. For this you need to know how to name ions.

Guide for Reading

Key Concepts

- How are the charges of Group A metal and nonmetal ions related to their positions in the periodic table?
- How are the charges of some transition metal ions determined?
- What are the two endings of the names of most polyatomic ions?

Vocabulary

monatomic ion

polyatomic ion

Reading Strategy

Outlining As you read, make an outline of the most important ideas in this section. Use the red headings as the main topics and the blue headings as subtopics. Add a sentence or a note after each heading to provide key information about the topic.

Monatomic Ions

Ionic compounds consist of a positive metal ion and a negative nonmetal ion combined in a proportion such that their charges add up to a net charge of zero. For example, the ionic compound sodium chloride (NaCl) consists of one sodium ion (Na^+) and one chloride ion (Cl^-). Probably you are already familiar with the name and formula of sodium chloride, which is common table salt. But it is important, in learning the language of chemistry, to be able to name and write the chemical formulas for all ionic compounds. The first step is to learn about the ions that form ionic compounds. Some ions, called **monatomic ions,** consist of a single atom with a positive or negative charge resulting from the loss or gain of one or more valence electrons, respectively.

Cations Recall that metallic elements tend to lose valence electrons. Lithium, sodium, and potassium in Group 1A lose one electron to form cations. All the Group 1A ions have a 1+ charge (Li^+, Na^+, K^+, Rb^+, and Cs^+). Magnesium and calcium are Group 2A metals. They tend to lose two electrons to form cations with a 2+ charge (Mg^{2+} and Ca^{2+}), as do all the other Group 2A metals. Aluminum is the only common Group 3A metal. As you might expect, it tends to lose three electrons to form a 3+ cation (Al^{3+}). **When the metals in Groups 1A, 2A, and 3A lose electrons, they form cations with positive charges equal to their group number.** Figure 9.1 shows some of the elements whose ionic charges can be obtained from their positions in the periodic table. The names of the cations of the Group 1A, Group 2A, and Group 3A metals are the same as the name of the metal, followed by the word *ion* or *cation*. Thus Na^+ is the sodium ion (or cation), Ca^{2+} is the calcium ion (or cation), and Al^{3+} is the aluminum ion (or cation).

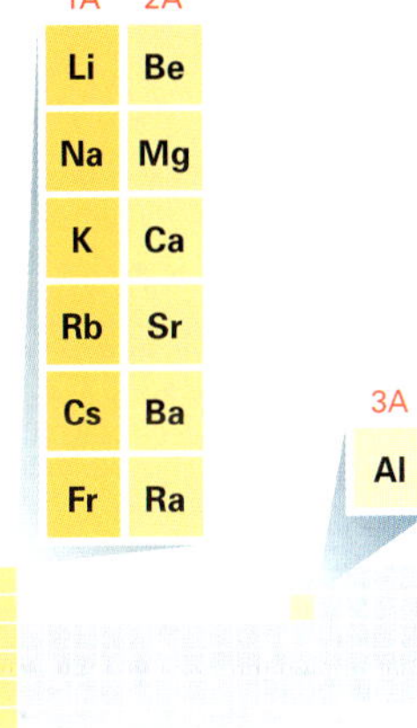

Figure 9.1 The representative elements shown form positive ions with charges equal to their group numbers. **Applying Concepts** ***Are the ions anions or cations?***

Section Resources

Print

- **Guided Reading and Study Workbook,** Section 9.1
- **Core Teaching Resources,** Section 9.1 Review, Interpreting Graphics
- **Transparencies,** T94–T95

Technology

- **Interactive Textbook with ChemASAP,** Problem-Solving 9.1, Assessment 9.1
- **Go Online,** Section 9.1

9.1

1 FOCUS

Objectives

9.1.1 **Identify** the charges on monatomic ions by using the periodic table, and **name** the ions.

9.1.2 **Define** a polyatomic ion and write the names and formulas of the most common polyatomic ions.

9.1.3 **Identify** the two common endings for the names of most polyatomic ions.

Guide for Reading

Build Vocabulary  L2

Word Parts Ask, **Can you think of words that begin with the prefixes *mono–* and *poly–*?** *(Students may respond with words such as monotone, monochrome, polygon, or polygamy.)* Discuss the meanings of these words and then ask, **How do you think monatomic ions and polyatomic ions might differ?** *(A monatomic ion has one atom; a polyatomic ion has more than one atom.)*

Reading Strategy  L2

Compare and Contrast As students read the section, have them make a chart that compares monatomic ions and polyatomic ions as to charge, composition, and name endings.

2 INSTRUCT

Connecting to Your World

Comment that you couldn't talk with people in a different country unless you knew their language. Emphasize that chemists everywhere use the same chemical language. Ask, **Why is it important that there be only one language of chemistry?** *(Chemists need to communicate with one another and having the same names for compounds makes this easier.)*

Answers to...

Figure 9.1 cations

Monatomic Ions

Use Visuals L1

Figures 9.1 and 9.2 Have students note that in each figure, the periodic table is shown in the background to help them locate the groups being discussed. Ask, **How do the ions formed by metals in Groups 1A, 2A, and 3A differ from the ions formed by nonmetals in Groups 5A, 6A, and 7A?** *(Groups 1A, 2A, and 3A form positive ions; Groups 5A, 6A, and 7A form negative ions.)* Ask, **Are the ions formed by the transition metals cations or anions?** *(They are cations.)*

CLASS Activity

Determining Ionic Charge L2

Purpose Students gain practice in using the periodic table to determine the charge on ions.

Materials A display-size periodic table

Procedure Have one student point, one-by-one and in no particular order, to any of the elements shown prominently in Figures 9.1 and 9.2. Have the other students vie to identify the charge of the ion formed by each element.

TEACHER Demo

Colorful Ions L2

Purpose Students see the various colors of solutions of transition metal ions and practice naming them.

Materials Beakers containing solutions of soluble metal salts, for example: $MnCl_2$, $FeCl_2$, $CoCl_2$, $NiCl_2$, and $CuCl_2$, or use the ions in Figure 9.3.

Procedure Write the symbols for the metal ions on the board. Have students describe the colors of the solutions. Ask them to use Table 9.2 to write the Stock and classical names of each cation. Ask students to suggest why chemists currently use only the Stock system for naming ions.

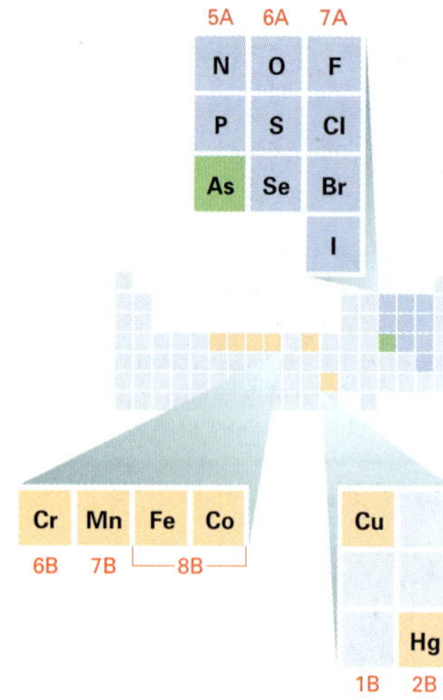

Figure 9.2 Note the positions of the nonmetals and the metalloid, arsenic. These elements form anions. Common transition elements that form more than one ion are also shown. **Applying Concepts** ***Do the transition metals form anions or cations?***

Table 9.1

Ionic Charges of Representative Elements

1A	2A	3A	4A	5A	6A	7A	8A
Li^+	Be^{2+}			N^{3-}	O^{2-}	F^-	
Na^+	Mg^{2+}	Al^{3+}		P^{3-}	S^{2-}	Cl^-	
K^+	Ca^{2+}			As^{3-}	Se^{2-}	Br^-	
Rb^+	Sr^{2+}					I^-	
Cs^+	Ba^{2+}						

Anions Nonmetals tend to gain electrons to form anions, so the charge of a nonmetallic ion is negative. **The charge of any ion of a Group A nonmetal is determined by subtracting 8 from the group number.** The elements in Group 7A form anions with a 1− charge (7 − 8 = −1). The name of an anion is not the same as the element's name. Anion names start with the stem of the element name and end in *-ide*. For example, two elements in Group 7A are fluorine and chlorine. The anions for these nonmetals are the fluor*ide* ion (F^-) and chlor*ide* ion (Cl^-). Anions of nonmetals in Group 6A have a 2− charge (6 − 8 = −2). Group 6A elements, oxygen and sulfur, form the ox*ide* anion (O^{2-}) and the sulf*ide* anion (S^{2-}), respectively. The three nonmetals in Group 5A—nitrogen, phosphorus, and arsenic—can form anions with a 3− charge (5 − 8 = −3). These have the symbols N^{3-}, P^{3-}, and As^{3-} and are called, respectively, nitride ion, phosphide ion, and arsenide ion. Figure 9.2 shows the Group A elements that form anions. Table 9.1 summarizes the ionic charges of representative elements that can be obtained from the periodic table.

The majority of the elements in the two remaining representative groups, 4A and 8A, usually do not form ions.

Checkpoint ***What is the name of the anion N^{3-}?***

Ions of Transition Metals The metals of Groups 1A, 2A, and 3A consistently form cations with charges of 1+, 2+, and 3+, respectively. Many of the transition metals (Groups 1B–8B) form more than one cation with different ionic charges. Some of these are shown in Figure 9.2. **The charges of the cations of many transition metal ions must be determined from the number of electrons lost.** For example, the transition metal iron forms two common cations, Fe^{2+} (two electrons lost) and Fe^{3+} (three electrons lost). Cations of tin and lead, the two metals in Group 4A, can also have more than one charge. Table 9.2 lists symbols for common ions of many of the metals that form more than one ion. Two methods are used to name these ions. The preferred method is called the Stock system. In the Stock system, a Roman numeral in parentheses is placed after the name of the element to indicate the numerical value of the charge. Table 9.2 shows that the cation Fe^{2+} is named iron(II) ion. Note that no space is left between the element name and the Roman numeral in parentheses. The name for Fe^{2+} is read "iron two ion." The Fe^{3+} ion is named iron(III) ion and is read "iron three ion." Colorful solutions of the transition metal ions Co^{3+}, Cr^{3+}, Fe^{3+}, Ni^{2+}, and Mn^{2+} are shown in Figure 9.3.

Figure 9.3 The ions of these transition metals produce an array of colors when dissolved in water. From left to right, the ions are Co^{3+}, Cr^{3+}, Fe^{3+}, Ni^{2+}, and Mn^{2+}.

An older, less useful method for naming these cations uses a root word with different suffixes at the end of the word. The older, or classical, name of the element is used to form the root name for the element. For example, *ferrum* is Latin for iron, so *ferr-* is the root name for iron. The suffix *-ous* is used to name the cation with the lower of the two ionic charges. The suffix *-ic* is used with the higher of the two ionic charges. Using this system, Fe^{2+} is the ferrous ion, and Fe^{3+} is the ferric ion, as shown in Table 9.2. Notice that you can usually identify an element from what may be an unfamiliar classical name by looking for the element's symbol in the name. Thus *fer*rous (Fe) is iron; *cu*prous (Cu) is copper; and *stan*nous (Sn) is tin. A major disadvantage of using classical names for ions is that they do not tell you the actual charge of the ion. A classical name tells you only that the cation has either the smaller (*-ous*) or the larger (*-ic*) charge of the pair of possible ions for that element.

A few transition metals have only one ionic charge. The names of these cations do not have a Roman numeral. These exceptions include silver, with cations that have a 1+ charge (Ag^{+}), as well as cadmium and zinc, with cations that have a 2+ charge (Cd^{2+} and Zn^{2+}). As Figure 9.4 shows, many transition metal compounds are colored and can be used as pigments. Pigments are compounds having intense colors that can be used to color other materials. For example, compounds of chromium are pigments used to make yellow, orange, red, or green paints. Various cadmium compounds range in color from yellow to red and maroon. Prussian blue is an important pigment composed of the transition element iron combined with carbon, hydrogen, and nitrogen.

Table 9.2

Symbols and Names of Common Metal Ions with More than One Ionic Charge

Symbol	Stock name	Classical name
Cu^{+}	Copper(I) ion	Cuprous ion
Cu^{2+}	Copper(II) ion	Cupric ion
Fe^{2+}	Iron(II) ion	Ferrous ion
Fe^{3+}	Iron(III) ion	Ferric ion
*Hg_2^{2+}	Mercury(I) ion	Mercurous ion
Hg^{2+}	Mercury(II) ion	Mercuric ion
Pb^{2+}	Lead(II) ion	Plumbous ion
Pb^{4+}	Lead(IV) ion	Plumbic ion
Sn^{2+}	Tin(II) ion	Stannous ion
Sn^{4+}	Tin(IV) ion	Stannic ion
Cr^{2+}	Chromium(II) ion	Chromous ion
Cr^{3+}	Chromium(III) ion	Chromic ion
Mn^{2+}	Manganese(II) ion	Manganous ion
Mn^{3+}	Manganese(III) ion	Manganic ion
Co^{2+}	Cobalt(II) ion	Cobaltous ion
Co^{3+}	Cobalt(III) ion	Cobaltic ion

*A diatomic elemental ion.

Figure 9.4 Many transition metals form brightly colored compounds that are used in making artists' paints.

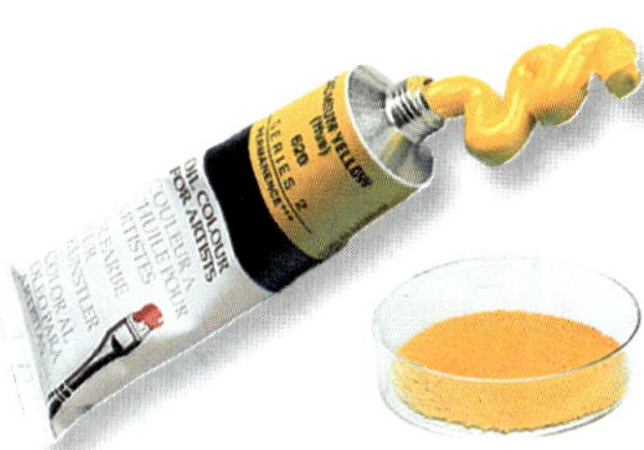

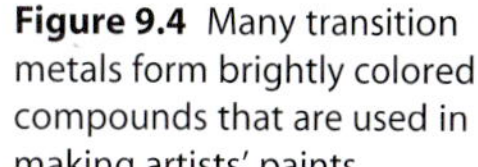

Symbols for Monatomic Ions L2

Purpose Students use the periodic table for writing symbols for monatomic ions.

Materials Blank photocopy of the periodic table for each pair of students.

Procedure Have students work in pairs. Allow them to refer to their textbooks or other resource materials. In the element blocks on the blank periodic table, have them write the symbols and names of the cations and anions of as many elements as possible. Have them use their tables to identify any trends in ionic charges. Encourage them to keep their tables for reference.

Relate 

Divide students into research teams to gather data about paint pigments from different sources. Possible assignments: (1) Examine the pigments available at an art-supply store and try to tell from the names what they contain. (2) Interview an art teacher about the pigments that he or she prefers. (3) Use the library or Internet to research which pigments were available and popular during earlier eras; look for information about natural sources of pigments. (4) Invite a curator or art restorer from a local museum to talk about using pigments to repair damaged art, authenticate art, or unmask a forgery.

Differentiated Instruction

Special Needs L1

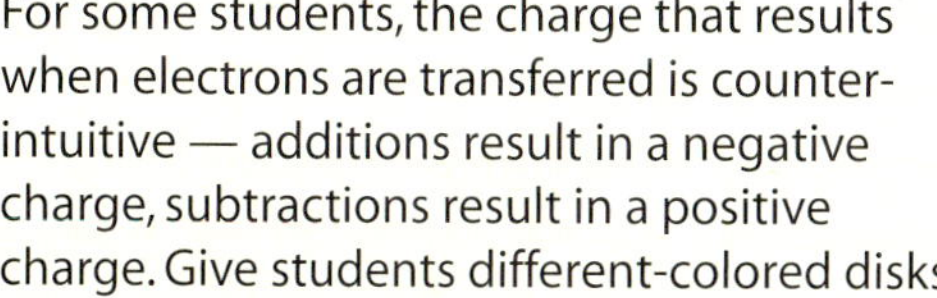

For some students, the charge that results when electrons are transferred is counterintuitive — additions result in a negative charge, subtractions result in a positive charge. Give students different-colored disks to represent protons and electrons in neutral atoms. Remove or add electrons and ask students to find the charge of the resulting "ion."

Answers to...

Figure 9.2 cations

 nitride ion

Careers *in* Chemistry

Pharmacist

Although pharmacists must often work in the evenings and on weekends, they are well paid and employment opportunities for pharmacists during the coming decade are expected to grow faster than for many other professions. The American Association of Colleges of Pharmacy (ASCP) operates the Pharmacy College Application Service to allow students to use a single web-based application for all applications to schools of pharmacy.

Have students research chemistry-related careers in the library or on the Internet. Students can then construct a table that describes the nature of the work, educational and training requirements, employment outlook, working conditions, and other necessary information.

CONCEPTUAL PROBLEM 9.1

Answers

1. **a.** selenide ion, anion
   b. barium ion, cation
   c. phosphide ion, anion
2. **a.** three electrons lost
   b. two electrons gained
   c. one electron lost

Practice Problems Plus L2

Write the symbol (including charge) for the ion formed by each element and classify each as an anion or a cation.

a. arsenic *(As^{3-}, anion)*
b. beryllium *(Be^{2+}, cation)*
c. astatine *(At^{-}, anion)*
d. gallium *(Ga^{3+}, cation)*

Careers *in* Chemistry

Pharmacist

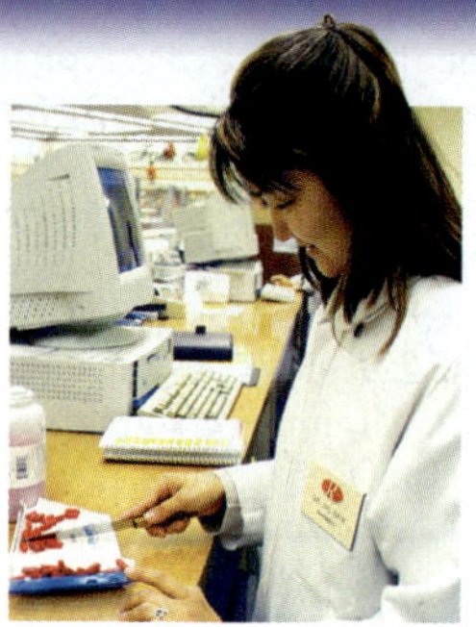

Doctors provide written instructions to pharmacists about medicines they wish to have dispensed to their patients. It is the responsibility of the pharmacist to prepare the medicine and to advise the patient about possible side effects. Pharmacists also make sure that physicians have not prescribed a medicine or dosage that could harm the patient.

A person with some knowledge of chemistry and biology could become a pharmacist's assistant with on-the-job training. Advancement to the position of pharmacist requires a college degree in pharmacy. This degree includes the study of chemistry, biology, mathematics, and statistics. Pharmacists must also learn about the biological effects of drugs and drug interactions (pharmacology). To obtain a license to dispense drugs, pharmacists are required to pass a state test and must work for a specified period of time under the supervision of another pharmacist.

For: Careers in Chemistry
Visit: PHSchool.com
Web Code: cdb-1091

CONCEPTUAL PROBLEM 9.1

Classifying and Naming Cations and Anions

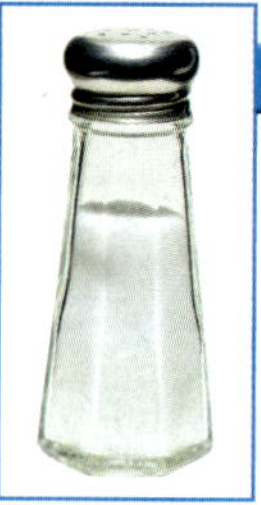

Write the symbol for the ion formed by each element. Classify the ions as cations or anions and name the ion. Potassium and iodine combine to form potassium iodide, an additive to table salt that protects the thyroid gland.

a. potassium **b.** iodine **c.** sulfur **d.** lead, 4 electrons lost

1 Analyze ***Identify the relevant concepts.***

a.–d. Use the periodic table or the electrons lost to write the symbol for the ion. Ions with positive charges are cations; ions with negative charges are anions. Apply the rules for naming cations and anions. The names of nonmetallic anions end in *-ide*. Metallic cations take the name of the metal. If the metal ion can have more than one ionic charge, use a Roman numeral in the Stock name or use the classical name with a suffix.

2 Solve ***Apply concepts to this situation.***

a. K^{+}: cation, potassium ion
b. I^{-}: anion, iodide ion
c. S^{2-}: anion, sulfide ion
d. Pb^{4+}: cation, lead(IV) or plumbic ion

The ions formed by metals are cations and the ions formed by nonmetals are anions. The rules for naming have been correctly applied.

Practice Problems

1. Name the ions formed by these elements and classify them as anions or cations.
   a. selenium **b.** barium **c.** phosphorus
2. How many electrons were lost or gained to form these ions?
   a. Fe^{3+} **b.** O^{2-} **c.** Cu^{+}

Problem-Solving 9.1 Solve Problem 1 with the help of an interactive guided tutorial.

with ChemASAP

256 Chapter 9

Facts and Figures

Potassium Iodide as Radiation Medicine

Potassium is stockpiled for use in the event of a nuclear accident that results in fallout of radioactive iodine-131. Whether radioactive or not, absorbed iodine accumulates in the thyroid gland. If a dose of nonradioactive KI is ingested in a timely way after a nuclear accident, iodine-131 is less likely to enter the thyroid gland. Iodine-131 has a relatively short half-life of eight days, but can cause thyroid cancer, especially in children.

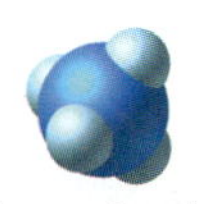

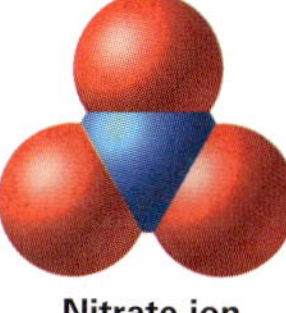

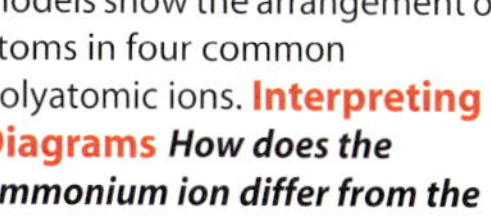

Figure 9.5 These molecular models show the arrangement of atoms in four common polyatomic ions. **Interpreting Diagrams** ***How does the ammonium ion differ from the other three?***

Polyatomic Ions

Some ions, called **polyatomic ions,** are composed of more than one atom. The sulfate anion consists of one sulfur atom and four oxygen atoms. These five atoms together comprise a single anion with an overall 2− charge. The formula is written SO_4^{2-}. Polyatomic ions, such as the sulfate ion, are tightly bound groups of atoms that behave as a unit and carry a charge. You can see the structures of four common polyatomic ions in Figure 9.5.

The names and formulas of some common polyatomic ions are shown in Table 9.3 grouped according to their charges. **The names of most polyatomic anions end in *-ite* or *-ate*.** For example, notice the endings of the names of the hypochlor*ite* ion (ClO^-) and the hydrogen carbon*ate* ion (HCO_3^-). Also notice that three important ions have different endings. The positively charged ammonium cation (NH_4^+) ends in *-ium*, and the cyanide ion (CN^-) and the hydroxide ion (OH^-) end in *-ide*. Use Table 9.3 as a reference until you have memorized its contents.

Sometimes the same two or three elements combine in different ratios to form different polyatomic ions. You can see examples in Table 9.3. Look for pairs of ions for which there is both an *-ite* and an *-ate* ending, for example, sulfite and sulfate. In the list below, examine the charge on each ion in the pair. Note the number of oxygen atoms and the endings on each name. You should be able to discern a pattern in the naming convention.

-ite	*-ate*
SO_3^{2-}, sulfite	SO_4^{2-}, sulfate
NO_2^-, nitrite	NO_3^-, nitrate
ClO_2^-, chlorite	ClO_3^-, chlorate

The charge on each polyatomic ion in a given pair is the same. The *-ite* ending indicates one less oxygen atom than the *-ate* ending. However, the ending does not tell you the actual number of oxygen atoms in the ion. For example, the nitrite ion has two oxygen atoms and the sulfite ion has three oxygen atoms. All anions with names ending in *-ite* or *-ate* contain oxygen.

 ***What is the formula for the hydrogen sulfite polyatomic ion?***

Table 9.3

Common Polyatomic Ions

Formula	Name
Charge = 1−	
$H_2PO_4^-$	Dihydrogen phosphate
$C_2H_3O_2^-$	Acetate
HSO_3^-	Hydrogen sulfite
HSO_4^-	Hydrogen sulfate
HCO_3^-	Hydrogen carbonate
NO_2^-	Nitrite
NO_3^-	Nitrate
CN^-	Cyanide
OH^-	Hydroxide
MnO_4^-	Permanganate
ClO^-	Hypochlorite
ClO_2^-	Chlorite
ClO_3^-	Chlorate
ClO_4^-	Perchlorate
Charge = 2−	
HPO_4^{2-}	Hydrogen phosphate
$C_2O_4^{2-}$	Oxalate
SO_3^{2-}	Sulfite
SO_4^{2-}	Sulfate
CO_3^{2-}	Carbonate
CrO_4^{2-}	Chromate
$Cr_2O_7^{2-}$	Dichromate
SiO_3^{2-}	Silicate
Charge = 3−	
PO_3^{3-}	Phosphite
PO_4^{3-}	Phosphate
Charge = 1+	
NH_4^+	Ammonium

Polyatomic Ions

Discuss

L2

Write the following formulas on the board: NaOH, H_2SO_4, NH_4NO_3, Na_2CO_3. Ask students what these compounds have in common. (*Students may note that they all contain oxygen.*) Tell students that this is an important observation, then explain that all of these compounds contain ions that are made up of more than one type of atom. In ions such as these, oxygen is almost always present.

Use Visuals

Table 9.3 Call attention to how the table is divided into sections according to charge. More than half of the ions have a charge of 1−. Only two ions have charges of 3−, and both of these contain the phosphorus atom. The ammonium ion is the only common polyatomic ion with a 1+ charge, and it does not contain oxygen. The remaining ions have charges of 2−. Have students pick out the *–ate* / *-ite* pairs and write their formulas on the board. Have students note that the members of each pair have the same charge. It is not the charge but the number of oxygen atoms in the members of an *-ate/-ite* pair that determines the endings of the names.

Discuss

Write on the board the names and formulas of at least ten different polyatomic ions. Purposely mismatch the names so that they do not appear with the correct formulas. Challenge students to see how quickly they can rearrange the names and formulas so that they are paired correctly.

Differentiated Instruction

English Learners

Have students prepare flash cards with the name of a polyatomic ion on one side and its formula on the other side. Pairs of students can use the flash cards to quiz each other on the names and formulas of these ions. Set aside a few minutes at the beginning or end of several class periods for this activity.

Facts and Figures

Two Kinds of Bonding

When an ionic compound forms between a metal ion and a polyatomic ion, the resulting compound contains both ionic bonds and covalent bonds. The atoms in all polyatomic ions are held together by covalent bonds.

Answers to...

Figure 9.5 The ammonium ion is a cation and does not contain oxygen.

 HSO_3^-

Section 9.1 (continued)

Shapes of Polyatomic Ions L2

Purpose Students visualize the geometry of some polyatomic ions.

Materials foam balls and wooden sticks

Procedure Pairs of students can research the three-dimensional structure of one of the polyatomic ions in Table 9.3. Have them prepare a ball-and-stick model of the ion. They may wish to differentiate atoms by size or color. Make sure that they pay attention to correct geometric placement of atoms.

3 ASSESS

Evaluate Understanding 

Write these symbols on the chalkboard: K, NH_3, CN, Mg, O, OH, Co, NH_4, Pb, PO_4, Ar, ClO, H_2O, and Zn. Ask students to identify which species can exist as ions by simply attaching a charge. *(K, CN, Mg, O, OH, Co, NH_4, Pb, PO_4, ClO, and Zn)* For these ions, ask students to identify the charges and name the ion.

Reteach 

Remind students that one of the most important properties of a compound is that it is electrically neutral. Because polyatomic ions carry a charge, they are always combined with other ions in compounds.

Writing Activity

Sodium is found in higher concentrations in extra-cellular fluids, such as plasma. Potassium is found in higher concentrations within cells. Both ions are involved in transmission of nerve impulses.

If your class subscribes to the Interactive Textbook, use it to review key concepts in Section 9.1.

with ChemASAP

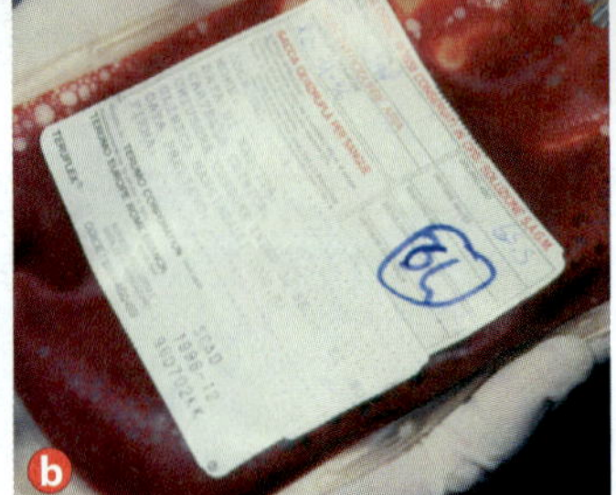

Figure 9.6 Hydrogen-containing polyatomic ions are part of many compounds that affect your daily life. **a** Sodium hydrogen carbonate, which contains the HCO_3^- ion, can relieve an upset stomach. **b** The presence of dissolved HCO_3^-, HPO_4^{2-}, and $H_2PO_4^-$ ions in your blood is critical for your health. **c** Crop dusters spread fertilizers containing HPO_4^{2-} and $H_2PO_4^-$ ions.

When the formula for a polyatomic ion begins with H (hydrogen), you can think of the H as representing a hydrogen ion (H^+) combined with another polyatomic ion. For example, HCO_3^- is a combination of H^+ and CO_3^{2-}. Note that the charge on the new ion is the algebraic sum of the ionic charges.

$$H^+ + \underset{\text{carbonate}}{CO_3^{2-}} \longrightarrow \underset{\text{hydrogen carbonate}}{HCO_3^-}$$

$$H^+ + \underset{\text{phosphate}}{PO_4^{3-}} \longrightarrow \underset{\text{hydrogen phosphate}}{HPO_4^{2-}}$$

$$H^+ + \underset{\text{hydrogen phosphate}}{HPO_4^{2-}} \longrightarrow \underset{\text{dihydrogen phosphate}}{H_2PO_4^-}$$

The hydrogen carbonate anion (HCO_3^-), the hydrogen phosphate anion (HPO_4^{2-}), and the dihydrogen phosphate anion ($H_2PO_4^-$) are essential components of living systems. In contrast, the cyanide ion (CN^-) is extremely poisonous to living systems because it blocks a cell's means of producing energy. Figure 9.6 shows some uses for compounds with hydrogen-containing polyatomic ions.

9.1 Section Assessment

3. **Key Concept** Explain how the charges of Group A metal and nonmetal ions are related to their positions in the periodic table.
4. **Key Concept** How are the charges of some transition metal ions determined?
5. **Key Concept** What are the usual endings for the names of polyatomic ions?
6. What are the charges on ions of Group 1A, Group 3A (aluminum), and Group 5A?
7. How does a polyatomic anion differ from a monatomic anion?
8. Write the symbol for the ion of each element. Classify the ion as an anion or a cation, and name the ion.
   - **a.** potassium **b.** oxygen
   - **c.** tin (2 electrons lost) **d.** bromine
   - **e.** beryllium **f.** cobalt (3 electrons lost)
9. Write the symbol or formula (including charge) for each of the following ions.
   - **a.** ammonium ion **b.** tin(II) ion
   - **c.** chromate ion **d.** nitrate ion

Writing Activity

Essay Sodium ions (Na^+) and potassium ions (K^+) are needed for the human body to function. Research where these ions are most likely to be found in the body and the roles they play. Write a brief essay describing your findings.

Assessment 9.1 Test yourself on the concepts in Section 9.1.

with ChemASAP

Section 9.1 Assessment

3. The positive charge of a group A metal is equal to its group number. The charge of a group A nonmetal is the group number minus 8.
4. from the number of electrons lost
5. *-ite or -ate*
6. Group 1A metals, 1+; Group 3A (aluminum), 3+, Group 5A nonmetals, 3–
7. A monatomic anion is a single atom with a negative charge; a polyatomic anion is two or more bound atoms with a negative charge.
8. **a.** K^+, cation, potassium ion **b.** O^{2-}, anion, oxide ion **c.** Sn^{2+}, cation, tin(II) ion **d.** Br^-, anion, bromide ion **e.** Be^{2+}, cation, beryllium ion **f.** Co^{3+}, cation, cobalt(III) ion
9. **a.** NH_4^+ **b.** Sn^{2+} **c.** CrO_4^{2-} **d.** NO_3^-

Technology & Society

Plasma TV

Today, a TV can be mounted on the wall just as if it were a picture in a frame. The cathode ray tube (CRT), which bulges from the back of a conventional TV, has been replaced by plasma. Plasma is a gaseous mixture of high-energy electrons and ions that activates tiny fluorescent tubes in chambers called pixels. A TV picture is created by hundreds of thousands of pixels, each contributing a single point of colored light. **Inferring** ***Explain why you don't see the points of light.***

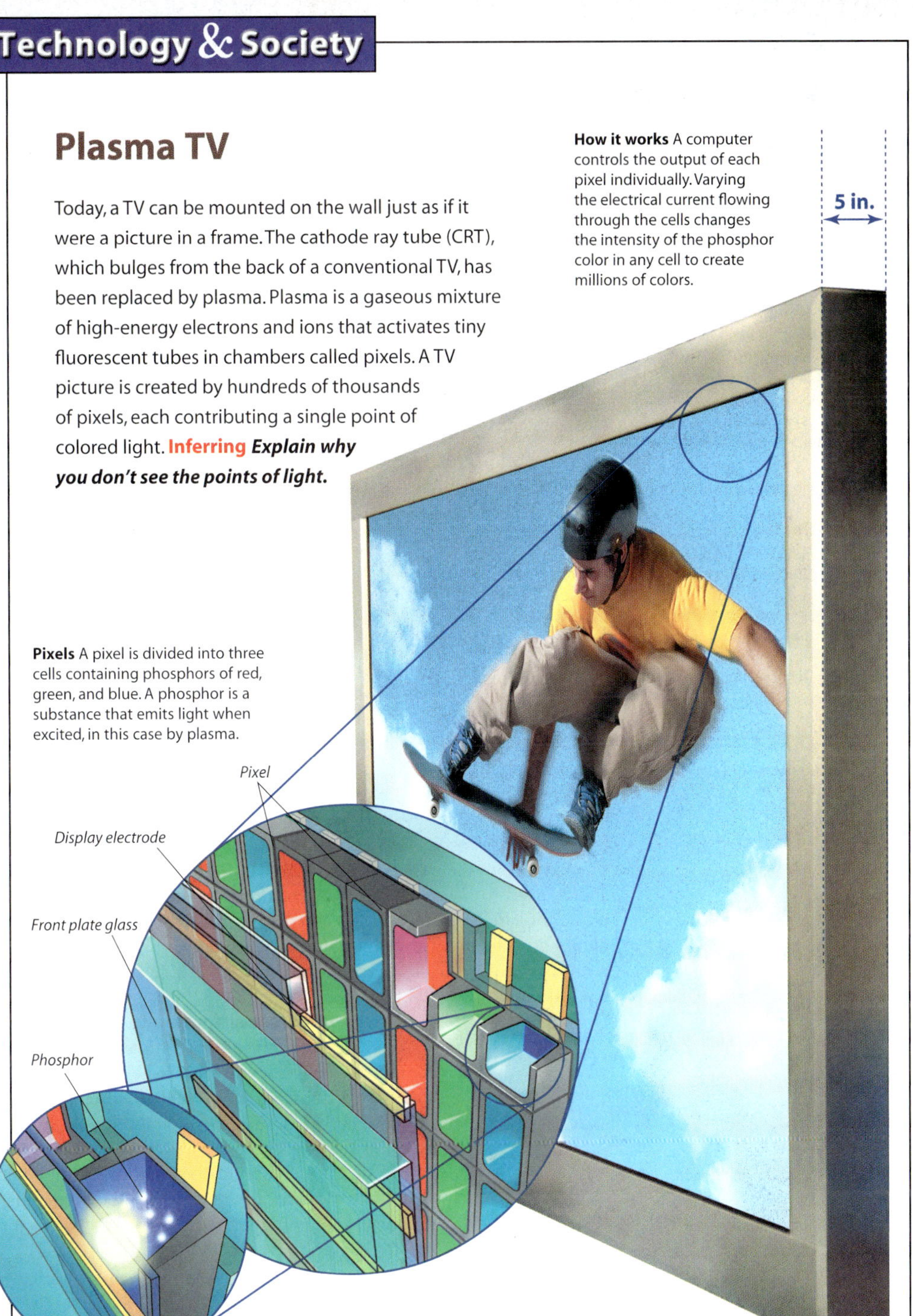

How it works A computer controls the output of each pixel individually. Varying the electrical current flowing through the cells changes the intensity of the phosphor color in any cell to create millions of colors.

Pixels A pixel is divided into three cells containing phosphors of red, green, and blue. A phosphor is a substance that emits light when excited, in this case by plasma.

Technology & Society

Plasma TV

Students may be interested in researching one or more of these topics:

- What is the composition of plasma? How is it produced? What is its role in creating a colored TV picture? Are there other ways in which plasma is used today?
- What are the properties of phosphors? What kinds of substances are they and how do they produce light?
- Color theory—how do red, green and blue light produce all imaginable colors of light?
- Plasma is considered a fourth state of matter. Explore the meaning of this statement.
- Explore the quality of plasma TV as compared with conventional TVs. Does the technology of plasma TV assure superior picture quality?

Answers to...

Inferring They are too small.

9.2

1 FOCUS

Objectives

9.2.1 **Apply the rules** for naming and writing formulas for binary ionic compounds.

9.2.2 **Apply the rules** for naming and writing formulas for compounds with polyatomic ions.

Guide for Reading

Build Vocabulary L2

Word Parts The word *compound* comes from the Latin *componere,* which means "to put." Parts are put together to make a whole. A compound always contains at least two parts. A compound with two parts is called a binary compound.

Reading Strategy L2

Anticipation Guide Have students page through the section and note the two major headings in red. Ask, **What are the two types of compounds you will be reading about?** *(binary ionic compounds and compounds with polyatomic ions)* Building on the vocabulary discussion above, ask, **What do you think is the difference between a binary compound and a compound with a polyatomic ion?** *(A binary compound contains only two elements. A compound with a polyatomic ion can contain three or more elements because a polyatomic ion can have two or more elements.)*

2 INSTRUCT

Connecting to Your World

Ask students to draw an analogy between a recipe that calls for 4 cups of tomato sauce and 1 teaspoon of salt and the formula for a chemical compound such as $CaCl_2$. *(Calcium and chlorine correspond to the ingredients tomato sauce and salt; the subscripts correspond to the quantities of the ingredients.)*

9.2 Naming and Writing Formulas for Ionic Compounds

Guide for Reading

Key Concepts

- How are the names of binary ionic compounds determined?
- How do you write the formulas for binary ionic compounds?
- How do you write the formulas and names of compounds containing polyatomic ions?

Vocabulary

binary compound

Reading Strategy

Predicting Before you read, preview the section by reading the Key Concepts, the headings, and the boldfaced sentences. Predict how you would write the formulas for binary ionic compounds. After you have read the section, check the accuracy of your prediction.

Connecting to Your World At festivals throughout the summer, contestants compete for blue ribbons for the best barbecue. Some cooks say the recipe for their barbecue sauce is the key to winning and they may hint at a secret ingredient. The recipe is the formula for the sauce—a complete list of ingredients and their proportions. With the recipe, anyone could reproduce a sauce, so a cook is likely to keep a prize-winning recipe a closely guarded secret. Chemistry also uses formulas, but without any secrets. Once you know the rules, you can write the formula for any chemical compound. In this section, you will learn how to write the formulas for ionic compounds.

Binary Ionic Compounds

In the days before the science of chemistry developed, the person who discovered a new compound often named it anything he or she wished. It was not uncommon for the name to describe some property of the substance or its source. For example, a common name for potassium carbonate (K_2CO_3) is *potash.* The name evolved because the compound was obtained by boiling wood ashes in iron pots. Baking soda ($NaHCO_3$) is added to batter to make cakes rise. Plaster of paris ($CaSO_4 \cdot \frac{1}{2}H_2O$) is the name for a substance used to make face masks like those shown in Figure 9.7. When plaster of Paris sets, it forms gypsum ($CaSO_4 \cdot 2H_2O$). Unfortunately, such names do not tell you anything about the chemical composition of a compound or give you any indication about how it is related to other compounds.

The French chemist Antoine-Laurent Lavoisier (1743–1794) determined the composition of many compounds in his experiments to show how chemical compounds form. As more and more compounds were identified, Lavoisier recognized that it was becoming impossible to memorize all the unrelated names of the compounds. He worked with other chemists to develop a systematic method for naming chemical compounds. Their work is the basis for naming compounds today.

Figure 9.7 The ionic compound in these masks has the common name gypsum ($CaSO_4 \cdot 2H_2O$). Gypsum is also used in wallboard.

Section Resources

Print

- **Small-Scale Chemistry Laboratory Manual,** Lab 12
- **Guided Reading and Study Workbook,** Section 9.2
- **Core Teaching Resources,** Section 9.2 Review
- **Transparencies,** T96–T97

Technology

- **Interactive Textbook with ChemASAP,** Simulation 9; Problem-Solving 9.11, 9.12; Assessment 9.2
- **Go Online,** Section 9.2

Naming Binary Ionic Compounds A **binary compound** is composed of two elements and can be either ionic or molecular. If you know the formula for a binary ionic compound, you can write its name. First you must verify that the compound is composed of a monatomic metallic cation and a monatomic nonmetallic anion. The compound Cs_2O is composed of the metal cesium and the nonmetal oxygen. Both cesium and oxygen are Group A elements, so each ion has only one charge. **To name any binary ionic compound, place the cation name first, followed by the anion name.** The name of Cs_2O, then, is cesium oxide. The name of NaBr is sodium bromide and the name of SrF_2 is strontium fluoride. But suppose you want to name the binary ionic compound CuO. Following the rule above, you would name this compound copper oxide. However, the name *copper oxide* is incomplete. Recall that copper commonly forms two cations: Cu^+ and Cu^{2+}. The names of these ions are copper(I) ion and copper(II) ion, respectively. How can you tell which of these cations forms the compound CuO? Working backward will help. The formula indicates that the copper cation and the oxide anion combine in a 1:1 ratio. You know that the oxide anion always has a 2− charge. Therefore, the charge of the copper cation must be 2+ in order to balance the 2− charge. The compound CuO must be copper(II) oxide. The formula for copper(I) oxide is Cu_2O.

Table 9.2 lists the symbols and names of the common metal ions that form more than one ion. Recall that the charges of monatomic anions can be determined from the periodic table; those of polyatomic anions are shown in Table 9.3. Using these sources, you can write the names of SnF_2 and SnS_2. Tin (Sn) forms cations with 2+ and 4+ charges. Fluorine is a Group 7A element, so the charge of the fluoride ion is 1−. In SnF_2, the ratio of cation to anion is 1:2. Therefore, the charge of the tin cation must be 2+ to balance the combined 2− charge of two fluoride ions. The name of SnF_2 is tin(II) fluoride or stannous fluoride. However, the name of SnS_2 is not tin(II) sulfide. Sulfur is a Group 6A element, so its charge is 2−. The charge of the tin cation must be 4+ to balance the combined charges of two sulfur anions. Thus the name of SnS_2 is tin(IV) sulfide or stannic sulfide. Figure 9.8 shows examples of uses of stannous fluoride and stannic sulfide.

Checkpoint *What are the charges of the two ions of copper?*

Word *Origins*

Binary comes from the Latin word *bini* meaning "two by two" or "twofold." Binary compounds consist of two elements. **Predict what a binary star might be and then check your prediction in the dictionary.**

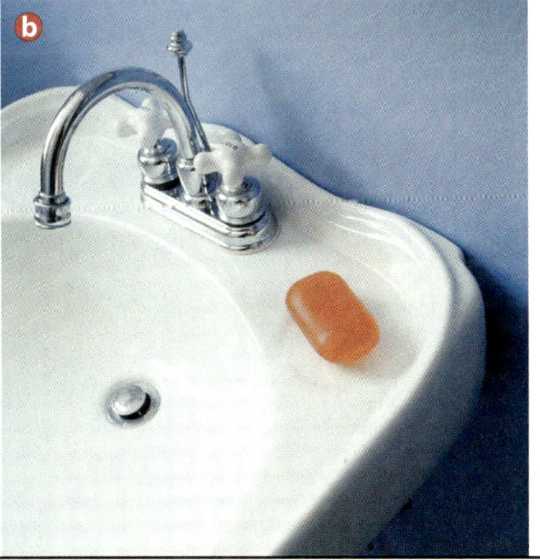

Figure 9.8 Tin(II) fluoride and tin(IV) sulfide have different compositions and uses. **a** Tin(II) fluoride is added to toothpastes to prevent cavities. **b** Tin(IV) sulfide is used in glazes for porcelain fixtures and dishes. **Inferring** ***What are the charges on the tin ions in the two compounds?***

Binary Ionic Compounds

Word *Origins* L2

A binary star is a system of two stars that revolve around each other.

Relate L1

Ask, **Do you or someone you know have a nickname? Does the nickname tell something about the person?** *(Students will have a variety of responses.)* Guide students to recognize that nicknames are somewhat like the common names used to describe compounds. Use sodium carbonate (Na_2CO_3) as an example. Its common name is soda or soda ash. Ask, **Which tells you more about the compound, soda or sodium carbonate?** *(sodium carbonate).*

CLASS Activity

Naming Ionic Binary Compounds L2

Purpose Students practice naming binary compounds.

Materials chalkboard, chalk

Procedure Divide the class into teams. Have each team write the names of ionic compounds on a sheet of paper as you write formulas on the board. Use chlorides, bromides, phosphides, and oxides of the elements of Groups 1A, 2A, and aluminum in 3A. Mix in sulfides, oxides, chlorides, and bromides of lead, tin, iron, cobalt, copper, and zinc. Have the teams compare their lists of names. When a discrepancy occurs, have them discuss it and come to agreement on the correct name. Save the lists of names for the activity on page 263.

Differentiated Instruction

Less Proficient Readers L1

Cut index cards to obtain pieces that are one-third and two- thirds of a card. Keep some cards whole. On some whole cards, mark three large positive signs; on others mark three large negative signs. Mark some of the one-third pieces with single positive signs; others with single negative signs. Mark the two-thirds pieces with two positive signs or two negative signs. On the opposite side of each card, write the formula for an ion having the appropriate charge. Have students match single positive and negative signs. Have them match double positive signs with double negative signs and with two single positive signs. Continue with all possible combinations. Then have students turn the cards over to see the formulas of the ionic compounds they produced.

Answers to...

Figure 9.8 2+ and 4+

1+ and 2+

TEACHER Demo

Making and Naming an Ionic Compound L2

Purpose Students see a reaction in which an ionic compound is formed and then name the compound and write its formula.

Materials 1 g powdered zinc, 4 g iodine (I_2), watch glass, eyedropper, 8 mL water

Safety Wear goggles and perform the demo in a fume hood. Place the zinc iodide in a plastic or cardboard container and dispose of it in an approved landfill site.

Procedure Wear safety goggles and use a fume hood. Mix 1 g of powdered zinc (Zn) and 4 g of iodine (I_2) on a watch glass. With an eyedropper, carefully add 8 mL of water, one drop at a time. After the reaction is complete, show students the zinc iodide that was formed. Write the formula unit for zinc iodide (ZnI_2) along with its name on the chalkboard. Ask, **What happened to elemental Zn and I_2 during the reaction?** *(Zinc and iodine reacted to form an ionic compound composed of Zn^{2+} and I^- ions.)* Use the criss-cross method to show how the formula unit for zinc iodide was derived. **Why is ZnI_2 named zinc iodide and not zinc(II) iodide?** *(Zinc forms only one ion.)*

Expected Outcome The masses of Zn and I_2 are the stoichiometic amounts. The reactants should be used up and white, crystalline ZnI_2 formed.

Download a worksheet on **Ionic Compounds** to complete, and find additional teacher support from NSTA SciLinks.

Figure 9.9 In the process for making steel, iron is extracted from hematite, an ore containing iron(III) oxide. **Applying Concepts** ***What is the formula for iron(III) oxide?***

For: Links on Ionic Compounds
Visit: www.SciLinks.org
Web Code: cdn-1092

Writing Formulas for Binary Ionic Compounds If you know the name of a binary ionic compound, you can write its formula. **Write the symbol of the cation and then the anion. Add whatever subscripts are needed to balance the charges.** The positive charge of the cation must balance the negative charge of the anion so that the net ionic charge of the formula is zero. The ionic compound potassium chloride is composed of potassium cations (K^+) and chloride anions (Cl^-), so potassium chloride is a binary ionic compound. The charge of each K^+ cation is balanced by the charge of each Cl^- anion, so in potassium chloride, the potassium and chloride ions combine in a 1:1 ratio. Thus the formula for potassium chloride is KCl. The net ionic charge of the formula unit is zero.

The binary ionic compound calcium bromide is composed of calcium cations (Ca^{2+}) and bromide anions (Br^-). The two ions do not have equal numerical charges. Thus each calcium ion with its 2+ charge must combine with (or be balanced by) two bromide ions, each with a 1− charge. That means that the ions must combine in a 1:2 ratio, so the formula for calcium bromide is $CaBr_2$. The net ionic charge of the formula unit is zero.

Figure 9.9 shows one step in the process of making steel from iron ore. Hematite, a common ore of iron, contains iron(III) oxide. What is the formula for this compound? Recall that a Roman numeral in the name of an ion shows the charge of the metal ion. Thus iron(III) oxide contains Fe^{3+} cations combined with oxide anions (O^{2-}). How can you balance a 3+ charge and a 2− charge? You must find the least common multiple of the charges, which is 6. Iron's three charges taken two times equals six ($3 \times 2 = 6$). Oxygen's two charges taken three times also equals six. Thus two Fe^{3+} cations (a 6+ charge) will balance three O^{2-} anions (a 6− charge). The balanced formula, then, is Fe_2O_3.

Another approach to writing a balanced formula for a compound is to use the crisscross method. In this method, the numerical value of the charge of each ion is crossed over and becomes the subscript for the other ion. Notice that the signs of the charges are dropped.

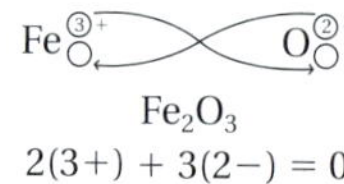

Fe_2O_3

$2(3+) + 3(2-) = 0$

The formula is correct because the overall charge of the formula is zero and the subscripts are in the lowest whole number ratio.

Simulation 9 Simulate combining ions and deriving the chemical formulas for several ionic compounds.

with ChemASAP

Facts and Figures

Lattice Energy in Ionic Bonding

Ionic compounds tend to be stable—energy in the form of heat or electricity is needed to decompose them. Thus, energy must be released when ionic bonds form. But the released energy is not primarily from the transfer of electrons. Forming a cation is an endothermic process. In addition, both the metal and the nonmetal must be vaporized—another endothermic process—before electron transfer can take place. Some energy (electron affinity) is released when the nonmetal gains one or more electrons, but this is not enough to make the entire process exothermic. It is lattice energy that makes the difference. Lattice energy is the energy released when gaseous cations and anions settle into the orderly crystal structure characteristic of a solid ionic compound.

If you use the crisscross method to write the formula for some compounds such as calcium sulfide (Ca^{2+} and S^{2-}), you will obtain the result Ca_2S_2. The 2:2 ratio of calcium and sulfide ions is not the lowest whole number ratio. The formula for calcium sulfide is CaS.

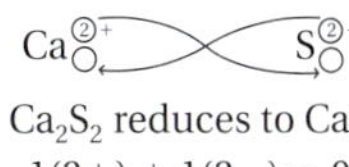

Ca_2S_2 reduces to CaS

$1(2+) + 1(2-) = 0$

Of course, if the magnitudes of the charges of the cation and anion are the same, as they are in this case, the ions combine in a 1:1 ratio and the charges are balanced.

Checkpoint ***Explain why the formula CaS is correct.***

CONCEPTUAL PROBLEM 9.2

Writing Formulas for Binary Ionic Compounds

Write formulas for these binary ionic compounds.

a. copper(II) sulfide, shown in the photo

b. potassium nitride

Analyze ***Identify the relevant concepts.***

Binary ionic compounds are composed of a monatomic cation and a monatomic anion. The ionic charges in an ionic compound must balance (add up to zero), and the ions must be in the lowest whole number ratio. The symbol for the cation appears first in the formula for the compound.

Solve ***Apply concepts to this situation.***

Write the symbol and charge for each ion in each compound.

a. Cu^{2+} and S^{2-} **b.** K^+ and N^{3-}

Balance the formula using appropriate subscripts.

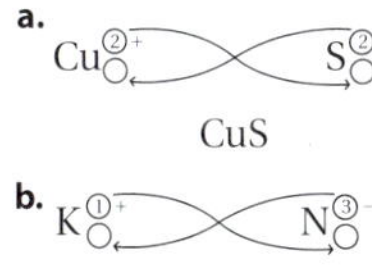

a. CuS

b. K_3N

The ions are in the lowest whole number ratio, and the net ionic charge is zero: $1(2+) + 1(2-) = 0$ and $3(1+) + 1(3-) = 0$.

Practice Problems

10. Write formulas for compounds formed from these pairs of ions.

a. Ba^{2+}, S^{2-} **b.** Li^+, O^{2-}

c. Ca^{2+}, N^{3-} **d.** Cu^{2+}, I^-

11. Write formulas for these compounds.

a. sodium iodide **b.** stannous chloride

c. potassium sulfide **d.** calcium iodide

Problem-Solving 9.11 Solve Problem 11 with the help of an interactive guided tutorial.

with **ChemASAP**

CLASS Activity

Formulas for Binary Ionic Compounds 

Purpose Students become more proficient in writing formulas.

Materials The lists of names of ionic compounds generated in the activity on p. 261.

Procedure Have teams of students write the formulas for the names on the lists. When the formulas on two or more lists do not agree, have the teams resolve the discrepancy by using the criss-cross method to write the formula on the board.

CONCEPTUAL PROBLEM 9.2

Answers

10. a. BaS **b.** Li_2O **c.** Ca_3N_2 **d.** CuI_2

11. a. NaI **b.** $SnCl_2$ **c.** K_2S **d.** CaI_2

Practice Problems Plus 

Write formulas for these compounds.

a. lithium fluoride *(LiF)*

b. aluminum chloride *($AlCl_3$)*

c. sodium nitride *(Na_3N)*

d. ferric oxide *(Fe_2O_3)*

Answers to...

Figure 9.9 Fe_2O_3

Checkpoint The charges are balanced and the subscripts are in the lowest whole number ratio.

Compounds with Polyatomic Ions

TEACHER Demo

Making and Naming Lead Carbonate L2

Purpose Students watch the formation of an ionic precipitate and postulate possible formulas.

Materials 50 mL 0.1*M* lead(II) nitrate, 50 mL 0.1*M* sodium carbonate, 100-mL beaker.

Procedure Wear goggles and gloves. Slowly mix the two solutions. Make sure students see the precipitate of lead carbonate that forms. Write on the board the symbols or formulas for the four ions that are in the two solutions: Pb^{2+}, NO_3^-, Na^+, CO_3^{2-}. Tell students that the two solutions contained these four ions and that the precipitate is an ionic compound formed by the combination of two of them. Ask students to list all possible formula units for the solid and to name each possibility. Identify the solid as lead(II) carbonate ($PbCO_3$).

Safety Wear goggles and gloves because lead(II) nitrate is toxic. For disposal, combine the reaction mixture with 10 g of solid NaCl in a beaker. Stir and allow to settle. Filter or decant to isolate the precipitate. Place the dry precipitate in a plastic or cardboard container and dispose of it in an approved landfill. Flush the filtrate down the drain with excess water.

Use Visuals L1

Figure 9.10 Tell students that the compounds mentioned in this caption are only a small fraction of the compounds with polyatomic ions that they may encounter daily. Ask, **What are the formulas for calcium carbonate, lead(II) sulfate, ammonium sulfate, and ammonium phosphate?** *($CaCO_3$, $PbSO_4$, $(NH_4)_2SO_4$, $(NH_4)_3PO_4$)*

Compounds with Polyatomic Ions

The pearl and the oyster shell shown in Figure 9.10 are both made of calcium carbonate ($CaCO_3$). Calcium carbonate is obviously not a binary compound because it contains more than two elements. Remember that an *-ate* or *-ite* ending on the name of a compound indicates that the compound contains a polyatomic anion that includes oxygen. This compound contains one monatomic ion (Ca^{2+}) and one polyatomic ion (CO_3^{2-}). Figure 9.10 also shows a typical automobile battery called a lead storage battery. The energy-producing reaction inside the battery uses the ionic compound lead(II) sulfate ($PbSO_4$), which consists of the monatomic ion Pb^{2+} and the polyatomic ion SO_4^{2-}. The fertilizer mixture also shown in the illustration could have been produced from such compounds as potassium hydrogen phosphate (K_2HPO_4), potassium sulfate (K_2SO_4), or sodium nitrate ($NaNO_3$). Each contains a polyatomic anion. How would you write the formula for an ionic compound with a polyatomic ion? You would do what you did for binary ionic compounds. **Write the symbol for the cation followed by the formula for the polyatomic ion and balance the charges.** For example, calcium nitrate is composed of a calcium cation (Ca^{2+}) and a polyatomic nitrate anion (NO_3^-). In calcium nitrate, two nitrate anions, each with a 1− charge, are needed to balance the 2+ charge of each calcium cation.

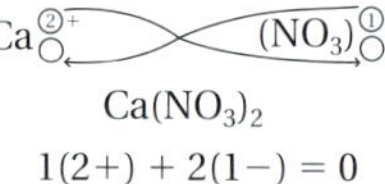

$$Ca(NO_3)_2$$

$$1(2+) + 2(1-) = 0$$

The charge is balanced and the ions are in the lowest whole number ratio, so the formula is correct. Parentheses are used around the nitrate ion in the formula because more than one nitrate anion is needed. The subscript 2 that follows the parentheses shows that the compound contains two nitrate anions. Use parentheses to set off the polyatomic ion in a formula only when the compound contains more than one polyatomic ion. The formula for strontium sulfate, $SrSO_4$, has no parentheses because one polyatomic sulfate anion balances the charge of the strontium cation.

Checkpoint ***When are parentheses used in a formula containing a polyatomic ion?***

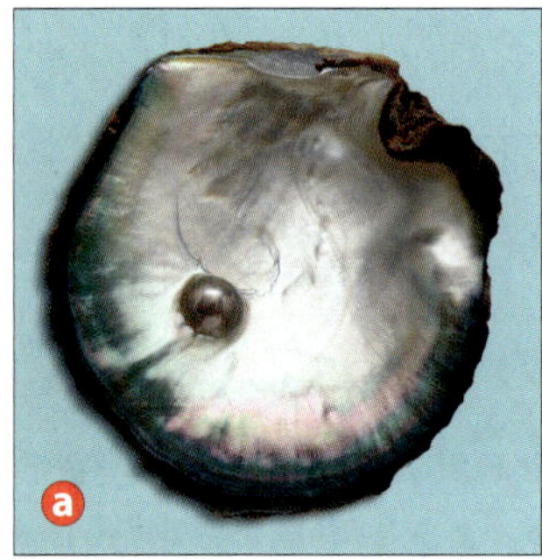

Figure 9.10 Some examples of ionic compounds containing polyatomic ions are shown. **a** Oysters produce calcium carbonate to form their shells and sometimes pearls. **b** Lead(II) sulfate is an important component of an automobile battery. **c** Compounds such as ammonium sulfate or ammonium phosphate are common fertilizers.

Differentiated Instruction

Gifted and Talented 

Have interested students research the possible mechanisms for the alleviation of symptoms of bipolar disorder by lithium ions. Challenge students to explain why other Group 1A ions, such as sodium and potassium, which are normally present in substantial concentrations in the body, do not have the same effects as lithium ions even though they have the same charge.

Lithium carbonate is a compound composed of lithium cations (Li^+) and polyatomic carbonate anions (CO_3^{2-}).

Li_2CO_3

In lithium carbonate, two lithium cations, each with a 1+ charge, are needed to balance the 2− charge of one carbonate anion. Parentheses are not needed to set off the polyatomic carbonate anion. Lithium carbonate can be prescribed for patients who have mood disorders, such as manic-depressive or bipolar disorder. A person with bipolar disorder experiences distressing mood swings, from elation to depression and back again. The exact mechanism is not known, but lithium ions may exert a mood-stabilizing effect on neurotransmission. Neurotransmission is the process by which "messages" are sent and received between nerve cells, including those in the brain.

CONCEPTUAL PROBLEM 9.3

Writing Formulas for Compounds with Polyatomic Ions

What are the formulas for these ionic compounds?

a. magnesium hydroxide, shown in the photo as milk of magnesia

b. potassium sulfate

1 Analyze ***Identify the relevant concepts.***

Write the formula for each ion in the order listed in the name. Use subscripts to balance the charges. If more than one polyatomic ion is needed to balance a formula, place the polyatomic ion formula in parentheses, followed by a subscript showing the number needed.

2 Solve ***Apply concepts to this situation.***

a.

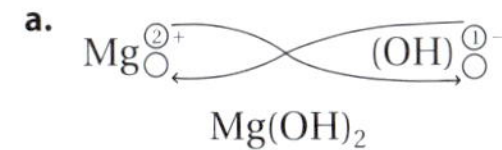

$Mg(OH)_2$

Two hydroxide anions with 1− charges are needed to balance the 2+ charge on one magnesium cation. The formula for magnesium hydroxide must make use of parentheses.

b.

K_2SO_4

Two potassium cations with 1+ charges are needed to balance the 2− charge on one sulfate anion. The formula for potassium sulfate is K_2SO_4.

Practice Problems

12. Write formulas for compounds formed from these pairs of ions.

a. NH_4^+, SO_3^{2-}

b. calcium ion, phosphate ion

13. Write formulas for these compounds.

a. lithium hydrogen sulfate

b. chromium(III) nitrite

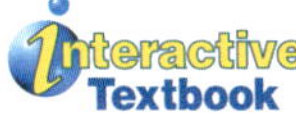

Problem-Solving 9.12 Solve Problem 12 with the help of an interactive guided tutorial.

with ChemASAP

CLASS Activity

Naming and Writing Formulas L2

Purpose Students gain experience in naming and writing formulas for compounds with polyatomic ions.

Procedure Divide the class into groups of two to four students. Have one student in each group randomly chose ions from Tables 9.2 and 9.3. The other member or members of the group must write the formula and name the compound. Rotate the position of "ion chooser" through the group. All students in the group should agree on the names and formulas.

Relate L1

Milk of magnesia, shown in the Conceptual Problem photo, is a suspension of magnesium hydroxide ($Mg(OH)_2$) in water and is useful as an antacid and laxative. In the stomach, the hydroxide ion reacts with excess acid that causes the discomfort of indigestion. In larger amounts, magnesium hydroxide also acts as a muscle relaxant and relieves constipation.

CONCEPTUAL PROBLEM 9.3

Answers

12. a. $(NH_4)_2SO_3$ **b.** $Ca_3(PO_4)_2$

13. a. $LiHSO_4$ **b.** $Cr(NO_2)_3$

Practice Problems Plus L2

Write formulas of compounds formed from these pairs of ions:

a. Pb^{2+}, NO_3^- *($Pb(NO_3)_2$)*

b. iron(III) ion and sulfate ion *($Fe_2(SO_4)_3$)*

c. Cr^{3+}, OH^- *($Cr(OH)_3$)*

d. sodium ion and hydrogen phosphate ion *(Na_2HPO_4)*

Facts and Figures

Lithium Carbonate as Medication

When John Cade, a doctor in the Victoria Department of Mental Health in Victoria, Australia, injected guinea pigs with a lithium salt of uric acid in 1949, he found that the usually hyperactive animals became calm and lethargic for a period of time. After further experimentation, Cade injected a severely disturbed patient with lithium carbonate. The results were so astonishingly positive for this patient, and for many others who followed, that within 10 years the treatment had spread throughout Europe and into the United States. Patients who had been confined to mental institutions for virtually their entire lives were able to return to normal living.

Answers to...

Parentheses are used in a formula when the formula contains more than one unit of the same polyatomic ion.

3 ASSESS

Evaluate Understanding L2

Write the following cations on one side of the board: NH_4^+, Mg^{2+}, Pb^{4+}. Write the following anions on the other side of the board: Br^-, S^{2-}, PO_4^{3-}. Ask students to write formula units for all the possible ionic compounds that these ions could form and name them. Ask them to explain how they arrived at their answers.

Reteach L1

Point out that three things must be considered when naming an ionic compound: (1) the identity of the ions, (2) the order of the names, and (3) the possibility that an element may form cations with more than one charge. (1) For binary compounds, ions can be identified from their symbols. The suffix *-ide* is used when naming the nonmetal ion. For compounds with polyatomic ions, the polyatomic ion can be obtained from Table 9.3 (2) The name of the cation always precedes the name of the anion. (3) For elements that form more than one cation, the correct charge is shown by a Roman numeral in parentheses directly after the cation name. Ask students to use the three-step naming procedure to name FeN and $Mg(NO_3)_2$. *(iron(III) nitride and magnesium nitrate)*

Writing Activity

Answers should mention that lithium carbonate is used because it is effective in regulating the patient's mental state.

If your class subscribes to the Interactive Textbook, use it to review key concepts in Section 9.2.

with ChemASAP

Figure 9.11 Sodium hypochlorite (NaClO) is often added to the water in swimming pools to prevent bacteria from exceeding safe levels.

Naming Compounds with Polyatomic Ions You have learned to write formulas for compounds containing polyatomic ions when you were given their names. Now, if you were given the formulas for these compounds, could you name them? When naming compounds containing polyatomic ions, follow these guidelines. First, recognize that the compound contains a polyatomic ion. If the ion is unfamiliar, find its name in Table 9.3. How would you name the compound LiCN? You could easily think that this compound is composed of individual atoms of lithium, carbon, and nitrogen. But you know that lithium is a Group 1A element and that this group forms 1+ ions. In addition, carbon is a Group 4A element that does not form a monatomic ion. From this you might suspect that the carbon and nitrogen atoms are part of a polyatomic ion with a 1− charge. Table 9.3 confirms your conclusion and tells you that the ion's name is cyanide ion. **To name a compound containing a polyatomic ion, state the cation first and then the anion just as you did in naming binary ionic compounds.** The name of LiCN is lithium cyanide.

The compound NaClO is used as a bleach and a disinfectant for swimming pools, as shown in Figure 9.11. The metallic cation in this compound is sodium (Na^+), a Group IA element that forms 1+ cations. The polyatomic ion must be ClO^-. This ion is called hypochlorite ion, so the name for NaClO is sodium hypochlorite.

Some ionic compounds containing polyatomic ions do not include a metal cation. Instead, the cation may be the polyatomic ammonium ion (NH_4^+). What is the name of $(NH_4)_2C_2O_4$? The ammonium cation has a charge of 1+. The anion must have a charge of 2− to balance the combined 2+ charge of two ammonium ions. Table 9.3 shows that the name of the anion $C_2O_4^{2-}$ is oxalate ion, so $(NH_4)_2C_2O_4$ is named ammonium oxalate.

9.2 Section Assessment

14. **Key Concept** Describe how to determine the names of binary ionic compounds.
15. **Key Concept** Describe how to write the formulas for binary ionic compounds.
16. **Key Concept** How do you write the formulas and the names of compounds with polyatomic ions?
17. Write the formula for these binary compounds.
    - **a.** beryllium chloride
    - **b.** cesium sulfide
    - **c.** sodium iodide
    - **d.** strontium oxide
18. Write the formula for these compounds containing polyatomic ions.
    - **a.** chromium(III) nitrite
    - **b.** sodium perchlorate
    - **c.** magnesium hydrogen carbonate
    - **d.** calcium acetate
19. Identify any incorrect formulas. Explain your answer.
    - **a.** $Mg_2(SO_4)_3$
    - **b.** Rb_3As
    - **c.** $BeCl_3$
    - **d.** NaF

Writing Activity

Report Investigate the role of lithium carbonate in the successful treatment of bipolar disorder. Write a brief report that includes information on bipolar disorder and why lithium carbonate is used to treat it.

Assessment 9.2 Test yourself on the concepts in Section 9.2.

with ChemASAP

Section 9.2 Assessment

14. Write the name of the cation followed by the name of the anion.
15. Write the symbol of the cation followed by the symbol of the anion, then use subscripts to balance the charges.
16. Write the symbol for the metal ion followed by the formula of the polyatomic ion and balance the charges. Name the cation first followed by the anion.
17. **a.** $BeCl_2$ **b.** Cs_2S **c.** NaI **d.** SrO
18. **a.** $Cr(NO_2)_3$ **b.** $NaClO_4$ **c.** $Mg(HCO_3)_2$ **d.** $Ca(C_2H_3O_2)_2$
19. **a.** incorrect; charges are not balanced, $MgSO_4$ **c.** incorrect; charges are not balanced, $BeCl_2$

Small-Scale LAB

Names and Formulas for Ionic Compounds

Purpose

To observe the formation of compounds, and to write their names and formulas.

Materials

- pencil
- paper
- ruler
- reaction surface
- chemicals shown in Figure A

Procedure   

On separate sheets of paper, draw two grids similar to Figure A. Make each square 2 cm on each side. Draw black Xs on one of the grids. Use the other grid as a data table to record your observations. Place a reaction surface over the grid with black Xs and add the chemicals as shown in Figure A.

Analyze

Using your experimental data, record the answers to the following questions below your data table.

1. Describe each precipitate (solid product) that forms. Use terms such as milky, grainy, cloudy, or gelatinous. Which mixture(s) did not form a precipitate?
2. Write the formulas and names of the chemical compounds produced in the mixings.

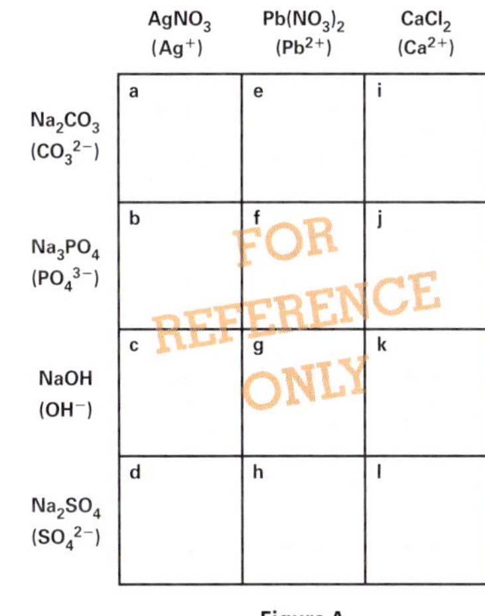

Figure A

You're The Chemist

The following small-scale activities allow you to develop your own procedures and analyze the results.

1. **Analyze It!** Repeat the experiment, using the chemicals in Figure B. Identify the precipitates, write their formulas, and name them.
2. **Explain It!** In ionic equations, the precipitate is written to the right of an arrow, and the ions that produced it are written to the left. Write ionic equations for the precipitates formed from the reactions related to Figure B. For example:

$$2Ag^+ + CO_3^{2-} \longrightarrow Ag_2CO_3$$

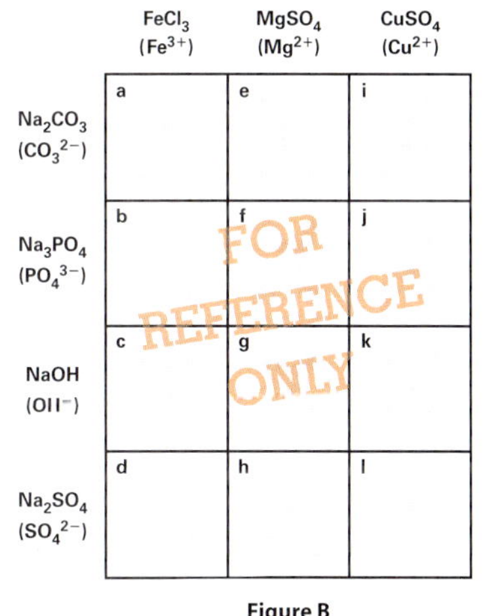

Figure B

Names and Formulas for Ionic Compounds L2

Objective After completing this activity, students will be able to:

- recognize and describe precipitates.
- write the names and formulas for the precipitates.

 **Prep Time** 1 hour

Solution	Preparation
0.05*M* $AgNO_3$	2.1 g in 250 mL
0.2*M* $Pb(NO_3)_2$	16.6 g in 250 mL
0.5*M* $CaCl_2$	13.9 g in 250 mL
1.0*M* Na_2CO_3	26.5 g in 250 mL
0.1*M* Na_3PO_4	9.5 g $Na_3PO_4 \cdot 12H_2O$ in 250 mL
0.5*M* NaOH	20.0 g in 1.0 L
0.2*M* Na_2SO_4	7.1 in 250 mL
0.2*M* $CuSO_4$	12.5 g $CuSO_4 \cdot 5H_2O$ in 250 mL
0.2*M* $MgSO_4$	6.0 g in 250 mL
0.1*M* $FeCl_3$	6.8 g $FeCl_3 \cdot 6H_2O$ in 25 mL of 1.0*M* NaCl (for stability); dilute to 250 mL

Class Time 30 minutes

Expected Outcome In Figure A, a, f, g, h, and j: milky white; b and k: cloudy white; e: cloudy tan; i and l: grainy white; c: muddy brown; d: no visible reaction. In Figure B, a, b, and c: orange; e, f, and g: white; i, j, and k: blue; d, h, and l: no visible reaction.

Analyze and Conclude

1. $AgNO_3$ and Na_2SO_4 did not form a precipitate.
2. **a.** Ag_2CO_3, silver carbonate **b.** Ag_3PO_4, silver phosphate **c.** AgOH, silver hydroxide (Note: The product is actually Ag_2O, silver oxide.)
   e. $PbCO_3$, lead(II) carbonate
   f. $Pb_3(PO_4)_2$, lead(II) phosphate
   g. $Pb(OH)_2$, lead(II) hydroxide
   h. $PbSO_4$, lead(II) sulfate
   i. $CaCO_3$, calcium carbonate
   j. $Ca_3(PO_4)_2$, calcium phosphate
   k. $Ca(OH)_2$, calcium hydroxide
   l. $CaSO_4$, calcium sulfate

You're The Chemist

1. **a.** $Fe_2(CO_3)_3$ iron(III) carbonate
   b. $FePO_4$, iron(III) phosphate
   c. $Fe(OH)_3$, iron(III) hydroxide
   e. $MgCO_3$, magnesium carbonate
   f. $Mg_3(PO_4)_2$, magnesium phosphate
   g. $Mg(OH)_2$, magnesium hydroxide
   i. $CuCO_3$, copper(II) carbonate
   j. $Cu_3(PO_4)_2$, copper(II) phosphate
   k. $Cu(OH)_2$, copper(II) hydroxide
2. **a.** $2Fe^{3+} + 3CO_3^{2-} \rightarrow Fe_2(CO_3)_3(s)$
   b. $Fe^{3+} + PO_4^{3-} \rightarrow FePO_4(s)$
   c. $Fe^{3+} + 3OH^- \rightarrow Fe(OH)_3(s)$
   e. $Mg^{2+} + CO_3^{2-} \rightarrow MgCO_3(s)$
   f. $3Mg^{2+} + 2PO_4^{3-} \rightarrow Mg_3(PO_4)_2(s)$
   g. $Mg^{2+} + 2OH^- \rightarrow Mg(OH)_2(s)$
   i. $Cu^{2+} + CO_3^{2-} \rightarrow CuCO_3(s)$
   j. $3Cu^{2+} + 2PO_4^{3-} \rightarrow Cu_3(PO_4)_2(s)$
   k. $Cu^{2+} + 2OH^- \rightarrow Cu(OH)_2(s)$

For Enrichment 

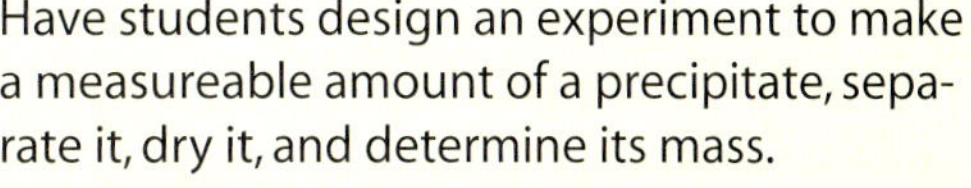

Have students design an experiment to make a measureable amount of a precipitate, separate it, dry it, and determine its mass.

9.3

1 FOCUS

Objectives

9.3.1 **Interpret** the prefixes in the names of molecular compounds in terms of their chemical formulas.

9.3.2 **Apply the rules** for naming and writing formulas for binary molecular compounds.

Guide for Reading

Build Vocabulary L2

Word Parts The word *prefix* comes from the Latin word *praefigere* meaning "to fasten before." Tell students they will learn to fasten a prefix such as *di-*, *tri-*, *tetra-* to the names of elements to indicate how many atoms of each element are in a formula of a molecular compound.

Reading Strategy L2

Visualize Encourage students to visualize the meaning of the prefixes used in binary molecular compounds. Write some names on the chalkboard, for example, carbon tetrachloride. Then draw one circle representing carbon and four circles representing chlorine. All the circles combined constitute the five atoms in CCl_4. Ask, **How many circles should I draw for fluorine in sulfur hexafluoride?** *(six)* How many should I draw for sulfur? *(one)*

2 INSTRUCT

Connecting to Your World

The prefix *milli-* means one-thousandth of something and the prefix *kilo-* means 1000 of something. Ask, **How many milligrams of gold are in one gram of gold?** *(1000)* **How many milligrams are in one kilogram of gold?** *(1,000,000)* **Based on $12,500 for a kilogram, is one cent the actual value for one milligram?** *(The price is actually one and one-quarter cent.)*

9.3 Naming and Writing Formulas for Molecular Compounds

Guide for Reading

Key Concepts

- What does a prefix in the name of a binary molecular compound tell you about the compound's composition?
- How do you write the formula for a binary molecular compound?

Reading Strategy

Monitoring Your Understanding Before you read, preview the Key Concepts, the headings, the boldfaced sentences, and the visuals. List two things you expect to learn. After reading, state what you learned about each item you listed.

Connecting to Your World

Gold was one of the first metals to attract human attention. When gold was discovered in California in the late 1840s, people from all over the world came to find it and make their fortune. Today, gold is still greatly prized and valued. Whereas one milligram of gold is worth only about one cent, one kilogram of gold is worth approximately $12,500. In this case, using the correct prefix (*milli-* or *kilo-*) makes quite a difference! Prefixes are important in chemistry, too. In this section, you will learn how prefixes in the name of a binary molecular compound tell you its composition.

Naming Binary Molecular Compounds

Recall that binary ionic compounds are composed of the ions of two elements, a metal and a nonmetal. Binary molecular compounds are also composed of two elements, but both elements are nonmetals and they are not ions. These differences affect the naming of these compounds and their formulas. Binary molecular compounds are composed of molecules, not ions, so ionic charges cannot be used to write formulas or to name them. In addition, when two nonmetallic elements combine, they often do so in more than one way. For example, the elements carbon and oxygen combine to form two invisible gaseous compounds, CO and CO_2. CO is illustrated in Figure 9.12. How would you name a binary compound formed by the combination of carbon and oxygen atoms? It might seem satisfactory to call it carbon oxide. However, the two carbon oxides, CO and CO_2, are very different compounds. Sitting in a room with small amounts of the carbon oxide CO_2 in the air would not present any problems. You exhale CO_2 as a product of your body chemistry. Thus it is normally present in the air you breathe. On the other hand, if the same amount of the other carbon oxide, CO, were in the room, you could die of asphyxiation. The binary compound CO is a poisonous gas that interferes with your blood's ability to carry oxygen to body cells. Obviously, a naming system that distinguishes between these two compounds is needed.

Section Resources

Print

- **Guided Reading and Study Workbook,** Section 9.3
- **Core Teaching Resources,** Section 9.3
- **Transparencies,** T98

Technology

- **Interactive Textbook with ChemASAP,** Assessment 9.3
- **Go Online,** Section 9.3

Table 9.4

Prefixes Used in Naming Binary Molecular Compounds										
Prefix	Mono-	Di-	Tri-	Tetra-	Penta-	Hexa-	Hepta-	Octa-	Nona-	Deca-
Number	1	2	3	4	5	6	7	8	9	10

Prefixes in the masses of gold samples distinguish between large and small samples. Prefixes in the names of binary molecular compounds help distinguish compounds containing different numbers of atoms. **A prefix in the name of a binary molecular compound tells how many atoms of an element are present in each molecule of the compound.** Table 9.4 lists the prefixes used to name binary molecular compounds. According to the table, the prefix *mono-* indicates the presence of one oxygen atom in CO. The prefix *di-* indicates the presence of the two oxygen atoms in CO_2. The two compounds of carbon and oxygen, CO and CO_2, are thus named carbon monoxide and carbon dioxide, respectively. *Laughing gas* is the common name for the gaseous compound dinitrogen monoxide (N_2O), which is used as an anesthetic. When inhaled, N_2O tends to make people laugh. Notice that the second element in the name ends with *-ide*. The names of all binary molecular compounds end in *-ide*. Also note that the vowel at the end of a prefix is sometimes dropped when the name of the element begins with a vowel. For CO, you would write carbon *mon*oxide, not carbon *mono*oxide. If just one atom of the first element is in the formula, omit the prefix *mono-* for that element.

Here are some guidelines for naming binary molecular compounds. First, confirm that the compound is a binary molecular compound—that is, a compound composed of two nonmetals. The name must identify the elements in the molecule and indicate the number of each atom of each element. Name the elements in the order listed in the formula. Use prefixes to indicate the number of each kind of atom. Omit the prefix *mono-* when the formula contains only one atom of the first element in the name. The suffix of the name of the second element is *-ide*. Now, apply these guidelines to naming N_2O. The formula shows that the compound consists of two nonmetals, so it is a binary molecular compound. Two atoms of nitrogen are combined with one atom of oxygen. Thus the prefix of nitrogen is *di-* and the prefix of oxygen is *mono-*. The name of the compound is dinitrogen monoxide. Using the same guidelines, the name of SF_6 is sulfur hexafluoride. Notice that it is not necessary to use the prefix *mono-* before sulfur. What about the compound Cl_2O_8? This binary molecular compound consists of two chlorine atoms (prefix *di-*) and eight oxygen atoms (prefix *octa-*). The name is dichlorine octoxide.

What suffix ends the names of all binary molecular compounds?

For: Links on Carbon Monoxide
Visit: www.SciLinks.org
Web Code: cdn-1093

Figure 9.12 Carbon monoxide is an invisible, gaseous compound of carbon and oxygen. It is a toxic product of incomplete burning, such as occurs in automobile engines and faulty furnaces.

Differentiated Instruction

Gifted and Talented 

Increasingly large quantities of the binary molecular compound CO_2 are released into the atmosphere annually by the burning of fossil fuels. Once in the atmosphere, carbon dioxide traps solar energy near Earth's surface. As a result, some scientists predict that over time global temperatures will rise. Ask students to learn more about global warming and prepare an oral or written report or a poster that explains it.

Naming Binary Molecular Compounds

Naming Binary Molecular Compounds L2

Purpose Students convert the subscripts in a formula to prefixes in the name.

Materials paper and pencil

Procedure Have students make a chart with three columns: *element name, number of atoms,* and *prefix*. Have them fill in the columns for N_2O and then name the compound. For example:

Nitrogen	2	di-
Oxygen	1	mono-

The name is dinitrogen monoxide. Write the molecular formulas for other compounds on the board and have students add them to their charts. Some possibilities are CCl_4, PBr_5, and P_4S_3.

Discuss 

Students may be interested in knowing how the order of the elements in the names of most molecular compounds is established. Ordinarily, the less electronegative element appears first in the name. For example, a compound containing carbon and sulfur is carbon disulfide (CS_2). Carbon is less electronegative than sulfur so it appears first. Electronegativity increases from left to right on the periodic table and decreases from top to bottom. Thus the first element in the name and formula is the element further to the left, for example, disulfur dichloride (S_2Cl_2). For elements in the same column, the element nearer the bottom of the table goes first, for example, iodine heptafluoride (IF_7).

Download a worksheet on **Carbon Monoxide** for students to complete, and find additional teacher support from NSTA SciLinks.

 *-ide*

Section 9.3 (continued)

Writing Formulas for Binary Molecular Compounds

Relate L1

Students may be interested in knowing that the hardness of substances such as silicon carbide have been quantified using the Mohs scale, which ranges from zero to 10. Tell students the graphite in their pencils has a rating of 0.5 Mohs. Diamond is the hardest mineral at 10 Mohs and silicon carbide is 9.3.

3 ASSESS

Evaluate Understanding L2

Challenge students to name the following pairs of compounds and to identify what each pair has in common: PBr_3 and $CrBr_3$, N_2O and Na_2O, Cl_4 and PbI_4, $P_2O_3$and Fe_2O_3 *(Each pair consists of a binary molecular compound and a binary ionic compound with one element in common.)*

Reteach L1

Have students write the names of these binary molecular compounds: NO, SiO_2, N_2O_4, ClF_3. *(nitrogen monoxide, silicon dioxide, dinitrogen tetroxide, chlorine trifluoride)* Have students write the formulas for these compounds: arsenic pentachloride, iodine tribromide, tetraphosphorus hexoxide. *($AsCl_5$, IBr_3, P_4O_6)*

Connecting Concepts

Yes, chlorine can form only one bond; silicon can form four. An electron dot structure shows silicon sharing one electron with each chlorine for a total of eight electrons around silicon. Each chlorine also is surrounded by eight electrons.

If your class subscribes to the Interactive Textbook, use it to review key concepts in Section 9.3.

with ChemASAP

Figure 9.13 A grinding wheel made of silicon carbide (SiC) can shape even the toughest materials. **Inferring** ***What causes the sparks?***

Writing Formulas for Binary Molecular Compounds

Suppose you know the name of a molecular compound and want to write the formula. **Use the prefixes in the name to tell you the subscript of each element in the formula. Then write the correct symbols for the two elements with the appropriate subscripts.** A simple example is silicon carbide. Silicon carbide is a hard material like diamond. It is used as an abrasive and for cutting and grinding, as shown in Figure 9.13. The name *silicon carbide* has no prefixes, so the subscripts of silicon and carbon must be one. Thus the formula for silicon carbide is SiC. The name of another binary molecular compound is dinitrogen tetroxide. The prefix *di-* before nitrogen tells you that the compound contains two nitrogen atoms; the prefix *tetra-* tells you that the molecule also contains four oxygen atoms. Thus the formula for dinitrogen tetroxide is N_2O_4.

9.3 Section Assessment

20. **Key Concept** What information do prefixes in the name of a binary molecular compound tell you about the composition of the compound?
21. **Key Concept** Describe how to write the formula of a binary molecular compound.
22. Write the names for these molecular compounds.
    a. NCl_3 **b.** BCl_3 **c.** NI_3
    d. SO_3 **e.** N_2H_4 **f.** N_2O_3
23. Write the formulas or names for these molecular compounds.
    a. CS_2 **b.** carbon tetrabromide
    c. Cl_2O_7 **d.** diphosphorus trioxide
24. Write the formulas for these binary molecular compounds.
    a. phosphorus pentachloride
    b. iodine heptafluoride
    c. chlorine trifluoride
    d. iodine dioxide
25. The name a student gives for the molecular compound $SiCl_4$ is monosilicon trichloride. Is this name correct? Explain.

Connecting Concepts

Covalent Bonds In Section 8.1, you learned about covalent bonds. Are the bonds between silicon and chlorine in silicon tetrachloride ($SiCl_4$) single bonds? Justify your answer by drawing an electron dot structure of silicon tetrachloride.

Assessment 9.3 Test yourself on the concepts in Section 9.3.

with ChemASAP

Section 9.3 Assessment

20. Prefixes indicate the number of atoms of each element in a molecule of the compound.
21. Write the symbols for each element with a subscript corresponding to the prefix before each element in the name.
22. **a.** nitrogen trichloride **b.** boron trichloride **c.** nitrogen triiodide **d.** sulfur trioxide **e.** dinitrogen tetrahydride **f.** dinitrogen trioxide
23. **a.** carbon disulfide **b.** CBr_4 **c.** dichlorine heptoxide **d.** P_2O_3
24. **a.** PCl_5 **b.** IF_7 **c.** ClF_3 **d.** IO_2
25. No; the correct name is silicon tetrachloride.

9.4 Naming and Writing Formulas for Acids and Bases

Connecting to Your World

Some ants can give painful stings when threatened or disturbed. Certain ant species called formicines have poison glands that produce venom containing formic acid. Formicines protect themselves by spraying this venom on their predators. Formic acid can stun or even kill the ants' most common enemies. A formicine attack on a human, however, is much less severe. The contact of formic acid with the skin usually results only in blistering. In this section, you will learn the names and formulas of some important acids such as formic acid.

Guide for Reading

Key Concepts

- What are the three rules for naming acids?
- How are the formulas of acids determined?
- How are bases named?

Vocabulary

acid

base

Reading Strategy

Comparing and Contrasting When you compare and contrast things, you examine how they are alike and how they are different. After you have read this section, list similarities and differences between acids and bases and how they are named.

Naming Acids

Acids are a group of ionic compounds with unique properties. As you will see in Chapter 19, acids can be defined in several ways. For now, it is enough to know that an **acid** is a compound that contains one or more hydrogen atoms and produces hydrogen ions (H^+) when dissolved in water. Acids have various uses, one of which is shown in Figure 9.14. When naming an acid, you can consider the acid to consist of an anion combined with as many hydrogen ions as are needed to make the molecule electrically neutral. Therefore, the chemical formulas of acids are in the general form H_nX where X is a monatomic or polyatomic anion and n is a subscript indicating the number of hydrogen ions that are combined with the anion.

Figure 9.14 To create designs such as this on glass, the glass is first coated with wax and the design is drawn through the wax. When the glass is dipped into hydrofluoric acid (HF), the acid etches (eats away) the glass wherever the wax has been removed.

Section Resources

Print

- **Guided Reading and Study Workbook,** Section 9.4
- **Core Teaching Resources,** Section 9.4
- **Transparencies,** T99

Technology

- **Interactive Textbook with ChemASAP,** Assessment 9.4
- **Go Online,** Section 9.4

9.4

1 FOCUS

Objectives

9.4.1 Apply three rules for naming acids.

9.4.2 Apply the rules in reverse to write formulas of acids.

9.4.3 Apply the rules for naming bases.

Guide for Reading

Build Vocabulary 

Connect the Terms Tell students that one of the properties of acids is a sour or sharp taste, which they can use to recognize foods containing acids, such as lemons. Then ask students to infer what it might mean to say a person has "an acid tongue."

Reading Strategy 

Summarize Encourage students to summarize the rules for naming acids in their own words.

2 INSTRUCT

Connecting to Your World

Ask, **Why is an attack by a formicine ant not usually serious for a human?** *(It would take more of the poison than a tiny ant could produce to be fatal for a person.)* **Can you imagine conditions under which a formicine attack might be serious for a person?** *(If many more than one ant attacked at once.)*

Naming Acids

Use Visuals 

Figure 9.14 Ask, **What inference can you make about the ability of glass to react with hydrofluoric acid compared to the ability of wax?** *(Apparently, glass reacts with hydrofluoric acid, but wax does not.)* **Would a glass bottle be an appropriate container for hydrofluoric acid?** *(No, the glass would react with the acid.)*

Section 9.4 (continued)

Use Visuals

L1

Table 9.5 Remind students that all acids have hydrogen as one component. It is the anion combined with hydrogen that identifies the acid. Call attention to the first column in which the endings of anions are given. Ask students to name some anions with the endings *-ide, -ite, and -ate. (They may mention chloride, bromide, nitrite, sulfite, carbonate, formate, among others)* Call attention to the third column. Ask, **When the anion ends in *-ide*, what does the acid name begin with?** *(hydro-)* **What does the acid name end with?** *(-ic)* Have students name an acid containing the bromide anion. *(hydrobromic acid)* Continue the questioning and naming through the other two rows of the table and encourage students to use the table until they are comfortable with naming acids.

Writing Formulas for Acids

Discuss

L2

The table for naming acids is also the table for writing formulas for acids, but it must be used in reverse. Give students an acid name, for example, perchloric acid. Ask, **What does the *-ic* ending tell you?** (*There is no* hydro- *in the name, so* -ic *means the anion ends in* -ate.) Ask, **What is the name and formula for the anion?** *(perchlorate, ClO_4^-)* Ask, **How many hydrogen atoms will combine with the perchlorate ion?** *(one, $HClO_4$)*

Download a worksheet on **Acids** for students to complete and find additional teacher support from NSTA SciLinks.

Table 9.5

Naming Common Acids			
Anion ending	**Example**	**Acid name**	**Example**
-ide	chlor*ide*, Cl^-	*hydro*-(stem)-*ic acid*	*hydro*chlor*ic acid*
-ite	sulf*ite*, SO_3^{2-}	(stem)-*ous acid*	sulfur*ous acid*
-ate	nitr*ate*, NO_3^-	(stem)-*ic acid*	nitr*ic acid*

Go Online SCI LINKS
For: Links on Acids
Visit: www.SciLinks.org
Web Code: cdn-1094

Three rules can help you name an acid with the general formula H_nX. Read the rules and the examples carefully. Notice that the naming system depends on the name of the anion. Each of the rules deals with an anion with a different suffix: *-ide*, *-ite*, and *-ate*.

1. **When the name of the anion (X) ends in *-ide*, the acid name begins with the prefix *hydro-*. The stem of the anion has the suffix *-ic* and is followed by the word *acid*.** Therefore, HCl(aq) (X = chloride) is named *hydro*chlor*ic acid*. H_2S(*aq*) (X = sulfide) is named *hydro*sulfur*ic acid*.
2. **When the anion name ends in *-ite*, the acid name is the stem of the anion with the suffix *-ous*, followed by the word *acid*.** Thus H_2SO_3 (*aq*) (X = sulfite) is named sulfur*ous acid*.
3. **When the anion name ends in *-ate*, the acid name is the stem of the anion with the suffix *-ic* followed by the word *acid*.** Thus HNO_3(*aq*) (X = nitrate) is named nitr*ic acid*.

The three rules are summarized in Table 9.5. Use the table to help you write acid names until you become an expert.

Writing Formulas for Acids

If you know the name of an acid, you can write its formula. **Use the rules for writing the names of acids in reverse to write the formulas for acids.** For example, what is the formula of hydrobromic acid? Following Rule 1, hydrobromic acid (*hydro-* prefix and *-ic* suffix) must be a combination of hydrogen ion (H^+) and bromide ion (Br^-). The formula of hydrobromic acid is HBr. How do you write the formula for phosphorous acid? Using Rule 2, hydrogen ion and phosphite ion (PO_3^{3-}) must be the components of phosphorous acid. The formula of phosphorous acid is H_3PO_3. (*Note:* Do not confuse *phosphorous* with *phosphorus*, the element name.) Finally, what is the formula for formic acid, the defensive weapon of the ants you read about in Connecting to Your World? According to Rule 3, formic acid (*-ic* ending) must be a combination of hydrogen ion (H^+) and form*ate* ion ($HCOO^-$). The formula for formic acid is HCOOH.

Many industrial processes, including steel and fertilizer manufacturing, use acids. In the laboratory, you will regularly use a few common acids such as those listed in Table 9.6. You should become familiar with their names and formulas.

Table 9.6

Common Acids	
Name	**Formula**
Hydrochloric acid	HCl
Sulfuric acid	H_2SO_4
Nitric acid	HNO_3
Acetic acid	CH_3COOH
Phosphoric acid	H_3PO_4
Carbonic acid	H_2CO_3

When does an acid name begin with the prefix **hydro-?**

Facts and Figures

A Toxic Environmental Gas

The binary molecular compound carbon monoxide is a colorless, odorless, flammable gas that is highly toxic to humans. Upon inhalation, carbon monoxide binds to the hemoglobin molecules in red blood cells, where it replaces oxygen. It binds to hemoglobin about 200 times more effectively than oxygen does, keeping the red blood cells from carrying oxygen to body tissues. Thus, even at low levels, carbon monoxide is a fast-acting poison that may cause serious illness or death. At peak traffic times, the level of carbon monoxide around roadways may reach as high as 100 parts per million (ppm) of air. In the U.S., automobiles must have catalytic converters that change toxic carbon monoxide to carbon dioxide. Carbon monoxide is also present in cigarette smoke. It takes several hours to replace the carbon monoxide in a smoker's blood after only one cigarette.

Names and Formulas for Bases

Another group of ionic compounds is the bases. A **base** is an ionic compound that produces hydroxide ions when dissolved in water. **Bases are named in the same way as other ionic compounds—the name of the cation is followed by the name of the anion.** For example, sodium hydroxide (NaOH) is a base used in making paper, cleaners, and soap, as shown in Figure 9.15. To write the formulas for bases, write the symbol for the metal cation followed by the formula for the hydroxide ion. Balance the ionic charges just as you do for any ionic compound. For example, aluminum hydroxide consists of the aluminum cation (Al^{3+}) and the hydroxide anion (OH^-). You need three hydroxide ions to balance the 3+ charge of the aluminum cation. Thus the formula for aluminum hydroxide is $Al(OH)_3$.

Figure 9.15 Sodium hydroxide is an important industrial and consumer product. **a** Recycled paper and wood are digested with NaOH to make pulp in the first step in making paper. **b** Cleaners containing NaOH cut through heavy grease. **c** An important use of sodium hydroxide is in making soap. **Inferring** ***Why is the woman in the second photo wearing gloves?***

9.4 Section Assessment

26. **Key Concept** List the rules for naming acids.
27. **Key Concept** How are the formulas for acids determined?
28. **Key Concept** How are bases named?
29. Give the names of these acids.
    a. HNO_2 **b.** $HMnO_4$
    c. HCN **d.** H_2S
30. Write the names of these bases.
    a. LiOH **b.** $Pb(OH)_2$
    c. $Mg(OH)_2$ **d.** $Al(OH)_3$
31. Identify each compound as an acid or a base.
    a. $Ba(OH)_2$ **b.** $HClO_4$
    c. $Fe(OH)_3$ **d.** KOH
32. Write the formulas for these compounds.
    a. carbonic acid
    b. sulfurous acid
    c. iron(III) hydroxide
    d. strontium hydroxide
33. What element generally appears in the formula of an acid? What ion generally appears in the formula of a base?

Elements Handbook

Sulfuric Acid Sulfuric acid is important to our economy. Refer to page R30 to learn more about H_2SO_4. Write a short report summarizing what you learn.

Assessment 9.4 Test yourself on the concepts in Section 9.4.

with **ChemASAP**

Section 9.4 Assessment

26. See the rules for naming on the SE facing page (page 272).
27. The rules for writing the names of acids are used in reverse to write the formulas for acids.
28. The name of the cation is followed by the name of the anion.
29. a. nitrous acid **b.** permanganic acid **c.** hydrocyanic acid **d.** hydrosulfuric acid
30. a. lithium hydroxide **b.** lead(II) hydroxide **c.** magnesium hydroxide **d.** aluminum hydroxide
31. a. base **b.** acid **c.** base **d.** base
32. a. H_2CO_3 **b.** H_2SO_3 **c.** $Fe(OH)_3$ **d.** $Sr(OH)_2$
33. acid, hydrogen; base, hydroxide ion

Names and Formulas for Bases

Relate L1

Sodium hydroxide is a powerful base. It can turn wood into the mushy pulp from which paper is made. It breaks up and dissolves the greasy matter that clogs drains. Care must be taken to protect the body when using strong bases.

3 ASSESS

Evaluate Understanding 

Divide students into an even number of groups, and assign to each a number of the polyatomic ions listed in Table 9.3. (Omit the hydroxide ion and ammonium ion.) Have the groups write the formulas for the acids corresponding to each assigned anion by adding the correct number of hydrogen atoms. Then ask students to name the acids on a separate piece of paper. Have them exchange papers with another group and write the formulas for the acid names they receive.

Reteach 

Use the names of the six common acids in Table 9.6 as models. Ask students to name the anion and write the formulas using Table 9.5.

Elements Handbook

Answers should include information on how sulfuric acid is produced. They should also include the various products requiring sulfuric acid in their preparation.

If your class subscribes to the Interactive Textbook, use it to review key concepts in Section 9.4.

with **ChemASAP**

Answers to...

Figure 9.15 She is protecting her hands from corrosive NaOH.

when the anion ends in *-ide.*

9.5

1 FOCUS

Objectives

9.5.1 **Define** the laws of definite proportions and multiple proportions.

9.5.2 **Apply the rules** for naming chemical compounds by using a flowchart.

9.5.3 **Apply the rules** for writing the formulas of chemical compounds by using a flowchart.

Guide for Reading

Build Vocabulary L2

List the Parts You Know A scientific law is a statement of a relationship that is invariable so far as it is known. Thus, the law of definite proportions and the law of multiple proportions are invariable statements governing the formation of compounds.

Reading Strategy L2

Think Aloud As you familiarize students with the flowcharts in Figures 9.20 and 9.22, verbalize the routes to a name or formula by asking and answering the questions posed at each intersection.

2 INSTRUCT

Connecting to Your World

Ask, **What would be the ratio of candles to birthday cake at a seventeenth birthday?** *(17:1)* **What would be the ratio of candles on two birthday cakes, one at a seventeenth birthday and the other at a fourth birthday?** *(17:4)*

The Laws of Definite and Multiple Proportions

Use Visuals L1

Figure 9.16 Ask students to use the models to determine the formulas for the two compounds. *(H_2O and H_2O_2)* Ask, **In what other way can you tell that the two compounds are different?** *(One bleaches cloth, the other does not.)*

9.5 The Laws Governing Formulas and Names

Guide for Reading

Key Concepts
- What are the two laws that describe how compounds form?
- How do you use a flowchart to write the name of a chemical compound?
- What four guidelines should you follow to write the formula of a chemical compound?

Vocabulary

law of definite proportions

law of multiple proportions

Reading Strategy

Relating Text and Visuals As you read, use Figure 9.20 and Figure 9.22 to help you become thoroughly familiar with writing the names and formulas for chemical compounds.

Connecting to Your World A birthday cake for a four-year-old has four candles. The ratio of candles to birthday cake is 4:1. A sixteen-year-old's birthday cake has 16 candles. The ratio of candles to cake is also a whole number ratio, 16:1. Is there a whole number ratio between the numbers of candles on one cake at two different birthdays? For the sixteenth and fourth birthdays, the ratio is 16:4 or 4:1. In chemistry, similar relationships exist among the masses of elements as they combine in compounds.

Figure 9.16 Water and hydrogen peroxide contain the same two elements, but they have different properties. (a) Water does not bleach dyes. (b) Hydrogen peroxide is a bleach.

The Laws of Definite and Multiple Proportions

The rules for naming and writing formulas for compounds are possible only because compounds form from the elements in predictable ways. **These ways are summed up in two laws: the law of definite proportions and the law of multiple proportions.**

The Law of Definite Proportions A chemical formula tells you, by means of subscripts, the ratio of atoms of each element in the compound. Ratios of atoms can also be expressed as ratios of masses. Magnesium sulfide (MgS) is composed of magnesium cations and sulfide anions. If you could take 100.00 g of magnesium sulfide and break it down into its elements, you will obtain 43.13 g of magnesium and 56.87 g of sulfur. The Mg:S ratio of these masses is 43.13/56.87 or 0.758:1. This mass ratio does not change no matter how the magnesium sulfide is formed or the size of the sample. Magnesium sulfide illustrates the **law of definite proportions,** which states that in samples of any chemical compound, the masses of the elements are always in the same proportions. Because atoms combine in simple whole-number ratios, it follows that their proportions by mass must always be the same.

The Law of Multiple Proportions Figure 9.16 shows two compounds, water (H_2O) and hydrogen peroxide (H_2O_2). Although these compounds are formed by the same two elements, they have different physical and chemical properties. Each compound obeys the law of definite proportions. In every sample of hydrogen peroxide, 16.0 g of oxygen are present for each 1.0 g of hydrogen. The mass ratio of oxygen to hydrogen is always 16:1. In every sample of water, the mass ratio of oxygen to hydrogen is always 8:1. If a sample of hydrogen peroxide has the same mass of hydrogen as a sample of water, the ratio of the mass of oxygen in the two compounds is exactly 2:1.

$$\frac{16 \text{ g O (in } H_2O_2 \text{ sample that has 1 g H)}}{8 \text{ g O (in } H_2O \text{ sample that has 1 g H)}} = \frac{16}{8} = \frac{2}{1} = 2:1$$

Section Resources

Print
- **Guided Reading and Study Workbook,** Section 9.5
- **Core Teaching Resources,** Section 9.5
- **Transparencies,** T100–T103

Technology
- **Interactive Textbook with ChemASAP,** Problem-Solving 9.34, Assessment 9.5

Using the results from studies of this kind, John Dalton stated the **law of multiple proportions:** Whenever the same two elements form more than one compound, the different masses of one element that combine with the same mass of the other element are in the ratio of small whole numbers. Figure 9.17 illustrates the law of multiple proportions.

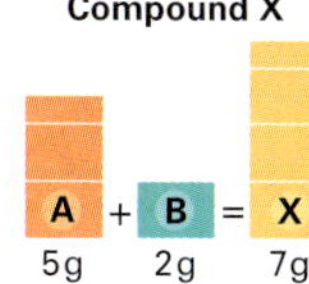

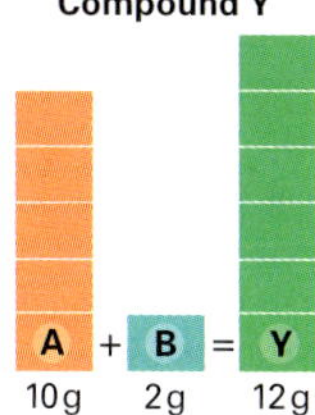

Figure 9.17 The diagram illustrates the law of multiple proportions. Two compounds, X and Y, contain equal masses of element B. The ratio of the masses of A in these compounds is 5:10 or 1:2 (a small whole number ratio). **Applying Concepts** ***Would the ratio be different if samples of X and Y contained 3 g of B?***

SAMPLE PROBLEM 9.1

Calculating Mass Ratios

Carbon reacts with oxygen to form two compounds. Compound A contains 2.41 g of carbon for each 3.22 g of oxygen. Compound B contains 6.71 g of carbon for each 17.9 g of oxygen. What is the lowest whole number mass ratio of carbon that combines with a given mass of oxygen?

1 Analyze ***List the knowns and the unknown.***

Knowns

- Compound A = 2.41 g C and 3.22 g O
- Compound B = 6.71 g C and 17.9 g O

Unknown

- Lowest whole number ratio of carbon per gram of oxygen in the two compounds = ?

Apply the law of multiple proportions to the two compounds. For each compound, find the grams of carbon that combine with 1.00 g of oxygen by dividing the mass of carbon by the mass of oxygen. Then find the ratio of the masses of carbon in the two compounds by dividing the larger value by the smaller. Confirm that the ratio is the lowest whole number ratio.

2 Calculate ***Solve for the unknown.***

- Compound A $\quad \frac{2.41 \text{ g C}}{3.22 \text{ g O}} = \frac{0.748 \text{ g C}}{1.00 \text{ g O}}$
- Compound B $\quad \frac{6.71 \text{ g C}}{17.9 \text{ g O}} = \frac{0.375 \text{ g C}}{1.00 \text{ g O}}$

Compare the masses of carbon per gram of oxygen in the compounds.

$$\frac{0.748 \text{ g C (in compound A)}}{0.375 \text{ g C (in compound B)}} = \frac{1.99}{1} = \text{roughly } \frac{2}{1} = 2:1$$

The mass ratio of carbon per gram of oxygen in the two compounds is 2:1.

3 Evaluate ***Does the result make sense?***

The ratio is a low whole number ratio, as expected. For a given mass of oxygen, compound A contains twice the mass of carbon as compound B.

Practice Problems

34. Lead forms two compounds with oxygen. One contains 2.98 g of lead and 0.461 g of oxygen. The other contains 9.89 g of lead and 0.763 g of oxygen. For a given mass of oxygen, what is the lowest whole number mass ratio of lead in the two compounds?

For help with using a calculator go to page R62.

Interactive Textbook

Problem-Solving 9.34 Solve Problem 34 with the help of an interactive guided tutorial.

with ChemASAP

Discuss L2

Students may be able to understand the law of multiple proportions more easily by thinking in terms of atoms. One atom of carbon in CO_2 has the same mass (12 amu) as one atom of carbon in CO. One atom of oxygen (16 amu) combines with 12 amu of carbon in CO. Two atoms of oxygen (32 amu) combine with 12 amu of carbon in CO_2. The ratio of the masses of oxygen in the two compounds is 32:16 or 2:1.

Use Visuals L1

Figure 9.17 Ask students, **How many grams of B are in compound X?** *(2 g)* **How many grams of B are in compound Y?** *(2 g)* Thus two compounds have the same masses of B. Ask, **What is the ratio between the masses of A in these two compounds that have equal masses of B?** *(10 g:5 g or 2:1)* Ask students to name the law illustrated by the figure and define it in their own words. *(the law of multiple proportions; When two compounds are formed from the same two elements, and you take a certain mass of one of the elements, the masses of the other element in the two compounds that combined with that certain mass are in a whole number ratio).*

Math Handbook

For a math refresher and practice, direct students to using a calculator, page R62.

Sample Problem 9.1

Answer

34. 2:1

Practice Problems Plus L2

In the compound iron oxide, also known as rust, the mass ratio of iron to oxygen is 7:3. A 33-g sample of a compound composed of iron and oxygen contains 10 g of oxygen. **Is the sample iron oxide? Explain your answer.** *(Yes, because the ratio in the compound is 2.3:1 and the ratio in iron oxide is 2.3:1)*

Facts and Figures

Proust's Law

In the early 1800s, many scientists were studying chemical reactions by carefully measuring the masses of reactants and products. The French scientist Joseph Proust (1766–1844) demonstrated that samples of copper carbonate always contain 5.3 parts by mass of copper to 4 parts of oxygen and 1 part of carbon. The principle embodied in this discovery became known as Proust's Law and later as the law of definite proportions. Proust's experiments inspired John Dalton to think in terms of atoms. With atoms in mind, Dalton designed experiments that soon led to the law of multiple proportions and Dalton's atomic theory.

Answers to...

Figure 9.17 No, as long as the two samples have the same mass of B.

Practicing Skills: Naming Chemical Compounds

Use Visuals L1

Figure 9.18 Have students read the first paragraph on this page and examine the photograph. Discuss the importance of naming conventions for compounds. Explain that chemists need to be able to understand papers written by colleagues and to reproduce experimental results. Focus on the problems that could occur if common names in multiple languages were the norm. Ask, **Why would knowing the correct name for a chemical be important to a poison-control center worker?** *(The worker needs to know the specific chemical ingested before offering a solution. Administering an antidote for the wrong chemical could cause further harm.)*

Relate L2

Ask students to think of other situations where specific names are important for accurate or fast results. *(When you are ordering items by telephone or on the internet)* Ask, **Why do many bureaucratic agencies use social security or driver's license numbers instead of people's names?** *(Many people could have the same name.)*

Figure 9.18 If someone ingests a poison, the poison control center can provide information about what immediate action to take. **Inferring** ***What information about the poison would be most helpful to the center?***

Practicing Skills: Naming Chemical Compounds

In the average home, you can probably find hundreds of chemicals, including cleaning products, drugs, and pesticides. Figure 9.18 shows a typical warning label on a product that tells about its possible dangers. Most people would not know what to do if some of these chemicals accidentally mixed together and began to react or if a small child ingested one. A phone call to a poison control center can provide lifesaving information to victims of such poisonings. But a poison control center can be much more effective if the caller can supply some information about the name or formula of the substance.

In this chapter, you learned two basic skills that could help you to deal with an emergency involving chemicals: writing chemical formulas and naming chemical compounds. If this is the first time you have tried to master these skills, you may feel a little overwhelmed. For example, you may find it difficult to know when you should or should not use prefixes and Roman numerals in a name. Or you may have trouble determining if a compound's name should end in *-ate*, *-ide*, or *-ite*. The flowchart in Figure 9.20 provides you with a sequence of questions for naming a compound when you know its formula. **Follow the arrows and answer the questions on the flowchart to write the correct name for a compound.** The sequence of questions can help you name chemicals you may have in your home as well as the colorful compounds that create the picturesque landscape of Gay Head, Massachusetts shown in Figure 9.19.

The flowchart shows the routes to the names of several compounds: HNO_3, N_2O_3, BaS, Li_2CO_3, $CuSO_4$, and $FeCl_2$. Apply the general formula Q_xR_y to each compound. Q and R can be atoms, monatomic ions, or polyatomic ions. For example, to name HNO_3, let H = Q and NO_3 = R. Follow the first arrow down to the question Q = H? The answer is yes, so the arrow to the right tells you that the compound is an acid. You can then follow the rules for naming acids. HNO_3 is nitric acid.

To name N_2O_3, let Q = N and R = O. The answer to the question Q = H? is no, so follow the arrow down. Does the compound have more than two elements? The answer is no, so follow the arrow to the left. The compound is binary and its name ends in *-ide*. Is Q a metal? The answer is no, so you must use prefixes in the name, which is *di*nitrogen *tri*oxide.

Figure 9.19 A variety of compounds create a colorful display of clay in the Gay Head cliffs on the island of Martha's Vineyard. Each colored compound can be named by the methods you are learning if you know the compound's formula.

Differentiated Instruction

Special Needs L1

Students who have difficulty with mathematics may need extra help with the laws of definite proportions and the law of multiple proportions. Assign such students to work with students with strong math skills. Encourage students with strong math skills to develop ways of sharing their problem-solving techniques with others.

Figure 9.20 This flowchart will help you name chemical compounds. Begin with the letters Q and R in the general formula Q_xR_y. Q and R can be atoms, monatomic ions, or polyatomic ions.

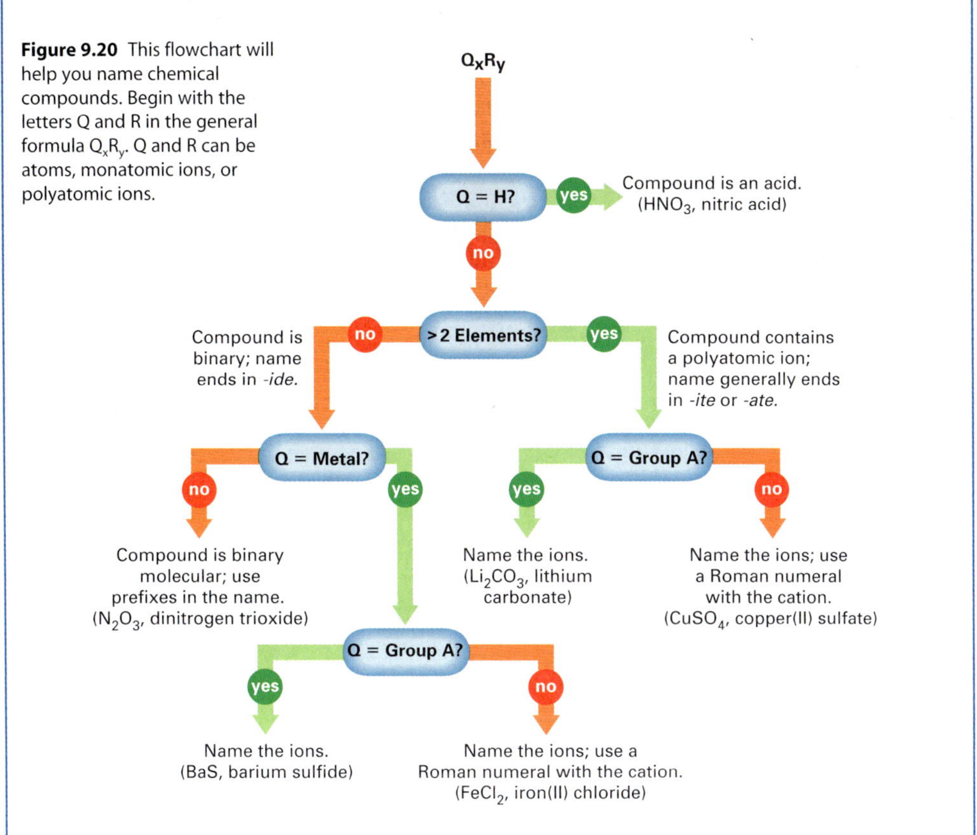

Another example shown on the flowchart is $CuSO_4$. In this case, Q = Cu and R = SO_4. Q does not equal H. The compound does have more than two elements, so it contains a polyatomic ion. Thus you should expect that the name will end in *-ite* or *-ate*. The answer to the next question, Q = Group A? is no, so you must name the ions and use a Roman numeral to identify the charge of the transition metal. The name is copper(II) sulfate. A sample of copper(II) sulfate is shown in Figure 9.21. Practice with the other compounds listed above, and then use the flowchart when doing naming exercises. Soon you won't need it anymore.

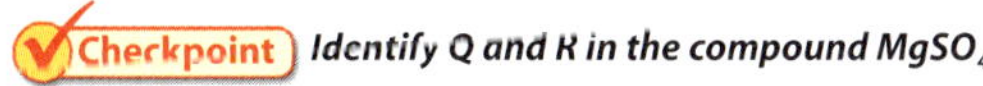
Checkpoint ***Identify Q and R in the compound $MgSO_4$.***

Figure 9.21 Blue copper(II) sulfate contains water in its crystal structure. When it is heated, it loses water and turns white. When the white solid absorbs water, it turns blue again.

Differentiated Instruction

Gifted and Talented L3

Challenge advanced students to use the flowchart as a model for writing a computer program for naming chemical compounds.

Use Visuals L1

Figure 9.20 Tell students that the flowchart is like a road map. At every intersection you need to make a decision and when you make the right decisions you are led to your destination (the correct name for the compound). Lead them through a few examples asking them to respond to the questions at each intersection. Then have volunteers from the class take the lead to continue the exercise.

Use Visuals L1

Figure 9.21 Students may ask about the two forms of copper sulfate (white and blue). Tell them that the blue form is called a hydrate because it has water molecules incorporated into its crystal structure. In the hydrate, each formula unit of $CuSO_4$ is accompanied by five molecules of water. Ask, **Would equal masses of the white and the blue compounds contain the same number of formula units of copper sulfate?** *(No, some of the mass of the hydrate is water.)*

Relate 

While students are involved in developing their skills in writing formulas for ionic and molecular compounds, you may have the opportunity to comment on the characteristics of representative ionic and molecular compounds found in Figure 9.20. For example, call attention to the high melting points of ionic compounds. Iron(II) chloride melts at 677°C; lithium carbonate at 723°C. In addition, a large amount of energy (heat of fusion) is required to melt these ionic compounds. By contrast, the molecular compound dinitrogen trioxide melts at −101°C and methane at −182°C.

Make the point that ionic and molecular compounds do not differ just in the way that they are named but fundamentally in the bonding that holds the atoms together in the compound.

Answers to...

Figure 9.18 the formula, the name, and possibly the concentration

Checkpoint Q is Mg^{2+}; R is SO_4^{2-}.

Practicing Skills: Writing Chemical Formulas

Discuss L2

Have students summarize the rules for writing formulas and naming compounds. Point out that naming compounds and writing formulas becomes easier with practice because chemical names and formulas are systematized. Ask students to consider how difficult the task would be if the names were assigned randomly. Because chemists adhere to an established system of naming, it is possible to quickly locate information on physical constants such as melting point, boiling point, and density in chemical handbooks such *Lange's Handbook of Chemistry, The Merck Index,* and the *CRC Handbook of Chemistry and Physics. The Merck Index* is particularly helpful to organic chemists interested in the synthesis and identification of a particular compound. Encourage students to look for these books in their local public library.

3 ASSESS

Evaluate Understanding L2

Give students formulas for compounds and have them use Figure 9.20 to determine their names. Then give them names for different compounds and have them use Figure 9.22 to write their formulas.

Reteach L1

Review the four guidelines for writing chemical formulas by asking students in turn what it means when each of the following is included in the name of a compound: an *-ide* ending, an *-ite* or *-ate* ending, a prefix, or a Roman numeral. Then give them the following names of compounds containing one of these components: hydrogen cyanide, potassium chromate, sulfur tetrafluoride, copper(I) chloride. Ask them to use Figure 9.22 to determine their formulas.

Practicing Skills: Writing Chemical Formulas

In writing a chemical formula from a chemical name, it is helpful to remember the following guidelines.

1. **An *-ide* ending generally indicates a binary compound.**
2. **An *-ite* or *-ate* ending means a polyatomic ion that includes oxygen is in the formula.**
3. **Prefixes in a name generally indicate that the compound is molecular.**
4. **A Roman numeral after the name of a cation shows the ionic charge of the cation.**

These guidelines and the questions in the flowchart in Figure 9.22 will help you write the formula for a compound when you know its name. For example, use the flowchart to write the formula for sodium chromate. The name does not contain prefixes, so it is ionic. The ions are sodium ion and chromate ion. Follow the arrows to the right and left. Sodium is a Group A element, so use the periodic table or Table 9.1 to obtain its ionic charge (1+). Chromate ion is a polyatomic ion, so use Table 9.3 to obtain its charge (2−). Balance the charges to obtain the formula Na_2CrO_4. Practice formula writing using the flowchart until you don't need it anymore.

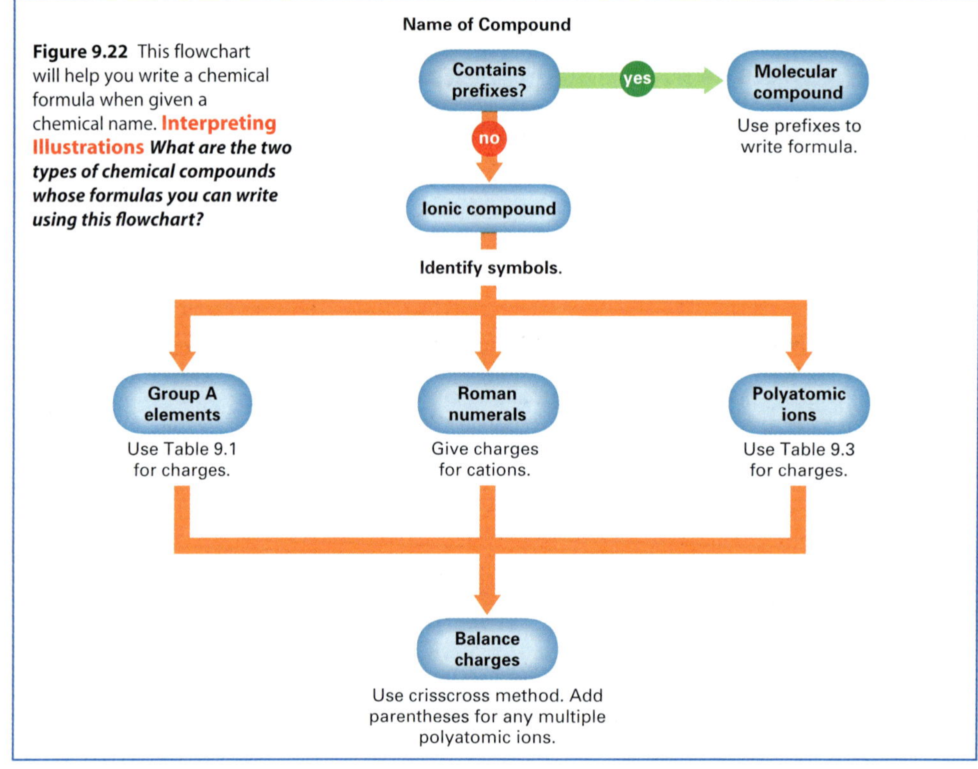

Figure 9.22 This flowchart will help you write a chemical formula when given a chemical name. **Interpreting Illustrations** ***What are the two types of chemical compounds whose formulas you can write using this flowchart?***

Facts and Figures

Other Molecular Compounds

The binary molecular compounds constitute only a small fraction of molecular compounds. An enormous number of compounds containing principally carbon, hydrogen, oxygen, nitrogen, and sulfur are classified as organic chemicals. A different, more complex but precise system of naming makes it possible for large and complex molecules to be precisely identified.

Quick LAB

Making Ionic Compounds

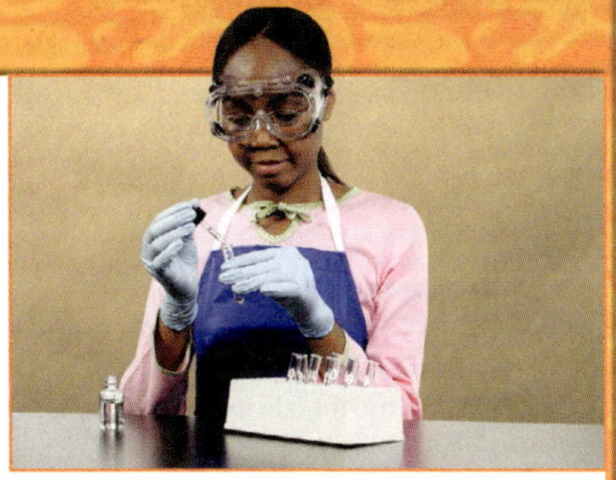

Purpose
To mix solutions containing cations and anions to make ionic compounds.

Materials

- 9 small test tubes
- test tube rack
- paper, pencil, ruler
- 6 solutions in plastic dropper bottles containing the following ions:

Solution A (Fe^{3+} ion)	Solution X (CO_3^{2-} ion)
Solution B (Ag^{+} ion)	Solution Y (OH^{-} ion)
Solution C (Pb^{2+} ion)	Solution Z (PO_4^{3-} ion)

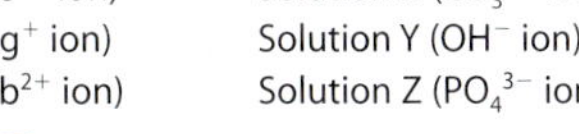

Procedure    

1. Label three test tubes A, three test tubes B, and three test tubes C.
2. Add 10 drops (approximately 0.5 mL) of solutions A, B, and C to appropriately labeled test tubes.
3. Add 10 drops of solution X to one test tube of A, 10 drops to one test tube of B, and 10 drops to one test tube of C. Observe each for the formation of a solid.
4. Make a 3-by-3 inch grid in which to record your observations. Label the rows A, B, and C. Label the columns X, Y, and Z. Describe any solid material you observe.
5. Repeat Step 3, adding 10 drops of solution Y to test tubes A, B, and C. Record your observations.
6. Repeat Step 3, adding 10 drops of solution Z to test tubes A, B, and C. Record your observations.

Analyze and Conclude

1. Some ionic compounds are insoluble (do not dissolve in water). Explain what you observed.
2. Write the formula for each ionic compound formed.
3. Name each ionic compound formed.
4. Will mixing any cation with any anion always lead to the formation of an insoluble ionic compound? Explain.

9.5 Section Assessment

35. **Key Concept** What two laws describe how chemical compounds form?
36. **Key Concept** How should you use a flowchart to name a chemical compound?
37. **Key Concept** What are four guidelines for writing the formulas of chemical compounds?
38. Two compounds containing copper and oxygen were found to contain the following masses:
    Compound A: 32.10 g Cu and 17.90 g Cl
    Compound B: 23.64 g Cu and 26.37 g Cl
    Are the compounds the same? If not, what is the lowest whole number mass ratio of copper that combines with a given mass of chlorine?
39. Name these compounds.
    a. $CaCO_3$ **b.** $PbCrO_4$ **c.** $SnCr_2O_7$
40. Write formulas for these compounds.
    a. tin(II) hydroxide **b.** barium fluoride
41. Identify the incorrect names or formulas.
    a. calcium(II) oxide **b.** aluminum oxide
    c. $Na_2C_2O_4$ **d.** $Mg(NH_4)_2$

Connecting Concepts

Ionic Bonds Review ionic bonds in Section 7.2 and show by means of electron dot structures why an ionic compound always has a charge of zero. Use magnesium bromide as an example.

Assessment 9.5 Test yourself on the concepts in Section 9.5.
with ChemASAP

Section 9.5 Assessment

35. law of definite proportions and law of multiple proportions
36. Follow the arrows and answer the questions on the flowchart.
37. See the four guidelines on the facing SE page (page 278).
38. no; 2:1
39. **a.** calcium carbonate **b.** lead(II) chromate **c.** tin(II) dichromate
40. **a.** $Sn(OH)_2$ **b.** BaF_2
41. **a.** incorrect; a Roman numeral is not used with a Group A metal, calcium oxide **b.** correct **c.** correct **d.** incorrect; ammonium ion, a polyatomic cation, will not form a compound with a monatomic cation.

Quick LAB

Making Ionic Compounds L2

Objective Students form precipitates and write their formulas and names.

Skills Focus Observing, applying rules

 **Prep Time** 30 minutes

Solution	Preparation
0.05*M* $AgNO_3$	2.1 g in 250 mL
0.2*M* $Pb(NO_3)_2$	16.6 g in 250 mL
1.0*M* $NaCO_3$	26.5 g in 250 mL
0.5*M* NaOH	5.0 g in 250 mL
0.1*M* $Na_3(PO_4)_2$	9.5 g $Na_3PO_4{\cdot}12H_2O$ in 250 mL
0.1*M* $FeCl_3$	6.8 g $FeCl_3{\cdot}6H_2O$ in 25 mL of 1.0*M* NaCl; dilute to 250 mL

Class Time 20 minutes

Analyze and Conclude

1. An insoluble compound formed in every tube.
2. $Fe_2(CO_3)_3$, $Fe(OH)_3$, $FePO_4$, Ag_2CO_3, AgOH, Ag_3PO_4, $PbCO_3$, $Pb(OH)_2$, $Pb_3(PO_4)_2$
3. iron(III) carbonate, iron(III) hydroxide, iron(III) phosphate, silver carbonate, silver hydroxide, silver phosphate, lead(II) carbonate, lead(II) hydroxide, lead(II) phosphate
4. No, NaCl, for example, is soluble

Connecting Concepts

Magnesium loses its two valence electrons. Two bromine atoms each gain one electron.

If your class subscribes to the Interactive Textbook, use it to review key concepts in Section 9.5.
with ChemASAP

Answers to...

Figure 9.22 ionic compounds and binary molecular compounds

CHAPTER 9

Study Guide

Study Tip

Identifying the Main Idea As you review each section, write down the key concepts that are found in bold print in each major division. Then add details and extensions until you have a complete outline of the topic.

If your class subscribes to the Interactive Textbook with ChemASAP, your students can go online to access an interactive version of the Student Edition and a self-test.

with **ChemASAP**

CHAPTER 9

Study Guide

Key Concepts

9.1 Naming Ions

- When the metals in Groups 1A, 2A, and 3A lose electrons, they form cations with positive charges equal to their group number.
- The charge of any ion of a Group A nonmetal is determined by subtracting 8 from the group number.
- The charges of cations of many transition metal ions must be determined from the number of electrons lost. When a cation can have more than one ionic charge, a Roman numeral is used in the name to indicate the charge.
- The names of most polyatomic anions end in *-ite* or *-ate*.

9.2 Naming and Writing Formulas for Ionic Compounds

- The name of a binary ionic compound is the cation name followed by the anion name.
- To write the formula for a binary ionic compound, write the symbol for the cation and then the anion. Then balance the charges.
- To write formulas for compounds containing polyatomic ions, write the symbol for the metal ion followed by the formula for the polyatomic ion and balance the charges.
- To name a compound containing a polyatomic ion, state the cation first and then the anion.

9.3 Naming and Writing Formulas for Molecular Compounds

- Prefixes show how many atoms of each element are present in a molecule of a binary compound.
- To write the formula for a binary molecular compound, write the symbols for the elements and use the prefixes to determine the subscripts. Omit *mono-* for a single atom.

9.4 Naming and Writing Formulas for Acids and Bases

- An acid is a combination of a monatomic or polyatomic anion with sufficient hydrogen atoms to make the compound electrically neutral. Acids are named as shown in Table 9.5.
- A base is a combination of a cation with as many hydroxide ions as are needed to make the compound electrically neutral. Bases are named in the same way as other ionic compounds.

9.5 The Laws Governing Formulas and Names

- The ways that compounds form are summed up in two laws: the law of definite proportions and the law of multiple proportions.
- To name a compound or write its formula, follow the flowcharts in Figures 9.20 and 9.22 to the correct name or formula.

Vocabulary

- acid (p. 271)
- base (p. 273)
- binary compound (p. 261)
- law of definite proportions (p. 274)
- law of multiple proportions (p. 275)
- monatomic ion (p. 253)
- polyatomic ion (p. 257)

Organizing Information

Use these terms to construct a concept map that organizes the major ideas of this chapter.

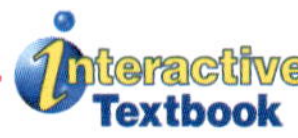

Concept Map 9 Solve the Concept Map with the help of an interactive guided tutorial.

with **ChemASAP**

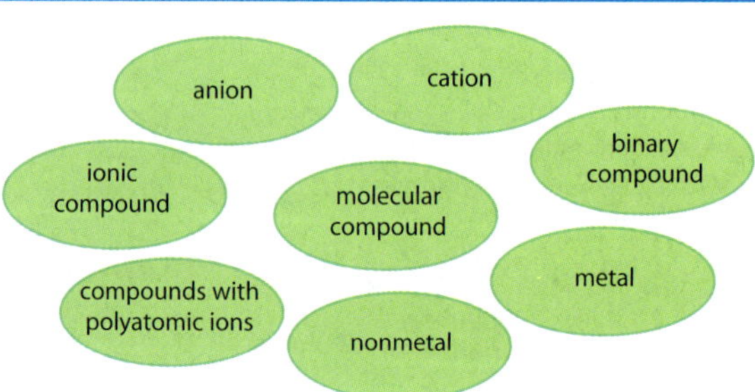

Chapter Resources

Print

- **Core Teaching Resources,** Chapter 9, Practice Problems, Vocabulary Review, Quiz, Chapter Test A, Chapter Test B

Technology

- **Computer Test Bank,** Chapter 9 Test
- **Interactive Textbook with ChemASAP,** Chapter 9

CHAPTER 9

Assessment

Reviewing Content

9.1 Naming Ions

42. Give the expected charges on the ions of elements of these groups of the periodic table.

a. Group 6A **b.** Group 1A
c. Group 7A **d.** Group 3A

43. Give the expected charge of the cations of these elements.

a. Sr **b.** Ca **c.** Al **d.** Cs

44. Name the following ions. Use Table 9.2 if necessary.

a. Ba^{2+} **b.** I^- **c.** Ag^+ **d.** Hg^{2+}

45. Write the names and formulas of the two polyatomic anions in Table 9.3 with names that do not end in *-ite* or *-ate*.

46. Name the following ions.

a. OH^- **b.** Pb^{4+} **c.** SO_4^{2-} **d.** O^{2-}

9.2 Naming and Writing Formulas for Ionic Compounds

47. What is the net ionic charge of every ionic compound? Explain.

48. How are formulas written for binary ionic compounds, given their names? How is the reverse done?

49. How do you determine the charge of a transition metal cation from the formula of an ionic compound containing that cation?

50. How are formulas written for ionic compounds with polyatomic ions, given their names? How is the reverse done?

51. Which of the following compounds are binary ionic compounds?

a. KBr **b.** sodium nitride
c. K_3PO_4 **d.** calcium sulfate

52. When must parentheses be used in a formula?

53. Complete the table by writing correct formulas for the compounds formed by combining positive and negative ions. Then name each compound.

	NO_3^-	CO_3^{2-}	CN^-	PO_4^{3-}
NH_4^+	a. ___	b. ___	c. ___	d. ___
Sn^{4+}	e. ___	f. ___	g. ___	h. ___
Fe^{3+}	i. ___	j. ___	k. ___	l. ___
Mg^{2+}	m. ___	n. ___	o. ___	p. ___

9.3 Naming and Writing Formulas for Molecular Compounds

54. What are the components of a binary molecular compound?

55. What prefix indicates each of the following numbers of atoms in the formula of a molecular compound?

a. 3 **b.** 1 **c.** 2 **d.** 6 **e.** 5 **f.** 4

56. How are formulas for binary molecular compounds written, given their names? How is the reverse done?

57. Write the formula or name for these compounds.

a. boron trichloride **b.** N_2O_5
c. dinitrogen tetrahydride **d.** CCl_4

9.4 Naming and Writing Formulas for Acids and Bases

58. Give the name or the formula for these acids.

a. HCl **b.** sulfuric acid
c. HNO_3 **d.** acetic acid

59. Is every molecular compound that contains hydrogen an acid? Explain.

60. Write formulas for these compounds.

a. nitrous acid
b. aluminum hydroxide
c. hydroselenic acid
d. strontium hydroxide
e. phosphoric acid

61. Write names or formulas for these compounds.

a. iron(II) hydroxide **b.** $Pb(OH)_2$
c. copper(II) hydroxide **d.** $Co(OH)_2$

9.5 The Laws Governing Formulas and Names

62. What is the law of definite proportions?

63. Describe the law of multiple proportions.

64. Nitrous oxide is known as laughing gas and is used as an anesthetic in dentistry. The mass ratio of nitrogen to oxygen is 7:4. A 68-g sample of a compound composed of nitrogen and oxygen contains 42 g of nitrogen. Is the sample nitrous oxide? Explain.

CHAPTER 9

Assessment

Reviewing Content

42. a. 2– **b.** 1+ **c.** 1– **d.** 3+

43. a. 2+ **b.** 2+ **c.** 3+ **d.** 1+

44. a. barium ion **b.** iodide ion **c.** silver ion **d.** mercury(II) ion

45. cyanide, CN^- and hydroxide, OH^-

46. a. hydroxide ion **b.** lead(IV) ion **c.** sulfate ion **d.** oxide ion

47. zero; A compound is electrically neutral.

48. The symbols for the cation and anion are written and the charges are balanced with subscripts. The name of the cation is followed by the name of the anion.

49. Determine the charge of the anion then work the formula backwards to find the charge of the transition metal cation needed to give a net charge of zero for the formula unit.

50. The symbols for the cation and anion are written and the charges are balanced with subscripts. Parentheses are used around the polyatomic ion if a subscript is needed. The name of the cation is followed by the name of the anion.

51. a and **b**

52. When more than a single polyatomic ion is needed to balance the formula

53. NH_4NO_3, ammonium nitrate; $(NH_4)_2CO_3$, ammonium carbonate; NH_4CN, ammonium cyanide; $(NH_4)_3PO_4$, ammonium phosphate; $Sn(NO_3)_4$, tin(IV) nitrate; $Sn(CO_3)_2$, tin(IV) carbonate; $Sn(CN)_4$, tin(IV) cyanide; $Sn_3(PO_4)_4$, tin(IV) phosphate; $Fe(NO_3)_3$, iron(III) nitrate; $Fe_2(CO_3)_3$, iron(III) carbonate; $Fe(CN)_3$, iron(III) cyanide; $FePO_4$, iron(III) phosphate; $Mg(NO_3)_2$, magnesium nitrate; $MgCO_3$, magnesium carbonate; $Mg(CN)_2$, magnesium cyanide; $Mg_3(PO_4)_2$, magnesium phosphate

54. two nonmetals

55. a. tri- **b.** mono- **c.** di- **d.** hexa- **e.** penta- **f.** tetra–

56. For formulas, write the correct symbols for each element with a subscript corresponding to the prefix before each element in the name. For naming, name each element in the order given. Use the subscript to determine the prefixes before each element in the name. The name ends in *-ide.*

57. a. BCl_3 **b.** dinitrogen pentoxide **c.** N_2H_4 **d.** carbon tetrachloride

58. a. hydrochloric acid **b.** H_2SO_4 **c.** nitric acid **d.** $HC_2H_3O_2$

59. No, to be an acid the compound must produce H^+ ions in water solution.

60. a. HNO_2 **b.** $Al(OH)_3$ **c.** H_2Se **d.** $Sr(OH)_2$ **e.** H_3PO_4

61. a. $Fe(OH)_2$ **b.** lead(II) hydroxide **c.** $Cu(OH)_2$ **d.** cobalt(II) hydroxide

62. In all samples of the same chemical compound, the masses of the elements are always in the same proportions.

63. Whenever two elements form more than one compound, the different masses of one element that combine with the same mass of the other element are in a ratio of small whole numbers.

64. no; The ratio of nitrogen to oxygen is 42:26, which is not a 7:4 ratio.

CHAPTER 9

Assessment continued

Understanding Concepts

65. **a.** $KMnO_4$ **b.** $Ca(HCO_3)_2$ **c.** Cl_2O_7 **d.** Si_3N_4 **e.** NaH_2PO_4 **f.** PBr_5 **g.** CCl_4

66. **a.** MgS **b.** Na_3PO_3 **c.** $Ba(OH)_2$ **d.** $Cu(NO_2)_2$ **e.** K_2SO_3 **f.** $CaCO_3$ **g.** $NaBr$ **h.** $Fe_2(SO_4)_3$

67. **a.** sodium chlorate
b. mercury(I) bromide
c. potassium chromate
d. perchloric acid
e. tin(IV) oxide
f. iron(III) acetate
g. potassium hydrogen sulfate
h. calcium hydroxide
i. barium sulfide

68. **a.** lithium perchlorate
b. dichlorine monoxide
c. mercury(II) fluoride
d. calcium oxide
e. barium phosphate
f. iodine
g. strontium sulfate
h. copper(I) acetate
i. silicon tetrachloride

69. **a.** magnesium permanganate
b. beryllium nitrate
c. potassium carbonate
d. dinitrogen tetrahydride
e. lithium hydroxide
f. barium fluoride
g. phosphorus triiodide
h. zinc oxide
i. phosphorous acid

70. **a.** $CaBr_2$ **b.** $AgCl$ **c.** Al_4C_3 **d.** NO_2 **e.** $Sn(CN)_4$ **f.** LiH **g.** $Sr(C_2H_3O_2)_2$ **h.** Na_2SiO_3

71. binary molecular compound

72. lithium carbonate, Li_2CO_3

73. $SnCl_4$

74. **a.** 2:1 **b.** PbI_2, lead(II) iodide and PbI_4, lead(IV) iodide

75. **a.** 9.85% **b.** nitrogen, oxygen, and chlorine; 54.9 billions of kg **c.** 34.7% **d.** H_2SO_4, N_2, O_2, NH_3, CaO, H_3PO_4, $NaOH$, Cl_2, Na_2CO_3, HNO_3

CHAPTER 9

Assessment continued

Understanding Concepts

65. Write formulas for these compounds.
a. potassium permanganate
b. calcium hydrogen carbonate
c. dichlorine heptoxide
d. trisilicon tetranitride
e. sodium dihydrogen phosphate
f. phosphorus pentabromide
g. carbon tetrachloride

66. Write formulas for these compounds.
a. magnesium sulfide **b.** sodium phosphite
c. barium hydroxide **d.** copper(II) nitrite
e. potassium sulfite **f.** calcium carbonate
g. sodium bromide **h.** ferric sulfate

67. Name these compounds.
a. $NaClO_3$ **b.** Hg_2Br_2 **c.** K_2CrO_4
d. $HClO_4$ **e.** SnO_2 **f.** $Fe(C_2H_3O_2)_3$
g. $KHSO_4$ **h.** $Ca(OH)_2$ **i.** BaS

68. Name each substance.
a. $LiClO_4$ **b.** Cl_2O **c.** HgF_2
d. CaO **e.** $Ba_3(PO_4)_2$ **f.** I_2
g. $SrSO_4$ **h.** $CuC_2H_3O_2$ **i.** $SiCl_4$

69. Name each compound.
a. $Mg(MnO_4)_2$ **b.** $Be(NO_3)_2$ **c.** K_2CO_3
d. N_2H_4 **e.** $LiOH$ **f.** BaF_2
g. PI_3 **h.** ZnO **i.** H_3PO_3

70. Write formulas for these compounds.
a. calcium bromide **b.** silver chloride
c. aluminum carbide **d.** nitrogen dioxide
e. tin(IV) cyanide **f.** lithium hydride
g. strontium acetate **h.** sodium silicate

71. A compound of general formula Q_xR_y contains no hydrogen, and Q and R are both elements. Neither Q nor R is a metal. Is Q_xR_y an acid, a binary ionic compound, an ionic compound containing a polyatomic anion, or a binary molecular compound?

72. A compound of general formula Q_xR_y contains no hydrogen, Q is the alkali metal of lowest atomic mass, and R contains the elements oxygen and carbon in a 3:1 ratio. Write the name and the formula of the compound.

73. Two compounds contain only tin and chlorine. The ratio of the masses of chlorine combined with 1.00 g of tin in the two compounds is 2:1. If one compound has the formula $SnCl_2$, what is the formula for the other compound?

74. Analysis of two compounds shows that they contain only lead and iodine in the following amounts:
Compound I: 22.48 g Pb and 27.52 g I
Compound II: 5.80 g Pb and 14.20 g I
a. Determine the ratio of lead contained in the two compounds for every 1 g of iodine.
b. Use your ratio and your knowledge of ionic charges to write the formulas and the names of the two compounds.

75. The United States produces thousands of different kinds of inorganic chemicals. Inorganic chemicals, for the most part, do not contain carbon. The table shows the amounts (in billions of kg) of the top ten inorganic chemicals produced in a recent year.

Chemical	Amount produced (billions of kg)
Sulfuric acid	39.4
Nitrogen	26.9
Oxygen	17.7
Ammonia	16.5
Lime	16.3
Phosphoric acid	11.2
Sodium hydroxide	11.0
Chlorine	10.3
Sodium carbonate	9.3
Nitric acid	6.8

a. What percentage of the total production of the top ten is lime (calcium oxide)?
b. Three diatomic gases are on the list. What are their names? What was the combined production of these gases in billions of kilograms?
c. What percentage of the total production of the top ten is the three acids?
d. Write formulas for the top ten inorganic chemicals.

Critical Thinking

76. Compare and contrast the information conveyed by a molecular formula with that given by a formula unit of a compound.

77. Where on the periodic table will you find the two elements in a binary molecular compound?

78. Why is it important for chemists to have a system of writing chemical names and formulas?

79. Criticize this statement: "The ionic charge of any metal can be determined from the position of the element in the periodic table."

80. Summarize the rules that chemists use for naming ionic compounds. What is the purpose for each rule?

81. Nitrogen and oxygen form a number of stable chemical compounds. In the models below, nitrogen is blue; oxygen is red. Write the chemical formula and name of each.

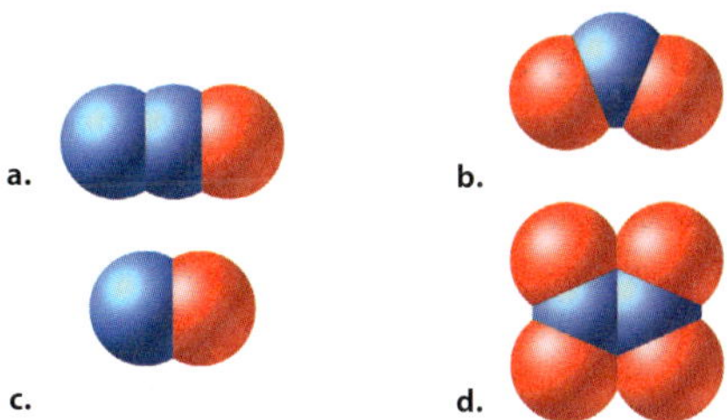

82. Examine the following names for ionic compounds. Show, by writing all possible formulas for the compounds, that the names are incomplete. Then, write each complete name.
a. copper sulfide **b.** iron sulfate **c.** lead oxide

83. Explain what is wrong with each formula.
a. $CsCl_2$ **b.** ZnO_2 **c.** LiNe **d.** Ba_2S_2

84. Separate the following compounds into five categories: binary ionic compounds, binary molecular compounds, compounds with polyatomic ions, acids, and bases. Some compounds may fit in more than one category.

a. CBr_4	**b.** HCN	**c.** NH_4OH
d. MgS	**e.** H_2SiO_3	**f.** ClBr
g. Al_2O_3	**h.** Na_2HPO_4	**i.** $KMnO_4$

Concept Challenge

85. *CRC Handbook of Chemistry and Physics* is a reference book that contains a wealth of information about elements and compounds. Two sections of this book you might use are "Physical Constants of Inorganic Compounds" and "Physical Constants of Organic Compounds." To familiarize yourself with this work, make a table with these headings: Name, Formula, Crystalline Form or Color, Density, Melting Point (°C), Boiling Point (°C), and Solubility in Water. Enter these substances in the body of the table: ammonium chloride, barium, barium sulfate, bromine, calcium carbonate, chlorine, copper(II) sulfate pentahydrate, iodine, iron(II) sulfate pentahydrate, mercury, potassium carbonate, and sulfur. Use the handbook to complete the table.

86. Use the table you prepared for Problem 85 to answer the following questions.
a. You have two unlabeled bottles, each containing a white powder. One of the substances is calcium carbonate, and the other is potassium carbonate. Describe a simple physical test you could carry out to distinguish between these two compounds.
b. How would you distinguish between samples of copper(II) sulfate pentahydrate and iron(II) sulfate pentahydrate?
c. A bottle contains a mixture of ammonium chloride and barium sulfate. How could you separate these two compounds?
d. List the elements in the table in order of increasing density. Identify the elements as metals or nonmetals.
e. List the compounds in the table in order of decreasing density.
f. Calculate the mass of 47.0 cm^3 of mercury.
g. Calculate the volume of 16.6 g of sulfur.
h. How would you distinguish among the Group 7A elements (halogens) listed in the table?

Critical Thinking

76. A molecular formula shows the number of each kind of atom in a molecule of the compound. The formula unit shows the lowest whole-number ratio of ions in a compound.

77. on the right side

78. Common names vary in different languages and are difficult to remember and convert to formulas.

79. The statement is true for the representative metals but not for the transition metals, which often have multiple charges.

80. Possible answers include: cations always come before anions; when a cation has more than one ionic charge, the charge is indicated by a Roman numeral; monatomic anions use an *–ide* ending. Each rule has a specific purpose; for example, an ionic charge is necessary information because it determines how many ions are in the formula unit of the compound.

81. **a.** N_2O, dinitrogen monoxide
b. NO_2, nitrogen dioxide
c. NO, nitrogen monoxide
d. N_2O_4, dinitrogen tetroxide

82. **a.** Cu_2S, copper(I) sulfide and CuS, copper(II) sulfide
b. $FeSO_4$, iron(II) sulfate and $Fe_2(SO_4)_3$, iron(III) sulfate
c. PbO, lead(II) oxide and PbO_2 lead(IV) oxide

83. **a.** The charges do not balance, CsCl.
b. The charges do not balance, ZnO.
c. Neon does not form compounds.
d. The subscripts are not the lowest whole-number ratio, BaS.

84. binary ionic compounds: *d* and *g;* binary molecular compounds: *a* and *f;* compounds with polyatomic ions: *b, c, e, h,* and *i;* acids: *b* and *e;* base: *c*

Concept Challenge

85. Spot check student answers.

86. **a.** Potassium carbonate has greater water solubility than $CaCO_3$.
b. The copper compound is blue; the iron compound is white.
c. Add water to dissolve the NH_4Cl; then filter out the insoluble $BaSO_4$.
d. chlorine (nonmetal), sulfur (nonmetal), bromine (nonmetal), barium (metal), iodine (nonmetal), mercury (metal)
e. barium sulfate, calcium carbonate, potassium carbonate, copper(II) sulfate pentahydrate, iron(II) sulfate pentahydrate, ammonium chloride
f. 639 g **g.** 7.54 cm^3 **h.** color, density, melting point, or boiling point

Cumulative Review

87. List five properties of the chair you are sitting on. Classify each as physical or chemical. (*Chapter 2*)

88. How many significant figures are in the following measurements? (*Chapter 3*)
a. 15.05 g **b.** 0.31 cm **c.** 890 mL
d. 300.0 cm^3 **e.** 3.0×10^5 kg **f.** 0.001 mm

89. Determine the sum of the following measurements to the correct number of significant figures. (*Chapter 3*)
1.55 cm + 0.235 cm + 3.4 cm

90. Make the following conversions. (*Chapter 3*)
a. 775 mL to microliters (μL)
b. −65°C to K
c. 8.32 mg Ag to centigrams of silver (cg Ag)

91. A student finds that 6.62 g of a substance occupies a volume of 12.3 cm^3. What is the density of the substance? (*Chapter 3*)

92. The diagrams show two models of the atom. (*Chapter 5*)
a. Which is the more accurate?
b. What do the positively charged particles represent?
c. What do the negatively charged particles represent?
d. What major subatomic particle is missing in both of these models?

a.

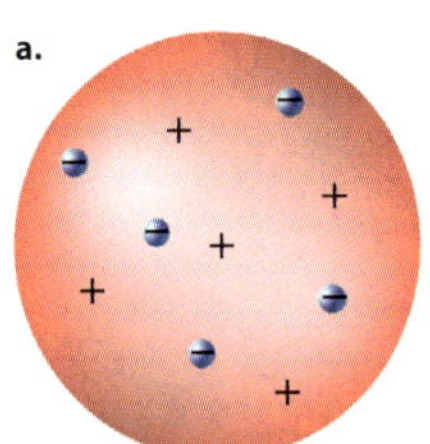

b.

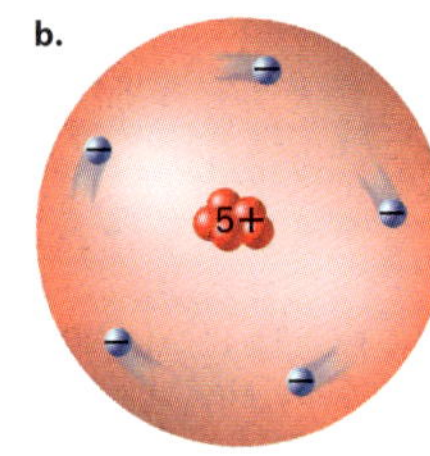

93. Compare and contrast neutrons and protons with respect to their charge, mass, and position in the atom. (*Chapter 4*)

94. What elements have these electron configurations? (*Chapter 5*)
a. $1s^22s^22p^6$ **b.** $1s^22s^22p^2$
c. $1s^22s^22p^1$ **d.** $1s^2$

95. How many valence electrons do atoms of the following elements have? (*Chapter 7*)
a. lithium **b.** sulfur **c.** phosphorus
d. calcium **e.** bromine **f.** neon

96. Where are the metalloids found on the periodic table? Compare the properties of the metalloids to metals and nonmetals. (*Chapter 6*)

97. Arrange the following groups of elements in order of increasing ionization energy. (*Chapter 6*)
a. potassium, cesium, lithium, sodium
b. fluorine, boron, lithium, carbon, neon

98. From the positions of the elements in the periodic table, choose the element in each pair with the higher electronegativity. (*Chapter 6*)
a. Cs and Li **b.** Sr and I **c.** S and Mg
d. O and Se **e.** Te and N **f.** C and F

99. The ions of the elements of Groups 1A and 2A have smaller radii than their neutral atoms, whereas the ions of Group 7A have larger radii than their neutral atoms. Explain. (*Chapter 6*)

100. Write the electron configuration for the element neon, then identify three ions that have the same electron configuration. (*Chapter 7*)

101. How many protons and electrons are in each ion? (*Chapter 7*)
a. magnesium ion **b.** bromide ion
c. strontium ion **d.** sulfide ion

102. Which of these compounds would you expect to contain covalent bonds? Why? (*Chapter 8*)
a. KCl **b.** PBr_3 **c.** ClBr **d.** NaI

103. Which of these substances would you expect to be polar? (*Chapter 8*)
a. Cl_2 **b.** CO **c.** CO_2 **d.** NH_3
e. CCl_4 **f.** H_2O **g.** CH_4

104. Draw electron dot structures for the substances in Question 103. (*Chapter 8*)

105. Explain what a hydrogen bond is and under what conditions a hydrogen bond will form. (*Chapter 8*)

106. Explain the difference between an ionic bond and a covalent bond. Use electron dot structures to illustrate your explanation. (*Chapter 8*)

Cumulative Review

87. Answers may include: color (physical), solid (physical), magnetic (physical), conducts electricity (physical), burns (chemical).

88. a. 4 **b.** 2 **c.** 2 **d.** 4 **e.** 2 **f.** 1

89. 5.2 cm

90. a. 7.75×10^5 μL **b.** 208 K **c.** 0.832 cg

91. 0.538 g/cm^3

92. a. b **b.** protons **c.** electrons **d.** neutrons

93. Both are in the nucleus and have a mass of about 1 amu. A proton is positively charged; a neutron has no charge.

94. a. neon **b.** carbon **c.** boron **d.** helium

95. a. 1 **b.** 6 **c.** 5 **d.** 2 **e.** 7 **f.** 8

96. The metalloids border a line separating the metals from the nonmetals. Their properties are intermediate between those of metals and nonmetals.

97. a. cesium, potassium, sodium, lithium **b.** lithium, boron, carbon, fluorine, neon

98. a. Li **b.** I **c.** S **d.** O **e.** N **f.** F

99. When metallic elements of Group 1A and 2A form ions, they lose all their outer shell electrons. This increases the attraction by the nucleus for the fewer remaining electrons and results in ions that are smaller than the neutral atoms. The electron that a Group 7A element gains in forming an ion enters the outer shell, resulting in a decrease in the effective nuclear attraction of the increased number of electrons. The anion is larger than the neutral atom.

100. $1s^22s^22p^6$; Possible answers are N^{3-}, O^{2-}, F^-, Na^+, Mg^{2+}, and Al^{3+}.

101. a. 12 p^+ and 10 e^- **b.** 35 p^+ and 36 e^- **c.** 38 p^+ and 36 e^- **d.** 16 p^+ and 18 e^-

102. b and c; Molecular compounds formed by two nonmetals have covalent bonds.

103. b, d, and f

104. a. :C̤l̤:C̤l̤: **b.** :C:::O: **c.** :Ö::C::Ö:

d. H:N̈:H with H below; **e.** :C̤l̤:C̤(:C̤l̤: above and below):C̤l̤: ; **f.** H:Ö:H ; **g.** H:C:H with H above and below

105. A hydrogen bond is an intermolecular force between a hydrogen atom covalently bonded to a very electronegative atom and an unshared pair of electrons on another electronegative atom.

106. ionic bond: electrons are transferred

Na· + ·F̤: → Na^+ :F̤:$^-$

covalent bond: electrons are shared

H· + ·H → H:H

Standardized Test Prep

Test-Taking Tip

Eliminate Wrong Answers If you don't know which response is correct, start by eliminating those you know are wrong. If you can rule out some choices, you'll have fewer left to consider and you'll increase your chances of choosing the correct answer.

Select the choice that best answers each question or completes each statement.

1. Identify the pair in which the formula does not match the name.
   a. sulfite, SO_3^{2-} b. dichromate, $Cr_2O_7^{2-}$
   c. hydroxide, OH^- d. nitrite, NO_3^-
   e. perchlorate, ClO_4^-
2. Which of these compounds are ionic?
   I. $CaSO_4$ II. N_2O_4 III. NH_4NO_3 IV. CaS
   a. I and II only b. II and III only
   c. III and IV only d. I, III, and IV only
   e. I, II, III, and IV
3. What is the name of $AlCl_3$?
   a. aluminum trichloride
   b. aluminum(III) chloride
   c. aluminum chlorite
   d. aluminum chlorate
   e. aluminum chloride
4. The Roman numeral in manganese(IV) sulfide indicates the
   a. group number on the periodic table.
   b. positive charge on the manganese ion.
   c. number of manganese ions in the formula.
   d. number of sulfide ions needed in the formula.
5. Which of these statements does not describe every binary molecular compound?
   a. Molecules of binary molecular compounds are composed of two atoms.
   b. The names of binary molecular compounds contain prefixes.
   c. The names of binary molecular compounds end in the suffix *-ide*.
   d. Binary molecular compounds are composed of two nonmetals.
6. What is the formula of ammonium carbonate?
   a. NH_4CO_3 b. NH_4C c. $(NH_4)_2CO_3$
   d. NH_3CO_4 e. $(NH_3)_2CO_4$ f. NH_4CO_2

The lettered choices below refer to Questions 7–11. A lettered choice may be used once, more than once, or not at all.

(A) QR (B) QR_2 (C) Q_2R (D) Q_2R_3

Which formula shows the correct ratio of ions in the compound formed by each pair of elements?

	Element Q	Element R
7.	aluminum	sulfur
8.	potassium	oxygen
9.	lithium	chlorine
10.	strontium	bromine
11.	sodium	sulfur

Use the data table to answer Questions 12–13. The table gives formulas for some of the ionic compounds formed when cations (M, N, P) combine with anions (A, B, C, D).

Cation	Anion			
	A	B	C	D
M	MA_2	(1)	(2)	MD
N	(3)	N_2B	(4)	(5)
P	PA_3	(6)	PC	$P_2(D)_3$

12. Use the given formulas to determine the ionic charge of each cation and anion.
13. Write formulas for compounds (1) through (6).

Use the atomic windows to answer Questions 14–15.

14. Identify the contents of each atomic window as a substance or a mixture.
15. Classify the contents as elements only, compounds only, or elements and compounds.

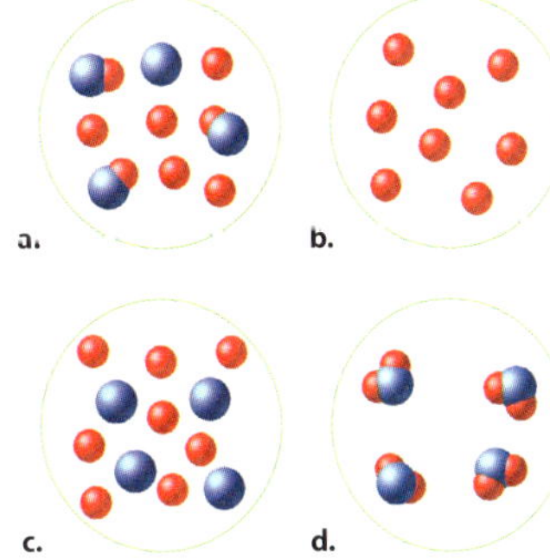

Standardized Test Prep

1. d
2. d
3. e
4. b
5. a
6. c
7. D
8. C
9. A
10. B
11. C
12. cations: M^{2+}, N^+, P^{3+}; anions: A^-, B^{2-}, C^{3-}, D^{2-}
13. (1) MB, (2) M_3C_2, (3) NA, (4) N_3C, (5) N_2D, (6) P_2B_3
14. **b** and **d** are substances; **a** and **c** are mixtures.
15. **a** contains elements and compounds, **b** and **c** contain only elements, and **d** contains only a compound.

CHAPTER 10

Chemical Quantities Planning Guide

SECTION OBJECTIVES	STANDARDS NATIONAL	STANDARDS STATE	ACTIVITIES and LABS
10.1 The Mole: A Measurement of Matter, pp. 287–296 1 block or 2 periods **10.1.1 Describe** methods of measuring the amount of something. **10.1.2 Define** Avogadro's number as it relates to a mole of a substance. **10.1.3 Distinguish** between the atomic mass of an element and its molar mass. **10.1.4 Describe** how the mass of a mole of a compound is calculated.	A-1, A-2, B-1, B-2		**SE** Inquiry Activity: Counting by Measuring Mass, p. 286 L2 **TE** Teacher Demo: Moles and Mass, p. 294 L1 **TE** Class Activity: Calculating Molar Mass, p. 295 L2 **SSLM** Lab13: Measuring Mass: A Means of Counting L2
10.2 Mole–Mass and Mole–Volume Relationships, pp. 297–304 1 block or 2 periods **10.2.1 Describe** how to convert the mass of a substance to the number of moles of a substance, and moles to mass. **10.2.2 Identify** the volume of a quantity of gas at STP.	A-1, A-2, B-2		**SE** Small-Scale Lab: Counting by Measuring Mass, p. 304 L2 **TE** Class Activity: Problem Solving, p. 298 L2 **TE** Teacher Demo: Molar Volume, p. 301 L2 **LM** Lab12: The Masses of Equal Volumes of Gases L2 **VCL** Lab 3: Counting by Measuring Mass L2
10.3 Percent Composition and Chemical Formulas, pp. 305–313 1 block or 2 periods **10.3.1 Describe** how to calculate the percent by mass of an element in a compound. **10.3.2 Interpret** an empirical formula. **10.3.3 Distinguish** between empirical and molecular formulas.	A-1, A-2, B-2, E-2		**SE** Quick Lab: Percent Composition, p. 308 L2 **TE** Class Activity: Empirical Formulas from Percent Composition, p. 309 L2 **TE** Teacher Demo: Paper Chromatography, p. 313 L1 **LM** Lab13: Empirical Formula Determination L2 **LP** Lab 3 Practical 10-1: Empirical Formulas L3

Ability Levels

- **L1** For students who need additional help
- **L2** For all students
- **L3** For students who need to be challenged

Components

SE	Student Edition	**LM**	Laboratory Manual	**GCP**	Graphing Calculator Problems	**T**	Transparencies
TE	Teacher's Edition	**SSLM**	Small-Scale Chemistry Lab Manual	**TTT**	Test-Taking Tips with Transparencies, Grades 9–12 Science	**CA**	Chem Alive and Chemistry Field Trips DVD
GRSW	Guided Reading and Study Workbook	**LP**	Laboratory Practicals	**STP**	Standardized Test Preparation Workbook	**iT**	Interactive Textbook with ChemASAP
CTR	Core Teaching Resources	**PLM**	Probeware Lab Manual			**GO**	Go Online Internet Resources
SSG	Spanish Study Guide	**VCL**	Virtual Chemistry Labs				
		CTB	Computer Test Bank				

RESOURCES PRINT and TECHNOLOGY	PROBLEMS and ASSESSMENT
GRSW Chapter 10, Section 1 **L1** T T104–107 **L2** **GO** Chapter 10, Section 1 **L2** **GO** Chapter 10, Section 1 **L2**	**SP** 10.1, Finding Mass from a Count **SP** 10.2, Converting Number of Atoms to Moles **SP** 10.3, Converting Moles to Number of Atoms **SP** 10.4, Finding the Molar Mass of a Compound **SE** Section 10.1 Assessment, p. 296 **iT** Section 10.1
GRSW Chapter 10, Section 2 **L1** T T108–109 **L2**	**SP** 10.5, Converting Moles to Mass **SP** 10.6, Converting Mass to Moles **SP** 10.7, Calculating the Volume of a Gas at STP **SP** 10.8, Calculating the Molar Mass of a Gas at STP **SE** Section 10.2 Assessment, p. 303 **iT** Section 10.2
GRSW Chapter 10, Section 3 **L1** T T110–112 **L2** **GO** Chapter 10, Section 3 **L2**	**SP** 10.9, Calculating Percent Composition from Mass Data **SP** 10.10, Calculating Percent Composition from a Formula **SP** 10.11, Determining the Empirical Formula of a Compound **SP** 10.12, Finding the Molecular Formula of a Compound **SE** Section 10.3 Assessment, p. 312 **iT** Section 10.3

Web Code: cdn-1101
Web Code: cdn-1104
Web Code: cdn-1103

MATERIALS List

Quantities for each group

STUDENT EDITION

Inquiry Activity, p. 286
100 paper clips of the same size, centigram balance

Small-Scale Lab, p. 304
1 teaspoon each of water, sodium chloride and calcium carbonate, plastic spoon, weighing paper, watchglass or small beaker, balance, paper, pencil, ruler

Quick Lab, p. 308
centigram balance, burner, 3 medium-sized test tubes, test-tube holder, test-tube rack, spatula, 2–3 g each of hydrated salts of copper(II) sulfate, calcium chloride, and sodium sulfate

TEACHER'S EDITION

Teacher Demo, p. 294
1 mol of each of several different substances, 1 sealed container per sample

Class Activity, p. 295
display from the Teacher Demo on p. 294

Teacher Demo, p. 301
50 g dry ice, towel, hammer, large plastic bag, duct tape, tongs, beaker, balance

Class Activity, p. 309
3 red marbles, 6 green marbles, 3 black marbles, 12 blue marbles

Teacher Demo, p. 313
coffee filter paper, water soluble marker, beaker

Chapter Assessment

CHAPTER ASSESSMENT

SE Chapter Assessment, pp. 315–318
CTB Chapter 10
iT Chapter 10

STANDARDIZED TEST PREP

SE Chapter 10, p. 319
STP Topic 3
iT Interactive Textbook with Assessment at PHSchool.com

CHAPTER 10

Chemical Quantities

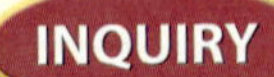
Activity

Counting by Measuring Mass

L2

Purpose

Students will develop methods of indirect measurement.

Prep Time 5 minutes
Class Time 15 minutes

Think About It

1. The number of paper clips determined by mass in Step 2 should equal the number of paper clips counted in Step 3. Any difference could be traced to a measurement error or some variation in the masses of individual paper clips.
2. To measure out 185 paper clips without counting, students should multiply the average mass of a paper clip (from Step 1) by 185. Then they can use the balance to find the corresponding mass of paper clips.
3. By using a larger sample in Step 1, students reduce the effect of any variation in the masses of individual paper clips on the average mass. However, more time is needed to count out the paper clips.

CHAPTER 10

Chemical Quantities

Chemists measure matter in a unit called the mole just as florists sell roses by the dozen.

INQUIRY Activity

Counting by Measuring Mass

Materials

100 paper clips of the same size, centigram balance

Procedure

1. Find the mass of 25 paper clips. Divide the total mass by 25 to find the average mass of a paper clip. Repeat this step using 25 different paper clips until your average masses agree.
2. Select about 75 percent of your paper clips and find their mass. Without counting, calculate the number of paper clips in your sample.
3. Count the number of paper clips in your sample.
4. Repeat Steps 2 and 3 with a different sample size.

Think About It

1. Did the number of paper clips you counted in the sample (Step 3) equal the number you calculated by mass (Step 2)?
2. Explain how you would use the balance to count out 185 paper clips.
3. What is the advantage of using a larger sample size in Step 1? What is a disadvantage?

10.1 The Mole: A Measurement of Matter

Connecting to Your World Every year, contestants from all over the world travel to Harrison Hot Springs in British Columbia, Canada, to compete in the world championship sand sculpture contest. Each contestant creates a beautiful work of art out of millions of tiny grains of sand. You could measure the amount of sand in a sculpture by counting the grains of sand. But wouldn't it be much easier to weigh the sand? In this section, you'll discover how chemists measure the amount of a substance using a unit called a mole.

Guide for Reading

Key Concepts

- What are three methods for measuring the amount of something?
- How is Avogadro's number related to a mole of any substance?
- How is the atomic mass of an element related to the molar mass of an element?
- How is the mass of a mole of a compound calculated?

Vocabulary

mole
Avogadro's number
representative particle
molar mass

Reading Strategy

Relating Text and Visuals As you read, look closely at Table 10.2. Explain how the information in the table helps you understand the basis for the molar masses of the elements.

Measuring Matter

You live in a quantitative world. The grade you got on your last exam, the number of times you heard your favorite song on the radio yesterday, and the cost of a bicycle you would like to own are all important quantities to you. These are quantities that answer questions such as "How much?" or "How many?" Scientists answer similar questions. How many kilograms of iron can be obtained from one hundred kilograms of iron ore? How many grams of hydrogen and nitrogen must be combined to make 200 grams of the fertilizer ammonia (NH_3)? Questions like these illustrate that chemistry is a quantitative science. In your study of chemistry, you will analyze the composition of samples of matter and perform chemical calculations that relate quantities of reactants in a chemical reaction to quantities of products. To solve these and other problems, you will have to be able to measure the amount of matter you have.

How do you measure matter? One way is to count how many of something you have. For example, you can count the CDs in your collection or the number of pins you knock down when bowling. Another way to measure matter is to determine its mass. You can buy potatoes by the kilogram or pound and gold by the gram or ounce. You can also measure matter by volume. For instance, people buy gasoline by the liter or the gallon and take cough medicine by the milliliter or the teaspoon. **You often measure the amount of something by one of three different methods—by count, by mass, and by volume.** For example, you can buy soda by the six-pack or by the liter. Figure 10.1 shows how you might measure the amount of grapes you want to buy.

Figure 10.1 A grocer's scale measures the weight of the grapes you buy.

Section Resources

Print

- **Guided Reading and Study Workbook,** Section 10.1
- **Core Teaching Resources,** Section 10.1 Review
- **Transparencies,** T104–T107

Technology

- **Interactive Textbook with ChemASAP,** Animation 11; Problem-Solving 10.1, 10.4, 10.5, 10.7; Assessment 10.1
- **Go Online,** Section 10.1

10.1

1 FOCUS

Objectives

10.1.1 Describe methods of measuring the amount of something.
10.1.2 Define Avogadro's number as it relates to a mole of a substance.
10.1.3 Distinguish between the atomic mass of an element and its molar mass.
10.1.4 Describe how the mass of a mole of a compound is calculated.

Guide for Reading

Build Vocabulary L2

Connect Terms Have students write two or three sentences, each of which relates a mole to one or two of the other vocabulary words. Answers might include the following: A mole contains Avogadro's number of particles. The mass of one mole of a substance is its molar mass, which is found by determining the mass in grams of its representative particle.

Reading Strategy L2

Visualize Students may more easily grasp the meaning of the mole by developing their own mental picture. For example, by analogy to a dozen, they may visualize a giant egg box with 6.02×10^{23} depressions for particles, or a giant sack bulging with Avogadro's number of particles.

2 INSTRUCT

Connecting to Your World

Ask, **Is it practical to count each grain of sand?** *(Students should realize that it is not practical to measure sand by counting individual grains.)* **How else might you measure, or quantify, the sand?** *(Find its mass or volume.)* Lead students to see that just as a small amount of sand contains millions of smaller particles, so also small amounts of chemical substances contain very large numbers of particles.

Measuring Matter

Relate 

Point out to students that they often use conversion factors. If a recipe calls for six eggs, this quantity can be considered half a dozen eggs (6 eggs × 1 doz/12 eggs). Four shoes equal two pairs (4 shoes × 1 pair/2 shoes). The concepts presented in this section involve the use of conversion factors in dimensional analysis or the factor-label method of problem solving. Ask students to write an everyday equality that is also a definition, such as 3 feet = 1 yard. Write their equalities on the board. Examples include 60 min = 1 h, 12 in. = 1 ft, 60 s = 1 min, 2.54 cm = 1 in., and 100 cm = 1 m. Point out that equalities can be written in either direction, so 60 min = 1 h can be written as 1 h = 60 min. Have them use these equalities to write conversion factors. Ask, **How many seconds are in 2.5 hours?** Show students how to use dimensional analysis to solve this problem. *(9000 s)*

Discuss 

Pass around bags, each of which contains a multiple of 12 beans. Ask students how they would go about determining the quantities of beans in the bags. They could count, weigh, or find the volume of the beans. With multiples of 12 beans in each bag, students may be guided to an understanding of how the word "dozen" can be used as a unit for the quantity of beans in each bag.

Some of the units used for measuring indicate a specific number of items. For example, a pair always means two. A pair of shoes is two shoes, and a pair of aces is two aces. Similarly, a dozen always means 12. A dozen eggs is 12 eggs and a dozen pens is 12 pens.

Apples are measured in three different ways. At a fruit stand, they are often sold by the count (3 for $2.40). In a supermarket, you usually buy apples by weight ($1.29/pound) or mass ($2.79/kg). At an orchard, you can buy apples by volume ($12.00/bushel). Each of these different ways to measure apples can be equated to a dozen apples.

By count: 1 dozen apples = 12 apples

For average-sized apples the following approximations can be used.

By mass: 1 dozen apples = 2.0 kg apples

By volume: 1 dozen apples = 0.20 bushel apples

Figure 10.2 shows other items sold by count, weight, and volume.

Knowing how the count, mass, and volume of apples relate to a dozen apples allows you to convert among these units. For example, based on the unit relationships given above, you could calculate the mass of a bushel of apples or the mass of 90 average-sized apples using conversion factors such as the following.

$$\frac{1 \text{ dozen apples}}{12 \text{ apples}} \quad \frac{2.0 \text{ kg apples}}{1 \text{ dozen apples}} \quad \frac{1 \text{ dozen apples}}{0.20 \text{ bushel apples}}$$

 ***What are three ways of measuring the amount of a substance?***

Figure 10.2 You can buy items by different types of measurements, such as a count, a weight or mass, or a volume. **Classifying** ***Which of these common items are being sold by weight? By volume? By count?***

Differentiated Instruction

English Learners 

For this chapter, team up a limited English proficient student with a student who has strong mathematical skills. Encourage the English learners to ask their partners for help with any difficulty they are having understanding the terms and how to solve the problems.

SAMPLE PROBLEM 10.1

Finding Mass from a Count

What is the mass of 90 average-sized apples if 1 dozen of the apples has a mass of 2.0 kg?

1 Analyze ***List the knowns and the unknown.***

Knowns
- number of apples = 90 apples
- 12 apples = 1 dozen apples
- 1 dozen apples = 2.0 kg apples

Unknown
- mass of 90 apples = ? kg

You can use dimensional analysis to convert the number of apples to the mass of apples. Carry out this conversion by performing the following sequence of conversions:

Number of apples ⟶ dozens of apples ⟶ mass of apples.

2 Calculate ***Solve for the unknown.***

The first conversion factor is $\frac{1 \text{ dozen apples}}{12 \text{ apples}}$.

The second conversion factor is $\frac{2.0 \text{ kg apples}}{1 \text{ dozen apples}}$.

Multiplying the original number of apples by these two conversion factors gives the answer in kilograms.

$$\text{mass of apples} = 90 \cancel{\text{apples}} \times \frac{1 \cancel{\text{dozen apples}}}{12 \cancel{\text{apples}}} \times \frac{2.0 \text{ kg apples}}{1 \cancel{\text{dozen apples}}}$$
$$= 15 \text{ kg apples}$$

The mass of 90 average-sized apples is 15 kg.

3 Evaluate ***Does the result make sense?***

Because a dozen apples has a mass of 2.0 kg, and 90 apples is less than 10 dozen apples, the mass should be less than 20 kg of apples (10 dozen × 2.0 kg/dozen).

Practice Problems

1. If 0.20 bushel is 1 dozen apples and a dozen apples has a mass of 2.0 kg, what is the mass of 0.50 bushel of apples?
2. Assume 2.0 kg of apples is 1 dozen and that each apple has 8 seeds. How many apple seeds are in 14 kg of apples?

CHEMath

Dimensional Analysis

Dimensional analysis is a tool for solving conversion problems—problems in which a measurement must be expressed in a different unit.

To solve a simple one-step conversion problem (for example, How many grams is 34 kg?), you must know the relationship between the unit of the known measurement (kg) and the unit of the desired answer (g). The relationship is: 1 kg = 1000 g.

Write this equality as a ratio (conversion factor). When you multiply the known measurement by the conversion factor, the unit kg cancels and the resulting answer has the unit g.

In more complex problems, you may need to use more than one conversion factor to obtain the answer, but the principle is the same.

Math Handbook

For help with dimensional analysis, go to page R66.

Interactive Textbook

Problem-Solving 10.1
Solve Problem 1 with the help of an interactive guided tutorial.

with **ChemASAP**

Sample Problem 10.1

Answers

1. 0.50 bushel × 1 dozen/0.20 bushel × 2.0 kg/1 dozen = 5.0 kg
2. 14 kg × 1 dozen/2.0 kg × 12 apples/1 dozen × 8 seeds/1 apple = 672 seeds

Practice Problems Plus L2

How many bushels of apples are in 1.0 kg of apples? *(0.10 bushels)*

CHEMath

Dimensional Analysis

Dimensional analysis uses conversion factors, which are ratios equal to one. Units that can be related by conversion factors can be changed from one unit to another. Chemistry problems involving dimensional analysis range from simple, one-factor conversions, such as changing m to cm, to problems that involve several conversion factors, such as changing a number of atoms to the mass of moles.

Math Handbook

For a math refresher and practice, direct students to dimensional analysis, page R66.

Differentiated Instruction

Special Needs 

Some students may experience difficulty with using dimensional analysis and should be given additional simple conversion problems for practice before they start doing problems involving the mole. Review the use of parentheses and the fraction bar as ways of grouping symbols to indicate the order of operations. Stress the importance of labels.

Answers to...

Figure 10.2 Weight, bird seed and screws; volume, paint; count, paint brush

Checkpoint by count, by mass, and by volume

What is a Mole?

Use Visuals L1

Figure 10.3 Have students study Figure 10.3 and read the text that discusses the number of particles in a mole. Point out that the mole represents a number of items just as *dozen, gross,* and *ream* all represent a number of items. Ask, **If you are counting the number of an extremely small item, should the number of items in a unit be small or large?** *(For a small item, a large number per counting unit is more convenient.)* A mole = 6.02×10^{23} items and is used to measure extremely small objects.

Relate 

Students usually need to review scientific notation and multiplication and division of exponents. Write several large and small numbers on the board using standard notation. Have students rewrite the numbers using scientific notation. Ask, **What are some measurements that should be written in scientific notation?** *(Students may mention extremely large or extremely small numbers such as the diameter of a virus or distances in space.)*

Download a worksheet on the **Mole** for students to complete, and find additional teacher support from NSTA SciLinks.

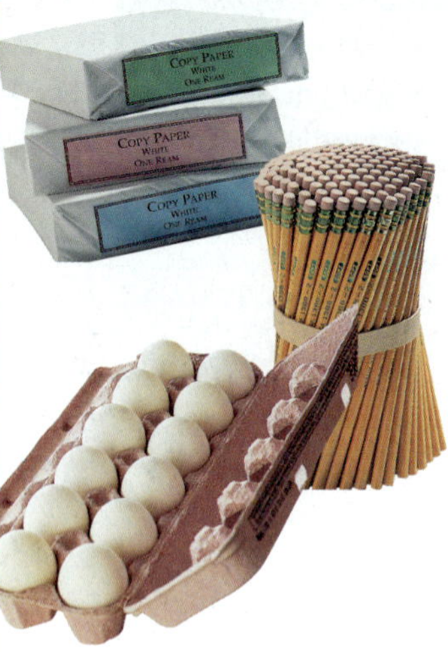

Figure 10.3 Words other than *mole* are used to describe a number of something—for example, a *ream* of paper (500 sheets), a *gross* of pencils (144), and a *dozen* eggs (12).

For: Links on the Mole
Visit: www.SciLinks.org
Web Code: cdn-1101

What Is a Mole?

Counting objects as big as apples is a reasonable way to measure the amount of apples. But imagine trying to count the grains of sand in a sand sculpture. That would be an endless job. Recall that matter is composed of atoms, molecules, and ions. These particles are much smaller than grains of sand and an extremely large number of them are in even a small sample of a substance. Obviously, counting particles one by one is not practical. However, think about counting eggs. It's easier when the eggs are grouped into dozens. A dozen is a specified number, 12, of things. Other common groupings of items are shown in Figure 10.3. Chemists also use a unit that is a specified number of particles. The unit is called a mole. Just as a dozen eggs is 12 eggs, a **mole** (mol) of a substance is 6.02×10^{23} representative particles of that substance and is the SI unit for measuring the amount of a substance. The number of representative particles in a mole, 6.02×10^{23}, is called **Avogadro's number.** It was named in honor of the Italian scientist Amedeo Avogadro di Quaregna (1776–1856) who helped clarify the difference between atoms and molecules.

The term **representative particle** refers to the species present in a substance: usually atoms, molecules, or formula units. The representative particle of most elements is the atom. Iron is composed of iron atoms. Helium is composed of helium atoms. Seven elements, however, normally exist as diatomic molecules (H_2, N_2, O_2, F_2, Cl_2, Br_2, and I_2). The representative particle of these elements and of all molecular compounds is the molecule. The molecular compounds water (H_2O) and sulfur dioxide (SO_2) are composed of H_2O and SO_2 molecules, respectively. For ionic compounds, such as calcium chloride, the representative particle is the formula unit $CaCl_2$. **A mole of any substance contains Avogadro's number of representative particles, or 6.02×10^{23} representative particles.** Table 10.1 summarizes the relationship between representative particles and moles of substances.

Converting Number of Particles to Moles The relationship, 1 mol = 6.02×10^{23} representative particles, is the basis for a conversion factor that you can use to convert numbers of representative particles to moles.

$$\text{moles} = \text{representative particles} \times \frac{1 \text{ mole}}{6.02 \times 10^{23} \text{ representative particles}}$$

 *How many representative particles are in one mole?*

Table 10.1

Representative Particles and Moles

Substance	Representative particle	Chemical formula	Representative particles in 1.00 mole
Atomic nitrogen	Atom	N	6.02×10^{23}
Nitrogen gas	Molecule	N_2	6.02×10^{23}
Water	Molecule	H_2O	6.02×10^{23}
Calcium ion	Ion	Ca^{2+}	6.02×10^{23}
Calcium fluoride	Formula unit	CaF_2	6.02×10^{23}
Sucrose	Molecule	$C_{12}H_{22}O_{11}$	6.02×10^{23}

Differentiated Instruction

Gifted and Talented 

Gifted students may wish to investigate different methods for determining Avogadro's number. They could present their findings to the class.

SAMPLE PROBLEM 10.2

Converting Number of Atoms to Moles

Magnesium is a light metal used in the manufacture of aircraft, automobile wheels, tools, and garden furniture. How many moles of magnesium is 1.25×10^{23} atoms of magnesium?

1 Analyze ***List the knowns and the unknown.***

Knowns
- number of atoms = 1.25×10^{23} atoms Mg
- 1 mol Mg = 6.02×10^{23} atoms Mg
- The desired conversion is: atoms $\longrightarrow$ moles

Unknown
- moles = ? mol Mg

2 Calculate ***Solve for the unknown.***

The conversion factor is $\frac{1 \text{ mol Mg}}{6.02 \times 10^{23} \text{ atoms Mg}}$.

Multiplying atoms of Mg by the conversion factor gives the answer.

$$\text{moles} = 1.25 \times 10^{23} \text{ atoms Mg} \times \frac{1 \text{ mol Mg}}{6.02 \times 10^{23} \text{ atoms Mg}}$$

$$\text{moles} = 2.08 \times 10^{-1} \text{ mol Mg} = 0.208 \text{ mol Mg}$$

3 Evaluate ***Does the result make sense?***

Because the given number of atoms is less than one-fourth of Avogadro's number, the answer should be less than one-fourth mole of atoms. The answer should have three significant figures.

Practice Problems

3. How many moles is 2.80×10^{24} atoms of silicon?

4. How many moles is 2.17×10^{23} representative particles of bromine?

For help with dimensional analysis, go to page R66.

Interactive Textbook

Problem-Solving 10.4
Solve Problem 4 with the help of an interactive guided tutorial.

with ChemASAP

Converting Moles to Number of Particles Now suppose you want to determine how many atoms are in a mole of a compound. To do this, you must know how many atoms are in a representative particle of the compound. This number is determined from the chemical formula. Figure 10.4 illustrates this idea with marbles (atoms) in cups (molecules). The number of marbles in a dozen cups is (6×12), or 72 marbles. In the formula of a molecule of carbon dioxide (CO_2), the subscripts show that carbon dioxide is composed of three atoms: one carbon atom and two oxygen atoms. A mole of carbon dioxide contains Avogadro's number of CO_2 molecules. But each molecule contains three atoms. Thus a mole of carbon dioxide contains three times Avogadro's number of atoms. A molecule of carbon monoxide (CO) consists of two atoms, so a mole of carbon monoxide contains two times Avogadro's number of atoms.

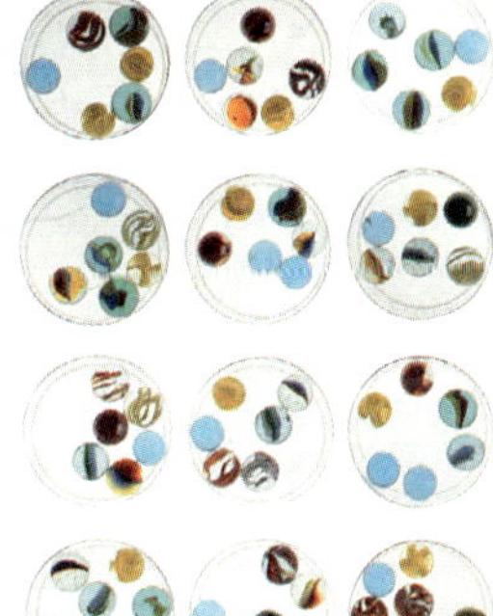

Figure 10.4 A dozen cups of marbles contain more than a dozen marbles. Similarly, a mole of molecules contains more than a mole of atoms. **Calculating** ***How many atoms are in one mole of molecules if each molecule consists of six atoms?***

Use Visuals 

Figure 10.4 Have students examine the photograph. Note that each cup contains six marbles. A dozen cups hold 72 marbles. Ask, **What are some everyday items that come in a "package" of more than one?** *(gum, 5 sticks; shoes, 2; batteries, 2, 4, 8)* Point out that a dozen packages of gum would be five dozen sticks of gum or 60 sticks of gum. Point out that this idea can be applied to molecules. For example, a mole of water (H_2O) consists of 2 mol H atoms and 1 mol O atoms. Thus a mole of water contains $3 \times 6.02 \times 10^{23}$ atoms or 1.8×10^{24} atoms.

Sample Problem 10.2

Answers

3. 2.80×10^{24} atoms Si $\times$ 1 mol/6.02×10^{23} atoms = 4.65 mol Si

4. 2.17×10^{23} representative particles $\times$ 1 mol/6.02×10^{23} representative particles = 0.360 mol Br_2

Practice Problems Plus L2

1. How many moles are equal to 2.41×10^{24} formula units of sodium chloride (NaCl)? *(4.00 moles)*

2. How many moles are equal to 9.03×10^{24} atoms of mercury (Hg)? *(15.0 moles)*

Math Handbook

For a math refresher and practice, direct students to dimensional analysis, page R66.

Answers to...

Figure 10.4 3.61×10^{24} atoms

6.02×10^{23} representative particles

Sample Problem 10.3

Answers

5. 1.14 mol × 6.02 × 10^{23} molecules/mol × 4 atoms/molecule = 2.75 × 10^{24} atoms
6. 4.65 × 10^{24} molecules NO_2 × 1 mol/6.02 × 10^{23} molecules = 7.72 mol NO_2

Practice Problems Plus 

1. **How many atoms are in 1.00 mole of sucrose, $C_{12}H_{22}O_{11}$?** *(2.71 × 10^{25} atoms)*
2. **How many atoms of C are in 2.00 moles of $C_{12}H_{22}O_{11}$?** *(1.44 × 10^{25} atoms)*
3. **How many atoms of H are in 2.00 moles of $C_{12}H_{22}O_{11}$?** *(2.65 × 10^{25} atoms)*
4. **How many atoms of O are in 3.65 moles of $C_{12}H_{22}O_{11}$?** *(2.42 × 10^{25} atoms)*

Math Handbook

For a math refresher and practice, direct students to dimensional analysis, page R66.

To find the number of atoms in a mole of a compound, you must first determine the number of representative particles. You can use the following conversion factor to convert a number of moles of a compound to the number of representative particles (molecules or formula units).

$$\text{representative particles} = \text{moles} \times \frac{6.02 \times 10^{23} \text{ representative particles}}{1 \text{ mole}}$$

The next step is to multiply the number of representative particles by the number of atoms in each molecule or formula unit.

Math Handbook

For help with dimensional analysis, go to page R66.

SAMPLE PROBLEM 10.3

Converting Moles to Number of Atoms

Propane is a gas used for cooking and heating. How many atoms are in 2.12 mol of propane (C_3H_8)?

1 Analyze ***List the knowns and the unknown.***

Knowns
- number of moles = 2.12 mol C_3H_8
- 1 mol C_3H_8 = 6.02 × 10^{23} molecules C_3H_8
- 1 molecule C_3H_8 = 11 atoms (3 carbon atoms and 8 hydrogen atoms)
- The desired conversion is: moles ⟶ molecules ⟶ atoms.

Unknown
- number of atoms = ? atoms

Use the relationships among units given above to write the desired conversion factors.

2 Calculate ***Solve for the unknown.***

The first conversion factor is $\frac{6.02 \times 10^{23} \text{ molecules } C_3H_8}{1 \text{ mol } C_3H_8}$.

The second conversion factor is $\frac{11 \text{ atoms}}{1 \text{ molecule } C_3H_8}$.

Multiply the moles of C_3H_8 by the proper conversion factors:

$$2.12 \cancel{\text{mol } C_3H_8} \times \frac{6.02 \times 10^{23} \cancel{\text{molecules } C_3H_8}}{1 \cancel{\text{mol } C_3H_8}} \times \frac{11 \text{ atoms}}{1 \cancel{\text{molecule } C_3H_8}}$$

$$= 1.4039 \times 10^{25} \text{ atoms} = 1.40 \times 10^{25} \text{ atoms}$$

3 Evaluate ***Does the result make sense?***

Because there are 11 atoms in each molecule of propane and more than 2 mol of propane, the answer should be more than 20 times Avogadro's number of propane molecules. The answer has three significant figures based on the three significant figures in the given measurement.

Practice Problems

5. How many atoms are in 1.14 mol SO_3?
6. How many moles are in 4.65 × 10^{24} molecules of NO_2?

Problem-Solving 10.5
Solve Problem 5 with the help of an interactive guided tutorial.

with ChemASAP

Differentiated Instruction

Special Needs 

Some students may experience difficulty with mathematical concepts. To help these students learn to do the many calculations in this chapter, have them write out, in their own words, a step-by-step procedure for doing each of the sample problems. Suggest that they ask a study partner to read their procedures and correct them, if necessary.

Perhaps you are wondering just how large a mole is. The SI unit, the mole, is not related to the small burrowing animal of the same name shown in Figure 10.5. But this little animal can help you appreciate the size of the number 6.02×10^{23}. Assume that an average animal-mole is 15 cm long, 5 cm tall, and has a mass of 150 g. Based on this information, the mass of 6.02×10^{23} animal-moles is 9.03×10^{22} kg. That means that the mass of Avogadro's number of animal-moles is equal to more than 60 times the combined mass of Earth's oceans. If spread over the entire surface of Earth, Avogadro's number of animal-moles would form a layer more than 8 million animal-moles thick. What about the length of 6.02×10^{23} animal-moles? If lined up end-to-end, 6.02×10^{23} animal-moles would stretch from Earth to the nearest star, Alpha Centauri, more than two million times. Are you beginning to understand how enormous Avogadro's number is?

Figure 10.5 An average animal-mole has a mass of 150 g. The mass of 6.02×10^{23} animal-moles is 9.03×10^{22} kg.

The Mass of a Mole of an Element

Remember that the atomic mass of an element (the mass of a single atom) is expressed in atomic mass units (amu). The atomic masses are relative values based on the mass of the most common isotope of carbon (carbon-12). Table 10.2 shows that an average carbon atom (C) with an atomic mass of 12.0 amu is 12 times heavier than an average hydrogen atom (H) with an atomic mass of 1.0 amu. Therefore, 100 carbon atoms are 12 times heavier than 100 hydrogen atoms. In fact, any number of carbon atoms is 12 times heavier than the same number of hydrogen atoms. The mass ratio of 12 carbon atoms to 1 hydrogen atom remains the same no matter what unit is used to express the masses. So 12.0 g of carbon atoms and 1.0 g of hydrogen atoms must contain the same number of atoms.

Animation 11 Find out how Avogadro's number is based on the relationship between the amu and the gram.

with ChemASAP

Table 10.2

CARBON ATOMS		HYDROGEN ATOMS		MASS RATIO
Number	Mass (amu)	Number	Mass (amu)	$\frac{\text{Mass carbon}}{\text{Mass hydrogen}}$
	12		1	$\frac{12 \text{ amu}}{1 \text{ amu}} = \frac{12}{1}$
	24 [2 × 12]		2 [2 × 1]	$\frac{24 \text{ amu}}{2 \text{ amu}} = \frac{12}{1}$
	120 [10 × 12]		10 [10 × 1]	$\frac{120 \text{ amu}}{10 \text{ amu}} = \frac{12}{1}$
	600 [50 × 12]		50 [50 × 1]	$\frac{600 \text{ amu}}{50 \text{ amu}} = \frac{12}{1}$
Avogadro's number	$(6.02 \times 10^{23}) \times (12)$	**Avogadro's number**	$(6.02 \times 10^{23}) \times (1)$	$\frac{(6.02 \times 10^{23}) \times (12)}{(6.02 \times 10^{23}) \times (1)} = \frac{12}{1}$

The Mass of a Mole of an Element

Discuss L2

Point out to students that the mass of a single atom can be expressed in atomic mass units, but it is not realistic to work with single atoms. Chemists work with large numbers of atoms for which the mass can be expressed in grams. In this text, the atomic masses are rounded to one place after the decimal point. Have students solve a given problem multiple times using a different rounding rule each time so they can see how rounding the atomic masses affects the answer.

Explain that the molar masses of all elements contain the same number of atoms because the atomic masses of the elements are relative values. Present this idea by telling the class that the mass of an atom of element X is twice as great as the mass of an atom of element Y. Ask, **If you have 10 grams of element X and 10 grams of element Y, would you expect both samples to contain the same number of atoms?** *(No, because atoms of element X are twice as massive as atoms of element Y. The sample of X would contain only half as many atoms as the sample of Y.)* **What would you have to do to get the same number of atoms in both samples?** *(Double the mass of element X to twice the mass of element Y.)*

Facts and Figures

History of Avogadro's Number

Avogadro's number was not developed by Avogadro. In the early 1900s, J. Perrin, a French scientist, first used the term "Avogadro's number" to describe the number of particles in a mole. Perrin used Brownian motion to determine the number.

TEACHER Demo

Moles and Mass L2

Purpose Students observe the difference in the mass and volume of 1 mol of different substances.

Materials 1 mol each of a variety of common chemicals, 1 sealed container per sample

Procedure Place 1 mol of at least two substances from each of the following categories in containers and seal them: molecular compounds—sucrose, water, paradichlorobenzene; ionic compounds—cobalt(II) chloride, potassium hydroxide, potassium dichromate; elements—sulfur, iron, carbon, mercury. On each container, mark the mass of each substance. Point out that each container holds 1 mol of a substance, no matter whether the representative particles are molecules, formula units, or atoms.

Expected Outcome Students observe that a mole of one substance has a different mass from a mole of another substance even though the same number of representative particles (6.02×10^{23}) are present.

Use Visuals 

Figure 10.6 Have students study the illustration and verify the stated values for molar masses by looking up the values for the atomic masses in the periodic table. Point out that the molar mass of an element is numerically equal to the atomic mass. Prepare a similar display of other elements in labeled containers for students to view. Ask students to write equations like those in Figure 10.6 for each of the elements displayed.

For: Links on Molar Mass
Visit: www.SciLinks.org
Web Code: cdn-1104

If you look at the atomic masses of the elements in the periodic table, you will notice that they are not whole numbers. For example, the atomic mass of carbon is not exactly 12 times the mass of hydrogen. Recall from Chapter 4 that this is because atomic masses are weighted average masses of the isotopes of each element.

Quantities measured in grams are convenient for working in the laboratory, so chemists have converted the relative scale of masses of the elements in amu to a relative scale of masses in grams. **The atomic mass of an element expressed in grams is the mass of a mole of the element.** The mass of a mole of an element is its **molar mass**. For carbon, the molar mass is 12.0 g. For atomic hydrogen, the molar mass is 1.0 g. Figure 10.6 shows one molar mass of mercury, carbon, iron, and sulfur. Compare the molar masses in the figure to the atomic masses in your periodic table. Notice that the molar masses were rounded off to one place after the decimal point. All the examples and problems in this text use molar masses that are rounded off in this way. If your teacher uses a different rounding rule for molar masses, your answers to problems may differ slightly from the answers given in the text.

If you were to compare 12.0 g of carbon atoms with 16.0 g of oxygen atoms, you would find they contain the same number of atoms. The molar masses of any two elements must contain the same number of atoms. How many atoms are contained in the molar mass of an element? You already know. The molar mass of any element contains 1 mol or 6.02×10^{23} atoms of that element.

The mole can now be further defined as the amount of substance that contains as many representative particles as the number of atoms in 12.0 g of carbon-12. You know that 12.0 g is the molar mass of carbon-12, so 12.0 g is 1 mol of carbon. The same relationship applies to hydrogen: 1.0 g of hydrogen is 1 mol of hydrogen atoms. Similarly, because 24.3 g is the molar mass of magnesium, 1 mol of magnesium (or 6.02×10^{23} atoms of magnesium) has a mass of 24.3 g. Molar mass is the mass of 1 mol of atoms of any element.

Checkpoint *How many atoms are in one mole of magnesium?*

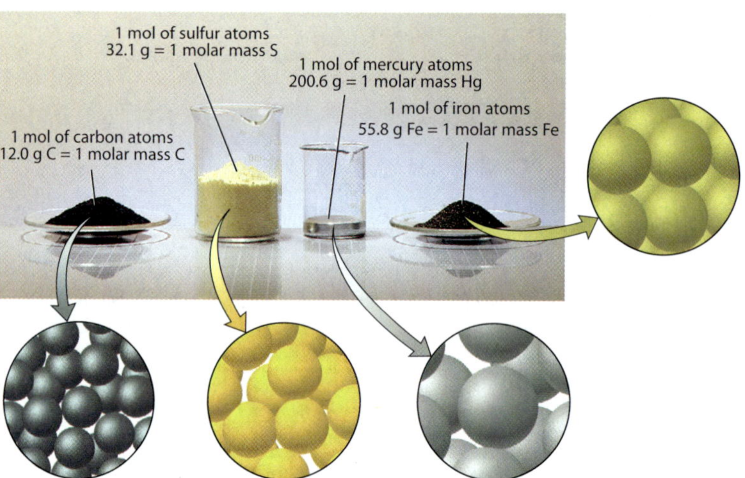

Figure 10.6 One molar mass of carbon, sulfur, mercury, and iron are shown. Each of the quantities contains one mole of the element. **Applying Concepts** *How many atoms of each element are shown?*

Differentiated Instruction

Special Needs 

For the Teacher Demo, mark the levels of the filled containers with tape on the outside of the containers so that sight-impaired students can tell by handling the displays that moles of different substances occupy different volumes and have different masses.

Go Online NSTA SCILINKS

Download a worksheet on **Molar Mass** for students to complete, and find additional teacher support from NSTA SciLinks.

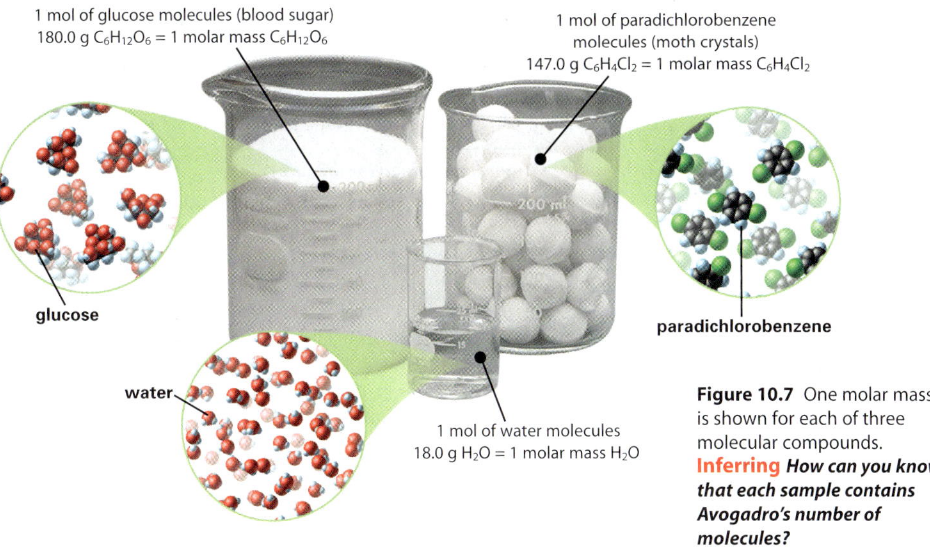

Figure 10.7 One molar mass is shown for each of three molecular compounds.
Inferring ***How can you know that each sample contains Avogadro's number of molecules?***

The Mass of a Mole of a Compound

To find the mass of a mole of a compound, you must know the formula of the compound. The formula of sulfur trioxide is SO_3. A molecule of SO_3 is composed of one atom of sulfur and three atoms of oxygen.

You can calculate the mass of a molecule of SO_3 by adding the atomic masses of the atoms making up the molecule. From the periodic table, the atomic mass of sulfur (S) is 32.1 amu. The mass of three atoms of oxygen is three times the atomic mass of a single oxygen atom (O): 3×16.0 amu = 48.0 amu. So, the molecular mass of SO_3 is 32.1 amu + 48.0 amu = 80.1 amu.

Now substitute the unit grams for atomic mass units to find the molar mass of SO_3. The molar mass (g/mol) of any compound is the mass in grams of 1 mol of that compound. Thus 1 mol of SO_3 has a mass of 80.1 g. This is the mass of 6.02×10^{23} molecules of SO_3.

To calculate the molar mass of a compound, find the number of grams of each element in one mole of the compound. Then add the masses of the elements in the compound. This method for calculating molar mass applies to any compound, molecular or ionic. The molar masses of glucose ($C_6H_{12}O_6$, 180.0 g), water (H_2O, 18.0 g), and paradichlorobenzene ($C_6H_4Cl_2$, 147.0 g) in Figure 10.7 were obtained in this way.

Differentiated Instruction

English Learners L2

The differences in the terms *mass*, *atomic mass*, *atomic mass unit*, and *molar mass* may be confusing to English learners. Have students compile a glossary in which they define each term in English and in their native language. Encourage students to include an example beside each definition. Before compiling their glossaries, students should be able to use the periodic table and calculate the mass of a mole of any substance.

The Mass of a Mole of a Compound

CLASS Activity

Calculating Molar Mass L2

Purpose Students learn to calculate the molar mass of a compound.

Materials the display prepared for the Teacher Demo on p. 294

Procedure Select a compound and show students how to calculate the mass of 1 mol using the atomic masses of the constituent elements from the periodic table. Have them calculate the molar masses of other compounds in the display. Point out that they have determined the mass of 6.02×10^{23} representative particles.

3 ASSESS

Evaluate Understanding L2

Ask, **What does a mole of a substance represent?** *(6.02×10^{23} representative particles of that substance)* **How can you convert the number of atoms or molecules of a substance to moles?** *(Multiply the number of atoms or molecules by the conversion factor 1 mole/ 6.02×10^{23} representative particles.)* **What is the molar mass of $Al_2(CO_3)_3$?** *(234.0 g)*

Reteach L1

Set up cooperative learning groups of three or four students with one student in each group who can help. Provide the groups with problems related to Section 10.1.

Answers to...

Figure 10.6 6.02×10^{23} atoms
Figure 10.7 The molar mass is the mass of Avogadro's number of molecules.

6.02×10^{23} atoms

Section 10.1 (continued)

Sample Problem 10.4

Answers

7. 1 mol P × 31.0 g P/1 mol P = 31.0 g P
   3 mol Cl × 35.5 g Cl/1 mol Cl = 106.5 g Cl
   31.0 g P + 106.5 g Cl = 137.5 g/mol
8. 1 mol Na × 23.0 g Na/1 mol Na = 23.0 g Na
   1 mol H × 1.0 g H/1 mol H = 1.0 g H
   1 mol C × 12.0 g C/1 mol C = 12.0 g C
   3 mol O × 16.0 g O/1 mol O = 48.0 g O
   23.0 g Na + 1.0 g H + 12.0 g C + 48.0 g O = 84.0 g

Practice Problems Plus L2

1. Determine the mass of one mole of each of the following compounds: CO_2 *(44.0 g)*, **SO_3** *(80.1 g)*, **Br_2** *(159.8 g)*, **H_2** *(2.0 g)*, **N_2** *(28.0 g)*, **NaOH** *(40.0 g)*, **$Al_2(SO_4)_3$** *(342.3 g)*, **and $Ba(NO_3)_2$** *(261.3 g)*.

2. What is the mass, in grams, of 1.72 mol $CaCl_2$? *(191 g)*

Math Handbook

For a math refresher and practice, direct students to significant figures, page R59.

Writing Activity

Answers might include that in the mid 1800s, the concept of the mole was based on the element oxygen.

If your class subscribes to the Interactive Textbook, use it to review key concepts in Section 10.1.

with ChemASAP

SAMPLE PROBLEM 10.4

Finding the Molar Mass of a Compound

The decomposition of hydrogen peroxide (H_2O_2) provides sufficient energy to launch a rocket. What is the molar mass of hydrogen peroxide?

1 Analyze *List the knowns and the unknown.*

Knowns
- molecular formula = H_2O_2
- 1 molar mass H = 1 mol H = 1.0 g H
- 1 molar mass O = 1 mol O = 16.0 g O

Unknown
- molar mass = ? g

One mol of hydrogen peroxide has 2 mol of hydrogen atoms and 2 mol of oxygen atoms. Convert moles of atoms to grams by using conversion factors (g/mol) based on the molar mass of each element. The sum of the masses of the elements is the molar mass.

2 Calculate *Solve for the unknown.*

Convert moles of hydrogen and oxygen to grams of hydrogen and oxygen. Then add the results.

$$2\ \cancel{\text{mol H}} \times \frac{1.0\text{ g H}}{1\ \cancel{\text{mol H}}} = 2.0\text{ g H}$$

$$2\ \cancel{\text{mol O}} \times \frac{16.0\text{ g O}}{1\ \cancel{\text{mol O}}} = 32.0\text{ g O}$$

$$\text{molar mass of } H_2O_2 = 34.0\text{ g}$$

3 Evaluate *Does the result make sense?*

The answer is the sum of two times the molar mass of hydrogen and oxygen. The answer is expressed to the tenth's place because the numbers being added are expressed to the tenth's place.

Practice Problems

7. Find the molar mass of PCl_3.
8. What is the mass of 1.00 mol of sodium hydrogen carbonate?

Math Handbook

For help with significant figures, go to page R59.

Interactive Textbook

Problem-Solving 10.7 Solve Problem 7 with the help of an interactive guided tutorial.

with ChemASAP

10.1 Section Assessment

9. **Key Concept** What are three ways to measure the amount of something?
10. **Key Concept** Describe the relationship between Avogadro's number and one mole of any substance.
11. **Key Concept** How is the atomic mass of an element related to the molar mass of the element?
12. **Key Concept** How can you calculate the mass of a mole of a compound?
13. How many moles is 1.50×10^{23} molecules NH_3?
14. How many atoms are in 1.75 mol $CHCl_3$?
15. What is the molar mass of $CaSO_4$?

Writing Activity

Report Research the history of Avogadro's number. What elements other than carbon have been used to define a mole? Write a report that summarizes your findings.

Assessment 10.1 Test yourself on the concepts in Section 10.1.

with ChemASAP

Section 10.1 Assessment

9. by count, by mass, and by volume
10. One mole of any substance contains Avogadro's number (6.02×10^{23}) of representative particles.
11. The molar mass of an element is its atomic mass expressed in grams.
12. Add together the masses, expressed in grams, of each element in one mole of the compound.
13. 2.49×10^{-1} mol NH_3
14. 5.27×10^{24} atoms
15. 136.2 g/mol

10.2 Mole–Mass and Mole–Volume Relationships

Connecting to Your World

Guess how many jelly beans are in the container and win a prize! You decide to enter the contest and you win. Was it just a lucky guess? Not exactly. You estimated the length and diameter of a jelly bean to find its approximate volume. Then you estimated the dimensions of the container to obtain its volume. You did the arithmetic and made your guess. In a similar way, chemists use the relationships between the mole and quantities such as mass, volume, and number of particles to solve chemistry problems. In this section you will find out how the mole and mass are related.

Guide for Reading

Key Concepts

- How do you convert the mass of a substance to the number of moles of the substance?
- What is the volume of a gas at STP?

Vocabulary

Avogadro's hypothesis
standard temperature and pressure (STP)
molar volume

Reading Strategy

Monitoring Your Understanding Before you read, preview the key concepts, the section heads, the boldfaced terms, and the visuals. List three things you expect to learn. After reading, state what you learned about each item you listed.

The Mole–Mass Relationship

In the previous section, you learned that the molar mass of any substance is the mass in grams of one mole of that substance. This definition applies to all substances—elements, molecular compounds, and ionic compounds. In some situations, however, the term molar mass may be unclear. For example, suppose you were asked what the molar mass of oxygen is? How you answer this question depends on what you assume to be the representative particle. If you assume the oxygen in the question is molecular oxygen (O_2), then the molar mass is 32.0 g (2 × 16.0 g). If you assume that the question is asking for the mass of a mole of oxygen atoms (O), then the answer is 16.0 g. You can avoid confusion such as this by using the formula of the substance, in this case, O_2 or O.

Suppose you need 3.00 mol of sodium chloride (NaCl) for a laboratory experiment. How can you measure this amount? It would be convenient to use a balance to measure the mass. But what mass in grams is 3.00 mol of NaCl? **Use the molar mass of an element or compound to convert between the mass of a substance and the moles of a substance.** The conversion factor for the calculation is based on the relationship: molar mass = 1 mol. Use the following equation to calculate the mass in grams of a given number of moles.

$$\text{mass (grams)} = \text{number of moles} \times \frac{\text{mass (grams)}}{1 \text{ mole}}$$

The molar mass of NaCl is 58.5 g/mol, so the mass of 3.00 mol NaCl is calculated in this way.

$$\text{mass of NaCl} = 3.00 \cancel{\text{mol}} \times \frac{58.5 \text{ g}}{1 \cancel{\text{mol}}} = 176 \text{ g}$$

When you measure 176 g of NaCl on a balance, you are measuring 3.00 moles of NaCl.

Section Resources

Print

- **Guided Reading and Study Workbook,** Section 10.2
- **Core Teaching Resources,** Section 10.2 Review
- **Laboratory Manual,** Lab 12
- **Small-Scale Chemistry Laboratory Manual,** Lab 13
- **Transparencies,** T108–T109

Technology

- **Virtual Chemistry Labs,** Lab 3
- **Interactive Textbook with ChemASAP,** Problem-Solving 10.16, 10.18, 10.20, 10.22; Simulation 10; Assessment 10.2

10.2

1 FOCUS

Objectives

10.2.1 Describe how to convert the mass of a substance to the number of moles of a substance, and moles to mass.

10.2.2 Identify the volume of a quantity of gas at STP.

Guide for Reading

Build Vocabulary 

Graphic Organizer Have students divide a piece of paper into three columns. In the first column, have students list what they know about each vocabulary term. In the second column, have them list what they want to know about each term. As they progress through the chapter, have them list in the third column what they learn about each term.

Reading Strategy 

Outline As students read this section, have them outline the concepts, using the section headings as the headings in the outline.

2 INSTRUCT

Connecting to Your World

Ask, **What method, other than estimating the volume of each individual candy, might you use to determine the number of jelly beans present?** *(Answers might include using the mass of one jelly bean and the total mass of all the jelly beans.)*

The Mole–Mass Relationship

Discuss 

Review the mathematical conversions of moles to number of particles and number of particles to moles. Stress that using dimensional analysis in problem solving allows students to perform these calculations without having to memorize the process. Emphasize the use of units when solving problems.

Sample Problem 10.5

Answers

16. 4.52×10^{-3} mol $C_{20}H_{42} \times 282.0$ g $C_{20}H_{42}$/1 mol $C_{20}H_{42} = 1.27$ g $C_{20}H_{42}$

17. 2.50 mol $Fe(OH)_2 \times 89.8$ g $Fe(OH)_2$/1 mol $Fe(OH)_2 = 225$ g $Fe(OH)_2$

Practice Problems Plus 

Calculate the mass in grams for 0.250 mol of each of the following compounds:

a. sucrose *(85.5 g)*
b. sodium chloride *(14.6 g)*
c. potassium permanganate *(39.5 g)*

Math Handbook

For a math refresher and practice, direct students to significant figures, page R59.

Relate 

In chemical manufacturing processes, reactants are purchased by mass, and products are sold by mass. However, each process is set up based on the knowledge of the ratio in which moles of reactants combine with each other to form moles of products. Ask, **If one mole of reactant produces one mole of product, how can you use the information on pp. 298 and 299 to find the mass of product if you know the mass of the reactant?** *(Change the mass of reactant to number of moles. Then, change that number of moles to mass of product.)* Tell students that they will learn more about conversions such as this in Chapter 12.

CLASS Activity

Problem Solving 

Students often can do one type of problem but have difficulty when two types of problems are interspersed. Have students do some practice problems for Sample Problems 10.5 and 10.6 individually. Then, have them do a mixture of the two types of problems.

Math Handbook

For help with significant figures go to page R59.

Interactive Textbook

Problem-Solving 10.16
Solve Problem 16 with the help of an interactive guided tutorial.
with ChemASAP

SAMPLE PROBLEM 10.5

Converting Moles to Mass

The aluminum satellite dishes in Figure 10.8 are resistant to corrosion because the aluminum reacts with oxygen in the air to form a coating of aluminum oxide (Al_2O_3). This tough, resistant coating prevents any further corrosion. What is the mass of 9.45 mol of aluminum oxide?

1 Analyze ***List the known and the unknown.***

Known
- number of moles = 9.45 mol Al_2O_3

Unknown
- mass = ? g Al_2O_3

The mass of the compound is calculated from the known number of moles of the compound. The desired conversion is moles ⟶ mass.

2 Calculate ***Solve for the unknown.***

Determine the molar mass of Al_2O_3: 1 mol Al_2O_3 = 102.0 g Al_2O_3

Multiply the given number of moles by the conversion factor relating moles of Al_2O_3 to grams of Al_2O_3.

$$\text{mass} = 9.45\ \cancel{\text{mol Al}_2\text{O}_3} \times \frac{102.0\ \text{g Al}_2\text{O}_3}{1\ \cancel{\text{mol Al}_2\text{O}_3}}$$
$$= 964\text{g Al}_2\text{O}_3$$

3 Evaluate ***Does the result make sense?***

The number of moles of Al_2O_3 is approximately 10, and each has a mass of approximately 100 g. The answer should be about 1000 g. The answer has been rounded to the correct number of significant figures.

Practice Problems

16. Find the mass, in grams, of 4.52×10^{-3} mol $C_{20}H_{42}$.

17. Calculate the mass, in grams, of 2.50 mol of iron(II) hydroxide.

Figure 10.8 These aluminum satellite dishes at the National Radio Astronomy Observatory near Soccoro, New Mexico are naturally protected from corrosion by the formation of a thin film of aluminum oxide (Al_2O_3).

In Sample Problem 10.5, you used a conversion factor based on the molar mass to convert moles to mass. Now suppose that in a laboratory experiment you obtain 10.0 g of sodium sulfate (Na_2SO_4). How many moles is this? You can calculate the number of moles using the same relationship you used in Sample Problem 10.5, 1 mol = molar mass, but this time the conversion factor is inverted. Use the following equation to convert your 10.0 g of Na_2SO_4 into moles.

$$\text{moles} = \text{mass (grams)} \times \frac{1 \text{ mole}}{\text{mass (grams)}}$$

The molar mass of Na_2SO_4 is 142.1 g/mol, so the number of moles of Na_2SO_4 is calculated this way.

$$\text{moles of } Na_2SO_4 = 10.0 \text{ g} \times \frac{1 \text{ mol}}{142.1 \text{ g}} = 7.04 \times 10^{-2} \text{ mol}$$

What conversion factor should you use to convert mass to moles?

Rust weakens an iron chain.

SAMPLE PROBLEM 10.6

Converting Mass to Moles

When iron is exposed to air, it corrodes to form red-brown rust. Rust is iron(III) oxide (Fe_2O_3). How many moles of iron(III) oxide are contained in 92.2 g of pure Fe_2O_3?

1 Analyze ***List the known and the unknown.***

Known
- mass = 92.2 g Fe_2O_3

Unknown
- number of moles = ? mol Fe_2O_3

The unknown number of moles of the compound is calculated from a known mass of a compound. The conversion is mass ⟶ moles.

2 Calculate ***Solve for the unknown.***

Determine the molar mass of Fe_2O_3: 1 mol = 159.6 g Fe_2O_3

Multiply the given mass by the conversion factor relating mass of Fe_2O_3 to moles of Fe_2O_3.

$$\text{moles} = 92.2 \text{ g } \cancel{Fe_2O_3} \times \frac{1 \text{ mol } Fe_2O_3}{159.6 \text{ g } \cancel{Fe_2O_3}}$$

$$= 0.578 \text{ mol } Fe_2O_3$$

3 Evaluate ***Does the result make sense?***

Because the given mass (about 90 g) is slightly larger than the mass of one-half mole of Fe_2O_3 (about 160 g), the answer should be slightly larger than one-half (0.5) mol.

Practice Problems

18. Find the number of moles in 3.70×10^{-1} g of boron.

19. Calculate the number of moles in 75.0 g of dinitrogen trioxide.

Math Handbook

For help with using a calculator, go to page R62.

Problem-Solving 10.18
Solve Problem 18 with the help of an interactive guided tutorial.
with ChemASAP

Sample Problem 10.6

Answers

18. 3.70×10^{-1} g B × 1 mol B/10.8 g B = 3.43×10^{-2} mol B

19. 75.0 g N_2O_3 × 1 mol N_2O_3/76.0 g N_2O_3 = 0.987 mol N_2O_3

Practice Problems Plus L2

Calculate the number of moles in 1.00×10^2 g of each of the following compounds.

a. sucrose *(0.292 mol)*
b. sodium chloride *(1.71 mol)*
c. potassium permanganate *(0.633 mol)*

Math Handbook

For a math refresher and practice, direct students to using a calculator, page R62.

Differentiated Instruction

Less Proficient Readers 

Provide students with several problems and have them analyze each problem for what information it contains, that is, help students identify the known(s) and the unknown(s). Also have them look for key words, such as *total, difference,* and *larger,* that show operations or relationships.

Answers to...

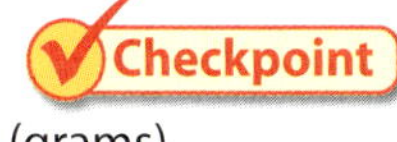

1 mole/mass (grams)

The Mole–Volume Relationship

Discuss L2

Ask, **What unit is used for the mass of a mole?** *(grams per mole, g/mol)* **What unit is used for the volume of a mole?** *(L per mole, L/mol)* Point out to students that, unlike solids and liquids, the molar volume of gases is predictable but is affected by temperature and pressure. Ask, **How does temperature affect the volume of a gas?** *(When temperature increases, volume increases. When temperature decreases, volume decreases.)* **How does pressure affect the volume of a gas?** *(When pressure increases, volume decreases. A decrease in pressure causes an increase in volume.)* Emphasize that when comparing the molar volumes of gases, it is necessary to have the gases at the same conditions of temperature and pressure.

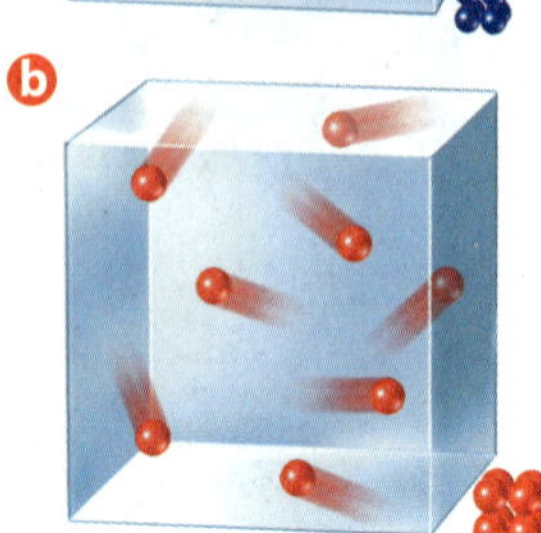

Figure 10.9 In each container, the volume occupied by the gas molecules is small compared with the container's volume, so the molecules are not tightly packed. ⓐ The molecules in this container are small. ⓑ This container can accommodate the same number of larger molecules.

The Mole–Volume Relationship

Look back at Figure 10.7. Notice that the volumes of one mole of different solid and liquid substances are not the same. For example, the volumes of one mole of glucose (blood sugar) and one mole of paradichlorobenzene (moth crystals) are much larger than the volume of one mole of water. What about the volumes of gases? Unlike liquids and solids, the volumes of moles of gases, measured under the same physical conditions, are much more predictable. Why should this be?

In 1811, Amedeo Avogadro proposed a groundbreaking explanation. **Avogadro's hypothesis** states that equal volumes of gases at the same temperature and pressure contain equal numbers of particles. The particles that make up different gases are not the same size. But the particles in all gases are so far apart that a collection of relatively large particles does not require much more space than the same number of relatively small particles. Whether the particles are large or small, large expanses of space exist between individual particles of gas, as shown in Figure 10.9.

If you buy a party balloon filled with helium and take it home on a cold day, you might notice that the balloon shrinks while it is outside. The volume of a gas varies with a change in temperature. The volume of a gas also varies with a change in pressure. In Figure 10.10, notice the changes in an empty water bottle when it is in the cabin of an airplane while in flight and after the plane has landed. The trapped air occupies the full volume of the bottle in the cabin where the air pressure is lower than it is on the ground. The increase in pressure when the plane lands causes the volume of the air in the bottle to decrease. Because of these variations due to temperature and pressure, the volume of a gas is usually measured at a standard temperature and pressure. **Standard temperature and pressure (STP)** means a temperature of 0°C and a pressure of 101.3 kPa, or 1 atmosphere (atm). **At STP, 1 mol or 6.02×10^{23} representative particles, of any gas occupies a volume of 22.4 L.** Figure 10.11 gives you an idea of the size of 22.4 L. The quantity, 22.4 L, is called the **molar volume** of a gas.

Checkpoint ***What is meant by standard temperature and pressure?***

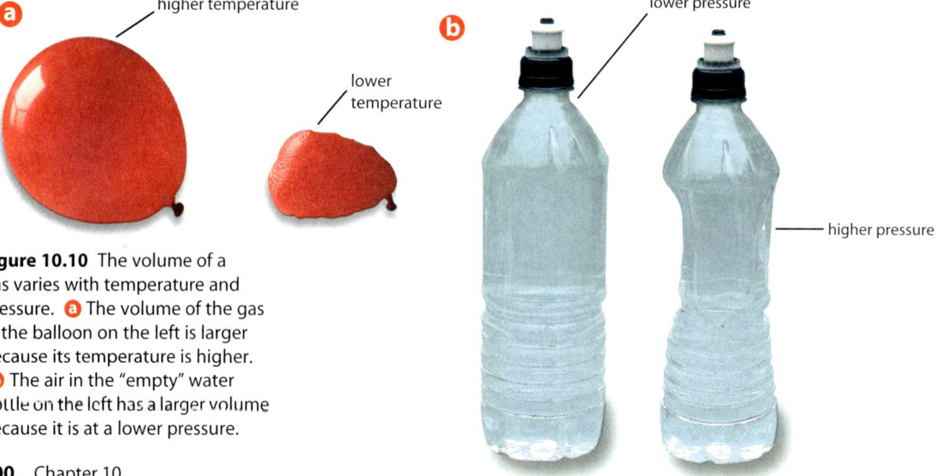

Figure 10.10 The volume of a gas varies with temperature and pressure. ⓐ The volume of the gas in the balloon on the left is larger because its temperature is higher. ⓑ The air in the "empty" water bottle on the left has a larger volume because it is at a lower pressure.

Differentiated Instruction

Gifted and Talented 

Have students apply their problem-solving skills to this question: Students heated a mixture of potassium chlorate and manganese dioxide, producing 0.377 L of oxygen gas at STP. **What was the mass of the gas collected?** *(0.539 g)*

Calculating Volume at STP The molar volume is used to convert a known number of moles of gas to the volume of the gas at STP. The relationship 22.4 L = 1 mol at STP provides the conversion factor.

$$\text{volume of gas} = \text{moles of gas} \times \frac{22.4\ \text{L}}{1\ \text{mol}}$$

Suppose you have 0.375 mol of oxygen gas and want to know what volume the gas will occupy at STP.

$$\text{volume of } O_2 = 0.375\ \cancel{\text{mol}} \times \frac{22.4\ \text{L}}{1\ \cancel{\text{mol}}} = 8.40\ \text{L}$$

Figure 10.11 This box, with a volume of 22.4 L, holds one mole of gas at STP.

SAMPLE PROBLEM 10.7

Calculating the Volume of a Gas at STP

Sulfur dioxide (SO_2) is a gas produced by burning coal. It is an air pollutant and one of the causes of acid rain. Determine the volume, in liters, of 0.60 mol SO_2 gas at STP.

1 Analyze *List the knowns and the unknown.*

Knowns
- moles = 0.60 mol SO_2
- 1 mol SO_2 = 22.4 L SO_2

Unknown
- volume = ? L SO_2

Use the relationship 1 mol SO_2 = 22.4 L SO_2 (at STP) to write the conversion factor needed to convert moles to volume.

The conversion factor is $\frac{22.4\ \text{L } SO_2}{1\ \text{mol } SO_2}$.

2 Calculate *Solve for the unknown.*

$$\text{volume} = 0.60\ \cancel{\text{mol } SO_2} \times \frac{22.4\ \text{L } SO_2}{1\ \cancel{\text{mol } SO_2}} = 13\ \text{L } SO_2$$

3 Evaluate *Does the result make sense?*

Because 1 mol of any gas at STP has a volume of 22.4 L, 0.60 mol should have a volume slightly larger than one half of a mole or 11.2 L. The answer should have two significant figures.

Practice Problems

20. What is the volume of these gases at STP?
    a. 3.20×10^{-3} mol CO_2
    b. 3.70 mol N_2

21. At STP, what volume do these gases occupy?
    a. 1.25 mol He
    b. 0.335 mol C_2H_6

Math Handbook

For help with dimensional analysis, go to page R66.

Problem-Solving 10.20
Solve Problem 20 with the help of an interactive guided tutorial.

with ChemASAP

The opposite conversion, from the volume of a gas at STP to the number of moles of gas, uses the same relationship: 22.4 L = 1 mol at STP. Suppose, in an experiment, you collect 0.200 liter of hydrogen gas at STP. You can calculate the number of moles of hydrogen in this way.

$$\text{moles} = 0.200\ \cancel{\text{L } H_2} \times \frac{1\ \text{mol } H_2}{22.4\ \cancel{\text{L } H_2}} = 8.93 \times 10^{-3}\ \text{mol } H_2$$

Differentiated Instruction

Gifted and Talented L3

Have students design a way to measure the volume of CO_2 produced in the Teacher Demo on this page. Methods might include measuring how much water the filled bag displaces.

Sample Problem 10.7

Answers

20. a. 3.20×10^{-3} mol $CO_2 \times 22.4$ L CO_2/1 mol $CO_2 = 7.17 \times 10^{-2}$ L CO_2
b. 3.70 mol $N_2 \times 22.4$ L N_2/1 mol N_2 = 82.9 L N_2

21. a. 1.25 mol He × 22.4 L He/1 mol He = 28.0 L He
b. 0.335 mol $C_2H_6 \times 22.4$ L C_2H_6/1 mol C_2H_6 = 7.50 L C_2H_6

Practice Problems Plus

At STP, what volume is occupied by each of the following gases?

a. 1.34 mol SO_2 *(30.0 L SO_2)*

b. 2.45×10^{-3} mol H_2S *(5.49×10^{-2} L H_2S)*

c. 6.7 mol H_2 *(1.5×10^2 L H_2)*

Math Handbook

For a math refresher and practice, direct students to dimensional analysis, page R66.

TEACHER Demo

Molar Volume

Purpose Students observe an approximation of molar volume.

Materials Dry ice, towel, hammer, large plastic bag, duct tape, tongs, beaker, balance

Procedure Wrap the dry ice in a towel and hit it with the hammer until it is in small pieces. Place 44 g (1 mol CO_2) of the small pieces in a beaker. Expel any air from the plastic bag, and tape the opening of the bag securely over the top of the beaker. As the dry ice sublimes, the bag will inflate.

Safety Wear goggles while crushing the dry ice, and do not allow dry ice to contact skin. Use tongs to handle the dry ice.

Expected Outcomes The volume of gas produced will not equal 22.4 L because conditions are not standard. However, the volume will be close to this value.

Answers to...

Checkpoint 0°C and 101.3 KPa or 1 atmosphere (atm)

Section 10.2 (continued)

Discuss 

Review the concept of density as a ratio of mass to volume. Discuss the units that go with density. *(g/mL, g/cm³, or g/L)* Ask, **If you had a mole of gas at STP, how could you calculate the density?** *(molar mass/22.4 L = density)*

Sample Problem 10.8

Answers

22. 3.58 g/1 L × 22.4 L/1 mol = 80.2 g/mol

23. 83.8 g/1 mol × 1 mol/22.4 L = 3.74 g/L

Practice Problems Plus 

1. A gas has a density of 0.902 g/L. What is the molar mass of this gas? *(20.2 g/mol)*

2. What is the density of oxygen gas at STP? *(1.43 g/L)*

Math Handbook

For a math refresher and practice, direct students to significant figures, page R59.

Discuss 

Emphasize to students that if the number of moles is known, the mass of the substance or the volume of a gas can be calculated. This concept will continue to be essential as students study mass–mass and other stoichiometric relationships in Chapter 12.

Calculating Molar Mass from Density A gas-filled balloon will either sink or float in the air depending on whether the density of the balloon's gas is greater or less than the density of the surrounding air. Different gases have different densities. Usually the density of a gas is measured in grams per liter (g/L) and at a specific temperature. The density of a gas at STP and the molar volume at STP (22.4 L/mol) can be used to calculate the molar mass of the gas.

molar mass = density at STP × molar volume at STP

$$\frac{\text{grams}}{\text{mole}} = \frac{\text{grams}}{\text{L}} \times \frac{22.4\ \text{L}}{1\ \text{mole}}$$

Checkpoint ***How is the density of a gas usually measured?***

Math Handbook

For help with significant figures, go to page R59.

SAMPLE PROBLEM 10.8

Calculating the Molar Mass of a Gas at STP

The density of a gaseous compound containing carbon and oxygen is found to be 1.964 g/L at STP. What is the molar mass of the compound?

1 Analyze ***List the knowns and the unknown.***

Knowns
- density = 1.964 g/L
- 1 mol (gas at STP) = 22.4 L

Unknown
- molar mass = ? g/mol

The conversion factor needed to convert density to molar mass is $\frac{22.4\ \text{L}}{1\ \text{mol}}$.

$$\text{molar mass} = \frac{\text{grams}}{\text{L}} \times \frac{22.4\ \text{L}}{1\ \text{mol}}$$

2 Calculate ***Solve for the unknown.***

$$\text{molar mass} = \frac{1.964\ \text{g}}{1\ \cancel{\text{L}}} \times \frac{22.4\ \cancel{\text{L}}}{1\ \text{mol}}$$
$$= 44.0\ \text{g/mol}$$

3 Evaluate ***Does the result make sense?***

The ratio of the calculated mass (44.0 g) to the volume (22.4 L) is about 2, which is close to the known density. The answer should have three significant figures.

Practice Problems

22. A gaseous compound composed of sulfur and oxygen, which is linked to the formation of acid rain, has a density of 3.58 g/L at STP. What is the molar mass of this gas?

23. What is the density of krypton gas at STP?

Problem-Solving 10.22
Solve Problem 22 with the help of an interactive guided tutorial.
with ChemASAP

Differentiated Instruction

Special Needs 

To help students learn to solve conversion problems, have them write each of the three pairs of conversion factors in Figure 10.12 on the two sides of a 3 × 5 card. For example, have them write

$$\frac{\text{molar mass (grams)}}{1.00\text{mol}}$$

on one side of a card and

$$\frac{1.00\text{mol}}{\text{molar mass (grams)}}$$

on the other. Also have them make a card with each of the units of the known and unknown quantities: *grams, moles, liters,* and *representative particles*. Students can first set up a problem using the cards in a way that will produce the desired unit for the answer and then fill in the actual numbers to solve the problem.

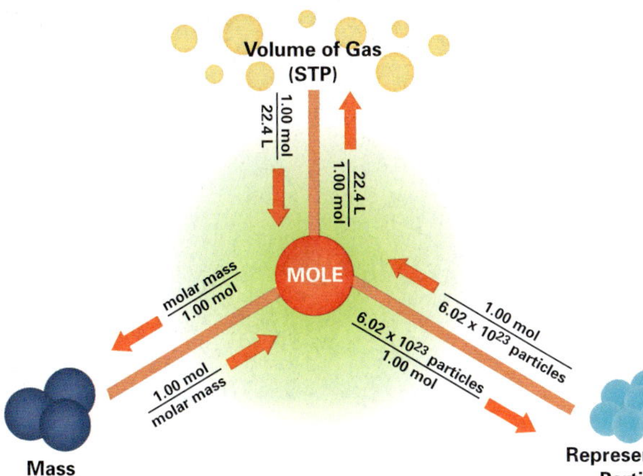

Figure 10.12 The map shows the conversion factors needed to convert among volume, mass, and number of particles.
Interpreting Diagrams ***How many conversion factors are needed to convert from the mass of a gas to the volume of a gas at STP?***

The Mole Road Map

You have now examined a mole in terms of particles, mass, and volume of gases at STP. Figure 10.12 summarizes these relationships and illustrates the importance of the mole. The mole is at the center of your chemical calculations. To convert from one unit to another, you must use the mole as an intermediate step. The form of the conversion factor depends on what you know and what you want to calculate.

Simulation 10 Use the mole road map to convert among mass, volume, and number of representative particles.

with **ChemASAP**

10.2 Section Assessment

24. **Key Concept** Describe how to convert between the mass and the number of moles of a substance.
25. **Key Concept** What is the volume of one mole of any gas at STP?
26. How many grams are in 5.66 mol of $CaCO_3$?
27. Find the number of moles in 508 g of ethanol (C_2H_6O).
28. Calculate the volume, in liters, of 1.50 mol Cl_2 at STP.
29. The density of an elemental gas is 1.7824 g/L at STP. What is the molar mass of the element?
30. The densities of gases A, B, and C at STP are 1.25 g/L, 2.86 g/L, and 0.714 g/L, respectively. Calculate the molar mass of each substance. Identify each substance as ammonia (NH_3), sulfur dioxide (SO_2), chlorine (Cl_2), nitrogen (N_2), or methane (CH_4).
31. Three balloons filled with three different gaseous compounds each have a volume of 22.4 L at STP. Would these balloons have the same mass or contain the same number of molecules? Explain.

Connecting Concepts

Density In Chapter 3 you learned that the densities of solids and liquids are measured in g/cm³ but the densities of gases are measured in g/L. Draw atomic diagrams of a solid and a gas that show why the two different units are practical.

Assessment 10.2 Test yourself on the concepts in Section 10.2.

with **ChemASAP**

The Mole Road Map

Use Visuals 

Figure 10.12 Have students study the figure. Guide them through examples of the various mole conversions. For example, start with 50.0 g of a compound or element and convert it to moles, and then to particles. Then, start with a certain volume of a gas and convert it to mass or particles.

3 ASSESS

Evaluate Understanding 

Have students work problems in which they use molar mass and molar volume to calculate the densities of gases. Have volunteers come to the board to show their calculations. Have the class determine whether the calculations are correct.

Reteach 

Review the concept of density as a ratio of mass to volume. Ask, **If you know the molar volume of a gas, how could density help you determine the molar mass?** *(Molar mass is the product of density and molar volume.)*

Connecting Concepts

Student answers should show diagrams of particles that are close enough to touch each other for liquids and solids. The diagrams of gases should show the particles far apart.

If your class subscribes to the Interactive Textbook, use it to review key concepts in Section 10.2.

with **ChemASAP**

Section 10.2 Assessment

24. To convert mass to moles, multiply the given mass by 1 mol/molar mass. To convert moles to mass, multiply the given number of moles by molar mass/1 mol.
25. 22.4 L
26. 567 g $CaCO_3$
27. 11.0 mol C_2H_6O
28. 33.6 L Cl_2
29. 39.9 g/mol
30. gas A: 28.0 g, nitrogen
gas B: 64.1 g, sulfur dioxide
gas C: 16.0 g, methane
31. The balloons have the same number of molecules. Each balloon is filled with one mole of gas, and one mole of any gas has the same number of molecules. The masses of the balloons will differ.

Answers to...

Figure 10.12 2

 The density of a gas is measured in grams per liter (g/L).

Counting by Measuring Mass L2

Objective After completing this activity, students will be able to:
- measure masses of chemicals and convert their data to moles and atoms.
- explore the quantitative chemical compositions of common substances

Skills Focus measuring, calculating

Prep Time 20 minutes
Class Time 40 minutes

Teaching Tips
- Explore with students ways of finding the mass of liquid and solid samples so that the container does not interfere with the measurement.
- If time allows, have students repeat the procedure and average the data.

Expected Outcome
See Data Table.

Analyze
Sample calculations using sample data:

1. 5.09 g NaCl × 1 mol NaCl/58.5 g NaCl = 0.0870 mol NaCl
2. See data table for answers.
3. See data table for answers.
4. 0.478 mol H × 6.02 × 10^{23} atoms/1 mol H = 2.88 × 10^{23} atoms H
5. water
6. water

You're the Chemist!
Sample answers are provided.

1. Find the mass of 100 drops of water, and then calculate the mass in grams per drop.
2. Find the mass of a piece of chalk. Write your name and find the mass of the chalk again. Convert the mass difference to moles and atoms.

For Enrichment 
Have students use their results from the lab to calculate the volume of one mole of each substance tested. Then, have them use a balance to measure one mole of each substance and a graduated cylinder to find its volume. Have them compare the calculated and experimental values and discuss any discrepancies.

Small-Scale LAB

Counting by Measuring Mass

Purpose
To determine the mass of several samples of chemical compounds and use the data to count atoms.

Materials
- chemicals shown in the table
- plastic spoon
- weighing paper
- watchglass or small beaker
- balance
- paper
- pencil
- ruler

Procedure 
Measure the mass of one level teaspoon of sodium chloride (NaCl), water (H_2O), and calcium carbonate ($CaCO_3$). Make a table similar to the one below.

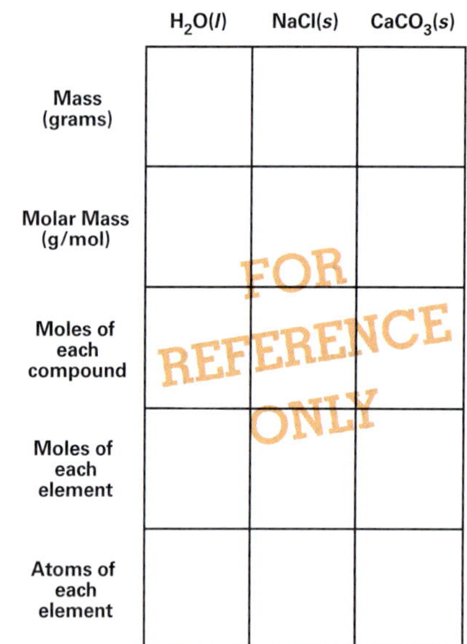

	$H_2O(l)$	$NaCl(s)$	$CaCO_3(s)$
Mass (grams)			
Molar Mass (g/mol)			
Moles of each compound			
Moles of each element			
Atoms of each element			

Analyze
Use your data to complete the following steps. Record your answers in or below your data table.

1. Calculate the moles of NaCl contained in one level teaspoon.

$$\text{moles of NaCl} = \text{g NaCl} \times \frac{1\ \text{mol NaCl}}{58.5\ \text{g}}$$

2. Repeat Step 1 for the remaining compounds. Use the periodic table to calculate the molar mass of water and calcium carbonate.
3. Calculate the number of moles of each element present in the teaspoon-sized sample of H_2O.

$$\text{moles of H} = \text{mol } H_2O \times \frac{2\ \text{mol H}}{1\ \text{mol } H_2O}$$

Repeat for the other compounds in your table.

4. Calculate the number of atoms of each element present in the teaspoon-sized sample of H_2O.

$$\text{atoms of H} = \text{mol H} \times \frac{6.02 \times 10^{23}\ \text{atoms H}}{1\ \text{mol } H_2O}$$

Repeat for the other compounds in your table.

5. Which of the three teaspoon-sized samples contains the greatest number of moles?
6. Which of the three compounds contains the most atoms?

You're the Chemist!
The following small-scale activities allow you to develop your own procedures and analyze the results.

1. **Design It!** Can you count by measuring volume? Design and carry out an experiment to do it!
2. **Design It!** Design an experiment that will determine the number of atoms of calcium, carbon, and oxygen it takes to write your name on the chalkboard with a piece of chalk. Assume chalk is 100 percent calcium carbonate, $CaCO_3$.

Sample Data

	$H_2O(l)$	$NaCl(s)$	$CaCO_3(s)$
Mass (g)	4.30	5.09	9.68
Molar mass (g/mol)	18.0	58.5	100.1
Moles of compound	0.239	0.0870	0.0967
Moles of elements	0.239 O 0.478 H	0.0870 Na 0.0870 Cl	0.0967 Ca 0.0967 C 0.290 O
Atoms of elements	1.44×10^{23} O 2.88×10^{23} H	5.24×10^{22} O 5.24×10^{22} Cl	5.82×10^{22} Ca 5.82×10^{22} C 1.75×10^{22} O

10.3 Percent Composition and Chemical Formulas

Connecting to Your World Is your shirt made of 100 percent cotton or wool, or is the fabric a combination of two or more fibers? A tag sewed into the seam of the shirt usually tells you what fibers were used to make the cloth and the percent of each. It helps to know the percents of the components in the shirt because they affect how warm it is, whether it will need to be ironed, and how it should be cleaned. In this section you will learn how the percents of the elements in a compound are important in chemistry.

Guide for Reading

Key Concepts

- How do you calculate the percent by mass of an element in a compound?
- What does the empirical formula of a compound show?
- How does the molecular formula of a compound compare with the empirical formula?

Vocabulary

percent composition

empirical formula

Reading Strategy

Comparing and Contrasting When you compare and contrast things, you examine how they are alike and different. As you read, list the similarities and differences between empirical and molecular formulas.

The Percent Composition of a Compound

If you have had experience with lawn care, you know that the relative amount, or the percent, of each nutrient in fertilizer is important. In spring, you may use a fertilizer that has a relatively high percent of nitrogen to "green" the grass. In fall, you may want to use a fertilizer with a higher percent of potassium to strengthen the root system. Knowing the relative amounts of the components of a mixture or compound is often useful.

The relative amounts of the elements in a compound are expressed as the **percent composition** or the percent by mass of each element in the compound. The percent composition of a compound consists of a percent value for each different element in the compound. As you can see in Figure 10.13, the percent composition of K_2CrO_4 is K = 40.3%, Cr = 26.8%, and O = 32.9%. These percents must total 100% (40.3% + 26.8% + 32.9% = 100%). **The percent by mass of an element in a compound is the number of grams of the element divided by the mass in grams of the compound, multiplied by 100%.**

$$\% \text{ mass of element} = \frac{\text{mass of element}}{\text{mass of compound}} \times 100\%$$

Percent Composition from Mass Data Imagine you are a chemist who has just finished the synthesis of a new compound. You have purified your product and stored the crystalline solid in a vial. Now you must verify the composition of your new compound and determine its molecular formula. You use analytical procedures to determine the relative masses of each element in the compound and calculate the percent composition.

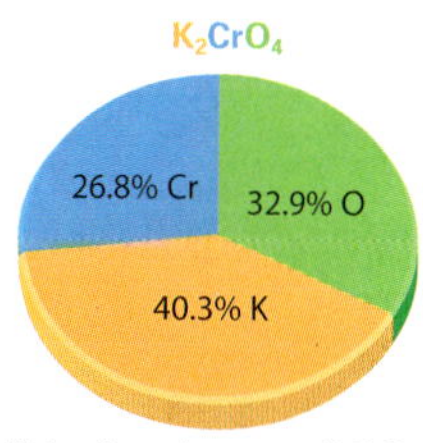

Potassium chromate, K_2CrO_4

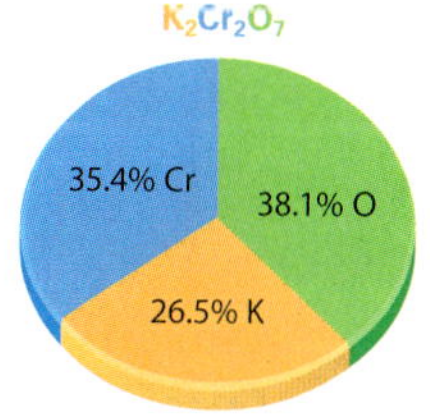

Potassium dichromate, $K_2Cr_2O_7$

Figure 10.13 Potassium chromate (K_2CrO_4) is composed of 40.3% potassium, 26.8% chromium, and 32.9% oxygen. **Interpreting Diagrams** *How does this percent composition differ from the percent composition of potassium dichromate ($K_2Cr_2O_7$), a compound composed of the same three elements?*

Section Resources

Print

- **Guided Reading and Study Workbook,** Section 10.3
- **Core Teaching Resources,** Section 10.3 Review, Interpreting Graphics
- **Laboratory Manual,** Lab 13
- **Transparencies,** T110–T112

Technology

- **Interactive Textbook with ChemASAP,** Problem-Solving 10.33, 10.34, 10.36, 10.38; Assessment 10.3
- **Go Online,** Section 10.3

10.3

1 FOCUS

Objectives

10.3.1 Describe how to calculate the percent by mass of an element in a compound.

10.3.2 Interpret an empirical formula

10.3.3 Distinguish between empirical and molecular formulas.

Guide for Reading

Build Vocabulary 

Paraphrase Have students read the definition of *percent composition* on this page. Then, have them define the term according to the definition in the dictionary. Their definitions should indicate that *percent composition* refers to the relative size of parts that make up 100 percent of something.

Reading Strategy 

Relate Text and Visuals As they read the chapter, students should examine each visual as it is referenced in the text. Have them read each caption and answer any question.

2 INSTRUCT

Connecting to Your World

Ask, **What should be the total of the percents listed on the label?** *(100%)* **If the shirt were 25% nylon, what percent would be cotton?** *(75%)*

The Percent Composition of a Compound

Use Visuals L1

Figure 10.13 Have students study the figure and read the text on percent composition. Point out that the three numbers in each circle graph add up to a total of 100%. Ask, **Which compound is a better source of potassium?** *(K_2CrO_4)*

Answers to...

Figure 10.13 The percent composition of $K_2Cr_2O_7$ is 26.5% K, 35.4% Cr, and 38.1% O.

Section 10.3 (continued)

Sample Problem 10.9

Answers

32. Mass of compound = 9.03 g + 3.48 g = 12.51 g; 9.03 g Mg/12.51 g compound × 100% = 72.2% Mg; 3.48 g N/12.51 g compound × 100% = 27.8% N

33. mass of O = 14.2 g − 13.2 g = 1.0 g O; 1.0 g O/14.2 g × 100% = 7.0% O; 13.2 g Hg/14.2 g × 100% = 93.0% Hg

Practice Problems Plus L2

1. What is the percent composition of the compound formed when 2.70 g of aluminum combine with oxygen to form 5.10 g of aluminum oxide? *(52.9% Al, 47.1% O)*

2. Interactive Textbook with ChemASAP contains the following problem: **Calculate the percent composition when 13.3 g Fe combine completely with 5.7 g O.** *(70% Fe, 30% O)*

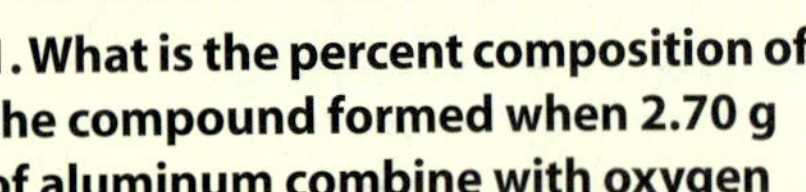

Math Handbook

For a math refresher and practice, direct students to percents, page R72.

Math Handbook

For help with percents go to page R72.

SAMPLE PROBLEM 10.9

Calculating Percent Composition from Mass Data

When a 13.60-g sample of a compound containing only magnesium and oxygen is decomposed, 5.40 g of oxygen is obtained. What is the percent composition of this compound?

1 Analyze ***List the knowns and the unknowns.***

Knowns
- mass of compound = 13.60 g
- mass of oxygen = 5.40 g O
- mass of magnesium = 13.60 g − 5.40 g = 8.20 g Mg

Unknowns
- percent Mg = ? % Mg
- percent O = ? % O

The percent by mass of an element in a compound is the mass of that element divided by the mass of the compound multiplied by 100%.

2 Calculate ***Solve for the unknown.***

$$\%\ \text{Mg} = \frac{\text{mass of Mg}}{\text{mass of compound}} \times 100\% = \frac{8.20\ \text{g}}{13.60\ \text{g}} \times 100\%$$
$$= 60.3\%$$
$$\%\ \text{O} = \frac{\text{mass of O}}{\text{mass of compound}} \times 100\% = \frac{5.40\ \text{g}}{13.60\ \text{g}} \times 100\%$$
$$= 39.7\%$$

3 Evaluate ***Does the result make sense?***

The percents of the elements add up to 100%:
60.3% + 39.7% = 100%.

Practice Problems

32. A compound is formed when 9.03 g Mg combines completely with 3.48 g N. What is the percent composition of this compound?

33. When a 14.2-g sample of mercury(II) oxide is decomposed into its elements by heating, 13.2 g Hg is obtained. What is the percent composition of the compound?

Interactive Textbook

Problem-Solving 10.33 Solve Problem 33 with the help of an interactive guided tutorial.

with ChemASAP

Figure 10.14 The percent composition of water is always the same regardless of the volume of the water sample. A sample of water is always 11.1% H and 88.9% O by mass.

Facts and Figures

Parts per Million and Parts per Billion

Percents are used to show relative parts of mixtures as well as the composition of a compound. But when an extremely small amount of a substance is present in a large amount of another substance, it might not be practical to use percents (parts per one hundred) to show the makeup of the mixture. Instead, concentrations of extremely dilute solutions are sometimes measured in units of parts per million (ppm) or parts per billion (ppb). For example, the composition of a mixture that consists of 1 gram of a substance per 10^6 grams of water (or 1 milligram of substance per liter of water) can be expressed as 1 ppm.

Percent Composition from the Chemical Formula You can also calculate the percent composition of a compound if you know only its chemical formula. The subscripts in the formula of the compound are used to calculate the mass of each element in a mole of that compound. The sum of these masses is the molar mass. Using the individual masses of the elements and the molar mass you can calculate the percent by mass of each element in one mole of the compound. Divide the mass of each element by the molar mass and multiply the result by 100%.

$$\% \text{ mass} = \frac{\text{mass of element in 1 mol compound}}{\text{molar mass of compound}} \times 100\%$$

The percent composition of a compound is always the same, as Figure 10.14 on the preceding page indicates.

How can you determine the percent by mass of an element in a compound if you know only the compound's formula?

Go Online NSTA SCILINKS

For: Links on Percent Composition
Visit: www.SciLinks.org
Web Code: cdn-1103

SAMPLE PROBLEM 10.10

Calculating Percent Composition from a Formula

Propane (C_3H_8), the fuel commonly used in gas grills, is one of the compounds obtained from petroleum. Calculate the percent composition of propane.

1 Analyze ***List the knowns and the unknowns.***

Knowns
- mass of C in 1 mol C_3H_8 = 36.0 g
- mass of H in 1 mol C_3H_8 = 8.0 g
- molar mass of C_3H_8 = 44.0 g/mol

Unknowns
- percent C = ? % C
- percent H = ? % H

Calculate the percent by mass of each element by dividing the mass of that element in one mole of the compound by the molar mass of the compound and multiplying by 100%.

2 Calculate ***Solve for the unknowns.***

$$\% \text{ C} = \frac{\text{mass of C}}{\text{mass of propane}} \times 100\% = \frac{36.0 \text{ g}}{44.0 \text{ g}} \times 100\% = 81.8\%$$

$$\% \text{ H} = \frac{\text{mass of H}}{\text{mass of propane}} \times 100\% = \frac{8.0 \text{ g}}{44.0 \text{ g}} \times 100\% = 18\%$$

3 Evaluate ***Does the result make sense?***
The percents of the elements add up to 100% when the answers are expressed to two significant figures.

Practice Problems

34. Calculate the percent composition of these compounds.
a. ethane (C_2H_6)
b. sodium hydrogen sulfate ($NaHSO_4$)

35. Calculate the percent nitrogen in these common fertilizers.
a. NH_3
b. NH_4NO_3

For help with significant figures go to page R59.

Interactive Textbook

Problem-Solving 10.35
Solve Problem 35 with the help of an interactive guided tutorial.
with **ChemASAP**

Sample Problem 10.10

Answers

34. a. 24.0 g C/30.0 g × 100% = 80.0% C
6.00 g H/30.0 g × 100% = 20.0% H
b. 23.0 g Na/120.1 g × 100% = 19.2% Na
1.0 g H/120.1 g × 100% = 0.83% H
32.1 g S/120.1 g × 100% = 26.7% S
64.0 g O/120.1 g × 100% = 53.3% O

35. a. 14.0 g N/17.0 g × 100% = 82.4% N
b. 28.0 g N/80.0 g × 100% = 35.0% N

Practice Problems Plus 

1. Determine the percent composition of the following oxides:
a. Fe_2O_3 *(69.9% Fe, 30.1%O)*
b. HgO *(92.6% Hg, 7.39%O)*
c. Ag_2O *(93.1% Ag, 6.90 %O)*
d. Na_2O *(74.2% Na, 25.8%O)*

2. Calculate the grams of oxygen in 90.0 g of Cl_2O. *(16.6 g)*

For a math refresher and practice, direct students to significant figures, page R59.

Go Online NSTA SCILINKS

Download a worksheet on **Percent Composition** for students to complete, and find additional teacher support from NSTA SciLinks.

Differentiated Instruction

Gifted and Talented 

Have students research the formulas of the three different oxides of iron. Ask, **Which of the oxides contains a higher percent of iron?** *(Of FeO (77.7% Fe), Fe_2O_3 (69.9% Fe), and Fe_3O_4 (72.3% Fe), FeO has the highest percent of iron.)*

Answers to...

Divide the mass of the element in one mole of the compound by the molar mass and multiply by 100%.

Percent Composition L2

Objective After completing this activity, students will be able to:
- determine the percent of water in a hydrate.

Skills Focus observing, calculating

 **Prep Time** 20 minutes

Class Time 30 minutes

Safety Students should wear safety goggles and tie back loose hair. Caution students that while heating test tubes, they should not aim the opening of the tube toward anyone. Tell them to move the test tube in the flame and not to heat one spot excessively. **CAUTION!** Be sure that students allow the tubes to cool completely before they touch them. Hot glass looks exactly like cold glass!

Teaching Tips For best results, students should do a second heating and cooling of each sample to determine whether all of the water has been driven off.

Expected Outcome See data table at the bottom of the page.

Think About It

1.–3. See data table.

4. The hydrated salt of sodium sulfate lost the greatest percent. The hydrated salt of calcium chloride lost the smallest percent.

For Enrichment 

Have students design and conduct a similar experiment to determine the percent of oxygen in potassium chlorate. Tell students that when potassium chlorate is heated, potassium chloride and oxygen are produced, $2KClO_3 \rightarrow 2KCl + 3O_2$. For classroom safety, no more than 5 g of potassium chlorate should be used. Results should show that potassium chlorate is approximately 39% oxygen.

Quick LAB

Percent Composition

Purpose

To measure the percent of water in a series of crystalline compounds called hydrates.

Materials
- centigram balance
- Bunsen burner
- 3 medium-sized test tubes
- test tube holder
- test tube rack
- spatula
- hydrated salts of copper(II) sulfate, calcium chloride, and sodium sulfate

Procedure
1. Label each test tube with the name of a salt. Measure and record the masses.
2. Add 2–3 g of salt (a good-sized spatula full) to the appropriately labeled test tube. Measure and record the mass of each test tube and salt.
3. Hold one of the tubes at a 45° angle and gently heat its contents over the burner, slowly passing it in and out of the flame. Note any change in the appearance of the solid salt.
4. As moisture begins to condense in the upper part of the test tube, gently heat the entire length of the tube. Continue heating until all of the moisture is driven from the tube. This may take 2–3 minutes. Repeat Steps 3 and 4 for the other two tubes.
5. Allow each tube to cool. Then measure and record the mass of each test tube and the heated salt.

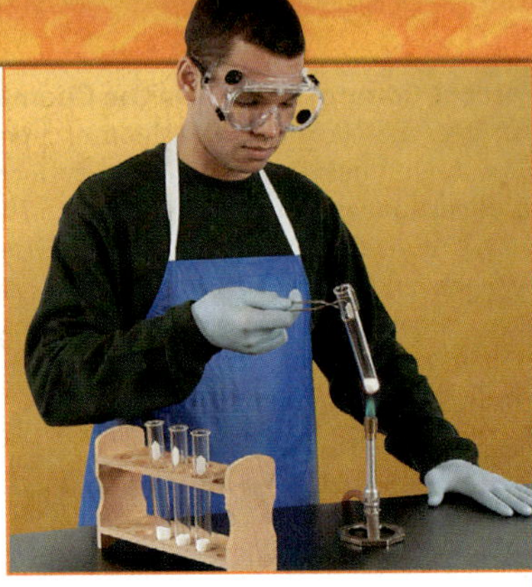

Analyze and Conclude
1. Set up a data table so that you can subtract the mass of the empty tube from the mass of the salt and the test tube, both before and after heating.
2. Calculate the difference between the mass of each salt before and after heating. This difference represents the amount of water lost by the hydrate on heating.
3. Calculate the percent by mass of water lost by each compound.
4. Which compound lost the greatest percent by mass of water? The smallest?

Percent Composition as a Conversion Factor You can use percent composition to calculate the number of grams of any element in a specific mass of a compound. To do this, multiply the mass of the compound by a conversion factor based on the percent composition of the element in the compound. Suppose you want to know how much carbon and hydrogen are contained in 82.0 g of propane. In Sample Problem 10.10, you found that propane is 81.8% carbon and 18% hydrogen. That means that in a 100-g sample of propane, you would have 81.8 g of carbon and 18 g of hydrogen. You can use the ratio 81.8 g C/100 g C_3H_8 to calculate the mass of carbon contained in 82.0 g of propane (C_3H_8).

$$82.0\ \text{g}\ \cancel{C_3H_8} \times \frac{81.8\ \text{g C}}{100\ \text{g}\ \cancel{C_3H_8}} = 67.1\ \text{g C}$$

Using the ratio 18 g H/100 g C_3H_8, you can calculate the mass of hydrogen.

$$82.0\ \text{g}\ \cancel{C_3H_8} \times \frac{18\ \text{g H}}{100\ \text{g}\ \cancel{C_3H_8}} = 15\ \text{g H}$$

The sum of the two masses equals 82 g, the sample size, to two significant figures (67.1 g C + 15 g H = 82 g C_3H_8).

 *How many grams of hydrogen are contained in a 100-g sample of propane?*

Data Table with Sample Data

	$CuSO_4 \cdot 5H_2O$	$CaCl_2 \cdot 2H_2O$	$Na_2SO_4 \cdot 10H_2O$
Test tube + hydrate (before heating)	23.88 g	23.60 g	23.92 g
Empty test tube	21.19 g	21.25 g	21.17 g
Mass of hydrate	2.69 g	2.35 g	2.75 g
Test tube + salt (after heating)	22.88 g	23.07 g	22.71 g
Empty test tube	21.19 g	21.25 g	21.17 g
Mass of anhydrous salt	1.69 g	1.82 g	1.54 g
Mass of water lost	1.00 g	0.53 g	1.21 g
Percent water (experimental)	37.2%	22.6%	44.0%
Percent water (theoretical)	36.1%	24.5%	55.9%

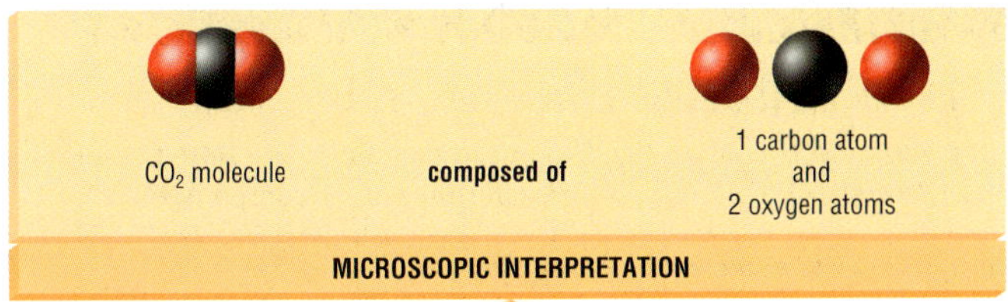

Figure 10.15 A formula can be interpreted on a microscopic level in terms of atoms or on a macroscopic level in terms of moles of atoms.

Word *Origins*

Empirical comes from the Latin word *empiricus* meaning a doctor relying on experience alone. An empirical formula must be obtained from experimental data. Thus, an empirical formula relies on experience. **Is a molecular formula also based on experimental data?**

Empirical Formulas

A useful formula for cooking rice is to use one cup of rice and two cups of water. If a larger amount of rice is needed, you could double or triple the amounts, for example, two cups of rice and four cups of water. The formulas for some compounds also show a basic ratio of elements. Multiplying that ratio by any factor can produce the formulas for other compounds.

The percent composition of your newly synthesized compound is the data you need to calculate the basic ratio of the elements contained in the compound. The basic ratio, called the **empirical formula,** gives the lowest whole-number ratio of the atoms of the elements in a compound. For example, a compound may have the empirical formula CO_2. The empirical formula shows the kinds and lowest relative count of atoms or moles of atoms in molecules or formula units of a compound. Figure 10.15 shows that empirical formulas may be interpreted at the microscopic (atomic) or macroscopic (molar) level.

An empirical formula may or may not be the same as a molecular formula. For example, the lowest ratio of hydrogen to oxygen in hydrogen peroxide is 1:1. Thus the empirical formula of hydrogen peroxide is HO. The actual molecular formula of hydrogen peroxide has twice the number of atoms as the empirical formula. The molecular formula is $(HO) \times 2$, or H_2O_2. But notice that the ratio of hydrogen to oxygen is still the same, 1:1. **The empirical formula of a compound shows the smallest whole-number ratio of the atoms in the compound.** The molecular formula tells the actual number of each kind of atom present in a molecule of the compound. For carbon dioxide, the empirical and molecular formulas are the same—CO_2. Figure 10.16 shows two compounds of carbon having the same empirical formula (CH) but different molecular formulas.

Figure 10.16 Ethyne (C_2H_2), also called acetylene, is a gas used in welder's torches. Styrene (C_8H_8) is used in making polystyrene. These two compounds have the same empirical formula. **Calculating** ***What is the empirical formula of ethyne and styrene?***

Empirical Formulas

Empirical Formulas from Percent Composition L2

Purpose Students are provided with an analogy that helps clarify the concepts of percent composition and empirical formulas.

Materials 3 red marbles, 6 green marbles, 3 black marbles, and 12 blue marbles

Procedure Provide pairs of students with sets of marbles. Have students express the number of different colored marbles as fractions and percents of the whole collection. Ask, **What percent of the collection do the red marbles represent?** *(12.5%)* Show them that the sums of the fractions and percents are equal to 1 and 100%, respectively. Ask, **What is the ratio of red: green:black:blue marbles in lowest terms?** *(1:2:1:4)* This activity can be extended if the different colored marbles are assumed to be atoms of different elements. Ask, **What is the empirical formula of a hypothetical "compound" that consists of 25% red marbles and 75% green marbles?** *(The ratio of red marbles to green marbles in the empirical formula would be 1:3.)*

Expected Outcomes Students express percent composition of the marbles and determine the "empirical formula" of a marble combination.

Word *Origins* L2

A molecular formula is based on experimental data in two different ways. The molar mass is determined experimentally, as is the empirical formula.

Differentiated Instruction

Less Proficient Readers L1

As students read about how to determine empirical and molecular formulas, have groups of students develop numbered lists of steps they would take to determine these formulas. They should have three lists: one for determining empirical formulas from percent composition; one for determining empirical formulas from mass data; and one for determining molecular formulas from the empirical formula and molar mass.

18 g H

Figure 10.16 CH

Sample Problem 10.11

Answers

36. a. 94.1 g O × 1 mol O/16.0 g O = 5.88 mol O
5.9 g H × 1 mol H/1.0 g H = 5.9 mol H
5.88 mol O/5.88 = 1.00 mol O
5.9 mol H/5.88 = 1.0 mol H
Empirical formula = HO
b. 67.6 g Hg × 1 mol Hg/200.6 g Hg = 0.337 mol Hg
10.8 g S × 1 mol S/32.1 g S = 0.336 mol S
21.6 g O × 1 mol O/16.0 g O = 1.35 mol O
0.337 mol Hg/0.336 = 1.00 mol Hg
0.336 mol S/0.336 = 1.00 mol S
1.35 mol O/0.336 = 4.02 mol O
Empirical formula = $HgSO_4$
37. 62.1 g C × 1 mol C/12.0 g C = 5.18 mol C
13.8 g H × 1 mol H/1.00 g H = 13.8 mol H
24.1 g N × 1 mol N/14.0 g N = 1.72 mol N
5.18 mol C/1.72 = 3.01 mol C
13.8 mol H/1.72 = 8.02 mol H
1.72 mol N/1.72 = 1.00 mol N
Empirical formula = C_3H_8N

Practice Problems Plus L2

What is the empirical formula of each of the following compounds?
a. 36.1% Ca, 63.9% Cl *($CaCl_2$)*
b. 40.0% C, 6.7% H, 53.3% O *(CH_2O)*
c. 3.7% H, 44.4% C, and 51.9% N *(HCN)*

Math Handbook
For a math refresher and practice, direct students to dimensional analysis, page R66.

SAMPLE PROBLEM 10.11

Determining the Empirical Formula of a Compound

A compound is analyzed and found to contain 25.9% nitrogen and 74.1% oxygen. What is the empirical formula of the compound?

1 Analyze *List the knowns and the unknown.*

Knowns
- percent of nitrogen = 25.9% N
- percent of oxygen = 74.1% O

Unknown
- Empirical formula = $N_?O_?$

The percent composition tells the ratio of the mass of nitrogen atoms to the mass of oxygen atoms in the compound. Change the ratio of masses to a ratio of moles by using conversion factors based on the molar mass of each element. Then reduce this ratio to the lowest whole-number ratio.

2 Calculate *Solve for the unknown.*

Because percent means parts per 100, you can assume that 100.0 g of the compound contains 25.9 g N and 74.1 g O. Use these values to convert to moles.

$$25.9\ \cancel{\text{g N}} \times \frac{1\ \text{mol N}}{14.0\ \cancel{\text{g N}}} = 1.85\ \text{mol N}$$

$$74.1\ \cancel{\text{g O}} \times \frac{1\ \text{mol O}}{16.0\ \cancel{\text{g O}}} = 4.63\ \text{mol O}$$

The mole ratio of nitrogen to oxygen is $N_{1.85}O_{4.63}$. But formulas must have whole-number subscripts. Divide each molar quantity by the smaller number of moles. This gives 1 mol for the element with the smaller number of moles.

$$\frac{1.85\ \text{mol N}}{1.85} = 1\ \text{mol N}; \quad \frac{4.63\ \text{mol O}}{1.85} = 2.50\ \text{mol O}$$

The result, $N_1O_{2.5}$, still has a subscript that is not a whole number. To obtain the lowest whole-number ratio, multiply each part of the ratio by the smallest whole number (in this case 2) that will convert both subscripts to whole numbers.

$$1\ \text{mol N} \times 2 = 2\ \text{mol N}$$
$$2.5\ \text{mol O} \times 2 = 5\ \text{mol O}$$

The empirical formula is N_2O_5.

3 Evaluate *Does the result make sense?*

The subscripts are whole numbers, and the percent composition of this empirical formula equals the percents given in the original problem.

Practice Problems

36. Calculate the empirical formula of each compound.
a. 94.1% O, 5.9% H
b. 67.6% Hg, 10.8% S, 21.6% O

37. 1,6-diaminohexane is used to make nylon. What is the empirical formula of this compound if it is 62.1% C, 13.8% H, and 24.1% N?

Math Handbook
For help with dimensional analysis go to page R66.

Interactive Textbook
Problem-Solving 10.37
Solve Problem 37 with the help of an interactive guided tutorial.
with ChemASAP

Facts and Figures

Computing Formulas

In using percent composition to determine empirical formula, 100.0 g of compound is arbitrarily chosen because it is easy to use. If an element comprises 28.5% of the mass of a compound, for example, it makes up 28.5 g of a 100.0-g sample. Any other mass of compound can be used but computation will be more difficult.

Table 10.3

Comparison of Empirical and Molecular Formulas

Formula (name)	Classification of formula	Molar mass
CH	Empirical	13
C_2H_2 (ethyne)	Molecular	26 (2 × 13)
C_6H_6 (benzene)	Molecular	78 (6 × 13)
CH_2O (methanal)	Empirical and Molecular	30
$C_2H_4O_2$ (ethanoic acid)	Molecular	60 (2 × 30)
$C_6H_{12}O_6$ (glucose)	Molecular	180 (6 × 30)

Molecular Formulas

Look at the compounds listed in Table 10.3. Ethyne and benzene have the same empirical formula—CH. Methanal, ethanoic acid, and glucose, shown in Figure 10.17 have the same empirical formula—CH_2O. But the compounds in these two groups have different molar masses. Their molar masses are simple whole-number multiples of the molar masses of the empirical formulas, CH and CH_2O. **The molecular formula of a compound is either the same as its experimentally determined empirical formula, or it is a simple whole-number multiple of its empirical formula.**

Once you have determined the empirical formula of your newly synthesized compound, you can determine its molecular formula, but you must know the compound's molar mass. A chemist often uses an instrument called a mass spectrometer to determine molar mass. The compound is broken into charged fragments (ions) that travel through a magnetic field. The magnetic field deflects the particles from their straight-line paths. The mass of the compound is determined from the amount of deflection experienced by the particles.

From the empirical formula, you can calculate the empirical formula mass (efm). This is simply the molar mass represented by the empirical formula. Then you can divide the experimentally determined molar mass by the empirical formula mass. This gives the number of empirical formula units in a molecule of the compound and is the multiplier to convert the empirical formula to the molecular formula. For example, recall that the empirical formula of hydrogen peroxide is HO. Its empirical formula mass is 17.0 g/mol. The molar mass of H_2O_2 is 34.0 g/mol.

$$\frac{34.0\ \text{g/}\cancel{\text{mol}}}{17.0\ \text{g/}\cancel{\text{mol}}} = 2$$

To obtain the molecular formula of hydrogen peroxide from its empirical formula, multiply the subscripts in the empirical formula by 2. $(HO) \times 2 = H_2O_2$.

 ***How does the molecular formula for a compound relate to its empirical formula?***

Figure 10.17 Methanal (formaldehyde), ethanoic acid (acetic acid), and glucose have the same empirical formula. **Applying Concepts** ***How could you easily obtain the molar mass of glucose using the molar mass of methanal?***

Molecular Formulas

Use Visuals 

Table 10.3 Have students examine the patterns shown in the table. Then have them add entries for these pairs of compounds: nitrogen dioxide (NO_2) and dinitrogen tetroxide (N_2O_4), diphosphorus pentoxide (P_2O_5) and tetraphosphorus decoxide (P_4O_{10}). Students can determine the empirical formula for each pair.

Facts and Figures

Some Carbohydrates Share the Same Empirical Formula

Several carbohydrates have the empirical formula CH_2O. Examples include glucose, which is abundant in plants and animals, and fructose, which is found in fruits and honey. Both of these simple structures have the molecular formula $C_6H_{12}O_6$, but they differ in structure.

Answers to...

Figure 10.17 Multiply the molar mass of methanal by 6.

 The molecular formula of a compound is the same as the empirical formula or it is a whole-number multiple of it.

Section 10.3 (continued)

Sample Problem 10.12

Answers

38. molar mass/efm = 62/31 = 2
molecular formula = $2(CH_3O)$ = $C_2H_6O_2$

39. a. same empirical formula (CH_2O)
b. different empirical formulas

Practice Problems Plus 

1. What is the molecular formula of a compound with the empirical formula CClN and a molar mass of 184.5? *($C_3Cl_3N_3$)*

2. What is the molecular formula of a compound that is 56.6% K, 8.7% C, and 34.7% O? *(K_2CO_3)*

3 ASSESS

Evaluate Understanding 

Have students list the steps they would take to calculate the molecular formula in each of the following situations:

- The empirical formula and molar mass are known.
- The percent composition and molar mass are known.

Reteach 

Point out to students that when they know the percent composition and molar mass of a compound, they must first use the percent composition to calculate the empirical formula. They can then calculate the empirical formula mass and compare it to the molar mass of the molecular compound to determine the molecular formula.

Elements Handbook

CaO: 71.5%
$CaCO_3$: 40.1%
$Ca(OH)_2$: 54.1%
$CaSO_4 \bullet 2H_2O$: 23.3%
$Ca_3(PO_4)_2$: 38.8%

If your class subscribes to the Interactive Textbook, use it to review key concepts in Section 10.3.

with ChemASAP

SAMPLE PROBLEM 10.12

Finding the Molecular Formula of a Compound

Calculate the molecular formula of a compound whose molar mass is 60.0 g/mol and empirical formula is CH_4N.

1 Analyze *List the knowns and the unknown.*

Knowns
- empirical formula = CH_4N
- molar mass = 60.0 g/mol

Unknown
- molecular formula = ?

2 Calculate *Solve for the unknown.*

First calculate the empirical formula mass. Then divide the molar mass by the empirical formula mass to obtain a whole number. To get the molecular formula, multiply the formula subscripts by this value.

Empirical formula	efm	Molar mass/efm	Molecular formula
CH_4N	30.0	60.0/30.0 = 2	$C_2H_8N_2$

3 Evaluate *Does the result make sense?*

The molecular formula has the molar mass of the compound.

Practice Problems

38. Find the molecular formula of ethylene glycol, which is used as antifreeze. The molar mass is 62 g/mol and the empirical formula is CH_3O.

39. Which pair of molecules has the same empirical formula?
a. $C_2H_4O_2$, $C_6H_{12}O_6$
b. $NaCrO_4$, $Na_2Cr_2O_7$

Interactive Textbook

Problem-Solving 10.38 Solve Problem 38 with the help of an interactive guided tutorial.

with ChemASAP

10.3 Section Assessment

40. Key Concept How do you calculate the percent by mass of an element in a compound?

41. Key Concept What information can you obtain from an empirical formula?

42. Key Concept How is the molecular formula of a compound related to its empirical formula?

43. Calculate the percent composition of the compound that forms when 222.6 g N combines completely with 77.4 g O.

44. Calculate the percent composition of calcium acetate ($Ca(C_2H_3O_2)_2$).

45. The compound methyl butanoate smells like apples. Its percent composition is 58.8% C, 9.8% H, and 31.4% O and its molar mass is 102 g/mol. What is its empirical formula? What is its molecular formula?

46. What is an empirical formula? Which of the following molecular formulas are also empirical formulas?
a. ribose ($C_5H_{10}O_5$)
b. ethyl butyrate ($C_6H_{12}O_2$)
c. chlorophyll ($C_{55}H_{72}MgN_4O_5$)
d. DEET ($C_{12}H_{17}ON$)

Elements Handbook

Calcium Select three important compounds that contain calcium from among those discussed on page R11 of the Elements Handbook. Determine the percent of calcium in each.

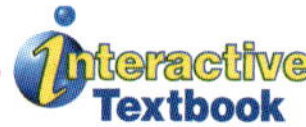

Assessment 10.3 Test yourself on the concepts in Section 10.3.

with ChemASAP

Section 10.3 Assessment

40. Divide the mass of an element in the compound by the mass of the compound; then multiply by 100%.

41. The empirical formula gives the lowest whole-number ratio of atoms in the compound.

42. The molecular formula of a compound is a simple whole-number multiple of the empirical formula.

43. 74.2% N, 25.8% O

44. 25.4% Ca, 30.4% C, 3.8% H, 40.5% O

45. $C_5H_{10}O_2$ is both its empirical and molecular formula.

46. An empirical formula has the lowest whole-number ratio of elements.
a. molecular **b.** molecular
c. molecular and empirical
d. molecular and empirical

Technology & Society

Drug Testing

A test to identify an abused substance in the body must be extremely accurate. A false-positive result could ruin a career. A false-negative result could endanger lives. The best method currently available to test for drug abuse is the gas chromatography/mass spectrometer system, or GC/MS. Gas chromatography separates a chemical mixture and identifies its components. Mass spectrometry uses masses to verify the identification. Used together, the GC/MS testing is nearly 100% reliable.

Interpreting Diagrams ***What is the purpose of the separation column?***

Testing athletes The results of a drug test could keep an athlete out of an upcoming event, and possibly off the team permanently.

1 The sample to be tested, such as urine or blood serum, is vaporized and mixed with an unreactive carrier gas such as helium. The resulting gas mixture is forced through a separation column.

2 The separation column is packed with a material that interacts physically with the vaporized sample. Components of the mixture move at different rates along the column and then pass through a detector.

3 Each peak on a gas chromatogram shows the retention time for a component. Retention time is used to identify compounds.

4 The separated component enters a mass spectrometer where its identity is confirmed.

Heated metal block

Detector

HELIUM

Facts and Figures

Drugs in the Blood

Current drug tests can detect even small traces of drugs in the blood or urine. For example, THC (the metabolic product of marijuana use) remains detectable in the body for at least 18 hours for an occasional user and up to 30 days for a habitual user. It can be detected in concentrations as small as 50 ng/mL of solution.

Technology & Society

Drug Testing

Discuss 

Chromatography Discuss with students the basic concepts of chromatography. Chromatography is based on the relative attraction of the material in the separation column (stationary phase) for the materials in the medium that moves through it (mobile phase). The amount of attraction is based on the bonding in the molecules in the different phases. If the bonds in one phase are polar and the bonds in the other phase are nonpolar, the phases have little attraction for each other, and the mobile phase passes quickly through the stationary phase. If the molecules in both phases are either polar or nonpolar, the attraction between the molecules in the phases causes this fraction of the mobile phase to move more slowly through the stationary phase.

TEACHER Demo

Paper Chromatography 

Purpose Students observe an example of a mixture being separated into its components.

Materials coffee filter paper, water soluble marker, beaker

Procedure Cut a strip of filter paper. About one third of the distance from the bottom of the paper, make a mark on the paper. Place water in the beaker to a depth of 2 cm. Place the end of the filter paper closest to the mark in the beaker, making sure the mark is above the water line. When the water rises almost to the top of the paper, remove the paper from the water, and observe what happened.

Expected Outcomes The components of the ink separate, based on the attraction of the components for the paper.

Answers to...

Interpreting Diagrams The column concentrates and separates the components of the mixture by causing each to move at a different rate.

CHAPTER 10 **Study Guide**

Study Tip

Organize New Information Different people learn in different ways. If you learn best visually, draw concept maps, outlines, and diagrams to help you organize the chapter content. If you learn best by listening, take turns reading the chapter aloud with a partner.

If your class subscribes to the Interactive Textbook with ChemASAP, your students can go online to access an interactive version of the Student Edition and a self-test.

with ChemASAP

CHAPTER 10 Study Guide

Key Concepts

10.1 The Mole: A Measurement of Matter

- Three methods for measuring the amount of a substance are by count, by mass, and by volume.
- A mole of any substance always contains Avogadro's number of representative particles, or 6.02×10^{23} representative particles.
- The atomic mass of an element expressed in grams is the mass of a mole of the element.
- To calculate the molar mass of a compound, find the number of grams of each element contained in one mole of the compound. Then add the masses of the elements in the compound.

10.2 Mole–Mass and Mole–Volume Relationships

- The molar mass of an element or compound is the conversion factor for converting between the mass and the number of moles of a substance.
- One mole of any gas occupies a volume of 22.4 L at standard temperature and pressure. One mole of any substance contains Avogadro's number of particles, so 22.4 L of any gas at STP contains 6.02×10^{23} representative particles of that gas.

10.3 Percent Composition and Chemical Formulas

- To determine the percent by mass of any element in a given compound, divide the element's mass by the mass of the compound and multiply by 100%.
- An empirical formula of a compound is the simplest whole-number ratio of atoms of the elements in the compound.
- The molecular formula of a compound is either the same as its experimentally determined empirical formula, or it is a simple whole-number multiple of it.

Vocabulary

- Avogadro's hypothesis (p. 300)
- Avogadro's number (p. 290)
- empirical formula (p. 309)
- molar mass (p. 294)
- molar volume (p. 300)
- mole (mol) (p. 290)
- percent composition (p. 305)
- representative particle (p. 290)
- standard temperature and pressure (STP) (p. 300)

Key Equations

- $\text{moles} = \text{representative particles} \times \dfrac{1 \text{ mole}}{6.02 \times 10^{23} \text{ representative particles}}$
- $\text{representative particles} = \text{moles} \times \dfrac{6.02 \times 10^{23} \text{ representative particles}}{1 \text{ mole}}$
- $\text{mass (grams)} = \text{number of moles} \times \dfrac{\text{mass (grams)}}{1 \text{ mole}}$
- $\text{moles} = \text{mass (grams)} \times \dfrac{1 \text{ mole}}{\text{mass (grams)}}$
- $\dfrac{\text{grams}}{\text{mole}} = \dfrac{\text{grams}}{\text{L}} \times \dfrac{22.4 \text{ L}}{1 \text{ mole}}$
- $\text{volume of gas} = \text{moles of gas} \times \dfrac{22.4 \text{ L}}{1 \text{ mol}}$
- $\%\text{ mass of element} = \dfrac{\text{mass of element}}{\text{mass of compound}} \times 100\%$

Organizing Information

Organize the major ideas of the chapter.

Concept Map 10 Solve the Concept Map with the help of an interactive tutorial.

with ChemASAP

- Avogadro's number
- Empirical formula
- Molar volume
- Molar mass
- Mole (mol)
- Percent composition
- Representative particle
- Standard temperature and pressure (STP)

Chapter Resources

Print

- **Core Teaching Resources,** Chapter 10, Practice Problems, Vocabulary Review, Quiz, Chapter Test A, Chapter Test B

Technology

- **Computer Test Bank,** Chapter 10 Test
- **Interactive Textbook with ChemASAP,** Chapter 10

CHAPTER 10

Assessment

Reviewing Content

10.1 The Mole: A Measurement of Matter

47. List three common ways that matter is measured. Give examples of each.

48. Name the representative particle (atom, molecule, or formula unit) of each substance.
a. oxygen gas **b.** sodium sulfide
c. sulfur dioxide **d.** potassium

49. How many hydrogen atoms are in a representative particle of each substance?
a. $Al(OH)_3$ **b.** $H_2C_2O_4$
c. $(NH_4)_2HPO_4$ **d.** $C_4H_{10}O$

50. Which contains more molecules: 1.00 mol H_2O_2, 1.00 mol C_2H_6, or 1.00 mol CO?

51. Which contains more atoms: 1.00 mol H_2O_2, 1.00 mol C_2H_6, or 1.00 mol CO?

52. Find the number of representative particles in each substance.
a. 3.00 mol Sn **b.** 0.400 mol KCl
c. 7.50 mol SO_2 **d.** 4.80×10^{-3} mol NaI

53. Calculate the molar mass of each substance.
a. H_3PO_4 **b.** N_2O_3 **c.** $CaCO_3$
d. $(NH_4)_2SO_4$ **e.** $C_4H_9O_2$ **f.** Br_2

54. Calculate the mass of 1.00 mol of each of these substances.
a. silicon dioxide (SiO_2)
b. diatomic nitrogen (N_2)
c. iron(III) hydroxide ($Fe(OH)_3$)
d. copper (Cu)

55. List the steps you would take to calculate the molar mass of any compound.

56. What is the molar mass of chlorine?

57. Construct a numerical problem to illustrate the size of Avogadro's number. Exchange problems with a classmate and then compare your answers.

10.2 Mole–Mass and Mole–Volume Relationships

58. How many moles is each of the following?
a. 15.5 g SiO_2 **b.** 0.0688 g AgCl
c. 79.3 g Cl_2 **d.** 5.96 g KOH
e. 937 g $Ca(C_2H_3O_2)_2$ **f.** 0.800 g Ca

59. Find the mass of each substance.
a. 1.50 mol C_5H_{12} **b.** 14.4 mol F_2
c. 0.780 mol $Ca(CN)_2$ **d.** 7.00 mol H_2O_2
e. 5.60 mol NaOH **f.** 3.21×10^{-2} mol Ni

60. Calculate the volume of each of the following gases at STP.
a. 7.6 mol Ar **b.** 0.44 mol C_2H_6

61. What is the density of each of the following gases at STP?
a. C_3H_8 **b.** Ne **c.** NO_2

62. Find each of the following quantities.
a. the volume, in liters, of 835 g SO_3 at STP
b. the mass, in grams, of a molecule of aspirin ($C_9H_8O_4$)
c. the number of atoms in 5.78 mol NH_4NO_3

10.3 Percent Composition and Chemical Formulas

63. Calculate the percent composition of each compound.
a. H_2S **b.** $(NH_4)_2C_2O_4$
c. $Mg(OH)_2$ **d.** Na_3PO_4

64. Using your answers from Problem 63, calculate the number of grams of these elements.
a. sulfur in 3.54 g H_2S
b. nitrogen in 25.0 g $(NH_4)_2C_2O_4$
c. magnesium in 97.4 g $Mg(OH)_2$
d. phosphorus in 804 g Na_3PO_4

65. Which of the following compounds has the highest iron content?
a. $FeCl_2$ **b.** $Fe(C_2H_3O_2)_3$
c. $Fe(OH)_2$ **d.** FeO

66. You find that 7.36 g of a compound has decomposed to give 6.93 g of oxygen. The only other element in the compound is hydrogen. If the molar mass of the compound is 34.0 g/mol, what is its molecular formula?

67. Which of the following can be classified as an empirical formula?
a. S_2Cl_2 **b.** $C_6H_{10}O_4$ **c.** Na_2SO_3

68. What is the molecular formula for each compound? Each compound's empirical formula and molar mass is given.
a. CH_2O, 90 g/mol **b.** HgCl, 472.2 g/mol

69. Determine the molecular formula for each compound.
a. 94.1% O and 5.9% H; molar mass = 34 g
b. 50.7% C, 4.2% H, and 45.1% O; molar mass = 142 g

CHAPTER 10

Assessment

Reviewing Content

47. Number, mass, or volume; examples will vary.

48. a. molecule
b. formula unit
c. molecule
d. atom

49. a. 3
b. 2
c. 9
d. 10

50. All contain 6.02×10^{23} molecules

51. 1.00 mol C_2H_6

52. a. 1.81×10^{24} atoms Sn
b. 2.41×10^{23} formula units KCl
c. 4.52×10^{24} molecules SO_2
d. 2.89×10^{21} formula units NaI

53. a. 98.0 g/mol
b. 76.0 g/mol
c. 100.1 g/mol
d. 132.1 g/mol
e. 89.0 g/mol
f. 159.8 g/mol

54. a. 60.1 g/mol
b. 28.0 g/mol
c. 106.8 g/mol
d. 63.5 g/mol

55. Answers will vary but should include
1. Determine the moles of each atom from the formula.
2. Look up the atomic mass of each element.
3. Multiply the number of moles of each atom by its molar mass.
4. Sum these products.

56. 71.0 g/mol Cl_2

57. Answers will vary.

58. a. 0.258 mol SiO_2
b. 4.80×10^{-4} mol AgCl
c. 1.12 mol Cl_2
d. 0.106 mol KOH
e. 5.93 mol $Ca(C_2H_3O_2)_2$
f. 2.00×10^{-2} mol Ca

59. a. 108 g C_5H_{12}
b. 547 g F_2
c. 71.8 g $Ca(CN)_2$
d. 238 g H_2O_2
e. 224 g NaOH
f. 1.88 g Ni

60. a. 1.7×10^2 L Ar
b. 9.9 L C_2H_6

61. a. 1.96 g/L
b. 0.902 g/L
c. 2.05 g/L

62. a. 234 L SO_3
b. 2.99×10^{-22} g $C_9H_8O_4$
c. 3.13×10^{25} atoms

63. a. 5.9% H, 94.1% S
b. 22.6% N, 6.5% H, 19.4% C, 51.6% O
c. 41.7% Mg, 54.9% O, 3.4% H
d. 42.1% Na, 18.9% P, 39.0% O

64. a. 3.33 g S
b. 5.65 g N
c. 40.6 g Mg
d. 152 g P

65. d. 77.7% Fe in FeO

66. H_2O_2

67. a. molecular
b. molecular
c. empirical

68. a. $C_3H_6O_3$
b. Hg_2Cl_2

69. a. H_2O_2
b. $C_6H_6O_4$

Understanding Concepts

70. You can measure the mass of 22.4 L of the compound at STP; this is the molar volume of the gas. The mass of the molar volume is the molar mass.

71. a. A, $C_2H_4O_2$; D, $C_5H_{10}O_5$; E, $C_6H_{12}O_6$
b. slope = 2.5/1, which is the ratio of the molar mass of the empirical formula to the mass of carbon in the empirical formula: 30/12 = 2.5/1.
c. mass of carbon = 36(*x*); molar mass = 90(*y*)
mass of carbon = 48(*x*); molar mass = 120(*y*)

72. a. A molecule is composed of two or more atoms.
b. There are 6.02×10^{23} molecules in 1 mol of a molecular substance.
c. A mole of CO_2 has 3 times Avogadro's number of atoms.

73. b. 0.842 mol C_2H_4

74. 24.5 g

75. a. CO
b. $C_2O_2NH_5$
c. Cl_2OC

76. a. 27 amu
b. aluminum

77. 3.01×10^{13} km

78. a. $CuBr_2$
b. CH_3

79. $C_3H_6O_3$

80. 3.34×10^{25} molecules H_2O

81. 2.73×10^{20} F atoms

82. 0.982 g He

CHAPTER 10 Assessment *continued*

Understanding Concepts

70. How can you determine the molar mass of a gaseous compound if you do not know its molecular formula?

71. A series of compounds has the empirical formula CH_2O. The graph shows the relationship between the molar mass of the compounds and the mass of carbon in each compound.

Molar mass vs. mass of carbon

Molar mass of compound (g): 0, 50, 100, 150, 200

Mass of C in compound (g): 20, 40, 60, 80

Points: A, D, E

a. What are the molecular formulas for the compounds represented by data points A, D, and E?

b. Find the slope of the line. Is this value consistent with the empirical formula? Explain.

c. Two other valid data points fall on the line between points A and D. What are the *x*, *y* values for these data points?

72. Explain what is wrong with each statement.

a. One mole of any substance contains the same number of atoms.

b. A mole and a molecule of a substance are identical in amount.

c. One molar mass of CO_2 contains Avogadro's number of atoms.

73. Which of the following contains the largest number of atoms?

a. 82.0 g Kr **b.** 0.842 mol C_2H_4
c. 36.0 g N_2

74. What is the total mass of a mixture of 3.50×10^{22} formula units Na_2SO_4, 0.500 mol H_2O, and 7.23 g AgCl?

75. Determine the empirical formulas of compounds with the following percent compositions.

a. 42.9% C and 57.1% O

b. 32.00% C, 42.66% O, 18.67% N, and 6.67% H

c. 71.72% Cl, 16.16% O, and 12.12% C

76. An imaginary "atomic balance" is shown below. Fifteen atoms of boron on the left side of the balance are balanced by six atoms of an unknown element E on the right side.

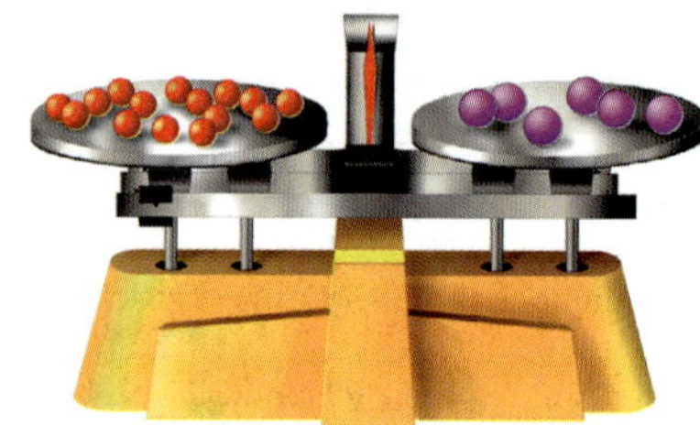

a. What is the atomic mass of element E?

b. What is the identity of element E?

77. A typical virus is 5×10^{-6} cm in diameter. If Avogadro's number of these virus particles were laid in a row, how many kilometers long would the line be?

78. Calculate the empirical formula for each compound.

a. compound consisting of 0.40 mol Cu and 0.80 mol Br

b. compound with 4 atoms of carbon for every 12 atoms of hydrogen

79. Muscle fatigue can result from the buildup of lactic acid resulting from overexercising. The percent composition of lactic acid is 40.0% C, 6.67% H, and 53.3% O. What is the molecular formula of lactic acid if its molar mass is 90.0 g/mol?

80. How many water molecules are in a 1.00-L bottle of water? The density of water is 1.00 g/mL.

81. The molecular formula of an antibacterial drug is $C_{17}H_{18}FN_3O_3$. How many fluorine atoms are in a 150-mg tablet of this drug?

82. What mass of helium is needed to inflate a balloon to a volume of 5.50 L at STP?

316 Chapter 10

Critical Thinking

83. What is the empirical formula of a compound that has three times as many hydrogen atoms as carbon atoms, but only half as many oxygen atoms as carbon atoms?

84. How are the empirical and molecular formulas of a compound related?

85. Why does one mole of carbon have a smaller mass than one mole of sulfur? How are the atomic structures of these elements different?

86. One mole of any gas at STP equals 22.4 L of that gas. It is also true that different elements have different atomic volumes, or diameters. How can you reconcile these two statements?

87. The graph shows the percent composition of phenylalanine.

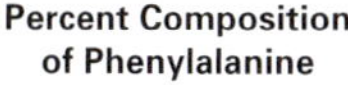

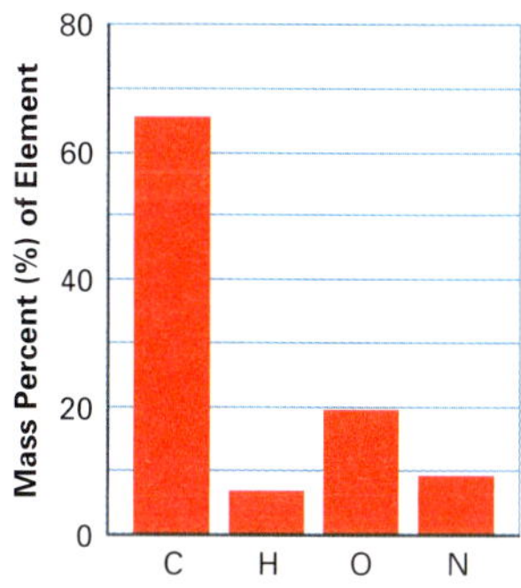

a. What is the empirical formula for phenylalanine?

b. If the molar mass of phenylalanine is 165.2 g/mol, what is its molecular formula?

Concept Challenge

88. Nitroglycerine contains 60% as many carbon atoms as hydrogen atoms, three times as many oxygen atoms as nitrogen atoms, and the same number of carbon and nitrogen atoms. The number of moles of nitroglycerine in 1.00 g is 0.00441. What is the molecular formula of nitroglycerine?

89. The density of nickel is 8.91 g/cm^3. How large a cube, in cm^3, would contain 2.00×10^{24} atoms of nickel?

90. Dry air is about 20.95% oxygen by volume. Assuming STP, how many oxygen molecules are in a 75.0-g sample of air? The density of air is 1.19 g/L.

91. The table below gives the molar mass and density of seven gases at STP.

Substance	Molar mass (g)	Density (g/l)
Oxygen	32.0	1.43
Carbon dioxide	44.0	1.96
Ethane	30.0	1.34
Hydrogen	2.0	0.089
Sulfur dioxide	64.1	2.86
Ammonia	17.0	0.759
Fluorine	38.0	1.70

a. Plot these data, with density on the *x*-axis.

b. What is the slope of the straight-line plot?

c. What is the molar mass of a gas at STP that has a density of 1.10 g/L?

d. A mole of a gas at STP has a mass of 56.0 g. Use the graph to determine its density.

92. A cubic meter of seawater contains 6×10^{-6} g gold. If the total mass of the water in Earth's oceans is 4×10^{20} kg, how many kilograms of gold are distributed throughout the oceans? (Assume that the density of seawater is 1 g/cm^3.) How many liters of seawater would have to be processed to recover 1 kg of gold (which has a value of about $12,500 at 2003 prices)? Do you think this recovery operation is feasible?

93. Avogadro's number has been determined by about 20 different methods. In one approach, the spacing between ions in an ionic substance is determined by using a technique called X-ray diffraction. X-ray diffraction studies of sodium chloride have shown that the distance between adjacent Na^+ and Cl^- ions is 2.819×10^{-8} cm. The density of solid NaCl is 2.165 g/cm^3. By calculating the molar mass to four significant figures, you can determine Avogadro's number. What value do you obtain?

Critical Thinking

83. C_2H_6O

84. A molecular formula is a whole-number multiple of its empirical formula.

85. Sulfur atoms have a greater atomic mass. The most abundant sulfur atom has 16 protons, 16 electrons, and 16 neutrons; carbon is composed of 6 protons, 6 electrons, and 6 neutrons. Therefore, 6.02×10^{23} sulfur atoms will have a greater mass than the same number of carbon atoms.

86. Gas molecules are separated by so much empty space their own volumes are insignificant in determining how much space a certain quantity of gas molecules occupies.

87. **a.** $C_9H_{11}O_2N$ **b.** $C_9H_{11}O_2N$

Concept Challenge

88. $C_3H_5O_9N_3$

89. 21.9 cm^3

90. 3.54×10^{23} O_2 molecules

91. **a.**

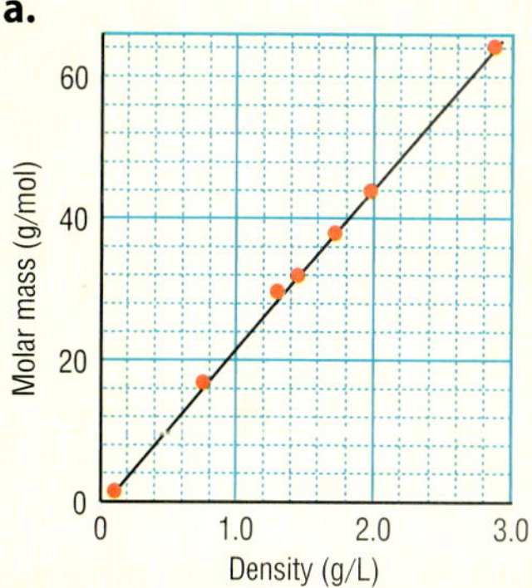

b. 22.4 L/mol

c. 24.6 g/mol

d. 2.5 g/L

92. 2.4×10^9 kg Au; 2×10^{11} L H_2O; not feasible

93. 6.025×10^{23} formula units/mol

CHAPTER 10

Assessment continued

Cumulative Review

94. chemical change: wax burning
physical change: wax melting
physical change: wax vaporizing

95. **a.** physical change **b.** chemical change **c.** chemical change **d.** physical change **e.** chemical change **f.** physical change

96. **a.** false **b.** true **c.** true **d.** false

97. No; the student has ignored the units. The density of sugar is 1.59 g/mL; the density of carbon dioxide is much less, 1.83 g/L.

98. The block will float. Its density, 0.848 g/mL, is less than the density of water.

99. **a.** 4.72×10^3 mg
b. 97 km/h
c. 4.4×10^{-2} dm

100. **a.** 40, 40, 50 **b.** 46, 46, 62 **c.** 35, 35, 46 **d.** 51, 51, 72

101. **a.** $1s^2 2s^2 2p^5$ **b.** $1s^2 2s^1$ **c.** [Kr] $5s^1$

102. Magnesium and barium are both in Group 2A and have 2 valence electrons.

103. Cr, Cd, Cu, and Co

104. For Group A elements, the group number equals the number of valence electrons.

105. A molecule is composed of two or more atoms.

106. **a.** pyramidal H:P̈:H (with H below P)
b. linear :O::C:
c. linear :S̈::C::S̈:
d. tetrahedral :F̈:C:F̈: (with :F̈: above and below C)

107. For single bond a single line connects the atoms (X—X). Atoms are connected by two lines in a double bond (X=X), and three lines in a triple bond (X≡X).

108. Answers will vary.
a. carbon monoxide (CO)
b. ozone (O_3)
c. nitrogen dioxide (NO_2)

109. Calculate the electronegativity difference between two atoms. If the difference is small (0.0 —0.4) the bond is nonpolar covalent. If the difference ≥ 2.0, the bond is most likely ionic. For values between 0.4 and 2.0, the bond is polar covalent.

110. **d.** CaS_2 **f.** Ba(OH)

111. **a.** iron(III) hydroxide **b.** ammonium iodide **c.** sodium carbonate **d.** carbon tetrachloride

112. **a.** KNO_3 **b.** CuO **c.** Mg_3N_2 **d.** AgF

CHAPTER 10

Assessment continued

Cumulative Review

94. Identify at least one chemical change and two physical changes that are occurring in the photo. (*Chapter 2*)

95. Classify each of the following as a physical change or a chemical change. (*Chapter 2*)
a. an aspirin tablet is crushed to a powder
b. a red rose turns brown
c. grape juice turns to wine
d. fingernail polish remover evaporates
e. a bean seed sprouts
f. a piece of copper is beaten into a thin sheet

96. Which of these statements are true about every solution? (*Chapter 2*)
a. Solutions are in the liquid state.
b. Solutions are homogeneous.
c. Solutions are mixtures.
d. Solutions are composed of at least two compounds.

97. A student writes down the density of table sugar as 1.59 and the density of carbon dioxide as 1.83. Can these values be correct? Explain. (*Chapter 3*)

98. A block of wood measuring 2.75 cm × 4.80 cm × 7.50 cm has a mass of 84.0 g. Will the block of wood sink or float in water? (*Chapter 3*)

99. Convert each of the following. (*Chapter 3*)
a. 4.72 g to mg
b. 2.7×10^3 cm/s to km/h
c. 4.4 mm to dm

100. How many protons, electrons, and neutrons are in each isotope? (*Chapter 4*)
a. zirconium-90 **b.** palladium-108
c. bromine-81 **d.** antimony-123

101. Write the complete electron configuration for each atom. (*Chapter 5*)
a. fluorine **b.** lithium **c.** rubidium

102. Why do the elements magnesium and barium have similar chemical and physical properties? (*Chapter 6*)

103. Which of the following are transition metals: Cr, Cd, Ca, Cu, Co, Cs, Ce? (*Chapter 6*)

104. How can the periodic table be used to infer the number of valence electrons in an atom? (*Chapter 7*)

105. How does a molecule differ from an atom? (*Chapter 8*)

106. Draw electron dot structures and predict the shapes of the following molecules. (*Chapter 8*)
a. PH_3 **b.** CO **c.** CS_2 **d.** CF_4

107. How are single, double, and triple bonds indicated in electron dot structures? (*Chapter 8*)

108. Give an example of each of the following. (*Chapter 8*)
a. coordinate covalent bonding
b. resonance structures
c. exceptions to the octet rule

109. Explain how you can use electronegativity values to classify a bond as nonpolar covalent, polar covalent, or ionic. (*Chapter 8*)

110. Identify any incorrect formulas among the following. (*Chapter 9*)
a. H_2O_2 **b.** $NaIO_4$ **c.** SrO
d. CaS_2 **e.** $CaHPO_4$ **f.** BaOH

111. Name these compounds. (*Chapter 9*)
a. $Fe(OH)_3$ **b.** NH_4I
c. Na_2CO_3 **d.** CCl_4

112. Write formulas for these compounds. (*Chapter 9*)
a. potassium nitrate **b.** copper(II) oxide
c. magnesium nitride **d.** silver fluoride

Standardized Test Prep

Test-Taking Tip

Wear a Watch. Be aware of how many questions you have to answer and how much time you have to answer them. Look at your watch or a clock frequently to keep track of your progress.

1. Choose the term that best completes the second relationship.

a. dozen: eggs
mole: ____
(A) atoms (B) 6.02×10^{23}
(C) size (D) grams

b. mole: Avogadro's number
molar volume: ____
(A) mole (B) water
(C) STP (D) 22.4 L

Select the choice that best answers each question or completes each statement.

2. Calculate the molar mass of ammonium phosphate, $(NH_4)_3PO_4$.
a. 149.0 g/mol **b.** 113.0 g/mol
c. 242.0 g/mol **d.** 121.0 g/mol

3. Based on the structural formula below, what is the empirical formula for tartaric acid, a compound found in grape juice?

```
HO — CH — COOH
      |
HO — CH — COOH
```

a. $C_2H_3O_3$ **b.** $C_4H_6O_6$
c. CHO **d.** $C_1H_{1.5}O_{1.5}$

4. How many hydrogen atoms are in six molecules of ethylene glycol, $HOCH_2CH_2OH$?
a. 6 **b.** 36
c. $6 \times 6.02 \times 10^{23}$ **d.** $36 \times 6.02 \times 10^{23}$

5. Which of these statements is true of a balloon filled with 1.00 mol $N_2(g)$ at STP?
I. The balloon has a volume of 22.4 L.
II. The contents of the balloon have a mass of 14.0 g.
III. The balloon contains 6.02×10^{23} molecules.
a. I only **b.** I and II only
c. I and III only **d.** II and III only
e. I, II, and III

6. Which of these compounds has the largest percent by mass of nitrogen?
a. N_2O **b.** NO **c.** NO_2
d. N_2O_3 **e.** N_2O_4

7. Allicin, $C_6H_{10}S_2O$, is the compound that gives garlic its odor. A sample of allicin contains 3.0×10^{21} atoms of carbon. How many atoms of hydrogen does this sample contain?
a. 1.8×10^{21} atoms **b.** 10 atoms
c. 5.0×10^{21} atoms **d.** 1.0×10^{21} atoms

The lettered choices below refer to Questions 8–12. A lettered choice may be used once, more than once, or not at all.

(A) CH (B) CH_2 (C) C_2H_5
(D) CH_3 (E) C_2H_3

Which of the formulas is the empirical formula for each of the following compounds?

8. C_8H_{12} **9.** C_6H_6 **10.** $C_{12}H_{24}$
11. C_2H_6 **12.** C_4H_{10}

Use the ball-and-stick models to answer Questions 13–15. In the models, carbon is black, hydrogen is light blue, oxygen is red, and nitrogen is dark blue. Write the molecular formula for each compound. Then calculate its molar mass.

13.

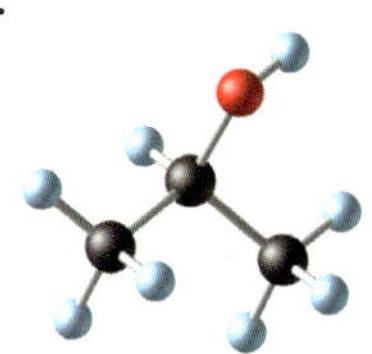

14.

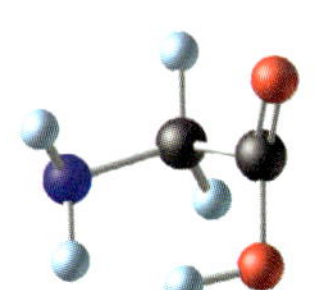

15.

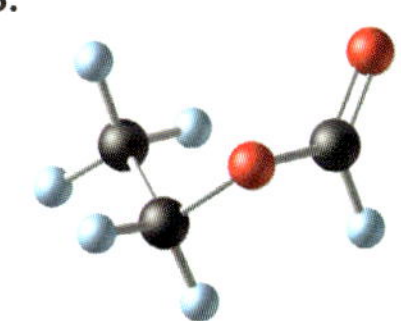

Standardized Test Prep

1. a. A **b.** D
2. a
3. a
4. b
5. c
6. a
7. c
8. E
9. A
10. B
11. D
12. C
13. C_3H_8O, 60.0 g/mol
14. $C_2H_5NO_2$, 75.0 g/mol
15. $C_3H_6O_2$, 74.0 g/mol

CHAPTER 11

Chemical Reactions Planning Guide

SECTION OBJECTIVES	STANDARDS NATIONAL	STANDARDS STATE	ACTIVITIES and LABS
11.1 Describing Chemical Reactions, pp. 321–329 1 block or 2 periods **11.1.1 Describe** how to write a word equation. **11.1.2 Describe** how to write a skeleton equation. **11.1.3 Describe** the steps for writing a balanced chemical equation.	A-1, A-2, B-3, G-1		**SE** Inquiry Activity: Modeling Chemical Reactions, p. 320 L2 **SE** Quick Lab: Removing Silver Tarnish, p. 326 L2 **TE** Teacher Demo: An Example of Chemical Change, p. 325 L2 **TE** Teacher Demo: Balance a Chemical Equation, p. 327 L2 **SSLM** Lab 14: Chemical Equations L2
11.2 Types of Chemical Reactions, pp. 330–341 1 block or 2 periods **11.2.1 Describe** the five general types of reactions. **11.2.2 Predict** the products of the five general types of reactions.	A-1, A-2, B-3, E-2, F-4, F-6		**TE** Teacher Demo: Single-Replacement Reactions, p. 333 L2 **TE** Teacher Demo: A Double-Replacement Reaction, p. 334 L1 **TE** Teacher Demo: Law of Conservation of Mass, p. 335 L2 **TE** Teacher Demo: Combustion of Iron, p. 337 L1 **TE** Teacher Demo: A Combination Reaction, p. 338 L2 **TE** Class Activity: Classifying Fires, p. 340 L2 **TE** Class Activity: Model a Fire Extinguisher, p. 341 L1 **LM** Lab 14: Types of Chemical Reactions L2 **LM** Lab 15: Reactivity of Metals L2 **SSLM** Lab 15: Balancing Chemical Equations L2 **LP** Lab Practical 11-1: Reactivity of Metals L2
11.3 Reactions in Aqueous Solution, pp. 342–345 1 block or 2 periods **11.3.1 Describe** the information found in a net ionic equation. **11.3.2 Predict** the formation of a precipitate in a double-replacement reaction.	A-1, A-2, B-3		**SE** Small-Scale Lab: Precipitation Reactions: Formation of Solids, p. 345 L2 **LM** Lab 16: Identification of Anions and Cations in Solution L2 **LM** Lab 17: Precipitation Reactions L2 **LM** Lab 18: Qualitative Analysis L2 **SSLM** Lab 16: Titration of Bleach L2 **SSLM** Lab 17: Halogen Ions in Solution L2 **LP** Lab Practical 11-2: Identification of Anions and Cations in Solution L2 **LP** Lab Practical 11-3: Precipitation Reactions L2 **LP** Lab Practical 11-4: Reaction Types L2 **LP** Lab Practical 11-5: Qualitative Analysis L3 **VCL** Labs 22–24 L2

Ability Levels

L1 For students who need additional help
L2 For all students
L3 For students who need to be challenged

Components

SE	Student Edition	LM	Laboratory Manual	GCP	Graphing Calculator Problems	T	Transparencies
TE	Teacher's Edition	SSLM	Small-Scale Chemistry Lab Manual	TTT	Test-Taking Tips with Transparencies, Grades 9–12 Science	CA	Chem Alive and Chemistry Field Trips DVD
GRSW	Guided Reading and Study Workbook	LP	Laboratory Practicals	STP	Standardized Test Preparation Workbook	iT	Interactive Textbook with ChemASAP
CTR	Core Teaching Resources	PLM	Probeware Lab Manual			GO	Go Online Internet Resources
SSG	Spanish Study Guide	VCL	Virtual Chemistry Labs				
		CTB	Computer Test Bank				

RESOURCES PRINT and TECHNOLOGY	PROBLEMS and ASSESSMENT
GRSW Chapter 11, Section 1 L1 T T113–T117 L2 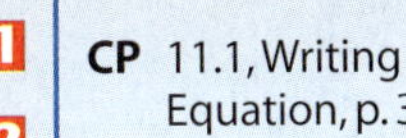 **GO** Chapter 11, Section 1 L2 **GO** Chapter 11, Section 1 L2	**CP** 11.1, Writing a Skeleton Equation, p. 324 **CP** 11.2, Writing a Balanced Chemical Equation, p. 327 **CP** 11.3, Balancing a Chemical Equation, p. 328 **SE** Section 11.1 Assessment, p. 329 **iT** Section 11.1
GRSW Chapter 11, Section 2 L1 T T118–T120 L2 **GO** Chapter 11, Section 2 L2	**CP** 11.4, Writing Equations for Combination Reactions, p. 331 **CP** 11.5, Writing the Equation for a Decomposition Reaction, p. 332 **CP** 11.6, Writing Equations for Single-Replacement Reactions, p. 334 **CP** 11.7, Writing Equations for Double-Replacement Reactions, p. 335 **CP** 11.8, Writing Equations for Combustion Reactions, p. 337 **SE** Section 11.2 Assessment, p. 339 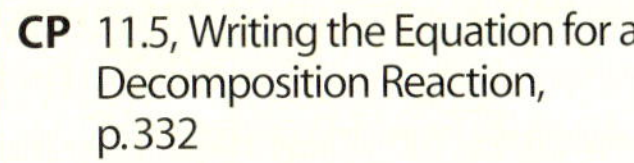 **iT** Section 11.2
GRSW Chapter 11, Section 3 L1  T T121 L2	**CP** 11.9, Writing and Balancing Net Ionic Equations, p. 343 **SE** Section 11.3 Assessment, p. 344 **iT** Section 11.3

Go Online

Web Code: cdn-1111
Web Code: cdn-1112
Web Code: cdn-1114

Web Code: cdb-1111

MATERIALS List

Quantities for each group

STUDENT EDITION

Inquiry Activity, p. 320
36 colored paper clips (12 each of 3 different colors)

Quick Lab, p. 326
aluminum foil, 20 cm × 20 cm, large beaker or glass pan, tarnished silver fork or spoon, sodium hydrogen carbonate, plastic tablespoon, hot water

Small-Scale Lab, p. 345
reaction surface, 0.05*M* $AgNO_3$, 0.2*M* $Pb(NO_3)_2$, 0.5*M* $CaCl_2$, 1.0*M* Na_2CO_3, 0.1*M* Na_3PO_4, 0.5*M* NaOH, 0.2*M* Na_2SO_4, 1.0*M* NaCl

TEACHER'S EDITION

Teacher Demo, p. 325
150-mL Pyrex beaker, table sugar, hood, concentrated sulfuric acid

Teacher Demo, p. 327
100-mL graduated cylinder, 0.1*M* copper(II) chloride solution, 400-mL beaker, metric ruler, scissors, aluminum foil

Teacher Demo, p. 333
magnesium ribbon, 10-mL graduated cylinder, 1*M* HCl, sodium, 2 beakers, tongs, water

Teacher Demo, p. 334
equimolar solutions of $BaCl_2$, Na_2SO_4, Na_3PO_4, $CaCl_2$, $Pb(NO_3)_2$, and KI

Teacher Demo, p. 335
50 mL 1.0*M* K_2CrO_4, 250-mL Erlenmeyer flask, 5 mL dilute $AgNO_3$ solution, small test tube, stopper, 50-mL graduated cylinder

Teacher Demo, p. 337
superfine steel wool, plastic sandwich bag, balance, ring stand, utility clamp, matches

Teacher Demo, p. 338
magnesium ribbon, large crucible, lab burner, cobalt blue glass or exposed photographic film, crucible tongs, matches or igniter

Class Activity, p. 340
research materials, graph paper

Teacher Demo, p. 341
calcium carbonate, dilute hydrochloric acid, 3 beakers, candle, matches

Chapter Assessment

CHAPTER ASSESSMENT

SE Chapter Assessment, pp. 347–350
CTB Chapter 11
 **iT** Chapter 11

STANDARDIZED TEST PREP

SE Chapter 11, p. 351
STP Topic 3
iT Interactive Textbook with Assessment at PHSchool.com

CHAPTER 11

Chemical Reactions

Modeling Chemical Reactions 

Purpose

Students describe what occurs during a chemical reaction

Prep Time 5 minutes
Class Time 15 minutes

Answers

Step 3: $2H_2 + O_2 \rightarrow 2H_2O$
Step 5: $CH_4 + 2O_2 \rightarrow CO_2 + 2H_2O$

Think About It

1. Dalton proposed that elements contain indivisible atoms that combine into compounds according to fixed ratios.
2. For the reactions modeled in Steps 2 and 4, the number of reactant atoms per element equals the number of product atoms per element. This conservation of atoms supports Dalton's theory that atoms are neither created nor destroyed during a reaction.

CHAPTER 11

Chemical Reactions

The brilliant colors of Autumn are evidence that chemical reactions have occurred.

INQUIRY Activity

Modeling Chemical Reactions

Materials

36 colored paper clips (12 each of 3 different colors)

Procedure

1. Each paper clip represents a single atom. Designate a different color of paper clip to represent atoms of oxygen (O), hydrogen (H), and carbon (C). Make two molecules each of hydrogen (H_2) and methane (CH_4) and six molecules of oxygen (O_2).
2. "React" one H_2 with one O_2 by splitting the molecules and joining one oxygen atom to two hydrogen atoms. Because there is an unreacted oxygen atom, you must react it with another hydrogen molecule to form a second water molecule.
3. Summarize what happened in this reaction by placing the number of each molecule reacted or formed in front of its formula.

   $___H_2 + ___O_2 \longrightarrow ___H_2O$
4. Now "react" methane (CH_4) with oxygen (O_2) to produce carbon dioxide (CO_2) and water (H_2O). Start with one molecule of methane and one molecule of oxygen. Continue reacting molecules of CH_4 and O_2 until all the reactant atoms have been used to form products.
5. Summarize what happened in this reaction.

   $___CH_4 + ___O_2 \longrightarrow ___CO_2 + ___H_2O$

Think About It

1. Summarize Dalton's atomic theory.
2. How might the numbers and kinds of atoms to the left and right of the arrow ($\longrightarrow$) in the model reactions you performed support this theory?

11.1 Describing Chemical Reactions

Connecting to Your World

On May 6, 1937, the huge airship *Hindenburg* was heading for its landing site in Lakehurst, New Jersey, after an uneventful trans-Atlantic crossing. Suddenly, to the horror of observers on the ground, the airship erupted into a fireball. Within a short time, 210,000 cubic meters of the airship's lifting gas, hydrogen, had burned and the airship was destroyed. The chemical reaction that occurred can be described as "hydrogen combines with oxygen to produce water." In this section, you will learn to represent this chemical reaction by a chemical equation.

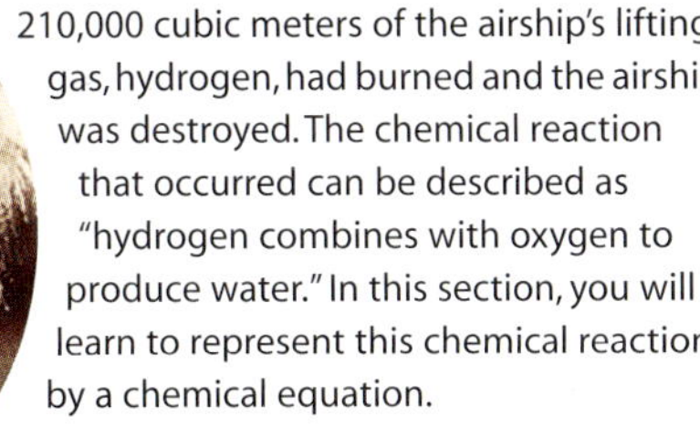

Guide for Reading

Key Concepts

- How do you write a word equation?
- How do you write a skeleton equation?
- What are the steps in writing a balanced chemical equation?

Vocabulary

chemical equation
skeleton equation
catalyst
coefficients
balanced equation

Reading Strategy

Relating Text and Visuals As you read this section, look at the illustrations of equations. In your notebook, explain how the illustrations demonstrate the difference between a balanced and unbalanced chemical equation.

Writing Chemical Equations

Every minute of the day chemical reactions take place—both inside you and around you. Not all are as dramatic as the explosion of the Hindenburg, but many are more complex. After a meal, a series of chemical reactions take place as your body digests food. Similarly, plants use sunlight to drive the photosynthetic processes needed to produce plant growth. Although the chemical reactions involved in photosynthesis and digestion are different, both chemical reactions are necessary to sustain life.

In a chemical reaction, one or more reactants change into one or more products. Figure 11.1a shows the ingredients for making leavened bread—flour, salt, yeast, and water. Chemical reactions take place when the ingredients are mixed together and heated in the oven. Figure 11.1b shows the product—a loaf of bread. Chemists use a chemical equation—a quick, shorthand notation—to convey as much information as possible about what happens in a chemical reaction.

Figure 11.1 Chemical changes occur when bread dough is mixed and baked. **a** Flour, salt, yeast, and water are the ingredients for making leavened bread. **b** Reactants in the ingredients undergo chemical changes to form the product (baked bread). **Observing** ***What evidence shows that chemical changes have occurred?***

Section Resources

Print

- **Guided Reading and Study Workbook,** Section 11.1
- **Core Teaching Resources,** Section 11.1 Review
- **Small-Scale Chemistry Laboratory Manual,** Lab 14
- **Transparencies,** T113–T117

Technology

- **Interactive Textbook with ChemASAP,** Animation 12; Simulation 11; Problem-Solving 11.2, 11.4, 11.6; Assessment 11.1
- **Go Online,** Section 11.1

11.1

1 FOCUS

Objectives

11.1.1 Describe how to write a word equation.

11.1.2 Describe how to write a skeleton equation

11.1.3 Describe the steps for writing a balanced chemical equation.

Guide for Reading

Build Vocabulary 

Paraphrase Write the following chemical equation on the board and have students provide synonymous words or phrases for each of the vocabulary words as they describe the equation:

$$2H_2O_2 \xrightarrow{MnO_2} 2H_2O + O_2$$

Reading Strategy 

Sequence As students read this section, have them write in their own words the sequence of steps that need to be taken to write and balance a chemical equation.

2 INSTRUCT

Connecting to Your World

Have students study the photograph and read the text. Explain that airships during the 1930s often contained hydrogen because it is the lightest element. However, when hydrogen is ignited in air, a vigorous reaction occurs. Have students describe the results as shown in the photograph. Write the word equation for the reaction of hydrogen with oxygen to produce water. Ask, **Do you think there is a simpler way to write chemical equations?** *(Yes, describe the reaction using chemical symbols of the elements and chemical formulas of compounds.)*

Answers to...

Figure 11.1 The separate ingredients of the bread are now combined into one product—the loaf of bread.

Writing Chemical Equations

Discuss L2

To assess students' prior knowledge about chemical compounds, ask, **What are the differences between molecular compounds and ionic compounds?** *(Molecular compounds are composed of nonmetallic elements. Ionic compounds are composed of a metal cation and an anion.)* **How are chemical compounds represented?** *(By chemical formulas that show the kinds and numbers of atoms in each molecule or formula unit.)* **What is the difference between a monatomic ion and a polyatomic ion?** *(A monatomic ion is formed when an atom gains or loses one or more electrons. A polyatomic ion is a tightly bound group of atoms that behaves as a unit and carries a charge.)*

Use Visuals L1

Figure 11.1 Have students study the photographs. Ask, **What are some indications that a chemical reaction has taken place?** *(increase in volume and change in color)* Explain that bakers use yeast to make bread dough rise. Yeasts are unicellular organisms that are able to extract energy from sugar in the absence of oxygen, a process known as fermentation. Point out that yeasts produce proteins that catalyze or facilitate the breakdown of carbohydrates in the dough, producing CO_2 and ethanol. The bubbles of trapped CO_2 cause the dough to rise. As the dough bakes, the ethanol evaporates.

Download a worksheet on **Chemical Equations** for students to complete, and find additional teacher support from NSTA SciLinks.

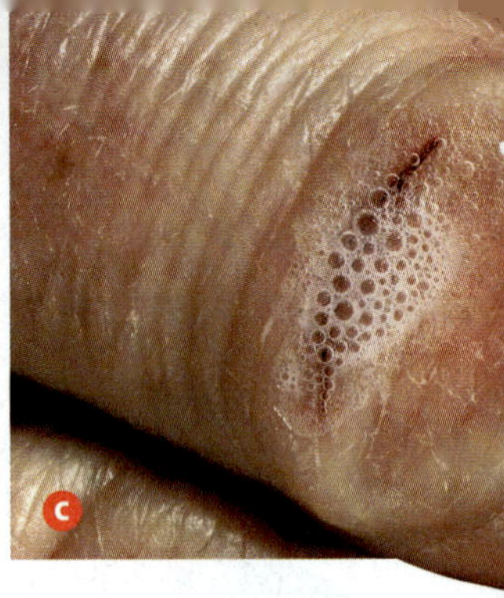

Figure 11.2 Three common chemical reactions are shown. **a** When methane gas burns, it combines with oxygen to form carbon dioxide and water. **b** Iron turns to red-brown rust (iron(III) oxide) in the presence of oxygen in the air. **c** Hydrogen peroxide decomposes to water and oxygen when used as an antiseptic.

Word Equations How do you describe what happens in a chemical reaction? Recall from Chapter 2 the shorthand method for writing a description of a chemical reaction. In this method, the reactants were written on the left and the products on the right. An arrow separated them. You read the arrow as *yields*, *gives*, or *reacts to produce*.

$$\text{Reactants} \longrightarrow \text{products}$$

How could you describe the rusting of iron shown in Figure 11.2b? You could say: "Iron reacts with oxygen to produce iron(III) oxide (rust)." That's a perfectly good description, but it might be quicker and easier to identify the reactants and product by means of a word equation.

$$\text{Iron} + \text{oxygen} \longrightarrow \text{iron(III) oxide}$$

To write a word equation, write the names of the reactants to the left of the arrow separated by plus signs; write the names of the products to the right of the arrow, also separated by plus signs. Notice that no plus sign is needed on the product side of this equation because iron(III) oxide is the only product.

Have you ever poured the antiseptic hydrogen peroxide on an open cut? Bubbles of oxygen gas form rapidly, as shown in Figure 11.2c. The production of a new substance, a gas, is evidence of a chemical change. Two new substances are produced in this reaction, oxygen gas and liquid water. You could describe this reaction by saying, "Hydrogen peroxide decomposes to form water and oxygen gas." You could also write a word equation.

$$\text{Hydrogen peroxide} \longrightarrow \text{water} + \text{oxygen}$$

When you light a burner on your stove, methane gas bursts into flames and produces the energy needed to heat your soup. Methane is the major component of natural gas, a common fuel for heating homes and cooking food. The burning of methane, as shown in Figure 11.2a, is a chemical reaction. How would you write the word equation for this reaction? Burning a substance typically requires oxygen, so methane and oxygen are the reactants. The products are water and carbon dioxide. Thus the word equation is this:

$$\text{Methane} + \text{oxygen} \longrightarrow \text{carbon dioxide} + \text{water}$$

What does the arrow ($\longrightarrow$) in a word equation mean?

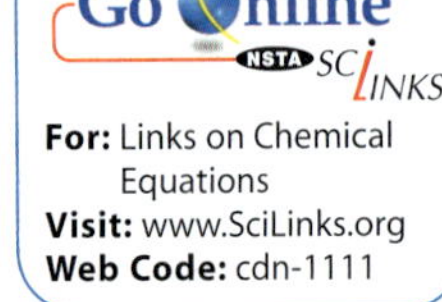

For: Links on Chemical Equations
Visit: www.SciLinks.org
Web Code: cdn-1111

Differentiated Instruction

English Learners L1

As they read through the chapter, Spanish-speaking students with limited English proficiency may find it helpful to consult the chapter summary and key terms in the Spanish supplement in the Teacher's Resource Package.

Chemical Equations Word equations adequately describe chemical reactions, but they are cumbersome. It's easier to use the formulas for the reactants and products to write chemical equations. A **chemical equation** is a representation of a chemical reaction; the formulas of the reactants (on the left) are connected by an arrow with the formulas of the products (on the right). Here is a chemical equation for rusting:

$$Fe + O_2 \longrightarrow Fe_2O_3$$

Equations that show just the formulas of the reactants and products are called skeleton equations. A **skeleton equation** is a chemical equation that does not indicate the relative amounts of the reactants and products. The first step in writing a complete chemical equation is to write the skeleton equation. **Write the formulas of the reactants to the left of the yields sign (arrow) and the formulas of the products to the right.**

To add more information to the equation, you can indicate the physical states of substances by putting a symbol after each formula. Use (*s*) for a solid, (*l*) for a liquid, (*g*) for a gas, and (*aq*) for a substance in aqueous solution (a substance dissolved in water). Here is the equation for rusting with symbols for the physical states added:

$$Fe(s) + O_2(g) \longrightarrow Fe_2O_3(s)$$

In many chemical reactions, a catalyst is added to the reaction mixture. A **catalyst** is a substance that speeds up the reaction but is not used up in the reaction. A catalyst is neither a reactant nor a product, so its formula is written above the arrow in a chemical equation. For example, Figure 11.3 shows that the compound manganese(IV) oxide ($MnO_2(s)$) catalyzes the decomposition of an aqueous solution of hydrogen peroxide ($H_2O_2(aq)$) to produce water and oxygen.

$$H_2O_2(aq) \xrightarrow{MnO_2} H_2O(l) + O_2(g)$$

Many of the symbols commonly used in writing chemical equations are listed in Table 11.1.

Table 11.1

Symbols Used in Chemical Equations

Symbol	Explanation
+	Used to separate two reactants or two products
$\longrightarrow$	"Yields," separates reactants from products
$\rightleftharpoons$	Used in place of $\longrightarrow$ for reversible reactions
(*s*)	Designates a reactant or product in the solid state; placed after the formula
(*l*)	Designates a reactant or product in the liquid state; placed after the formula
(*g*)	Designates a reactant or product in the gaseous state; placed after the formula
(*aq*)	Designates an aqueous solution; the substance is dissolved in water; placed after the formula
$\xrightarrow{\Delta}$ $\xrightarrow{heat}$	Indicates that heat is supplied to the reaction
$\xrightarrow{Pt}$	A formula written above or below the yield sign indicates its use as a catalyst (in this example, platinum).

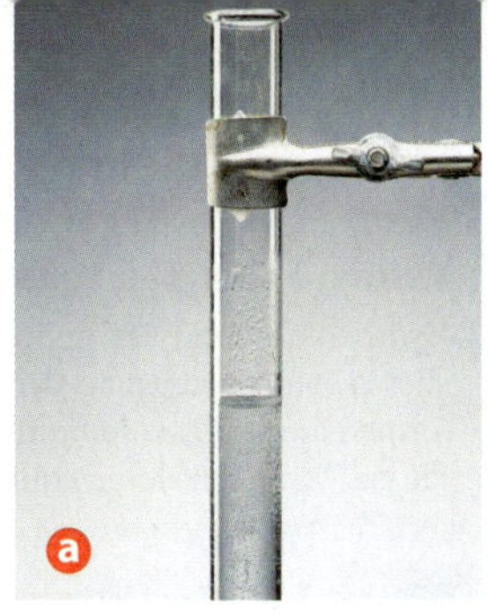

Figure 11.3 Hydrogen peroxide decomposes to form water and oxygen gas. **a** Bubbles of oxygen appear slowly as decomposition proceeds. **b** With the addition of the catalyst manganese(IV) oxide (MnO_2), decomposition speeds up. The white "smoke" is condensed water vapor.

Animation 12 Relate chemical symbols and formulas to the information they communicate.

with ChemASAP

Use Visuals L1

Table 11.1 Use an overhead projector to display Table 11.1. Make sure students understand the meaning and role of each symbol. Point out that the phase symbols provide important clues about reactions. Note that the items placed above the yield arrow represent conditions that must be met before the reaction can take place at a reasonable pace.

Use Visuals L1

Figure 11.3 Point out that although hydrogen peroxide contains the same kinds of elements as water, it has very different properties. Hydrogen peroxide is much less stable than water. Have students write the balanced chemical equation for the reaction described in the caption. Students should indicate the MnO_2 catalyst above the reaction arrow. Write the equation on the board and explain that this is an example of a decomposition reaction—a single compound is broken down into two or more products. Explain that although hydrogen peroxide is an unstable compound that readily decomposes to produce water and oxygen, it does so slowly at room temperature. Therefore, hydrogen peroxide must come in contact with a catalyst in order for the decomposition to proceed rapidly, as shown in **b**.

Relate L2

Tell students that frequently in nature the products of one reaction become the reactants of a subsequent reaction. An example is the combustion of octane (C_8H_{18}), the primary ingredient of gasoline, which produces carbon dioxide and water vapor. The carbon dioxide can combine with atmospheric water vapor to produce carbonic acid (H_2CO_3), which is a component of acid rain. Ask students if they can think of other such processes.

Facts and Figures

Hydrogen Peroxide

Challenge students to propose a hypothesis about why our bodies produce an enzyme that helps catalyze the decomposition of hydrogen peroxide. *(to prevent the accumulation in the body of hydrogen peroxide, a toxic by-product of reactions involving oxygen.)* Students may be interested to learn that hydrogen peroxide is sometimes used to restore the clarity of old paintings. Lead-based paints darken with time (PbS). Hydrogen peroxide converts PbS to $PbSO_4$.

Answers to...

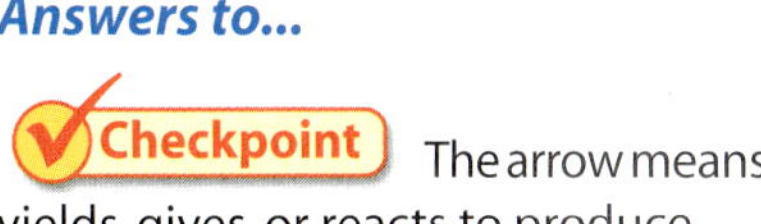

The arrow means yields, gives, or reacts to produce.

CONCEPTUAL PROBLEM 11.1

Answers

1. When solid sodium is dropped in water, hydrogen gas and aqueous sodium hydroxide are produced.
2. $S(s) + O_2(g) \rightarrow SO_2(g)$

Balancing Chemical Equations

Discuss L2

Review with students the law of conservation of mass. Emphasize that mass, or matter, cannot be created or destroyed. Explain that it is because of this law that a chemical equation is not complete until it is balanced.

Address Misconceptions

Students often think that they can balance an equation by changing the subscripts in one or more of the formulas. Use the following example to show why this approach is incorrect. $H_2(g) + O_2(g) \rightarrow H_2O(l)$ could be balanced by changing the formula of the product to H_2O_2. But H_2O_2 (hydrogen peroxide) is not the same substance as water. Therefore, the equation would describe a different reaction. To help students overcome this misconception, have them draw a box around the formulas for the reactants and products before they start to balance the equation. Tell them that the boxes are "off-limits" or "out of bounds." They cannot change any number that appears inside a box.

Relate L2

Glass is often etched to provide a design. Glass contains calcium silicate, $CaSiO_3$. When hydrofluoric acid, HF, is placed on the glass, it reacts with the calcium silicate, etching the glass by producing aqueous calcium fluoride, silicon tetrafluoride, and water. Have students write a balanced chemical equation for this reaction. *($CaSiO_3 + 6HF \rightarrow CaF_2 + SiF_4 + 3H_2O$)*

CONCEPTUAL PROBLEM 11.1

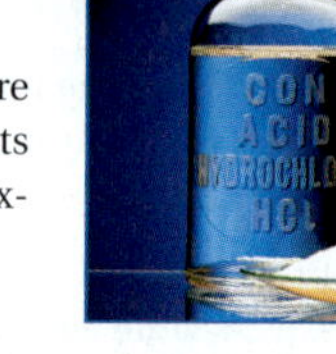

Writing a Skeleton Equation

Hydrochloric acid and solid sodium hydrogen carbonate are shown before being placed in the beaker to react. The products formed are aqueous sodium chloride, water, and carbon dioxide gas. Write a skeleton equation for this chemical reaction.

 **Analyze** ***Identify the relevant concepts.***
Write the correct formula for each substance in the reaction. Separate the reactants from the products by means of an arrow. Indicate the state of each substance.

2 **Solve** ***Apply concepts to this situation.***
solid sodium hydrogen carbonate: $NaHCO_3(s)$
hydrochloric acid: $HCl(aq)$
aqueous sodium chloride: $NaCl(aq)$
water: $H_2O(l)$
carbon dioxide gas: $CO_2(g)$
$NaHCO_3(s) + HCl(aq) \longrightarrow NaCl(aq) + H_2O(l) + CO_2(g)$

Practice Problems

1. Write a sentence that describes this chemical reaction.
$Na(s) + H_2O(l) \longrightarrow NaOH(aq) + H_2(g)$
2. Sulfur burns in oxygen to form sulfur dioxide. Write a skeleton equation for this chemical reaction. Include appropriate symbols from Table 11.1.

Interactive Textbook

Problem-Solving 11.2
Solve Problem 2 with the help of an interactive guided tutorial.
with ChemASAP

Balancing Chemical Equations

How would you write a word equation for the manufacture of bicycles? Simplify your task by limiting yourself to four major components: frames, wheels, handlebars, and pedals. Your word equation for making a bicycle could read like this.

$$\underset{\text{(reactants)}}{\text{Frame} + \text{wheel} + \text{handlebar} + \text{pedal}} \longrightarrow \underset{\text{(product)}}{\text{bicycle}}$$

Your word equation shows the reactants (the kinds of parts) and the product (a bicycle). But if you were responsible for ordering parts to make a bicycle, this word equation would be inadequate because it does not indicate the quantity of each part needed to make one bicycle.

A standard bicycle is composed of one frame (F), two wheels (W), one handlebar (H), and two pedals (P). Using these symbols, the formula for a bicycle would be FW_2HP_2. The skeleton equation for bicycle assembly would be this:

$$F + W + H + P \longrightarrow FW_2HP_2$$

This is an unbalanced equation. An unbalanced equation does not indicate the quantity of the reactants needed to make the product. A complete description of the reaction must include not only the kinds of parts involved but also the quantities of parts required.

Differentiated Instruction

Special Needs L1

Have students, working in pairs, take turns writing equations for each other to balance. Two pairs of students may enjoy getting together for a "doubles match" in which each pair works as a team to balance equations that the opposing team creates.

324 Chapter 11

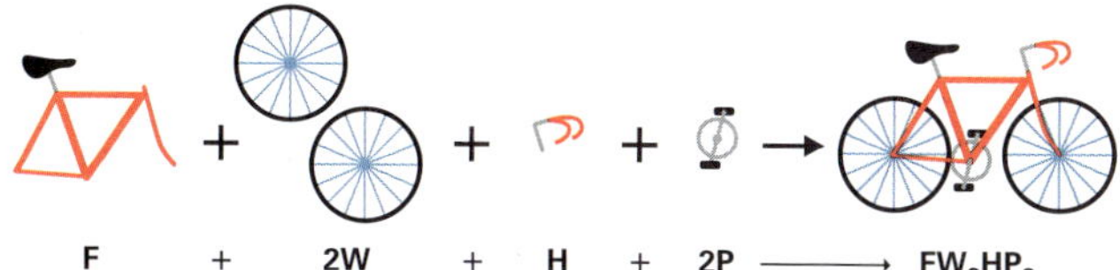

This is a balanced equation for making a bicycle. It tells you that one frame, two wheels, one handlebar, and two pedals produce one bicycle. To balance the equation, the number 2 was placed before wheels and pedals. The number 1 is understood to be in front of frame, handlebar, and bicycle. These numbers are called **coefficients**—small whole numbers that are placed in front of the formulas in an equation in order to balance it. In this balanced equation, the number of each bicycle part on the reactant side is the same as the number of those parts on the product side. A chemical reaction is also described by a **balanced equation** in which each side of the equation has the same number of atoms of each element and mass is conserved. Real bicycles are being assembled in Figure 11.4.

Recall that John Dalton's atomic theory states that as reactants are converted to products, the bonds holding the atoms together are broken and new bonds are formed. The atoms themselves are neither created nor destroyed; they are merely rearranged. This part of Dalton's theory explains the law of conservation of mass: In any chemical change, mass is conserved. The atoms in the products are the same atoms that were in the reactants—they are just rearranged.

Representing a chemical reaction by a balanced chemical equation is a two-step process. **To write a balanced chemical equation, first write the skeleton equation. Then use coefficients to balance the equation so that it obeys the law of conservation of mass.** In every balanced equation, each side of the equation has the same number of atoms of each element.

Sometimes a skeleton equation may already be balanced. For example, carbon burns in the presence of oxygen to produce carbon dioxide.

$$C(s) + O_2(g) \longrightarrow CO_2(g)$$

Reactants 1 carbon atom, 2 oxygen atoms — **Product** 1 carbon atom, 2 oxygen atoms

This equation is balanced. One carbon atom and two oxygen atoms are on each side of the equation. You do not need to change the coefficients; they are all understood to be 1.

Checkpoint ***What is the law of conservation of mass?***

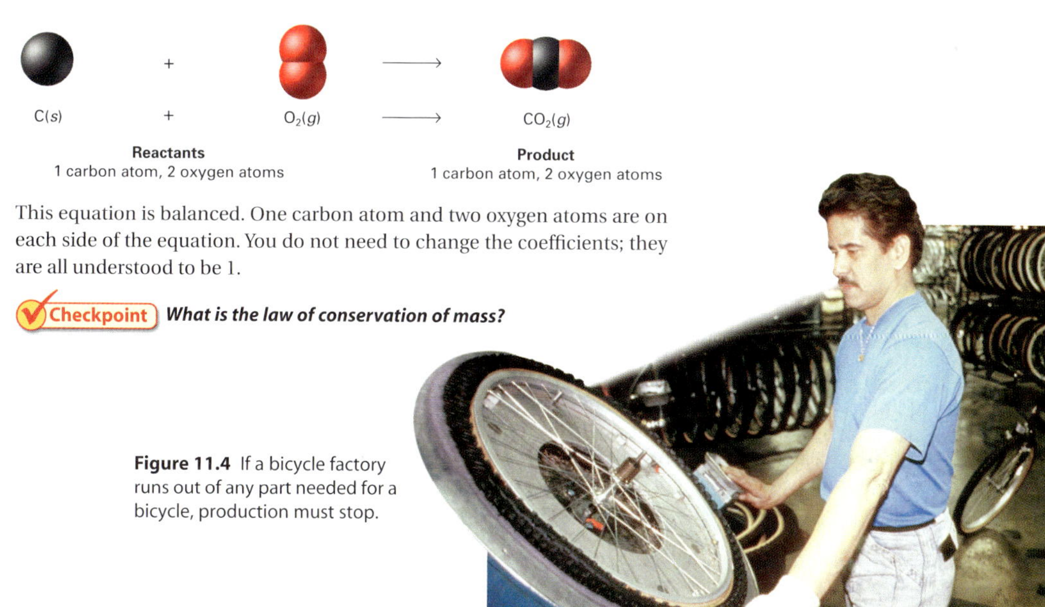

Figure 11.4 If a bicycle factory runs out of any part needed for a bicycle, production must stop.

Differentiated Instruction

Gifted and Talented L3

Point out to students that not all chemical equations are easily balanced. Methods other than those they currently use must be used to balance some equations, such as those they will encounter in Chapter 21.

Challenge students to balance the following equation.

$HCl + HNO_3 \rightarrow HOCl + NO + H_2O$

($3HCl + 2HNO_3 \rightarrow 3HOCl + 2NO + H_2O$)

TEACHER Demo

An Example of Chemical Change L2

Purpose Students observe a dramatic example of chemical change.

Materials 150-mL Pyrex beaker, table sugar, hood, concentrated sulfuric acid

Procedure Fill a 150-mL Pyrex beaker to the 30-mL mark with table sugar. Place the beaker in a hood, and add 30 mL of concentrated sulfuric acid, which will dehydrate the sugar.

Safety and Disposal The hood will remove any sulfur dioxide and carbon monoxide, which are possible by-products of the reaction. Wear goggles, a face shield, a lab apron, and protective gloves. Use a plexiglas shield to separate the students from the reaction. As a further precaution, have them wear goggles. While wearing protective gear, place the beaker and contents into a 500-mL beaker half full of water. Swirl the small beaker in the water and then neutralize using a base with a molarity of less than 1*M*. Decant the liquid and fill the large beaker halfway with water once again. Let stand overnight in a protected location, test for pH, and neutralize. Repeat if necessary. Wrap the carbon in newspaper before placing in a wastebasket.

Expected Outcome The carbon left behind will expand to form a porous, foam cylinder as the water vapor escapes.

Discuss L2

Tell students that it is not necessary for coefficients to be the same on both sides of the equation in order for the number of atoms of each type to balance. Use the diagram of the formation of CO_2 on the student page to illustrate this concept. The atoms balance, but the coefficients total 2 on the left side of the equation and 1 on the right side. If the sums of the coefficients on both sides of a balanced chemical equation are the same number, it is coincidental.

Answers to...

Checkpoint In any physical or chemical change, mass is conserved.

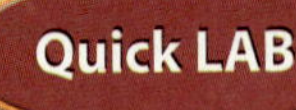

Removing Silver Tarnish L2

Objective After completing this activity, students will be able to:

- describe and write a balanced chemical equation for a single-replacement reaction.

Skills Focus Observing, Inferring, Communicating Results

Prep Time 10 minutes
Class Time 20 minutes

Expected Outcome Aluminum replaces the silver in the tarnish on the fork or spoon.

Analyze and Conclude

1. After the reaction, the fork looks silvery, and the tarnish is gone.
2. The aluminum becomes darkened.
3. Al_2S_3
4. $3Ag_2S(s) + 2Al(s) \rightarrow 6Ag(s) + Al_2S_3(s)$

For Enrichment

Have students repeat the Quick Lab using a piece of magnesium instead of aluminum foil. Have them compare the results and the speed of the reaction. Students will find that the reaction occurs more quickly with magnesium. After students learn about single-replacement reactions on page 333, have them explain why the reaction proceeded more rapidly with magnesium. *(Magnesium is more active than aluminum.)*

Relate

Tell students that one way of purifying air in space vehicles is to react lithium peroxide with carbon dioxide, producing lithium carbonate and oxygen. Ask, **What is the balanced chemical equation for this reaction?** *($2Li_2O_2 + 2CO_2 \rightarrow 2Li_2CO_3 + O_2$)*

Quick LAB

Removing Silver Tarnish

Materials

- aluminum foil, 20 cm × 20 cm
- large beaker or glass pan
- tarnished silver fork or spoon
- sodium hydrogen carbonate
- plastic tablespoon
- hot water

Procedure

1. Fill the beaker about three-quarters full of hot water and add 2 tablespoons of sodium hydrogen carbonate ($NaHCO_3$).
2. Crush the aluminum foil into a loose ball and place it in the beaker.
3. Write a brief description of the tarnished silver fork, then place it in the beaker so that it is touching the aluminum ball.
4. Allow the beaker to stand undisturbed for 30 minutes.
5. Remove the fork and aluminum ball and rinse them with water.

Analyze and Conclude

1. Compare the silver fork with your observations before placing the fork in the water. What changes do you observe?
2. Did a chemical reaction occur? How do you know?
3. The tarnish on the silver fork is silver sulfide (Ag_2S). Silver becomes tarnished when it is exposed to air, egg yolk, or rubber bands. Each of these substances contains sulfur. Look carefully for a pale yellow precipitate of aluminum sulfide on the bottom of the beaker. Write the formula for aluminum sulfide.
4. The unbalanced equation for the reaction is
   $$Ag_2S(s) + Al(s) \longrightarrow Al_2S_3(s) + Ag(s)$$
   Balance the equation.

What about the equation for the reaction of hydrogen gas and oxygen gas? This is the reaction that occurred in the Hindenburg disaster, which you read about in *Connecting to Your World.* When hydrogen and oxygen are mixed, a spark will initiate a rapid reaction. The product of the reaction is water. This is the equation for the burning of hydrogen:

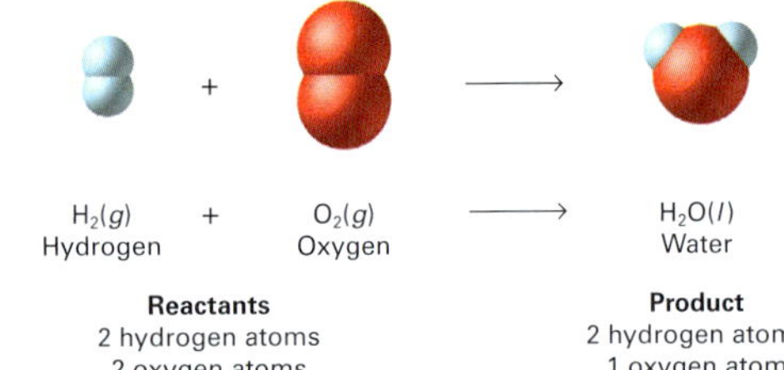

The formulas for all the reactants and product are correct, but this equation is not balanced. Count the atoms on both sides of the equation. Two oxygen atoms are on the reactant (left) side of the equation and only one oxygen atom is on the product (right) side. As written, the equation does not obey the law of conservation of mass and so it does not describe what really happens. What can you do to balance it? A few guidelines for writing and balancing equations will help.

Interactive Textbook
Simulation 11 Sharpen your skills by balancing chemical equations.
with ChemASAP

326 Chapter 11

Differentiated Instruction

Special Needs

Polystyrene models can help students understand the meaning of subscripts and coefficients when balancing chemical reactions. They can also help illustrate how atoms are rearranged during a chemical reaction. Use a knife or razor blade to cut polystyrene balls of various sizes in half. Paint the hemispheres different colors to stand for various elements. Glue a small, flat magnet to the flat side of each of the hemispheres. Using the spheres, show on a metal board how formulas of reactants become formulas of products in various types of reactions.

Rules for Writing and Balancing Equations

1. Determine the correct formulas for all the reactants and products.
2. Write the skeleton equation by placing the formulas for the reactants on the left and the formulas for the products on the right with a yields sign ($\longrightarrow$) in between. If two or more reactants or products are involved, separate their formulas with plus signs.
3. Determine the number of atoms of each element in the reactants and products. Count a polyatomic ion as a single unit if it appears unchanged on both sides of the equation.
4. Balance the elements one at a time by using coefficients. When no coefficient is written, it is assumed to be 1. Begin by balancing elements that appear only once on each side of the equation. Never balance an equation by changing the subscripts in a chemical formula. Each substance has only one correct formula.
5. Check each atom or polyatomic ion to be sure they are equal on both sides of the equation.
6. Make sure all the coefficients are in the lowest possible ratio.

CONCEPTUAL PROBLEM 11.2

Writing a Balanced Chemical Equation

Hydrogen and oxygen react to form water. The reaction releases enough energy to launch a rocket. Write a balanced equation for the reaction.

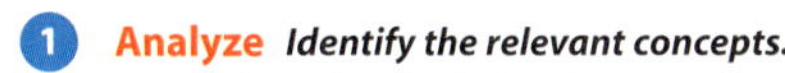

1 Analyze ***Identify the relevant concepts.*** Apply the rules for balancing equations to the skeleton equation describing the reaction.

2 Solve ***Apply concepts to this situation.*** Write correct formulas to give the skeleton equation.

$$H_2(g) + O_2(g) \longrightarrow H_2O(l)$$

Count the number of each kind of atom. Hydrogen is balanced but oxygen is not. If you put the coefficient 2 in front of H_2O, the oxygen will be balanced.

$$H_2(g) + O_2(g) \longrightarrow 2H_2O(l)$$

Now twice as many hydrogen atoms are in the product as are in the reactants. To correct this, put the coefficient 2 in front of H_2. Four hydrogen atoms and two oxygen atoms are on each side of the chemical equation. The equation is now balanced.

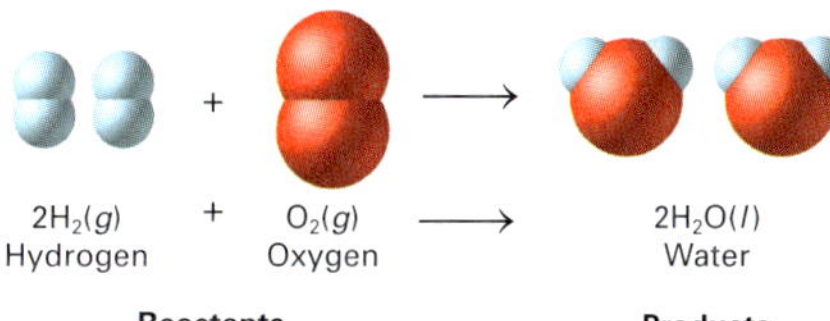

Practice Problems

3. Balance each equation.
   a. $AgNO_3 + H_2S \longrightarrow Ag_2S + HNO_3$
   b. $Zn(OH)_2 + H_3PO_4 \longrightarrow Zn_3(PO_4)_2 + H_2O$
4. Rewrite these word equations as balanced chemical equations.
   a. hydrogen + sulfur ⟶ hydrogen sulfide
   b. iron(III) chloride + calcium hydroxide ⟶ iron(III) hydroxide + calcium chloride

Problem-Solving 11.4 Solve Problem 4 with the help of an interactive guided tutorial.

with **ChemASAP**

Differentiated Instruction

English Learners

Have students name different elements in their native languages and in English. Some periodic table Internet sites list the names of elements in several languages other than English. Then have them list the symbols for several elements in their native languages. Point out that the names might differ, but chemical symbols are the same in any language. Balancing equations uses only universal symbols and numbers and is not language dependent.

CONCEPTUAL PROBLEM 11.2

Answers

3. a. $2AgNO_3 + H_2S \rightarrow Ag_2S + 2HNO_3$

b. $3Zn(OH)_2 + 2H_3PO_4 \rightarrow Zn_3(PO_4)_2 + 6H_2O$

4. a. $H_2 + S \rightarrow H_2S$

b. $2FeCl_3 + 3Ca(OH)_2 \rightarrow 2Fe(OH)_3 + 3CaCl_2$

Practice Problems Plus

1. Rewrite the following word equation as a balanced chemical equation: aluminum sulfate + calcium hydroxide → aluminum hydroxide + calcium sulfate. *($Al_2(SO_4)_3 + 3Ca(OH)_2 \rightarrow 2Al(OH)_3 + 3CaSO_4$)*

2. Rewrite the following word equation as a balanced chemical equation: phosphoric acid + sodium hydroxide → sodium phosphate + water. *($H_3PO_4 + 3NaOH \rightarrow Na_3PO_4 + 3H_2O$)*

TEACHER Demo

Balance a Chemical Equation

Purpose Students observe a chemical reaction and write skeleton and balanced chemical equations for the reaction.

Materials 100-mL graduated cylinder, 0.1*M* copper(II) chloride solution, 400-mL beaker, metric ruler, scissors, aluminum foil

Safety $CuCl_2$ is an irritant.

Procedure Add 200 mL of 0.1*M* copper(II) chloride solution to a 400-mL beaker. Cut a 5-cm square piece of aluminum foil. Crumple the piece of aluminum foil into a loose ball and place it in the copper solution. Ask students to note any evidence of chemical change. Write a sentence on the board that describes the chemical reaction. Have students write a skeleton equation for the reaction in their notebooks. Then have students balance the equation. Decant the liquid and flush down the drain with excess water. Dry the copper residue and use it again.

Expected Outcome The solution will begin to produce gas and a red-brown precipitate of copper will fall to the bottom of the beaker. The skeleton equation is $CuCl_2(aq) + Al(s) \rightarrow AlCl_3(aq) + Cu(s)$. The balanced chemical equation is $3CuCl_2(aq) + 2Al(s) \rightarrow 2AlCl_3(aq) + 3Cu(s)$.

Section 11.1 (continued)

CONCEPTUAL PROBLEM 11.3

Answers

5. a. $FeCl_3 + 3NaOH \rightarrow Fe(OH)_3 + 3NaCl$

b. $CS_2 + 3Cl_2 \rightarrow CCl_4 + S_2Cl_2$

6. $Ca(OH)_2 + H_2SO_4 \rightarrow CaSO_4 + 2H_2O$

Discuss L2

Emphasize to students that in order to write a correct balanced chemical equation, the formulas for all reactants and products must be written correctly. Write an unbalanced chemical equation on the board, with all the formulas written correctly. Rewrite the equation with one formula written incorrectly. Identify the incorrect formula so that students consistently recognize that it is not correct. Have students balance the two equations and note the differences. Often an equation with an incorrect formula cannot be balanced.

Relate L2

Point out that balanced chemical equations can help manufacturers who use chemical processes to know how much of each reactant should be purchased and how much of each product will be made. For example, a welder might use oxygen and acetylene (C_2H_2) in a torch. Have students write a balanced chemical equation for the reaction that occurs in the torch. Water and carbon dioxide are produced. *($2C_2H_2 + 5O_2 \rightarrow 4CO_2 + 2H_2O$)* Ask, **From this balanced equation, what would be an effective acetylene: oxygen ratio in the torch?** *(2:5)*

Have students research chemistry-related careers in the library or on the Internet. Students can then construct a table that describes the nature of the work, educational and training requirements, employment outlook, working conditions, and other necessary information.

The reaction of copper metal with an aqueous solution of silver nitrate is described by this skeleton equation. How can it be balanced?

$$Cu(s) + AgNO_3(aq) \longrightarrow Cu(NO_3)_2(aq) + Ag(s)$$

The nitrate ion appears unchanged on both sides of the equation, so this ion can be balanced as a unit. Do this by placing a 2 in front of $AgNO_3$.

$$Cu(s) + 2AgNO_3(aq) \longrightarrow Cu(NO_3)_2(aq) + Ag(s)$$

But now the atoms of silver are not balanced. You must add a 2 in front of Ag on the product side to balance the atoms of silver.

$$Cu(s) + 2AgNO_3(aq) \longrightarrow Cu(NO_3)_2(aq) + 2Ag(s)$$

The equation is now balanced because the same number of each kind of atom are on both sides of the equation.

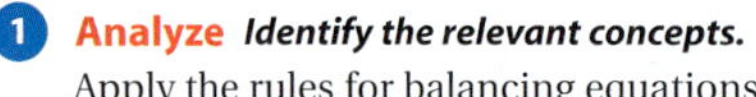

CONCEPTUAL PROBLEM 11.3

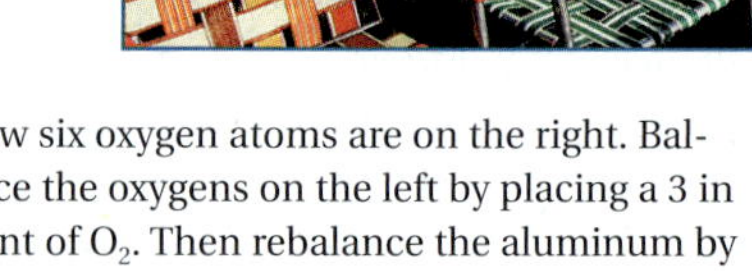

Balancing a Chemical Equation

Aluminum is a good choice for outdoor furniture because it reacts with oxygen in the air to form a thin protective coat of aluminum oxide. Balance the equation for this reaction.

$$Al(s) + O_2(g) \longrightarrow Al_2O_3(s)$$

1 Analyze ***Identify the relevant concepts.***

Apply the rules for balancing equations.

2 Solve ***Apply concepts to this situation.***

First balance the aluminum by placing the coefficient 2 in front of $Al(s)$.

$$2Al(s) + O_2(g) \longrightarrow Al_2O_3(s)$$

How can the odd number of oxygen atoms in the product (right side) balance the even number of oxygen atoms on the left? Any whole-number coefficient placed in front of the O_2 will give an even number of oxygen atoms on the left. This is because the coefficient is always being multiplied by the subscript 2. The solution is to multiply the formula with the odd number of oxygen atoms by 2.

$$2Al(s) + O_2(g) \longrightarrow 2Al_2O_3(s)$$

Now six oxygen atoms are on the right. Balance the oxygens on the left by placing a 3 in front of O_2. Then rebalance the aluminum by changing the coefficient of $Al(s)$ from 2 to 4.

$$4Al(s) + 3O_2(g) \longrightarrow 2Al_2O_3(s)$$

Suppose the equation for the formation of aluminum oxide was written this way.

$$8Al(s) + 6O_2(g) \longrightarrow 4Al_2O_3(s)$$

Because this equation obeys the law of conservation of mass, it is correct. However, equations are normally written with coefficients in their lowest possible ratio. Each of the coefficients can be divided by 2 to give the previous equation, which has the lowest whole-number ratio of coefficients.

Practice Problems

5. Balance each equation.

a. $FeCl_3 + NaOH \longrightarrow Fe(OH)_3 + NaCl$

b. $CS_2 + Cl_2 \longrightarrow CCl_4 + S_2Cl_2$

6. Write and balance this equation.

calcium hydroxide + sulfuric acid $\longrightarrow$ calcium sulfate + water

Problem-Solving 11.6
Solve Problem 6 with the help of an interactive guided tutorial.
with ChemASAP

Differentiated Instruction

Special Needs L1

Provide students with molecular model kits or Styrofoam balls of different sizes and colors. Using the balls to represent different types of atoms, have students model balancing a simple equation. Provide students with several additional unbalanced chemical equations, and have them balance the equations using the models. Caution students to keep "atoms" in compounds together so that they use multiples of the entire unit and not multiples of just one atom in the unit.

Careers in Chemistry

Hazardous Materials Specialist

Hazardous materials specialists do the vital work of keeping the environment safe from harmful substances. Backed by regulations devised primarily by the Environmental Protection Agency (EPA), these specialists are responsible for the safe handling, treatment, storage, and transportation of hazardous materials. They work at all levels of government and are employed by industries and universities where dangerous materials are generated or used. Part of their jobs may be to educate those who need to know, including the public, about the rules and regulations applied to hazardous materials.

Hazardous materials specialists often work in the field inspecting, testing, overseeing cleanup work, and determining whether storage facilities are in compliance with regulations. First on the scene when a chemical spill occurs, these specialists must be able to identify the spilled chemical substance and know how to take emergency action to neutralize its effect and thus protect the public. Then they must devise a cleanup procedure and oversee its implementation.

Entry into the field usually requires a degree from a four-year college with a major in chemistry or other physical science. Other areas of concentration might be industrial hygiene, environmental health, or engineering.

For: Careers in Chemistry
Visit: PHSchool.com
Web Code: cdb-1111

11.1 Section Assessment

7. **Key Concept** How do you write a word equation?
8. **Key Concept** How do you write a skeleton equation?
9. **Key Concept** Describe the steps in writing a balanced chemical equation.
10. Write skeleton equations for these reactions.
    a. Heating copper(II) sulfide in the presence of diatomic oxygen produces pure copper and sulfur dioxide gas.
    b. When heated, baking soda (sodium hydrogen carbonate) decomposes to form the products sodium carbonate, carbon dioxide, and water.
11. Write and balance equations for the following reactions.
    a. Iron metal and chlorine gas react to form solid iron(III) chloride.
    b. Solid aluminum carbonate decomposes to form solid aluminum oxide and carbon dioxide gas.
    c. Solid magnesium reacts with aqueous silver nitrate to form solid silver and aqueous magnesium nitrate.
12. Balance the following equations.
    a. $SO_2 + O_2 \longrightarrow SO_3$
    b. $Fe_2O_3 + H_2 \longrightarrow Fe + H_2O$
    c. $P + O_2 \longrightarrow P_4O_{10}$
    d. $Al + N_2 \longrightarrow AlN$

Writing Activity

Paragraph Some products are marketed as biodegradable. What does *biodegradable* mean? Identify three biodegradable products. How do these products benefit the environment?

Assessment 11.1 Test yourself on the concepts in Section 11.1.

with ChemASAP

Careers in Chemistry

Hazardous Materials Specialist

Materials are classified as hazardous when the chemical reactions they undergo are potentially dangerous to humans or the environment. Being a hazardous materials specialist is an ever-changing career because the list of materials that are considered hazardous is constantly changing.

3 ASSESS

Evaluate Understanding 

To evaluate students' understanding of how to write and interpret chemical equations, write some word equations on the board and have students produce skeleton equations. Write some unbalanced chemical equations on the board; have students balance them and describe them in words. Write some equations on the board and begin to balance them by changing subscripts. When students object, ask why this approach is incorrect.

Reteach 

Review with students the key steps in writing balanced equations. Have them make a flow chart to describe the best way to subdivide the task.

Writing Activity

Student answers will vary but should reflect an understanding of the term *biodegradable* as applied to materials that can and will decompose through natural organic processes.

If your class subscribes to the Interactive Textbook, use it to review key concepts in Section 11.1.

with ChemASAP

Section 11.1 Assessment

7. Write the names of the reactants separated by a plus sign followed by an arrow followed by the names of the products separated by a plus sign.
8. Write the formulas of the reactants to the left of the yields sign and the formulas of the products to the right.
9. Write the skeleton equation with the correct formulas of the reactants on the left and the correct formulas of the products on the right. Then balance the equation using coefficients so that it obeys the law of conservation of mass.
10. a. $CuS + O_2 \xrightarrow{heat} Cu + SO_2$
    b. $NaHCO_3 \rightarrow Na_2CO_3 + CO_2 + H_2O$
11. a. $2Fe(s) + 3Cl_2(g) \rightarrow 2FeCl_3(s)$
    b. $Al_2(CO_3)_3(s) \rightarrow Al_2O_3(s) + 3CO_2(g)$
    c. $Mg(s) + 2AgNO_3(aq) \rightarrow 2Ag(s) + Mg(NO_3)_2(aq)$
12. a. $2SO_2 + O_2 \rightarrow 2SO_3$ b. $Fe_2O_3 + 3H_2 \rightarrow 2Fe + 3H_2O$ c. $4P + 5O_2 \rightarrow P_4O_{10}$
    d. $2Al + N_2 \rightarrow 2AlN$

11.2

1 FOCUS

Objectives

11.2.1 Describe the five general types of reactions.

11.2.2 Predict the products of the five general types of reactions.

Guide for Reading

Build Vocabulary L2

Graphic Organizers Have students draw concept maps entitled "Types of chemical reactions." Have them include all the vocabulary terms in the concept map. Ask them to explain how the name of each reaction tells what occurs in the reaction.

Reading Strategy L2

Visualize As students read the section, have them visualize the process that occurs in each type of chemical reaction. Suggest that they sketch their visualizations.

2 INSTRUCT

Connecting to Your World

Have students study the photograph and read the text. Explain that an important product of combustion reactions is energy, which can be in the form of heat and/or light. **Where else can you observe combustion reactions on a daily basis?** *(Examples include natural gas stoves and heaters, butane lighters, acetylene torches, and automobile engines.)*

Download a worksheet on **Reaction Types** for students to complete, and find additional teacher support from NSTA SciLinks.

11.2 Types of Chemical Reactions

Guide for Reading

Key Concepts

- What are the five general types of reactions?
- How can you predict the products of the five general types of reactions?

Vocabulary

combination reaction
decomposition reaction
single-replacement reaction
activity series
double-replacement reaction
combustion reaction

Reading Strategy

Outlining As you read, make an outline of the most important ideas in this section. Use the red headings as the main topics and the blue headings as subtopics. Add a sentence or a note after each heading to provide key information about each topic.

Connecting to Your World Often charcoal briquettes provide the heat for barbeque grills through the burning of carbon. Have you ever felt the heat and smelled the smoke coming from a burning charcoal grill? The heat and smoke are the products of a combustion reaction. Combustion is one of the five general types of chemical reactions. In this chapter, you will learn that if you can recognize a reaction as being a particular type, you may be able to predict the products of the reaction.

Classifying Reactions

The five general types of reaction are combination, decomposition, single-replacement, double-replacement, and combustion. Not all chemical reactions fit uniquely into only one category. Occasionally, a reaction may fit equally well into two categories. Nevertheless, recognizing a reaction as a particular type is useful. Patterns of chemical behavior will become apparent and allow you to predict the products of reactions.

Combination Reactions The first type of reaction is the combination, or synthesis, reaction. A **combination reaction** is a chemical change in which two or more substances react to form a single new substance. As shown in Figure 11.5, magnesium metal and oxygen gas combine to form the compound magnesium oxide.

$$2Mg(s) + O_2(g) \longrightarrow 2MgO(s)$$

Notice that in this reaction, as in all combination reactions, the product is a single substance (MgO), which is a compound. The reactants in this combination reaction (Mg and O_2) are two elements. This is often the case, but two compounds may also combine to form a single substance.

When a Group A metal and a nonmetal react, the product is a compound consisting of the metal cation and the nonmetal anion.

$$2K(s) + Cl_2(g) \longrightarrow 2KCl(s)$$

When two nonmetals react in a combination reaction, more than one product is often possible.

$$S(s) + O_2(g) \longrightarrow SO_2(g) \text{ sulfur dioxide}$$
$$2S(s) + 3O_2(g) \longrightarrow 2SO_3(g) \text{ sulfur trioxide}$$

More than one product may also result from the combination reaction of a transition metal and a nonmetal.

$$Fe(s) + S(s) \longrightarrow FeS(s) \text{ iron(II) sulfide}$$
$$2Fe(s) + 3S(s) \longrightarrow Fe_2S_3(s) \text{ iron(III) sulfide}$$

For: Links on Reaction Types
Visit: www.SciLinks.org
Web Code: cdn-1112

Section Resources

Print

- **Guided Reading and Study Workbook,** Section 11.2
- **Core Teaching Resources,** Section 11.2 Review
- **Laboratory Manual,** Labs 14, 15
- **Small-Scale Chemistry Laboratory Manual,** Lab 15
- **Transparencies,** T118–T120
- **Lab Practical** 11–1

Technology

- **Interactive Textbook with ChemASAP,** Simulation 12; Problem-Solving 11.14, 11.15, 11.17, 11.18, 11.21; Assessment 11.2
- **Go Online,** Section 11.2

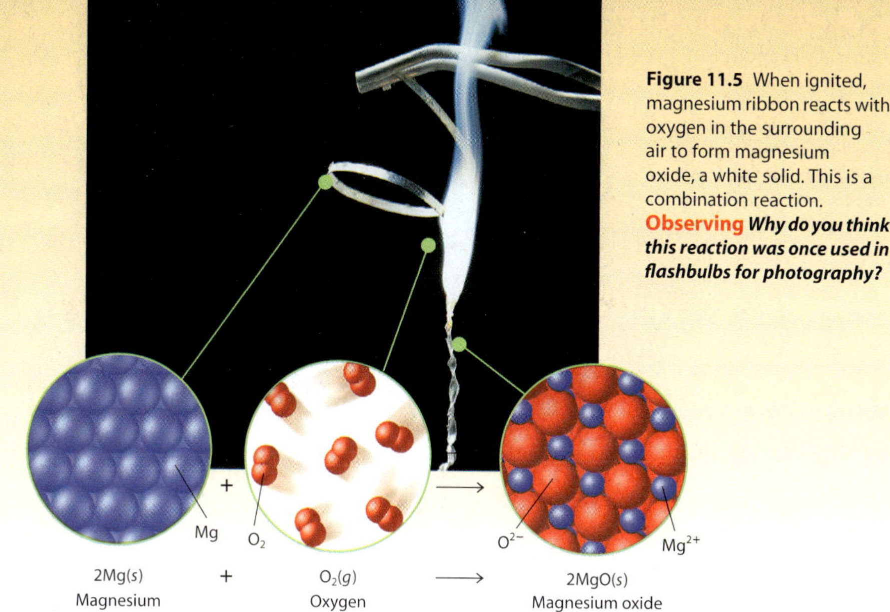

Figure 11.5 When ignited, magnesium ribbon reacts with oxygen in the surrounding air to form magnesium oxide, a white solid. This is a combination reaction. **Observing** ***Why do you think this reaction was once used in flashbulbs for photography?***

CONCEPTUAL PROBLEM 11.4

Writing Equations for Combination Reactions

Copper and sulfur, shown in the photo, are the reactants in a combination reaction. Complete the equation for the reaction.

$Cu(s) + S(s) \longrightarrow$ (two reactions possible)

1 Analyze ***Identify the relevant concepts.***
Two reactions are possible because copper is a transition metal and has more than one common ionic charge (Cu^+ and Cu^{2+}). Determine the formulas for the two products. Balance the two possible equations.

2 Solve ***Apply concepts to this situation.***
Write the skeleton equation first, then apply the rules for balancing equations.

For copper(II):

$Cu(s) + S(s) \longrightarrow CuS(s)$ (balanced)

For copper(I):

$Cu(s) + S(s) \longrightarrow Cu_2S(s)$

$2Cu(s) + S(s) \longrightarrow Cu_2S(s)$ (balanced)

Practice Problems

13. Complete and balance this equation for a combination reaction.

$Be + O_2 \longrightarrow$

14. Write and balance the equation for the formation of magnesium nitride (Mg_3N_2) from its elements.

Interactive Textbook

Problem-Solving 11.14 Solve Problem 14 with the help of an interactive guided tutorial.

with ChemASAP

Differentiated Instruction

Gifted and Talented  L3

Combination reactions are sometimes referred to as synthesis reactions, but synthesis reactions and combination reactions are no longer considered to be synonymous. Have students research the meaning of *synthesis*. Have them investigate the synthesis of polymers and the process of photosynthesis. Point out that synthesis involves making a final product from other substances, but not all synthesis processes—such as those in photosynthesis—are true combination reactions.

Classifying Reactions

CONCEPTUAL PROBLEM 11.4

Answers

13. $2Be + O_2 \rightarrow 2BeO$

14. $3Mg + N_2 \rightarrow Mg_3N_2$

Practice Problems Plus  L2

Write and balance an equation for the formation of aluminum chloride ($AlCl_3$) from its elements.
($2Al + 3Cl_2 \rightarrow 2AlCl_3$)

Discuss L2

Write a number of equations for combination reactions on the board and have students practice balancing them. Point out that many combination reactions release large amounts of energy. Rewrite some of the equations with the product on the reactant side and the reactants on the product side. Ask, **Is it possible to reverse the process?** *(yes)* **What is the name of the reverse process?** *(a decomposition reaction)* **Why is energy usually needed for a decomposition reaction to occur?** *(Decomposition involves the breaking of bonds, which requires energy.)*

Answers to...

Figure 11.5 The reaction produces bright, white light.

Section 11.2 (continued)

CONCEPTUAL PROBLEM 11.5

Answers

15. $2HI \rightarrow H_2 + I_2$

16. HBr

Use Visuals L1

Conceptual Problem 11.5 Have students study the photograph in the problem. Explain that the explosive properties of dynamite are due to the rapid production of large amounts of gases. A related reaction is the decomposition of trinitrotoluene (TNT). Write the equation for the decomposition of TNT on the board: $2C_7H_5N_3O_6(s) \rightarrow 3N_2(g) + 7CO(g) + 5H_2O(g) + 7C(s)$. Point out to students that for every 2 mol of TNT that decompose, 15 mol of hot, expanding gases are produced.

Discuss L2

Help students create a list of criteria for identifying decomposition and combination reactions. For example, decomposition reactions are characterized by one molecule on the reactant side and smaller molecules or elements on the product side. Students should supplement their lists later, when they study single-replacement, double-replacement, and combustion reactions.

Discuss L2

Emphasize to students that decomposition reactions consist of one substance forming two or more different substances. Students tend to limit decomposition reactions to the decomposition of a compound into its component elements. A compound can also break down into an element and a compound or two or more compounds. Provide students with the following chemical equations as examples of decomposition reactions that result in products other than elements: $2H_2O_2 \rightarrow 2H_2O + O_2$ and $H_2CO_3 \rightarrow H_2O + CO_2$.

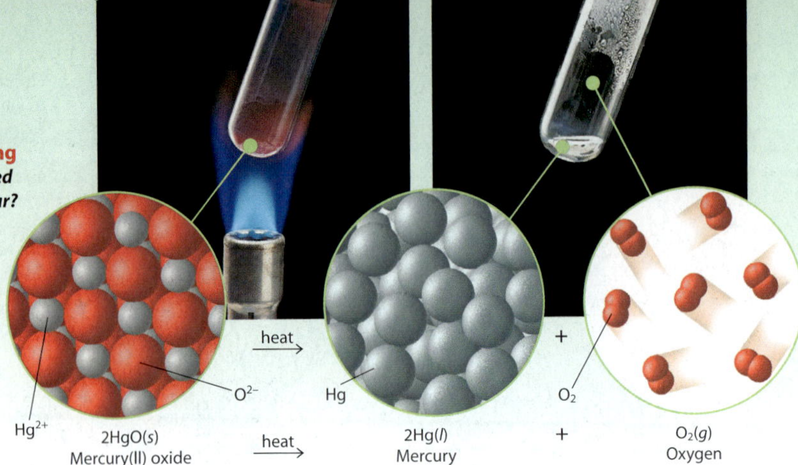

Figure 11.6 When orange-colored mercury(II) oxide is heated, it decomposes into its constituent elements: liquid mercury and gaseous oxygen. **Comparing and Contrasting** ***How are the reactions pictured in Figures 11.5 and 11.6 similar? How are they different?***

Decomposition Reactions When mercury(II) oxide is heated, it decomposes or breaks down into two simpler compounds, as shown in Figure 11.6.

$$2HgO(s) \longrightarrow 2Hg(l) + O_2(g)$$

A **decomposition reaction** is a chemical change in which a single compound breaks down into two or more simpler products. Decomposition reactions involve only one reactant and two or more products. The products can be any combination of elements and compounds. It is usually difficult to predict the products of decomposition reactions. However, when a simple binary compound such as HgO breaks down, you know that the products must be the constituent elements Hg and O_2. Most decomposition reactions require energy in the form of heat, light, or electricity.

CONCEPTUAL PROBLEM 11.5

Writing the Equation for a Decomposition Reaction

Decomposition reactions that produce gases and heat are sometimes explosive, as the photo shows. Write a balanced equation for the following decomposition reaction.

$$H_2O(l) \xrightarrow{\text{electricity}}$$

1 Analyze ***Identify the relevant concepts.***
Water, a binary compound, breaks down into its elements. Balance the equation, remembering that hydrogen and oxygen are both diatomic molecules.

2 Solve ***Apply concepts to this situation.***
Write the skeleton equation, then apply the rules for balancing equations.

$$H_2O(l) \xrightarrow{\text{electricity}} H_2(g) + O_2(g)$$
$$2H_2O(l) \xrightarrow{\text{electricity}} 2H_2(g) + O_2(g) \text{ (balanced)}$$

Practice Problems

15. Complete and balance this decomposition reaction.
$HI \longrightarrow$

16. Write the formula for the binary compound that decomposes to the products H_2 and Br_2.

Problem-Solving 11.15 Solve Problem 15 with the help of an interactive guided tutorial.
with ChemASAP

Single-Replacement Reactions Dropping a small piece of potassium into a beaker of water creates the vigorous reaction shown in Figure 11.7. The reaction produces hydrogen gas and a large quantity of heat. The released hydrogen gas can ignite explosively.

$$2K(s) + 2H_2O(l) \longrightarrow 2KOH(aq) + H_2(g)$$

Similar but less spectacular reactions can occur. For example, if you drop a piece of zinc into a solution of copper nitrate, this reaction occurs:

$$Zn(s) + Cu(NO_3)_2(aq) \longrightarrow Cu(s) + Zn(NO_3)_2(aq)$$

These equations describe two examples of single-replacement reactions. A **single-replacement reaction** is a chemical change in which one element replaces a second element in a compound. You can identify a single-replacement reaction by noting that both the reactants and the products consist of an element and a compound. In the equation above, zinc and copper change places. The reacting element Zn replaces copper in the reactant compound $Cu(NO_3)_2$. The products are the element Cu and the compound $Zn(NO_3)_2$.

Whether one metal will displace another metal from a compound depends upon the relative reactivities of the two metals. The **activity series** of metals, given in Table 11.2, lists metals in order of decreasing reactivity. A reactive metal will replace any metal listed below it in the activity series. Thus iron will displace copper from a copper compound in solution, but iron does not similarly displace zinc or calcium.

A halogen can also replace another halogen from a compound. The activity of the halogens decreases as you go down Group 7A of the periodic table—fluorine, chlorine, bromine, and iodine. Bromine is more active than iodine, so this reaction occurs:

$$Br_2(aq) + NaI(aq) \longrightarrow NaBr(aq) + I_2(aq)$$

But bromine is less active than chlorine, so this reaction does not occur:

$$Br_2(aq) + NaCl(aq) \longrightarrow \text{No reaction}$$

Table 11.2

Activity Series of Metals

Decreasing reactivity ↓	Name	Symbol
	Lithium	Li
	Potassium	K
	Calcium	Ca
	Sodium	Na
	Magnesium	Mg
	Aluminum	Al
	Zinc	Zn
	Iron	Fe
	Lead	Pb
	(Hydrogen)	(H)*
	Copper	Cu
	Mercury	Hg
	Silver	Ag

*Metals from Li to Na will replace H from acids and water; from Mg to Pb they will replace H from acids only.

Figure 11.7 The alkali metal potassium displaces hydrogen from water and forms a solution of potassium hydroxide in a single-replacement reaction. The heat of the reaction is often sufficient to ignite the hydrogen. **Inferring** ***Why are alkali metals stored under mineral oil or kerosene?***

Differentiated Instruction

Gifted and Talented L3

Encourage students to devise general statements to represent each of the first four types of chemical reactions discussed in this section. For example: X + Y → XY, XY → X + Y, X + CD → XD + C, AB + CD → AD + CB.

Use Visuals L1

Table 11.2 With Table 11.2 displayed on an overhead projector, review the definition of a single-replacement reaction on the board. Explain that single-replacement reactions can be compared to partners cutting in on each other at a dance. A person who is alone approaches a couple and cuts in. The person replaces one member of the couple, who is now left alone. In a chemical reaction, however, only certain substances can replace other substances in a given compound. The activity series of metals tells us which metals can replace other metals in a given compound. Review the reactions on this page. Show students how to use Table 11.2 to predict whether a reaction will occur.

TEACHER Demo

Single-Replacement Reactions L2

Purpose Students observe single-replacement reactions.

Materials magnesium ribbon, 10-mL graduated cylinder, 1*M* HCl, sodium, 2 beakers, tongs, water

Safety HCl(*aq*) is corrosive and can cause severe burns. The piece of sodium should be no larger than a match head. Wear plastic gloves and use tongs to avoid contact between sodium and your skin.

Procedure Place a 2-cm piece of magnesium ribbon in 10 mL of 1*M* HCl. Using tongs, place a small piece of sodium in 250 mL of cold water.

Expected Outcomes In both reactions, the metal replaces hydrogen in a compound. Hydrogen gas is released. The reaction using sodium is much more dramatic and rapid than the reaction involving magnesium.

Answers to...

Figure 11.6 Both involve two elements and one compound. In Figure 11.5, a compound is being formed from the elements; in Figure 11.6, a compound is being decomposed to its elements.

Figure 11.7 to prevent their reaction with water vapor and oxygen in the air

CONCEPTUAL PROBLEM 11.6

Answers

17. a. $Fe(s) + Pb(NO_3)_2(aq) \rightarrow Fe(NO_3)_2(aq) + Pb(s)$

b. $Cl_2(aq) + 2NaI(aq) \rightarrow 2NaCl(aq) + I_2(aq)$

c. $Ca(s) + 2H_2O(l) \rightarrow Ca(OH)_2(aq) + H_2(g)$

Practice Problems Plus L2

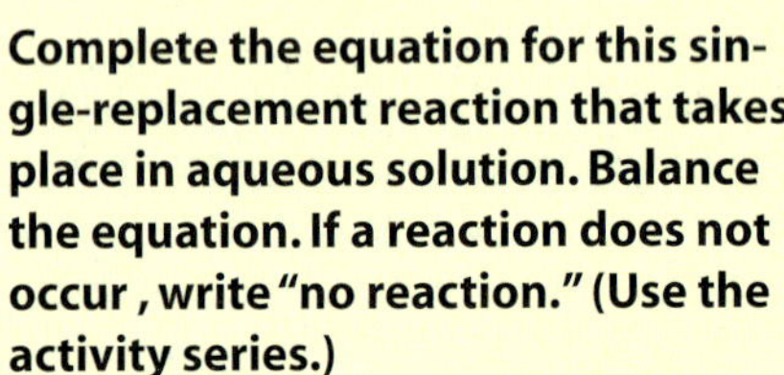

Complete the equation for this single-replacement reaction that takes place in aqueous solution. Balance the equation. If a reaction does not occur, write "no reaction." (Use the activity series.)

$K(s) + H_2SO_4(aq) \rightarrow$

($2K(s) + H_2SO_4(aq) \rightarrow K_2SO_4(aq) + H_2(g)$)

Discuss L2

Explain to students that a double-replacement reaction always involves two ionic compounds in aqueous solution. In addition, one of the products must be a precipitate, a gas, or a molecular compound.

TEACHER Demo

A Double-Replacement Reaction L1

Purpose Students observe double-replacement reactions.

Materials equimolar solutions of $BaCl_2$, Na_2SO_4, Na_3PO_4, $CaCl_2$, $Pb(NO_3)_2$, and KI

Procedure Mix equimolar solutions of the following ionic compounds: $BaCl_2$ and Na_2SO_4; Na_3PO_4 and $CaCl_2$; and $Pb(NO_3)_2$ and KI. Ask, **What evidence of a double-replacement reaction do you observe?** *(a precipitate)* Write the formula of each precipitate formed in the reactions.

Expected Outcome Precipitates of $BaSO_4$, $Ca_3(PO_4)_2$, and PbI_2 form.

CONCEPTUAL PROBLEM 11.6

Writing Equations for Single-Replacement Reactions

The photo shows the reaction between $Zn(s)$ and $H_2SO_4(aq)$. Write a balanced chemical equation for each single-replacement reaction. The reactions take place in aqueous solution.

a. $Zn(s) + H_2SO_4(aq) \longrightarrow$

b. $Cl_2(aq) + NaBr(aq) \longrightarrow$

1 Analyze *Identify the relevant concepts.*

a. According to the activity series of metals, zinc displaces hydrogen from an acid and takes its place. Balance the equation, remembering that elemental hydrogen is diatomic.

b. Chlorine is more reactive than bromine and displaces bromine from its compounds. Balance the equation. Bromine is diatomic.

2 Solve *Apply concepts to this situation.*

Write the skeleton equation first, then apply the rules for balancing equations.

a. $Zn(s) + H_2SO_4(aq) \longrightarrow ZnSO_4(aq) + H_2(g)$ (balanced)

b. $Cl_2(aq) + NaBr(aq) \longrightarrow NaCl(aq) + Br_2(aq)$

$Cl_2(aq) + 2NaBr(aq) \longrightarrow 2NaCl(aq) + Br_2(aq)$ (balanced)

Practice Problem

17. Complete the equations for these single-replacement reactions in aqueous solution. Balance each equation. Write "no reaction" if a reaction does not occur.

a. $Fe(s) + Pb(NO_3)_2(aq) \longrightarrow$

b. $Cl_2(aq) + NaI(aq) \longrightarrow$

c. $Ca(s) + H_2O(l) \longrightarrow$

Problem-Solving 11.17 Solve Problem 17 with the help of an interactive guided tutorial.

with ChemASAP

Double-Replacement Reactions Sometimes, when two solutions of ionic compounds are mixed, nothing happens. At other times, the ions in the two solutions react. Figure 11.8 shows that mixing aqueous solutions of potassium carbonate and barium chloride results in a chemical reaction. A white precipitate of solid barium carbonate is formed. Potassium chloride, the other product of the reaction, remains in solution. This is an example of a **double-replacement reaction,** which is a chemical change involving an exchange of positive ions between two compounds. Double-replacement reactions are also referred to as double-displacement reactions. They generally take place in aqueous solution and often produce a precipitate, a gas, or a molecular compound such as water. For a double-replacement reaction to occur, one of the following is usually true.

1. One of the products is only slightly soluble and precipitates from solution. For example, the reaction of aqueous solutions of sodium sulfide and cadmium nitrate produces a yellow precipitate of cadmium sulfide.

$$Na_2S(aq) + Cd(NO_3)_2(aq) \longrightarrow CdS(s) + 2NaNO_3(aq)$$

2. One of the products is a gas. Poisonous hydrogen cyanide gas is produced when aqueous sodium cyanide is mixed with sulfuric acid.

$$2NaCN(aq) + H_2SO_4(aq) \longrightarrow 2HCN(g) + Na_2SO_4(aq)$$

3. One product is a molecular compound such as water. Combining solutions of calcium hydroxide and hydrochloric acid produces water.

$$Ca(OH)_2(aq) + 2HCl(aq) \longrightarrow CaCl_2(aq) + 2H_2O(l)$$

Differentiated Instruction

Gifted and Talented L3

When hydrochloric acid (HCl) is dropped on calcium carbonate ($CaCO_3$) a double-replacement reaction occurs. Ask, **What are the products of this reaction?** *($CaCl_2$ and H_2CO_3)* Neither of these products is a gas, but bubbles of gas are released during this reaction. Ask students to infer the identity of the gas. *(CO_2)* Ask, **What type of reaction could produce this gas?** *(The decomposition reaction $H_2CO_3 \rightarrow H_2O + CO_2$ releases carbon dioxide gas.)*

334 Chapter 11

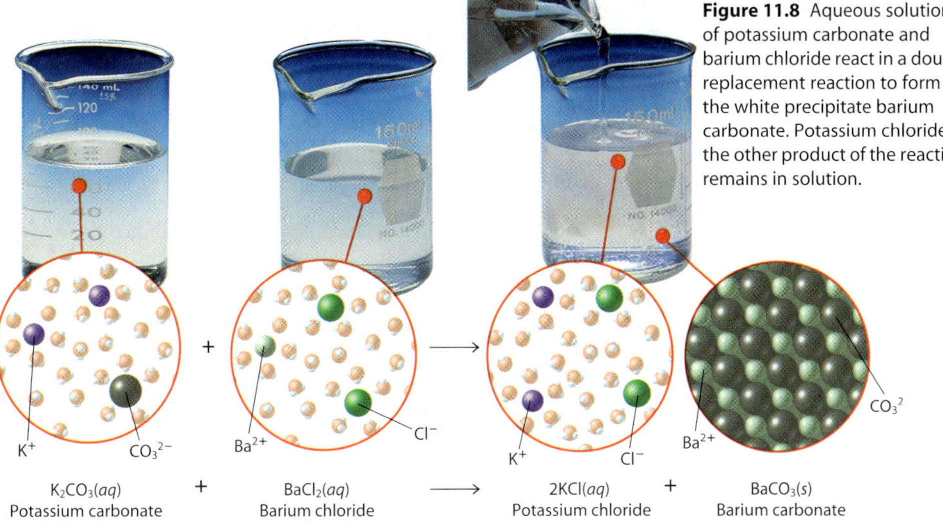

$K_2CO_3(aq)$ + $BaCl_2(aq)$ ⟶ $2KCl(aq)$ + $BaCO_3(s)$

Potassium carbonate + Barium chloride ⟶ Potassium chloride + Barium carbonate

Figure 11.8 Aqueous solutions of potassium carbonate and barium chloride react in a double-replacement reaction to form the white precipitate barium carbonate. Potassium chloride, the other product of the reaction, remains in solution.

CONCEPTUAL PROBLEM 11.7

Writing Equations for Double-Replacement Reactions

Write a balanced chemical equation for each double-replacement reaction.

a. $CaBr_2(aq) + AgNO_3(aq) \longrightarrow$ (A precipitate of silver bromide is formed.)

b. $FeS(s) + HCl(aq) \longrightarrow$ (Hydrogen sulfide gas (H_2S) is formed.)

1 Analyze *Identify the relevant concepts.*

a. The driving force behind the reaction is the formation of a precipitate, which is shown in the photo. Write correct formulas of the products using ionic charges. Then balance the equation.

b. A gas is formed. Use ionic charges to write the correct formula of the other product. Then balance the equation.

2 Solve *Apply concepts to this situation.*

For each reaction, write the skeleton equation first, then apply the rules for balancing equations.

a. $CaBr_2(aq) + AgNO_3(aq) \longrightarrow AgBr(s) + Ca(NO_3)_2(aq)$

$CaBr_2(aq) + 2AgNO_3(aq) \longrightarrow 2AgBr(s) + Ca(NO_3)_2(aq)$ (balanced)

b. $FeS(s) + HCl(aq) \longrightarrow H_2S(g) + FeCl_2(aq)$

$FeS(s) + 2HCl(aq) \longrightarrow H_2S(g) + FeCl_2(aq)$ (balanced)

Practice Problems

18. Write the products of these double-replacement reactions. Then balance each equation.

a. $NaOH(aq) + Fe(NO_3)_3(aq) \longrightarrow$ (Iron(III) hydroxide is a precipitate.)

b. $Ba(NO_3)_2(aq) + H_3PO_4(aq) \longrightarrow$ (Barium phosphate is a precipitate.)

19. Write a balanced equation for each reaction.

a. $KOH(aq) + H_3PO_4(aq) \longrightarrow$

b. $H_2SO_4(aq) + Al(OH)_3(aq) \longrightarrow$

Interactive Textbook

Problem-Solving 11.18 Solve Problem 18 with the help of an interactive guided tutorial.

with ChemASAP

CONCEPTUAL PROBLEM 11.7

Answers

18. a. $3NaOH(aq) + Fe(NO_3)_3(aq) \rightarrow Fe(OH)_3(s) + 3NaNO_3(aq)$

b. $3Ba(NO_3)_2(aq) + 2H_3PO_4(aq) \rightarrow Ba_3(PO_4)_2(s) + 6HNO_3(aq)$

19. a. $3KOH(aq) + H_3PO_4(aq) \rightarrow K_3PO_4(aq) + 3H_2O(l)$

b. $3H_2SO_4(aq) + 2Al(OH)_3(aq) \rightarrow Al_2(SO_4)_3(aq) + 6H_2O(l)$

Practice Problems Plus L2

Write the products for this double-replacement reaction. Then balance the equation. $Pb(NO_3)_2(aq) + NaI(aq) \rightarrow$ (lead iodide is a precipitate)

($Pb(NO_3)_2(aq) + 2NaI(aq) \rightarrow PbI_2(s) + 2NaNO_3(aq)$)

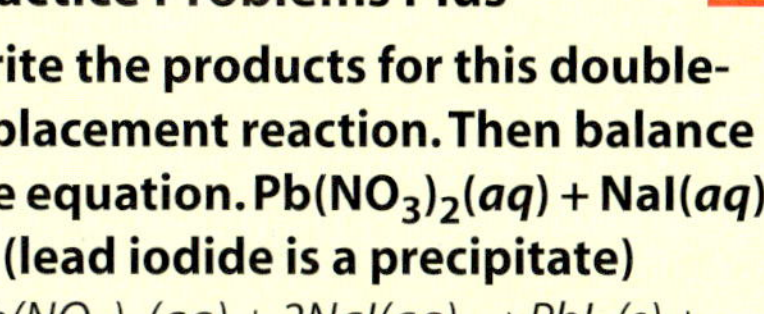

TEACHER Demo

Law of Conservation of Mass L2

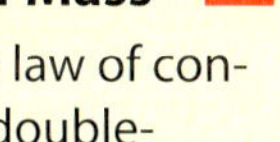

Purpose Demonstrate the law of conservation of mass using a double-replacement reaction.

Materials 50 mL 1.0*M* K_2CrO_4, 250-mL Erlenmeyer flask, 5 mL dilute $AgNO_3$ solution, small test tube, stopper, 50-mL graduated cylinder

Safety Avoid skin contact with silver nitrate.

Procedure Place 50 mL of 1.0*M* potassium chromate (K_2CrO_4) in a 250-mL Erlenmeyer flask. Place about 5 mL of dilute silver nitrate solution ($AgNO_3$) in a small test tube that will fit inside the flask. Stopper the flask and determine the mass of the system. Invert the flask sufficiently so that the two solutions mix. Ask students to note any evidence of chemical changes. Measure the mass of the flask after the reaction. Flush the products down the drain with excess water.

Expected Outcomes A bright red precipitate forms. The mass has not changed.

Facts and Figures

Launching the Space Shuttle

Decomposition and displacement reactions supply the energy needed for launching the space shuttle into space. Have students research what fuels are used to propel the space shuttle orbiter. Have volunteers describe the reactions that occur and how the products provide the thrust necessary to launch vehicles.

Discuss

L2

Point out to students that an important product of most combustion reactions is energy, which is usually in the form of heat or light. Emphasize that one of the reactants must be oxygen. Relate this fact to everyday experience by recalling how removing oxygen from a combustion reaction—for example, snuffing out a candle—causes the reaction to stop.

Relate

L2

Ask students to infer why it is important that combustion reactions, such as those used to heat a home or run an automobile, take place in properly ventilated areas. *(Without proper ventilation and enough available oxygen, the combustion may be incomplete, and poisonous carbon monoxide may be produced.)*

Use Visuals

L1

Figure 11.9 Have students study the figure. Remind them that the complete combustion of a hydrocarbon, such as methane, always produces water, carbon dioxide, heat, and light. Use a disposable lighter to show students the combustion of butane. Ask students to write the balanced equation for the combustion of butane (C_4H_8). Have them use the diagram and chemical equation on this page as an aid.

Discuss

L2

Explain that some combustion reactions are also combination reactions in which an element or a compound combines with oxygen to form a single product plus energy. For example: $2Mg(s) + O_2(g) \rightarrow 2MgO(s)$, and $4Fe(s) + 3O_2(g) \rightarrow 2Fe_2O_3(s)$.

Download a worksheet on **Combustion** for students to complete, and find additional teacher support from NSTA SciLinks.

For: Links on Combustion
Visit: www.SciLinks.org
Web Code: cdn-1114

Combustion Reactions The flames of a campfire or a gas grill are evidence that a combustion reaction is taking place. A **combustion reaction** is a chemical change in which an element or a compound reacts with oxygen, often producing energy in the form of heat and light. A combustion reaction always involves oxygen as a reactant. Often the other reactant is a hydrocarbon, which is a compound composed of hydrogen and carbon. The complete combustion of a hydrocarbon produces carbon dioxide and water. But if the supply of oxygen is limited during a reaction, the combustion will not be complete. Elemental carbon (soot) and toxic carbon monoxide gas may be additional products. The complete combustion of a hydrocarbon releases a large amount of energy as heat. That's why hydrocarbons such as methane (CH_4), propane (C_3H_8), and butane (C_4H_{10}) are important fuels. The combustion reaction for methane is shown in Figure 11.9. Gasoline is a mixture of hydrocarbons that can be approximately represented by the formula C_8H_{18}. The complete combustion of gasoline in a car engine is shown by this equation.

$$2C_8H_{18}(l) + 25O_2(g) \longrightarrow 16CO_2(g) + 18H_2O(l)$$

The reactions between oxygen and some elements other than carbon are also examples of combustion reactions. For example, both magnesium and sulfur will burn in the presence of oxygen. As you look at these combustion equations, notice that the reactions could also be classified as combination reactions.

$$2Mg(s) + O_2(g) \longrightarrow 2MgO(s)$$

$$S(s) + O_2(g) \longrightarrow SO_2(g)$$

Checkpoint ***What are the products of the combustion of a hydrocarbon?***

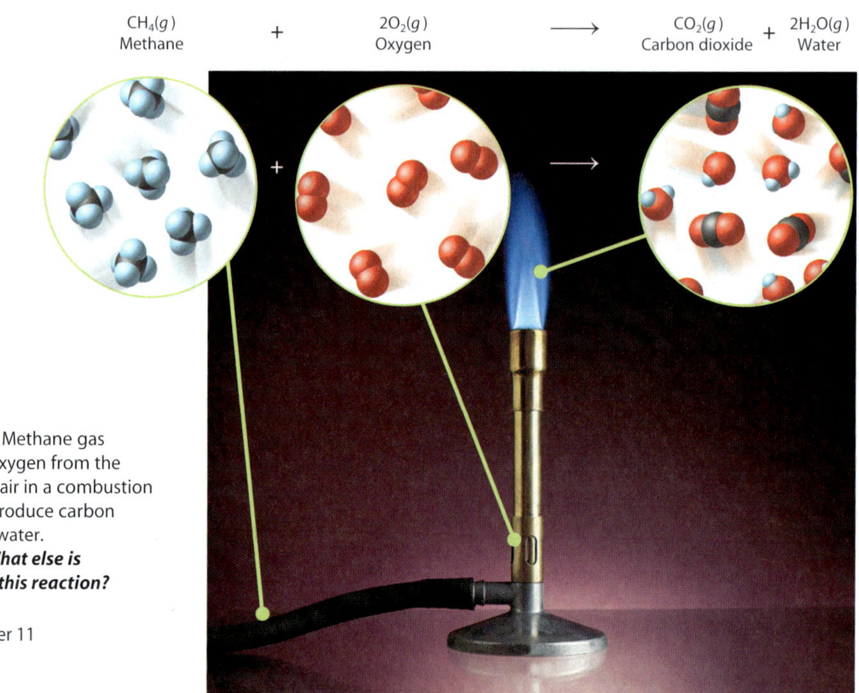

Figure 11.9 Methane gas reacts with oxygen from the surrounding air in a combustion reaction to produce carbon dioxide and water. **Inferring** ***What else is produced in this reaction?***

336 Chapter 11

Differentiated Instruction

English Learners

L1

Make sure that students understand clearly the five terms that describe general types of chemical reactions: combination, decomposition, single-replacement, double-replacement, and combustion. Help them to spell and pronounce each term as well as to define it. Encourage them to ask questions about any term they do not fully understand.

CONCEPTUAL PROBLEM 11.8

Writing Equations for Combustion Reactions

An alcohol lamp often uses ethanol as its fuel. Write balanced equations for the complete combustion of these compounds.

a. benzene ($C_6H_6(l)$) **b.** ethanol ($CH_3CH_2OH(l)$)

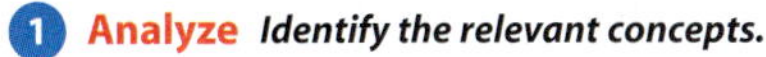

1 Analyze ***Identify the relevant concepts.***

Oxygen is the other reactant in these combustion reactions. The products are CO_2 and H_2O. Write the skeleton equation for each reaction, then balance the equation.

2 Solve ***Apply concepts to this situation.***

For each reaction, write the skeleton equation, then apply the rules for balancing equations.

a. $C_6H_6(l) + O_2(g) \longrightarrow CO_2(g) + H_2O(g)$

$2C_6H_6(l) + 15O_2(g) \longrightarrow 12CO_2(g) + 6H_2O(g)$ (balanced)

b. $CH_3CH_2OH(l) + O_2(g) \longrightarrow CO_2(g) + H_2O(g)$

$CH_3CH_2OH(l) + 3O_2(g) \longrightarrow 2CO_2(g) + 3H_2O(g)$ (balanced)

Practice Problems

20. Write a balanced equation for the complete combustion of each compound.

a. formic acid (HCOOH)

b. heptane (C_7H_{16})

21. Write a balanced equation for the complete combustion of glucose ($C_6H_{12}O_6$).

Problem-Solving 11.21 Solve Problem 21 with the help of an interactive guided tutorial.

with ChemASAP

Predicting the Products of a Chemical Reaction

Now that you have learned about some of the basic reaction types, you can predict the products of many reactions. **The number of elements and/or compounds reacting is a good indicator of possible reaction type and thus possible products.** For example, in a combination reaction, two or more reactants (elements or compounds) combine to form a single product. In a decomposition reaction, a single compound is the reactant; two or more substances are the products. An element and a compound are the reactants in a single-replacement reaction. A different element and a new compound are the products. In a double-replacement reaction, two ionic compounds are the reactants; two new compounds are the products. The reactants in a combustion reaction are oxygen and usually a hydrocarbon. The products of most combustion reactions are carbon dioxide and water.

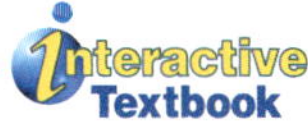

Simulation 12 Practice classifying reactions according to reaction type.

with ChemASAP

Differentiated Instruction

Gifted and Talented L3

Have students choose a particular chemical compound, such as NaOH, H_2SO_4, or CH_3OH, and research how it is produced commercially. A report or poster should be presented describing in as much detail as possible the reactions that are necessary to produce the compound.

CONCEPTUAL PROBLEM 11.8

Practice Problems Answers

20. a. $2HCOOH + O_2 \rightarrow 2CO_2 + 2H_2O$

b. $C_7H_{16} + 11O_2 \rightarrow 7CO_2 + 8H_2O$

21. $C_6H_{12}O_6 + 6O_2 \rightarrow 6CO_2 + 6H_2O$

Practice Problems Plus 

Write a balanced equation for the complete combustion of ethanol (C_2H_5OH). *($C_2H_5OH + 3O_2 \rightarrow 2CO_2 + 3H_2O$)*

Predicting the Products of a Chemical Reaction

TEACHER Demo

Combustion of Iron L1

Purpose Students observe the combustion of iron.

Materials superfine steel wool, plastic sandwich bag, balance, ring stand, utility clamp, matches

Safety Steel wool burns readily.

Procedure Place a superfine steel wool pad in a plastic sandwich bag, measure the mass, and write the value on the board. Take the steel wool out of the bag and unfold it to full length. Clamp it to a ring stand with a utility clamp. Set the steel wool on fire. Ask students to predict what effect the burning will have on the mass of the sample. As soon as the steel wool stops burning and is cool, place the remains in the same plastic sandwich bag and measure the mass again. Write the new value on the board next to the original mass. Discuss why the sample gained mass.

Expected Outcomes Two elements, $Fe(s)$ and $O_2(g)$, combined in a combustion reaction to form a binary compound, Fe_2O_3. The sample gained mass because initially only the mass of the iron was determined. During the reaction, oxygen from the air combined with the iron.

Answers to...

Figure 11.9 energy in the form of heat and light

Checkpoint carbon dioxide and water, plus energy

TEACHER Demo

A Combination Reaction L2

Purpose Students observe a combination reaction.

Materials magnesium ribbon, large crucible, lab burner, cobalt blue glass or exposed photographic film, crucible tongs, matches or igniter

Safety Students should not look directly at burning magnesium. Have them observe through pieces of cobalt blue glass or exposed photographic film. Another option is to conduct the reaction inside a large, metal can.

Procedure Explain to students that in a combination reaction, the reactants combine to make one new product. Measure a 5- to 7-cm strip of magnesium ribbon. Light the burner. Hold one end of the magnesium ribbon with a pair of crucible tongs. Ignite the magnesium and hold it above the crucible. Ask students to note any evidence of chemical change. Have students note the condition of the residue compared to the original magnesium. The product may be disposed of in the trash.

Expected Outcomes Metallic magnesium and oxygen gas from the air form a white powder, magnesium oxide. Other evidence of chemical change includes release of energy as heat and light. Tell students that small amounts of magnesium nitride also form.

For Enrichment L3

Extend the demonstration on this page. Ask, **What are the reactants in this combination reaction?** *(Mg and O_2)* **What is the product of this reaction?** *(MgO)* **Does this reaction obey the law of conservation of mass? Explain.** *(Yes, the sum of the masses of magnesium and oxygen is equal to the mass of the magnesium oxide formed.)* Have students write the balanced equation for the reaction. *($2Mg(s) + O_2(g) \rightarrow 2MgO(s)$)* Challenge students to write the balanced equation for the formation of the small amount of magnesium nitride that forms during this reaction. *($3Mg(s) + N_2(g) \rightarrow Mg_3N_2(s)$)*.

Figure 11.10 The five types of chemical reactions discussed in this chapter are summarized here.

1 Combination Reaction

General Equation: $R + S \longrightarrow RS$

Reactants: Generally two elements, or two compounds (where at least one compound is a molecular compound)
Probable Products: A single compound
Example: Burning magnesium in air

$2Mg(s) + O_2(g) \longrightarrow 2MgO(s)$

2 Decomposition Reaction

General Equation: $RS \longrightarrow R + S$

Reactants: Generally a single binary compound or a compound with a polyatomic ion
Probable Products: Two elements (for a binary compound), or two or more elements and/or compounds (for a compound with a polyatomic ion)
Example: Heating mercury(II) oxide

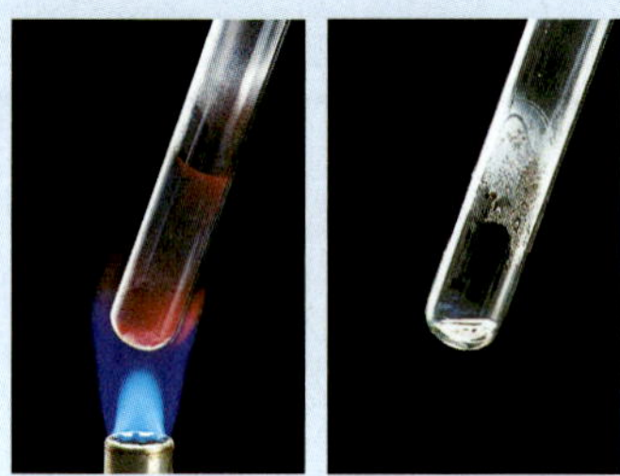

$2HgO(s) \longrightarrow 2Hg(l) + O_2(g)$

3 Single-Replacement Reaction

General Equation: $T + RS \longrightarrow TS + R$

Reactants: An element and a compound
In a single-replacement reaction, an element replaces another element from a compound in aqueous solution. For a single-replacement reaction to occur, the element that is displaced must be less active than the element that is doing the displacing.
Probable Products: A different element and a new compound
Example: Potassium in water

$2K(s) + 2H_2O(l) \longrightarrow 2KOH(aq) + H_2(g)$

Facts and Figures

Magnesium from Sea Water

Magnesium metal is an important component of alloys used to make consumer materials. The main commercial source of Mg(s) is seawater. The Mg^{2+} ion is the third most abundant dissolved ion in the oceans. A process for isolating magnesium from seawater depends on the fact that because of a double-replacement reaction, Mg^{2+} will precipitate when OH^- is added. Students can research the commercial process.

4 Double-Replacement Reaction

General Equation: $R^+S^- + T^+U^- \longrightarrow R + U^- + T^+S^-$

Reactants: Two ionic compounds
In a double-replacement reaction, two ionic compounds react by exchanging cations to form two different compounds.
Probable Products: Two new compounds
Double-replacement reactions are driven by the formation of a precipitate, a gaseous product, or water.
Example: Reaction of aqueous solutions of barium chloride and potassium carbonate

$K_2CO_3(aq) + BaCl_2(aq) \longrightarrow 2KCl(aq) + BaCO_3(s)$

5 Combustion Reaction

General Equation: $C_xH_y + (x + y/4)\,O_2 \longrightarrow xCO_2 + (y/2)H_2O$

Reactants: Oxygen and a compound of C, H, (O)
When oxygen reacts with an element or compound, combustion may occur.
Probable Products: CO_2 and H_2O
With incomplete combustion, C and CO may also be products.
Example: The combustion of methane gas in air

$CH_4(g) + 2O_2(g) \longrightarrow CO_2(g) + 2H_2O(g)$

11.2 Section Assessment

22. **Key Concept** What are the five types of chemical reactions?
23. **Key Concept** What are the keys to predicting the products of the five general types of reactions?
24. Classify each reaction and balance the equations.
    a. $C_3H_6 + O_2 \longrightarrow CO_2 + H_2O$
    b. $Al(OH)_3 \longrightarrow Al_2O_3 + H_2O$
    c. $Li + O_2 \longrightarrow Li_2O$
    d. $Zn + AgNO_3 \longrightarrow Ag + Zn(NO_3)_2$
25. Which of the five general types of reaction would most likely occur, given each set of reactants? What are the probable products?
    a. an aqueous solution of two ionic compounds
    b. a single compound
    c. two elements
    d. oxygen and a compound of carbon and hydrogen
26. Complete and balance an equation for each reaction.
    a. $CaI_2 + Hg(NO_3)_2 \longrightarrow$ (HgI_2 precipitates.)
    b. $Al + Cl_2 \longrightarrow$
    c. $Ag + HCl \longrightarrow$
    d. $C_2H_2 + O_2 \longrightarrow$
    e. $MgCl_2 \longrightarrow$
27. What are the three types of products that result from double-replacement reactions?

Connecting Concepts

Molecular Compounds Hydrogen peroxide is an antiseptic that undergoes a decomposition reaction in the presence of living cells. Refer to Section 8.1 and write a paragraph giving evidence that hydrogen peroxide is a molecular compound.

Assessment 11.2 Test yourself on the concepts in Section 11.2.

with ChemASAP

Section 11.2 Assessment

22. combination, decomposition, single-replacement, double-replacement, and combustion
23. the number of elements and/or compounds reacting
24. **a.** combustion; $2C_3H_6 + 9O_2 \rightarrow 6CO_2 + 6H_2O$ **b.** decomposition; $2Al(OH)_3 \rightarrow Al_2O_3 + 3H_2O$ **c.** combination; $4Li + O_2 \rightarrow 2Li_2O$ **d.** single-replacement; $Zn + 2AgNO_3 \rightarrow 2Ag + Zn(NO_3)_2$
25. **a.** double replacement; two different compounds **b.** decomposition; two or more elements and/or compounds **c.** combination; a compound **d.** combustion; carbon dioxide and water
26. **a.** $CaI_2 + Hg(NO_3)_2 \rightarrow HgI_2 + Ca(NO_3)_2$ **b.** $2Al + 3Cl_2 \rightarrow 2AlCl_3$ **c.** no reaction **d.** $2C_2H_2 + 5O_2 \rightarrow 4CO_2 + 2H_2O$ **e.** $MgCl_2 \xrightarrow{\text{electricity}} Mg + Cl_2$
27. gas, precipitate, and molecular compound

3 ASSESS

Evaluate Understanding L2

Ask students to give an example of each type of reaction discussed in this section. Write several chemical equations on the board and ask students to classify them. In addition, write only the products of combination, decomposition, and displacement reactions, and challenge students to fill in the reactants. Have students refer to Table 11.2 to answer the following questions. **What would happen if a piece of iron was placed in a solution of lead(II) nitrate?** *(Iron will displace lead.)* **What would happen if a piece of aluminum was placed in a solution of calcium chloride?** *(Nothing will happen because Al will not displace Ca.)*

Reteach L1

Help students develop a branched flowchart of chemical reactions similar to those used in qualitative analysis. Start by asking if there is a single reactant. If so, the reaction is a decomposition. If not, proceed to the next step. Ask if oxygen is one of the reactants. If so, the reaction is either combination or combustion. If not, proceed to the next step. Continue through the five general types of reactions.

Connecting Concepts

Paragraphs should include the information that hydrogen atoms and oxygen atoms tend to share electrons to become stable. Units formed when atoms share electrons are molecular compounds.

If your class subscribes to the Interactive Textbook, use it to review key concepts in Section 11.2.

with ChemASAP

Technology & Society

Discuss 

After students have read the article, ask, **What three things are necessary for a fire to burn?** *(fuel, oxygen, and energy to initiate combustion)* **Why is it not safe to use a single kind of fire extinguisher on all fires?** *(A fire extinguisher that controls one type of fire may actually enhance other types of combustion reactions. For example, water is not sprayed on burning magnesium because the intense heat can decompose water, producing flammable hydrogen and oxygen gases.)*

CLASS Activity

Classifying Fires 

Purpose Students classify fires according to type.

Materials research materials, graph paper

Procedure Have students find out what the letters A, B, C, and D refer to in classifying fires. Have them contact their local fire department to obtain statistics on how many class A, B, C, and D fires have occurred in their area during the past year or six months. As a class, have students graph the data and discuss any conclusions that can be drawn from the data.

Expected Outcomes Class A fires are those where ordinary combustibles, such as wood or plastic, are burning. Class B fires involve flammable liquids, such as gasoline. Class C fires are electrical. Class D fires are metal fires.

Relate 

Discuss with students the proper use of a fire extinguisher. Tell them to remember PASS. **P**ull the pin. **A**im the nozzle at the base of the fire. **S**queeze the handle. **S**weep the contents of the extinguisher from side to side until the fire is out. Then, they should shut off the fire extinguisher and watch to see that the fire does not rekindle.

Technology & Society

Combating Combustion

A fire has three requirements: oxygen, fuel, and a temperature high enough to initiate and sustain combustion. Firefighters put out fires by eliminating one or more of these requirements. When water is sprayed on a typical building fire, it stops the fire by lowering the temperature of the burning material and soaking it with noncombustible water. Steam from the vaporizing water also tends to displace air from around the fuel, which denies oxygen to the fuel. To improve the ability of water to saturate the fuel, for example, upholstered furniture and rugs, a substance called a surfactant is added to the water. **Inferring** ***How can it help to roll on the ground if your clothes are on fire?***

Water Water is the most important tool for firefighters. Water-based foams are more effective, but they are also more expensive.

Facts and Figures

Fire Extinguisher Labels

Fire extinguishers indicate on their labels what types of fires they are designed to extinguish. Some extinguishers might be labeled "AB," for example, indicating that they are effective on more than one type of fire. Many fire extinguishers now are labeled with pictures that show the type of fire. For example, a type A extinguisher might show a picture of burning wood. The extinguisher might also show what type of fire the extinguisher should not be used on by showing a picture of that type of fire in a circle with a black line through it.

Forest fires Firefighters combat forest fires from the air by spreading substances that coat the surfaces of the trees to prevent burning. They can also cut the fire off from its fuel by using bulldozers to cut a clear path through the trees or by setting a controlled blaze.

Grease fires Water sprayed on a grease fire can spread the flames. A carbon-dioxide fire extinguisher produces a cloud of heavier-than-air CO_2 that blankets the fire and cuts off the oxygen supply.

Electrical fires Chemicals such as monoammonium phosphate (MAP), blown from a dry-chemical extinguisher, cover an electrical fire and cut off oxygen.

Relate (L2)

Fire extinguishers can be quite heavy. Purchasers of fire extinguishers should be sure that they can handle and operate an extinguisher effectively. Extinguishers must be recharged after each use, even if they are not entirely empty. Periodically, they should be checked to make sure they are operable and effective.

Discuss with students the requirements of a gas that might be used to put out a fire. The gas cannot itself burn or support the combustion of another material. It must be heavier than air so that it will settle on a fire, depriving it of oxygen. Discuss some specific gases in terms of these factors. For example, nitrogen and helium are relatively nonreactive, but they are not heavy enough. Hydrogen and methane will burn. Oxygen supports the burning of the fuel. Carbon dioxide will not burn or support combustion and is heavier than air. Ask, **Is a piece of paper combustible? Is it flammable?** *(Paper is combustible but not flammable.)* Explain that these two terms are often used synonymously, but combustible means that the material will burn, and flammable means a material may easily burst into flame.

Model a Fire Extinguisher  (L1)

Purpose Students observe the effect of carbon dioxide on a flame.

Materials calcium carbonate, dilute hydrochloric acid, 3 beakers, candle, matches

Safety Exert caution when using open flames or acid.

Procedure Place some calcium carbonate in a beaker. Add hydrochloric acid, and allow several minutes for the reaction to produce collectable amounts of carbon dioxide. Collect the carbon dioxide in a beaker. This gas will stay in the beaker because it is heavier than air. Light a candle, and place it in the third beaker. Pour the carbon dioxide over the flame, and observe what happens.

Expected Outcomes The carbon dioxide extinguishes the flame.

Differentiated Instruction

Gifted and Talented  (L3)

Have students check the Internet to learn more about careers in fire science. Prepare a class list of each career, its description, and its requirements. Careers might include firefighter, arson investigator, or fire extinguisher manufacturer.

Answers to...

Inferring Rolling on the ground deprives the burning clothes of oxygen.

11.3

1 FOCUS

Objectives

11.3.1 Describe the information found in a net ionic equation.

11.3.2 Predict the formation of a precipitate in a double-replacement reaction.

Guide for Reading

Build Vocabulary L2

List the Parts You Know Help students appreciate the meaning of "net ionic equation" by exploring the meaning of phrases such as net income and net worth.

Reading Strategy L2

Anticipation Guide Ask, **In a double-replacement reaction, two ions form a solid precipitate, a molecular compound, or a gas. What happens to the other ions?** *(They remain dissolved in the solution.)* **What would you need to know to predict whether a precipitate forms in a double-replacement reaction?** *(The solubility of all compounds that could form.)*

2 INSTRUCT

Connecting to Your World

Point out that the formation of stalactites and stalagmites is a precipitation reaction. Ask, **How do limestone caverns form?** *(Calcium carbonate reacts with carbon dioxide dissolved in water to form soluble calcium hydrogen carbonate. Carbon dioxide then converts calcium hydrogen carbonate back to calcium carbonate, which precipitates and forms stalactites and stalagmites.)*

Net Ionic Equations

Discuss L2

Ask, **What type of reaction is a precipitation reaction?** *(double-replacement)* **What evidence shows that a double-replacement reaction has occurred?** *(formation of a precipitate, a gas, or a molecular compound)*

11.3 Reactions in Aqueous Solution

Guide for Reading

Key Concepts

- What does a net ionic equation show?
- How can you predict the formation of a precipitate in a double-replacement reaction?

Vocabulary

complete ionic equation

spectator ion

net ionic equation

Reading Strategy

Comparing and Contrasting When you compare and contrast things, you examine how they are alike and different. As you read, list the ways that complete ionic equations and net ionic equations are the same and how they are different.

Connecting to Your World The beauty of a limestone cavern is the result of chemical reactions involving water. Limestone caverns form as calcium carbonate reacts with carbon dioxide dissolved in water and forms soluble calcium hydrogen carbonate. Additional carbon dioxide then converts the calcium hydrogen carbonate back into calcium carbonate. The calcium carbonate precipitates and forms dramatic stalactites and stalagmites. In this section, you will learn to predict the formation of precipitates and write equations to describe the reactions that produce them.

Net Ionic Equations

Your world is water-based. More than 70% of Earth's surface is covered by water, and about 66% of the adult human body is water. It is not surprising, then, that many important chemical reactions take place in water—that is, in aqueous solution.

The reaction of aqueous solutions of silver nitrate with sodium chloride to form solid silver chloride and aqueous sodium nitrate is a double-replacement reaction. The reaction is shown in Figure 11.11.

$$AgNO_3(aq) + NaCl(aq) \longrightarrow AgCl(s) + NaNO_3(aq)$$

This is the way you have been writing equations involving aqueous solutions of ionic compounds. However, the equation does not show that like most ionic compounds, the reactants and one of the products dissociate, or separate, into cations and anions when they dissolve in water. For example, when sodium chloride dissolves in water, it separates into sodium ions ($Na^+(aq)$) and chloride ions ($Cl^-(aq)$). Similarly, silver nitrate dissociates into silver ions ($Ag^+(aq)$) and nitrate ions ($NO_3^-(aq)$). You can use these ions to write a **complete ionic equation,** an equation that shows dissolved ionic compounds as dissociated free ions.

$$Ag^+(aq) + NO_3^-(aq) + Na^+(aq) + Cl^-(aq) \longrightarrow AgCl(s) + Na^+(aq) + NO_3^-(aq)$$

Notice that the nitrate ion and the sodium ion appear unchanged on both sides of the equation. The equation can be simplified by eliminating these ions because they don't participate in the reaction.

$$Ag^+(aq) + \cancel{NO_3^-(aq)} + \cancel{Na^+(aq)} + Cl^-(aq) \longrightarrow AgCl(s) + \cancel{Na^+(aq)} + \cancel{NO_3^-(aq)}$$

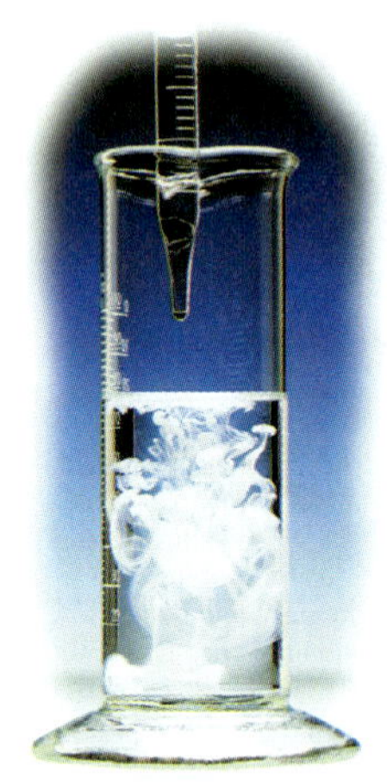

Figure 11.11 A precipitate of silver chloride forms when aqueous solutions of silver nitrate and sodium chloride are mixed. **Inferring** ***Which ions do not participate in the reaction?***

Section Resources

Print

- **Guided Reading and Study Workbook,** Section 11.3
- **Core Teaching Resources,** Section 11.3 Review, Interpreting Graphics
- **Laboratory Manual,** Labs 16, 17, 18
- **Small-Scale Chemistry Laboratory Manual,** Lab 16, 17
- **Transparencies,** T121
- **Lab Practical,** 11-2, 11-3, 11-4, 11-5

Technology

- **Interactive Textbook with ChemASAP,** Problem-Solving 11.28, Assessment 11.3

An ion that appears on both sides of an equation and is not directly involved in the reaction is called a **spectator ion.** When you rewrite an equation leaving out the spectator ions, you have the net ionic equation. The **net ionic equation** is an equation for a reaction in solution that shows only those particles that are directly involved in the chemical change.

$$Ag^{+}(aq) + Cl^{-}(aq) \longrightarrow AgCl(s)$$

In writing balanced net ionic equations, you must make sure that the ionic charge is balanced. For the previous reaction, the net ionic charge on each side of the equation is zero and is therefore balanced. But consider the skeleton equation for the reaction of lead with silver nitrate.

$$Pb(s) + AgNO_3(aq) \longrightarrow Ag(s) + Pb(NO_3)_2(aq)$$

The nitrate ion is the spectator ion in this reaction. The net ionic equation is this.

$$Pb(s) + Ag^{+}(aq) \longrightarrow Ag(s) + Pb^{2+}(aq) \text{ (unbalanced)}$$

Why is this equation unbalanced? Notice that a single unit of positive charge is on the reactant side of the equation. Two units of positive charge are on the product side. Placing the coefficient 2 in front of $Ag^{+}(aq)$ balances the charge. A coefficient of 2 in front of $Ag(s)$ rebalances the atoms.

$$Pb(s) + 2Ag^{+}(aq) \longrightarrow 2Ag(s) + Pb^{2+}(aq) \text{ (balanced)}$$

A net ionic equation shows only those particles involved in the reaction and is balanced with respect to both mass and charge.

Word Origins

Spectator comes from the Latin verb *spectare,* meaning "to watch." Thus a spectator ion can be thought of as only watching a reaction, not participating. **During a football game, what analogy can you draw to the people in the seats and the football players on the field?**

CONCEPTUAL PROBLEM 11.9

Writing and Balancing Net Ionic Equations

In the photograph, aqueous solutions of iron(III) chloride and potassium hydroxide are mixed. A precipitate of iron(III) hydroxide forms. Identify the spectator ions and write a balanced net ionic equation for the reaction.

 **Analyze** ***Identify the relevant concepts.***

Write the complete ionic equation for the reaction, showing any soluble ionic compounds as individual ions. Eliminate aqueous ions that appear as both reactants and products. Balance the net ionic equation.

 **Solve** ***Apply concepts to this situation.***

$$Fe^{3+}(aq) + 3Cl^{-}(aq) + 3K^{+}(aq) + 3OH^{-}(aq) \longrightarrow Fe(OH)_3(s) + 3K^{+}(aq) + 3Cl^{-}(aq)$$

The spectator ions are K^{+} and Cl^{-}. The balanced net ionic equation is

$$Fe^{3+}(aq) + 3OH^{-}(aq) \longrightarrow Fe(OH)_3(s)$$

Practice Problems

28. Write the balanced net ionic equation for this reaction.

$$Ca^{2+}(aq) + OH^{-}(aq) + H^{+}(aq) + PO_4^{3-}(aq) \longrightarrow Ca^{2+}(aq) + PO_4^{3-}(aq) + H_2O(l)$$

29. Write the complete ionic equation and net ionic equation for the reaction of aqueous calcium hydroxide with phosphoric acid. The products are calcium phosphate and water.

Problem-Solving 11.28 Solve Problem 28 with the help of an interactive guided tutorial.

with ChemASAP

Differentiated Instruction

Less Proficient Readers L1

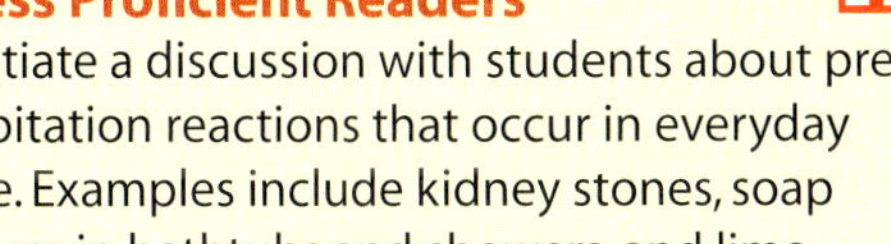

Initiate a discussion with students about precipitation reactions that occur in everyday life. Examples include kidney stones, soap scum in bathtubs and showers, and limestone deposits. Explain the value of being able to write net ionic equations as a direct and simple way to describe chemical change. Relate the concept of net ionic equations to net income. Spectator ions can be compared to two people who go to a dance on a double date and then watch while their partners pair up to dance.

CONCEPTUAL PROBLEM 11.9

Answers

28. $OH^{-}(aq) + H^{+}(aq) \rightarrow H_2O(l)$

29. complete ionic equation:

$$3Ca^{2+}(aq) + 6OH^{-}(aq) + 6H^{+}(aq) + 2PO_4^{3-}(aq) \rightarrow Ca_3(PO_4)_2(s) + 6H_2O(l)$$

net ionic equation: same as complete ionic equation

Practice Problems Plus L2

Write a balanced net ionic equation for the reaction between sulfuric acid (H_2SO_4) and sodium hydroxide (NaOH). *($H^{+}(aq) + OH^{-}(aq) \rightarrow H_2O(l)$)*

Word Origins L2

The spectators at a football game are present at the game, but they don't participate in it.

Relate 

If any students have visited underground caves such as Carlsbad Caverns, have them describe what they saw. Explain that the caverns form when carbonic acid dissolves the calcium carbonate (calcite) in limestone. The icicle-like deposits on the roof of the cavern are called **stalactites.** The deposits on the floor are called **stalagmites.** Both form when dissolved calcium carbonate precipitates from groundwater.

Discuss 

Explain to students that a net ionic equation differentiates between ions that react to form a solid precipitate, a gas, or water, and ions that simply remain in aqueous solution. It is important to note, however, that spectator ions are not completely unaffected by the reaction. They end up paired with different anions or cations than they were paired with when the reaction began.

Answers to...

Figure 11.11 Na^{+} and NO_3^{-}

Predicting the Formation of a Precipitate

Discuss 

On the board, write the products only of several double-replacement reactions; challenge students to fill in the reactants. Examples include: $\rightarrow AgBr(s) + KNO_3(aq)$, $\rightarrow BaSO_4(s) + 2NaCl(aq)$, and $\rightarrow 2KCl(aq) + CaSO_4(s)$. *($KBr(aq) + AgNO_3(aq)$, $BaCl_2(aq) + Na_2SO_4(aq)$, and $CaCl_2(aq) + K_2SO_4(aq)$)* Remind students that in a double-replacement reaction, the positive ions (cations) of one compound trade places with the positive ions of another compound.

3 ASSESS

Evaluate Understanding 

To evaluate students' understanding of complete ionic equations, net ionic equations, and the formation of precipitates, write the reactants in some precipitation reactions on the board. Include at least one example for which no reaction occurs. Have students write complete and net ionic equations.

Reteach 

Review with students the writing and balancing of complete and net ionic equations and the use of solubility rules to predict the outcome of double-replacement reactions. Stress that it is important to note the physical states of reactants and products in precipitation reactions.

Elements Handbook

In the formation of limestone caves, calcium carbonate dissolves in carbonic acid (H_2CO_3). Another reaction occurs when stalagmites form: $Ca^{2+}(aq) + 2HCO_3^{-}(aq) \rightarrow CaCO_3(s) + CO_2(g) + H_2O(l)$

If your class subscribes to the Interactive Textbook, use it to review key concepts in Section 11.3.

with ChemASAP

Table 11.3
Solubility Rules for Ionic Compounds

Compounds	Solubility
Salts of alkali metals and ammonia	Soluble
Nitrate salts and chlorate salts	Soluble
Sulfate salts, except compounds with Pb^{2+}, Ag^{+}, Hg_2^{2+}, Ba^{2+}, Sr^{2+}, and Ca^{2+}	Soluble
Chloride salts, except compound with Ag^{+}, Pb^{2+}, and Hg_2^{2+}	Soluble
Carbonates, phosphates, chromates, sulfides, and hydroxides	Most are insoluble

Predicting the Formation of a Precipitate

You have seen that mixing solutions of two ionic compounds can sometimes result in the formation of an insoluble salt called a precipitate. Some combinations of solutions produce precipitates, while others do not. Whether or not a precipitate forms depends upon the solubility of the new compounds that form. **You can predict the formation of a precipitate by using the general rules for solubility of ionic compounds.** These rules are shown in Table 11.3. Will a precipitate form when aqueous solutions of $Na_2CO_3(aq)$ and $Ba(NO_3)_2(aq)$ are mixed?

$$2Na^{+}(aq) + CO_3^{2-}(aq) + Ba^{2+}(aq) + 2NO_3^{-}(aq) \longrightarrow ?$$

When these four ions are mixed, the cations could change partners. If they did, the two new compounds that would form are $NaNO_3$ and $BaCO_3$. These are the only new combinations of cation and anion possible. To find out if an exchange will occur, refer to Table 11.3, which gives guidelines for determining whether ion combinations are soluble. Recall that sodium is an alkali metal. Rows 1 and 2 tell you that sodium nitrate will not form a precipitate because alkali metal salts and nitrate salts are soluble. Row 5 indicates that carbonates in general are insoluble. Barium carbonate will precipitate. In this reaction Na^{+} and NO_3^{-} are spectator ions. The net ionic equation for this reaction is as follows.

$$Ba^{2+}(aq) + CO_3^{2-}(aq) \longrightarrow BaCO_3(s)$$

11.3 Section Assessment

30. **Key Concept** What is a net ionic equation?

31. **Key Concept** How can you predict the formation of a precipitate in a double-replacement reaction?

32. Write a balanced net ionic equation for each reaction.
- **a.** $Pb(NO_3)_2(aq) + H_2SO_4(aq) \longrightarrow PbSO_4(s) + HNO_3(aq)$
- **b.** $Pb(C_2H_3O_2)_2(aq) + HCl(aq) \longrightarrow PbCl_2(s) + HC_2H_3O_2(aq)$
- **c.** $Na_3PO_4(aq) + FeCl_3(aq) \longrightarrow NaCl(aq) + FePO_4(s)$
- **d.** $(NH_4)_2S(aq) + Co(NO_3)_2(aq) \longrightarrow CoS(s) + NH_4NO_3(aq)$

33. Write a balanced net ionic equation for each reaction. Identify the spectator ions in each reaction.
- **a.** $HCl(aq) + AgNO_3(aq) \longrightarrow$
- **b.** $Pb(C_2H_3O_2)_2(aq) + LiCl(aq) \longrightarrow$
- **c.** $Na_3PO_4(aq) + CrCl_3(aq) \longrightarrow$

34. Identify the precipitate formed when solutions of these ionic compounds are mixed.
- **a.** $H_2SO_4 + BaCl_2 \longrightarrow$
- **b.** $Al_2(SO_4)_3 + NH_4OH \longrightarrow$
- **c.** $AgNO_3 + H_2S \longrightarrow$
- **d.** $CaCl_2 + Pb(NO_3)_2 \longrightarrow$
- **e.** $Ca(NO_3)_2 + Na_2CO_3 \longrightarrow$

35. Will a precipitate form when the following aqueous solutions of ionic compounds are mixed?
- **a.** $AgNO_3$ and Na_2SO_4
- **b.** NH_4Cl and $Ba(NO_3)_2$
- **c.** $CaCl_2$ and K_2SO_4
- **d.** $Pb(NO_3)_2$ and HCl

Elements Handbook

Limestone Caves Refer to page R13 in the Elements Handbook to learn more about the formation of limestone caves. What part does acid (hydrogen ion) play in the dissolving process? What is the reaction that deposits stalactites and stalagmites?

Assessment 11.3 Test yourself on the concepts in Section 11.3.

with ChemASAP

Section 11.3 Assessment

30. A net ionic equation shows only those particles involved in the reaction and is balanced with respect to both mass and charge.

31. Use the general rules for solubility of ionic compounds (Table 11.3).

32. **a.** $Pb^{2+}(aq) + SO_4^{2-}(aq) \rightarrow PbSO_4(s)$
b. $Pb^{2+}(aq) + 2Cl^{-}(aq) \rightarrow PbCl_2(s)$
c. $Fe^{3+}(aq) + PO_4^{3-}(aq) \rightarrow FePO_4(s)$
d. $Co^{2+}(aq) + S^{2-}(aq) \rightarrow CoS(s)$

33. **a.** $Ag^{+}(aq) + Cl^{-}(aq) \rightarrow AgCl(s)$; spectator ions are $H^{+}(aq)$ and $NO_3^{-}(aq)$
b. $Pb^{2+}(aq) + 2Cl^{-}(aq) \rightarrow PbCl_2(s)$; spectator ions are $Li^{+}(aq)$ and $C_2H_3O_2^{-}(aq)$
c. $Cr^{3+}(aq) + PO_4^{3-}(aq) \rightarrow CrPO_4(s)$; spectator ions are $Na^{+}(aq)$ and $Cl^{-}(aq)$

34. **a.** $BaSO_4(s)$ **b.** $Al(OH)_3(s)$ **c.** $Ag_2S(s)$ **d.** $PbCl_2(s)$ **e.** $CaCO_3(s)$

35. **a.** yes; $Ag_2SO_4(s)$ **b.** no **c.** yes; $CaSO_4(s)$ **d.** yes; $PbCl_2$

Small-Scale LAB

Precipitation Reactions: Formation of Solids

Purpose

To observe, identify, and write balanced equations for precipitation reactions.

Materials

- pencil
- paper
- ruler
- reaction surface
- chemicals shown in the grid below

	$AgNO_3$ (Ag^+)	$Pb(NO_3)_2$ (Pb^{2+})	$CaCl_2$ (Ca^{2+})
Na_2CO_3 (CO_3^{2-})	a	f	k
Na_3PO_4 (PO_4^{3-})	b	g	l
NaOH (OH^-)	c	h	m
Na_2SO_4 (SO_4^{2-})	d	i	n
NaCl (Cl^-)	e	j	o

Procedure   

Copy the grid on two sheets of paper. Make each square 2 cm on each side. Draw large black Xs on one of the grids. Place a reaction surface over the grid with black Xs and add the chemicals as shown. Use the other grid as a data table to record your observations for each solution.

Analyze

Using your experimental data, record your answers to the following in the space below your data table.

1. Translate the following word equations into balanced chemical equations and explain how the equations represent what happens in grid spaces *a* and *g*.
   a. In grid space *a*, sodium carbonate reacts with silver nitrate to produce sodium nitrate and solid silver carbonate.
   b. In grid space *g*, sodium phosphate reacts with lead(II) nitrate to produce sodium nitrate and solid lead(II) phosphate.
2. Write a word equation to represent what happens in grid space *m*.
3. What happens in grid space *d*? Which other mixings gave similar results? Is it necessary to write an equation when no reaction occurs? Explain.
4. Write balanced equations for the other precipitation reactions you observed.
5. Write balanced net ionic equations for the other precipitation reactions you observed.

You're The Chemist

The following small-scale activities allow you to develop your own procedures and analyze the results.

1. **Explain It!** Mix a solution of potassium iodide (KI) with silver nitrate. Then mix potassium iodide solution with lead(II) nitrate. Describe your results. Write balanced equations and net ionic equations for each reaction.
2. **Design It!** Table salt is mostly sodium chloride. Design and carry out an experiment to find out if table salt will form a precipitate with either lead(II) nitrate or silver nitrate. Interpret your results.
3. **Design It!** Design and carry out an experiment to show that iodized table salt contains potassium iodide.

Small-Scale LAB

Precipitation Reactions: Formation of Solids 

Objective Students observe, identify, and write balanced equations for precipitation reactions.

Prep Time 30 minutes
Class Time 40 minutes

Solution	Preparation
0.05*M* $AgNO_3$	2.1 g in 250 mL
0.2*M* $Pb(NO_3)_2$	16.6 g in 250 mL
0.5*M* $CaCl_2$	13.9 g in 250 mL
1.0*M* Na_2CO_3	26.5 g in 250 mL
0.1*M* Na_3PO_4	9.5 g $Na_3PO_4 \cdot 12H_2O$ in 250 mL
0.5*M* NaOH	20.0 g in 1.0 L
0.2*M* Na_2SO_4	7.1 g in 250 mL
1.0*M* NaCl	14.6 g in 250 mL

Disposal Flush all chemicals down the drain with excess water.

Analyze

1. $Na_2CO_3 + 2AgNO_3 \rightarrow 2NaNO_3 + Ag_2CO_3(s)$
   $2Na_3PO_4 + 3Pb(NO_3)_2 \rightarrow 6NaNO_3 + Pb_3(PO_4)_2(s)$
2. Sodium hydroxide and calcium chloride form sodium chloride and solid calcium hydroxide.
3. Mixings d, n, and o do not react.
4. $Na_3PO_4 + 3AgNO_3 \rightarrow 3NaNO_3 + Ag_3PO_4(s)$
   $NaOH + AgNO_3 \rightarrow NaNO_3 + AgOH(s)$
   (actual products: $NaNO_3 + Ag_2O(s) + H_2O$)
   $NaCl + AgNO_3 \rightarrow NaNO_3 + AgCl(s)$
   $Na_2CO_3 + Pb(NO_3)_2 \rightarrow 2NaNO_3 + PbCO_3(s)$
   $2NaOH + Pb(NO_3)_2 \rightarrow 2NaNO_3 + Pb(OH)_2(s)$
   $Na_2SO_4 + Pb(NO_3)_2 \rightarrow 2NaNO_3 + PbSO_4(s)$
   $2NaCl + Pb(NO_3)_2 \rightarrow 2NaNO_3 + PbCl_2(s)$
   $Na_2CO_3 + CaCl_2 \rightarrow 2NaCl + CaCO_3(s)$
   $2Na_3PO_4 + 3CaCl_2 \rightarrow 6NaCl + Ca_3(PO_4)_2(s)$
   $2NaOH + CaCl_2 \rightarrow 2NaCl + Ca(OH)_2(s)$
5. $2Ag^+ + CO_3^{2-} \rightarrow Ag_2CO_3(s)$
   $3Ag^+ + PO_4^{3-} \rightarrow Ag_3PO_4(s)$
   $Ag^+ + OH^- \rightarrow AgOH(s)$
   $Ag^+ + Cl^- \rightarrow AgCl(s)$
   $Pb^{2+} + CO_3^{2-} \rightarrow PbCO_3(s)$
   $3Pb^{2+} + 2PO_4^{3-} \rightarrow Pb_3(PO_4)_2(s)$
   $Pb^{2+} + 2OH^- \rightarrow Pb(OH)_2(s)$
   $Pb^{2+} + SO_4^{2-} \rightarrow PbSO_4(s)$
   $Pb^{2+} + 2Cl^- \rightarrow PbCl_2(s)$
   $Ca^{2+} + CO_3^{2-} \rightarrow CaCO_3(s)$
   $3Ca^{2+} + 2PO_4^{3-} \rightarrow Ca_3(PO_4)_2(s)$
   $Ca^{2+} + 2OH^- \rightarrow Ca(OH)_2(s)$

You're The Chemist

1. $KI + AgNO_3 \rightarrow KNO_3 + AgI(s)$
   $Ag^+ + I^- \rightarrow AgI(s)$
   $2KI + Pb(NO_3)_2 \rightarrow 2KNO_3 + PbI_2(s)$
   $Pb^{2+} + 2I^- \rightarrow PbI_2(s)$
2. Add a drop of $Pb(NO_3)_2$ or $AgNO_3$ to a few grains of salt. Look for white crystals.
3. Place a drop of $Pb(NO_3)_2$ on a small pile of dry table salt. Keep part of the pile dry. Look for yellow lead iodide.

CHAPTER 11 Study Guide

Study Tip

Use Physical Models to Clarify Concepts For example, when balancing chemical equations or classifying a reaction according to its type, physical models of atoms can aid visual learners. Molecular model kits, Styrofoam balls, gumdrops, or modeling clay can be used to represent different types of atoms.

If your class subscribes to the Interactive Textbook with ChemASAP, your students can go online to access an interactive version of the Student Edition and a self-test.

with ChemASAP

CHAPTER 11 Study Guide

Key Concepts

11.1 Describing Chemical Reactions

- To write a word equation, write the names of the reactants to the left of the arrow separated by plus signs; write the names of the products to the right of the arrow, also separated by plus signs.
- To write a skeleton equation, write the formulas for the reactants to the left of the yields sign and the formulas for the products to the right.
- After writing the skeleton equation, use coefficients to balance the equation so that it obeys the law of conservation of mass.

11.2 Types of Chemical Reactions

- The five general types of reactions are combination, decomposition, single-replacement, double-replacement, and combustion.
- The number of elements and/or compounds reacting is a good indicator of possible reaction type and thus possible products.
- In a combination reaction, there is always a single product.
- A decomposition reaction involves the breakdown of a single compound into two or more simpler substances.
- In a single-replacement reaction, both the reactants and the products are an element and a compound.
- A double-replacement reaction generally takes place between two ionic compounds in aqueous solution.
- A combustion reaction always involves oxygen as a reactant.

11.3 Reactions in Aqueous Solution

- A net ionic equation shows only those particles involved in the reaction and is balanced with respect to mass and charge.
- You can predict the formation of a precipitate by using the general rules for solubility of ionic compounds.

Vocabulary

- activity series (p. 333)
- balanced equation (p. 325)
- catalyst (p. 323)
- chemical equation (p. 323)
- coefficients (p. 325)
- combination reaction (p. 330)
- combustion reaction (p. 336)
- complete ionic equation (p. 342)
- decomposition reaction (p. 332)
- double-replacement reaction (p. 334)
- net ionic equation (p. 343)
- single-replacement reaction (p. 333)
- skeleton equation (p. 323)
- spectator ion (p. 343)

Organizing Information

Use these terms to construct a concept map that organizes the major ideas of this chapter.

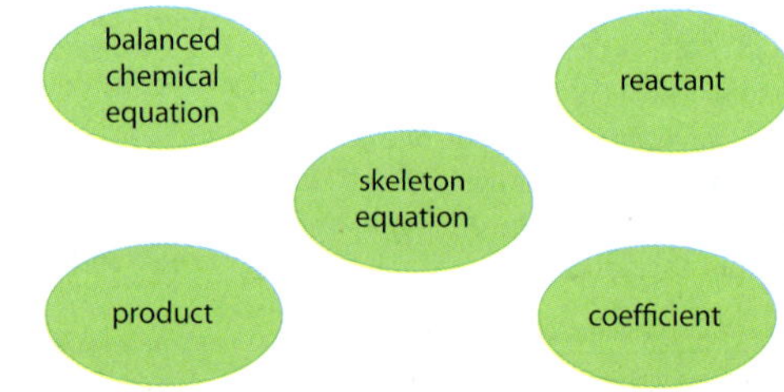

Concept Map 11 Solve the Concept Map with the help of an interactive tutorial.

with ChemASAP

Chapter Resources

Print

- **Core Teaching Resources,** Chapter 11, Practice Problems Vocabulary Review, Quiz, Chapter Test A, Chapter Test B

Technology

- **Computer Test Bank,** Chapter 11 Test
- **Interactive Textbook with ChemASAP,** Chapter 11

CHAPTER 11

Assessment

Reviewing Content

11.1 Describing Chemical Reactions

36. Identify the reactants and products in each chemical reaction.

a. Hydrogen gas and sodium hydroxide are formed when sodium is dropped into water.

b. In photosynthesis, carbon dioxide and water react to form oxygen gas and glucose.

37. How did John Dalton explain a chemical reaction using his atomic theory?

38. What is the function of an arrow ($\longrightarrow$) in a chemical equation? A plus sign (+)?

39. Write sentences that completely describe each of the chemical reactions shown in these skeleton equations.

a. $NH_3(g) + O_2(g) \xrightarrow{Pt} NO(g) + H_2O(g)$

b. $H_2SO_4(aq) + BaCl_2(aq) \longrightarrow BaSO_4(s) + HCl(aq)$

c. $N_2O_3(g) + H_2O(l) \longrightarrow HNO_2(aq)$

40. What is the purpose of a catalyst?

41. Balance equations for each item. The formula for each product (object) is given.

a. a basketball team

center + forward + guard $\longrightarrow$ team

$C + F + G \longrightarrow CF_2G_2$

b. a tricycle

frame + wheel + seat + pedal $\longrightarrow$ tricycle

$F + W + S + P \longrightarrow FW_3SP_2$

42. The equation for the formation of water from its elements, $H_2(g) + O_2(g) \longrightarrow H_2O(l)$, can be "balanced" by changing the formula of the product to H_2O_2. Explain why this is incorrect.

43. Balance the following equations.

a. $PbO_2 \longrightarrow PbO + O_2$

b. $Fe(OH)_3 \longrightarrow Fe_2O_3 + H_2O$

c. $(NH_4)_2CO_3 \longrightarrow NH_3 + H_2O + CO_2$

d. $NaCl + H_2SO_4 \longrightarrow Na_2SO_4 + HCl$

11.2 Types of Chemical Reactions

44. What is a characteristic of every combination reaction?

45. Write balanced chemical equations for the following combination reactions.

a. $Mg + O_2 \longrightarrow$

b. $P + O_2 \longrightarrow$ diphosphorus pentoxide

c. $Ca + S \longrightarrow$

46. What is a distinguishing feature of every decomposition reaction?

47. Write a balanced chemical equation for each decomposition reaction.

a. $Ag_2O(s) \xrightarrow{\Delta}$?

b. ammonium nitrate $\xrightarrow{\Delta}$ dinitrogen monoxide + water

48. Use the activity series of metals to write a balanced chemical equation for each single-replacement reaction.

a. $Au(s) + KNO_3(aq) \longrightarrow$

b. $Zn(s) + AgNO_3(aq) \longrightarrow$

c. $Al(s) + H_2SO_4(aq) \longrightarrow$

49. Write a balanced equation for each of the following double-replacement reactions.

a. $H_2C_2O_4(aq) + KOH(aq) \longrightarrow$

b. $CdBr_2(aq) + Na_2S(aq) \longrightarrow$ (Cadmium sulfide is a precipitate.)

50. What substance is common to all combustion reactions?

51. Write a balanced equation for the complete combustion of each compound.

a. butene (C_4H_8)

b. acetone (C_3H_6O)

52. Balance each equation and identify its type.

a. $Hf + N_2 \longrightarrow Hf_3N_4$

b. $Mg + H_2SO_4 \longrightarrow MgSO_4 + H_2$

c. $C_2H_6 + O_2 \longrightarrow CO_2 + H_2O$

d. $Pb(NO_3)_2 + NaI \longrightarrow PbI_2 + NaNO_3$

e. $Fe + O_2 \longrightarrow Fe_3O_4$

11.3 Reactions in Aqueous Solution

53. What is a spectator ion?

54. Write a balanced net ionic equation for the following reactions.

a. $HCl(aq) + Ca(OH)_2(aq) \longrightarrow$

b. $AgNO_3(aq) + AlCl_3(aq) \longrightarrow$ (Silver chloride is a precipitate.)

55. Complete each equation and then write a net ionic equation.

a. $Al(s) + H_2SO_4(aq) \longrightarrow$

b. $HCl(aq) + Ba(OH)_2(aq) \longrightarrow$

c. $Au(s) + HCl(aq) \longrightarrow$

CHAPTER 11

Assessment

Reviewing Content

36. a. reactants: sodium and water; products: hydrogen and sodium hydroxide

b. reactants: carbon dioxide and water; products: oxygen and glucose

37. Dalton said that the atoms of reactants are rearranged to form new substances as products.

38. The arrow separates the reactants from the products and indicates a reaction that progresses in the forward direction. A plus sign separates individual reactants and individual products from one another.

39. a. Gaseous ammonia and oxygen react in the presence of a platinum catalyst to produce nitrogen monoxide gas and water vapor.

b. Aqueous solutions of sulfuric acid and barium chloride are mixed to produce a precipitate of barium sulfate and aqueous hydrochloric acid.

c. The gas dinitrogen trioxide reacts with water to produce an aqueous solution of nitrous acid.

40. A catalyst speeds up a chemical reaction.

41. a. $C + 2F + 2G \rightarrow CF_2G_2$

b. $F + 3W + S + 2P \rightarrow FW_3SP_2$

42. A formula is a unique identifier of a substance. A different formula would indicate a different substance, not the one that is taking part in the reaction you are trying to balance.

43. a. $2PbO_2 \rightarrow 2PbO + O_2$

b. $2Fe(OH)_3 \rightarrow Fe_2O_3 + 3H_2O$

c. $(NH_4)_2CO_3 \rightarrow 2NH_3 + H_2O + CO_2$

c. $2NaCl + H_2SO_4 \rightarrow Na_2SO_4 + 2HCl$

44. a single product

45. a. $2Mg + O_2 \rightarrow 2MgO$

b. $4P + 5O_2 \rightarrow 2P_2O_5$

c. $Ca + S \rightarrow CaS$

46. a single reactant

47. a. $2Ag_2O \xrightarrow{\Delta} 4Ag + O_2$

b. $NH_4NO_3 \xrightarrow{\Delta} N_2O + 2H_2O$

48. a. no reaction

b. $Zn(s) + 2AgNO_3(aq) \rightarrow Zn(NO_3)_2(aq) + 2Ag(s)$

c. $2Al(s) + 3H_2SO_4(aq) \rightarrow Al_2(SO_4)_3(aq) + 3H_2(g)$

49. a. $H_2C_2O_4(aq) + 2KOH(aq) \rightarrow K_2C_2O_4(aq) + 2H_2O(l)$

b. $CdBr_2(aq) + Na_2S(aq) \rightarrow CdS(s) + 2NaBr(aq)$

50. oxygen

51. a. $C_4H_8 + 6O_2 \rightarrow 4CO_2 + 4H_2O$

b. $C_3H_6O + 4O_2 \rightarrow 3CO_2 + 3H_2O$

52. a. $3Hf + 2N_2 \rightarrow Hf_3N_4$; combination

b. $Mg + H_2SO_4 \rightarrow MgSO_4 + H_2$; single replacement

c. $2C_2H_6 + 7O_2 \rightarrow 4CO_2 + 6H_2O$; combustion

d. $Pb(NO_3)_2 + 2NaI \rightarrow PbI_2 + 2NaNO_3$; double replacement

e. $3Fe + 2O_2 \rightarrow Fe_3O_4$; combination

53. an ion that does not participate in the reaction

54. a. $H^+(aq) + OH^-(aq) \rightarrow H_2O(l)$

b. $Ag^+(aq) + Cl^-(aq) \rightarrow AgCl(s)$

55. a. $2Al(s) + 6H^+(aq) \rightarrow 2Al^{3+}(aq) + 3H_2(g)$

b. $H^+(aq) + OH^-(aq) \rightarrow H_2O(l)$

c. no reaction

Understanding Concepts

56. **a.** $Cl_2 + 2NaI \rightarrow 2NaCl + I_2$
b. $2NH_3 \rightarrow N_2 + 3H_2$
c. $4Na + O_2 \rightarrow 2Na_2O$

57. **a.** $Cl_2(g) + 2KI(aq) \rightarrow I_2(aq) + 2KCl(aq)$
b. $2Fe(s) + 6HCl(aq) \rightarrow 2FeCl_3(aq) + 3H_2(g)$
c. $P_4O_{10}(s) + 6H_2O(l) \rightarrow 4H_3PO_4(aq)$

58. **a.** $ZnS(aq) + H_2SO_4(aq) \rightarrow H_2S(g) + ZnSO_4(aq)$
b. $NaOH(aq) + HNO_3(aq) \rightarrow H_2O(l) + NaNO_3(aq)$
c. $2KF(aq) + Ca(NO_3)_2(aq) \rightarrow CaF_2(s) + 2KNO_3(aq)$

59. **a.** $Na_2O(s) + H_2O(l) \rightarrow 2NaOH(aq)$
b. $H_2(g) + Br_2(g) \rightarrow 2HBr(g)$
c. $Cl_2O_7(l) + H_2O(l) \rightarrow 2HClO_4(aq)$

60. **a.** $Fe(s) + H_2SO_4(aq) \rightarrow FeSO_4(aq) + H_2(g)$
b. no reaction
c. $Br_2(l) + BaI_2(aq) \rightarrow BaBr_2(aq) + I_2(aq)$

61. **a.** tube A
b. $2Na(s) + 2H_2O(l) \rightarrow 2NaOH(aq) + H_2(g)$
single-replacement

62. **a.** $2C_8H_{18} + 25O_2 \rightarrow 16CO_2 + 18H_2O$
b. $C_6H_{12}O_6 + 6O_2 \rightarrow 6CO_2 + 6H_2O$
c. $HC_2H_3O_2 + 2O_2 \rightarrow 2CO_2 + 2H_2O$

63 **a.** $2Al_2O_3 \xrightarrow{\text{energy}} 4Al + 3O_2$
b. $Sn(OH)_4 \xrightarrow{\Delta} SnO_2 + 2H_2O$
c. $Ag_2CO_3 \xrightarrow{\Delta} Ag_2O + CO_2$

64. **a.** $H^+(aq) + OH^-(aq) \rightarrow H_2O(l)$
b. $S^{2-}(aq) + 2H^+(aq) \rightarrow H_2S(g)$
c. $3OH^-(aq) + Fe^{3+}(aq) \rightarrow Fe(OH)_3(s)$

65. **a.** $CdS(s)$
b. $Na^+(aq)$ and $NO_3^-(aq)$
c. $Cd^{2+}(aq) + S^{2-}(aq) \rightarrow CdS(s)$

66. **a.** $K_2O(s) + H_2O(l) \rightarrow 2KOH(aq)$; combination
b. $C_{19}H_{40}(s) + 29O_2(g) \rightarrow 19CO_2(g) + 20H_2O(l)$
c. $2Rb(s) + 2H_2O(l) \rightarrow H_2(g) + 2RbOH(aq)$
single-replacement

Understanding Concepts

56. Each equation is incorrect. Find the errors, then rewrite and balance each equation.
a. $Cl_2 + NaI \longrightarrow NaCl_2 + I$
b. $NH_3 \longrightarrow N + H_3$
c. $Na + O_2 \longrightarrow NaO_2$

57. Write a balanced chemical equation for each reaction. Use the necessary symbols from Table 11.1 to describe the reaction completely.
a. Bubbling chlorine gas through a solution of potassium iodide gives elemental iodine and a solution of potassium chloride.
b. Bubbles of hydrogen gas and aqueous iron(III) chloride are produced when metallic iron is dropped into hydrochloric acid.
c. Solid tetraphosphorus decoxide reacts with water to produce phosphoric acid.

58. Write balanced chemical equations for these double-replacement reactions that occur in aqueous solution.
a. Zinc sulfide is added to sulfuric acid.
b. Sodium hydroxide reacts with nitric acid.
c. Solutions of potassium fluoride and calcium nitrate are mixed.

59. Write a balanced chemical equation for each combination reaction.
a. sodium oxide + water $\longrightarrow$
b. hydrogen + bromine $\longrightarrow$
c. dichlorine heptoxide + water $\longrightarrow$

60. Write a balanced chemical equation for each single-replacement reaction that takes place in aqueous solution. Write "no reaction" if a reaction does not occur.
a. Steel wool (iron) is placed in sulfuric acid.
b. Mercury is poured into an aqueous solution of zinc nitrate.
c. Bromine reacts with aqueous barium iodide.

61. Pieces of sodium and magnesium are dropped into separate water-filled test tubes (A and B). There is vigorous bubbling in Tube A, but not in Tube B.
a. Which tube contains the sodium metal?
b. Write an equation for the reaction in the tube containing the sodium metal. What type of reaction is occurring in this tube?

62. Write a balanced equation for the complete combustion of each compound. Assume that the products are carbon dioxide and water.
a. octane (C_8H_{18})
b. glucose ($C_6H_{12}O_6$)
c. acetic acid ($HC_2H_3O_2$)

63. Write balanced chemical equations for these decomposition reactions.
a. Aluminum is obtained from aluminum oxide with the addition of a large amount of electrical energy.
b. Heating tin(IV) hydroxide gives tin(IV) oxide and water.
c. Silver carbonate decomposes into silver oxide and carbon dioxide when it is heated.

64. Write a balanced net ionic equation for each reaction. The product that is not ionized is given.
a. $H_2C_2O_4 + KOH \longrightarrow [H_2O]$
b. $Na_2S + HCl \longrightarrow [H_2S]$
c. $NaOH + Fe(NO_3)_3 \longrightarrow [Fe(OH)_3]$

65. Aqueous solutions of sodium sulfide and cadmium nitrate were mixed in a beaker.

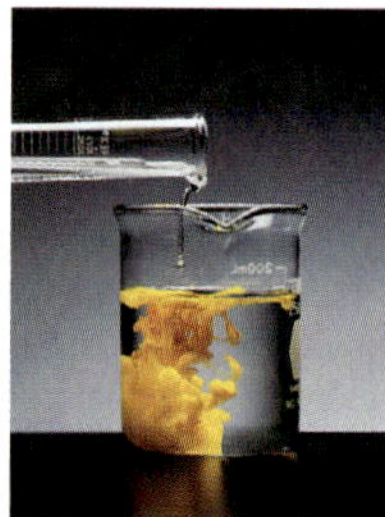

a. Write the formula of the yellow precipitate.
b. Identify the spectator ions in the solution.
c. Write the net ionic equation for the reaction.

66. Each of these equations has water as either a reactant or a product. Balance each equation and identify the type of those reactions in which water is a reactant.
a. $K_2O(s) + H_2O(l) \longrightarrow KOH(aq)$
b. $C_{19}H_{40}(s) + O_2(g) \longrightarrow CO_2(g) + H_2O(l)$
c. $Rb(s) + H_2O(l) \longrightarrow H_2(g) + RbOH(aq)$

Critical Thinking

67. Fill in the missing reactant and then balance each equation.

a. $K(s) + ____ \longrightarrow KOH(aq) + H_2(g)$

b. $C_2H_5OH(l) + ____ \longrightarrow CO_2(g) + H_2O(g)$

c. $Bi(NO_3)_3(aq) + ____ \longrightarrow Bi_2S_3(s) + HNO_3(aq)$

d. $Al(s) + ____ \longrightarrow AlBr_3(s)$

e. $HNO_3(aq) + ____ \longrightarrow Ba(NO_3)_2(aq) + H_2O(l)$

68. Why is smoking not permitted near an oxygen source? What would happen if a match were struck in a room filled with oxygen?

69. Alkanes are hydrocarbon molecules that have the general formula C_nH_{2n+2}. The graph shows the number of oxygen, carbon dioxide, and water molecules needed to balance the equations for the complete combustion of every alkane having from one to ten carbon atoms.

$$C_nH_{2n+2} + ____ O_2 \longrightarrow ____ CO_2 + ____ H_2O$$

Coefficients in Combustion Equations

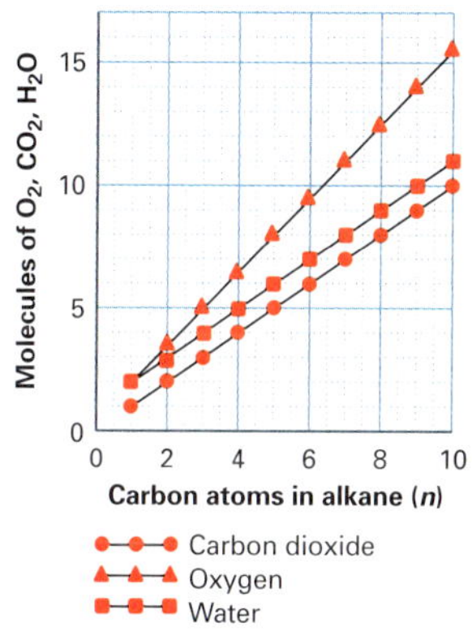

a. Use the graph to write balanced equations for the combustion of C_5H_{12} and C_9H_{20}.

b. Extrapolate the graph and write balanced equations for the combustion of $C_{12}H_{26}$ and $C_{17}H_{36}$.

c. The coefficient for O_2 in the general equation is as follows.

$$n + \left(\frac{n+1}{2}\right)$$

What are the coefficients for CO_2 and H_2O?

Concept Challenge

70. Write a balanced chemical equation for each reaction. Classify each as to type.

a. Sodium iodide reacts with phosphoric acid.

b. Potassium oxide reacts with water.

c. Heating sulfuric acid produces water, oxygen, and sulfur dioxide.

d. Aluminum reacts with sulfuric acid.

e. Pentane (C_5H_{12}) reacts with oxygen.

71. The photos show various types of reactions.

(1) (2)

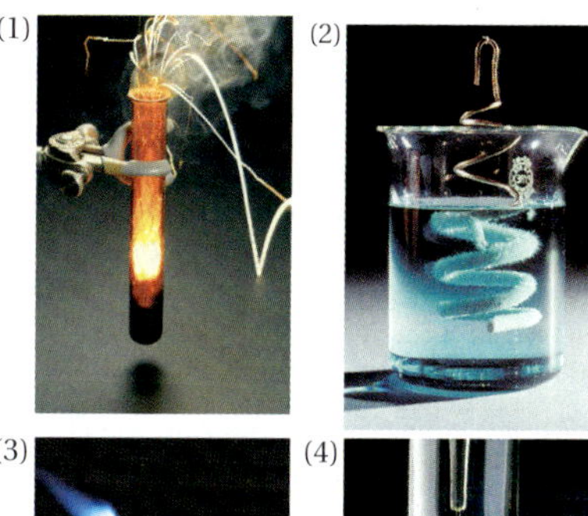

(3)

(4)

(1) Aluminum reacting with bromine

(2) The reaction of copper with aqueous silver nitrate

(3) Propane (C_3H_8) reacting with oxygen

(4) The reaction of lead(II) nitrate with potassium iodide

a. Identify each type of reaction.

b. Write the equation for each reaction.

72. When pale yellow chlorine gas is bubbled through a clear, colorless solution of sodium iodide, the solution turns brown.

a. What type of reaction is taking place?

b. Write the net ionic equation.

Critical Thinking

67. a. $2K(s) + 2H_2O(l) \rightarrow 2KOH(aq) + H_2(g)$

b. $C_2H_5OH(l) + 3O_2(g) \rightarrow 2CO_2(g) + 3H_2O(g)$

c. $2Bi(NO_3)_3(aq) + 3H_2S(g) \rightarrow Bi_2S_3(s) + 6HNO_3(aq)$

d. $2Al(s) + 3Br_2(l) \rightarrow 2AlBr_3(s)$

e. $2HNO_3(aq) + Ba(OH)_2(aq) \rightarrow Ba(NO_3)_2(aq) + 2H_2O(l)$

68. Smoking is not permitted near an oxygen source because a fire will burn faster in an area of high oxygen concentration. However, if a match were struck in a room full of oxygen and isolated from combustible material, it would only burn more vigorously.

69. a. $C_5H_{12} + 8O_2 \rightarrow 5CO_2 + 6H_2O$

$C_9H_{20} + 14O_2 \rightarrow 9CO_2 + 10H_2O$

b. $2C_{12}H_{26} + 37O_2 \rightarrow 24CO_2 + 26H_2O$

$C_{17}H_{36} + 26O_2 \rightarrow 17CO_2 + 18H_2O$

c. $n = CO_2$; $(n + 1) = H_2O$

Concept Challenge

70. a. $3NaI + H_3PO_4 \rightarrow 3HI + Na_3PO_4$; double-replacement

b. $K_2O + H_2O \rightarrow 2KOH$; combination

c. $2H_2SO_4 \xrightarrow{\Delta} 2H_2O + O_2 + 2SO_2$; decomposition

d. $2Al + 3H_2SO_4 \rightarrow 3H_2 + Al_2(SO_4)_3$; single-replacement

e. $C_5H_{12} + 8O_2 \rightarrow 5CO_2 + 6H_2O$; combustion

71. a. (1) combination

(2) single-replacement

(3) combustion

(4) double-replacement

b. (1) $2Al(s) + 3Br_2(l) \rightarrow 2AlBr_3(s)$

(2) $Cu(s) + 2AgNO_3(aq) \rightarrow Cu(NO_3)_2(aq) + 2Ag(s)$

(3) $C_3H_8(g) + 5O_2(g) \rightarrow 3CO_2(g) + 4H_2O(g)$

(4) $Pb(NO_3)_2(aq) + 2KI(aq) \rightarrow PbI_2(s) + 2KNO_3(aq)$

72. a. single-replacement

b. $Cl_2(g) + 2I^-(aq) \rightarrow I_2(aq) + 2Cl^-(aq)$

Assessment continued

Cumulative Review

73. a. water
b. water vapor in the air
c. physical change
74. Element: gold; compounds: sodium chloride, ice with water; homogeneous mixtures, salt water, air; heterogeneous mixture: salt and sand; substance: sodium chloride, gold, water with ice
75. 36.6 kg
76. 22 protons, 28 neutrons, and 22 electrons
77. a. $1s^22s^22p^63s^23p^64s^23d^{10}4p^6$
b. $1s^22s^22p^63s^23p^6$
c. $1s^22s^22p^63s^23p^63d^{10}$
d. $1s^22s^22p^63s^23p^63d^{10}$
78. Electronegativity is the tendency for an atom to attract bonded electrons to itself. Electronegativity values increase from left to right in a period.
79. a. incorrect; KBr
b. correct
c. incorrect; Ca_3N_2
d. correct
80. a. K_2CrO_4
b. $NaHSO_3$
c. permanganic acid
d. potassium oxalate
81. a. 2.41 mol
b. 6.91×10^{-2} mol
c. 0.934 mol
d. 7.09 mol
82. a. compound 1: $FeSO_4$, compound 2: $FeSO_3$
b. $FeSO_4$, iron(II) sulfate; $FeSO_3$, iron(III) sulfite
83. $C_8H_{10}O_2N_4$
84. a.

$CaCl_2$ (mol)	H_2O (mol)
a. 0.156	b. 0.312
c. 0.439	d. 0.878
e. 1.12	f. 2.24
g. 3.03	h. 6.06

CHAPTER 11 Assessment continued

Cumulative Review

73. When you take a glass of cold liquid outside on a warm, humid day, drops of liquid soon form on the outside of the glass. *(Chapter 2)*
a. What is the liquid?
b. Where did the liquid come from?
c. Did a chemical or physical change occur?

74. Classify each of the following as an element, a compound, a homogeneous mixture, a heterogeneous mixture, or a substance. Some may fit in more than one category. *(Chapter 2)*
a. salt water **b.** salt and sand
c. sodium chloride **d.** gold
e. air **f.** water with ice

75. A block of ice measures 25.0 cm × 42.0 cm × 38.0 cm. What is the mass of the ice in kilograms? The density of ice is 0.917 g/cm^3. *(Chapter 3)*

76. List the number of protons, neutrons, and electrons in this isotope of titanium: $^{50}_{22}Ti$. *(Chapter 4)*

77. Write electron configurations for the following ions. *(Chapter 5)*
a. Sr^{2+} **b.** S^{2-} **c.** Ga^{3+} **d.** Cu^{+}

78. Explain what is meant by electronegativity. How do electronegativity values change across a row of representative elements? *(Chapter 6)*

79. Are any of the following formula units for ionic compounds incorrect? If so, write the correct formulas. *(Chapter 7)*
a. K_2Br **b.** Na_2S
c. CaN_2 **d.** Al_2O_3

80. Give the name or formula for the following compounds. *(Chapter 9)*
a. Potassium chromate
b. Sodium hydrogen sulfite
c. $HMnO_4$
d. $K_2C_2O_4$

81. Calculate the number of moles in each substance. *(Chapter 10)*
a. 54.0 L of nitrogen dioxide (at STP)
b. 1.68 g of magnesium ions
c. 69.6 g of sodium hypochlorite
d. 4.27×10^{24} molecules of carbon monoxide

82. The graphs show the percent composition of two different compounds formed by the elements iron, oxygen, and sulfur. *(Chapter 10)*

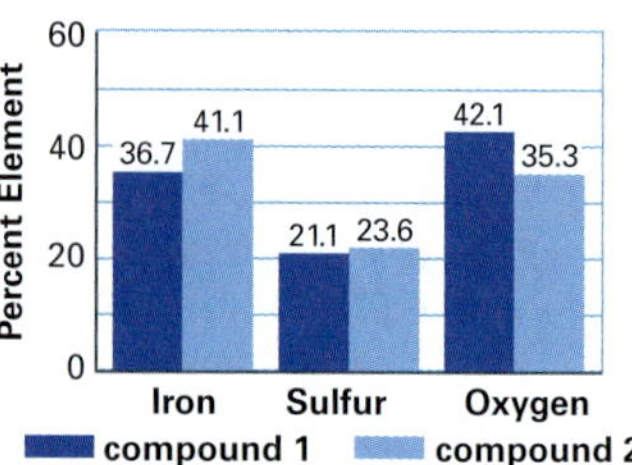

a. Using the data on the graphs, calculate the empirical formula of each compound.
b. Name each compound.

83. Many coffees and colas contain the stimulant caffeine. The percent composition of caffeine is 49.5% C, 5.20% H, 16.5% O, and 28.9% N. What is the molecular formula of caffeine if its molar mass is 194.1 g/mol? *(Chapter 10)*

84. Calcium chloride ($CaCl_2$) is a white solid used as a drying agent. The maximum amount of water absorbed by different quantities of $CaCl_2$ is given in the table below. *(Chapter 10)*

$CaCl_2$ (g)	$CaCl_2$ (mol)	H_2O (g)	H_2O (mol)
17.3	a. ____	5.62	b. ____
48.8	c. ____	15.8	d. ____
124	e. ____	40.3	f. ____
337	g. ____	109	h. ____

a. Complete the table.
b. Plot the moles of water absorbed (*y*-axis) versus the moles of $CaCl_2$.
c. Based on your graph, how many molecules of water does each formula unit of $CaCl_2$ absorb?

b.

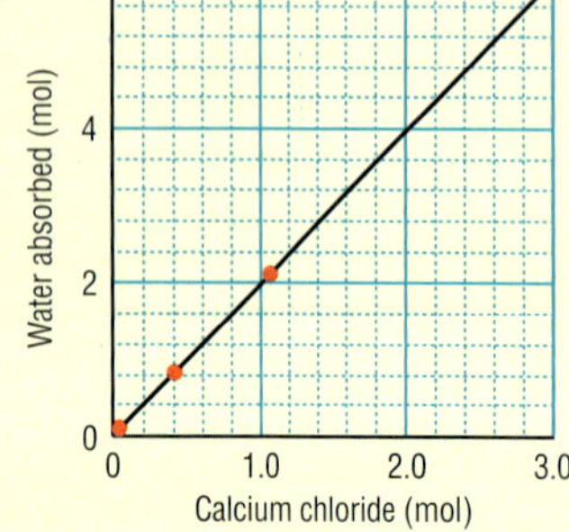

c. Two molecules of water.

Standardized Test Prep

Test-Taking Tip

Before you answer questions about a diagram, study the diagram carefully. Read all captions and labels. Look at all the information in the diagram and think about how it all relates together. Ask yourself some questions. *What is this diagram showing? What does this diagram tell me?*

Select the choice that best answers each question or completes each statement.

1. When the equation $Fe_2O_3 + H_2 \longrightarrow Fe + H_2O$ is balanced using whole-number coefficients, what is the coefficient of H_2?
   a. 6 **b.** 3 **c.** 2 **d.** 1

2. Identify the spectator ion in this reaction.
   $Ba(OH)_2(aq) + H_2SO_4(aq) \longrightarrow BaSO_4(s) + H_2O(l)$
   a. Ba^{2+}
   b. SO_4^{2-}
   c. OH^-
   d. H^+
   e. There is no spectator ion.

3. Magnesium ribbon reacts with an aqueous solution of copper(II) chloride in a single replacement reaction. Which are the products of the balanced net ionic equation for the reaction?
   a. $Mg^{2+}(aq) + 2Cl^-(aq) + Cu(s)$
   b. $Mg^+(aq) + Cl^-(aq) + Cu^+(aq)$
   c. $Mg^{2+}(aq) + Cu(s)$
   d. $Cu(s) + 2Cl^-(aq)$

Use the following description and data table to answer Questions 4–6.

Dropper bottles labeled P, Q, and R contain one of three aqueous solutions: potassium carbonate, K_2CO_3; hydrochloric acid, HCl; and calcium nitrate, $Ca(NO_3)_2$. The table shows what happens when pairs of solutions are mixed.

Solution	P	Q	R
P	—	precipitate	no reaction
Q	precipitate	—	gas forms
R	no reaction	gas forms	—

4. Identify the contents of each dropper bottle.

5. Write the net ionic equation for the formation of the precipitate.

6. Write the complete ionic equation for the formation of the gas.

7. Which are the expected products of the decomposition reaction of potassium oxide, K_2O?
   a. $K^+(s)$ and $O^{2-}(g)$
   b. $K^+(s)$ and $O_2(g)$
   c. $K(s)$ and $O_2^{2-}(g)$
   d. $K(s)$ and $O_2(g)$

Use the diagram to answer Questions 8–11.

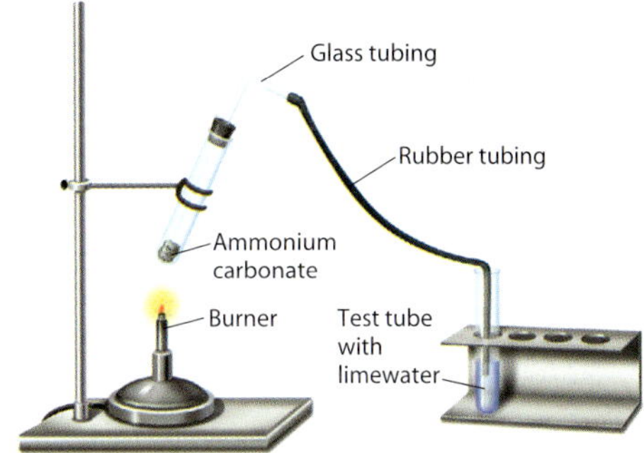

8. When ammonium carbonate is heated, water, ammonia, and carbon dioxide are produced. What type of chemical reaction is occurring?

9. Write formulas for the reaction products.

10. Write a balanced equation for the reaction. Include states for reactants and products.

11. What purpose does the limewater serve?

Use the atomic windows to answer Question 12. Oxygen atoms are red; nitrogen atoms are blue.

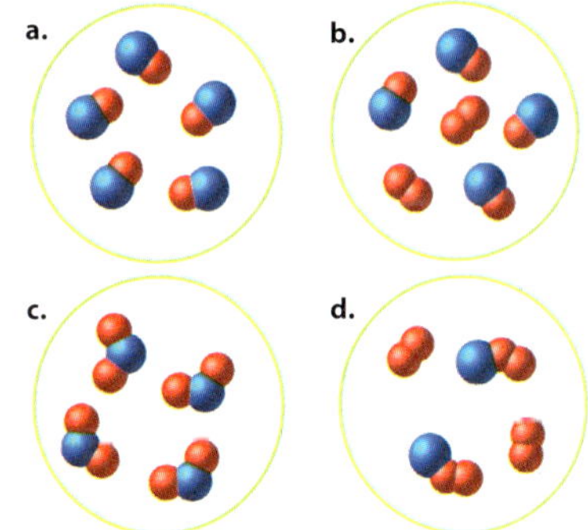

12. Which windows represent (1) the reactants and (2) the products for the following reaction?
   $$2NO(g) + O_2(g) \longrightarrow 2NO_2(g)$$

Standardized Test Prep

1. b
2. e
3. c
4. P is calcium nitrate, Q is potassium carbonate, and R is hydrochloric acid.
5. $Ca^{2+}(aq) + CO_3^{2-}(aq) \rightarrow CaCO_3(s)$
6. $2K^+(aq) + CO_3^{2-}(aq) + 2H^+(aq) + 2Cl^-(aq) \rightarrow 2K^+(aq) + 2Cl^-(aq) + H_2O(l) + CO_2(g)$
7. d
8. decomposition reaction
9. H_2O, NH_3, and CO_2
10. $(NH_4)_2CO_3(s) \rightarrow 2NH_3(g) + CO_2(g) + H_2O(g)$
11. to react with and identify the carbon dioxide
12. (b) represents the reactants (c) represents the products

CHAPTER 12

Stoichiometry Planning Guide

SECTION OBJECTIVES	STANDARDS NATIONAL	STANDARDS STATE	ACTIVITIES and LABS	
12.1 The Arithmetic of Equations, pp. 353–358 1 block or 2 periods **12.1.1** **Explain** how balanced equations apply to both chemistry and everyday life. **12.1.2** **Interpret** balanced chemical equations in terms of moles, representative particles, mass, and gas volume at STP. **12.1.3** **Identify** the quantities that are always conserved in chemical reactions.	A-2, B-3		**SE** Inquiry Activity: How Many Can You Make?, p. 352 **TE** Teacher Demo: Interpreting a Chemical Equation, p. 357 **LM** Lab 19: Quantitative Analysis **LP** Lab Practical 12-1: Stoichiometry in a Reaction	L2 L2 L2 L2
12.2 Chemical Calculations, pp. 359–367 1 block or 2 periods **12.2.1** **Construct** mole ratios from balanced chemical equations and **apply** these ratios in stoichiometric calculations. **12.2.2** **Calculate** stoichiometric quantities from balanced chemical equations using units of moles, mass, representative particles, and volumes of gases at STP.	A-2, B-3		**SE** Small-Scale Lab: Analysis of Baking Soda, p. 367 **TE** Teacher Demo: Interpreting a Chemical Equation, p. 361 **TE** Class Activity: Stoichiometric Flash Cards, p. 365 **LM** Lab 19: Quantitative Analysis **SSLM** Lab 18: Titration: Determining How Much Acid is in a Solution **SSLM** Lab 19: Mass Titrations: Measuring Molar Concentrations **PLM** Analysis of Baking Soda **LP** Lab Practical 12-2: Limiting Reagent **VCL** Lab 28: Analysis of Baking Soda	L2 L2 L2 L2 L2 L2 L2 L2 L2
12.3 Limiting Reagent and Percent Yield, pp. 368–377 1 block or 2 periods **12.3.1** **Identify** the limiting reagent in a reaction. **12.3.2** **Calculate** theoretical yield, actual yield, or percent yield given appropriate information.	A-2, B-3		**SE** Quick Lab: Limiting Reagents, p. 372 **TE** Teacher Demo: Limiting Factor, p. 369 **TE** Class Activity: Actual Yield and Heat, p. 373 **LM** Lab 20: Balanced Chemical Equations	L2 L2 L2 L2

Ability Levels	
L1	For students who need additional help
L2	For all students
L3	For students who need to be challenged

Components

SE	Student Edition	**LM**	Laboratory Manual	**GCP**	Graphing Calculator Problems	**T**	Transparencies
TE	Teacher's Edition	**SSLM**	Small-Scale Chemistry Lab Manual	**TTT**	Test-Taking Tips with Transparencies, Grades 9–12 Science	**CA**	Chem Alive and Chemistry Field Trips DVD
GRSW	Guided Reading and Study Workbook	**LP**	Laboratory Practicals			**iT**	Interactive Textbook with ChemASAP
CTR	Core Teaching Resources	**PLM**	Probeware Lab Manual	**STP**	Standardized Test Preparation Workbook	**GO**	Go Online Internet Resources
SSG	Spanish Study Guide	**VCL**	Virtual Chemistry Labs				
		CTB	Computer Test Bank				

RESOURCES PRINT and TECHNOLOGY	PROBLEMS and ASSESSMENT
GRSW Chapter 12, Section 1 L1 **T** T122-T125 L2 **GO** Chapter 12, Section 1 L2	**CP** 12.1, Interpreting a Balanced Chemical Equation **SP** 12.1, Using a Balanced Equation as a Recipe **SE** Section 12.1 Assessment, p. 358 **iT** Section 12.1
GRSW Chapter 12, Section 2 L1 **T** T126-T132 L2	**SP** 12.2, Calculating Moles of a Product **SP** 12.3, Calculating the Mass of a Product **SP** 12.4, Calculating Moles of a Product **SP** 12.5, Volume-Volume Stoichiometric Calculations **SP** 12.6, Finding the Volume of a Gas Needed for a Reaction **SE** Section 12.2 Assessment, p. 366 **iT** Section 12.2
GRSW Chapter 12, Section 12.3 L1 **T** T133-T138 L2 **GO** Chapter 12, Section 3 L2	**SP** 12.7, Determining the Limiting Reagent in a Reaction **SP** 12.8, Using a Limiting Reagent to Find the Quantity of a Product **SP** 12.9, Calculating the Theoretical Yield of a Reaction **SP** 12.10, Calculating the Percent Yield of a Reaction **SE** Section 12.3 Assessment, p. 375 **iT** Section 12.3

Go Online

Web Code: cdn-1121
Web Code: cdn-1123

MATERIALS List

Quantities for each group

STUDENT EDITION

Inquiry Activity, p. 352
20 metal paper clips (symbol M), 20 identically colored vinyl-coated paper clips (symbol C), plastic sandwich bag

Small-Scale Lab, p. 367
baking soda, 3 plastic cups, soda straw, balance, pipets of HCl, NaOH, and thymol blue, pH sensor (optional)

Quick Lab, p. 372
graduated cylinder, balance, 3 250-mL Erlenmeyer flasks, 3 rubber balloons, 4.2 g magnesium ribbon, 300 mL 1.0*M* hydrochloric acid

TEACHER'S EDITION

Teacher Demo, p. 357
2.5–3.5-cm strip of magnesium, 50 mL 1*M* HCl, 100-mL beaker, baking soda

Teacher Demo, p. 362
0.1*M* KI, 0.1*M* $Pb(NO_3)_2$, 250-mL beakers

Class Activity, p. 365
8 index cards, 1 colored index card, paper punch, 2 brass paper fasteners

Teacher Demo, p. 369
15 plastic bottles, 30 plastic caps to fit the bottles, 6 containers to hold 5 caps each

Class Activity, p. 373
3 foam cups, a thermometer, 100 mL of 1.0*M* HCl, approximately 200 mL of 1.0*M* NaOH

Chapter Assessment

CHAPTER ASSESSMENT
SE Chapter Assessment, pp. 379–382
CTB Chapter 12
iT Chapter 12

STANDARDIZED TEST PREP
SE Chapter 12, p. 383
STP Topic 3
iT Interactive Textbook with Assessment at PHSchool.com

CHAPTER 12

Stoichiometry

INQUIRY Activity

How Many Can You Make?

Purpose

Students will use simple molecular models to demonstrate the concept of a limiting reagent in a chemical reaction.

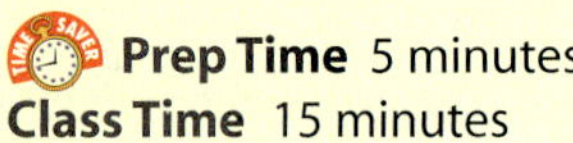

Prep Time 5 minutes
Class Time 15 minutes

Think About It

1. Acceptable answers include $n_{M_2} = 5\text{–}10, n_{C_2} = 15 - n_{M_2}$.
2. Answers will depend on n_{M_2} and n_{C_2}.
3. Answers will depend on n_{M_2} and n_{C_2}.
4. Answers will depend on n_{M_2} and n_{C_2}.
5. Students should note that the number of MC_3 molecules formed and the number and type of reactant molecules remaining depend on the initial number of M_2 and C_2 molecules and the ratio of M to C atoms in the molecular formula.

CHAPTER 12

Stoichiometry

Like a chemical equation, a recipe tells you the amount of ingredients (your reactants) needed to make a pie (your product).

INQUIRY Activity

How Many Can You Make?

Materials

20 metal paper clips (symbol M), 20 identically colored vinyl-coated paper clips (symbol C), plastic sandwich bag.

Procedure

1. Join together pairs of paper clips of the same color to form models representing 10 diatomic molecules of each reactant. Place these molecules in the plastic bag.
2. Without looking, remove 15 molecules from the plastic bag.
3. Line up the M_2 and C_2 molecules in two adjacent vertical rows.
4. Pair up reactant molecules in the 1:3 M_2-to-C_2 ratio as shown in the equation $M_2 + 3C_2 \longrightarrow 2MC_3$.
5. Make the molecules "react" by taking them apart and forming two molecules of the product.
6. Continue making M_2 and C_2 react in a 1:3 ratio until you run out of one of the reactants.

Think About It

1. List the number of each type of reactant molecule that you removed from the bag.
2. How many molecules of the product did you form?
3. Which reactant molecule did you run out of first?
4. How many molecules of each reactant remained at the completion of the reaction?
5. Repeat the experiment and compare the results.

12.1 The Arithmetic of Equations

Connecting to Your World

Silk is one of the most beautiful and luxurious of all fabrics. It is spun from the cocoons of silkworms. Silk manufacturers know from experience that to produce enough silk to make just one elegant Japanese kimono they will need over 3000 cocoons. In a similar fashion, chemists need to know how much reactant is needed to make a certain amount of product. The answer lies in chemical equations. From a balanced chemical equation, you can determine the quantities of reactants and products in a reaction.

Guide for Reading

Key Concepts

- How is a balanced equation like a recipe?
- How do chemists use balanced chemical equations?
- In terms of what quantities can you interpret a balanced chemical equation?
- What quantities are conserved in every chemical reaction?

Vocabulary

stoichiometry

Reading Strategy

Using Prior Knowledge Before you read, jot down three things you know about balanced chemical equations. When you have read the section, explain how what you already knew helped you learn something new.

Using Everyday Equations

When you bake cookies, you probably use a recipe. A cookie recipe tells you the precise amounts of ingredients to mix to make a certain number of cookies, as shown in Figure 12.1. If you need a larger number of cookies than the recipe provides, you can double or triple the amounts of ingredients. **A balanced chemical equation provides the same kind of quantitative information that a recipe does.** In a cookie recipe, you can think of the ingredients as the reactants, and the cookies as the products.

Here is another example. Imagine you are in charge of manufacturing for the Tiny Tyke Tricycle Company. The business plan for Tiny Tyke requires the production of 640 custom-made tricycles each week. One of your responsibilities is to be sure that there are enough parts available at the start of each workweek to make these tricycles. How can you determine the number of parts you need per week?

Figure 12.1 A cookie recipe tells you the number of cookies that you can expect to make from the listed amounts of ingredients. **Using Models** ***How can you express a cookie recipe as a balanced equation?***

Section Resources

Print

- **Guided Reading and Study Workbook,** Section 12.1
- **Core Teaching Resources,** Section 12.1 Review
- **Transparencies,** T122–T125
- **Laboratory Manual,** Lab 19

Technology

- **Interactive Textbook with ChemASAP,** Problem Solving 12.1, 12.4, Assessment 12.1
- **Go Online,** Section 12.1

12.1

1 FOCUS

Objectives

12.1.1 Explain how balanced equations apply to both chemistry and everyday situations.

12.1.2 Interpret balanced chemical equations in terms of moles, representative particles, mass, and gas volume at STP.

12.1.3 Identify the quantities that are always conserved in chemical reactions.

Guide for Reading

Build Vocabulary 

Paraphrase Introduce the term stoichiometry in your own words. Stress that stoichiometry allows students to calculate the amounts of chemical substances involved in chemical reactions using information obtained from balanced chemical equations.

Reading Strategy 

Relate Text and Visuals Have students construct tables similar to Figure 12.3 to interpret balanced chemical equations for stoichiometric problems. Students need to use only those levels of interpretation—particles, molecules, moles, mass, or gas volumes—appropriate to the problem.

2 INSTRUCT

Connecting to Your World

Have students study the photograph and read the text that opens the section. Ask, **How many cocoons would be required to produce enough silk for two Japanese kimonos?** *(twice as many, or 6000 cocoons)* **How did you calculate the number of cocoons?** *(by multiplying the number needed for one kimono by two)*

Answers to...

Figure 12.1 Acceptable answers will list ingredients as reactants and cookies as the product.

Using Everyday Equations

Discuss L2

Write this statement on the board: "A frame, a seat, wheels, a handlebar, and pedals are needed to assemble a complete tricycle." Have students look at the visual representation of how a tricycle is assembled and ask, **Does this statement correctly describe the process?** *(no)* **What is the correct statement?** *(A frame, a seat, three wheels, a handlebar, and two pedals are needed to assemble a complete tricycle.)*

Using Balanced Chemical Equations

Discuss L2

Remind students that during a chemical reaction, atoms are rearranged into new combinations and groupings. It is somewhat similar to changing silk cocoons into a kimono. Just as a chemical reaction has reactants and products, the cocoon is the reactant and the kimono is the product. Much like having the proper supply of silk cocoons, chemists must have an adequate supply of reactants for a chemical reaction.

Word *Origins* 

The term *spectrometry* means "the study of light or spectrum."

To simplify this discussion, assume that the major components of the tricycle are the frame (F), the seat (S), the wheels (W), the handlebars (H), and the pedals (P), in other words, your reactants. The figure below illustrates how an equation can represent the manufacturing of a single tricycle.

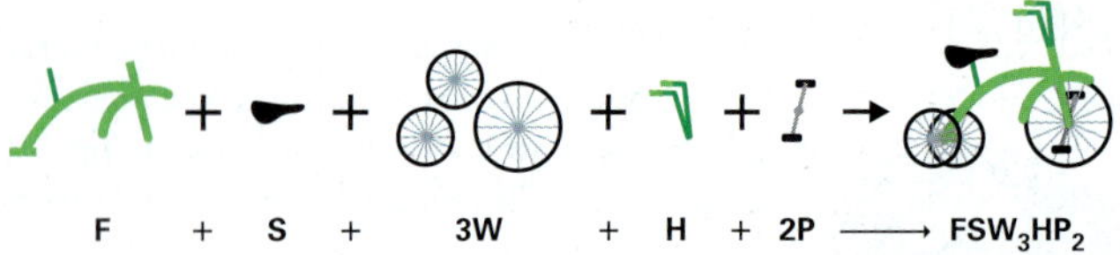

The finished tricycle, your product, has a "formula" of FSW_3HP_2. The balanced equation for making a single tricycle is

$$F + S + 3W + H + 2P \longrightarrow FSW_3HP_2$$

This balanced equation is a "recipe" to make a single tricycle: Making a tricycle requires assembling one frame, one seat, three wheels, one handlebar, and two pedals. Now look at Sample Problem 12.1. It shows you how to use the balanced equation to calculate the number of parts needed to manufacture a given number of tricycles.

Using Balanced Chemical Equations

Nearly everything you use is manufactured from chemicals—soaps, shampoos and conditioners, CDs, cosmetics, medicines, and clothes. In manufacturing such items, the cost of making them cannot be greater than the price they are sold at. Otherwise, the manufacturer will not make a profit. Therefore, the chemical processes used in manufacturing must be carried out economically. This is where balanced equations help.

A balanced chemical equation tells you what amounts of reactants to mix and what amounts of product to expect. **Chemists use balanced chemical equations as a basis to calculate how much reactant is needed or product is formed in a reaction.** When you know the quantity of one substance in a reaction, you can calculate the quantity of any other substance consumed or created in the reaction. Quantity usually means the amount of a substance expressed in grams or moles. However, quantity could just as well be in liters, tons, or molecules.

The calculation of quantities in chemical reactions is a subject of chemistry called **stoichiometry.** Calculations using balanced equations are called stoichiometric calculations. For chemists, stoichiometry is a form of bookkeeping. For example, accountants can track income, expenditures, and profits for a small business by tallying each in dollars and cents. Chemists can track reactants and products in a reaction by stoichiometry. It allows chemists to tally the amounts of reactants and products using ratios of moles or representative particles.

 *How is stoichiometry similar to bookkeeping?*

Word *Origins*

Stoichiometry comes from the combination of the Greek words *stoikheioin*, meaning "element," and *metron*, meaning "to measure." Stoichiometry is the calculation of amounts of substances involved in chemical reactions. **What do think the term *spectrometry* means?**

354 Chapter 12

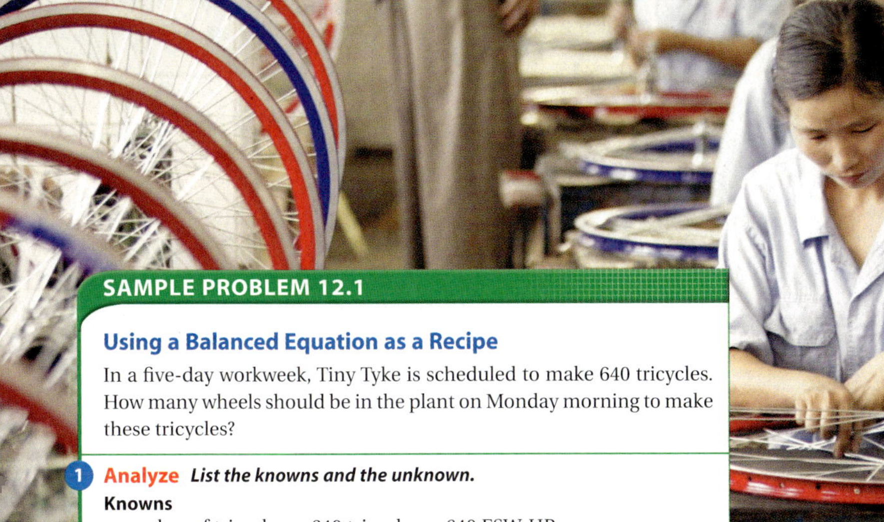

SAMPLE PROBLEM 12.1

Using a Balanced Equation as a Recipe

In a five-day workweek, Tiny Tyke is scheduled to make 640 tricycles. How many wheels should be in the plant on Monday morning to make these tricycles?

1 Analyze ***List the knowns and the unknown.***

Knowns

- number of tricycles = 640 tricycles = 640 FSW_3HP_2
- $F + S + 3W + H + 2P \longrightarrow FSW_3HP_2$

Unknown

- number of wheels = ? wheels

The desired conversion is tricycles (FSW_3HP_2) $\longrightarrow$ wheels (W). The balanced equation tells you that each tricycle has three wheels, or $1\ FSW_3HP_2 = 3\ W$. The problem can be solved by using the proper conversion factor derived from this expression.

2 Calculate ***Solve for the unknown.***

You can write two conversion factors relating wheels to tricycles.

$$\frac{3\ W}{1\ FSW_3HP_2} \quad \text{and} \quad \frac{1\ FSW_3HP_2}{3\ W}$$

The desired unit is W, so use the conversion factor on the left—the one that has W in the numerator. Multiply the number of tricycles by the conversion factor.

$$640\ \cancel{FSW_3HP_2} \times \frac{3W}{1\ \cancel{FSW_3HP_2}} = 1920\ W$$

3 Evaluate ***Does the result make sense?***

If three wheels are required for each tricycle, and a total of more than 600 tricycles are being made, then a number of wheels in excess of 1800 is a logical answer. The unit of the known cancels with the unit in the denominator of the conversion factor, and the answer is in the unit of the unknown.

Practice Problems

1. Tiny Tike has decided to make 288 tricycles each day. How many tricycle seats, wheels, and pedals are needed?
2. Write an equation that gives your own "recipe" for making a skateboard.

Math Handbook

For help with dimensional analysis, go to page R66.

Interactive Textbook

Problem Solving 12.1 Solve Problem 1 with the help of an interactive guided tutorial.

with ChemASAP

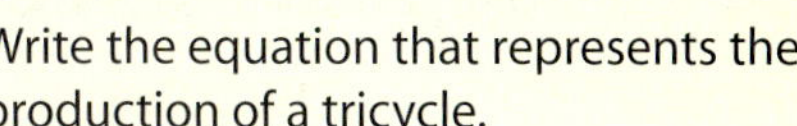

Discuss **L2**

Write the equation that represents the production of a tricycle.

$F + S + 3W + H + 2P \rightarrow FSW_3HP_2$

Point out that the balanced equation contains information that not only relates reactant to products but relates one reactants to another.

Ask, **How many handlebars are needed to complete tricycles if 24 pedals are available?** *(12)* Show students how to set up the proportions to solve this problem.

$?\ H = 24\ \cancel{P} \times (1\ H/2\cancel{P}) = 12\ H$

Ask, **How many handlebars are needed to complete tricycles if 24 seats are available?** *(24)* **24 wheels?** *(8)*

Sample Problem 12.1

Answers

1. 288 seats, 864 wheels, 576 pedals
2. Answers will vary but should include the correct number of "parts" to make the product.

Practice Problems Plus **L2**

Chapter 21 Assessment problem 57 is related to Sample Problem 12.1.

Math Handbook

For a math refresher and preview, direct students to dimensional analysis, page R66.

Answers to...

Checkpoint In stoichiometry, the amount of product formed by a chemical reaction can be calculated based on the amounts of the reactants. In bookkeeping, the profits of a business can be calculated based on income and expenditures.

Interpreting Chemical Equations

Discuss L2

Review balancing chemical reactions by writing several unbalanced equations on the board. For example:

$CuO(s) + NH_3(aq) \rightarrow Cu(s) + H_2O(l) + N_2(g)$

$NH_3(g) + O_2(g) \rightarrow NO(g) + H_2O(g)$

$KClO_3(s) \rightarrow KCl(s) + O_2(g)$.

Have students balance the equations as shown.

$3CuO(s) + 2NH_3(aq) \rightarrow 3Cu(s) + 3H_2O(l) + N_2(g)$

$4NH_3(g) + 5O_2(g) \rightarrow 4NO(g) + 6H_2O(g)$

$2KClO_3(s) \rightarrow 2KCl(s) + 3O_2(g)$

Ask, **Why is it not correct to balance an equation by changing the subscripts in one or more of the formulas?** *(Changing the subscripts in a formula changes the chemical identity of the substance.)*

Mass Conservation in Chemical Reactions

Use Visuals L1

Figure 12.3 Remind students that the term STP represents "standard temperature and pressure." Ask, **What are the values of STP?** *(0°C and 101.3 kPa)* **Why is the volume of a gas usually measured at STP?** *(The volume of a gas is usually measured at STP because its volume varies with temperature and pressure.)* **What is the molar volume of any gas at STP?** *(22.4 L/mol)* **How many particles does it contain?** *(22.4 L of any ideal gas at STP contains 6.02×10^{23} particles of that gas.)*

Figure 12.2 Gardeners use ammonium salts as fertilizer. The nitrogen in these salts is essential to plant growth.

Interpreting Chemical Equations

In gardens such as the one shown in Figure 12.2, fertilizers are often used to improve the growth of flowers. As you may recall from Chapter 10, ammonia is widely used as a fertilizer. Ammonia is produced industrially by the reaction of nitrogen with hydrogen.

$$N_2(g) + 3H_2(g) \longrightarrow 2\,NH_3(g)$$

The balanced chemical equation tells you the relative amounts of reactants and product in the reaction. However, your interpretation of the equation depends on how you quantify the reactants and products. **A balanced chemical equation can be interpreted in terms of different quantities, including numbers of atoms, molecules, or moles; mass; and volume.** As you study stoichiometry, you will learn how to read a chemical equation in terms of any of these quantities.

Number of Atoms At the atomic level, a balanced equation indicates that the number and type of each atom that makes up each reactant also makes up each product. Thus, both the number and types of atoms are not changed in a reaction. In the synthesis of ammonia, the reactants are composed of two atoms of nitrogen and six atoms of hydrogen. These eight atoms are recombined in the product.

Number of Molecules The balanced equation indicates that one molecule of nitrogen reacts with three molecules of hydrogen. Nitrogen and hydrogen will always react to form ammonia in a 1:3:2 ratio of molecules. If you could make 10 molecules of nitrogen react with 30 molecules of hydrogen, you would expect to get 20 molecules of ammonia. Of course, it is not practical to count such small numbers of molecules and allow them to react. You could, however, take Avogadro's number of nitrogen molecules and make them react with three times Avogadro's number of hydrogen molecules. This would be the same 1:3 ratio of molecules of reactants. The reaction would form two times Avogadro's number of ammonia molecules.

Moles A balanced chemical equation also tells you the number of moles of reactants and products. The coefficients of a balanced chemical equation indicate the relative numbers of moles of reactants and products in a chemical reaction. This is the most important information that a balanced chemical equation provides. Using this information, you can calculate the amounts of reactants and products. In the synthesis of ammonia, one mole of nitrogen molecules reacts with three moles of hydrogen molecules to form two moles of ammonia molecules. As you can see from this reaction, the total number of moles of reactants does not equal the total number of moles of product.

Checkpoint ***What do the coefficients of a balanced chemical equation indicate?***

Facts and Figures

Ammonia in the Nitrogen Cycle

Earth's atmosphere contains 0.01 parts per million of ammonia, and small amounts of ammonia occur in volcanic gases. Most ammonia cycles through the living world without returning to the atmosphere. Ammonia plays a role in several stages of the nitrogen cycle. Nitrogen-fixing bacteria form nodules, or swellings, on the roots of plants in the legume family, such as beans and clover plants. These bacteria change atmospheric nitrogen into ammonia molecules or ammonium ions. Other bacteria break down the nitrogenous material in dead plants and animals into ammonia molecules. Certain soil bacteria oxidize these molecules into nitrate ions, the form readily absorbed by plant roots. When a plant dies, this cycle begins again.

356 Chapter 12

Mass A balanced chemical equation obeys the law of conservation of mass. This law states that mass can be neither created nor destroyed in an ordinary chemical or physical process. As you recall, the number and type of atoms does not change in a chemical reaction. Therefore, the total mass of the atoms in the reaction does not change. Using the mole relationship, you can relate mass to the number of atoms in the chemical equation. The mass of 1 mol of N_2 (28.0 g) plus the mass of 3 mol of H_2 (6.0 g) equals the mass of 2 mol of NH_3 (34 g). Although the number of moles of reactants does not equal the number of moles of product, the total number of grams of reactants does equal the total number of grams of product.

Volume If you assume standard temperature and pressure, the equation also tells you about the volumes of gases. Recall that 1 mol of any gas at STP occupies a volume of 22.4 L. The equation indicates that 22.4 L of N_2 reacts with 67.2 L (3 × 22.4 L) of H_2. This reaction forms 44.8 L (2 × 22.4 L) of NH_3.

For: Links on Conservation of Mass
Visit: www.SciLinks.org
Web Code: cdn-1121

Mass Conservation in Chemical Reactions

Figure 12.3 summarizes the information derived from the balanced chemical equation for the formation of ammonia. As you can see, the mass of the reactants equals the mass of the products. In addition, the number of atoms of each type in the reactants equals the number atoms of each type in the product. **Mass and atoms are conserved in every chemical reaction.** However, molecules, formula units, moles, and volumes are not necessarily conserved—although they may be. Consider, for example, the formation of hydrogen iodide,

$$H_2(g) + I_2(g) \longrightarrow 2HI(g)$$

In this reaction, molecules, moles, and volume are all conserved. But in the majority of chemical reactions, they are not.

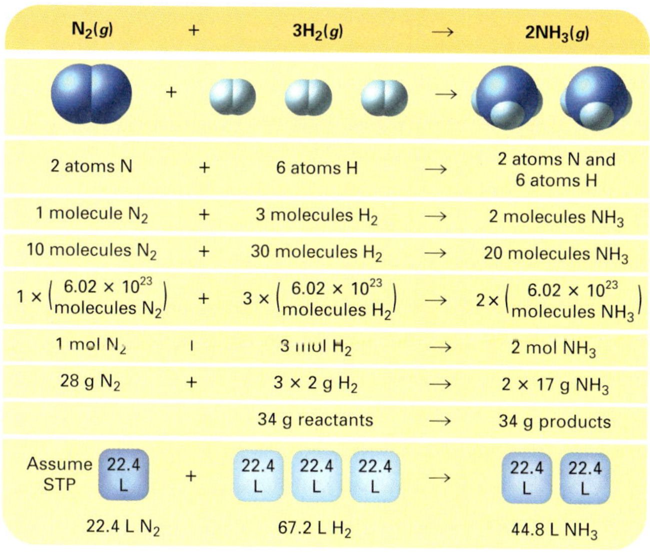

$N_2(g)$	+	$3H_2(g)$	→	$2NH_3(g)$
	+		→	
2 atoms N	+	6 atoms H	→	2 atoms N and 6 atoms H
1 molecule N_2	+	3 molecules H_2	→	2 molecules NH_3
10 molecules N_2	+	30 molecules H_2	→	20 molecules NH_3
$1 \times \left(\begin{matrix}6.02 \times 10^{23} \\ \text{molecules } N_2\end{matrix}\right)$	+	$3 \times \left(\begin{matrix}6.02 \times 10^{23} \\ \text{molecules } H_2\end{matrix}\right)$	→	$2 \times \left(\begin{matrix}6.02 \times 10^{23} \\ \text{molecules } NH_3\end{matrix}\right)$
1 mol N_2	+	3 mol H_2	→	2 mol NH_3
28 g N_2	+	3 × 2 g H_2	→	2 × 17 g NH_3
		34 g reactants	→	34 g products
Assume STP 22.4 L	+	22.4 L 22.4 L 22.4 L	→	22.4 L 22.4 L
22.4 L N_2		67.2 L H_2		44.8 L NH_3

Figure 12.3 The balanced chemical equation for the formation of ammonia can be interpreted in several ways.
Predicting ***How many molecules of NH_3 could be made from 5 molecules of N_2 and 15 molecules of H_2?***

Download a worksheet on **Conservation of Mass** for students to complete and find additional support for NSTA SciLinks.

TEACHER Demo

Interpreting a Chemical Equation

Purpose Students interpret a balanced equation of the reaction of magnesium and hydrochloric acid.

Materials 2.5–3.5-cm strip of magnesium, 50 mL 1*M* HCl(*aq*) in a 100-mL beaker, baking soda

Safety Wear safety glasses and an apron. Neutralize remaining HCl(*aq*) with baking soda before flushing down the drain.

Procedure Identify the two reactants as magnesium and hydrochloric acid. Have students observe the reaction as you carefully add the magnesium strip to the acid. Ask students to write a balanced chemical equation for the reaction of magnesium and hydrochloric acid. *[$Mg(s) + 2HCl(aq) \rightarrow MgCl_2(aq) + H_2(g)$]* Have students interpret the equation in terms of particles, moles, and molar masses.

Expected Outcome Students should express the balanced equation at the particle level as one atom of magnesium reacts with two molecules of hydrogen chloride to produce one formula unit of magnesium chloride and one molecule of hydrogen gas. Similarly, one mole of magnesium reacts with two moles of hydrogen chloride to produce one mole of magnesium chloride and one mole of hydrogen gas. Finally, 24.31 g Mg + 72.92 g HCl produces 95.21 g $MgCl_2$ + 2.02 g H_2.

Differentiated Instruction

Less Proficient Readers L1

Have students construct a table like the one shown in Figure 12.3 for the reaction of hydrogen gas with oxygen gas to form water. Students should begin by writing the balanced equation. Encourage students to draw pictures as shown in Figure 12.3 to represent reactants and products. Have students use the completed table to answer questions such as, "How many moles of water are produced by reacting 4 moles of hydrogen gas with excess oxygen?" Other reactions for students to analyze:

$CO(g) + 2H_2(g) \rightarrow CH_3OH(l)$

$CH_4(g) + 2O_2(g) \rightarrow CO_2(g) + 2H_2O(g)$

Answers to...

Figure 12.3 10 molecules NH_3

the relative numbers of representative particles, the relative numbers of moles, and, for gases, the relative volumes

Section 12.1 (continued)

CONCEPTUAL PROBLEM 12.1

Answers

3. 2 molecules H_2 + 1 molecule $O_2 \rightarrow$ 2 molecules H_2O
   2 mol H_2 + 1 mol $O_2 \rightarrow$ 2 mol H_2O
   44.8 L H_2 + 22.4 L $O_2 \rightarrow$ 44.8 L H_2O
4. 2 mol C_2H_2 + 5 mol $O_2 \rightarrow$ 4 mol CO_2 + 2 mol H_2O
   44.8 L C_2H_2 + 112 L $O_2 \rightarrow$ 89.6 L CO_2 + 44.8 L H_2O
   212 g reactants $\rightarrow$ 212 g products

Practice Problems Plus L2

Balance the following equation:

$C_5H_{12}(g) + O_2(g) \rightarrow CO_2(g) + H_2O(g)$

Interpret the balanced equation in terms of relative number of moles, volumes of gas at STP, and masses of reactants and products. *[1 mol $C_5H_{12}(g)$ + 8 mol $O_2(g) \rightarrow$ 5 mol $CO_2(g)$ + 6 mol $H_2O(g)$; 22.4 L $C_5H_{12}(g)$ + 179 L $O_2(g) \rightarrow$ 112 L $CO_2(g)$ + 134 L $H_2O(g)$; 328 g reactants $\rightarrow$ 328 g products]*

3 ASSESS

Evaluate Understanding L2

Have pairs of students write a balanced chemical equation for a simple reaction. Have pairs exchange equations and write quantitative relationships between reactants and products in terms of mass, moles, particles, and, where appropriate, volumes.

Reteach L1

Remind students that the coefficients in a balanced chemical equation state the relationships among substances involved in the reaction.

Writing Activity

Moles of reactants and products are conserved for some reactions, but this is generally not the case.

If your class subscribes to the Interactive Textbook, use it to review key concepts in Section 12.1.

with ChemASAP

CONCEPTUAL PROBLEM 12.1

Interpreting a Balanced Chemical Equation

Hydrogen sulfide, which smells like rotten eggs, is found in volcanic gases. The balanced equation for the burning of hydrogen sulfide is:

$$2H_2S(g) + 3O_2(g) \longrightarrow 2SO_2(g) + 2H_2O(g)$$

Interpret this equation in terms of
a. numbers of representative particles and moles.
b. masses of reactants and products.

1 Analyze *Identify the relevant concepts.*
a. The coefficients in the balanced equation give the relative number of molecules or moles of reactants and products.
b. A balanced chemical equation obeys the law of conservation of mass.

2 Solve *Apply concepts to this situation.*
a. 2 molecules H_2S + 3 molecules $O_2 \longrightarrow$ 2 molecules SO_2 + 2 molecules H_2O
   2 mol H_2S + 3 mol $O_2 \longrightarrow$ 2 mol SO_2 + 2 mol H_2O

b. Multiply the number of moles of each reactant and product by its molar mass:
2 mol H_2S + 3 mol $O_2 \longrightarrow$ 2 mol SO_2 + 2 mol H_2O.

$$\left(2\ \cancel{mol} \times 34.1\frac{g}{\cancel{mol}}\right) + \left(3\ \cancel{mol} \times 32.0\frac{g}{\cancel{mol}}\right) \longrightarrow \left(2\ \cancel{mol} \times 64.1\frac{g}{\cancel{mol}}\right) + \left(2\ \cancel{mol} \times 18.0\frac{g}{\cancel{mol}}\right)$$

68.2 g H_2S + 96.0 g $O_2 \longrightarrow$ 128.2 g SO_2 + 36.0 g H_2O
164.2 g = 164.2 g

Practice Problems

3. Interpret the equation for the formation of water from its elements in terms of numbers of molecules and moles, and volumes of gases at STP.
$$2H_2(g) + O_2(g) \longrightarrow 2H_2O(g)$$
4. Balance the following equation.
$$C_2H_2(g) + O_2(g) \longrightarrow CO_2(g) + H_2O(g)$$
Interpret the balanced equation in terms of relative numbers of moles, volumes of gas at STP, and masses of reactants and products.

Problem-Solving 12.4 Solve Problem 4 with the help of an interactive guided tutorial.

with ChemASAP

12.1 Section Assessment

5. **Key Concept** How is a balanced equation similar to a recipe?
6. **Key Concept** How do chemists use balanced equations?
7. **Key Concept** Chemical reactions can be described in terms of what quantities?
8. **Key Concept** What quantities are always conserved in chemical reactions?
9. Interpret the given equation in terms of relative numbers of representative particles, numbers of moles, and masses of reactants and products.
$$2K(s) + 2H_2O(l) \longrightarrow 2KOH(aq) + H_2(g)$$
10. Balance this equation: $C_2H_5OH(l) + O_2(g) \longrightarrow CO_2(g) + H_2O(g)$. Show that the balanced equation obeys the law of conservation of mass.

Writing Activity

Explanatory Paragraph Explain this statement: "Mass and atoms are conserved in every chemical reaction, but moles are not necessarily conserved."

Assessment 12.1 Test yourself on the concepts in Section 12.1.

with ChemASAP

Section 12.1 Assessment

5. Both a balanced equation and a recipe give quantitative information about the starting and end materials.
6. as a basis to calculate how much reactant is needed or product is formed in a reaction
7. numbers of atoms, molecules, or moles; mass; and volumes
8. mass and atoms
9. 2 atoms K + 2 molecules $H_2O \rightarrow$ 2 formula units KOH + 1 molecule H_2
   2 mol K + 2 mol $H_2O \rightarrow$ 2 mol KOH + 1 mol H_2
   78.2 g K + 36.0 g $H_2O \rightarrow$ 112.2 g KOH + 2.0 g H_2
10. $C_2H_5OH + 3O_2 \rightarrow 2CO_2 + 3H_2O$
   46.0 g C_2H_5OH + 96.0 g $O_2 \rightarrow$ 88.0 g CO_2 + 54.0 g H_2O
   142.0 g reactants $\rightarrow$ 142.0 g products

12.2 Chemical Calculations

Connecting to Your World Air bags inflate almost instantaneously upon a car's impact. The effectiveness of air bags is based on the rapid conversion of a small mass of sodium azide into a large volume of gas. The gas fills an air bag, preventing the driver from hitting the steering wheel or dashboard. The entire reaction occurs in less than a second. In this section you will learn how to use a balanced chemical equation to calculate the amount of product formed in a chemical reaction.

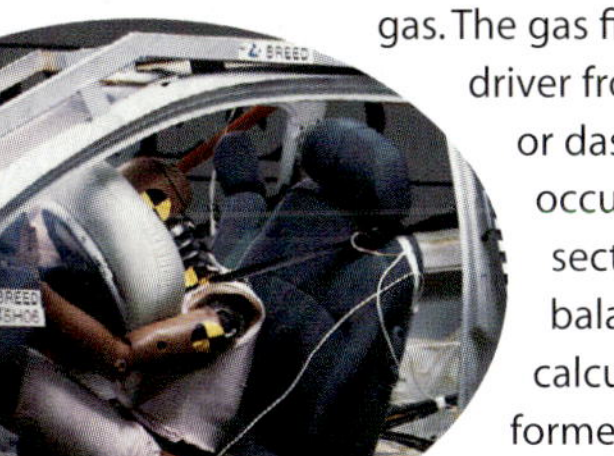

Guide for Reading

Key Concepts
- How are mole ratios used in chemical calculations?
- What is the general procedure for solving a stoichiometric problem?

Vocabulary
mole ratio

Reading Strategy
Relating Text and Visuals As you read, look closely at Figure 12.8. Explain how this illustration helps you understand the relationship between known and unknown quantities in a stoichiometric problem.

Writing and Using Mole Ratios

As you just learned, a balanced chemical equation provides a great deal of quantitative information. It relates particles (atoms, molecules, formula units), moles of substances, and masses. A balanced chemical equation also is essential for all calculations involving amounts of reactants and products. For example, suppose you know the number of moles of one substance. The balanced chemical equation allows you to determine the number of moles of all other substances in the reaction.

Look again at the balanced equation for the production of ammonia from nitrogen and hydrogen:

$$N_2(g) + 3H_2(g) \longrightarrow 2NH_3(g)$$

The most important interpretation of this equation is that 1 mol of nitrogen reacts with 3 mol of hydrogen to form 2 mol of ammonia. Based on this interpretation, you can write ratios that relate moles of reactants to moles of product. A **mole ratio** is a conversion factor derived from the coefficients of a balanced chemical equation interpreted in terms of moles. **In chemical calculations, mole ratios are used to convert between moles of reactant and moles of product, between moles of reactants, or between moles of products.** Three mole ratios derived from the balanced equation above are:

$$\frac{1 \text{ mol } N_2}{3 \text{ mol } H_2} \qquad \frac{2 \text{ mol } NH_3}{1 \text{ mol } N_2} \qquad \frac{3 \text{ mol } H_2}{2 \text{ mol } NH_3}$$

Mole-Mole Calculations In the mole ratio below, W is the unknown quantity and G is the given quantity. The values of a and b are the coefficients from the balanced equation. Thus a general solution for a mole-mole problem, such as Sample Problem 12.2, is given by

$$\underbrace{x \cancel{\text{mol } G}}_{\text{Given}} \times \underbrace{\frac{b \text{ mol } W}{a \cancel{\text{mol } G}}}_{\text{Mole ratio}} = \underbrace{\frac{xb}{a} \text{ mol } W}_{\text{Calculated}}$$

Figure 12.4 Manufacturing plants produce ammonia by combining nitrogen with hydrogen. Ammonia is used in cleaning products, fertilizers, and in the manufacture of other chemicals.

12.2

1 FOCUS

Objectives

12.2.1 Construct mole ratios from balanced chemical equations and **apply** these ratios in stoichiometric calculations.

12.2.2 Calculate stoichiometric quantities from balanced chemical equations using units of moles, mass, representative particles, and volumes of gases at STP.

Guide for Reading

Build Vocabulary 

Paraphrase Have students work with a partner to define the term *mole ratio* in their own words. They may do so by reading this section and by using what they have already learned about balanced chemical equations. Have student pairs read their definitions to the class.

Reading Strategy 

Identify Main Ideas/Details As you read the material under the heading *Mass-Mass Calculations,* identify and list the main ideas presented by the text.

2 INSTRUCT

Connecting to Your World

Have students study the photograph and read the text that opens the section. Write the equation for the decomposition of sodium azide with heat as one of the products $2NaN_3(s) \rightarrow 2Na(s) + 3N_2(g)$ + heat. Ask, **How can stoichiometry be used to calculate the volume of a gas produced in this reaction?** *(The number of moles (and volume) of nitrogen gas formed by this reaction depends on the number of moles of sodium azide that decompose.)*

Section Resources

Print
- **Guided Reading and Study Workbook,** Section 12.2
- **Core Teaching Resources,** Section 12.2 Review
- **Transparencies,** T126–T132
- **Laboratory Manual,** Lab 19
- **Small-Scale Chemistry Laboratory Manual,** Labs 18, 19
- **Probeware Laboratory Manual,** Section 12.2

Technology
- **Interactive Textbook with ChemASAP,** Simulation 13, Problem-Solving 12.12, 12.13, 12.15, 12.19, Assessment 12.2
- **Virtual Chemistry Lab** 28

Section 12.2 (continued)

Writing and Using Mole Ratios

Using Visuals L2

Figure 12.4 Have students consider the sizes of the containers shown relative to the sizes of containers used in a classroom laboratory. Have them imagine they are managing the manufacturing facility pictured. Ask, **What factors would they need to consider to meet demands for ammonia?** *(Acceptable answers include the number of customers, the number of cylinders per customer, the amount of ammonia per cylinder, and the amount of H_2 and N_2 needed to produce that quantity of NH_3.)*

Sample Problem 12.2

Answers

11. a. $\frac{4 \text{ mol Al}}{3 \text{ mol } O_2}$ $\frac{3 \text{ mol } O_2}{4 \text{ mol Al}}$ $\frac{4 \text{ mol Al}}{2 \text{ mol } Al_2O_3}$

$\frac{2 \text{ mol } Al_2O_3}{4 \text{ mol Al}}$ $\frac{3 \text{ mol } O_2}{2 \text{ mol } Al_2O_3}$ $\frac{2 \text{ mol } Al_2O_3}{3 \text{ mol } O_2}$

b. 7.4 mol

12. a. 11.1 mol

b. 0.52 mol

Practice Problems Plus L2

Chapter 12 Assessment problem 38 is related to Sample Problem 12.2.

Math Handbook

For a math refresher and practice, direct students to dimensional analysis, page R66.

Discuss L2

Point out that heat is produced in the decomposition of sodium azide, used in air safety bags. Explain the heat produced by a reaction can also be measured and related to the amount of reactant(s) consumed. In this case, the amount of heat produced depends on the mass of sodium azide (the reactant) that decomposes. It is possible to relate grams of reactant to moles of reactant to heat produced.

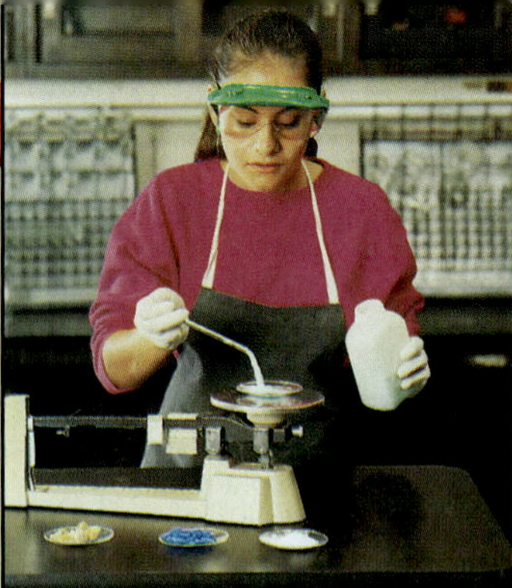

Figure 12.5 To determine the number of moles in a sample of a compound, first measure the mass of the sample. Then use the molar mass to calculate the number of moles in that mass.

Math Handbook

For help with dimensional analysis, go to page R66.

Interactive Textbook

Problem-Solving 12.12 Solve Problem 12 with the help of an interactive guided tutorial.

with ChemASAP

SAMPLE PROBLEM 12.2

Calculating Moles of a Product

How many moles of ammonia are produced when 0.60 mol of nitrogen reacts with hydrogen?

1 Analyze ***List the known and the unknown.***

Known

- moles of nitrogen = 0.60 mol N_2

Unknown

- moles of ammonia = ? mol NH_3

The conversion is mol $N_2 \longrightarrow$ mol NH_3. According to the balanced equation, 1 mol N_2 combines with 3 mol H_2 to produce 2 mol NH_3. To determine the number of moles of NH_3, the given quantity of N_2 is multiplied by the form of the mole ratio from the balanced equation that allows the given unit to cancel. This mole ratio is 2 mol NH_3/1 mol N_2.

2 Calculate ***Solve for the unknown.***

$$0.60 \text{ mol } N_2 \times \frac{2 \text{ mol } NH_3}{1 \text{ mol } N_2} = 1.2 \text{ mol } NH_3$$

3 Evaluate ***Does the result make sense?***

The ratio of 1.2 mol NH_3 to 0.60 mol N_2 is 2:1, as predicted by the balanced equation.

Practice Problems

11. This equation shows the formation of aluminum oxide, which is found on the surface of aluminum objects exposed to the air.

$$4Al(s) + 3O_2(g) \longrightarrow 2Al_2O_3(s)$$

a. Write the six mole ratios that can be derived from this equation.

b. How many moles of aluminum are needed to form 3.7 mol Al_2O_3?

12. According to the equation in Problem 11:

a. How many moles of oxygen are required to react completely with 14.8 mol Al?

b. How many moles of Al_2O_3 are formed when 0.78 mol O_2 reacts with aluminum?

Mass-Mass Calculations No laboratory balance can measure substances directly in moles. Instead, the amount of a substance is usually determined by measuring its mass in grams, as shown in Figure 12.5. From the mass of a reactant or product, the mass of any other reactant or product in a given chemical equation can be calculated. The mole interpretation of a balanced equation is the basis for this conversion. If the given sample is measured in grams, the mass can be converted to moles by using the molar mass. Then the mole ratio from the balanced equation can be used to calculate the number of moles of the unknown. If it is the mass of the unknown that needs to be determined, the number of moles of the unknown can be multiplied by the molar mass. As in mole-mole calculations, the unknown can be either a reactant or a product.

Differentiated Instruction

Less Proficient Readers L1

Encourage students to find a method of problem solving that capitalizes on their strengths. For example, a visual learner might draw pictures of the reactants and products, instead of just writing the symbols. A kinesthetic learner may prefer to manipulate molecular models of the reactants and products.

SAMPLE PROBLEM 12.3

Calculating the Mass of a Product

Calculate the number of grams of NH_3 produced by the reaction of 5.40 g of hydrogen with an excess of nitrogen. The balanced equation is

$$N_2(g) + 3H_2(g) \longrightarrow 2NH_3(g)$$

1 Analyze ***List the knowns and the unknown.***

Knowns

- mass of hydrogen = 5.40 g H_2
- 3 mol H_2 = 2 mol NH_3 (from balanced equation)
- 1 mol H_2 = 2.0 g H_2 (molar mass)
- 1 mol NH_3 = 17.0 g NH_3 (molar mass)

Unknown

- mass of ammonia = ? g NH_3

The mass in grams of hydrogen will be used to find the mass in grams of ammonia:

$$\text{g } H_2 \longrightarrow \text{g } NH_3$$

The following steps are necessary to determine the mass of ammonia:

$$\text{g } H_2 \longrightarrow \text{mol } H_2 \longrightarrow \text{mol } NH_3 \longrightarrow \text{g } NH_3$$

The coefficients of the balanced equation show that 3 mol H_2 reacts with 1 mol N_2 to produce 2 mol NH_3. The mole ratio relating mol NH_3 to mol H_2 is 2 mol NH_3/3 mol H_2.

2 Calculate ***Solve for the unknown.***

This following series of calculations can be combined:

$$\text{g } H_2 \longrightarrow \text{mol } H_2 \longrightarrow \text{mol } NH_3 \longrightarrow \text{g } NH_3$$

$$\underset{\text{Given quantity}}{5.40 \text{ g } \cancel{H_2}} \times \underset{\text{Change given unit to moles}}{\frac{1 \cancel{\text{mol } H_2}}{2.0 \cancel{\text{g } H_2}}} \times \underset{\text{Mole ratio}}{\frac{2 \cancel{\text{mol } NH_3}}{3 \cancel{\text{mol } H_2}}} \times \underset{\text{Change moles to grams}}{\frac{17.0 \text{ g } NH_3}{1 \cancel{\text{mol } NH_3}}} = 31 \text{ g } NH_3$$

3 Evaluate ***Does the result make sense?***

Because there are three conversion factors involved in this solution, it is more difficult to estimate an answer. However, because the molar mass of NH_3 is substantially greater than the molar mass of H_2, the answer should have a larger mass than the given mass. The answer should have two significant figures.

Practice Problems

13. Acetylene gas (C_2H_2) is produced by adding water to calcium carbide (CaC_2).

$$CaC_2(s) + 2H_2O(l) \longrightarrow C_2H_2(g) + Ca(OH)_2(aq)$$

How many grams of acetylene are produced by adding water to 5.00 g CaC_2?

14. Using the same equation, determine how many moles of CaC_2 are needed to react completely with 49.0 g H_2O.

CHEMath

Significant Figures

The significant figures in a measurement are all the digits known with certainty plus one estimated digit. The number of significant figures in the measurements used in a calculation determines how you round the answer.

When multiplying and dividing measurements, the rounded answer can have no more significant figures than the least number of significant figures in any measurement in the calculation.

The product of 3.6 m × 2.48 m = 8.928 m² is rounded to 8.9 m² (2 significant figures).

When adding and subtracting measurements, the answer can have no more decimal places than the least number of decimal places in any measurement in the problem. The difference of 8.78 cm − 2.2 cm = 6.58 cm is rounded to 6.6 cm (one decimal place).

Math Handbook

For help with significant figures, go to page R59.

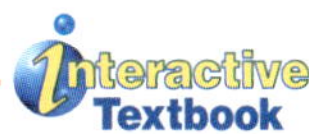

Problem-Solving 12.13 Solve Problem 13 with the help of an interactive guided tutorial.

with ChemASAP

Sample Problem 12.3

Answers

13. 2.03 g C_2H_2

14. 1.36 mol CaC_2

Practice Problems Plus L2

Rust (Fe_2O_3) is produced when iron (Fe) reacts with oxygen (O_2).

$$4Fe(s) + 3O_2(g) \rightarrow 2Fe_2O_3(s)$$

How many grams of Fe_2O_3 are produced when 12.0 g of iron rusts? *(17.2 g)*

CHEMath

Significant Figures

Point out the concept of significant figures applies only to measured quantities. Review the following condensed rules for determining which digits are significant.

1. Nonzero digits are significant.

2. a. A zero is significant only if it as at the right end of a number and after a decimal point, or
b. between digits that are significant according to rules 1 and 2a.

3. If a quantity is exact, it has an unlimited number of significant digits.

4. All digits of a quantity written in scientific notation are significant. Ask, **How many significant figures are in each of the following measurements?**

a. 30, 400 s *(3)*
b. 150.0 cm *(4)*
c. 2401 km *(4)*
d. 168.030 m *(6)*
e. 0.058 m *(2)*
f. 3.010×10^8 s *(4)*

Math Handbook

For a math refresher and practice, direct students to significant figures, page R59.

TEACHER Demo

Interpreting a Chemical Equation L2

Purpose Students interpret a balanced equation in terms of moles and mass.

Materials Prior to the demonstration prepare 0.1*M* solutions of potassium iodide and lead(II) nitrate. Measure 50.0 mL of $Pb(NO_3)_2$ and 150 mL of KI into separate 250-mL beakers.

Safety Wear safety glasses and apron.

Procedure Tell students that you are going to mix 0.005 moles of lead(II) nitrate with excess potassium iodide. Have student observe as you combine both solutions in the 250-mL beaker. Have students write a balanced chemical equation for the observed reaction. *[$2KI(aq) + Pb(NO_3)_2(aq) \rightarrow 2KNO_3(aq) + PbI_2(s)$]* Have students predict the number of moles of product produced. *(0.005 moles PbI_2 assuming the reaction was complete)* Note that, in an actual reaction, the amounts of reactants often are not present in the mole ratios predicted by the coefficients in a balanced equation. Explain the importance of the mole ratios in an equation for calculating relative quantities. Ask, **What is the mass of lead(II) nitrate reacted and the mass of lead(II) iodide produced?** *(1.66 g $Pb(NO_3)_2$ and 2.30 g PbI_2)*

Expected Outcome A bright yellow precipitate will form.

Discuss L2

Students sometimes try to do mass-mass conversions by incorrectly using the mole ratio as a mass ratio. (That is, they use grams instead of moles as the units in the mole ratio, and then skip the mass-mole conversion step.) Stress that because the number of grams in one mole of a substance varies with its molar mass, a mass-mole conversion is a necessary intermediate step in mass-mass stoichiometric problems.

Figure 12.6 In this Hubble Space Telescope image, clouds of condensed ammonia are visible covering the surface of Saturn.

If the law of conservation of mass is true, how is it possible to make 31 g NH_3 from only 5.40 g H_2? Looking back at the equation for the reaction, you will see that hydrogen is not the only reactant. Another reactant, nitrogen, is also involved. If you were to calculate the number of grams of nitrogen needed to produce 31 g NH_3 and then compare the total masses of reactants and products, you would have an answer to this question. Go ahead and try it!

Mass-mass problems are solved in basically the same way as mole-mole problems. Figure 12.7 reviews the steps for the mass-mass conversion of any given mass (G) and any wanted mass (W).

Steps in Solving a Mass-Mass Problem

1. Change the mass of G to moles of G (mass $G \longrightarrow$ mol G) by using the molar mass of G.
$$\text{mass } G \times \frac{1 \text{ mol } G}{\text{molar mass } G} = \text{mol } G$$
2. Change the moles of G to moles of W (mol G $\longrightarrow$ mol W) by using the mole ratio from the balanced equation.
$$\text{mol } G \times \frac{b \text{ mol } W}{a \text{ mol } G} = \text{mol } W$$
3. Change the moles of W to grams of W (mol W $\longrightarrow$ mass W) by using the molar mass of W.
$$\text{mol } W \times \frac{\text{molar mass } W}{1 \text{ mol } W} = \text{mass } W$$

Figure 12.7 also shows the steps for doing mole-mass and mass-mole stoichiometric calculations. For a mole-mass problem, the first conversion (from mass to moles) is skipped. For a mass-mole problem, the last conversion (from moles to mass) is skipped. You can use parts of the three-step process shown in Figure 12.7 as they are appropriate to the problem you are solving.

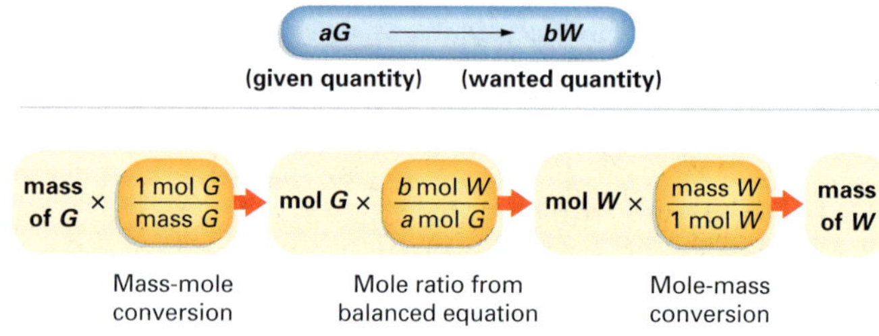

Figure 12.7 This general solution diagram indicates the steps necessary to solve a mass-mass stoichiometry problem: convert mass to moles, use the mole ratio, and then convert moles to mass. **Inferring** ***Is the given always a reactant?***

Facts and Figures

Atmospheric Ammonia

Ammonia is found in trace amounts in the atmospheres of three Jovian planets—Jupiter, Saturn, and Uranus. In Jupiter's atmosphere, the clouds of ammonia consist of frozen ammonia droplets changing to liquid ammonia droplets nearer the planet's surface. Because of colder temperatures, the ammonia clouds in the atmosphere of Saturn and Uranus consist of frozen ammonia droplets.

$aG \longrightarrow bW$

(given quantity) (wanted quantity)

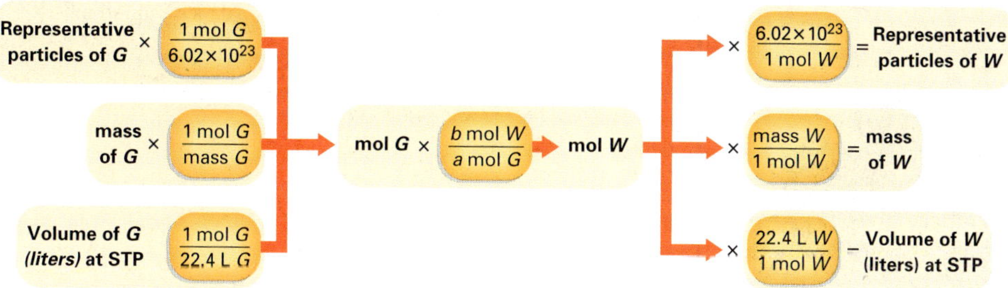

Figure 12.8 With your knowledge of conversion factors and this problem-solving approach, you can solve a variety of stoichiometric problems. **Identifying** ***What conversion factor is used to convert moles to representative particles?***

Other Stoichiometric Calculations

As you already know, you can obtain mole ratios from a balanced chemical equation. From the mole ratios, you can calculate any measurement unit that is related to the mole. The given quantity can be expressed in numbers of representative particles, units of mass, or volumes of gases at STP. The problems can include mass-volume, particle-mass and volume-volume calculations. For example, you can use stoichiometry to relate volumes of reactants and products in the reaction shown in Figure 12.8. **In a typical stoichiometric problem, the given quantity is first converted to moles. Then the mole ratio from the balanced equation is used to calculate the number of moles of the wanted substance. Finally, the moles are converted to any other unit of measurement related to the unit mole, as the problem requires.**

Thus far, you have learned how to use the relationship between moles and mass (1 mol = molar mass) in solving mass-mass, mass-mole, and mole-mass stoichiometric problems. The mole-mass relationship gives you two conversion factors.

$$\frac{1\text{ mol}}{\text{molar mass}} \quad \text{and} \quad \frac{\text{molar mass}}{1\text{ mol}}$$

Recall from Chapter 10 that the mole can be related to other quantities as well. For example, 1 mol = 6.02×10^{23} representative particles, and 1 mol of a gas = 22.4 L at STP. These two relationships provide four more conversion factors that you can use in stoichiometric calculations.

$$\frac{1\text{ mol}}{6.02 \times 10^{23}\text{ particles}} \quad \text{and} \quad \frac{6.02 \times 10^{23}\text{ particles}}{1\text{ mol}}$$

$$\frac{1\text{ mol}}{22.4\text{ L}} \quad \text{and} \quad \frac{22.4\text{ L}}{1\text{ mol}}$$

Figure 12.8 summarizes the steps for a typical stoichiometric problem. Notice that the units of the given quantity will not necessarily be the same as the units of the wanted quantity. For example, given the mass of *G*, you might be asked to calculate the volume of *W* at STP.

What conversion factors can you write based on the mole-mass and mole-volume relationships?

Simulation 13 Strengthen your analytical skills by solving stoichiometric problems.

with **ChemASAP**

Discuss

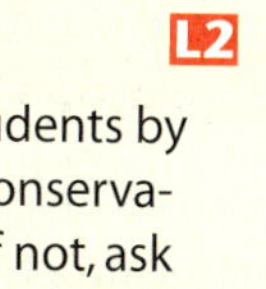

Initiate a discussion with students by asking whether the law of conservation of mass is always true. If not, ask them to give an example. (Nuclear reactions are the only cases where this law does not hold true.) Ask, **Why isn't there a "law of conservation of moles"?** *(In reactions involving rearrangements of atoms, reactants can combine or decompose to produce fewer or greater numbers of moles of product. Although the total mass of reactants and products is constant, the number of moles of particles can increase or decrease depending on the final grouping of atoms.)* **Give an example of a reaction in which the number of moles of products is greater than the number of moles of reactants.**

[$2H_2O(l) \rightarrow 2H_2(g) + O_2(g)$]

Give an example of a reaction in which the number of moles of products is less than the number of moles of reactants.

[$2Mg(s) + O_2(g) \rightarrow 2MgO(s)$]

Other Stoichiometric Calculations

Discuss

L2

On the board, write equations for reactions in which the reactants are both gases or are a gas and a solid. Ask students how the reactants and products in each reaction would most likely be measured. Have students relate these measurements to the concept of a mole.

Answers to...

Figure 12.7 No; the given could be a product.

Figure 12.8 6.02×10^{23} representative particles/1 mol

$$\frac{1\text{ mol}}{\text{molar mass}} \qquad \frac{\text{molar mass}}{1\text{ mol}}$$

$$\frac{22.4\text{ L}}{1\text{ mol}} \qquad \frac{1\text{ mol}}{22.4\text{ L}}$$

Discuss 

Construct a diagram on the board or overhead projector showing the relationships that are useful for solving stoichiometry problems. One simple model reaction is A → B. Use double-headed arrows to connect the terms: *Particles of A, Moles of A, Grams of A, Moles of B, Particles of B, and Grams of B.* Above the appropriate arrows, write: *Avogadro's number, Coefficients,* and *Molar mass.* Explain that the only "transitions" allowed are between quantities connected by arrows. Point out that the required conversion factor to make a "transition" is written above each arrow. Have students refer to the diagram when working practice problems.

Sample Problem 12.4

Answers

15. 4.82×10^{22} molecules O_2

16. 11.5 g NO_2

Practice Problems Plus 

Hydrogen gas can be made by reacting methane (CH_4) with high-temperature steam:

$CH_4(g) + H_2O(g) \rightarrow CO(g) + 3H_2(g)$

How many hydrogen molecules are produced when 158 g of methane reacts with steam? (*1.78×10^{25} hydrogen molecules*)

Math Handbook

For a math refresher and practice, direct students to dimensional analysis, page R66.

Hydrogen (H_2) Oxygen (O_2)

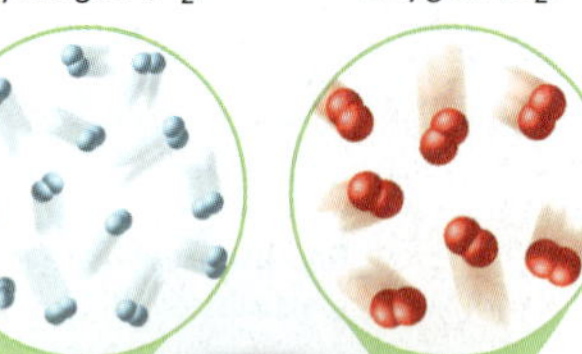

The electrolysis of water causes it to decompose into hydrogen and oxygen.

Math Handbook

For help with dimensional analysis, go to page R66.

Problem-Solving 12.15 Solve Problem 15 with the help of an interactive guided tutorial.

with ChemASAP

SAMPLE PROBLEM 12.4

Calculating Molecules of a Product

How many molecules of oxygen are produced when 29.2 g of water is decomposed by electrolysis according to this balanced equation?

$$2H_2O(l) \xrightarrow{\text{electricity}} 2H_2(g) + O_2(g)$$

1 Analyze ***List the knowns and the unknown.***

Knowns

- mass of water = 29.2 g H_2O
- 2 mol H_2O = 1 mol O_2 (from balanced equation)
- 1 mol H_2O = 18.0 g H_2O (molar mass)
- 1 mol O_2 = 6.02×10^{23} molecules O_2

Unknown

- molecules of oxygen = ? molecules O_2

The following calculations need to be done:

$$\text{g } H_2O \longrightarrow \text{mol } H_2O \longrightarrow \text{mol } O_2 \longrightarrow \text{molecules } O_2$$

The appropriate mole ratio relating mol O_2 to mol H_2O from the balanced equation is 1 mole O_2/2 mol H_2O.

2 Calculate ***Solve for the unknown.***

$$\underset{\text{Given quantity}}{29.2 \text{ g } H_2O} \times \underset{\text{Change to moles}}{\frac{1 \text{ mol } H_2O}{18.0 \text{ g } H_2O}} \times \underset{\text{Mole ratio}}{\frac{1 \text{ mol } O_2}{2 \text{ mol } H_2O}} \times \underset{\text{Change to molecules}}{\frac{6.02 \times 10^{23} \text{ molecules } O_2}{1 \text{ mol } O_2}}$$

$$= 4.88 \times 10^{23} \text{ molecules } O_2$$

3 Evaluate ***Does the result make sense?***

The given mass of water should produce a little less than 1 mol of oxygen, or a little less than Avogadro's number of molecules. The answer should have three significant figures.

Practice Problems

15. How many molecules of oxygen are produced by the decomposition of 6.54 g of potassium chlorate ($KClO_3$)?

$$2KClO_3(s) \longrightarrow 2KCl(s) + 3O_2(g)$$

16. The last step in the production of nitric acid is the reaction of nitrogen dioxide with water.

$$3NO_2(g) + H_2O(l) \longrightarrow 2HNO_3(aq) + NO(g)$$

How many grams of nitrogen dioxide must react with water to produce 5.00×10^{22} molecules of nitrogen monoxide?

The coefficients in a chemical equation indicate the relative number of particles and the relative number of moles of reactants and products. For a reaction involving gaseous reactants or products, the coefficients also indicate relative amounts of each gas. As a result, you can use volume ratios in the same way you have used mole ratios.

Differentiated Instruction

Gifted and Talented 

Have students use the calculations in the sample problems as general algorithms to write computer programs that solve stoichiometric problems. Have students demonstrate and explain their programs to interested students.

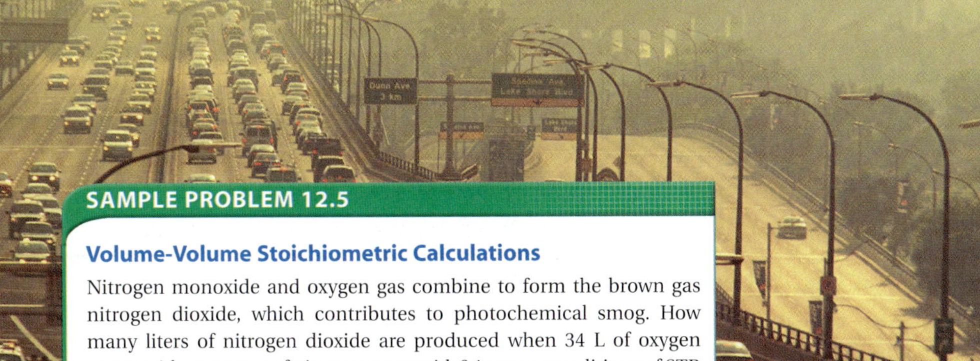

SAMPLE PROBLEM 12.5

Volume-Volume Stoichiometric Calculations

Nitrogen monoxide and oxygen gas combine to form the brown gas nitrogen dioxide, which contributes to photochemical smog. How many liters of nitrogen dioxide are produced when 34 L of oxygen reacts with an excess of nitrogen monoxide? Assume conditions of STP.

$$2NO(g) + O_2(g) \longrightarrow 2NO_2(g)$$

1 Analyze ***List the knowns and the unknown.***

Knowns
- volume of oxygen = 34 L O_2
- 2 mol NO_2/1 mol O_2 (mole ratio from balanced equation)
- 1 mol O_2 = 22.4 L O_2 (at STP)
- 1 mol NO_2 = 22.4 L NO_2 (at STP)

Unknown
- volume of nitrogen dioxide = ? L NO_2

2 Calculate ***Solve for the unknown.***

$$\underbrace{34\ \cancel{L\ O_2}}_{\text{Given quantity}} \times \underbrace{\frac{1\ \cancel{mol\ O_2}}{22.4\ \cancel{L\ O_2}}}_{\text{Change to moles}} \times \underbrace{\frac{2\ \cancel{mol\ NO_2}}{1\ \cancel{mol\ O_2}}}_{\text{Mole ratio}} \times \underbrace{\frac{22.4\ L\ NO_2}{1\ \cancel{mol\ NO_2}}}_{\text{Change to liters}} = 68\ L\ NO_2$$

3 Evaluate ***Does the result make sense?***

Because 2 mol NO_2 is produced for each 1 mol O_2 that reacts, the volume of NO_2 should be twice the given volume of O_2. The answer should have two significant figures.

Practice Problems

17. The equation for the combustion of carbon monoxide is

$$2CO(g) + O_2(g) \longrightarrow 2CO_2(g)$$

How many liters of oxygen are required to burn 3.86 L of carbon monoxide?

18. Phosphorus and hydrogen can be combined to form phosphine (PH_3).

$$P_4(s) + 6H_2(g) \longrightarrow 4PH_3(g)$$

How many liters of phosphine are formed when 0.42 L of hydrogen reacts with phosphorus?

Problem-Solving 12.18 Solve Problem 18 with the help of an interactive guided tutorial.

with ChemASAP

Did you notice that in Sample Problem 12.5 the 22.4 L/mol factors canceled out? This will always be true in a volume-volume problem. Remember that coefficients in a balanced chemical equation indicate the relative numbers of moles. The coefficients also indicate the relative volumes of interacting gases.

Sample Problem 12.5

Answers

17. 1.93 L O_2
18. 0.28 L PH_3

Practice Problems Plus L2

Ammonia (NH_3) reacts with oxygen (O_2) to produce nitrogen monoxide (NO) and water.

$$4NH_3(g) + 5O_2(g) \rightarrow 4NO(g) + 6H_2O(l)$$

How many liters of NO are produced when 1.40 L of oxygen reacts with ammonia? *(1.12 L)*

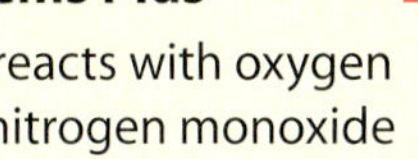

CLASS Activity

Stoichiometric Flash Cards L2

Purpose To aid students in sequencing the steps in solving stoichiometric problems

Materials 8 index cards, 1 colored index card, paper punch, 2 brass paper fasteners

Procedure Distribute the cards to the students. Have them place the eight index cards on two piles of four cards each. On the first card of the first pile, have them write *Converting a given measured quantity to moles.* On the each of the three remaining cards, have students write the conversion factors for converting mass to moles, representative particles to moles, and volume to moles, respectively. For the second set of cards, have students label the first card *Changing moles of wanted substances to a measured quantity.* On each of the reaming cards, have them write the appropriate conversion factor. On the colored card have students write *Converting moles of given to moles of wanted using mole ratio from balanced chemical equation* b *mol W/a mol* G. Have the students use the paper pinch to punch each of the two sets of cards and attach them with a brass paper fastener. Allow students to practice using the cards to solve the Practice Problems.

Expected Outcome The cards should aid in sequencing the steps in solving stoichiometric problems.

Sample Problem 12.6

Answers L2

19. 18.6 mL SO_2
20. 1.9 dL CO_2

Math Handbook

For a math refresher and practice, direct students to dimensional analysis, page R66.

3 ASSESS

Evaluate Understanding L2

Write a balanced equation on the board, such as $H_2(g) + I_2(g) \rightarrow 2HI(g)$. Have students write all the different mole ratios for the reaction. Then pose a problem, such as, **How many moles of hydrogen iodide are formed when 0.75 mol I_2 gas is reacted with excess hydrogen gas?** Have students choose the correct mole ratio for the problem. Repeat with other problems and reactions, including one with gases.

Reteach 

Use molecular models to review the importance of mole ratios. Illustrate how the mole ratios from the balanced chemical equation are related to the individual atoms, formula units, and molecules of the reactants and products as described by the equation.

Connecting Concepts

Figure 10.12, the "mole road map," describes how the mass, volume, and number of representative particles of a substance can be converted into moles (and vice versa). These relationships apply to six of the calculations shown in Figure 12.8; the seventh is a mole-mole conversion, based on a mole ratio.

If your class subscribes to the Interactive Textbook, use it to review key concepts in Section 12.2.

with ChemASAP

SAMPLE PROBLEM 12.6

Finding the Volume of a Gas Needed for a Reaction

Assuming STP, how many milliliters of oxygen are needed to produce 20.4 mL SO_3 according to this balanced equation?

$$2SO_2(g) + O_2(g) \longrightarrow 2SO_3(g)$$

1 Analyze ***List the knowns and the unknown.***

Knowns
- volume of sulfur trioxide = 20.4 mL
- 2 mL SO_3/1 mL O_2 (volume ratio from balanced equation)

Unknown
- volume of oxygen = ? mL O_2

2 Calculate ***Solve for the unknown.***

$$20.4\ \cancel{\text{mL SO}_3} \times \frac{1\ \text{mL O}_2}{2\ \cancel{\text{mL SO}_3}} = 10.2\ \text{mL O}_2$$

3 Evaluate ***Does the result make sense?***

Because the volume ratio is 2 volumes SO_3 to 1 volume O_2, the volume of O_2 should be half the volume of SO_3. The answer should have three significant figures.

Practice Problems

Consider this equation:

$$CS_2(l) + 3O_2(g) \longrightarrow CO_2(g) + 2SO_2(g)$$

19. Calculate the volume of sulfur dioxide produced when 27.9 mL O_2 reacts with carbon disulfide.

20. How many deciliters of carbon dioxide are produced when 0.38 L SO_2 is formed?

Math Handbook

For help with dimensional analysis, go to page R66.

Interactive Textbook

Problem-Solving 12.19 Solve Problem 19 with the help of an interactive guided tutorial.

with ChemASAP

12.2 Section Assessment

21. **Key Concept** How are mole ratios used in chemical calculations?

22. **Key Concept** Outline the sequence of steps needed to solve a typical stoichiometric problem.

23. Write the 12 mole ratios that can be derived from the equation for the combustion of isopropyl alcohol.

$$2C_3H_7OH(l) + 9O_2(g) \longrightarrow 6CO_2(g) + 8H_2O(g)$$

24. The combustion of acetylene gas is represented by this equation:

$$2C_2H_2(g) + 5O_2(g) \longrightarrow 4CO_2(g) + 2H_2O(g)$$

How many grams of CO_2 and grams of H_2O are produced when 52.0 g C_2H_2 burns in oxygen?

Connecting Concepts

Chemical Quantities Review the "mole road map" at the end of Section 10.2. Explain how this road map ties into the summary of steps for stoichiometric problems shown in Figure 12.8.

Interactive Textbook

Assessment 12.2 Test yourself on the concepts in Section 12.2.

with ChemASAP

366 Chapter 12

Section 12.2 Assessment

21. Mole ratios are written using the coefficients from a balanced chemical equation. They are used to relate moles of reactants and products in stoichiometric calculations.

22. Convert the given quantity to moles; use the mole ratio from the equations to find the moles of the wanted; convert moles of wanted to the desired unit.

23. $\frac{2\ \text{mol C}_3\text{H}_7\text{OH}}{9\ \text{mol O}_2}$ $\frac{2\ \text{mol C}_3\text{H}_7\text{OH}}{6\ \text{mol CO}_2}$ $\frac{2\ \text{mol C}_3\text{H}_7\text{OH}}{8\ \text{mol H}_2\text{O}}$ $\frac{9\ \text{mol O}_2}{6\ \text{mol CO}_2}$

$\frac{9\ \text{mol O}_2}{8\ \text{mol H}_2\text{O}}$ $\frac{6\ \text{mol CO}_2}{8\ \text{mol H}_2\text{O}}$ $\frac{9\ \text{mol O}_2}{2\ \text{mol C}_3\text{H}_7\text{OH}}$ $\frac{6\ \text{mol CO}_2}{2\ \text{mol C}_3\text{H}_7\text{OH}}$

$\frac{8\ \text{mol H}_2\text{O}}{2\ \text{mol C}_3\text{H}_7\text{OH}}$ $\frac{6\ \text{mol CO}_2}{9\ \text{mol O}_2}$ $\frac{8\ \text{mol H}_2\text{O}}{9\ \text{mol O}_2}$ $\frac{8\ \text{mol H}_2\text{O}}{6\ \text{mol CO}_2}$

24. 176 g CO_2, 36.0 g H_2O

Analysis of Baking Soda

Purpose
To determine the mass of sodium hydrogen carbonate in a sample of baking soda using stoichiometry.

Materials
- baking soda
- 3 plastic cups
- soda straw
- balance
- pipets of HCl, NaOH, and thymol blue
- pH sensor (optional)

Procedure

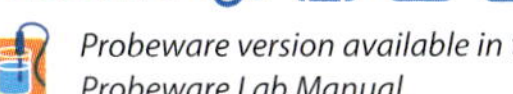
Probeware version available in the Probeware Lab Manual.

A. Measure the mass of a clean, dry plastic cup.
B. Using the straw as a scoop, fill one end with baking soda to a depth of about 1 cm. Add the sample to the cup and measure its mass again.
C. Place two HCl pipets that are about 3/4 full into a clean cup and measure the mass of the system.
D. Transfer the contents of both HCl pipets to the cup containing baking soda. Swirl until the fizzing stops. Wait 5–10 minutes to be sure the reaction is complete. Measure the mass of the two empty HCl pipets in their cup again.
E. Add 5 drops of thymol blue to the plastic cup.
F. Place two full NaOH pipets in a clean cup and measure the mass of the system.
G. Add NaOH slowly to the baking soda/HCl mixture until the pink color just disappears. Measure the mass of the NaOH pipets in their cup again.

Analyze
Using your experimental data, record the answers to the following questions below your data table.

1. Write a balanced equation for the reaction between baking soda ($NaHCO_3$) and HCl.
2. Calculate the mass in grams of the baking soda.
   (Step B − Step A)
3. Calculate the total mmol of 1*M* HCl.
   Note: Every gram of HCl contains 1 mmol.
   (Step C − Step D) × 1.00 mmol/g
4. Calculate the total mmol of 0.5*M* NaOH.
   Note: Every gram of NaOH contains 0.5 mmol.
   (Step F − Step G) × 0.500 mmol/g
5. Calculate the mmol of HCl that reacted with the baking soda. *Note:* The NaOH measures the amount of HCl that did not react.
   (Step 3 − Step 4)
6. Calculate the mass of the baking soda from the reaction data.
   (0.084 g/mmol × Step 5)
7. Calculate the percent error of the experiment.
   $$\frac{(\text{Step 2} - \text{Step 6})}{\text{Step 2}} \times 100\%$$

You're the Chemist
The following small-scale activities allow you to develop your own procedures and analyze the results.

1. **Analyze It!** For each calculation you did, substitute each quantity (number and unit) into the equation and cancel the units to explain why each step gives the quantity desired.
2. **Design It!** Baking powder consists of a mixture of baking soda, sodium hydrogen carbonate, and a solid acid, usually calcium dihydrogen phosphate ($Ca(H_2PO_4)_2$). Design and carry out an experiment to determine the percentage of baking soda in baking powder.

Analysis of Baking Soda 

Objective
Students calculate the mass of $NaHCO_3$ in a sample using stoichiometry.

 **Prep Time** 1 hour

Materials Baking soda; plastic cups; soda straws; mass balances; pipets of HCl, NaOH, and thymol blue; pH sensor (optional)

Advance Prep

Solution	Preparation
0.5*M* NaOH	20.0 g in 1.0 L
1.0*M* HCl	82 mL of 12*M* in 1.0 L **Caution** *Always add acid to water carefully and slowly.*
0.04% TB	100 mg in 21.5 mL of 01*M* NaOH; dilute to 250 mL

Class Time 30 minutes
Safety Have students wear safety glasses and follow the standard safety procedures.

Teaching Tips
- Stress that the procedure measures the amount of excess HCl that is not reacted with the baking soda (Step 4). Because this excess HCl reacts with the NaOH in a 1:1 mole ratio, the moles of NaOH equal the moles of HCl in excess. Subtracting the excess moles of HCl from the total moles used in the experiment (Step 5) yields the moles reacted with the baking soda, which is 100% $NaHCO_3$.
- If the mixture does not turn red when thymol blue is added, the student should find the mass of a third pipet and add just enough HCl to turn the mixture cherry red. Then the student should find the mass of the half-empty pipet so the mass of HCl added can be calculated and added to the total mass used.

Expected Outcome
Sample data: Step A. 2.83 g, B. 3.28 g, C. 10.70 g, D. 4.29 g, F. 10.53 g, G. 8.78 g

Analyze
1. $HCl + NaHCO_3(s) \rightarrow CO_2(g) + H_2O + NaCl$
2. 3.28 g – 2.83 g = 0.45 g
3. (10.70 – 4.29)g × 1.00 mmol/g = 6.41 mmol
4. (10.53 – 8.78)g × 0.500 mmol/g = 0.875 mmol (0.875 mmol HCl unreacted)
5. 6.41 mmol total – 0.875 mmol unreacted = 5.53 mmol (5.53 mmol $NaHCO_3$)
6. (0.0840 g/mmol) × 5.53 mmol = 0.46 g
7. (0.46 – 0.45) g/0.45g × 100% = 2% error (assuming baking soda is one hundred percent sodium hydrogen carbonate)

You're the Chemist
1. See Steps 2–7.
2. Repeat Steps A–G and 1–7 using baking powder instead of baking soda. The percent error is the percent of baking soda in baking powder, assuming no other errors.

For Enrichment 
Ask students to predict how much baking soda and 1*M* HCl are needed to produce enough CO_2 to fill a 1-L plastic bag. Have them write a procedure and then carry out the experiment.

12.3

1 FOCUS

Objectives

12.3.1 Identify the limiting reagent in a reaction.

12.3.2 Calculate theoretical yield, actual yield, or percent yield given appropriate information.

Guide for Reading

Build Vocabulary L2

LINCS Have students use the LINCS strategy for the terms *theoretical yield, actual yield, and percent yield.* Students should **L**ist the parts of a term they know; **I**magine a picture of the term; **N**ote a "sound-alike word" for the term; **C**onnect the terms; and **S**elf-test the terms.

Reading Strategy 

Compare and Contrast Compare and contrast the terms *limiting reagent* and *excess reagent.*

2 INSTRUCT

Connecting to Your World

Have students study the photograph and read the text. Ask, **If the carpenter decided to construct a built-in table using two legs for support, what would limit the number of tables constructed?** *(tabletops)* **What and how many unused parts would remain** *(3 legs)*

Limiting and Excess Reagents

Use Visuals L1

Figure 12.9 Ask, **Where else do you see examples of processes or activities being "limited" on an everyday basis?** *(Acceptable answers include the number of people you can fit into a car, and the number of cookies that can be made from a given amount of dough.)*

12.3 Limiting Reagent and Percent Yield

Guide for Reading

 **Key Concepts**

- How is the amount of product in a reaction affected by an insufficient quantity of any of the reactants?
- What does the percent yield of a reaction measure?

Vocabulary

limiting reagent
excess reagent
theoretical yield
actual yield
percent yield

Reading Strategy

Building Vocabulary After you have read the section, explain the differences among *theoretical yield, actual yield,* and *percent yield.*

Connecting to Your World

If a carpenter had two tabletops and seven table legs, he would have difficulty building more than one functional four-legged table. The first table would require four of the legs, leaving just three legs for the second table. In this case, the number of table legs is the limiting factor in the construction of four-legged tables. A similar concept applies in chemistry. The amount of product made in a chemical reaction may be limited by the amount of one or more of the reactants.

Limiting and Excess Reagents

Many cooks follow a recipe when making a new dish. They know that sufficient quantities of all the ingredients must be available. Suppose, for example, that you are preparing to make lasagna and you have more than enough meat, tomato sauce, ricotta cheese, eggs, mozzarella cheese, spinach, and seasoning on hand. However, you have only half a box of lasagna noodles. The amount of lasagna you can make will be limited by the quantity of noodles you have. Thus, the noodles are the limiting ingredient in this baking venture. Figure 12.9 illustrates another example of a limiting ingredient in the kitchen. A chemist often faces a similar situation. **In a chemical reaction, an insufficient quantity of any of the reactants will limit the amount of product that forms.**

Figure 12.9 The amount of product is determined by the quantity of the limiting reagent. In this example, the rolls are the limiting reagent. No matter how much of the other ingredients you have, with two rolls you can make only two sandwiches.

Section Resources

Print

- **Guided Reading and Study Workbook,** Section 12.3
- **Core Teaching Resources,** Section 12.3 Review, Interpreting Graphics
- **Transparencies,** T133–T138
- **Laboratory Manual,** Lab 20

Technology

- **Interactive Textbook with ChemASAP,** Animation 13, Problem-Solving 12.25, 12.28, 12.29, 12.31, Assessment 12.3
- **Go Online,** Section 12.3

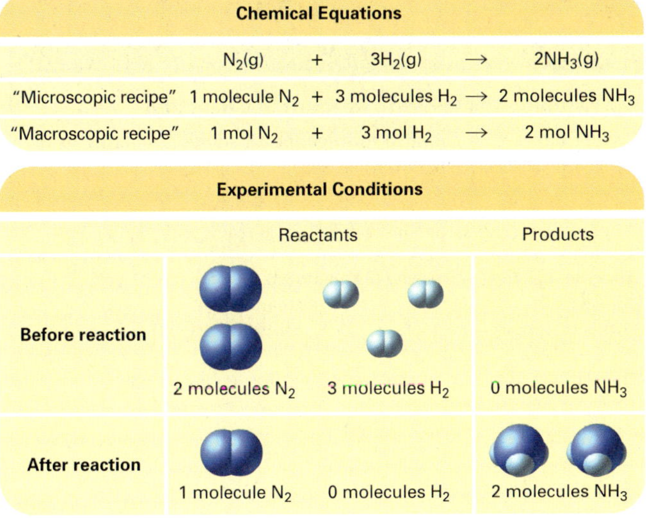

Chemical Equations

	$N_2(g)$	+	$3H_2(g)$	$\longrightarrow$	$2NH_3(g)$
"Microscopic recipe"	1 molecule N_2	+	3 molecules H_2	$\longrightarrow$	2 molecules NH_3
"Macroscopic recipe"	1 mol N_2	+	3 mol H_2	$\longrightarrow$	2 mol NH_3

Experimental Conditions

	Reactants		Products
Before reaction	2 molecules N_2	3 molecules H_2	0 molecules NH_3
After reaction	1 molecule N_2	0 molecules H_2	2 molecules NH_3

Figure 12.10 The "recipe" calls for 3 molecules of H_2 for every 1 molecule of N_2. In this particular experiment, H_2 is the limiting reagent and N_2 is in excess.
Inferring ***How would the amount of products formed change if you started with four molecules of N_2 and three molecules of H_2?***

As you know, a balanced chemical equation is a chemist's recipe. You can interpret the recipe on a microscopic scale (interacting particles) or on a macroscopic scale (interacting moles). The coefficients used to write the balanced equation give both the ratio of representative particles and the mole ratio. Recall the equation for the preparation of ammonia:

$$N_2(g) + 3H_2(g) \longrightarrow 2NH_3(g)$$

When one molecule (mole) of N_2 reacts with three molecules (moles) of H_2, two molecules (moles) of NH_3 are produced. What would happen if two molecules (moles) of N_2 reacted with three molecules (moles) of H_2? Would more than two molecules (moles) of NH_3 be formed? Figure 12.10 shows both the particle and the mole interpretations of this problem.

Before the reaction takes place, nitrogen and hydrogen are present in a 2:3 molecule (mole) ratio. The reaction takes place according to the balanced equation. One molecule (mole) of N_2 reacts with three molecules (moles) of H_2 to produce two molecules (moles) of NH_3. At this point, all the hydrogen has been used up, and the reaction stops. One molecule (mole) of unreacted nitrogen is left in addition to the two molecules (moles) of NH_3 that have been produced by the reaction.

In this reaction, only the hydrogen is completely used up. It is the **limiting reagent,** or the reagent that determines the amount of product that can be formed by a reaction. The reaction occurs only until the limiting reagent is used up. By contrast, the reactant that is not completely used up in a reaction is called the **excess reagent.** In this example, nitrogen is the excess reagent because some nitrogen will remain unreacted.

Sometimes in stoichiometric problems, the given quantities of reactants are expressed in units other than moles. In such cases, the first step in the solution is to convert each reactant to moles. Then the limiting reagent can be identified. The amount of product formed in a reaction can be determined from the given amount of limiting reagent.

Checkpoint *How do limiting and excess reagents differ?*

For: Links on Hydrogen
Visit: www.SciLinks.org
Web Code: cdn-1123

Animation 13 Apply the limiting reagent concept to the production of iron from iron ore.
with ChemASAP

TEACHER Demo

Limiting Factor 

Purpose Students model the concept of a limiting reagent.

Materials 15 plastic bottles, 30 plastic caps to fit the bottles, 6 containers to hold 5 caps each

Procedure Before the demonstration, place five caps in each container and put them out of the sight of the students. Set out all 15 plastic bottles. Bring out one container of caps. Ask, **How many closed plastic bottles can we have?** *(5)* **What limits the number?** *(caps)* Bring out another container of caps. Again, ask the students what will happen. Continue bringing out caps and matching them to the bottles.

Expected Outcome With two containers of caps, student should indicate that the caps are the limiting factor. With three containers, students should realize that there is a one-to-one correspondence and therefore no limiting factor. (Or, students may argue that both the bottles and caps are equally limiting factors.) With the remaining containers, the caps are in excess and students should indicate that the bottles are the limiting factor.

Download a worksheet on **Hydrogen** for students to complete and find additional support for NSTA SciLinks.

Differentiated Instruction

Less Proficient Readers 

Help students understand the concept of a limiting reagent by comparing it to everyday situations. For example, suppose a person is planning a dinner party. He has 6 glasses, 8 plates, and 4 chairs. How many people can he invite? Or, how many letters can you send if you have 24 pieces of stationery, 12 envelopes, and 10 stamps?

Answers to...

Figure 12.10 The amount would remain the same; two molecules of NH_3 would form.

Checkpoint The limiting reagent is completely used up in a reaction. The excess reagent is not completely used up; some of it remains after the reaction takes place.

Section 12.3 (continued)

Discuss 

Write the balanced chemical equation for the following reaction on the board: $2Na(s) + Cl_2(g) \rightarrow 2NaCl(s)$. Ask a series of questions that require students to predict the amount of product formed based on molar quantities of reactant supplied by you. Use small whole numbers that students can easily manipulate without calculators. As students solve the problems, create a list of molar quantities under each reactant and product showing how they are related for each set of conditions. In each case, ask students to state which reactant is the limiting reagent and circle the value in the list to designate it as the limiting quantity.

Sample Problem 12.7

Answers

25. O_2 is the limiting reagent.
26. HCl is the limiting reagent.

Practice Problems Plus 

Phosphoric acid reacts with sodium hydroxide according to the equation:

$$H_3PO_4(aq) + 3NaOH(aq) \rightarrow Na_3PO_4(aq) + 3H_2O(l)$$

If 1.75 mol H_3PO_4 is made to react with 5.00 mol NaOH, identify the limiting reagent. *(5.25 mol NaOH required; NaOH is the limiting reagent)*

Math Handbook

For a math refresher and practice, direct students to dimensional analysis, page R66.

Sulfur (S)

Copper (Cu)

Copper(I) sulfide (Cu_2S)

Math Handbook

For help with dimensional analysis, go to page R66.

Interactive Textbook

Problem-Solving 12.25 Solve Problem 25 with the help of an interactive guided tutorial.

with ChemASAP

SAMPLE PROBLEM 12.7

Determining the Limiting Reagent in a Reaction

Copper reacts with sulfur to form copper(I) sulfide according to the following balanced equation.

$$2Cu(s) + S(s) \longrightarrow Cu_2S(s)$$

What is the limiting reagent when 80.0 g Cu reacts with 25.0 g S?

1 Analyze ***List the knowns and the unknown.***

Knowns	Unknown
• mass of copper = 80.0 g Cu	• limiting reagent = ?
• mass of sulfur = 25.0 g S	

The number of moles of each reactant must first be found:

$$g\ Cu \longrightarrow mol\ Cu$$

$$g\ S \longrightarrow mol\ S$$

The balanced equation is used to calculate the number of moles of one reactant needed to react with the given amount of the other reactant:

$$mol\ Cu \longrightarrow mol\ S$$

The mole ratio relating mol S to mol Cu from the balanced chemical equation is 1 mol S/2 mol Cu.

2 Calculate ***Solve for the unknown.***

$$80.0\ \cancel{g\ Cu} \times \frac{1\ mol\ Cu}{63.5\ \cancel{g\ Cu}} = 1.26\ mol\ Cu$$

$$25.0\ \cancel{g\ S} \times \frac{1\ mol\ S}{32.1\ \cancel{g\ S}} = 0.779\ mol\ S$$

$$1.26\ \cancel{mol\ Cu} \times \frac{1\ mol\ S}{2\ \cancel{mol\ Cu}} = 0.630\ mol\ S$$

Given quantity · Mole ratio · Needed amount

Comparing the amount of sulfur needed (0.630 mol S) with the given amount (0.779 mol S) indicates that sulfur is in excess. Thus copper is the limiting reagent.

3 Evaluate ***Do the results make sense?***

Since the ratio of the given mol Cu to mol S was less than the ratio (2:1) from the balanced equation, copper should be the limiting reagent.

Practice Problems

25. The equation for the complete combustion of ethene (C_2H_4) is

$$C_2H_4(g) + 3O_2(g) \longrightarrow 2CO_2(g) + 2H_2O(g)$$

If 2.70 mol C_2H_4 is reacted with 6.30 mol O_2, identify the limiting reagent.

26. Hydrogen gas can be produced by the reaction of magnesium metal with hydrochloric acid.

$$Mg(s) + 2HCl(aq) \longrightarrow MgCl_2(aq) + H_2(g)$$

Identify the limiting reagent when 6.00 g HCl reacts with 5.00 g Mg.

370 Chapter 12

In Sample Problem 12.7, you may have noticed that even though the mass of copper used in the reaction is greater than the mass of sulfur, copper is the limiting reagent. The reactant that is present in the smaller amount by mass or volume is not necessarily the limiting reagent.

SAMPLE PROBLEM 12.8

Using a Limiting Reagent to Find the Quantity of a Product

What is the maximum number of grams of Cu_2S that can be formed when 80.0 g Cu reacts with 25.0 g S?

$$2Cu(s) + S(s) \longrightarrow Cu_2S(s)$$

1 Analyze ***List the knowns and the unknown.***

Knowns

- limiting reagent = 1.26 mol Cu (from Sample Problem 12.7)
- 1 mol Cu_2S = 159.1 g Cu_2S (molar mass)

Unknown

- mass copper(I) sulfide = ? g Cu_2S

The limiting reagent, which was determined in the previous sample problem, is used to calculate the maximum amount of Cu_2S formed:

$$\text{mol Cu} \longrightarrow \text{mol } Cu_2S \longrightarrow \text{g } Cu_2S$$

2 The equation yields the appropriate mole ratio: 1 mol Cu_2S/2 mol Cu.

Calculate ***Solve for the unknown.***

$$1.26 \text{ mol Cu} \times \frac{1 \text{ mol } Cu_2S}{2 \text{ mol Cu}} \times \frac{159.1 \text{ g } Cu_2S}{1 \text{ mol } Cu_2S} = 1.00 \times 10^2 \text{ g } Cu_2S$$

The given quantity of copper, 80.0 g, could have been used for this step instead of the moles of copper, which were calculated in Sample Problem 12.7.

3 Evaluate ***Do the results make sense?***

Copper is the limiting reagent in this reaction. The maximum number of grams of Cu_2S produced should be more than the amount of copper that initially reacted because copper is combining with sulfur. However, the mass of Cu_2S produced should be less than the total mass of the reactants (105.0 g) because sulfur was in excess.

Practice Problems

27. The equation below shows the incomplete combustion of ethene.

$$C_2H_4(g) + 2O_2(g) \longrightarrow 2CO(g) + 2H_2O(g)$$

If 2.70 mol C_2H_4 is reacted with 6.30 mol O_2,

a. identify the limiting reagent.

b. calculate the moles of water produced.

28. The heat from an acetylene torch is produced by burning acetylene (C_2H_2) in oxygen.

$$2C_2H_2(g) + 5O_2(g) \longrightarrow 4CO_2(g) + 2H_2O(g)$$

How many grams of water can be produced by the reaction of 2.40 mol C_2H_2 with 7.40 mol O_2?

Problem-Solving 12.28 Solve Problem 28 with the help of an interactive guided tutorial.

with ChemASAP

Relate **L2**

Have students reread the Inquiry Activity. Ask, **What is the maximum number of product molecules that can be formed using the procedure described in the activity?** *(Given that (a) one molecule of M_2 reacts with three molecules of C_2, (b) there are only ten molecules of each reactant available, and (c) fifteen molecules are chosen, a maximum of three M_2 molecules can react with nine C_2 molecules to form a total of six MC_3 molecules.)*

Sample Problem 12.8

Answers

27. a. 5.40 mol O_2 required; C_2H_4 is the limiting reactant.

b. 5.40 mol H_2O

28. 43.2 g H_2O

Practice Problems Plus **L2**

Chapter 12 Assessment problem 51 is related to Sample Problem 12.8.

Facts and Figures

Comparative Torch Temperatures

- Acetylene welding torches produce flame temperature of 3300–3400°C.
- Propane torches reach flame temperatures of about 1400°C.
- The flame of a match has a temperature range of 600–800°C.

Quick LAB

Limiting Reagents L2

Objective After completing this activity, students will be able to

- compare the amount of hydrogen gas produced from the reaction of magnesium and hydrochloric acid with the amounts of magnesium.
- infer the limiting reagent in reactions of magnesium and hydrochloric acid.

Skills Focus Observing, measuring, calculating

 **Prep Time** 15 minutes

Materials graduated cylinders, mass balances, 250-mL Erlenmeyer flasks, rubber balloons, magnesium ribbons, 1.0*M* hydrochloric acid

Class Time 20 minutes

Safety The balloons contain hydrogen gas and should be kept away from heat and open flames during and after the experiment. Caution students about the corrosive nature of HCl. Students should wear safety glasses and aprons. Once the reaction in a flask is complete, carefully remove the balloon. Disperse the gas in a fume hood or outside. Neutralize remaining $HCl(aq)$ with baking soda before flushing down the drain.

Teaching Tips

- Have students carefully balance the equation for the reaction and then look at the mole ratio from the balanced equation and the molar amounts of the two reactants present before they make their prediction in Step 4.

Expected Outcome The approximate volumes of the balloons are 0.60 g Mg, 0.5 L H_2; 1.2 g Mg, 1 L H_2; 2.4 g Mg, 1 L H_2.

Analyze and Conclude

1. Those students who failed to balance the equation, obtain the mole ratio, and compare the molar amounts of reactants, likely predicted a doubling of the volume with each doubling of the mass of magnesium.
2. $Mg + 2HCl \rightarrow MgCl_2 + H_2$
3. A mass of 0.60 g Mg is 0.025 mol Mg. Because Mg and HCl react in a 1:2 mol ratio, 0.10 mol HCl is in excess and Mg is limiting. According to the balanced equation, 0.025 mol Mg should produce 0.025 mol H_2. A mass of 1.2 g Mg is 0.050 mol Mg. According to the balanced equation, 0.050 mol Mg will react with 0.1 mol HCl to produce 0.05 mol H_2.
4. A mass of 2.4 g Mg is 0.10 mol Mg. Because there is only 0.1 mol HCl in the flask and, according to the equation, 0.20 mol HCl is needed to react with 0.10 mol Mg, HCl is limiting. Therefore, only 0.05 mol H_2 is produced.

For Enrichment 

Ask, **What other clue could be used to determine if HCl was the limiting reagent?** *(If HCl were the limiting reagent, remaining magnesium metal would be present after the reaction.)*

Quick LAB

Limiting Reagents

Purpose

To illustrate the concept of a limiting reagent in a chemical reaction.

Materials

- graduated cylinder
- balance
- 3 250-mL Erlenmeyer flasks
- 3 rubber balloons
- 4.2 g magnesium ribbon
- 300 mL 1.0*M* hydrochloric acid

Procedure    

1. Add 100 mL of the hydrochloric acid solution to each flask.
2. Weigh out 0.6 g, 1.2 g, and 2.4 g of magnesium ribbon, and place each sample into its own balloon.
3. Stretch the end of each balloon over the mouth of each flask. Do not allow the magnesium ribbon in the balloon to fall into the flask.
4. Magnesium reacts with hydrochloric acid to form hydrogen gas. When you mix the magnesium with the hydrochloric acid in the next step, you will generate a certain volume of hydrogen gas. How do you think the volume of hydrogen produced in each flask will compare?
5. Lift up on each balloon and shake the magnesium into each flask. Observe the volume of gas produced until the reaction in each flask is completed.

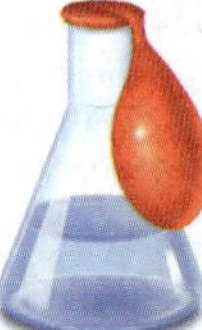

Analyze and Conclude

1. How did the volumes of hydrogen gas produced, as measured by the size of the balloons, compare? Did the results agree with your prediction?
2. Write a balanced equation for the reaction you observed.
3. The 100 mL of hydrochloric acid contained 0.10 mol HCl. Show by calculation why the balloon with 1.2 g Mg inflated to about twice the size of the balloon with 0.60 g Mg.
4. Show by calculation why the balloons with 1.2 g and 2.4 g Mg inflated to approximately the same volume. What was the limiting reagent when 2.4 g Mg was added to the acid?

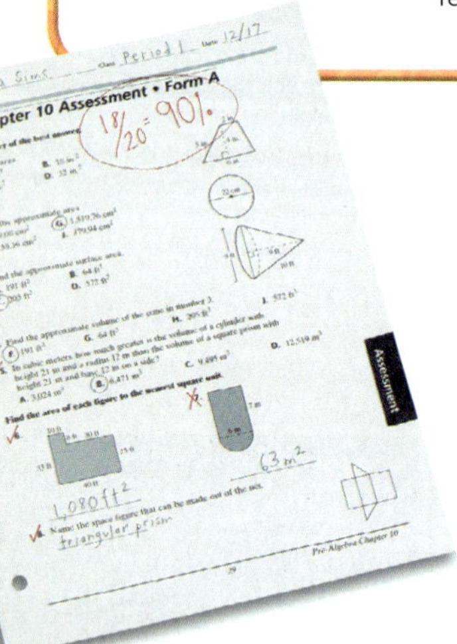

Figure 12.11 Calculating the ratio of the number of correct answers to the number of questions on the exam is a measure of how well the student performed on the exam.

Percent Yield

In theory, when a teacher gives an exam to the class, every student should get a grade of 100%. This generally does not occur, as shown in Figure 12.11. Instead, the performance of the class is usually spread over a range of grades. Your exam grade, expressed as a percentage, is a ratio of two items. The first item is the number of questions you answered correctly. The second is the total number of questions. The grade compares how well you performed with how well you could have performed if you had answered all the questions correctly. Chemists perform similar calculations in the laboratory when the product from a chemical reaction is less than expected, based on the balanced chemical equation.

When an equation is used to calculate the amount of product that will form during a reaction, the calculated value represents the theoretical yield. The **theoretical yield** is the maximum amount of product that could be formed from given amounts of reactants. In contrast, the amount of product that actually forms when the reaction is carried out in the laboratory is called the **actual yield.** The **percent yield** is the ratio of the actual yield to the theoretical yield expressed as a percent.

$$\text{Percent yield} = \frac{\text{actual yield}}{\text{theoretical yield}} \times 100\%$$

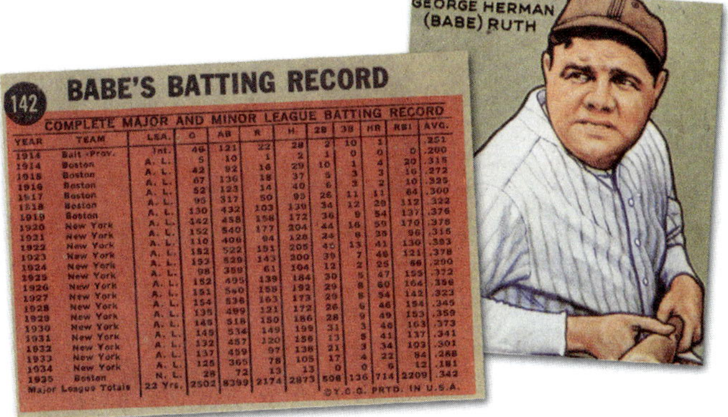

Figure 12.12 A batting average is actually a percent yield.

Because the actual yield of a chemical reaction is often less than the theoretical yield, the percent yield is often less than 100%. **The percent yield is a measure of the efficiency of a reaction carried out in the laboratory.** This is similar to an exam score measuring your efficiency of learning, or a batting average measuring your efficiency of hitting a baseball.

A percent yield should not normally be larger than 100%. Many factors cause percent yields to be less than 100%. Reactions do not always go to completion; when this occurs, less than the calculated amount of product is formed. Impure reactants and competing side reactions may cause unwanted products to form. Actual yield can also be lower than the theoretical yield due to a loss of product during filtration or in transferring between containers. Moreover, if reactants or products have not been carefully measured, a percent yield of 100% is unlikely.

An actual yield is an experimental value. Figure 12.13 shows a typical laboratory procedure for determining the actual yield of a product of a decomposition reaction. For reactions in which percent yields have been determined, you can calculate and therefore predict an actual yield if the reaction conditions remain the same.

What factors can cause the actual yield to be less than the theoretical yield?

Figure 12.13 Sodium hydrogen carbonate ($NaHCO_3$) will decompose when heated. (a) The mass of $NaHCO_3$, the reactant, is measured. (b) The reactant is heated. (c) The mass of one of the products, sodium carbonate (Na_2CO_3), the actual yield, is measured. The percent yield is calculated once the actual yield is determined. **Predicting** ***What are the other products of this reaction?***

Percent Yield

Actual Yield and Heat

Purpose Students assess actual yield of a reaction based on temperature changes

Materials 3 foam cups, a thermometer, 100 mL of 1.0*M* HCl, and approximately 200 mL of 1.0*M* NaOH

Safety Students should wear safety glasses and aprons. HCl and NaOH are corrosive. Caution students to avoid skin contact with these chemicals. Neutralized solutions can be flushed down the drain with excess water.

Procedure Write the balanced chemical equation on the board for students to use as a reference: $HCl(aq) + NaOH(aq) \rightarrow H_2O(l) + NaCl(aq) + \text{heat}$. Have students interpret the balanced equation as 1 mole of HCl reacts with 1 mole of NaOH to form 1 mole of NaCl and 1 mole of water. Have students transfer 30 mL of the HCl to a styrofoam cup and measure the temperature. Next, have them add 10 mL of NaOH and gently stir the contents with the thermometer. Students should record the highest temperature reached as the reaction proceeds. Using a new cup each time, repeat the measurement with 30 mL of HCl and 30 mL of NaOH, and with 30 mL of HCl and 60 mL of NaOH.

Expected Outcomes The second trial, in which reactants are present in a 1:1 mole ratio, produces the greatest temperature change. A slightly lower temperature change for the third trial can be accounted for by the greater mass of reagents. Ask, **Which trial had the highest yield?** *(the one that produced the greatest temperature change)*

Answers to...

Figure 12.13 H_2O and CO_2

Reactions do not always go to completion; side reactions may occur; poor laboratory technique.

Discuss 

Point out the importance of the yield in a chemical reaction. Note that actual yields are usually less than theoretical yields. For industrial chemists or chemical engineers, the goal is to find cost-effective methods for converting reactants into products. Industrial chemists and chemical engineers want to achieve the maximum product yield at the lowest cost. One way to control costs is to minimize the amount of excess reagent by calculating stoichiometric quantities of the reactants.

Sample Problem 12.9

Answers

29. 59.3 g Fe

30. 17.0 g Ag

Practice Problems Plus 

Chapter 12 Assessment problem 49 is related to Sample Problem 12.9.

Math Handbook

For a math refresher and practice, direct students to dimensional analysis, page R66.

Sample Problem 12.10

Answers

31. 83.5%

32. 57.7%

Practice Problems Plus 

If 75.0 g of siderite ore ($FeCO_3$) is heated with an excess of oxygen, 45.0 g of ferric oxide (Fe_2O_3) is produced.

$$4FeCO_3(s) + O_2(g) \rightarrow 2Fe_2O_3(s) + 4CO_2(g)$$

What is the percent yield of this reaction? *(87.0%)*

SAMPLE PROBLEM 12.9

Calculating the Theoretical Yield of a Reaction

Calcium carbonate, which is found in seashells, is decomposed by heating. The balanced equation for this reaction is:

$$CaCO_3(s) \xrightarrow{\Delta} CaO(s) + CO_2(g)$$

What is the theoretical yield of CaO if 24.8 g $CaCO_3$ is heated?

1 Analyze ***List the knowns and the unknown.***

Knowns

- mass of calcium carbonate = 24.8 g $CaCO_3$
- 1 mol $CaCO_3$ = 100.1 g $CaCO_3$ (molar mass)
- 1 mol CaO = 56.1 g CaO (molar mass)

Unknown

- theoretical yield of calcium oxide = ? g CaO

Calculate the theoretical yield using the mass of the reactant:

$$\text{g } CaCO_3 \longrightarrow \text{mol } CaCO_3 \longrightarrow \text{mol CaO} \longrightarrow \text{g CaO}$$

The appropriate mole ratio is 1 mol CaO/1 mol $CaCO_3$.

2 Calculate ***Solve for the unknown.***

$$24.8 \text{ g } \cancel{CaCO_3} \times \frac{1 \cancel{\text{ mol } CaCO_3}}{100.1 \text{ g } \cancel{CaCO_3}} \times \frac{1 \cancel{\text{ mol CaO}}}{1 \cancel{\text{ mol } CaCO_3}} \times \frac{56.1 \text{ g CaO}}{1 \cancel{\text{ mol CaO}}} = 13.9 \text{ g CaO}$$

3 Evaluate ***Does the result make sense?***

The mole ratio of CaO to $CaCO_3$ is 1:1. The ratio of their masses in the reaction should be the same as the ratio of their molar masses, which is slightly greater than 1:2. The result of the calculations shows that the mass of CaO is slightly greater than half the mass of $CaCO_3$.

Math Handbook

For help with dimensional analysis, go to page R66.

Practice Problems

29. When 84.8 g of iron(III) oxide reacts with an excess of carbon monoxide, iron is produced.

$$Fe_2O_3(s) + 3CO(g) \longrightarrow 2Fe(s) + 3CO_2(g)$$

What is the theoretical yield of iron?

30. When 5.00 g of copper reacts with excess silver nitrate, silver metal and copper(II) nitrate are produced. What is the theoretical yield of silver in this reaction?

Interactive Textbook

Problem-Solving 12.29 Solve Problem 29 with the help of an interactive guided tutorial.

with ChemASAP

Recall that the percent yield is calculated by multiplying the ratio of the actual yield to theoretical yield by 100%. Therefore, you must have values of both the theoretical yield and the actual yield to calculate the percent yield.

SAMPLE PROBLEM 12.10

Calculating the Percent Yield of a Reaction

What is the percent yield if 13.1 g CaO is actually produced when 24.8 g $CaCO_3$ is heated?

$$CaCO_3(s) \xrightarrow{\Delta} CaO(s) + CO_2(g)$$

Analyze ***List the knowns and the unknown.***

Knowns

- actual yield = 13.1 g CaO
- theoretical yield = 13.9 g CaO (from Sample Problem 12.9)
- percent yield = $\frac{\text{actual yield}}{\text{theoretical yield}} \times 100\%$

Unknown

- percent yield = ? %

2 Calculate ***Solve for the unknown.***

$$\text{percent yield} = \frac{13.1\ \text{g CaO}}{13.9\ \text{g CaO}} \times 100\% = 94.2\%$$

Evaluate ***Does the result make sense?***

In this example, the actual yield is slightly less than theoretical yield. Therefore, the percent yield should be slightly less than 100%. The answer should have three significant figures.

Practice Problems

31. If 50.0 g of silicon dioxide is heated with an excess of carbon, 27.9 g of silicon carbide is produced.

$$SiO_2(s) + 3C(s) \longrightarrow SiC(s) + 2CO(g)$$

What is the percent yield of this reaction?

32. If 15.0 g of nitrogen reacts with 15.0 g of hydrogen, 10.5 g of ammonia is produced. What is the percent yield of this reaction?

Math Handbook

For help with percents, go to page R72.

Interactive Textbook

Problem-Solving 12.31 Solve Problem 31 with the help of an interactive guided tutorial.

with **ChemASAP**

12.3 Section Assessment

33. **Key Concept** In a chemical reaction, how does an insufficient quantity of a reactant affect the amount of product formed?

34. **Key Concept** How can you gauge the efficiency of a reaction carried out in the laboratory?

35. What is the percent yield if 4.65 g of copper is produced when 1.87 g of aluminum reacts with an excess of copper(II) sulfate?

$$2Al(s) + 3CuSO_4(aq) \longrightarrow Al_2(SO_4)_3(aq) + 3Cu(s)$$

Elements Handbook

Haber Process Read about ammonia on page R26. Examine the flow chart summarizing the Haber process. What experimental data would you need to determine the percent yield of the Haber process?

Assessment 12.3 Test yourself on the concepts in Section 12.3.

with **ChemASAP**

Section 12.3 Assessment

33. In a chemical reaction, an insufficient quantity of any of the reactants will limit the amount of product that forms.

34. The efficiency of a reaction carried out in a laboratory can be measured by calculating the percent yield.

35. 70.5%

3 ASSESS

Evaluate Understanding L2

Limiting Reaction

Following the safety precaution listed in the Quick Lab on page 372, place a small piece of zinc in a large beaker of dilute hydrochloric acid. Explain the reaction to the students.

$$Zn(s) + 2HCl(aq) \rightarrow ZnCl_2(aq) + H_2(g)$$

Ask, **Which of the two reactants is the limiting reagent?** *(If the zinc completely disappears after some time, students may conclude that the zinc is used up first, and is, therefore, the limiting reagent.)* After students respond, ask, **How can you test your hypothesis?** *(Acceptable answers include measure the volume of H_2 produced for various amounts of zinc added to a fixed volume of HCl. Determine the stoichiometric quantity of zinc, the amount that reacts to give the greatest volume of H_2.)*

Reteach 

Emphasize the importance of having a correctly balanced equation in order to properly calculate the theoretical yield for a reaction. Explain that the way to decide which substance, if any, is the limiting reagent, is to compare the mole ratios of the given substances to the required mole ratio, as shown in the balanced chemical equation.

Elements Handbook

the amounts of N_2 and H_2 used (to calculate the theoretical yield), and the amount of NH_3 produced (i.e., the actual yield)

If your class subscribes to the Interactive Textbook, use it to review key concepts in Section 12.3.

with **ChemASAP**

Technology & Society

Just the Right Volume of Gas

Write the following reactions on the board.

Reaction I $2NaN_3(s) \rightarrow 2Na(s) + 3N_2(g)$

Reaction II $10Na(s) + 2KNO_3(s) \rightarrow K_2O(s) + 5Na_2O(s) + N_2(g) + \text{heat}$

Point out that the proper inflation of the air bag requires two reactions. Explain that an electrical current produced by the igniter causes the decomposition of sodium azide into sodium metal and nitrogen gas. Note that the sodium metal produced is dangerously reactive. In a second reaction, potassium nitrate reacts with the elemental sodium and forms potassium oxide, sodium oxide, and additional nitrogen gas. The heat causes all the solid products to fuse with SiO_2, powdered sand, which is also part of the reaction mixture. The fused product is a safe, unreactive glass.

Discuss

L2

Ask, **How many moles of potassium nitrate must be included in the reaction mixture to consume the sodium produced by the decomposition of one mole of sodium azide?** *(0.2 mol KNO_3)* **How many liters of N_2 are produced at STP if 1.0 mole of sodium azide and 0.20 mole of potassium nitrate react?** *(36 L)* Have students speculate how the pressure of the gas inside the air bag depends on the number of moles of nitrogen produced and the temperature inside the air bag. *(Acceptable answers should indicate that because gas pressure depends on the number of gas particles present, pressure depends on the number of moles of gas particles present. The heat released by this reaction raises the temperature of the gaseous products, helping the bag inflate even faster.)*

Technology & Society

Just the Right Volume of Gas

In a front-end collision, proper inflation of an air bag may save your life. Engineers use stoichiometry to determine the exact quantity of each reactant in the air bag's inflation system. **Interpreting Diagrams** ***What is the source of the gas that fills an air bag?***

Car Facts

17.1 million cars and light trucks were sold in 2002 in the United States.

Monaco has the highest number of vehicles in relation to its road network. In 1996 (most recent figures), it had 480 vehicles for each kilometer of road. If they were required to park behind one another on the streets, half would have nowhere to park!

At night, headlights illuminate 160 ft in front of your car. If you are driving 40 mi/h at night, your reaction distance is 88 ft and your braking distance is 101 ft. Is your stopping distance less than or greater than 160 ft?

Reaction distance + **Braking distance** = **Stopping distance**

Facts and Figures

Automobile Restraint Systems

History	
1947	Tucker Automobile Company makes safety belts available.
1949	Nash Motor Company provides lap-safety belts.
1958	Swedish engineer, Nils Bohlin, patents chest/lap-safety belt.
1963	Volvo makes chest/lap-safety belts standard equipment on its U.S. models.
1988	Chrysler Motor Company makes air-bag restraint systems standard equipment.

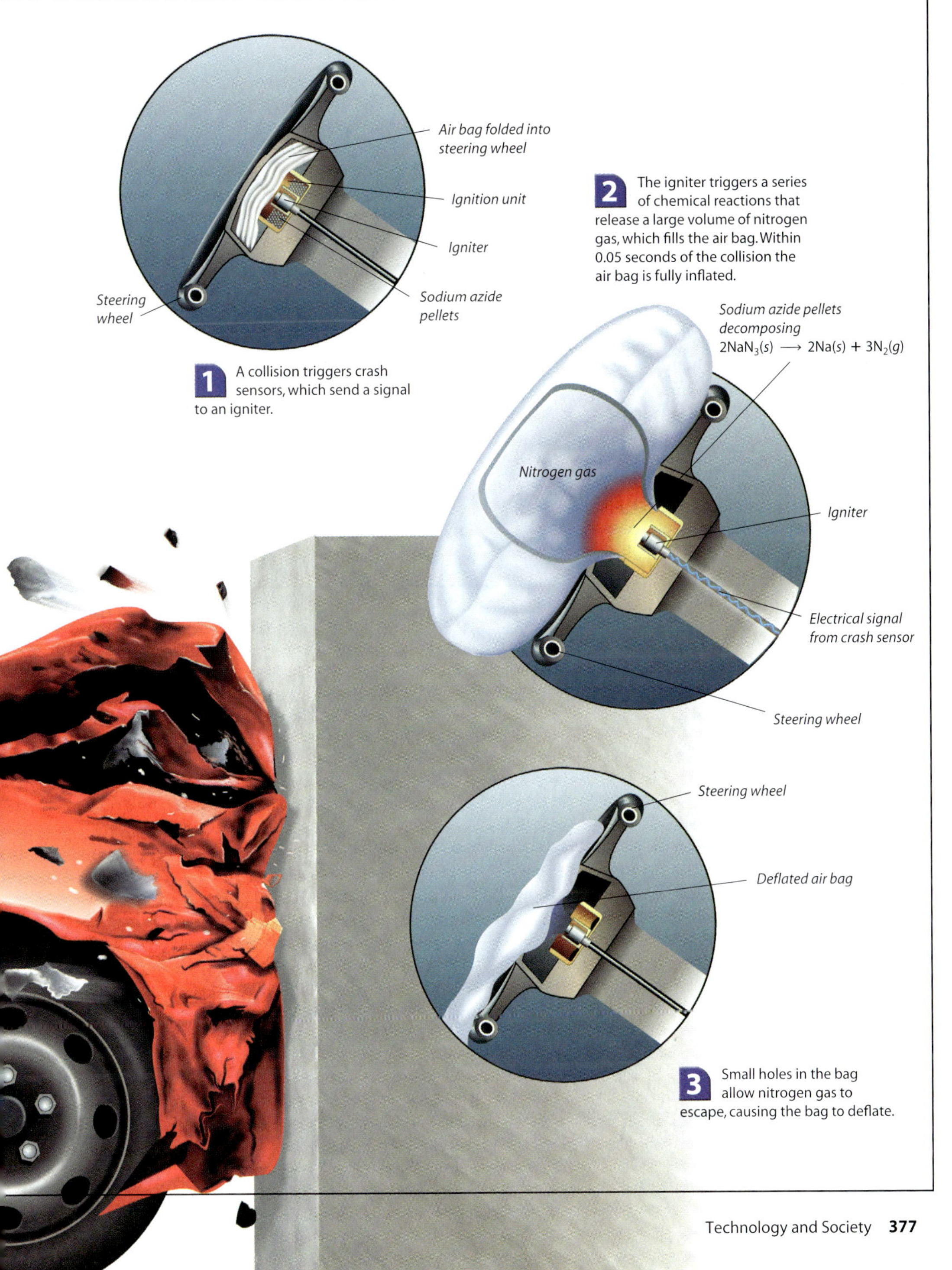

Technology and Society 377

Discuss L2

Have students research the safety restrictions on the use of air bags with infants and young children. Have students research new designs and applications of safety air bags.

Facts and Figures

Automobile Restraint Systems

Statistic The use of air-bag restraint systems (seatbelt/air bags) reduces the risk of fatalities in accidents by about 70%.

Answers to...

Interpreting Diagrams

The sources of the N_2 gas are the decomposition reaction of sodium azide (NaN_3) and the reaction of sodium with potassium nitrate.

CHAPTER 12 Study Guide

Study Tip

Prioritize Schedule your time realistically. Stick to your deadlines.

If your class subscribes to the Interactive Textbook with ChemASAP, your students can go online to access an interactive version of the Student Edition and a self-test.

with ChemASAP

CHAPTER 12 Study Guide

Key Concepts

12.1 The Arithmetic of Equations

- A balanced chemical equation provides the same kind of quantitative information that a recipe does.
- Chemists use balanced chemical equations as a basis to calculate how much reactant is needed or product is formed in a reaction.
- A balanced chemical equation can be interpreted in terms of different quantities, including numbers of atoms, molecules, or moles; mass; and volume.
- Mass and atoms are conserved in every chemical reaction.

12.2 Chemical Calculations

- In chemical calculations, mole ratios are used to convert between moles of reactant and moles of product, between moles of reactants, or between moles of products.
- In a typical stoichiometric problem, the given quantity is first converted to moles. Then the mole ratio from the balanced equation is used to calculate the moles of the wanted substance. Finally, the moles are converted to any other unit of measurement related to the unit mole.

12.3 Limiting Reagent and Percent Yield

- In a chemical reaction, an insufficient quantity of any of the reactants will limit the amount of product that forms.
- The percent yield is a measure of the efficiency of a reaction performed in the laboratory.

Vocabulary

- actual yield (p. 372)
- excess reagent (p. 369)
- limiting reagent (p. 369)
- mole ratio (p. 359)
- percent yield (p. 372)
- stoichiometry (p. 354)
- theoretical yield (p. 372)

Key Equations

- mole-mole relationship for $aG \longrightarrow bW$:

$$x \text{ mol } G \times \frac{b \text{ mol } W}{a \text{ mol } G} = \frac{xb}{a} \text{ mol } W$$

- percent yield $= \frac{\text{actual yield}}{\text{theoretical yield}} \times 100\%$

Organizing Information

Use these terms to construct a concept map that organizes the major ideas of this chapter.

mole ratio · percent yield · chemical equation · actual yield · theoretical yield · excess reagent · stoichiometry · limiting reagent

Concept Map 12 Solve the Concept Map with the help of an interactive guided tutorial.

with ChemASAP

Chapter Resources

Print

- **Core Teaching Resources,** Chapter 12, *Practice Problems, Vocabulary Review, Quiz, Chapter Test A, Chapter Test B*

Technology

- **Computer Test Bank,** Chapter 12 Test
- **Interactive Textbook with ChemASAP,** Chapter 12

378 Chapter 12

CHAPTER 12 Assessment

Reviewing Content

12.1 The Arithmetic of Equations

36. Interpret each chemical equation in terms of interacting particles.

a. $2KClO_3(s) \longrightarrow 2KCl(s) + 3O_2(g)$

b. $4NH_3(g) + 6NO(g) \longrightarrow 5N_2(g) + 6H_2O(g)$

c. $4K(s) + O_2(g) \longrightarrow 2K_2O(s)$

37. Interpret each equation in Problem 36 in terms of interacting numbers of moles of reactants and products.

38. Calculate and compare the mass of the reactants with the mass of the products for each equation in Problem 36. Show that each balanced equation obeys the law of conservation of mass.

12.2 Chemical Calculations

39. Explain the term *mole ratio* in your own words. When would you use this term?

40. Carbon disulfide is an important industrial solvent. It is prepared by the reaction of coke with sulfur dioxide.

$$5C(s) + 2SO_2(g) \longrightarrow CS_2(l) + 4CO(g)$$

a. How many moles of CS_2 form when 2.7 mol C reacts?

b. How many moles of carbon are needed to react with 5.44 mol SO_2?

c. How many moles of carbon monoxide form at the same time that 0.246 mol CS_2 forms?

d. How many mol SO_2 are required to make 118 mol CS_2?

41. Methanol (CH_3OH) is used in the production of many chemicals. Methanol is made by reacting carbon monoxide and hydrogen at high temperature and pressure.

$$CO(g) + 2H_2(g) \longrightarrow CH_3OH(g)$$

a. How many moles of each reactant are needed to produce 3.60×10^2 g CH_3OH?

b. Calculate the number of grams of each reactant needed to produce 4.00 mol CH_3OH.

c. How many grams of hydrogen are necessary to react with 2.85 mol CO?

42. The reaction of fluorine with ammonia produces dinitrogen tetrafluoride and hydrogen fluoride.

$$5F_2(g) + 2NH_3(g) \longrightarrow N_2F_4(g) + 6HF(g)$$

a. If you have 66.6 g NH_3, how many grams of F_2 are required for complete reaction?

b. How many grams of NH_3 are required to produce 4.65 g HF?

c. How many grams of N_2F_4 can be produced from 225 g F_2?

43. What information about a chemical reaction is derived from the coefficients in a balanced equation?

44. Lithium nitride reacts with water to form ammonia and aqueous lithium hydroxide.

$$Li_3N(s) + 3H_2O(l) \longrightarrow NH_3(g) + 3LiOH(aq)$$

a. What mass of water is needed to react with 32.9 g Li_3N?

b. When the above reaction takes place, how many molecules of NH_3 are produced?

c. Calculate the number of grams of Li_3N that must be added to an excess of water to produce 15.0 L NH_3 (at STP).

12.3 Limiting Reagent and Percent Yield

45. What is the significance of the limiting reagent in a reaction? What happens to the amount of any reagent that is present in an excess?

46. How would you identify a limiting reagent in a chemical reaction?

47. In a reaction chamber, 3.0 mol of aluminum is mixed with 5.3 mol Cl_2 and reacts. The reaction is described by the following balanced chemical equation.

$$2Al + 3Cl_2 \longrightarrow 2AlCl_3$$

a. Identify the limiting reagent for the reaction.

b. Calculate the number of moles of product formed.

c. Calculate the number of moles of excess reagent remaining after the reaction.

48. Heating an ore of antimony (Sb_2S_3) in the presence of iron gives the element antimony and iron(II) sulfide.

$$Sb_2S_3(s) + 3Fe(s) \longrightarrow 2Sb(s) + 3FeS(s)$$

When 15.0 g Sb_2S_3 reacts with an excess of Fe, 9.84 g Sb is produced. What is the percent yield of this reaction?

CHAPTER 12 Assessment

Reviewing Content

36. a. Two formula units $KClO_3$ decompose to form two formula units KCl and three molecules O_2.
b. Four molecules NH_3 react with six molecules NO to form five molecules N_2 and six molecules H_2O.
c. Four atoms K react with one molecule O_2 to form two formula units K_2O.

37. a. Two mol $KClO_3$ decompose to form two mol KCl and three mol O_2.
b. Four mol NH_3 react with six mol NO to form five mol N_2 and six mol H_2O.
c. Four mol K react with one mol O_2 to form two mol K_2O.

38. a. 245.2 g **b.** 248.0 g
c. 188.4 g
All obey the law of conservation of mass.

39. Acceptable answers include the idea of writing a ratio using the coefficients of two substances from a balanced equation as the number of moles of each substance reacting or being formed.

40. a. 0.54 mol **b.** 13.6 mol
c. 0.984 mol **d.** 236 mol

41. a. 11.3 mol CO, 22.5 mol H_2
b. 112 g CO, 16.0 g H_2
c. 11.4 g H_2

42. a. 372 g F_2
b. 1.32 g NH_3
c. 123 g N_2F_4

43. The coefficients indicate the relative number of moles (or particles) of reactants and products.

44. a. 51.2 g H_2O
b. 5.71×10^{23} molecules NH_3
c. 23.2 g Li_3N

45. The amount of the limiting reagent determines the maximum amount of product that can be formed. The excess reagent is only partially consumed in the reaction.

46. To identify the limiting reagent, express quantities of reactants as moles; compare to the mole ratios from the balanced equation.

47. a. Al
b. 3.0 mol $AlCl_3$
c. 0.8 mol Cl_2

48. 91.5%

CHAPTER 12 Assessment *continued*

Understanding Concepts

49. Calcium carbonate reacts with phosphoric acid to produce calcium phosphate, carbon dioxide, and water.

$$3CaCO_3(s) + 2H_3PO_4(aq) \longrightarrow Ca_3(PO_4)_2(aq) + 3CO_2(g) + 3H_2O(l)$$

a. How many grams of phosphoric acid react with excess calcium carbonate to produce 3.74 g $Ca_3(PO_4)_2$?

b. Calculate the number of grams of CO_2 formed when 0.773 g H_2O is produced.

50. Nitric acid and zinc react to form zinc nitrate, ammonium nitrate, and water.

$$4Zn(s) + 10HNO_3(aq) \longrightarrow 4Zn(NO_3)_2(aq) + NH_4NO_3(aq) + 3H_2O(l)$$

a. How many atoms of zinc react with 1.49 g HNO_3?

b. Calculate the number of grams of zinc that must react with an excess of HNO_3 to form 29.1 g NH_4NO_3.

51. Hydrazine (N_2H_4) is used as rocket fuel. It reacts with oxygen to form nitrogen and water.

$$N_2H_4(l) + O_2(g) \longrightarrow N_2(g) + 2H_2O(g)$$

a. How many liters of N_2 (at STP) form when 1.0 kg N_2H_4 reacts with 1.0 kg O_2?

b. How many grams of the excess reagent remain after the reaction?

52. When 50.0 g of silicon dioxide is heated with an excess of carbon, 32.2 g of silicon carbide is produced.

$$SiO_2(s) + 3C(s) \longrightarrow SiC(s) + 2CO(g)$$

a. What is the percent yield of this reaction?

b. How many grams of CO gas are made?

53. If the reaction below proceeds with a 96.8% yield, how many kilograms of $CaSO_4$ are formed when 5.24 kg SO_2 reacts with an excess of $CaCO_3$ and O_2?

$$2CaCO_3(s) + 2SO_2(g) + O_2(g) \longrightarrow 2CaSO_4(s) + 2CO_2(g)$$

54. Ammonium nitrate will decompose explosively at high temperatures to form nitrogen, oxygen, and water vapor.

$$2NH_4NO_3(s) \longrightarrow 2N_2(g) + 4H_2O(g) + O_2(g)$$

What is the total number of liters of gas formed when 228 g NH_4NO_3 is decomposed? (Assume STP.)

55. In an experiment, varying masses of sodium metal are reacted with a fixed initial mass of chlorine gas. The amounts of sodium used and the amounts of sodium chloride formed are shown on the following graph.

NaCl Produced by Reacting Sodium With Chlorine

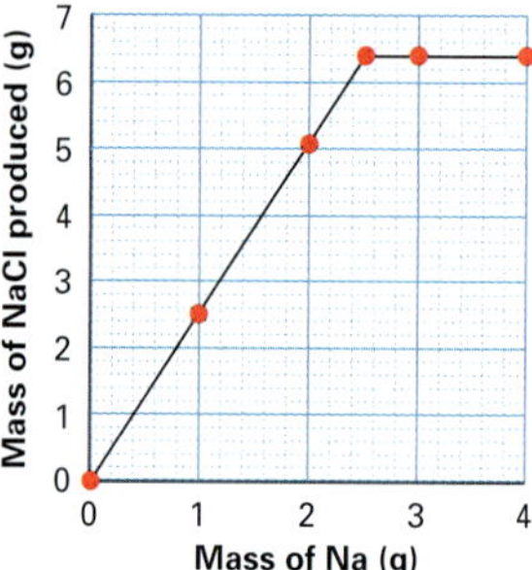

a. Explain the general shape of the graph.

b. Estimate the amount of chlorine gas used in this experiment at the point where the curve becomes horizontal.

56. The manufacture of compound F requires five separate chemical reactions. The initial reactant, compound A, is converted to compound B, compound B is converted to compound C, and so forth. The diagram below summarizes the stepwise manufacture of compound F, including the percent yield for each step. Provide the missing quantities or missing percent yields. Assume that the reactant and product in each step react in a one-to-one mole ratio.

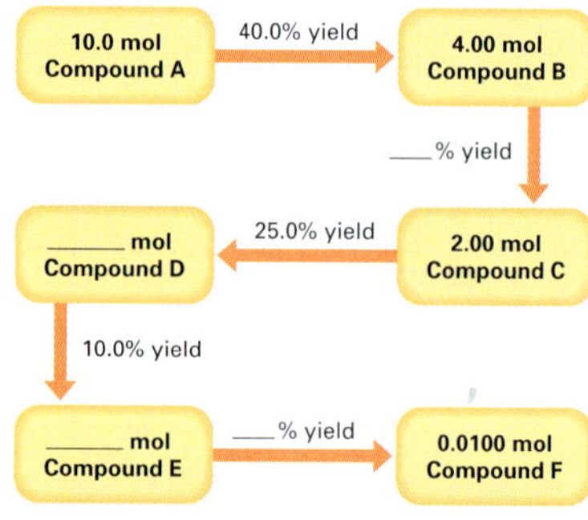

CHAPTER 12 Assessment *continued*

Understanding Concepts

49. a. 2.36 g H_3PO_4
b. 1.89 g CO_2

50. a. 5.70×10^{21} atoms Zn
b. 95.2 g Zn

51. a. 7.0×10^2 L N_2
b. no reagent in excess

52. a. 96.4%
b. 45.0 g

53. 10.7 kg $CaSO_4$

54. 224 L gas

55. a. Initially, the amount of NaCl formed increases as the amount of Na used increases. For this part of the curve, sodium is the limiting reagent. Beyond a mass of about 2.5 g of Na, the amount of product formed remains constant because chlorine is now the limiting reagent.
b. Chlorine becomes the limiting reagent when the mass of sodium exceeds 2.5 g. This corresponds to a mass of about 3.9 g chlorine.

56. 50.0% yield; 0.500 mol; 0.0500 mol; 20.0% yield

380 Chapter 12

Critical Thinking

57. Given a certain quantity of reactant, you calculate that a particular reaction should produce 55 g of a product. When you perform the reaction, you find that you have produced 63 g of product. What is your percent yield? What could have caused a percent yield greater than 100%?

58. Would the law of conservation of mass hold in a net ionic equation? Explain.

59. A bicycle-built-for-three has a frame, two wheels, six pedals, and three seats. The balanced equation for this bicycle is

$$F + 2W + 6P + 3S \longrightarrow FW_2P_6S_3$$

How many of each part are needed to make 29 bicycles-built-for-three?

a. frames **b.** wheels
c. pedals **d.** seats

60. A car gets 9.2 kilometers to a liter of gasoline. Assuming that gasoline is 100% octane (C_8H_{18}), which has a density of 0.69 g/cm^3, how many liters of air (21% oxygen by volume at STP) will be required to burn the gasoline for a 1250-km trip? Assume complete combustion.

61. Ethyl alcohol (C_2H_5OH) can be produced by the fermentation of glucose ($C_6H_{12}O_6$). If it takes 5.0 h to produce 8.0 kg of alcohol, how many days will it take to consume 1.0×10^3 kg of glucose? (An enzyme is used as a catalyst.)

$$C_6H_{12}O_6 \xrightarrow{\text{enzyme}} 2C_2H_5OH + 2CO_2$$

Concept Challenge

62. A 1004.0-g sample of $CaCO_3$ that is 95.0% pure gives 225 L CO_2 at STP when reacted with an excess of hydrochloric acid.

$$CaCO_3 + 2HCl \longrightarrow CaCl_2 + CO_2 + H_2O$$

What is the density (in g/L) of the CO_2?

63. The white limestone cliffs of Dover, England, contain a large percentage of calcium carbonate ($CaCO_3$). A sample of limestone with a mass of 84.4 g reacts with an excess of hydrochloric acid to form calcium chloride.

$$CaCO_3 + 2HCl \longrightarrow CaCO_3 + CaCl_2 + H_2O + CO_2$$

The mass of calcium chloride formed is 81.8 g. What is the percentage of calcium carbonate in the limestone?

64. For the reaction below there is a 100.0 g of each reactant available. Which reagent is the limiting reagent?

$$2MnO_2 + 4KOH + O_2 + Cl_2 \longrightarrow 2KMnO_4 + 2KCl + 2H_2O$$

65. The equation for one of the reactions in the process of reducing iron ore to the metal is

$$Fe_2O_3(s) + 3\,CO(g) \longrightarrow 2\,Fe(s) + 3\,CO_2(g)$$

a. What is the maximum mass of iron, in grams, that can be obtained from 454 g (1.00 lb) of iron(III) oxide?

b. What mass of CO is required to reduce the iron(III) oxide to iron metal?

66. SO_3 can be produced in the following two-step process:

$$FeS_2 + O_2 \longrightarrow Fe_2O_3 + SO_2$$
$$SO_2 + O_2 \longrightarrow SO_3$$

Assuming that all the FeS_2 reacts, how many grams of SO_3 are produced when 20.0 g of the FeS_2 reacts with 16.0 g of O_2?

Critical Thinking

57. The percent yield is 115%; such a yield could be attributed to experimenter error, or to unreacted starting material, or to outside materials contaminating the product.

58. Yes, a net ionic equation is balanced and thus obeys the law of conservation of mass.

59. **a.** 29 frames
b. 58 wheels
c. 174 pedals
d. 87 seats

60. 1.1×10^6 L air

61. 13 days

Concept Challenge

62. 1.86 g/L

63. 87.4% $CaCO_3$

64. KOH is the limiting reagent.

65. **a.** 347 g Fe
b. 239 g CO

66. 6.51 g SO_3

CHAPTER 12 Assessment continued

Cumulative Review

67. **a.** 22, 22, 25
b. 50, 50, 70
c. 8, 8, 10
d. 12, 12, 14

68. **a.** ultraviolet
b. ultraviolet
c. ultraviolet

69. **a.** sodium
b. arsenic
c. cesium

70. **a.** Cs·
b. :B̈r· (with dots above and below)
c. ·Ca·
d. ·P̈· (with dots above and below)

71. c and d

72. single bond, one pair shared electrons; double bond, two pairs of shared electrons; triple bond, three pairs of shared electrons

73. Yes, an ionic compound with at least one polyatomic ion has covalent bonds.

74. A cation has a positive charge, and an anion has a negative charge.

75. **a.** phosphate ion
b. aluminum ion
c. selenide ion
d. ammonium ion

76. **a.** silicon dioxide
b. potassium sulfate
c. carbonic acid
d. magnesium sulfide

77. **a.** $Al_2(CO_3)_3$
b. NO_2
c. K_2S
d. $MnCrO_4$
e. NaBr

78. 1.30×10^{-22} g

79. 7.38 g Be

80. $C_2H_2O_4$

81. **a.** 0.473 mol KNO_3
b. 9.91×10^{-2} mol SO_2
c. 3.74×10^{-2} mol PCl_3

82. **a.** $2Pb(NO_3)_2 \rightarrow 2PbO + 4NO_2 + O_2$
b. $2C_3H_7OH + 9O_2 \rightarrow 6CO_2 + 8H_2O$
c. $2Al + 3FeO \rightarrow 3Fe + Al_2O_3$

83. **a.** 1, 1, 1, 2
b. 1, 3, 3, 1
c. 1, 1, 1, 2

CHAPTER 12 Assessment continued

Cumulative Review

67. How many electrons, protons, and neutrons are in an atom of each isotope? *(Chapter 4)*
a. titanium-47
b. tin-120
c. oxygen-18
d. magnesium-26

68. When comparing ultraviolet and visible electromagnetic radiation, which has *(Chapter 5)*
a. a higher frequency
b. a higher energy
c. a shorter wavelength

69. Identify the larger atom of each pair. *(Chapter 6)*
a. sodium and chlorine
b. arsenic and nitrogen
c. fluorine and cesium

70. Write electron dot formulas for the following atoms. *(Chapter 7)*
a. Cs
b. Br
c. Ca
d. P

71. Which of these elements form ions with a 2+ charge? *(Chapter 7)*
a. potassium
b. sulfur
c. barium
d. magnesium

72. Distinguish among single, double, and triple covalent bonds. *(Chapter 8)*

73. Can a compound have both ionic and covalent bonds? Explain your answer. *(Chapter 8)*

74. How do you distinguish between a cation and an anion? *(Chapter 9)*

75. Name these ions. *(Chapter 9)*
a. PO_4^{3-}
b. Al^{3+}
c. Se^{2-}
d. NH_4^+

76. Name each substance. *(Chapter 9)*
a. SiO_2
b. K_2SO_4
c. H_2CO_3
d. MgS

77. Write the formula for each compound. *(Chapter 9)*
a. aluminum carbonate
b. nitrogen dioxide
c. potassium sulfide
d. manganese(II) chromate
e. sodium bromide

78. What is the mass, in grams, of a molecule of benzene (C_6H_6)? *(Chapter 10)*

79. How many grams of beryllium are in 147 g of the mineral beryl ($Be_3Al_2Si_6O_{18}$)? *(Chapter 10)*

80. What is the molecular formula of oxalic acid, molar mass 90 g/mol? Its percent composition is 26.7% C, 2.2% H, and 71.1% O. *(Chapter 10)*

81. How many moles is each of the following? *(Chapter 10)*
a. 47.8 g KNO_3
b. 2.22 L SO_2 (at STP)
c. 2.25×10^{22} molecules PCl_3

82. Write a balanced chemical equation for each reaction. *(Chapter 11)*
a. When heated, lead(II) nitrate decomposes to form lead(II) oxide, nitrogen dioxide, and molecular oxygen.
b. The complete combustion of isopropyl alcohol (C_3H_7OH) produces carbon dioxide and water vapor.
c. When a mixture of aluminum and iron(II) oxide is heated, metallic iron and aluminum oxide are produced.

83. Balance each equation. *(Chapter 11)*
a. $Ba(NO_3)_2(aq) + Na_2SO_4(aq) \longrightarrow BaSO_4(s) + NaNO_3(aq)$
b. $AlCl_3(aq) + AgNO_3(aq) \longrightarrow AgCl(s) + Al(NO_3)_3(aq)$
c. $H_2SO_4(aq) + Mg(OH)_2(aq) \longrightarrow MgSO_4(aq) + H_2O(l)$

84. Write a net ionic equation for each reaction in Problem 83. *(Chapter 11)*

85. Identify the spectator ions in each reaction in Problem 83. *(Chapter 11)*

86. Write a balanced chemical equation for the complete combustion of ribose, $C_5H_{10}O_5$. *(Chapter 11)*

84. **a.** $Ba^{2+}(aq) + SO_4^{2-}(aq) \rightarrow BaSO_4(s)$
b. $Ag^+(aq) + Cl^-(aq) \rightarrow AgCl(s)$
c. $H^+(aq) + OH^-(aq) \rightarrow H_2O(l)$

85. **a.** sodium ion and nitrate ion
b. aluminum ion and nitrate ion
c. magnesium ion and sulfate ion

86. $C_5H_{10}O_5 + 5O_2 \rightarrow 5CO_2 + 5H_2O$

Standardized Test Prep

Test-Taking Tip

Anticipate the answer. Use what you know to predict what you think the answer should be. Then look to see if your answer, or one much like it, is given as an option.

Select the choice that best answers each question or completes each statement.

1. Nitric acid is formed by the reaction of nitrogen dioxide with water.

   $$3NO_2(g) + H_2O(l) \longrightarrow NO(g) + 2HNO_3(aq)$$

   How many moles of water are needed to react with 8.4 mol NO_2?
   - **a.** 2.8 mol
   - **b.** 3.0 mol
   - **c.** 8.4 mol
   - **d.** 25 mol

2. Phosphorus trifluoride is formed from its elements.

   $$P_4(s) + 6F_2(g) \longrightarrow 4PF_3(g)$$

   How many grams of fluorine are needed to react with 6.20 g of phosphorus?
   - **a.** 2.85 g
   - **b.** 5.70 g
   - **c.** 11.4 g
   - **d.** 37.2 g

3. Magnesium nitride is formed in the reaction of magnesium metal with nitrogen gas.

   $$3Mg(s) + N_2(g) \longrightarrow Mg_3N_2(s)$$

   The reaction of 4.0 mol of nitrogen with 6.0 mol of magnesium produces
   - **a.** 2.0 mol of Mg_3N_2 and 2.0 mol of excess N_2.
   - **b.** 4.0 mol of Mg_3N_2 and 1.0 mol of excess Mg.
   - **c.** 6.0 mol of Mg_3N_2 and 3.0 mol of excess N_2.
   - **d.** no product because the reactants are not in the correct mole ratio.

Questions 4 and 5 involve the reaction between diatomic element P and diatomic element Q to form the compound P_3Q.

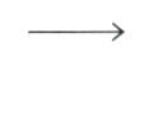

4. Write a balanced equation for the reaction between element P and element Q.
5. Based on the atomic windows below, identify the limiting reagent.

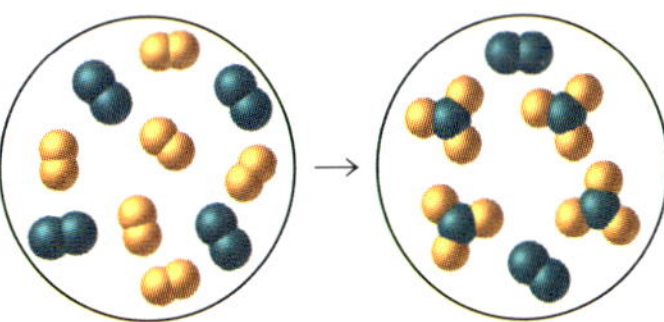

For each question there are two statements. Decide whether each statement is true or false. Then decide whether Statement II is a correct explanation for Statement I.

	Statement I		Statement II
6.	Every stoichiometry calculation uses a balanced equation.	BECAUSE	Every chemical reaction obeys the law of conservation of mass.
7.	A percent yield is always greater than 0% and less than 100%.	BECAUSE	The actual yield in a reaction is never more than the theoretical yield.
8.	The amount of the limiting reagent left after a reaction is zero.	BECAUSE	The limiting reagent is completely used up in a reaction.
9.	The coefficients in a balanced equation represent the relative masses of the reactants and products.	BECAUSE	The mass of the reactants must equal the mass of the products in a chemical reaction.
10.	A mole ratio is always written with the larger number in the numerator.	BECAUSE	A mole ratio will always be greater than 1.

Standardized Test Prep

1. a
2. c
3. a
4. $3P_2 + Q_2 \rightarrow 2P_3Q$
5. P_2 is the limiting reagent.
6. True, True, correct explanation
7. False, True
8. True, True, correct explanation
9. False, True
10. False, False

CHAPTER 13

States of Matter Planning Guide

SECTION OBJECTIVES	STANDARDS NATIONAL	STANDARDS STATE	ACTIVITIES and LABS	
13.1 The Nature of Gases, pp. 385–389 1 block or 2 periods **13.1.1 Describe** the assumptions of the kinetic theory as it applies to gases. **13.1.2 Interpret** gas pressure in terms of kinetic theory. **13.1.3 Define** the relationship between Kelvin temperature and average kinetic energy.	A-1, A-2, B-2, B-5		**SE** Inquiry Activity: Observing Gas Pressure, p. 384 **TE** Teacher Demo: Elastic Collisions, p. 386 **TE** Teacher Demo: Air Pressure, p. 387 **TE** Class Activity: Particle Motion and Pressure, p. 388	L2 L2 L2 L2
13.2 The Nature of Liquids, pp. 390–395 1 block or 2 periods **13.2.1 Identify** factors that determine physical properties of a liquid. **13.2.2 Define** evaporation in terms of kinetic energy. **13.2.3 Describe** the equilibrium between a liquid and its vapor. **13.2.4 Identify** the conditions at which boiling occurs.	A-2, B-2, B-5		**TE** Teacher Demo: Vapor Pressure, p. 392 **TE** Class Activity: Water Versus Alcohol, p. 393 **TE** Class Activity: Temperature and Boiling, p. 394 **SSLM** Lab 20: Absorption of Water by Paper Towels: A Consumer Lab	L2 L2 L2 L2
13.3 The Nature of Solids, pp. 396–400 1 block or 2 periods **13.3.1 Evaluate** how the way particles are organized explains the properties of solids. **13.3.2 Identify** the factors that determine the shape of a crystal. **13.3.3 Explain** how allotropes of an element are different.	A-1, A-2, B-2, B-3, E-2, G-1		**SE** Small-Scale Lab: The Behavior of Liquids and Solids, p. 400 **TE** Teacher Demo: Crystalline Solid Model, p. 397 **LM** Lab 21: Allotropic Forms of Sulfur	L2 L2 L2
13.4 Changes of State, pp. 401–405 1 block or 2 periods **13.4.1 Identify** the conditions necessary for sublimation. **13.4.2 Describe** how equilibrium conditions are represented in a phase diagram.	A-1, A-2, B-2, B-3, E-2, G-1		**SE** Quick Lab: Sublimation, p. 402 **LM** Lab 22: Changes of Physical State **LP** Lab Practical 13-1: Changes of Physical State	L2 L2 L2

Ability Levels

- **L1** For students who need additional help
- **L2** For all students
- **L3** For students who need to be challenged

Components

SE	Student Edition	**LM**	Laboratory Manual	**GCP**	Graphing Calculator Problems	**T**	Transparencies
TE	Teacher's Edition	**SSLM**	Small-Scale Chemistry Lab Manual	**TTT**	Test-Taking Tips with Transparencies, Grades 9–12 Science	**CA**	Chem Alive and Chemistry Field Trips DVD
GRSW	Guided Reading and Study Workbook	**LP**	Laboratory Practicals	**STP**	Standardized Test Preparation Workbook	**iT**	Interactive Textbook with ChemASAP
CTR	Core Teaching Resources	**PLM**	Probeware Lab Manual			**GO**	Go Online Internet Resources
SSG	Spanish Study Guide	**VCL**	Virtual Chemistry Labs				
		CTB	Computer Test Bank				

RESOURCES PRINT and TECHNOLOGY	PROBLEMS and ASSESSMENT
GRSW Chapter 13, Section 1 **L1** **T** T139–T141 **L2** **GO** Chapter 13, Section 1 **L2**	**SP** 13.1, Converting between Units of Pressure **SE** Section 13.1 Assessment, p. 389 **iT** Section 13.1
GRSW Chapter 11, Section 2 **L1** **T** T142–T144 **L2**	**SE** Section 13.2 Assessment, p. 395 **iT** Section 13.2
GRSW Chapter 13, Section 3 **L1** **T** T145–T147 **L2** **GO** Chapter 13, Section 3 **L2**	**SE** Section 13.3 Assessment, p. 399 **iT** Section 13.3
GRSW Chapter 13, Section 4 **L1** **T** T148–T149 **L2** **GO** Chapter 13, Section 4 **L2** **GO** Chapter 13, Section 4 **L2**	**SE** Section 13.4 Assessment, p. 404 **iT** Section 13.4

Web Code: cdn-1131
Web Code: cdn-1133
Web Code: cdn-1134

Web Code: cdb-1134

MATERIALS List

Quantities for each group

STUDENT EDITION

Inquiry Activity, p. 384
small glass with a smooth, even rim; water; index card

Small-Scale Lab, p. 400
plastic Petri dish, water, ice, rubbing alcohol, calcium chloride, graph paper (1-cm cycle), bromthymol blue solution (BTB), vinegar, aqueous ammonia

Quick Lab, p. 402
small pieces of solid air freshener, small shallow container, 2 clear plastic cups (8 oz each), hot tap water, ice, 3 thick cardboard strips

TEACHER'S EDITION

Teacher Demo, p. 386
Newtonian cradle

Teacher Demo, p. 387
empty aluminum beverage can, water, hot plate, foil or other material for sealing opening of can

Class Activity, p. 388
cardboard containers (such as half-pint milk cartons), ball bearings

Teacher Demo, p. 392
manometer, 1-L flask with 2-hole stopper, separatory funnel, short piece of glass tubing bent at a 90° angle, rubber hose, 50 mL acetone, ice water

Class Activity, p. 393
2 cotton balls, water, rubbing alcohol

Class Activity, p. 394
thermometer, beaker, water, ring stand, thermometer clamp, hot plate

Class Activity, p. 397
samples of wallpaper with repeating patterns

Teacher Demo, p. 398
dishwashing detergent, watch glass, overhead projector, drinking straw

Chapter Assessment

CHAPTER ASSESSMENT

SE	Chapter Assessment, pp. 407–410
CTB	Chapter 13
iT	Chapter 13

STANDARDIZED TEST PREP

SE	Chapter 13, p. 411
STP	Topic 4

iT Interactive Textbook with Assessment at PHSchool.com

CHAPTER 13

States of Matter

INQUIRY Activity

Observing Gas Pressure 

Purpose

Students will recognize that air exerts pressure.

Prep Time 5 minutes
Class Time 10 minutes

Think About It

1. The index card sticks to the glass and the water remains in the inverted glass.
2. The number of grams of water equals the number of milliliters of water in the glass.
3. The external atmospheric pressure is greater than the pressure of the water in the inverted glass.

Extensions

- Have students repeat the activity, substituting a saucer or pan of water for the index card. Compare the assembly to a barometer.
- Ask students to calculate the total mass of water that could be held in the inverted glass. Explain that atmospheric pressure at sea level is about 14.7 psi (pounds per square inch) or 1.03 kilograms per square centimeter. Have students express their results in both psi and kg/cm^2.

After students read Section 13.1, ask them to evaluate and revise their answers to Question 3.

CHAPTER 13

States Of Matter

On a hot summer day ice cream turns quickly into a liquid.

INQUIRY Activity

Observing Gas Pressure

Materials

small glass with a smooth, even rim; water; index card

Procedure

1. Fill the glass to its rim with water.
2. Place the index card on top of the glass.
3. Working over a sink, use one hand to press the index card firmly to the top of the glass. Then quickly invert the glass, keeping your hand in place.
   CAUTION *Wipe up any spill immediately.*
4. Remove your hand from the index card.

Think About It

1. What did you observe when you removed your hand?
2. Measure the volume of water in the glass. How many grams of water are in the glass? (Remember that 1 mL H_2O = 1 g H_2O.)
3. What keeps the water inside the glass?

13.1 The Nature of Gases

Connecting to Your World

You are walking your dog in the woods. Suddenly your dog begins to bark and run toward what you think is a black cat. But before you realize that the "cat" is not a cat, the damage is done. The skunk has released its spray! Within seconds you smell that all-too-familiar foul odor. In this section, you will discover some general characteristics of gases that help explain how odors travel through the air, even on a windless day.

Kinetic Theory and a Model for Gases

The word *kinetic* refers to motion. The energy an object has because of its motion is called **kinetic energy.** According to the **kinetic theory,** all matter consists of tiny particles that are in constant motion. The particles in a gas are usually molecules or atoms. The kinetic theory as it applies to gases includes the following fundamental assumptions about gases.

The particles in a gas are considered to be small, hard spheres with an insignificant volume. Within a gas, the particles are relatively far apart compared with the distance between particles in a liquid or solid. Between the particles, there is empty space. No attractive or repulsive forces exist between the particles. The motion of one particle in a gas is independent of the motion of all the other particles.

The motion of the particles in a gas is rapid, constant, and random. As a result, gases fill their containers regardless of the shape and volume of the containers. An uncontained gas can spread out into space without limit. The particles travel in straight-line paths until they collide with another particle, or another object, such as the wall of their container. The particles change direction only when they rebound from collisions with one another or with other objects.

Measurements indicate that the average speed of oxygen molecules in air at 20°C is an amazing 1700 km/h! At these high speeds, the odor from a hot cheese pizza in Washington, D.C., should reach Mexico City in about 115 minutes. That does not happen, however, because the molecules responsible for the odor are constantly striking molecules in air and rebounding in other directions. Their path of uninterrupted travel in a straight line is very short. The aimless path the molecules take is called a random walk.

All collisions between particles in a gas are perfectly elastic. During an elastic collision, kinetic energy is transferred without loss from one particle to another, and the total kinetic energy remains constant. The diagrams in Figure 13.1 on the next page illustrate the assumptions of kinetic theory as applied to gases.

Guide for Reading

Key Concepts

- What are the three assumptions of the kinetic theory as it applies to gases?
- How does kinetic theory explain gas pressure?
- What is the relationship between the temperature in kelvins and the average kinetic energy of particles?

Vocabulary

kinetic energy
kinetic theory
gas pressure
vacuum
atmospheric pressure
barometer
pascal (Pa)
standard atmosphere (atm)

Reading Strategy

Relating Text and Visuals As you read, look closely at Figure 13.3. Explain how this illustration helps you understand how kinetic energy is distributed among the particles of a gas at two different temperatures.

Word *Origins*

Kinetic comes from the Greek word *kinetos,* meaning "to move." Kinetic energy is the energy an object has because of its motion. **Some sculptures are kinetic. What characteristic do they share?**

Section Resources

Print

- **Guided Reading and Study Workbook,** Section 13.1
- **Core Teaching Resources,** Section 13.1 Review
- **Transparencies,** T139–T141

Technology

- **Interactive Textbook with ChemASAP,** Problem-Solving 13.1, Animation 14, Assessment 13.1
- **Go Online,** Section 13.1

13.1

1 FOCUS

Objectives

13.1.1 Describe the assumptions of the kinetic theory as it applies to gases.

13.1.2 Interpret gas pressure in terms of kinetic theory.

13.1.3 Define the relationship between Kelvin temperature and average kinetic energy.

Guide for Reading

Build Vocabulary L2

Word Forms The word *vacuum* comes from the Latin *vacare,* meaning "to be empty." Discuss what happens in a vacuum cleaner *(Suction occurs as matter rushes in to fill a vacuum.)*

Reading Strategy L2

Active Comprehension Read the first paragraph of *Kinetic Energy and Temperature.* Ask students what more they would like to know about this topic. Students can look for answers to their questions as they read the rest of *Kinetic Energy and Temperature.*

2 INSTRUCT

Connecting to Your World

Ask, **How can you smell a skunk's odor?** *(Molecules emitted by the skunk travel through the air to receptors in your nose.)*

Kinetic Theory and a Model for Gases

Discuss L2

Because multiple units are introduced for pressure, you many want to review the use of conversion factors.

Word *Origins* L2

A kinetic sculpture has moving parts. Some are free-hanging mobiles; some are stabiles with motors.

Gas Pressure

TEACHER Demo

Elastic Collisions 

Purpose Students will differentiate elastic colllisions from perfectly elastic collisions.

Materials Newtonian cradle (a device in which small steel balls are suspended by thin nylon tethers to horizontal wooden sticks)

Procedure Explain that in an elastic collision, kinetic energy is transferred between the objects that collide. In a perfectly elastic collision, the total amount of kinetic energy remains constant. Then pull back one of the balls and let it fall into the other balls. After the balls come to rest, ask, **Were the collisions between the balls elastic?** *(Yes, because kinetic energy was transferred with each collision.)* **Why did the balls eventually stop moving?** *(The collisions were not perfectly elastic; some kinetic energy was lost as heat during each collision.)*

Remind students that the kinetic theory assumes that all collisions between particles in a gas are perfectly elastic. (In Chapter 14, students will learn that the assumptions of the kinetic theory do not hold true for real gases under all conditions.)

Use Visuals 

Figure 13.2 Point out the parts of the barometer and explain how the barometer is used to measure atmospheric pressure. Ask, **What is the air pressure on top of Mount Everest in kPa?** *(33.7 kPa)*

FYI

In Chapter 14, students will study factors that affect gas pressure—amount of gas, temperature, and volume.

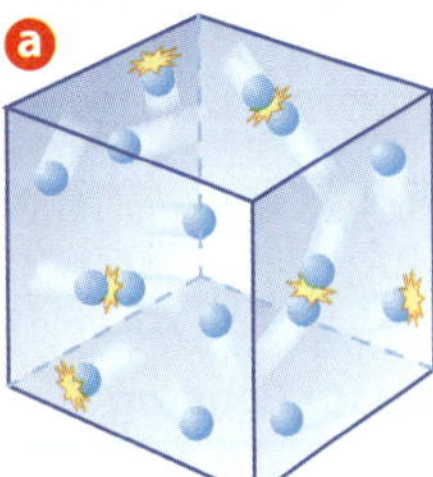
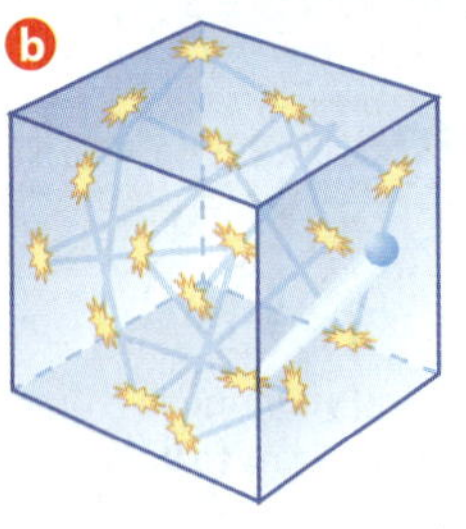

Figure 13.1 Gases share some general characteristics. **a** The rapid, constant motion of particles in a gas causes them to collide with one another and with the walls of their container. **b** The particles travel in straight-line paths between collisions. **c** A gas fills all the available space in its container.

Gas Pressure

A helium-filled balloon maintains its shape because of the pressure of the gas within it. **Gas pressure** results from the force exerted by a gas per unit surface area of an object. What causes this force? Moving bodies exert a force when they collide with other bodies. Although a single particle in a gas is a moving body, the force it exerts is extremely small. Yet it is not hard to imagine that simultaneous collisions involving many particles would produce a measurable force on an object. **Gas pressure is the result of simultaneous collisions of billions of rapidly moving particles in a gas with an object.** If there are no particles, there cannot be collisions. Consequently, there is no pressure. An empty space with no particles and no pressure is called a **vacuum.**

A gas pressure that you are familiar with is that caused by a mixture of gases—air. Air exerts pressure on Earth because gravity holds the particles in air in Earth's atmosphere. **Atmospheric pressure** results from the collisions of atoms and molecules in air with objects. Atmospheric pressure decreases as you climb a mountain because the density of Earth's atmosphere decreases as the elevation increases.

A **barometer** is a device that is used to measure atmospheric pressure. Figure 13.2 shows an early type of mercury barometer. The height of the mercury column in the tube depends on the pressure exerted by particles in air colliding with the surface of the mercury in the dish. Atmospheric pressure depends on weather and on altitude. In fair weather at sea level, the atmospheric pressure is sufficient to support a mercury column about 760 mm high.

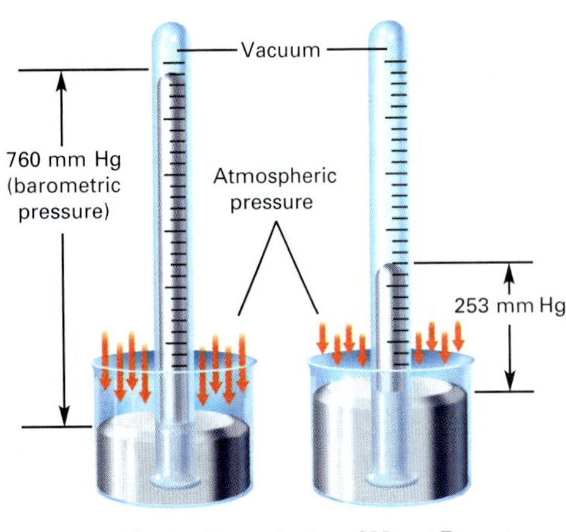

Figure 13.2 At sea level, air exerts enough pressure to support a 760-mm column of mercury. On top of Mount Everest, at 9000 m, the air exerts only enough pressure to support a 253-mm column of mercury.
Calculating ***What is the decrease in pressure from sea level to the top of Mount Everest?***

Differentiated Instruction

Gifted and Talented 

Encourage students to examine critically how barometric pressure is reported in local newspapers and on television. Ask, **What units are commonly used in weather reports to express barometric pressure?** *(Unfortunately, units are rarely stated in the reports. Pressure is usually reported in inches of mercury. Students may infer the unit based on the numerical values of the stated pressures.)* Bring in the weather sections of newspapers and have students determine today's pressure in mm Hg, in kPa, and in atm. Students may wish to track the daily pressure and display the changes on a graph.

The SI unit of pressure is the **pascal (Pa).** It represents a very small amount of pressure. For example, normal atmospheric pressure is about 100,000 Pa, that is, 100 kilopascals (kPa). Two older units of pressure are still commonly used. These units are millimeters of mercury (mm Hg) and atmospheres. One **standard atmosphere (atm)** is the pressure required to support 760 mm of mercury in a mercury barometer at 25°C. The numerical relationship among the three units is given below.

$$1 \text{ atm} = 760 \text{ mm Hg} = 101.3 \text{ kPa}$$

In the case of gases, it is important to be able to relate measured values to standards. Recall that the standard temperature and pressure (STP) are defined as a temperature of 0°C and a pressure of 101.3 kPa, or 1 atm.

SAMPLE PROBLEM 13.1

Converting Between Units of Pressure

A pressure gauge records a pressure of 450 kPa. What is this measurement expressed in atmospheres and millimeters of mercury?

1 Analyze ***List the knowns and the unknowns.***

Knowns
- pressure = 450 kPa
- 1 atm = 101.3 kPa
- 1 atm = 760 mm Hg

Unknowns
- pressure = ? atm
- pressure = ? mm Hg

For converting kPa ⟶ atm, the appropriate conversion factor is

$$\frac{1 \text{ atm}}{101.3 \text{ kPa}}$$

For converting kPa ⟶ mm Hg, the appropriate conversion factor is

$$\frac{760 \text{ mm Hg}}{101.3 \text{ kPa}}$$

2 Calculate ***Solve for the unknowns.***

$$450 \text{ kPa} \times \frac{1 \text{ atm}}{101.3 \text{ kPa}} = 4.4 \text{ atm}$$

$$450 \text{ kPa} \times \frac{760 \text{ mm Hg}}{101.3 \text{ kPa}} = 3400 \text{ mm Hg} = 3.4 \times 10^3 \text{ mm Hg}$$

3 Evaluate ***Do the results make sense?***

Because the first conversion factor is much less than 1 and the second much greater than 1, it makes sense that the values expressed in atm and mm Hg are respectively smaller and larger than the value expressed in kPa.

Practice Problems

1. What pressure, in kilopascals and in atmospheres, does a gas exert at 385 mm Hg?
2. The pressure at the top of Mount Everest is 33.7 kPa. Is that pressure greater or less than 0.25 atm?

CHEMath

Using a Calculator

After you analyze a sample problem, you can use a calculator to solve for the unknown. A calculator provides the four basic arithmetic functions of addition (+), subtraction (−), multiplication (×), and division (÷). You can also raise to a power (x^2), take the square root ($\sqrt{x}$), and take the logarithm (*log*). You will need to be able to enter measurements written in scientific notation. On many calculators, you will use the *EE* key or *EXP* key to enter such measurements.

Math Handbook

For help using a calculator, go to page R62.

Problem-Solving 13.1 Solve Problem 1 with the help of an interactive guided tutorial.

with **ChemASAP**

Sample Problem 13.1

Answers

1. 51.3 kPa, 0.507 atm
2. 33.7 kPa is greater than 0.25 atm

Practice Problems Plus 

What pressure, in mm Hg and atm, does a sample of neon gas exert at 75.0 kPa? *(563 mm Hg, 0.740 atm)*

CHEMath

Using a Calculator

When students use a calculator to convert between units, they should evaluate the answer to make sure that they entered the data correctly and that they used an appropriate conversion factor.

Math Handbook

For a math refresher and practice, direct students to using a calculator, page R62.

TEACHER Demo

Air Pressure 

Purpose Students will become aware of the tremendous pressure exerted by Earth's atmosphere.

Materials empty aluminum beverage can, water, hot plate, foil or other material for sealing opening of can

Procedure Explain to students that they live at the bottom of a very deep "ocean" of air. Fill an empty aluminum beverage container with water to a depth of about 1 to 2 cm. Set the can on a hot plate and bring the water to a boil. Allow the water to boil vigorously for several minutes. Remove the can from the heat source, seal the opening, and allow the can to cool at room temperature or invert the can in a pan of cool water.

Expected Outcome Once the water vapor inside the can condenses, the atmospheric pressure will crush the can. Ask students to interpret their observations.

Facts and Figures

Aneroid Barometers

Unlike traditional barometers, aneroid barometers do not balance atmospheric pressure against a liquid of known density. In fact, they contain no liquid and are independent of gravity. If possible, bring one to class and explain its components. Have interested students research the advantages of aneroid barometers. *(They are compact, self enclosed, and easily transported. They can operate under conditions not suited to mercury barometers.)* Interested students may wish to construct an aneroid barometer and demonstrate its use.

Answers to...

Figure 13.2 507 mm Hg

Kinetic Energy and Temperature

Interpreting Graphs 

a. a point near the peak of the curve **b.** The curves have the same overall shape, but the curve for hot water is wider with a lower peak. **c.** At an even higher temperature, the graph would be wider than the red curve with a lower peak; at an even lower temperature, the graph would be narrower than the blue curve with a higher peak.

Enrichment Question 

Compare the kinetic energy of the particles represented by the points on the red and blue curves where the curves intersect. *(The particles have the same kinetic energy.)*

CLASS Activity

Kinetic Energy and Frequency of Collisions 

Purpose Students will be able to visualize what is happening in a gas as the average kinetic energy increases.

Materials cardboard containers (e.g., half-pint milk cartons), ball bearings

Procedure Give students cardboard boxes containing the same number of ball bearings. Have students shake the boxes slowly at first. Then have them gradually increase the rate of shaking and observe what happens. Tell students that the shaking is analogous to adding energy to the particles in a gas. The greater the rate of the shaking, the more energy is added. Then ask how an increase in energy affects the frequency of collisions.

Expected Outcome Student should hear and feel the increase in number of collisions as the rate of shaking (or the amount of energy added) increases.

Download a worksheet on **Kinetic Theory** for students to complete, and find additional teacher support from NSTA SciLinks.

For: Links on Kinetic Theory
Visit: www.SciLinks.org
Web Code: cdn-1131

Kinetic Energy and Temperature

As a substance is heated, its particles absorb energy, some of which is stored within the particles. This stored portion of the energy, or potential energy, does not raise the temperature of the substance. The remaining absorbed energy speeds up the particles—that is, increases their kinetic energy. This increase in kinetic energy results in an increase in temperature.

Average Kinetic Energy The particles in any collection of atoms or molecules at a given temperature have a wide range of kinetic energies. Most of the particles have kinetic energies somewhere in the middle of this range. Therefore, average kinetic energy is used when discussing the kinetic energy of a collection of particles in a substance. At any given temperature the particles of all substances, regardless of physical state, have the same average kinetic energy. For example, the ions in table salt, the molecules in water, and the atoms in helium all have the same average kinetic energy at room temperature even though the three substances are in different physical states.

Figure 13.3 shows the distribution of kinetic energies of water molecules at two different temperatures. The blue curve shows the distribution of kinetic energy among the water molecules in cold water. The red curve shows the distribution of kinetic energy among the water molecules in hot water. In both cases, most of the molecules have intermediate kinetic energies, which are close to the average value. Notice that at the higher temperature there is a wider range of kinetic energies.

There is a relationship between the average kinetic energy of the particles in a substance and the substance's temperature. An increase in the average kinetic energy of the particles causes the temperature of a substance to rise. As a substance cools, the particles tend to move more slowly, and their average kinetic energy declines.

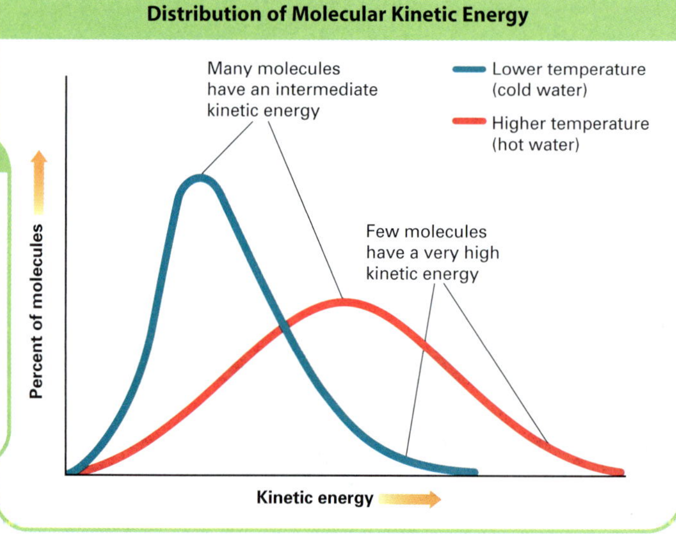

Figure 13.3 The red and blue curves show the kinetic energy distributions of a typical collection of molecules at two different temperatures.

INTERPRETING GRAPHS

a. Inferring Which point on each curve represents the average kinetic energy?
b. Analyzing Data Compare the shapes of the curves for cold water and hot water.
c. Predicting What would happen to the shape of the curve if the water temperature were even higher? Even lower?

Figure 13.4 In this vacuum chamber, scientists cooled sodium vapor to nearly absolute zero. To keep the atoms from sticking to the walls of the chamber, the scientists used magnetism and gravity to trap the atoms 0.5 cm above the coil in the center of the chamber. The coil is shown at about two times its actual size.

You could reasonably expect the particles of all substances to stop moving at some very low temperature. The particles would have no kinetic energy at that temperature because they would have no motion. Absolute zero (0 K, or −273.15°C) is the temperature at which the motion of particles theoretically ceases. There is no temperature lower than absolute zero. Absolute zero has never been produced in the laboratory. However, near-zero temperatures of about 0.0000000005 K (0.5×10^{-9} K), which is 0.5 nanokelvin, have been achieved in the vacuum chamber shown in Figure 13.4.

Average Kinetic Energy and Kelvin Temperature The Kelvin temperature scale reflects the relationship between temperature and average kinetic energy. **The Kelvin temperature of a substance is directly proportional to the average kinetic energy of the particles of the substance.** For example, the particles in helium gas at 200 K have twice the average kinetic energy as the particles in helium gas at 100 K. The effects of temperature on particle motion in liquids and solids are more complex than in gases.

Animation 14 Observe particles in motion and discover the connection between temperature and kinetic energy.

with ChemASAP

What happens to the temperature of a substance when the average kinetic energy of its particles decreases?

13.1 Section Assessment

3. **Key Concept** Briefly describe the assumptions of kinetic theory as applied to gases.
4. **Key Concept** Use kinetic theory to explain what causes gas pressure.
5. **Key Concept** How is the Kelvin temperature of a substance related to the average kinetic energy of its particles?
6. Convert the following pressures to kilopascals.
   a. 0.95 atm **b.** 45 mm Hg
7. A cylinder of oxygen gas is cooled from 300 K (27°C) to 150 K (−123°C). By what factor does the average kinetic energy of the oxygen molecules in the cylinder decrease?

Writing Activity

Describing Motion in a Gas Write a paragraph describing the behavior of an oxygen molecule in a sealed container of air. Include what happens when the molecule collides with another molecule or the walls of the container.

Assessment 13.1 Test yourself on the concepts in Section 13.1.

with ChemASAP

3 ASSESS

Evaluate Understanding L2

To assess students' comprehension of kinetic theory as it applies to gases, ask, **What is kinetic energy?** *(energy due to the motion of an object)* **How is the average kinetic energy of a collection of particles related to temperature?** *(Average kinetic energy is directly proportional to the Kelvin temperature. Higher temperatures reflect a greater average kinetic energy.)*

Reteach L1

Have students look at Figure 13.3. Ask, **What does the blue curve represent?** *(The distribution of kinetic energy in cold water.)* **What does the red curve represent?** *(The distribution of kinetic energy in hot water.)* **In which sample is the average kinetic energy of the particles higher?** *(hot water).*

Writing Activity

Student paragraphs should describe straight-line, random motion until particles collide with each other and with the sides of the container. After a collision, the direction of motion changes.

If your class subscribes to the Interactive Textbook, use it to review key concepts in Section 13.1.

with ChemASAP

Section 13.1 Assessment

3. A gas is composed of tiny particles whose motion is rapid, constant, and random. Collisions between particles are perfectly elastic.
4. Gas pressure is the result of simultaneous collisions of billions of rapidly moving particles with an object.
5. The Kelvin temperature is directly proportional to the average kinetic energy of the particles.
6. **a.** 96 kPa
   b. 6.0 kPa
7. by one-half

Answers to...

 The temperature decreases.

13.2

1 FOCUS

Objectives

13.2.1 Identify factors that determine physical properties of a liquid.

13.2.2 Define evaporation in terms of kinetic energy.

13.2.3 Describe the equilibrium between a liquid and its vapor.

13.2.4 Identify the conditions at which boiling occurs.

Guide for Reading

Build Vocabulary 

Concept Map After students read *Vapor Pressure*, have them make a concept map using the terms *evaporation, condensation, dynamic equilibrium,* and *vapor pressure.*

Reading Strategy 

Compare and Contrast Have students write a short paragraph comparing and contrasting evaporation and boiling.

2 INSTRUCT

Connecting to Your World

Have students read the introduction and study the photograph of a lava flow. Ask, **Is lava a liquid or a solid?** *(a liquid)*

A Model for Liquids

Use Visuals L1

Figure 13.5 After students study the figure ask, **How are the colored water and the bromine vapor similar?** *(They can both flow and take the shape of their beakers.)* **How are water and bromine vapor different?** *(Water has a definite volume. The vapor doesn't have a definite volume.)* **According to kinetic theory, why doesn't the water spontaneously flow out of the beaker?** *(Attractions among water molecules hold the molecules together.)*

13.2 The Nature of Liquids

Guide for Reading

 **Key Concepts**

- What factors determine the physical properties of a liquid?
- What is the relationship between evaporation and kinetic energy?
- When can a dynamic equilibrium exist between a liquid and its vapor?
- Under what conditions does boiling occur?

Vocabulary

vaporization
evaporation
vapor pressure
boiling point
normal boiling point

Reading Strategy

Using Prior Knowledge Before you read, write a definition for the term *boiling point.* After you read this section, compare and contrast the definition of *boiling point* in the text with your original definition.

Connecting to Your World

The Kilauea volcano in Hawaii is the most active volcano in the world. It has been erupting for centuries. The hot lava oozes and flows, scorching everything in its path, occasionally overrunning nearby houses. When the lava cools, it solidifies into rock. The properties of liquids are related to intermolecular interactions. In this section you will learn about some of the properties of liquids.

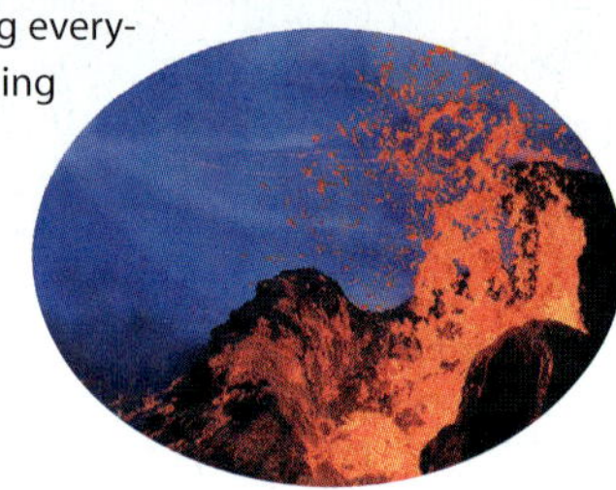

A Model for Liquids

According to the kinetic theory, both the particles in gases and the particles in liquids have kinetic energy. This energy allows the particles in gases and liquids to flow past one another, as shown in Figure 13.5. Substances that can flow are referred to as fluids. The ability of gases and liquids to flow allows them to conform to the shape of their containers.

There is a key difference between gases and liquids. According to the kinetic theory, there are no attractions between the particles in a gas. But the particles in a liquid are attracted to each other. These intermolecular attractions keep the particles in a liquid close together, which is why liquids have a definite volume. **The interplay between the disruptive motions of particles in a liquid and the attractions among the particles determines the physical properties of liquids.**

Intermolecular attractions also reduce the amount of space between the particles in a liquid. Thus liquids are much more dense than gases. Increasing the pressure on a liquid has hardly any effect on its volume. The same is true of solids. For that reason, liquids and solids are known as condensed states of matter.

Figure 13.5 Both liquids and gases can flow. The liquid on the left is colored water. The gas on the right is bromine vapor. If a gas is denser than air, it can be poured from one container into another. These pictures were taken in a fume hood because bromine is both toxic and corrosive.
Predicting ***Over time, what will happen to the gas in the uncovered beaker? Explain.***

Section Resources

Print

- **Guided Reading and Study Workbook,** Section 13.2
- **Core Teaching Resources,** Section 13.2 Review
- **Transparencies,** T142–T144
- **Small-Scale Chemistry Laboratory Manual,** Lab 20

Technology

- **Interactive Textbook with ChemASAP,** Animation 15, 16; Assessment 13.2

Evaporation

As you probably know from experience, water in an open container like the one in Figure 13.6a eventually moves into the air. The conversion of a liquid to a gas or vapor is called **vaporization.** When such a conversion occurs at the surface of a liquid that is not boiling, the process is called **evaporation.** Most of the molecules in a liquid don't have enough kinetic energy to overcome the attractive forces and escape into the gaseous state. **During evaporation, only those molecules with a certain minimum kinetic energy can escape from the surface of the liquid.** Even some of the particles that do escape collide with molecules in the air and rebound into the liquid.

You may have noticed that a liquid evaporates faster when heated. This occurs because heating the liquid increases the average kinetic energy of its particles. The added energy enables more particles to overcome the attractive forces keeping them in the liquid state. As evaporation occurs, the particles with the highest kinetic energy tend to escape first. The particles left in the liquid have a lower average kinetic energy than the particles that have escaped. The process is similar to removing the fastest runner from a race. The remaining runners have a lower average speed. As evaporation takes place, the liquid's temperature decreases. Therefore, evaporation is a cooling process.

You have observed the effects of evaporative cooling on hot days. When you perspire, water molecules in your perspiration absorb heat from your body and evaporate from your skin's surface. This evaporation leaves the remaining perspiration cooler. The perspiration that remains cools you further by absorbing more body heat.

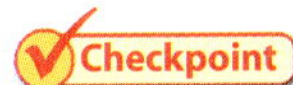 ***What happens to the rate of evaporation when a liquid is heated?***

Figure 13.6 The process of evaporation has a different outcome in an open system, such as a lake, than in a closed system, such as a terrarium. a In an open container, molecules that evaporate can escape from the container. b In a closed container, the molecules cannot escape. They collect as a vapor above the liquid. Some molecules condense back into a liquid.

Animation 15
Observe the phenomenon of evaporation from a molecular perspective.
with ChemASAP

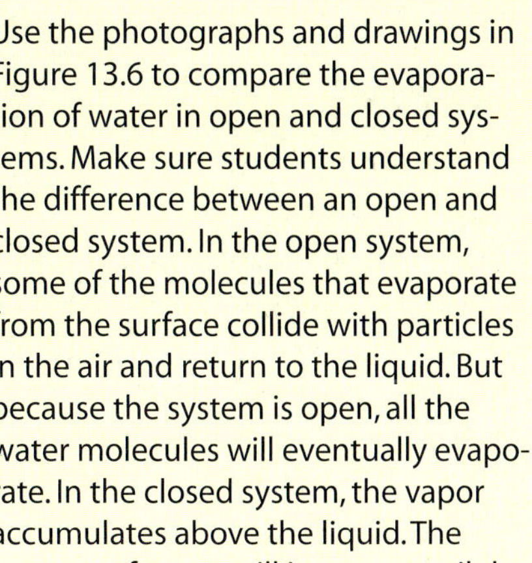

Evaporation

Discuss L2

Use the photographs and drawings in Figure 13.6 to compare the evaporation of water in open and closed systems. Make sure students understand the difference between an open and closed system. In the open system, some of the molecules that evaporate from the surface collide with particles in the air and return to the liquid. But because the system is open, all the water molecules will eventually evaporate. In the closed system, the vapor accumulates above the liquid. The amount of vapor will increase until the rates of evaporation and condensation are equal.

Answers to...

Figure 13.5 It will disperse in the surrounding air.

 It increases.

Vapor Pressure

Vapor Pressure L2

Purpose Students will observe how the vapor pressure of a liquid changes with temperature.

Materials manometer, 1-L flask with 2-hole stopper, separatory funnel, short piece of glass tubing bent at a 90° angle, rubber hose, 50 mL acetone, ice water

Safety To insert the glass tubing in the stopper, use glycerol. Wear gloves or hold the tubing in a towel to protect your hands. Be sure there are no open flames in the room because acetone is flammable.

Procedure You will need to obtain or construct a manometer and assemble the following apparatus. Insert a two-hole stopper into a 1-L flask. Into one hole of the stopper, insert a separatory funnel capable of containing 50 mL of acetone. Insert the bent glass tubing into the other hole. Use a rubber hose to connect the glass tubing to one end of the manometer. Be sure the hose fits both connections tightly. Use the manometer to measure the initial pressure. Then open the valve on the separatory funnel, and let the acetone drain into the flask. Have students observe the pressure in the flask. The maximum pressure observed is the vapor pressure of acetone at room temperature. Next, place the flask in ice water, and have students observe the change in pressure.

Expected Outcomes The pressure rises immediately as the acetone flows into the flask. The pressure falls when the flask is cooled.

Vapor Pressure

The evaporation of a liquid in a closed container differs from evaporation in an open container. No particles can escape into the outside air from the closed container in Figure 13.6b. When a partially filled container of liquid is sealed, some of the particles at the surface of the liquid vaporize. These particles collide with the walls of the sealed container and produce a vapor pressure. **Vapor pressure** is a measure of the force exerted by a gas above a liquid. Over time, the number of particles entering the vapor increases and some of the particles condense and return to the liquid state. The following equation summarizes the process.

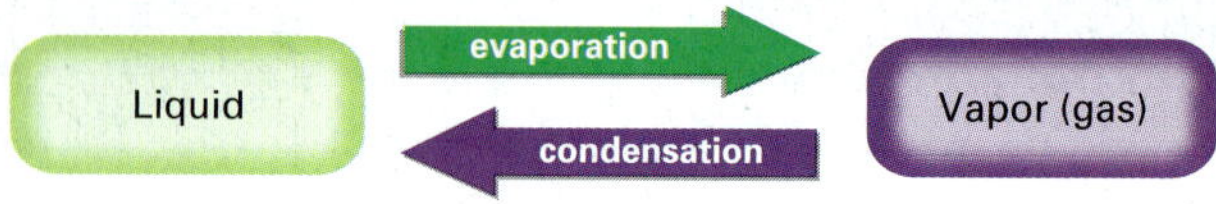

Eventually, the number of particles condensing will equal the number of particles vaporizing. The vapor pressure will then remain constant. **In a system at constant vapor pressure, a dynamic equilibrium exists between the vapor and the liquid. The system is in equilibrium because the rate of evaporation of liquid equals the rate of condensation of vapor.**

At equilibrium, the particles in the system continue to evaporate and condense, but there is no net change in the number of particles in the liquid or vapor. The sealed terrarium in Figure 13.6b is an example of a closed container at equilibrium. The moisture on the inner walls of the terrarium is a sign that equilibrium has been established. Particles that once evaporated are condensing, but other particles are evaporating to take their place.

Vapor Pressure and Temperature Change An increase in the temperature of a contained liquid increases the vapor pressure. This happens because the particles in the warmed liquid have increased kinetic energy. As a result, more of the particles will have the minimum kinetic energy necessary to escape the surface of the liquid. The particles escape the liquid and collide with the walls of the container at a greater frequency. Table 13.1 gives the vapor pressures of some common liquids at various temperatures. The vapor pressure data indicates how volatile a given liquid is, or how easily it evaporates. Of the three liquids shown, diethyl ether is the most volatile and water is the least volatile.

Table 13.1
Vapor Pressure (in kPa) of Three Substances at Different Temperatures

	0°C	20°C	40°C	60°C	80°C	100°C
Water	0.61	2.33	7.37	19.92	47.34	101.33
Ethanol	1.63	5.85	18.04	47.02	108.34	225.75
Diethyl ether	24.70	58.96	122.80	230.65	399.11	647.87

Differentiated Instruction

Special Needs

If students are having difficulty grasping the concept of a dynamic equilibrium, try this analogy. There are three students on one side of a room and three students on the other side. One group has 30 foam balls. The other group has none. The balls represent water molecules. Students in the first group begin to toss the balls toward the other group. This action represents evaporation. At first the only process occurring is evaporation. But soon students in the second group begin to toss the balls back. This action represents condensation. When the number of balls being tossed in each direction at any given moment is equal, a dynamic equilibrium exists.

Vapor Pressure Measurements The vapor pressure of a liquid can be determined with a device called a manometer. Figure 13.7 shows how a simple manometer works. One end of a U-shaped glass tube containing mercury is attached to a container. The other end of the tube is open to the surrounding atmosphere. When there is only air in the container, the pressure is the same on both sides of the tube and the mercury level is the same in each arm of the tube. When a liquid is added to the container, the pressure in the container increases due to the vapor pressure of the liquid. The vapor pressure of the liquid pushes the mercury on the container side of the U-tube. The levels of mercury in the U-tube are no longer the same. You can determine the vapor pressure in mm of Hg by measuring the difference between the two levels of mercury. As the vapor pressure increases, so does the difference between the two levels.

 ***What two physical changes occur in a partially filled, sealed container of liquid?***

Boiling Point

The rate of evaporation of a liquid from an open container increases as the liquid is heated. Heating allows a greater number of particles at the liquid's surface to overcome the attractive forces that keep them in the liquid state. The remaining particles in the liquid move faster and faster as they absorb the added energy. Thus, the average kinetic energy of the particles in the liquid increases and the temperature of the liquid increases. **When a liquid is heated to a temperature at which particles throughout the liquid have enough kinetic energy to vaporize, the liquid begins to boil.** Bubbles of vapor form throughout the liquid, rise to the surface, and escape into the air. The temperature at which the vapor pressure of the liquid is just equal to the external pressure on the liquid is the **boiling point** (bp).

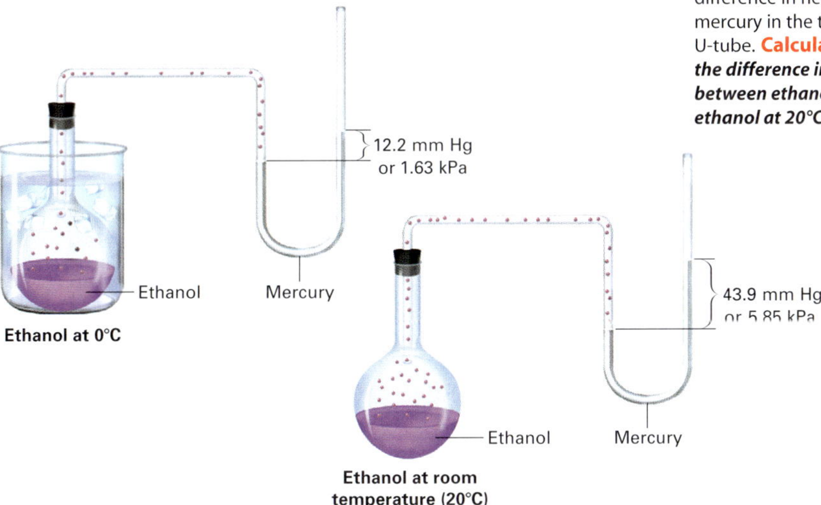

Figure 13.7 The vapor pressure of a contained liquid can be measured in a manometer. The vapor pressure is equal to the difference in height of the mercury in the two arms of the U-tube. **Calculating** ***What is the difference in vapor pressure between ethanol at 0°C and ethanol at 20°C?***

Water versus Alcohol L2

Purpose Students will infer the relative vapor pressures of two liquids.

Materials 2 cotton balls, water, rubbing alcohol

Procedure Use cotton balls to simultaneously dab spots of water and rubbing alcohol onto the chalkboard. Have the class observe what happens to the spots. Ask students to infer which liquid has the greater vapor pressure at room temperature and explain their reasoning.

Boiling Point

Discuss L2

Explain that boiling is different from evaporation because evaporation can occur at any temperature. Boiling takes place at specific temperatures and pressures. Atmospheric pressure opposes the formation of bubbles in a liquid. Bubbles form only when the vapor pressure inside a bubble is equal to atmospheric pressure.

Answers to...

Figure 13.7 4.22 kPa or 31.7 mm Hg

 evaporation of the liquid and condensation of the vapor

CLASS Activity

Temperature and Boiling L1

Purpose To dispel the misconception that the temperature of a liquid increases as it boils

Materials thermometer, beaker, ring stand, thermometer clamp, water, hot plate

Safety Students must wear safety goggles, an apron, and mitts to protect them from the boiling water and steam. **CAUTION** *Steam can cause a more severe burn than boiling water can.*

Procedure Have students measure the temperature every 30 seconds for several minutes as water is heated and then boiled. Have students use their data to construct a graph of temperature versus time. Ask students to write a short paragraph explaining their results.

Relate L2

Discuss the effect of altitude on cooking time. Ask, **Why does it take longer to cook food in water at high altitudes?** *(Because atmospheric pressure decreases with altitude, water boils at a lower temperature and the food doesn't cook as quickly.)* **Why does food take less time to cook in a pressure cooker?** *(In a pressure cooker the pressure can be greater than atmospheric pressure. Thus, the boiling point of water can be higher, and the food takes less time to cook.)*

Interpreting Graphs L2

a. 60°C **b.** about 20 kPa **c.** about 30 kPa

Enrichment Question L3

Ask students to construct a similar graph for diethyl ether using the data in Table 13.1. Have them use the graph to determine the normal boiling point of diethyl ether. *(34.5 °C)*

Sea Level Atmospheric pressure at the surface of water at 70°C is greater than its vapor pressure. Bubbles of vapor cannot form in the water, and it does not boil.

Sea Level At the boiling point, the vapor pressure is equal to atmospheric pressure. Bubbles of vapor form in the water, and it boils.

Atop Mount Everest At higher altitudes, the atmospheric pressure is lower than it is at sea level. Thus the water boils at a lower temperature.

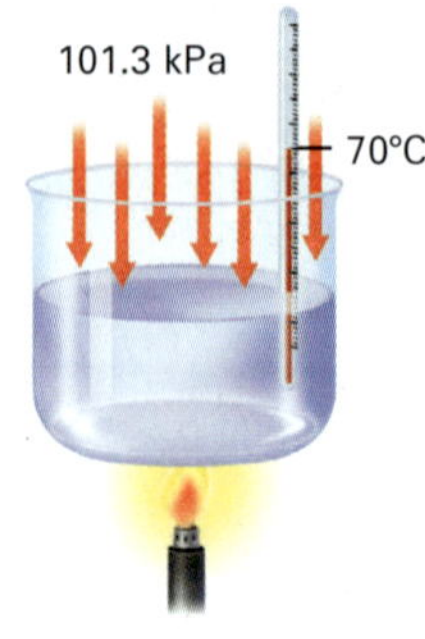

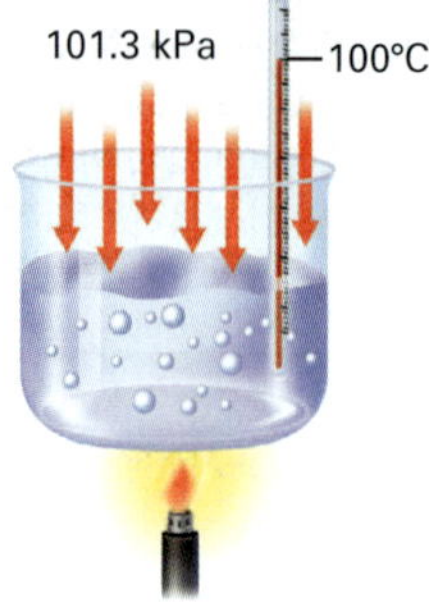

34 kPa
70°C

Figure 13.8 A liquid boils when the vapor pressure of particles within the liquid equals the atmospheric pressure. The boiling point varies with altitude.

Boiling Point and Pressure Changes Because a liquid boils when its vapor pressure is equal to the external pressure, liquids don't always boil at the same temperature. Figure 13.8 shows how a change in altitude affects the boiling point of water. Because atmospheric pressure is lower at higher altitudes, boiling points decrease at higher altitudes. In Denver, which is 1600 m above sea level, the average atmospheric pressure is 85.3 kPa. So water boils at about 95°C. In a pressure cooker, the vapor cannot escape and the vapor pressure increases. So water boils at a temperature above 100°C and food can cook more quickly.

Look at the vapor pressure versus temperature graph in Figure 13.9. You can use the graph to show how the boiling point of a liquid is related to vapor pressure. At a lower external pressure, the boiling point decreases. The particles in the liquid need less kinetic energy to escape from the liquid. At a higher external pressure, the boiling point increases. The particles in the liquid need more kinetic energy to escape from the liquid.

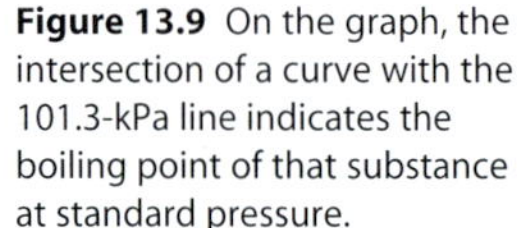

Figure 13.9 On the graph, the intersection of a curve with the 101.3-kPa line indicates the boiling point of that substance at standard pressure.

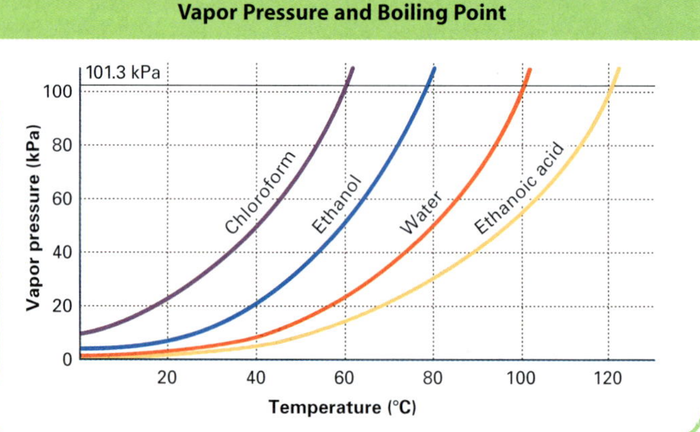

INTERPRETING GRAPHS

a. Analyzing Data What is the boiling point of chloroform at 101.3 kPa?

b. Analyzing Data What is the vapor pressure of ethanol at 40°C?

c. Analyzing Data What would atmospheric pressure need to be for ethanoic acid to boil at 80°C?

394 Chapter 13

Differentiated Instruction

Gifted and Talented L3

Ask students to use the data in Table 13.1 to construct a graph of vapor pressure versus temperature. They will need to plot log P versus $1/T$, where P is the vapor pressure in mm Hg and T is the absolute temperature.

Table 13.2

The Normal Boiling Points of Several Substances

Name and formula	Boiling Point (°C)
Carbon disulfide (CS_2)	46.0
Chloroform ($CHCl_3$)	61.7
Methanol (CH_4O)	64.7
Tetrachloromethane (CCl_4)	76.8
Ethanol (C_2H_6O)	78.5
Water (H_2O)	100.0

Boiling is a cooling process similar to evaporation. During boiling, the particles with the highest kinetic energy escape first when the liquid is at the boiling point. Turning off the source of external heat drops the liquid's temperature below its boiling point. Supplying more heat allows more particles to acquire enough kinetic energy to escape. However, the temperature of the boiling liquid never rises above its boiling point. If heat is supplied at a greater rate, the liquid only boils faster. The vapor produced is at the same temperature as that of the boiling liquid. Although the vapor has the same average kinetic energy as the liquid, its potential (or stored energy) is much higher. Thus, a burn from steam is more severe than one from an equal mass of boiling water at the same temperature.

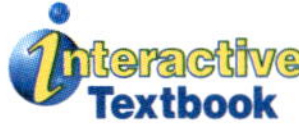

Animation 16 Relate vapor pressure and boiling point to intermolecular attractive forces.

with ChemASAP

Normal Boiling Point Because a liquid can have various boiling points depending on pressure, the **normal boiling point** is defined as the boiling point of a liquid at a pressure of 101.3 kPa. For example, the normal boiling point of water is 100°C. Table 13.2 lists the normal boiling points of six molecular compounds.

13.2 Section Assessment

8. **Key Concept** What factors help determine the physical properties of liquids?
9. **Key Concept** In terms of kinetic energy, explain how a molecule in a liquid evaporates.
10. **Key Concept** A liquid is in a closed container and has a constant vapor pressure. What is the relationship between the rate of evaporation of the liquid and the rate of condensation of the vapor in the container?
11. **Key Concept** What condition must exist for a liquid to boil?
12. Use Figure 13.9 to determine the boiling point of each liquid.
    - **a.** ethanoic acid at 27 kPa
    - **b.** chloroform at 80 kPa
    - **c.** ethanol at 50 kPa
13. Explain why the boiling point of a liquid varies with atmospheric pressure.
14. Explain how evaporation lowers the temperature of a liquid.

Writing Activity

Contrast Paragraph Look for high-altitude cooking directions on packages of cake, brownie, or muffin mixes. Contrast them with standard directions.

Assessment 13.2 Test yourself on the concepts in Section 13.2.

with ChemASAP

3 ASSESS

Evaluate Understanding 

To summarize section content, ask, **What are some of the general characteristics that distinguish a liquid from a gas?** *(Gases can expand to fill their containers. Gases are easily compressed. Liquids conform to the shape of their containers, but do not expand to fill them. Liquids are not as easy to compress as gases because the spaces between particles in a liquid are much smaller than those in a gas.)*

Reteach 

Remind students that the temperature at which the vapor pressure of a liquid is equal to the atmospheric pressure is the boiling point of the liquid. The exact temperature at which boiling occurs depends on the strength of the intermolecular attractions between molecules in the liquid and the magnitude of the atmospheric pressure. Explain that when the atmospheric pressure is reduced below the vapor pressure of water, water will boil at room temperature.

Writing Activity

Examples of recommended adjustments include adding more flour, greasing nonstick muffin tins, reducing the amount of leavening, increasing the amount of liquids, increasing the cooking time, and increasing the baking temperature.

If your class subscribes to the Interactive Textbook, use it to review key concepts in Section 13.2.

with ChemASAP

Section 13.2 Assessment

8. the interplay between the disruptive motions of particles in a liquid and the attractions among the particles
9. A molecule with a certain minimum kinetic energy can escape from the surface of the liquid and vaporize.
10. Rate of evaporation of the liquid equals the rate of condensation of the vapor.
11. Particles throughout the liquid must have enough kinetic energy to vaporize.
12. **a.** about 76°C **b.** about 52°C **c.** about 62°C
13. Boiling occurs when the vapor pressure of a liquid equals the external pressure. If the atmospheric pressure changes, the boiling point will change.
14. When the molecules with the highest kinetic energy escape from the liquid, the average kinetic energy of the remaining particles is lower and the temperature decreases.

13.3

1 FOCUS

Objectives

13.3.1 Evaluate how the way particles are organized explains the properties of solids.

13.3.2 Identify the factors that determine the shape of a crystal.

13.3.3 Explain how allotropes of an element are different.

Guide for Reading

Build Vocabulary L2

Word Parts The word *amorphous* comes from the Greek prefix *a-*, meaning "not," and the Greek root *morph*, meaning "form." An amorphous substance has no form. Morphing is a special effect used in films or videos to change the shape or form of an object. The branch of biology that deals with form and structure is morphology.

Reading Strategy L2

Relate Text and Visuals Have students use Figure 13.11 to make a table summarizing the characteristics of the seven crystal systems.

2 INSTRUCT

Connecting to Your World

Explain that for some elements, the atoms can be arranged in more than one way. Ask, **Which form of carbon was discovered in 1985?** *(buckminsterfullerene)*

A Model for Solids

Discuss L2

Check students' knowledge of types of compounds by asking them to compare the structures of molecular and ionic compounds. *(Ionic compounds are composed of ions and are usually formed from a metal and a nonmetal. Molecular compounds are composed of molecules formed from two or more nonmetallic elements.)*

13.3 The Nature of Solids

Guide for Reading

Key Concepts

- How are the structure and properties of solids related?
- What determines the shape of a crystal?

Vocabulary

melting point
crystal
unit cell
allotropes
amorphous solid
glass

Reading Strategy

Building Vocabulary After you have read the section, give an example of a substance that is and one that is not an *amorphous solid.*

Connecting to Your World

In 1985, scientists discovered a new form of carbon. They called this form of carbon buckminsterfullerene, or buckyball for short. The molecules in buckyball are hollow spheres. The carbon atoms are arranged so that the pattern on the surface of the sphere resembles the surface of a soccer ball. In this section, you will learn how the arrangement of particles in solids determines some general properties of solids.

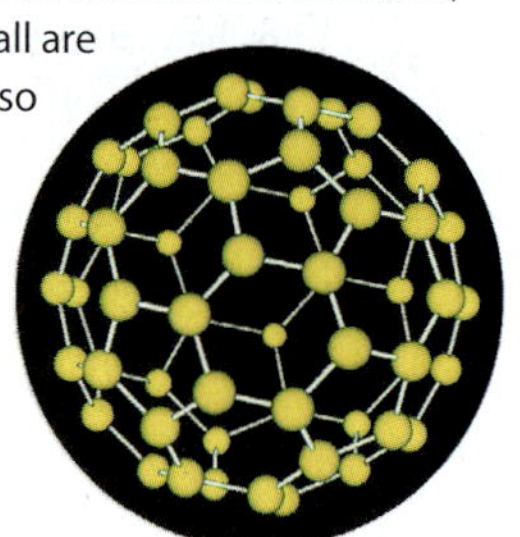

A Model for Solids

The general properties of solids reflect the orderly arrangement of their particles and the fixed locations of their particles. In most solids, the atoms, ions, or molecules are packed tightly together. These solids are dense and not easy to compress. Because, the particles in solids tend to vibrate about fixed points, solids do not flow.

When you heat a solid, its particles vibrate more rapidly as their kinetic energy increases. The organization of particles within the solid breaks down, and eventually the solid melts. The **melting point** (mp) is the temperature at which a solid changes into a liquid. At this temperature, the disruptive vibrations of the particles are strong enough to overcome the attractions that hold them in fixed positions. The melting and freezing points of a substance are at the same temperature. At that temperature, the liquid and solid phases are in equilibrium.

Crystal Structure and Unit Cells

Most solid substances are crystalline. In a **crystal** the particles are arranged in an orderly, repeating, three-dimensional pattern called a crystal lattice. Figure 13.10 shows part of the crystal lattice in sodium chloride. **The shape of a crystal reflects the arrangement of the particles within the solid.**

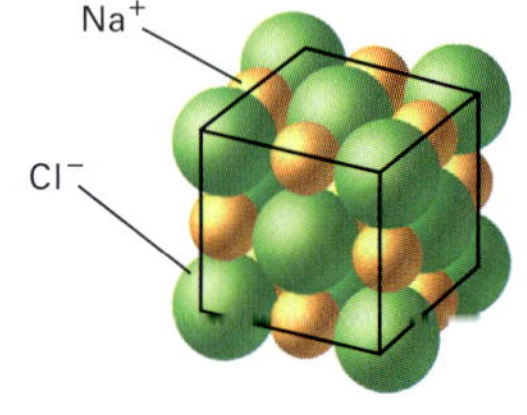

Figure 13.10 The orderly arrangement of sodium and chloride ions within a sodium chloride crystal determines the shape of the crystal. The closely packed ions vibrate about fixed points on the crystal.

Section Resources

Print

- **Guided Reading and Study Workbook,** Section 13.3
- **Core Teaching Resources,** Section 13.3 Review
- **Transparencies,** T145–T147
- **Laboratory Manual,** Lab 21

Technology

- **Interactive Textbook with ChemASAP,** Assessment 13.3
- **Go Online,** Section 13.3

The type of bonding that exists between particles in crystals determines their melting points. In general, ionic solids have high melting points because relatively strong forces hold them together. By contrast, molecular solids have relatively low melting points. Not all solids melt, however. Wood and cane sugar, for example, decompose when heated.

Crystal Systems A crystal has sides, or faces. The angles at which the faces of a crystal intersect are always the same for a given substance and are characteristic of that substance. Crystals are classified into seven groups, or crystal systems, which have the characteristic shapes shown in Figure 13.11. The edges are labeled a, b, and c. The angles are labeled α, β, and γ. The seven crystal systems differ in terms of the angles between the faces and in the number of edges of equal length on each face.

Figure 13.11 Crystals are classified into seven crystal systems. **Classifying** ***In which of the systems are all three angles equal to 90°?***

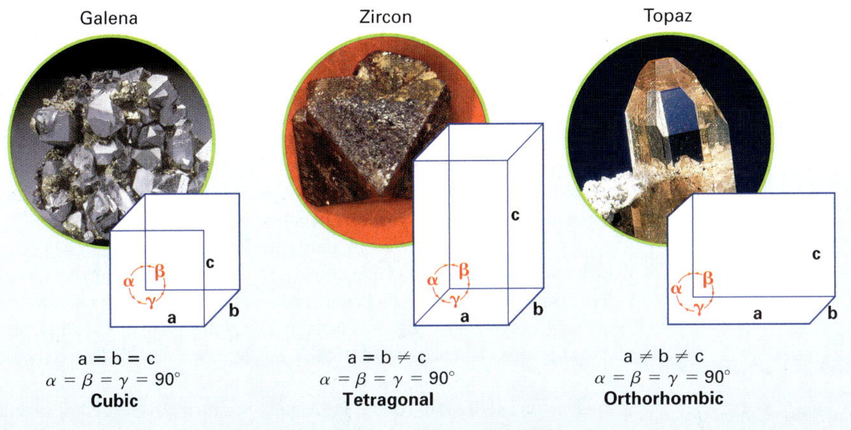

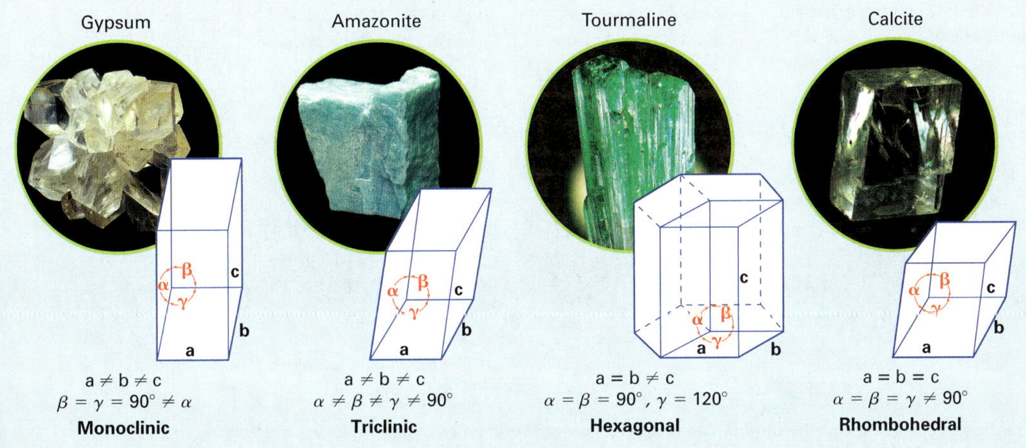

Crystal Structure and Unit Cells

CLASS Activity

Wallpaper Lattices 

Purpose Students explore an analogy for the crystal lattices in solids.

Materials samples of wallpaper with repeating patterns

Procedure Explain that wallpaper patterns often consist of a small group of images that are repeated to obtain the overall effect. The group of images is like a unit cell in a crystal lattice. Ask students to identify the "unit cells" in wallpaper samples. Show students how to obtain a set of "lattice points" by choosing the same point in each unit of the repeating pattern. The collection of lattice points shows the fundamental arrangement of the units in the pattern. Ask students if they could arrange the components of each unit differently to produce a different "unit cell."

Expected Outcome Students observe two-dimensional arrangements at the macroscopic level that are analogous to the three-dimensional arrangements in crystal lattices.

Differentiated Instruction

Gifted and Talented L3

Have students do research on the crystalline structure and formation of natural, synthetic, and simulated diamonds. Simulated diamonds are known as YAGs (yttrium aluminum garnet). Have students write a short paper comparing the three kinds of crystalline structures.

Answers to...

Figure 13.11 cubic, tetragonal, and orthorhombic

TEACHER Demo

Crystalline Solid Model

L2

Materials dishwashing detergent, watch glass, overhead projector, drinking straw

Procedure You can use an aqueous solution of detergent to make a model of a crystalline solid. Place the solution in a watch glass on an overhead projector and use a straw to blow bubbles into the solution. The bubbles form a regular, repeating pattern that is analogous to the arrangement of unit cells in a crystalline solid.

Download a worksheet on **Allotropes** for students to complete, and find additional teacher support from NSTA SciLinks.

The shape of a crystal depends on the arrangement of the particles within it. The smallest group of particles within a crystal that retains the geometric shape of the crystal is known as a **unit cell.** A crystal lattice is a repeating array of any one of fourteen kinds of unit cells. There are from one to four types of unit cells that can be associated with each crystal system. Figure 13.12 shows the three kinds of unit cells that can make up a cubic crystal system.

For: Links on Allotropes
Visit: www.SciLinks.org
Web Code: cdn-1133

Allotropes Some solid substances can exist in more than one form. A good example is the element carbon. Diamond is one crystalline form of carbon. It forms when carbon crystallizes under tremendous pressure (thousands of atmospheres). A different crystalline form of carbon is graphite. The lead in a pencil is not the element lead; it is graphite. In graphite, the carbon atoms are packed in sheets rather than in the extended three-dimensional array characteristic of diamond.

In 1985, a third form of carbon was discovered in ordinary soot. This form of carbon is called buckminsterfullerene, or buckyball. The 60 carbon atoms in molecules of buckyball are bonded together to form a hollow sphere, or cage. The atoms are arranged in a pattern of hexagons and pentagons on the surface of the cage, similar to the pattern on the surface of a soccer ball. Since 1985, other molecules of carbon with hollow cages have been discovered. The one with 70 carbon atoms is shaped like a football. As a group, these forms of carbon are called fullerenes.

The physical properties of diamond, graphite, and fullerenes are quite different. Diamond has a high density and is very hard. Graphite has a relatively low density and is soft and slippery. The hollow cages in fullerenes give them great strength and rigidity. Diamond, graphite, and buckyballs are allotropes of carbon. **Allotropes** are two or more different molecular forms of the same element in the same physical state. Although allotropes are composed of atoms of the same element, they have different properties because their structures are different. Figure 13.13 compares the structures of carbon allotropes. Only a few elements have allotropes. In addition to carbon, these include the nonmetals phosphorus, sulfur, and oxygen (O_2 and O_3), and the metalloids boron and antimony.

Figure 13.12 The unit cell in a cubic crystal system may be simple cubic, body-centered cubic, or face-centered cubic. In the space-filling models and line drawings, the spheres represent atoms or ions.

Simple cubic

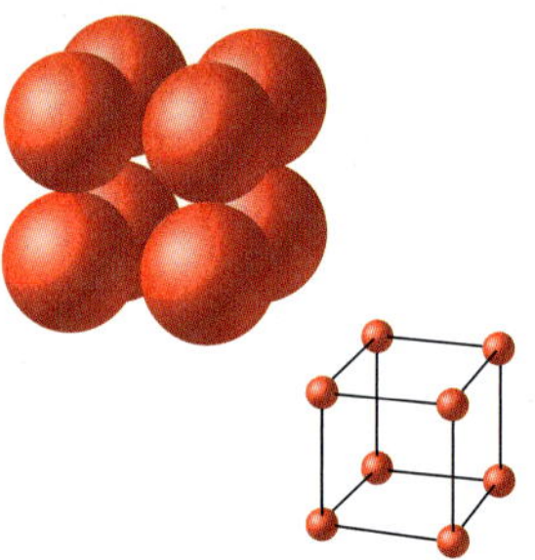

In a simple cubic unit cell, the atoms or ions are arranged at the corners of an imaginary cube.

Body-centered

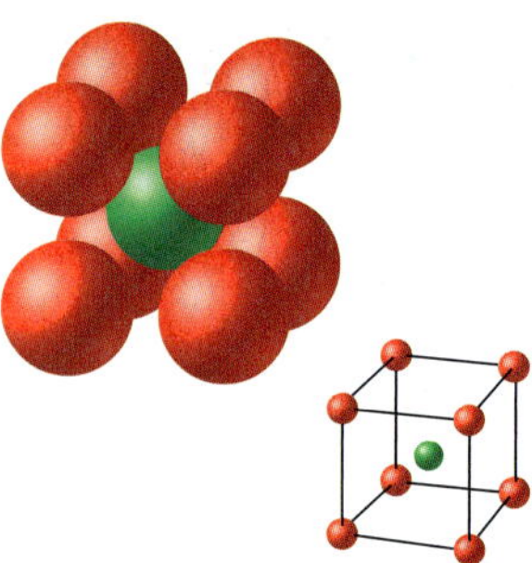

In a body-centered cubic unit cell, the atoms or ions are at the corners and in the center of an imaginary cube.

Face-centered

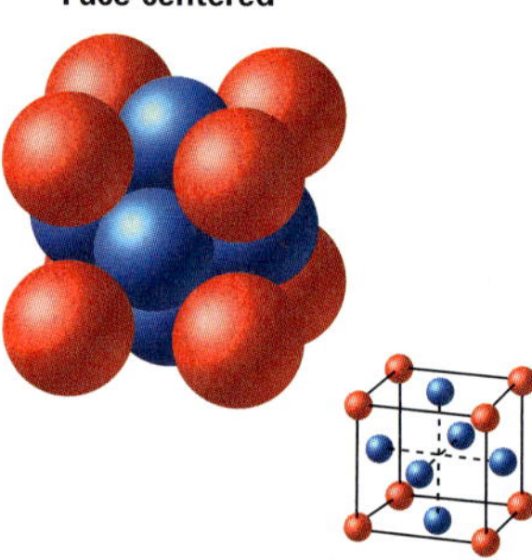

In a face-centered cubic unit cell, there are atoms or ions at the corners and in the center of each face of the imaginary cube.

Facts and Figures

LCDs

Liquid-crystal displays (LCDs) are used in calculators and wristwatches. Liquid crystals exhibit properties both of liquids and solids. Interested students may want to find out how liquid crystal displays work, with emphasis on their physical properties.

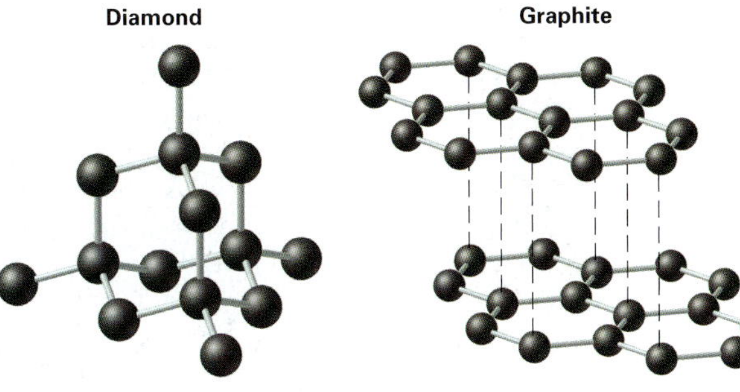

In diamond, each carbon atom in the interior of the diamond is strongly bonded to four others. The array is rigid and compact.

In graphite, the carbon atoms are linked in widely spaced layers of hexagonal (six-sided) arrays.

In buckminsterfullerene, 60 carbon atoms form a hollow sphere. The carbons are arranged in pentagons and hexagons.

Figure 13.13 Diamond, graphite, and fullerenes are allotropes of carbon. **Classifying** ***Based on the arrangements of their atoms, explain why the properties of fullerenes are closer to those of diamond than of graphite?***

Non-Crystalline Solids Not all solids are crystalline in form; some solids are amorphous. An **amorphous solid** lacks an ordered internal structure. Rubber, plastic, and asphalt are amorphous solids. Their atoms are randomly arranged. Other examples of amorphous solids are glasses. A **glass** is a transparent fusion product of inorganic substances that have cooled to a rigid state without crystallizing. Glasses are sometimes called supercooled liquids. The irregular internal structures of glasses are intermediate between those of a crystalline solid and those of a free-flowing liquid. Glasses do not melt at a definite temperature. Instead, they gradually soften when heated. This softening with temperature is critical to the glassblower's art. When a crystalline solid is shattered, the fragments tend to have the same surface angles as the original solid. By contrast, when an amorphous solid, such as glass, is shattered, the fragments have irregular angles and jagged edges.

Checkpoint ***What property of glass is important to glassblowers?***

13.3 Section Assessment

15. **Key Concept** In general, how are the particles arranged in solids?
16. **Key Concept** What does the shape of a crystal tell you about the structure of a crystal?
17. How do allotropes of an element differ?
18. What phases are in equilibrium at a substance's melting point?
19. How do the melting points of ionic solids generally compare with those of molecular solids?
20. What is the difference between a crystal lattice and a unit cell?

Elements Handbook

Glass Read about optical glass on page R20. What are the general physical properties of glass? How does optical glass differ from other types of glass? What are some uses of optical glass?

Interactive Textbook

Assessment 13.3 Test yourself on the concepts in Section 13.3.

with ChemASAP

Discuss 

Ask students if they know of any elements besides carbon that exist in different allotropic forms. *(Elemental oxygen occurs as O_2 and O_3, both of which are gases; sulfur can exist as monoclinic or orthorhombic crystals; phosphorus can exist as red or white phosphorus, which are solids at room temperature.)*

3 ASSESS

Evaluate Understanding 

Ask students to describe the distinguishing characteristics of crystalline solids and amorphous solids. *(Crystalline solids are characterized by an orderly, repeating three-dimensional arrangement of atoms, ions, or molecules. All crystals have a regular shape that reflects the arrangement of particles in the solid. Amorphous solids lack a well-defined arrangement of particles.)*

Reteach 

Describe how a higher degree of organization and stronger intermolecular attractions between particles distinguish solids from gases and liquids.

Elements Handbook

Glass has the disordered structure of a liquid, but the hardness of a solid. Optical glass is purer than window glass and transmits more light. Optical glass is used in eyeglasses, lenses, and optical fibers.

Interactive Textbook

If your class subscribes to the Interactive Textbook, use it to review key concepts in Section 13.3.

with ChemASAP

Section 13.3 Assessment

15. Particles in solids are packed tightly together in an orderly arrangement. The locations of the particles are fixed.
16. The shape of a crystal reflects the arrangement of the particles within the solid.
17. Allotropes are different molecular forms of the same element in the same physical state.
18. The liquid and solid states are in equilibrium.
19. Ionic solids generally have higher melting points than do molecular solids.
20. A crystal lattice is a repeating array of unit cells.

Answers to...

Figure 13.13 The hollow cages have a structure that is closer to diamond's rigid structure than to the widely spaced layers in graphite.

Checkpoint Glass gradually softens when heated.

The Behavior of Liquids and Solids

Objective After completing this activity, students will be able to:

- explain some behaviors of liquids and solids.

Skills Focus Observing, inferring, drawing conclusions

Prep Time 20 minutes

Materials plastic Petri dish, water, ice, rubbing alcohol, calcium chloride, graph paper (1-cm), bromothymol blue solution (BTB), vinegar, aqueous ammonia

Advance Prep

- Purchase vinegar, aqueous ammonia, and rubbing alcohol.
- If you purchase BTB as a powder, prepare a 0.04% solution by dissolving 100 mg BTB powder in 16.0 mL 0.01*M* NaOH. Dilute to 250 mL.

Class Time 40 minutes

Safety Be sure students wear goggles and aprons. BTB can stain clothing and skin. The room should be well ventilated and there should be no flames.

Expected Outcomes The answers to the Analyze questions include a description of the expected outcome for each experiment.

Analyze

1. Water in the dish evaporates and condenses into a cloud when it contacts the cold surface under the ice.
2. The drop of water on top of the dish provides enough cooling to cause cloud formation.
3. Water beads up and alcohol spreads because the intermolecular attractions in water are stronger.
4. Calcium chloride absorbs water from the environment in the dish.
5. The many pieces of calcium chloride effectively dry the atmosphere, leaving no water vapor in the dish.

Small-Scale LAB

The Behavior of Liquids and Solids

Purpose

To explore and explain some behaviors of liquids and solids.

Materials

- plastic Petri dish
- water
- ice
- rubbing alcohol
- graph paper, 1-cm
- calcium chloride

Procedure

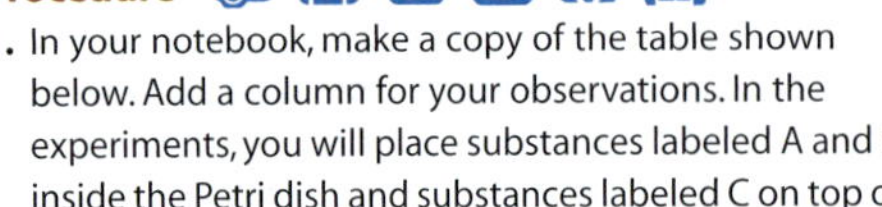

1. In your notebook, make a copy of the table shown below. Add a column for your observations. In the experiments, you will place substances labeled A and B inside the Petri dish and substances labeled C on top of the dish.
2. For Experiment 1, place one drop of water in the Petri dish. Replace the cover and place a small piece of ice on top of the cover.
3. After a few minutes, observe the interior surface of the Petri dish cover and the contents of the dish. Record your observations. Clean and dry the Petri dish and cover.
4. Repeat Steps 2 and 3 for Experiments 2–5, using the materials listed in the table. For Experiment 4, place the Petri dish on the graph paper so that you can place the water and the calcium chloride about 3 cm apart.

Experiment	Substance A	Substance B	Substance C
1	drop of water		ice cube
2	drop of water		drop of water
3	drop of rubbing alcohol		drop of water
4	drop of water	piece of $CaCl_2$	
5		several pieces of $CaCl_2$	ice cube

Analyze

Using your experimental data, record the answers to the following questions beneath your data table.

1. Explain your observations in Experiment 1 in terms of the behavior of liquids.
2. Why is ice not needed for cloud formation in Experiment 2?
3. What differences do you observe about the behavior of rubbing alcohol in Experiment 3 and the behavior of water in the previous experiments? Explain.
4. What happens to solid calcium chloride in a humid environment?
5. Propose an explanation for no cloud formation in Experiment 5.

You're The Chemist

1. **Analyze It!** Place a drop of water and a drop of rubbing alcohol about 3 cm apart in a Petri dish. Cover the dish and place it on a piece of graph paper. Be careful not to mix the contents. Observe what happens to the size of the water drop over time.
2. **Analyze It!** Add a drop of bromthymol blue (BTB) to a drop of vinegar. What happens?
3. **Design It!** Vinegar is a solution of water and ethanoic acid, CH_3COOH. Design and carry out an experiment to see if ethanoic acid will evaporate from a drop of vinegar. Does ethanoic acid evaporate?
4. **Design It!** Design and carry out an experiment to see if ammonia will evaporate from a drop of aqueous ammonia.

You're the Chemist

1. The water drop increases in diameter over time as the alcohol evaporates and is "captured" by the water drop. The attractions in the resulting mixture are weaker overall.
2. The BTB turns from green to yellow in the presence of vinegar.
3. Place a drop of vinegar and a drop of BTB about 3 cm apart in a Petri dish. Cover and observe. The BTB slowly changes from green to yellow. Ethanoic acid that evaporates is "captured" by the BTB.
4. Place a drop of ammonia and a drop of BTB about 3 cm apart in a Petri dish. Cover and observe. The BTB slowly changes from green to blue. Ammonia that evaporates is "captured" by the BTB.

13.4 Changes of State

Connecting to Your World Familiar weather events can remind you that water exists on Earth as a liquid, a solid, and a vapor. A spring shower brings liquid raindrops and a winter blizzard delivers solid snowflakes. On a humid summer day, you may be uncomfortable because there is a high concentration of water vapor in the air. As water cycles through the atmosphere, the oceans, and Earth's crust, it undergoes repeated changes of state. In this section, you will learn what conditions can control the state of a substance.

Guide for Reading

Key Concepts
- When can sublimation occur?
- How are the conditions at which phases are in equilibrium represented on a phase diagram?

Vocabulary

sublimation

phase diagram

triple point

Reading Strategy

Predicting In your notebook, predict what causes ice cubes that are left in the freezer for a long time to get smaller.

Sublimation

If you hang wet laundry on a clothesline on a very cold day, the water in the clothes quickly freezes to ice. Eventually, however, the clothes become dry although the ice never thaws. The ice changes directly to water vapor without melting and passing through the liquid state. The change of a substance from a solid to a vapor without passing through the liquid state is called **sublimation.** Sublimation can occur because solids, like liquids, have a vapor pressure. **Sublimation occurs in solids with vapor pressures that exceed atmospheric pressure at or near room temperature.**

Iodine is another example of a substance that undergoes sublimation. This violet-black solid ordinarily changes into a purple vapor without passing through a liquid state. Notice in Figure 13.14 how dark crystals of iodine deposit on the underside of a watch glass placed on top of a beaker containing solid iodine that is being heated. The iodine vapor sublimes from iodine crystals in the bottom of the beaker and condenses to form crystals on the watch glass.

Sublimation has many useful applications. If freshly brewed coffee is frozen and the water vapor is removed with a vacuum pump, the result is freeze-dried coffee. Solid carbon dioxide (dry ice) is often used as a coolant for goods, such as ice cream, that must remain frozen during shipment. Dry ice has a low temperature of −78°C. Because it sublimes, it does not produce a liquid as ordinary ice does when it melts. Solid air fresheners contain a variety of substances that sublime at room temperature. Sublimation is also useful for separating substances. Organic chemists use sublimation to separate mixtures and to purify compounds.

Figure 13.14 When solid iodine is heated, the crystals sublime, going directly from the solid to the gaseous state. When the vapor cools, it goes directly from the gaseous to the solid state.

13.4

1 FOCUS

Objectives

13.4.1 Identify the conditions necessary for sublimation.

13.4.2 Describe how equilibrium conditions are represented in a phase diagram.

Guide for Reading

Build Vocabulary 

Word Forms The word *sublimation* comes from the Latin word *sublimare*, meaning "to elevate." The meaning of *sublimare* was expanded by the Middle Ages to include "to refine" or purify. The process of sublimation can be used to remove impurities from a substance.

Reading Strategy 

Use Prior Knowledge After students read this section, have them write a paragraph comparing sublimation to evaporation.

2 INSTRUCT

Connecting to Your World

Explain that rain is the condensation product of water vapor in the atmosphere. Ask students to describe the water cycle in terms of phase changes. *(They should focus on evaporation, condensation, and precipitation.)* Use the discussion to introduce sublimation.

Sublimation

Relate 

The sublimation of iodine can be used to develop fingerprints found on paper. When iodine vapor is absorbed by fatty acids in sweat, the iodine imparts a yellow-brown color to the ridge pattern. Because the fingerprints tend to fade, they must be photographed or fixed with a reagent such as starch solution.

Section Resources

Print
- **Guided Reading and Study Workbook,** Section 13.4
- **Core Teaching Resources,** Section 13.4 Review, Interpreting Graphics
- **Transparencies,** T148–T149
- **Laboratory Manual,** Lab 22
- **Laboratory Practicals,** 13-1

Technology
- **Interactive Textbook with ChemASAP,** Simulation 14, Assessment 13.4
- **Go Online,** Section 13.4

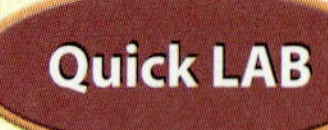

Sublimation L2

Objective After completing this activity, students will be able to:
- describe the process of sublimation.

Skills Focus Observing, inferring, predicting, communicating

Prep Time 15 minutes
Class Time 20 minutes

Safety Perform this activity in a well-ventilated room. Some people are allergic to substances in air fresheners.

Teaching Tips

- Students should not fill the shallow containers to the point where the sublimator floats, to prevent it from tipping over.
- If a colored air freshener is used, the condensed solid will be white.

Expected Outcome The heat from the hot water causes the air freshener to sublime. When the vapor reaches the cold surface of the upper cup, it condenses back into a solid.

Analyze and Conclude

1. the vaporization of a solid
2. At room temperature, there would be less sublimation; at the boiling point, there would be more sublimation.
3. When a substance sublimes in a mixture, other components remain as residue.

For Enrichment 

If you used a colored air freshener, ask students to explain why the condensed solid is white. *(The dye does not sublime and remains behind at the bottom of the sublimator.)*

Download a worksheet on **Phase Diagrams** for students to complete, and find additional teacher support from NSTA SciLinks.

Quick LAB

Sublimation

Purpose
To observe the sublimation of air freshener.

Materials
- small pieces of solid air freshener
- small shallow container
- 2 clear 8 oz plastic cups
- hot tap water
- ice
- 3 thick cardboard strips

Procedure   

1. Place a few pieces of air freshener in one of the cups. **CAUTION** *Work in a well-ventilated room.*
2. Bend the cardboard strips and place them over the rim of the cup that has the air freshener pieces.
3. Place the second cup inside the first. The base of the second cup should not touch the air freshener. Adjust the cardboard as necessary. This assembly is your sublimator.
4. Fill the top cup with ice. Do not get any ice or water in the bottom cup.
5. Fill the shallow container about one-third full with hot tap water.
6. Carefully place your sublimator in the hot water. Observe what happens.

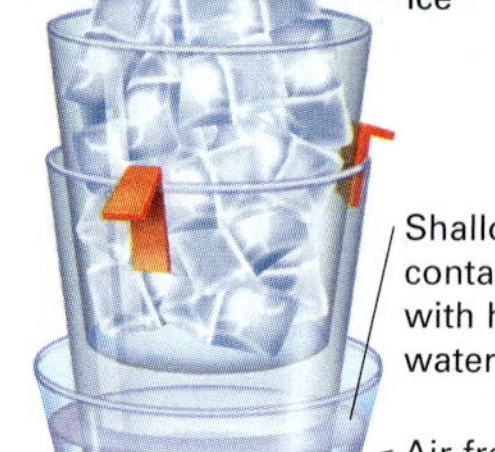

Analyze and Conclude
1. Define sublimation.
2. What do you think would happen if the water in the shallow container were at room temperature? If it were boiling?
3. Why is it possible to separate the substances in some mixtures by sublimation?

Phase Diagrams

The relationships among the solid, liquid, and vapor states (or phases) of a substance in a sealed container can be represented in a single graph. The graph is called a **phase diagram.** A phase diagram gives the conditions of temperature and pressure at which a substance exists as solid, liquid, and gas (vapor). **The conditions of pressure and temperature at which two phases exist in equilibrium are indicated on a phase diagram by a line separating the phases.**

Figure 13.15 shows the phase diagram for water. In each of the colored regions on the phase diagram, water is in a single phase. The curving line that separates water's vapor phase from its liquid phase are the equilibrium conditions for liquid and vapor. The line also illustrates how the vapor pressure of water varies with temperature. The other two curving lines give the conditions for equilibrium between liquid water and ice and between water vapor and ice. The point on the diagram at which all three curves meet is called the triple point. The **triple point** describes the only set of conditions at which all three phases can exist in equilibrium with one another. For water, the triple point is a temperature of 0.016°C and a pressure of 0.61 kPa (0.0060 atm). Figure 13.16 shows water at its triple point.

By referring to Figure 13.15, you can determine what happens if you melt ice or boil water at pressures less than 101.3 kPa. A decrease in pressure lowers the boiling point and raises the melting point. An increase in pressure will raise the boiling point and lower the melting point.

 *What are the variables that are plotted on a phase diagram?*

Interactive Textbook

Simulation 14 Predict the physical states present at different points on a phase diagram.

with ChemASAP

For: Links on Phase Diagrams
Visit: www.SciLinks.org
Web Code: cdn-1134

Differentiated Instruction

Less Proficient Readers 

The ability to interpret data in graphical format is fundamental to a course in chemistry. To reinforce this skill, have students compare the graph in Figure 13.9 to the graph in Figure 13.15. Ask students which line on the phase diagram corresponds to the curve for water in Figure 13.9. *(the line between liquid water and water vapor)*

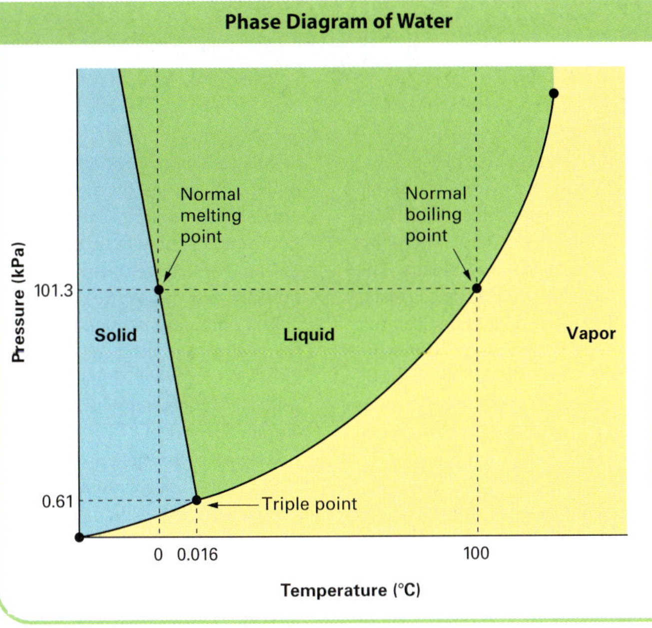

Figure 13.15 The phase diagram of water shows the relationship among pressure, temperature, and physical states of water.

INTERPRETING GRAPHS

a. Analyzing Data At the triple point of water, what are the values of temperature and pressure?
b. Inferring What states of matter are present at the triple point of water?
c. Analyzing Data Assuming standard pressure, at what temperature is there an equilibrium between water vapor and liquid water? Between liquid water and ice?

Look at Figure 13.15. Follow the equilibrium line between liquid water and water vapor to the triple point. Below the triple point, the vapor and liquid cannot exist in equilibrium. Increasing the pressure won't change the vapor to a liquid. The solid and the vapor are in equilibrium at temperatures below 0.016°C. With an increase in pressure, the vapor begins to behave more like a solid. For example, it is no longer easily compressed.

Figure 13.15 also illustrates how an increase in pressure affects the melting point of ice. For years, the accepted hypothesis for how ice skaters move along the ice was the following. The blades of the skates exert pressure, which lowers the melting point of the ice. The ice melts and a film of water forms under the blades of the skates. This film acts as a lubricant, enabling the skaters to glide gracefully over the ice. This hypothesis fails to explain why skiers also glide along very nicely on another solid form of water—snow. Wide skis exert much less pressure per unit area of snow than narrow skate blades exert on ice. Recent research shows that the surface of ice has a slippery, water-like surface layer that exists well below ice's melting point. Even ice that is at −129°C has this layer. A new hypothesis proposes that the liquid-like surface layer provides the lubrication needed for smooth skating and skiing.

Figure 13.16 At the triple point, ice, liquid water, and water vapor can exist at equilibrium. Freezing, melting, boiling, and condensation can all occur at the same time, as shown in the flask.

Phase Diagrams

Interpreting Graphs L2

a. 0.016°C and 0.61 kPa
b. solid, liquid, and vapor
c. 100°C; 0°C

Enrichment Question L3

Show students the phase diagram for carbon on R18. Ask, **How many phases of carbon are shown in this diagram?** *(4)* **Which phases are in equilibrium at the two triple points?** *(diamond, graphite, and liquid carbon; graphite, liquid carbon, and carbon vapor)*

Use Visuals L1

Figure 13.15 Display the figure on an overhead projector. Explain that a phase diagram is a convenient way to summarize the conditions of temperature and pressure at which the solid, liquid, and gas phases of a substance are most stable. Point out that each line in the diagram represents the set of possible temperature-pressure values at which different phases are in dynamic equilibrium with one another. For example, in a closed system at 100°C and 101.3 kPa, liquid water is in dynamic equilibrium with water vapor. (Note that the *x*- and *y*-axes are not drawn to scale so that the triple point and normal boiling point can fit on the graph.) Ask, **At what temperature and pressure are the liquid, solid, and vapor phases of water in dynamic equilibrium?** *(0.016°C, 0.61 kPa)*

FYI

There is fourth phase of matter. This phase is called the plasma phase. In the plasma phase, matter consists of positive ions and free electrons. The plasma phase exists only at high temperatures. The glowing gas in a fluorescent lamp, the sun, and other stars are in the plasma phase.

Facts and Figures

Freeze-Dried Foods

Astronauts and backpackers use freeze-dried foods because they are lightweight and easily reconstituted by adding water. Freeze-dried foods don't require refrigeration because bacteria cannot multiply in the absence of moisture. The sublimation of ice at low pressure makes the process of freeze-drying possible. Food is frozen and placed in a chamber attached to a vacuum pump. Then the pressure in the chamber is lowered until the ice crystals sublime. Freeze-dried food tends to maintain its flavor because the substances in the food that impart flavor remain in the food as the ice sublimes.

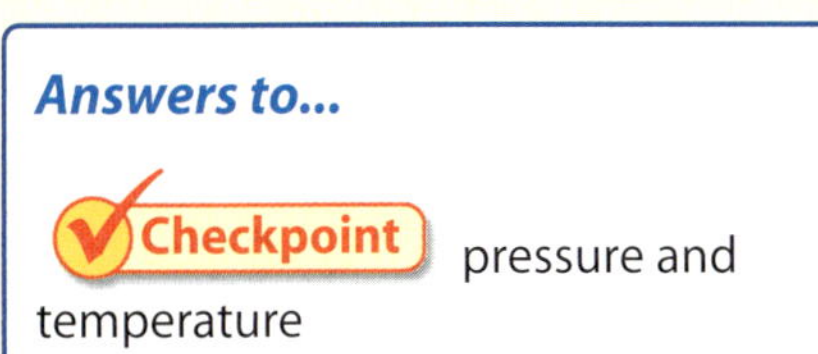

pressure and temperature

Solid State Chemist

Go Online PHSchool.com

Have students research chemistry-related careers in the library or on the Internet. Students can then construct a table that describes the nature of the work, educational and training requirements, employment outlook, working conditions, and other necessary information.

3 ASSESS

Evaluate Understanding L2

List a series of temperature–pressure values on the board. Have students state which phase of water is most stable under each set of conditions.

Reteach L1

On the board, draw a stoppered, sidearm flask containing water and a chunk of ice. Note that the sidearm is connected to a vacuum pump. The diagram illustrates ice, liquid water, and water vapor in equilibrium. Have a student add an arrow and label to the diagram to represent a phase change. Continue until there are six arrows on the diagram.

Writing Activity

The substance is likely molecular because the bonds between ions are typically stronger than the attractions between molecules.

If your class subscribes to the Interactive Textbook, use it to review key concepts in Section 13.4.

with ChemASAP

Careers *in* Chemistry

Solid State Chemist

By now you are familiar with chemical reactions in aqueous solutions and the gaseous state. But did you know that chemical reactions can also take place in solids, without a liquid or a gas?

Reactions that take place in the solid state are extremely useful. These reactions occur during the manufacture of semiconductors for the electronics industry. The device that ignites the gas in a gas grill is made of a type of ceramic called a piezoelectric. This type of ceramic produces a voltage when subjected to mechanical pressure. There is a class of light-sensitive solids called photopolymers. These solids are used to make plates for printing presses that use dry inks instead of water-based inks. A solid-state reaction takes place in lenses that are designed to darken when exposed to light.

Solid state chemists often study crystal structures, superconductors, and semiconductors. They may also study the electrical, magnetic, and optical properties of solid compounds.

Solid state chemists often work at major universities, teaching and doing basic research. Corporations employ them to do pure research or to develop applications for new discoveries in this field. Companies in the petroleum, electronic, and computer industries are a few of the places a solid state chemist might be employed.

To work in this field usually requires an undergraduate or graduate degree in chemistry or chemical engineering. Courses studied include chemistry, physics, and mathematics.

For: Careers in Chemistry
Visit: www.PHSchool.com
Web Code: cdb-1134

13.4 Section Assessment

21. **Key Concept** What properties must a solid have to undergo sublimation?
22. **Key Concept** What do the curved lines on a phase diagram represent?
23. Describe one practical use of sublimation.
24. What does the triple point on a phase diagram describe?
25. Using Figure 13.15, estimate the boiling point of water at a pressure of 50 kPa.

Connecting Concepts

Types of Bonding Would you expect a substance that sublimates at or near room temperature to be a molecular substance or an ionic substance? Use what you learned about bonding in Chapters 7 and 8 to explain your answer.

Assessment 13.4 Test yourself on the concepts in Section 13.4.

with ChemASAP

404 Chapter 13

Section 13.4 Assessment

21. Sublimation occurs in solids that have vapor pressures that exceed atmospheric pressure at or near room temperature.
22. The lines show the conditions of temperature and pressure at which two phases exist in equilibrium.
23. freeze-dried coffee, dry ice as a coolant, air fresheners, separating mixtures, and purifying substances
24. The triple point describes the only set of conditions at which three phases can exist in equilibrium.
25. about 60°C

Technology & Society

Thin Films

A thin film consists of layers of atoms deposited on a surface at a uniform depth of 10^{-9} m to 10^{-6} m. Engineers can produce thin films with specific electrical properties by varying the element deposited and the thickness of the deposit. **Interpreting** ***Thin films are sometimes referred to as super crystals. What does this imply about their structure?***

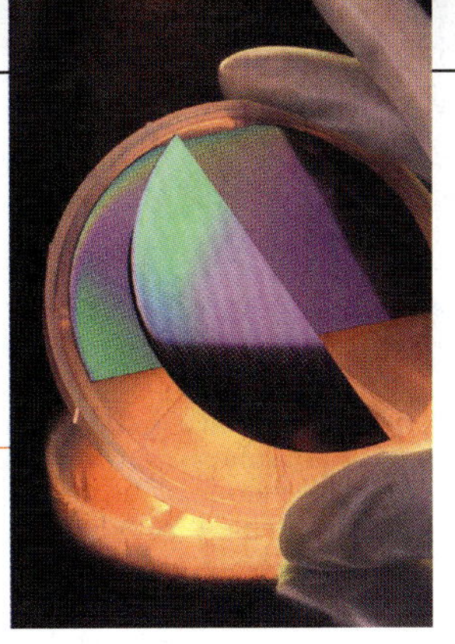

Semiconductor microchips Thin films of conducting materials, insulating materials, and semiconducting materials are layered and then etched to make a chip that contains thousands of microcircuits.

Smart windows Layers of transparent thin films, sandwiched between glass, change tint when electrically charged. The original tint returns when the electrical charge is removed.

CDs and DVDs The reflective surface of CDs are bands of thin films that carry digital information as notches on the surface of the film. DVDs increase information storage by using two or more thin-film layers.

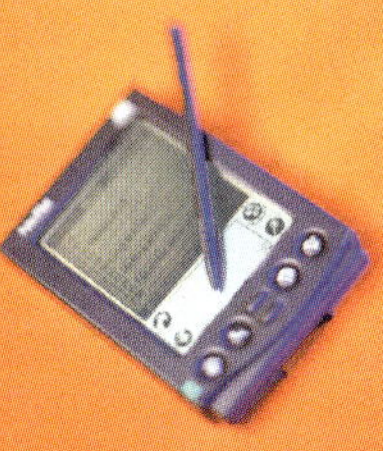

PDAs The display of a PDA contains thousands of liquid crystals attached to transparent, thin-film transistors (TFT). Images are produced as each TFT sends a charge to its linked liquid crystal.

Computer monitors Using thin-film transistors and a liquid-crystal display, flatscreen monitors provide sharp images, save space, and are about 60% more efficient than cathode ray monitors.

Technology & Society

Thin Films

L3

Purpose

This feature builds on the discussion of crystal structure in Section 13.3.

Background

New research in thin films has centered on the development of single-crystal substrates. A substrate is an underlying template that lines up the atoms of a crystalline thin film grown on it, similar to a waffle iron that confers a grid pattern on batter poured into it while hot. A substrate must have a crystalline structure that matches that of the thin film and that aligns the film's structure so that electrons or light can be conducted through the film. Single-crystal MgO is used with barium titanate as a substrate for optical modulators and switches; with yttrium-barium-copper oxides (YBCO) for superconducting thin films; and with other oxides as a substrate for optical waveguide films. These new substrates are useful because of the important applications of devices made using the thin films that are grown on the substrates. High-temperature superconducting films are currently being developed for a variety of electronic devices such as magnetic sensors for medical, geological, and industrial applications; microwave components for radar and communication technologies; and ultrafast switches, interconnects, and current leads.

Answers to...

Interpreting A thin film is composed of a single crystal.

CHAPTER 13 Study Guide

Study Tip

Learn By Teaching One excellent way to evaluate your knowledge of a key concept is to try to explain the concept to another person. The questions the other person asks will help you assess the depth of your knowledge.

If your class subscribes to the Interactive Textbook with ChemASAP, your students can go online to access an interactive version of the Student Edition and a self-test.

with ChemASAP

CHAPTER 13 Study Guide

Key Concepts

13.1 The Nature of Gases

- Particles in a gas are considered to be small, hard spheres with an insignificant volume. The motion of the particles in a gas is rapid, constant, and random. All collisions between particles in a gas are perfectly elastic.
- Gas pressure is the result of simultaneous collisions of billions of rapidly moving particles in a gas with an object.
- The Kelvin temperature of a substance is directly proportional to the average kinetic energy of the particles of the substance.

13.2 The Nature of Liquids

- The interplay between the disruptive motions of particles in a liquid and the attractions among the particles determines the physical properties of liquids.
- During evaporation, only those molecules with a certain minimum kinetic energy can escape from the surface.
- In a system at constant vapor pressure, a dynamic equilibrium exists between the vapor and the liquid. The rates of evaporation and condensation are equal.
- At a temperature at which particles throughout a liquid have enough kinetic energy to vaporize, the liquid begins to boil.

13.3 The Nature of Solids

- The general properties of solids and the shapes of crystals reflect the orderly arrangement and the fixed locations of particles within the solids.

13.4 Changes of State

- Sublimation occurs in solids that have vapor pressures that exceed atmospheric pressure at or near room temperature.
- Conditions of pressure and temperature at which two phases exist in equilibrium are indicated on a phase diagram by a line separating the two regions representing the phases.

Vocabulary

- allotrope (p. 398)
- amorphous solid (p. 399)
- atmospheric pressure (p. 386)
- barometer (p. 386)
- boiling point (p. 393)
- crystal (p. 396)
- evaporation (p. 391)
- gas pressure (p. 386)
- glass (p. 399)
- kinetic energy (p. 385)
- kinetic theory (p. 385)
- melting point (p. 396)
- normal boiling point (p. 395)
- pascal (Pa) (p. 387)
- phase diagram (p. 402)
- standard atmosphere (atm) (p. 387)
- sublimation (p. 401)
- triple point (p. 402)
- unit cell (p. 398)
- vacuum (p. 386)
- vaporization (p. 391)
- vapor pressure (p. 392)

Key Equation

- 1 atm = 760 mm Hg = 101.3 kPa

Organizing Information

Use these terms to construct a concept map that organizes the major ideas of this chapter.

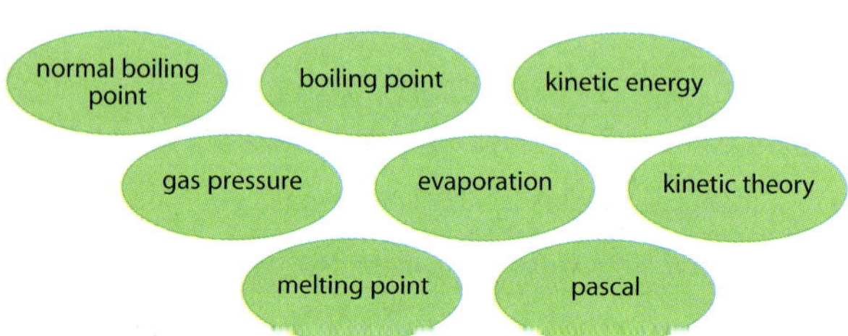

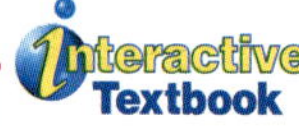

Concept Map 13 Solve the Concept Map with the help of an interactive guided tutorial.

with ChemASAP

Chapter Resources

Print

- **Core Teaching Resources,** Chapter 13, *Practice Problems, Vocabulary Review, Quiz, Chapter Test A, Chapter Test B*

Technology

- **Computer Test Bank,** Chapter 13 Test
- **Interactive Textbook with ChemASAP,** Chapter 13

CHAPTER 13

Assessment

Reviewing Content

13.1 The Nature of Gases

26. What is meant by an elastic collision?

27. Which of these statements are characteristic of matter in the gaseous state?

- **a.** Gases fill their containers completely.
- **b.** Gases exert pressure.
- **c.** Gases have mass.
- **d.** The pressure of a gas is independent of its temperature.
- **e.** Gases are compressible.
- **f.** The distances between particles in a gas are relatively large.

28. List the various units used to measure pressure, and identify the SI unit.

29. Change 1656 kPa to atm.

30. Convert 190 mm Hg to the following.

- **a.** kilopascals
- **b.** atmospheres of pressure

31. Explain the relationship between the Kelvin temperature of a substance and the kinetic energy of its particles.

32. How is the average kinetic energy of water molecules affected when you pour hot water from a kettle into cups at the same temperature as the water?

33. What does the abbreviation STP represent?

34. What is significant about the temperature absolute zero?

35. By what factor does the average kinetic energy of the molecules of gas in an aerosol container increase when the temperature is raised from 27°C (300 K) to 627°C (900 K)?

13.2 The Nature of Liquids

36. Explain why liquids and gases differ in density and the ability to be compressed.

37. Compare the evaporation of a liquid in a closed container with that of liquid in an open container.

38. Describe what is happening at the molecular level when a dynamic equilibrium occurs.

39. Explain why increasing the temperature of a liquid increases its rate of evaporation.

40. Would you expect a dynamic equilibrium in a liquid in an open container? Explain your answer.

41. Describe the effect that increasing temperature has on the vapor pressure of a liquid.

42. Distinguish between the boiling point and the normal boiling point of a liquid.

43. Use the graph to answer each question.

Vapor Pressure vs. Temperature for Water

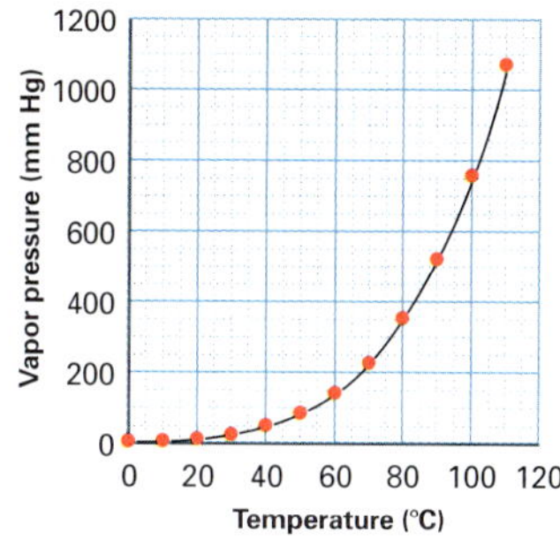

- **a.** What is the vapor pressure of water at 40°C?
- **b.** At what temperature is the vapor pressure of water 600 mm Hg?
- **c.** What is the significance of the vapor pressure of water at 100°C?

44. Explain how boiling is a cooling process.

13.3 The Nature of Solids

45. Name at least one physical property that would permit you to distinguish a molecular solid from an ionic solid.

46. Describe what happens when a solid is heated to its melting point.

47. Explain why molecular solids usually have lower melting points than ionic solids.

13.4 Changes of State

48. When you remove the lid from a food container that has been left in a freezer for several months, you discover a large collection of ice crystals on the underside of the lid. Explain what has happened.

49. Explain why a liquid stays at a constant temperature while it is boiling.

CHAPTER 13 Assessment

Reviewing Content

26. In an elastic collision energy is transferred between particles.

27. a, b, c, e, and f

28. pascal (Pa), SI; millimeter of mercury (mm Hg); atmosphere (atm)

29. 16.35 atm

30. **a.** 25 kPa **b.** 0.25 atm

31. The Kelvin temperature is directly proportional to the average kinetic energy.

32. If the temperature does not change, the average kinetic energy is not affected.

33. STP stands for standard temperature (0°C) and pressure (101.3 kPa or 1 atm).

34. At absolute zero, the motion of particles would theoretically cease.

35. The average kinetic energy triples.

36. The particles in a gas are relatively far apart compared to the particles in a liquid. Because of the extra space between particles, a gas is less dense and easier to compress.

37. In both cases, particles with sufficient kinetic energy move from the liquid to the vapor phase. In a closed container, a dynamic equilibrium is set up between the contained liquid and its vapor.

38. Two opposing processes are occurring at identical rates.

39. More molecules have enough energy to escape the attractions within the liquid.

40. No. In an open container, most of the particles that escape from the surface of the liquid as vapor do not condense back to liquid.

41. The average kinetic energy increases, which allows more vapor to form above the liquid.

42. The boiling point is the temperature at which the vapor pressure equals the external pressure. At the normal boiling point, the external pressure is 101.3 kPa.

43. **a.** about 50 mm Hg **b.** about 94°C **c.** 760 mm Hg is standard pressure

44. Escaping molecules have more kinetic energy than the average. Thus, the average kinetic energy and temperature of the remaining molecules are lower.

45. Ionic compounds generally have higher melting points than molecular solids do.

46. The particles have sufficient kinetic energy to overcome the attractive forces holding them in place.

47. The intermolecular attractions between molecules are weaker than the attractions between ions.

48. Water from the food sublimed and then condensed on the lid.

49. The temperature remains constant while the liquid boils because the energy that is added is used to vaporize the molecules.

CHAPTER 13

Assessment continued

Understanding Concepts

50. The average kinetic energy of the molecules is greater because, by definition, a fever is a state of increased body temperature.

51. a. 121°C
b. chloroform
c. chloroform
d. The external pressure on ethanol would have to increase; the external pressure on ethanoic acid would have to decrease.

52. Evaporation is the conversion of a liquid to a gas or vapor when the liquid is below its boiling point. The vapor pressure is the force per unit area exerted by the vaporized particles on the walls of a sealed container. The boiling point is the temperature at which the vapor pressure of the liquid is equal to the external pressure.

53. Although the net amounts of vapor and liquid remain constant, some molecules are evaporating while an equal number of molecules are condensing.

54. a. 80°C
b. 100°C

55. decrease; As the attractions become stronger, it becomes more difficult for molecules to overcome the attractions and vaporize.

56. a. liquid **b.** vapor **c.** liquid **d.** vapor **e.** solid **f.** solid

57. about 77°C

58. Atmospheric pressure results from the collisions of particles in air with objects. There are fewer particles in a given volume of air at the top of a mountain than at sea level.

59. As the temperature drops to −196°C, the average kinetic energy of particles in the air decreases drastically as does the pressure. So the volume of the balloon, which is a flexible container, decreases. As the balloon warms back to room temperature, the average kinetic energy of the particles increases and the balloon expands to its previous volume.

CHAPTER 13

Assessment continued

Understanding Concepts

50. What happens to the average kinetic energy of the water molecules in your body when you have a fever?

51. Refer to Figure 13.9 to answer these questions about chloroform, ethanoic acid, water, and ethanol.
a. What is the normal boiling point of ethanoic acid?
b. Which liquid has the highest vapor pressure at 40°C?
c. At standard atmospheric pressure, which of the substances are in the gaseous state at 70°C?
d. Water boils at 100°C at standard pressure. How would the pressure on ethanol and on ethanoic acid have to change for these liquids to boil at 100°C?

52. Describe evaporation, vapor pressure, and boiling point.

53. Why is the equilibrium that exists between a liquid and its vapor in a closed container called a dynamic equilibrium?

54. The table gives the vapor pressure of isopropyl alcohol at various temperatures. Graph the data. Use a smooth curve to connect the data points.

Temperature (°C)	Vapor pressure (kPa)
0	1.11
25	6.02
50	23.9
75	75.3
100	198
125	452

a. What is the estimated normal boiling point of isopropyl alcohol?
b. What is the boiling point of isopropyl alcohol when the external pressure is increased to twice standard pressure?

55. In a series of liquids, as the intermolecular forces of attraction strengthen, would you expect the vapor pressure to increase or decrease? Explain.

56. Predict the physical state of each of these substances at the indicated temperature. Use the melting point and boiling point data from the table below.
a. phenol at 99°C
b. ammonia at −25°C
c. methanol in an ice-water bath
d. methanol in a boiling-water bath
e. ammonia at −100°C
f. phenol at 25°C

Substance	Melting Point (°C)	Boiling Point (°C)
ammonia	−77.7	−33.4
methanol	−97.7	64.7
water	0	100
phenol	40.9	181.9

57. Mount McKinley in Alaska is the tallest peak in North America at 6194 m. The atmospheric pressure at its peak is 44 kPa. Use Figure 13.9 to find the boiling point of water at the peak of Mount McKinley.

58. What causes atmospheric pressure, and why is it much lower on the top of a mountain than it is at sea level?

59. Pouring liquid nitrogen onto a balloon decreases the volume of the balloon dramatically, as shown in the photograph on the left. In time, the balloon reinflates as shown in the photograph on the right. Use kinetic theory to explain this sequence of events. The temperature of liquid nitrogen is −196°C.

Critical Thinking

60. What role does atmospheric pressure play when someone is drinking a liquid through a straw?
61. Your lab partner measures the boiling point of water in an open beaker as 108.2°C. Even though you know that water can be made to boil at this temperature, you ask your partner to repeat the measurement. Explain.
62. What everyday evidence suggests that all matter is in constant motion?
63. Is the average kinetic energy of the particles in a block of ice at 0°C the same as or different from the average kinetic energy of the particles in a gas-filled weather balloon at 0°C? Explain.
64. Are perfectly elastic collisions possible between objects that are large enough for you to see?
65. How does perspiration help cool your body on a hot day?
66. Why do different liquids have different normal boiling points?
67. There is a liquid–vapor equilibrium in a container. Explain why the vapor pressure in the container is not affected when the volume of the container is changed.
68. A teacher wants to demonstrate that unheated water can boil at room temperature in a beaker within a bell jar connected to a vacuum pump. However, the vacuum pump is faulty and can reduce pressures only to 15 kPa. Can the teacher use this pump to perform the demonstration successfully? Explain your answer.
69. You have two similar sealed jars of water at the same temperature. In the first jar, there is a large amount of water. In the second jar, there is a small amount of water. Explain how the vapor pressure can be the same in both jars even though many more water molecules are evaporating in the first jar.
70. Why are pressure cookers recommended for cooking at high-altitude?
71. A mixture of gases contains oxygen, nitrogen, and water vapor. What physical process could you use to remove the water vapor from the sample?

Concept Challenge

72. The ions in sodium chloride are arranged in a face-centered cubic pattern. Sketch a layer of the ions in a crystal of sodium chloride.
73. Using Figure 13.11, identify the crystal system described by these characteristics.
    a. three unequal edges meet at right angles
    b. three equal edges with three equal angles that are not right angles
    c. two equal edges and one unequal edge meet at right angles
    d. three unequal edges do not meet at right angles
    e. three equal edges meet at right angles
74. Use this drawing to answer the questions.

    a. What type of unit cell is in a lattice of cesium chloride?
    b. What is the coordination number of Cs^+?
    c. Based on the diagram, what is the formula of cesium chloride? Explain your answer.
75. Relative humidity is defined by the following equation

$$\textit{Relative humidity} = \frac{(a)}{(b)} \times 100\%$$

where (a) is the pressure of water vapor in the air and (b) is the equilibrium vapor pressure of water in air at the same temperature. Can the relative humidity ever exceed 100%? Explain.
76. The solid-liquid equilibrium line in the phase diagram of a given substance slants to the right. How is the substance's freezing point affected by increased pressure?

Critical Thinking

60. When you draw on a straw, the pressure inside the straw is less than the pressure (atmospheric) on the liquid in the container. So liquid is pushed up the straw.
61. Possible answer: Since the beaker is an open container, the water should boil at 100°C at or close to sea level. Your partner probably misread the thermometer and should recheck the value.
62. Possible answers: Odors will travel through a room; ink will spread throughout a beaker of water.
63. The average kinetic energy is the same, because the temperature is the same.
64. No. Collisions between large objects involve some loss of kinetic energy to heat.
65. Because the evaporation of perspiration is an endothermic process the skin is cooled.
66. The intermolecular attractions in some compounds are stronger than in others.
67. The vapor pressure depends only on the kinetic energy of the escaping molecules.
68. No. At 15 kPa water would boil at a temperature of about 50°C, which is much higher than room temperature.
69. The kinetic energy of the molecules in the vapor is the same in both cases; so the vapor pressure is the same.
70. At high altitude, the boiling point of water is less than 100°C because the atmospheric pressure is lower. The increased pressure in the pressure cooker increases the temperature at which water boils.
71. condensation of the water vapor on a cold surface

Concept Challenge

72. $Na^+Cl^-Na^+Cl^-$
$Cl^-Na^+Cl^-Na^+$
$Na^+Cl^-Na^+Cl^-$
$Cl^-Na^+Cl^-Na^+$

73. **a.** orthorhombic
b. rhombohedral
c. tetragonal
d. triclinic
e. cubic

74. **a.** body-centered cubic
b. 8
c. CsCl (one Cl^- ion and 8 × 1/8 equals one Cs^+ ion)

75. No; if (a) = (b) then water vapor will condense at the same rate that the liquid evaporates.

76. The freezing point increases.

Cumulative Review

77. inversely

78. a. S
b. K

79. a. $1s^2 2s^2 2p^6 3s^2 3p^6$
b. $1s^2 2s^2 2p^6 3s^2 3p^6$
c. $1s^2$

80. a. 2
b. 2

81. dispersion forces, dipole interactions, hydrogen bonds

82. a. Cu_2SO_3
b. HNO_2

83. a. CO
b. PBr_3

84. a. Fe^{3+}
b. Cd^{2+}

85. a. 53.7% Fe
b. 34.6% Al

86. a. 13.9 mol SO_2
b. 0.0472 mol NH_3

87. a. 51.2 g Cl_2O_7
b. 30.6 mL H_2O

88. a. 0.38 mol CO_2
b. 3.73 mol NH_3

89. $H_2S(aq) + Cd(NO_3)_2(aq) \rightarrow 2HNO_3(aq) + CdS(s)$

90. a. $V_2O_5 + 2H_2 \rightarrow V_2O_3 + 2H_2O$
b. $(NH_4)_2Cr_2O_7 \rightarrow Cr_2O_3 + N_2 + 4H_2O$

91. a. Mg
b. Li

92. a. combination
b. combustion

93. a. 198 g H_2O
b. 23 mol
c. 144 g C

94. C_2H_4

95. 39.4 g FeS (0.448 mol)

96. 41.9%

CHAPTER 13 Assessment continued

Cumulative Review

77. How are the frequency and wavelength of light waves related? (*Chapter 5*)

78. Which atom in each pair has the larger atomic radius? (*Chapter 6*)
a. O and S **b.** K and Br

79. Write the complete electron configuration of each ion. (*Chapter 7*)
a. Ca^{2+} **b.** S^{2-} **c.** Li^{+}

80. How many unshared pairs of electrons are in each molecule? (*Chapter 8*)
a. H_2O **b.** CO

81. List the intermolecular attractions between molecules in order of increasing strength. (*Chapter 8*)
a. dispersion forces
b. hydrogen bonds
c. dipole interactions

82. Write a correct formula for each compound. (*Chapter 9*)
a. copper(I) sulfite **b.** nitrous acid

83. Identify the binary molecular compound in each pair of substances. (*Chapter 9*)
a. NaCl or CO **b.** PBr_3 or LiOH

84. Write formulas for these ions. (*Chapter 9*)
a. iron(III) **b.** cadmium

85. Calculate the percent by mass of the metal in each compound. (*Chapter 10*)
a. Fe_2S_3 **b.** $Al(OH)_3$

86. How many moles are there in each sample? (*Chapter 10*)
a. 888 g of sulfur dioxide
b. 2.84×10^{22} molecules of ammonia

87. Perchloric acid forms by the reaction of water with dichlorine heptoxide. (*Chapter 10*)

$$Cl_2O_7 + H_2O \longrightarrow 2HClO_4$$

a. How many grams of Cl_2O_7 react with an excess of H_2O to form 56.2 g of $HClO_4$?
b. How many mL of water are needed to form 3.40 mol $HClO_4$?

88. How many moles are there in each sample? (*Chapter 10*)
a. 8.6 L CO_2 (at STP) **b.** 63.4 g NH_3

89. When hydrogen sulfide gas is bubbled into a solution of cadmium nitrate in water, the products are nitric acid and a precipitate of cadmium sulfide. Write a balanced equation for the reaction. Include physical states for all reactants and products. (Hydrogen sulfide gas is soluble in water.) (*Chapter 11*)

90. Balance these equations. (*Chapter 11*)
a. $V_2O_5 + H_2 \longrightarrow V_2O_3 + H_2O$
b. $(NH_4)_2Cr_2O_7 \longrightarrow Cr_2O_3 + N_2 + H_2O$

91. List the metal that ranks higher in the activity series of metals. (*Chapter 11*)
a. magnesium or mercury
b. potassium or lithium

92. Classify each reaction as a combination, decomposition, single-replacement, double-replacement, or combustion. (*Chapter 11*)
a. $2Li(s) + Br_2(l) \longrightarrow 2LiBr(s)$
b. $2C_2H_6(g) + 7O_2(g) \longrightarrow 4CO_2(g) + 6H_2O(g)$

93. The complete decomposition of sucrose (table sugar) caused by strong heating may be represented by this equation. (*Chapter 12*)

$$C_{12}H_{22}O_{11}(s) \longrightarrow 11H_2O(l) + 12C(s)$$

For the decomposition of 1.00 mol of sucrose:
a. How many grams of H_2O are produced?
b. What is the total number of moles of products produced?
c. How many grams of C are produced?

94. Hydrogen reacts with ethene (C_2H_4) to form ethane (C_2H_6). (*Chapter 12*)

$$C_2H_4 + H_2 \longrightarrow C_2H_6$$

What is the limiting reagent when 40.0 g C_2H_4 reacts with 3.0 g H_2?

95. Iron(II) sulfide is produced when iron is heated with sulfur. (*Chapter 12*)

$$Fe(s) + S(s) \xrightarrow{\Delta} FeS(s)$$

What is the theoretical yield of FeS if 25.0 g Fe is heated with 32.0 g S?

96. What is the percent yield in Question 95 if 16.5 g of FeS is produced? (*Chapter 12*)

Standardized Test Prep

Test-Taking Tip

Interpreting Graphs A line graph helps you see the relationship between two variables. Before you answer a question about a graph, identify the variables and the general relationship between the variables based on the shape of the curve.

Use the graph to answer Questions 1 and 2.

Vapor Pressure Graph of Three Substances

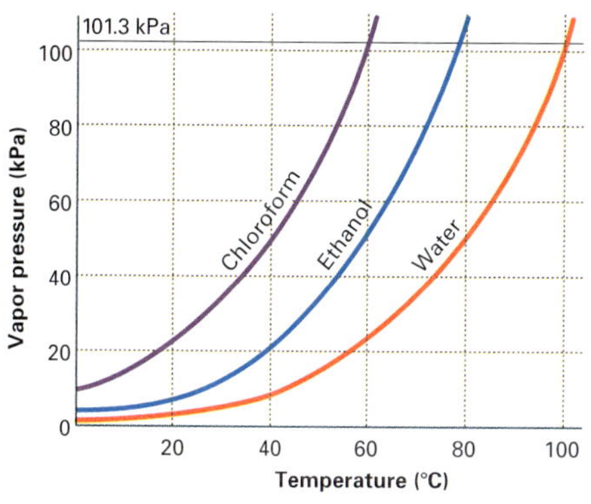

1. What is the normal boiling point of ethanol?
2. Can chloroform be heated to 90°C in an open container?
3. Which sequence has the states of CH_3OH correctly ordered in terms of increasing average kinetic energy?
   a. $CH_3OH(s)$, $CH_3OH(g)$, $CH_3OH(l)$
   b. $CH_3OH(g)$, $CH_3OH(l)$, $CH_3OH(s)$
   c. $CH_3OH(l)$, $CH_3OH(g)$, $CH_3OH(s)$
   d. $CH_3OH(s)$, $CH_3OH(l)$, $CH_3OH(g)$

Use the drawing to answer Questions 4–6. The same liquid is in each flask.

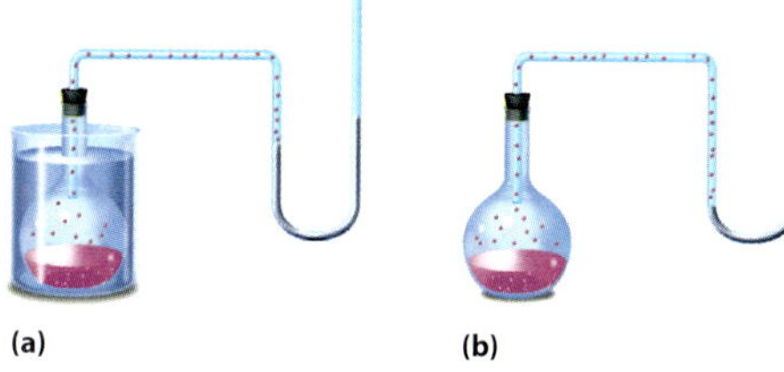

4. In which flask is the vapor pressure lower? Give a reason for your answer.
5. In which flask is the liquid at the higher temperature? Explain your answer.
6. How can the vapor pressure in each flask be determined?

For each question there are two statements. Decide whether each statement is true or false. Then decide whether Statement II is a correct explanation for Statement I.

	Statement I		Statement II
7.	In an open container, the rate of evaporation of a liquid always equals the rate of condensation.	BECAUSE	A dynamic equilibrium exists between the liquid and its vapor in an open container.
8.	Water boils at a temperature below 100°C on top of a mountain.	BECAUSE	Atmospheric pressure decreases with an increase in altitude.
9.	The temperature of a substance always increases as heat is added to the substance.	BECAUSE	The average kinetic energy of the particles in a substance increase with an increase in temperature.
10.	Solids have a fixed volume.	BECAUSE	Particles in a solid cannot move.
11.	Gases are more compressible than liquids.	BECAUSE	There is more space between particles in a gas than between particles in a liquid.

Standardized Test Prep

1. about 78°C
2. no
3. d
4. Vapor pressure is lower in flask B because the level of mercury in the left arm of B is higher, which indicates less pressure on the mercury.
5. Temperature is higher in flask A because vapor increases with temperature.
6. by measuring the difference in heights between the two arms
7. False, False
8. True, True, correct explanation
9. False, True
10. True, False
11. True, True, correct explanation

CHAPTER 14

The Behavior of Gases Planning Guide

SECTION OBJECTIVES	STANDARDS NATIONAL	STANDARDS STATE	ACTIVITIES and LABS
14.1 Properties of Gases, pp. 413–417 1 block or 2 periods **14.1.1 Explain** why gases are easier to compress than solids or liquids are. **14.1.2 Describe** the three factors that affect gas pressure.	B-2, B-5		**SE** Inquiry Activity: Observing Volume Changes, p. 412 L2 **TE** Teacher Demo: Pressure and Particle Size, p. 415 L2 **VCL** Lab 15: Investigation of Gas Pressure and Mass L2
14.2 The Gas Laws, pp. 418–425 1 block or 2 periods **14.2.1 Describe** the relationships among the temperature, pressure, and volume of a gas. **14.2.2 Use** the combined gas law to solve problems.	A-1, A-2, B-2, B-5		**TE** Teacher Demo: Pressure and Volume, p. 419 L2 **TE** Class Activity: Observing the Effect of Pressure on Temperature, p. 422 L1 **LM** Lab 23: Pressure–Volume Relationships for Gases L2 **LM** Lab 24: Temperature–Volume Relationships for Gases L2 **PLM** Probeware Lab: Analysis of Baking Soda L2 **LP** Lab Practical 14-1: Boyle's Law L2 **LP** Lab Practical 14-2: Charle's Law L2 **VCL** Lab 11: Pressure-Volume Relationship for Gases L2 **VCL** Lab 12: Temperature-Volume Relationship for Gases L2
14.3 Ideal Gases, pp. 426–431 1 block or 2 periods **14.3.1 Compute** the value of an unknown using the ideal gas law. **14.3.2 Compare and contrast** real and ideal gases.	A-1, A-2, B-2, E-2, F-1		**SE** Quick Lab: Carbon Dioxide from Antacid Tablets, p. 428 L2 **PLM** Probeware Lab: Carbon Dioxide from Antacid Tablets L2 **VCL** Lab 13: Derivation of the Ideal Gas Law L2 **VCL** Lab 14: Ideal vs Real Gases L2
14.4 Gases: Mixtures and Movements, pp. 432–437 1 block or 2 periods **14.4.1 Relate** the total pressure of a mixture of gases to the partial pressures of the component gases. **14.4.2 Explain** how the molar mass of a gas affects the rate at which the gas diffuses and effuses.	A-1, A-2, B-2		**SE** Small-Scale Lab: Diffusion, p. 437 L2 **TE** Class Activity: Model Partial Pressure, p. 433 L2 **TE** Class Activity: Effusion, p. 435 L1 **LM** Lab 25: Diffusion of Gases L2 **SSLM** Lab 21: Synthesis and Qualitative Analysis of Gases L2

Ability Levels

- **L1** For students who need additional help
- **L2** For all students
- **L3** For students who need to be challenged

Components

SE	Student Edition	**LM**	Laboratory Manual
TE	Teacher's Edition	**SSLM**	Small-Scale Chemistry Lab Manual
GRSW	Guided Reading and Study Workbook	**LP**	Laboratory Practicals
CTR	Core Teaching Resources	**PLM**	Probeware Lab Manual
SSG	Spanish Study Guide	**VCL**	Virtual Chemistry Labs
		CTB	Computer Test Bank
GCP	Graphing Calculator Problems	**T**	Transparencies
TTT	Test-Taking Tips with Transparencies, Grades 9–12 Science	**CA**	Chem Alive and Chemistry Field Trips DVD
STP	Standardized Test Preparation Workbook	**iT**	Interactive Textbook with ChemASAP
		GO	Go Online Internet Resources

RESOURCES PRINT and TECHNOLOGY	PROBLEMS and ASSESSMENT
GRSW Chapter 14, Section 1 **L1** **T** T150–T151 **L2** **GO** Chapter 14, Section 1 **L2**	**SE** Section 14.1 Assessment, p. 417 **iT** Section 14.1
GRSW Chapter 14, Section 2 **L1** **T** T152–T155 **L2** **GO** Chapter 14, Section 2 **L2**	**SE** Section 14.2 Assessment, p. 425 **iT** Section 14.2 **SP** 14.1, Using Boyle's Law, p. 419 **SP** 14.2, Using Charles's Law, p. 421 **SP** 14.3, Using Gay–Lussac's Law, p. 423 **SP** 14.4, Using the Combined Gas Law, p. 424
GRSW Chapter 14, Section 3 **L1** **T** T156–T157 **L2**	**SE** Section 14.3 Assessment, p. 429 **iT** Section 14.3 **SP** 14.5, Using the Ideal Gas Law to Find the Amount of a Gas, p. 427
GRSW Chapter 14, Section 4 **L1** **T** T158–T159 **L2** **GO** Chapter 14, Section 4 **L2**	**SE** Section 14.4 Assessment, p. 436 **iT** Section 14.4 **SP** 14.6, Using Dalton's Law of Partial Pressures, p. 434

Web Code: cdn-1141
Web Code: cdn-1142
Web Code: cdn-1144

MATERIALS List

Quantities for each group

STUDENT EDITION

Inquiry Activity, p. 412
round non-latex balloon, marking pen, freezer, metric tape measure, sunny window

Quick Lab, p. 428
6 effervescent antacid tablets, 3 rubber balloons (spherical), plastic medicine dropper, water, clock or watch, metric tape measure, graph paper, pressure sensor (optional)

Small-Scale Lab, p. 437
clear plastic cup or Petri dish; reaction surface; dropper bottles containing bromthymol blue, hydrochloric acid, and sodium hydrogen sulfite solutions; ruler; cotton swab; NaOH, NH_4Cl, KI, and $NaNO_2$ (optional)

TEACHER'S EDITION

Teacher Demo, p. 415
2 identical glass containers, 2 different sizes of beads

Class Activity, p. 419
vacuum pump, bell jar, marshmallows

Class Activity, p. 422
2 large vats, ice water, hot water, inflated bicycle tire

Class Activity, p. 433
balance, 4 marbles, 2 buttons, 3 pennies, 4 nickels

Class Activity, p. 435
2 round, identical balloons; helium; metric tape measure

Chapter Assessment

CHAPTER ASSESSMENT

- **SE** Chapter Assessment, pp. 439–442
- **CTB** Chapter 14
- **IT** Chapter 14

STANDARDIZED TEST PREP

- **SE** Chapter 14, p. 443
- **STP** Topic 4
- **IT** Interactive Textbook with Assessment at PHSchool.com

CHAPTER 14

The Behavior of Gases

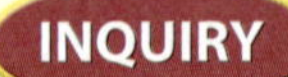

Observing Volume Changes 

Purpose

In this activity, students will:

- observe the effect of temperature on the volume of a gas.

Prep Time 10 minutes
Class Time 20 minutes actual class time, 60 minutes wait time

Safety If only latex balloons are available, don't let students who are allergic to latex handle the balloons.

Think About It

1. The volume decreased.
2. The volume increased.
3. As temperature decreases, the volume of a gas decreases. As temperature increases, the volume of a gas increases.

After students study Charles's law in Section 14.2, ask them to explain the relationship between temperature and volume they observed in this activity.

CHAPTER 14

The Behavior of Gases

The tank of compressed gas allows this diver to observe the sea turtle in its underwater environment.

INQUIRY Activity

Observing Volume Changes

Materials

round non-latex balloon, marking pen, freezer, metric tape measure, sunny window

Procedure 

1. Inflate the balloon and tie it closed. Use the marking pen to draw a line around the middle of the balloon. This line is the circumference of the balloon. You will use the circumference to track any change in the volume of the balloon.
2. Place the balloon in the freezer for half an hour. Remove it and quickly measure and record the circumference of the balloon in centimeters.
3. Place the balloon in a sunny window for half an hour. Measure and record the circumference of the balloon in centimeters.

Think About It

1. How did the volume of the balloon change when you placed it in the freezer?
2. What happened to the volume of the balloon when you placed it in a sunny window?
3. Use your results to suggest a relationship between the temperature and volume of a gas.

14.1 Properties of Gases

Connecting to Your World In organized soccer, there are rules about equipment. For international competitions, the ball's mass must be not more than 450 grams and not less than 410 grams. The pressure of the air inside the ball must be no lower than 0.6 atmospheres and no higher than 1.1 atmospheres at sea level. A ball that is properly inflated will rebound faster and travel farther than a ball that is under-inflated. If the pressure is too high, the ball may burst when it is kicked. In this section, you will study variables that affect the pressure of a gas.

Guide for Reading

Key Concepts
- Why are gases easier to compress than solids or liquids are?
- What are the three factors that affect gas pressure?

Vocabulary
compressibility

Reading Strategy
Relating Cause and Effect A cause makes something happen. As you read, write down three possible causes for an increase in gas pressure.

Compressibility

When a car comes to a sudden stop, the people in the car will continue to move forward unless they are restrained. The driver and any passengers are more likely to survive a collision if they are wearing seat belts to restrict their forward movement. Cars also contain air bags as a second line of defense. A sudden reduction in speed triggers a chemical reaction inside an air bag. One product of the reaction is nitrogen gas, which causes the bag to inflate. An inflated air bag keeps the driver from colliding with the steering wheel, as shown in Figure 14.1. An inflated air bag keeps a passenger from colliding with the dashboard or windshield.

Recall that a gas can expand to fill its container, unlike a solid or liquid. The reverse is also true. Gases are easily compressed, or squeezed into a smaller volume. **Compressibility** is a measure of how much the volume of matter decreases under pressure. Why does a collision with an inflated air bag cause much less damage than a collision with a steering wheel or dashboard? When a person collides with an inflated air bag, the impact forces the molecules of gas in the bag closer together. The compression of the gas absorbs the energy of the impact.

Figure 14.1 A crash dummy can be used to test the effectiveness of an air bag. Because gases can be compressed, the air bag absorbs some of the energy from the impact of a collision. Air bags work best when combined with seat belts.

Section Resources

Print
- **Guided Reading and Study Workbook,** Section 14.1
- **Core Teaching Resources,** Section 14.1 Review
- **Transparencies,** T150–T151

Technology
- **Interactive Textbook with ChemASAP,** Assessment 14.1
- **Go Online,** Section 14.1
- **Virtual Chemistry Labs,** Lab 15

14.1

1 FOCUS

Objectives

14.1.1 Explain why gases are easier to compress than solids or liquids are.

14.1.2 Describe the three factors that affect gas pressure.

Guide for Reading

Build Vocabulary 

Word Parts Have students look up the meanings of the prefix *com-* (together), the word *press* (to act on with a steady force or weight), and the suffix *-ibility* (ability or tendency). Have them use these meanings to define *compressibility*.

Reading Strategy 

Identify Main Ideas For each paragraph in this section, have students identify the main idea. They can use their list as a study guide.

2 INSTRUCT

Connecting to Your World

Ask students to examine the photograph that opens the section. Ask, **What factors affect the pressure of the air inside the ball?** *(the temperature of the air in the ball; and the volume of the ball)*

Compressibility

Discuss 

Point out that in this chapter students will build upon their knowledge of kinetic theory to describe the behavior of gases. Ask, **What is kinetic energy?** *(energy of motion)* Ask, **How are temperature and kinetic energy related?** *(Temperature is a measure of average kinetic energy.)*

FYI

You may want to use the T&S feature on air bags in Chapter 12 on p. 376.

Section 14.1 (continued)

Use Visuals L1

Figure 14.2 This figure illustrates the amount of space in a gas under ordinary conditions. Ask, **Estimate the amount of space occupied by the molecules in the figure.** *(Students are likely to say less than 5%.)* Explain that, for practical reasons, in most drawings, the particles will be too close together for their depicted size. Remind students how small the actual particles are, and how many particles there are in a small volume of gas.

Factors Affecting Gas Pressure

Discuss 

To assess students' knowledge of the kinetic theory, ask, **How can you describe gas pressure?** *(Gas pressure is the result of collisions of particles with their container.)* Ask, **How does increasing the number of particles in a gas affect the pressure?** *(The more particles colliding with the walls of a container, the higher the pressure.)*

Relate 

Lead the class in a discussion of the four variables—volume, temperature, pressure, and number of moles—used to describe a gas. Discuss how the properties of gases lead to some unique uses of them. For example, because the particles in a gas are relatively far apart, gases are poor thermal conductors and are used for insulation between the glass panes in double-paned windows.

FYI

In this section, students look at the relationships between the variables *P*, *V*, *T*, and *n* qualitatively before addressing these relationships quantitatively in Section 14.2.

Download a worksheet on **Gases** for students to complete, and find additional teacher support from NSTA SciLinks.

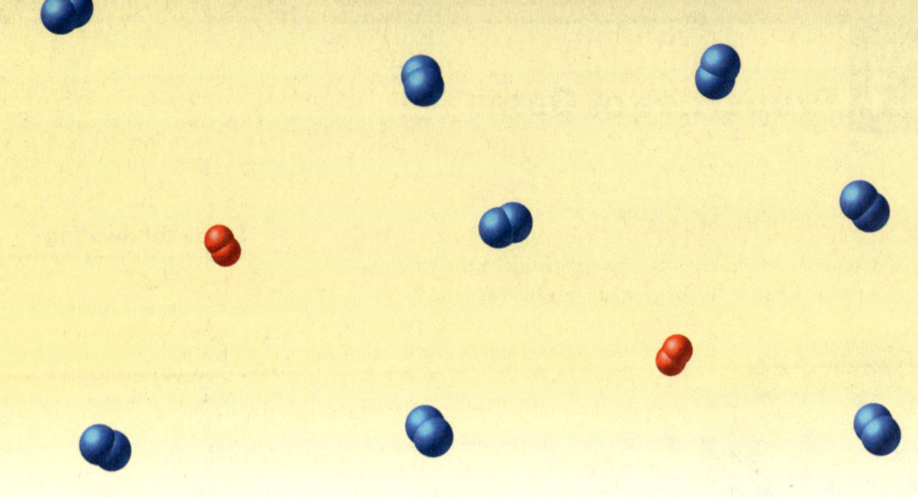

Figure 14.2 There are only a few nitrogen and oxygen molecules in this model of air. At room temperature, the distance between molecules in a container of air at standard pressure is about 10 times the diameter of a molecule.

Kinetic theory can explain why gases are compressed more easily than liquids or solids. **Gases are easily compressed because of the space between the particles in a gas.** Remember that the volume of the particles in a gas is small compared to the overall volume of the gas. So the distance between particles in a gas is much greater than the distance between particles in a liquid or solid. Under pressure, the particles in a gas are forced closer together, or compressed.

At room temperature, the distance between particles in an enclosed gas is about 10 times the diameter of a particle. Figure 14.2 is a model of an air sample at room temperature. Only oxygen and nitrogen—the two main gases in air—are represented. Note how few particles can fit in this relatively large space when the distance between particles is reasonable for the size of the particles. It isn't practical to represent the actual distances between particles in all the molecular drawings of gases in this book. The drawings would be too large to easily fit on a page. For the drawings to fit on a page, the particles must be drawn closer together.

 ***How does the volume of the particles in a gas compare to the overall volume of the gas?***

Factors Affecting Gas Pressure

Kinetic theory can help explain other properties of gases, such as their ability to expand and take the shape and volume of their containers. Recall these assumptions about the particles in a gas. The particles move along straight-line paths until they collide with other particles or the walls of their container. The motion of the particles is constant and random. Because there are no significant forces of attraction or repulsion among particles in a gas, particles in a gas can move freely.

Four variables are generally used to describe a gas. The variables and their common units are pressure (*P*) in kilopascals, volume (*V*) in liters, temperature (*T*) in kelvins, and the number of moles (*n*). **The amount of gas, volume, and temperature are factors that affect gas pressure.**

Go Online NSTA SciLinks

For: Links on Gases
Visit: www.SciLinks.org
Web Code: cdn-1141

Differentiated Instruction

Less Proficient Readers 

Help students make a list of assumptions about gases in the kinetic theory. Review the list and ask for a behavior of gases related to each assumption.

Amount of Gas An air-filled raft blasts through a narrow opening between rocks and plummets over a short waterfall into churning white water below. The raft bends and twists, absorbing some of the pounding energy of the river. The strength and flexibility of the raft are impressive and rely on the pressure of the gas inside the raft. The raft must be made of a material that is strong enough to withstand the pressure of the air inside the raft. The material must also keep air from leaking out of the raft. The volume of the inflated raft in Figure 14.3 is dramatically larger than the volume of the raft before it is inflated. As air is added, the raft expands to its intended volume. The pressure of the air inside the raft keeps the raft inflated.

Figure 14.3 The volume of this air-filled raft is much larger than its volume before it was inflated. Using a pump to force air into a raft increases the pressure of the air inside the raft. The increased pressure causes the raft to inflate to its intended size.

You can use kinetic theory to predict and explain how gases will respond to a change of conditions. If you inflate an air raft, for example, the pressure inside the raft will increase. Collisions of particles with the inside walls of the raft result in the pressure that is exerted by the enclosed gas. By adding gas, you increase the number of particles. Increasing the number of particles increases the number of collisions, which explains why the gas pressure increases.

Figure 14.4 shows what happens when gas is added to an enclosed, rigid container. Because the container is rigid, the volume of the gas is constant. Assume also that the temperature of the gas does not change. Under these conditions, doubling the number of particles of gas doubles the pressure. Tripling the number of particles triples the pressure, and so on. With a powerful pump and a strong container, you can generate very high pressures by adding more and more gas. However, once the pressure exceeds the strength of the container, the container will burst. Removing gas from a rigid container has the opposite effect. As the amount of gas is reduced, the pressure inside the container is reduced.

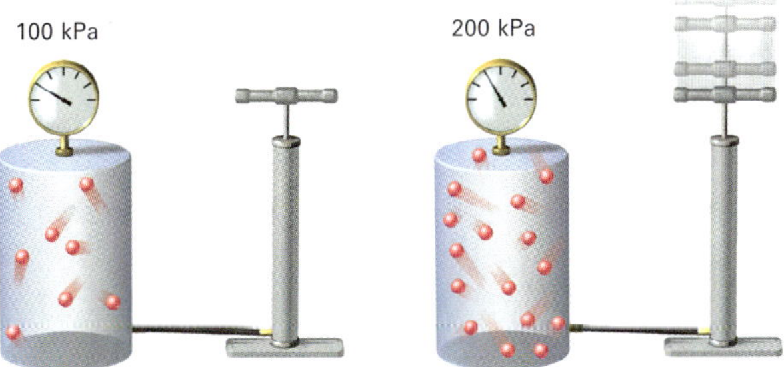

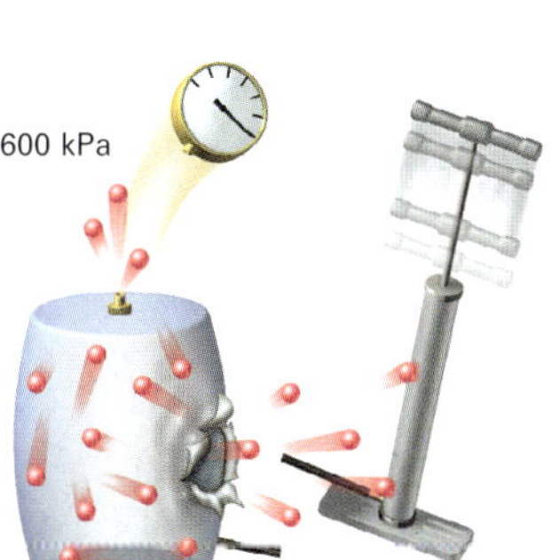

Figure 14.4 When a gas is pumped into a closed rigid container, the pressure increases as more particles are added. If the number of particles is doubled, the pressure will double. **Predicting** ***What would happen to the pressure in the container if the number of particles were tripled? If the number of particles were cut in half?***

Use Visuals L1

Figures 14.3 and 14.4 Make sure students understand the difference between a container with a volume that can vary (within limits), such as a balloon or raft, and a rigid container with a constant volume.

TEACHER Demo

Pressure and Particle Size L2

Purpose Students use an analogy to infer the effect of particle size on gas pressure.

Materials 2 identical glass containers, 2 different sizes of beads

Procedure Add small beads to a glass container until it is approximately one-eighth full. Add the same number of larger beads to an identical container. Have the students assume that the beads represent particles of gases in sealed containers at the same temperature. Ask students to compare the pressures in the two containers.

Expected Outcome The pressures will be identical because both gases contain the same number of particles with the same average kinetic energy in the same volume.

Differentiated Instruction

Special Needs 

Inflate a deflated object with a pump. During the process, ask students to describe and explain what is happening. Ask students to predict what will happen when you remove air from the inflated object. Then deflate the object.

Answers to...

Figure 14.4 The pressure would triple; the pressure would be cut in half.

 The volume of the particles is much smaller than the overall volume of the gas.

Discuss 

Ask students to think about why the instructions on an aerosol say to hold the can upright while spraying. Ask, **If the can is not held upright, what material will come out of the can?** *(the propellant)* **How would the loss of propellant affect the use of the can when it is later held upright and sprayed?** *(The amount of propellant is reduced, so there is less pressure to force the paint out of the can.)*

Use Visuals 

Figure 14.6 Figure 14.6 is the first of the drawings that show a piston and cylinder. If possible, show students an actual piston and cylinder. Explain how the tight fit of the piston in the cylinder keeps the gas contained and, thus, able to be compressed. Ask, **What will happen to the pressure in the cylinder as the volume is decreased?** *(Pressure will increase.)* Point out that the number of particles in both diagrams is the same. Then ask, **Why does the pressure of a contained gas double when the volume is reduced by one-half?** *(Reducing the volume by one-half doubles the number of simultaneous collisions of particles with the walls of the piston, doubling the pressure.)*

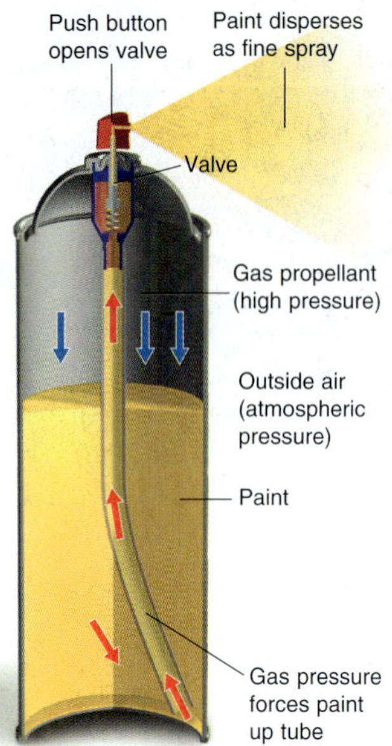

Figure 14.5 The pressure of the gas inside a new can of spray paint is greater than the air pressure outside the can. When gas rushes though an opening in the top of the can, it propels, or forces, paint out of the can. As the can is used, the pressure of the propellant decreases.
Relating Cause and Effect ***What happens when the pressure of the propellant equals the air pressure outside the can?***

If the pressure of the gas in a sealed container is lower than the outside air pressure, air will rush into the container when the container is opened. This movement causes the whoosh you hear when you open a vacuum-packed container. When the pressure of a gas in a sealed container is higher than the outside air pressure, the gas will flow out of the container when the container is unsealed.

Aerosol cans depend on the movement of a gas from a region of high pressure to a region of lower pressure. Aerosol cans may contain whipped cream, hair mousse, or spray paint. Figure 14.5 shows how a can of spray paint works. The can contains a gas stored at high pressure. The air outside the can is at a lower pressure. Pushing the spray button creates an opening between the inside of the can and the air outside. The gas flows through the opening to the lower pressure region outside. The movement of the gas propels, or forces, the paint out of the can. As the gas is depleted, the pressure inside the can decreases until the gas can no longer propel paint from the can.

Volume You can raise the pressure exerted by a contained gas by reducing its volume. The more the gas is compressed, the greater is the pressure that the gas exerts inside the container. When gas is in a cylinder, as in an automobile engine, a piston can be used to reduce its volume. The snug-fitting piston keeps gas from escaping as the cylinder moves down and up.

Figure 14.6 shows a cylinder of gas under two different conditions. When the cylinder has a volume of 1 L, the gas exerts a pressure of 100 kPa. When the volume is halved to 0.5 L, the pressure is doubled to 200 kPa. Increasing the volume of the contained gas has the opposite effect. If the volume is doubled, the particles can expand into a volume that is twice the original volume. With the same number of particles in twice the volume, the pressure of the gas is cut in half.

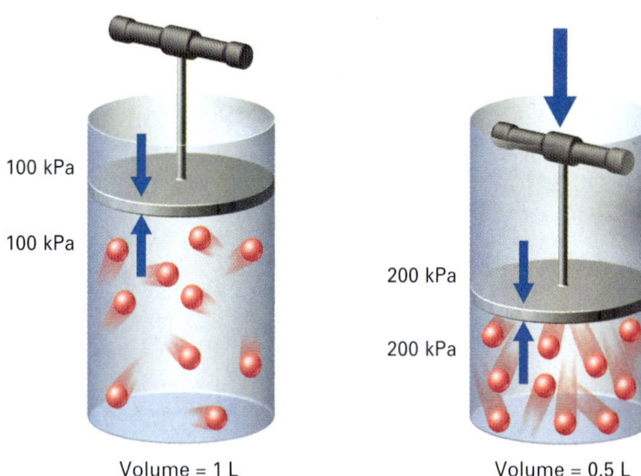

Figure 14.6 A piston can be used to force a gas in a cylinder into a smaller volume. When the volume is decreased, the pressure the gas exerts is increased.
Interpreting Diagrams ***What happens to the gas pressure when the volume is reduced from 1 L to 0.5 L?***

Facts and Figures

Gas and Momentum

The particles in a gas have momentum, the product of their mass and velocity (*mv*). The pressure exerted by a gas in a given volume is proportional to the average kinetic energy of its molecules ($1/2\ mv^2$). The mass of the particles is constant, so any change in momentum is due to a change in velocity. A rise in temperature reflects a rise in average kinetic energy, corresponding to an increase in momentum and pressure.

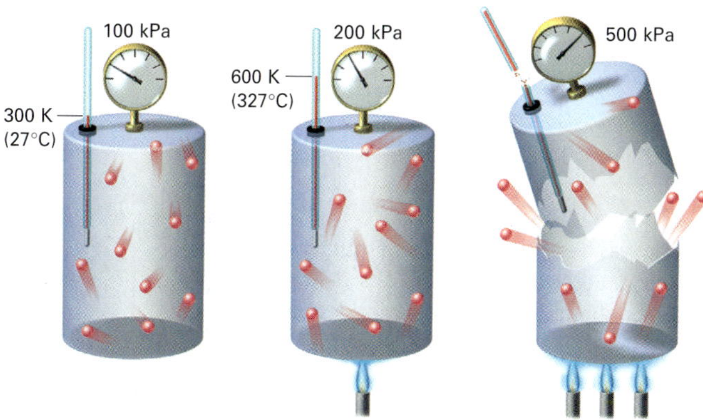

Figure 14.7 An increase in temperature causes an increase in the pressure of an enclosed gas. The container can explode if there is too great an increase in the pressure.

Temperature A sealed bag of potato chips bulges at the seams when placed in a sunny location. The bag bulges because an increase in the temperature of an enclosed gas causes an increase in its pressure. You can use kinetic theory to explain what happens. As a gas is heated, the temperature increases and the average kinetic energy of the particles in the gas increases. Faster-moving particles strike the walls of their container with more energy.

Look at Figure 14.7. The volume of the container and the amount of gas is constant. When the Kelvin temperature of the enclosed gas doubles from 300 K to 600 K, the pressure of the enclosed gas doubles from 100 kPa to 200 kPa. A gas in a sealed container may generate enormous pressure when heated. For that reason, an aerosol can, even an "empty" one, may explode if thrown onto a fire. By contrast, as the temperature of an enclosed gas decreases, the pressure decreases. The particles, on average, move more slowly and have less kinetic energy. They strike the container walls with less force. Halving the Kelvin temperature of a gas in a rigid container decreases the gas pressure by half.

14.1 Section Assessment

1. **Key Concept** Why is a gas easy to compress?
2. **Key Concept** List three factors that can affect gas pressure.
3. Why does a collision with an air bag cause less damage than a collision with a steering wheel?
4. How does a decrease in temperature affect the pressure of a contained gas?
5. If the temperature is constant, what change in volume would cause the pressure of an enclosed gas to be reduced to one quarter of its original value?
6. Assuming the gas in a container remains at a constant temperature, how could you increase the gas pressure in the container a hundredfold?

Writing Activity

Explaining Aerosol Cans Write a paragraph explaining how the compressed gas in an aerosol can force paint out of the can. Describe how the gas pressure inside the can changes as the paint is sprayed, and identify the variable that causes this change. Refer to Figure 14.5.

Assessment 14.1 Test yourself on the concepts in Section 14.1.

with ChemASAP

Section 14.1 Assessment

1. Gases are easily compressed because of the space between particles in a gas.
2. temperature, pressure, and amount of gas
3. Because the gas in the inflated air bag can be compressed, the bag absorbs some of the energy from the impact of a collision. The solid steering wheel cannot.
4. If the temperature decreases, the pressure will also decrease.
5. The volume would need to increase by a factor of four.
6. Increase the amount of gas in the container one hundredfold.

3 ASSESS

Evaluate Understanding L2

Determine students' understanding of gas pressure. Ask, **What effect would tripling the number of particles of a gas in a closed container have on the pressure exerted?** *(The pressure would triple.)* **What effect would doubling the volume of an enclosed gas have on the pressure?** *(The pressure would decrease by one half.)* **How does the pressure of an enclosed gas change with increasing temperature?** *(The number and force of collisions increase with temperature, and the pressure increases.)*

Reteach L1

Help students make a table to summarize how gas pressure changes as variables increase or decrease. Label the rows *Volume, Temperature,* and *Number of Particles.* Label the columns *Increase* and *Decrease.* Have students use up or down arrows to complete the table.

Writing Activity

The pressure of the gas inside the can is greater than atmospheric pressure. As the gas is used to propel the paint out the can, the number of particles of gas in the can decreases and the pressure of the gas decreases.

If your class subscribes to the Interactive Textbook, use it to review key concepts in Section 14.1.

with ChemASAP

Answers to...

Figure 14.5 The gas can no longer propel the product out of the can.
Figure 14.6 The pressure is doubled from 100 kPa to 200 kPa.

14.2

1 FOCUS

Objectives

14.2.1 Describe the relationships among the temperature, pressure, and volume of a gas.

14.2.2 Use the combined gas law to solve problems.

Guide for Reading

Build Vocabulary L2

Graphic Organizer Have students make a compare/contrast table for the gas laws with variables, constants, and the classification "direct" or "inverse."

Reading Strategy L2

Preview Have students look at Figures 14.8 and 14.10, and use the graphs to write preliminary versions of Boyle's law and Charles's law.

2 INSTRUCT

Connecting to Your World

Have students study the photograph and read the text that opens the section. Ask, **What is the effect of heating a gas at constant pressure?** *(The density of the gas decreases. As the density of the gas inside the balloon is lowered relative to the density of the gas outside the balloon, the balloon rises.)*

Boyle's Law: Pressure and Volume

Interpreting Graphs L2

a. 50 kPa

b. about 33 kPa

c. As either pressure or volume increases, the other variable decreases.

Enrichment Question L3

Pick two points on the graph, and use them to show mathematically that Boyle's Law describes an inverse relationship. *(For this graph, PV = 100 kPa·L for any point on the line.)*

14.2 The Gas Laws

Guide for Reading

Key Concepts

- How are the pressure, volume, and temperature of a gas related?
- When is the combined gas law used to solve problems?

Vocabulary

Boyle's law
Charles's law
Gay-Lussac's law
combined gas law

Reading Strategy

Relating Text and Visuals As you read, look closely at Figure 14.9. Explain how a combination of photographs and drawings helps you understand the relationship between the temperature and volume of a gas.

Connecting to Your World

This hot air balloon was designed to carry a passenger around the world. Because warm air is less dense than cooler air, the pilot heats the air inside the balloon to make the balloon rise. The pilot releases hot air through a vent in the top of the balloon to make the balloon descend. In this section, you will study some laws that will allow you to predict gas behavior under specific conditions, such as in a hot air balloon.

Boyle's Law: Pressure and Volume

How would an increase in pressure affect the volume of a contained gas? **If the temperature is constant, as the pressure of a gas increases, the volume decreases.** In turn, as the pressure decreases, the volume increases. Robert Boyle was the first person to study this pressure-volume relationship in a systematic way. In 1662, Boyle proposed a law to describe the relationship. **Boyle's law** states that for a given mass of gas at constant temperature, the volume of the gas varies inversely with pressure.

Look at Figure 14.8. A gas with a volume of 1.0 L (V_1) is at a pressure of 100 kPa (P_1). As the volume increases to 2.0 L (V_2), the pressure decreases to 50 kPa (P_2). The product $P_1 \times V_1$ (100 kPa $\times$ 1.0 L = 100 kPa·L) is the same as the product $P_2 \times V_2$ (50 kPa $\times$ 2.0 L = 100 kPa·L). As the volume decreases to 0.5 L (V_3), the pressure increases to 200 kPa (P_3). Again, the product of the pressure and the volume equals 100 kPa·L.

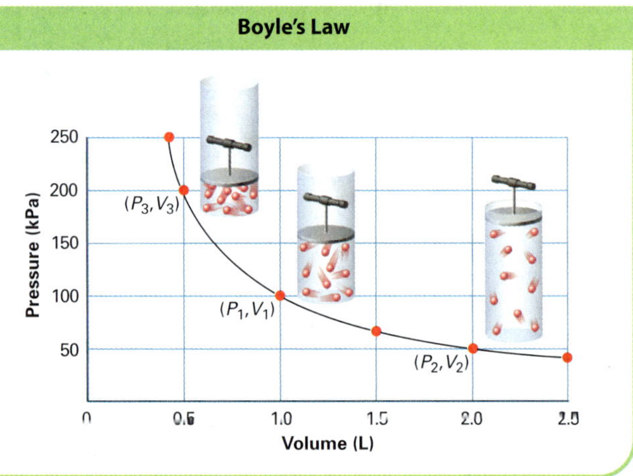

Figure 14.8 The pressure of a gas changes as the volume changes.

INTERPRETING GRAPHS

a. Observing When the volume is 2.0 L, what is the pressure?
b. Predicting What would the pressure be if the volume were increased to 3.0 L?
c. Drawing Conclusions Based on the shape of the graph, describe the general pressure-volume relationship.

Section Resources

Print

- **Guided Reading and Study Workbook,** Section 14.2
- **Core Teaching Resources,** Section 14.2 Review
- **Transparencies,** T152–T155
- **Laboratory Manual,** Labs 23–24
- **Probeware Lab Manual,** Section 14.2
- **Laboratory Practicals,** 23–24

Technology

- **Interactive Textbook with ChemASAP,** Simulation 15, 16, 17; Problem-Solving 14.8, 14.10, 14.12, 1414; Assessment 14.2
- **Go Online,** Section 14.2
- **Virtual Chemistry Labs,** 11, 12

In an inverse relationship, the product of the two variable quantities is constant. So the product of pressure and volume at any two sets of pressure and volume conditions is always constant at a given temperature. The mathematical expression of Boyle's law is as follows.

$$P_1 \times V_1 = P_2 \times V_2$$

The graph of an inverse relationship is always a curve, as in Figure 14.8.

Interactive Textbook

Simulation 15 Examine the relationship between gas volume and pressure.

with **ChemASAP**

SAMPLE PROBLEM 14.1

Using Boyle's Law

A balloon contains 30.0 L of helium gas at 103 kPa. What is the volume of the helium when the balloon rises to an altitude where the pressure is only 25.0 kPa? (Assume that the temperature remains constant.)

1 Analyze ***List the knowns and the unknown.***

Knowns
- $P_1 = 103$ kPa
- $V_1 = 30.0$ L
- $P_2 = 25.0$ kPa

Unknown
- $V_2 = ?$ L

Use Boyle's law ($P_1 \times V_1 = P_2 \times V_2$) to calculate the unknown value (V_2).

2 Calculate ***Solve for the unknown.***

Rearrange Boyle's law to isolate V_2.

$$V_2 = \frac{V_1 \times P_1}{P_2}$$

Substitute the known values for P_1, V_1, and P_2 into the equation and solve.

$$V_2 = \frac{30.0\ \text{L} \times 103\ \text{kPa}}{25.0\ \text{kPa}}$$

$$= 1.24 \times 10^2\ \text{L}$$

3 Evaluate ***Does the result make sense?***

A decrease in pressure at constant temperature must correspond to a proportional increase in volume. The calculated result agrees with both kinetic theory and the pressure-volume relationship. Also, the units have canceled correctly and the answer is expressed to the proper number of significant figures.

Practice Problems

7. Nitrous oxide (N_2O) is used as an anesthetic. The pressure on 2.50 L of N_2O changes from 105 kPa to 40.5 kPa. If the temperature does not change, what will the new volume be?

8. A gas with a volume of 4.00 L at a pressure of 205 kPa is allowed to expand to a volume of 12.0 L. What is the pressure in the container if the temperature remains constant?

Math Handbook

For help with algebraic equations, go to page R69.

Interactive Textbook

Problem Solving 14.8 Solve Problem 8 with the help of an interactive guided tutorial.

with **ChemASAP**

Sample Problem 14.1

Answers

7. $105\ \text{kPa} \times 2.50\ \text{L} = 40.5\ \text{kPa} \times V_2$

$V_2 = 105\ \text{kPa} \times 2.50\ \text{L}/40.5\ \text{kPa} = 6.48\ \text{L}$

8. $205\ \text{kPa} \times 4.00\ \text{L} = P_2 \times 12.0\ \text{L}$

$P_2 = 205\ \text{kPa} \times 4.00\ \text{L}/12.0\ \text{L} = 68.3\ \text{kPa}$

Practice Problems Plus L2

The volume of a gas at 99.6 kPa and 24°C is 4.23 L. **What volume will it occupy at 93.3 kPa and 24°C?** *(4.52 L)*

Math Handbook

For a math refresher and practice, direct students to algebraic equations, page R69.

Discuss L2

Have students consider what will happen when a helium-filled balloon is released into the sky. Assume the temperature remains constant. Remind students that, as the elevation increases, the atmospheric pressure decreases. Ask, **If the balloon contains 30 L of gas at 100 kPa, what would its volume be at 25 kPa?** *(120 L)*

TEACHER Demo

Pressure and Volume L2

Purpose Students observe the effect that changing pressure has on the volume of a gas.

Materials vacuum pump, bell jar, marshmallows

Procedure Explain that marshmallows contain trapped air. Place several marshmallows in the bell jar, and then pull a vacuum in the jar. Ask students to explain why the marshmallows increase in size.

Expected Outcome The removal of air surrounding the marshmallows reduces the pressure on them. Air trapped inside the marshmallows can expand to a greater volume.

FYI

See *Explosives* on R26 of the Elements Handbook for an example of what can happen when a confined gas under pressure is allowed to expand.

Differentiated Instruction

Special Needs L1

Consider having students work in pairs to solve the practice problems in this chapter. Match up students who have a mastery of algebraic equations with students who need more practice solving for an unknown.

Charles's Law: Temperature and Volume

Discuss

L2

Point out that when the pressure and amount of a gas are unchanged, the ratio of the volume of the gas to the absolute temperature of the gas is a constant. Mathematically, this constant can be expressed as $V_1/T_1 = V_2/T_2$ and is known as Charles's law.

Interpreting Graphs

L2

a. kelvin
b. The volume increases
c. The volume would be 0.

Enrichment Question

L3

Point out the ΔV and ΔT labels. Ask, **What does the ratio $\Delta V/\Delta T$ represent?** *(slope)*

Relate

L2

Jacques Charles was one of three passengers in the second balloon ascension that carried humans. The hydrogen-filled balloon was launched on December 1, 1783, in Paris, France. More than 400,000 curious onlookers attended. Have interested students research the history of ballooning. Have them prepare a report for the class, which includes the types of gases used and the challenges involved.

Figure 14.9 When the gas in the blue balloon is cooled at constant pressure, the volume of the gas decreases. When the gas is heated at constant pressure, the volume increases.
Calculating ***What is the ratio of volume to temperature for each set of conditions? Round your answer to two significant figures.***

Charles's Law: Temperature and Volume

Figure 14.9 compares the volume of a balloon in a mixture of ice, salt, and water to the volume of the same balloon in hot water. The amount of gas and the pressure are constant. **As the temperature of an enclosed gas increases, the volume increases, if the pressure is constant.**

In 1787, the French physicist Jacques Charles studied the effect of temperature on the volume of a gas at constant pressure. When he graphed his data, Charles observed that a graph of gas volume versus temperature (in °C) is a straight line for any gas. When he extrapolated, or extended, the line to zero volume ($V = 0$), the line always intersected the temperature axis at −273.15°C. This value is equal to 0 on the Kelvin temperature scale.

The observations that Charles made are summarized in Charles's law. **Charles's law** states that the volume of a fixed mass of gas is directly proportional to its Kelvin temperature if the pressure is kept constant. Look at the graph in Figure 14.10. When the temperature is 300 K, the volume is 1.0 L. When the temperature is 900 K, the volume is 3.0 L. In both cases, the ratio of V to T is 0.0033.

Figure 14.10 This graph shows how the volume changes as the temperature of a gas changes.

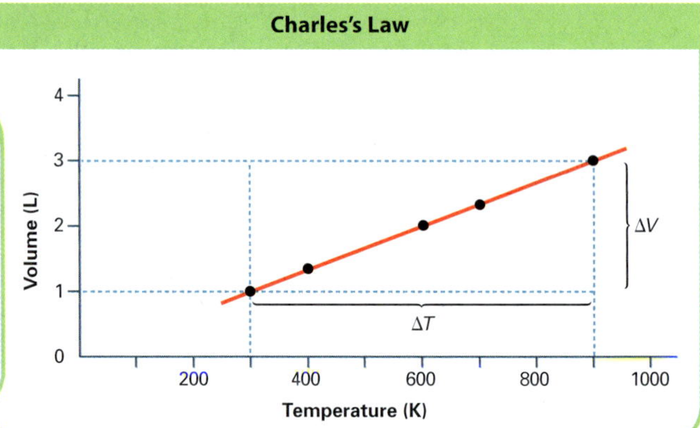

INTERPRETING GRAPHS

a. Observing What is the unit of temperature?
b. Drawing Conclusions What happens to the volume as the temperature rises?
c. Predicting If the temperature of a gas were 0 K, what would the volume of the gas be?

Differentiated Instruction

Gifted and Talented

L3

Explain to students that Charles summarized his observations of the relationship between the volume and temperature of a gas in the following equation: $V = V_0(1 + aT)$ where V_0 is the volume of the gas at 0°C, T is its temperature expressed in °C, and a is a constant for all gases. Have students show that the numerical value of a is approximately 1/273.

The ratio V_1/T_1 is equal to the ratio V_2/T_2. Because this ratio is constant at all conditions of temperature and volume, when the pressure is constant, you can write Charles's law as follows.

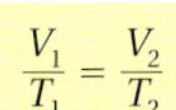

$$\frac{V_1}{T_1} = \frac{V_2}{T_2}$$

The ratio of the variables is always a constant in a direct relationship, and the graph is always a straight line. It is not a direct relationship if the temperatures are expressed in degrees Celsius. So when you solve gas law problems, the temperature must always be expressed in kelvins.

Interactive Textbook

Simulation 16 Examine the relationship between gas volume and temperature.

with **ChemASAP**

SAMPLE PROBLEM 14.2

Using Charles's Law

A balloon inflated in a room at 24°C has a volume of 4.00 L. The balloon is then heated to a temperature of 58°C. What is the new volume if the pressure remains constant?

1 Analyze ***List the knowns and the unknown.***

Knowns
- $V_1 = 4.00$ L
- $T_1 = 24$°C
- $T_2 = 58$°C

Unknown
- $V_2 = ?$ L

Use Charles's law ($V_1/T_1 = V_2/T_2$) to calculate the unknown value (V_2).

2 Calculate ***Solve for the unknown.***

Because you will use a gas law, express the temperatures in kelvins.

$$T_1 = 24°\text{C} + 273 = 297 \text{ K}$$

$$T_2 = 58°\text{C} + 273 = 331 \text{ K}$$

Rearrange Charles's law to isolate V_2.

$$V_2 = \frac{V_1 \times T_2}{T_1}$$

Substitute the known values for T_1, V_1, and T_2 into the equation and solve.

$$V_2 = \frac{4.00 \text{ L} \times 331 \cancel{\text{K}}}{297 \cancel{\text{K}}}$$

$$= 4.46 \text{ L}$$

3 Evaluate ***Does the result make sense?***

The volume increases as the temperature increases. This result agrees with both the kinetic theory and Charles's law.

Practice Problems

9. If a sample of gas occupies 6.80 L at 325°C, what will its volume be at 25°C if the pressure does not change?

10. Exactly 5.00 L of air at −50.0°C is warmed to 100.0°C. What is the new volume if the pressure remains constant?

Problem Solving 14.10 Solve Problem 10 with the help of an interactive guided tutorial.

with **ChemASAP**

Sample Problem 14.2

Answers

9. $T_1 = 325°\text{C} + 273 = 598$ K
$T_2 = 25°\text{C} + 273 = 298$ K
6.80 L/598 K = V_2/298 K
$V_2 = 298 \text{ K} \times 6.80 \text{ L}/598 \text{ K} = 3.39$ L

10. $T_1 = -50.0°\text{C} + 273 = 223$ K
$T_2 = 100.0°\text{C} + 273 = 373$ K
5.00 L/223 K = V_2/373 K
$V_2 = 373 \text{ K} \times 5.00 \text{ L}/223 \text{ K} = 8.36$ L

Practice Problems Plus 

The volume of a gas is 0.80 L at 101.3 kPa and 0°C. **What volume will it occupy at 101.3 kPa and 24°C?** *(0.87 L)*

Discuss 

While solving Charles's law problems, students should remember that the new volume of a gas is equal to its original volume times a quotient. The value of the quotient indicates whether the gas is heated or cooled. If the gas is heated ($T_2 > T_1$), the new volume is greater because the gas expands. So, the quotient has to be greater than one (T_2/T_1). If the gas is cooled ($T_2 < T_1$), the new volume is smaller because the gas contracts. So, the quotient must be less than one (T_1/T_2).

Answers to...

Figure 14.9 0.0038

Gay-Lussac's Law: Pressure and Temperature

Relate L2

Use tire pressure to discuss the relationship between gas pressure and Kelvin temperature. Ask, **Why do auto tire manufacturers recommend checking for proper inflation before driving the car more than a mile?** *(The tires get warm as the car moves, increasing the pressure inside the tires.)*

CLASS Activity

Observing the Effect of Pressure on Temperature L1

Purpose Students observe the relationship between the pressure and temperature of a gas.

Materials 2 large vats, ice water, hot water, inflated bicycle tire

Procedure Fill a large vat with ice water and a second vat with hot water. First, have students squeeze an inflated bicycle tire to assess its firmness. Next, immerse the tire in ice water and have them feel its firmness. Finally, immerse the tire in hot water and assess the firmness. Have the students describe the relationship between pressure and temperature at constant volume.

Expected Outcome A direct relationship exists between pressure and temperature at a constant volume.

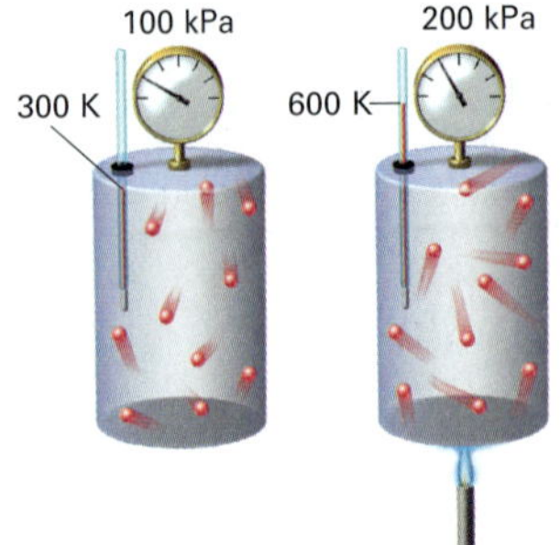

Figure 14.11 When a gas is heated at constant volume, the pressure increases.
Interpreting Diagrams ***How can you tell from the drawings that there is a fixed amount of gas in the cylinders?***

Gay-Lussac's Law: Pressure and Temperature

When tires are not inflated to the recommended pressure, fuel efficiency and traction decrease. Treads can wear down faster. Most importantly, improper inflation can lead to tire failure. A driver should not check tire pressure after driving a long distance because the air in a tire heats up during a drive. **As the temperature of an enclosed gas increases, the pressure increases, if the volume is constant.**

Joseph Gay-Lussac (1778–1850), a French chemist, discovered the relationship between the pressure and temperature of a gas in 1802. His name is on the gas law that describes the relationship. **Gay-Lussac's law** states that the pressure of a gas is directly proportional to the Kelvin temperature if the volume remains constant. Look at Figure 14.11. When the temperature is 300 K, the pressure is 100 kPa. When the temperature is doubled to 600 K, the pressure doubles to 200 kPa. Because Gay-Lussac's law involves direct proportions, the ratios P_1/T_1 and P_2/T_2 are equal at constant volume. You can write Gay-Lussac's law as follows.

$$\frac{P_1}{T_1} = \frac{P_2}{T_2}$$

Gay-Lussac's law can be applied to reduce the time it takes to cook food. One cooking method involves placing food above a layer of water and heating the water. The water vapor, or steam, that is produced cooks the food. Steam that escapes from the pot is at a temperature of about 100°C when the pressure is near one atmosphere. In a pressure cooker, like the one shown in Figure 14.12, steam is trapped inside the cooker. The temperature of the steam reaches about 120°C. The food cooks faster at this higher temperature, but the pressure rises, which increases the risk of an explosion. A pressure cooker has a valve that allows some vapor to escape when the pressure exceeds the set value.

Checkpoint ***How does a pressure cooker affect the time needed to cook food?***

Figure 14.12 In a pressure cooker, food cooks faster than in an ordinary pot with a lid.

Differentiated Instruction

English Learners 

Have English learners make a list of terms in this section that they don't understand. Pair them with other students who can use paraphrasing to explain the terms.

SAMPLE PROBLEM 14.3

Using Gay-Lussac's Law

Aerosol cans carry warnings on their labels that say not to incinerate (burn) them or store the cans above a certain temperature. This problem will show why it is dangerous to dispose of aerosol cans in a fire. The gas in a used aerosol can is at a pressure of 103 kPa at 25°C. If the can is thrown onto a fire, what will the pressure be when the temperature reaches 928°C?

1 Analyze *List the knowns and the unknown.*

Knowns	Unknown
• P_1 = 103 kPa	• P_2 = ? kPa
• T_1 = 25°C	
• T_2 = 928°C	

Use Gay-Lussac's law ($P_1/T_1 = P_2/T_2$) to calculate the unknown (P_2). Remember, because this problem involves temperatures and a gas law, the temperatures must be expressed in kelvins.

2 Calculate *Solve for the unknown.*

First convert degrees Celsius to kelvins.

$$T_1 = 25°C + 273 = 298\ K$$

$$T_2 = 928°C + 273 = 1201\ K$$

Rearrange Gay-Lussac's law to isolate P_2.

$$P_2 = \frac{P_1 \times T_2}{T_1}$$

Substitute the known values for P_1, T_2, and T_1 into the equation and solve.

$$P_2 = \frac{103\ kPa \times 1201\ \cancel{K}}{298\ \cancel{K}}$$

$$= 415\ kPa$$

$$= 4.15 \times 10^2\ kPa$$

3 Evaluate *Does the result make sense?*

From the kinetic theory, one would expect the increase in temperature of a gas to produce an increase in pressure if the volume remains constant. The calculated value does show such an increase.

Practice Problems

11. A sample of nitrogen gas has a pressure of 6.58 kPa at 539 K. If the volume does not change, what will the pressure be at 211 K?

12. The pressure in a car tire is 198 kPa at 27°C. After a long drive, the pressure is 225 kPa. What is the temperature of the air in the tire? Assume that the volume is constant.

Interactive Textbook

Simulation 17 Examine the relationship between gas pressure and temperature.

with ChemASAP

Problem Solving 14.12 Solve Problem 12 with the help of an interactive guided tutorial.

with ChemASAP

Sample Problem 14.3

Answers

11. 6.58 kPa/539 K = P_2/211 K
P_2 = 211 K × 6.58 kPa/539 K = 2.58 kPa

12. T_1 = 27°C + 273 = 300 K
198 kPa/300 K = 225 kPa/T_2
T_2 = 225 kPa × 300 K/198 kPa = 341 K (68°C)

Practice Problems Plus 

The pressure in a sealed plastic container is 108 kPa at 41°C. **What is the pressure when the temperature drops to22°C?** Assume that the volume has not changed. *(101 kPa)*

Discuss 

The direct relationship between pressure and temperature at constant volume can be expressed as $P_1/T_1 = P_2/T_2$. In Gay-Lussac's law problems, the new pressure of the gas is equal to its original pressure times a quotient. The value of the quotient indicates whether the gas is heated or cooled. If the gas is heated ($T_2 > T_1$), the new pressure is greater. So, the quotient has to be greater than one (T_2/T_1). If the gas is cooled ($T_2 < T_1$), the new pressure is less. So, the quotient must be less than one (T_1/T_2).

Differentiated Instruction

Gifted and Talented 

Gases such as oxygen, nitrogen, helium, and hydrogen can be liquefied by lowering the temperature and increasing the pressure. But different gases must be cooled to different temperatures before they condense, no matter how high the pressure is raised. This temperature is called the "critical temperature."

Have students look up the critical temperatures for oxygen, nitrogen, helium, and hydrogen. Ask, **What do these temperatures indicate about the relative strengths of the intermolecular attractions for these gases?** *(The lower the critical temperature, the lower the intermolecular attractive forces.)*

Answers to...

Figure 14.11 The number of particles in each container is the same.

A pressure cooker reduces cooking time.

The Combined Gas Law

Sample Problem 14.4

Answers

13. $T_1 = 25°C + 273 = 298\ K$
$T_2 = 125°C + 273 = 398\ K$
$155\ kPa \times 1.00\ L/298\ K = 605\ kPa \times V_2/398\ K$
$V_2 = 398\ K \times 155\ kPa \times 1.00\ L/298\ K \times 605\ kPa = 0.342\ L$

14. $T_1 = -50°C + 273 = 223\ K$
$T_2 = 102°C + 273 = 375\ K$
$107\ kPa \times 5.00\ L/223\ K = P_2 \times 7.00\ L/375\ K$
$P_2 = 375\ K \times 107\ kPa \times 5.00\ L/223\ K \times 7.00\ L = 1.29 \times 10^2\ kPa$

Practice Problems Plus L2

The volume of a gas at 26°C and 75 kPa is 10.5 L. The pressure is increased to 116 kPa. **What final temperature would be required to reduce the volume to 9.5 L?** *(418 K or 145°C)*

Discuss L2

Write the equation for the combined gas law on the board. Ask, **What variable that is used to describe a gas is missing from this equation?** (*n, the number of moles of gas*)

Download a worksheet on **The Gas Laws** for students to complete, and find additional teacher support from NSTA SciLinks.

FYI

The ascent of a weather balloon is representative of many events that occur outside a laboratory. Instead of a controlled experiment with manipulated and responding variables (with all other variables controlled), multiple variables are changing simultaneously.

Go Online NSTA SciLinks

For: Links on The Gas Laws
Visit: www.SciLinks.org
Web Code: cdn-1142

The Combined Gas Law

There is a single expression that combines Boyle's, Charles's, and Gay-Lussac's laws. The **combined gas law** describes the relationship among the pressure, temperature, and volume of an enclosed gas. **The combined gas law allows you to do calculations for situations in which only the amount of gas is constant.**

$$\frac{P_1 \times V_1}{T_1} = \frac{P_2 \times V_2}{T_2}$$

SAMPLE PROBLEM 14.4

Using the Combined Gas Law

The volume of a gas-filled balloon is 30.0 L at 313 K and 153 kPa pressure. What would the volume be at standard temperature and pressure (STP)?

1 Analyze *List the knowns and the unknown.*

Knowns
- $V_1 = 30.0\ L$
- $T_1 = 313\ K$
- $P_1 = 153\ kPa$
- $T_2 = 273\ K$ (standard temperature)
- $P_2 = 101.3\ kPa$ (standard pressure)

Unknown
- $V_2 = ?\ L$

Use the known values and the combined gas law to calculate the unknown (V_2).

2 Calculate *Solve for the unknown.*

Rearrange the combined gas law to isolate V_2.

$$V_2 = \frac{V_1 \times P_1 \times T_2}{P_2 \times T_1}$$

Substitute the known quantities into the equation and solve.

$$V_2 = \frac{30.0\ L \times 153\ \cancel{kPa} \times 273\ \cancel{K}}{101.3\ \cancel{kPa} \times 313\ \cancel{K}} = 39.5\ L$$

3 Evaluate *Does the result make sense?*

A decrease in temperature and a decrease in pressure have opposite effects on the volume of gas in a balloon. To evaluate the increase in volume, multiply the original volume (30.0 L) by the ratio of P_1 to P_2 (1.51) and the ratio of T_2 to T_1 (0.872). The result is 39.5 L.

Practice Problems

13. A gas at 155 kPa and 25°C has an initial volume of 1.00 L. The pressure of the gas increases to 605 kPa as the temperature is raised to 125°C. What is the new volume?

14. A 5.00-L air sample has a pressure of 107 kPa at a temperature of −50.0°C. If the temperature is raised to 102°C and the volume expands to 7.00 L, what will the new pressure be?

Interactive Textbook

Problem Solving 14.14
Solve Problem 14 with the help of an interactive guided tutorial.
with ChemASAP

Differentiated Instruction

Gifted and Talented L3

Have students use the combined gas law to derive the equations for Charles's law and Gay-Lussac's law. *(The variables n and P are constant for Charles's law, and n and V are constant for Gay-Lussac's law.)*

Weather balloons, like the one in Figure 14.13, carry a package of data-gathering instruments up into the atmosphere. At an altitude of about 27,000 meters, the balloon bursts. The combined gas law can help to explain this situation. Both outside temperature and pressure drop as the balloon rises. These changes have opposite effects on the volume of the weather balloon. A drop in temperature causes the volume of an enclosed gas to decrease. A drop in outside pressure causes the volume to increase. Given that the balloon bursts, the drop in pressure must affect the volume more than the drop in temperature does.

The combined gas law can also help you solve gas problems when only two variables are changing. It may seem challenging to remember four different expressions for the gas laws. But you actually only need to remember one expression—the combined gas law. You can derive the other laws from the combined gas law by holding one variable constant.

To illustrate, suppose you hold the temperature constant ($T_1 = T_2$). Rearrange the combined gas law so that the two temperature terms are on the same side of the equation. Because $T_1 = T_2$, the ratio of T_1 to T_2 is equal to one. Multiplying by 1 does not change a value in an equation. So when the temperature is constant, you can delete the temperature ratio from the rearranged combined gas law. What you are left with is the mathematical expression of Boyle's law.

$$P_1 \times V_1 = P_2 \times V_2 \times \frac{\cancel{T_1}}{\cancel{T_2}}$$

$$P_1 \times V_1 = P_2 \times V_2$$

A similar process yields Charles's law when pressure remains constant and Gay-Lussac's law when volume remains constant.

Figure 14.13 Weather balloons carry instruments that can gather data about Earth's atmosphere. **Predicting** ***Explain why helium is more likely than air to be used in weather balloons.***

14.2 Section Assessment

15. **Key Concept** How are the pressure and volume of a gas related at constant temperature?
16. **Key Concept** If pressure is constant, how does a change in temperature affect the volume of a gas?
17. **Key Concept** What is the relationship between the temperature and pressure of a contained gas at constant volume?
18. **Key Concept** In what situations is the combined gas law useful?
19. Write the mathematical equation for Boyle's law and explain the symbols.
20. A given mass of air has a volume of 6.00 L at 101 kPa. What volume will it occupy at 25.0 kPa if the temperature does not change?
21. Explain how Charles's law can be derived from the combined gas law.
22. The volume of a weather balloon increases as the balloon rises in the atmosphere. Why doesn't the drop in temperature at higher altitudes cause the volume to decrease?

Connecting Concepts

Phase Changes Use what you learned about phase changes in Section 13.4 to try to explain this statement: Scientists cannot collect temperature and volume data for an enclosed gas at temperatures near absolute zero.

Assessment 14.2 Test yourself on the concepts in Section 14.2.

with **ChemASAP**

3 ASSESS

Evaluate Understanding L2

Sketch two balloons (one twice as large as the other) on the board. Label the smaller balloon V_1 and the other V_2, and indicate that $V_2 = 2V_1$. Copy the following table onto the chalkboard, listing only those values in boldface type. Have students complete the table.

T_1	T_2
(256 K)	**512 K**
40°C	*(353°C)*
(T_1)	**$2T_1$**
(−136°C)	**1°C**

Reteach L1

In many situations, a sample of gas is simultaneously subjected to pressure and temperature changes that have opposite effects on volume. Consider a weather balloon rising through the atmosphere. The higher it rises, the colder the temperature and the lower the volume. At the same time, atmospheric pressure decreases, allowing the gas to expand. The combined gas law allows students to determine which variable has the greater effect on the volume.

Connecting Concepts

As temperatures decrease toward absolute zero, particles in the gas slow down, and attractions between particles increase. A gas would liquefy, then solidify, as it is cooled to temperatures near absolute zero.

If your class subscribes to the Interactive Textbook, use it to review key concepts in Section 14.2.

with **ChemASAP**

Section 14.2 Assessment

15. If the temperature is constant, the volume of a gas decreases as the pressure increases.
16. If the pressure is constant, as the temperature increases, the volume increases.
17. If the volume is constant, as the temperature increases, the pressure increases.
18. The combined gas law allows you to do calculations when the only constant is the amount of gas
19. $P_1 \times V_1 = P_2 \times V_2$; P = pressure; V = volume; the subscript 1 represents the starting conditions; the subscript 2, the final conditions.
20. 24.2 L
21. When the pressure is constant, $P_1 = P_2$, the pressure terms cancel, leaving an equation for Charles' law.
22. The outside pressure decreases, causing a greater increase in the balloon's volume.

Answers to...

Figure 14.13 Helium is less dense than air.

14.3

1 FOCUS

Objectives

14.3.1 Compute the value of an unknown using the ideal gas law.

14.3.2 Compare and contrast real and ideal gases.

Guide for Reading

Build Vocabulary L2

Word Parts Have students look up the meanings of *ideal* and *real*. Have them write definitions that include the terms *theoretical* and *actual*.

Reading Strategy L2

Make Inferences Have students use their understanding of the terms *real* and *ideal* to infer the differences between real and ideal gases. As the students read the section, have them evaluate and revise their inference.

2 INSTRUCT

Connecting to Your World

Have students study the photograph and read the text that opens the section. Ask, **How does the existence of dry ice violate the assumptions of the kinetic theory?** *(The kinetic theory assumes that there are no attractions among the particles in a gas. But there must be attractions between carbon dioxide molecules for the gas to solidify.)*

Ideal Gas Law

Discuss L2

Sketch two containers with identical volumes on the board. Tell students one container is filled with neon gas and the other with helium at the same temperature and pressure. Ask students to use the ideal gas law to find the equation for the number of moles in each container. Students should determine that $n_{He} = n_{Ne}$.

14.3 Ideal Gases

Guide for Reading

Key Concepts

- What is needed to calculate the amount of gas in a sample at given conditions of volume, temperature, and pressure?
- Under what conditions are real gases most likely to differ from ideal gases?

Vocabulary

ideal gas constant

ideal gas law

Reading Strategy

Building Vocabulary After you read this section, explain the difference between *ideal* and *real* as these terms are applied to gases.

Connecting to Your World

Solid carbon dioxide, or dry ice, is used to protect products that need to be kept cold during shipping. The adjective *dry* refers to a key advantage of shipping with dry ice. Dry ice doesn't melt. It sublimes. Dry ice can exist because gases don't obey the assumptions of kinetic theory at all conditions. In this section, you will learn how real gases differ from the ideal gases on which the gas laws are based.

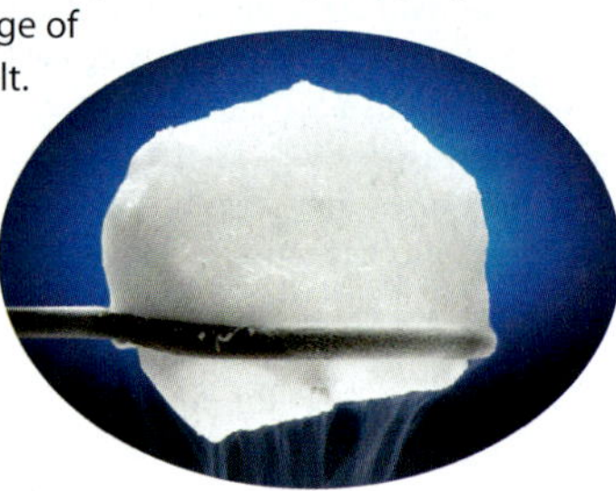

Ideal Gas Law

With the combined gas law, you can solve problems with three variables: pressure, volume, and temperature. The combined gas law assumes that the amount of gas does not vary. You cannot use the combined gas law to calculate the number of moles of a gas in a fixed volume at a known temperature and pressure. **To calculate the number of moles of a contained gas requires an expression that contains the variable n.** The combined gas law can be modified to include the number of moles.

The number of moles of gas is directly proportional to the number of particles. The volume occupied by a gas at a specified temperature and pressure also must depend on the number of particles. So moles must be directly proportional to volume as well. You can introduce moles into the combined gas law by dividing each side of the equation by n.

$$\frac{P_1 \times V_1}{T_1 \times n_1} = \frac{P_2 \times V_2}{T_2 \times n_2}$$

This equation shows that $(P \times V)/(T \times n)$ is a constant. This constant holds for ideal gases—gases that conform to the gas laws.

If you know the values for P, V, T, and n for one set of conditions, you can calculate a value for the constant. Recall that 1 mol of every gas occupies 22.4 L at STP (101.3 kPa and 273 K). You can use these values to find the value of the constant, which has the symbol R and is called the ideal gas constant. Insert the values of P, V, T, and n into $(P \times V)/(T \times n)$.

$$R = \frac{P \times V}{T \times n} = \frac{101.3\text{ kPa} \times 22.4\text{ L}}{273\text{ K} \times 1\text{ mol}} = 8.31\text{ (L·kPa)/(K·mol)}$$

The **ideal gas constant** (R) has the value 8.31 (L·kPa)/(K·mol). The gas law that includes all four variables—P, V, T, and n—is called the **ideal gas law**. It is usually written as follows.

$$P \times V = n \times R \times T \text{ or } PV = nRT$$

Section Resources

Print

- **Guided Reading and Study Workbook,** Section 14.3
- **Core Teaching Resources,** Section 14.3 Review, Interpreting Graphics
- **Transparencies,** T156–T157
- **Probeware Laboratory Manual,** Section 14.3

Technology

- **Interactive Textbook with ChemASAP,** Problem Solving 14.24, Assessment 14.3
- **Virtual Chemistry Labs,** 13, 14

SAMPLE PROBLEM 14.5

Using the Ideal Gas Law to Find the Amount of a Gas

A deep underground cavern contains 2.24×10^6 L of methane gas (CH_4) at a pressure of 1.50×10^3 kPa and a temperature of 315 K. How many kilograms of CH_4 does the cavern contain?

1 Analyze ***List the knowns and the unknown.***

Knowns
- $P = 1.50 \times 10^3$ kPa
- $V = 2.24 \times 10^6$ L
- $T = 315$ K
- $R = 8.31$ (L·kPa)/(K·mol)
- molar mass$_{CH_4}$ = 16.0 g

Unknown
- ? kg CH_4

Calculate the number of moles (n) using the ideal gas law. Use the molar mass to convert moles to grams. Then convert grams to kilograms.

2 Calculate ***Solve for the unknown.***

Rearrange the equation for the ideal gas law to isolate n.

$$n = \frac{P \times V}{R \times T}$$

Substitute the known quantities into the equation to find the number of moles of methane.

$$n = \frac{(1.50 \times 10^3 \text{ kPa}) \times (2.24 \times 10^6 \text{ L})}{8.31 \frac{\text{L} \times \text{kPa}}{\text{K} \times \text{mol}} \times 315 \text{ K}} = 1.28 \times 10^6 \text{ mol } CH_4$$

A mole-mass conversion gives the number of grams of methane.

$$1.28 \times 10^6 \text{ mol } CH_4 \times \frac{16.0 \text{ g } CH_4}{1 \text{ mol } CH_4} = 20.5 \times 10^6 \text{ g } CH_4$$
$$= 2.05 \times 10^7 \text{ g } CH_4$$

Convert this answer to kilograms.

$$2.05 \times 10^7 \text{ g } CH_4 \times \frac{1 \text{ kg}}{10^3 \text{ g}} = 2.05 \times 10^4 \text{ kg } CH_4$$

3 Evaluate ***Does the result make sense?***

Although the methane is compressed, its volume is still very large. So it is reasonable that the cavern contains a large mass of methane.

Practice Problems

23. When the temperature of a rigid hollow sphere containing 685 L of helium gas is held at 621 K, the pressure of the gas is 1.89×10^3 kPa. How many moles of helium does the sphere contain?

24. A child's lungs can hold 2.20 L. How many grams of air do her lungs hold at a pressure of 102 kPa and a body temperature of 37°C? Use a molar mass of 29 g for air, which is about 20% O_2 (32 g/mol) and 80% N_2 (28 g/mol).

Engineers use drilling rods to explore for natural gas in the crust below the ocean floor.

Math Handbook

For help with conversion problems, go to page R66.

Problem Solving 14.24
Solve Problem 24 with the help of an interactive guided tutorial.
with ChemASAP

Sample Problem 14.5

Answers

23. 251 or 2.51×10^2 mol He*(g)*
24. 2.5 g air

Practice Problems Plus 

At 34.0°C, the pressure inside a nitrogen-filled tennis ball with a volume of 148 cm^3 is 212 kPa. **How many moles of N_2 are in the tennis ball?** *(1.23×10^{-2} mol N_2)*

A helium-filled balloon contains 0.16 mol He at 101 kPa and a temperature of 23°C. **What is the volume of the gas in the balloon?** *(3.9 L)*

Math Handbook

For a math refresher and practice, direct students to conversion factors, page R66.

Discuss 

Explain that Avogadro's hypothesis makes it possible to relate the molar quantity of a gas to its temperature, volume, and pressure. The hypothesis assumes that as long as the particles are not tightly packed, equal volumes of gases at the same temperature and pressure contain equal numbers of particles. Emphasize that Avogadro's hypothesis works only for gases because a large portion of their volume is empty space. At low pressure, the volumes of individual molecules are negligible compared to the volume of the container holding the gas. The volume of a gas depends on the number of particles present, not their size. Ask, **How would you determine the mass of a balloon full of helium gas at STP without making any mass measurements?** (According to Avogadro's hypothesis, one mole of a gas has a volume of 22.4 L at STP. Calculate the volume of the balloon in liters at STP. Then, convert from liters to moles. To find the mass, multiply the number of moles by the molar mass of He.)

Ideal Gases and Real Gases

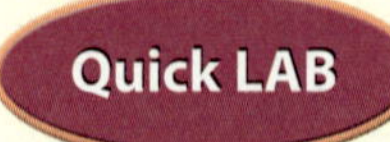

Quick LAB

Carbon Dioxide from Antacid Tablets L2

Objective After completing this activity, students will be able to:

- measure the amount of carbon dioxide gas given off when antacid tablets dissolve in water.

Skills Focus Observing, Calculating, Measuring

Prep Time 10 minutes

Class Time 40 minutes

Safety If you use latex balloons, check to see if any students are allergic to latex.

Expected Outcome The volumes of CO_2 produced will reflect the amount of antacid each balloon contains.

Analyze and Conclude

1. The volume of the balloon is directly proportional to the number of tablets.
2. Answers will vary, but the masses and numbers of moles should be in ratios of 1:2:3 for the three balloons.
3. Possible answer: 2.0 g of $NaHCO_3$ (molar mass = 84.01 g) should yield about 1.2×10^{-2} mol of CO_2.

For Enrichment

Have students use a similar procedure to compare different brands of effervescent antacids instead of different amounts of the same antacid.

Quick LAB

Carbon Dioxide from Antacid Tablets

Purpose

To measure the amount of carbon dioxide gas given off when antacid tablets dissolve in water.

Materials

- 6 effervescent antacid tablets
- 3 rubber balloons (spherical)
- plastic medicine dropper
- water
- clock or watch
- metric tape measure
- graph paper
- pressure sensor (optional)

Procedure

Sensor version available in the Probeware Lab Manual.

1. Break six antacid tablets into small pieces. Keep the pieces from each tablet in a separate pile. Put the pieces from one tablet into the first balloon. Put the pieces from two tablets into a second balloon. Put the pieces from three tablets into a third balloon. **CAUTION** *If you are allergic to latex, do not handle the balloons.*
2. After you use the medicine dropper to squirt about 5 mL of cold water into each balloon, immediately tie off each balloon.
3. Shake the balloons to mix the contents. Allow the contents to warm to room temperature.
4. Measure and record the circumference of each balloon several times during the next 20 minutes.
5. Use the maximum circumference of each balloon to calculate its volume. (*Hint:* Volume of a sphere $= \frac{4\pi r^3}{3}$ and r = circumference/2π.)

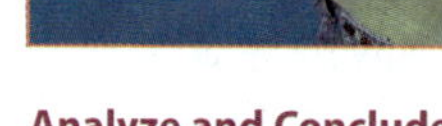

Analyze and Conclude

1. Make a graph of volume versus number of tablets. Use your graph to describe the relationship between the number of tablets used and the volume of the balloon.
2. Assume that the balloon is filled with carbon dioxide gas at 20°C and standard pressure. Calculate the mass and the number of moles of CO_2 in each balloon at maximum inflation.
3. If a typical antacid tablet contains 2.0 g of sodium hydrogen carbonate, how many moles of CO_2 should one tablet yield? Compare this theoretical value with your results.

Figure 14.14 In this flask used to store liquid nitrogen, there are two walls with a vacuum in between.

Ideal Gases and Real Gases

An ideal gas is one that follows the gas laws at all conditions of pressure and temperature. Such a gas would have to conform precisely to the assumptions of kinetic theory. Its particles could have no volume, and there could be no attraction between particles in the gas. As you probably suspect, there is no gas for which these assumptions are true. So an ideal gas does not exist. Nevertheless, at many conditions of temperature and pressure, real gases behave very much like an ideal gas.

The particles in a real gas do have volume, and there are attractions between the particles. Because of these attractions, a gas can condense, or even solidify, when it is compressed or cooled. For example, if water vapor is cooled below 100°C at standard atmospheric pressure, it condenses to a liquid. The behavior of other real gases is similar, although lower temperatures and greater pressures may be required. Such conditions are required to produce the liquid nitrogen in Figure 14.14. **Real gases differ most from an ideal gas at low temperatures and high pressures.**

Real Gases Deviate From the Ideal

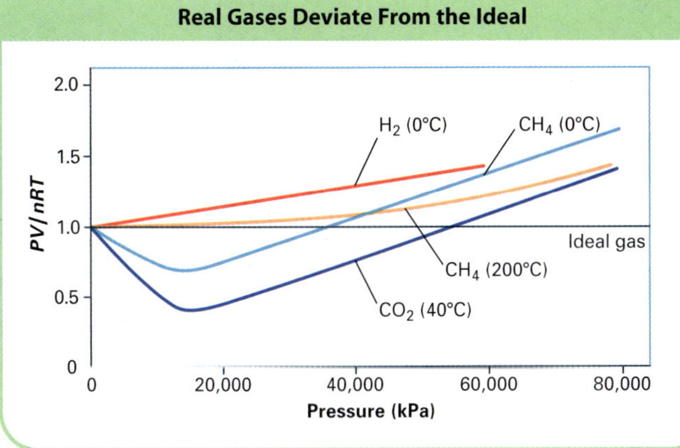

Figure 14.15 This graph shows how real gases deviate from the ideal gas law at high pressures.

INTERPRETING GRAPHS

a. Observing What are the values of $(PV)/(nRT)$ for an ideal gas at 20,000 and 60,000 kPa?
b. Comparing What variable is responsible for the differences between the two (CH_4) curves?
c. Making Generalizations How does an increase in pressure affect the $(PV)/(nRT)$ ratio for real gases?

Figure 14.15 shows how the value of the ratio (PV/nRT) changes as pressure increases. For an ideal gas, the result is a horizontal line because the ratio is always equal to 1. For real gases at high pressure, the ratio may deviate, or depart, from the ideal. When the ratio is greater than 1, the curve rises above the ideal gas line. When the ratio is less than 1, the curve drops below the line. The deviations can be explained by two factors. As attractive forces reduce the distance between particles, a gas occupies less volume than expected, causing the ratio to be less than 1. But the actual volume of the molecules causes the ratio to be greater than 1.

In portions of the curves below the line, intermolecular attractions dominate. In portions of the curves above the line, molecular volume dominates. Look at the curves for methane (CH_4) at 0°C and at 200°C. At 200°C, the molecules have more kinetic energy to overcome intermolecular attractions. So the curve for CH_4 at 200°C never drops below the line.

14.3 Section Assessment

25. **Key Concept** What do you need to calculate the amount of gas in a sample at given conditions of temperature, pressure, and volume?
26. **Key Concept** Under what conditions do real gases deviate most from ideal behavior?
27. What is an ideal gas?
28. Determine the volume occupied by 0.582 mol of a gas at 15°C if the pressure is 81.8 kPa.
29. What pressure is exerted by 0.450 mol of a gas at 25°C if the gas is in a 0.650-L container?
30. Use the kinetic theory of gases to explain this statement: No gas exhibits ideal behavior at all temperatures and pressures.

Connecting Concepts

Polarity At standard pressure, ammonia will condense at −33.3°C. At the same pressure, nitrogen does not condense until −195.79°C. Use what you learned about intermolecular attractions and polarity in Section 8.3 to explain this difference.

Assessment 14.3 Test yourself on the concepts in Section 14.3.

with ChemASAP

Interpreting Graphs 

a. 1.0 in both cases
b. temperature
c. The ratio increases.

Enrichment Question L3

How would a curve for CO_2 at 200°C differ from the curve at 40°C? *(At a higher temperature, the behavior of a gas is closer to the ideal; so the curve might not dip below the straight line.)*

Use Visuals L1

Figure 14.15 Point out that the straight line represents ideal conditions. For an ideal gas, by definition $PV = nRT$, and the ratio of PV to nRT is 1. When the volume of a gas is greater than expected, the ratio tends to be greater than 1. When the volume is less than expected, the ratio tends to be less than 1. The main factors that affect the volume are intermolecular attractions and the actual volume of the particles.

3 ASSESS

Evaluate Understanding L2

Have students apply the ideal gas law to a sample of gas in a balloon. Ask them to explain why in a balloon n and R are constants and P, V, and T are variables.

Reteach L1

Point out to students that one advantage of the ideal gas law is that it enables them to find the number of moles of a gas by measuring its temperature, pressure, and volume.

Connecting Concepts

The nitrogen molecule is nonpolar and the ammonia molecule is polar. So there are stronger intermolecular attractions in ammonia.

Section 14.3 Assessment

25. an expression that contains the variable n
26. Real gases deviate from ideal behavior at low temperatures and high pressures.
27. An ideal gas is a gas that follows the gas laws at all conditions of pressure and temperature.
28. 17.0 L
29. 1.71×10^3 kPa
30. In real gases, there are attractions between molecules, and the molecules have volume. At low temperatures, attractions between molecules pull them together and reduce the volume. At high pressures, the volume occupied by the molecules is a significant part of the total volume.

If your class subscribes to the Interactive Textbook, use it to review key concepts in Section 14.3.

with ChemASAP

Technology & Society

Diving In

Decompression sickness is an application of Henry's law (which is discussed in Chapter 16): At a given temperature, the solubility of a gas in a liquid is directly proportional to the pressure of the gas. Dissolved nitrogen is more problematic than dissolved oxygen because oxygen released during decompression can be removed from the blood and used by the cells. Nitrogen is not used up by the cells and must be excreted through the lungs. Varying the composition of the compressed gas and using dive charts are two strategies for combating decompression sickness.

Discuss L2

Discuss the content of the article in the context of Dalton's law of partial pressures. Remind students that, according to Dalton's law of partial pressures, the total pressure exerted by a mixture of gases is equal to the sum of the partial pressures exerted by each of the different gases in the mixture. The fractional contribution to pressure exerted by each gas does not change as the temperature, pressure, or volume changes as long as the composition of the mixture is constant. Ask students to use Dalton's law of partial pressures to explain how changing the mixture of gases in the tanks used by divers can help prevent problems.

CLASS Activity

Effect of Depth on Partial Pressures L2

With Table 14.1 displayed on an overhead projector, point out that the air at sea level is about 21% oxygen and 78% nitrogen. The values for the partial pressures given in Table 14.1 reflect conditions at 1 atm or 101.3 kPa. Have students calculate the partial pressures of each gas that a scuba diver, breathing the same air mixture, would experience at depths of 100 feet (approximately 4 atm) and 300 feet (approximately 10 atm).

Technology & Society

Diving In

Divers who depend on the air in their lungs can stay under water for only a few minutes. But with tanks of compressed air, scuba divers can stay under water for hours. They can descend to great depths to explore a coral reef or salvage a sunken ship. Compressed air allows engineers to build, inspect, and repair ships, bridges, and oil platforms. **Interpreting Diagrams** ***Which of the two main components of air must be in a diver's tank?***

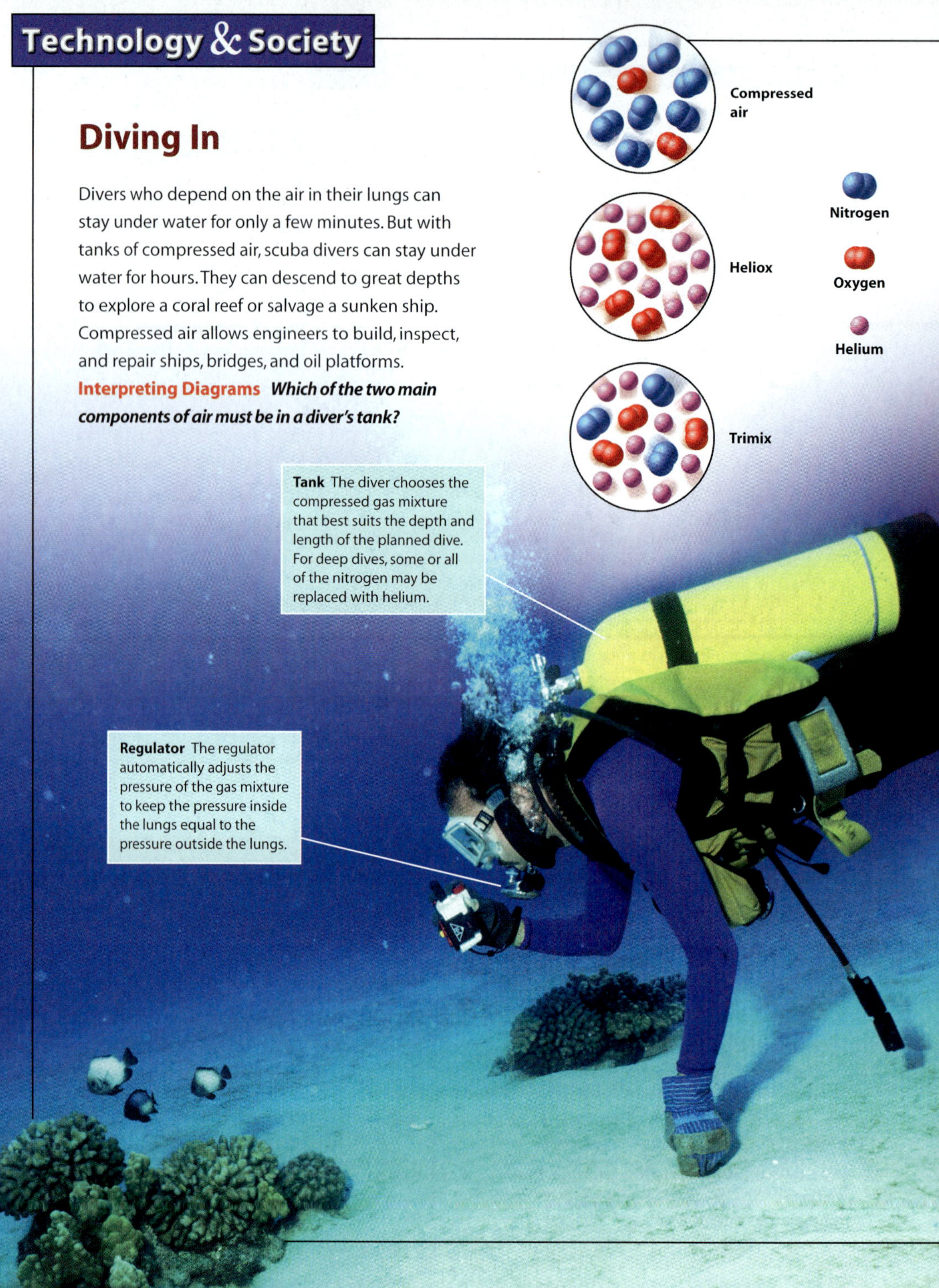

Facts and Figures

Decompression Sickness

The invention of compressed air in the 1840s allowed people to work under water, with one major drawback. Decompression sickness was initially called caisson disease because workers constructing the Brooklyn Bridge worked in water-tight containers called caissons. The pressure of the air in the caissons needed to be greater than atmospheric pressure to withstand the pressure of the surrounding water. The condition was also called "the bends". In 1878, Paul Bert stated that workers could avoid the bends if they ascended gradually to the surface. Bert referred to the work of Robert Boyle. In 1667, Boyle observed a bubble form in the eye of a viper that was placed in a compressed atmosphere and then removed. (Boyle reported that the viper appeared distressed by the experience.)

Dive tables help a diver avoid decompression sickness. Divers use the data to control the length and frequency of their dives, and the rate at which they return to the surface.

Table 1 End-of-dive lettergroup At the end of a dive, the diver matches the time and depth of the dive to a letter group. Letter A represents the least amount of nitrogen left in the blood after the dive. The circled numbers show the maximum time a diver can spend at a given depth without having to make a decompression stop during the ascent.

Table 2 Surface interval time The longer a diver spends at the surface, the more nitrogen is excreted through the lungs. A diver can move from Group H to Group A after 8 hours at the surface.

NAUI DIVE TABLES

TABLE 1 - END-OF-DIVE LETTER GROUP

TABLE 2 - SURFACE INTERVAL TIME (SIT) TABLE

TABLE 3 - REPETITIVE DIVE TIMETABLE

Table 3 Repetitive dive timetable Divers use this table to determine an adjusted maximum dive time before doing a second dive. They also use the table to determine a letter group at the end of the second dive.

Dissolved nitrogen As a diver descends, the water exerts greater pressure. More gas pressure is required to keep the lungs expanded. This increased gas pressure causes more nitrogen to dissolve in the diver's blood.

Nitrogen narcosis Below a depth of about 30 meters, dissolved nitrogen interferes with the transmission of nerve impulses. The effects are similar to those of alcohol and include dizziness, slowed reaction time, and an inability to think clearly.

Decompression sickness As the diver returns to the surface, pressure decreases and dissolved nitrogen is released from the blood. Bubbles of nitrogen can block small blood vessels and reduce the supply of oxygen to cells, causing severe pain in the joints, dizziness, and vomiting.

Use Visuals L1

Discuss the purpose of dive tables and explain how to use them. Further explanation of the tables is provided at the NAUI website. Have students use the indicated table to answer the following questions.

Table 1

1. How long can a diver stay at 24 m without needing to make a decompression stop? *(35 minutes)*

2. A diver stays at a depth of 33 m for 20 minutes. **Will she have to make a decompression stop when returning to the surface? Explain.** *(Yes, the maximum time with no decompression at this depth is 15 minutes.)*

Table 2

Explain that after a dive, a diver is assigned a letter based on the depth and time of the dive. This letter determines how soon the diver can dive again. The higher the letter, the longer it takes for nitrogen to be removed from the blood.

1. A diver assigned to Group G has been out of the water for 4 hours. **What group is he now in?** *(Group C)*

2. After how many hours has all nitrogen that might be harmful been eliminated from the blood? *(24 hours)*

Differentiated Instruction

Gifted and Talented L3

Have students find out how to use the repetitive dive table (Table 3). Have them teach other students how to use the table and provide problems for the students to solve.

Answers to...

Interpreting Diagrams Oxygen is essential; nitrogen is optional.

14.4

1 FOCUS

Objectives

14.4.1 Relate the total pressure of a mixture of gases to the partial pressures of the component gases.

14.4.2 Explain how the molar mass of a gas affects the rate at which the gas diffuses and effuses.

Guide for Reading

Build Vocabulary L2

Imagine a Picture As students learn about *effusion* and *diffusion*, have them visualize each process at the microscopic level. Then have them describe what is happening.

Reading Strategy L2

Monitor Your Understanding If students are having difficulty reading *Graham's Law*, have them think about how they are reading and how they might deal with the difficulty.

2 INSTRUCT

Connecting to Your World

Have students study the photograph and read the text that opens the section. Ask students to try to explain why high-altitude climbers carry a supply of oxygen. *(Students may recall that atmospheric pressure decreases with altitude. They may conclude that the drop in pressure is related to a decrease in the concentration of the gases in air, including oxygen.)*

Dalton's Law

Discuss L2

Assess students' knowledge about gases by asking them to compare the number of particles in two identical, sealed containers of propane and helium gas at the same temperature and pressure. *(Equal volumes of gases at the same temperature and pressure contain equal numbers of particles.)*

14.4 Gases: Mixtures and Movements

Guide for Reading

Key Concepts

- How is the total pressure of a mixture of gases related to the partial pressures of the component gases?
- How does the molar mass of a gas affect the rate at which the gas effuses or diffuses?

Vocabulary

partial pressure
Dalton's law of partial pressures
diffusion
effusion
Graham's law of effusion

Reading Strategy

Previewing Before you read, look at the equation on this page that is highlighted in yellow. In your own words, describe the relationship that the equation is expressing.

Connecting to Your World

The top of Mount Everest is more than 29,000 feet above sea level. A list of gear for an expedition to Mount Everest includes climbing equipment such as an ice axe and a climbing harness. It includes ski goggles and a down parka with a hood. All the items on the list are important, but none is as important as the compressed-gas cylinders of oxygen. In this section, you will find out why a supply of oxygen is essential at higher altitudes.

Dalton's Law

Gas pressure results from collisions of particles in a gas with an object. If the number of particles increases in a given volume, more collisions occur. If the average kinetic energy of the particles increases, more collisions occur. In both cases, the pressure increases. Gas pressure depends only on the number of particles in a given volume and on their average kinetic energy. Particles in a mixture of gases at the same temperature have the same average kinetic energy. So the kind of particle is not important.

Table 14.1 shows the composition of dry air, air that does not contain any water vapor. The contribution each gas in a mixture makes to the total pressure is called the **partial pressure** exerted by that gas. In dry air, the partial pressure of nitrogen is 79.11 kPa. **In a mixture of gases, the total pressure is the sum of the partial pressures of the gases.**

$$P_{total} = P_1 + P_2 + P_3 + \ldots$$

This equation is a mathematical expression of a law proposed by Dalton. **Dalton's law of partial pressures** states that, at constant volume and temperature, the total pressure exerted by a mixture of gases is equal to the sum of the partial pressures of the component gases.

Animation 17 Observe the behavior of a mixture of nonreacting gases.

with ChemASAP

Table 14.1

Composition of Dry Air

Component	Volume (%)	Partial pressure (kPa)
Nitrogen	78.08	79.11
Oxygen	20.95	21.22
Carbon dioxide	0.04	0.04
Argon and others	0.93	0.95
Total	100.00	101.32

Section Resources

Print

- **Guided Reading and Study Workbook,** Section 14.4
- **Core Teaching Resources,** Section 14.4 Review
- **Transparencies,** T158–T159
- **Laboratory Manual,** Lab 25
- **Small-Scale Chemistry Laboratory Manual,** Lab 21

Technology

- **Interactive Textbook with ChemASAP,** Animation 17, 18; Problem-Solving 14.32; Assessment 14.4
- **Go Online,** Section 14.4

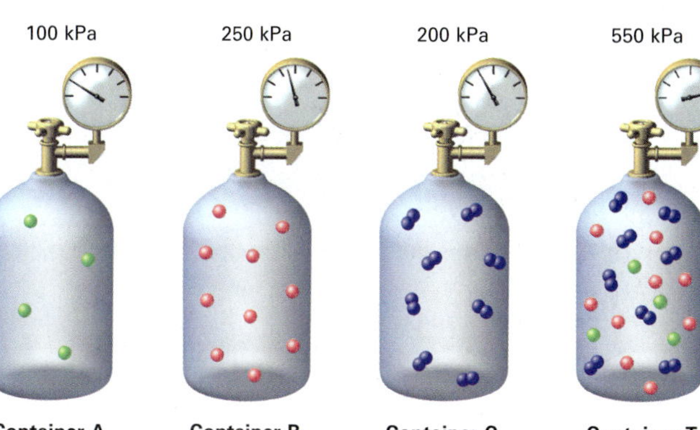

Figure 14.16 Three gases are combined in container T. The pressure that each gas exerts is independent of the pressure exerted by the other two gases. The pressure in container T is the sum of the pressures in containers A, B, and C. **Interpreting Diagrams** ***What is the relationship between the number of particles in containers A and C and the partial pressures of the gases in A and C?***

Look at Figure 14.16. Containers A, B, C, and T (for total) have the same volume and are at the same temperature. The gases in containers A, B, and C are combined in container T. Because the volumes are identical, each gas in the mixture exerts the pressure it exerted before the gases were mixed in container T. So the pressure in container T (550 kPa) is the sum of the pressures in containers A, B, and C (100 + 250 + 200 kPa).

If the percent composition of a mixture of gases does not change, the fraction of the pressure exerted by a gas does not change as the total pressure changes. This fact is important for people who must operate at high altitudes. For example, at the top of Mount Everest, the total atmospheric pressure is 33.73 kPa. This is about one-third of its value at sea level. The partial pressure of oxygen is also reduced by one third, to 7.06 kPa. The partial pressure of oxygen must be 10.67 kPa or higher to support respiration in humans. The climber in Figure 14.17 needs an oxygen mask and a cylinder of compressed oxygen to survive.

Figure 14.17 This climber is using a tank of compressed gas to supplement the supply of oxygen available at high altitudes.

433

CLASS Activity

Model Partial Pressure L2

Purpose Students make a model that is analogous to Dalton's law.

Materials balance, 4 marbles, 2 buttons, 3 pennies, 4 nickels

Procedure Have students find the total mass of the marbles and record it in a data table. Have them repeat this process for the buttons and the pennies. Then, have them add all the marbles, buttons, and pennies together and find the overall mass. Ask, **How does the total mass of the items compare to the sum of the masses of the individual items?** *(The sum of the individual masses should equal the total mass.)* **If the nickels are added to the mixture, how could you find the mass of the nickels?** *(The mass of the nickels could be calculated by finding a new total mass and subtracting from it the masses of the marbles, the buttons, and the pennies.)*

Expected Outcome Students should recognize that mass is being used to represent pressure, and that the contribution of each type of item to the total mass is independent of the other types of items in the mixture.

FYI

You may want to compare the use of a compressed gas at high altitudes and the use of a compressed gas during deep underwater dives.

Facts and Figures

Relative Partial Pressure

The relative partial pressure exerted by a gas in a mixture of gases does not vary with temperature, pressure, or volume of the mixture. The ratio of n_1/n (the relative amount of the gas in the mixture) remains the same. Because the value of n_1/n remains constant, the relative partial pressure (P_1/P) exerted by the gas remains the same.

Answers to...

Figure 14.16 The ratio of the particles in Containers C and A (2:1) is equal to the ratio of their partial pressures.

Sample Problem 14.6

Answers

31. 20.0 kPa + 46.7 kPa + 26.7 kPa = 93.4 kPa or 9.34×10^1 kPa

32. 32.9 kPa – 6.6 kPa – 23.0 kPa = 3.3 kPa

Practice Problems Plus L2

The pressure in an automobile tire filled with air is 245.0 kPa. The P_{O_2} = 51.3 kPa, P_{CO_2} = 0.10 kPa, and P_{others} = 2.3 kPa. **What is the P_{N2}?** (191.3 kPa)

Use Visuals L1

Have students examine the visual on this page and read the caption. Ask, **How does the amount of oxygen in the tank of compressed air compare to the amount of oxygen in the air you normally breathe?** *(Air is usually about 21% oxygen, so the air in the tanks contains about the same percentage of oxygen.)*

Firefighters carry tanks of compressed air. The tanks contain from 19.5% to 23.5% oxygen by volume.

SAMPLE PROBLEM 14.6

Using Dalton's Law of Partial Pressures

Air contains oxygen, nitrogen, carbon dioxide, and trace amounts of other gases. What is the partial pressure of oxygen (P_{O_2}) at 101.30 kPa of total pressure if the partial pressures of nitrogen, carbon dioxide, and other gases are 79.10 kPa, 0.040 kPa, and 0.94 kPa, respectively?

1 Analyze *List the knowns and the unknown.*

Knowns
- P_{N_2} = 79.10 kPa
- P_{CO_2} = 0.040 kPa
- P_{others} = 0.94 kPa
- P_{total} = 101.30 kPa

Unknown
- P_{O_2} = ? kPa

Use Dalton's law of partial pressures ($P_{total} = P_{O_2} + P_{N_2} + P_{CO_2} + P_{others}$) to calculate the unknown value (P_{O_2}).

2 Calculate *Solve for the unknown.*

Rearrange Dalton's law to isolate P_{O_2}. Substitute the values for the partial pressures and solve the equation.

$$P_{O_2} = P_{total} - (P_{N_2} + P_{CO_2} + P_{others})$$
$$= 101.30 \text{ kPa} - (79.10 \text{ kPa} + 0.040 \text{ kPa} + 0.94 \text{ kPa})$$
$$= 21.22 \text{ kPa}$$

3 Evaluate *Does the result make sense?*

The partial pressure of oxygen must be smaller than that of nitrogen because P_{total} is only 101.30 kPa. The other partial pressures are small, so an answer of 21.22 kPa seems reasonable.

Practice Problems

31. Determine the total pressure of a gas mixture that contains oxygen, nitrogen, and helium. The partial pressures are: P_{O_2} = 20.0 kPa, P_{N_2} = 46.7 kPa, and P_{He} = 26.7 kPa.

32. A gas mixture containing oxygen, nitrogen, and carbon dioxide has a total pressure of 32.9 kPa. If P_{O_2} = 6.6 kPa and P_{N_2} = 23.0 kPa, what is P_{CO_2}?

Problem Solving 14.32
Solve Problem 32 with the help of an interactive guided tutorial.

with ChemASAP

434 Chapter 14

Graham's Law

Suppose you open a perfume bottle in one corner of a room. At some point a person standing in the opposite corner will be able to smell the perfume. Molecules in the perfume evaporate and diffuse, or spread out, through the air in the room. **Diffusion** is the tendency of molecules to move toward areas of lower concentration until the concentration is uniform throughout. In Figure 14.18A, bromine vapor is diffusing through the air in a graduated cylinder. The bromine vapor in the bottom of the cylinder has started to move upward toward the area where there is a lower concentration of bromine. In Figure 14.18B, the bromine has diffused almost to the top of the cylinder. If the process is allowed to continue, the bromine vapor will spill out of the cylinder.

There is another process that involves the movement of molecules in a gas. This process is called effusion. During **effusion**, a gas escapes through a tiny hole in its container. With effusion and diffusion, the type of particle is important. **Gases of lower molar mass diffuse and effuse faster than gases of higher molar mass.**

Thomas Graham's Contribution The Scottish chemist Thomas Graham studied rates of effusion during the 1840s. From his observations, he proposed a law. **Graham's law of effusion** states that the rate of effusion of a gas is inversely proportional to the square root of the gas's molar mass. This law can also be applied to the diffusion of gases.

Graham's law makes sense if you know how the mass, velocity, and kinetic energy of a moving object are related. The expression that relates the mass (m) and the velocity (v) of an object to its kinetic energy (KE) is $KE = 1/2mv^2$. For the kinetic energy to be constant, any increase in mass must be balanced by a decrease in velocity. For example, a ball with a mass of 2 g must travel at 5 m/s to have the same kinetic energy as a ball with a mass of 1 g traveling at 7 m/s. There is an important principle here. If two objects with different masses have the same kinetic energy, the lighter object must move faster.

Word *Origins*

Diffusion and **effusion** come from the Latin *fundere* meaning "to pour." They differ only in their prefixes. The prefix *dis-* means "apart." The prefix *ex-* means "out." **How do these prefixes help to contrast what happens to a gas during diffusion and effusion?**

For: Links on Diffusion/Effusion
Visit: www.SciLinks.org
Web Code: cdn-1144

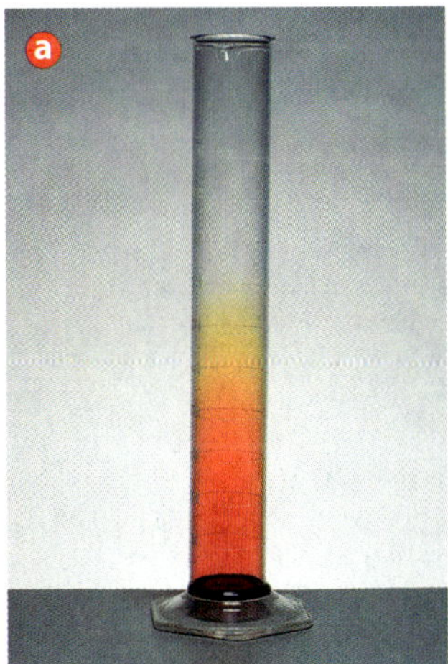

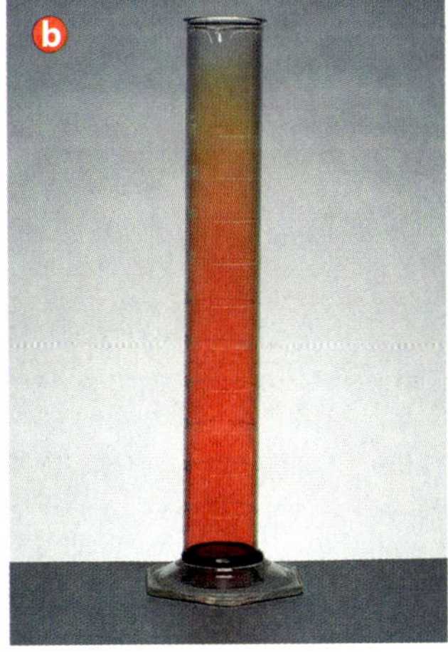

Figure 14.18 The diffusion of one substance through another is a relatively slow process. **a** Bromine vapor is diffusing upward through the air in a graduated cylinder. **b** After several hours, bromine vapors are near the top of the cylinder. **Predicting** ***What will happen as the bromine continues to diffuse?***

Graham's Law

Word *Origins* L2

During diffusion, the particles move apart. During effusion, the particles move out of a tiny hole. (Students may have trouble with the scientific definitions of *effusion* and *diffusion* because of the multiple, often overlapping, meanings of these terms in common usage. For example, in many dictionaries one definition of *effuse* is "to spread out; diffuse; radiate.")

Use Visuals 

Figure 14.18 Have students study the photograph. Explain that diffusion is a relatively slow process whose rate is inversely proportional to the square root of the molar mass of the gas.

CLASS Activity

Effusion L1

Purpose Students compare effusion rates of helium and air.

Materials 2 round, identical balloons; helium; metric tape measure.

Safety Wear safety goggles when handling balloons.

Procedure Fill two identical, round non-latex balloons with equal volumes of helium and air. Have volunteers measure the circumference of the balloons. One day later, measure the circumferences again.

Expected Outcome Helium effused from the balloon at a faster rate.

Download a worksheet on **Diffusion/Effusion** for students to complete, and find additional teacher support from NSTA SciLinks.

Differentiated Instruction

Less Proficient Readers L1

Have students work in groups to rank the relative rates of diffusion at a constant temperature for carbon dioxide, helium, and nitrogen. *(helium > nitrogen > carbon dioxide)*

Answers to...

Figure 14.18 The bromine will spill out of the container.

3 ASSESS

Evaluate Understanding L2

Tell students that the partial pressures of oxygen and hydrogen gases in a container are both 100 kPa. Ask, **In which sample are there more molecules? In which sample do the molecules have greater average kinetic energy?** *(Both gases have the same number of molecules and same average kinetic energy.)* Ask, **About how much faster does helium diffuse compared to oxygen?** *(Helium diffuses almost three times faster than oxygen.)*

Reteach L1

Remind students that *diffusion* is a general term that applies to molecules moving away from a region of high concentration. Effusion is a specific example of diffusion in which molecules pass through a narrow opening. Graham's law applies to both.

Elements Handbook

Both tables report data for dry air. Both show that air is mainly nitrogen and oxygen. Table 14.1 uses percents and includes a compound. The table on R4 uses ppm and provides specific data for more elements .

If your class subscribes to the Interactive Textbook, use it to review key concepts in Section 14.4.

with ChemASAP

Figure 14.19 The character balloons used in parades are filled with helium gas so that they will float.

Comparing Effusion Rates The balloon in Figure 14.19 is used in holiday parades. It is filled with helium so that it will float above the crowd along the parade route. There is a drawback to using helium in a balloon. Both helium atoms and the molecules in air can pass freely through the tiny pores in a balloon. But a helium-filled balloon will deflate sooner than an air-filled balloon. Kinetic theory can explain this difference.

If the balloons are at the same temperature, the particles in each balloon have the same average kinetic energy. But helium atoms are less massive than oxygen or nitrogen molecules. So the molecules in air move more slowly than helium atoms with the same kinetic energy. Because the rate of effusion is related only to a particle's speed, Graham's law can be written as follows for two gases, A and B.

$$\frac{\text{Rate}_A}{\text{Rate}_B} = \sqrt{\frac{\text{molar mass}_B}{\text{molar mass}_A}}$$

The rates of effusion of two gases are inversely proportional to the square roots of their molar masses. You can use the expression to compare the rates of effusion of nitrogen (molar mass = 28.0 g) and helium (molar mass = 4.0 g). Helium effuses (and diffuses) nearly three times faster than nitrogen at the same temperature.

$$\frac{\text{Rate}_{He}}{\text{Rate}_{N_2}} = \sqrt{\frac{28.0\text{ g}}{4.0\text{ g}}} = \sqrt{7.0} = 2.7$$

Animation 18 Observe the processes of gas effusion and diffusion.

with ChemASAP

14.4 Section Assessment

33. **Key Concept** In a mixture of gases, how is the total pressure determined?
34. **Key Concept** What is the effect of molar mass on rates of diffusion and effusion?
35. How is the partial pressure of a gas in a mixture calculated?
36. What distinguishes effusion from diffusion? How are these processes similar?
37. How can you compare the rates of effusion of two gases in a mixture?
38. Explain why the rates of diffusion of nitrogen gas and carbon monoxide are almost identical at the same temperature.

Elements Handbook

Table 14.1 on page 432 and the *Elements in the Atmosphere* table on page R4 provide data on the composition of air. Look at the data included in each table. Identify two ways in which the tables are similar. Describe at least three differences.

Assessment 14.4 Test yourself on the concepts in Section 14.4.

with ChemASAP

Section 14.4 Assessment

33. Total pressure is equal to the sum of the partial pressures of the components.
34. Gases with lower molar masses diffuse and effuse faster than gases with higher molar masses.
35. Subtract the partial pressures of the other gases from the total pressure.
36. During effusion, a gas escapes through a tiny hole in its container. In both cases, the rate depends on the molar mass.
37. The rate of effusion of two gases in a mixture is inversely proportional to the square roots of their molar masses.
38. Carbon monoxide and nitrogen have almost identical molar masses when the masses are rounded to two significant figures (28 g).

436 Chapter 14

Small-Scale LAB

Diffusion

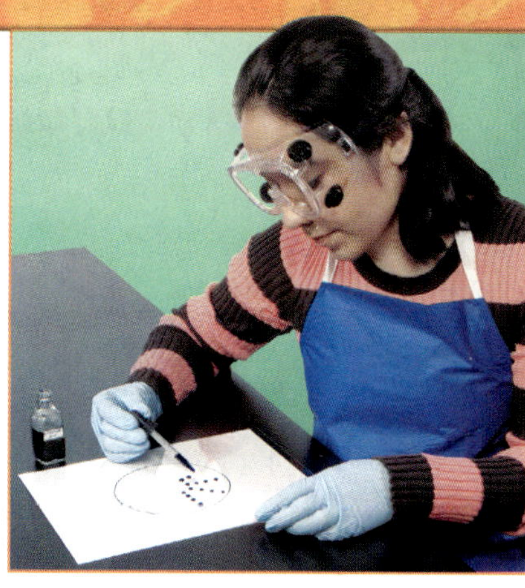

Purpose

To infer diffusion of a gas by observing color changes during chemical reactions.

Materials

- clear plastic cup or Petri dish
- reaction surface
- dropper bottles containing bromthymol blue, hydrochloric acid, and sodium hydrogen sulfite
- ruler
- cotton swab
- NaOH, NH_4Cl, KI, and $NaNO_2$ (optional)

Procedure   

1. Use the plastic cup or Petri dish to draw the large circle shown below on a sheet of paper.

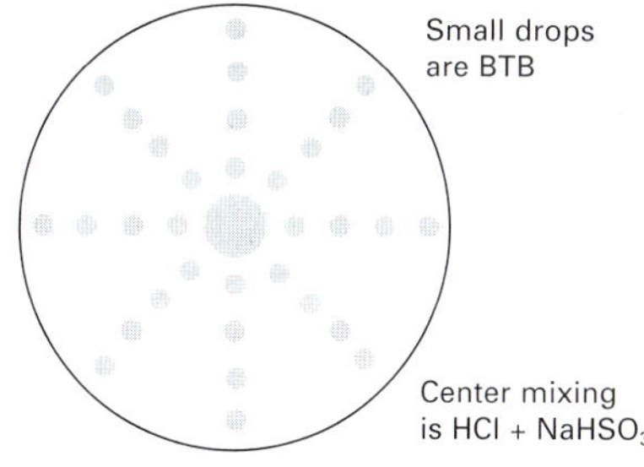

2. Place a reaction surface over the grid and add small drops of bromthymol blue (BTB) in the pattern shown by the small circles. Make sure the drops do not touch one another.
3. Mix one drop each of hydrochloric acid (HCl) and sodium hydrogen sulfite ($NaHSO_3$) in the center of the pattern.
4. Place the cup or Petri dish over the grid and observe what happens.
5. If you plan to do You're The Chemist Activity 1, don't dispose of your materials yet.

Analyze

Using your experimental data, record the answers to the following questions below your data table.

1. Describe in detail the changes you observed in the drops of BTB over time. Draw pictures to illustrate the changes.
2. Draw a series of pictures showing how one of the BTB drops might look over time if you could view the drop from the side.
3. The BTB changed even though you added nothing to it. If the mixture in the center circle produced a gas, would this explain the change in the drops of BTB? Use kinetic theory to explain your answer.
4. Translate the following word equation into a balanced chemical equation: Sodium hydrogen sulfite reacts with hydrochloric acid to produce sulfur dioxide gas, water, and sodium chloride.

You're The Chemist

The following small-scale activities allow you to develop your own procedures and analyze the results.

1. **Analyze It!** Carefully absorb the center mixture of the original experiment onto a cotton swab and replace it with one drop of NaOH and one drop of NH_4Cl. Describe what happens and explain in terms of kinetic theory. Ammonium chloride reacts with sodium hydroxide to produce ammonia gas, water, and sodium chloride. Write and balance a chemical equation to describe this reaction.
2. **Design It!** Design an experiment to observe the effect of the size of the BTB drops on the rate at which they change. Explain your results in terms of kinetic theory.
3. **Analyze It!** Repeat the original experiment, using KI in place of BTB and mixing sodium nitrite ($NaNO_2$) with hydrochloric acid at the center. Record your results. Write and balance an equation for the reaction. $NaNO_2$ reacts with HCl to produce nitrogen monoxide gas, water, sodium nitrate, and sodium chloride.

Analyze and Conclude

1. The drops near the center change immediately. As the gas diffuses, all the drops change color. The color change begins at the outer edge of each drop.

    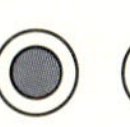   

2. 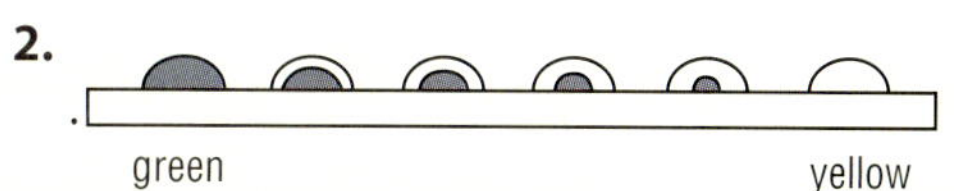

3. The particles of gas produced are in motion. As the particles diffuse from the center, they collide and react with molecules of BTB.
4. $NaHSO_3 + HCl \rightarrow SO_2 + H_2O + NaCl$

Diffusion 

Objective After completing this activity, students will be able to:

- infer diffusion of a gas by observing color changes during reactions.

Skills Focus Inferring, Drawing Conclusions

Prep Time 30 minutes
Class Time 40 minutes

Advance Preparation

Solution	Preparation
1.0*M* HCl	82 mL of 12*M* in 1.0 L
0.04% BTB	100 mg in 16.0 mL 0.01*M* NaOH, dilute to 250 mL
0.5*M* NaOH	20.0 g in 1.0 L
1.0*M* NH_4Cl	13.4 g in 250 mL
0.4*M* $NaNO_2$	6.9 g in 250 mL
0.1*M* KI	4.2 g in 250 mL
1.0*M* $NaHSO_3$	26 g in 250 mL

Safety Always slowly add acid to water.

Teaching Tips

- Before class, add a drop of BTB to the paper to be sure the paper is not acidic. The BTB should remain blue.

Expected Outcome

The drops of BTB turn yellow, starting with those closest to the center.

You're the Chemist!

1. As ammonia diffuses, BTB changes from yellow to blue.
   $NH_4Cl + NaOH \rightarrow NH_3 + H_2O + NaCl$
2. Vary the size of the BTB drops from "pin-heads" to "puddles." Tiny drops are better able to detect small quantities of gases.
3. The KI turned orange in the same manner as the BTB turned yellow.
   $3NaNO_2 + 2HCl \rightarrow 2NO + H_2O + NaNO_3 + 2NaCl$

For Enrichment 

Have students design an experiment that shows the effect of temperature on rate of diffusion. The drops change color more quickly at higher temperatures.

CHAPTER 14 Study Guide

Study Tip

Color Code Variables Suggest that students practice solving numeric problems by assigning a different color to each variable when they write the mathematical expression they will use to solve the problem. Students should continue using the assigned colors as they rearrange the expression to solve for the unknown and as they substitute values for the variables.

If your class subscribes to the Interactive Textbook with ChemASAP, your students can go online to access an interactive version of the Student Edition and a self-test.

with ChemASAP

CHAPTER 14 Study Guide

Key Concepts

14.1 Properties of Gases

- Gases are easily compressed because of the space between particles in a gas.
- The amount of gas (n), volume (V), and temperature (T) are factors that affect gas pressure (P).

14.2 The Gas Laws

- As the pressure of a gas increases, the volume decreases, if the temperature is constant.
- As the temperature of an enclosed gas increases, the volume increases, if the pressure is constant.
- As the temperature of an enclosed gas increases, the pressure increases, if the volume is constant.
- The combined gas law allows you to do calculations for situations in which only the amount of gas is constant.

14.3 Ideal Gases

- To calculate the number of moles of a contained gas requires an expression that contains the variable n.
- Real gases differ most from an ideal gas at low temperatures and high pressures.

14.4 Gases: Mixtures and Movements

- In a mixture of gases, the total pressure is the sum of the partial pressures of the gases.
- Gases of lower molar mass diffuse and effuse faster than gases of higher molar mass.

Vocabulary

- Boyle's law (p. 418)
- Charles's law (p. 420)
- combined gas law (p. 424)
- compressibility (p. 413)
- Dalton's law of partial pressures (p. 432)
- diffusion (p. 435)
- effusion (p. 435)
- Gay-Lussac's law (p. 422)
- Graham's law of effusion (p. 435)
- ideal gas constant (p. 426)
- ideal gas law (p. 426)
- partial pressure (p. 432)

Key Equations

- Boyle's law: $P_1 \times V_1 = P_2 \times V_2$
- Charles's law: $\frac{V_1}{T_1} = \frac{V_2}{T_2}$
- Gay-Lussac's law: $\frac{P_1}{T_1} = \frac{P_2}{T_2}$
- Combined gas law: $\frac{P_1 \times V_1}{T_1} = \frac{P_2 \times V_2}{T_2}$
- Ideal gas law: $P \times V = n \times R \times T$ or $PV = nRT$
- Dalton's law: $P_{total} = P_1 + P_2 + P_3 + \ldots$
- Graham's law: $\frac{\text{Rate}_A}{\text{Rate}_B} = \sqrt{\frac{\text{molar mass}_B}{\text{molar mass}_A}}$

Organizing Information

Use these terms to construct a concept map that organizes the major ideas of this chapter.

Boyle's law | Charles's law | partial pressure | combined gas law | diffusion

Concept Map 14 Solve the Concept Map with the help of an interactive guided tutorial.

with ChemASAP

Chapter Resources

Print

- **Core Teaching Resources,** Chapter 14
  Practice Problems, Vocabulary Review, Quiz, Chapter Test A, Chapter Test B

Technology

- **Computer Test Bank,** Chapter 14 Test
- **Interactive Textbook with ChemASAP,** Chapter 14

CHAPTER 14

Assessment

Reviewing Content

14.1 Properties of Gases

39. What happens to the particles in a gas when the gas is compressed?

40. Explain why heating a contained gas that is held at a constant volume increases its pressure.

41. Describe what happens to the volume of a balloon when it is taken outside on a cold winter day. Explain why the observed change happens.

42. A metal cylinder contains 1 mol of nitrogen gas. What will happen to the pressure if another mole of gas is added to the cylinder, but the temperature and volume do not change?

43. If a gas is compressed from 4 L to 1 L and the temperature remains constant, what happens to the pressure?

44. Use the drawing to help explain why gas pressure decreases when gas is removed from a container with a fixed volume.

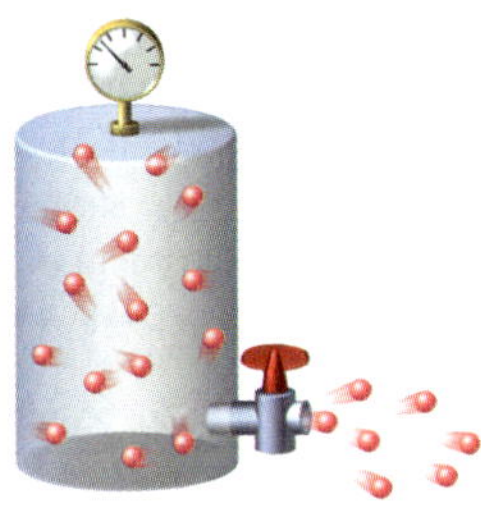

14.2 The Gas Laws

45. Write the mathematical equation for Charles's law and explain the symbols.

46. The gas in a closed container has a pressure of 3.00×10^2 kPa at 30°C (303 K). What will the pressure be if the temperature is lowered to −172°C (101 K)?

47. Calculate the volume of a gas (in L) at a pressure of 1.00×10^2 kPa if its volume at 1.20×10^2 kPa is 1.50×10^3 mL.

48. A gas with a volume of 4.0 L at 90.0 kPa expands until the pressure drops to 20.0 kPa. What is its new volume if the temperature doesn't change?

49. A gas with a volume of 3.00×10^2 mL at 150.0°C is heated until its volume is 6.00×10^2 mL. What is the new temperature of the gas if the pressure remains constant during the heating process?

50. Write the mathematical expression for the combined gas law.

51. A sealed cylinder of gas contains nitrogen gas at 1.00×10^3 kPa pressure and a temperature of 20°C. When the cylinder is left in the sun, the temperature of the gas increases to 50°C. What is the new pressure in the cylinder?

52. Show how Gay-Lussac's law can be derived from the combined gas law.

14.3 Ideal Gases

53. Describe an ideal gas.

54. Explain why it is impossible for an ideal gas to exist.

55. What is the volume occupied by 1.24 mol of a gas at 35°C if the pressure is 96.2 kPa?

56. What volume will 12.0 g of oxygen gas (O_2) occupy at 25°C and a pressure of 52.7 kPa?

57. If 4.50 g of methane gas (CH_4) is in a 2.00-L container at 35°C, what is the pressure in the container?

58. A helium-filled weather balloon has a volume of 2.4×10^2 L at 99 kPa pressure and a temperature of 0°C. What is the mass of the helium in the balloon?

14.4 Gases: Mixtures and Movements

59. In your own words, state Dalton's law of partial pressure.

60. Which gas effuses faster: hydrogen or chlorine? How much faster?

61. Which gas effuses faster at the same temperature: molecular oxygen or atomic argon?

62. Calculate the ratio of the velocity of helium atoms to the velocity of neon atoms at the same temperature.

63. Calculate the ratio of the velocity of helium atoms to the velocity of fluorine molecules at the same temperature.

CHAPTER 14

Assessment

Reviewing Content

39. The space between particles is reduced.

40. The increased kinetic energy of the particles causes collisions to occur more frequently and with more force.

41. The volume decreases. The molecules have less kinetic energy and cause less pressure on the inside of the balloon.

42. The pressure doubles.

43. The pressure quadruples.

44. As particles are removed from the container, there is a decrease in the number of collisions.

45. $V_1/T_1 = V_2/T_2$
V_1 and V_2 are the initial and final volumes; T_1 and T_2 are the initial and final temperatures.

46. 1.00×10^2 kPa

47. 1.80 L

48. 18 L

49. 846 K (573°C)

50. $(P_1V_1)/T_1 = (P_2V_2)/T_2$

51. 1.10×10^3 kPa

52. When the volume is constant, $V_1 = V_2$, so the volume terms cancel, leaving the equation for Gay-Lussac's law.

53. Its particles have no volume, there are no attractions between them, and collisions are elastic. An ideal gas follows the gas laws at all temperatures and pressures.

54. Particles in a real gas have a finite volume and are attracted to one another at low temperatures and high pressures.

55. 33.0 L

56. 17.6 L

57. 3.60×10^2 kPa

58. 42 g

59. The total pressure of a gaseous mixture is equal to the sum of the individual pressures of each gas.

60. Hydrogen gas diffuses faster than chlorine gas by an approximate factor of six.

61. molecular oxygen

62. 2.25:1

63. 3.08:1

Understanding Concepts

64. How does kinetic theory explain the compressibility of gases?

Use this description to answer Questions 65 and 66.

A teacher adds 1 mL of water to an empty metal soda can. The teacher heats the can over a burner until the water boils and then quickly plunges the can upside down in an ice-water bath. The can immediately collapses inward as though crushed in a trash compactor.

65. Use kinetic theory to explain why the can collapsed inward.

66. If the experiment were done with a dry can, would the results be similar? Explain.

67. Why do aerosol containers display the warning, "Do not incinerate"?

68. The manufacturer of an aerosol deodorant packaged in a 150-mL container plans to produce a container of the same size that will hold twice as much gas. How will the pressure of the gas in the new product compare with that of the gas in the original container?

69. Why must Kelvin temperatures be used in calculations that involve gases?

70. Explain how using a pressure cooker reduces the time required to cook food.

71. The ratio of two variables is always a constant. What can you conclude about the relationship between the two variables?

72. A 3.50-L gas sample at 20°C and a pressure of 86.7 kPa expands to a volume of 8.00 L. The final pressure of the gas is 56.7 kPa. What is the final temperature of the gas, in degrees Celsius?

73. Explain the reasons why real gases deviate from ideal behavior.

74. How would the number of particles of two gases compare if their partial pressures in a container were identical?

75. Why does a balloon filled with helium deflate more quickly than a balloon filled with air?

76. A certain gas effuses four times as fast as oxygen (O_2). What is the molar mass of the gas?

77. During an effusion experiment, a certain number of moles of an unknown gas passed through a tiny hole in 75 seconds. Under the same conditions, the same number of moles of oxygen gas passed through the hole in 30 seconds. What is the molar mass of the unknown gas?

78. The photograph shows a tube with cotton balls at each end. The cotton ball at the left was soaked with hydrochloric acid. The cotton ball on the right was soaked with a solution of ammonia. When these compounds react, they form a white solid, ammonium chloride. Based on the location of the ammonium chloride in the tube, which gas diffuses at a faster rate, hydrogen chloride or ammonia? Explain.

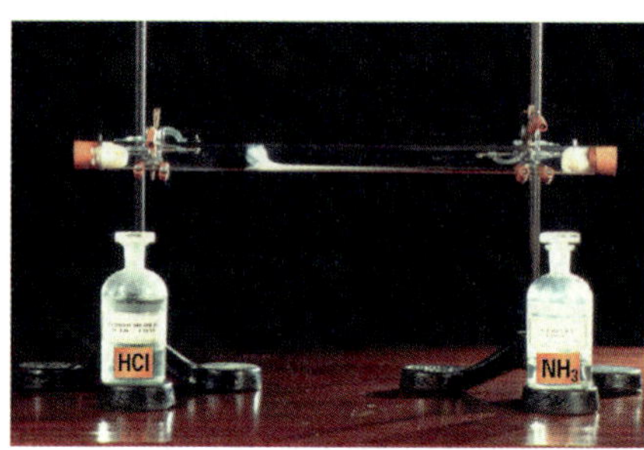

CHAPTER 14 Assessment continued

Understanding Concepts

64. Gases are easily compressed because there is a lot of space between the particles.

65. Boiling the water fills the can with steam. When the can is plunged upside down into ice water the steam is trapped and rapidly condenses, reducing gas pressure inside the can. The walls of the can are not strong enough to withstand the comparatively high atmospheric pressure, which crushes the can.

66. The results would be much less dramatic. The change in volume (and in internal pressure) from heated air to cold air is much less dramatic than when steam (a vapor) condenses to water (a liquid).

67. High temperatures increase the pressure of the contents of the container and may cause it to explode.

68. The pressure will double.

69. Temperatures measured on the Kelvin scale are directly proportional to the average kinetic energy of the particles. Celsius temperatures are not.

70. The water boils at a higher temperature, which speeds the cooking process.

71. The variables are directly proportional.

72. 165°C

73. The particles in a real gas have a finite volume and are attracted to one another.

74. The number of particles would be equal.

75. Helium atoms have a smaller molar mass than oxygen and nitrogen molecules and effuse faster through pores in the balloon.

76. 2.0 g

77. 2.0×10^2 g

78. Ammonia diffuses faster than hydrogen chloride. Based on the location of the reaction product, the ammonia molecules travel about twice the distance of the hydrogen chloride molecules, in the same time.

440 Chapter 14

Critical Thinking

79. How does the vacuum in the flask used to store liquid nitrogen prevent heat transfer?

80. Gases will diffuse from a region of higher concentration to a region of lower concentration. Why don't the gases in Earth's atmosphere escape into the near-vacuum of space?

81. What real gas comes closest to having the characteristics of an ideal gas? Explain your answer.

82. Death Valley in California is at 86 m below sea level. Will the partial pressure of oxygen in Death Valley be the same, lower, or higher than the partial pressure of oxygen at sea level? Give a reason for your answer.

83. The following reaction takes place in a sealed 40.0-L container at a temperature of 120°C.

$$4NH_3(g) + 5O_2(g) \longrightarrow 4NO(g) + 6H_2O(g)$$

a. When 34.0 g of NH_3 reacts with 96.0 g of O_2, what is the partial pressure of NO in the sealed container?

b. What is the total pressure in the container?

84. The graph shows the direct relationship between volume and temperature for three different gas samples. Offer at least one explanation for why the graphs are not identical for the three samples. (*Hint:* What variables other than temperature and volume can be used to describe a gas?)

Volume vs. Temperature
Data for Three Gases

Gas A
Gas B
Gas C
Volume (L)
Temperature (°C)

Concept Challenge

85. Oxygen is produced in the laboratory by heating potassium nitrate (KNO_3). The data table below gives the volume of oxygen produced at STP from different quantities of KNO_3. Use the data to determine the mole ratio by which KNO_3 and O_2 react.

Mass of KNO_3 (g)	Volume of O_2 (cL)
0.84	9.3
1.36	15.1
2.77	30.7
4.82	53.5
6.96	77.3

86. A mixture of ethyne gas (C_2H_2) and methane gas (CH_4) occupied a certain volume at a total pressure of 16.8 kPa. When the sample burned, the products were CO_2 gas and H_2O vapor. The CO_2 was collected and its pressure found to be 25.2 kPa in the same volume and at the same temperature as the original mixture. What percentage of the original mixture was methane?

87. A 0.10-L container holds 3.0×10^{20} molecules of H_2 at 100 kPa and 0°C.

a. If the volume of a hydrogen molecule is 6.7×10^{-24} mL, what percentage of the volume of the gas is occupied by its molecules?

b. If the pressure is increased to 100,000 kPa, the volume of the gas is 1×10^{-4} L. What fraction of the total volume do the hydrogen molecules now occupy?

88. Many gases that have small molecules, such as N_2 and O_2, have the expected molar volume of 22.41 L at STP. However, other gases behave in a very non-ideal manner, even if extreme pressures and temperatures are not involved. The molar volumes of CH_4, CO_2, and NH_3 at STP are 22.37 L, 22.26 L, and 22.06 L, respectively. Explain the reasons for these large departures from the ideal.

Critical Thinking

79. A vacuum contains no matter to allow the transfer of kinetic energy between molecules.

80. The gases that make up the atmosphere, just like any other form of matter, are held near Earth by the force of gravity.

81. Helium gas is composed of small atoms with little attraction for each other.

82. The partial pressure of oxygen would be greater than at sea level, because the atmospheric pressure would be greater than at sea level.

83. a. 1.63×10^2 kPa

b. 4.48×10^2 kPa

84. There could be different amounts of each gas or the gases could be at different pressures.

Concept Challenge

85. 2 mol KNO_2 for each 1 mol O_2

86. 46% CH_4

87. a. 2.0×10^{-3}%

b. 2.0%

88. Because attractions between molecules in gases such as nitrogen and oxygen are insignificant, these gases have the molar volume of an ideal gas, 22.4 L at STP. Based on their molar volumes at STP, there are attractions between molecules of CH_4, CO_2, and NH_3. These attractions increase in strength from methane to carbon dioxide to ammonia.

CHAPTER 14

Assessment continued

Cumulative Review

89. K = °C + 273
90. 10.6 g/cm^3
91. 82 protons, 82 electrons, 124 neutrons
92. nickel, Ni
93. a. tungsten
94. c. selenium
95. b. SO_2
96. **a.** tin(II) bromide
b. barium sulfate
c. magnesium hydroxide
d. iodine pentafluoride
97. 206 g
98. **a.** 158 g
b. 98 g
c. 342 g
d. 331 g
99. It is the volume occupied by 1 mol of a gas at STP.
100. **a.** $C_4H_8O_2$
b. C_8H_8
c. $C_3H_6O_3$
101. **a.** single-replacement
b. decomposition
102. **a.** $P_4O_{10} + 6H_2O \rightarrow 4H_3PO_4$
b. $Al_2S_3 + 6H_2O \rightarrow 2Al(OH)_3 + 3H_2S$
103. 60.0% C, 13.3% H, 26.7% O
104. **a.** $4Al + 3O_2 \rightarrow 2Al_2O_3$
b. 3.09×10^2 g Al; 2.75×10^2 g O_2
105. The motion of particles in a gas is constant, random, and rapid.
106. Gas pressure is the result of the collisions of gas particles with the container wall.

CHAPTER 14

Assessment *continued*

Cumulative Review

89. What is the mathematical relationship between the Kelvin and Celsius temperature scales? *(Chapter 3)*

90. A metal sample has a mass of 9.92 g and measures 4.5 cm × 1.3 cm × 1.6 mm. What is the density of the metal? *(Chapter 3)*

91. How many electrons, protons, and neutrons are there in an atom of lead-206? *(Chapter 4)*

92. Which element has the following electron configuration? *(Chapter 5)*

1s	2s	2p	3s	3p	4s	3d
↑↓	↑↓	↑↓	↑↓	↑↓	↑↓	↑↓
		↑↓		↑↓		↑↓
		↑↓		↑↓		↑↓
						↑
						↑

93. Which of these elements are metals? *(Chapter 6)*
a. arsenic **b.** tungsten **c.** xenon

94. Which element is most likely to form a compound with strontium? *(Chapter 7)*
a. neon **b.** tin **c.** selenium

95. Which compound contains at least one double bond? *(Chapter 8)*
a. H_2Se **b.** SO_2 **c.** PCl_3

96. Name each compound. *(Chapter 9)*
a. $SnBr_2$ **b.** $BaSO_4$
c. $Mg(OH)_2$ **d.** IF_5

97. An atom of lead has a mass 17.16 times greater than the mass of an atom of carbon-12. What is the molar mass of this isotope of lead? *(Chapter 10)*

98. Calculate the molar mass of each substance. *(Chapter 10)*
a. $Ca(CH_3CO_2)_2$
b. H_3PO_4
c. $C_{12}H_{22}O_{11}$
d. $Pb(NO_3)_2$

99. What is the significance of the volume 22.4 L? *(Chapter 10)*

100. Calculate the molecular formula of each of the following compounds. *(Chapter 10)*
a. The empirical formula is C_2H_4O and the molar mass is 88 g.
b. The empirical formula is CH and the molar mass is 104 g.
c. The molar mass is 90 g. The percent composition is 26.7% C, 71.1% O, and 2.2% H.

101. What type of reaction is each of the following? *(Chapter 11)*
a. Calcium reacts with water to form calcium hydroxide and hydrogen gas.
b. Mercury and oxygen are prepared by heating mercury(II) oxide.

102. Write a balanced equation for each chemical reaction. *(Chapter 12)*
a. Tetraphosphorus decoxide reacts with water to form phosphoric acid.
b. Aluminum hydroxide and hydrogen sulfide form when aluminum sulfide reacts with water.

103. Calculate the percent composition of 2-propanol (C_3H_7OH). *(Chapter 12)*

104. Aluminum oxide is formed from its elements. *(Chapter 12)*

$$Al(s) + O_2(g) \longrightarrow Al_2O_3(s)$$

a. Balance the equation.
b. How many grams of each reactant are needed to form 583 g $Al_2O_3(s)$?

105. Explain why a gas expands until it takes the shape and volume of its container. *(Chapter 13)*

106. Use the drawings to explain how gas pressure is produced. *(Chapter 13)*

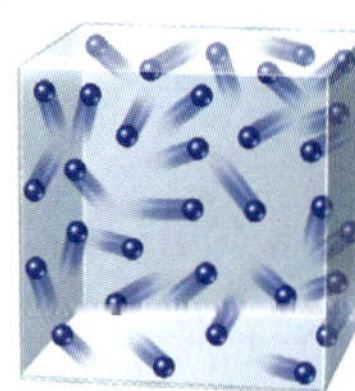

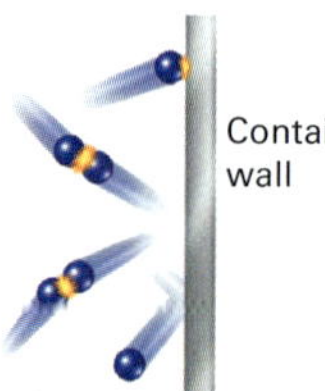

Standardized Test Prep

Test-Taking Tip

Constructing a Diagram You may be asked to draw a diagram or make changes to an existing drawing. Sketch lightly at first so you can erase easily if you need to, or do a sketch on a separate piece of paper. Once you are sure of your answer, draw the final diagram.

Select the choice that best answers each question or completes each statement.

1. A gas in a balloon at constant pressure has a volume of 120.0 mL at −123°C. What is its volume at 27.0°C?

a. 60.0 mL **b.** 240.0 mL
c. 26.5 mL **d.** 546 mL

2. If the Kelvin temperature of a gas is tripled and the volume is doubled, the new pressure will be

a. 1/6 the original pressure.
b. 2/3 the original pressure.
c. 3/2 the original pressure.
d. 5 times the original pressure.

3. Which of these gases effuses fastest?

a. Cl_2 **b.** NO_2 **c.** NH_3 **d.** N_2

4. All the oxygen gas from a 10.0-L container at a pressure of 202 kPa is added to a 20.0-L container of hydrogen at a pressure of 505 kPa. After the transfer, what are the partial pressures of oxygen and hydrogen?

a. Oxygen is 101 kPa; hydrogen is 505 kPa.
b. Oxygen is 202 kPa; hydrogen is 505 kPa.
c. Oxygen is 101 kPa; hydrogen is 253 kPa.
d. Oxygen is 202 kPa; hydrogen is 253 kPa.

5. Which of the following changes would increase the pressure of a gas in a closed container?

I. Part of the gas is removed.
II. The container size is decreased.
III. Temperature is increased.

a. I and II only
b. II and III only
c. I and III only
d. I, II, and III

6. A real gas behaves most nearly like an ideal gas

a. at high pressure and low temperature.
b. at low pressure and high temperature.
c. at low pressure and low temperature.
d. at high pressure and high temperature.

Use the graphs to answer Questions 7–10. A graph may be used once, more than once, or not at all.

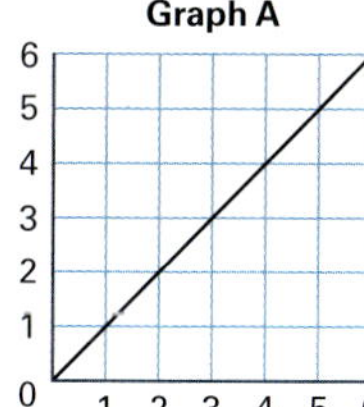

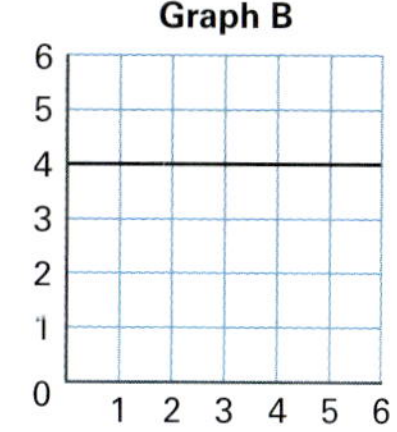

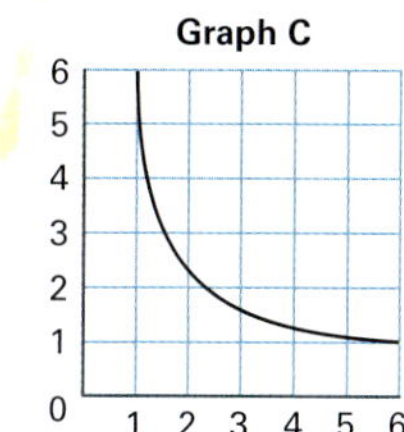

Which graph shows each of the following?

7. directly proportional relationship
8. graph with slope = 0
9. inversely proportional relationship
10. graph with a constant slope

Use the drawing to answer Questions 11 and 12.

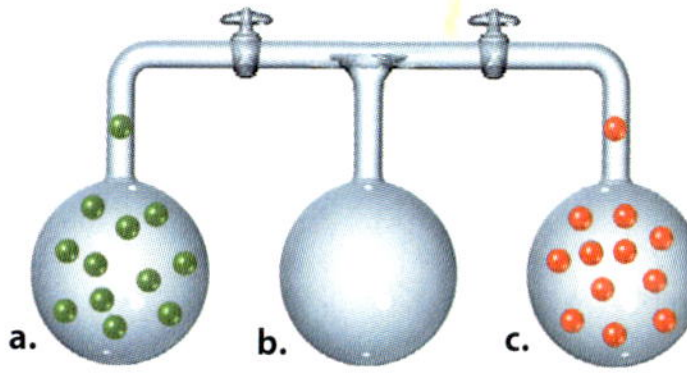

11. Bulb A and bulb C contain different gases. Bulb B contains no gas. If the valves between the bulbs are opened, how will the particles of gas be distributed when the system reaches equilibrium? Assume none of the particles are in the tubes that connect the bulbs.

12. Make a three-bulb drawing with 6 blue spheres in bulb A, 9 green spheres in bulb B, and 12 red spheres in bulb C. Then draw the setup to represent the distribution of gases after the valves are opened and the system reaches equilibrium.

Standardized Test Prep

1. b
2. c
3. c
4. a
5. b
6. b
7. A
8. B
9. C
10. A and B
11. There will be four green spheres and four red spheres in each bulb.
12. Each bulb will contain two blue, three green, and four red spheres.

CHAPTER 15

Water and Aqueous Systems Planning Guide

SECTION OBJECTIVES	STANDARDS NATIONAL	STANDARDS STATE	ACTIVITIES and LABS	
15.1 Water and Its Properties, pp. 445–449 1 block or 2 periods **15.1.1 Explain** the high surface tension and low vapor pressure of water in terms of the structure of the water molecule and hydrogen bonding. **15.1.2 Describe** the structure of ice.	A-1, B-2		**SE** Inquiry Activity: Observing Surface Tension, p. 444 **SE** Quick Lab: Surfactants, p. 448	L2 L2
15.2 Homogeneous Aqueous Systems, pp. 450–458 1 block or 2 periods **15.2.1 Distinguish** between a solvent and a solute. **15.2.2 Describe** what happens in the solution process. **15.2.3 Explain** why all ionic compounds are electrolytes. **15.2.4 Demonstrate** how the formula for a hydrate is written.	A-1, B-2, G-1		**SE** Small-Scale Lab: Electrolytes p. 458 **TE** Teacher Demo: Electrolytes, p. 453 **TE** Teacher Demo: Magic Writing, p. 454 **LM** Lab 26: Distillation **LM** Lab 27: The Solvent Properties of Water **LM** Lab 28: Water of Hydration **LM** Lab 29: Electrolytes and Nonelectrolytes **SSLM** Lab 22: Reactions of Aqueous Ionic Compounds **SSLM** Lab 23: Identification of Eight Unknown Solutions **SSLM** Lab 24: Electrolytes **LP** Lab Practical 15-1: Electrolytes and Nonelectrolytes **PLM** Probeware Lab: Electrolytes **VCL** Lab 21: Electrolytes	L2 L2 L1 L2 L2 L2 L2 L2 L3 L2 L2 L2 L2
15.3 Heterogeneous Aqueous Systems, pp. 459–463 1 block or 2 periods **15.3.1 Distinguish** between a suspension and a solution. **15.3.2 Identify** the distinguishing characteristic of a colloid.	E-2		**TE** Teacher Demo: Motion of Colloidal Particles, p. 461 **SSLM** Lab 25: Hard and Soft Water	L1 L2

Ability Levels	
L1	For students who need additional help
L2	For all students
L3	For students who need to be challenged

Components

SE	Student Edition	**LM**	Laboratory Manual	**GCP**	Graphing Calculator Problems	**T**	Transparencies
TE	Teacher's Edition	**SSLM**	Small-Scale Chemistry Lab Manual	**TTT**	Test-Taking Tips with Transparencies, Grades 9–12 Science	**CA**	Chem Alive and Chemistry Field Trips DVD
GRSW	Guided Reading and Study Workbook	**LP**	Laboratory Practicals	**STP**	Standardized Test Preparation Workbook	**iT**	Interactive Textbook with ChemASAP
CTR	Core Teaching Resources	**PLM**	Probeware Lab Manual			**GO**	Go Online Internet Resources
SSG	Spanish Study Guide	**VCL**	Virtual Chemistry Labs				
		CTB	Computer Test Bank				

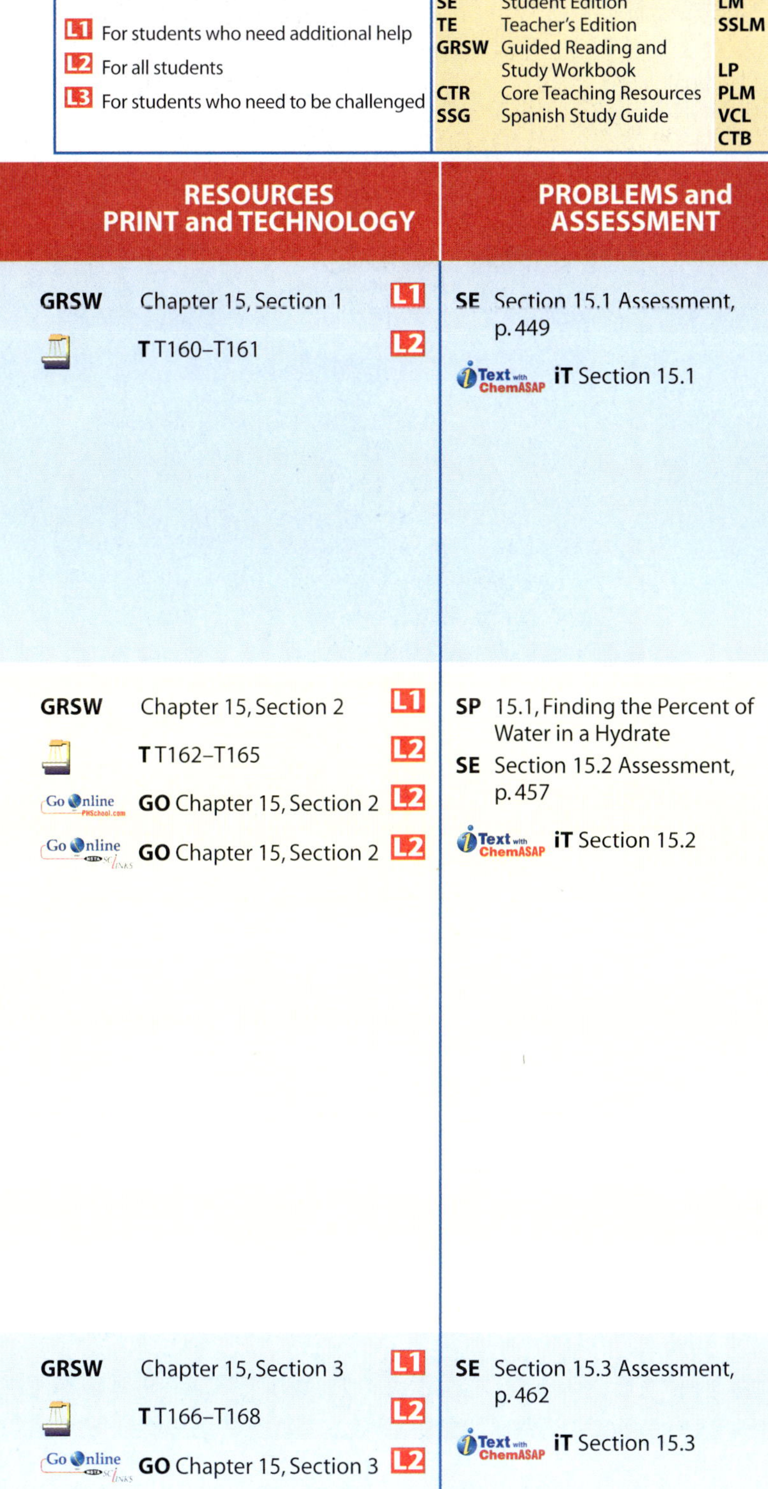

RESOURCES PRINT and TECHNOLOGY			PROBLEMS and ASSESSMENT
GRSW	Chapter 15, Section 1	L1	**SE** Section 15.1 Assessment, p. 449
	T T160–T161	L2	**iT** Section 15.1
GRSW	Chapter 15, Section 2	L1	**SP** 15.1, Finding the Percent of Water in a Hydrate
	T T162–T165	L2	**SE** Section 15.2 Assessment, p. 457
	GO Chapter 15, Section 2	L2	**iT** Section 15.2
	GO Chapter 15, Section 2	L2	
GRSW	Chapter 15, Section 3	L1	**SE** Section 15.3 Assessment, p. 462
	T T166–T168	L2	**iT** Section 15.3
	GO Chapter 15, Section 3	L2	

Web Code: cdn-1152
Web Code: cdn-1153

Web Code: cdb-1152

MATERIALS List

Quantities for each group

STUDENT EDITION

Inquiry Activity, p. 444
waxed paper, ruler, teaspoon, tap water, cup, liquid dish detergent

Quick Lab, p. 448
shallow dish or Petri dish, water, paper clip, rubber band (approximately 2 inches in diameter), 2 micropipets or droppers, vegetable oil, liquid dish detergent

Small-Scale Lab, p. 458
paper; pencil; ruler; reaction surface; conductivity tester; water; micropipet or dropper; conductivity probe (optional); NaCl(s); Na_2CO_3(s); $NaHCO_3$(s); KCl(s); $MgSO_4$(s), table sugar, cornstarch, KI(s); aqueous solutions of HCl, H_2SO_4, HNO_3, CH_3COOH, NH_3, and NaOH; rubbing alcohol; distilled water; various liquid foods

TEACHER'S EDITION

Teacher Demo, p. 453
light bulb in a porcelain socket; 9V or lantern battery; 2 copper metal strips; lamp cord; alligator clips; 0.1*M* solutions of glucose, alanine, glycine, ascorbic acid, malonic acid, citric acid, acetic acid, hydrochloric acid; 8 beakers

Teacher Demo, p. 454
$CoCl_2 \cdot 6H_2O$ solution, cotton swab, white paper, hot plate, misting bottle, water

Teacher Demo, p. 461
whole milk, water, beaker, stirring rod, projector or laser pointer

Chapter Assessment

CHAPTER ASSESSMENT

SE Chapter Assessment, pp. 465–468
CTB Chapter 15
iT Chapter 15

STANDARDIZED TEST PREP

SE Chapter 15, p. 469
iT Interactive Textbook with Assessment at PHSchool.com

CHAPTER 15

Water and Aqueous Systems

Observing Surface Tension

Purpose

Students will observe that water forms nearly spherical-shaped drops on nonpolar surfaces, due to the surface tension resulting from intermolecular hydrogen bonding between water molecules. Students will recognize that the surface properties of water are different from those of a detergent–water mixture.

Prep Time 5 minutes
Class Time 10 minutes

Think About It

1. The water forms spherical-shaped drops.
2. The water spreads out more evenly on the surface of the waxed paper.
3. The detergent may interfere with hydrogen bonding between water molecules, causing the drop of water to spread out.

CHAPTER 15

Water and Aqueous Systems

Water is an amazing liquid that makes plant and animal life on Earth possible.

INQUIRY Activity

Observing Surface Tension

Materials

waxed paper, ruler, teaspoon, tap water, cup, liquid dish detergent

Procedure

1. Place a 15-cm square of waxed paper on a flat surface.
2. Drop 1/3 teaspoon of water from a height of about 20 cm onto the center of the paper.
3. Add 1 drop of detergent to 1/2 cup of water and stir.
4. Repeat Steps 1 and 2 using the diluted detergent in place of water.
5. You may wish to repeat this activity with squares of aluminum foil, clear wrap, and writing paper.

Think About It

1. What do you observe when you drop water onto the waxed paper surface?
2. What do you observe when you use diluted detergent instead of water?
3. Use your results to suggest how the detergent may have changed the physical properties of water.

15.1 Water and Its Properties

Connecting to Your World In December of 1968, when the Apollo 8 astronauts first saw their home planet from a distance of thousands of kilometers, they affectionately called it the big blue marble. Water covers about three-quarters of Earth's surface. In addition to making up Earth's oceans, water forms the polar ice caps and cycles through the atmosphere. All life forms that are known to exist are made mostly of water. In this section, you will learn about the properties of water and what makes this unique substance essential to life on Earth.

Guide for Reading

Key Concepts

- How can you account for the high surface tension and low vapor pressure of water?
- How would you describe the structure of ice?

Vocabulary

surface tension

surfactant

Reading Strategy

Monitoring Your Understanding Before you read, list what you already know about water and aqueous systems. After you read, write what you have learned. Is what you already knew correct? If not, how do you need to revise that information?

Water in the Liquid State

When you turn on the faucet, you expect that water will stream out to supply all your needs. You couldn't live without water, nor could all the plants and animals, like those in Figure 15.1 that share space on the "big blue marble." Besides the water visible on Earth's surface, immense reserves of water exist deep underground. Water in the form of ice and snow dominates the polar regions of Earth. Icebergs drift in the oceans, and snow blankets the temperate zones in winter. Water vapor from the evaporation of surface water and from steam spouted from geysers and volcanoes is always present in Earth's atmosphere.

Figure 15.1 Water is vital to life. Animals that live on the grasslands depend on watering holes, such as the one shown here.

Section Resources

Print

- **Guided Reading and Study Workbook,** Section 15.1
- **Core Teaching Resources,** Section 15.1 Review
- **Transparencies,** T160–T161

Technology

- **Interactive Textbook with ChemASAP,** Animation 19, 20; Assessment 15.1

15.1

1 FOCUS

Objectives

15.1.1 Explain the high surface tension and low vapor pressure of water in terms of the structure of the water molecule and hydrogen bonding.

15.1.2 Describe the structure of ice.

Guide for Reading

Build Vocabulary L2

Word Parts Tension comes from the Latin *tendere*, which means "to stretch." Thus, surface tension causes the surface of a liquid to appear stretched and taut. Ask students to use the derivation of tension to explain the expression "nervous tension". *(a condition of stress or tautness in an individual brought about by some external cause)*

Reading Strategy L2

Use Prior Knowledge Before students begin to read this chapter, encourage them to review the concepts of polarity and hydrogen bonding in earlier chapters.

2 INSTRUCT

Connecting to Your World

Have students study the photograph and read the text. Ask, **What are the white swirls in the photo?** *(clouds)* **How are the white swirls relevant to the subject matter of this chapter?** *(Clouds are composed of water vapor, water in its gaseous state.)*

Water in the Liquid State

Discuss L2

Tell students that, although three-quarters of Earth's surface is water, only a small fraction of it can be used for drinking. The majority of Earth's surface water is in the oceans and contains too high a concentration of salts for the human body to tolerate.

Use Visuals 

Figures 15.2, 15.3 Have students study the figures. Ask, **What is meant by the term *polarity*?** *(Polarity refers to the net molecular dipole resulting from electronegativity differences between covalently bonded atoms.)* **What part of the water molecule has a partial negative charge?** *(the oxygen atom)* **What part has a partial positive charge?** *(the hydrogen atoms)* **What kind of substances would you expect to dissolve in water?** *(polar substances, ionic compounds)*

Discuss 

Write the structural formula of water on the board. Using VSEPR theory, show students how the oxygen's non-bonding electron pairs reduce the bond angle in a water molecule to 105°C. Point out how the net molecular dipole and hydrogen bonding properties of water are due to water's bent shape and the large electronegativity differences between H and O. Note that hydrogen bonding contributes to water's unusually low vapor pressure and high surface tension.

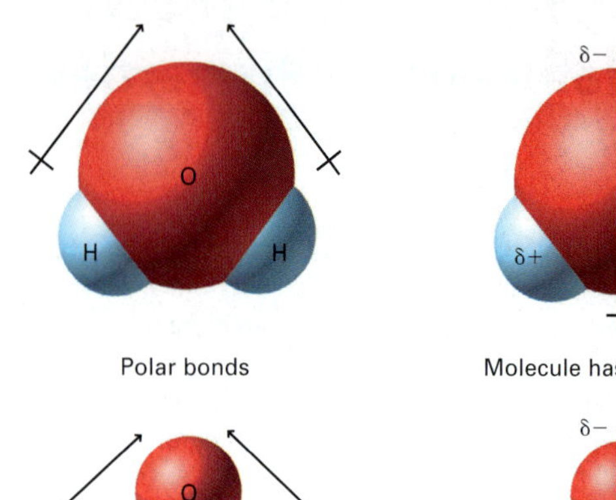

Figure 15.2 In a water molecule, the bond polarities are equal, but the two poles do not cancel each other because a water molecule is bent. The molecule as a whole is polar. **Applying Concepts** ***Which element in water has the higher electronegativity?***

Animation 19 See how hydrogen bonding results in the unique properties of water.

with **ChemASAP**

Recall that water is a simple triatomic molecule, H_2O. The oxygen atom forms a covalent bond with each of the hydrogen atoms. Because of its greater electronegativity, oxygen attracts the electron pair of the covalent O—H bond to a greater extent than hydrogen. As a result, the oxygen atom acquires a partial negative charge ($\delta-$). The less electronegative hydrogen atoms acquire partial positive charges ($\delta+$). Thus, the O—H bonds are highly polar. How do the polarities of the two O—H bonds affect the polarity of the molecule? The shape of the molecule is the determining factor. The bond angle of the water molecule is approximately 105°, which gives the molecule a bent shape. The two O—H bond polarities do not cancel, so the water molecule as a whole is polar. The net polarity of the water molecule is illustrated in Figure 15.2.

In general, polar molecules are attracted to one another by dipole interactions. The negative end of one molecule attracts the positive end of another molecule. Figure 15.3 shows how this intermolecular attraction among water molecules results in the formation of hydrogen bonds. **Many unique and important properties of water—including its high surface tension and low vapor pressure—result from hydrogen bonding.**

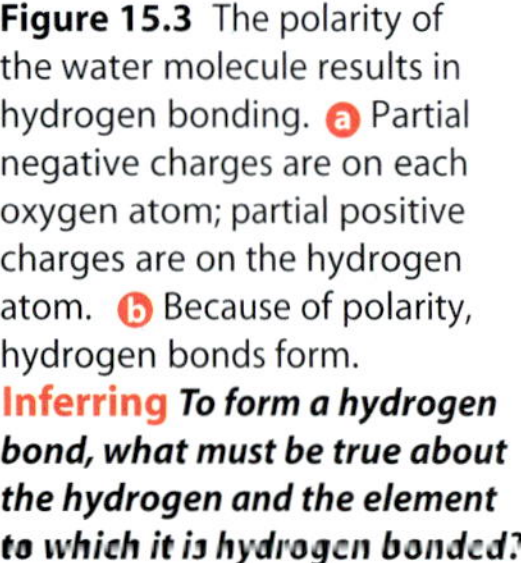

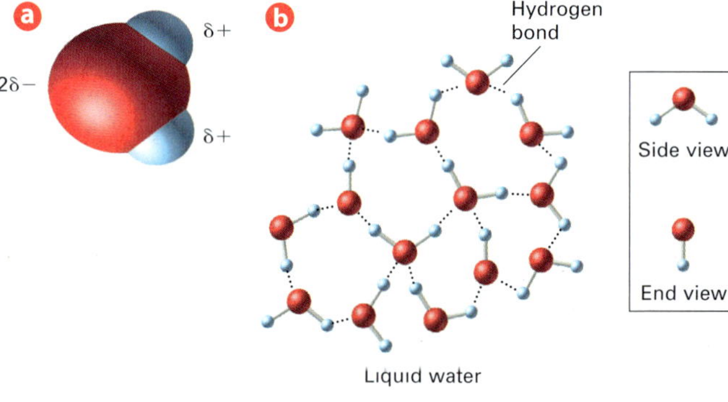

Figure 15.3 The polarity of the water molecule results in hydrogen bonding. **a** Partial negative charges are on each oxygen atom; partial positive charges are on the hydrogen atom. **b** Because of polarity, hydrogen bonds form. **Inferring** ***To form a hydrogen bond, what must be true about the hydrogen and the element to which it is hydrogen bonded?***

Facts and Figures

Availability of Water

Have students compare areas of population density and the distribution of water. Give them two different maps. One map should show the locations of cities and towns. The other should be a regional contour map showing water courses, rivers, lakes, and reservoirs. Ask students why most large cities or even historic sites are located near water sources. What happens when they're not? Have them consider the vast irrigation networks and water channels that crisscross the nation's agricultural areas to bring water to arid regions and to population centers with insufficient fresh water.

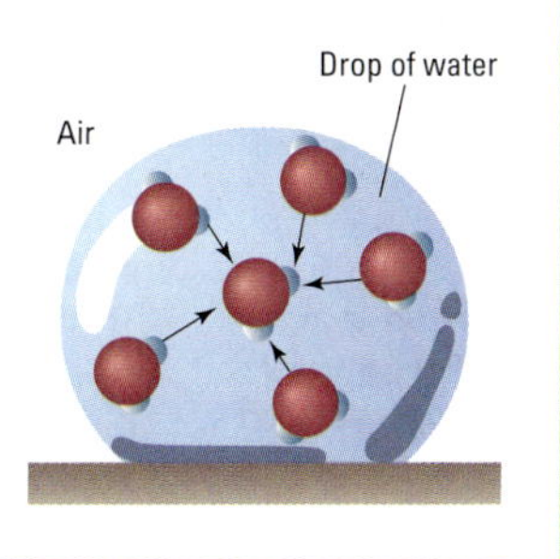

Figure 15.4 Surface tension makes it possible for some insects, such as this water strider, to walk on water. Water molecules at the surface of the water drop above cannot form hydrogen bonds with air molecules, so they are drawn into the body of the liquid, producing surface tension.

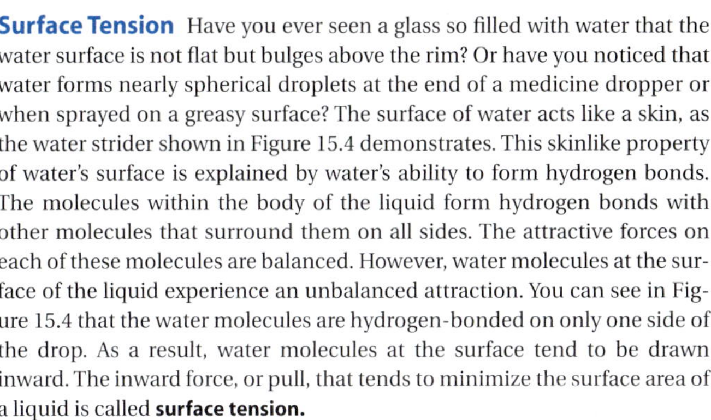

Surface Tension Have you ever seen a glass so filled with water that the water surface is not flat but bulges above the rim? Or have you noticed that water forms nearly spherical droplets at the end of a medicine dropper or when sprayed on a greasy surface? The surface of water acts like a skin, as the water strider shown in Figure 15.4 demonstrates. This skinlike property of water's surface is explained by water's ability to form hydrogen bonds. The molecules within the body of the liquid form hydrogen bonds with other molecules that surround them on all sides. The attractive forces on each of these molecules are balanced. However, water molecules at the surface of the liquid experience an unbalanced attraction. You can see in Figure 15.4 that the water molecules are hydrogen-bonded on only one side of the drop. As a result, water molecules at the surface tend to be drawn inward. The inward force, or pull, that tends to minimize the surface area of a liquid is called **surface tension.**

All liquids have a surface tension, but water's surface tension is higher than most. The surface tension of water tends to hold a drop of liquid in a spherical shape. The drop is not a perfect sphere because the force of gravity tends to pull it down, causing it to flatten. This is why, on some surfaces, water tends to bead up rather than spread out. It is possible to decrease the surface tension of water by adding a surfactant. A **surfactant** is any substance that interferes with the hydrogen bonding between water molecules and thereby reduces surface tension. Soaps and detergents are surfactants. Adding a detergent to beads of water on a greasy surface reduces surface tension, causing the beads of water to collapse and spread out.

Vapor Pressure Hydrogen bonding between water molecules also explains water's unusually low vapor pressure. Remember that the vapor pressure of a liquid is the result of molecules escaping from the surface of the liquid and entering the vapor phase. Because hydrogen bonds hold water molecules to one another, the tendency of these molecules to escape is low and evaporation is slow. Imagine what would happen if it were not. All the lakes and oceans, with their large surface areas, would tend to evaporate!

Checkpoint *What is surface tension?*

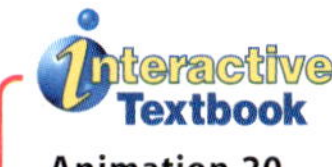
Interactive Textbook

Animation 20
Discover how some insects can walk on water.

with ChemASAP

Discuss 

To help students understand why liquid water assumes a spherical shape on many surfaces, explain that nature tends to find the path of least resistance. Moving molecules takes work. A spherical shape provides the minimum surface area for a given volume; molecules expend the least energy possible to move into a spherical arrangement while maximizing their interactions with one another in the bulk liquid. This energy efficiency creates the surface tension. In order to adopt any other shape, more work would have to be done. The work "saved" is the surface tension.

Use Visuals 

Figure 15.4 Point out that the skinlike qualities of water are due to an exceptionally high surface tension that is created by an extensive network of hydrogen bonds. As an analogy, describe the following scene. A crowd of autograph seekers surrounds a celebrity. As each person approaches as closely as possible, an impenetrable circular barrier forms. This barrier will remain intact as long as there is a net attraction toward the circle's center. The process is dynamic. As one person wiggles closer to the celebrity, another is forced to retreat, but the overall shape doesn't change. Similarly, within a drop of water, attractions between individual molecules may shift, but the overall shape remains constant as molecules continue to be drawn toward a central focal point.

Answers to...

Figure 15.2 oxygen

Figure 15.3 The hydrogen atom must be covalently bonded to a very electronegative atom with one or more lone pairs of electrons.

Checkpoint the inward pull that tends to minimize the surface area of a liquid

Quick LAB

Surfactants

L2

Objective After completing this activity, students will be able to:

- identify a property of water that results from hydrogen bonding.
- describe the effect of a surfactant on surface tension.

Address Misconceptions

Students may believe that the surface contains a different kind of water molecule than the interior of a volume of water. Remind students that surface tension still exists even after you mix the water sample to redistribute the molecules.

Skills Focus Observing, inferring, drawing conclusions

 **Prep Time** 10 minutes

Advance Prep

- Gather paper clips, rubber bands, detergent, and vegetable oil.

Class Time 20 minutes

Teaching Tips

- Have students dry their hands before touching the paper clips.

Expected Outcome The paper clip floats. The rubber band takes on a circular shape when filled with oil. Detergent destroys the surface tension and causes the rubber band to return to its original, non-circular shape.

Analyze and Conclude

1. The paper clip floated. The surface tension of the water supported the paper clip.
2. No. The surface tension is broken, and the paper clip sinks because it is made of a metal that is more dense than water.
3. The rubber band took a circular shape. The surface tension of the surrounding water pulling the rubber band radially outward is not balanced by the surface tension of the oil.
4. The oil layer was broken, and the rubber band returned to its original shape.

Quick LAB

Surfactants

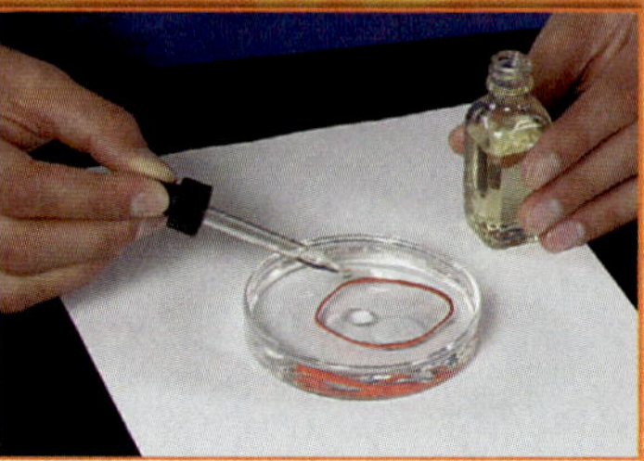

Purpose
To observe an unusual surface property of water that results from hydrogen bonding.

Materials

- shallow dish or Petri dish
- water
- paper clip
- rubber band, approximately 5 cm in diameter
- micropipets or droppers (2)
- vegetable oil
- liquid dish detergent

Procedure

1. Thoroughly clean and dry the dish.
2. Fill the dish almost full with water. Dry your hands.
3. Being careful not to break the surface, gently place the paper clip on the water. Observe what happens.
4. Repeat Steps 1 and 2.
5. Gently place the open rubber band on the water.
6. Slowly add oil drop by drop onto the water encircled by the rubber band until that water is covered with a layer of oil. Observe for 15 seconds.
7. Allow one drop of dish detergent to fall onto the center of the oil layer. Observe the system for 15 seconds.

Analyze and Conclude

1. What happened to the paper clip in Step 3? Why?
2. If a paper clip becomes wet, does it float? Explain your answer.
3. What shape did the rubber band take when the water inside it was covered with oil? Why did it take the observed shape?
4. Describe what happened when dish detergent was dropped onto the layer of oil.

Water in the Solid State

You have seen that water in the liquid state exhibits some unique properties. The same is true for water in the solid state. For example, ice cubes float in your glass of iced tea because solid water has a lower density than liquid water. This is not usual for liquids. As a typical liquid cools, it begins to contract and its density increases gradually. Increasing density means that the molecules of the liquid move closer together so that a given volume of the liquid contains more molecules and thus more mass. If the cooling continues, the liquid eventually solidifies with a density greater than the density of the liquid. Because the density of a typical solid is greater than that of the corresponding liquid, the solid sinks in its own liquid.

As water begins to cool, it behaves initially like a typical liquid. It contracts slightly and its density gradually increases, as shown in Table 15.1. Notice that at 4°C, the density of water is at its maximum of 1.000 g/cm^3. When the temperature of the water falls below 4°C, the density of water actually starts to decrease. Below 4°C, water no longer behaves like a typical liquid. Ice, which forms at 0°C, has about a 10 percent lower density than water at 0°C. You may have noticed that ice begins to form at the surface of a pond when the temperature reaches 0°C, but the ice does not sink. It floats at the surface, making ice skating and ice fishing possible. Ice is one of only a few solids that float in their own liquid.

Table 15.1
Density of Liquid Water and Ice

Temperature (°C)	Density (g/cm^3)
100 (liquid water)	0.9584
50	0.9881
25	0.9971
10	0.9997
4	1.000*
0 (liquid water)	0.9998
0 (ice)	0.9168

*Most dense

 ***At what temperature does water have its maximum density?***

For Enrichment

L3

Water striders depend on the surface tension of water for moving around and capturing food. Tell students that detergents are increasingly found as pollutants in bodies of water in nature. Have students design an experiment to determine the concentration of detergent at which water can no longer support the weight of a water strider.

Differentiated Instruction

Gifted and Talented L3

The structure of ice was determined using X-ray diffraction, a method that was also used by Rosalind Franklin in her work with DNA. This work contributed to the discovery of the helical structure of DNA by James Watson and Francis Crick. Have students research how X-ray diffraction is used to determine the structure of a substance.

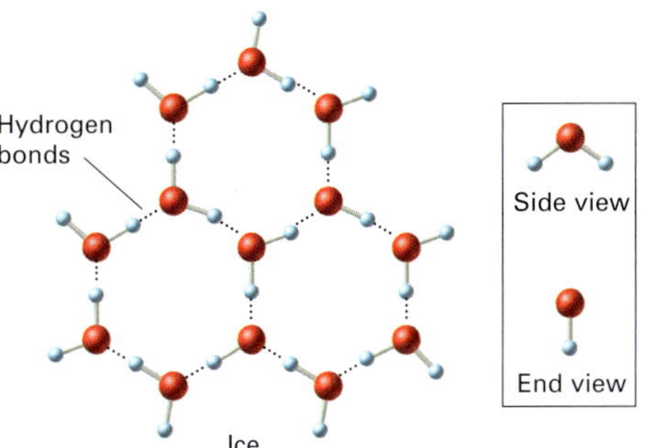

Figure 15.5 Extensive hydrogen bonding in ice holds the water molecules farther apart in a more ordered arrangement than in liquid water. The hexagonal symmetry of a snowflake reflects the structure of the ice crystal.

Why is ice less dense than liquid water? As you can see in Figure 15.5, hydrogen bonds hold the water molecules in place in the solid phase. **The structure of ice is a regular open framework of water molecules arranged like a honeycomb.** When ice melts, the framework collapses and the water molecules pack closer together, making liquid water more dense than ice.

The fact that ice floats has important consequences for organisms. A layer of ice on the top of a pond acts as an insulator for the water beneath, preventing it from freezing solid except under extreme conditions. Because the liquid water at the bottom of an otherwise frozen pond is warmer than 0°C, fish and other aquatic life are better able to survive. If ice were denser than liquid water, bodies of water would tend to freeze solid during the winter months, destroying many types of organisms.

Ice melts at 0°C. This is a high melting temperature for a molecule with such a low molar mass. A considerable amount of energy is required to return water molecules in the solid state to the liquid state. The heat absorbed when 1 g of water changes from a solid to a liquid is 334 J. This same amount of energy is needed to raise the temperature of 1 g of liquid water from 0°C to 80°C.

15.1 Section Assessment

1. **Key Concept** What causes the high surface tension and low vapor pressure of water?
2. **Key Concept** How would you describe the structure of ice?
3. What effect does a surfactant have on the surface tension of water?
4. What are two factors that determine how spherical a drop of a liquid will be?
5. The molecules water (H_2O) and methane (CH_4) have similar masses, but methane changes from a gas to a liquid at −161°C. Water becomes a gas at 100°C. What could account for the difference?

Writing Activity

Short Story Use your imagination to write a short children's story about a water strider whose "magic" ability to walk on water saved him from a predator bird. For your young readers, be sure to include an explanation of the life saving magic.

Assessment 15.1 Test yourself on the concepts in Section 15.1.

with ChemASAP

Section 15.1 Assessment

1. Water molecules are hydrogen-bonded to each other, but not to air molecules. Net attraction is inward, minimizing the water surface area. Hydrogen bonding makes it more difficult for water molecules to escape from the liquid phase to the vapor phase.
2. Ice has a honeycomb-like structure of hydrogen-bonded water molecules.
3. Surfactants lower the surface tension of water by interfering with hydrogen bonding.
4. The surface tension of a liquid tends to hold a drop of liquid in a spherical shape; gravity tends to flatten the drop.
5. Water has intermolecular hydrogen bonding between its molecules; methane does not.

Water in the Solid State

Use Visuals L2

Table 15.1 Have students note that the density of liquid water increases with decreasing temperature. Ask, **Why doesn't this trend continue below 4°C?** *(Below 4°C, the kinetic energy of the molecules is insufficient to overcome hydrogen bonding, which holds the water molecules in fixed positions.)*

3 ASSESS

Evaluate Understanding 

Ask, **Why is the surface tension of water so high compared to that of other liquids?** *(Water molecules form a large number of hydrogen bonds, in addition to dipole-dipole forces between molecules.)* **Why does water form a meniscus in a narrow tube?** *(Water molecules have a greater attraction to the molecules on the surface of the glass than they do to each other.)*

Reteach 

Have students explain what property of water allows ice to float in its liquid phase. *(Ice has a lower density than liquid water. Because there is more space between water molecules in the solid state than in the liquid state, there are fewer molecules in any given volume. Less mass per volume results in a lower relative density of ice as compared to water.)*

Writing Activity

Stories will vary in imaginative scope, but should include a simplified explanation of surface tension.

If your class subscribes to the Interactive Textbook, use it to review key concepts in Section 15.1.

with ChemASAP

Answers to...

Checkpoint 4°C

15.2

1 FOCUS

Objectives

15.2.1 Distinguish between a solvent and a solute.

15.2.2 Describe what happens in the solution process.

15.2.3 Explain why all ionic compounds are electrolytes.

15.2.4 Demonstrate how the formula for a hydrate is written.

Guide for Reading

Build Vocabulary L2

Concept Map Have students make a concept map using the following terms: *solvation, solute, aqueous solution, solvent.*

Reading Strategy L2

Out!ine Before students read this section, encourage them to make an outline using the red and blue headings as divisions and subdivisions. As they read, have them write a short summary under each heading.

2 INSTRUCT

Connecting to Your World

Have students study the photograph and read the text. Ask, **What is it about the pickle that allows it to produce light?** *(It conducts electricity.)* Ask students if they know of other materials that conduct electricity. *(Students will probably answer that metals conduct electricity. They may also know that ionic solutions conduct.)*

Solvents and Solutes

Discuss L2

Remind students that an anion is any negatively charged atom or group of atoms. A cation is any positively charged atom or group of atoms. Have students suggest a memory aid to help them remember. *(For example, the first two letters of anion, a and n, relate anion to negative.)*

15.2 Homogeneous Aqueous Systems

Guide for Reading

Key Concepts

- What is the difference between a solvent and a solute?
- What happens in the solution process?
- Why are all ionic compounds electrolytes?
- How do you write the formula for a hydrate?

Vocabulary

aqueous solution
solvent
solute
solvation
electrolyte
nonelectrolyte
strong electrolyte
weak electrolyte
hydrate

Reading Strategy

Relating Text and Visuals As you read, look carefully at Figure 15.7. In your notebook, explain in your own words how this illustration helps clarify the process of solvation.

Connecting to Your World

Is it possible to read by the light of a glowing pickle? Although it sounds absurd, an ordinary dill pickle from the deli can be a source of light! Iron or copper electrodes are inserted into the ends of the pickle and connected to a source of alternating electric current. After a time, during which the pickle becomes hot and produces water vapor, the pickle begins to glow. The mechanism by which the light is generated is not fully understood, but it is clear that conduction of electricity by the pickle is an important factor. In this section you will learn what kind of solution conducts electricity.

Solvents and Solutes

Water dissolves so many of the substances that it comes in contact with that you won't find chemically pure water in nature. Even the tap water you drink is a solution that contains varying amounts of dissolved minerals and gases. An **aqueous solution** is water that contains dissolved substances. In a solution, the dissolving medium is the **solvent,** and the dissolved particles are the **solute. A solvent dissolves the solute. The solute becomes dispersed in the solvent.** Solvents and solutes may be gases, liquids, or solids.

Recall that solutions are homogeneous mixtures. They are also stable mixtures. For example, sodium chloride does not settle out when its solutions are allowed to stand, provided other conditions, such as temperature, remain constant. Solute particles can be atoms, ions, or molecules, and their average diameters are usually less than 1 nm (10^{-9} m). Therefore, if you filter a solution through filter paper, both the solute and the solvent pass through the filter, as Figure 15.6 shows.

Substances that dissolve most readily in water include ionic compounds and polar covalent molecules. Nonpolar covalent molecules, such as methane, and compounds found in oil, grease, and gasoline, do not dissolve in water. However, oil and grease will dissolve in gasoline. To understand this difference, you must know more about the structures of the solvent and the solute and what attractions exist between them.

Figure 15.6 A solution cannot be separated by filtration. The small size of the solute particles allows them to pass through filter paper.

Section Resources

Print

- **Guided Reading and Study Workbook,** Section 15.2
- **Core Teaching Resources,** Section 15.2 Review
- **Laboratory Manual,** Lab 26, 27, 28, 29
- **Small-Scale Chemistry Laboratory Manual,** Labs 22, 23, 24
- **Laboratory Practicals,** 15.1
- **Probeware Lab Manual,** Section 15.2
- **Transparencies,** T162–T165

Technology

- **Virtual Chemistry Labs,** Lab 21
- **Interactive Textbook with ChemASAP,** Simulation 18, 19, Problem-Solving 15.6, Assessment 15.2
- **Go Online,** Section 15.2

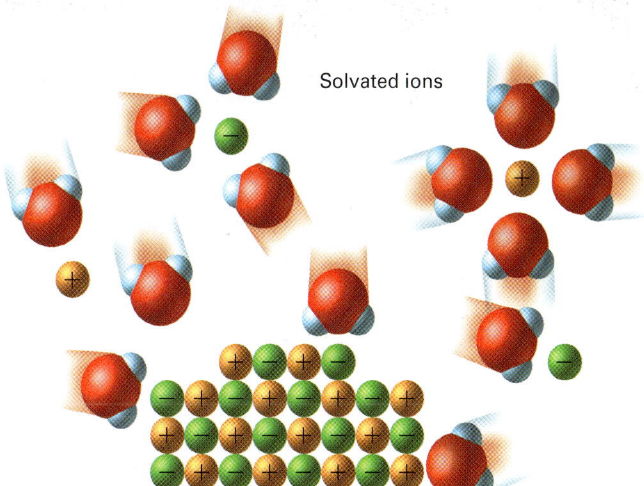

Figure 15.7 When an ionic solid dissolves, the ions become solvated or surrounded by solvent molecules. **Inferring** ***Why do the water molecules orient themselves differently around the anions and the cations?***

The Solution Process

Water molecules are in continuous motion because of their kinetic energy. When a crystal of sodium chloride is placed in water, the water molecules collide with it. Remember that a water molecule is polar, with a partial negative charge on the oxygen atom and partial positive charges on the hydrogen atoms. The polar solvent molecules (H_2O) attract the solute ions (Na^+, Cl^-). **As individual solute ions break away from the crystal, the negatively and positively charged ions become surrounded by solvent molecules and the ionic crystal dissolves.** The process by which the positive and negative ions of an ionic solid become surrounded by solvent molecules is called **solvation.** Figure 15.7 shows a model of the solvation of an ionic solid such as sodium chloride.

In some ionic compounds, the attractions among the ions in the crystals are stronger than the attractions exerted by water. These compounds cannot be solvated to any significant extent and are therefore nearly insoluble. Barium sulfate ($BaSO_4$) and calcium carbonate ($CaCO_3$) are examples of nearly insoluble ionic compounds.

Figure 15.8 shows that oil and water do not mix. But what about oil in gasoline? Both oil and gasoline are composed of nonpolar molecules. The attractive forces that hold two oil molecules together are similar in magnitude to the forces that hold two gasoline molecules together. Oil molecules can easily separate and replace gasoline molecules to form a solution. As a rule, polar solvents such as water dissolve ionic compounds and polar compounds; nonpolar solvents such as gasoline dissolve nonpolar compounds. This relationship can be summed up in the expression "like dissolves like."

What is meant by solvation?

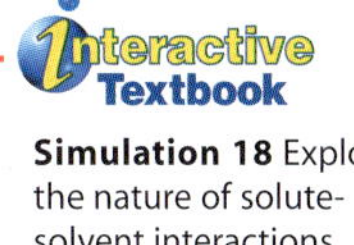

Simulation 18 Explore the nature of solute-solvent interactions.

with ChemASAP

Figure 15.8 Oil and water do not mix. Oil is less dense than water, so it floats on top. The colors result from the bending of light rays by the thin film of oil.

The Solution Process

Use Visuals L1

Figure 15.7 Have students study the figure. Stress the idea that when a solid dissolves in water, it breaks into small pieces. Ask, **What must happen for an ionic solid to dissolve?** *(The molecules of the solvent must be able to overcome the attractive forces between ions that hold the solid together.)* **What part of a water molecule is attracted to a negatively charged solute ion?** *(the hydrogen atoms)*

Discuss L2

Remind students that solutions are homogeneous mixtures containing a solvent and one or more solutes. Usually the solvent is defined as the component in the system that is present in the greatest amount. Point out that a water-soluble solute can be a solid, liquid, or gas. Ask students to name some of the solutes in blood. *(Solutes in blood are typically ions such as sodium, potassium, calcium, chloride, hydrogen carbonate, and phosphate. Glucose, a covalent compound, also is dissolved in blood. Dissolved gases such as oxygen and carbon dioxide are also present.)*

Differentiated Instruction

Less Proficient Readers L1

Have students preview the section by looking for vocabulary and other unfamiliar terms. Encourage students to write the terms, their phonetic respellings, and their definitions in their notebooks.

Answers to...

Figure 15.7 The positive hydrogen end of the water molecule orients itself toward the anion. The negative-oxygen end of the water molecule orients itself toward the cation.

the process by which the ions of an ionic solid become surrounded by solvent molecules

Careers *in* Chemistry

Wastewater Engineer

Disinfection is the last step in the process of making wastewater safe for drinking or for release back into the environment. Chlorine is a safe, inexpensive, and effective disinfectant against waterborne bacteria and viruses. Chlorine remains in the water and continues to provide protection against these organisms. Other disinfectants include ozone and ultraviolet light. Although ozone and UV are both powerful disinfectants, they do not remain in the water to provide residual protection. Chloramines, compounds formed by the reaction of chlorine and ammonia or other nitrogen compounds, provide residual protection, but they do not have the disinfecting power of chlorine. Chlorine dioxide is a powerful disinfectant and offers residual protection; however, the technology is relatively new and complex.

Have students research chemistry-related careers in the library or on the Internet. Students can then construct a table that describes the nature of the work, educational and training requirements, employment outlook, working conditions, and other necessary information.

Careers *in* Chemistry

Wastewater Engineer

Wastewater must be physically and chemically treated before it is returned to the environment or recycled for human use. A wastewater engineer is responsible for monitoring this process. These engineers control the amount of water treated, the level of treatment, and the quality of water produced. When water is received at a treatment plant, a wastewater engineer oversees primary treatment that involves filtering the water for solids and debris. The engineer then adds microorganisms to the water that convert the remaining dissolved organic matter into solids that can be removed as sludge. If this water is to be released into a river or stream, the engineer then tests the water for levels of remaining organic compounds. If amounts of these compounds fall below recommended levels, the water is safe to be released.

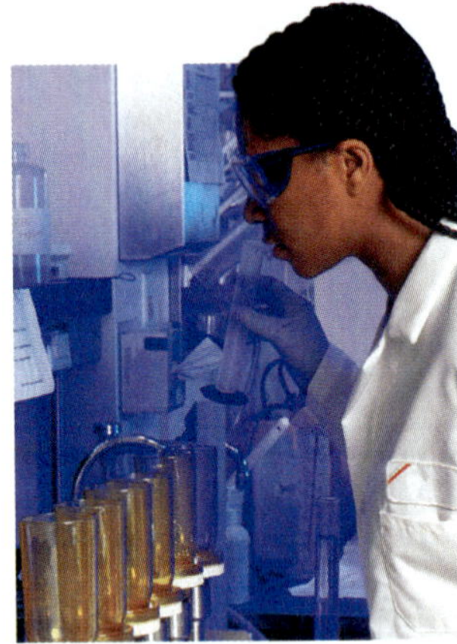

Production of drinking water often requires additional steps. The engineer must test the water for nitrogen and phosphorus compounds that would need to be removed. The engineer then disinfects the water using chlorine, ozone, or other disinfectants, and might add fluoride to strengthen consumers' teeth.

Interpreting the results of tests requires experience and an understanding of the entire chemical, and often biological, environment. Wastewater engineers usually have a degree in engineering. They must take courses in water toxicology, organic chemistry, and environmental biology.

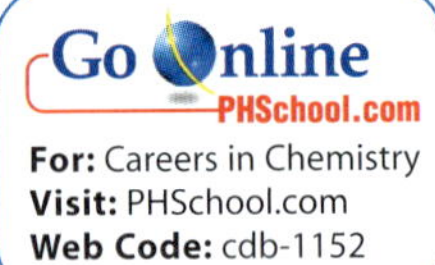

For: Careers in Chemistry
Visit: PHSchool.com
Web Code: cdb-1152

Electrolytes and Nonelectrolytes

Remember the glowing pickle that you read about in *Connecting to Your World?* The pickle contained an electrolyte. An **electrolyte** is a compound that conducts an electric current when it is in an aqueous solution or in the molten state. Conduction of electricity requires ions that are mobile and thus able to carry an electrical current. **All ionic compounds are electrolytes because they dissociate into ions.** Sodium chloride, copper(II) sulfate, and sodium hydroxide are typical water-soluble electrolytes. Barium sulfate is an ionic compound that cannot conduct electricity in aqueous solution because it is insoluble, but it can conduct in the molten state.

A compound that does not conduct an electric current in either aqueous solution or the molten state is called a **nonelectrolyte.** Many molecular compounds are nonelectrolytes because they are not composed of ions. Most compounds of carbon, such as table sugar (sucrose) and the alcohol in rubbing alcohol (2-propanol), are nonelectrolytes.

Checkpoint *How do electrolytes and nonelectrolytes differ?*

Facts and Figures

Wastewater Treatment

Cleaning wastewater protects the environment from the harmful substances it carries. Wastewater treatment also helps recycle water, a resource that is in increasingly short supply. Wastewater treatment begins with the used water from your home, your school, or industry. The water collects in underground pipes called sewers. Gravity moves the water through the sewers to the treatment plant. Once at the plant, the wastewater undergoes a number of purification stages. The primary treatment stage is mostly a physical process involving filtration and sedimentation. In secondary treatment, large populations of microorganisms convert the remaining dissolved organic matter into solids that separate as a sludge. Sometimes tertiary treatment is needed as well. At this level of treatment, nutrients such as nitrogen or phosphorus compounds that produce unwanted growth of algae are removed.

Some polar molecular compounds are nonelectrolytes in the pure state, but become electrolytes when they dissolve in water. This occurs because such compounds ionize in solution. For example, neither ammonia ($NH_3(g)$) nor hydrogen chloride ($HCl(g)$) is an electrolyte in the pure state. Yet an aqueous solution of ammonia conducts electricity because ammonium ions (NH_4^+) and hydroxide ions (OH^-) form when ammonia dissolves in water.

$$NH_3(g) + H_2O(l) \longrightarrow NH_4^+(aq) + OH^-(aq)$$

Similarly, in aqueous solution, hydrogen chloride produces hydronium ions (H_3O^+) and chloride ions (Cl^-). An aqueous solution of hydrogen chloride conducts electricity and is therefore an electrolyte.

$$HCl(g) + H_2O(l) \longrightarrow H_3O^+(aq) + Cl^-(aq)$$

Not all electrolytes conduct an electric current to the same degree. In the simple conductivity test shown in Figure 15.9, a bulb glows brightly when electrodes attached to it are immersed in a sodium chloride solution. The bright glow shows that sodium chloride is a **strong electrolyte** because nearly all the dissolved sodium chloride exists as separate Na^+ and Cl^- ions. The ions move in solution and conduct an electric current. Most soluble salts, inorganic acids, and inorganic bases are strong electrolytes.

The bulb glows dimly when the electrodes are immersed in a mercury(II) chloride solution because mercury(II) chloride is a weak electrolyte. A **weak electrolyte** conducts electricity poorly because only a fraction of the solute in the solution exists as ions. Other weak electrolytes are ammonia (NH_3) and organic acids and bases. In a solution of glucose, the bulb does not glow. Glucose ($C_6H_{12}O_6$) is a molecular compound. It does not form ions, so it is a nonelectrolyte.

Checkpoint ***Name an electrolyte and a nonelectrolyte.***

Interactive Textbook

Simulation 19 Simulate the behavior of electrolytes and nonelectrolytes in a circuit and at the atomic level.

with **ChemASAP**

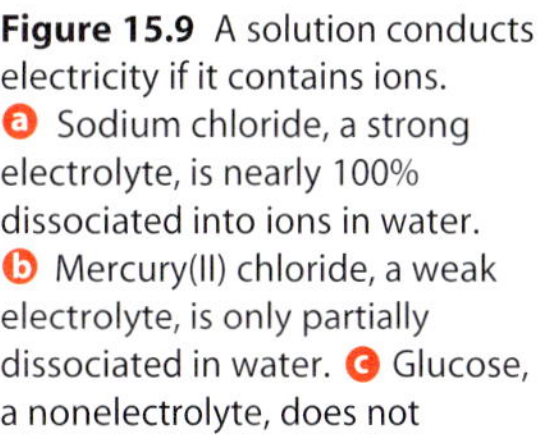

Figure 15.9 A solution conducts electricity if it contains ions. **a** Sodium chloride, a strong electrolyte, is nearly 100% dissociated into ions in water. **b** Mercury(II) chloride, a weak electrolyte, is only partially dissociated in water. **c** Glucose, a nonelectrolyte, does not dissociate in water.

Electrolytes and Nonelectrolytes

TEACHER Demo

Electrolytes

L2

Purpose Students compare electrolytes using a conductivity tester

Materials light bulb in a porcelain socket; 9V or lantern battery; 2 copper metal strips; lamp cord; alligator clips; 0.1*M* solutions of glucose ($C_6H_{12}O_6$), alanine ($HC_3H_6O_2N$), glycine ($HC_2H_4O_2N$), ascorbic acid ($HC_6H_7O_6$), malonic acid ($H_2C_3H_2O_4$), citric acid ($H_3C_6H_5O_7$), acetic acid ($HC_2H_3O_2$), and hydrochloric acid (HCl); 8 beakers labeled 1–8; marking pen

The complete circuit consists of a light bulb (in a porcelain socket), a 9V or lantern battery, and two copper metal strips to be immersed in the test solution. Connect the leads of the battery and lightbulb sockets using lamp cord and alligator clips. If possible, prepare 8 such testers so that students can directly compare the brightness of the lightbulbs.

Procedure Write the names and numbers of the beakers on the board. Pour an equal volume of each solution into a labeled beaker. Remind students that all the solutions have the same concentration (0.1*M*). Immerse a set of electrodes in each solution and have students compare the brightness of the light bulbs. Ask students to list the substances in order from strongest to weakest electrolyte. Remind them that a strong electrolyte is a substance that exists in solution almost entirely as ions. A weak electrolyte is a substance that only partly dissociates into ions in solution.

Expected Outcomes Students infer that the different substances are dissociated, or ionized, to different extents.

Answers to...

Checkpoint An electrolyte conducts electricity; a nonelectrolyte does not.

Checkpoint electrolyte: sodium chloride; nonelectrolyte: glucose

Facts and Figures

Electrolytes

Electrolytes are essential to all metabolic processes. Sodium and potassium ions control nerve impulse transmission and muscle contraction. If renal function is impaired or malabsorption from the gut disturbs optimum sodium and potassium levels, then serious nervous-system problems arise. Loss of consciousness or difficulty in maintaining muscle coordination could result. Any condition that causes prolonged bouts of diarrhea can be life threatening because of the dramatic loss of electrolytes. Electrolytes are excreted through the skin via sweat, and they must be replenished or cramps and heat stroke may occur. Sports drinks are a good source of electrolytes; they contain Na^+, K^+, and Ca^{2+}.

Hydrates

Use Visuals L1

Figure 15.10 Have students study the photographs. Ask, **What is a hydrate?** *(a compound that contains weakly bound water molecules as part of its crystal structure)* **What does the dot between the $CuSO_4$ and the $5H_2O$ in the compound formula mean?** *(It indicates that water is an integral part of the representative unit of the compound.)* **How can the presence of water in the structure of the crystal lattice be inferred?** *(by the change in the appearance of the hydrated substance when the water is driven off)*

TEACHER Demo

Magic Writing 

Purpose Students observe the color change when a hydrate gains and loses water.

Materials $CoCl_2{\cdot}6H_2O$ solution, cotton swab, white paper, hot plate, misting bottle, water

Procedure Use a cotton swab to draw a picture or write a message on a piece of white paper using a solution of hydrated cobalt(II) chloride. Place the paper on a warm (not hot) hot plate until the writing dries. Use a misting bottle to mist water on the paper. The paper can be used as a weather predictor.

Safety and Disposal Wear gloves. Cobalt chloride is toxic if ingested. Collect cobalt chloride solution, evaporate to dryness, place cooled residue in bag, and label for disposal.

Expected Outcomes The picture or message will appear blue when the paper dries and pink when the paper is misted with water.

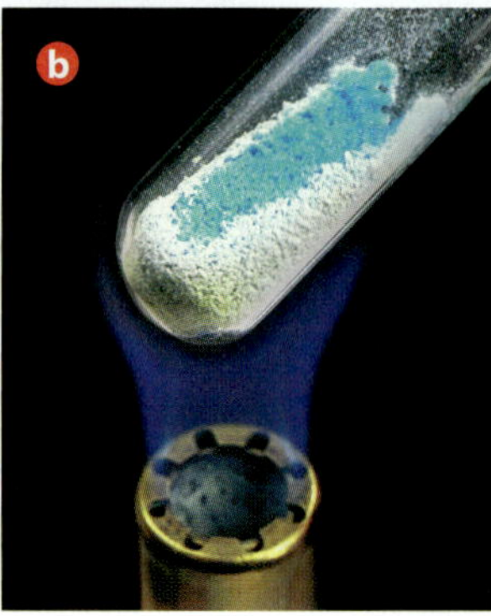

Figure 15.10 Water can be driven from a hydrate by heating. **a** Heating of a sample of blue $CuSO_4{\cdot}5H_2O$ begins. **b** After a time, much of the blue hydrate has been converted to white anhydrous $CuSO_4$.

Hydrates

When an aqueous solution of copper(II) sulfate is allowed to evaporate, deep-blue crystals of copper(II) sulfate pentahydrate are deposited. The chemical formula for this compound is $CuSO_4{\cdot}5H_2O$. Water molecules are an integral part of the crystal structure of copper(II) sulfate pentahydrate and many other substances. The water contained in a crystal is called the water of hydration or water of crystallization. A compound that contains water of hydration is called a **hydrate.** **In writing the formula of a hydrate, use a dot to connect the formula of the compound and the number of water molecules per formula unit.** Crystals of copper(II) sulfate pentahydrate always contain five molecules of water for each copper and sulfate ion pair. The deep-blue crystals are dry to the touch. They are unchanged in composition or appearance in normally moist air. But when heated above 100°C, the crystals lose their water of hydration. Figure 15.10 shows how the blue crystals of $CuSO_4{\cdot}5H_2O$ crumble to a white anhydrous powder that has the formula $CuSO_4$. If anhydrous copper(II) sulfate is treated with water, the blue pentahydrate is regenerated.

$$CuSO_4{\cdot}5H_2O(s) \underset{-\text{heat}}{\overset{+\text{heat}}{\rightleftharpoons}} CuSO_4(s) + 5H_2O(g)$$

Another compound that changes color in the presence of moisture is cobalt(II) chloride. A piece of filter paper that has been dipped in an aqueous solution of cobalt(II) chloride and then dried is blue in color (anhydrous $CoCl_2$). But as you can see in Figure 15.11, when the paper is exposed to moist air, it turns pink because of the formation of the hydrate cobalt(II) chloride hexahydrate ($CoCl_2{\cdot}6H_2O$). The blue paper could be used to test for the presence of water.

Some familiar hydrates are listed in Table 15.2 and shown in Figure 15.12. Each one contains a fixed quantity of water and has a definite composition.

Figure 15.11 Paper treated with anhydrous cobalt(II) chloride is blue. In the presence of moisture the paper turns pink. **Inferring** ***How could you change the pink paper back to blue?***

454 Chapter 15

Table 15.2

Some Common Hydrates		
Formula	**Chemical name**	**Common name**
$MgSO_4 \cdot 7H_2O$	magnesium sulfate heptahydrate	Epsom salt
$Ba(OH)_2 \cdot 8H_2O$	barium hydroxide octahydrate	
$CaCl_2 \cdot 2H_2O$	calcium chloride dihydrate	
$CuSO_4 \cdot 5H_2O$	copper(II) sulfate pentahydrate	blue vitriol
$Na_2SO_4 \cdot 10H_2O$	sodium sulfate decahydrate	Glauber's salt
$KAl(SO_4)_2 \cdot 12H_2O$	potassium aluminum sulfate dodecahydrate	alum
$Na_2B_4O_7 \cdot 10H_2O$	sodium tetraborate decahydrate	borax
$FeSO_4 \cdot 7H_2O$	iron(II) sulfate heptahydrate	green vitriol
$H_2SO_4 \cdot H_2O$	sulfuric acid hydrate (mp 8.6°C)	

Figure 15.12 Epsom salts can be used as a mild laxative. Borax is a strong cleaner, water softener, and fireproofing agent.

Efflorescent Hydrates The forces holding the water molecules in hydrates are not very strong, so the water is easily lost and regained. Because the water molecules are held by weak forces, hydrates often have an appreciable vapor pressure. If a hydrate has a vapor pressure higher than the pressure of water vapor in the air, the hydrate will lose its water of hydration or effloresce. For example, $CuSO_4 \cdot 5H_2O$ has a vapor pressure of about 1.0 kPa at room temperature. The average pressure of water vapor at room temperature is about 1.3 kPa. Copper(II) sulfate pentahydrate is stable until the humidity decreases. When the vapor pressure drops below 1.0 kPa, the hydrate effloresces. Washing soda, or sodium carbonate decahydrate ($Na_2CO_3 \cdot 10H_2O$), is efflorescent. As the crystals lose water of hydration, they effloresce and become coated with a white powder of anhydrous sodium carbonate (Na_2CO_3).

Hygroscopic Hydrates Hydrated salts that have a low vapor pressure remove water from moist air to form higher hydrates. These hydrates and other compounds that remove moisture from air are called hygroscopic. For example, calcium chloride monohydrate spontaneously absorbs a second molecule of water when exposed to moist air.

$$CaCl_2 \cdot H_2O(s) \xrightarrow{\text{moist air}} CaCl_2 \cdot 2H_2O(s)$$

Calcium chloride is used as a desiccant in the laboratory. A desiccant is a substance used to absorb moisture from the air and create a dry atmosphere. For example, anhydrous $CaCl_2$ can be placed in the bottom of a tightly sealed container called a desiccator. Substances that must be kept dry are stored inside. A solid desiccant such as calcium sulfate ($CaSO_4$) can also be added to a liquid solvent to keep it dry. A bottle labeled "dry ethanol" may have solid calcium sulfate at the bottom. The calcium sulfate does not dissolve appreciably in the solvent but absorbs water from the ethanol and keeps it dry. When a desiccant has absorbed all the water it can hold, the salt can be returned to its anhydrous state by heating.

What is a hygroscopic hydrate?

Discuss L2

Many of the terms used to describe the properties of hydrates may be unfamiliar to students. Review the definition and pronunciation of the terms *effloresce, hygroscopic, desiccant,* and *deliquescent.* Have students write the definition for each term in their notebooks. Demonstrate some of these phenomena by setting out watch glasses containing fresh NaOH pellets, anhydrous calcium chloride, and anhydrous copper sulfate. Advise students not to touch the compounds. Point out the compounds that are hygroscopic and those that are deliquescent. Many students may have seen rice in salt shakers at restaurants. Ask them what purpose the rice serves. (*Rice acts as a desiccant.*) Explain why hygroscopic agents are preferred over deliquescent agents for use as desiccants or drying agents. Emphasize the reversible properties of hydration. Desiccants can be recycled by heating to drive off absorbed water. If possible, show students a desiccator and explain why desiccators are used in the laboratory to store hygroscopic chemicals.

Use Visuals 

Table 15.2 Display the table on an overhead projector. Review the nomenclature used to name salts and the use of prefixes to indicate the number of water molecules in the crystalline hydrate. Ask, **Do you know the uses for any of the compounds in the table?** (*Epsom salt is used as a cathartic; Glauber's salt is used as a cathartic or diuretic; alum is used as an astringent, as an emetic, and in the manufacture of baking powder, dyes, and paper; borax is used in the manufacture of glass, enamel, artificial gems, soaps, and antiseptics.*)

Facts and Figures

Desiccants

Cameras and electronic equipment made in Japan and other Asian countries are usually shipped to the United States in the holds of ships. Because ocean air contains so much moisture, the packaging for this equipment usually contains small packets of a desiccant. The desiccant absorbs the water vapor from the air, preventing it from condensing on the delicate wiring and circuit boards of the equipment, which could cause corrosion and short circuits.

Answers to...

Figure 15.11 Heat it gently to drive off the water.

a hydrate that removes water from moist air to form a higher hydrate

Section 15.2 (continued)

Sample Problem 15.1

Answers

6. 36.1%
7. 49.3%

Practice Problems Plus L2

Magnesium sulfate forms a heptahydrate.

a. Write the equation for the formation of this hydrate from the anhydrous salt. *($MgSO_4 + 7H_2O \rightarrow MgSO_4 \cdot 7H_2O$)*

b. Calculate the percent by mass of water. *(51.2%)*

Relate 

After students have studied Sample Problem 15.1, have them work in groups to find the percent by mass of water in the compounds in Table 15.2.

Math Handbook

For a math refresher and practice, direct students to percents, page R72.

Download a worksheet on **Deliquescent Compounds** for students to complete, and find additional teacher support from NSTA SciLinks.

Suppose you are measuring a mass of Na_2CO_3 for a chemical reaction. You want to use the hydrate of the compound because it is less expensive than the anhydrous compound. To determine what percent of the hydrate is water, first determine the mass of the number of moles of water in one mole of hydrate. Then determine the total mass of the hydrate. The percent by mass of water can be calculated using this equation.

$$\text{percent } H_2O = \frac{\text{mass of water}}{\text{mass of hydrate}} \times 100\%$$

SAMPLE PROBLEM 15.1

Finding the Percent of Water in a Hydrate

Calculate the percent by mass of water in washing soda, sodium carbonate decahydrate ($Na_2CO_3 \cdot 10H_2O$).

1 Analyze ***List the known and the unknown.***

Known	**Unknown**
• Formula of hydrate = $Na_2CO_3 \cdot 10H_2O$	• percent H_2O = ? %

To determine the percent by mass, determine the mass of 10 moles of water and the mass of one mole of the hydrated compound. Substitute these values into the following equation and solve.

$$\text{Percent } H_2O = \frac{\text{mass of water}}{\text{mass of hydrate}} \times 100\%$$

2 Calculate ***Solve for the unknown.***

mass of 10 moles H_2O = 180 g

molar mass of $Na_2CO_3 \cdot 10H_2O$ = 286.0 g

$$\text{percent } H_2O = \frac{1.80 \times 10^2 \text{ g}}{286.0 \text{ g}} \times 100\% = 62.9\%$$

3 Evaluate ***Does the result make sense?***

Because the mass of the water accounts for more than half the molar mass of the compound, a percentage greater than 50% should be expected. The answer should have three significant figures.

Practice Problems

6. What is the percent by mass of water in $CuSO_4 \cdot 5H_2O$?

7. Calculate the percent by mass of water in calcium chloride hexahydrate ($CaCl_2 \cdot 6H_2O$).

Math Handbook

For help with percents go to page R72.

Interactive Textbook

Problem-Solving 15.6
Solve Problem 6 with the help of an interactive guided tutorial.
with ChemASAP

If you need 5.00 g of anyhydrous Na_2CO_3 for your reaction, how many grams of $Na_2CO_3 \cdot 10H_2O$ could you use instead? You know now that 62.9% of the hydrate is water, so 37.1 g out of every 100 g is Na_2CO_3 (100.0 g − 62.9 g = 37.1 g). Calculate how much hydrate you need as follows.

$$5.00 \text{ g } \cancel{Na_2CO_3} \times \frac{(100 \text{ g } Na_2CO_3 \cdot 10H_2O)}{37.1 \text{ g } \cancel{Na_2CO_3}} = 13.5 \text{ g } Na_2CO_3 \cdot 10H_2O$$

456 Chapter 15

Figure 15.13 Deliquescent substances can remove water from the air. ⓐ Sodium hydroxide pellets absorb moisture from the air. ⓑ Eventually a solution is formed. **Applying Concepts** ***Identify the solvent and the solute.***

Deliquescent Compounds Have you ever noticed the small packets of silica gel that are often packaged with electronic equipment and leather goods? Although the structure of silica gel is not the same as a hydrated salt, it is a hygroscopic substance used to absorb moisture from the air to prevent damage to sensitive equipment and materials. Some compounds are so hygroscopic that they become wet when exposed to normally moist air. These compounds are deliquescent, which means that they remove sufficient water from the air to dissolve completely and form solutions. Figure 15.13 shows that pellets of sodium hydroxide are deliquescent. For this reason, containers of NaOH and other chemicals should always be tightly stoppered and the chemicals should never be touched by fingers. The solution formed by a deliquescent substance has a lower vapor pressure than that of the water in the air.

For: Links on Deliquescent Compounds
Visit: www.SciLinks.org
Web Code: cdn-1152

15.2 Section Assessment

8. **Key Concept** In the formation of a solution, how does the solvent differ from the solute?
9. **Key Concept** Describe what happens to the solute and the solvent when an ionic compound dissolves in water.
10. **Key Concept** Why are all ionic compounds electrolytes?
11. **Key Concept** How do you write the formula for a hydrate?
12. Which of the following substances dissolve to a significant extent in water? Explain your answer in terms of polarity.
    a. CH_4 **b.** KCl **c.** He
    d. $MgSO_4$ **e.** Sucrose **f.** $NaHCO_3$
13. Identify the solvent and the solute in vinegar, a dilute aqueous solution of acetic acid.
14. Distinguish between efflorescent and hygroscopic substances.
15. Calculate the percent by mass of water in magnesium sulfate heptahydrate ($MgSO_4 \cdot 7H_2O$).

Connecting Concepts

Percent Composition Review Sample Problem 10.10 in Chapter 10 and compare it with Sample Problem 15.1. How are the procedures the same? Is the percent of copper in $CuSO_4 \cdot 5H_2O$ the same as in $CuSO_4$? Explain.

Assessment 15.2 Test yourself on the concepts in Section 15.2.

with ChemASAP

Section 15.2 Assessment

8. The dissolving medium is the solvent, and the dissolved particles are the solute.
9. As individual solute ions break away from the crystal, the negatively and positively charged ions become surrounded by solvent molecules and the ionic crystal dissolves.
10. Because they dissociate into ions.
11. In writing the formula of a hydrate, a dot is used to connect the formula of the compound with the number of water molecules per formula unit.
12. **a.** insoluble, nonpolar **b.** soluble, ionic **c.** insoluble, nonpolar **d.** soluble, ionic **e.** soluble, polar **f.** soluble, ionic
13. Water is the solvent; acetic acid is the solute.
14. Efflorescent compounds such as certain hydrates lose water to the air. Hygroscopic compounds remove moisture from air, sometimes forming hydrates.
15. 51.2% water

3 ASSESS

Evaluate Understanding 

To assess students' understanding of the properties of aqueous solutions, ask, **What types of substances will dissolve in water to form aqueous solutions?** *(ionic compounds and polar covalent molecules)* Dissolve a small amount of nickel(II) chloride in 500 mL of distilled water and ask, **In this solution, what is the solute and what is the solvent?** *(Water is the solvent and nickel(II) chloride is the solute.)* **What type of electrolyte is nickel(II) chloride? Explain.** *(Nickel(II) chloride is an ionic compound and is soluble in water; therefore, it is a strong electrolyte.)*

Reteach 

Project models of water molecules, cations, and anions on an overhead projector to show how water molecules orient their dipoles to solvate ions. Point out that cations and anions are attracted to different ends of the water molecule. Stress that the classification of water-soluble substances as strong electrolytes, weak electrolytes, or nonelectrolytes is determined by the relative number of ions in solution. Nonelectrolytes do not produce ions.

Connecting 

In both procedures, the mass of a component of a compound is compared to the mass of the compound. The percent of copper in copper sulfate and copper sulfate pentahydrate is different because the formula masses of the two compounds are different.

If your class subscribes to the Interactive Textbook, use it to review key concepts in Section 15.2.

with ChemASAP

Answers to...

Figure 15.13 The solute is sodium hydroxide and the solvent is water.

Electrolytes

L2

Objectives After completing this activity, students will be able to:

- classify compounds as electrolytes or nonelectrolytes by testing them for conductivity.

Address Misconceptions

Students often think that a substance will conduct electricity if it dissolves in water. Tell students that some covalent compounds dissolve in water but do not conduct electricity.

Prep Time 45 minutes

Advance Prep

- Make up reagents as follows:

Solution	Preparation
1.0*M* HCl	82 mL of 12*M* in 1.0 L
1.0*M* HNO_3	63 mL of 15.8*M* in 1.0 L
1.0*M* H_2SO_4	56 mL of 18*M* in 1.0 L

CAUTION: *Always add acid to water.*

1.0*M* NH_3	67 mL of 15*M* in 1.0 L

CAUTION: *Using proper ventilation, add ammonia to water.*

CH_3COOH	Use white vinegar.
0.5*M* NaOH	20.0 g in 1.0 L

- Obtain various liquids such as rubbing alcohol, distilled water, soft drinks, orange juice, pickle juice, and coffee.

Class Time 20 minutes

Safety Be sure students wear goggles and aprons.

Teaching Tips

- Ask students to bring in small samples of orange juice and other liquid foods.

Expected Outcome None of the compounds conduct electricity when in the solid state. Aqueous solutions of NaCl, $MgSO_4$, Na_2CO_3, $NaHCO_3$, KCl, and KI conduct electricity and are thus electrolytes. Cornstarch and table sugar do not conduct electricity and are nonelectrolytes.

Analyze

1. Electrolytes: NaCl, $MgSO_4$, Na_2CO_3, $NaHCO_3$, KCl, KI; nonelectrolytes: sugar, cornstarch
2. No, because the ions are locked in a crystal lattice and cannot move.
3. Table sugar and cornstarch are covalent compounds. NaCl, $MgSO_4$, Na_2CO_3, $NaHCO_3$, KCl, and KI are ionic compounds. An electrolyte is a compound that dissociates into ions in solution.

You're The Chemist

1. $NaHCO_3(s) \rightarrow Na^+(aq) + HCO_3^-(aq)$
   $KCl(s) \rightarrow K^+(aq) + Cl^-(aq)$
   $MgSO_4(s) \rightarrow Mg^{2+}(aq) + SO_4^{2-}(aq)$
   $KI(s) \rightarrow K^+(aq) + I^-(aq)$
   Diagrams should be similar to Figure 15.9.
2. Test each solution. Strong electrolytes: HCl, H_2SO_4, HNO_3, NaOH; Weak electrolytes: CH_3COOH, NH_3; Nonelectrolytes: rubbing alcohol, distilled water
3. Strong electrolytes: soft drinks, pickle juice Weak electrolytes: orange juice, coffee

For Enrichment

Have students design and carry out an experiment to determine whether the conductivity of an electrolyte depends on its concentration.

Small-Scale LAB

Electrolytes

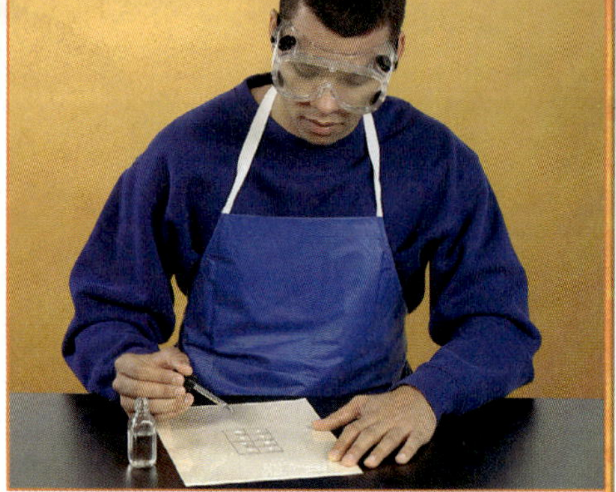

Purpose

To classify compounds as electrolytes by testing their conductivity in aqueous solution.

Materials

- pencil
- paper
- ruler
- reaction surface
- chemicals shown in the grid below
- conductivity tester
- water
- micropipet or dropper
- conductivity probe (optional)

Procedure

Probeware version available in the Probeware Lab Manual.

On separate sheets of paper, draw two grids similar to the one below. Make each square 2 cm on each side. Place a reaction surface over one of the grids and place a few grains of each solid in the indicated places. Test each solid for conductivity. Then add 1 drop of water to each solid and test the wet mixture for conductivity. Be sure to clean and dry the conductivity leads between each test.

NaCl(*s*)	$MgSO_4$(*s*)
Na_2CO_3(*s*)	Table Sugar $C_{12}H_{22}O_{11}$
$NaHCO_3$(*s*)	Cornstarch $(C_6H_{12}O_6)n$
KCl(*s*)	KI(*s*)

Analyze

Using your experimental data, record the answers to the following questions in the space below your data table.

1. Electrolytes are compounds that conduct electric current in aqueous solution. Which compounds in your table are electrolytes? Which are not electrolytes?
2. Do any of these electrolytes conduct electric current in the solid form? Explain.
3. Are these ionic or covalent compounds? Classify each compound in the grid as ionic or covalent. For a compound to be an electrolyte, what must happen when it dissolves in water?

You're The Chemist

The following small-scale activities allow you to develop your own procedures and analyze the results.

1. **Analyze It!** When an ionic solid dissolves in water, water molecules attract the ions, causing them to come apart, or dissociate. The resulting dissolved ions are electrically charged particles that allow the solution to conduct electric current. The following chemical equations represent this phenomenon.

   $$NaCl(s) \longrightarrow Na^+(aq) + Cl^-(aq)$$
   $$Na_2CO_3(s) \longrightarrow 2Na^+(aq) + CO_3^{2-}(aq)$$

   Write a similar chemical equation for each electrolyte you tested. Draw diagrams to explain how the ions conduct electric current.
2. **Design It!** Obtain the following aqueous solutions: HCl, H_2SO_4, HNO_3, CH_3COOH, NH_3, NaOH, rubbing alcohol, and distilled water. Design and carry out an experiment to test their conductivity. Use your data to classify each substance as a strong electrolyte, weak electrolyte, or nonelectrolyte.
3. **Design It!** Test various liquids for conductivity. Try soft drinks, orange juice, pickle juice, and coffee. Which liquids are electrolytes?

458 Chapter 15

15.3 Heterogeneous Aqueous Systems

Connecting to Your World It wiggles and jiggles. It comes in many colors and flavors. When you pop it in your mouth, it dissolves. It is gelatin, one of the most popular desserts in the United States. In fact, more than a million packages of gelatin are purchased or eaten every day. Gelatin has even traveled into space. In 1996, American astronaut Shannon Lucid shared a gelatin dessert with her Russian crewmates. Gelatin is a heterogeneous mixture called a colloid. In this section, you will learn more about the characteristics of colloids and a related mixture called a suspension.

Guide for Reading

Key Concepts

- What is the difference between a suspension and a solution?
- What distinguishes a colloid from a suspension and a solution?

Vocabulary

suspension
colloid
Tyndall effect
Brownian motion
emulsion

Reading Strategy

Previewing Before you read, rewrite the headings of this section as questions. As you read, write answers to your questions.

Suspensions

So far in this chapter, you have learned about homogeneous mixtures that are formed when compounds such as inorganic acids, bases, and ionic salts mix with water. These mixtures are classified as solutions. In contrast, heterogeneous mixtures are not solutions.

If you shake a piece of clay with water, the clay breaks into fine particles. The water becomes cloudy because the clay particles are suspended in the water. But if you stop shaking, the particles begin to settle out. A **suspension** is a mixture from which particles settle out upon standing. **A suspension differs from a solution because the particles of a suspension are much larger and do not stay suspended indefinitely.** The particles in a typical suspension have an average diameter greater than 1000 nm. By contrast, the particle size in a solution is usually about 1 nm. The larger size of suspended particles means that gravity plays a larger role in causing them to settle out of the mixture. Cooks use suspensions of flour or cornstarch in water to thicken sauces and gravies. These mixtures must be shaken or stirred immediately before use because the suspended particles quickly settle out.

Suspensions are heterogeneous because at least two substances can be clearly identified. In the example of clay particles mixed with water, you can clearly see the dispersed phase (clay) in the dispersion medium (water). Figure 15.14 shows that if muddy water is filtered, the filter traps the suspended clay particles and clear water passes through.

Figure 15.14 A suspension is a heterogeneous mixture. Suspended particles can be removed by filtration. **Comparing and Contrasting** ***How does the filtration of a suspension compare with the filtration of a solution?***

Section Resources

Print

- **Guided Reading and Study Workbook,** Section 15.3
- **Core Teaching Resources,** Section 15.3 Review
- **Transparencies,** T166–T168
- **Small-Scale Chemistry Laboratory Manual,** Lab 24

Technology

- **Interactive Textbook with ChemASAP,** Assessment 15.3
- **Go Online,** Section 15.3

15.3

1 FOCUS

Objectives

15.3.1 Distinguish between a suspension and a solution.

15.3.2 Identify the distinguishing characteristic of a colloid.

Guide for Reading

Build Vocabulary 

Venn Diagram Have students make a Venn diagram using the following vocabulary terms: *solution, suspension, colloid, Tyndall effect, Brownian motion.*

Reading Strategy 

Predict Before students begin to read this chapter, have them preview the section and predict how the properties of suspensions and colloids differ from those of solutions. Then, as they read, have them correct any misconceptions they may have.

2 INSTRUCT

Connecting to Your World

Have students study the photograph and read the text. Comment that gelatin is referred to as a heterogeneous mixture called a colloid. Ask, **What does that tell you about the composition of gelatin?** *(Its composition is not uniform throughout.)* **If a colloid such as gelatin differs from a solution with respect to the size of its particles, would you expect a colloid to have smaller or larger particles than a solution?** *(larger)*

Answers to...

Figure 15.14 When a solution is filtered, nothing is removed. When a suspension is filtered, suspended particles are removed.

Suspensions

Discuss 

Refer students to the Career feature on wastewater treatment. Explain that sedimentation and filtration play important roles in the treatment process. First, large particles are removed from the water in large settling tanks by sedimentation. Lime and alum are added to help coagulate the particles. Filtration through sandy loams (diatomaceous earth) then removes the suspended matter and may absorb colloidal material as well.

Colloids

Word *Origins* L2

Egg white, jellies, and other sols and gels often have a gluelike appearance.

Use Visuals 

Table 15.3 Display Table 15.3 on an overhead projector. Explain that colloids are characterized by the physical state of the *dispersed phase* and of the *continuous phase*—also called *dispersion medium*. Ask, **What is the dispersion medium in fog?** *(air)* **What is dispersed in fog?** *(water droplets)* Help students understand this two-part system by comparing it to a solution, which, instead of a dispersed phase and a dispersion medium, consists of a solute and a solvent.

Discuss 

Point out that the main difference between solutions, suspensions, and colloids is particle size. Solution particles are typically less than 1 nm in diameter. Colloid particles are between 1 nm and 1000 nm in diameter. Suspension particles are typically larger than 1000 nm. Smaller particles are less susceptible to the effects of gravity and are influenced more by the effects of Brownian movement. The collisions of molecules with extremely small colloidal particles are sufficiently energetic to move colloidal particles in a random fashion that prevents their settling to the bottom.

Word *Origins*

Colloid comes from the Greek word *kolla* meaning "glue." The word *colloid* is based on the glue-like appearance of many colloids and was coined by the English scientist Thomas Graham in 1861. **As you read, identify examples of colloids that meet Graham's original definition.**

Colloids

You read in *Connecting to Your World* that gelatin is a type of mixture called a colloid. A **colloid** is a heterogeneous mixture containing particles that range in size from 1 nm to 1000 nm. The particles are spread throughout the dispersion medium, which can be a solid, liquid, or gas. The first substances to be identified as colloids were glues. Other colloids include such mixtures as gelatin, paint, aerosol sprays, and smoke. Table 15.3 lists some common colloidal systems and gives examples of familiar colloids.

How do the properties of colloids differ from those of suspensions and solutions? Like suspensions, many colloids are cloudy or milky in appearance when they are concentrated. Like solutions, colloids may look clear or almost clear when they are dilute. The important difference between colloids and solutions and suspensions is in the size of the particles. **Colloids have particles smaller than those in suspensions and larger than those in solutions.** These intermediate-sized particles cannot be retained by filter paper as are the larger particles of a suspension, and they do not settle out with time. Colloids can be distinguished by the Tyndall effect and by the observation of Brownian motion. They are also subject to coagulation or clumping together, and they can be emulsified or made stable.

Table 15.3

Some Colloidal Systems

System			
Dispersed phase	**Dispersion medium**	**Type**	**Example**
gas	liquid	foam	whipped cream
gas	solid	foam	marshmallow
liquid	liquid	emulsion	milk, mayonnaise
liquid	gas	aerosol	fog, aerosols
solid	gas	smoke	dust in air
solid	liquid	sols and gels	egg white, jellies, paint, blood, colloidal gold, starch in water, gelatin

Differentiated Instruction

English Learners 

Have students use a dictionary to look up the various uses of the word *disperse* or *dispersion*. Ask students to explain how the seeds of a plant being dispersed by the wind are related to the dispersed phase and dispersion medium of a colloid.

Figure 15.15 The path of light is visible only when the light is scattered by particles. (a) Fog or mist is a colloid and thus exhibits the Tyndall effect. (b) Particles in colloids and suspensions reflect or scatter light in all directions. Solutions do not scatter light.

The Tyndall Effect Ordinarily you can't see a beam of sunlight unless the light passes through particles of water (mist) or dust in the air. These particles scatter the sunlight. Similarly, a beam of light is visible as it passes through a colloid. The scattering of visible light by colloidal particles is called the **Tyndall effect.** Suspensions also exhibit the Tyndall effect, but solutions do not. The particles in solutions are too small to scatter light. Figure 15.15 shows how the Tyndall effect can differentiate solutions from colloids and suspensions.

Brownian Motion Flashes of light, or scintillations, are seen when colloids are studied under a microscope. Colloids scintillate because the particles reflecting and scattering the light move erratically. The chaotic movement of colloidal particles, which was first observed by the Scottish botanist Robert Brown (1773–1858), is called **Brownian motion.** Brownian motion is caused by collisions of the molecules of the dispersion medium with the small, dispersed colloidal particles. These collisions help prevent the colloidal particles from settling.

Go Online NSTA SciLinks
For: Links on Brownian Motion
Visit: www.SciLinks.org
Web Code: cdn-1153

Coagulation Colloidal particles also tend to stay suspended because they become charged by adsorbing ions from the dispersing medium onto their surface. Some colloidal particles become positively charged by absorbing positively charged ions. Other colloidal particles become negatively charged by absorbing negatively charged ions. All the colloidal particles in a particular colloidal system will have the same charge, although the colloidal system is neutral. The repulsion between the like-charged particles prevents the particles from forming heavier aggregates that would have a greater tendency to settle out. Thus, a colloidal system can be destroyed or coagulated by the addition of ions having a charge opposite to that of the colloidal particles. The added ions neutralize the charged colloidal particles. The particles can clump together to form heavier aggregates and precipitate from the dispersion.

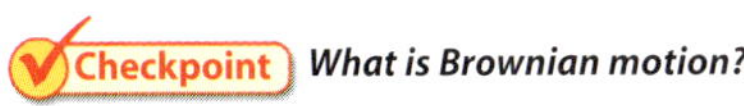
Checkpoint *What is Brownian motion?*

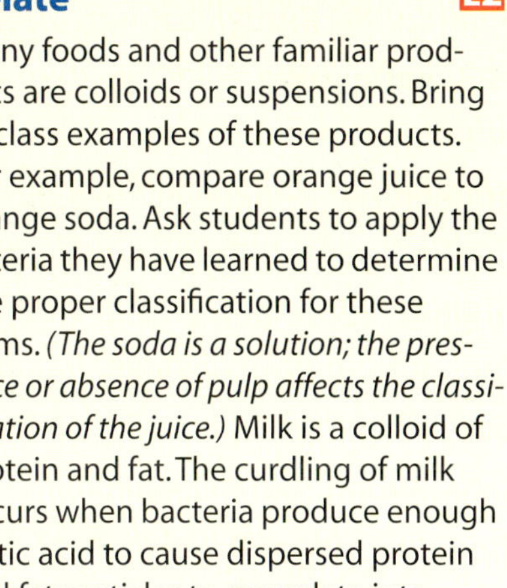

Relate L2

Many foods and other familiar products are colloids or suspensions. Bring to class examples of these products. For example, compare orange juice to orange soda. Ask students to apply the criteria they have learned to determine the proper classification for these items. *(The soda is a solution; the presence or absence of pulp affects the classification of the juice.)* Milk is a colloid of protein and fat. The curdling of milk occurs when bacteria produce enough lactic acid to cause dispersed protein and fat particles to coagulate into larger particles, which separate from the rest of the mixture. This process is comparable to the clotting of blood.

Motion of Colloidal Particles L1

Purpose Students observe the Tyndall effect.

Materials whole milk, water, beaker, stirring rod, projector or laser pointer

Procedure Add a small amount of whole milk to some water in a beaker, and stir to mix. In a darkened room, shine a light from a projector or a laser pointer through the beaker.

Expected Outcome When viewed from the side, the path of the light beam is observed. This is the Tyndall effect. Students also can observe Brownian motion of airborne dust particles in the light beam. **CAUTION:** *If a laser pointer is used, warn students not to look directly at it.*

Relate L2

Ask, **Why it is usually recommended that drivers use low beams when driving under foggy conditions at night?** *(Fog is a colloid that produces the Tyndall effect. Bright lights produce a higher degree of light scattering in all directions, including straight back into the driver's eyes.)*

Facts and Figures

Paint

Students may be interested in knowing why paint must be stirred before it is used. Explain that latex paint is a complex mixture of binders, pigments, and drying agents. Some ingredients are water-soluble; others are present as dispersed particles. For uniform color and composition, the paint must be stirred well before being used.

Download a worksheet on **Brownian Motion** for students to complete, and find additional teacher support from NSTA SciLinks.

Answers to...

Checkpoint the random movement of colloidal particles caused by the collision of molecules of the dispersion medium with the small, dispersed colloidal particles

Section 15.3 (continued)

3 ASSESS

Evaluate Understanding L2

Ask, **In what way are colloids similar to solutions?** *(In both, dispersed particles are small enough to pass through standard filter paper and to withstand the pull of gravity.)* **In what way are colloids similar to suspensions?** *(Both types of mixtures produce the Tyndall effect.)*

Reteach 

On the board, draw the relative sizes of solute, colloid, and suspension particles by making an analogy to golf balls, baseballs, and basketballs. Ask students which "particles" would get caught in a sieve a bit smaller than a basketball hoop and which would pass through? Remind students that solution particles are typically less than 1 nm in diameter, whereas colloid particles are ten to one thousand times larger and those of suspensions exceed the largest colloid particles.

Elements Handbook

Students' paragraphs should mention that electrolytes are excreted from the body in urine and sweat and can be restored by drinking water, juices, or sports drinks.

If your class subscribes to the Interactive Textbook, use it to review key concepts in Section 15.3.

with ChemASAP

Figure 15.16 The addition of an egg yolk to a mixture of oil and vinegar produces mayonnaise, a stable emulsion.

Emulsions An **emulsion** is a colloidal dispersion of a liquid in a liquid. An emulsifying agent is essential for the formation of an emulsion and for maintaining the emulsion's stability. For example, oils and greases are not soluble in water. However, they readily form a colloidal dispersion if soap or detergent is added to the water. Soaps and detergents are emulsifying agents. One end of a large soap or detergent molecule is polar and is attracted to water molecules. The other end of the soap or detergent molecule is nonpolar and is soluble in oil or grease. Soaps and other emulsifying agents thus allow the formation of colloidal dispersions between liquids that do not ordinarily mix. Figure 15.16 shows a familiar example of an emulsion—mayonnaise. Mayonnaise is a heterogeneous mixture of oil and vinegar. Such a mixture would quickly separate without the presence of egg yolk, which is the emulsifying agent. Other foods such as milk, margarine, and butter are also emulsions. Cosmetics, shampoos, and lotions are formulated with emulsifiers to maintain consistent quality. Table 15.4 summarizes the properties of solutions, colloids, and suspensions.

Table 15.4

Properties of Solutions, Colloids, and Suspensions

	System		
Property	**Solution**	**Colloid**	**Suspension**
Particle type	ions, atoms, small molecules	large molecules or particles	large particles or aggregates
Particle size	0.1–1 nm	1–1000 nm	1000 nm and larger
Effect of light	no scattering	exhibits Tyndall effect	exhibits Tyndall effect
Effect of gravity	stable, does not separate	stable, does not separate	unstable, sediment forms
Filtration	particles not retained on filter	particles not retained on filter	particles retained on filter
Uniformity	homogeneous	heterogeneous	heterogeneous

15.3 Section Assessment

16. **Key Concept** How does a suspension differ from a solution?
17. **Key Concept** What distinguishes a colloid from a suspension and a solution?
18. How can you determine through observation that a mixture is a suspension?
19. Could you separate a colloid by filtering? Explain.
20. How can the Tyndall effect be used to distinguish between a colloid and a solution?
21. What causes Brownian motion? Can the presence of Brownian motion distinguish between a solution and a colloid? Explain.

Elements Handbook

Electrolytes Go to page R8 in the Elements Handbook and read about the importance of electrolytes in the body. Write a paragraph explaining why the concentration of these ions may decline and how they can be restored.

Assessment 15.3 Test yourself on the concepts in Section 15.3.

with ChemASAP

Section 15.3 Assessment

16. The particles of a suspension are much larger and do not stay suspended indefinitely.
17. Colloids have particles smaller than those in suspensions and larger than those in solutions.
18. The particles in a suspension will settle out over time.
19. Particles in a colloid such as gelatin are smaller than the holes in filter paper and cannot be removed by filtering.
20. A beam of light is visible as it passes through a colloid; a beam of light is invisible as it passes through a solution.
21. Brownian motion is caused by collisions of the molecules of the dispersion medium with the small, dispersed colloidal particles. Flashes of light, or scintillations, are seen when colloids are studied under a microscope. Colloids scintillate because the particles reflecting and scattering the light move erratically. The particles in a solution are too small to be seen under a microscope and do not cause scintillations.

Technology & Society

Water Worth Drinking

When you turn on a faucet, you probably don't think about the substances other than water that may end up in your glass. To ensure the safety of the water you drink, the Environmental Protection Agency (EPA) has established drinking water standards that state and local governments must follow. These standards set upper limits on potentially harmful substances that could be present in water. The standards are based on studies of how these substances affect human health. **Inferring** ***What might happen if water were not treated to kill bacteria and other pathogens?***

1 **Laboratory analysis** identifies natural or synthetic compounds that may be harmful when present in drinking water. Standards are set for these substances in parts per million (ppm) or parts per billion (ppb).

2 If periodic **water samples** indicate too high a level of coliform bacteria, disinfectants such as chlorine must be added to destroy these and other potentially dangerous microorganisms.

3 When water must be **treated with chlorine,** frequent tests ensure that levels of this disinfectant do not fall below 0.5 ppm.

Local water providers test their public water supplies regularly to monitor levels of minerals, chlorine, and microorganisms such as bacteria.

Facts and Figures

Parts per million

Parts per million, or ppm, is a unit of concentration. It is commonly used when measuring levels of pollutants in air and water. One ppm is 1 part in 1,000,000. A solution whose concentration is 1 ppm contains 1 g of solute for each million (10^6) grams of solution. Because the density of water is 1 g/mL, 1 kg of a dilute aqueous solution will have a volume very close to 1 L. Thus, 1 ppm also corresponds to 1 mg of solute per liter of solution. The maximum allowable concentration of arsenic is 0.01 ppm or 0.01mg of arsenic per liter of water.

Technology & Society

Water Worth Drinking

Purpose

This feature builds on the discussion of water in the liquid state in section 15.1 and on some of the substances and materials that might be dissolved or suspended in the water.

Background

- 1999 marked the 25th year of public health protection under the Safe Drinking Water Act. The Act, passed in 1974 and amended in 1986 and 1996, gives the Environmental Protection Agency (EPA) the authority to set drinking water standards. Drinking water standards apply to public water systems, i. e., systems that provide water for human consumption through at least 15 service connections, or regularly serve at least 25 individuals. Public water systems include municipal water companies, homeowner associations, schools, businesses, campgrounds and shopping malls.
- Standards for safety are constantly being examined by the EPA and may change as new data become available. For example, the standard for arsenic was recently changed from 50 ppb to 10 ppb.
- Coliform bacteria are part of the normal bacterial flora of the intestinal tract of humans and other animals. *E. coli* is the best known of these bacteria, but there are many other species. The presence of coliform bacteria in the water supply is a strong indication of contamination by human or animal wastes. Microbes in these wastes can cause diarrhea, cramps, nausea, headaches, or other symptoms.
- *Cryptosporidium* is a parasite that enters lakes and rivers through sewage and animal waste. It causes cryptosporidiosis, a mild gastrointestinal disease. However, the disease can be severe or fatal for people with severely weakened immune systems.

Answers to...

Inferring A serious epidemic of typhoid, cholera, or other pathogen-caused disease might occur.

CHAPTER 15 Study Guide

Study Tip

Organize New Information Suggest that students review the chapter contents by using each of the vocabulary terms correctly in a sentence. Check their sentences for errors. Then, have students rewrite their sentences as fill-in-the-blank questions by leaving out the vocabulary terms. Students can use the questions to quiz a classmate.

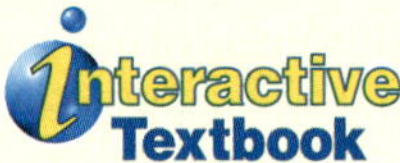

If your class subscribes to the Interactive Textbook with ChemASAP, your students can go online to access an interactive version of the Student Edition and a self-test.

with ChemASAP

CHAPTER 15 Study Guide

Key Concepts

 **15.1 Water and Its Properties**

- The high surface tension and low vapor pressure of water are the results of hydrogen bonding.
- The structure of ice is a regular open framework of water molecules held together by hydrogen bonds and arranged like a honeycomb.
- Molecules in liquid water are packed closer together than they are in ice.

 **15.2 Homogeneous Aqueous Systems**

- In a solution, a solvent dissolves the solute. The solute becomes dispersed in the solvent.
- In the dissolving process, individual solute ions break away from the crystal. Solvent molecules surround the negatively and positively charged ions in a process called solvation and the ionic crystal dissolves.
- All ionic compounds are electrolytes because they dissociate into ions and thus can conduct electricity.
- The formula of a hydrate consists of the formula of the ionic salt followed by a dot and the number of water molecules associated with one formula unit of the salt.

 **15.3 Heterogeneous Aqueous Systems**

- A suspension differs from a solution because the component particles of a suspension are much larger and do not stay suspended indefinitely.
- Colloids have particles smaller than those in suspensions and larger than those in solutions.

Vocabulary

- aqueous solution (p. 450)
- Brownian motion (p. 461)
- colloid (p. 460)
- electrolyte (p. 452)
- emulsion (p. 462)
- hydrate (p. 454)
- nonelectrolyte (p. 452)
- solute (p. 450)
- solvation (p. 451)
- solvent (p. 450)
- strong electrolyte (p. 453)
- surfactant (p. 447)
- suspension (p. 459)
- surface tension (p. 447)
- Tyndall effect (p. 461)
- weak electrolyte (p. 453)

Key Equation

- Percent $H_2O = \frac{\text{mass of water}}{\text{mass of hydrate}} \times 100\%$

Organizing Information

Use these terms to construct a concept map that organizes the major ideas of this chapter.

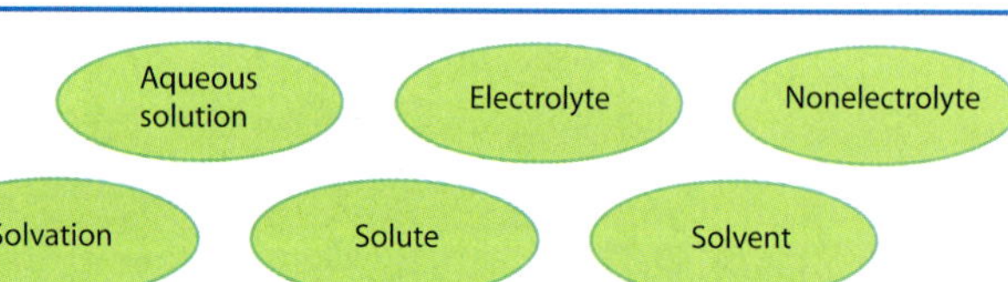

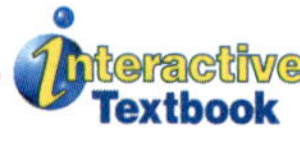

Concept Map 15 Solve the Concept Map with the help of an interactive guided tutorial.

with ChemASAP

Chapter Resources

Print

- **Core Teaching Resources,** Chapter 15, *Practice Problems, Vocabulary Review, Quiz, Chapter Test A, Chapter Test B*

Technology

- **Computer Test Bank,** Chapter 15 Test
- **Interactive Textbook with ChemASAP,** Chapter 15

CHAPTER 15 Assessment

Reviewing Content

15.1 Water and Its Properties

22. Why does water have high surface tension?
23. Why do the particles at the surface of a liquid behave differently from those in the bulk of the liquid?
24. Describe some observable effects that are produced by the surface tension of a liquid.
25. What is a surfactant? Explain how it works.
26. How can the unusually low vapor pressure of water be explained?
27. Explain why bodies of water with large surface areas such as lakes and oceans do not evaporate rapidly.
28. How does the structure of ice differ from the structure of water?
29. What would be some of the consequences if ice were denser than water?
30. Explain the role of hydrogen bonds in ice.

15.2 Homogeneous Aqueous Systems

31. Distinguish between a solution in general and an aqueous solution.
32. Identify the solvent and the solute in a solution of table sugar in water.
33. Why is water an excellent solvent for most ionic and polar covalent compounds but not for nonpolar compounds?
34. Suppose an aqueous solution contains both sugar and salt. Can you separate either of these solutes from the water by filtration? Explain your reasoning.
35. Describe the process of solvation.
36. Which of the following substances dissolve appreciably in water? Give reasons for your choice.
    a. HCl b. NaI
    c. NH_3 d. $MgSO_4$
    e. CH_4 f. $CaCO_3$
37. Explain why gasoline does not dissolve in water.
38. What particles must be present in a solution if it is to conduct electricity?
39. Why does molten sodium chloride conduct electricity?
40. What is the main distinction between an aqueous solution of a strong electrolyte and an aqueous solution of a weak electrolyte?
41. What is meant by a substance's water of hydration?
42. Write formulas for these hydrates.
    a. sodium sulfate decahydrate
    b. calcium chloride dihydrate
    c. barium hydroxide octahydrate
43. Name each hydrate.
    a. $SnCl_4 \cdot 5H_2O$
    b. $FeSO_4 \cdot 7H_2O$
    c. $BaBr_2 \cdot 4H_2O$
    d. $FePO_4 \cdot 4H_2O$
44. Epsom salt ($MgSO_4 \cdot 7H_2O$) changes to the monohydrate form at 150°C. Write an equation for this change.
45. Explain why a hygroscopic substance can be used as a desiccant.
46. Why is it important to keep some hygroscopic substances in tightly sealed containers?
47. Some hydrates are efflorescent. Explain what that means. Under what conditions will a hydrate effloresce?

15.3 Heterogeneous Aqueous Systems

48. How can you distinguish between a suspension and a solution?
49. Arrange colloids, suspensions, and solutions in order of increasing particle size.
50. What is the Tyndall effect?
51. Why don't solutions demonstrate the Tyndall effect?
52. What causes Brownian motion?
53. What are two circumstances that help keep colloidal particles in suspension?
54. How can a colloid be destroyed?
55. What makes a colloidal dispersion stable?

CHAPTER 15 Assessment

Reviewing Content

22. Strong hydrogen bonding causes an inward pull that tends to minimize the surface area.
23. Surface molecules are attracted to the liquid molecules below but not to the air above.
24. Drops are spherical; objects denser than a liquid can float on its surface.
25. A wetting agent such as a soap or detergent; a surfactant interferes with hydrogen bonding between water molecules and reduces surface tension.
26. Hydrogen bonds hold water molecules to each other, so the tendency for them to escape the solution is low.
27. Water has a low vapor pressure.
28. Ice is a regular open framework of hydrogen-bonded water molecules arranged like a honeycomb. This structure collapses in liquid water.
29. Bodies of water would freeze from the bottom up. This would kill many forms of aquatic life.
30. Hydrogen bonds between water molecules hold the molecules in a regular, open structure.
31. An aqueous solution is a solution that has water as the solvent.
32. Water: solvent; sugar: solute
33. Polar water molecules electrostatically attract ions and polar covalent molecules, but nonpolar compounds are unaffected because they have no charges.
34. No; the molecules and ions are smaller than the pores of the filter.
35. Cations and anions become surrounded by solvent molecules.
36. **a.** HCl (polar) dissolves. **b.** NaI (ionic) dissolves. **c.** NH_3 (polar) dissolves. **d.** $MgSO_4$ (ionic) dissolves. **e.** CH_4 (nonpolar) will not dissolve. **f.** $CaCO_3$ (strong ionic forces) will not dissolve.
37. Water is polar, and gasoline is nonpolar.
38. ions
39. Its ions are free to move toward an electrode.
40. A strong electrolyte is almost totally ionized.
41. water in the crystal structure of a substance
42. **a.** $Na_2SO_4 \cdot 10H_2O$ **b.** $CaCl_2 \cdot 2H_2O$ **c.** $Ba(OH)_2 \cdot 8H_2O$
43. **a.** tin(IV) chloride pentahydrate **b.** iron(II) sulfate heptahydrate **c.** barium bromide tetrahydrate **d.** iron(III) phosphate tetrahydrate
44. $MgSO_4 \cdot 7H_2O \rightarrow MgSO_4 \cdot H_2O + 6H_2O$
45. They absorb water vapor from the air.
46. They absorb water vapor from air.
47. Efflorescence is the loss of water of hydration that occurs when the hydrate has a higher vapor pressure than that of the water vapor in air.
48. A suspension has large particles that settle to the bottom. The particles in a solution stay suspended.
49. solutions, colloids, suspensions
50. scattering of visible light by colloids and suspensions
51. The molecules or ions are too small to have reflective surfaces.

52. collisions of the molecules of the dispersing medium with the small, dispersed colloidal particles
53. Brownian motion and repulsion between like-charged ions adsorbed on the surfaces of colloidal particles
54. Add ions with a charge opposite that of the colloidal particles.
55. the addition of an emulsifier

Understanding Concepts

56. hexane, ethanol, water
57. Water molecules have maximum density at 4°C. Below 4°C the molecules arrange in a regular network because of the attractions between them. As a result, ice has a lower density than water.
58. **a.** 1.00 g/mL **b.** 4°C
 c. No; the density of ice is 0.917 g/mL at 0°C. There would be a break in the curve at 0°C as liquid water at 0°C changes to ice at 0°C.
59. **a.** Water expands when it freezes.
 b. Water is polar and wax is nonpolar, and water has a higher surface tension.
 c. Water has a lower vapor pressure than alcohol.
60. The container would break because water expands as it freezes.
61. **a.** gasoline **b.** water
 c. gasoline **d.** water
62. Anions are attracted to the hydrogen atoms of the water molecule because they have a partial positive charge. Cations are attracted to the oxygen atom in the water molecule because it has a partial negative charge.
63. **a.** No, they both dissolve in water.
 b. drying to examine the crystals, testing for electrical conductivity, doing a flame test
64. Ethanol has both a polar hydroxyl end (–OH) that dissolves in water and a nonpolar hydrocarbon end (C_2H_5–) that dissolves in gasoline.
65. No; nonpolar molecules do not dissolve in polar solvents.
66. **a.** $NH_4Cl(s) \rightarrow NH_4^+(aq) + Cl^-(aq)$
 b. $Cu(NO_3)_2(s) \rightarrow Cu^{2+}(aq) + 2NO_3^-(aq)$
 c. $HC_2H_3O_2(s) \rightarrow H^+(aq) + C_2H_3O_2^-(aq)$
 d. $HgCl_2(s) \rightarrow Hg^{2+}(aq) + 2Cl^-(aq)$
67. **a.** sodium carbonate monohydrate, 14.5 % H_2O **b.** magnesium sulfate heptahydrate, 51.2% H_2O
68. 15.6 g $CuSO_4$
69. $CaCl_2{\cdot}2H_2O$
70. **a.** (1), (3), (6), (8) **b.** (1), (2), (5), (6), (7), (9)
 c. (2), (4), (5)

CHAPTER 15 Assessment continued

Understanding Concepts

56. From your knowledge of intermolecular forces, arrange these liquids in order of increasing surface tension: water (H_2O), hexane (C_6H_{14}), ethanol (C_2H_5OH).

57. Water has its maximum density at 4°C. Discuss the consequences of this fact.

58. The graph below shows the density of water over the temperature range 0°C to 20°C.

Density vs. Temperature for Liquid Water

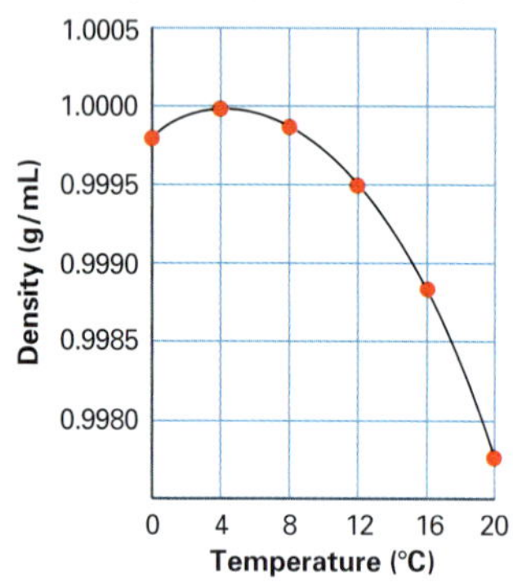

 a. What is the maximum density of water?
 b. At what temperature does the maximum density of water occur?
 c. Would it be meaningful to expand the smooth curve of the graph to the left to temperatures below 0°C?

59. Explain which properties of water are responsible for these occurrences.
 a. Water in tiny cracks in rocks helps break up the rocks when it freezes.
 b. Water beads up on a newly waxed car.
 c. A longer time is needed for a teaspoon of water to evaporate than a teaspoon of alcohol.

60. Describe what might happen if you put a sealed glass container full of water into a freezer.

61. Water is a polar solvent; gasoline is a nonpolar solvent. Decide which compounds are more likely to dissolve in water and which are more likely to dissolve in gasoline.
 a. CCl_4 **b.** Na_2SO_4
 c. methane (CH_4) **d.** KCl

62. Explain why ions become solvated in aqueous solution.

63. You have a solution containing either sugar or salt dissolved in water.
 a. Can you tell which it is by visual inspection? Explain.
 b. Give two ways by which you could easily tell which it is.

64. Explain why ethanol (C_2H_5OH) will dissolve in both gasoline and water.

65. Are all liquids soluble in each other? Explain.

66. Write equations to show how these substances ionize or dissociate in water.
 a. NH_4Cl **b.** $Cu(NO_3)_2$
 c. $HC_2H_3O_2$ **d.** $HgCl_2$

67. Name these hydrates and determine the percent by mass of water in each.
 a. $Na_2CO_3{\cdot}H_2O$ **b.** $MgSO_4{\cdot}7H_2O$

68. How many grams of copper(II) sulfate pentahydrate would you need to measure in order to have 10.0 g of anhydrous copper(II) sulfate?

69. A hydrate has the following percent composition: 75.5% $CaCl_2$ and 24.5% H_2O. What is the formula for this hydrate?

70. Match each term with the following descriptions. A description may apply to more than one term.
 a. solution **b.** colloid **c.** suspension
 (1) does not settle out on standing
 (2) heterogeneous mixture
 (3) particle size less than 1 nm
 (4) particles can be filtered out
 (5) demonstrates Tyndall effect
 (6) particles are invisible to the unaided eye
 (7) homogenized milk
 (8) salt water
 (9) jelly

Critical Thinking

71. When ethyl alcohol (C_2H_6O) dissolves in water, the volume of the final solution is less than the separate volumes of the water and alcohol added together. Can you explain this result? Do you think that it might be possible to mix two different liquids and get a mixture volume that is larger than the sum of the volumes of the two components? Explain.

72. Describe what would happen to a pond at 0°C if the density of ice were greater than the density of water. Do you think the pond would freeze more quickly? Explain.

73. When the humidity is low and the temperature high, humans must take in large quantities of water or face serious dehydration. Why do you think water is so important for the proper functioning of your body?

74. Describe as specifically as possible what would happen if a nonpolar molecular liquid were added to water. What would form if you shook this mixture vigorously?

75. Why is the midday sky blue while the evening sky in the west is often an orange or red color? (*Hint:* There are colloidal-size particles in the atmosphere.)

Concept Challenge

76. Make a drawing to show how the oxygen atom in one molecule of water can be connected to as many as four other molecules of water by hydrogen bonds. Write an explanation of your drawing.

77. A problem for firefighters is that much of the water they spray on a fire doesn't soak in but runs off carrying debris and pollution into the environment. Explain how the addition of a surfactant to water used to fight fires could help put out the fire more rapidly and protect the environment.

78. When spring comes, ice melting at the surface of a pond begins a beneficial process that stirs up the water of the pond. Explain why the pond water begins to mix when the ice melts. (*Hint:* Consider the changes in density.)

79. After a winter of alternate periods of freezing and thawing, some roads have broken pavement and potholes. Using what you know of the properties of water, explain why potholes form.

80. What relationships exist between the following volumes?
    a. 1 g of ice at 0°C and 1 g of liquid water at 0°C
    b. 1 g of liquid water at 100°C and 1 g of steam at 100°C

81. Cobalt chloride test paper is blue. This paper is made by soaking strips of paper in an aqueous solution of $CoCl_2{\cdot}6H_2O$. The paper strips are then dried in an oven.

$$CoCl_2{\cdot}6H_2O\ (\text{pink}) + \text{heat} \longrightarrow CoCl_2\ (\text{blue}) + 6H_2O$$

    a. When cobalt(II) chloride hexahydrate is dissolved in water, what is the color of the solution?
    b. What is the color of wet cobalt chloride paper?
    c. What is the color of dry cobalt chloride paper?
    d. What is the percent by mass of water in the hexahydrate?
    e. What does cobalt chloride test paper test for?

Critical Thinking

71. The hydrogen-bonded structures in water are disrupted when ethyl alcohol is added because the alcohol competes for hydrogen bonds with water molecules and the water structure collapses. Thus, mixtures of water and ethyl alcohol have less volume than the sum of the volumes of the components. Mixing of two liquids could result in a greater volume than the sum of the volumes of the components if there is greater structural ordering in the mixture than in the separated components.

72. Ice would form on top and sink to the bottom. The pond would freeze more quickly because no surface ice would insulate the water below from the freezing air.

73. Most of the important chemical reactions of life take place in aqueous solutions inside cells.

74. The liquid would form a layer on top of the water if its density was less than that of water and on the bottom if its density was greater. The mixture would form a temporary emulsion.

75. Sunlight is scattered by particles in the atmosphere. The shorter wavelengths of visible light (blue and violet) are more visible at midday when the sun is closest to Earth. The longer wavelengths of visible light (yellow and red) are more visible at sunrise and sunset, when the sun is at its greatest distance from Earth.

Concept Challenge

76.

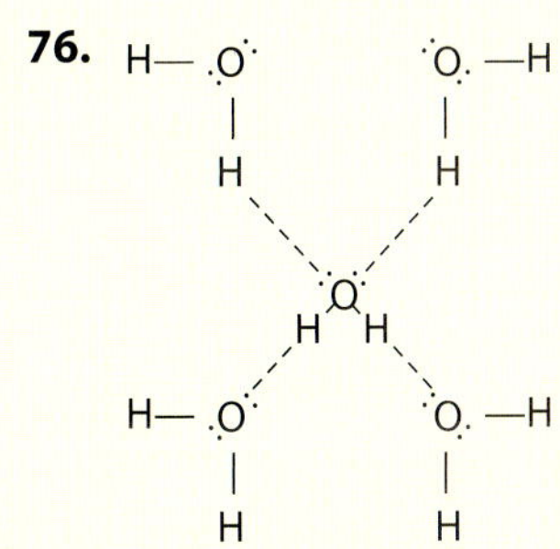

77. A surfactant helps to wet the burning material, so less water rolls off carrying pollutants into the environment.

78. In spring, when the ice melts and the temperature of the surface water increases to 4°C, it becomes heavier than the water below it and sinks. The downward movement of surface water forces water in the deeper parts of the lake upward, where it is warmed.

79. Water enters cracks in pavement and expands when it freezes, creating larger cracks. Continuous freeze-thaw cycles cause pavement to break up.

80. a. The volume of 1 g of ice at 0°C is greater. **b.** The volume of 1 g of liquid water at 100°C is less.

81. a. pink **b.** pink **c.** blue **d.** 45.4% **e.** water or water vapor

CHAPTER 15 Assessment continued

Cumulative Review

82. a. 195 g **b.** 1.95×10^5 mg
c. 0.195 kg

83. a. 5 **b.** 2 **c.** 2 **d.** 4

84. $1s^2 2s^2 2p^6$, neon

85. $H^+ + H{:}\ddot{\underset{..}{O}}{:}H \longrightarrow H{:}\ddot{\underset{..}{O}}{:}H^+$ (with a third H bonded below the O)

86. a. 22.4 L. **b.** mass He: 4.00 g; mass CH_4: 16.0 g; mass O_2: 32.0 g
c. density He: 0.179 g/L; density CH_4: 0.716 g/L; density O_2: 1.43 g/L
d. Balloons containing He and CH_4 will rise; O_2 balloon will sink.

87. a. $6CO_2 + 6H_2O \rightarrow C_6H_{12}O_6 + 6O_2$
b. $2Na + 2H_2O \rightarrow 2Na^+ + 2OH^- + H_2$

88. 9.07 g H_2; 72.0 g O_2

89. 3.60×10^{-2} g H_2O, 2.24×10^{-2} L O_2

90. 25.7 g H_2O

91. 636 g C_2H_4O

92. a. 0.300 mol O_2
b. 6.72 L O_2

93. a. hydrogen **b.** 0.048 g H_2O
c. oxygen **d.** 0.010 L

94. a. raises the boiling point
b. lowers the boiling point

95. 1.27 atm

CHAPTER 15 Assessment continued

Cumulative Review

82. A cylindrical vessel, 28.0 cm in height and 3.00 cm in diameter, is filled with water at 50°C. The density of water is 0.988 g/cm^3 at this temperature. Express the mass of water in the vessel in the following units. *(Chapter 3)*
a. grams **b.** milligrams **c.** kilograms

83. How many significant figures are in each measurement? *(Chapter 3)*
a. 56.003 g **b.** 0.0056 cm
c. 750 mL **d.** 0.4005 dg

84. Write the correct electron configuration for the oxide ion. Which noble gas has the same electron configuration? *(Chapter 7)*

85. When a proton is attracted to the unshared electron pair of a water molecule, the polyatomic hydronium ion (H_3O^+) is formed. Draw electron dot structures to show the formation of this ion. *(Chapter 8)*

86. The balloons contain 1 mol of He, CH_4, and O_2 at STP. *(Chapter 10)*

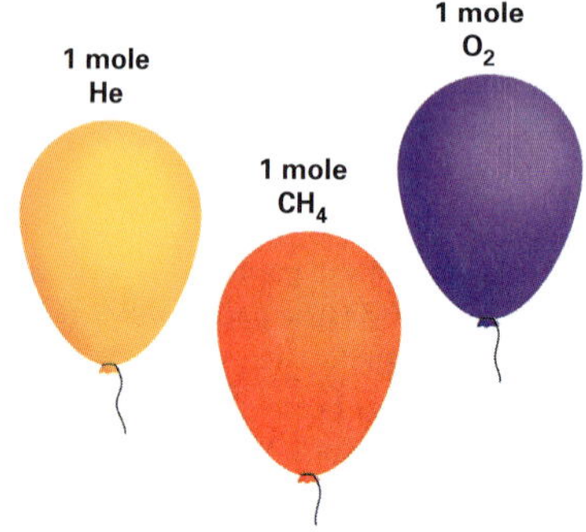

a. What is the volume of each balloon?
b. What is the mass of each balloon?
c. Calculate the density of the gas in each balloon.
d. The density of air at room temperature is about 1.2 g/mL. Predict whether each balloon will rise or sink when released.

87. Balance these equations. *(Chapter 11)*
a. $CO_2 + H_2O \longrightarrow C_6H_{12}O_6 + O_2$
b. $Na + H_2O \longrightarrow Na^+ + OH^- + H_2$

88. How many grams each of hydrogen gas and oxygen gas are required to produce 4.50 mol of water? *(Chapter 12)*

89. The decomposition of hydrogen peroxide is given by this equation.
$$2H_2O_2(l) \longrightarrow 2H_2O(l) + O_2(g)$$
Calculate the mass of water (in grams) and the volume of oxygen at STP formed when 2.00×10^{-3} mol of hydrogen peroxide is decomposed. *(Chapter 12)*

90. Calculate the mass of water produced in the complete combustion of 8.00 L of propane (C_3H_8) at STP, given this unbalanced equation. *(Chapter 12)*
$$C_3H_8 + O_2 \longrightarrow CO_2 + H_2O$$

91. Acetaldehyde (C_2H_4O) is produced commercially by the reaction of acetylene (C_2H_2) with water, as shown by this equation.
$$C_2H_2 + H_2O \longrightarrow C_2H_4O$$
How many grams of C_2H_4O can be produced from 2.60×10^2 g H_2O, assuming sufficient C_2H_2 is present? *(Chapter 12)*

92. Hydrogen reacts with oxygen to form water. *(Chapter 12)*
$$2H_2 + O_2 \longrightarrow 2H_2O$$
a. How many moles of oxygen are required to produce 10.8 g H_2O?
b. How many liters of oxygen is this at STP?

93. A mixture of 40 cm^3 of oxygen gas and 60 cm^3 of hydrogen gas at STP is ignited. *(Chapter 12)*
a. Which gas is the limiting reagent?
b. What is the mass of water produced?
c. Which gas remains after reaction?
d. What is the volume, at STP, of the remaining gas?

94. Explain how the following changes in the pressure on the surface of water affect the water's boiling point. *(Chapter 13)*
a. an increase in pressure
b. a decrease in pressure

95. The temperature of 1 L of steam at constant volume and 1.00 atm pressure is increased from 100°C to 200°C. Calculate the final pressure of the steam in atmospheres, assuming the volume does not change. *(Chapter 14)*

Standardized Test Prep

Test-Taking Tip

Constructed Response You will probably answer most constructed response questions by writing a sentence or a paragraph. Put as much information into each sentence as possible, but avoid unnecessary words. Your answer should address all the specific points asked for in the question. For longer responses, you may find it helpful to prepare a brief outline.

Select the choice that best answers each question or completes each statement.

1. When a sugar cube completely dissolves in a glass of water, it forms
   a. a colloid. **b.** a suspension.
   c. an emulsion. **d.** a solution.
2. How many water molecules are tied up per formula unit of a compound that is an octahydrate?
   a. nine **b.** eight **c.** seven **d.** six
3. Which property is characteristic of water?
   a. relatively high surface tension
   b. relatively high vapor pressure
   c. relatively low solvent ability
   d. relatively low polarity

Use the description below and the data table in the next column to answer Questions 4–6.

A student used a conductivity meter to measure the conductivity of several aqueous solutions. The magnitude of the conductivity value is proportional to the number of ions in the solution. The SI conductivity unit is the microsiemens/cm (μS/cm). The table gives the results reported by the student who tested six solutions plus distilled water. Each solution had a concentration of 0.02*M*.

Solution	Conductivity (μS/cm)
potassium chloride, KCl	2050
aluminum chloride, $AlCl_3$	4500
calcium chloride, $CaCl_2$	3540
sodium hydroxide, NaOH	2180
ethanol, C_2H_5OH	0
magnesium bromide, $MgBr_2$	3490
distilled water	0

4. Why do distilled water and the ethanol solution have zero conductivity?
5. Explain why two pairs of conducting solutions have similar conductivities.
6. The $AlCl_3$ solution has a conductivity that is about twice that of the KCl solution. Explain.

Use the atomic windows to answer Question 7.

7. Atomic window (a) represents solute particles in a given volume of solution. Which window represents the solute particles in the same volume of solution when the amount of solvent is doubled?

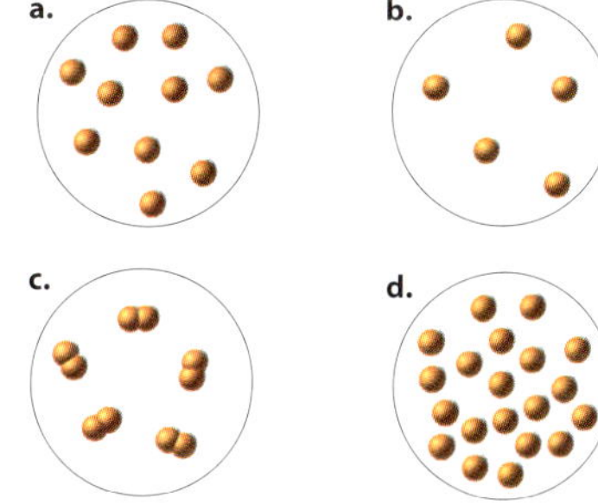

For each question there are two statements. Decide whether each statement is true or false. Then decide whether Statement II is a correct explanation for Statement I.

	Statement I		Statement II
8.	Water has a relatively high surface tension.	BECAUSE	Water molecules form strong hydrogen bonds with other water molecules.
9.	Methanol, CH_3OH, is a strong electrolyte.	BECAUSE	Methanol molecules can form hydrogen bonds with water.
10.	Particles in a colloid settle out faster than particles in a solution.	BECAUSE	Particles in a colloid are larger than particles in a solution.
11.	Water molecules are polar.	BECAUSE	The bond between hydrogen and oxygen atoms in a water molecule is polar.

Standardized Test Prep

1. d
2. b
3. a
4. When water is purified by distillation, dissolved ions are removed. Ethanol is a molecular compound that does not ionize in water.
5. KCl and NaOH have similar conductivities because they each provide two moles of ions per mole of compound. $CaCl_2$ and $MgBr_2$ have similar conductivities because they each provide three moles of ions per mole of compound.
6. Aluminum chloride provides twice as many ions per mole as potassium chloride does.
7. b
8. True, True, correct explanation
9. False, True
10. False, True
11. True, True, correct explanation

CHAPTER 16

Solutions Planning Guide

SECTION OBJECTIVES	STANDARDS NATIONAL	STANDARDS STATE	ACTIVITIES and LABS
16.1 Properties of Solutions, pp. 471–479 1 block or 2 periods **16.1.1 Identify** the factors that determine the rate at which a solute dissolves. **16.1.2 Identify** the units usually used to express the solubility of a solute. **16.1.3 Identify** the factors that determine the mass of solute that will dissolve in a given mass of solute.	A-1, B-5, C-1, E-2, F-4, F-6		**SE** Inquiry Activity: Salt and the Freezing Point of Water, p. 470 L2 **TE** Teacher Demo: Solubility of Gases, p. 475 L1 **SSLM** Lab 26: Solubility Rules L2 **LM** Lab 30: Factors Affecting Solution Formation L2 **LM** Lab 31: Supersaturation L2
16.2 Concentrations of Solutions, pp. 480–486 1 block or 2 periods **16.2.1 Solve** problems involving the molarity of a solution. **16.2.2 Describe** the effect of dilution on the total moles of solute in solution. **16.2.3 Define** percent by volume and percent by mass solutions.	A-1, A-2		**TE** Class Activity: Preparing Solutions, p. 482 L2 **TE** Class Activity: Solution Calculations, p. 483 L2 **TE** Teacher Demo: Serial Dilutions, p. 485 L2 **LM** Lab 32: Introduction to Chromatography L2 **LP** Lab Practical 16-1: Chromotography L2
16.3 Colligative Properties of Solutions, pp. 487–490 1 block or 2 periods **16.3.1 Identify** three colligative properties of solutions. **16.3.2 Explain** why the vapor pressure, freezing point, and boiling point of a solution differ from those properties of the pure solvent.	A-1, A-2		**SE** Quick Lab: Solutions and Colloids, p. 489 L2 **TE** Class Activity: Freezing Point Depression, p. 489 L1
16.4 Calculations Involving Colligative Properties, pp. 491–497 1 block or 2 periods **16.4.1 Solve Problems** related to the molality and mole fraction of a solution. **16.4.2 Describe** how freezing-point depression and boiling-point elevation are related to molality.	A-1, A-2		**SE** Small-Scale Lab: Making a Solution, p. 497 L2 **TE** Class Activity: Diagramming Methods of Concentration Calculation, p. 493 L3 **LM** Lab 33: Freezing Point L2

Ability Levels	
L1	For students who need additional help
L2	For all students
L3	For students who need to be challenged

Components

SE	Student Edition	LM	Laboratory Manual	GCP	Graphing Calculator Problems	T	Transparencies
TE	Teacher's Edition	SSLM	Small-Scale Chemistry Lab Manual	TTT	Test-Taking Tips with Transparencies, Grades 9–12 Science	CA	Chem Alive and Chemistry Field Trips DVD
GRSW	Guided Reading and Study Workbook	LP	Laboratory Practicals			iT	Interactive Textbook with ChemASAP
CTR	Core Teaching Resources	PLM	Probeware Lab Manual				
SSG	Spanish Study Guide	VCL	Virtual Chemistry Labs	STP	Standardized Test Preparation Workbook	GO	Go Online Internet Resources
		CTB	Computer Test Bank				

RESOURCES PRINT and TECHNOLOGY	PROBLEMS and ASSESSMENT
GRSW Chapter 16, Section 1 L1 **T** T169–T170 L2 **GO** Chapter 16, Section 1 L2	**SP** 16.1, Calculating the Solubility of a Gas **SE** Section 16.1 Assessment, p. 477 **iT** Section 16.1
GRSW Chapter 16, Section 2 L1 **T** T171–T174 L2	**SP** 16.2, Calculating the Molarity of a Solution **SP** 16.3, Finding the Moles of Solute in a Solution **SP** 16.4, Preparing a Dilute Solution **SP** 16.5, Calculating Percent (Volume/Volume) **SE** Section 16.2 Assessment, p. 486 **iT** Section 16.2
GRSW Chapter 16, Section 3 L1 **T** T175–T176 L2 **GO** Chapter 16, Section 3 L2	**SE** Section 16.3 Assessment, p. 490 **iT** Section 16.3
GRSW Chapter 16, Section 4 L1 **T** T177–T179 L2	**SP** 16.6, Using Solution Molality **SP** 16.7, Calculating Mole Fractions **SP** 16.8, Calculating the Freezing-Point Depression of a Solution **SP** 16.9, Calculating the Boiling Point of an Ionic Solution **SE** Section 16.4 Assessment, p. 496 **iT** Section 16.4

Go Online

Web Code: cdn-1161
Web Code: cdn-1163

MATERIALS List

Quantities for each group

STUDENT EDITION

Inquiry Activity, p. 470
plastic plate, 20-cm piece of string or narrow ribbon, water, an ice cube, table salt (sodium chloride)

Quick Lab, p. 489
sodium hydrogen carbonate, cornstarch, stirring rod, distilled water, flashlight, masking tape, 3 jars with parallel sides, teaspoon, cup

Small-Scale Lab, p. 497
NaCl (lab grade), water, 50-mL volumetric flask, balance

TEACHER'S EDITION

Teacher Demo, p. 475
bottle of warm soda, bottle of cold soda, 2 other bottles of soda

Class Activity, p. 482
water, NaCl, sucrose; 50-mL, 100-mL, and 1-L volumetric flasks; balance; graph paper

Class Activity, p. 483
calculator

Teacher Demo, p. 485
four 20-mL test tubes, 10 mL of 0.5*M* sucrose, 15 mL water, 10-mL pipet

Class Activity, p. 489
ice, water, rock salt, thermometer, foam cup

Chapter Assessment

CHAPTER ASSESSMENT

SE	Chapter Assessment, pp. 499-502
CTB	Chapter 16
iT	Chapter 16

STANDARDIZED TEST PREP

SE	Chapter 16, p. 503
STP	Topic 6
iT	Interactive Textbook with Assessment at PHSchool.com

CHAPTER 16

Solutions

INQUIRY Activity

Salt and the Freezing Point of Water

L2

Purpose

To observe the effect of salt on the freezing point of water

 **Prep Time** 10 minutes
Class Time 20 minutes

Think About It

1. Students should observe that sprinkling salt along the string causes the ice beneath the string to melt. (The salt lowers the freezing point of the ice. Because the ice absorbs heat from the wet string, the string freezes to the ice.)
2. Students may have seen rock salt used to freeze homemade ice cream. In areas that experience heavy snow falls, salt, sand, and gravel mixtures are spread on roads to prevent the formation of "black ice."
3. Both sugar and baking soda lower the freezing point of water, but to different degrees.

CHAPTER 16

Solutions

Water has shaped these cliffs in Grand Canyon National Park and carried away some of the rock in solution.

INQUIRY Activity

Salt and the Freezing Point of Water

Materials

plastic plate, a 20-cm piece of string or narrow ribbon, water, an ice cube, and some table salt (sodium chloride)

Procedure

1. Remove an ice cube from the freezer and place it on the plate.
2. Moisten the string or ribbon with water and place it in a straight line across the ice cube.
3. Sprinkle some salt along the length of the string on the ice.
4. Wait one to two minutes and then slowly lift the ends of the string. The ice cube should lift off the plate.
5. If you are not successful in lifting the cube with the string, repeat Steps 3 and 4.

Think About It

1. What effect does the salt have on the water/ice?
2. Can you name some practical instances of salt being used to affect the freezing point of water?
3. Do you think sugar or baking soda would work equally well? Try it!

470

16.1 Properties of Solutions

Connecting to Your World

It was there one minute and gone the next! An entire house was swallowed up by Earth. A victim of moving groundwater, the house had disappeared into a sinkhole! Groundwater, the rain and melted snow that soaks into the ground, can dissolve huge amounts of rock over time and create beautiful limestone caves. A sinkhole forms when the roof of a cave weakens from being dissolved and suddenly collapses. One recorded sinkhole swallowed a house, several other buildings, five cars, and a swimming pool! In this section, you will learn how the solution process occurs and the factors that influence the process.

Guide for Reading

Key Concepts
- What factors determine the rate at which a substance dissolves?
- How is solubility usually expressed?
- What conditions determine the amount of solute that will dissolve in a given solvent?

Vocabulary

saturated solution
solubility
unsaturated solution
miscible
immiscible
supersaturated solution
Henry's law

Reading Strategy

Outlining As you read, make an outline of the most important ideas in this section. Use the red headings as the main topics and the blue headings as subtopics. Add a sentence or a note after each heading to provide key information about each topic.

Solution Formation

Have you noticed, when making tea, that granulated sugar dissolves faster than sugar cubes, and that both granulated sugar and sugar cubes dissolve faster in hot tea or when you stir? Figure 16.1 illustrates these observations. You will be able to explain these observations, as well as the formation of a sinkhole, once you have gained an understanding of the properties of solutions.

Recall that solutions are homogeneous mixtures that may be solid, liquid, or gaseous. **The compositions of the solvent and the solute determine whether a substance will dissolve. Stirring (agitation), temperature, and the surface area of the dissolving particles determine how fast the substance will dissolve.** These three factors involve the contact of the solute with the solvent.

Figure 16.1 Stirring and heating increase the rate at which a solute dissolves. (a) A cube of sugar in cold tea dissolves slowly. (b) Granulated sugar dissolves in cold water more quickly than a sugar cube, especially with stirring. (c) Granulated sugar dissolves very quickly in hot tea.

Section Resources

Print
- **Guided Reading and Study Workbook,** Section 16.1
- **Core Teaching Resources,** Section 16.1 Review, Interpreting Graphics
- **Small-Scale Chemistry Laboratory Manual,** Lab 26
- **Laboratory Manual,** Labs 30, 31
- **Transparencies,** T169–T170

Technology
- **Interactive Textbook with ChemASAP,** Simulation 20, Problem-Solving 16.2, Assessment 16.1
- **Go Online,** Section 16.1

16.1

1 FOCUS

Objectives

16.1.1 Identify the factors that determine the rate at which a solute dissolves.

16.1.2 Identify the units usually used to express the solubility of a solute.

16.1.3 Identify the factors that determine the mass of solute that will dissolve in a given mass of solute.

Guide for Reading

Build Vocabulary L2

Graphic Organizer Students can compare and contrast the three types of solutions—*saturated solution*, *unsaturated solution*, and *supersaturated solution*—using a table. Have them include definitions and examples of each solution type.

Reading Strategy L2

Relate Text and Visuals Have students study Figures 16.2 and 16.6. Have them make simple sketches that show how the three types of solutions differ.

2 INSTRUCT

Connecting to Your World

Ask, **What causes the formation of limestone caves?** *(groundwater, rain, and melted snow that soaks into the ground and dissolves a significant amount of limestone)*

Solution Formation

Relate L1

Prepare a bulletin board display using pictures that illustrate the importance of solution processes in nature. Examples include a farmer applying fertilizer to a field, a close-up of a plant, a person eating food, fish swimming, a volcanic eruption, and natural crystals. Encourage students to contribute to the display by posting pictures they find.

Download a worksheet on **Solubility** for students to complete, and find additional teacher support from NSTA SciLinks.

Solubility

Relate L2

Explain that solubility is somewhat like population density. Both terms express the concentration of objects. With solubility, the objects are molecules or ions dissolved in a given quantity of solvent. With population density, the objects are organisms per unit area. However, point out that the precise amount of solute that a particular solvent holds under given conditions is fixed. Population density is not limited in the same way.

For: Links on Solubility
Visit: www.SciLinks.org
Web Code: cdn-1161

Stirring and Solution Formation If a teaspoon of granulated sugar (sucrose) is placed in a glass of tea, the crystals dissolve slowly. If the contents of the glass are stirred, however, the crystals dissolve more quickly. The dissolving process occurs at the surface of the sugar crystals. Stirring speeds up the process because fresh solvent (the water in tea) is continually brought into contact with the surface of the solute (sugar). It's important to realize, however, that agitation (stirring or shaking) affects only the rate at which a solid solute dissolves. It does not influence the amount of solute that will dissolve. An insoluble substance remains undissolved regardless of how vigorously or for how long the solvent/solute system is agitated.

Temperature and Solution Formation Temperature also influences the rate at which a solute dissolves. Sugar dissolves much more rapidly in hot tea than in iced tea. At higher temperatures, the kinetic energy of water molecules is greater than at lower temperatures so they move faster. The more rapid motion of the solvent molecules leads to an increase in the frequency and the force of the collisions between water molecules and the surfaces of the sugar crystals.

Particle Size and Solution Formation The rate at which a solute dissolves also depends upon the size of the solute particles. A spoonful of granulated sugar dissolves more quickly than a sugar cube because the smaller particles in granulated sugar expose a much greater surface area to the colliding water molecules. Remember, the dissolving process is a surface phenomenon. The more surface area of the solute that is exposed, the faster the rate of dissolving.

Solubility

If you add 36.0 g of sodium chloride to 100 g of water at 25°C, all of the 36.0 g of salt dissolves. But if you add one more gram of salt and stir, no matter how vigorously or for how long, only 0.2 g of the last portion will dissolve. Why does the remaining 0.8 g of salt remain undissolved? According to the kinetic theory, water molecules are in continuous motion. Therefore, they should continue to bombard the excess solid, removing and solvating the ions. As ions are solvated, they dissolve in the water. Based on this information, you might expect all of the sodium chloride to dissolve eventually. That does not happen, however, because an exchange process is occurring. New particles from the solid are solvated and enter into solution, as shown in Figure 16.2. At the same time an equal number of already dissolved particles crystalize. These particles come out of solution and are deposited as a solid. The mass of undissolved crystals remains constant.

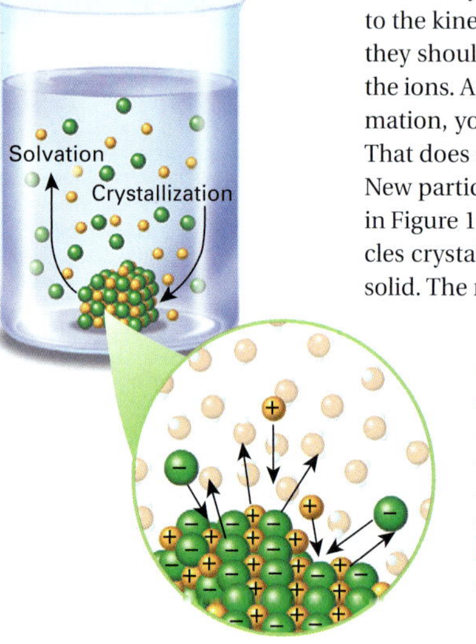

Figure 16.2 In a saturated solution, a state of dynamic equilibrium exists between the solution and the excess solute. The rate of solvation (dissolving) equals the rate of crystallization, so the total amount of dissolved solute remains constant. **Inferring** ***What would happen if you added more solute?***

Facts and Figures

Fertilizer Runoff

Most farmers use fertilizers that contain salts of one or more of three elements essential to plant growth: potassium, nitrogen, and phosphorus. When more fertilizer is applied than the soil can absorb, rain washes off the excess salts. Not only is this economically wasteful, it is hazardous to the environment. The water containing these dissolved salts flows into streams and rivers, where the salts contribute to the eutrophication of the water. Eutrophied waters are rich in nutrients but deficient in dissolved oxygen. Signs of eutrophication include overgrowth of water plants, algal blooms, and bad odors resulting from the growth of bacteria that do not need oxygen. In recent years, the problem of fertilizer runoff has lessened with the introduction of slow-release nitrogenous fertilizers. Many farmers have also adopted new methods to reduce runoff.

Figure 16.3 Liquids that are insoluble in one another are immiscible. ⓐ A thin film of oil spreads over a water surface. Light rays, bent by the film, create patterns of color. ⓑ Vinegar, which is water-based, and oil are immiscible.

What is happening? Particles move from the solid into the solution. Other dissolved particles move from the solution back to the solid. Because these two processes occur at the same rate, no net change occurs in the overall system. As Figure 16.2 illustrates, a state of dynamic equilibrium exists between the solution and the undissolved solute. The system will remain the same as long as the temperature remains constant. Such a solution is said to be saturated. A **saturated solution** contains the maximum amount of solute for a given quantity of solvent at a constant temperature and pressure. For example, 36.2 g of sodium chloride dissolved in 100 g of water is a saturated solution at 25°C. If additional solute is added to this solution, it will not dissolve. The **solubility** of a substance is the amount of solute that dissolves in a given quantity of a solvent at a specified temperature and pressure to produce a saturated solution. **Solubility is often expressed in grams of solute per 100 g of solvent.** Sometimes the solubility of a gas is expressed in grams per liter of solution (g/L). A solution that contains less solute than a saturated solution at a given temperature and pressure is an **unsaturated solution.** If additional solute is added to an unsaturated solution, the solute will dissolve until the solution is saturated.

Some liquids—for example, water and ethanol—are infinitely soluble in each other. Any amount of ethanol will dissolve in a given volume of water, and vice versa. Similarly, ethylene glycol and water mix in all proportions. Pairs of liquids such as these are said to be completely miscible. Two liquids are **miscible** if they dissolve in each other in all proportions. In such a solution, the liquid that is present in the larger amount is usually considered the solvent. Liquids that are slightly soluble in each other—for example, water and diethyl ether—are partially miscible. Liquids that are insoluble in one another are **immiscible.** As you can see in Figure 16.3, oil and vinegar are immiscible, as are oil and water.

 Checkpoint *What is a saturated solution?*

Word *Origins*

Miscible comes from the Latin word *miscere,* meaning "to mix." Completely miscible liquids dissolve in each other in all proportions. **If the prefix *im-* means "not," what would you call two liquids that are insoluble in each other?**

Word *Origins* L2

Two liquids that are insoluble in each other are called immiscible. Similarly, the word *immense* means transcending ordinary means of measurement, or not capable of normal measurement. The word *immobile* means incapable of being moved.

Use Visuals 

Figure 16.3 Have students study the photographs. Ask, **Why do the two liquids not mix?** *(For nonpolar liquids to mix with water, hydrogen bonds must be broken and replaced by much weaker forces between water and the nonpolar compound. In the cases shown, the forces of attraction between molecules are maximized if the liquids remain unmixed.)*

Differentiated Instruction

English Learners L1

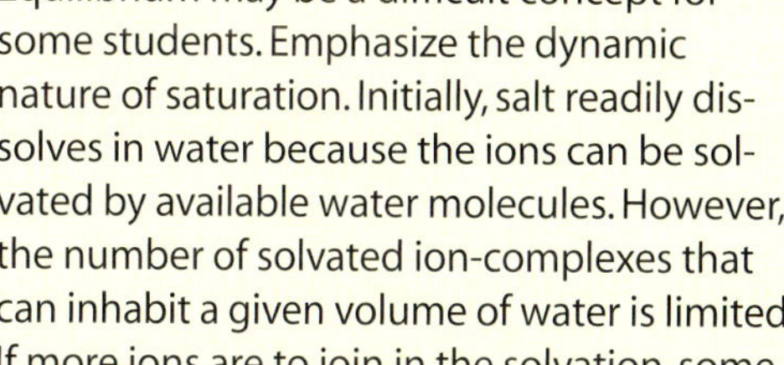

Equilibrium may be a difficult concept for some students. Emphasize the dynamic nature of saturation. Initially, salt readily dissolves in water because the ions can be solvated by available water molecules. However, the number of solvated ion-complexes that can inhabit a given volume of water is limited. If more ions are to join in the solvation, some ions must give up their places. Use substitutions in a basketball game as an analogy for the dynamic nature of saturation. Tell students that only five players from each team can be on the court at one time. Players on the bench must wait until a player comes off the court. In this analogy, the players on the sidelines represent ions in a crystal; and those on the court represent dissolved ions.

Answers to...

Figure 16.2 nothing

 Checkpoint a solution that contains the maximum amount of solute for a given quantity of solvent at a constant temperature and pressure

Section 16.1 (continued)

Factors Affecting Solubility

Interpreting Graphs

L2

a. The solubility of KNO_3 increases as the temperature increases.
b. $Yb_2(SO_4)_3$ shows a decrease in solubility as temperature increases. NaCl shows the least change in solubility.
c. Only a negligible amount of NaCl would go into solution, if any.

Enrichment Question

L3

Ask students which of the compounds in Figure 16.4 they would predict would have negative molar heats of solution. Have students look up the ΔH_{soln} for each compound in the graph (*CRC Handbook of Chemistry and Physics*). Students should state how the solubility curve for a compound is related to its ΔH_{soln}. *(Students should find that with the exception of $Yb_2(SO_4)_3$, for which data is not given, all of the compounds in the graph have positive ΔH_{soln} values at 25°C. Thus, heat is absorbed in the solution process for these compounds. Students may predict, then, that providing more heat by raising the temperature of a solution will increase solubility. Students may also note that NaCl, which shows the least increase in solubility with temperature, has the lowest positive ΔH_{soln} among the compounds discussed.)*

Figure 16.4 Changing the temperature usually affects the solubility of a substance.

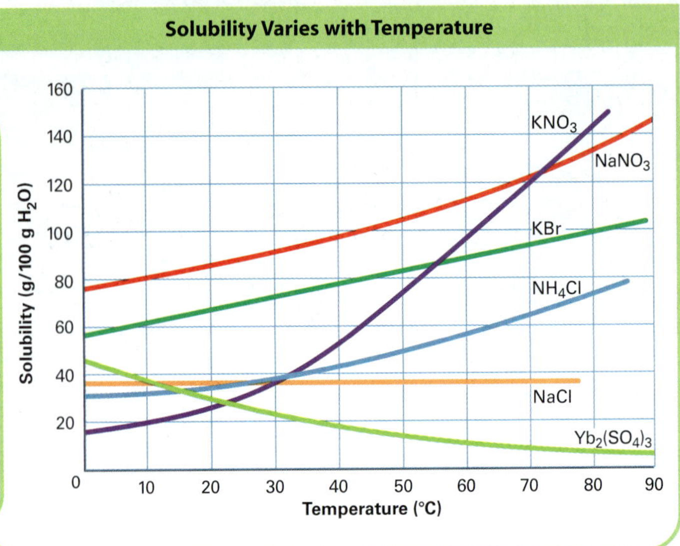

INTERPRETING GRAPHS

a. Describe What happens to the solubility of KNO_3 as the temperature increases?
b. Identify Which substance shows a decrease in solubility as temperature increases? Which substance exhibits the least change in solubility?
c. Apply Concepts Suppose you added some solid sodium chloride (NaCl) to a saturated solution of sodium chloride at 20°C and warmed the mixture to 40°C. What would happen to the added sodium chloride?

Factors Affecting Solubility

You have read that solubility is defined as the mass of solute that dissolves in a given mass of a solvent at a specified temperature. **Temperature affects the solubility of solid, liquid, and gaseous solutes in a solvent; both temperature and pressure affect the solubility of gaseous solutes.**

Temperature The solubility of most solid substances increases as the temperature of the solvent increases. Figure 16.4 shows how the solubility of several substances changes as temperature increases. The mineral deposits around hot springs, such as the one shown in Figure 16.5, result from the cooling of the hot, saturated solution of minerals emerging from the spring. As the solution cools in air, it cannot contain the same concentration of minerals as it did at a higher temperature, so some of the minerals precipitate.

For a few substances, solubility decreases with temperature. For example, the solubility of ytterbium sulfate ($Yb_2(SO_4)_3$) in water drops from 44.2 g per 100 g of water at 0°C to 5.8 g per 100 g of water at 90°C. Table 16.1 lists the solubilities of some common substances at various temperatures.

Suppose you make a saturated solution of sodium ethanoate (sodium acetate) at 30°C and let the solution stand undisturbed as it cools to 25°C. Because the solubility of this compound is greater at 30°C than at 25°C, you expect that solid sodium ethanoate will crystallize from the solution as the temperature drops. But no crystals form. You have made a supersaturated solution. A **supersaturated solution** contains more solute than it can theoretically hold at a given temperature. The crystallization of a supersaturated solution can be initiated if a very small crystal, called a seed crystal, of the solute is added. The rate at which excess solute deposits upon the surface of a seed crystal can be very rapid, as shown in Figure 16.6. Crystallization can also occur if the inside of the container is scratched.

Figure 16.5 Mineral deposits form around the edges of this hot spring because the hot water is saturated with minerals. As the water cools, some of the minerals crystallize because they are less soluble at the lower temperature.

Differentiated Instruction

English Learners and Less Proficient Readers

L1

Have students draw and label a diagram illustrating the dynamic equilibrium of a saturated solution with arrows describing the motion of particles between the crystalline and solvated states.

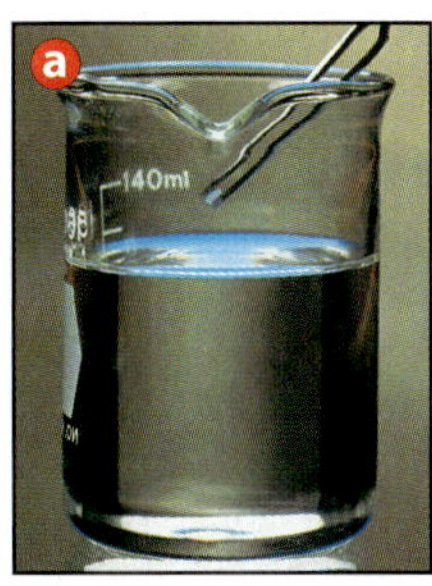
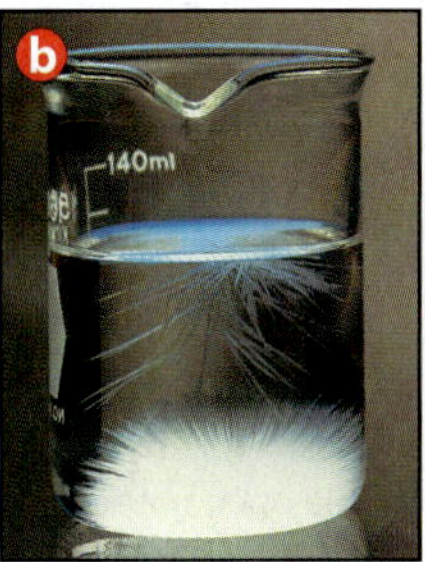
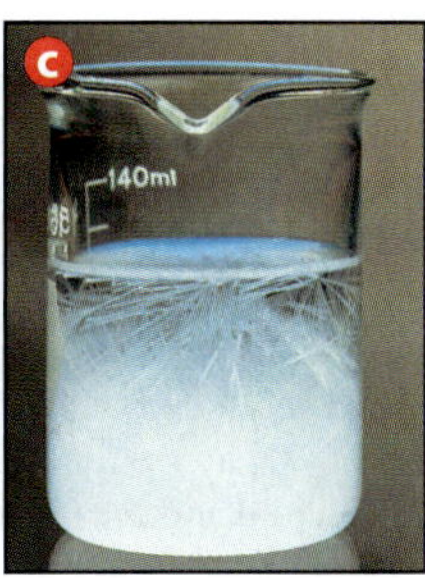

Figure 16.6 A supersaturated solution crystallizes rapidly when disturbed. **a** The solution is clear before a seed crystal is added. **b** Crystals begin to form in the solution immediately after the addition of a seed crystal. **c** Excess solute crystallizes rapidly. **Applying Concepts** ***When the crystallization has ceased, will the solution be saturated or unsaturated?***

Another example of crystallization in a supersaturated solution is the production of rock candy. A solution is supersaturated with sugar. Seed crystals cause the sugar to crystallize out of solution onto a string for you to enjoy!

The effect of temperature on the solubility of gases in liquid solvents is opposite that of solids. The solubilities of most gases are greater in cold water than in hot. For example, Table 16.1 shows that the most important component of air for living beings—oxygen—becomes less soluble in water as the temperature of the solution rises. This fact has some important consequences. When an industrial plant takes water from a lake to use for cooling and then dumps the resulting heated water back into the lake, the temperature of the entire lake increases. Such a change in temperature is known as thermal pollution. Aquatic animal and plant life can be severely affected because the increase in temperature lowers the concentration of dissolved oxygen in the lake water.

Simulation 20 Observe the effect of temperature on the solubility of solids and gases in water.

with ChemASAP

Table 16.1

Solubilities of Some Substances in Water at Various Temperatures

Substance	Formula	Solubility ($g/100\ g\ H_2O$) 0°C	20°C	50°C	100°C
Barium hydroxide	$Ba(OH)_2$	1.67	31.89	—	—
Barium sulfate	$BaSO_4$	0.00019	0.00025	0.00034	—
Calcium hydroxide	$Ca(OH)_2$	0.189	0.173	—	0.07
Lead(II) chloride	$PbCl_2$	0.60	0.99	1.70	—
Lithium carbonate	Li_2CO_3	1.5	1.3	1.1	0.70
Potassium chlorate	$KClO_3$	4.0	7.4	19.3	56.0
Potassium chloride	KCl	27.6	34.0	42.6	57.6
Sodium chloride	$NaCl$	35.7	36.0	37.0	39.2
Sodium nitrate	$NaNO_3$	74	88.0	114.0	182
Aluminum chloride	$AlCl_3$	30.84	31.03	31.60	33.32
Silver nitrate	$AgNO_3$	122	222.0	455.0	733
Lithium bromide	$LiBr$	143.0	166	203	266.0
Sucrose (cane sugar)	$C_{12}H_{22}O_{11}$	179	230.9	260.4	487
Hydrogen*	H_2	0.00019	0.00016	0.00013	0.0
Oxygen*	O_2	0.0070	0.0043	0.0026	0.0
Carbon dioxide*	CO_2	0.335	0.169	0.076	0.0

*Gas at 101 kPa (one atmosphere) total pressure

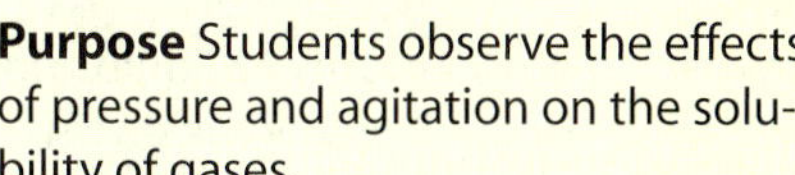

Solubility of Gases L1

Purpose Students observe the effects of pressure and agitation on the solubility of gases.

Materials bottle of warm soda, bottle of cold soda, 2 other bottles of soda

Procedure Open a bottle of warm soda and a bottle of cold soda. Ask students to observe and compare the results. *(Carbon dioxide gas comes out of solution much more quickly in the warm soda.)* The rush of bubbles in a warm bottle of soda can be a model for the "bends," a malady that afflicts divers when dissolved nitrogen forms bubbles in the bloodstream. The model also shows how the escape of carbon dioxide from rising magma can trigger a volcanic eruption. In each case, high pressure keeps a gas in solution; reduced pressure allows it to escape. Repeat the demonstration, this time using a shaken and unshaken bottle of soda. Have students compare and contrast the effect of agitation on the solubility of gases and solids. **CAUTION:** *Be sure to open the shaken bottle in a sink or other partially enclosed area, to avoid spilling.*

Facts and Figures

Rainmakers

U.S. scientists Vincent Schaefer, Irving Langmuir, and Bernard Vonnegut carried out the first successful rainmaking experiments. They knew that water vapor in the atmosphere condenses around tiny nuclei, such as particles of soot. In 1946, Schaefer produced snow by dropping dry ice (solid carbon dioxide) pellets into clouds. A year later, Vonnegut discovered that silver iodide particles also serve as "precipitation nuclei." Since that time, seeding clouds with silver iodide has been the standard method.

Answers to...

Figure 16.6 saturated

Section 16.1 (continued)

Relate

L2

Explain that the action of a gas pressing down on the surface of a liquid and increasing the solubility of the gas in the liquid is like the action of a hammer driving a nail into a piece of wood. When you hit the nail harder, more of the nail is driven into the wood. The hammer is analogous to pressure, the nail to solute gas, and the wood to solvent liquid.

Sample Problem 16.1

Answers

1. $S_2 = S_1 \times P_2/P_1 = 0.16$ g/L $\times 288$ kPa/ 104 kPa $= 4.4 \times 10^{-1}$ g/ L
2. $P_2 = P_1 \times S_2/S_1 = 1.0$ atm $\times 9.5$ g/L /3.6 g/L $= 2.6$ atm

Practice Problems Plus

The solubility of a gas is 1.04 g/L at 98.0 kPa and 25°C. **What is the solubility of the same gas at 125 kPa and 25°C?** *(1.33 g/L)*

Math Handbook

For a math refresher and practice, direct students to algebraic equations, page R69.

Pressure Changes in pressure have little effect on the solubility of solids and liquids, but pressure strongly influences the solubility of gases. Gas solubility increases as the partial pressure of the gas above the solution increases. Carbonated beverages are a good example. These drinks contain large amounts of carbon dioxide (CO_2) dissolved in water. Dissolved CO_2 makes the liquid fizz and your mouth tingle. The drinks are bottled under a high pressure of CO_2 gas, which forces large amounts of the gas into solution. When a carbonated-beverage container is opened, the partial pressure of CO_2 above the liquid decreases. Immediately, bubbles of CO_2 form in the liquid and escape from the open bottle, as shown in Figure 16.7. As a result, the concentration of dissolved CO_2 decreases. If the bottle is left open, the drink becomes "flat" as the solution loses its CO_2.

How is the partial pressure of carbon dioxide gas related to the solubility of CO_2 in a carbonated beverage? The relationship is described by **Henry's law,** which states that at a given temperature, the solubility (S) of a gas in a liquid is directly proportional to the pressure (P) of the gas above the liquid. In other words, as the pressure of the gas above the liquid increases, the solubility of the gas increases. Similarly, as the pressure of the gas decreases, the solubility of the gas decreases. You can write the relationship in the form of an equation.

$$\frac{S_1}{P_1} = \frac{S_2}{P_2}$$

S_1 is the solubility of a gas at one pressure, P_1; S_2 is the solubility at another pressure, P_2.

Checkpoint ***How is the solubility of a gas affected by pressure?***

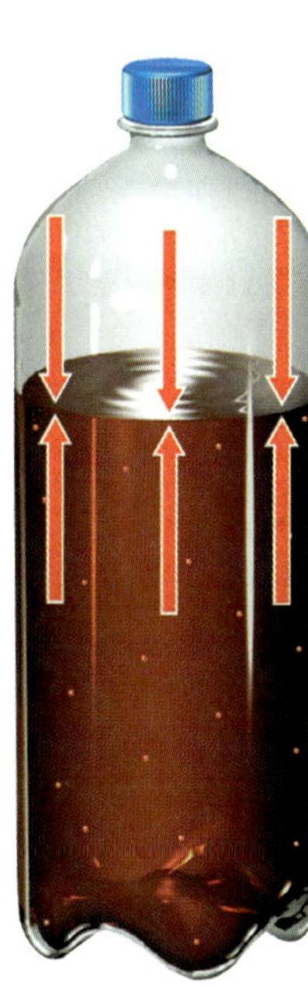

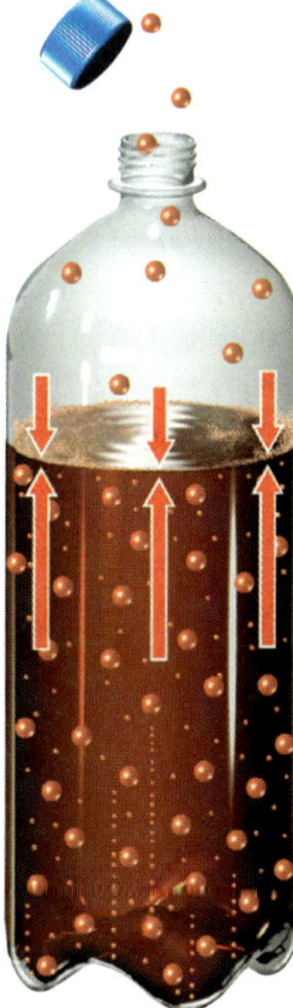

Figure 16.7 When a carbonated-beverage bottle is sealed, the pressure of CO_2 above the liquid is high and the concentration of CO_2 in the liquid is also high. When the cap is removed, the pressure of CO_2 gas above the liquid decreases and carbon dioxide bubbles out of the liquid.

SAMPLE PROBLEM 16.1

Calculating the Solubility of a Gas

If the solubility of a gas in water is 0.77 g/L at 3.5 atm of pressure, what is its solubility (in g/L) at 1.0 atm of pressure? (The temperature is held constant at 25°C.)

1 Analyze ***List the knowns and the unknown.***

Knowns

- $P_1 = 3.5$ atm
- $S_1 = 0.77$ g/L
- $P_2 = 1.0$ atm
- Henry's law: $\frac{S_1}{P_1} = \frac{S_2}{P_2}$

Unknown

- $S_2 = ?$ g/L

2 Calculate ***Solve for the unknown.***

Solve Henry's law for S_2. Substitute the known values and calculate.

$$S_2 = \frac{S_1 \times P_2}{P_1} = \frac{0.77 \text{ g/L} \times 1.0 \cancel{\text{atm}}}{3.5 \cancel{\text{atm}}} = 0.22 \text{ g/L}$$

3 Evaluate ***Does the result make sense?***

The new pressure is approximately one-third of the original pressure, so the new solubility should be approximately one-third of the original. The answer is correctly expressed to two significant figures.

Practice Problems

1. The solubility of a gas in water is 0.16 g/L at 104 kPa. What is the solubility when the pressure of the gas is increased to 288 kPa? Assume the temperature remains constant.
2. A gas has a solubility in water at 0°C of 3.6 g/L at a pressure of 1.0 atm. What pressure is needed to produce an aqueous solution containing 9.5 g/L of the same gas at 0°C?

Math Handbook

For help with algebraic equations, go to page R69.

Problem-Solving 16.2 Solve Problem 2 with the help of an interactive guided tutorial.

with ChemASAP

16.1 Section Assessment

3. **Key Concept** What determines whether a substance will dissolve? What determines how fast a substance will dissolve?
4. **Key Concept** What units are usually used to express the solubility of a solute?
5. **Key Concept** What are two conditions that determine the mass of solute that will dissolve in a given mass of solvent?
6. What would you do to change
   a. a saturated solid/liquid solution to an unsaturated solution?
   b. a saturated gas/liquid solution to an unsaturated solution?
7. The solubility of a gas is 0.58 g/L at a pressure of 104 kPa. What is its solubility if the pressure increases to 250 kPa at the same temperature?

Writing Activity

Procedure Research how to grow crystals. Then write a stepwise procedure for growing rock candy, which is crystallized sugar, or sucrose ($C_{12}H_{22}O_{11}$).

Assessment 16.1 Test yourself on the concepts in Section 16.1.

with ChemASAP

Section 16.1 Assessment

3. chemical composition of the solute and solvent; agitation, temperature, particle size of the solute
4. grams of solute per 100 g of solvent
5. temperature and pressure (if the solute is a gas)
6. **a.** Add solvent.
   b. Increase the pressure.
7. 1.4 g/L

3 ASSESS

Evaluate Understanding L2

Have students use the concept of solvation to explain how agitation, temperature, and particle size affect the rate of solution formation. Ask, **Which of these factors affect the solubility of a substance?** *(temperature only)* **How is the solubility of a gas affected by the pressure of the gas above the liquid?** *(The solubility of the gas increases with increasing pressure.)*

Reteach L1

Review the factors that affect the aqueous solubilities of solid and gaseous substances. *(temperature and pressure)* Make sure that students are able to distinguish between the factors that affect the solubility and those that affect the rate of solution, such as agitation and particle size. Tell students that solubility is the amount of a substance that will dissolve in a given amount of solvent, whereas rate of solution is the amount of solute entering a solution per unit time. Temperature is the only factor discussed that affects both the rate of solution and the solubility.

Writing Activity

Student procedures should include making a saturated aqueous solution of table sugar, heating it, and suspending a thread or string in the solution. The loosely covered solution is allowed to stand undisturbed for about a week.

If your class subscribes to the Interactive Textbook use it to review key concepts in Section 16.1.

with ChemASAP

Answers to...

Checkpoint The solubility of a gas increases with increasing pressure.

Technology & Society

A Solution for Kidney Failure

Comparing and Contrasting

Both kidneys and hemodialysis "clean" blood by removing waste products using semipermeable membranes for filtering. Kidneys undergo this internal process continually, filtering the body's entire blood supply approximately every 45 minutes. Hemodialysis must be performed three times a week and takes several hours to complete a session.

Kidney Function

As part of their normal functioning, the body's cells continually produce waste products and dump these wastes into the blood. If they are not removed from the blood, the dissolved waste products build up and become toxic to the body. The work of filtering waste products from blood is done by thousands of small nephrons within the kidneys. Blood enters a nephron through a tiny blood vessel. Everything except blood cells and very large molecules can then pass out of the blood through a semipermeable membrane into a long collecting duct. As the fluid travels through this tube, all the necessary components of your blood—water, ions, and other solutes—pass through the walls of the tube and back into the blood. Waste products, however, remain in the tube. By the time the tube exits the nephron, it contains only urine. Blood that leaves the nephron has all its necessary components—minus the waste products in the urine.

Technology & Society

A Solution for Kidney Failure

Your blood transports oxygen and other nutrients to cells throughout your body. It also picks up discarded waste materials from cells and carries them to your kidneys. Kidneys have the important job of filtering potentially toxic materials from blood and excreting them in urine. The body's entire blood supply passes through its two kidneys approximately every 45 minutes. Should the kidneys fail to function, life-threatening poisons would build up in the body. In that case, the treatment is usually hemodialysis, a procedure for cleansing the blood outside the body. **Comparing and Contrasting** ***What are some similarities and differences between the way kidneys work and hemodialysis?***

A A tube connects a patient to a dialysis machine. The machine pumps blood from a vein, circulates it through a dialyzing bath, and returns it to another vein.

Nephrons Millions of tiny processing units called nephrons filter blood through a network of capillaries. The collecting duct carries toxic materials such as urea to the ureter for excretion.

Facts and Figures

The Makeup of Blood

Blood is a solution, a colloid, and a suspension all at the same time. Blood contains gases, ions, and nutrient molecules in solution. It also contains biologically important particles—protein molecules—in colloidal dispersion, and cells in suspension.

For your body to function properly, solutes in blood must be kept at the right concentrations. Many of your organs participate in this process. The liver and pancreas are responsible for regulating the concentration of sugar in your body, while kidneys keep the concentration of dissolved waste products in your blood at a minimum.

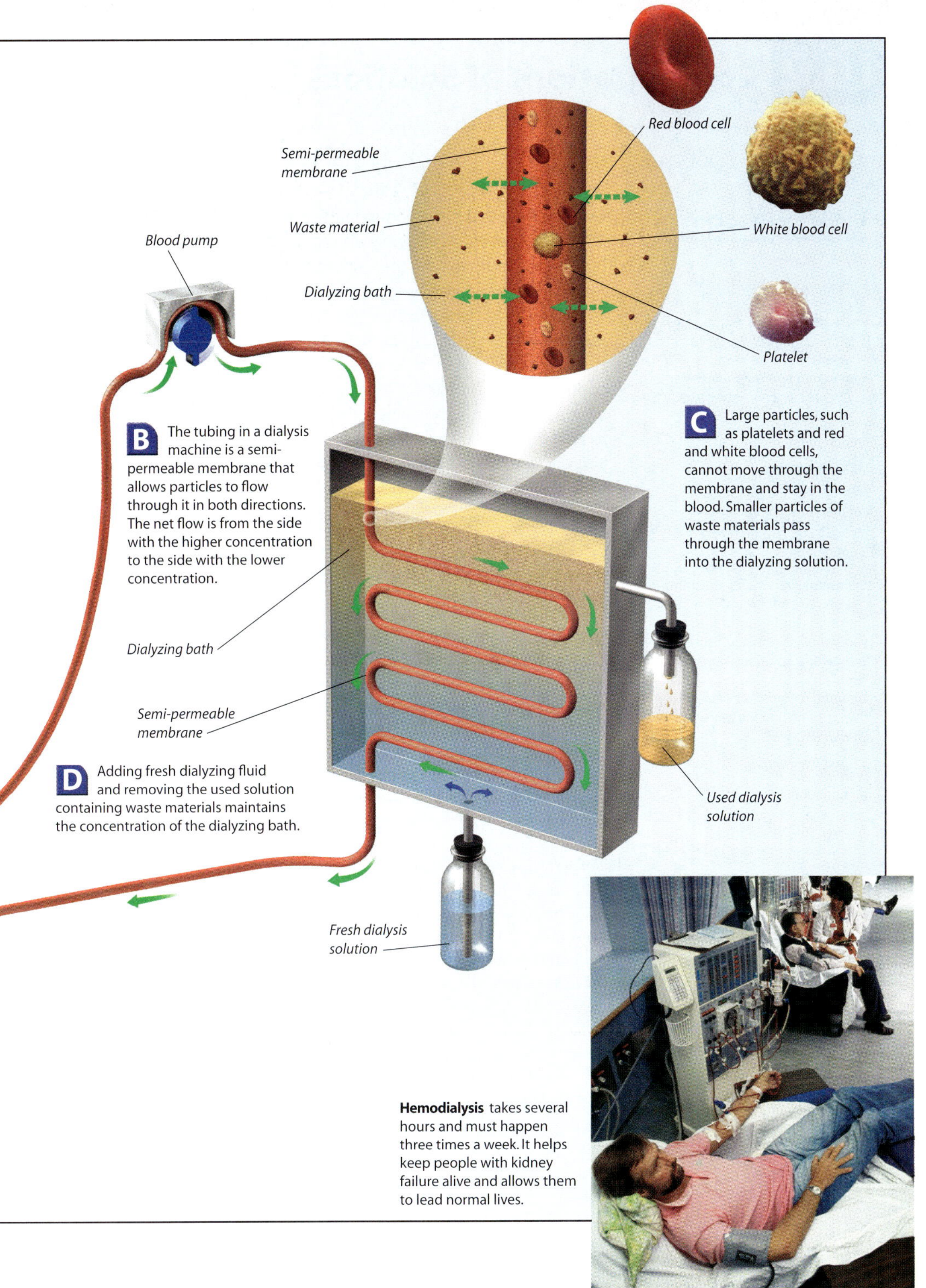

Hemodialysis takes several hours and must happen three times a week. It helps keep people with kidney failure alive and allows them to lead normal lives.

Blood Composition and Health

After students have had a chance to read the article, ask them to name some of the molecular and ionic solutes found in the bloodstream. Create a table on the chalkboard that lists the different components in the bloodstream, including water, proteins, Na^+, K^+, Mg^{2+}, Ca^{2+}, Cl^-, HCO_3^-, SO_4^{2-}, HPO_4^{2-}, O_2, CO_2, red and white blood cells, and platelets. Create additional columns with the headings *concentration, possible causes of elevated levels,* and *possible causes of depressed levels.* Explain that changes in the composition of blood, for example in the concentration of one or more of the ions, can be an indication of a disease or disorder. Doctors frequently use blood tests to check on patients' health. Have students do research to find the concentrations of each of the components in "normal" blood and to find out what types of diseases are linked to changes in the concentrations of certain ions. For example, elevated levels of Na^+ are associated with overactive adrenal glands. Ask students which components of blood are suspended and which are dissolved.

Answers to...

Comparing and Contrasting

Both use a semipermeable membrane to filter waste products from the blood. The process must occur at regular intervals—every 45 minutes in the kidneys and three times a week with hemodialysis. In hemodialysis, the cleansing of the blood takes place outside the body

16.2

1 FOCUS

Objectives

16.2.1 Solve problems involving the molarity of a solution.

16.2.2 Describe the effect of dilution on the total moles of solute in solution.

16.2.3 Define percent by volume and percent by mass solutions.

Guide for Reading

Build Vocabulary L2

Graphic Organizer Students can make a web diagram showing how *dilute solution, concentrated solution,* and *molarity* are related to *concentration.*

Reading Strategy L2

Identify Main Ideas/Details Have students identify the main ideas of this section by listing the key concepts under each heading.

2 INSTRUCT

Connecting to Your World

Have students study the photograph and read the text. Ask, **What substances may be contaminants in drinking water?** *(metals, pesticides, bacteria, water treatment by-products)* **How is the quality of drinking water assessed?** *(Federal and state governments set standards that limit the amount of contaminants allowed in drinking water.)*

16.2 Concentrations of Solutions

Guide for Reading

Key Concepts

- How do you calculate the molarity of a solution?
- What effect does dilution have on the total moles of solute in solution?
- What are two ways to express the percent concentration of a solution?

Vocabulary

concentration
dilute solution
concentrated solution
molarity (*M*)

Reading Strategy

Summarizing When you summarize, you restate the key ideas in your own words. As you read about molarity, making dilutions, and percent solutions, summarize the main ideas in the text. In your summary, be sure to include all the key terms and the sentences in boldfaced type.

Connecting to Your World A supply of clean drinking water is important for all communities. What constitutes clean water? The federal government, along with state governments, has set standards limiting the amount of contaminants allowed in drinking water. These contaminants include metals, pesticides, bacteria, and even the by-products of water treatment. Water must be tested continually to ensure that the concentrations of these contaminants do not exceed established limits. In this section, you will learn how solution concentrations are calculated.

Molarity

You have learned that a substance can dissolve to some extent in a particular solvent to form a solution. In this section, you will learn how to express the actual extent of dissolving, that is, the concentration of a solution. The **concentration** of a solution is a measure of the amount of solute that is dissolved in a given quantity of solvent. A **dilute solution** is one that contains a small amount of solute. By contrast, a **concentrated solution** contains a large amount of solute. An aqueous solution of sodium chloride containing 1 g NaCl per 100 g H_2O might be described as dilute when compared with a sodium chloride solution containing 30 g NaCl per 100 g H_2O. But the same solution might be described as concentrated when compared with a solution containing only 0.01 g NaCl per 100 g H_2O. You can see that the terms *concentrated* and *dilute* are only qualitative descriptions of the amount of a solute in solution.

How can concentration be expressed quantitatively? In chemistry, the most important unit of concentration is molarity. **Molarity (*M*)** is the number of moles of solute dissolved in one liter of solution. **To calculate the molarity of a solution, divide the moles of solute by the volume of the solution.**

$$\text{Molarity } (M) = \frac{\text{moles of solute}}{\text{liters of solution}}$$

Note that the volume involved is the total volume of the resulting solution, not the volume of the solvent alone. Molarity is also known as molar concentration. When the symbol *M* is accompanied by a numerical value, it is read as "molar." For example, a solution labeled 3*M* NaCl is read as three molar sodium chloride solution. Figure 16.8 demonstrates how to make a 0.5-molar solution, that is, a solution with a molarity of 0.5.

Section Resources

Print

- **Guided Reading and Study Workbook,** Section 16.2
- **Core Teaching Resources,** Section 16.2 Review
- **Laboratory Manual,** Lab 32
- **Lab Practical** 16-1
- **Transparencies,** T171–T174

Technology

- **Interactive Textbook with ChemASAP,** Problem Solving 16.8, 16.11, 16.12, 16.14; Assessment, 16.2

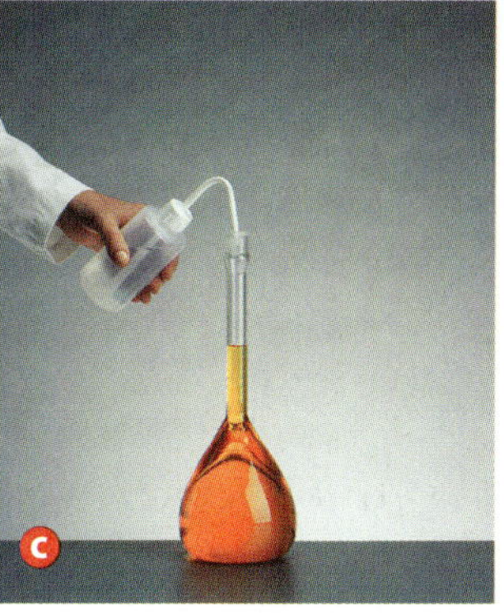

Figure 16.8 The photos show how to make a 0.5-molar (0.5*M*) solution. ⓐ Add 0.5 mol of solute to a 1-L volumetric flask half filled with distilled water. ⓑ Swirl the flask carefully to dissolve the solute. ⓒ Fill the flask with water exactly to the 1-L mark.

SAMPLE PROBLEM 16.2

Calculating the Molarity of a Solution

Intravenous (IV) saline solutions are often administered to patients in the hospital. One saline solution contains 0.90 g NaCl in exactly 100 mL of solution. What is the molarity of the solution?

1 Analyze ***List the knowns and the unknown.***

Knowns

- solution concentration = 0.90 g NaCl/100 mL
- molar mass NaCl = 58.5 g/mol

Unknown

- solution concentration = ?*M*

Convert the concentration from g/100 mL to mol/L. The sequence is g/100 mL → mol/100 mL → mol/L.

2 Calculate ***Solve for the unknown.***

Use the molar mass to convert g NaCl/100 mL to mol NaCl/100 mL. Then use the conversion factor between milliliters and liters to convert to mol/L, which is molarity.

$$\text{solution concentration} = \frac{0.90\ \text{g NaCl}}{100\ \text{mL}} \times \frac{1\ \text{mol NaCl}}{58.5\ \text{g NaCl}} \times \frac{1000\ \text{mL}}{1\ \text{L}}$$
$$= 0.15\ \text{mol/L} = 0.15M$$

3 Evaluate ***Does the result make sense?***

The answer should be less than 1*M* because a concentration of 0.90 g/100 mL is the same as 9.0 g/1000 mL (9.0 g/1 L), and 9.0 g is less than 1 mol NaCl. The answer is correctly expressed to two significant figures.

Practice Problems

8. A solution has a volume of 2.0 L and contains 36.0 g of glucose ($C_6H_{12}O_6$). If the molar mass of glucose is 180 g/mol, what is the molarity of the solution?

9. A solution has a volume of 250 mL and contains 0.70 mol NaCl. What is its molarity?

Math Handbook

For help with dimensional analysis, go to page R66.

Interactive Textbook

Problem-Solving 16.8 Solve Problem 8 with the help of an interactive guided tutorial.

with ChemASAP

Molarity

Discuss

L2

To assess students' knowledge about chemical quantities ask, **How many moles are in each quantity of the following substances: 12.0 g NaCl** (0.205 mol NaCl)**; 53.8 g KNO_3** *(0.532 mol KNO_3)* **Find the mass in grams of each of these amounts of substances: 1.50 mol NaOH** (60.0 g)**; 0.575 mol $NaHCO_3$** (48.3 g) **Which contains more molecules: 1.00 mol SO_2 or 1.00 mol SO_3?** *(numbers are equal)* **Which contains more mass?** *(1.00 mol SO_3)*

Use Visuals

L1

Figure 16.8 Students may think that solutions of given molarity can be correctly prepared by addition of the solute to a premeasured volume of solvent. They often forget that the solute will take up some of the available space in the volumetric flask. Explain that the sequence of steps is important when making up standard solutions because the solute plus the solvent may take up more volume than the solvent alone. Ask, **Why is it important that the solute be added to some of the solvent and dissolved, and then solvent be added up to the 1 L mark on the volumetric flask?** *(The total volume of the mixture is likely to exceed the desired volume.)*

Sample Problem 16.2

Answers

8. 36.0 g/2.0 L solution × 1 mol/180 g = $1.0 \times 10^{-1}M$

9. 0.70 mol/250 mL × 1000 mL/ 1 L = 2.8*M*

Math Handbook

For a math refresher and practice, direct students to dimensional analysis, page R66.

Facts and Figures

IV Solutions

A patient in a hospital is often administered an intravenous (IV) drip containing an aqueous solution. The patient may need the solution to prevent dehydration or to administer nutrients or medicines. Great care must be taken when fluids are introduced into the bloodstream. The balance of dissolved electrolytes inside and outside of the tissue cells could be disturbed. The effects of an imbalance in electrolytes and water can be harmful or could even cause the death of a patient. Aqueous solutions containing 0.85% - 0.9% (mass/mass) of sodium chloride or 5% (mass/mass) of glucose prevent undesired changes in the electrolyte balance.

Sample Problem 16.3

Answers

10. 335 mL × 1L/1000 mL × 0.425 mol/1L = 1.42×10^{-1} mol NH_4NO_3

11. 0.2 mol/L × 1L/1000 mL × 250 mL = 5.0×10^{-1} mol $CaCl_2$; 5.6×10^{1} g $CaCl_2$

Practice Problems Plus L2

How many grams of solute are in 2.40 L of 0.650*M* $HClO_2$? *(107 g)*

Math Handbook

For a math refresher and practice, direct students to conversion problems, page R66.

CLASS Activity

Preparing Solutions L2

Purpose Students learn how to prepare molar solutions.

Materials water; NaCl; sucrose; 50-mL, 100-mL, and 1-L volumetric flasks; balance; graph paper

Procedure Have students determine the mass of a 50-mL, 100-mL, and a 1-L volumetric flask. Review the definition of 1*M* and how to calculate the mass of 1 mol of NaCl (58.5 g). Add 500 mL of water to the 1-L volumetric flask and add 1 mol of NaCl. Swirl the flask until the NaCl dissolves. Add water up to the 1-L mark. Ask, **Does the flask contain a liter of water?** *(No, because the dissolved NaCl takes up some volume.)* Have students prepare 0.0625*M*, 0.125*M*, 0.250*M*, and 0.500*M* solutions of sucrose using the 100-mL and 50-mL volumetric flasks. Have students determine the mass of each solution and calculate the densities of the solutions. *[Density = (mass of flask + solution − mass of flask)/mL of solution.]* Have students make a graph of density versus concentration; include the density of pure water (1 g/mL).

Expected Outcome The graph shows that density is directly proportional to concentration.

Sometimes, you may need to determine the number of moles of solute dissolved in a given volume of solution. You can do this if the molarity of the solution is known. For example, how many moles are in 2.00 L of 2.5*M* lithium chloride (LiCl)? Rearrange the formula for molarity to solve for the number of moles.

$$\text{Molarity } M = \frac{\text{moles of solute}}{\text{liters of solution}}$$

$$\text{Moles of solute} = \text{molarity } (M) \times \text{liters of solution } (V)$$

$$\text{Moles of LiCl} = 2.5\,M \times 2.00\text{ L} = \left(\frac{2.5\text{ mol}}{1\text{ \cancel{L}}}\right) \times 2.00\text{ \cancel{L}}$$

$$= 5.0\text{ mol}$$

Thus 2.00 L of 2.5 *M* lithium chloride solution contains 5.0 mol LiCl.

SAMPLE PROBLEM 16.3

Finding the Moles of Solute in a Solution

Household laundry bleach is a dilute aqueous solution of sodium hypochlorite (NaClO). How many moles of solute are present in 1.5 L of 0.70*M* NaClO?

1 Analyze ***List the knowns and the unknown.***

Knowns
- volume of solution = 1.5 L
- solution concentration = 0.70*M* NaClO
- moles solute = $M \times \text{L} = \frac{\text{mol}}{\text{L}} \times \text{L}$

Unknown
- moles solute = ? mol

The conversion is volume of solution ⟶ moles of solute. Molarity has the units mol/L and is a conversion factor between moles of solute and volume of solution. Multiply the given volume by the molarity expressed in mol/L.

2 Calculate ***Solve for the unknown.***

$$\text{moles solute} = \frac{0.70\text{ mol NaClO}}{1\text{ \cancel{L}}} \times 1.5\text{ \cancel{L}}$$

$$= 1.1\text{ mol NaClO}$$

3 Evaluate ***Does the result make sense?***

The answer should be greater than 1 mol but less than 1.5 mol because the solution concentration is less than 0.75 mol/L and the volume is less than 2 L. The answer is correctly expressed to two significant figures.

Practice Problems

10. How many moles of ammonium nitrate are in 335 mL of 0.425*M* NH_4NO_3?

11. How many moles of solute are in 250 mL of 2.0*M* $CaCl_2$? How many grams of $CaCl_2$ is this?

Math Handbook

For help with conversion problems, go to page R66.

Interactive Textbook

Problem-Solving 16.11 Solve Problem 11 with the help of an interactive guided tutorial.

with ChemASAP

Differentiated Instruction

Gifted and Talented L3

Have students imagine they are writing a recipe for someone else to follow who does not know much about chemistry. Have them write out the steps necessary to make a 0.5*M* aqueous solution of KOH and a 0.2*M* aqueous solution of $MgCl_2$. Have them include tips to ensure success.

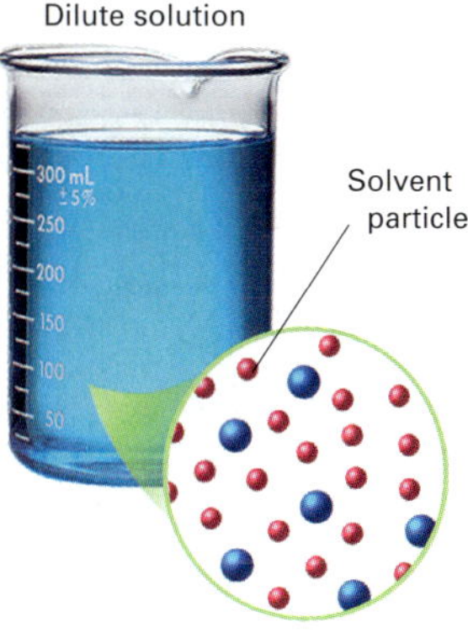

Figure 16.9 Adding solvent to a concentrated solution lowers the concentration, but the total number of moles of solute present remains the same.

Making Dilutions

Both solutions in Figure 16.9 contain the same amount of solute. You can tell by the color of solution A that it is more concentrated than solution B; that is, solution A has the greater molarity. The more dilute solution B was made from solution A by adding more solvent. The procedure for diluting a solution is shown in Figure 16.10. **Diluting a solution reduces the number of moles of solute per unit volume, but the total number of moles of solute in solution does not change.**

Moles of solute before dilution = moles of solute after dilution

Recall the definition of molarity.

$$\text{Molarity}\ (M) = \frac{\text{moles of solute}}{\text{liters of solution}}$$

Rearranging the equation gives an expression for moles of solute.

$$\text{Moles of solute} = \text{molarity}\ (M) \times \text{liters of solution}\ (V)$$

The total number of moles of solute remains unchanged upon dilution, so you can write this equation.

$$\text{Moles of solute} = M_1 \times V_1 = M_2 \times V_2$$

M_1 and V_1 are the molarity and volume of the initial solution, and M_2 and V_2 are the molarity and volume of the diluted solution. Volumes can be in liters or milliliters, as long as the same units are used for both V_1 and V_2.

Figure 16.10 The student is preparing 100 mL of 0.40*M* $MgSO_4$ from a stock solution of 2.0*M* $MgSO_4$. (a) She measures 20 mL of the stock solution with a 20-mL pipet. (b) She transfers the 20 mL to a 100-mL volumetric flask. (c) She carefully adds water to the mark to make 100 mL of solution. **Inferring** ***How many significant figures does the new molarity have?***

Differentiated Instruction

Less Proficient Readers L1

Have students work in groups of three to explore dilution. Have each group bring in a shoe box and put 20 red beads into it. These red beads represent solute molecules. Then add 40 green beads. These beads represent solvent molecules. Students should see that the box contains twice as many green beads as red beads; thus, the concentration is 0.5 red/green. Now add another 40 green beads. This is analogous to a dilution from 0.5 to 0.25 red/green. Although the concentration of red beads has been halved, the total number of red beads has not changed. If the 20 red beads represent moles of solute, 20 moles of solute would still be in solution.

Making Dilutions

Discuss

L2

Remind students that when substances react, they do so as particles and that moles of different substances contain the same number of representative particles. Because many reactions occur in solution, a unit of concentration based on moles (such as molarity) is needed to provide information about the number of reactant particles in solution.

CLASS Activity

Solution Calculations

L2

Purpose Students practice calculating the variables in the definition of molarity.

Materials calculator, paper, pencil

Procedure Divide students into groups of three. One student should name a solute; the other two should each specify one of three quantities: *solute mass*, *solution volume*, or *molarity*. The unspecified quantity should then be calculated with input from each member. For example, the first student may say "sodium chloride," the second may say "12 grams," and the third may say "2.5 liters." The group would then have to calculate the molarity of the resulting solution. Each group should carry out at least one calculation involving each of the possible unknown quantities.

Answers to...

Figure 16.10 two

Sample Problem 16.4

Answers

12. $V_1 = M_2 \times V_2 / M_1 = 0.760M \times 250.0\ \text{mL} / 4.00M = 4.75 \times 10^1\ \text{mL}$

13. Use a pipet to transfer 5.0×10^1 mL of the 1.0*M* solution to a 250-ml volumetric flask. Then add distilled water up to the mark.

Practice Problems Plus  L2

What volume of 12.00*M* sulfuric acid is required to prepare 1.00 L of 0.400*M* sulfuric acid? *(33.3 mL)*

Math Handbook

For a math refresher and practice, direct students to algebraic equations, page R69.

Discuss  L2

If students are having difficulty with the concepts of molarity and dilution, have them consider diluting a container of juice concentrate (frozen or liquid). No matter how much water is added to the concentrate, the initial quantity of concentrate does not change. What changes is the amount of concentrate in a given volume of the diluted solution. Then have students consider using only a fraction of the original concentrate. For example, a gallon of juice prepared from half a can will contain half as much concentrate as a gallon prepared from the entire can. Thus, the molarity would be cut in half.

Discuss  L2

Write the dilution equation on the board. ($M_1 \times V_1 = M_2 \times V_2$) Show students that when mol/*V* is substituted for *M*, the volumes cancel out, leaving moles. Emphasize that, when diluting a solution, the number of moles you start with is the same as the number of moles you end with. Work through several examples using the dilution equation. Ask, **How many milliliters of a 6.00*M* stock solution of $NH_3(aq)$ would you need to prepare 100.0 mL of 0.30*M* $NH_3(aq)$?** *(5.0 mL)*

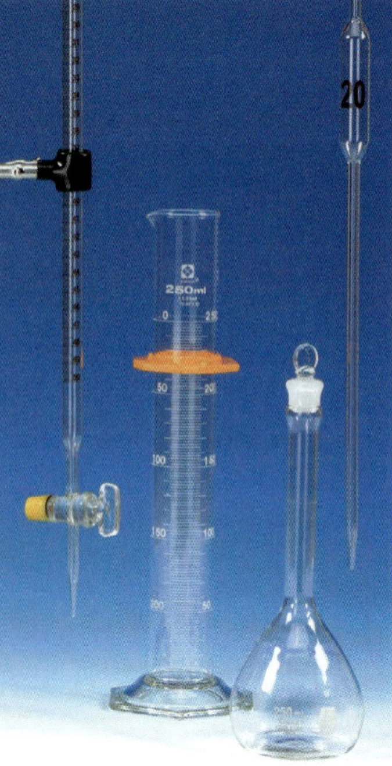

Figure 16.11 The photo shows a buret, a graduated cylinder, a volumetric flask, and a volumetric pipet. These are just some examples of volume-measuring devices.

Math Handbook

For help with algebraic equations, go to p. R69.

Problem-Solving 16.12 Solve Problem 12 with the help of an interactive guided tutorial.

with ChemASAP

SAMPLE PROBLEM 16.4

Preparing a Dilute Solution

How many milliliters of aqueous 2.00*M* $MgSO_4$ solution must be diluted with water to prepare 100.0 mL of aqueous 0.400*M* $MgSO_4$?

1 Analyze ***List the knowns and the unknown.***

Knowns

- $M_1 = 2.00M\ MgSO_4$
- $M_2 = 0.400M\ MgSO_4$
- $V_2 = 100.0\ \text{mL of}\ 0.400M\ MgSO_4$
- $M_1 \times V_1 = M_2 \times V_2$

Unknown

- $V_1 = ?\ \text{mL of}\ 2.00M\ MgSO_4$

2 Calculate ***Solve for the unknown.***

Solving for V_1 yields:

$$V_1 = \frac{M_2 \times V_2}{M_1} = \frac{0.400\cancel{M} \times 100.0\ \text{mL}}{2.00\cancel{M}}$$

$$= 20.0\ \text{mL}$$

Thus 20.0 mL of the initial solution must be diluted by adding enough water to increase the volume to 100.0 mL.

3 Evaluate ***Does the result make sense?***

The concentration of the initial solution is five times larger than the concentration of the diluted solution. Because the moles of solute in each solution are the same, the volume of the solution to be diluted (20.0 mL) should be one-fifth the final volume of the diluted solution. The answer is correctly expressed to three significant figures.

Practice Problems

12. How many milliliters of a solution of 4.00*M* KI are needed to prepare 250.0 mL of 0.760*M* KI?

13. How could you prepare 250 mL of 0.20*M* NaCl using only a solution of 1.0*M* NaCl and water?

Which of the volume-measuring devices shown in Figure 16.11 should you use to make a dilution? Your choice depends upon how precise you want the concentration of the solution to be. The dilution described in Sample Problem 16.4 requires a molarity with three significant figures. For this you would need to measure 20.0 mL of the 2.00*M* $MgSO_4$ solution with a 20-mL volumetric pipet or a buret. A graduated cylinder could not provide the required precision. You would transfer the solution to a 100-mL volumetric flask and add distilled water to the flask exactly up to the etched line. The contents would then be exactly 100.0 mL of 0.400*M* $MgSO_4$. For measurements that require less precision, graduated cylinders are appropriate for measuring volumes.

Percent Solutions

Another way to describe the concentration of a solution is by the percent of a solute in the solvent. **The concentration of a solution in percent can be expressed in two ways: as the ratio of the volume of the solute to the volume of the solution or as the ratio of the mass of the solute to the mass of the solution.**

Concentration in Percent (Volume/Volume) If both the solute and the solvent are liquids, a convenient way to make a solution is to measure the volumes of the solute and the solution. The concentration of the solute is then expressed as a percent of the solution by volume. For example, isopropyl alcohol (2-propanol) is sold as a 91% solution, as shown in Figure 16.12. This solution consists of 91 mL of isopropyl alcohol mixed with enough water to make 100 mL of solution. The concentration can be expressed as 91 percent (volume/volume), or 91% (v/v). The relationship between percent by volume and the volumes of solute and solution is

$$\text{Percent by volume } (\%\ (v/v)) = \frac{\text{volume of solute}}{\text{volume of solution}} \times 100\%$$

Figure 16.12 The label clearly distinguishes this solution of isopropyl alcohol from rubbing alcohol which is a 70% solution of isopropyl alcohol. **Applying Concepts** ***How many milliliters of isopropyl alcohol are in 100 mL of 91% alcohol?***

SAMPLE PROBLEM 16.5

Calculating Percent (Volume/Volume)

What is the percent by volume of ethanol (C_2H_6O, or ethyl alcohol) in the final solution when 85 mL of ethanol is diluted to a volume of 250 mL with water?

1 Analyze ***List the knowns and the unknown.***

Knowns
- volume of ethanol = 85 mL
- volume of solution = 250 mL
- $\text{Percent by volume } (\%\ (v/v)) = \frac{\text{volume of solute}}{\text{volume of solution}} \times 100\%$

Unknown
- % ethanol (v/v) = ? %

2 Calculate ***Solve for the unknown.***

Substitute the known values into the equation and solve.

$$\%\ (v/v) = \frac{85 \text{ mL ethanol}}{250 \text{ mL}} \times 100\%$$
$$= 34\% \text{ ethanol } (v/v)$$

3 Evaluate ***Does the result make sense?***

The volume of the solute is about one-third the volume of the solution, so the answer is reasonable. The answer is correctly expressed to two significant figures.

Practice Problems

14. If 10 mL of propanone (or acetone (C_3H_6O)) is diluted with water to a total solution volume of 200 mL, what is the percent by volume of propanone in the solution?

15. A bottle of the antiseptic hydrogen peroxide (H_2O_2) is labeled 3.0% (v/v). How many mL H_2O_2 are in a 400.0-mL bottle of this solution?

For help with percents, go to page R72.

Interactive Textbook

Problem-Solving 16.15 Solve Problem 15 with the help of an interactive guided tutorial.

with ChemASAP

Percent Solutions

Sample Problem 16.5

Answers

14. %(v/v) = 10 mL propanone/200 mL × 100% = 5.0 % v/v

15. mL H_2O_2 = 400.0 mL × 3% / 100% = 1.2×10^1 mL

Practice Problems Plus 

Ethanol is mixed with gasoline to make gasohol. **What is the percent by volume of ethanol in gasohol when 95 mL of ethanol is added to sufficient gasoline to make 1.0 L of gasohol?** *(9.5% (v/v))*

Math Handbook

For a math refresher and practice, direct students to percents, page R72.

TEACHER Demo

Serial Dilutions L2

Purpose Students learn how to prepare standard solutions efficiently.

Materials four 20-mL test tubes, 10 mL of 0.5*M* surcose, 15 mL water, 10-mL pipet.

Procedure Remind students of the activity in which they prepared molar solutions of sucrose. Each solution was prepared independently. Ask, **Is there a more efficient way to prepare this same set of molar solutions?** *(Yes; make serial dilutions.)* Place 10 mL of 0.5*M* sucrose in one test tube and 5 mL water in each of three others. Using a 10-mL pipet, transfer 5 mL of the 0.5*M* solution to the second test tube; mix by pipeting up and down. Repeat this process for the third and fourth test tubes. Have students calculate the molarities of each of the resulting solutions. Tell students that serial dilutions are used to prepare extremely dilute solutions, such as 0.000001*M* sucrose. The procedure eliminates the error that would arise from having to measure out very small volumes of stock solutions.

Facts and Figures

Calculating Percent by Volume

Percent by volume is easy to calculate because there is no need to calculate molar masses. However, because a volume change can occur when two liquids mix, a percent by volume can be misleading. When 20 mL of ethanol is mixed with 80 mL of water, the resulting volume is less than 100 mL; more water must be added.

Answers to...

Figure 16.12 91 mL

Section 16.2 (continued)

3 ASSESS

Evaluate Understanding L2

Ask students to compare percent by mass and percent by volume with molarity as ways of expressing the concentration of a solution. Have students write out a numbered list of steps they would follow to prepare a 1*M* aqueous solution of KCl.

Reteach L1

Remind students that when making dilute solutions, the accuracy of the resulting solution's concentration is dependent upon the measuring devises used. Using a graduated cylinder to measure is much less accurate than using a volumetric pipet or buret. Review how to use all of these measuring devises.

Writing Activity

Students may find it hard to tell which percent method is used on labels. Have students consider whether one or the other method might present the product in a more favorable light. Even the name "fruit drink" can be misleading.

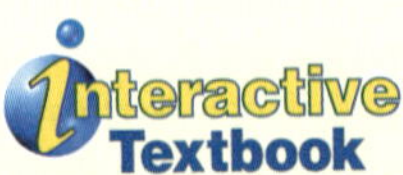

If your class subscribes to the Interactive Textbook, use it to review key concepts in Section 16.2.

with ChemASAP

Concentration in Percent (Mass/Mass) Another way to express the concentration of a solution is as a percent (mass/mass), which is the number of grams of solute in 100 grams of solution. Percent by mass is sometimes a convenient unit of concentration when the solute is a solid. For example, a solution containing 7 g of sodium chloride in 100 grams of solution is 7 percent (mass/mass), or 7% (m/m).

$$\text{Percent by mass (\% (m/m))} = \frac{\text{mass of solute}}{\text{mass of solution}} \times 100\%$$

Suppose you want to make 2000 g of a solution of glucose in water that has a 2.8% (m/m) concentration of glucose. How much glucose should you use? In a 2.8 percent solution, each 100 g of solution contains 2.8 g of solute. Thus, to find the grams of solute, you can multiply the mass of the solution by the ratio of grams of solute to grams of solution.

$$2000 \text{ g } \cancel{\text{solution}} \times \frac{2.8 \text{ g glucose}}{100 \text{ g } \cancel{\text{solution}}} = 56 \text{ g glucose}$$

How much solvent (water) should be used? The mass of the solvent equals the mass of the solution minus the mass of the solute or 1944 g (2000 g − 56 g). Thus a 2.8% (m/m) glucose solution contains 56 g of glucose dissolved in 1944 g of water.

Information is often expressed as percent composition on food labels. For example, the label on a fruit drink or on maple-flavored pancake syrup should show the percent of fruit juice or maple syrup contained in the product. Such information can be misleading unless the units are given. When you use percentages to express concentration, be sure to state the units (v/v) or (m/m).

16.2 Section Assessment

16. **Key Concept** How do you calculate the molarity of a solution?
17. **Key Concept** Compare the number of moles of solute before dilution with the number of moles of solute after dilution.
18. **Key Concept** What are two ways of expressing the concentration of a solution as a percent?
19. Calculate the molarity of a solution containing 400 g $CuSO_4$ in 4.00 L of solution.
20. How many moles of solute are present in 50.0 mL of 0.20*M* KNO_3?
21. How many milliliters of a stock solution of 2.00*M* KNO_3 would you need to prepare 100.0 mL of 0.150*M* KNO_3?
22. What is the concentration, in percent (v/v), of a solution containing 50 mL of diethyl ether ($C_4H_{10}O$) in 2.5 L of solution?
23. How many grams of K_2SO_4 would you need to prepare 1500 g of 5.0% K_2SO_4 (m/m) solution?

Writing Activity

Report Write a report comparing the concentration of fruit juice in different types of juice drinks. Is the percent of fruit juice given by % (v/v), or % (m/m)? Explain why these concentrations can be confusing or misleading.

Assessment 16.2 Test yourself on the concepts in Section 16.2.

with ChemASAP

Section 16.2 Assessment

16. If the number of moles and the volume of a solution is known, its molarity is determined by dividing the moles of solute by the volume of the solution.
17. Diluting a solution reduces the number of moles of solute per unit volume, but the total number of moles of solute in solution does not change.
18. Express the concentration as the ratio of the volume of the solute to the volume of the solution (v/v) or as the ratio of the mass of the solute to the mass of the solution (m/m)
19. $6.27 \times 10^{-1} M$ $CuSO_4$
20. 1.00×10^{-2} mol KNO_3
21. 7.50 mL
22. 2.0% (v/v) diethyl ether
23. 7.5×10^{1} g K_2SO_4

16.3 Colligative Properties of Solutions

Connecting to Your World

The wood frog is a remarkable creature because it can survive being frozen. Scientists believe that a substance in the cells of this frog acts as a natural antifreeze, which prevents the cells from freezing. Although fluids surrounding the frog's cells may freeze, the cells themselves do not. In this section, you will discover how a solute can change the freezing point of a solution.

Guide for Reading

Key Concepts

- What are three colligative properties of solutions?
- What factor determines the amount by which a solution's vapor pressure, freezing point, and boiling point differ from those properties of the solvent?

Vocabulary

colligative property
freezing-point depression
boiling-point elevation

Reading Strategy

Relating Text and Visuals As you read, look carefully at Figure 16.14. In your notebook, explain how this visual explains why equal molar solutions of different substances can have different freezing point depressions and boiling point elevations.

Vapor-Pressure Lowering

You probably won't be surprised to learn that the physical properties of a solution differ from those of the pure solvent used to make the solution. After all, tea is not the same as pure water. But it might surprise you to learn that some of these differences in properties have little to do with the specific identity of the solute. Instead, they depend upon the number of solute particles in the solution. A property that depends only upon the number of solute particles, and not upon their identity is called a **colligative property.** **Three important colligative properties of solutions are vapor-pressure lowering, boiling-point elevation, and freezing-point depression.**

Recall that vapor pressure is the pressure exerted by a vapor that is in dynamic equilibrium with its liquid in a closed system. A solution that contains a solute that is nonvolatile (not easily vaporized) always has a lower vapor pressure than the pure solvent, as shown in Figure 16.13. Glucose, a molecular compound, and sodium chloride, an ionic compound, are examples of nonvolatile solutes. When glucose or sodium chloride is dissolved in a solvent, the vapor pressure of the solution is lower than the vapor pressure of the pure solvent. Why is this true?

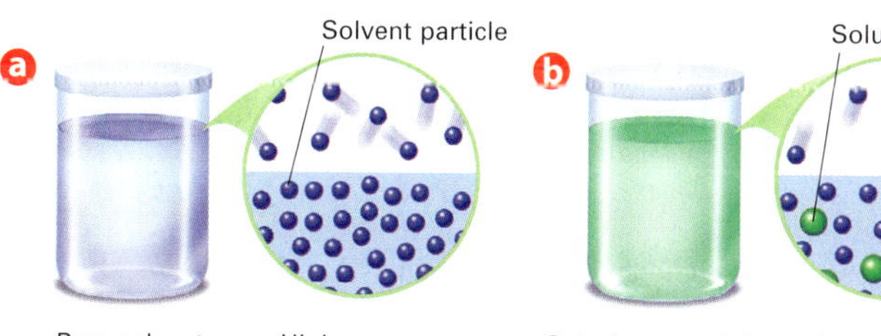
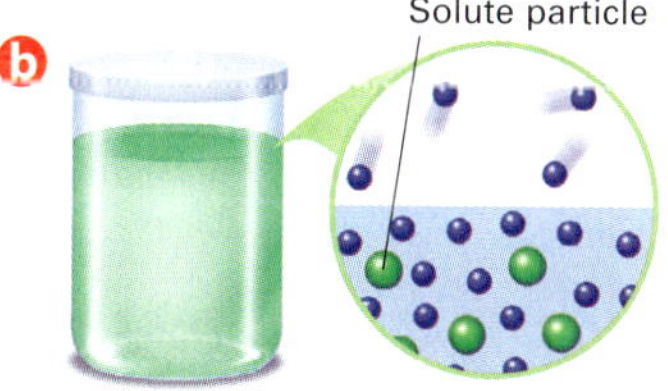

Figure 16.13 The vapor pressure of a solution of a nonvolatile solute is less than the vapor pressure of a pure solvent. **a** In a pure solvent, equilibrium is established between the liquid and the vapor. **b** In a solution, solute particles reduce the number of free solvent particles able to escape the liquid. Equilibrium is established at a lower vapor pressure. **Interpreting Diagrams** ***How is decreased vapor pressure represented in the diagram?***

16.3

1 FOCUS

Objectives

16.3.1 Identify three colligative properties of solutions.

16.3.2 Explain why the vapor pressure, freezing point, and boiling point of a solution differ from those properties of the pure solvent.

Guide for Reading

Build Vocabulary L2

Word Parts The adjective colligative comes from the Latin *colligare* meaning "to tie." A colligative property is one in which particles are seemingly tied together in their actions and act as a group rather than as individuals.

Reading Strategy L2

Use Prior Knowledge Relate the section to prior knowledge by having students write the chemical equations for the following examples:

- hydrochloric acid in water ($HCl(g) \rightarrow H^+(aq) + Cl^-(aq)$)
- magnesium chloride in water ($MgCl_2(s) \rightarrow Mg^{2+}(aq) + 2Cl^-(aq)$)
- Glucose in water ($C_6H_{12}O_6(s) \rightarrow C_6H_{12}O_6(aq)$)

2 INSTRUCT

Connecting to Your World

Ask, **What is a possible explanation for the wood frog's ability to survive being frozen?** *(The frog's cells contain a natural antifreeze.)* Mention to students that some fish and reindeer also benefit from so-called "natural antifreeze." When water freezes, the water molecules form a crystalline lattice work. **How might a solute change the freezing point of a solid?** *(It might slow down the formation of the crystal lattice.)*

Answers to...

Figure 16.13 by fewer solvent particles in the space above the liquid

Section Resources

Print

- **Guided Reading and Study Workbook,** Section 16.3
- **Core Teaching Resources,** Section 16.3 Review
- **Transparencies,** T175–T176

Technology

- **Interactive Textbook with ChemASAP,** Assessment 16.3
- **Go Online,** Section 16.3

Section 16.3 (continued)

Vapor-Pressure Lowering

Discuss L2

Remind students that ionic compounds and certain molecular compounds, such as HCl, produce two or more particles when they dissolve in water. Most molecular compounds, such as glucose, do not dissociate when they dissolve in water. For each formula unit of $MgCl_2$ that dissolves, three particles are formed in solution. Ask, **How many particles are formed when $FeCl_3$ dissolves in water?** *(4)*

Freezing-Point Depression

Use Visuals L2

Figure 16.14 Emphasize that colligative properties do not depend on the kind of particles, but on their concentration. For colligative properties, a mole of one kind of particle has the same effect as a mole of any other kind of particle. Ask, **Which produces a greater change in colligative properties—an ionic solid or a molecular solid?** *(An ionic solid produces a greater change because it will produce two or more moles of ions for every mole of solid that dissolves.)*

Download a worksheeet on **Solutions** for students to complete, and find additional teacher support from NSTA SciLinks.

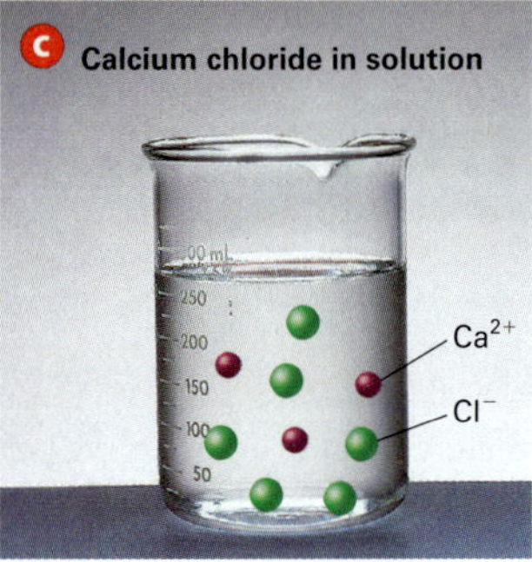

Figure 16.14 Particle concentrations differ for dissolved covalent and ionic compounds in water. **a** Three moles of glucose dissolved in water produce 3 mol of particles because glucose does not dissociate. **b** Three moles of sodium chloride dissolved in water produce 6 mol of particles because each formula unit of NaCl dissociates into two ions. **c** Three moles of calcium chloride dissolved in water produce 9 mol of particles because each formula unit of $CaCl_2$ dissociates into three ions.

In an aqueous solution of sodium chloride, sodium ions and chloride ions are dispersed throughout the liquid water. Both within the liquid and at the surface, the ions are surrounded by layers of associated water molecules, or shells of water of solvation. The formation of these shells of water of solvation reduces the number of solvent molecules that have enough kinetic energy to escape as vapor. Thus, the solution has a lower vapor pressure than the pure solvent (water) would have at the same temperature.

Ionic solutes that dissociate, such as sodium chloride and calcium chloride, have greater effects on the vapor pressure than does a nondissociating solute such as glucose. Recall that each formula unit of the ionic compound sodium chloride (NaCl) produces two particles in solution, a sodium ion and a chloride ion. Each formula unit of calcium chloride ($CaCl_2$) produces three particles, a calcium ion and two chloride ions. When glucose dissolves, the molecules do not dissociate. Figure 16.14 compares the number of particles in three solutions of the same concentration. The vapor-pressure lowering caused by 0.1 mol of sodium chloride in 1000 g of water is twice that caused by 0.1 mol of glucose in the same quantity of water. In the same way, 0.1 mol $CaCl_2$ in 1000 g of water produces three times the vapor-pressure lowering as 0.1 mol of glucose in the same quantity of water. **The decrease in a solution's vapor pressure is proportional to the number of particles the solute makes in solution.**

Which compound affects the vapor pressure of a solution the least: glucose, sodium chloride, or calcium chloride?

Freezing-Point Depression

When a substance freezes, the particles of the solid take on an orderly pattern. The presence of a solute in water disrupts the formation of this pattern because of the shells of water of solvation. As a result, more kinetic energy must be withdrawn from a solution than from the pure solvent to cause the solution to solidify. The freezing point of a solution is lower than the freezing point of the pure solvent. The difference in temperature between the freezing point of a solution and the freezing point of the pure solvent is the **freezing-point depression.**

For: Links on Solutions
Visit: www.SciLinks.org
Web Code: cdn-1163

488 Chapter 16

Facts and Figures

Osmotic Pressure

Another colligative property, which is important in many biological processes, is osmotic pressure: the tendency for solvent to flow through a semipermeable membrane from a region of low solute concentration to a region of higher solute concentration. Osmotic pressure helps regulate the movement of fluids across cell membranes. Cells remain healthy as long as they are continually bathed in an isotonic medium, which maintains the correct osmotic balance between the inside and outside of the cell membrane. Physicians make use of the osmotic effect to treat patients with kidney failure. Afflicted individuals undergo dialysis, a process in which the blood is circulated through a machine equipped with a semipermeable membrane. In the machine, osmotic pressure is used to separate waste materials from the blood.

Quick LAB

Solutions and Colloids

Purpose
To classify mixtures as solutions or colloids using the Tyndall effect.

Materials
- sodium hydrogen carbonate
- cornstarch
- stirring rod
- distilled water (or tap water)
- flashlight
- masking tape
- 3 jars with parallel sides
- teaspoon
- cup

Procedure  
1. In a cup, make a paste by mixing $\frac{1}{2}$ teaspoon cornstarch with 4 teaspoons water.
2. Fill one jar with water. Add one half teaspoon sodium hydrogen carbonate to a second jar and fill with water. Stir to mix. Add the cornstarch paste to the third jar and fill with water. Stir to mix.
3. Turn out the lights in the room. Shine the beam of light from the flashlight at each of the jars and record your observations.

Analyze and Conclude
1. In which of the jars was it possible to see the path of the beam of light?
2. What made the light beam visible?
3. If a system that made the light beam visible were filtered, would the light beam be visible in the filtrate? Explain.
4. Predict what you would observe if you were to replace the sodium hydrogen carbonate with sucrose (cane sugar) or sodium chloride (table salt).
5. Predict what you would observe if you were to replace the cornstarch with flour or diluted milk.
6. Explain how you could use this method to distinguish a colloid from a suspension.

Freezing-point depression is another colligative property. **The magnitude of the freezing-point depression is proportional to the number of solute particles dissolved in the solvent and does not depend upon their identity.** The addition of 1 mol of solute particles to 1000 g of water lowers the freezing point by 1.86°C. For example, if you add 1 mol (180 g) of glucose to 1000 g of water, the solution freezes at –1.86°C. However, if you add 1 mol (58.5 g) of sodium chloride to 1000 g of water, the solution freezes at –3.72°C, double the change for glucose. This is because 1 mol NaCl produces 2 mol of particles, and thus doubles the freezing-point depression. The freezing-point depression of aqueous solutions makes walks and driveways safer when people, like the man in Figure 16.15, sprinkle salt on icy surfaces to make the ice melt. The melted ice forms a solution with a lower freezing point than that of pure water. Similarly, ethylene glycol ($C_2H_6O_2$, antifreeze) is added to the water in automobile cooling systems to depress the freezing point of the water below 0°C. Automobiles can thus withstand subfreezing temperatures without freezing up.

Figure 16.15 Surfaces can be free of ice even at temperatures below freezing if salt is applied. **Inferring** ***Why would calcium chloride ($CaCl_2$) be a better salt for this purpose than sodium chloride (NaCl)?***

Quick LAB

Solutions and Colloids 

Objective After completing this activity, students will be able to
- Classify mixtures as solutions or colloids using the Tyndall effect.

Skills Focus observing, inferring, predicting

Prep Time 30 minutes
Class Time 20 minutes

Expected Outcome The beam of light is visible in the cornstarch/water mixture but not in water or sodium hydrogen carbonate solution.

Analyze and Conclude
1. undissolved cornstarch
2. Cornstarch particles reflect light.
3. yes for a colloid; no for a suspension
4. Because two solutions form, the beam of light would not be visible.
5. Because flour and milk do not dissolve, the beam of light will be visible.
6. Cannot; both reflect light.

For Enrichment 
Have students test and classify various household mixtures using the Tyndall effect.

CLASS Activity

Freezing Point Depression L1

Purpose To observe the freezing point depression of ice by the addition of rock salt (NaCl) to a slurry of ice-water

Materials thermometer, foam cup, rock salt, water, and ice.

Procedure Have students make an ice-water slurry in their cups. Ask them to measure the initial temperature of the ice-water slurry and the lowest temperature reached after the addition of the rock salt. Explain that the observed difference in temperature is the freezing point depression for this solution of NaCl in water.

Answers to...
Figure 16.15 $CaCl_2$ produces three ions in solution; NaCl produces two.

 glucose

Section 16.2 (continued)

Boiling-Point Elevation

3 ASSESS

Evaluate Understanding L2

Have students explain how the addition of solute particles to a solvent causes the vapor pressure and freezing point of the solution to decrease relative to the pure solvent and the boiling point to increase.

Reteach L1

Remind students that colligative properties are those physical properties of solutions that depend only on the number of particles of solute in solution and not on the chemical composition of the particles. The magnitudes of the observed physical changes are proportional to the quantity of solute particles in solution.

Connecting Concepts

Volatile solutes quickly evaporate at higher temperatures and so would not be present to cause the elevation of the solvent's boiling point.

If your class subscribes to the Interactive Textbook, use it to review key concepts in Section 16.3 with ChemASAP

Figure 16.16 Cooling the boiling mixture, when the temperature on the candy thermometer reaches a certain value, gives you solid fudge.

Boiling-Point Elevation

The boiling point of a substance is the temperature at which the vapor pressure of the liquid phase equals atmospheric pressure. As you just learned, adding a nonvolatile solute to a liquid solvent decreases the vapor pressure of the solvent. Because of the decrease in vapor pressure, additional kinetic energy must be added to raise the vapor pressure of the liquid phase of the solution to atmospheric pressure and initiate boiling. Thus the boiling point of a solution is higher than the boiling point of the pure solvent. The difference in temperature between the boiling point of a solution and the boiling point of the pure solvent is the **boiling-point elevation.** The same antifreeze, added to automobile engines to prevent freeze-ups in winter, protects the engine from boiling over in summer.

Remember that boiling-point elevation is a colligative property; it depends on the concentration of particles, not on their identity. Thus you can think about boiling-point elevation in terms of particles. It takes additional kinetic energy for the solvent particles to overcome the attractive forces that keep them in the liquid. Thus the presence of a solute elevates the boiling point of the solvent. **The magnitude of the boiling-point elevation is proportional to the number of solute particles dissolved in the solvent.** The boiling point of water increases by 0.512°C for every mole of particles that the solute forms when dissolved in 1000 g of water.

Apparent applications of boiling point elevation are sometimes erroneous or misleading. For example, to make fudge, a lot of sugar and some flavorings are mixed with water and the solution is boiled, as shown in Figure 16.16. As the water slowly boils away, the concentration of sugar in the solution increases. This may appear to be a case of boiling point elevation caused by an increased concentration of the solute, sugar. The real reason for the boiling point elevation is that the sugar decomposes into different substances as it is heated and this new mixture has a higher boiling point. The composition of the mixture will, when it is cooled, give you a delicious tasting treat.

16.3 Section Assessment

24. **Key Concept** What are three colligative properties of solutions?
25. **Key Concept** What factor determines how much the vapor pressure, freezing point, and boiling point of a solution differ from those properties of the pure solvent?
26. Would a dilute or a concentrated sodium fluoride solution have a higher boiling point? Explain.
27. An equal number of moles of KI and MgI_2 are dissolved in equal volumes of water. Which solution has the higher
    a. boiling point?
    b. vapor pressure?
    c. freezing point?
28. Explain why the vapor pressure, boiling point, and freezing point of an aqueous solution of a nonvolatile solute are not the same as those of the pure solvent.

Connecting Concepts

Vapor Pressure Review what you learned in Section 13.2 about the relationship between the vapor pressure of liquids and their boiling points. Explain why only nonvolatile solutes cause the elevation of the solvent's boiling point.

Assessment 16.3 Test yourself on the concepts in Section 16.3. with ChemASAP

Section 16.3 Assessment

24. vapor-pressure lowering, boiling-point elevation, and freezing-point depression

25. the number of solute particles dissolved in the solvent

26. concentrated sodium fluoride, because the magnitude of the boiling-point elevation is proportional to the number of solute particles dissolved in the solvent

27. a. MgI_2 solution **b.** KI solution **c.** KI solution

28. Formation of solvation shells around solute particles reduces the number of water molecules with sufficient kinetic energy to escape the solution. Therefore, the vapor pressure is lower relative to pure solvent. More energy must be supplied to reach the boiling point; therefore the boiling point is elevated relative to pure solvent. Solvation shells interfere with the formation of hydrogen-bonded ice structure. This results in depression of the freezing point relative to the pure solvent.

16.4 Calculations Involving Colligative Properties

Connecting to Your World Cooking instructions for a wide variety of foods, from dried pasta to packaged beans to frozen fruits to fresh vegetables, often call for the addition of a small amount of salt to the cooking water. Most people like the flavor of food cooked with salt. But adding salt can have another effect on the cooking process. Recall that dissolved salt elevates the boiling point of water. Suppose you added a teaspoon of salt to two liters of water. A teaspoon of salt has a mass of about 20 g. Would the resulting boiling point increase be enough to shorten the time required for cooking? In this section, you will learn how to calculate the amount the boiling point of the cooking water would rise.

Guide for Reading

Key Concepts

- What are two ways of expressing the concentration of a solution?
- How are freezing-point depression and boiling-point elevation related to molality?

Vocabulary

molality (m)

mole fraction

molal freezing-point depression constant (K_f)

molal boiling-point elevation constant (K_b)

Reading Strategy

Before you read, make a list of the vocabulary terms above. As you read, write the symbols or formulas that apply to each term and describe the symbols or formulas using words.

Molality and Mole Fraction

Recall that colligative properties depend only upon solute concentration. **The unit molality and mole fractions are two additional ways in which chemists express the concentration of a solution.** The unit **molality (m)** is the number of moles of solute dissolved in 1 kilogram (1000 g) of solvent. Molality is also known as molal concentration.

$$\text{Molality} = \frac{\text{moles of solute}}{\text{kilogram of solvent}}$$

Note that molality is not the same as molarity. Molality refers to moles of solute per kilogram of solvent rather than moles of solute per liter of solution. In the case of water as the solvent, 1 kg or 1000 g equals a volume of 1000 mL, or 1 L.

You can prepare a solution that is 1.00 molal ($1m$) in glucose, for example, by adding 1.00 mol (180 g) of glucose to 1000 g of water. Figure 16.17 shows how a 0.500 molal ($0.500m$) solution in sodium chloride is prepared by dissolving 0.500 mol (29.3 g) of NaCl in 1.000 kg (1000 g) of water.

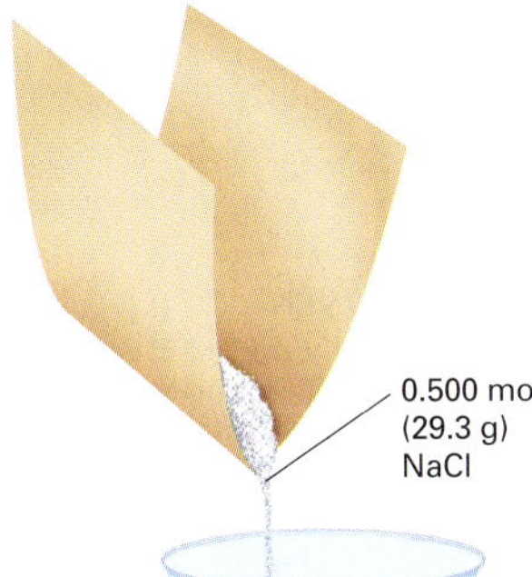

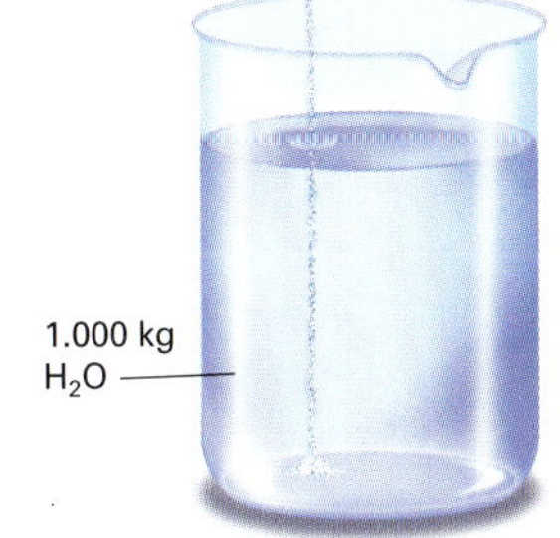

Figure 16.17 To make a $0.500m$ solution of NaCl, use a balance to measure 1.000 kg of water and add 0.500 mol (29.3 g) NaCl. **Calculating** ***What would be the molality if only 0.500 kg of water were used?***

Section Resources

Print

- **Guided Reading and Study Workbook,** Section 16.4
- **Core Teaching Resources,** Section 16.3 Review
- **Transparencies,** T177–T179
- **Laboratory Manual,** Lab 33

Technology

- **Interactive Textbook with ChemASAP,** Problem-Solving 16.29, 16.31, 16.33, 16.36; Simulation 21; Assessment 16.4

16.4

1 FOCUS

Objectives

16.4.1 Solve problems related to the molality and mole fraction of a solution

16.4.2 Describe how freezing-point depression and boiling-point elevation are related to molality.

Guide to Reading

Build Vocabulary  L2

Graphic Organizers Use a chart to organize the definitions and the mathematical formulas associated with each vocabulary term.

Reading Strategy  L2

Preview Have students preview this section by reading the key concepts and skimming the headings, visuals, and boldfaced materials.

2 INSTRUCT

Connecting to Your World

Ask, **What effect does adding salt have on the cooking process?** *(Adding salt increases the boiling point of the cooking water.)* Ask, **Do you think the resulting boiling point increase would be enough to significantly shorten the time required for cooking?** *(Students are likely to predict, correctly, that the change would be negligible.)*

Molality and Mole Fraction

Use Visuals L1

Figure 16.17 Display the figure on an overhead projector. Ask students to write the definition of molality in their notebooks. Show the step-by-step procedure a chemist would use to prepare a $0.500m$ solution of NaCl. Have students confirm your calculations and the data given in the figure.

Answers to...

Figure 16.17 $1.00m$

Sample Problem 16.6

Answers

29. 750 g water × 0.400 mol NaF / 1000 g water × 42.0 g NaF / 1 mol NaF = 1.26×10^{1} g NaF

30. 10.0 g NaCl / 600 g water × 1 mol NaCl / 58.5 g NaCl × 1000 g water/ 1 kg water = $2.85 \times 10^{-1} m$ NaCl

Practice Problems Plus L2

How many grams of lithium bromide must be dissolved in 444 g of water to prepare a 0.140*m* LiBr solution?
(5.40 g LiBr)

Discuss L2

Write the expressions defining molarity and molality on the board. Compare the chemical quantities in each expression. Point out that molarity is denoted by *M* and molality by *m*. Explain that the molality of a solution does not vary with temperature because the mass of the solvent does not change. In contrast, the molarity of a solution does vary with temperature because the liquid can expand and contract. When studying colligative properties such as boiling-point elevation and freezing-point depression, it is preferable to use a concentration that does not depend on temperature.

Math Handbook
For a math refresher and practice, direct students to dimensional analysis, page R66.

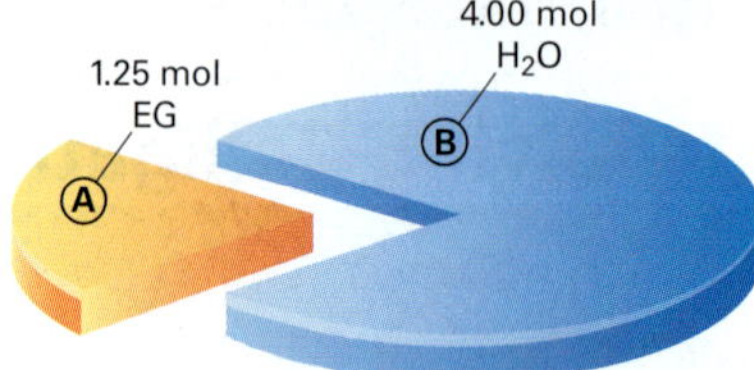

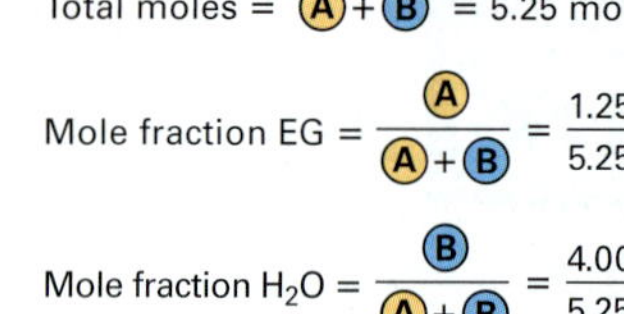

Figure 16.18 Ethylene glycol (EG) is added to water as antifreeze in the proportions shown. A mole fraction is the ratio of the number of moles of one substance to the total number of moles of all substances in the solution. **Inferring** ***What is the sum of all mole fractions in a solution?***

SAMPLE PROBLEM 16.6

Using Solution Molality

How many grams of potassium iodide must be dissolved in 500.0 g of water to produce a 0.060 molal KI solution?

1 Analyze ***List the knowns and the unknown.***

Knowns
- mass of water = 500.0g = 0.5000 kg
- solution concentration = 0.060*m*
- molar mass KI = 166.0 g/mol

Unknown
- mass of solute = ? g KI

According to the definition of molal, the final solution must contain 0.060 mol KI per 1000 g H_2O. Use the molality as a conversion factor to convert from mass of water to moles of the solute (KI). Then use the molar mass of KI to convert from mol KI to g KI. The steps are

$$\text{mass of } H_2O \longrightarrow \text{mol KI} \longrightarrow \text{g KI.}$$

2 Calculate ***Solve for the unknown.***

$$0.5000\ \cancel{\text{kg } H_2O} \times \frac{0.060\ \cancel{\text{mol KI}}}{1.000\ \cancel{\text{kg } H_2O}} \times \frac{166.0 \text{ g KI}}{1\ \cancel{\text{mol KI}}} = 5.0 \text{ g KI}$$

Math Handbook
For help with dimensional analysis, go to page R66.

3 Evaluate ***Does the result make sense?***

A 1 molal KI solution is one molar mass of KI (166.0 g) dissolved in 1000 g of water. The desired molal concentration (0.060*m*) is about $\frac{1}{20}$ of that value, so the mass of KI should be much less than the molar mass. The answer is correctly expressed to two significant figures.

Practice Problems

29. How many grams of sodium fluoride are needed to prepare a 0.400*m* NaF solution that contains 750 g of water?

30. Calculate the molality of a solution prepared by dissolving 10.0 g NaCl in 600 g of water.

Interactive Textbook
Problem-Solving 16.29 Solve Problem 29 with the help of an interactive guided tutorial.
with ChemASAP

The concentration of a solution also can be expressed as a mole fraction, as shown in Figure 16.18. The **mole fraction** of a solute in a solution is the ratio of the moles of that solute to the total number of moles of solvent and solute. In a solution containing n_A mol of solute A and n_B mol of solvent B, the mole fraction of solute A (X_A) and the mole fraction of solvent B (X_B) can be expressed as follows.

$$X_A = \frac{n_A}{n_A + n_B} \qquad X_B = \frac{n_B}{n_A + n_B}$$

Checkpoint ***What is a mole fraction?***

Differentiated Instruction

Less Proficient Readers L1

Have students write sentences using the words molarity and molality, and have them circle the letter in each word that makes them distinct. The mnemonic, 'r' for molarity and liter may help.

Facts and Figures

Mole Fraction Uses

Explain that the mole fraction compares the number of moles of a solute to the total number of moles in the solution. Organic chemists, who frequently work with nonaqueous solvent systems, often use this method of expressing concentration. The mole fraction is also used when calculating the vapor pressure of a solution.

492 Chapter 16

SAMPLE PROBLEM 16.7

Calculating Mole Fractions

Ethylene glycol ($C_2H_6O_2$) is added to automobile cooling systems to protect against cold weather. What is the mole fraction of each component in a solution containing 1.25 mol of ethylene glycol (EG) and 4.00 mol of water?

1 Analyze *List the knowns and the unknowns.*

Knowns
- moles of ethylene glycol (n_{EG}) = 1.25 mol EG
- moles of water (n_{H_2O}) = 4.00 mol H_2O

Unknowns
- mole fraction EG (X_{EG}) = ?
- mole fraction H_2O (X_{H_2O}) = ?

The mole fraction of ethylene glycol (X_{EG}) in the solution is the number of moles of ethylene glycol divided by the total number of moles in the solution:

$$X_{EG} = \frac{n_{EG}}{n_{EG} + n_{H_2O}}$$

Similarly, the mole fraction of water (X_{H_2O}) in the solution is the number of moles of water divided by the total number of moles in the solution:

$$X_{H_2O} = \frac{n_{H_2O}}{n_{EG} + n_{H_2O}}$$

2 Calculate *Solve for the unknowns.*

$$X_{EG} = \frac{n_{EG}}{n_{EG} + n_{H_2O}} = \frac{1.25\ \cancel{mol}}{1.25\ \cancel{mol} + 4.00\ \cancel{mol}} = 0.238$$

$$X_{H_2O} = \frac{n_{H_2O}}{n_{EG} + n_{H_2O}} = \frac{4.00\ \cancel{mol}}{1.25\ \cancel{mol} + 4.00\ \cancel{mol}} = 0.762$$

3 Evaluate *Does the result make sense?*

The mole fraction is a dimensionless quantity. The sum of the mole fractions of all the components in a solution must equal 1. Note that $X_{EG} + X_{H_2O} = 1.000$. Each answer is correctly expressed to three significant figures.

Practice Problems

31. What is the mole fraction of each component in a solution made by mixing 300 g of ethanol (C_2H_5OH) and 500 g of water?

32. A solution contains 50.0 g of carbon tetrachloride (CCl_4) and 50.0 g of chloroform ($CHCl_3$). Calculate the mole fraction of each component in the solution.

Figure 16.19 The concentration of antifreeze used in an automobile cooling system can be described by mole fractions.

Math Handbook

For help with using a calculator, go to page R62.

Interactive Textbook

Problem-Solving 16.32 Solve Problem 32 with the help of an interactive guided tutorial.

with ChemASAP

Sample Problem 16.7

Answers

31. $X_{C_2H_5OH} = 0.190; X_{H_2O} = 0.810$

32. $X_{CCl_4} = 0.437; X_{CHCl_3} = 0.563$

Practice Problems Plus L2

Calculate the mole fraction of each component in a solution of 42 g CH_3OH, 35 g C_2H_5OH, and 50 g C_3H_7OH. *($CH_3OH = 0.45$; $C_2H_5OH = 0.26$; $C_3H_7OH = 0.29$)*

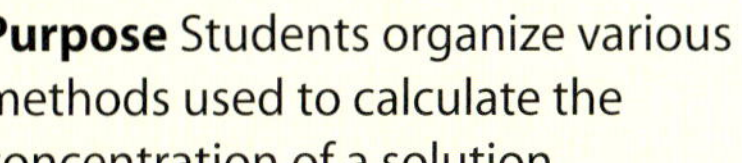

Math Handbook

For a math refresher and practice, direct students to using a calculator, page R62.

CLASS Activity

Diagramming Methods of Concentration Calculation L3

Purpose Students organize various methods used to calculate the concentration of a solution.

Procedure Have students design a diagram that includes definition of each type of concentration unit. More than one design is possible, but each should include the terms *concentration of a solution, percent by mass, percent by volume, molarity, molality,* and *mole fraction*. Students may wish to include additional detail by showing interconversions between *mass of solute* and *moles of solute*. Underneath each concentration unit, students should provide examples of when the unit would be used in the laboratory and in everyday life.

Answers to...

Figure 16.18 one

Checkpoint Mole fraction is the ratio of the number of moles of one component of a mixture to the total number of moles of all the components of the mixture.

Freezing-Point Depression and Boiling-Point Elevation

Interpreting Graphs L2

a. 0°C; 100°C
b. The freezing point of the solution is lower than that of pure water and the boiling point is higher than that of pure water.
c. Adding a solute to water allows it to remain as a liquid over a longer temperature range because the solution changes to a solid at a lower temperature and changes to a vapor at a higher temperature.

Enrichment Question 

Which aqueous solution would have the larger boiling point elevation and freezing point depression: 1*m* KCl or 1*m* $CaBr_2$? *($CaBr_2$)*

Use Visuals 

Figure 16.20 Display Figure 16.20 on an overhead projector. Choose an arbitrary concentration for an aqueous solution of NaCl or ethylene glycol. Calculate the boiling-point elevation and freezing-point depression. Then write the temperatures on the horizontal axis. Show students how to read the data for the boiling points of the pure solvent and of the solution. Then have students read the corresponding values for the freezing points. Ask, **What are ΔT_b and ΔT_f for this solution?** *(Point out that the solute affects both the freezing point and boiling point of a liquid.)* **Why would knowing the boiling-point elevation and freezing-point depression be important when choosing antifreeze for car radiators?** *(Despite its name, antifreeze protects against both freezing and overheating.)*

Table 16.2
K_f Values for Some Common Solvents

Solvent	K_f (°C/*m*)
Water	1.86
Acetic acid	3.90
Benzene	5.12
Nitrobenzene	7.00
Phenol	7.40
Cyclohexane	20.2
Camphor	37.7

Freezing-Point Depression and Boiling-Point Elevation

The graph in Figure 16.20 shows that the freezing point of a solvent is lowered and its boiling point is raised by the addition of a nonvolatile solute. **The magnitudes of the freezing-point depression (ΔT_f) and the boiling-point elevation (ΔT_b) of a solution are directly proportional to the molal concentration (*m*), when the solute is molecular, not ionic.**

$$\Delta T_f \propto m \quad \Delta T_b \propto m$$

The change in the freezing temperature (ΔT_f) is the difference between the freezing point of the solution and the freezing point of the pure solvent. Similarly, the change in the boiling temperature (ΔT_b) is the difference between the boiling point of the solution and the boiling point of the pure solvent. The term *m* is the molal concentration of the solution.

With the addition of a constant, the proportionality between the freezing point depression (ΔT_f) and the molality *m* can be expressed as an equation.

$$\Delta T_f = K_f \times m$$

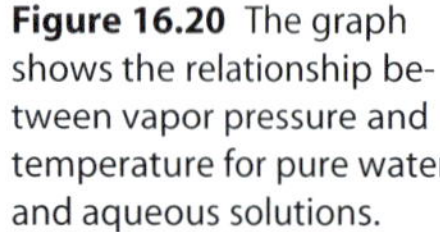

Simulation 21 Discover the principle underlying the colligative properties of solutions.

with ChemASAP

The constant, K_f, is the **molal freezing-point depression constant,** which is equal to the change in freezing point for a 1-molal solution of a nonvolatile molecular solute. The value of K_f depends upon the solvent. Its units are °C/*m*. Table 16.2 lists the K_f values for water and some other solvents.

The boiling-point elevation of a solution can also be expressed as an equation.

$$\Delta T_b = K_b \times m$$

The constant, K_b, is the **molal boiling-point elevation constant,** which is equal to the change in boiling point for a 1-molal solution of a nonvolatile molecular solute. The value of K_b depends upon the solvent. Its units are °C/*m*. Table 16.3 lists the K_b values for water and some other solvents.

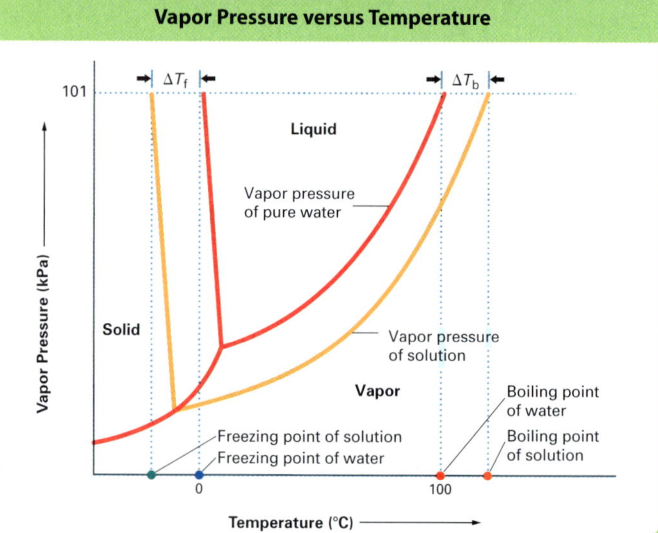

Figure 16.20 The graph shows the relationship between vapor pressure and temperature for pure water and aqueous solutions.

INTERPRETING GRAPHS

a. Identify What is the freezing point of water? What is the boiling point?
b. Describe How do the freezing and boiling points of the solution compare to those of pure water?
c. Apply Concepts Does adding a solute to water allow it to remain as a liquid over a longer or shorter temperature range? Explain.

494 Chapter 16

Sample Problem 16.8 shows how these equations can be used for calculating ΔT_f and ΔT_b of solutions if the solute is a molecular compound. For ionic compounds, both the freezing-point depression and the boiling-point elevation depend upon the number of ions produced by each formula unit. This number is used to calculate an effective molality, as shown in Sample Problem 16.9.

SAMPLE PROBLEM 16.8

Calculating the Freezing-Point Depression of a Solution

Antifreeze protects a car from freezing. It also protects it from overheating. Calculate the freezing-point depression and the freezing point of a solution containing 100 g of ethylene glycol ($C_2H_6O_2$) antifreeze in 0.500 kg of water.

1 Analyze ***List the knowns and the unknowns.***

Knowns
- mass of solute = 100 g $C_2H_6O_2$
- mass of solvent = 0.500 kg H_2O
- K_f for H_2O = 1.86°C/m
- $\Delta T_f = K_f \times m$

Unknown
- ΔT_f = ?°C
- freezing point = ?°C

Calculate the number of moles of solute and the molality. Then calculate the freezing-point depression and freezing point.

2 Calculate ***Solve for the unknown.***

$$\text{moles } C_2H_6O_2 = 100 \text{ g } \cancel{C_2H_6O_2} \times \frac{1 \text{ mol}}{62.0 \text{ g } \cancel{C_2H_6O_2}} = 1.61 \text{ mol}$$

$$m = \frac{\text{mol solute}}{\text{kg solvent}} = \frac{1.61 \text{ mol}}{0.500 \text{ kg}} = 3.22m$$

$$\Delta T_f = K_f \times m = 1.86°\text{C}/m \times 3.22m = 5.99°\text{C}$$

The freezing point of the solution is 0.00°C − 5.99°C = −5.99°C.

3 Evaluate ***Does the result make sense?***

A 1-molal solution reduces the freezing temperature by 1.86°C, so a decrease of 5.99°C for an approximately 3-molal solution is reasonable. The answer is correctly expressed with three significant figures.

Practice Problems

33. What is the freezing point depression of an aqueous solution of 10.0 g of glucose ($C_6H_{12}O_6$) in 50.0 g H_2O?

34. Calculate the freezing-point depression of a benzene solution containing 400 g of benzene and 200 g of the molecular compound acetone (C_3H_6O). K_f for benzene is 5.12°C/m.

Table 16.3

K_b Values for Some Common Solvents

Solvent	K_b (°C/m)
Water	0.512
Ethanol	1.19
Benzene	2.53
Cyclohexane	2.79
Acetic acid	3.07
Phenol	3.56
Nitrobenzene	5.24
Camphor	5.95

Math Handbook

For help with algebraic equations, go to page R69.

Problem-Solving 16.33 Solve Problem 33 with the help of an interactive guided tutorial.

with ChemASAP

Sample Problem 16.8

Answers

33. mol $C_6H_{12}O_6$ = 10.0 g glucose × 1 mol / 180.2 g = 0.0555 mol; m = 0.0555 mol / 50.0 g H_2O × 1000g / 1 kg = 1.11m; $\Delta T_f = K_f \times m$ = 1.86°C/m × 1.11m = 2.06°C

34. mol C_3H_6O = 200 g C_3H_6O × 1 mol/ 58.0 g C_3H_6O = 3.45 mol C_3H_6O; m = 3.45 mol C_3H_6O/400 g × 1000 g/1 kg = 8.63m; $K_f \times m$ = 5.12°C/m × 8.63m = 44.2°C

Math Handbook

For a math refresher and practice, direct students to algebraic equations, page R69.

Discuss 

Some students may think that freezing points and boiling points can be depressed or elevated without end. Explain that as the concentration of a solute, such as ethylene glycol, increases, there comes a point when the quantity of solute exceeds the quantity of solvent. Ethylene glycol becomes the solvent and water becomes the solute. The trends in colligative properties begin to reflect ethylene glycol instead of water. If the solute is a solid, such as NaCl, eventually the solution becomes saturated. But even before this saturation point, the magnitude of certain colligative properties may reach a maximum.

3 ASSESS

Evaluate Understanding 

Ask, **Which solution has a higher boiling point, 1 mol of $Al(NO_3)_3$ in 1000 g of water or 1.5 mol of KCl in 1000 g of water?** Have students explain their answers. *(The solution of $Al(NO_3)_3$ has a higher boiling point because $Al(NO_3)_3$ dissociates into a larger number of particles.)* **Why is it important to distinguish between nonvolatile and volatile compounds when discussing certain colligative properties?** *(Volatile solutes would quickly evaporate at higher temperatures, which would change the molal concentration of the solution.)*

Sample Problem 16.9

Answers

35. $m = 1.25 \text{ mol } CaCl_2 / 1400 \text{ g} \times 1000 \text{ g} / 1 \text{ kg} = 0.893m\ CaCl_2$; Each formula unit of $CaCl_2$ dissociates into three particles; $0.893m\ CaCl_2 \times 3 = 2.68m$; $\Delta T_b = K_b \times m = 0.512°C/m \times 2.68m = 1.37°C$; $100°C + 1.37°C = 101.37°C$

36. $2.00°C / 0.512°C/m = 3.91m$; Each formula unit of NaCl dissociates into two particles; $3.91m / 2 = 1.96m$; In 1 kg of water, 1.96 mol $NaCl \times 58.5 \text{ g} / 1 \text{ mol} = 115 \text{ g NaCl}$

Practice Problems Plus

L2

What is the boiling point of a solution containing 96.7 g of sucrose ($C_{12}H_{22}O_{11}$) in 250.0 g water at 1 atm? *(100.579°C)*

Math Handbook

For a math refresher and practice, direct students to significant figures, page R59.

Reteach

Remind students that when they do calculations involving ionic solids, they must find the molality of the solution in terms of the total number of particles. Work through several examples to be sure they understand this important point.

Elements Handbook

Given the list of dissolved elements and their ppm, students should infer that the freezing point of ocean water is less than 0°C at 1 atm and that the rate of evaporation of ocean waters is slower because of lowered vapor pressure.

If your class subscribes to the Interactive Textbook, use it to review key concepts in Section 16.4.

with ChemASAP

SAMPLE PROBLEM 16.9

Calculating the Boiling Point of a Solution

What is the boiling point of a 1.50m NaCl solution?

1 Analyze *List the knowns and the unknown.*

Knowns
- concentration = 1.50m NaCl
- K_b for H_2O = 0.512°C/m
- $\Delta T_b = K_b \times m$

Unknown
- boiling point = ?°C

Each formula unit of NaCl dissociates into two particles, Na^+ and Cl^-. Based on the total number of dissociated particles, the effective molality is $2 \times 1.50m = 3.00m$. Calculate the boiling-point elevation and then add it to 100°C.

2 Calculate *Solve for the unknown.*

$$\Delta T_b = K_b \times m = \frac{0.512°C}{m} \times 3.00m = 1.54°C$$

The boiling point of the solution is 100°C + 1.54°C = 101.54°C

3 Evaluate *Does the result make sense?*

The boiling point increases about 0.5°C for each mole of solute particles, so the total change is reasonable. Because the boiling point of water is defined as exactly 100°C, this value does not limit the number of significant figures in the solution of the problem.

Math Handbook

For help with significant figures, go to page R59.

Practice Problems

35. What is the boiling point of a solution that contains 1.25 mol $CaCl_2$ in 1400 g of water?

36. What mass of NaCl would have to be dissolved in 1.000 kg of water to raise the boiling point by 2.00°C?

Interactive Textbook

Problem-Solving 16.36 Solve Problem 36 with the help of an interactive guided tutorial.

with ChemASAP

16.4 Section Assessment

37. Key Concept What are two ways of expressing the ratio of solute particles to solvent particles?

38. Key Concept How are freezing-point depression and boiling-point elevation related to molality?

39. How many grams of sodium bromide must be dissolved in 400.0 g of water to produce a 0.500 molal solution?

40. Calculate the mole fraction of each component in a solution of 2.50 mol ethanoic acid (CH_3COOH) in 10.00 mol of water.

41. What is the freezing point of a solution of 12.0 g of CCl_4 dissolved in 750.0 g of benzene? The freezing point of benzene is 5.48°C; K_f is 5.12°C/m.

Elements Handbook

Element Distribution Look at the table on page R4 of the Elements Handbook showing the distribution of elements in the oceans. What generalization can you make about the temperature at which ocean water will freeze? What effect does the presence of dissolved elements in the ocean have on the rate of evaporation of ocean water?

Assessment 16.4 Test yourself on the concepts in Section 16.4.

with ChemASAP

Section 16.4 Assessment

37. Molality and mole fractions are two convenient ways of expressing the ratio of solute particles to solvent particles.

38. The magnitudes of the freezing-point depression (ΔT_f) and the boiling-point elevation (ΔT_b) of a solution are directly proportional to the molal concentration (m), when the solute is molecular, not ionic.

39. 20.6 g NaBr

40. $X_{CH_3COOH} = 0.200$ $X_{H_2O} = 0.800$

41. 4.95°C

Small-Scale LAB

Making a Solution

Purpose

To make a solution and use carefully measured data to calculate the solution's concentration.

Materials

- solid NaCl
- water
- 50-mL volumetric flask
- balance

Procedure 

Measure the mass of a clean, dry, volumetric flask. Add enough solid NaCl to approximately fill one-tenth of the volume of the flask. Measure the mass of the flask again. Half fill the flask with water and shake it gently until all the NaCl dissolves. Fill the flask with water to the 50-mL mark and measure the mass again.

Analyze

Using your experimental data, record the answers to the following questions below your data table.

1. Percent by mass tells how many grams of solute are present in 100 g of solution.

$$\text{\% by mass} = \frac{\text{mass of solute}}{\text{mass of solute} + \text{solvent}} \times 100\%$$

a. Calculate the mass of the solute (NaCl).

b. Calculate the mass of the solvent (water).

c. Calculate the percent by mass of NaCl in the solution.

2. Mole fraction tells how many moles of solute are present for every 1 mol of total solution.

$$\text{Mole fraction} = \frac{\text{mol NaCl}}{\text{mol NaCl} + \text{mol } H_2O}$$

a. Calculate the moles of NaCl solute.
Molar mass NaCl = 58.5 g/mol

b. Calculate the moles of water.
Molar mass H_2O = 18.0 g/mol

c. Calculate the mole fraction of your solution.

3. Molality (*m*) tells how many moles of solute are present in 1 kg of solvent.

$$m = \frac{\text{mol NaCl}}{\text{kg } H_2O}$$

Calculate the molality of your solution.

4. Molarity (*M*) tells how many moles of solute are dissolved in 1 L of solution.

$$M = \frac{\text{mol NaCl}}{\text{L solution}}$$

a. Calculate the liters of solution. 1000 mL = 1 L

b. Calculate the molarity of the NaCl solution.

5. Density tells how many grams of solution are present in 1 mL of solution.

$$\text{Density} = \frac{\text{g solution}}{\text{mL solution}}$$

Calculate the density of the solution.

You're The Chemist

The following small-scale activities allow you to develop your own procedures and analyze the results.

1. Analyze It Use a small-scale pipet to extract a sample of your NaCl solution and deliver it to a massed empty volumetric flask. Measure the mass of the flask and fill it with water to the 50-mL line. Measure the mass of the flask again. Calculate the concentration of this dilute solution using the same units you used to calculate the concentration of the NaCl solution. Are the results you obtained reasonable?

2. Design It! Design and carry out an experiment to make a solution of table sugar quantitatively. Calculate the concentration of the table sugar solution using the same units you used to calculate the concentration of the NaCl solution. Is the effective molality of the table sugar solution the same as the effective molality of a sodium chloride solution of the same concentration? Recall that effective molality is the concentration value used to calculate boiling-point elevation and freezing-point depression.

Making a Solution 

Objective

After completing this activity, students will be able to:

- make a solution
- calculate the solution's molarity, molality, percent by mass, and mole fraction

 **Prep Time** 20 minutes
Class Time 20 minutes
Safety Wear safety goggles and a lab apron.

Teaching Tips

Household table salt contains a small amount of desiccant, such as sodium silicate, which gives the salt solution an opaque appearance. Use lab grade NaCl instead. Remind students about significant figures. Point out that the balance is probably more accurate than the volumetric flask. Calculations based only on measurements using the balance will have a larger number of significant figures.

Expected Outcome Students make a solution and calculate the solution's concentration in various units.

Analyze

Sample data:
dry bottle = 15.98 g
flask + NaCl = 22.88 g
flask + NaCl + water = 69.09 g
1. a. 6.90 g **b.** 46.21 g **c.** 13.0%
2. a. 0.118 mol NaCl **b.** 2.57 mol H_2O
c. 0.0439
3. 2.55*m*
4. a. 0.050 L **b.** 2.4*M*
5. 1.1 g/mL

You're the Chemist

1. Sample data:
dry flask = 15.98 g
flask + NaCl solution = 22.88 g
flask + NaCl solution + water = 69.09 g
mass of NaCl solution = 2.09 g
mass of water = 47.77 g
mass of NaCl = 0.271 g
percent mass of NaCl = 0.544%
moles NaCl = 4.64×10^{-3} mol
mass of water = 49.59 g
moles of water = 2.76 mol
mole fraction = 1.68×10^{-3}
molality = 0.0936*m*
density = 1.0 g mL

2. Sample data:
mass of dry flask = 16.72 g
mass of flask + sugar = 20.85 g
mass of flask + sugar + water = 69.53 g
mass of sugar = 4.13 g
mass of solvent = 48.68 g
percent mass of sugar = 7.82%
moles of sugar = 0.0121 mol
moles of water = 2.70 mol
mole fraction = 4.46×10^{-3}
molality = 0.249*m*
molarity = 0.242*M*
density = 1.1 g/mL
No, the effective molaliy of the NaCl solution is twice the effective molality of the sugar solution.

For Enrichment 

Students could compare the freezing point depressions of the sugar and salt solutions. *(One way to do this is to partially freeze the solutions and determine the temperature of the mixtures of liquid and solid.)*

CHAPTER 16 Study Guide

Study Tip

Organize New Information Create flashcards to help you visualize relationships and formulas. Try to integrate what you're studying with what you already know.

If your class subscribes to the Interactive Textbook with ChemASAP, your students can go online to access an interactive version of the Student Edition and a self-test.

with ChemASAP

CHAPTER 16 Study Guide

Key Concepts

16.1 Properties of Solutions

- Whether or not a substance dissolves depends upon the nature of the solvent and the solute. Factors that determine how fast a substance dissolves are stirring, temperature, and surface area.
- Solubility is usually expressed in grams of solute per 100 g of solvent.
- The amount of solute that dissolves in a given solvent depends upon the temperature and pressure.

16.2 Concentrations of Solutions

- To calculate the molarity of a solution, divide the moles of solute by the volume of the solution in liters.
- Diluting a solution does not change the total number of moles of solute in solution.
- The concentration of a solution in percent is the ratio of the volume of the solute to the volume of the solution times 100% or the ratio of the mass of the solute to the mass of the solution times 100%.

16.3 Colligative Properties of Solutions

- Three colligative properties of solutions are vapor-pressure lowering, freezing-point depression, and boiling-point elevation.
- The magnitude of each colligative property is directly proportional to the number of solute molecules or ions present.

16.4 Calculations Involving Colligative Properties

- Molality and mole fractions are two additional ways of expressing the concentration of a solution.
- Freezing-point depression and boiling-point elevation are proportional to molality.

Vocabulary

- boiling-point elevation (p. 490)
- concentrated solution (p. 480)
- concentration (p. 480)
- colligative property (p. 487)
- dilute solution (p. 480)
- freezing-point depression (p. 488)
- Henry's law (p. 476)
- immiscible (p. 473)
- miscible (p. 473)
- molal freezing-point depression constant (K_f) (p. 494)
- molal boiling-point elevation constant (K_b) (p. 494)
- molality (m) (p. 491)
- molarity (M) (p. 480)
- mole fraction (p. 492)
- saturated solution (p. 473)
- solubility (p. 473)
- supersaturated solution (p. 474)
- unsaturated solution (p. 473)

Key Equations

- Henry's law: $\frac{S_1}{P_1} = \frac{S_2}{P_2}$
- Molarity (M) $= \frac{\text{moles of solute}}{\text{liters of solution}}$
- $M_1 \times V_1 = M_2 \times V_2$
- Percent by volume $= \frac{\text{volume of solute}}{\text{volume of solution}} \times 100\%$
- Percent by mass $= \frac{\text{mass of solute}}{\text{mass of solution}} \times 100\%$
- Molality $= \frac{\text{moles of solute}}{\text{kilogram of solvent}}$
- Mole fractions: $X_A = \frac{n_A}{n_A + n_B}$ $\quad X_B = \frac{n_B}{n_A + n_B}$
- $\Delta T_f = K_f \times m$
- $\Delta T_b = K_b \times m$

Chapter Resources

Print

- **Core Teaching Resources,** Chapter 16, *Practice Problems, Vocabulary Review, Quiz, Chapter Test A, Chapter Test B*

Technology

- **Computer Test Bank,** Chapter 16 Test
- **Interactive Textbook with ChemASAP,** Chapter 16

CHAPTER 16

Assessment

Reviewing Content

16.1 Properties of Solutions

42. Name and distinguish between the two components of a solution.

43. Explain why the dissolved component does not settle out of a solution.

44. Define the following terms: solubility, saturated solution, unsaturated solution, miscible, and immiscible.

45. If a saturated solution of sodium nitrate is cooled, what change might you observe?

46. Can a solution with undissolved solute be supersaturated? Explain.

47. What mass of $AgNO_3$ can be dissolved in 250 g of water at 20°C? Use Table 16.1.

48. What is the effect of pressure on the solubility of gases in liquids?

49. The solubility of methane, the major component of natural gas, in water at 20°C and 1.00 atm pressure is 0.026 g/L. If the temperature remains constant, what will be the solubility of this gas at the following pressures?
a. 0.60 atm **b.** 1.80 atm

16.2 Concentrations of Solutions

50. Knowing the molarity of a solution is more meaningful than knowing whether a solution is dilute or concentrated. Explain.

51. Define molarity, then calculate the molarity of each solution.
a. 1.0 mol KCl in 750 mL of solution
b. 0.50 mol $MgCl_2$ in 1.5 L of solution

52. How many milliliters of 0.500*M* KCl solution would you need to dilute to make 100.0 mL of 0.100*M* KCl?

53. Calculate the moles and grams of solute in each solution.
a. 1.0 L of 0.50*M* NaCl
b. 5.0×10^2 mL of 2.0*M* KNO_3
c. 250 mL of 0.10*M* $CaCl_2$

54. Calculate the grams of solute required to make the following solutions.
a. 2500 g of saline solution (0.90% NaCl (m/m))
b. 0.050 kg of 4.0% (m/m) $MgCl_2$

55. What is the concentration (in % (v/v)) of the following solutions?
a. 25 mL of ethanol (C_2H_5OH) is diluted to a volume of 150 mL with water.
b. 175 mL of isopropyl alcohol (C_3H_7OH) is diluted with water to a total volume of 275 mL.

16.3 Colligative Properties of Solutions

56. What are colligative properties? Identify three colligative properties and explain why each occurs.

57. Which has the higher boiling point:
a. seawater or distilled water?
b. 1.0*M* KNO_3 or 1.5*M* KNO_3?
c. 0.100*M* KCl or 0.100*M* $MgCl_2$?

58. Why does a 1*m* solution of calcium nitrate have a lower freezing point than a 1*m* solution of sodium nitrate?

59. Explain how a decrease in the vapor pressure of a solution results in an increase in its boiling point.

60. In old-fashioned ice cream makers, a mixture of rock salt (NaCl) and ice is used to cool the creamy mixture as it is stirred. What is the purpose of the salt?

16.4 Calculations Involving Colligative Properties

61. Distinguish between a 1*M* solution and a 1*m* solution.

62. Describe how you would make an aqueous solution of methanol (CH_3OH) in which the mole fraction of methanol is 0.40.

63. What is the boiling point of each solution?
a. 0.50 mol glucose in 1000 g H_2O
b. 1.50 mol NaCl in 1000 g H_2O

64. What is the freezing point of each solution?
a. 1.40 mol Na_2SO_4 in 1750 g H_2O
b. 0.060 mol $MgSO_4$ in 100 g H_2O

65. Determine the freezing points of each 0.20*m* aqueous solution.
a. K_2SO_4 **b.** $CsNO_3$ **c.** $Al(NO_3)_3$

CHAPTER 16 Assessment

Reviewing Content

42. The solvent is the substance in which the solute is dissolved.

43. Random collisions of the solvent molecules with the solute particles provide enough force to overcome gravity.

44. solubility: the amount of a substance that dissolves in a given quantity of solvent at specified conditions of temperature and pressure to produce a saturated solution; saturated solution: a solution containing the maximum amount of solute for a given amount of solvent at a constant temperature and pressure; unsaturated solution: a solution that contains less solute than a saturated solution at a given temperature and pressure; miscible: describes liquids that dissolve in each other; immiscible: describes liquids that are insoluble in each other

45. Particles of solute crystallize.

46. No; if there were undissolved solute, the excess solute would come out of a supersaturated solution.

47. 5.6×10^2 g $AgNO_3$

48. Solubility increases with pressure.

49. **a.** 1.6×10^{-2} g/L **b.** 4.7×10^{-2} g/L

50. *Dilute* and *concentrated* are relative terms and are not quantitative. Molarity provides the exact number of moles of solute per liter of solution.

51. Molarity is the number of moles of solute dissolved in one liter of solution. **a.** 1.3*M* KCl
b. 3.3×10^{-1}*M* $MgCl_2$

52. 2.00×10^1 mL

53. **a.** 5.0×10^{-1} mol NaCl, 29 g NaCl
b. 1.0 mol KNO_3, 1.0×10^2 g KNO_3
c. 2.5×10^{-2} mol $CaCl_2$, 2.8 g $CaCl_2$

54. **a.** 2.3×10^1 g NaCl **b.** 2.0 g $MgCl_2$

55. **a.** 16% (v/v) ethanol
b. 63.6% (v/v) isopropyl alcohol

56. Colligative properties are properties of a solution that depend only on the number of solute particles; boiling-point elevation, freezing-point depression, and vapor-pressure lowering. Boiling points are elevated because solvent molecules surround solute particles, reducing the number of solvent molecules that have sufficient energy to escape the solution; relative to the pure solvent, the amount of energy required to cause vaporization or boiling increases. Solutes disrupt the ordering of the solvent structure, so more kinetic energy must be withdrawn from a solution for it to solidify. This lowers the freezing point of the solution.

57. **a.** sea water **b.** 1.5*M* KNO_3
c. 0.100*M* $MgCl_2$

58. The effective molality of the $Ca(NO_3)_2$ solution is 3*m*. The effective molality of the $NaNO_3$ solution is 2*m*.

59. When vapor pressure is lowered relative to pure solvent, more energy must be supplied to reach the boiling point; thus the boiling point is increased relative to pure solvent.

60. The salt lowers the freezing point of the ice-water cooling mixture.

61. 1*M* solution: 1 mol of solute in 1 L of solution; 1*m* solution: 1 mol of solute in 1000 g of solvent

CHAPTER 16

Assessment *continued*

62. Add 27.0 g H_2O to 32.0 g CH_3OH.
63. a. 100.26°C **b.** 101.54°C
64. a. –4.46°C **b.** –2.2°C
65. a. –1.1°C **b.** –0.74°C **c.** –1.5°C

Understanding Concepts

66. a. The freezing-point depression is twice as great for solute B; solute B must provide twice as many particles in solution.
b. Solute A probably forms a saturated solution.

67. $\Delta T_f = -9.60°C$; $\Delta T_b = +4.74°C$

68. Each gram of acetone requires 0.93 g of water.

69. The mole fraction of $NaHCO_3$ is 0.020; of water is 0.98. The solution is 1.1*m*.

70. The mole fraction of NaCl is 2.69×10^{-3}; the mole fraction of H_2O is 9.97×10^{-1}.

71. Add one crystal of KNO_3. If the solution is supersaturated, crystallization occurs. If it is saturated, the crystal does not dissolve; if unsaturated, the crystal dissolves.

72.

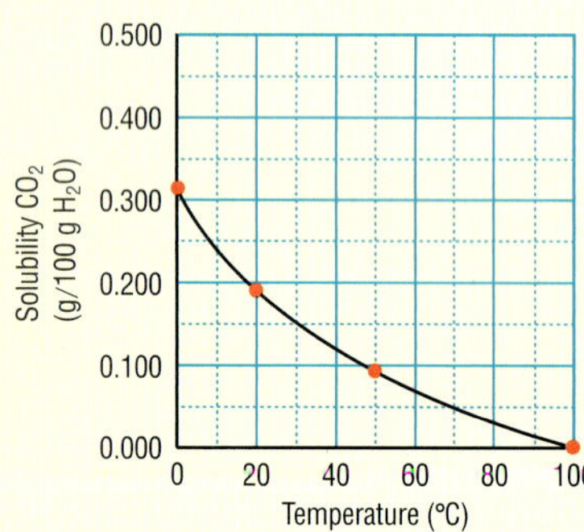

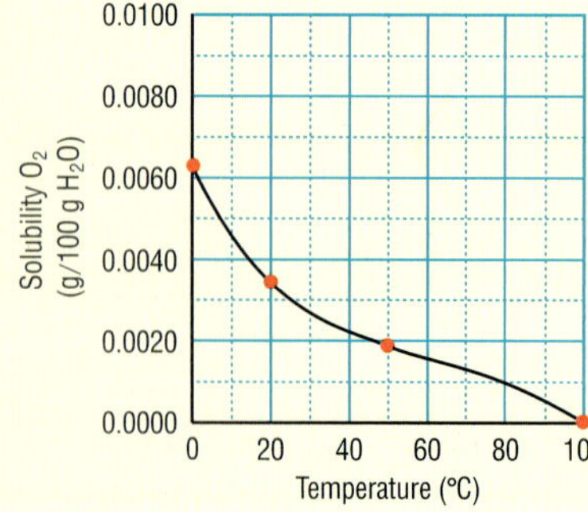

CHAPTER 16

Assessment *continued*

Understanding Concepts

66. Different numbers of moles of two different solutes, A and B, were added to identical quantities of water. The graph shows the freezing point of each of the solutions formed.

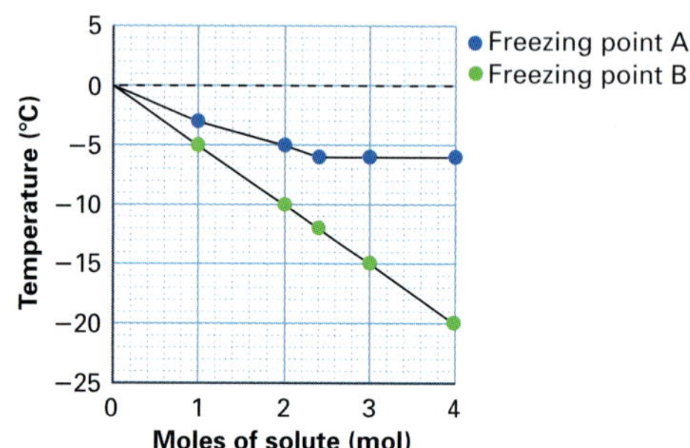

a. Explain the relative slopes of the two lines between 0 and 2 mol of solute added.
b. Why does the freezing point for solution A not continue to drop as amounts of solute A are added beyond 2.4 mol?

67. Calculate the freezing- and boiling-point changes for a solution containing 12.0 g of naphthalene ($C_{10}H_8$) in 50.0 g of benzene.

68. Describe how you would prepare an aqueous solution of acetone (CH_3COCH_3) in which the mole fraction of acetone is 0.25.

69. The solubility of sodium hydrogen carbonate ($NaHCO_3$) in water at 20°C is 9.6 g/100 g H_2O. What is the mole fraction of $NaHCO_3$ in a saturated solution? What is the molality of the solution?

70. A solution is labeled 0.150*m* NaCl. What are the mole fractions of the solute and solvent in this solution?

71. You are given a clear aqueous solution containing KNO_3. How would you determine experimentally if the solution is unsaturated, saturated, or supersaturated?

72. Plot a graph of solubility versus temperature for the three gases listed in Table 16.1.

73. A mixture of ethylene glycol (EG) and water is used as antifreeze in automobile engines. The freezing point and density of the mixture vary with the percent by mass of (EG) in the mixture. On the following graph, point A represents 20% (EG) by mass; point B, 40%; and point C, 60%.

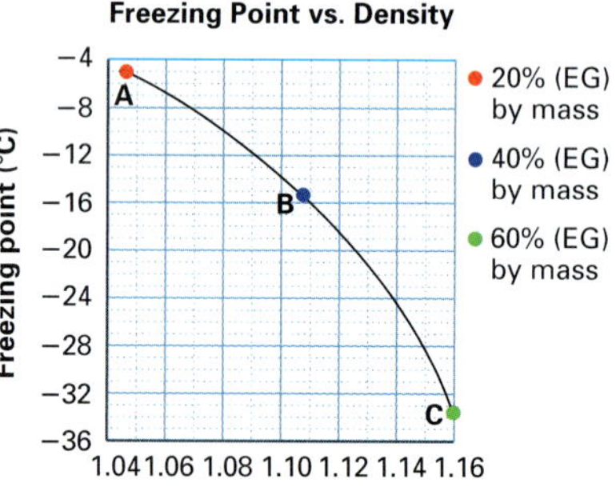

a. What is the density of the antifreeze mixture that freezes at –25°C?
b. What is the freezing point of a mixture that has a density of 1.06?
c. Estimate the freezing point of a mixture that is 30% by mass (EG).

74. Calculate the freezing point and the boiling point of a solution that contains 15.0 g of urea (CH_4N_2O) in 250 g of water. Urea is a covalently bonded compound.

75. Calculate the mole fractions in a solution that is 25.0 g of ethanol (C_2H_5OH) and 40.0 g of water.

76. Estimate the freezing point of an aqueous solution of 20.0 g of glucose ($C_6H_{12}O_6$) dissolved in 500.0 g of water.

77. The solubility of KCl in water is 34.0 g KCl/100 g H_2O at 20°C. A warm solution containing 50.0 g KCl in 130 g H_2O is cooled to 20°C.
a. How many grams of KCl remain dissolved?
b. How many grams came out of solution?

78. How many moles of ions are present when 0.10 mol of each compound is dissolved in water?
a. K_2SO_4 **b.** $Fe(NO_3)_3$ **c.** $Al_2(SO_4)_3$ **d.** $NiSO_4$

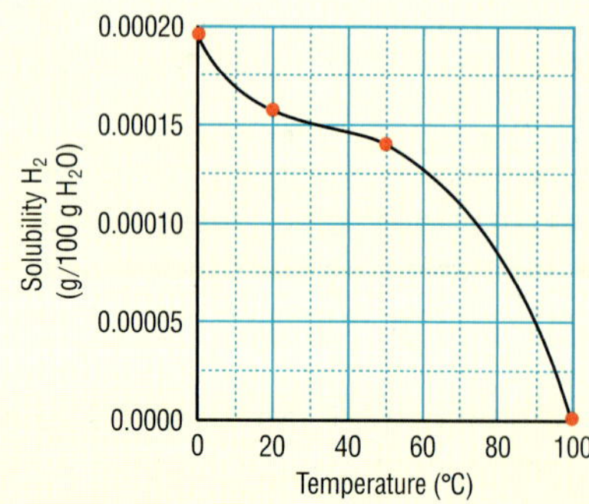

73. a. about 1.14 g/cm^3
b. about –7.2°C
c. about –9.5°C

74. fp = –1.86°C; bp = 100.512°C

75. $X_{C_2H_5OH} = 0.20$; $X_{H_2O} = 0.80$

76. –0.413°C

77. a. 44.2 g KCl
b. 5.8 g KCl

78. a. 0.30 mol
b. 0.40 mol
c. 0.50 mol
d. 0.20 mol

Critical Thinking

79. A solution contains 26.5 g NaCl in 75.0 g H_2O at 20°C. Determine if the solution is unsaturated, saturated, or supersaturated. (The solubility of NaCl at 20°C is 36.0 g/100 g H_2O.)

80. An aqueous solution freezes at –2.47°C. What is its boiling point?

81. Hydrogen peroxide is often sold commercially as a 3.0% (m/v) aqueous solution.
   a. If you buy a 250-mL bottle of 3.0% H_2O_2 (m/v), how many grams of hydrogen peroxide have you purchased?
   b. What is the molarity of this solution?

82. How many grams of $NaNO_3$ will precipitate if a saturated solution of $NaNO_3$ in 200 g H_2O at 50°C is cooled to 20°C?

83. What is the molar mass of a nondissociating compound if 5.76 g of the compound in 750 g of benzene gives a freezing-point depression of 0.460°C?

84. The molality of an aqueous solution of sugar ($C_{12}H_{22}O_{11}$) is 1.62*m*. Calculate the mole fractions of sugar and water.

85. Why might calcium chloride spread on icy roads be more effective at melting ice than an equal amount of sodium chloride?

86. The following table lists the molar concentrations of the most abundant monatomic ions in seawater. Calculate the mass in grams of each ion contained in 5.00 L of seawater. The density of sea water is 1.024 g/mL.

Ion	Molarity (*M*)
Chloride	0.546
Sodium	0.470
Magnesium	0.053
Calcium	0.0103
Potassium	0.0102

87. Which will have a greater boiling point elevation: 3.00 g $Ca(NO_3)_2$ in 60.0 g of water, or 6.00 g $Ca(NO_3)_2$ in 30.0 g of water?

Concept Challenge

88. When an excess of zinc is added to 800 mL of a hydrochloric acid solution, the solution evolves 1.21 L of hydrogen gas measured over water at 21°C and 747.5 mm Hg. What was the molarity of the acid? The vapor pressure of water at 21°C is 18.6 mm Hg.

89. How many milliliters of 1.50*M* HNO_3 contain enough nitric acid to dissolve an old copper penny with a mass of 3.94 g?

$$3Cu + 8HNO_3 \longrightarrow 3Cu(NO_3)_2 + 2NO + 4H_2O$$

90. One way to express the solubility of a compound is in terms of moles of compound that will dissolve in 1 kg of water. Solubility depends on temperature. Plot a graph of the solubility of potassium nitrate (KNO_3) from the following data.

Temperature (C°)	Solubility (mol/kg)
0	1.61
20	2.80
40	5.78
60	11.20
80	16.76
100	24.50

From your graph estimate
   a. the solubility of KNO_3 at 76°C and at 33°C.
   b. the temperature at which its solubility is 17.6 mol/kg of water.
   c. the temperature at which the solubility is 4.24 mol/kg of water.

91. A 250-mL sample of Na_2SO_4 is reacted with an excess of $BaCl_2$. If 5.28 g $BaSO_4$ is precipitated, what is the molarity of the Na_2SO_4 solution?

92. Suppose you have an unknown compound and want to identify it by means of its molar mass. Design an experiment that uses the concept of freezing-point depression to obtain the molar mass. What laboratory measurements would you need to make? What calculations would be needed?

Critical Thinking

79. unsaturated
80. 100.680°C
81. **a.** 7.5 g H_2O_2 **b.** $8.8 \times 10^{-1} M$
82. 5.2×10^1 g $NaNO_3$
83. 8.55×10^1 g/mol
84. $X_{H_2O} = 0.972; X_{C_{12}H_{22}O_{11}} = 0.028$
85. $CaCl_2$ produces three particles upon dissolving; NaCl produces only two particles. Freezing-point depression depends on the number of solute particles in the solvent.
86. chloride: 96.9 g
sodium: 54.1 g
magnesium: 6.44 g
calcium: 2.07 g
potassium: 1.99 g
87. The solution with the higher concentration of ions will have the greater boiling point elevation; 6.00 g $Ca(NO_3)_2$ in 30 g of water.

Concept Challenge

88. $1.2 \times 10^{-1} M$ HCl
89. 1.10×10^2 mL HNO_3
90. **a.** 76°C: 15 mol/kg; 33°C: 5 mol/kg
b. 82°C
c. 30°C
91. $9.0 \times 10^{-2} M$ Na_2SO_4

92. Determine the freezing point of a suitable solvent. Dissolve a known mass of the unknown molecular compound in a known mass of the solvent. Determine the freezing point of the solution. This gives the freezing point depression. Use $\Delta T_f = K_f \times m$ to find the molality of the solution. Use the molality to find the moles of solute. Use the moles of solute and the measured mass of solute to calculate the molar mass.

$$\frac{\text{g solute}}{\cancel{\text{g solvent}}} \times \frac{\cancel{1000 \text{ g solvent}}}{\cancel{1 \text{ kg solvent}}} \times \frac{\cancel{1 \text{ kg solvent}}}{1 \text{ mol solute}} = \frac{\text{g solute}}{1 \text{ mol solute}}$$

The molar mass obtained is valid only for an undissociating molecular solute.

CHAPTER 16 **Assessment** *continued*

Cumulative Review

93. a. 1.98×10^2 g H_2O
b. 1.98×10^5 mg H_2O
c. 1.98×10^{-1} kg H_2O

94. a. 3.47×10^{-1} kg **b.** 7.3×10^{-5} kg
c. 9.43×10^{-6} kg **d.** 8.77×10^{-4} kg

95. Rutherford's model contains a nucleus.

96. a. manganese, Mn **b.** indium, In **c.** francium, Fr **d.** polonium, Po

97. Calcium permanganate is $Ca(MnO_4)_2$. Four formula units contain 4 Ca atoms, 8 Mn atoms, and 32 O atoms.

98. $C_8H_6O_4$

99. a. 5.58×10^1 g Fe, 6.35×10^1 g Cu, 2.01×10^2 g Hg, 3.21×10^1 g S
b. Each sample contains 6.02×10^{23} atoms.
c. 4.48×10^{-1} mol Fe, 3.93×10^{-1} mol Cu, 1.25×10^{-1} mol Hg, 7.80×10^{-1} mol S

100. 1.7×10^4 L

101. a. combination **b.** decomposition **c.** single-replacement **d.** combustion **e.** single-replacement **f.** double-replacement

102. $2H^+(aq) + S^{2-}(aq) \rightarrow H_2S(g)$

103. a. $NH_4Cl(s) \rightarrow NH_4^+(aq) + Cl^-(aq)$
b. $Cu(NO_3)_2(s) \rightarrow Cu^{2+}(aq) + 2NO_3^-(aq)$
c. $HNO_3(aq) \rightarrow H^+(aq) + NO_3^-(aq)$
d. $HC_2H_3O_2(l) \rightarrow H^+(aq) + C_2H_3O_2^-(aq)$
e. $Na_2SO_4(s) \rightarrow 2Na^+(aq) + SO_4^{2-}(aq)$
f. $HgCl_2(s) \rightarrow Hg^{2+}(aq) + 2Cl^-(aq)$

104. 4.9×10^1 L O_2

105. a. :I· **b.** ·Te· **c.** :Sb· **d.** ·Sr·

106. 1.08×10^2 kPa

107. The particles in an ideal gas have no volume and there are no attractions between particles.

108. Unbalanced intermolecular attractions between molecules at the surface of the liquid and those below the surface create an inward pull, or force, that minimizes the surface area to create surface tension.

109. The very polar hydrogen chloride molecule produces hydronium ions (H_3O^+) and chloride ions (Cl^-) that are stabilized by becoming surrounded by water molecules in aqueous solution. Hydrogen chloride does not dissociate in nonpolar benzene. Polar compounds generally have low solubility in nonpolar solvents.

110. a solution, if the soap mixture is very dilute; More concentrated mixtures of soap in water form colloids.

CHAPTER 16 **Assessment** *continued*

Cumulative Review

93. A cylindrical vessel, 28.0 cm in height and 3.00 cm in diameter, is filled with water at 50°C. The density of water is 0.988 g/cm^3 at this temperature. Express the mass of water in the vessel in the following units. (*Chapter 3*)
a. grams
b. milligrams
c. kilograms

94. Convert each of the following mass measurements to its equivalent in kilograms. (*Chapter 3*)
a. 347 g **b.** 73 mg
c. 9.43 mg **d.** 877 mg

95. What is the most significant difference between the Thomson model of the atom and the Rutherford model? (*Chapter 5*)

96. Name and give the symbol for the element in the following positions in the periodic table. (*Chapter 6*)
a. Group 7B, period 4 **b.** Group 3A, period 5
c. Group 1A, period 7 **d.** Group 6A, period 6

97. How many atoms of each element are present in four formula units of calcium permanganate? (*Chapter 6*)

98. Terephthalic acid is an organic compound used in the synthesis of polyesters. Terephthalic acid contains 57.8 percent C, 3.64 percent H, and 38.5 percent O. The molar mass is approximately 166 g/mol. What is the molecular formula of terephthalic acid? (*Chapter 10*)

99. The photograph shows one mole each of iron, copper, mercury, and sulfur. (*Chapter 10*)
a. What is the mass of each element?
b. How many atoms are in each sample?
c. How many moles is 25.0 g of each element?

100. What is the volume occupied by 1500 g of hydrogen gas (H_2) at STP? (*Chapter 10*)

101. Identify the type of chemical reaction. (*Chapter 11*)
a. $H_2(g) + Cl_2(g) \longrightarrow 2HCl(g)$
b. $2H_2O(l) \longrightarrow O_2(g) + 2H_2(g)$
c. $2K(s) + 2H_2O(l) \longrightarrow 2KOH(aq) + H_2(g)$
d. $C_2H_6O(l) + 3O_2(g) \longrightarrow 2CO_2(g) + 3H_2O(l)$
e. $Cl_2(aq) + 2KBr(aq) \longrightarrow 2KCl(aq) + Br_2(aq)$
f. $Pb(NO_3)_2(aq) + 2NaCl(aq) \longrightarrow PbCl_2(s) + 2NaNO_3(aq)$

102. Write the net ionic equation for the following reaction. (*Chapter 11*)

$$2HI(aq) + Na_2S(aq) \longrightarrow H_2S(g) + 2NaI(aq)$$

103. Indicate by simple equations how the following substances ionize or dissociate in water. (*Chapter 11*)
a. NH_4Cl **b.** $Cu(NO_3)_2$ **c.** HNO_3
d. $HC_2H_3O_2$ **e.** Na_2SO_4 **f.** $HgCl_2$

104. The equation for the combustion of methanol (CH_3OH) is the following:

$$2CH_3OH(l) + 3O_2(g) \longrightarrow 2CO_2(g) + 4H_2O(l)$$

What volume of oxygen, measured at STP, is required to completely burn 35.0 g of methanol? (*Chapter 12*)

105. Draw electron dot structures for the following atoms. (*Chapter 7*)
a. I **b.** Te **c.** Sb **d.** Sr

106. A cylinder of nitrogen gas at 25°C and 101.3 kPa is heated to 45°C. What is the new pressure of the gas? (*Chapter 14*)

107. Why does an ideal gas not exist? (*Chapter 14*)

108. What relationship exists between surface tension and intermolecular attractions in a liquid? (*Chapter 15*)

109. The solubility of hydrogen chloride gas in the polar solvent water is much greater than its solubility in the nonpolar solvent benzene. Why? (*Chapter 15*)

110. When soap is shaken with water, is a solution, a suspension, or a colloid formed? Explain. (*Chapter 15*)

502 Chapter 16

Standardized Test Prep

Test-Taking Tip

Interpreting Data Tables Tables present a large amount of data in a small space. They allow you to make comparisons and to analyze the information present. Start by reading the title (if there is one). Then read the headings. Try to figure out the relationship between the different columns and rows of information. Ask yourself some questions: *What information is related in the table? How are the relationships represented?*

Select the choice that best answers each question or completes each statement.

1. An aqueous solution is 65% (v/v) rubbing alcohol. How many milliliters of water are in a 95-mL sample of this solution?
   a. 62 mL **b.** 1.5 mL **c.** 33 mL **d.** 30 mL

2. Which of these actions will cause more sugar to dissolve in a saturated sugar water solution?
   I. Add more sugar while stirring.
   II. Add more sugar and heat the solution.
   III. Grind the sugar to a powder; then add while stirring.
   a. I only
   b. II only
   c. III only
   d. I and II only
   e. II and III only

3. When 2.0 mol of methanol is dissolved in 45 g of water, the mole fraction of methanol is
   a. 0.44. **b.** 0.043. **c.** 2.25. **d.** 0.55.

The lettered choices below refer to Questions 4–7.

A lettered choice may be used once, more than once, or not at all.

A. moles/liter of solution
B. grams/mole
C. moles/kilogram of solvent
D. °C/molal
E. no units

Which of the above units is appropriate for each measurement?

4. molality
5. mole fraction
6. molar mass
7. molarity

Use the description and the data table to answer Questions 8–11.

A student measured the freezing points of three different aqueous solutions at five different concentrations. The data table summarizes the data.

Molarity (*M*)	Freezing Point Depression (°C)		
	NaCl	$CaCl_2$	C_2H_5OH
0.5	1.7	2.6	0.95
1.0	3.5	5.6	2.0
1.5	5.3	8.3	3.0
2.0	7.2	11.2	4.1
2.5	9.4	14.0	5.3

8. Graph the data for all three solutes on the same graph, using molarity as the independent variable.
9. Summarize the relationship between molarity and freezing-point depression.
10. Compare the slopes of the three lines and explain any difference.
11. If you collected similar data for KOH and added a fourth line to your graph, which existing line would the new line approximate?

Use the atomic windows to answer Questions 12–14. The windows show water and two aqueous solutions with different concentrations. The red spheres represent the solute particles; the blue spheres represent water.

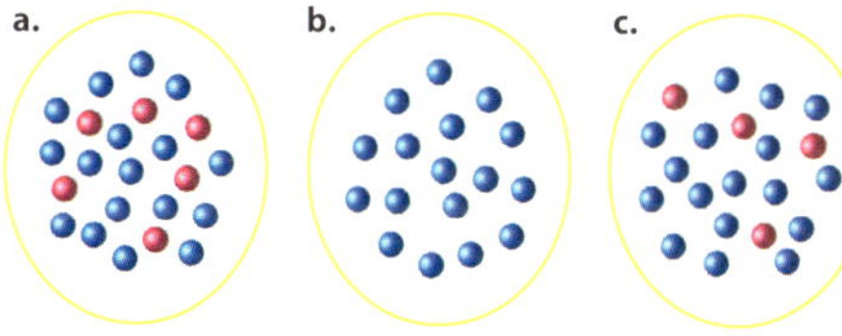

12. Which solution has the highest vapor pressure?
13. Which solution has the lowest vapor pressure?
14. Which solution has the lowest boiling point?

Write a brief essay to answer Question 15.

15. Describe how you would prepare 100 mL of 0.50*M* KCl starting with a stock solution that is 2.0*M* KCl.

Standardized Test Prep

1. c
2. b
3. a
4. C
5. E
6. B
7. A
8.

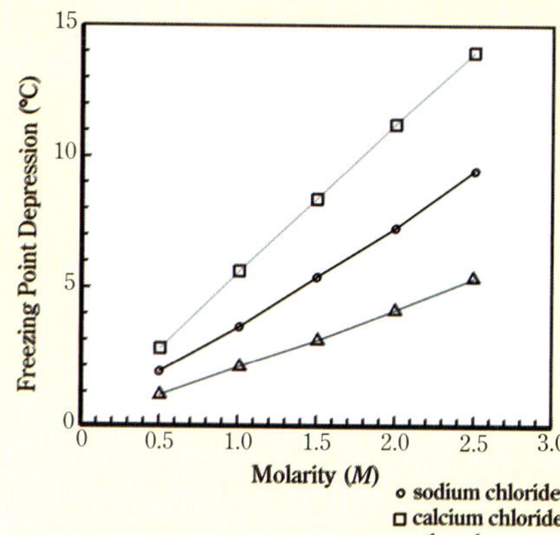

9. As the molarity increases, the freezing-point depression increases.
10. The slopes are an approximate 1:2:3 ratio that reflects the relative number of particles per mole of each solute in solution.
11. NaCl
12. b
13. a
14. b
15. Transfer 25 mL of the stock solution to a 100-ml volumetric flask and add water to make 100 mL of solution.

CHAPTER 17

Thermochemistry Planning Guide

SECTION OBJECTIVES	STANDARDS NATIONAL	STANDARDS STATE	ACTIVITIES and LABS
17.1 The Flow of Energy—Heat and Work, pp. 505–510 1 block or 2 periods **17.1.1 Explain** how energy, heat, and work are related. **17.1.2 Classify** processes as either exothermic or endothermic. **17.1.3 Identify** the units used to measure heat transfer. **17.1.4 Distinguish** between heat capacity and specific heat.	A-1, A-2, B-3, B-5		**SE** Inquiry Activity: Observing Heat Flow, p. 504 L2 **TE** Teacher Demo: An Endothermic Reaction, p. 506 L2 **TE** Class Activity: Heat Transfer, p. 508 L2 **LM** Lab 34: The Specific Heat of a Metal L2 **LP** Lab practical 17-1: Specific Heat L2 **VCL** Lab 16: The Specific Heat of a Metal L2
17.2 Measuring and Expressing Enthalpy Changes, pp. 511–519 1 block or 2 periods **17.2.1 Describe** how calorimeters are used to measure heat flow. **17.2.2 Construct** thermochemical equations. **17.2.3 Solve** for enthalpy changes in chemical reactions by using heats of reaction.	A-1, A-2, B-3, E-2, F-6, G-1		**TE** Class Activity: Calorimetry Measurements, p. 512 L2 **TE** Teacher Demo: An Exothermic Reaction, p. 515 L2 **LM** Lab 33: The Specific Heat of Metal L2 **LM** Lab 34: The Specific Heat of a Metal L2 **LM** Lab 35: Heats of Reaction L2 **LP** Lab Practical 17-1: Specific Heat L2 **LP** Lab Practical 17-2: Heat of Reaction L2 **VCL** Lab 17: Heat of Fusion of Water L2 **VCL** Lab 19: Heat of Combustion L2
17.3 Heat in Changes of State, pp. 520–526 1 block or 2 periods **17.3.1 Classify** the enthalpy change that occurs when a substance melts, freezes, boils, condenses, or dissolves. **17.3.2 Solve** for the enthalpy change that occurs when a substance melts, freezes, boils, condenses, or dissolves.	A-1, A-2, B-3		**SE** Quick Lab: Heat of Fusion of Ice, p. 522 L2 **TE** Class Activity: Melting and Boiling, p. 521 L2 **TE** Teacher Demo: Exothermic Solidification, p. 523 L2 **TE** Class Activity: Heating Curve for Ethanol, p. 524 L2 **TE** Class Activity: Hot and Cold Packs, p. 525 L2 **SSLM** Lab 27: Heat of Fusion of Ice L2 **PLM** Probeware Lab: Heat of Fusion of Ice L2
17.4 Calculating Heats of Reaction, pp. 527–533 1 block or 2 periods **17.4.1 State** Hess's law of heat summation and **describe** how it is used in chemistry. **17.4.2 Solve** for enthalpy changes by using Hess' law or standard heats of formation.	A-1, A-2, B-3		**SE** Small-Scale Lab: Heat of Combustion of a Candle, p. 533 L2 **LM** Lab 35: Heats of Reaction L2 **PLM** Probeware Lab: Heat of Combustion of a Candle L2 **LP** Lab Practical 17-2: Heat of Reaction L2 **VCL** Lab 18: Heats of Reaction L2

Ability Levels

- **L1** For students who need additional help
- **L2** For all students
- **L3** For students who need to be challenged

Components

SE	Student Edition	**LM**	Laboratory Manual	**GCP**	Graphing Calculator Problems	**T**	Transparencies
TE	Teacher's Edition	**SSLM**	Small-Scale Chemistry Lab Manual	**TTT**	Test-Taking Tips with Transparencies, Grades 9–12 Science	**CA**	Chem Alive and Chemistry Field Trips DVD
GRSW	Guided Reading and Study Workbook	**LP**	Laboratory Practicals	**STP**	Standardized Test Preparation Workbook	**iT**	Interactive Textbook with ChemASAP
CTR	Core Teaching Resources	**PLM**	Probeware Lab Manual			**GO**	Go Online Internet Resources
SSG	Spanish Study Guide	**VCL**	Virtual Chemistry Labs				
		CTB	Computer Test Bank				

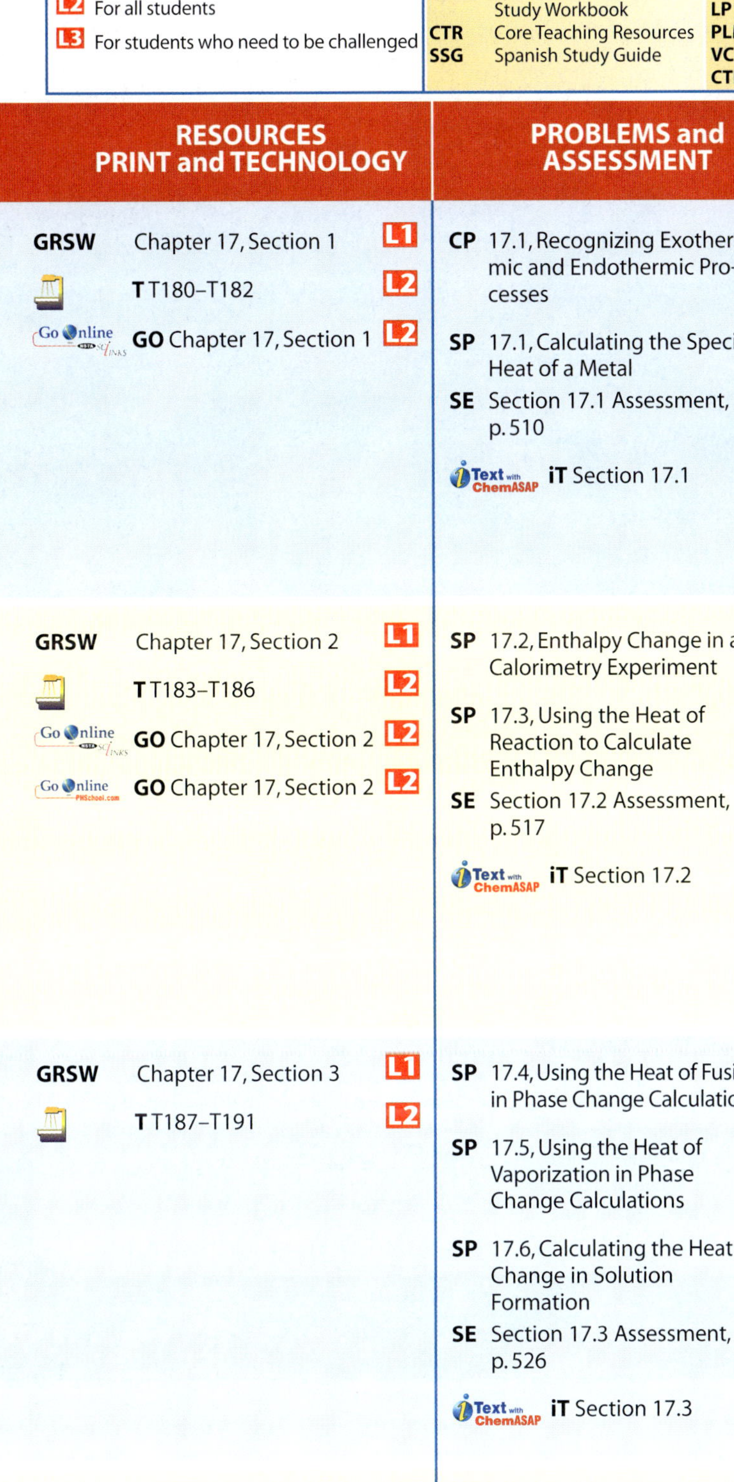

RESOURCES PRINT and TECHNOLOGY	PROBLEMS and ASSESSMENT
GRSW Chapter 17, Section 1 **L1** **T** T180–T182 **L2** **GO** Chapter 17, Section 1 **L2**	**CP** 17.1, Recognizing Exothermic and Endothermic Processes **SP** 17.1, Calculating the Specific Heat of a Metal **SE** Section 17.1 Assessment, p. 510 **iT** Section 17.1
GRSW Chapter 17, Section 2 **L1** **T** T183–T186 **L2** **GO** Chapter 17, Section 2 **L2** **GO** Chapter 17, Section 2 **L2**	**SP** 17.2, Enthalpy Change in a Calorimetry Experiment **SP** 17.3, Using the Heat of Reaction to Calculate Enthalpy Change **SE** Section 17.2 Assessment, p. 517 **iT** Section 17.2
GRSW Chapter 17, Section 3 **L1** **T** T187–T191 **L2**	**SP** 17.4, Using the Heat of Fusion in Phase Change Calculations **SP** 17.5, Using the Heat of Vaporization in Phase Change Calculations **SP** 17.6, Calculating the Heat Change in Solution Formation **SE** Section 17.3 Assessment, p. 526 **iT** Section 17.3
GRSW Chapter 17, Section 4 **L1** **T** T192–T195 **L2** **GO** Chapter 17, Section 4 **L2**	**SP** 17.7, Calculating the Standard Heat of Reaction **SE** Section 17.4 Assessment, p. 532 **iT** Section 17.4

Go Online

Web Code: cdn-1171
Web Code: cdn-1172
Web Code: cdn-1174

Web Code: cdb-1172

MATERIALS List

Quantities for each group

STUDENT EDITION

Inquiry Activity, p. 504
clean, medium-sized rubber band

Quick Lab, p. 522
100-mL graduated cylinder, hot water, foam cup, thermometer, temperature probe (optional), ice

Small-Scale Lab, p. 533
candle, metric ruler, aluminum foil, balance, safety matches, clock or watch, temperature probe (optional)

TEACHER'S EDITION

Teacher Demo, p. 506
barium hydroxide octahydrate, ammonium chloride. 250-mL Erlenmeyer flask with stopper

Class Activity, p. 508
sheet metal (13 cm × 18 cm), foam (13 cm × 18 cm), wood (40 cm × 20 cm), glue

Class Activity, p. 512
1*M* HCl, foam cup, thermometer, mossy zinc, magnesium turnings

Teacher Demo, p. 515
CaO, metal tray, thin aluminum pan, water, egg (raw)

Class Activity, p. 521
beaker, water, boiling chips, hot plate, thermometer, ice cubes, foam cup

Teacher Demo, p. 523
3 wide-mouth test tubes, sodium thiosulfate pentahydrate, hot water bath, 30°C water bath, thermometers, stirring rods

Class Activity, p. 525
hot and cold packs

Chapter Assessment

CHAPTER ASSESSMENT

SE	Chapter Assessment, pp. 535–538
CTB	Chapter 17
iT	Chapter 17

STANDARDIZED TEST PREP

SE	Chapter 17, p. 539
STP	Topic 5

iT Interactive Textbook with Assessment at PHSchool.com

CHAPTER 17

Thermochemistry

Observing Heat Flow

Purpose

Students will begin to
- infer that heat and temperature changes are related.
- understand that heat flows from a warm object to a cooler object.

Prep Time 5 minutes
Class Time 10 minutes

Think About It

1. The rubber band felt warmer after it was stretched.
2. It felt cooler after it was allowed to relax.
3. Students may infer that heat is related to the observed temperature changes. From their experiences, students may realize that heat flows from a warm object to a cool object. The stretching of rubber is exothermic. The reverse process (relaxation of the rubber) is endothermic.

CHAPTER 17

Thermochemistry

A marshmallow absorbs heat as it is toasted over a fire.

INQUIRY Activity

Observing Heat Flow

Materials

a clean, medium-sized rubber band

Procedure

1. Hook your index fingers through each end of the rubber band. Without stretching the rubber band, place it against your upper lip or forehead. Note the temperature of the rubber band. **CAUTION** *If you are allergic to latex, do not handle rubber bands.*
2. Move the rubber band away from your skin, quickly stretch and hold it, and then place it back against your skin. Note any temperature change.
3. Fully stretch the rubber band, and then allow it to return to its original shape. Place it against your skin and note any temperature change.
4. Repeat Steps 2 and 3 until you are certain of the temperature change in each step.

Think About It

1. Did the rubber band feel cooler or warmer after it was stretched in Step 2?
2. Did the rubber band feel cooler or warmer after it returned to its original shape in Step 3?
3. Think about the temperature changes you observed and form some initial answers to the following questions. What is heat? In what direction does heat flow?

17.1 The Flow of Energy—Heat and Work

Connecting to Your World Lava flowing out of an erupting volcano is very hot. Its temperature ranges from 550°C to 1400°C. As lava flows down the side of a volcano, it loses heat and begins to cool slowly. In some instances, the lava may flow into the ocean, where it cools more rapidly. In this section, you will learn about heat flow and why some substances cool down or heat up more quickly than others.

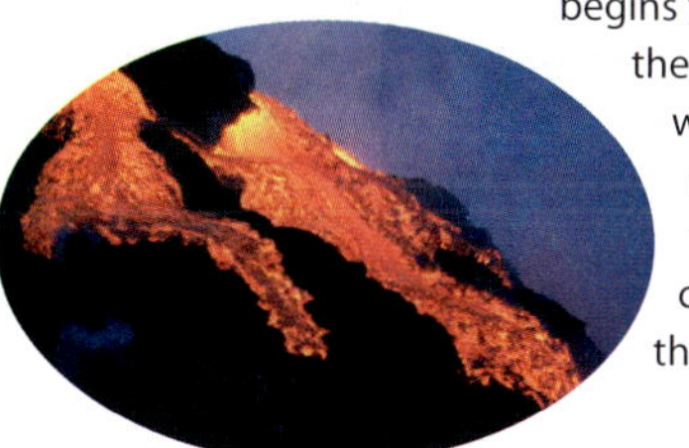

Guide for Reading

Key Concepts
- In what direction does heat flow?
- What happens in endothermic and exothermic processes?
- In what units is heat flow measured?
- On what two factors does the heat capacity of an object depend?

Vocabulary
thermochemistry
chemical potential energy
heat
system
surroundings
law of conservation of energy
endothermic process
exothermic process
heat capacity
specific heat

Reading Strategy
Building Vocabulary As you read the section, write a definition of each vocabulary term in your own words.

Energy Transformations

Energy is the capacity for doing work or supplying heat. Unlike matter, energy has neither mass nor volume. Energy is detected only because of its effects—for example, the motion of the race car in Figure 17.1.

Thermochemistry is the study of energy changes that occur during chemical reactions and changes in state. Every substance has a certain amount of energy stored inside it. The energy stored in the chemical bonds of a substance is called **chemical potential energy.** The kinds of atoms and their arrangement in the substance determine the amount of energy stored in the substance.

During a chemical reaction, a substance is transformed into another substance with a different amount of chemical potential energy. When you buy gasoline, you are actually buying the stored potential energy it contains. The controlled explosions of the gasoline in a car's engine transform the potential energy into useful work, which can be used to propel the car. At the same time, however, heat is also produced, making the car's engine extremely hot. Energy changes occur as either heat transfer or work, or a combination of both.

Heat, represented by q, is energy that transfers from one object to another because of a temperature difference between them. One of the effects of adding heat to an object is an increase in its temperature. It is the radiant heat of the sun's rays that makes a summer day hot. In this example, the air is the object that absorbs heat and increases in temperature. **Heat always flows from a warmer object to a cooler object.** If two objects remain in contact, heat will flow from the warmer object to the cooler object until the temperature of both objects is the same.

Figure 17.1 When fuel is burned in a car engine, chemical potential energy is released and is used to do work.

Section Resources

Print
- **Guided Reading and Study Workbook,** Section 17.1
- **Core Teaching Resources,** Section 17.1 Review
- **Transparencies,** T180–T182
- **Laboratory Manual,** Lab 34

Technology
- **Interactive Textbook with ChemASAP,** Problem-Solving 17.2, 17.4, Assessment 17.1
- **Go Online,** Section 17.1
- **Virtual Chemistry Labs,** Lab 16

17.1

1 FOCUS

Objectives

17.1.1 Explain how energy, heat, and work are related.
17.1.2 Classify processes as either exothermic or endothermic.
17.1.3 Identify the units used to measure heat transfer.
17.1.4 Distinguish between heat capacity and specific heat.

Guide for Reading

Build Vocabulary L2

Paraphrase Ask students to write a definition in their own words for each vocabulary term, and then write a complete sentence with the term.

Reading Strategy L2

Use Prior Knowledge To assess whether students know the difference between temperature and energy, ask, **What does a thermometer measure?** *(average kinetic energy)* **Does a thermometer measure heat?** *(no)*

2 INSTRUCT

Connecting to Your World

Ask, **What happens to the heat in lava once it flows out of a volcano?** *(The heat is released.)* **Why does lava cool more quickly in water than on land?** *(Water has a greater capacity to absorb heat than does air; the temperature difference between the lava and water may be greater than between the lava and the air.)*

Energy Transformations

Use Visuals L1

Figure 17.1 Have students study the photograph of the race car. Ask, **What work is done as the chemical potential energy of the fuel is released?** *(The race car moves around the track.)* **In addition to work, what other type of energy change occurs as the fuel is burned?** *(heat is released)*

Section 17.1 (continued)

Exothermic and Endothermic Processes

Discuss

L2

Discuss with students the distinction between kinetic and potential energy. Kinetic energy is the energy associated with an object because of its motion. Potential energy is the energy associated with an object because of its position in a field of force or due to its particular chemical composition. Explain that the potential energy of a reactant or a product in a chemical reaction is determined by the strengths of the attractive and repulsive forces between atoms. In a chemical reaction, atoms are rearranged into new groupings that have different relative potential energies. The change in potential energy is either the result of absorption of energy from the surroundings (endothermic reaction) or the release of energy to the surroundings (exothermic reaction).

TEACHER Demo

An Endothermic Reaction

L2

Purpose Students will observe an endothermic reaction between two solids.

Materials barium hydroxide octahydrate, ammonium chloride, 250-mL Erlenmeyer flask with stopper

Safety Barium salts are poisonous if ingested. Inhalation of concentrated ammonia vapors can be dangerous. Perform this demo in a fume hood or a well-ventilated room. Disposal: Flush the solution down the drain with excess water.

Procedure Mix 32 g of barium hydroxide octahydrate ($Ba(OH)_2 \cdot 8H_2O$) with 11 g of ammonium chloride (NH_4Cl) in a 250-mL Erlenmeyer flask fitted with a stopper. Swirl to mix.

Expected Outcome The flask becomes extremely cold. Place the flask on a wet piece of wood and the flask will freeze to the wood.

(Reference: *Chemical Demonstrations* by Bassam Z. Shakhashiri, The University Press, 1983)

Word *Origins*

Exothermic comes from the Greek words *exo*, meaning "outside of," and *therme*, meaning "heat." **Based on the characteristics of endothermic reactions, what do you think the Greek prefix *endo-* means?**

Exothermic and Endothermic Processes

Chemical reactions and changes in physical state generally involve either the release or the absorption of heat. In studying energy changes, you can define a **system** as the part of the universe on which you focus your attention. The **surroundings** include everything else in the universe. In thermochemical experiments, you can consider the region in the immediate vicinity of the system as the surroundings. Together, the system and its surroundings make up the universe. A major goal of thermochemistry is to examine the flow of heat between the system and its surroundings. The **law of conservation of energy** states that in any chemical or physical process, energy is neither created nor destroyed. If the energy of the system decreases during that process, the energy of the surroundings must increase by the same amount so that the total energy of the universe remains unchanged.

In thermochemical calculations, the direction of the heat flow is given from the point of view of the system. An **endothermic process** is one that absorbs heat from the surroundings. **In an endothermic process, the system gains heat as the surroundings cool down.** In Figure 17.2a, the system (the person) gains heat from its surroundings (the fire). Heat flowing into a system from its surroundings is defined as positive; q has a positive value. An **exothermic process** is one that releases heat to its surroundings. **In an exothermic process, the system loses heat as the surroundings heat up.** In Figure 17.2b, the system (the body) loses heat to the surroundings (the perspiration on the skin, and the air). Heat flowing out of a system into its surroundings is defined as negative; q has a negative value because the system is losing heat.

Checkpoint ***What does the law of conservation of energy state?***

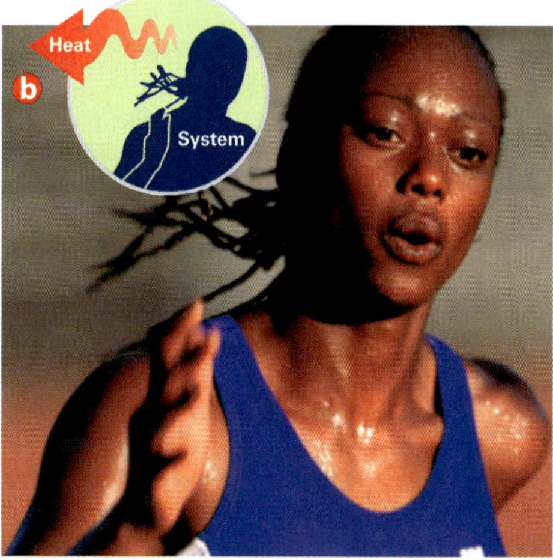

Figure 17.2 Heat flow is defined from the point of view of the system. **a** In an endothermic process, heat flows into the system from the surroundings. **b** In an exothermic process, heat flows from the system to the surroundings. In both cases, energy is conserved.
Interpreting Diagrams ***In which process does q have a negative value?***

Word *Origins*

L2

The Greek prefix *endo*-means "within." Some examples of words that use this prefix are *endoskeleton*, *endomorph*, and *endoderm*.

Differentiated Instruction

Less Proficient Readers L1

Have students write definitions for each of the key terms in this chapter. Special attention should be given to those terms that are defined by mathematical relationships. Example calculations showing how to apply the formulas should be included in their notes.

CONCEPTUAL PROBLEM 17.1

Recognizing Exothermic and Endothermic Processes

On a sunny winter day, the snow on a rooftop begins to melt. As the meltwater drips from the roof, it refreezes into icicles. Describe the direction of heat flow as the water freezes. Is this process endothermic or exothermic?

1 Analyze ***Identify the relevant concepts.***

Heat always flows from a warmer object to a cooler object. An endothermic process absorbs heat from the surroundings. An exothermic process releases heat to the surroundings.

2 Solve ***Apply concepts to this situation.***

In order for water to freeze, its temperature must decrease. So heat must flow out of the water (the system). Because heat is released from the system to the surroundings (the air), the process is exothermic.

Practice Problems

1. A container of melted paraffin wax is allowed to stand at room temperature until the wax solidifies. What is the direction of heat flow as the liquid wax solidifies? Is the process exothermic or endothermic?

2. When solid barium hydroxide octahydrate ($Ba(OH)_2 \cdot 8H_2O$) is mixed in a beaker with solid ammonium thiocyanate (NH_4SCN), a reaction occurs. The beaker quickly becomes very cold. Is the reaction exothermic or endothermic?

Problem-Solving 17.1 Solve Problem 1 with the help of an interactive guided tutorial.

with ChemASAP

Units for Measuring Heat Flow

Describing the amount of heat flow requires units different than those used to describe temperature.  **Heat flow is measured in two common units, the calorie and the joule.**

You have probably heard of someone exercising to "burn calories." During exercise your body breaks down sugars and fats into carbon dioxide and water, and this process releases heat. Although there is not an actual fire burning the sugars and fats within your body, chemical reactions accomplish the same result. In breaking down 10 g of sugar, for example, your body releases a certain amount of heat. The same amount of heat would be released if 10 g of sugar were completely burned in a fire.

A calorie (cal) is defined as the quantity of heat needed to raise the temperature of 1 g of pure water 1°C. The word *calorie* is written with a small c except when referring to the energy contained in food. The dietary Calorie, written with a capital C, always refers to the energy in food. One dietary Calorie is actually equal to one kilocalorie, or 1000 calories.

$$1 \text{ Calorie} = 1 \text{ kilocalorie} = 1000 \text{ calories}$$

The statement "10 g of sugar has 41 Calories" means that 10 g of sugar releases 41 kilocalories of heat when completely burned.

The joule is the SI unit of energy. One joule of heat raises the temperature of 1 g of pure water 0.2390°C. You can convert between calories and joules using the following relationships.

$$1 \text{ J} = 0.2390 \text{ cal} \qquad 4.184 \text{ J} = 1 \text{ cal}$$

CONCEPTUAL PROBLEM 17.1

Answers

1. Heat flows from the system (paraffin) to the surroundings (air). The process is exothermic.
2. Since the beaker becomes cold, heat is absorbed by the system (chemicals within the beaker) from the surroundings (beaker and surrounding air). The process is endothermic.

Practice Problems Plus 

Chapter 17 Assessment problem 47 is related to Conceptual Problem 17.1.

Units for Measuring Heat Flow

Discuss 

Make sure students understand the terms introduced in this section and how to convert from one to another. Students may have difficulty with the distinction between *calorie* and *Calorie* because everyday usage of the terms is often not accurate. The distinction between *heat capacity* and *specific heat* can be hard to remember because the terms are so closely related and sound so similar. Tell students to remember that *specific* heats of different materials can be compared because the quantity of matter involved (1 g) is *specified.*

Facts and Figures

Significant Figures

Point out that the concept of significant figures applies only to measured quantities. Students may wonder why an estimated digit can be considered significant. Tell them a significant figure is one that is known to be reasonably reliable. A careful estimate fits this criterion. On the chalkboard, illustrate different examples—such as a liquid in a graduated cylinder or beaker—and discuss where the estimated digit can occur. Go over the rules for determining which digits are significant and give examples of each. Stress that zeros that serve only as placeholders are not significant. For example, the numbers 3200 and 0.0032 both have 2 significant figures.

Answers to...

Figure 17.2 In the process shown in Figure 17.2b, the system loses heat to its surroundings, so q is negative.

The law of conservation of energy states that in any chemical or physical process, energy is neither created nor destroyed.

Heat Capacity and Specific Heat

Use Visuals

Table 17.1 Have students study Table 17.1. Ask, **Which has a higher specific heat capacity—water or steam?** *(water)* **What factors do you think affect the specific heat of each substance?** *(Specific heat is affected by the amount of heat and the change in temperature.)*

Download a worksheet on **Specific Heat** for students to complete, and find additional teacher support from NSTA SciLinks.

CLASS Activity

Heat Transfer

Purpose Students will compare heat transfer of different materials.

Materials sheet metal (13 cm × 18 cm), foam (13 cm × 18 cm), wood (40 cm × 20 cm), glue

Procedure Glue a 13 cm × 18 cm piece of sheet metal and a 13 cm × 18 cm piece of foam to a piece of wood that is about 40 cm × 20 cm. Pass the wood around the classroom, and ask students to put their hands on each of the three surfaces and describe the temperature of each.

Expected Outcome Usually students indicate that the metal feels the coldest and the foam feels the same temperature as their hands. Explain that all of the surfaces are colder than their hand, and that heat is being transferred from their hand to those surfaces. Metals conduct heat away from the hand more efficiently than foam, so they feel colder than foam.

For: Links on specific heat
Visit: www.SciLinks.org
Web Code: cdn-1171

Heat Capacity and Specific Heat

The amount of heat needed to increase the temperature of an object exactly 1°C is the **heat capacity** of that object. **The heat capacity of an object depends on both its mass and its chemical composition.** The greater the mass of the object, the greater its heat capacity. One of the massive steel girders in Figure 17.3, for example, requires much more heat to raise its temperature 1°C than a small steel nail does. Similarly, a cup of water has a much greater heat capacity than a drop of water.

Different substances with the same mass may have different heat capacities. On a sunny day, a 20-kg puddle of water may be cool, while a nearby 20-kg iron sewer cover may be too hot to touch. This situation illustrates how different heat capacities affect the temperature of objects. Assuming that both the water and the iron absorb the same amount of radiant energy from the sun, the temperature of the water changes less than the temperature of the iron because the specific heat capacity of water is larger.

The specific heat capacity, or simply the **specific heat,** of a substance is the amount of heat it takes to raise the temperature of 1 g of the substance 1°C. Table 17.1 gives specific heats for some common substances. Water has a very high specific heat compared with the other substances in the table. You can see from the table that one calorie of heat raises the temperature of 1 g of water 1°C. Metals, however, have low specific heats. One calorie of heat raises the temperature of 1 g of iron 9°C. So water has a specific heat nine times that of iron. The same amount of heat affects the temperature of objects with a high specific heat much less than the temperature of those with a low specific heat.

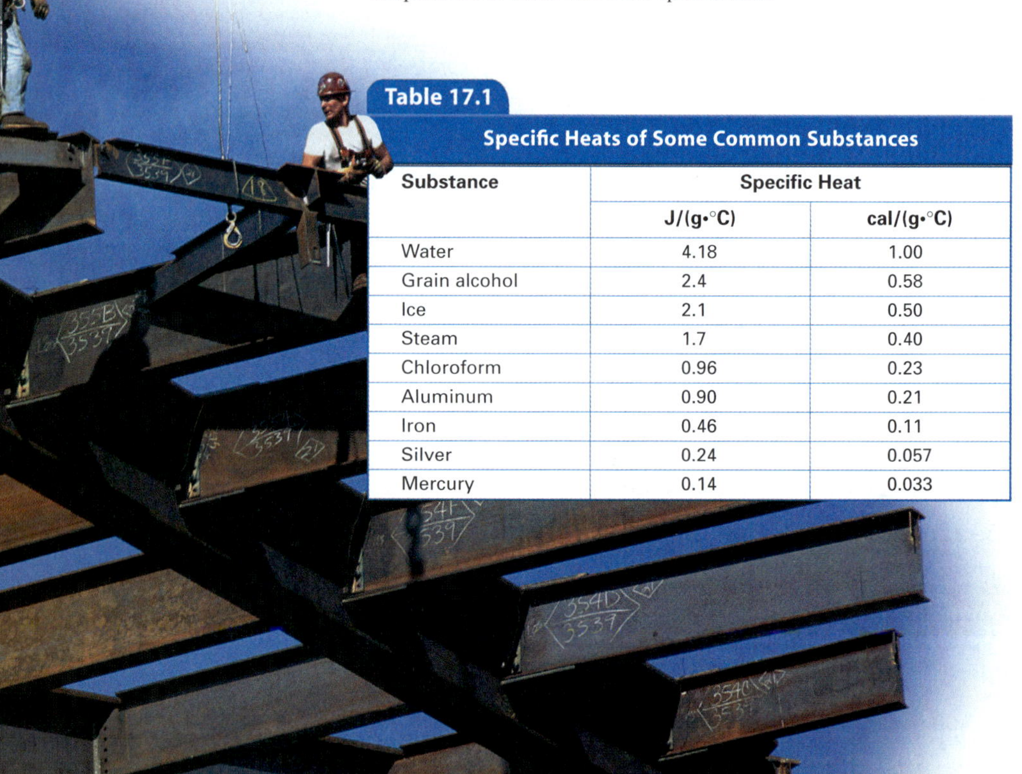

Figure 17.3 A massive steel girder has a higher heat capacity than a steel nail. **Inferring** ***What factors affect how quickly the different areas of a construction site will heat up during the day?***

Table 17.1

Specific Heats of Some Common Substances

Substance	Specific Heat	
	J/(g·°C)	cal/(g·°C)
Water	4.18	1.00
Grain alcohol	2.4	0.58
Ice	2.1	0.50
Steam	1.7	0.40
Chloroform	0.96	0.23
Aluminum	0.90	0.21
Iron	0.46	0.11
Silver	0.24	0.057
Mercury	0.14	0.033

Figure 17.4 Water releases a lot of heat as it cools. (a) During freezing weather, farmers protect citrus crops by spraying them with water. The ice that forms has a protective effect as long as its temperature does not drop much below 0°C. (b) Because it is mostly water, the filling of a hot apple pie is much more likely to burn your tongue than the crust.

Just as it takes a lot of heat to raise the temperature of water, water also releases a lot of heat as it cools. Water in lakes and oceans absorbs heat from the air on hot days and releases it back into the air on cool days. This property of water is responsible for moderate climates in coastal areas. Figure 17.4 illustrates two other common effects associated with the high specific heat of water. The orange trees in Figure 17.4a have been sprayed with water to protect the fruit from frost damage during icy weather. As the water freezes, it releases heat, which helps prevent the fruit from freezing. In Figure 17.4b, the label on a box of apple pie warns that the "filling is hot." When a freshly baked apple pie comes out of the oven, both the filling and crust are at the same temperature. However, the filling, which is mostly water, has a higher specific heat than the crust. In order to cool down, the filling must give off a lot of heat. This is why you have to be careful not to burn your tongue when eating hot apple pie.

To calculate the specific heat (C) of a substance, you divide the heat input by the temperature change times the mass of the substance.

$$C = \frac{q}{m \times \Delta T} = \frac{\text{heat (joules or calories)}}{\text{mass (g)} \times \text{change in temperature (°C)}}$$

In the equation above, q is heat and m is mass. The symbol ΔT (read "delta T") represents the change in temperature. ΔT is calculated from the equation $\Delta T = T_f - T_i$, where T_f is the final temperature and T_i is the initial temperature. As you can see from the equation, specific heat may be expressed in terms of joules or calories. Therefore, the units of specific heat are either J/(g·°C) or cal/(g·°C).

 Checkpoint ***What is the relationship between specific heat and heat capacity?***

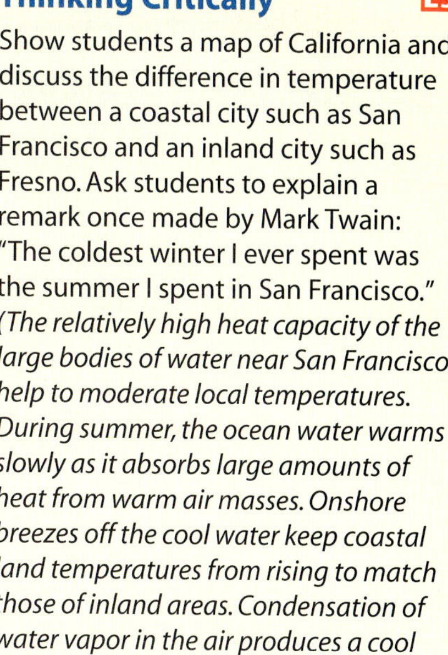

Thinking Critically L3

Show students a map of California and discuss the difference in temperature between a coastal city such as San Francisco and an inland city such as Fresno. Ask students to explain a remark once made by Mark Twain: "The coldest winter I ever spent was the summer I spent in San Francisco." *(The relatively high heat capacity of the large bodies of water near San Francisco help to moderate local temperatures. During summer, the ocean water warms slowly as it absorbs large amounts of heat from warm air masses. Onshore breezes off the cool water keep coastal land temperatures from rising to match those of inland areas. Condensation of water vapor in the air produces a cool fog, which probably enhanced Mark Twain's impression of the cool of San Francisco summers.)*

Use Visuals L1

Figure 17.4 Have students refer to Figure 17.4. Discuss how spraying fruit and flooding orchards with a few inches of water can keep the fruit from freezing. *(As water cools and freezes, it releases energy that can warm the air and protect the fruit.)*

Facts and Figures

Dietary Calories

Your proper caloric intake depends on your level of physical activity. In an eight-hour day at a desk, you burn about 800 Calories. This is about the number of Calories in two helpings of spaghetti. When exercising, however, you become a relative biochemical blast furnace. In vigorous activities such as running and jumping, you expend 7–10 Calories per minute, or 420–600 Calories per hour. At these rates, a runner who covers a 26-mile marathon course in 3 hours might expend 1800 Calories, or the equivalent of 4.5 helpings of spaghetti. Have students do research to compare the energy expended (in Calories per minute) during several types of sports related and non-sports related activities.

Answers to...

 **Checkpoint** The specific heat is the heat capacity of an object divided by its mass in grams.

Sample Problem 17.1

Answers

3. 2.0 J/(g•°C)
4. 1.8 kJ

Practice Problems Plus 

How much heat is required to raise the temperature of 400.0 g of silver 45°C? *(4.3×10^3 J)*

Math Handbook

For a math refresher and practice, direct students to algebraic equations, page R69.

3 ASSESS

Evaluate Understanding 

Place an ice cube in a beaker of cool water. Have students discuss the flow of heat when the ice and water are the system and surroundings, respectively, and when the ice/water mixture and a 37°C room are the system and surroundings, respectively.

Reteach 

Emphasize that substances vary in their response to an input of heat. A given amount of heat raises the temperature of some substances (such as metals) far more than others (such as nonmetals). Point out that specific heat is a property of a substance. It is a measure of the ability of a substance to store heat.

Writing Activity

The water has a higher heat capacity than the concrete. Thus, the sun's heat raises the temperature of the concrete more than that of the water.

If your class subscribes to the Interactive Textbook, use it to review key concepts in Section 17.1.

with ChemASAP

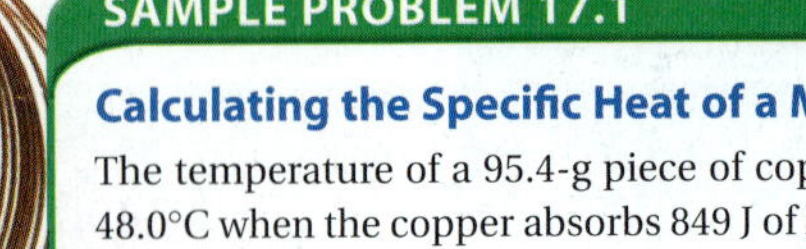

SAMPLE PROBLEM 17.1

Calculating the Specific Heat of a Metal

The temperature of a 95.4-g piece of copper increases from 25.0°C to 48.0°C when the copper absorbs 849 J of heat. What is the specific heat of copper?

1 Analyze ***List the knowns and the unknown.***

Knowns

- $m_{Cu} = 95.4 g$
- $\Delta T = (48.0°C - 25.0°C) = 23.0°C$
- $q = 849$ J

Unknown

- $C_{Cu} = ?$ J/(g·°C)

2 Calculate ***Solve for the unknown.***

Use the known values and the definition of specific heat, $C = \frac{q}{m \times \Delta T}$, to calculate the unknown value C_{Cu}.

$$C_{Cu} = \frac{q}{m \times \Delta T} = \frac{849 \text{ J}}{95.4 \text{ g} \times 23.0°\text{C}} = 0.387 \text{ J/(g·°C)}$$

3 Evaluate ***Does the result make sense?***

Water has a very high specific heat (4.18 J/(g·°C)). Metals, however, have low specific heats, so the calculated value of 0.387 J/(g·°C) seems reasonable.

Practice Problems

3. When 435 J of heat is added to 3.4 g of olive oil at 21°C, the temperature increases to 85°C. What is the specific heat of the olive oil?
4. How much heat is required to raise the temperature of 250.0 g of mercury 52°C?

Math Handbook

For help with algebraic equations, go to page R69.

Interactive Textbook

Problem-Solving 17.4 Solve Problem 4 with the help of an interactive guided tutorial.

with ChemASAP

17.1 Section Assessment

5. **Key Concept** In what direction does heat flow between two objects?
6. **Key Concept** How do endothermic processes differ from exothermic processes?
7. **Key Concept** What units are used to measure heat flow?
8. **Key Concept** On what factors does the heat capacity of an object depend?
9. Using calories, calculate how much heat 32.0 g of water absorbs when it is heated from 25.0°C to 80.0°C. How many joules is this?
10. A chunk of silver has a heat capacity of 42.8 J/°C and a mass of 181 g. Calculate the specific heat of silver.
11. How many kilojoules of heat are absorbed when 1.00 L of water is heated from 18°C to 85°C?

Writing Activity

Explanatory Paragraph Use the concept of heat capacity to explain why on a sunny day the concrete deck around an outdoor swimming pool becomes hot, while the water stays cool.

Assessment 17.1 Test yourself on the concepts in Section 17.1.

with ChemASAP

Section 17.1 Assessment

5. Heat flows from the object of higher temperature to the object of lower temperature.
6. Endothermic processes absorb heat from the surroundings; exothermic processes release heat to the surroundings.
7. calories and joules
8. mass and chemical composition
9. 1.76×10^3 cal (1.76 kcal); 7.36×10^3 J (7.36 kJ)
10. 2.36×10^{-1} J/(g•°C)
11. 2.8×10^2 kJ

17.2 Measuring and Expressing Enthalpy Changes

Connecting to Your World As you know, a burning match gives off heat. When you strike a match, heat is released to the surroundings in all directions. You can classify this reaction as exothermic. In addition to describing the direction of heat flow, you may also want to determine the quantity of heat that is transferred. How much heat does a burning match release to the surroundings? In this section, you will learn how you can measure heat flow in chemical and physical processes by applying the concept of specific heat.

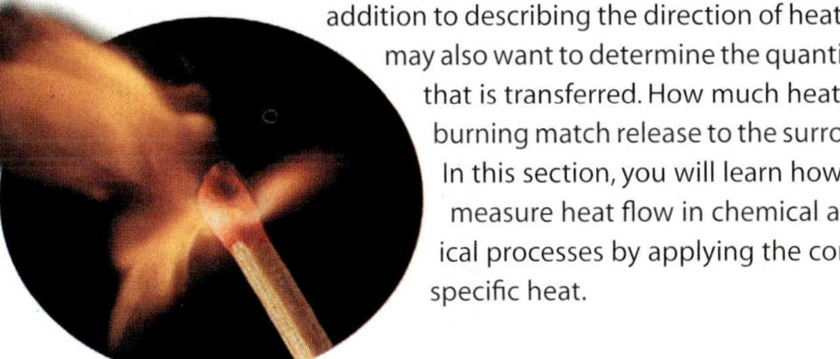

Guide for Reading

Key Concepts
- What basic concepts apply to calorimetry?
- How can you express the enthalpy change for a reaction in a chemical equation?

Vocabulary
calorimetry
calorimeter
enthalpy
thermochemical equation
heat of reaction
heat of combustion

Reading Strategy
Relating Text and Visuals As you read, look at Figure 17.7. Explain how these diagrams help you understand exothermic and endothermic processes.

Calorimetry

Heat that is released or absorbed during many chemical reactions can be measured by a technique called calorimetry. **Calorimetry** is the precise measurement of the heat flow into or out of a system for chemical and physical processes. **In calorimetry, the heat released by the system is equal to the heat absorbed by its surroundings. Conversely, the heat absorbed by a system is equal to the heat released by its surroundings.** The insulated device used to measure the absorption or release of heat in chemical or physical processes is called a **calorimeter.**

Constant-Pressure Calorimeters Foam cups are excellent heat insulators. Because they do not let much heat in or out, they can be used as simple calorimeters. In fact, the heat flows for many chemical reactions can be measured in a constant-pressure calorimeter similar to the one shown in Figure 17.5. Because most chemical reactions and physical changes carried out in the laboratory are open to the atmosphere, these changes occur at constant pressure. The heat content of a system at constant pressure is the same as a property called the **enthalpy** (H) of the system. The heat released or absorbed by a reaction at constant pressure is the same as the change in enthalpy, symbolized as ΔH. Because the reactions presented in this textbook occur at constant pressure, the terms *heat* and *enthalpy change* are used interchangeably. In other words, $q = \Delta H$.

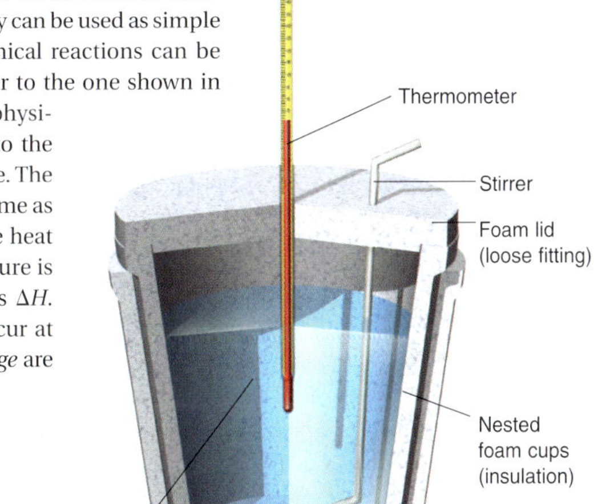

Figure 17.5 In a simple constant-pressure calorimeter, a thermometer records the temperature change as chemicals react in water. The substances reacting in solution constitute the system. The water constitutes the surroundings.

Section Resources

Print
- **Guided Reading and Study Workbook,** Section 17.2
- **Core Teaching Resources,** Section 17.2 Review, Interpreting Graphics
- **Transparencies,** T183-T186
- **Laboratory Manual,** Labs 34, 35

Technology
- **Interactive Textbook with ChemASAP,** Problem-Solving 17.13, 17.15, Simulation 22, Assessment 17.2
- **Go Online,** Section 17.2
- **Virtual Chemistry Labs,** 17, 19

17.2

1 FOCUS

Objectives

17.2.1 Describe how calorimeters are used to measure heat flow.

17.2.2 Construct thermochemical equations.

17.2.3 Solve for enthalpy changes in chemical reactions by using heats of reaction.

Guide for Reading

Build Vocabulary L2 

Concept Map Have students make a concept map using the vocabulary terms in this section.

Reading Strategy L2 

Directed Reading/Thinking Activity Ask students if there is a way to measure how much heat is released from a burning match. *(Yes, but only indirectly. If the reaction were confined, then any temperature changes in the surroundings could be attributed to heat transfer from the reaction.)*

2 INSTRUCT

Connecting to Your World

Tell students that a match won't ignite unless you strike it and add heat to it. Ask, **Where does the heat that is added to the match come from?** *(friction)* **Is the burning of a match an endothermic reaction? Explain.** *(No; the reaction releases more energy, in the form of heat and light, than the amount it absorbs to start.)*

Calorimetry

Relate L2 

Show students typical utility bills for electricity, gas, or oil. Ask, **Why would the size of the bill vary from one household to another?** *(floor space, number of occupants, thermostat setting, insulation, type of fuel used)* **Why might the bill vary from month to month?** *(changes in the gradient between internal and external temperatures, fluctuations in fuel prices)*

CLASS Activity

Calorimetry Measurements L2

Purpose Students will make calorimetry measurements and calculations. Two separate but related oxidation-reduction reactions will be studied by different groups of students. The results can then be pooled and compared to provide insight into how calorimetry can be used as a tool to investigate the physical and chemical properties of substances.

Materials 1*M* HCl, foam cup, thermometer, mossy zinc, magnesium turnings

Safety Students must wear protective goggles and aprons. Remind them of the safety precautions needed for handling HCl. Use a fume hood if possible. Students should not be allowed to dispense concentrated HCl themselves.

Procedure Pour 100 mL of 1*M* HCl into a foam cup and determine the temperature to the nearest tenth of a degree Celsius. Add 0.5 g of mossy zinc or 0.5 g of magnesium turnings.

Expected Outcome For calculations, ignore the mass of the solids added. Assume that HCl is primarily water and has the same specific heat and density of water.

$\Delta H = 100\ g \times 4.18\ J/(g \cdot °C) \times \Delta T$

$\Delta H_{rxn}Zn = -150\ kJ/mol$

$\Delta H_{rxn}Mg = -460\ kJ/mol$

Discuss sources of error such as heat lost to surroundings.

References:

1. Shakhashiri, B.Z. *Chemical Demonstrations*, Vol. 1, University Press, 1983, 1.10

2. Alyea, H.N. *J.Chem. Educ.* 1970, 47, A-387

Go Online NSTA SciLINKS

Download a worksheet on **Enthalpy** for students to complete, and find additional teacher support from NSTA SciLinks.

For: Links on enthalpy
Visit: www.SciLinks.org
Web Code: cdn-1172

To measure the enthalpy change for a reaction in aqueous solution in a foam cup calorimeter, you dissolve the reacting chemicals (the system) in known volumes of water (the surroundings). Then measure the initial temperature of each solution and mix the solutions in the foam cup. After the reaction is complete, measure the final temperature of the mixed solutions. Because you know the initial and final temperatures and the heat capacity of water, you can calculate the heat absorbed by the surroundings (q_{surr}) using the formula for specific heat.

$$q_{surr} = m \times C \times \Delta T$$

In this expression, m is the mass of the water; C is the specific heat of water; and $\Delta T = T_f - T_i$. Because the heat absorbed by the surroundings is equal to, but has the opposite sign of, the heat released by the system, the enthalpy change for the reaction (ΔH) can be written as follows.

$$q_{sys} = \Delta H = -q_{surr} = -m \times C \times \Delta T$$

The sign of ΔH is negative for an exothermic reaction and positive for an endothermic reaction.

Constant-Volume Calorimeters Calorimetry experiments can also be performed at constant volume using a device called a bomb calorimeter. In a bomb calorimeter, shown in Figure 17.6, a sample of a compound is burned in a constant-volume chamber in the presence of oxygen at high pressure. The heat that is released warms the water surrounding the chamber. By measuring the temperature increase of the water, it is possible to calculate the quantity of heat released during the combustion reaction.

Checkpoint ***What formula is used to calculate the enthalpy change in a constant-pressure calorimeter?***

Figure 17.6 Nutritionists use bomb calorimeters to measure the energy content of the foods you eat. To see some of their data (expressed in Calories per serving), you can look at a nutrition label. **Calculating** ***According to the nutrition label below, each serving contains 140 Calories. How many kilojoules is this?***

Nutrition Facts
Serving Size 1 Cookie (26g / 0.9oz.)
Servings Per Container 8

Amount Per Serving	
Calories 140	Calories from Fat 70
	% Daily Value*
Total Fat 8g	**12%**
Saturated Fat 2.5g	**13%**
Polyunsaturated Fat 0g	
Monounsaturated Fat 3g	
Cholesterol 10mg	**3%**
Sodium 80mg	**3%**
Total Carbohydrate 16g	**5%**
Dietary Fiber 0g	0%

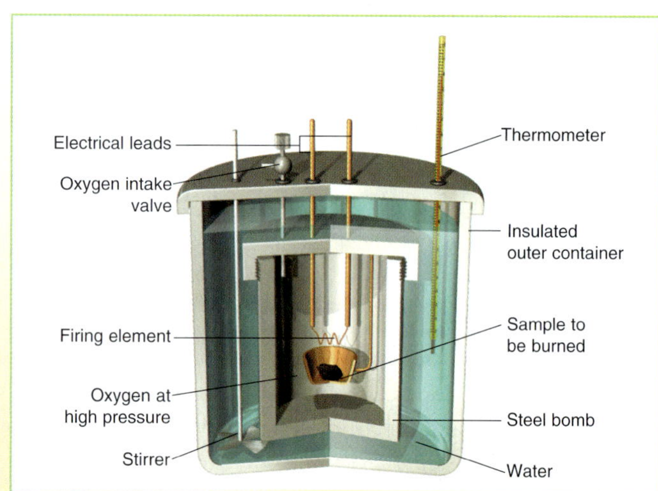

Differentiated Instruction

Gifted and Talented L3

A large percentage of the electricity used in the United States is produced by the burning of fossil fuels. Power plants generate electricity by transferring heat to water during the combustion of fossil fuels. The heated water turns to steam, which drives turbines that generate electricity. In order to condense and recycle the steam, a source of coolant water is also needed. Consequently, power plants (including nuclear power plants) are often built near a source of water, such as an ocean, lake, or river. However, when the coolant water is returned to its source, it is usually warmer. This increase in temperature is called thermal pollution. Have students do research to find out if this label is justified.

SAMPLE PROBLEM 17.2

Enthalpy Change in a Calorimetry Experiment

When 25.0 mL of water containing 0.025 mol HCl at 25.0°C is added to 25.0 mL of water containing 0.025 mol NaOH at 25.0°C in a foam cup calorimeter, a reaction occurs. Calculate the enthalpy change (in kJ) during this reaction if the highest temperature observed is 32.0°C. Assume the densities of the solutions are 1.00 g/mL.

1 Analyze ***List the knowns and the unknown.***

Knowns

- $C_{water} = 4.18\ J/(g\cdot°C)$
- $V_{final} = V_{HCl} + V_{NaOH} = 25.0\ mL + 25.0\ mL = 50.0\ mL$
- $T_i = 25.0°C$
- $T_f = 32.0°C$
- $Density_{solution} = 1.00\ g/mL$

Unknown

- $\Delta H = ?\ kJ$

Use dimensional analysis to determine the mass of the water. You must also calculate ΔT. Use $\Delta H = -q_{surr} = -m \times C \times \Delta T$ to solve for ΔH.

2 Calculate ***Solve for the unknown.***

First, calculate the total mass of the water.

$$m = (50.0\ \cancel{mL}) \times \left(\frac{1.00\ g}{\cancel{mL}}\right) = 50.0\ g$$

Now calculate ΔT.

$$\Delta T = T_f - T_i = 32.0°C - 25.0°C = 7.0°C$$

Use the values for m, C_{water}, and ΔT to calculate ΔH.

$$\Delta H = -m \times C \times \Delta T = -(50.0\ \cancel{g})(4.18\ J/(\cancel{g} \times \cancel{°C}))(7.0\cancel{°C})$$
$$= -1463\ J = -1.5 \times 10^3\ J = -1.5\ kJ$$

3 Evaluate ***Does the result make sense?***

The sign of ΔH is negative; the reaction releases 1.5 kJ of heat to the water. The heat released by the system (the reaction) equals the heat absorbed by the surroundings (the water). About 4 J of heat raises the temperature of 1 g of water 1°C; 50 g of water requires about 200 J. To heat 50 g of water 7°C requires about 1400 J, or 1.4 kJ. This estimated answer is very close to the calculated value of ΔH.

Practice Problems

12. When 50.0 mL of water containing 0.50 mol HCl at 22.5°C is mixed with 50.0 mL of water containing 0.50 mol NaOH at 22.5°C in a calorimeter, the temperature of the solution increases to 26.0°C. How much heat (in kJ) was released by this reaction?

13. A small pebble is heated and placed in a foam cup calorimeter containing 25.0 mL of water at 25.0°C. The water reaches a maximum temperature of 26.4°C. How many joules of heat were released by the pebble?

Math Handbook

For help with dimensional analysis, go to page R66.

Problem-Solving 17.13
Solve Problem 13 with the help of an interactive guided tutorial.
with ChemASAP

Sample Problem 17.2

Answers

12. 1.46 kJ
13. 146 J

Practice Problems Plus L2

A lead mass is heated and placed in a foam cup calorimeter containing 40.0 mL of water at 17.0°C. The water reaches a temperature of 20.0°C. **How many joules of heat were released by the lead?** *(502 J)*

Math Handbook

For a math refresher and practice, direct students to dimensional analysis, page R66.

Discuss 

If possible, show photographs or actual examples of the types of calorimeter(s) used in laboratories today. Discuss the use of calorimetry as an analytical tool; for example, in investigating the caloric content of foods. Make sure students understand all the terms in the equation $q_{sys} = \Delta H = -m \times C \times \Delta T$. Emphasize that the delta notation (Δ) represents an algebraic operation involving two separate measurements. When determining the algebraic value of ΔT, the algebraic signs of the final and initial temperatures must be included in the operation.

Answers to...

Figure 17.6 586 kJ

$q_{sys} = \Delta H = -q_{surr} = -m \times C \times \Delta T$

Careers *in* Chemistry

Firefighter

Firefighters must have a working knowledge of chemistry because they are often called to identify and safely remove hazardous wastes. Knowledge of mechanics, first-aid, physics, and record-keeping is also useful.

Firefighters may work in industry, nuclear power plants, natural resources, airports, colleges, and in rural or urban firefighting services.

Have students research chemistry-related careers in the library or on the Internet. Students can then construct a table that describes the nature of the work, training requirements, working conditions, and other necessary information.

Thermochemical Equations

Discuss L2

To illustrate writing a thermochemical equation, refer to two familiar reactions. The combustion of methane in a Bunsen burner adds energy to the surroundings. Thus, the burning of methane has a negative ΔH.

$$CH_4(g) + 2O_2(g) \rightarrow CO_2(g) + 2H_2O(g) + 890\text{ kJ}$$

$$\Delta H = -890\text{ kJ/mol}$$

During the electrolysis of water, electrical energy is used to decompose the water into hydrogen gas and oxygen gas. Energy is added to the system from the surroundings. Thus, the electrolysis of water has a positive ΔH.

$$241.8\text{ kJ} + 2H_2O(l) \rightarrow 2H_2(g) + O_2(g)$$

$$\Delta H = +241.8\text{ kJ/mol}$$

Remind students that the sign of ΔH indicates whether the reaction is endothermic (+) or exothermic (–). Although by convention, an exothermic reaction has a negative ΔH, when heat is shown as a "product" in an equation, it is expressed as an absolute value.

Careers *in* Chemistry

Firefighter

If you are courageous, team-oriented, and have a strong sense of public service, you may wish to consider a career as a firefighter. Firefighters respond to fires, medical emergencies, and hazardous chemical spills. In responding to a fire, firefighters must know what types of chemicals can be used on different types of fires. Having knowledge of the requirements of different chemical fires and spills enables firefighters to put out rather than feed fires. Regular reviews of fire science literature keep firefighters up to date on current technology developments and policy changes. Firefighters also engage in physical fitness activities to build stamina and improve agility.

Although the job is hazardous and includes unpredictable hours, competition for jobs is relatively high. To become a firefighter, you must be at least 18 years of age with a high school education or equivalent. Selection for positions in fire departments often depends on passing a written test as well as a medical examination. Once hired, training consists of instruction in firefighting and rescue techniques, emergency medical procedures, basic chemistry of fires and firefighting, and instruction in local building codes and fire prevention methods.

For: Careers in Chemistry
Visit: PHSchool.com
Web Code: cdb-1172

Thermochemical Equations

If you mix calcium oxide with water, the water in the mixture becomes warm. This exothermic reaction occurs when cement, which contains calcium oxide, is mixed with water to make concrete. When 1 mol of calcium oxide reacts with 1 mol of water, 1 mol of calcium hydroxide forms and 65.2 kJ of heat is released. **In a chemical equation, the enthalpy change for the reaction can be written as either a reactant or a product.** In the equation describing the exothermic reaction of calcium oxide and water, the enthalpy change can be considered a product.

$$CaO(s) + H_2O(l) \longrightarrow Ca(OH)_2(s) + 65.2\text{ kJ}$$

This equation is presented visually in Figure 17.7a. A chemical equation that includes the enthalpy change is called a **thermochemical equation.**

The **heat of reaction** is the enthalpy change for the chemical equation exactly as it is written. You will usually see heats of reaction reported as ΔH, which is equal to the heat flow at constant pressure. The physical state of the reactants and products must also be given. The standard conditions are that the reaction is carried out at 101.3 kPa (1 atmosphere) and that the reactants and products are in their usual physical states at 25°C. The heat of reaction, or ΔH, in the above example is −65.2 kJ. Each mole of calcium oxide and water that react to form calcium hydroxide produces 65.2 kJ of heat.

$$CaO(s) + H_2O(l) \longrightarrow Ca(OH)_2(s) \qquad \Delta H = -65.2\text{ kJ}$$

Differentiated Instruction

Gifted and Talented L3

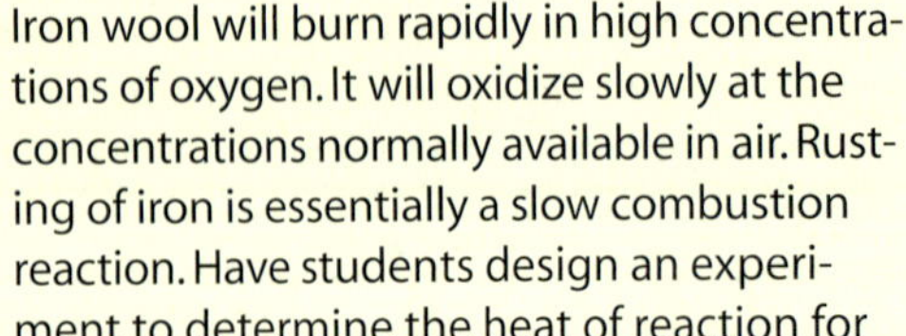

Iron wool will burn rapidly in high concentrations of oxygen. It will oxidize slowly at the concentrations normally available in air. Rusting of iron is essentially a slow combustion reaction. Have students design an experiment to determine the heat of reaction for the rusting of iron wool. *(Wrapping moist iron wool around the bulb of a thermometer demonstrates the heat released by this reaction. A foam cup calorimeter can be used to quantify the thermodynamic parameters of the reaction.)*

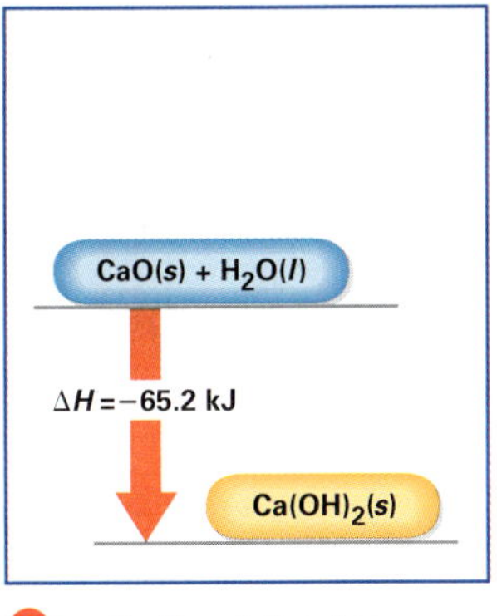

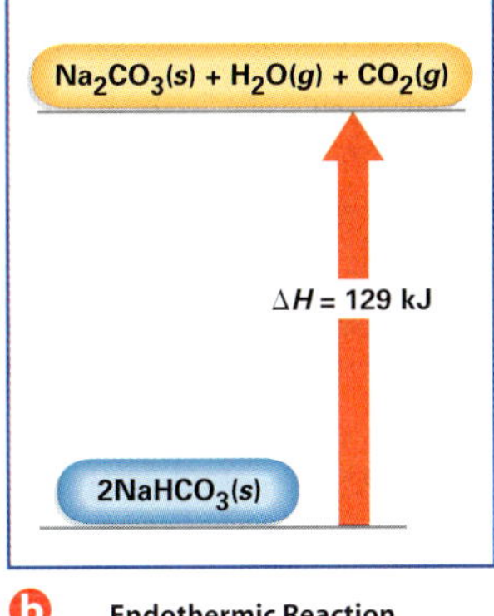

 Exothermic Reaction

ⓑ Endothermic Reaction

Figure 17.7 These enthalpy diagrams show exothermic and endothermic processes: ⓐ the reaction of calcium oxide and water and ⓑ the decomposition of sodium bicarbonate.
Identifying ***In which case is the enthalpy of the reactant(s) higher than that of the product(s)?***

Other reactions absorb heat from the surroundings. For example, baking soda (sodium bicarbonate) decomposes when it is heated. The carbon dioxide released in the reaction causes a cake to rise while baking. This process is endothermic.

$$2NaHCO_3(s) + 129\ kJ \longrightarrow Na_2CO_3(s) + H_2O(g) + CO_2(g)$$

Remember that ΔH is positive for endothermic reactions. Therefore, you can write the reaction as follows.

$$2NaHCO_3(s) \longrightarrow Na_2CO_3(s) + H_2O(g) + CO_2(g) \qquad \Delta H = 129\ kJ$$

Figure 17.7b shows the enthalpy diagram for this reaction.

Chemistry problems involving enthalpy changes are similar to stoichiometry problems. The amount of heat released or absorbed during a reaction depends on the number of moles of the reactants involved. The decomposition of 2 mol of sodium bicarbonate, for example, requires 129 kJ of heat. Therefore, the decomposition of 4 mol of the same substance would require twice as much heat, or 258 kJ. In this and other endothermic processes, the potential energy of the product(s) is higher than the potential energy of the reactant(s).

The physical state of the reactants and products in a thermochemical reaction must also be stated. To see why, compare the following two equations for the decomposition of 1 mol H_2O.

$$H_2O(l) \longrightarrow H_2(g) + \tfrac{1}{2}O_2(g) \qquad \Delta H = 285.8\ kJ$$
$$H_2O(g) \longrightarrow H_2(g) + \tfrac{1}{2}O_2(g) \qquad \Delta H = 241.8\ kJ$$
$$\text{difference} = 44.0\ kJ$$

Although the two equations are very similar, the different physical states of H_2O result in different ΔH values. In one case, the reactant is a liquid; in the other case, the reactant is a gas. The vaporization of 1 mole of liquid water to water vapor at 25°C requires an extra 44.0 kJ of heat.

$$H_2O(l) \longrightarrow H_2O(g) \qquad \Delta H = 44.0\ kJ$$

 ***Why must the physical states of the reactants and products be stated in a thermochemical equation?***

Interactive Textbook

Simulation 22 Simulate a combustion reaction and compare the ΔH results for several compounds.

with ChemASAP

Use Visuals 

Figure 17.7 Ask students to draw similar diagrams for the burning of CH_4 and the electrolysis of H_2O. For each reaction, ask, **Is the potential energy of the reactants higher or lower than the products?** *(higher for CH_4, lower for H_2O)* **In what direction does heat move in diagram 17.7a?** *(into the surroundings)* **In what direction does heat move in diagram 17.7b?** *(absorbed from the surroundings by the reaction)*

TEACHER Demo

An Exothermic Reaction 

Purpose Students will observe an exothermic reaction and write a thermochemical equation.

Materials CaO, metal tray, thin aluminum pan, water, egg (raw)

Safety CaO is corrosive.

Procedure Place 100 g of CaO in a metal tray. Place a thin aluminum pan or pie tin in the tray so that its bottom is completely in contact with the calcium oxide. Pour just enough water into the tray to cover the calcium oxide. Then break a fresh egg into the aluminum pan. Ask, **Is this an exothermic or endothermic reaction?** *(Exothermic; 65.2 kJ of heat is released per mole of CaO.)* Have students write a thermochemical equation to describe the reaction. Check to see that students express the enthalpy of reaction as a negative value. Have students use the thermochemical equation to answer stoichiometric questions such as: **How much heat is released when 100.0 g of calcium oxide reacts with excess water?** *(116 kJ)*

Expected Outcome Students should observe that enough heat is liberated by the reaction to cook the egg.

Differentiated Instruction

Gifted and Talented 

Many metals react with oxygen to produce metal oxides. Have students research the combustion reactions of a series of selected metals (e.g., Mg, Ca, Fe). Students should write thermochemical equations for each reaction and compare the relative amounts of heat released in each case. Ask students to describe any trend(s) they observe.

Less Proficient Readers 

Special attention should be given to vocabulary terms in the chapter that are defined by mathematical relationships. Have students write definitions of these terms. Sample calculations showing how to apply the formulas should be included in students' notes.

Answers to...

Figure 17.7 for the exothermic reaction: $CaO(s) + H_2O(l) \rightarrow Ca(OH)_2(s)$

 Different physical states of the same substance have different ΔH values.

Sample Problem 17.3

Answers

14. 6.63 kJ
15. 89.4 kJ

Practice Problems Plus 

The burning of magnesium is a highly exothermic reaction.

$2Mg(s) + O_2(g) \rightarrow 2MgO(s) + 1204\ kJ$

How many kilojoules of heat are released when 0.75 mol of Mg burn in an excess of O_2?

(450 kJ)

CHEMath

Conversion Problems

A conversion factor relating kilojoules of heat and moles of sodium bicarbonate is needed to solve the Sample Problem. It is important to label the units of the conversion factor so that the answer contains the correct units.

Math Handbook

For a math refresher and practice, direct students to conversion problems, page R66.

Relate 

The decomposition of sodium bicarbonate to form carbon dioxide gas is essential for successful baking. Sodium bicarbonate, also called baking soda, is the source of the carbon dioxide that causes cakes, muffins, and quick breads to rise in the oven. Without this substance, these foods would be flat and heavy.

CHEMath

Conversion Problems

A conversion factor is a ratio of two quantities that are equal to one another. When doing conversions, write the conversion factors so that the unit of a given measurement cancels, leaving the correct unit for your answer. Note that the equalities needed to write a particular conversion factor may be given in the problem. In other cases, you will need to know or look up the necessary equalities.

Math Handbook

For help with conversion problems, go to page R66.

SAMPLE PROBLEM 17.3

Using the Heat of Reaction to Calculate Enthalpy Change

Using the thermochemical equation in Figure 17.7b on page 515, calculate the amount of heat (in kJ) required to decompose 2.24 mol $NaHCO_3(s)$.

1 Analyze ***List the knowns and the unknown.***

Knowns

- 2.24 mol $NaHCO_3(s)$ decomposes
- $\Delta H = 129$ kJ (for 2 mol $NaHCO_3$)

Unknown

- $\Delta H = ?$ kJ

Use the thermochemical equation,

$$2NaHCO_3(s) + 129\ kJ \longrightarrow Na_2CO_3(s) + H_2O(g) + CO_2(g),$$

to write a conversion factor relating kilojoules of heat and moles of $NaHCO_3$. Then use the conversion factor to determine ΔH for 2.24 mol $NaHCO_3$.

2 Calculate ***Solve for the unknown.***

The thermochemical equation indicates that 129 kJ are needed to decompose 2 mol $NaHCO_3(s)$. Use this relationship to write the following conversion factor.

$$\frac{129\ kJ}{2\ mol\ NaHCO_3(s)}$$

Using dimensional analysis, solve for ΔH.

$$\Delta H = 2.24\ \cancel{mol\ NaHCO_3(s)} \times \frac{129\ kJ}{2\ \cancel{mol\ NaHCO_3(s)}} = 144\ kJ$$

3 Evaluate ***Does the result make sense?***

Because the $\Delta H = 129$ kJ refers to the decomposition of 2 mol $NaHCO_3(s)$, the decomposition of 2.24 mol should absorb about 10% more heat than 129 kJ, or slightly more than 142 kJ. The answer of 144 kJ is consistent with this estimate.

Practice Problems

14. When carbon disulfide is formed from its elements, heat is absorbed. Calculate the amount of heat (in kJ) absorbed when 5.66 g of carbon disulfide is formed.

$$C(s) + 2S(s) \longrightarrow CS_2(l) \quad \Delta H = 89.3\ kJ$$

15. The production of iron and carbon dioxide from iron(III) oxide and carbon monoxide is an exothermic reaction. How many kilojoules of heat are produced when 3.40 mol Fe_2O_3 reacts with an excess of CO?

$$Fe_2O_3(s) + 3CO(g) \longrightarrow 2Fe(s) + 3CO_2(g) + 26.3\ kJ$$

Interactive Textbook

Problem-Solving 17.15
Solve Problem 15 with the help of an interactive guided tutorial.

with ChemASAP

Facts and Figures

Keeping Warm

How do animals such as polar bears and seals survive the cold land and water temperatures where they live? A good coat of fur helps, but it is not enough to keep them warm. These animals also have unique fat cells that help generate heat. These fat cells are in tissue called brown fat. The cells of brown fat are unlike other fat cells in these animals' bodies. Most other cells store chemical energy released during the breakdown of carbohydrates and fatty acids in the bonds of adenosine triphosphate (ATP). ATP acts as the central source of energy and growth of all organisms. Heat is mostly a waste product in ATP-producing cells. The heat generated by the brown fat tissue, however, helps the animal keep relatively comfortable even at subzero temperatures.

Table 17.2

Heats of Combustion at 25°C		
Substance	**Formula**	**ΔH (kJ/mol)**
Hydrogen	$H_2(g)$	−286
Carbon	$C(s)$, graphite	−394
Methane	$CH_4(g)$	−890
Acetylene	$C_2H_2(g)$	−1300
Ethanol	$C_2H_5OH(l)$	−1368
Propane	$C_3H_8(g)$	−2220
Glucose	$C_6H_{12}O_6(s)$	−2808
Octane	$C_8H_{18}(l)$	−5471
Sucrose	$C_{12}H_{22}O_{11}(s)$	−5645

Figure 17.8 The combustion of natural gas is an exothermic reaction. As bonds in methane (the main component of natural gas) and oxygen are broken and bonds in carbon dioxide and water are formed, large amounts of energy are released.

Table 17.2 lists heats of combustion for some common substances. The **heat of combustion** is the heat of reaction for the complete burning of one mole of a substance. The combustion of natural gas, which is mostly methane, is an exothermic reaction used to heat many homes around the country.

$$CH_4(g) + 2O_2(g) \longrightarrow CO_2(g) + 2H_2O(l) + 890\text{ kJ}$$

You can also write this equation as follows.

$$CH_4(g) + 2O_2(g) \longrightarrow CO_2(g) + 2H_2O(l) \qquad \Delta H = -890\text{ kJ}$$

Burning 1 mol of methane releases 890 kJ of heat. The heat of combustion (ΔH) for this reaction is −890 kJ per mole of carbon burned.

Like other heats of reaction, heats of combustion are reported as the enthalpy changes when the reactions are carried out at 101.3 kPa of pressure and the reactants and products are in their physical states at 25°C.

17.2 Section Assessment

16. **Key Concept** Calorimetry is based on what basic concepts?
17. **Key Concept** How are enthalpy changes treated in chemical equations?
18. When 2 mol of solid magnesium (Mg) combines with 1 mole of oxygen gas (O_2), 2 mol of solid magnesium oxide (MgO) is formed and 1204 kJ of heat is released. Write the thermochemical equation for this combustion reaction.
19. Gasohol contains ethanol (C_2H_5OH)(l), which when burned reacts with oxygen to produce $CO_2(g)$ and $H_2O(g)$. How much heat is released when 12.5 g of ethanol burns?

$$C_2H_5OH(l) + 3O_2(g) \longrightarrow 2CO_2(g) + 3H_2O(l)$$
$$\Delta H = -1368\text{ kJ}$$

20. Explain the term *heat of combustion.*

Elements Handbook

Group 6A Look at the important chemical reactions involving Group 6A elements on page R29. Write two examples of thermochemical equations—one describing an exothermic reaction and another describing an endothermic reaction.

Assessment 17.2 Test yourself on the concepts in Section 17.2.

with ChemASAP

3 ASSESS

Evaluate Understanding L2

To determine students' knowledge about measuring and expressing heat changes, ask, **Explain the difference between H and ΔH.** *(H is enthalpy or heat content; ΔH represents a change in heat content.)* **Use the data in Table 17.2 to write the thermochemical equation for the combustion of propane.** *($C_3H_8(g) + 5O_2(g) \rightarrow 3CO_2(g) + 4H_2O(l)$; $\Delta H = -2220$ kJ)* **Draw an enthalpy diagram for this reaction like those in Figure 17.7. Calculate the amount of heat produced when 64.0 grams of propane burns completely.** *(3.23×10^3 kJ)*

Reteach L1

Select a substance from Table 17.2, and write the thermochemical equation on the chalkboard. Discuss the ΔH for the reaction, why it is negative or positive, and draw the enthalpy diagram for the reaction. Remind students that heat is the energy that flows into or out of a thermodynamic system because of differences in temperature between the system and its surroundings. Remind students that the total energy (potential and kinetic) of the system and the surroundings must remain the same in any chemical or physical process.

Elements Handbook

Example exothermic reactions:

$4K(s) + O_2(g) \rightarrow 2K_2O(s) + 363.2$ kJ/mol

$C_6H_{12}O_6(s) + 6O_2(g) \rightarrow 6CO_2(g) + 6H_2O(g) + 2808$ kJ/mol

Example endothermic reaction:

$3O_2(g) + 285$ kJ/mol $\rightarrow 2O_3(g)$

If your class subscribes to the Interactive Textbook, use it to review key concepts in Section 17.2.

with ChemASAP

Section 17.2 Assessment

16. The heat released by the system is equal to the heat absorbed by the surroundings. Conversely, the heat absorbed by a system is equal to the heat released by its surroundings.
17. The enthalpy change in a chemical reaction can be written as either a reactant or a product.
18. $2Mg(s) + O_2(g) \rightarrow 2MgO(s) + 1204$ kJ, or $2Mg(s) + O_2(g) \rightarrow 2MgO(s)$ $\Delta H = -1204$ kJ
19. 3.72×10^2 kJ
20. Heat of combustion is the heat of reaction for the complete burning of one mole of a substance.

Technology & Society

Solar Power Plants

Background

- Solar One, an early demonstration solar energy plant in the Mojave Desert in California, used water as the fluid in the receiver. The receiver in the tower generated steam to drive a turbine to generate electricity. The plant was later converted to Solar Two, which used molten salt as the fluid. The hot salt can be stored, then used when needed to boil water into steam that drives the turbine.
- CSP is the least expensive solar electricity technology for large-scale power generation and has the potential to make solar power available at a very competitive rate. As a result, government, industry, and utilities are forming partnerships with the goal of reducing the manufacturing cost of CSP technologies.

CLASS Activity

Energy Conversions L2

Have students construct a concept map or flow chart depicting the conversion of solar energy into other forms of energy. For example, solar energy can be used to heat water, which can be used to generate electricity, which in turn can be converted into light (e.g., a reading lamp) or chemical energy (e.g., recharging a battery).

Technology & Society

Solar Power Plants

When you consider energy sources other than coal, oil, or natural gas, you may think of the sun. Solar energy, or energy from the sun, is free, clean, plentiful, and readily available. The daily amount of solar energy received by Earth is about 200,000 times the total world electrical-generating capacity. However, the process of collecting solar energy and using it to generate electricity can be expensive, complex, and space-consuming. Large-scale solar power plants use concentrating solar power (CSP) technologies, which convert solar energy into heat and then electrical energy. **Interpreting Diagrams** ***In which part of a power tower is solar energy transformed into heat?***

From sun to socket Because a heliostat field collects solar energy faster than the system can use it, some of the energy is stored (as a molten salt) for later use. Energy storage enables a power tower to supply electricity during cloudy weather or even at night.

Solar power by the tower
A power tower is a CSP technology that uses a vast field of sun-tracking mirrors, called heliostats, to concentrate sunlight on a receiver that sits on top of a tower. Before the solar energy can be used to do work, it must first be transformed into heat by means of a fluid such as steam or a molten salt.

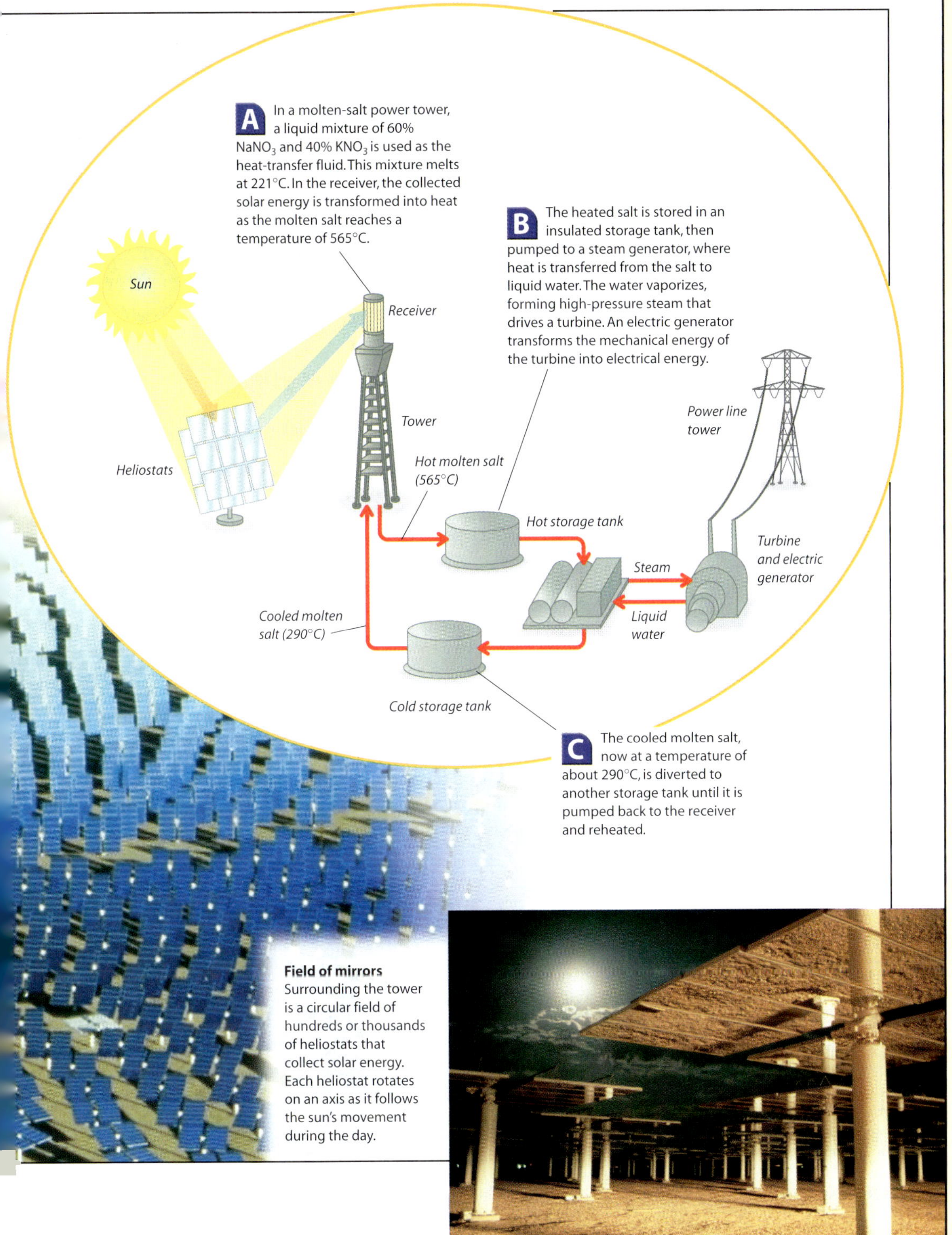

A In a molten-salt power tower, a liquid mixture of 60% $NaNO_3$ and 40% KNO_3 is used as the heat-transfer fluid. This mixture melts at 221°C. In the receiver, the collected solar energy is transformed into heat as the molten salt reaches a temperature of 565°C.

B The heated salt is stored in an insulated storage tank, then pumped to a steam generator, where heat is transferred from the salt to liquid water. The water vaporizes, forming high-pressure steam that drives a turbine. An electric generator transforms the mechanical energy of the turbine into electrical energy.

C The cooled molten salt, now at a temperature of about 290°C, is diverted to another storage tank until it is pumped back to the receiver and reheated.

Field of mirrors Surrounding the tower is a circular field of hundreds or thousands of heliostats that collect solar energy. Each heliostat rotates on an axis as it follows the sun's movement during the day.

CLASS Activity

Developments in Solar Power L2

Have students research and write a report on a topic related to solar power. Possible topics include:

- hybrid solar power plants that combine solar energy collection with fossil fuel combustion
- mirror technologies used in solar power plants
- heat-transfer fluids used in CSP technologies (e.g., air, steam, molten salt, oil)
- houses and other buildings that have solar-powered heating and hot-water systems
- photovoltaic technologies used in solar power plants, solar cars, and solar houses

Answers to...

Interpreting Diagrams The receiver on the power tower converts solar energy to heat.

17.3

1 FOCUS

Objectives

17.3.1 Classify the enthalpy change that occurs when a substance melts, freezes, boils, condenses, or dissolves.

17.3.2 Solve for the enthalpy change that occurs when a substance melts, freezes, boils, condenses, or dissolves.

Guide for Reading

Build Vocabulary L2

Paraphrase Have students write a short paragraph explaining in their own words what each vocabulary term means.

Reading Strategy L2

Relate Text and Visuals After students read about the enthalpy change that occurs during a particular phase change, they should examine Figure 17.9 and find the arrow that applies to that phase change. This will help them to visualize the process.

2 INSTRUCT

Connecting to Your World

Remind students that evaporation is a physical change from a liquid state to a gas (vapor). Ask, **On a molecular level, what is the difference between liquid water and water vapor?** *(In the vapor phase, the water molecules are farther apart and not as strongly attracted to each other.)* Ask, **Because of the attraction between molecules in a liquid, what must happen for a liquid to change into a gas?** *(Energy must be added to overcome the attraction.)* **Is evaporation endothermic or exothermic?** *(endothermic)*

17.3 Heat in Changes of State

Guide for Reading

Key Concepts

- How does the quantity of heat absorbed by a melting solid compare to the quantity of heat released when the liquid solidifies?
- How does the quantity of heat absorbed by a vaporizing liquid compare to the quantity of heat released when the vapor condenses?
- What thermochemical changes can occur when a solution forms?

Vocabulary

molar heat of fusion
molar heat of solidification
molar heat of vaporization
molar heat of condensation
molar heat of solution

Reading Strategy

Summarizing After you read this section, construct a flowchart that summarizes the enthalpy changes involved as a substance undergoes phase changes.

Connecting to Your World An athlete can burn a lot of calories during a race. These calories are either used to do work or are released as heat. When your body heats up, you start to sweat. The evaporation of sweat is your body's way of cooling itself to a normal temperature. This section will help you to understand how the evaporation of sweat from your skin helps to rid your body of excess heat.

Heats of Fusion and Solidification

What happens if you place an ice cube on a table in a warm room? The ice cube is the system, and the table and air around it are the surroundings. The ice absorbs heat from its surroundings and begins to melt. The temperature of the ice and the water produced remains at 0°C until all of the ice has melted. The temperature of the water begins to increase only after all of the ice has melted.

Like ice cubes, all solids absorb heat as they melt to become liquids. The gain of heat causes a change of state instead of a change in temperature. Whenever a change of state occurs by a gain or loss of heat, the temperature of the substance undergoing the change remains constant. The heat absorbed by one mole of a solid substance as it melts to a liquid at a constant temperature is the **molar heat of fusion** (ΔH_{fus}). The **molar heat of solidification** (ΔH_{solid}) is the heat lost when one mole of a liquid solidifies at a constant temperature. **The quantity of heat absorbed by a melting solid is exactly the same as the quantity of heat released when the liquid solidifies; that is, $\Delta H_{fus} = -\Delta H_{solid}$.** This relationship is shown in Figure 17.9.

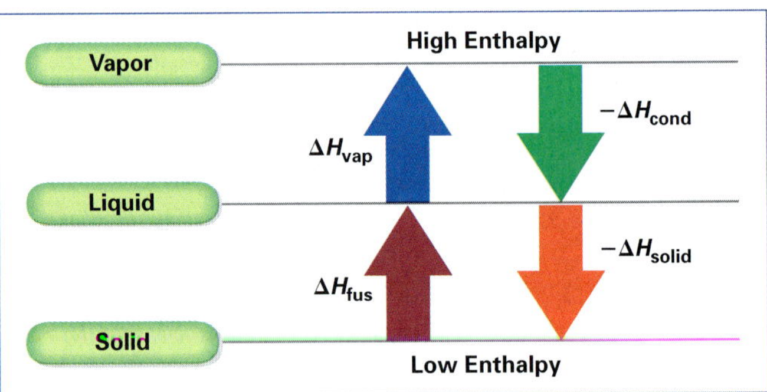

Figure 17.9 Enthalpy changes accompany changes in state. Fusion and vaporization are endothermic processes. Solidification and condensation are exothermic processes.
Interpreting Diagrams *Which arrows represent processes that release heat to the surroundings?*

Section Resources

Print

- **Guided Reading and Study Workbook,** Section 17.3
- **Core Teaching Resources,** Section 17.3 Review
- **Transparencies,** T187–T191
- **Small-Scale Chemistry Lab Manual,** Lab 27
- **Probeware Laboratory Manual,** Section 17.3

Technology

- **Interactive Textbook with ChemASAP,** Problem-Solving 17.21, 17.24, 17.26, Animation 21, Assessment 17.3
- **Go Online,** Section 17.3

The melting of 1 mol of ice at 0°C to 1 mol of water at 0°C requires the absorption of 6.01 kJ of heat. This quantity of heat is the molar heat of fusion of water. Likewise, the conversion of 1 mol of water at 0°C to 1 mol of ice at 0°C releases 6.01 kJ. This quantity of heat is the molar heat of solidification of water.

$$H_2O(s) \longrightarrow H_2O(l) \qquad \Delta H_{fus} = 6.01\ \text{kJ/mol}$$

$$H_2O(l) \longrightarrow H_2O(s) \qquad \Delta H_{solid} = -6.01\ \text{kJ/mol}$$

SAMPLE PROBLEM 17.4

Using the Heat of Fusion in Phase-Change Calculations

How many grams of ice at 0°C will melt if 2.25 kJ of heat are added?

1 Analyze ***List the knowns and the unknown.***

Knowns
- Initial and final temperatures are 0°C
- ΔH_{fus} = 6.01 kJ/mol
- ΔH = 2.25 kJ

Unknown
- m_{ice} = ? g

Use the thermochemical equation

$$H_2O(s) + 6.01\ \text{kJ} \longrightarrow H_2O(l)$$

to find the number of moles of ice that can be melted by the addition of 2.25 kJ of heat. Convert moles of ice to grams of ice.

2 Calculate ***Solve for the unknown.***

Express ΔH_{fus} and the molar mass of ice as conversion factors.

$$\frac{1\ \text{mol ice}}{6.01\ \text{kJ}} \quad \text{and} \quad \frac{18.0\ \text{g ice}}{1\ \text{mol ice}}$$

Multiply the known enthalpy change (2.25 kJ) by the conversion factors

$$m_{ice} = 2.25\ \cancel{\text{kJ}} \times \frac{1\ \cancel{\text{mol ice}}}{6.01\ \cancel{\text{kJ}}} \times \frac{18.0\ \text{g ice}}{1\ \cancel{\text{mol ice}}}$$
$$= 6.74\ \text{g ice}$$

3 Evaluate ***Does the result make sense?***

6.01 kJ is required to melt 1 mol of ice. Because only about one third of this amount of heat (roughly 2 kJ) is available, only about one-third mol of ice, or 18.0 g/3 = 6 g, should melt. This estimate is close to the calculated answer.

Practice Problems

21. How many kilojoules of heat are required to melt a 10.0-g popsicle at 0°C? Assume the popsicle has the same molar mass and heat of fusion as water.

22. How many grams of ice at 0°C could be melted by the addition of 0.400 kJ of heat?

Math Handbook

For help with conversion problems, go to page R66.

Interactive Textbook

Problem-Solving 17.21 Solve Problem 21 with the help of an interactive guide tutorial.

with ChemASAP

Heats of Fusion and Solidification

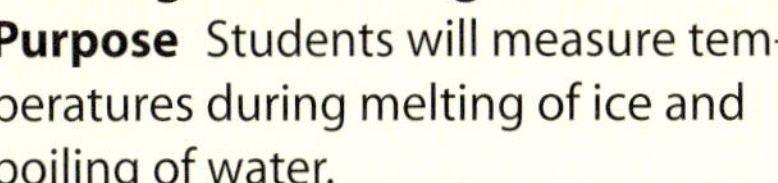

Melting and Boiling L2

Purpose Students will measure temperatures during melting of ice and boiling of water.

Materials beaker, water, boiling chips, hot plate, thermometer, ice cubes, foam cup

Procedure Have students record the temperature of gently boiling water in a beaker (use boiling chips) and of melting ice in a foam cup. Students should record the temperatures every 30 seconds.

Expected Outcomes Even if the thermometers do not record exactly 100°C or 0°C, students should note that the temperature does not change during boiling and melting.

Sample Problem 17.4

Answers

21. 3.34 kJ

22. 1.20 g ice

Practice Problems Plus 

How much heat must be removed to freeze a tray of ice cubes if the water has a mass of 50.0 g? *(16.7 kJ)*

Math Handbook

For a math refresher and practice, direct students to conversion problems, page R66.

Answers to...

Figure 17.9 solidification and condensation

Heat of Fusion of Ice L2

Objective Students will melt an ice cube and calculate the heat of fusion of ice.

Skills Focus Observing, measuring, calculating, designing experiments

 **Prep Time** 10 minutes

Materials 100-mL graduated cylinder, hot water, foam cup, thermometer, temperature probe (optional), ice

Advance Prep Provide students with a source of water between 50°C and 60°C.

Class Time 20 minutes

Safety Students should wear safety goggles. Warn students not to use the thermometer as a stirring rod and discuss what students should do if the thermometer breaks.

Teaching Tips

- Remind students to make all temperature measurements to the nearest tenth of a degree and all volume measurements to the nearest milliliter.
- Small ice cubes or crushed ice work best.

Expected Outcome Students should calculate values of approximately 6.01 kJ/mol for the heat of fusion of ice.

Analyze and Conclude

1. The mass of ice will be equal to the final volume of water minus the initial volume of water, expressed in grams.
2. See Expected Outcome.
3. See Expected Outcome. If the experiment is not done quickly, loss of heat to the classroom may cause ΔT to be too large, resulting in a larger ΔH_{fus}. Careful reading of the thermometer and graduated cylinder is crucial. Some water may adhere to the thermometer and cup; this may affect volume measurements.
4. Acceptable answers include the following: To determine the mass of warm water cooled and the mass of ice melted, measure the mass of (1) the empty cup, (2) the cup plus the hot water, and (3) the cup after the ice has melted. Use a covered calorimeter to decrease the influence of room temperature.

Quick LAB

Heat of Fusion of Ice

Purpose

To estimate the heat of fusion of ice.

Materials

- ice
- foam cup
- 100-ml graduated cylinder
- thermometer
- hot water
- temperature probe (optional)

Procedure   

 *Probe version available in the Probeware Lab Manual.*

1. Fill the graduated cylinder with hot tap water and let stand for 1 minute. Pour the water into the sink.
2. Use the graduated cylinder to measure 70 mL of hot water. Pour the water into the foam cup. Measure the temperature of the water.
3. Add an ice cube to the cup of water. Gently swirl the cup. Measure the temperature of the water as soon as the ice cube has completely melted.
4. Pour the water into the graduated cylinder and measure the volume.

Analyze and Conclude

1. Calculate the mass of the ice. (*Hint:* The mass of ice melted is the same as the increase in the volume of the water.) Convert this into moles.
2. Calculate ΔH_{fus} of ice (kJ/mol) by dividing the heat transferred from the water by the moles of ice melted.
3. Compare your experimental value of ΔH_{fus} of ice with the accepted value of 6.01 kJ/mol. Account for any error.
4. How might you revise the procedure to achieve more accurate results?

Heats of Vaporization and Condensation

When liquids absorb heat at their boiling points, they become vapors. The amount of heat necessary to vaporize one mole of a given liquid is called its **molar heat of vaporization** (ΔH_{vap}). Table 17.3 lists the molar heats of vaporization for several substances at their normal boiling points.

The molar heat of vaporization of water is 40.7 kJ/mol. This means that it takes 40.7 kJ of energy to convert 1 mol of water molecules in the liquid state to 1 mol of water molecules in the vapor state at the normal boiling point of water (100°C at 101.3 kPa). This process is described in the thermochemical equation below.

$$H_2O(l) \longrightarrow H_2O(g) \qquad \Delta H_{vap} = 40.7 \text{ kJ/mol}$$

Table 17.3

Heats of Physical Change

Substance	ΔH_{fus} (kJ/mol)	ΔH_{vap} (kJ/mol)
Ammonia (NH_3)	5.65	23.4
Ethanol (C_2H_5OH)	4.60	43.5
Hydrogen (H_2)	0.12	0.90
Methanol (CH_3OH)	3.16	35.3
Oxygen (O_2)	0.44	6.82
Water (H_2O)	6.01	40.7

For Enrichment 

Have students design a new experiment that incorporates their suggestions for revising the procedure (their answers to Question 4). If students perform the new experiment, they should compare the results of the two experiments and calculate the respective percent errors.

Diethyl ether ($C_4H_{10}O$) is a low-boiling-point liquid (bp = 34.6°C) that is a good solvent and was formerly used as an anesthetic. If diethyl ether is poured into a beaker on a warm, humid day, the ether will absorb heat from the beaker walls and evaporate very rapidly. If the beaker loses enough heat, the water vapor in the air may condense and freeze on the beaker walls. If so, a coating of frost will form on the outside of the beaker. Diethyl ether has a molar heat of vaporization (ΔH_{vap}) of 15.7 kJ/mol.

$$C_4H_{10}O(l) \longrightarrow C_4H_{10}O(g) \qquad \Delta H_{vap} = 15.7 \text{ kJ/mol}$$

Condensation is the exact opposite of vaporization. When a vapor condenses, heat is released. The amount of heat released when 1 mol of vapor condenses at the normal boiling point is called its **molar heat of condensation** (ΔH_{cond}). This value is numerically the same as the corresponding molar heat of vaporization, however the value has the opposite sign. **The quantity of heat absorbed by a vaporizing liquid is exactly the same as the quantity of heat released when the vapor condenses; that is, $\Delta H_{vap} = -\Delta H_{cond}$.** Figure 17.10 summarizes the enthalpy changes that occur as a solid is heated to a liquid and then to a vapor. You should be able to identify certain trends regarding the temperature during changes of state and the energy requirements that accompany these changes from the graph. The large values for ΔH_{vap} and ΔH_{cond} are the reason hot vapors such as steam can be very dangerous. You can receive a scalding burn from steam when the heat of condensation is released as the steam touches your skin.

$$H_2O(g) \longrightarrow H_2O(l) \qquad \Delta H_{cond} = -40.7 \text{ kJ/mol}$$

Checkpoint ***What is the molar heat of vaporization?***

Interactive Textbook

Animation 21 Observe the phase changes as ice is converted to steam when heat is added.

with **ChemASAP**

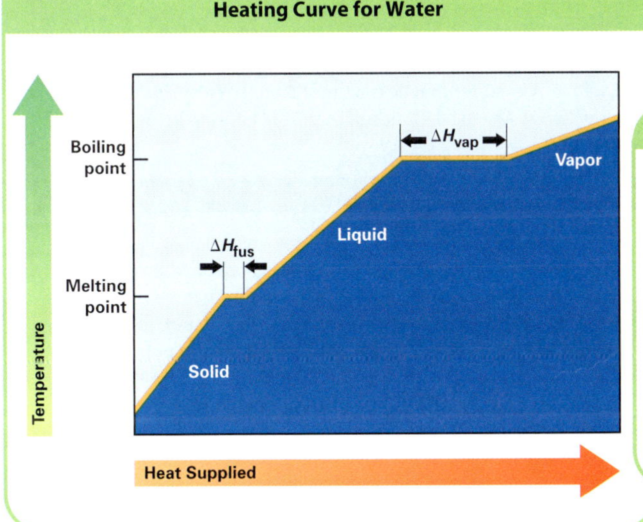

Figure 17.10 A heating curve graphically describes the enthalpy changes that take place during phase changes.

INTERPRETING GRAPHS

a. Identify In which region(s) of the graph is temperature constant?

b. Describe How does the amount of energy required to melt a given mass of ice compare to the energy required to vaporize the same mass of water? Explain.

c. Apply Concepts Which region of the graph represents the coexistence of solid and liquid? Liquid and vapor?

Facts and Figures

Ether

Before the mid-1840s, patients undergoing surgery received no anesthesia, and they had to endure intense pain. Then the anesthetic properties of diethyl ether (commonly called ether) and chloroform were discovered. By the Civil War, use of these two substances was common; when available, they were used in surgeries on the battlefield. Anesthetics were generally applied by pouring the agent on linen, holding it at some distance from the patient's nose and mouth for the first inhalation, and then gradually moving the cloth closer until the desired effects were produced. Ether is no longer used as an anesthetic because it is flammable and causes many undesirable side reactions in patients.

Heats of Vaporization and Condensation

Interpreting Graphs L2

a. at the melting point and the boiling point

b. It takes much less energy to melt a given mass of ice than to vaporize the same mass of water. The ΔH_{fus} is much smaller than ΔH_{vap}.

c. the plateau of the curve between solid and liquid; the plateau between liquid and vapor

Enrichment Question 

How would the rate of heating affect the heating curve? *(The faster the rate of heating is, the steeper the slope of the curve will be.)*

TEACHER Demo

Exothermic Solidification 

Purpose Students will observe the release of heat when a liquid solidifies.

Materials 3 wide-mouth test tubes, sodium thiosulfate pentahydrate, hot water bath, 30°C water bath, thermometers, stirring rods

Procedure Set three wide-mouth test tubes each containing about 20 grams of sodium thiosulfate pentahydrate ($Na_2S_2O_3 \cdot 5H_2O$) in a hot water bath until the chemical is completely melted. Transfer the tubes to a cool water bath at about 30°C. Let the test tubes sit undisturbed until they are close to 30°C and feel lukewarm to touch. Give the tubes to three student volunteers at different locations in the classroom so that many students can see and feel what happens. Give each volunteer a thermometer and a seed crystal. Ask them to quickly take the temperature of the liquid, drop in the seed crystal, and stir gently.

Expected Outcomes The temperature of the test tube will rise. Students will observe the change by touching the test tube and reading the thermometer.

Answers to...

Checkpoint the amount of heat needed to vaporize one mole of a given liquid

Sample Problem 17.5

Answers

23. 1.44×10^2 kJ

24. 1.9×10^{-1} kJ

Practice Problems Plus — L2

How many kilojoules of heat are required to vaporize 50.0 g of ethanol, C_2H_5OH? The boiling point of ethanol is 78.5°C. Its molar heat of vaporization is 43.5 kJ/mol. *(47.3 kJ)*

Math Handbook

For a math refresher and practice, direct students to using a calculator, page R62.

CLASS Activity

Heating Curve for Ethanol — L2

Have students look up the molar heat of fusion, the molar heat of vaporization, and the specific heat of ethanol (grain alcohol) in Tables 17.1 and 17.3. Tell students that ethanol freezes at 158.7 K and boils at 351.5 K. Have students summarize the enthalpy changes that occur as solid ethanol is heated to a liquid and then to a gas by constructing a heating curve like the one shown in Figure 17.10 for water. Ask students to find the amount of heat necessary to convert 1 mole of solid ethanol to ethanol vapor.

SAMPLE PROBLEM 17.5

Using the Heat of Vaporization in Phase-Change Calculations

How much heat (in kJ) is absorbed when 24.8 g $H_2O(l)$ at 100°C and 101.3 kPa is converted to steam at 100°C?

1 Analyze ***List the knowns and the unknown.***

Knowns

- Initial and final conditions are 100°C and 101.3 kPa
- mass of water converted to steam = 24.8 g
- ΔH_{vap} = 40.7 kJ/mol

Unknown

- ΔH = ? kJ

Refer to the following thermochemical equation.

$$H_2O(l) + 40.7\ \text{kJ/mol} \longrightarrow H_2O(g)$$

ΔH_{vap} is given in kJ/mol, but the quantity of water is given in grams. You must first convert grams of water to moles of water. Then multiply by ΔH_{vap}.

Math Handbook

For help with using a calculator, go to page R62.

2 Calculate ***Solve for the unknown.***

The required conversion factors come from ΔH_{vap} and the molar mass of water.

$$\frac{1\ \text{mol}\ H_2O(l)}{18.0\ \text{g}\ H_2O(l)} \quad \text{and} \quad \frac{40.7\ \text{kJ}}{1\ \text{mol}\ H_2O(l)}$$

Multiply the mass of water in grams by the conversion factors.

$$\Delta H = 24.8\ \text{g}\ \cancel{H_2O(l)} \times \frac{1\ \cancel{\text{mol}\ H_2O(l)}}{18.0\ \text{g}\ \cancel{H_2O(l)}} \times \frac{40.7\ \text{kJ}}{1\ \cancel{\text{mol}\ H_2O(l)}}$$
$$= 56.1\ \text{kJ}$$

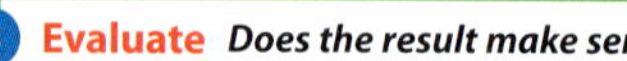

3 Evaluate ***Does the result make sense?***

Knowing the molar mass of water is 18.0 g/mol, 24.8 g $H_2O(l)$ can be estimated to be somewhat less than 1.5 mol H_2O. The calculated enthalpy change should be a little less than 1.5 $\cancel{\text{mol}}$ × 40 kJ/$\cancel{\text{mol}}$ = 60 kJ, and it is.

Practice Problems

23. How much heat is absorbed when 63.7g $H_2O(l)$ at 100°C and 101.3 kPa is converted to steam at 100°C? Express your answer in kJ.

24. How many kilojoules of heat are absorbed when 0.46 g of chloroethane (C_2H_5Cl, bp 12.3°C) vaporizes at its normal boiling point? The molar heat of vaporization of chloroethane is 26.4 kJ/mol.

Interactive Textbook

Problem-Solving 17.24 Solve Problem 24 with the help of an interactive guided tutorial.

with ChemASAP

524 Chapter 17

Heat of Solution

If you've ever used a hot pack or a cold pack, then you have felt the enthalpy changes that occur when a solute dissolves in a solvent. **During the formation of a solution, heat is either released or absorbed.** The enthalpy change caused by dissolution of one mole of substance is the **molar heat of solution** (ΔH_{soln}). Sodium hydroxide provides a good example of an exothermic molar heat of solution. When 1 mol of sodium hydroxide (NaOH)(*s*) is dissolved in water, the solution can become so hot that it steams. The heat from this process is released as the sodium ions and the hydroxide ions interact with the water. The temperature of the solution increases, releasing 445.1 kJ of heat as the molar heat of solution.

$$NaOH(s) \xrightarrow{H_2O(l)} Na^+(aq) + OH^-(aq)$$
$$\Delta H_{soln} = -445.1 \text{ kJ/mol}$$

A practical application of exothermic dissolving is a hot pack. A hot pack mixes calcium chloride ($CaCl_2$) and water, which produces the heat characteristic of an exothermic reaction.

$$CaCl_2(s) \xrightarrow{H_2O(l)} Ca^{2+}(aq) + 2Cl^-(aq)$$
$$\Delta H_{soln} = -82.8 \text{ kJ/mol}$$

The dissolution of ammonium nitrate (NH_4NO_3)(*s*) is an example of an endothermic process. When ammonium nitrate dissolves in water, the solution becomes so cold that frost may form on the outside of the container. Instant cold packs, which are used to treat muscle aches and sore joints, work by endothermic dissolving. The cold pack in Figure 17.11 contains solid ammonium nitrate crystals and water. Once the solute dissolves in the solvent, the pack becomes cold. In this case, the solution process absorbs energy from the surroundings.

$$NH_4NO_3(s) \xrightarrow{H_2O(l)} NH_4^+(aq) + NO_3^-(aq)$$
$$\Delta H_{soln} = 25.7 \text{ kJ/mol}$$

Checkpoint ***What is the molar heat of solution?***

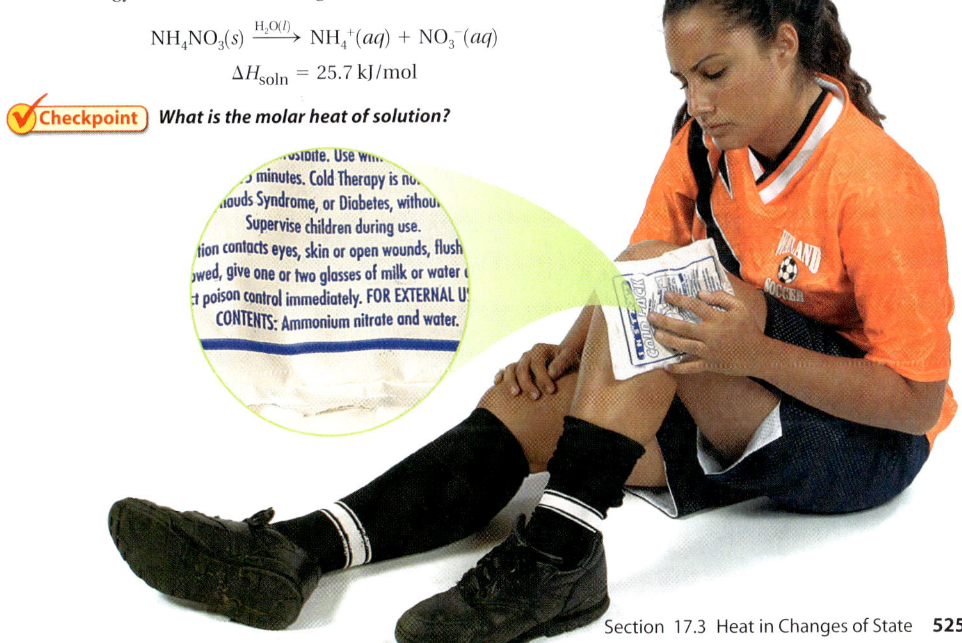

Figure 17.11 The cold pack shown has two sealed plastic bags, one inside the other. The outer bag contains ammonium nitrate crystals. The inner bag contains liquid water. When the pack is squeezed, the inner bag breaks, allowing the ammonium nitrate and water to mix. **Inferring** ***How would you define the system and the surroundings in this process?***

Heat of Solution

Use Visuals L1

Figure 17.11 Have students look at the photo of a cold pack. Ask, **Is the process that occurs in a cold pack endothermic or exothermic? How do you know?** *(Endothermic; heat is absorbed and the temperature of the surroundings decreases.)* **In which direction is heat flowing?** *(from the surroundings to the system)* **Is the process in a hot pack endothermic or exothermic? How do you know?** *(Exothermic; heat is released and the temperature of the surroundings increases.)* **In which direction is heat flowing?** *(from the system to the surroundings)*

CLASS Activity

Hot and Cold Packs

- Purchase a number of hot and cold packs used by athletes and show the class how they work. Most instant cold and hot packs operate by utilizing the heat released or absorbed when certain substances dissolve in water. Hot packs usually contain calcium chloride, $CaCl_2(s)$, while cold packs usually contain ammonium nitrate, $NH_4NO_3(s)$. If the labels specify the amount of salt contained in each package, have students use the molar heats of solution provided in the text to calculate the amount of heat absorbed and released in each case.
- Have students design and perform an experiment that measures the temperature change in a hot or cold pack. Also, you could change the solute to something not mentioned in the text (e.g., sodium acetate), and then have students conclude whether the solution process is endothermic or exothermic.

Differentiated Instruction

Less Proficient Readers L1

Provide at least two brands of cold or hot packs. Have students activate the packs at exactly the same time and then measure the maximum or minimum temperature achieved. Have students determine how long the packs remain effective. Ask students to decide, based on the data and the cost of each pack, which brand provided the best value.

Answers to...

Figure 17.11 The system is the cold pack; the athlete and the air around the cold pack constitute the surroundings.

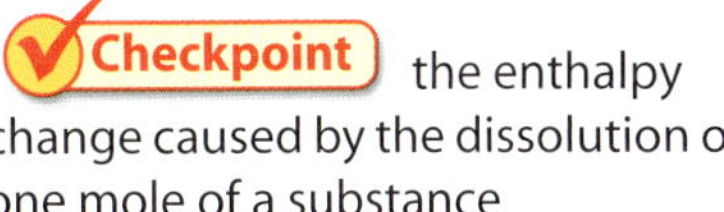
Checkpoint the enthalpy change caused by the dissolution of one mole of a substance

Sample Problem 17.6

Answers

25. -3.01×10^2 kJ

26. 3.42 mol $NH_4NO_3(s)$

Practice Problems Plus L2

How many kilojoules of heat are released when 25.0 g of NaOH are dissolved in water? *(–278 kJ)*

Math Handbook

For a math refresher and practice, direct students to significant figures, page R59.

3 ASSESS

Evaluate Understanding L2

Ask students to sketch a heating curve for 1 mol of ice being heated from –10°C to 110°C. Have students write brief explanations of their graphs. Ask students to calculate how much energy would be required to change 54.0 g of ice at –10.0°C to water vapor at 110°C. *(165 kJ)*

Reteach 

Display Figure 17.10 on an overhead projector. Remind students that during endothermic phase changes (melting or evaporation), energy absorbed does not increase the temperature because the energy is being used to overcome attractions between particles. The energy can be released back to the surroundings during the opposing exothermic phase change (condensation or freezing).

Connecting Concepts

Water molecules are very polar, and many hydrogen bonds form among the molecules. It takes a great deal of energy to break the hydrogen bonds and pull the water molecules apart.

If your class subscribes to the Interactive Textbook, use it to review key concepts in Section 17.3.

with ChemASAP

SAMPLE PROBLEM 17.6

Calculating the Enthalpy Change in Solution Formation

How much heat (in kJ) is released when 2.500 mol NaOH(s) is dissolved in water?

1 Analyze *List the knowns and the unknown.*

Knowns
- $\Delta H_{soln} = -445.1$ kJ/mol
- amount of NaOH(s) dissolved = 2.500 mol

Unknown
- $\Delta H = ?$ kJ

Use the heat of solution from the following chemical equation to solve for the amount of heat released (ΔH).

$$NaOH(s) \xrightarrow{H_2O(l)} Na^+(aq) + OH^-(aq) + 445.1 \text{ kJ/mol}$$

2 Calculate *Solve for the unknown.*

Multiply the number of moles of NaOH by ΔH_{soln}.

$$\Delta H = 2.500 \cancel{\text{mol NaOH}(s)} \times \frac{-445.1 \text{ kJ}}{1 \cancel{\text{mol NaOH}(aq)}} = -1113 \text{ kJ}$$

3 Evaluate *Does the result make sense?*

ΔH is 2.5 times greater than ΔH_{soln}, as it should be. Also, ΔH should be negative, as the dissolution of NaOH in water is exothermic.

Math Handbook

For help with significant figures go to page R59.

Interactive Textbook

Problem-Solving 17.26 Solve Problem 26 with the help of an interactive guided tutorial.

with ChemASAP

Practice Problems

25. How much heat (in kJ) is released when 0.677 mol NaOH(s) is dissolved in water?

26. How many moles of $NH_4NO_3(s)$ must be dissolved in water so that 88.0 kJ of heat is absorbed from the water?

17.3 Section Assessment

27. **Key Concept** How does the molar heat of fusion of a substance compare to its molar heat of solidification?

28. **Key Concept** How does the molar heat of vaporization of a substance compare to its molar heat of condensation?

29. **Key Concept** What enthalpy changes occur when a solute dissolves in a solvent?

30. Identify each enthalpy change by name and classify each change as exothermic or endothermic.
- **a.** 1 mol $C_3H_8(l) \longrightarrow$ 1 mol $C_3H_8(g)$
- **b.** 1 mol $Hg(l) \longrightarrow$ 1 mol $Hg(s)$
- **c.** 1 mol $NH_3(g) \longrightarrow$ 1 mol $NH_3(l)$
- **d.** 1 mol $NaCl(s)$ + 3.88 kJ/mol $\longrightarrow$ 1 mol $NaCl(aq)$
- **e.** 1 mol $NaCl(s) \longrightarrow$ 1 mol $NaCl(l)$

31. Why is a burn from steam potentially far more serious than a burn from very hot water?

Connecting Concepts

Hydrogen Bonding Reread the description of intermolecular attractions in Section 8.3. Use what you know about hydrogen bonding to explain why water absorbs such a large amount of heat as it vaporizes.

Assessment 17.3 Test yourself on the concepts in Section 17.3.

with ChemASAP

Section 17.3 Assessment

27. Molar heat of fusion and molar heat of solidification have an identical numerical value but are of opposite sign.

28. Molar heat of vaporization and molar heat of condensation have an identical numerical value but are of opposite sign.

29. Heat is either released or absorbed in the formation of a solution.

30. **a.** molar heat of vaporization, endothermic
b. molar heat of solidification, exothermic
c. molar heat of condensation, exothermic
d. molar heat of solution, endothermic
e. molar heat of fusion, endothermic

31. When a mole of steam condenses on your skin, it releases a substantial amount of heat—its molar heat of condensation.

17.4 Calculating Heats of Reaction

Connecting to Your World Emeralds are beautiful gemstones composed of the elements chromium, aluminum, silicon, oxygen, and beryllium. It is possible to break down an emerald into its component elements and measure the resulting enthalpy change—but then the gem would be destroyed. What if you wanted to determine the heat of reaction without actually performing the reaction? In this section you will see how you can calculate heats of reaction from known thermochemical equations and enthalpy data.

Guide for Reading

Key Concepts

- What are two ways that you can determine the heat of reaction when it cannot be directly measured?

Vocabulary

Hess's law of heat summation

standard heat of formation

Reading Strategy

Outlining As you read, make an outline of the most important ideas in this section.

Hess's Law

Sometimes it is hard to measure the enthalpy change for a reaction. For example, the reaction might take place too slowly to actually measure the enthalpy change. Or, the reaction might be an intermediate step in a series of reactions. Or, as mentioned above, you might not want to destroy the material that undergoes the reaction. Fortunately, it is possible to measure a heat of reaction indirectly. **Hess's law of heat summation** states that if you add two or more thermochemical equations to give a final equation, then you can also add the heats of reaction to give the final heat of reaction. **Hess's law allows you to determine the heat of reaction indirectly.**

Figure 17.12 shows a diamond mine. Diamonds are a form of carbon that exists at 25°C. Another is graphite. Because graphite is more stable than diamond, you might predict the following reaction.

$$C(s, \text{diamond}) \longrightarrow C(s, \text{graphite})$$

Fortunately for people who own diamonds, the conversion of diamond to graphite takes millions and millions of years. This enthalpy change cannot be measured directly because the reaction is far too slow. Hess's law, however, provides a way to calculate the heat of reaction.

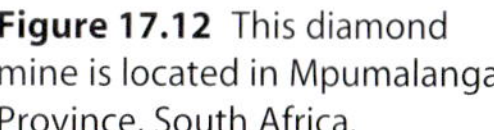

Figure 17.12 This diamond mine is located in Mpumalanga Province, South Africa.

Section Resources

Print

- **Guided Reading and Study Workbook,** Section 17.4
- **Core Teaching Resources,** Section 17.4 Review
- **Transparencies,** T192–T195
- **Laboratory Manual,** Lab 35
- **Probeware Lab Manual,** Section 17.4

Technology

- **Interactive Textbook with ChemASAP,** Problem-Solving 17.32, Assessment 17.4
- **Go Online,** Section 17.4
- **Virtual Chemistry Labs,** Lab 18

17.4

1 FOCUS

Objectives

17.4.1 State Hess's law of heat summation and **describe** how it is used in chemistry.

17.4.2 Solve for enthalpy changes by using Hess's law or standard heats of formation.

Guide for Reading

Build Vocabulary  L2

Paraphrase Ask students why they think Hess's law is called a law of heat summation. What does the word *summation* imply?

Reading Strategy  L2

Using Prior Knowledge Have students recall how the overall enthalpy change for converting one mole of ice to water vapor can be calculated by summing the enthalpy changes for each step in the process. A similar strategy can be used for determining the enthalpy change for a chemical reaction.

2 INSTRUCT

Connecting to Your World

Have students examine the photograph of emeralds. Ask, **Given that an emerald is too valuable to be destroyed, is there a way to determine the heat of reaction without actually performing the reaction?** *(Yes. The heat of reaction might be determined indirectly by studying the enthalpy changes for related reactions. These reactions could involve less valuable forms of the structures found in an emerald crystal.)*

Hess's Law

Use Visuals L1

Figure 17.13 Have students study the diagram of enthalpy change. Ask, **What is the enthalpy change of the reverse reaction—from graphite to diamond?** *($\Delta H = 1.9$ kJ)*

Discuss L2

Provide the following analogy. A group of hikers is attempting to reach the summit of a mountain. Two routes are possible: one is a short, steep path that goes directly to the top; the other is a long path that is less steep but involves a number of changes in direction. Although the distances covered by hikers using the two routes would be different, the final altitude would be the same. The change in altitude is analogous to the enthalpy change for a reaction. The distances traveled by the hikers on each path is analogous to the number of steps required to form a given set of products. No matter how one goes from a given set of reactants to a given set of products, in one step or in several, the overall enthalpy change is the same.

Relate L2

Made almost entirely of carbon, diamonds are the world's oldest substance. At 3.3 billion years of age, they are almost as old as Earth itself. Diamonds only form under extremely intense temperature and pressure. The only place on Earth these conditions are found naturally is 100 miles underground. Volcanic eruptions lift diamonds and other minerals to the surface.

Figure 17.13 Hess's law is used to determine the enthalpy change for the conversion of diamond to graphite.
Interpreting Diagrams ***Which of the reactions shown is an example of decomposition?***

You can use Hess's law to find the enthalpy change for the conversion of diamond to graphite by using the following combustion reactions and Figure 17.13.

a. $C(s, \text{graphite}) + O_2(g) \longrightarrow CO_2(g) \qquad \Delta H = -393.5 \text{ kJ}$

b. $C(s, \text{diamond}) + O_2(g) \longrightarrow CO_2(g) \qquad \Delta H = -395.4 \text{ kJ}$

Write equation **a** in reverse to give:

c. $CO_2(g) \longrightarrow C(s, \text{graphite}) + O_2(g) \qquad \Delta H = 393.5 \text{ kJ}$

When you write a reverse reaction, you must also change the sign of ΔH. If you now add equations **b** and **c,** you get the equation for the conversion of diamond to graphite. The $CO_2(g)$ and $O_2(g)$ terms on both sides of the summed equations cancel, just as they do in algebra. Now if you also add the values of ΔH for equations **b** and **c,** you get the heat of reaction for this conversion.

$$C(s, \text{diamond}) + \cancel{O_2(g)} \longrightarrow \cancel{CO_2(g)} \qquad \Delta H = -395.4 \text{ kJ}$$
$$\cancel{CO_2(g)} \longrightarrow C(s, \text{graphite}) + \cancel{O_2(g)} \qquad \Delta H = 393.5 \text{ kJ}$$
$$C(s, \text{diamond}) \longrightarrow C(s, \text{graphite}) \qquad \Delta H = -1.9 \text{ kJ}$$

The conversion of diamond to graphite is an exothermic process, so its heat of reaction has a negative sign. Conversely, the change of graphite to diamond is an endothermic process.

Facts and Figures

The 1867 discovery of a diamond in South Africa radically changed the diamond industry. Until then, India and Brazil had been the world's leading diamond producers, but by the 1880s, South Africa was producing 95% of the world's diamonds. Today, South Africa is third in production in terms of value. Canada, Australia, Brazil, China and Russia now collectively account for more than half of world diamond mining.

Another case in which Hess's law is useful is when reactions yield products in addition to the product of interest. Suppose you want to determine the enthalpy change for the formation of carbon monoxide from its elements. You can write the following equation for this reaction.

$$C(s, \text{graphite}) + \tfrac{1}{2}O_2(g) \longrightarrow CO(g) \qquad \Delta H = ?$$

Although it is easy to write the equation, carrying out the reaction in the laboratory as written is virtually impossible. Carbon dioxide (a "side product") is produced along with carbon monoxide (the "desired product"). Therefore, any measured heat of reaction is related to the formation of both $CO(g)$ and $CO_2(g)$, and not $CO(g)$ alone. However, you can calculate the desired enthalpy change by using Hess's law and the following two reactions that can be carried out in the laboratory.

a. $C(s, \text{graphite}) + O_2(g) \longrightarrow CO_2(g) \qquad \Delta H = -393.5\ \text{kJ}$

b. $CO(g) + \tfrac{1}{2}O_2(g) \longrightarrow CO_2(g) \qquad \Delta H = -283.0\ \text{kJ}$

Writing the reverse of equation **b** and changing the sign of ΔH yields equation **c.**

c. $CO_2(g) \longrightarrow CO(g) + \tfrac{1}{2}O_2(g) \qquad \Delta H = 283.0\ \text{kJ}$

Adding equations **a** and **c** gives the expression for the formation of $CO(g)$ from its elements. The enthalpy diagram for this heat summation is shown in Figure 17.14. Notice that only $\tfrac{1}{2}O_2(g)$ cancels from each equation.

$$C(s, \text{graphite}) + O_2(g) \longrightarrow \cancel{CO_2(g)} \qquad \Delta H = -393.5\ \text{kJ}$$

$$\cancel{CO_2(g)} \longrightarrow CO(g) + \tfrac{1}{2}O_2(g) \qquad \Delta H = 283.0\ \text{kJ}$$

$$C(s, \text{graphite}) + \tfrac{1}{2}O_2(g) \longrightarrow CO(g) \qquad \Delta H = -110.5\ \text{kJ}$$

The formation of $CO(g)$ is exothermic; 110.5kJ of heat is given off when 1 mol $CO(g)$ is formed from its elements.

Why can't you directly measure the heat of reaction for the formation of carbon monoxide from its elements?

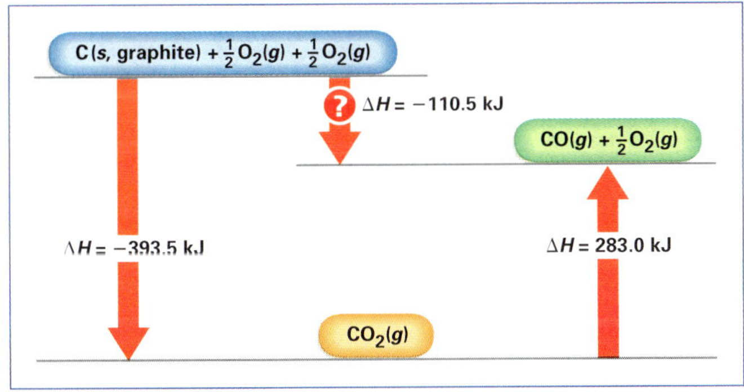

Figure 17.14 Hess's law is used to determine the enthalpy change for the formation of $CO(g)$ from its elements. **Interpreting Diagrams** ***How does the diagram represent endothermic and exothermic reactions differently?***

For: Links on diamond and graphite
Visit: www.SciLinks.org
Web Code: cdn-1174

Differentiated Instruction

Less Proficient Readers 

If students are having trouble visualizing the application of Hess's law, give them the following simplified equations.

$1 + 2 \rightarrow 3$

$3 + 4 \rightarrow 5 + 2$

Show students how 2 and 3 can be cancelled from both sides of the equations. The two equations can then be added together to produce $1 + 4 \rightarrow 5$.

Download a worksheet on **Diamond and Graphite** for students to complete, and find additional teacher support from NSTA SciLinks.

Use Visuals 

Figure 17.14 Have students study the diagram of enthalpy change for the formation of CO. Ask: **How can you tell from the diagram that the formation of CO is exothermic?** *(The arrow points downward and ΔH is negative.)*

Answers to...

Figure 17.13 The endothermic reaction $CO_2(g) \rightarrow C(s, \text{graphite}) + O_2(g)$ is a decomposition reaction.

Figure 17.14 The arrow points downward for an exothermic reaction and upward for an endothermic reaction.

 The reaction in the laboratory is impossible because carbon dioxide is produced along with carbon monoxide.

Standard Heats of Formation

Use Visuals L1

Table 17.4 With Table 17.4 displayed on an overhead projector, define *standard heat of formation* (the change in enthalpy that accompanies the formation of 1 mole of a compound from its elements with all substances in their standard states at 25°C and 101.3 kPa.) Point out that the standard heat of formation of a free element in its standard state is defined as 0. Demonstrate how to use standard heats of formation to calculate heats of reaction at standard conditions.

Use Visuals 

Figure 17.15 Have students study the enthalpy diagram. Ask, **Does water have a lower or higher enthalpy than the elements from which it is formed?** *(lower)* **On what other basis can you account for your answer?** *(It takes energy to decompose water.)*

Table 17.4

Standard Heats of Formation (ΔH_f^0) at 25°C and 101.3 kPa

Substance	ΔH_f^0 (kJ/mol)	Substance	ΔH_f^0 (kJ/mol)
$Al_2O_3(s)$	−1676.0	$H_2O_2(l)$	−187.8
$Br_2(g)$	30.91	$I_2(g)$	62.4
$Br_2(l)$	0.0	$I_2(s)$	0.0
C(*s*, diamond)	1.9	$N_2(g)$	0.0
C(*s*, graphite)	0.0	$NH_3(g)$	−46.19
$CH_4(g)$	−74.86	$NO(g)$	90.37
$CO(g)$	−110.5	$NO_2(g)$	33.85
$CO_2(g)$	−393.5	$NaCl(s)$	−411.2
$CaCO_3(s)$	−1207.0	$O_2(g)$	0.0
$CaO(s)$	−635.1	$O_3(g)$	142.0
$Cl_2(g)$	0.0	P(*s*, white)	0.0
$Fe(s)$	0.0	P(*s*, red)	−18.4
$Fe_2O_3(s)$	−822.1	S(*s*, rhombic)	0.0
$H_2(g)$	0.0	S(*s*, monoclinic)	0.30
$H_2O(g)$	−241.8	$SO_2(g)$	−296.8
$H_2O(l)$	−285.8	$SO_3(g)$	−395.7

Standard Heats of Formation

Enthalpy changes generally depend on conditions of the process. In order to compare enthalpy changes, scientists specify a common set of conditions as a reference point. These conditions, called the standard state, refer to the stable form of a substance at 25°C and 101.3 kPa. The **standard heat of formation** (ΔH_f^0) of a compound is the change in enthalpy that accompanies the formation of one mole of a compound from its elements with all substances in their standard states at 25°C. The ΔH_f^0 of a free element in its standard state is arbitrarily set at zero. Thus, $\Delta H_f^0 = 0$ for the diatomic molecules $H_2(g)$, $N_2(g)$, $O_2(g)$, $F_2(g)$, $Cl_2(g)$, $Br_2(l)$, and $I_2(s)$. Similarly, $\Delta H_f^0 = 0$ for the graphite form of carbon, C(*s*, graphite). Table 17.4 lists ΔH_f^0 values for some common substances.

Standard heats of formation provide an alternative to Hess's law in determining heats of reaction indirectly. **For a reaction that occurs at standard conditions, you can calculate the heat of reaction by using standard heats of formation.** Such an enthalpy change is called the standard heat of reaction (ΔH^0). The standard heat of reaction is the difference between the standard heats of formation of all the reactants and products.

$$\Delta H^0 = \Delta H_f^0(\text{products}) - \Delta H_f^0(\text{reactants})$$

Figure 17.15 is an enthalpy diagram for the formation of water from its elements at standard conditions. The enthalpy difference between the reactants and products, −285.8 kJ/mol, is the standard heat of formation of liquid water from the gases hydrogen and oxygen. Notice that water has a lower enthalpy than the elements from which it is formed.

Figure 17.15 This enthalpy diagram shows the standard heat of formation of water. **Classifying** ***Is this reaction endothermic or exothermic?***

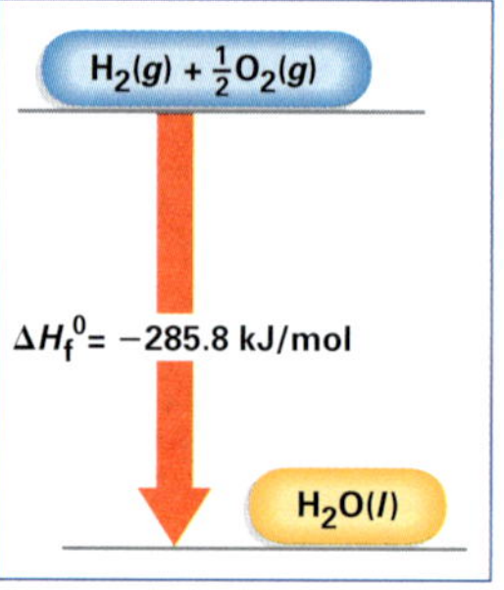

Facts and Figures

Thermite Reaction

In this reaction, iron(III) oxide is reduced to metallic iron by aluminum.

$$Fe_2O_3(s) + 2Al(s) \rightarrow 2Fe(s) + Al_2O_3(s)$$
$$\Delta H = -853.9 \text{ kJ}$$

The reaction reaches a temperature of about 3000°C, meaning both reaction products are molten ($mp_{Fe} = 1530°C$; $mp_{Al_2O_3} = 2030°C$).

SAMPLE PROBLEM 17.7

Calculating the Standard Heat of Reaction

What is the standard heat of reaction (ΔH^0) for the reaction of $CO(g)$ with $O_2(g)$ to form $CO_2(g)$?

1 Analyze *List the knowns and the unknown.*

Knowns

(from Table 17.4)

- $\Delta H_f^0 O_2(g) = 0$ kJ/mol (free element)
- $\Delta H_f^0 CO(g) = -110.5$ kJ/mol
- $\Delta H_f^0 CO_2(g) = -393.5$ kJ/mol

Unknown

- $\Delta H^0 = ?$ kJ

Balance the equation of the reaction of $CO(g)$ with $O_2(g)$ to form $CO_2(g)$. Then determine ΔH^0 using the standard heats of formation of the reactants and products.

2 Calculate *Solve for the unknown.*

First, write the balanced equation.

$$2CO(g) + O_2(g) \longrightarrow 2CO_2(g)$$

Next, find and add the ΔH_f^0 of all of the reactants, taking into account the number of moles of each.

$$\Delta H_f^0(\text{reactants}) = 2\ \text{mol}\ \cancel{CO(g)} \times \frac{-110.5\ \text{kJ}}{1\ \text{mol}\ \cancel{CO(g)}} + 0\ \text{kJ}$$
$$= -221.0\ \text{kJ}$$

Then, find the ΔH_f^0 of the product in a similar way.

$$\Delta H_f^0(\text{product}) = 2\ \text{mol}\ \cancel{CO_2(g)} \times \frac{-393.5\ \text{kJ}}{1\ \text{mol}\ \cancel{CO_2(g)}}$$
$$= -787.0\ \text{kJ}$$

Finally, solve for the unknown

$$\Delta H^0 = \Delta H_f^0(\text{products}) - \Delta H_f^0(\text{reactants})$$
$$\Delta H^0 = (-787.0\ \text{kJ}) - (-221.0\ \text{kJ})$$
$$\Delta H^0 = -566.0\ \text{kJ}$$

3 Evaluate *Does the result make sense?*

The ΔH^0 is negative. Therefore, the reaction is exothermic. This makes sense because combustion reactions always release heat.

Practice Problems

32. Calculate ΔH^0 for the following reactions.

a. $Br_2(g) \longrightarrow Br_2(l)$

b. $CaCO_3(s) \longrightarrow CaO(s) + CO_2(g)$

c. $2NO(g) + O_2(g) \longrightarrow 2NO_2(g)$

33. With one exception, the standard heats of formation of $Na(s)$, $O_2(g)$, $Br_2(l)$, $CO(g)$, $Fe(s)$, and $He(g)$ are identical. What is the exception? Explain.

Carbon Monoxide Detector

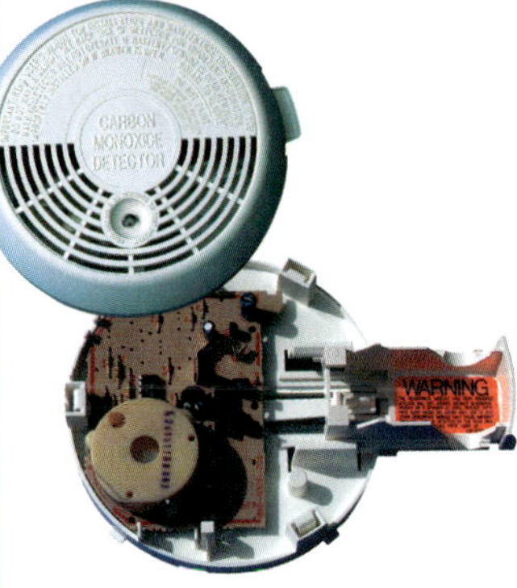

Math Handbook

For help with significant figures, go to page R59.

Problem-Solving 17.32

Solve Problem 32 with the help of an interactive guided tutorial.

with ChemASAP

Sample Problem 17.7

Answers

32. a. -3.091×10^1 kJ

b. 1.784×10^2 kJ

c. -1.130×10^2 kJ

33. CO is a compound and not an element in its standard state.

Practice Problems Plus

L2

Calculate the standard heat of reaction for the following reaction: $2SO_2(g) + O_2(g) \rightarrow 2SO_3(g)$ (-197.8 kJ)

Math Handbook

For a math refresher and practice, direct students to significant figures, page R59.

Answers to...

Figure 17.15 exothermic

3 ASSESS

Evaluate Understanding L2

Have students explain two ways that scientists can determine the heat of a reaction indirectly. *(1. Add two or more thermochemical equations to give the desired equation. 2. Find the difference between the standard heats of formation for all reactants and the standard heats of formation for all products in the reaction.)*

Reteach L1

Work out a heat-of-reaction problem using both of the methods described above.

Elements Handbook

$H_2O(g) + CH_4(g) \longrightarrow CO(g) + 3H_2(g)$ $\Delta H = 206$ kJ

$H_2O(g) + CO(g) \longrightarrow CO_2(g) + H_2(g)$ $\Delta H = -41$ kJ

$2H_2O(g) + CH_4(g) \longrightarrow CO_2(g) + 4H_2(g)$ $\Delta H = 165$ kJ

If your class subscribes to the Interactive Textbook, use it to review key concepts in Section 17.4.

with ChemASAP

Figure 17.16 Standard heats of formation are used to calculate the enthalpy change for the reaction of carbon monoxide and oxygen. **Interpreting Diagrams** ***How does this diagram also demonstrate Hess's law?***

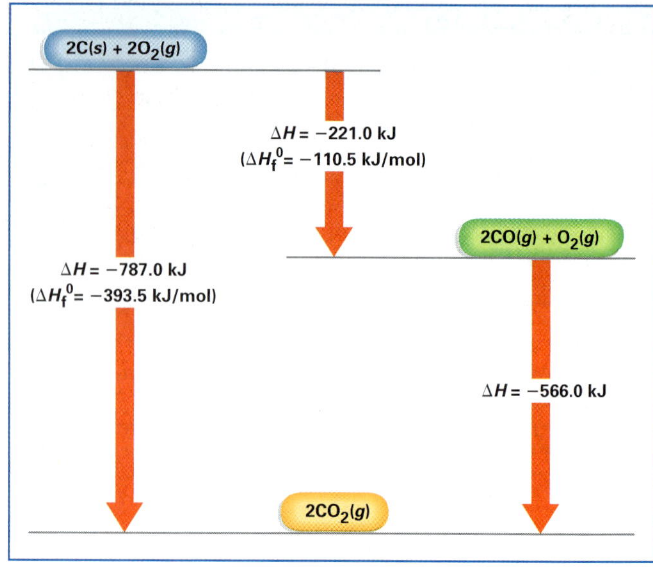

Figure 17.16 is an enthalpy diagram that shows how the standard heat of reaction was calculated in Sample Problem 17.7.

$$2CO(g) + O_2(g) \longrightarrow 2CO_2(g)$$

The standard heat of formation of the product, $CO_2(g)$, is -393.5 kJ/mol. The standard heats of formation of the reactants, $CO(g)$ and $O_2(g)$, are -110.5 kJ/mol and 0 kJ/mol, respectively. The diagram shows the difference between ΔH_f^0(product) and ΔH_f^0(reactants) after taking into account the number of moles of each.

17.4 Section Assessment

34. **Key Concept** What are two ways that the heat of reaction can be determined when it cannot be directly measured?

35. Calculate the enthalpy change (ΔH) in kJ for the following reaction.

$$2Al(s) + Fe_2O_3(s) \longrightarrow 2Fe(s) + Al_2O_3(s)$$

Use the enthalpy changes for the combustion of aluminum and iron:

$$2Al(s) + \tfrac{3}{2}O_2(g) \longrightarrow Al_2O_3(s) \quad \Delta H = -1669.8 \text{ kJ}$$

$$2Fe(s) + \tfrac{3}{2}O_2(g) \longrightarrow Fe_2O_3(s) \quad \Delta H = -824.2 \text{ kJ}$$

36. What is the formula for calculating the standard heat of reaction?

37. What is the standard heat of reaction (ΔH^0) for the decomposition of hydrogen peroxide?

$$2H_2O_2(l) \longrightarrow 2H_2O(l) + O_2(g)$$

Elements Handbook

Sources of Hydrogen Use Hess's law and two thermochemical equations on page R38 to calculate ΔH for the following reaction.

$$2H_2O(g) + CH_4(g) \longrightarrow CO_2(g) + 4H_2(g)$$

Assessment 17.4 Test yourself on the concepts in Section 17.4.

with ChemASAP

Section 17.4 Assessment

34. Use Hess's law of heat summation or use standard heats of formation.

35. -8.456×10^2 kJ

36. $\Delta H^0 = \Delta H_f^0(\text{products}) - \Delta H_f^0(\text{reactants})$

37. -1.960×10^2 kJ

Answers to...

Figure 17.16 It shows the summation of thermochemical equations to give a final equation. It also shows the enthalpy change for each step.

Heat of Combustion of a Candle

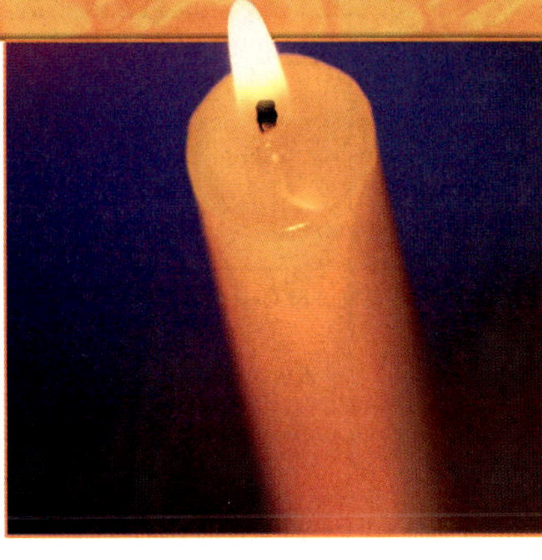

Purpose

To observe a burning candle and calculate the heat associated with the combustion reaction.

Materials

- candle
- aluminum foil
- safety matches
- ruler
- balance
- temperature probe (optional)

Procedure   

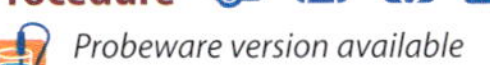

Probeware version available in the Probeware Lab Manual.

Measure and record the length of a candle in centimeters. Place the candle on a small piece of aluminum foil and measure the mass of the foil-candle system. Note the time as you light the candle. Let the candle burn for about five minutes. **CAUTION** *Keep clothing away from the flame.* While you wait, begin answering the Analyze questions. After about 5 minutes, extinguish the candle and record the time. Measure the mass of the foil-candle system again. **DO NOT** try to measure the mass while the candle is burning.

Analyze

Using your experimental data, answer the following questions.

1. Observe the candle burn and draw a picture of what you see.
2. Examine the flame closely. Is it the wax or the wick that burns?
3. If you said the wax, how does the wax burn without touching the flame? If you said the wick, what is the function of the wax?
4. If you could measure the temperature near the flame, you would find that the air is much hotter above the flame than it is beside it. Why? Explain.
5. Scientists have often wondered if a candle would burn well in zero gravity. How would zero gravity change the shape of the flame?
6. How much length and mass did the candle lose? Are these data more consistent with the wax or the wick burning?
7. Keeping in mind that *wick* is also a verb, explain how a candle works.
8. The formula for candle wax can be approximated as $C_{20}H_{42}$. Write and balance an equation for the complete combustion of the candle wax.
9. Calculate the number of moles of candle wax burned in the experiment.
10. Calculate the heat of combustion of candle wax in kJ/mol. The standard heat of formation of candle wax ($C_{20}H_{42}$) is −2230 kJ/mol. The standard heats of formation of carbon dioxide and water are −394 kJ/mol and −242 kJ/mol, respectively. The heat of combustion of candle wax equals the sum of the heats of formation of the products minus the sum of the heats of formation of the reactants.
11. Calculate the amount of heat (in kJ) released in your reaction. (*Hint:* Multiply the number of moles of candle wax burned in the experiment by the heat of combustion of candle wax.)

You're the Chemist

The following small-scale activities allow you to develop your own procedures and analyze the results.

1. **Design It!** Design an experiment to show that the candle wax does not burn with complete combustion.
2. **Design It!** Design an experiment to show that water is a product of the combustion of a candle.

Small-Scale LAB

Heat of Combustion of a Candle 

Objective After completing this activity, students will be able to:

- explore the way a candle burns.
- apply basic principles of thermochemistry to the combustion reaction.
- calculate the heat of combustion.

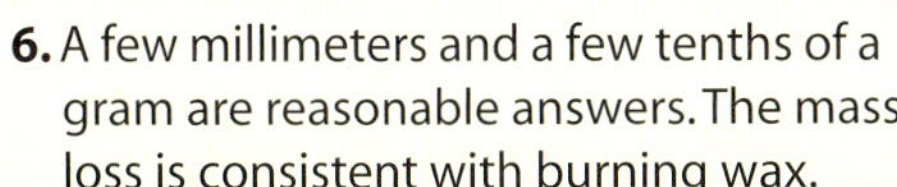

Address Misconceptions

Students have many misconceptions about the way candles burn. Questions 1 and 2 prompt students to wrongly conclude that the wick burns and the wax just slows the rate of burning. Questions 5 and 6 focus students on the key experiment, the determination of the mass loss of the candle, and allow them to revise their opinions.

 **Prep Time** 10 minutes

Materials candle, metric ruler, aluminum foil, balance, safety matches, clock or watch, temperature probe (optional)

Advance Prep Purchase or collect candles and a roll of aluminum foil.

Class Time 20 minutes

Safety Remind students to wear safety glasses and to keep the flame away from any combustible material.

Teaching Tips

- For best results and greater safety, use short, stout votive candles.
- The day before the lab, remind students to tie long hair back and to avoid wearing loose-sleeved clothes.

Expected Outcome The candle loses mass and length as it burns.

Analyze

1. Acceptable drawings should show the wax melting.
2. Many students incorrectly choose the wick.
3. The wick draws melted wax to the flame. Those who think the wick burns may suggest that the wax slows the rate of burning.
4. The hot gases expand and rise.
5. The flame might be round in zero gravity.
6. A few millimeters and a few tenths of a gram are reasonable answers. The mass loss is consistent with burning wax.
7. Heat from the combustion melts wax, which is drawn up by the wick, evaporated, and burned.
8. $C_{20}H_{42} + \frac{61}{2}O_2 \longrightarrow 20CO_2 + 21H_2O$
9. Sample answer: 0.0018 mol
10. Sample answer: −10,700 kJ/mol
11. Sample answer: 19 kJ

You're the Chemist

1. Soot will appear on a glass Petri dish held over the flame.
2. Water will condense on the bottom of an ice-filled, glass Petri dish held over the flame.

For Enrichment 

Have students design an experiment to determine the heat of combustion of table sugar. This is a reaction that you might wish to demonstrate for the class.

CHAPTER 17 Study Guide

Study Tip

Study with Friends Divide the class into small groups, and ask each group to write a review question for each Key Concept and each vocabulary term in one of the sections of the chapter. When students have finished writing their questions, have the groups place their lists in a central location. Then, ask each group to pick up a list of questions for one of the other three sections. After groups have had a chance to collaborate on answering the questions, repeat the process, with groups answering the questions for a third and fourth section.

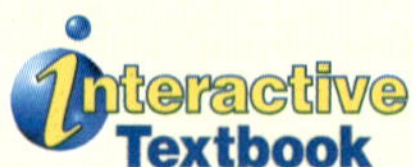

If your class subscribes to the Interactive Textbook with ChemASAP, your students can go online to access an interactive version of the Student Edition and a self-test.

with ChemASAP

CHAPTER 17 Study Guide

Key Concepts

17.1 The Flow of Energy—Heat and Work

- Heat always flows from a warmer object to a cooler object.
- A system gains heat in an endothermic process, and loses heat in an exothermic process.
- Heat flow is measured with two common units, the calorie and the joule.
- The heat capacity of an object depends on both its mass and its chemical composition.

17.2 Measuring and Expressing Enthalpy Changes

- In calorimetry, the heat released by a system equals the heat absorbed by its surroundings. Conversely, the heat absorbed by a system equals the heat released by its surroundings.
- The enthalpy change for a reaction can be treated like any other reactant or product.

Heat in Changes of State

- The heat absorbed by a melting solid is exactly the same as the heat lost when the liquid solidifies; that is, $\Delta H_{fus} = -\Delta H_{solid}$.
- The heat absorbed by a vaporizing liquid is exactly the same as the heat lost when the vapor condenses; that is, $\Delta H_{vap} = -\Delta H_{cond}$.
- Heat is either released or absorbed during the formation of a solution.

17.4 Calculating Heats of Reaction

- You can calculate the heat of a reaction by applying Hess's law of heat summation or by using standard heats of formation.

Vocabulary

- calorimeter (p. 511)
- calorimetry (p. 511)
- chemical potential energy (p. 505)
- endothermic process (p. 506)
- enthalpy (p. 511)
- exothermic process (p. 506)
- heat (p. 505)
- heat capacity (p. 508)
- heat of combustion (p. 517)
- heat of reaction (p. 514)
- Hess's law of heat summation (p. 527)
- law of conservation of energy (p. 506)
- molar heat of condensation (p. 523)
- molar heat of fusion (p. 520)
- molar heat of solidification (p. 520)
- molar heat of solution (p. 525)
- molar heat of vaporization (p. 522)
- specific heat (p. 508)
- standard heat of formation (p. 530)
- surroundings (p. 506)
- system (p. 506)
- thermochemical equation (p. 514)
- thermochemistry (p. 505)

Key Equations

- 1 Calorie = 1 kilocalorie = 1000 calories
- $C = \dfrac{q}{m \times \Delta T}$
- $q_{sys} = \Delta H = -q_{surr} = -m \times C \times \Delta T$
- $\Delta H^0 = \Delta H_f^0(\text{products}) - \Delta H_f^0(\text{reactants})$

Organizing Information

Use these terms to construct a concept map that organizes the major ideas of this chapter.

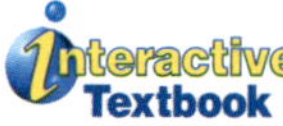

Concept Map 17 Solve the Concept Map with the help of an interactive guided tutorial.

with ChemASAP

- specific heat
- standard heats of formation
- heats of changes of state
- endothermic and exothermic reactions
- thermochemistry
- heat capacity
- thermochemical equations
- calorimetry
- Hess's law

Chapter Resources

Print

- **Core Teaching Resources,** Chapter 17 *Practice Problems, Vocabulary Review, Quiz, Chapter Test A, Chapter Test B*

Technology

- **Computer Test Bank,** Chapter 17 Test
- **Interactive Textbook with ChemASAP,** Chapter 17

CHAPTER 17

Assessment

Reviewing Content

17.1 The Flow of Energy—Heat

38. Explain in your own words the law of conservation of energy.

39. What always happens when two objects of different temperatures come in contact? Give an example from your own experience.

40. Define potential energy in terms of chemistry.

41. What factors determine heat capacity?

42. What is the relationship between a calorie and a Calorie?

43. Make the following conversions.

a. 8.50×10^2 cal to Calories

b. 444 cal to joules

c. 1.8 kJ to joules

d. 4.5×10^{-1} kJ to calories

44. Why do you think it is important to define the system and the surroundings?

45. Describe the sign convention that is used in thermochemical calculations.

46. Two substances in a glass beaker chemically react, and the beaker becomes too hot to touch.

a. Is the reaction exothermic or endothermic?

b. If the two substances are defined as the system, what constitutes the surroundings?

47. Classify these processes as exothermic or endothermic.

a. condensing steam **b.** evaporating alcohol

c. burning alcohol **d.** baking a potato

17.2 Measuring and Expressing Enthalpy Changes

48. What name is given to a heat change at constant pressure?

49. What is the function of a calorimeter?

50. There are some obvious sources of error in experiments that use foam cups as calorimeters. Name at least three.

51. What device would you use to measure the heat released at constant volume?

52. Give the standard conditions for heat of combustion.

53. What information is given in a thermochemical equation?

17.3 Heat in Changes of State

54. Explain why ice melts at 0°C without an increase of temperature, even though heat flows from the surroundings to the system (the ice).

55. Calculate the quantity of heat gained or lost in the following changes.

a. 3.50 mol of water freezes at 0°C.

b. 0.44 mol of steam condenses at 100°C.

c. 1.25 mol NaOH(*s*) dissolves in water.

d. 0.15 mol $C_2H_5OH(l)$ vaporizes at 78.3°C.

56. Sodium acetate dissolves readily in water according to the following equation.

$$NaC_2H_3O_2(s) \longrightarrow NaC_2H_3O_2(aq)$$
$$\Delta H = -17.3 \text{ kJ/mol}$$

Would this process increase or decrease the temperature of the water?

17.4 Calculating Heats of Reaction

57. Explain Hess's law of heat summation.

58. A considerable amount of heat is required for the decomposition of aluminum oxide.

$$2Al_2O_3(s) \longrightarrow 4Al(s) + 3O_2(g)$$
$$\Delta H = 3352 \text{ kJ}$$

a. What is the enthalpy change for the formation of 1 mol of aluminum oxide from its elements?

b. Is the reaction exothermic or endothermic?

59. Calculate the enthalpy change for the formation of lead(IV) chloride by the reaction of lead(II) chloride with chlorine.

$$PbCl_2(s) + Cl_2(g) \longrightarrow PbCl_4(l)$$
$$\Delta H = ?$$

Use the following thermochemical equations.

$$Pb(s) + 2Cl_2(g) \longrightarrow PbCl_4(l) \quad \Delta H = -329.2 \text{ kJ}$$
$$Pb(s) + Cl_2(g) \longrightarrow PbCl_2(s) \quad \Delta H = -359.4 \text{ kJ}$$

60. What is the standard heat of formation of a free element in its standard state?

61. Consider the statement, "the more negative the value of ΔH_f^0, the more stable the compound." Is this statement true or false? Explain.

CHAPTER 17

Assessment

Reviewing Content

38. Answers will vary, but should include the idea that energy is conserved in every physical and chemical process.

39. Heat flows from the object at the higher temperature to the object at the lower temperature. For example, holding a cold soft-drink can makes your hand cold.

40. Potential energy is energy stored in a substance because of its chemical composition.

41. the chemical composition of the substance and its mass

42. 1 Cal = 1000 cal = 1 kcal

43. **a.** 8.50×10^{-1} Calorie **b.** 1.86×10^3 J **c.** 1.8×10^3 J **d.** 1.1×10^2 cal

44. Answers will vary, but should mention that thermochemistry measures heat flow across the boundary between the system and the surroundings.

45. A negative sign is given to heat flow from the system to the surroundings. A positive sign is given to heat flow to the system from the surroundings.

46. **a.** exothermic

b. The immediate surroundings are the glass beaker and the air. If one or more of the substances is in water, the water is also considered part of the surroundings.

47. **a.** exothermic **b.** endothermic **c.** exothermic **d.** endothermic

48. enthalpy

49. A calorimeter is an instrument used to measure enthalpy changes in physical or chemical processes.

50. The foam cup will absorb heat. Some heat will be lost to the air. If the reactants are not completely mixed, temperature measurements will not be accurate.

51. bomb calorimeter

52. one atmosphere pressure (101.3 kPa); all reactants and products in their normal physical state at 25°C.

53. amount of heat released or absorbed in a chemical change at constant pressure

54. Heat is being used to melt the ice.

55. **a.** -2.10×10^1 kJ **b.** -1.8×10^1 kJ **c.** -5.56×10^2 kJ **d.** 6.5 kJ

56. increase

57. Hess's law allows the calculation of the enthalpy change of a reaction from the known enthalpy changes of two or more other reactions.

58. **a.** -1.676×10^3 kJ **b.** exothermic

59. 3.02×10^1 kJ

60. zero

61. The statement is true, since stability implies lower energy. The greater the release of heat, the more stable is the compound relative to its elements (all of which have $\Delta H_f^{\,0} = 0$).

CHAPTER 17

Assessment continued

Understanding Concepts

62. Substance B; For equal masses, the substance with the greater heat capacity undergoes the smaller temperature change.

63. 4.00×10^1 g water; 9.60×10^2 g ice

64. **a.** Graphs should show the mass on the *x*-axis and heat (in calories) on the *y*-axis.

b. about 54 cal/g

c. The two values are essentially the same.

65. 2.44×10^4 cal; 1.02×10^5 J

66. -7.50×10^2 kJ

67. **a.** -8.902×10^2 kJ

b. -5.660×10^2 kJ

68. The standard heat of formation (ΔH_f^0) of a compound is the change in enthalpy that accompanies the formation of one mole of a compound from its elements with all substances in their standard states at 25°C.

69. **a.** -2.21×10^2 kJ

b. -1.96×10^2 kJ

c. -9.046×10^2 kJ

70. **a.** 3.19×10^3 cal, 3.19 kcal, 1.34×10^4 J

b. 1.28×10^2 g H_2O

71. 2.36×10^1 kJ

72. 1.42×10^3 g

73. 2.38×10^2 kJ

74. 1.8×10^2 kJ

75. 6.71×10^1 kJ

CHAPTER 17

Assessment continued

Understanding Concepts

62. Equal masses of two substances absorb the same amount of heat. The temperature of substance A increases twice as much as the temperature of substance B. Which substance has the higher specific heat? Explain.

63. If 3.20 kcal of heat is added to 1.00 kg of ice at 0°C, how much water at 0°C is produced, and how much ice remains?

64. The amounts of heat required to change different quantities of carbon tetrachloride (CCl_4)(*l*) into vapor are given in the table.

Mass of CCl_4 (g)	Heat (J)	Heat (cal)
2.90	652	156
7.50	1689	404
17.0	3825	915
26.2	5894	1410
39.8	8945	2140
51.0	11453	2740

a. Graph the data, using heat as the dependent variable.

b. What is the slope of the line?

c. The heat of vaporization of $CCl_4(l)$ is 53.8 cal/g. How does this value compare with the slope of the line?

65. Calculate the enthalpy change in calories when 45.2 g of steam at 100°C condenses to water at the same temperature. What is the enthalpy change in joules?

66. Find the enthalpy change for the formation of phosphorus pentachloride from its elements.

$$2P(s) + 5Cl_2(g) \longrightarrow 2PCl_5(s)$$

Use the following thermochemical equations.

$$PCl_5(s) \longrightarrow PCl_3(g) + Cl_2(g) \quad \Delta H = 87.9 \text{ kJ}$$

$$2P(s) + 3Cl_2(g) \longrightarrow 2PCl_3(g) \quad \Delta H = -574 \text{ kJ}$$

67. Calculate the change in enthalpy (in kJ) for the following reactions.

a. $CH_4(g) + 2O_2(g) \longrightarrow CO_2(g) + 2H_2O(l)$

b. $2CO(g) + O_2(g) \longrightarrow 2CO_2(g)$

68. What is the standard heat of formation of a compound?

69. Use standard heats of formation (ΔH_f^0) to calculate the change in enthalpy for these reactions.

a. $2C(s, \text{graphite}) + O_2(g) \longrightarrow 2CO(g)$

b. $2H_2O_2(l) \longrightarrow 2H_2O(l) + O_2(g)$

c. $4NH_3(g) + 5O_2(g) \longrightarrow 4NO(g) + 6H_2O(g)$

70. An ice cube with a mass of 40.0 g melts in water originally at 25.0°C.

a. How much heat does the ice cube absorb from the water when it melts? Report your answer in calories, kilocalories, and joules.

b. Calculate the number of grams of water that can be cooled to 0°C by the melting ice cube.

71. The molar heat of vaporization of ethanol ($C_2H_5OH(l)$) is 43.5 kJ/mol. Calculate the heat required to vaporize 25.0 g of ethanol at its boiling point.

72. An orange contains 445 kJ of energy. What mass of water could this same amount of energy raise from 25.0°C to the boiling point?

73. The combustion of ethane (C_2H_4) is an exothermic reaction.

$$C_2H_4(g) + 3O_2(g) \longrightarrow 2CO_2(g) + 2H_2O(l) \quad \Delta H = -1.39 \times 10^3 \text{ kJ}$$

Calculate the amount of heat liberated when 4.79 g C_2H_4 reacts with excess oxygen.

74. Calculate the enthalpy change (ΔH) for the formation of nitrogen monoxide from its elements.

$$N_2(g) + O_2(g) \longrightarrow 2NO(g)$$

Use these thermochemical equations.

$$4NH_3(g) + 3O_2(g) \longrightarrow 2N_2(g) + 6H_2O(l) \quad \Delta H = -1.53 \times 10^3 \text{ kJ}$$

$$4NH_3(g) + 5O_2(g) \longrightarrow 4NO(g) + 6H_2O(l) \quad \Delta H = -1.17 \times 10^3 \text{ kJ}$$

75. How much heat must be removed from a 45.0-g sample of naphthalene ($C_{10}H_8$) at its freezing point to bring about solidification? The heat of fusion of naphthalene is 191.2 kJ/mol.

536 Chapter 17

Critical Thinking

76. Your fingers quickly begin to feel cold when you touch an ice cube. What important thermochemical principle does this illustrate?

77. You place a bottle containing 2.0 L of mineral water at 25°C into a refrigerator to cool to 7°C.

a. How many kJ of heat is lost by the water?

b. How many kJ of heat is absorbed by the refrigerator?

c. What assumptions did you make in your calculations?

78. When 1.000 mol of $N_2(g)$ reacts completely with 3.000 mol of $H_2(g)$, 2.000 mol of $NH_3(g)$ and 92.38 kJ of heat are produced.

$$N_2(g) + 3H_2(g) \longrightarrow 2NH_3(g) + 92.38\ \text{kJ}$$

Use this thermochemical equation to calculate ΔH for the following reactions.

a. $2N_2(g) + 6H_2(g) \longrightarrow 4NH_3(g)$

b. $\frac{3}{2}N_2(g) + \frac{9}{2}H_2(g) \longrightarrow 3NH_3(g)$

c. $\frac{1}{2}N_2(g) + \frac{3}{2}H_2(g) \longrightarrow NH_3(g)$

79. Explain why fusion is an endothermic process but freezing is an exothermic process.

80. Evaluate this statement: The energy content of a substance is higher in the liquid phase than in the vapor phase at the same temperature.

81. Using the following equations:

$$Ca(s) + 2C(s) \longrightarrow CaC_2(s) \quad \Delta H = -62.8\ \text{kJ}$$

$$CO_2(g) \longrightarrow C(s) + O_2(g) \quad \Delta H = 393.5\ \text{kJ}$$

$$CaCO_3(s) + CO_2(g) \longrightarrow CaC_2(s) + 2\tfrac{1}{2}O_2(g) \quad \Delta H = 1538\ \text{kJ}$$

determine the heat of reaction (in kJ) for:

$$Ca(s) + C(s) + \tfrac{3}{2}O_2(g) \longrightarrow CaCO_3(s)$$

82. The sugar glucose ($C_6H_{12}O_6$) is an important nutrient for living organisms to meet their energy needs. The standard heat of formation (ΔH_f^0) of glucose is –1260 kJ/mol. Calculate how much heat (in kJ/mol) is released at standard conditions if 1 mol of glucose undergoes the following reaction.

$$C_6H_{12}O_6(s) + 6O_2(g) \longrightarrow 6CO_2(g) + 6H_2O(l)$$

Concept Challenge

83. The temperature of a person with an extremely high fever can be lowered with a sponge bath of isopropyl alcohol (C_3H_7OH). The heat of vaporization of this alcohol is 11.1 kcal/mol.

a. How many kilocalories of heat are removed from a person's skin when 175 g of isopropyl alcohol evaporates? How many kilojoules?

b. How many kilograms of water would this energy loss cool in lowering the temperature from 40.0°C to 36.0°C?

84. Ethane ($C_2H_6(g)$) can be formed by the reaction of ethene ($C_2H_4(g)$) with hydrogen gas.

$$C_2H_4(g) + H_2(g) \longrightarrow C_2H_6(g)$$

Use the heats of combustion for the following reactions to calculate the heat change for the formation of ethane from ethene and hydrogen.

$$2H_2(g) + O_2(g) \longrightarrow 2H_2O(l) \quad \Delta H = -5.72 \times 10^2\ \text{kJ}$$

$$C_2H_4(g) + 3O_2(g) \longrightarrow 2H_2O(l) + 2CO_2(g) \quad \Delta H = -1.401 \times 10^3\ \text{kJ}$$

$$2C_2H_6(g) + 7O_2(g) \longrightarrow 6H_2O(l) + 4CO_2(g) \quad \Delta H = -3.100 \times 10^3\ \text{kJ}$$

85. An ice cube at 0°C was dropped into 30.0 g of water in a cup at 45.0°C. At the instant that all of the ice was melted, the temperature of the water in the cup was 19.5°C. What was the mass of the ice cube?

86. The molar heat of vaporization of water at various temperatures is given in the graph. Estimate the amount of heat required to convert 1 L of water to steam on the summit of Mount Everest (8850 m), where the boiling temperature of water is 70°C.

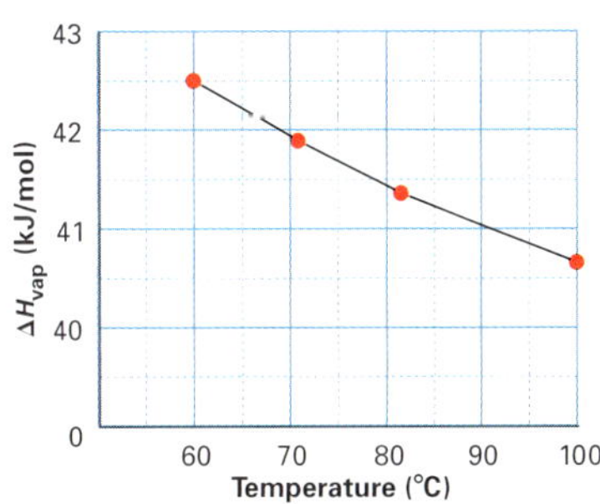

Critical Thinking

76. Your fingers feel cold because heat always passes from a warmer object (fingers) to a cooler object (ice).

77. a. 1.5×10^2 kJ

b. The refrigerator absorbs 1.5×10^2 kJ of heat.

c. assumes the mineral water has the same specific heat as chemically pure water, that no heat is lost by the refrigerator, and the volume of the water is exactly 2 L

78. a. –184.8 kJ

b. –138.6 kJ

c. –46.19 kJ

79. When a solid reaches its melting point, additional heat must be absorbed to convert it to a liquid. Therefore, fusion of a solid is endothermic. This heat of fusion is released when a liquid freezes, so freezing is exothermic.

80. When the vapor condenses to a liquid, the heat of vaporization is released. Therefore, the energy content of the vapor must have been higher than that of the liquid. The statement is incorrect.

81. –1207 kJ

82. –2816 kJ

Concept Challenge

83. a. 3.24×10^1 kcal, 1.36×10^2 kJ

b. 8.13 kg

84. -1.37×10^2 kJ

85. 9.6 g

86. ΔH_{vap} for water at 70°C is approximately 42 kJ/mol. 1 L of water (1000 mL) has a mass of 1000 g and contains 55.6 mol water. Therefore, the amount of heat required is 42 kJ/mol × 55.6 mol = 2.34×10^3 kJ

CHAPTER 17

Assessment continued

Cumulative Review

87. The manipulated variable is the variable you change during an experiment. The responding variable is the variable you observe during an experiment.

88. **a.** Cr
b. Cu
c. C
d. Ca
e. Cs
f. Cl

89. **a.** 6.99
b. 10.68
c. 3.6×10^2
d. 4.44

90. See table below.

91. 32.2 m

92. phosphorus, arsenic, germanium

93. **a.** 2
b. 2
c. 3
d. 1

94. A covalent bond is nonpolar when the electrons of the bonding atoms are shared equally. A covalent bond is polar when the electrons of the bonding atoms are shared unequally. The bond between oxygen atoms in the oxygen molecule (O_2) is nonpolar. The bond between a carbon atom and an oxygen atom in the carbon monoxide molecule is polar

95. **a.** K_3N
b. Al_2S_3
c. $Ca(NO_3)_2$
d. $CaSO_4$

96. 1.20×10^{24} H_2 molecules

97. $Ag^+(aq) + Cl^-(aq) \rightarrow AgCl(s)$

98. $N_2(g) + O_2(g) \rightarrow 2NO(g)$
$2NO(g) + O_2(g) \rightarrow 2NO_2(g)$

99. 1.18×10^1 g O_2

100. 0.40

101. 11.1 L

102. Water is a very good solvent and dissolves at least a small part of most of what it comes in contact with in nature.

103. solutions; suspensions

CHAPTER 17

Assessment continued

Cumulative Review

87. Explain the difference between a manipulated variable and a responding variable. (*Chapter 1*)

88. Write the correct chemical symbol for each element. (*Chapter 2*)
a. chromium **b.** copper
c. carbon **d.** calcium
e. cesium **f.** chlorine

89. Express the results of the following calculations with the correct number of significant figures. (*Chapter 3*)
a. 6.723×1.04
b. $8.934 + 0.2005 + 1.55$
c. $864 \div 2.4$
d. $9.258 - 4.82$

90. List three kinds of subatomic particles in an atom. Describe each kind in terms of charge, relative mass, and location with respect to the nucleus. (*Chapter 4*)

91. Calculate the wavelength of a radio wave with a frequency of $93.1 \times 10^6\ s^{-1}$. (*Chapter 5*)

92. List the following atoms in order of increasing atomic radius: phosphorus, germanium, arsenic. (*Chapter 6*)

93. How many chloride ions would be required to react with these cations to make an electrically neutral particle? (*Chapter 7*)
a. strontium cation
b. calcium cation
c. aluminum cation
d. lithium cation

94. How does a polar covalent bond differ from a nonpolar covalent bond? Which type of bond is found in molecular oxygen (O_2)? In carbon monoxide (CO)? (*Chapter 8*)

95. Write formulas for the following compounds. (*Chapter 9*)
a. potassium nitride
b. aluminum sulfide
c. calcium nitrate
d. calcium sulfate

96. How many hydrogen molecules are in 44.8 L $H_2(g)$ at STP? (*Chapter 10*)

97. Write the net ionic equation for the reaction of aqueous solutions of sodium chloride and silver acetate. (*Chapter 11*)

98. When lightning flashes, nitrogen and oxygen combine to form nitrogen monoxide. The nitrogen monoxide reacts with oxygen to form nitrogen dioxide. Write equations for these two reactions. (*Chapter 11*)

99. How many grams of oxygen are formed by the decomposition of 25.0 g of hydrogen peroxide? (*Chapter 12*)

$$2H_2O_2(l) \longrightarrow 2H_2O(l) + O_2(g)$$

100. What fraction of the average kinetic energy of hydrogen gas at 100 K does hydrogen gas have at 40 K? (*Chapter 13*)

101. A gas has a volume of 8.57 L at 273 K. What will be the volume at 355 K if its pressure does not change? (*Chapter 14*)

102. What property of water makes it impossible to find pure water in nature? (*Chapter 15*)

103. Do colloids, suspensions, or solutions contain the smallest particles? Which contain the largest particles? (*Chapter 15*)

538 Chapter 17

Answer to question 90:

Particle	Charge	Relative Mass	Location
proton	1+	1	nucleus
neutron	0	1	nucleus
electron	1–	1/1840	outside nucleus

Standardized Test Prep

Test-Taking Tip

Eliminate Wrong Answers If you don't know which response is correct, start by eliminating those you know are wrong. If you can rule out some choices, you'll have fewer left to consider and you'll increase your chances of choosing the correct answer.

Select the choice that best answers each question or completes each statement.

1. The ΔH_{fus} of ethanol (C_2H_5OH) is 4.60 kJ/mol. How many kilojoules are required to melt 24.5 g of ethanol at its freezing point?
   a. 2.45 kJ **b.** 5.33 kJ
   c. 245 kJ **d.** 18.8 kJ
2. How much heat, in kilojoules, must be added to 178 g of water to increase the temperature of the water by 5.0°C?
   a. 890 kJ **b.** 36 kJ
   c. 3.7 kJ **d.** 0.093 kJ
3. The standard heat of formation of a free element in its standard state is always
   a. zero. **b.** positive.
   c. negative. **d.** higher for solids than gases.
4. If ΔH for the reaction $2HgO(s) \longrightarrow 2Hg(l) + O_2(g)$ is 181.66 kJ, then ΔH for the reaction $Hg(l) + \frac{1}{2}O_2(g) \longrightarrow HgO(s)$ is
   a. 90.83 kJ. **b.** −90.83 kJ.
   c. 181.66 kJ. **d.** −181.66 kJ.
5. The specific heat capacity of grain alcohol is ten times larger than the specific heat capacity of silver. A hot bar of silver with a mass of 55 g is dropped into an equal mass of cool alcohol. If the temperature of the silver bar drops 45 °C, the temperature of the alcohol
   a. increases 45°C. **b.** decreases 4.5°C.
   c. increases 4.5°C. **d.** decreases 45°C.
6. Hydrogen gas and fluorine gas react to form hydrogen fluoride, HF. Calculate the enthalpy change (in kJ) for the conversion of 15.0 g of $H_2(g)$ to $HF(g)$ at constant pressure.

$$H_2(g) + F_2(g) \longrightarrow 2HF(g)$$
$$\Delta H = -536 \text{ kJ}$$

The lettered choices below refer to Questions 7–10. A lettered choice may be used once, more than once, or not at all.

(A) kJ/mol
(B) (J·°C)/g
(C) J/(g·°C)
(D) kJ
(E) kJ/°C

Which unit is appropriate for each of the following measurements?

7. heat of reaction
8. heat capacity
9. molar heat of fusion
10. specific heat

Use the graph and table to answer Questions 11–14. Assume 1.00 mol of substance in each container.

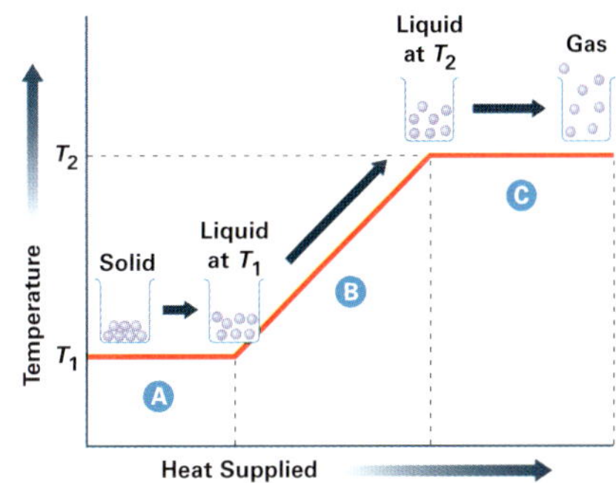

Substance	Freezing point (K)	ΔH_{fus} (kJ/mol)	Boiling point (K)	ΔH_{vap} (kJ/mol)
Ammonia	195.3	5.65	239.7	23.4
Benzene	278.7	9.87	353.3	30.8
Methanol	175.5	3.16	337.2	35.3
Neon	24.5	0.33	27.1	1.76

11. Calculate heat absorbed in region A for neon.
12. Calculate heat absorbed in region C for benzene.
13. Calculate heat absorbed in regions B and C for methanol. [specific heat = 81.6 J/(g·°C)]
14. Calculate heat absorbed in regions A, B, and C for ammonia. [specific heat = 35.1 J/(g·°C)]

Standardized Test Prep

1. a
2. c
3. a
4. b
5. c
6. -3.98×10^3 kJ
7. D
8. E
9. A
10. C
11. 0.33 kJ
12. 30.8 kJ
13. region B: 422 kJ; region C: 35.3 kJ
14. region A: 5.65 kJ; region B: 26.5 kJ; region C: 23.4 kJ

CHAPTER 18

Reaction Rates and Equilibrium Planning Guide

SECTION OBJECTIVES	STANDARDS NATIONAL	STANDARDS STATE	ACTIVITIES and LABS	
18.1 Rates of Reaction, pp. 541–548 1 block or 2 periods **18.1.1 Describe** how to express the rate of a chemical reaction. **18.1.2 Identify** four factors that influence the rate of a chemical reaction.	A-1, A-2, B-3, E-2		**SE** Inquiry Activity: Temperature and Reaction Rates, p. 540 **SE** Quick Lab: Does Steel Burn? p. 544 **TE** Teacher Demo: Use of Heat and Catalyst in a Reaction, p. 546 **LM** Lab 36: Factors Affecting Reaction Rates **LM** Lab 37: The Clock Reaction **SSLM** Lab 28: Factors Affecting the Rate of a Chemical Reaction	L2 L2 L2 L2 L3 L2
18.2 Reversible Reactions and Equilibrium, pp. 549–559 1 block or 2 periods **18.2.1 Describe** how the amounts of reactants and products change in a chemical system at equilibrium. **18.2.2 Identify** three stresses that can change the equilibrium position of a chemical system. **18.2.3 Explain** what the value of K_{eq} indicates about the position of equilibrium.	A-1, A-2, B-3		**TE** Teacher Demo: Revisiting Banana Oil Demonstration, p. 552 **TE** Teacher Demo: Temperature and Equilibrium, p. 555 **LM** Lab 38: Disturbing Equilibrium **SSLM** Lab 29: Le Chatelier's Principle and Chemical Equilibrium	L1 L2 L2 L2
18.3 Solubility Equilibrium, pp. 560–565 1 block or 2 periods **18.3.1 Describe** the relationship between the solubility product constant and the solubility of a compound. **18.3.2 Predict** whether precipitation will occur when two salt solutions are mixed.	A-1, A-2, B-2		**TE** Class Activity: Solubility Tables, p. 561 **TE** Teacher Demo: Common Ion Effect, p. 562 **LM** Lab 39: A Solubility Product Constant	L2 L2 L2
18.4 Entropy and Free Energy, pp. 566–574 1 block or 2 periods **18.4.1 Identify** two characteristics of spontaneous reactions. **18.4.2 Describe** the role of entropy in chemical reactions. **18.4.3 Identify** two factors that determine the spontaneity of a reaction. **18.4.4 Define** Gibbs free-energy change.	A-1, A-2, B-2, B-3, B-5		**SE** Small-Scale Lab: Enthalpy and Entropy, p. 574 **TE** Class Activity: Research Photosynthesis, p. 568 **TE** Teacher Demo: The Entropy of Water, p. 570 **TE** Teacher Demo: Observing Spontaneous Reaction, p. 572 **VCL** Lab 20: Enthalpy and Entropy	L2 L2 L2 L2 L2
18.5 The Progress of Chemical Reactions, pp. 575–579 1 block or 2 periods **18.5.1 Describe** the general relationship between the value of the specific rate constant, *k*, and the speed of a chemical reaction. **18.5.2 Interpret** the hills and valleys in a reaction progress curve.	A-1, A-2, B-3, G-1, G-2		**LP** Lab Practical 18-1: Reaction Rate	L2

Ability Levels	
L1	For students who need additional help
L2	For all students
L3	For students who need to be challenged

Components

SE	Student Edition	**LM**	Laboratory Manual	**GCP**	Graphing Calculator Problems	**T**	Transparencies
TE	Teacher's Edition	**SSLM**	Small-Scale Chemistry Lab Manual	**TTT**	Test-Taking Tips with Transparencies, Grades 9–12 Science	**CA**	Chem Alive and Chemistry Field Trips DVD
GRSW	Guided Reading and Study Workbook	**LP**	Laboratory Practicals	**STP**	Standardized Test Preparation Workbook	**iT**	Interactive Textbook with ChemASAP
CTR	Core Teaching Resources	**PLM**	Probeware Lab Manual			**GO**	Go Online Internet Resources
SSG	Spanish Study Guide	**VCL**	Virtual Chemistry Labs				
		CTB	Computer Test Bank				

RESOURCES PRINT and TECHNOLOGY			PROBLEMS and ASSESSMENT
GRSW	Chapter 18, Section 1	L1	**SE** Section 18.1 Assessment, p. 547
	T T196–T198	L1	**iT** Section 18.1
	GO Chapter 18, Section 1	L2	
GRSW	Chapter 18, Section 2	L1	**CP** 18.1, Applying Le Châtelier's Principle
	T T199–T202	L2	**SP** 18.1, Expressing and Calculating K_{eq}
			SP 18.2, Finding the Equilibrium Constant
			SE Section 18.2 Assessment, p. 559
			iT Section 18.2
GRSW	Chapter 18, Section 3	L1	**SP** 18.3, Finding the Ion Concentrations in a Saturated Solution
	T T203–T205	L2	**SP** 18.4, Finding Equilibrium Ion Concentrations in the Presence of a Common Ion
	GO Chapter 18, Section 3	L2	**SE** Section 18.3 Assessment, p. 565
			iT Section 18.3
GRSW	Chapter 18, Section 4	L1	**SE** Section 18.4 Assessment, p. 573
	T T206–T208	L2	**iT** Section 18.4
	GO Chapter 18, Section 4	L2	
GRSW	Chapter 18, Section 5	L1	**CP** 18.2, Finding the Order of a Reaction from Experimental Data
	T T209–T212	L2	**SE** Section 18.5 Assessment, p. 579
	GO Chapter 18, Section 5	L2	**iT** Section 18.5

Go Online

Web Code: cdn-1181
Web Code: cdn-1182
Web Code: cdn-1184

Web Code: cdb-1185

MATERIALS List

Quantities for each group

STUDENT EDITION

Inquiry Activity, p. 540
masking tape, 4 plastic cups, hot and cold tap water, ice, thermometer, 4 (or 8) effervescent antacid tablets, clock or watch with a second hand, graph paper

Quick Lab, p. 544
#0000 steel-wool pad, tissue paper, tongs, Bunsen burner, heat-resistant pad

Small-Scale Lab, p. 574
NaCl, NH_4Cl, $CaCl_2$, thermometer, 4 1-oz. plastic cups, plastic spoon, crushed ice, water, stirring rod. Optional solids: KCl, $NaHCO_3$, Na_3PO_4

TEACHER'S EDITION

Teacher Demo, p. 546
10 mL ethanoic acid, 10 mL isoamyl alcohol, 2 test tubes, anhydrous calcium sulfate, sulfuric acid, water

Teacher Demo, p. 552
10 mL ethanoic acid, 10 mL isoamyl alcohol, 2 test tubes, anhydrous calcium sulfate, sulfuric acid, water

Teacher Demo, p. 555
3 g copper turnings, 3 heat-resistant test tubes, 10 mL of 6*M* nitric acid, one-hole stopper, 2 solid stoppers, glass delivery tube, rubber tubing, warm water bath, ice water bath

Teacher Demo, p. 562
saturated solutions of AgCl and $AgNO_3$, 2 test tubes, dropper

Teacher Demo, p. 570
block of ice, 2 beakers, water, hot plate

Teacher Demo, p. 572
0.1*M* solution of $KMnO_4$ in 0.1*M* HCl, graduated cylinder, 1000-mL beaker, thermometer, 20 mL of 30% H_2O_2, watch glass, splint, matches

Chapter Assessment

CHAPTER ASSESSMENT

SE	Chapter Assessment, pp. 581–584
CTB	Chapter 18
iT	Chapter 18

STANDARDIZED TEST PREP

SE	Chapter 18, p. 585
STP	Topic 7

iT Interactive Textbook with Assessment at PHSchool.com

CHAPTER 18

Reaction Rates and Equilibrium

INQUIRY Activity

Temperature and Reaction Rates

L2

Purpose

To observe the effect of temperature on chemical reaction rates.

 **Prep Time** 10 minutes
Class Time 15 minutes

Think About It

1. Yes; the increased temperatures provide increased energy for the reaction.
2. Increasing the surface area by crushing the tablets makes each reaction go faster, but the overall pattern is the same.

CHAPTER 18

Reaction Rates and Equilibrium

Chemical reactions, like stock cars, have measurable speeds.

INQUIRY Activity

Temperature and Reaction Rates

Materials

masking tape, four plastic cups, hot and cold tap water, ice, a thermometer, four effervescent antacid tablets, a clock or watch with a second hand, graph paper, and a pen or pencil.

Procedure

1. Mark an A, B, C, or D on four separate pieces of masking tape. Label each plastic cup with one of the lettered pieces of tape.
2. Fill each cup three-quarters full as follows: (A) cold tap water plus some ice, (B) cold tap water, (C) half-and-half mixture of cold tap water and hot tap water, (D) hot tap water.
3. Measure and record the temperature of the water in cup A.
4. Drop a tablet into cup A and measure the time it takes for the reaction (bubble production) to end. Repeat Steps 3 and 4 using the remaining three cups.
5. Draw a line or bar graph of temperature versus reaction time.

Think About It

1. Do the tablets react faster at higher temperatures? Explain why or why not.
2. What results would you expect if you crushed the tablets into a powder before adding them to the water? Explain your prediction. If time permits, check this prediction experimentally.

540

18.1 Rates of Reaction

Connecting to Your World

Metal objects can be damaged by corrosion, but this process can be useful to hungry soldiers, truck drivers, and others who want a hot meal but do not have a place to cook it. Products now on the market use the heat given off by the corrosion reaction of an iron-magnesium alloy with salt water to produce a hot meal. Normally the reaction takes place at such a slow rate that the heat released from the process cannot be used. But when the rate is increased by the addition of salt water, heat is produced rapidly. In this section, you will learn some ways in which the rate of a reaction can be increased.

Guide for Reading

 **Key Concepts**

- How is the rate of a chemical change expressed?
- What four factors influence the rate of a chemical reaction?

Vocabulary

rate
collision theory
activation energy
activated complex
transition state
inhibitor

Reading Strategy

Outlining As you read, make an outline of the most important ideas in this section. Use the red headings as the main topics and the blue as subtopics. Add a sentence or a note after each heading to provide key information about each topic.

Collision Theory

When you strike a match, it erupts into flame almost instantly and burns quickly. Some other reactions occur more slowly. For example, millions of years were required for dead plants beneath Earth's surface to be converted to coal. These examples show that the speed of chemical reactions can vary from instantaneous to extremely slow.

The concept of speed is already familiar to you. The winning sprinter in Figure 18.1 covers 100 meters in 11.5 seconds. She runs at a speed of 8.70 m/s. Another sprinter takes 15.0 seconds to cover the same distance. She runs at a speed of 6.67m/s. Both 8.70 m/s and 6.67 m/s express rates of travel.

Figure 18.1 Speed is measured as a change in distance in a given interval of time. A world-class sprinter might cover 100 meters in 11.5 seconds; the speed, or rate, is 100 m/11.5 s = 8.70 m/s.

Section Resources

Print

- **Guided Reading and Study Workbook,** Section 18.1
- **Core Teaching Resources,** Section 18.1 Review
- **Transparencies,** T196–T198
- **Laboratory Manual,** Labs 36, 37
- **Small-Scale Chemistry Laboratory Manual,** Lab 28

Technology

- **Interactive Textbook with ChemASAP,** Animation 22, Simulation 23, Assessment 18.1
- **Go Online,** Section 18.1

18.1

1 FOCUS

Objectives

18.1.1 Describe how to express the rate of a chemical reaction.

18.1.2 Identify four factors that influence the rate of a chemical reaction.

Guide for Reading

Build Vocabulary 

Graphic Organizers Have students draw a concept map of the vocabulary terms as they read the section.

Reading Strategy 

Relate Cause and Effect For the factors that affect reaction rates, ask students to make flow charts that show the factor, its direct effect, and its effect on rate of reaction.

2 INSTRUCT

Connecting to Your World

Explain that all chemical reactions occur at rates specific to the nature of the reactants. Ask, **How can the rate of the corrosion of iron-magnesium alloy be increased?** *(by adding salt water)* **Name some situations where salt affects the rate of corrosion of metals.** *(possible answers: autos on salted roads in winter, metals on boats, and houses near the ocean)*

Collision Theory

Discuss 

Tell students that the time required for a chemical change to take place may be $<10^{-6}$ seconds or $>10^{6}$ years. Draw three vertical columns on the board and label them "very fast," "moderate," and "very slow." Ask students to suggest chemical changes that fit into each category. If they have trouble thinking of reactions that fit in the moderate category, suggest reactions that take place over several hours or days, such as food spoilage or the rusting of iron.

Use Visuals 

Figure 18.3 Explain that the rates of chemical reactions are often measured by the decrease in the concentration of one of the reactants or the increase in the concentration of one of the products. Ask, **What happens to the reaction rate?** *(It slows down.)*

Discuss 

Students may think that particles always react when they collide. Stress that particles form bonds when they collide only if the particles have sufficient kinetic energy. In any given reaction, only a certain fraction of the particles have enough energy to react upon collision.

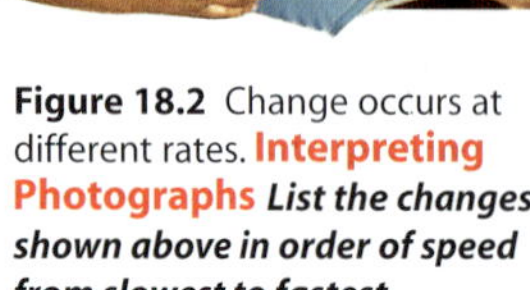

Figure 18.2 Change occurs at different rates. **Interpreting Photographs** ***List the changes shown above in order of speed from slowest to fastest.***

A **rate** is a measure of the speed of any change that occurs within an interval of time. The interval of time may range from fractions of a second to centuries. Figure 18.2 shows some familiar examples of rates of change. If one half of a 1-mole piece of iron turns to rust in one year, the rate at which iron rusts might be expressed as 0.5 mol/yr. **In chemistry, the rate of chemical change or the reaction rate is usually expressed as the amount of reactant changing per unit time.** The progress of a typical reaction is illustrated in Figure 18.3.

Rates of chemical reactions are related to the properties of atoms, ions, and molecules through a model called collision theory. According to **collision theory,** atoms, ions, and molecules can react to form products when they collide with one another, provided that the colliding particles have enough kinetic energy. Particles lacking the necessary kinetic energy to react bounce apart unchanged when they collide. Figure 18.4 is a model for the reaction of hydrogen and oxygen molecules to form water.

You can use two balls of soft modeling clay to illustrate collision theory. If you throw the balls of clay together gently, they don't stick to one another. This is analogous to colliding particles of low energy that fail to react. The same balls of clay thrown together with great force, however, stick tightly to each other. This is analogous to the collision of two high-energy particles that results in the formation of a product.

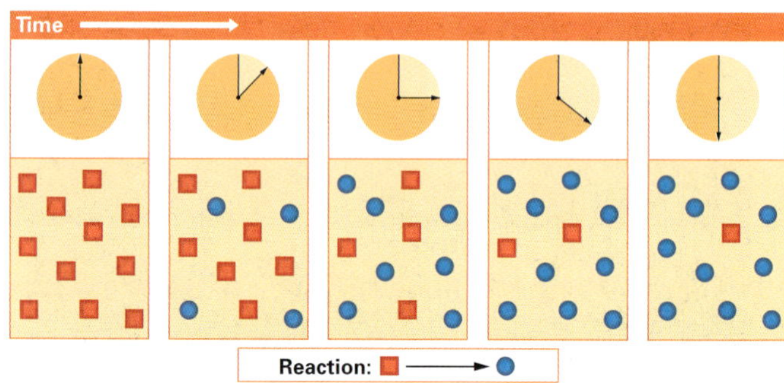

Figure 18.3 As time passes, the amount of reactant (red squares) decreases and the amount of product (blue spheres) increases. Rates of chemical reactions are often measured as a change in the number of moles during an interval of time. **Interpreting Diagrams** ***Assuming equal time intervals between the boxes, how can you tell that the rate of conversion of reactant to product is not constant throughout this reaction?***

Differentiated Instruction

Special Needs 

Because the rest of the concepts in Chapter 18 are based on collision theory, mastery of this theory is crucial. Create a Velcro model of colliding molecules with fuzzy, lightweight balls and a hooked backboard. If students throw the balls hard enough, the "molecules" will stick to the backboard or "react." If they miss or throw with too little kinetic energy, the molecules won't "react."

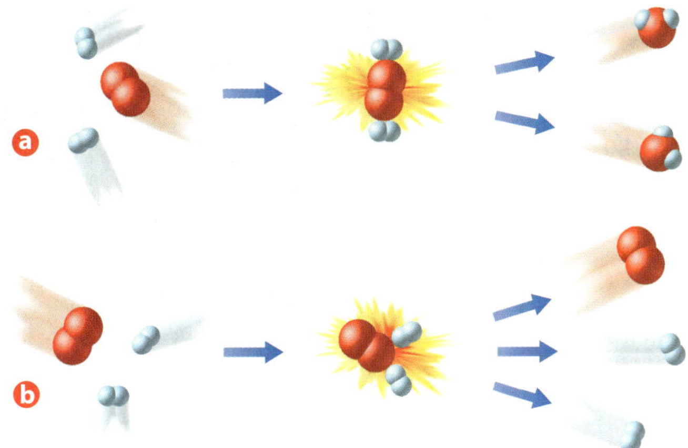

Figure 18.4 If colliding particles have enough kinetic energy and collide at the right orientation, they can react to form a new product. (a) An effective collision of reactant molecules produces product molecules. (b) An ineffective collision of reactant molecules produces no reaction, and the reactants bounce apart unchanged.

You can also use modeling clay to illustrate another point about chemical reactions. If you roll clay into a rope and begin to shake one end more and more vigorously, you will come to a point where the clay rope will break. Similarly, if enough energy is applied to a molecule, the bonds holding the molecule together can break apart. Substances, when supplied with sufficient energy, decompose to simpler substances or reorganize themselves into new substances. The minimum energy that colliding particles must have in order to react is called the **activation energy.** In a sense, the activation energy for a chemical reaction is a barrier that reactants must cross to be converted to products. You can see in the energy diagram in Figure 18.5 that when two reactant particles with the necessary activation energy collide, a new entity called the activated complex may form.

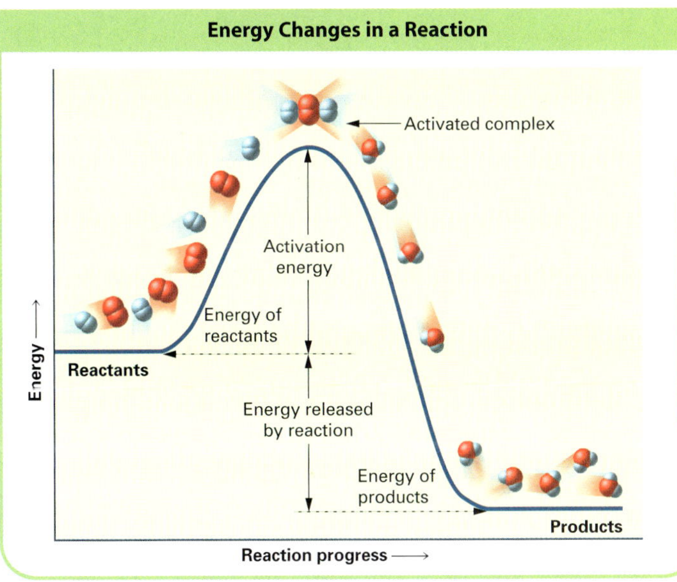

Figure 18.5 The activation-energy barrier must be crossed before reactants are converted to products.

INTERPRETING GRAPHS

a. Navigate Which are at a higher energy, the reactants or products?
b. Read Is energy absorbed or released in progressing from the reactants to the activated complex?
c. Interpret Once the activated complex is formed, will it always proceed to form products? Explain.

Use Visuals  L1

Figure 18.5 Explain that the diagram graphically depicts the energy changes that occur during a reaction. Point out that the activation-energy barrier must be crossed before reactants can be converted to products. Explain that the activation energy determines how rapidly a reaction occurs at a given temperature. Ask, **How does a large activation energy affect reaction time?** *(It causes the reaction to be slow.)* **Why?** *(Few molecules collide with enough energy to form an activated complex.)*

Interpreting Graphs  L2

a. reactants
b. absorbed
c. No; it could also revert to the reactants.

Enrichment Question  L3

Ask, **Will changing the temperature of the reactants change the activation energy? Explain.** *(No; it will only increase the number of reactants having the activation energy needed to react.)*

Differentiated Instruction

Gifted and Talented  L3

Tell students that collision theory uses two-particle collisions to explain reaction mechanisms because the probability of three or more particles colliding simultaneously is relatively remote. Have students research typical reactions that require the collision of three or more particles and summarize their findings in a report that they present to the class.

Answers to...

Figure 18.2 aging, ripening, burning
Figure 18.3 The number of additional blue balls (products) decreases in each successive time interval.

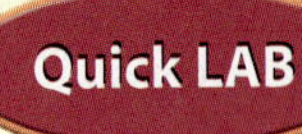

Does Steel Burn? L2

Objective After completing this activity, students will be able to:

- relate particle size or surface area to the rate of a reaction.
- write a balanced equation for the reaction.

Skills Focus Observing, writing equations, inferring

 **Prep Time** 20 minutes

Class Time 20 minutes

Safety Steel wool burns readily. Use proper precautions.

Expected Outcome

Tight Ball: except for the glowing of a few loose steel fibers on the surface, there are no apparent changes; Loose ball: glows; Loose fibers: burn readily.

Analyze and Conclude

1. Except for the glowing of a few loose steel fibers on the surface, no apparent changes occurred when the tight ball was heated. The loose ball glowed. The loose fibers burst into flame almost immediately. The ball acts like a single large particle of steel that is in contact with atmospheric oxygen only at its relatively small surface. The loose fibers act like many small particles, with larger total surface area, and each small particle is exposed to atmospheric oxygen.
2. $4Fe(s) + 3O_2(g) \rightarrow 2Fe_2O_3(s)$
3. The combustion of the steel wool is faster because the high temperature provides the energy for the iron and oxygen to overcome the activation-energy barrier. The small particle size exposes more of the iron surface to atmospheric oxygen.
4. The steel wool may ignite.

For Enrichment 

Have students predict whether you can use a nine-volt battery and steel wool to start a fire. Provide a battery, tongs, and steel wool. Have students design an emergency fire-starter kit that would be safe to carry in a backpack.

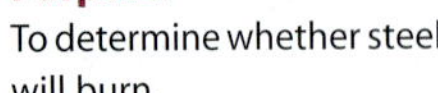

Does Steel Burn?

Purpose

To determine whether steel will burn.

Materials

- #0000 steel wool pad
- tongs
- Bunsen burner
- heat-resistant pad
- pencil and paper

Procedure    

1. Roll a small piece of steel wool into a very tight, pea-sized ball.
2. Holding the ball with tongs, heat the steel wool in the blue-tip flame of the burner for no longer than 10 seconds. In this and subsequent steps, place pieces of heated steel wool on the heat-resistant pad to cool. Observe all appropriate safety precautions when working with the burner and the heated materials. Record your observations.
3. Gently roll a second piece of steel wool into a loose, pea-sized ball. Holding the loose ball with the tongs, heat the wool in the burner flame for no longer than 10 seconds. Record your observations.
4. Pull a few individual fibers of steel wool from the pad. Hold one end of the loose fibers with the tongs and heat them in the flame of the burner for no longer than 10 seconds. Again record your observations.

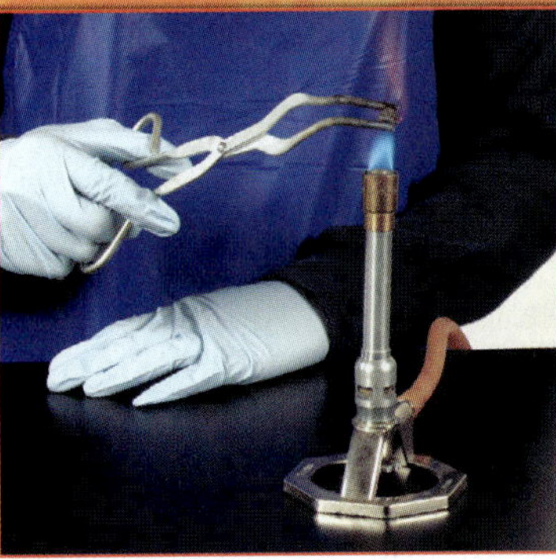

Analyze and Conclude

1. What differences did you observe when the tight ball, the loose ball, and the loose fibers were heated in the flame? Give a reason for any differences you observed.
2. Write the balanced equation for any chemical reaction you may have observed (assume that the steel wool is composed mainly of iron).
3. How do your results differ from those observed in the rusting of an automobile body?
4. Explain why steel wool is a hazard in shops where there are hot plates, open flames, or sparking motors.

An **activated complex** is an unstable arrangement of atoms that forms momentarily at the peak of the activation-energy barrier. An activated complex forms only if the colliding particles have sufficient energy and if the atoms are oriented properly. The lifetime of an activated complex is typically about 10^{-13} s. Its brief existence ends with the re-formation of the reactants or with the formation of products. The two outcomes are equally likely. Thus the activated complex is sometimes called the **transition state.**

Collision theory explains why some naturally occurring reactions are immeasurably slow at room temperature. Carbon and oxygen react when charcoal burns, but this reaction has a high activation energy. At room temperature, the collisions of oxygen and carbon molecules are not energetic enough to break the O—O and C—C bonds. These bonds must be broken to form the activated complex. Thus the reaction rate of carbon with oxygen at room temperature is essentially zero.

 ***What are two events that might occur after the formation of an activated complex?***

Factors Affecting Reaction Rates

Every chemical reaction proceeds at its own rate. Some reactions are naturally fast and some are naturally slow under the same conditions. However, by varying the conditions of the reaction, the rate of almost any reaction can be modified. **The rate of a chemical reaction depends upon temperature, concentration, particle size, and the use of a catalyst.** Collision theory helps explain why changing one or more of these factors may affect the rate of a chemical reaction.

Animation 22 Explore several factors that control the speed of a reaction.

with ChemASAP

Temperature Usually, raising the temperature speeds up reactions, while lowering the temperature slows down reactions. At higher temperatures, the motions of the reactant particles are faster and more chaotic than they are at lower temperatures. Increasing the temperature increases both the frequency of collisions and the number of particles that have enough kinetic energy to slip over the activation-energy barrier to become products. An increase in temperature, therefore, causes products to form faster.

A familiar example of the effect of temperature on reaction rate is the burning of charcoal. At room temperature, a bag of charcoal in contact with air does not burn. But when a starter flame touches the charcoal, atoms of the reactants (carbon and oxygen) collide with higher energy and greater frequency. Some collisions are at a high enough energy to form the product (carbon dioxide). The heat released by the reaction then supplies enough energy to get more carbon and oxygen over the activation-energy barrier. When the starter flame is removed, the reaction continues.

For: Links on Reaction Rates
Visit: www.SciLinks.org
Web Code: cdn-1181

Concentration In a crowded room where people are moving about rapidly, you may find yourself bumping into people more frequently than if you were one of a few occupants. Similarly, the number of particles in a given volume affects the rate at which reactions occur. Cramming more particles into a fixed volume increases the concentration of reactants and consequently the frequency of collision. Increased collision frequency leads to a higher reaction rate. The lighted splint in Figure 18.6 only glows in air, which is 20% oxygen. It may soon be extinguished. But when the glowing splint is plunged into pure oxygen, it immediately bursts into flame. The increased concentration of oxygen greatly speeds up the combustion reaction. This is why flames are forbidden in areas where bottled oxygen is in use.

Figure 18.6 The rate of a reaction depends upon the concentrations of the reactants. **a** In air, a lighted splint glows and soon goes out. **b** When placed in pure oxygen, the splint bursts into flame. **Inferring** ***What accounts for the difference in reactivity?***

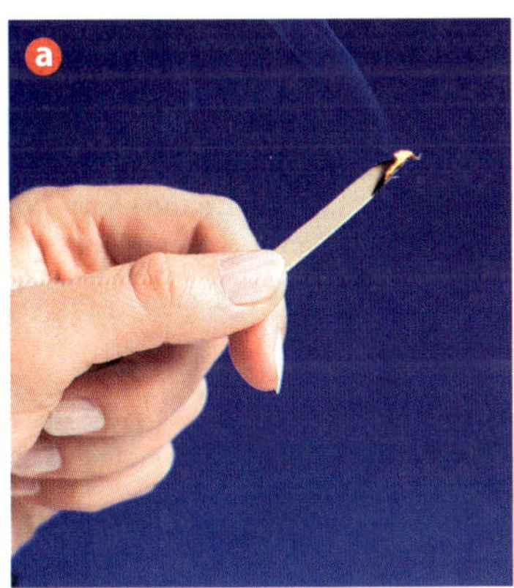

Factors Affecting Reaction Rates

Discuss L2

Point out that an increase in temperature causes an increase in average kinetic energy. The number of collisions increases because molecules are moving faster. More important, more collisions are effective. Ask, **How do refrigerators prevent food from spoiling?** *(The kinetic energy of reactants is reduced; fewer particles possess sufficient energy to react.)*

Download a worksheet on **Reaction Rates** for students to complete, and find additional teacher support from NSTA SciLinks.

Facts and Figures

Northern Lights

Generally, the spontaneity of a reaction is increased by heating the reactants. In certain reactions, however, other types of energy are more effective than heat. During periods of intense solar activity, for example, bombardment by charged particles causes a small fraction of nitrogen molecules in the upper atmosphere to dissociate, yielding excited nitrogen atoms. The green light of the aurora borealis results from the fluorescence of molecular nitrogen, which is produced when two excited nitrogen atoms recombine. Scientists who have heated samples of air in the laboratory have found that no amount of thermal energy is sufficient to reproduce the reaction that causes the green emission.

Answers to...

Figure 18.6 The concentration of oxygen (a reactant) is increased.

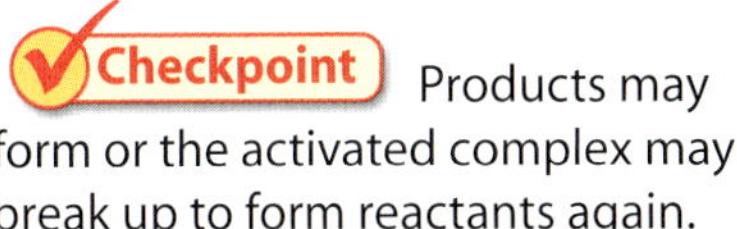

Products may form or the activated complex may break up to form reactants again.

Section 18.1 (continued)

TEACHER Demo

Use of Heat and Catalyst in a Reaction L2

Purpose Students observe the effect of heat and a catalyst on the rate of a chemical reaction.

Materials 10 mL ethanoic acid, 10 mL isoamyl alcohol, 2 test tubes, anhydrous calcium sulfate, concentrated sulfuric acid, water

Procedure Mix 5 mL ethanoic acid (acetic acid) and 5 mL isoamyl alcohol in each of two test tubes. Add a few granules of anhydrous calcium sulfate to each test tube. To one test tube, add three drops of concentrated sulfuric acid and heat this test tube in a boiling water bath for a few minutes. Write the general chemical equation for the formation of an ester on the board. (organic acid + alcohol → ester + water) Inform students that esters are responsible for the characteristic odors of many fruits and may be prepared by reacting an organic acid with an alcohol. Have students smell the odor emanating from both test tubes by carefully wafting the odor towards their noses. Ask them for their observations? (*The solution containing sulfuric acid that was heated smells like bananas while the other solution has a strong odor of ethanoic acid (vinegar); a change in odor is a sign of a reaction.*) **Ask, what purpose does the sulfuric acid and heat serve in this reaction?** (*Both are used to speed up the reaction rate. The sulfuric acid is a catalyst for the reaction.*)

Expected Outcome The solution containing sulfuric acid smells like bananas after being heated, whereas the other solution smells like vinegar, or ethanoic acid.

Figure 18.7 An explosion destroyed this grain elevator. The minute size of the reactant particles (grain dust), and the mixture of the grain dust with oxygen in the air caused the reaction to be explosive.

Particle Size If you put a bundle of sticks on a campfire, they burn quickly. A log, however, burns more slowly. Small pieces of wood have more surface area than a log, and surface area plays an important role in determining the rate of reaction. The same is true for other chemical reactions. The total surface area of a solid or liquid reactant affects the reaction rate. The smaller the particle size, the larger is the surface area for a given mass of particles. An increase in surface area increases the amount of the reactant exposed for reaction, which further increases the collision frequency and the reaction rate.

One way to increase the surface area of solid reactants is to dissolve them. In a solution, particles are separated and more accessible to other reactants. You can also increase the surface area of a solid by grinding it into a fine powder. Small dustlike particles, however, can be dangerous when suspended in air. As coal miners know, large chunks of coal pose little danger, but coal dust mixed with air is an explosive hazard because of its large surface area. The same is true of dust in a flour mill or a grain elevator, as the dramatic photograph in Figure 18.7 shows.

Catalysts Increasing the temperature is not always the best way to increase the rate of a reaction. A catalyst is often better. Recall that a catalyst is a substance that increases the rate of a reaction without being used up during the reaction. Catalysts permit reactions to proceed along a lower energy path, as Figure 18.8 shows. Notice that the activation energy barrier for the catalyzed reaction is lower than that of the uncatalyzed reaction. With a lower activation-energy barrier, more reactants have the energy to form products within a given time. For instance, the rate of the combination reaction of hydrogen and oxygen at room temperature is negligible, but with a trace of finely divided platinum (Pt) as a catalyst, the reaction is rapid.

$$2H_2(g) + O_2(g) \xrightarrow{Pt} 2H_2O(l)$$

Because a catalyst is not consumed during a reaction, it does not appear as a reactant or product in the chemical equation. Instead, the catalyst is often written above the yield arrow, as in the equation above.

Interactive Textbook

Simulation 23 Explore the effects of concentration, temperature, and a catalyst on reaction rate.

with ChemASAP

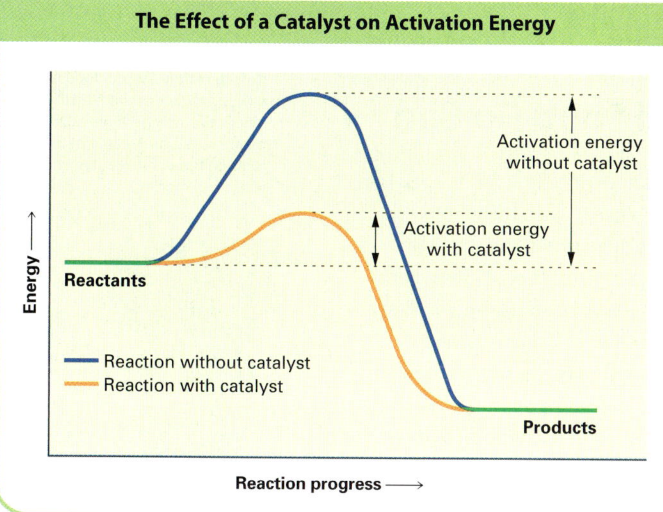

Figure 18.8 A catalyst increases the rate of a reaction by lowering the activation-energy barrier.

INTERPRETING GRAPHS

a. Navigate How does the catalyst affect the magnitude of the activation energy?
b. Read Does the catalyst change the amount of energy released in the reaction?
c. Interpret Along which of the two reaction paths are reactants converted more rapidly to products?

Catalysts are crucial for many life processes. Your body temperature is only 37°C and cannot be raised significantly without danger. Without catalysts, few reactions in the body would proceed fast enough. Enzymes are biological catalysts that increase the rates of biological reactions. When you eat a meal containing protein, enzymes in your digestive tract break down the protein molecules in a few hours. Without enzymes, the digestion of protein at 37°C would take many years!

An **inhibitor** is a substance that interferes with the action of a catalyst. Some inhibitor molecules work by reacting with, or "poisoning," the catalyst itself. Thus the inhibitor reduces the amount of functional catalyst available. Reactions slow or even stop when a catalyst is poisoned.

18.1 Section Assessment

1. **Key Concept** How is the rate of a chemical reaction expressed?
2. **Key Concept** What are four factors that affect the rate of a chemical reaction?
3. Suppose a thin sheet of zinc containing 0.2 mol of the metal is completely converted to zinc oxide (ZnO) in one month. How would you express the rate of conversion of the zinc?
4. Does every collision between reacting particles lead to products? Explain.
5. Refrigerated food stays fresh for long periods. The same food stored at room temperature quickly spoils. Why?

Writing Activity

Write a Report Catalytic converters are required on all cars manufactured in the United States. Find out what substances are used as the catalysts. What chemical reactions does the catalyst facilitate? Write a report summarizing your findings.

Assessment 18.1 Test yourself on the concepts in Section 18.1.

with ChemASAP

Section 18.1 Assessment

1. The rate of a chemical reaction is the amount of reactant changing per unit of time.
2. temperature, concentration, particle size, use of a catalyst
3. Rate = 0.2 mol Zn/month
4. No; the collision must have sufficient energy to break and form bonds.
5. Chemical reactions involved in food spoilage occur faster at higher temperatures because more energy is available.

Interpreting Graphs 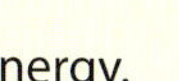

a. It lowers the activation energy.
b. no
c. the line for reaction with a catalyst

Enrichment Question 

Inhibitors poison catalysts. **How would adding an inhibitor affect both graphs?** *(The activation energy would be higher than the graph for the reaction with a catalyst, but no higher than the activation energy without a catalyst.)*

3 ASSESS

Evaluate Understanding L2

Have students use the activated complex theory (Figure 18.5) to explain why hydrogen and oxygen do not react at room temperature but do combine explosively in the presence of a spark or flame. *(The explosive reaction indicates that the product (water) has less energy than the reactants. The reaction does not occur at room temperature because the reaction has a very high activation energy.)*

Reteach 

Remind students that chemical changes occur as a result of individual reactions between particles. Therefore, factors that affect collisions between particles are the factors that influence the rate of reaction. Have students write explanations of how temperature, concentration, and surface area affect the number of collisions.

Writing Activity

The catalysts are platinum, rhodium, and palladium. They facilitate these reactions: $2NO \rightarrow N_2 + O_2$; $2NO_2 \rightarrow N_2 + 2O_2$

Interactive Textbook

If your class subscribes to the Interactive Textbook, use it to review key concepts in Section 18.1.

with ChemASAP

Technology & Society

Don't Let Good Food Go Bad

Because oxygen has six valence electrons in its highest occupied energy level, it has a strong tendency to combine with atoms that can donate or share electrons. Antioxidants interfere with this tendency because they are more easily oxidized than the substances they protect. In effect, they serve as competitive inhibitors. Carotenes, vitamin C, and vitamin E are naturally occurring antioxidants. Ask students to name foods that are rich in these substances. *(broccoli, carrots, corn, squash, and tomatoes)*

Encourage students to find as many products as they can that contain the antioxidants described in this article. Have them make a table listing the products and the antioxidants they contain. Ask students to state any similarities they see in the products. Finally, have students research the pros and cons of using chemical preservatives in food.

Answers to...

Inferring to minimize exposure to oxygen

Technology & Society

Don't Let Good Food Go Bad

For thousands of years, people have dried fruits and vegetables and salted or smoked meat and fish to keep them from spoiling. These methods slow oxidation and the action of microorganisms. When food oxidizes, or reacts with oxygen, it becomes rancid and turns brown. Microorganisms such as bacteria, yeast, and molds destroy or contaminate food, making it unfit to eat. Today, methods for preserving food also include refrigeration and the addition of preservatives—antioxidants and antimicrobials. **Inferring** ***Why is it a good practice to keep food in closed containers?***

Refrigeration Storing foods in a refrigerator keeps them fresh longer. Low temperatures slow microbial action. Meat kept at freezing temperatures has a lifetime of months.

Antioxidants The preservatives BHT (butylated hydroxytoluene) and BHA (butylated hydroxyanisole), added to packaging or to the food inside, keep the food from becoming stale quickly.

Oxygen-free packaging Many food packages contain nitrogen or another nonreactive gas rather than air. Excluding oxygen prolongs shelf life.

Sulfites and sulfur dioxide These compounds are antioxidants and antimicrobials used in drying fruits and preserving fruit juices. Canning and pickling slow the action of microbes and limit contact with air.

18.2 Reversible Reactions and Equilibrium

Connecting to Your World For years, scientists tried to produce nitrogen compounds for fertilizers to increase the amount of available foodstuffs. Unfortunately, none of these efforts proved to be commercially successful. Finally, in the early 1900s, two German chemists, Fritz Haber and Karl Bosch, refined the process of making ammonia from elemental nitrogen and hydrogen. Their success came from controlling the temperature and pressure under which the two gases are reacted. In this section, you will learn how changing the reaction conditions can influence the yield of a chemical reaction.

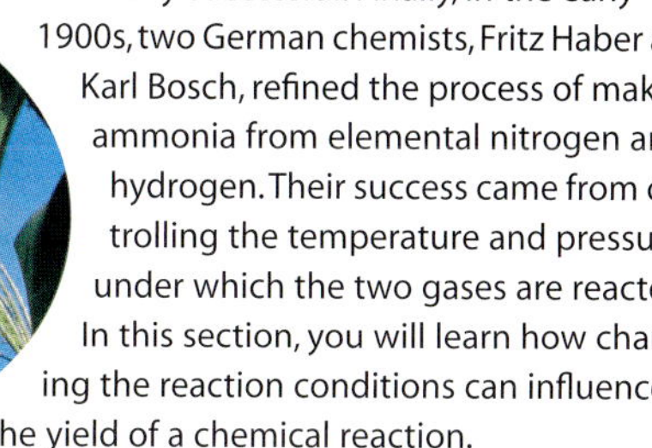

Guide for Reading

Key Concepts

- How do the amounts of reactants and products change in a chemical system at equilibrium?
- What three stresses can cause a change in the equilibrium position of a chemical system?
- What does the value of K_{eq} indicate about the equilibrium position of a reaction?

Vocabulary

reversible reaction
chemical equilibrium
equilibrium position
Le Châtelier's principle
equilibrium constant

Reading Strategy

Previewing Before you read this section, rewrite the headings as how, why, and what questions about reversible reactions and equilibrium. As you read, write answers to the questions.

Reversible Reactions

From what you have already learned, you may have inferred that chemical reactions go completely from reactants to products, just as the chemical equation indicates. This is not usually the case. Some reactions are reversible. A **reversible reaction** is one in which the conversion of reactants to products and the conversion of products to reactants occur simultaneously. One example of a reversible reaction is the following:

$$\text{Forward reaction: } 2SO_2(g) + O_2(g) \longrightarrow 2SO_3(g)$$

$$\text{Reverse reaction: } 2SO_2(g) + O_2(g) \longleftarrow 2SO_3(g)$$

In the first reaction, which is read from left to right, sulfur dioxide and oxygen produce sulfur trioxide. In the second reaction, which is read from right to left, sulfur trioxide decomposes into oxygen and sulfur dioxide. The first reaction is called the forward reaction. The second is called the reverse reaction. The two equations can be combined into one using a double arrow. The double arrow tells you that this reaction is reversible.

$$\underset{\text{Sulfur dioxide}}{2SO_2(g)} + \underset{\text{Oxygen}}{O_2(g)} \rightleftharpoons \underset{\text{Sulfur trioxide}}{2SO_3(g)}$$

Figure 18.9 shows that this equation represents two opposite reactions.

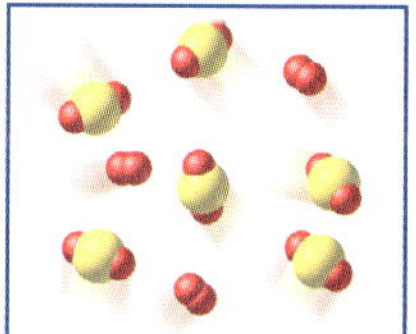

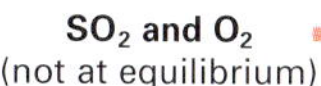

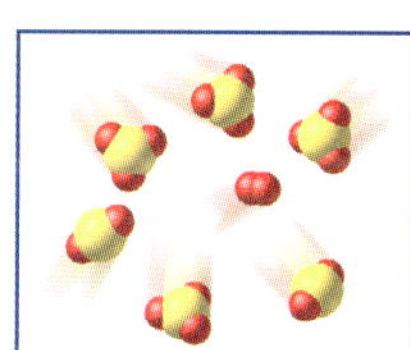

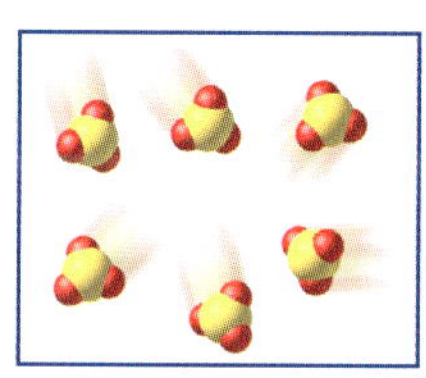

SO_2 and O_2 (not at equilibrium) → $2SO_2 + O_2 \rightleftharpoons 2SO_3$ (at equilibrium) ← SO_3 (not at equilibrium)

Figure 18.9 Molecules of SO_2 and O_2 react to give SO_3. Molecules of SO_3 decompose to give SO_2 and O_2. At equilibrium, all three types of molecules are present in the mixture.

18.2

1 FOCUS

Objectives

18.2.1 Describe how the amounts of reactants and products change in a chemical system at equilibrium.

18.2.2 Identify three stresses that can change the equilibrium position of a chemical system.

18.2.3 Explain what the value of K_{eq} indicates about the position of equilibrium.

Guide for Reading

Build Vocabulary L2

LINCS Have students use the LINCS strategy to learn what Le Châtelier's principle means. In LINCS exercises, students list what they know about each term, imagine a picture that describes the term, note a reminding "sound-alike" word, connect the terms to the sound-alike word by making up a short story, and then perform a brief self-test.

Reading Strategy L2

Monitor Your Understanding Ask students if the flow chart relating cause and effect helped them read Section 18.1. Ask them to consider where they would use that technique in this section.

2 INSTRUCT

Connecting to Your World

Point out that the two elements involved in production of ammonia are gases. Ask, **Do you think that controlling temperature and pressure of liquids or solids in a similar reaction would have had a similar affect? Why?** *(Temperature might, but liquids and solids aren't sensitive to pressure difference the way gases are.)*

Section Resources

Print

- **Guided Reading and Study Workbook,** Section 18.2
- **Core Teaching Resources,** Section 18.2 Review
- **Transparencies,** T199–T202
- **Laboratory Manual,** Lab 38
- **Small-Scale Chemistry Laboratory Manual,** Lab 29

Technology

- **Interactive Textbook with ChemASAP,** Animation 23; Simulation 24; Problem-Solving 18.6, 18.7, 18.9; Assessment 18.2

Reversible Reactions

Use Visuals L1

Figure 18.9 Explain that chemical reactions do not always go completely to products because they are often reversible. Once the product(s) form, they may decompose to form the reactants. Ask, **Does the reaction continue when the products and reactants are at equilibrium?** *(Yes; reactants change into products, and products reform into reactants, but their net amounts stay the same.)*

Interpreting Graphs L2

a. Initial concentrations are at the far left, and equilibrium concentrations are at the far right.

b. SO_3

c. SO_3 is much higher than SO_2 and O_2; O_2 is lowest.

Enrichment Question 

Would these graphs change if you changed volume, pressure, or temperature? Explain. *(Yes, you can change the concentration of the gas by changing the volume of its container, pressure, and temperature.)*

Word *Origins* L2

Disequilibrium for a chemical reaction means the opposite of being in balance. Therefore, the amount of reactant and product is still changing.

Changes in Concentrations of Reactants and Products

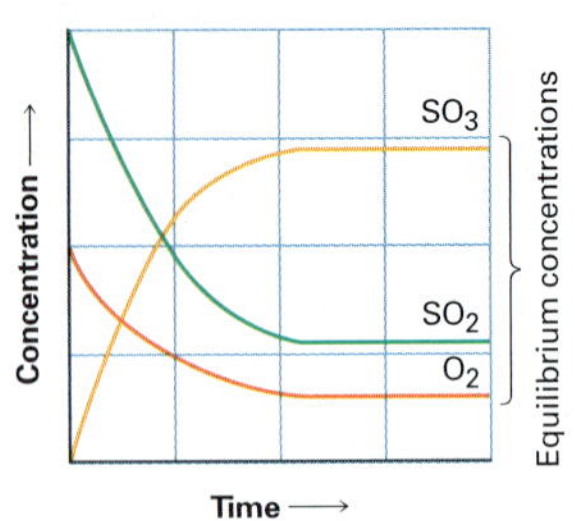

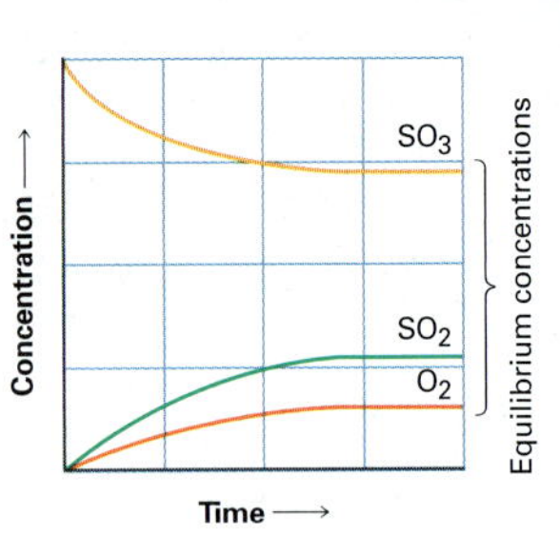

Figure 18.10 These graphs show how the concentrations of O_2, SO_2, and SO_3 vary with time. Left: Initially, SO_2 and O_2 are present. Right: Initially, only SO_3 is present.

INTERPRETING GRAPHS

a. Navigate Where on the graphs can you find the initial concentrations of the reactants and products? The equilibrium concentrations?
b. Read Which gas is most abundant at equilibrium?
c. Interpret How do the equilibrium concentrations of O_2, SO_2, and SO_3 compare?

What actually happens when sulfur dioxide and oxygen gases are mixed in a sealed chamber? The two reactants begin to react to form sulfur trioxide at a particular rate. Because no sulfur trioxide is present at the beginning of the reaction, the initial rate of the reverse reaction is zero. As sulfur trioxide begins to form, however, the decomposition of sulfur trioxide begins. This reverse reaction proceeds slowly at first, but its rate increases as the concentration of sulfur trioxide increases. Simultaneously, the rate of the forward reaction decreases because sulfur dioxide and oxygen are being used up. Eventually sulfur trioxide is decomposing to sulfur dioxide and oxygen as fast as sulfur dioxide and oxygen are forming sulfur trioxide. When the rates of the forward and reverse reactions are equal, the reaction has reached a state of balance called **chemical equilibrium.** Changes in concentrations of the three components during the course of the reaction are shown in the graphs in Figure 18.10. The graph on the left shows the progress of a reaction that starts with specific concentrations of SO_2 and O_2, but with zero concentration of SO_3. The graph on the right shows concentrations for a reaction that begins with an initial concentration of SO_3 and zero concentrations for SO_2 and O_2. Notice that after a certain time, all concentrations remain constant. **At chemical equilibrium, no net change occurs in the actual amounts of the components of the system.** The amount of SO_3 in the equilibrium mixture is the maximum amount that can be produced by this reaction under the conditions of the reaction.

Word *Origins*

Equilibrium comes from the Latin word *aequilibrium*, meaning "in balance." In a chemical system at equilibrium, the rate of the forward reaction is balanced by the rate of the reverse reaction. **If the prefix *dis*-means "opposite or absence of," what does a state of disequilibrium indicate?**

The unchanging amounts of SO_2, O_2, and SO_3 in the reaction mixture at equilibrium might cause you to think that both reactions have stopped. This is not the case. Chemical equilibrium is a dynamic state. Both the forward and reverse reactions continue, but because their rates are equal, no net change occurs in their concentrations. The escalators in Figure 18.11 are like the double arrows in a dynamic equilibrium equation. The number of people using the up escalator must equal the number of people using the down escalator for the number of people on both floors to remain constant.

 **Checkpoint** ***Why is chemical equilibrium called a dynamic state?***

Although the rates of the forward and reverse reactions are equal at chemical equilibrium, the concentrations of the components on both sides of the chemical equation are not necessarily the same. In fact, they can be dramatically different. Figure 18.10 shows the equilibrium concentrations of SO_2, O_2, and SO_3. The concentration of SO_3 is significantly greater than the concentrations of SO_2 and O_2. The relative concentrations of the reactants and products at equilibrium constitute the **equilibrium position** of a reaction. The equilibrium position indicates whether the reactants or products are favored in a reversible reaction. If A reacts to give B and the equilibrium mixture contains significantly more of B—say 1% A and 99% B—then the formation of B is said to be favored.

$$\underset{1\%}{A} \rightleftharpoons \underset{99\%}{B}$$

On the other hand, if the mixture contains 99% A and 1% B at equilibrium, then the formation of A is favored.

$$\underset{99\%}{A} \rightleftharpoons \underset{1\%}{B}$$

Notice that the equilibrium arrows are not of equal length; the longer of the two arrows indicates the favored direction of a reaction.

In principle, almost all reactions are reversible to some extent under the right conditions. In practice, one set of components is often so favored at equilibrium that the other set cannot be detected. If one set of components (reactants) is completely converted to new substances (products), you can say that the reaction has gone to completion, or is irreversible. When you mix chemicals expecting to get a reaction but no products can be detected, you can say that there is no reaction. Reversible reactions occupy a middle ground between the theoretical extremes of irreversibility and no reaction.

A catalyst speeds up both the forward and the reverse reactions equally because the reverse reaction is exactly the opposite of the forward reaction. The catalyst lowers the activation energy of the reaction by the same amount in both the forward and reverse directions. Catalysts do not affect the amounts of reactants and products present at equilibrium; they simply decrease the time it takes to establish equilibrium.

What is chemical equilibrium?

Animation 23 Take a close look at a generalized reversible reaction.

with ChemASAP

Figure 18.11 If the rate at which shoppers move from the first floor to the second is equal to the rate at which shoppers move from the second floor to the first, then the number of shoppers on each floor remains constant. **Applying Concepts** ***Is it necessary that an equal number of shoppers be on each floor? Explain.***

Discuss

L2

Point out that, in many cases, the concept of reactants turning into products, as described by a chemical equation, is an oversimplification. Many reactions reach an equilibrium state in which the reaction mixture contains both reactant and product particles. The percent of reactants converted to products varies considerably. When hydrogen and oxygen combine to form water, for example, nearly all of the molecules react. In the reaction of hydrogen and nitrogen to form ammonia, the percent of reactants converted to product is much smaller. Stress that equilibrium refers to the rates of the forward or reverse reactions, not to the quantities of reactants and products.

Answers to...

Figure 18.11 No, the numbers can be different but the number of shoppers moving between floors must be the same.

because both the forward and reverse reactions continue

a situation in which the forward and reverse reactions occur at the same rate

Factors Affecting Equilibrium: Le Châtelier's Principle

Discuss

L1

Refer to the formation of banana oil, which was described on page 546. Write on the board the general word equation for the formation of an ester (organic acid + alcohol → ester + water). Add a reverse arrow to show the possibility of the opposite reaction. Remind students that the reaction was effective in the test tube to which anhydrous calcium sulfate had been added. Calcium sulfate reacted with water to form a hydrate, which limited the reverse reaction by removing a product.

TEACHER Demo

Revisiting Banana Oil Demonstration 

You may want to repeat the banana oil demonstration on page 546. This time, add concentrated sulfuric acid to both test tubes before heating. However, add the anhydrous calcium sulfate to only one tube. Ask students to predict which reaction will yield the most product. Then compare the results qualitatively by strength of odor or quantitatively if you have access to gas phase chromatography (GC) equipment.

Expected Outcome The tube containing anhydrous calcium sulfate yields the larger amount of product.

Factors Affecting Equilibrium: Le Châtelier's Principle

A delicate balance exists in a system at equilibrium. Changes of almost any kind can disrupt this balance. When the equilibrium of a system is disturbed, the system makes adjustments to restore equilibrium. However, the equilibrium position of the restored equilibrium is different from the original equilibrium position; that is, the amounts of products and reactants may have increased or decreased. Such a change is called a shift in the equilibrium position.

The French chemist Henri Le Châtelier (1850–1936) studied how the equilibrium position shifts as a result of changing conditions. He proposed what has come to be called **Le Châtelier's principle:** If a stress is applied to a system in dynamic equilibrium, the system changes in a way that relieves the stress. **Stresses that upset the equilibrium of a chemical system include changes in the concentration of reactants or products, changes in temperature, and changes in pressure.**

The following examples of applications of Le Châtelier's principle all involve reversible reactions. For simplicity and clarity, the components to the left of the reaction arrow will be considered the reactants and the components to the right of the reaction arrow will be considered the products. Blue arrows indicate the shifts resulting from additions to or removals from the system. The arrows always point in the direction of the resulting shift in the equilibrium position—that is, toward the favored side.

Concentration Changing the amount, or concentration, of any reactant or product in a system at equilibrium disturbs the equilibrium. The system adjusts to minimize the effects of the change. Consider the decomposition of carbonic acid (H_2CO_3) in aqueous solution to form the products carbon dioxide and water. The system has reached equilibrium. At equilibrium the amount of carbonic acid is less than 1%.

$$\underset{<1\%}{H_2CO_3(aq)} \underset{\substack{\text{Remove } CO_2 \\ \text{Direction of shift} \rightarrow}}{\overset{\substack{\text{Add } CO_2 \\ \leftarrow \text{Direction of shift}}}{\rightleftharpoons}} \underset{>99\%}{CO_2(aq) + H_2O(l)}$$

Adding more carbon dioxide disturbs the equilibrium. It increases the concentration of CO_2 in the mixture and causes the rate of the reverse reaction to increase. As more reactant (H_2CO_3) is formed, the rate of the forward reaction also begins to increase. In time, the rates of the forward and reverse reactions again become equal, and a new equilibrium is established with a higher concentration of reactant (H_2CO_3). Adding a product to a reaction at equilibrium pushes a reversible reaction in the direction of reactants.

If, on the other hand, carbon dioxide is removed, the concentration of CO_2 decreases. This causes the rate of the reverse reaction to decrease. As less reactant (H_2CO_3) is being formed, the rate of the forward reaction also begins to decrease. When the rates of the forward and reverse reactions again become equal, equilibrium is restored but at a different equilibrium position. Removing a product always pushes a reversible reaction in the direction of products.

Farmers use this technique to increase the yield of eggs laid by hens. Hens lay eggs and then proceed to hatch them. If the eggs are removed after they are laid (removing the product), the hen will lay more eggs (increasing the yield). Similarly as products are removed from a reaction mixture, the system continually changes to restore equilibrium by producing more products. But because the products are being removed, the reaction can never reestablish equilibrium. The reaction continues to produce products until the reactants are used up. Another example of this concept is found in the body. Blood contains dissolved carbonic acid in equilibrium with carbon dioxide and water. The body uses the removal of products to keep the concentration of carbonic acid within a safe range. When the athletes in Figure 18.12 exhale carbon dioxide, the equilibrium shifts toward carbon dioxide and water, thus reducing the amount of carbonic acid. The same principle applies to adding or removing reactants. When a reactant is added to a system at equilibrium, the reaction shifts in the direction of the formation of products. When a reactant is removed, the reaction shifts in the direction of formation of reactants.

Checkpoint ***According to Le Châtelier's principle, how does a system at equilibrium respond to a stress?***

Interactive Textbook

Simulation 24 Simulate Le Châtelier's principle for the synthesis of ammonia.

with **ChemASAP**

Figure 18.12 The rapid exhalation of CO_2 during and after vigorous exercise helps reestablish the body's correct CO_2:H_2CO_3 equilibrium. This keeps the acid concentration in the blood within a safe range.

553

Relate **L2**

Write the equation for the dissociation of H_2CO_3 on the board. Explain that our bodies expel CO_2 to maintain the correct equilibrium between CO_2 and H_2CO_3. During and after vigorous exercise, more CO_2 than usual needs to be removed to maintain equilibrium. Ask, **What are some signs that the body is trying to reestablish equilibrium after exercising?** *(a faster heartbeat and breathing rate)*

Facts and Figures

Carbon Monoxide

Molecules such as carbon monoxide (CO) that are structurally similar to the oxygen molecule can bind to the hemoglobin molecule. Therefore, they compete with oxygen for the same binding locations. This competition is the cause of carbon monoxide poisoning. Because their affinity for the heme group is greater, carbon monoxide molecules can displace oxygen molecules from hemoglobin.

$$HbO_2 + CO \rightleftharpoons HbCO + O_2$$

The equilibrium in this reaction lies far to the right because the affinity of carbon monoxide for heme is approximately 150 times stronger than that of oxygen.

Answers to...

Checkpoint It moves in the direction that relieves the stress.

Section 18.2 (continued)

Use Visuals 

Figure 18.13 Ask students to study the diagrams and explain how equilibrium in **a** relates to equilibrium in **c**. *(Increasing the pressure on the system in* **a** *caused the reaction to shift to the right producing more molecules of product in* **c**.*)* Ask, **What kind of substances does a change in pressure affect?** *(gases)*

Relate 

The industrial production of ammonia provides a practical application of Le Châtelier's principle. Write the equilibrium equation on the board. Ask, **Based on the equilibrium equation, select ideal temperature and pressure conditions for the highest possible yield of ammonia.** *(high pressure and low temperature)* Tell students that the industrial process devised by Carl Bosch actually uses both high temperature and pressure plus an iron-based catalyst. Ask, **Why doesn't Bosch's process use the conditions predicted by Le Châtelier's principle?** *(A higher-than-ideal temperature (plus the catalyst) is needed to increase the reaction rate.)* Point out that an efficient industrial process must often strike a balance between multiple goals, in this case, yield and reaction time.

CONCEPTUAL PROBLEM 18.1

Answers

6. **a.** favors reactants
b. favors reactants
c. favors products
d. favors products

Practice Problems Plus 

The gases dinitrogen tetroxide and nitrogen dioxide exist in equilibrium according to the equation

$$N_2O_4(g) + 58\text{ kJ} \rightleftharpoons 2NO_2(g)$$

What effect do the following changes have on the equilibrium position?

a. addition of heat *(favors products)*
b. decrease in pressure *(favors products)*
c. addition of NO_2 *(favors reactants)*
d. removal of N_2O_4 *(favors reactants)*

Temperature Increasing the temperature causes the equilibrium position of a reaction to shift in the direction that absorbs heat. The heat absorption reduces the applied temperature stress. For example, consider the following exothermic reaction that occurs when SO_3 is produced from the reaction of SO_2 and O_2.

$$2SO_2(g) + O_2(g) \underset{\text{Remove heat (cool)} \atop \text{Direction of shift} \rightarrow}{\overset{\text{Add heat} \atop \leftarrow \text{Direction of shift}}{\rightleftharpoons}} 2SO_3(g) + \text{heat}$$

Heat can be considered to be a product, just like SO_3. Heating the reaction mixture at equilibrium pushes the equilibrium position to the left, which favors the reactants. As a result, the product yield decreases. Cooling, or removing heat, pulls the equilibrium to the right, and the product yield increases.

Pressure A change in the pressure on a system affects only gaseous equilibria that have an unequal number of moles of reactants and products. An example is the reaction you read about in *Connecting to Your World:* hydrogen and nitrogen react to form ammonia. Imagine that the three gases are at equilibrium in a cylinder that has a piston attached to a plunger—similar to a bicycle pump but with the hose sealed. A catalyst has been included to speed up the reaction. What happens to the pressure when you push the plunger down? The pressure on the gases momentarily increases because the same number of molecules is contained in a smaller volume. The system immediately relieves some of the pressure increase by reducing the number of gas molecules. For every two molecules of ammonia made, four molecules of the reactants are used up (three molecules of hydrogen and one of nitrogen). Therefore, the equilibrium position shifts to make more ammonia. There are then fewer molecules in the system. The pressure decreases, although it will not decrease all the way to the original pressure. As you can see in Figure 18.13, increasing the pressure on the system results in a shift in the equilibrium position that favors the formation of product.

$$N_2(g) + 3H_2(g) \underset{\text{Reduce pressure} \atop \leftarrow \text{Direction of shift}}{\overset{\text{Increase pressure} \atop \text{Direction of shift} \rightarrow}{\rightleftharpoons}} 2NH_3(g)$$

Figure 18.13 Pressure affects a mixture of nitrogen, hydrogen, and ammonia at equilibrium. **a** The system is at equilibrium. **b** Equilibrium is disturbed by an increase in pressure. **c** A new equilibrium position is established with fewer gas molecules. **Interpreting Diagrams** ***What effect does a decrease in volume have on the number of gas molecules?***

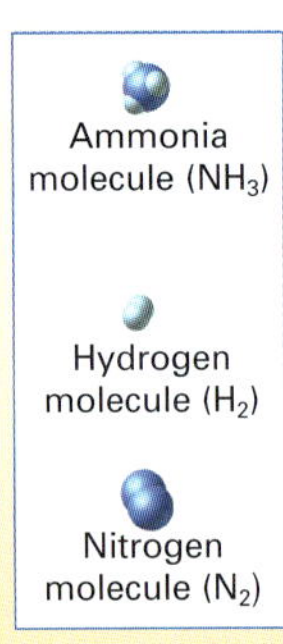

a Initial equilibrium condition (11 gas molecules)

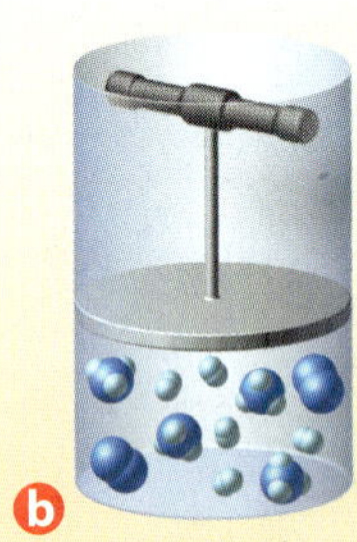

b Pressure increased, equilibrium disturbed

c New equilibrium condition at increased pressure (9 gas molecules)

The equilibrium position for this reaction can be made to favor the reactants instead of the product. Imagine pulling the plunger of the piston device back up so the volume containing the gases increases. This increase in volume decreases the pressure on the system. To restore the higher starting pressure, the system can produce more gas molecules by the decomposition of some ammonia molecules. Decomposition of two molecules of gaseous NH_3 produces four molecules of reactants (three H_2 and one N_2). Pressure at the new equilibrium is higher than when the pressure was first decreased, but not as high as it was at the starting equilibrium. Lowering the pressure on the system thus results in a shift of the equilibrium to favor the reactants.

A stress on this physical equilibrium could result in a loss of balance.

CONCEPTUAL PROBLEM 18.1

Applying Le Châtelier's Principle

What effect do each of the following changes have on the equilibrium position for this reversible reaction?

$$PCl_5(g) + \text{heat} \rightleftharpoons PCl_3(g) + Cl_2(g)$$

a. addition of Cl_2 **b.** increase in pressure
c. removal of heat **d.** removal of PCl_3 as it is formed

1 Analyze ***Identify the relevant concepts.***
a.–d. The stress placed on each system is known. The effect of each stress is unknown. By Le Châtelier's principle, the equilibrium system will shift in a direction that minimizes the imposed stress. Analyze the effect of each change on the reaction.

2 Solve ***Apply concepts to this situation.***

a. The addition of Cl_2, a product, shifts the equilibrium to the left, forming more PCl_5.

b. The equation shows 2 mol of gaseous product and 1 mol of gaseous reactant. The increase in pressure is relieved if the equilibrium shifts to the left, because a decrease in the number of moles of gaseous substances produces a decrease in pressure.

c. The removal of heat causes the equilibrium to shift to the left, because the reverse reaction is heat-producing.

d. The removal of PCl_3 causes the equilibrium to shift to the right to produce more PCl_3.

Practice Problem

6. How is the equilibrium position of this reaction affected by the following changes?
$C(s) + H_2O(g) + \text{heat} \rightleftharpoons CO(g) + H_2(g)$

a. lowering the temperature
b. increasing the pressure
c. removing hydrogen
d. adding water vapor

Interactive Textbook

Problem-Solving 18.6 Solve a similar problem with the help of an interactive guided tutorial.

TEACHER Demo

Temperature and Equilibrium L2

Purpose Students observe the effect of temperature changes on the equilibrium position of an exothermic reaction.

Materials 3 g copper turnings, 3 heat-resistant test tubes, 10 mL 6*M* nitric acid, one-hole stopper, 2 solid stoppers, glass delivery tube, rubber tubing, warm water bath, ice water bath

Preparation Place a small amount (3 g or less) of copper turnings in a test tube, and add 10 mL of 6*M* nitric acid (HNO_3). Insert a one-hole stopper, fitted with a glass delivery tube and rubber tubing, into the test tube. Fill two test tubes with the brown gas that forms. When the color is the same as in the reacting test tube, stopper the tubes. Once sealed, the tubes can be saved indefinitely for repeated use, or discharged in a fume hood.

Safety Use a fume hood. Nitric acid is corrosive, and NO_2 is a toxic gas.

Procedure Set up an ice bath and a warm water bath (70–80°C). Hold up the tubes for students to see and tell them that each tube contains NO_2 gas. Place one tube in warm water and the other in cold water. After a few minutes, remove the tubes and have students compare their colors. Now reverse the positions of the tubes in the water baths and after a few minutes, have students again observe the colors.

Write the reaction equation on the board:

$$2NO_2(g) \rightleftharpoons N_2O_4(g) + \text{heat}$$

Explain that nitrogen dioxide is brown and dinitrogen tetroxide is colorless. Ask students to explain the effect of changing temperature. *(Adding heat shifts the equilibrium in the direction that absorbs heat, to the left. Removing heat shifts the equilibrium in the direction that produces heat, to the right.)*

Answers to...

Figure 18.13 Decreasing the volume shifts the equilibrium in the direction of fewer gas molecules.

Equilibrium Constants

Discuss L2

Point out that whenever a chemical system reaches equilibrium, the reactants and products have a fixed numerical relationship. This relationship is stated in the equilibrium constant expression. The value of the equilibrium constant can be used to predict the position of equilibrium. Because the equilibrium constant refers to an expression in which products are divided by reactants, a K_{eq} value greater than 1 favors the products. Conversely, a K_{eq} value less than 1 favors the reactants.

Equilibrium Constants

Chemists express the position of equilibrium in terms of numerical values. These values relate the amounts of reactants to products at equilibrium. In a general reaction, *a* mol of reactant A and *b* mol of reactant B react to give *c* mol of product C and *d* mol of product D at equilibrium.

$$aA + bB \rightleftharpoons cC + dD$$

The **equilibrium constant** (K_{eq}) is the ratio of product concentrations to reactant concentrations at equilibrium, with each concentration raised to a power equal to the number of moles of that substance in the balanced chemical equation. The expression for the equilibrium constant can be written this way.

$$K_{eq} = \frac{[C]^c \times [D]^d}{[A]^a \times [B]^b}$$

The exponents in the equilibrium-constant expression are the coefficients in the balanced chemical equation. The square brackets indicate the concentrations of substances in moles per liter (mol/L). The value of K_{eq} depends on the temperature of the reaction. If the temperature changes, the value of K_{eq} also changes.

The size of the equilibrium constant shows whether products or reactants are favored at equilibrium. **A value of K_{eq} greater than 1 means that products are favored over reactants; a value of K_{eq} less than 1 means that reactants are favored over products.**

$K_{eq} > 1$, products favored at equilibrium

$K_{eq} < 1$, reactants favored at equilibrium

When the numerical value of an equilibrium constant is calculated, the cancellation of units may or may not lead to a unit for the constant. Chemists have agreed to report all equilibrium constants without stated units. Sample Problem 18.1 shows how to calculate the equilibrium constant for the reaction illustrated in Figure 18.14.

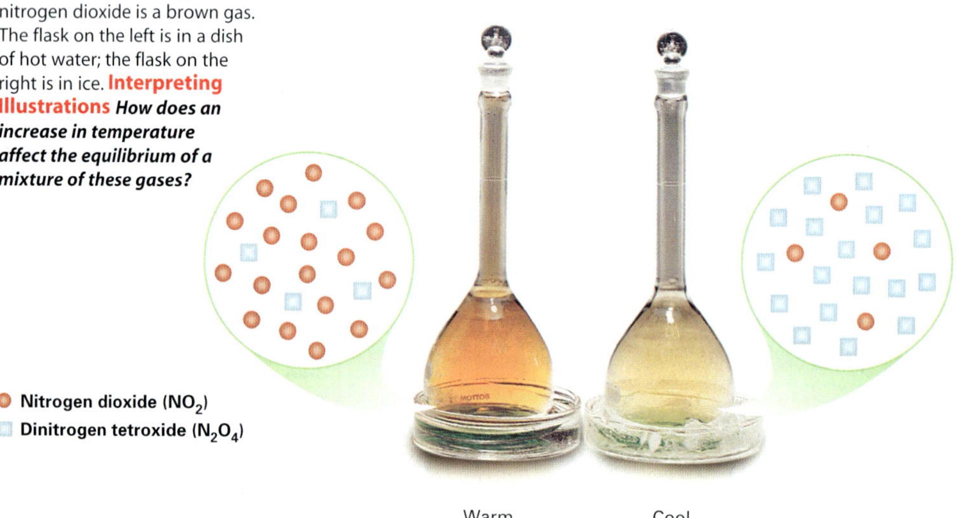

Figure 18.14 Dinitrogen tetroxide is a colorless gas; nitrogen dioxide is a brown gas. The flask on the left is in a dish of hot water; the flask on the right is in ice. **Interpreting Illustrations** ***How does an increase in temperature affect the equilibrium of a mixture of these gases?***

SAMPLE PROBLEM 18.1

Expressing and Calculating K_{eq}

The colorless gas dinitrogen tetroxide (N_2O_4) and the dark brown gas nitrogen dioxide (NO_2) exist in equilibrium with each other.

$$N_2O_4(g) \rightleftharpoons 2NO_2(g)$$

A liter of a gas mixture at equilibrium at 10°C contains 0.0045 mol of N_2O_4 and 0.030 mol of NO_2. Write the expression for the equilibrium constant and calculate the equilibrium constant (K_{eq}) for the reaction.

1 Analyze ***List the knowns and the unknowns.***

Knowns

- $[N_2O_4] = 0.0045$ mol/L
- $[NO_2] = 0.030$ mol/L
- $K_{eq} = \dfrac{[C]^c \times [D]^d}{[A]^a \times [B]^b}$

Unknowns

- K_{eq} (algebraic expression) = ?
- K_{eq} (numerical value) = ?

2 Calculate ***Solve for the unknowns.***

To write the equilibrium constant expressions, place the concentration of the product in the numerator and the concentration of the reactant in the denominator. Raise each to the power equal to its coefficient in the chemical equation. Substitute the given concentrations and calculate K_{eq}.

$$K_{eq} = \frac{[NO_2]^2}{[N_2O_4]} = \frac{(0.030 \text{ mol/L})^2}{0.0045 \text{ mol/L}}$$

$$= \frac{(0.030 \text{ mol/L} \times 0.030 \text{ mol/L})}{0.0045 \text{ mol/L}} = 0.20$$

3 Evaluate ***Does the result make sense?***

Each concentration is raised to the correct power. The numerical value of the constant is correctly expressed to two significant figures.

Practice Problems

7. The reversible reaction $N_2(g) + 3H_2(g) \rightleftharpoons 2NH_3(g)$ produces ammonia, which is a fertilizer. At equilibrium, a 1-L flask contains 0.15 mol H_2, 0.25 mol N_2, and 0.10 mol NH_3. Calculate K_{eq} for the reaction.

8. For the same mixture, under the same conditions described in Problem 7, calculate K_{eq} for $2NH_3(g) \rightleftharpoons N_2(g) + 3H_2(g)$. How is the K_{eq} for a forward reaction related to the K_{eq} for a reverse reaction?

Math Handbook

For help with using a calculator, go to page R62.

Interactive Textbook

Problem Solving 18.7 Solve Problem 7 with the help of an interactive guided tutorial.

with ChemASAP

Ammonia (NH_3), produced by a reversible reaction, is used extensively as a fertilizer.

Sample Problem 18.1

Answers

7. $K_{eq} = 12$

8. $K_{eq} = 8.3 \times 10^{-2}$; one is the inverse of the other.

Practice Problems Plus 

Analysis of an equilibrium mixture of hydrogen, iodine, and hydrogen iodide contained in a 10.0-L flask at a certain temperature gives the following results: hydrogen, 0.15 mol; iodine, 0.15 mol; hydrogen iodide, 0.87 mol. **Calculate K_{eq} for the reaction:**

$$H_2(g) + I_2(g) \rightleftharpoons 2HI(g)$$

(K_{eq} = 34)

Math Handbook

For a math refresher and practice, direct students to using a calculator, page R62.

Answers to...

Figure 18.14 An increase in temperature causes the equilibrium to shift toward the formation of more NO_2.

Discuss 

Show students how to set up and solve problems involving equilibrium constants. Stress the need to clearly establish what is given and what is being asked. Break down the process into easily followed steps. Point out that the only quantities that can go into the brackets used in the equilibrium constant expression are concentrations, or quantities like partial pressures, which are related to concentrations. Warn students not to forget to subtract the number of moles that react from the number of moles originally present in order to determine the number remaining at equilibrium. Point out that it is this value for the concentration of a reactant that is substituted in the denominator of the equation for K_{eq}.

Sample Problem 18.2

Answers

9. $K_{eq} = 1.6 \times 10^{-3}$
10. $K_{eq} = 0.79$

Math Handbook

For a math refresher and practice, direct students to using a calculator, page R62.

SAMPLE PROBLEM 18.2

Finding the Equilibrium Constant

One mol of colorless hydrogen gas and 1.00 mol of violet iodine vapor are sealed in a 1-L flask and allowed to react at 450°C. At equilibrium, 1.56 mol of colorless hydrogen iodide is present, together with some of the reactant gases. Calculate K_{eq} for the reaction.

$$H_2(g) + I_2(g) \rightleftharpoons 2HI(g)$$

1 Analyze ***List the knowns and the unknowns.***

Knowns

- $[H_2]$ (initial) = 1.00 mol/L
- $[I_2]$ (initial) = 1.00 mol/L
- $[HI]$ (equilibrium) = 1.56 mol/L
- $K_{eq} = \dfrac{[C]^c \times [D]^d}{[A]^a \times [B]^b}$

Unknowns

- K_{eq} (algebraic expression) = ?
- K_{eq} (numerical value) = ?

2 Calculate ***Solve for the unknowns.***

The balanced equation indicates that 1.00 mol H_2 and 1.00 mol I_2 are needed to form 2.00 mol HI. Making 1.56 mol of HI, therefore, consumes $\frac{1}{2} \times 1.56$ mol of each or 0.78 mol of H_2 and 0.78 mol of I_2. Calculate how much H_2 and I_2 remain in the flask at equilibrium.

$$\text{mol } H_2 = \text{mol } I_2 = (1.00 \text{ mol} - 0.78 \text{ mol}) = 0.22 \text{ mol}$$

Write the expression for K_{eq}: $K_{eq} = \dfrac{[HI]^2}{[H_2] \times [I_2]}$

Substitute the equilibrium concentrations of the reactants and products into the equation and calculate.

$$K_{eq} = \frac{(1.56 \text{ mol/L})^2}{0.22 \text{ mol/L} \times 0.22 \text{ mol/L}} = \frac{1.56 \cancel{\text{mol/L}} \times 1.56 \cancel{\text{mol/L}}}{0.22 \cancel{\text{mol/L}} \times 0.22 \cancel{\text{mol/L}}} = 50$$

3 Evaluate ***Does the result make sense?***

Each concentration is raised to the correct power. The numerical value of the constant is correctly rounded to two significant figures.

Practice Problems

9. Suppose the following system reaches equilibrium.

$$N_2(g) + O_2(g) \rightleftharpoons 2NO(g)$$

Analysis of the equilibrium mixture in a 1-L flask gives the following results: N_2 = 0.50 mol, O_2 = 0.50 mol, and NO = 0.020 mol. Calculate K_{eq} for the reaction.

10. At 750°C the following reaction reaches equilibrium in a 1-L flask.

$$H_2(g) + CO_2(g) \rightleftharpoons H_2O(g) + CO(g)$$

Analysis of the equilibrium mixture gives the following results: H_2 = 0.053 mol, CO_2 = 0.053 mol, H_2O = 0.047 mol, and CO = 0.047 mol. Calculate K_{eq} for the reaction.

Math Handbook

For help with using a calculator, go to page R62.

Interactive Textbook

Problem Solving 18.9 Solve Problem 9 with the help of an interactive guided tutorial.

with ChemASAP

You can also calculate the equilibrium concentrations of the reactants and products if you know the equilibrium constant and if you have some information about the initial or final concentration of one of the constituents. For example, bromine chloride (BrCl) decomposes to form bromine and chlorine according to this equation.

$$2BrCl(g) \rightleftharpoons Br_2(g) + Cl_2(g)$$

At a certain temperature, the equilibrium constant for this reaction is 11.1. Suppose that pure BrCl is placed in a 1-L container and allowed to come to equilibrium. Analysis then shows that the reaction mixture contains 4.00 moles of Cl_2. How many moles of Br_2 and BrCl are also present in the equilibrium mixture?

According to the equation, when BrCl decomposes, equal numbers of moles of Br_2 and Cl_2 are formed. At equilibrium, 4.00 mol of Cl_2 is present, so 4.00 mol of Br_2 must also be present. Because the volume of the container is 1 L, the concentrations of Br_2 and Cl_2 are each 4.00 mol/L. These concentrations can be used in the equilibrium constant expression to calculate the concentration of BrCl.

$$11.1 = \frac{[Br_2] \times [Cl_2]}{[BrCl]^2}$$

Rearrange the equation to solve for $[BrCl]^2$ and substitute the known values.

$$[BrCl]^2 = \frac{[Br_2] \times [Cl_2]}{11.1} = \frac{(4.00 \text{ mol/L}) \times (4.00 \text{ mol/L})}{11.1} = 1.44 \text{ mol}^2/\text{L}^2$$

$$[BrCl] = \sqrt{1.44 \text{ mol}^2/\text{L}^2} = 1.20 \text{ mol/L}$$

The concentration of the reactant (BrCl) is less than the concentrations of the products, so the products (Br_2 and Cl_2) are slightly favored at equilibrium. This is to be expected because $K_{eq} > 1$. Yellow chlorine gas and brown liquid bromine with its vapor are shown in Figure 18.15.

Figure 18.15 Chlorine and bromine are two of the halogens. **Comparing and Contrasting** ***How are they similar? In what ways are they different?***

18.2 Section Assessment

11. **Key Concept** How do the amounts of reactants and products change after a reaction has reached chemical equilibrium?
12. **Key Concept** What are three stresses that can upset the equilibrium of a chemical system?
13. **Key Concept** What does the value of the equilibrium constant tell you about the amounts of reactants and products present at equilbrium?
14. How can a balanced chemical equation be used to write an equilibrium-constant expression?
15. Can a pressure change shift the equilibrium position in every reversible reaction? Explain.
16. Using the following equilibrium constants for several reactions, determine in which reactions the products are favored. Why?
    a. $K_{eq} = 1 \times 10^2$ **b.** $K_{eq} = 0.003$ **c.** $K_{eq} = 3.5$

Connecting Concepts

Percent Yield Recall the concept of percent yield of a chemical reaction in Section 12.3. Could the fact that a reaction is reversible and establishes equilibrium affect the percent yield of a reaction? Explain.

Assessment 18.2 Test yourself on the concepts in Section 18.2.

with **ChemASAP**

3 ASSESS

Evaluate Understanding L2

Write several reversible chemical equations on the board. Have students use Le Châtelier's principle to predict how the yield of a reaction product could be increased. Ask them to write the equation for calculating an equilibrium constant using the generalized expression

$$aA + bB \rightleftharpoons cC + dD$$

$$K_{eq} = \frac{[C]^c \times [D]^d}{[A]^a \times [B]^b}$$

Reteach L1

If students have trouble writing the correct equilibrium constant expression from a balanced equation, list the steps on the board. Use as an example the equilibrium process for the production and decomposition of ozone in the stratosphere. Explain that stratospheric ozone screens out about 95% of the UV radiation from the sun. Within the ozone layer, ozone is continually formed and consumed through an equilibrium process:

$$3O_2(g) \rightleftharpoons 2O_3(g)$$

Connecting Concepts

Yes; 100% yield is obtained only when all reactants are converted to products. If a reversible reaction stops before all reactants are used up, it can never reach 100% yield.

If your class subscribes to the Interactive Textbook, use it to review key concepts in Section 18.2.

with **ChemASAP**

Section 18.2 Assessment

11. They do not change.
12. changes in concentrations of reactants and products, changes in temperature, and changes in pressure
13. $K_{eq} > 1$ favors products; $K_{eq} < 1$ favors reactants.
14. An equilibrium constant expression is a ratio. In the numerator, product concentrations are multiplied. In the denominator, reactant concentrations are multiplied. Each concentration is raised to a power equal to the coefficient for that species in the balanced equation.
15. no; only in reversible reactions in which the mole ratios of gaseous reactants and products are unequal
16. Products are favored in reactions a and c because $K_{eq} > 1$.

Answers to...

Figure 18.15 Both are in Group 7A, form ions with 1– charge, and exist as diatomic molecules. Chlorine is a yellow gas; bromine is a brown liquid at ordinary temperatures.

18.3

1 FOCUS

Objectives

18.3.1 Describe the relationship between the solubility product constant and the solubility of a compound.

18.3.2 Predict whether precipitation will occur when two salt solutions are mixed.

Guide for Reading

Build Vocabulary L2

Paraphrase Make sure students understand the difference between the terms *solubility* and *solubility product*. Have them write the definitions in their own words.

Reading Strategy L2

Summarize After reading this section, ask students to summarize what they learned about solubility product constants and the common ion effect. Suggest that they use an equation to illustrate their summary.

2 INSTRUCT

Connecting to Your World

Ask, **Why is $BaSO_4$ ingested by a patient before having an X-ray taken of the stomach?** *(Soft tissues don't show up well in X-rays.)* **Why is the digestive system well suited to this type of image enhancement?** *(Chemicals are easily brought into the system and easily removed.)*

The Solubility Product Constant

Discuss L2

Any ionic solid placed in water establishes an equilibrium between its dissociated ions in solution and the solid that is undissolved. The solubility product constant is an equilibrium constant that describes the equilibrium between a solid and its ions in solution.

18.3 Solubility Equilibrium

Guide for Reading

Key Concepts

- What is the relationship between the solubility product constant and the solubility of a compound?
- How can you predict whether precipitation will occur when two salt solutions are mixed?

Vocabulary

solubility product constant

common ion

common ion effect

Reading Strategy

Predicting Based on what you know about equilibrium and the dissociation of ionic compounds in aqueous solution, predict the form of the equilibrium constant expression for the dissolving of an ionic compound in water. After you read the section, check the accuracy of your prediction and clarify any misconceptions you may have had.

Connecting to Your World X-rays allow physicians to make observations of bones and internal organs without invasive surgery. Sometimes chemical substances are used to improve the clarity of the X-ray image. Barium sulfate is ingested by a patient before X-ray images of the digestive tract are taken. Barium sulfate absorbs the X-rays, thereby producing light areas on the developed X-ray film. However, barium salts are usually toxic. In this section, you will learn why patients can ingest this poisonous substance without harm.

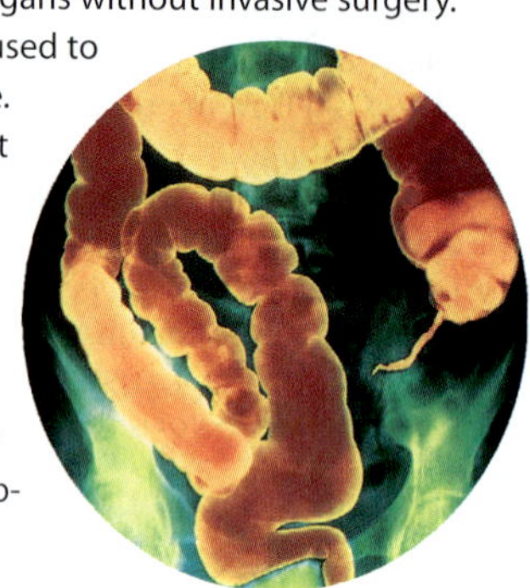

The Solubility Product Constant

Ionic compounds differ in their solubilities. Most salts of the alkali metals are soluble in water. More than 35 g of sodium chloride will dissolve in only 100 g of water. Many classes of ionic compounds, however, are insoluble. For example, compounds that contain phosphate, sulfide, sulfite, or carbonate ions are often insoluble. Exceptions are compounds in which these ions are combined with ammonium ions or alkali metal ions. Table 18.1 summarizes the solubilities of many ionic compounds in water.

Most insoluble salts will actually dissolve to some extent in water. These salts are said to be slightly, or sparingly, soluble in water. For example, Figure 18.16 shows what happens when the "insoluble" salt silver chloride is mixed with water. A very small amount of silver chloride actually dissolves in the water.

$$AgCl(s) \rightleftharpoons Ag^+(aq) + Cl^-(aq)$$

You can write an equilibrium expression for this process.

$$K_{eq} = \frac{[Ag^+] \times [Cl^-]}{[AgCl]}$$

As long as some undissolved AgCl (solid) is present, the concentration of the AgCl is a constant. Thus the concentration of AgCl, a constant, can be combined with the equilibrium constant to form a new constant.

$$K_{eq} \times [AgCl] = [Ag^+] \times [Cl^-] = K_{sp}$$

Figure 18.16 Silver chloride is slightly soluble in water. **Predicting** ***Would adding solid silver chloride to this test tube disturb the equilibrium?***

Section Resources

Print

- **Guided Reading and Study Workbook,** Section 18.3
- **Core Teaching Resources,** Section 18.3 Review
- **Transparencies,** T203–T205
- **Laboratory Manual,** Lab 39

Technology

- **Interactive Textbook with ChemASAP,** Problem-Solving 18.17, 18.19; Assessment 18.3
- **Go Online,** Section 18.3

Table 18.1

Solubilities of Ionic Compounds in Water

Compounds	Solubility	Exceptions
Salts of Group 1A metals and ammonia	Soluble	Some lithium compounds
Ethanoates, nitrates, chlorates, and perchlorates	Soluble	Few exceptions
Sulfates	Soluble	Compounds of Pb, Ag, Hg, Ba, Sr, and Ca
Chlorides, bromides, and iodides	Soluble	Compounds of Ag and some compounds of Hg and Pb
Sulfides and hydroxides	Most are insoluble	Alkali metal sulfides and hydroxides are soluble. Compounds of Ba, Sr, and Ca are slightly soluble.
Carbonates, phosphates, and sulfites	Insoluble	Compounds of the alkali metals and of ammonium ions

This new constant, called the **solubility product constant** (K_{sp}), equals the product of the concentrations of the ions each raised to a power equal to the coefficient of the ion in the dissociation equation. The coefficients for the dissociation of silver chloride are each 1. The value of K_{sp} for silver chloride at 25°C is 1.8×10^{-10}.

$$K_{sp} = [Ag^+] \times [Cl^-]$$
$$K_{sp} = 1.8 \times 10^{-10}$$

What does the size of the solubility product constant tell you about the solubility of the compound? **The smaller the numerical value of the solubility product constant, the lower the solubility of the compound.** The solubility product constants for some common sparingly soluble salts are given in Table 18.2 on the next page. The mineral deposits around sink drains, such as the one shown in Figure 18.17, are composed of compounds such as calcium carbonate ($K_{sp} = 4.5 \times 10^{-9}$). The low solubility of such compounds makes them difficult to remove.

What does the solubility product constant expression consist of?

Figure 18.17 Scale, formed by the precipitation of slightly soluble salts, often builds up around kitchen and bathroom faucets. **Applying Concepts** ***Why are insoluble substances hard to remove?***

Solubility Tables 

Purpose Students illustrate the solubilities of salts containing various anions.

Materials Calculator, paper, pencil

Procedure Divide the class into groups of four students. Ask the groups to create color-coded tables illustrating the solubilities of salts containing various anions. For example, students can design separate tables for nitrates, sulfates, chlorides, sulfides, hydroxides, and carbonates. Have them use different colors in each table to show which elements form soluble compounds containing a given anion and which form insoluble compounds. Students should divide the tasks of looking up solubilities, designing the tables, and drawing the final tables.

Use Visuals 

Table 18.1 Have students reread the Connecting to Your World on page 560. Ask them if Table 18.1 helps them understand why barium sulfate can be used in humans for X-ray imaging. *(because it isn't soluble in human tissues)*

Discuss 

Explain that the solubility product constant is closely related to the equilibrium constant that students learned about in Section 18.2. Point out similarities in the equations for these two quantities. Write the example problems on the board and carry out the solutions for the solubility product, K_{sp}. Focus on how to set up the algebraic expression.

Answers to...

Figure 18.16 No, because the concentration of AgCl is constant.

Figure 18.17 because they don't dissolve in a solvent such as water.

The product of the concentrations of the ions, each raised to a power equal to the coefficient of the ion in the dissociation equation.

TEACHER Demo

Common Ion Effect **L2**

Purpose Students hypothesize the result of mixing silver nitrate and silver chloride

Materials silver chloride, silver nitrate, 2 test tubes, dropper

Procedure Prepare saturated solutions of silver chloride and silver nitrate in two test tubes. Write the solubility product constant expression for silver chloride on the board.

$K_{sp} = [Ag^+] \times [Cl^-] = 1.8 \times 10^{-10}$

Have students hypothesize what would occur if a few drops of the silver nitrate solution were added to the test tube containing the silver chloride. Add the drops of silver nitrate solution and ask students what they observe. *(A precipitate, silver chloride, comes out of solution.)* Explain that the value of $[Ag^+]$ increases in the K_{sp} expression as the silver nitrate is added, so that silver chloride must precipitate until the solubility product again equals K_{sp}.

Sample Problem 18.3

Answers

17. $2 \times 10^{-14} M$

18. $6.7 \times 10^{-5} M$

Practice Problems Plus 

What is the concentration of barium and sulfate ions in a saturated barium sulfate solution at 25°C?
$K_{sp} = 1.1 \times 10^{-10}$
($1.0 \times 10^{-5} M$)

Table 18.2

Solubility Product Constants (K_{sp}) at 25°C

Salt	K_{sp}	Salt	K_{sp}	Salt	K_{sp}
Halides		**Sulfates**		**Hydroxides**	
AgCl	1.8×10^{-10}	$PbSO_4$	6.3×10^{-7}	$Al(OH)_3$	3.0×10^{-34}
AgBr	5.0×10^{-13}	$BaSO_4$	1.1×10^{-10}	$Zn(OH)_2$	3.0×10^{-16}
AgI	8.3×10^{-17}	$CaSO_4$	2.4×10^{-5}	$Ca(OH)_2$	6.5×10^{-6}
$PbCl_2$	1.7×10^{-5}	**Sulfides**		$Mg(OH)_2$	7.1×10^{-12}
$PbBr_2$	2.1×10^{-6}	NiS	4.0×10^{-20}	$Fe(OH)_2$	7.9×10^{-16}
PbI_2	7.9×10^{-9}	CuS	8.0×10^{-37}	**Carbonates**	
PbF_2	3.6×10^{-8}	Ag_2S	8.0×10^{-51}	$CaCO_3$	4.5×10^{-9}
CaF_2	3.9×10^{-11}	ZnS	3.0×10^{-23}	$SrCO_3$	9.3×10^{-10}
Chromates		FeS	8.0×10^{-19}	$ZnCO_3$	1.0×10^{-10}
$PbCrO_4$	1.8×10^{-14}	CdS	1.0×10^{-27}	Ag_2CO_3	8.1×10^{-12}
Ag_2CrO_4	1.2×10^{-12}	PbS	3.0×10^{-28}	$BaCO_3$	5.0×10^{-9}

The chromate ion is responsible for the brilliant yellow color of lead(II) chromate.

SAMPLE PROBLEM 18.3

Finding the Ion Concentrations in a Saturated Solution

What is the concentration of lead ions and chromate ions in a saturated lead chromate solution at 25°C? ($K_{sp} = 1.8 \times 10^{-14}$)

1 Analyze ***List the knowns and the unknowns.***

Knowns

- $K_{sp} = 1.8 \times 10^{-14}$
- $K_{sp} = [Pb^{2+}] \times [CrO_4^{2-}]$
- $PbCrO_4(s) \rightleftharpoons Pb^{2+}(aq) + CrO_4^{2-}(aq)$

Unknowns

- $[Pb^{2+}] = ?\,M$
- $[CrO_4^{2-}] = ?\,M$

For each Pb^{2+} ion formed, one CrO_4^{2-} ion is formed.

2 Calculate ***Solve for the unknowns.***

$$K_{sp} = [Pb^{2+}] \times [CrO_4^{2-}] = 1.8 \times 10^{-14}$$

At equilibrium $[Pb^{2+}] = [CrO_4^{2-}]$. Substitute $[Pb^{2+}]$ for $[CrO_4^{2-}]$ in the expression for K_{sp} to get an equation with one unknown.

$$K_{sp} = [Pb^{2+}] \times [Pb^{2+}] = [Pb^{2+}]^2 = 1.8 \times 10^{-14}$$

Solve for $[Pb^{2+}]$.

$$[Pb^{2+}] = [CrO_4^{2-}] = \sqrt{1.8 \times 10^{-14}} = 1.3 \times 10^{-7} M$$

3 Evaluate ***Does the result make sense?***

The product $[Pb^{2+}] \times [CrO_4^{2-}]$ is close to the value of K_{sp}. The answer should have two significant figures.

Practice Problems

17. Lead(II) sulfide (PbS) has a K_{sp} of 3.0×10^{-28}. What is the concentration of lead(II) ions in a saturated solution of PbS?

18. What is the concentration of calcium ions in a saturated calcium carbonate solution at 25°C? ($K_{sp} = 4.5 \times 10^{-9}$)

Problem Solving 18.17 Solve Problem 17 with the help of an interactive guided tutorial.

with ChemASAP

Figure 18.18 Lead(II) chromate is slightly soluble in water. **a** A saturated solution of $PbCrO_4$ is pale yellow. **b** When a few drops of lead nitrate ($Pb(NO_3)_2$) are added to the solution, it becomes cloudy as more lead(II) chromate precipitates.
Applying Concepts ***Explain what happened in b by considering the addition of lead nitrate as a stress on the $PbCrO_4$ equilibrium.***

The Common Ion Effect

In a saturated solution of lead(II) chromate, an equilibrium is established between the solid lead(II) chromate and its ions in solution.

$$PbCrO_4(s) \rightleftharpoons Pb^{2+}(aq) + CrO_4^{2-}(aq) \quad K_{sp} = 1.8 \times 10^{-14}$$

What do you think would happen if you added some lead nitrate to this solution? The concentration of lead ion would increase. Immediately, the product of $[Pb^{2+}]$ and $[CrO_4^{2-}]$ would be greater than K_{sp}. Applying Le Châtelier's principle, the stress of the additional Pb^{2+} can be relieved if the reaction shifts to the left. Figure 18.18 demonstrates that lead ions combine with chromate ions in solution to form additional solid $PbCrO_4$. Lead chromate continues to precipitate from the solution until the product of $[Pb^{2+}]$ and $[CrO_4^{2-}]$ once again equals 1.8×10^{-14}.

$$K_{sp} \text{ for the original solution: } [Pb^{2+}][CrO_4^{2-}] = 1.8 \times 10^{-14}$$

$$K_{sp} \text{ after addition of } Pb(NO_3)_2\text{: } \mathbf{[Pb^{2+}]}[CrO_4^{2-}] = 1.8 \times 10^{-14}$$

In this example, the lead ion is called a common ion. A **common ion** is an ion that is found in both salts in a solution. Adding lead nitrate to a saturated solution of $PbCrO_4$ causes the solubility of $PbCrO_4$ to decrease. The lowering of the solubility of an ionic compound as a result of the addition of a common ion is called the **common ion effect.** Adding sodium chromate to the solution of $PbCrO_4$ would also produce the common ion effect. The additional chromate ion, a different common ion, would similarly cause the solubility equilibrium to shift to the left and produce more $PbCrO_4$.

Recall from *Connecting to Your World* that patients must ingest toxic barium sulfate ($BaSO_4$) when having an X-ray taken. Table 18.2 shows that the K_{sp} of barium sulfate is 1.1×10^{-10}. Considering this low value, it is understandable that barium cannot harm the patient because so little of it is in solution and able to enter the bloodstream. The solubility can be further reduced by adding sodium sulfate (Na_2SO_4), a nontoxic soluble compound that provides the common ion SO_4^{2-}.

Checkpoint ***What is the common ion effect?***

For: Links on the Common Ion Effect
Visit: www.SciLinks.org
Web Code: cdn-1182

The Common Ion Effect

Discuss L2

Point out that the common ion effect is not a new concept, but an example of Le Châtelier's principle at work. Explain that adding an ion to an equilibrium system containing that ion will cause the equilibrium to shift in the direction that uses up the ion.

Download a worksheet on the **Common Ion Effect** for students to complete, and find additional teacher support from NSTA SciLinks.

Facts and Figures

Other Compounds in Radiology

Barium sulfate is just one of many compounds used as contrast materials in radiology. Barium compounds are used to look at gastrointestinal areas. Water-soluble iodinated compounds are contrast materials injected to make blood vessels, urinary tract, and spinal canal visible in X-rays. Naturally placed or injected air also functions as a contrast material.

Answers to...

Figure 18.18 An increase in $Pb^{2+}(aq)$ shifts the equilibrium to the left creating more $PbCrO_4$ (*s*).

Checkpoint The lowering of the solubility of an ionic compound as a result of the addition of one of the ions in the compound

Relate 

Tiny crystals or grains of silver bromide are immersed in a gelatinous emulsion on the surface of unexposed photographic film. When photons enter a camera and strike the film, the energy of the photons causes some bromide ions to lose an electron and some silver ions to gain an electron. These grains of silver bromide are said to be activated. In the developing process, only silver ions in activated grains are converted to silver atoms to form the image.

Sample Problem 18.4

Answers

19. $2 \times 10^{-17} M$

20. $3.2 \times 10^{-6} M$

Math Handbook

For a math refresher and practice, direct students to algebraic equations, page R69.

SAMPLE PROBLEM 18.4

Finding Equilibrium Ion Concentrations in the Presence of a Common Ion

Photographic film is covered with a light-sensitive emulsion containing silver bromide. The K_{sp} of silver bromide is 5.0×10^{-13}. What is the bromide-ion concentration of a 1.00-L saturated solution of AgBr to which 0.020 mol of $AgNO_3$ is added?

1 Analyze *List the knowns and the unknown.*

Knowns

- $K_{sp} = 5.0 \times 10^{-13}$
- moles of $AgNO_3$ added = 0.020 mol
- volume of solution = 1 L
- $AgBr(s) \rightleftharpoons Ag^+(aq) + Br^-(aq)$
- $K_{sp} = [Ag^+] \times [Br^-]$

Unknown

- $[Br^-] = ?\ M$

Express the concentrations of the two ions in one unknown. Let the equilibrium concentration of bromide ion from the dissociation be x. Then the equilibrium concentration of silver ion is $x + 0.020$.

2 Calculate *Solve for the unknown.*

Because of the small value of K_{sp}, you can make a simplifying assumption: x will be negligibly small compared to 0.020. Thus the $[Ag^+]$ at equilibrium is approximately equal to $0.020M$. Solve for x and substitute these values into the K_{sp} expression.

$$K_{sp} = [Ag^+] \times [Br^-] = [Ag^+] \times x$$

$$x = \frac{K_{sp}}{[Ag^+]} = \frac{(5.0 \times 10^{-13})}{[Ag^+]}$$

$$= \frac{(5.0 \times 10^{-13})}{0.020}$$

$$= [Br^-] = 2.5 \times 10^{-11} M$$

The equilibrium concentration of bromide ion is $2.5 \times 10^{-11} M$.

3 Evaluate *Does the result make sense?*

Because the $[Ag^+]$ is so high, $[Br^-]$ should be very low, approaching the numerical value for K_{sp}. The answer should have two significant figures, and the simplifying assumption was justified.

Practice Problems

19. What is the concentration of sulfide ion in a 1.0-L solution of iron(II) sulfide to which 0.04 mol of iron(II) nitrate is added? The K_{sp} of FeS is 8×10^{-19}.

20. The K_{sp} of $SrSO_4$ is 3.2×10^{-7}. What is the equilibrium concentration of sulfate ion in a 1.0-L solution of strontium sulfate to which 0.10 mol of $Sr(CH_3CO_2)_2$ has been added?

Math Handbook

For help with algebraic equations, go to page R69.

Interactive Textbook

Problem Solving 18.19 Solve Problem 19 with the help of an interactive guided tutorial.

with ChemASAP

Differentiated Instruction

Less Proficient Readers 

Pair students who have trouble with word problems with those who are proficient at algebra and problem solving. Have each pair work through the practice problems, taking turns being the lead solver. The lead student should explain the process being used to arrive at the solution.

The solubility product (K_{sp}) can be used to predict whether a precipitate will form when solutions are mixed. **If the product of the concentrations of two ions in the mixture is greater than the K_{sp} of the compound formed from the ions, a precipitate will form.** After precipitation, the solution will be saturated with the precipitated compound. If the product of the concentrations is less than the K_{sp}, no precipitate will form and the solution is unsaturated.

Suppose you have 0.50 L of 0.002*M* $Ba(NO_3)_2$ and mix it with 0.50 L of 0.008 M Na_2SO_4. You then have one liter of solution. The insoluble compound that could precipitate is barium sulfate. To predict whether a precipitate of $BaSO_4$ will form, you need to determine the concentration of the ions after the dilution that occurs on mixing. Precipitation will occur if the product of the concentrations of the two ions (Ba^{2+} and SO_4^{2-}) exceeds the K_{sp} of $BaSO_4$, which is 1.1×10^{-10}. Because each solution was diluted with an equal volume of the other solution, the concentrations of both Ba^{2+} and SO_4^{2-} after mixing will be half of their original concentrations. Thus in the combined solution, $[Ba^{2+}] = 0.001M$ and $[SO_4^{2-}] = 0.004M$. These concentrations can now be multiplied together as a trial product and the result compared with the K_{sp}.

$$[Ba^{2+}] \times [SO_4^{2-}] = (0.001M) \times (0.004M) = 4 \times 10^{-6}$$

The K_{sp} of barium sulfate is 1.1×10^{-10}, so the trial product is larger than the K_{sp}. A precipitate will form. Barium sulfate will precipitate until the product of the concentrations of the ions remaining in solution equals 1.1×10^{-10}. Figure 18.19 shows the mixing of the solutions and the precipitation of barium sulfate.

Figure 18.19 A precipitate of barium sulfate forms as barium nitrate ($Ba(NO_3)_2$) and sodium sulfate (Na_2SO_4) solutions are mixed.
Applying Concepts ***What is the product of the concentrations of barium ion and sulfate ion after precipitation is complete?***

18.3 Section Assessment

21. **Key Concept** What is the relationship between the solubility product constant and the solubility of a compound?
22. **Key Concept** How can you predict whether a precipitate will form when two salt solutions are mixed?
23. Write the solubility product expression for Ag_2CO_3.
24. What is the concentration of lead ions and sulfide ions in a saturated solution of lead sulfide (PbS) solution at 25°C? ($K_{sp} = 3.0 \times 10^{-28}$)
25. The K_{sp} of barium sulfate is 1.1×10^{-10}. What is the sulfate ion concentration of a 1.00 L saturated solution of $BaSO_4$ to which 0.015 mol of $Ba(NO_3)_2$ is added?
26. Would precipitation occur when 500 mL of a 0.02*M* solution of $AgNO_3$ is mixed with 500 mL of a 0.001*M* solution of NaCl? Explain.
27. Which compound, FeS ($K_{sp} = 8.0 \times 10^{-19}$) or CuS ($K_{sp} = 8.0 \times 10^{-37}$), has the higher solubility?
28. What is the K_{sp} of nickel(II) sulfide if the equilibrium concentrations of Ni^{2+} and S^{2-} in a saturated solution of NiS are each $1 \times 10^{-10}M$?

Connecting Concepts

Spectator Ions Refer to the discussion of spectator ions and net ionic equations in Section 11.3. In a precipitation reaction, would the addition of a spectator ion lead to the formation of more precipitate because of the common ion effect? Explain.

Assessment 18.3 Test yourself on the concepts in Section 18.3.

with ChemASAP

Section 18.3 Assessment

21. the smaller the K_{sp}, the lower the solubility of the compound
22. If the product of the concentrations of two ions in a mixture is greater than the K_{sp} of the compound formed from the ions, a precipitate will form.
23. $[Ag^+]^2 \times [CO_3^{2-}] = K_{sp}$
24. $2 \times 10^{-14}M$
25. $7.3 \times 10^{-9}M$
26. yes; the product $[Ag^+] \times [Cl^-] = 0.01M \times 0.0005M = 5 \times 10^{-6} >> K_{sp}$ AgCl $= 1.8 \times 10^{-10}$
27. FeS
28. 4×10^{-20}

Discuss

Have students hypothesize how the common ion effect could be used to identify ions in solution. Suggest that they use Table 18.1 to find a common property of salts listed in Table 18.2 (insolubility). Explain that saturated solutions of many of the salts in Table 18.2 can be used to test for the presence of different ions in solution. If a precipitate forms when a few drops of an unknown solution is added to a known saturated solution, the unknown solution must have an ion in common with the known solution.

3 ASSESS

Evaluate Understanding

Have students classify the salts in Table 18.2 on the basis of solubility. Ask, **What is the relationship between the solubility product constant and the relative solubility of a salt?** *(the larger the solubility product constant, the more soluble the salt)*

Reteach

Project a transparency of Table 18.1. Point out some salient features, for example, water is the solvent, the last column is for exceptions to the first column. Ask students to look in Table 18.2 for examples of ions described as low-solubility in Table 18.1. Sketch a graph of the relationship between solubility product constants and solubility.

Connecting Concepts

Spectator ions don't participate in the reaction. They stay in solution and don't precipitate.

If your class subscribes to the Interactive Textbook, use it to review key concepts in Section 18.3.

with ChemASAP

Answers to...

Figure 18.19 1.1×10^{-10} (K_{sp})

18.4

1 FOCUS

Objectives

18.4.1 Identify two characteristics of spontaneous reactions.

18.4.2 Describe the role of entropy in chemical reactions.

18.4.3 Identify two factors that determine the spontaneity of a reaction.

18.4.4 Define Gibbs free-energy change.

Guide for Reading

Build Vocabulary L2

Paraphrase Have students write definitions for *spontaneous* and *nonspontaneous* in their own words. Ask them to consider the meaning of the terms as they relate to human behavior.

Reading Strategy L2

Identify Details In this section, students need to acquire a firm grasp of the characteristics of spontaneous and nonspontaneous reactions to build on. As they read, have them create a table or chart describing the two types of reactions and how they differ.

2 INSTRUCT

Connecting to Your World

Ask, **What conditions produce a spontaneous chemical reaction?** *(Based on prior knowledge, students may suggest reactions that have a low activation energy, reactions that involve active elements such as alkali metals, combustion reactions above the kindling temperature, or reactions in solution that produce a precipitate.)*

18.4 Entropy and Free Energy

Guide for Reading

Key Concepts

- What are two characteristics of spontaneous reactions?
- What part does entropy play in chemical reactions?
- What two factors determine the spontaneity of a reaction?
- Is the Gibbs free-energy change positive or negative in a spontaneous process?

Vocabulary

free energy
spontaneous reaction
nonspontaneous reaction
entropy
law of disorder
Gibbs free-energy change

Reading Strategy

Relating Text and Visuals As you read about spontaneous and nonspontaneous reactions, study Figure 18.25. Summarize three ways in which the energy terms of the free energy equation result in a spontaneous reaction and three ways in which they result in a nonspontaneous reaction.

Connecting to Your World

Fires in homes, barns, and other places sometimes seem to start on their own, without any application of heat from an external source. These fires could be caused by spontaneous combustion. Inside a pile of oily rags or a stack of hay that has not been thoroughly dried, decomposition causes heat to build up. If the existing conditions do not allow the heat to escape, the temperature can build up enough to cause a fire. In this section, you will learn about the conditions that will produce a spontaneous chemical reaction.

Free Energy and Spontaneous Reactions

Many chemical and physical processes release energy that can be used to bring about other changes. For example, some of the energy liberated in a chemical reaction can be harnessed to do work, such as driving the pistons of an internal-combustion engine. **Free energy** is energy that is available to do work. Just because free energy is available to do work, however, does not mean that it can be used efficiently. The internal-combustion engine in a car is only about 30% efficient; that is, only about 30% of the free energy released by burning gasoline is used to propel the car. The remaining 70% is lost as friction and waste heat. However, no process can be made 100 percent efficient. Even in living things, which are among the most efficient users of free energy, processes are seldom more than 70% efficient.

Another complication, in addition to energy efficiency, is that energy can be obtained from a reaction only if the reaction actually, rather than theoretically, occurs. In other words, although a balanced equation can be written for a chemical reaction, the reaction may not actually take place. For example, you can write an equation for the decomposition of carbon dioxide to carbon and oxygen.

$$CO_2(g) \longrightarrow C(s) + O_2(g)$$

This equation, which represents the reverse of combustion, is balanced. However, experience tells you that the reaction represented by the equation does not tend to occur. Carbon and oxygen react to form carbon dioxide, not the reverse. So the world of balanced chemical equations is really divided into two groups. One group contains equations representing reactions that actually occur; the other contains equations representing reactions that do not tend to occur, or at least not efficiently.

Section Resources

Print

- **Guided Reading and Study Workbook,** Section 18.4
- **Core Teaching Resources,** Section 18.4 Review, Interpreting Graphics
- **Transparencies,** T206–T208

Technology

- **Interactive Textbook with ChemASAP,** Assessment 18.4
- **Go Online,** Section 18.4
- **Virtual Chemistry Labs,** Lab 20

The first, and more important group involves processes that are spontaneous. A **spontaneous reaction** occurs naturally and favors the formation of products at the specified conditions. **Spontaneous reactions produce substantial amounts of products at equilibrium and release free energy.** The colorful fireworks display shown in Figure 18.20 is an example of a spontaneous reaction. In contrast, a **nonspontaneous reaction** is a reaction that does not favor the formation of products at the specified conditions. Nonspontaneous reactions do not give substantial amounts of products at equilibrium.

Figure 18.20 Fireworks displays are the result of highly favored spontaneous reactions. A large quantity of free energy is released.

Consider again the reversible decomposition of carbonic acid in water.

$$\underset{<1\%}{H_2CO_3(aq)} \rightleftharpoons \underset{>99\%}{CO_2(g) + H_2O(l)}$$

Carbonic acid is the reactant in the forward reaction. If you could start with pure carbonic acid in water and let the system come to equilibrium, more than 99% of the reactant would be converted to the products carbon dioxide and water. These products are highly favored at equilibrium. The natural tendency is for carbonic acid to decompose to carbon dioxide and water. Thus the forward reaction is spontaneous and releases free energy. In the reverse reaction, carbon dioxide and water are the reactants, and carbonic acid is the product. If you permit a solution of carbon dioxide in water to come to equilibrium, less than 1% of the reactants combines to form carbonic acid. The reactants show little natural tendency to go to products. Thus the combination of carbon dioxide and water to form carbonic acid is nonspontaneous. In nearly all reversible reactions, one reaction is favored over the other.

Another example, shown in Figure 18.21, is the spontaneous reaction of aqueous cadmium nitrate with aqueous sodium sulfide to produce aqueous sodium nitrate and solid yellow cadmium sulfide. Cadmium sulfide is highly favored in this equilibrium. The reverse reaction, the production of cadmium nitrate and sodium sulfide from cadmium sulfide and sodium nitrate, is nonspontaneous.

$$Cd(NO_3)_2(aq) + Na_2S(aq) \rightleftharpoons CdS(s) + 2NaNO_3(aq)$$

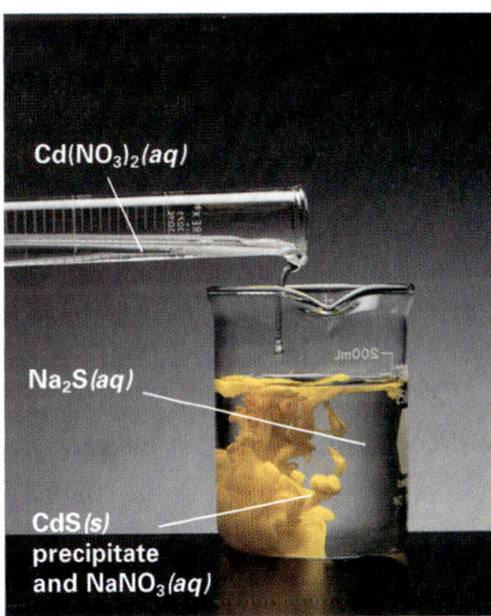

Figure 18.21 A precipitate of cadmium sulfide (CdS) forms spontaneously when solutions of sodium sulfide (Na_2S) and cadmium nitrate ($Cd(NO_3)_2$) are mixed. **Inferring** ***Is free energy released in this reaction?***

Free Energy and Spontaneous Reactions

Discuss L2

Discuss what is meant by work. Point out that work is done on a gas when a force is applied to reduce its volume. Work is done by rapidly expanding gases such as those produced in an explosion. Ask, **Is work done on molecules when the temperature is raised or when chemical bonds are broken?** *(yes, in both cases)* **How is energy related to work?** *(Work and energy have the same units.)*

Use Visuals L1

Figure 18.20 Explain that a fireworks display is the result of highly spontaneous reactions that release a large quantity of free energy. Also explain that this reaction requires a very high initiation temperature to produce the fast reaction rate observed.

Relate L2

Commercially, chemical reactions are usually carried out to manufacture a chemical substance or to produce energy. The commercial value of a chemical reaction depends on the conditions under which it occurs spontaneously and on the reaction rate. Emphasize that all spontaneous reactions release free energy.

Discuss L2

Students may think that a nonspontaneous reaction is one in which no products are formed. Remind them that, in a reversible reaction under a given set of conditions, one reaction is always spontaneous and the other reaction is nonspontaneous. Therefore, the difference between spontaneous and nonspontaneous reactions is the *amount* of products at equilibrium.

Answers to...

Figure 18.21 Free energy is released because the reaction is spontaneous.

Relate 

Have students consider the combustion of glucose to produce carbon dioxide and water. Point out that the reverse of this reaction is the overall reaction for photosynthesis, the process by which plants make glucose using carbon dioxide and water. Ask students to consider the magnitude of the free-energy value for the photosynthesis reaction, +2868 kJ/mol. They should realize that the reaction is nonspontaneous under ordinary conditions.

CLASS Activity

Research Photosynthesis 

Have students research the process of photosynthesis and prepare a report explaining how it occurs. *(Bright light provides the energy needed for this highly endothermic reaction. The reaction also requires chlorophyll, a molecule that can capture light energy.)*

Entropy

Download a worksheet on **Entropy** for students to complete, and find additional teacher support from NSTA SciLinks.

Figure 18.22 Many nonspontaneous reactions are required for a plant to grow. Nonspontaneous reactions can occur when they are coupled with spontaneous reactions. **Inferring** ***What is another requirement for this photosynthesis reaction?***

$$6CO_2 + 6H_2O \xrightarrow{\text{light energy}} C_6H_{12}O_6 + 6O_2$$

It is important to note that the terms *spontaneous* and *nonspontaneous* do not refer to how fast reactants go to products. Some spontaneous reactions go so slowly that they appear to be nonspontaneous. The reaction of sugar and oxygen, for example, produces carbon dioxide and water. However, a bowl of sugar on a table does nothing, and you might assume that the equilibrium among sugar, oxygen, carbon dioxide, and water greatly favors the sugar and oxygen. The equilibrium does favor the products, but at room temperature the reaction is so slow that it takes thousands of years to reach equilibrium. When you supply energy in the form of heat, however, the reaction is fast. Then it is obvious that the formation of carbon dioxide and water is highly favored.

Some reactions that are nonspontaneous at one set of conditions may be spontaneous at other conditions. Changing the temperature or pressure, for example, may determine whether or not a reaction is spontaneous. The nonspontaneous photosynthesis reaction summarized in Figure 18.22 could not be driven to completion without the energy supplied by sunlight.

Sometimes a nonspontaneous reaction can be made to occur if it is coupled to a spontaneous reaction—a reaction that releases free energy. Coupled reactions are a common feature of the complex biological processes that take place in living organisms. Within cells, a series of spontaneous reactions release the energy stored in glucose. As you will learn in Chapter 24, molecules in cells can capture and transfer free energy to nonspontaneous reactions, such as the formation of proteins.

 ***If a reaction releases free energy, is the reaction spontaneous or nonspontaneous?***

For: Links on Entropy
Visit: www.SciLinks.org
Web Code: cdn-1184

Entropy

Recall that heat (enthalpy) changes accompany most chemical and physical processes. The combustion of carbon, for example, is exothermic and spontaneous. The heat released during this reaction is 393.5 kJ for each mole of carbon (graphite) burned.

$$C(s, \text{graphite}) + O_2(g) \longrightarrow CO_2(g) + 393.5 \text{ kJ/mol}$$

568 Chapter 18

You might expect that only exothermic reactions would be spontaneous. Some processes, however, may be spontaneous even though they absorb heat. For example, consider the physical process of melting ice to water. As it turns from a solid to a liquid, 1 mol of ice at 25°C absorbs 6.0 kJ of heat from its surroundings.

$$H_2O(s) + 6.0\ \text{kJ/mol} \longrightarrow H_2O(l)$$

Ice melts spontaneously at 25°C, even though the water produced contains more heat than the ice from which it forms. Considering only enthalpy changes, the energy of the water seems higher than the energy of the ice. That would seem to violate the rule that in spontaneous processes, the direction of the change in energy is from higher energy to lower energy; that is, free energy is released. Yet this process occurs anyway. Some factor other than the enthalpy change must help determine whether a physical or chemical process is spontaneous.

The other factor is related to order. You are probably familiar with everyday ideas about order and disorder. For example, a handful of marbles is relatively ordered in the sense that all the marbles are collected in one place. It is improbable that when permitted to fall, the marbles will end up in the same neat arrangement. Instead, the marbles scatter on the ground. They become disordered. Scattered marbles have a higher entropy than a handful of marbles. **Entropy** is a measure of the disorder of a system.

The concept that physical and chemical systems attain the lowest possible energy has a companion idea called the **law of disorder,** which states that the natural tendency is for systems to move in the direction of maximum disorder or randomness. You already know something about this natural tendency toward disorder. Maybe the photographs in Figure 18.23 look familiar to you. More than likely, your bedroom is neat and clean at the beginning of the week. But unless you put energy (work) into maintaining it, your room becomes messy by the end of the week.

The law of disorder (entropy change) also operates at the level of atoms and molecules and so it is a factor in determining the direction of chemical reactions. **An increase in entropy favors the spontaneous chemical reaction; a decrease favors the nonspontaneous reaction.** Figure 18.24 on the following page illustrates some generalities that will help you to predict the course of many reactions.

Figure 18.23 The entropy of a recently cleaned and orderly room is low. However, over time a room tends to become disorderly. Entropy has a tendency to increase.
Applying Concepts ***Which has the greater entropy, an assembled jigsaw puzzle, or the pieces in the box?***

Use Visuals 

Figure 18.23 Explain that entropy is a measure of the disorder in a system. Ask, **Which bedroom has higher entropy?** *(the messy room)* Ask students to describe other human-made or natural environments that can become disordered over time. *(possible examples: weeds in a garden, books on library shelves, a table of sweaters on a sale day, sand castles or ice sculptures, the contents of a house after a flood or earthquake)*

Discuss 

Explain that in a game of chance, the probable event can happen in many ways. The same is true of order and disorder. An ordered arrangement conforms to some chosen requirements; all other arrangements are called disordered. Because far more arrangements are possible in the latter category, disorder is the natural condition of things.

Facts and Figures

Origin of *Entropy*

The term *entropy* was coined by the German physicist Rudolf Clausius in 1865. He used the term to describe the portion of free energy that is unavoidably lost to the environment as waste heat. Clausius pointed out that in any process involving the flow of energy, there is always some loss, so that the entropy of the universe is continually increasing.

Answers to...

Figure 18.22 light energy
Figure 18.23 the pieces in the box

 spontaneous

Section 18.4 (continued)

Discuss 

Review the different situations that increase entropy. Ask students for everyday examples in which entropy increases or decreases, for example, fans entering a stadium and fans seated during a game.

TEACHER Demo

The Entropy of Water 

Purpose Students infer the relative magnitude of the entropy of water in different states.

Materials block of ice, beaker of water, rapidly boiling water, hot plate

Procedure Place a block of ice, a beaker of water, and a beaker of rapidly boiling water in front of the classroom. Ask, **Which item has the least entropy?** *(block of ice)* **The most?** *(boiling water)* **Why?** *(In solid ice the water molecules are highly ordered in a lattice. In liquid water, despite the hydrogen bonding between molecules, the molecules are relatively free to move around. In the water vapor produced by boiling water, randomness is greatly increased because the water molecules are able to move freely.)*

Figure 18.24 Entropy is a measure of the disorder of a system.

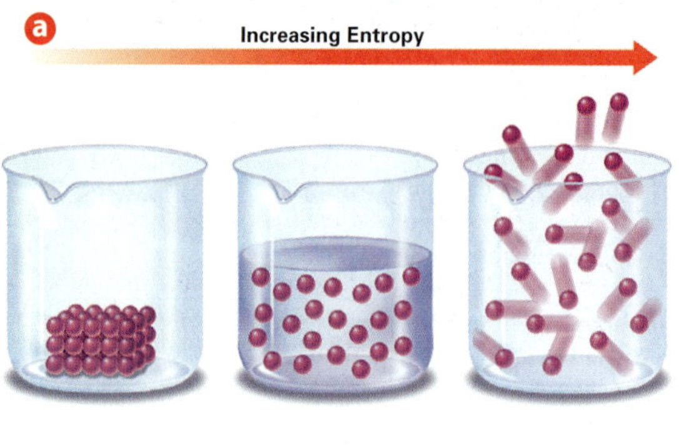

a For a given substance, the entropy of the gas is greater than the entropy of the liquid or the solid. Similarly, the entropy of the liquid is greater than that of the solid. Thus entropy increases in reactions in which solid reactants form liquid or gaseous products. Entropy also increases when liquid reactants form gaseous products.

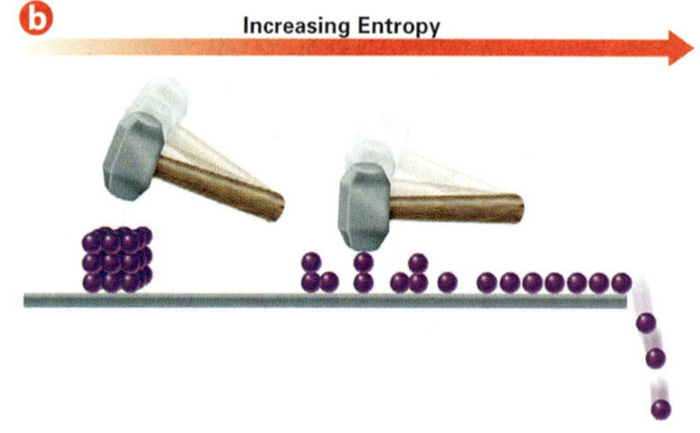

b Entropy increases when a substance is divided into parts. For instance, entropy increases when a crystalline ionic compound, such as sodium chloride, dissolves in water. This is because the solute particles—sodium ions and chloride ions—are more separated in solution than they are in the crystal form.

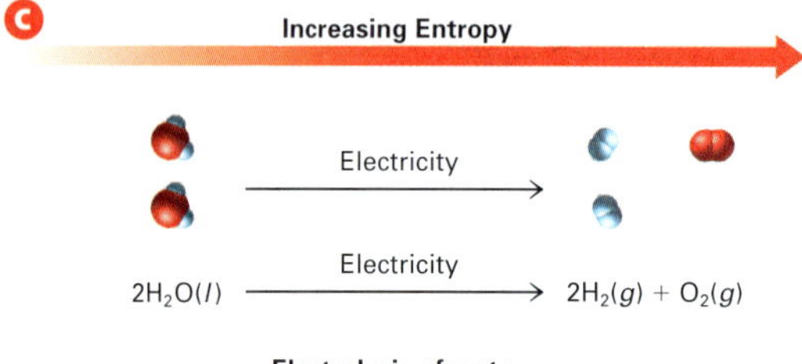

c Entropy tends to increase in chemical reactions in which the total number of product molecules is greater than the total number of reactant molecules.

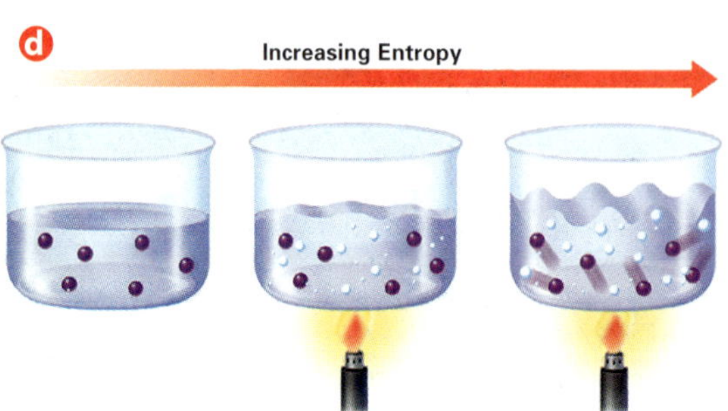

d Entropy tends to increase when temperature increases. As the temperature increases, the molecules move faster and faster, which increases the disorder.

Differentiated Instruction

Gifted and Talented 

Pose the following problem: Entropy decreases in the reaction of iron, air, and water to form rust. Why is this reaction spontaneous? *(An ordered solid (Fe) and a less-ordered gas (O_2) release enough energy to balance the production of a more-ordered solid (Fe_2O_3) in the presence of a catalyst (H_2O).)*

Table 18.3

How Changes in Enthalpy and Entropy Affect Reaction Spontaneity		
Enthalpy change	**Entropy**	**Spontaneous reaction?**
Decreases (exothermic)	Increases (more disorder in products than in reactants)	Yes
Increases (endothermic)	Increases	Only if unfavorable enthalpy change is offset by favorable entropy change
Decreases (exothermic)	Decreases (less disorder in products than in reactants)	Only if unfavorable entropy change is offset by favorable enthalpy change
Increases (endothermic)	Decreases	No

Enthalpy, Entropy, and Free Energy

In every chemical reaction, heat is either released or absorbed and entropy or randomness either increases or decreases. How do these two energy factors determine the course of a reaction? **The size and direction of enthalpy changes and entropy changes together determine whether a reaction is spontaneous; that is, whether it favors products and releases free energy.** An exothermic reaction accompanied by an increase in entropy is definitely spontaneous because both factors are favorable. The combustion of carbon, for example, is exothermic. Entropy also increases as solid carbon is converted to gaseous carbon dioxide. Because both factors are favorable, the reaction must be spontaneous. The reverse reaction—production of carbon and oxygen from carbon dioxide—is nonspontaneous; neither the enthalpy nor the entropy change is favorable.

A reaction is also spontaneous if a decrease in entropy is offset by a large release of heat. Similarly, an endothermic, or heat-absorbing, reaction is spontaneous if an entropy increase offsets the heat absorption. For example, enthalpy change and entropy change work in opposition when ice melts. The melting of ice is endothermic, but the process is still spontaneous above 0°C. At such temperatures, the absorption of heat is more than offset by a favorable entropy change.

Table 18.3 summarizes how entropy and enthalpy changes affect the spontaneity of chemical reactions. Notice that either of the two variables, but not both, can be unfavorable for a spontaneous process.

Nonspontaneous reactions, those in which the products are not favored, can have enthalpy changes, entropy changes, or both working against them. For example, a reaction might be highly endothermic with a large decrease in entropy. In that case, both changes work against the formation of products and the reaction is nonspontaneous. In another case, a reaction may be exothermic but involve a decrease in entropy large enough to offset the favorable enthalpy change. This reaction will also be nonspontaneous. Alternatively, an endothermic reaction could have an increase in entropy too small to overcome the unfavorable enthalpy change. Figure 18.25 on the following page summarizes the relationship among enthalpy, entropy, and free-energy changes for spontaneous reactions.

Checkpoint *What two energy factors determine the course of a reaction?*

Interactive Textbook

Simulation 25 Simulate how enthalpy and entropy changes combine to determine the spontaneity of a reaction.

with ChemASAP

Enthalpy, Entropy, and Free Energy

Relate 

Evidence suggests that Earth is being warmed because of the greenhouse effect, which is largely due to the release into the atmosphere of CO_2 produced by combustion. Another aspect of the greenhouse effect that relates to the topic of spontaneity, is the increase in the quantity of atmospheric CO_2 gas that would occur if the oceans became warmer. CO_2 is quite soluble in water. The oceans currently contain a great deal of dissolved CO_2. If the oceans become warmer, more CO_2 would be released into the atmosphere because CO_2, like all other gases, is less soluble in warm ocean water than in cool ocean water. The process whereby CO_2 dissolves in ocean water would become nonspontaneous, and the reverse process, the release of CO_2 gas, would become spontaneous. The increased temperature would release more CO_2, which would further increase temperature, and so on.

Answers to...

Checkpoint enthalpy and entropy

Gibbs Free Energy

Discuss L2

Free energy refers to more than just the enthalpy change, ΔH, that occurs in a chemical reaction. Free energy combines ΔH with the change in entropy, ΔS. ΔH and ΔS are usually assumed to be constant for the same reactions at different temperatures. Therefore, the contribution to the free energy from the $T\Delta S$ term comes mostly from the temperature itself.

TEACHER Demo

Observing a Spontaneous Reaction 

Purpose Students observe a spontaneous reaction with a negative ΔH and a positive ΔS.

Materials 0.1*M* solution of $KMnO_4$ in 0.1*M* HCl, graduated cylinder, 1000-mL beaker, thermometer, 20 mL of 30% H_2O_2, watch glass, splint, matches

Safety Wear gloves and safety goggles. Potassium permanganate is a powerful oxidizing agent and a strong skin irritant. The manganese ion is considered hazardous waste and must be disposed of in compliance with local regulations.

Procedure Transfer 100 mL of the $KMnO_4$ solution to a 1000-mL beaker. Record its temperature. With a graduated cylinder, measure 20 mL of 30% H_2O_2. (Use a more dilute solution for a less vigorous reaction.) Record its temperature. Add the peroxide to the permanganate and cover the beaker with a watch glass. Record the temperature when the most vigorous reaction has subsided. Test the gas produced with a glowing splint. Note that the reaction is spontaneous, so ΔG must be negative. The entropy change must be positive because a gas is produced. From the free-energy equation ($\Delta G = \Delta H - T\Delta S$) and the temperature change, it is obvious that ΔH must be negative at room temperature. The reaction is

$$2MnO_4^-(aq) + 5H_2O_2(l) + 6H^+(aq) \rightarrow 2Mn^{2+}(aq) + 8H_2O(l) + 5O_2(g).$$

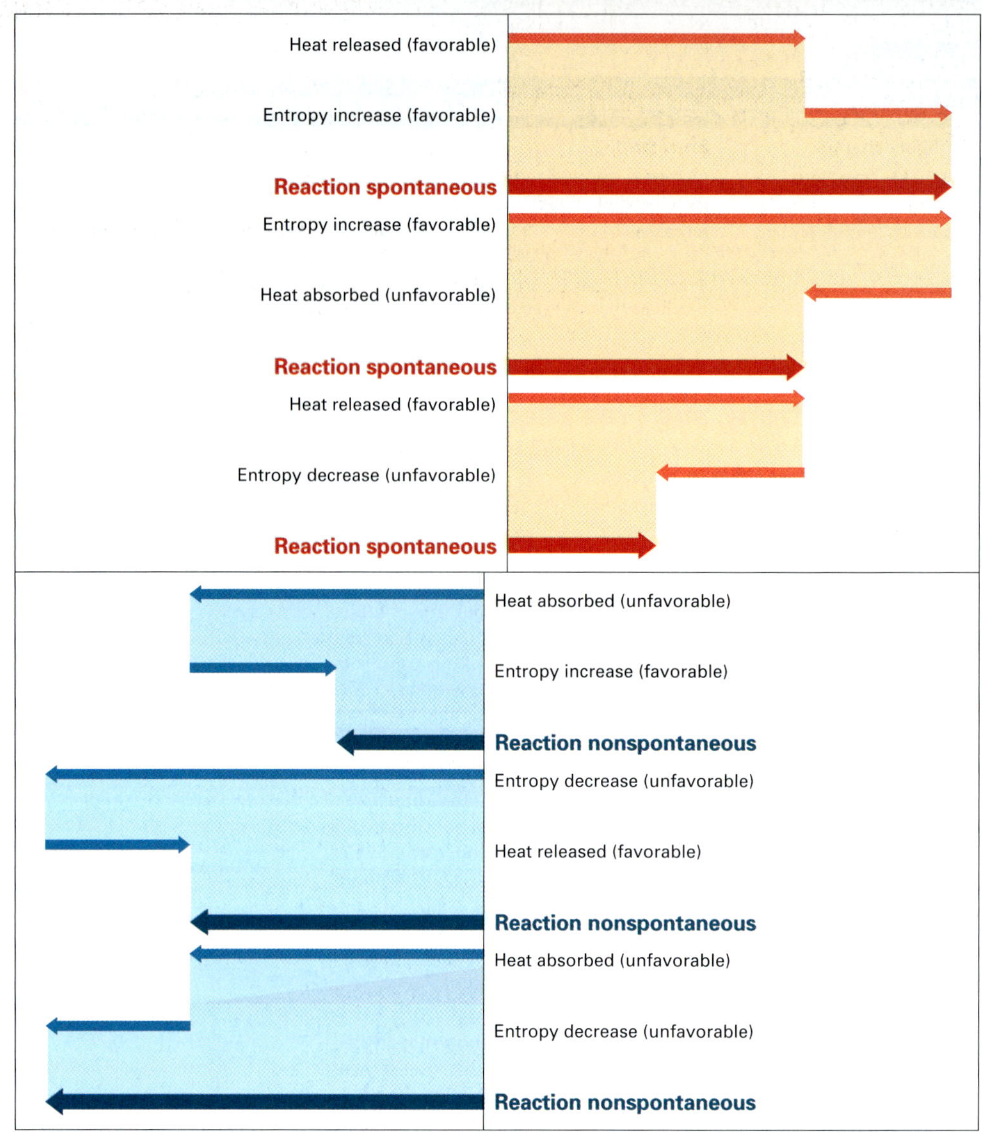

Figure 18.25 The combination of the enthalpy change and the change in entropy for a reaction determines whether or not the reaction is spontaneous.

Gibbs Free Energy

Recall that in every spontaneous process, some energy becomes available to do work. This energy, called the **Gibbs free-energy change,** is the maximum amount of energy that can be coupled to another process to do useful work. The change in Gibbs free energy is related to the change in entropy (ΔS) and the change in enthalpy (ΔH) of the system by the free-energy equation.

$$\Delta G = \Delta H - T\Delta S$$

The temperature (T) is in kelvins.

All spontaneous processes release free energy. **The numerical value of ΔG is negative in spontaneous processes because the system loses free energy.** Nonspontaneous processes require that work be expended to make them go forward at the specified conditions. Therefore, the numerical value of ΔG is positive for a nonspontaneous process. However, a reaction that is nonspontaneous under one set of conditions may be spontaneous under another set of conditions.

To see the effect of temperature on the spontaneity of a reaction, consider this chemical reaction. Solid calcium carbonate decomposes to give calcium oxide and carbon dioxide.

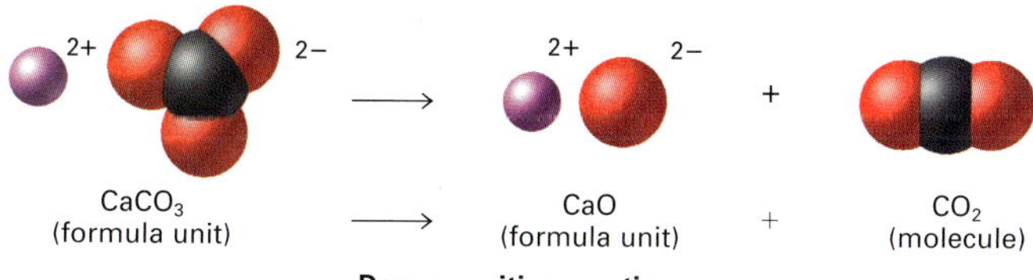

$CaCO_3$ (formula unit) ⟶ CaO (formula unit) + CO_2 (molecule)

Decomposition reaction

In this reaction, the entropy increases because one of the products formed from the solid reactant is a gas. However, the reaction is endothermic and the entropy increase is not great enough to offset the unfavorable enthalpy change. Thus the reaction is nonspontaneous at ordinary temperatures. The effect of an entropy increase, however, is magnified as the temperature increases. At temperatures above 850°C, the favorable entropy-temperature term $T\Delta S$ in the equation $\Delta G = \Delta H - T\Delta S$ outweighs the unfavorable enthalpy term ΔH. Under these conditions, the decomposition of calcium carbonate becomes spontaneous. Calcium oxide obtained from this spontaneous reaction is used to neutralize acid soil, as shown in Figure 18.26.

Figure 18.26 Soil that becomes too acidic for proper plant growth can be neutralized by the addition of lime, which is the common name for calcium oxide (CaO).

18.4 Section Assessment

29. **Key Concept** What are two characteristics of spontaneous reactions?
30. **Key Concept** What part does entropy play in determining whether a chemical reaction is spontaneous or nonspontaneous?
31. **Key Concept** What two factors determine the spontaneity of a reaction?
32. **Key Concept** What is the sign (+ or −) of the Gibbs free-energy change in a spontaneous process?
33. Where does released free energy typically end up? Does free energy lost as heat ever serve a useful function? Explain.
34. Suppose the products in a spontaneous process are more ordered than the reactants. Is the entropy change favorable or unfavorable?
35. Explain why, under certain circumstances, a nonspontaneous reaction becomes spontaneous.

Elements Handbook

Explosives Among this class of compounds are the reactants in rapid, violent reactions. Read about some common explosives on page R26 of the Elements Handbook. Choose a reaction and explain why it fits the definition of a spontaneous reaction.

Assessment 18.4 Test yourself on the concepts in Section 18.4.

with ChemASAP

3 ASSESS

Evaluate Understanding L2

Have students pair up to review what they have learned. One student should explain the significance of positive and negative changes in entropy. Together they should discuss any defects in the explanation. Then the other student should explain how the values of ΔG, ΔH, and ΔS determine whether a reaction is spontaneous. Again the students should evaluate the explanation.

Reteach L1

Have students compare Table 18.3 and Figure 18.25. These visuals contain the information that students need to explain why some reactions are spontaneous and others are not. Ask, **What changes can occur during a reaction to favor spontaneity?** *(changes in which heat is released and entropy increases.)* **Under what conditions is a reaction least likely to be spontaneous?** *(when heat is absorbed and entropy decreases.)* **What happens when one change favors spontaneity and one change does not?** *(The relative strengths of the opposing changes determine the outcome.)* For reactions that lead to equilibrium, the magnitudes of the enthalpy and entropy changes determine the point at which the opposing reactions are balanced.

Elements Handbook

The reaction of nitroglycerine releases 1427kJ of heat, and in the process, converts 1 mol of a liquid to 29 mol of four different gases. Thus both enthalpy and entropy changes favor spontaneity.

Section 18.4 Assessment

29. They produce substantial amounts of products at equilibrium and release free energy.
30. An increase favors a spontaneous reaction; a decrease favors a nonspontaneous reaction.
31. The size and direction of enthalpy and entropy changes together determine whether a reaction is spontaneous.
32. negative (–) sign
33. As heat; it can be useful; for example, waste heat from chemical reactions in the body maintains body temperature at 37°C, and waste heat from industrial processes and electrical generation can be used to heat water and buildings.
34. unfavorable
35. A change in state of a reactant, due to change in temperature or pressure, may change the balance between ΔS and ΔH.

Interactive Textbook

If your class subscribes to the Interactive Textbook, use it to review key concepts in Section 18.4.

with ChemASAP

Enthalpy and Entropy L2

Objective After completing this lab, students will be able to:

- investigate the energy associated with chemical reactions
- relate changes to exothermic and endothermic reactions

Class Time 40 minutes

Materials NaCl, NH_4Cl, $CaCl_2$, alcohol thermometer, 4 1-oz. plastic cups, plastic spoon, crushed ice, water, stirring rod. Optional solids: KCl, $NaHCO_3$, Na_3PO_4

Safety Use small alcohol thermometers. If students are careful, they can use the thermometers to gently stir the contents of the plastic cup. Remind students of the hazard posed by broken glass. Never use mercury thermometers. They could break, releasing their poisonous contents. Look for a research lab that will trade alcohol thermometers for mercury thermometers.

Analyze

1. Sample data for ΔT: **a.** 0°C; **b.** −16°C; **c.** +32°C
2. $NH_4Cl + H_2O$ is endothermic. ΔH is positive. $CaCl_2 + H_2O$ is exothermic. ΔH is negative.
3. $NaCl + H_2O(l)$ ΔH is close to 0.
4. heat + $NH_4Cl(s) \rightarrow NH_4^+(aq) + Cl^-(aq)$
   $CaCl_2(s) \rightarrow Ca^{2+}(aq) + 2Cl^-(aq)$ + heat
5. All of the solids dissolved rapidly. Entropy usually increases during solvation. ΔS is positive in each case.
6. $\Delta G = \Delta H - T\Delta S$
   $\Delta G = (0) - T(+)$ for $NaCl(s)$; ΔG is −.
   $\Delta G = (+) - T(+)$ for $NH_4Cl(s)$; ΔG is + or −.
   $\Delta G = (-) - T(+)$ for $CaCl_2(s)$; ΔG is −.

You're the Chemist

1. The temperature can drop as much as 15°C.
2. The melting of ice is endothermic. Endothermic processes absorb heat and cool the environment.
3. Both $CaCl_2$ and NH_4Cl depress the freezing point and cause a drop in temperature.
4. Salts such as KCl, $NaHCO_3$, and Na_3PO_4 dissolve endothermically, that is, with little or no change in temperature.

Small-Scale LAB

Enthalpy and Entropy

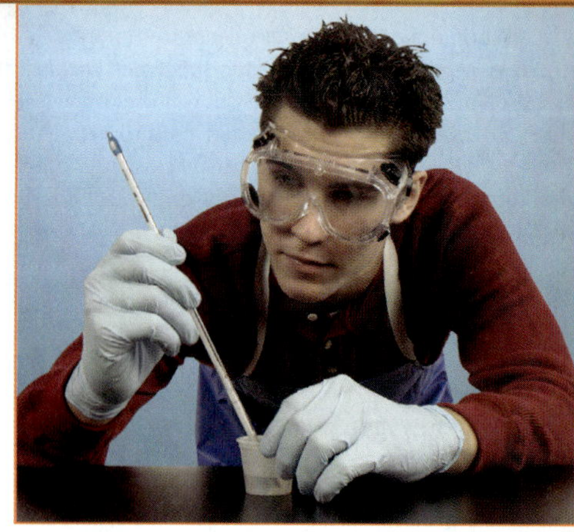

Purpose

To observe and measure energy changes during the formation of a solution and to describe and explain these changes in terms of entropy and enthalpy.

Materials

- alcohol thermometer
- four 1-oz plastic cups
- plastic spoon
- solid chemicals shown in the table
- crushed ice
- water
- stirring rod

Procedure   

Place two level spoonfuls of water in a plastic cup and measure the temperature (T_1) of the water. Add one level spoonful of solid NaCl to the cup, stir gently, and measure the highest or lowest temperature (T_2). Record this information on a separate sheet of paper in a table similar to the one below. Repeat the experiment two more times, using NH_4Cl and $CaCl_2$. Record the temperatures.

Mixture	T_1	T_2	ΔT
a. $NaCl(s) + H_2O(l)$			
b. $NH_4Cl(s) + H_2O(l)$			
c. $CaCl_2(s) + H_2O(l)$			

Analyze

Using your experimental data, record the answers to the following questions in or below your data table.

1. Calculate ΔT for each mixture. $\Delta T = T_2 - T_1$
2. An exothermic process releases heat. An endothermic process absorbs heat. Which solutions are endothermic and which are exothermic? What is the sign of ΔH in each case?
3. Which solution(s) had little or no change in temperature?
4. When sodium chloride dissolves in water, the ions dissociate.

$$NaCl(s) \longrightarrow Na^+(aq) + Cl^-(aq)$$

Write ionic equations, similar to the one above, that describe how NH_4Cl and $CaCl_2$ each dissociate as they dissolve in water. Include heat as a reactant or product in each equation.

5. Which solids in this experiment rapidly dissolved in water? Does the dissolving process usually occur with an increase or decrease in entropy? What is the sign of ΔS in each case?
6. Consider the equation $\Delta G = \Delta H - T\Delta S$. For each dissolving process, substitute the sign of ΔS and ΔH into the equation and determine the sign of ΔG. For which process might ΔG be either positive or negative?

You're The Chemist

The following small-scale activities allow you to develop your own procedures and analyze the results.

1. **Analyze It!** Mix a tablespoon of crushed ice with a tablespoon of solid NaCl. Stir gently, then measure and record the lowest temperature reached. Compare this mixture with the mixture of solid NaCl and liquid water. What is the difference in temperature?
2. **Analyze It!** Is the process of melting ice exothermic or endothermic? Does this explain the difference in temperature in Activity 1? Explain.
3. **Design It!** What effect does mixing other salts with crushed ice have on temperature? Try substituting different salts in Activity 1.
4. **Design It!** What effect does dissolving other salts in liquid water have on temperature? Try dissolving the following salts in liquid water: potassium chloride (KCl), sodium hydrogen carbonate ($NaHCO_3$), and sodium phosphate (Na_3PO_4).

For Enrichment 

Have students use and then dissect a chemical heat pack that uses sodium acetate and stainless steel. Ask, What is the exothermic process here? *(crystallization of sodium acetate)* What does the steel do? *(acts as a catalyst)*

18.5 The Progress of Chemical Reactions

Connecting to Your World

The Tour de France is one of the most famous bicycle races in the world. It is held from mid-July to early August. During the course of the race, cyclists travel almost 4000 kilometers and cross steep mountains at heights of 1600 meters or more. To cross these high mountains requires extra amounts of energy. A chemical reaction can be thought of as a process involving peaks and valleys also. In this section, you will learn about the energy requirements for a chemical reaction to progress from initial reactants to final products.

Guide for Reading

Key Concepts

- What is the general relationship between the value of the specific rate constant, *k*, and the speed of a chemical reaction?
- What do the hills and valleys in a reaction progress curve represent?

Vocabulary

rate law
specific rate constant
first-order reaction
elementary reaction
reaction mechanism
intermediate

Reading Strategy

Monitoring Your Understanding After you read the section, identify what you don't yet understand. Reread the section or ask for help, then write your clarified understanding.

Rate Laws

The rate of a reaction depends in part on the concentrations of the reactants. For a reaction in which reactant A forms product B in one step, you can write a simple equation.

$$\mathrm{A} \longrightarrow \mathrm{B}$$

The rate at which A forms B can be expressed as the change in A (ΔA) with time, where concentration A_1 is the initial concentration of A at time t_1 and concentration A_2 is the concentration of A at a later time, t_2.

$$\text{Rate} = -\frac{\Delta \mathrm{A}}{\Delta t} = -\frac{\text{concentration } \mathrm{A}_2 - \text{concentration } \mathrm{A}_1}{t_2 - t_1}$$

Because A, the reactant, is decreasing, the concentration of A is smaller at a later time than initially and so ΔA will always be negative. The negative sign in the expression is needed to make the rate positive, as all rates must be. The rate of disappearance of A is proportional to the concentration of A.

$$-\frac{\Delta \mathrm{A}}{\Delta t} \propto [\mathrm{A}]$$

The proportionality becomes an equation with the insertion of a constant (k).

$$\text{Rate} = -\frac{\Delta \mathrm{A}}{\Delta t} = k \times [\mathrm{A}]$$

This equation, called a **rate law,** is an expression for the rate of a reaction in terms of the concentration of reactants. The **specific rate constant** (k) for a reaction is a proportionality constant relating the concentrations of reactants to the rate of the reaction. The magnitude of the specific rate constant depends on the conditions of the reaction and is determined experimentally. **The value of the specific rate constant, *k*, is large if the products form quickly; the value is small if the products form slowly.**

Animation 24 Observe the characteristics of a first-order reaction.

with ChemASAP

Section Resources

Print

- **Guided Reading and Study Workbook,** Section 18.5
- **Core Teaching Resources,** Section 18.5 Review
- **Transparencies,** T209–T212
- **Lab Practicals,** Lab 18-1

Technology

- **Interactive Textbook with ChemASAP,** Animation 24, Problem-Solving 18.37, Assessment 18.5
- **Go Online,** Section 18.5

18.5

1 FOCUS

Objectives

18.5.1 Describe the general relationship between the value of the specific rate constant, *k*, and the speed of a chemical reaction.

18.5.2 Interpret the hills and valleys in a reaction progress curve.

Guide for Reading

Build Vocabulary 

Word Parts Write the word *intermediate* on the board. Have students look up the prefix *inter-* (*between*) and the stem *mediate* (*middle*). Then have them write a definition for the word in their own terms. Compare students definitions to the dictionary definition.

Reading Strategy 

Use Prior Knowledge Have students recall the four factors that can influence the rate of a chemical reaction. *(temperature, concentration, particle size, and the presence of a catalyst)* Ask them how the concentration of reactants affects the rate of a chemical reaction. Tell them that they will use this knowledge as they read this section.

2 INSTRUCT

Connecting to Your World

Ask, **How is the Tour de France bicycle race like the reaction-progress curve for a complex reaction?** *(A series of chemical reactions may require more energy at some points [i.e., activation energy] and release energy at other points.)*

Rate Laws

Use Visuals L2

Figure 18.27 Explain that the slope of a straight line is the ratio of the rise (*y*) to the run (*x*). Demonstrate how to calculate the slope for several straight lines. Explain that the slope of the tangent to a curve at any point is the slope of the curve at that point. Discuss how to calculate the rate of reaction by determining the slope of the tangent line for the two points shown on the graph. Have students calculate the initial rate of reaction. *(rate = (10 – 7.5)/(0.5 – 0) = 5 concentration/time units)* Next have students calculate the rate at time = 2.5. *(rate = (5.5 – 2.5)/(3.7 – 1.7) = 1.5 concentration/time units)* Point out that the rate decreases, as expected, as the concentration of reactant A decreases.

Discuss how the effect of concentration on the reaction rate is expressed as a quantitative relationship. Each rate expression contains a rate constant, *k*, which is specific to the reaction. A knowledge of this dependence not only allows prediction of the actual rate, but also gives clues to the step-by-step mechanism by which the reaction occurs. Demonstrate how to set up the rate equation for a chemical reaction such as the formation of water from hydrogen and oxygen.

Interpreting Graphs L2

a. relative concentration of A
b. It decreases.
c. initial rate

Enrichment Question L3

How would the graph look for a first-order reaction that is reversible? *(It would never reach zero concentration.)*

Figure 18.27 The graph shows how the concentration of a reactant changes as a reaction proceeds.

INTERPRETING GRAPHS

a. Navigate What is the dependent variable?
b. Read What happens to the relative concentration of reactant A as time progresses?
c. Interpret The short colored lines illustrate the reaction rates at two distinct points of time. Is the *initial rate* or the *subsequent rate* faster?

The Rate of a First-Order Reaction

Relative concentration of A (0–10); Relative time (1–12); Initial rate; Subsequent rate; ΔA; Δt

The order of a reaction is the power to which the concentration of a reactant must be raised to give the experimentally observed relationship between concentration and rate. In a **first-order reaction,** the reaction rate is directly proportional to the concentration of only one reactant. The conversion of A to B in a one-step reaction is an example of a first-order reaction; the reaction rate is proportional to the concentration of A raised to the first power: $[A]^1 = [A]$. As a first-order reaction progresses, the rate of reaction decreases, as shown in Figure 18.27. This decrease occurs because the concentration of reactant is decreasing. On a graph, the rate ($\Delta A/\Delta t$) at any point equals the slope of the tangent to the curve at that point. For a first-order reaction, a reduction of [A] by one-half reduces the reaction rate by one-half.

In some kinds of reactions, such as double-replacement, two substances react to give products. The coefficients in the general equation for such a reaction are represented by lowercase letters.

$$aA + bB \longrightarrow cC + dD$$

For a one-step reaction of A with B, the rate of reaction is dependent on the concentrations of both A and B.

$$\text{Rate} = k[A]^a[B]^b$$

When each of the exponents *a* and *b* in the rate law equals 1, the reaction is said to be first-order in A, first-order in B, and second-order overall. The overall order of a reaction is the sum of the exponents for the individual reactants.

For any one-step reaction, the experimentally determined exponents in the rate law are the same as the coefficients *a* and *b* in the chemical equation. However, the exponents in the rate law and the coefficients in the equation do not correspond in most real reactions because most reactions are more complex than the one-step reactions used in the examples. The actual order of a reaction must be determined by experiment.

The rate of blooming for the century plant *(Agave americana)* is one time per 10 to 15 years.

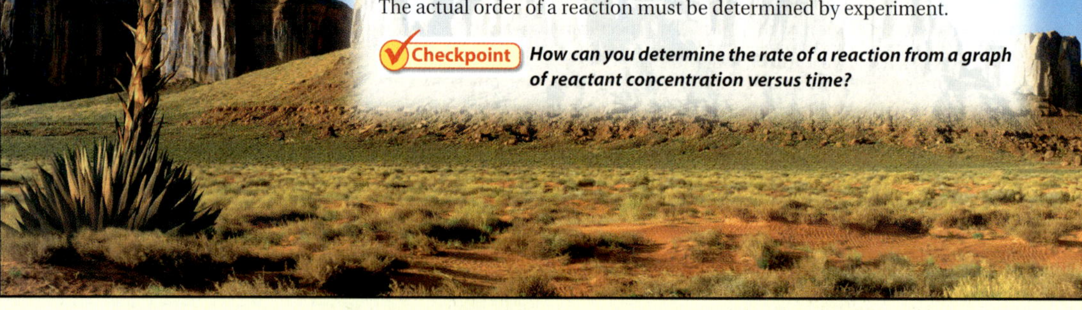

Checkpoint ***How can you determine the rate of a reaction from a graph of reactant concentration versus time?***

CONCEPTUAL PROBLEM 18.2

Finding the Order of a Reaction from Experimental Data

The rate law for the one-step reaction $aA \longrightarrow B$ is of the form: Rate = $k[A]^a$. From the data in the following table, find the order of the reaction with respect to A and the overall order of the reaction.

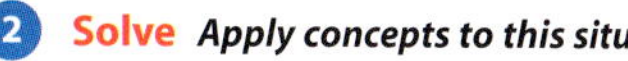

Initial concentration of A (mol/L)	Initial rate (mol/L·s)
0.050	3.0×10^{-4}
0.10	12×10^{-4}
0.20	48×10^{-4}

1 Analyze ***Identify the relevant concepts.***
The initial concentration, the initial rate data, and the form of the rate law are given. This is a one-step reaction. Use these data to find the desired orders by inspection.

2 Solve ***Apply concepts to this situation.***
If a reaction is first-order in A, then $a = 1$ in the rate equation. In that case, the initial rate is directly proportional to the initial [A]. However, the reaction cannot be first-order in A because a doubling of A causes the rate to increase by four times. This suggests that $a = 2$, because $2^2 = 4$. Increasing [A] fourfold should, therefore, increase the rate sixteenfold ($4^2 = 16$). When the initial concentration of A is increased from 0.050 mol/L to 0.20 mol/L, the initial rate increases from 3.0×10^{-4} mol/L·s to 48×10^{-4} mol/L·s, which is a sixteenfold increase. The reaction is thus second-order in A and the reaction must be second-order overall because A is the only reactant.

Practice Problems

36. Show that the unit of k for a first-order reaction is a reciprocal unit of time, such as a reciprocal second (s^{-1}).

37. Suppose a first-order reaction initially proceeds at a rate of 0.5 mol/L·s. What is the rate when half the starting material remains? When one-fourth of the starting material remains?

Interactive Textbook

Problem Solving 18.36 Solve Problem 36 with the help of an interactive guided tutorial.

with ChemASAP

Discuss L2

Point out that most reactions are more complex than the one-step reactions used as examples in the text. In most cases, reaction order cannot be predicted solely on the basis of a balanced equation; it must be determined experimentally. This is because the rate depends primarily on the slowest step in the reaction mechanism.

Conceptual Problem 18.2

Answers

36. Rate = k[A]; rate is moles per liter per second. [A] is moles per liter.
k = rate/[A].
$k = 1/s = s^{-1}$

37. 0.25 mol/L•s; 0.125 mol/L•s

Answers to...

Checkpoint Calculate the slope of the tangent to the curve at any point.

Section 18.5 (continued)

Reaction Mechanisms

Discuss 

Explain that even complex reactions occur as a series of simple steps. Each step usually involves the collision of no more than two different particles. Use the reaction between hydrogen and bromine as an example. The overall reaction $H_2 + Br_2 \rightarrow 2HBr$ occurs as a sequence of three elementary steps.

$Br_2 \rightarrow 2Br$
$Br + H_2 \rightarrow HBr + H$
$H + Br \rightarrow HBr$

Interpreting Graphs L2

a. four
b. three
c. It would have no effect.

Enrichment Question 

How would a catalyst affect this graph? (*It would lower the activation energy for the slowest step in the reaction mechanism.*)

Math Handbook

For a math refresher and practice, direct students to graphing, page R74.

Use Visuals 

Figure 18.28 The energy of a reaction (y-axis), determined experimentally, is plotted versus time or reaction progress (x-axis). The graph shows that the slope changes from positive to negative frequently during the course of the reaction. The usual way to construct such a graph is to measure the energy at several points, most likely at the tops of the energy barriers and at the positions of the intermediate. The remaining points can be interpolated. Displaying the data in a graph provides a more complete picture than a table of data.

CHEMath

Graphing

Relationships between variables are easier to see when data is displayed in a graph. The independent variable is plotted on the *x*-axis (horizontal axis). The dependent variable is plotted on the *y*-axis (vertical axis).

A graph may show a trend. For example, as the value of the independent variable increases, the value of the dependent variable may either increase or decrease.

A straight-line graph with a positive slope indicates a direct proportion between the variables. A smooth curve dropping from left to right shows that two variables are inversely proportional.

Math Handbook

For help with making and interpreting graphs, go to page R74.

Reaction Mechanisms

If enough information were available, you could graph all the energy changes that occur as reactants are converted to products in a chemical reaction. Such a graph would constitute a reaction progress curve. The simplest reaction progress curve is for an elementary reaction, such as the one shown in Figure 18.5. An **elementary reaction** is a reaction in which reactants are converted to products in a single step. Such a reaction has only one activation-energy peak between reactants and products and thus only one activated complex. Most chemical reactions consist of a number of elementary reactions. The series of elementary reactions or steps that take place during the course of a complex reaction is called a **reaction mechanism.** The reaction progress curve for a complex reaction resembles a series of peaks and valleys, as you can see in Figure 18.28. **The peaks correspond to the energies of the activated complexes. Each valley corresponds to the energy of an intermediate.** An **intermediate** is a product of one of the steps in the reaction mechanism. It becomes a reactant in the next step. Intermediates have a significant lifetime compared with an activated complex. They have real ionic or molecular structures and some stability. They are reactive enough, however, to take part in the next step of the reaction mechanism.

Intermediates do not appear in the overall chemical equation for a reaction. The reaction mechanism for the decomposition of nitrous oxide (N_2O), for example, is believed to occur in two elementary steps:

$$N_2O(g) \longrightarrow N_2(g) + O(g)$$
$$N_2O(g) + O(g) \longrightarrow N_2(g) + O_2(g)$$
$$2N_2O(g) \longrightarrow 2N_2(g) + O_2(g)$$

Notice that in the first elementary step, oxygen atoms are formed. These are an intermediate product in this step of the mechanism. They are a reactant in the second step, and they disappear when the two elementary reactions are summed to give the final chemical equation. This illustrates an important general point: The overall chemical equation for a complex reaction gives no information about the reaction mechanism. Reaction mechanisms must be determined experimentally.

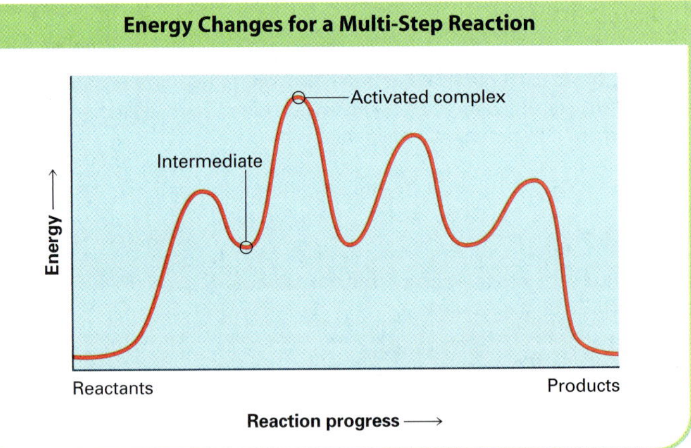

Figure 18.28 A reaction progress curve shows an activation-energy peak for each elementary reaction. Valleys indicate the formation of intermediates.

INTERPRETING GRAPHS

a. Navigate How many elementary reactions are part of this reaction?
b. Read How many intermediates are formed?
c. Interpret How would a catalyst affect the energy of the reactants and the energy of the products of this reaction?

Facts and Figures

Ozone

Ozone is an essential gas in the stratosphere. It absorbs UV radiation from the sun, keeping the UV levels at Earth's surface below a lethal level. The Chapman cycle describes the equilibrium of O_2 and O_3 in the stratosphere. In the first step, O_2 + solar radiation $\rightarrow$ 2O. Then, O + O_2 combine to make O_3. Natural levels of nitrogen oxides react with O and O_3 to keep O_3 at an equilibrium of 6 ppm. Human-generated chemicals such as chlorinated fluorocarbons change the equilibrium, so that less ozone is produced. UV radiation levels at Earth's surface go up as a result.

Careers *in* Chemistry

Professor of Chemistry

If you study chemistry in college, your teacher probably will be a Professor of Chemistry. Chemistry professors train the next generation of chemists. In addition to their classroom teaching, they often undertake fundamental research projects to learn more about the how and why of chemical interactions.

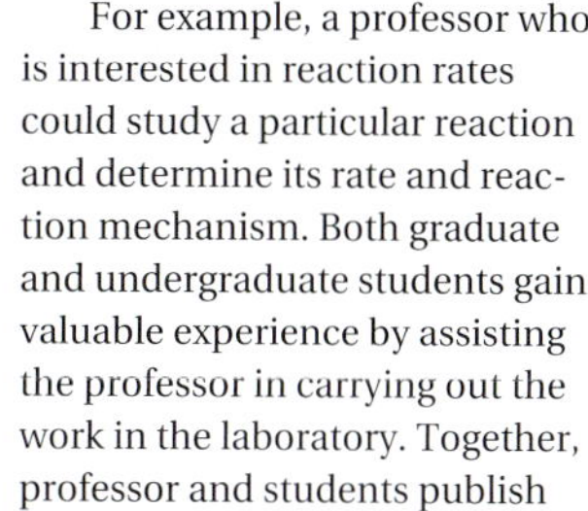

For example, a professor who is interested in reaction rates could study a particular reaction and determine its rate and reaction mechanism. Both graduate and undergraduate students gain valuable experience by assisting the professor in carrying out the work in the laboratory. Together, professor and students publish the results of their research in scientific journals. Much of the basic research reported in journals is done in university laboratories.

Many chemistry professors specialize in one kind of chemistry, for example, inorganic chemistry, organic chemistry, biochemistry, or analytical chemistry. As experts in their particular field, they may share what they know by writing books.

A new member of the Chemistry faculty might start at the level of Instructor. The usual steps in advancement are Assistant Professor, Associate Professor, and then Professor. The minimum qualification for Instructor is usually a PhD in chemistry. Often candidates for a position also have completed a number of years of post-doctoral research at a university.

Go Online PHSchool.com

For: Careers in Chemistry
Visit: PHSchool.com
Web Code: cdb-1185

18.5 Section Assessment

38. **Key Concept** How does the size of the specific rate constant (k) for a reaction relate to the speed of the reaction?
39. **Key Concept** What do the hills and valleys in a reaction progress curve represent?
40. Define each of the following terms as applied to chemical reactions.
    a. elementary reaction
    b. intermediate
    c. reaction mechanism
    d. activation energy
41. Show that the unit of k for a reaction that is second-order in A and second-order overall can be expressed in L/(mol·s).
42. The rate law for this reaction is first-order in NO and O_3, and second-order overall.
$$NO(g) + O_3(g) \longrightarrow NO_2(g) + O_2(g)$$
Write the complete rate law for this reaction.

Writing Activity

Write a Report Because of its potential impact on human health and agriculture, ozone depletion in the stratosphere is an area of continuing scientific interest and study. Research and write a report on the reaction mechanisms for the depletion of ozone in the stratosphere caused by the presence of CFCs.

Interactive Textbook

Assessment 18.5 Test yourself on the concepts in Section 18.5.

with ChemASAP

Section 18.5 Assessment

38. the larger the specific rate constant k, the faster the reaction

39. Peaks correspond to energies of activated complexes. Valleys correspond to energies of intermediates.

40. **a.** converts reactants to products in a single step
b. a product of one reaction that becomes the reactant in a second reaction
c. the sequence of elementary reactions of a complex reaction
d. minimum energy required for conversion of a reactant to product (or product to reactant) in an elementary reaction

41. Rate = k[A][B] and rates are moles per liter per second. [A] and [B] are moles per liter. k = rate/[A][B]
$k = 1/M$•s or L/mol•s

42. rate = $k[NO][O_3]$

Careers *in* Chemistry

Professor of Chemistry

Different universities and colleges emphasize different parts of a professor's job, although all schools expect a professor to teach. Point out that department web sites at universities often report on the research professors are doing and what their graduate students are working on.

3 ASSESS

Evaluate Understanding L2

Ask students to construct a reaction rate equation for the production of NO_2F from NO_2 and F_2. The two-step reaction mechanism is

$NO_2 + F_2 = NO_2F + F$ (slow)
$NO_2 + F = NO_2F$ (fast)
$2\,NO_2 + F_2 = 2NO_2F$

Because the first step is the slower, it is the rate-determining step. The rate law is rate = $k[NO_2][F_2]$.

Reteach L1

Have students note the difference between rate determination and the calculation of an equilibrium constant, which can be determined from the overall reaction. Stress that there is no correlation between the equilibrium constant and the rate of a chemical reaction. Remind students that many familiar reactions have multiple steps. Ask them to suggest possible mechanisms for the following reactions:

$C + O_2 \rightarrow CO_2$
$2H_2 + O_2 \rightarrow 2H_2O$
$CH_4 + 2O_2 \rightarrow CO_2 + 2H_2O$

Writing Activity

Solar radiation in the stratosphere breaks CF_2Cl_2 into active substances Cl and CCl_2F. These react in several ways with O_3 to produce O_2.

Interactive Textbook

If your class subscribes to the Interactive Textbook, use it to review key concepts in Section 18.5.

with ChemASAP

CHAPTER 18 Study Guide

Study Tip

Switch Tasks When you feel yourself losing focus, switch the type of task you are working on, the subject that you are studying, or the environment that you are in. Take a break and walk around a bit. Stop studying when you are no longer being productive.

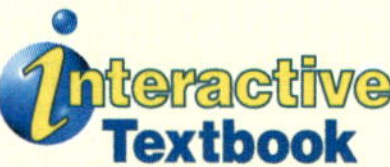

If your class subscribes to the Interactive Textbook, your students can go online to address an interactive version of the Student Edition and a self-test.

with ChemASAP

CHAPTER 18 Study Guide

Key Concepts

18.1 Rates of Reaction

- The rate of a chemical reaction is usually expressed as the change in the amount of a reactant per unit time.
- Factors that influence the rate of a chemical reaction are temperature, concentration, particle size, and the use of a catalyst.

18.2 Reversible Reactions and Equilibrium

- Chemical equilibrium is a state in which the forward and reverse reactions take place at the same rate. At chemical equilibrium, no net change occurs in the system.
- Stresses that can alter the equilibrium position of a reaction are changes in the concentrations of reactants or products, changes in temperature, and changes in pressure.
- A K_{eq} value greater than 1 means that products are favored over reactants; a K_{eq} less than 1 means that reactants are favored over products.

18.3 Solubility Equilibrium

- The smaller the numerical value of the solubility product constant, the lower the solubility of the compound.
- A precipitate will form if the product of the concentrations of two ions in a mixture is greater than the K_{sp} of the compound formed from the ions.

18.4 Entropy and Free Energy

- Spontaneous reactions produce substantial amounts of product and release free energy.
- An increase in entropy favors the spontaneous reaction; a decrease favors the nonspontaneous reaction.
- The size and direction of enthalpy changes and entropy changes together determine whether a reaction is spontaneous.
- The numerical value of ΔG is negative in spontaneous processes because the system loses free energy.

18.5 The Progress of Chemical Reactions

- The value of the specific rate constant, k, in the rate equation is large if products form quickly from reactants.
- Graphs of the progress of a reaction show peaks that correspond to the energies of activated complexes and valleys that correspond to the energies of intermediates and products.

Vocabulary

- activated complex (p. 544)
- activation energy (p. 543)
- chemical equilibrium (p. 550)
- collision theory (p. 542)
- common ion (p. 563)
- common ion effect (p. 563)
- elementary reaction (p. 578)
- entropy (p. 569)
- equilibrium constant (p. 556)
- equilibrium position (p. 551)
- first-order reaction (p. 576)
- free energy (p. 566)
- Gibbs free-energy change (p. 572)
- inhibitor (p. 547)
- intermediate (p. 578)
- law of disorder (p. 569)
- Le Châtelier's principle (p. 552)
- nonspontaneous reaction (p. 567)
- rate (p. 542)
- rate law (p. 575)
- reaction mechanism (p. 578)
- reversible reaction (p. 549)
- solubility product constant (p. 561)
- specific rate constant (p. 575)
- spontaneous reaction (p. 567)
- transition state (p. 544)

Key Equations

- $K_{eq} = \frac{[C]^c \times [D]^d}{[A]^a \times [B]^b}$
- $\Delta G = \Delta H - T\Delta S$
- $\text{rate} = k[A]^a[B]^b$

Chapter Resources

Print

- **Core Teaching Resources,** Chapter 18 *Practice Problems, Vocabulary Review, Quiz, Chapter Test A, Chapter Test B*

Technology

- **Computer Test Bank,** Chapter 18 Test
- **Interactive Textbook with ChemASAP,** Chapter 18

CHAPTER 18

Assessment

Reviewing Content

18.1 Rates of Reaction

43. Explain the collision theory of reactions.

44. How is the activation energy of a reaction like a wall or barrier?

45. How is the rate of a reaction influenced by a catalyst? How do catalysts make this possible?

46. Which of these statements is true?
- **a.** Chemical reactions can be slowed down by increasing the temperature.
- **b.** Once a chemical reaction gets started, the reacting particles no longer have to collide for products to form.
- **c.** Enzymes are biological catalysts.

47. When the gas to a stove is turned on, the gas does not burn unless lit by a flame. Once lit, however, the gas burns until turned off. Explain these observations in terms of the effect of temperature on reaction rate.

18.2 Reversible Reactions and Equilibrium

48. In your own words, define a reversible reaction.

49. How do the rates of the forward and reverse reactions compare at a state of dynamic chemical equilibrium?

50. What is Le Châtelier's principle? Use it to explain why carbonated drinks go flat when their containers are left open.

51. Write the expression for the equilibrium constant for each reaction.
- **a.** $4H_2(g) + CS_2(g) \rightleftharpoons CH_4(g) + 2H_2S(g)$
- **b.** $PCl_5(g) \rightleftharpoons PCl_3(g) + Cl_2(g)$

52. Comment on the favorability of product formation in each reaction.
- **a.** $H_2(g) + F_2(g) \rightleftharpoons 2HF(g)\ K_{eq} = 1 \times 10^{13}$
- **b.** $SO_2(g) + NO_2(g) \rightleftharpoons NO(g) + SO_3(g)\ K_{eq} = 1 \times 10^{2}$
- **c.** $2H_2O(g) \rightleftharpoons 2H_2(g) + O_2(g)\ K_{eq} = 6 \times 10^{-28}$

18.3 Solubility Equilibrium

53. Write the solubility product expression for each salt.
- **a.** NiS
- **b.** $BaCO_3$

54. What does the solubility product constant (K_{sp}) represent?

55. Use Table 18.2 to rank these salts from most soluble to least soluble.
a. CuS **b.** $BaSO_4$ **c.** $SrCO_3$ **d.** AgI

56. How does the addition of a common ion affect the solubility of another substance?

18.4 Entropy and Free Energy

57. How does the free energy of a reaction help predict whether the reaction will be spontaneous?

58. What is the meaning of entropy?

59. The products in a spontaneous process are more ordered than the reactants. Is this entropy change favorable or unfavorable?

60. Which system has the lower entropy?
- **a.** 50 ml of liquid water or 50 ml of ice
- **b.** 10 g of sodium chloride crystals or a solution containing 10 g of sodium chloride

61. Predict the direction of the entropy change in each reaction.
- **a.** $CaCO_3(s) \longrightarrow CaO(s) + CO_2(g)$
- **b.** $NH_3(g) + HCl(g) \longrightarrow NH_4Cl(s)$

62. Is it true that all spontaneous processes are exothermic? Explain your answer.

63. At normal atmospheric pressure, steam condenses to liquid water even though this is an unfavorable entropy change. Explain.

64. What two factors together determine whether a reaction is spontaneous?

18.5 The Progress of Chemical Reactions

65. What is meant by each term?
- **a.** specific rate constant
- **b.** first-order reaction
- **c.** rate law

66. Half of the reactant in a first-order reaction disappears in 50 minutes. How many minutes are required for the reaction to be 75% complete?

67. Sketch a reaction progress curve for the overall reaction with the following mechanism.

$$2NO(g) \longrightarrow N_2O_2(g)\ \text{(fast)}$$
$$N_2O_2(g) + O_2(g) \longrightarrow 2NO_2(g)\ \text{(slow)}$$

Write the balanced equation for the overall reaction.

CHAPTER 18

Assessment

Reviewing Content

43. Chemical reactions require collisions with sufficient energy to break and form bonds.

44. Reactant particles must have a certain minimum amount of energy to react to form product, just as it takes a certain amount of energy to climb over a wall or barrier.

45. A catalyst increases the rate of reactions by providing an alternative reaction mechanism with a lower activation energy.

46. c

47. Gas molecules and oxygen molecules do not have enough energy to react at room temperature. The flame raises the temperature and the energy of collisions, so the reaction begins. The heat released by the reaction maintains the high temperature, and the reaction continues spontaneously.

48. In a reversible reaction, reactants are continually forming products and products are continually forming reactants.

49. They are equal.

50. A system in dynamic equilibrium changes to relieve a stress applied to it. Carbonated drinks in closed containers are in a state of dynamic equilibrium between the CO_2 in the liquid and above it. When containers are opened, CO_2 gas above the liquid escapes. The gas bubbles out of the liquid in an attempt to reestablish equilibrium.

51. **a.** $K_{eq} = \dfrac{[H_2S]^2 \times [CH_4]}{[H_2]^4 \times [CS_2]}$ **b.** $K_{eq} = \dfrac{[PCl_3] \times [Cl_2]}{[PCl_5]}$

52. **a.** highly favorable **b.** favorable **c.** highly unfavorable

53. **a.** $[Ni^{2+}][S^{2-}]$ **b.** $[Ba^{2+}][CO_3^{\,2-}]$

54. the product of the ion concentrations raised to the power of their coefficients

55. c, b, d, a

56. lowers the solubility

57. A spontaneous reaction has a negative free energy.

58. a measure of disorder in a system

59. unfavorable

60. **a.** ice **b.** sodium chloride crystals

61. **a.** entropy increases **b.** entropy decreases

62. No; some endothermic processes are spontaneous because of their favorable change in entropy.

63. The favorable exothermic change of the condensation process offsets the unfavorable entropy change.

64. enthalpy change and entropy change

65. **a.** proportionality constant that relates concentrations of reactants to rate of reaction **b.** reaction rate directly proportional to concentration of one reactant **c.** expression of rate of reaction in terms of concentrations of reactants

66. 100 minutes

67.

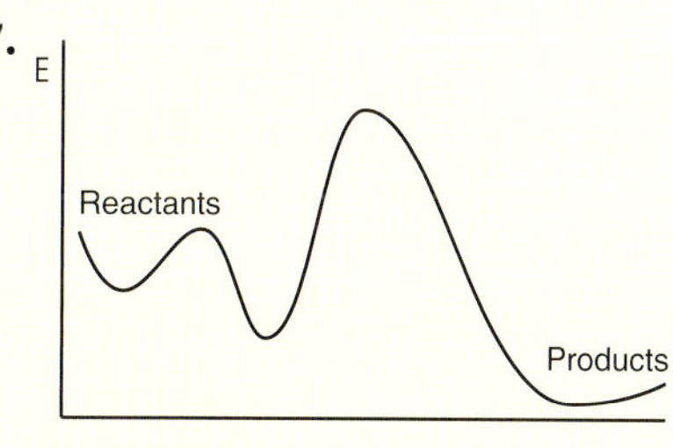

$2NO + O_2 \rightleftharpoons 2NO_2$

CHAPTER 18 Assessment continued

Understanding Concepts

68. a. 1.22×10^{-4} mol/(L•s)
b. 1.94×10^{-4} mol/(L•s)

69. c

70. 25 kJ

71. The change from **a** to **b** is spontaneous, favored by an increase in entropy. The change from **b** to **c** will not occur, because it would result in a decrease in entropy, causing the process to be nonspontaneous.

72. a. yes; a solid becomes a gas
b. yes; A single solid becomes two products, one of them a gas.
c. no; Two particles become one.
d. yes; One particle becomes two.

73. increase in products

74. $K_{eq} = \dfrac{[SO_3]^2}{[SO_2]^2 \times [O_2]}$

75. $K_{eq} = 6.59 \times 10^{-1}$

76. Increasing pressure tends to reduce volume and increase density, so the system responds by favoring production of liquid, which is more dense.

77. The product of the concentrations of the ions must be greater than the ion-product constant (K_{sp}).

78. $3.0 \times 10^{-5} M$

79.

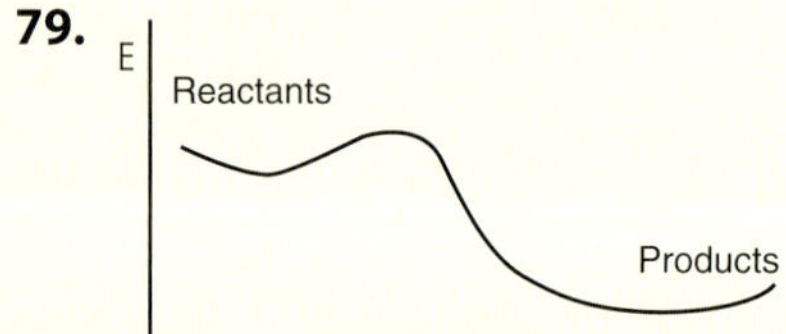

80. A catalyst will establish equilibrium more quickly, but it does not change the equilibrium position.

81. a. IO^-
b. No; the I^- is changed in the reaction. A catalyst would not appear in the reaction as a reactant, an intermediate, or a product.
c. two
d. the slow reaction
e. $2H_2O_2 \rightarrow 2H_2O + O_2$

82. $2.0 \times 10^{-8} M$

CHAPTER 18 Assessment continued

Understanding Concepts

68. Consider the decomposition of N_2O_5 in carbon tetrachloride (CCl_4) at 45°C.

$$2N_2O_5(soln) \longrightarrow 4NO_2(g) + O_2(g)$$

The reaction is first order in N_2O_5, with the specific rate constant 6.08×10^{-4}/s. Calculate the reaction rate at these conditions.
a. $[N_2O_5] = 0.200$ mole/liter
b. $[N_2O_5] = 0.319$ mole/liter

69. Which of the following tells you that a reaction is spontaneous?
a. The reaction is exothermic.
b. Entropy is increased in the reaction.
c. Free energy is released in the reaction.

70. For the reaction $A + B \rightleftharpoons C$, the activation energy of the forward reaction is 5 kJ and the total energy change is −20 kJ. What is the activation energy of the reverse reaction?

71. A large box is divided into two compartments with a door between them. Equal quantities of two different monatomic gases are placed in the compartments, as shown in (a). The door between the compartments is opened and the gas particles start to mix, as in (b). Why would it be highly unlikely for the situation in (b) to progress to the situation shown in (c)?

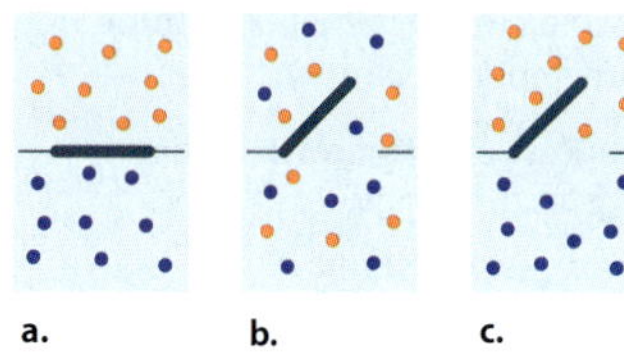

72. Would you expect the entropy to increase in each of the following reactions? Explain.
a. $C(s) + O_2(g) \longrightarrow CO_2(g)$
b. $2Al_2O_3(s) \longrightarrow 4Al(s) + 3O_2(g)$
c. $2N(g) \longrightarrow N_2(g)$
d. $N_2(g) \longrightarrow 2N(g)$

73. What would be the effect on the equilibrium position if the volume is decreased in this reaction?

$$4HCl(g) + O_2(g) \rightleftharpoons 2Cl_2(g) + 2H_2O(g)$$

74. Write the equilibrium-constant expression for this reaction.

$$2SO_2(g) + O_2(g) \rightleftharpoons 2SO_3(g)$$

75. A mixture at equilibrium at 827°C contains 0.552 mol CO_2, 0.552 mol H_2, 0.448 mol CO, and 0.448 mol H_2O. The balanced equation is shown below.

$$CO_2(g) + H_2(g) \rightleftharpoons CO(g) + H_2O(g)$$

What is the value of K_{eq}?

76. The freezing of liquid water at 0°C can be represented as follows.

$$H_2O(l, d = 1.00\ g/cm^3) \rightleftharpoons H_2O(s, d = 0.92\ g/cm^3)$$

Explain why the application of pressure causes ice to melt.

77. What must be true about the concentration of two ions if precipitation occurs when solutions of the two ions are mixed?

78. What is the concentration of carbonate ions in a saturated solution of $SrCO_3$? $K_{sp} = 9.3 \times 10^{-10}$

79. Sketch a reaction progress curve for a reaction that has an activation energy of 22 kJ, and the total energy change is −103 kJ.

80. Predict what will happen if a catalyst is added to a slow reversible reaction. What happens to the equilibrium position?

81. The decomposition of hydrogen peroxide is catalyzed by iodide ions. The mechanism is thought to be as follows.

$$H_2O_2(aq) + I^-(aq) \longrightarrow H_2O(l) + IO^-(aq) \text{ (slow)}$$
$$IO^-(aq) + H_2O_2(aq) \longrightarrow H_2O(l) + O_2(g) + I^-(aq) \text{ (fast)}$$

a. What is the reactive intermediate?
b. Does I^- qualify as a catalyst? Explain.
c. What is the minimum number of activated complexes needed to describe the reaction?
d. Which of the two reactions has the smaller specific rate constant?
e. Write the balanced equation for the reaction.

82. What is the equilibrium concentration of barium ion in a 1.0-L saturated solution of barium carbonate to which 0.25 mol K_2CO_3 has been added?

Critical Thinking

83. Suppose equilibrium exists for the following reaction at 425 K.

$$Fe_3O_4(s) + 4H_2(g) \rightleftharpoons 3Fe(s) + 4H_2O(g)$$

How would the equilibrium concentration of H_2O be affected by these actions?

a. adding more H_2 to the mixture
b. adding more $Fe(s)$
c. removing $H_2(g)$
d. adding a catalyst

84. An increase in temperature raises the energy of the collisions between reactant molecules. An increase in the concentration of reactants increases the number of collisions. What is the effect of a catalyst on the collisions between molecules?

85. Make a list of five things you did today that resulted in an increase in entropy.

86. Explain each of the following

a. A campfire is "fanned" to help get it going.
b. An explosion at a grain elevator is blamed on dust.
c. A pinch of powdered manganese dioxide causes hydrogen peroxide to explode even though the manganese dioxide is not changed.

87. Ammonium ions and nitrite ions react in water to form nitrogen gas.

$$NO_2^-(aq) + NH_4^+(aq) \longrightarrow N_2(g) + 2H_2O(l)$$

From the following data, decide the order of the reaction with respect to NH_4^+ and NO_2^-, and the overall order of the reaction.

Initial concentration of NO_2^- (mol/L)	Initial concentration of NH_4^+ (mol/L)	Initial rate (mol/L·s)
0.0100	0.200	5.4×10^{-7}
0.0200	0.200	10.8×10^{-7}
0.0400	0.200	21.5×10^{-7}
0.0600	0.200	32.3×10^{-7}
0.200	0.0202	10.8×10^{-7}
0.200	0.0404	21.6×10^{-7}
0.200	0.0606	32.4×10^{-7}
0.200	0.0808	43.3×10^{-7}

Concept Challenge

88. The following data were collected for the decomposition of compound AB into its elements. The reaction is first-order in AB.

[AB] (mol/L)	Times (s)
0.300	0
0.246	50
0.201	100
0.165	150
0.135	200
0.111	250
0.090	300
0.075	350

a. Make a graph of concentration (*y*-axis) versus time (*x*-axis).
b. Determine the rate of this reaction at $t = 100$ seconds and $t = 250$ seconds.

89. When table sugar, or sucrose, is dissolved in an acidic solution, the sucrose slowly decomposes into two simpler sugars: fructose and glucose. Use the graph to answer these questions.

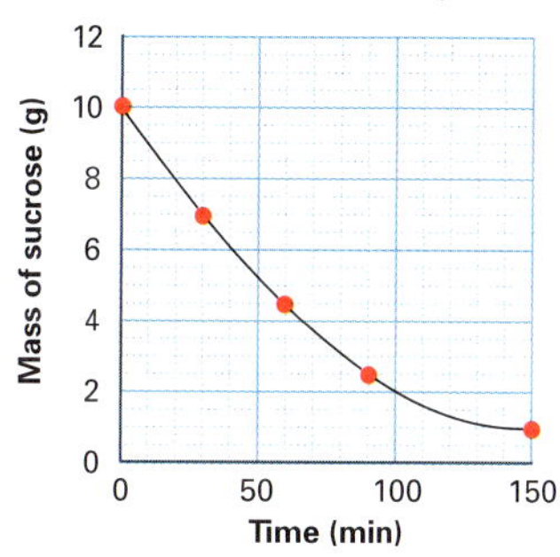

a. How many grams of sucrose decompose in the first 30 minutes?
b. How many grams of sucrose decompose in the 30-minute interval between 90 and 120 minutes?
c. In general, what happens to the rate of decomposition with time?

Critical Thinking

83. a. favors products
b. favors reactants
c. favors reactants
d. no effect

84. A catalyst increases the efficiency of the collisions; more collisions result in the formation of the product.

85. Possible answers: using a blow dryer, flushing a toilet, mowing the lawn, cooking breakfast, driving a car, and breathing

86. a. Fanning brings more oxygen into contact with the campfire.
b. Dust particles have a large surface area to react with oxygen with explosive rapidity.
c. MnO_2 is a catalyst; the reaction occurs with explosive rapidity.

87. first order in NO_2, first order in NH_4^+, second order overall

Concept Challenge

88. a.

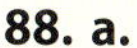

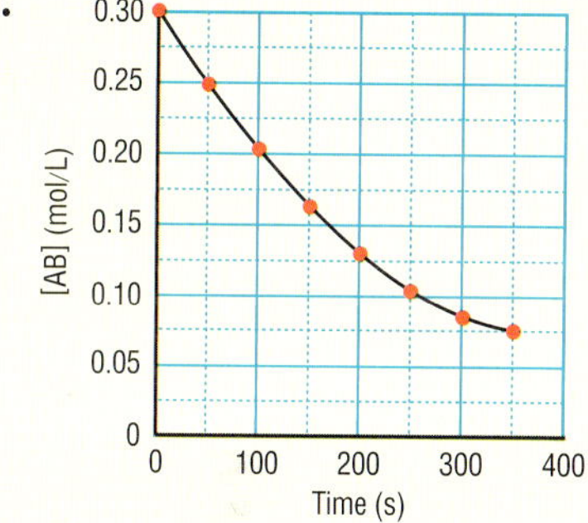

b. The rate when $t = 100$ s is 8×10^{-4} mol/L·s. The rate when $t = 250$ s is 4×10^{-4} mol/L·s.

89. a. 3 g
b. 1.3 g
c. The rate decreases.

CHAPTER 18

Assessment *continued*

Cumulative Review

90. Write electron configurations and draw electron dot structures for the following elements. *(Chapter 5)*

a. Ge **b.** Ca **c.** O
d. Ar **e.** Cl **f.** P

91. What is wrong with saying that solid potassium chloride is composed of KCl molecules? *(Chapter 7)*

92. Name each ion and identify it as an anion or a cation. *(Chapter 9)*

a. F^- **b.** Cu^{2+} **c.** P^{3-} **d.** H^+
e. Na^+ **f.** I^- **g.** O^{2-} **h.** Mg^{2+}

93. Name the following compounds and give the charge on the anion for each. *(Chapter 9)*

a. $NaClO_4$ **b.** $KMnO_4$ **c.** $Ca_3(PO_4)_2$
d. $MgCO_3$ **e.** Na_2SO_4 **f.** $K_2Cr_2O_7$

94. Which atoms from the following list would you expect to form positive ions, and which would you expect to form negative ions? *(Chapter 9)*

a. Cl **b.** Ca **c.** P **d.** Se
e. Cu **f.** Sn **g.** K **h.** Fe

95. Find the mass in grams of each quantity. *(Chapter 10)*

a. 4.50 mol Fe
b. 36.8 L CO (at STP)
c. 1 molecule of glucose, $C_6H_{12}O_6$
d. 0.0642 mol ammonium phosphate

96. Aqueous silver nitrate reacts with aqueous potassium iodide to form a precipitate of silver iodide. *(Chapter 11)*

a. Write the complete ionic equation.
b. What are the spectator ions?
c. Write the net ionic equation.

97. When heated, potassium chlorate decomposes into potassium chloride and oxygen gas. *(Chapter 12)*

a. Write the balanced equation for this reaction.
b. How many grams of oxygen are formed when 4.88 g $KClO_3$ decompose?

98. Give the names and abbreviations of three units of pressure. *(Chapter 13)*

99. Is the boiling point of a liquid compound a constant? Explain. *(Chapter 13)*

100. What happens to the pressure of a contained gas in each instance? *(Chapter 14)*

a. more gas particles are added
b. the temperature of the gas is decreased
c. the volume of the container is reduced

101. What volume will 24.5 g of carbon dioxide gas occupy at 55°C and a pressure of 88.8 kPa? *(Chapter 14)*

102. Which of these should readily dissolve in water? *(Chapter 15)*

a. $KI(s)$ **b.** $C_2H_6(g)$
c. $NH_4Cl(s)$ **d.** $Na_3PO_4(s)$

103. Calculate the percent by mass of water in barium bromide tetrahydrate. *(Chapter 15)*

104. The water solubility of which of these substances would most likely decrease with an increase in temperature? *(Chapter 16)*

a. $NH_4NO_3(s)$ **b.** $NH_3(g)$
c. $KI(s)$ **d.** $NaCl(s)$

105. How many moles of solute are in 2.40 L of 0.66*M* KCl? *(Chapter 16)*

106. How many liters of a stock solution of 6.00*M* HCl would you need to prepare 15.0 L of 0.500*M* HCl? *(Chapter 16)*

107. A small amount of ethanol (C_2H_5OH) is dissolved in a large beaker of water. *(Chapter 16)*

a. Identify the solute and the solvent.
b. Is the freezing point of the solution above or below 0°C?

108. How much heat is evolved when 12.4 g of steam at 100°C condenses to water at 100°C? *(Chapter 17)*

109. When solid sodium hydroxide is dissolved in water, the solution gets warm. Is this an exothermic or endothermic process? *(Chapter 17)*

110. The equation for the complete combustion of ethene is the following. *(Chapter 17)*

$$C_2H_4(g) + 3O_2(g) \longrightarrow 2CO_2(g) + 2H_2O(g) \quad \Delta H = -1411 \text{ kJ}$$

How many kilojoules of heat are released when 32.8 g of ethene are burned?

CHAPTER 18 Assessment continued

Cumulative Review

90. **a.** $1s^22s^22p^63s^23p^63d^{10}4s^24p^2$ ·Ġe·
b. $1s^22s^22p^63s^23p^64s^2$ ·Ca·
c. $1s^22s^22p^4$:Ȯ·
d. $1s^22s^22p^63s^23p^6$:Ar:
e. $1s^22s^22p^63s^23p^5$:Cl·
f. $1s^22s^22p^63s^23p^3$:Ṗ·

91. Potassium chloride is an ionic compound, not a molecular compound.

92. **a.** fluoride anion
b. copper(II) cation
c. phosphide anion
d. hydrogen cation
e. sodium cation
f. iodide anion
g. oxide anion
h. magnesium cation

93. **a.** sodium perchlorate, ClO_4^-
b. potassium permanganate, MnO_4^-
c. calcium phosphate, PO_4^{3-}
d. magnesium carbonate, CO_3^{2-}
e. sodium sulfate, SO_4^{2-}
f. potassium dichromate, $Cr_2O_7^{2-}$

94. positive: b, d–h; negative: a, c

95. **a.** 2.51×10^2 g
b. 4.59×10^1 g
c. 2.99×10^{-22} g
d. 9.57 g

96. **a.** $Ag^+(aq) + NO_3^-(aq) + K^+(aq) + I^-(aq) \rightarrow AgI(s) + K^+(aq) + NO_3^-(aq)$
b. potassium ions and nitrate ions
c. $Ag^+(aq) + I^-(aq) \rightarrow AgI(s)$

97. **a.** $2KClO_3(s) + \text{heat} \rightarrow 2KCl(s) + 3O_2(g)$
b. 1.91 g O_2

98. Pascal (Pa), atmosphere (atm), millimeters of mercury (mm Hg)

99. No; the boiling point is the temperature at which the vapor pressure of the liquid equals the atmospheric pressure; it changes if the atmospheric pressure changes.

100. **a.** increases
b. decreases
c. increases

101. 1.71×10^1 L

102. a, c, d

103. 19.5%

104. b

105. 1.58 mol

106. 1.25 L

107. **a.** The solute is ethanol; the solvent is water.
b. below

108. -2.80×10^1 kJ

109. exothermic

110. -1.65×10^3 kJ

584 Chapter 18

Standardized Test Prep

Test-Taking Tip

Multiple Parts Sometimes two phrases in a true/false question are connected by a word such as *because* or *therefore*, or by a group of words like *and so*. These words imply a relationship between one part of the sentence and another. They usually imply that one thing caused another thing to happen. Statements that include such words can be false even if both parts of the statement are true by themselves.

Select the choice that best answers each question or completes each statement.

1. Which reaction is represented by the following expression for an equilibrium constant?

$$K_{eq} = \frac{[CO]^2 \times [O_2]}{[CO_2]^2}$$

a. $2CO_2 \rightleftharpoons O_2 + 2CO$
b. $CO_2^2 \rightleftharpoons O_2 + 2CO^2$
c. $O_2 + 2CO \rightleftharpoons 2CO_2$
d. $O_2 + CO_2 \rightleftharpoons CO_2^2$

2. At 25°C, zinc sulfide has a K_{sp} of 3.0×10^{-23}, zinc carbonate has a K_{sp} of 1.0×10^{-10}, and silver iodide has a K_{sp} of 8.3×10^{-17}. Order these salts from most soluble to least soluble.
a. zinc carbonate, zinc sulfide, silver iodide
b. silver iodide, zinc carbonate, zinc sulfide
c. zinc carbonate, silver iodide, zinc sulfide
d. zinc sulfide, silver iodide, zinc carbonate

Use the table to answer Questions 3 and 4.

ΔS	ΔH	ΔG	Spontaneous?
+	−	(a)	Yes
+	(b)	+ or −	At high *T*
(c)	+	+	No
−	−	(d)	At low *T*

3. The value of ΔG depends on the enthalpy (ΔH) and entropy (ΔS) terms for a reaction. The value of ΔG also varies as a function of temperature. Use the data in the table to identify the missing entries (a), (b), (c), and (d).

4. Which of these reactions would you expect to be spontaneous at relatively low temperatures? At relatively high temperatures?
a. $H_2O(l) \longrightarrow H_2O(g)$
b. $H_2O(g) \longrightarrow H_2O(l)$
c. $H_2O(s) \longrightarrow H_2O(l)$

5. The atomic windows below represent different degrees of entropy. Arrange the windows in order of increasing entropy.

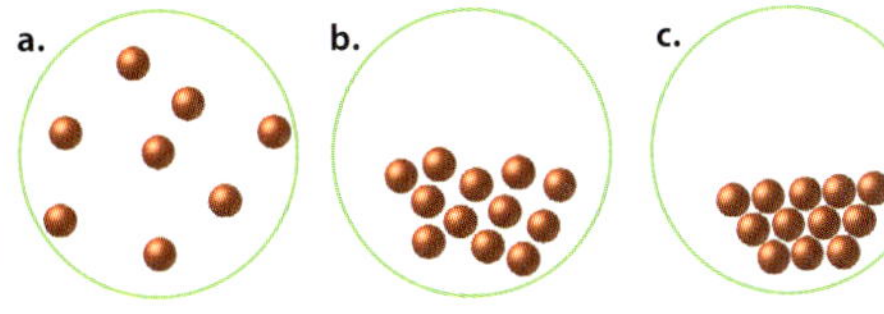

For each question below there are two statements. Decide whether each statement is true or false. Then decide whether Statement II is a correct explanation for Statement I.

	Statement I		Statement II
6.	A catalyst lowers the activation energy for a reaction.	BECAUSE	A catalyst makes a reaction more exothermic.
7.	An exothermic reaction is always a spontaneous reaction.	BECAUSE	Exothermic reactions release heat to the surroundings.
8.	The entropy of ice is greater than the entropy of steam.	BECAUSE	The density of ice is greater than the density of steam.
9.	The rate of a chemical reaction is affected by a change in temperature.	BECAUSE	The kinetic energy of particles is related to the temperature.
10.	A large value for an equilibrium constant indicates that products are favored at equilibrium.	BECAUSE	The ratio of products to reactants at equilibrium is always > 1.

Standardized Test Prep

1. a
2. c
3. (a). −
   (b). +
   (c). −
   (d). + or −
4. a and b at low temperatures; c at high temperatures
5. c, b, a
6. true, false
7. false, true
8. false, true
9. true, true, correct explanation
10. true, false

CHAPTER 19

Acids, Bases, and Salts
Planning Guide

SECTION OBJECTIVES	STANDARDS NATIONAL	STANDARDS STATE	ACTIVITIES and LABS
19.1 Acid-Base Theories, pp. 587–593 1 block or 2 periods **19.1.1 Define** the properties of acids and bases. **19.1.2 Compare and contrast** the theories of Arrhenius, Brønsted-Lowry, and Lewis.	A-1, B-3, G-2, G-3		**SE** Inquiry Activity: Effect of Foods on Baking Soda, p. 586 L2 **TE** Teacher Demo: Reactive Acids, p. 588 L2
19.2 Hydrogen Ions and Acidity, pp. 594–604 1 block or 2 periods **19.2.1 Describe** how $[H^+]$ and $[OH^-]$ are related in an aqueous solution. **19.2.2 Classify** a solution as neutral, acidic, or basic given the hydrogen-ion or hydroxide-ion concentration. **19.2.3 Convert** hydrogen-ion concentrations into pH values and hydroxide-ion concentrations into pOH values. **19.2.4 Describe** the purpose of an acid-base pH indicator.	A-1, A-2, B-3		**SE** Quick Lab: Indicators from Natural Sources, p. 604 L2 **TE** Class Activity: Using a pH Meter, p. 595 L2 **TE** Teacher Demo: pH Indicators, p. 600 L2 **TE** Teacher Demo: Observing pH Change, p. 601 L2 **TE** Class Activity: Comparing pH Indicators and pH Meters p. 603 L2 **LM** Lab 40: Estimation of pH L2 **SSLM** Lab 30: A Small-Scale Colorimetric pH Meter L2 **LP** Lab Practical 19-1: Determining pH L2
19.3 Strengths of Acids and Bases, pp. 605–611 1 block or 2 periods **19.3.1 Define** strong acids and weak acids. **19.3.2 Describe** how an acid's strength is related to the value of its acid dissociation constant. **19.3.3 Calculate** an acid dissociation constant (K_a) from concentration and pOH measurements. **19.3.4 Order** acids by strength according to their acid dissociation constants (K_a). **19.3.5 Order** bases by strength according to their base dissociation constants (K_b).	A-1, A-2, B-3		**TE** Class Activity: Shampoo Survey, p. 608 L1 **VCL** Lab 27: Ionizaton Constants of Weak Acids L2
19.4 Neutralization Reactions, pp. 612–617 1 block or 2 periods **19.4.1 Define** the products of an acid-base reaction. **19.4.2 Explain** how acid-base titration is used to calculate the concentration of an acid or a base. **19.4.3 Explain** the concept of equivalence in neutralization reactions. **19.4.4 Describe** the relationship between equivalence point and the end point of a titration.	A-1, A-2, B-3		**SE** Small-Scale Lab: Ionization Constants of Weak Acids, p. 617 L2 **TE** Teacher Demo, Titration Using Indicators, p. 613 L2 **TE** Teacher Demo, Titration Using a pH Meter p. 615 L2 **LM** Lab 41: Reactions of Acids L2 **LM** Lab 42: Neutralization Reactions L2 **LM** Lab 43: Acid–Base Titrations L2 **SSLM** Lab 31: Titration Curves L2 **LP** Lab Practical 19-2: Acids and Bases L2 **LP** Lab Practical 19-3: Titration L2 **PLM** The Neutralizing Power of Antacids L2 **PLM** Small-Scale Titrations L2
19.5 Salts in Solution, pp. 618–623 1 block or 2 periods **19.5.1 Describe** when a solution of a salt is acidic or basic. **19.5.2 Demonstrate** with equations how buffers resist changes in pH.	A-1, B-3, E-2		**TE** Teacher Demo: Predicting pH of Solutions, p. 619 L2 **TE** Teacher Demo: Comparing Commercial Buffers, p. 620 L2 **LM** Lab 44: Salt Hydrolysis L2 **LM** Lab 45: Buffers L2 **SSLM** Lab 32: Strong and Weak Acids and Bases L2 **SSLM** Lab 33: Buffers L2 **VCL** Labs 25, 26, and 29 L2

Ability Levels

L1 For students who need additional help
L2 For all students
L3 For students who need to be challenged

Components

SE	Student Edition	**LM**	Laboratory Manual	**GCP**	Graphing Calculator Problems	**T**	Transparencies
TE	Teacher's Edition	**SSLM**	Small-Scale Chemistry Lab Manual	**TTT**	Test-Taking Tips with Transparencies, Grades 9–12 Science	**CA**	Chem Alive and Chemistry Field Trips DVD
GRSW	Guided Reading and Study Workbook	**LP**	Laboratory Practicals	**STP**	Standardized Test Preparation Workbook	**iT**	Interactive Textbook with ChemASAP
CTR	Core Teaching Resources	**PLM**	Probeware Lab Manual			**GO**	Go Online Internet Resources
SSG	Spanish Study Guide	**VCL**	Virtual Chemistry Labs				
		CTB	Computer Test Bank				

RESOURCES PRINT and TECHNOLOGY			PROBLEMS and ASSESSMENT
GRSW	Chapter 19, Section 1	L1	**CP** 19.1, Identifying Lewis Acids and Bases **SE** Section 19.1 Assessment, p. 593 **iT** Section 18.2
	T T213–T214	L2	
	GO Chapter 19, Section 1	L2	
GRSW	Chapter 19, Section 2	L1	**SP** 19.1, Finding the [OH^-] of a Solution **SP** 19.2, Calculating pH from [H^+] **SP** 19.3, Using pH to Find [H^+] **SP** 19.4, Calculating pH from [OH^-] **SE** Section 19.2 Assessment, p. 604 **iT** Section 19.2
	T T215–T222	L2	
	GO Chapter 19, Section 2	L2	
GRSW	Chapter 19, Section 3	L1	**SP** 19.5, Calculating a Dissociation Constant **SE** Section 19.3 Assessment, p. 611 **iT** Section 19.3
	T T223–224	L2	
	GO Chapter 19, Section 3	L2	
GRSW	Chapter 19, Section 4	L1	**SP** 19.6, Finding the Number of Moles of an Acid in Neutralization **SP** 19.7, Determining the Concentration of an Acid by Titration **SE** Section 19.4 Assessment, p. 616 **iT** Section 19.4
	T T225–T226	L2	
GRSW	Chapter 19, Section 5	L1	**CP** 19.2, Using Equations to Illustrate the Action of a Buffer **SE** Section 19.5 Assessment, p. 622 **iT** Section 19.5
	T T227–T229	L2	

Go Online

Web Code: cdn-1191
Web Code: cdn-1192

Web Code: cdb-1193

MATERIALS List

Quantities for each group

STUDENT EDITION

Inquiry Activity, p. 586
variety of fruits and vegetables, knife, paper towels, large paper plate, baking soda

Quick Lab, p. 604
See SE p. 604.

Small-Scale Lab, p. 617
reaction surface, metric ruler, pH buffer, 0.04% solutions of BCG, BTB, BPB, MCP, and TB; 0.05% MO, 0.02% AYR, 0.1% phenolphthalein

TEACHER'S EDITION

Teacher Demo, p. 588
5-10 mL of 1*M* HCl, test tube, small pieces of zinc, safety match

Class Activity, p. 595
several test tubes, 0.1*M* HCl, 0.1*M* NaOH, household products such as lemon juice, vinegar, shampoo, liquid detergent, pH meter

Teacher Demo, p. 600
0.1% solutions of thymol blue, methyl red, bromthymol blue, phenolphthalein, aqueous buffers spanning pH 4 to pH 10, test tubes, pH meter

Teacher Demo, p. 601
lemon juice, tea, antacid tablets, glass of 10% grape juice, household ammonia

Class Activity, p. 603
chemical supply catalogs

ClassActivity, p. 608
Different brands and types of shampoo, 10-mL graduated cylinder, beakers, universal indicator paper

Teacher Demo, p. 613
equimolar solutions of HCl and NaOH, phenolphthalein, electric stirrer with a magnetic stir bar, beaker, table salt

Teacher Demo, p. 615
equimolar solutions of HCl and NaOH, phenolphthalein, beaker, buret, pH meter, graph paper

Teacher Demo, p. 619
1*M* solutions of NH_4NO_3, KCl, $NaHCO_3$, Na_2SO_4; pH meter or universal pH paper

Teacher Demo, p. 620
antacid tablets, flasks and burets, 0.5*M* HCl and NaOH, distilled water, methyl red, bromthymol blue

Chapter Assessment

CHAPTER ASSESSMENT
SE Chapter Assessment, pp. 625–628
CTB Chapter 19
iT Chapter 19

STANDARDIZED TEST PREP
SE Chapter 19, p. 629
STP Topic 8
iT Interactive Textbook with Assessment at PHSchool.com

CHAPTER 19

Acids, Bases, and Salts

INQUIRY Activity

Effect of Foods on Baking Soda

L2

Purpose Students observe that fruits and vegetables with sour or tart tastes react with baking soda to produce bubbles of (carbon dioxide) gas.

Prep Time 30 minutes
Class Time 15 minutes
Safety Use care when using sharp knives.

Think About It

1. Bubbling occurs due to a gas (carbon dioxide) being given off.
2. Reacting foods include lemon, orange, grapefruit, and tomato. Nonreacting foods include celery, banana, grape, and apple.

CHAPTER 19

Acids, Bases, and Salts

Tomatoes contain ascorbic acid (vitamin C), an important nutrient and antioxidant.

INQUIRY Activity

Effect of Foods on Baking Soda

Materials

variety of fruits and vegetables (e.g., a celery stalk, a banana, a grape, a tomato, a lemon, an apple, an orange, a grapefruit), knife, paper towels, large paper plate, and baking soda (sodium hydrogen carbonate, $NaHCO_3$)

Procedure

1. Carefully cut the fruits and vegetables into small pieces and place them on the plate. **Caution** *Never taste anything in the laboratory, including food products.* Make sure the pieces are well separated from each other. Be sure to wipe any juice off the knife after cutting each fruit and vegetable.
2. Sprinkle a pinch of baking soda on each sample.

Think About It

1. What do you observe?
2. Which fruits and vegetables showed a reaction with the baking soda and which ones did not?

586

19.1 Acid-Base Theories

Connecting to Your World Bracken Cave, near San Antonio, Texas, is home to twenty to forty million bats, which is probably the largest colony of mammals in the world. Visitors to the cave must wear protective goggles and respirators to protect themselves from the dangerous levels of ammonia in the cave. Ammonia is a by-product of the bats' urine. In this section, you will learn why ammonia is considered a base.

Guide for Reading

Key Concepts

- What are the properties of acids and bases?
- How did Arrhenius define an acid and a base?
- What distinguishes an acid from a base in the Brønsted-Lowry theory?
- How did Lewis define an acid and a base?

Vocabulary

monoprotic acids
diprotic acids
triprotic acids
conjugate acid
conjugate base
conjugate acid-base pair
hydronium ion (H_3O^+)
amphoteric
Lewis acid
Lewis base

Reading Strategy

Building Vocabulary As you read the section, write a definition for each vocabulary term. Include an example with a formula for each term.

Properties of Acids and Bases

Acids and bases play a central role in much of the chemistry that affects your daily life. Your body needs acids and bases to function properly. Vinegar, carbonated drinks, and foods such as citrus fruits contain acids. The electrolyte in a car battery is an acid. Most manufacturing processes use acids or bases. Bases are present in many commercial products, including antacids and household cleaning agents. Figure 19.1 shows some of the many products that contain acids and bases.

Acids Acids have several distinctive properties with which you are probably familiar. Acidic compounds give foods a tart or sour taste. For example, vinegar imparts a tart taste to salad dressing. Vinegar contains ethanoic acid, sometimes called acetic acid. Lemons, which taste sour enough to make your mouth pucker, contain citric acid.

Aqueous solutions of acids are electrolytes. Recall from Chapter 15 that electrolytes conduct electricity. Some acid solutions are strong electrolytes, and others are weak electrolytes. Acids cause certain chemical dyes, called indicators, to change color. Many metals, such as zinc and magnesium, react with aqueous solutions of acids to produce hydrogen gas. Acids react with compounds containing hydroxide ions to form water and a salt. **Acids taste sour, will change the color of an acid-base indicator, and can be strong or weak electrolytes in aqueous solution.**

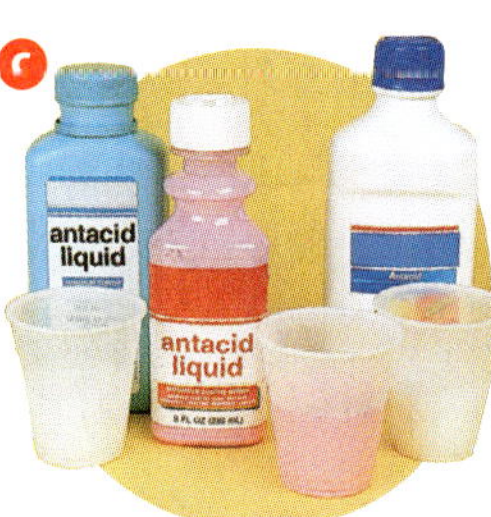

Figure 19.1 Many items contain acids or bases, or produce acids and bases when dissolved in water. **a** Citrus fruit contain citric acid. **b** Tea contains tannic acid. **c** Antacids use bases to neutralize excess stomach acid. **d** The base calcium hydroxide is a component of mortar.

1 FOCUS

Objectives

19.1.1 Define the properties of acids and bases.

19.1.2 Compare and contrast acids and bases as defined by the theories of Arrhenius, Brønsted-Lowry, and Lewis.

Guide for Reading

Build Vocabulary  L2

Word Parts Explain that the prefixes *mono, di,* and *tri* have Greek origins, meaning *one, two,* and *three,* respectively. Have students think of other words that use these prefixes. (*monopoly, monolog; dipole; triangle, triarchy*)

Reading Strategy  L2

Predict Have students make a two-column chart with the headings *Acid* and *Base***.** Have them list substances they think contain either acids or bases. Ask students how they identified each substance as an acid or base.

2 INSTRUCT

Connecting to Your World

Have students study the photograph and read the text that opens the section. Ask, **Why do visitors to Bracken Cave in Texas need to wear protective goggles and respirators?** (*Students will likely suggest that ammonia has a strong, pungent odor and that high levels of ammonia are dangerous.*)

Section Resources

Print

- **Guided Reading and Study Workbook,** Section 19.1
- **Core Teaching Resources,** Section 19.1 Review
- **Transparencies,** T213–T214

Technology

- **Interactive Textbook with ChemASAP,** Animation 25, Problem-Solving 19.1, Assessment 19.1
- **Go Online,** Section 19.1

Properties of Acids and Bases

Reactive Acids L2

Purpose Demonstrate that many metals and acids react to produce hydrogen gas.

Materials 5–10 mL of 1*M* HCl, test tube, small pieces of zinc, match

Safety Wear safety goggles, gloves, and a lab apron while performing the demonstration.

Procedure Add 5–10 mL 1*M* HCl to a test tube and add a few small pieces of zinc metal.

Expected Outcome Gas bubbles are liberated from the solution. Explain that many metals react with aqueous acids to produce hydrogen gas. Remind students that hydrogen is a highly flammable gas. Demonstrate the liberation of the gas by placing a lit match at the mouth of the test tube to ignite the hydrogen.

Arrhenius Acids and Bases

Use Visuals 

Table 19.1 Note that the acids in the table are listed in order of decreasing tendency to yield hydrogen ions. Ask, **What element do all the acids in this table have in common?** *(hydrogen)* **Do diprotic acids always produce more hydrogen ions in solution than monoprotic acids?** *(They may, but there is no direct relationship between the number of hydrogen atoms in a formula and the number of hydrogen ions produced.)*

Table 19.1

Some Common Acids

Name	Formula
Hydrochloric acid	HCl
Nitric acid	HNO_3
Sulfuric acid	H_2SO_4
Phosphoric acid	H_3PO_4
Ethanoic acid	CH_3COOH
Carbonic acid	H_2CO_3

Bases Bases have properties with which you are less familiar. Bases have a bitter taste, but tasting most bases is hazardous. Soap is a familiar material that has the properties of a base. If you have tasted soap, you know that it has a bitter taste. The slippery feel of soap is another property of bases.

Like acids, aqueous solutions of bases are electrolytes and will cause an indicator to change color. Water and a salt are formed when a base that contains hydroxide ions reacts with an acid. With a few exceptions, none of the foods you eat are bases. Milk of magnesia (a suspension of magnesium hydroxide in water) is a base used to treat the problem of excess stomach acid. **Bases taste bitter, feel slippery, will change the color of an acid-base indicator, and can be strong or weak electrolytes in aqueous solution.**

Checkpoint ***Give an example of a base.***

Arrhenius Acids and Bases

Although chemists had recognized the properties of acids and bases for many years, they were not able to propose a theory to explain this behavior. Then, in 1887, the Swedish chemist Svante Arrhenius (1859–1927) proposed a revolutionary way of defining and thinking about acids and bases. **Arrhenius said that acids are hydrogen-containing compounds that ionize to yield hydrogen ions (H^+) in aqueous solution. He also said that bases are compounds that ionize to yield hydroxide ions (OH^-) in aqueous solution.**

Arrhenius Acids Table 19.1 lists some common acids. Acids that contain one ionizable hydrogen, such as nitric acid (HNO_3), are called **monoprotic acids.** Acids that contain two ionizable hydrogens, such as sulfuric acid (H_2SO_4), are called **diprotic acids.** Acids that contain three ionizable hydrogens, such as phosphoric acid (H_3PO_4), are called **triprotic acids.** Not all compounds that contain hydrogen are acids, however. Also, not all the hydrogens in an acid may be released as hydrogen ions. Only the hydrogens in very polar bonds are ionizable. In such bonds, hydrogen is joined to a very electronegative element. When a compound that contains such bonds dissolves in water, it releases hydrogen ions because the hydrogen ions are stabilized by solvation. An example is the hydrogen chloride molecule, shown in Figure 19.2. Hydrogen chloride is a polar covalent molecule. It ionizes to form an aqueous solution of hydronium ions and chloride ions.

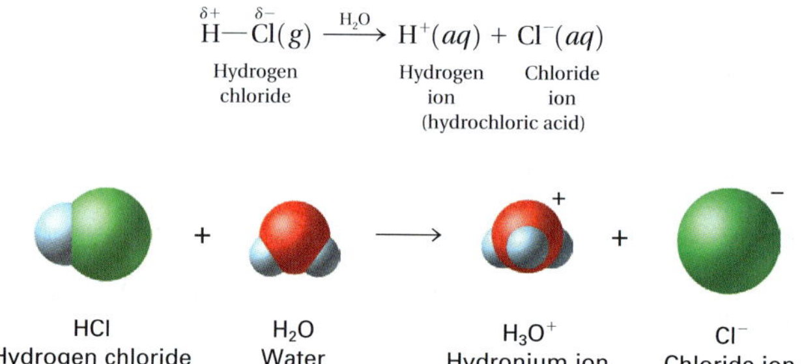

Figure 19.2 Hydrochloric acid is actually an aqueous solution of hydrogen chloride. Hydrogen chloride forms hydronium ions, making this compound an acid.

Facts and Figures

Acid Names and Uses

Acids can often be purchased at hardware and home supply stores, but they are often labeled with "old names." For example, muriatic acid, a term no longer used in chemistry, is hydrochloric acid. Among other uses, this acid is purchased to clean lime deposits from tile and porcelain and to clean mortar stains from brick and tile.

In contrast, the four hydrogens in methane (CH_4) are attached by weakly polar C—H bonds. Methane has no ionizable hydrogens and is not an acid. Ethanoic (acetic) acid (CH_3COOH), used in the manufacture of plastics, pharmaceuticals, and photographic chemicals, is different. Although this molecule contains four hydrogens, ethanoic acid is a monoprotic acid. The structural formula shows why.

```
     H    O
     |    ||
H —  C —  C — O — H
     |
     H
```

Ethanoic acid (CH_3COOH)

The three hydrogens attached to the carbon are in weakly polar bonds. They do not ionize. Only the hydrogen bonded to the highly electronegative oxygen can be ionized. As you gain more experience looking at written formulas for acids, you will be able to recognize which hydrogen atoms can be ionized.

Arrhenius Bases Table 19.2 lists some common bases. The base with which you are perhaps most familiar is sodium hydroxide (NaOH). Sodium hydroxide is commonly known as lye. Sodium hydroxide is an ionic solid. It dissociates into sodium ions and hydroxide ions in aqueous solution.

$$\underset{\text{Sodium hydroxide}}{NaOH(s)} \xrightarrow{H_2O} \underset{\text{Sodium ion}}{Na^+(aq)} + \underset{\text{Hydroxide ion}}{OH^-(aq)}$$

Because of its extremely caustic nature, sodium hydroxide is a major component of consumer products used to clean clogged drains.

Potassium hydroxide (KOH) is another ionic solid that dissociates to form potassium ions and hydroxide ions in aqueous solution.

$$\underset{\text{Potassium hydroxide}}{KOH(s)} \xrightarrow{H_2O} \underset{\text{Potassium ion}}{K^+(aq)} + \underset{\text{Hydroxide ion}}{OH^-(aq)}$$

Sodium and potassium are Group 1A elements. The elements in Group 1A, the alkali metals, react with water to produce solutions that are basic. Sodium metal reacts violently with water to form sodium hydroxide and hydrogen gas. The following equation illustrates the reaction of sodium metal with water.

$$\underset{\text{Sodium metal}}{2Na(s)} + \underset{\text{Water}}{2H_2O(l)} \longrightarrow \underset{\text{Sodium hydroxide}}{2NaOH(aq)} + \underset{\text{Hydrogen}}{H_2(g)}$$

Table 19.2

Some Common Bases

Name	Formula	Solubility in water
Potassium hydroxide	KOH	High
Sodium hydroxide	NaOH	High
Calcium hydroxide	$Ca(OH)_2$	Very low
Magnesium hydroxide	$Mg(OH)_2$	Very low

Go Online NSTA SCILINKS
For: Links on Acids and Bases in the home
Visit: www.SciLinks.org
Web Code: cdn-1191

Use Visuals

Table 19.2 Have students examine the table. Ask them to examine the metal cations in each of the bases, noting that the bases containing Group 1A elements appear to be highly soluble in water while those containing Group 2A elements are not. Ask, **Based on your generalizations, is LiOH likely to be soluble in water?** *(LiOH is probably soluble in water.)*

Explain that hydroxides involving Group 2A elements increase in solubility going down the periodic table column. Ask, **Do you think $Be(OH)_2$ is soluble in water?** *(insoluble)* **$Sr(OH)_2$?** *(quite soluble)*

Discuss

Acid–base concepts have been a part of chemistry for more than 300 years. Over time, theories have been developed to explain the observed behavior of acids and bases. Arrhenius's theory addresses compounds whose formulas contain obvious clues to their acidic or basic nature.

Download a worksheet on **Acids and Bases in the Home** for students to complete, and find additional teacher support from NSTA SciLinks.

Answers to...

Soap and milk of magnesia are examples of bases.

Brønsted-Lowry Acids and Bases

Discuss 

Brønsted arrived at his theory of acids and bases through his work in kinetics and thermodynamics. Brønsted noticed that a great many compounds that would not be classified as acids and bases by the Arrhenius definition behaved like acids and bases in his experiments. He found that these compounds were capable of hydrogen-ion transfer reactions. This led him to define acids as hydrogen-ion donors and bases as hydrogen-ion acceptors in chemical reactions. Some students may think the conjugate base of an acid in a given reaction appears on the same side of the equation as the acid. Explain that an acid does not react with its conjugate base but instead produces it or is produced by it in a reaction.

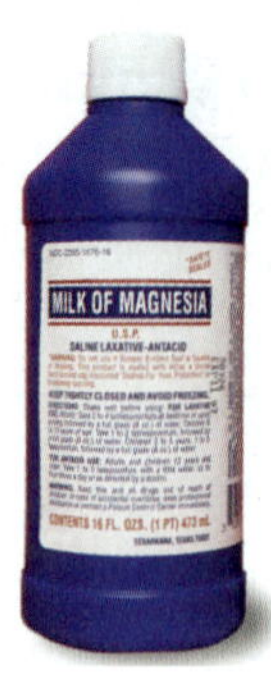

Figure 19.3 Milk of magnesia is a base used as an antacid. Bases are usually hazardous when taken internally, but the low solubility of milk of magnesia makes it safe to use.

Sodium hydroxide and potassium hydroxide are very soluble in water. Concentrated solutions of these compounds can be readily prepared. Such solutions, like other basic solutions, would have a bitter taste and slippery feel. However, they are extremely caustic to the skin and can cause deep, painful, slow-healing wounds if not immediately washed off.

Calcium hydroxide ($Ca(OH)_2$) and magnesium hydroxide ($Mg(OH)_2$) are compounds of Group 2A metals. These compounds are not very soluble in water. Their solutions are always very dilute, even when saturated. The concentration of hydroxide ions in such solutions is correspondingly low. A saturated solution of calcium hydroxide contains only 0.165 g $Ca(OH)_2$ per 100 g of water. Magnesium hydroxide is much less soluble than calcium hydroxide. A saturated solution contains only 0.0009 g $Mg(OH)_2$ per 100 g of water. Suspensions of magnesium hydroxide in water contain low concentrations of hydroxide ion. People take these suspensions internally as milk of magnesia, shown in Figure 19.3, which is an antacid and a mild laxative.

Brønsted-Lowry Acids and Bases

The Arrhenius definition of acids and bases is not a very comprehensive one. It defines acids and bases rather narrowly and does not include certain substances that have acidic or basic properties. For example, aqueous solutions of sodium carbonate ($Na_2CO_3(aq)$) and ammonia ($NH_3(aq)$) are basic. Neither of these compounds is a hydroxide, however, and neither would be classified as a base under the Arrhenius definition. In 1923, the Danish chemist Johannes Brønsted (1879–1947) and the English chemist Thomas Lowry (1874–1936) independently proposed a new definition. **The Brønsted-Lowry theory defines an acid as a hydrogen-ion donor, and a base as a hydrogen-ion acceptor.** All the acids and bases included in the Arrhenius theory are also acids and bases according to the Brønsted-Lowry theory. Some compounds not included in the Arrhenius theory are classified as bases in the Brønsted-Lowry theory.

Figure 19.4 Sodium carbonate decahydrate ($Na_2CO_3 \cdot 10H_2O$) is used as a laundry aid. Its common name is washing soda.

Why Ammonia is a Base The behavior of ammonia as a base can be understood by using the Brønsted-Lowry theory. Ammonia gas is very soluble in water. When ammonia dissolves in water, it acts as a base because it accepts a hydrogen ion from water.

$$NH_3(aq) + H_2O(l) \rightleftharpoons NH_4^+(aq) + OH^-(aq)$$

$NH_3(aq)$	$H_2O(l)$	$NH_4^+(aq)$	$OH^-(aq)$
Ammonia (hydrogen-ion acceptor, Brønsted-Lowry base)	Water (hydrogen-ion donor, Brønsted-Lowry acid)	Ammonium ion	Hydroxide ion (makes the solution basic)

In this reaction, ammonia is the hydrogen-ion acceptor and therefore is a Brønsted-Lowry base. Water, the hydrogen-ion donor, is a Brønsted-Lowry acid. Hydrogen ions are transferred from water to ammonia, as is shown in Figure 19.5. This causes the hydroxide-ion concentration to be greater than it is in pure water. As a result, solutions of ammonia are basic.

 ***Why is ammonia considered to be a Brønsted-Lowry base?***

590 Chapter 19

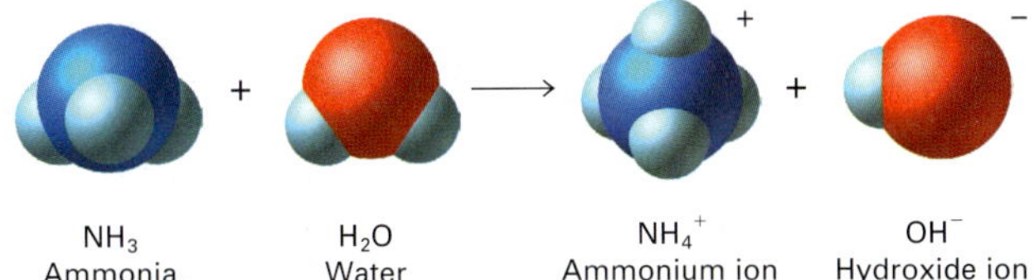

Figure 19.5 Ammonia dissolves in water to form ammonium ions and hydroxide ions. In this reaction, the water molecule donates a hydrogen ion to the ammonia molecule.
Explaining ***Why is ammonia not classified as an Arrhenius base?***

Conjugate Acids and Bases Because all gases become less soluble in water as temperature increases, increasing the temperature of an aqueous solution of ammonia releases ammonia gas. As ammonia gas leaves the solution, the equilibrium in the equation shifts to the left. The ammonium ion (NH_4^+) reacts with OH^- to form NH_3 and H_2O. When the reaction goes from right to left, NH_4^+ gives up a hydrogen ion; it acts as a Brønsted-Lowry acid. The hydroxide ion accepts an H^+; it acts as a Brønsted-Lowry base. Overall, then, this equilibrium has two acids and two bases.

$$\underset{\text{Base}}{NH_3(aq)} + \underset{\text{Acid}}{H_2O(l)} \rightleftharpoons \underset{\text{Conjugate acid}}{NH_4^+(aq)} + \underset{\text{Conjugate base}}{OH^-(aq)}$$

When ammonia dissolves and then reacts with water, NH_4^+ is the conjugate acid of the base NH_3. A **conjugate acid** is the particle formed when a base gains a hydrogen ion. Similarly, OH^- is the conjugate base of the acid water. A **conjugate base** is the particle that remains when an acid has donated a hydrogen ion. Conjugate acids and bases are always paired with a base or an acid, respectively. A **conjugate acid-base pair** consists of two substances related by the loss or gain of a single hydrogen ion. The ammonia molecule and ammonium ion are a conjugate acid-base pair. The water molecule and hydroxide ion are also a conjugate acid-base pair.

$$\underset{\text{Base}}{NH_3(aq)} + \underset{\text{Acid}}{H_2O(l)} \rightleftharpoons \underset{\text{Conjugate acid}}{NH_4^+(aq)} + \underset{\text{Conjugate base}}{OH^-(aq)}$$

The Brønsted-Lowry theory also applies to acids. Consider the dissociation of hydrogen chloride in water.

$$\underset{\text{Acid}}{HCl(g)} + \underset{\text{Base}}{H_2O(l)} \rightleftharpoons \underset{\text{Conjugate acid}}{H_3O^+(aq)} + \underset{\text{Conjugate base}}{Cl^-(aq)}$$

In this reaction, hydrogen chloride is the hydrogen-ion donor. Thus it is a Brønsted-Lowry acid. Water is the hydrogen-ion acceptor and therefore water is a Brønsted-Lowry base. A water molecule that gains a hydrogen ion becomes a positively charged **hydronium ion (H_3O^+).** The chloride ion is the conjugate base of the acid HCl. The hydronium ion is the conjugate acid of the base water.

Table 19.3

Several Conjugate Acid-Base Pairs

Acid	Base
HCl	Cl^-
H_2SO_4	HSO_4^-
H_3O^+	H_2O
HSO_4^-	SO_4^{2-}
CH_3COOH	CH_3COO^-
H_2CO_3	HCO_3^-
HCO_3^-	CO_3^{2-}
NH_4^+	NH_3
H_2O	OH^-

Use Visuals L1

Table 19.3 Have students study the Brønsted-Lowry acid-base conjugate pairs. Ask, **How are hydrogen ions related to Brønsted-Lowry acids?** *(Acids lose hydrogen ions.)* **How are hydrogen ions related to Brønsted-Lowry bases?** *(Bases gain hydrogen ions.)*

Answers to...

Figure 19.5 It is not a hydroxide.

Ammonia is considered to be a Brønsted-Lowry base because it is a hydrogen-ion acceptor.

Word Origins L2

An amphibious airplane is designed to take off from and land on either land or water. The prefix, *amph-*, is also used in *amphibian* which is an animal that has aquatic larvae, with gills that grow into air-breathing adults with lungs.

Lewis Acids and Bases

Discuss L2

Point out that Lewis was the first scientist to discuss the significance of electron pairs in bonding (Lewis electron-dot diagrams). His theory of acids and bases was an extension of his concept of electron pairs. A Lewis acid accepts a pair of electrons to form a covalent bond and a Lewis base donates a pair of electrons to form a covalent bond.

Answers to...

Figure 19.6 hydronium ion; hydrogen sulfate ion

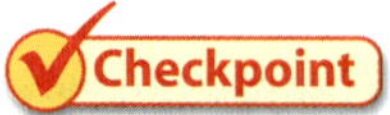

A Lewis acid is an electron-pair acceptor; a Lewis base is an electron-pair donor.

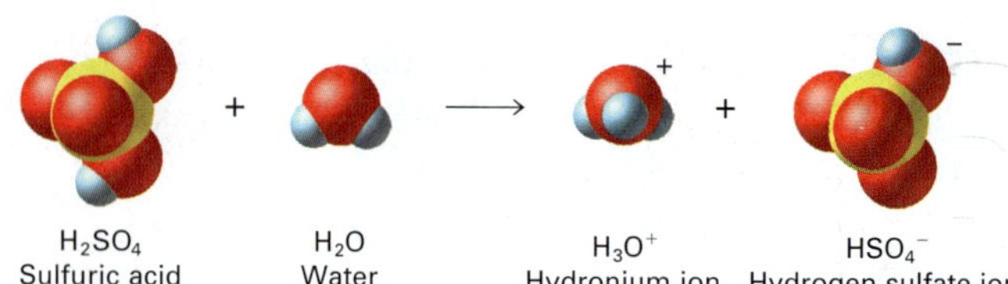

Figure 19.6 When sulfuric acid dissolves in water, it forms hydronium ions and hydrogen sulfate ions. **Identify** ***Which ion is the conjugate acid and which is the conjugate base?***

Figure 19.6 shows the reaction that occurs when sulfuric acid dissolves in water. The solution formed consists of hydronium ions and hydrogen sulfate ions.

Sometimes water accepts a hydrogen ion. At other times, it donates a hydrogen ion. A substance that can act as both an acid and a base is said to be **amphoteric.** Water is amphoteric. In the reaction with HCl, water accepts a proton and is therefore a base.

Word Origins

Amphoteric comes from the Greek word *amphoteros*, meaning "partly one and partly the other." An amphoteric substance can act as both an acid and a base. **What makes an amphibious airplane different from other airplanes?**

Lewis Acids and Bases

A third theory of acids and bases was proposed by Gilbert Lewis (1875–1946). **Lewis proposed that an acid accepts a pair of electrons during a reaction, while a base donates a pair of electrons.** This concept is more general than either the Arrhenius theory or the Brønsted-Lowry theory. A **Lewis acid** is a substance that can accept a pair of electrons to form a covalent bond. A **Lewis base** is a substance that can donate a pair of electrons to form a covalent bond. A hydrogen ion (Brønsted-Lowry acid) can accept a pair of electrons in forming a bond. A hydrogen ion, therefore, is also a Lewis acid. A Brønsted-Lowry base, or a substance that accepts a hydrogen ion, must have a pair of electrons available and, therefore, is also a Lewis base. Consider the reaction of H^+ and OH^-.

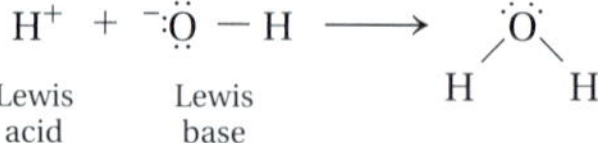

In this reaction, a hydroxide ion is a Lewis base. It is also a Brønsted-Lowry base. The hydrogen ion is both a Lewis acid and a Brønsted-Lowry acid. The Lewis definition also includes some compounds not classified as Brønsted-Lowry acids or bases.

Ammonia dissolved in water is another example of a Lewis acid and Lewis base. In this reaction, the hydrogen ion from the dissociation of water is an electron-pair acceptor and is a Lewis acid. Ammonia is an electron-pair donor and is a Lewis base.

Animation 25 Compare the three important definitions of acids and bases.

with ChemASAP

Table 19.4

Acid-Base Definitions		
Type	**Acid**	**Base**
Arrhenius	H^+ producer	OH^- producer
Brønsted-Lowry	H^+ donor	H^+ acceptor
Lewis	electron-pair acceptor	electron-pair donor

Checkpoint ***What is a Lewis acid and a Lewis base?***

CONCEPTUAL PROBLEM 19.1

Identifying Lewis Acids and Bases

Ammonia is widely used in fertilizers, plastics, and explosives. Identify the Lewis acid and the Lewis base in this reaction involving ammonia.

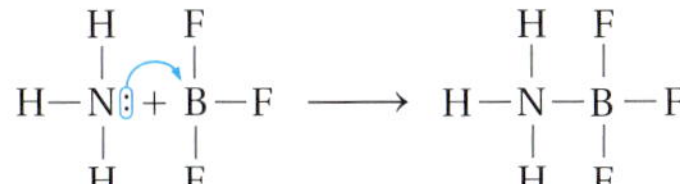

$$H_3N{:} + BF_3 \longrightarrow H_3N{-}BF_3$$

1 Analyze ***Identify the relevant concepts.*** The Lewis acid–Lewis base definitions, which are to be used in solving the problem, are based on the acceptance and donation of a pair of electrons.

2 Solve ***Apply concepts to this situation.*** Ammonia is donating a pair of electrons. Boron trifluoride is accepting a pair of electrons. Lewis bases donate electrons, so ammonia is acting as a Lewis base. Lewis acids accept a pair of electrons, so boron trifluoride is acting as a Lewis acid.

Practice Problems

1. Identify the Lewis acid and Lewis base in each reaction.
   a. $H^+ + H_2\ddot{O}{:} \longrightarrow H_3O^+$
   b. $AlCl_3 + Cl^- \longrightarrow AlCl_4^-$

2. Would you predict PCl_3 to be a Lewis acid or a Lewis base in typical reactions? Explain your prediction.

Interactive Textbook

Problem-Solving 19.1 Solve Problem 1 with the help of an interactive guided tutorial.

with ChemASAP

19.1 Section Assessment

3. **Key Concept** What are the properties of acids and bases?
4. **Key Concept** How did Arrhenius define an acid and a base?
5. **Key Concept** How are acids and bases defined by the Brønsted-Lowry theory?
6. **Key Concept** What is the Lewis theory of acids and bases?
7. **a.** What is a conjugate acid-base pair?
   b. Write equations for the ionization of HNO_3 in water and the reaction of CO_3^{2-} with water. For each equation, identify the hydrogen-ion donor and hydrogen-ion acceptor. Then label the conjugate acid-base pairs in each equation.
8. Identify the following acids as monoprotic, diprotic, or triprotic. Explain your reasoning.
   a. H_2CO_3 **b.** H_3PO_4 **c.** HCl **d.** H_2SO_4

Writing Activity

Write a Report Household drain cleaners contain pellets of sodium hydroxide (NaOH) and small metal particles. Use the library or Internet to find out how drain cleaners work. Include the identity of the metal particles in your written report.

Interactive Textbook

Assessment 19.1 Test yourself on the concepts in Section 19.1.

with ChemASAP

Section 19.1 Assessment

3. Both are electrolytes and change the color of an acid-base indicator. Acids have a sour taste; bases taste bitter.
4. An acid gives hydrogen ions. A base gives hydroxide ions.
5. Acids are hydrogen-ion donors and bases are hydrogen-ion acceptors.
6. A Lewis acid is an electron-pair acceptor; a Lewis base is an electron-pair donor.
7. **a.** two substances related by the loss or gain of a single hydrogen ion
   b. $HNO_3 + H_2O \rightarrow H_3O^+ + NO_3^-$; HNO_3 is the hydrogen-ion donor; its conjugate base is NO_3^-. H_2O is the hydrogen-ion acceptor; its conjugate acid is H_3O^+.
   $CO_3^{2-} + H_2O \rightleftharpoons HCO_3^- + OH^-$
   CO_3^{2-} is the hydrogen-ion acceptor; its conjugate acid is HCO_3^-. H_2O is the hydrogen-ion donor; its conjugate base is OH^-.
8. **a.** diprotic; two ionizable hydrogens
   b. triprotic; three ionizable hydrogens
   c. monoprotic; one ionizable hydrogen
   d. diprotic; two ionizable hydrogens

CONCEPTUAL PROBLEM 19.1

Answers

1. **a.** H^+ is the Lewis acid; H_2O is the Lewis base.
   b. $AlCl_3$ is the Lewis acid; Cl^- is the Lewis base.
2. A Lewis base; it has a nonbonding pair of electrons that it can donate.

Practice Problems Plus L2

Identify each reactant as an acid or base.

a. $KOH + HBr \rightarrow KBr + H_2O$
(KOH is the base; HBr is the acid)

b. $HCl + H_2O \rightarrow H_3O^+ + Cl^-$
(HCl is the acid; H_2O is the base)

3 ASSESS

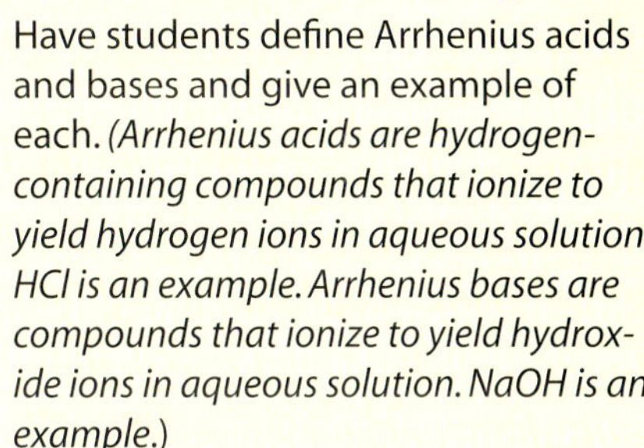

Evaluate Understanding L2

Have students define Arrhenius acids and bases and give an example of each. *(Arrhenius acids are hydrogen-containing compounds that ionize to yield hydrogen ions in aqueous solution. HCl is an example. Arrhenius bases are compounds that ionize to yield hydroxide ions in aqueous solution. NaOH is an example.)*

Writing Activity

Drain cleaners contain caustic NaOH. Reaction with aluminum generates heat, which softens greases and oils, and also generates hydrogen, which agitates the mixture.

Interactive Textbook

If your class subscribes to the Interactive Textbook, use it to review key concepts in Section 19.1.

with ChemASAP

19.2

1 FOCUS

Objectives

19.2.1 Describe how [H⁺] and [OH] are related in an aqueous solution.

19.2.2 Classify a solution as neutral, acidic, or basic given the hydrogen-ion or hydroxide-ion concentration.

19.2.3 Convert hydrogen-ion concentrations into pH values and hydroxide-ion concentrations into pOH values.

19.2.4 Describe the purpose of an acid-base indicator.

Guide for Reading

Build Vocabulary L2

Graphic Organizers Students can compare and contrast the properties of a *neutral solution,* an *acid solution,* and a *basic solution* using a table.

Reading Strategy L2

Identify Main Idea/Details Have students write down the titles of the red heads in the section. Have them follow each title with a sentence that expresses the section's main idea.

2 INSTRUCT

Connecting to Your World

Have students study the photograph and read the text that opens the section. Ask, **What scale is commonly used to express the acidity of a solution?** (pH scale)

Hydrogen Ions from Water

Use Visuals L1

Figure 19.7 Ask students to examine the figure. Illustrate the reaction on the board: Show how two water molecules can react to yield a hydronium ion and a hydroxide ion. Ask, **Which element is donating a pair of electons?** *(oxygen)* **Which element is accepting a pair of electrons?** (hydrogen)

19.2 Hydrogen Ions and Acidity

Guide for Reading

Key Concepts

- How are [H⁺] and [OH⁻] related in an aqueous solution?
- How is the hydrogen-ion concentration used to classify a solution as neutral, acidic, or basic?
- What is the most important characteristic of an acid-base indicator?

Vocabulary

self-ionization
neutral solution
ion-product constant for water (K_w)
acidic solution
basic solution
alkaline solutions
pH

Reading Strategy

Relating Text and Visuals
As you read about pH, look carefully at the diagram in Figure 19.10. Make sure that you can explain why the differences in [H⁺] and [OH⁻] exist in acidic, neutral, and basic solutions.

Connecting to Your World

A patient is brought to a hospital unconscious and with a fruity odor on his breath. The doctor suspects the patient has fallen into a diabetic coma. To confirm her diagnosis, she orders several tests, including one of the acidity of the patient's blood. The results from this test will be expressed in units of pH, not molar concentration. In this section, you will learn how the pH scale is used to indicate the acidity of a solution and why the pH scale is used.

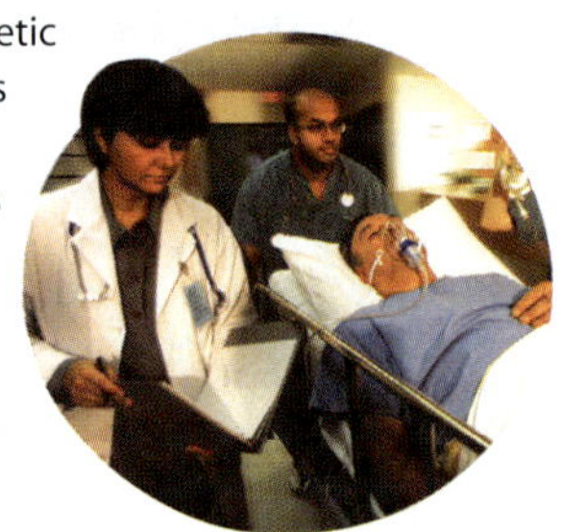

Hydrogen Ions from Water

As you already know, water molecules are highly polar and are in continuous motion, even at room temperature. Occasionally, the collisions between water molecules are energetic enough to transfer a hydrogen ion from one water molecule to another. A water molecule that loses a hydrogen ion becomes a negatively charged hydroxide ion (OH^-). A water molecule that gains a hydrogen ion becomes a positively charged hydronium ion (H_3O^+).

The reaction in which water molecules produce ions is called the **self-ionization** of water. This reaction can be written as a simple dissociation.

$$H_2O(l) \rightleftharpoons \underset{\text{Hydrogen ion}}{H^+(aq)} + \underset{\text{Hydroxide ion}}{OH^-(aq)}$$

In water or aqueous solution, hydrogen ions (H^+) are always joined to water molecules as hydronium ions (H_3O^+). Hydrogen ions in aqueous solution have several names. Some chemists call them protons. Others prefer to call them hydrogen ions or hydronium ions. In this textbook, either H^+ or H_3O^+ is used to represent hydrogen ions in aqueous solution. Figure 19.7 shows how two water molecules react to form one hydronium ion and one hydroxide ion.

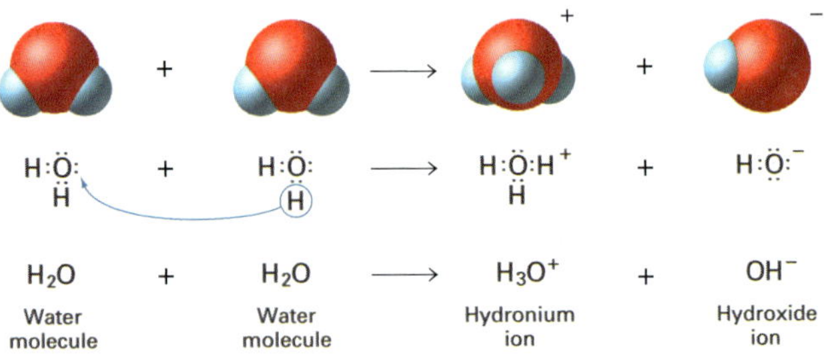

Figure 19.7 The self-ionization of water. A proton (hydrogen ion) transfers from one water molecule to another water molecule. The result is one hydronium ion (H_3O^+) and one hydroxide ion (OH^-).

Section Resources

Print

- **Guided Reading and Study Workbook,** Section 19.2
- **Core Teaching Resources,** Section 19.2 Review
- **Transparencies,** T215–T222
- **Laboratory Manual,** Lab 40
- **Small-Scale Chemistry Laboratory Manual,** Lab 30

Technology

- **Interactive Textbook with ChemASAP,** Problem-Solving 19.10, 19.12, 19.14, 19.15, Assessment 19.2
- **GoOnline,** Section 19.2

The self-ionization of water occurs to a very small extent. In pure water at 25°C, the equilibrium concentration of hydrogen ions ($[H^+]$) and the equilibrium concentration of hydroxide ions ($[OH^-]$) are each only $1 \times 10^{-7}M$. This means that the concentrations of H^+ and OH^- are equal in pure water. Any aqueous solution in which $[H^+]$ and $[OH^-]$ are equal is described as a **neutral solution.**

Ion Product Constant for Water

In any aqueous solution, when $[H^+]$ increases, $[OH^-]$ decreases. When $[H^+]$ decreases, $[OH^-]$ increases. Le Châtelier's principle, which you learned about in Chapter 18, applies here. If additional ions (either hydrogen ions or hydroxide ions) are added to a solution, the equilibrium shifts. The concentration of the other type of ion decreases. More water molecules are formed in the process.

$$H^+(aq) + OH^-(aq) \rightleftharpoons H_2O(l)$$

For aqueous solutions, the product of the hydrogen-ion concentration and the hydroxide-ion concentration equals 1.0×10^{-14}.

$$[H^+] \times [OH^-] = 1.0 \times 10^{-14}$$

This equation is true for all dilute aqueous solutions at 25°C. As you will see, the concentrations of H^+ and OH^- may change when substances are added to water. However, the product of $[H^+]$ and $[OH^-]$ is always 1×10^{-14}.

The product of the concentrations of the hydrogen ions and hydroxide ions in water is called the **ion-product constant for water (K_w).**

$$K_w = [H^+] \times [OH^-] = 1.0 \times 10^{-14}$$

Not all solutions are neutral. When some substances dissolve in water, they release hydrogen ions. For example, when hydrogen chloride dissolves in water, it forms hydrochloric acid.

$$HCl(g) \xrightarrow{H_2O} H^+(aq) + Cl^-(aq)$$

In such a solution, the hydrogen-ion concentration is greater than the hydroxide-ion concentration. The hydroxide ions are present from the self-ionization of water. An **acidic solution** is one in which $[H^+]$ is greater than $[OH^-]$. The $[H^+]$ of an acidic solution is greater than $1 \times 10^{-7}M$.

When sodium hydroxide dissolves in water, it forms hydroxide ions in solution.

$$NaOH(s) \xrightarrow{H_2O} Na^+(aq) + OH^-(aq)$$

In such a solution, the hydrogen-ion concentration is less than the hydroxide-ion concentration. Remember, the hydrogen ions are present from the self-ionization of water. A **basic solution** is one in which $[H^+]$ is less than $[OH^-]$. The $[H^+]$ of a basic solution is less than $1 \times 10^{-7}M$. Basic solutions are also known as **alkaline solutions.** Some uses for acids and bases are shown in Figure 19.8.

What is a basic solution?

Figure 19.8 Acids and bases have many uses in the home and in industry. **a** Unrefined hydrochloric acid, commonly called muriatic acid, is used to clean stone buildings and swimming pools. **b** Sodium hydroxide, or lye, is commonly used as a drain cleaner.
Predicting ***How will each chemical affect the hydrogen-ion and hydroxide-ion concentration of an aqueous solution?***

Ion Product Constant for Water

CLASS Activity

Using a pH Meter L1

Purpose Students observe how a pH meter measures the PH of solutions.

Materials several test tubes; $0.1M$ HCl; $0.1M$ NaOH; household products such as lemon juice, vinegar, shampoo; and liquid detergent; pH meter

Procedure Set up test tubes with acids and bases of varying strength. In addition to HCl and NaOH, include household products such as lemon juice, vinegar, shampoo, and liquid detergent. Use a pH meter to measure the pH of each solution. Explain that the reading on the pH meter depends upon the concentration of hydrogen ions in the solutions.

Differentiated Instruction

English Learners L2

Students who do not have a good background in scientific notation may have trouble understanding the relationship between hydrogen- and hydroxide-ion concentrations. Draw diagrams to help students visualize the shift in equilibrium that occurs when one or the other kind of ion is added or removed.

Answers to...

Figure 19.8 Hydrochloric acid will increase $[H^+]$ and decrease $[OH^-]$; sodium hydroxide will increase $[OH^-]$ and decrease $[H^+]$.

A basic solution is one in which $[H^+]$ is less than $[OH^-]$

Sample Problem 19.1

Answers

9. a. basic
   b. basic
   c. acidic
   d. neutral
10. $1.0 \times 10^{-11} M$; basic

Practice Problems Plus 

The $[OH^-]$ of a carbonated soft drink is $1.0 \times 10^{-11} M$. **What is the $[H^+]$ of this solution? Is the solution acidic or basic?** *($1.0 \times 10^{-3} M$; acidic)*

Math Handbook

For a math refresher and practice, direct students to scientific notation, page R56.

The pH Concept

Discuss 

Pure water self-ionizes to form hydrogen and hydroxide ions, yet it does not conduct electric current well. Ask, **Why does this phenomenon occur?** *(Pure water is a poor conductor because the concentrations of the ions are low.)*

Math Handbook

For help with scientific notation, go to page R56.

Interactive Textbook

Problem-Solving 19.10 Solve Problem 10 with the help of an interactive guided tutorial.

with ChemASAP

SAMPLE PROBLEM 19.1

Finding the $[OH^-]$ of a Solution

Colas are slightly acidic. If the $[H^+]$ in a solution is $1.0 \times 10^{-5} M$, is the solution acidic, basic, or neutral? What is the $[OH^-]$ of this solution?

1 Analyze *List the knowns and the unknowns.*

Knowns
- $[H^+] = 1.0 \times 10^{-5} M$
- Ion-product constant for water:

$$K_w = [H^+] \times [OH^-] = 1 \times 10^{-14}$$

Unknowns
- solution = acidic, basic, or neutral?
- $[OH^-] = ?M$

2 Calculate *Solve for the unknowns.*

$[H^+]$ is $1.0 \times 10^{-5} M$. Because this is greater than $1.0 \times 10^{-7} M$, the solution is acidic.

By definition $K_w = [H^+] \times [OH^-]$. Therefore, $[OH^-] = \frac{K_w}{[H^+]}$.

Substituting the known numerical values, compute $[OH^-]$ as follows.

$$[OH^-] = \frac{1.0 \times 10^{-14}}{1.0 \times 10^{-5}} = 1.0 \times 10^{-9} M$$

3 Evaluate *Do the results make sense?*

If $[H^+]$ is greater than $1.0 \times 10^{-7} M$, the $[OH^-]$ must be less than $1.0 \times 10^{-7} M$. At $1 \times 10^{-9} M$, $[OH^-]$ is less than $1 \times 10^{-7} M$.

Practice Problems

9. Classify each solution as acidic, basic, or neutral.
   a. $[H^+] = 6.0 \times 10^{-10} M$
   b. $[OH^-] = 3.0 \times 10^{-2} M$
   c. $[H^+] = 2.0 \times 10^{-7} M$
   d. $[OH^-] = 1.0 \times 10^{-7} M$

10. If the hydroxide-ion concentration of an aqueous solution is $1 \times 10^{-3} M$, what is the $[H^+]$ in the solution? Is the solution acidic, basic, or neutral?

The pH Concept

Expressing hydrogen-ion concentration in molarity is cumbersome. A more widely used system for expressing $[H^+]$ is the pH scale, proposed in 1909 by the Danish scientist Søren Sørensen (1868–1939). On the pH scale, which ranges from 0 to 14, neutral solutions have a pH of 7. A pH of 0 is strongly acidic. A solution with a pH of 14 is strongly basic.

Calculating pH The **pH** of a solution is the negative logarithm of the hydrogen-ion concentration. The pH may be represented mathematically using the following equation.

$$pH = -\log[H^+]$$

Facts and Figures

Ions in Water

Explain to the students that all aqueous systems contain both hydrogen and hydroxide ions due to self-ionization of water. At 25°C, about 1 molecule of water out of 550 000 000 will dissociate. This amounts to 1 g of hydrogen ions and 17 g of hydroxide ions in 10 000 000 L of water. No matter how small the concentrations, both ions are always present in a water solution.

The relationship between the concentrations of H^+ and OH^- in aqueous solutions at constant temperature is similar to the relationship between the pressure and volume of a gas at constant temperature. In both cases, the relationship is inverse. The product of the two quantities is a constant—as one quantity increases, the other decreases.

In a neutral solution, the $[H^+] = 1 \times 10^{-7}M$. The pH of a neutral solution is 7.

$$\begin{aligned} pH &= -\log(1 \times 10^{-7}) \\ &= -(\log 1 + \log 10^{-7}) \\ &= -(0.0 + (-7.0)) \\ &= 7.0 \end{aligned}$$

You can calculate the pH of a solution using the log function key on a calculator. (You can review finding the logarithm of a number on a calculator in the Math Handbook on page R78.)

Figure 19.9 shows how the hydrogen-ion concentration of a solution is used to classify the solution as neutral acidic, or basic. **A solution in which $[H^+]$ is greater than $1 \times 10^{-7}M$ has a pH less than 7.0 and is acidic. The pH of pure water or a neutral aqueous solution is 7.0. A solution with a pH greater than 7 is basic and has a $[H^+]$ of less than $1 \times 10^{-7}M$.**

- Acidic solution: $pH < 7.0$ $[H^+]$ greater than $1 \times 10^{-7}M$
- Neutral solution: $pH = 7.0$ $[H^+]$ equals $1 \times 10^{-7}M$
- Basic solution: $pH > 7.0$ $[H^+]$ less than $1 \times 10^{-7}M$

For: Links on pH
Visit: www.SciLinks.org
Web Code: cdn-1192

$[H_3O^+]$ and $[OH^-]$ in Acidic, Neutral, and Basic Solutions

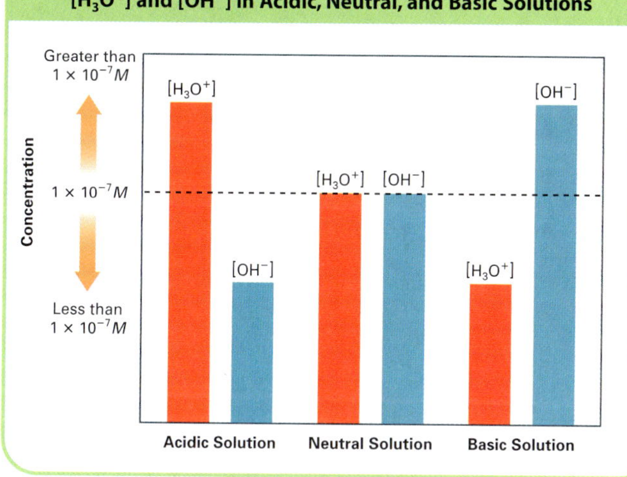

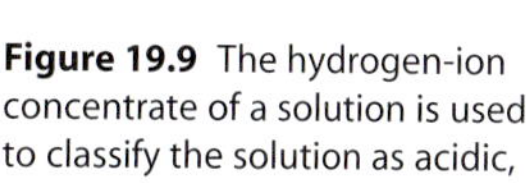

Figure 19.9 The hydrogen-ion concentrate of a solution is used to classify the solution as acidic, neutral, or basic.

INTERPRETING GRAPHS

a. Identify What is $[H_3O^+]$ in a neutral solution?
b. Describe How does $[H_3O^+]$ compare with $[OH^-]$ in an acidic solution?
c. Compare and Contrast In terms of ion concentrations, how are basic solutions different from acidic solutions?

The pH values of several common aqueous solutions are listed in Table 19.5. The table also summarizes the relationship among $[H^+]$, $[OH^-]$, and pH. You may notice that pH can sometimes be read from the value of $[H^+]$. If $[H^+]$ is written in scientific notation and has a coefficient of 1, then the pH of the solution equals the exponent, with the sign changed from minus to plus. For example, a solution with $[H^+] = 1 \times 10^{-2}M$ has a pH of 2.0 and a solution with $[H^+] = 1 \times 10^{-13}M$ has a pH of 13.0.

If the pH is an integer number, it is also possible to directly write the value of $[H^+]$. A solution with a pH of 9.0 has a $[H^+] = 1 \times 10^{-9}M$. A pH of 4 indicates a $[H^+]$ of $1 \times 10^{-4}M$.

Download a worksheet on **pH** for students to complete, and find additional teacher support from NSTA SciLinks.

Use Visuals 

Figure 19.9 Have students study the figure and the text that follows. Explain that the pH scale shows the relationship between pH and the hydrogen-ion concentration. Write the expression for the ion-product constant for water on the board, and remind students about the relationship between the concentrations of hydronium ion and hydroxide ion in an aqueous solution. Point out that the product of these concentrations in aqueous solutions is always 1×10^{-14} at 25°C.

Arbitrarily assign concentration values to each of the hydronium-ion bars in Figure 19.9 for the acidic and basic solutions (for the neutral solution, $[H^+]$ is $1 \times 10^{-7}M$); then have students calculate the pH of those solutions. Ask, **What is $[OH^-]$ for these solutions?** *(Have students refer to Table 19.5 to find examples of aqueous systems with these $[H^+]$ and $[OH^-]$.)*

Interpreting Graphs 

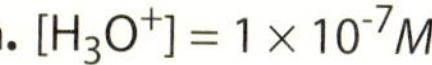

a. $[H_3O^+] = 1 \times 10^{-7}M$
b. $[H_3O^+] = [OH^-]$
c. basic: $[OH^-] > [H_3O^+]$
acidic: $[H_3O^+] > [OH^-]$

Differentiated Instruction

Gifted and Talented 

Have students identify occupations in which people need to measure pH. Interested students may select a local person in an occupation from the list and ask to interview that person about his or her use of pH measurement. Have students write up their interviews as if they were preparing them for a newspaper article.

Discuss 

Introduce pH as a simpler way to express hydrogen-ion concentration. Point out that it is easier to say the pH of a solution is 3.00 than to say that the hydrogen ion concentration is equal to 1.0×10^{-3} moles per liter. For students who have no concept of logarithms, explain that pH is found by taking the negative of the power (exponent) of 10 that expresses the hydrogen ion concentration. Show students this example: $[H] = 0.00010 = 1.0 \times 10^{-4}M$; $pH = -\log_{10}(1.0 \times 10^{-4}) = 4.00$. Most pH values are positive numbers, but negative values are also possible. Assuming 100% ionization, for example, the pH of a 10*M* HCl solution is –1.00.

Discuss 

Have students compare the pH and pOH of a solution by posing a series of questions such as the following: **In an acidic solution with a pH of 3.25, are there any OH^- ions?** *(Yes, there are always some OH^- ions in an aqueous solution.)* **How do you know this?** *(The product of the hydrogen ion concentration and the hydroxide ion concentration must always equal K_w.)* **What is the pOH of this solution?** *($pH + pOH = 14$; $pOH = 14 - 3.25 = 10.75$)*

Table 19.5

Relationship among $[H^+]$, $[OH^-]$, and pH

	$[H^+]$ (mol/L)	$[OH^-]$ (mol/L)	pH	Aqueous system
	1×10^{0}	1×10^{-14}	0.0	1*M* HCl
Increasing acidity ↑	1×10^{-1}	1×10^{-13}	1.0	0.1*M* HCl
	1×10^{-2}	1×10^{-12}	2.0	Gastric juice; Lemon juice
	1×10^{-3}	1×10^{-11}	3.0	
	1×10^{-4}	1×10^{-10}	4.0	Tomato juice
	1×10^{-5}	1×10^{-9}	5.0	Black coffee
	1×10^{-6}	1×10^{-8}	6.0	Milk
Neutral	1×10^{-7}	1×10^{-7}	7.0	Pure water
	1×10^{-8}	1×10^{-6}	8.0	Blood
Increasing basicity ↓	1×10^{-9}	1×10^{-5}	9.0	Sodium bicarbonate, sea water
	1×10^{-10}	1×10^{-4}	10.0	
	1×10^{-11}	1×10^{-3}	11.0	Milk of magnesia; Household ammonia
	1×10^{-12}	1×10^{-2}	12.0	Washing soda
	1×10^{-13}	1×10^{-1}	13.0	0.1*M* NaOH
	1×10^{-14}	1×10^{0}	14.0	1*M* NaOH

Calculating pOH The pOH of a solution equals the negative logarithm of the hydroxide-ion concentration.

$$pOH = -\log[OH^-]$$

A neutral solution has a pOH of 7. A solution with a pOH less than 7 is basic. A solution with a pOH greater than 7 is acidic. A simple relationship between pH and pOH allows you to find either one when the other is known.

$$pH + pOH = 14$$

$$pH = 14 - pOH$$

$$pOH = 14 - pH$$

pH and Significant Figures For pH calculations, you should express the hydrogen-ion concentration in scientific notation. For example, a hydrogen-ion concentration of 0.0010*M*, rewritten as $1.0 \times 10^{-3}M$ in scientific notation, has two significant figures. The pH of this solution is 3.00, with the two numbers to the right of the decimal point representing the two significant figures in the concentration. A solution with a pH of 3.00 is acidic, as shown in Figure 19.10.

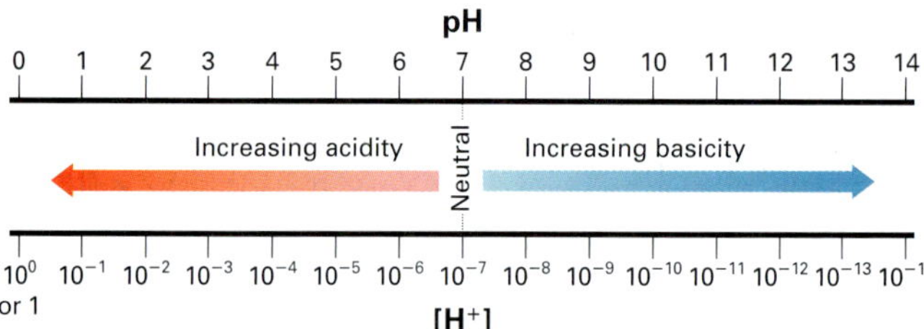

Figure 19.10 The pH scale shows the relationship between pH and the hydrogen-ion concentration. **Interpreting Diagrams** ***What happens to $[H^+]$ as pH increases?***

598 Chapter 19

Most pH values are not whole numbers. For example, milk of magnesia has a pH of 10.5. Using the definition of pH, this means that $[H^+]$ must equal $1 \times 10^{-10.5}M$. Thus $[H^+]$ must be less than $1 \times 10^{-10}M$ (pH 10.0), but greater than $1 \times 10^{-11}M$ (pH 11.0). If $[H^+]$ is written in scientific notation but its coefficient is not 1, then you use a calculator with a log function key to calculate pH.

SAMPLE PROBLEM 19.2

Calculating pH from $[H^+]$

What is the pH of a solution with a hydrogen-ion concentration of $4.2 \times 10^{-10}M$?

1 Analyze ***List the knowns and the unknown.***

Knowns
- $[H^+] = 4.2 \times 10^{-10}M$
- $pH = -\log[H^+]$

Unknown
- pH = ?

2 Calculate ***Solve for the unknown.***

$pH = -\log[H^+]$

Using a calculator, find $\log[H^+]$.

$\log(4.2 \times 10^{-10}) = -9.37675$

$pH = -(-9.38)$

$= 9.38$

3 Evaluate ***Do the results make sense?***

The calculated pH is between 9 ($[H^+] = 1 \times 10^{-9}M$) and 10 ($[H^+] = 1 \times 10^{-10}M$). The pH is rounded to two decimal places because the hydrogen-ion concentration has two significant figures.

Practice Problems

11. Find the pH of each solution.
a. $[H^+] = 1 \times 10^{-4}M$
b. $[H^+] = 0.0015M$

12. What are the pH values of the following solutions, based on their hydrogen-ion concentrations?
a. $[H^+] = 1.0 \times 10^{-12}M$
b. $[H^+] = 0.045M$

You can calculate the hydrogen-ion concentration of a solution if you know the pH. For example, if the solution has a pH of 3.00, then $[H^+] = 1 \times 10^{-3}M$. When the pH is not a whole number, you will need a calculator with a y^x function key to calculate the hydrogen-ion concentration. For example, if the pH is 3.70, the hydrogen-ion concentration is greater than $1 \times 10^{-4}M$ (pH 4.0) and less than $1 \times 10^{-3}M$ (pH 3.0). To get an accurate value, use a calculator as shown in the following example.

CHEMath

Logarithms

The common logarithm (log) of a number (N) is the exponent (*x*) to which the base 10 must be raised to yield the number. If $N = 10^x$, then $\log N = x$.

The use of logarithms allows a large range of values to be conveniently expressed as small, nonexponential numbers. For example, $\log 10^3 = 3$ and $\log 10^6 = 6$. Although there is a range of three orders of magnitude ($10 \times 10 \times 10$ or 1000) between the numbers 10^3 and 10^6, the range of the log values is only 3.

The concentration of hydrogen ions in most aqueous solutions, although small, can vary over many orders of magnitude. Scientists use pH, a logarithmic scale that ranges between 0 and 14, to more conveniently express the hydrogen ion concentration of a solution.

Math Handbook

For help calculating logarithms, go to page R78.

Interactive Textbook

Problem-Solving 19.12 Solve Problem 12 with the help of an interactive guided tutorial.

with **ChemASAP**

Sample Problem 19.2

Answers

11. a. $pH = -\log[H^+] = -\log(1 \times 10^{-4}) = 4.0$

b. $pH = -\log[H^+] = -\log(1.5 \times 10^{-3}) = 2.82$

12. a. $pH = -\log[H^+] = -\log(1.0 \times 10^{-12}) = 12.00$

b. $pH = -\log[H^+] = -\log(4.5 \times 10^{-2}) = 1.35$

Practice Problems Plus L2

What are the pH values of the following three solutions, based on their hydrogen ion concentrations?

a. $[H^+] = 1.0 \times 10^{-11}M$ *(11.00)*
b. $[H^+] = 1.0 \times 10^{-8}M$ *(8.00)*
c. $[H^+] = 0.00001M$ *(5.0)*

CHEMath

Logarithms

The CHEMath feature will help students who are struggling with logarithms. In Section 19.2, focus on solutions of strong acids and bases. Save calculating dissociation constants for Section 19.3.

Math Handbook

For a math refresher and practice, direct students to logarithms, page R78.

Answers to...

Figure 19.10 $[H^+]$ decreases

Section 19.2 (continued)

Sample Problem 19.3

Answers

13. a. $-\log[H^+] = 5.00$; $\log[H^+] = -5.00$; $[H^+] = 1.0 \times 10^{-5}M$
b. $-\log[H^+] = 12.83$; $\log[H^+] = -12.83$; $[H^+] = 1.5 \times 10^{-13}M$

14. a. $-\log[H^+] = 4.00$; $\log[H^+] = -4.00$; $[H^+] = 1.0 \times 10^{-4}M$
b. $-\log[H^+] = 11.55$; $\log[H^+] = -11.55$; $[H^+] = 2.8 \times 10^{-12}M$

Practice Problems Plus L2

What are the hydroxide-ion concentrations for solutions with the following pH values?

a. 4.00 *($1.0 \times 10^{-10}M$)*
b. 8.00 *($1.0 \times 10^{-6}M$)*
c. 12.00 *($1.0 \times 10^{-2}M$)*

A soft drink has a pH of 3.80. **What is the hydrogen-ion concentration in the drink?** *($1.6 \times 10^{-4}M$)*

Measuring pH

TEACHER Demo

pH Indicators L2

Purpose Students observe how pH indicators react to the acidity of their environment.

Materials 0.1% solutions of thymol blue, methyl red, bromthymol blue, and phenolphthalein; test tubes; aqueous buffers spanning pH 4 to pH 10 (Either create a range of buffers spanning pH 4 to pH 10 or higher, or purchase standard buffers with different pH values.)

Procedure Explain that many natural and synthetic pigments are weak acids that change color with varying pH. Transfer a small volume of each buffer to a test tube. Label the tubes with their respective pH values. Add 5 drops of indicator to each test tube. Set up a separate rack of tubes for each indicator and ask students to infer the pH range for which each indicator is most suitable. Students should note the colors and pH values of the solutions in tubes. In addition, add 5 drops of the indicators to two or three tubes containing a solution with an "unknown" pH and have students estimate its pH value using the reference tubes. Ask, **Why would an investigator prefer to use a pH meter to measure pH?** *(Individual indicators are only responsive to pH changes in narrow ranges. Many different indicators are needed to span the entire pH range. A pH meter allows an investigator to collect quantitative values on a continuous basis throughout the entire pH range. The precision and accuracy of pH meters are superior to standard indicators. pH meters can be calibrated and used reliably at different temperatures.)*

Math Handbook

For help calculating logarithms, go to page R78.

SAMPLE PROBLEM 19.3

Using pH to Find $[H^+]$

The pH of an unknown solution is 6.35. What is its hydrogen-ion concentration?

1 Analyze *List the knowns and the unknown.*

Knowns	Unknown
• pH = 6.35	• $[H^+] = ?M$
• $pH = -\log[H^+]$	

2 Calculate *Solve for the unknown.*

First, rearrange the equation for the definition of pH to solve for the unknown.

$$-\log[H^+] = pH$$

Next, substitute the value of pH.

$$-\log[H^+] = 6.35$$

Change the signs on both sides of the equation.

$$\log[H^+] = -6.35$$

Finally, determine the number that has a log of −6.35, the antilog of −6.35, using the 10^x key on the calculator. The antilog of −6.35 is 4.5×10^{-7}. Therefore, $[H^+] = 4.5 \times 10^{-7}M$.

3 Evaluate *Do the results make sense?*

The hydrogen ion concentration must be between $1 \times 10^{-6}M$ (pH = 6) and $1 \times 10^{-7}M$ (pH = 7). The answer is rounded to two significant figures because the pH was measured to two decimal places.

Practice Problems

13. Calculate $[H^+]$ for each solution.
a. pH = 5.00
b. pH = 12.83

14. What are the hydrogen-ion concentrations for solutions with the following pH values?
a. 4.00
b. 11.55

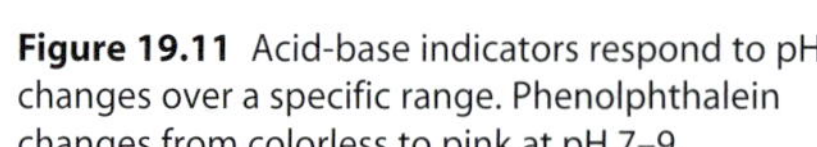

Interactive Textbook

Problem-Solving 19.14 Solve Problem 14 with the help of an interactive guided tutorial.

with ChemASAP

Measuring pH

People need to be able to measure the pH of the solutions they use. From maintaining the correct acid-base balance in a swimming pool, to creating soil conditions ideal for plant growth, to making medical diagnoses, pH measurements have valuable applications. For preliminary pH measurements and for small-volume samples, an indicator such as the one shown in Figure 19.11 is often used. For precise and continuous measurements, a pH meter is preferred.

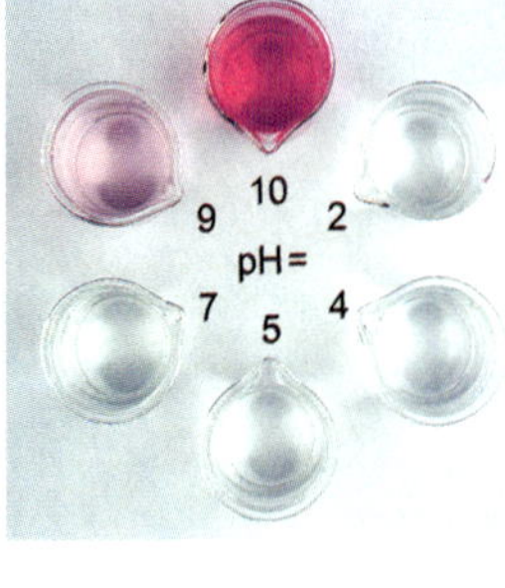

Figure 19.11 Acid-base indicators respond to pH changes over a specific range. Phenolphthalein changes from colorless to pink at pH 7–9.

If you know the $[OH^-]$ of a solution, you can find its pH. The ion-product for water defines the relationship between $[H^+]$ and $[OH^-]$. Therefore, you can determine $[H^+]$ by dividing K_w by the known $[OH^-]$.

SAMPLE PROBLEM 19.4

Calculating pH from $[OH^-]$

What is the pH of a solution if $[OH^-] = 4.0 \times 10^{-11}M$?

1 Analyze ***List the knowns and the unknown.***

Knowns
- $[OH^-] = 4.0 \times 10^{-11}M$
- $K_w = [OH^-] \times [H^+] = 1 \times 10^{-14}$
- $pH = -\log[H^+]$

Unknown
- pH = ?

2 Calculate ***Solve for the unknown.***

To calculate pH, first calculate $[H^+]$ by using the definition of K_w.

$$K_w = [OH^-] \times [H^+]$$

$$[H^+] = \frac{K_w}{[OH^-]} = \frac{1.0 \times 10^{-14}}{4.0 \times 10^{-11}} = 0.25 \times 10^{-3}M$$

$$= 2.5 \times 10^{-4}M$$

With the value of $[H^+]$ determined, use the definition of pH to solve for the pH.

$$pH = -\log[H^+] = -\log(2.5 \times 10^{-4})$$

A calculator indicates that $\log 2.5 \times 10^{-4}M = -3.60205$, therefore

$$pH = -(-3.60)$$
$$= 3.60$$

3 Evaluate ***Do the results make sense?***

A solution in which $[OH^-]$ is less than $1 \times 10^{-7}M$ would be acidic because $[H^+]$ would be greater than $1 \times 10^{-7}M$. The pH is expressed to 2 decimal places because $[OH^-]$ is expressed to 2 significant figures.

Practice Problems

15. Calculate the pH of each solution.
- **a.** $[OH^-] = 4.3 \times 10^{-5}M$
- **b.** $[OH^-] = 4.5 \times 10^{-11}M$

16. Calculate the pH of each solution.
- **a.** $[H^+] = 5.0 \times 10^{-5}M$
- **b.** $[H^+] = 8.3 \times 10^{-10}M$

Acid-Base Indicators An indicator (HIn) is an acid or a base that undergoes dissociation in a known pH range. **An indicator is a valuable tool for measuring pH because its acid form and base form have different colors in solution.** The following generalized equation represents the dissociation of an indicator (HIn).

$$\underset{\text{Acid form}}{HIn(aq)} \underset{H^+}{\overset{OH^-}{\rightleftharpoons}} H^+(aq) + \underset{\text{Base form}}{In^-(aq)}$$

The acid form dominates the dissociation equilibrium at low pH (high $[H^+]$), and the base form dominates the equilibrium at high pH (high $[OH^-]$).

Problem-Solving 19.15 Solve Problem 15 with the help of an interactive guided tutorial.

with **ChemASAP**

Sample Problem 19.4

Answers

15. a. $[H^+] = K_w/[OH^-] = (1.0 \times 10^{-14})/(4.3 \times 10^{-5}) = 2.3 \times 10^{-10}M$; $pH = -\log(2.3 \times 10^{-10}) = 9.63$

b. $[H^+] = K_w/[OH^-] = (1.0 \times 10^{-14})/(4.5 \times 10^{-11}) = 2.2 \times 10^{-4}M$; $pH = -\log(2.2 \times 10^{-4}) = 3.65$

16. a. $pH = -\log(5.0 \times 10^{-5}) = 4.30$

b. $pH = -\log(8.3 \times 10^{-10}) = 9.08$

Practice Problems Plus 

Calculate the pH for each solution.

a. $[OH^-] = 2.5 \times 10^{-11}M$ *(3.40)*

b. $[OH^-] = 6.5 \times 10^{-8}M$ *(6.81)*

c. $[OH^-] = 8.7 \times 10^{-5}M$ *(9.94)*

Math Handbook

For a math refresher and practice, direct students to logarithms, page R78.

TEACHER Demo

Observing pH Change L2

Purpose Students observe a color change resulting from a change in pH.

Materials Lemon juice, tea, antacid tablet, glass of 10% grape juice, household ammonia

Procedure Without explanation, add lemon juice to tea, or an antacid tablet to grape juice, or a few drops of ammonia to grape juice. Have the students try to explain the changes they observe.

Expected Outcome Color changes occur when lemon juice, an antacid tablet, or ammonia are added. The pigments in tea and grape juice are weak acids that change color with varying pH.

Relate 

Point out to students that the pH scale can be compared to the Richter scale, which measures the strength of earthquakes. On each scale, a change of one unit represents a tenfold change in the value being measured. On the Richter scale, for example, a tremor measuring 4.0 is ten times stronger than one measuring 3.0. On the pH scale, the hydrogen-ion concentration of a solution with a pH of 3.0 is ten times greater than that of a solution with a pH of 4.0.

Interpreting Graphs L2

a. thymol blue

b. The indicators change color over a very limited range of pH. This allows for a more accurate reading of pH change.

c. bromphenol blue or bromcresol green

Enrichment Question L3

Which indicator(s) would best show a change in pH from 4.6 to 4.9?

(Using both bromcresol green and methyl red would give the best results.)

Figure 19.12 Indicators change color at a different pH.

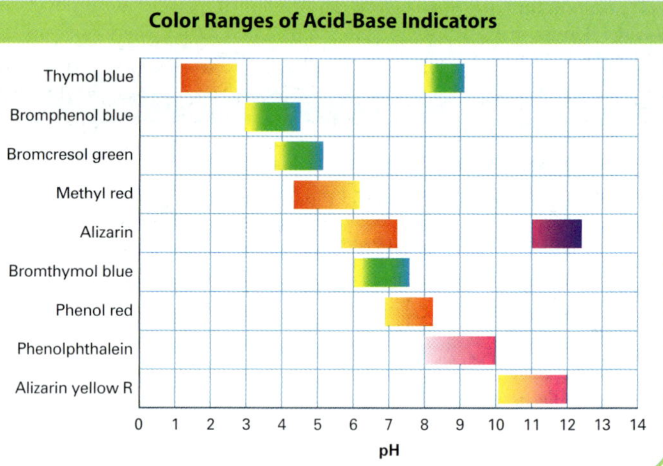

INTERPRETING GRAPHS

a. Identify Which indicator changes color in a solution with a pH of 2?

b. Compare and Contrast What do you notice about the range over which each indicator changes color?

c. Apply Concepts Which indicator would you choose to show that a solution has changed from pH 3 to pH 5?

For each indicator, the change from dominating acid form to dominating base form occurs in a narrow range of approximately two pH units. Within this range, the color of the solution is a mixture of the colors of the acid and the base forms. Knowing the pH range over which this color change occurs can give you a rough estimate of the pH of a solution. At all pH values below this range, you would see only the color of the acid form. At all pH values above this range, you would see only the color of the base form. You could get a more precise estimate of the pH of the solution by repeating the experiment with indicators that have different pH ranges for their color changes. Many different indicators are needed to span the entire pH spectrum. Figure 19.12 shows the pH ranges of some commonly used indicators.

Indicators have certain characteristics that limit their usefulness. The listed pH values of indicators are usually given for 25°C. At other temperatures, an indicator may change color at a different pH. If the solution being tested is not colorless, the color of the indicator may be distorted. Dissolved salts in a solution may also affect the indicator's dissociation. Using indicator strips can help overcome these problems. An indicator strip is a piece of paper or plastic impregnated with an indicator. The paper is dipped into an unknown solution and compared with a color chart to measure the pH. Some indicator paper is impregnated with multiple indicators. The colors that result, which cover a wide pH range, are shown in Figure 19.13.

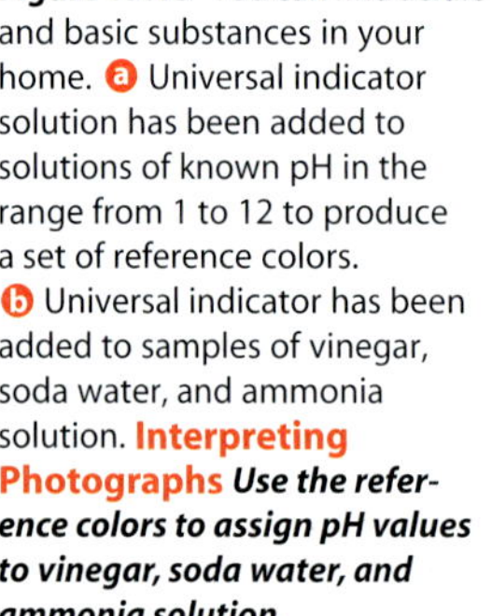

Figure 19.13 You can find acidic and basic substances in your home. **a** Universal indicator solution has been added to solutions of known pH in the range from 1 to 12 to produce a set of reference colors. **b** Universal indicator has been added to samples of vinegar, soda water, and ammonia solution. **Interpreting Photographs** ***Use the reference colors to assign pH values to vinegar, soda water, and ammonia solution.***

Figure 19.14 Altering soil pH can affect the development of plants. (a) In acidic soils, hydrangeas produce blue flowers. (b) In basic soils, hydrangeas produce pink flowers. Evergreen plants (c) suffer from chlorosis, a yellowing of the foliage (d) if soil pH is too basic.

pH Meters A pH meter makes rapid, accurate pH measurements. Most chemistry laboratories have a pH meter. A pH meter connected to a computer or chart recorder can be used to make a continuous recording of pH changes. As you can see in Figure 19.15, the pH meter gives a direct readout of pH.

A pH meter is often easier to use than liquid indicators or indicator strips. Measurements of pH obtained with a pH meter are typically accurate to within 0.01 pH unit of the true pH. The color and cloudiness of the unknown solution do not affect the accuracy of the pH value obtained. Hospitals use pH meters to find small but meaningful changes of pH in blood and other body fluids. Sewage, industrial effluents, and soil pH are also easily monitored with a pH meter.

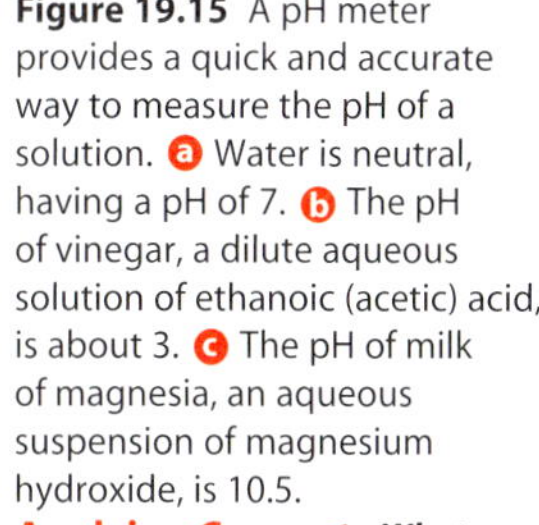

Figure 19.15 A pH meter provides a quick and accurate way to measure the pH of a solution. (a) Water is neutral, having a pH of 7. (b) The pH of vinegar, a dilute aqueous solution of ethanoic (acetic) acid, is about 3. (c) The pH of milk of magnesia, an aqueous suspension of magnesium hydroxide, is 10.5.
Applying Concepts ***What are some advantages of using a pH meter rather than an indicator?***

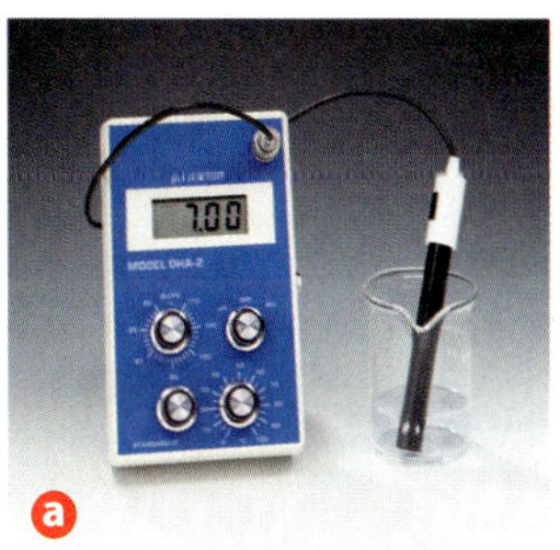

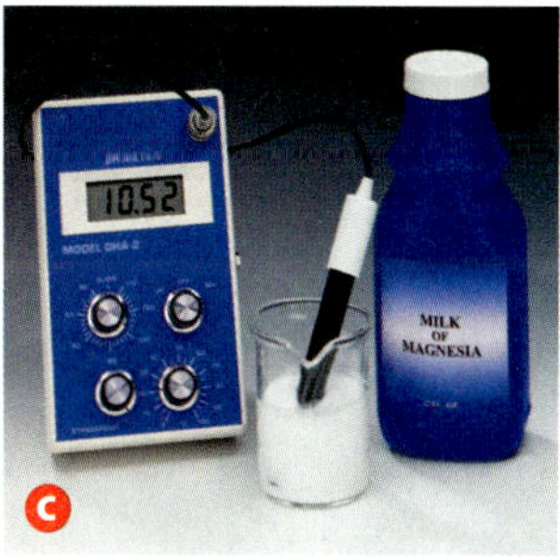

CLASS Activity

Comparing pH Indicators and pH Meters L2

Purpose Students compare pH indicators and pH meters.

Materials chemical supply catalogs

Procedure Allow students to compare the costs and capabilities of pH meters versus other quantitative acid-base indicators. Some pH meters are hand held devices, while others are larger, bench instruments, which can be connected to computer hardware and are intended for more detailed analyses. Students should compare the resolution and precision of each of the instruments and pH indicators. Ask, **What are some advantages and uses of hand held devices? Why would a chemist want to have a pH meter that can be calibrated at different temperatures?** *(Hand-held pH meters are portable and, therefore, convenient to use in the field. Outdoor temperatures may vary widely.)*

Answers to...

Figure 19.13 vinegar: 2; soda water: 4; ammonia solution: 10 to 11

Figure 19.15 A pH meter is more accurate, faster, and provides continuous readings.

3 ASSESS

Evaluate Understanding L2

Have students use equations to describe the relationship between the concentrations of hydrogen and hydroxide ions in pure water and to show what happens to the equilibrium when HCl is added to water. Repeat the procedure for NaOH. *(At 25°C, $[H^+] \times [OH^-] = 1.0 \times 10^{-14}$. When acid is added, the $[H^+]$ increases and the $[OH^-]$ decreases. When base is added the $[OH^-]$ increases and the $[H^+]$ decreases.)*

Quick LAB

Indicators from Natural Sources L2

Objective Students will measure the pH of various household materials using a natural indicator and an indicator chart.

Prep Time 20 minutes

Class Time If the period allotted to lab work is short, spread the work over two days.

Safety Perform this lab in a well-ventilated room.

Expected Outcome Students will use a natural indicator to determine if tested materials are acidic, basic, or neutral.

Analyze and Conclude

1. The initial color is purple. In an acidic solution, the indicator is red; in a neutral solution, it is blue-purple; and in a basic solution, it is green.
2. Changes in the relative number of H^+ and OH^- ions present are responsible for the color changes.
3. Answers will vary depending on the materials chosen.
4. Personal hygiene items mainly test neutral; cleaning materials, such as soap, mainly test basic.

Elements Handbook

Students will note that the pH affects the relative concentrations of ClO^- and HClO in pool water.

Quick LAB

Indicators from Natural Sources

Purpose

To measure the pH of various household materials by using a natural indicator to make an indicator chart.

Materials

- knife
- red cabbage leaves
- 1-cup measure
- hot water
- 2 jars
- clean white cloth
- teaspoon
- tape
- 3 sheets of plain white paper
- pencil
- ruler
- 10 clear plastic cups
- white vinegar (CH_3COOH)
- baking soda ($NaHCO_3$)
- household ammonia
- dropper
- various household items

Procedure

1. Put $\frac{1}{2}$ cup of finely chopped red cabbage leaves in a jar and add $\frac{1}{2}$ cup of hot water. Stir and crush the leaves with a spoon. Continue the extraction until the water is distinctly colored.
2. Strain the extract through a piece of cloth into a clean jar. This liquid is your natural indicator.
3. Tape three sheets of paper end to end. Draw a line along the center and label it at 5-cm intervals with the numbers 1 to 14. This is your pH scale.
4. Pour your indicator to about 1-cm depth into each of three plastic cups. To one cup, add several drops of vinegar, to the second add a pinch of baking soda, and to the third add several drops of ammonia. The resulting colors indicate pH values of about 3, 9, and 11, respectively. Place these colored positions on your pH scale.
5. Repeat Step 4 for household items such as table salt, borax, milk, lemon juice, laundry detergent, dish detergent, milk of magnesia, mouthwash, toothpaste, shampoo, and carbonated beverages.

Analyze and Conclude

1. What was the color of the indicator at acidic, neutral, and basic conditions?
2. What chemical changes were responsible for the color changes?
3. Label the materials you tested as acidic, basic, or neutral.
4. Which group contains items used for cleaning or for personal hygiene?

19.2 Section Assessment

17. **Key Concept** What is the relationship between $[H^+]$ and $[OH^-]$ in an aqueous solution?
18. **Key Concept** What is true about the relative concentrations of hydrogen ions and hydroxide ions in each kind of solution?
    a. basic **b.** acidic **c.** neutral
19. **Key Concept** What is true about the colors of a pH indicator?
20. Determine the pH of each solution.
    a. $[H^+] = 1 \times 10^{-6}M$ **b.** $[H^+] = 0.00010M$
    c. $[OH^-] = 1 \times 10^{-2}M$ **d.** $[OH^-] = 1 \times 10^{-11}M$
21. What are the hydroxide-ion concentrations for solutions with the following pH values?
    a. 6.00 **b.** 9.00 **c.** 12.00

Elements Handbook

Group 7A Elements Hypochlorite salts are used to disinfect swimming pools. Use page R34 of the Elements Handbook to find out how the pH of swimming pool water is regulated to maintain the necessary concentration of hypochlorous acid, HOCl.

Assessment 19.2 Test yourself on the concepts in Section 19.2.

with ChemASAP

Interactive Textbook

If your class subscribes to the Interactive Textbook, use it to review key concepts in Section 19.2.

with ChemASAP

Section 19.2 Assessment

17. $[H^+] \times [OH^-] = 1.0 \times 10^{-14}$; when $[H^+]$ in a solution increases, the $[OH^-]$ decreases.
18. **a.** Hydroxide ion concentration is greater.
    b. Hydrogen ion concentration is greater.
    c. The concentrations are equal.
19. The color of HIn (aq) must be different than the color of In^- (aq)
20. **a.** 6.0 **21. b.** 4.00 **22. c.** 12.0 **23. d.** 3.0
21. **a.** $1.0 \times 10^{-8}M$ **22. b.** $1.0 \times 10^{-5}M$
    c. $1.0 \times 10^{-2}M$

19.3 Strengths of Acids and Bases

Connecting to Your World

Lemons and grapefruits have a sour taste because they contain citric acid. When you make lemonade, or cut up a grapefruit, you probably do not wear safety goggles or chemical-resistant clothing even though you are working with an acid. However, some acids require such precautions. For example, sulfuric acid is a widely used industrial chemical that can quickly cause severe burns if it comes into contact with skin. In this section, you will learn what makes some acids weak acids and other acids strong acids.

Strong and Weak Acids and Bases

Acids are classified as strong or weak depending on the degree to which they ionize in water. In general, **strong acids** are completely ionized in aqueous solution. Hydrochloric acid and sulfuric acid are strong acids.

$$HCl(g) + H_2O(l) \longrightarrow H_3O^+(aq) + Cl^-(aq) \text{ (100\% ionized)}$$

Weak acids ionize only slightly in aqueous solution. The ionization of ethanoic acid (acetic acid), a typical weak acid, is not complete.

$$\underset{\text{Ethanoic acid}}{CH_3COOH(aq)} + \underset{\text{Water}}{H_2O(l)} \rightleftharpoons \underset{\text{Hydronium ion}}{H_3O^+(aq)} + \underset{\text{Ethanoate ion}}{CH_3COO^-(aq)}$$

Table 19.6 shows the relative strengths of some common acids and bases.

Table 19.6

Relative Strengths of Common Acids and Bases

Substance	Formula	Relative Strength
Hydrochloric acid	HCl	Strong Acid
Nitric acid	HNO_3	
Sulfuric acid	H_2SO_4	
Phosphoric acid	H_3PO_4	↑ Increasing strength of acid
Ethanoic acid	CH_3COOH	
Carbonic acid	H_2CO_3	
Hypochlorous acid	HClO	
Neutral Solution		Neutral Solution
Ammonia	NH_3	↓ Increasing strength of base
Sodium silicate	Na_2SiO_3	
Calcium hydroxide	$Ca(OH)_2$	Strong Base
Sodium hydroxide	NaOH	
Potassium hydroxide	KOH	

Guide for Reading

Key Concepts

- How does the value of an acid dissociation constant relate to the strength of an acid?
- How can you calculate an acid dissociation constant (K_a) of a weak acid?

Vocabulary

strong acids
weak acids
acid dissociation constant (K_a)
strong bases
weak bases
base dissociation constant (K_b)

Reading Strategy

Comparing and Contrasting Compare the concentrations of H^+ and OH^- in solutions of strong acids and weak acids. Explain these differences. Do the same for strong bases and weak bases.

19.3

1 FOCUS

Objectives

19.3.1 Define strong acids and weak acids.

19.3.2 Describe how an acid's strength is related to the value of its acid dissociation constant.

19.3.3 Calculate an acid dissociation constant (K_a) from concentration and pH measurements.

19.3.4 Order acids by strength according to their acid dissociation constants (K_a).

19.3.5 Order bases by strength according to their base dissociation constants (K_b).

Guide for Reading

Build Vocabulary 

Graphic Organizer Students can make two columns labeled *Acid* and *Base*. Under each column, write the related vocabulary words and their definitions.

Reading Strategy 

Preview Students can preview the information in this section by skimming the headings, visuals, and boldfaced material.

2 INSTRUCT

Connecting to Your World

Have students study the photograph and read the text that opens the section. Ask, **What is an example of a weak acid?** *(lemon or grapefruit)* **What is an example of a strong acid?** *(sulfuric acid)*

Strong and Weak Acids and Bases

Use Visuals 

Table 19.6 Have students study the table. Ask, **Which is the weakest acid in the table?** *(hypochlorous acid)* **Which is the weakest base?** *(ammonia)*

Section Resources

Print

- **Guided Reading and Study Workbook,** Section 19.3
- **Core Teaching Resources,** Section 19.3 Review
- **Transparencies,** T223–T224

Technology

- **Interactive Textbook with ChemASAP,** Problem-Solving 19.23, Assessment 19.3
- **Go Online,** Section 19.3
- **Virtual Chemistry Labs,** Lab 27

Use Visuals 

Figure 19.16. Explain to students that the bar graphs compare the extent of ionization or dissociation of strong to weak acids. Ask, **What is the primary difference between strong and weak acids?** *(Strong acids are nearly 100% dissociated, while weak acids are only partially dissociated.)*

Interpreting Graphs 

a. The strong acid dissociates completely, forming equal amounts of ions.

b. The amount of acid that dissociates is equal to the amount of the two resulting ions.

c. The bar graph for the first dissociation of oxalic acid should show low concentrations of H^+ and $HOOCCOO^-$. The bar graph for the second dissociation should show even lower concentrations of H^+ and $OOCCOO^{2-}$. Perceptive students may note that $[H^+]$ in the second bar graph actually should be much higher than $[OOCCOO^{2-}]$ because of the H^+ ions produced in the first dissociation.

Enrichment Question 

Relate the amount of dissociation with the properties of strong and weak acids. *(With increasing amount of dissociation, the acid becomes stronger.)*

Acid Dissociation Constant In an aqueous solution of ethanoic acid, fewer than 1% of ethanoic acid molecules are ionized at any instant. Therefore, ethanoic acid is considered a weak acid. Figure 19.16 compares the extent of dissociation of strong, weak, and very weak acids. A strong acid completely dissociates in water. As a result, $[H_3O^+]$ is high. Hydrochloric acid and sulfuric acid are examples of strong acids. On the other hand, weak acids, such as boric acid and carbonic acid, remain largely undissociated. The $[H_3O^+]$ of a weak acid is low.

You can write the equilibrium-constant expression from the balanced chemical equation. The equilibrium-constant expression for ethanoic acid is shown below.

$$K_{eq} = \frac{[H_3O^+] \times [CH_3COO^-]}{[CH_3COOH] \times [H_2O]}$$

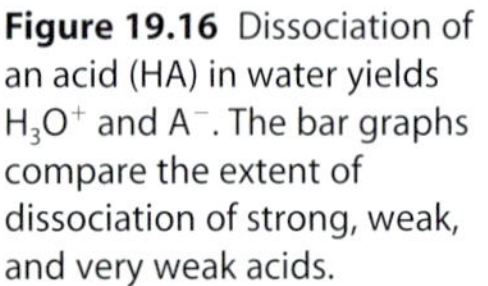

Figure 19.16 Dissociation of an acid (HA) in water yields H_3O^+ and A^-. The bar graphs compare the extent of dissociation of strong, weak, and very weak acids.

INTERPRETING GRAPHS

a. Explain In the graph for the strong acid, why are the heights of the H_3O^+ and A^- bars the same as the height of HA bar?

b. Inferring In the graph for the weak acid, why is the height of the H_3O^+ bar the same as the distance from the top of the second HA bar to the dotted line?

c. Apply Concepts Draw a bar graph for the dissociation of the weak diprotic acid, oxalic acid. Be sure to include the first and second dissociation.

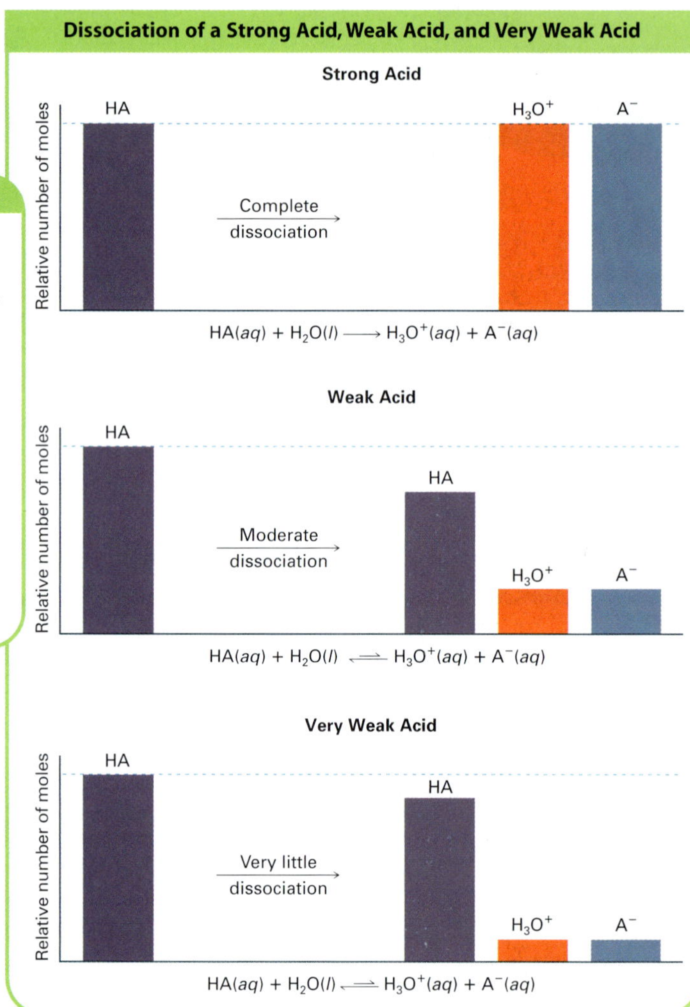

Table 19.7

Dissociation Constants of Weak Acids

Acid	Ionization	K_a (25°C)
Oxalic acid	$HOOCCOOH(aq) \rightleftharpoons H^+(aq) + HOOCCOO^-(aq)$	5.6×10^{-2}
	$HOOCCOO^-(aq) \rightleftharpoons H^+(aq) + OOCCOO^{2-}(aq)$	5.1×10^{-5}
Phosphoric acid	$H_3PO_4(aq) \rightleftharpoons H^+(aq) + H_2PO_4^-(aq)$	7.5×10^{-3}
	$H_2PO_4^-(aq) \rightleftharpoons H^+(aq) + HPO_4^{2-}(aq)$	6.2×10^{-8}
	$HPO_4^{2-}(aq) \rightleftharpoons H^+(aq) + PO_4^{3-}(aq)$	4.8×10^{-13}
Methanoic acid	$HCOOH(aq) \rightleftharpoons H^+(aq) + HCOO^-(aq)$	1.8×10^{-4}
Benzoic acid	$C_6H_5COOH(aq) \rightleftharpoons H^+(aq) + C_6H_5COO^-(aq)$	6.3×10^{-5}
Ethanoic acid	$CH_3COOH(aq) \rightleftharpoons H^+(aq) + CH_3COO^-(aq)$	1.8×10^{-5}
Carbonic acid	$H_2CO_3(aq) \rightleftharpoons H^+(aq) + HCO_3^-(aq)$	4.3×10^{-7}
	$HCO_3^-(aq) \rightleftharpoons H^+(aq) + CO_3^{2-}(aq)$	4.8×10^{-11}

For dilute solutions, the concentration of water is a constant. It can be combined with K_{eq} to give an acid dissociation constant. An **acid dissociation constant (K_a)** is the ratio of the concentration of the dissociated (or ionized) form of an acid to the concentration of the undissociated (nonionized) form. The dissociated form includes both the H_3O^+ and the anion.

$$K_{eq} \times [H_2O] = K_a = \frac{[H_3O^+] \times [CH_3COO^-]}{[CH_3COOH]}$$

The acid dissociation constant (K_a) reflects the fraction of an acid in the ionized form. For this reason, dissociation constants are sometimes called ionization constants. If the value of the dissociation constant is small, then the degree of dissociation or ionization of the acid in the solution is small. **Weak acids have small K_a values. The stronger an acid is, the larger is its K_a value.** A larger value of K_a means the dissociation or ionization of the acid is more complete. For example, nitrous acid (HNO_2) has a K_a of 4.4×10^{-4}, whereas ethanoic acid (acetic acid) has a K_a of 1.8×10^{-5}. This means that nitrous acid is more ionized in solution than ethanoic acid. Nitrous acid is a stronger acid than ethanoic acid.

Therefore, a strong acid has a higher $[H_3O^+]$ and a large dissociation constant. Conversely, a weak acid has a low $[H_3O^+]$ and a small dissociation constant. Diprotic and triprotic acids lose their hydrogens one at a time. Each ionization reaction has a separate dissociation constant. Thus, phosphoric acid (H_3PO_4) has three dissociation constants to go with its three ionizable hydrogens. Table 19.7 shows the ionization reactions and dissociation constants of some common weak acids, ranked by the value of the first dissociation constant of each acid.

What distinguishes a strong acid from a weak acid?

Figure 19.17 A solution of phosphoric acid is used to remove lime deposits from plumbing fixtures. Lime deposits are calcium carbonate.

Facts and Figures

Losing Ions

Diprotic and triprotic acids lose their hydrogen ions one at a time. Each successive ion is more difficult to lose from the negative ion. Phosphoric acid (H_3PO_4) loses one hydrogen ion relatively easily, leaving an $H_2PO_4^-$ ion. This negatively charged ion loses fewer hydrogen ions than phosphoric acid, and even fewer hydrogen ions leave HPO_4^{2-}. Although phosphoric acid contains three acidic hydrogens, it does not have a high degree of ionization and is, therefore, a relatively weak acid.

Discuss **L2**

Remind students that, in solution, the negative ion of an acid is a base. Explain that strong acids, such as nitric acid, dissociate almost completely because the negative ion is a very weak base. That is, its tendency to combine with a hydrogen ion is slight. In contrast, the dissociation of a weak acid, such as carbonic acid, is limited, because its negative ion is a relatively strong base. These ions combine with most of the available hydrogen ions to form the undissociated form of the acid again.

The terms strong and weak, as applied to acids, are often confused with the concept of concentration. Stress that the dissociation of an acid or base into ions involves the establishment of equilibrium. The terms strong and weak refer to the position of the equilibrium. When a strong acid or base dissolves, the equilibrium favors the products. When a weak acid or base dissolves, the equilibrium favors the reactants.

Show students that the extent to which products or reactants are favored can be determined from the values of K_a or K_b. Also, discuss how values of K_a and K_b can be used to compare the strengths of acids and bases.

Answers to...

Stronger acids have larger K_a values and ionize more completely to produce H_3O^+ ions.

CLASS Activity

Shampoo Survey 

Purpose Students compare shampoos for normal, dry, or oily hair.

Materials different brands and types of shampoo; 10-mL graduated cylinder; beakers; universal indicator paper

Procedure Have groups of students conduct a survey of shampoo products. Each group should gather data on a different brand and type of shampoo. Decide as a class what information should be included in the survey and how the data will be reported. To assure uniform pH testing, have groups prepare 1% shampoo solutions and use universal indicator paper.

Expected Outcome Shampoos labeled for use on normal, dry, or oily hair are formulated by controlling the strength and amount of the synthetic detergent. The quantity of the active ingredient controls the "defatting" action, which removes oil from the hair.

Base Dissociation Constant Just as there are strong acids and weak acids, there are also strong bases and weak bases. **Strong bases** dissociate completely into metal ions and hydroxide ions in aqueous solution. Some strong bases, such as calcium hydroxide and magnesium hydroxide, are not very soluble in water. The small amounts of these bases that do dissolve dissociate completely.

Weak bases react with water to form the hydroxide ion and the conjugate acid of the base. Ammonia is an example of a weak base.

$$\underset{\text{Ammonia}}{NH_3(aq)} + \underset{\text{Water}}{H_2O(l)} \rightleftharpoons \underset{\text{Ammonium ion}}{NH_4^+(aq)} + \underset{\text{Hydroxide ion}}{OH^-(aq)}$$

The equilibrium of this equation greatly favors the reverse reaction. Only about 1% of the ammonia is present as NH_4^+, the conjugate acid of NH_3. The concentrations of NH_4^+ and OH^- are low and equal. The equilibrium-constant expression for the reaction of ammonia with water is

$$K_{eq} = \frac{[NH_4^+] \times [OH^-]}{[NH_3] \times [H_2O]}$$

The concentration of water is constant in dilute solutions. It can be combined with K_{eq} to give a base dissociation constant (K_b).

$$K_{eq} \times [H_2O] = K_b = \frac{[NH_4^+] \times [OH^-]}{[NH_3]}$$

In general, the **base dissociation constant (K_b)** is the ratio of the concentration of the conjugate acid times the concentration of the hydroxide ion to the concentration of the base. The general form of this equation is as follows.

$$K_b = \frac{[\text{conjugate acid}] \times [OH^-]}{[\text{base}]}$$

The magnitude of K_b indicates the ability of a weak base to compete with the very strong base OH^- for hydrogen ions. Because bases such as ammonia are weak relative to the hydroxide ion, K_b for such bases is usually small. The K_b for ammonia is 1.8×10^{-5}. The smaller the value of K_b, the weaker is the base.

Checkpoint ***Which is larger, the K_a of a weak acid or the K_a of a strong acid?***

Figure 19.18 Window cleaners use an ammonia solution to clean glass because it is a weak base. Being a weak base also makes ammonia relatively safe to use.

Table 19.8

Concentration of Some Common Laboratory Acids and Bases

Acid or base	Concentration: Moles/liter (molarity)	Concentration: Grams/liter
Concentrated hydrochloric acid	12	438
Dilute hydrochloric acid	6	219
Concentrated sulfuric acid	18	1764
Dilute sulfuric acid	6	588
Concentrated phosphoric acid	15	1470
Concentrated nitric acid	16	1008
Dilute nitric acid	6	378
Ethanoic acid, glacial	17	1020
Ethanoic acid, dilute	6	360
Dilute sodium hydroxide	6	240
Concentrated aqueous ammonia	15	255
Dilute aqueous ammonia	6	102

Concentration and Strength The words *concentrated* and *dilute* indicate how much of an acid or base is dissolved in solution. These terms refer to the number of moles of the acid or base in a given volume. The words *strong* and *weak* refer to the extent of ionization or dissociation of an acid or base. They indicate how many of the particles ionize or dissociate into ions. Hydrochloric acid (HCl)(*aq*) is a strong acid; it is completely dissociated into ions. Gastric juice in the stomach is a dilute solution of hydrochloric acid. A relatively small number of HCl molecules are present in a given volume of gastric juice, but they are all dissociated into ions. A sample of hydrochloric acid added to a large volume of water becomes more dilute, but it is still a strong acid. Vinegar is a dilute solution of a weak acid, ethanoic acid. Pure ethanoic acid (glacial acetic acid) is still a weak acid, even though it is highly concentrated. Solutions of ammonia can be dilute or concentrated, depending on the amount of ammonia dissolved in a given volume of water. In any solution of ammonia, however, whether concentrated or dilute, ammonia will be a weak base because the amount of ionization will be small. Table 19.8 lists the concentrations of acids and bases commonly found in the laboratory.

Calculating Dissociation Constants

You can calculate the acid dissociation constant (K_a) of a weak acid or the base dissociation constant (K_b) of a weak base from experimental data. **To find the K_a of a weak acid or the K_b of a weak base, substitute the measured concentrations of all the substances present at equilibrium into the expression for K_a or K_b.** For a weak acid, you can determine these concentrations experimentally if you know the initial molar concentration of the acid and the pH (or $[H_3O^+]$) of the solution at equilibrium.

Calculating Dissociation Constants

Discuss

L2

Explain how the dissociation constant of an acid or base can be determined experimentally from the concentration of the solution and its pH. The calculation is based on the assumption that the acid molecule forms an equal number of hydrogen ions and negative ions when it dissociates. This is true only if there is no additional source of either hydrogen ions or the negative ions.

Point out that calculating K_a is much more complicated in the case of a polyprotic acid. The dissociation of H_3PO_4, for example, involves three separate ionizations, each with its own K_a.

Answers to...

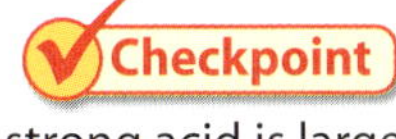

Checkpoint The K_a of a strong acid is larger.

Sample Problem 19.5

Answers

22. $K_a = (4.2 \times 10^{-3})(4.2 \times 10^{-3})/(9.58 \times 10^{-2}) = 1.8 \times 10^{-4}$

23. $K_a = (9.86 \times 10^{-4})(9.86 \times 10^{-4})/(2.0 \times 10^{-1}) = 4.89 \times 10^{-6}$

Practice Problems Plus 

A solution of a weak acid, exactly 0.500*M*, has a $[H^+] = 5.77 \times 10^{-6}M$

a. What is the pH of this solution? *(5.239)*

b. What is the value of K_a for this acid? *(6.66×10^{-11})*

Math Handbook

For a math refresher and practice, direct students to scientific notation, page R56.

Careers in Chemistry

Stone Conservator

Have students do research about the sources and causes of acid rain. *(Acid rain is thought to be due principally to the release of sulfur oxides and nitrogen oxides into the atmosphere.)* Ask, **How do weather patterns affect the distribution of environmental damage due to acid rain?** *(Prevailing winds generally carry pollutants from west to east.)* **What types of vegetation, if any, are most resistant to acid rain, or tend to thrive in acidic soils?** *(Students may wish to speak to staff at nurseries for help answering this question.)* **What types of conservation efforts are used to adjust the pH of lakes and rivers that become too acidic?** *(A special slurry of $CaCO_3$, or limestone, is sometimes added to lakes and rivers to control pH.)*

Have students research chemistry-related careers in the library or on the Internet. Students can then construct a table that describes the nature of the work, educational and training requirements, employment outlook, working conditions, and other necessary information.

In general, for an acid in water you can find K_a by substituting the concentrations of the acid, [HA], the negative ion from the dissociation of the acid, $[A^-]$, and the hydrogen ion, $[H^+]$, into the equation below.

$$K_a = \frac{[H^+][A^-]}{[HA]}$$

SAMPLE PROBLEM 19.5

Calculating a Dissociation Constant

A 0.1000*M* solution of ethanoic acid is only partially ionized. From measurements of the pH of the solution, $[H^+]$ is determined to be $1.34 \times 10^{-3}M$. What is the acid dissociation constant (K_a) of ethanoic acid?

1 Analyze ***List the knowns and the unknown.***

Knowns

- [ethanoic acid] = 0.1000*M*
- $[H^+] = 1.34 \times 10^{-3}M$
- $CH_3COOH(aq) + H_2O(l) \rightleftharpoons H_3O^+(aq) + CH_3COO^-(aq)$
- $K_a = \dfrac{[H^+] \times [CH_3COO^-]}{[CH_3COOH]}$

Unknown

- $K_a = ?$

2 Calculate ***Solve for the unknown.***

Each molecule of CH_3COOH that ionizes gives an H^+ and a CH_3COO^- ion. Therefore, at equilibrium $[H^+] = [CH_3COO^-] = 1.34 \times 10^{-3}M$. The equilibrium concentration of CH_3COOH is the initial concentration minus the concentration of the ionized acid or $(0.1000 - 0.00134)M = 0.0987M$.

Concentration	$[CH_3COOH]$	$[H^+]$	$[CH_3COO^-]$
Initial	0.1000	0	0
Change	-1.34×10^{-3}	1.34×10^{-3}	1.34×10^{-3}
Equilibrium	0.0987	1.34×10^{-3}	1.34×10^{-3}

Substitute the equilibrium values into the expression for K_a.

$$K_a = \frac{[H^+] \times [CH_3COO^-]}{[CH_3COOH]} = \frac{(1.34 \times 10^{-3}) \times (1.34 \times 10^{-3})}{0.0987}$$

$$= 1.82 \times 10^{-5}$$

3 Evaluate ***Does the result make sense?***

The value of K_a is consistent with that of a weak acid.

Practice Problems

22. In an exactly 0.1*M* solution of methanoic acid, $[H^+] = 4.2 \times 10^{-3}M$. Calculate the K_a of methanoic acid.

23. In an exactly 0.2*M* solution of a monoprotic weak acid, $[H^+] = 9.86 \times 10^{-4}M$. What is the K_a for this acid?

Math Handbook

For help with scientific notation, go to page R56.

Interactive Textbook

Problem-Solving 19.23 Solve Problem 23 with the help of an interactive guided tutorial.

with ChemASAP

Careers *in* Chemistry

Stone Conservator

As more is learned about the damaging effects of airborne pollutants, such as acid rain, there is a growing concern to preserve historic buildings and pieces of sculpture from damage. Stone conservators work to prevent and repair the damage to stone used in buildings and sculptures.

Stone conservators must clean statues properly to remove pollution deposits. One method they use is to apply a thin, clay mudpack to the stone's surface to pull out the deposits. Lasers are also used to remove pollution from stone. When making repairs, stone conservators sometimes use surgical microscopes to examine the surfaces of statues. If the stone was originally painted, the original paint is chemically analyzed to determine its composition. In addition, conservators research and apply methods to preserve the stone once it has been repaired. One preservation technique is to seal the stone to prevent water, which carries dissolved gases and salts, from seeping into the pores of the stone.

Stone conservators often work for museums, universities, or private companies that specialize in the cleaning and restoration of stone objects. They must have knowledge of both chemistry and art because they often work closely with both chemists and museum personnel. A stone conservator's level of education may range from a bachelor's degree to a doctorate.

For: Careers in Chemistry
Visit: PHSchool.com
Web Code: cdb-1193

19.3 Section Assessment

24. **Key Concept** Compare a strong acid and a weak acid in terms of the acid dissociation constant.
25. **Key Concept** How do you determine the K_a of a weak acid or the K_b of a weak base?
26. Which acid in Table 19.6 would you expect to have the lowest ionization constant?
27. Acid HX has a very small value of K_a. How do the relative amounts of H^+ and HX compare at equilibrium?
28. Write the equations for the ionization or dissociation of the following acids and bases in water.
    a. nitric acid
    b. ethanoic acid
    c. ammonia
    d. magnesium hydroxide
29. Compare the terms *strong/weak* and *concentrated/dilute* as they pertain to acids and bases.

Writing Activity

Cause and Effect Paragraph The chief cause of tooth decay is the weak acid called lactic acid ($C_3H_6O_3$). Lactic acid is formed in the mouth by the action of specific bacteria, such as *Streptococcus mutans*, on sugars present in sticky plaque on tooth surfaces. Starting with the information on page R34, research current efforts to thwart tooth decay. Write a report summarizing your findings.

Assessment 19.3 Test yourself on the concepts in Section 19.3.

with ChemASAP

Section 19.3 Assessment

24. A strong acid is completely ionized in aqueous solution and has a large K_a. A weak acid is ionized only slightly in aqueous solution and has a small K_a.

25. Substitute the measured concentrations of all the substances present at equilibrium into the expressions for K_a or K_b.

26. hypochlorous acid

27. The [HX] is much greater than the $[H^+]$.

28. **a.** $HNO_3 + H_2O \rightarrow H_3O^+ + NO_3^-$
b. $CH_3COOH + H_2O \rightleftharpoons CH_3COO^- + H_3O^+$
c. $NH_3 + H_2O \rightleftharpoons NH_4^+ + OH^-$
d. $Mg(OH)_2 \rightarrow Mg^{2+} + 2OH^-$

29. Strong acids and bases are completely ionized in aqueous solution. Weak acids and bases ionize only slightly in aqueous solution. Concentrated acid or base solutions contain large amounts (high concentrations) of acid or base. Dilute acid or base solutions contain small amounts (low concentrations) of acid or base.

3 ASSESS

Evaluate Understanding L2

Ask, **List the two conditions needed to calculate the dissociation constant of a weak acid.** *(You must know the initial molar concentration of the acid and the pH or $[H^+]$ of the solution at equilibrium.)*

Reteach L1

Table 19.7, may pose difficulties for some students. Use an example to explain the information in the table. Point out that the K_a for each acid is equal to the product of the concentrations of the ions on the right of the equation, divided by the concentration of the non-ionized acid. Also explain that the K_a value is a quantitative indication of acid strength. A lower K_a value means the reactants are favored in the equilibrium; thus, the acid is weaker.

As a class activity, have students prepare a numbered list of steps for calculating dissociation constants. Give them a different example than those in Sample Problem 19.5, and have them use the numbered procedure to work through the problem.

Writing Activity

To thwart tooth decay, reduce sugar intake; brush and floss regularly, use a fluoride toothpaste or rinse; and have regular dental check-ups to catch minor decay early.

If your class subscribes to the Interactive Textbook, use it to review key concepts in Section 19.3.

with ChemASAP

19.4

1 FOCUS

Objectives

19.4.1 Define the products of an acid-base reaction.

19.4.2 Explain how acid-base titration is used to calculate the concentration of an acid or a base.

19.4.3 Explain the concept of equivalence in neutralization reactions.

19.4.4 Describe the relationship between equivalence point and the end point of a titration.

Guide for Reading

Build Vocabulary L2

Graphic Organizers Have students write definitions of *standard solution, titration,* and *end point* in the first column of a two column table. Have them describe how these three terms are related in the second column.

Reading Strategy L2

Relate Text and Visuals Have students reference the visuals as they learn about titration.

2 INSTRUCT

Connecting to Your World

Have students study the photograph and read the text that opens the section. Ask, **What is a neutralization reaction?** *(the reaction between an acid and a base, producing a salt and water)*

Why not use a stronger base to provide quicker relief from excess stomach acid? *(The stronger base would be too corrosive for the digestive tract.)*

Acid-Base Reactions

Discuss L2

Ask students if they have witnessed or read a news report about people trying to clean up a chemical spill on a highway or at a factory. Ask, **If a spill involves an acid, what type of treatment is possible?** *(A weak base in powder form, such as sodium carbonate, can be spread over the spill to absorb, react with, and neutralize the acid.)*

19.4 Neutralization Reactions

Guide for Reading

Key Concepts

- What are the products of the reaction of an acid with a base?
- What is the endpoint of a titration?

Vocabulary

neutralization reactions
equivalence point
titration
standard solution
end point

Reading Strategy

Identifying a Sequence
A sequence is the order in which a series of events occurs. As you read about acid-base titrations, list the steps that should be used in carrying out a precise titration. Include the reactants and how they are measured, the indicator, and what to look for as the titration nears its end point.

Connecting to Your World Nearly all of the adult population suffers from acid indigestion at some time. Although hydrochloric acid is always present in the stomach, an excess can cause heartburn and a feeling of nausea. A common way to relieve the pain of acid indigestion is to take antacids to neutralize the stomach acid. The active ingredient in many antacids is sodium hydrogen carbonate, aluminum hydroxide, or magnesium hydroxide. In this section, you will learn what a neutralization reaction is.

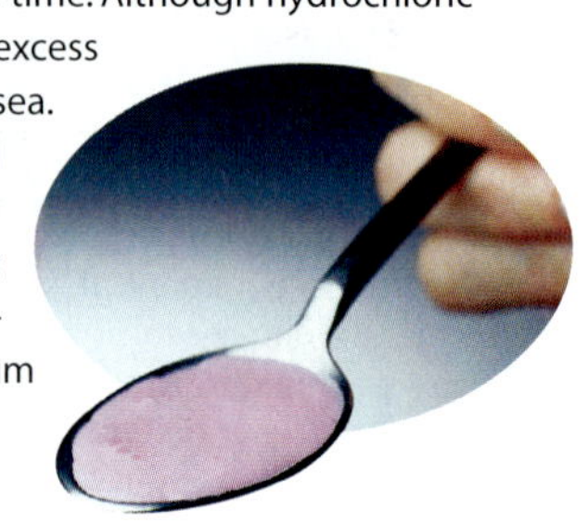

Acid-Base Reactions

If you mix a solution of a strong acid containing hydronium (hydrogen) ions with a solution of a strong base that has an equal number of hydroxide ions, a neutral solution results. The final solution has properties that are characteristic of neither an acidic nor a basic solution. Consider these examples:

$$HCl(aq) + NaOH(aq) \longrightarrow NaCl(aq) + H_2O(l)$$

$$H_2SO_4(aq) + 2KOH(aq) \longrightarrow K_2SO_4(aq) + 2H_2O(l)$$

In each example, a strong acid reacts with a strong base. If solutions of these substances are mixed in the mole ratios specified by the balanced equation, neutral solutions will result. Similar reactions of weak acids and/or weak bases do not usually produce neutral solutions. In general, however, reactions in which an acid and a base react in an aqueous solution to produce a salt and water are called **neutralization reactions.** The formation of water in a neutralization reaction is shown in Figure 19.20.

Neutralization reactions are one way to prepare pure samples of salts. You could prepare potassium chloride, for example, by mixing equal molar quantities of hydrochloric acid and potassium hydroxide. An aqueous solution of potassium chloride would result. You could heat the solution to evaporate the water, leaving the salt potassium chloride. Table 19.9 lists some common salts and their applications.

In general, the reaction of an acid with a base produces water and one of a class of compounds called salts. When you hear the word *salt,* you may think of the substance that flavors your French fries or scrambled eggs. Table salt (sodium chloride) is one example of a salt, but there are many more. Salts are compounds consisting of an anion from an acid and a cation from a base.

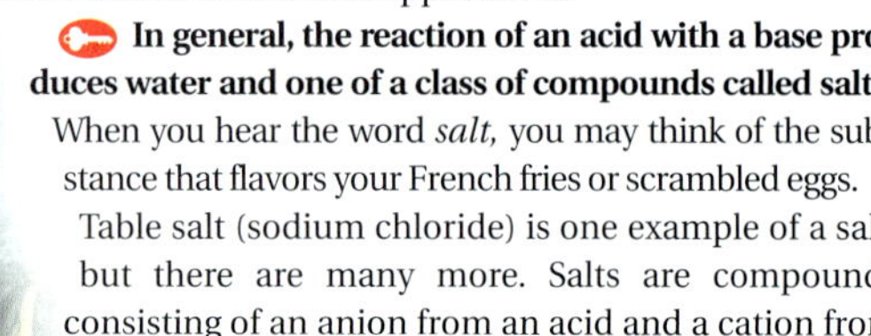

Figure 19.19 For fish and other aquatic animals to survive, the water in which they live must be maintained at the proper pH.

Section Resources

Print

- **Guided Reading and Study Workbook,** Section 19.4
- **Core Teaching Resources,** Section 19.4 Review
- **Transparencies,** T225–T226
- **Laboratory Manual,** Labs 41–43
- **Small-Scale Chemistry Laboratory Manual,** Lab 31
- **Probeware Lab Manual,** The Neutralizing Power of Antacids, Small-Scale Titrations

Technology

- **Interactive Textbook with ChemASAP,** Problem-Solving 19.30, 19.33, Simulation 26, Assessment 19.4

Table 19.9

Some Salts and Their Applications		
Name	**Formula**	**Applications**
Ammonium sulfate	$(NH_4)_2SO_4$	Fertilizer
Barium sulfate	$BaSO_4$	Gastrointestinal studies; white pigment
Calcium chloride	$CaCl_2$	De-icing roadways and sidewalks
Potassium chloride	KCl	Sodium-free salt substitute
Silver bromide	$AgBr$	Photographic emulsions
Sodium hydrogen carbonate (baking soda)	$NaHCO_3$	Antacid
Sodium carbonate decahydrate (washing soda)	$Na_2CO_3 \cdot 10H_2O$	Glass manufacture; water softener
Sodium chloride (table salt)	$NaCl$	Body electrolyte; chlorine manufacture

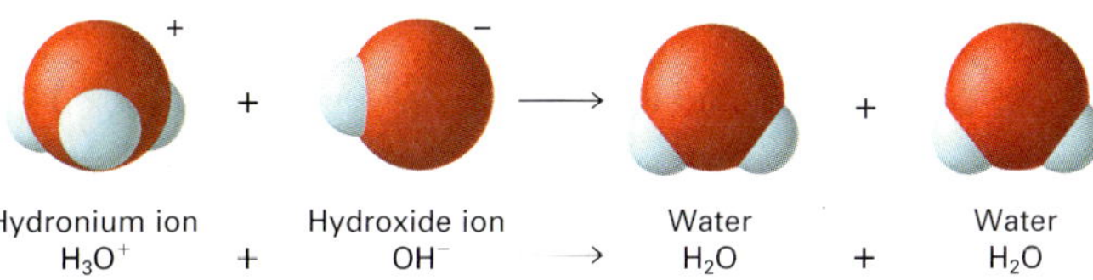

Figure 19.20 In a neutralization reaction, hydronium ions (H_3O^+) combine with hydroxide ions (OH^-) to form neutral water.

The properties of acids, bases, and salts help explain many diverse phenomena. The usefulness of antacids, for example, depends on the process of acid-base neutralization. Farmers use a similar process to control the pH of soil.

Titration

Acids and bases sometimes, but not always, react in a 1:1 mole ratio.

$$\underset{1\text{ mol}}{HCl(aq)} + \underset{1\text{ mol}}{NaOH(aq)} \longrightarrow \underset{1\text{ mol}}{NaCl(aq)} + \underset{1\text{ mol}}{H_2O(l)}$$

When sulfuric acid reacts with sodium hydroxide, however, the ratio is 1:2. Two moles of the base sodium hydroxide are required to neutralize one mole of H_2SO_4.

$$\underset{1\text{ mol}}{H_2SO_4(aq)} + \underset{2\text{ mol}}{2NaOH(aq)} \longrightarrow \underset{1\text{ mol}}{Na_2SO_4(aq)} + \underset{2\text{ mol}}{2H_2O(l)}$$

Similarly, hydrochloric acid and calcium hydroxide react in a 2:1 ratio.

$$\underset{2\text{ mol}}{2HCl(aq)} + \underset{1\text{ mol}}{Ca(OH)_2(aq)} \longrightarrow \underset{1\text{ mol}}{CaCl_2(aq)} + \underset{2\text{ mol}}{2H_2O(l)}$$

You will notice in the preceding examples that the number of moles of hydrogen ions provided by the acid are equivalent to the number of hydroxide ions provided by the base. When an acid and base are mixed, the **equivalence point** is when the number of moles of hydrogen ions equals the number of moles of hydroxide ions.

What is the equivalence point of a reaction between an acid and a base?

Titration

Discuss L2

Point out that neutralization is a process that occurs whenever an acid reacts with a base in the mole ratios specified by the balanced equation. Explain that not all neutralization reactions produce neutral solutions. The relative strengths of the reactants determines whether the solution will be acidic, basic, or neutral.

TEACHER Demo

Titration Using Indicators L2

Purpose Students observe properties of titration.

Materials equimolar solutions of HCl and NaOH, phenolphthalein, electric stirrer with a magnetic stir bar, beaker, table salt

Safety Wear protective goggles and laboratory apron.

Procedure Write the following equation on the board: $HCl(aq) + NaOH(aq) \rightarrow NaCl(aq) + H_2O(l)$

Prepare equimolar solutions of HCl and NaOH. Add a few drops of phenolphthalein indicator to the NaOH solution. Slowly add the HCl to the NaOH while carefully stirring. If possible, use an electric stirrer with a magnetic stir bar. Ask, **What happens to the color of the solution as the acid is added?** *(The deep pink color slowly fades until the solution becomes colorless.)* Remind students that phenolphthalein is an acid-base indicator that is colorless at pH lower than 8. Have a few volunteers come up and gently touch the outside of the reaction beaker. Ask them to describe their observations to the class.

Ask, **Is the neutralization reaction endothermic or exothermic?** *(exothermic)*

Display a quantity of table salt equivalent to the amount of salt that could be retrieved if the neutral solution were evaporated.

Answers to...

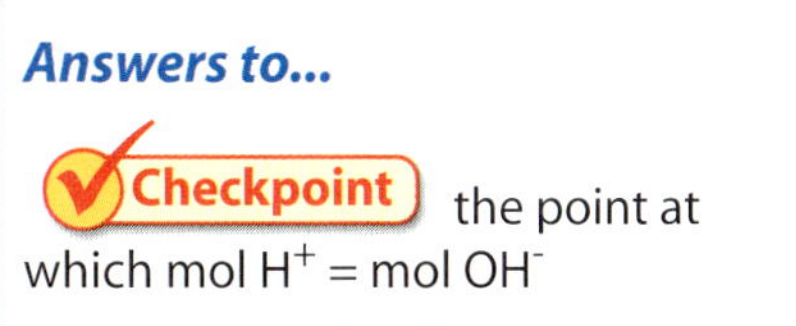
the point at which mol H^+ = mol OH^-

Sample Problem 19.6

Answers

30. $H_3PO_4 + 3KOH \rightarrow K_3PO_4 + 3H_2O$; 1.56 mol H_3PO_4 × 3 mol KOH/1 mol H_3PO_4 = 4.68 mol KOH

31. $HNO_3 + NaOH \rightarrow NaNO_3 + H_2O$; 0.20 mol HNO_3 × 1 mol NaOH/1 mol HNO_3 = 0.20 mol NaOH

Practice Problems Plus L2

How many moles of aluminum hydroxide are needed to neutralize 2.30 mol of sulfuric acid? *(1.53 moles)*

Math Handbook

For a math refresher and practice, direct students to dimensional analysis, page R66.

Use Visuals 

Table 19.5 and Figure 19.21 Have students study Table 19.5 and Figure 19.21. Red cabbage, when used as an acid-base indicator, turns red between pH 2 and 3, pink between pH 3 and 4, and green between pH 8-12. Ask, **What color will emerge when testing household ammonia?** *(green)* **Lemon juice?** *(red)* **Seawater?** *(green)*

Math Handbook

For help with dimensional analysis, go to page R66.

Problem-Solving 19.30 Solve Problem 30 with the help of an interactive guided tutorial.

with ChemASAP

SAMPLE PROBLEM 19.6

Finding the Number of Moles of an Acid in Neutralization

How many moles of sulfuric acid are required to neutralize 0.50 mol of sodium hydroxide?

1 Analyze ***List the knowns and the unknown.***

Knowns

- mol NaOH = 0.50 mol
- $H_2SO_4(aq) + 2NaOH(aq) \longrightarrow Na_2SO_4(aq) + 2H_2O(l)$
- $\dfrac{\text{mol } H_2SO_4}{\text{mol NaOH}} = \dfrac{1}{2}$

Unknown

- moles H_2SO_4 = ? mol

2 Calculate ***Solve for the unknown.***

The mole ratio of H_2SO_4 to NaOH is 1:2. The necessary number of moles of H_2SO_4 is calculated using this ratio.

$$0.50 \, \cancel{\text{mol NaOH}} \times \frac{1 \text{ mol } H_2SO_4}{2 \, \cancel{\text{mol NaOH}}} = 0.25 \text{ mol } H_2SO_4$$

3 Evaluate ***Does the result make sense?***

Because the mole ratio of H_2SO_4 to NaOH is 1:2, the expected number of moles of H_2SO_4 should be half the given number of moles of NaOH. The answer should have two significant figures.

Practice Problems

30. How many moles of potassium hydroxide are needed to completely neutralize 1.56 mol of phosphoric acid?

31. How many moles of sodium hydroxide are required to neutralize 0.20 mol of nitric acid?

You can determine the concentration of acid (or base) in a solution by performing a neutralization reaction. You must use an appropriate acid-base indicator to show when neutralization has occurred. As you can see in Figure 19.21, the juice of the red cabbage is an acid-base indicator. In the laboratory, phenolphthalein is often the preferred indicator for acid-base neutralization reactions. Solutions that contain phenolphthalein turn from colorless to deep pink as the pH of the solution changes from acidic to basic. In slightly basic solutions, the indicator is very faintly pink.

Figure 19.21 Red cabbage juice is used as an acid-base indicator. As the solution changes from highly acidic to basic, the color changes from red to violet to green to yellow. **Predicting** ***Would the yellow solution have a high or low pH?***

Facts and Figures

Acids, Bases, and Curly Hair

Curly hair from a permanent wave involves an acid-base neutralization reaction. The waving solution is basic and has a pH of about 9.20. The active ingredient in the waving solution is often an ammonium compound that is applied once the hair has been rolled. After a specified time, the hair is rinsed to remove the excess base. Then a neutralizer is applied. The neutralizer is an acid solution that has a pH range of 2 to 6. If the hair is exposed to the waving solution for too long a period of time, the hair's texture and appearance are damaged.

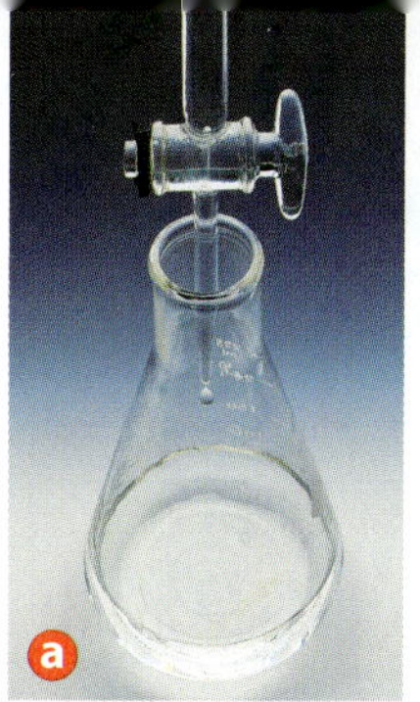
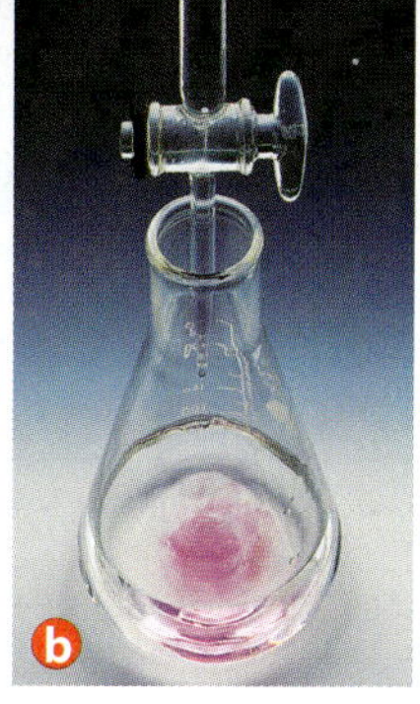
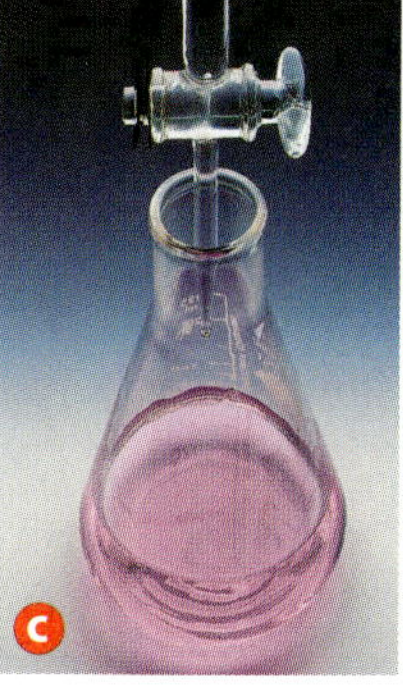

Figure 19.22 The titration of an acid with a base is shown here. ⓐ A known volume of an acid (plus a few drops of phenolphthalein indicator) in a flask is placed beneath a buret filled with a base of known concentration. ⓑ Base is slowly added from the buret to the acid while the flask is gently swirled. ⓒ A change in the color of the indicator signals that neutralization has occurred.

The steps in a neutralization reaction are as follows.

1. A measured volume of an acid solution of unknown concentration is added to a flask.
2. Several drops of the indicator are added to the solution while the flask is gently swirled.
3. Measured volumes of a base of known concentration are mixed into the acid until the indicator just barely changes color.

The process of adding a known amount of solution of known concentration to determine the concentration of another solution is called **titration.** The solution of known concentration is called the **standard solution.** Use a buret to add the standard solution. A titration is continued until the indicator shows that neutralization has just occurred. The point at which the indicator changes color is the **end point** of the titration. The titration of an acid of unknown concentration with a standard base is shown in Figure 19.22. You can use a similar procedure to find the concentration of a base using a standard acid.

Figure 19.23 shows how the pH of a solution changes during the titration of a strong acid (HCl) with a strong base (NaOH). The pH of the initial acid solution is low. As the base is added, the pH increases because some of the acid is neutralized. As the titration approaches the point of neutralization, at a pH of 7, the pH increases dramatically as hydrogen ions are used up. Once past the point of neutralization, additional base produces a further increase of pH. **The point of neutralization is the end point of the titration.** At this point, the contents of the beaker consist of only H_2O and NaCl, which is the salt derived from the strong acid HCl and the strong base NaOH, plus a trace of indicator.

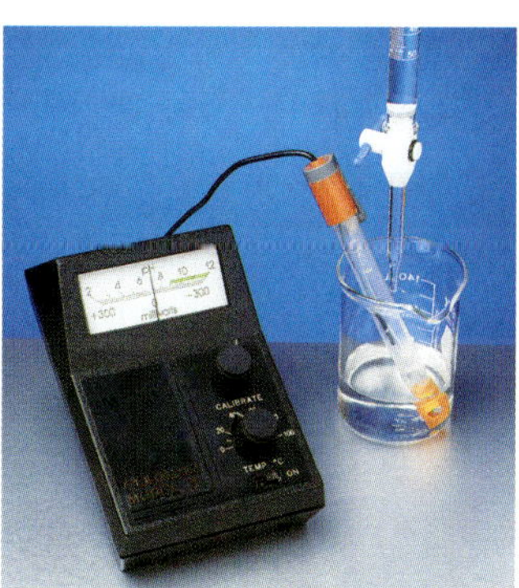

Titration of a Strong Acid with a Strong Base

pH: 0, 2, 4, 6, 8, 10, 12, 14

0.10*M* NaOH added (mL): 50, 100

Equivalence point

Figure 19.23 In this titration of a strong acid with a strong base, 0.10*M* NaOH is slowly added from a buret to 50.0 mL of 0.10*M* HCl in the beaker. The equivalence point, the midpoint on the vertical portion of the pH titration curve, occurs at 50.0 mL of NaOH added. **Interpreting Diagrams** ***What is true concerning [H⁺] and [OH⁻] at the equivalence point?***

Interactive Textbook

Simulation 26 Simulate the titration of several acids and bases and observe patterns in the pH at equivalence.

with ChemASAP

Titration Using a pH Meter L2

Purpose Students observe properties of the equivalence point.

Materials equimolar solutions of HCl and NaOH, phenolphthalein, beaker, buret, pH meter, graph paper

Procedure Repeat the demo described on page 613. This time place the base in a beaker and the acid in a buret. Use a pH meter to monitor the titration. Construct a graph similar to the one in Figure 19.23

Expected Outcome Compared with Figure 19.23, the graph will be reversed because an acidic solution is being added to a basic solution.

Ask, **What happens to the graph as the titration nears its equivalence point?** *(The slope becomes very steep.)* **What would happen to the pH if a small amount of base were added to the beaker after the equivalence point is reached?** *(The pH would increase dramatically.)*

Add some base to the beaker and observe the pH. Ask, **What can be done to regain the equivalence point?** *(Add acid from the buret.)*

Answers to...

Figure 19.21 high

Figure 19.23 They are equal for this titration.

Section 19.4 (continued)

Sample Problem 19.7

Answers

32. 25.0 mL KOH × 1.00 mol KOH/1000mL KOH × 1 mol HCl/1 mol KOH × 1000 mL HCl/0.45 mol HCl = 56 mL HCl

33. 38.5 mL NaOH × 0.150 mol NaOH/1000 mL NaOH × 1 mol H_3PO_4/3 mol NaOH = 0.00193 mol H_3PO_4; 0.001923 mol H_3PO_4/0.0150 L H_3PO_4 = 0.129*M* H_3PO_4

Practice Problems Plus 

What is the molarity of sulfuric acid if 25.0 mL of the solution is neutralized by 25.5 mL of 0.50M KOH? *(0.26M)*

Math Handbook

For a math refresher and practice, direct students to dimensional analysis, page R66.

3 ASSESS

Evaluate Understanding 

Ask students to summarize the titration process in their own words. *(During a titration, equivalent volumes of an acid and of a base are determined. One is a standard solution with a known concentration; the concentration of the other solution is not known.)*

Reteach 

Have students use Figures 19.22 and 19.23 to describe how titration works.

Connecting Concepts

Acid-base neutralizations are double displacement reactions. The positive ions are exchanged between the acid and base reactants; the products are a salt and water.

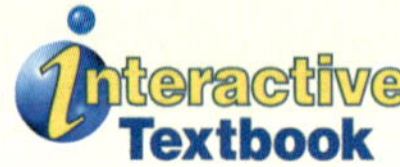

If your class subscribes to the Interactive Textbook, use it to review key concepts in Section 19.4.

with ChemASAP

SAMPLE PROBLEM 19.7

Determining the Concentration of an Acid by Titration

A 25-mL solution of H_2SO_4 is completely neutralized by 18 mL of 1.0*M* NaOH. What is the concentration of the H_2SO_4 solution?

1 Analyze ***List the knowns and the unknown.***

Knowns

- molarity base = 1.0*M* NaOH
- volume base = 18 mL = 0.018 L
- volume acid = 25 mL = 0.025 L
- $H_2SO_4(aq) + 2NaOH(aq) \longrightarrow Na_2SO_4(aq) + 2H_2O(l)$

Unknown

- molarity acid = ?*M* H_2SO_4

Use the molarity to convert the volume of base to moles of base. Use the mole ratio to find the number of moles of H_2SO_4. Calculate the molarity by dividing the number of moles of H_2SO_4 by the number of liters of H_2SO_4.

The conversion steps are as follows:

$$\text{L NaOH} \longrightarrow \text{mol NaOH} \longrightarrow \text{mol } H_2SO_4 \longrightarrow M\ H_2SO_4$$

2 Calculate ***Solve for the unknown.***

$$0.018\ \cancel{\text{L NaOH}} \times \frac{1.0\ \cancel{\text{mol NaOH}}}{1\ \cancel{\text{L NaOH}}} \times \frac{1\ \text{mol } H_2SO_4}{2\ \cancel{\text{mol NaOH}}} = 0.0090\ \text{mol } H_2SO_4$$

$$\text{molarity} = \frac{\text{moles}}{\text{liters}} = \frac{0.0090\ \text{mol}}{0.025\ \text{L}} = 0.36\ M$$

The $[H_2SO_4]$ is 0.36*M*.

3 Evaluate ***Does the result make sense?***

Because the volume of acid was greater than the volume of base, the concentration is less than 1.0*M*. The answer has two significant figures.

Math Handbook

For help with dimensional analysis, go to page R66.

Interactive Textbook

Problem-Solving 19.33 Solve Problem 33 with the help of an interactive guided tutorial.

with ChemASAP

Practice Problems

32. How many milliliters of 0.45*M* HCl will neutralize 25.0 mL of 1.00*M* KOH?

33. What is the molarity of H_3PO_4 if 15.0 mL is completely neutralized by 38.5 mL of 0.150*M* NaOH?

19.4 Section Assessment

34. **Key Concept** What are the products of a reaction between an acid and a base?

35. **Key Concept** What occurs at the endpoint of a titration?

36. How many moles of HCl are required to neutralize aqueous solutions of these bases?

a. 2 mol NH_3 **b.** 0.1 mol $Ca(OH)_2$

37. Write complete balanced equations for the following acid-base reactions?

a. $H_2SO_4(aq) + KOH(aq) \longrightarrow$

b. $H_3PO_4(aq) + Ca(OH)_2(aq) \longrightarrow$

c. $HNO_3(aq) + Mg(OH)_2(aq) \longrightarrow$

Connecting Concepts

Types of Reactions Reread the information on types of chemical reactions in Section 11.2. Which type of reaction are all acid-base neutralizations? Explain your answer.

Assessment 19.4 Test yourself on the concepts in Section 19.4.

with ChemASAP

616 Chapter 19

Section 19.4 Assessment

34. a salt and water

35. neutralization

36. a. 2 mol **b.** 0.2 mol

37. a. $H_2SO_4 + 2KOH \rightarrow 2H_2O + K_2SO_4$

b. $2H_3PO_4 + 3Ca(OH)_2 \rightarrow 6H_2O + Ca_3(PO_4)_2$

c. $2HNO_3 + Mg(OH)_2 \rightarrow 2H_2O + Mg(NO_3)_2$

Ionization Constants of Weak Acids

Purpose

To measure ionization constants of weak acids such as bromocresol green (BCG).

Materials

- ruler
- reaction surface
- pH buffer and BCG

Procedure  

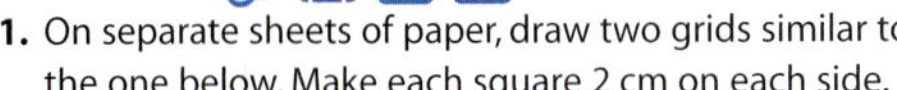

1. On separate sheets of paper, draw two grids similar to the one below. Make each square 2 cm on each side.
2. Place a reaction surface over one of the grids and place one drop of BCG in each square.
3. Place one drop of pH buffer in each square corresponding to its pH value.
4. Use the second grid as a data table to record your observations for each solution.

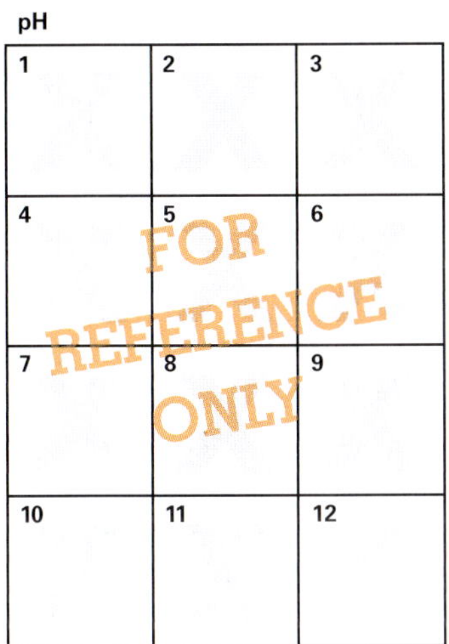

Analyze

Using your experimental data, record the answers to the following questions below your data table.

1. What is the color of the lowest-pH solutions?
2. What is the color of the highest-pH solutions?
3. At which pH does the bromcresol green change from one color to the other? At which pH does an intermediate color exist?

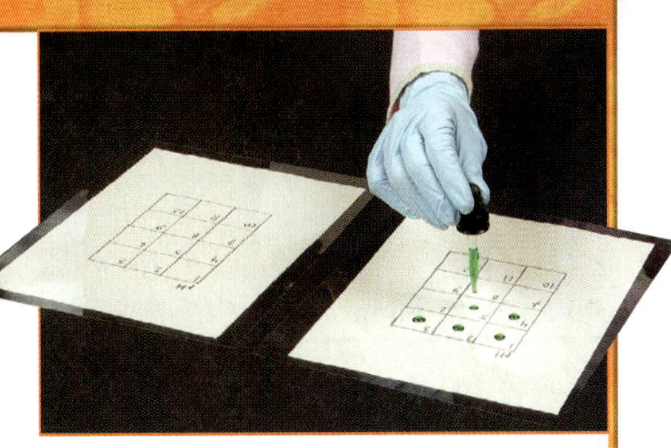

An acid-base indicator is usually a weak acid with a characteristic color. Because bromcresol green is an acid, it is convenient to represent its rather complex formula as HBCG. HBCG ionizes in water according to the following equation.

$$\underset{\text{(yellow)}}{HBCG} + H_2O \rightleftharpoons \underset{\text{(blue)}}{BCG^-} + H_3O^+$$

The K_a expression is

$$K_a = \frac{[BCG^-] \times [H_3O^+]}{[HBCG]}$$

When $[BCG^-] = [HBCG]$, $K_a = [H_3O^+]$.

4. What color is the conjugate acid of BCG^-?
5. What color is the conjugate base of HBCG?
6. What color is an equal mixture of the conjugate acid and conjugate base of bromcresol green? At what pH does this equal mixture occur?

You're The Chemist

The following small-scale activities allow you to develop your own procedures and analyze the results.

1. **Design It!** Design and carry out an experiment to measure the K_a of several more acid-base indicators. Record the colors of the conjugate acids (low pH) and the conjugate bases (high pH). Determine the K_a for each acid.
2. **Analyze It!** Explain in your own words how to measure the K_a of a colored acid-base indicator. Explain what you would do and how you would interpret your results.

Ionization Constants of Weak Acids 

Objective Students measure ionization constants of weak acids.

Prep Time 1 hour
Class Time 30 minutes

Solution	Preparation
*0.04% BTB	100 mg in 16.0 mL 0.01*M* NaOH, dilute to 250 mL
*0.04% BCG	100 mg in 14.3 mL 0.01*M* NaOH, dilute to 250 mL
*0.04% BPB	100 mg in 14.9 mL 0.01*M* NaOH, dilute to 250 mL
*0.04% MCP	100 mg in 26.2 mL 0.01*M* NaOH, dilute to 250 mL
*0.04% TB	100 mg in 21.5 mL 0.01*M* NaOH, dilute to 250 mL
0.1% phenolphthalein	250 mg in 250 mL of 70% 2-propanol
0.05% MO	125 mg in 250 mL
0.02% AYR	50 mg in 250 mL
Solution A	14.9 g boric acid and 12.6 g citric acid in 1200 mL
Solution B	45.6 g $Na_2SO_4 \cdot 12H_2O$ in 1200 mL

*Can dissolve sodium salt directly in 250 mL of distilled water.

pH*	Solution A	Solution B
2	195 mL	5 mL
3	176 mL	24 mL
4	155 mL	45 mL
5	134 mL	66 mL
6	118 mL	82 mL
7	99 mL	101 mL
8	85 mL	115 mL
9	69 mL	131 mL
10	54 mL	146 mL
11	44 mL	156 mL
12	17 mL	183 mL

*Use 1.0*M* H_3PO_4 for pH 1.

Analyze

1. pH solutions 1 to 3 are yellow.
2. pH solutions 5 to 12 are blue.
3. pH solution 4 is green (an intermediate between yellow and blue).
4. The conjugate acid, HBCG, is yellow.
5. The conjugate base, BCG^-, is blue.
6. The equal mixture of HBCG and BCG^- is green at pH = 4.

You're the Chemist

1. Results will vary depending on the indicators chosen.
2. To measure the K_a of a colored acid-base indicator, mix one drop of the indicator solution with one drop of each pH 1–12 buffer solution. Look for the pH of the color change. This pH is close to the K_a of the acid.

19.5

1 FOCUS

Objectives

19.5.1 Describe when a solution of a salt is acidic or basic.

19.5.2 Demonstrate with equations how buffers resist changes in pH.

Guide for Reading

Build Vocabulary L2

Paraphrase Students can look up the definition of a buffer in the dictionary. Based on this definition, they can define *buffer capacity* in their own words.

Reading Strategy L2

KWL Have students construct a KWL chart about buffers, with the three column titles, *What I Know about Buffers; What I Want to Know about Buffers; and What I Learned about Buffers.* Students will write in the appropriate columns before, during, and after reading this section.

2 INSTRUCT

Connecting to Your World

Have students read the opening paragraph. Ask, **What is the normal pH of human blood?** *(7.4)*

What are the pH limits of human blood before causing death? *(between 6.8 and 7.8)*

Salt Hydrolysis

Interpreting Graphs L2

a. (a) 8.7, (b) 7

b. The number of equivalents of acid and base are equal.

c. In (a), CH_3COO^- ions react with water to produce OH^- ions. In (b), Cl^- ions do not react to produce either H_3O^+ or OH^- ions.

Enrichment Question L3

Are there limits to the pH of salt solutions? *(In general, salt solutions would not have pH values in the range of strong acids or bases.)*

19.5 Salts in Solution

Guide for Reading

Key Concepts

- When is the solution of a salt acidic or basic?
- What are the components of a buffer?

Vocabulary

salt hydrolysis

buffer

buffer capacity

Reading Strategy

Predicting Based on what you have learned about acid-base neutralization reactions, predict the possible pH range of solutions of salts of strong acids and weak bases. Account for this pH range. Do the same for salts of weak acids and strong bases.

Connecting to Your World The pH inside most living cells is close to 7. Because the chemical processes of the cell are very sensitive to pH, even a slight change in pH can be harmful. Human blood, for example, is normally maintained at a pH very close to 7.4. A person cannot survive for more than a few minutes if the blood pH drops to 6.8 or rises to 7.8. In this section, you will learn about chemical processes that ensure that the pH of blood is kept near 7.4.

Salt Hydrolysis

A salt consists of an anion from an acid and a cation from a base. It forms as a result of a neutralization reaction. Although solutions of many salts are neutral, some are acidic and others are basic. Solutions of sodium chloride and of potassium sulfate are neutral. A solution of ammonium chloride is acidic. A solution of sodium ethanoate (sodium acetate) is basic. Figure 19.24 shows a titration curve obtained by adding a solution of sodium hydroxide, a strong base, to a solution of ethanoic (acetic) acid, a weak acid. An aqueous solution of sodium ethanoate exists at the equivalence point.

$$\underset{\text{Ethanoic acid}}{CH_3COOH(aq)} + \underset{\text{Sodium hydroxide}}{NaOH(aq)} \longrightarrow \underset{\text{Sodium ethanoate}}{CH_3COONa(aq)} + \underset{\text{Water}}{H_2O(l)}$$

The pH at the equivalence point is 8.7—basic.

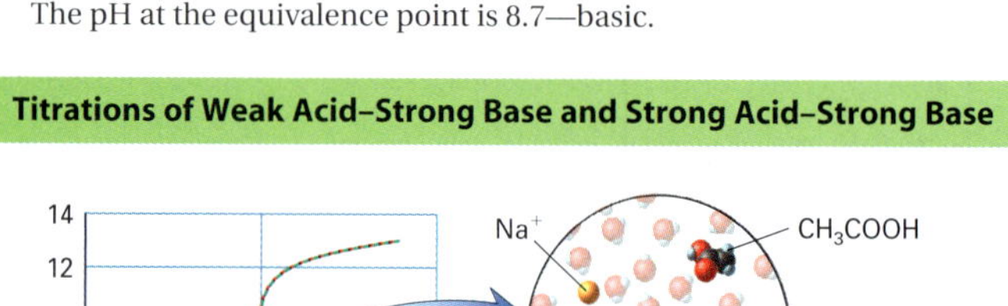
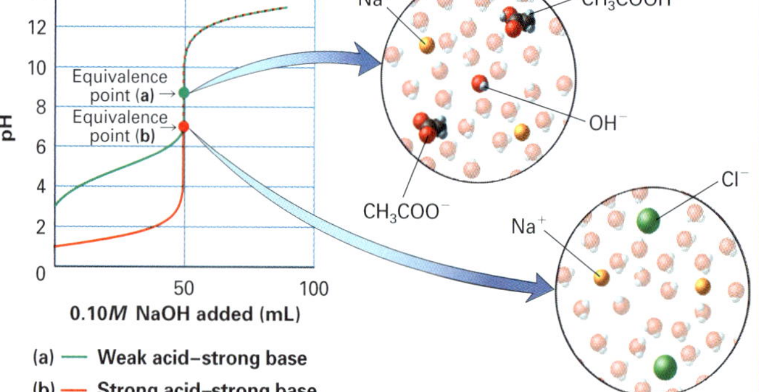

Figure 19.24 The titration curve for a weak acid and a strong base is compared with the titration curve for a strong acid and a strong base.

INTERPRETING GRAPHS

a. Identify What is the pH of the equivalence point of each titration?

b. Describe Why is the same amount of base used in each titration?

c. Apply Concepts Explain why the equivalence points of the two titrations are different.

Section Resources

Print

- **Guided Reading and Study Workbook,** Section 19.5
- **Core Teaching Resources,** Section 19.5 Review, Interpreting Graphics
- **Transparencies,** T227–T229
- **Laboratory Manual,** Labs 44–45
- **Small-Scale Chemistry Laboratory Manual,** Labs 32–33

Technology

- **Interactive Textbook with ChemASAP,** Animation 26, Problem-Solving 19.39, Assessment 19.5
- **Virtual Chemistry Labs,** 25, 26, 29

For a strong acid–strong base titration, the pH at the equivalence point is 7, or neutral. This difference exists because some salts promote hydrolysis. In **salt hydrolysis,** the cations or anions of a dissociated salt remove hydrogen ions from or donate hydrogen ions to water. Depending on the direction of the hydrogen-ion transfer, solutions containing hydrolyzing salts may be either acidic or basic. Hydrolyzing salts are usually derived from a strong acid and a weak base, or from a weak acid and a strong base. Sodium carbonate, washing soda, is the salt of the strong base sodium hydroxide and carbonic acid, a weak acid. Ammonium nitrate, used in fertilizers, is the salt of the weak base ammonia and nitric acid, a strong acid. Soap is the salt of a strong base, usually sodium hydroxide, and stearic acid, a weak acid present in fats. **In general, salts that produce acidic solutions contain positive ions that release protons to water. Salts that produce basic solutions contain negative ions that attract protons from water.**

Sodium ethanoate (CH_3COONa) is the salt of a weak acid (ethanoic acid, CH_3COOH) and a strong base (sodium hydroxide, NaOH). In solution, the salt is completely ionized.

$$\underset{\text{Sodium ethanoate}}{CH_3COONa(aq)} \longrightarrow \underset{\text{Ethanoate ion}}{CH_3COO^-(aq)} + \underset{\text{Sodium ion}}{Na^+(aq)}$$

The ethanoate ion is a Brønsted-Lowry base, which means it is a hydrogen-ion acceptor. It establishes an equilibrium with water, forming electrically neutral ethanoic acid and negative hydroxide ions.

$$\underset{\substack{\text{(hydrogen-ion}\\\text{acceptor, Brønsted-}\\\text{Lowry base)}}}{CH_3COO^-(aq)} + \underset{\substack{\text{(hydrogen-ion}\\\text{donor, Brønsted-}\\\text{Lowry acid)}}}{H_2O(l)} \rightleftharpoons CH_3COOH(aq) + \underset{\substack{\text{(makes the}\\\text{solution}\\\text{basic)}}}{OH^-(aq)}$$

This process is called hydrolysis because it splits a hydrogen ion off a water molecule. The resulting solution contains a hydroxide-ion concentration greater than the hydrogen-ion concentration. Thus the solution is basic.

Ammonium chloride (NH_4Cl) is the salt of a strong acid (hydrochloric acid, HCl) and a weak base (ammonia, NH_3). It is completely ionized in solution.

$$NH_4Cl(aq) \longrightarrow NH_4^+(aq) + Cl^-(aq)$$

Figure 19.25 Vapors of the strong acid $HCl(aq)$ and the weak base $NH_3(aq)$ combine to form the acidic white salt ammonium chloride (NH_4Cl).

Figure 19.26 A few drops of universal indicator solution have been added to each of these 0.10*M* aqueous salt solutions. The color of the indicator shows the following. (a) Ammonium chloride (NH_4Cl), is acidic (pH about 5.3). (b) Sodium chloride (NaCl) is neutral (pH 7). (c) Sodium ethanoate (CH_3COONa) is basic (pH about 8.7).

Predicting pH of Solutions L2

Purpose Students predict the pH of different salt solutions and and give reasons for their predictions.

Materials 1*M* solutions of NH_4NO_3, KCl, $NaHCO_3$, Na_2SO_4; pH meter *or* universal pH paper

Procedure Have students speculate about the pH of these solutions and explain their reasoning. After some discussion, check the pH of the solutions with a pH meter or universal pH paper. After you obtain the experimental results, have students try to explain the data. *(NH_4NO_3 is acidic because the salt results from a strong acid and a weak base; KCl is neutral because the salt results from a strong acid and a strong base; $NaHCO_3$ is basic because the salt results from a strong base and a weak acid; Na_2SO_4 is neutral because the salt results from a strong acid and a strong base.)*

Discuss L2

Point out that the salts that give non-neutral solutions in the demo above and in the illustrations are classified as Brønsted-Lowry acids and bases because the cations and anions of these salts accept or donate hydrogen ions from water (hydrolysis). Have students learn the following:

Strong acid + Strong base → Neutral solution

Strong acid + Weak base → Acidic solution

Weak acid + Strong base → Basic solution

Use Visuals L1

Figures 19.24 and 19.26 Have students compare the figures. Explain that the equivalence point is not always at pH 7. The salts that form may hydrolyze, which results in a reversible reaction that produces a higher concentration of hydrogen or hydroxide ions. Ask, **What is different about neutralization reactions that produce a neutral solution at the equivalence point and those that produce acidic or basic solutions?** *(The strengths of the reactants; when a strong acid and base react, the solution is neutral; when a strong base and a weak acid react, the solution is basic; when a strong acid and a weak base react, the solution is acidic.)*

Buffers

Word Origins L2

A buffer zone is a neutral area set up around a site to prevent any changes or to the site being investigated. In order for the correct conclusions to be drawn, no evidence can be corrupted.

TEACHER Demo

Comparing Commerical Buffers L2

Purpose To demonstrate that the neutralizing capacities of commercial buffers differ, students compare antacid tablets, buffered aspirin, and aspirin.

Materials Antacid tablets, flasks and burets, 0.5*M* HCl and NaOH, distilled water, methyl red, bromothymol blue

Procedure Set up pairs of flasks containing 100 mL of distilled water to test each product. Dissolve equal masses of the tablets in each flask. Add a few drops of methyl red to one of the paired flasks and titrate with 0.5*M* NaOH. To the other flask, add a few drops of bromothymol blue and titrate with 0.5*M* HCl. Titrate until the end point is reached. Explain that the buffering capacity is directly related to the amount of acid or base required to reach the end point. Have students compare the buffering capacities of the tablets tested.

Use Visuals L1

Figure 19.27 Explain to students that buffers are solutions in which the pH stays relatively constant when small amounts of either acid or base are added. Point out that in the buffered solution on the left, there is little change in pH.

The ammonium ion (NH_4^+) is a strong enough acid to donate a hydrogen ion to a water molecule, although the equilibrium is strongly to the left.

$$\underset{\text{(hydrogen-ion donor, Brønsted-Lowry acid)}}{NH_4^+(aq)} + \underset{\text{(hydrogen-ion acceptor, Brønsted-Lowry base)}}{H_2O(l)} \rightleftharpoons NH_3(aq) + \underset{\text{(makes the solution acidic)}}{H_3O^+(aq)}$$

This process is also called hydrolysis. It results in the formation of ammonia molecules and hydronium (hydrogen) ions. The $[H_3O^+]$ is greater than the $[OH^-]$. Thus, a solution of ammonium chloride is acidic. To determine if a salt solution is acidic or basic, remember the following rules:

Strong acid + **Strong base** ⟶ Neutral solution

Strong acid + Weak base ⟶ **Acidic** solution

Weak acid + **Strong base** ⟶ **Basic** solution

Word Origins

Buffer comes from the Old English word *buff*, meaning "firmly or sturdily." A buffer solution resists changes in its pH even when acidic or basic solutions are added to it. **What does it mean if authorities set up a buffer zone around a fire investigation site?**

Buffers

The addition of 10 mL of 0.10*M* sodium hydroxide to 1 L of pure water increases the pH by 4.0 pH units (from 7.0 to 11.0). A solution containing 0.20 mol/L each of ethanoic acid and sodium ethanoate has a pH of 4.76. When moderate amounts of either acid or base are added to this solution, however, the pH changes little. The addition of 10 mL of 0.10*M* sodium hydroxide to 1 L of this solution, for example, increases the pH by only 0.01 pH unit, from 4.76 to 4.77. Figure 19.27 shows what happens when 1.0 mL of 0.01*M* HCl solution is added to an unbuffered solution.

The solution of ethanoic acid and sodium ethanoate is an example of a typical buffer. A **buffer** is a solution in which the pH remains relatively constant when small amounts of acid or base are added. **A buffer is a solution of a weak acid and one of its salts, or a solution of a weak base and one of its salts.**

A buffer solution is better able to resist drastic changes in pH than is pure water. Figure 19.28 illustrates how a buffer works. Ethanoic acid (CH_3COOH) and its anion (CH_3COO^-) act as reservoirs of neutralizing power. They react with any hydroxide ions or hydrogen ions added to the solution. For example, consider the buffer solution in which the sodium ethanoate (CH_3COONa) is completely ionized.

$$\underset{\text{Sodium ethanoate}}{CH_3COONa(aq)} \longrightarrow \underset{\text{Ethanoate ion}}{CH_3COO^-(aq)} + \underset{\text{Sodium ion}}{Na^+(aq)}$$

a

b

Figure 19.27 A buffer is a solution in which the pH remains relatively constant. **a** The indicator shows that the buffered solution on the left and the unbuffered solution on the right are basic—pH about 8. **b** After the addition of 1.0 mL of 0.01*M* HCl solution, the pH of the buffered solution shows no visible change. The pH of the unbuffered solution, however, is now about 3 and the solution is acidic.

Facts and Figures

Blood Buffers

Your body functions properly only when the pH of your blood lies between 7.35 and 7.45. A blood pH outside these narrow limits can be life-threatening! To keep your blood at the proper pH, your blood contains buffers. The most important blood buffer consists of carbonic acid and hydrogen carbonate ions. Hydrogen ions produced by chemical reactions in your body are the main threat to your blood's pH. This buffer maintains the desired blood pH by removing excess hydrogen ions.

$$\underset{\text{Hydrogen carbonate ion}}{HCO_3^-(aq)} + \underset{\text{Hydrogen ion}}{H^+(aq)} \rightleftharpoons \underset{\text{Carbonic acid}}{H_2CO_3(aq)}$$

As long as there are hydrogen carbonate ions available, the excess hydrogen ions are removed, and the pH of the blood changes very little.

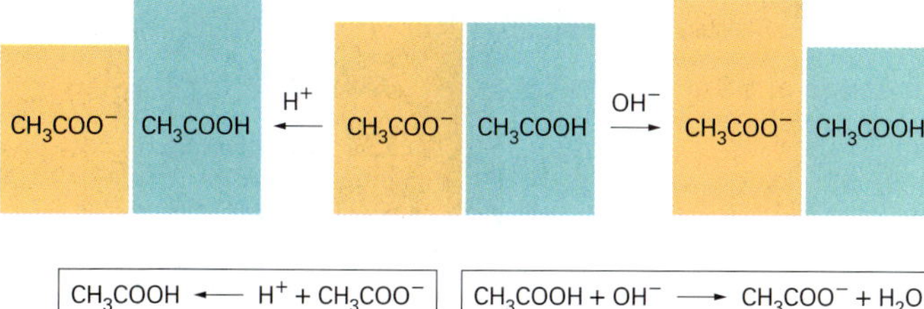

Figure 19.28 The buffer described here is made of ethanoic acid (CH_3COOH) and sodium ethanoate, which is the source of ethanoate ions (CH_3COO^-). To begin, the concentrations of ethanoic acid and ethanoate ions are equal. When either H^+ or OH^- is added, the buffer produces additional ethanoic acid or ethanoate ions. In both situations, the ratio of $[CH_3COOH]$ to $[CH_3COO^-]$ and consequently, the pH, changes very little.

When an acid is added to the solution, the ethanoate ions (CH_3COO^-) act as a hydrogen-ion "sponge." This creates ethanoic acid, which does not ionize extensively in water; thus the pH does not change appreciably.

$$\underset{\text{Ethanoate ion}}{CH_3COO^-(aq)} + \underset{\text{Hydrogen ion}}{H^+(aq)} \rightleftharpoons \underset{\text{Ethanoic acid}}{CH_3COOH(aq)}$$

A base is a source of hydroxide ions. When a base is added to the solution, the ethanoic acid and the hydroxide ions react to produce water and the ethanoate ion.

$$\underset{\text{Ethanoic acid}}{CH_3COOH(aq)} + \underset{\text{Hydroxide ion}}{OH^-(aq)} \rightleftharpoons \underset{\text{Ethanoate ion}}{CH_3COO^-(aq)} + \underset{\text{Water}}{H_2O(l)}$$

The ethanoate ion is not a strong enough base to accept hydrogen ions from water extensively. Again, the pH does not change very much.

An ethanoate buffer solution cannot control the pH when too much acid is added, because no more ethanoate ions are present to accept hydrogen ions. The ethanoate buffer also becomes ineffective when too much base is added. Do you know why this is so? When too much base is added, no more ethanoic acid molecules are present to donate hydrogen ions. When too much acid or base is added, the buffer capacity of a solution is exceeded. The **buffer capacity** is the amount of acid or base that can be added to a buffer solution before a significant change in pH occurs.

Two buffer systems are crucial in maintaining human blood pH within a very narrow range (pH 7.35–7.45). One is the carbonic acid–hydrogen carbonate buffer system. The other is the dihydrogen phosphate–monohydrogen phosphate buffer system. Table 19.10 lists several important buffer systems.

Animation 26 Discover the chemistry behind buffer action.

with ChemASAP

Table 19.10

Important Buffer Systems

Buffer name	Buffer species	Buffer pH (components 0.1*M*)
Ethanoic acid–ethanoate ion	CH_3COOH/CH_3COO^-	4.76
Dihydrogen phosphate ion–hydrogenphosphate ion	$H_2PO_4^-/HPO_4^{2-}$	7.20
Carbonic acid–hydrogen carbonate ion (solution saturated with CO_2)	H_2CO_3/HCO_3^-	6.46
Ammonium ion–ammonia	NH_4^+/NH_3	9.25

Discuss L2

Explain that buffers present a practical application of concepts in this chapter. The limited dissociation of weak acids and bases gives these substances the ability to act as buffers. This ability is greatly expanded if the salt of the acid or base is added to the solution. Buffer action utilizes the equilibrium established when a weak acid or base is combined with its salt. When a base is added to a buffered solution, the acidic form removes hydroxide ions from the solution. When an acid is added to a buffered solution, the basic form removes hydrogen ions from the solution.

Section 19.5 (continued)

CONCEPTUAL PROBLEM 19.2

Answers

38. a. $HPO_4^{2-} + H^+ \rightarrow H_2PO_4^-$

b. $H_2PO_4^- + OH^- \rightarrow HPO_4^{2-} + H_2O$

39. $CH_3COO^- + H^+ \rightarrow CH_3COOH$

3 ASSESS

Evaluate Understanding L2

Have students explain in their own words what determines whether a solution containing a hydrolyzing salt is acidic or basic. *(Salts formed from a strong acid and strong base are neutral. Those formed from a strong base and weak acid are basic; those formed from a strong acid and weak base are acidic.)* Ask students to explain how ethanoate ions can effectively remove hydrogen ions from solution. *(The ethanoate ions are a Brønsted–Lowry base that accept the hydrogen ions to form ethanoic acid.)*

Reteach L1

Point out that a hydrolysis reaction always uses one part of the water molecule and leaves behind the other. Use equations to show how the salt of a weak acid always reacts with water to produce hydroxide ions and the salt of a weak base always reacts to produce hydrogen ions. Remind students that chemical buffers are equilibrium systems that tend to resist change due to external influences.

Writing Activity

The most important blood buffer consists of carbonic acid and hydrogen carbonate ions. $HCO_3^-(aq) + H^+(aq) \rightleftharpoons H_2CO_3(aq)$ The buffer maintains the desired blood pH by removing or providing hydrogen ions.

CONCEPTUAL PROBLEM 19.2

Using Equations to Illustrate the Action of a Buffer

Show how the carbonic acid–hydrogen carbonate buffer can "mop up" added hydrogen ions and hydroxide ions.

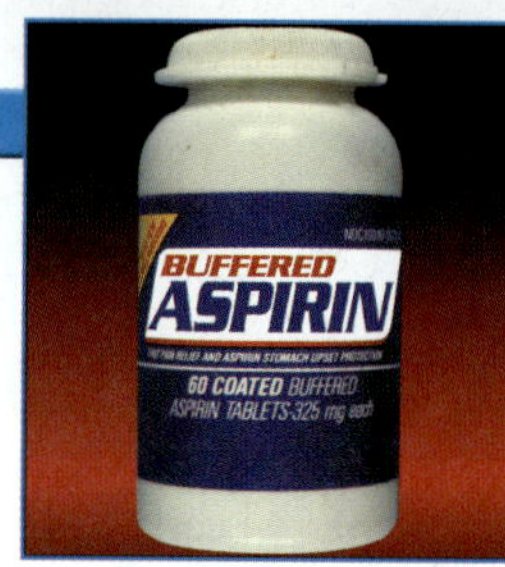

1 Analyze ***Identify the relevant concepts.***

The definition of a buffer, in terms of its ability to act as a reservoir of neutralizing power, should be illustrated. One component in a buffer can react with H^+; this component is a proton acceptor. Another component can react with OH^-; this component is a proton donor.

2 Solve ***Apply concepts to this situation.***

The carbonic acid–hydrogen carbonate buffer is a solution of carbonic acid (H_2CO_3) and hydrogen carbonate ions (HCO_3^-). When a base is added to this buffer, it reacts with H_2CO_3 (a proton donor) to produce neutral water. The pH changes very little.

$$\underset{\text{Carbonic acid}}{H_2CO_3(aq)} + \underset{\text{Hydroxide ion}}{(OH^-)(aq)} \rightleftharpoons \underset{\text{Hydrogen carbonate ion}}{HCO_3^-(aq)} + \underset{\text{Water}}{H_2O(l)}$$

When an acid is added to the buffer, it reacts with HCO_3^- (a proton acceptor) to produce un-ionized carbonic acid. Again the pH changes very little.

$$\underset{\text{Hydrogen carbonate ion}}{HCO_3^-(aq)} + \underset{\text{Hydrogen ion}}{H^+(aq)} \rightleftharpoons \underset{\text{Carbonic acid}}{H_2CO_3(aq)}$$

Practice Problems

38. Write reactions to show what happens when the following occur.

a. Acid is added to a solution of HPO_4^{2-}.

b. Base is added to a solution of $H_2PO_4^-$.

39. Write an equation that shows what happens when acid is added to the ethanoic acid–ethanoate buffer.

Problem-Solving 19.39 Solve Problem 39 with the help of an interactive guided tutorial.

with ChemASAP

19.5 Section Assessment

40. Key Concept What type of salt produces an acidic solution? A basic solution?

41. Key Concept What substances are combined to make a buffer?

42. Which of these salts would form an acidic aqueous solution?

a. $KC_2H_3O_2$ **b.** LiCl **c.** $NaHCO_3$ **d.** $(NH_4)_2SO_4$

43. Using equations, show what happens when acid is added to an ammonium ion–ammonia buffer. What happens when base is added?

Writing Activity

Explanatory Paragraph To maintain a proper pH, your blood contains buffers. Research the buffer systems in your blood and write a paragraph on how one of them works.

Assessment 19.5 Test yourself on the concepts in Section 19.5.

with ChemASAP

Interactive Textbook

If your class subscribes to the Interactive Textbook, use it to review key concepts in Section 19.5.

with ChemASAP

Section 19.5 Assessment

40. acidic solution: salt of a strong acid and weak base; basic solution: salt of a weak acid and a strong base

41. a weak acid and one of its salts or a weak base and one of its salts

42. d

43. $NH_3 + H^+ \rightarrow NH_4^+$

$NH_4^+ + OH^- \rightarrow NH_3 + H_2O$

Technology & Society

Rescuing Crumbling Books

Millions of books printed since the mid-nineteenth century slowly decay as they sit on library shelves. The cause is the acidity of the paper, which comes from alum (aluminum sulfate), used to prevent inks from soaking into paper. Libraries around the world are working with chemists to find ways to remove the acid from these books. One such method involves placing a book in a vacuum chamber, removing all the moisture from the pages, and applying a gas to neutralize the acid. **Inferring** ***Why might a library decide to use the method shown below?***

Deterioration The pages of this book are so fragile that they crumble at a touch. Deacidifying the paper will extend the life of the book.

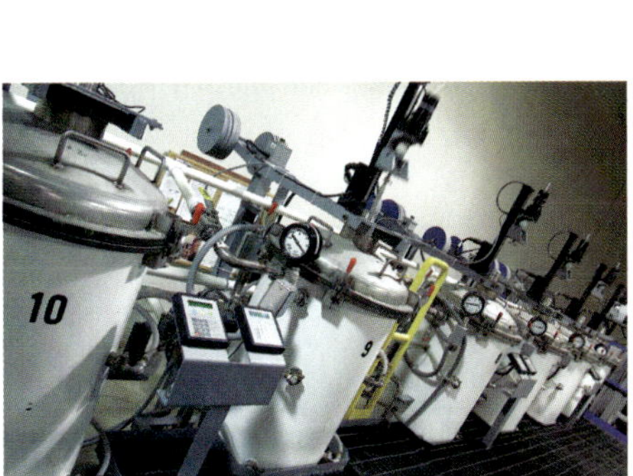

Deacidification In the tank system above, many books are treated at the same time. The machine moves the books gently in deacidification fluid, bathing each page. After the fluid is pumped away, a vacuum dries the treated books.

Restoration King James of England issued this charter to establish a colony in Virginia and find a water passage to the Pacific Ocean. It was severely damaged from prolonged exposure to moisture and a poor storage environment, after which chemists washed and deacidified each page.

Technology & Society

Rescuing Crumbling Books

Explain that when aluminum sulfate $[Al_2(SO_4)_3]$ combines with water vapor in the atmosphere, small amounts of sulfuric acid form on the pages of books, causing the paper to turn yellow and brittle. Humidity has an important influence on how much acid forms and how quickly books deteriorate. Therefore, it is important for libraries to have adequate climate control systems.

After students have read the feature, lead them in a discussion about the advantages and disadvantages of having the reference materials in your library available as bound books, on microfilm, and on audiotapes.

Facts and Figures

Treating Books in Mass

For an estimated cost of \$6 to \$10 per book, hundreds of volumes can be treated at once. If books had to be treated page by page, deacidification could cost \$1000 per book.

Answers to...

Inferring

Library books are handled by a lot of people over time. It is important that the books are preserved. Also, it would cost less to do many books at one time, as done in the deacidification method described, than individual treatments using a different method.

CHAPTER 19 Study Guide

Study Tip

Get into the habit of actively studying each day. Take time each night to review the day's work. Make notes of difficult concepts to ask your teacher about or discuss with a study friend.

If your class subscribes to the Interactive Textbook with ChemASAP, your students can go online to access an interactive version of the Student Edition and a self-test.

with ChemASAP

CHAPTER 19 Study Guide

Key Concepts

19.1 Acid-Base Theories

- Acids taste sour, bases taste bitter and feel slippery. Both are electrolytes and cause indicators to change colors.
- In an aqueous solution, an Arrhenius acid yields hydrogen ions and an Arrhenius base yields hydroxide ions.
- A Brønsted-Lowry acid is a hydrogen-ion donor; a Brønsted-Lowry base is a hydrogen-ion acceptor.
- A Lewis acid is an electron-pair acceptor; a Lewis base is an electron-pair donor.

19.2 Hydrogen Ions and Acidity

- For an aqueous solution, the product of $[H^+]$ and $[OH^-]$ equals 1×10^{-14}.
- On the pH scale, 0 is strongly acidic, 7 is neutral, and 14 is strongly basic.
- The acid and base form of an indicator have different colors in solution.

19.3 Strengths of Acids and Bases

- The stronger an acid is, the larger its K_a value.
- To find K_a of a weak acid or K_b of a weak base, substitute the concentrations of the substances into the equilibrium expression.

19.4 Neutralization Reactions

- An acid and a base react to produce a salt and water.
- The point of neutralization is the endpoint of a titration.

19.5 Salts in Solution

- Salts that produce acidic solutions contain positive ions that release protons to water; salts that produce basic solutions have negative ions that attract protons from water.
- A buffer is a solution of a weak acid or weak base and one of its salts.

Vocabulary

- acid dissociation constant (K_a) (p. 607)
- acidic solution (p. 595)
- alkaline solution (p. 595)
- amphoteric (p. 592)
- base dissociation constant (K_b) (p. 608)
- basic solution (p. 595)
- buffers (p. 620)
- buffer capacity (p. 621)
- conjugate acid (p. 591)
- conjugate acid-base pair (p. 591)
- conjugate base (p. 591)
- diprotic acid (p. 588)
- end point (p. 615)
- equivalence point (p. 613)
- hydronium ion (H_3O^+) (p. 591)
- ion-product constant for water (K_w) (p. 595)
- Lewis acid (p. 592)
- Lewis base (p. 592)
- monoprotic acids (p. 588)
- neutral solution (p. 595)
- neutralization reaction (p. 612)
- pH (p. 596)
- salt hydrolysis (p. 619)
- self-ionization (p. 594)
- standard solution (p. 615)
- strong acid (p. 605)
- strong base (p. 608)
- titration (p. 615)
- triprotic acid (p. 588)
- weak acid (p. 605)
- weak base (p. 608)

Key Equations

- $K_w = [H^+] \times [OH^-] = 1.0 \times 10^{-14}$
- $pH = -\log[H^+]$
- $pOH = -\log[OH^-]$
- $K_a = \dfrac{[H^+][A^-]}{[HA]}$

Organizing Information

Use these terms to construct a concept map that organizes the major ideas of this chapter.

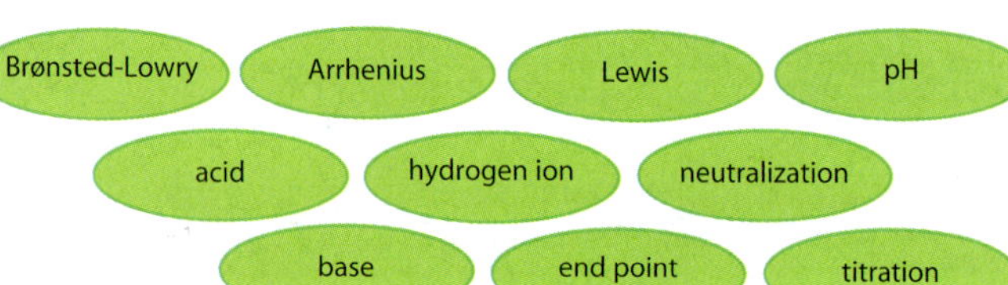

Concept Map 19 Create your Concept Map using the computer.

with ChemASAP

Chapter Resources

Print

- **Core Teaching Resources,** Chapter 19, *Practice Problems, Vocabulary Review, Quiz, Chapter Test A, Chapter Test B*

Technology

- **Computer Test Bank,** Chapter 19 Test
- **Interactive Textbook with ChemASAP,** Chapter 19

CHAPTER 19

Assessment

Reviewing Content

19.1 Acid-Base Theories

44. How did Arrhenius describe acids and bases?

45. Classify each compound as an Arrhenius acid or an Arrhenius base.
a. $Ca(OH)_2$ **b.** HNO_3 **c.** KOH
d. C_2H_5COOH **e.** HBr **f.** H_2SO_4

46. Write an equation for the dissociation of each compound in water.
a. potassium hydroxide
b. magnesium hydroxide

47. Write balanced equations for the reaction of each metal with water.
a. lithium **b.** barium

48. Identify each reactant in the following equations as a hydrogen-ion donor (acid) or a hydrogen-ion acceptor (base).
a. $HNO_3 + H_2O \longrightarrow H_3O^+ + NO_3^-$
b. $CH_3COOH + H_2O \rightleftharpoons H_3O^+ + CH_3COO^-$
c. $NH_3 + H_2O \rightleftharpoons NH_4^+ + OH^-$
d. $H_2O + CH_3COO^- \rightleftharpoons CH_3COOH + OH^-$

49. Label the conjugate acid-base pairs in each equation in Problem 48.

50. What is a Lewis acid? A Lewis base? In what sense is the Lewis theory more general than the Arrhenius and Brønsted-Lowry theories?

19.2 Hydrogen Ions and Acidity

51. Write an equation showing the ionization of water.

52. What are the concentrations of H^+ and OH^- in pure water at 25°C?

53. How is the pH of a solution calculated?

54. Why is the pH of pure water at 25°C equal to 7.00?

55. Calculate the pH for the following solutions and indicate whether each solution is acidic or basic.
a. $[H^+] = 1 \times 10^{-2}M$
b. $[OH^-] = 1 \times 10^{-2}M$
c. $[OH^-] = 1 \times 10^{-8}M$

56. What are the hydroxide-ion concentrations for solutions with the following pH values?
a. 4.00 **b.** 8.00 **c.** 12.00

57. Calculate the pH or $[H^+]$ for each solution.
a. $[H^+] = 2.4 \times 10^{-6}M$ **b.** pH = 13.20

19.3 Strengths of Acids and Bases

58. Identify each compound as a strong or weak acid or base.
a. NaOH **b.** HCl **c.** NH_3 **d.** H_2SO_4

59. Would a strong acid have a large or a small K_a? Explain.

60. Why are $Mg(OH)_2$ and $Ca(OH)_2$ considered to be strong bases even though their saturated solutions are only mildly basic?

61. Write the expression for K_a for each acid. Assume only one hydrogen is ionized.
a. HF **b.** H_2CO_3

19.4 Neutralization Reactions

62. Write a general word equation for a neutralization reaction.

63. Identify the products and write balanced equations for each neutralization reaction.
a. $HNO_3(aq) + KOH(aq) \longrightarrow$
b. $HCl(aq) + Ca(OH)_2(aq) \longrightarrow$
c. $H_2SO_4(aq) + NaOH(aq) \longrightarrow$

64. What is characteristic of the end point of a titration?

65. What is the molarity of sodium hydroxide if 20.0 mL of the solution is neutralized by each of the following 1.00*M* solutions?
a. 28.0 mL of HCl
b. 17.4 mL of H_3PO_4

19.5 Salts in Solution

66. What kinds of salts hydrolyze water?

67. Write an equation showing why an aqueous solution of sodium hydrogen carbonate is basic.

68. Explain why solutions of salts that hydrolyze water do not have a pH of 7.

69. Predict whether an aqueous solution of each salt will be acidic, basic, or neutral.
a. $NaHCO_3$ **b.** NH_4NO_3
c. KCl **d.** Na_2CO_3
e. Na_2SO_4 **f.** NH_4Cl

70. A buffered solution cannot absorb an unlimited amount of acid or base. Explain.

CHAPTER 19

Assessment

Reviewing Content

44. Acids ionize to give hydrogen ions in aqueous solution. Bases ionize to give hydroxide ions in aqueous solution.

45. a. base **b.** acid **c.** base
d. acid **e.** acid **f.** acid

46. a. $KOH \rightarrow K^+ + OH^-$
b. $Mg(OH)_2 \rightarrow Mg^{2+} + 2OH^-$

47. a. $2Li + 2H_2O \rightarrow 2LiOH + H_2$
b. $Ba + 2H_2O \rightarrow Ba(OH)_2 + H_2$

48. a. HNO_3, acid; H_2O, base
b. CH_3COOH, acid; H_2O, base
c. H_2O, acid; NH_3, base
d. H_2O, acid; CH_3COO^-, base

49. a. HNO_3 with NO_3^-, H_2O with H_3O^+
b. CH_3COOH with CH_3COO^-, H_2O with H_3O^+
c. H_2O with OH^-, NH_3 with NH_4^+
d. H_2O with OH^-, CH_3COO^- with CH_3COOH

50. A Lewis acid accepts a pair of electrons to form a covalent bond. A Lewis base donates a pair of electrons to form a covalent bond. The Lewis theory explains the behavior of compounds that act like bases without accepting hydrogen ions or that act like acids without donating hydrogen ions.

51. $H_2O \rightleftharpoons H^+ + OH^-$

52. $1.0 \times 10^{-7}M$ for both H^+ and OH^- at 25° C

53. the negative logarithm of the $[H^+]$

54. The hydrogen ion concentration in pure water at 25° C is $1 \times 10^{-7}M$. The negative logarithm or pH of this concentration is 7.0.

55. a. pH = 2.00, acidic **b.** pH = 12.00, basic **c.** pH = 6.00, acidic

56. a. $1.0 \times 10^{-10}M$ **b.** $1.0 \times 10^{-6}M$
c. $1.0 \times 10^{-2}M$

57. a. 5.62 **b.** $6.3 \times 10^{-14}M$

58. a. strong base **b.** strong acid
c. weak base **d.** strong acid

59. A strong acid is completely dissociated; K_a must be large.

60. They have high K_b values. Their concentration in saturated solution is low because of their low solubility.

61. a. $K_a = \frac{[H^+][F^-]}{[HF]}$ **b.** $K_a = \frac{[H^+][HCO_3^-]}{[H_2CO_3]}$

62. Acid + base → a salt + water

63. a. $HNO_3 + KOH \rightarrow KNO_3 + H_2O$
b. $2HCl + Ca(OH)_2 \rightarrow CaCl_2 + 2H_2O$
c. $H_2SO_4 + 2NaOH \rightarrow Na_2SO_4 + 2H_2O$

64. The indicator changes color.

65. a. 1.40*M* **b.** 2.61*M*

66. salts with a cation from a weak base and an anion from a strong acid, or with a cation from a strong base and an anion from a weak acid

67. $HCO_3^-(aq) + H_2O(l) \rightarrow H_2CO_3(aq) + OH^-(aq)$

68. Weak-acid anions accept protons from water, increasing the pH of the solution. Weak-base cations donate protons to water, decreasing the pH.

69. a. basic **b.** acidic **c.** neutral
d. basic **e.** neutral **f.** acidic

70. Eventually the buffer capacity of the buffer is exceeded and the pH will change significantly with the addition of a strong acid or base.

Understanding Concepts

71. Yes, acids like acetic acid dissolve well but ionize poorly.
72. $HPO_4^{2-} \rightarrow H^+ + PO_4^{3-}$ (acting as an acid); $HPO_4^{2-} + H^+ \rightarrow H_2PO_4^-$ (acting as a base)
73. 4.6×10^{-4}
74. **a.** CO_3^{2-}, carbonate ion
b. I^-, iodide ion
c. NH_3, ammonia
d. HSO_3^-, hydrogen sulfite ion
75. **a.** $HClO_2$, chlorous acid
b. H_3PO_4, phosphoric acid
c. H_3O^+, hydronium ion
d. NH_4^+, ammonium ion
76. **a.** $[OH^-] = 4.0 \times 10^{-10}M$
b. pH = 12.26
c. $[OH^-] = 2.0 \times 10^{-5}M$
d. pH = 5.86
77. $H_3PO_4 \rightleftharpoons H^+ + H_2PO_4^-$; $H_2PO_4^- \rightleftharpoons H^+ + H_2PO_4^{2-}$; $HPO_4^{2-} \rightleftharpoons H^+ + PO_4^{3-}$
78. **a.** KOH is the base; HBr is the acid.
b. HCl is the acid; H_2O is the base.
79. **a.** HSO_4^-
b. CN^-
c. OH^-
d. NH_3
80. $H_2PO_4^- + OH^- \rightleftharpoons H_2O + HPO_4^{2-}$
81. **a.** $2HCl + Mg(OH)_2 \rightarrow MgCl_2 + 2H_2O$
b. $2HCl + CaCO_3 \rightarrow H_2O + CO_2 + CaCl_2$
c. $Al(OH)_3 + 3HCl \rightarrow AlCl_3 + 3H_2O$
82. **a.** shift to the left
b. shift to the right
83. **a.** 8.73
b. phenolphthalein or phenol red
84. $NaC_2H_3O_2 + H_2O \rightarrow Na^+ + HC_2H_3O_2 + OH^-$
85. b, c, d, a
86. **a.** $NaOH + HCl \rightarrow NaCl + H_2O$
b. $0.72M$ NaOH

Understanding Concepts

71. Is it possible to have a concentrated weak acid? Explain.

72. Write equations showing that the hydrogen phosphate ion (HPO_4^{2-}) is amphoteric.

73. The pH of a $0.50M$ HNO_2 solution is 1.83. What is the K_a of this acid?

74. Write the formula and name of the conjugate base of each Brønsted-Lowry acid.
a. HCO_3^- **b.** HI
c. NH_4^+ **d.** H_2SO_3

75. Write the formula and name of the conjugate acid of each Brønsted-Lowry base.
a. ClO_2^- **b.** $H_2PO_4^-$
c. H_2O **d.** NH_3

76. Calculate the $[OH^-]$ or pH of each solution.
a. pH = 4.60 **b.** $[OH^-] = 1.8 \times 10^{-2}M$
c. pH = 9.30 **d.** $[OH^-] = 7.3 \times 10^{-9}M$

77. Write the three equations for the stepwise ionization of phosphoric acid.

78. Use the Brønsted-Lowry and Lewis definitions of acids and bases to identify each reactant as an acid or a base.
a. $KOH + HBr \longrightarrow KBr + H_2O$
b. $HCl + H_2O \longrightarrow Cl^- + H_3O^+$

79. Write the formula for the conjugate base of each acid.
a. H_2SO_4 **b.** HCN
c. H_2O **d.** NH_4^+

80. Use the phosphate buffer ($H_2PO_4^-/HPO_4^{2-}$) to illustrate how a buffer system works. Show, by means of equations, how the pH of a solution can be kept almost constant when small amounts of acid or base are added.

81. Write an equation for the reaction of each antacid with hydrochloric acid.
a. magnesium hydroxide
b. calcium carbonate
c. aluminum hydroxide

82. How would the equilibrium between hypochlorous acid and the hypochlorite ion be affected by the addition of each?
$HOCl(aq) + OH^-(aq) \rightleftharpoons OCl^-(aq) + H_2O(l)$
a. HCl **b.** NaOH

83. The following data were collected from a titration of 50.00 mL of ethanoic acid (CH_3COOH) of unknown concentration with $0.100M$ NaOH. Plot these data (pH on the y-axis) to obtain a titration curve.

Volume of NaOH (mL)	pH	Volume of NaOH (mL)	pH
0		50.00	8.73
10.00	4.15	50.01	8.89
25.00	4.76	51.00	11.00
40.00	5.36	60.00	11.96
49.00	6.45	75.00	12.30
49.99	8.55	100.00	12.52

a. What is the pH at the end point of this titration?
b. Use Figure 19.12 to identify one or more indicators that could be used to determine the end point in this titration.

84. Write an equation to show that an aqueous solution of sodium ethanoate will be basic.

85. Arrange these solutions in order of decreasing acidity.
a. $0.1M$ NaOH
b. $0.1M$ HCl
c. $0.1M$ ammonium chloride
d. $0.1M$ sodium ethanoate

86. The graph shows the number of millimoles (mmol) of water formed by the drop-by-drop addition of $1.0M$ HCl to a 25.0-mL sample of NaOH of unknown concentration.

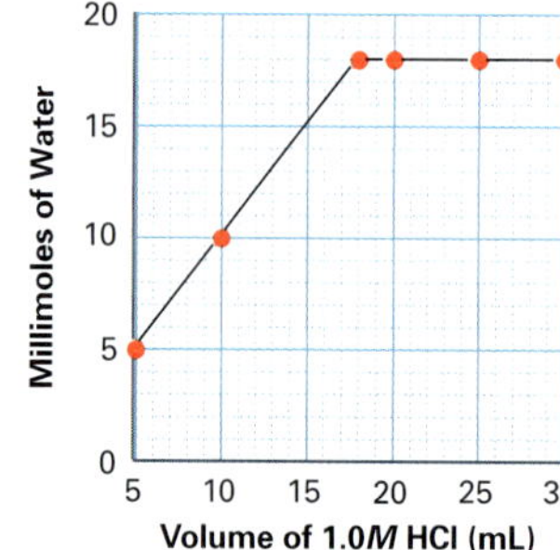

a. Write an equation for the reaction.
b. Estimate the concentration of the NaOH.

Critical Thinking

87. Arrhenius, Brønsted-Lowry, and Lewis all developed theories to explain acids and bases.
   a. Which theory is easiest for you to understand?
   b. How can all three theories be accepted by chemists?

88. Demonstrate the relationship pH + pOH = 14. (*Hint:* Use the expression for K_w.)

89. Which quantity might correspond to the *y*-axis on this graph: $[H^+]$, pH, $[OH^-]$, or pOH? Explain.

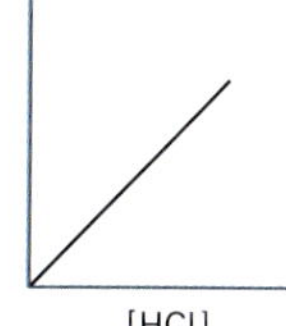

90. The following equilibria are involved in the solubility of carbon dioxide in water.

$$CO_2(g) \rightleftharpoons CO_2(aq)$$
$$CO_2(aq) + H_2O(l) \rightleftharpoons H_2CO_3(aq)$$
$$H_2CO_3(aq) \rightleftharpoons H^+(aq) + HCO_3^-(aq)$$
$$HCO_3^-(aq) \rightleftharpoons H^+(aq) + CO_3^{2-}(aq)$$

If sea water is slightly alkaline, would you expect the concentration of dissolved CO_2 to be higher or lower than in pure water? Explain.

91. Critique the accuracy of each of these statements.
   a. Indicators such as methyl red provide accurate and precise measurements of pH.
   b. According to the Arrhenius definition of acids and bases, ammonia qualifies as a base.
   c. The strength of an acid or base changes as its concentration changes.

92. The hydrogen carbonate ion–carbonic acid buffer system is an important buffer system in the blood. This system is represented by the following equation.

$$H_2O(l) + CO_2(g) \rightleftharpoons H_2CO_3(aq) \rightleftharpoons H^+(aq) + HCO_3^-(aq)$$

Explain how abnormal breathing patterns can lead to acid-base imbalances in the blood that are called respiratory acidosis (abnormally low pH) and respiratory alkalosis (abnormally high pH). Too-rapid and too-deep breathing also lead to respiratory alkalosis. Explain.

Concept Challenge

93. Calculate the pH of a 0.010*M* solution of sodium cyanide (NaCN). The K_b of CN^- is 2.1×10^{-5}.

94. Show that for any conjugate acid-base pair $K_a \times K_b = K_w$.

95. The K_w of water varies with temperature, as indicated in the table below.

Temperature (°C)	K_w	pH
0	1.137×10^{-15}	a.______
10	2.917×10^{-15}	b.______
20	6.807×10^{-15}	c.______
30	1.469×10^{-14}	d.______
40	2.917×10^{-14}	e.______
50	5.470×10^{-14}	f.______

   a. Find the pH of water for each temperature in the table. Use these data to prepare a graph of pH versus temperature.
   b. Using your graph, estimate the pH of water at 5°C.
   c. At what temperature is the pH of water approximately 6.85?

96. Use the cyanate buffer $HOCN/OCN^-$ to illustrate how a buffer system works. Show, by means of equations, how the pH of a solution can be kept almost constant when small amounts of acid or base are added.

97. Suppose you slowly add 0.1*M* NaOH to 50.0 mL of 0.1*M* HCl. What volume of NaOH must be added for the resulting solution to become neutral? Explain your reasoning.

98. What is the molarity of an H_2SO_4 solution if 80.0 mL of the solution reacts completely with 0.424 g Na_2CO_3?

$$H_2SO_4(aq) + Na_2CO_3(aq) \longrightarrow H_2O(l) + CO_2(aq) + Na_2SO_4(aq)$$

99. Household bleach is a solution of sodium hypochlorite. It is considered a hazardous solution because it is basic and because it contains the active bleaching ingredient, ClO^-. What is the $[OH^-]$ in an aqueous solution that is 5.0% NaClO by mass? What is the pH of the solution? (The density of the solution is 1.0 g/mL and $K_a = 3.5 \times 10^{-8}$.)

Critical Thinking

87. a. Answers will vary; students are likely to consider the Arrhenius theory the easiest to understand, and the Lewis theory the best because it is the most general. All three theories provide definitions and describe accepted behavior of a certain group of compounds. The Brønsted-Lowry theory includes a greater number of compounds than the Arrhenius theory because it is more general, and the Lewis theory includes the greatest number of compounds because it is the most general.
b. Each theory has advantages in certain circumstances.

88. Because log (a × b) = log a + log b, and $K_w = [H^+] \times [OH^-] = 1.0 \times 10^{-14}$, taking the log of each concentration term gives: $\log[H^+] + \log[OH^-] = -14; -\log[H^+] - \log[OH^-] = 14$; pH + pOH = 14.

89. The *y*-axis might correspond to $[H^+]$ because HCl is a strong acid.

90. CO_2 concentration is higher in pure water. Less CO_2 becomes carbonate because pure water does not have the OH^- ions needed to reduce H^+ concentration.

91. a. False; an indicator determines a range of pH values.
b. False; an Arrhenius base dissociates to give hydroxide ions in aqueous solution. Ammonia does not do this.
c. False; strength is a measure of dissociation or ionization, not concentration.

92. Hyperventilation releases CO_2. The equilibrium shift causes H_2CO_3 and H^+ concentrations to decrease. The loss of H^+ increases the basicity of the blood and alkalosis results. Hypoventilation does not release enough CO_2. The CO_2 build up increases the H_2CO_3 and H^+ concentrations. The H^+ concentration increases the acidity of the blood and acidosis results.

Concept Challenge

93. pH = 10.66

94. $K_w = K_aK_b = \frac{[H^+][A^-]}{[HA]} \times \frac{[HA][OH^-]}{[A^-]} = [H^+][OH^-]$

95. a. 7.4721, 7.2675, 7.0835, 6.9165, 6.7675, 6.6310

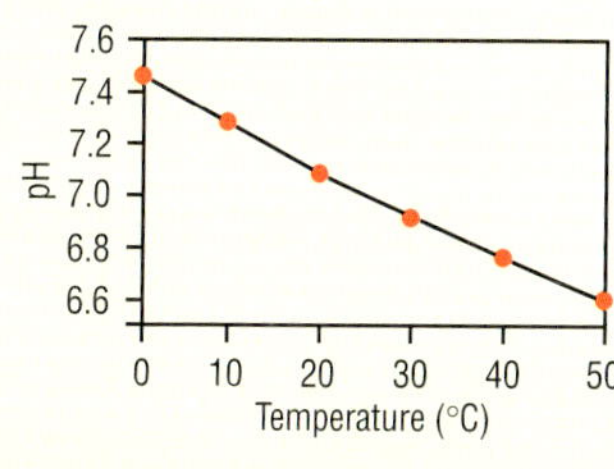

b. 7.37 **c.** 35°C

96. $HOCN + OH^- \rightleftharpoons H_2O + OCN^-$
$OCN^- + H^+ \rightleftharpoons HOCN$

97. 50.0 mL; The pH = 7 when $[H^+] = [OH^-]$. Because HCl is a strong acid that supplies one hydrogen ion per formula unit and NaOH is a strong base that supplies one hydroxide ion per formula unit, $[H^+] = [OH^-]$ when equal volumes of solutions of the same molarity are combined.

98. $5.00 \times 10^{-2}M$

99. $[OH^-] = 4.4 \times 10^{-4}M$
pH = 10.66

CHAPTER 19

Assessment continued

Cumulative Review

100. **a.** $K_2O(s)$
b. $CaS(s)$
c. $AlF_3(s)$

101. 131 g O_2

102. liquid

103. The total pressure in a mixture of gases is equal to the sum of the partial pressures of each gas in the mixture.

104. **b.** Boyle's law

105. suspension

106. **c.** $SiO_2(s)$

107. hydrogen bond

108. 2.25 g KCl

109. Dissolve 0.272 mol KOH(*s*) in water and add sufficient water to give 400.0 mL of solution.

110. 0.49 L

111. b, c, and d

112. 1.8×10^4 J

113. **a.** 144 J
b. 1.0×10^3 kJ
c. 82.9 cal

114. 2.8 kJ

115. The product of the concentrations of the two ions must be greater than the solubility product.

116. **a.** $K_{eq} = \dfrac{[CO]^2 \times [O_2]}{[CO_2]^2}$

b. $K_{eq} = \dfrac{[NH_3]^2}{[N_2] \times [N_2]^3}$

117. $2.0 \times 10^{-8}M$

118. **a.** NaCl(*aq*)
b. $CO_2(g)$
c. hot water

119. **a.** shift right
b. no change
c. shift right
d. shift right
e. shift right

120. A: first order
B: second order
C: zero order
overall: third order

CHAPTER 19

Assessment continued

Cumulative Review

100. Write the product of each of these combination reactions. (*Chapter 11*)
a. $K(s) + O_2(g) \longrightarrow$
b. $Ca(s) + S(s) \longrightarrow$
c. $F_2(g) + Al(s) \longrightarrow$

101. How many grams of oxygen are needed to completely burn 87.4 g of sulfur to form sulfur trioxide? (*Chapter 12*)

$$S(s) + O_2(g) \longrightarrow SO_3(g)$$

102. Which state of matter is not part of the process of sublimation? (*Chapter 13*)

103. State Dalton's law of partial pressures. (*Chapter 14*)

104. Which of these laws describes an inverse relationship? (*Chapter 14*)
a. Charles' law
b. Boyle's law
c. Gay-Lussac's law

105. Which has the largest particles, a solution, a colloid, or a suspension? (*Chapter 15*)

106. Which of these is not an electrolyte? (*Chapter 15*)
a. NaCl(*l*) **b.** $KNO_2(aq)$ **c.** $SiO_2(s)$ **d.** NaCl(*aq*)

107. What type of bond is responsible for water's high surface tension? (*Chapter 15*)

108. How many grams of potassium chloride are in 45.0 mL of a 5.00% (by mass) solution? (*Chapter 16*)

109. How would you prepare 400.0 mL of a 0.680*M* KOH solution? (*Chapter 16*)

110. How many liters of 8.0*M* HCl are needed to prepare 1.50 L of 2.5*M* HCl? (*Chapter 16*)

111. Which of these is an endothermic process? (*Chapter 17*)
a. burning wax **b.** evaporating water
c. melting wax **d.** roasting a marshmallow

112. How many joules of heat are required to melt a 55.0-g ice cube at 0°C? (*Chapter 17*)

113. Make the following conversions. (*Chapter 17*)
a. 34.5 cal to joules
b. 250 Cal to kilojoules
c. 0.347 kJ to calories

114. The specific heat capacity of iron is 0.46 J/(g·°C). How many kilojoules of energy are needed to raise the temperature of a 432-g iron bar 14°C? (*Chapter 17*)

115. What must be true about the concentration of two ions if precipitation occurs when solutions of the two ions are mixed? (*Chapter 18*)

116. Write an equilibrium-constant expression for each equation. (*Chapter 18*)
a. $2CO_2(g) \rightleftharpoons 2CO(g) + O_2(g)$
b. $N_2(g) + 3H_2(g) \rightleftharpoons 2NH_3(g)$

117. What is the equilibrium concentration of barium ion in a 1.0-L saturated solution of barium carbonate to which 0.25 mol K_2CO_3 has been added? (*Chapter 18*)

118. In each pair, which has the higher entropy? (*Chapter 18*)
a. NaCl(*s*) or NaCl(*aq*)?
b. $CO_2(s)$ or $CO_2(g)$?
c. hot water or cold water?

119. How would each change affect the position of equilibrium of this reaction? (*Chapter 18*)

$$2H_2(g) + O_2(g) \rightleftharpoons 2H_2O(g) + \text{heat}$$

a. increasing the pressure
b. adding a catalyst
c. increasing the concentration of $H_2(g)$
d. cooling the reaction mixture
e. removing water vapor from the container

120. For the reaction $A(g) + B(g) + C(g) \longrightarrow D(g)$ the following data were obtained at a constant temperature. From the data decide the kinetic order of reaction with respect to A, B, and C, and the overall order of reaction. (*Chapter 18*)

Initial [A] (mol/L)	Initial [B] (mol/L)	Initial [C] (mol/L)	Initial Rate (mol/L·min)
0.0500	0.0500	0.0100	6.25×10^{-3}
0.1000	0.0500	0.0100	1.25×10^{-2}
0.1000	0.1000	0.0100	5.00×10^{-2}
0.0500	0.0500	0.0200	6.25×10^{-3}

Standardized Test Prep

Test-Taking Tip

Eliminate Wrong Answers If you don't know which response is correct, start by eliminating those you know are wrong. If you can rule out some choices, you'll have fewer left to consider and you'll increase your chances of choosing the correct answer.

Select the choice that best answers each question or completes each statement.

1. An acid has a measured K_a of 3×10^{-6}.
   - **a.** The acid is a strong acid.
   - **b.** An aqueous solution of the acid would have a pH < 7.
   - **c.** The acid is a strong electrolyte.
   - **d.** All of the above are correct.
2. The pH of a sample of orange juice is 3.5. A sample of tomato juice has a pH of 4.5. Compared to the $[H^+]$ of orange juice, the $[H^+]$ of tomato juice is
   - **a.** 1.0 times higher. **b.** 10 times lower.
   - **c.** 10 times higher. **d.** 1.0 times lower.
3. Which species is the conjugate base of an ammonium ion, NH_4^+?
   - **a.** H_2O **b.** OH^- **c.** NH_3 **d.** H_3O^+
4. How many moles of NaOH are required to neutralize 2.4 mol H_2SO_4?
   - **a.** 1.2 mol **b.** 2.4 mol **c.** 3.6 mol **d.** 4.8 mol
5. A solution with a hydrogen ion concentration of $2.3 \times 10^{-8}M$ has a pH between
   - **a.** 2 and 3. **b.** 3 and 4. **c.** 7 and 8. **d.** 8 and 9.
6. The net ionic equation for the neutralization reaction between solutions of potassium hydroxide and hydrochloric acid is
   - **a.** $H^+(aq) + OH^-(aq) \longrightarrow H_2O(l)$.
   - **b.** $KOH(aq) + HCl(aq) \longrightarrow H_2O(l) + KCl(aq)$.
   - **c.** $K^+(aq) + Cl^-(aq) \longrightarrow KCl(aq)$.
   - **d.** $K^+(aq) + OH^-(aq) + H^+(aq) + Cl^-(aq) \longrightarrow KCl(aq) + H_2O(l)$.
7. Calculate the molarity of an HCl solution if 25.0 mL of the solution is neutralized by 15.5 mL of 0.800M NaOH.
   - **a.** 0.248M **b.** 0.496M **c.** 1.29M **d.** 0.645M
8. Which combination of compound and ion would not make a useful buffer solution?
   - **a.** ammonium ion and ammonia
   - **b.** hydrogen carbonate ion and carbonic acid
   - **c.** sulfate ion and sulfuric acid
   - **d.** ethanoate ion and ethanoic acid

The lettered choices below refer to Questions 9–11. A lettered choice may be used once, more than once, or not at all.

(A) PQ (B) P_2Q_3 (C) PQ_3 (D) P_3Q (E) PQ_2

Which of the choices is the general formula for the salt formed in each of the following neutralization reactions? P is a cation; Q is an anion.

9. $H_3PO_4 + NaOH \longrightarrow$
10. $H_2SO_4 + Mg(OH)_2 \longrightarrow$
11. $HNO_3 + Al(OH)_3 \longrightarrow$

Use the drawings below to answer Questions 12 and 13. Water molecules have been omitted from the solution windows.

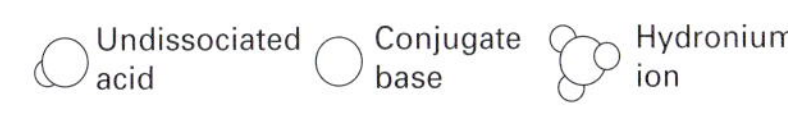

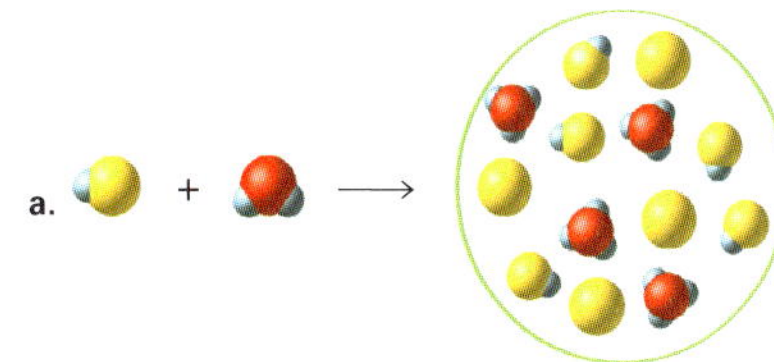

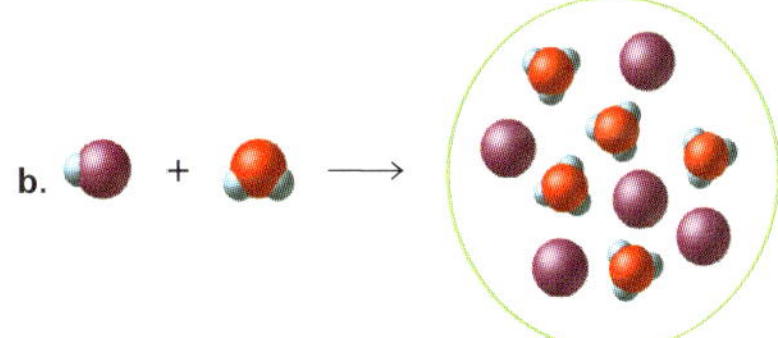

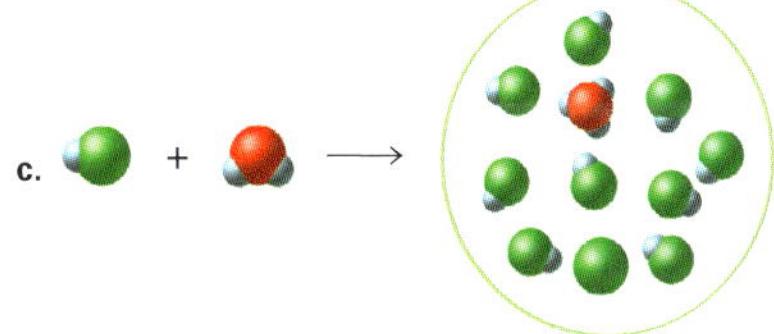

12. Put the acids in order of increasing strength.
13. How many of the acids are strong acids?

Standardized Test Prep

1. b
2. b
3. c
4. d
5. c
6. a
7. b
8. c
9. (D)
10. (A)
11. (C)
12. c, a, b
13. one

CHAPTER 20

Oxidation-Reduction Reactions Planning Guide

Use these planning tools
Lesson Plan Book
Teacher Express

SECTION OBJECTIVES	STANDARDS NATIONAL	STANDARDS STATE	ACTIVITIES and LABS	
20.1 The Meaning of Oxidation and Reduction, pp. 631–638 1 block or 2 periods **20.1.1 Define** oxidation and reduction in terms of the loss or gain of oxygen and the loss or gain of electrons. **20.1.2 State** the characteristics of a redox reaction and **identify** the oxidizing agent and reducing agent. **20.1.3 Describe** what happens to iron when it corrodes.	A-1, B-2, B-3		**SE** Inquiry Activity: Rusting, p. 630	L2
			TE Teacher Demo: Oxidation and Reduction, p. 632	L2
			TE Class Activity: Born-Haber Cycles, p. 633	L2
			TE Class Activity: Identifying Oxidizing and Reducing Agents, p. 634	L2
			TE Class Activity: Metal Corrosion, p. 636	L2
			LM Lab 46: Oxidation–Reduction Reactions	L2
			LM Lab 47: Corrosion	L2
			SSLM Lab 34: Determination of an Activity Series	L2
			LP Lab Practical 20-1: Reduction of Metals	L2
20.2 Oxidation Equations, pp. 639–644 1 block or 2 periods **20.2.1 Determine** the oxidation number of an atom of any element in a pure substance. **20.2.2 Define** oxidation and reduction in terms of a change in oxidation number and identify atoms being oxidized or reduced in redox reactions.	A-1, B-2, B-3, E-2		**TE** Teacher Demo: Redox Reactions, p. 640	L2
			TE Class Activity: Assigning Oxidation Numbers, p. 641	L2
			TE Class Activity: Identifying Redox Reactions, p. 642	L2
20.3 Balancing Redox Equations, pp. 645–655 1 block or 2 periods **20.3.1 Describe** how oxidation numbers are used to identify redox reactions. **20.3.2 Balance** a redox equation using the oxidation-number-change method. **20.3.3 Balance** a redox equation by breaking the equation into oxidation and reduction half-reactions, and then using the half-reaction method.	A-1, A-2, B-3, G-1		**SE** Quick Lab: Bleach It! Oxidize the Color Away, p. 653	L2
			SE Small-Scale Lab: Half-Reactions, p. 655	L2
			TE Teacher Demo: Color Changes in Redox Reactions, p. 648	L2
			TE Class Activity: Preventing Oxidation, p. 651	L1
			SSLM Lab 35: Oxidation–Reduction Reactions	L2
			PLM Probeware Lab: Bleach It! Oxide the Color Away	L2
			VCL Lab 30: Redox Titrations: Determination of Iron	L2

Ability Levels

L1 For students who need additional help

L2 For all students

L3 For students who need to be challenged

Components

SE	Student Edition	**LM**	Laboratory Manual	**GCP**	Graphing Calculator Problems	**T**	Transparencies
TE	Teacher's Edition	**SSLM**	Small-Scale Chemistry Lab Manual	**TTT**	Test-Taking Tips with Transparencies, Grades 9–12 Science	**CA**	Chem Alive and Chemistry Field Trips DVD
GRSW	Guided Reading and Study Workbook	**LP**	Laboratory Practicals	**STP**	Standardized Test Preparation Workbook	**iT**	Interactive Textbook with ChemASAP
CTR	Core Teaching Resources	**PLM**	Probeware Lab Manual			**GO**	Go Online Internet Resources
SSG	Spanish Study Guide	**VCL**	Virtual Chemistry Labs				
		CTB	Computer Test Bank				

RESOURCES PRINT and TECHNOLOGY	PROBLEMS and ASSESSMENT
GRSW Chapter 20, Section 1  L1	**CP** 20.1, Identifying Oxidized and Reduced Reactants
T T230–T233 L2	**SE** Section 20.1 Assessment, p. 638
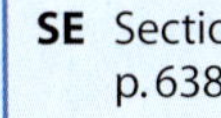 **GO** Chapter 20, Section 1 L2	 **iT** Section 20.1
GO Chapter 20, Section 1 L2	
 **GRSW** Chapter 20, Section 2  L1	**CP** 20.2, Assigning Oxidation Numbers to Atoms
T T234–T237  L2	**CP** 20.3, Identifying Atoms Oxidized and Reduced
	SE Section 20.2 Assessment, p. 643
	 **iT** Section 20.2
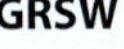 **GRSW** Chapter 20, Section 3  L1	**CP** 20.4, Identifying Redox Equations
T T238–T241 L2	**CP** 20.5, Balancing Redox Equations by Oxidation-Number Change
 **GO** Chapter 20, Section 3 L2	**CP** 20.6, Balancing Redox Equations by Half-Reactions
 **GO** Chapter 20, Section 3 L2	**SE** Section 20.3 Assessment, p. 654
	 **iT** Section 20.3

Web Code: cdn-1201
Web Code: cdn-1202
Web Code: cdn-1203

Web Code: cdb-1203

MATERIALS List

Quantities for each group

STUDENT EDITION

Inquiry Activity, p. 630
iron finishing nails, pliers, scissors, copper wire, zinc strips, sandpaper, plastic wrap, saucer, water, table salt, petroleum jelly, paper towels

Quick Lab, p. 653
spot plates, medicine droppers, liquid and powder bleach, 3% H_2O_2, 1% oxalic acid, 0.2*M* $Na_2S_2O_3$, I_2 in KI(*aq*), $KMnO_4$(*aq*), grape juice, rusty water, colored fabric, grass-stained fabric, flower petals

Small-Scale Lab, p. 655
reaction surfaces, 1.0*M* HCl, HNO_3, and H_2SO_4, pieces of Zn, Mg, Cu, and Fe

TEACHER'S EDITION

Teacher Demo, p. 632
3 g copper(II) oxide (CuO), 3 g charcoal (C), crucible with cover, Bunsen burner, tongs, glass plate

Teacher Demo, p. 640
100 mL of 0.1*M* lead(II) acetate $[Pb(CH_3COO)_2]$, 150-mL beaker, 1-cm × 3-cm zinc strip, sodium sulfide, 3*M* NaOH, 1*M* iron(III) chloride, stirring rod, plastic container

Class Activity, p. 642
3 small glass test tubes, zinc, copper, magnesium, 0.1*M* HCl, medicine dropper or pipet

Teacher Demo, p. 648
6 crystals of potassium permanganate ($KMnO_4$), 540 mL of water, large beaker, 1 g sodium hydrogen sulfite ($NaHSO_3$), two 50-mL beakers, 1 g barium chloride dihydrate ($BaCl_2 \cdot 2H_2O$), 3*M* H_2SO_4, stirring rod, plastic container

Class Activity, p. 651
potatoes, apples, beakers, water, boiled water, lemon juice, carbonated beverage, sugar, salt, and vinegar solutions, paper towels

Chapter Assessment

CHAPTER ASSESSMENT

SE Chapter Assessment, pp. 657–660

CTB Chapter 20

 **iT** Chapter 20

STANDARDIZED TEST PREP

SE Chapter 20, p. 661

STP Topic 9

iT Interactive Textbook with Assessment at PHSchool.com

CHAPTER 20

Oxidation-Reduction Reactions

INQUIRY Activity

Rusting

L2

Purpose

Students observe how rust occurs and how to prevent it.

Prep Time 20 minutes
Class Time 20 minutes on the first day and 20 minutes the following day

Safety

Be careful not to puncture skin with any sharp nails.

Think About It

1. Nails 1, 2, 5, and 6 show evidence of rusting whereas nails 3, 4, and 7 do not.
2. The nails that were not protected showed signs of rust.
3. Students should suggest a relationship between evidence of rusting and the presence of water, air, salt, and more-or-less active metals.

CHAPTER 20

Oxidation-Reduction Reactions

Rust is the result of the iron in this ship reacting with oxygen and water.

INQUIRY Activity

Rusting

Materials

fine sandpaper, 7 iron finishing nails (approximately 6 cm long), paper towels, scissors, water, saucer, pliers, copper wire, zinc strip, petroleum jelly, table salt, plastic wrap

Procedure

1. Use sandpaper to polish seven nails. Wipe them clean with a paper towel.
2. Place two wet paper towels on the saucer.
3. Nail 1: Using pliers, bend into a U shape. Nail 2: Wrap one end with copper wire. Nail 3: Wrap one end with a strip of zinc. Nail 4: Cover the entire nail with a thin coat of petroleum jelly. Nail 5: Moisten with water and sprinkle with salt. Nail 6: Leave untreated. Nail 7: Leave untreated.
4. Place nails 1 through 6 on the wet paper towel. Make sure the nails do not touch. Cover them with a piece of plastic wrap. Place nail 7 (the control) on top of the plastic wrap. During a 24-hour period, record your observations in a table.

Think About It

1. What changes took place with the nails over 24 hours?
2. How can you account for these changes?
3. How can you account for differences in the appearance of the nails after 24 hours?

20.1 The Meaning of Oxidation and Reduction

Connecting to Your World During winter in cold climates, salt is often spread on roads to lower the freezing point of water and thereby prevent the buildup of slippery ice. Salt may make driving safer, but the salt that clings to the metallic parts of cars can cause them to corrode or rust relatively quickly. This corrosion is one example of a chemical reaction called oxidation-reduction. In this section, you will learn about oxidation-reduction reactions.

Guide for Reading

Key Concepts

- How is the gain or loss of oxygen in a chemical reaction linked to oxidation and reduction?
- How is the gain or loss of electrons linked to oxidation and reduction?
- What happens to iron when it corrodes?

Vocabulary

oxidation-reduction reactions
redox reactions
oxidation
reduction
reducing agent
oxidizing agent

Reading Strategy

Comparing and Contrasting As you read, compare the older definitions of oxidation and reduction with the current definitions.

What Are Oxidation and Reduction?

The combustion of gasoline in an automobile engine and the burning of wood in a fireplace are reactions that require oxygen as they release energy. The reactions that break down food in your body and release energy use oxygen from the air you breathe.

Oxygen and Redox Early chemists saw oxidation only as the combination of an element with oxygen to produce an oxide. The burning of a fuel is also an oxidation reaction that uses oxygen. For example, when methane (CH_4), the main component of natural gas, burns in air, it oxidizes and forms oxides of carbon and hydrogen, as shown in Figure 20.1. One oxide of carbon is carbon dioxide, CO_2.

Not all oxidation processes that use oxygen involve burning. For example, when elemental iron turns to rust, it slowly oxidizes to compounds such as iron(III) oxide (Fe_2O_3). Bleaching stains in fabrics is an example of oxidation that does not involve burning. Common liquid household bleach contains sodium hypochlorite (NaClO), a substance that releases oxygen, which oxidizes stains to a colorless form. Powder bleaches may contain calcium hypochlorite ($Ca(ClO)_2$), sodium perborate ($NaBO_3$), or sodium percarbonate ($2Na_2CO_3 \cdot 3H_2O_2$). Hydrogen peroxide (H_2O_2) also releases oxygen when it decomposes. It is both a bleach and a mild antiseptic that kills bacteria by oxidizing them.

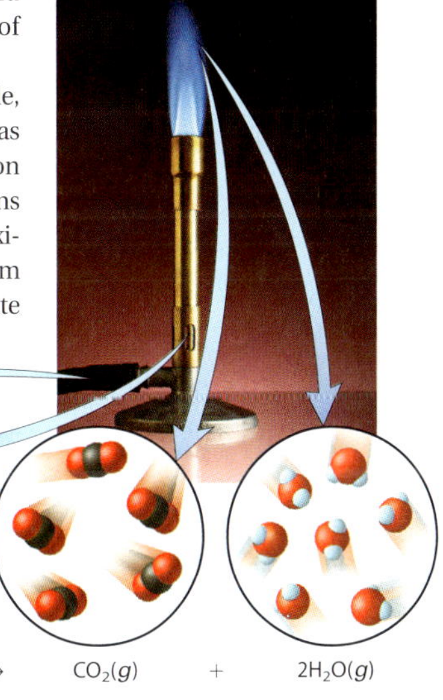

Figure 20.1 A Bunsen burner oxidizes the methane in natural gas to carbon dioxide and water.

20.1

1 FOCUS

Objectives

20.1.1 Define oxidation and reduction in terms of the loss or gain of oxygen and the loss or gain of electrons.

20.1.2 State the characteristics of a redox reaction and **identify** the oxidizing agent and reducing agent.

20.1.3 Describe what happens to iron when it corrodes.

Guide for Reading

Build Vocabulary L2

Graphic Organizer Use a web diagram to connect this section's vocabulary terms. Place *oxidation-reduction reactions* and *redox reactions* in the center and then place the four remaining terms around the center, along with an explanation of how the terms relate.

Reading Strategy L2

Outline Have students preview the section's content by making an outline using the B and C headings.

2 INSTRUCT

Connecting to Your World

Have students read the opening paragraph. Ask, **How might cars be protected against corrosion?** *(Put a protective undercoat on the car or use plastic car parts.)*

Section Resources

Print

- **Guided Reading and Study Workbook,** Section 20.1
- **Core Teaching Resources,** Section 20.1 Review
- **Laboratory Manual,** Labs 46, 47
- **Small-Scale Chemistry Laboratory Manual,** Lab 34
- **Transparencies,** T230–T233

Technology

- **Interactive Textbook with ChemASAP,** Problem-Solving 20.1, Assessment 20.1
- **Go Online,** Section 20.1

What Are Oxidation and Reduction?

Discuss L2

To assess students' prior knowledge about electron transfer, ask, **What happens to magnesium and oxygen when they react to form magnesium oxide, MgO?** *(Mg loses electrons to form Mg^{2+} and oxygen gains electrons to form O^{2-}.)*

TEACHER Demo

Oxidation and Reduction L2

Purpose Students will observe how pure metals are obtained from their oxide ores by reduction (lost of oxygen).

Materials 3 g copper(II) oxide (CuO), 3 g charcoal (C), crucible with cover, Bunsen burner, tongs, glass plate

Procedure Mix 3 g copper(II) oxide (CuO) and 3 g charcoal (C) in a crucible. Cover the crucible. Then heat strongly and evenly over a Bunsen burner for 10 minutes. Cool the crucible and remove the copper with tongs. Pass the copper around on a glass plate for the students to examine. Pass around a bit of the original CuO as well so that the students can see the change in appearance. Write the reaction that occurred on the board: $2CuO(s) + C(s) \rightarrow 2Cu(s) + CO_2(g)$. **Ask, what species were oxidized and reduced based on loss or gain of oxygen?** *(C was oxidized, and CuO was reduced.)*

Expected Outcome Copper metal is formed.

Figure 20.2 When charcoal, which is mostly carbon, is burned in air, carbon dioxide and heat are produced.

The oxide of hydrogen is water, H_2O. Charcoal, shown in Figure 20.2, also oxidizes when it burns, forming CO_2.

A process called reduction is the opposite of oxidation. Originally, reduction meant the loss of oxygen from a compound. The reduction of iron ore to metallic iron involves the removal of oxygen from iron(III) oxide. The reduction is accomplished by heating the ore with carbon, usually in the form of coke. The equation for the reduction of iron ore is shown below.

$$\underset{\text{Iron (III) oxide}}{2Fe_2O_3(s)} + \underset{\text{Carbon}}{3C(s)} \longrightarrow \underset{\text{Iron}}{4Fe(s)} + \underset{\text{Carbon dioxide}}{3CO_2(g)}$$

The reduction of iron also includes an oxidation process. As iron(III) oxide is reduced to iron by losing oxygen, carbon oxidizes to carbon dioxide by gaining oxygen. Oxidation and reduction always occur simultaneously. **The substance gaining oxygen is oxidized, while the substance losing oxygen is reduced.** No oxidation occurs without reduction, and no reduction occurs without oxidation. Reactions that involve these processes are therefore called oxidation-reduction reactions. **Oxidation-reduction reactions** are also known as **redox reactions.**

Checkpoint *What is another name for redox reactions?*

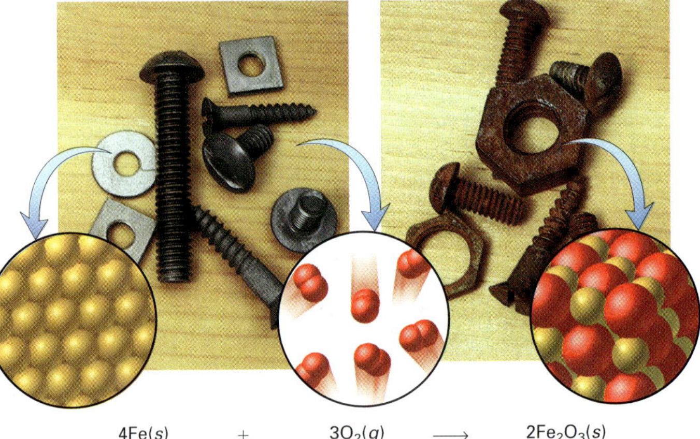

Figure 20.3 When items made of iron are exposed to moist air, the Fe atoms react with O_2 molecules. The iron rusts; it is oxidized to compounds such as iron(III) oxide (Fe_2O_3).

Differentiated Instruction

Gifted and Talented 

Have advanced students do research on redox reactions involving oxygen that are important in nature. Examples are the reduction of carbon in carbon dioxide to form carbohydrates during photosynthesis and the oxidation of glucose during cell respiration to release energy. Have students prepare reports or posters based on their research.

Electron Shift in Redox Reactions The modern concepts of oxidation and reduction have been extended to include many reactions that do not even involve oxygen. You learned in Chapter 6 that, with the exception of fluorine, oxygen is the most electronegative element. As a result, when oxygen bonds with an atom of a different element (other than fluorine), electrons from that atom shift toward oxygen. Redox reactions are currently understood to involve any shift of electrons between reactants. **Oxidation** is now defined to mean complete or partial loss of electrons or gain of oxygen. **Reduction** is now defined to mean complete or partial gain of electrons or loss of oxygen.

Oxidation	Reduction
Loss of electrons Gain of oxygen	Gain of electrons Loss of oxygen

Redox Reactions That Form Ions During a reaction between a metal and a nonmetal, electrons are transferred from atoms of the metal to atoms of the nonmetal. For example, when magnesium metal is heated with the nonmetal sulfur, the ionic compound magnesium sulfide is produced as shown in Figure 20.4. Two electrons are transferred from a magnesium atom to a sulfur atom. The magnesium atoms become more stable by the loss of electrons. The sulfur atoms become more stable by the gain of electrons.

$$\cdot Mg \cdot + \cdot \ddot{S}: \longrightarrow Mg^{2+} + :\ddot{S}:^{2-}$$

Magnesium atom, Sulfur atom, Magnesium ion, Sulfide ion

Because it loses electrons, the magnesium atom is said to be oxidized to a magnesium ion. Simultaneously, the sulfur atom gains two electrons and is reduced to a sulfide ion. The overall process is represented as the two component processes below.

Oxidation: $\cdot Mg \cdot \longrightarrow Mg^{2+} + 2e^-$ (loss of electrons)

Reduction: $\cdot \ddot{S}: + 2e^- \longrightarrow :\ddot{S}:^{2-}$ (gain of electrons)

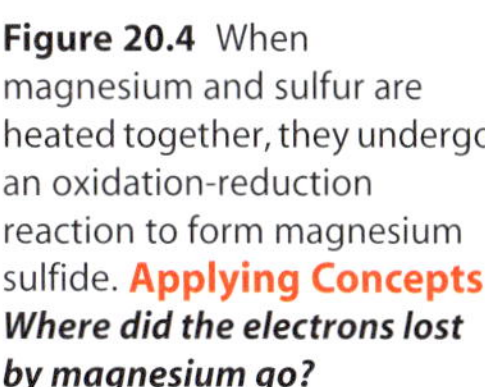

Word Origins

Reduction comes from the Latin word *reducere*, meaning "to lead back." Reduction of metal oxides by the removal of oxygen leads to a smaller mass of the metal compared to the original metal oxide. **Sometimes when a company reorganizes, it undergoes a "reduction in force." What does this mean?**

Figure 20.4 When magnesium and sulfur are heated together, they undergo an oxidation-reduction reaction to form magnesium sulfide. **Applying Concepts** ***Where did the electrons lost by magnesium go?***

Word Origins L2

A "reduction in force" occurs when a company reorganizes its structure, leading to a smaller work force and people losing their jobs. Similarly, a "reduction in weight" refers to a weight loss.

Discuss L2

Ask students to recall the electron structure of an oxygen atom and how oxygen atoms bond with other atoms. Be sure they identify the source of the electrons an oxygen atom gains during bonding. Point out that when a metal combines with oxygen, it loses electrons, and when oxygen is removed from the oxide of a metal, the metal gains electrons. Explain to students that this knowledge led to a broader definition of oxidation and reduction as an exchange of electrons.

CLASS Activity

Born-Haber Cycles L3

Show students how to construct Born-Haber cycles for the redox reactions of sodium and magnesium with oxygen by finding tabulated energy values for the heats of formation of MgO and Na_2O; heats of vaporization of Mg(*s*) and Na(*s*); heat of forming O atoms from O_2; ionization energies of gaseous Mg and Na; and the electron affinity of O *(From their diagrams, the students should be able to determine the lattice energies of MgO and Na_2O. Large lattice energies indicate that these redox reactions are highly exothermic.)*

Facts and Figures

The History of the Name "Copper"

Reduction processes were used to produce some metals more than 5000 years ago. Copper oxide and copper sulfide ores were roasted with charcoal to form copper, carbon dioxide, and sulfur dioxide. Large amounts of copper ore were mined on the island of Cyprus, off the coast of the area now known as Turkey. The Romans, who eventually imported this valuable ore, called it aes cyprium, or "ore of Cyprus." The name was later shortened to cyprium and then corrupted to cuprum. The English word "copper" derives from this corruption of the Latin word for Cyprus.

Answers to...

Figure 20.4 They were gained by sulfur.

Checkpoint oxidation-reduction reactions.

Section 20.1 (continued)

CLASS Activity

Identifying Oxidizing and Reducing Agents **L2**

Purpose Students will write a reaction equation and identify the oxidizing and reducing agents.

Materials paper, pencil, periodic table

Procedure Separate students into groups of two or three. Have one student in each group select an active metal and a nonmetal from the periodic table. Have this student ask the other student(s) to identify the ionic compound that forms when electron transfer occurs between the metal and the nonmetal. Have students write the equation for the reaction and identify the oxidizing and reducing agents.

CONCEPTUAL PROBLEM 20.1

Answers

1. **a.** Na: oxidized (reducing agent)
   S: reduced (oxidizing agent)
   b. Al: oxidized (reducing agent)
   O_2: reduced (oxidizing agent)
2. **a.** oxidation
   b. reduction

Practice Problems Plus 

Determine what is oxidized and what is reduced in the following reaction. Also identify the oxidizing agent and reducing agent.

$2Ba(s) + O_2(g) \rightarrow 2BaO(s)$

(Ba is oxidized and is the reducing agent; O is reduced and is the oxidizing agent.)

Losing electrons is oxidation. Gaining electrons is reduction. The substance that loses electrons is called the **reducing agent.** By losing electrons to sulfur, magnesium reduces the sulfur. Magnesium is thus the reducing agent. The substance that accepts electrons is called the **oxidizing agent.** By accepting electrons from magnesium, sulfur oxidizes the magnesium. Sulfur is the oxidizing agent. Another way of identifying oxidizing and reducing agents is to remember that the species that is reduced is the oxidizing agent and the species oxidized is the reducing agent.

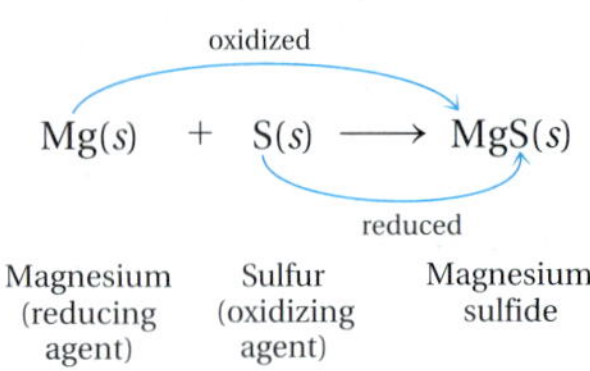

You might find it helpful to use the following mnemonic device to remember the definitions of oxidation and reduction: "*LEO* the lion goes *GER*." LEO stands for Losing Electrons is Oxidation; GER stands for Gaining Electrons is Reduction.

CONCEPTUAL PROBLEM 20.1

Identifying Oxidized and Reduced Reactants

Silver nitrate reacts with copper to form copper nitrate and silver. From the equation below, determine what is oxidized and what is reduced. Identify the oxidizing agent and the reducing agent.

$$2AgNO_3(aq) + Cu(s) \longrightarrow Cu(NO_3)_2(aq) + 2Ag(s)$$

1 Analyze ***Identify the relevant concepts.***

The species that loses electrons is oxidized and is the reducing agent. The species that gains electrons is reduced and is the oxidizing agent.

2 Solve ***Apply concepts to this situation.***

Rewrite the equation in ionic form. It is then easier to analyze the reaction.

$$2Ag^+ + 2NO_3^- + Cu \longrightarrow Cu^{2+} + 2NO_3^- + 2Ag$$

In this reaction, two electrons are lost from a copper atom (Cu) when it becomes a Cu^{2+} ion. These electrons are gained by two silver ions (Ag^+), which become neutral silver atoms.

Oxidation: $Cu \longrightarrow Cu^{2+} + 2e^-$ (loss of electrons)

Reduction: $2Ag^+ + 2e^- \longrightarrow 2Ag$ (gain of electrons)

The Cu is the reducing agent. The Ag^+ is the oxidizing agent.

Practice Problems

1. Determine what is oxidized and what is reduced in each reaction. Identify the oxidizing agent and reducing agent in each case.
   a. $2Na(s) + S(s) \longrightarrow Na_2S(s)$
   b. $4Al(s) + 3O_2(g) \longrightarrow 2Al_2O_3(s)$
2. Identify these processes as either oxidation or reduction.
   a. $2I^- \longrightarrow I_2 + 2e^-$ **b.** $Zn^{2+} + 2e^- \longrightarrow Zn$

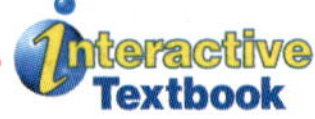

Problem-Solving 20.1 Solve Problem 1 with the help of an interactive guided tutorial.

with **ChemASAP**

Table 20.1	
Processes Leading to Oxidation and Reduction	
Oxidation	**Reduction**
Complete loss of electrons (ionic reactions)	Complete gain of electrons (ionic reactions)
Shift of electrons *away* from an atom in a covalent bond	Shift of electrons *toward* an atom in a covalent bond
Gain of oxygen	Loss of oxygen
Loss of hydrogen by a covalent compound	Gain of hydrogen by a covalent compound
Increase in oxidation number	Decrease in oxidation number

Go Online
NSTA SCiLINKS
For: Links on Oxidation and Reduction
Visit: www.SciLinks.org
Web Code: cdn-1202

Redox With Covalent Compounds When a metal and a nonmetal react and form ions, it is easy to identify complete transfers of electrons. But some reactions involve covalent compounds, that is, compounds in which complete electron transfer does not occur. One such example is the reaction of hydrogen with oxygen.

$$2H_2(g) + O_2(g) \longrightarrow 2H_2O(l)$$

Consider what happens to the bonding electrons in the formation of a water molecule. In each reactant hydrogen molecule, the bonding electrons are shared equally between the hydrogen atoms. In water, however, the bonding electrons are pulled toward oxygen because it is much more electronegative than hydrogen. The result is a shift of bonding electrons away from hydrogen, even though there is not a complete transfer. Hydrogen is oxidized because it undergoes a partial loss of electrons.

In oxygen, the other reactant, the bonding electrons are shared equally between oxygen atoms in the reactant oxygen molecule. However, when oxygen bonds to hydrogen in the water molecule, there is a shift of electrons toward oxygen. Oxygen is thus reduced because it undergoes a partial gain of electrons.

H—H electrons shared equally

O—O electrons shared equally

H—O (with H below O) shift of bonding electrons away from hydrogen and toward oxygen

In the reaction of hydrogen and oxygen to produce water, hydrogen is the reducing agent because it is oxidized. Oxygen is the oxidizing agent because it is reduced. This redox reaction is highly exothermic—that is, it releases a great deal of energy, as shown in Figure 20.5.

In some reactions involving covalent reactants or products, the partial electron shifts are less obvious. Some general guidelines are thus helpful. For example, for carbon compounds, the addition of oxygen or the removal of hydrogen is always oxidation. Table 20.1 above lists processes that constitute oxidation and reduction. The last entry in the table refers to oxidation numbers which are introduced in Section 20.2.

What happens to the bonding electrons in the formation of water?

Figure 20.5 This welder is using an oxyhydrogen torch to cut and weld steel. When hydrogen burns in oxygen, the redox reaction generates temperatures of about 2600°C.

Discuss L2

Give examples of redox reactions and identify oxidizing and reducing agents based on the processes listed in Table 20.1. Emphasize that, for a given reaction, one or more of the processes may be in effect. Tell students that the issue of oxidation numbers will be addressed in Section 20.2.

Use Visuals 

Figure 20.5 Have the students examine the photograph in the figure and read the caption. On the board, write the balanced redox reaction for hydrogen burning in oxygen. Identify the reducing and oxidizing agents, and explain the identities on the basis of partial loss or gain of electrons. State that another gas, acetylene (C_2H_2) is also used in torches because it produces a large amount of heat when it burns.

Have Students write the balanced chemical equation for this redox reaction.

($2C_2H_2(g) + 5O_2(g) \rightarrow 4CO_2(g) + 2H_2O(g)$)

Then, have students identify the oxidizing and reducing agents. Ask them to explain their answers.

Download a worksheet on **Oxidation and Reduction** for students to complete, and find additional teacher support from NSTA SciLinks.

Facts and Figures

Developing Film

Black-and-white photography involves oxidation-reduction reactions. Exposing black-and-white film to light activates very fine grains of silver bromide in the film. The film is developed by placing it in a developing solution that is actually a reducing agent. The developer, usually an organic chemical, such as hydroquinone ($C_6H_4(OH)_2$), reduces the activated silver bromide to finely divided, black metallic silver. Any silver bromide that remains unactivated is removed from the film by using a solvent called a fixer. Sodium thiosulfate, commonly called hypo, is used for this purpose. The areas of the film exposed to the most light appear darkest because they have the highest concentration of metallic silver. The reversed image, called a negative, is used to produce a positive print, the black-and-white photograph, of the image.

Answers to...

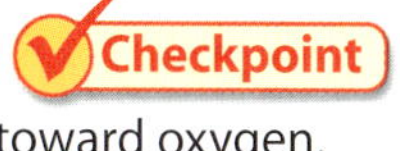 They shift toward oxygen.

Corrosion

Download a worksheet on **Corrosion** for students to complete, and find additional teacher support from NSTA SciLinks.

CLASS Activity

Metal Corrosion

L2

Purpose Students will order metals based on their ease of corrosion.

Procedure Have students try to prepare a list of metals ordered by their ease of corrosion. Have students base their lists on known behaviors of specific metals. Suggest that students consider how different metals are used as a clue to their behavior. After students compile their lists, refer them to Table 11.2, the activity series for metals, on page 333. Have students write a paragraph discussing how closely their lists correspond to the actual activity series. Ask, **Why might there be discrepancies between their lists and the activity series?** *(Reasons might include unavailability of data for some metals or difficulty in rating metals whose behaviors and uses are similar.)*

Figure 20.6 Oxidation-reduction reactions cause corrosion. The copper on this roof reacted with water vapor, carbon dioxide, and other substances in the air to form a patina. This patina consists of a pale-green film of basic copper(II) carbonate. Because patinas enhance the surface appearance of objects, they are valued by architects and artists.

For: Links on Corrosion
Visit: www.SciLinks.org
Web Code: cdn-1201

Corrosion

Billions of dollars are spent yearly to prevent and to repair damage caused by the corrosion of metals. **Iron, a common construction metal often used in the form of the alloy steel, corrodes by being oxidized to ions of iron by oxygen.** Water in the environment accelerates the rate of corrosion. Oxygen, the oxidizing agent, is reduced to oxide ions (in compounds such as Fe_2O_3) or to hydroxide ions. The following equations describe the corrosion of iron to iron hydroxides in moist conditions.

$$2Fe(s) + O_2(g) + 2H_2O(l) \longrightarrow 2Fe(OH)_2(s)$$

$$4Fe(OH)_2(s) + O_2(g) + 2H_2O(l) \longrightarrow 4Fe(OH)_3(s)$$

Corrosion occurs more rapidly in the presence of salts and acids. These substances produce electrically conducting solutions that make electron transfer easier. The corrosion of some metals can be a desirable feature, as Figure 20.6 shows.

Resistance to Corrosion Not all metals corrode easily. Gold and platinum are called noble metals because they are very resistant to losing their electrons by corrosion. Other metals lose electrons easily but are protected from extensive corrosion by the oxide coating formed on their surface. For example, aluminum oxidizes quickly in air to form a coating of very tightly packed aluminum oxide particles. This coating protects the aluminum object from further corrosion, as shown in Figure 20.7. Iron also forms a coating when it corrodes, but the coating of iron oxide that forms is not tightly packed. Water and air can penetrate the coating and attack the iron metal below it. The corrosion continues until the iron object becomes only a pile of rust.

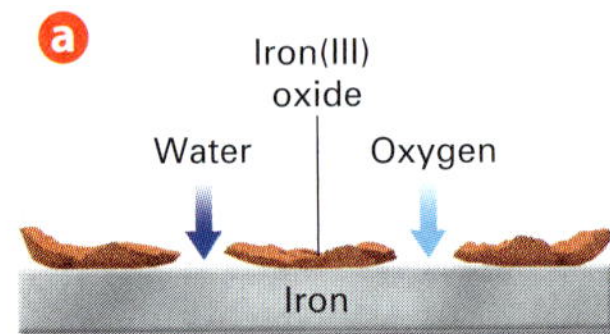

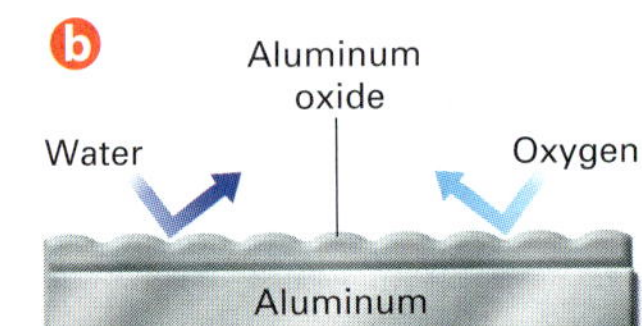

Figure 20.7 Oxidation causes the complete corrosion of some metals. **a** Iron reacts with water and oxygen to form iron(III) oxide, or rust. **b** Aluminum, however, resists such corrosion because it forms a protective coating of aluminum oxide.
Applying Concepts ***How does the aluminum oxide on aluminum differ from the iron(III) oxide formed on corroding iron?***

Controlling Corrosion The corrosion of objects such as shovels or knives is a common problem but not usually a serious one. In contrast, the corrosion of a steel support pillar of a bridge, as in Figure 20.8, or the hull of an oil tanker is much more serious and costly. To prevent corrosion in such cases, the metal surface may be coated with oil, paint, plastic, or another metal, as shown in Figure 20.9. These coatings exclude air and water from the surface, thus preventing corrosion. If the coating is scratched or worn away, however, the exposed metal will begin to corrode.

In another method of corrosion control, one metal is "sacrificed," or allowed to corrode, in order to save a second metal. For example, to protect an iron object, a piece of magnesium (or another active metal) may be placed in electrical contact with the iron. When oxygen and water attack the iron object, the iron atoms lose electrons as the iron begins to be oxidized. However, because magnesium is a better reducing agent than iron, the magnesium immediately transfers electrons to the iron atoms, preventing their oxidation.

Figure 20.8 Bridges like this one become unsafe because of corrosion.

Figure 20.9 Painting a surface (left) protects it from the effects of the environment. Chromium metal also serves as a protective coating and imparts an attractive, mirrorlike finish (right). Like aluminum, chromium forms a corrosion-resistant oxide film on its surface.

Discuss 

Lead a class discussion on metal corrosion. Ask, **Can you identify the common chemical characteristic of all metal corrosion?** *(the transfer of electrons from metals to oxidizing agents)* Have students cite examples of desirable and undesirable metal corrosion. Ask students to suggest ways in which the rate of corrosion could be decreased or increased.

Differentiated Instruction

English Learners 

Tell students that the steel cans used for commercial canning are coated with a thin layer of tin. Ask them to infer the reason for this procedure and to predict whether or not aluminum cans have a similar lining. Have students explain their reasoning.

Answers to...

Figure 20.7 Unlike iron(III) oxide, the tough aluminum oxide coating on the surface of the metal actually protects the aluminum object from further corrosion.

Section 20.1 (continued)

3 ASSESS

Evaluate Understanding L2

On the board, write several equations for redox reactions. Ask students to identify the oxidizing and reducing agents. Make sure to assess the logic students use to arrive at their answers.

Reteach L2

To reinforce the dual nature of redox reactions, compare redox reactions to acid-base reactions. Explain that the transfer of electrons from a reducing agent to an oxidizing agent is analogous to the transfer of protons from acids to bases. In neither case, can the donor act independent of an acceptor.

Writing Activity

Silver is more easily oxidized than is gold. Silver tarnish is mainly silver sulfide. Tarnish can be removed if the silver is placed in a dilute NaOH bath and a weak current is passed through the solution.

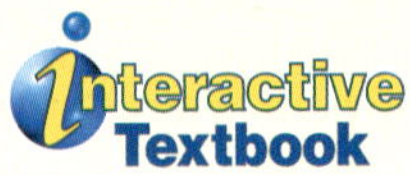

If your class subscribes to the Interactive Textbook, use it to review key concepts in Section 20.1.

with ChemASAP

Answers to...

coating a surface and "sacrificing" one metal for another

Figure 20.10 Zinc blocks, the white rectangles in the photograph, are attached to the steel (iron) hull of this ship. The zinc blocks oxidize (corrode) instead of the iron, preventing the hull from corroding.

Sacrificial zinc and magnesium blocks are sometimes attached to piers and ship hulls to prevent corrosion damage in areas submerged in water. Figure 20.10 shows zinc blocks attached to the steel hull of a ship. The blocks corrode instead of the iron. Underground pipelines and storage tanks may be connected to magnesium blocks for protection. It is easier and cheaper to replace a block of magnesium or zinc than to replace a bridge or a pipeline.

Checkpoint ***What are two ways to control corrosion?***

20.1 Section Assessment

3. **Key Concept** Define oxidation and reduction in terms of the gain or loss of oxygen.
4. **Key Concept** Define oxidation and reduction in terms of the gain or loss of electrons.
5. **Key Concept** What happens to the atoms in an iron nail that corrodes?
6. How do you identify the oxidizing agent and the reducing agent in a redox reaction?
7. Use electron transfer or electron shift to identify what is oxidized and what is reduced in each reaction. Use the electronegativity values in Table 6.2 in Chapter 6, for molecular compounds.
   - **a.** $2Na(s) + Br_2(l) \longrightarrow 2NaBr(s)$
   - **b.** $H_2(g) + Cl_2(g) \longrightarrow 2HCl(g)$
   - **c.** $2Li(s) + F_2(g) \longrightarrow 2LiF(s)$
   - **d.** $S(s) + Cl_2(g) \longrightarrow SCl_2(g)$
   - **e.** $N_2(g) + 2O_2(g) \longrightarrow 2NO_2(g)$
   - **f.** $Mg(s) + Cu(NO_3)_2(aq) \longrightarrow Mg(NO_3)_2(aq) + Cu(s)$
8. Identify the reducing agent and the oxidizing agent for each reaction in Problem 7.

Writing Activity

Write a Report Shipwrecks of Spanish galleons sometimes contain gold or silver treasures. Why is the recovered gold hardly ever changed while the silver has turned black? Research and write a report on how the thick layers of tarnish are removed from the silver artifacts.

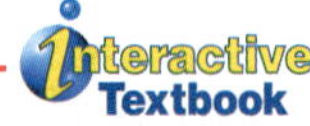

Assessment 20.1 Check your understanding of the important ideas and concepts in Section 20.1.

with ChemASAP

Section 20.1 Assessment

3. Oxidation is the gain of oxygen. Reduction is the loss of oxygen.
4. Oxidation is the loss of electrons. Reduction is the gain of electrons.
5. Iron atoms are oxidized when iron corrodes.
6. The species reduced is the oxidizing agent. The species oxidized is the reducing agent.
7. **a.** Na oxidized; Br_2 reduced **b.** H_2 oxidized ; Cl_2 reduced **c.** Li oxidized; F_2 reduced **d.** S oxidized; Cl_2 reduced **e.** N_2 oxidized; O_2 reduced **f.** Mg oxidized; Cu reduced
8. **a.** reducing agent, Na; oxidizing agent, Br_2 **b.** reducing agent, H_2; oxidizing agent, Cl_2 **c.** reducing agent, Li; oxidizing agent, F_2 **d.** reducing agent, S; oxidizing agent, Cl_2 **e.** reducing agent, N_2; oxidizing agent, O_2 **f.** reducing agent, Mg; oxidizing agent, Cu

20.2 Oxidation Numbers

Connecting to Your World

The bursts of bright white light produced by fireworks are the result of metals being burned. Powdered or flaked aluminum or magnesium are included in the fireworks. When these metals are heated to high temperatures in the explosion, they burn with an intense white light. The fireworks shown here are called stars. As elements burn, their oxidation numbers change. In this section, you will learn about how oxidation and reduction are defined in terms of a change in oxidation number.

Guide for Reading

Key Concepts

- What is the general rule for assigning oxidation numbers?
- How are oxidation and reduction defined in terms of a change in oxidation number?

Vocabulary

oxidation number

Reading Strategy

Using Prior Knowledge Before you read, predict how oxidation and reduction affect the charge on an ion. Check your predictions as you read.

Assigning Oxidation Numbers

An **oxidation number** is a positive or negative number assigned to an atom to indicate its degree of oxidation or reduction. **As a general rule, a bonded atom's oxidation number is the charge that it would have if the electrons in the bond were assigned to the atom of the more electronegative element.** The following set of rules should help you determine oxidation numbers.

Rules for Assigning Oxidation Numbers

1. The oxidation number of a monatomic ion is equal in magnitude and sign to its ionic charge. For example, the oxidation number of the bromide ion (Br^{1-}) is −1; that of the Fe^{3+} ion is +3.
2. The oxidation number of hydrogen in a compound is +1, except in metal hydrides, such as NaH, where it is −1.
3. The oxidation number of oxygen in a compound is −2, except in peroxides, such as H_2O_2, where it is −1, and in compounds with the more electronegative fluorine, where it is positive.
4. The oxidation number of an atom in uncombined (elemental) form is 0. For example, the oxidation number of the potassium atoms in potassium metal (K) or of the nitrogen atoms in nitrogen gas (N_2) is 0.
5. For any neutral compound, the sum of the oxidation numbers of the atoms in the compound must equal 0.
6. For a polyatomic ion, the sum of the oxidation numbers must equal the ionic charge of the ion.

Section Resources

Print

- **Guided Reading and Study Workbook,** Section 20.2
- **Core Teaching Resources,** Section 20.2 Review
- **Transparencies,** T234–T237

Technology

- **Interactive Textbook with ChemASAP,** Problem-Solving 20.9, Assessment 20.2

20.2

1 FOCUS

Objectives

20.2.1 Determine the oxidation number of an atom of any element in a pure substance.

20.2.2 Define oxidation and reduction in terms of a change in oxidation number and **identify** atoms being oxidized or reduced in redox reactions.

Guide for Reading

Build Vocabulary 

Word Forms Students can relate the vocabulary term, *oxidation number*, with other similar terms used in the chapter.

Reading Strategy 

Outline Have students preview the section's content by making an outline of the content.

2 INSTRUCT

Connecting to Your World

Have students read the opening paragraph. Ask, **What other colors can be seen in fireworks?** *(yellow, green, blue-green, red)* **Which elements are responsible for these colors?** *(yellow — sodium; green — barium; blue-green — copper; crimson — strontium)*

Assigning Oxidation Numbers

Relate 

Explain that determining oxidation numbers of elements in compounds is a way for chemists to keep track of electron transfer during redox reactions. Ask, **What are other examples where items are numbered to keep track of movement?** *(The numbers on sports players' jerseys and the area codes assigned to telephone numbers in different regions are two examples.)*

TEACHER Demo

Redox Reactions

L2

Purpose Students will observe a redox reaction and write a balanced equation describing it.

Materials 100 mL of 0.1*M* lead(II) acetate [$Pb(CH_3COO)_2$], 150-mL beaker, 1-cm × 3-cm zinc strip, sodium sulfide, 3*M* NaOH, 1*M* iron(III) chloride, stirring rod, plastic container

Safety Lead(II) acetate is a suspected carcinogen and mutagen. Perform activity in a well-ventilated area. Wear safety goggles, aprons, and disposable gloves.

Procedure Pour 100 mL of 0.1*M* lead(II) acetate [$Pb(CH_3COO)_2$] into a 150-mL beaker and add a 1-cm × 3-cm zinc strip. **CAUTION!** Lead(II) acetate is a suspected carcinogen and mutagen.

Expected Outcome A "tree" of lead that forms over the course of a day.

Ask students to write a balanced equation for the redox reaction that occurred. Have them assign oxidation numbers for Zn and Pb. Based on changes in oxidation number, have students identify the elements that were oxidized and reduced.

Answers $Zn(s) + Pb(CH_3COO)_2(aq) \rightarrow Pb(s) + Zn(CH_3COO)_2(aq)$ Reactants: Zn^0, Pb^{+2}; Products: Zn^{+2}, Pb^0 Zn was oxidized. Pb was reduced.

Disposal Dispose of the reaction mixture by adding a threefold molar excess of sodium sulfide to the contents of the beaker and stirring occasionally for one hour. Adjust the pH to neutral with 3*M* NaOH. Filter or decant the mixture. Allow the residue to dry. Add a threefold excess of 1*M* iron(III) chloride to the filtrate, stirring slowly. Allow the precipitate to settle, then filter or decant. Put the dry precipitate and filtered residue in a plastic container and bury in an approved landfill site. Flush the neutral filtrate down the drain with excess water.

Figure 20.11 The oxidation number of any element in the free or uncombined state is zero. The elements shown here (left to right) are white phosphorus (stored under water), sulfur, potassium (stored under oil), carbon, and bromine (liquid and vapor). The potassium and phosphorus are stored under a liquid to prevent them from reacting with oxygen in the air.

In binary ionic compounds, such as NaCl and $CaCl_2$, the oxidation numbers of the atoms equal their ionic charges (Rule 1). The compound sodium chloride is composed of sodium ions (Na^{1+}) and chloride ions (Cl^{1-}). Thus the oxidation number of sodium is +1, and that of chlorine is −1. Notice that the sign is put before the oxidation number.

Because water is a molecular compound, no ionic charges are associated with its atoms. However, oxygen is reduced in the formation of water. Oxygen is more electronegative than hydrogen. So, in water, the two shared electrons in the H—O bond are shifted toward oxygen and away from hydrogen. Imagine that the electrons contributed by the two hydrogen atoms are completely transferred to the oxygen. The charges that would result from this transfer are the oxidation numbers of the bonded elements. The oxidation number of oxygen is −2, and the oxidation number of each hydrogen is +1 (Rules 2 and 3). Oxidation numbers are often written above the chemical symbols in a formula. For example, water can be represented as

$$\overset{+1}{H_2}\overset{-2}{O}$$

Many elements can have several different oxidation numbers. Use rules 5 and 6 to determine the oxidation number of atoms of these elements, plus other elements not covered in the first four rules. The substances in Figure 20.12 are compounds of chromium, but chromium has a different oxidation number in each compound.

Figure 20.12 Yellow potassium chromate (K_2CrO_4) and green chromium(III) oxide (Cr_2O_3) both are compounds of chromium. **Inferring** ***What is the oxidation number of chromium in each compound?***

640 Chapter 20

Differentiated Instruction

Gifted and Talented 

Have students do research and report to the class why carbon monoxide is extremely poisonous to living systems whereas carbon dioxide is relatively harmless. Point out that carbon in carbon monoxide has an oxidation number that is lower than that in carbon dioxide. Therefore, one difference is that carbon monoxide is a better reducing agent than carbon dioxide.

CONCEPTUAL PROBLEM 20.2

Assigning Oxidation Numbers to Atoms

What is the oxidation number of each kind of atom in the following ions and compounds?

a. SO_2 **b.** CO_3^{2-} **c.** Na_2SO_4 **d.** $(NH_4)_2S$

1 Analyze ***Identify the relevant concepts.***

Use the set of rules you just learned to assign and calculate oxidation numbers.

2 Solve ***Apply concepts to this situation.***

a. There are two oxygen atoms and the oxidation number of each oxygen is −2 (Rule 3). The sum of the oxidation numbers for the neutral compound must be 0 (Rule 5). Therefore, the oxidation number of sulfur is +4, because $+4 + (2 \times (-2)) = 0$.

$$\overset{+4}{S}\overset{-2}{O_2}$$

b. The oxidation number of oxygen is −2 (Rule 3).

$$\overset{?}{C}\overset{-2}{O_3}{}^{2-}$$

The sum of the oxidation numbers of the carbon and oxygen atoms must equal the ionic charge, −2 (Rule 6). The oxidation number of carbon must be +4, because $+4 + (3 \times (-2)) = -2$.

$$\overset{+4}{C}\overset{-2}{O_3}{}^{2-}$$

c. The oxidation number of each sodium ion, Na^+, is the same as its ionic charge, +1 (Rule 1). The oxidation number of oxygen is −2 (Rule 3).

$$\overset{+1}{Na_2}\overset{?}{S}\overset{-2}{O_4}$$

For the sum of the oxidation numbers in the compound to be 0 (rule 5), the oxidation number of sulfur must be +6, because $(2 \times (+1)) + (+6) + (4 \times (-2)) = 0$.

$$\overset{+1}{Na_2}\overset{+6}{S}\overset{-2}{O_4}$$

d. Ammonium ions, NH_4^+, have an ionic charge of +1, so the sum of the oxidation numbers of the atoms in the ammonium ion must be +1. The oxidation number of hydrogen is +1 in this ion.

$$\overset{?}{N}\overset{+1}{H_4}{}^{+}$$

$$? + 4(+1) = +1$$

So, the oxidation number of nitrogen must be −3 because $-3 + (+4) = +1$.

Two ammonium ions have a total charge of +2. Since the compound $(NH_4)_2S$ is neutral, sulfur must have a balancing oxidation number of −2.

$$(\overset{-3}{N}\overset{+1}{H_4})_2\overset{-2}{S}$$

Practice Problems

9. Determine the oxidation number of each element in the following.

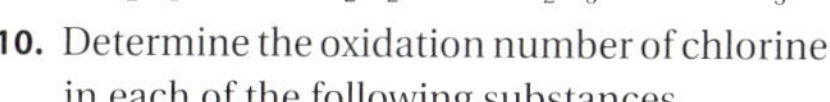

a. S_2O_3 **b.** Na_2O_2 **c.** P_2O_5 **d.** NO_3^-

10. Determine the oxidation number of chlorine in each of the following substances.

a. $KClO_3$ **b.** Cl_2 **c.** $Ca(ClO_4)_2$ **d.** Cl_2O

Interactive Textbook

Problem-Solving 20.9 Solve Problem 9 with the help of an interactive guided tutorial.

with **ChemASAP**

Discuss L2

Lead a class discussion on determining oxidation numbers for elements in compounds by applying the rules for assigning oxidation numbers. Go over many examples until students feel comfortable with determining these numbers. Also explain that the same element can have several oxidation numbers, depending on the compound. Ask, **What are the oxidation numbers of iodine in HIO_4, HIO_3, HIO, I_2, and HI?** *(+7, +5, +1, 0, −1, respectively)*

CONCEPTUAL PROBLEM 20.2

Answers

9. a. S, +3; O, −2
b. Na, +1; O, −1
c. P, +5; O, −2
d. N, +5; O, −2

10. a. +5
b. 0
c. +7
d. +1

CLASS Activity

Assigning Oxidation Numbers L2

Purpose Students will work cooperatively to assign oxidation numbers for various elements.

Procedure Write formulas for ionic and molecular compounds on the chalkboard. Divide the class into groups of two and have students work together to assign oxidation numbers to all of the elements in each compound. Students can use the problem-solving strategy outlined in Conceptual Problem 20.2.

Facts and Figures

How a Breathalyzer Works

A breath analyzer, or Breathalyzer, is used to estimate the level of alcohol in a person's bloodstream. The chemical potassium dichromate is used in this process. If any ethanol is present in the person's breath when breathing into the apparatus, the Cr^{+6} in the potassium dichromate (orange) is reduced to one of its lower states, such as Cr^{3+} (green). The color change can be related to the amount of alcohol in the bloodstream.

Answers to...

Figure 20.12 K_2CrO_4, Cr +6; Cr_2O_3, Cr +3.

Oxidation-Number Changes in Chemical Reactions

CLASS Activity

Identifying Redox Reactions L2

Purpose Students will identify and write equations for redox reactions.

Materials 3 small, glass test tubes; zinc; copper; magesium; 0.1*M* HCl; medicine dropper or pipet

Procedure Have students work in pairs. Give each pair a set of three small, glass test tubes containing a small portion of zinc, copper, and magnesium. Have students add approximately 10 drops of 0.1*M* HCl to the metals and note signs of reaction. They should see bubbles form in the tubes containing zinc and magnesium, but not in the tube containing copper. Ask, **Can you guess the identity of the gas in the tubes containing zinc and magnesium?** ($H_2(g)$) Have students write equations for the reactions of zinc and magnesium with HCl. Have students explain why these reactions are redox reactions.

Use Visuals L1

Figure 20.14 Have students examine the figure 20.14 and ask, **What would happen if a copper nail were added to a solution of iron(II) sulfate instead?** *(Nothing would happen because copper is below iron in the activity series of metals.)*

Figure 20.13 Copper reacts with silver nitrate. **a** A copper wire is placed in a silver nitrate solution. **b** Crystals of silver coat the wire and the solution slowly turns blue as a result of the formation of copper(II) nitrate. **Drawing Conclusions** ***What change occurs in the oxidation number of silver? How does the oxidation number of copper change?***

Oxidation-Number Changes in Chemical Reactions

Figure 20.13 shows what happens when copper wire is placed in a solution of silver nitrate. In this reaction, the oxidation number of silver decreases from +1 to 0 as each silver ion (Ag^{1+}) gains an electron and is reduced to silver metal (Ag^0). Copper's oxidation number increases from 0 to +2 as each atom of copper metal (Cu^0) loses two electrons and is oxidized to a copper(II) ion (Cu^{2+}). Here is the equation with oxidation numbers added:

$$\overset{+1\,+5-2}{2AgNO_3(aq)} + \overset{0}{Cu(s)} \longrightarrow \overset{+2\,+5-2}{Cu(NO_3)_2(aq)} + \overset{0}{2Ag(s)}$$

Figure 20.14 illustrates a redox reaction that shows what occurs when a shiny iron nail is dipped into a solution of copper(II) sulfate.

You can define oxidation and reduction in terms of a change in oxidation number. **An increase in the oxidation number of an atom or ion indicates oxidation. A decrease in the oxidation number of an atom or ion indicates reduction.**

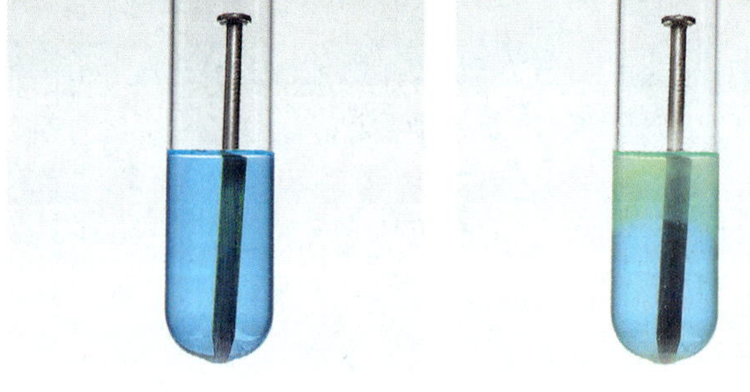

Figure 20.14 A redox reaction occurs between iron and copper. **a** An iron nail is placed in a copper(II) sulfate solution. **b** The iron reduces Cu^{2+} ions in solution and is simultaneously oxidized to Fe^{2+}. The iron becomes coated with metallic copper.

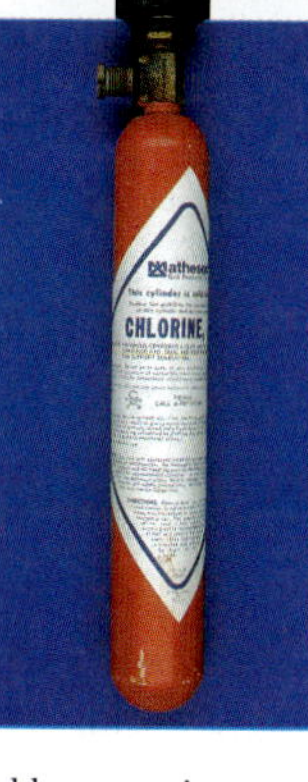

CONCEPTUAL PROBLEM 20.3

Identifying Oxidized and Reduced Atoms

Use changes in oxidation number to identify which atoms are oxidized and which are reduced in the following reaction. Also identify the oxidizing agent and the reducing agent.

$$Cl_2(g) + 2HBr(aq) \longrightarrow 2HCl(aq) + Br_2(l)$$

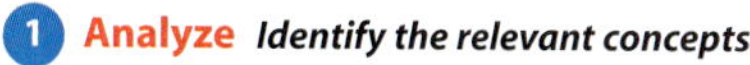

1 Analyze ***Identify the relevant concepts.***

Use the rules to assign oxidation numbers to each atom in the equation. An increase in oxidation number indicates oxidation. A decrease in oxidation number indicates reduction. The substance that is oxidized in a redox reaction is the reducing agent. The substance that is reduced is the oxidizing agent.

2 Solve ***Apply concepts to this situation.***

$$\overset{0}{Cl_2}(g) + 2\overset{+1-1}{HBr}(aq) \longrightarrow 2\overset{+1-1}{HCl}(aq) + \overset{0}{Br_2}(l)$$

The element chlorine is reduced because its oxidation number decreases (0 to −1). The bromide ion from HBr(*aq*) is oxidized because its oxidation number increases (−1 to 0). Chlorine is reduced, so Cl_2 is the oxidizing agent. The bromide ion from HBr(*aq*) is oxidized, so Br^- is the reducing agent.

Practice Problems

11. Identify which atoms are oxidized and which are reduced in each reaction.

a. $2H_2(g) + O_2(g) \longrightarrow 2H_2O(l)$

b. $2KNO_3(s) \longrightarrow 2KNO_2(s) + O_2(g)$

12. Identify the oxidizing agent and the reducing agent in each equation in Practice Problem 11.

Problem-Solving 20.11 Solve Problem 11 with the help of an interactive guided tutorial.

with **ChemASAP**

20.2 Section Assessment

13. **Key Concept** What is the general rule for assigning oxidation numbers?

14. **Key Concept** How is a change in oxidation number related to the process of oxidation and reduction?

15. Use the changes in oxidation numbers to identify which atoms are oxidized and which are reduced in each reaction.

a. $2Na(s) + Cl_2(g) \longrightarrow 2NaCl(s)$

b. $2HNO_3(aq) + 6HI(aq) \longrightarrow 2NO(g) + 3I_2(s) + 4H_2O(l)$

16. Identify the oxidizing agent and the reducing agent in each reaction in Problem 15.

Connecting Concepts

Types of Reactions Reread Section 11.2 and identify which of the five types of reactions are most likely redox reactions.

Assessment 20.2 Test yourself on the concepts in Section 20.2.

with **ChemASAP**

Section 20.2 Assessment

13. The oxidation number is the charge a bonded atom would have if the electrons in the bond were assigned to the more electronegative element.

14. An increase in oxidation number indicates oxidation; a decrease in oxidation number indicates reduction.

15. a. Na, oxidized. Cl_2, reduced **b.** I, oxidized; N, reduced

16. a. Na reducing agent, Cl_2 oxidizing agent **b.** HNO_3 oxidizing agent, HI reducing agent

CONCEPTUAL PROBLEM 20.3

Answers

11. a. H_2 oxidized, O_2 reduced **b.** N reduced, O oxidized

12. a. H_2 reducing agent, O_2 oxidizing agent **b.** N oxidizing agent, O reducing agent

3 ASSESS

Evaluate Understanding  L2

Ask, **What are the oxidation numbers of the elements in the following compounds? KNO_3**$(K +1, N +5, O -2)$; **Na_2SO_3**$(Na +1, S +4, O -2)$; **P_2O_5** $(P +5, O -2)$; **MgH_2**$(Mg +2, H -1)$; **ClF_3** $(Cl +3, F -1)$; **HIO** $(H +1, I +1, O -2)$

What does the sign and magnitude of the oxidation number of an element in a compound indicate? *(The sign indicates loss (+) or gain (–) of electrons compared to the isolated element; the magnitude indicates the degree of the gain or loss.)* **If fluorine is converted to F^- from F_2, what is the change in oxidation number? Is this oxidation or reduction?** *(change is 0 to –1; reduction)*

Reteach  L1

Stress that oxidation numbers refer to the combining capacity of single atoms. Thus, a change in oxidation number must be calculated on a per-atom basis. Use examples such as Mn_2O_3 and $Cr_2O_7^{2-}$ to explain to students that they must consider the number of atoms involved.

Connecting Concepts

combination, decomposition, single-replacement, combustion.

If your class subscribes to the Interactive Textbook, use it to review key concepts in Section 20.2.

with **ChemASAP**

Answers to...

Figure 20.13 The oxidation number of silver changes from +1 to 0; the oxidation number of copper changes from 0 to +2.

Technology & Society

Cold Water Makes a Hot Meal

Developing the FRH was not as simple as packaging a few grams of magnesium in a box. When exposed to air, magnesium oxidizes to magnesium oxide (MgO), which forms a thin film on the metal's surface. This film prevents further oxidation of the metal, either by air or by water. So the researchers had to find a way of letting water get through the MgO film.

They discovered that adding common table salt (NaCl) and powdered iron to the magnesium would do the trick. When water is added to this mixture, the chloride ions (Cl^-) react with the MgO film to form MgOHCl, which eats away at the remaining MgO.

Once the MgO film is gone, the water can react directly with the magnesium metal. The powdered iron seems to help this reaction take place more rapidly than it normally would. The researchers do not understand exactly how the iron helps, but they theorize that it assists in the transfer of electrons between the water and the magnesium. In any case, with the iron present, the reaction occurs rapidly and vigorously. As a result of putting corrosion to work, the energy stored in the chemical bonds of the reactants is released in just a few minutes, and a hot meal is ready.

After students have read the article, ask, **Why is water needed to activate the flameless ration heater?** *(It serves as an oxidizing agent.)*

Ask, **What is the reducing agent in this reaction?** *(Mg)*

Name other metals that might react with water to generate heat and hydrogen gas. *(Students answers will vary, but may include Ca, Sr, or Ba.)*

Applying Concepts

$Mg(s) + 2H_2O(l) \rightarrow Mg(OH)_2 + H_2(g)$

Technology & Society

Cold Water Makes a Hot Meal

Researchers have solved a long-standing problem: how to provide astronauts and military personnel with hot meals when there are no cooking facilities or time to cook. They developed the Flameless Ration Heater (FRH). An FRH is very small and weighs only one and a half ounces. It heats precooked, packaged meal rations with no flame or electricity. Simply place an MRE (Meal Ready to Eat) pouch in an FRH and add water. After about 15 minutes, the food is a warm 60°C. **Applying Concepts** ***What is the chemical equation for the reaction of magnesium and water?***

How it works An FRH uses corrosion to generate heat. Water oxidizes magnesium to form magnesium hydroxide, hydrogen, and heat. Just 24 grams of magnesium releases 355,000 joules of heat, enough to boil about 1 L of water.

Eating in space Astronauts N. Jan Davis and Stephen K. Robinson on the space shuttle Discovery eat Japanese rice. Astronauts use the FRH to heat packaged meals.

Facts and Figures

Preventing Metal Corrosion

People spend a considerable amount of time and energy trying to prevent or inhibit the simple redox reaction called corrosion. Many metals oxidize when exposed to the oxygen in air. To keep them from corroding or rusting, the metals can be painted, plated with chromium, or galvanized with zinc. But corrosion is not all bad. Because the oxidation of metal is generally an exothermic reaction, it can be used to generate useful heat—as long as the oxidation takes place quickly enough to release the heat energy in a short span of time.

20.3 Balancing Redox Equations

Connecting to Your World One of the long pursued goals of alchemists was to change common metals, such as lead or copper, into gold. One tool that the alchemists employed in their quest was *aqua regia*, which means "royal water." *Aqua regia* is a mixture of concentrated hydrochloric (HCl) and nitric (HNO_3) acids. When metallic gold is added to *aqua regia*, oxidation and reduction reactions produce gaseous nitrogen monoxide (NO) and soluble, stable $AuCl_4^-$ ions. In this section, you will learn how to write and balance chemical equations for redox reactions such as this one.

Guide for Reading

Key Concepts

- How are oxidation numbers used to identify redox reactions?
- How are changes in oxidation numbers used to balance a redox equation?
- What is the procedure for balancing a redox equation using half-reactions?

Vocabulary

oxidation-number-change method

half-reaction

half-reaction method

Reading Strategy

Identifying a Sequence As you read, list the steps used to balance a redox equation by the oxidation-number change method and the half-reaction method.

Identifying Redox Reactions

In general, all chemical reactions can be assigned to one of two classes. One class is oxidation-reduction (redox) reactions, in which electrons are transferred from one reacting species to another. Many single-replacement reactions, combination reactions, decomposition reactions, and combustion reactions are redox reactions. Two examples of redox reactions are shown in Figure 20.15. Figure 20.15a shows what happens when potassium metal reacts with water. Figure 20.15b shows the reaction of zinc with hydrochloric acid. The other class includes all other reactions, in which no electron transfer occurs. For example, double-replacement reactions and acid-base reactions are not redox reactions.

Figure 20.15 Many single replacement reactions are redox reactions. **a** Potassium metal reacts violently with water to produce hydrogen gas (which ignites) and potassium hydroxide. **b** Zinc metal reacts vigorously with hydrochloric acid to produce hydrogen gas and zinc chloride. **Applying Concepts** ***Explain why each reaction is a redox reaction.***

Section Resources

Print

- **Guided Reading and Study Workbook,** Section 20.3
- **Core Teaching Resources,** Interpreting Graphics, Section 20.3 Review
- **Small-Scale Chemistry Laboratory Manual,** Lab 35
- **Probeware Laboratory Manual,** Section 20.3
- **Transparencies,** T238–T241

Technology

- **Interactive Textbook with ChemASAP,** Problem-Solving 20.18, Assessment 20.3
- **Go Online**, Section 20.3
- **Virtual Chemistry Labs**, Lab 30

20.3

1 FOCUS

Objectives

20.3.1 Describe how oxidation numbers are used to identify redox reactions.

20.3.2 Balance a redox equation using the oxidation-number-change method.

20.3.3 Balance a redox equation by breaking the equation into oxidation and reduction half-reactions, and then using the half-reaction method.

Guide for Reading

Build Vocabulary L2

Paraphrase Explain the meaning of the new terms using examples.

Reading Strategy L2

Use Prior Knowledge Have students relate balancing redox equations with the law of conservation of mass.

2 INSTRUCT

Connecting to Your World

Have students read the opening paragraph. Ask, **Were the alchemists making gold? Explain.** *(No, the alchemists were only combining elemental gold into a gold compound.)*

Identifying Redox Reactions

Discuss L2

Ask students to recall the types of chemical reactions they studied in previous chapters. Write an example of each kind of equation on the board or overhead projector. Ask students to decide which equations represent redox reactions and which do not.

Answers to...

Figure 20.15 The oxidation number of potassium changes from 0 to +1 so this is a redox reaction. The oxidation number of zinc changes from 0 to +2 so this is a redox reaction.

Relate L2

Remind students of the concepts that underlie the need to balance equations. For all equations, mass must be conserved by balancing the number of atoms and electrical neutrality by balancing charges. The balancing of redox reactions must also reflect the electrons gained and lost by oxidizing and reducing agents. In other words, electrons cannot be created or destroyed.

Ask students for other examples in which "balancing" is required to achieve conservation. One example is the first law of thermodynamics, which states that energy must be conserved. To conserve energy, any energy change in a system must be balanced by an equal and opposite energy change in the surroundings. For a non-scientific example, consider a checking account. To maintain sufficient funds to cover checks written, money withdrawn would eventually need to be balanced by deposits.

Figure 20.16 The redox reaction between nitrogen and oxygen that takes place when lightning bolts heat the air to extreme temperatures forms nitrogen monoxide.

During an electrical storm, as shown in Figure 20.16, oxygen molecules and nitrogen molecules in the air react to form nitrogen monoxide. This is an example of a combination reaction. The equation for the reaction is shown below.

$$N_2(g) + O_2(g) \longrightarrow 2NO(g)$$

How can you tell if this is a redox reaction? **If the oxidation number of an element in a reacting species changes, then that element has undergone either oxidation or reduction. Therefore, the reaction as a whole must be a redox reaction.** In the example above, the oxidation number of nitrogen increases from 0 to +2 while the oxidation number of oxygen decreases from 0 to −2. Therefore, the reaction between nitrogen and oxygen to form nitrogen monoxide is a redox reaction.

Many reactions in which color changes occur are redox reactions. One example is shown in Figure 20.17. Written in ionic form, the unbalanced equation for this reaction is

$$\underset{\text{Permanganate ion (purple)}}{MnO_4^-(aq)} + \underset{\text{Bromide ion (colorless)}}{Br^-(aq)} \longrightarrow \underset{\text{Manganese(II) ion (colorless)}}{Mn^{2+}(aq)} + \underset{\text{Bromine (brown)}}{Br_2(aq)}$$

Figure 20.17 A color change can signal a redox reaction. When a colorless solution containing bromide ions (Br^-) is added to a solution containing permanganate ions (MnO_4^-), the distinctive purple color of the permanganate ion is replaced by the pale brown color of bromine.

646 Chapter 20

CONCEPTUAL PROBLEM 20.4

Identifying Redox Reactions

Use the change in oxidation number to identify whether each reaction is a redox reaction or a reaction of some other type. If a reaction is a redox reaction, identify the element reduced, the element oxidized, the reducing agent, and the oxidizing agent.

a. $Cl_2(g) + 2NaBr(aq) \longrightarrow 2NaCl(aq) + Br_2(g)$

b. $2NaOH(aq) + H_2SO_4(aq) \longrightarrow Na_2SO_4(aq) + 2H_2O(l)$

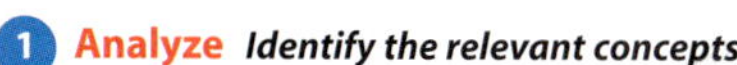

1 Analyze ***Identify the relevant concepts.***

Assign oxidation numbers to show whether or not oxidation numbers change. If changes occur, the reaction is a redox reaction. The element whose oxidation number increases is oxidized and is the reducing agent. The element whose oxidation number decreases is reduced and is the oxidizing agent.

2 Solve ***Apply concepts to this situation.***

a. $\overset{0}{Cl_2}(g) + 2\overset{+1\,-1}{NaBr}(aq) \longrightarrow 2\overset{+1\,-1}{NaCl}(aq) + \overset{0}{Br_2}(g)$

This is a single-replacement reaction. The chlorine is reduced. The bromide ion is oxidized. This is a redox reaction. Chlorine is the oxidizing agent; bromide ion is the reducing agent.

b. $2\overset{+1-2+1}{NaOH}(aq) + \overset{+1+6-2}{H_2SO_4}(aq) \longrightarrow \overset{+1+6-2}{Na_2SO_4}(aq) + 2\overset{+1\,-2}{H_2O}(l)$

This is an acid-base (neutralization) reaction. None of the elements changes in oxidation number. This is not a redox reaction.

Practice Problems

17. Identify which of the following are oxidation-reduction reactions. If a reaction is a redox reaction, name the element oxidized and the element reduced.

a. $Mg(s) + Br_2(l) \longrightarrow MgBr_2(s)$

b. $H_2CO_3(aq) \longrightarrow H_2O(l) + CO_2(g)$

18. Identify which of the following are oxidation-reduction reactions. If a reaction is a redox reaction, name the element oxidized and the element reduced.

a. $CaCO_3(s) + 2HCl(aq) \longrightarrow CaCl_2(aq) + H_2O(l) + CO_2(g)$

b. $CuO(s) + H_2(g) \longrightarrow Cu(s) + H_2O(l)$

Interactive Textbook

Problem-Solving 20.18 Solve Problem 18 with the help of an interactive guided tutorial.

with ChemASAP

Two Ways to Balance Redox Equations

Chemists and chemical engineers must be able to write a correctly balanced equation that accurately represents what happens in a chemical reaction. Many oxidation-reduction reactions are too complex to be balanced by trial and error. Fortunately, two systematic methods are available. These two methods are based on the fact that the total number of electrons gained in reduction must equal the total number of electrons lost in oxidation. One method uses oxidation number changes, and the other uses half reactions.

Using Oxidation-Number Changes You can use oxidation numbers to keep track of electron transfers—a type of chemical bookkeeping. In the **oxidation-number-change method,** you balance a redox equation by comparing the increases and decreases in oxidation numbers.

CONCEPTUAL PROBLEM 20.4

Answers

17. a. redox reaction: Mg is oxidized, Br_2 is reduced.

b. not a redox reaction

18. a. not a redox reaction

b. redox reaction: H_2 is oxidized, Cu is reduced.

Two Ways to Balance Redox Equations

Discuss L2

When you present the half-reaction method for balancing redox equations, students may question why hydrogen ions, hydroxide ions, or water molecules can be added to an equation during the process. Remind them that the aqueous solutions in which these reactions take place contain an abundance of these species.

TEACHER Demo

Color Changes in Redox Reactions L2

Purpose Students observe that redox reactions can sometimes be identified as a result of color changes.

Materials 6 crystals of potassium permanganate ($KMnO_4$), 540 mL of water, large beaker, 1 g sodium hydrogen sulfite ($NaHSO_3$), two 50-mL beakers, 1 g barium chloride dihydrate ($BaCl_2 \cdot 2H_2O$), 3*M* H_2SO_4, stirring rod, plastic container

CAUTION! Potassium permanganate is an irritant and suspected mutagen. Barium chloride is toxic.

Procedure Dissolve 6 crystals of potassium permanganate ($KMnO_4$) in 500 mL of water in a large beaker. Dissolve 1 g sodium hydrogen sulfite ($NaHSO_3$) in 20 mL of water in a 50-mL beaker. Add this solution to the $KMnO_4$ solution and stir. Dissolve 1 g barium chloride dihydrate ($BaCl_2 \cdot 2H_2O$) in 20 mL of water in another 50-mL beaker. Add this solution to the large beaker and stir.

Expected Outcomes In the first reaction, the purple solution becomes colorless. When the $BaCl_2$ solution is added, the reaction mixture turns cloudy-white. The first reaction involves reduction of MnO_4^- to colorless Mn^{2+} and oxidation of HSO_3^- to HSO_4^-. The second reaction is not a redox reaction; it involves the formation of insoluble $BaSO_4$.

Discuss Ask students to use the oxidation number change method to balance this redox equation.

$Cu(s) + HNO_3(aq) \rightarrow Cu(NO_3)_2(aq) + NO(g) + H_2O(l)$

[$3Cu(s) + 8HNO_3(aq) \rightarrow 3Cu(NO_3)_2(aq) + 2NO(g) + 4H_2O(l)$]

Disposal Dispose of the waste solution by adding 3*M* H_2SO_4 until precipitation of $BaSO_4$ appears to be complete. Let stand overnight. Filter or decant the liquid. Dry the residue, place in a plastic container, and bury in an approved landfill site. Neutralize the filtrate, and flush down the drain with excess water.

Figure 20.18 In a blast furnace like this one, air is blown through a combination of iron ore and coke. The carbon monoxide produced from the oxidation of coke reduces the Fe^{3+} ions to metallic iron.

To use the oxidation-number-change method, start with the skeleton equation for the redox reaction. As an example, look at the process used to obtain metallic iron from iron ore in a blast furnace shown in Figure 20.18.

$$Fe_2O_3(s) + CO(g) \longrightarrow Fe(s) + CO_2(g) \text{ (unbalanced)}$$

Step 1 *Assign oxidation numbers to all the atoms in the equation.* Write the numbers above the atoms.

$$\overset{+3\ -2}{Fe_2O_3}(s) + \overset{+2-2}{CO}(g) \longrightarrow \overset{0}{Fe}(s) + \overset{+4-2}{CO_2}(g)$$

Note that the oxidation number is stated per atom. So although the total positive charge of Fe ions in Fe_2O_3 is 6+, the oxidation number of each Fe ion is +3.

Step 2 *Identify which atoms are oxidized and which are reduced.* In this reaction, iron decreases in oxidation number from +3 to 0, a change of −3. Therefore, iron is reduced. Carbon increases in oxidation number from +2 to +4, a change of +2. Thus, carbon is oxidized.

Step 3 *Use one bracketing line to connect the atoms that undergo oxidation and another such line to connect those that undergo reduction.* Write the oxidation-number change at the midpoint of each line.

+2 (oxidation)

$$\overset{+3\ -2}{Fe_2O_3}(s) + \overset{+2-2}{CO}(g) \longrightarrow \overset{0}{Fe}(s) + \overset{+4-2}{CO_2}(g)$$

−3 (reduction)

Remember that a change in oxidation number represents the number of electrons transferred. Each carbon atom in CO loses 2 electrons in oxidation, and each iron atom in Fe_2O_3 accepts 3 electrons in reduction. As the equation is written, the number of electrons transferred in oxidation does not equal the number transferred in reduction. **In a balanced redox equation, the total increase in oxidation number of the species oxidized must be balanced by the total decrease in the oxidation number of the species reduced.** Step 4 will make the oxidation-number changes equal.

Step 4 *Make the total increase in oxidation number equal to the total decrease in oxidation number by using appropriate coefficients.* In this example, the oxidation-number increase should be multiplied by 3 and the oxidation-number decrease should be multiplied by 2, which gives an increase of +6 and a decrease of −6. This can be done in the equation by placing the coefficient 2 in front of Fe on the right side and the coefficient 3 in front of both CO and CO_2. The formula Fe_2O_3 does not need a coefficient because the formula already indicates 2 Fe.

3 × (+2) = +6

$$Fe_2O_3(s) + 3CO(g) \longrightarrow 2Fe(s) + 3CO_2(g)$$

2 × (−3) = −6

Step 5 *Finally, make sure that the equation is balanced for both atoms and charge.* If necessary, finish balancing the equation by inspection.

$$Fe_2O_3(s) + 3CO(g) \longrightarrow 2Fe(s) + 3CO_2(g)$$

CONCEPTUAL PROBLEM 20.5

Balancing Redox Equations by Oxidation-Number Change

Balance this redox equation by using the oxidation-number-change method.

$$K_2Cr_2O_7(aq) + H_2O(l) + S(s) \longrightarrow KOH(aq) + Cr_2O_3(s) + SO_2(g)$$

1 Analyze ***Identify the relevant concepts.***

You can balance redox equations by determining changes in oxidation numbers and applying the five steps.

2 Solve ***Apply concepts to this situation.***

Assign oxidation numbers.

$$\overset{+1\,+6\,-2}{K_2Cr_2O_7}(aq) + \overset{+1\,-2}{H_2O}(l) + \overset{0}{S}(s) \longrightarrow \overset{+1\,-2\,+1}{KOH}(aq) + \overset{+3\,-2}{Cr_2O_3}(s) + \overset{+4\,-2}{SO_2}(g)$$

Identify the atoms that are oxidized and reduced. Cr is reduced. S is oxidized.

Connect the atoms that change in oxidation number. Indicate the signs and magnitudes of the changes.

$$K_2\overset{+6}{Cr_2}O_7(aq) + H_2O(l) + \overset{0}{S}(s) \longrightarrow KOH(aq) + \overset{+3}{Cr_2}O_3(s) + \overset{+4}{S}O_2(g)$$

(Cr: −3; S: +4)

Balance the increase and decrease in oxidation numbers. Four chromium atoms must be reduced $(4 \times (-3) = -12$ decrease) for each three sulfur atoms that are oxidized $(3 \times (+4) = +12$ increase). Put the coefficient 3 in front of S and SO_2, and the coefficient 2 in front of $K_2Cr_2O_7$ and Cr_2O_3.

$$2K_2\overset{+6}{Cr_2}O_7(aq) + H_2O(l) + 3\overset{0}{S}(s) \longrightarrow KOH(aq) + 2\overset{+3}{Cr_2}O_3(s) + 3\overset{+4}{S}O_2(g)$$

(Cr: $(4)(-3) = -12$; S: $(3)(+4) = +12$)

Check the equation and balance by inspection if necessary. The coefficient 4 in front of KOH balances potassium. The coefficient 2 in front of H_2O balances hydrogen and oxygen. The final equation is

$$2K_2Cr_2O_7(aq) + 2H_2O(l) + 3S(s) \longrightarrow 4KOH(aq) + 2Cr_2O_3(s) + 3SO_2(g).$$

Practice Problems

19. Balance each redox equation using the oxidation-number change method.

a. $KClO_3(s) \longrightarrow KCl(s) + O_2(g)$

b. $HNO_2(aq) + HI(aq) \longrightarrow NO(g) + I_2(s) + H_2O(l)$

20. Balance each redox equation using the oxidation-number change method.

a. $Bi_2S_3(s) + HNO_3(aq) \longrightarrow Bi(NO_3)_3(aq) + NO(g) + S(s) + H_2O(l)$

b. $SbCl_5(aq) + KI(aq) \longrightarrow SbCl_3(aq) + KCl(aq) + I_2(s)$

Problem-Solving 20.20 Solve Problem 20 with the help of an interactive guided tutorial.

with ChemASAP

CONCEPTUAL PROBLEM 20.5

Answers

19. a. 2, 2, 3
b. 2, 2, 2, 1, 2
20. a. 1, 8, 2, 2, 3, 4
b. 1, 2, 1, 2, 1

Practice Problems Plus

Write chemical equations for the following redox reactions. Balance each equation using the oxidation-number change method.

Solid barium chlorate decomposes when heated, yielding solid barium chloride and gaseous oxygen.
($Ba(ClO_3)_2(s) \rightarrow BaCl_2(s) + 3O_2(g)$)

Solid lead(II) sulfide reacts with gaseous oxygen to produce solid lead(II) oxide and gaseous sulfur dioxide.
($2PbS(s) + 3O_2(g) \rightarrow 2PbO(s) + 2SO_2(g)$)

Facts and Figures

Redox Reaction and Rockets

Evidence exists that the use of rockets—and of redox reactions to power them—began in the thirteenth century when Chinese armies used "arrows of flying fire" to defend their cities. These rockets were leather cylinders filled with gunpowder and mounted on a stick. It is generally thought that the gunpowder mixture originated in ninth century China.

The essential components of a rocket propellant are a source of energy (the fuel) and an oxidizing agent to react with the fuel. When the mixture is ignited, the fuel is oxidized rapidly and the oxidizing agent is reduced. Gunpowder is a mixture of charcoal (carbon), sulfur, and potassium nitrate. It reacts explosively to produce potassium sulfide, gaseous carbon dioxide, and nitrogen. In the redox reaction, carbon is the fuel and other components are oxidizing agents.

Use Visuals 

Table 20.2 Have students study Table 20.2 and ask, **What are the oxidation numbers of all the elements in H_2SO_4?** *(H is +1, S is +6, and O is −2.)* **What are the oxidation numbers of all the elements in $Na_2S_2O_3$?** *(Na is +1, S is +2, and O is −2.)*

Discuss 

Point out that the underlying assumption when using half-reactions is that any redox reaction can always be considered as two separate processes occurring simultaneously. Therefore, the "whole" reaction can be thought of as the sum of two halves—an oxidation reaction and a reduction reaction. Go through the half-reaction method with a specific example, clearly explaining each step. Emphasize that oxidation half-reactions are always written with the electrons on the right-hand side of the equation. Reduction half-reactions are always written with the electrons on the left-hand side of the equation.

Discuss L2

In Step 3, H_2O and either H^+ or OH^- can be added to either side of an equation to balance O and H atoms. **Explain why these ions can be added to either side of the equation.** *(The reaction itself is not changed by doing so; rather the medium of the reaction is simply being defined. In an acidic solution, H^+ ions are present in excess. In a basic solution, OH^- ions are present in excess.)*

Table 20.2

Oxidation Numbers of Sulfur in Different Substances

Substance	Oxidation number
H_2SO_4	+6
SO_3	+6
H_2SO_3	+4
SO_2	+4
$Na_2S_2O_3$	+2
SCl_2	+2
S_2Cl_2	+1
S	0
H_2S	−2

Using Half-Reactions A second method for balancing redox equations involves the use of half-reactions. A **half-reaction** is an equation showing just the oxidation or just the reduction that takes place in a redox reaction. In the **half-reaction method,** you write and balance the oxidation and reduction half-reactions separately before combining them into a balanced redox equation. The outcome is the same as with the oxidation-number-change method. **To balance a redox reaction using half-reactions, write separate half-reactions for the oxidation and the reduction. After you balance atoms in each half-reaction, balance electrons gained in the reduction with electrons lost in the oxidation.** The half-reaction method is particularly useful in balancing equations for ionic reactions.

Sulfur is an element that can have several different oxidation numbers, as you can see in Table 20.2. The oxidation of sulfur by nitric acid in aqueous solution is one example of a redox reaction that can be balanced by the half-reaction method.

$$S(s) + HNO_3(aq) \longrightarrow SO_2(g) + NO(g) + H_2O(l) \text{ (unbalanced)}$$

The following steps show how to balance this equation using the half-reaction method.

Step 1 *Write the unbalanced equation in ionic form.* In this case, only HNO_3 is ionized. The products are covalent compounds.

$$S(s) + H^+(aq) + NO_3^-(aq) \longrightarrow SO_2(g) + NO(g) + H_2O(l)$$

Step 2 *Write separate half-reactions for the oxidation and reduction processes.* Sulfur is oxidized in this reaction because its oxidation number increases from 0 to +4. Nitrogen is reduced because its oxidation number decreases from 5 to +2.

Oxidation half-reaction: $\overset{0}{S}(s) \longrightarrow \overset{+4}{S}O_2(g)$

Reduction half-reaction: $\overset{+5}{N}O_3^-(aq) \longrightarrow \overset{+2}{N}O(g)$

Notice that H^+ ions and H_2O are not included in the half-reactions because they are neither oxidized nor reduced. However, they will be used in balancing the half-reactions.

Step 3 *Balance the atoms in the half-reactions.*

a. Balance the oxidation half-reaction. Sulfur is already balanced in the half-reaction, but oxygen is not. This reaction takes place in acid solution, so H_2O and $H^+(aq)$ are present and can be used to balance oxygen and hydrogen as needed. Add two molecules of H_2O on the left to balance the oxygen in the half-reaction.

$$2H_2O(l) + S(s) \longrightarrow SO_2(g)$$

Oxygen is now balanced, but four hydrogen ions ($4H^+$) must be added to the right to balance the hydrogen on the left.

$$2H_2O(l) + S(s) \longrightarrow SO_2(g) + 4H^+(aq)$$

This half-reaction is now balanced in terms of atoms.

Facts and Figures

Bioluminescence

Some organisms produce light by means of oxidation-reductions in a process known as bioluminescence. Light is given off when one of a class of compounds called luciferins is oxidized by combining with oxygen. A product of these reactions is energy in the form of light. Bioluminescence serves different functions in different organisms. For example, fireflies use bioluminescence to attract mates, while some squids release a glowing cloud of "ink" to confuse and escape from their predators.

b. Balance the reduction half-reaction. Nitrogen is already balanced. Add two molecules of H_2O on the right to balance the oxygen.

$$NO_3^-(aq) \longrightarrow NO(g) + 2H_2O(l)$$

Oxygen is balanced, but four hydrogen ions ($4H^+$) must be added to the left to balance hydrogen.

$$4H^+(aq) + NO_3^-(aq) \longrightarrow NO(g) + 2H_2O(l)$$

This half-reaction is now balanced in terms of atoms.

Step 4 *Add enough electrons to one side of each half-reaction to balance the charges.* Note that neither half-reaction is balanced for charge. Four electrons are needed on the right side in the oxidation half-reaction.

$$\text{Oxidation: } 2H_2O(l) + S(s) \longrightarrow SO_2(g) + 4H^+(aq) + 4e^-$$

Three electrons are needed on the left side in the reduction half-reaction.

$$\text{Reduction: } 4H^+(aq) + NO_3^-(aq) + 3e^- \longrightarrow NO(g) + 2H_2O(l)$$

Each half-reaction is now balanced with respect to both atoms and charge.

Step 5 *Multiply each half-reaction by an appropriate number to make the numbers of electrons equal in both.* The number of electrons lost in oxidation must equal the number of electrons gained in reduction. In this case, the oxidation half-reaction is multiplied by 3 and the reduction half-reaction is multiplied by 4. Therefore the number of electrons lost in oxidation and the number of electrons gained in reduction both equal 12.

$$\text{Oxidation: } 6H_2O(l) + 3S(s) \longrightarrow 3SO_2(g) + 12H^+(aq) + 12e^-$$

$$\text{Reduction: } 16H^+(aq) + 4NO_3^-(aq) + 12e^- \longrightarrow 4NO(g) + 8H_2O(l)$$

Step 6 *Add the balanced half-reactions to show an overall equation.*

$$6H_2O(l) + 3S(s) + 16H^+(aq) + 4NO_3^-(aq) + 12e^- \longrightarrow 3SO_2(g) + 12H^+(aq) + 12e^- + 4NO(g) + 8H_2O(l)$$

Then subtract terms that appear on both sides of the equation.

$$3S(s) + 4H^+(aq) + 4NO_3^-(aq) \longrightarrow 3SO_2(g) + 4NO(g) + 2H_2O(l)$$

Step 7 *Add the spectator ions and balance the equation.* Recall that spectator ions are present but do not participate in or change during a reaction. Because none of the ions in the reactants appear in the products, there are no spectator ions in this particular example. The balanced equation is correct. However, it can be written to show the HNO_3 as not ionized.

$$3S(s) + 4HNO_3(aq) \longrightarrow 3SO_2(g) + 4NO(g) + 2H_2O(l)$$

For: Links on Redox Reactions
Visit: www.SciLinks.org
Web Code: cdn-1203

Balancing a Redox Equation Using the Half-Reaction Method

Step 1 Write the unbalanced equation in ionic form.

Step 2 Write separate half-reactions for the oxidation and reduction processes.

Step 3 Balance the atoms in the half-reactions.

Step 4 Add enough electrons to one side of each half-reaction to balance the charges. (*Hint:* The electrons will be added to opposite sides of the half-reactions.)

Step 5 Multiply each half-reaction by an appropriate number to make the numbers of electrons equal in both.

Step 6 Add the half-reactions to show an overall equation.

Step 7 Add the spectator ions and balance the equation.

Differentiated Instruction

Less Proficient Readers L1

Balancing redox reactions by the half-reaction method is a lengthy process that requires good organizational skills and attention to detail. Students who are impatient or easily frustrated may have difficulty with these problems. Working in small groups can help them organize their efforts. Supervise the work closely to be sure each student is contributing.

Download a worksheet on Redox Reactions for students to complete, and find additional teacher support from NSTA SciLinks.

CLASS Activity

Preventing Oxidation L1

Purpose Students observe the oxidation process in everyday foods.

Materials potatoes; apples; beakers; water; boiled water; lemon juice; carbonated beverage; sugar, salt and vinegar solutions; paper towels

Procedure Students should be familiar with the "browning" of sliced potatoes or apples upon exposure to oxygen. The oxidation is catalyzed by enzymes in food. Explain that oxidation can be prevented or slowed down. The outer skin provides a natural physical barrier to oxygen. Chemical protection can be provided by antioxidants, which act as reducing agents or by inhibiting enzymes. Liquids such as lemon juice contain antioxidants. Have students form groups of two or three and give each group a potato and an apple. Provide beakers filled with tap water, water that has been boiled to exclude most dissolved oxygen, lemon juice, a carbonated beverage, plus sugar, salt, and vinegar solutions. Have students quickly cut the potato and apple into slices and immediately place a slice in each beaker. Have them label locations on paper towels to match the labels on the beakers. Have them remove the slices after they have soaked for several minutes and place them on the paper towels. Have them make periodic observations of the slices to determine the degree of "browning." On the basis of their observations, have students decide which liquids offer the best protection. Have them try to explain why the liquids are good or bad protectants. They can present their findings in the form of a poster or a written/oral report.

CONCEPTUAL PROBLEM 20.6

Answers

21. $4ZN + NO_3^- + 6H_2O + 7OH^- \rightarrow 4Zn(OH)_4^{2-} + NH_3$

Practice Problems Plus L2

The following reaction takes place in basic solution. Use the half-reaction method to write a balanced ionic equation.

$Zn(s) + As_2O_3(aq) \rightarrow AsH_3(aq) + Zn^{2+}(aq)$

$[6Zn(s) + As_2O_3(aq) + 9H_2O \rightarrow 6Zn^{2+}(aq) + 2AsH_3(aq) + 12OH^-(aq)]$

CONCEPTUAL PROBLEM 20.6

Balancing Redox Equations by Half-Reactions

Balance this redox equation using the half-reaction method.

$$KMnO_4(aq) + HCl(aq) \longrightarrow MnCl_2(aq) + Cl_2(g) + H_2O(l) + KCl(aq)$$

1 Analyze ***Identify the relevant concepts.***

You can use the seven steps of the half-reaction method.

2 Solve ***Apply concepts to this situation.***

Write the equation in ionic form.

$$K^+(aq) + MnO_4^-(aq) + H^+(aq) + Cl^-(aq) \longrightarrow Mn^{2+}(aq) + 2Cl^-(aq) + Cl_2(g) + H_2O(l) + K^+(aq) + Cl^-(aq)$$

Write half-reactions. Determine the oxidation and reduction process.

Oxidation: $\overset{-1}{Cl^-} \longrightarrow \overset{0}{Cl_2}$

Reduction: $\overset{+7}{MnO_4^-} \longrightarrow \overset{+2}{Mn^{2+}}$

Balance the atoms. The solution is acidic, so use H_2O and H^+ to balance the oxygen and hydrogen. (If the solution is basic, H_2O and OH^- are used.)

Oxidation:

$2Cl^-(aq) \longrightarrow Cl_2(g)$ (atoms balanced)

Reduction:

$MnO_4^-(aq) + 8H^+(aq) \longrightarrow Mn^{2+}(aq) + 4H_2O(l)$ (atoms balanced)

Balance the charges.

Oxidation: $2Cl^-(aq) \longrightarrow Cl_2(g) + 2e^-$ (charges balanced)

Reduction:

$MnO_4^-(aq) + 8H^+(aq) + 5e^- \longrightarrow Mn^{2+}(aq) + 4H_2O(l)$ (charges balanced)

Make the numbers of electrons equal. Multiply the oxidation half-reaction by 5 and the reduction half-reaction by 2.

Oxidation: $10Cl^-(aq) \longrightarrow 5Cl_2(g) + 10e^-$

Reduction:

$2MnO_4^-(aq) + 16H^+(aq) + 10e^- \longrightarrow 2Mn^{2+}(aq) + 8H_2O(l)$

Add the half-reactions.

$$10Cl^-(aq) + 2MnO_4^-(aq) + 16H^+(aq) + \cancel{10e^-} \longrightarrow 5Cl_2(g) + \cancel{10e^-} + 2Mn^{2+}(aq) + 8H_2O(l)$$

The equation, after subtracting terms that appear on both sides:

$$10Cl^-(aq) + 2MnO_4^-(aq) + 16H^+(aq) \longrightarrow 5Cl_2(g) + 2Mn^{2+}(aq) + 8H_2O(l)$$

Add the spectator ions.

$$10Cl^- + 2MnO_4^- + 2K^+ + 16H^+ + 6Cl^- \longrightarrow 5Cl_2 + 2Mn^{2+} + 4Cl^- + 8H_2O + 2K^+ + 2Cl^-$$

Adding spectator and nonspectator Cl^- on each side gives

$$16Cl^-(aq) + 2MnO_4^-(aq) + 2K^+(aq) + 16H^+(aq) \longrightarrow 5Cl_2(g) + 2Mn^{2+}(aq) + 6Cl^-(aq) + 8H_2O(l) + 2K^+(aq)$$

The equation is balanced for atoms and charge. To show it balanced for the substances given in the question (rather than for ions), rewrite it as

$$2KMnO_4(aq) + 16HCl(aq) \longrightarrow 2MnCl_2(aq) + 5Cl_2(g) + 8H_2O(l) + 2KCl(aq)$$

Practice Problem

21. The following reaction takes place in basic solution. Balance the equation using the half-reaction method.

$Zn(s) + NO_3^-(aq) \longrightarrow NH_3(aq) + Zn(OH)_4^{2-}(aq)$

Interactive Textbook

Problem-Solving 20.21 Solve Problem 21 with the help of an interactive guided tutorial.

with **ChemASAP**

Choosing a Balancing Method

Some redox equations, such as the following example, are simple enough to balance by inspection.

$$Zn(s) + 2AgNO_3(aq) \longrightarrow 2Ag(s) + Zn(NO_3)_2(aq)$$

Balancing by oxidation number change usually works well if the oxidized and reduced species appear only once on each side of the equation and no acids or bases are involved. However, in some redox reactions, the same element is both oxidized and reduced. In the reaction of bromine with a base such as KOH, elemental bromine is simultaneously oxidized from 0 in Br_2 to +5 in $KBrO_3$ and reduced from 0 in Br_2 to −1 in KBr.

$$3Br_2(l) + 6KOH(aq) \longrightarrow 5KBr(aq) + KBrO_3(aq) + 3H_2O(l)$$

You can tell from the coefficients in the balanced equation that it is not easy to balance. Equations such as this one and equations for reactions that take place in acidic or alkaline solution are best balanced by the half-reaction method.

Quick LAB

Bleach It! Oxidize the Color Away

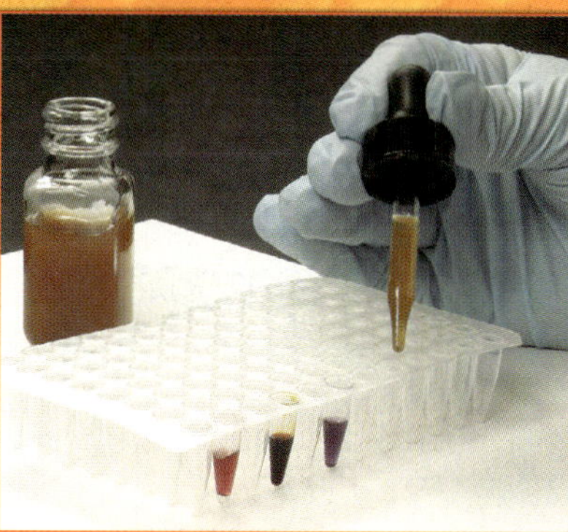

Purpose

To test the effect of oxidizing agents on stains and dyes.

Materials

- spot plate
- dropper
- water
- colorimeter (optional)

Oxidizing agents

- liquid chlorine bleach (5% (m/v) sodium hypochlorite)
- powder bleach
- oxalic acid solution (1% (m/v))
- sodium thiosulfate solution (hypo) (0.2*M* $Na_2S_2O_3$)
- hydrogen peroxide (3% (v/v) H_2O_2)

Samples

- iodine solution (1% I_2 in 2% (m/v) KI)
- potassium permanganate solution (0.05*M* $KMnO_4$)
- grape juice
- rusty water
- piece of colored fabric
- colored flower petals
- grass stain on piece of white fabric

Procedure    

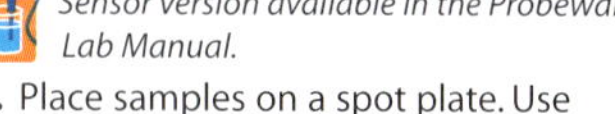
Sensor version available in the Probeware Lab Manual.

1. Place samples on a spot plate. Use 4 drops of each liquid or a small piece of each solid.
2. Describe the color and appearance of each sample in Step 1.
3. Add a few drops of the first oxidizing agent to each sample.
4. Describe any immediate change in appearance and any further change after 15 minutes.
5. Repeat Steps 1–4 with each oxidizing agent.

Analyze and Conclude

1. Make a grid and record your observations.
2. Compare the oxidizing power of the oxidizing agents.
3. How do you know that chemical changes have occurred?

Choosing a Balancing Method

Discuss 

A reaction in which the same compound is both oxidized and reduced is called *disproportional*. When NO_2 is dissolved in water, it disproportionates to form nitric acid (HNO_3) and nitric oxide (NO). The balanced equation for this reaction is $3NO_2(g) + H_2O(l) \rightarrow 2H^+(aq) + 2NO_3^-(aq) + NO(g)$

Quick LAB

Bleach It! Oxidize the Color Away L2

Objective After completing this activity, students will be able to:

- Test the effect of oxidizing agents on stains and dyes.

Skills Focus observing, comparing, making conclusions

 **Prep Time** 20 minutes

Materials spot plates, medicine droppers, liquid and powder bleach, 3% H_2O_2, 1% oxalic acid, 0.2*M* $Na_2S_2O_3$, I_2 in KI(*aq*), $KMnO_4(aq)$, grape juice, rusty water, colored fabric, grass-stained fabric, flower petals

Advance Prep Prepare the oxidizing agents and other solutions before class. Arrange to bring items such as grape juice and flower petals to class on the day of the lab.

Safety The oxidizing agents in this lab pose a slight contact hazard and should be handled carefully. Safety goggles should be worn at all times. Students should also be careful with splashing and spillage because the oxidizing solutions can bleach/destroy clothing; lab coats are recommended.

Expected Outcome Various color changes occur.

Analyze and Conclude

1. Grids contain student observations.
2. Rankings depend on observations.
3. Color changes occur.

For Enrichment 

Have students extend this lab by bringing in stained or old clothing from home. Have them test the effect of the oxidizing agents on these materials.

Careers *in* Chemistry

Quality Control Chemist

Quality control chemists are an important part of the food and drug industries. Strict regulations within the industry and also imposed by the government serve to protect the nation's health. These regulations cover all aspects of the industry.

3 ASSESS

Evaluate Understanding L2

In order to evaluate the students' understanding of this section, ask, **Copper(II) hydroxide ($Cu(OH)_2$) is converted to copper(II) oxide (CuO) and water upon heating. Is this a redox reaction? Why?** *(No, none of the elements undergoes a change in oxidation number.)* **Copper(II) oxide (CuO) is converted to copper metal upon heating in the presence of carbon. Is this a redox reaction? Why?** *(Yes, copper is reduced; its oxidation number decreases from +2 to 0. Carbon is oxidized; its oxidation number increases from 0 to +4 when it is converted to CO_2.)*

Reteach L1

Point out that the distinguishing feature of redox reactions is the change in oxidation numbers that occurs. To decide whether a given reaction is a redox reaction, students need only compare the oxidation numbers of the reactants to the oxidation numbers of the products. If a change has occurred, the reaction is a redox reaction.

Elements Handbook

oxidizing agent, Na^+; reducing agent, Cl^-

oxidizing agent, Cl_2; reducing agent, Br^-

oxidizing agent, IO_3^- or I^{+5}; reducing agent, HSO_3^- or S^{+4}

If your class subscribes to the Interactive Textbook, use it to review key concepts in Section 20.3.

with ChemASAP

Careers *in* Chemistry

Quality Control Chemist

A quality control chemist is responsible for seeing that products meet manufacturing standards. The consistent quality of many products you use is due to the work of a quality control chemist. They work in a wide variety of industries, including those that manufacture air bags for automobiles. They are involved in every aspect of producing the product. They contribute to the original research and development of the product and help refine its manufacturing process. They then develop methods to test the product or to determine why it has failed to perform for the consumer.

Quality control chemists must know how to make and interpret measurements. Experience in analytical chemistry and the ability to develop new analytical methods are particularly important. The specific requirements for the job depend on the industry and company that employs the quality control chemist. A good quality control chemist assures that his or her company's customers receive the highest-quality products.

For: Careers in Chemistry
Visit: PHSchool.com
Web Code: cdb-1203

20.3 Section Assessment

22. **Key Concept** What happens to the oxidation number of a species reacting in a redox reaction?
23. **Key Concept** In any redox reaction, how does the increase in oxidation number of the species oxidized compare to the decrease in the oxidation number of the species reduced?
24. **Key Concept** In balancing a redox reaction using half-reactions, what must be true concerning the number of electrons lost in the oxidation half-reaction compared to the number of electrons gained in the reduction half-reaction?
25. Balance each redox equation, using the oxidation-number-change method.
    a. $ClO_3^-(aq) + I^-(aq) \longrightarrow Cl^-(aq) + I_2(aq)$ [acidic solution]
    b. $C_2O_4^{2-}(aq) + MnO_4^-(aq) \longrightarrow Mn^{2+}(aq) + CO_2(g)$ [acidic solution]
    c. $Br_2(l) + SO_2(g) \longrightarrow Br^-(aq) + SO_4^{2-}(aq)$ [acidic solution]

Elements Handbook

Sodium, Iodine, and Bromine Review the equations for the production of sodium, bromine, and iodine found on pages R6 and R32 of the Elements Handbook. Identify the oxidizing agent and the reducing agent in each reaction.

Assessment 20.3 Test yourself on the concepts in Section 20.3.

with ChemASAP

Section 20.3 Assessment

22. The oxidation number changes.

23. The total increase in oxidation number of the species oxidized equals the total decrease in oxidation number of the species reduced.

24. The number of electrons lost in oxidation equals the number of electrons gained in reduction.

25. a. $ClO_3^-(aq) + 6I^-(aq) + 6H^+(aq) \rightarrow Cl^-(aq) + 3I_2(aq) + 3H_2O(l)$

b. $5C_2O_4^{2-}(aq) + 2MnO_4^-(aq) + 16H^+(aq) \rightarrow 2Mn^{2+}(aq) + 10CO_2(g) + 8H_2O(l)$

c. $Br_2(l) + SO_2(g) + 2H_2O(l) \rightarrow 2Br^-(aq) + SO_4^{2-}(aq) + 4H^+(aq)$

Small-Scale LAB

Half-Reactions

Purpose
To observe redox reactions and to write half-reactions that describe them.

Materials
- pencil
- paper
- ruler
- reaction surface
- chemicals listed in the grid

Procedure   
On separate sheets of paper, draw two grids similar to the one below.

	HCl	HNO_3	H_2SO_4
Zn			
Mg			
Cu			
Fe			

Make each square 2 cm on each side. Place a reaction surface over one of the grids and add one drop of each acid solution to one piece of each metal, as shown above. Use the second grid as a data table to record your observations for each solution.

Analyze
Using your experimental data, answer the following questions.

1. Which metal is the most reactive? Explain. Which metal did not react with any of the acids? List the metals in order of decreasing reactivity.
2. What is the chemical formula of the gas produced in each reaction?
3. An active metal reacts with an acid to produce hydrogen gas and a salt. Write equations and net ionic equations to describe the reactions you observed. Are all of these redox reactions? Explain.
4. The half-reaction for the oxidation of Zn is shown below.

   $$Zn(s) \longrightarrow Zn^{2+}(aq) + 2e^-$$

   Write the oxidation half-reaction for the other metals that react.
5. The half-reaction for the reduction of H^+ from the acid is shown below.

   $$2H^+ + 2e^- \longrightarrow H_2(g)$$

   Notice that this half-reaction is the same for all the acids. Demonstrate how adding this half-reaction to each oxidation half-reaction results in the overall net ionic equations.

You're The Chemist
The following small-scale activities allow you to develop your own procedures and analyze the results.

1. **Analyze It!** Pennies minted after 1982 are made of zinc with a thin copper coating. Use a penny that has been damaged so that a portion of the zinc shows through to compare the reactivity of the zinc and the copper toward various acids.
2. **Design It!** Many household products, such as toilet-bowl cleaners and vinegar, contain acids. Design and carry out experiments to find out if these products also react with metals.

Half-Reactions 

Objective After completing this activity, students will be able to:
- Write half-reactions to describe observed redox reactions between acids and metals.

 **Prep Time** 45 minutes

Advance Prep Prepare the following solutions.

Solution	Preparation
1.0*M* HCl	82 mL of 12*M* in 1.0 L
1.0*M* HNO_3	63 mL of 15.8*M* in 1.0 L
1.0*M* H_2SO_4	56 mL of 18*M* in 1.0 L

Class Time 45 minutes

Safety Always add acid to water carefully and slowly.

Teaching Tips
Emphasize that a half-reaction shows what happens to electrons in redox reactions. Use magnesium turnings or cut magnesium ribbon into 0.5-cm pieces. Use galvanized nails for zinc, staples from a stapler for iron, and pennies for copper. Recycle all the metals except the staples.

Analyze
1. Mg is most reactive because it bubbles most vigorously. Cu does not react. The order of reactivity is Mg > Zn > Fe > Cu.
2. $H_2(g)$ is produced in all the reactions.
3. $Mg(s) + 2HCl(aq) \rightarrow H_2(g) + MgCl_2(aq)$

   $Mg(s) + 2H^+(aq) \rightarrow H_2(g) + Mg^{2+}(aq)$

   Reactions of HCl with Zn and Fe are similar.

   All are redox reactions because the oxidation numbers of reactants change.
4. $Mg(s) \rightarrow Mg^{2+}(aq) + 2e^-$

   $Zn(s) \rightarrow Zn^{2+}(aq) + 2e^-$

   $Fe(s) \rightarrow Fe^{2+}(aq) + 2e^-$
5. $2H^+ + 2e^- \rightarrow H_2(g)$

   $Mg(s) \longrightarrow Mg^{2+}(aq) + 2e^-$

   $Mg(s) + 2H^+(aq) \longrightarrow H_2(g) + Mg^{2+}(aq)$

You're The Chemist
1. Add a drop of any acid to the damaged part of the penny and notice that only the zinc interior reacts.
2. Vinegar and many toilet-bowl cleaners dissolve metals. Keep products containing acids away from metal pipes and fixtures.

For Enrichment L3
Have students research the redox reactions that take place in the black-and-white photography process. Have them write half-reactions and complete redox equations for the reactions.

CHAPTER 20 Study Guide

Study Tip

Organize Information Tell students that organizing information will help them remember it more easily. Suggest that students write key information from the chapter on index cards or large self-stick notes and use them to create large concept maps.

If your class subscribes to the Interactive Textbook with ChemASAP, your students can go online to access an interactive version of the Student Edition and a self-test.

with ChemASAP

CHAPTER 20 Study Guide

Key Concepts

20.1 The Meaning of Oxidation and Reduction

- Oxidation and reduction always occur simultaneously. The substance gaining oxygen is oxidized, while the substance losing oxygen is reduced.
- Losing electrons is oxidation. Gaining electrons is reduction.
- Iron, often used in the form of the alloy steel, corrodes by being oxidized to ions of iron by oxygen.

20.2 Oxidation Numbers

- As a general rule, a bonded atom's oxidation number is the charge that it would have if the electrons in the bond were assigned to the atom of the more electronegative element.
- An increase in the oxidation number of an atom or ion indicates oxidation. A decrease in the oxidation number of an atom or ion indicates reduction.

20.3 Balancing Redox Equations

- If the oxidation number of an element in a reacting species changes, then that element has undergone either oxidation or reduction. Therefore, the reaction as a whole must be a redox reaction.
- In a balanced redox equation, the total increase in oxidation number of the species oxidized must be balanced by the total decrease in the oxidation number of the species reduced.
- To balance a redox reaction using half-reactions, write separate half-reactions for the oxidation and the reduction. After you balance atoms in each half-reaction, balance electrons gained in the reduction with electrons lost in the oxidation.

Vocabulary

- half-reaction (p. 650)
- half-reaction method (p. 650)
- oxidation (p. 633)
- oxidation number (p. 639)
- oxidation-number-change method (p. 647)
- oxidation-reduction reaction (p. 632)
- oxidizing agent (p. 634)
- redox reaction (p. 632)
- reducing agent (p. 634)
- reduction (p. 633)

Organizing Information

Use these terms to construct a concept map that organizes the major ideas of this chapter.

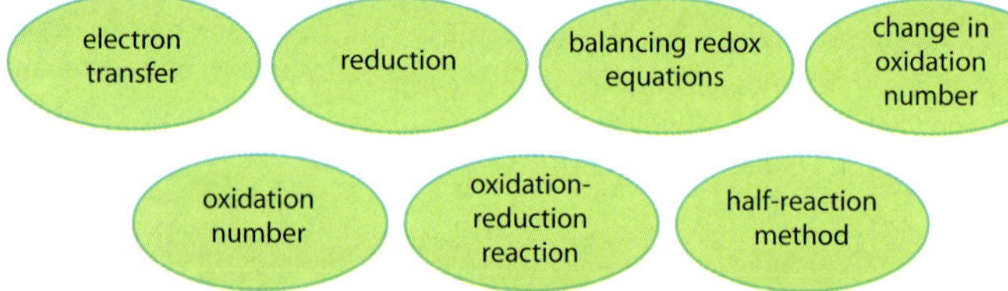

Concept Map 20 Create your concept map using the computer.

with ChemASAP

Chapter Resources

Print

- **Core Teaching Resources,** Chapter 20, *Practice Problems, Vocabulary Review, Quiz, Chapter Test A, Chapter Test B*

Technology

- **Computer Test Bank,** Chapter 20 Test
- **Interactive Textbook with ChemASAP,** Chapter 20

CHAPTER 20

Assessment

Reviewing Content

20.1 The Meaning of Oxidation and Reduction

26. What chemical process must always accompany a reduction process?

27. What happens to an oxidizing agent during a redox reaction?

28. Balance each redox equation and identify whether the first substance was oxidized or reduced.

a. $Ba(s) + O_2(g) \longrightarrow BaO(s)$
b. $CuO(s) + H_2(g) \longrightarrow Cu(s) + H_2O(l)$
c. $C_2H_4(g) + O_2(g) \longrightarrow CO_2(g) + H_2O(l)$
d. $CaO(s) + Al(s) \longrightarrow Al_2O_3(s) + Ca(s)$

29. Identify each process as either oxidation or reduction.

a. $Al \longrightarrow Al^{3+} + 3e^-$ **b.** $2Cl^- \longrightarrow Cl_2 + 2e^-$
c. $S^{2-} \longrightarrow S + 2e^-$ **d.** $Sr \longrightarrow Sr^{2+} + 2e^-$

30. Which of the following would most likely be oxidizing agents and which would most likely be reducing agents? (*Hint:* Think in terms of tendencies to lose or gain electrons.)

a. Cl_2 **b.** K **c.** Ag^+ **d.** Zn^{2+}

31. Refer to the electronegativity values in Table 6.2 to determine which reactant is oxidized and which reactant is reduced in each reaction.

a. $H_2(g) + S(s) \longrightarrow H_2S(g)$
b. $N_2(g) + 3H_2(g) \longrightarrow 2NH_3(g)$
c. $S(s) + O_2(g) \longrightarrow SO_2(g)$
d. $2H_2(g) + O_2(g) \longrightarrow 2H_2O(l)$

32. Identify the oxidizing agent and the reducing agent for each reaction in Problem 31.

20.2 Oxidation Numbers

33. What is an oxidation number?

34. Which of these statements is false?

a. The oxidation number of an uncombined element is zero.
b. The sum of the oxidation numbers of the atoms in a polyatomic ion must equal the charge of the ion.
c. Every element has a single oxidation number.
d. The oxidation number of oxygen in a compound or a polyatomic ion is almost always −2.

35. Determine the oxidation number of each metal atom.

a. Ca^{2+} **b.** Al_2S_3 **c.** Na_2CrO_4 **d.** V_2O_5 **e.** MnO_4^-

36. Assign oxidation numbers to the atoms in the following ions:

a. OH^- **b.** PO_4^{3-} **c.** IO_3^- **d.** $H_2PO_4^-$ **e.** HSO_4^-

20.3 Balancing Redox Equations

37. Use the changes in oxidation numbers to identify which atoms are oxidized and which are reduced in each reaction.

a. $Al(s) + MnO_2(s) \longrightarrow Al_2O_3(s) + Mn(s)$
b. $K(s) + H_2O(l) \longrightarrow KOH(aq) + H_2(g)$
c. $HgO(s) \longrightarrow Hg(l) + O_2(g)$
d. $P_4(s) + O_2(g) \longrightarrow P_4O_{10}(s)$

38. Balance each redox equation.

a. $Al(s) + Cl_2(g) \longrightarrow AlCl_3(s)$
b. $Al(s) + Fe_2O_3(s) \longrightarrow Al_2O_3(s) + Fe(s)$
c. $Cl_2(g) + KOH(aq) \longrightarrow KClO_3(aq) + KCl(aq) + H_2O(l)$
d. $HNO_3(aq) + H_2S(g) \longrightarrow S(s) + NO(g) + H_2O(l)$
e. $KIO_4(aq) + KI(aq) + HCl(aq) \longrightarrow KCl(aq) + I_2(s) + H_2O(l)$

39. Identify which of these unbalanced equations represent redox reactions.

a. $Li(s) + H_2O(l) \longrightarrow LiOH(aq) + H_2(g)$
b. $K_2Cr_2O_7(aq) + HCl(aq) \longrightarrow KCl(aq) + CrCl_3(aq) + H_2O(l) + Cl_2(g)$
c. $Al(s) + HCl(aq) \longrightarrow AlCl_3(aq) + H_2(g)$
d. $Cl_2(g) + H_2O(l) \longrightarrow HCl(aq) + HClO(aq)$
e. $I_2O_5(s) + CO(g) \longrightarrow I_2(s) + CO_2(g)$
f. $H_2O(l) + SO_3(g) \longrightarrow H_2SO_4(aq)$

40. Use the half-reaction method to write a balanced ionic equation for each reaction. All are basic solutions.

a. $MnO_4^-(aq) + ClO_2^-(aq) \longrightarrow MnO_2(s) + ClO_4^-(aq)$
b. $Cr^{3+}(aq) + ClO^-(aq) \longrightarrow CrO_4^{2-}(aq) + Cl^-(aq)$
c. $Mn^{3+}(aq) + I^-(aq) \longrightarrow Mn^{2+}(aq) + IO_3^-(aq)$

CHAPTER 20

Assessment

Reviewing Content

26. oxidation

27. The oxidizing agent is reduced.

28. a. $2Ba(s) + O_2(g) \rightarrow 2BaO(s)$; barium is oxidized
b. $CuO(s) + H_2(g) \rightarrow Cu(s) + H_2O(l)$; copper is reduced
c. $C_2H_4(g) + 3O_2(g) \rightarrow 2CO_2(g) + 2H_2O(l)$; carbon is oxidized
d. $3CaO(s) + 2Al(s) \rightarrow Al_2O_3(s) + 3Ca(s)$; calcium is reduced

29. a. oxidation **b.** oxidation **c.** oxidation **d.** oxidation

30. a. oxidizing agent **b.** reducing agent **c.** oxidizing agent **d.** oxidizing agent

31. a. H_2 is oxidized; S is reduced.
b. N_2 is reduced; H_2 is oxidized.
c. S is oxidized; O_2 is reduced.
d. H_2 is oxidized; O_2 is reduced.

32. a. H_2 is the reducing agent; S is the oxidizing agent.
b. N_2 is the oxidizing agent; H_2 is the reducing agent.
c. S is the reducing agent; O_2 is the oxidizing agent.
d. H_2 is the reducing agent; O_2 is the oxidizing agent.

33. An oxidation number is the charge an atom would have if the electrons in each bond were assigned to the atoms of the more electronegative element.

34. c is false.

35. a. +2 **b.** +3 **c.** Na, +1; Cr, +6 **d.** +5 **e.** +7

36. a. O, −2; H, +1 **b.** P, +5; O, −2 **c.** I, +5; O, −2 **d.** H, +1; P, +5; O, −2 **e.** H, +1; S, +6; O, −2

37. a. Al is oxidized; Mn is reduced.
b. K is oxidized; H is reduced.
c. Hg is reduced; O is oxidized.
d. P is oxidized; O is reduced.

38. a. $2Al(s) + 3Cl_2(g) \rightarrow 2AlCl_3(s)$
b. $2Al(s) + Fe_2O_3(s) \rightarrow Al_2O_3(s) + 2Fe(s)$
c. $3Cl_2(g) + 6KOH(aq) \rightarrow KClO_3(aq) + 5KCl(aq) + 3H_2O(l)$
d. $2HNO_3(aq) + 3H_2S(aq) \rightarrow 3S(s) + 2NO(g) + 4H_2O(l)$
e. $KIO_4(aq) + 7KI(aq) + 8HCl(aq) \rightarrow 8KCl(aq) + 4I_2(s) + 4H_2O(l)$

39. redox: a, b, c, d, e

40. a. $4MnO_4^-(aq) + 3ClO_2^-(aq) + 2H_2O(l) \rightarrow 4MnO_2(s) + 3ClO_4^-(aq) + 4OH^-(aq)$
b. $2Cr^{3+}(aq) + 3ClO^-(aq) + 10OH^-(aq) \rightarrow 2CrO_4^{2-}(aq) + 3Cl^-(aq) + 5H_2O(l)$
c. $6Mn^{3+}(aq) + I^-(aq) + 6OH^-(aq) \rightarrow 6Mn^{2+}(aq) + IO_3^-(aq) + 3H_2O(l)$

CHAPTER 20

Assessment continued

Understanding Concepts

41. a. $4Al(s) + 3MnO_2(s) \rightarrow 2Al_2O_3(s) + 3Mn(s)$
b. $2K(s) + 2H_2O(l) \rightarrow 2KOH(aq) + H_2(g)$
c. $2HgO(s) \rightarrow 2Hg(l) + O_2(g)$
d. $P_4(s) + 5O_2(g) \rightarrow P_4O_{10}(s)$

42. a. $2Li(s) + 2H_2O(l) \rightarrow 2LiOH(aq) + H_2(g)$
b. $K_2Cr_2O_7(aq) + 14HCl(aq) \rightarrow 2KCl(aq) + 2CrCl_3(aq) + 7H_2O(l) + 3Cl_2(g)$
c. $2Al(s) + 6HCl(aq) \rightarrow 2AlCl_3(aq) + 3H_2(g)$
d. $Cl_2(g) + H_2O(l) \rightarrow HCl(aq) + HClO(aq)$
e. $I_2O_5(s) + 5CO(g) \rightarrow I_2(s) + 5CO_2(g)$
f. $H_2O(l) + SO_3(g) \rightarrow H_2SO_4(aq)$

43. a. +4 **b.** +5 **c.** +5 **d.** +3 **e.** +5 **f.** +3

44. K_2CrO_4, Cr +6; $K_2Cr_2O_7$, Cr +6

45. a. Cl oxidized, Mn reduced, Mn oxidizing agent, Cl reducing agent
b. Cu oxidized, N reduced, N oxidizing agent, Cu reducing agent
c. P oxidized, N reduced, N oxidizing agent, P reducing agent
d. Sn oxidized, Bi reduced, Bi oxidizing agent, Sn reducing agent

46. a. $MnO_2(s) + 4HCl(aq) \rightarrow MnCl_2(aq) + Cl_2(g) + 2H_2O(l)$
b. $Cu(s) + 4HNO_3(aq) \rightarrow Cu(NO_3)_2(aq) + 2NO_2(g) + 2H_2O(l)$
c. $3P(s) + 5HNO_3(aq) + 2H_2O(l) \rightarrow 5NO(g) + 3H_3PO_4(aq)$
d. $2Bi(OH)_3(s) + 3Na_2SnO_2(aq) \rightarrow 2Bi(s) + 3Na_2SnO_3(aq) + 3H_2O(l)$

47. a. $16H^+(aq) + 2Cr_2O_7^{2-}(aq) + C_2H_5OH(aq) \rightarrow 4Cr^{3+}(aq) + 2CO_2(g) + 11H_2O(l)$
b. oxidizing agent

48. a. $WO_3(s) + 3H_2(g) \rightarrow W(s) + 3H_2O(g)$ **b.** H_2 **c.** H

49. a. oxidized
b. H is the oxidizing agent; Ag is the reducing agent.
c. $2Ag(s) + H_2S(s) \rightarrow Ag_2S(s) + H_2(g)$

50. Lead atoms are oxidized by losing 2 electrons to form Pb^{2+} ions. Oxygen atoms are reduced by gaining 2 electrons to form O^{2-} ions.

CHAPTER 20

Assessment continued

Understanding Concepts

41. Balance the equations in Problem 37 by an appropriate method.

42. Balance the equations in Problem 39 by an appropriate method.

43. Determine the oxidation number of phosphorus in each substance.

a. P_4O_8 **b.** PO_4^{3-} **c.** P_2O_5
d. P_4O_6 **e.** $H_2PO_4^-$ **f.** PO_3^{3-}

44. What is the oxidation number for chromium in each of the compounds shown below?

45. Identify the element oxidized, the element reduced, the oxidizing agent, and the reducing agent in each unbalanced redox equation.

a. $MnO_2(s) + HCl(aq) \longrightarrow MnCl_2(aq) + Cl_2(g) + H_2O(l)$
b. $Cu(s) + HNO_3(aq) \longrightarrow Cu(NO_3)_2(aq) + NO_2(g) + H_2O(l)$
c. $P(s) + HNO_3(aq) + H_2O(l) \longrightarrow NO(g) + H_3PO_4(aq)$
d. $Bi(OH)_3(s) + Na_2SnO_2(aq) \longrightarrow Bi(s) + Na_2SnO_3(aq) + H_2O(l)$

46. Balance each redox equation in Problem 45 by using the oxidation-number-change method.

47. A device called a Breathalyzer is used to test a person's breath for the alcohol ethanol, C_2H_5OH. In this test ethanol reacts with an acidic solution of orange dichromate ion to form green chromium(III) ion.

$$Cr_2O_7^{2-}(aq) + C_2H_5OH(aq) \longrightarrow Cr^{3+}(aq) + CO_2(g)$$

The amount of color change is proportional to the amount of ethanol in the exhaled breath.

a. Balance this equation by the half-reaction method.
b. Is dichromate ion an oxidizing agent or a reducing agent?

48. The metallic element tungsten, used as a filament in incandescent light bulbs, is obtained by heating tungsten(VI) oxide with hydrogen.

$$WO_3(s) + H_2(g) \longrightarrow W(s) + H_2O(g)$$

a. Balance the equation.
b. What is the reducing agent in this reaction?
c. Which element undergoes an increase in oxidation number?

49. Silver tarnishes when it reacts with hydrogen sulfide in the air.

$$Ag(s) + H_2S(g) \longrightarrow Ag_2S(s) + H_2(g)$$

a. Is silver oxidized or reduced in this reaction?
b. Identify the oxidizing agent and the reducing agent.
c. Balance the equation.

50. The following equation represents an oxidation-reduction reaction that uses oxygen. Show how this reaction can also be defined as an oxidation-reduction reaction in terms of electron transfer.

$$Pb(s) + O_2(g) \longrightarrow PbO(s)$$

51. Does each of the following equations represent a redox reaction? Explain how you know.

a. $Bi_2O_3(s) + 3C(s) \longrightarrow 2Bi(s) + 3CO(g)$
b. $Cr_2O_3(s) + 3H_2S(g) \longrightarrow Cr_2S_3(s) + 3H_2O(l)$
c. $BCl_3(g) + 3H_2O(l) \longrightarrow H_3BO_3(s) + 3HCl(g)$

52. The following unbalanced equation represents a reaction that can occur when dinitrogen tetroxide, N_2O_4, is combined with hydrazine, N_2H_4.

$$N_2O_4(l) + N_2H_4(l) \longrightarrow N_2(g) + H_2O(g)$$

Balance the equation and describe in words the electron transfer that takes place.

53. Examine the following hypothetical redox equation.

$$2X + 3H_2Y \longrightarrow X_2Y_3 + 3H_2$$

a. What is the oxidation number of element X on each side of the equation?
b. What is the oxidation number of element Y on each side of the equation?
c. What is oxidized in this equation?
d. What is reduced in this equation?

51. a. Yes, the oxidation number of bismuth changes from +3 to zero; the oxidation number of carbon changes from zero to +2.
b. No, there is no change in oxidation number of any of the atoms in this reaction.
c. No, there is no change in oxidation number of any of the atoms in this reaction.

52. $N_2O_4(l) + 2N_2H_4(l) \rightarrow 3N_2(g) + 4H_2O(g)$ Based on oxidation number changes, nitrogen atoms from N_2O_4 gain electrons in forming N_2 and nitrogen atoms from N_2H_4 lose electrons in forming N_2.

53. a. reactant, 0; product, +3
b. reactant, –2; product, –2
c. X
d. H

Critical Thinking

54. Explain why the number of electrons lost must equal the number of electrons gained in every redox reaction.

55. The highest possible oxidation number that chlorine exhibits in any compound is +7, whereas its most negative oxidation number is −1. Write the electron configuration of chlorine and explain why these are the limiting oxidation numbers for chlorine.

56. Explain why a sodium atom is a reducing agent but a sodium ion is not.

57. Many decomposition, single-replacement, combination, and combustion reactions are also redox reactions. Why is a double-replacement reaction never a redox reaction?

58. Why must every redox reaction have a reducing agent and an oxidizing agent?

59. Which substance in each pair is more likely to be an oxidizing agent?
- **a.** S^{2-} or SO_4^{2-}
- **b.** H_2O or H_2O_2
- **c.** NO_2^- or NO_3^-
- **d.** $Cr_2O_7^{2-}$ or Cr^{3+}
- **e.** H_2 or H_2O

60. Which is more likely to be a strong reducing agent, a group 1A metal or a group 7A nonmetal? Explain.

61. Predict the product(s) and write the balanced equation for each of these redox reactions. Identify the oxidizing agent in each reaction.
- **a.** rubidium + iodine $\longrightarrow$
- **b.** barium + water $\longrightarrow$
- **c.** aluminum + iron(II) sulfate $\longrightarrow$
- **d.** butene (C_4H_8) + oxygen $\longrightarrow$
- **e.** zinc + hydrobromic acid $\longrightarrow$
- **f.** magnesium + bromine $\longrightarrow$

62. The electronegativity of rhenium, Re, is 1.9, and the electronegativity of selenium is 2.4. If rhenium were to react with selenium to form a compound, which element would be oxidized and which element would be reduced? Explain.

63. Which of the following ions is most likely to be an oxidizing agent? Explain your choice.

$$MnO_4^-, MnO_4^{2-}, Mn^{+2}$$

Concept Challenge

64. How many grams of copper are needed to reduce completely the silver ions in 85.0 mL of 0.150*M* $AgNO_3(aq)$ solution?

65. How many milliliters of 0.280*M* $K_2Cr_2O_7(aq)$ solution are needed to oxidize 1.40 g of sulfur? First balance the equation.

$$K_2Cr_2O_7(aq) + H_2O(l) + S(s) \longrightarrow SO_2(g) + KOH(aq) + Cr_2O_3(aq)$$

66. Carbon monoxide can be removed from the air by passing it over solid diiodine pentoxide.

$$CO(g) + I_2O_5(s) \longrightarrow I_2(s) + CO_2(g)$$

- **a.** Balance the equation.
- **b.** Identify the element being oxidized and the element being reduced.
- **c.** How many grams of carbon monoxide can be removed from the air by 0.55 g of diiodine pentoxide (I_2O_5)?

67. What is the oxidation number of nitrogen in each of theses species?

a. HNO_3		**b.** NH_3	
c. N_2O_3		**d.** NO_2^-	
e. N_2O		**f.** NH_4Cl	
g. NO		**h.** NO_2	

68. The elements fluorine and oxygen can react to form fluorine monoxide, F_2O. Write the balanced chemical equation for this reaction. Check electronegativity values and then identify the elements oxidized and reduced.

69. The oxidation number of nitrogen can range from a minimum of −3 to a maximum of +5. Use this information to explain why the nitride ion, N^{3-}, can act only as a reducing agent, and why the nitrate ion, NO_3^-, can act only as an oxidizing agent.

70. Combine each of the following pairs of half-reactions into a complete, balanced, ionic redox equation.
- **a.** $Hg^{2+} + 2e^- \longrightarrow Hg$
  $Al \longrightarrow Al^{3+} + 3e^-$
- **b.** $MnO_2 + 4H^+ + 2e^- \longrightarrow Mn^{2+} + 2H_2O$
  $Fe \longrightarrow Fe^{2+} + 2e^-$
- **c.** $Fe^{3+} + e^- \longrightarrow Fe^{2+}$
  $Cd \longrightarrow Cd^{2+} + 2e^-$

Critical Thinking

54. When one reactant loses electrons another reactant must gain them.

55. $1s^2 2s^2 2p^6 3s^2 3p^5$
A chlorine atom can "lose" its seven valence electrons or it can gain one electron to fill the last 3*p* orbital.

56. A sodium atom achieves a stable electron configuration by losing its one valence electron, making it a reducing agent. A sodium ion has a stable electron configuration.

57. Double-replacement reactions never involve the transfer of electrons; instead they involve the transfer of ions in aqueous solution.

58. In every redox reaction one species loses one or more electrons and is a reducing agent. Another substance gains one or more electrons and is an oxidizing agent.

59. **a.** SO_4^{2-} **b.** H_2O_2 **c.** NO_3^- **d.** $Cr_2O_7^{2-}$ **e.** H_2O

60. Group 1A metals; a reducing agent must give up electrons. Group 1A metals achieve stability by giving up electrons and forming positively charged ions.

61. **a.** $Rb(s) + I_2(s) \rightarrow RbI_2(s)$; oxidizing agent is I
b. $Ba(s) + 2H_2O(l) \rightarrow Ba(OH)_2(aq) + H_2(g)$; oxidizing agent is H
c. $2Al(s) + 3FeSO_4(aq) \rightarrow Al_2(SO_4)_3(aq) + 3Fe(s)$; oxidizing agent is Fe
d. $C_4H_8(g) + 6O_2(g) \rightarrow 4CO_2(g) + 4H_2O(l)$; oxidizing agent is O
e. $Zn(s) + 2HBr(aq) \rightarrow ZnBr_2(aq) + H_2(g)$; oxidizing agent is H
f. $Mg(s) + Br_2(l) \rightarrow MgBr_2(s)$, oxidizing agent is Br

62. Rhenium is oxidized and selenium is reduced, because the bonded electrons would shift toward the more electronegative element, selenium.

63. MnO_4^-, because the manganese is at its highest oxidation state.

Concept Challenge

64. 0.406 g Cu

65. 104 mL $K_2Cr_2O_7$

66. **a.** $5CO + I_2O_5 \rightarrow I_2 + 5CO_2$ **b.** C is oxidized. I is reduced. **c.** 0.22 g CO

67. **a.** +5 **b.** −3 **c.** +3 **d.** +3 **e.** +1 **f.** −3 **g.** +2 **h.** +4

68. $2F_2 + O_2 \rightarrow 2F_2O$; fluorine is reduced; oxygen is oxidized

69. Nitride ion has the minimum oxidation number of −3, therefore it cannot gain additional electrons and be an oxidizing agent. It can lose electrons, however, and be a reducing agent. Nitrogen in the nitrate ion has the maximum oxidation number of +5, therefore it cannot lose additional electrons and enable NO_3^- be a reducing agent. It can gain electrons, however, and be an oxidizing agent.

70. **a.** $3Hg^{2+} + 2Al \rightarrow 3\ Hg + 2Al^{3+}$
b. $MnO_2 + 4H^+ + Fe \rightarrow Mn^{2+} + 2H_2O + Fe^{2+}$
c. $2Fe^{3+} + Cd \rightarrow 2Fe^{2+} + Cd^{2+}$

Cumulative Review

71. a.

Alkane burned	O_2 used (mol)	CO_2 produced (mol)	H_2O produced (mol)
CH_4	2	1	2
C_2H_6	3.5	2	3
C_3H_8	5	3	4
C_4H_{10}	6.5	4	5
C_5H_{12}	8	5	6
C_6H_{14}	9.5	6	7

b. $C_xH_y + [x + (y/4)]O_2 \rightarrow xCO_2 + (y/2)H_2O$

72. sublimation

73. 1.8×10^2 kPa

74. suspension; particles settle out of a suspension; not a solution

75. a, c, and d

76. Dilute 110 mL of 6.0*M* HCl to 440 mL total volume.

77. $Ca(NO_3)_2$; boiling point elevation is a colligative property that depends on the number of particles in solution. $Ca(NO_3)_2$ gives three particles per formula unit; LiF gives two particles per formula unit.

78. 0.379*M* H_3PO_4

79. Solubility $PbBr_2 = 8.1 \times 10^{-3}M$

80. Test tube B has the NaCl added to it. Due to the common ion effect, the addition of either sulfate ion or barium ion to a saturated solution of $BaSO_4$ will cause the solubility product of $BaSO_4$ to be exceeded and barium sulfate will precipitate as shown in test tubes A and C.

81. a. $1.0 \times 10^{-2}M$ **b.** $1.0 \times 10^{-11}M$ **c.** $1.6 \times 10^{-9}M$

82. 56.3 mL KOH

83. a. NH_4^+ and NH_3; H_2O and H_3O^+ **b.** H_2SO_3 and HSO_3^-; NH_2^- and NH_3 **c.** HNO_3 and NO_3^-; I^- and HI

84. a. acidic **b.** basic **c.** basic **d.** acidic

85. a. 5.00 **b.** 10.00 **c.** 13.00 **d.** 6.52

86. large

CHAPTER 20 Assessment continued

Cumulative Review

71. The complete combustion of hydrocarbons involves the oxidation of both carbon and hydrogen atoms by oxygen. The following table lists the moles of O_2 used and the moles of CO_2 and moles of H_2O produced when a series of hydrocarbons called alkanes are burned. *(Chapter 12)*

Alkane burned	O_2 used (mol)	CO_2 produced (mol)	H_2O produced (mol)
CH_4	2	1	2
C_2H_6	3.5	2	3
C_3H_8	5	3	4
C_4H_{10}	a. ___	b. ___	c. ___
C_5H_{12}	d. ___	e. ___	f. ___
C_6H_{14}	g. ___	h. ___	i. ___

a. Complete the table.

b. Based on the data, write a balanced generalized equation for the complete oxidation of any alkane. Use the following form, and write the coefficients in terms of *x* and *y*:

$$C_xH_y + ___ O_2 \longrightarrow ___ CO_2 + ___ H_2O$$

72. Name a change of state that does not involve a liquid. *(Chapter 13)*

73. A gas cylinder has a volume of 6.8 L and is filled with 13.8 g of N_2. Calculate the pressure of N_2 at 25°C. *(Chapter 14)*

74. A particular paint must be stirred before using. Is the stirred paint a solution or a suspension? Explain. *(Chapter 15)*

75. Which of these are nonelectrolytes? *(Chapter 15)*

a. $S(s)$

b. $NH_4Cl(aq)$

c. $SiO_2(s)$

d. $F_2(g)$

76. How would you make 440 mL of 1.5*M* HCl solution from a stock solution of 6.0*M* HCl? *(Chapter 16)*

77. One mole of LiF and $Ca(NO_3)_2$ are each dissolved in 1.0 L of water. Which solution has the higher boiling point? Explain. *(Chapter 16)*

78. What is the molarity of the solution prepared by dissolving 46.4 g H_3PO_4 in enough water to make 1.25 L of solution? *(Chapter 16)*

79. The K_{sp} of lead(II) bromide ($PbBr_2$) at 25°C is 2.1×10^{-6}. What is the solubility of $PbBr_2$ (in mol/L) at this temperature? *(Chapter 18)*

80. Bottles containing 0.1*M* solutions of Na_2SO_4, $BaCl_2$, and NaCl have had their labels accidentally switched. To discover which bottle contains the NaCl, you place a clear saturated solution of $BaSO_4$ ($K_{sp} = 1.1 \times 10^{-10}$) into three test tubes. To each test tube you add a few drops of each mislabeled solution. The results are shown below. To which tube was NaCl added? Explain. *(Chapter 18)*

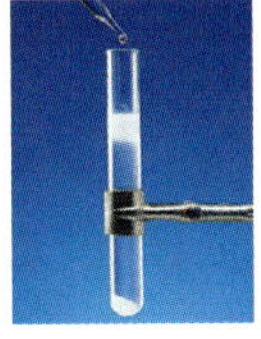
Test tube A

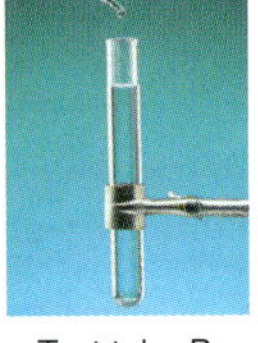
Test tube B

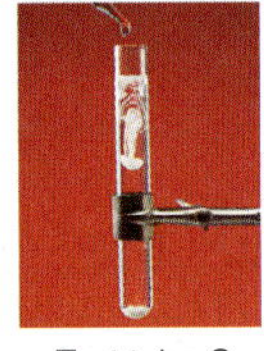
Test tube C

81. What is the hydrogen-ion concentration of solutions with the following pH? *(Chapter 19)*

a. 2.00 **b.** 11.00 **c.** 8.80

82. How many milliliters of a 4.00*M* KOH solution are needed to neutralize 45.0 mL of 2.50*M* H_2SO_4 solution? *(Chapter 19)*

83. Identify the conjugate acid-base pairs in each equation. *(Chapter 19)*

a. $NH_4^+(aq) + H_2O(l) \longrightarrow NH_3(aq) + H_3O^+(aq)$

b. $H_2SO_3(aq) + NH_2^-(aq) \longrightarrow HSO_3^-(aq) + NH_3(aq)$

c. $HNO_3(aq) + I^-(aq) \longrightarrow HI(aq) + NO_3^-(aq)$

84. Classify each solution in Problem 85 as acidic, basic, or neutral. *(Chapter 19)*

85. Calculate the pH of solutions with the following hydrogen-ion or hydroxide-ion concentrations. *(Chapter 19)*

a. $[H^+] = 0.000\ 010M$

b. $[OH^-] = 1.0 \times 10^{-4}M$

c. $[OH^-] = 1.0 \times 10^{-1}M$

d. $[H^+] = 3.0 \times 10^{-7}M$

86. A reaction goes essentially to completion. Would you expect the value of *K* to be large or small? *(Chapter 19)*

660 Chapter 20

Standardized Test Prep

Test-Taking Tip

Interpreting Data Tables Tables present a large amount of data in a small space. They allow you to make comparisons and to analyze the information present. To interpret the content in a table, start by reading the title (if there is one). Then read the headings. Try to figure out the relationship between the different columns and rows of information. Ask yourself some questions: *What information is related in the table? How are the relationships represented?*

Select the choice that best answers each question or completes each statement.

1. Which of these processes is not an oxidation?
a. a decrease in oxidation number
b. a complete loss of electrons
c. a gain of oxygen
d. a loss of hydrogen by a covalent molecule

2. In which of these pairs of nitrogen-containing ions and compounds is the oxidation number of nitrogen in the ion higher than in the nitrogen compound?

I. N_2H_4 and NH_4^+
II. NO_3^- and N_2O_4
III. N_2O and NO_2^-

a. I only **b.** I and II only
c. I and III only **d.** II and III only
e. I, II, and III

3. Identify the elements oxidized and reduced in this reaction.

$$2ClO^- + H_2 + 2e^- \longrightarrow 2Cl^- + 2OH^-$$

a. Cl is oxidized; H is reduced
b. H is oxidized; Cl is reduced
c. Cl is oxidized; O is reduced
d. O is oxidized; Cl is reduced

4. Which of these half-reactions represents a reduction?

I. $Fe^{2+} \longrightarrow Fe^{3+}$
II. $Cr_2O_7^{2-} \longrightarrow Cr^{3+}$
III. $MnO_4^- \longrightarrow Mn^{2+}$

a. I and II only
b. II and III only
c. I and III only
d. I, II, and III

5. Which of these general types of reactions is not a redox reaction?
a. single replacement **b.** double replacement
c. combustion **d.** combination

6. What is the reducing agent in this reaction?

$$MnO_4^- + SO_2 \longrightarrow Mn^{2+} + SO_4^{2-}$$

a. SO_2 **b.** SO_4^{2-} **c.** Mn^{2+} **d.** MnO_4^-

Use the table to answer Questions 7–11.

Arrow	Group	Metal	Metal ion	Arrow
	3	K	K^+	
	3	Ca	Ca^{2+}	
	3	Na	Na^+	
	4	Mg	Mg^{2+}	
	4	Al	Al^{3+}	
	4	Zn	Zn^{2+}	
1	4	Fe	Fe^{2+}	2
	4	Ni	Ni^{2+}	
	4	Sn	Sn^{2+}	
	4	Pb	Pb^{2+}	
	5	Cu	Cu^{2+}	
	5	Ag	Ag^+	

7. Which arrow indicates increasing ease of oxidation? Of reduction?

8. Which numbered group of metals are the strongest reducing agents?

9. Which numbered group of metals are the most difficult to oxidize?

10. Which is a stronger oxidizing agent, Na or Fe?

11. Which is a stronger reducing agent, Mg or Ag?

Use this diagram to answer Questions 12 and 13. It shows the formation of an ion from an atom.

12. Does the diagram represent oxidation or reduction? Does the oxidation number increase or decrease when the ion forms?

13. Draw a diagram showing the formation of a sulfide ion from a sulfur atom. Make the relative sizes of the atom and ion realistic. Does your drawing represent an oxidation or a reduction?

Standardized Test Prep

1. a
2. d
3. b
4. b
5. b
6. a
7. arrow 1; arrow 2
8. group 3
9. group 5
10. iron
11. magnesium
12. oxidation; oxidation number increases
13. sulfide ion should be larger than sulfur atom; reduction

CHAPTER 21

Electrochemistry Planning Guide

SECTION OBJECTIVES	STANDARDS NATIONAL	STANDARDS STATE	ACTIVITIES and LABS
21.1 Electrochemical Cells, pp. 663–670 2 block or 4 periods **21.1.1 Interpret** an activity series and **identify** the elements that are most easily oxidized and those that are least easily oxidized. **21.1.2 Name** the types of reactions involved in elecrochemical processes. **21.1.3 Describe** how a voltaic cell produces electrical energy. **21.1.4 Describe** current technologies that use electrochemical processes to produce energy.	B-3, E-2		**SE** Inquiry Activity: A Lemon Battery, p. 662 L2 **TE** Teacher Demo: A Redox Reaction, p. 664 L2 **TE** Teacher Demo: Inside a Dry Cell, p. 667 L2 **TE** Teacher Demo: Making a Lead Cell, p. 668 L2 **LM** Lab 47: Corrosion L2 **SSLM** Lab 36: Small-Scale Voltaic Cells L2
21.2 Half-Cells and Cell Potentials, pp. 671–677 2 block or 4 periods **21.2.1 Identify** what causes the electrical potential of an electrochemical cell. **21.2.2 Define** the standard hydrogen electrode. **21.2.3 Describe** how the standard reduction potential of a half-cell is determined. **21.2.4 Interpret** the meaning of the sign of the standard cell potential.	A-1, A-2, B-3		**TE** Teacher Demo: The Corrosion of Iron, p. 673 L2 **TE** Class Activity: Determining Cell Voltages, p. 675 L2 **LM** Lab 47: Corrosion L2
21.3 Electrolytic Cells, pp. 678–685 1 block or 2 periods **21.3.1 Distinguish** between electrolytic and voltaic cells. **21.3.2 Identify** the products of the electrolysis of water. **21.3.3 Describe** chemical changes that take place during the electrolysis of brine. **21.3.4 Name** three ways that electrolysis is used in metal processing.	A-1, A-2, B-3, E-2		**SE** Small-Scale Lab: Electrolysis of Water, p. 684 L2 **TE** Teacher Demo: The Electrolysis of Water, p. 680 L2 **LM** Lab 48: Electrochemistry L2 **LP** Lab Practical 21-1: Electroplating L2

Ability Levels	
L1	For students who need additional help
L2	For all students
L3	For students who need to be challenged

Components

SE	Student Edition	**LM**	Laboratory Manual	**GCP**	Graphing Calculator Problems	**T**	Transparencies
TE	Teacher's Edition	**SSLM**	Small-Scale Chemistry Lab Manual	**TTT**	Test-Taking Tips with Transparencies, Grades 9–12 Science	**CA**	Chem Alive and Chemistry Field Trips DVD
GRSW	Guided Reading and Study Workbook	**LP**	Laboratory Practicals	**STP**	Standardized Test Preparation Workbook	**iT**	Interactive Textbook with ChemASAP
CTR	Core Teaching Resources	**PLM**	Probeware Lab Manual			**GO**	Go Online Internet Resources
SSG	Spanish Study Guide	**VCL**	Virtual Chemistry Labs				
		CTB	Computer Test Bank				

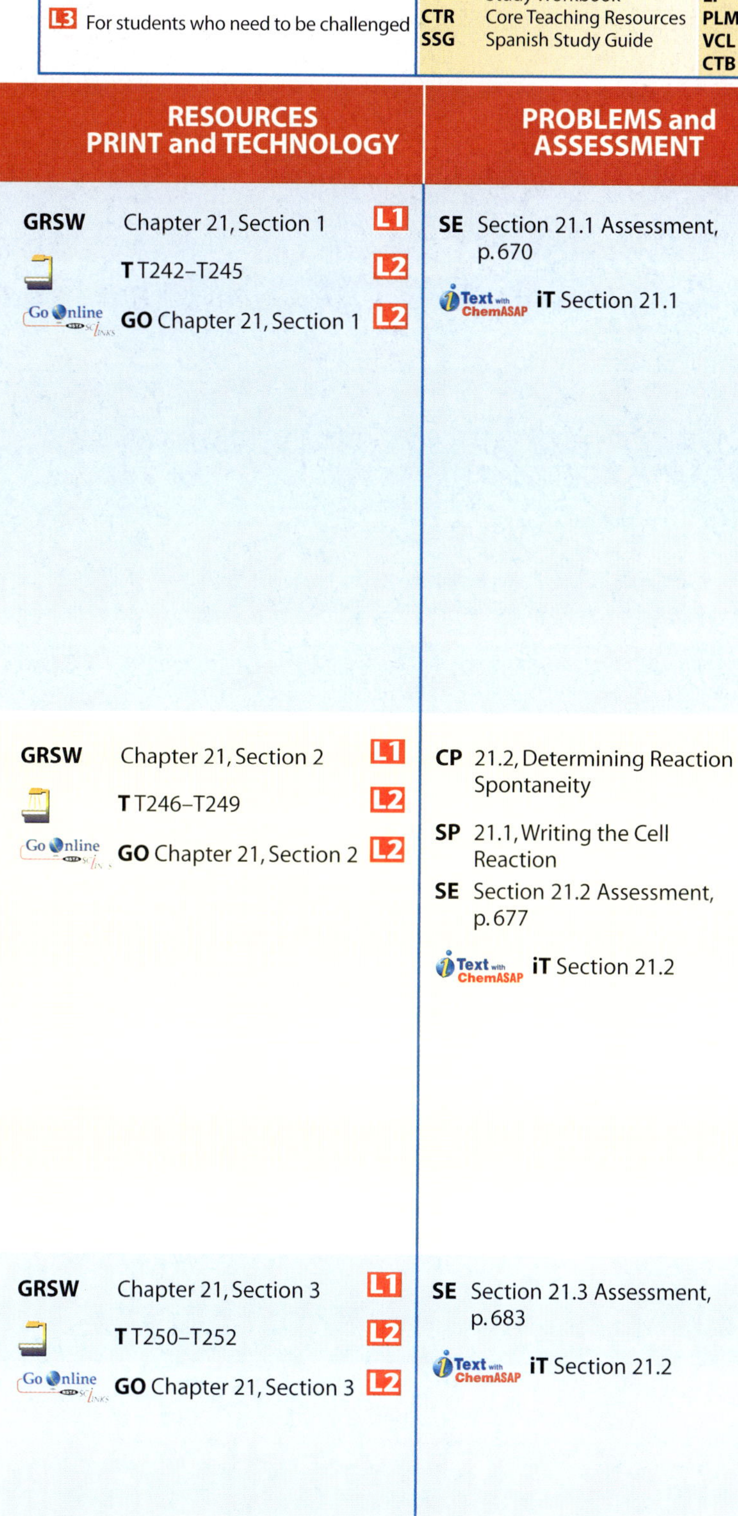

RESOURCES PRINT and TECHNOLOGY	PROBLEMS and ASSESSMENT
GRSW Chapter 21, Section 1 L1 **T** T242–T245 L2 **GO** Chapter 21, Section 1 L2	**SE** Section 21.1 Assessment, p. 670 **iT** Section 21.1
GRSW Chapter 21, Section 2 L1 **T** T246–T249 L2 **GO** Chapter 21, Section 2 L2	**CP** 21.2, Determining Reaction Spontaneity **SP** 21.1, Writing the Cell Reaction **SE** Section 21.2 Assessment, p. 677 **iT** Section 21.2
GRSW Chapter 21, Section 3 L1 **T** T250–T252 L2 **GO** Chapter 21, Section 3 L2	**SE** Section 21.3 Assessment, p. 683 **iT** Section 21.2

Web Code: cdn-1211
Web Code: cdn-1213

MATERIALS List

Quantities for each group

STUDENT EDITION

Inquiry Activity, p. 662
copper strip, zinc strip, fine sandpaper, a whole lemon, voltmeter

Quick Lab, p. 682
9-V battery, sodium sulfate solution, copper penny, nickel coin, iron nail, filter paper, aluminum foil, reaction surface

Small-Scale Lab, p. 684
reaction surface, electrolysis device, 0.2*M* Na_2SO_4, 0.1*M* KI, 0.2*M* $CuSO_4$, 0.04% BTB, 1.0*M* NaCl, 0.2*M* KBr, liquid starch

TEACHER'S EDITION

Teacher Demo, p. 664
200 mL of 0.1M silver nitrate ($AgNO_3$), 250 mL beaker, glass stirring rod, strip of polished copper metal

Teacher Demo, p. 667
Dry cell, voltmeter, 2 wire leads, zinc electrode, carbon electrode, manganese dioxide/ammonium chloride paste.

Teacher Demo, p. 668
2 strips of lead, wooden rod, 2 connecting wires, 250-mL beaker, dilute sulfuric acid (H_2SO_4), 6-V DC power supply, doorbell

Teacher Demo, p. 673
2 pennies, metal wires from twist ties, table salt, paper towel

Teacher Demo, p. 679
Hoffman apparatus or apparatus constructed using iron nails as electrodes, 2 test tubes, 6M NaOH, 250- or 500-mL beaker, 2 insulated wire leads

Chapter Assessment

CHAPTER ASSESSMENT

SE	Chapter Assessment, pp. 687–690
CTB	Chapter 21
iT	Chapter 21

STANDARDIZED TEST PREP

SE	Chapter 21, p. 691
STP	Topic 9
iT	Interactive Textbook with Assessment at PHSchool.com

CHAPTER 21

Electrochemistry

A Lemon Battery

L2

Purpose

Students experience the transfer of electrons.

Prep Time 20 minutes
Class Time 20 minutes

Safety

Be sure the edges of the metal strips are smooth. Have students rinse and dry the strips after use to remove saliva.

Think About It

1. a tingling sensation on the tongue
2. No, with two strips of the same metal, there was no tingling.
3. An orange is like a lemon, so an orange would probably produce tingling.
4. Any 12 V battery could start a car, but it probably wouldn't be rechargeable.

CHAPTER 21

Electrochemistry

An electrochemical process was used to produce the shiny chrome finish on this motorcycle.

INQUIRY Activity

A Lemon Battery

Materials

copper strip (1 cm × 3 cm), zinc strip (1 cm × 3 cm), fine sandpaper, a whole lemon, voltmeter

Procedure

1. Polish the copper and zinc strips with sandpaper.
2. Clean the strips with hot, soapy water and dry them.
3. Push the zinc strip into the lemon. Leave about 1.5 cm of the zinc strip sticking out of the lemon.
4. Push the copper strip into the lemon about 1 cm away from the zinc strip. It should not make contact with the zinc strip. Leave about 1.5 cm of the copper strip sticking out of the lemon.
5. Briefly touch both metal strips at the same time with your tongue and observe what happens. Do not share the lemon battery with another student.
6. Measure the voltage of your lemon battery.

Think About It

1. Describe the sensation you felt on your tongue.
2. Do you think you would feel the same sensation using two copper strips or two zinc strips? Try it!
3. What would you expect to happen if, in place of the lemon, you were to use an orange?
4. If you connected 12 lemon batteries together in series, you would generate a voltage of about 12 V. Could you use this assembly to start a car? Explain.

21.1 Electrochemical Cells

Connecting to Your World

On a summer evening, fireflies glow to attract their mates. In the ocean depths, anglerfish emit light to attract prey. Luminous shrimp, squid, jellyfish, and even bacteria also exist. These organisms, and others, are able to give off energy in the form of light as a result of redox reactions. In this section, you will discover that the transfer of electrons in a redox reaction produces energy.

Guide for Reading

Key Concepts

- For any two metals in an activity series, which metal is more readily oxidized?
- What type of chemical reaction is involved in all electrochemical processes?
- How does a voltaic cell produce electrical energy?
- What current technologies use electrochemical processes to produce electrical energy?

Vocabulary

electrochemical process
electrochemical cell
voltaic cells
half-cell
salt bridge
electrode
anode
cathode
dry cell
battery
fuel cells

Reading Strategy

Building Vocabulary As you read the section, write a definition of each vocabulary term in your own words.

Electrochemical Processes

As you have learned, chemical processes can either release energy or absorb energy. The energy can sometimes be in the form of electricity. Electrochemistry has many applications in the home as well as in industry. Flashlight and automobile batteries are familiar examples of devices used to generate electricity. The manufacture of sodium and aluminum metals and the silverplating of tableware involve the use of electricity. Biological systems also use electrochemistry to carry nerve impulses. In this chapter, you will learn about the relationship between redox reactions and electrochemistry.

A Spontaneous Redox Reaction When a strip of zinc metal is dipped into an aqueous solution of blue copper sulfate, the zinc becomes copper-plated, as shown in Figure 21.1. The net ionic equation involves only zinc and copper.

$$Zn(s) + Cu^{2+}(aq) \longrightarrow Zn^{2+}(aq) + Cu(s)$$

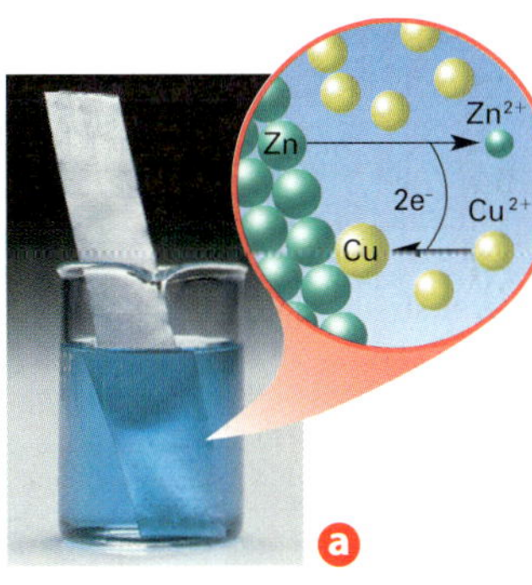

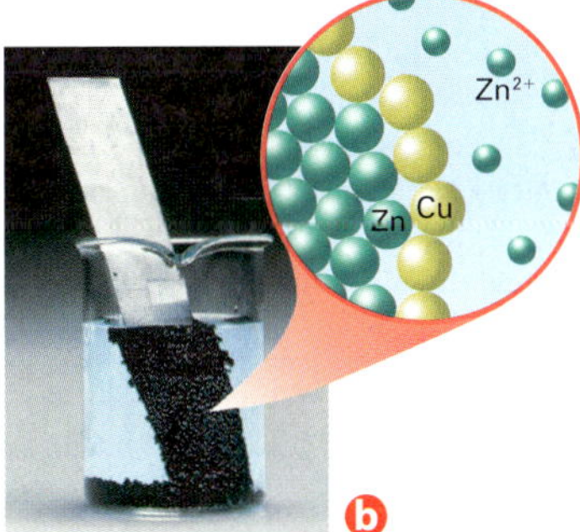

Figure 21.1 Zinc metal oxidizes spontaneously in a copper-ion solution. (a) A zinc strip is immersed in a solution of copper sulfate. (b) As the redox reaction proceeds, copper plates out onto the zinc, and the blue copper sulfate solution is replaced by a solution of zinc sulfate.
Classifying ***How can you classify this reaction?***

Section Resources

Print

- **Guided Reading and Study Workbook,** Section 21.1
- **Core Teaching Resources,** Section 21.1 Review
- **Transparencies,** T242–T245
- **Laboratory Manual,** Lab 47

Technology

- **Interactive Textbook with ChemASAP,** Assessment 21.1
- **GoOnline,** Section 21.1

21.1

1 FOCUS

Objectives

21.1.1 Interpret an activity series and **identify** the elements that are most easily oxidized and those that are least easily oxidized.

21.1.2 Name the type of reactions involved in electrochemical processes.

21.1.3 Describe how a voltaic cell produces electrical energy.

21.1.4 Describe current technologies that use electrochemical processes to produce energy.

Guide for Reading

Build Vocabulary L2

Word Parts Cathode comes from the Greek *kathodos* meaning way down. Anode also comes from the Greek *anodos* meaning way up. Thus the two words, cathode and anode, are opposites.

Reading Strategy L2

Outline Show students how to make an outline as they read this section. Have them use the red headings as their main entries and the blue headings as secondary entries. Ask them to include, under each heading, a sentence summarizing the important idea.

2 INSTRUCT

Connecting to Your World

Ask, **What have you learned that shows that energy can be released in the form of light?** *(When electrons drop from higher to lower energy levels in an atom they emit energy in the form of light.)*

Answers to...

Figure 21.1 It is a redox reaction. Zinc is oxidized, and copper is reduced.

Electrochemical Processes

Use Visuals L1

Figure 21.1 Ask, **What is the color of the copper sulfate solution in Figure 21.2a?** *(blue)* **Is the color different in Figure 21.1b?** *(It's still blue but lighter in color.)* **What is the significance of the lighter color?** *(Blue copper ions have been removed from the solution and changed into the copper atoms on the zinc.)*

A Redox Reaction 

Purpose Students observe a redox reaction.

Materials 200 mL 0.1*M* silver nitrate ($AgNO_3$) in a 250 mL beaker, glass stirring rod, strip of polished copper metal

Safety Wear gloves, apron, and safety goggles.

Procedure Put a glass stirring rod across the top of the beaker and suspend a polished copper strip from the rod so that the strip dips into the $AgNO_3$ solution. The metal surface will darken and appear "fuzzy" as silver metal is deposited. Over time, a layer of silver will form on the copper. Ask, **What is the significance of the blue color of the solution?** *(Blue copper(II) ions are being produced.)* Write on the board the net reaction for the electrochemical process:

$$2Ag^+(aq) + Cu(s) \rightarrow 2Ag(s) + Cu^{2+}(aq)$$

Point out that the reaction involves a transfer of electrons from copper atoms to silver ions. Ask, **What was oxidized? What was reduced?** *(Copper was oxidized; silver ions were reduced.)* Combine liquid wastes and add a 50% molar excess of NaCl. Filter or decant and dry the AgCl residue. Put in a plastic container and bury in an approved landfill. Flush the filtrate down the drain with excess water.

Expected Outcome Silver crystals form on the copper wire. The solution turns blue because of dissolved copper(II) ions.

For: Links on Activity Series
Visit: www.SciLinks.org
Web Code: cdn-1211

Table 21.1
Activity Series of Metals, with Half-Reactions for Oxidation Process

	Element	Oxidation half-reactions
Most active and most easily oxidized	Lithium	$Li(s) \longrightarrow Li^+(aq) + e^-$
	Potassium	$K(s) \longrightarrow K^+(aq) + e^-$
	Barium	$Ba(s) \longrightarrow Ba^{2+}(aq) + 2e^-$
	Calcium	$Ca(s) \longrightarrow Ca^{2+}(aq) + 2e^-$
	Sodium	$Na(s) \longrightarrow Na^+(aq) + e^-$
Decreasing activity ↓	Magnesium	$Mg(s) \longrightarrow Mg^{2+}(aq) + 2e^-$
	Aluminum	$Al(s) \longrightarrow Al^{3+}(aq) + 3e^-$
	Zinc	$Zn(s) \longrightarrow Zn^{2+}(aq) + 2e^-$
	Iron	$Fe(s) \longrightarrow Fe^{2+}(aq) + 2e^-$
	Nickel	$Ni(s) \longrightarrow Ni^{2+}(aq) + 2e^-$
	Tin	$Sn(s) \longrightarrow Sn^{2+}(aq) + 2e^-$
	Lead	$Pb(s) \longrightarrow Pb^{2+}(aq) + 2e^-$
	Hydrogen*	$H_2(g) \longrightarrow 2H^+(aq) + 2e^-$
	Copper	$Cu(s) \longrightarrow Cu^{2+}(aq) + 2e^-$
	Mercury	$Hg(s) \longrightarrow Hg^{2+}(aq) + 2e^-$
Least active and least easily oxidized	Silver	$Ag(s) \longrightarrow Ag^+(aq) + e^-$
	Gold	$Au(s) \longrightarrow Au^{3+}(aq) + 3e^-$

*Hydrogen is included for reference purposes.

Electrons are transferred from zinc atoms to copper ions. This is a redox reaction that occurs spontaneously. As the reaction proceeds, zinc atoms lose electrons as they are oxidized to zinc ions. The zinc metal slowly dissolves. At the same time, copper ions in solution gain the electrons lost by the zinc. They are reduced to copper atoms and are deposited as metallic copper. As the copper ions in solution are gradually replaced by zinc ions, the blue color of the solution fades. Balanced half-reactions for this redox reaction can be written as follows.

Oxidation: $Zn(s) \longrightarrow Zn^{2+}(aq) + 2e^-$

Reduction: $Cu^{2+}(aq) + 2e^- \longrightarrow Cu(s)$

If you look at the activity series of metals in Table 21.1, you will see that zinc is higher on the list than copper. **For any two metals in an activity series, the more active metal is the more readily oxidized.** As Figure 21.1 shows, zinc is more readily oxidized than copper because when dipped into a copper sulfate solution, zinc becomes plated with copper. In contrast, when a copper strip is dipped into a solution of zinc sulfate, the copper does not spontaneously become zinc-plated. This is because copper metal is not oxidized by zinc ions. When a zinc strip is dipped into a copper sulfate solution, electrons are transferred from zinc metal to copper ions.

 ***Why doesn't a copper strip that is dipped in zinc sulfate solution become spontaneously zinc-plated?***

Download a worksheet on **The Activity Series** for students to complete, and find additional teacher support from NSTA SciLinks.

Differentiated Instruction

Gifted and Talented 

Challenge students to research the electrochemical nature of the nervous system. Students may want to focus on different aspects of the system and combine their findings into a visual display and/or oral report. Possible areas to be addressed include the role of sodium and potassium ions, how signals are transmitted across synapses, what an EEG measures, and the value of squid for nervous system research.

Redox Reactions and Electrochemistry An **electrochemical process** is any conversion between chemical energy and electrical energy. **All electrochemical processes involve redox reactions.** If a redox reaction is to be used as a source of electrical energy, the two half-reactions must be physically separated. In the case of the zinc-metal–copper-ion reaction, the electrons released by zinc must pass through an external circuit to reach the copper ions if useful electrical energy is to be produced. In that situation, the system serves as an electrochemical cell. Alternatively, an electric current can be used to produce a chemical change. That system, too, serves as an electrochemical cell. An **electrochemical cell** is any device that converts chemical energy into electrical energy or electrical energy into chemical energy. Redox reactions occur in all electrochemical cells.

Voltaic Cells

In 1800, the Italian physicist Alessandro Volta built the first electrochemical cell that could be used to generate a direct (DC) electric current. Figure 21.2 shows a photograph of Volta's illustration of one of his early cells. **Voltaic cells** (named after their inventor) are electrochemical cells used to convert chemical energy into electrical energy. **Electrical energy is produced in a voltaic cell by spontaneous redox reactions within the cell.** You use a voltaic cell every time you turn on a flashlight or a battery-powered calculator.

Constructing a Voltaic Cell A **half-cell** is one part of a voltaic cell in which either oxidation or reduction occurs. A typical half-cell consists of a piece of metal immersed in a solution of its ions. Figure 21.3 on the following page shows a voltaic cell that makes use of the zinc–copper reaction. In this cell, one half-cell is a zinc rod immersed in a solution of zinc sulfate. The other half-cell is a copper rod immersed in a solution of copper sulfate.

The half-cells are connected by a **salt bridge**—a tube containing a strong electrolyte, often potassium sulfate (K_2SO_4). Salt bridges usually are made of agar, a gelatinous substance. A porous plate may be used instead of a salt bridge. The porous plate allows ions to pass from one half-cell to the other but prevents the solutions from mixing completely. A wire carries the electrons in the external circuit from the zinc rod to the copper rod. A voltmeter or light bulb can be connected in the circuit. The driving force of such a voltaic cell is the spontaneous redox reaction between zinc metal and copper(II) ions in solution.

The zinc and copper rods in this voltaic cell are the electrodes. An **electrode** is a conductor in a circuit that carries electrons to or from a substance other than a metal. The reaction at the electrode determines whether the electrode is labeled as an anode or a cathode. The electrode at which oxidation occurs is called the **anode.** Because electrons are produced at the anode, it is labeled the negative electrode. The electrode at which reduction occurs is called the **cathode.** Electrons are consumed at the cathode. As a result, the cathode is labeled the positive electrode. Neither electrode is really charged, however. All parts of the voltaic cell remain balanced in terms of charge at all times. The moving electrons balance any charge that might build up as oxidation and reduction occur.

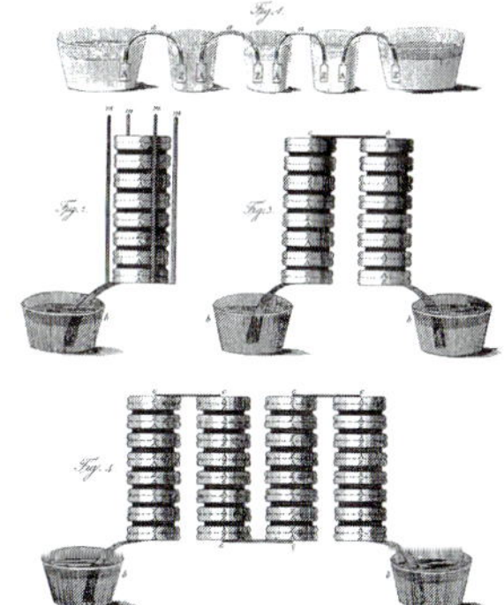

Figure 21.2 Volta built his electrochemical cell using piles of silver and zinc plates separated by cardboard soaked in salt water. He used his cell to obtain an electrical current.

Discuss 

Explain that redox reactions make it possible for energy interconversion between electrical energy and chemical energy. Review oxidation and reduction half-cell reactions. Stress that the electrons must be balanced. Show how these reactions are combined to form the net ionic equation for an electrochemical process such as the one just demonstrated. The net equation summarizes the transfer of electrons from the species being oxidized to the species being reduced.

Voltaic Cells

Discuss 

Remind students of the lemon battery they made in the Inquiry Activity. Explain that the lemon battery is an example of a voltaic cell. Use its parts to illustrate terms such as cathode, anode, salt bridge, and half-cell. Introduce the shorthand method for representing an electrochemical cell.

Relate 

Italian physicist Alesandro Vólta had already been experimenting with static electricity when, in 1780, his friend Luigi Galvani demonstrated that a current of electricity could be generated if two different metals were placed in contact with the muscle of a frog. Volta later showed that animal tissue was not necessary current flowed with only the two dissimilar metals. A controversy arose between scientists who believed in animal-generated electricity and those who believed in metallic electricity. Volta settled the matter when, in 1801, he demonstrated his first voltaic cell for Napoleon. The unit of electrical potential, the volt, was named in Volta's honor.

Differentiated Instruction

Special Needs 

Have students think about the half-reactions in Table 21.1 in terms of where the element is located in the periodic table. Remind them of what they learned in Chapter 6 about periodic trends in chemical properties. Point out that the elements on the left side of the periodic table are more easily oxidized (lose electrons) than the elements on the right side of the periodic table. Group 1A and 2A metals, listed at the top of the activity series, are some of the most reactive elements in nature. They readily form ions by giving up one or two electrons. Au, Ag, and Cu, represent some of the most stable elements in nature. They are not readily oxidized and so are listed at the bottom of the activity series.

Answers to...

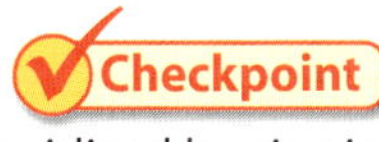

Checkpoint Copper is not oxidized by zinc ions.

Use Visuals L1

Figure 21.3 Explain that voltaic cells can be used as sources of electrical energy because the two half-reactions are physically separated. The reaction in the illustration could take place in a single beaker, but it would not be possible to produce a stream of electrons. **At which electrode does oxidation (loss of electrons) take place?** *(Oxidation occurs at the anode (negative electrode).)* **Where does reduction (gain of electrons) take place?** *(Reduction occurs at the cathode (positive electrode).)* **What path do the electrons given up by zinc follow?** *(They go through the wire and the electric light to the copper electrode.)* **What happens to the electrons at the copper electrode?** *(They reduce copper ions to copper.)*

Relate 

In discussing the need for a salt bridge in the Zn/Cu voltaic cell, explain that as zinc is oxidized at the anode, Zn^{2+} ions enter the solution. They have no negative ions to balance their charges, so a positive charge tends to build up around the anode. Similarly, at the cathode, Cu^{2+} ions are reduced to Cu and taken out of the solution leaving behind unbalanced negative ions. Thus a negative charge tends to develop around the cathode. The salt bridge allows negative ions, such as SO_4^{2-}, to be drawn to the anode compartment to balance the growing positive charge. Positive ions, such as K^+, are drawn from the salt bridge to balance the growing negative charge at the cathode.

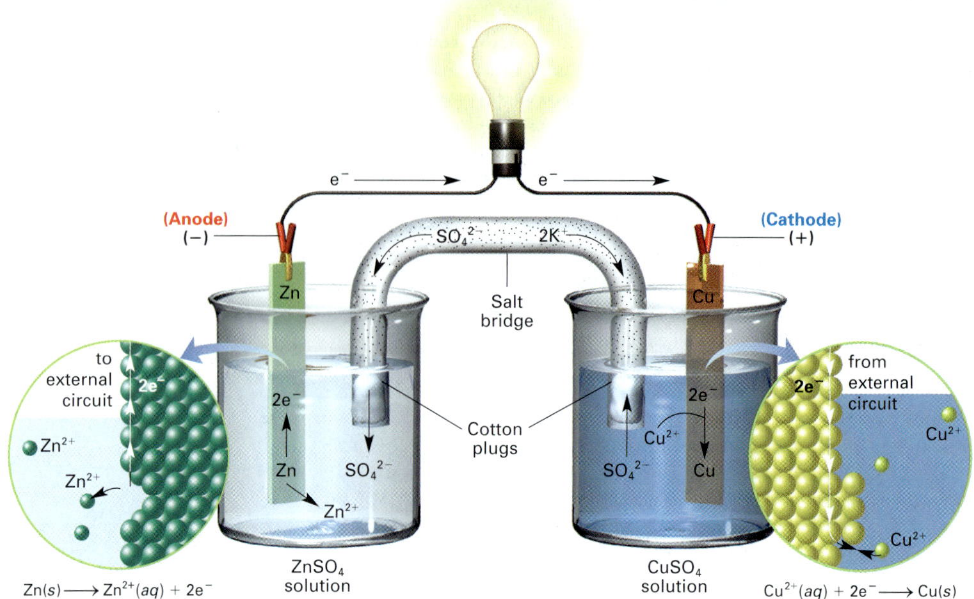

Figure 21.3 In this voltaic cell, the electrons generated from the oxidation of Zn to Zn^{2+} flow through the external circuit (the wire) into the copper strip. These electrons reduce the surrounding Cu^{2+} to Cu. To maintain neutrality in the electrolytes, anions flow through the salt bridge.

How a Voltaic Cell Works The electrochemical process that occurs in a zinc–copper voltaic cell can best be described in a number of steps. These steps actually occur at the same time.

1. Electrons are produced at the zinc rod according to the oxidation half-reaction.

$$Zn(s) \longrightarrow Zn^{2+}(aq) + 2e^-$$

Because it is oxidized, the zinc rod is the anode, or negative electrode.

2. The electrons leave the zinc anode and pass through the external circuit to the copper rod. (If a bulb is in the circuit, the electron flow will cause it to light. If a voltmeter is present, it will indicate a voltage.)
3. Electrons enter the copper rod and interact with copper ions in solution. There the following reduction half-reaction occurs.

$$Cu^{2+}(aq) + 2e^- \longrightarrow Cu(s)$$

Because copper ions are reduced at the copper rod, the copper rod is the cathode, or positive electrode, in the voltaic cell.

4. To complete the circuit, both positive and negative ions move through the aqueous solutions via the salt bridge. The two half-reactions can be summed to show the overall cell reaction. Note that the electrons in the overall reaction must cancel.

$$\begin{array}{l} Zn(s) \longrightarrow Zn^{2+}(aq) + 2e^- \\ Cu^{2+}(aq) + 2e^- \longrightarrow Cu(s) \\ \hline Zn(s) + Cu^{2+}(aq) \longrightarrow Zn^{2+}(aq) + Cu(s) \end{array}$$

When the zinc sulfate and copper(II) sulfate solutions in the voltaic half-cells are both 1.0*M*, the cell generates an electrical potential of 1.10 volts (V). If different metals are used for the electrodes or if different solution concentrations are used, the voltage will differ.

Differentiated Instruction

Less Proficient Readers 

Students may have difficulty remembering that oxidation occurs at the anode of a voltaic cell and reduction occurs at the cathode. Tell them that consonants go together and vowels go together. Both cathode and reduction begin with a consonant and anode and oxidation begin with a vowel.

Representing Electrochemical Cells You can represent the zinc–copper voltaic cell by using the following shorthand form.

$$Zn(s) \mid ZnSO_4(aq) \parallel CuSO_4(aq) \mid Cu(s)$$

The single vertical lines indicate boundaries of phases that are in contact. The zinc rod ($Zn(s)$) and the zinc sulfate solution ($ZnSO_4(aq)$), for example, are separate phases in physical contact. The double vertical lines represent the salt bridge or porous partition that separates the anode compartment from the cathode compartment. The half-cell that undergoes oxidation (the anode) is written first, to the left of the double vertical lines.

Which half-cell is written first in the shorthand representation of an electrochemical cell?

For: Links on Electrochemical Cells
Visit: www.SciLinks.org
Web Code: cdn-1210

Using Voltaic Cells as Energy Sources

Although the zinc-copper voltaic cell is of historical importance, it is no longer used commercially. **Current technologies that use electrochemical processes to produce electrical energy include dry cells, lead storage batteries, and fuel cells.**

Dry Cells When a compact, portable electrical energy source is required, a dry cell is usually chosen. A **dry cell** is a voltaic cell in which the electrolyte is a paste. A type of dry cell that is very familiar to you is the common flashlight battery, which, despite the name, is not a true battery. In such a dry cell, a zinc container is filled with a thick, moist electrolyte paste of manganese(IV) oxide (MnO_2), zinc chloride ($ZnCl_2$), ammonium chloride (NH_4Cl), and water (H_2O). As shown in Figure 21.4a, a graphite rod is embedded in the paste. The zinc container is the anode and the graphite rod is the cathode. The thick paste and its surrounding paper liner prevent the contents of the cell from freely mixing, so a salt bridge is not needed. The half-reactions for this cell are shown below.

Oxidation: $Zn(s) \longrightarrow Zn^{2+}(aq) + 2e^-$ (anode reaction)

Reduction: $2MnO_2(s) + 2NH_4^+(aq) + 2e^- \longrightarrow$
$Mn_2O_3(s) + 2NH_3(aq) + H_2O(l)$ (cathode reaction)

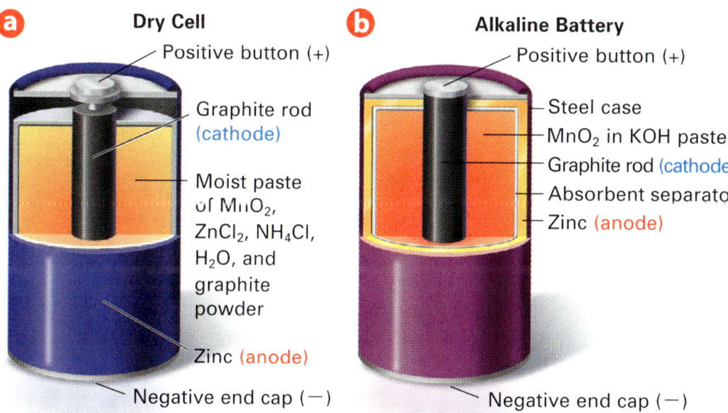

Figure 21.4 Both dry cells and alkaline batteries are single electrochemical cells that produce about 1.5 V. **a** The dry cell is inexpensive, has a short shelf life, and suffers from voltage drop when in use. **b** The alkaline battery costs more than the dry cell, has a longer shelf life, and does not suffer from voltage drop. **Inferring** ***What is oxidized in these cells and what is reduced?***

Using Voltaic Cells as Energy Sources

TEACHER Demo

Inside a Dry Cell

Purpose Students observe the construction of a dry cell.

Materials dry cell, voltmeter, 2 wire leads, zinc electrode, carbon electrode, manganese dioxide/ammonium chloride paste (Prepare the paste by adding saturated $NH_4Cl(aq)$ to powdered MnO_2 until thick.)

Procedure Before class, cut through a dry cell with a hacksaw from the top down on the right side of the central carbon electrode. In class, have students identify the three parts of the cell. *(The graphite rod is the cathode, the zinc container is the anode, and the manganese dioxide/ammonium chloride paste is the electrolyte.)* Construct a dry cell by placing a zinc electrode and a carbon electrode into a manganese dioxide/ammonium chloride paste. Connect the cell to a voltmeter and measure the potential.

Expected Outcome Students should be able to identify the anode, cathode, and electrolyte in each cell.

Discuss

Explain to students that, by itself, a dry cell does not provide a complete circuit, that is, electrons cannot flow from the anode to the cathode. When devices using dry cells are turned on, an external circuit is completed, allowing the flow of electrons from the anode to the cathode.

Use Visuals L1

Figure 21.4 Describe the difference between the common dry cell and the alkaline battery. *(In the alkaline dry cell, the electrolyte is a basic KOH paste.)*

Answers to...

Figure 21.4 In both the alkaline battery and the dry cell, zinc is oxidized and MnO_2 is reduced.

The half cell that undergoes oxidation is written first.

Section 21.1 (continued)

Use Visuals L1

Figure 21.5 Call attention to the grids, two containing Pb and two containing PbO_2. Ask, **What are the oxidation numbers of the metal lead (Pb) and the lead in lead(IV) oxide (PbO_2)?** *(oxidation number of elemental Pb is 0; oxidation number of Pb in PbO_2 is +4.)* **Which is the anode and which is the cathode?** *(Pb is the anode and PbO_2 is the cathode.)* **How do you know?** *(Pb can only lose electrons and be oxidized, and Pb^{4+} can gain electrons and be reduced.)* Write on the board the overall reaction for the battery when it is discharging. Call attention to the products ($PbSO_4$ and H_2O) and note that Pb is oxidized to form $PbSO_4$ and Pb^{4+} (in PbO_2) is reduced to also form $PbSO_4$. **What is the oxidation number of the lead in $PbSO_4$?** *(+2; Pb is oxidized to Pb^{2+} and Pb^{4+} is reduced to Pb^{2+}.)*

TEACHER Demo

Making a Lead Cell L2

Purpose Students observe a simple version of the lead storage battery.

Materials 2 strips of lead, wooden rod, 2 connecting wires, 250-mL beaker, dilute sulfuric acid (H_2SO_4), 6-V DC power supply, doorbell

Safety Sulfuric acid is corrosive. Wear safety goggles, gloves, and lab apron.

Procedure Attach two lead strips to a wooden rod so that the strips hang vertically in a 250-mL beaker. Place the strips 4 cm apart. Pour sufficient dilute sulfuric acid into the beaker to cover two-thirds of each strip. Connect a 6-V DC power supply to the strips with wires, and charge the battery for a few minutes. Then connect the cell to a doorbell.

Expected Outcome Students observe that after charging the battery, the bell rings. Discuss the half-cell reactions that occur. Ask students to write the shorthand notation for the electrochemical cell. *[Pb(s) I $PbSO_4$(s) II PbO_2(s) I $PbSO_4$(s)]*

In an ordinary dry cell, the graphite rod serves only as a conductor and does not undergo reduction, even though it is the cathode. The manganese in MnO_2 is the species that is actually reduced. The electrical potential of this cell starts out at 1.5 V but decreases steadily during use to about 0.8 V. Dry cells of this type are not rechargeable because the cathode reaction is not reversable.

The alkaline battery, shown in Figure 21.4b on the previous page, is an improved dry cell used for the same purposes. In the alkaline battery, the reactions are similar to those in the common dry cell, but the electrolyte is a basic KOH paste. This change in design eliminates the buildup of ammonia gas and maintains the Zn electrode, which corrodes more slowly under alkaline conditions.

Checkpoint ***What electrolyte is used in an alkaline battery?***

Lead Storage Batteries People depend on lead storage batteries to start their cars. A **battery** is a group of cells connected together. A 12-V car battery consists of six voltaic cells connected together. Each cell produces about 2 V and consists of lead grids, as shown in Figure 21.5. One set of grids, the anode, is packed with spongy lead. The other set, the cathode, is packed with lead(IV) oxide (PbO_2). The electrolyte for both half-cells in a lead storage battery is concentrated sulfuric acid. Using the same electrolyte for both half-cells allows the cell to operate without a salt bridge or porous separator. The half-reactions are as follows.

Oxidation: $Pb(s) + SO_4^{2-}(aq) \longrightarrow PbSO_4(s) + 2e^-$

Reduction: $PbO_2(s) + 4H^+(aq) + SO_4^{2-}(aq) + 2e^- \longrightarrow PbSO_4(s) + 2H_2O(l)$

When a lead storage battery discharges, it produces the electrical energy needed to start a car. The overall spontaneous redox reaction that occurs is the sum of the oxidation and reduction half-reactions.

$$Pb(s) + PbO_2(s) + 2H_2SO_4(aq) \longrightarrow 2PbSO_4(s) + 2H_2O(l)$$

This equation shows that lead sulfate forms during discharge. The sulfate slowly builds up on the plates, and the concentration of sulfuric acid decreases.

The reverse reaction occurs when a lead storage battery is recharged. This reaction occurs whenever the car's generator is working properly.

$$2PbSO_4(s) + 2H_2O(l) \longrightarrow Pb(s) + PbO_2(s) + 2H_2SO_4(aq)$$

This is not a spontaneous reaction. To make the reaction proceed as written, a direct current must pass through the cell in a direction opposite that of the current flow during discharge. In theory, a lead storage battery can be discharged and recharged indefinitely, but in practice its lifespan is limited. This is because small amounts of lead sulfate fall from the electrodes and collect on the bottom of the cell. Eventually, the electrodes lose so much lead sulfate that the recharging process is ineffective or the cell is shorted out. The battery must then be replaced. The processes that occur during the discharge and recharge of a lead-acid battery are summarized in Figure 21.6.

Figure 21.5 One cell of a 12-V lead storage battery is illustrated here. Current is produced when lead at the anode and lead(IV) oxide at the cathode are both converted to lead sulfate. These processes decrease the sulfuric acid concentration in the battery. Reversing the reaction recharges the battery.

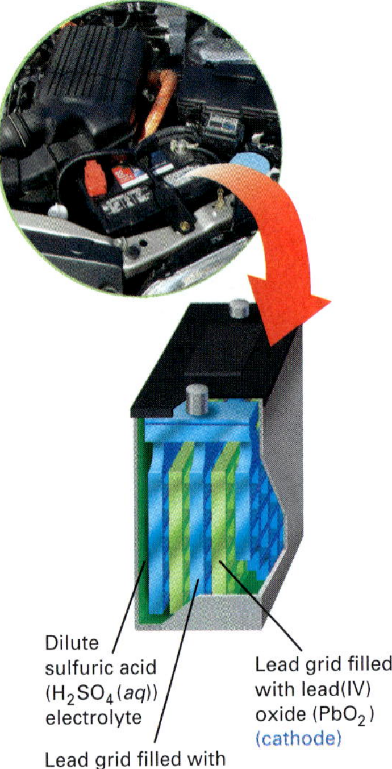

Facts and Figures

Car Battery

The forerunner of today's car battery was built by the French physicist Gaston Plant in 1859. Plant's first battery consisted of a single cell containing two sheets of lead separated by rubber strips and immersed in a 10% sulfuric acid solution. A year later he presented to Acadmie des Sciences a battery consisting of nine cells enclosed in box with outside terminals. Plant's invention was significant because it provided the first means of storing electricity.

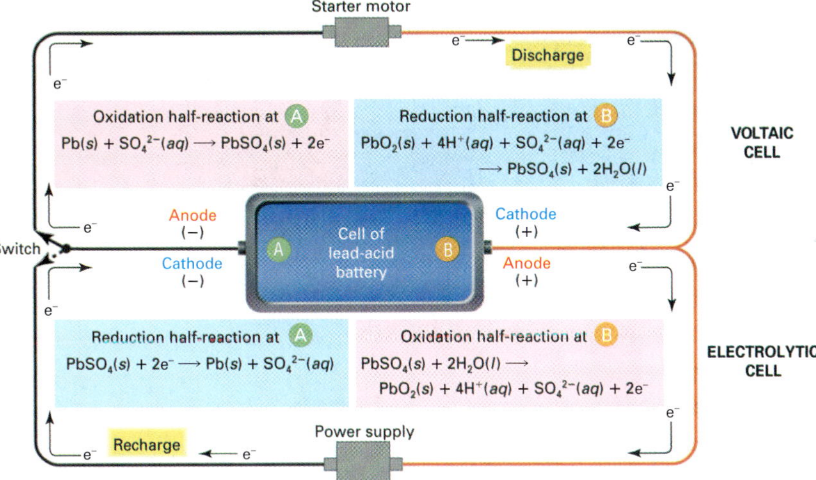

Figure 21.6 The lead-acid battery in an automobile acts as a voltaic cell (top) when it supplies current to start the engine. Some of the power from the running engine is used to recharge the battery which then acts as an electrolytic cell (bottom). You will learn more about electrolytic cells in Section 21.3.

Fuel Cells To overcome the disadvantages associated with lead storage batteries, cells with renewable electrodes have been developed. Such cells, called **fuel cells,** are voltaic cells in which a fuel substance undergoes oxidation and from which electrical energy is continuously obtained. Fuel cells do not have to be recharged. They can be designed to emit no air pollutants and to operate more quietly and more cost-effectively than a conventional electrical generator.

Perhaps the simplest fuel cell is the hydrogen–oxygen fuel cell, which is shown in Figure 21.7 on the following page. In this fuel cell, there are three compartments separated from one another by two electrodes made of porous carbon. Oxygen (the oxidizer) is fed into the cathode compartment. Hydrogen (the fuel) is fed into the anode compartment. The gases diffuse slowly through the electrodes. The electrolyte in the central compartment is a hot, concentrated solution of potassium hydroxide. Electrons from the oxidation half-reaction at the anode pass through an external circuit to enter the reduction half-reaction at the cathode.

Oxidation: $2H_2(g) + 4OH^-(aq) \longrightarrow 4H_2O(l) + 4e^-$ (anode)

Reduction: $O_2(g) + 2H_2O(l) + 4e^- \longrightarrow 4OH^-(aq)$ (cathode)

The overall reaction in the hydrogen–oxygen fuel cell is the oxidation of hydrogen to form water.

$$2H_2(g) + O_2(g) \longrightarrow 2H_2O(l)$$

Other fuels, such as methane (CH_4) and ammonia (NH_3), can be used in place of hydrogen. Other oxidizers, such as chlorine (Cl_2) and ozone (O_3), can be used in place of oxygen.

Discuss L2

Point out that fuel cells were developed for space travel where lightweight, reliable power systems are needed. Fuel cells differ from lead storage batteries in that they are not self-contained. Operation depends on a steady flow of fuel and oxygen into the cell—where combustion takes place—and the flow of the combustion product out of the cell. In the case of the hydrogen fuel cell, the product is pure water. Both the electricity generated and the water produced are consumed in space flights. Fuel cells convert 75% of the available energy into electricity, in contrast with a conventional electric power plant that converts from 35 to 40% of the energy of coal to electricity. Fuels other than hydrogen can also be used, for example, ammonia (NH_3), hydrazine (N_2H_4), and methane (CH_4). Students may be interested in the equations for the reactions that take place in these cells.

$4NH_3(g) + 3O_2(g) \rightarrow 2N_2(g) + 6H_2O(g)$

$2N_2H_4(g) + 2O_2(g) \rightarrow 2N_2(g) + 4H_2O(g)$

$CH_4(g) + 2O_2(g) \rightarrow CO_2(g) + 2H_2O(g)$

In each case, the products are gases and water vapor which are normally found in Earth's atmosphere.

Relate 

Divide the class into groups of three or four students. Have them research the use of fuel cells in the space shuttle and by utility companies. Ask them to prepare posters showing different types of fuel cells and their applications. Where possible, have students include the half-cell and overall reactions for each fuel cell. Have them describe the advantages gained by using a fuel cells rather than more conventional sources of electrical energy.

Differentiated Instruction

Gifted and Talented 

Electric cars are quiet and nonpolluting—no noisy engines spew noxious gases into the atmosphere. So why have car makers not built, promoted, and sold many electric vehicles? Have interested students research the current status of electric cars in the marketplace. Have them find out what obstacles stand in the way of mass usage of these cars. Students could choose to write a report or prepare an oral presentation. *(One obstacle that students may discover is the need for inexpensive, lightweight batteries for efficient storing of electrical energy to lengthen the distance of travel without recharging.)*

Answers to...

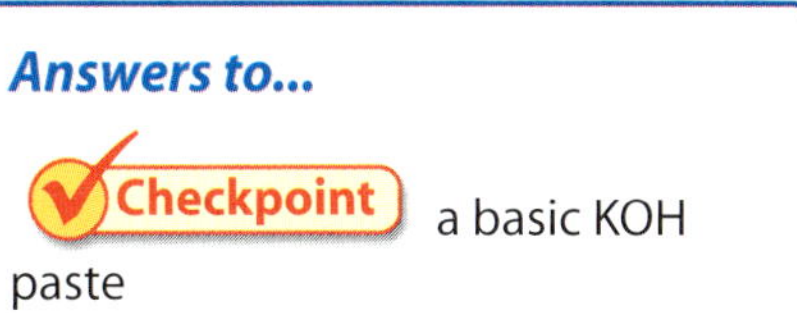

a basic KOH paste

Section 21.1 (continued)

3 ASSESS

Evaluate Understanding L2

Have students make a sketch of a tin/lead voltaic cell (*Sn| SnSO$_4$||PbSO$_4$|Pb*). Have them label the cathode and anode, and indicate the direction of electron flow. *(Tin is the anode; lead is the cathode. The electrons flow from tin to lead.)* Ask them to write the equations for the half-reactions.

$[Sn(s) \rightarrow Sn^{2+}(aq) + 2e^-$

$Pb^{2+}(aq) + 2e^- \rightarrow Pb(s)]$

Reteach L1

Emphasize that a chemical reaction can produce a flow of electrons or a flow of electrons can cause a chemical reaction to occur. Note that reduction always occurs at the cathode, and oxidation always occurs at the anode. Discuss the half-reactions for the charging process in a lead cell and compare them to the half-reactions when the cell is producing electric current.

Writing Activity

In a zinc–copper voltaic cell, zinc is oxidized and copper ions are reduced.
Anode: $Zn(s) \rightarrow Zn^{2+}(aq) + 2e^-$
Cathode: $Cu^{2+}(aq) + 2e^- \rightarrow Cu(s)$
A zinc rod serves as the anode; a copper rod serves as the cathode. The half-cells are connected by a salt bridge through which both positive and negative ions migrate. The overall reaction is as follows:
$Zn(s) + Cu^{2+}(aq) \rightarrow Zn^{2+}(aq) + Cu(s)$

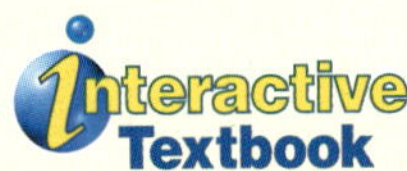

If your class subscribes to the Interactive textbook, use it to review key concepts in Section 21.1

with ChemASAP

Answers to...

Figure 21.7 Water is the only product.

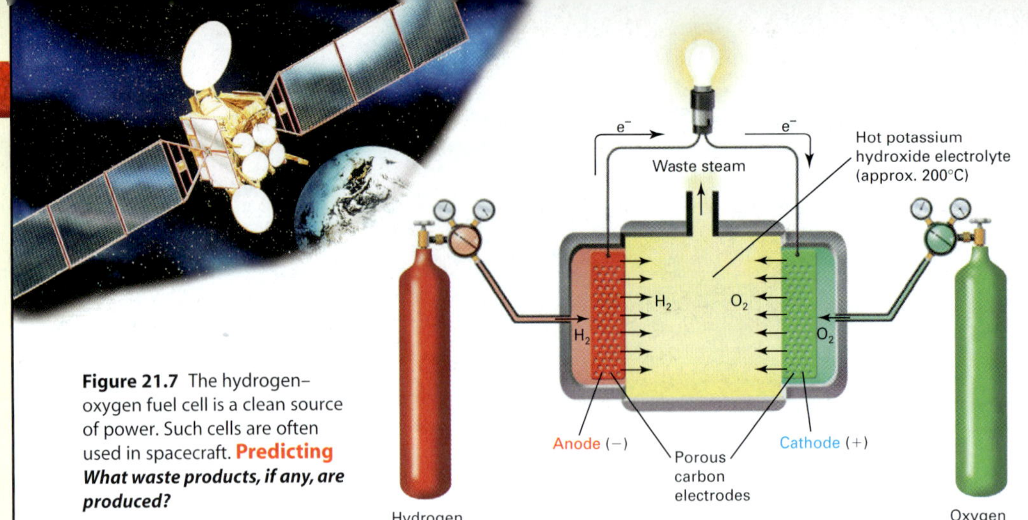

Figure 21.7 The hydrogen–oxygen fuel cell is a clean source of power. Such cells are often used in spacecraft. **Predicting** ***What waste products, if any, are produced?***

Since the 1960s, astronauts have used fuel cells as an energy source aboard spacecraft. Hydrogen–oxygen fuel cells with a mass of approximately 100 kg each were used in the Apollo spacecraft missions. Fuel cells are well suited for extended space missions because they offer a continuous energy source that releases no pollutants. On space shuttle missions, for example, astronauts drink the water produced by onboard hydrogen–oxygen fuel cells. Fuel cells are also used as auxiliary power sources for submarines and other military vehicles. At present, however, they are too expensive for general use.

21.1 Section Assessment

1. **Key Concept** If the relative activities of two metals are known, which metal is more easily oxidized?
2. **Key Concept** What type of reaction occurs during an electrochemical process?
3. **Key Concept** What is the source of electrical energy produced in a voltaic cell?
4. **Key Concept** What are three examples of technologies that use electrochemical processes to supply electrical energy?
5. Describe the structure of a dry cell. What substance is oxidized? What substance is reduced?
6. What is the electrolyte in a lead storage battery? Write the half-reactions for such a battery.
7. Write the overall reaction that takes place in a hydrogen–oxygen fuel cell. What product(s) are formed? Describe the half-reactions in this cell.
8. Predict the result when a strip of copper is dipped into a solution of iron(II) sulfate.

Writing Activity

Explanatory Paragraph Write a paragraph explaining how a zinc–copper voltaic cell works. Make sure to mention the half-reactions and the overall reaction in your explanation. (*Hint:* Use Figure 21.3 on page 666 as a reference.)

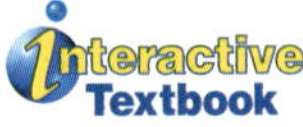

Assessment 21.1 Test yourself on the concepts in Section 21.1.

with ChemASAP

Section 21.1 Assessment

1. the metal with the higher activity
2. a redox reaction
3. spontaneous redox reactions within the cell
4. fuel cells, lead storage batteries, and dry cells
5. A zinc container (anode) filled with electrolyte paste; the cathode is a graphite rod embedded in the paste. Zn is oxidized; MnO_2 is reduced.
6. concentrated sulfuric acid;
Anode: $Pb(s) + SO_4^{2-}(aq) \rightarrow PbSO_4(s) + 2e^-$
Cathode:
$PbO_2(s) + 4H^+(aq) + SO_4^{2-}(aq) + 2e^- \rightarrow PbSO4(s) + 2H_2O(l)$
7. $2H_2(g) + O_2(g) \rightarrow 2H_2O(l)$; water is the product. H_2 is oxidized at the anode; O_2 is reduced at the cathode
8. no reaction

21.2 Half-Cells and Cell Potentials

Connecting to Your World

Batteries provide current to power lights and many kinds of electronic devices—such as the personal digital assistant shown here. One type of battery that you're probably familiar with is a car battery. The electrical potential between the negative and positive terminals of a car battery is 12 V. Other common batteries are the ones used to power flashlights, radios, and CD players. They are commonly known as AAA, AA, C, and D batteries. The electrical potential in these batteries is 1.5 V. In this section, you will learn how to calculate electrical potential.

Guide for Reading

Key Concepts

- What causes the electrical potential of an electrochemical cell?
- What value is assigned to the electrical potential of a standard hydrogen electrode?
- How can you determine the standard reduction potential of a half-cell?
- How can you interpret the cell potential of a redox reaction in terms of the spontaneity of the reaction?

Vocabulary

electrical potential
reduction potential
cell potential
standard cell potential
standard hydrogen electrode

Reading Strategy

Monitoring Your Understanding After you read the section, identify what you don't yet understand about half-cells and cell potentials. Reread the section or ask for help. Then write your clarified understanding.

Electrical Potential

The **electrical potential** of a voltaic cell is a measure of the cell's ability to produce an electric current. Electrical potential is usually measured in volts (V). The potential of an isolated half-cell cannot be measured. For example, you cannot measure the electrical potential of a zinc half-cell or of a copper half-cell separately. When these two half-cells are connected to form a voltaic cell, however, the difference in potential can be measured. The electrical potential of a 1*M* zinc–copper cell is +1.10 V.

The electrical potential of a cell results from a competition for electrons between two half-cells. The half-cell that has a greater tendency to acquire electrons will be the one in which reduction occurs. Oxidation occurs in the other half-cell as there is a loss of electrons. The tendency of a given half-reaction to occur as a reduction is called the **reduction potential.** The half-cell in which reduction occurs has a greater reduction potential than the half-cell in which oxidation occurs. The difference between the reduction potentials of the two half-cells is called the **cell potential.**

$$\text{cell potential} = \begin{pmatrix}\text{reduction potential}\\ \text{of half-cell in which}\\ \text{reduction occurs}\end{pmatrix} - \begin{pmatrix}\text{reduction potential}\\ \text{of half-cell in which}\\ \text{oxidation occurs}\end{pmatrix}$$

$$\text{or} \quad E^0_{cell} = E^0_{red} - E^0_{oxid}$$

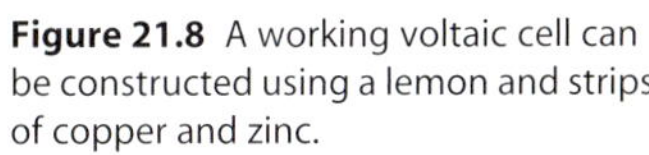
Figure 21.8 A working voltaic cell can be constructed using a lemon and strips of copper and zinc.

Section Resources

Print

- **Guided Reading and Study Workbook,** Section 21.2
- **Core Teaching Resources,** Section 21.2 Review
- **Transparencies,** T246–T249
- **Laboratory Manual,** Lab 47

Technology

- **Interactive Textbook with ChemASAP,** Simulation 27, Problem-Solving 21.9, 21.12, Assessment 21.2

21.2

1 FOCUS

Objectives

21.2.1 Identify what causes the electrical potential of an electrochemical cell.

21.2.2 Define the standard hydrogen electrode.

21.2.3 Describe how the standard reduction potential of a half-cell is determined.

21.2.4 Interpret the meaning of the sign of the standard cell potential.

Guide for Reading

Build Vocabulary 

Word Parts Potential comes from the Latin *potens* meaning "power." Electrical potential is the ability of a cell to produce an electric current. Ask students, **What does a weather forecast mean when it says that there is a potential for rain?** *(It could possibly rain.)*

Reading Strategy 

Summarize Read the first page of the text and, with the participation of the class, summarize the contents in a sentence or two. Ask students to write the summary in their notebooks. Encourage them to continue to summarize the principal ideas under each heading.

2 INSTRUCT

Connecting to Your World

Ask, **Which of the batteries mentioned is rechargeable?** *(the car battery)* **Do you use other rechargeable batteries?** *(Students may have experienced charging batteries in items such as cell phones, electric toothbrushes, and small vacuum cleaners.)*

Electrical Potential

Use Visuals 

Figure 21.8 Check student understanding of electrochemical, or voltaic, cells by ask students to identify the anode and cathode in the lemon battery. *(Zinc is the anode and copper is the cathode.)* **What process goes on at the anode?** *(oxidation)* **What process goes on the cathode?** *(reduction)* **In which process are electrons lost?** *(oxidation)* **What role does the lemon play in the battery?** *(It is the salt bridge.)*

Standard Cell Potential

Use Visuals 

Figure 21.9 Have students study the diagram. Ask, **What is the significance of the standard hydrogen electrode? How is it different from other electrodes?** *(It has been assigned a standard reduction potential of 0.00 V at 25°C.)* **What does the standard hydrogen half-cell consist of?** *(A platinum electrode with hydrogen gas at atmospheric pressure bubbling over its surface; the electrode is immersed in 1.00M acid solution at 25°C.)*

Discuss L2

Students may be confused by the arbitrary assignment of 0.00 V to the hydrogen half-cell. Use the Celsius temperature scale as an analogy. When the Celsius temperature scale was invented, 0°C was arbitrarily assigned to the temperature at which water freezes. All measured temperatures are relative to that assigned value. Temperatures higher than the freezing point of water are positive temperatures. Temperatures lower than 0°C are negative.

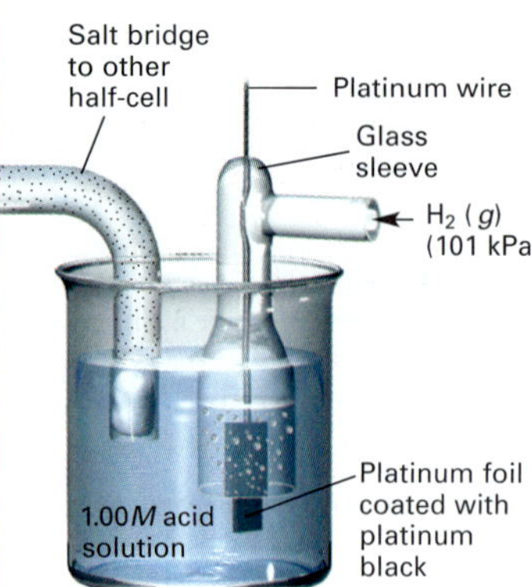

Figure 21.9 The standard hydrogen electrode is arbitrarily assigned a standard reduction potential of 0.00 V at 25°C.

Standard Cell Potential

The **standard cell potential** (E^0_{cell}) is the measured cell potential when the ion concentrations in the half-cells are 1*M*, any gases are at a pressure of 101 kPa, and the temperature is 25°C. The symbols E^0_{red} and E^0_{oxid} represent the standard reduction potentials for the reduction and oxidation half-cells, respectively. The relationship between these values follows the general relationship for cell potential given on the previous page.

$$E^0_{cell} = E^0_{red} - E^0_{oxid}$$

Because half-cell potentials cannot be measured directly, scientists have chosen an arbitrary electrode to serve as a reference. The **standard hydrogen electrode** is used with other electrodes so the reduction potentials of the other cells can be measured.  **The standard reduction potential of the hydrogen electrode has been assigned a value of 0.00 V.** The standard hydrogen electrode, which is illustrated in Figure 21.9, consists of a platinum electrode immersed in a solution with a hydrogen-ion concentration of 1*M*. The solution is at 25°C . The electrode itself is a small square of platinum foil coated with finely divided platinum, known as platinum black. Hydrogen gas at a pressure of 101 kPa is bubbled around the platinum electrode. The half-cell reaction that occurs at the platinum black surface is as follows.

$$2H^+(aq, 1M) + 2e^- \rightleftharpoons H_2(g, 101\text{ kPa}) \qquad E^0_{H^+} = 0.00\text{ V}$$

The double arrows in the equation indicate that the reaction is reversible. The symbol $E^0_{H^+}$ represents the standard reduction potential of H^+. The standard reduction potential of H^+ is the tendency of H^+ ions to acquire electrons and be reduced to $H_2(g)$. Whether this half-cell reaction occurs as a reduction or as an oxidation is determined by the reduction potential of the half-cell to which the standard hydrogen electrode is connected.

 Checkpoint ***What components and conditions are required for a standard hydrogen electrode?***

Standard Reduction Potentials

A voltaic cell can be made by connecting a standard hydrogen half-cell to a standard zinc half-cell, as shown in Figure 21.10. To determine the overall reaction for this cell, first identify the half-cell in which reduction takes place. In all electrochemical cells, reduction takes place at the cathode and oxidation takes place at the anode. A voltmeter gives a reading of +0.76 V when the zinc electrode is connected to the negative terminal and the hydrogen electrode is connected to the positive terminal. The zinc is oxidized, which means that it is the anode. Hydrogen ions are reduced, which means that the hydrogen electrode is the cathode. You can now write the half-reactions and the overall cell reaction.

Oxidation: $Zn(s) \longrightarrow Zn^{2+}(aq) + 2e^-$ (at anode)

Reduction: $2H^+(aq) + 2e^- \longrightarrow H_2(g)$ (at cathode)

Cell reaction: $Zn(s) + 2H^+(aq) \longrightarrow Zn^{2+}(aq) + H_2(g)$

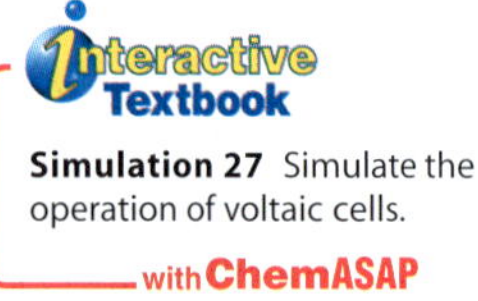

Simulation 27 Simulate the operation of voltaic cells.

with ChemASAP

Differentiated Instruction

Special Needs L1

Give students small cards with e^- printed on them in large letters. Divide students into pairs. Let one student role play being an anode and the other a cathode. The anode should hand one or more cards to the cathode

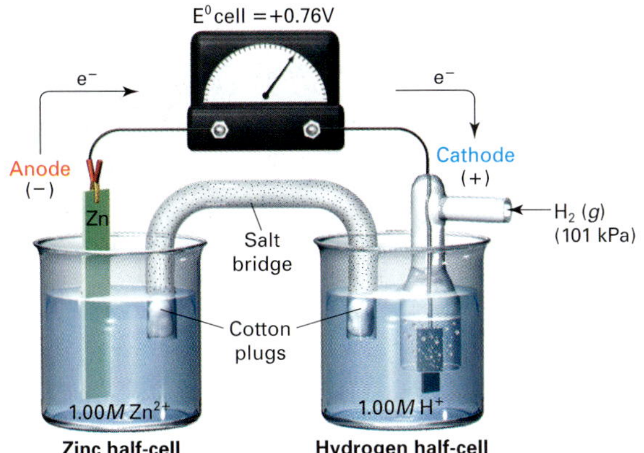

Figure 21.10 This voltaic cell consists of zinc and hydrogen half-cells. **Interpreting Diagrams** ***Where does reduction occur, and what is reduced in this cell?***

You can determine the standard reduction potential of a half-cell by using a standard hydrogen electrode and the equation for standard cell potential. In the zinc–hydrogen cell, zinc was oxidized and hydrogen ions were reduced. Let $E^0_{red} = E^0_{H^+}$ and $E^0_{oxid} = E^0_{Zn^{2+}}$ in the standard cell potential equation.

$$E^0_{cell} = E^0_{red} - E^0_{oxid}$$
$$E^0_{cell} = E^0_{H^+} - E^0_{Zn^{2+}}$$

The cell potential (E^0_{cell}) was measured at +0.76 V. The reduction potential of the hydrogen half-cell is a defined standard: $E^0_{H^+}$ always equals 0.00 V. Substituting these values into the preceding equation will give the standard reduction potential for the zinc half-cell.

$$+0.76\ V = 0.00\ V - E^0_{Zn^{2+}}$$
$$E^0_{Zn^{2+}} = -0.76\ V$$

The standard reduction potential for the zinc half-cell is −0.76 V. The value is negative because the tendency of zinc ions to be reduced to zinc metal in this cell is less than the tendency of hydrogen ions to be reduced to hydrogen gas (H_2). Consequently, the zinc ions are not reduced. Instead, the opposite occurs: Zinc metal is oxidized to zinc ions.

Many different half-cells can be paired with the hydrogen half-cell in a similar manner. Using this method, the standard reduction potential for each half-cell can be obtained. For a standard copper half-cell, for example, the measured standard cell potential is +0.34 V. Copper is the cathode, and Cu^{2+} ions are reduced to Cu metal when the cell operates. The hydrogen half-cell is the anode, and H_2 gas is oxidized to H^+ ions. You can calculate the standard reduction potential for copper as follows.

$$E^0_{cell} = E^0_{red} - E^0_{oxid}$$
$$E^0_{cell} = E^0_{Cu^{2+}} - E^0_{H^+}$$
$$+0.34\ V = E^0_{Cu^{2+}} - 0.00$$
$$E^0_{Cu^{2+}} = +0.34\ V$$

This value is positive because the tendency for copper ions to be reduced in the cell is greater than the tendency of hydrogen ions to be reduced.

Standard Reduction Potentials

The Corrosion of Iron 

Purpose Students observe the galvanic corrosion of iron compared to copper.

Materials 2 pennies, metal wires from twist ties, table salt, paper towel

Procedure Thoroughly clean the pennies and the wires. Moisten a paper towel with tap water and sprinkle it with salt. Wrap one wire around a penny and place it on the paper towel. Place an unwrapped penny and a coil of wire on the paper towel as a control. Observe what happens after at least 24 hours.

Expected Outcomes Iron corroded; copper did not. Ask students to explain. *(In galvanic corrosion, the corrosion rate (oxidation) of one metal accelerates while that of the other is decreased. Because iron is more active than copper, its corrosion was accelerated.)*

Differentiated Instruction

Less Proficient Readers 

Have pairs of students look at the diagram of the voltaic cell in Figure 21.10. Have them trace the movement of the electrons from one electrode to the other. They should be able to explain that the electrons travel up from the zinc anode into the external wire, through the voltmeter, and down through the wire into the hydrogen/platinum cathode. Have them identify where oxidation is taking place and where reduction is taking place. Refer them to Table 21.2 to find the appropriate half-cell reactions and have them write the overall reaction for the cell.

Answers to...

Figure 21.10 Reduction occurs at the cathode. In this cell hydrogen ions are reduced to hydrogen gas.

 The standard hydrogen electrode consists of a platinum electrode immersed in a 1*M* acid solution at 25°C.

Discuss 

Explain that the two half-cells of a voltaic cell are competing for electrons and that oxidation or reduction could occur in either cell. The half-cell with the more positive reduction potential will win the competition and undergo reduction; oxidation will occur in the other half-cell. The potential produced by the electrochemical cell is the difference in the reduction potentials of the two half-cell reactions. Explain that the quantitative value of any half-cell potential is obtained by measuring it against the standard hydrogen electrode. Point out that Table 21.2 contains tabulated values for some half-cells. These can be used to calculate cell potentials.

Table 21.2

Reduction Potentials at 25°C with 1*M* Concentrations of Aqueous Species

Electrode	Half-reaction	E^0 (V)
Li^+/Li	$Li^+ + e^- \longrightarrow Li$	−3.05
K^+/K	$K^+ + e^- \longrightarrow K$	−2.93
Ba^{2+}/Ba	$Ba^{2+} + 2e^- \longrightarrow Ba$	−2.90
Ca^{2+}/Ca	$Ca^{2+} + 2e^- \longrightarrow Ca$	−2.87
Na^+/Na	$Na^+ + e^- \longrightarrow Na$	−2.71
Mg^{2+}/Mg	$Mg^{2+} + 2e^- \longrightarrow Mg$	−2.37
Al^{3+}/Al	$Al^{3+} + 3e^- \longrightarrow Al$	−1.66
H_2O/H_2	$2H_2O + 2e^- \longrightarrow H_2 + 2OH^-$	−0.83
Zn^{2+}/Zn	$Zn^{2+} + 2e^- \longrightarrow Zn$	−0.76
Cr^{3+}/Cr	$Cr^{3+} + 3e^- \longrightarrow Cr$	−0.74
Fe^{2+}/Fe	$Fe^{2+} + 2e^- \longrightarrow Fe$	−0.44
H_2O/H_2 (pH 7)	$2H_2O + 2e^- \longrightarrow H_2 + 2OH^-$	−0.42
Cd^{2+}/Cd	$Cd^{2+} + 2e^- \longrightarrow Cd$	−0.40
$PbSO_4/Pb$	$PbSO_4 + 2e^- \longrightarrow Pb + SO_4^{2-}$	−0.36
Co^{2+}/Co	$Co^{2+} + 2e^- \longrightarrow Co$	−0.28
Ni^{2+}/Ni	$Ni^{2+} + 2e^- \longrightarrow Ni$	−0.25
Sn^{2+}/Sn	$Sn^{2+} + 2e^- \longrightarrow Sn$	−0.14
Pb^{2+}/Pb	$Pb^{2+} + 2e^- \longrightarrow Pb$	−0.13
Fe^{3+}/Fe	$Fe^{3+} + 3e^- \longrightarrow Fe$	−0.036
H^+/H_2	$2H^+ + 2e^- \longrightarrow H_2$	0.000
$AgCl/Ag$	$AgCl + e^- \longrightarrow Ag + Cl^-$	+0.22
Hg_2Cl_2/Hg	$Hg_2Cl_2 + 2e^- \longrightarrow 2Hg + 2Cl^-$	+0.27
Cu^{2+}/Cu	$Cu^{2+} + 2e^- \longrightarrow Cu$	+0.34
O_2/OH^-	$O_2 + 2H_2O + 4e^- \longrightarrow 4OH^-$	+0.40
Cu^+/Cu	$Cu^+ + e^- \longrightarrow Cu$	+0.52
I_2/I^-	$I_2 + 2e^- \longrightarrow 2I^-$	+0.54
Fe^{3+}/Fe^{2+}	$Fe^{3+} + e^- \longrightarrow Fe^{2+}$	+0.77
Hg_2^{2+}/Hg	$Hg_2^{2+} + 2e^- \longrightarrow 2Hg$	+0.79
Ag^+/Ag	$Ag^+ + e^- \longrightarrow Ag$	+0.80
O_2/H_2O (pH 7)	$O_2 + 4H^+ + 4e^- \longrightarrow 2H_2O$	+0.82
Hg^{2+}/Hg	$Hg^{2+} + 2e^- \longrightarrow Hg$	+0.85
Br_2/Br^-	$Br_2 + 2e^- \longrightarrow 2Br^-$	+1.07
O_2/H_2O	$O_2 + 4H^+ + 4e^- \longrightarrow 2H_2O$	+1.23
MnO_2/Mn^{2+}	$MnO_2 + 4H^+ + 2e^- \longrightarrow Mn^{2+} + 2H_2O$	+1.28
$Cr_2O_7^{2-}/Cr^{3+}$	$Cr_2O_7^{2-} + 14H^+ + 6e^- \longrightarrow 2Cr^{3+} + 7H_2O$	+1.33
Cl_2/Cl^-	$Cl_2 + 2e^- \longrightarrow 2Cl^-$	+1.36
PbO_2/Pb^{2+}	$PbO_2 + 4H^+ + 2e^- \longrightarrow Pb^{2+} + 2H_2O$	+1.46
MnO_4^-/Mn^{2+}	$MnO_4^- + 8H^+ + 5e^- \longrightarrow Mn^{2+} + 4H_2O$	+1.51
$PbO_2/PbSO_4$	$PbO_2 + 4H^+ + SO_4^{2-} + 2e^- \longrightarrow PbSO_4 + 2H_2O$	+1.69
F_2/F^-	$F_2 + 2e^- \longrightarrow 2F^-$	+2.87

Differentiated Instruction

Special Needs 

Some students may have trouble remembering which reactions in Table 21.2 have the greatest tendency to occur as reductions. Have them compare the table with the activity series of metals in Table 21.1. Point out that the most active metals are at the top of Table 21.1; ions of these same metals are at the top of Table 21.2. Because active metals lose electrons easily, they are most easily oxidized. By the same reasoning, their ions are least likely to be reduced.

Table 21.2 on the preceding page lists some standard reduction potentials at 25°C. The half-reactions are arranged in increasing order of their tendency to occur in the forward direction—that is, as a reduction. Thus, the half-reactions at the top of the table have the least tendency to occur as reductions. The half-reactions at the bottom of the table have the greatest tendency to occur as reductions. For example, lithium ions (at the top) have very little tendency to be reduced to lithium metal.

Calculating Standard Cell Potentials

To function, a cell must be constructed of two half-cells. The half-cell reaction having the more positive (or less negative) reduction potential occurs as a reduction in the cell. With this in mind, it is possible to write cell reactions and calculate cell potentials for cells without actually assembling them. You can simply use the known standard reduction potentials for the various half-cells (from Table 21.2) to predict the half-cells in which reduction and oxidation will occur. This information can then be used to find the resulting E^0_{cell} value. **If the cell potential for a given redox reaction is positive, then the reaction is spontaneous as written. If the cell potential is negative, then the reaction is nonspontaneous.** This latter reaction will be spontaneous in the reverse direction, however, and the cell potential will then have a numerically equal but positive value.

CONCEPTUAL PROBLEM 21.1

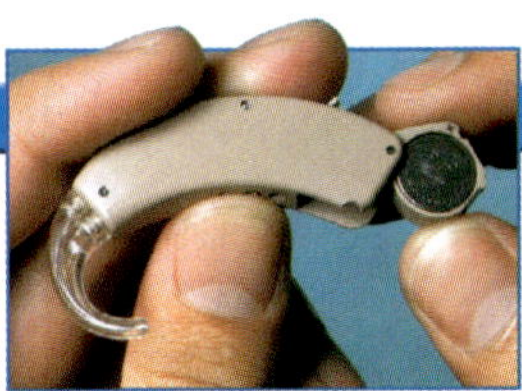

Determining Reaction Spontaneity

Many hearing aid batteries contain zinc and silver oxide. Show that the following redox reaction between Zn and Ag^+ is spontaneous.

$$Zn(s) + 2Ag^+(aq) \longrightarrow Zn^{2+}(aq) + 2Ag(s)$$

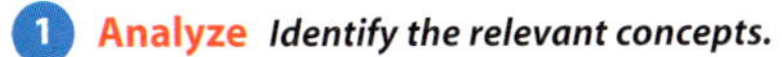

1 Analyze ***Identify the relevant concepts.*** Identify the half-reactions, and calculate the standard cell potential ($E^0_{cell} = E^0_{red} - E^0_{oxid}$). If E^0_{cell} is positive, the reaction is spontaneous.

2 Solve ***Apply concepts to this situation.*** The half-reactions from the equation are:

Oxidation: $Zn(s) \longrightarrow Zn^{2+}(aq) + 2e^-$

Reduction: $Ag^+(aq) + e^- \longrightarrow Ag(s)$

Use Table 21.2 to look up the standard reduction potentials for the half-cells.

$Zn^{2+}(aq) + 2e^- \longrightarrow Zn(s)$ $\quad E^0_{Zn^{2+}} = -0.76\ V$

$Ag^+(aq) + e^- \longrightarrow Ag(s)$ $\quad E^0_{Ag^+} = +0.80\ V$

Calculate the standard cell potential.

$$E^0_{cell} = E^0_{red} - E^0_{oxid} = E^0_{Ag^+} - E^0_{Zn^{2+}}$$
$$= 0.80\ V - (-0.76\ V) = +1.56\ V$$

$E^0_{cell} > 0$, so the reaction is spontaneous.

Practice Problems

9. Determine whether the following redox reaction will occur spontaneously.

$$3Zn^{2+}(aq) + 2Cr(s) \longrightarrow 3Zn(s) + 2Cr^{3+}(aq)$$

10. Is this redox reaction spontaneous as written?

$$Co^{2+}(aq) + Fe(s) \longrightarrow Fe^{2+}(aq) + Co(s)$$

Interactive Textbook

Problem-Solving 21.9 Solve Problem 9 with the help of an interactive guided tutorial.

with ChemASAP

Calculating Standard Cell Potentials

CLASS Activity

Determining Cell Voltages L2

Purpose Students become familiar with combining reduction half-reactions.

Procedure Divide the class into groups of three or four. Have them use Table 21.2 to combine half-reactions and create cells having potentials that fall in the following ranges: >+2.5 V, +2.4 V to +1.4 V, +1.0 V to +0.5 V, and +0.4 V to +0.1 V.

Expected Outcome Several cells could be created in each category. Typical cells might be Na/Na^+ with Sn^{2+}/Sn, +2.57 V; Mg^{2+}/Mg with Pb^{2+}/Pb, +2.24 V; Ca^{2+}/Ca with Al^{3+}/Al, 1.21 V; $PbSO_4/Pb$ with Ni^{2+}/Ni, +0.11 V.

CONCEPTUAL PROBLEM 21.1

Answers

9. nonspontaneous; $E^0_{cell} = -0.02\ V$

10. yes; $E^0_{cell} = +0.16\ V$

Practice Problems Plus 

Chapter Assessment problem 41 is related to Conceptual problem 21.1.

Differentiated Instruction

Gifted L3

Have students investigate the practice of attaching large pieces of magnesium to the hulls of ocean going ships having steel (iron) hulls. Ask them to explain the reason for this practice and provide any supporting figures they can. *(The more active metal, magnesium, is oxidized rather than the iron. The reduction potential for magnesium (Mg^{2+}/Mg) is −2.37 V and the reduction potential for iron (Fe^{3+}/Fe) is −0.036 V).*

Discuss

L2

As students work Sample Problems 21.1 and 21.2, they may think that when a half-cell reaction must be multiplied by a factor such as 2 or 3, the E^0 value must also be multiplied by that factor. Point out that even though two or three times as many electrons are present, the tendency for the electrons to flow is not two or three times greater. The tendency, which is measured by the E^0 value, remains the same.

Discuss

L2

Point out that until now it has been assumed that reactions in standard half-cells always go in the spontaneous direction. For example, a Ag/Cu cell can be constructed in such a way that the only substance available for oxidation is silver metal and the only substance available for reduction is copper ion. In this case, the reaction is nonspontaneous and the cell potential is negative. Thus, the sign of the cell potential is an indicator of reaction spontaneity. Students will learn in the next section that nonspontaneous reactions can be driven by an outside source of energy.

Sample Problem 21.1

Answers

11. $2Al(s) + 3Cu^{2+}(aq) \rightarrow 2Al^{3+}(aq) + 3Cu(s)$

12. $Cu(s) + 2Ag^{+}(aq) \rightarrow Cu^{2+}(aq) + 2Ag(s)$

Practice Problems Plus

L2

A voltaic cell is constructed using the following half-reactions:

$Li^{+}(aq) + e^{-} \rightarrow Li(s)$ $E^0_{Li^+} = -3.05$ V

$Mg^{2+}(aq) + 2e^{-} \rightarrow Mg(s)$ $E^0_{Mg^{2+}} = -2.37$ V

Determine the standard cell reaction.
[$2Li(s) + Mg^{2+}(aq) \rightarrow Mg(s) + 2Li^{+}(aq)$]

Math Handbook

For a math refresher and practice, direct students to significant figures, page R59.

SAMPLE PROBLEM 21.1

Writing the Cell Reaction

Determine the cell reaction for a voltaic cell composed of the following half-cells.

$$Fe^{3+}(aq) + e^{-} \longrightarrow Fe^{2+}(aq) \qquad E^0_{Fe^{3+}} = +0.77\ V$$
$$Ni^{2+}(aq) + 2e^{-} \longrightarrow Ni(s) \qquad E^0_{Ni^{2+}} = -0.25\ V$$

1 Analyze *List the knowns and the unknowns.*

Knowns
- $E^0_{Fe^{3+}} = +0.77$ V
- $E^0_{Ni^{2+}} = -0.25$ V

Unknowns
- cell reaction = ?
- E^0_{cell} = ? V

The half-cell with the more positive reduction potential is the one in which reduction occurs (the cathode). That means that the oxidation reaction occurs at the anode. Add the half-reactions, making certain that the number of electrons lost equals the number of electrons gained.

2 Calculate *Solve for the unknowns.*

The Fe^{3+} half-cell, which has the more positive reduction potential, is the cathode. So in this voltaic cell, Fe^{3+} is reduced and Ni is oxidized. The half-cell reactions, written in the direction in which they actually occur, are as follows.

Oxidation: $Ni(s) \longrightarrow Ni^{2+}(aq) + 2e^{-}$ (at anode)

Reduction: $Fe^{3+}(aq) + e^{-} \longrightarrow Fe^{2+}(aq)$ (at cathode)

Before adding the half-reactions, be sure that the electrons cancel—that is, that they are present in equal number on both sides of the equation. Because Ni loses two electrons as it is oxidized, while Fe^{3+} gains only one electron as it is reduced, you must multiply the Fe^{3+} half-cell equation by 2.

$$Ni(s) \longrightarrow Ni^{2+}(aq) + 2e^{-}$$
$$2[Fe^{3+}(aq) + e^{-} \longrightarrow Fe^{2+}(aq)]$$
$$\overline{Ni(s) + 2Fe^{3+}(aq) \longrightarrow Ni^{2+}(aq) + 2Fe^{2+}(aq)}$$

3 Evaluate *Does the result make sense?*

The electrons cancelled when the half-reactions were added. The equation for the cell reaction is correctly written.

Practice Problems

11. A voltaic cell is constructed using the following half-reactions.

$Cu^{2+}(aq) + 2e^{-} \longrightarrow Cu(s)$
$E^0_{Cu^{2+}} = +0.34$ V
$Al^{3+}(aq) + 3e^{-} \longrightarrow Al(s)$
$E^0_{Al^{3+}} = -1.66$ V

Determine the cell reaction.

12. A voltaic cell is constructed using the following half-reactions.

$Ag^{+}(aq) + e^{-} \longrightarrow Ag(s)$
$E^0_{Ag^{+}} = +0.80$ V
$Cu^{2+}(aq) + 2e^{-} \longrightarrow Cu(s)$
$E^0_{Cu^{2+}} = +0.34$ V

Determine the cell reaction.

Interactive Textbook

Problem-Solving 21.12 Solve Problem 12 with the help of an interactive guided tutorial.

with ChemASAP

SAMPLE PROBLEM 21.2

Calculating the Standard Cell Potential

Calculate the standard cell potential for the voltaic cell described in Sample Problem 21.1. The half-reactions are as follows.

$$Fe^{3+}(aq) + e^- \longrightarrow Fe^{2+}(aq) \quad E^0_{Fe^{3+}} = +0.77\ V$$
$$Ni^{2+}(aq) + 2e^- \longrightarrow Ni(s) \quad E^0_{Ni^{2+}} = -0.25\ V$$

1 Analyze ***List the knowns and the unknown.***

Knowns
- $E^0_{Fe^{3+}} = +0.77\ V$
- $E^0_{Ni^{2+}} = -0.25\ V$
- cathode: Fe^{3+} half-cell

Unknown
- $E^0_{cell} = ?$

Use the equation $E^0_{cell} = E^0_{red} - E^0_{oxid}$ to calculate standard cell potential.

2 Calculate ***Solve for the unknown.***

$$E^0_{cell} = E^0_{red} - E^0_{oxid} = E^0_{Fe^{3+}} - E^0_{Ni^{2+}}$$
$$= +0.77\ V - (-0.25\ V) = +1.02\ V$$

3 Evaluate ***Does the result make sense?***

If the reduction potential of the reduction is positive, and the reduction potential of the oxidation is negative, E^0_{cell} must be positive.

Practice Problems

13. A voltaic cell is constructed using the following half-reactions.

$$Cu^{2+}(aq) + 2e^- \longrightarrow Cu(s) \quad E^0_{Cu^{2+}} = +0.34\ V$$
$$Al^{3+}(aq) + 3e^- \longrightarrow Al(s) \quad E^0_{Al^{3+}} = -1.66\ V$$

Calculate the standard cell potential.

14. A voltaic cell is constructed using the following half-reactions.

$$Ag^+(aq) + e^- \longrightarrow Ag(s) \quad E^0_{Ag^+} = +0.80\ V$$
$$Cu^{2+}(aq) + 2e^- \longrightarrow Cu(s) \quad E^0_{Cu^{2+}} = +0.34\ V$$

Calculate the standard cell potential.

Math Handbook

For help with significant figures, go to page R59.

21.2 Section Assessment

15. Key Concept What causes the electrical potential of a cell?

16. Key Concept What is the electrical potential of a standard hydrogen electrode?

17. Key Concept How can you find the standard reduction potential of a half-cell?

18. Key Concept What cell potential values indicate a spontaneous reaction? A nonspontaneous reaction?

19. The standard reduction potential for a cadmium half-cell is −0.40 V. What does this mean?

Connecting Concepts

Batteries Reread the discussion on batteries in Section 21.1. Use Figure 21.6 and Table 21.2 to calculate the standard cell potential of a voltaic cell in a lead–acid battery.

Assessment 21.2 Test yourself on the concepts in Section 21.2.

with ChemASAP

Section 21.2 Assessment

15. Electrical potential of cell results from competition for electrons between two half-cells.

16. The standard reduction potential of the hydrogen electrode has been assigned a value of 0.00 V at 25°C.

17. by connecting it to a standard hydrogen electrode and measuring the cell potential

18. If the cell potential for a redox reaction is positive, the reaction is spontaneous; if it is negative, the reaction is nonspontaneous.

19. The cadmium half-cell has a tendency to undergo oxidation.

Sample Problem 21.2

Answers

13. $E^0_{cell} = +2.00\ V$

14. $E^0_{cell} = +0.46\ V$

Practice Problems Plus 

Calculate the standard cell potential of this redox reaction to determine whether the reaction will occur spontaneously.

$2Al^{3+}(aq) + 3Mg(s) \rightarrow 2Al(s) + 3Mg^{2+}(aq)$

(spontaneous; $E^0_{cell} = +0.71\ V$)

3 ASSESS

Evaluate Understanding 

Ask students to use Table 21.2 to create an electrochemical cell that will operate spontaneously. Have them calculate the cell potential and write the equations for the two half-reactions and the overall cell reaction. Finally have them use the shorthand method to represent the cell.

Reteach L1

Write a nonspontaneous redox reaction on the board, for example: $Cd^{2+}(aq) + Ag(s) \rightarrow Cd(s) + Ag^+(aq)$ Ask, **What element is being reduced?** *(cadmium)* Have students direct you in writing the half-reaction for the reduction. [*$Cd^{2+}(aq) + 2e^- \rightarrow Cd(s)$*] Ask, **What element is being oxidized?** *(silver)* Write the half-reaction with student input. [*$Ag(s) \rightarrow Ag^+(aq) + e^-$*] Guide students through the process of calculating the cell potential *(−1.20 V)* and ask whether the cell operates spontaneously. *(No, it has a negative potential.)*

Connecting Concepts

$Pb + PbO_2 + 2H_2SO_4^{2-} \rightarrow 2PbSO_4 + 2H_2O; E^0_{cell} = +2.05\ V$

If your class subscribes to the Interactive Textbook, use it to review key concepts in Section 21.2.

with ChemASAP

21.3

1 FOCUS

Objectives

21.3.1 Distinguish between electrolytic and voltaic cells.

21.3.2 Identify the products of the electrolysis of water.

21.3.3 Describe the chemical changes that take place during the electrolysis of brine.

21.3.4 Name three ways that electrolysis is used in metal processing.

Guide for Reading

Build Vocabulary L2

Word Parts Electr- or electro- are combining forms from the New Latin word *electricus.* Ask students what words they know outside of chemistry that contain these combining forms. *(electricity, electrocardiogram, electronic, electromagnet)*

Reading Strategy 

Preview Before beginning this section, have students leaf through the pages and read the headings. Then have them read the boldface sentences under each heading. Ask them if they now have an idea of what this section is about.

2 INSTRUCT

Connecting to Your World

Ask students about the importance of electricity in making and using DVDs. *(Electricity is needed to deposit metal on the master and also to play a DVD. Students might also recall that the laser used to burn the DVD requires electricity.)*

21.3 Electrolytic Cells

Guide for Reading

 **Key Concepts**

- How do voltaic and electrolytic cells differ?
- What products are formed by the electrolysis of water?
- What oxidation and reduction reactions occur during the electrolysis of brine?
- How are electrolytic cells used in metal processing?

Vocabulary

electrolysis

electrolytic cell

Reading Strategy

Relating Text and Visuals As you read about electrolytic cells, look at Figure 21.12. Make sure that you can explain the differences between a voltaic cell and an electrolytic cell.

Connecting to Your World DVDs are used to store all types of data, from multimedia computer programs to the latest popular movie. DVDs have made it possible to store large amounts of data in a very small space. To manufacture a DVD, a laser is used to transfer all the necessary data to the master DVD. Then, metal is deposited on the master using electricity. Once the metal disc is removed from the master, it is used to stamp out the duplicates sold to the public. In this section, you will read about electrochemical processes that require (rather than produce) electrical energy.

Electrolytic vs. Voltaic Cells

In Section 21.1, you learned how a spontaneous chemical reaction can be used to generate a flow of electrons (an electric current). In this section, you will learn how an electric current can be used to make a nonspontaneous redox reaction go forward. The process in which electrical energy is used to bring about such a chemical change is called **electrolysis.** Although you may not have realized it, you are already familiar with some results of electrolysis: silver-plated dishes and utensils, gold-plated jewelry, and chrome-plated automobile parts are a few examples.

The apparatus in which electrolysis is carried out is an electrolytic cell. An **electrolytic cell** is an electrochemical cell used to cause a chemical change through the application of electrical energy. An electrolytic cell uses electrical energy (direct current) to make a nonspontaneous redox reaction proceed to completion. The vast array of electrolytic cells shown in Figure 21.11 are used for the commercial production of chlorine and sodium hydroxide from brine, a concentrated aqueous solution of sodium chloride.

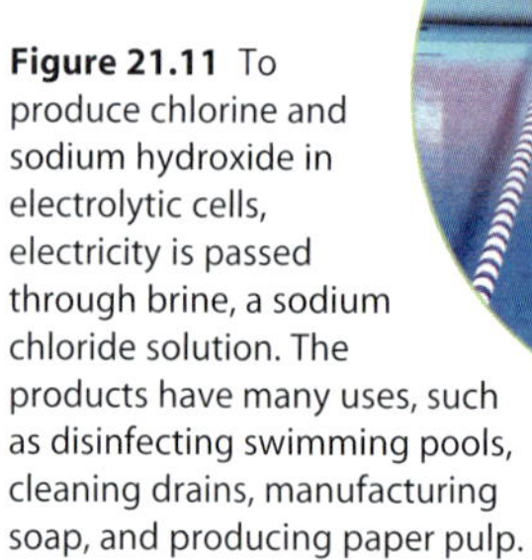

Figure 21.11 To produce chlorine and sodium hydroxide in electrolytic cells, electricity is passed through brine, a sodium chloride solution. The products have many uses, such as disinfecting swimming pools, cleaning drains, manufacturing soap, and producing paper pulp.

Section Resources

Print

- **Guided Reading and Study Workbook,** Section 21.3
- **Core Teaching Resources,** Section 21.3 Review, Interpreting Graphics
- **Transparencies,** T250–T252
- **Laboratory Manual,** Lab 48
- **Small-Scale Chemistry Laboratory Manual,** Labs 35, 36

Technology

- **Interactive Textbook with ChemASAP,** Animation 27, Assessment 21.3
- **Go Online,** Section 21.3

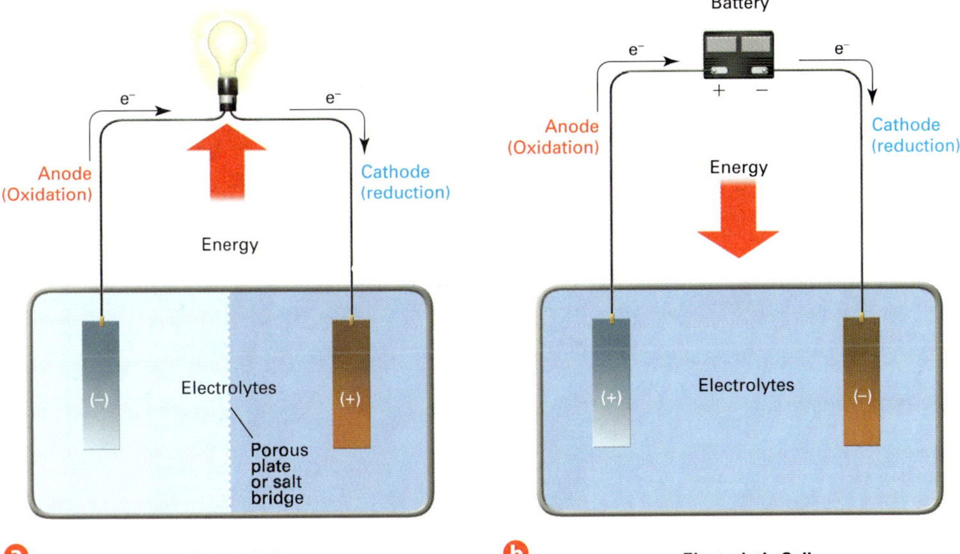

a Voltaic Cell **b** Electrolytic Cell

Figure 21.12 Electrochemical cells can be classified as voltaic or electrolytic. **a** In a voltaic cell, energy is released from a spontaneous redox reaction. The system (cell) does work on the surroundings (lightbulb). **b** In an electrolytic cell, energy is absorbed to drive a non-spontaneous reaction. The surroundings (battery or power supply) do work on the system (cell). **Comparing and Contrasting** ***How are voltaic and electrolytic cells similar? How are they different?***

In both voltaic and electrolytic cells, electrons flow from the anode to the cathode in the external circuit. As shown in Figure 21.12, for both types of cells, the electrode at which reduction occurs is the cathode. The electrode at which oxidation occurs is the anode.

The key difference between voltaic and electrolytic cells is that in a voltaic cell, the flow of electrons is the result of a spontaneous redox reaction, whereas in an electrolytic cell, electrons are pushed by an outside power source, such as a battery. The redox process in the voltaic cell is spontaneous; in the electrolytic cell, the redox process is nonspontaneous. Electrolytic and voltaic cells also differ in the assignment of charge to the electrodes. In an electrolytic cell, the cathode is considered to be the negative electrode. This is because it is connected to the negative electrode of the battery. (Remember that in a voltaic cell, the anode is the negative electrode and the cathode is the positive electrode.) The anode in the electrolytic cell is considered to be the positive electrode because it is connected to the positive electrode of the battery. It is important to remember these conventions about the two kinds of cells.

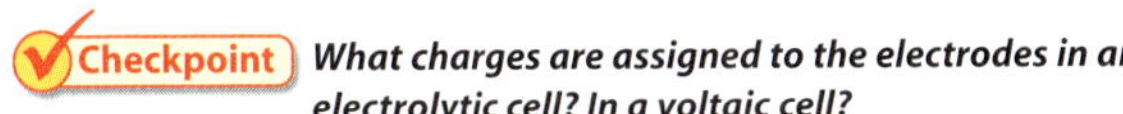

Checkpoint ***What charges are assigned to the electrodes in an electrolytic cell? In a voltaic cell?***

Differentiated Instruction

Gifted

Explain that electrolytic cells are used in the recovery of low-grade ores in mining. Have students select a metal and research how it is recovered and purified by electrolysis.

Electrolytic vs. Voltaic Cells

Relate L2

The spontaneous cell reaction in a lead-storage battery provides immediate energy to start a car. Ask students to describe what happens in the battery. *(Lead and lead(IV) oxide undergo a redox reaction in which lead(II) sulfate is formed.)* Students may have experienced a dead battery. Often, recharging can revive a dead battery. Ask what reaction occurs during recharging. *(Lead(II) ions in lead(II) sulfate are oxidized to lead(IV) and reduced to lead.)* Note that this is a nonspontaneous reaction that requires energy from another source.

Discuss

Electrolytic processes are used to separate active metals such as aluminum, magnesium, and sodium from their salts. The same process is used to recover metals from ores. Compare electrolytic cells to voltaic cells. In both, oxidation occurs at the anode and reduction at the cathode. In the voltaic cell, a spontaneous redox reaction occurs. In the electrolytic cell, external electrical energy makes a nonspontaneous redox reaction occur.

Answers to...

Figure 21.12 In both the voltaic cell and in the electrolytic cell, electrons flow from the anode to the cathode in the external circuit. In the voltaic cell, energy is released from a spontaneous redox reaction. In the electrolytic cell, energy is absorbed to drive a nonspontaneous reaction.

Checkpoint In an electrolytic cell, the anode is assigned a positive charge, and the cathode is assigned a negative charge. In a voltaic cell, the anode is negative, and the cathode is positive.

Electrolysis of Water

Use Visuals

Figure 21.13 Call attention to the volumes of hydrogen and oxygen that have been collected in the photograph. Ask, **Why is one volume larger than the other?** *(When water is decomposed, two molecules of hydrogen are formed for every one molecule of oxygen.)* For emphasis, write the equation for the decomposition of water on the board. **Why is it necessary to add an acid to the water in the cell?** *(Water is a molecular compound; it does not form a significant number of ions. The liquid in the cell must be an electrolyte, that is, it must contain ions.)*

The Electrolysis of Water L2

Purpose Students observe the electrolysis of water and note that energy is required for the process.

Materials Hoffman apparatus or construct a substitute using iron nails as electrodes, 2 test tubes, 6*M* NaOH, 250- or 500-mL beaker, 2 insulated wires

Safety Wear gloves, apron and safety goggles. NaOH is corrosive.

Procedure If you have sufficient class time, assemble the apparatus in front of the class. Otherwise prepare the apparatus in advance. Insert the iron electrodes into the lower ends of the test tubes. Use insulated wires to connect the electrodes to a power supply to avoid contact with the electrolyte. Fill the test tubes with the NaOH solution and invert them in a beaker that is one-third filled with the solution. When all is ready, turn on the power supply.

Expected Outcomes Hydrogen gas is collected at the negative electrode (cathode); oxygen gas is collected at the positive electrode (anode). Demonstrate with a lit match (H_2) and a glowing splint (O_2) that one test tube contains hydrogen and the other oxygen.

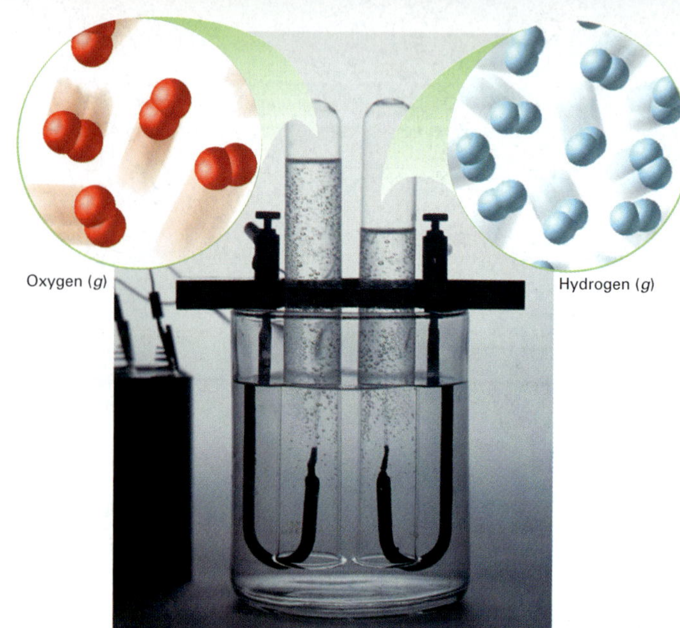

Figure 21.13 When an electric current is passed through water, the water decomposes into oxygen gas and hydrogen gas. **Interpreting Photographs** ***Which electrode in the photograph is the cathode? The anode?***

Electrolysis of Water

When a current is applied to two electrodes immersed in pure water, nothing happens. There is no current flow and no electrolysis. However, when an electrolyte such as H_2SO_4 or KNO_3 in low concentration is added to the pure water, the solution conducts electricity and electrolysis occurs. This process is illustrated in Figure 21.13. **The products of the electrolysis of water are hydrogen gas and oxygen gas.** Water is reduced to hydrogen at the cathode according to the following reduction half-reaction.

$$\text{Reduction: } 2H_2O(l) + 2e^- \longrightarrow H_2(g) + 2OH^-(aq) \text{ (at cathode)}$$

Water is oxidized at the anode according to the following oxidation half-reaction.

$$\text{Oxidation: } 2H_2O(l) \longrightarrow O_2(g) + 4H^+(aq) + 4e^- \text{ (at anode)}$$

The region around the cathode turns basic due to the production of OH^- ions. The region around the anode turns acidic due to an increase in H^+ ions. The overall cell reaction is obtained by adding the half-reactions (after doubling the first one to balance electrons).

$$\begin{aligned} \text{Reduction: } & 2[2H_2O(l) + 2e^- \longrightarrow H_2(g) + 2OH^-(aq)] \\ \text{Oxidation: } & 2H_2O(l) \longrightarrow O_2(g) + 4H^+(aq) + 4e^- \\ \hline \text{Overall cell reaction: } & 6H_2O(l) \longrightarrow 2H_2(g) + O_2(g) + 4H^+(aq) + 4OH^-(aq) \end{aligned}$$

The ions produced tend to recombine to form water.

$$4H^+(aq) + 4OH^-(aq) \longrightarrow 4H_2O(l)$$

For that reason they are not included in the net reaction.

$$2H_2O(l) \xrightarrow{\text{electrolysis}} 2H_2(g) + O_2(g)$$

Differentiated Instruction

English Learners L1

Encourage students to review the chapter looking for the vocabulary words in bold print. Have them compile a glossary in which they define each term in their own words. Have them also look for the key concepts, which are printed in bold type. After carefully reading each sentence have them write it in their own words.

Download a worksheet on **Electrolysis** for students to complete, and find additional teacher support from NSTA SciLinks.

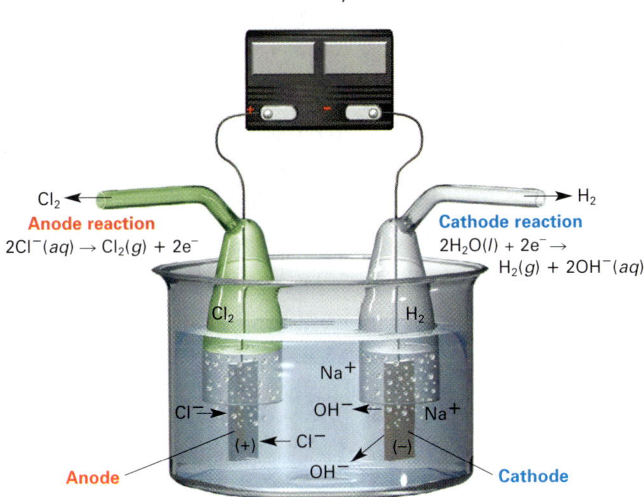

Figure 21.14 The electrolytic cell used for the electrolysis of brine. **Interpreting Diagrams** ***Which substances are produced by oxidation? By reduction?***

Electrolysis of Brine

If the electrolyte in an aqueous solution is more easily oxidized or reduced than water, then the products of electrolysis will be substances other than hydrogen and oxygen. An example is the electrolysis of brine, a concentrated aqueous solution of sodium chloride, which simultaneously produces chlorine gas, hydrogen gas, and sodium hydroxide. The electrolytic cell for this process is shown in Figure 21.14.

During electrolysis of brine, chloride ions are oxidized to produce chlorine gas at the anode. Water is reduced to produce hydrogen gas at the cathode. Sodium ions are not reduced to sodium metal in the process because water molecules are more easily reduced than are sodium ions. The reduction of water produces hydroxide ions as well as hydrogen gas. Thus the electrolyte in solution becomes sodium hydroxide (NaOH). The half-reactions are

Oxidation: $2Cl^-(aq) \longrightarrow Cl_2(g) + 2e^-$ (at anode)

Reduction: $2H_2O(l) + 2e^- \longrightarrow H_2(g) + 2OH^-(aq)$ (at cathode)

The overall ionic equation is the sum of the two half-reactions.

$$2Cl^-(aq) + 2H_2O(l) \longrightarrow Cl_2(g) + H_2(g) + 2OH^-(aq)$$

The spectator ion Na^+ can be included in the equation (as part of NaCl and of NaOH) to show the formation of sodium hydroxide during the electrolytic process.

$$2NaCl(aq) + 2H_2O(l) \longrightarrow Cl_2(g) + H_2(g) + 2NaOH(aq)$$

When the sodium hydroxide solution is about 10% (m/v), it is removed from the cell and processed further.

What is the overall ionic equation for the electrolysis of brine?

For: Links on Electrolysis
Visit: www.SciLinks.org
Web Code: cdn-1213

Electrolysis of Brine

Use Visuals L1

Figure 21.14 Have students study the diagram. Ask, **What is the purpose of the battery?** *(The battery is a voltaic cell that supplies the electrical energy needed to drive the nonspontaneous redox reaction in the electrolytic cell.)* **What substance is being oxidized in the electrolytic cell?** *(Cl^-)* **What substance is being reduced?** *(H_2O)* Make a sketch of the cell on the board. Use different colors to label the cathode (1) and anode (2) on the battery. Show the flow of electrons from the anode (negative terminal) of the battery to the cathode (negative electrode) of the electrolytic cell, and from the anode (positive electrode) of the electrolytic cell to the cathode (positive terminal) of the battery. Show students that in both the voltaic and electrolytic cells, the definitions of a cathode and an anode are the same: reduction always takes place at the cathode and oxidation always takes place at the anode.

Why is a DC current, and not an AC current, used to run the reaction? *(DC devices maintain a constant potential difference. The voltage potential from an AC generator alternates in sign.)* Write the overall equation for the reaction on the board and ask students to use the standard reduction potentials in Table 21.2 to show why this reaction is nonspontaneous.

[$Cl_2(g) + 2e^- \rightarrow 2Cl^-(aq); E^0 = +1.36\ V$

$2H_2O(l) + 2e^- \rightarrow H_2(g) + 2OH^-(aq); E^0 = -0.42\ V$

$E^0_{cell} = E^0_{red} - E^0_{oxid} = -0.42\ V - 1.36\ V = -1.78 V$

The potential is negative; therefore, energy is needed to make this reaction occur.]

Facts and Figures

Anodizing

Anodizing is an electrolytic process for applying a corrosion-resistant oxide coating to aluminum and certain other metals. When an electric current flows through an electrolytic cell containing an aluminum anode, a thin coat of aluminum oxide forms on the surface of the aluminum. Immersion of the aluminum object in pure boiling water permanently seals the film against corrosion. Dyeing of anodized aluminum can produce metallic red, metallic blue, and a host of other metallic colors on the metal surface. Also, the high porosity of the oxide coating permits it to absorb pigments and lubricants.

Answers to...

Figure 21.13 The electrode on the right is the cathode; the electrode on the left is the anode.

Figure 21.14 Oxidation produces chlorine gas. Reduction produces hydrogen gas and OH^- ions.

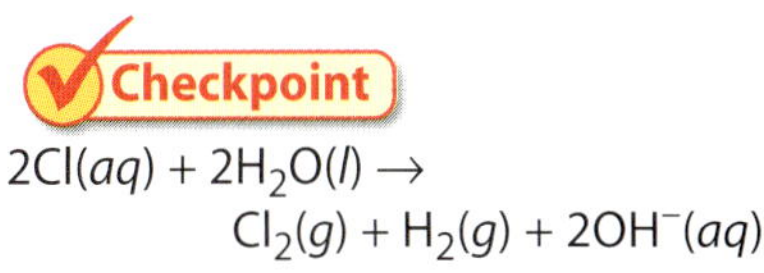

$2Cl(aq) + 2H_2O(l) \rightarrow Cl_2(g) + H_2(g) + 2OH^-(aq)$.

Using Electrolysis in Metal Processing

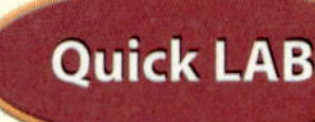

Electrochemical Analysis of Metals 

Objective Students electrochemically oxidize metals and identify the products.

Skills Focus Observing, inferring

 **Prep Time** 15 minutes

Materials 9-volt battery, copper penny, nickel coin, iron nail, filter paper, aluminum foil, reaction surface, Na_2SO_4

Advance Prep

Solution	Preparation
0.2*M* Na_2SO_4	7.1 g in 250 mL water

Class Time 20 minutes

Expected Outcomes In Step 3, a light blue image of the penny appears on the filter paper. In Step 4, a blue-green image of the nickel appears. In Step 5, a brownish-red image of the nail appears.

Analyze and Conclude

1. See Expected Outcomes.
2. Each oxidation occurs at the anode:
   $Cu(s) \rightarrow Cu^{2+}(aq) + 2e^-$
   $Ni(s) \rightarrow Ni^{2+}(aq) + 2e^-$
   $Fe(s) \rightarrow Fe^{2+}(aq) + 2e^-$
3. The battery oxidized each metal and the resulting colored metal ion migrated to the filter paper.
4. $H_2O(l) + 2e^- \rightarrow \frac{1}{2}\ H_2O(g) + OH^-(aq)$
5. $H_2O(l) + 2e^- \rightarrow \frac{1}{2} H_2(g) + OH^-(aq)$
   $Cu(s) \rightarrow Cu^{2+}(aq) + 2e^-$
   $H_2O(l) + Cu(s) \rightarrow \frac{1}{2} H_2(g) + OH^-(aq) + Cu^{2+}(aq)$

For Enrichment 

Have students repeat Steps 1–3 but replace $Na_2SO_4(aq)$ with one drop of 0.1*M* KI and one drop of 20% (v/v) starch solution. Explain that the battery oxidized the iodide ions to form iodine. Ask students to write a half-reaction for the oxidation and explain the color change. *($2I^- \rightarrow I_2 + 2e^-$; I^- ions are oxidized to I_2, which reacts with starch to form a blue-black substance.)*

Quick LAB

Electrochemical Analysis of Metals

Purpose

To electrochemically oxidize metals and identify the products.

Materials

- 9-volt battery
- sodium sulfate solution
- copper penny
- nickel coin
- iron nail
- filter paper
- aluminum foil
- reaction surface

Procedure  

1. Stack the following in order on a reaction surface: a 3-cm square of aluminum foil, a 2-cm square of filter paper, 1 drop of Na_2SO_4 solution, and a penny. The penny should be roughly centered on the filter paper, which should be roughly centered on the foil.
2. Apply the negative (−) terminal of the 9-volt battery to the aluminum foil and the positive (+) terminal to the penny for no more than three seconds.
3. Remove the penny and observe the filter paper.
4. Repeat Steps 1–3, replacing the penny with the nickel coin.
5. Repeat Steps 1–3, replacing the penny with the iron nail.

Analyze and Conclude

1. What colors formed on the filter paper for each object?
2. For each metal object you tested, the battery oxidized the metal to form metal cations with a 2+ charge. Write a half-reaction for each metal oxidation you observed. Did these reactions take place at the anode or the cathode?
3. Explain in your own words why the colors formed in the filter paper.
4. The aluminum foil serves as the cathode, where the reduction of water takes place. Write the half-reaction for the reduction of water.
5. Combine the half-reaction for the oxidation of copper with the half-reaction for the reduction of water to form an overall equation.

Animation 27 Take an atomic-level look at how electricity can bring about a chemical change.

with ChemASAP

Using Electrolysis in Metal Processing

Electrolysis has many important applications in the field of metallurgy. **Electrolytic cells are commonly used in the plating, purifying, and refining of metals.** Many of the shiny, metallic objects you see every day—such as chrome-plated fixtures or nickel-plated coins—were manufactured with the help of electrolytic processes.

Electroplating and Electroforming Electroplating is the deposition of a thin layer of a metal on an object in an electrolytic cell. An object may be electroplated to protect the surface of the base metal from corrosion or to make it more attractive. An object that is to be silver-plated is made the cathode in an electrolytic cell. The anode is the metallic silver that is to be deposited, and the electrolyte is a solution of a silver salt, such as silver cyanide. When a direct current is applied, silver ions move from the anode to the object to be plated.

$$\text{Reduction:}\quad Ag^+(aq) + e^- \longrightarrow Ag(s)\ \text{(at cathode)}$$

Facts and Figures

Plating Jewelry

Gold and silver are the most common metals used for plating jewelry. A thin layer of the precious metal is deposited over much less expensive base metals such as copper. Nonmetallic objects may be used as the base ma[illegible]l as long as their surface is first coated with a conductor such as graphite. Plating solutions contain a suitable electrolyte (an acid or base) to adjust pH. Small amounts of other substances may be added to vary the color of gold jewelry. Nickel, silver, copper, cobalt, and iridium may be deposited along with the gold.

The net result is that silver transfers from the silver electrode to the object being plated. Figure 21.15 shows a silver-plated tea set. Many factors contribute to the quality of the metal coating that forms. In the plating solution, the concentration of the cations to be reduced must be carefully controlled. The solution must also contain compounds to control the acidity and to increase the conductivity. Other compounds may be used to make the metal coating brighter or smoother.

Electroforming is a process in which an object is reproduced by making a metal mold of it at the cathode of a cell. For example, a phonograph record can be coated with metal so it will conduct a current. It is then electroplated with a thick coating of metal. This coating can be stripped off and used as a mold to produce copies of the record.

Figure 21.15 Pure silver was electroplated onto steel to produce this silver-plated tea set.

Electrowinning In another process, called electrowinning, impure metals can be purified in electrolytic cells. The cations of molten salts or aqueous solutions are reduced at the cathode to give very pure metals. A common use of electrowinning is in the extraction of aluminum from its ore, bauxite. Bauxite is impure alumina (Al_2O_3). In a method known as the Hall-Heroult process, purified alumina is dissolved in molten cryolite (Na_3AlF_6), and heated to above 1000°C in a carbon-lined tank. The carbon lining, connected to a direct current, serves as the cathode. The anode consists of carbon rods dipped into the tank. At the cathode, Al^{3+} ions are reduced, forming molten aluminum. At the anode, carbon is oxidized, forming carbon dioxide gas.

$$2Al_2O_3(l) + 3C(s) \longrightarrow 4Al(l) + 3CO_2(g)$$

Electrorefining In the process of electrorefining, a piece of impure metal is made the anode of the cell. It is oxidized to the cation and then reduced to the pure metal at the cathode. This technique is used to obtain ultrapure silver, lead, and copper.

Other Processes Other electrolytic processes are centered on the anode rather than the cathode. In electropolishing, for example, the surface of an object at the anode is dissolved selectively to give it a high polish. In electromachining, a piece of metal at the anode is partially dissolved until the remaining portion is an exact copy of the object at the cathode.

21.3 Section Assessment

20. **Key Concept** What is the difference between an electrolytic cell and a voltaic cell?
21. **Key Concept** What products form during the electrolysis of water?
22. **Key Concept** What chemical changes occur during the electrolysis of brine?
23. **Key Concept** What are some applications of electrolysis in the field of metallurgy?
24. What is the charge on the anode of an electrolytic cell? Of a voltaic cell?
25. Which process, oxidation or reduction, always occurs at the cathode of an electrolytic cell?

Elements Handbook

Manufacturing Aluminum Read about the Hall-Heroult process on page R17. Write the half-reactions for the electrolysis that takes place, and explain the function of cryolite in the process.

Interactive Textbook

Assessment 21.3 Test yourself on the concepts in Section 21.3.

with ChemASAP

Section 21.3 Assessment

20. A voltaic cell uses an electrochemical reaction to produce electrical energy; an electrolytic cell uses electrical energy to bring about a chemical change.
21. $H_2(g)$ and $O_2(g)$
22. Chloride ions are oxidized to produce chlorine gas, and water is reduced to produce hydrogen gas.
23. Electroplating (deposition of a thin layer of metal on an object), electrorefining (purification of metals), and electrowinning (extraction of metals).
24. electrolytic cell anode (+); voltaic cell anode (–)
25. Reduction always occurs at the cathode.

Relate L2

Ask, **Can metallic sodium be obtained by electrolyzing brine?** *(No; the products are chlorine gas, hydrogen gas, and sodium hydroxide.)* Tell students that sodium is obtained by electrolysis of molten NaCl in the Downs cell. The cell operates at 801°C to keep the NaCl melted. The anode and cathode of the cell are separated to prevent recombination of sodium and chlorine. Reduction of Na^+ occurs at an iron cathode; oxidation of Cl^- occurs at a graphite anode. Liquid sodium rises to the top of the molten NaCl and is drawn off. Ask students to write the equation for the overall reaction. *[$2NaCl(l) \rightarrow 2Na(l) + Cl_2(g)$]*

3 ASSESS

Evaluate Understanding L2

Ask students to list ways in which voltaic and electrolytic cells are the same and ways in which they are different.

Reteach L1

Make a sketch of a simple electrolytic cell on the board. Label the cathode, anode, and the positive and negative terminals of the battery. Indicate the path of electrons from the anode to the cathode.

Elements Handbook

Anode: $C(s) + 2O^{2-}(l) \rightarrow CO_2(g) + 4e^-$
Cathode: $3e^- + Al^{3+}(l) \rightarrow Al(l)$
Adding cryolite to aluminum oxide results in a mixture that melts at a much lower temperature (1012°C) than pure aluminum oxide (2045°C).

Interactive Textbook

If your class subscribes to the Interactive textbook, use it to review key concepts in Section 21.3.

with ChemASAP

Electrolysis of Water L2

Objective Students electrolyze solutions and interpret their observations.

Skills Focus Observing, drawing conclusions

Materials Reaction surface, electrolysis device, reagents listed below.

 **Prep Time** 30 minutes

Advance Prep

Solution	Preparation
0.2*M* Na_2SO_4	7.1 g in 250 mL
0.1*M* KI	4.2 g in 250 mL
0.2*M* $CuSO_4$	12.5 g $CuSO_4 \cdot 5H_2O$ in 250 mL
starch	50 mL liquid starch in 200 mL
0.04% BTB	100 mg in 16.0 mL of 0.01M NaOH; dilute to 250 mL
1.0*M* NaCl	14.6 g in 250 mL
0.2*M* KBr	6.0 g in 250 mL

Class Time 30 minutes

Expected Outcomes

Figure A *H_2O:* no visible reaction; *Na_2SO_4:* bubbles at anode and cathode; *Na_2SO_4 + BTB:* anode yellow, cathode blue.

Figure B *KI:* bubbles at cathode, yellow soln at anode; *KI + starch:* bubbles at cathode, black soln at anode; *KI + BTB:* bubbles and blue soln at cathode, yellow soln at anode.

Analyze

1. Pure water has too few ions to carry an electric current.
2. In solution, sodium sulfate dissociates into ions, which carry an electric current.
3. The bubbles are $H_2(g)$, and the blue BTB solution indicates the presence of OH^- ions.
4. The bubbles are $O_2(g)$, and H^+ ions in solution impart the yellow color to the BTB solution.
5. $H_2O \rightarrow H_2(g) + \frac{1}{2}O_2(g)$

You're the Chemist

1. Bubbles are $H_2(g)$; blue solution indicates presence of OH^- ions; yellow solution is $I_2(aq)$, which is black in presence of starch.
2. **NaCl** Cathode: $2H_2O + 2e^- \rightarrow H_2(g) + 2OH^-$ (bubbles, blue BTB); Anode: $2Cl^- \rightarrow Cl_2(aq) + 2e^-$ (yellow solution)
KBr Cathode: $2H_2O + 2e^- \rightarrow H_2(g) + 2OH^-$ (bubbles, blue BTB); Anode: $2Br^- \rightarrow Br_2(aq) + 2e^-$ (yellow solution)
$CuSO_4$ Cathode: $Cu^{2+} + 2e^- \rightarrow H_2(g) + 2OH^-$ (copper plates out); Anode: $H_2O \rightarrow \frac{1}{2}O_2(g) + 2H^+ + 2e^-$ (bubbles, yellow BTB)
3. I^- is more likely to oxidize than H_2O because it has a more positive E^0 value:

$$H_2O \rightarrow \frac{1}{2}O_2(g) + 2H^+ + 2e^-;\ E^0 = -0.82\text{ V}$$

$$2I^- \rightarrow I_2(aq) + 2e^-;\ E^0 = -0.54\text{ V}$$

For Enrichment 

In 1985, treasure hunter Mel Fisher salvaged tons of silver coins from the shipwrecked galleon Atocha, which sank in 1622. Have students research the electrochemistry Fisher used to rescue the treasure from corrosion.

Small-Scale LAB

Electrolysis of Water

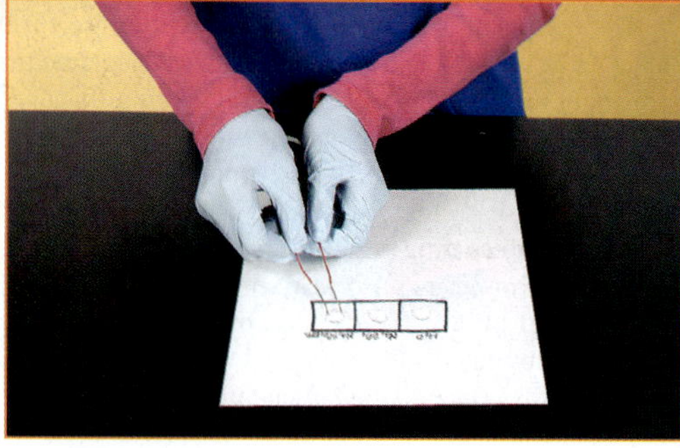

Purpose

To electrolyze solutions and interpret your observations in terms of chemical reactions and equations.

Materials

- pencil
- paper
- ruler
- reaction surface
- electrolysis device
- chemicals shown in Figure A

Procedure    

1. On separate sheets of paper, draw two grids similar to Figure A. Make each square 2 cm on each side.
2. Place a reaction surface over one of the grids and add one drop of each solution as shown in Figure A.
3. Apply the leads of the electrolysis device to each solution. Be sure to clean the leads between each experiment. Look carefully at the cathode (negative lead) and the anode (positive lead).
4. Use the second grid as a data table to record your observations for each solution.

H_2O Na_2SO_4 Na_2SO_4 + BTB

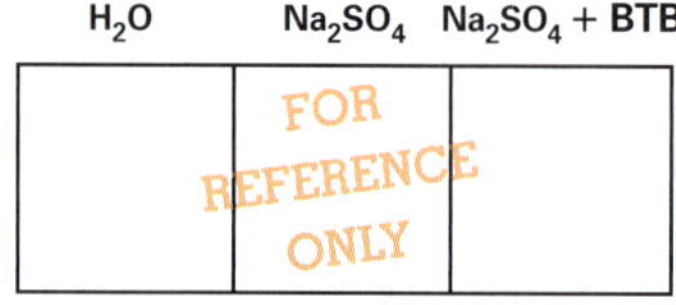

Figure A

Analyze

Using your experimental data, record the answers to the following questions below your data table.

1. Explain why pure water does not conduct electricity and does not undergo electrolysis.
2. Explain why water with sodium sulfate conducts electricity and undergoes electrolysis.
3. The cathode (negative lead) provides electrons to water and the following half-reaction occurs.

$$2H_2O + 2e^- \longrightarrow H_2(g) + 2OH^-$$

Explain how your observations correspond to the products shown in this reaction.

4. The anode (positive lead) takes away electrons from water and the following half-reaction occurs.

$$H_2O \longrightarrow \tfrac{1}{2}O_2(g) + 2H^+ + 2e^-$$

Explain how your observations correspond to the products shown in this reaction.

5. Add the two half-reactions to obtain the overall reaction for the electrolysis of water. Simplify the result by adding the OH^- and H^+ to get HOH, and then canceling anything that appears on both sides of the equation.

You're the Chemist

The following small-scale activities allow you to develop your own procedures and analyze the results.

1. **Analyze It!** Perform the above experiment using the chemicals shown in Figure B. Record your results. The cathode and anode reactions are

$$2H_2O + 2e^- \longrightarrow H_2(g) + 2OH^-\ \text{(at cathode)}$$
$$2I^- \longrightarrow I_2(aq) + 2e^-\ \text{(at anode)}$$

Explain how your observations correspond to the products shown in these half-reactions.

KI KI + starch KI + BTB

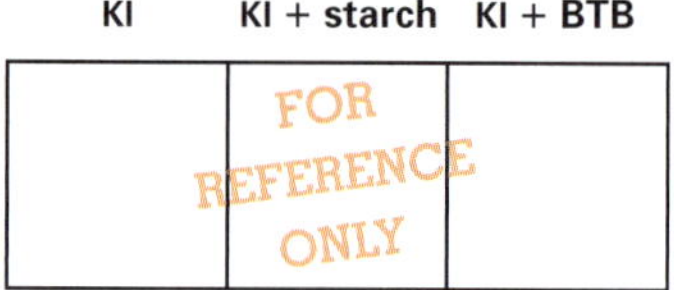

Figure B

2. **Design It!** Design an experiment to explore what happens when you electrolyze NaCl, KBr, and $CuSO_4$ with and without BTB. Use half-reactions to predict your results.
3. **Analyze It!** For each half-reaction listed above, look up the E^0 values in Section 21.2. Show that the E^0 values are consistent with what happens in each case.

Technology & Society

Electroplating

Electroplating is the use of electrolysis to deposit a thin coating of metal on an object, often to enhance its appearance or value. Metals commonly used for electroplating include gold, silver, copper, nickel, and chromium. The layer of the deposited metal is very thin, usually from 5×10^{-5} cm to 1×10^{-3} cm thick. In addition to jewelry and tableware, common electroplated objects include electronic connectors, circuit boards, and contacts coated with elements such as gold, platinum, ruthenium, or osmium.

Applying Concepts ***What are the half-reactions for the electrolytic cell shown below?***

Plating Metals Gold and silver are often used to plate jewelry. Chrome and nickel are used to plate faucets, kitchen appliances, and automotive trim. Zinc coatings are used to protect steel objects from corrosion.

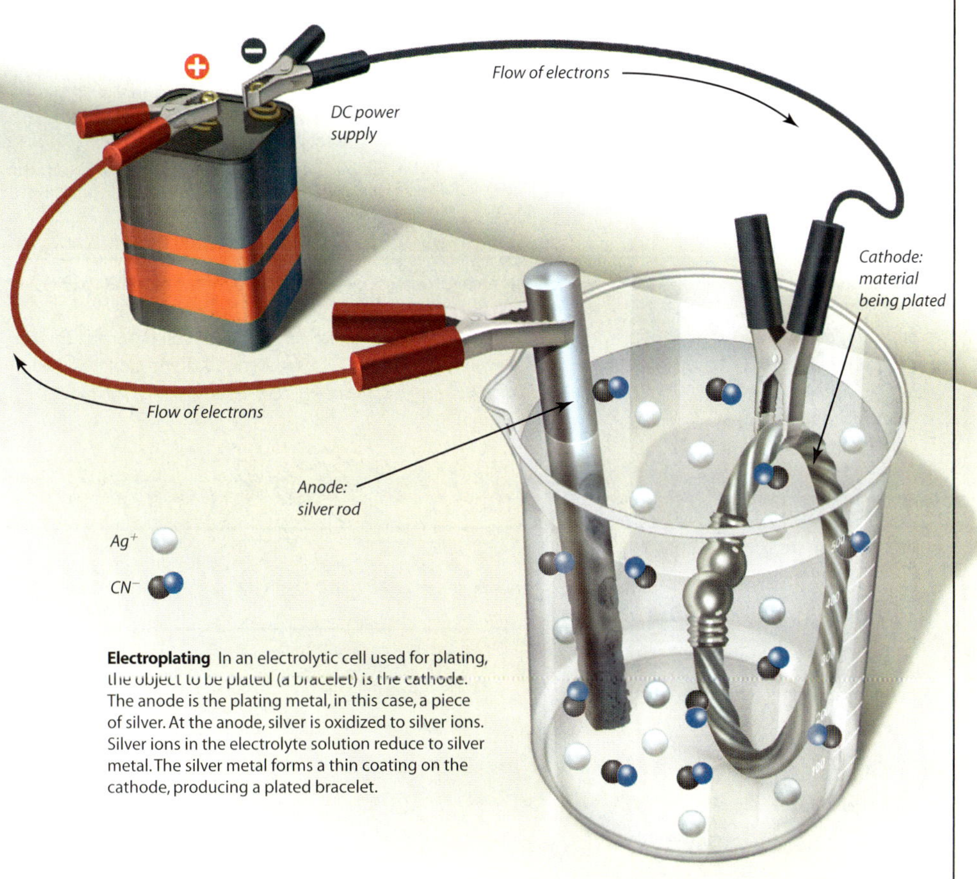

Electroplating In an electrolytic cell used for plating, the object to be plated (a bracelet) is the cathode. The anode is the plating metal, in this case, a piece of silver. At the anode, silver is oxidized to silver ions. Silver ions in the electrolyte solution reduce to silver metal. The silver metal forms a thin coating on the cathode, producing a plated bracelet.

Technology & Society

Electroplating

The cathode in an electroplating apparatus must be able to conduct electricity. Thus you might think that nonmetallic substances such as wood, plastic, and leather could not be electroplated. However, chromium-plated plastic automobile parts enhance the appearance of most cars. Ask students to research the pretreatment of plastics that makes electroplating possible.

Students could research how a base metal can be electroplated with an alloy, such as brass or bronze.

Students may be interested in knowing that the modern era of electroplating began in 1800 with Alesandro Volta's discovery of the voltaic pile, which was the world's first battery. The voltaic pile provided the first significant supply of continuous direct-current electricity. Today, electricity is abundant, but its cost is an important factor in determining whether an electrolytic process is economically viable. Electrochemical industries often position their plants in areas where the cost of electricity is lowest.

Charles Hall and Paul Heroult independently developed the process for electrolytically extracting aluminum from its ore. Students may enjoy researching the amazing coincidences in the lives of these two men.

Answers to...

Applying Concepts

Cathode: $Ag^+(aq) + e^- \rightarrow Ag(s)$

Anode: $Ag(s) \rightarrow Ag^+(aq) + e^-$

CHAPTER 21 **Study Guide**

Study Tip

Study for short focused blocks of time. Take a break when you feel tired. When you are no longer being productive, take a short walk or stretch.

If your class subscribes to the Interactive Textbook with ChemASAP, your students can go online to access an interactive version of the Student Edition and a self-test.

with **ChemASAP**

CHAPTER 21 Study Guide

Key Concepts

21.1 Electrochemical Cells

- For any two metals in an activity series, the more active metal is the more readily oxidized.
- All electrochemical processes involve redox reactions.
- Energy is produced in a voltaic cell by spontaneous redox reactions within the cell.
- Current electrochemical technologies used to produce electrical energy include dry cells, lead storage batteries, and fuel cells.

21.2 Half-Cells and Cell Potentials

- The electrical potential of a cell results from a competition for electrons between two half-cells.
- The standard reduction potential of the hydrogen electrode has been assigned a value of 0.00 V.
- You can determine the standard reduction potential of a half-cell by using a standard hydrogen electrode and the equation for standard cell potential.
- If the cell potential for a given redox reaction is positive, then the reaction is spontaneous. If the cell potential is negative, then the reaction is nonspontaneous.

21.3 Electrolytic Cells

- The key difference between voltaic and electrolytic cells is that in a voltaic cell, the flow of electrons is the result of a spontaneous redox reaction, whereas in an electrolytic cell, electrons are pushed by an outside power source, such as a battery.
- The products of the electrolysis of water are hydrogen gas and oxygen gas.
- During electrolysis of brine, chloride ions are oxidized to produce chlorine gas at the anode; water is reduced at the cathode to produce hydrogen gas.
- Electrolytic cells are commonly used in the plating, purifying, and refining of metals.

Vocabulary

- anode (p. 665)
- battery (p. 668)
- cathode (p. 665)
- cell potential (p. 671)
- dry cell (p. 667)
- electrical potential (p. 671)
- electrochemical cell (p. 665)
- electrochemical process (p. 665)
- electrode (p. 665)
- electrolysis (p. 678)
- electrolytic cell (p. 678)
- fuel cell (p. 669)
- half-cell (p. 665)
- reduction potential (p. 671)
- salt bridge (p. 665)
- standard cell potential (p. 672)
- standard hydrogen electrode (p. 672)
- voltaic cell (p. 665)

Key Equation

- $E^0_{cell} = E^0_{red} - E^0_{oxid}$

Organizing Information

Use these terms to construct a concept map that organizes the major ideas of this chapter.

cell potential · fuel cell · battery · voltaic cell · electrochemical process · electrolytic cell

Concept Map 21 Solve the Concept Map with the help of an interactive guided tutorial.

with **ChemASAP**

Chapter Resources

Print

- **Core Teaching Resources,** Chapter 21, *Practice Problems, Vocabulary Review, Quiz, Chapter Test A, Chapter Test B*

Technology

- **Computer Test Bank,** Chapter 21 Test
- **Interactive Textbook with ChemASAP,** Chapter 21

CHAPTER 21

Assessment

Reviewing Content

21.1 Electrochemical Cells

26. What is meant by the term *half-reaction*?

27. Write the half-reactions that occur when a strip of aluminum is dipped into a solution of copper(II) sulfate.

28. What would you expect to happen when a strip of lead is placed in an aqueous solution of magnesium nitrate?

29. For each pair of metals listed below, decide which metal is more readily oxidized.

a. Hg, Cu **b.** Ca, Al **c.** Ni, Mg
d. Sn, Ag **e.** Pb, Zn **f.** Cu, Al

30. At which electrode in a voltaic cell does reduction always occur?

31. Explain the function of the salt bridge in a voltaic cell.

32. In a typical flashlight battery, what material is used for the anode? For the cathode?

33. Explain why the specific gravity of the electrolyte in a lead storage battery decreases during the discharge process.

34. Use the shorthand method to represent the electrochemical reaction in a car battery.

35. Fuel cells can be designed to generate electrical energy while emitting no air pollutants, yet they are not widely used. Explain.

36. List the advantages of a fuel cell over a lead storage battery.

21.2 Half-Cells and Cell Potentials

37. How was the standard reduction potential of the hydrogen electrode determined?

38. What is the electric potential of a cell?

39. How does the order of the metals in Table 21.1 on page 664 compare with the order in Table 21.2? Why?

40. Explain how to determine the standard reduction potential for the aluminum half-cell.

41. Determine whether these redox reactions will occur spontaneously. Calculate the standard cell potential in each case.

a. $Cu(s) + 2H^+(aq) \longrightarrow Cu^{2+}(aq) + H_2(g)$
b. $2Ag(s) + Fe^{2+}(aq) \longrightarrow 2Ag^+(aq) + Fe(s)$

42. Use the information in Table 21.2 on page 674 to calculate standard cell potentials for these voltaic cells.

a. $Ni\,|\,Ni^{2+}\,||\,Cl_2\,|\,Cl^-$
b. $Sn\,|\,Sn^{2+}\,||\,Ag^+\,|\,Ag$

21.3 Electrolytic Cells

43. Why is direct current, not alternating current, used in the electroplating of metals?

44. Describe briefly how you would electroplate a teaspoon with silver.

Use the diagram to answer Questions 45–47.

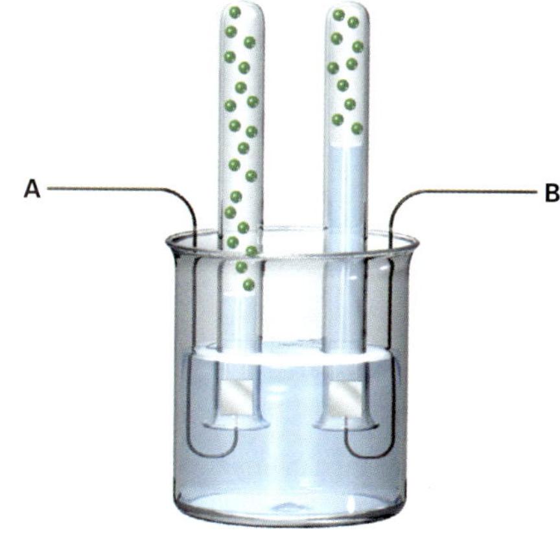

45. Write an equation for the decomposition of water by electrolysis.

46. At which electrode, A or B, is hydrogen produced?

47. The equation for the electrolysis of brine is

$$2NaCl(aq) + 2H_2O(l) \longrightarrow Cl_2(g) + H_2(g) + 2NaOH(aq)$$

How would you modify the electrolysis diagram to make it quantitatively represent the formation of hydrogen and chlorine?

48. Distinguish between voltaic and electrolytic cells.

49. Why is it not possible to measure the potential of an isolated half-cell?

CHAPTER 21

Assessment

Reviewing Content

26. an equation that represents the oxidation or the reduction in a redox reaction

27. Oxidation: $Al(s) \rightarrow Al^{3+}(aq) + 3e^-$
Reduction: $Cu^{2+}(aq) + 2e^- \rightarrow Cu(s)$

28. Nothing

29. **a.** Cu **b.** Ca **c.** Mg **d.** Sn **e.** Zn **f.** Al

30. cathode

31. The salt bridge allows ions to pass from one half-cell to the other but prevents the solutions from mixing.

32. anode: zinc, cathode: carbon (graphite)

33. Water is produced by the redox reaction and sulfuric acid is used up; water has a lower density than sulfuric acid.

34. $Pb(s)|PbSO_4(aq)||PbO_2(s)|PbSO_4(s)$

35. Fuel cells cannot generate electricity as economically as more conventional forms of electrical generation.

36. A fuel cell needs no recharging; it does not produce toxic wastes; and if the fuel is hydrogen gas, it is lighter.

37. It was arbitrarily set at zero.

38. the ability of a voltaic cell to produce a current

39. The relative order is the same because both tables rank the elements according to their tendency to undergo oxidation/reduction.

40. Connect the aluminum half-cell to a standard hydrogen half-cell and use a voltmeter. The indicated voltage is the standard reduction potential for the aluminum half-cell.

41. **a.** nonspontaneous, –0.34 V
b. nonspontaneous, –1.24 V

42. **a.** 1.61 V **b.** 0.94 V

43. A direct current flows in one direction only.

44. The teaspoon is the cathode in an electrolytic cell with silver cyanide as the electrolyte. When the DC current flows, the silver ions deposit as silver on the teaspoon.

45. $2H_2O(l) \rightarrow O_2(g) + 2H_2(g)$

46. electrode A

47. The small spheres representing molecules of H_2 and Cl_2 would be in 1:1 ratio.

48. Voltaic cells convert chemical energy into electrical energy. Electrolytic cells use electrical energy to cause a chemical reaction.

49. Two half-cells are needed because oxidation or reduction cannot occur in isolation. One half-cell gains electrons and one loses them, producing an electric current.

Understanding Concepts

50. Describe the composition of the anode, cathode, and electrolyte in a fully discharged lead storage battery.
51. Predict what will happen, if anything, when an iron nail is dipped into a solution of copper sulfate. Write the oxidation and reduction half-reactions for this process and the balanced equation for the overall reaction.
52. Calculate E^0_{cell} and write the overall cell reaction for these cells.
    a. $Sn|Sn^{2+}||Pb^{2+}|Pb$
    b. $H_2|H^+||Br_2|Br^-$
53. What property do lead(II) sulfate and lead dioxide have that makes salt bridges unnecessary in a lead storage battery?
54. Complete this data table for the electrolysis of water.

H_2O used	H_2 formed	O_2 formed
a. 2.0 mol	____ mol	____ mol
b. ____ g	____ g	16.0 g
c. ____ mL	10.0 g	____ g
d. 44.4 g	____ g	____ g
e. ____ g	8.80 L (STP)	____ L (STP)
f. 66.0 mL	____ g	____ L (STP)

55. In one process used to produce aluminum metal, ore containing aluminum oxide is converted to aluminum chloride. Aluminum metal is then produced by the electrolysis of molten aluminum chloride ($AlCl_3$).
    a. Write the half-reactions that take place at each electrode.
    b. Write the equation for the overall cell reaction.
    c. Identify the products produced at each electrode.
56. What relationship does the voltage produced by a redox reaction have to the spontaneity of the reaction?
57. The reactions that take place in voltaic cells produce electric current and the reactions in electrolytic cells can be made to take place when an electric current is applied. What common feature do these redox reactions share?
58. Use the information in Table 21.2 to determine which of the following cell reactions will proceed spontaneously.
    a. $Zn + Pb^{2+} \longrightarrow ?$
    b. $Cu + Fe^{2+} \longrightarrow ?$
    c. $Ag + Cu^{2+} \longrightarrow ?$
    d. $H_2 + Cu \longrightarrow ?$
    e. $Fe + Pb^{2+} \longrightarrow ?$
    f. $Na + Cl_2 \longrightarrow ?$
59. For each spontaneous reaction in Question 58, write the half-reaction that takes place at the anode. Be sure to include the electrons.
60. Write the overall balanced equation for each reaction in Question 58 that proceeds spontaneously.
61. Determine the standard cell potential for each spontaneous reaction in Question 58.
62. In certain cases, more than one reaction is possible at an electrode. How can you determine which reaction actually takes place?
63. Answer the following questions for the electrolysis of brine (concentrated sodium chloride solution).
    a. Write the equations for the two possible reactions that can take place at the anode?
    b. Write the equations for the two possible reactions that can take place at the cathode?
    c. Which reaction actually takes place at the anode? Explain why this reaction takes place in preference to the other possible reaction.
    d. Which reaction actually takes place at the cathode? Explain why this reaction takes place in preference to the other possible reaction.
64. Identify the better oxidizing agent in each of the following pairs.
    a. Li^+, Ca^{2+}
    b. Fe^{3+}, Hg_2^{2+}
    c. Cu^{2+}, Cu^+
    d. Hg^{2+}, I_2
65. Gold is not included in Table 21.2 on page 674. In what part of the table does gold belong?
66. Lead storage batteries can be recharged. Why are dry cells not rechargeable?

CHAPTER 21 Assessment continued

Understanding Concepts

50. The anode and cathode grids are both packed with $PbSO_4$. The electrolyte is very dilute sulfuric acid.
51. Some of the iron dissolves and the nail becomes coated with copper.
    Oxidation: $Fe \rightarrow Fe^{2+} + 2e^-$
    Reduction: $Cu^{2+} + 2e^- \rightarrow Cu$
    $Fe(s) + CuSO_4(aq) \rightarrow FeSO_4(aq) + Cu(s)$
52. **a.** $Sn(s) + Pb^{2+}(aq) \rightarrow Sn^{2+}(aq) + Pb(s)$
    $E^0_{cell} = +0.01$ V
    b. $H_2(g) + Br_2(l) \rightarrow 2H^+(aq) + 2Br^-(aq)$
    $E^0_{cell} = +1.07$ V
53. Lead(II) sulfate and lead dioxide are very insoluble in sulfuric acid.
54. **a.** 2.0, 1.0
    b. 18.0, 2.0
    c. 90, 80
    d. 4.9, 39.5
    e. 7.07, 4.40
    f. 7.3, 41.1
55. **a.** oxidation: $6Cl^-(l) \rightarrow 3Cl_2(g) + 6e^-$ (anode)
    reduction: $2Al^{3+}(l) + 6e^- \rightarrow 2Al(l)$ (cathode)
    b. overall reaction: $2AlCl_3(l) \rightarrow 2Al(l) + 3Cl_2(g)$
    c. anode, chlorine gas; cathode, liquid aluminum
56. If the cell potential for a redox reaction is positive, the reaction is spontaneous as written.
57. In each type of cell, oxidation occurs at the anode and reduction occurs at the cathode.
58. a, e, and f are spontaneous; b, c, and d are nonspontaneous.
59. **a.** $Zn \rightarrow Zn^{2+} + 2e^-$
    e. $Fe \rightarrow Fe^{2+} + 2e^-$
    f. $Na \rightarrow Na^+ + e^-$
60. **a.** $Zn + Pb^{2+} \rightarrow Zn^{2+} + Pb$
    e. $Fe + Pb^{2+} \rightarrow Fe^{2+} + Pb$
    f. $2Na + Cl_2 \rightarrow 2Na^+ + 2Cl^-$
61. **a.** + 0.63 V **e.** + 0.21 V **f.** + 4.07 V
62. It will be the reaction that has the larger cell potential, or more readily oxidized or reduced.
63. **a.** possible oxidation reactions at anode:
    (i) $2Cl^-(aq) \rightarrow Cl_2(g) + 2e^-$
    (ii) $2H_2O(l) \rightarrow O_2(g) + 4H^+(aq) + 4e^-$
    b. possible reduction reactions at cathode:
    (i) $Na^+(aq) + e^- \rightarrow Na(s)$
    (ii) $2H_2O(l) + 2e^- \rightarrow H_2(g) + 2OH^-(aq)$
    c. (i) Chloride ions are more readily oxidized to chlorine gas than water molecules are oxidized to oxygen.
    d. (ii) Water molecules are more easily reduced than sodium ions.
64. **a.** Li^+ **b.** Fe^{3+} **c.** Cu^{2+} **d.** I_2
65. Gold belongs near the bottom, below silver, because it is one of the least active metals.
66. The paste in a dry cell allows for the movement of electrons but not ions.

Critical Thinking

67. For any voltaic cell, chemists consider the electrode that produces electrons to be negative, and they call it the anode. Most dictionaries, however, define the anode as the positively charged electrode. Explain.

68. Describe the process that is occurring in the following illustration.

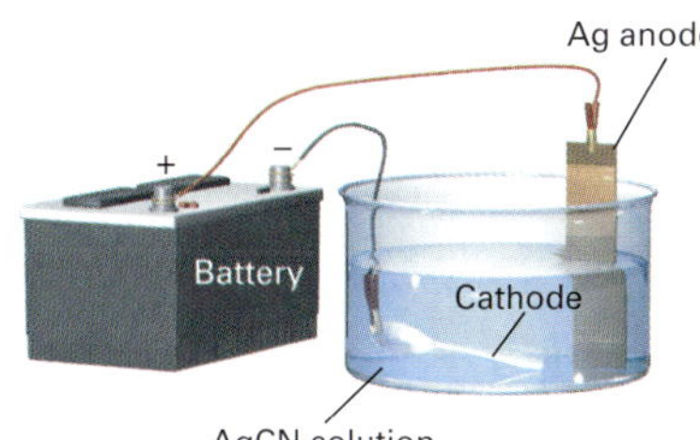

69. Which plot is characteristic of a dry cell? Explain your answer.

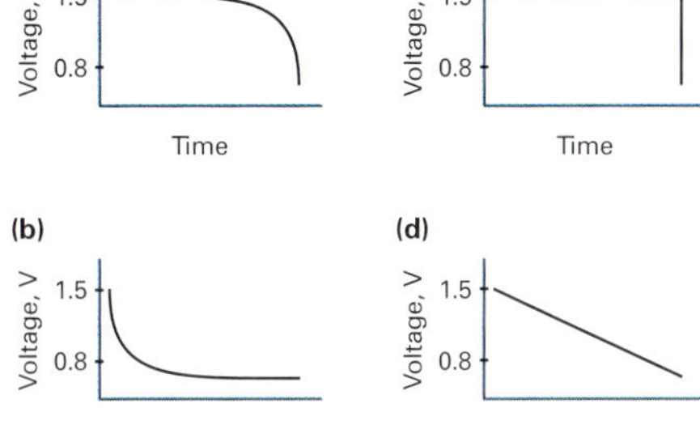

70. An engineer has proposed a new battery design that uses silver as the electrode in both half-cells. As a chemist, what would you tell the engineer about the proposal?

71. In most voltaic cells, the half-cells are connected by a salt bridge or a porous barrier instead of a piece of wire made of copper or some other metal. Why is a metal wire not suitable for connecting the half-cells of a voltaic cell? Explain your answer.

Concept Challenge

72. Write the overall cell reactions and calculate E^0_{cell} for voltaic cells composed of the following sets of half-reactions.

a. $AgCl(s) + e^- \longrightarrow Ag(s) + Cl^-(aq)$
$Ni^{2+}(aq) + 2e^- \longrightarrow Ni(s)$

b. $Al^{3+}(aq) + 3e^- \longrightarrow Al(s)$
$Cl_2(g) + 2e^- \longrightarrow 2Cl^-(aq)$

73. Impure copper is purified in an electrolytic cell. Design an electrolytic cell, with H_2SO_4 as electrolyte, that will allow you to carry out this process. Give the oxidation and reduction half-reactions and a balanced equation for the overall reaction.

74. This spontaneous redox reaction occurs in the voltaic cell illustrated below.

$$Ni^{2+}(aq) + Fe(s) \longrightarrow Ni(s) + Fe^{2+}(aq)$$

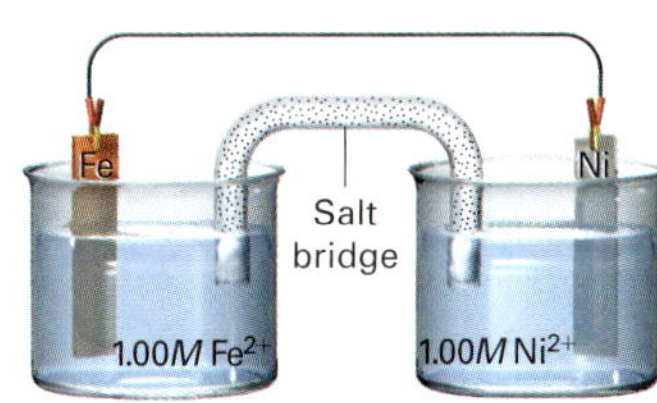

a. Identify the anode and the cathode.
b. Assign charges to the electrodes.
c. Write the half-reactions.
d. Calculate the standard cell potential when the half-cells are at standard conditions.

75. The cells in an automobile battery are separated from each other. However the electrolyte is the same in all the cells. What would be the consequence of placing all the electrodes into a single container of electrolyte solution rather than into separate compartments?

Critical Thinking

67. The chemists' definition focuses on the electrons that are produced by oxidation at the anode of a voltaic cell; the dictionary definition is probably based on an electrolytic cell, whose electrodes are defined by the battery terminals to which they are attached.

68. The spoon is being plated with silver.

69. d; the voltage falls steadily.

70. It will not work as a battery.

71. As electrons flow from the anode to the cathode in the external circuit, anions must flow from the cathode compartment to the anode compartment to maintain neutrality in the electrolytes. Anions cannot flow through wire made of copper or any other metal. The cell will not function if the salt bridge is replaced with a metal wire.

72. a. $2AgCl + Ni \rightarrow 2Ag + NiCl_2$
$E^0_{cell} = +0.47\ V$
b. $3Cl_2 + 2Al \rightarrow 2AlCl_3$
$E^0_{cell} = +3.02\ V$

Concept Challenge

73. Oxidation:
$2Cu(impure) + 2H_2SO_4 \rightarrow 2Cu^{2+} + 2H_2 + SO_4^{2+}$
Reduction:
$2Cu^{2+} + 2SO_4^{2+} + 2H_2O \rightarrow 2Cu(pure) + 2H_2SO_4 + O_2$
Overall reaction:
$2Cu(impure) + 2H_2O \rightarrow 2Cu(pure) + 2H_2 + O_2$

74. a. The iron electrode is the anode; the nickel electrode is the cathode.
b. The anode is negative; the cathode is positive.
c. Anode: $Fe(s) \rightarrow Fe^{2+}(aq) + 2e^-$
Cathode: $Ni^{2+}(aq) + 2e^- \rightarrow Ni(s)$
d. $E^0_{cell} = +0.19\ V$

75. The battery output would not be 12V.

CHAPTER 21 **Assessment** *continued*

Cumulative Review

76. a. 3, 2, 3, 2, 4
b. 2, 1, 1, 2
c. 3, 6, 5, 1, 3

77. 467 mL

78. a. $SnCl_4 \cdot 5H_2O$ **b.** $MgSO_4 \cdot 7H_2O$
c. $FePO_4 \cdot 4H_2O$ **d.** $CaCl_2 \cdot 2H_2O$

79. a. 0.0125g NaCl **b.** 101g KNO_3

80. Dilute 31 mL 16*M* HNO_3 to 500 mL with water.

81. a. 4.32×10^2 kJ **b.** 2.55×10^5 cal
c. 2.70×10^3 J

82. a. -2.78×10^2 kJ **b.** −7.03 kJ
c. 13 kJ **d.** −27 kJ

83. 267 kJ

84. a and c; $K_{eq} < 1$

85. $\dfrac{[N_2] \times [H_2]^3}{[NH_3]^2}$

86. a. pH = 8.00 **b.** pH = 5.00
c. pH = 10.00 **d.** pH = 5.00

87. a. $[OH^-] = 1 \times 10^{-7}M$
b. $[OH^-] = 1 \times 10^{-10}M$
c. $[OH^-] = 1 \times 10^{-6}M$

88. a. $2Na(s) + 2H_2O(l) \rightarrow 2NaOH(aq) + H_2(g)$
b. $Ca(s) + 2H_2O(l) \rightarrow Ca(OH)_2(aq) + H_2(g)$

89. a. +6 **b.** −2 **c.** +4 **d.** +2
e. 0 **f.** +4

90. a. Ca, +2; Cr, +6; O, −2
b. K, +1; Mn, +7; O, −2
c. Ca, +2; N, +5; O, −2
d. Al, +3; O, −2; H +1

91. b; Ca is oxidized, Cl_2 is reduced.
d; Ca is oxidized, H is reduced.

92. a. 3, 6, 1, 5, 3
b. 1, 3, 2, 3

93. a. Cr, +6 **b.** I, +5 **c.** Mn, +7
d. Fe, +3

CHAPTER 21 **Assessment** *continued*

Cumulative Review

76. Balance each equation. *(Chapter 11)*
a. $H_2S(g) + HNO_3(aq) \longrightarrow S(s) + NO(g) + H_2O(l)$
b. $AgNO_3(aq) + Pb(s) \longrightarrow Pb(NO_3)_2(aq) + Ag(s)$
c. $Cl_2(g) + NaOH(aq) \longrightarrow NaCl(aq) + NaClO_3(aq) + H_2O(l)$

77. A sample of oxygen gas has a volume of 425 mL at 30°C. What is the new volume of the gas if the temperature is raised to 60°C while the pressure is kept constant? *(Chapter 14)*

78. Write formulas for these hydrates. *(Chapter 15)*
a. Tin(IV) chloride pentahydrate
b. Magnesium sulfate heptahydrate
c. Iron(III) phosphate tetrahydrate
d. Calcium chloride dihydrate

79. Calculate the grams of solute required to make the following solutions. *(Chapter 16)*
a. 250 mL of 0.50% NaCl (m/v)
b. 500 mL 2.0*M* KNO_3

80. Concentrated nitric acid is 16*M*. How would you prepare 500 mL of 1.0*M* HNO_3? *(Chapter 16)*

81. Convert the following. *(Chapter 17)*
a. 4.32×10^5 joules to kilojoules
b. 255 Calories to calories
c. 645 calories to joules

82. Calculate the quantity of heat lost or gained in the following changes. *(Chapter 17)*
a. 0.625 mol NaOH(*s*) dissolves in water.
b. 1.17 mol water freezes at 0°C.
c. 0.30 mol $C_2H_5OH(l)$ vaporizes.
d. 0.66 mol of steam condenses at 100°C.

83. The combustion of natural gas, methane (CH_4), is an exothermic reaction. *(Chapter 17)*

$$CH_4(g) + 2O_2(g) \longrightarrow CO_2(g) + 2H_2O(l) \quad \Delta H = -890 \text{ kJ}$$

Calculate the amount of heat liberated when 4.80 g CH_4 reacts with an excess of oxygen.

84. Four reactions have the following equilibrium constants. Identify in which of these reactions the reactants are favored over products. Why? *(Chapter 18)*
a. $K_{eq} = 0.006$ **b.** $K_{eq} = 5.3$
c. $K_{eq} = 8 \times 10^{-4}$ **d.** $K_{eq} = 2 \times 10^3$

85. Give the equilibrium-constant expression for the decomposition of ammonia to nitrogen and hydrogen. *(Chapter 18)*

$$2NH_3(g) \rightleftharpoons N_2(g) + 3H_2(g)$$

86. Determine the pH for each solution. *(Chapter 19)*
a. $[H^+] = 1.0 \times 10^{-8}M$
b. $[H^+] = 0.000010M$
c. $[OH^-] = 1.0 \times 10^{-4}M$
d. $[OH^-] = 1.0 \times 10^{-9}M$

87. Three solutions have the following pH values. What are the hydroxide-ion concentrations of these solutions? *(Chapter 19)*
a. pH = 7.0 **b.** pH = 4.0 **c.** pH = 9.0

88. Write a balanced equation for the reaction of each of the following metals with water. *(Chapter 19)*
a. sodium **b.** calcium

89. Determine the oxidation number of sulfur in each of the following. *(Chapter 20)*
a. H_2SO_4 **b.** H_2S **c.** SO_2
d. $Na_2S_2O_3$ **e.** S **f.** SO_3^{2-}

90. Determine the oxidation number of each element in these substances. *(Chapter 20)*
a. $CaCr_2O_7$ **b.** $KMnO_4$
c. $Ca(NO_3)_2$ **d.** $Al(OH)_3$

91. Identify which of the following are oxidation-reduction reactions. If a reaction is a redox reaction, name the element oxidized and the element reduced. *(Chapter 20)*
a. $CaCO_3(s) \longrightarrow CaO(s) + CO_2(g)$
b. $Ca(s) + Cl_2(g) \longrightarrow CaCl_2(s)$
c. $CaCl_2(aq) + 2AgNO_3(aq) \longrightarrow Ca(NO_3)_2(aq) + 2AgCl(s)$
d. $Ca(s) + 2H_2O(l) \longrightarrow Ca(OH)_2(aq) + 2H_2(g)$

92. Balance each redox equation. *(Chapter 20)*
a. $Br_2(g) + NaOH(aq) \longrightarrow NaBrO_3(aq) + NaBr(aq) + H_2O(l)$
b. $Fe_2O_3(s) + H_2(g) \longrightarrow Fe(s) + H_2O(l)$

93. What is the oxidation number of the italicized element in each formula? *(Chapter 20)*
a. $K_2\mathit{Cr}_2O_7$ **b.** $K\mathit{I}O_3$
c. $\mathit{Mn}O_4^-$ **d.** $\mathit{Fe}Cl_3$

Standardized Test Prep

Test-Taking Tip

Eliminate Wrong Answers If you don't know which response is correct, start by eliminating those you know are wrong. If you can rule out some choices, you'll have fewer left to consider and you'll increase your chances of choosing the correct answer.

Select the choice that best answers each question or completes each statement.

1. Which statement describes electrolysis?
   - **a.** Reduction occurs at the anode.
   - **b.** Energy is produced.
   - **c.** Oxidation occurs at the cathode.
   - **d.** Positive ions move to the cathode.
2. Which of these statements about rusting is true?
   I. Iron is oxidized.
   II. Oxygen is the oxidizing agent.
   III. Iron atoms gain electrons.
   - **a.** I and II only
   - **b.** II and III only
   - **c.** I only
   - **d.** I, II, and III
3. The energy source for an ordinary flashlight is a
   - **a.** dry cell.
   - **b.** voltaic cell.
   - **c.** battery.
   - **d.** all of the above
4. Magnesium metal is prepared by the electrolysis of molten $MgCl_2$. One half-reaction is $Mg^{2+}(l) + 2e^- \longrightarrow Mg(l)$.
   - **a.** This half-reaction occurs at the cathode.
   - **b.** Magnesium ions are oxidized.
   - **c.** Chloride ions are reduced at the anode.
   - **d.** Chloride ions gain electrons during this process.
5. If the cell potential for a redox reaction is positive,
   - **a.** the redox reaction is spontaneous.
   - **b.** the redox reaction is not spontaneous.
   - **c.** the reaction only occurs during electrolysis.
   - **d.** More than one statement is correct.

Use the data table to answer Questions 6–12. Hydrogen is included as a reference point for the metals.

Activity Series of Selected Metals

Element	Oxidation half-reaction
Lithium	$Li(s) \longrightarrow Li^+(aq) + e^-$
Potassium	$K(s) \longrightarrow K^+(aq) + e^-$
Sodium	$Na(s) \longrightarrow Na^+(aq) + e^-$
Aluminum	$Al(s) \longrightarrow Al^{3+}(aq) + 3e^-$
Zinc	$Zn(s) \longrightarrow Zn^{2+}(aq) + 2e^-$
Iron	$Fe(s) \longrightarrow Fe^{2+}(aq) + 2e^-$
Hydrogen	$H_2(g) \longrightarrow 2H^+(aq) + 2e^-$
Copper	$Cu(s) \longrightarrow Cu^{2+}(aq) + 2e^-$

6. Which metal will more easily lose an electron, sodium or potassium?
7. Which metal is more easily oxidized, copper or aluminum?
8. What is the relationship between ease of oxidation and the activity of a metal?
9. Describe what would happen if you placed a clean strip of aluminum in a solution of copper(II) sulfate. Explain your answer.
10. Would a copper strip placed in a solution containing zinc ions react spontaneously with the zinc ions? Explain your reasoning.
11. Based on the positions of zinc and iron in the table, explain how attaching zinc blocks to a steel ship hull protects the steel from corrosion.
12. Write the half-reaction for the reduction of aluminum ions.
13. An electrolytic cell is shown below. Draw this cell on a separate piece of paper and label the anode, cathode, and direction of electron flow.

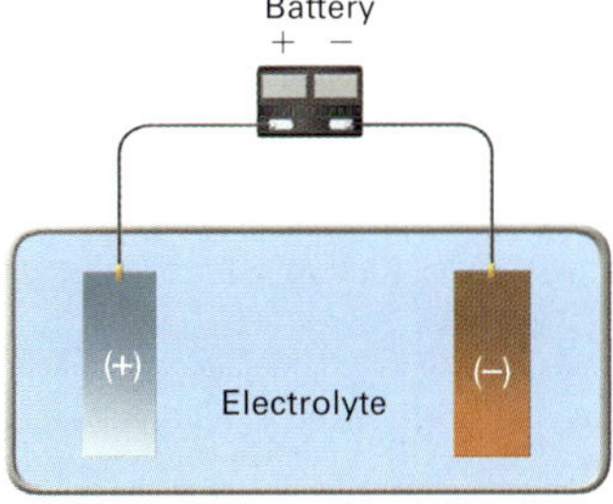

Standardized Test Prep

1. d
2. a
3. d
4. a
5. a
6. potassium
7. aluminum
8. The more active the metal, the more easily it is oxidized.
9. Aluminum would be oxidized and copper ions would be reduced. Copper would plate out on the aluminum.
10. No, because zinc is more easily oxidized than copper, copper cannot reduce the zinc ions in solution.
11. Zinc is more readily oxidized than iron. If iron atoms lose electrons through oxidation, zinc can act as a reducing agent to replace those lost electrons.
12. $Al^{3+}(aq) + 3e^- \rightarrow Al(s)$
13. Answers should look like the labels on Figure 21.12b on page 679.

CHAPTER 22

Hydrocarbon Compounds Planning Guide

SECTION OBJECTIVES	STANDARDS NATIONAL	STANDARDS STATE	ACTIVITIES and LABS
22.1 Hydrocarbons, pp. 693–701 1 block or 2 periods **22.1.1 Describe** the relationship betwen the number of valence electrons and bonding in carbon. **22.1.2 Define** and **describe** alkanes. **22.1.3 Relate** the polarity of hydrocarbons to their solubility.	A-1, A-2, B-2, G-1, G-2		**SE** Inquiry Activity: What Dissolves What? p. 692 L2 **TE** Teacher Demo: Methane Shapes, p. 694 L2 **TE** Class Activity: Model of Ethane, p. 695 L2 **TE** Teacher Demo: Alkane Structures, p. 696 L2 **TE** Teacher Demo: Properties of Alkanes, p. 697 L2 **TE** Teacher Demo: Mixing Oil and Water, p. 700 L2
22.2 Unsaturated Hydrocarbons, pp. 702–703 1 block or 2 periods **22.2.1 Describe** the difference between unsaturated and saturated hydrocarbons. **22.2.2 Distinguish** between the structures of alkenes and alkynes.	B-2		**TE** Class Activity: Double Bond Rigidity, p. 702 L2
22.3 Isomers, pp. 704–708 1 block or 2 periods **22.3.1 Explain** why structural isomers have different properties. **22.3.2 Describe** the conditions under which geometric isomers are possible. **22.3.3 Identify** optical isomers.	A-1, B-2, C-1		**TE** Class Activity: Modeling Isomers, p. 704 L2 **SE** Quick Lab: Structural Isomers of Heptane, p. 706 L2 **SE** Small-Scale Lab: Hydrocarbon Isomers, p. 708 L2 **SSLM** Lab 37: Molecular Structure of Hydrocarbons L2
22.4 Hydrocarbon Rings, pp. 709–711 1 block or 2 periods **22.4.1 Identify** cyclic ring structures. **22.4.2 Describe** bonding in benzene.	B-2		**LM** Lab 49: Hydrocarbons: A Structural Study L2 **LP** Lab Practical 22-1: Model Building with Organic Compounds L2
22.5 Hydrocarbons From Earth's Crust, pp. 712–717 1 block or 2 periods **22.5.1 Identify** three important fossil fuels and describe their origins. **22.5.2 Describe** the composition of natural gas, petroleum, and coal. **22.5.3 Describe** what happens when petroleum is refined.	B-2, E-2		**TE** Class Activity: Crude Oil, p. 713 L2

Ability Levels	
L1	For students who need additional help
L2	For all students
L3	For students who need to be challenged

Components

SE	Student Edition	LM	Laboratory Manual	GCP	Graphing Calculator Problems	T	Transparencies
TE	Teacher's Edition	SSLM	Small-Scale Chemistry Lab Manual	TTT	Test-Taking Tips with Transparencies, Grades 9–12 Science	CA	Chem Alive and Chemistry Field Trips DVD
GRSW	Guided Reading and Study Workbook	LP	Laboratory Practicals	STP	Standardized Test Preparation Workbook	iT	Interactive Textbook with ChemASAP
CTR	Core Teaching Resources	PLM	Probeware Lab Manual			GO	Go Online Internet Resources
SSG	Spanish Study Guide	VCL	Virtual Chemistry Labs				
		CTB	Computer Test Bank				

RESOURCES PRINT and TECHNOLOGY	PROBLEMS and ASSESSMENT
GRSW Chapter 22, Section 1 L1 T T253–T255 L2 GO Chapter 22, Section 1 L2 GO Chapter 22, Section 1 L2	**CP** 22.1, Drawing Structural Formulas for Alkanes **CP** 22.2, Naming Branched-Chain Alkanes **CP** 22.3, Drawing Structural Formulas for Branched-Chain Alkanes **SE** Section 22.1 Assessment, p. 701 **iT** Section 22.1
GRSW Chapter 22, Section 2 L1 T T256 L2	**SE** Section 22.2 Assessment, p. 703 **iT** Section 22.2
GRSW Chapter 22, Section 3 L1 T T257–T258 L2	**CP** 22.4, Identifying Asymmetric Carbon Atoms **SE** Section 22.3 Assessment, p. 707 **iT** Section 22.3
GRSW Chapter 22, Section 4 L1 T T259 L2 GO Chapter 22, Section 4 L2	**SE** Section 22.4 Assessment, p. 711 **iT** Section 22.4
GRSW Chapter 22, Section 5 L1 T T260–T261 L2 GO Chapter 22, Section 5 L2	**SE** Section 22.5 Assessment, p. 715 **iT** Section 22.5

Web Code: cdn-1221
Web Code: cdn-1224
Web Code: cdn-1225

Web Code: cdb-1221

MATERIALS List

Quantities for each group

STUDENT EDITION

Inquiry Activity, p. 692
candle wax, vegetable oil, petroleum jelly, rubbing alcohol, mineral oil, water, 3 square glass plates, toothpicks

Quick Lab, p. 706
ball-and-stick molecular model kits

Small-Scale Lab, p. 708
toothpicks, modeling clay

TEACHER'S EDITION

Teacher Demo, p. 694
colored foam balls, applicator sticks

Class Activity, p. 695
foam balls and applicator sticks or molecular model sets

Teacher Demo, p. 696
foam balls and applicator sticks or molecular model sets

Teacher Demo, p. 697
samples of gas (methane), liquid (hexane), and solid (candle wax) alkanes, Table 1

Class Activity, p. 698
different colored markers

Teacher Demo, p. 700
large test tube with stopper, water, mineral or baby oil, food coloring

Class Activity, p. 704
molecular model sets

Teacher Demo, p. 705
two ball-and-stick models of CHFClBr

Chapter Assessment

CHAPTER ASSESSMENT
SE Chapter Assessment, pp. 719–722
CTB Chapter 22
iT Chapter 22

STANDARDIZED TEST PREP
SE Chapter 22, p. 723
STP Topic 10
iT Interactive Textbook with Assessment at PHSchool.com

CHAPTER 22

Hydrocarbon Compounds

What Dissolves What? L2

Purpose

Students identify general solubility trends in hydrocarbons.

 **Prep Time** 10 minutes

Class Time 20 minutes

Safety Make sure there are no open flames in the classroom and that the room is well ventilated.

Think About It

1. Petroleum jelly and vegetable oil are soluble in mineral oil. Candle wax does not dissolve in mineral oil. None of the materials is soluble in water.
2. The solubility of these materials in rubbing alcohol is more like that in water.
3. Materials that dissolve in mineral oil will probably not dissolve in water or rubbing alcohol.

After students study Section 22.1, you may want to have them review their answers to Question 3, revise the answers (if needed), and offer explanations for the general statement about solubility.

CHAPTER 22

Hydrocarbon Compounds

An oil rig drills deep into the ocean floor in search of petroleum, a valuable mixture of hydrocarbons.

INQUIRY Activity

What Dissolves What?

Materials

candle wax, vegetable oil, petroleum jelly, rubbing alcohol, mineral oil, water, three square glass plates, nine toothpicks

Procedure

1. Place 3 pea-sized portions of petroleum jelly 10 cm apart on a glass plate.
2. Add 5 drops of rubbing alcohol to one portion of the petroleum jelly, 5 drops of mineral oil to another, and 5 drops of water to the third.
3. Use a clean toothpick to mix each portion as completely as possible. Observe whether the petroleum jelly appears to dissolve.
4. On a second glass plate repeat Steps 1 through 3 using pea-sized portions of candle wax.
5. On a third glass plate, repeat Steps 1 through 3 using pea-sized portions of vegetable oil.

Think About It

1. How did the solubility of these three materials (petroleum jelly, candle wax, and vegetable oil) in water compare with their solubility in mineral oil?
2. Was the solubility of these materials in rubbing alcohol more like that in water or in mineral oil?
3. Use your observations to develop a general statement about the solubility of the materials you tested.

692

22.1 Hydrocarbons

Connecting to Your World The gasoline used in cars and lawnmowers, the diesel fuel used in trucks and buses, and the kerosene used in lanterns and camping stoves are examples of liquid fuels. Often, a barbecue grill uses a gaseous fuel. A solid fuel, coal, produced the steam for the locomotives that pulled old-time trains. All these fuels are mixtures of compounds called hydrocarbons. In this section, you will learn about the structure and properties of hydrocarbons.

Organic Chemistry and Hydrocarbons

Fewer than 200 years ago, it was thought that only living organisms could synthesize the carbon compounds found in their cells. So these compounds were classified as organic compounds and the study of these compounds as organic chemistry. Many people thought that a mysterious vital force directed the formation of carbon compounds. A German chemist, Friedrich Wöhler (1800–1882), refuted this idea in 1828. He was able to use inorganic substances to synthesize urea—a carbon compound found in urine. Today, organic chemistry includes the chemistry of almost all carbon compounds, regardless of their origin.

In a reference book that lists properties of common compounds, the list of organic compounds is much longer than the list of inorganic compounds. In fact, there are more than a million organic compounds. By definition, a compound contains at least two elements. The simplest organic compounds contain only carbon and hydrogen and are called **hydrocarbons.** The two simplest hydrocarbons are methane and ethane.

Methane (CH_4) is the major component of natural gas. It is sometimes called marsh gas because it is formed by the action of bacteria on decaying plants in swamps and other marshy areas. Livestock and termites also emit substantial quantities of methane as a product of digestion. Recall that a carbon atom has four valence electrons and a hydrogen atom has one valence electron. So one carbon atom can form a single covalent bond with four hydrogen atoms.

$$\cdot\dot{\underset{\cdot}{C}}\cdot \quad + \quad 4H\cdot \quad \longrightarrow \quad H:\overset{H}{\underset{H}{\ddot{\underset{..}{C}}}}:H$$

Carbon atom — Hydrogen atoms — Methane molecule

Because carbon has four valence electrons, a carbon atom always forms four covalent bonds. Remembering this principle will help you to write correct structures for organic compounds.

Guide for Reading

Key Concepts
- How is the number of valence electrons in carbon atoms related to the bonds that carbon atoms form?
- What are two possible arrangements of carbon atoms in an alkane?
- In terms of their polarity, what type of molecules are alkanes?

Vocabulary
hydrocarbons
alkane
straight-chain alkanes
homologous series
condensed structural formula
substituent
alkyl group
branched-chain alkane

Reading Strategy
Relating Text and Visuals As you read, look at Figure 22.1. Explain how the formulas and models can help you understand the structure of methane and ethane molecules.

Animation 28 Get a glimpse of the staggering variety of hydrocarbon compounds.
with ChemASAP

Section Resources

Print
- **Guided Reading and Study Workbook,** Section 22.1
- **Core Teaching Resources,** Section 22.1 Review
- **Transparencies,** T253–T255

Technology
- **Interactive Textbook with ChemASAP,** Animation 28; Problem-Solving 22.1, 22.3, 22.6; Assessment 22.1
- **Go Online,** Section 22.1

22.1

1 FOCUS

Objectives

22.1.1 Describe the relationship between number of valence electrons and bonding in carbon.
22.1.2 Define and describe alkanes.
22.1.3 Relate the polarity of hydrocarbons to their solubility.

Guide for Reading

Build Vocabulary 

Paraphrase Have students use simpler words or phrases to rewrite the definition of *homologous series*.

Reading Strategy 

Sequence As students read the rules for naming branched-chain alkanes, have them make a graphic organizer to summarize the steps.

2 INSTRUCT

Connecting to Your World

Have students read the introduction. Then ask, **What do many fuels have in common?** *(They are mixtures of compounds called hydrocarbons.)* **Based on its name, predict the composition of a hydrocarbon.** *(Hydrocarbons are compounds that contain only hydrogen and carbon.)*

Organic Chemistry and Hydrocarbons

Discuss 

To review bonding ask, **What is the difference between an ionic bond and a covalent bond?** *(Electrons are transferred between atoms in ionic bonds and shared in covalent bonds.)* **What type of bonding occurs in carbon?** *(covalent bonding)* **What are valence electrons?** *(electrons in the highest occupied energy level of an atom)*

TEACHER Demo

Methane's Shape L2

Purpose Students observe the three-dimensional shape of the simplest hydrocarbon, methane.

Materials colored foam balls, applicator sticks.

Procedure Use colored foam balls and applicator sticks to construct models that are large enough for all the students to see. Review the tetravalency of carbon as well as the octet rule. To assess students' knowledge about molecular shape, ask, **Why does methane have a tetrahedral shape?** *(VSEPR theory states that, because electron pairs repel each other, electron pairs are as far apart as possible in molecules.)* Explain that one of the key ideas of organic chemistry is "structure dictates function." The arrangement of atoms is as important as the number and type of atoms.

Alkanes

Use Visuals 

Table 22.1 Use an overhead transparency to display Table 22.1. Explain the IUPAC system for naming alkanes. Suggest that students memorize the names of the first ten alkanes. Ask, **Why are all of these compounds considered straight-chain alkanes?** *(The carbon atoms form an unbroken line. Each carbon atom within the chain is bonded to no more than two other carbon atoms.)*

Figure 22.1 Different types of formulas and models can be used to represent hydrocarbons such as methane and ethane. **Interpreting Diagrams** ***What does a stick in a ball-and-stick model represent?***

Structural formula	Ball-and-stick model	Space-filling model
Methane (CH_4) — Line represents shared electrons		
Ethane (C_2H_6)		

Methane is not typical of the vast majority of organic compounds, because there isn't a bond between carbon atoms in a methane molecule. There is a carbon-carbon bond in ethane. Like methane, ethane (C_2H_6) is a gas at standard temperature and pressure. In an ethane molecule, two carbon atoms share a pair of electrons. The remaining six valence electrons form bonding pairs with the electrons from six hydrogen atoms.

$$2\cdot\dot{\text{C}}\cdot + 6\text{H}\cdot \longrightarrow \cdot\dot{\text{C}}\text{:}\dot{\text{C}}\cdot + 6\text{H}\cdot \longrightarrow \text{H}_3\text{C:CH}_3$$

Carbon atoms, Hydrogen atoms, Ethane molecule

The ability of carbon to form stable carbon-carbon bonds is one reason that carbon can form so many different compounds.

Figure 22.1 shows the molecular formulas, structural formulas, ball-and-stick models, and space-filling models for methane and ethane. Structural formulas are a convenient way to show the arrangement of atoms in a molecule. But two-dimensional structural formulas do not provide accurate information about how the atoms in a molecule are arranged in space. Three-dimensional molecular models represent the shapes of molecules more accurately. Throughout this chapter and the next, ball-and-stick models and space-filling models will be used along with structural formulas to represent organic molecules. The bonding theories you studied in Section 8.3—hybrid orbital theory and VSEPR theory—are used to predict the molecular shapes.

 ***What is the main advantage of a molecular model as compared to a structural formula?***

Alkanes

Methane and ethane are examples of alkanes. An **alkane** is a hydrocarbon in which there are only single covalent bonds. In any alkane, all the carbon-carbon bonds are single covalent bonds, and all the other bonds are carbon-hydrogen bonds. **The carbon atoms in an alkane can be arranged in a straight chain or in a chain that has branches.**

Differentiated Instruction

English Learners 

Encourage students to preview the chapter looking for words in bold print. Have students compile a glossary in which they define each term in English and in their native language. Suggest that students also include illustrations when appropriate.

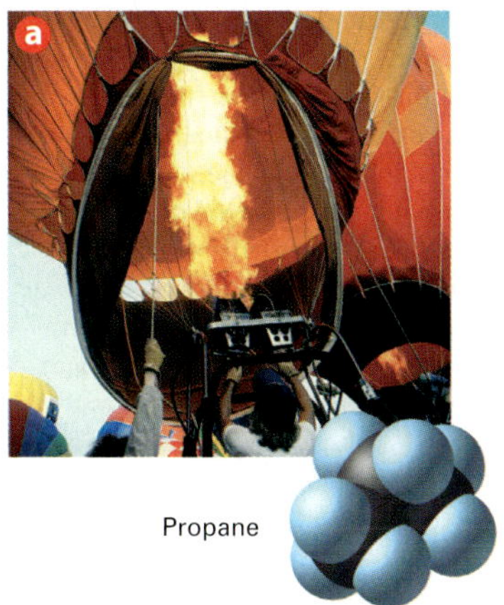

Figure 22.2 Hydrocarbons are used as fuels. **a** Pressurized tanks of propane are used to fuel the burners in hot-air balloons. **b** Butane serves as the fuel for many lighters.

Straight-Chain Alkanes Ethane is the simplest of the **straight-chain alkanes,** which contain any number of carbon atoms, one after the other, in a chain. Propane (C_3H_8) has three carbon atoms bonded in a chain with eight electrons shared with eight hydrogen atoms. Butane (C_4H_{10}) has a chain of four carbons and ten hydrogens. Figure 22.2 shows how propane and butane are used.

Table 22.1 shows the straight-chain alkanes containing up to ten carbons. The straight-chain alkanes are an example of a homologous series. A group of compounds forms a **homologous series** if there is a constant increment of change in molecular structure from one compound in the series to the next. A CH_2 group is the increment of change in straight-chain alkanes. This change can be traced in the structural formulas in Table 22.1. Notice in Table 22.1 that the boiling points of the straight-chain alkanes increase as the number of carbons in the chain increase. The melting points also increase in the same way. The change reflects the increase in molar mass from methane to decane.

Word *Origins*

Homologous comes from the Greek prefix *homo,* meaning "same," and *logos,* meaning "word." The homologous series of alkanes have the same general formula C_nH_{2n+2} where *n* is any integer. **What is the formula for a 14-carbon alkane?**

Table 22.1

The First Ten Straight-Chain Alkanes

Name	Molecular formula	Structural formula	Boiling point (°C)
Methane	CH_4	CH_4	−161.0
Ethane	C_2H_6	CH_3CH_3	−88.5
Propane	C_3H_8	$CH_3CH_2CH_3$	−42.0
Butane	C_4H_{10}	$CH_3CH_2CH_2CH_3$	0.5
Pentane	C_5H_{12}	$CH_3CH_2CH_2CH_2CH_3$	36.0
Hexane	C_6H_{14}	$CH_3CH_2CH_2CH_2CH_2CH_3$	68.7
Heptane	C_7H_{16}	$CH_3CH_2CH_2CH_2CH_2CH_2CH_3$	98.5
Octane	C_8H_{18}	$CH_3CH_2CH_2CH_2CH_2CH_2CH_2CH_3$	125.6
Nonane	C_9H_{20}	$CH_3CH_2CH_2CH_2CH_2CH_2CH_2CH_2CH_3$	150.7
Decane	$C_{10}H_{22}$	$CH_3CH_2CH_2CH_2CH_2CH_2CH_2CH_2CH_2CH_3$	174.1

Word *Origins* L2

($C_{14}H_{30}$) To help students remember the definition of a homologous series, explain that *logical* is related to *logos*. One meaning of *logical* is "necessary or to be expected because of what has gone before."

CLASS Activity

Model of Ethane 

Purpose students construct a ball-and-stick model of ethane.

Materials molecular model sets or foam balls and applicator sticks

Procedure Have students construct a ball-and-stick model of ethane. Check the models for tetrahedral geometry at each carbon. Ask, **How many covalent bonds are there in ethane?** *(7)* **Why are three-dimensional models a more accurate representation of organic molecules than are two-dimensional models?** *(Three-dimensional models convey the spatial relationships among the atoms.)*

Differentiated Instruction

Less Proficient Readers 

Have students list the section objectives in their notebooks, leaving space after each objective. As they read the section, they should list terms and phrases that provide the information requested for each objective. Check their entries for accuracy.

Answers to...

Figure 22.1 shared electron pair in covalent bond

A molecular model shows the three-dimensional shape of a molecule as well as the arrangement of its atoms.

Download a worksheet on **Alkanes** for students to complete, and find additional teacher support from NSTA SciLinks.

Alkane Structures L2

Purpose Students compare molecular models and structural formulas of alkanes.

Materials foam balls and applicator sticks or molecular model sets

Procedure Write the complete structural formulas for methane, ethane, propane, and butane on the board. Ask students how the structures are similar and how they are different. Then show students molecular models for each of the compounds. Pull apart an ethane model and show how a CH_2 fragment can be inserted between the CH_3 (methyl) sections to make larger alkanes. Use the models of propane and butane to show how the tetrahedral arrangement of the atoms around each carbon results in a "zig-zag" shape for the carbon backbone. Illustrate the rotational freedom of the atoms in a saturated hydrocarbon by twisting the models of propane and butane about the carbon-carbon axis. Once students understand how the atoms are joined together in the larger alkanes, write the condensed formulas under the complete structural formulas for ethane, propane, and butane. Explain that the condensed formulas are a convenient shorthand notation.

For: Links on Alkanes
Visit: www.SciLinks.org
Web Code: cdn-1221

The names of the alkanes listed in Table 22.1 follow rules established by the International Union of Pure and Applied Chemistry (IUPAC). Every alkane has a name that ends with the suffix *-ane*. For the straight-chain alkanes with one to four carbon atoms, the official names and the common names are the same. They are methane, ethane, propane, and butane, respectively. A mixture of Latin and Greek prefixes are used to name the hydrocarbons having straight chains longer than four carbon atoms. The prefixes are *pent-* for 5, *hex-* for 6, *hept-* for 7, *oct-* for 8, and so on.

To draw a structural formula for a straight-chain alkane, write the symbol for carbon as many times as necessary to get the proper chain length. Then complete the formula with hydrogens and lines representing covalent bonds. Complete structural formulas show all the atoms and bonds in a molecule. Sometimes, however, shorthand or condensed structural formulas work just as well. In a **condensed structural formula,** some bonds and/or atoms are left out of the structural formula. Although the bonds and atoms do not appear, you must understand that they are there. Table 22.2 shows several ways to draw condensed structural formulas for butane.

 ***What suffix is used in the name of an alkane?***

Table 22.2

Formulas for Butane

Formula	Description
C_4H_{10}	Molecular formula
$\begin{array}{ccccccccc} & & H & & H & & H & & H & \\ & & \vert & & \vert & & \vert & & \vert & \\ H & - & C & - & C & - & C & - & C & - & H \\ & & \vert & & \vert & & \vert & & \vert & \\ & & H & & H & & H & & H & \end{array}$	Complete structural formula
$CH_3-CH_2-CH_2-CH_3$	Condensed structural formula; C — H bonds understood
$CH_3CH_2CH_2CH_3$	Condensed structural formula; C — H and — C — C — bonds understood
$CH_3(CH_2)_2CH_3$ Subscript Methylene unit	Condensed structural formula; all bonds understood; parentheses indicate CH_2 units are linked together in a continuous chain (the — CH_2 — unit is called a methylene group); subscript 2 to the right of parenthesis indicates two methylene groups are linked together
C — C — C — C	Carbon skeleton; all hydrogens and C — H bonds understood
╱╲╱	Line-angle formula; all carbons and hydrogens understood; carbon atoms are located at each intersection and at the ends of lines

CONCEPTUAL PROBLEM 22.1

Drawing Structural Formulas for Alkanes

Draw complete structural formulas for the straight-chain alkanes that have three and four carbons.

 **Analyze** ***Identify the relevant concepts.*** In an alkane, each carbon atom forms four covalent bonds to hydrogen or to other carbon atoms. Because these are straight-chain alkanes, the appropriate number of carbons will be written in a straight line and then connected to each other by single bonds. The chain will then be filled in with hydrogen atoms so that each carbon forms four covalent bonds.

 **Solve** ***Apply concepts to this situation.*** A three-carbon straight-chain alkane has three carbon atoms in a straight line. The center carbon bonds to two hydrogen atoms. Each of the two end carbons bonds to three hydrogen atoms. A four-carbon straight-chain alkane has four carbon atoms in a straight line. Each of the two center carbons bonds to two hydrogen atoms, and each of the two end carbons bond to three hydrogen atoms.

```
    H   H   H             H   H   H   H
    |   |   |             |   |   |   |
H - C - C - C - H     H - C - C - C - C - H
    |   |   |             |   |   |   |
    H   H   H             H   H   H   H
```

Practice Problems

1. Draw complete structural formulas for the straight-chain alkanes with five and six carbons.
2. How many single bonds are there in a propane molecule?

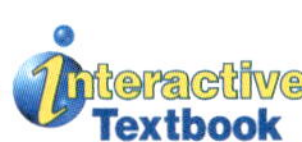

Problem-Solving 22.1 Solve Problem 1 with the help of an interactive guided tutorial.

with **ChemASAP**

Branched-Chain Alkanes Alkanes and other hydrocarbons do not always have carbon atoms bonded in straight chains. Because a carbon atom forms four covalent bonds, it can bond not only to one or two other carbon atoms, but also to three or even four other carbons, resulting in branched chains. In organic chemistry, branches on a hydrocarbon chain are discussed as if they were substituted for a hydrogen atom on the chain. An atom or group of atoms that can take the place of a hydrogen atom on a parent hydrocarbon molecule is called a **substituent.** Look at the diagrams below. The longest continuous carbon chain of a branched-chain hydrocarbon is called the *parent* alkane. All other carbon atoms or groups of carbon atoms are regarded as substituents. In Chapter 23, you will study compounds in which atoms of halogens, oxygen, nitrogen, sulfur, and phosphorus can take the place of a hydrogen atom on the carbon chain.

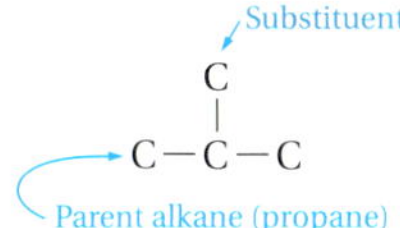

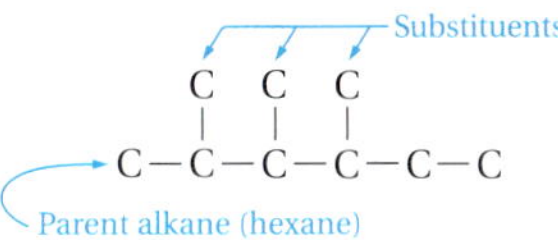

CONCEPTUAL PROBLEM 22.1

Answers

1.

```
    H   H   H   H   H
    |   |   |   |   |
H - C - C - C - C - C - H
    |   |   |   |   |
    H   H   H   H   H

    H   H   H   H   H   H
    |   |   |   |   |   |
H - C - C - C - C - C - C - H
    |   |   |   |   |   |
    H   H   H   H   H   H
```

2. 10

TEACHER Demo

Properties of Alkanes 

Purpose Students compare some properties of alkanes.

Materials samples of methane gas, liquid hexane, and solid candle wax, Table 22.1 transparency

Procedure If possible, bring samples of methane, hexane, and candle wax to class. With the condensed formulas on the board and Table 22.1 on the overhead projector, ask, **What trend do you see regarding the number of carbon atoms and the boiling point of the alkanes?** *(the longer the carbon atom chain, the higher the boiling point)* **What can you conclude about the attraction between molecules as the number of carbon atoms increases?** *(The intermolecular attractions increase.)*

Differentiated Instruction

Gifted and Talented 

Have students research sources of methane and methane's relationship to the greenhouse effect and global warming. Students can start with the discussion of greenhouse gases on R22. Questions to be answered could include:

- What is the greenhouse effect, and how does methane contribute to it?
- What are the major sources of methane in Earth's atmosphere?
- Is the methane content of Earth's atmosphere increasing or decreasing?

Answers to...

Checkpoint *-ane*

Section 22.1 (continued)

Discuss L2

Discuss how branched-chain alkanes differ from straight-chain alkanes. Use ball-and-stick models to show how straight-chain butane (C_4H_{10}) can be rearranged as a branched-chain structure (2-methylpropane) with the same molecular formula. Stress that each arrangement of atoms represents a different compound with different physical and chemical properties. Show models of several other branched-chain alkanes. Use an overhead transparency to display the IUPAC rules for naming branched-chain alkanes, and use the rules to name the branched-chain structures you have built.

CLASS Activity

Distinguishing Parent Structures From Substituents L1

Purpose Students distinguish the parent structures in a branched-chain hydrocarbon from the substituents.

Materials different colored markers

Procedure Write structural formulas for branched-chain hydrocarbons on the board in order from simplest to most complex. Include formulas in which the longest chain is not drawn in a straight line. Have students use one color to circle the carbons in the parent structure, another to circle the methyl groups, and so on.

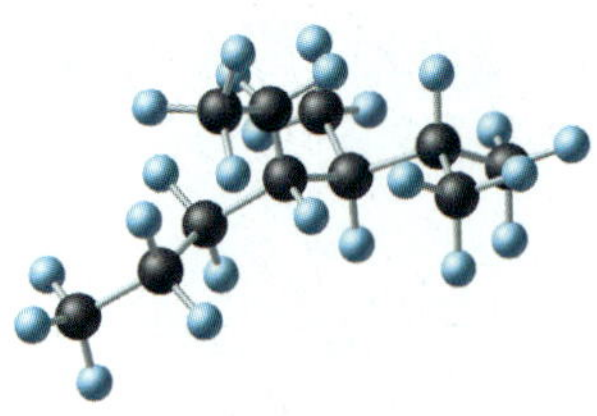

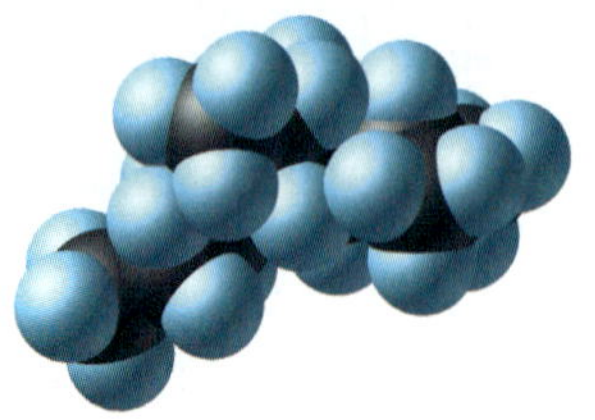

Figure 22.3 Ball-and-stick and space-filling models show the arrangement of atoms in 4-ethyl-2,3-dimethylheptane. **Interpreting Diagrams** ***Based on the ball-and-stick model, how many carbons are there in the longest straight chain?***

A hydrocarbon substituent is called an **alkyl group.** An alkyl group can be one carbon or several carbons long. Alkyl groups are named by removing the *-ane* ending from the parent hydrocarbon name and adding *-yl.* The three smallest alkyl groups are the methyl group (CH_3—), the ethyl group (CH_3CH_2—), and the propyl group ($CH_3CH_2CH_2$—). An alkyl group consists of an alkane with one hydrogen removed.

When a substituent alkyl group is attached to a straight-chain hydrocarbon, branches are formed. An alkane with one or more alkyl groups is called a **branched-chain alkane.** The IUPAC rules for naming branched-chain alkanes are quite straightforward. The name of a branched-chain alkane is based on the name of the longest continuous carbon chain. Each alkyl substituent is named according to the length of its chain and numbered according to its position on the main chain. The compound with the following structural formula can be used as an example. Figure 22.3 shows models of the compound.

$$\begin{array}{ccccccccccccc}
7 & & 6 & & 5 & & 4 & & 3 & & 2 & & 1 \\
CH_3 & - & CH_2 & - & CH_2 & - & CH & - & CH & - & CH & - & CH_3 \\
 & & & & & & | & & | & & | & & \\
 & & & & & & CH_2 & & CH_3 & & CH_3 & & \\
 & & & & & & | & & & & & & \\
 & & & & & & CH_3 & & & & & &
\end{array}$$

1. Find the longest chain of carbons in the molecule. This chain is considered the parent structure. In the example, the longest chain contains seven carbon atoms. Therefore, the parent hydrocarbon structure is heptane.
2. Number the carbons in the main chain in sequence. To do this, start at the end that will give the groups attached to the chain the smallest numbers. This has already been done in the example. As you can see, numbering the chain from right to left places the substituent groups at carbon atoms 2, 3, and 4. If the chain were numbered from left to right, the groups would be at positions 4, 5, and 6. These higher numbers would violate the rule.
3. Add numbers to the names of the substituent groups to identify their positions on the chain. These numbers become prefixes to the name of the alkyl group. In this example the substituents and positions are 2-methyl, 3-methyl, and 4-ethyl.
4. Use prefixes to indicate the appearance of the same group more than once in the structural formula. Common prefixes are *di-* (twice), *tri-* (three times), *tetra-* (four times), and *penta-* (five times). This example has two methyl substituents. Thus, *dimethyl* will be part of the complete name.
5. List the names of alkyl substituents in alphabetical order. For purposes of alphabetizing, ignore the prefixes *di-*, *tri-*, and so on. In this example, the 4-ethyl group is listed before the 2-methyl and 3-methyl groups (which are combined as 2,3-dimethyl in the name).
6. Use proper punctuation. This is very important in writing the names of organic compounds in the IUPAC system. Commas are used to separate numbers. Hyphens are used to separate numbers and words. The entire name is written without any spaces.

Facts and Figures

Carbocations

George Olah won the 1994 Nobel Prize in Chemistry for his work with hydrocarbons. Olah showed that some chemical reactions involving hydrocarbons proceed through short-lived, highly reactive, intermediate compounds which he called *carbocations.* Olah was able to produce, isolate, and preserve carbocations so that they could be studied in detail.

According to the IUPAC rules, the name of the compound in the example is 4-ethyl-2,3-dimethylheptane. Note that the name of the parent alkane, heptane, follows directly after the final prefix, dimethyl. (It would be incorrect to write the name 4-ethyl-2,3-dimethyl heptane.)

CONCEPTUAL PROBLEM 22.2

Naming Branched-Chain Alkanes

Name this compound using the IUPAC system. Notice that the longest chain is not written in a straight line.

```
              CH3
              |
CH3 — CH2 — C — CH3
              |
              CH2
              |
              CH2
              |
              CH3
```

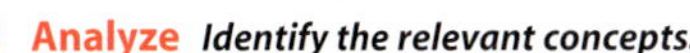

1 Analyze ***Identify the relevant concepts.***
The parent structure is the longest chain of carbons. All other groups are substituents. The carbons are numbered to give the first substituent the lowest possible number.

These location numbers become part of the name as prefixes. The names of the substituents are listed in alphabetical order with numbers separated by commas and numbers and words separated by hyphens.

2 Solve ***Apply concepts to this situation.***
The longest carbon chain in the molecule is hexane (six carbons). There are two methyl substituents on carbon 3, so the prefix is 3,3-dimethyl. The correct IUPAC name is 3,3-dimethylhexane.

Practice Problems

3. Name these compounds according to the IUPAC system.

a.
```
CH2 — CH2 — CH — CH2 — CH3
|            |
CH3          CH2
             |
             CH3
```

b.
```
CH3 — CH2 — CH — CH3
             |
             CH3
```

4. Name these compounds according to the IUPAC system.

a.
```
        H    CH3  H
        |    |    |
CH3 — C — C — C — CH3
        |    |    |
        CH3  H    H
```

b.
```
CH3 — CH — CH2 — CH2 — CH3
        |
        CH3
```

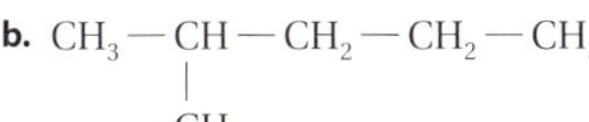

Problem-Solving 22.3 Solve Problem 3 with the help of an interactive guided tutorial.
with **ChemASAP**

With an alkane name and knowledge of the IUPAC rules, it is easy to reconstruct the structural formula.

1. Find the root word (ending in *-ane*) in the hydrocarbon name. Then write the longest carbon chain to create the parent structure.
2. Number the carbons on this parent carbon chain.
3. Identify the substituent groups. Attach the substituents to the numbered parent chain at the proper positions.
4. Add hydrogens as needed.

CONCEPTUAL PROBLEM 22.2

Answers

3. a. 3-ethylhexane
b. 2-methylbutane
4. a. 2,3-dimethylpentane
b. 2-methylpentane

Practice Problems Plus L2

Name the following compounds according to IUPAC system.

a.
```
CH3 — CH — CH2 — CH2 — CH3
       |
       CH3
```
(2-methylpentane)

b.
```
CH2 — CH2 — CH2 — CH — CH2
|                  |     |
CH3                CH2   CH3
                   |
                   CH3
```
(3-ethylheptane)

Answers to...
Figure 22.3 7

CONCEPTUAL PROBLEM 22.2

Answers

5.
$$\begin{array}{c}\qquad CH_3\\ \qquad |\\ CH_3CHCHCH_2CH_2CH_3\\ \quad |\\ \quad CH_3\end{array}$$

6.
$$\begin{array}{cccccccccccccccc} & & CH_3 & & CH_3 & & CH_3 \\ & & | & & | & & | \\ CH_3 & - & CH & - & CH & - & C & - & CH_2 & - & CH_2 & - & CH_2 & - & CH_3 \\ & & & & & & | \\ & & & & & & CH_2 \\ & & & & & & | \\ & & & & & & CH_3 \end{array}$$

Practice Problems Plus **L2**

Draw structural formulas for these compounds.

a. 3-methylhexane
$$\begin{array}{ccccccc} CH_3 & - & CH & - & CH_2 & - & CH_2 \\ & & | & & & & | \\ & & CH_2 & & & & CH_3 \\ & & | \\ & & CH_3 \end{array}$$

b. 3-ethyl-3,4-dimethyloctane
$$\begin{array}{ccccccccccccccc} & & & & CH_2CH_3 \\ & & & & | \\ CH_3 & - & CH_2 & - & C & - & CH & - & CH_2 & - & CH_2 & - & CH_2 & - & CH_3 \\ & & & & | & & | \\ & & & & CH_3 & & CH_3 \end{array}$$

Properties of Alkanes

TEACHER Demo

Mixing Oil and Water **L1**

Purpose Students observe the inability of water and oil to mix.

Materials large test tube with stopper, water, mineral oil or baby oil, food coloring

Procedure Prepare a large test tube with equal volumes of water and either mineral oil or baby oil. A small amount of food coloring can be added to the water to more clearly demarcate the polar and nonpolar phases. Shake the test tube vigorously and allow the phases to separate. Discuss the rule "like dissolves like." Ask, **Why don't the oil and the water form a solution?** *(Water is a polar compound, oil is a nonpolar compound.)* Review the role that molecular shape plays in determining if a molecule is polar or nonpolar.

CONCEPTUAL PROBLEM 22.3

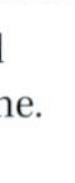

Drawing Structural Formulas for Branched-Chain Alkanes

The compound 2,2,4-trimethylpentane (isooctane) is found in gasoline. Draw a complete structural formula for isooctane.

Analyze ***Identify the relevant concepts.*** Prefixes indicate the types of substituents, the number of times each appears, and their locations on the parent chain. The part of the name that ends in *-ane* indicates the parent structure. The carbons on the parent chain are numbered so that the carbons to which alkyl substituents are attached have the lowest numbers. Hydrogens are added as needed.

2 Solve ***Apply concepts to this situation.*** The parent structure is pentane, which has five carbon atoms. There are two methyl groups on carbon 2 and one on carbon 4. Nine hydrogens are added to complete the structure.

$$\begin{array}{ccccccccc} & & CH_3 & & & & CH_3 \\ & & | & & & & | \\ CH_3 & - & C & - & CH_2 & - & CH & - & CH_3 \\ 1 & & 2| & & 3 & & 4 & & 5 \\ & & CH_3 \end{array}$$

Practice Problems

5. Draw a structural formula for 2,3-dimethylhexane.
6. Draw the structural formula for 4-ethyl-2,3,4-trimethyloctane.

Problem-Solving 22.5 Solve Problem 5 with the help of an interactive guided tutorial.

with ChemASAP

Properties of Alkanes

The electron pair in a carbon-hydrogen or a carbon-carbon bond is shared almost equally by the nuclei of the atoms forming the bond. **Molecules of hydrocarbons, such as alkanes, are nonpolar molecules.** The attractions between nonpolar molecules are weak van der Waals forces. So alkanes of low molar mass tend to be gases or liquids that boil at a low temperature.

Recall the general rule "like dissolves like." Two nonpolar compounds will form a solution, as will two polar compounds. But a nonpolar compound and a polar compound will not form a solution. For example, because oil is a mixture of hydrocarbons, oil and water do not mix. So the oil from an oil spill will float on top of the water, as shown in Figure 22.4.

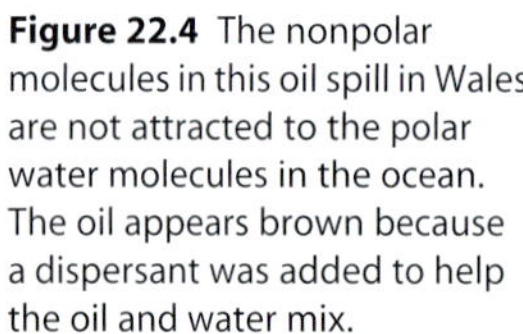

Figure 22.4 The nonpolar molecules in this oil spill in Wales are not attracted to the polar water molecules in the ocean. The oil appears brown because a dispersant was added to help the oil and water mix.

Careers in Chemistry

Organic Chemist

Infrared (IR) spectrophotometry determines the frequencies of infrared radiation absorbed by a molecule. The resulting spectrum is a "fingerprint" of the compound. With nuclear magnetic resonance (NMR) spectrometry, many atoms can be identified by the characteristic radiofrequencies that they absorb when placed in a magnetic field.

Careers in Chemistry

Organic Chemist

An organic chemist might be best described as part inventor and part detective. Most organic chemists use existing carbon-based molecules to synthesize new organic compounds. An organic chemist uses scientific theories and scientific methods to design and build a molecule with a specific set of new properties.

To confirm the identity of new molecules, organic chemists use infrared (IR) spectrophotometry and nuclear magnetic resonance (NMR) spectrometry to analyze their structures.

Companies that manufacture medicines, dyes, cosmetics, or foods are among the companies that employ organic chemists.

To become an organic chemist, you will need at least an undergraduate education in chemistry. In addition to courses in organic chemistry, you must have a firm background in physical, inorganic, and analytical chemistry.

If you are interested in the computer-aided design of bioactive molecules, such as new drugs, biochemistry and computer training are also important. An organic chemist must know how to use laboratory equipment and how to handle chemicals safely.

For: Careers in Chemistry
Visit: PHSchool.com
Web Code: cdb-1221

22.1 Section Assessment

7. **Key Concept** Explain why carbon atoms form four covalent bonds.
8. **Key Concept** What are two ways that carbon atoms can be arranged in an alkane?
9. **Key Concept** Describe the polarity of alkane molecules.
10. Write complete structural formulas for the following alkanes.
    a. propane **b.** pentane
11. Draw a condensed structural formula for 2,2-dimethylbutane.
12. Explain why mineral oil, which is a mixture of hydrocarbons, does not dissolve in water.

Connecting Concepts

Bond Dissociation Energy Except for combustion reactions, alkanes usually do not react easily. Use what you learned about bond dissociation energy in Section 8.2 to explain why alkane molecules tend to be unreactive.

Assessment 22.1 Test yourself on the concepts in Section 22.1.

with ChemASAP

Have students research chemistry-related careers in the library or on the Internet. Students can then construct a table that describes the nature of their work, educational and training requirements, employment outlook, working conditions, and other necessary information.

3 ASSESS

Evaluate Understanding L2

To assess students' knowledge of alkanes, ask, **Why are straight-chain alkanes considered a homologous series?** *(There is a constant increment of change in molecular structure from one compound in the series to the next.)* **What is the difference between straight-chain and branched-chain alkanes?** *(In branched-chain alkanes, substituents are attached to carbons on the parent chain.)*

Reteach L1

Quiz students orally to ensure memorization of the first ten prefixes used in hydrocarbon nomenclature. Then write the following names on the board: 3-ethylpentane, 2,3,4-trimethylhexane, 2,2-dimethylbutane. Have students write structural formulas for these compounds. Remind students to (1) write the alkane carbon skeleton, (2) attach the substituent groups, and (3) add hydrogen atoms to all the unfilled bonding sites.

Connecting Concepts

The high bond dissociation energy indicates a strong covalent bond, one that is hard to break. The bond dissociation energy of a carbon-carbon single bond is 347 kJ/mol.

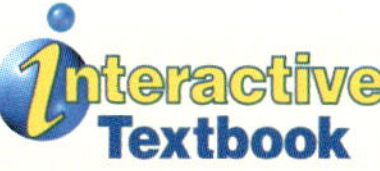

If your class subscribes to the Interactive Textbook, use it to review key concepts in Section 22.1.

with ChemASAP

Section 22.1 Assessment

7. because a carbon atom contains four valence electrons
8. in a straight chain or in a chain with branches
9. Alkane molecules are nonpolar
10. a.
```
    H   H   H
    |   |   |
H — C — C — C — H
    |   |   |
    H   H   H
```
b.
```
    H   H   H   H   H
    |   |   |   |   |
H — C — C — C — C — C — H
    |   |   |   |   |
    H   H   H   H   H
```
11.
```
         CH3
         |
CH3 — C — CH2 — CH3
         |
         CH3
```
12. The nonpolar hydrocarbons in mineral oil will not form a solution with a polar compound such as water.

22.2

1 FOCUS

Objectives

22.2.1 Describe the difference between unsaturated and saturated hydrocarbons.

22.2.2 Distinguish the structures of alkenes and alkynes.

Guide for Reading

Build Vocabulary L2

Graphic Organizer Have students make a compare/contrast table for saturated and unsaturated hydrocarbons.

Reading Strategy L2

Use Prior Knowledge Assess students' prior knowledge of multiple covalent bonds. Ask, **What kind of bonding occurs between carbon and oxygen in a molecule of CO_2 and between nitrogen atoms in a molecule of N_2?** *(a double bond; a triple bond)*

2 INSTRUCT

Connecting to Your World

Tell students that alkanes are saturated hydrocarbons. Ask, **Based on the bonding in alkanes, predict the bonding in an unsaturated hydrocarbon.** *(Unsaturated hydrocarbons might contain double or triple bonds.)*

Alkenes

CLASS Activity

Double Bond Rigidity L1

Purpose Students model the rigidity of a double bond.

Procedure Have students touch just the tips of the first fingers of both hands. With the fingers touching, have them rotate their hands in opposite directions and observe the position of their thumbs. Repeat the procedure, but with two fingers of each hand touching. Rotation about the "double bond" is impossible.

22.2 Unsaturated Hydrocarbons

Guide for Reading

Key Concepts

- What are the structural characteristics of alkenes?
- What are the structural characteristics of alkynes?

Vocabulary

saturated compounds
unsaturated compounds
alkenes
alkynes
aliphatic hydrocarbons

Reading Strategy

Comparing and Contrasting After you read the section, write a paragraph about alkenes and alkynes. How are these compounds similar? How are they different?

Connecting to Your World

A cockroach can transmit bacterial diseases when it comes in contact with food. Roach traps are baited with pheromones, which roaches produce to attract mates. The structure of many pheromones is based on unsaturated hydrocarbons. In this section, you will study two types of unsaturated hydrocarbons—alkenes and alkynes.

Alkenes

Organic compounds that contain the maximum number of hydrogen atoms per carbon atom are called **saturated compounds.** Alkanes are saturated compounds because the only bonds in alkanes are single covalent bonds. Compounds that contain double or triple carbon-carbon bonds are called **unsaturated compounds.** The ratio of hydrogen atoms to carbon atoms is lower in an unsaturated compound than in a saturated compound.

Alkenes are hydrocarbons that contain one or more carbon-carbon double covalent bonds. A carbon-carbon double bond is shown in structural formulas as two parallel lines. **At least one carbon-carbon bond in an alkene is a double covalent bond. Other bonds may be single carbon-carbon bonds and carbon-hydrogen bonds.**

Ethene (C_2H_4) is the simplest alkene. It is often called by the common name ethylene. Figure 22.5 shows the ball-and-stick and space-filling models of ethene. To name an alkene by the IUPAC system, find the longest chain in the molecule that contains the double bond. This chain is the parent alkene. It has the root name of the alkane with the same number of carbons plus the ending *-ene.* The chain is numbered so that the carbon atoms of the double bond have the lowest possible numbers. Substituents on the chain are named and numbered in the same way they are for the alkanes. Some examples of the structures and IUPAC names of simple alkenes are shown below.

$H_2C{=}CH_2$ — Ethene (ethylene)

$CH_3{-}CH{=}CH_2$ — Propene (propylene)

$CH_2{=}CH{-}CH_2{-}CH_3$ — 1-butene

$CH_3{-}CH{=}CH{-}CH_3$ — 2-butene

$CH_3{-}CH(CH_3){-}CH{=}CH{-}CH_3$ — 4-methyl-2-pentene

Ethene

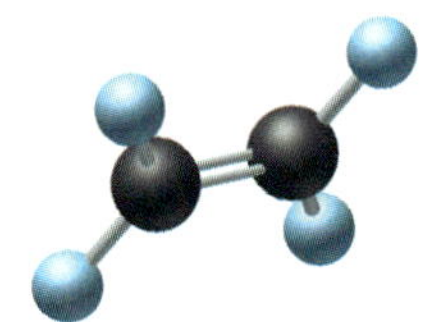

Figure 22.5 Because rotation is restricted around the double bond, atoms in ethene lie in one plane. **Calculating** ***What is the ratio of hydrogen to carbon in ethene?***

Section Resources

Print

- **Guided Reading and Study Workbook,** Section 22.2
- **Core Teaching Resources,** Section 22.2 Review
- **Transparencies,** T256

Technology

- **Interactive Textbook with ChemASAP,** Assessment 22.2

Table 22.3

Boiling Points of Homologous Alkanes, Alkenes, and Alkynes

Name	Molecular structure	Boiling point (°C)
C_2		
Ethane	CH_3-CH_3	−88.5
Ethene	$CH_2=CH_2$	−103.9
Ethyne	$CH\equiv CH$	−81.8
C_3		
Propane	$CH_3CH_2CH_3$	−42.0
Propene	$CH_3CH=CH_2$	−47.0
Propyne	$CH_3C\equiv CH$	−23.3

Alkynes

Hydrocarbons that contain one or more carbon-carbon triple covalent bonds are called **alkynes.** A carbon-carbon triple bond is shown in structural formulas as three parallel lines. **At least one carbon-carbon bond in an alkyne is a triple covalent bond. Other bonds may be single or double carbon-carbon bonds and single carbon-hydrogen bonds.** Like alkenes, alkynes are unsaturated compounds.

$$-C\equiv C-$$

Alkynes are not plentiful in nature. The simplest alkyne is the gas ethyne (C_2H_2), which has the common name acetylene. Acetylene is the fuel burned in oxyacetylene torches, which are used in welding. Figure 22.6 shows that the single bonds that extend from the carbons in the carbon-carbon triple bond of ethyne are separated by the maximum angle of 180°. This makes ethyne a linear molecule.

Straight-chain and branched-chain alkanes, alkenes, and alkynes are **aliphatic hydrocarbons.** The major attractions between aliphatic molecules are weak van der Waals forces. As a result, the introduction of a double or triple bond into a hydrocarbon does not have a dramatic effect on physical properties such as boiling point. Compare the boiling points in Table 22.3.

Figure 22.6 The triple bond restricts rotation in an ethyne molecule, which has a linear shape. **Calculating** ***What is the ratio of hydrogen to carbon in ethyne?***

22.2 Section Assessment

13. **Key Concept** Describe the bonding between atoms in an alkene.
14. **Key Concept** What types of bonds are present in an alkyne?
15. What is the difference between saturated and unsaturated hydrocarbons?
16. Write electron dot structures for ethene and ethyne. Describe the shape of each molecule.
17. Draw and name all the alkenes with the molecular formula C_4H_8.

Connecting Concepts

Saturated and Unsaturated Write a paragraph explaining what saturated and unsaturated hydrocarbons have in common with saturated and unsaturated solutions. Refer to Section 16.1.

Assessment 22.2 Test yourself on the concepts in Section 22.2.

with **ChemASAP**

Section 22.2 Assessment

13. At least one carbon-carbon bond in an alkene is a double bond. Other bonds may be single C–C bonds or C–H bonds.
14. An alkyne contains at least one carbon-carbon triple bond.
15. Saturated hydrocarbons have the maximum number of hydrogen atoms per carbon. Unsaturated hydrocarbons have fewer hydrogen atoms per carbon.
16. H:C::C:H (each C also bearing an H above) ethene; H:C:::C:H ethyne
17. $CH_2=CH-CH_2-CH_3$ 1-butene; $CH_3-C(H)=C(H)-CH_3$ 2-butene; $CH_2=C(CH_3)-CH_3$ methylpropene

Alkynes

Use Visuals L1

Figure 22.6 Have students compare the models of ethene and ethyne. Ask, **What is similiar about these mol–ecules?** *(Symmetrical carbon atoms can be superimposed upon one another.)* **What is different?** *(In ethene, the carbon atoms are joined by a double bond; in ethyne, by a triple bond. There are two fewer hydrogen atoms in ethyne and ethyne is linear.)*

3 ASSESS

Evaluate Understanding 

Have students explain why compounds with double or triple bonds are called unsaturated compounds.

Reteach L1

Have students construct ball-and-stick models of ethane, ethene, and ethyne. Ask students to write the complete structural formula for each molecule. Point out the lack of rotation about the double and triple bonds. Have students describe the molecular geometry of each molecule. *(tetrahedral, planar, and linear, respectively)*

Connecting Concepts

There is a higher concentration of solute in a saturated solution than in an unsaturated solution. There are more hydrogen atoms per carbon atom in a saturated hydrocarbon than in an unsaturated hydrocarbon.

If your class subscribes to the Interactive Textbook, use it to review key concepts in Section 22.2.

with **ChemASAP**

Answers to...

Figure 22.5 2:1

Figure 22.6 1:1

22.3

1 FOCUS

Objectives

22.3.1 Explain why structural isomers have different properties.

22.3.2 Describe the conditions under which geometric isomers are possible.

22.3.3 Identify optical isomers.

Guide for Reading

Build Vocabulary L2

Word Parts Explain that the term *isomer* derives from Greek terms meaning "same parts." Ask, **What is another chemical term that begins with the prefix *iso-* and what does it mean?** *(Isotopes are atoms of the same elements with different masses.)*

Reading Strategy L2

Identify Main Ideas Have students use headings and key concepts to identify the main ideas of this section.

2 INSTRUCT

Connecting to Your World

Have students read the introduction. Ask, **What do three-dimensional structures of molecules reveal?** *(how atoms are arranged in space)*

Structural Isomers

CLASS Activity

Modeling Isomers L2

Purpose Students use models to identify isomers

Procedure Have students use model kits to construct isomers of pentane (C_5H_{12}) and hexane (C_6H_{12}). Have them draw the structural formulas for each isomer. Ask, **How does the number of pentane isomers compare with the number of hexane isomers?** *(There are more hexane isomers.)*

22.3 Isomers

Guide for Reading

 **Key Concepts**

- How do the properties of structural isomers differ?
- What are the two types of stereoisomers?

Vocabulary

isomers
structural isomers
stereoisomers
geometric isomers
trans configuration
cis configuration
asymmetric carbon
optical isomers

Reading Strategy

Building Vocabulary As you read, write definitions of the four types of isomers described. Be sure your definitions differentiate among the types.

Connecting to Your World The retinal molecule in the rod and cone cells of your eye has a hydrocarbon skeleton. The first step in the process of vision occurs when light enters your eye and strikes a cell containing retinal. The light causes a change in the three-dimensional structure of the retinal molecule. The structures before and after the light strikes are examples of isomers. In this section, you will study different types of isomers.

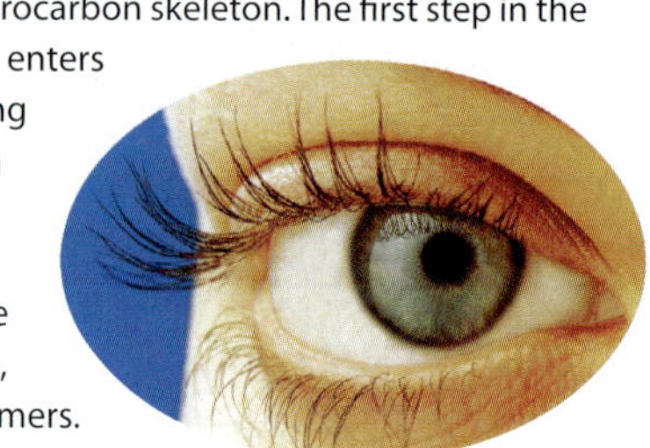

Structural Isomers

You may have noticed that the structures of some hydrocarbons differ only in the positions of substituent groups or of multiple bonds in their molecules. Look at the structural formulas for butane and 2-methylpropane shown below and at their ball-and-stick models in Figure 22.7. Even though both compounds have the formula C_4H_{10}, their boiling points and other properties differ. Because their structures are different, they are different substances. Compounds that have the same molecular formula but different molecular structures are called **isomers.**

$CH_3—CH_2—CH_2—CH_3$

Butane (C_4H_{10}) (bp −0.5°C)

$CH_3—CH(CH_3)—CH_3$

2-methylpropane (C_4H_{10}) (bp −10.2°C)

Butane and 2-methylpropane represent a category of isomers called structural isomers. **Structural isomers** are compounds that have the same molecular formula, but the atoms are joined together in a different order. **Structural isomers differ in physical properties such as boiling point and melting point. They also have different chemical reactivities.** In general, the more highly branched the hydrocarbon structure, the lower the boiling point of the isomer compared with less branched isomers. For example, 2-methylpropane has a lower boiling point than butane.

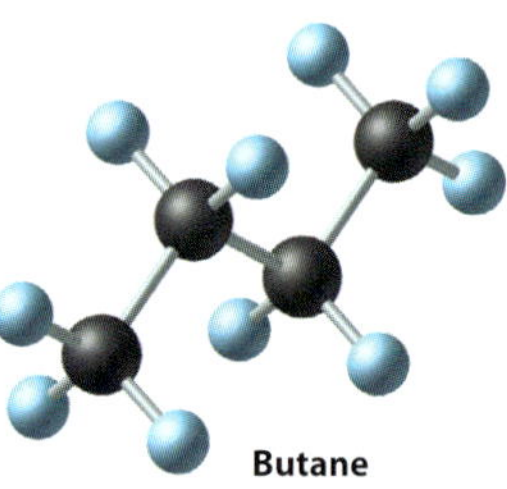

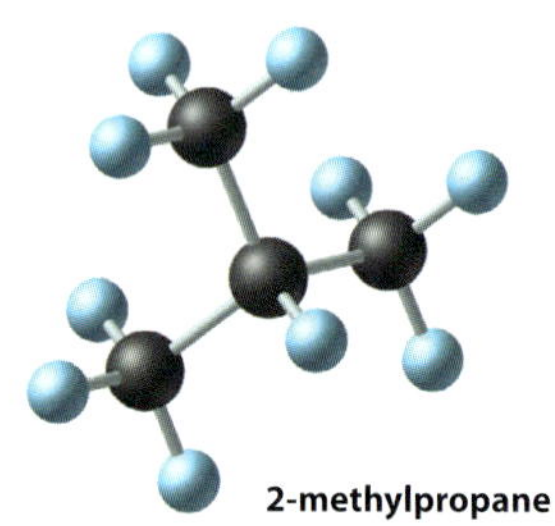

Figure 22.7 Both butane and 2-methylpropane have the molecular formula C_4H_{10}. But the atoms in their molecules are arranged in a different order. So these compounds are structural isomers.

Section Resources

Print

- **Guided Reading and Study Workbook,** Section 22.3
- **Core Teaching Resources,** Section 22.3 Review
- **Transparencies,** T257–T258
- **Small-Scale Chemistry Laboratory Manual,** Lab 37

Technology

- **Interactive Textbook with ChemASAP,** Simulation 28, Problem-Solving 22.18, Assessment 22.3

Stereoisomers

Remember that molecules are three-dimensional structures. So molecules with the same molecular formula and with atoms joined in exactly the same order may still be isomers. **Stereoisomers** are molecules in which the atoms are joined in the same order, but the positions of the atoms in space are different. **Two types of stereoisomers are geometric isomers and optical isomers.**

Geometric Isomers The first category of stereoisomers is based on the presence of a double bond in a molecule. A double bond between two carbon atoms prevents them from rotating with respect to each other. Because of this lack of rotation, groups on either side of the double bond can have different orientations in space. **Geometric isomers** have atoms joined in the same order, but differ in the orientation of groups around a double bond. Look at the models of 2-butene in Figure 22.8. Two arrangements are possible for the methyl groups with respect to the rigid double bond. In the ***trans* configuration,** the methyl groups are on opposite sides of the double bond. In the ***cis* configuration,** the methyl groups are on the same side of the double bond. *Trans*-2-butene and *cis*-2-butene have different physical and chemical properties. For example, the density of *cis*-2-butene is 0.616 g/cm^3, and the density of *trans*-2-butene is 0.599 g/cm^3. The groups attached to the carbons of the double bond do not need to be the same. Geometric isomerism is possible whenever each carbon of the double bond has at least one substituent.

CH_3 H
C=C
H CH_2CH_3
trans-2-pentene

CH_3 CH_2CH_3
C=C
H   H
cis-2-pentene

CH_3 H
C=C
CH_3CH_2 H
2-methyl-1-butene
(no *cis, trans* isomers)

Optical Isomers The second category of stereoisomers occurs whenever a carbon atom has four different atoms or groups attached. A carbon with four different atoms or groups attached is an **asymmetric carbon.** Look at the models in Figure 22.9. Because H, F, Cl, and Br atoms are attached to a single carbon atom, the carbon is asymmetric. The relationship between the molecules is similar to the relationship between right and left hands.

Think about an object placed in front of a mirror. If the object is symmetrical, like a ball, then its mirror image can be superimposed. That is, the appearance of the ball and its reflection are indistinguishable. By contrast, a pair of hands is distinguishable even though the hands consist of identical parts. The right hand reflects as a left hand and the left hand reflects as a right hand. The images cannot be superimposed. Many ordinary objects, such as ears, feet, shoes, and bird wings, are similarly related.

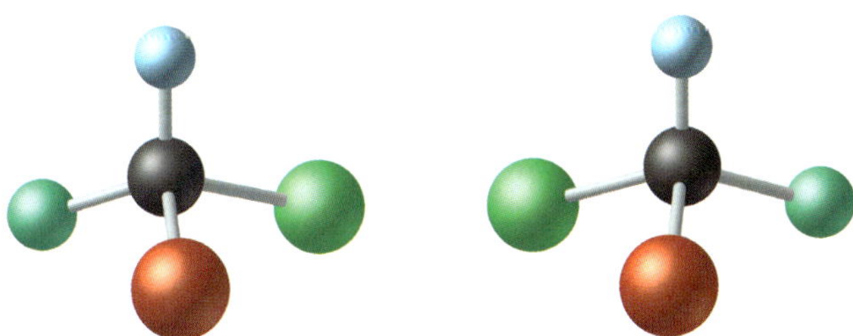

Figure 22.9 In these models, four different atoms are attached to the same carbon atom, making the carbon atom an asymmetric carbon.

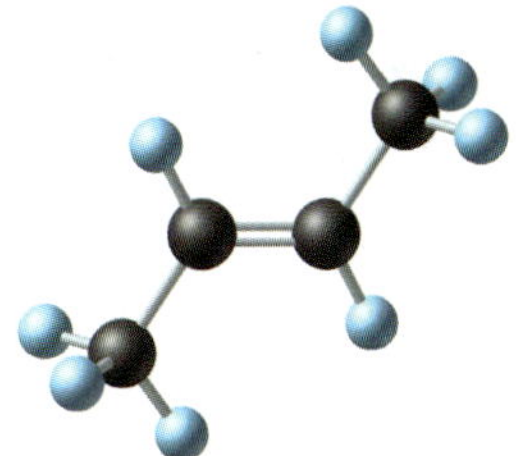

Trans configuration

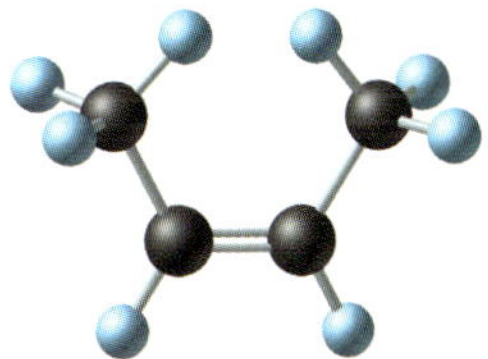

Cis configuration

Figure 22.8 There is a *trans* and a *cis* configuration of 2-butene because a methyl group is attached to each carbon of the double bond.
Comparing and Contrasting
How are the trans and cis configurations different?

Discuss L2

Point out that the number of possible structural isomers for an alkane increases dramatically with increasing numbers of carbon atoms. With more carbon atoms there are more ways to arrange the atoms in space. Remind students that each isomer has a unique set of physical and chemical properties. Name the three isomers of pentane. Have students name the five isomers of hexane they constructed in the *Modeling Isomers* class activity. Point out that the existence of isomers is a major reason that so many different organic compounds exist.

Stereoisomers

Discuss L2

Point out that the number of possible isomers for an alkene can be much greater than for an alkane. Draw the complete structural formulas of 1-butene, *cis*-2-butene, and *trans*-2-butene on the board to show the three possible isomers for butene (C_4H_8). Introduce the term *geometric isomers* and explain how *cis*-2-butene and *trans*-2-butene are different molecules due to the lack of free rotation about the carbon-carbon double bond.

TEACHER Demo

Optical Isomers L2

Purpose Students observe three-dimensional models of optical isomers.

Materials two ball-and-stick models of CHFClBr

Procedure Demonstrate how it is not possible to superimpose one model on the other.

Answers to...

Figure 22.8 The groups are on opposite sides of the double bond in the *trans* configuration; the groups are on the same side of the bond in the *cis* configuration.

CONCEPTUAL PROBLEM 22.3

Answers

18. a.

$$CH_3-\overset{*}{\underset{\underset{Cl}{|}}{\overset{\overset{H}{|}}{C}}}-CHO$$

b. no asymmetric carbon

19. a.

$$CH_3CH_2-\overset{*}{\underset{\underset{F}{|}}{\overset{\overset{CH_3}{|}}{C}}}-Br$$

b. no asymmetric carbon

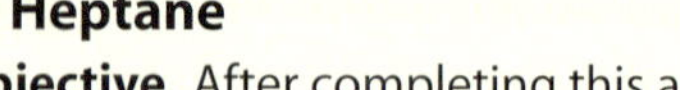

Structural Isomers of Heptane L2

Objective After completing this activity, students will be able to:

- build ball-and-stick models of heptane isomers and name them.

Skills Focus Using models, observing, analyzing

Prep Time 5 minutes
Class Time 25 minutes

Expected Outcome Students will model and name nine structural isomers of heptane.

Analyze and Conclude

1. heptane, 2-methylhexane, 3-methylhexane, 2,3-dimethylpentane, 2,4-dimethylpentane, 2,2-dimethylpentane, 3,3-dimethylpentane, 3-ethylpentane, 2,2,3-trimethylbutane
2. 4
3. Each is a unique compound.

For Enrichment L3

Students can model, draw, and name the 18 structural isomers of octane.

Simulation 28 Play the isomer game of "Pick the Pairs."

with ChemASAP

Pairs of molecules that differ only in the way that four different groups are arranged around a central carbon atom are called **optical isomers.** The term *optical* is used to describe this type of isomer because the term is related to the concept of objects and mirror images. The molecules shown in Figure 22.9 are optical isomers. The models cannot be superimposed because they are mirror images of each other. Try to visualize the models in three dimensions. Imagine rotating one model 180°. At first, it might appear that the models are now identical. However, the direction of each bond in space is not the same. For example, the carbon-bromine bond would be pointing away from you rather than toward you. No matter how you turn the models, you cannot make the models look exactly alike. Unless bonds are broken, these molecules cannot be made to look alike.

How does the presence of a carbon-carbon double bond in a molecule result in geometric isomers?

CONCEPTUAL PROBLEM 22.4

Identifying Asymmetric Carbon Atoms

Compounds with the following formulas are alcohols. The alcohol represented by the formula in **a.** is the alcohol in rubbing alcohol. Which compound has an asymmetric carbon?

a. $CH_3\underset{\underset{OH}{|}}{C}HCH_3$ **b.** $CH_3\underset{\underset{OH}{|}}{C}HCH_2CH_3$

1 Analyze ***Identify the relevant concepts.***
An asymmetric carbon atom has four different groups attached.

2 Solve ***Apply concepts to this situation.***
Draw the structures in a more complete form to determine the presence of a carbon atom with four different groups attached.

a. The central carbon has two CH_3 groups attached. It is not asymmetric.

b. The central carbon has one H, one OH, one CH_3, and one CH_2CH_3 group attached. Because these four groups are different, the central carbon is asymmetric. In a structural formula, an asymmetric carbon is often marked with an asterisk.

Practice Problems

18. Identify the asymmetric carbon, if any, in each of the following structures.

a. $CH_3\underset{\underset{Cl}{|}}{C}HCHO$ **b.** $CH_3\underset{\underset{CH_3}{|}}{C}HOH$

19. Identify the asymmetric carbon, if any, in each of the following structures.

a.

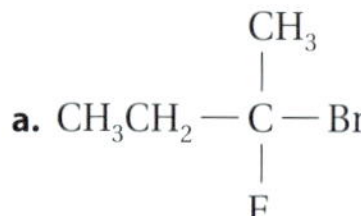

b. CH_2Cl_2

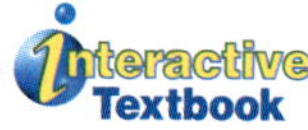

Problem-Solving 22.18 Solve Problem 18 with the help of an interactive guided tutorial.

with ChemASAP

Facts and Figures

Mirror Images

Tell students that mirror images exist in nature. For example, 99% of the shells of some snails are right-handed spirals and 1% are left-handed spirals. The right- and left-handed spirals are mirror images. Other species have shells that are normally left-handed.

Quick LAB

Structural Isomers of Heptane

Purpose

To build ball-and-stick models and name the nine structural isomers of heptane (C_7H_{16}).

Materials

- ball-and-stick molecular model kit (Colors used to represent elements in the kit may not match colors used to represent elements in this book.)

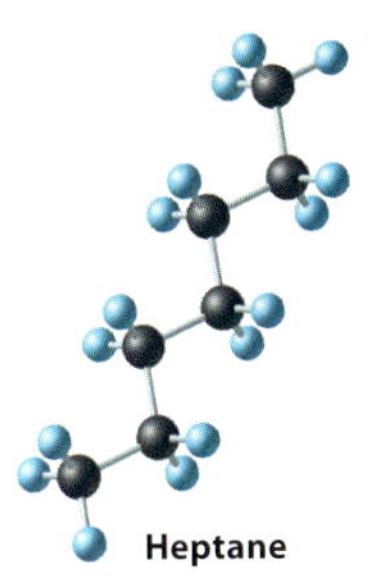
Heptane

Procedure

1. Build a model for the straight-chain isomer of C_7H_{16}. Draw the structural formula and write the IUPAC name for this isomer.
2. Move one carbon atom from the end of the chain and add a methyl substituent to the chain. Draw the structural formula and name this isomer.
3. Move the methyl group to a new position on the chain. Then draw and name this third isomer. Is there a third position that the methyl group can be moved to on the chain of six carbons to form yet another different isomer?
4. Make other structural isomers by shortening the longest straight chain and using the removed carbons as substituents. Draw the structural formulas and name each isomer.

Analyze and Conclude

1. What are the names of the nine structural isomers of C_7H_{16}?
2. What is the shortest possible straight carbon chain in the isomers?
3. Why does each structural isomer have its own unique name?

22.3 Section Assessment

20. **Key Concept** Explain why you would expect two structural isomers to have different boiling points.
21. **Key Concept** Name the two types of stereoisomers.
22. What structure must be present in a molecule for geometric isomers to exist?
23. How can an asymmetric carbon be identified?
24. What is the relationship between two molecules that are optical isomers?
25. Draw structural formulas for the following alkenes. If a compound has geometric isomers, draw both the *cis* and *trans* forms.
    a. 1-pentene
    b. 2-hexene
    c. 2-methyl-2-hexene

Writing Activity

Writing Instructions Write instructions that could be used to decide whether two compounds are isomers and, if so, what type of isomers they might be. (1) How can you tell if two compounds are isomers? (2) How can you tell if two compounds are structural isomers? (3) How can you tell if two compounds can be geometric isomers? (4) How can you tell if two compounds can be optical isomers?

Assessment 22.3 Test yourself on the concepts in Section 22.3.

with ChemASAP

3 ASSESS

Evaluate Understanding L2

Have students explain what the term *isomer* means and then describe the differences among structural isomers, geometric isomers, and optical isomers. Have them draw examples of each type of isomer.

Reteach

Use ball-and-stick models to review the requirements for classifying two molecules with the same molecular formula as structural, geometric, or optical isomers.

Writing Activity

Answers should include (1) same molecular formula, different structures; (2) atoms joined in a different order; (3) presence of a double bond; and (4) presence of an asymmetric carbon.

If your class subscribes to the Interactive Textbook, use it to review key concepts in Section 22.3.

with ChemASAP

Section 22.3 Assessment

20. Structural isomers are different compounds with different physical properties.
21. geometric isomers and optical isomers
22. when each carbon in a double bond has at least one substituent
23. It has four different substituents.
24. The relationship is similiar to the relationship between right and left hands, which can't be superimposed.
25. a. H, H–C=C–$CH_2CH_2CH_3$, H (H and H on first carbon; $CH_2CH_2CH_3$ and H on second carbon)
    1-pentene
    b. CH_3, H on first carbon; H, $CH_2CH_2CH_3$ on second carbon (C=C)
    trans-2-hexene
    CH_3, H on first carbon; $CH_2CH_2CH_3$, H on second carbon (C=C)
    cis-2-hexene
    c. CH_3, CH_3 on first carbon; H, $CH_2CH_2CH_3$ on second carbon (C=C)
    2-methyl-2-hexene

Answers to...

Checkpoint The double bond prevents rotation around the carbon atoms, which allows for different spatial arrangements of substituents around the bond.

Hydrocarbon Isomers L2

Objective After completing this activity, students will be able to:
- write line-angle formulas for hydrocarbons.

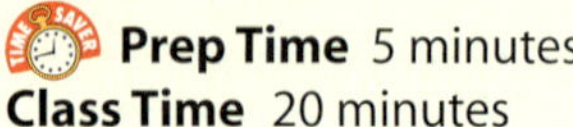

Prep Time 5 minutes
Class Time 20 minutes

Expected Outcome

Students will learn to write line-angle formulas for hydrocarbons.

Analyze

1. Students should draw formulas for pentane, 2-methylbutane, and 2,2-dimethylpropane.
2. Subtract from four the number of lines drawn to any point.
3. $CH_3(CH_2)_2CH_3$ butane; $(CH_3)_2CHCH_3$ methylpropane

For Enrichment L3

There are eight isomers of C_7H_{16}. Draw their line-angle formulas and name them.

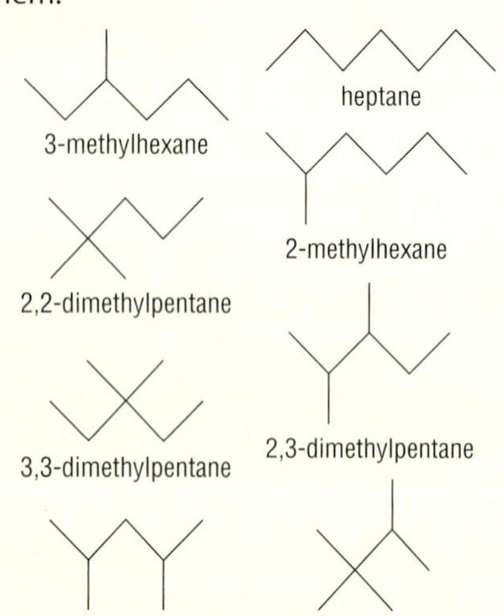

You're the Chemist

1.

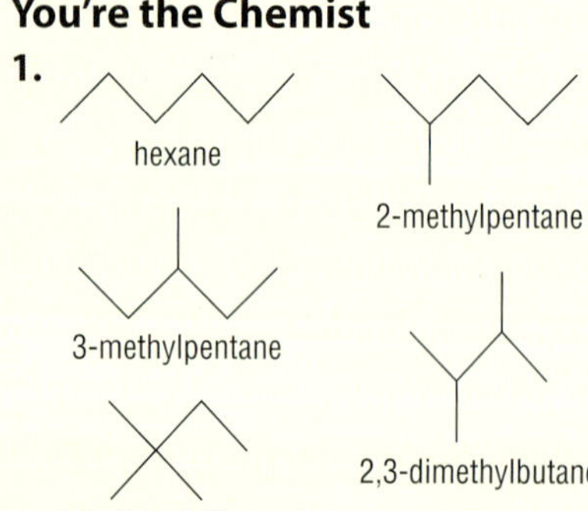

Small-Scale LAB

Hydrocarbon Isomers

Purpose

To draw line-angle formulas and name some of the isomers in gasoline.

Materials

- toothpicks
- modeling clay

Procedure

Gasoline is a complex mixture of hydrocarbon molecules. Each molecule contains between five and ten carbon atoms. Many of the components of gasoline are isomers with the same molecular formula. These components include the isomers of pentane. Study the formulas and names of the isomers of C_5H_{12} in the chart below. Make a model of each isomer using toothpicks and modeling clay. Compare the models and make accurate drawings on your paper.

Formulas Representing Isomers of C_5H_{12}

Condensed	Line-angle	Space-filling
$CH_3CH_2CH_2CH_2CH_3$ pentane		
$CH_3CHCH_2CH_3$ with CH_3 branch; 2-methylbutane		
CH_3CCH_3 with two CH_3 branches; 2,2-dimethylpropane		

Analyze

Using your experimental data, answer the following questions.

1. Complete structural formulas include all the atoms and all the chemical bonds in a molecule. Draw the complete structural formula for each isomer of C_5H_{12} in the chart.
2. In a line-angle formula, each line represents a carbon-carbon bond. Each end of a line, as well as the intersection of lines, represents a carbon atom. All hydrogen atoms are understood. Knowing that carbon always forms four bonds in organic compounds, explain how to determine the number of hydrogen atoms bonded to each carbon in a line-angle formula.
3. Because butane can vaporize readily, it is used in the formulations of gasolines in cold climates during winter. Draw condensed structural formulas and line-angle formulas for the two isomers of butane (C_4H_{10}). Make models of each isomer.

You're The Chemist

The following small-scale activities allow you to develop your own procedures and analyze the results.

1. **Analyze It!** Gasoline contains isomers of hexane. Draw the line-angle formulas and name the five isomers of C_6H_{14}. Make a model of one isomer and convert that model into the other four.
2. **Design It!** Gasoline also contains small amounts of the six isomers of pentene. Two of the isomers are *cis* and *trans* configurations of the same structural isomer. Experiment with your models to make these isomers. Use two toothpicks to represent a double bond. Draw structural formulas for the *cis* and *trans* isomers, and line-angle formulas for the others.

708 Chapter 22

2. $(H)(H)C{=}C(H)(CH_2CH_2CH_3)$ 1-pentene; $(CH_3)(H)C{=}C(CH_2CH_3)(H)$ *cis*-2-pentene; $(CH_3)(H)C{=}C(H)(CH_2CH_3)$ *trans*-2-pentene

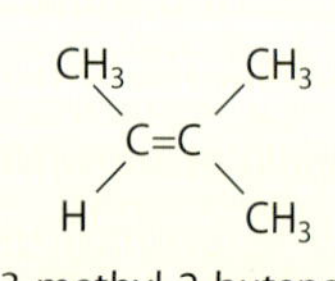

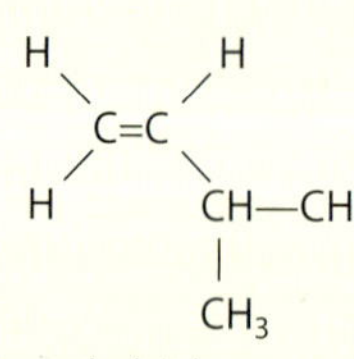

$(H)(H)C{=}C(CH_3)(CH_2CH_3)$ 2-methyl-1-butene

22.4 Hydrocarbon Rings

Connecting to Your World

Not everyone loves carrots, but you have probably heard about how good they are for you. One reason carrots are a healthful food choice is that they contain beta-carotene, an important nutrient for vision. Beta-carotene also gives carrots their orange color. The hydrocarbon skeleton of beta-carotene contains hydrocarbon rings. In this section, you will learn about hydrocarbon rings and the properties of compounds formed from these rings.

Guide for Reading

Key Concepts

- What is the general structure of cyclic hydrocarbons?
- What is the most accurate description of the bonding in a benzene ring?

Vocabulary

cyclic hydrocarbons

aromatic compound

Reading Strategy

Relating Text and Visuals As you read, look at the benzene diagrams on page 710. What is the value of the diagram with a circle in a hexagon?

Cyclic Hydrocarbons

Saturated and unsaturated straight-chain and branched-chain hydrocarbons are not the only types of hydrocarbons. **In some hydrocarbon compounds, the carbon chain is in the form of a ring.** Figure 20.10 shows the structures of some of these hydrocarbons. Compounds that contain a hydrocarbon ring are called **cyclic hydrocarbons.**

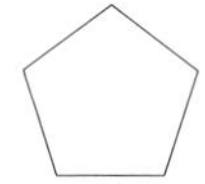
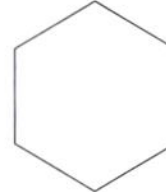

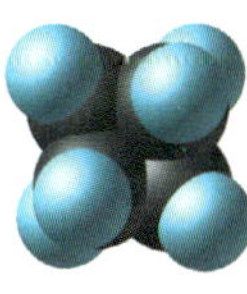

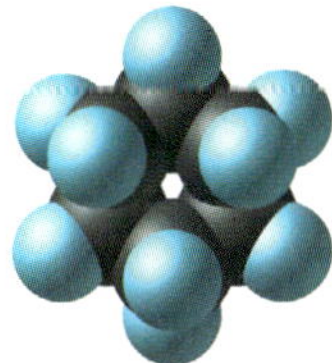

Cyclopropane (bp –34.4°C) | Cyclobutane (bp –13°C) | Cyclopentane (bp 49.5°C) | Cyclohexane (bp 81.4°C)

Figure 22.10 The simplest cyclic hydrocarbon has a 3-carbon ring. Rings containing up to 20 carbon atoms are found in nature. Rings with 5 or 6 carbons are the most abundant.

Section Resources

Print

- **Guided Reading and Study Workbook,** Section 22.4
- **Core Teaching Resources,** Section 22.4 Review, Interpreting Graphics
- **Transparencies,** T259
- **Laboratory Manual,** Lab 49
- **Laboratory Practicals,** 22-1

Technology

- **Interactive Textbook with ChemASAP,** Assessment 22.4
- **Go Online,** Section 22.4

22.4

1 FOCUS

Objectives

22.4.1 Identify cyclic ring structures.
22.4.2 Describe bonding in benzene.

Guide for Reading

Build Vocabulary 

Word Forms Explain that *aromatic* comes from a Greek word meaning "sweet spice." *Aromatherapy* is the use of aromatic oils for their supposed therapeutic effects when applied to the skin or inhaled.

Reading Strategy 

Compare and Contrast Have students make a Venn diagram for aliphatic cyclic hydrocarbons and aromatic cyclic hydrocarbons.

2 INSTRUCT

Connecting to Your World

Have students read the introduction. Then ask, **Predict how the carbon atoms in hexane could form a ring.** *(The two carbons at the end of the chain could bond to one another.)*

Cyclic Hydrocarbons

Use Visuals 

Figure 22.11 Ask, **Why does cyclohexane have two fewer hydrogen atoms than hexane?** *(When carbon atoms at the end of a chain bond to form a ring, those atoms can form only two additional bonds instead of three.)*

Discuss 

The formation of a ring causes bond strain in all cycloalkanes with rings that contain fewer than six carbons. The bond angle in these cycloalkanes tends to be less than the normal carbon-carbon bond angle of 109.5°. Bond strain explains the increased chemical reactivity of small-ring cycloalkanes.

Aromatic Hydrocarbons

Download a worksheet on **Aromatic Hydrocarbons** for students to complete and find additional teacher support from NSTA SciLinks.

Discuss 

Using the diagrams in the text on page 711, discuss the nomenclature of aromatic compounds.

For: Links on Aromatic Hydrocarbons
Visit: www.SciLinks.org
Web Code: cdn-1224

Aromatic Hydrocarbons

There is a class of organic compounds that are responsible for the aromas of spices such as vanilla, cinnamon, cloves, and ginger. These compounds were originally called aromatic compounds because they have distinct, pleasant aromas. Not all compounds currently classified as aromatic have an odor. Molecules of aromatic compounds contain a single ring or a group of rings. Benzene (C_6H_6) is the simplest example of an aromatic compound. An **aromatic compound** is an organic compound that contains a benzene ring or other ring in which the bonding is like that of benzene. Another name for an aromatic compound is an arene. Because of the structure of benzene, the properties of aromatic compounds are quite different from those of aliphatic compounds.

The Structure of Benzene Friedrich Kekulé (1829–1896) made one major contribution to chemistry. He was the first to describe the structure of a benzene molecule. Look at the space-filling model of benzene in Figure 22.11. The benzene molecule is a six-membered carbon ring with a hydrogen atom attached to each carbon. This arrangement leaves one electron from each carbon free to participate in a double bond. Two different structures with alternating double bonds can be written for benzene.

Single bond

Double bond

These structural formulas show only the extremes in electron sharing between any two adjacent carbons in benzene. One extreme is a normal single bond. The other extreme is a normal double bond. Recall that when two or more equally valid structures can be drawn for a molecule, resonance occurs. The actual bonding in a benzene ring doesn't alternate between the two extreme resonance structures. Rather, all the bonds in the ring are identical hybrids of single and double bonds. **In a benzene molecule, the bonding electrons between carbon atoms are shared evenly around the ring.** Benzene and other molecules that exhibit resonance are more stable than similar molecules that do not exhibit resonance. Thus, benzene is not as reactive as six-carbon alkenes.

Figure 22.11 This is a space-filling model of benzene. **Applying Concepts** ***What is the molecular formula of benzene?***

Because of the way electrons are distributed, drawing a solid or dashed circle inside a hexagon is a good way to represent benzene. However, it does not show the number of electrons involved. For this reason, the traditional structure, which is the structure shown on the right, is used in this textbook. Remember, though, that each bond in the ring is identical.

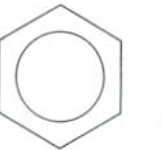  

What is the name of the simplest aromatic hydrocarbon and what is its formula?

Facts and Figures

The Benzene Problem

August Kekulé (1829–1896) proposed that carbon atoms link together to form chains. Later he pondered the problem of benzene, which seemed to have too few hydrogen atoms to exist in chain form. After working on the benzene problem one evening, Kekulé fell asleep in front of his fireplace and dreamed of snakes twisting and turning. One of the snakes grasped its tail in its mouth and formed itself into a ring. Kekulé awoke at once and worked out his ring structure for benzene that night. Afterwards, when other scientists were confronted with difficult problems, Kekulé told them "Let us learn to dream, gentlemen, and then perhaps we shall learn the truth."

Substituted Aromatic Compounds The dyes used to produce the intense colors shown in Figure 22.12 are substituted aromatic compounds. Compounds containing substituents attached to a benzene ring are named as derivatives of benzene. When the benzene ring is a substituent on an alkane, the C_6H_5 group is called a phenyl group.

CH_3 (on benzene ring) — Methylbenzene (toluene)

$CH_3-CH_2-CH-CH_2-CH_2-CH_3$ (phenyl on CH) — 3-phenylhexane

CH_2CH_3 (on benzene ring) — Ethylbenzene

Some derivatives of benzene have two substituents. These derivatives are called disubstituted benzenes. There are three structural isomers for the liquid aromatic compound dimethylbenzene ($C_6H_4(CH_3)_2$). Dimethylbenzenes are also called xylenes. The boiling points of the xylenes are a reminder that structural isomers have different physical properties.

1,2-dimethylbenzene (*o*-xylene) (bp 144°C)

1,3-dimethylbenzene (*m*-xylene) (bp 139°C)

1,4-dimethylbenzene (*p*-xylene) (bp 139°C)

In the IUPAC naming system, the possible positions of two substituents in disubstituted benzene are designated as 1,2; 1,3; or 1,4. Common names for disubstituted benzenes use the terms *ortho*, *meta*, and *para* (abbreviated as *o*, *m*, and *p*) in place of numbers.

Figure 22.12 The molecules of many dyes include benzene rings. Dyes that contain aromatic compounds are called aniline dyes.

22.4 Section Assessment

26. **Key Concept** What is a cyclic hydrocarbon?

27. **Key Concept** Describe the bonding between carbon atoms in benzene.

28. The molecules of both cyclohexane and benzene have six carbon atoms bonded in a ring. What is the difference between these two compounds?

29. Name the following compounds.

a. CH_2CH_3 (on benzene ring)

b. CH_2CH_3 and $CH_2CH_2CH_3$ (on benzene ring)

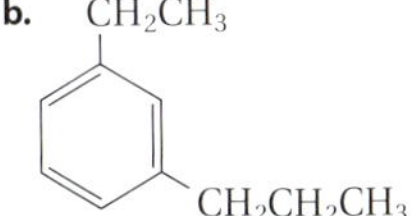

Writing Activity

Explanatory Paragraph The alternate name for an aromatic compound is arene. Write a paragraph explaining why *arene* is not an entirely accurate choice for an aromatic compound. (*Hint:* What does the suffix *-ene* indicate when used with aliphatic compounds?)

Assessment 22.4 Test yourself on the concepts in Section 22.4.

with ChemASAP

Section 22.4 Assessment

26. In a cyclic hydrocarbon, the carbon chain is in the form of a ring.

27. In benzene, the bonding electrons between carbon atoms are shared evenly around the ring.

28. Cyclohexane is a saturated hydrocarbon with a carbon chain in the form of a ring. Benzene is an unsaturated aromatic hydrocarbon.

29. a. ethylbenzene
b. 1-ethyl-3-propylbenzene

3 ASSESS

Evaluate Understanding L2

Ask students to compare and contrast the following terms as they relate to hydrocarbons.

1. *aromatic* and *aliphatic*
2. *chain* and *cyclic*
3. *saturated* and *unsaturated*

Reteach L1

Review the terms *alkane, alkene, alkyne, cycloalkane, aliphatic,* and *aromatic*. Note that both groups of unsaturated hydrocarbons have a name ending in *-ene*. Explain that it is not necessary for a compound to have a distinctive odor for it to be an aromatic compound. *Aromatic* refers strictly to the presence of a benzene ring in the structure.

Writing Activity

The suffix *-ene* indicates the presence of a double bond, but the bonds in the rings in aromatic compounds are hybrid bonds, not alternating single and double bonds.

If your class subscribes to the Interactive Textbook, use it to review key concepts in Section 22.4.

with ChemASAP

Answers to....

Figure 22.11 C_6H_6

benzene

22.5

1 FOCUS

Objectives

22.5.1 Identify three important fossil fuels and describe their origins.

22.5.2 Describe the composition of natural gas, petroleum, and coal.

25.5.3 Describe What happens when petroleum is refined.

Guide for Reading

Build Vocabulary 

Word Forms Relate the chemical process of cracking to the physical process when an objects cracks or breaks apart.

Reading Strategy L2

Relating Text and Visuals Have students relate the description of fractions in Table 22.4 to the diagram of the fractionating column in Figure 22.14.

2 INSTRUCT

Connecting to Your World

Have students read the introduction. Then ask, **Why are hydrocarbons harder to obtain on Earth than they would be on Titan?** *(On Earth, the hydrocarbons are in the crust, not in the atmosphere.)*

Natural Gas

Relate 

It was not until the invention of the kerosene lamp that people valued petroleum. By the time electric lights reduced the need for kerosene, automobiles required gasoline, which previously had been discarded as a useless by-product of kerosene production. Before the petroleum era began, fats from domesticated animals, whale oil, wood, and coal were used as energy sources.

22.5 Hydrocarbons From Earth's Crust

Guide for Reading

 **Key Concepts**

- What type of hydrocarbons are in natural gas?
- What is the first step in the refining of petroleum?
- How is coal classified?
- What is the chemical composition of coal?

Vocabulary

cracking

Reading Strategy

Previewing Before you read, jot down five things you know about petroleum. As you read, explain how what you learn relates to what you already know.

Connecting to Your World

The photograph shows a painting in which the artist imagines Saturn's moon Titan as a refueling station in space. A traveler would refill a fuel tank with hydrocarbons from Titan's atmosphere. On Earth, hydrocarbons are also used as fuels. However, they are more difficult to obtain because they are found in Earth's crust, not in Earth's atmosphere. In this section, you will learn about three fossil fuels—natural gas, petroleum, and coal.

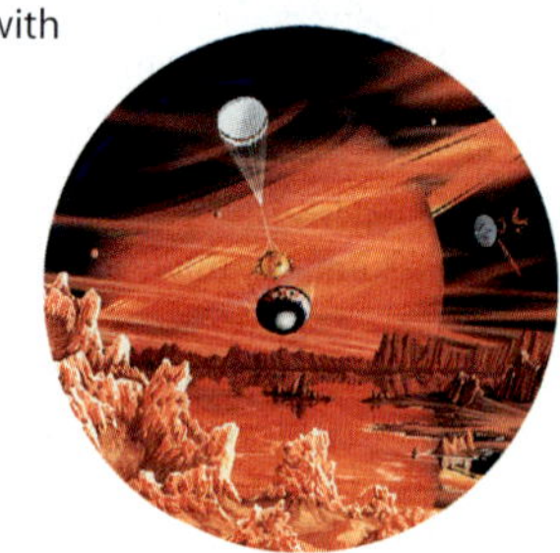

Natural Gas

Much of the world's energy is supplied by burning fossil fuels. Fossil fuels are carbon-based because they are derived from the decay of organisms. Millions of years ago, marine organisms died, settled on the ocean floor, and were buried in ocean sediments. Heat, pressure, and bacteria changed the residue into petroleum and natural gas, which contain mostly aliphatic hydrocarbons. Figure 22.13 shows one possible location for these fuels.

Natural gas is an important source of alkanes of low molar mass. Typically, natural gas is composed of about 80% methane, 10% ethane, 4% propane, and 2% butane. The remaining 4% consists of nitrogen and hydrocarbons of higher molar mass. Natural gas also contains a small amount of the noble gas helium. In fact, natural gas is a major source of helium. Methane, the major constituent of natural gas, is especially prized for combustion because it burns with a hot, clean flame.

$$CH_4(g) + 2O_2(g) \longrightarrow CO_2(g) + 2H_2O(g) + \text{heat}$$

Propane and butane are separated from the other gases in natural gas by liquefaction. These heating fuels are sold in liquid form in pressurized tanks as liquid petroleum gas (LPG).

Oxygen is necessary for the efficient combustion of a hydrocarbon. If there is not enough oxygen available, the combustion is incomplete. Complete combustion of a hydrocarbon gives a blue flame. Incomplete combustion gives a yellow flame. This is due to the formation of small, glowing carbon particles that are deposited as soot when they cool. Carbon monoxide, a toxic gas, is also formed along with carbon dioxide and water during incomplete combustion.

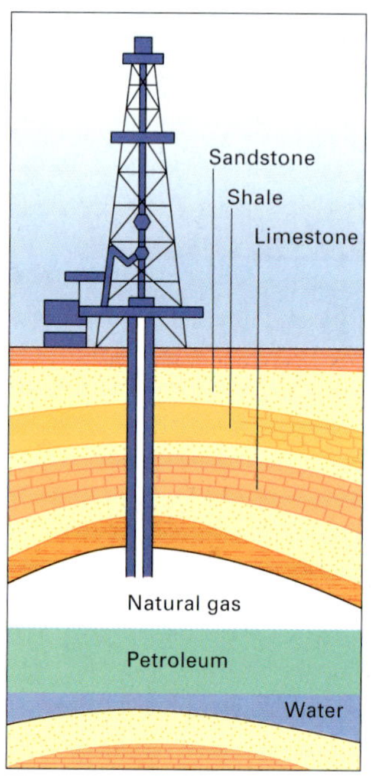

Figure 22.13 Wells are drilled to reach natural gas and petroleum. Pressure from the trapped gas may force petroleum up a well pipe, but pumping is usually required.

Section Resources

Print

- **Guided Reading and Study Workbook,** Section 22.5
- **Core Teaching Resources,** Section 22.5 Review
- **Transparencies,** T260–T261

Technology

- **Interactive Textbook with ChemASAP,** Assessment 22.5
- **Go Online,** Section 22.5

Petroleum

The organic compounds found in petroleum, or crude oil, are more complex than those in natural gas. Most of the hydrocarbons in petroleum are straight-chain and branched-chain alkanes. Petroleum also contains small amounts of aromatic compounds and sulfur-, oxygen-, and nitrogen-containing organic compounds.

Humans have known about petroleum for centuries; ancient peoples found it seeping from the ground in certain areas. In the late 1850s, a vast deposit of petroleum was discovered in Pennsylvania when a well was drilled to obtain petroleum for use as a fuel. Within decades, petroleum deposits had also been found in the Middle East, Europe, and the East Indies. Petroleum has since been found in other parts of the world as well.

Petroleum is a mixture of hydrocarbons having from one to more than 40 carbon atoms. Without further treatment, petroleum is not very useful. The mixture must be separated, or refined, into parts called fractions, which have many commercial uses. **The refining process starts with the distillation of petroleum (crude oil) into fractions according to boiling point.** A schematic of a petroleum refining distillation tower is shown in Figure 22.14. Each distillation fraction contains several different hydrocarbons. The fractions and their composition are listed in Table 22.4.

The amounts of products obtained by fractional distillation are not in proportion to the demand of the market. Gasoline is by far the most commonly used product, so other processes are used to make the supply meet the demand. **Cracking** is a controlled process by which hydrocarbons are broken down or rearranged into smaller, more useful molecules. For example, fractions containing compounds of higher molar mass are "cracked" to produce the more useful short-chain components of gasoline and kerosene. Hydrocarbons are cracked with the aid of a catalyst and with heat. This process also produces low-molar-mass alkanes, which are used to manufacture paints and plastics. Other catalytic processes besides cracking are used to increase the amounts of components that improve the performance of gasoline.

 ***When and where was the first petroleum well drilled?***

Fractionating Column

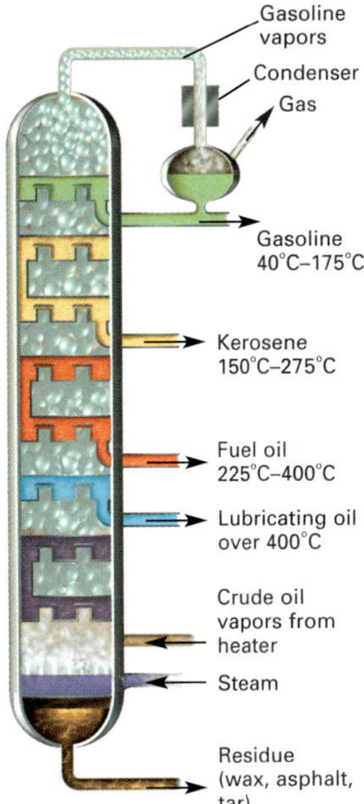

Figure 22.14 In fractional distillation, the crude oil is heated so that it vaporizes and rises through the fractionating column. The column is hotter at the bottom and cooler at the top. Compounds with the highest boiling points condense near the bottom. Compounds with the lowest boiling points condense near the top.

Table 22.4

Fractions Obtained from Crude Oils

Fraction	Composition of carbon chains	Boiling range (°C)	Percent of crude oil
Natural gas	C_1 to C_4	Below 20	10%
Petroleum ether (solvent)	C_5 to C_6	30 to 60	
Naphtha (solvent)	C_7 to C_8	60 to 90	
Gasoline	C_5 to C_{12}	40 to 175	40%
Kerosene	C_{12} to C_{15}	150 to 275	10%
Fuel oils, mineral oil	C_{15} to C_{18}	225 to 400	30%
Lubricating oil, petroleum jelly greases, paraffin wax, asphalt	C_{16} to C_{24}	Over 400	10%

Petroleum

Discuss 

Review the balanced equation for combustion of methane and explain the conditions that lead to the formation of CO. Refer students to *Carbon Monoxide* on page R23 of the Elements Handbook .

CLASS Activity

Fractions of Crude Oil 

Divide the class into five groups. Have each group create a display on one of the fractions obtained from crude oil. Each group should use Table 22.4 to make a pie graph of the percents of the fractions in crude oil, highlighting the slice that represents their fraction. Displays should also include information on the composition of the fraction and uses of the products derived from that fraction. Students can illustrate their displays with original art or magazine photographs.

Differentiated Instruction

Gifted and Talented 

Have students research the energy sources used in the United States. Have them summarize their findings in a bar chart showing to what extent the United States depends on each type of energy source. The bar chart should profile the energy sources for the years 1850, 1900, 1950, 1990, and 2000.

Gifted and Talented 

Have students research one or more specific examples in which bacteria have been used to clean up an oil spill. Was the method successful? Why or why not? Challenge the students to explain how such organisms are able to metabolize hydrocarbons. What are the products of the bacterial digestion of oil?

Answers to...

Checkpoint In Pennsylvania in the late 1850s

Coal

Discuss 

Have students study Figure 22.15 and read the caption. If possible, show the students samples of peat, lignite, bituminous coal, and anthracite coal. Explain that coal can be converted to methane by a process called coal gasification in which coal and hydrogen react under high temperature and pressure. Synthetic crude oil can be made by a process called coal liquefaction in which coal is dissolved in a solvent at high pressure and temperature. The product, called solvent-refined coal, is then purified and heated. The hard, brittle residue that remains can be melted and used as a liquid fuel.

Download a worksheet on **Natural Resources** for students to complete, and find additional teacher support from NSTA SciLinks.

For: Links on Natural Resources
Visit: www.SciLinks.org
Web Code: cdn-1225

Coal

Geologists think that coal had its origin some 300 million years ago when huge tree ferns and mosses grew abundantly in swampy tropical regions. When the plants died, they formed thick layers of decaying vegetation. Layer after layer of soil and rock eventually covered the decaying vegetation, which caused a buildup of intense pressure. This pressure, together with heat from Earth's interior, slowly turned the plant remains into coal.

Coal Formation The first stage in the formation of coal is an intermediate material known as peat. Peat, shown in Figure 22.15, is a soft, brown, spongy, fibrous material. When first dug out of a bog, peat has a very high water content. After it has been allowed to dry, it produces a low-cost but smoky fuel. If peat is left in the ground, it continues to change. After a long period of time, peat loses most of its fibrous texture and becomes lignite, or brown coal. Lignite is much harder than peat and has a higher carbon content (about 50%). The water content, however, is still high. Continued pressure and heat slowly change lignite into bituminous, or soft coal. Bituminous coal has a lower water content and higher carbon content (70–80%) than lignite. In some regions of Earth's crust, even greater pressures have been exerted. In those places, such as eastern Pennsylvania, soft coal has been changed into anthracite, or hard coal. Anthracite has a carbon content that exceeds 80%, making it an excellent fuel source. **Coal is classified by its hardness and carbon content.**

Coal, which is usually found in seams from 1 to 3 meters thick, is obtained from both underground and surface mines. In North America, coal mines are usually less than 100 meters underground. Much of the coal is so close to the surface that it is strip-mined, as shown in Figure 22.16. By contrast, many coal mines in Europe and other parts of the world extend 1000 to 1500 meters below Earth's surface.

Figure 22.15 Coal formed when tree ferns and mosses died. The layers of decaying organic material were compressed over millions of years between layers of soil and rock. The first stage in coal formation is peat. Continued pressure and heat transform peat into lignite, bituminous coal, and anthracite coal.

Facts and Figures

Cleaning up Oil

Spilled oil from oil tankers can destroy fragile marine ecosystems. Because cleaning up oil spills is difficult and expensive, researchers are developing new cleanup technologies and safer transportation methods. Most current cleanup methods involve containing the oil and either pumping it away or soaking it up with absorbent materials. Some of the new developments include the genetic engineering of bacteria to "eat" up the oil spill and the design of double-hulled tankers that are less likely to spill their oil in a collision.

Figure 22.16 Coal is mined from both surface mines, shown here, and underground mines.

Composition of Coal **Coal consists largely of condensed aromatic compounds of extremely high molar mass. These compounds have a high proportion of carbon compared with hydrogen.** Due to the high proportion of aromatic compounds, coal leaves more soot upon burning than do the more aliphatic fuels obtained from petroleum. The majority of the coal that was once burned in North America contained about 7% sulfur, which burns to form the major air pollutants SO_2 and SO_3.

Coal may be distilled to obtain a variety of products: coke, coal tar, coal gas, and ammonia. Coke is the solid material left after coal distillation. It is used as a fuel in many industrial processes and is the crucial reducing agent in the smelting of iron ore. Because it is almost pure carbon, coke produces intense heat and little or no smoke when it burns. Coal gas consists mainly of hydrogen, methane, and carbon monoxide, all of which are flammable. Coal tar can be distilled further into benzene, toluene, naphthalene, phenol, and pitch. The ammonia from distilled coal is converted to ammonium sulfate for use as a fertilizer.

22.5 Section Assessment

30. **Key Concept** Describe the hydrocarbons found in natural gas.
31. **Key Concept** Describe the first process used in the refining of petroleum.
32. **Key Concept** What are the two variables used to classify coal?
33. **Key Concept** Describe the chemical composition of coal.
34. How did the three major fossil fuels form?
35. What are the principal sources of aliphatic hydrocarbons? What is the principal source of aromatic hydrocarbons?
36. Explain why cracking is a necessary step in petroleum refining.

Elements Handbook

Catalytic Converters When fossil fuels burn in an internal combustion engine, the exhaust contains more than carbon dioxide gas and water vapor. Read about catalytic converters on page R42. Write a paragraph explaining what happens to pollutants in a catalytic converter.

Assessment 22.5 Test yourself on the concepts in Section 22.5.

with ChemASAP

Section 22.5 Assessment

30. alkanes of low molar mass
31. The refining process starts with the distillation of petroleum into fractions according to boiling point.
32. hardness and carbon content
33. Coal consists largely of condensed ring compounds of extremely high molar mass. These compounds have a high ratio of carbon to hydrogen.
34. Heat, pressure, and bacteria changed marine organisms buried in ocean sediments into petroleum and natural gas. Heat and pressure changed buried layers of vegetation into coal.
35. natural gas and petroleum, coal
36. Amounts of products obtained by fractional distillation don't match demands. Cracking breaks down hydrocarbons into smaller, more useful components.

3 ASSESS

Evaluate Understanding 

Have students compare the formation of coal, natural gas, and petroleum. *(Petroleum and natural gas are derived from marine organisms buried under ocean sediments; coal, from buried wetland plants. Natural gas and petroleum contain mostly aliphatic compounds; coal contains a large percentage of aromatic compounds.)*

Reteach 

Display Figure 22.14 on an overhead projector. If possible, bring a sample of crude oil to class so that students can observe its color and odor. Explain that petroleum is a complex mixture of hydrocarbons whose molecules range in size from 5 to 30 carbon atoms. To be used effectively, the crude oil must be refined by fractional distillation, which separates compounds on the basis of their boiling points. Explain that in general, low molar mass compounds have low boiling points and high molar mass compounds have high boiling points.

Elements Handbook

Transition metals, such as rhodium and platinum, act as catalysts for reactions that change nitrogen oxides to nitrogen, and hydro–carbons and carbon monoxide to carbon dioxide and water.

If your class subscribes to the Interactive Textbook, use it to review key concepts in Section 22.5.

with ChemASAP

Technology & Society

A Number You Can't Knock

Causes of Knocking

Knocking is influenced by two main variables. The first variable is engine compression—the degree to which the gas–air mixture in the cylinder is compressed prior to ignition by the spark plug (see Step 2). The higher the compression, the more power delivered by the ignition of the fuel. Higher compression also means greater likelihood of engine knock. The second variable is the chemical structure of the hydrocarbons that make up the fuel. Straight-chain alkanes are more prone to knocking than are branched-chain alkanes, cycloalkanes, and aromatic hydrocarbons. In practice, these two variables mean that engine compression—and thus performance—is limited by the composition of the gasoline used.

Octane Ratings

The designers of the octane system developed a way of testing fuels in a special one-cylinder engine with a variable compression ratio. For each fuel tested, they could increase the compression until the engine began to knock. Then they chose two pure fuels as standards against which to measure others—heptane and isooctane (2,2,4-trimethylpentane). These two hydrocarbons had similar physical properties, but isooctane was very resistant to knocking, and heptane was very prone to it. The designers arbitrarily assigned heptane an octane rating of 0 and isooctane a rating of 100. Every fuel tested could then be assigned an octane number relative to these hydrocarbons. An octane rating of 89 meant that fuel performed the same in the special test engine as a mixture that was 89 parts isooctane and 11 parts heptane.

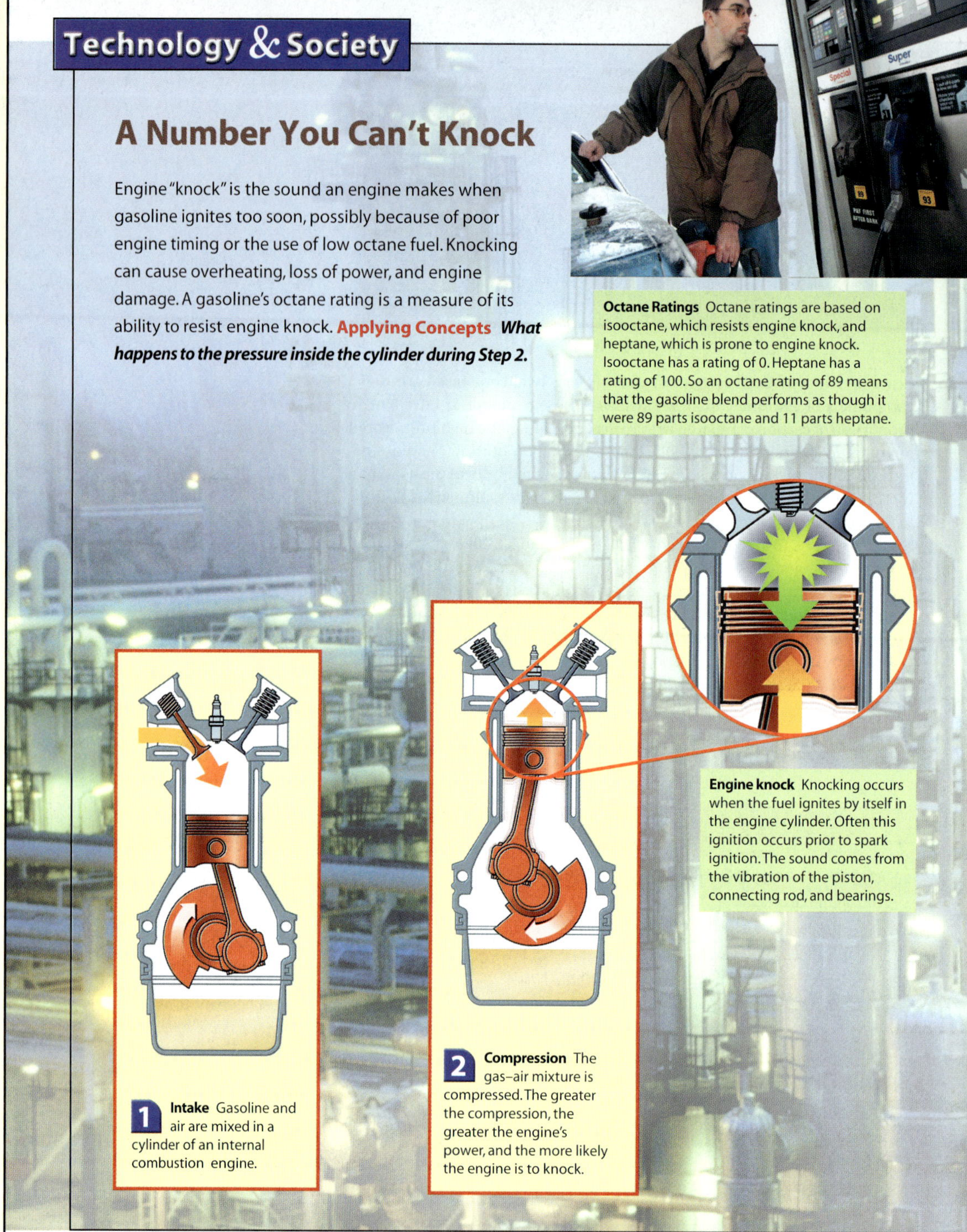

Technology & Society

A Number You Can't Knock

Engine "knock" is the sound an engine makes when gasoline ignites too soon, possibly because of poor engine timing or the use of low octane fuel. Knocking can cause overheating, loss of power, and engine damage. A gasoline's octane rating is a measure of its ability to resist engine knock. **Applying Concepts** ***What happens to the pressure inside the cylinder during Step 2.***

Octane Ratings Octane ratings are based on isooctane, which resists engine knock, and heptane, which is prone to engine knock. Isooctane has a rating of 0. Heptane has a rating of 100. So an octane rating of 89 means that the gasoline blend performs as though it were 89 parts isooctane and 11 parts heptane.

Engine knock Knocking occurs when the fuel ignites by itself in the engine cylinder. Often this ignition occurs prior to spark ignition. The sound comes from the vibration of the piston, connecting rod, and bearings.

1 Intake Gasoline and air are mixed in a cylinder of an internal combustion engine.

2 Compression The gas–air mixture is compressed. The greater the compression, the greater the engine's power, and the more likely the engine is to knock.

716 Chapter 22

This fuel rating system is still in use today. The major modification is that two somewhat different testing methods are used. One, called the motor octane rating (M), tests the fuel under conditions that approximate high-speed driving with heavy loads. The other, called the research octane rating (R), uses conditions that simulate more normal driving. Usually, the octane number assigned to gasolines is an average of the numbers given by these two methods. This is indicated on gasoline pumps as (R + M)/2.

Discuss L2

Review the refining processes used to convert large molar mass hydrocarbons to smaller branched-chain hydrocarbons. Point out that the petroleum industry uses cracking, reforming, and isomerization to produce gasolines with the desired octane rating. Cracking is the process by which large hydrocarbons are broken down into smaller molecules with a catalyst and heat. Reforming and isomerization are the processes used to convert straight-chain alkanes, such as *n*-octane, to branched-chain alkanes, such as isooctane.

Answers to...

Applying Concepts The pressure inside the cylinder increases during Step 2.

CHAPTER 22 Study Guide

Study Tip

Organize new information. Use flashcards to help you visualize the different types of hydrocarbons. Try to integrate what you're studying with what you already know. Find a study partner and quiz one another using the flashcards.

If your class subscribes to the Interactive Textbook with ChemASAP, your students can go online to access an interactive version of the Student Edition and a self-test.

with ChemASAP

CHAPTER 22 Study Guide

Key Concepts

22.1 Hydrocarbons

- Because carbon has four valence electrons, carbon atoms always form four covalent bonds.
- The carbon atoms in an alkane can be arranged in a straight chain or in a chain that has branches.
- Molecules of hydrocarbons, such as alkanes, are nonpolar molecules.

22.2 Unsaturated Hydrocarbons

- At least one carbon-carbon bond in an alkene is a double bond. Other bonds may be single carbon-carbon and carbon-hydrogen bonds.
- At least one carbon-carbon bond in an alkyne is a triple bond. Other bonds may be single or double carbon-carbon bonds and single carbon-hydrogen bonds.

22.3 Isomers

- Structural isomers differ in physical properties and have different chemical reactivites.
- Two types of stereoisomers are geometric isomers and optical isomers.

22.4 Hydrocarbon Rings

- Some hydrocarbon compounds have a carbon chain that is in the form of a ring.
- In a benzene molecule, the bonding electrons between carbon atoms are shared evenly around the ring.

22.5 Hydrocarbons From Earth's Crust

- Natural gas is an important source of alkanes of low molar mass.
- The refining of petroleum starts with the distillation of petroleum into fractions according to boiling point.
- Coal consists largely of condensed aromatic compounds of extremely high molar mass. These compounds have a high proportion of carbon compared with hydrogen.

Vocabulary

- aliphatic hydrocarbons (p. 703)
- alkanes (p. 694)
- alkenes (p. 702)
- alkyl group (p. 698)
- alkynes (p. 703)
- aromatic compound (p. 710)
- asymmetric carbon (p. 705)
- branched-chain alkane (p. 698)
- *cis* configuration (p. 705)
- condensed structural formulas (p. 696)
- cracking (p. 713)
- cyclic hydrocarbon (p. 709)
- homologous series (p. 695)
- hydrocarbons (p. 693)
- geometric isomers (p. 705)
- isomers (p. 704)
- optical isomers (p. 706)
- saturated compounds (p. 702)
- stereoisomers (p. 705)
- straight-chain alkanes (p. 695)
- structural isomers (p. 704)
- substituent (p. 697)
- *trans* configuration (p. 705)
- unsaturated compounds (p. 702)

Organizing Information

Use these terms to construct a concept map that organizes the major ideas of this chapter.

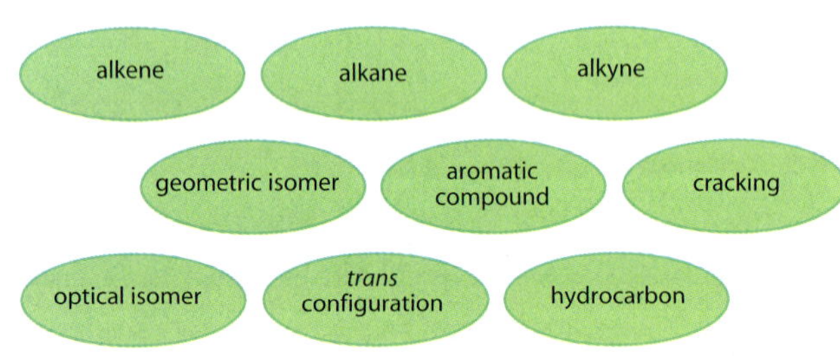

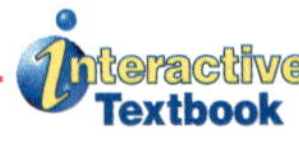

Concept Map 22 Solve the Concept Map with the help of an interactive guided tutorial.

with ChemASAP

Chapter Resources

Print

- **Core Teaching Resources,** Chapter 22, *Practice Problems, Vocabulary Review, Quiz, Chapter Test A, Chapter Test B*

Technology

- **Computer Test Bank,** Chapter 22 Test
- **Interactive Textbook with ChemASAP,** Chapter 22

CHAPTER 22

Assessment

Reviewing Content

22.1 Hydrocarbons

37. Draw condensed structural formulas for pentane and hexane. Assume that the C—H and C—C bonds are understood.

38. Name the alkanes that have the following molecular or structural formulas.
- **a.** $CH_3CH_2CH_3$
- **b.** $CH_3(CH_2)_6CH_3$
- **c.** H—C—C—C—C—C—H (each carbon bonded to an H above and an H below)

39. Draw structures for the alkyl groups derived from methane, ethane, and propane.

40. Give the IUPAC name for each compound.
- **a.** $CH_3—CH—CH_2$ with CH_3 on CH and CH_3 on CH_2
- **b.** $CH_3—CH—CH—CH_3$ with CH_3 on each CH
- **c.** $CH_3—CH—CH_2—CH_2$ with $CH_2—CH_3$ on CH and CH_3 on the last CH_2

41. Why are alkane molecules nonpolar?

22.2 Unsaturated Hydrocarbons

42. Give a systematic name for these alkenes.
- **a.** $CH_3CH{=}CH_2$
- **b.** CH_3 and H on one C, H and CH_2CH_3 on the other (C=C)
- **c.** $CH_3CHCH_2CH{=}CH_2$ with CH_3 on the second carbon
- **d.** CH_3 and CH_3 on one C, CH_2CH_3 and CH_2CH_3 on the other (C=C)

43. Name and draw a structural formula for each alkene with the molecular formula C_5H_{10}.

22.3 Isomers

44. Draw and name all the structural isomers with the molecular formula C_6H_{14}. (You may wish to draw only the carbon skeletons.)

45. Draw one structural isomer of each compound.
- **a.** $CH_3—C—CH_3$ with CH_3 above and CH_3 below the central C
- **b.** $CH_3—CH—CH—CH_3$ with CH_3 on the first CH and $CH_2—CH_3$ on the second CH

46. Draw a structural formula or carbon skeleton for each of the following alkenes. If *cis* and *trans* forms are present, include both forms.
- **a.** 2-pentene
- **b.** 2-methyl-2-pentene
- **c.** 3-ethyl-2-pentene

47. Do all molecules have optical isomers? Explain.

48. Can you draw a structural isomer of hexane (C_6H_{14}) that has an asymmetric carbon? Explain.

22.4 Hydrocarbon Rings

49. Draw a structural formula for each compound.
- **a.** 1,4-diethylbenzene
- **b.** 2-methyl-3-phenylpentane
- **c.** 1,3-dimethylbenzene

50. Explain why both of these structures represent 1,2-diethylbenzene.

(Two benzene ring structures, each with CH_2CH_3 and CH_2CH_3 on adjacent carbons, drawn with different double-bond positions)

22.5 Hydrocarbons From Earth's Crust

51. How are catalysts used in petroleum refining?

52. Rank these materials in order of increasing hardness: bituminous coal, peat, lignite, and anthracite coal.

53. What happens to the sulfur when coal burns?

CHAPTER 22

Assessment

Reviewing Content

37. pentane: $CH_3CH_2CH_2CH_2CH_3$
hexane: $CH_3CH_2CH_2CH_2CH_2CH_3$

38. **a.** propane **b.** octane **c.** pentane

39. methyl: H—C— (with H above and below); ethyl: H—C—C— (each C with H above and below); propyl: H—C—C—C— (each C with H above and below)

40. **a.** 2-methylbutane
b. 2,3-dimethylbutane
c. 3-methylhexane

41. The carbon-carbon bonds are nonpolar and the carbon-hydrogen bonds are very weakly polar..

42. **a.** propene **b.** *trans*-2-pentene
c. 4-methyl-1-pentene
d. 3-ethyl-2-methyl-2-pentene

43. $CH_2{=}CHCH_2CH_2CH_3$ 1-pentene
$CH_3CH{=}CHCH_2CH_3$ 2-pentene
$CH_3—C{=}CH—CH_3$ (with CH_3 on C) 2-methyl-2-butene

44. Five structural isomers with the molecular formula C_6H_{14} exist.
- C—C—C—C—C—C hexane
- C—C—C—C (with C above and C below the second C) 2, 2-dimethylbutane
- C—C—C—C—C (with C on the second C) 2-methylpentane
- C—C—C—C—C (with C on the third C) 3-methylpentane
- C—C—C—C (with C on the second and third C) 2, 3-dimethylbutane

45. **a.** Accept any isomer with 5 carbons and 12 hydrogens. **b.** Accept any isomer with 7 carbons and 16 hydrogens.

46. **a.** CH_3 and H on one C, H and CH_2CH_3 on the other (C=C): *trans*-2-pentene; CH_3 and H on one C, CH_2CH_3 and H on the other (C=C): *cis*-2-pentene
b. CH_3 and CH_3 on one C, H and CH_2CH_3 on the other (C=C)
2-methyl-2-pentene
c. CH_3 and H on one C, CH_2CH_3 and CH_2CH_3 on the other (C=C)
3-ethyl-2-pentene

47. No, Only molecules with at least one asymmetric carbon have optical isomers.

48. No, Hexane contains no substituted groups.

49. **a.** benzene ring with CH_2CH_3 at top and CH_2CH_3 at bottom **b.** $CH_3—CH—CH—CH_2—CH_3$ with CH_3 on the first CH and a benzene ring on the second CH
c. benzene ring with CH_3 and CH_3 in meta positions

50. Two different structural formulas are possible because a benzene ring exhibits resonance.

51. During cracking, catalysts are used to produce more short-chain components.

52. peat, lignite, bituminous coal, anthracite coal

53. When coal that contains a high percent of sulfur burns, the major air pollutants SO_2 and SO_3 are by-products.

CHAPTER 22

Assessment *continued*

Understanding Concepts

54. a. The *di-* indicates two methyl groups, but only one location is given; 2,2-dimethlypentane.
b. Alkyl groups on end carbons are part of the chain; pentane.
c. The chain wasn't numbered so the substituent has the lowest possible number; 2-methylbutane.
d. The methyl group on carbon 4 is part of the chain; 3-methylpentane

55. a. Ethyne has one triple C–C bond and two single C–H bonds.
b. All bonds in propane are single bonds. **c.** In methylbenzene, there are hybrid bonds within the ring and single bonds within substituents and between substituents and the ring.

56. a. $CH{\equiv}C{-}CH_3$ **b.** cyclohexane ring with two H on each carbon **c.** $CH_3{-}CH{-}CH_3$ with a benzene ring on the middle carbon

d. $CH_3{-}C(CH_3)_2{-}CH_2{-}CH(CH_3){-}CH_3$

57. propane, butane, pentane

58. Geometric isomers differ in the arrangement of substituent groups attached to each carbon in a double bond. In optical isomers, there is at least one asymmetric carbon.

59. a. H:C::C:H with one H above and one H below each C **b.** H:C:C:C:H with one H above and one H below each C

c. H:C:::C:H **d.** a ring of four C atoms with two H on each C

60. $2C_8H_{18} + 25O_2 \rightarrow 16CO_2 + 18H_2O$

61. The middle structure is most stable due to resonance within the ring.

62. a. 9.6 billion **b.** 81%

63. No, the structures are identical; one has been flipped over.

64. a. (3) **b.** (1) **c.** (5) **d.** (4) **e.** (2)

CHAPTER 22

Assessment *continued*

Understanding Concepts

54. Why are the following names incorrect? What are the correct names?
a. 2-dimethylpentane
b. 1,3-dimethylpropane
c. 3-methylbutane
d. 3,4-dimethylbutane

55. For each hydrocarbon shown, identify the type of covalent bonds and name the compound.

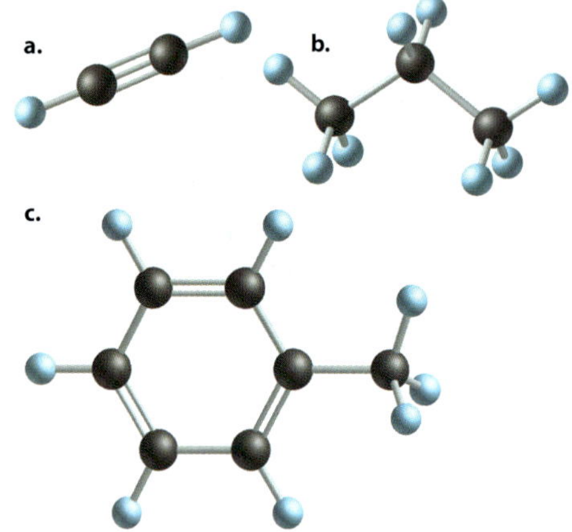

56. Write structural formulas for these compounds.
a. propyne
b. cyclohexane
c. 2-phenylpropane
d. 2,2,4-trimethylpentane

57. Name the next three higher homologs of ethane.

58. Compare geometric and optical isomers.

59. Draw electron dot structures for each compound.
a. ethene **b.** propane
c. ethyne **d.** cyclobutane

60. Write an equation for the combustion of octane.

61. Compare these three molecular structures. Which would you expect to be most stable? Why?

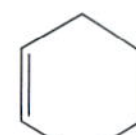 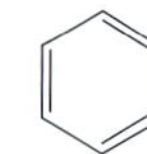 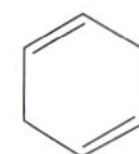

62. The seven organic chemicals produced in the largest amounts in the United States in a recent year are listed in the table below. Answer the following questions based on the data given.

Chemical	Amount produced (billions of kg)
Ethylene	15.9
Propylene	8.4
Urea	6.8
Ethylene dichloride	6.3
Benzene	5.3
Ethyl benzene	4.3
Vinyl chloride	3.7

a. How many billion kilograms of aromatic compounds were produced?
b. Of the total mass of all seven compounds produced, what percent by mass was made up of aliphatic compounds?

63. Are these two structures geometric isomers? Explain your answer.

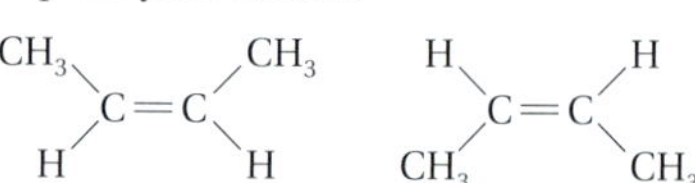

64. Use the labeled features in the molecular structure to answer the following questions.

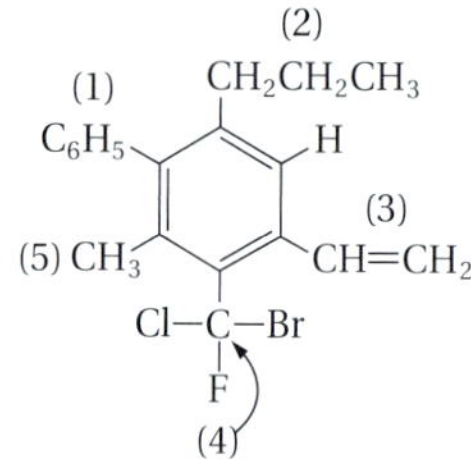

a. Which label identifies a double bond?
b. Which label identifies a phenyl group?
c. Which label identifies a methyl group?
d. Which label identifies an asymmetric carbon?
e. Which label identifies a propyl group?

Critical Thinking

65. Methane (CH_4), a widely used fuel, has a heat of combustion (ΔH) of −890 kJ/mol. The ΔH for benzene (C_6H_6) is much higher, −3268 kJ/mol, yet benzene alone is never used as a fuel. Suggest some reasons why benzene is a less desirable fuel than methane.

66. Explain why you cannot draw a structural formula for methene.

67. Use the isomers of 2-pentene to show how lack of rotation about a carbon-carbon double bond leads to geometric isomers.

68. Most cyclic hydrocarbons have higher boiling points than alkanes with the same number of carbons. Suggest a possible explanation for this general difference in boiling points.

69. Alkadienes are hydrocarbons with two double bonds. Draw the structural formula of the alkadiene with the molecular formula C_3H_4.

70. The molecular formula C_4H_6 could represent an alkyne, a cycloalkene, or a hydrocarbon with two double bonds. Write the condensed structural formulas for each. Which compound do you think is the least stable, and why?

71. Draw structural formulas for the following compounds.

a. *cis*-2,3-dimethyl-3-hexene

b. 1-ethyl-2-methylcyclopentane

72. Draw the correct structure for any of the ones below that are incorrect.

a. $CH_3—CH=CH—CH_2—CH_3$

b. (cyclopentene ring with CH_3 and CH_3 on the same carbon)

c. $CH_3—C≡CH—CH_2—CH_3$

d. $CH_3=CH—CH_2—CH_3$

73. What structural feature is associated with each of these hydrocarbons: an alkane; an alkene; an aromatic hydrocarbon; a cycloalkane.

74. Alkenes can undergo an addition reaction in which substances are added to the carbons in the carbon-carbon double bond. Predict whether an alkane or an alkyne is more likely to undergo an addition reaction. Explain your answer.

Concept Challenge

75. Use the data in Table 22.1 to make a graph of boiling point versus number of carbons for the first ten straight-chain alkanes. Is the graph a straight line? Use the graph to predict the boiling point of undecane, the straight-chain alkane containing eleven carbons. Use a chemistry handbook to find the actual boiling point of undecane. Compare the actual boiling point with your prediction.

76. Fossil fuels such as oil and natural gas are the raw materials for many consumer products. Should this information affect the decision to develop energy sources other than fossil fuels? Explain.

77. The graph shows the number of structural isomers for alkanes with three through ten carbon atoms.

Number of Possible Structural Isomers

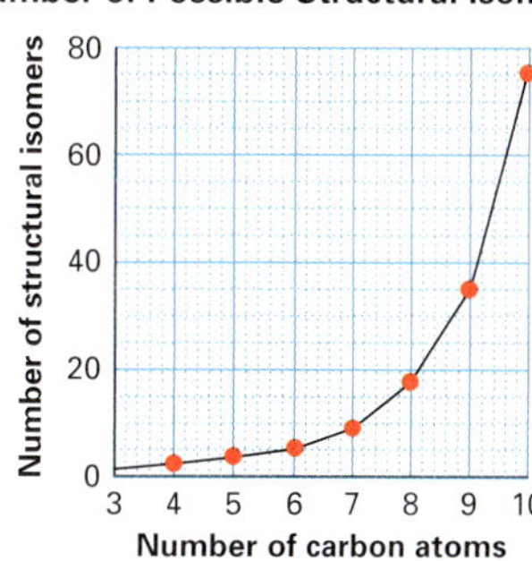

a. How many structural isomers are there for the C_6, C_7, C_8, C_9, and C_{10} alkanes?

b. The difference between the number of isomers for C_7 and C_8 is 9. The difference between the number of isomers for C_9 and C_{10} is 40. In each case, one additional carbon atom is added to the molecule. Why is the change in the number of isomers so different?

78. Correct each of the following names and draw the correct structural formulas.

a. 4-methylhexane

b. 1,4-diethyl cyclopentane

c. 3,3methyl-4-ethyloctane

Critical Thinking

65. The amount of heat per carbon is higher for methane (–890 kJ/mol of carbon) than for benzene (–545 kJ/mol of carbon). Methane undergoes complete combustion if sufficient air is present. Burning aromatic compounds produces more soot.

66. *Meth-* implies one carbon atom; *ene-* implies a double bond, which requires two carbon atoms.

67. The methyl and ethyl groups can be on the same side of the bond or opposite sides.

$CH_3(H)C=C(H)CH_2CH_3$ with CH_3 and CH_2CH_3 on opposite sides — *trans*-2-pentene

$CH_3(H)C=C(H)CH_2CH_3$ with CH_3 and CH_2CH_3 on the same side — *cis*-2-pentene

68. Students may infer that the ring structures in cyclic hydrocarbons produce stronger van der Waals attractions, which increase the energy required for vaporization.

69. $H_2C=C=CH_2$

70. The cycloalkene would be most unstable because the bond angles are 90° instead of the 120° predicted by VSEPR theory.

$C≡C—C—C$ (alkyne)

□ (cycloalkene)

$C=C—C=C$ (alkadiene)

71. a. $(CH_3)(CH_3CH_2)C=C(CH_3)(CH_2CH_3)$ — CH_3 groups on the same side, CH_3CH_2 groups on the same side

b. cyclopentane ring with CH_2CH_3 and CH_3 on adjacent carbons

72. b. cyclopentene ring with CH_3 and CH_3 on the same carbon

c. $CH_3—C≡C—CH_2—CH_3$

d. $CH_2=CH—CH_2—CH_3$

73. Alkanes contain only single bonds. Alkenes contain at least one double bond. Aromatic hydrocarbons contain a benzene ring or a similar ring. Cycloalkanes contain aliphatic chains linked end-to-end.

74. an alkyne, because it contains a triple bond

Concept Challenge

75. The graph isn't a straight line. The estimated bp should be greater than 150°C. The actual bp is 196°C.

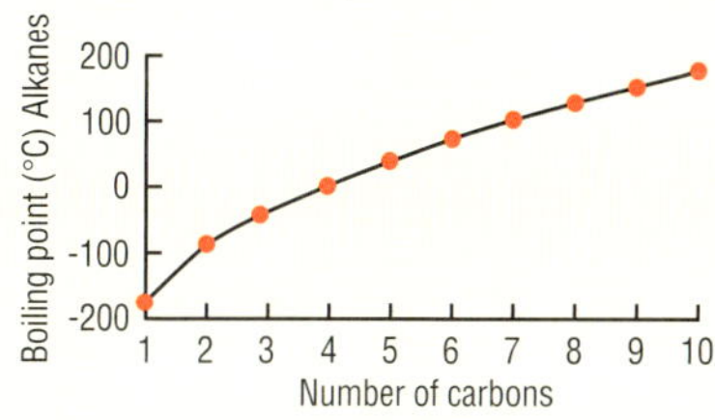

76. Answers will vary, but should reflect that fossil fuels are not renewable.

77. a. $C_6 = 5, C_7 = 9, C_8 = 18, C_9 = 35, C_{10} = 75$

b. the more carbon atoms, the more possible ways to arrange them

78. a. $CH_3CH_2CH(CH_3)CH_2CH_2CH_3$

3-methylhexane

b. cyclopentane ring with two CH_2CH_3 groups

1,3-diethylcyclopentane

c. $CH_3CH_2—C(CH_3)_2—CH(CH_2CH_3)CH_2CH_2CH_2CH_3$

3,3-dimethyl-4-ethyloctane

CHAPTER 22 Assessment continued

Cumulative Review

79. a. 13.9 L **b.** 1 L **c.** 20 kPa
80. 3.04×10^2 calories
81. 1.13 mol KNO_3; 1.14×10^2 g KNO_3
82. 1.67×10^2 J
83. 1 cal = 4.184 J; 4.184×10^3 J
84. a. Smaller particle size speeds up the reaction. **b.** Higher temperature usually speeds up the reaction.
85. a. favors reactants **b.** favors products
86. a. $K_{eq} = \dfrac{[ICl]^2}{[I_2][Cl_2]}$

b. $K_{eq} = \dfrac{[H_2][Br_2]}{[HBr]^2}$

c. $K_{eq} = \dfrac{[HCl]^4[S]^3[SO_2]}{[S_2Cl_2]^2[H_2O]^2}$

d. $K_{eq} = \dfrac{[NH_3]^2}{[N_2][H_2]^2}$

87. a. 10.00
b. 7.59
c. 12.00
d. 11.70
88. a. ~8 **b.** Use a pH meter
89. a. H_3PO_4 **b.** CsOH
c. H_2CO_3 **d.** $Be(OH)_2$
90. a. $NaOH \rightarrow Na^+ + OH^-$
b. $Ba(OH)_2 \rightarrow Ba^{2+} + 2OH^-$
91. a. Ca, +2; C, +4; O, −2 **b.** Cl, 0
c. Li, +1; I, +5; O, −2
d. Na, +1; S, +4; O, −2
92. a. reduction **b.** reduction
c. reduction **d.** oxidation
93. a. +4 **b.** +4 **c.** +3 **d.** +5 **e.** +5 **f.** +2
94. a. The coefficients are 2, 9, 6, 8.
b. The coefficients are 3, 2, 1, 3.
95. It is the cell potential when the ion concentrations in the half-cells are 1*M*, the temperature is 25°C, and the pressure of any gases present is 101.3 kPa.
96. Reduction occurs in the half-cell with the more positive, or less negative, reduction potential. Ni^{2+} is reduced and Al is oxidized.
Cell reaction: $3Ni^{2+}(aq) + 2Al(s) \rightarrow 3Ni(s) + 2Al^{+3}(aq)$; $E^0_{cell} = +1.66$ V
97. The reaction is nonspontaneous.
98. reduction; reduction

CHAPTER 22 Assessment continued

Cumulative Review

79. Calculate the following quantities. (*Chapter 14*)
a. The number of liters occupied at STP by 6.20×10^{-1} mol $Cl_2(g)$.
b. The volume of a gas at 3 kPa of pressure if the same gas has a volume of 6 L at 0.5 kPa and the temperature is constant.
c. The partial pressure of gas X (P_x) in a mixture of three gases, X, Y, and Z, if the total pressure (P_{total}) is 50 kPa and the sum of the partial pressures of Y and Z is 30 kPa.

80. How many calories are absorbed when 56.0 g of liquid water at 100°C is vaporized to steam at 100°C? (*Chapter 15*)

81. How many moles of solute are in 750 mL of 1.50*M* KNO_3? How many grams of KNO_3 is this? (*Chapter 16*)

82. A silver dollar is heated and placed in a foam cup calorimeter containing 50.0 mL of water at 26.5°C. The water reached a maximum temperature of 27.3°C. How many joules of heat were released by the silver dollar? (*Chapter 17*)

83. What is the relationship between a calorie and a joule? How many joules is 1 kcal? (*Chapter 17*)

84. How does (**a**) particle size and (**b**) temperature affect the rate of a chemical reaction? (*Chapter 18*)

85. Explain how the equilibrium position of this reaction is affected by (**a**) decreasing the temperature and (**b**) removing CO_2. (*Chapter 18*)

$CaCO_3(s) + heat \rightleftharpoons CaO(s) + CO_2(g)$

86. Write equilibrium constant expressions for the following reactions. (*Chapter 18*)
a. $Cl_2(g) + I_2(g) \rightleftharpoons 2ICl(g)$
b. $2HBr(g) \rightleftharpoons H_2(g) + Br_2(g)$
c. $2S_2Cl_2(g) + 2H_2O(g) \rightleftharpoons 4HCl(g) + 3S(g) + SO_2(g)$
d. $N_2(g) + 3H_2(g) \rightleftharpoons 2NH_3(g)$

87. What are the pH values for aqueous solutions containing each of the following hydroxide-ion concentrations? (*Chapter 19*)
a. $1.0 \times 10^{-4}M$ **b.** $3.9 \times 10^{-7}M$
c. 0.010*M* **d.** 0.0050*M*

88. A colorless solution of unknown pH turns blue when tested with the acid-base indicator bromthymol blue. It remains colorless when tested with phenolphthalein. (*Chapter 19*)
a. What is the approximate pH of the solution?
b. How could you determine the pH more accurately?

89. Write the formula for each acid or base. (*Chapter 19*)
a. phosphoric acid **b.** cesium hydroxide
c. carbonic acid **d.** beryllium hydroxide

90. Write the reaction for the dissociation of each of the following compounds in water. (*Chapter 19*)
a. sodium hydroxide **b.** barium hydroxide

91. Give the oxidation number of each element in the following substances. (*Chapter 20*)
a. $CaCO_3$ **b.** Cl_2
c. $LiIO_3$ **d.** Na_2SO_3

92. Identify these processes as either oxidation or reduction. (*Chapter 20*)
a. $Fe^{3+} + e^- \longrightarrow Fe^{2+}$ **b.** $Cl_2 + 2e^- \longrightarrow 2Cl^-$
c. $Fe^{3+} + 3e^- \longrightarrow Fe$ **d.** $Zn \longrightarrow Zn^{2+} + 2e^-$

93. Determine the oxidation number of nitrogen in the following substances and ions. (*Chapter 20*)
a. N_2O_4 **b.** NO_2 **c.** NH_3
d. NO_3^- **e.** NH_4^+ **f.** NO

94. Balance these redox equations. (*Chapter 20*)
a. $C_3H_7OH(l) + O_2(g) \longrightarrow CO_2(g) + H_2O(l)$
b. $BaO(s) + Al(s) \longrightarrow Al_2O_3(s) + Ba(s)$

95. Explain the term *Standard Cell Potential*. (*Chapter 21*)

96. A voltaic cell is made of the following half-cells. Determine the cell reaction and calculate the standard cell potential. (*Chapter 21*)

$Al^{3+}(aq) + 3e^- \longrightarrow Al(s)\ E^0_{Al^{3+}} = -1.66V$
$Ni^{2+}(aq) + 2e^- \longrightarrow Ni(s)\ E^0_{Ni^{2+}} = -0.25V$

97. The calculated standard cell potential for a redox reaction is a negative number. What does this tell you about the reaction? (*Chapter 21*)

98. What process always occurs at the cathode of an electrolytic cell? At the cathode of a voltaic cell? (*Chapter 21*)

Standardized Test Prep

Test-Taking Tip

Eliminate Wrong Answers If you don't know which response to a question is correct, start by eliminating those you know are wrong. If you can rule out some choices, you'll have fewer choices left to consider and you'll increase your chances of choosing the correct answer.

Select the choice that best answers each question or completes each statement.

1. What is the name of the compound with the following structural formula?

```
         CH3  H   CH3
          |   |    |
CH3 — C — C — C — CH3
          |   |    |
          H   H    H
```

   a. 1,2,3,3-tetramethylpropane
   b. heptane
   c. 2,4-dimethylpentane
   d. 1,5-dimethylbutane

2. Which of these are characteristic of all alkenes?
   I. unsaturated
   II. carbon-carbon double bond
   III. optical isomers
   a. I and II only
   b. II and III only
   c. I and III only
   d. I, II, and III

3. How many carbon atoms are in a molecule of 4,5-diethyloctane?
   a. 10
   b. 12
   c. 14
   d. 16

4. *Cis-trans* geometric isomerism is possible in
   a. 2-pentene.
   b. 2-butane.
   c. propyne.
   d. benzene.

5. A structural isomer of heptane is
   a. methylbenzene.
   b. 3,3-dimethylpentane.
   c. cycloheptane.
   d. 3-methylhexene.

6. Which molecule has optical isomers?
   a. CH_4
   b. CF_2H_2
   c. CFClBrI
   d. CF_2ClH

Use the space-filling models of pentane isomers to answer Questions 7 and 8.

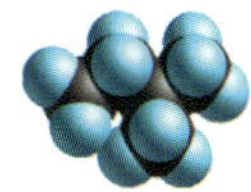
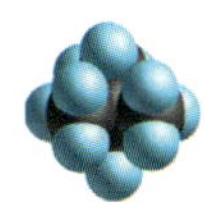

7. Write structural formulas for the three structural isomers of pentane, C_5H_{12}. Name each isomer.

8. Write structural formulas for the four structural isomers of cyclopentane. Draw ball-and-stick models of the isomers.

The lettered choices below refer to Questions 9–12. A lettered choice may be used once, more than once, or not at all.

(A) alkene
(B) arene
(C) alkyne
(D) alkane

To which of the above classes of hydrocarbons does each of the following compounds belong?

9. C_7H_{16}
10. C_5H_8
11. C_6H_6
12. C_8H_{16}

Use the molecular structures below to answer Questions 13–16. A molecular structure may be used once, more than once, or not at all.

a. 

b. $CH_3 — CH_2 — CH_2 — CH_3$

c. d. 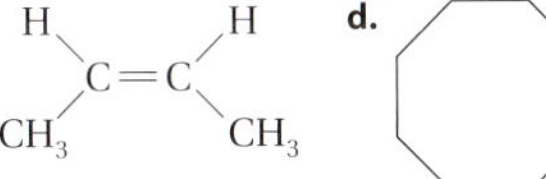

13. Which structure is a cycloalkane?
14. Which structure is a saturated hydrocarbon?
15. Which structure is a *cis*-isomer?
16. Which structure is a *trans*-isomer?

Standardized Test Prep

1. c
2. a
3. b
4. a
5. b
6. c
7.

```
    H   H   H   H   H
    |   |   |   |   |
H — C — C — C — C — C — H
    |   |   |   |   |
    H   H   H   H   H
```

pentane

```
    H   H   H   H
    |   |   |   |
H — C — C — C — C — H
    |   |   |   |
    H   |   H   H
     H— C —H
        |
        H
```

2-methylbutane

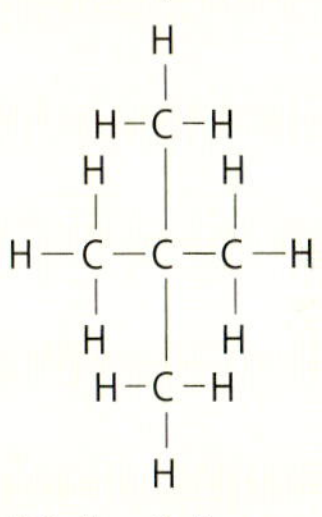

2,2-dimethylpropane

8. In the ball-and-stick models, the black spheres represent carbon and the white spheres represent hydrogen.

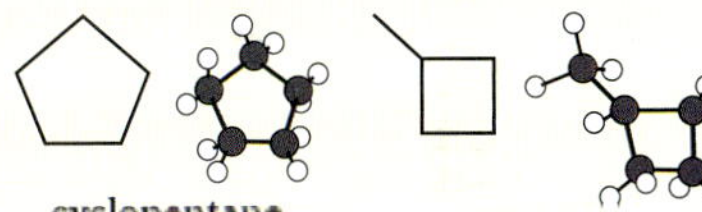

cyclopentane

methylcyclobutane

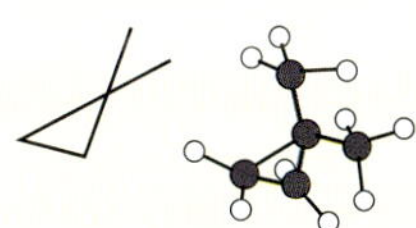

1,1-dimethylcyclopropane

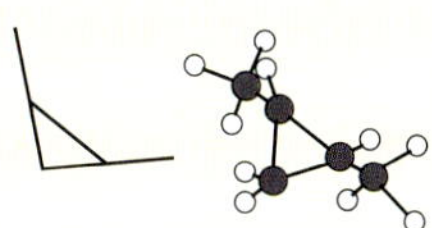

1,2-dimethylcyclopropane

9. D
10. C
11. B
12. A
13. d
14. b, d
15. c
16. a

CHAPTER 23

Functional Groups Planning Guide

SECTION OBJECTIVES	STANDARDS NATIONAL	STANDARDS STATE	ACTIVITIES and LABS	
23.1 Introduction to Functional Groups, pp. 725–729 1 block or 2 periods **23.1.1 Explain** how organic compounds are classified. **23.1.2 Identify** halocarbons and the IUPAC rules for naming halocarbons. **23.1.3 Describe** how halocarbons can be prepared.	A-2, B-2, B-3		**SE** Inquiry Activity: Making the Slimiest Polymer, p 724.	L2
			TE Class Activity: Odors of Functional Groups, p. 726	L1
			TE Class Activity: Halocarbon Structures and Names, p. 727	L2
			TE Teacher Demo: Testing for Halocarbons, p. 728	L2
23.2 Alcohols and Ethers, pp. 730–736 1 block or 2 periods **23.2.1 Identify** how alcohols are classified and named. **23.2.2 Predict** how the solubility of an alcohol varies with the length of its carbon chain. **23.2.3 Name** the reactions of alkenes that may be used to introduce functional groups. **23.2.4 Construct** the general structure of an ether and **describe** how ethers are named.	B-2, B-3, G-2		**TE** Teacher Demo: Are Alcohols Basic? p. 732	L1
			TE Class Activity: Household Products with Alcohol, p. 733	L2
			TE Teacher Demo: Aqueous Solubility of Alcohols, p. 733	L2
			TE Teacher Demo: Test for Unsaturation, p. 734	L2
23.3 Carbonyl Compounds, pp. 737–746 2 blocks or 4 periods **23.3.1 Identify** the structure of a carbonyl group as found in aldehydes and ketones. **23.3.2 Construct** the general formula for carboxylic acids and **explain** how they are named. **23.3.3 Describe** an ester. **23.3.4 Explain** why dehydrogenation is an oxidation reaction.	A-1, B-2, B-3		**SE** Quick Lab: Testing for an Aldehyde, p. 746	L2
			TE Class Activity: Model Building, p. 738	L2
			TE Teacher Demo: Making an Ester, p. 742	L2
			TE Teacher Demo: Oxidation of Alcohols, p. 744	L2
			LM Lab 50: Esters of Carboxylic Acids	L2
			SSLM Lab 38: Vitamin C in Tablets	L2
			SSLM Lab 39: Vitamin C in Drinks	L2
			LP Lab Practical 23-1: Esters of Carboxylic Acids	L2
23.4 Polymerization, pp. 747–755 1 block or 2 periods **23.4.1 Describe** how addition polymers are formed. **23.4.2 Describe** how condensation polymers are formed.	A-1, B-2, B-3, E-2, G-3		**SE** Small-Scale Lab: Polymers, p. 753	L2
			TE Teacher Demo: Building a Polymer Model, p. 748	L2
			TE Teacher Demo: Making Nylon, p. 751	L2
			TE Class Activity: Pigment Extraction, p. 755	L1

Ability Levels

L1 For students who need additional help
L2 For all students
L3 For students who need to be challenged

Components

SE	Student Edition	**LM**	Laboratory Manual	**GCP**	Graphing Calculator Problems	**T**	Transparencies
TE	Teacher's Edition	**SSLM**	Small-Scale Chemistry Lab Manual	**TTT**	Test-Taking Tips with Transparencies, Grades 9–12 Science	**CA**	Chem Alive and Chemistry Field Trips DVD
GRSW	Guided Reading and Study Workbook	**LP**	Laboratory Practicals	**STP**	Standardized Test Preparation Workbook	**iT**	Interactive Textbook with ChemASAP
CTR	Core Teaching Resources	**PLM**	Probeware Lab Manual			**GO**	Go Online Internet Resources
SSG	Spanish Study Guide	**VCL**	Virtual Chemistry Labs				
		CTB	Computer Test Bank				

RESOURCES PRINT and TECHNOLOGY	PROBLEMS and ASSESSMENT
GRSW Chapter 23, Section 1 **L1** T262–T264 **L2** **GO** Chapter 23, Section 1 **L2**	**SE** Section 23.1 Assessment, p. 729 **iT** Section 23.1
GRSW Chapter 23, Section 2 **L1** T265–T267 **L2** **GO** Chapter 23, Section 2 **L2**	**SE** Section 23.2 Assessment, p. 736 **iT** Section 23.2
GRSW Chapter 23, Section 3 **L1** T268–T269 **L2** **GO** Chapter 23, Section 3 **L2**	**SE** Section 23.3 Assessment, p. 746 **iT** Section 23.3
GRSW Chapter 23, Section 4 **L1** T270 **L2** **GO** Chapter 23, Section 4 **L2**	**SE** Section 23.4 Assessment, p. 752 **iT** Section 23.4

Go Online

Web Code: cdn-1231
Web Code: cdn-1232
Web Code: cdn-1233
Web Code: cdn-1234

MATERIALS List

Quantities for each group

STUDENT EDITION

Inquiry Activity, p. 724
white paper glue, 100-mL graduated cylinder, 2 large glasses, 4 paper or foam cups, borax, balance, spoon, 10-mL graduated cylinder

Quick Lab, p. 746
See SE p. 746.

Small-Scale Lab, p. 753
3.5-oz cup, soda straw, powdered guar gum, 4% borax solution, pipet

TEACHER'S EDITION

Class Activity, p. 726
2-propanol, propanoic or butanoic acid, butyl amine, vanillin or cinnamaldehyde, octyl acetate or ethyl butyrate

Teacher Demo, p. 728
copper wire, Bunsen burner, forceps, bromobenzene, benzoic acid

Teacher Demo, p. 732
several alcohols, litmus paper

Class Activity, p. 733
common household products that contain alcohols

Teacher Demo, p. 733
10 or 15 test tubes with stoppers, distilled water, ethanol, 1-butanol, 1-hexanol, 1-octanol, a hexanediol

Teacher Demo, p. 734
$KMnO_4$, NaOH, distilled water, cyclohexane, cyclohexene, 2 test tubes with stoppers

Class Activity, p. 738
model-building kits

Teacher Demo, p. 742
water bath, hot plate, test tubes, several carboxylic acids, several corresponding alcohols, concentrated sulfuric acid, filter paper, watch glass

Teacher Demo, p. 744
CrO_3, concentrated sulfuric acid, water, 3 test tubes, acetone, ethanol, isopropanol, *t*-butyl alcohol

Teacher Demo, p. 748
metal-link chain, paper clips

Teacher Demo, p. 751
5% aqueous solution of 1,6-hexanediamine, small beaker, 20% NaOH, dropper, 5% solution of adipoyl chloride in cyclohexane, forceps or tongs, paper towel

Chapter Assessment

CHAPTER ASSESSMENT
SE Chapter Assessment, pp. 757–760
CTB Chapter 23
iT Chapter 23

STANDARDIZED TEST PREP
SE Chapter 23, p. 761
STP Topic 10
iT Interactive Textbook with Assessment at PHSchool.com

CHAPTER
23

Functional Groups

INQUIRY Activity

Making the Slimiest Polymer 

Purpose

Students recognize that the properties of a polymer can vary by changing the quantities of its components.

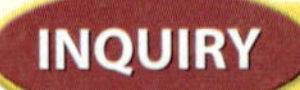 **Prep Time** 15 minutes
Class Time 30 minutes
Safety Note that borax is toxic if ingested and can irritate the skin if left in contact for long periods. Have students wash their hands thoroughly after performing this activity.

Teaching Tips

- Have students work in pairs and summarize their observations in a table.
- Ask students to compare the properties of the product to those of the materials with which they started. Have them propose a hypothesis for what may have happened at the molecular level. *(The properties of the product are those of a deformable solid rather than those of the liquids from which it was made. The borax causes interlinking between the polymers in the glue.)*

Think About It

1. the smallest amount of borax, 5 mL
2. The consistency of the slime would probably be too liquid with less than 5 mL of borax and too solid with more than 20 mL.
3. The slime would be thick and rigid without water. It would not be very slimy.

CHAPTER 23 Functional Groups

The sweet smell of lavender comes, in part, from linalool, an alcohol.

INQUIRY Activity

Making the Slimiest Polymer

Materials
White paper glue, 100-mL graduated cylinder, two large glasses, four paper or foam cups, borax, balance, spoon, 10-mL graduated cylinder

Procedure

1. Mix 60 mL (about 2 oz) of white paper glue with 60 mL of water in a glass.
2. Pour 30 mL of this mixture into each of the four cups.
3. Make a borax solution by stirring 5 g of borax into 100 mL of water in the other glass.
4. Add 5 mL of borax solution to the first cup of glue, 10 mL to the second, 15 mL to the third, and 20 mL to the fourth. Stir the mixture in each cup and knead it with your fingers until it is smooth.
5. Take the contents of the cup with 5 mL of borax and slowly stretch it into a long strand. Repeat for the other three cups.

Think About It

1. Which amount of borax allowed you to pull the longest strand?
2. Do you think adding less than 5 mL or more than 20 mL of the borax solution would allow you to pull a longer strand?
3. Would the results be different if you did not add water to the glue initially?

23.1 Introduction to Functional Groups

Connecting to Your World The lights dim and the audience chatter subsides as the musicians enter the hall. All the musicians wear black, and from a distance they look nearly identical. But their differences become apparent when they pick up their instruments. As the conductor signals the beginning of the piece, a single flute is heard. The warm sounds of the stringed instruments join in, and soon each musician contributes a unique sound to the music. In a similar way, one hydrocarbon is nearly identical to another until it picks up a functional group. In this section, you will learn how functional groups determine the character of organic compounds.

Guide for Reading

Key Concepts

- How are organic compounds classified?
- What is a halocarbon?
- How may halocarbons be prepared?

Vocabulary

functional group
halocarbons
alkyl halides
aryl halides
substitution reaction

Reading Strategy

Making and Using Tables As you read, examine the tables carefully. Make a table in your notebook that organizes the material in this section.

Functional Groups

In Chapter 22, you learned about hydrocarbon chains and rings—the essential components of every organic compound. Yet in most chemical reactions involving organic molecules, the saturated hydrocarbon skeletons of the molecules are chemically inert. How, then, can there be hundreds of different kinds of organic reactions?

Most organic chemistry involves substituents, which are groups attached to hydrocarbon chains. The substituents of organic molecules often contain oxygen, nitrogen, sulfur, and/or phosphorus. They are called functional groups because they are the chemically functional parts of the molecules. A **functional group** is a specific arrangement of atoms in an organic compound that is capable of characteristic chemical reactions. Most organic chemistry is functional-group chemistry. **Organic compounds can be classified according to their functional groups.** Figure 23.1 shows several consumer products that contain various functional groups. The symbol R represents any carbon chains or rings attached to the functional group. Because the double and triple bonds of alkenes and alkynes are chemically reactive, they are also considered functional groups. Table 23.1 on the following page lists the functional groups that you will learn about in this chapter and in Chapter 24. You will find it helpful to refer to this table throughout the chapter.

Figure 23.1 Many consumer products contain hydrocarbon derivatives. The hydrocarbon skeletons in these products are chemically similar. Functional groups give each product unique properties and uses.

Section Resources

Print

- **Guided Reading and Study Workbook,** Section 23.1
- **Core Teaching Resources,** Section 23.1 Review
- **Transparencies,** T262–264

Technology

- **Interactive Textbook with ChemASAP,** Assessment 23.1
- **Go Online,** Section 23.1

23.1

1 FOCUS

Objectives

23.1.1 Explain how organic compounds are classified.

23.1.2 Identify halocarbons and the IUPAC rules for naming halocarbons.

23.1.3 Describe how halocarbons can be prepared.

Guide for Reading

Build Vocabulary 

Paraphrase Have students explain in their own words the meaning of the term *substitution reaction.*

Reading Strategy 

Use Prior Knowledge Review Chapter 22. A thorough knowledge of hydrocarbons is needed before students learn about substituents. A review of intermolecular forces also will be helpful.

2 INSTRUCT

Connecting to Your World

Ask, **How do functional groups determine the character of organic compounds?** *(Each functional group has certain chemical properties that affect the behavior of any compound that contains the functional group.*

Functional Groups

Discuss 

Stress the importance of structure in the chemistry of organic molecules. Molecular shape is dictated by bonding functional groups. Ask students to hypothesize about the structure of a molecule related to its taste or smell. *(Both senses rely on the ability of chemoreceptor molecules on cell surfaces, which recognize a molecule by its shape.)*

Use Visuals L1

Table 23.1 Have students study the table. Encourage them to memorize the general compound structure for each functional group. It may be helpful to list specific examples of compounds for each functional group so that students become more familiar with these groups in larger molecules. Include examples of alkenes and alkynes. Ask, **Why are the carbon–carbon double and triple bonds considered to be functional groups?** *(They are chemically reactive.)* **What do most of the functional groups in the table have in common?** *(Most functional groups contain oxygen.)*

CLASS Activity

Odors of Functional Groups 

Purpose Students will learn that compounds containing the same functional group have similar chemical and physical properties.

Materials 2-propanol, propanoic or butanoic acid, butyl amine, vanillin or cinnamaldehyde, octyl acetate or ethyl butyrate

Safety Have students carefully waft vapors of the compounds in order to detect their odors. Remind students that normally it is not safe to smell chemicals. An exception is being made for these known compounds.

Procedure Bring samples of organic compounds to class. Allow students to observe the characteristic odors associated with alcohols (2-propanol), acids (propanoic or butanoic acid), amines (butyl amine), aldehydes (vanillin or cinnamaldehyde), and esters (the orange odor of octyl acetate and the pineapple odor of ethyl butyrate). Have students record their observations, and then make a summary table on the board. Ask students to consider what other properties molecules with the same functional groups might have in common.

Table 23.1

Organic Compounds Classified by Functional Group

Compound type	Compound structure	Functional group
Halocarbon	R—X (X = F, Cl, Br, or I)	Halogen
Alcohol	R—OH	Hydroxyl
Ether	R—O—R	Ether
Aldehyde	$R{-}\overset{O}{\overset{\Vert}{C}}{-}H$	Carbonyl
Ketone	$R{-}\overset{O}{\overset{\Vert}{C}}{-}R$	Carbonyl
Carboxylic acid	$R{-}\overset{O}{\overset{\Vert}{C}}{-}OH$	Carboxyl
Ester	$R{-}\overset{O}{\overset{\Vert}{C}}{-}O{-}R$	Ester
Amine	$R{-}NH_2$	Amino
Amide	$R{-}\overset{O}{\overset{\Vert}{C}}{-}\overset{H}{\overset{\vert}{N}}{-}R$	Amide

Halogen Substituents

Halocarbons are a class of organic compounds containing covalently bonded fluorine, chlorine, bromine, or iodine. **A halocarbon is a carbon-containing compound with a halogen substituent.** The IUPAC rules for naming halocarbons are based on the name of the parent hydrocarbon. The halogen groups are named as substituents. Examples of IUPAC names for several simple halocarbons along with their structural formulas and space-filling models are given in Figure 23.2. Common names are in parentheses.

Common names of halocarbons consist of two parts. The first part names the hydrocarbon portion of the molecule as an alkyl group, such as *methyl-* or *ethyl-*. The second part gives the halogen with an *-ide* ending.

Figure 23.2 The IUPAC rules for naming halocarbons are based on the name of the parent hydrocarbon. Structural formulas and space-filling models of halocarbons. (a) The common name for chloromethane is methyl chloride. (b) The common name for chloroethene is vinyl chloride. (c) The common name for chlorobenzene is phenyl chloride.

Differentiated Instruction

Gifted and Talented 

Have students research examples in which the structure of a chemical plays an important role in a biological process. Suggestions for research include neurotransmitters, such as acetylcholine, and hormones, such as estradiol and testosterone. Have students research how environmental pollutants can interfere with the action of these natural chemicals.

Table 23.2

Names of Some Common Alkyl Groups		
Name	**Alkyl group**	**Remarks**
Isopropyl	CH_3—C(CH_3)(H)—	The prefix *iso-* is used when there is a methyl group on the carbon second from the unsubstituted end of the longest chain.
Isobutyl	CH_3—CH(CH_3)—CH_2— (primary carbon)	The carbon joining this alkyl group to another group is bonded to one other carbon; it is a primary carbon.
Secondary butyl (*sec*-butyl)	CH_3—CH_2—CH(—)—CH_3 (secondary carbon)	The carbon joining this alkyl group to another group is bonded to two other carbons; it is a secondary carbon.
Tertiary butyl (*tert*-butyl)	CH_3—C(CH_3)(CH_3)— (tertiary carbon)	The carbon joining this alkyl group to another group is bonded to three carbons; it is a tertiary carbon.
Vinyl	$H_2C{=}CH$—	When used as an alkyl group in giving compounds common names, this group is called vinyl.
Phenyl	C_6H_5— (benzene ring)	Phenyl is derived from benzene.

On the basis of their common names, halocarbons in which a halogen is attached to a carbon of an aliphatic chain are called **alkyl halides.** The number of carbon atoms attached to the carbon that is bonded to the halogen determines whether the carbon is primary, secondary, or tertiary. Table 23.2 lists the names of alkyl groups other than methyl, ethyl, and propyl. Halocarbons in which a halogen is attached to a carbon of an arene ring are called **aryl halides.**

The attractions between halocarbon molecules are primarily the result of the weak van der Waals interactions called dispersion forces. These attractions increase with the degree of halogen substitution. Thus, more highly halogenated organic compounds have higher boiling points, as illustrated in Table 23.3 on the next page.

Very few halocarbons are found in nature, but they can be readily prepared and used for many purposes. For example, halothane (2-bromo-2-chloro-1,1,1-trifluoroethane) is used as an anesthetic. Hydrofluorocarbons are used as refrigerants in automobile air-conditioning systems.

Give an example of a halocarbon.

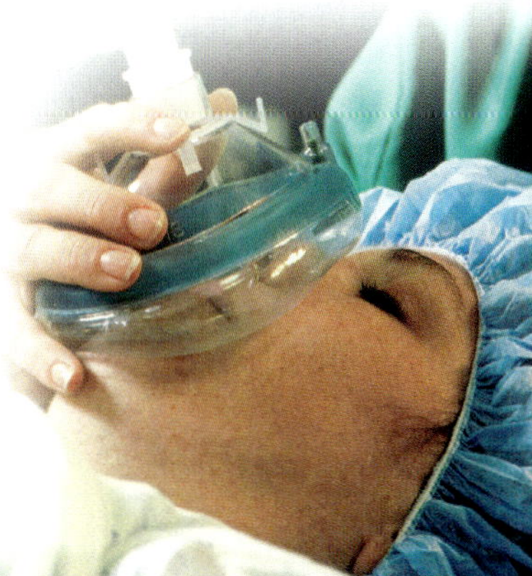

Figure 23.3 Halothane is a halocarbon that is used as an anesthetic.

Halogen Substituents

Use Visuals L1

Table 23.2 Display the table using a transparency and overhead projector and discuss the rules for naming alkyl halides. Write examples of primary, secondary, tertiary, vinyl, and aryl halides on the board. Help students to recognize the alkyl portion of the molecule by circling it in a different color chalk. Write the structure of the original parent hydrocarbon next to the alkyl halide. After students have learned the meaning of the terms primary, secondary, and tertiary, point out the derivation of the common names: *sec*-butylbromide and *tert*-butylbromide.

CLASS Activity

Halocarbon Structures and Names L2

Divide the class into groups of two or three students. Have each group practice naming halocarbons from structural formulas and writing structural formulas from their names. Students in each group should take turns devising questions and giving answers.

Discuss L2

Use examples from this section to show how a chemist can start with a simple alkane, change it to an alkyl halide, then change the halide to an alcohol. Point out that this is one way organic chemists synthesize molecules.

Facts and Figures

Anesthetics

Contrary to popular belief, some anesthetics were used during the Civil War. Opium was sometimes added to wine or liquor, if it was available. Chloroform and ether were not always available on the battlefield. And most surgeries, especially amputations, were done without anesthesia.

Differentiated Instruction

English Learners L2

Encourage students to practice writing and then naming the structures of 1-, 2-, and 3-carbon alkyl halides. Have them build a model of one of these structures and discuss how it is produced from the corresponding alkane.

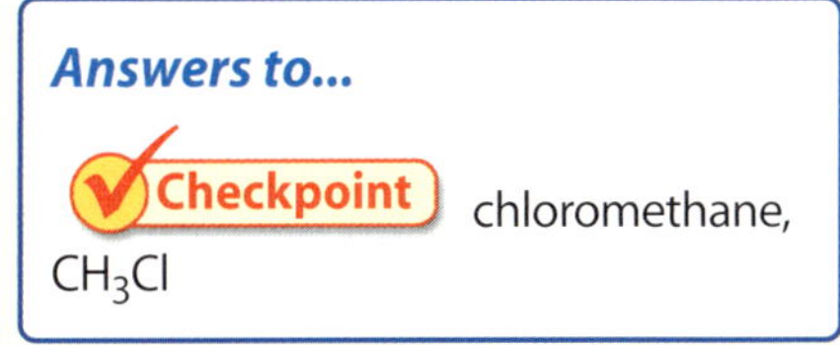

Answers to...

Checkpoint chloromethane, CH_3Cl

Use Visuals

L1

Table 23.3 Have students study the table. Point out the trend in boiling-point temperatures. Ask, **What can you infer about the cause of this pattern in boiling-point temperatures?** *(The more halogen substituents a molecule has, the greater will be the attraction of the molecule to other halocarbon molecules.)*

Substitution Reactions

TEACHER Demo

Testing for Halocarbons

L2

Purpose Students will observe the Beilstein test for determining the presence of halogen functionalities in organic compounds. Point out that organic chemists often perform qualitative chemical tests to identify unknown organic compounds.

Safety Wear safety goggles and a laboratory apron while performing the demo.

Materials copper wire, Bunsen burner, forceps, bromobenzene, benzoic acid

Procedure Prepare a piece of copper wire with a small loop in one end. Heat the loop in a Bunsen burner flame and allow the wire to cool. Next, dip the loop into a sample of aryl halide (bromobenzene). Heat the wire in the flame again. Repeat the test with a negative control, benzoic acid.

Expected Outcomes A green flame will be produced with bromobenzene, indicating the presence of a halogen. The flame will not be green with benzoic acid.

Download a worksheet on **Substitution Reactions** for students to complete, and find additional teacher support from NSTA SciLinks.

Table 23.3

Molecular Masses and Boiling Points of Chloromethanes

Molecular structure	Name	Molar mass	Boiling point (°C)
CH_4*	Methane	16.0	−161
CH_3Cl	Chloromethane (methyl chloride)	50.5	−24
CH_2Cl_2	Dichloromethane (methylene chloride)	85.0	40
$CHCl_3$	Trichloromethane (chloroform)	119.5	61
CCl_4	Tetrachloromethane (carbon tetrachloride)	154.0	74

* Included for purposes of comparison.

Substitution Reactions

Organic reactions often proceed more slowly than inorganic reactions. This is because organic reactions commonly involve the breaking of relatively strong covalent bonds. Catalysts are often needed. Many organic reactions are complex, often producing a mixture of products. The desired product must then be separated by distillation, crystallization, or other means. A common type of organic reaction is a **substitution reaction,** in which an atom, or a group of atoms, replaces another atom or group of atoms.

A halogen can replace a hydrogen atom on an alkane to produce a halocarbon. The symbol X stands for a halogen in this generalized equation.

$$\underset{\text{Alkane}}{R{-}H} + \underset{\text{Halogen}}{X_2} \longrightarrow \underset{\text{Halocarbon}}{R{-}X} + \underset{\text{Hydrogen halide}}{HX}$$

Sunlight or another source of ultraviolet radiation usually serves as a catalyst. From the generalized equation, you can write a specific one.

$$\underset{\text{Methane}}{CH_4} + \underset{\text{Chlorine}}{Cl_2} \xrightarrow[\text{light}]{\text{UV}} \underset{\text{Chloromethane}}{CH_3Cl} + \underset{\text{Hydrogen chloride}}{HCl}$$

Even under controlled conditions, this simple halogenation reaction produces a mixture of mono-, di-, tri-, and tetrachloromethanes.

Checkpoint ***What is a substitution reaction?***

Go Online NSTA SCILINKS
For: Links on Substitution Reactions
Visit: www.SciLinks.org
Web Code: cdn-1231

Facts and Figures

Perc

Perchloroethylene (PCE, or perc) is the solvent used in about 95 percent of dry-cleaning businesses in the United States. This halocarbon has been the subject of close government and public scrutiny for more than 20 years because there is some evidence that it has adverse health effects, including cancer, in humans. Because government agencies in the United States and around the world have not agreed about the potential of environmental exposure to PCE to cause health problems, use of the compound has not been restricted.

Treating benzene with a halogen in the presence of a catalyst causes the substitution of a hydrogen atom in the ring. Iron compounds are often used as catalysts for aromatic substitution reactions. For example, a rusty nail dropped in the reaction flask can act as a catalyst.

$$C_6H_6 + Br_2 \xrightarrow{\text{catalyst}} C_6H_5Br + HBr$$

Halogens on carbon chains are readily displaced by hydroxide ions to produce an alcohol and a salt. The general reaction is as follows.

$$\underset{\text{Halocarbon}}{R{-}X} + \underset{\text{Hydroxide ion}}{OH^-} \xrightarrow[100°C]{H_2O} \underset{\text{Alcohol}}{R{-}OH} + \underset{\text{Halide ion}}{X^-}$$

Chemists usually use aqueous solutions of sodium or potassium hydroxide as the source of hydroxide ions. The chemical equations for two specific examples are shown below.

$$\underset{\substack{\text{Iodomethane}\\\text{(methyl iodide)}}}{CH_3{-}I(l)} + \underset{\substack{\text{Potassium}\\\text{hydroxide}}}{KOH(aq)} \xrightarrow[100°C]{H_2O} \underset{\text{Methanol}}{CH_3{-}OH(l)} + \underset{\substack{\text{Potassium}\\\text{iodide}}}{KI(aq)}$$

$$\underset{\substack{\text{Bromoethane}\\\text{(ethyl bromide)}}}{CH_3CH_2Br(l)} + \underset{\substack{\text{Sodium}\\\text{hydroxide}}}{NaOH(aq)} \xrightarrow[100°C]{H_2O} \underset{\text{Ethanol}}{CH_3CH_2OH(l)} + \underset{\substack{\text{Sodium}\\\text{bromide}}}{NaBr(aq)}$$

Fluoro groups are not easily displaced. Thus, fluorocarbons are seldom, if ever, used to make alcohols.

23.1 Section Assessment

1. **Key Concept** How are organic compounds classified?
2. **Key Concept** What is a halocarbon?
3. **Key Concept** How can a halocarbon be prepared?
4. Identify the functional group in each structure.
   - **a.** $CH_3{-}OH$
   - **b.** $CH_3{-}CH_2{-}NH_2$
   - **c.** $C_6H_5{-}\overset{\displaystyle O}{\overset{\|}{C}}{-}OH$
   - **d.** $CH_3{-}CH_2{-}CH_2{-}Br$
   - **e.** $CH_3{-}CH_2{-}O{-}CH_2{-}CH_3$
5. Give the structural formula for each compound.
   - **a.** isopropyl chloride
   - **b.** 1-iodo-2,2-dimethylpentane
   - **c.** *p*-bromotoluene
6. Write the names of all possible dichloropropanes that could form from the chlorination of propane.

Elements Handbook

Halogens Learn more about the chemistry of halogens on pages R32–R35 of the Elements Handbook. Write a short report about the importance of halogens as functional groups in organic chemistry.

Assessment 23.1 Check your understanding of the important ideas and concepts in Section 23.1.

with ChemASAP

3 ASSESS

Evaluate Understanding L2

Write the structural formulas for ethene, ethyl chloride, and ethanol on the board. Ask students to name the compounds and identify their functional groups. *(double bond, halogen, hydroxyl group)* Ask students to choose two of the compounds to illustrate a substitution reaction. *(Hydroxide ions can replace the chlorine in ethyl chloride to form ethanol.)* Have students explain the terms *primary, secondary,* and *tertiary. (They are used to characterize the carbon atom that joins an alkyl group to other groups.)*

Reteach L1

Work with students to construct a concept map showing the connection between alkanes and their derivatives: alkenes, alkynes, halocarbons, and alcohols. Use propane as a specific example and start with the heading *alkanes.* Write the structural formula for the propyl derivative in each case.

Elements Handbook

Student paragraphs should describe some of the many uses of halocarbons. In addition to their uses as anesthetics and cleaning agents, they are used for non-stick surfaces (Teflon); pipes and other plastic items (PVC); and refrigerants (chlorofluorocarbons and hydrofluorocarbons).

If your class subscribes to the Interactive Textbook, use it to review key concepts in Section 23.1.

with ChemASAP

Section 23.1 Assessment

1. according to their functional groups
2. Halocarbons are a class of organic compounds containing covalently bonded fluorine, chlorine, bromine, or iodine.
3. by reacting an alkane with a halogen, catalyzed by UV light
4. **a.** —OH, hydroxyl **b.** —NH_2, amino **c.** —COOH, carboxyl **d.** —Br, halogen **e.** —C—O—C—, ether
5. **a.** $CH_3{-}C(CH_3)(H){-}Cl$ **b.** $I{-}CH_2C(CH_3)_2CH_2CH_2CH_3$ **c.** $CH_3{-}C_6H_4{-}Br$ (para)
6. 1,1-dichloropropane; 1,2-dichloropropane; 2,2-dichloropropane; 1,3-dichloropropane

Answers to...

Checkpoint a reaction in which an atom or group of atoms replaces another atom or group of atoms

23.2

1 FOCUS

Objectives

23.2.1 Identify how alcohols are classified and named.

23.2.2 Predict how the solubility of an alcohol varies with the length of its carbon chain.

23.2.3 Name the reactions of alkenes that may be used to introduce functional groups.

23.2.4 Construct the general structure of an ether and **describe** how ethers are named.

Guide for Reading

Build Vocabulary L2

Word Forms To help students relate vocabulary to concepts discussed in this section, have them distinguish between hydration and hydrogenation. Although both terms come from the same root word, hydration is the addition of water, whereas hydrogenation is the addition of hydrogen.

Reading Strategy L2

Compare and Contrast Have students write a short paragraph comparing structural formulas and space-filling models. Students should explain what information each representation best shows about a compound and what each representation does not show.

2 INSTRUCT

Connecting to Your World

Have students read the opening paragraph of the section. Ask, **Why do you think ethyl ether is no longer used as an anesthetic?** *(It causes nausea, is highly flammable, and it is difficult to control the dosage of an inhaled vapor.)*

23.2 Alcohols and Ethers

Guide for Reading

Key Concepts

- How are alcohols classified and named?
- How does the solubility of an alcohol vary with the length of its carbon chain?
- What reactions of alkenes may be used to introduce functional groups into organic molecules?
- What is the general structure of an ether and how are their alkyl groups named?

Vocabulary

alcohol
hydroxyl group
fermentation
denatured alcohol
addition reaction
hydration reaction
hydrogenation reaction
ether

Reading Strategy

Relating Text and Visuals As you read, look carefully at the structural formulas and the space-filling models in the section. In your notebook, explain how each type of representation of a compound helps you to understand organic compounds and their functional groups.

Connecting to Your World Prior to the 1840s, patients had to endure surgery while they were fully conscious. Today, when having major surgery, a patient often receives a general anesthetic. In addition to causing the patient to lose consciousness, a general anesthetic also causes the patient's muscles to relax. The major benefit of a general anesthetic is that the patient does not experience pain during surgery. The earliest anesthetics, used during the Civil War, belonged to a class of chemical compounds called ethers. In this section, you will read about the chemical characteristics of ethers that make them good anesthetics.

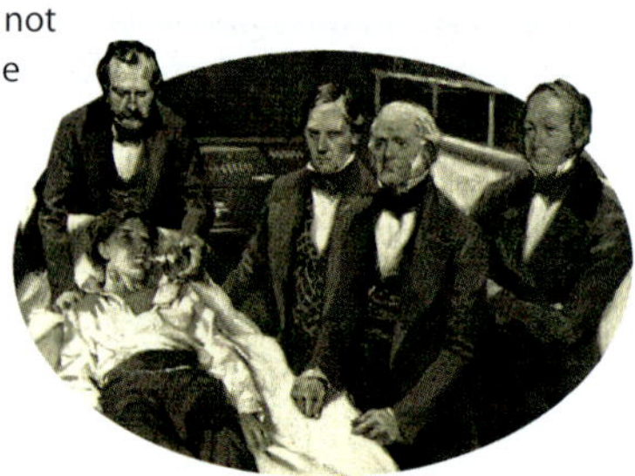

Alcohols

What do mouthwash, perfume, and hairspray have in common? They all contain an alcohol of some type. An **alcohol** is an organic compound with an —OH group.

R—Ö—H

Alcohol molecule

The —OH functional group in alcohols is called a **hydroxyl group** or hydroxy function. **Aliphatic alcohols can be classified into structural categories according to the number of R groups attached to the carbon with the hydroxyl group.** If one R group is attached, the alcohol is a primary alcohol; if two R groups, a secondary alcohol; if three R groups, a tertiary alcohol. This nomenclature is summarized below.

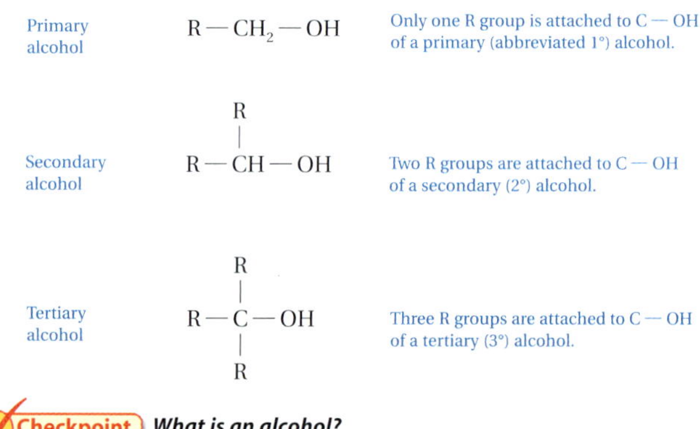

Checkpoint *What is an alcohol?*

730 Chapter 23

Section Resources

Print

- **Guided Reading and Study Workbook,** Section 23.2
- **Core Teaching Resources,** Section 23.2 Review
- **Transparencies,** T265–T267

Technology

- **Interactive Textbook with ChemASAP,** Assessment 23.2
- **Go Online,** Section 23.2

Both IUPAC and common names are used for alcohols. **When using the IUPAC system to name continuous-chain and substituted alcohols, drop the *-e* ending of the parent alkane name and add the ending *-ol.*** The parent alkane is the longest continuous chain of carbons that includes the carbon attached to the hydroxyl group. In numbering the longest continuous chain, the position of the hydroxyl group is given the lowest possible number. Alcohols containing two, three, and four —OH substituents are named diols, triols, and tetrols, respectively.

Common names of aliphatic alcohols are written in the same way as those of the halocarbons. The alkyl group ethyl, for example, is named and followed by the word alcohol, as in ethyl alcohol. Figure 23.4 shows several products that contain ethyl alcohol. Compounds with more than one —OH substituent are called glycols. The structural formulas for some simple aliphatic alcohols along with their IUPAC and common names are shown below.

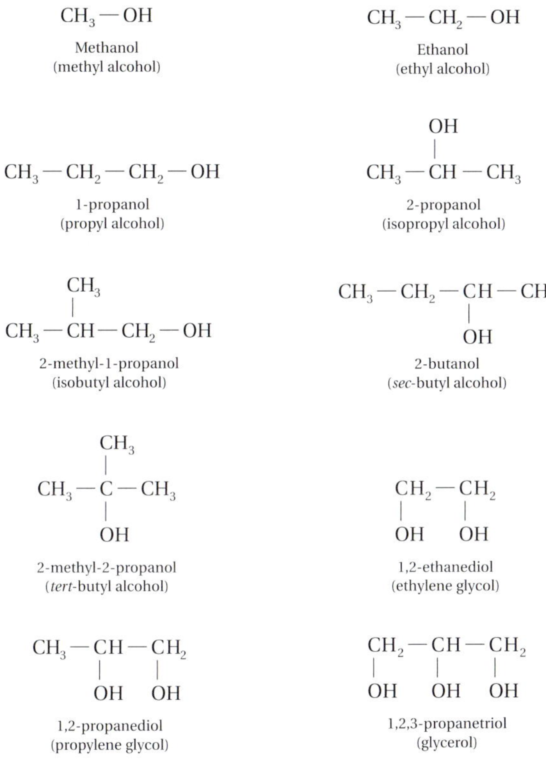

Phenols are compounds in which a hydroxyl group is attached directly to an aromatic ring. Phenol is the parent compound. Cresol is the common name for the *o, m,* and *p* structural isomers of methylphenol. Phenols are commonly used to make plastics, fibers, and drugs.

For: Links on Alcohols
Visit: www.SciLinks.org
Web Code: cdn-1232

Figure 23.4 Ethanol (ethyl alcohol) is a common component of many household products.

Alcohols

Discuss

L2

Explain that many biochemically active molecules contain carbon atoms bonded to oxygen atoms. Point out that the presence of a single oxygen atom significantly alters the chemical and physical properties of a hydrocarbon. Explain that alcohols and ethers are two important classes of organic compounds that contain oxygen.

Discuss

L2

Discuss the IUPAC conventions for naming aliphatic alcohols. Write the condensed structural formulas of several primary, secondary, and tertiary alcohols on the board, and help students to name them. Remind students how to distinguish primary, secondary, and tertiary carbons. Write the structures of methanol, ethanol, and isopropyl alcohol on the board and point out that students already know a lot about the uses and properties of these alcohols. Help students name the structures shown and challenge them to cite everyday examples where these alcohols are used. *(Methanol is a good fuel for internal combustion engines; it is sometimes used for race cars. Methanol is also an excellent solvent, and is used for paints, shellacs, and varnishes. Ethanol is found in alcoholic beverages and is sometimes used as a gasoline additive to produce cleaner burning fuels. Isopropyl alcohol is used as an antiseptic—rubbing alcohol.)* Ask, **How does a hydroxyl group differ from a hydroxide ion?** *(The bonding between carbon and the hydroxyl group in alcohols is covalent. The bonding between a metal and the hydroxide ion in bases is ionic.)*

Download a worksheet on **Alcohols** for students to complete, and find additional teacher support from NSTA SciLinks.

Answers to...

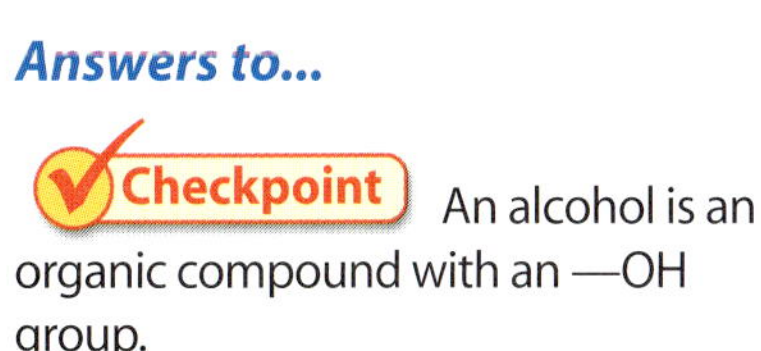

An alcohol is an organic compound with an —OH group.

Properties of Alcohols

Discuss

L2

Write the structure of water and the general structures of alcohols and ethers on the board. Point out that alcohols and ethers can be viewed as derivatives of water. Allow students to handle ball-and-stick and space-filling models of water, ethanol, and diethyl ether. Write the condensed structural formula for each on the board. Ask students to describe the structural similarities between all three compounds. (Students should identify the bent geometry around the oxygen atom.) Ask, **What significance does this bent shape have for the overall polarity of the molecules?** *(Each O—H and O—C bond is polar. Because the molecules are bent, the bond polarities do not cancel, and the molecules as a whole are polar. The region around the oxygen atom has a slight negative charge.)* Point out that the polar character of alcohols and ethers is tempered by the alkyl groups, which are nonpolar. Challenge students to name the important intermolecular forces holding like molecules together in each case. Ask, **How are water, ethanol, and diethyl ether ranked in order of increasing boiling point?** *(Ether, ethanol, water; ether molecules are attracted to one another by intermolecular dipolar attractions, while ethanol and water molecules have strong intermolecular hydrogen bonding.)*

Are Alcohols Basic?

L1

Purpose Students will observe that, although alcohols contain —OH groups, they are not hydroxides.

Materials Several alcohols, litmus paper

Procedure Test several alcohols with litmus paper to show that they have no basic properties.

CH_3-OH

Methanol (methyl alcohol)

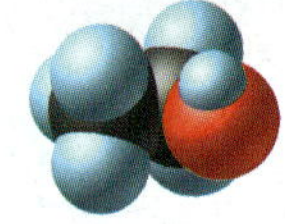

CH_3-CH_2-OH

Ethanol (ethyl alcohol)

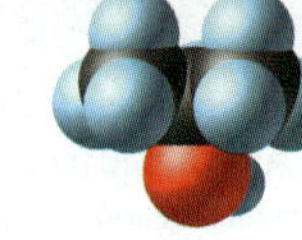

$CH_3-CH(OH)-CH_3$

2-propanol (isopropyl alcohol)

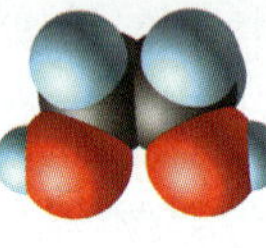

$CH_2(OH)-CH_2(OH)$

1,2-ethanediol (ethylene glycol)

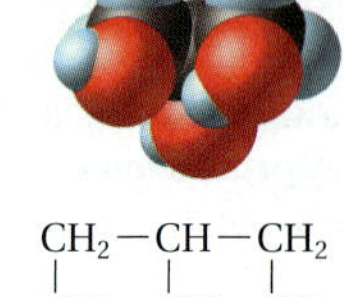

$CH_2(OH)-CH(OH)-CH_2(OH)$

1,2,3-propanetriol (glycerol)

Figure 23.5 These alcohols contain one, two, or three hydroxyl groups. **Classifying** ***Which space-filling model depicts a diol? A triol?***

Properties of Alcohols

Alcohols are capable of intermolecular hydrogen bonding. Therefore, they boil at higher temperatures than alkanes and halocarbons containing comparable numbers of atoms.

Because alcohols are derivatives of water (the hydroxyl group is part of a water molecule), they are somewhat soluble. **Alcohols of up to four carbons are soluble in water in all proportions. The solubility of alcohols with four or more carbons in the chain is usually much lower.** This is because alcohols consist of two parts: the carbon chain and the hydroxyl group. The carbon chain is nonpolar and is not attracted to water. The hydroxyl group is polar and strongly interacts with water through hydrogen bonding. For alcohols of up to four carbons, the polarity of the hydroxyl group is more significant than the nonpolarity of the carbon chain. Thus, these alcohols are soluble in water. As the number of carbon atoms increases above four, however, the nonpolarity of the chain becomes more significant, and the solubility decreases.

Many aliphatic alcohols are used in laboratories, clinics, and industry as shown in Figure 23.6. Isopropyl alcohol (IUPAC: 2-propanol), which is more familiarly known as rubbing alcohol, is a colorless, somewhat odorless liquid (bp 82°C) often used as an antiseptic. It is also used as a base for perfumes, creams, lotions, and other cosmetics. Ethylene glycol (IUPAC: 1,2-ethanediol) is the principal ingredient of certain antifreezes. Its boiling point is 197°C. Its advantages over other liquids with high boiling points are its solubility in water and its freezing point of −17.4°C. If water is added to ethylene glycol, the mixture freezes at an even lower temperature. For example, a 50% (v/v) aqueous solution of ethylene glycol freezes at −36°C. Another aliphatic alcohol, glycerol (IUPAC: 1,2,3-propanetriol), is a viscous, sweet-tasting, water-soluble liquid used as a moistening agent in cosmetics, foods, and drugs. Glycerol is also an important component of fats and oils, which you will learn about in Chapter 24.

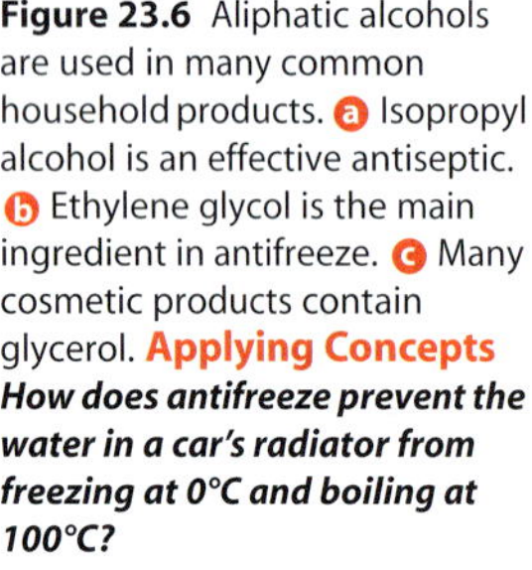

Figure 23.6 Aliphatic alcohols are used in many common household products. **a** Isopropyl alcohol is an effective antiseptic. **b** Ethylene glycol is the main ingredient in antifreeze. **c** Many cosmetic products contain glycerol. **Applying Concepts** ***How does antifreeze prevent the water in a car's radiator from freezing at 0°C and boiling at 100°C?***

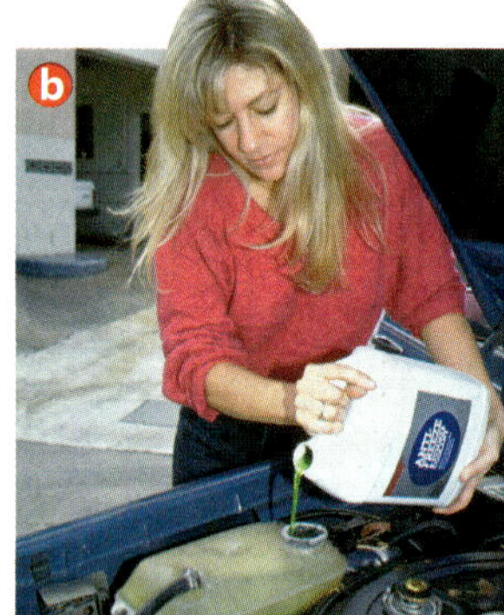

Facts and Figures

Antifreeze Poisoning

Every year, about 10,000 dogs and cats are victims of accidental poisoning by automobile antifreeze. A pet does not have to drink a lot of antifreeze to be poisoned. Most brands of commercial antifreeze consist of 95 percent ethylene glycol, an extremely toxic chemical. Even a few licks of this sweet-tasting liquid can be fatal to a cat or dog. For a medium-sized dog, ingestion of about 3 to 4 tablespoons of antifreeze is toxic. For cats, as little as 1 to 2 teaspoons can be lethal. Antifreeze poisoning commonly occurs in spring and fall when car owners replace the old antifreeze in their car radiators with fresh antifreeze. However, poisoning can happen anytime, particularly when a car boils over or when a hose leaks, releasing the antifreeze.

Figure 23.7 Cracker and bread dough are both made primarily of flour and water, but bread dough also contains yeast. Carbon dioxide, a product of sugar fermentation by yeast, causes bread to rise. **Inferring** ***What happens to the ethanol that is also produced?***

Ethyl alcohol (IUPAC: ethanol), which has a boiling point of 78.5°C, is also called grain alcohol. It is an important industrial chemical. Most ethanol is still produced by yeast fermentation of sugar. **Fermentation** is the production of ethanol from sugars by the action of yeast or bacteria. The bread shown in Figure 23.7 is different from the crackers because of fermentation. The enzymes of the yeast or bacteria serve as catalysts for the transformation. The breakdown of the sugar glucose ($C_6H_{12}O_6$) is an important fermentation reaction.

$$\underset{\text{Glucose}}{C_6H_{12}O_6(aq)} \longrightarrow \underset{\text{Ethanol}}{2CH_3CH_2OH(aq)} + \underset{\text{Carbon dioxide}}{2CO_2(g)}$$

Ethanol is the intoxicating substance in alcoholic beverages. It is a depressant that can be fatal if taken in large doses at once. Over time, continuous abuse of alcoholic beverages can damage the liver, which can lead to death.

The ethanol used in industrial applications is denatured. **Denatured alcohol** is ethanol with an added substance to make it toxic (poisonous). That added substance, or denaturant, is often methyl alcohol (IUPAC: methanol). Methyl alcohol is sometimes called wood alcohol because, prior to 1925, it was prepared by the distillation of wood. Wood alcohol is extremely toxic. As little as 10 mL has been reported to cause permanent blindness, and as little as 30 mL has been known to cause death.

Addition Reactions

The carbon–carbon single bonds in alkanes are not easy to break. In an alkene, however, one of the bonds in the double bond is somewhat weaker and thus is easier to break than a carbon–carbon single bond. So it is sometimes possible for a compound of general structure X—Y to add to a double bond. In an **addition reaction,** a substance is added at the double or triple bond of an alkene or alkyne. **Addition reactions of alkenes are an important method of introducing new functional groups into organic molecules.** In the general reaction shown below, X and Y represent the two parts of the compound that are added.

$$\rangle C{=}C\langle + X{-}Y \longrightarrow -\overset{X}{\underset{|}{\overset{|}{C}}}-\overset{Y}{\underset{|}{\overset{|}{C}}}-$$

Household Products with Alcohols L2

Bring household products that contain alcohols to class. Read the name of the alcohol given on each label, and help students to write the structure for each alcohol. Explain why alcohols are used in so many commercial products. *(Because alcohols have both polar and nonpolar characteristics, and are capable of forming hydrogen bonds, they make excellent solvents for a wide variety of polar and nonpolar compounds.)*

TEACHER Demo

Aqueous Solubility of Alcohols 

Purpose Students will observe the aqueous solubilities of alcohols and glycols.

Materials 10 or 15 test tubes with stoppers, distilled water, ethanol, 1-butanol, 1-hexanol, 1-octanol, a hexanediol

Procedure Prepare two or three sets of 5 test tubes, labeled 1 to 5, each containing a small volume of distilled water and an equal volume of ethanol, 1-butanol, 1-hexanol, 1-octanol, and a hexanediol. Stopper the tubes securely. Instruct students to mix the contents of each tube and allow the mixture to settle. Have students record their observations and describe any trends. Students should generalize about the aqueous solubility of straight-chain alcohols and about the role that —OH groups play in this solubility.

Expected Outcome Students should observe that the solubilities of the alcohols decrease as the alkyl chains increase in length.

Differentiated Instruction

Gifted and Talented L3

Have students research the role of microorganisms in fermentation processes and fermentation as a cellular process that extracts energy from glucose in the absence of oxygen. Have students design a simple experiment to show that fermentation can take place when the starting ingredient is table sugar (sucrose) rather than glucose.

Less Proficient Readers L1

Students may not understand why breaking one of the bonds in a double bond opens up two bonding sites. Have them count the number of bonds each carbon atom can form after the double bond has been broken. Remind them that each carbon atom can form single bonds with four other atoms. Therefore, one atom may be added to each carbon involved.

Answers to...

Figure 23.5 Ethylene glycol is a diol; glycerol is a triol.

Figure 23.6 A nonvolatile solute depresses water's freezing point and elevates its boiling point.

Figure 23.7 It evaporates.

Addition Reactions

Use Visuals L1

Figure 23.8 Direct students to study the figure. Ask, **If the orange color remains, what would you conclude about the sample?** *(The sample is saturated.)* Remind students that a double or triple bond is more reactive than a single covalent bond. Write the equation for the addition of Br_2 to 2-butene on the board. Explain that when a molecule containing a double bond reacts, one of the bonds in the double bond is broken. This reaction opens up two bonding sites, one on each of the carbon atoms that previously shared the double bond. Explain that the addition of molecules to alkenes makes it possible for chemists to synthesize a large variety of compounds, including polymers.

TEACHER Demo

Test for Unsaturation L2

Purpose Students will observe the reactions of alkenes and alkynes with oxidizing agents such as potassium permanganate. This reaction can be used to test for unsaturated compounds.

Materials $KMnO_4$, NaOH, distilled water, cyclohexane, cyclohexene, 2 test tubes with stoppers

Procedure Prepare an alkaline solution of potassium permanganate by dissolving 0.1 g $KMnO_4$ and 15 g NaOH in 100 mL distilled water. Put 12 mL cyclohexane in one test tube and 12 mL cyclohexene in a second test tube. Add 1 mL of the alkaline permanganate solution to each. Stopper and shake. The general equation for the reaction of permanganate with an unsaturated hydrocarbon is

$$\text{R—RCH=CH—R} + MnO_4^- \rightarrow \text{R—CHOH—CHOH—R} + MnO_2$$

Expected Outcomes The decolorization of cyclohexene indicates the presence of a double or triple bond. Cyclohexane does not decolorize.

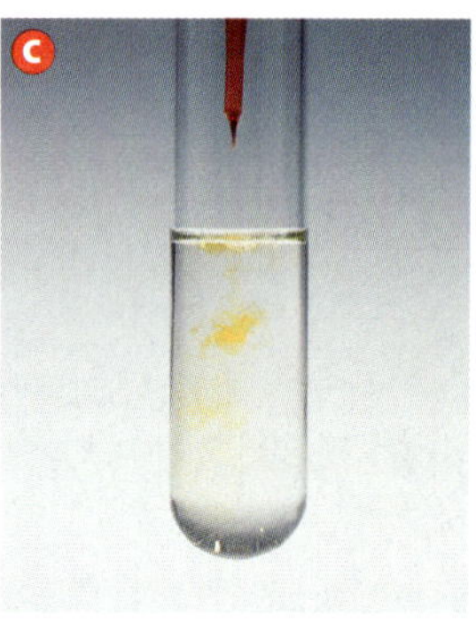

Figure 23.8 Bromine can be used to identify unsaturated compounds. **a** Bromine has a brownish-orange color. **b** A few drops of bromine solution are added to an unsaturated organic compound. **c** The bromine reacts to form a colorless halocarbon.

The addition of water to an alkene is a **hydration reaction.** Hydration reactions usually occur when the alkene and water are heated to about 100°C in the presence of a trace of strong acid. The acid, usually hydrochloric acid or sulfuric acid, serves as a catalyst for the reaction. The addition of water to ethene, the equation for which is shown below, is a typical hydration reaction. The parts of ethanol that come from the addition of water are shown in blue.

$$\text{H}_2\text{C}{=}\text{CH}_2 + \text{H}{-}\text{OH} \xrightarrow[100^\circ\text{C}]{\text{H}^+} \text{H}{-}\underset{\text{H}}{\overset{\text{H}}{\text{C}}}{-}\underset{\text{H}}{\overset{\text{OH}}{\text{C}}}{-}\text{H}$$

Ethene — Water — Ethanol

When the reagent X—Y is a halogen molecule such as chlorine or bromine, the product of the reaction is a disubstituted halocarbon. The addition of bromine to ethene to form the disubstituted halocarbon 1,2-dibromoethane is an example.

$$\text{H}_2\text{C}{=}\text{CH}_2 + \text{Br}{-}\text{Br} \longrightarrow \text{H}{-}\underset{\text{H}}{\overset{\text{Br}}{\text{C}}}{-}\underset{\text{H}}{\overset{\text{Br}}{\text{C}}}{-}\text{H}$$

Ethene (colorless) — Bromine (brownish orange) — 1,2-dibromoethane (colorless)

The addition of bromine to carbon–carbon multiple bonds is often used as a chemical test for unsaturation in an organic molecule. Bromine has a brownish-orange color, but most organic compounds of bromine are colorless. The test for unsaturation is performed by adding a few drops of a 1% solution of bromine in carbon tetrachloride (CCl_4) to the suspected alkene. As Figure 23.8 shows, the loss of the orange color is a positive test for unsaturation. If the orange color remains, the sample is completely saturated.

Hydrogen halides, such as HBr or HCl, also can add to a double bond. Because the product contains only one substituent, it is called a monosubstituted halocarbon. The addition of hydrogen chloride to ethene is an example.

$$\text{H}_2\text{C}{=}\text{CH}_2 + \text{H}{-}\text{Cl} \longrightarrow \text{H}{-}\underset{\text{H}}{\overset{\text{H}}{\text{C}}}{-}\underset{\text{H}}{\overset{\text{Cl}}{\text{C}}}{-}\text{H}$$

Ethene (ethylene) — Hydrogen chloride — Chloroethane (ethyl chloride)

The addition of hydrogen to a carbon–carbon double bond to produce an alkane is called a **hydrogenation reaction.** Hydrogenation reactions usually require a catalyst. Finely divided platinum (Pt) or palladium (Pd) is often used. The manufacture of margarine from unsaturated vegetable oils is a common application of a hydrogenation reaction. Adding hydrogen to unsaturated oils results in the formation of saturated fats that have higher melting points than the unsaturated oils and remain solid at room temperature.

Facts and Figures

Trans Fats

Trans fats are produced through hydrogenation, the chemical process by which hydrogen is added to unsaturated fatty acids in vegetable oil. Hydrogenation converts the unsaturated bonds in the oil to saturated bonds, creating margarine—a solid, spreadable fat with increased shelf life. Although hydrogenation eliminates some double bonds, it transforms others from the natural *cis* configuration to the *trans* configuration. Research indicates that eating trans fats is associated with an increased risk for heart disease.

The hydrogenation of a double bond is a reduction reaction. In the examples below, ethene is reduced to ethane, and cyclohexene is reduced to cyclohexane.

$$CH_2{=}CH_2 + H{-}H \xrightarrow{Pt} CH_3{-}CH_3$$

Ethene (ethylene) Hydrogen Ethane

$$C_6H_{10} + H{-}H \xrightarrow{Pt} C_6H_{12}$$

Cyclohexene Hydrogen Cyclohexane

Under normal conditions, benzene resists hydrogenation. It also resists the addition of a halogen or a hydrogen halide. Under conditions of high temperatures and high pressures of hydrogen, and with certain catalysts however, three molecules of hydrogen gas can reduce one molecule of benzene to form one molecule of cyclohexane.

$$C_6H_6 + 3H{-}H \xrightarrow{Pt} C_6H_{12}$$

Benzene Hydrogen Cyclohexane

Ethers

Another class of organic compounds may sound familiar to you—ethers. An **ether** is a compound in which oxygen is bonded to two carbon groups. **The general structure of an ether is R—O—R. The alkyl groups attached to the ether linkage are named in alphabetical order and are followed by the word *ether*.**

R—Ö—R (Ether molecule) $CH_3CH_2{-}O{-}CH_3$ (Ethylmethyl ether) $CH_3{-}O{-}C_6H_5$ (Methylphenyl ether (anisole))

Some ethers, such as ethylmethyl ether and methylphenyl ether, are nonsymmetric. This is because the R groups attached to the ether oxygen are different. When both R groups are the same, the ether is symmetric. Symmetric ethers are named by using the prefix *di-*. Sometimes, however, the prefix *di-* is dropped and a compound such as diethyl ether is simply called ethyl ether. Figure 23.9 shows examples of a nonsymmetric ether and a symmetric ether.

Checkpoint *What is an ether?*

a

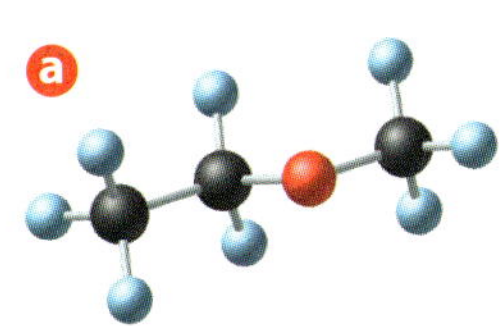

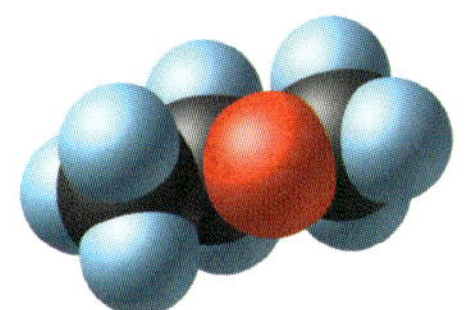

b

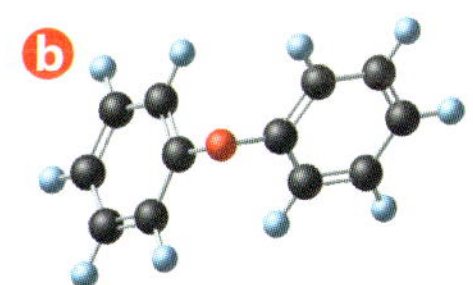

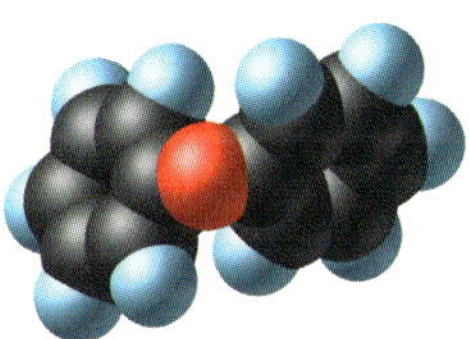

Figure 23.9 Ethers are either nonsymmetric or symmetric. **a** Ethylmethyl ether is an example of a nonsymmetric ether. **b** Diphenyl ether is a symmetric ether. Both space-filling and ball-and-stick models of each ether are shown. **Interpreting Visuals** ***Describe the differences between a nonsymmetric ether and a symmetric ether.***

Discuss L2

Write several examples of addition reactions on the board and help students predict the structure of the expected product in each case.

Ethers

Use Visuals L1

Figure 23.9 Have students study the models in the figure. Write the structures for these molecules on the board and describe the molecular structure of ethers. Point out that ethers are slightly polar due to the bent geometry around the oxygen atom. Ask, **Are ethers more or less soluble in water than alkanes of similar size?** *(Ethers are generally more soluble due to their ability to participate in hydrogen bonding with water.)* **Would you expect an ether named butyl ether to be symmetric or nonsymmetric? How do you know?** *(Symmetric, the prefix* di- *is often dropped when naming symmetric ethers.)* Discuss the naming of ethers using a number of specific examples. Explain that because ethers contain nonpolar and polar regions and because they are very inert to most chemical reagents, they make excellent solvents for reactions involving nonpolar and polar organic compounds.

Facts and Figures

Ethers as Solvents

Ethers are used as nonpolar solvents to extract natural products such as lipids. Low molar-mass ethers are partially soluble in water due to the ability of the oxygen atoms in the ether functional groups to form hydrogen bonds with water molecules. Solubility in water decreases as the mass of the carbon groups increases.

Answers to...

Figure 23.9 A nonsymmetric ether contains two different alkyl groups; a symmetric ether contains two identical alkyl groups.

Checkpoint a compound in which oxygen is bonded to two carbon groups.

Section 23.2 (continued)

Discuss **L2**

Ask students whether it is possible to have a compound that contains both ether and alcohol functional groups. *(Yes, 4-methoxy-2-butanol is an example.)*

3 ASSESS

Evaluate Understanding **L2**

Have students draw the structural formulas for 3-pentanol, 2-methyl-2-butanol, 1-heptanol, ethylpropyl ether, and diphenyl ether. Have them classify each alcohol as primary, secondary, or tertiary. Have students name and describe a chemical test for unsaturation in an organic molecule. Have them write the chemical equation for the reaction. Write the structure for 2-butanol on the board and ask, **What reactants could a chemist use to prepare this compound?** *(2-butene and aqueous acid)*

Reteach **L1**

Using the reactions given in this section, point out that the two atoms being added to double-bonded carbons in an alkene can be identical or they can be different. In hydrogenation, the atoms added are the same. When HCl reacts with an alkene to produce an alkyl chloride or when water and an alkene react to produce an alcohol, the atoms added are different.

Writing Activity

Engines knock when gasoline burns too soon or too quickly. MTBE replaced tetraethyl lead as an antiknocking additive. MTBE from leaking storage tanks and fuel spills contaminates groundwater.

If your class subscribes to the Interactive Textbook, use it to review key concepts in Section 23.2.

with ChemASAP

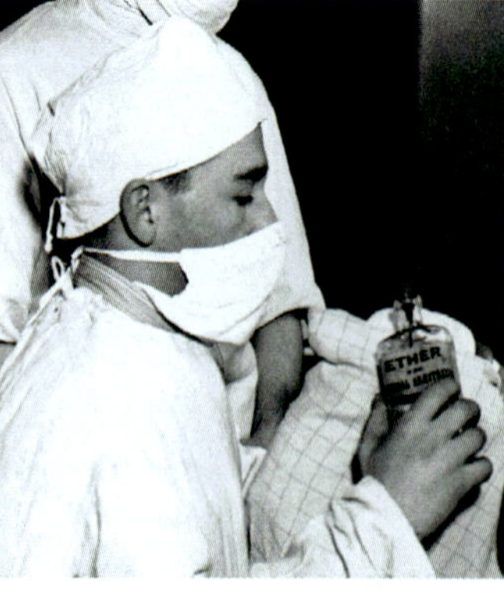

Figure 23.10 Diethyl ether is a dangerous, highly flammable liquid. It is usually stored in metal cans to keep it away from direct sunlight.

Diethyl ether, a volatile liquid (bp 35°C) shown in Figure 23.10, was the first reliable general anesthetic. Originally reported in 1849 by Crawford W. Long, an American physician, diethyl ether was used by doctors for more than a century.

$$CH_3CH_2—O—CH_2CH_3$$

Diethyl ether (ethyl ether)

Diphenyl ether (phenyl ether)

Diethyl ether has been replaced by other anesthetics such as halothane because it is highly flammable and often causes nausea. Diphenyl ether is used in the manufacture of perfumes and soaps.

Ethers usually have lower boiling points than alcohols of comparable molar mass. They have higher boiling points than comparable hydrocarbons and halocarbons. Ethers are more soluble in water than hydrocarbons and halocarbons, but less soluble than alcohols. This is because the oxygen atom in an ether is a hydrogen acceptor. Ethers have no hydroxyl hydrogen atoms to donate in hydrogen bonding. Ethers, therefore, form more hydrogen bonds than hydrocarbons and halocarbons but fewer hydrogen bonds than alcohols.

23.2 Section Assessment

7. **Key Concept** How are alcohols classified and named?
8. **Key Concept** How does the solubility of alcohols vary with the length of the carbon chains?
9. **Key Concept** How can functional groups be introduced into organic molecules?
10. **Key Concept** Write the general structure for an ether and explain how ethers are named.
11. Give the structure for the expected organic product from each of the following reactions.
    a. $(CH_3)(H)C{=}C(H)(CH_3) + HBr \longrightarrow$
    b. + Cl_2 $\xrightarrow{\text{catalyst}}$
    c. + $3H_2$ $\xrightarrow[\text{pressure}]{\text{catalyst}}$
12. Write the common names for the following compounds.
    a. $CH_3CH_2CHCH_3$ (with OH on the third carbon)
    b. CH_3CHCH_2OH (with CH_3 on the second carbon)
    c. $CH_3CH_2OCH_2CH_3$
    d. $CH_3CH_2CH_2OCH_2CH_2CH_2CH_3$

Writing Activity

Portfolio Project Research and write a report on methyl-*tert*-butyl ether (MTBE). Find out why MTBE was added to gasoline and why its use is now limited.

Assessment 23.2 Check your understanding of the important ideas and concepts in Section 23.2.

with ChemASAP

Section 23.2 Assessment

7. Alcohols are classified as primary (1°), secondary (2°), or tertiary (3°). They are named by dropping the *-e* ending of the parent alkane name and adding the ending *-ol*.
8. Alcohols are soluble in water when the carbon chain of the alcohol contains four or fewer carbons. The solubility of longer chain alcohols is much lower.
9. by addition or substitution reactions
10. R—O—R; list the alkyl groups attached to the ether linkage in alphabetical order and add *ether*.
11. a. $H—C(CH_3)(H)—C(CH_3)(Br)—H$ b. (chlorobenzene, Cl) c. (cyclohexane)
12. a. *sec*-butyl alcohol b. isobutyl alcohol c. diethyl ether (ethyl ether) d. butylpropyl ether

23.3 Carbonyl Compounds

Connecting to Your World What is your favorite flavor, benzaldehyde or vanillin? Although these two flavors might sound like they belong in the foods of a science-fiction movie, the chances are good that you have eaten these organic molecules, called aldehydes, in ice cream or cookies. Aldehydes and other carbonyl compounds have important everyday uses—even though they may have unfamiliar names. In this section, you will read about the properties that are associated with carbonyl compounds.

Guide for Reading

Key Concepts

- What is the structure of a carbonyl group found in aldehydes and ketones?
- What is the general formula for a carboxylic acid?
- What is the general structure of an ester?
- Why is dehydrogenation an oxidation reaction?

Vocabulary

carbonyl group
aldehyde
ketone
carboxylic acid
carboxyl group
fatty acids
esters
dehydrogenation reaction

Reading Strategy

Summarizing As you read about carbonyl compounds, summarize the main ideas in the text that follow each red and blue heading.

Aldehydes and Ketones

Remember that in an alcohol, an oxygen atom is bonded to a carbon group and a hydrogen atom. In an ether, an oxygen atom is bonded to two carbon groups. An oxygen atom can also be bonded to a single carbon atom by a double covalent bond. Such an arrangement is called a carbonyl group. A **carbonyl group** is a functional group with the general structure C=O. **The C=O functional group is present in aldehydes and ketones.** The structural formulas and space-filling models of some carbonyl-group compounds are shown in Figure 23.11.

An **aldehyde** is an organic compound in which the carbon of the carbonyl group is always joined to at least one hydrogen. The general formula for an aldehyde is RCHO.

A **ketone** is an organic compound in which the carbon of the carbonyl group is joined to two other carbons. The general formula for a ketone is RCOR.

$$\begin{array}{c} O \\ \| \\ R-C-H \end{array} \leftarrow \text{Carbonyl group} \qquad \begin{array}{c} O \\ \| \\ R-C-R \end{array} \leftarrow \text{Carbonyl group}$$

Aldehyde   Ketone

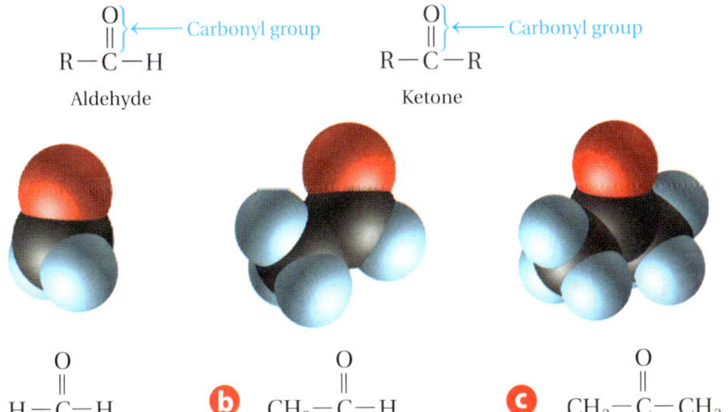

a $\begin{array}{c} O \\ \| \\ H-C-H \end{array}$ b $\begin{array}{c} O \\ \| \\ CH_3-C-H \end{array}$ c $\begin{array}{c} O \\ \| \\ CH_3-C-CH_3 \end{array}$

Figure 23.11 These low molar-mass carbonyl group compounds are completely soluble in water. (a) Methanal is commonly known as formaldehyde. (b) Ethanal is commonly known as acetaldehyde. (c) Propanone is commonly known as acetone. **Interpreting Diagrams** ***Use the structures to explain why these compounds are soluble in water.***

Section Resources

Print

- **Laboratory Manual,** Lab 50
- **Small-Scale Chemistry Laboratory Manual,** Labs 38–39
- **Guided Reading and Study Workbook,** Section 23.3
- **Core Teaching Resources,** Section 23.3 Review
- **Transparencies,** T268–T269

Technology

- **Interactive Textbook with ChemASAP,** Assessment 23.3
- **Go Online,** Section 23.3

23.3

1 FOCUS

Objectives

23.3.1 Identify the structure of a carbonyl group as found in aldehydes and ketones.
23.3.2 Construct the general formula for carboxylic acids and **explain** how they are named.
23.3.3 Describe an ester.
23.3.4 Explain why dehydrogenation is an oxidation reaction.

Guide for Reading

Build Vocabulary  L2

Graphics Organizer Use the following terms to make a concept map: *carbonyl group, aldehydes, ketones, carboxylic acids, carboxyl group, and fatty acids.)*

Reading Strategy  L2

Compare and Contrast Have students write the general formula for an alcohol and an ether. As they read about aldehydes, ketones, and carboxylic acids, have them write those general structures and compare them with the structures of alcohols and ethers. *(Students should note that all these structures contain oxygen but that the bonding is different.)*

2 INSTRUCT

Connecting to Your World

How do you think pure vanilla extract compares chemically with artificial vanilla flavoring, which is vanillin? *(Vanilla extract is a mixture that contains other substances in addition to vanillin. Artificial vanilla flavoring is a pure substance—vanillin.)*

Answers to...

Figure 23.11 The polar oxygen atom makes up a relatively large proportion of the molecule, so the molecule is soluble in water.

Aldehydes and Ketones

Relate L2

Have students gently sniff the contents of labeled test tubes containing small amounts of vanillin, isoamyl acetate, and benzaldehyde. Ask: **Which odor is that of vanilla ice cream?** *(vanillin)* **Bananas?** *(isoamyl acetate)* **Maraschino cherries?** *(benzaldehyde)* Write structures for the compounds on the board. Refer to Table 23.4 for benzaldehyde and vanillin; isoamyl acetate is $CH_3COOCH_2CH_2CH(CH_3)_2$. Explain that compounds with carbonyl groups account for the odor/taste of many foods.

Use Visuals L1

Table 23.4 Write several examples of aldehydes and ketones on the board and discuss the naming of these compounds. Explain to students that the difference between an aldehyde and a ketone lies in the position of the carbonyl group. Display the table using a transparency and overhead projector and point out that in ketones, as in ethers, the functional group lies between two R groups. Except for methanal, the carbonyl group in aldehydes is attached to one R group and one hydrogen atom.

CLASS Activity

Model Building L2

Many students may find it helpful in understanding the differences between aldehydes and ketones if they build models of compounds containing these functional groups. Have students make models of propanal, propanone, and propanoic acid. Have them examine their models and make comparisons. If model-building kits are not available, toothpicks and gumdrops or marshmallows can be used. Warn students not to eat gumdrops or marshmallows used in model building.

Table 23.4

Some Common Aldehydes and Ketones

Condensed formula	Structural formula	IUPAC name	Common name
Aldehydes			
HCHO	O ‖ H—C—H	Methanal	Formaldehyde
CH_3CHO	O ‖ CH_3—C—H	Ethanal	Acetaldehyde
C_6H_5CHO	(benzene ring)—C(=O)—H	Benzaldehyde	Benzaldehyde
$C_6H_5CH{=}CHCHO$	(benzene ring)—CH=CH—C(=O)—H	3-phenyl-2-propenal	Cinnamaldehyde
$CH_3O(OH)C_6H_3CHO$	(benzene ring with HO, OCH_3)—C(=O)—H	4-hydroxy-3-methoxybenzaldehyde	Vanillin
Ketones			
CH_3COCH_3	O ‖ CH_3—C—CH_3	Propanone	Acetone (dimethyl ketone)
$C_6H_5COC_6H_5$	(benzene ring)—C(=O)—(benzene ring)	Diphenylmethanone	Benzophenone (diphenyl ketone)

The IUPAC system may be used for naming aldehydes and ketones. For either class of compounds, first identify the longest hydrocarbon chain that contains the carbonyl group. Replace the *-e* ending of the hydrocarbon by *-al* to designate an aldehyde. In the IUPAC system, the continuous-chain aldehydes are named methanal, ethanal, propanal, butanal, and so forth.

Ketones are named by changing the ending of the longest continuous carbon chain that contains the carbonyl group from *-e* to *-one.* Table 23.4 illustrates the naming of some common aldehydes and ketones. If the carbonyl group of a ketone could occur at more than one place on the chain, then its position is designated by the lowest possible number.

What name ending designates a ketone?

738 Chapter 23

Differentiated Instruction

Gifted and Talented L3

Have students study the structure of cinnamaldehyde and other carbonyl-containing compounds that contain carbon-carbon bonds conjugated with a carbon-oxygen double bond. Have students consider how the location of a carbonyl group next to an alkene group affects the stability of the molecule through resonance.

Table 23.5

Boiling Points of Some Compounds with One and Two Carbons

Compound	Formula	Molar mass	Boiling point (°C)	Primary intermolecular interactions
One carbon				
Methane	CH_4	16	−161	No hydrogen-bonding or polar-polar interactions
Methanal	HCHO	30	−21	Polar-polar interactions
Methanol	CH_3OH	32	65	Hydrogen bonding
Two carbons				
Ethane	C_2H_6	30	−89	No hydrogen-bonding or polar-polar interactions
Ethanal	CH_3CHO	44	21	Polar-polar interactions
Ethanol	CH_3CH_2OH	46	78	Hydrogen bonding

Properties of Aldehydes and Ketones Aldehydes and ketones can form weak hydrogen bonds between the carbonyl oxygen and the hydrogen atoms of water. The lower members of the series—methanal (formaldehyde), ethanal (acetaldehyde), and propanone (acetone)—are soluble in water in all proportions. As the length of the hydrocarbon chain increases, however, water solubility decreases. When the carbon chain exceeds five or six carbons, solubility of both aldehydes and ketones is very low. As might be expected, all aldehydes and ketones are soluble in nonpolar solvents.

Aldehydes and ketones cannot form intermolecular hydrogen bonds because they lack hydroxyl (—OH) groups. Consequently, they have boiling points that are lower than those of the corresponding alcohols. Aldehydes and ketones can attract each other, however, through polar–polar interactions of their carbonyl groups. As a result, their boiling points are higher than those of the corresponding alkanes. These attractive forces account for the fact that nearly all aldehydes and ketones are either liquids or solids at room temperature. The exception is methanal, which is an irritating, pungent gas. Table 23.5 compares the boiling points of alkanes, aldehydes, and alcohols of similar molar mass.

Uses of Aldehydes and Ketones A wide variety of aldehydes and ketones have been isolated from plants and animals. Many of them, particularly those with high molar masses, have fragrant or penetrating odors. They are usually known by their common names, which can indicate their natural sources or perhaps a characteristic property. Aromatic aldehydes are often used as flavoring agents. Benzaldehyde is the simplest aromatic aldehyde. Also known as oil of bitter almond, benzaldehyde is a constituent of almonds. It is a colorless liquid with a pleasant almond odor. Cinnamaldehyde imparts the characteristic odor of oil of cinnamon.

Discuss 

Explain that many familiar smells and flavors come from organic compounds. For example, many ketones and aldehydes, particularly those with high molar masses, have fragrant odors. Ask, **What are some functional groups other than aldehydes and ketones that can have an effect on the particular odor of an organic compound?** *(alcohols, esters, amines)* Have students name some of their favorite scents. Ask, **What are the natural sources of these perfumes?** *(animal oils, plant extracts)* Explain that perfume makers usually combine many different compounds to create a complex scent. Although these compounds may come from animal oils and plant extracts, they can be synthesized artificially. Citral, found in roses and citrus fruits, is an aldehyde. Civetone, the odor of musk, is a ketone.

Use Visuals 

Table 23.5 Ask students to think of some examples of simple organic molecules that contain carbonyl groups. Have them write the structures and build models of these molecules. Ask, **What are some structural features of carbonyl-containing molecules?** *(Carbonyl groups display trigonal planar geometry.)* Point out that a carbonyl carbon represents the positive center of aldehydes and ketones and that an oxygen atom represents the negative center. Discuss the types of intermolecular forces operating between the functional groups in aldehydes and ketones. Ask, **How does the shape of a molecule affect the strength of intermolecular attractions?** *(If the shape causes bond polarities within the molecule to be unbalanced, the molecule is usually polar, and intermolecular forces will be stronger than if the molecule were nonpolar.)*

Facts and Figures

Vanilla Beans

Vanilla flavoring comes from the seed capsules, called "beans," of several species of orchid of the genus Vanilla. The capsules are collected, boiled, dried, and cured to produce the long, thin, brown bean-like pods that are familiar to cooks everywhere. The vanilla orchid was first used as a flavoring by the Aztecs and the Mayans in Mexico and Central America for their *xocolatl* (chocolate) drinks. Due to advances in pollination techniques, most of the natural vanilla produced today comes from Madagascar, Reunion, and Tahiti rather than from Central America.

Answers to...

-one

Carboxylic Acids

Discuss L2

Write the structural formula for a carboxylic acid on the board or overhead projector. Ask, **What functional groups are present and what characteristics would you expect the molecule to have?** *(Students should recognize the hydroxyl and carbonyl groups and may suggest that the molecule would have properties of both an alcohol and an aldehyde.)* **Why are carboxylic acids considered to be acids?** *(They contain an ionizable proton.)* Explain that the presence of the double-bonded oxygen weakens the bond between the other oxygen atom and the hydrogen atom bonded to it, so that the hydrogen can ionize easily. Point out that the overall functional group is called a carboxyl group and the molecule containing it is called a carboxylic acid.

Use Visuals L1

Figure 23.14 Have students study the photograph. Explain that carboxylic acids are found in many natural products. Ask students to name the acids in the food products shown. Write the structural formulas for the acids on the board. Point out the carboxyl functional group in each. Have students name other examples of food products that contain acids and describe the tastes and odors associated with carboxylic acids.

Figure 23.12 Vanilla beans, the seed pods of the vanilla orchid, are the natural source of this vanilla flavoring for ice cream and other foods.

Figure 23.13 Many types of nail polish remover contain acetone, a common ketone.

Vanilla beans, which are responsible for the popular vanilla flavor, are shown in Figure 23.12.

The simplest aldehyde is methanal, also called formaldehyde. Methanal is very important industrially, having its greatest use in the manufacture of synthetic resins, but it is hard to handle in the gaseous state. Methanal is usually available as a 40% aqueous solution, known as formalin. Formalin can be used to preserve biological specimens. The methanal in solution combines with protein in tissues to make the tissues hard and insoluble in water, preventing the specimen from decaying. Today, formalin is used infrequently as a preservative because it causes cancer.

The most common industrial ketone is propanone, also called acetone. Propanone is a colorless, volatile liquid that boils at 56°C. It is miscible with water in all proportions. Propanone is used as a solvent for resins, plastics, and varnishes and is often found in nail-polish removers as shown in Figure 23.13.

Figure 23.14 Carboxylic acids give a variety of foods—spoiled as well as fresh—a distinctive sour taste. **Predicting** ***What would you expect the pH range of aqueous solutions containing carboxylic acids to be?***

Carboxylic Acids

A **carboxylic acid** is a compound with a carboxyl group. A **carboxyl group** consists of a carbonyl group attached to a hydroxyl group.

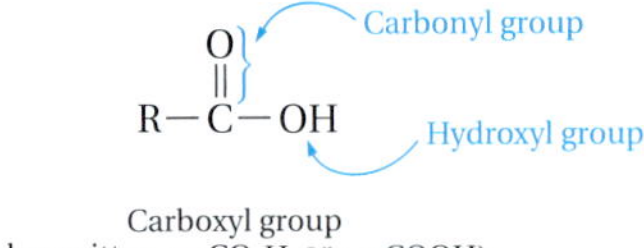

The general formula for a carboxylic acid is RCOOH. Carboxylic acids are weak acids because they ionize slightly in solution to give a carboxylate ion and a hydrogen ion.

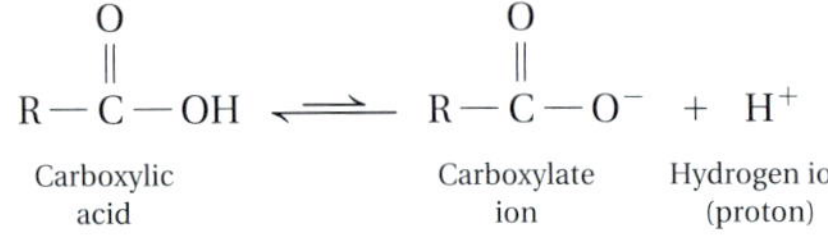

In the IUPAC system, carboxylic acids are named by replacing the *-e* ending of the parent alkane with the ending *-oic acid*. Remember, the parent alkane of a carboxylic acid is the hydrocarbon with the longest continuous carbon chain containing the carboxyl group.

Differentiated Instruction

Less Proficient Readers

Write the general structures of an aldehyde, a ketone, a carboxylic acid, and an ester on the board. Have students copy the structures in their notebooks. Have them label the carbonyl groups, hydroxyl groups, aliphatic, and aryl groups. Have them indicate whether each structure is an aldehyde, ketone, ester, or carboxylic acid.

740 Chapter 23

Table 23.6

Saturated Aliphatic Carboxylic Acids				
Formula	**Carbon atoms**	**IUPAC name**	**Common name**	**Melting point (°C)**
HCOOH	1	Methanoic acid	Formic acid	8
CH_3COOH	2	Ethanoic acid	Acetic acid	17
CH_3CH_2COOH	3	Propanoic acid	Propionic acid	−22
$CH_3(CH_2)_2COOH$	4	Butanoic acid	Butyric acid	−6
$CH_3(CH_2)_4COOH$	6	Hexanoic acid	Caproic acid	−3
$CH_3(CH_2)_6COOH$	8	Octanoic acid	Caprylic acid	16
$CH_3(CH_2)_8COOH$	10	Decanoic acid	Capric acid	31
$CH_3(CH_2)_{10}COOH$	12	Dodecanoic acid	Lauric acid	44
$CH_3(CH_2)_{12}COOH$	14	Tetradecanoic acid	Myristic acid	58
$CH_3(CH_2)_{14}COOH$	16	Hexadecanoic acid	Palmitic acid	63
$CH_3(CH_2)_{16}COOH$	18	Octadecanoic acid	Stearic acid	70

Carboxylic acids are abundant and widely distributed in nature. Many have common names derived from a Greek or Latin word that describes their natural sources. For example, the common name for ethanoic acid is acetic acid, which comes from the Latin word *acetum*, meaning vinegar. Common household vinegar contains about 5% (v/v) acetic acid. The formation of acetic acid in wine causes the wine to turn sour, become vinegar, and develop a pungent aroma. One type of vinegar made from wine is shown in Figure 23.14. Many continuous-chain carboxylic acids were first isolated from fats and are called **fatty acids.** Propionic acid, the three-carbon acid, literally means first fatty acid. Common names are used more often than IUPAC names for carboxylic acids. Table 23.6 lists the names and formulas of some common aliphatic carboxylic acids.

The low-molar-mass members of the aliphatic carboxylic acid series are colorless, volatile liquids. They have sharp, unpleasant odors. The higher members of the series are nonvolatile, waxy, odorless solids with low melting points. Stearic acid, an 18-carbon acid obtained from beef fat, is used to make inexpensive wax candles.

Like alcohols, carboxylic acids form intermolecular hydrogen bonds. Thus, carboxylic acids have higher boiling and melting points than other compounds of similar molar mass. All aromatic carboxylic acids are crystalline solids at room temperature.

The carboxyl group in carboxylic acids is polar and readily forms hydrogen bonds with water molecules. Figure 23.15 shows a model of ethanoic acid. Methanoic, ethanoic, propanoic, and butanoic acids are completely miscible with water. The solubility of carboxylic acids of higher molar mass drops sharply, however. Most carboxylic acids dissolve in organic solvents such as ethanol or propanone.

What is a fatty acid?

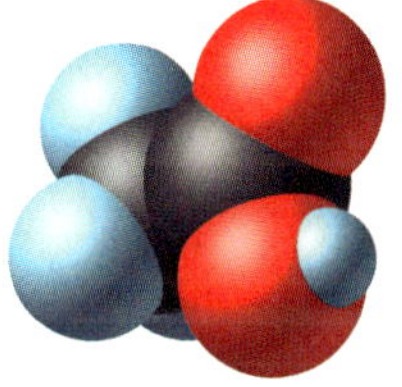

CH_3COOH

Figure 23.15 Ethanoic acid, shown in this formula and space-filling model, is a colorless, volatile liquid. **Applying Concepts** ***Is ethanoic acid soluble in water? Why?***

Discuss L2

Ask students why carboxylic acids are considered weak acids. Write the chemical equation for the ionization of dilute aqueous ethanoic acid on the board. Point out that the acid is only partly ionized at equilibrium. Challenge students to write the equilibrium constant expression for the reaction shown. Have them write the chemical equation for the ionization of propanoic acid.

Use Visuals L1

Table 23.6 Use an overhead projector to display a transparency of Table 23.6. Discuss the IUPAC rules for naming carboxylic acids. Remind students that the systematic names are derived from the parent alkanes. Discuss the basis for the observed trends in physical properties for the homologous series of aliphatic carboxylic acids. Ask, **What is the name of an aliphatic carboxylic acid that is a solid at room temperature?** *(Answers will vary, but should be restricted to those acids with melting points higher than 25°C.)* **What is the name of an aliphatic carboxylic acid that is a liquid at room temperature?** *(Answers will vary, but should be restricted to those acids with melting points lower than 25°C.)* **What is the common name of an 18-carbon aliphatic carboxylic acid? What is a source of this substance?** *(stearic acid; found in beef fat)*

Answers to...

Figure 23.14 below pH 7

Figure 23.15 Yes. The carboxyl group is polar and can form hydrogen bonds with water molecules. The combined van der Waals forces between the single methyl groups of ethanoic acid are not great enough to prevent water miscibility.

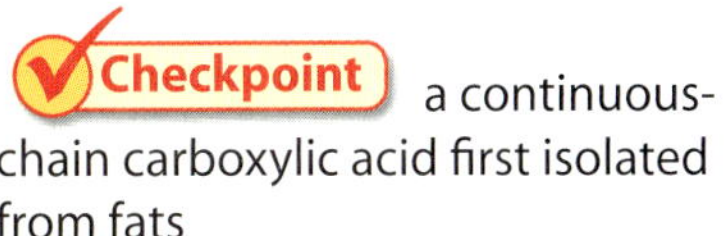

a continuous-chain carboxylic acid first isolated from fats

Esters

TEACHER Demo

Making an Ester

L2

Purpose Students will observe the synthesis of an ester from an acid and an alcohol.

Materials water bath, hot plate, test tubes, several carboxylic acids, several corresponding alcohols, concentrated sulfuric acid, filter paper, watch glass

Safety and Disposal Sulfuric acid is caustic and can burn skin and clothing. Neutralize the combined reaction mixtures and flush them down the drain.

Procedure Explain that esters are derivatives of carboxylic acids in which the —OH group of the carboxyl group has been replaced by an —OR group from an alcohol. The synthesis of an ester from an acid and an alcohol is called esterification. Synthesize a variety of fragrant esters in class and give students a chance to smell the products. Set up a water bath on a hot plate to maintain the temperature at about 80°C. To synthesize the ester, mix 1–2 mL (or 0.2–0.6 g if the acid is a solid) of each carboxylic acid with 3–5 mL of the corresponding alcohol in a test tube. Add 1-2 drops of concentrated sulfuric acid and heat for 15 minutes in the water bath. Dip a piece of filter paper into the test tube, and place it on a clean watch glass. Have students record the esters you prepare and their associated odors. Ask them to write the chemical equation for each reaction. Point out that the only difference in these reactions is in the R group attached to the carboxyl and alcohol functional groups.

Expected Outcome

Acid	Alcohol	Odor of Ester
acetic	ethyl	apple
acetic	isoamyl	banana
acetic	amyl	apricot
acetic	octyl	orange
butyric	ethyl	pineapple
salicylic	methyl	wintergreen

Figure 23.16 Esters impart the characteristic aromas and flavors of many flowers and fruits. Marigolds, raspberries, and bananas all contain esters.

Esters

Esters are probably the most pleasant and delicious organic compounds one can study. Many esters have pleasant, fruity odors. Esters give blueberries, pineapples, apples, pears, bananas, and many other fruits their characteristic aromas. They give many perfumes their fragrances. **Esters** are derivatives of carboxylic acids in which the —OH of the carboxyl group has been replaced by an —OR from an alcohol. **Esters contain a carbonyl group and an ether link to the carbonyl carbon. The general formula for an ester is RCOOR.** The R groups can be short-chain or long-chain aliphatic (alkyl) or aromatic (aryl) groups, saturated or unsaturated.

$$\mathrm{R{-}C({=}O){-}O{-}R}$$

Carbonyl group (from the acid)
Alkyl or aryl group (from the alcohol)

Figure 23.16 shows some of the flowers and fruits that contain esters. Simple esters are neutral substances. Although the molecules are polar, they cannot form hydrogen bonds with one another because they do not contain hydrogen attached to oxygen or another electronegative atom. As a result, only weak attractions hold ester molecules to one another. As you might expect, esters have much lower boiling points than the strongly hydrogen-bonded carboxylic acids from which they are derived. The low-formula-mass esters are somewhat soluble in water, but esters containing more than four or five carbons have very limited solubility.

Esters may be prepared from a carboxylic acid and an alcohol. The process is called esterification. The reactants, usually a carboxylic acid and a primary or secondary alcohol, are heated with a mineral acid as a catalyst. The reaction is reversible.

$$\underset{\text{Carboxylic acid}}{\mathrm{R{-}C({=}O){-}OH}} + \underset{\text{Alcohol}}{\mathrm{RO{-}H}} \underset{}{\overset{H^+}{\rightleftharpoons}} \underset{\text{Carboxylate ester}}{\mathrm{R{-}C({=}O){-}OR}} + \underset{\text{Water}}{\mathrm{H{-}OH}}$$

742 Chapter 23

Facts and Figures

Scientists Growing Flavors

Scientists are learning how to produce flavor molecules and other naturally occurring compounds in plant cell cultures and genetically altered yeast cells. For example, Japanese scientists are using a tissue culture technique called root culture to produce ginseng for teas and medicines and for the red pigment used in cosmetics. In the United States, scientists are obtaining vanilla flavor from cells surgically removed from a vanilla orchid and cultured in a glass bioreactor. Prior to any commercial production, a patent application is usually filed, and manufacturers must seek approval from the Food and Drug Administration to use the prepared flavor molecules in food products.

The formation of the ester ethyl ethanoate from ethanoic acid and ethanol is an example of esterification. The product of this reaction, ethyl ethanoate, is shown in Figure 23.17.

$$\underset{\text{Ethanoic acid}}{CH_3{-}C({=}O){-}OH} + \underset{\text{Ethanol}}{CH_3CH_2O{-}H} \xrightleftharpoons{H^+} \underset{\text{Ethyl ethanoate}}{CH_3{-}C({=}O){-}OCH_2CH_3} + \underset{\text{Water}}{H_2O}$$

If an ester is heated with water for several hours, usually very little happens. In strong acid or base solutions, however, the ester breaks down. An ester is hydrolyzed by the addition of water to produce a carboxylic acid and an alcohol. The reaction is rapid in acidic solution.

$$\underset{\text{Ethyl ethanoate}}{CH_3{-}\overset{O}{\overset{\|}{C}}{-}OCH_2CH_3} + H{-}OH \xrightleftharpoons{H^+} \underset{\text{Ethanoic acid}}{CH_3{-}\overset{O}{\overset{\|}{C}}{-}OH} + \underset{\text{Ethanol}}{HOCH_2CH_3}$$

Hydroxide ions also promote this reaction. The usual agent for ester hydrolysis is an aqueous solution of sodium hydroxide or potassium hydroxide. Because many esters do not dissolve in water, a solvent such as ethanol is added to make the solution homogeneous. The reaction mixture is usually heated. All of the ester is converted to products. The carboxylic acid product is in solution as its sodium or potassium salt.

$$\underset{\text{Ethyl ethanoate}}{CH_3{-}\overset{O}{\overset{\|}{C}}{-}OCH_2CH_3} + NaOH \longrightarrow \underset{\text{Sodium ethanoate}}{CH_3{-}\overset{O}{\overset{\|}{C}}{-}O^-Na^+} + \underset{\text{Ethanol}}{HOCH_2CH_3}$$

If the reaction mixture is acidified, the carboxylic acid forms.

$$\underset{\text{Sodium ethanoate}}{CH_3{-}\overset{O}{\overset{\|}{C}}{-}O^-Na^+} + HCl \longrightarrow \underset{\text{Ethanoic acid}}{CH_3{-}\overset{O}{\overset{\|}{C}}{-}OH} + NaCl$$

Oxidation-Reduction Reactions

All the classes of organic compounds you have just studied are related by oxidation and reduction reactions. Recall from Chapter 20 that oxidation is the gain of oxygen, loss of hydrogen, or loss of electrons and reduction is the loss of oxygen, gain of hydrogen, or gain of electrons. Also remember that one does not occur without the other.

In organic chemistry, the number of oxygen atoms and hydrogen atoms attached to carbon indicates the degree of oxidation of a compound. The fewer hydrogens on a carbon–carbon bond, the more oxidized the bond. Thus, a triple bond (an alkyne) is more oxidized than a double bond (an alkene), which is more oxidized than a single bond (an alkane). In other words, an alkane can be oxidized to an alkene and then to an alkyne; ethane (an alkane) can be oxidized to ethene (an alkene), then to ethyne (an alkyne).

$$\underset{\text{Ethyl ethanoate}}{CH_3{-}C({=}O){-}OCH_2CH_3}$$

Figure 23.17 Ethyl ethanoate, shown with its structural formula and space-filling model, is a low-molar-mass ester.

Differentiated Instruction

English Learners/Less Proficient Readers L1

Students may find it difficult to understand that oxidation and reduction can be defined in terms of the loss or gain of hydrogen. Pair up students who are having trouble with this concept. Have students write out each series of oxidation-reduction reactions on this and the next few pages of the text, noting the number of hydrogen and oxygen atoms attached to carbon at each step.

Relate 

Explain that the formation of an ester from a carboxylic acid and an alcohol is an excellent example of dynamic equilibrium. Under the proper conditions, the equilibrium can be shifted greatly in favor of the products. Other conditions favor the decomposition of the ester into the carboxylic acid and alcohol. Have students review Le Chtelier's principle and use it to predict conditions that would favor the reactants and conditions that would favor the products. Ask, **How can a chemist improve the yield of an ester from an esterification reaction?** *(Use an excess of a reactant, or remove the water or ester by distillation as it is formed.)*

Discuss 

Discuss the base-promoted hydrolysis of an ester. Write a number of examples of reactions on the chalkboard or overhead projector. Explain which part of the ester yields the alcohol and which part yields carboxylic acid products. Ask students about the difference in volatility of the products and how a chemist might separate them once they are formed. Write the molecular structure for propyl ethanoate on the board. Ask, **What products would be formed from the base-promoted hydrolysis of propyl ethanoate?** *(propanol (propyl alcohol) and ethanoic (acetic) acid)*

Oxidation–Reduction Reactions

Discuss 

Go through the sequence of oxidations that occurs for alkanes. Explain that carbon dioxide is the most oxidized form of carbon, and an alkane is the most reduced form of carbon. Discuss each step as being an example of an oxidation or reduction. Remind students that the loss of hydrogen is oxidation, and the gain of hydrogen is reduction. In any series of oxidations or reductions involving hydrocarbons, an alkane is the least oxidized and an alkyne is most oxidized.

TEACHER Demo

Oxidation of Alcohols L2

Purpose Students observe the oxidation of primary and secondary alcohols to carboxylic acids and ketones respectively by chromic acid.

Materials CrO_3; concentrated sulfuric acid; water; 3 test tubes, acetone, ethanol, isopropanol, and *t*-butyl alcohol

Procedure Explain how the loss of a molecule of hydrogen by an alcohol can lead to the formation of a carbonyl group. Prepare an acidic solution of CrO_3 by dissolving 0.25 g of CrO_3 in 0.75 mL of water and adding 0.25 mL of concentrated sulfuric acid. A clear orange solution should be obtained. Set up three test tubes each containing 1 mL of acetone. Dissolve about 5 drops of ethanol, isopropanol, and *t*-butyl alcohol in separate tubes. Add 1–2 drops of chromic acid reagent to each tube and gently tap the tubes to mix. Note any changes that occur. A positive test for a primary or secondary alcohol is indicated by a change in color of the solution from orange to blue-green. No change should occur in the case of *t*-butyl alcohol. Have students record their observations and write the chemical equations for any reactions they observed. Ask, **Why was no reaction observed for the test tube containing t-butyl alcohol?** *(There is no hydrogen on the carbon bearing the hydroxyl group.)*

Figure 23.18 Oxidation reactions occur in many daily activities. **a** These basketball players are energized by oxidation reactions taking place within the cells of their bodies. **b** Much of the world relies on hydrocarbon combustion as a source of energy for vehicles.

The loss of hydrogen is a **dehydrogenation reaction.** Strong heating and a catalyst are usually necessary to make dehydrogenation reactions occur. **Dehydrogenation is an oxidation reaction because the loss of each molecule of hydrogen involves the loss of two electrons from the organic molecule.** The remaining carbon electrons pair to make a second or third bond.

$$H_3C-CH_3 \xrightarrow[\text{oxidation}]{\text{loss of hydrogen (dehydrogenation)}} H_2C=CH_2 \xrightarrow[\text{oxidation}]{\text{loss of hydrogen (dehydrogenation)}} H-C\equiv C-H$$

Least oxidized (most reduced) → Most oxidized (least reduced)

These reactions are reversible. Alkynes can be reduced to alkenes, and alkenes can be reduced to alkanes by addition of hydrogen to a double bond.

$$H-C\equiv C-H \xrightarrow[\text{reduction}]{\text{gain of hydrogen (hydrogenation)}} H_2C=CH_2 \xrightarrow[\text{reduction}]{\text{gain of hydrogen (hydrogenation)}} H_3C-CH_3$$

Most oxidized (least reduced) → Most reduced (least oxidized)

Oxidation in organic chemistry also involves the number and degree of oxidation of oxygen atoms attached to carbon. For example, methane, a saturated hydrocarbon, can be oxidized in steps to carbon dioxide. This occurs if it alternately gains oxygen atoms and loses hydrogen atoms. Methane is oxidized to methanol, then to methanal, then to methanoic acid, and finally to carbon dioxide. The same sequence of oxidations occurs for other alkanes. Each series consists of an alkane, alcohol, aldehyde (or ketone), carboxylic acid, and carbon dioxide. The carbon dioxide is most oxidized or least reduced, and the alkane is least oxidized or most reduced.

$$CH_4 \xrightarrow[\text{oxidation}]{\text{gain of oxygen}} H_3C-OH \xrightarrow[\text{oxidation}]{\text{loss of hydrogen}} H-\overset{O}{\overset{\|}{C}}-H \xrightarrow[\text{oxidation}]{\text{gain of oxygen}} H-\overset{O}{\overset{\|}{C}}-OH \xrightarrow[\text{oxidation}]{\text{loss of hydrogen}} O=C=O$$

Methane (most energetic molecule); Methanol (methyl alcohol); Methanal (formaldehyde); Methanoic acid (formic acid); Carbon dioxide (least energetic molecule)

The more reduced a carbon compound is, the more energy it can release upon its complete oxidation to carbon dioxide. The oxidation of organic compounds is exothermic. The energy-releasing properties of oxidation reactions are extremely important for the production of energy in living systems. They also explain why the combustion of hydrocarbons such as methane is a good source of energy. Figure 23.18 shows how energy production by oxidation reactions is used in both living and nonliving systems.

Primary alcohols can be oxidized to aldehydes, and secondary alcohols can be oxidized to ketones, as shown below.

$$R-\underset{H}{\underset{|}{\overset{OH}{\overset{|}{C}}}}-H \xrightarrow[-2H]{\text{oxidation}} R-\overset{O}{\overset{\|}{C}}-H \qquad R-\underset{H}{\underset{|}{\overset{OH}{\overset{|}{C}}}}-R \xrightarrow[-2H]{\text{oxidation}} R-\overset{O}{\overset{\|}{C}}-R$$

Primary alcohol → Aldehyde; Secondary alcohol → Ketone

Differentiated Instruction

Gifted and Talented 

Have students assign oxidation numbers to carbon for each step in the sequence of oxidation reactions shown in the text in which methane is converted to carbon dioxide. Ask, **How many molecules of hydrogen are given up by methane?** *(2)* ***How many electrons are lost?*** *(4)* ***How many molecules of H_2 would ethane yield?*** *(3)*

Tertiary alcohols, however, cannot be oxidized because there is no hydrogen atom present on the carbon bearing the hydroxyl group. A comparison of the structure of primary, secondary, and tertiary alcohols follows.

$$\text{H}-\underset{\text{H}}{\overset{\text{H}}{\text{C}}}-\underset{\text{H}}{\overset{\text{H}}{\text{C}}}-\underset{\text{H}}{\overset{\text{H}}{\text{C}}}-\text{OH} \qquad \text{H}-\underset{\text{H}}{\overset{\text{H}}{\text{C}}}-\underset{\text{OH}}{\overset{\text{H}}{\text{C}}}-\underset{\text{H}}{\overset{\text{H}}{\text{C}}}-\text{H} \qquad \text{H}-\underset{\text{H}}{\overset{\text{H}}{\text{C}}}-\underset{\text{OH}}{\overset{\text{CH}_3}{\text{C}}}-\underset{\text{H}}{\overset{\text{H}}{\text{C}}}-\text{H}$$

Primary alcohol — Secondary alcohol — Tertiary alcohol

The primary alcohols methanol and ethanol can be oxidized to aldehydes by warming them at about 50°C with acidified potassium dichromate ($K_2Cr_2O_7$). In these reactions, methanol produces formaldehyde, and ethanol produces acetaldehyde.

$$\text{H}-\underset{\text{H}}{\overset{\text{OH}}{\text{C}}}-\text{H} \xrightarrow[\text{H}_2\text{SO}_4]{\text{K}_2\text{Cr}_2\text{O}_7} \text{H}-\overset{\text{O}}{\overset{\|}{\text{C}}}-\text{H} \qquad \text{CH}_3-\underset{\text{H}}{\overset{\text{OH}}{\text{C}}}-\text{H} \xrightarrow[\text{H}_2\text{SO}_4]{\text{K}_2\text{Cr}_2\text{O}_7} \text{CH}_3-\overset{\text{O}}{\overset{\|}{\text{C}}}-\text{H}$$

Methanol (methyl alcohol) (bp 65°C) — Methanal (formaldehyde) (bp −21°C) — Ethanol (ethyl alcohol) (bp 78°C) — Ethanal (acetaldehyde) (bp 21°C)

Preparing an aldehyde by this method is often a problem because aldehydes are easily oxidized further to carboxylic acids.

$$\text{R}-\overset{\text{O}}{\overset{\|}{\text{C}}}-\text{H} \xrightarrow[\text{H}_2\text{SO}_4]{\text{K}_2\text{Cr}_2\text{O}_7} \text{R}-\overset{\text{O}}{\overset{\|}{\text{C}}}-\text{OH}$$

Aldehyde — Carboxylic acid

Oxidation of the secondary alcohol 2-propanol by warming it with acidified potassium dichromate produces acetone.

$$\text{CH}_3-\underset{\text{H}}{\overset{\text{OH}}{\text{C}}}-\text{CH}_3 \xrightarrow[\text{H}_2\text{SO}_4]{\text{K}_2\text{Cr}_2\text{O}_7} \text{CH}_3-\overset{\text{O}}{\overset{\|}{\text{C}}}-\text{CH}_3$$

2-Propanol (isopropyl alcohol) — Propanone (acetone)

Unlike aldehydes, ketones are resistant to further oxidation, so there is no need to remove them from the reaction mixture during the reaction.

Tests for aldehydes make use of the ease with which these compounds are oxidized. Benedict's and Fehling's reagents are deep-blue alkaline solutions of copper(II) sulfate. Figure 23.19 illustrates Fehling's test for an aldehyde. When an aldehyde is oxidized with Benedict's or Fehling's reagent, a red precipitate of copper(I) oxide (Cu_2O) is formed. The aldehyde is oxidized to its acid, and copper(II) ions (Cu^{2+}) are reduced to copper(I) ions (Cu^{+}).

Go Online SCI*LINKS*

For: Links on Oxidation and Reduction
Visit: www.SciLinks.org
Web Code: cdn-1233

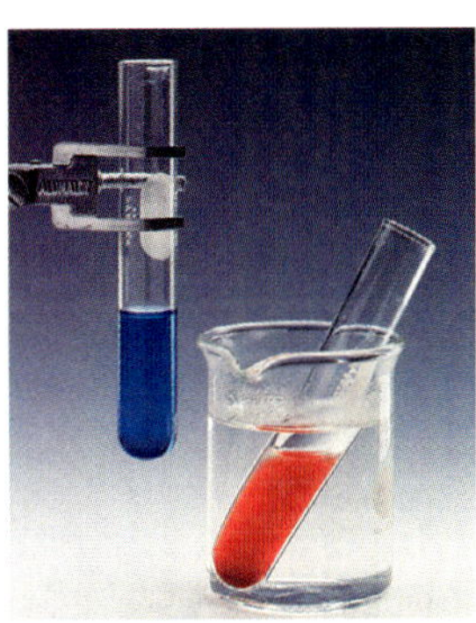

Figure 23.19 When an aldehyde is mixed with Fehling's reagent (left test tube) and heated, the blue copper(II) ions in Fehling's reagent are reduced to form Cu_2O, a red precipitate (right test tube). **Inferring** ***What is the oxidation state of copper in the product?***

Go Online SCI*LINKS*

Download a worksheet on **Oxidation and Reduction** for students to complete, and find additional teacher support from NSTA SciLinks.

3 ASSESS

Evaluate Understanding 

Write the molecular structures of vanillin, 3-heptanone, ethyl butanoate, and pentanoic acid on the board. Have students classify each compound as an aldehyde, ketone, ester, or carboxylic acid.

Answers to...

Figure 23.19 +1

Quick LAB

Testing for an Aldehyde L2

Objective After completing this activity, students will be able to describe a positive test for an aldehyde.

Teaching Tips It may be necessary to warm the test tube in a bath of warm water.

Class Time 30 minutes

Safety Do not store Tollens' reagent for extended periods of time. It decomposes on standing and yields an explosive mixture. Avoid skin contact with these chemicals.

Expected Outcome The deposition of a silver mirror on the inner walls of the test tube is a positive indicator for the presence of an aldehyde functional group. (The silver mirror can be removed from the inside of the test tube with nitric acid.)

Analyze and Conclude

1. A layer of reflective, silver metal is deposited on the inner walls of test tube 1. No reaction is observed for test tubes 2 and 3.

2. In an alkaline solution of silver nitrate, the aldehyde is oxidized to a carboxylic acid according to the following reaction:

$$RCHO(aq) + 2Ag(NH_3)_2OH(aq) \rightarrow 2Ag(s) + RCOO^-(aq) + NH_4^+(aq) + H_2O(l) + 3NH_3(aq)$$

3. The test is useful for distinguishing aldehydes from alcohols and ketones.

Connecting Concepts

Students' comparisons should include descriptions of how the bonding in carboxylic acids causes them to be weak acids.

If your class subscribes to the Interactive Textbook, use it to review key concepts in Section 23.3.

with ChemASAP

Quick LAB

Testing for an Aldehyde

Purpose

To distinguish an aldehyde from an alcohol or a ketone using Tollens's reagent.

Materials

- 1*M* sodium hydroxide
- 5% silver nitrate
- 6*M* aqueous ammonia
- 4 small test tubes
- test tube rack
- plastic droppers
- glucose solution
- propanone
- ethanol

Procedure    

1. Add 1 drop of 1*M* sodium hydroxide to 2 mL of 5% silver nitrate in a test tube. Add 6*M* aqueous ammonia drop by drop, gently agitating the tube after each addition until the brownish precipitate dissolves. This will be your Tollens's reagent.
2. Place 10 drops of Tollens's reagent in each of three clean, labeled test tubes.
3. To test tube 1, add 2 drops of glucose solution. To test tube 2, add 2 drops of propanone. To test tube 3, add 2 drops of ethanol. Gently agitate each test tube to mix the contents.
4. Observe the test tubes, leaving them undisturbed for at least 5 minutes.

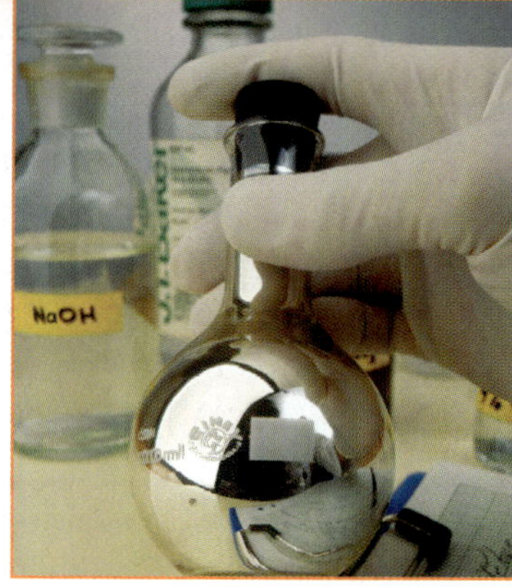

Analyze and Conclude

1. What evidence of a chemical reaction did you observe in test tube 1? In test tube 2? In test tube 3?
2. Write the equation for any chemical reaction you observed.
3. If you observed a chemical reaction in one or more of the test tubes, what practical uses might the reaction have?

23.3 Section Assessment

13. Key Concept Describe the structure of the carbonyl groups that are characteristic of aldehydes and ketones.

14. Key Concept What is the general formula for a carboxylic acid?

15. Key Concept What is the general structure of an ester?

16. Key Concept Explain why dehydrogenation is an oxidation reaction.

17. What products are expected when the following compounds are oxidized?

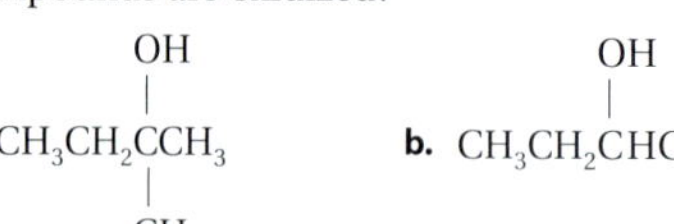

a. $CH_3CH_2C(OH)(CH_3)CH_3$ **b.** $CH_3CH_2CH(OH)CH_3$

18. Give the IUPAC name for the aldehyde and the ketone.

a. CH_3CH_2CHO **b.** $CH_3CH_2CH_2C(=O)CH_2CH_3$

Connecting Concepts

Acids Review inorganic acids in Chapter 19. Write a paragraph comparing the properties of inorganic acids with those of carboxylic acids. Are carboxylic acids weak or strong acids? Explain.

Assessment 23.3 Check your understanding of the important ideas and concepts in Section 23.3.

with ChemASAP

Section 23.3 Assessment

13. A carbonyl group is a carbon atom double bonded to an oxygen atom. Aldehyde: carbon of the carbonyl group is always joined to at least one hydrogen (RCHO); ketone: carbon is joined to two other carbons (RCOR). Carboxylic acid: carbon is attached to a hydroxyl group (RCOOH). Ester: a derivative of a carboxylic acid in which the —OH of the carboxyl group has been replaced by an —OR from an alcohol (RCOOR)

14. RCOOH or RCO_2H

15. a compound with the molecular formula RCOOR or RCO_2R

16. The loss of hydrogen (H_2) also results in the loss of two electrons.

17. a. no reaction **b.** 2-butanone

18. a. propanal **b.** 3-hexanone

23.4 Polymerization

Connecting to Your World

Snap beads are a favorite toy for toddlers. These colorful plastic shapes of many different colors can be snapped together end-to-end to form chains of different lengths. They can also be snapped together to form loops of various sizes. Chemical compounds called monomers and polymers resemble snap beads. Monomers are like individual snap beads. When monomers are joined end-to-end to other monomers they form long chains called polymers. In this section, you will learn about some characteristics of monomers and polymers.

Guide for Reading

Key Concepts

- How does an addition polymer form?
- How are condensation polymers formed?

Vocabulary

polymer

monomers

Reading Strategy

Predicting In your notebook, write a prediction about how polymers are formed. As you read the section, correct your prediction, if necessary. After you finish the section, write a short summary of the two main ways polymers are formed.

Addition Polymers

Most of the reactions that you have learned about so far involve reactants and products of low molar mass. Some of the most important organic compounds that exist, however, are giant molecules called polymers. Each day, you see many different polymers. For example, the materials you know as plastics are polymers. The kinds and uses of plastics are numerous indeed!

A **polymer** is a large molecule formed by the covalent bonding of repeating smaller molecules. The smaller molecules that combine to form a polymer are called **monomers.** Some polymers contain only one type of monomer. Others contain two or more types of monomers. The reaction that joins monomers to form a polymer is called polymerization. Most polymerization reactions require a catalyst.

An addition polymer forms when unsaturated monomers react to form a polymer. Ethene undergoes addition polymerization. The ethene molecules bond to one another to form the long-chain polymer polyethylene.

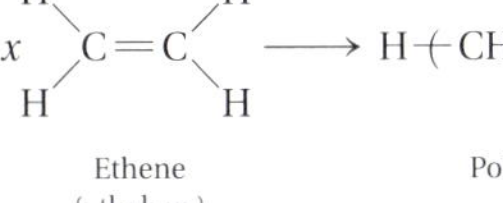

$$x\,H_2C{=}CH_2 \longrightarrow H{+}CH_2{-}CH_2{+}_xH$$

x is number of ethylene units that combine to form a long chain.

Ethene (ethylene)

Polyethylene

x is number of repeating $-CH_2-CH_2-$ units in polymer; parentheses identify the repeating unit.

Polyethylene, which is chemically resistant and easy to clean, is an important industrial product. It is used to make familiar items such as the plastic products pictured in Figure 23.20.

Figure 23.20 Polyethylene is used for many familiar household items, including plastic bottles, bags, and food containers.

23.4

1 FOCUS

Objectives

23.4.1 Describe how addition polymers are formed.

23.4.2 Describe how condensation polymers are formed.

Guide for Reading

Build Vocabulary 

Word Parts The word *polymer* comes from the Greek word *poly*, meaning "many" and the Greek word *meros*, meaning "parts." A polymer is a compound that is formed by the covalent bonding of repeating small parts, called monomers. Nylon, plastics, and many other large molecules are polymers.

Reading Strategy 

Relate Text and Visuals Students may have trouble understanding the concept of a polymer as an array of repeating units. Drawing a segment containing 3 or 4 monomers of each polymer will help students visualize the polymer.

2 INSTRUCT

Connecting to Your World

Relate the construction of a snap-bead chain to the formation of a polymer. Ask, **What is the repeating unit, or monomer, in the chain?** *(each bead of the chain)* **What do you think are some of the characteristics of monomers?** *(Answers will vary. Students may suggest that monomers must have the ability to be joined together, must be small molecules, or must be chemically reactive.)*

Section Resources

Print

- **Guided Reading and Study Workbook,** Section 23.4
- **Core Teaching Resources,** Section 23.4 Review, Interpreting Graphics
- **Transparency,** T270

Technology

- **Interactive Textbook with ChemASAP** Assessment 23.4
- **Go Online,** Section 23.4

Addition Polymers

Discuss L2

Discuss the similarities between the addition of hydrogen and halogens to alkenes and the formation of addition polymers. Explain that chemists have also found that when certain catalysts are used, alkenes will add to one another to form a long chain of hydrocarbon units. Ask, **What is a catalyst?** *(A catalyst is a substance that increases the rate of a reaction without being used up itself in the reaction.)*

Discuss L2

Write the formula for the simplest repeating unit of a polymer on the board. Ask students to identify the polymer and to write out the structure for the polymer showing how three successive monomers are linked together. Ask, **Why can't an exact formula for a polymer be written?** *(The length of the carbon chain varies from polymer molecule to polymer molecule. The best that can be done is to write the formula for the simplest repeating unit, the monomer, in parentheses. The subscript x indicates a very large number.)*

TEACHER Demo

Building a Polymer Model L2

Purpose Students wil better understand how monomers link together to form polymers.

Procedure Bring a length of metal-link chain to class. Use the metal chain as a model for a polymer. Explain that a polymerization reaction is a reaction that joins monomers to form a polymer. Point out the features in the chain that are similar to a polymer molecule. In addition, model a polymerization reaction using paperclips. Each paperclip represents a monomer. Explain that each end of the paperclip represents a reaction site. Build a chain by successively connecting one paperclip to another. The chain formed is analogous to a polymer. Point out that the main differences between polymers in organic chemistry are due to the type of monomers used to form the chains.

The physical properties of polyethylene can be controlled by shortening or lengthening the carbon chains. Polyethylene that contains relatively short chains ($x = 100$) has the consistency of paraffin wax. Polyethylene with long chains ($x = 1000$) is harder and more rigid.

Polymers of substituted ethenes also are useful. Polypropylene is prepared by the polymerization of propene.

$$x\mathrm{CH_2{=}CH(CH_3)} \longrightarrow \mathrm{+CH_2-CH(CH_3)+_x}$$

Propene (propylene) → Polypropylene

Polypropylene, a stiffer polymer than polyethylene, is used extensively in utensils and containers, as well as items such as the whistle shown in Figure 23.21a. Polystyrene is prepared by the polymerization of styrene. Polystyrene, in the form of a rigid foam, is a poor heat conductor, which makes it useful for insulating homes and for manufacturing molded items such as coffee cups and picnic coolers.

$$x\mathrm{CH_2{=}CH(C_6H_5)} \longrightarrow \mathrm{+CH_2-CH(C_6H_5)+_x}$$

Styrene (vinyl benzene) → Polystyrene

Many halocarbon polymers, including polyvinyl chloride (PVC), have useful properties. Vinyl chloride is the monomer of polyvinyl chloride.

$$x\mathrm{CH_2{=}CH(Cl)} \longrightarrow \mathrm{+CH_2-CH(Cl)+_x}$$

Chloroethene (vinyl chloride) → Polyvinyl chloride (PVC)

Polyvinyl chloride (PVC) is used for the plumbing pipes shown in Figure 23.21b. It is also produced in sheets, sometimes with a fabric backing, for use as a tough plastic upholstery covering.

Figure 23.21 These common polymers are used for various applications. **a** Polypropylene is used in the manufacture of a variety of items, such as this whistle. **b** Polyvinyl chloride is used for pipes in plumbing.

Facts and Figures

The First Plastic—Celluloid

One of the first plastics, developed by an Englishman named Alexander Parkes, was a synthetic ivory named pyroxlin. Parkes marketed it under the trade name "Parkesine," and it won a bronze medal at the 1862 World's Fair in London. Parkesine was made from cellulose treated with nitric acid and a solvent. An American amateur inventor named John Wesley Hyatt later improved the product. Because cellulose was the main constituent used in the synthesis of his new material, Hyatt named it celluloid. A large variety of objects were made of celluloid, and it was a huge success. However, celluloid had a dangerous defect—it burned easily and tended to explode. Hyatt liked to tell stories about celluloid billiard balls exploding when struck very hard. Today, celluloid is used only to make table-tennis balls.

Figure 23.22 PTFE is used to coat cookware and to insulate wires, cables, motors, and generators. It also is suspended in motor oils as a friction-reducing agent.

Polytetrafluoroethene (Teflon™ or PTFE) is the product of the polymerization of tetrafluoroethene monomers. PTFE is very resistant to heat and chemical corrosion. You are probably familiar with this polymer as a coating on the nonstick cookware shown in Figure 23.22. Because PTFE is very durable and slick, it is formed into bearings and bushings used in chemical reactors.

$$\underset{\text{Tetrafluoroethene}}{x\mathrm{CF_2{=}CF_2}} \longrightarrow \underset{\text{Teflon (PTFE)}}{\mathrm{+CF_2{-}CF_2+_{x}}}$$

Polyisoprene, harvested from tropical plants such as the rubber tree shown in Figure 23.23, is the polymer that constitutes natural rubber. It is used to make tires, rubber bands, soles of athletic shoes, and many other common items.

$$\underset{\text{Isoprene}}{x\mathrm{CH_2{=}\underset{\underset{CH_3}{|}}{C}H{=}CH_2}} \longrightarrow \underset{\text{Polyisoprene}}{\mathrm{+CH_2{-}C(CH_3){=}C(H){-}CH_2+_{x}}}$$

 **Checkpoint** ***What is polyisoprene?***

Figure 23.23 Rubber is harvested from tropical plants and is used in a variety of products.
Interpreting Photographs
What is the source of natural polyisoprene?

Use Visuals 

Figure 23.22 Ask, **Why do you think PTFE is used for cooking utensils?** *(Food does not stick to it, and it is very resistant to heat.)* **What is the common name for PTFE?** *(Teflon)*

Discuss 

Have students study the chemical equations for the formation of the addition polymers discussed in this section. Write the equations on the board and help students identify the repeating units. Discuss the features of an addition polymerization and how the name of a polymer is derived from the monomer units. Bring to class samples of items that are composed of addition polymers. Have students describe and compare their physical characteristics. Ask, **What are some of the advantages of synthetic polymers over metals and other natural materials?** *(They are moldable, lightweight, chemically resistant, and easy to clean.)*

Differentiated Instruction

Less Proficient Readers L1

Have students collect several household items composed of different polymer materials. Ask the students to create a table listing the items and the plastic container codes found on the bottom of each item. Have students do research to find out how the codes are involved in recycling.

Answers to...

Figure 23.23 tropical plants such as the rubber tree

 **Checkpoint** the polymer of isoprene that constitutes natural rubber

Condensation Polymers

Discuss L2

Explain that the key to polymerization is the ability of molecules to form repeating chains. For this to happen, each reacting molecule must have at least two reactive sites. In addition polymerization, the two active sites become available as double bonds are opened up. In condensation polymerization, the sites become available as two groups (such as H groups and OH groups) are split off the molecules.

Relate L2

Have students research the use of synthetic polymers in medicine. Some applications include prosthetics, skin substitutes, artificial artery grafts, and heart valve replacements.

Download a worksheet on **Polymers** for students to complete, and find additional teacher support from NSTA SciLinks.

For: Links on Polymers
Visit: www.SciLinks.org
Web Code: cdn-1234

Condensation Polymers

Condensation polymers are formed by the head-to-tail joining of monomer units. This is usually accompanied by the formation of water as a reaction product, thus the name condensation polymers. The formation of a polyester is an example of condensation polymerization. Polyesters are polymers that consist of many repeating units of dicarboxylic acids and dihydroxy alcohols joined by ester bonds. The formation of a polyester can be represented by a block diagram, which shows only the functional groups involved in the polymerization reaction. The squares and circles represent unreactive parts of the organic molecules.

Condensation polymerization requires that there be two functional groups on each monomer molecule.

$$x\,\text{HO}-\overset{\text{O}}{\overset{\|}{\text{C}}}-\square-\overset{\text{O}}{\overset{\|}{\text{C}}}-\text{OH} + x\,\text{HO}-\bigcirc-\text{OH} \longrightarrow \left(\overset{\text{O}}{\overset{\|}{\text{C}}}-\square-\overset{\text{O}}{\overset{\|}{\text{C}}}-\text{O}-\bigcirc-\text{O}\right)_x + 2x\,\text{H}_2\text{O}$$

Dicarboxylic acid — Dihydroxy alcohol — Representative polymer unit of a polyester

The polyester polyethylene terephthalate (PET) is formed from terephthalic acid and ethylene glycol.

$$\text{HO}-\overset{\text{O}}{\overset{\|}{\text{C}}}-\text{C}_6\text{H}_4-\overset{\text{O}}{\overset{\|}{\text{C}}}-\text{OH} + \text{HO}-\text{CH}_2\text{CH}_2-\text{OH} \longrightarrow \left(\overset{\text{O}}{\overset{\|}{\text{C}}}-\text{C}_6\text{H}_4-\overset{\text{O}}{\overset{\|}{\text{C}}}-\text{O}-\text{CH}_2\text{CH}_2-\text{O}\right)_x + 2x\,\text{H}_2\text{O}$$

Terephthalic acid — Ethylene glycol — Representative polymer unit of polyethylene terephthalate (PET)

PET fibers form when the compound is melted and forced through tiny holes in devices called spinnerettes. The fibers, sold as Dacron™, Fortrel®, or Terylene (depending on the manufacturer), are used for tire cord and permanent-press clothing. Figure 23.24 shows how woven Dacron tubing is used to replace major blood vessels. PET fibers are often blended with cotton to make clothes that are more comfortable on hot, humid days than those containing 100% polymer. These clothes retain the wrinkle resistance of 100% polyester. PET melts may also be forced through a narrow slit to produce sheets of Mylar™, a polymer used extensively as magnetic tape for tape recorders and computers.

Checkpoint ***How are PET fibers made?***

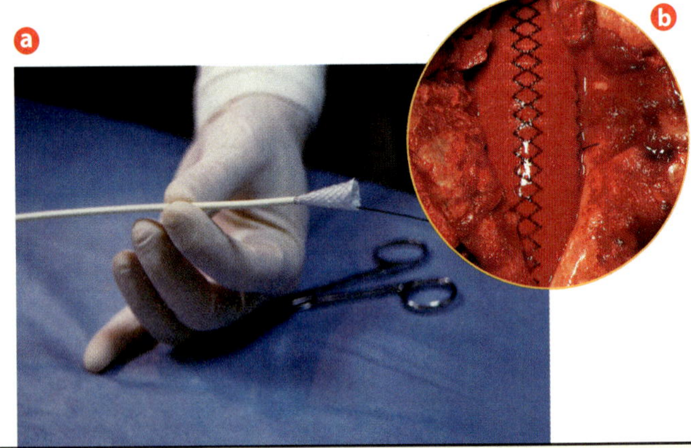

Figure 23.24 Woven Dacron (PET fibers) tubing can be used to replace major blood vessels. **a** The blood vessel replacement before surgery. **b** The blood vessel after surgery.
Applying Concepts ***What properties of Dacron make it useful for this purpose?***

Facts and Figures

Super-Absorbent Polymers

Sodium polyacrylate is a super-absorbent polymer that is used in disposable diapers, alkaline batteries, fuel filters, and potting soil. Diapers contain a layer of sodium polyacrylate granules that become a gel-like material when mixed with water. Hydrogen bonding between the water and the polymer allows sodium polyacrylate to absorb 800 times its weight in distilled water. The water moves through a membrane because of osmotic pressure as the polymer and water become a gel. The concentration of sodium ions is higher inside the membrane. Students may enjoy finding the mass of a diaper and then adding a measured volume of water until the diaper can absorb no more. **CAUTION** *Do not remove the polymer granules from the diaper. Inhaling them can cause adverse health reactions.*

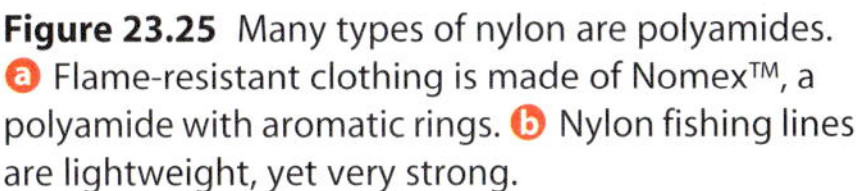

Figure 23.25 Many types of nylon are polyamides. **a** Flame-resistant clothing is made of Nomex™, a polyamide with aromatic rings. **b** Nylon fishing lines are lightweight, yet very strong.

Many important polymers are formed by the reaction of carboxylic acids and amines. The amines used to make polymers generally contain the amino functional group $-NH_2$. The condensation of a carboxylic acid and an amine produces an amide.

$$\underset{\text{Carboxylic acid}}{R-\overset{\overset{\displaystyle O}{\|}}{C}-OH} + \underset{\text{Amine}}{H-\overset{\overset{\displaystyle H}{|}}{N}-R} \longrightarrow \underset{\text{Amide}}{R-\overset{\overset{\displaystyle O}{\|}}{C}-\overset{\overset{\displaystyle H}{|}}{N}-R} + \underset{\text{Water}}{HO-H}$$

Polyamides are polymers in which the carboxylic acid and amine monomer units are linked by amide bonds. The many types of nylon are polyamides. You are probably familiar with a range of nylon products; more nylon products are shown in Figure 23.25. Nylon polymer has a molar mass of about 1×10^4 g/mol and a melting point of 250°C. The representative polymer unit of nylon is derived from 6-aminohexanoic acid, a compound that contains both carboxyl and amino functional groups. The long polymer chain is formed by the successive attachment of the carboxyl group of one molecule of the monomer to the amino group of the next monomer by the formation of an amide bond.

$$\underset{\text{6-Aminohexanoic acid}}{xH_2N-CH_2{+}CH_2{+}_4\overset{\overset{\displaystyle O}{\|}}{C}-OH} \xrightarrow{\text{heat}} \underset{\text{Representative polymer unit of nylon}}{\left(CH_2{+}CH_2{+}_4\overset{\overset{\displaystyle O}{\|}}{C}-\overset{\overset{\displaystyle H}{|}}{N}\right)_x} + xH_2O$$

The melted polymer can be spun into very fine, yet very strong, fibers. Nylon fibers are used for carpeting, tire cord, fishing lines, sheer hosiery, and textiles. Nylon is also molded into gears, bearings, and zippers.

Checkpoint ***What are polyamides?***

TEACHER Demo

Making Nylon L2

Purpose Students observe the synthesis of nylon.

Materials 5% aqueous solution of 1,6-hexanediamine, small beaker, 20% sodium hydroxide, dropper, 5% solution of adipoyl chloride in cyclohexane, forceps or tongs, water, paper towel

Safety Wear safety goggles, a chemical-resistant apron, and chemical-resistant gloves while preparing solutions and performing the demonstration.

Procedure Pour 5 mL of a 5% aqueous solution of 1,6-hexanediamine into a small beaker. Add 5 drops of 20% sodium hydroxide. Slowly pour 5 mL of a 5% solution of adipoyl chloride in cyclohexane down the wall of the beaker. A two-phase system will form, and a polymer film will appear at the interface. Use forceps or tongs to slowly draw out a long strand of nylon. Rinse thoroughly with water and set it on a paper towel to dry. Have students examine the color, hardness, and strength. Discuss the uses of nylon, and explain that it is a polyamide and a condensation polymer.

Discuss L2

Discuss the features of condensation polymerization. Write specific examples of condensation polymers on the board. Relate the structure of polyamides to the structure of proteins. Have students identify some other kinds of naturally occurring polymers. *(polysaccharides such as cellulose, starch, and chitin; nucleic acids such as DNA; proteins)*

Facts and Figures

Development of Nylon

Nylon was developed by Wallace Carothers and his co-workers at DuPont during the 1930s. DuPont planned to make nylon stockings as the first product made from nylon because silk stockings were becoming expensive. However, problems developed in the processes of making the nylon filaments and weaving the material, and the production of nylon hosiery was delayed. The filaments that were produced while the manufacturing processes were being refined were not discarded; instead, they were used to make the first commercial product that contained nylon—the bristles in toothbrushes. Nylon hosiery was not sold until 1940 and was an immediate success.

Answers to...

Figure 23.24 It is nonreactive and non-allergenic, and it can be formed into thin fibers and tubes.

Checkpoint PET is melted and extruded through tiny holes or slits to form fibers and sheets.

Checkpoint polymers of amines and carboxylic acids or polymers of amino-substituted carboxylic acids

Section 23.4 (continued)

3 ASSESS

Evaluate Understanding L2

Have students write sentences in their own words describing what takes place during the formation of a polymer. Ask, **What is the difference between an addition polymer and a condensation polymer? Give examples of each.** *(An addition polymer is formed when unsaturated monomers react to form a polymer. One example is the reaction of ethene to form polyethylene. Condensation polymers are formed by the head-to-tail joining of monomer units. The other product of the condensation reaction is a small molecule such as water.)* Have students describe the uses of some addition and condensation polymers.

Reteach L1

Remind students that the area of chemistry devoted to the study of polymers involves more than just plastics. Synthetic fibers are used to create clothing, automobile tires, and materials used in the medical and dental fields. The abundance of synthetic polymers can be traced to the unique bonding ability of carbon. To show how small changes in the structures of monomers can lead to dramatic differences in physical properties, compare polymers based on ethene and vinyl chloride.

Writing Activity

Students' analogies might include putting together a toddler's plastic pop beads, stringing a bead necklace, or making a paperclip, or popcorn chain.

If your class subscribes to the Interactive Textbook, use it to review key concepts in Section 23.4.

with ChemASAP

Polyamides that contain aromatic rings are extremely tough and flame resistant. The aromatic rings make the resulting fiber stiffer and tougher. Nomex™ is a polyamide with a carbon skeleton consisting of aromatic rings derived from isophthalic acid and *m*-phenylenediamine.

$$\text{HO}-\overset{\text{O}}{\overset{\|}{\text{C}}}-\text{C}_6\text{H}_4-\overset{\text{O}}{\overset{\|}{\text{C}}}-\text{OH} + \text{H}_2\text{N}-\text{C}_6\text{H}_4-\text{NH}_2 \longrightarrow \left(\overset{\text{O}}{\overset{\|}{\text{C}}}-\text{C}_6\text{H}_4-\overset{\text{O}}{\overset{\|}{\text{C}}}-\underset{\text{H}}{\underset{|}{\text{N}}}-\text{C}_6\text{H}_4-\underset{\text{H}}{\underset{|}{\text{N}}}\right)_x + 2x\text{H}_2\text{O}$$

Isophthalic acid — *m*-Phenylenediamine — Representative unit of Nomex

Like nylon, Nomex is a poor conductor of electricity. Because it is more rigid than nylon, it is used to make parts for electrical fixtures. Nomex is also used in the manufacture of flame-resistant clothing for race-car drivers and firefighters and in the fabrication of flame-resistant building materials.

Kevlar™ has a structure similar to that of Nomex. It is a polyamide made from terephthalic acid and *p*-phenylenediamine.

$$\text{HO}-\overset{\text{O}}{\overset{\|}{\text{C}}}-\text{C}_6\text{H}_4-\overset{\text{O}}{\overset{\|}{\text{C}}}-\text{OH} + \text{H}_2\text{N}-\text{C}_6\text{H}_4-\text{NH}_2 \longrightarrow \left(\overset{\text{O}}{\overset{\|}{\text{C}}}-\text{C}_6\text{H}_4-\overset{\text{O}}{\overset{\|}{\text{C}}}-\overset{\text{H}}{\overset{|}{\text{N}}}-\text{C}_6\text{H}_4-\overset{\text{H}}{\overset{|}{\text{N}}}\right)_x + 2x\text{H}_2\text{O}$$

Terephthalic acid — *p*-Phenylenediamine — Representative unit of Kevlar

Figure 23.26 Bulletproof vests are made of Kevlar.

Kevlar is used extensively where strength and flame resistance are needed. A properly constructed vest made of Kevlar, shown in Figure 23.26, is strong enough to stop high-speed bullets, yet is light and flexible enough to be worn under normal clothing.

Proteins, which are polyamides of naturally occurring amino acids, rank among the most important of all biological molecules. You will learn about these essential polymers in more detail in Chapter 24.

23.4 Section Assessment

19. **Key Concept** Describe how addition polymers form.
20. **Key Concept** Describe how condensation polymers form.
21. What is a polymer? A monomer?
22. Give names and uses for three types of polymers.
23. How is water involved in the formation of condensation polymers?
24. What is formed when a carboxylic acid and an amine combine? Give an example of the type of polymer that is formed by this reaction.
25. What structure must a monomer have if it is to undergo addition polymerization?

Writing Activity

Write an Analogy Analogies are often used to help explain concepts. Write an analogy that you can use to describe to a middle-school student how a polymer is constructed from monomers. Use diagrams as well as words to describe your analogy.

Assessment 23.4 Check your understanding of the important ideas and concepts in Section 23.4.

with ChemASAP

Section 23.4 Assessment

19. formed when unsaturated monomers react and link to one another
20. formed by the head-to-tail joining of monomer units, usually with loss of water
21. A polymer is a large molecule formed by the covalent bonding of repeating small molecules. Monomers are the smaller molecules that combine to form the repeating unit of a polymer.
22. Answers will vary. Polyethylene: milk bottles and other containers; polyester: a textile for permanent press clothing; Kevlar: bullet-proof vests
23. Water is a product of most condensation polymerizations.
24. an amide bond; nylon is a polyamide.
25. at least one carbon-carbon double bond.

Small-Scale LAB

Polymers

Purpose

To cross-link some polymers and examine their properties.

Materials

- $3\frac{1}{2}$-oz plastic cup
- soda straw
- powdered guar gum
- plastic spoon
- 4% borax solution
- pipet

Procedure   

1. Half fill a $3\frac{1}{2}$-oz cup with water.
2. Use a soda straw as a measuring scoop to obtain approximately 2 cm of powdered guar gum. **Caution:** *Do not use your mouth to draw up the guar gum into the straw.* Gently sprinkle the guar gum powder into the water while stirring with a plastic spoon. Add the guar gum powder slowly to prevent it from clumping. Stir the mixture well.
3. While stirring, add one full pipet (about 4 mL) of borax solution. Continue to stir until a change occurs.

Analyze

Using your experimental data, record the answers to the following questions below your observations.

1. Describe the polymer you just made. Is it a liquid or a solid? What special characteristics does it have?
2. Guar gum is a carbohydrate, a polymer with many repeating alcohol functional groups (—OH). Draw a zigzag line to represent a crude polymer chain.
3. Add —OH groups along the chain to represent the alcohol functional groups.
4. Borate ions combine with alcohol to form water and borate complexes of the alcohol.

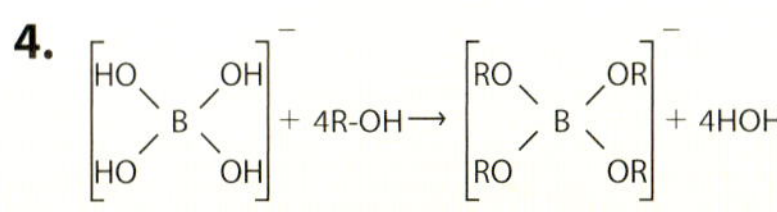

Write a similar equation that replaces all of the —OH groups on the borate with —OR groups.

5. If two polymer chains each contain two nearby —OH groups, borate will cross-link the polymer chains by forming a complex with two alcohols on each chain. Draw a structure similar to the one you drew for Question 4, but replace your four R groups with two polymer chains.

You're the Chemist

The following small-scale activities allow you to develop your own procedures and analyze the results.

1. **Design It!** Try using borax to cross-link other common carbohydrate polymers, such as cornstarch or liquid laundry starch. For each polymer, half fill a $3\frac{1}{2}$-oz cup with the chosen carbohydrate polymer and add enough water to bring the liquid to within about 1 cm of the rim. Stir carefully and thoroughly. Add one full pipet (about 4 mL) of borax solution while stirring. Describe the similarities and differences between this cross-linked polymer and the polymer you made previously. Compare the properties of these polymers.
2. **Analyze It!** Cut a 1 cm × 15 cm strip of paper. Use a drop of glue or a stapler to fasten one end of the paper strip to the other end to form a ring. Now cut out some identical-sized strips of paper, and glue or staple them together into an interlocking chain of paper rings. Explain how this chain is like a polymer. Make another paper ring that cross-links your chain to one of your classmates' chains. Explain how these linked chains are like cross-linked polymers.

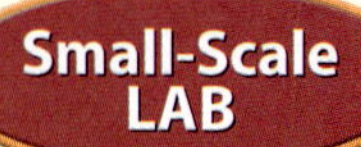

Polymers 

Objective After completing this activity, students will be able to

- use a borax solution to cross-link carbohydrate polymers.
- write structural formulas for cross-linked polymers.
- relate the properties of some polymers to their structures.

Students tend to think of polymerization as a large-scale industrial process. This lab shows that polymers can be synthesized on a much smaller scale.

Skills Focus Observing, inferring, drawing conclusions, interpreting diagrams, comparing and contrasting

Prep Time 30 minutes

Materials $3\frac{1}{2}$-oz plastic cups, soda straw, powdered guar gum, plastic spoons, 4% borax solution, pipet, corn starch or liquid laundry starch, paper, metric ruler, white school glue

Advance Prep Prepare the solution.

Solution	Preparation
4% borax	10 g $Na_2B_4O_7{\cdot}10H_2O$ in 250 mL

Class Time 30 minutes

Safety and Disposal Students should not take home the polymers as the mixtures tend to develop mold in a few days. Dispose of these polymers in the waste basket. While the cross–linked polymers are non-toxic, borax is toxic if ingested. Remind students to wash their hands thoroughly when finished.

Teaching Tips

- School glue adheres to the stirring spoon as it cross-links. Decant any excess liquid that does not adhere.
- Wooden craft sticks make great stirring sticks.

Expected Outcome Students will make a polymer that has some liquid properties and some solid properties.

Analyze

1. The polymer is a viscous gel that oozes. It has properties of both a liquid and a solid.

2.–3.

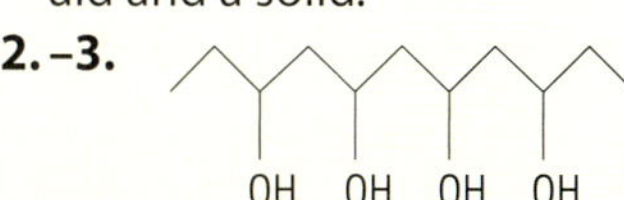

4. [HO, HO, OH, OH around B]⁻ + 4R-OH ⟶ [RO, RO, OR, OR around B]⁻ + 4HOH

5.

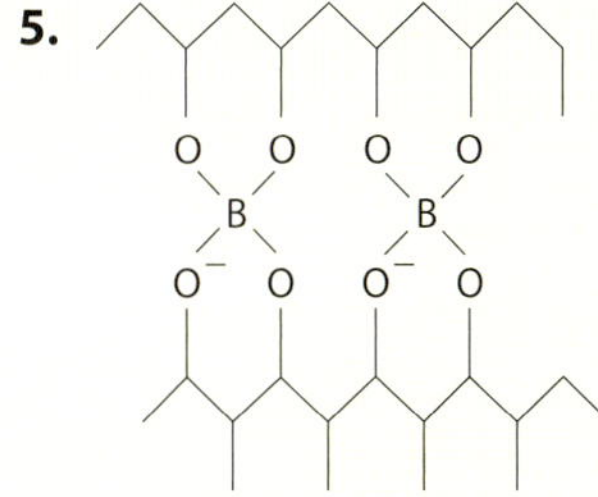

For Enrichment L3

Have students quantify the synthesis of the polymer by varying the amount of water used in the reaction.

You're the Chemist

1. Through experimentation, students are able to produce an amazing variety of polymers with different properties. Students should compare and contrast the properties of the various polymers they make.
2. The chain is similar to a polymer because it contains many repeating units linked end to end.

Technology & Society

Organic Dyes

Purpose

Students will identify the various plants and insects that are used to make organic dyes.

Background

Most dye molecules contain one or more benzene rings to which various functional groups are attached. Some of the more common functional groups in dyes are amino groups, carbonyl groups, carbon-carbon double bonds, carbon-sulfur double bonds, and nitrogen-nitrogen double bonds. In addition, carboxyl, hydroxyl, and amino groups in the molecule help the dye adhere to the fibers of a fabric.

Tyrian purple was one of the most expensive dyes in history, and its use was usually reserved for royalty, leading to the term *royal purple.* Tyrian purple was prepared from the secretions of sea snails, specifically *Murex brandaris* and *Purpura haemostoma,* which are found on the Mediterranean and Atlantic coasts. Huge quantities of these mollusks were used for dyeing the fabrics of kings, emperors, and chief magistrates. It took the secretions of 10,000 snails to produce 1 g of Tyrian purple dye.

The sea snails that secrete tyrian purple were used in the preparation of the purple ink and in dyeing parchments upon which the codices of Byzantium were written. Because the snails are also found on the coasts of the British Isles, the ink was used to write some of the early English, Irish, and French manuscripts. The color went out of use about the eighth century, although it may have been used occasionally up until the eleventh century.

Technology & Society

Organic Dyes

Organic dyes are made from plants or insects and generally produce colors that are soft and natural. Using organic dyes requires two steps: extracting the dyestuff, and fixing the dye to a surface or material. The earliest recorded use of organic dyes to color fabrics dates from 2600 B.C. in China.

Interpreting Photographs ***What plant yields a dye that produces salmon pinks to deep reds?***

Dying fabric Long lengths of fabric are bathed in large vats of dye before being wound into bolts and sold.

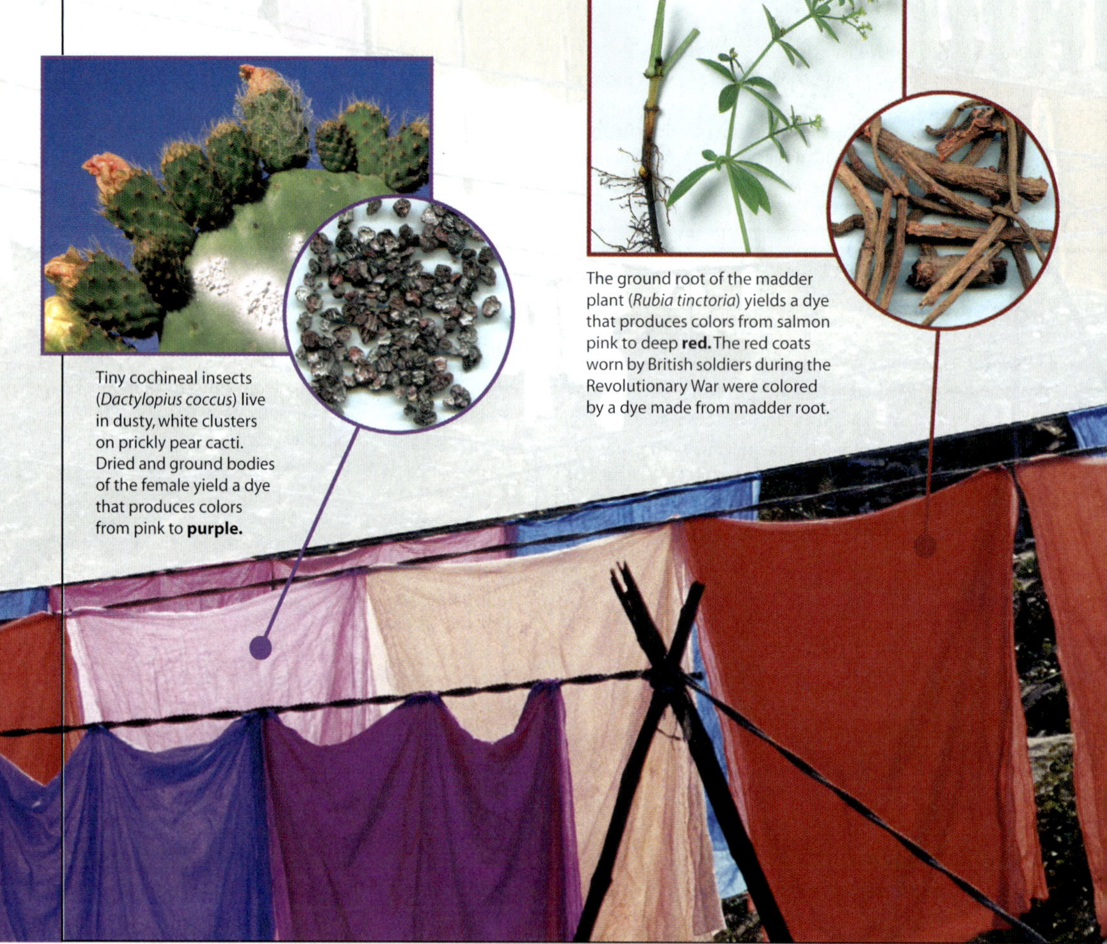

Tiny cochineal insects (*Dactylopius coccus*) live in dusty, white clusters on prickly pear cacti. Dried and ground bodies of the female yield a dye that produces colors from pink to **purple.**

The ground root of the madder plant (*Rubia tinctoria*) yields a dye that produces colors from salmon pink to deep **red.** The red coats worn by British soldiers during the Revolutionary War were colored by a dye made from madder root.

The shredded wood and root bark of the Osage orange tree (*Maclura pomifera*) yield a dye that can produce colors ranging from yellow to tan to **green.** During World War I, dye from Osage orange was used to color U.S. uniforms khaki.

The wood of the smoke tree (*Cotinus coggygria*) yields fustic, which produces colors from yellow to gold to **orange.**

Made from the indigo plant (*Indigofera tinctoria*), indigo dye produces a wide variety of blues from pale sky **blue** to dark navy.

Discuss

Discuss the color spectrum. Bring in a prism to demonstrate that white light is made up of many colors. Remind students that an object appears to be a particular color because wavelengths of that color light are reflected, while the other colors of white light are absorbed. Ask, **Why does a black fabric appear black?** *(All the wavelengths of white light are absorbed by the dyes in the fabric.)*

Research Projects

- Have students research the use of mordants to make dyes colorfast. Mordants are metallic compounds that react with the organic dyes in a fabric to produce a precipitate. This insoluble compound remains "anchored" in the fabric and does not wash out, making the dye colorfast.
- Have students research laser dyes. A laser dye is a complex fluorescent organic dye dissolved in an organic solvent. Two examples of laser dyes are coumarin and rhodamine.

CLASS Activity

Pigment Extraction L1

Have students bring in naturally colored material and try to extract the pigment using various solvents. For example, leaves from birch trees yield a green dye. Students can then use the pigment extracts to dye cotton fabrics or clothing such as T-shirts.

Answers to...

Interpreting Photographs the ground root of the madder plant

CHAPTER 23 Study Guide

Study Tip

Identify Key Concepts and Vocabulary For each section of the chapter, read the Key Concepts that are listed in the Study Guide. Then review the chapter text for any concepts that you do not fully understand. Next, define each vocabulary term in your own words and check definitions against the text's definitions. When you check the text, read all the related text, not just the sentence that defines the boldfaced vocabulary term.

If your class subscribes to the Interactive Textbook with ChemASAP, your students can go online to access an interactive version of the Student Edition and a self-test.

with ChemASAP

CHAPTER 23 Study Guide

Key Concepts

23.1 Introduction to Functional Groups

- Organic compounds can be classified according to their functional groups.
- A halocarbon is a carbon-containing compound with a halogen substituent.
- A halogen can replace a hydrogen atom on an alkane to produce a halocarbon.

23.2 Alcohols and Ethers

- Aliphatic alcohols can be arranged into structural categories according to the number of R groups attached to the carbon with the hydroxyl group.
- When using the IUPAC system to name continuous-chain and substituted alcohols, drop the *-e* ending of the parent alkane name and add the ending *-ol.*
- Alcohols of up to four carbons are soluble in water in all proportions. The solubility of alcohols with four or more carbons in the chain is usually much lower.
- Addition reactions of alkenes are an important method of introducing new functional groups into organic molecules.
- The general structure of an ether is R—O—R. The alkyl groups attached to the ether linkage are named in alphabetical order and are followed by the word *ether.*

23.3 Carbonyl Compounds

- A carbonyl group, with the general structure C=O, is the functional group in aldehydes and ketones.
- The general formula for a carboxylic acid is RCOOH.
- Esters contain a carbonyl group and an ether link to the carbonyl carbon. The general formula for an ester is RCOOR.
- Dehydrogenation is an oxidation reaction because the loss of each molecule of hydrogen involves the loss of two electrons from the organic molecule.

23.4 Polymerization

- An addition polymer forms when unsaturated monomers react to form a polymer.
- Condensation polymers are formed by the head-to-tail joining of monomer units.

Vocabulary

- addition reaction (p. 733)
- alcohol (p. 730)
- aldehyde (p. 737)
- alkyl halide (p. 727)
- aryl halide (p. 727)
- carbonyl group (p. 737)
- carboxyl group (p. 740)
- carboxylic acid (p. 740)
- dehydrogenation reaction (p. 743)
- denatured alcohol (p. 733)
- ester (p. 741)
- ether (p. 735)
- fatty acid (p. 740)
- fermentation (p. 733)
- functional group (p. 725)
- halocarbon (p. 726)
- hydration reaction (p. 734)
- hydrogenation reaction (p. 734)
- hydroxyl group (p. 730)
- ketone (p. 737)
- monomer (p. 747)
- polymer (p. 747)
- substitution reaction (p. 728)

Organizing Information

Use these terms to construct a concept map that organizes the major ideas of this chapter.

ether | functional group | ketone
halocarbon | carboxylic acid | polymer
alcohol | ester | aldehyde

Concept Map 23 Create your Concept Map using the computer.

with ChemASAP

Chapter Resources

Print

- **Core Teaching Resources,** Chapter 23 Practice Problems, Vocabulary Review, Quiz, Chapter Test A, Chapter Test B

Technology

- **Computer Test Bank,** Chapter 23 Test
- **Interactive Textbook with ChemASAP,** Chapter 23

CHAPTER 23

Assessment

Reviewing Content

23.1 Introduction to Functional Groups

26. What does R in the formula R—CH_2Cl represent?

27. Write a structural formula for each compound.
a. 1,2,2-trichlorobutane
b. 1,3,5-tribromobenzene
c. 1,2-dichlorocyclohexane

28. Name the following halocarbons.
a. $CH_2{=}CHCH_2Cl$
b. $CH_3CH(CH_3)CH_2CH(Cl)CH_2Cl$
c. benzene ring with Br substituents (meta)

29. Write structural formulas and give IUPAC names for all the isomers of the following compounds.
a. $C_3H_6Cl_2$ **b.** C_4H_9Br

30. What organic products are formed in the following reactions?
a. $C_6H_5Br + NaOH \xrightarrow{heat}$ ____ $+ NaBr$
b. $C_6H_{11}Cl + NaOH \longrightarrow$ ____ $+ NaCl$
c. $CH_3CH(CH_3)Cl + NaOH \longrightarrow$ ____ $+ NaCl$
d. $C_6H_6 + Br_2 \xrightarrow{catalyst}$ ____ $+ HBr$

23.2 Alcohols and Ethers

31. Give the IUPAC names for the following alcohols.
a. $CH_3{-}CH(OH){-}CH_3$
b. $CH_2(OH){-}CH(OH){-}CH_3$

32. Write the structure for the expected product from each reaction.
a. $CH_2{=}CHCH_2CH_3 + Br_2 \longrightarrow$
b. $CH_3CH{=}CHCH_3 + I_2 \longrightarrow$
c. $CH_3CH{=}CHCH_3 + H_2 \longrightarrow$
d. cyclohexene $+ Cl_2 \longrightarrow$

33. Write structures and names of the products obtained upon addition of each of the following reagents to ethene.
a. HBr **b.** Cl_2 **c.** H_2O **d.** H_2 **e.** HCl

34. Name the following ethers.
a. $CH_3OCH_2CH_3$
b. $C_6H_5{-}O{-}CH_2CH_3$
c. $CH_2{=}CHOCH{=}CH_2$
d. $CH_3CH(CH_3)OCH(CH_3)CH_3$

23.3 Carbonyl Compounds

35. Name these aldehydes and ketones.
a. $CH_3C(=O)CH_3$
b. $CH_3CH(CH_3)CH_2CHO$
c. $C_6H_5{-}CH_2CHO$
d. CH_3CHO
e. $C_6H_5{-}C(=O){-}C_6H_5$
f. $CH_3CH_2C(=O)CH_2CH_2CH_3$

23.4 Polymerization

36. Different samples of a polymer such as polyethylene can have different properties. Explain.

37. Draw the structure of the repeating units in a polymer that has the following monomers.
a. 1-butene **b.** 1,2-dichloroethene

CHAPTER 23

Assessment

Reviewing Content

26. a carbon chain or ring

27. a. $ClCH_2CCl_2CH_2CH_3$
b. 1,3,5-tribromobenzene (benzene ring with Br at 1, 3, 5)
c. cyclohexane ring with Cl on adjacent carbons

28. a. 3-chloropropene
b. 1,2-dichloro-4-methylpentane
c. 1,3-dibromobenzene

29. a. $CHCl_2{-}CH_2{-}CH_3$
1, 1-dichloropropane
$CH_2Cl{-}CHCl{-}CH_3$
1, 2-dichloropropane
$CH_2Cl{-}CH_2{-}CH_2Cl$
1, 3-dichloropropane
$CH_3{-}CCl_2{-}CH_3$
2, 2-dichloropropane

b. $CH_3{-}CH_2{-}CH_2{-}CH_2Br$
1-bromobutane
$CH_3{-}CH_2{-}CHBr{-}CH_3$
2-bromobutane
$CH_3{-}CH(CH_3){-}CH_2Br$
1-bromo-2-methylpropane
$CH_3{-}CBr(CH_3){-}CH_3$
2-bromo-2-methylpropane

30. a. $C_6H_5{-}OH$ **b.** $C_6H_{11}{-}OH$
c. $CH_3{-}CH(CH_3){-}OH$ **d.** $C_6H_5{-}Br$

31. a. 2-propanol
b. 1,2-propanediol

32. a. $CH_2BrCHBrCH_2CH_3$ **b.** $CH_3CHICHICH_3$
c. $CH_3CH_2CH_2CH_3$ (H H added) **d.** 1,2-dichlorocyclohexane

33. a. $CH_2(H){-}CH_2Br$ bromoethane **b.** $CH_2Cl{-}CH_2Cl$ 1, 2-dichloroethane
c. $CH_2(H){-}CH_2OH$ ethanol **d.** $CH_2(H){-}CH_2(H)$ ethane **e.** $CH_2(H){-}CH_2Cl$ chloroethane

34. a. ethylmethyl ether
b. ethylphenyl ether
c. divinyl ether or vinyl ether
d. diisopropyl ether or isopropyl ether

35. a. propanone or acetone
b. 3-methylbutanal
c. 2-phenylethanal
d. ethanal or acetaldehyde
e. diphenylmethanone or diphenyl ketone or benzophenone
f. 3-hexanone or ethylpropylketone

CHAPTER 23

Assessment continued

36. The properties of polyethylene vary with the length of the chains.

37. a. $-CH_2-CH(CH_2CH_3)-$ **b.** $-CH(Cl)-CH(Cl)-$

Understanding Concepts

38. a. R–X **b.** $R-C(=O)-R$

c. $R-C(=O)-O-R$ **d.** $R-C(=O)-N(H)-R$

39. c, a, b

40. a. $CH_3-C(=O)-OCH_3 + H_2O$

b. $CH_3CH_2CH_2-C(=O)-O^-Na^+ + CH_3CH_2OH$

c. $CH_3-C(=O)-H$

41. Both atoms in a carbon-carbon double bond have the same electronegativity, so the bond is nonpolar. Because oxygen is more electronegative than carbon, a carbon-oxygen bond is very polar.

42. a. $HCOO^-K^+ + H_2O$

b. $CH_3CH_2COO^-Na^+ + H_2O$

c. $CH_3COO^-Na^+ + H_2O$

43. a. phenol

b. ether

c. alcohol

d. phenol

e. alcohol

44. a. $CH_3CH_2CH(Cl)-CH_2(Cl)$

b. $CH_3CH_2CH(Br)-CH_2(Br)$

c. cyclohexane ring with H and Br on adjacent carbons

45. a. carboxyl group, ethanoic acid (acetic acid)

b. ketone (carbonyl group), propanone (acetone)

c. ether, diethyl ether (ethyl ether)

d. alcohol (hydroxyl group), ethanol (ethyl alcohol)

CHAPTER 23

Assessment continued

Understanding Concepts

38. Write a general structure for each type of compound.

a. halocarbon **b.** ketone

c. ester **d.** amide

39. Place the following compounds in order from lowest boiling point to highest boiling point. Molar masses are given in parentheses.

a. CH_3CHO (44) **b.** CH_3CH_2OH (46)

c. $CH_3CH_2CH_3$ (44)

40. Write the structure and name of the expected products for each reaction.

a. $CH_3COOH + CH_3OH \xrightarrow{H^+}$

b. $CH_3CH_2CH_2COOCH_2CH_3 + H_2O \xrightarrow{NaOH}$

c. $CH_3CH_2OH \xrightarrow{K_2Cr_2O_7}$

41. Explain why a carbon–carbon double bond is nonpolar, but a carbon–oxygen double bond is very polar.

42. What are the products of each reaction?

a. $HCOOH + KOH \longrightarrow$

b. $CH_3CH_2COOH + NaOH \longrightarrow$

c. $CH_3COOH + NaOH \longrightarrow$

43. Classify each compound as an alcohol, a phenol, or an ether.

a. naphthalene ring–OH

b. benzene ring–O–benzene ring

c. benzene ring–CH_2OH

d. benzene ring–OH

e. $CH_3CH_2CH(OH)CH_3$

44. Write the structural formulas for the products of these reactions.

a. $CH_3CH_2CH{=}CH_2 + Cl_2 \longrightarrow$

b. $CH_3CH_2CH{=}CH_2 + Br_2 \longrightarrow$

c. cyclohexene + HBr $\longrightarrow$

45. For each compound pictured, identify the functional group and name the compound. The red atoms represent oxygen.

a.

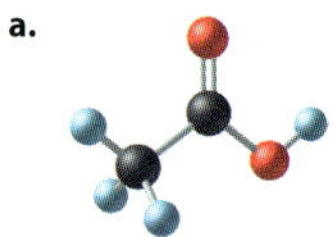

b.

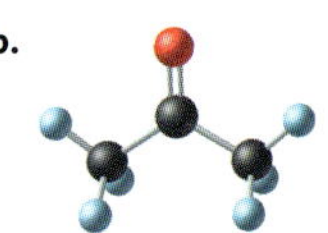

c.

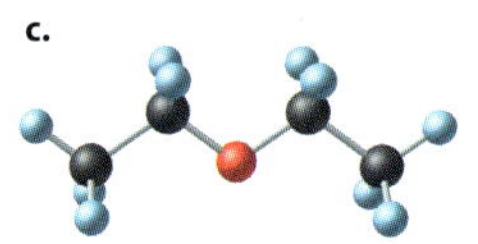

d.

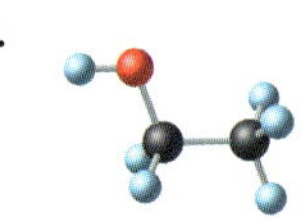

46. Write the name and structure for the alcohol that must be oxidized to make each carbonyl compound.

a. HCHO **b.** $CH_3C(=O)CH_3$

c. $CH_3CH(CH_3)CHO$ **d.** cyclohexanone

47. Write the structures of the expected products for these reactions.

a. $CH_3CH_2COOCH_2CH_3 \xrightarrow{NaOH}$

b. CH_3COO–benzene ring $\xrightarrow{KOH}$

c. $CH_3CH_2COOCH_2CH(CH_3)CH_3 \xrightarrow{HCl}$

46. a. CH_3OH methanol **b.** $CH_3-CH(OH)-CH_3$ 2-propanol

c. $CH_3-CH(CH_3)-C(H)(H)-OH$ 2-methyl-1-propanol **d.** cyclohexane ring with OH and H on one carbon, cyclohexanol

47. a. $CH_3CH_2COO^-Na^+$, CH_3CH_2OH

b. $CH_3COO^-K^+$, benzene ring–OH

c. CH_3CH_2COOH, $CH_3CH(CH_3)CH_2OH$

Critical Thinking

48. Complete the following reactions by writing the structures of the expected products and by naming the reactants and products.

a. $CH_3COOCH_3 + H_2O \xrightarrow{HCl}$

b. $CH_3CH_2CH_2COOCH_2CH_2CH_3 + H_2O \xrightarrow{NaOH}$

c. $HCOOCH_2CH_3 + H_2O \xrightarrow{KOH}$

49. Benzene is poisonous and a proven carcinogen. Yet many compounds containing benzene rings, such as benzaldehyde, are common in the foods you eat. Why are some organic compounds with phenyl groups safe to eat?

50. Explain why diethyl ether is more soluble in water than dihexyl ether. Would you expect propane or diethyl ether to be more soluble in water? Why?

51. Explain why 1-butanol has a higher boiling point than diethyl ether. Which compound would you expect to be more soluble in water? Why?

52. Propane ($CH_3CH_2CH_3$) and acetaldehyde (CH_3CHO) have the same molar mass, but propane boils at −42°C and acetaldehyde boils at 21°C. Account for this difference.

53. How would you expect the water solubility of ethanoic and decanoic acids to compare?

54. The processes used to synthesize many organic compounds often use compounds that contain double or triple bonds as reactants. Explain why the use of unsaturated reactants might be an advantage over using saturated reactants.

55. Cadaverine (1,5-diaminopentane) and putrescine (1,4-diaminobutane) are unpleasant-smelling compounds that are formed by bacteria in rotting flesh. Draw their structures. What class of organic compounds are these two chemicals?

56. For the following compounds write a chemical equation showing how to produce the compound.

a.

b.

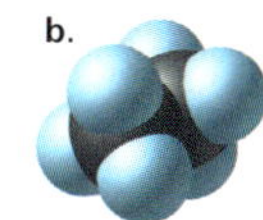

Concept Challenge

57. Cholesterol is a compound in your diet and is also synthesized in the liver. Sometimes it is deposited on the inner walls of blood vessels, causing hardening of the arteries. Describe the structural features and functional groups of this important molecule.

58. Hydrocarbons from petroleum are an important source of raw material for the chemical industry. Using reactions covered in this chapter and any required inorganic chemicals, propose a scheme for the manufacture of ethylene glycol, a major component of antifreeze, from petrochemical ethene.

59. Human hair is composed of long-chain polymers. Some of the monomers in these polymers contain sulfur atoms. When two sulfur atoms are adjacent to one another, they form strong, covalent disulfide (S—S) bonds that can link two polymer molecules together. The location of the cross links between the hair polymers affects how curly or straight the hair is and helps to hold the polymers in their shape. Waving lotion that is used in permanent waves is a reducing agent, and neutralizing agent is an oxidizing agent. Using this information, explain in terms of chemistry how a permanent wave can change the shape of hair. Describe what chemical steps should be taken to change the shape of someone's hair from straight to curly.

60. Which of the following monomers would be used to produce the addition polymer $-(CF_2)_x-$: $CH_2{=}CF_2$, $CH_2{=}CHF$, CF_4, $CF_2{=}CF_2$, or $CHF{=}CHF$?

Critical Thinking

48. a. $CH_3COOCH_3 + H_2O \xrightarrow{HCl}$
methyl ethanoate (methyl acetate), water

$CH_3-C(=O)-OH + CH_3OH$
ethanoic acid (acetic acid), methanol

b. $CH_3CH_2CH_2COOCH_2CH_2CH_3 + H_2O \xrightarrow{NaOH}$
propyl butanoate (propyl butyrate), water

$CH_3CH_2CH_2-C(=O)-O^-Na^+ + CH_3CH_2CH_2OH$
sodium butanoate (sodium butyrate), 1-propanol

c. $HCOOCH_2CH_3 + H_2O \xrightarrow{KOH}$
ethyl methanoate (ethyl formate), water

$H-C(=O)-O^-K^+ + CH_3CH_2OH$
potassium methanoate (potassium formate), ethanol

49. The chemical properties (and toxicity) of organic compounds are determined by the compound as a whole. As a substituent in a molecule, a phenyl group ring does not have the same properties as benzene.

50. The oxygen atom in diethyl ether polarizes the small molecule. This enables diethyl ether to dissolve in water, which is also polar. The large dihexyl ether molecule has large nonpolar parts and does not dissolve. Propane is less soluble in water than is diethyl ether because propane is nonpolar.

51. The alcohol molecules form hydrogen bonds with one another, resulting in a higher boiling point. They also form hydrogen bonds with water molecules, causing 1-butanol to be more soluble than diethyl ether. (Although diethyl ether is polar, 1-butanol has greater polarity.)

52. Acetaldehyde is polarized by its carbonyl oxygen forming stronger intermolecular attractions. Nonpolar propane has weak intermolecular attractions. Thus, propane molecules are more easily liberated from the liquid state.

53. The short-chain ethanoic acid has a higher water solubility.

54. Substitution of an alkane by a halogen, for example, usually gives a mixture of products. Addition to a double bond or a triple can give a single addition product.

55. $H_2NCH_2(CH_2)_3CH_2NH_2$ cadaverine; $H_2NCH_2(CH_2)_2CH_2NH_2$ putrescine; both compounds are amines.

56. a. $CH_4 + Cl_2 \xrightarrow{\text{UV light}} CH_3Cl + HCl$

b. $CH_2{=}CH_2 + H_2 \xrightarrow{Pt} CH_3CH_3$

Concept Challenge

57. Cholesterol is an alcohol with a hydroxyl group on a cycloalkane. It has four nonaromatic rings. It has a double bond in one of its rings, as well as a large alkyl group, making it nonpolar.

58. $CH_2CH_2(g) + Br_2(l) \rightarrow CH_2BrCH_2Br(l)$
$CH_2BrCH_2Br(l) + 2NaOH(aq) \rightarrow CH_2OHCH_2OH(l) + 2NaBr(aq)$

59. Waving lotion reduces —S—S— bonds to —SH bonds. Hair can be placed in curlers to form the hair in the desired shape. The neutralizing agent is an oxidizing agent that reforms —S—S— bonds, locking the hair into its curly shape. Similar steps could be used to straighten curly hair.

60. $CF_2{=}CF_2$

Cumulative Review

61. b. 3

62. **a.** $:\ddot{F}\cdot + e^{-} \longrightarrow :\ddot{F}:^{-}$

b. $H\cdot + \cdot\ddot{O}: + e^{-} \longrightarrow (:\ddot{O}:H)^{-}$

63. 2.86 g SO_2

64. 1.15×10^2 kPa

65. Anhydrous $Na_2CO_3(s)$ is the better value; the decahydrate is 63.0% water.

66. The amount of dissolved oxygen is greater entering the plant; solubilities of gases tend to decrease with increasing temperature.

67. 0.117*M* $Ca(NO_3)_2$

68. At any given moment, the rate of dissolving of solute is equal to the rate of precipitation of solute. As a result, the concentration of the solution remains constant.

69. 71 kJ

70. **a.** shift toward products
b. shift toward reactants
c. shift toward reactants
d. shift toward reactants

71. a, c, d, b

72. **a.** Na +1; N +3; O −2
b. Co +2; S +6; O −2
c. Se +4; O -2
d. Zn +2; O -2; H +1
e. K +1; Pt +2; Cl −1

73. Oxidized: H of BH_4 Reduced: H of H_2O unaffected: Na, B, O

74. spontaneous redox reactions

75. Reduction always occurs at the cathode. In the electrolytic cell, the cathode is the negative electrode.

Cumulative Review

61. What is the maximum number of orbitals in the *p* sublevel of an atom? (*Chapter 5*)
a. 1 **b.** 3 **c.** 5 **d.** 9

62. Using electron dot structures, illustrate the formation of F^- from a fluorine atom and a hydroxide ion from atoms of hydrogen and oxygen. (*Chapter 7*)

63. Calculate the mass, in grams, of one liter of SO_2 at standard temperature and pressure. (*Chapter 12*)

64. A sample of 1.40 L of nitrogen gas in a sealed container at 25.0°C and a pressure of 1.00×10^2 kPa is heated to 68.7°C. What is the new pressure? (*Chapter 14*)

65. Sodium carbonate is often sold as the anhydrous compound [$Na_2CO_3(s)$] or as the decahydrate [$Na_2CO_3{\cdot}10H_2O(s)$]. If the price per kilogram of the anhydrous and decahydrate forms are the same, which compound is the better value? (*Chapter 15*)

66. Assume that water enters a power plant at 20°C and leaves at 30°C. Will the amount of dissolved oxygen in the water be greater entering or leaving the plant? (*Chapter 16*)

67. A solution is made by diluting 250 mL of 0.210*M* $Ca(NO_3)_2$ solution with water to a final volume of 450 mL. Calculate the molarity of $Ca(NO_3)_2$ in the diluted solution. (*Chapter 16*)

68. In a saturated solution containing undissolved solute, the solute is continually dissolving, but the solution concentration remains constant. Explain. (*Chapter 16*)

69. A 500-g aluminum tray at 22°C is heated to 180°C in an oven. How many kJ of heat does the pan absorb if the specific heat of aluminum is 0.90 J/(g·°C)? (*Chapter 17*)

70. Predict the direction of shift in the equilibrium position for each change in conditions. (*Chapter 18*)

$$2NO_2(g) \rightleftharpoons 2NO(g) + O_2(g)$$

a. O_2 partial pressure decreased
b. total pressure increased
c. O_2 partial pressure increased
d. NO partial pressure increased

71. List these K_a values for weak acids in order of increasing acid strength. (*Chapter 19*)
a. 3.5×10^{-6}
b. 2.7×10^{-3}
c. 1.5×10^{-5}
d. 6.6×10^{-5}

72. Assign an oxidation number to each atom in these compounds. (*Chapter 20*)
a. $NaNO_2$
b. $CoSO_4$
c. SeO_2
d. $Zn(OH)_2$
e. K_2PtCl_4

73. Solid sodium borohydride ($NaBH_4$) is being studied as a possible source of hydrogen fuel for hydrogen-powered vehicles. The borohydride reacts with water to produce hydrogen gas and sodium metaborate. Identify which atoms of the reactants are oxidized, which are reduced, and which are unaffected in this reaction. (*Chapter 20*)

$$NaBH_4(s) + 2H_2O(l) \longrightarrow 4H_2(g) + NaBO_2(aq)$$

74. What is the source of the electrical energy produced in a voltaic cell? (*Chapter 21*)

75. At which electrode in an electrolytic cell does reduction always occur? What is the charge on this electrode? (*Chapter 21*)

76. Draw a condensed structural formula for each compound. (*Chapter 22*)
a. 1,2-dimethylcyclobutane
b. 2-methyl-2-pentene
c. 2-butene
d. 2-pentyne

77. Is petroleum or coal most likely to be a good source of aromatic compounds? (*Chapter 22*)

78. Which of these statements applies to ethene? (*Chapter 22*)
a. saturated hydrocarbon
b. H—C—H bond angle of 120°
c. alkene
d. aromatic compound

76. **a.**

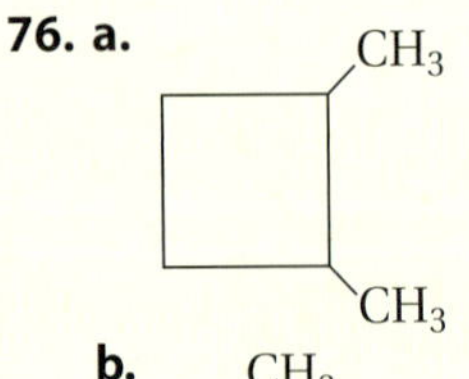

b.

$$CH_3C(CH_3){=}CHCH_2CH_3$$

c. $CH_3CH{=}CHCH_3$

d. $CH_3C{\equiv}CCH_2CH_3$

77. coal

78. b, c

Standardized Test Prep

Test-Taking Tip

Eliminate Wrong Answers If you don't know which response is correct, start by eliminating those you know are wrong. If you can rule out some choices, you'll have fewer left to consider and you'll increase your chances of choosing the correct answer.

Select the choice that best answers each question or completes each statement.

1. The acid-catalyzed hydrolysis of an ester gives a carboxylic acid and
   a. an amine. **b.** an ether.
   c. an alcohol. **d.** an alkene.
2. Ethane, methanal, and methanol have similar molar masses. Which series lists the compounds in order of increasing boiling point?
   a. ethane, methanal, methanol
   b. methanal, methanol, ethane
   c. methanol, methanal, ethane
   d. ethane, methanol, methanal
3. A carbonyl group is characterized by a
   a. carbon–carbon double bond.
   b. carbon–oxygen double bond.
   c. carbon–nitrogen single bond.
   d. carbon–oxygen single bond.

The lettered choices below refer to Questions 4–7. A lettered choice may be used once, more than once, or not at all.

(A) alcohol (D) ether
(B) ketone (E) aldehyde
(C) carboxylic acid

To which class of organic compounds does each of the following compounds belong?

4. CH_3CH_2COOH
5. $CH_3CH_2CH_2OH$
6. $CH_3CH_2OCH_3$
7. CH_3COCH_3

Use the space-filling models with Questions 8 and 9.

a.

b.

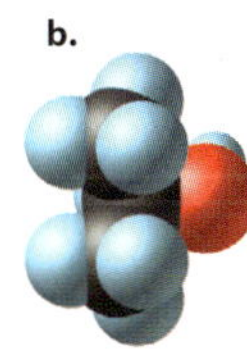

c.

8. Using models a, b, and c as guides, draw ball-and-stick models of the three compounds that have the formula C_3H_8O. Name each compound.
9. There are two compounds with a carbonyl group that have the molecular formula C_3H_6O. Write a complete structural formula and draw a ball-and-stick model for each compound. Name each compound.

Characterize the reactions in Questions 10–14 as addition, esterification, oxidation, polymerization, or substitution reactions.

10. $CH_3CHO \xrightarrow[H_2SO_4]{K_2Cr_2O_7} CH_3COOH$
11. $CH_2{=}CH_2 + HCl \longrightarrow CH_3CH_2Cl$
12. $CH_3CO_2H + CH_3CH_2OH \xrightarrow{H^+} CH_3COOCH_2CH_3 + H_2O$
13. $x CH_2{=}CH_2 \longrightarrow H{+}CH_2-CH_2{+}_x H$
14. benzene $+ Br_2 \xrightarrow{\text{catalyst}}$ bromobenzene (ring with Br) $+ HBr$

For each question there are two statements. Decide whether each statement is true or false. Then decide whether Statement II is a correct explanation for Statement I.

	Statement I		Statement II
15.	The addition of hydrogen to an alkene is a reduction reaction.	BECAUSE	The addition of hydrogen to any molecule is a reduction reaction.
16.	Aldehydes are easily oxidized.	BECAUSE	Oxidation of aldehydes produces alcohols.
17.	Ethanol (CH_3CH_2OH) is immiscible in water in all proportions.	BECAUSE	Ethanol molecules can form hydrogen bonds with other ethanol molecules.
18.	The hydrogenation of benzene is easier than the hydrogenation of ethene.	BECAUSE	The benzene ring is stabilized by resonance.

Standardized Test Prep

1. c
2. a
3. b
4. C
5. A
6. D
7. B
8. a.

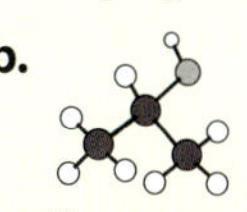

1-propanol

b.

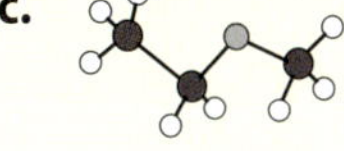

2-propanol

c.

ethylmethyl ether

9. Propanal and propanone are also known as dimethyl ketone and acetone. In the models, the black spheres represent carbon, the gray spheres represent oxygen, and the white spheres represent hydrogen.

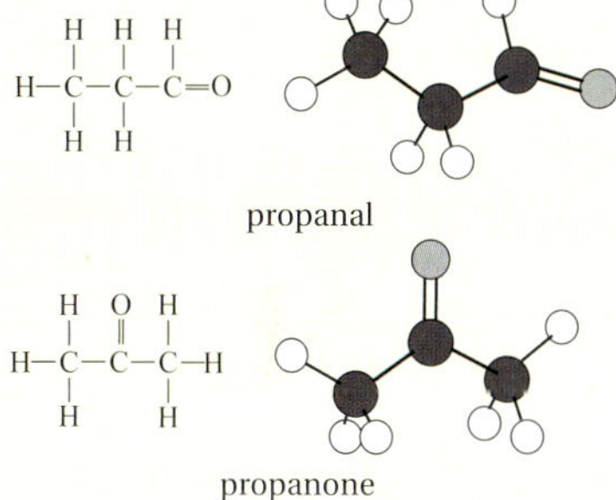

propanal

propanone

10. oxidation
11. addition
12. esterification
13. polymerization
14. substitution
15. True, True, correct explanation
16. True, False
17. False, True
18. False, True

CHAPTER 24

The Chemistry of Life Planning Guide

SECTION OBJECTIVES	STANDARDS NATIONAL	STANDARDS STATE	ACTIVITIES and LABS	
24.1 A Strategy for Life, pp. 763–765 1 block or 2 periods **24.1.1 Identify** the two major cell types that occur in nature. **24.1.2 Describe** the chemical changes that occur during photosynthesis.	C-1, C-3, C-4		**SE** Inquiry Activity: Biological Catalysis, p. 762 **TE** Class Activity: Living and Nonliving Things, p. 764	L2 L2
24.2 Carbohydrates, pp. 766–768 1 block or 2 periods **24.2.1 Identify** where glucose is found in nature. **24.2.2 Describe** how two simple sugars can be linked.	B-2, C-1		**TE** Teacher Demo: Glucose and Fructose Models, p. 767 **TE** Teacher Demo: Benedict's Test, p. 767	L2 L2
24.3 Amino Acids and Their Polymers, pp. 769–774 1 block or 2 periods **24.3.1 Diagram** the structure of an amino acid. **24.3.2 Describe** how peptide bonds form and **identify** what determines the properties of peptides and proteins. **24.3.3 Describe** how enzymes affect biochemical reactions.	A-1, B-2, B-3, C-1		**SE** Small-Scale Lab: The Egg: A Biochemical Storehouse, p. 774 **TE** Teacher Demo: Peptide Models, p. 770 **TE** Teacher Demo: Protein in Hard Tissue, p. 771 **TE** Teacher Demo: Enzymes, p. 772	L2 L2 L2 L2
24.4 Lipids, pp. 775–777 1 block or 2 periods **24.4.1 Identify** the physical property that distinguishes lipids from other biological molecules. **24.4.2 Describe** the structure of a lipid bilayer.	B-2, B -3, C-1		**TE** Class Activity: Better Butter **LM** Lab 51: Preparation of Soap **SSLM** Lab 40: Reactions of Biomolecules	L2 L2 L2
24.5 Nucleic Acids, pp. 778–785 1 block or 2 periods **24.5.1 Identify** the functions of DNA and RNA. **24.5.2 Describe** how information is sorted in genetic material and how genes can mutate. **24.5.3 Describe** how DNA fingerprinting and recombinant DNA technology are used.	A-1, A-2, C-1, C-2, C-3, F-6		**SE** Quick Lab: A Model of DNA, p. 780 **TE** Class Activity: Genetic Code, p. 782	L2 L2
24.6 Metabolism, pp. 786–791 1 block or 2 periods **24.6.1 Describe** the function of ATP in cells. **24.6.2 Distinguish** between catabolism and anabolism. **24.6.3 Describe** the nitrogen cycle.	A-2, B-3, C-1, C-4, C-5, E-2, F-4			

Ability Levels	
L1	For students who need additional help
L2	For all students
L3	For students who need to be challenged

Components							
SE	Student Edition	**LM**	Laboratory Manual	**GCP**	Graphing Calculator Problems	**T**	Transparencies
TE	Teacher's Edition	**SSLM**	Small-Scale Chemistry Lab Manual	**TTT**	Test-Taking Tips with Transparencies, Grades 9–12 Science	**CA**	Chem Alive and Chemistry Field Trips DVD
GRSW	Guided Reading and Study Workbook	**LP**	Laboratory Practicals	**STP**	Standardized Test Preparation Workbook	**iT**	Interactive Textbook with ChemASAP
CTR	Core Teaching Resources	**PLM**	Probeware Lab Manual			**GO**	Go Online Internet Resources
SSG	Spanish Study Guide	**VCL**	Virtual Chemistry Labs				
		CTB	Computer Test Bank				

RESOURCES PRINT and TECHNOLOGY			PROBLEMS and ASSESSMENT
GRSW	Chapter 24, Section 1	L1	**SE** Section 24.1 Assessment, p. 765
	T T271–T272	L2	**iT** Section 24.1
GRSW	Chapter 24, Section 2	L2	**SE** Section 24.2 Assessment, p. 768
	T T273–T274	L2	**iT** Section 24.2
Go Online	**GO** Chapter 24, Section 2	L2	
GRSW	Chapter 24, Section 3	L1	**SE** Section 24.3 Assessment, p. 773
	T T275–T278	L2	**iT** Section 24.3
Go Online	**GO** Chapter 24, Section 3	L2	
GRSW	Chapter 24, Section 4	L1	**SE** Section 24.4 Assessment, p. 777
	T T279–T280	L2	**iT** Section 24.4
Go Online	**GO** Chapter 24, Section 4	L2	
GRSW	Chapter 24, Section 5	L1	**SE** Section 24.5 Assessment, p. 785
	T T281–T284	L2	**iT** Section 24.5
GRSW	Chapter 24, Section 6	L1	**SE** Section 24.6 Assessment, p. 790
	T T285	L2	**iT** Section 24.6

Web Code: cdn-1242
Web Code: cdn-1243
Web Code: cdn-1244

MATERIALS List

Quantities for each group

STUDENT EDITION

Inquiry Activity, p. 762
2 small pieces of red meat, such as ground beef, one raw and one cooked; 2 paper plates; small amount of 3% hydrogen peroxide; dropper

Small-Scale Lab, p. 774
chicken egg, ruler, balance

Quick Lab, p. 780
cardboard tube from paper-towel roll; felt-tip markers, two colors; metric ruler; thumbtack; 10 toothpicks

TEACHER'S EDITION

Class Activity, p. 764
photographs of living and nonliving things

Teacher Demo, p. 767
foam balls and toothpicks or molecule kit

Teacher Demo, p. 767
hot-water bath, 2–5 mL of 1% carbohydrate solution, 2 test tubes, distilled water, 2–3 mL of Benedict's reagent, dropper, samples of mono- and polysaccharides

Teacher Demo, p. 770
models of amino acids

Teacher Demo, p. 771
thigh bone from cooked poultry, 1*M* HCl, beaker

Teacher Demo, p. 772
2 cartons of fresh milk, Lactaid®, water, dropper, spoons or small cups

Class Activity, p. 776
butter, regular margarine, low-calorie margarine, 3 test tubes, balance, warm-water bath

Chapter Assessment

CHAPTER ASSESSMENT

SE Chapter assessment, pp. 793–796
CTB Chapter 24
iT Chapter 24

STANDARDIZED TEST PREP

SE Chapter 24, p. 797
STP Topic 10
iT Interactive Textbook with Assessment at PHSchool.com

CHAPTER 24

The Chemistry of Life

Biological Catalysts L2

Purpose

Students infer the action of a catalyst that is destroyed by heat.

Prep Time 20 minutes
Class Time 10 minutes

Think About It

1. Students should observe bubbles of gas when H_2O_2 is added to the raw meat, but no bubbles when H_2O_2 is added to the cooked meat.
2. Students may recall from earlier chapters that the rate of decomposition of H_2O_2 into O_2 and H_2O increases greatly in the presence of a catalyst.
3. A catalyst exists in raw meat that is destroyed when the meat is cooked. (The catalyst is called catalase.)

CHAPTER 24

The Chemistry of Life

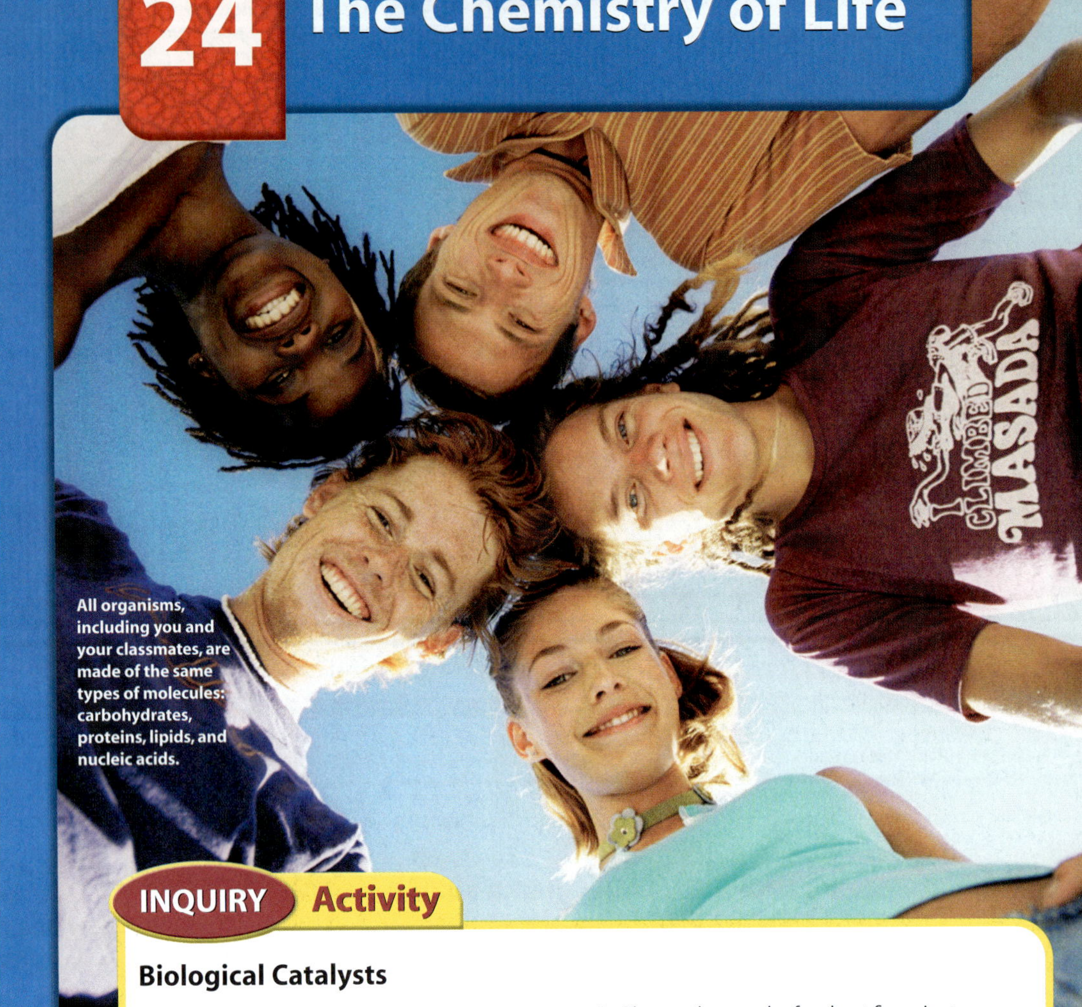

All organisms, including you and your classmates, are made of the same types of molecules: carbohydrates, proteins, lipids, and nucleic acids.

INQUIRY Activity

Biological Catalysts

Materials

2 small pieces of red meat, such as ground beef, one raw and one cooked; 2 paper plates; small amount of 3% hydrogen peroxide; dropper

Procedure  

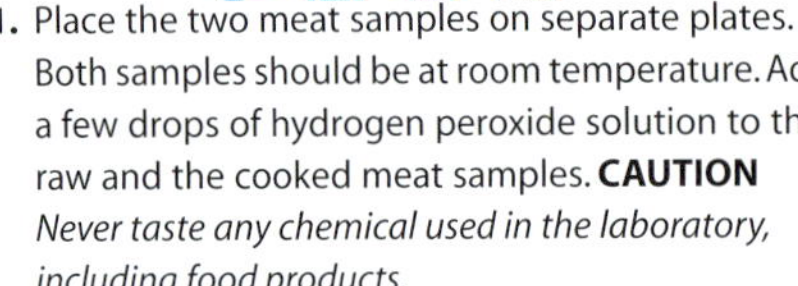

1. Place the two meat samples on separate plates. Both samples should be at room temperature. Add a few drops of hydrogen peroxide solution to the raw and the cooked meat samples. **CAUTION** *Never taste any chemical used in the laboratory, including food products.*
2. Observe the samples for about five minutes.

Think About It

1. Do you observe any difference in the appearance of the hydrogen peroxide placed on the raw meat and on the cooked meat?
2. What do you think causes the difference in the appearance of the hydrogen peroxide on the raw and cooked meats?
3. What can you conclude from this experiment?

24.1 A Strategy for Life

Connecting to Your World The air you breathe is composed mainly of nitrogen (N_2) and oxygen (O_2). Earth's early atmosphere may have been very different and most inhospitable to life. It is thought that the atmosphere changed over time as a result of photosynthesis, a process carried out by green organisms such as these cyanobacteria. One of the products of photosynthesis is oxygen, which accumulated in the atmosphere and changed the composition of air. In this section, you will learn about the characteristics of living things, including the organisms in which photosynthesis occurs.

Guide for Reading

 **Key Concepts**

- What are the two major types of cells that occur in nature?
- What compound is reduced during photosynthesis? What compounds are formed?

Vocabulary

photosynthesis

Reading Strategy

Relating Text and Visuals As you read, look carefully at Figure 24.3. In your notebook, explain how this illustration helps you understand the relationship between photosynthesis and the carbon compounds used by all organisms.

The Structure of Cells

Life! You are certainly familiar with it, but what does it really mean? Until recently, life was defined as the ability of an organism to grow and to reproduce its own kind. However, recent discoveries made at the fringes of life seem to blur this simple definition. As difficult as it is to define life, you can generally regard tiny structures called cells as the fundamental units of life. You will learn about the construction of cells and the chemical processes that cells undergo to obtain the energy they need to survive.

Organisms are composed of as few as one cell or as many as billions of cells. **Two major cell types occur in nature: the cells of bacteria, known as prokaryotic cells, and the cells of all other organisms, known as eukaryotic cells.** The prokaryotic cell is the more ancient of the two. Microscopic examination of fossilized remains shows that prokaryotic cells were present on Earth at least 3 billion years ago. Eukaryotic cells did not appear until about 1 billion years ago. Figure 24.1 shows both types of cells.

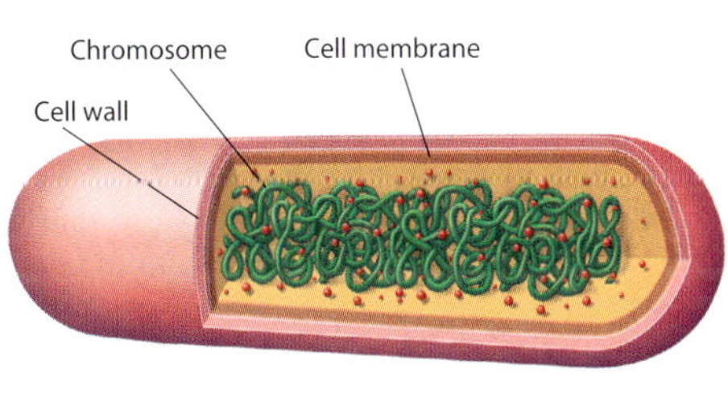

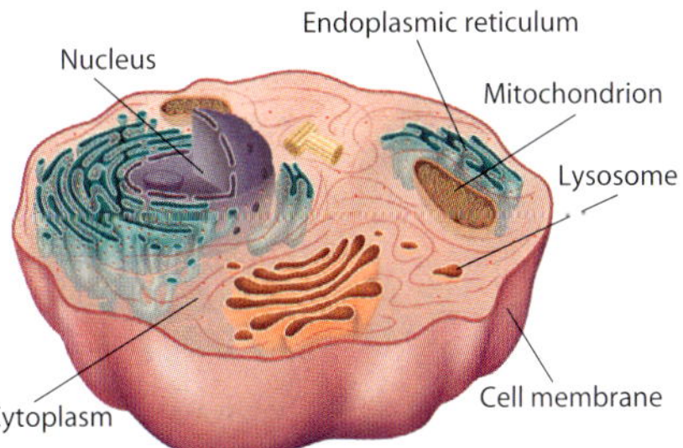

Figure 24.1 Typical prokaryotic and eukaryotic cells are shown here. Note that only the eukaryotic cell has a nucleus.
Comparing and Contrasting ***How do prokaryotic and eukaryotic cells compare in size?***

Section Resources

Print

- **Guided Reading and Study Workbook,** Section 24.1
- **Core Teaching Resources,** Section 24.1
- **Transparencies,** T271–T272

Technology

- **Interactive Textbook with ChemASAP,** Animation 29, Assessment 24.1

24.1

1 FOCUS

Objectives

24.1.1 Identify the two major cell types that occur in nature.

24.1.2 Describe the chemical changes that occur during photosynthesis.

Guide for Reading

Build Vocabulary 

Word Parts The word *photosynthesis* has two Greek roots. *Phot* means "light," and *suntithenae* means "to put together." Photosynthesis uses light to combine carbon dioxide and water into glucose.

Reading Strategy 

Active Comprehension Read Connecting to Your World aloud. Ask, **What more would you like to know about the process of photosynthesis?** Write down students' responses. Tell students to consider the questions as they read the section.

2 INSTRUCT

Connecting to Your World

Ask students to speculate about what types of gases may have dominated before the appearance of photosynthetic organisms. (Studies suggest the following: CO_2, CH_4, H_2, NH_3, N_2, and H_2O.) Ask, **What life-supporting gas does photosynthesis produce?** *(oxygen)* **What type of organisms photosynthesize?** *(many types of bacteria and almost all plant species)*

Answers to...

Figure 24.1 Eukaryotic cells are larger than prokaryotic cells.

The Structure of Cells

Use Visuals L1

Figure 24.1 Point out that eukaryotic cells can vary greatly in size and shape, but they are always larger than prokaryotic cells. Point out that many metabolic reactions in eukaryotic cells are compartmentalized. Ask, **What possible advantages would organelles provide for a cell?** *(Many of the chemical activities that occur in cells involve similar reactants and products. By nature, they are incompatible. If these processes are physically separated, they can proceed concurrently without conflict.)*

CLASS Activity

Living and NonLiving 

Have teams of students create a list of characteristics that distinguish living from nonliving systems. *(Living things have one or more cells; they obtain and consume energy; they respond to changing conditions; they grow and develop, and they can reproduce.)* Present photographs of living and nonliving things and help students use their list to classify each subject as living or nonliving.

Energy and Carbon Cycle

Relate L2

The carbon cycle plays a large part in controlling the amount of CO_2 in the atmosphere. CO_2 is a greenhouse gas that traps infrared radiation in Earth's atmosphere. One concern of climatologists is the rise in CO_2 levels in the atmosphere caused by fossil fuel burning, deforestation, and forest fires.

Both eukaryotic and prokaryotic cells contain all the chemicals necessary for life, encased in a cell membrane. The cell membrane is a sack that holds the contents of a cell and acts as a selective barrier for the passage of substances into and out of the cell. Eukaryotic cells are considerably larger and more complex than prokaryotic cells, but the chemical processes carried out by both types of cells are very similar.

One major feature that distinguishes eukaryotic cells from prokaryotic cells is the presence of membrane-enclosed organelles. Organelles, meaning little organs, are small structures suspended in the interior cellular fluid, or cytoplasm. The organelles are the sites of many specialized functions in eukaryotic cells. The nucleus, a structure that is important in eukaryotic cell reproduction, is not present in prokaryotic cells. Mitochondria (singular: mitochondrion) are the source of cellular energy in eukaryotic cells that use oxygen. Mitochondria are often referred to as the powerhouses of the cell. Lysosomes are the sites for the digestion of substances taken into a cell. Yet another membrane-enclosed structure in eukaryotic cells is the highly folded, netlike endoplasmic reticulum (ER). Among its various functions, the ER serves as an attachment site for ribosomes. The ribosomes, small organelles that are not membrane-enclosed, are the sites where essential substances called proteins are made.

Which organelle is involved in cell reproduction?

The Energy and Carbon Cycle

Interactive Textbook

Animation 29 Explore the complexity of chemicals essential to all life.

with ChemASAP

Organisms must have energy to survive. The ultimate source of this energy is the sun. Cells of green plants and certain algae contain organelles called chloroplasts that are able to capture solar energy and make food. Figure 24.2 shows a chloroplast from the cell of a corn leaf. Within the chloroplast is a light-capturing system that converts light energy into chemical energy by a process called **photosynthesis.** In addition to sunlight, photosynthetic organisms require carbon dioxide and water. **Photosynthesis uses the energy from sunlight to reduce carbon dioxide to compounds that contain C—H bonds, mainly in the form of glucose ($C_6H_{12}O_6$).** The following equation describes the process.

$$\underset{\text{Carbon dioxide (carbon in more oxidized state)}}{6CO_2} + \underset{\text{Water}}{6H_2O} + \underset{\text{from sunlight}}{\text{Energy}} \longrightarrow \underset{\text{Glucose (carbon in more reduced state)}}{C_6H_{12}O_6} + \underset{\text{Oxygen}}{6O_2}$$

Figure 24.2 The reactions of photosynthesis take place on the inner membranes and in the spaces between membranes of a chloroplast.

Figure 24.3 illustrates the relationship between photosynthesis and the carbon compounds used by all organisms. In the energy and carbon cycle, photosynthetic organisms produce necessary carbon compounds. Animals, which do not carry out photosynthesis, get these carbon compounds by eating plants or by eating animals that feed on plants. Both plants and animals get energy by unleashing the energy stored in the chemical bonds of these carbon compounds. The nutrients are oxidized back to carbon dioxide and water in the process.

$$\underset{\text{Glucose (carbon in more reduced state)}}{C_6H_{12}O_6} + \underset{\text{Oxygen}}{6O_2} \longrightarrow \underset{\text{Carbon dioxide (carbon in more oxidized state)}}{6CO_2} + \underset{\text{Water}}{6H_2O} + \text{Energy}$$

Facts and Figures

Studying Photosynthesis

Melvin Calvin (1911–1997) and others in his laboratory were important participants in tracing the chemical reactions that take place in photosynthesis. Their work was made possible by two techniques: using radioactive carbon-14 to follow carbon's path through the complex process, and chromatography that identified the products at each step.

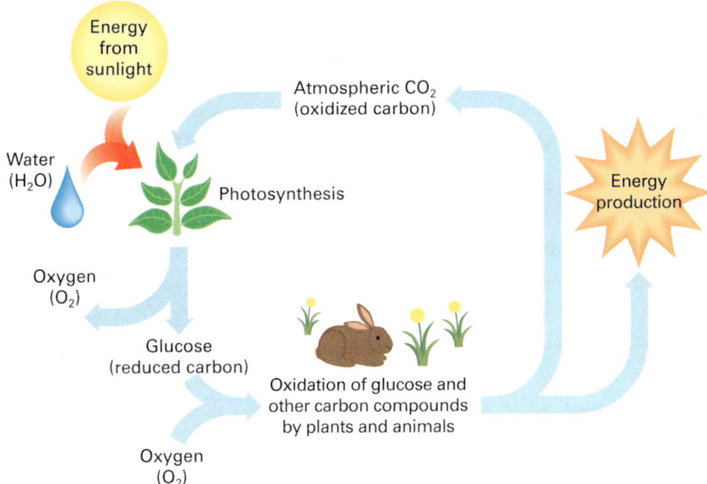

Figure 24.3 In the energy and carbon cycle, photosynthesis and the oxidation of glucose are responsible for the major transformations and movements of carbon. Plants, such as rain forest trees, release oxygen into the atmosphere through photosynthesis. **Interpreting Diagrams** ***In which part of the cycle is oxygen released?***

Thus, although plant life could survive without animals, animal life could never survive without plants. Without photosynthesis, the supply of carbon compounds that animals need to get energy would not exist.

Oxygen also is an important product of photosynthesis. Photosynthetic organisms produce the oxygen found in Earth's atmosphere. The oxygen is required for most organisms to sustain life. The importance of photosynthetic organisms in producing carbon compounds and oxygen is a major reason for the concern about the loss of such organisms through the destruction of rain forests.

All biological processes, including photosynthesis, are based on certain essential kinds of chemical substances. Surprisingly, the great complexity of life arises from just a few types of biological molecules. In the remainder of this chapter, you will learn about the molecular structures of the great classes of biological molecules and the roles they play in the body.

24.1 Section Assessment

1. **Key Concept** What two types of cells occur in nature?
2. **Key Concept** What chemical changes occur during photosynthesis?
3. What are the fundamental units of life?
4. Describe the structure of a eukaryotic cell.
5. What is the function of chloroplasts in green plants and algae?
6. Write an equation that describes the oxidation of glucose.
7. Explain how carbon moves through the environment.

Elements Handbook

Greenhouse Gases Read the feature on greenhouse gases on page R22. Describe how the energy and carbon cycle normally keeps atmospheric CO_2 levels balanced. What factors have disrupted this balance and caused atmospheric CO_2 levels to rise?

Interactive Textbook

Assessment 24.1 Test yourself on the concepts in Section 24.1.

with **ChemASAP**

3 ASSESS

Evaluate Understanding 

Write the equation for photosynthesis on the board. Ask, **Should this reaction be classified as acid–base, redox, or combination?** *(redox; CO_2 is reduced and H_2O is oxidized.)* **Is photosynthesis endothermic or exothermic?** *(endothermic)* **Which part of the plant cell specializes in the light-harvesting reactions of photosynthesis?** *(the chloroplast)*

Reteach 

Have students examine Figure 24.3. Explain that the diagram represents the flow of energy and nutrients through Earth's biosphere, the regions on Earth where living organisms exist. Point out that energy is required to maintain the highly organized chemical environments found in living cells. Ask, **Where does the energy come from that drives the cycle?** *(the Sun)* **Why are plants and other photosynthetic organisms called "producers" and animals called "consumers"?** *(Photosynthetic organisms produce high-energy forms of carbon compounds that are consumed by animals.)*

Elements Handbook

CO_2 is released into the air by cellular respiration and removed from the air during photosynthesis. The burning of fossil fuels and the clearing of forests for agriculture have contributed to increased CO_2 levels in the atmosphere.

Interactive Textbook

If your class subscribes to the Interactive Textbook, use it to review key concepts in Section 24.1.

with **ChemASAP**

Section 24.1 Assessment

1. eukaryotic cells and prokaryotic cells
2. CO_2 is reduced to glucose and oxygen is released.
3. cells
4. A eukaryotic cell has a nucleus and other membrane-enclosed organelles.
5. Chloroplasts capture solar energy and make food.
6. $6CO_2 + 6H_2O + \text{Sunlight} \rightarrow C_6H_{12}O_6 + 6O_2$
7. Carbon moves through the environment in the carbon and energy cycle: Through photosynthesis, plants convert carbon dioxide to carbon compounds used by for energy and growth; animals ingest plant products for energy, producing carbon dioxide in the process, which returns to the atmosphere.

Answers to...

Figure 24.3 photosynthesis

Checkpoint The nucleus.

24.2

1 FOCUS

Objectives

24.2.1 Identify where glucose is found in nature.

24.2.2 Describe how two simple sugars can be linked.

Guide for Reading

Build Vocabulary L2

Graphic Organizers As you learn the new vocabulary, make a concept map of the relationships among carbohydrates, monosaccharides, disaccharides, and polysaccharides.

Reading Strategy L2

Use Prior Knowledge Remind students of what they know about polymers. As they read this section, they should use this prior knowledge to confirm differences among different types of polysaccharides.

2 INSTRUCT

Connecting to Your World

Chitin is a tough yet lightweight covering that protects many insects. **What is a polymer?** *(a high molar mass molecule formed by the covalent bonding of repeating smaller units, called monomers)* **Where does a cicada get the chemicals to build chitin and other polymers it needs?** *(by eating plant products)*

Monosaccharides

Use Visuals L1

Figure 24.4 Point out that plants are mostly composed of a carbohydrate known as cellulose. Ask, **What are carbohydrates, and what is the general formula of these compounds?** *(Carbohydrates are a group of organic compounds that includes simple and complex sugars. Most carbohydrates have the general formula $C_n[H_2O]_n$.)*

24.2 Carbohydrates

Guide for Reading

Key Concepts

- Where is glucose found abundantly in nature?
- How can the cyclic forms of two simple sugars be linked?

Vocabulary

carbohydrates
monosaccharides
disaccharide
polysaccharides

Reading Strategy

Previewing Before you read this section, skim the section to find out the types of carbohydrates you will be learning about. List the types of carbohydrates and, as you read, write a description of each one.

Connecting to Your World

This cicada is molting—shedding its old exoskeleton and forming a new one. An exoskeleton is the hard protective covering of insects, lobsters, and other arthropods. As an arthropod grows, it must molt to make room for its bigger body. An arthropod's exoskeleton is made of a polymer called chitin, which belongs to a class of organic molecules known as carbohydrates. In this section, you will learn about the structures and functions of carbohydrates.

Monosaccharides

Long-distance runners often prepare for a big race by eating a great deal of bread and pasta, a process known as carbohydrate loading. Breads and pastas are excellent sources of the family of important molecules called carbohydrates. **Carbohydrates** are monomers and polymers of aldehydes and ketones that have numerous hydroxyl groups attached; they are made up of carbon, hydrogen, and oxygen. The name carbohydrate comes from the early observation that many of these compounds have the general formula $C_n(H_2O)_n$ and, as a result, appear to be hydrates of carbon. They are not, in fact, true hydrates.

You would probably call a diet based mainly on bread and pasta a starchy diet. You would be correct. Such foods contain certain carbohydrates called starches. In this section, you will learn about the similarities and differences among some well-known types of carbohydrates.

Carbohydrates are present in most foods, as Figure 24.4 shows. The simplest carbohydrate molecules are called simple sugars, or **monosaccharides.** Glucose and fructose are examples of simple sugars. **Glucose is abundant in plants and animals.** Depending on the source, glucose has also been called corn sugar, grape sugar, or blood sugar. Fructose occurs in a large number of fruits and in honey. Glucose and fructose both have the molecular formula $C_6H_{12}O_6$. However, glucose has an aldehyde functional group, whereas fructose has a ketone functional group. Therefore, they are structural isomers. Both undergo many of the same reactions as ordinary aldehydes and ketones.

Figure 24.4 Carbohydrates are the most abundant sources of energy in food. Fruits contain simple carbohydrates called sugars. Bread and pasta are good sources of complex carbohydrates called starches.

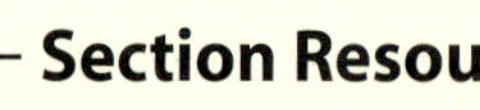

Section Resources

Print

- **Guided Reading and Study Workbook,** Section 24.2
- **Core Teaching Resources,** Section 24.2
- **Transparencies,** T273–T274

Technology

- **Interactive Textbook with ChemASAP,** Assessment 24.2
- **Go Online,** Section 24.2

In aqueous solution, simple sugars such as glucose and fructose exist in a dynamic equilibrium between straight-chain and cyclic forms. The cyclic form predominates. The structures for each sugar in both forms are:

```
 CHO                 CH2OH
  |                   |
H-C-OH           H    C---O    H
  |               \  /|    \  /
HO-C-H      <=>     C  H     C
  |                / \OH H  / \
H-C-OH           HO   C---C    OH
  |                   |   |
H-C-OH                H   OH
  |
 CH2OH
```

Straight-chain and cyclic forms of glucose

```
 CH2OH
  |
  C=O              HOCH2    O     OH
  |                    \  /   \  /
HO-C-H      <=>          C      C
  |                     / \H HO/ \
H-C-OH                 H   C---C  CH2OH
  |                        |   |
H-C-OH                     OH  H
  |
 CH2OH
```

Straight-chain and cyclic forms of fructose

Note the aldehyde functional group (—CHO) on the straight-chain form of glucose and the ketone functional group ($-\overset{\displaystyle O}{\overset{\|}{C}}-$) on the straight-chain form of fructose.

Which form of a simple sugar is favored at equilibrium in aqueous solution?

Disaccharides and Polysaccharides

Simple sugars form the building blocks of more complex carbohydrates. **The cyclic forms of two simple sugars can be linked by means of a condensation reaction.** For example, the linking of glucose and fructose with the loss of a water molecule produces sucrose—common table sugar. A sugar such as sucrose that forms from two monosaccharides in this way is known as a **disaccharide.** The reaction by which it forms is as follows.

```
Glucose (cyclic, with OH)
   +                 -H2O
Fructose (cyclic, with OH)   ----->   Sucrose (glucose-O-fructose)
```

Glucose

Fructose

Sucrose

Sucrose is obtained commercially mainly from the juice of sugar cane and sugar beets. The world's production from these sources exceeds 7×10^9 metric tons per year.

The formation of a disaccharide is sometimes the first step in a condensation polymerization reaction that produces extremely large molecules. The polymers produced by the linkage of many monosaccharide monomers are called **polysaccharides.** Starches, the major storage form of glucose in plants, are polysaccharide polymers that consist of glucose monomers. Figure 24.5 shows a portion of a starch molecule.

Word *Origins*

Polysaccharide comes from the Greek words *polys*, meaning "full" or "many," and *sakcharon*, meaning "sugar." A polysaccharide is a compound that is formed by the linking together of many simple sugar molecules. Starches, cellulose, and glycogen are all polysaccharides. **What are the names of some other organic molecules that contain the prefix *poly-*?**

Facts and Figures

Glucose Testing

The ability to detect mono- and polysaccharides is an important function of clinical laboratories and at-home tests for those afflicted with diabetes mellitus. The Benedict's test is one way to test for monosaccharides. It is not specific to glucose, however, but tests the total reducing substance. Available test strips allow more precise measurement of glucose. They combine enzymes that are specific to glucose and indicators that produce colors that vary depending on the concentration of glucose.

Disaccharides and Polysaccharides

TEACHER Demo

Glucose and Fructose Models L2

Purpose To build models of sugars.

Materials foam balls and toothpicks or molecule kit

Procedure Construct ball-and-stick models of glucose and fructose. Point out the functional groups. Demonstrate the cyclization of the straight-chain forms. Connect the two cyclic forms to form a disaccharide and note the loss of a water molecule. Point out that if this condensation reaction is allowed to continue, many monosaccharide units could join to form a polysaccharide.

TEACHER Demo

Benedict's Test L2

Purpose Students classify sugars a[s] reducing or nonreducing.

Materials hot-water bath, 2–5 mL of 1% carbohydrate solution, 2 test tube[s], distilled water, 2–3 mL of Benedict's reagent, dropper, samples of mono- and polysaccharides

Procedure Place 2–5 mL of 1% carbohydrate solution (corn syrup) in one test tube; place distilled water in another test tube as a negative control. Add 2–3 mL of Benedict's reagent to each tube and heat in hot-water bath for a few minutes. The Cu^{2+} ions react with reducing sugars to form a brick-red cuprous oxide (Cu_2O) precipitate. The amount of precipitate formed depends on the amount of sugar present. Have students observe the formation of a red precipitate. Test a number of mono- and polysaccharides and have students classify them as reducing or nonreducing.

Expected Outcome The precipitate indicates presence of a simple sugar.

Word *Origins* L2

Polypropylene, polystyrene, and polyvinyl chloride (PVC).

Answers to....

the cyclic form

Download a worksheet on **Carbohydrates** for students to complete, and find additional teacher support from NSTA SciLinks.

3 ASSESS

Evaluate Understanding L2

Have students describe the structure of mono-, di-, and polysaccharides and give a common example of each. Ask students to specify the types of elements, functional groups, and linkages found in carbohydrates.

Reteach L1

...dents study the structures for ... of sucrose (page 767). ...**ppens in this condensa-** ...**n?** *(Fructose and glucose ... make sucrose, losing one ...ecule.)* Have students ...how polymers form by con- ...on.

...necting Concepts

...yester polyethylene terephthalate ...ET) is a condensation polymer of ...erephthalic acid and ethylene glycol. Nylon is a condensation polymer of 6-aminohexanoic acid.

If your class subscribes to the Interactive Textbook, use it to review key concepts in Section 24.2.

with **ChemASAP**

Answers to...

Figure 24.5 Besides writing structural formulas, physical models (such as a ball-and-stick model) could be constructed to show the differences between starch and cellulose.

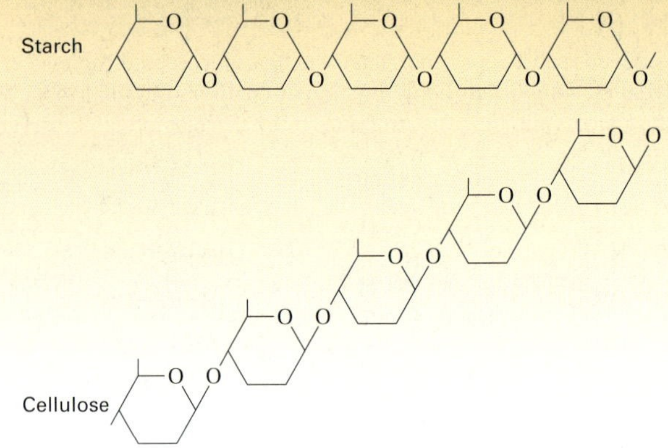

Figure 24.5 Starch and cellulose are similar polymers made up of hundreds of glucose monomers. They differ in the orientation of the bond between the glucose units. Because of this difference, starch is readily digestible, but cellulose is indigestible by most organisms. **Using Models** ***What other types of models could you use to show the structural differences between starch and cellulose?***

A typical linear starch molecule contains hundreds of glucose monomers. Other starches are branched molecules, each branch containing about a dozen glucose units. Glycogen, the energy source stored in the liver and muscle cells of animals, is a more highly branched molecule than plant starches. Glycogen, too, consists of glucose monomers.

Cellulose is probably the most abundant biological molecule on Earth. As you can see in Figure 24.5, cellulose also is a polymer of glucose. The orientation of the bond that links the glucose monomers in cellulose is different, however, from the bond orientation in starch and in glycogen. Starch can be digested by most organisms and is partially soluble in water. Cellulose, however, can be digested by only a few kinds of microorganisms, such as those that live in the digestive tracts of cattle and termites. Cellulose is insoluble in water and is an important structural polysaccharide that provides form, hardness, and rigidity in plants. Plant cell walls are made of cellulose. Cotton is about 80% cellulose.

For: Links on Carbohydrates
Visit: www.SciLinks.org
Web Code: cdn-1242

24.2 Section Assessment

8. **Key Concept** Where is glucose found in nature?
9. **Key Concept** How can the cyclic forms of two simple sugars be combined?
10. Distinguish between the important structural features of sucrose, glucose, and fructose.
11. Describe the main characteristics of monosaccharides, disaccharides, and polysaccharides.
12. Starch and cellulose have different properties, but both are composed of glucose units. Explain what makes them different.
13. Name a source for each polysaccharide:
    a. starch **b.** cellulose **c.** glycogen
14. What is the most abundant carbohydrate on Earth and where is it found?

Connecting Concepts

Polymers Starch is a condensation polymer of glucose. Reread Section 23.4, and describe some other examples of molecules that form by condensation polymerization.

Assessment 24.2 Test yourself on the concepts in Section 24.2.

with **ChemASAP**

Section 24.2 Assessment

8. in plants and animals
9. via a condensation reaction
10. *sucrose:* disaccharide of glucose and fructose; *glucose:* monosaccharide with aldehyde functional group; *fructose:* monosaccharide with ketone functional group.
11. Monosaccharides (simple sugars) can exist in straight-chain or cyclic form. Disaccharides form when two simple sugars condense. Polysaccharides are polymers of monosaccharides.
12. The orientation of the bonds that link the glucose monomers is different.
13. a. plants **b.** plant cell walls **c.** liver and muscle cells of animals
14. Cellulose; it is found in plant cell walls.

24.3 Amino Acids and Their Polymers

Connecting to Your World Many people are lactose intolerant, which means that they cannot digest milk or products containing milk. These people cannot digest milk products because their bodies do not produce enough of the enzyme lactase to digest lactose, the sugar found in milk. The undigested lactose causes bloating and stomach upset. Some people with lactose intolerance can enjoy milk products if they take a pill containing lactase before eating. In this section, you will learn what enzymes are and what function they serve in the body.

Guide for Reading

 **Key Concepts**

- What is the general structure of an amino acid?
- Which functional groups are always involved in amide bonds between amino acids?
- What determines the differences in the chemical and physiological properties of peptides and proteins?
- How do enzymes affect the rates of reactions in living things?

Vocabulary

amino acid
peptide
peptide bond
protein
enzymes
substrates
active site

Reading Strategy

Building Vocabulary As you read the section, write a definition of each vocabulary term in your own words.

Amino Acids

Many biological compounds contain nitrogen in addition to carbon, oxygen, and hydrogen. Some of the most important nitrogen-containing molecules in organisms are amino acids. In fact, the polymers of amino acids, which you will also learn about, make up more than one half of the dry weight of your body.

An **amino acid** is any compound that contains an amino group ($—NH_2$) and a carboxyl group (—COOH) in the same molecule. For chemists and biochemists, however, the term is usually reserved for the 20 common amino acids that are formed and used by living organisms. **Amino acids have a skeleton that consists of a carboxyl group and an amino group, both of which are covalently bonded to a central carbon atom. The remaining two groups on the central carbon atom are hydrogen and an R group that constitutes the amino acid side chain.**

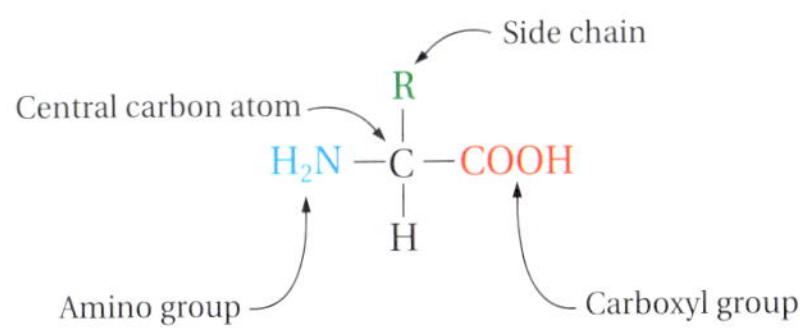

The chemical nature of the side-chain group accounts for the differences in properties among the 20 amino acids. In some amino acids, the side chains are nonpolar aliphatic or aromatic hydrocarbons. In other amino acids, the side chains are neutral but polar. In still others, the side chains are acidic or basic.

Because the central carbon of amino acids is asymmetric, these compounds can exist as optical isomers. As you may recall from Section 22.3, optical isomers may be right- or left-handed. Nearly all the amino acids found in nature are of the left-handed, or L, form.

Section Resources

Print
- **Guided Reading and Study Workbook,** Section 24.3
- **Core Teaching Resources,** Section 24.3
- **Transparencies,** T275–T278

Technology
- **Interactive Textbook with ChemASAP,** Assessment 24.3
- **Go Online,** Section 24.3

24.3

1 FOCUS

Objectives

24.3.1 Diagram the structure of an amino acid.

24.3.2 Describe how peptide bonds form and **identify** what determines the properties of peptides and proteins.

24.3.3 Describe how enzymes affect biochemical reactions.

Guide for Reading

Build Vocabulary  L2

Graphic Organizers Have students use a Venn diagram to compare peptides to proteins. They should include the vocabulary for the section and ask themselves, What do peptides and proteins have in common? What is unique to each?

Reading Strategy  L2

Relate Text and Visuals Have students skim this section, paying attention to the figures. Ask, **What are some of the three-dimensional shapes formed by peptide molecules?** *(straight chains, helical chains, and pleated sheets)*

2 INSTRUCT

Connecting to Your World

Point out that the prevalence of lactose intolerance in a population depends on ethnic origins. In the United States it is least prevalent in people of northern European descent and most prevalent in Asian Americans. Ask, **What does the enzyme lactase do for people?** *(helps them digest milk)*

Amino Acids

Discuss L2

Discuss how to calculate the number of possible arrangements of amino acids in proteins. Ask, **How many different ways can eight different amino acids be arranged to form a peptide?** *($8! = 8 \times 7 \times 6 \times 5 \times 4 \times 3 \times 2 \times 1 = 40{,}320$)* Ask, **How many combinations are possible for a peptide made of 10 different amino acids?** *($10! = 3{,}628{,}800$)* Point out to students that an average protein usually contains more than 100 amino acids.

Peptides

TEACHER Demo

Peptide Models L2

Purpose Students observe how amino acids link through amide bonds to form peptides.

Procedure Create models of peptides by linking amino acids with amide bonds. Point out that peptide formation is another example of a condensation polymerization. Ask students to describe how continuing the polymerization process produces protein molecules.

Expected Outcome Students should be able to describe peptide formation by manipulating the models.

Proteins

Discuss 

Point out that the types of amino acids in proteins have a large influence on the physical and chemical properties of the protein. The twenty common amino acids can be grouped into two major categories: polar or nonpolar. The ratio of polar to nonpolar amino acids profoundly affects the shape of a protein. Discuss why this should be. When a protein folds, nonpolar amino acids are sequestered to the interior of the protein, away from the aqueous environment. Polar amino acid residues often predominate on the surface of proteins.

Table 24.1 gives the names of the amino acids with their three-letter abbreviations. Examine the abbreviations. You will use them as shortcuts when you read or write about protein structure.

Peptides

A **peptide** is any combination of amino acids in which the amino group of one amino acid is united with the carboxyl group of another amino acid. The amide bond between the carboxyl group of one amino acid and the nitrogen in the amino group of the next amino acid in the peptide chain is called a **peptide bond.**

$$H_2N{-}\underset{H}{\overset{R}{C}}{-}\overset{O}{\overset{\|}{C}}{-}OH \;+\; H{-}\underset{H}{N}{-}\underset{H}{\overset{R}{C}}{-}\overset{O}{\overset{\|}{C}}{-}OH \longrightarrow H_2N{-}\underset{H}{\overset{R}{C}}{-}\overset{O}{\overset{\|}{C}}{-}\underset{H}{N}{-}\underset{H}{\overset{R}{C}}{-}\overset{O}{\overset{\|}{C}}{-}OH + H_2O$$

Amino acid — Amino acid — Peptide (Peptide bond)

The amide bonds between amino acids always involve the central amino and central carboxyl groups. The side chains are not involved in the bonding.

Note that a free amino group remains at one end of the resulting peptide. The convention is to write the formula of the peptide so that this free amino group is at the left end. There is also a free carboxyl group, which appears at the right end.

More amino acids may be added to the peptide in the same fashion to form long chains by condensation polymerization. The order in which the amino acids of a peptide molecule are linked is called the amino acid sequence of that molecule. The amino acid sequence of a peptide is conveniently expressed using the three-letter abbreviations for the amino acids. For example, Asp—Glu—Gly represents a peptide containing three amino acids. This tripeptide contains aspartic acid, glutamic acid, and glycine, in that order, with the free amino group assumed to be on the left end (on the Asp) and the free carboxyl group on the right end (on the Gly). Note that Asp—Glu—Gly is a different peptide from Gly—Glu—Asp because the order of amino acids is reversed, and thus the free amino group and free carboxyl group are on different amino acids.

Table 24.1 Abbreviations for Amino Acids

Amino acid	Abbreviation
Alanine	Ala
Arginine	Arg
Asparagine	Asn
Aspartic acid	Asp
Cysteine	Cys
Glutamine	Gln
Glutamic acid	Glu
Glycine	Gly
Histidine	His
Isoleucine	Ile
Leucine	Leu
Lysine	Lys
Methionine	Met
Phenylalanine	Phe
Proline	Pro
Serine	Ser
Threonine	Thr
Tryptophan	Trp
Tyrosine	Tyr
Valine	Val

By what process can amino acids be added to peptides to form long chains?

Proteins

In theory, the process of adding amino acids to a peptide chain may continue indefinitely. Any peptide with more than ten amino acids is called a polypeptide. A peptide with more than about 100 amino acids is called a **protein.** On average, a molecule of 100 amino acids has a molar mass of about 10,000 amu. Proteins are an important class of biomolecules. Your skin, hair, nails, muscles, and the hemoglobin molecules in your blood are made of protein. Proteins are needed for almost all chemical reactions that take place in the body.

Facts and Figures

Ethnobiologists

Often exploring the remotest jungles and forests, ethnobiologists search for information about plants, fungi, and even insects that people use as medicine. These scientists carefully interview local specialists, who administer or use the remedies; collect samples of the remedies, and send these samples to laboratories for analysis of their chemical components. Ethnobiologists carry out an important task—fewer than 3% of the flowering plants on Earth have been examined for medicinal value, but about 40% of medicines in use today come from plant sources.

770 Chapter 24

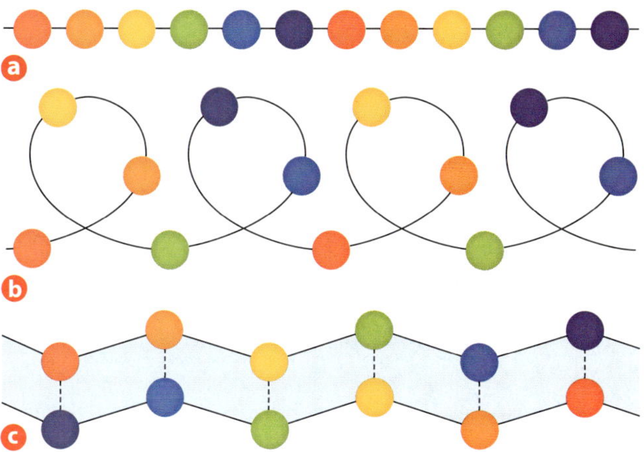

Figure 24.6 Peptides form three-dimensional shapes. **a** This is a representation of amino acids in a peptide chain. **b** The chain may coil into a helix. **c** Two peptide chains may become arranged in a pleated, sheet-like structure.
Applying Concepts ***What types of bonds determine the three-dimensional shape of a protein?***

Differences in the chemical and physiological properties of peptides and proteins result from differences in the amino acid sequence. 20 amino acids can be linked in an enormous number of ways in a protein molecule. As many as 20^{100} amino acid sequences are possible for a protein of 100 amino acids containing a combination of the 20 different amino acids.

Protein molecules are folded into relatively stable three-dimensional shapes. Figure 24.6a represents a long peptide chain of a protein. Figure 24.6b shows how sections of peptide chain may coil into a regular spiral, known as a helix. Peptide chains may also be arranged side by side to form a pleated sheet, as shown in Figure 24.6c. Irregular folding of the chains also can occur. The three-dimensional shape of a protein is determined by interactions among the amino acids in its peptide chains. Protein shape is partly maintained by hydrogen bonds between adjacent folded chains. Covalent bonds also form between sulfur atoms of cysteine side chains that are folded near each other. In that way, separate polypeptide chains may be joined into a single protein. Figure 24.7 traces the shape of myoglobin, a protein that stores oxygen in muscle cells. As you can see, the peptide chains of most of the myoglobin molecule are twisted into helixes.

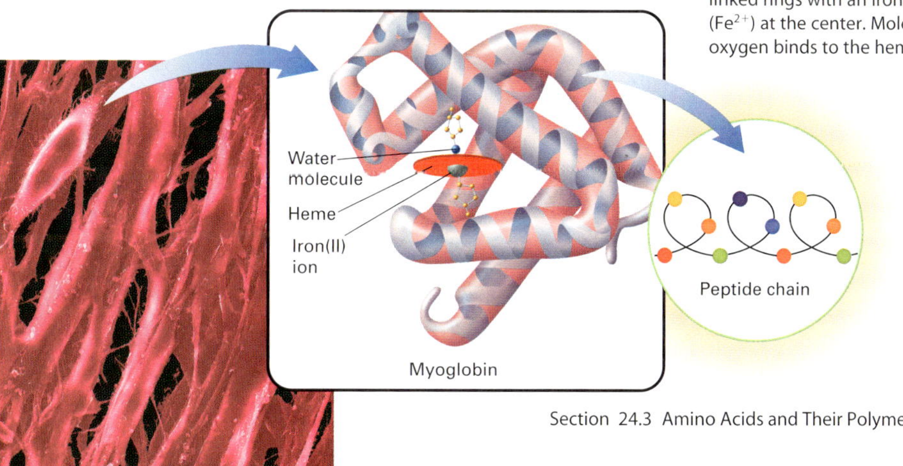

Figure 24.7 The three-dimensional structure of myoglobin, the oxygen storage protein of this muscle tissue, is shown here. Most of the peptide chain of myoglobin is wound into helixes. Myoglobin also contains a nonprotein structure called heme. The heme group is shown as a disk in the myoglobin structure. Heme contains four linked rings with an iron(II) ion (Fe^{2+}) at the center. Molecular oxygen binds to the heme iron.

TEACHER Demo

Protein in Hard Tissue L2

Purpose Students learn that some proteins play structural roles in hard tissue.

Materials thigh bone from cooked poultry, 1*M* HCl, beaker

Procedure Students may be surprised to learn that protein is a major structural constituent of bone. Illustrate this fact by soaking a thigh bone from cooked poultry in 1*M* HCl overnight. The next day, rinse the bone with water.

Expected Outcome Upon visual inspection it may seem as if no reaction has occurred at all; demonstrate that the bone can be bent and twisted quite easily. Point out that the mineral salts of the bone were dissolved in the acid. The acid-insoluble portion remaining consists of a special protein called collagen. Ask **What possible roles does collagen have in the formation and maintenance of healthy bone tissue?** *(It provides a framework for the deposition of calcium phosphate, the mineral that makes up bone, and for the growth of bone marrow cells.)*

Differentiated Instruction

Gifted and Talented 

Scientists have long sought a reliable, general method for predicting the three-dimensional shape of any protein based on its amino acid sequence. Because a protein's shape is so closely associated with its substrate binding properties, a code for predicting protein structure would allow scientists to design new proteins and modify existing ones for a variety of purposes.

Have students find out how the amino acid sequence of a protein dictates its secondary and tertiary structure. Ask them to draw a simple protein (or part of one) that shows some of the bonds affecting its shape.

Answers to...

Figure 24.6 Hydrogen bonds

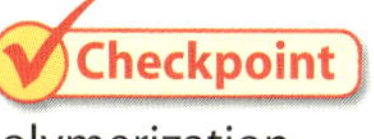

condensation polymerization

Enzymes

Use Visuals 

Figure 24.8 Compare the binding of a substrate and an enzyme to the way in which a key fits into a lock. Ask, **How does the enzyme exclude all but the correct substrate?** *(The substrate has a matching, or complementary, shape.)* **What would happen to the system if you heated it until the enzyme shape changed?** *(The substrate would no longer fit.)*

Enzymes 

Purpose Students observe that enzymes speed up chemical reactions.

Materials Lactaid®, dropper, water, 2 cartons of fresh milk, spoons or small cups

Procedure Treat one container of milk with Lactaid and one with an equal number of water dops. Store the milk cartons in the refrigerator for one to two days. Then have student volunteers do a comparative test.

Expected Outcome The container with added lactase should taste sour; the other container should still be fresh.

Go Online NSTA SciLinks

Download a worksheet on **Proteins/ Enzymes** for students to complete, and find additional teacher support from NSTA SciLinks.

Relate 

Coenzymes are often vitamins; for example, B vitamins help release energy from foods. Lack of B vitamins usually produces malnourishment of some kind, since food can't be digested properly without them. Niacin and B_6 deficiencies, for example, produce skin and nervous system disorders, such as headaches, depression, and memory loss.

For: Links on Proteins/ Enzymes
Visit: www.SciLinks.org
Web Code: cdn-1243

Enzymes

Enzymes are proteins that act as biological catalysts. **Enzymes increase the rates of chemical reactions in living things.** In 1926, the American chemist James B. Sumner reported the first isolation and crystallization of an enzyme. The enzyme he isolated was urease, which hydrolyzes urea, a constituent of urine, into ammonia and carbon dioxide. The strong ammonia smell of wet diapers that are allowed to stand for a long time is the result of the action of bacteria that contain this enzyme. The reaction is shown below.

$$\underset{\text{Urea}}{H_2N-\overset{\overset{\displaystyle O}{\|}}{C}-NH_2(aq)} + \underset{\text{Water}}{H_2O(l)} \xrightarrow{\text{urease}} \underset{\text{Ammonia}}{2NH_3(g)} + \underset{\text{Carbon dioxide}}{CO_2(g)}$$

Since the discovery of urease, thousands of enzymes have been isolated and structurally characterized as proteins.

In addition to being able to promote reactions, enzymes have two other properties of true catalysts. First, they are unchanged by the reaction they catalyze. Second, they do not change the normal equilibrium position of a chemical system. The same amount of product is eventually formed whether or not an enzyme is present. Few reactions in cells ever reach equilibrium, however. The products tend to convert rapidly to another substance in a subsequent enzyme-catalyzed reaction. According to Le Châtelier's principle, such removal of a product pulls the reaction toward completion.

How Enzymes Work Enzymes catalyze most of the chemical changes that occur in the cell. **Substrates** are the molecules on which an enzyme acts. In a typical enzymatic reaction, diagrammed in Figure 24.8, the substrate interacts with side chains of the amino acids on the enzyme. These interactions cause the making and breaking of bonds. A substrate molecule must make contact with and bind to an enzyme molecule before the substrate can be transformed into product. The place on an enzyme where a substrate binds is called the **active site.** An active site is usually a pocket or crevice formed by folds in the peptide chains of the enzyme protein. The peptide chain of an enzyme is folded in a unique way to accommodate the substrate at the active site. The diagram in Figure 24.9 shows this folding for an enzyme called HIV protease, which is produced by the virus that causes AIDS.

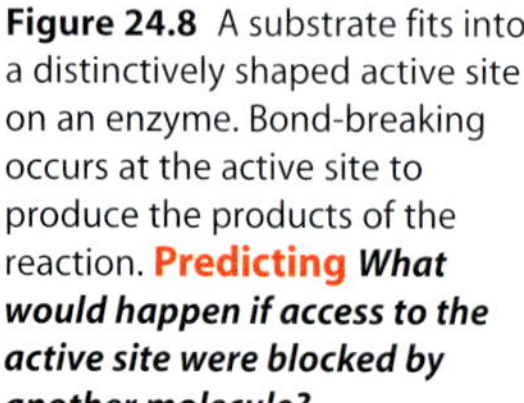

Figure 24.8 A substrate fits into a distinctively shaped active site on an enzyme. Bond-breaking occurs at the active site to produce the products of the reaction. **Predicting** ***What would happen if access to the active site were blocked by another molecule?***

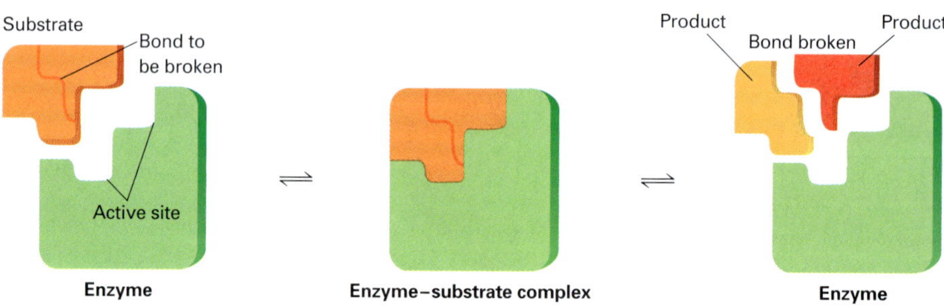

Differentiated Instruction

Less Proficient Readers 

Many laundry detergents contain enzymes that are supposed to help remove stains. Ask students to read the labels of a number of laundry detergents and find out if enzymes are listed among the ingredients. Ask them to hypothesize how enzymes improve the cleaning performance of a detergent.

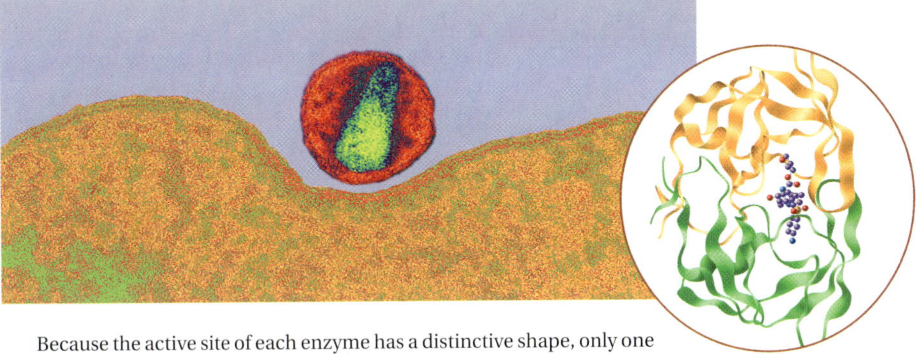

Because the active site of each enzyme has a distinctive shape, only one specific substrate molecule can fit into the enzyme, in the same way that only one key will fit into a certain lock. Each enzyme catalyzes only one chemical reaction, with only one substrate. An enzyme-substrate complex is formed when an enzyme molecule and a substrate molecule are joined.

To see the efficiency of enzymes, consider an enzyme called carbonic anhydrase. It catalyzes the reversible breakdown of carbonic acid to carbon dioxide and water. One molecule of carbonic anhydrase can catalyze the breakdown of 36 million molecules of carbonic acid in one minute!

$$\underset{\text{Carbonic acid}}{H_2CO_3(aq)} \xrightleftharpoons{\text{carbonic anhydrase}} \underset{\text{Carbon dioxide}}{CO_2(g)} + \underset{\text{Water}}{H_2O(l)}$$

Figure 24.9 This color-enhanced scanning electron micrograph shows HIV (red particle) infecting a human white blood cell (orange). The diagram models the enzyme HIV protease. The green and yellow ribbons trace the two peptide chains of the enzyme. A substrate molecule (purple) is embedded in the active site formed between the two peptide chains.

Coenzymes Some enzymes can directly catalyze the transformation of biological substrates without assistance from other substances. Other enzymes need nonprotein coenzymes, also called cofactors, to assist the transformation. Coenzymes are metal ions or small organic molecules that must be present for an enzyme-catalyzed reaction to occur. Many water-soluble vitamins, such as B vitamins, are coenzymes. Metal ions that act as coenzymes include the cations of magnesium, potassium, iron, and zinc. The enzyme catalase includes an iron(III) ion in its structure. Catalase catalyzes the breakdown of hydrogen peroxide to water and oxygen.

$$2H_2O_2(aq) \xrightarrow{\text{catalase}} 2H_2O(l) + O_2(g)$$

24.3 Section Assessment

15. Key Concept What are the four groups that surround the central carbon atom in an amino acid?
16. Key Concept Which functional groups are always involved in amide bonds?
17. Key Concept What determines the differences in the properties of peptides and proteins?
18. Key Concept How do enzymes affect the reaction rates in living things?
19. What is meant by the amino acid sequence of a protein?
20. Describe three properties of enzymes.

Connecting Concepts

Reversible Reactions Reread Section 18.2 on reversible reactions and equilibrium. Use Le Châtelier's principle to explain why few reactions in cells ever reach equilibrium.

Interactive Textbook

Assessment 24.3 Test yourself on the concepts in Section 24.3.

with ChemASAP

3 ASSESS

Evaluate Understanding L2

Have students sketch the general structure of a tripeptide—such as Gly-Val-Trp—using letter abbreviations in place of the R group. Ask them to label the peptide bonds and the carboxyl and amino ends. Ask, **Which element do proteins contain that carbohydrates do not?** *(nitrogen)* **What is an enzyme?** *(Enzymes are proteins that function as biological catalysts.)* Have students describe one enzyme discussed and its function. *(One example is urease, which catalyzes the hydrolysis of urea.)*

Reteach 

Use dryer duct tubing or slinky toys to help illustrate the secondary and tertiary structure of proteins. Point out that many of the amino acids in an enzyme are not directly involved in catalysis because they lie far from the active site. However, they do participate in intermolecular forces, which maintain the folded shape of the protein, a prerequisite for its biological activity.

Connecting Concepts

In cellular reactions, the products tend to convert rapidly to another substance in a subsequent enzyme-catalyzed reaction. According to Le Châtelier's principle, if the products of a reaction mixture are continually removed, the reaction will tend to go to completion rather than establish equilibrium.

Interactive Textbook

If your class subscribes to the Interactive Textbook, use it to review key concepts in Section 24.3.

with ChemASAP

Section 24.3 Assessment

15. amino, carboxyl, R, and hydrogen
16. The amino group of one amino acid combines with the carboxylic acid group of another to form an amide linkage
17. differences in the amino acid sequence
18. Enzymes speed up the rate of reaction but do not affect the position of equilibrium.
19. the order of amino acids in the protein molecule
20. catalysts (they remain unchanged in a reaction); usually proteins; specific to the reactions they catalyze

Answers to...

Figure 24.8 Transformation of the substrate into product could not occur.

The Egg: A Biochemical Storehouse L2

Objective Students examine the physical and chemical properties of a chicken egg.

Skills Focus Measuring, calculating, designing experiments

Prep Time 20 minutes

Materials Chicken egg, metric ruler, mass balance, NaOH, $CuSO_4$, HCl

Advance Prep Make up reagents as follows:

Solution	Preparation
0.5*M* NaOH	20.0 g in 1.0 L
0.2*M* $CuSO_4$	12.5 g $CuSO_4{\cdot}5H_2O$ in 250 mL
1.0*M* HCl	82 mL of 12*M* in 1.0 L

Class Time 20 minutes

Safety Always add acid to water carefully and slowly!

Teaching Tips

- For each class, obtain at least a dozen eggs of two or three different sizes: medium, large, and extra large. Some markets also sell small or jumbo eggs. This will allow students to obtain data from classmates about eggs that are a different size than their own.
- Have students write their names on their eggs so they can identify them the next day.
- Store some eggs in a refrigerator and some at room temperature if students are going to do activities 1, 2, and 8 in the You're the Chemist section.

Small-Scale LAB

The Egg: A Biochemical Storehouse

Purpose

To explore some physical and chemical properties of a chicken egg

Materials

- chicken egg
- ruler
- balance
- pencil
- paper

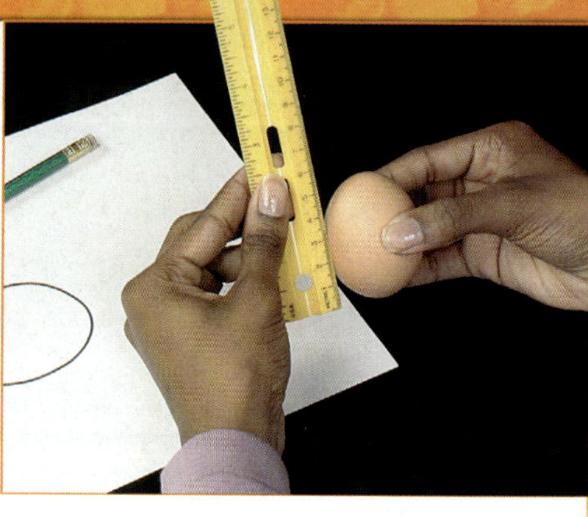

Procedure   

Obtain a chicken egg. Examine the egg's shape and measure its length and width in centimeters. Measure the mass of the egg. Make an accurate, life-size sketch of your egg and record all your data on the sketch.

Analyze

Using your experimental data, record the answers to the following questions below your drawing.

1. A common way to compare the shapes of eggs is by using a shape index. The shape index is the width of an egg expressed as a percentage of its length.

$$\text{Shape index} = \frac{\text{width}}{\text{length}} \times 100\%$$

Calculate the shape index of your egg.

2. The volume, original mass (when freshly laid), and surface area of an egg can easily be estimated by using the following equations.

$V = (0.5236)(\ell w^2)$ $\quad m = (0.5632)(\ell w^2)$

$A = (3.138)(\ell w^2)^{2/3}$

V = volume $\quad m$ = original mass

A = surface area $\quad \ell$ = length $\quad w$ = width

Use your data to calculate the volume, original mass, and surface area of your egg. Show your work, and record your results.

3. Compare your measured mass of the egg with your calculated mass. Which is greater? Suggest why the mass of an egg might change over time.
4. Using your measured mass and your calculated volume, calculate the density of your egg. Compare this value with the density of a freshly laid egg (d = 1.075 g/cm^3).

You're the Chemist

The following small-scale activities allow you to develop your own procedures and analyze the results.

1. **Design It!** Design an experiment to answer the following question: Does the mass of an egg change over time?
2. **Analyze It!** Using your measured mass, your calculated original mass, and your experiments on the mass loss of an egg over time, estimate the age of your egg. What assumptions must you make?
3. **Design It!** Design and carry out an experiment to measure the volume of your egg. Write down what you did and what you found.
4. **Design It!** Carry out a series of experiments, or consult with your classmates and use their data, to determine if and how the shape index varies with the size of the egg (small, medium, large, extra large, jumbo).
5. **Analyze It!** Determine how the mass of an egg varies with its size (small, medium, large, extra large, jumbo).
6. **Analyze It!** An eggshell contains a calcium carbonate matrix with a protein cuticle. Place one drop of HCl on an eggshell and observe what happens. Write a chemical equation for this reaction. **CAUTION** *HCl is caustic and can burn skin.*
7. **Analyze It!** Proteins can be detected by adding aqueous solutions of copper(II) sulfate and sodium hydroxide to a sample. A violet color indicates the presence of protein. Test powdered milk and an eggshell for protein. What are your results?
8. **Design It!** Design and carry out an experiment to answer the following question. Does temperature affect the mass of an egg over time?

Expected Outcome

Sample data: length = 5.90 cm; width = 4.55 cm; mass = 62.42 g.

Analyze

1. shape index = $\frac{4.55 \text{ cm}}{5.90 \text{ cm} \times 100\%}$ =77.1
2. $V = (0.5236)(5.90 \text{ cm})(4.55 \text{ cm})^2 = 64.0 \text{ cm}^3$

   $M = (0.5632)(5.90 \text{ cm})(4.55 \text{ cm})^2 = 68.8 \text{ cm}^3$

   $A = 3.138[(5.90 \text{ cm})(4.55 \text{ cm})^2]^{2/3} = 77.3 \text{ cm}^3$
3. The measured mass of 62.42 g is less than the calculated mass of 68.8 g by 6.4 g. The egg may have lost water.
4. 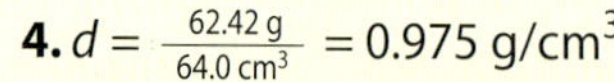 $d = \frac{62.42 \text{ g}}{64.0 \text{ cm}^3} = 0.975 \text{ g/cm}^3$

   This is less than the density of a freshly laid egg.

You're the Chemist

1. Measure the mass each day for two or three days. Eggs lose 0.2 to 0.5 grams per day.
2. Assuming the egg loses 0.20 g per day, the age of the egg is:

   $(68.8\text{g} - 62.42\text{g}) \times \frac{1 \text{ day}}{0.20\text{g}} = 32 \text{ days old}$
3. Measure the volume by water displacement.
4. Larger eggs have smaller shape indexes.
5. Extra large eggs are usually more than 70 g, and medium eggs are less than 50 g.
6. $2HCl + CaCO_3 \rightarrow CO_2\,(g) + H_2O + CaCl_2$
7. Both powdered milk and the eggshell produce a violet color.
8. Measure the mass of a refrigerated egg for three days. Measure its mass for three more days while storing it at room temperature. The warmer the temperature, the greater the mass loss.

24.4 Lipids

Connecting to Your World

Candles were an early invention of ancient civilizations. Candlesticks that date to at least 5000 years ago have been found with other artifacts from the civilizations of ancient Egypt and Crete. Before the invention of electric lighting, wax candles were the major source of lighting in homes. In this section, you will read more about waxes and similar compounds that make up the class of biomolecules known as lipids.

Guide for Reading

Key Concepts

- What physical property distinguishes lipids from other classes of biological molecules?
- How do phospholipid molecules arrange themselves in water?

Vocabulary

lipids
triglyceride
saponification
phospholipids
waxes

Reading Strategy

Building Vocabulary As you read the section, write a definition of each vocabulary term in your own words.

Triglycerides

Fats, oils, and other water-insoluble compounds are called **lipids.** The difference between fats and oils is simply that fats are solid at room temperature and oils are liquid. Most fats, such as butter, are obtained from animals. The fats from palm kernels and coconuts, however, are exceptions. Most oils, such as olive oil, are plant products.

Although excessive dietary fat is harmful, you do need some lipids in your diet to stay healthy. Experts recommend that your lipid intake make up less than 30% of your daily caloric intake of food. Lipids such as the butter shown in Figure 24.10 provide an efficient way for your body to store energy. They are also needed to keep your cell membranes healthy. In this section, you will learn about the chemical composition and biological uses of lipids.

Lipids tend to dissolve readily in organic solvents, such as ether and chloroform, rather than in highly polar solvents such as water. This property sets them apart from most biological substances such as carbohydrates and proteins. Natural fats and oils exist as triesters of glycerol with fatty acids, which are long-chain carboxylic acids (C_{12} through C_{24}). This form of lipid is known as a **triglyceride.** Triglycerides are important as the long-term storage form of energy in the human body. The following equation shows the general reaction for the formation of triglycerides.

$$\begin{array}{c} CH_2OH \\ | \\ CHOH \\ | \\ CH_2OH \\ \text{Glycerol} \end{array} + \begin{array}{c} HO-\overset{O}{\overset{\|}{C}}-R \\ HO-\overset{O}{\overset{\|}{C}}-R \\ HO-\overset{O}{\overset{\|}{C}}-R \\ \text{3 Fatty acid molecules} \end{array} \longrightarrow \begin{array}{c} CH_2-O-\overset{O}{\overset{\|}{C}}-R \\ | \\ CH-O-\overset{O}{\overset{\|}{C}}-R \\ | \\ CH_2-O-\overset{O}{\overset{\|}{C}}-R \\ \text{Triglyceride (triester of glycerol)} \end{array} + \begin{array}{c} 3H_2O \\ \text{Water} \end{array}$$

Figure 24.10 Moderate levels of dietary fats and oils are essential to health.

Section Resources

Print

- **Guided Reading and Study Workbook,** Section 24.4
- **Core Teaching Resources,** Section 24.4
- **Transparencies,** T279–T280
- **Laboratory Manual,** Lab 51
- **Small-Scale Chemistry Laboratory Manual,** Lab 40

Technology

- **Interactive Textbook with ChemASAP,** Assessment 24.4
- **Go Online,** Section 24.4

24.4

1 FOCUS

Objectives

24.4.1 Identify the physical property that distinguishes lipids from other biological molecules.

24.4.2 Describe the structure of a lipid bilayer.

Guide for Reading

Build Vocabulary 

Word Parts Have students dissect the words *hydrophilic* and *hydrophobic*. Ask them for examples of words that use the same roots, to help distinguish between the meanings.

Reading Strategy 

Predicting Have students look ahead through the section and predict what the properties of lipids are and what types exist. They should check their predictions after they read the section.

2 INSTRUCT

Connecting to Your World

Point out that candlepower and foot–candles were once used as units of illumination. Ask students to imagine today's towns lit only by candles. Ask, **How would your life be different if candles were the only way to get light at night?** *(Answers will vary.)*

Triglycerides

Discuss 

Explain that lipids contain identifiable functional groups, but their most distinguishing property is their insolubility in water. Discuss why they are insoluble. *(They are nonpolar.)* Ask, **What utility does this characteristic give lipids?** *(They can be used to contain water-soluble products and processes.)*

CLASS Activity

Better Butter? 

Have students compare the composition of butter, regular margarine, and low-calorie margarine by melting the same mass of each kind of spread in separate test tubes. Put the test tubes in warm water to liquefy the samples. Then allow the samples to settle. The liquid will form two layers: the top layer is a lipid (butterfat or partially hydrogenated vegetable oil) and the bottom layer is water. Students can compare the samples quantitatively and qualitatively and calculate the percentage of water in each fat.

Phospholipids

Relate 

Lipid bilayers are an important component of cells today and were probably essential to the origin of cellular life. Before DNA and RNA existed to direct the formation of complex molecules, molecules had to be able to self-assemble. Bipolar molecules like phospholipids can assemble themselves into membranes. A membrane-bound environment was an important step in providing a place for complex biological chemistry.

Download a worksheet on **Fats** for students to complete, and find additional teacher support from NSTA SciLinks.

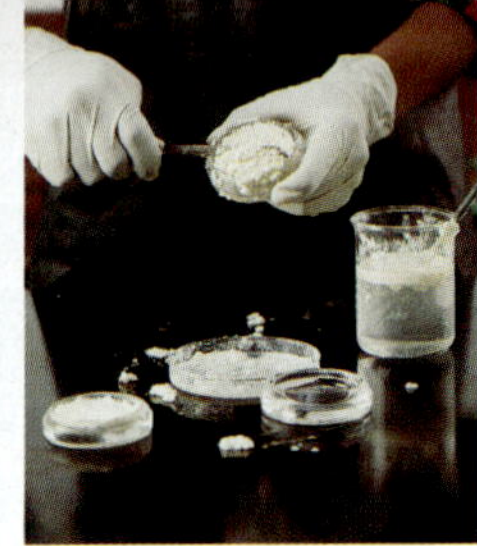

Figure 24.11 These photographs illustrate soapmaking. Once the soap is formed, it is poured into molds. Later it may be milled, or shredded, with scent or color added, and then remolded to produce a finished product.

Like other esters, fats and oils are easily hydrolyzed in the presence of acids and bases. The hydrolysis of oils or fats by boiling with an aqueous solution of an alkali-metal hydroxide is called **saponification.** Saponification is used to make soap. Soaps are thus the alkali metal (Na, K, or Li) salts of fatty acids. A typical saponification reaction is shown below.

$$\begin{array}{l} CH_2-O-\overset{O}{\overset{\|}{C}}-(CH_2)_{16}CH_3 \\ | \\ CH-O-\overset{O}{\overset{\|}{C}}-(CH_2)_{16}CH_3 \\ | \\ CH_2-O-\overset{O}{\overset{\|}{C}}-(CH_2)_{16}CH_3 \end{array} + 3NaOH \longrightarrow \begin{array}{l} CH_2OH \\ | \\ CHOH \\ | \\ CH_2OH \end{array} + 3CH_3(CH_2)_{16}-\overset{O}{\overset{\|}{C}}-O^-Na^+$$

Tristearin (triester of glycerol and stearic acid) — Glycerol — Sodium stearate (a soap)

As shown in Figure 24.11, soap can be made by heating a fat, such as beef tallow or coconut oil, with an excess of sodium hydroxide. When sodium chloride is added to the saponified mixture, the sodium salts of the fatty acids separate as a thick curd of crude soap. The crude soap is then purified. Glycerol is an important by-product of saponification reactions. It is recovered by evaporating the water layer.

 Checkpoint ***What are the products of a saponification reaction?***

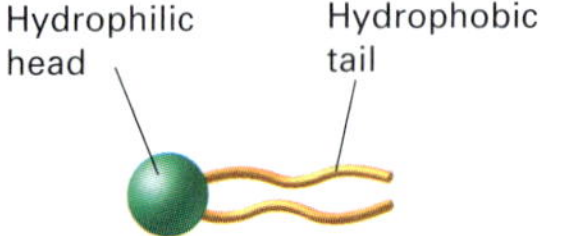

Figure 24.12 A space-filling model of lecithin is shown here. In the simplified representation above, the hydrophilic head is shown as a sphere and the hydrophobic tails as wavy lines.

Phospholipids

Phospholipids, or lipids that contain phosphate groups, are abundant in cells. Figure 24.12 shows a typical phospholipid molecule, lecithin. The lecithin molecule has a hydrophilic (water-loving) ionic head and oily or hydrophobic (water-hating) hydrocarbon tails. You know the solubility rule, "Like dissolves like." Lecithin, which is partly hydrophobic and partly hydrophilic, behaves in water like both an insoluble hydrocarbon and a soluble ionic compound. In water, the hydrophobic carbon chains aggregate to exclude water. The hydrophilic part is drawn to water, which can solvate it.  **In water, phospholipids spontaneously form a spherical double layer, called a lipid bilayer, in which the hydrophobic tails of phospholipid molecules are sandwiched between two layers of hydrophilic heads.**

Differentiated Instruction

Special Needs L1

Create a model of a cell membrane by floating a layer of table-tennis balls in a tub of water. Add a few tennis balls. Explain that the water underneath the balls represents the cytoplasm of the cell, the table-tennis balls represent the phospholipids, and the tennis balls represent the proteins.

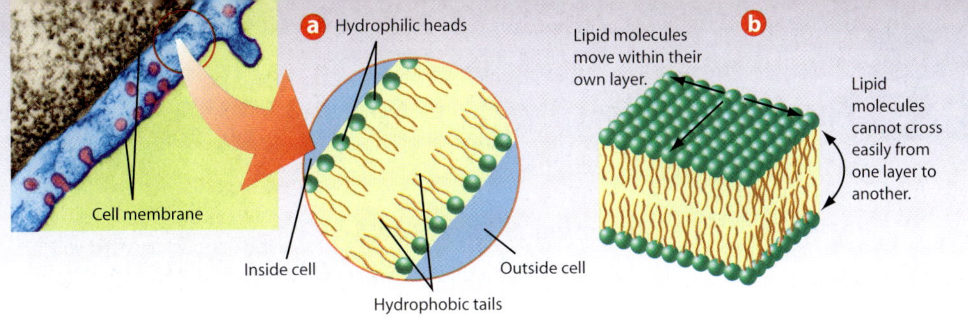

Figure 24.13 A cell membrane has a lipid bilayer structure. **a** The hydrophilic heads are in contact with water, but the hydrophobic tails are not. **b** The lipid molecules move easily within their own layer but do not readily move to the other layer. **Applying Concepts** ***What prevents a lipid molecule from crossing to the opposite side of the bilayer?***

Cell membranes, such as the one shown in Figure 24.13, consist primarily of lipid bilayers. The lipid bilayer of a cell membrane acts as a barrier against the passage of molecules and ions into and out of the cell. However, cells do need to take in certain ions and molecules, such as nutrients, while excluding other materials. Selective absorption is accomplished by the protrusion of protein molecules through the lipid bilayer. These proteins form channels through which specific ions and molecules can selectively pass. Not all membrane proteins extend all the way through the membrane. Proteins, such as enzymes, may be bound to the interior surface of the membrane. Many membrane proteins have attached carbohydrate molecules. The carbohydrate portion is on the exterior of the lipid bilayer, where it can hydrogen-bond with water. The protein portion is on the interior of the lipid bilayer, so it does not contact the water.

Waxes

Waxes are also part of the lipid family. **Waxes** are esters of long-chain fatty acids and long-chain alcohols. The hydrocarbon chains for both the acid and the alcohol usually contain from 10 to 30 carbon atoms. Waxes are low-melting, stable solids. In many plants, a wax coat protects the surfaces of leaves from water loss and attack by microorganisms. For example, carnauba wax, a major ingredient in car wax and floor polish, is found on the leaves of a South American palm tree. In animals, waxes coat the skin, hair, and feathers and help keep these structures pliable and waterproof.

For: Links on Fats
Visit: www.SciLinks.org
Web Code: cdn-1244

24.4 Section Assessment

21. **Key Concept** What physical property sets lipids apart from biological substances such as carbohydrates and proteins?
22. **Key Concept** How do phospholipids behave in water?
23. Compare the molecular structures of the three main types of lipids.
24. What role do phospholipids and proteins play in cell membranes?
25. What two classes of organic compounds combine to form a wax?

Writing Activity

Explain a Concept Write a paragraph explaining how cells are able to selectively absorb certain ions and molecules while excluding other materials.

Assessment 24.4 Test yourself on the concepts in Section 24.4.

with ChemASAP

Use Visuals 

Figure 27.13 Ask, **Why are the phospholipid tails facing each other?** *(They are hydrophobic and the membrane is surrounded by water.)*

Waxes

Discuss 

Point out that an important difference between waxes and oils is the length of the hydrocarbon backbone. Waxes are derivatives of high molar mass carboxylic acids. Remind students that carboxylic acids react with alcohols to form esters. Waxes are essentially high molar mass esterification products of carboxylic acids.

3 ASSESS

Evaluate Understanding L2

Display samples of different types of lipids and ask students to decide what type they are, why they think so, and what that implies about their structure.

Reteach 

Review the structures on pages 775–776. Point out that fats and oils hydrolyze in the presence of acids or bases because lipids are esters.

Writing Activity

Protein molecules protruding through the cell membrane form channels that allow specific ions and molecules to pass into the cell, but exclude other materials.

Interactive Textbook

If your class subscribes to the Interactive Textbook, use it to review key concepts in Section 24.4.

with ChemASAP

Section 24.4 Assessment

21. Lipids do not dissolve in highly polar solvents such as water.
22. They spontaneously form a lipid bilayer, with hydrophobic tails of the phospholipids in the middle.
23. Triglycerides are triesters of glycerol with long chains of fatty acids attached. Phospholipids are lipids that contain phosphate groups; they have a hydrophobic and hydrophilic end. Waxes are esters of long-chain fatty acids and long-chain alcohols.
24. Phospholipids make up the lipid bilayer of the cell membrane. Protein molecules protrude from this layer, forming channels through which nutrients pass.
25. long-chain fatty acids and long-chain alcohols

Answers to...

Figure 24.13 the hydrophobic middle of a lipid bilayer

soap and glycerol

24.5

1 FOCUS

Objectives

24.5.1 Identify the functions of DNA and RNA.

24.5.2 Describe how information is sorted in genetic material and how genes can mutate.

24.5.3 Describe how DNA fingerprinting and recombinant DNA technology are used.

Guide for Reading

Build Vocabulary L2

Word Parts The root of many key words in this section relates to the nucleus of the cell. Nucleic acid, for example, was first found in the nucleus of eukaryotic cells. Nucleotides are the compounds that make up these nucleus-inhabiting acids.

Reading Strategy L2

Identifying the Main Idea Have students identify the main idea under each red heading. Ask them to note details as they read that support the main idea for that portion of the text.

2 INSTRUCT

Connecting to Your World

Point out that some physical traits are inherited, while others are acquired during life. Ask, **What are some inherited traits?** *(hair color, eye color, blood type)* **What are some acquired traits?** *(muscle mass, hair length, skills)*

24.5 Nucleic Acids

Guide for Reading

Key Concepts

- What are the functions of DNA and RNA?
- How long a base sequence of DNA is required to specify one amino acid in a peptide chain?
- What are gene mutations?
- What forms the basis for a method of identifying a person from biological samples?
- What is recombinant DNA technology?

Vocabulary

nucleic acids
nucleotides
gene

Reading Strategy

Summarizing As you read the section, summarize each part that begins with a red head in a few sentences.

Connecting to Your World Maybe people have told you that you have your mother's eyes or your father's nose. Although this is not literally true (your eyes and nose are your own), you do inherit the instructions for assembling the proteins of your body from your parents. In this section, you will learn what these instructions are called and how they code for proteins.

DNA and RNA

More than 100 years ago, a Swiss biochemist discovered a class of nitrogen-containing compounds in the nuclei of cells. The nuclei were first obtained from dead white blood cells in the pus of infected wounds. The eventual understanding of the biological role of the compounds has led to a revolution in biochemistry.

These nitrogen-containing compounds, called **nucleic acids,** are polymers that are found primarily in cell nuclei. They are indispensable components of every living thing. Two kinds of nucleic acids are found in cells—*d*eoxyribo*n*ucleic *a*cid (DNA) and *r*ibo*n*ucleic *a*cid (RNA). **DNA stores the information needed to make proteins and governs the reproduction and growth of cells and new organisms. RNA has a key role in the transmission of the information stored in DNA and in the synthesis of proteins.**

The monomers that make up the DNA and RNA polymers are called **nucleotides.** Nucleic acids are therefore polynucleotides. As shown below, each nucleotide consists of a phosphate group, a five-carbon sugar, and a nitrogen-containing unit called a nitrogen base.

Phosphate + Sugar + Nitrogen base ⟶ Phosphate — Sugar — Nitrogen base

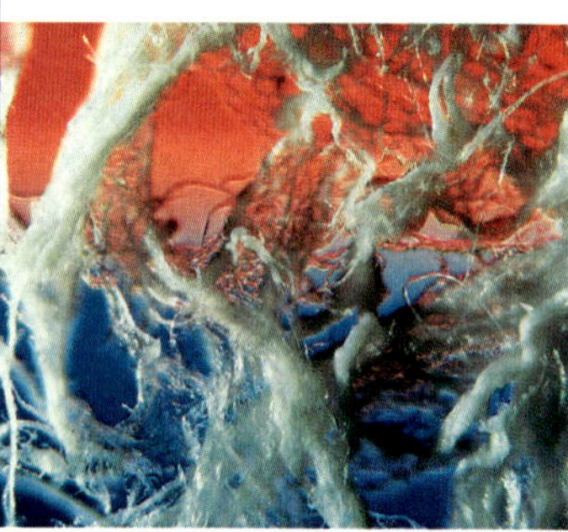

Figure 24.14 This photo shows strands of DNA extracted from cellular material.

The sugar unit in the nucleotides of DNA is the five-carbon monosaccharide known as deoxyribose. There are four different nitrogen bases in DNA—adenine, guanine, thymine, and cytosine. These four bases are abbreviated A, G, T, and C, respectively, and are shown in a short segment of a DNA molecule in Figure 24.15. Notice that adenine and guanine each contains a double ring and that thymine and cytosine contain a single ring. Ribose, which has one more oxygen atom than deoxyribose, is the sugar found in the nucleotide monomers of RNA. The base thymine is never found in RNA. Instead, it is replaced by a fifth nitrogen base, called uracil, which is abbreviated U.

778 Chapter 24

Section Resources

Print

- **Guided Reading and Study Workbook,** Section 24.5
- **Core Teaching Resources,** Section 24.5
- **Transparencies,** T281–T284

Technology

- **Interactive Textbook with ChemASAP,** Simulation 29, Assessment 24.5

Figure 24.15 The nucleotide monomers of DNA are linked together through their phosphate groups.

Nitrogen base
Simple sugar
Phosphate group

Adenine
Cytosine
Guanine
Thymine

Chemists studying nucleic acids discovered that the amount of adenine in DNA always equals the amount of thymine (A = T). Similarly, the amount of guanine always equals the amount of cytosine (G = C). The significance of these facts was not apparent until 1953, when James Watson and Francis Crick proposed that the structure of DNA consists of two polynucleotide chains wrapped into a spiral shape. This is the famous double helix of DNA. For the nitrogen bases to fit neatly into the double helix, every double-ringed base on one strand must be paired with a single-ringed base on the opposing strand. The pairing of A with T and G with C not only provides the best possible fit; it also allows the maximum number of hydrogen bonds to form between the opposing bases, as Figure 24.16 shows. Thus, the pairing of A and T (with two hydrogen bonds between the opposing bases) and of G and C (with three hydrogen bonds) makes for the most stable arrangement in the double helix.

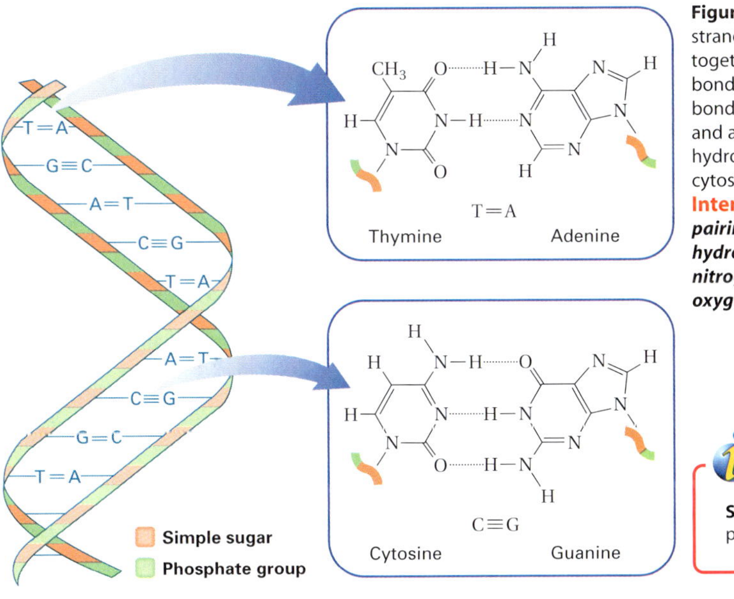

Figure 24.16 The two DNA strands in a double helix are held together by many hydrogen bonds; there are two hydrogen bonds between each thymine (T) and adenine (A) and three hydrogen bonds between each cytosine (C) and guanine (G). **Interpreting Diagrams** ***In the pairing of C with G, how many hydrogen bonds involve nitrogen? How many involve oxygen?***

Interactive Textbook

Simulation 29 Construct a portion of a DNA molecule.

with ChemASAP

DNA and RNA

Relate 

Have students read the autobiography of the Nobel prize-winning chemist, Arthur Kornberg, *For the Love of Enzymes: The Odyssey of a Biochemist*. Kornberg provides an exceptionally readable historical account of the discovery and characterization of DNA and nucleic acids. This book documents many of the important biological discoveries made in the twentieth century. Encourage students to make a time line of the history of DNA.

Use Visuals 

Figure 24.16 Point out the sequence complementarity between strands. Ask, **How does the experimental data that moles A = moles T, and moles G = moles C support the concept of base-pairing?** *(A and G are always hydrogen-bonded to T and C, respectively, in the opposite strand. Thus, there is a one-to-one mole correspondence.)* **What about the structure of the bases makes this arrangement the best fit?** *(two hydrogen bonds in the T-A connection, three for C-G, and length of bonds)*

Answers to...

Figure 24.16 nitrogen, 1; oxygen, 2

Quick LAB

A Model of DNA L2

Objective Students construct a model of double-stranded DNA.

Skills Focus Using models, inferring

 **Prep Time** 10 minutes

Materials Cardboard tube from paper-towel roll; felt-tip markers, two colors; metric ruler; thumbtack; 10 toothpicks

Class Time 20 minutes

Teaching Tips To get parallel strands, have students cut a thin strip of paper that just wraps around the tube (following the spiral seam); then measure the length of the paper, marking points along it every 5 cm. Students can use the paper to mark points on the tube and connect points to draw the parallel strand.

Expected Outcome Models should have an orderly column of bases on the inside, with phosphate and sugar groups on the outside. Ask, **What forces hold the two strands of DNA together?** *(Primarily hydrogen bonding; dispersion forces also act between stacked bases.)* **What forces the two strands apart?** *(Repulsion of negative charges on phosphate groups; two strands have greater entropy than a single helical structure.)* Point out that DNA is held together by many reinforcing bonds. Ask students to speculate about the influence of pH on the stability of DNA. *(Extremes in pH could disrupt the hydrogen bonding between bases in opposite strands due to ionization of nitrogen.).*

Analyze and Conclude

1. Base pairs are neatly stacked inside the double helix.
2. the exterior
3. the exterior

For Enrichment 

Ask students what has to happen to replicate base pairs of DNA. *(The two strands must separate.)* Have students alter their model so that it can be used to demonstrate DNA replication.

Quick LAB

A Model of DNA

Purpose

To construct a model of double-stranded DNA

Materials

- cardboard tube from paper-towel roll
- felt-tip markers (two colors)
- metric ruler
- thumbtack
- 10 toothpicks

Procedure 

1. The typical cardboard tube has a seam that, when viewed from one end, describes a spiral that moves away from the observer. This spiral is a helix. Outline the spiral seam with a colored marker.
2. Using a different-colored marker, draw a second spiral midway between the lines of the first. These two spirals represent the two strands of double-stranded DNA.
3. Measure along the tube, and mark a dot on each spiral every 5 cm. Label each dot with the letter S to indicate a sugar unit. Make a hole in the spirals at each S mark with the thumbtack. Move down each spiral and mark a letter P to indicate a phosphate group halfway between each two S dots.
4. Color each toothpick along half its length with a marker. A toothpick represents a base pair in the DNA molecule.

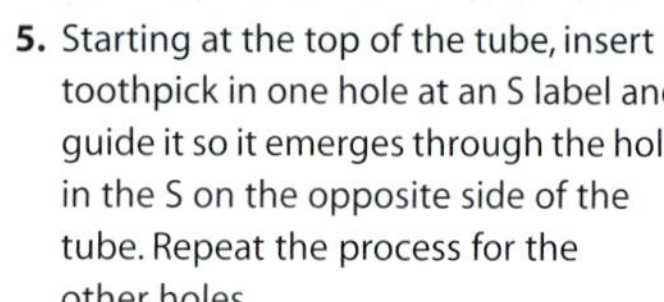

5. Starting at the top of the tube, insert a toothpick in one hole at an S label and guide it so it emerges through the hole in the S on the opposite side of the tube. Repeat the process for the other holes.

Analyze and Conclude

1. Are the bases on the interior or the exterior of the double helix? Are they randomly arranged or neatly stacked?
2. Are the phosphate groups on the exterior or the interior of the DNA structure?
3. Are the sugar groups on the interior or the exterior of the DNA molecule?

The Genetic Code

An organism contains many proteins that are characteristic of that particular organism. The proteins of earthworms are different from the proteins of pine trees, which are different from the proteins of humans. How do cells in a given kind of organism know which proteins to make? The cells use the instructions contained in the organism's DNA. A **gene** is a segment of DNA that carries the instructions for making one peptide chain. Thus the products of genes are the peptides and proteins found in an organism.

You can think of DNA as a reference manual that stores the instructions for building proteins. The instructions are written in a simple language that has 4 "letters"—the bases A, T, G, and C. Experimental data show that each "word" in a DNA manual is exactly three letters in length. Each three-letter base sequence, or triplet, codes for one of the 20 common amino acids. The code words are strung together in the DNA molecule to form genes, which specify the order of amino acids in peptides and proteins. **Three bases of DNA arranged in a specific sequence are required to specify one amino acid in a peptide or protein chain.**

 ***How do cells in a given kind of organism know which proteins to make?***

Table 24.2 provides the DNA code words for the 20 common amino acids. For example, you can see that the DNA code word AAA specifies the amino acid phenylalanine and that the DNA code word CGA specifies the amino acid alanine. Note that most amino acids are specified by more than one code word, but a code word never specifies more than one amino acid. With DNA code words of three letters, 900 bases arranged in a specific sequence would be required to code for a peptide chain made up of 300 amino acids arranged in a specific sequence.

Three code words (ATT, ATC, and ACT) are reserved as end, or termination, code words. The translation of a base sequence of DNA in a gene into the amino acid sequence of a peptide begins after one termination code word and runs continuously until another termination code word is reached. The termination code word signals a stop to the addition of amino acids in the production of the peptide, which is thus completed. You can think of a termination code as being similar to the period at the end of a sentence.

The molar masses of DNA molecules reach into the millions and possibly billions. Even with only four bases, the number of possible sequences of nucleotides in a DNA chain is enormous. The sequence of the nitrogen bases A, T, G, and C in the DNA of an organism constitutes the genetic plan, or blueprint, for that organism. This genetic plan is inherited from parents and passed to offspring. Differences in the number and sequence of the bases in DNA ultimately are responsible for the great diversity of living creatures found on Earth.

Table 24.2

Three-Letter DNA Code Words for the Amino Acids

First Letter in Code Word	Second Letter in Code Word: A	G	T	C	Third Letter in Code Word
A	AAA Phe	AGA Ser	ATA Tyr	ACA Cys	A
	AAG Phe	AGG Ser	ATG Tyr	ACG Cys	G
	AAT Leu	AGT Ser	ATT End	ACT End	T
	AAC Leu	AGC Ser	ATC End	ACC Trp	C
G	GAA Leu	GGA Pro	GTA His	GCA Arg	A
	GAG Leu	GGG Pro	GTG His	GCG Arg	G
	GAT Leu	GGT Pro	GTT Gln	GCT Arg	T
	GAC Leu	GGC Pro	GTC Gln	GCC Arg	C
T	TAA Ile	TGA Thr	TTA Asn	TCA Ser	A
	TAG Ile	TGG Thr	TTG Asn	TCG Ser	G
	TAT Ile	TGT Thr	TTT Lys	TCT Arg	T
	TAC Met	TGC Thr	TTC Lys	TCC Arg	C
C	CAA Val	CGA Ala	CTA Asp	CCA Gly	A
	CAG Val	CGG Ala	CTG Asp	CCG Gly	G
	CAT Val	CGT Ala	CTT Glu	CCT Gly	T
	CAC Val	CGC Ala	CTC Glu	CCC Gly	C

Figure 24.17 This space-filling model shows only a tiny segment of a DNA molecule and gives you an idea of the enormous mass of the entire molecule.

The Genetic Code

Relate 

The fact that millions of different sequences are possible from the combination of only four nucleotides can be compared to the fact that it is possible to form all the words of the English language from only two symbols—the dot and dash of Morse code. Moreover, both systems use patterns to communicate information. Point out, however, that Morse code is a relatively simple human invention that uses arbitrary combinations of dots and dashes to represent letters of the alphabet. The genetic code is a complex natural communication system that directs and controls all life processes.

Discuss L2

Cracking the genetic code was one of the greatest achievements of modern science. Point out that the genetic code is (nearly) universal. The three-letter codes used to translate DNA to protein are the same in humans and in bacteria. Discuss the implications of this for evolution of life on Earth. *(It supports the argument that life on Earth evolved only once.)*

Use Visuals 

Table 24.2 Ask, **How many three-letter codes are there?** *(64)* **Are all possible combinations of bases used?** *(yes)* **Why are there more codes than amino acids?** *(Some amino acids have more than one code.)* **Is there an advantage to redundant codes?** *(Redundancy might protect against harmful mutations.)*

Differentiated Instruction

Gifted and Talented 

Challenge students to write a computer program that converts any sequence of DNA bases into an amino-acid sequence using the three-letter codes in Table 24.2.

Answers to...

 The cells use instructions contained in the organism's DNA.

Gene Mutations

Genetic Code

L2

Have students practice reading the genetic code. Write a number of short DNA sequences on the board and ask students to use Table 24.2 to translate the genetic information into an amino–acid sequence. (Example: *A-C-G-G-C-A-A-G-A-G-A-A-T-G-G-A-C-T = Cys-Arg-Ser-Leu-Thr-End)* Next, ask students to interpret the effects of substitutions, additions, and deletions on the translation of the DNA sequence. Ask, **How would the sequence be affected if a thymine were substituted for the first guanine?** *(This mutation would insert a stop codon. The protein synthesis would prematurely terminate, and there would most likely be no biological activity.)*

Gene Mutations

When a change occurs in a DNA code word, the result is a mutation. **Substitutions, additions, or deletions of one or more nucleotides in the DNA molecule are called gene mutations.** The effect of the deletion of a single base from a gene can be illustrated by the following analogy. Suppose a string of letters of the alphabet goes as follows:

PATTHEBADCAT

The letters may not make sense at first glance. However, if you separate them into three-letter words, they form a perfectly sensible statement:

PAT THE BAD CAT

Now delete the first letter, and again separate the string into three-letter segments:

ATT HEB ADC AT

This last sequence is nonsensical. Similarly, the deletion of a base in the DNA base sequence can turn the information into nonsense. A sequence that once may have coded for the proper sequence of amino acids in a necessary protein may be replaced by a sequence that produces a useless or damaging amino acid sequence. The same sort of harmful effect may be obtained by mutations involving substitutions or additions of nucleotides.

Such mutations might result in the production of a faulty protein or of no protein at all. Diseases that result from gene mutations are called genetic disorders. Thousands of genetic disorders have been identified. Sickle cell anemia, which affects mainly people of African descent, results from a mutation in one of the peptide chains of hemoglobin, the oxygen-carrying protein of blood. Figure 24.18 shows the three-dimensional structure of normal hemoglobin. The defective hemoglobin made by the sickle cell gene reduces the oxygen supply to the tissues, triggering the painful episodes of the disease. In methemoglobinemia, another genetic disorder, iron in some of the hemoglobin is in the oxidized iron(III) state rather than the normal reduced iron(II) state. A faulty hemoglobin chain permits the undesirable oxidation of the iron(II) to occur.

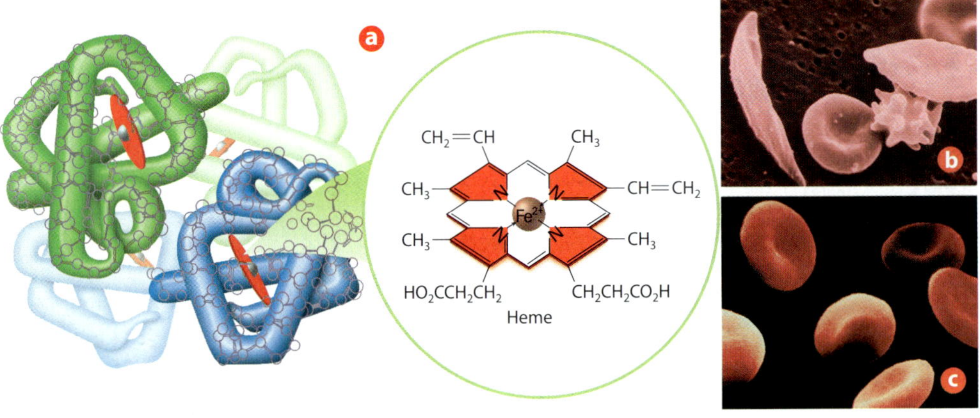

Figure 24.18 The abbreviated molecular structure of hemoglobin, the oxygen-transporting protein of blood, is shown here. **a** The hemoglobin molecule consists of four peptide chains, each containing a heme group. **b** In the genetic disorder sickle cell anemia, defective hemoglobin causes red blood cells to take on a distorted shape. **c** Normal red blood cells have a doughnut-like shape. **Interpreting Diagrams** ***How many nitrogen atoms does a heme group contain?***

Facts and Figures

Caffeine Mutations?

Many people have looked for health risks associated with the consumption of caffeinated beverages. One concern raised is the potential mutagenic effects due to the misincorporation of caffeine into nucleic acids. Caffeine is a nitrogen compound that is structurally related to purine bases, especially guanine. Its resemblance may allow it to bind to DNA-synthesizing enzymes, which would incorporate it into a polynucleotide. However, caffeine lacks hydrogen for proper base-pairing, which would destabilize the DNA helix.

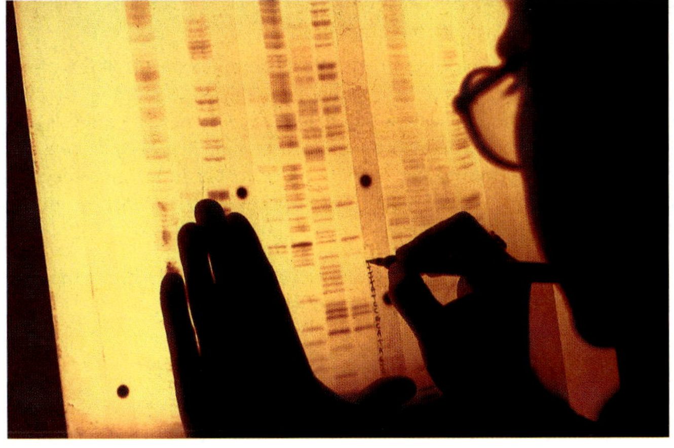

Figure 24.19 This scientist is comparing DNA fingerprints. DNA fingerprinting is an important tool in the identification of people. A person's DNA fingerprint contains DNA fragments that are characteristic of his or her father and mother but that would not be identical to those of the father or the mother.

Not all gene mutations are harmful. Occasionally, a mutation can result in the synthesis of a protein that is more efficient than the version that previously existed. Such a mutation could thus be beneficial to the survival of an organism.

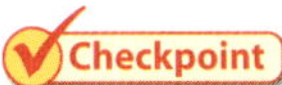 ***What are diseases that result from gene mutations called?***

DNA Fingerprinting

Only about 5% of a human's DNA is used for coding the information needed for protein synthesis. The remaining 95% consists of repeating, noncoding base sequences that separate or sometimes interrupt gene coding sequences. The role of these stretches of noncoding DNA is unclear. The noncoding sequences are similar for members of the same family but are slightly different for almost every individual. Differences also exist in the coding portions of DNA. The base sequences of DNA are slightly different for different individuals, except for identical twins. Identical twins are exactly alike because they have identical DNA.

The variation in the DNA of individuals forms the basis for a method of identifying a person from samples of his or her hair, skin cells, or body fluid. Because DNA sequences, like fingerprints, are unique for each individual, the DNA method is called DNA fingerprinting. To construct a DNA fingerprint, scientists first isolate the DNA in a sample. Only a tiny sample is needed. Enzymes that cut DNA chains between specific base sequences are used to cleave the DNA in the sample, thus providing a large number of DNA fragments. The fragments are of different lengths and base compositions for different individuals and can be separated and visualized. The pattern, or DNA fingerprint, that emerges can then be compared with a sample of DNA from a known individual, as Figure 24.19 shows.

If the DNA fingerprints are identical, it can be stated with a high degree of certainty that the DNA in the unknown sample is from the known individual. The chance in favor of a DNA fingerprint being unique to one individual can be as high as 1×10^{19} to 1—a high degree of certainty, given that the world's population is only about 6×10^9 people. In practice, however, the chance of positively identifying an individual from a sample of DNA is generally lower than predicted and is often subject to interpretation.

DNA Fingerprinting

Relate

L2

DNA fingerprinting has been controversial in terms of its reliability. The more matches between two samples of DNA, the more likely that the two samples are from one individual. The price of the test goes up with each region scanned, however, so the number of scans has a practical limit, and some risk of misidentification remains. The reliability of DNA fingerprinting varies, depending on the number of regions scanned and interpretation. However, it is much higher than the reliability of an eyewitness, which is about 50:50.

Answers to...

Figure 24.18 four

 genetic disorders

Recombinant DNA Technology

Discuss L2

Point out that the treatment of human diseases with recombinant proteins, though effective, is a replacement therapy: Patients must have a continuous supply of the protein to remain asymptomatic. Ultimately, genetic engineers would like to develop methods for treating the cause of a genetically linked disorder by providing the patient's own cells with the ability to produce a missing or faulty protein, an area of research referred to as gene therapy. Have students research this approach and report on a successful application of gene therapy. Suggested topics include cystic fibrosis and hemophilia.

Questions about the certainty of identification by means of DNA fingerprints are often at issue in criminal cases. DNA fingerprints can be obtained from evidence, such as blood, left at the scene of a crime. However, there is no standard for absolute determination of innocence or guilt from DNA fingerprinting, nor is there likely to be. Juries must decide whether the evidence is convincing. Evidence based on DNA fingerprints is permitted in more than half of the states in the United States.

How do scientists construct a DNA fingerprint from a sample of hair, skin cells, or body fluid?

Recombinant DNA Technology

Scientists have learned to manipulate genes by various methods. **Recombinant DNA technology consists of methods for cleaving a DNA chain, inserting a new piece of DNA into the gap created by the cleavage, and resealing the chain.** Figure 24.20 illustrates such a method. The altered DNA formed by this method is known as recombinant DNA.

Applications in Medicine The first practical application of recombinant DNA technology was to insert the gene for making human insulin into bacteria. Most people naturally make insulin, a polypeptide that controls levels of blood sugar, but insufficient insulin production results in diabetes. The symptoms of diabetes can often be controlled by insulin injections. In the past, human insulin was not available for this purpose. Pig insulin, which is quite similar in structure to human insulin, was used as a substitute. Some patients, however, were allergic to pig insulin. Today, diabetic patients use the human form of insulin produced by bacteria that have been altered by recombinant DNA technology. Use of this insulin removes the need for the potentially dangerous use of pig insulin.

Other proteins produced by recombinant DNA technology are used as medicinal drugs. For example, an enzyme called tissue plasminogen activator (TPA) is used to dissolve blood clots in patients who have suffered heart attacks. Another protein, interferon, is thought to relieve or delay some of the debilitating effects of multiple sclerosis. Recombinant DNA technology is also being applied to the cure of genetic disorders in a treatment known as gene therapy.

Applications in Agriculture In agriculture, new recombinant DNA techniques can make plants resistant to pests and weed killers and produce fruits and vegetables that are better suited for shipping and storage. Figure 24.21 shows genetically altered tomatoes that have a longer ripening period, which prevents spoilage and allows for their year-round availability. Genetically altered organisms have many potential benefits, but some people have concerns about their safety.

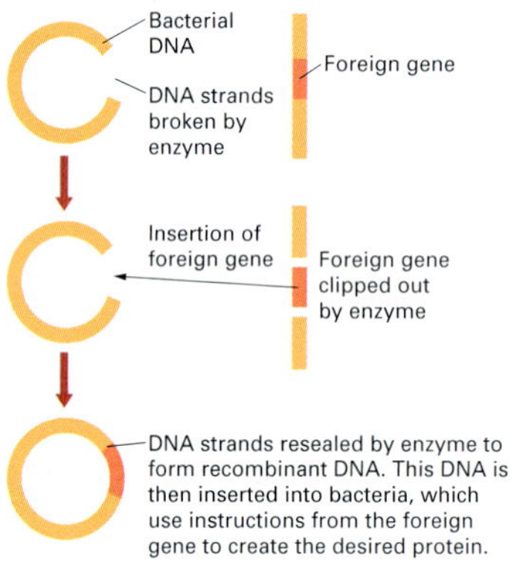

Figure 24.20 Included here are the elements of an experiment involving recombinant DNA. In this experiment, DNA from one organism is inserted into the DNA of a different organism.

Facts and Figures

Recombinant DNA

Recombinant DNA technology uses small circular DNA molecules called plasmids that occur naturally in bacteria and in some yeast. Once a human gene has been combined with the plasmid DNA, the recombinant molecule is reintroduced into bacteria. Because human DNA and bacterial DNA use the same bases, and because the genetic code is universal, the bacteria translate the human gene in the recombinant DNA into protein exactly as they would their own genes. When grown in large incubators, the bacteria can produce massive quantities of the recombinant protein.

Figure 24.21 These genetically engineered tomatoes were altered to increase the length of time it takes them to ripen.

Cloning Ethical concerns were raised in 1997 when Scottish scientists announced the birth of a lamb named Dolly, shown in Figure 24.22. In normal animal reproduction, an offspring is a genetic mixture of the characteristics of both parents. Dolly, however, was a clone—an offspring of a single individual. A clone is an exact genetic copy of its parent because it is formed using the DNA of only that parent. The birth of Dolly has raised the question of whether humans might eventually be cloned. Many people are concerned about some of the possible outcomes of cloning identical individuals. These reactions are one aspect of more general concerns about the uniqueness of life. Human cloning experiments have been outlawed in several countries.

Figure 24.22 Dolly the sheep had no father. She was cloned from a single cell taken from her mother. Dolly died in February, 2003. **Inferring** ***How did Dolly's DNA compare with the DNA in her mother's cells?***

24.5 Section Assessment

26. **Key Concept** What functions are performed by DNA and RNA?
27. **Key Concept** What does a three-letter base sequence of DNA specify?
28. **Key Concept** What are three types of gene mutations?
29. **Key Concept** What is the basis for a method of identifying a person from samples of hair, skin cells, or body fluid?
30. **Key Concept** What methods are used in recombinant DNA technology?
31. Why do you think cloning is controversial? What are your personal thoughts on this matter?

Connecting Concepts

Scientific Methods Write a paragraph describing how scientific methods might be applied to the process of DNA fingerprinting.

Assessment 24.5 Test yourself on the concepts in Section 24.5.

with ChemASAP

3 ASSESS

Evaluate Understanding L2

Have students review the DNA models on pages 779–781 and write a short paragraph describing the structure of DNA.

Reteach L1

Display Figure 24.16 on an overhead projector. Use it to discuss the chemical and functional differences between DNA and RNA. *(RNA contains a different sugar and one different base. RNA transmits genetic information from the nucleus to the cytoplasm in a eukaryotic cell.)*

Connecting Concepts

Forensic investigators use scientific methods when trying to solve a crime. At the crime scene, they make observations and collect evidence, including DNA samples. The investigators make hypotheses about how the crime was committed and who may be responsible. They can test these hypotheses by questioning suspects and comparing DNA fingerprints of the suspects with those left at the crime scene.

If your class subscribes to the Interactive Textbook, use it to review key concepts in Section 24.5.

with ChemASAP

Section 24.5 Assessment

26. DNA stores genetic information; RNA transmits genetic information.
27. an amino acid
28. substitution, addition, and deletion of a nucleotide
29. variation in DNA
30. Answers might include the ideas of cleaving a DNA chain, inserting a new piece of DNA into the gap created by the cleavage, resealing the chain, and expressing the new gene product.
31. People have different opinions about issues such as the dignity of human life and interfering with nature.

Answers to...

Figure 24.22 Their DNA was the same.

 isolate DNA from a sample; fragment it; separate and visualize the fragments

24.6

1 FOCUS

Objectives

24.6.1 Describe the function of ATP in cells.

24.6.2 Distinguish between catabolism and anabolism.

24.6.3 Describe the nitrogen cycle.

Guide for Reading

Build Vocabulary L2

Paraphrase Have students write definitions of *metabolism, catabolism,* and *anabolism* in their own words.

Reading Strategy L2

Monitor Your Understanding After students read about catabolism and anabolism, ask them to create a diagram that shows how carbon moves from one type of metabolism to the other. They should show where energy (ATP) comes into play.

2 INSTRUCT

Connecting to Your World

Ask, **What is metabolic rate?** *(how fast the body's chemical reactions are)* **Why is iodine important in controlling human metabolism?** *(It is a component of the hormone thyroxine, which controls metabolic rate.)*

ATP

Discuss L2

Have students compare the free energy changes associated with spontaneous and nonspontaneous processes. Energy released by a reaction that is available to do work is called free energy. Spontaneous processes, which favor formation of products, release free energy. Point out that it is the phosphate–oxygen anhydride bonds in ATP that store the chemical energy in ATP. When broken by water, these bonds release energy that is used to form new types of chemical bonds in other molecules.

24.6 Metabolism

Guide for Reading

Key Concepts

- What is the function of ATP in living cells?
- How does a cell obtain the energy and building blocks needed for the construction of new biological compounds?
- What happens in a cell during anabolism?
- How do nitrogen-fixing bacteria provide plants with a useable form of nitrogen?

Vocabulary

adenosine triphosphate (ATP)
metabolism
catabolism
anabolism

Reading Strategy

Comparing and Contrasting When you compare and contrast things, you examine how they are alike and different. As you read, list the similarities and differences between catabolism and anabolism.

Connecting to Your World

The body's metabolic rate is controlled by a hormone called thyroxine that is released by the thyroid gland. When a person's diet does not contain adequate iodine, the thyroid gland is unable to produce and release thyroxine. The gland begins to swell, forming a goiter. In this section, you will learn where some of the nutrients in a person's diet come from and how they are involved in metabolic reactions.

ATP

All living things need energy to function properly. **Adenosine triphosphate (ATP),** shown in Figure 24.23, is a molecule that transmits this energy in the cells of living organisms. The function of ATP can be compared to a belt connecting an electric motor to a pump. The motor generates energy capable of operating the pump. But if a belt does not connect the motor to the pump, the energy produced by the motor is wasted. You can think of ATP as the belt that couples the production and use of energy by cells. **In living cells, ATP is the energy carrier between the spontaneous reactions that release energy and nonspontaneous reactions that use energy.**

Recall that oxidation reactions, such as the combustion of methane in a furnace or the oxidation of glucose in a living cell, are spontaneous reactions that release energy. This energy can be captured when adenosine diphosphate (ADP) condenses with an inorganic phosphate group to become ATP. The addition of a phosphate group, called phosphorylation, occurs during certain biochemical oxidation reactions.

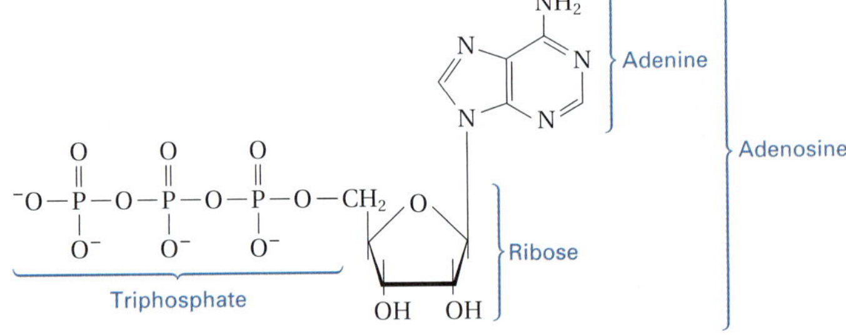

Figure 24.23 ATP is ADP that has been phosphorylated. ATP provides energy to muscles for moving the body. **Comparing and Contrasting** ***How is the structure of ATP similar to that of a DNA nucleotide?***

Section Resources

Print

- **Guided Reading and Study Workbook,** Section 24.6
- **Core Teaching Resources,** Section 24.6 Review, Interpreting Graphics
- **Transparencies,** T285

Technology

- **Interactive Textbook with ChemASAP,** Assessment 24.6
- **Go Online,** Section 24.6

$$\text{Adenosine}-\overset{\overset{\Large O}{\|}}{\underset{\underset{\Large OH}{|}}{P}}-O-\overset{\overset{\Large O}{\|}}{\underset{\underset{\Large OH}{|}}{P}}-OH + HO-\overset{\overset{\Large O}{\|}}{\underset{\underset{\Large OH}{|}}{P}}-OH \longrightarrow \text{Adenosine}-\overset{\overset{\Large O}{\|}}{\underset{\underset{\Large OH}{|}}{P}}-O-\overset{\overset{\Large O}{\|}}{\underset{\underset{\Large OH}{|}}{P}}-O-\overset{\overset{\Large O}{\|}}{\underset{\underset{\Large OH}{|}}{P}}-OH + H_2O$$

Adenosine diphosphate (ADP) | Inorganic phosphate (P_i) | Adenosine triphosphate (ATP) | Water

The formation of ATP efficiently captures energy produced by the oxidation reactions in living cells. Every mole of ATP produced by the phosphorylation of ADP stores about 30.5 kJ of energy. The reverse happens when ATP is hydrolyzed back to ADP: Every mole of ATP that is hydrolyzed back to ADP releases about 30.5 kJ of energy. Cells use this released energy to drive processes that would ordinarily be nonspontaneous. Because of its ability to capture energy from one process and transmit it to another, ATP is sometimes referred to as a high-energy compound; however, the energy produced by the breakdown of ATP to ADP is not particularly high for the breaking of a covalent bond. ATP is important because it occupies an intermediate position in the energetics of the cell. It can be formed by using the energy obtained from a few higher-energy oxidation reactions. The energy that is contained in the bonds of ATP can then be used to drive other cellular processes.

Catabolism

Thousands of chemical reactions take place in the cells of a living organism. The entire set of chemical reactions carried out by an organism is known as the organism's **metabolism.** In metabolism, unneeded cellular components and the nutrients in food are broken down into simpler compounds by chemical reactions collectively called **catabolism.** Catabolic reactions release energy as well as produce simple compounds. **The degradation of complex biological molecules such as carbohydrates, lipids, proteins, and nucleic acids during catabolism provides the energy and the building blocks for the construction of new biological compounds needed by the cell.** Through the formation of ATP, catabolic reactions provide the energy for such needs as body motion and the transport of nutrients to cells where they are required. The oxidation reactions of catabolism also provide energy in the form of heat. These reactions help keep your body temperature constant at 37°C. The need for energy and building blocks is the reason why all organisms, such as the field mouse shown in Figure 24.24, require food.

The complete oxidation of glucose to carbon dioxide and water is one of the most important energy-yielding processes of catabolism. Study Figure 24.25, which summarizes the major steps in the degradation of one glucose molecule to six molecules of carbon dioxide. The complete oxidation actually involves many reactions, which are not shown. As you can see in the figure, the major carbon-containing reactants and products are named, and they are also referred to according to the number of carbons they contain.

Figure 24.24 Organisms such as this mouse use the energy stored in the chemical bonds of food molecules to power their body processes.

Facts and Figures

Free Energy

All living things require a continuous input of free energy for muscle contraction, transport of nutrients, and synthesis of biomolecules. More than 50% of the free energy released by cellular reactions is used to cause other chemical changes. The rest is lost as heat. Compare this efficiency to that of the internal combustion engine. At most, 30% of the free energy of combustion is used to propel a car.

Discuss 

Use Hess's law of heat summation to show how the free energy of ATP hydrolysis can be harnessed by the cell to drive a nonspontaneous reaction. Consider the reversible, nonspontaneous reaction:

$$X \rightleftharpoons Y \quad \Delta G^\circ = +21 \text{ kJ} = \text{mol}$$

Normally, K_{eq} for this reaction is much less than 1, but if the reaction is coupled to the hydrolysis of ATP:

$$ATP + H_2O \rightarrow ADP + P + H^+ \quad \Delta G^\circ = -30.51 \text{ kJ/mol}$$

$$X + ATP + H_2O \rightarrow YP + ADP + H^+ \quad \Delta G^\circ = -9.5 \text{ kJ/mol}$$

Conversion of X to Y is driven forward by high intracellular concentrations of ATP.

Catabolism

Relate 

In the human body, glycogen is stored in skeletal muscles and the liver. Liver glycogen is used to maintain normal blood sugar level. Muscle glycogen, however, is used solely for muscle work. Large amounts of carbohydrates, including glycogen, are used by the body during athletic events that require power or speed. The carbohydrates provide the necessary energy. Interval training, a technique that alternates rest and work phases, is designed as much to train biochemical energy pathways as to train the muscles used in an athletic activity. Proper training and diet can increase the amount of glycogen stored after a training session by more than a thousandfold.

Download a worksheet on **food as fuel** for students to complete, and find additional teacher support from NSTA SciLinks.

Answers to...

Figure 24.23 Like the nucleotides of DNA and RNA, ATP has a basic structure consisting of a phosphate group, a five-carbon sugar, and a nitrogen base. The sugar unit in both ATP and RNA is ribose.

Anabolism

Use Visuals L1

Figure 24.26 Ask, **Are there any inputs (created by catabolism) that aren't reflected in outputs?** *(no)* Discuss why this is so for an animal, for example. *(We need all these things to live, and there has to be a way to break them down and use them to build a body.)*

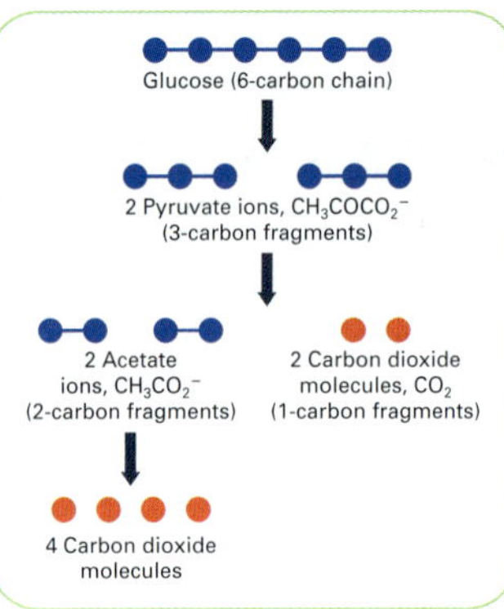

Figure 24.25 The breakdown of glucose to carbon dioxide and water is one of the most important energy-yielding processes of catabolism. **Making Generalizations** ***What happens to the number of carbon-carbon bonds from one step to the next?***

The combustion of 1 mole of glucose to 6 moles of carbon dioxide and 6 moles of water, either by fire or by oxidation in a living cell, produces 2.82×10^3 kJ of energy. Cells that use oxygen may produce up to 38 moles of ATP by capturing the energy released by the complete oxidation of a single mole of glucose! The large amount of ATP produced from the oxidation of glucose makes it the likeliest mode of energy production for most kinds of cells. In fact, if glucose is available, brain cells use no other source of carbon compounds for energy production.

Checkpoint ***How much energy is released by the complete combustion of 1 mole of glucose?***

Anabolism

Some of the simple compounds produced by catabolism are used to synthesize more-complex biological molecules—carbohydrates, lipids, proteins, and nucleic acids—necessary for the health and growth of an organism. The synthesis reactions of metabolism are called **anabolism.** Unlike catabolism, which releases energy, anabolism uses energy.

Figure 24.26 gives an overview of the relationship between catabolism and anabolism. Nutrients and unneeded cell components are degraded to simpler components by the reactions of catabolism. The oxidative reactions of catabolism yield energy captured in the formation of ATP.  **In anabolism, the products and the energy of catabolism are used to make new cell parts and compounds needed for cellular life and growth.** You already know that energy produced by physical and chemical processes is of little value unless the energy can be captured to do work. If it is not captured, the energy is lost as heat. The chemical energy produced by catabolism must have some means of being used for the chemical work of anabolism. The ATP molecule is that means of transmitting energy.

Figure 24.26 Simple compounds produced by catabolism are used in the synthesis reactions of anabolism. **Applying Concepts** ***What part of metabolism releases energy? What part uses energy?***

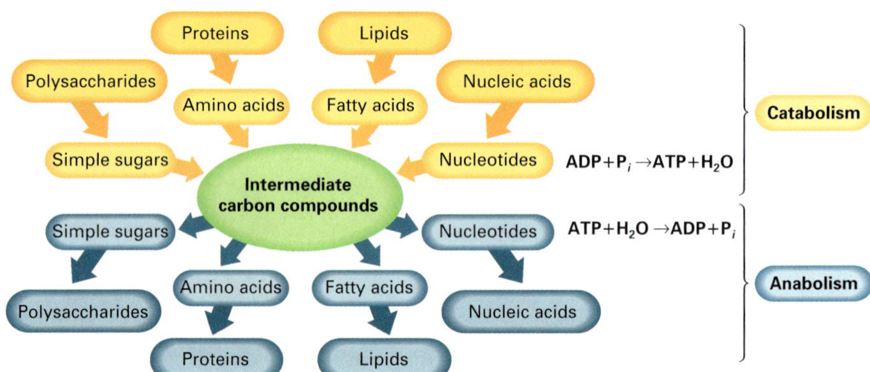

Differentiated Instruction

Less Proficient Readers L1

Have students create cards for the steps in glucose breakdown, shown in Figure 24.26. They should expand on the shorthand version in the visual, using words or chemical structures, so that each step makes sense to them. They can shuffle their cards and have another student put them into the correct sequence.

788 Chapter 24

The Nitrogen Cycle

You have learned that the biological molecules taken into an organism's body as nutrients in food are broken down during catabolism. Food contains carbohydrates, proteins, lipids, nucleic acids, vitamins, and minerals. These nutrients are composed mainly of carbon, hydrogen, and oxygen atoms. Many biological compounds, such as proteins, contain nitrogen as well. Although Earth's atmosphere is 78% nitrogen gas, no animals and only a few plants can use this form of nitrogen to make nitrogen-containing compounds. However, certain bacteria can convert nitrogen gas into useable forms in a process called nitrogen fixation. **Nitrogen-fixing bacteria reduce atmospheric nitrogen ($N_2(g)$) to ammonia ($NH_3(g)$), a water-soluble form of nitrogen that can be used by plants.** In soil and biological fluids, most ammonia is present as ammonium ions.

Plants incorporate ammonia into biological nitrogen compounds such as proteins, nucleic acids, and ATP. Because animals cannot synthesize these compounds, they get them by eating plants or other animals that eat plants. When these plants and animals die, they decay with the aid of bacteria. Decaying matter returns nitrogen to the soil as ammonia, nitrite ions (NO_2^-), or nitrate ions (NO_3^-). Moreover, some nitrogen gas is returned to the atmosphere. This flow of nitrogen between the atmosphere and Earth and its living creatures is the nitrogen cycle, shown in Figure 24.27.

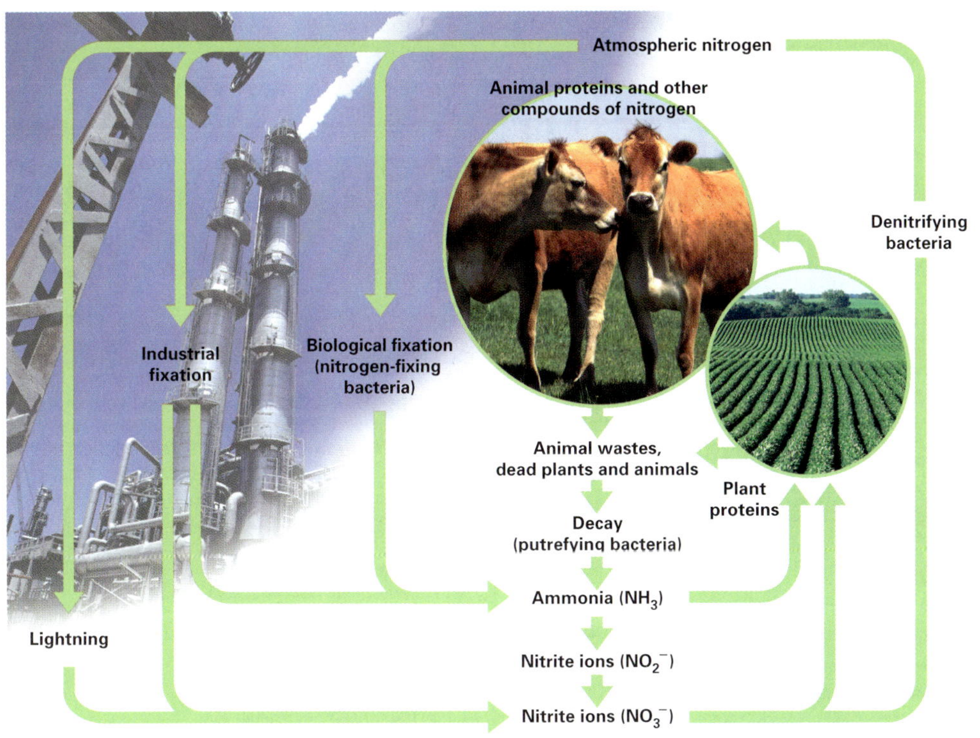

Figure 24.27 Nitrogen moves between the atmosphere and the biosphere in the nitrogen cycle.

The Nitrogen Cycle

Relate

Nitrogen often is the limiting factor in how many plants can grow to maturity in a system. Some plants specialize in low-nitrogen settings, which prevent faster growing plants from taking over the habitat. Release of extraordinary nitrogen from fertilizers allows these other plants to outcompete the low-nitrogen specialists and may eliminate them from the habitat.

Answers to...

Figure 24.25 The number of C—C bonds decreases.

Figure 24.26 Catabolism releases energy; anabolism uses energy.

2.82×10^3 kJ

Section 24.6 (continued)

3 ASSESS

Evaluate Understanding L2

Ask students to compare the carbon cycle to the nitrogen cycle. Ask, **Where does energy come into play? What is its source?**

Reteach L1

Draw a schematic diagram on the board that illustrates the exchange of energy between anabolic and catabolic reactions. Connect the terms ATP and ADP by upper and lower arrows. Label the upper arrow "anabolism". In anabolic reactions, small molecules are put together to form larger ones. These nonspontaneous processes must be coupled to the free energy of ATP hydrolysis; thus, ADP is produced during anabolic reactions. Label the lower arrow "catabolism". In catabolic reactions, large molecules are broken down. Catabolic reactions most often release energy that can be used to drive the synthesis of ATP.

Writing Activity

The nitrogen cycle is the flow of nitrogen between the atmosphere and biosphere. Atmospheric nitrogen is converted to ammonia and ammonium ions by nitrogen-fixing bacteria. Plants, and then plant-eating organisms, synthesize nitrogen-containing compounds. The decay of these plants and animals returns nitrogen to the atmosphere.

If your class subscribes to the Interactive Textbook, use it to review key concepts in Section 24.6.

with ChemASAP

Figure 24.28 The bumps on these plant roots contain bacteria, which live in a symbiotic relationship with the plant. The plant gets nitrogen in a form it can use, and the bacteria get food in the form of sugars that the plant makes during photosynthesis.

Biological Nitrogen Fixation Nitrogen-fixing bacteria are of two types: free-living and symbiotic. Free-living bacteria lead an independent existence in soil. Symbiotic bacteria, such as *Rhizobium,* live in a mutually beneficial arrangement with plants. Symbiotic bacteria live in nodules on the roots of legumes, such as alfalfa, clover, peas, and beans. These root nodules are shown in Figure 24.28. Soil fertility can be improved by plowing nitrogen-rich legumes back into the ground instead of harvesting them.

Industrial Nitrogen Fixation Modern agriculture uses an enormous quantity of nitrogen, which plays a role in the nitrogen cycle. For the past several years, the daily amount of atmospheric nitrogen fixed by industrial processes in the production of fertilizers has probably exceeded the amount fixed by living organisms in Earth's forests and oceans. Nitrogen fertilizers enter the biosphere when they are taken up by plants. In addition, a small amount of atmospheric nitrogen is fixed by lightning discharges, which produce the soluble nitrogen oxides (NO, NO_2, N_2O_4, N_2O_5).

24.6 Section Assessment

32. **Key Concept** What is the role of ATP in energy production and energy use in living cells?
33. **Key Concept** What is the function of catabolism in the cells of living organisms?
34. **Key Concept** How does anabolism make use of the products of catabolism?
35. **Key Concept** What form of nitrogen is supplied to plants by nitrogen-fixing bacteria?
36. How many moles of ATP are formed from the complete oxidation of 1 mol of glucose in a cell that uses oxygen?

Writing Activity

Explain a Process Review the diagram of the nitrogen cycle in Figure 24.27. Write a paragraph that describes how nitrogen moves between the atmosphere and the biosphere. (*Hint:* Use atmospheric nitrogen as the starting point in your description of the process.)

Assessment 24.6 Test yourself on the concepts in Section 24.6.

with ChemASAP

Section 24.6 Assessment

32. ATP transmits energy in the body. Energy is released and captured in the bonds of ATP when the bonds of ATP are broken.
33. Catabolism is the subset of metabolic reactions that involve degradation of complex molecules and production of energy.
34. Catabolic reactions provide the energy and the building blocks for the construction of new compounds needed by the cell in anabolism.
35. Atmospheric nitrogen is converted to ammonia and ammonium ions.
36. as many as 38 moles of ATP

Technology & Society

Biomedical Implants

Science, medicine, and technology have combined to create artificial limbs, artificial organs, and other parts for the human body. Some are mechanical devices, some have batteries, and some even contain computers. Scientists are developing replacement wrists and shoulders, researching ways to extend the life of implants, and creating sophisticated implants using robotics. **Inferring** ***What do you think would be considered desirable properties for the materials used to make a knee implant?***

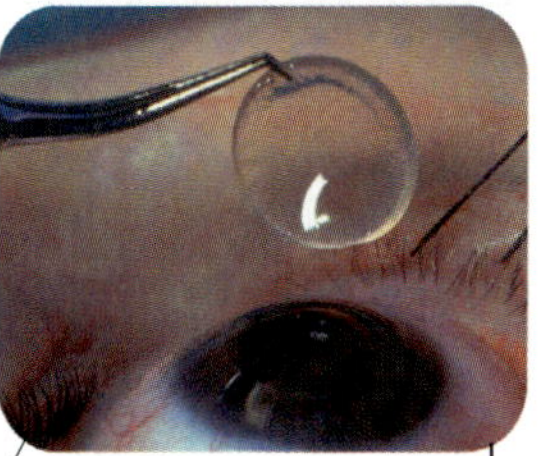

Eye implants Small acrylic disks or rings reshape the eye to correct nearsightedness. Plastic or acrylic lens implants are used to treat patients with cataracts.

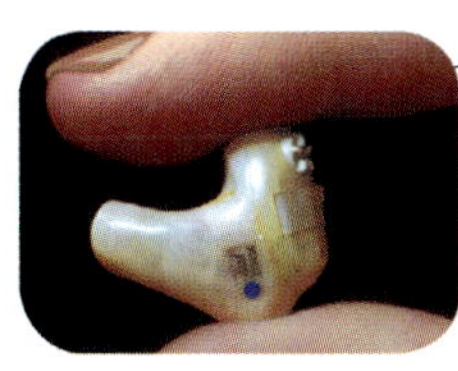

Cochlear implants An external microphone picks up sound, and a mini-computer translates the sound into signals received by a device implanted in the inner ear.

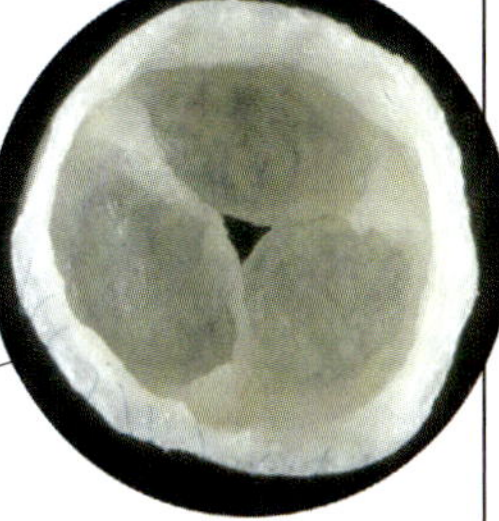

Heart valve implants Heart valves control blood flow through the heart. Implants replace valves damaged by age or disease.

Knee implants Made of metal and plastic, the first implants used a hinge design. Newer models more closely emulate the complex movements of a knee.

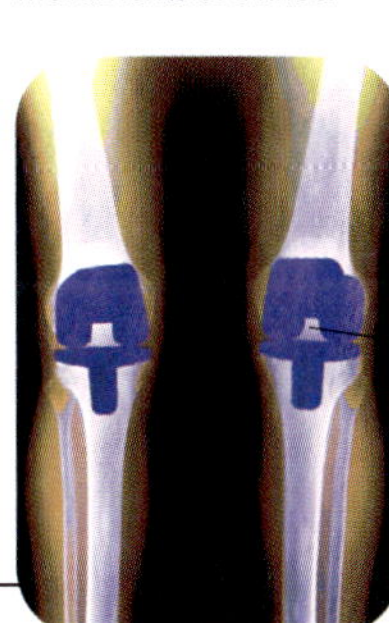

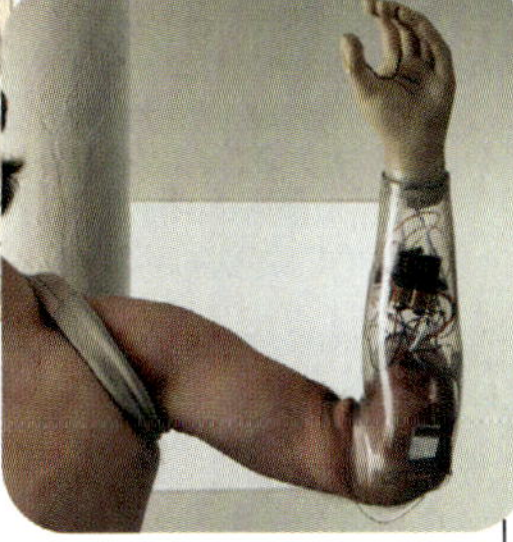

Arm implants Signals from muscles switch motors on or off to move the hand, wrist, or elbow. Researchers hope to develop arm implants that will respond directly to signals from the brain.

Technology & Society

Biomedical Implants

Implants require the efforts of many professionals, including materials scientists, mechanical engineers, surgeons, and biomedical engineers. After implant surgery, follow-up is essential to make it work, which involves therapists working in a variety of fields. The cochlear implant, for example, requires the help of speech and hearing specialists for many hours to train people to recognize speech and other sounds. Implants don't last forever. For example, knees and hips produce debris during wear. The debris level builds up and causes joint loosening, discomfort, and the need for repair or replacement surgery.

Students can find a variety of Web sites about biomedical implants. The sites range from information for surgeons, to how to make a decision on whether to get an implant, to support groups for people living with an implant.

Answers to...

Inferring They should have strength, good range of motion, and smooth joint surfaces.

CHAPTER 24 Study Guide

Study Tip

Study with a friend. Quiz each other, compare notes, discuss discrepancies. Organize the types of biomolecules you learned about in this chapter. You could make a chart that compares the different types of chemicals, or a flow-chart of processes. Work together to make sure you include all the important points.

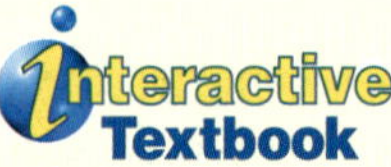

If your class subscribes to the Interactive Textbook with ChemASAP, your students can go online to address an interactive version of the Student Edition and a self-test.

with ChemASAP

CHAPTER 24 Study Guide

Key Concepts

24.1 A Strategy for Life

- Two major cell types occur in nature: prokaryotic cells and eukaryotic cells.
- Photosynthesis uses sunlight to reduce CO_2 to compounds that contain C—H bonds, mainly in the form of glucose.

24.2 Carbohydrates

- Glucose is abundant in plants and animals.
- The cyclic forms of two simple sugars can be linked by means of a condensation reaction.

24.3 Amino Acids and Their Polymers

- An amino acid has a carboxyl group, an amino group, a hydrogen atom, and an R group bonded to a central carbon atom.
- Amide bonds between amino acids involve the amino and carboxyl groups, not side groups.
- Differences in the amino acid sequence result in differences in the properties of peptides.
- Enzymes increase reaction rates.

24.4 Lipids

- Unlike other biological molecules, lipids dissolve readily in organic solvents.
- In water, phospholipids spontaneously form a spherical double layer called a lipid bilayer.

24.5 Nucleic Acids

- DNA stores information needed to make proteins and governs the reproduction of cells. RNA transmits information stored in DNA during protein synthesis.
- A sequence of three bases of DNA is required to specify one amino acid in a peptide.
- Gene mutations occur when nucleotides in DNA are substituted, added, or deleted.
- The variation in DNA forms the basis for identifying a person from biological samples.
- In recombinant DNA technology, DNA is cleaved and a new piece of DNA is inserted.

24.6 Metabolism

- In living cells, ATP is the energy carrier between reactions that release and use energy.
- The degradation of biological molecules during catabolism provides the energy and the building blocks for making new compounds.
- In anabolism, new compounds needed for cellular life and growth are made from the products of catabolism.
- Nitrogen-fixing bacteria reduce atmospheric nitrogen to ammonia.

Vocabulary

- active site (p. 772)
- adenosine triphosphate (p. 786)
- amino acid (p. 769)
- anabolism (p. 788)
- carbohydrates (p. 766)
- catabolism (p. 787)
- disaccharide (p. 767)
- enzymes (p. 772)
- gene (p. 780)
- lipids (p. 775)
- metabolism (p. 787)
- monosaccharides (p. 766)
- nucleic acids (p. 778)
- nucleotides (p. 778)
- peptide (p. 770)
- peptide bond (p. 770)
- phospholipids (p. 776)
- photosynthesis (p. 764)
- polysaccharides (p. 767)
- protein (p. 770)
- saponification (p. 776)
- substrates (p. 772)
- triglyceride (p. 775)
- waxes (p. 777)

Key Equations

- Photosynthesis:
  $6CO_2 + 6H_2O + \text{Energy} \longrightarrow C_6H_{12}O_6 + 6O_2$
- Oxidation of glucose:
  $C_6H_{12}O_6 + 6O_2 \longrightarrow 6CO_2 + 6H_2O + \text{Energy}$

Organizing Information

Use these terms to construct a concept map that organizes the major ideas of this chapter.

Concept Map 24 Solve the Concept Map with the help of an interactive guided tutorial.

with ChemASAP

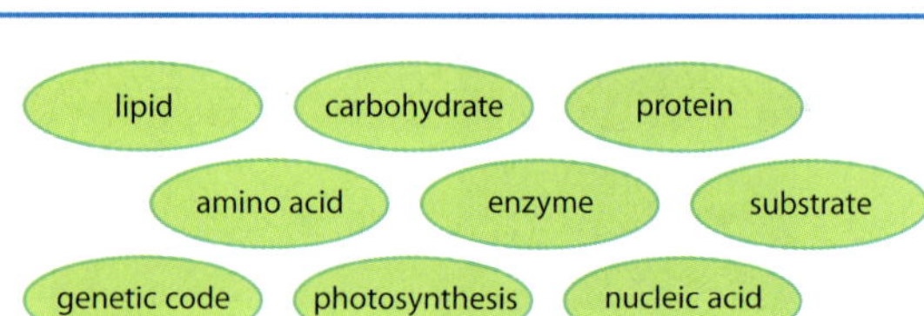

Chapter Resources

Print

- **Core Teaching Resources,** Chapter 24, *Practice Problems, Vocabulary Review, Quiz, Chapter Test A, Chapter Test B*

Technology

- **Computer Test Bank,** Chapter 24 Test
- **Interactive Textbook with ChemASAP,** Chapter 24

CHAPTER 24

Assessment

Reviewing Content

24.1 A Strategy for Life

37. What is the main difference between a prokaryotic and a eukaryotic cell?

38. Explain what happens in photosynthesis.

39. Write a balanced equation for the complete oxidation of glucose.

40. Describe three organelles found in eukaryotic cells. Give a function of each organelle.

24.2 Carbohydrates

41. Name two important monosaccharides.

42. Where in nature are glucose and fructose found?

43. How does the carbonyl functional group differ in glucose and fructose?

44. Which monosaccharides combine to form the disaccharide sucrose?

45. What is the product of the complete hydrolysis of each of the following polysaccharides?
a. starch
b. glycogen

46. What product is formed when cellulose is broken down?

24.3 Amino Acids and Their Polymers

47. What is the name given to the bond connecting two amino acids in a peptide chain?

48. How many peptide bonds does the tripeptide Ser—Gly—Phe have?

49. Describe two common patterns found in the folding of protein chains.

50. Are the structures of the following two tripeptides the same? Explain.
a. Ala—Ser—Gly
b. Gly—Ser—Ala

51. Describe the function of an enzyme.

52. What is an enzyme-substrate complex? How does it form?

24.4 Lipids

53. Distinguish between a fat and an oil.

54. What is a triglyceride?

55. What is a soap?

56. Draw structural formulas for the products of the complete hydrolysis of tristearin.

57. Draw a simple representation of a lipid bilayer.

58. What two types of compounds combine to form a wax?

59. What is the function of waxes in plants? In animals?

24.5 Nucleic Acids

60. What two types of nucleic acids do cells contain?

61. What are the components of a nucleotide?

62. What is the structural difference between the sugar unit in RNA and the sugar unit in DNA?

63. What type of bonding helps hold a DNA double helix together?

64. Which of the following base pairs are found in a DNA molecule: A—A, A—T, C—G, G—A, A—U, T—U?

65. How many bases specify an amino acid in the genetic code?

66. What are the consequences of a substitution of one base for another in DNA?

67. What is the basis for identifying an individual by DNA fingerprinting?

68. What is recombinant DNA?

24.6 Metabolism

69. Write an abbreviated, balanced equation for the hydrolysis of ATP to ADP.

70. Where do the complex biomolecules your body degrades during catabolism come from?

71. Compare and contrast catabolism and anabolism.

72. Describe the nitrogen cycle in your own words.

73. What is the source of raw materials used in anabolic reactions?

74. What is meant by industrial nitrogen fixation? What is produced during this process?

CHAPTER 24

Assessment

Reviewing Content

37. A eukaryotic cell has a cell membrane and organelles. Prokaryotic cells have a cell membrane, but lack organelles.

38. Photosynthetic organisms use the sun's energy to synthesize carbon compounds from CO_2 and H_2O

39. $C_6H_{12}O_6 + 6O_2 \rightarrow 6CO_2 + 6H_2O +$ energy

40. sample answer: nucleus (cellular reproduction); mitochondria (cellular energy production); lysosomes (digestion of substances)

41. glucose and fructose

42. Glucose is found in blood, corn, and grapes. Fructose is found in honey and many fruits.

43. Glucose is an aldehyde; fructose is a ketone.

44. glucose and fructose

45. **a.** glucose **b.** glucose

46. glucose

47. peptide bond

48. two

49. Peptide chains fold into helixes or into sheets in which peptide chains lie side by side.

50. No; Both peptides contain alanine, serine, and glycine, but the sequences are different.

51. Enzymes catalyze biological reactions.

52. complex formed when an enzyme and substrate bind to each other

53. At room temperature, animal fats are solid; plant oils are liquid.

54. triester of glycerol and fatty acids

55. alkali metal salt of a fatty acid

56.

$$\begin{array}{l} CH_2OH \\ | \\ CHOH \\ | \\ CH_2OH \end{array} + 3CH_3(CH_2)_{16}-\overset{\overset{\displaystyle O}{\|}}{C}-O^-Na^+$$

Glycerol   Sodium stearate

57.

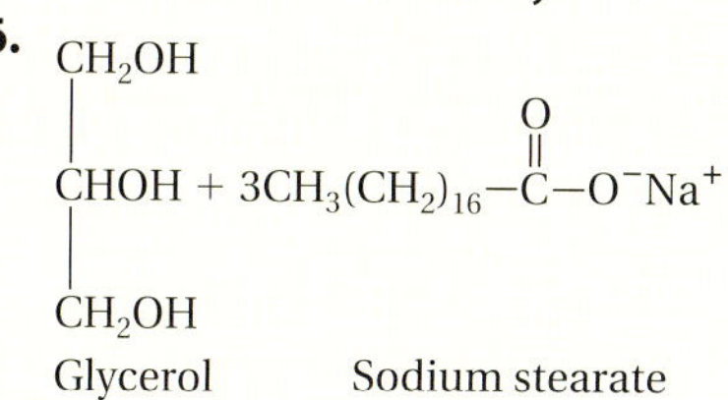

58. long-chain fatty acid and long-chain alcohol

59. protection from water loss and microorganisms in plants; pliability and waterproofing in animals

60. DNA and RNA

61. phosphate group, sugar unit, nitrogen base

62. Ribose has one more oxygen.

63. hydrogen bonding

64. A–T and C–G

65. three

66. a possible gene mutation

67. Each individual's DNA is unique.

68. altered DNA that contains a foreign gene

69. $ATP + H_2O \rightarrow ADP + P_i$

70. foods or existing body tissues

71. In catabolism, molecules are broken down and energy is captured. These products are used to make biomolecules in anabolism.

72. N_2 is converted to NH_3, which is taken up by plants, which are eaten by animals. Decaying organisms return N_2 to the air.

73. products of catabolism

74. Atmospheric nitrogen is reduced to ammonia by industrial processes.

CHAPTER 24 Assessment continued

Understanding Concepts

75. Catabolism of 1 mol glucose yields 2.82×10^3 kJ of energy.

76.

$$CH_3(CH_2)_{14}-\overset{\overset{\displaystyle O}{\|}}{C}-O^-Na^+$$

sodium palmitate

77. The negatively charged hydrophilic phosphate heads interact favorably with water.

78. a. Arg-Gly-Cys-Asn
b. Arg-Gly-Cys-Asn

79. a. triglyceride
b. glucose
b. dipeptide

80. C-G-x-C-C-x-T-C-A(or G); There are multiple codes for the same amino acid.

81. G-C-T-A-G-G-T

82. a. amino acid
b. monosaccharide (simple sugar)
c. nucleotide

83. permit passage of nutrients into cells

84. No; three code words specify the termination of a peptide chain.

85. many applications in medicine and agriculture

86. More oxygen is needed for energy-producing processes; it rids the body of excess CO_2.

87. $30.5 \text{ kJ/mol} \times 38 \text{ mol}/2.82 \times 10^3 \text{ kJ/mol} \times 100\% = 41.1\%$

88. about 275 mol ATP

89. The shape of the cellulose and starch molecules are different; humans lack enzymes necessary to cleave cellulose to glucose monomers.

90. A substrate molecule or molecules binds to the active site of the enzyme to form an enzyme-substrate complex.

91. Coenzymes must be present for an enzyme-catalyzed reaction to occur.

92. evidence from fossilized bacteria

93. a. iron(II) ion (Fe^{2+})
b. iron(III) ion. The consequence is methemoglobinemia.

94. Mutations could occur that would change how the organism constructs a protein.

95. Photosynthesis provides the carbon compounds that plants and animals need to exist.

96. monosaccharides: single simple sugar; glucose, fructose disaccharides: two linked monosaccharide units; sucrose polysaccharide: many linked monosaccharide units; starch, glycogen, cellulose

97. A carboxylic acid group of one amino acid and amino group of another amino acid undergo condensation polymerization to form an amide (peptide) bond.

98. hydrogen bonding between A-T (A-U in RNA) and G-C in DNA; helps hold DNA in double helix

---

CHAPTER 24 Assessment continued

Understanding Concepts

75. Some experts recommend that complex carbohydrates provide 50–55% of dietary Calories. What is the relationship between energy production and complex carbohydrates?

76. The formula for palmitic acid is $CH_3(CH_2)_{14}CO_2H$. A popular soap is mostly sodium palmitate. Draw a structural formula for sodium palmitate.

77. Why are the hydrophilic heads located on the outsides of the cell membrane?

78. Consider the following sequence of DNA: GCC–CCA–ACG–TTA.
a. Using the code words for amino acids in Table 24.2 on page 781, write the amino acid sequence formed by translation of the DNA sequence into a peptide.
b. What amino acid sequence would result from the substitution of adenine (A) for the second cytosine (C)?

79. Identify or classify each of the following biological molecules.

a.

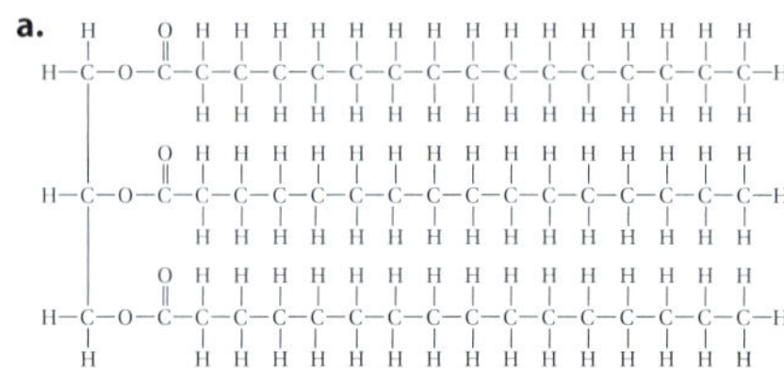

b.

c.

80. Use Table 24.2 on page 781 to write a base sequence for DNA that codes for the tripeptide Ala—Gly—Ser. Why might your answer be different from the answers of your classmates?

81. A segment of a DNA strand has the following base sequence: CGATCCA. Write the base sequence that would be found on the other strand in the double helix.

82. Which type of monomer produces each of the following polymers?
a. protein
b. polysaccharide
c. nucleic acid

83. What is one function of membrane proteins?

84. Does every code word in DNA specify an amino acid in protein synthesis? Explain.

85. What are some of the outcomes of recombinant DNA research?

86. Why is it necessary to breathe deeply during vigorous exercise?

87. The complete oxidation of glucose releases 2.82×10^3 kJ/mol of energy, and the formation of ATP from ADP requires 30.5 kJ/mol. What percent of the energy released in the complete oxidation of glucose is captured in the formation of ATP?

88. An average adult expends about 8400 kJ of energy every day. How many moles of ATP must be converted to ADP to provide this amount of energy?

89. Why can't humans digest cellulose, considering that it is made of the same monomers as starch?

90. How is an enzyme-substrate complex formed in the enzyme's active site?

91. What role do coenzymes play in metabolism?

92. Suggest a reason why prokaryotic cells are thought to be more ancient than eukaryotic cells.

93. What is the oxidation state of iron in each of the following?
a. normal hemoglobin **b.** methemoglobon
What is the consequence of this change in oxidation state?

94. What are the possible consequences of an error in DNA sequence?

95. Explain why photosynthesis might be considered the most important chemical process on Earth.

96. Describe the differences among monosaccharides, disaccharides, and polysaccharides. Give examples of each type of carbohydrate.

97. Describe how amino acids join to form a peptide bond.

98. What is base-pairing? How does base-pairing relate to the structure of DNA?

Critical Thinking

99. You may know that freshly picked corn is sweeter than corn that has been picked and stored for several days. This is because the stored corn contains less sugar than the freshly picked corn. About half of the sugar in fresh-picked corn has been converted to starch only one day after the corn has been picked. The sweetness of corn can be preserved by "blanching" the ears—dipping the ears of corn into boiling water for a few minutes, then cooling them in cold water. Corn that has been blanched and then frozen maintains its sweetness for a long time. Propose a biochemical reason why blanching corn maintains its sweetness.

100. Interpret this statement: "Carbon dioxide is an energy-poor molecule, but glucose is an energy-rich molecule."

101. In the DNA double helix, where are the base pairs located in relation to the backbone structure: inside the double helix, or outside the double helix? What must happen before the protein-making machinery of the cell can "read" the code words formed by the DNA bases?

102. Which type of gene mutation do you think will do more damage to an organism: a substitution mutation in which one base is substituted for another base, or an addition mutation in which a base is added to a sequence of bases? Explain.

103. Suggest a reason why a bean plant might not grow well if planted in sterilized soil.

104. What class of polymer is formed from each of the following monomers?

a. amino acids **b.** monosaccharides **c.** nucleotides

105. Complete the following equation by drawing the structural formulas of the products.

(glucose, ring structure with CH_2OH, H, OH, HO groups) + (fructose, ring structure with CH_2OH, H, OH, HO, CH_2OH groups)

Concept Challenge

106. Compare the structure of a DNA nucleotide with an RNA nucleotide.

107. Write a possible DNA sequence that codes for a polypeptide with the following amino acids: histidine-cysteine-glycine-arginine-proline. Is it possible for a mutation to occur in which there is a substitution of one of the last three bases with no effect on the amino acids present in the chain?

108. Describe the structural features all amino acids have in common. What structural features differ among the amino acids?

109. Using structural formulas, write a chemical equation for the formation of a dipeptide from two amino acids. What functional group does the reaction create?

110. Explain why cell growth stops when the dietary intake of nutrients is insufficient.

111. The following compound is hydrolyzed by boiling with sodium hydroxide. What are the saponification products?

$$CH_2-O-\overset{O}{\overset{\|}{C}}-(CH_2)_{14}CH_3$$
$$CH-O-\overset{O}{\overset{\|}{C}}-(CH_2)_{12}CH_3$$
$$CH_2-O-\overset{O}{\overset{\|}{C}}-(CH_2)_{16}CH_3$$

112. What causes the spontaneous formation of a lipid bilayer?

113. Peptide chains fold and bend into three-dimensional shapes. Suggest how a peptide chain is held in this 3-D shape.

114. A sequence of nine bases in a gene codes for the amino acid sequence Trp-Met-Met. What is the sequence of bases in this DNA fragment? Use Table 24.2 on page 781 to help you. Could you determine the base sequence for certain if the amino acid sequence were Trp-Met-Leu? Why or why not?

Critical Thinking

99. Blanching the corn destroys most of the enzymes responsible for the conversion of glucose to starch. Freezing the corn slows down the action of any remaining starch-producing enzymes and prevents spoilage.

100. CO_2 cannot produce energy through oxidation. Glucose can be oxidized to CO_2 and produce energy.

101. Inside the helix; strands must unwind

102. Substitution of one base could result in no change in the amino acid sequence of a protein (if there are multiple code words for the amino acid), or a different amino acid, which could be damaging or helpful to the organism. A base added to the start of the gene would result in a nonsense amino acid sequence for the protein produced by the gene. A base added to the end of the gene could make the termination codon untranslatable, and the protein product might not terminate.

103. Bacteria necessary for nitrogen fixation might be killed by sterilization of the soil.

104. a. peptides and proteins
b. polysaccharides
c. polynucleotides (RNA and DNA)

105. (sucrose structure: glucose ring linked through O to fructose ring with $HOCH_2$, CH_2OH groups) $+ H_2O$

Concept Challenge

106. A nucleotide consists of a phosphate group, a five-carbon sugar unit, and a nitrogen base. In DNA, the sugar is deoxyribose; in RNA, the sugar is ribose. In DNA, the bases may be A, G, T, or C; in RNA, the bases may be A, G, U, or C.

107. Sample answer: GTA-ACA-CCA-GCA-GGA; yes, because GGA, GGG, GGT, and GGC all code for proline.

108. Every amino acid consists of a central carbon atom bonded to an amine group, a carboxyl group, a hydrogen atom, and a side chain. Different amino acids have different side chains.

109. An amide goup is formed by the reaction.

$$H_2N-\underset{H}{\overset{R}{C}}-\overset{O}{\overset{\|}{C}}-OH + H-\underset{H}{N}-\underset{H}{\overset{R}{C}}-\overset{O}{\overset{\|}{C}}-OH \longrightarrow$$
$$H_2N-\underset{H}{\overset{R}{C}}-\overset{O}{\overset{\|}{C}}-\underset{H}{N}-\underset{H}{\overset{R}{C}}-\overset{O}{\overset{\|}{C}}-OH + H_2O$$

110. Nutrients supply carbon compounds, ions, and energy needed for growth. The (insufficient) food primarily is used for energy production necessary to sustain the cell.

111. glycerol + a soap

112. Phospholipids have hydrophobic tails that exclude water and polar heads that dissolve in water.

113. hydrogen bonds between adjacent parts of folded chains and covalent bonds between side-chain groups of cysteine

114. A-C-C-T-A-C-T-A-C; no, Leu has 6 possible code words.

Cumulative Review

115. The particles in real gases have volumes and are attracted to one another.

116. a and b are electrolytes. c and d are nonelectrolytes.

117. a. 0.519°C
b. 0.209°C
c. 0.238°C
d. 0.227°C

118. 1.71×10^{1} kJ heat absorbed

119. Besides being composed of highly combustible material, the needles have a large surface area that increases their rate of combustion.

120. The equivalents of acid and base must be equal.

121. a. 4.15
b. 5.26
c. 12.79
d. 10.36

122. a. fluorine
b. oxygen
c. chlorine

123. Oxidation always occurs at the anode. In the voltaic cell the anode is the negative electrode.

124. The nickel wire would be silver-plated.

125. a. Cu **b.** Ni **c.** Ag
d. Fe **e.** Cd **f.** Cu

126. a. $CH_3—CH_2—CH_2—CH_2—CH_2—CH_2—CH_3$

b. $CH_3—CH(CH_3)—CH=CH—CH_2—CH_3$

c. $CH_3—CH(C_6H_5)—CH_2—CH_3$

d. $C_6H_4(CH_2CH_3)_2$ (CH_2CH_3 groups at positions 1 and 3)

127. a. cyclopentane
b. 2-methyl-2-propanol
c. 3-pentanone

CHAPTER 24 Assessment continued

Cumulative Review

115. Describe two factors that cause real gases to depart from the ideal gas law. *(Chapter 14)*

116. Characterize these compounds as electrolytes or nonelectrolytes. *(Chapter 15)*
a. NaCl
b. $CuSO_4$
c. CCl_4
d. H_2O

117. Calculate the boiling-point elevation for these aqueous solutions. *(Chapter 16)*
a. 0.507*m* NaCl
b. 0.204*m* NH_4Cl
c. 0.155*m* $CaCl_2$
d. 0.222*m* $NaHSO_4$

118. How much heat (in kJ) is released or absorbed when 0.265 mol of sodium bicarbonate is decomposed according to the following reaction? *(Chapter 17)*

$$2NaHCO_3(s) \longrightarrow Na_2CO_3(s) + H_2O(g) + CO_2(g) \quad \Delta H = 129 \text{ kJ}$$

119. Explain why the needles on a dried out fir tree can burn with almost explosive rapidity. *(Chapter 18)*

120. What must be true at the end point of an acid-base titration? *(Chapter 19)*

121. Calculate the pH of each of the following solutions. *(Chapter 19)*
a. $[H^+] = 7.0 \times 10^{-5}M$
b. $[OH^-] = 1.8 \times 10^{-9}M$
c. $[OH^-] = 6.1 \times 10^{-2}M$
d. $[H^+] = 4.4 \times 10^{-11}M$

122. Identify the oxidizing agent in each reaction. *(Chapter 20)*
a. xenon + fluorine ⟶ xenon tetrafluoride
b. sulfur + oxygen ⟶ sulfur trioxide
c. gaseous chlorine + aqueous sodium bromide ⟶ aqueous bromine + aqueous sodium chloride

123. At which electrode in a voltaic cell does oxidation always occur? What is the charge on this electrode? *(Chapter 21)*

124. What would you observe when a length of nickel wire is immersed in an aqueous solution of silver nitrate? *(Chapter 21)*

125. For each pair of metals shown below, decide which metal is more readily reduced. *(Chapter 21)*
a. Cu, Mg **b.** Cd, Ni **c.** Ag, Sn
d. Zn, Fe **e.** Ni, Cd **f.** Al, Cu

126. Write a molecular structure for each compound. *(Chapter 22)*
a. heptane **b.** 2-methyl-3-hexene
c. 2-phenylbutane **d.** 1,3-diethylbenzene

127. Give the IUPAC name for these compounds. *(Chapter 23)*

a. cyclic CH_2, CH_2, H_2C, $H_2C—CH_2$ (five-membered ring)

c. $H—CH_2—CH_2—C(=O)—CH_2—CH_2—H$ (H—C—C—C—C—C—H with H's on each carbon and O double-bonded to the middle C)

b.

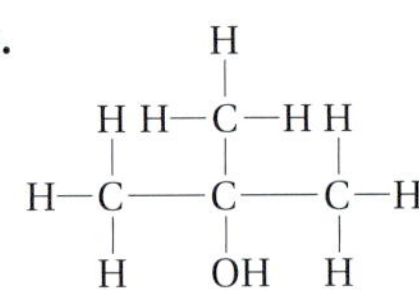

128. Name the next highest homolog of each of these compounds. *(Chapter 22)*
a. 1-butene **b.** cyclooctane
c. pentane **d.** nonane

129. Name the four structural isomers of C_4H_9Cl. *(Chapter 23)*

130. Write a molecular formula for each compound. *(Chapter 23)*
a. methyl acetate **b.** butanal
c. 2-hydroxypropanoic acid **d.** 2-butanone

131. Name each polymer and state at least one of its uses. *(Chapter 23)*
a. $—(CF_2—CF_2)_x—$ **b.** $H—(CH_2—CH_2)_x—H$
c. $—(CH_2—CHCl)_x—$ **d.** $—(CH_2—CH(C_6H_5))_x—$

132. Capsaicin, shown below, is the major contributor to the heat of chili peppers. Circle and name the functional groups in capsaicin. *(Chapter 23)*

CH_3O, HO on benzene ring $—CH_2NHC(=O)(CH_2)_4CH=CHCH(CH_3)_2$

796 Chapter 24

128. a. 1-pentene **b.** cyclononane
c. hexane **d.** decane

129. 1-chlorobutane, 2-chlorobutane, 1-chloro-2-methylpropane, 2-chloro-2-methylpropane

130. a. $CH_3CO_2CH_3$
b. $CH_3(CH_2)_2COH$
c. $CH_3CHOHCOOH$
d. $CH_3COC_2H_5$

131. a. polytetrafluoroethene (non-stick cookware)
b. polyethylene (plastic bags and containers)
c. polyvinyl chloride (PVC pipe)
d. polystyrene (electric insulation)

132.

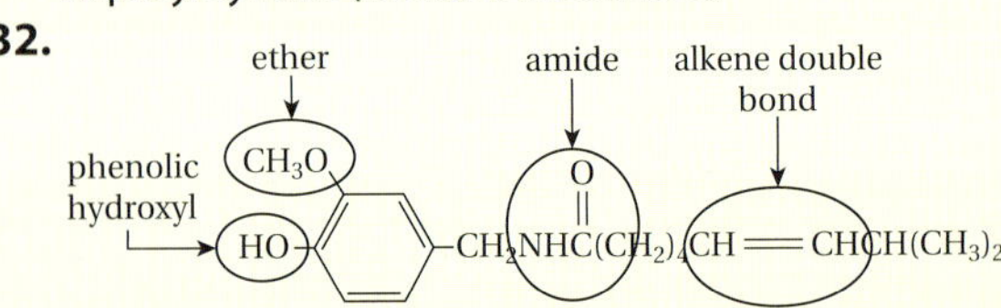

Standardized Test Prep

Test-Taking Tip

Anticipating the Answer You can sometimes figure out the answer to a question before you look at the answer choices. Try to answer the question in your mind before you look at the answer choices. Then read the choices and choose the choice that agrees with the answer in your mind.

Select the choice that best answers each question or completes each statement.

1. What phrase best describes ATP?
   a. energy producer **b.** energy consumer
   c. energy transmitter **d.** energy pump

2. Which element is not found in amino acids?
   a. phosphorus **b.** nitrogen **c.** carbon
   d. oxygen **e.** hydrogen

3. For any enzyme to function, the substrate must bind to the
   a. product. **b.** cofactor.
   c. active site. **d.** peptide.

Use the paragraph to answer Questions 4–6.

Because an amino acid contains both a carboxyl group and an amino group, it is amphoteric, that is, it can act as either an acid or a base. Crystalline amino acids have some properties—relatively high melting points and high water solubilities—that are more characteristic of ionic substances than of molecular substances.

4. Write an equation showing glycine acting as an acid in a reaction with water. (Glycine is the simplest amino acid. Its side chain R = H.)

5. Write an equation showing glycine acting as a base in a reaction with water.

6. It is possible for glycine to undergo an internal Brønsted-Lowry acid-base neutralization reaction. Write the resulting structural formula.

Explain how this reaction would account for the ionic properties of glycine.

Use the photo to answer Question 7.

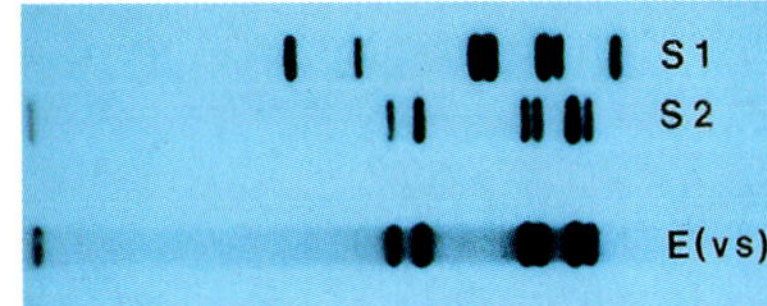

7. The photo shows three DNA fingerprints. The DNA in sample E was taken from a crime scene. The DNA in samples S1 and S2 were taken from suspects. Based on the DNA evidence, which suspect did not commit the crime?

For Questions 8–11, name the category of organic compounds most closely identified with each biological molecule.

8. proteins **9.** nucleic acids
10. lipids **11.** carbohydrates

Use these models to answer Question 12.

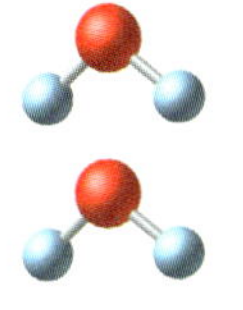
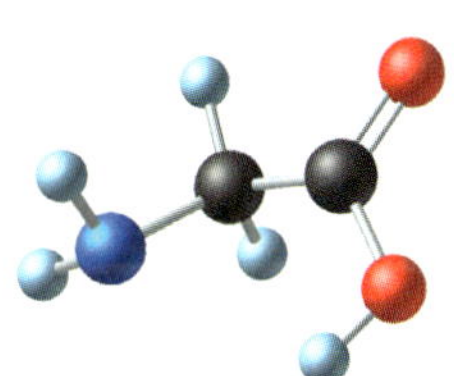

12. Glycine can form hydrogen bonds with water. Draw a glycine molecule with at least two water molecules bonded to each end.

For each question there are two statements. Decide whether each statement is true or false. Then decide whether Statement II is a correct explanation for Statement I.

	Statement I		Statement II
13.	Lipids tend to be insoluble in water.	BECAUSE	Lipids have mainly nonpolar bonds.
14.	Starch and cellulose are both digestible by most organisms.	BECAUSE	Glucose is the monomer in both starch and cellulose.
15.	Many of the reactions in catabolism are oxidation reactions.	BECAUSE	Oxidation reactions tend to be energy-producing reactions.

Standardized Test Prep

1. c
2. a
3. c
4. $H_2NCH_2COOH(aq) + H_2O(l) \rightarrow H_2NCH_2COO^-(aq) + H_3O^+(aq)$
5. $H_2NCH_2COOH(aq) + H_2O(l) \rightarrow H_3N^+CH_2COOH(aq) + OH^-(aq)$
6. The carboxylic acid group can donate a proton to the amino group to form $H_3N^+CH_2COO^-$.
7. suspect S1
8. amino acids
9. nucleotides
10. fatty acids
11. monosaccharides
12. See the drawing below. In the drawing, black spheres represent carbon, gray spheres represent oxygen, and white spheres represent hydrogen.

13. True, True, correct explanation
14. False, True
15. True, True, correct explanation

Nuclear Chemistry Planning Guide

SECTION OBJECTIVES	STANDARDS NATIONAL	STANDARDS STATE	ACTIVITIES and LABS
25.1 Nuclear Radiation, pp. 799–802 1 block or 2 periods **25.1.1 Explain** how an unstable nucleus releases energy. **25.1.2 Describe** the three main types of nuclear radiation.	B-1, B-3, G-3		**SE** Inquiry Activity: Simulating Radioactive Decay, p.798 L2 **TE** Teacher Demo: An Effect of Radiation, p. 800 L1
25.2 Nuclear Transformations, pp. 803–809 1 block or 2 periods **25.2.1 Describe** the type of decay a radioisotope undergoes. **25.2.2 Solve problems** that involve half-life. **25.2.3 Identify** the two ways transmutation can occur.	A-1, B-1		**SE** Small-Scale Lab: Radioactivity and Half-Lives, p. 809 L2 **TE** Class Activity: Particle Accelerators, p.807 L2 **SSLM** Lab 41: Half-Lives and Reaction Rates L2 **LP** Lab Practical 25-2: Half-Life Simulation L2
25.3 Fission and Fusion of Atomic Nuclei, pp. 810–815 1 block or 2 periods **25.3.1 Describe** what happens in a nuclear chain reaction. **25.3.2 Explain** the role of water in the storage of spent fuel rods. **25.3.3 Distinguish** fission reactions from fusion reactions.	B-1, E-2, F-6, G-1, G-3		**TE** Teacher Demo: Model a Chain Reaction, p. 811 L2 **TE** Class Activity: Nuclear Fussion, p. 812 L2
25.4 Radiation in Your Life, pp. 816–819 1 block or 2 periods **25.4.1 Identify** three devices that are used to detect radiation. **25.4.2 Describe** how radioisotopes are used in medicine.	A-1, A-2, B-1, E-2, F-1		**SE** Quick Lab: Studying Inverse-Square Relationships, p. 818 L2 **TE** Teacher Demo: Background Radiation, p. 817 L2 **LM** Lab 52: Radioactivity and Radiation L2 **PLM** Probeware Lab: Studying Inverse-Square Relationships L2 **LP** Lab Practical 25-1: Radioactivity and Radiation L2

Ability Levels	
L1	For students who need additional help
L2	For all students
L3	For students who need to be challenged

Components

SE	Student Edition	LM	Laboratory Manual	GCP	Graphing Calculator Problems	T	Transparencies
TE	Teacher's Edition	SSLM	Small-Scale Chemistry Lab Manual	TTT	Test-Taking Tips with Transparencies, Grades 9–12 Science	CA	Chem Alive and Chemistry Field Trips DVD
GRSW	Guided Reading and Study Workbook	LP	Laboratory Practicals	STP	Standardized Test Preparation Workbook	iT	Interactive Textbook with ChemASAP
CTR	Core Teaching Resources	PLM	Probeware Lab Manual			GO	Go Online Internet Resources
SSG	Spanish Study Guide	VCL	Virtual Chemistry Labs				
		CTB	Computer Test Bank				

RESOURCES PRINT and TECHNOLOGY	PROBLEMS and ASSESSMENT
GRSW Chapter 25, Section 1 L1 **T** T286–T288 L2	**SE** Section 25.1 Assessment, p. 802 **iT** Section 25.1
GRSW Chapter 25, Section 2 L1 **T** T289–T292 L2 **GO** Chapter 25, Section 2 L2	**SP** 25.1, Using Half-lives in Calculations **SE** Section 25.2 Assessment, p. 808 **iT** Section 25.2
GRSW Chapter 25, Section 3 L1 **T** T293 L2	**SE** Section 25.3 Assessment, p. 813 **iT** Section 25.3
GRSW Chapter 25, Section 4 L1 **T** T294 L2 **GO** Chapter 25, Section 4 L2	**SE** Section 25.4 Assessment, p. 819 **iT** Section 25.4

Web Code: cdn-1252
Web Code: cdn-1254

MATERIALS List

Quantities for each group

STUDENT EDITION

Inquiry Activity, p. 798
a container, a lid, 128 pennies, a flat table or desk surface, a scientific calculator, loose-leaf paper, graph paper, pen, or pencil.

Small-Scale Lab, p. 809
pencil, paper, ruler, penny, graph paper

Quick Lab, p. 818
flashlight, strips of duct tape, scissors, poster board, white (50 cm × 50 cm) meter ruler or tape measure flat surface, long enough to hold the meter ruler, loose-leaf paper, graph paper, pen or pencil, light sensor (optional)

TEACHER'S EDITION

Teacher Demo, p. 800
photographic film, radiation source, key

Class Activity, p. 807
reference materials and Internet access, miscellaneous materials for models

Teacher Demo, p. 811
wooden matchsticks

Class Activity, p. 812
reference materials and Internet access

Teacher Demo, p. 817
Geiger counter

Chapter Assessment

CHAPTER ASSESSMENT

SE Chapter Assessment, pp. 821–824
CTB Chapter 25
iT Chapter 25

STANDARDIZED TEST PREP

SE Chapter 25, p. 825
STP Topic 11
iT Interactive Textbook with Assessment at PHSchool.com

CHAPTER 25

Nuclear Chemistry

INQUIRY Activity

Simulating Radioactivity L2

Purpose

Students will model radioactive decay.

Prep Time 5 minutes
Class Time 15 minutes

Think About It

1. After studying Section 25.2, students should realize that the curves on the graphs they prepared by plotting "number of heads" versus "trial number" resemble exponential decay curves.
2. When students plot "log of number of heads" versus "trial number," the graph is linear. With careful study of the data, some students may realize that the number of heads appearing after n trials is approximately $128 \times (^1/_2)^n$.

CHAPTER 25

Nuclear Chemistry

A nuclear submarine uses energy released by nuclear reactions.

INQUIRY Activity

Simulating Radioactive Decay

Materials

a container, a lid, 128 pennies, a flat table or desk surface, a scientific calculator, loose-leaf paper, graph paper, pen or pencil

Procedure

1. Make a two-column data table with the headings "Trial" and "Number of heads."
2. Put 128 pennies in the container and shake them.
3. Carefully pour the pennies onto a flat surface.
4. Pick up, count, and set aside those pennies that are turned heads up. In the table, enter the number of heads for Trial 1.
5. Return the remaining pennies (those that were tails) to the container and repeat Steps 3–5 for a total of five trials.
6. If time permits, repeat the entire experiment four more times and find the average of your data or the data collected by the class.

Think About It

1. Prepare a graph of the number of heads (y-axis) versus the trial number (x-axis). Is the result a straight line or a curve? Add a third column titled "Log of number of heads" to the data table, and use your calculator to complete the table.
2. Prepare a graph of the data in Column 3 (y-axis) versus trial number (x-axis). Is the result a straight line or a curve?

25.1 Nuclear Radiation

Connecting to Your World

Marie Curie was a Polish scientist whose research led to many discoveries about radiation and radioactive elements. In 1903 she and her husband Pierre, along with Antoine Henri Becquerel, won the Nobel Prize in physics for their work on radioactivity. She was also awarded the Nobel Prize in chemistry in 1911 for her research on radioactive elements. Marie Curie's research was essential to understanding and using newly discovered radioactive elements. In 1934 she died from leukemia caused by her long-term exposure to radiation. In this section, you will learn about the various types of radiation and their effects.

Guide for Reading

 **Key Concepts**

- How does an unstable nucleus release energy?
- What are the three main types of nuclear radiation?

Vocabulary

radioactivity
radiation
radioisotopes
alpha particle
beta particle
gamma ray

Reading Strategy

Relating Text and Visuals As you read about types of radiation, look at Figure 25.1. List how the positive, negative, or neutral electric charge of each type of radiation, as described in the text, is related to how the radiation is deflected as it passes by the charged plates.

Radioactivity

In 1896, the French chemist Antoine Henri Becquerel (1852–1908) made an accidental discovery. He was studying the ability of uranium salts that had been exposed to sunlight to fog photographic film plates. During bad weather, Becquerel could not expose the sample to sunlight, but happened to leave it on top of the photographic plate. When he developed the plate, he discovered that the uranium salt still fogged the plate. At that time, two of Becquerel's associates were Marie Curie (1867–1934) and Pierre Curie (1859–1906). The Curies were able to show that rays emitted by the uranium atoms caused the fogging of the plates. Marie Curie named the process by which materials give off such rays **radioactivity.** The penetrating rays and particles emitted by a radioactive source are called **radiation.**

Nuclear reactions, which account for radioactivity, differ from chemical reactions in a number of important ways. In chemical reactions, atoms tend to attain stable electron configurations by losing electrons or sharing electrons. In nuclear reactions, the nuclei of unstable isotopes, called **radioisotopes,** gain stability by undergoing changes. These changes are always accompanied by the emission of large amounts of energy. Unlike chemical reactions, nuclear reactions are not affected by changes in temperature, pressure, or the presence of catalysts. They are also unaffected by the compounds in which the unstable isotopes are present. The nuclear reactions of a given radioisotope cannot be speeded up, slowed down, or turned off.

 ***What is radiation?***

Section Resources

Print

- **Guided Reading and Study Workbook,** Section 25.1
- **Core Teaching Resources,** Section 25.1 Review
- **Transparencies,** T286–288

Technology

- **Interactive Textbook with ChemASAP,** Assessment 25.1

25.1

1 FOCUS

Objectives

25.1.1 Explain how an unstable nucleus releases energy.

25.1.2 Describe the three main types of nuclear radiation.

Guide for Reading

Build Vocabulary 

Word Parts Have students look up the meanings of the first three vocabulary terms. From these meanings, have them infer the meaning of the prefix used in all three terms. Students should infer that the prefix *radio-* means "to send out in all directions."

Reading Strategy 

Using Context Clues Have students use context clues to infer the definition of *decay,* as it refers to radioactivity. Inferences should include that *decay* means "breaking down into smaller parts."

2 INSTRUCT

Connecting to Your World

Have students examine the periodic table. Ask, **What element was named in honor of the Curies?** *(curium)* **Is this element radioactive?** *(yes)*

Radioactivity

Discuss 

Ask, **What particles in an atom are involved in chemical reactions?** *(valence electrons)* **What is the difference between a chemical reaction and a nuclear reaction?** *(Chemical reactions involve the transfer or sharing of electrons; the nuclei remain unchanged. Nuclear reactions involve changes in the nuclei of atoms.)*

Answers to...

 the penetrating rays and particles emitted by a radioactive source

Types of Radiation

Discuss L2

Explain that the nuclei of a radioactive element spontaneously decompose. Nuclear chemistry is the study of changes in matter that originate in atomic nuclei. Ask, **What types of radiation exist, and how harmful are they?** *(The three most common types of radiation emitted by unstable nuclei are alpha (α), beta (β), and gamma (γ) radiation. Although all forms of radiation are somewhat harmful, gamma rays are particularly dangerous because they penetrate body tissues.)* Explain that radioactivity reflects the tendency of atomic nuclei to achieve stability. Ask, **What makes a nucleus unstable?** *(It has too many or too few neutrons relative to the number of protons.)* Point out that the energy produced during nuclear decay can be released as electromagnetic radiation, such as gamma rays, or as kinetic energy, which is carried by particles, such as alpha particles, that are emitted from the nucleus.

TEACHER Demo

An Effect of Radiation L1

Purpose Students observe the effects of radiation on photographic film.

Materials photographic film, radiation source, key

Safety Use only safely packaged radiation sources. Do not directly touch any radioactive material.

Procedure Explain that radiation reduces silver in photographic film. This property led to the discovery of radioactivity by Becquerel. Place a metal key on top of unexposed photographic film that has been wrapped in black paper. Place a radiation source on top of the key. After one week, develop the film.

Expected Outcome The area of the film that was beneath the key is unexposed, unlike the rest of the film.

Point out that the film badges worn by persons working with radioactivity operate by the same mechanism. They serve as important radiation detectors, safeguarding against overexposure to ionizing radiation.

The discovery of radioactivity disproved Dalton's assumption that atoms are indivisible. A radioactive atom, or radioisotope, undergoes drastic changes as it emits radiation. These radioisotopes have unstable nuclei. The stability of a nucleus depends on the ratio of neutrons to protons in the nucleus, and on the overall size of the nucleus. Too many or too few neutrons relative to the number of protons makes a nucleus unstable. **An unstable nucleus releases energy by emitting radiation during the process of radioactive decay.** Unstable radioisotopes of one element are transformed into stable (nonradioactive) isotopes of a different element. Radioactive decay is spontaneous and does not require any input of energy.

Types of Radiation

Radiation is emitted during radioactive decay. **The three main types of nuclear radiation are alpha radiation, beta radiation, and gamma radiation.** Table 25.1 summarizes the characteristics of these three types of radiation. The different types of radiation can be separated by an electric field as shown in Figure 25.1.

Alpha Radiation The type of radiation called alpha radiation consists of helium nuclei that have been emitted from a radioactive source. Each of these emitted particles, called an **alpha particle,** contains two protons and two neutrons and has a double positive charge. In nuclear equations, an alpha particle is written ${}^{4}_{2}He$ or α. The electric charge symbol is generally omitted.

The radioisotope uranium-238 releases alpha radiation and is transformed into another radioisotope, thorium-234. Figure 25.2a illustrates this process.

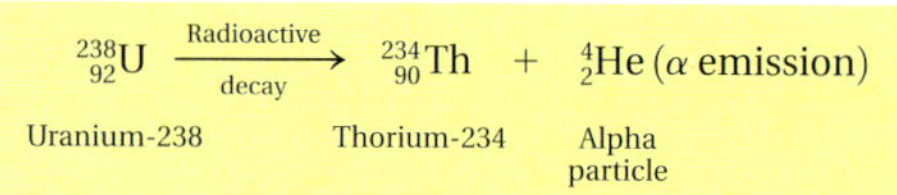

$$ {}^{238}_{92}U \xrightarrow[\text{decay}]{\text{Radioactive}} {}^{234}_{90}Th + {}^{4}_{2}He\ (\alpha \text{ emission}) $$

Note that this nuclear equation is balanced. The sum of the mass numbers (superscripts) on the right equals the sum on the left. The same is true for the atomic numbers (subscripts).

When an atom loses an alpha particle, the atomic number of the product atom is lower by two and its mass number is lower by four. Because of their large mass and charge, alpha particles do not tend to travel very far and are not very penetrating. A sheet of paper or the surface of your skin easily stops them. However, radioisotopes that emit alpha particles are dangerous when ingested. The particles do not have to travel far to penetrate soft tissue and cause damage.

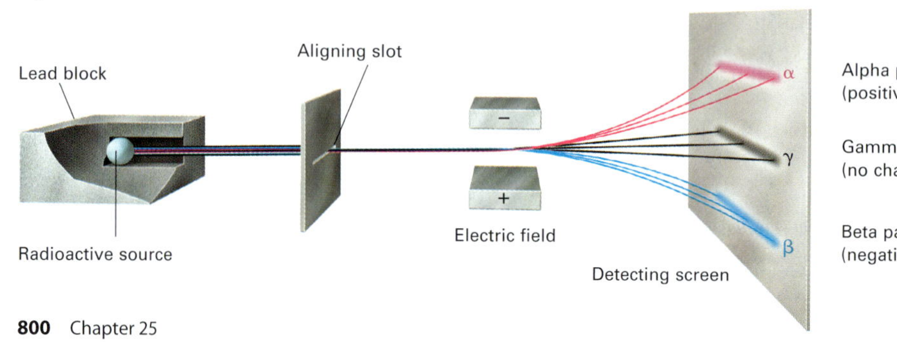

Figure 25.1 These three types of radiation deflect differently as they pass between a pair of electrically charged plates. Alpha particles (α) and beta particles (β) are deflected in opposite directions—alpha particles toward the negative plate and beta particles toward the positive plate. Gamma rays are undeflected. **Applying Concepts** ***Why are gamma rays not deflected?***

800 Chapter 25

Facts and Figures

Nuclear Decay

Nuclear decay, unlike chemical reactions, is unaffected by temperature, pressure, or the presence of catalysts. Spontaneous nuclear decay cannot be slowed down, speeded up, or stopped. Nuclear decay illustrates the interconnected nature of matter and energy; matter is related to energy according to the equation $E = mc^2$, where c is a constant equal to the speed of light.

Table 25.1

Characteristics of Some Types of Radiation

Property	Alpha radiation	Beta radiation	Gamma radiation
Composition	Alpha particle (helium nucleus)	Beta particle (electron)	High-energy electromagnetic radiation
Symbol	α, ${}^4_2He$	β, ${}^{0}_{-1}e$	γ
Charge	2+	1−	0
Mass (amu)	4	1/1837	0
Common source	Radium-226	Carbon-14	Cobalt-60
Penetrating power	Low (0.05 mm body tissue)	Moderate (4 mm body tissue)	Very high (penetrates body easily)
Shielding	Paper, clothing	Metal foil	Lead, concrete (incompletely shields)

Beta Radiation An electron resulting from the breaking apart of a neutron in an atom is called a **beta particle.** The neutron breaks apart into a proton, which remains in the nucleus, and a fast-moving electron, which is released.

$$\underset{\text{Neutron}}{{}^1_0n} \longrightarrow \underset{\text{Proton}}{{}^1_1H} + \underset{\substack{\text{Electron}\\\text{(beta particle)}}}{{}^{0}_{-1}e}$$

The symbol for the electron has a subscript of −1 where the atomic number would be written. This represents the electron's negative charge. The superscript 0 where a mass number would be written represents the extremely small mass of the electron compared to that of a proton.

Carbon-14 emits a beta particle as it undergoes radioactive decay to form nitrogen-14. Figure 25.2b shows this beta emission.

$$\underset{\substack{\text{Carbon-14}\\\text{(radioactive)}}}{{}^{14}_{6}C} \longrightarrow \underset{\substack{\text{Nitrogen-14}\\\text{(stable)}}}{{}^{14}_{7}N} + \underset{\substack{\text{Beta}\\\text{particle}}}{{}^{0}_{-1}e} \ (\beta \text{ emission})$$

The nitrogen-14 atom has the same atomic mass number as carbon-14, but its atomic number has increased by 1. It contains an additional proton and one fewer neutron. The nuclear equation is balanced.

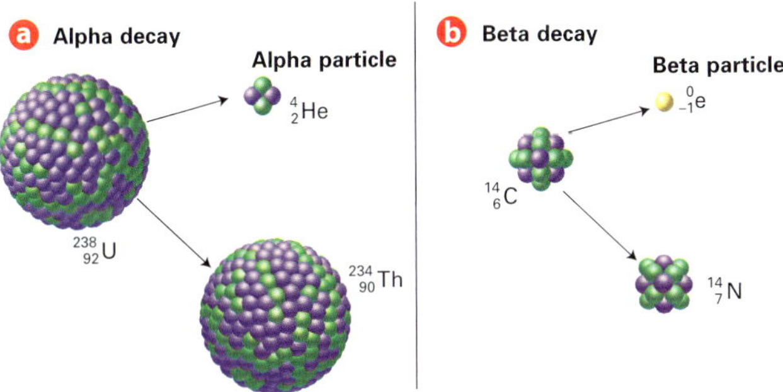

Figure 25.2 Radiation is emitted during radioactive decay. **a** Uranium-238 undergoes alpha decay to form thorium-234. **b** Carbon-14 undergoes beta decay to form nitrogen-14.
Interpreting Diagrams ***What particle is emitted in each decay process?***

Use Visuals L1

Table 25.1 Have students examine the table. Encourage them to memorize the names and symbols used to represent each type of radiation. Point out the differences in charge and mass. Students may wish to recreate Table 25.1 on an index card and use the information to help them write and balance nuclear equations. Point out the relative penetrating power of each type of radiation and the precautions necessary to effectively block different types of radiation.

Discuss L2

Tell students that beta decay can result in either the emission of an electron (a β^- particle) or a positron (a β^+ particle). Present students with the partial equation for the decay of chlorine-32 into sulfur-32. Ask, **Which kind of particle would need to be emitted to balance the equation?** *(a positron)* Explain that during positron emission, a proton is converted into a neutron. Ask, **What kind of unstable nuclei would be likely to emit positrons?** *(one with too many protons for the number of neutrons)* Explain that emission of a positron also results in a neutrino, while emission of an electron is accompanied by emission of an anti-neutrino. Neutrinos and anti-neutrinos have no charge and virtually no mass. The emission of neutrinos and anti-neutrinos are not included in the student text because they are not chemically significant and do not react significantly with matter. However, the laws of energy-mass conservation and momentum conservation are satisfied by the presence of neutrinos and anti-neutrinos

Facts and Figures

Isotopes vs. Radioisotopes

Radioisotopes of an element are not chemically different from other isotopes of that element. Chemical reactivity depends on the electron structure of an atom, not its nuclear makeup.

Answers to...

Figure 25.1 They have no charge and so are unaffected by the electric field.

Figure 25.2 alpha particle (helium nucleus); beta particle (electron)

Section 25.1 (continued)

Relate 

Point out that radio waves, microwaves, visible light, ultraviolet light, and X-rays are all forms of electromagnetic radiation. Ask, **How does a gamma ray differ from other types of electromagnetic radiation?** *(These other types are not directly produced by the radioactive decay of a nucleus.)*

3 ASSESS

Evaluate Understanding 

Write several partial equations for nuclear decay involving alpha and beta particles. Include parent and daughter nuclei only in the equations. For each equation, have students compare the parent and daughter nuclei to identify the type of particle emitted. Ask students to justify their conclusions. Ask, **What other type of emission is possible during nuclear decay?** *(gamma emission)*

Reteach 

Remind students that when writing a nuclear equation, the sums of the mass numbers and atomic numbers of the reactants must equal the sums of the mass and atomic numbers of the products. Students should check their equations by comparing the sums of the superscripts and subscripts on each side of the equation.

Connecting Concepts

Atomic numbers (charge) and mass numbers must be balanced.

If your class subscribes to the Interactive Textbook, use it to review key concepts in Section 25.1.

with ChemASAP

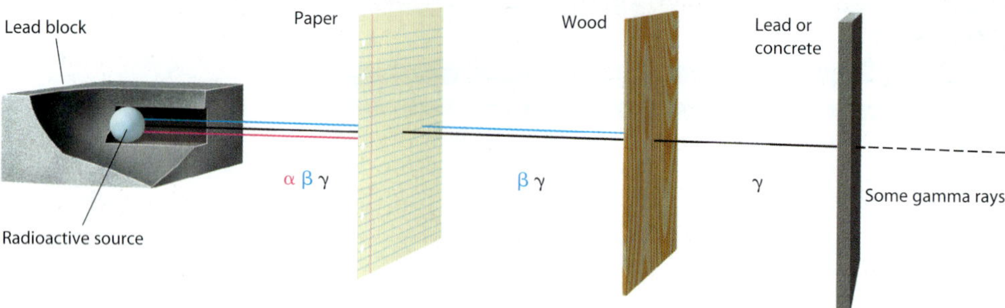

Figure 25.3 Because of their large mass and charge, alpha particles (red) are the least penetrating of the three main types of radiation. Gamma rays (black) have no mass or charge and are the most penetrating.

A beta particle has less charge than an alpha particle and much less mass than an alpha particle. Consequently, beta particles are more penetrating. Beta particles can pass through paper but are stopped by aluminum foil or thin pieces of wood.

Gamma Radiation A high-energy photon emitted by a radioisotope is called a **gamma ray.** The high-energy photons are electromagnetic radiation. Nuclei often emit gamma rays along with alpha or beta particles during radioactive decay. The following example demonstrates this process.

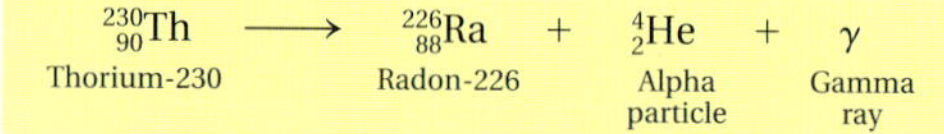

$$\underset{\text{Thorium-230}}{{}^{230}_{90}\text{Th}} \longrightarrow \underset{\text{Radon-226}}{{}^{226}_{88}\text{Ra}} + \underset{\text{Alpha particle}}{{}^{4}_{2}\text{He}} + \underset{\text{Gamma ray}}{\gamma}$$

Because gamma rays have no mass and no electrical charge, the emission of gamma radiation does not alter the atomic number or mass number of an atom. Gamma rays are extremely penetrating and can be very dangerous. Gamma rays pass easily through paper, wood, and the human body. They can be stopped, although not completely, by several meters of concrete or several centimeters of lead, as shown in Figure 25.3.

25.1 Section Assessment

1. **Key Concept** How does an unstable nucleus release energy?
2. **Key Concept** What are the three main types of nuclear radiation?
3. What part of an atom undergoes change during radioactive decay?
4. How is the atomic number of a nucleus changed by alpha decay? By beta decay? By gamma decay?
5. How is the atomic mass number of a nucleus changed by alpha decay? By beta decay? By gamma decay?
6. Which of the three kinds of radiation described in this section is the most penetrating?

Connecting Concepts

Connecting Concepts Refer back to the rules for balancing chemical equations introduced in Section 11.1. An equation for radioactive decay has different nuclei appearing on each side of the yields sign so the atoms cannot be balanced. What two items must be balanced to achieve a balanced equation for radioactive decay?

Assessment 25.1 Test yourself on the concepts in Section 25.1.

with ChemASAP

Section 25.1 Assessment

1. An unstable nucleus releases energy by emitting radiation during radioactive decay.
2. alpha radiation, beta radiation, and gamma radiation
3. The nucleus undergoes change.
4. In alpha decay, the atomic number decreases by two. In beta decay, the atomic number increases by one. The emission of gamma radiation does not change the atomic number.
5. In alpha decay, the mass number decreases by four. In beta decay, the mass number does not change. The emission of gamma radiation does not change the mass number.
6. Gamma radiation is most penetrating.

25.2 Nuclear Transformations

Connecting to Your World

Weather stripping and insulation help lower heating and cooling bills by conserving energy. They can, however, reduce the exchange of indoor and outdoor air. As a result, radioactive substances such as radon gas can accumulate indoors and pose a health risk. Radon-222 is a radioactive isotope that is present naturally in the soil in some areas. It has a constant rate of decay. In this section, you will learn about decay rates of radioactive substances.

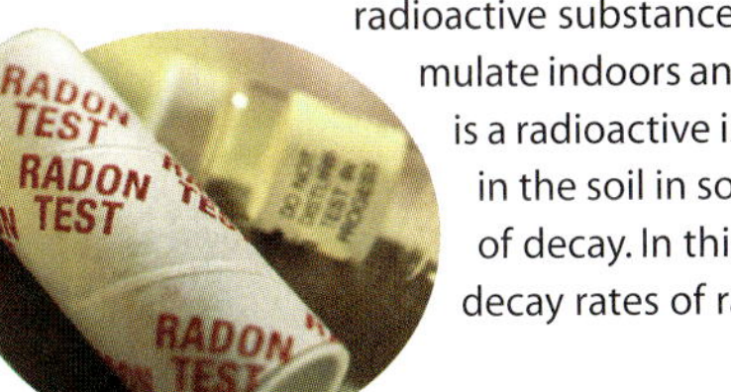

Guide for Reading

Key Concepts

- What determines the type of decay a radioisotope undergoes?
- How much of a sample of a radioisotope remains after each half-life?
- What are two ways that transmutation can occur?

Vocabulary

nuclear force
band of stability
positron
half-life
transmutation
transuranium elements

Reading Strategy

Building Vocabulary Before you read, make a list of the vocabulary terms above. As you read the section, write a definition of each vocabulary term in your own words.

Nuclear Stability and Decay

All nuclei, except those of hydrogen atoms, consist of neutrons and two or more protons. If a force did not hold these subatomic particles together, the like-charged protons would repel one another and fly apart. The **nuclear force** is an attractive force that acts between *all* nuclear particles that are extremely close together, such as protons and neutrons in a nucleus. At these short distances, the nuclear force dominates over electromagnetic repulsions and holds the nucleus together.

More than 1,500 different nuclei are known. Of those, only 264 are stable and do not decay or change with time. The stability of a nucleus depends on the neutron-to-proton ratio. Figure 25.4 shows a graph of number of neutrons vs. number of protons for all known stable nuclei. These nuclei are in a region called the **band of stability.** For elements of low atomic number (below about 20), this ratio is about 1. Above atomic number 20, stable nuclei have more neutrons than protons.

Number of Neutrons vs. Number of Protons for Stable Nuclei

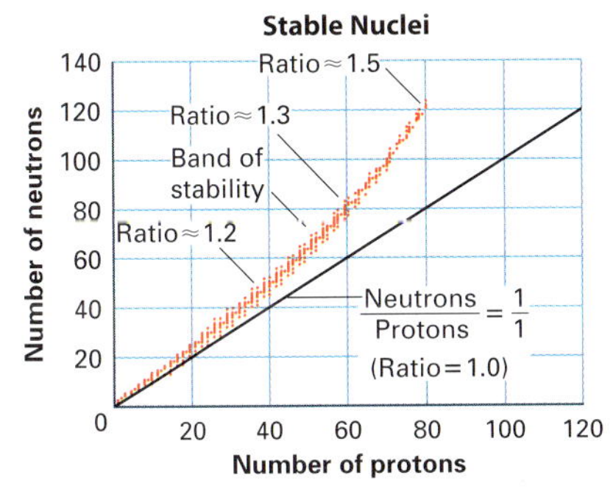

Figure 25.4 A neutron-versus-proton plot of all stable nuclei forms a pattern called the band of stability (shown in red).

INTERPRETING GRAPHS

a. Identify What do the dots on the graph represent?
b. Apply Concepts What is the approximate ratio of neutrons to protons for neodymium, whose atomic number is 60?
c. Describe How does the neutron-to-proton ratio change as the number of protons increases?

Section Resources

Print

- **Guided Reading and Study Workbook,** Section 25.2
- **Core Teaching Resources,** Section 25.2 Review, Interpreting Graphics
- **Transparencies,** T289–T292
- **Small-Scale Chemistry Lab Manual,** Lab 41

Technology

- **Interactive Textbook with ChemASAP,** Simulation 30, Problem-Solving 25.7, Assessment 25.2
- **Go Online,** Section 25.2

25.2

1 FOCUS

Objectives

25.2.1 Describe the type of decay a radioisotope undergoes.
25.2.2 Solve problems that involve half-life.
25.2.3 Identify two ways transmutations can occur.

Guide for Reading

Build Vocabulary

Paraphrase Have students look up the definitions of the vocabulary words. Then, have them paraphrase each of these definitions.

Reading Strategy

Relate Cause and Effect As students read through the section, have them list examples of cause and effect. For example, nuclei are unstable because their neutron-to-proton ratio is greater than one.

2 INSTRUCT

Connecting to Your World

Have students study the photograph and read the text that opens the section. Ask, **Which type of radiation does radon emit?** *(alpha particles)* **How do scientists express the decay rates of radioactive isotopes?** *(as a half-life—the time it takes for half of the nuclei in the sample to decay)*

Nuclear Stability and Decay

Interpreting Graphs

a. The dots represent stable nuclei.
b. 1.3
c. It also increases.

Enrichment

Ask, **In terms of stability, what purpose do the neutrons serve in an atom?** *The neutrons separate the positive charges in a nucleus and contribute to the **strong force**, an attractive force between all nucleons.*

Section 25.2 (continued)

Half-Life

Discuss L2

Explain that, for each element, there exists only a small range of neutron-to-proton ratios that produce stable nuclei. If a nucleus does not reflect a stable ratio, it spontaneously decays until a stable ratio of neutrons to protons results.

Relate 

Explain that the nuclear stability that results from a proper ratio of neutrons to protons in an atom is like the structural stability that results from a proper ratio of mortar to bricks in a building. In the building, gravitational and adhesive forces are balanced; in the nucleus, forces of repulsion (between protons) and attraction (the strong force) are balanced.

Interpreting Graphs 

a. 50%
b. 25%
c. three half-lives

Enrichment Question 

Will the curve on the graph ever reach zero? *(Theoretically, it could reach zero if a final atom decays. However, it would not be practical to graph the number of half-lives this event would take, and the amount of radioisotope present would be undectable.)*

Use Visuals 

Figure 25.5 After students examine the graph, have them consider the role neutrons play in stabilizing the nuclei of atoms. Encourage students to research the strong nuclear force, the force of attraction that holds nucleons together in a stable nucleus.

Table 25.2

Decay Processes			
Beta Emission			
${}^{66}_{29}\text{Cu}$	$\longrightarrow$	${}^{66}_{30}\text{Zn} + {}^{0}_{-1}\text{e}$	
${}^{14}_{6}\text{C}$	$\longrightarrow$	${}^{14}_{7}\text{N} + {}^{0}_{-1}\text{e}$	
Electron Capture			
${}^{59}_{28}\text{Ni} + {}^{0}_{-1}\text{e}$	$\longrightarrow$	${}^{59}_{27}\text{Co}$	
${}^{37}_{18}\text{Ar} + {}^{0}_{-1}\text{e}$	$\longrightarrow$	${}^{37}_{17}\text{Cl}$	
Positron Emission			
${}^{8}_{5}\text{B}$	$\longrightarrow$	${}^{8}_{4}\text{Be} + {}^{0}_{+1}\text{e}$	
${}^{15}_{8}\text{O}$	$\longrightarrow$	${}^{15}_{7}\text{N} + {}^{0}_{+1}\text{e}$	
Alpha Emission			
${}^{226}_{88}\text{Ra}$	$\longrightarrow$	${}^{222}_{86}\text{Rn} + {}^{4}_{2}\text{He}$	
${}^{232}_{90}\text{Th}$	$\longrightarrow$	${}^{228}_{88}\text{Ra} + {}^{4}_{2}\text{He}$	

A nucleus may be unstable and undergo spontaneous radioactive decay for several reasons. **The neutron-to-proton ratio determines the type of decay that occurs.** Some nuclei have too many neutrons relative to the number of protons. These nuclei decay by turning a neutron into a proton to emit a beta particle (an electron) from the nucleus.

$${}^{1}_{0}\text{n} \longrightarrow {}^{1}_{1}\text{H} + {}^{0}_{-1}\text{e}$$

This process is known as beta emission. It increases the number of protons while decreasing the number of neutrons. Table 25.2 shows examples of beta emission.

Other nuclei are unstable because they have too few neutrons relative to the number of protons. These nuclei increase their stability by converting a proton to a neutron. An electron is captured by a nucleus during this process of electron capture.

A **positron** is a particle with the mass of an electron but a positive charge. During positron emission, a proton changes to a neutron.

All nuclei that have an atomic number greater than 83 are radioactive. These nuclei have both too many neutrons and too many protons to be stable. Therefore they undergo radioactive decay. Most of them emit alpha particles. Alpha emission increases the neutron-to-proton ratio, which tends to increase the stability of the nucleus. In alpha emission the mass number decreases by four and the atomic number decreases by two.

If all the masses in a nuclear reaction were measured accurately enough, you would find that mass is not conserved. An extremely small quantity of mass is converted into energy released in radioactive decay.

Half-Life

Every radioisotope has a characteristic rate of decay measured by its half-life. A **half-life** ($t_{1/2}$) is the time required for one-half of the nuclei of a radioisotope sample to decay to products, as shown in Figure 25.5. **After each half-life, half of the existing radioactive atoms have decayed into atoms of a new element.**

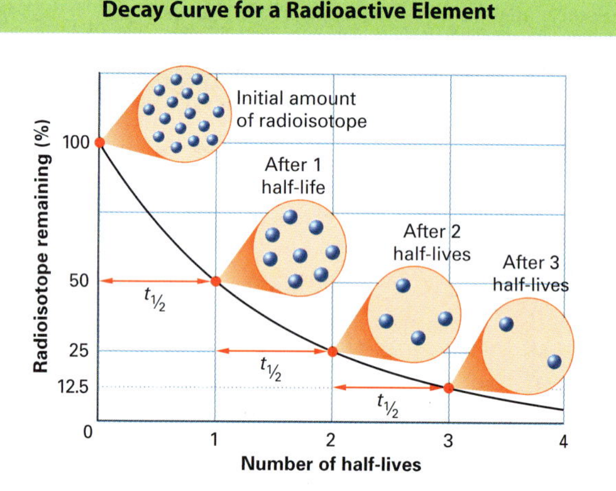

Figure 25.5 This decay curve shows that during each half-life, half of the radioactive atoms decay into atoms of another element.

INTERPRETING GRAPHS

a. Identify What percent of the atoms remains after 1 half-life?
b. Describe What percent of the atoms remains after two half-lives?
c. Apply Concepts Approximately how many half-lives does it take for 12.5% of the radioisotope to remain?

Facts and Figures

Unstable Matter

Nearly 85% of all known nuclei are unstable. This fact does not mean that 85% of all matter is unstable. The distribution of stable and unstable isotopes is not even. Because neutrons help bind protons together in the nucleus, the ratio of protons to neutrons is a major factor in determining stability.

Table 25.3

Half-Lives and Radiation of Some Naturally Occurring Radioisotopes

Isotope	Half-life	Radiation emitted
Carbon-14	5.73×10^3 years	β
Potassium-40	1.25×10^9 years	β, γ
Radon-222	3.8 days	α
Radium-226	1.6×10^3 years	α, γ
Thorium-234	24.1 days	β, γ
Uranium-235	7.0×10^8 years	α, γ
Uranium-238	4.46×10^9 years	α

Go Online SCI LINKS

For: Links on Radioactive Dating
Visit: www.scilinks.org
Web Code: cdn-1252

Half-lives can be as short as a fraction of a second or as long as billions of years. Table 25.3 shows the half-lives of some radioisotopes that occur in nature. Scientists use the half-lives of some radioisotopes found in nature to determine the age of ancient artifacts. Many artificially produced radioisotopes have short half-lives, which is useful in nuclear medicine. The short-lived isotopes are not a long-term radiation hazard to the patient.

One isotope that has a long half-life is uranium-238. Uranium-238 decays through a complex series of radioactive isotopes to the stable isotope lead-206. Figure 25.6 illustrates this process. The age of uranium-containing minerals can be estimated by measuring the ratio of uranium-238 to lead-206. Because the half-life of uranium-238 is 4.5×10^9 years, it is possible to use its half-life to date rocks as old as the solar system.

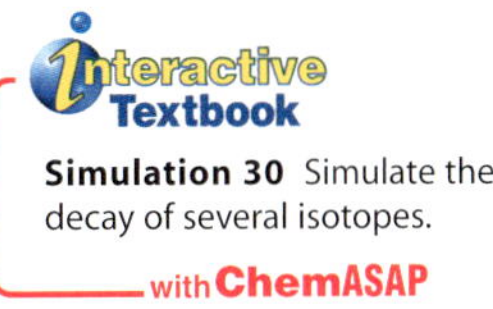

Simulation 30 Simulate the decay of several isotopes.

with ChemASAP

 ***What is the half-life of carbon-14?***

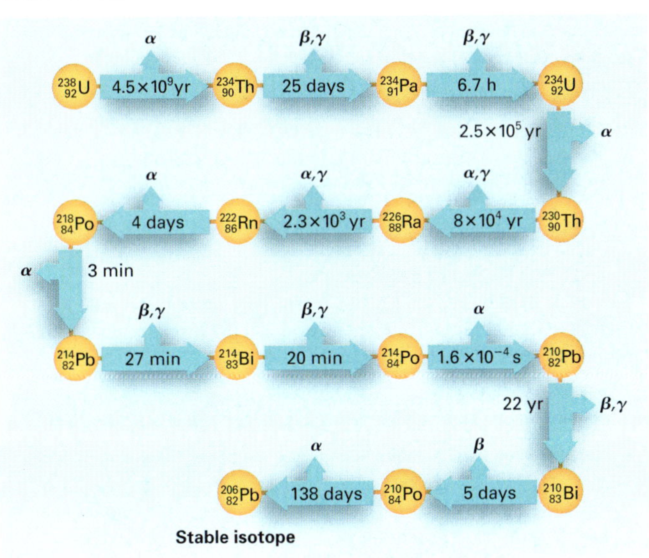

Figure 25.6 Uranium-238 decays through a complex series of radioactive intermediates, including radon (Rn) gas. **Interpreting Diagrams** ***What is the stable end product of this series?***

Use Visuals L1

Table 25.3 Refer students to the table and have them note the range of half-life values. Remind them that these values do not indicate how long a given atom of an isotope will exist, but only how long it takes for half of the atoms in a sample to undergo radioactive decay.

Download a worksheet on **Radioactive Dating** for students to complete, and find additional teacher support from NSTA Sci Links.

Discuss 

Emphasize that the rate of disintegration of the nuclei of an isotope is unaffected by factors such as heat, pressure, or chemical reactions. Note that the half-life for a particular radioactive isotope is unique to that isotope. No two isotopes have exactly the same half-life.

Use Visuals 

Figure 25.6 Have students study Figure 25.6, which charts the radioactive decay of uranium-238. Explain that many radioisotopes go through a complex series of nuclear reactions before a stable product is formed.

Differentiated Instruction

Less Proficient Readers 

Have students who have difficulty reading draw inferences from tables and numbers. For example, have students compare and contrast the change in radioactivity levels over time for samples of the isotopes potassium-40 and thorium-234. Using Table 25.3, they should be able to infer that the radiation emitted by a sample of K-40 would diminish much less rapidly than the radiation from a sample of Th-234 because of the much longer half-life of K-40.

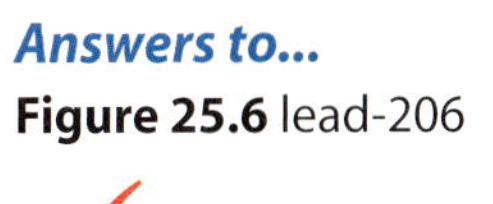

Answers to...

Figure 25.6 lead-206

 5.73×10^3 years

Sample Problem 25.1

Answers

7. 10.4 h/2.6 h/half-life = 4 half-lives
1.0 mg/2^4 = 0.063 mg Mn-56

8. 48.2 da/24.1 da/half-life = 2 half-lives
No, $(^1/_2)^2$, or $^1/_4$, of the sample will remain.

Practice Problems Plus L2

1. **How much of a 0.74-mg sample of uranium-235 will remain after 2.8×10^9 years?** *(4.6×10^{-2} mg)*

2. A 0.456-mg sample of hydrogen-3 was collected. After 24.52 years, 0.114 mg of the sample remains. **What is the half-life of hydrogen-3?** *(12.26 years)*

3. Strontium-90 is a beta emitter with a half-life of 29 years. **What is the mass of strontium-90 in a 5.0-g sample of the isotope at the end of 87 years?** *(0.63 g)*

Math Handbook

For a math refresher and practice, direct students to dimensional analysis, page R66.

Figure 25.7 This archaeologist is digging for artifacts. The age of an artifact can often be determined from its measured carbon-14 content.

Scientists often find the age of an object that was once part of a living system by measuring the amount of carbon-14 ($^{14}_{6}C$) it contains. Carbon-14 has a half-life of 5730 years. Most of Earth's carbon, however, consists of the more stable isotopes $^{12}_{6}C$ and $^{13}_{6}C$. The ratio of $^{14}_{6}C$ to the other carbon isotopes in the environment is fairly constant because high-energy cosmic rays from space constantly produce $^{14}_{6}C$ in the upper atmosphere. Plants grow by producing sugars, cellulose, and other compounds from carbon dioxide in the atmosphere. Animals grow by eating the plants, and sometimes other animals. This keeps the ratio of carbon-14 to other carbon isotopes constant during any living organism's lifetime. When an organism dies, it stops exchanging carbon with the environment and its radioactive $^{14}_{6}C$ atoms decay without being replaced. Therefore, the ratio of $^{14}_{6}C$ to stable carbon in the remains of an organism changes in a predictable way that enables the archaeologist in Figure 25.7 to obtain an estimate of its age.

Math Handbook

For help with scientific notation, go to page R56 of the Math Handbook.

SAMPLE PROBLEM 25.1

Using Half-lives in Calculations

Carbon-14 emits beta radiation and decays with a half-life ($t_{1/2}$) of 5730 years. Assume you start with a mass of 2.00×10^{-12} g of carbon-14.

a. How long is three half-lives?

b. How many grams of the isotope remain at the end of three half-lives?

1 Analyze *List the knowns and the unknowns.*

Knowns
- $t_{1/2}$ = 5730 years
- initial mass 2.00×10^{-12} g
- number of half-lives = 3

Unknowns
- 3 half-lives = ? years
- mass after 3 half-lives = ? g

First, calculate the time for three half-lives by multiplying the length of each half-life by three. Find the mass remaining by multiplying the original mass by $\frac{1}{2}$ three times.

2 Calculate *Solve for the unknowns.*

a. 3 half-lives = 3 × 5730 years = 17,190 years

b. The initial mass of carbon-14 is reduced by one half for each of the three half-lives, so for three half-lives

$$\text{Remaining mass} = 2.00 \times 10^{-12}\text{g} \times \frac{1}{2} \times \frac{1}{2} \times \frac{1}{2} = 0.250 \times 10^{-12}\text{g}$$

3 Evaluate *Do the results make sense?*

The mass of carbon-14 after three half-lives should be much lower than the original mass. The final answer has the proper units and the proper number of significant figures.

Practice Problems

7. Manganese-56 is a beta emitter with a half-life of 2.6 h. What is the mass of manganese-56 in a 1.0-mg sample of the isotope at the end of 10.4 h?

8. A sample of thorium-234 has a half-life of 24.1 days. Will all the thorium undergo radioactive decay in 48.2 days? Explain.

Problem-Solving 25.7
Solve Problem 7 with the help of an interactive guided tutorial.

with ChemASAP

Differentiated Instruction

Gifted and Talented L3

Review exponential notation and rules of exponents as they apply to logarithms. Review the rules for manipulating logs, such as $\log xy = \log x + \log y$ and $\log x^a = a \log x$. Assign practice problems that require students to apply these rules. Challenge students to apply logarithms to solving half-life problems that involve fractions of half-lives.

Transmutation Reactions

The conversion of an atom of one element to an atom of another element is called **transmutation.** There are at least two ways transmutation occurs. **Transmutation can occur by radioactive decay. Transmutation can also occur when particles bombard the nucleus of an atom.** The bombarding particles may be protons, neutrons, or alpha particles.

Many transmutations occur in nature. The production of carbon-14 from naturally occurring nitrogen-14, for example, takes place in the upper atmosphere. Another naturally occurring isotope, uranium-238, undergoes 14 transmutations before reaching a stable isotope. Many other transmutations are done in laboratories or in nuclear reactors. The earliest artificial transmutation was performed in 1919 by Ernest Rutherford (1871–1937). He bombarded nitrogen gas with alpha particles to produce an unstable isotope of fluorine. The results of this reaction are shown in Figure 25.8. The first step of the reaction forms fluorine-18.

$$\underset{\text{Nitrogen-14}}{{}^{14}_{7}\mathrm{N}} + \underset{\text{Alpha particle}}{{}^{4}_{2}\mathrm{He}} \longrightarrow \underset{\text{Fluorine-18}}{{}^{18}_{9}\mathrm{F}}$$

The fluorine isotope quickly decomposes to a stable isotope of oxygen and a proton.

$$\underset{\text{Fluorine-18}}{{}^{18}_{9}\mathrm{F}} \longrightarrow \underset{\text{Oxygen-17}}{{}^{17}_{8}\mathrm{O}} + \underset{\text{Proton}}{{}^{1}_{1}\mathrm{H}}$$

Rutherford's experiment eventually led to the discovery of the proton. James Chadwick's discovery of the neutron in 1932 also involved a transmutation experiment. Neutrons were produced when beryllium-9 was bombarded with alpha particles.

$$\underset{\text{Beryllium-9}}{{}^{9}_{4}\mathrm{Be}} + \underset{\text{Alpha particle}}{{}^{4}_{2}\mathrm{He}} \longrightarrow \underset{\text{Carbon-12}}{{}^{12}_{6}\mathrm{C}} + \underset{\text{Neutron}}{{}^{1}_{0}\mathrm{n}}$$

What is transmutation?

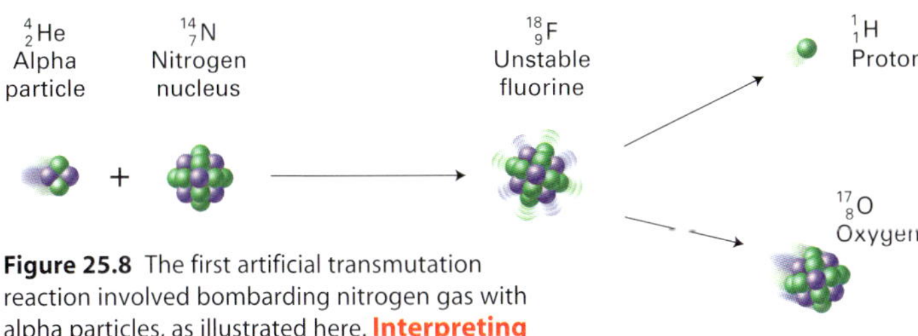

Figure 25.8 The first artificial transmutation reaction involved bombarding nitrogen gas with alpha particles, as illustrated here. **Interpreting Diagrams** ***What particles were formed?***

Transmutation Reactions

Discuss L2

Explain that one of the main goals of medieval alchemists was the conversion of common metals to precious metals. No chemical reaction can achieve this goal. However, through transmutations, modern chemists can change one element into another. Point out that transmutation reactions also allow chemists to produce elements that do not occur naturally.

CLASS Activity

Particle Accelerators L2

Purpose Students learn about and model an accelerator.

Materials reference materials and Internet access, materials for making models

Procedure Particle accelerators, such as linear accelerators, cyclotrons, and synchrotrons, are used in transmutation experiments. Divide the class into groups of four to five. Have each group choose one type of accelerator and conduct library research on its design and function. The group should create a model of the accelerator, which students can display and explain.

Expected Outcome Models and explanations should show understanding of how transmutation is accomplished.

Discuss 

Write equations for a number of transmutation reactions on the board. Point out that the mass numbers and atomic numbers for reactants and products in transmutation reactions are balanced. Write partial equations such as

$${}^{239}_{94}\mathrm{Pu} + ? \longrightarrow {}^{242}_{96}? + {}^{1}_{0}\mathrm{n}$$

$${}^{238}_{92}\mathrm{U} + {}^{?}_{?}\mathrm{C} \longrightarrow {}^{246}_{98}\mathrm{Cf} + 4\,{}^{1}_{0}\mathrm{n}$$

on the board. Then have students supply the missing information.

Differentiated Instruction

Special Needs L1

To reinforce the concept of half-life, have students consider the following problem: Imagine winning a $1000 prize but the conditions of the award require that half of the remainder of the prize is spent each month. Ask, **After how many months would you be left with less than $1? What is the half-life of this prize?** *(10 months; 1 month)* As an extension, have students plot a decay curve, like the one in Figure 25.5, except the *x*-axis should represent time (days or years), for one of the elements in Table 25.3. Tell students that the initial amount of radioisotope is 100 g.

Answers to...

Figure 25.8 ultimately, oxygen-17 atoms and protons

 the conversion of an atom of one element to an atom of another element

Section 25.2 (continued)

3 ASSESS

Evaluate Understanding L2

Ask students how the ratio of neutrons to protons changes in nuclei that undergo beta particle and alpha particle emission. Ask students to explain the concept of half-life using a numerical example. (For example: original state = 256 atoms; one half-life = 128 atoms; two half-lives = 64 atoms, and so on.) Write several balanced and unbalanced nuclear equations on the board and have students identify and correct the unbalanced equations.

Reteach L1

Point out the three natural processes that can result in a nucleus attaining a stable ratio of neutrons to protons. A beta emission increases the number of protons and decreases the number of neutrons. A positron emission decreases the number of protons and increases the number of neutrons. In heavy elements, the nucleus may emit an alpha particle, which decreases the numbers of both neutrons and protons.

Writing Activity

C-14 is used to date once-living artifacts such as bones or coral. Its range is limited to 5.00×10^4 years by its relatively short half-life. The oldest rocks are dated with U-238. Its half-life is about 4.5 billion years. The decay product is Pb-207. K-40 has a half-life of 1.25×10^9 years. It decays to Ar-40. Items made of clay, such as pottery, can be dated using a property called thermoluminescence.

If your class subscribes to the Interactive Textbook, use it to review key concepts in Section 25.2.

with ChemASAP

Figure 25.9 Fermilab is a major accelerator center located in Batavia, Illinois. The main accelerator is a ring that has a radius of 1.0 km.

The elements in the periodic table with atomic numbers above 92, the atomic number of uranium, are called the **transuranium elements.** All the transuranium elements undergo transmutation. None of them occurs in nature, and all of them are radioactive. These elements have been synthesized in nuclear reactors and nuclear accelerators. Accelerators like the one in Figure 25.9 accelerate bombarding particles to very high speeds, while reactors produce beams of low-energy particles. When uranium-238 is bombarded with relatively slow neutrons from a nuclear reactor, some uranium nuclei capture neutrons to produce uranium-239.

$$^{238}_{92}\text{U} + ^{1}_{0}\text{n} \longrightarrow ^{239}_{92}\text{U}$$

Uranium-239 is radioactive and emits a beta particle. The product is an isotope of the artificial radioactive element neptunium (atomic number 93).

$$^{239}_{92}\text{U} \longrightarrow ^{239}_{93}\text{Np} + ^{0}_{-1}\text{e}$$

Neptunium is unstable and decays, emitting a beta particle, to produce a second artificial element, plutonium (atomic number 94).

$$^{239}_{93}\text{Np} \longrightarrow ^{239}_{94}\text{Pu} + ^{0}_{-1}\text{e}$$

Plutonium and neptunium are both transuranium elements and do not occur in nature. Scientists in Berkeley, California synthesized these first artificial elements in 1940. Since that time, more than 20 additional transuranium elements have been produced artificially.

25.2 Section Assessment

9. **Key Concept** What determines the type of decay a radioisotope will undergo?
10. **Key Concept** How much of a sample of radioisotope remains after one half-life? After two half-lives?
11. **Key Concept** What are two ways that transmutation can occur?
12. Complete and balance the equations for the following nuclear reactions.
    - **a.** $^{27}_{13}\text{Al} + ^{4}_{2}\text{He} \longrightarrow ^{30}_{14}\text{Si} + ?$
    - **b.** $^{214}_{83}\text{Bi} \longrightarrow ^{4}_{2}\text{He} + ?$
    - **c.** $^{27}_{14}\text{Si} \longrightarrow ^{0}_{-1}\text{e} + ?$
    - **d.** $^{66}_{29}\text{Cu} \longrightarrow ^{66}_{30}\text{Zn} + ?$
13. A radioisotope has a half-life of 4 days. How much of a 20-gram sample of this radioisotope remains at the end of each time period?
    - **a.** 4 days
    - **b.** 8 days
14. The mass of cobalt-60 in a sample is found to have decreased from 0.800 g to 0.200 g in a period of 10.5 years. From this information, calculate the half-life of cobalt-60.

Writing Activity

Research the methods used to date materials such as pottery, coral, and stone. Prepare a written report that summarizes your findings on the radioisotopes used, their half-lives, and their limitations.

Assessment 25.2 Test yourself on the concepts in Section 25.2.

with ChemASAP

Section 25.2 Assessment

9. the neutron-to-proton ratio
10. 50% of the sample remains after one half-life; 25% remains after two half-lives.
11. radioactive decay and particle bombardment of a nucleus
12. **a.** $^{27}_{13}\text{Al} + ^{4}_{2}\text{He} \longrightarrow ^{30}_{14}\text{Si} + ^{1}_{1}\text{H}$

    b. $^{214}_{83}\text{Bi} \longrightarrow ^{4}_{2}\text{He} + ^{210}_{81}\text{Tl}$

    c. $^{27}_{14}\text{Si} \longrightarrow ^{0}_{-1}\text{e} + ^{27}_{15}\text{P}$

    d. $^{66}_{29}\text{Cu} \longrightarrow ^{66}_{30}\text{Zn} + ^{0}_{-1}\text{e}$
13. **a.** 10 g
    b. 5 g
14. 5.25 years

Small-Scale LAB

Radioactivity and Half-Lives

Purpose

To simulate the transformation of a radioactive isotope over time and to graph the data and relate it to radioactive decay and half-lives.

Materials

- pencil
- ruler
- penny
- paper
- graph paper

Procedure

On a sheet of paper, make a data table similar to the one below. For trial number 1, flip a penny 100 times and, in your grid, record the total number of heads that result. Now flip the penny the same number of times as the number of heads that you obtained in the first 100 flips. Record the total number of flips and the number of heads that result. Continue this procedure until you obtain no more heads.

Trial #	Number of flips	Number of heads
1	100	
2		
3		
4		
5		
6		
7		

Analyze

Using your experimental data, record the answers to the following questions below your data table.

1. Use graph paper to plot the number of flips (*y*-axis) versus the trial number (*x*-axis). Draw a smooth line through the points.
2. Examine your graph. Is the rate of the number of heads produced over time linear or nonlinear? Is the rate constant over time or does it change?
3. Why does each trial reduce the number of heads by approximately one-half?
4. A half-life is the time required for one-half of the atoms of a radioisotope to emit radiation and to decay to products. What value represents one half-life for the process of flipping coins?

You're The Chemist

The following small-scale activities allow you to develop your own procedures and analyze the results.

1. **Design It!** Design and carry out an experiment using a single die to model radioactive decay. Plot your data.
2. **Analyze It!** Many radioisotopes undergo alpha decay. They emit an alpha particle (helium nucleus $^{4}_{2}He$). For example,

$$^{222}_{86}Rn \longrightarrow ^{218}_{84}Po + ^{4}_{2}He$$

Find the half-life of radon-222 in Table 25.3 and determine how long it takes for only one eighth of a sample of radon-222 to remain.

3. **Analyze It!** Other radioisotopes undergo beta decay, emitting a beta particle (electron $^{0}_{-1}e$). For example,

$$^{14}_{6}C \longrightarrow ^{14}_{7}N + ^{0}_{-1}e$$

Find the half-life of carbon-14 in Table 25.3 and determine what fraction of the carbon-14 in a sample has not yet decayed by beta emission after 11,460 years.

Radioactivity and Half-Lives L2

Objective After completing this activity, students will be able to simulate the transformation of a radioactive isotope and relate simulated data to radioactive decay and half-life.

Prep Time 5 minutes
Class Time 20 minutes

Teaching Tips

- Emphasize that the trials involving flipping a coin to simulate radioactivity. The appearance and removal of a "head" represents the decay of an unstable nucleus. The rate of removal is analogous to the half-life of a radioactive isotope—around 50%. Although the number of heads decreases over time, the percent of heads produced remains relatively steady until the sample becomes too small.
- Point out that in a sample as small as 100, the likelihood of producing a number of heads other than 50 is high. If the sample size were increased, the relative error would decrease. As the sample size approaches zero, the results are no longer statistically reliable.
- Remind students that the probabilities relate to the overall sample, not to any individual coin or atom. Predictions for a mole of atoms, which provides a large sample, can be quite accurate.

Expected Outcome For each trial, the number of heads is approximately half the number of flips.

Analyze

1. Student graphs should resemble the graph in Figure 25.5. **2.** Nonlinear; the rate decreases over time. **3.** For each flip, the probability of a head is 0.50. **4.** one trial

You're the Chemist

1. Count the total number of even numbers that result in 100 rolls of the die. Roll the die again a number of times equal to the number obtained on the first trial. Do trials until the number of events equals zero. Plot number of evens versus trial.
2. After 3.8 days, half the sample remains. After 7.6 days, one-fourth remains, and, after 11.4 days, one-eighth remains.
3. This time period is two half-lives (11,460 years/5730 years = 2) of carbon-14. After two half-lives, one-fourth of the sample remains.

For Enrichment L3

Repeat the procedure four more times, using a new table for each trial. Average the results in each cell of the table for the five trials. Ask, **Which is closer to the expected values, the results from a single trial or the averaged results?** *(The average values should be closer.)* Ask students to discuss the advantages of using multiple trials in an experiment.

25.3

1 FOCUS

Objectives

25.3.1 Describe what happens in a nuclear chain reaction.

25.3.2 Explain the role of water in the storage of spent fuel rods.

25.3.3 Distinguish fission reactions from fusion reactions.

Guide for Reading

Build Vocabulary L2

Graphic Organizers Have students draw a concept map that correctly relates the vocabulary terms for this section.

Reading Strategy L2

Preview Before students read the section in detail, have them preview the section headings, visuals, and boldfaced material.

2 INSTRUCT

Connecting to Your World

Have students study the photograph and read the text that opens the section. Explain that the sun contains hydrogen nuclei. Ask, **In which state of matter do these hydrogen nuclei exist?** *(plasma)* **Describe a hydrogen nucleus.** *(a proton)* Explain that protons can combine to produce helium nuclei. Ask, **How is the energy of the sun produced?** *(Energy is released during the formation of helium nuclei.)*

Nuclear Fission

Discuss L2

Ask, **What is the general meaning of fission?** *(An object splits into smaller parts.)* **What is the general meaning of fusion?** *(Objects combine into a larger whole.)* Explain that these terms have more precise meanings when applied to nuclear reactions but still include the concepts of fragmentation and merging.

25.3 Fission and Fusion of Atomic Nuclei

Guide for Reading

Key Concepts

- What happens in a nuclear chain reaction?
- Why are spent fuel rods from a nuclear reaction stored in water?
- How do fission reactions and fusion reactions differ?

Vocabulary

fission
neutron moderation
neutron absorption
fusion

Reading Strategy

Identifying Details In the three paragraphs following the heading *Nuclear Fission,* the sentence describing a chain reaction states the main idea. List the details about energy changes in fission and chain reactions that are related to the main idea.

Connecting to Your World

The sun appears as a fiery ball in the sky—so bright it should never be viewed with unprotected eyes. Although its surface temperature is about 5800 K, the sun is not actually burning. If the energy given off by the sun were the product of a combustion reaction, the sun would have burned out approximately 2000 years after it was formed, long before today. In this section, you will learn how energy is produced in the sun.

Nuclear Fission

When the nuclei of certain isotopes are bombarded with neutrons, they undergo **fission,** the splitting of a nucleus into smaller fragments. Uranium-235 and plutonium-239 are the only fissionable isotopes. Figure 25.10 shows how a fissionable atom, such as uranium-235, breaks into two fragments of roughly the same size when struck by a slow-moving neutron. At the same time, more neutrons are released by the fission. These neutrons strike the nuclei of other uranium-235 atoms, continuing the fission by a chain reaction. **In a chain reaction, some of the neutrons produced react with other fissionable atoms, producing more neutrons which react with still more fissionable atoms.**

Nuclear fission can release enormous amounts of energy. The fission of 1 kg of uranium-235, for example, yields an amount of energy equal to that generated in the explosion of 20,000 tons of dynamite. In an uncontrolled nuclear chain reaction, the total energy release takes only fractions of a second. Atomic bombs are devices that start uncontrolled nuclear chain reactions.

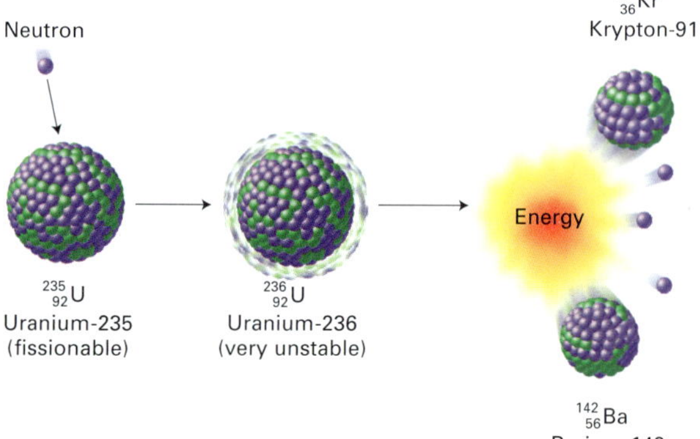

Figure 25.10 In nuclear fission, a uranium-235 nucleus breaks into two smaller nuclei and releases neutrons. **Predicting** ***What happens when the released neutrons strike other uranium-235 nuclei?***

Section Resources

Print

- **Guided Reading and Study Workbook,** Section 25.3
- **Core Teaching Resources,** Section 25.3 Review
- **Transparency,** T293

Technology

- **Interactive Textbook with ChemASAP,** Animation 30, Assessment 25.3

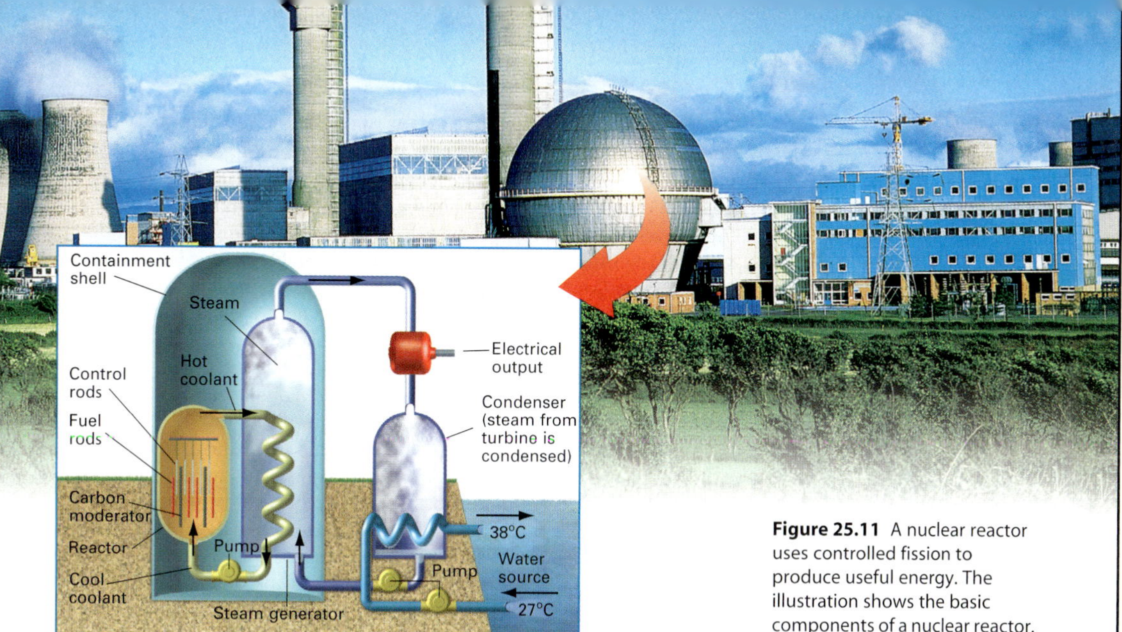

Figure 25.11 A nuclear reactor uses controlled fission to produce useful energy. The illustration shows the basic components of a nuclear reactor. Energy from the fission process heats the circulating coolant. The heated coolant is used to produce steam that turns a steam-driven turbine. The turbine drives a generator to produce electrical energy.

Fission can be controlled so energy is released more slowly. Nuclear reactors, such as the one illustrated in Figure 25.11, use controlled fission to produce useful energy. In the controlled fission reaction within a nuclear reactor, much of the energy generated is in the form of heat. A coolant fluid, usually liquid sodium or water, removes the heat from the reactor core. The heat is used to generate steam, which drives a turbine that in turn generates electricity. The control of fission in a nuclear reactor involves two steps, neutron moderation and neutron absorption.

Neutron Moderation **Neutron moderation** is a process that slows down neutrons so the reactor fuel (uranium-235 or plutonium-239) captures them to continue the chain reaction. Moderation is necessary because most of the neutrons produced move so fast that they will pass right through a nucleus without being absorbed. Water and carbon in the form of graphite are good moderators.

Neutron Absorption To prevent the chain reaction from going too fast, some of the slowed neutrons must be trapped before they hit fissionable atoms. **Neutron absorption** is a process that decreases the number of slow-moving neutrons. Control rods, made of a material such as cadmium, are used to absorb neutrons. When the control rods extend almost all the way into the reactor core, they absorb many neutrons, and fission occurs slowly. As the rods are pulled out, they absorb fewer neutrons and the fission process speeds up. If the chain reaction were to go too fast, heat might be produced faster than the coolant could remove it. In this case, the reactor core would overheat, which could lead to mechanical failure and release of radioactive materials into the atmosphere. Ultimately, a meltdown of the reactor core might occur.

Checkpoint ***Why must neutrons in a reactor be slowed down?***

Interactive Textbook

Animation 30 Take a close look at a nuclear fission chain reaction.

with ChemASAP

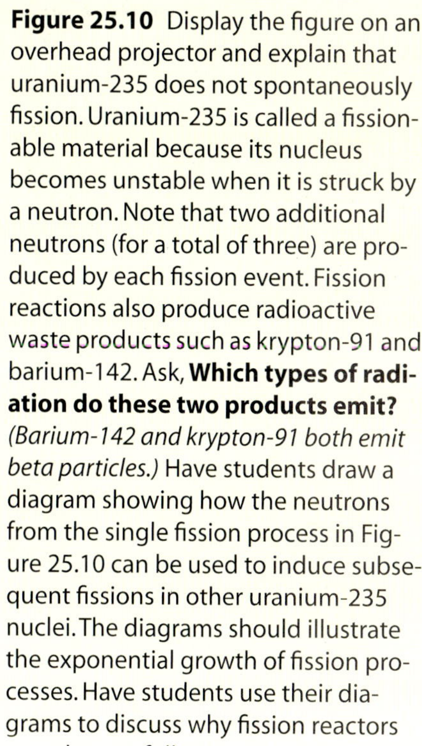

Use Visuals L1

Figure 25.10 Display the figure on an overhead projector and explain that uranium-235 does not spontaneously fission. Uranium-235 is called a fissionable material because its nucleus becomes unstable when it is struck by a neutron. Note that two additional neutrons (for a total of three) are produced by each fission event. Fission reactions also produce radioactive waste products such as krypton-91 and barium-142. Ask, **Which types of radiation do these two products emit?** *(Barium-142 and krypton-91 both emit beta particles.)* Have students draw a diagram showing how the neutrons from the single fission process in Figure 25.10 can be used to induce subsequent fissions in other uranium-235 nuclei. The diagrams should illustrate the exponential growth of fission processes. Have students use their diagrams to discuss why fission reactors must be carefully monitored and controlled.

TEACHER Demo

Model a Chain Reaction L2

Purpose Students observe a chain reaction.

Materials wooden matchsticks

Safety Keep all flammable materials away from the demonstration area. Have a fire extinguisher ready.

Procedure Cut some wooden matchsticks in half and arrange the heads in branching chains on a noncombustible surface. Construct the chain so that the end of each matchstick is touching the heads of two other matchsticks, forming a "Y." Ignite the first matchstick head.

Expected Outcome The matches ignite in a chain reaction.

Differentiated Instruction

English Learners L2

Have students write the definitions of fission and fusion in their notebooks in their native language and in English. Have them prepare a table, complete with sample reactions, describing the characteristics of fission and fusion processes in nuclear chemistry.

Differentiated Instruction

Gifted and Talented L3

The difference between the mass of a nucleus and the masses of the separated protons and neutrons is called the mass defect. Challenge students to use the mass defect and Albert Einstein's famous equation, $E = mc^2$, to calculate the energy released during a chemical reaction and during a nuclear fission reaction. Have students explain why so much more energy is released during a nuclear fission reaction.

Answers to...

Figure 25.10 They react and cause the production of more neutrons.

Checkpoint so the reactor fuel can capture them to continue the chain reaction

Section 25.3 (continued)

Nuclear Waste

Discuss

L2

Have students compare and contrast nuclear fission and nuclear fusion. Ask students to describe the advantages and disadvantages of each type of process as a means to meet future energy needs. Ask, **What types of technical hurdles remain to be solved in order to utilize the energy produced by nuclear fusion?** *(Accept any reasonable answer, such as containment of the reaction.)*

Relate

L2

Explain that the wastes produced in fission reactors contain isotopes with half-lives measured in the thousands or hundreds of thousands of years. Many proposals for storing or disposing of these wastes involve methods and materials that may be highly unsuitable. For example, the placement of wastes in thick drums that would then be sunk in the oceans—one suggested method—may contain the waste for only decades or a few centuries. The containers would begin to leak long before the contents were safe. The disposal problem has led some people to propose permanent abandonment of nuclear reactors as sources of energy. After students read the text, lead a discussion on the pros and cons of using nuclear energy from fission reactors.

CLASS Activity

Nuclear Fission

L2

After the neutron was discovered in the 1930's, scientists set out to create elements heavier than uranium in the laboratory. The culmination of this work was the discovery of nuclear fission. Have students research the contributions of Lise Meitner and Enrico Fermi.

Nuclear Waste

Fuel rods from nuclear power plants are one major source of nuclear waste. The fuel rods are made from a fissionable isotope, either uranium-235 or plutonium-239. The fuel rods are long and narrow—typically 3 meters long with a 0.5-cm diameter. Three hundred fuel rods are bundled together to form an assembly, and one hundred assemblies are arranged to form the reactor core. During fission, the amount of fissionable isotope in each fuel rod decreases. Eventually there is no longer enough fuel in the rods to ensure that the output of the power station remains constant. The isotope-depleted, or spent, fuel rods must be removed and replaced with new fuel rods.

Spent fuel rods are classified as high-level nuclear waste. They contain a mixture of highly radioactive isotopes, including both the fission products and what remains of the nuclear fuel. Some of these fission products have very short half-lives, on the order of fractions of seconds. Others have half-lives of hundreds or thousands of years. All nuclear power plants have holding tanks, or "swimming pools," for spent fuel rods. **Water cools the spent rods, and also acts as a radiation shield to reduce the radiation levels.** The pools, like the one shown in Figure 25.12, are typically 12 meters deep, and are filled with water. Storage racks at the bottom of these pools are designed to hold the spent fuel assemblies. The rods continue to produce heat for years after their removal from the core.

The assemblies of spent fuel rods may spend a decade or more in a holding tank. In the past, plant operators expected used fuel rods to be reprocessed to recover the remaining fissionable isotope, which would be recycled in the manufacture of new fuel rods. However, with large deposits of uranium ore available—many in the United States—it is less expensive to mine new fuel than to reprocess depleted fuel. At some nuclear plants, there is no space left in the storage pool. In order to keep these plants open, their fuel rods must be moved to off-site storage facilities.

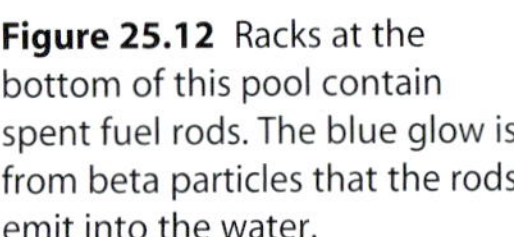

Figure 25.12 Racks at the bottom of this pool contain spent fuel rods. The blue glow is from beta particles that the rods emit into the water.

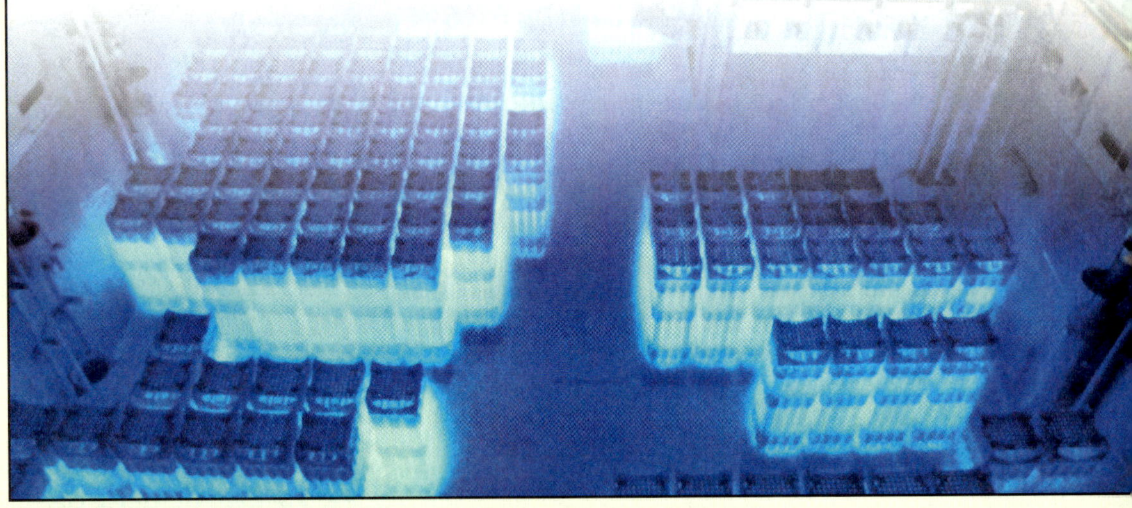

Facts and Figures

Nuclear Waste and Nuclear Power

Concerns about nuclear wastes are magnified by the timeframe during which the wastes will be hazardous. As of 1997, 440 nuclear power plants were in operation worldwide and about 30 countries got some of their electricity from nuclear power stations. In France, for example, 75% of the electricity is generated by nuclear power. By contrast, only about 20% of electricity generated in the United States comes from nuclear power plants. The DOE is responsible for cleaning up 130 nuclear sites and safely managing their waste. The sites include locations where uranium was milled, research labs, and former nuclear weapons production facilities. Have students research how the cleanups are progressing.

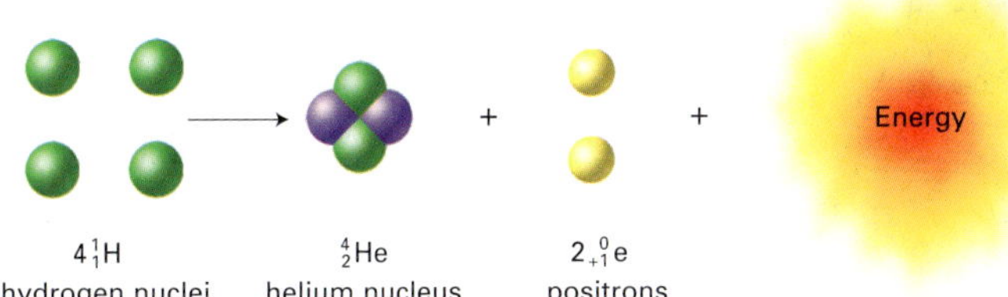

Figure 25.13 In solar fusion, hydrogen nuclei fuse to produce helium nuclei. **Interpreting Diagrams** ***What other particles are produced by this reaction?***

Nuclear Fusion

The sun, directly and indirectly, is the source of most energy used on Earth. The energy released by the sun results from nuclear fusion. **Fusion** occurs when nuclei combine to produce a nucleus of greater mass. In solar fusion, hydrogen nuclei (protons) fuse to make helium nuclei. Figure 25.13 shows that the reaction also produces two positrons. **Fusion reactions, in which small nuclei combine, release much more energy than fission reactions, in which large nuclei split.** However, fusion reactions occur only at very high temperatures—in excess of 40,000,000°C.

The use of controlled nuclear fusion as an energy source on Earth is appealing. The potential fuels are inexpensive and readily available. One reaction that scientists are studying is the combination of a deuterium (hydrogen-2) nucleus and a tritium (hydrogen-3) nucleus to form a helium nucleus.

$$^2_1H + ^3_1H \longrightarrow ^4_2He + ^1_0n + \text{energy}$$

The problems with fusion lie in achieving the high temperatures necessary to start the reaction and in containing the reaction once it has started. The high temperatures required to initiate fusion reactions have been achieved by using a fission bomb. Such a bomb is the triggering device used for setting off a hydrogen bomb, which is an uncontrolled-fusion device. Such a process is clearly of no use, however, as a controlled generator of power.

Word *Origins*

Fusion comes from the Latin word *fusus* meaning "melt together." Fusion is the combination of two low-mass nuclei to form a nucleus of larger mass, accompanied by the release of a large amount of energy. **In politics, what might be the goal of a fusionist?**

25.3 Section Assessment

15. **Key Concept** Explain what happens in a nuclear chain reaction.
16. **Key Concept** Why are spent fuel rods from a nuclear reaction stored in water?
17. **Key Concept** How are fusion reactions different from fission reactions?
18. What does nuclear moderation accomplish in a nuclear reactor?
19. What is the source of the radioactive nuclei present in spent fuel rods?
20. Assuming technical problems could be overcome, what are some advantages to using a fusion reactor to produce electricity?

Elements Handbook

Heavy Water Reactors Read the paragraph on Heavy Water Reactors on page R39 of the Elements Handbook. Use what you have learned about nuclear reactors to explain what a neutron moderator does, and then explain the advantages of using heavy water instead of ordinary water as a neutron moderator.

Assessment 25.3 Test yourself on the concepts in Section 25.3.

with **ChemASAP**

Nuclear Fusion

Word *Origins* L2

A fusionist might bring together different factions into one cohesive group.

3 ASSESS

Evaluate Understanding L2

Ask students to write nuclear equations describing fission and fusion. Ask, **How might a meltdown of a nuclear reactor occur?** *(Rapid removal of the control rods could allow too many neutrons to react with fissionable nuclei.)* **Describe the conditions under which nuclear fusion will occur.** *(temperatures in excess of 4.00×10^7 °C, methods to confine and control the dense plasmas are needed)*

Reteach L1

Explain that both nuclear fission and nuclear fusion produce energy by the conversion of matter to energy. In fission, heavy nuclei are split into lighter nuclei. In fusion, light nuclei combine to form heavier nuclei. Fission reactions are relatively easy to control but produce radioactive wastes. Fusion reactions are difficult to initiate and control but produce little radioactive waste.

Elements Handbook

A neutron moderator slows neutrons so that reactor-fuel nuclei may capture them and sustain the chain reaction. D_2O is a more efficient moderator than H_2O because D_2O absorbs fewer neutrons.

Section 25.3 Assessment

15. Neutrons produced by fissionable atoms react with other fissionable atoms, producing more neutrons that react with other fissionable atoms.
16. Water cools spent fuel rods and provides a radiation shield.
17. Fission reactions involve splitting nuclei. In fusion reactions, small nuclei combine and release much more energy.
18. slows down neutrons
19. unused nuclear fuel and fission products
20. Potential fuels are inexpensive and readily available.

If your class subscribes to the Interactive Textbook, use it to review key concepts in Section 25.3.

with **ChemASAP**

Answers to...

Figure 25.13 positrons

Technology & Society

Dating a Fossil

Discuss 

Discuss with students the limits of using carbon-14 to date fossils. First, some organic material must be present, as the material tested must be from an organism that once lived. If only an imprint remains and all organic material is gone, therefore, carbon-14 cannot be used in dating the fossil. Second, after a certain time, so little carbon-14 remains that it cannot be detected. Experts disagree as to what this age is, but it is commonly accepted that carbon-14 dating cannot be used reliably on objects that are more than 75,000 years old. Ask, **To the nearest whole number, how many half-lives of carbon-14 is 75,000 years?** *(75,000 years/5730 years per half-life = 13 half-lives.)* **What percent of the carbon-14 remains after this number of half lives?** *(0.0012%)*

Technology & Society

Dating a Fossil

Direct information about earlier life on Earth comes from fossils and artifacts. The age of an artifact of biological origin, such as bone, wood, cloth, and plant fibers, from 200 to about 50,000 years old can be determined by carbon-14 dating. All living organisms contain carbon-12 and carbon-14 in a fixed ratio. After an organism dies, the ratio of carbon-12 to carbon-14 changes. Carbon-14 is radioactive with a half-life of 5730 years and decays to nitrogen-14, while carbon-12 remains constant. **Calculating** ***An unearthed wooden tool was found to have only 50% of the carbon-14 content of a sample of living wood. How old is the wooden tool?***

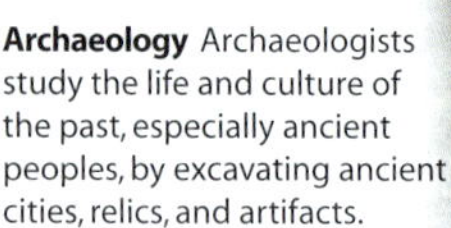

Archaeology Archaeologists study the life and culture of the past, especially ancient peoples, by excavating ancient cities, relics, and artifacts.

The Wooly Mammoth The fossil remains of a woolly mammoth are found in Pleistocene deposits (the Pleistocene epoch was from 2,500,000 to 10,000 years ago). The abundance of well-preserved carcasses in the permanently frozen ground of Siberia have provided much information about these extinct animals.

C. 23,000 BC At the time of death, the carbon-12 to carbon-14 ratio of the woolly mammoth was very similar to that of an animal living today.

C. 17,300 BC The carbon-14 in the dead mammoth was one-half of its value in 23,000 BC.

C. 11,600 BC The carbon-14 was 25% of its initial value.

C. 5,900 BC After another 5,730 years, the carbon-14 is now 12.5% of its 23,000 BC value.

Facts and Figures

Relative Dating

Fossils can also be dated by relating them to known geological events. For example, during a large volcanic eruption, vast amounts of volcanic ash are spread over large areas of the surrounding area. This ash will settle and eventually become a distinctive clay layer. Any fossil in or near this layer can be dated according to the date of the event that formed the layer.

Mummies This mummy from Peru was dated between the late 13th and early 14th century by carbon-14 dating of the decorative cloth around the remains.

c. 200 BC

The carbon-14 is 6.25% of its initial value.

Today, AD 2005

In what year will the carbon-14 level have decreased to 3.125% of its 23,000 BC value?

Woolly Mammoth This well-preserved frozen skin of a baby mammoth was discovered in frozen ground in Siberia in 1900. It had been completely refrigerated for about 25,000 years.

Use Visuals 

Note in the large photo that the fish fossil is in a distinct rock layer. Ask, **What methods other than carbon-14 dating might be used to date the fossil?** *(Answers will vary but might include looking for other fossils in the layer and dating them.)* Discuss how organisms found in the same rock layer probably formed at the same time. Some of these fossils might be dated by using carbon-14, and others might be from organisms that existed only during a known, short period of time.

Relate 

Ask students to relate the half-lives in Table 25.3 on page 805 to the usefulness of isotopes in dating materials. For example, uranium-238 is used to date rocks that are quite old. If U-238 were used to date a sample that is only 50,000 years old, not enough of the U-238 would have decayed for the change to be noticed. Ask, **Which of the radioisotopes listed in the table might be used to date a rock that is 20 million years old?** *(Answers will vary, but students might suggest potassium-40.)*

Differentiated Instruction

Gifted and Talented 

Have students write down the exact procedure used to work problems involving whole-number half-lives. Then have them use scientific calculators to expand this procedure to include partial half-lives. Provide students with problems to work that involve partial half-lives. Examples might include problems such as the following: A fossil was found to contain 60% of its original carbon-14. How old is the fossil? *($k = \ln 2/t_{1/2} = 0.693/5{,}730\ yr = 1.21 \times 10^{-4} yr^{-1}$; $t = 1/k \ln A_o/A_t = (1/1.21 \times 10^{-4} yr^{-1}) \ln(1/0.60) = 4{,}200\ yr$)*

Answers to...

Calculating 5,730 years.

25.4

1 FOCUS

Objectives

25.4.1 Identify three devices that are used to detect radiation.

25.4.2 Describe how radioisotopes are used in medicine.

Guide for Reading

Build Vocabulary 

Graphic Organizers Have students prepare a table to compare and contrast the structures, uses, and other aspects of the different types of radiation detectors listed as vocabulary terms.

Reading Strategy 

Reciprocal Teaching Encourage students to take a leadership role in reading this section. Teacher and students take turns in reading and in conducting discussions about what was read.

2 INSTRUCT

Connecting to Your World

Have students study the photograph and read the text that opens the section. Am-241 undergoes α decay. Ask, **What is the other product of the decay of an Am-241 nucleus?** *(Np-237)* **Why do you have to exert special care in disposing of these smoke detectors?** *(The detectors still contain radioactive materials.)*

Detecting Radiation

Use Visuals L1

Explain that a cloud chamber can be used to detect ionizing radiation. Ask, **What are some other devices used to detect radiation and how do they function?** *(Radiation can expose a photographic plate, cause the sealed gas in a Geiger counter to be ionized so that it conducts electricity, or produce a flash of light on the phosphor-coated surface of a scintillation counter.)*

25.4 Radiation in Your Life

Guide for Reading

 **Key Concepts**

- What are three devices used to detect radiation?
- How are radioisotopes used in medicine?

Vocabulary

ionizing radiation
Geiger counter
scintillation counter
film badge
neutron activation analysis

Reading Strategy

Summarizing When you summarize, you review and state, in the correct order, the main points you have read. As you read about using radiation, write a brief summary of the text following each use that is discussed. Your summary should include only the most important information.

Connecting to Your World

A common household smoke detector uses the radiation from a radioisotope. The smoke detector contains a small amount of americium, $^{241}_{95}\text{Am}$, in the form AmO_2. Radiation from the $^{241}_{95}\text{Am}$ nuclei ionizes the nitrogen and oxygen in smoke-free air, allowing a current to flow. When smoke particles get in the way, they are ionized instead. The drop in current is detected by an electronic circuit, causing it to sound an alarm. In this section, you will learn about some of the other practical uses of radiation.

Detecting Radiation

Radiation emitted by radioisotopes is called ionizing radiation. **Ionizing radiation** is radiation with enough energy to knock electrons off some atoms of the bombarded substance to produce ions. Radiation cannot be seen, heard, felt, or smelled. Thus radiation detection instruments must be used to alert people to the presence of radiation and to monitor its level. **Devices such as Geiger counters, scintillation counters, and film badges are commonly used to detect radiation.** These devices operate because of the effects of the radiation when it strikes atoms or molecules. For example, the radiation can produce ions, which can then be detected, or it can expose a photographic plate and produce images such as the one shown in Figure 25.14. When the plate is developed, its darkened areas show where the plate has been exposed to radiation.

 *What is ionizing radiation?*

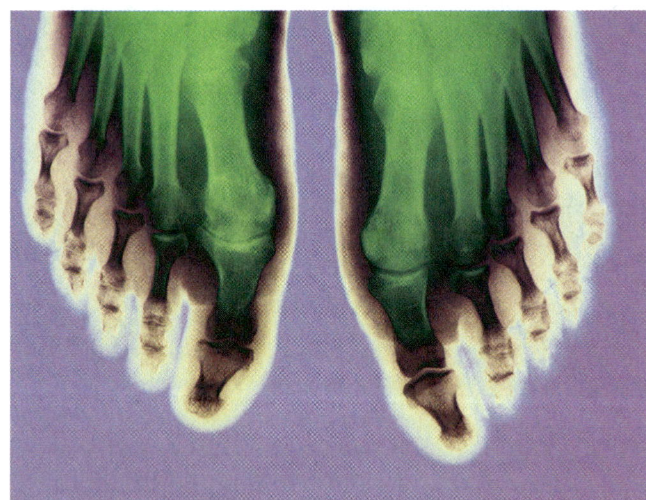

Figure 25.14 Radiation can expose a photographic plate.

Section Resources

Print

- **Laboratory Manual,** Lab 52
- **Probeware Laboratory Manual,** Section 25.4
- **Guided Reading and Study Workbook,** Section 25.4
- **Core Teaching Resources,** Section 25.4 Review
- **Transparency,** T294

Technology

- **Interactive Textbook with ChemASAP,** Assessment 25.4
- **Go Online,** Section 25.4

Geiger Counter A **Geiger counter** uses a gas-filled metal tube to detect radiation. The tube has a central wire electrode that is connected to a power supply. When ionizing radiation penetrates a thin window at one end of tube, the gas inside the tube becomes ionized. Because of the ions and free electrons produced, the gas becomes an electrical conductor. Each time a Geiger tube is exposed to radiation, current flows. The bursts of current drive electronic counters or cause audible clicks from a built-in speaker. Geiger counters can detect alpha, beta, and gamma radiation.

Geiger counter

Scintillation counter

Scintillation Counter A **scintillation counter** uses a phosphor-coated surface to detect radiation. Ionizing radiation striking the phosphor surface produces bright flashes of light, or scintillations. The number of flashes and their respective energies are detected electronically. The information is then converted into electronic pulses, which are measured and recorded. Scintillation counters have been designed to detect all types of ionizing radiation.

Film Badge A **film badge** consists of several layers of photographic film covered with black lightproof paper, all encased in a plastic or metal holder. The film badge is an important radiation detector for persons who work near any type of radiation source. The badge is worn the entire time the person is at work. At specific intervals, with the frequency depending on the type of work involved, the film is removed and developed. The strength and type of radiation exposure are determined from the darkening of the film. Records are kept of the results. Film badges do not protect a person from radiation exposure, but they do monitor the degree of exposure to radiation. Protection against radiation is achieved by keeping a safe distance from the source and by using adequate shielding.

Film badge

Facts and Figures

Gold-Foil Experiment

Hans Geiger (1882–1945) was a student of Ernest Rutherford. It was Geiger, along with another student named Marsden, who carried out the gold-foil, alpha-scattering experiment at Rutherford's direction. Geiger gained experience with the problem of detecting radiation, which ultimately led to his invention of the Geiger counter. Ask students to review the gold-foil experiment and describe the detection method used to measure the scattering of alpha particles. This method was a precursor of the scintillation counter.

Discuss

Explain that the most commonly used radiation detector, the Geiger counter, is primarily used for the detection of beta radiation, because alpha particles cannot penetrate the tube and most gamma and X rays pass through the gas without causing many ionizations. Scintillation counters, though not generally portable, are designed to detect all types of radiation. They rely on a phosphor surface, which produces a bright flash when struck by ionizing radiation. The flashes and their energies can be detected electronically. The process is analogous to electrons striking the phosphor screen inside a television set. Explain that the third type of radiation detector, the film badge, is a modernday version of Becquerel's photographic plate.

TEACHER Demo

Background Radiation

Purpose Students observe background radiation.

Materials Geiger counter

Procedure Using a Geiger counter equipped with a rate meter and a loudspeaker, have students listen to the random clicking noise. Be sure that there are no radioactive sources nearby. Point out that the counter is recording background radiation, which is always present. Ask students about the possible sources of background radiation. *(cosmic, soil, and interior of Earth)* Explain that background radiation varies from location to location. To make an accurate measurement of radiation from a given source, you must take local background radiation into account. Next, place the Geiger counter near some objects containing radioactive elements, such as the luminous dial of a watch or a smoke detector.

Expected Outcome The Geiger counter indicates small amounts of background radiation and larger amounts near a source of radioactivity.

Answers to...

radiation with enough energy to knock electrons off some atoms of the bombarded substance to produce ions

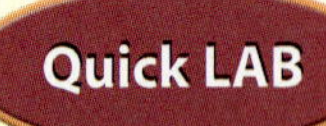

Section 25.4 (continued)

Using Radiation

Quick LAB

Studying Inverse-Square Relationships  L2

Objective After completing this activity, students will be able to:

- demonstrate the relationship between radiation intensity and the distance from the radiation source.

Skills Focus observing, measuring, using graphs, calculating

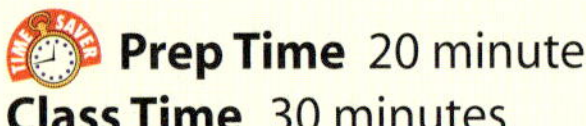

Prep Time 20 minutes

Class Time 30 minutes

Expected Outcome As distance increases, intensity decreases and area increases.

Analyze and Conclude

1. Intensity is inversely related to the square of the distance.
2. When distance is doubled, area increases by a factor of four, and intensity is one-fourth as much. When distance is tripled, area increases by a factor of nine, and intensity is one-ninth as much.

For Enrichment  L3

Have students repeat the Quick Lab two times, changing the size of the hole from what it is in the original lab. On one trial, the hole should be slightly larger. On the other trial, the hole should be slightly smaller. Ask, **Does the size of the hole affect the results?** *(No, the results are the same.)*

Go Online NSTA SCiLINKS

Download a worksheet on **Radioactive Tracers** for students to complete, and find additional teacher support from NSTA Sci Links.

Quick LAB

Studying Inverse-Square Relationships

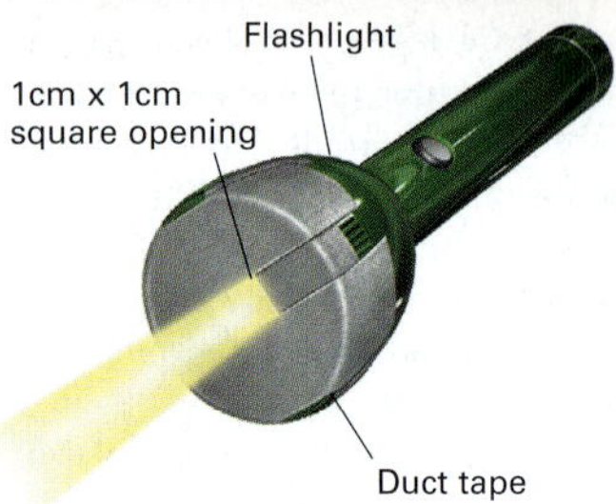

Purpose

To demonstrate the relationship between radiation intensity and the distance from the radiation source.

Materials

- flashlight
- strips of duct tape
- scissors
- poster board, white (50 cm × 50 cm)
- meter ruler or tape measure
- flat surface, long enough to hold the meter ruler
- loose-leaf paper
- graph paper
- pen or pencil
- light sensor (optional)

Procedure    

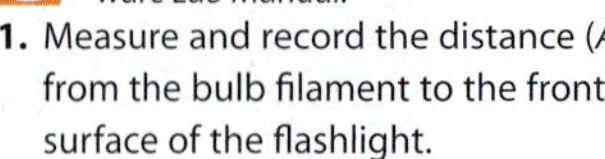

Probeware version available in the Probeware Lab Manual.

1. Measure and record the distance (*A*) from the bulb filament to the front surface of the flashlight.
2. Cover the end of a flashlight with tape. Leave a 1 cm × 1 cm square hole in the center.
3. Place the flashlight on its side on a flat, horizontal surface. Turn on the flashlight. Darken the room.
4. Mount a large piece of white poster board in front of the flashlight, perpendicular to the horizontal surface.
5. Move the flashlight away from the board in short increments. At each position, record the distance (*B*) from the flashlight to the board and the length (*L*) of one side of the square image on the board.
6. On a sheet of graph paper, plot *L* on the *y*-axis versus $A + B$ on the *x*-axis. On another sheet, plot L^2 on the *y*-axis versus $A + B$ on the *x*-axis.

Analyze and Conclude

1. As the flashlight is moved away from the board, what do you notice about the intensity of the light in the illuminated square? Use your graphs to describe the relationship between intensity and distance.
2. When the distance of the flashlight from the board is doubled and tripled, what happens to the areas and intensities of the illuminated squares?

Using Radiation

Although radiation can be harmful, it can be used safely and is important in many scientific procedures. **Neutron activation analysis** is a procedure used to detect trace amounts of elements in samples. A sample of the material to be studied is bombarded with neutrons from a radioactive source. Some atoms in the sample become radioactive. The half-life and type of radiation emitted by the radioisotopes are detected, and a computer processes this information. Because this information is characteristic for each element, scientists can determine what radioisotopes were produced and what elements were originally in the sample. Neutron activation analysis is used by museums to detect art forgeries, and by crime laboratories to analyze gunpowder residues.

Radioisotopes called tracers are used in agriculture to test the effects of herbicides, pesticides, and fertilizers. The tracer is introduced into the substance being tested to make the substance radioactive. Next, the plants are treated with the radioactively labeled substance. Then, the radioactivity of the plants is measured to find the locations of the substance. Often, the tracer is also monitored in animals that consume the plants, in water, and in soil.

Go Online NSTA SCiLINKS

For: Links on Radioactive Tracers
Visit: www.scilinks.org
Web Code: cdn-1254

Facts and Figures

Radioisotopes

Many of the radioactive isotopes used in medicine and agriculture are not naturally occurring. They are produced artificially by methods such as neutron bombardment.

Radioisotopes are used by physicians to diagnose and treat some forms of cancer. Actively dividing cells are usually more sensitive to radiation than are normal cells.

Radioisotopes can be used to diagnose medical problems and, in some cases, to treat diseases. Iodine-131, for example, is used to detect thyroid problems. The thyroid gland extracts iodide ions from the bloodstream and uses them to make the hormone thyroxine. To diagnose thyroid disease, the patient is given a drink containing a small amount of the radioisotope iodine-131. After about two hours, the amount of iodide uptake is measured by scanning the patient's throat with a radiation detector. Figure 25.15 shows the results of such a scan. In a similar way, the radioisotope technetium-99m is used to detect brain tumors and liver disorders. Phosphorus-32 is used to detect skin cancer.

Radiation has become a routine part of the treatment of some cancers. Cancer is a disease in which abnormal cells in the body are produced at a rate far beyond the rate for normal cells. The mass of cancerous tissue resulting from this runaway growth is called a tumor. Radiation therapy is often used to treat cancer because the fast-growing cancer cells are more susceptible to damage by high-energy radiation such as gamma rays than are the healthy cells. The cancerous area can be treated with radiation to kill the cancer cells. Some normal cells are also killed, however, and cancer cells at the center of the tumor may be resistant to the radiation. Therefore, the benefits of the treatment and the risks to the patient must be carefully evaluated before radiation treatment begins.

In a technique called teletherapy, a narrow beam of high-intensity gamma radiation is directed at cancerous tissue. Salts of radioisotopes can also be sealed in gold tubes and directly implanted in tumors. These seeds emit beta and gamma rays that kill the surrounding cancer cells. Because the radioisotope is in a sealed container, it is prevented from traveling throughout the body.

Pharmaceuticals containing radioisotopes of gold, iodine, or phosphorus are sometimes given in radiation therapy. For example, a dose of iodine-131 larger than that used simply to detect thyroid diseases can be given to treat the diseased thyroid. The radioactive iodine accumulates in the thyroid and emits beta and gamma rays to provide therapy.

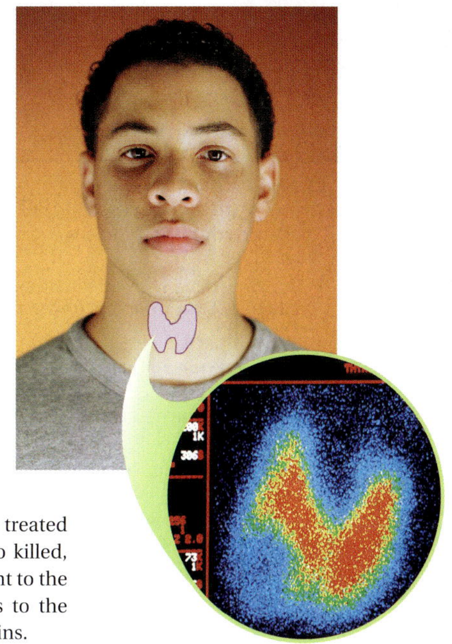

Figure 25.15 This scanned image of a thyroid gland shows where radioactive iodine-131 has been absorbed. Doctors use these images to identify thyroid disorders.

25.4 Section Assessment

21. **Key Concept** Describe three methods of detecting radiation.
22. **Key Concept** Describe two applications of radioisotopes in medicine.
23. If you work regularly near a radiation source, why might your employer want to monitor your exposure to radiation by having you use a film badge rather than a Geiger counter?
24. What is an advantage of using a radioactive seed, consisting of a radioisotope in a small gold tube, to treat a cancerous tumor?

Writing Activity

Write a Report Research and report how technetium-99m is produced, what the letter *m* at the end of its name means, and how the isotope is used in bone imaging and other methods of medical diagnosis.

Assessment 25.4 Test yourself on the concepts in Section 25.4.

with **ChemASAP**

Section 25.4 Assessment

21. Geiger counter, scintillation counter, and film badge
22. medical diagnosis and treatment of disease
23. A film badge monitors the degree and type of exposure to all types of radiation for an individual.
24. The seeds emit beta particles and gamma rays to kill the surrounding cancer cells. The seed cannot move from its inserted location.

3 ASSESS

Evaluate Understanding L2

Have students compare and contrast detection of radiation by a film badge and detection of radiation by a scintillation counter. Ask, **What are two medical uses of radioactive isotopes?** *(as tracers in disease diagnosis and in the treatment of cancer)*

Reteach L1

Explain that the most common home smoke detector is based on the same principles as a radiation detector. The smoke detector contains a radioisotope that ionizes the air in a small chamber. The ionized air allows a current to flow in the chamber. When smoke particles enter the chamber, the ionization decreases, which decreases the current. The decreased current is detected by a secondary circuit, which triggers an alarm.

Writing Activity

Tc-99m is made from the decay of molybdenum-99, which also has a relatively short half-life. The *m* means "metastable," which indicates that the isotope might become more or less stable. Medical uses might include the following examples. To pinpoint a liver problem, doctors inject a mixture containing Tc-99m, which should be absorbed by the liver. If gamma radiation emitted by the Tc-99m atoms in the mixture is detected in the kidneys, there is likely a problem with the liver. If the mixture reaches the liver but does not reach the gallbladder, the duct connecting the organs is probably blocked. Tc-99m is also used to detect the presence of cancerous cells in bone marrow.

If your class subscribes to the Interactive Textbook, use it to review key concepts in Section 25.4.

with **ChemASAP**

CHAPTER 25 Study Guide

Study Tip

Show all your calculations when working problems. Writing down the steps of a calculation helps you think through each step in the thought process of problem solving. Also, if you don't show your calculations, you have more difficulty finding the source of your error if you made a mistake.

If your class subscribes to the Interactive Textbook with ChemASAP, your students can go online to access an interactive version of the Student Edition and a self-test.

with ChemASAP

CHAPTER 25 Study Guide

Key Concepts

25.1 Nuclear Radiation

- An unstable nucleus loses energy by emitting radiation during the process of radioactive decay.
- The three main types of radiation are alpha radiation, beta radiation, and gamma radiation.

25.2 Nuclear Transformations

- The type of radioactive decay that occurs depends on the neutron-to-proton ratio of the unstable nucleus.
- After each half-life, half of the existing atoms have decayed into atoms of a new element.
- Transmutation can occur by radioactive decay, and can also occur when particles bombard the nucleus of an atom.

25.3 Fission and Fusion of Atomic Nuclei

- In a chain reaction, some of the neutrons produced react with other fissionable atoms, producing more neutrons which react with still more fissionable atoms.
- Spent fuel rods are stored in on-site pools of water to absorb the heat they continue generating and to help reduce radiation levels.
- In nuclear fusion, light nuclei fuse to make more massive nuclei. In fission, massive nuclei split apart to form less massive nuclei.

25.4 Radiation in Your Life

- Radiation may be detected with a Geiger counter or a scintillation counter. A film badge monitors radiation exposure of individuals who work with radioactive materials.
- Radioactivity and radiation are used in medical diagnosis, and in the treatment of some diseases, including some forms of cancer.

Vocabulary

- alpha particle (p. 800)
- band of stability (p. 803)
- beta particle (p. 801)
- film badge (p. 817)
- fission (p. 810)
- fusion (p. 813)
- gamma ray (p. 802)
- Geiger counter (p. 817)
- half-life (p. 804)
- ionizing radiation (p. 816)
- neutron absorption (p. 811)
- neutron activation analysis (p. 818)
- neutron moderation (p. 811)
- nuclear force (p. 803)
- positron (p. 804)
- radiation (p. 799)
- radioactivity (p. 799)
- radioisotopes (p. 799)
- scintillation counter (p. 817)
- transmutation (p. 807)
- transuranium elements (p. 808)

Key Equations

- $^{238}_{92}\text{U} \xrightarrow[\text{decay}]{\text{Radioactive}} {}^{234}_{90}\text{Th} + {}^{4}_{2}\text{He}$
- $^{1}_{0}\text{n} \longrightarrow {}^{1}_{1}\text{H} + {}^{0}_{-1}\text{e}$
- $^{14}_{6}\text{C} \longrightarrow {}^{14}_{7}\text{N} + {}^{0}_{-1}\text{e}$
- $^{230}_{90}\text{Th} \longrightarrow {}^{226}_{88}\text{Ra} + {}^{4}_{2}\text{He} + \gamma$

Organizing Information

Use these terms to construct a concept map that organizes the major ideas of this chapter.

Concept Map 25 Solve the Concept Map with the help of an interactive guided tutorial.

with ChemASAP

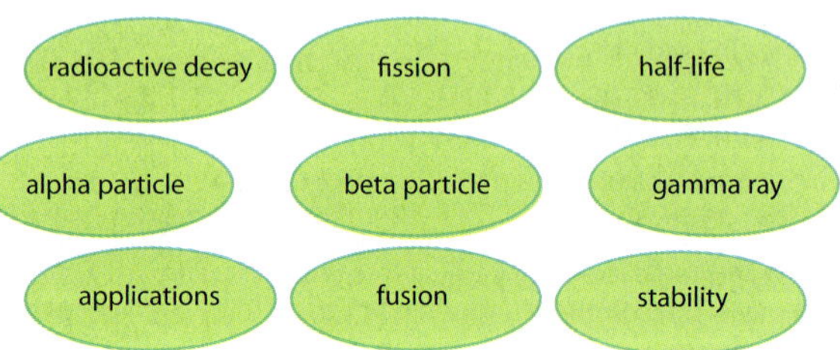

Chapter Resources

Print

- **Core Teaching Resources,** Chapter 25, Practice Problems, Vocabulary Review, Quiz, Chapter Test A, Chapter Test B

Technology

- **Computer Test Bank,** Chapter 25 Test
- **Interactive Textbook with ChemASAP,** Chapter 25

820 Chapter 25

CHAPTER 25

Assessment

Reviewing Content

25.1 Nuclear Radiation

25. Explain the difference between an isotope and a radioisotope.

26. The disintegration of the radioisotope radium-226 produces an isotope of the element radon and alpha radiation. The atomic number of radium (Ra) is 88; the atomic number of radon (Rn) is 86. Write a balanced equation for this transformation.

27. A radioisotope of the element lead (Pb) decays to an isotope of the element bismuth (Bi) by emission of a beta particle. Complete the equation for the decay process by supplying the missing atomic number and mass number.

$$ {}^{210}_{?}\text{Pb} \longrightarrow {}^{?}_{83}\text{Bi} + {}^{0}_{-1}\text{e} $$

28. Write the symbol and charge for each.

a. alpha particle
b. beta particle
c. gamma ray

29. Alpha radiation is emitted during the disintegration of the following isotopes. Write balanced nuclear equations for their decay processes. Name the element produced in each case.

a. uranium-238 (${}^{238}_{92}\text{U}$) **b.** thorium-230 (${}^{230}_{90}\text{Th}$)
c. uranium-235 (${}^{235}_{92}\text{U}$) **d.** radon-222 (${}^{222}_{86}\text{Rn}$)

30. The following radioisotopes are beta emitters. Write balanced nuclear equations for their decay processes.

a. carbon-14 (${}^{14}_{6}\text{C}$) **b.** strontium-90 (${}^{90}_{38}\text{Sr}$)
c. potassium-40 (${}^{40}_{19}\text{K}$) **d.** nitrogen-13 (${}^{13}_{7}\text{N}$)

31. How are the mass number and atomic number of a nucleus affected by the loss of the following?

a. beta particle
b. alpha particle
c. gamma ray

32. The following radioactive nuclei decay by emitting alpha particles. Write the product of the decay process for each.

a. ${}^{238}_{94}\text{Pu}$
b. ${}^{210}_{83}\text{Bi}$
c. ${}^{210}_{84}\text{Po}$
d. ${}^{230}_{90}\text{Th}$

25.2 Nuclear Transformations

33. What happens to an atom with a nucleus that falls outside the band of stability?

34. Write an equation for the radioactive decay of fluorine-17 by positron emission.

35. Identify the more stable isotope in each pair.

a. ${}^{14}_{6}\text{C}$, ${}^{13}_{6}\text{C}$
b. ${}^{3}_{1}\text{H}$, ${}^{1}_{1}\text{H}$
c. ${}^{16}_{8}\text{O}$, ${}^{18}_{8}\text{O}$
d. ${}^{14}_{7}\text{N}$, ${}^{15}_{7}\text{N}$

36. Explain half-life.

37. Why is it important that radioactive isotopes used internally for diagnosis or treatment have relatively short half-lives?

38. A patient is administered 20 mg of iodine-131. How much of this isotope will remain in the body after 40 days if the half-life for iodine-131 is 8 days?

39. What is the difference between natural and artificial radioactivity?

40. What are the transuranium elements? Why are they unusual?

25.3 Fission and Fusion of Atomic Nuclei

41. Describe the process of nuclear fission, and define a nuclear chain reaction.

42. Why are spent fuel rods removed from a reactor core? What do they contain? What happens to them after they are removed?

43. Fusion reactions produce enormous amounts of energy. Why is fusion not used to generate electrical power?

25.4 Radiation in Your Life

44. Why are x-rays and the radiation emitted by radioisotopes called ionizing radiation?

45. What is the purpose of wearing a film badge when working with ionizing radiation sources?

46. Explain how iodine-131 is used in both the diagnosis and treatment of thyroid disease.

CHAPTER 25 Assessment

Reviewing Content

25. Each isotope of an element has the same atomic number but a different atomic mass. A radioisotope is an isotope that is radioactive.

26. ${}^{226}_{88}\text{Ra} \longrightarrow {}^{222}_{86}\text{Rn} + {}^{4}_{2}\text{He}$

27. ${}^{210}_{82}\text{Pb} \longrightarrow {}^{210}_{83}\text{Bi} + {}^{0}_{-1}\text{e}$

28. a. α, +2 **b.** β, -1 **c.** γ, 0

29. a. ${}^{238}_{92}\text{U} \longrightarrow {}^{234}_{90}\text{Th} + {}^{4}_{2}\text{He}$; thorium–234

b. ${}^{230}_{90}\text{Th} \longrightarrow {}^{226}_{88}\text{Ra} + {}^{4}_{2}\text{He}$; radium–226

c. ${}^{235}_{92}\text{U} \longrightarrow {}^{231}_{90}\text{Th} + {}^{4}_{2}\text{He}$; thorium–231

d. ${}^{222}_{86}\text{Rn} \longrightarrow {}^{218}_{84}\text{Po} + {}^{4}_{2}\text{He}$; polonium–218

30. a. ${}^{14}_{6}\text{C} \longrightarrow {}^{14}_{7}\text{N} + {}^{0}_{-1}\text{e}$

b. ${}^{90}_{38}\text{Sr} \longrightarrow {}^{90}_{39}\text{Y} + {}^{0}_{-1}\text{e}$

c. ${}^{40}_{19}\text{K} \longrightarrow {}^{40}_{20}\text{Ca} + {}^{0}_{-1}\text{e}$

d. ${}^{13}_{7}\text{N} \longrightarrow {}^{13}_{8}\text{O} + {}^{0}_{-1}\text{e}$

31. a. mass number: unchanged; atomic number: increases by 1
b. mass number: decreases by 4; atomic number: decreases by 2
c. Mass number and atomic number are both unchanged.

32. a. ${}^{234}_{92}\text{U}$ **b.** ${}^{206}_{81}\text{Tl}$ **c.** ${}^{206}_{82}\text{Pb}$ **d.** ${}^{226}_{88}\text{Ra}$

33. It undergoes radioactive decay.

34. ${}^{17}_{9}\text{F} \longrightarrow {}^{17}_{8}\text{O} + {}^{0}_{+1}\text{e}$

35. a. ${}^{13}_{6}\text{C}$ **b.** ${}^{1}_{1}\text{H}$ **c.** ${}^{16}_{8}\text{O}$ **d.** ${}^{14}_{7}\text{N}$

36. One half-life is the time required for one-half of the atoms of a radioisotope to emit radiation and decay.

37. so the person is exposed to radioactivity for a limited time

38. 6.3×10^{-1} mg

39. Natural radioactivity comes from elements in nature. Artificial radioactivity comes from elements created in nuclear reactors and accelerators.

40. The elements with an atomic number greater than 92; none occur in nature and all are radioactive.

41. The nuclei of certain isotopes are bombarded with neutrons. The nuclei break into two fragments and release more neutrons. Released neutrons hit other nuclei to start a chain reaction that releases large amounts of energy.

42. A power plant cannot maintain a constant output of electricity with spent fuel rods, which contain depleted fissionable isotopes and fission products. The rods are stored in pools of water.

43. Fusion requires extremely high temperatures, making it difficult to start or contain the reaction.

44. Ionizing radiation, such as x-rays and gamma radiation, has sufficient energy to remove electrons from the atoms it hits.

45. The film badge measures radiation exposure; an exposed film badge indicates how much radiation a worker has received.

46. In diagnosis, the amount of iodine uptake in the thyroid is measured and studied; in treatment, the radioactive iodine-131 is concentrated in and by the thyroid.

CHAPTER 25

Assessment *continued*

Understanding Concepts

47. a. ${}^{30}_{15}P + {}^{0}_{-1}e \longrightarrow {}^{30}_{14}Si$

b. ${}^{13}_{6}C + {}^{1}_{0}n \longrightarrow {}^{14}_{6}C$

c. ${}^{131}_{53}I \longrightarrow {}^{131}_{54}Xe + {}^{0}_{-1}e$

48. Nuclear fusion takes place in the sun. A nuclear reactor utilizes nuclear fission.

49. a. ${}^{32}_{16}S$ **b.** ${}^{14}_{6}C$ **c.** ${}^{4}_{2}He$ **d.** ${}^{141}_{57}La$ **e.** ${}^{185}_{79}Au$

50. a. ${}^{90}_{38}Sr \longrightarrow {}^{90}_{39}Y + {}^{0}_{-1}e$

b. ${}^{14}_{6}C \longrightarrow {}^{14}_{7}N + {}^{0}_{-1}e$

c. ${}^{137}_{55}Cs \longrightarrow {}^{137}_{56}Ba + {}^{0}_{-1}e$

d. ${}^{239}_{93}Np \longrightarrow {}^{239}_{94}Pu + {}^{0}_{-1}e$

e. ${}^{50}_{22}Ti \longrightarrow {}^{50}_{23}V + {}^{0}_{-1}e$

51. a. about 20% **b.** about 85 g **c.** about 83 days **d.** about 25 days

52. a. ${}^{222}_{86}Rn \longrightarrow {}^{218}_{84}Po + {}^{4}_{2}He$

b. ${}^{234}_{90}Th \longrightarrow {}^{230}_{88}Ra + {}^{4}_{2}He$

c. ${}^{210}_{84}Po \longrightarrow {}^{206}_{82}Pb + {}^{4}_{2}He$

53. a. named radioactivity and discovered several radioactive elements
b. discovered natural radioactivity from uranium ores
c. discovered the neutron
d. artificially transmuted elements

54. 1 million

55. ${}^{215}_{85}At$

56. a. 27 protons and 33 neutrons
b. 82 protons and 124 neutrons
c. 90 protons and 143 neutrons
d. 1 proton and 2 neutrons

57. 11,460 years

58. a. ${}^{0}_{-1}e$ **b.** ${}^{238}_{92}U$ **c.** ${}^{68}_{32}Ge$ **d.** ${}^{0}_{+1}e$

59. a. ${}^{231}_{91}Pa \longrightarrow {}^{227}_{89}Ac + {}^{4}_{2}He$

b. ${}^{241}_{95}Am \longrightarrow {}^{237}_{93}Np + {}^{4}_{2}He$

c. ${}^{226}_{88}Ra \longrightarrow {}^{222}_{86}Rn + {}^{4}_{2}He$

d. ${}^{252}_{99}Es \longrightarrow {}^{248}_{97}Bk + {}^{4}_{2}He$

CHAPTER 25

Assessment *continued*

Understanding Concepts

47. Write nuclear equations for these conversions.
a. ${}^{30}_{15}P$ to ${}^{30}_{14}Si$
b. ${}^{13}_{6}C$ to ${}^{14}_{6}C$
c. ${}^{131}_{53}I$ to ${}^{131}_{54}Xe$

48. What is the difference between the nuclear reactions taking place in the sun and the nuclear reactions taking place in a nuclear reactor?

49. Complete these nuclear equations.
a. ${}^{32}_{15}P \longrightarrow \square + {}^{0}_{-1}e$
b. $\square \longrightarrow {}^{14}_{7}N + {}^{0}_{-1}e$
c. ${}^{238}_{92}U \longrightarrow {}^{234}_{90}Th + \square$
d. ${}^{141}_{56}Ba \longrightarrow \square + {}^{0}_{-1}e$
e. $\square \longrightarrow {}^{181}_{77}Ir + {}^{4}_{2}He$

50. Write nuclear equations for the beta decay of the following isotopes.
a. ${}^{90}_{38}Sr$ **b.** ${}^{14}_{6}C$ **c.** ${}^{137}_{55}Cs$ **d.** ${}^{239}_{93}Np$ **e.** ${}^{50}_{22}Ti$

51. The following graph shows the radioactive decay curve for thorium-234. Use the graph to answer the questions.

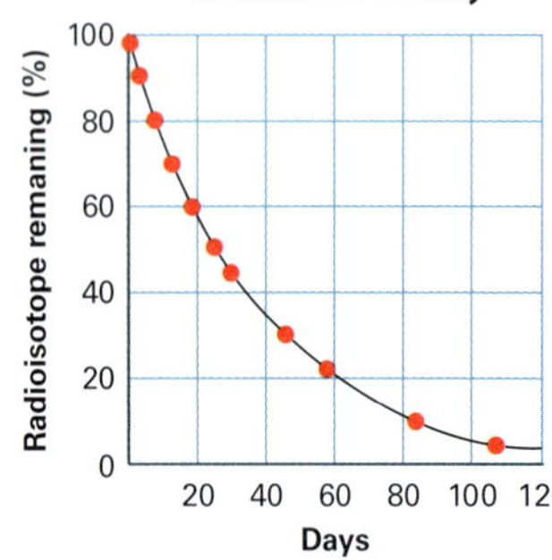

a. What percent of the isotope remains after 60 days?
b. How many grams of a 250-g sample of thorium-234 would remain after 40 days had passed?
c. How many days would pass while 44 g of thorium-234 decayed to 4.4 g of thorium-234?
d. What is the half-life of thorium-234?

52. Write a nuclear equation for each word equation.
a. Radon-222 emits an alpha particle to form polonium-218.
b. Radium-230 is produced when thorium-234 emits an alpha particle.
c. When polonium-210 emits an alpha particle, the product is lead-206.

53. Describe the various contributions the following people made to the fields of nuclear and radiation chemistry.
a. Marie Curie **b.** Antoine Henri Becquerel
c. James Chadwick **d.** Ernest Rutherford

54. If you started with 32 million radioactive atoms, how many would you have left after five half-lives?

55. A radioactive nucleus decays to give an alpha particle and a bismuth-211 (${}^{211}_{83}Bi$) nucleus. What was the original nucleus?

56. How many protons and how many neutrons are there in each of the following nuclei?
a. ${}^{60}_{27}Co$ **b.** ${}^{206}_{82}Pb$ **c.** ${}^{233}_{90}Th$ **d.** ${}^{3}_{1}H$

57. The carbon-14 in an artifact gives 4 counts per minute per gram of carbon. Living matter has a carbon-14 content that gives 16 counts per minute per gram of carbon. What is the age of the artifact?

58. Complete the following nuclear reaction equations.
a. ${}^{38}_{19}K \longrightarrow {}^{38}_{20}Ca + ?$
b. ${}^{242}_{94}Pu \longrightarrow ? + {}^{4}_{2}He$
c. ${}^{68}_{31}Ga \longrightarrow ? + {}^{0}_{-1}e$
d. ${}^{68}_{32}Ge \longrightarrow {}^{68}_{31}Ga + ?$

59. Write balanced nuclear equations for alpha emission by each of the following isotopes.
a. ${}^{231}_{91}Pa$ **b.** ${}^{241}_{95}Am$ **c.** ${}^{226}_{88}Ra$ **d.** ${}^{252}_{99}Es$

60. Write balanced nuclear equations for beta emission by each of the following isotopes.
a. ${}^{3}_{1}H$ **b.** ${}^{28}_{12}Mg$ **c.** ${}^{131}_{53}I$ **d.** ${}^{75}_{34}Se$

61. The ratio of carbon-14 to carbon-12 in a chunk of charcoal from an archeological dig is determined to be one-half the ratio in a piece of freshly cut wood. How old is the charcoal from the dig site?

60. a. ${}^{3}_{1}H \longrightarrow {}^{3}_{2}He + {}^{0}_{-1}e$

b. ${}^{28}_{12}Mg \longrightarrow {}^{28}_{13}Al + {}^{0}_{-1}e$

c. ${}^{131}_{53}I \longrightarrow {}^{131}_{54}Xe + {}^{0}_{-1}e$

d. ${}^{75}_{34}Se \longrightarrow {}^{75}_{35}Br + {}^{0}_{-1}e$

61. 5730 years

Critical Thinking

62. Name the elements represented by the following symbols and indicate which of them would have no stable isotopes.

a. Pt **b.** Th **c.** Fr **d.** Ti
e. Xe **f.** Cf **g.** V **h.** Pd

63. Why would it be advantageous for radioactive tracers used in the study of living organisms to have a relatively short half-life?

64. Compare the half-life of an element to a single-elimination sports tournament.

65. Why does the relatively large mass and charge of an alpha particle make it less penetrating than beta and gamma radiation?

66. Why might radioisotopes of C, N, and O be especially harmful to living creatures?

67. A target of bismuth-209 was bombarded with iron-58 for several days. One atom of meitnerium ($^{266}_{109}Mt$) was produced. How many neutrons were released in the process?

$$^{209}_{83}Bi + ^{58}_{26}Fe \longrightarrow ^{266}_{109}Mt + ?\,^{1}_{0}n$$

68. When neutrons bombard magnesium-24 ($^{24}_{12}Mg$), a neutron is captured and photons are ejected. What new element is formed?

69. Plutonium-239 emits alpha particles and is therefore especially hazardous when inhaled or ingested. What new element is formed by this alpha emission? Write a balanced nuclear equation.

70. Tritium (hydrogen-3) has a half-life of 12.3 years. How old is a bottle of wine if the tritium activity is determined to be 25% that of new wine?

71. A 0.25 mg sample of californium-249 ($^{249}_{98}Cf$) was used as the target in the synthesis of seaborgium ($^{236}_{106}Sg$). Four neutrons were emitted for each transformed $^{249}_{98}Cf$ and the result was a nucleus with 106 protons and a mass of 263 amu. What was the bombarding particle?

$$^{249}_{98}Cf + ? \longrightarrow 4\,^{1}_{0}n + ^{263}_{106}Sg$$

Concept Challenge

72. The radioisotope cesium-137 has a half-life of 30 years. A sample decayed at the rate of 544 counts per minute in the year 1985. In what year will the decay rate be 17 counts per minute?

73. Describe the process depicted in the following graph.

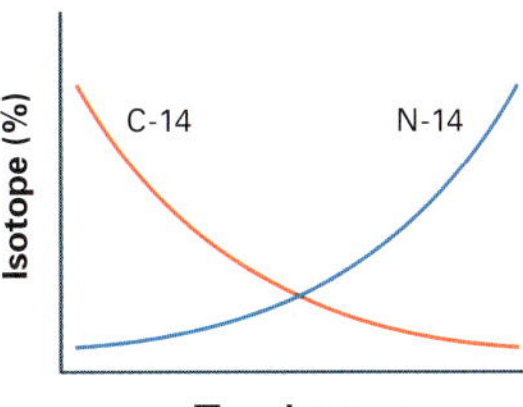

74. Bismuth-211 is a radioisotope. It decays by alpha emission to yield another radioisotope, which emits beta radiation as it decays to a stable isotope. Write equations for the nuclear reactions and name the decay products.

75. What isotope remains after three beta particles and five alpha particles are lost from a thorium-234 isotope? (Refer to the uranium-238 decay series to check your answer.)

76. Write a paragraph analyzing the overall logic of the reasoning in the following argument.

(1) Radiation kills fast-growing cells.
(2) Cancer cells are fast-growing.
(3) Therefore, radiation kills only cancer cells.

77. Uranium has a density of 19 g/cm^3. What volume does a mass of 8.0 kg of uranium occupy?

78. Element 107 (Bh) is formed when nuclei of element 109 (Mt) each emit an alpha particle. Nuclei of element 107, in turn, also each emit an alpha particle, forming an atom with a mass number of 262. Write balanced equations for these two nuclear reactions.

Critical Thinking

62. **a.** platinum **b.** thorium **c.** francium **d.** titanium **e.** xenon **f.** californium **g.** vanadium **h.** palladium. Thorium (b), francium (c), and californium (f) have no stable isotopes.

63. The organism would be exposed to less harmful radiation.

64. In every round of the tournament, one-half the teams are eliminated; in every half-life, one-half the nuclei decay. In the tournament a single team eventually emerges as the winner and the tournament stops, but radioactive decay continues (almost) indefinitely.

65. An alpha particle is much more likely than other kinds of radiation to collide with another particle and be stopped. At the atomic level, the larger the size of a particle, the greater is the chance of it striking another particle. The greater the magnitude of a particle's charge, the more strongly it will be attracted to particles of opposite charge.

66. Radioactive isotopes of these elements can be incorporated into the body tissue of organisms. When the isotopes decay, they can damage tissue very easily.

67. one neutron

68. $^{25}_{12}Mg$

69. uranium; $^{239}_{94}Pu \longrightarrow ^{235}_{92}U + ^{4}_{2}He$

70. 24.6 years

71. $^{18}_{8}O$

Concept Challenge

72. 2135

73. the radioactive (beta) decay of carbon-14, along with the increase of the nitrogen product

74. $^{211}_{83}Bi \longrightarrow ^{207}_{81}Tl + ^{4}_{2}He$; thallium – 207

$^{207}_{81}Tl \longrightarrow ^{207}_{82}Pb + ^{0}_{-1}e$; lead – 207

75. Bismuth-214 remains.

76. The reasoning is not sound. Cells other than cancer cells may be fast-growing and therefore killed by radiation as well.

77. 4.2×10^2 cm^3

78. $^{268}_{109}Mt \longrightarrow ^{264}_{107}Bh + ^{4}_{2}He$

$^{264}_{107}Bh \longrightarrow ^{262}_{105}Db + ^{4}_{2}He$

CHAPTER 25

Assessment *continued*

Cumulative Review

79. a. 26 protons and 33 neutrons
b. 92 protons and 143 neutrons
c. 24 protons and 28 neturons

80. The Pauli exclusion principle states that no two electrons in an atom can have the same quantum numbers. Hund's rule states that electrons that occupy orbitals of equal energy distribute with unpaired spins as much as possible.

81. a. covalent **b.** ionic **c.** covalent **d.** ionic

82. a. $Ca(OH)_2 + 2HCl \rightarrow CaCl_2 + 2H_2O$
b. $Fe_2O_3 + 3H_2 \rightarrow 2Fe + 3H_2O$
c. $2NaHCO_3 + H_2SO_4 \rightarrow Na_2SO_4 + 2CO_2 + 2H_2O$
d. $2C_2H_6 + 7O_2 \rightarrow 4CO_2 + 6H_2O$

83. $9.22 \times 10^3\ cm^3\ H_2$; $0.412\ mol\ H_2$

84. The oxygen atom carries a partial negative charge, and the hydrogen atoms carry partial positive charges. The water molecule's bent shape results in a net polarity, and the negative end of one molecule attracts the positive end of another molecule.

85. 6.7 mL

86. a.

```
        CH3
        |
CH3—C—CH2—CH2—CH2—CH3
        |
        CH3
```

b.

```
   CH3
   |
  /\
 /  \—CH3
 \__/   (cyclopentane ring)
```

c.

```
      CH3
      |
CH3—C=CH—CH2—CH2—CH2—CH3
```

d. $CH_3—C{\equiv}C—CH_3$

e. CH_3–(benzene ring)–CH_3

f.

```
            CH2CH3
            |
CH3—CH2—CH—CH2—CH2—CH2—CH2—CH3
```

CHAPTER 25

Assessment *continued*

Cumulative Review

79. How many protons, neutrons, and electrons are in an atom of each isotope? *(Chapter 4)*
a. iron-59
b. uranium-235
c. chromium-52

80. What is the Pauli exclusion principle? What is Hund's rule? *(Chapter 5)*

81. Identify the bonds between each pair of atoms as ionic or covalent. *(Chapter 8)*
a. carbon and silicon
b. calcium and fluorine
c. sulfur and nitrogen
d. bromine and cesium

82. Balance the following equations. *(Chapter 11)*
a. $Ca(OH)_2 + HCl \longrightarrow CaCl_2 + H_2O$
b. $Fe_2O_3 + H_2 \longrightarrow Fe + H_2O$
c. $NaHCO_3 + H_2SO_4 \longrightarrow Na_2SO_4 + CO_2 + H_2O$
d. $C_2H_6 + O_2 \longrightarrow CO_2 + H_2O$

83. A piece of magnesium with a mass of 10.00 g is added to sulfuric acid. How many cubic centimeters of hydrogen gas (at STP) will be produced if the magnesium reacts completely? How many moles is this? *(Chapter 12)*

84. The diagram below shows a water molecule. Identify the location of any partial positive and partial negative charges on the molecule. Then explain how the partial charges and their locations produce an attraction between different water molecules. *(Chapter 15)*

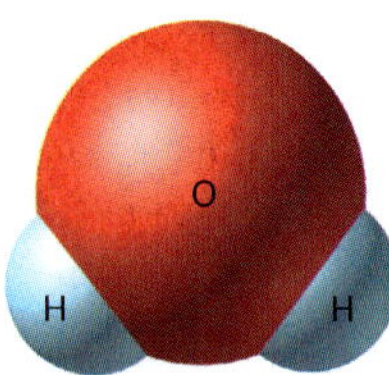

Polar bonds

85. You have a $0.30M$ solution of sodium sulfate. What volume (in mL) must be measured to give 0.0020 mol of sodium sulfate? *(Chapter 16)*

86. Draw the structural formula for each compound. *(Chapter 22)*
a. 2,2-dimethylhexane
b. 1,2-dimethylcyclopentane
c. 2-methyl-2-heptene
d. 2-butyne
e. 1,4-dimethylbenzene
f. 3-ethyloctane

87. Name each compound. *(Chapter 23)*
a. CH_3CH_2COOH **b.** CH_3CH_2CHO
c. $CH_3CH_2CH_2OH$ **d.** $CH_3CH_2CH_2NH_2$
e. $CH_3CH_2CH_2Cl$ **f.** $CH_3CH_2OCH_3$

88. For each pair of compounds, which is the more highly oxidized? *(Chapter 23)*
a. ethanol and ethanal
b. ethane and ethene
c. ethanoic acid and ethanal
d. ethyne and ethane

89. What two compounds result from the acid-catalyzed hydrolysis of propyl ethanoate? *(Chapter 23)*

90. Which of these classes of compounds does not contain a carbon–oxygen double bond? *(Chapter 23)*
a. amide **b.** ketone
c. aldehyde **d.** carboxylic acid

91. Match a numbered item with each term or phrase. *(Chapter 24)*
a. amino acid **(1)** carbohydrate
b. fat **(2)** nucleic acid
c. monosaccharide **(3)** lipid
d. peptide bond **(4)** protein
e. sugar
f. DNA
g. saponification
h. genetic code
i. enzyme
j. triglyceride

87. a. propanoic acid **b.** propanal
c. 1-propanol **d.** 1-aminopropane
e. 1-chloropropane **f.** ethylmethyl ether

88. a. ethanal **b.** ethene
c. ethanoic acid **d.** ethyne

89. 1-propanol and ethanoic acid

90. They all contain a carbon-oxygen double bond.

91. a. (4) **b.** (3) **c.** (1) **d.** (4) **e.** (1) **f.** (2) **g.** (3) **h.** (2) **i.** (4) **j.** (3)

Standardized Test Prep

Test-Taking Tip

Anticipate the Answer Use what you know to decide what you think the answer should be. Then look to see if your answer, or one much like it, is given as an option.

Select the choice that best answers each question.

1. If a radioisotope undergoes beta emission,
   - **a.** the atomic number changes.
   - **b.** the number of neutrons remains constant.
   - **c.** the isotope loses a proton.
   - **d.** the mass number changes.
2. The radioisotope radon-222 has a half-life of 3.8 days. How much of an initial 20.0-g sample of radon-222 would remain after 15.2 days?
   a. 5.00 g **b.** 12.5 g **c.** 1.25 g **d.** 2.50 g
3. Spent fuel rods
   - **a.** are no longer radioactive.
   - **b.** are stored under water for at least a decade.
   - **c.** contain only one isotope of uranium, ^{238}U.
   - **d.** remain radioactive for less than 100 years.
4. What particle is needed to balance this equation?
   $$^{27}_{13}Al + ^{4}_{2}He \longrightarrow ______ + ^{30}_{15}P$$

For each nuclear equation in Questions 5–8, name the particle that is being emitted or captured.

5. $^{59}_{26}Fe \longrightarrow ^{59}_{27}Co + ^{0}_{-1}e$
6. $^{185}_{79}Au \longrightarrow ^{181}_{77}Ir + ^{4}_{2}He$
7. $^{59}_{27}Co + ^{1}_{0}n \longrightarrow ^{60}_{27}Co$
8. $^{118}_{54}Xe \longrightarrow ^{118}_{53}I + ^{0}_{+1}e$
9. Use the graph below to find whether neon-21, zirconium-90, and neodymium-130 have stable nuclei.

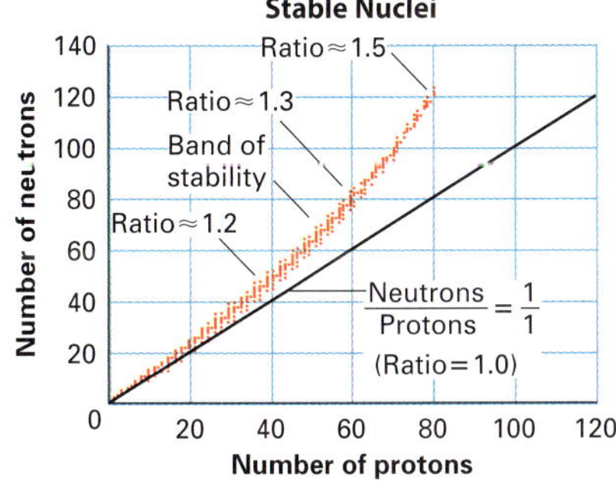

Use the graph to answer Questions 10–12.

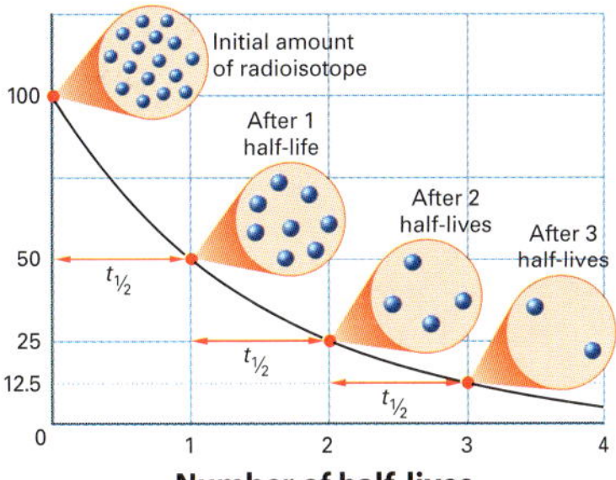

Estimate the percent remaining of the radioisotope after the given number of half-lives.

10. $0.5\ t_{1/2}$ **11.** $1.25\ t_{1/2}$ **12.** $3.75\ t_{1/2}$

Use the drawings of atomic nuclei to answer Questions 13 and 14.

13. Write the name and symbol for each isotope.

a. 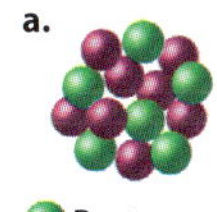 **b.** 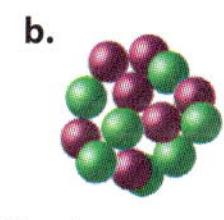 **c.** 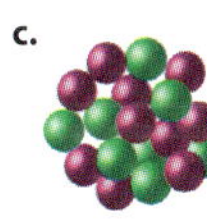

Proton Neutron

14. Which isotope is radioactive?

The lettered choices below refer to Questions 15–19. A lettered choice may be used once, more than once, or not at all.

(A) film badge
(B) radioactive tracer
(C) radiation therapy
(D) neutron activation analysis
(E) Geiger counter

Which of the above items or processes is best described by each of the following applications?

15. treating some cancers
16. detecting ionizing radiation
17. monitoring exposure to radiation
18. diagnosing some diseases
19. detecting art forgeries

Standardized Test Prep

1. a
2. c
3. b
4. neutron, $^{1}_{0}n$
5. beta emission
6. alpha emission
7. neutron capture
8. positron emission
9. Zirconium-90 and neon-21 nuclei are stable; neodymium-130 nuclei are unstable.
10. 71%
11. 42%
12. 7.5%
13. a. carbon-14, $^{14}_{6}C$
b. nitrogen-14, $^{14}_{7}N$
c. oxygen-18, $^{18}_{8}O$
14. Carbon-14 is radioactive.
15. (C)
16. (E)
17. (A)
18. (B)
19. (D)

Appendices

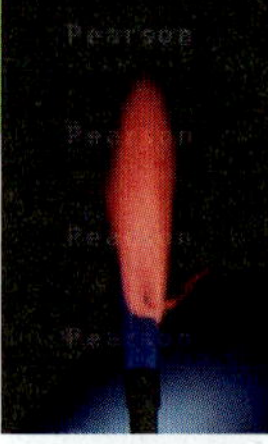
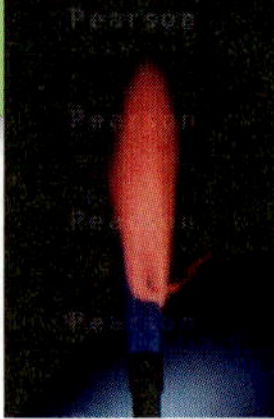

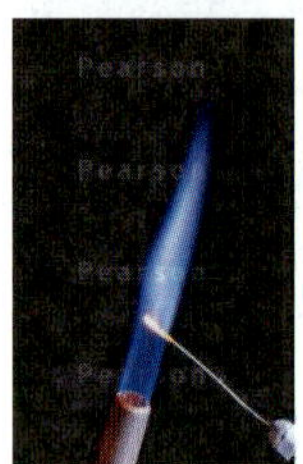

Contents

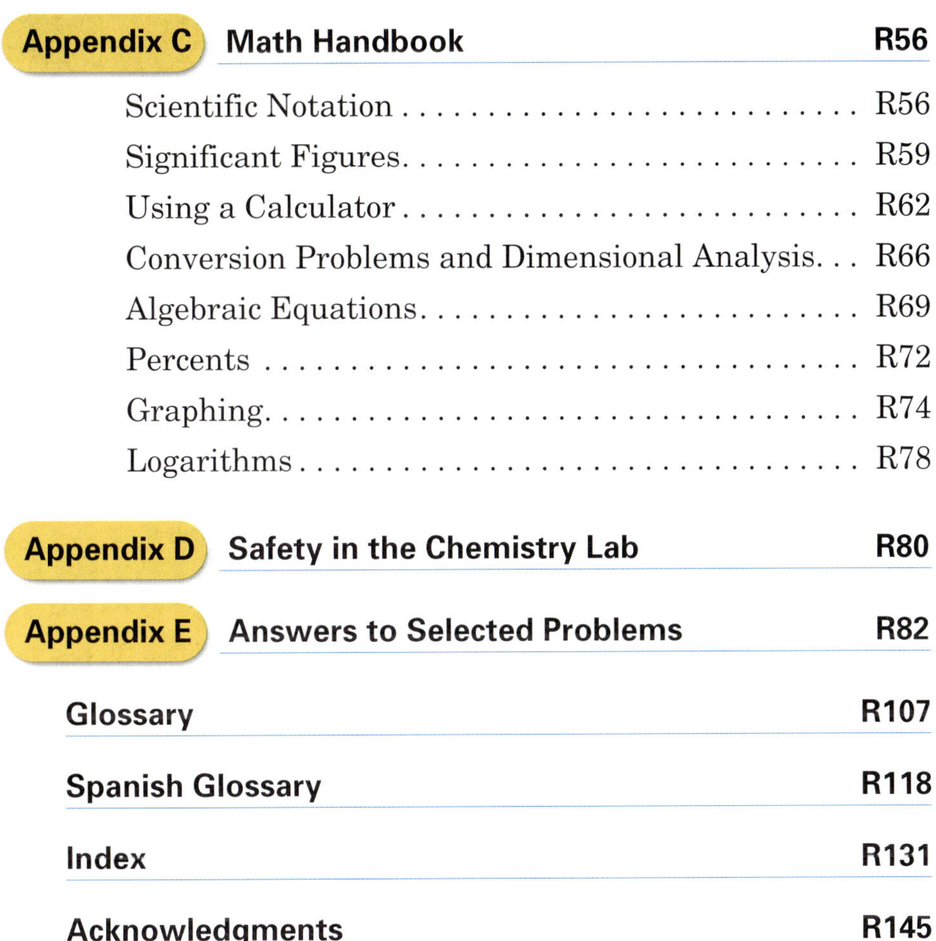

Elements Handbook

Contents

The Elements Handbook provides information that has traditionally been included within the text in a descriptive chemistry chapter. Beginning on page R6, the information in the handbook is organized by groups of elements. For most groups, there is a two-page spread that contains bulleted lists of properties. The data is divided into categories for easy access. A property spread also supplies information on sources of elements, a reminder of where a group fits into the periodic table, and data on the discovery of elements.

For most groups, there is a two-page spread that contains short articles about specific elements. The articles assume knowledge of the appropriate concepts from the text. How much background students will need depends on the topic. Students could read *Gold* on page R44 early in the year and understand most of the content, but *Catalytic Converters* on page R42 would be more challenging. The experience would be like reading an article in a newspaper that contains some unfamiliar terms and concepts. The articles are meant to introduce topics without attempting to cover all possible aspects of a topic. For many articles, you will find additional background in the TE wrap.

Appendix A: Elements Handbook

Contents

How to Use the Elements Handbook

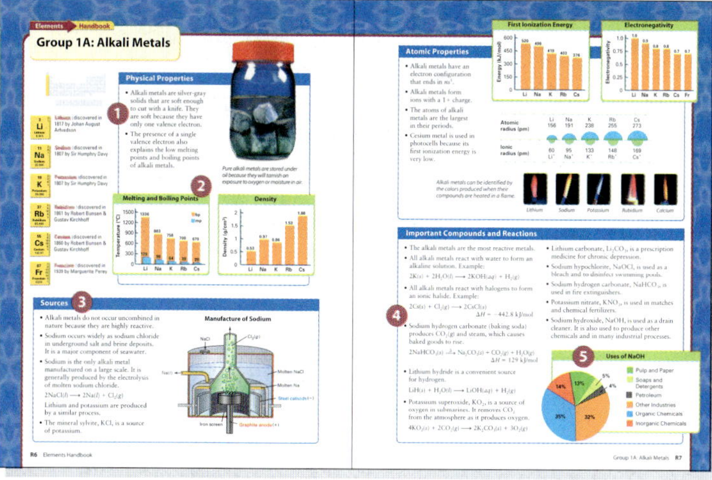

1. Look for data that describes what the elements in a group have in common.
2. Interpret graphs to reveal trends in properties within a group.
3. Learn where elements exist in nature and how they are processed.
4. Find examples of typical chemical reactions.
5. Find out which compounds have important industrial uses.

6. Make connections between chemistry and technology.
7. Explore the impact of chemical reactions on the environment.
8. Find out the effects of certain chemicals on your health.
9. Discover some fun facts about elements.

How to Use the Elements Handbook

There are a number of ways to use the Elements Handbook.

- One section assessment in each chapter includes an activity that requires students to use information from a specific location in the Elements Handbook.
- In the TE wrap for the Elements Handbook, there are suggestions for ways that students can analyze data in the handbook or research topics.
- You may want to encourage students to browse through the Elements Handbook now and then to see how their understanding of the content increases with their knowledge of chemistry.
- Of course, as you become familiar with the Elements Handbook, you will find your own ways to use it with your students.

Elements Around You

There are thousands of lines in the visible region of the absorption spectrum of the sun. Scientists have used these lines to identify 67 different chemical elements. There may be others. However, two elements comprise 99.9% of all the atoms in the sun. Students will be able to explain this distribution after they study fusion in Chapter 25.

To Do

If students total the data in the tables, they will discover that the parts per million for elements in Earth's crust, elements dissolved in the oceans, and elements in the atmosphere add up to 977,800 ppm, 33,315 ppm, and 999,725 ppm, respectively. Students should be able to figure out that the discrepancy in the first and last categories are due to missing data on additional elements. The large discrepancy in the data for the oceans may seem puzzling at first, until students recall that a solution requires a solvent. The percentages of elements in the sun add up to 100.0461%, which is more than 100%. Ask students how the rules for rounding could produce such a result.

Elements Around You

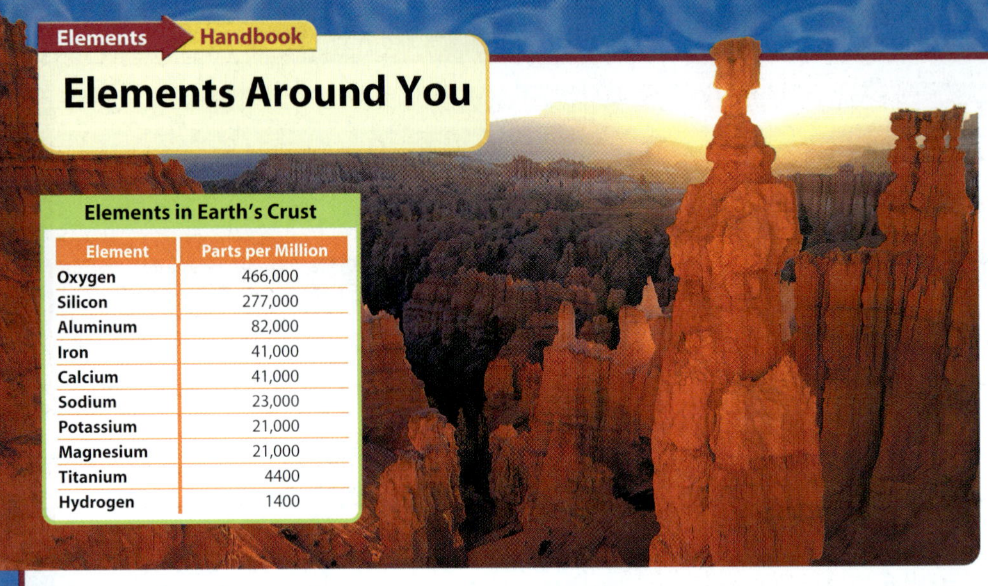

Elements in Earth's Crust

Element	Parts per Million
Oxygen	466,000
Silicon	277,000
Aluminum	82,000
Iron	41,000
Calcium	41,000
Sodium	23,000
Potassium	21,000
Magnesium	21,000
Titanium	4400
Hydrogen	1400

Elements Dissolved in the Oceans

Element	Parts per Million
Chlorine	19,400
Sodium	10,800
Magnesium	1300
Sulfur	904
Calcium	411
Potassium	392
Bromine	67
Carbon	28
Strontium	8
Boron	5

Elements in the Atmosphere

Element	Parts per Million*
Nitrogen	780,900
Oxygen	209,500
Argon	9300
Neon	18
Helium	5.2
Krypton	1.14
Hydrogen	0.5
Xenon	0.086
Radon	Traces

*Data is for dry air.

Elements in the Sun

Element	Percent of Total Atoms
Hydrogen	92.1
Helium	7.8
Oxygen	0.078
Carbon	0.043
Nitrogen	0.0088
Silicon	0.0045
Magnesium	0.0038
Neon	0.0035
Iron	0.0030
Sulfur	0.0015

Elements Within You

Did You Know...

The mass of an adult's brain is about 2% of total body mass. Yet brain cells consume about 20% of the available **oxygen** when the body is at rest.

Did You Know...

Bacteria in the mouth produce **sulfur** compounds that cause bad breath. The culprits are hydrogen sulfide, H_2S, methyl mercaptan, CH_3SH, and dimethyl sulfide, $(CH_3)_2S$.

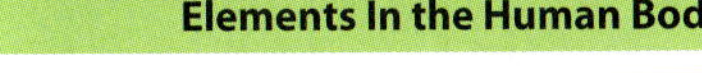

Did You Know...

The human body contains more **calcium** than any other metal. Your bones and teeth contain 99.2% of the calcium by mass.

Did You Know...

Although air is about 78% **nitrogen**, your body must get the nitrogen it needs from food, not from the air you inhale.

Elements In the Human Body

Element	Percent of Total Body Mass
Oxygen	61
Carbon	23
Hydrogen	10
Nitrogen	2.6
Calcium	1.4
Phosphorus	1.1
Sulfur	0.2
Potassium	0.2
Sodium	0.14
Chlorine	0.12

Did You Know...

There is evidence that white blood cells are able to produce toxic **chlorine** gas. The gas helps to protect against infection.

Did You Know...

Phosphorus is an important element for growth. So blood levels of the phosphate ion (PO_4^{3-}) in children are double what they are in adults.

Elements Within You

After students study Chapter 24, they should be able to offer an informed explanation for the distribution of elements in the human body. For example, they can note the presence of nitrogen in proteins and nucleic acids, and the presence of phosphorus in nucleic acids, phospholipids, and the molecules that capture and store energy.

Group 1A: Alkali Metals

The meaning of the term *discovery* can vary because scientists sometimes knew that an element existed, and had named the element, long before the element could be isolated from its compounds. For example, Arfvedson discovered lithium during an analysis of ore containing $LiAl(Si_2O_5)_2$, but lithium was not isolated until W.T. Brande and Sir Humphrey Davy performed an electrolysis of lithium oxide.

Physical Properties

It isn't always easy to establish the melting point or boiling point for a substance. Energy added to a sample to cause a change of state can cause other changes. For example, a melting point cannot be determined for sucrose because it decomposes when heated. It is possible to melt iodine, but not at atmospheric pressure because at that pressure iodine will sublime. To determine the melting point of diamond, the sample must be heated in an inert atmosphere. Otherwise it reacts with oxygen in the air. Even in an inert atmosphere, it is not possible to determine a melting point for graphite because the graphite sublimes.

To Do

To see how physical properties change across the periods in the periodic table, students can compare the ranges on the *y*-axes of the melting and boiling point graphs and the density graphs for Groups 1A–8A.

Elements Handbook

Group 1A: Alkali Metals

Lithium | discovered in 1817 by Johan August Arfvedson

Sodium | discovered in 1807 by Sir Humphry Davy

Potassium | discovered in 1807 by Sir Humphry Davy

Rubidium | discovered in 1861 by Robert Bunsen & Gustav Kirchhoff

Cesium | discovered in 1860 by Robert Bunsen & Gustav Kirchhoff

Francium | discovered in 1939 by Marguerite Perey

Physical Properties

- Alkali metals are silver-gray solids that are soft enough to cut with a knife. They are soft because they have only one valence electron.
- The presence of a single valence electron also explains the low melting points and boiling points of alkali metals.

Pure alkali metals are stored under oil because they will tarnish on exposure to oxygen or moisture in air.

Melting and Boiling Points

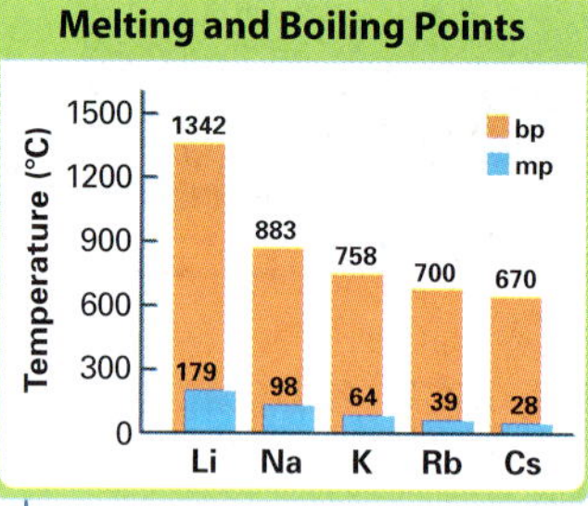

Density

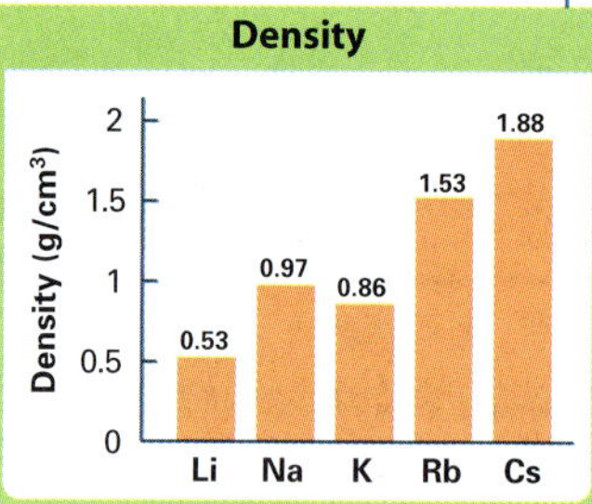

Sources

- Alkali metals do not occur uncombined in nature because they are highly reactive.
- Sodium occurs widely as sodium chloride in underground salt and brine deposits. It is a major component of seawater.
- Sodium is the only alkali metal manufactured on a large scale. It is generally produced by the electrolysis of molten sodium chloride.

  $$2NaCl(l) \longrightarrow 2Na(l) + Cl_2(g)$$

  Lithium and potassium are produced by a similar process.
- The mineral sylvite, KCl, is a source of potassium.

Manufacture of Sodium

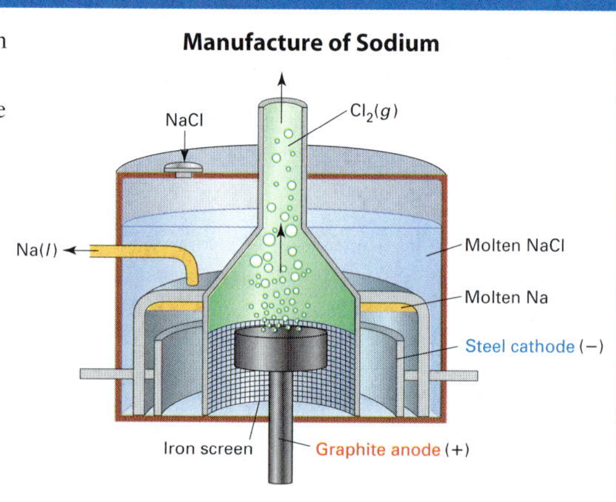

Sources

As new sources are discovered, the relative abundance of an element may need to be revised. For example, rubidium is now considered more abundant in Earth's crust than zinc or copper. However, there is little market demand for rubidium. Uses for rubidium include fiber optics, night-vision devices, PET scans, and photoelectric cells. Some rubidium is produced as a by-product of processing lithium and cesium ores.

To Do

Some students may want to research how rubidium atoms have been used to produce a Bose-Einstein condensate at a temperature only a few billionths of a degree above absolute zero.

Atomic Properties

- Alkali metals have an electron configuration that ends in ns^1.
- Alkali metals form ions with a 1+ charge.
- The atoms of alkali metals are the largest in their periods.
- Cesium metal is used in photocells because its first ionization energy is very low.

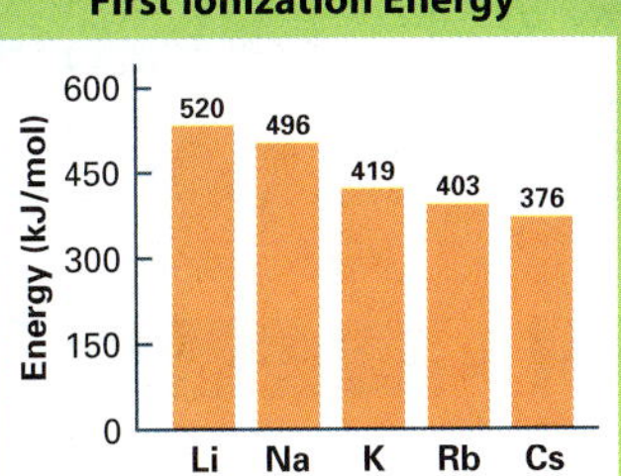

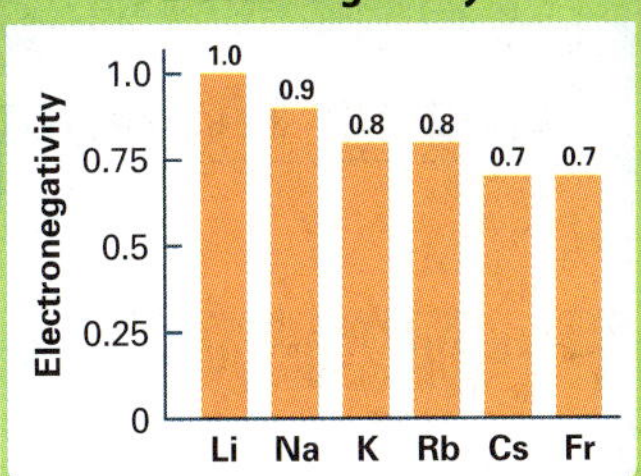

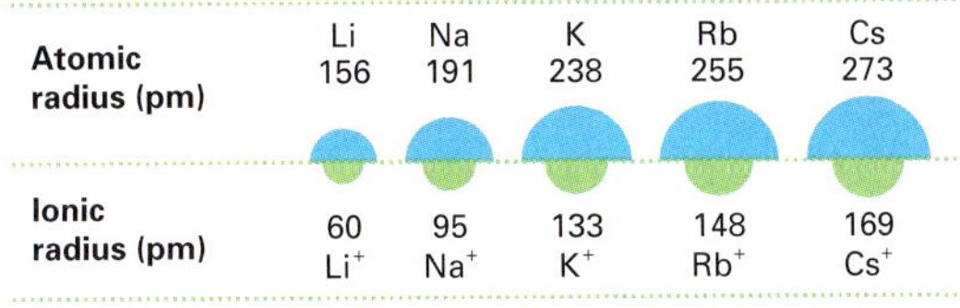

Alkali metals can be identified by the colors produced when their compounds are heated in a flame.

Lithium

Sodium

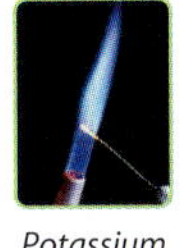
Potassium

Rubidium

Cesium

Important Compounds and Reactions

- The alkali metals are the most reactive metals.
- All alkali metals react with water to form an alkaline solution. Example:

$$2K(s) + 2H_2O(l) \longrightarrow 2KOH(aq) + H_2(g)$$

- All alkali metals react with halogens to form an ionic halide. Example:

$$2Cs(s) + Cl_2(g) \longrightarrow 2CsCl(s) \quad \Delta H = -442.8 \text{ kJ/mol}$$

- Sodium hydrogen carbonate (baking soda) produces $CO_2(g)$ and steam, which causes baked goods to rise.

$$2NaHCO_3(s) \xrightarrow{\Delta} Na_2CO_3(s) + CO_2(g) + H_2O(g) \quad \Delta H = 129 \text{ kJ/mol}$$

- Lithium hydride is a convenient source for hydrogen.

$$LiH(s) + H_2O(l) \longrightarrow LiOH(aq) + H_2(g)$$

- Potassium superoxide, KO_2, is a source of oxygen in submarines. It removes CO_2 from the atmosphere as it produces oxygen.

$$4KO_2(s) + 2CO_2(g) \longrightarrow 2K_2CO_3(s) + 3O_2(g)$$

- Lithium carbonate, Li_2CO_3, is a prescription medicine for chronic depression.
- Sodium hypochlorite, NaClO, is used as a bleach and to disinfect swimming pools.
- Sodium hydrogen carbonate, $NaHCO_3$, is used in fire extinguishers.
- Potassium nitrate, KNO_3, is used in matches and chemical fertilizers.
- Sodium hydroxide, NaOH, is used as a drain cleaner. It is also used to produce other chemicals and in many industrial processes.

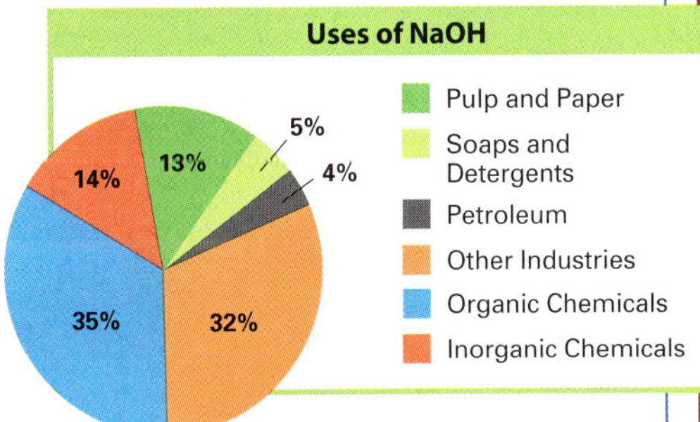

Atomic Properties

The source for most of the atomic radii data is *Bonding and Structure* by N. W. Alcock (Ellis Horwood, 1999).

Of all elements with atomic numbers between 1 and 100, atoms of francium are the least stable. The longest-lived isotope of francium, francium-233, has a half-life of only 22 minutes.

To Do

Comparing the ranges on the *y*-axes of the first ionization energy graphs for Groups 1A–8A provides another way to emphasize the variation of properties across periods in the periodic table.

Important Compounds and Reactions

To Do

The U.S. chemical industry produces large amounts annually (by mass) of sulfuric acid, phosphoric acid, ammonia, hydrogen chloride, sodium hydroxide, nitric acid, ammonium nitrate, and ammonium sulfate. As a long-term class project, students could collect data from the Elements Handbook and other sources to explain this production.

When students study organic chemistry, they could do similar research on ethylene, propylene, ethylene dichloride, urea, ethylbenzene, styrene, ethylene oxide, cumene, and 1,3-butadiene.

Group 1A: Alkali Metals

Vapor Lamps

Adding mercury to a sodium vapor lamp helps with the problem of color recognition because of the strong blue and green lines in mercury's emission spectrum. However, because the tubes contain mercury, used lamps must be handled according to the rules of disposal for hazardous wastes or the tubes must be recycled.

The sodium vapor in a lamp can be at low pressure or at high pressure. Low-pressure sodium vapor lamps that do not contain mercury cause the least light pollution and are less likely to interfere with astronomical observations.

Because sodium vapor can etch glass, the tube in a sodium vapor lamp is made from a very pure form of aluminum oxide (Al_2O_3) called polycrystalline alumina (PCA). PCA doesn't react with the sodium, and it allows light to pass through.

To Do

Students could collect data on the background colors and the color of lettering used for traffic signs (e.g., stop, yield, wrong way, and do not enter signs; speed limits and mileposts; parking and construction signs; and information on local tourist attractions). Then students could consider how a sodium vapor lamp might affect the visibility of the signs.

Group 1A: Alkali Metals

Na Vapor Lamps

When streetlights have a golden glow, the source of the light is probably sodium vapor. Inside the lamp is a sealed tube containing a sodium-mercury alloy and a starter gas, such as xenon. Electrodes at each end of the tube are connected to an electric circuit. When the lamp is on, a continuous spark, or arc, forms between the electrodes. The arc produces enough heat to vaporize the sodium and mercury atoms. Within the arc, atoms ionize. Outside the arc, ions recombine with electrons and light is emitted—yellow for sodium and blue-green for mercury.

A sodium vapor lamp uses less energy than most other light sources and costs less to operate. But the lamp isn't a perfect light source. The color of an object is visible only when light of that color is reflected off the object. So yellow lines on the pavement appear yellow under a sodium vapor lamp, but a red stop sign appears gray. ■

Na K Restoring Electrolytes

"Don't sweat it" may be good advice for handling stress, but not for maintaining a healthy body. The sweat you produce on a hot day or during exercise cools your body as it evaporates. Sweat consists mainly of water, sodium chloride, and small amounts of other inorganic salts. The salts are electrolytes that help keep the volume of body fluids constant. Electrolytes produce ions when they dissolve in water.

Body fluids contain sodium ions and potassium ions. Potassium ions are the principal cations inside cells. Sodium ions are the principal cations in the fluids outside of cells. The transmission of nerve impulses depends on the movement of sodium and potassium ions across the membranes of nerve cells. Potassium ions cause the heart muscle to relax between heartbeats.

Replacing the water lost during exercise is important, but not sufficient. Electrolytes must be replaced too. Some signs of electrolyte depletion are muscle cramps, nausea, and an inability to think clearly. Many athletes use sports drinks to replace electrolytes. Some experts recommend these drinks for people who lose more than 8 liters of sweat daily or exercise continuously for more than 60 minutes.

Table salt is the chief source of sodium in the diet. But large amounts are also found in unexpected places, such as eggs. For a healthy adult, the recommended daily intake of sodium chloride is about 5 grams per day—about half the amount many people consume. The daily recommended amount of potassium is about 1 gram. The word *daily* is important because your kidneys excrete potassium even when the supply is low. Eating foods high in potassium and low in sodium is ideal. Such foods include bananas, chicken, and orange juice. ■

Restoring Electrolytes

To Do

Relying more on fresh foods than processed foods can reduce the total amount of sodium chloride in a person's diet. Students could look at the labels for some processed foods and compare the sodium chloride content.

Students might also want to estimate how much sodium chloride they ingest daily from processed foods. Remind them that this quantity is in addition to the sodium chloride that is present in unprocessed foods.

Cs Cesium Atomic Clock

Many watches contain a quartz crystal that vibrates at a constant rate. The vibration provides the "beat" that is translated into the time you see displayed. A clock with a quartz crystal is more accurate than a mechanical clock, which has moving parts that can be worn down by friction. But a quartz crystal isn't accurate enough for modern communication and navigation systems.

For greater accuracy, you need an atomic clock, which may gain or lose only one second in 20 million years! In most atomic clocks, cesium-133 atoms provide the "beat." Unlike quartz crystals, all cesium-133 atoms are identical and they don't wear out with use. The clock is designed so that the atoms repeatedly absorb and emit radiation. The emitted radiation has a frequency of exactly 9,192,631,770 cycles per second. The cycles are counted and translated into seconds, minutes, and hours. Since 1967, the second had been defined based on the cesium clock.

There is a standard cesium atomic clock at the National Institute of Standards and Technology in Fort Collins, Colorado. A short-wave radio station transmits accurate time signals from this clock to most of North America. Clocks and watches advertised as "atomic" or "radio controlled" contain a tiny antenna and receiver that pick up the signal and decode the information. The owner sets the time zone. The radio signal sets the time. ■

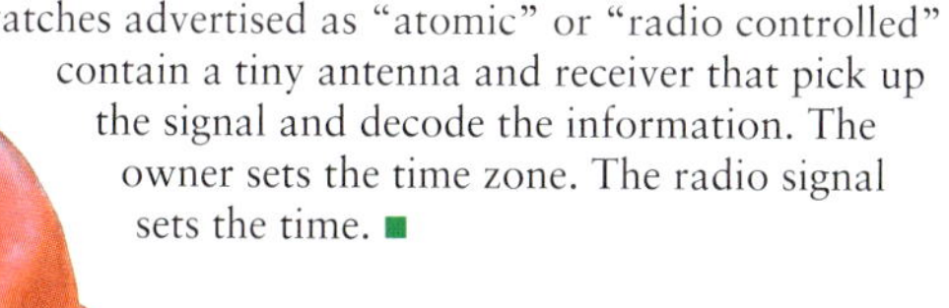

Global Positioning System (GPS) receivers can pinpoint any location on Earth to within a few meters. They depend on accurate time signals from atomic clocks in 24 Earth-orbiting satellites.

Na Salt of the Earth

In the ancient world, table salt (NaCl) was an extremely valuable commodity. Before refrigeration, salt was used to preserve foods such as meats and fish. As salt draws the water out of bacteria, they shrivel up and die. Sailors ("salty dogs") especially depended on salt pork and fish to survive long voyages.

Trade routes were established and roads were built to transport salt. In ancient China, coins were made of salt and taxes were levied on salt. Roman soldiers were paid an allotment called a *salarium argentum*, or "salt silver," from which the word *salary* is derived. ■

Li Bipolar Disorder

About two million people in the United States experience the extreme mood swings of bipolar disorder. During a manic phase, they think they can conquer the world. During a depression, they may feel hopeless. The ionic compound lithium carbonate often is used to control these symptoms. Lithium ions probably have an effect on the transmission of messages between brain cells. ■

Did You Know...

Most alkali metals are stored under oil. But **lithium** is so light that it floats on oil. So a coat of petroleum jelly is applied to lithium before it is stored.

Cesium Atomic Clock

A quartz crystal can keep accurate time to about one second every ten years. Over time, the vibrations of the crystal are affected by changes in temperature, impurities in the crystal, and the accumulated effects of the vibrations.

In a cesium clock, the valence electron in a cesium-133 atom oscillates between two hyperfine states. In one state, the magnetic field of the valence electron is in the same direction as the magnetic field of the nucleus. In the other state, the magnetic field is in the opposite direction.

GPS satellites transmit data that indicates their location and current time. 24 satellites are operating at all times with 3 satellites in reserve. Signals that are released simultaneously from different satellites will arrive at a receiver at slightly different times. The receiver compares data from four satellites to pinpoint its position.

To Do

Ask students to interpret the following statement. A person with one watch knows the time very well, but a person with two watches is not so certain.

Salt of the Earth

A good person is sometimes described as the "salt of the earth" and an unworthy person as "not worth his salt," or not worth the wages he receives.

In non-industrial countries, the principal use of NaCl is still food preparation and preservation. In the United States, 40% of NaCl is used for the production of chlorine and sodium hydroxide, and another 40% in de-icing roads. Table salt accounts for only 1% of usage.

Bipolar Disorder

Dr. John Cade discovered the mood-altering effect of lithium ions in 1949. His discovery had a major impact on the field of psychiatry. Up to that point, the accepted view was that mental health problems were always psychological. Talk therapy was the approved method of treatment. But many patients did not respond to this approach. Dr. Cade's work demonstrated that a mental illness could be caused by the wrong amount of a chemical in the body.

Today, both chemical and psychological approaches are used to treat mental illnesses. Lithium carbonate is initially given during a manic phase. Once the patient is stabilized, continued use of the medication can also help control the depression. The disorder may begin during adolescence or early adulthood. It can easily be misdiagnosed.

Group 2A: Alkaline Earth Metals

Vauquelin discovered an element he called gluinium as an oxide in beryl. The name reflected the sweet taste of the oxide. When Friedrich Wöhler and A. Bussy isolated the element in 1828, Wohler renamed it beryllium. Adair Crawford noted the existence of strontium in minerals in 1790.

Physical Properties

Beryllium's strength and rigidity make it an ideal material for mid-range and tweeter diaphragms. In most loud speakers, a current alternating at the frequency of the sound that generated it flows through a coil. As the coil is attracted to and repelled by a permanent magnet, the coil and the cone-shaped diaphragm attached to the coil vibrate. The vibration of the diaphragm reproduces the original sound. A beryllium diaphragm can reproduce frequencies that cannot be heard when other materials are used.

Sources

To Do

Help students write balanced equations and classify the reactions involved in the production of magnesium.

(1) Decomposition

$CaCO_3(s) \xrightarrow{\Delta} CaO(s) + CO_2(g)$

(2) Combination

$CaO(s) + H_2O(l) \longrightarrow Ca(OH)_2(aq)$

(3) Double-replacement

$Ca(OH)_2(aq) + MgCl_2(aq) \longrightarrow Mg(OH)_2(s) + CaCl_2(aq)$

(4) Double-replacement/neutralization

$Mg(OH)_2(s) + 2HCl(aq) \longrightarrow MgCl_2(aq) + 2H_2O(l)$

(5) Decomposition

$MgCl_2(aq) \xrightarrow{elec.} Mg(s) + Cl_2(g)$

(6) Combination

$Cl_2(g) + H_2(g) \longrightarrow 2HCl(l)$

Students may wonder what the point is of reactions (3) and (4). The precipitation of $Mg(OH)_2$ is used to concentrate the magnesium. If you add 250 mL of 0.1*M* $Ca(OH)_2$ to 250 mL of seawater, students can observe the precipitate, which can be separated from the remaining solution by decanting or filtration.

Group 2A: Alkaline Earth Metals

Beryllium | discovered in 1798 by Nicholas Vauquelin

Magnesium | isolated in 1808 by Sir Humphry Davy

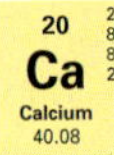

Calcium | discovered in 1808 by Sir Humphry Davy

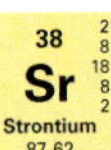

Strontium | discovered in 1808 by Sir Humphry Davy

Barium | discovered in 1808 by Sir Humphry Davy

88 Ra Radium (226) 2 8 18 32 18 8 2

Radium | discovered in 1898 by Pierre Curie & Marie Curie

Physical Properties

- Alkaline earth metals are relatively soft, but harder than alkali metals.
- Alkaline earth metals have a gray white luster when freshly cut. When exposed to air, they quickly form a tough, thin oxide coating.
- Densities, melting points, and boiling points tend to be higher than for the alkali metal in the same period.

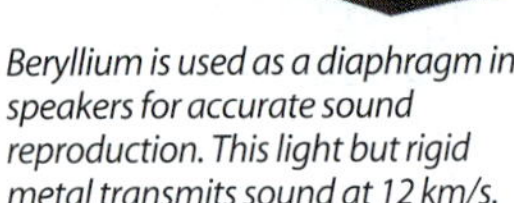

Diaphragm

Beryllium is used as a diaphragm in speakers for accurate sound reproduction. This light but rigid metal transmits sound at 12 km/s.

- Strong, lightweight magnesium alloys are used in cameras, lawnmowers, aircraft, and automobiles.

Melting and Boiling Points

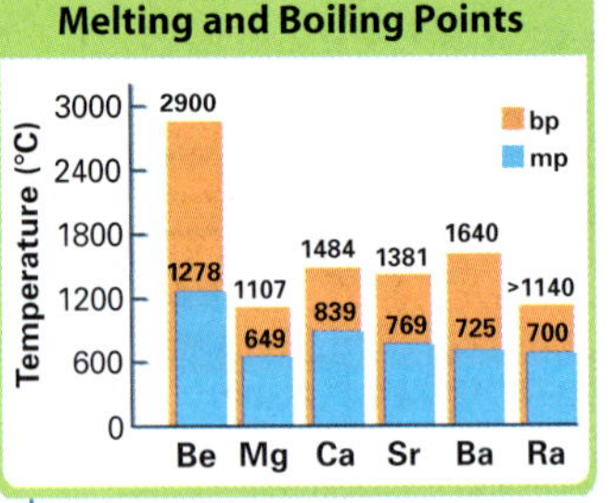

Density

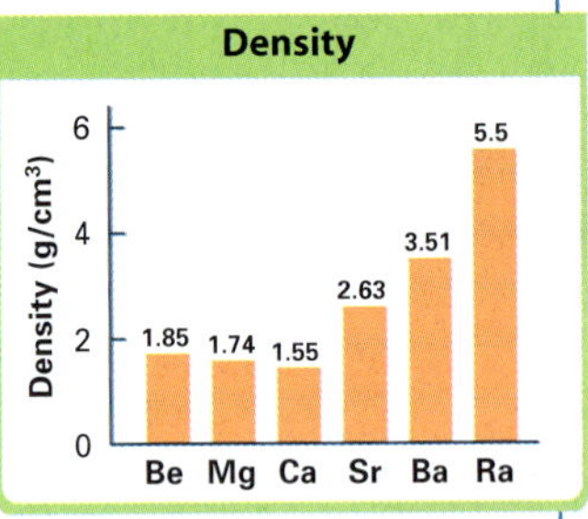

Sources

- Alkaline earth metals are not found in nature in the elemental state.
- Many mountain ranges contain alkaline earth carbonates—limestone ($CaCO_3$) and dolomite ($CaCO_3 \cdot MgCO_3$).
- Oyster shells containing $CaCO_3$ are used to extract magnesium. The chlorine gas produced as a by-product is fed back into the process.
- Barium is made by reduction of its oxide with aluminum at high temperature.

  $3BaO(s) + 2Al(s) \longrightarrow 3Ba(l) + Al_2O_3(s)$
- Salts of highly radioactive radium are a by-product of uranium refining.

Production of Magnesium

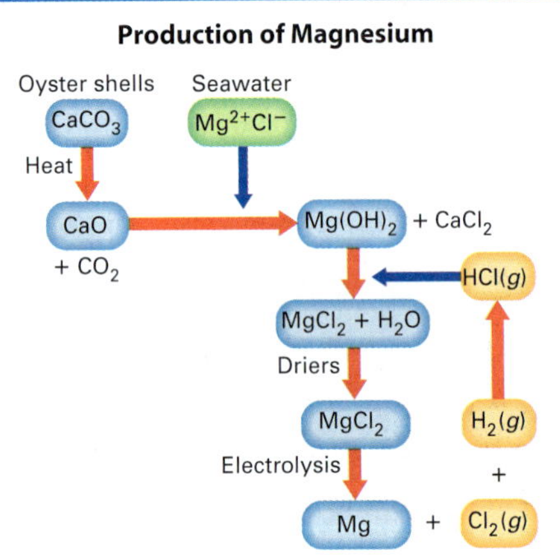

Atomic Properties

- Alkaline earth metals have an electron configuration that ends in ns^2.
- The alkaline earth metals are strong reducing agents, losing 2 electrons and forming ions with a 2+ charge.
- Because radium is luminous, it was once used to make the hands and numbers on watches glow in the dark.
- The ratio of ^{87}Sr to ^{86}Sr varies with location. This data can be used to solve problems, such as the source of materials.

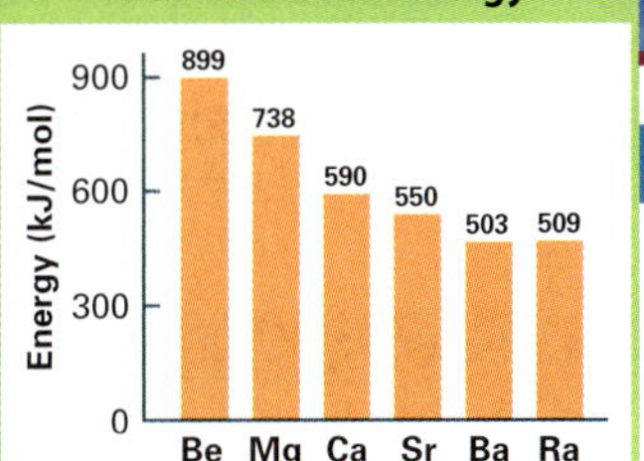

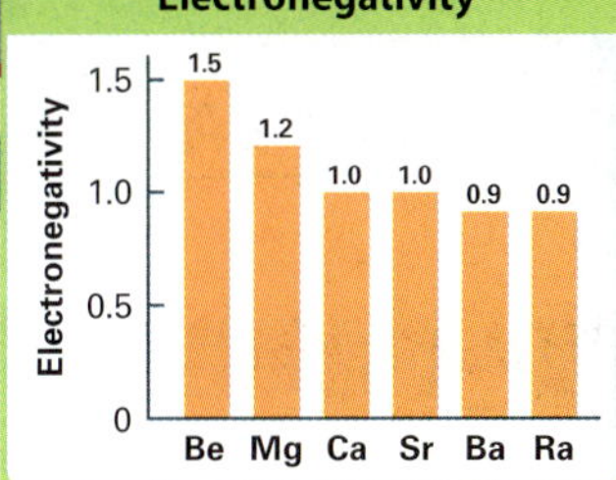

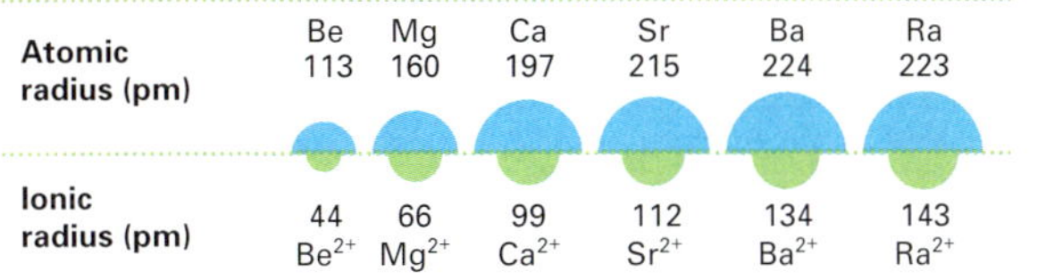

Calcium, strontium, and barium can be identified by the colors produced when their compounds are heated in a flame.

Calcium

Strontium

Barium

Important Compounds and Reactions

- Alkaline earth metals are less reactive than alkali metals.
- Alkaline earth metals react with halogens to form ionic halides. Example:

  $Mg(s) + Br_2(l) \longrightarrow MgBr_2(s)$
- All Group 2A metals (except Be) react with water to form an alkaline solution. Example:

  $Sr(s) + 2H_2O(l) \longrightarrow Sr(OH)_2(aq) + H_2(g)$
- Alkaline earth metals react with oxygen to form binary oxides. Example:

  $2Ca(s) + O_2(g) \longrightarrow 2CaO(s)$ $\quad \Delta H = -635.1$ kJ/mol
- Heating limestone produces lime, CaO.

  $CaCO_3(s) \xrightarrow{\Delta} CaO(s) + CO_2(g)$ $\quad \Delta H = 176$ kJ/mol
- Slaked lime, $Ca(OH)_2$, reacts with carbon dioxide to form limestone.

  $Ca(OH)_2(s) + CO_2(g) \longrightarrow CaCO_3(s) + H_2O(g)$
- Barium peroxide is used as a dry powdered bleach. It reacts with water to form the bleaching agent, hydrogen peroxide.

  $BaO_2(s) + 2H_2O(l) \longrightarrow H_2O_2(aq) + Ba(OH)_2(aq)$
- Gypsum, calcium sulfate dihydrate, $CaSO_4{\cdot}2H_2O$, is used to make plasterboard.
- Calcium phosphate, $Ca_3(PO_4)_2$, is the major component of bone and teeth.

Slaked lime is an ingredient in plaster, cement, and the mortar used in this stone wall.

Atomic Properties

One of the problems that the ratio of ^{87}Sr to ^{86}Sr was used to solve was the source of timber for the prehistoric great houses in Chaco Canyon, New Mexico. Of the more than 200,000 trees that were used to build the houses, at least one-fifth were spruce or fir trees that were carried from mountainous sites 75–100 km away. Scientists can identify specific sites by comparing $^{87}Sr/^{86}Sr$ ratios in beams to ratios in trees from the sites. The ratios show that the timber for construction beams came from the Chuska and San Mateo mountains, but not from the San Pedro Mountains.

Important Compounds and Reactions

To Do

Students could look at the relationship between solubility and the properties and uses of calcium oxide, calcium carbonate, calcium sulfate, and calcium phosphate. Refer to Table B.9 on page R54.

Group 2A: Alkaline Earth Metals

Chlorophyll

Chlorophyll is an example of a chelate, a substance in which a central metal ion is bonded to a large organic molecule. There are two types of chlorophyll. In chlorophyll *b*, CHO replaces the methyl group at the upper right in the formula. Chlorophyll absorbs light in the red and blue-violet regions of the visible spectrum. In chlorophyll *a* and chlorophyll *b*, the peaks are at different wavelengths. Thus, chlorophyll absorbs a broader range of wavelengths than it would if there were only one type.

Fireworks

The first fireworks were probably invented in China more than a thousand years ago. They were a mixture of sulfur, charcoal, and potassium nitrate, which burned quickly with a flash. This black powder is still a major component of modern fireworks. When iron filings or copper or zinc were added to this gunpowder, early versions of fireworks with limited colors (orange, yellow, white) were produced. The rest of the colors were not added to the palette until the 1800s.

The pattern produced by a shell depends on how the stars are arranged within the shell. Shells are named for their patterns—*chrysanthemum, weeping willow, or strobe.* Strobes are clusters of flickering silvery lights that drift slowing to the ground. The gold sparks in a display come from burning iron filings or charcoal. Brilliant white flames are produced when magnesium, titanium, and aluminum burn at very high temperatures.

To Do

Because fireworks displays are often a central part of public celebrations, students may not realize that fireworks are classified as hazardous materials. Suggest that students investigate the hazards associated with the misuse of fireworks, and the regulations that are in place in your state to prevent injuries.

Group 2A: Alkaline Earth Metals

Mg Chlorophyll

A plant that is deprived of magnesium will turn yellow and eventually die. The yellow color is a sign that the plant is not producing enough of the green pigment chlorophyll, which is found in structures called chloroplasts. There are about half a million chloroplasts in one square millimeter of leaf surface. In chloroplasts, light energy is changed to chemical energy.

Chlorophyll molecules absorb wavelengths of blue light and red light, and reflect green light. There is a magnesium ion at the center of each chlorophyll molecule. It is embedded like a jewel in a crown-like ring. The magnesium ion forms a bond with each nitrogen atom.

Chlorophyll a

Chlorophyll is typical of a family of compounds called porphyrins, which contain a central metal ion. In heme, the ion is Fe^{2+}. There are four heme molecules in hemoglobin, which transports oxygen in the blood. ■

Sr Ba Fireworks

Producing an aerial fireworks display requires skill and knowledge of chemistry. Shells are stored in steel pipes anchored in sand. A typical shell contains two charges, one to launch the shell and one to cause the shell to burst. When the lifting charge explodes, the shell is blown out of the pipe and the fuse attached to the bursting charge ignites. The shell travels for a few seconds before it bursts.

The name of this effect is chrysanthemum.

The explosions that launch the shells and cause them to burst are exothermic redox reactions. The reducing agent, or fuel, may be aluminum, magnesium, or sulfur. The oxidizing agent may be a nitrate, a chlorate, or a perchlorate, such as potassium perchlorate ($KClO_4$).

When a shell bursts, it releases tiny pellets called "stars," which produce the colors. Only a few elements are needed to produce these colors. Two are alkaline earth metals, strontium for bright red and barium for light green. These same colors are produced when strontium and barium compounds are heated in a flame. Flame tests can identify elements because an element emits characteristic wavelengths of light when its atoms absorb heat and then return to the ground state. ■

Labels for shells used in a fireworks display describe the color and the effect produced by the shell.

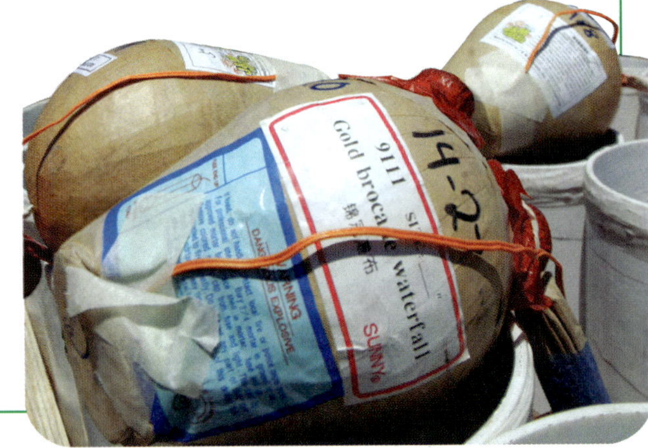

Ca Limestone Caves

The chemical reactions that form a limestone cave are simple. Carbon dioxide in the air dissolves in rain to form weak carbonic acid, H_2CO_3. As rain passes through soil, it dissolves carbon dioxide produced by decaying plants and becomes even more acidic. The rainwater seeps into limestone, $CaCO_3$, beneath the soil. The $CaCO_3$ dissolves in the carbonic acid, forming a solution of calcium hydrogen carbonate, $Ca(HCO_3)_2$.

$$CaCO_3(s) + H_2CO_3(aq) \longrightarrow Ca^{2+}(aq) + 2HCO_3^-(aq)$$

Over millions of years, as more and more limestone dissolves, a cave forms and slowly grows in size. Once the cave stops growing, another process may occur. Calcium hydrogen carbonate solution drips through the cave's roof. Carbon dioxide is released from solution, leaving behind a tiny deposit of solid calcium carbonate.

$$Ca^{2+}(aq) + 2HCO_3^-(aq) \longrightarrow CaCO_3(s) + CO_2(g) + H_2O(l)$$

The deposit slowly grows into a stalactite, which hangs from the ceiling like an icicle. Drops of solution that fall from the ceiling form stalagmites on the cave floor. Calcium carbonate is white. So any color in the deposits is due to traces of metal ions, such as copper (blue-green) and iron (red-brown). The growth rate of the formations depends mainly on the volume of water that drips through the roof and the concentration of calcium bicarbonate in the water. In many caves, the rate of growth is measured in centimeters per hundreds or thousands of years. ■

Stalactites and stalagmites may join together to form columns.

Ca Building Strong Bones

As a young adult, you can have a major effect on the health of your bones later in life. You acquire 90% of the bone tissue in your skeleton before age 18 as females and age 20 as males. Physical activity and the amount of calcium in your diet affect the buildup of bone mass.

To maintain a constant level of calcium in your blood, your body may release calcium from bone tissue.

If enough calcium is lost from bones, they become brittle and tend to break easily. This condition, known as osteoporosis, is most common in older people. However, it can occur in young and middle-aged adults.

The recommended daily intake of calcium from age 9 to age 18 is 1.3 g per day. This is the amount in a liter of milk.

Calcium-fortified foods, such as orange juice, are a good choice for those who cannot tolerate the lactose in milk. Vitamin D in orange juice helps you absorb calcium. ■

Did You Know...

Calcium oxide (lime) emits a bright white light when it is heated. Before electric lights, theaters used lime to focus light on a single actor. So the expression *in the limelight* describes a person in a prominent position.

Limestone Caves

Limestone caves tend to be longer and more complex than other types of caves. Limestone is a sedimentary rock, which generally forms in warm, shallow seas. When the water table drops in a limestone cave, dry spaces that can be explored are left. It is in these dry spaces that stalactites and stalagmites form.

Building Strong Bones

In recent years there has been a debate about how the increased consumption of soft drinks by teens has affected the calcium content of their bones. One hypothesis is that acids in soft drinks interfere with absorption of calcium. Another is that the consumption of soft drinks reduces the consumption of calcium sources such as milk.

Group 2A: Alkaline Earth Metals **R13**

Group 3A

Physical Properties

To Do

To celebrate its 80th anniversary, *Chemical & Engineering News* published an issue on the elements (September 8, 2003). The articles can be accessed on the Internet. Students are likely to enjoy the article on gallium by Oliver Sacks.

Sources

In the Elements Handbook, students will often see references to minerals and ores. A mineral is a naturally occurring, inorganic, crystalline solid with a characteristic chemical composition. Some properties used to identify minerals are crystal structure, color, luster, density, hardness, and how its crystals fracture. An ore is a material from which a valuable mineral can be profitably extracted.

Group 3A

Boron | discovered in 1808 by Sir Humphry Davy and by Joseph-Louis Gay-Lussac & Louis-Jacques Thénard

Aluminum | discovered in 1825 by Hans Christian Oersted

Gallium | discovered in 1875 by Paul-Emile Lecoq de Boisbaudran

Indium | discovered in 1863 by Ferdinand Reich & Hieronymus T. Richter

Thallium | discovered in 1861 by Sir William Crookes

Physical Properties

- Boron is a metalloid. The rest of the Group 3A elements are metals.
- Aluminum is a valuable structural material because of its strength, especially in alloys with silicon or iron. These alloys have a low density and resist corrosion.
- Gallium has an extremely wide liquid temperature range (30°C to 2204°C). Solid gallium floats in liquid gallium, which is unusual for a metal.

Boron is black, lustrous, and extremely hard, but brittle.

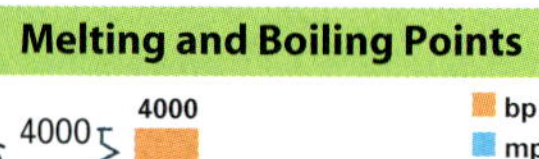

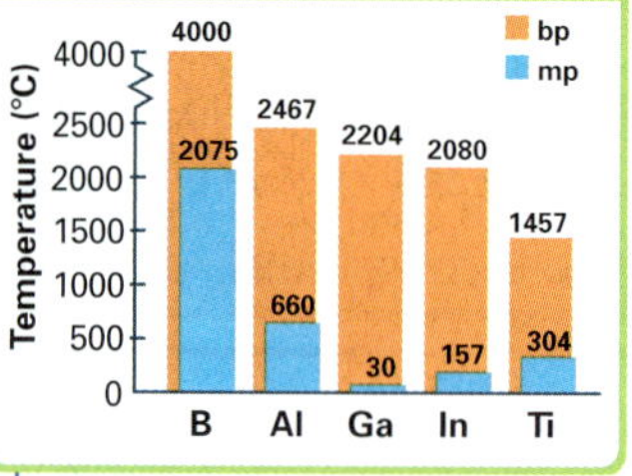

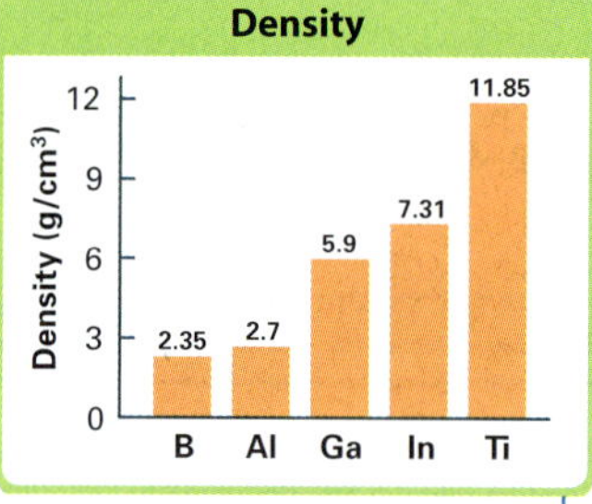

Sources

- Boron is always combined with oxygen in nature. Boron can be prepared by the reaction of its oxide with magnesium metal.

  $B_2O_3(s) + 3Mg(s) \longrightarrow 2B(s) + 3MgO(s)$
- Bauxite is a common ore of aluminum. The primary mineral in bauxite is alumina, Al_2O_3, which is reduced to aluminum by electrolysis.
- Gallium, indium, and thorium are quite rare. They are typically extracted from ores being processed to extract other metals.

The conveyor belt is being used to unload bauxite from a train. Bauxite is a major source of aluminum. It is also a source of gallium.

Atomic Properties

- Group 3A elements have an electron configuration that ends in ns^2np^1.
- The most common oxidation number for boron, aluminum, gallium and indium is +3. For thallium, it is +1.
- Group 3A elements become more metallic from top to bottom within the group.
- Radioactive thallium-201 is injected into patients taking the stress test used in the diagnosis of heart disease.

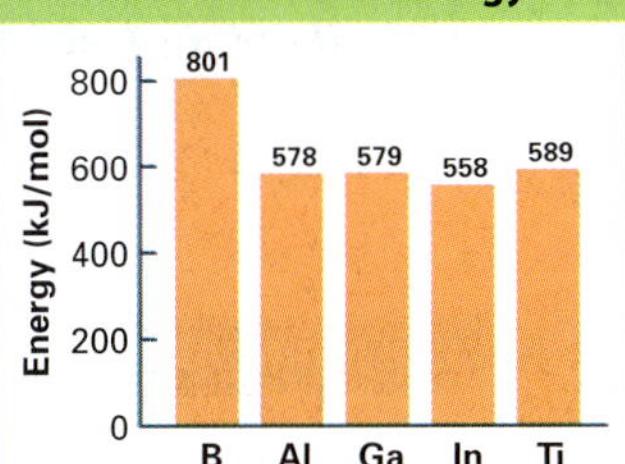

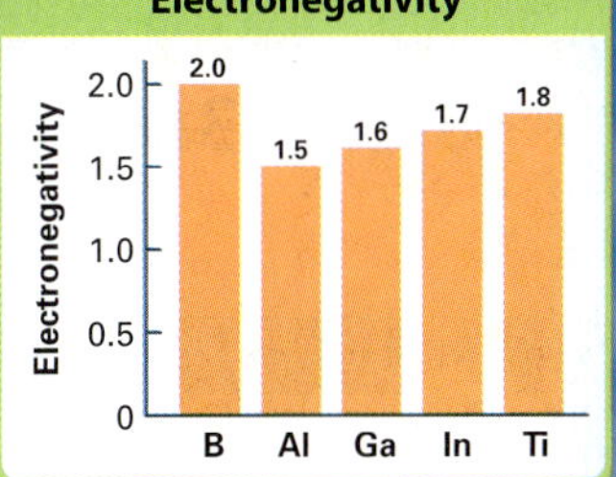

	B	Al	Ga	In	Tl
Atomic radius (pm)	83	143	141	166	172

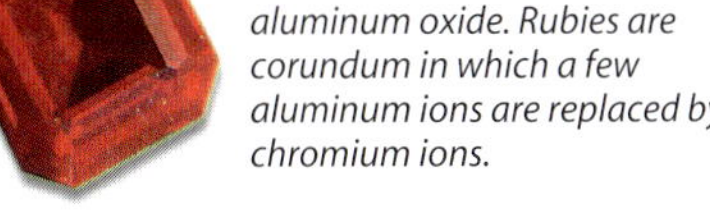

Ionic radius (pm)	23	51	62	81	95
	B^{3+}	Al^{3+}	Ga^{3+}	In^{3+}	Ti^{3+}

Important Compounds and Reactions

- Group 3A elements react with halogens to form halides. Example:

  $2Al(s) + 3Cl_2(g) \longrightarrow 2AlCl_3(s)$

  Aluminum chloride is used as a catalyst in organic reactions.
- Group 3A elements react with oxygen to form oxides. Example:

  $4Al(s) + 3O_2(g) \longrightarrow 2Al_2O_3(s)$ $\Delta H = -1676$ kJ/mol

  Because this reaction is so exothermic, powdered aluminum is a component of some explosives, fireworks, and rocket fuels.
- Aluminum sulfate (alum), $Al_2(SO_4)_3{\cdot}18H_2O$, is used in water treatment plants.
- Gallium arsenide, GaAs, converts electric current to light in light-emitting diodes (LEDs). It is produced as follows.

  $(CH_3)_3Ga(g) + AsH_3(g) \longrightarrow GaAs(s) + 3CH_4(g)$

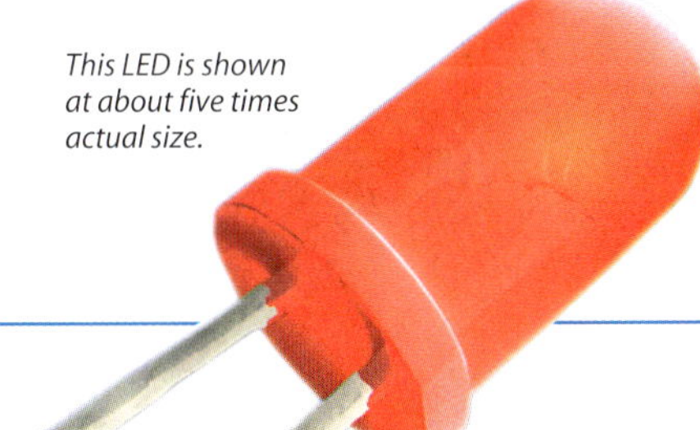

This LED is shown at about five times actual size.

Corundum is a mineral form of aluminum oxide. Rubies are corundum in which a few aluminum ions are replaced by chromium ions.

- Borax, $Na_2B_4O_7{\cdot}10H_2O$, is used to soften water and is found in glasses and glazes. When a mixture of borax and hydrochloric acid is heated, boric acid is produced.

  $Na_2B_4O_7(aq) + 2HCl(aq) + 5H_2O(l) \longrightarrow 4H_3BO_3(aq) + 2NaCl(aq)$
- Boric acid is poisonous if ingested. The solid is used as an insecticide against cockroaches. A dilute solution of boric acid is an eyewash.
- Boron carbide, B_4C, is almost as hard as diamond. It is used for items that must resist wear, such as cutting tools.
- Laboratory glassware is made of heat-resistant glass containing 12–15% boric oxide.

Atomic Properties

To Do

Challenge students to figure out why thallium commonly has an oxidation number of +1 in addition to +3 when other elements in the group do not. Thallium can lose an electron from a 6*p* orbital more easily than the other elements can lose an electron from their highest occupied *p* orbital because the energy difference between the 6*s* and 6*p* orbitals is greater than the difference between the *s* and *p* orbitals in the lower principal energy levels. The occupied 4*f* sublevel in thallium is responsible for the difference.

Important Compounds and Reactions

The compositions of sapphires and rubies are almost identical. They are mainly crystals of colorless aluminum oxide (Al_2O_3). Traces of cobalt make sapphires blue and traces of chromium make rubies red.

Crystalline boron nitride is an industrial abrasive, which is used to cut and polish metals. But when finely ground boron nitride is added to face powder, it provides a soft, silky texture.

Group 3A

Uses of Indium

Traditionally, solder was an alloy of tin and lead, which melts at about 180°C. Nowadays many solders are lead-free and some low-temperature solders do not even contain any tin. One mixture used to solder glass and ceramics is 97% indium and 3% silver. Often a flux is used when metals are being soldered. A flux is used to remove oxides from the surface of metals before they are joined.

To Do

You might want to arrange for a demonstration of soldering by a teacher in your school or another expert. Ask the person doing the demonstration to discuss how the composition of solder affects its properties and its uses.

Recycling Aluminum

When old bridges are repaired, the steel or concrete decks may be replaced with aluminum decks. The aluminum decks require much less maintenance because a thin oxide layer protects the underlying layers of aluminum. The aluminum decks can carry equal or greater loads than steel decks because aluminum has a high strength to weight ratio. Strength is defined as the ability of a material to withstand forces of compression, tension, and shear. At the limit of its compressive strength, a material will tend to compact, i.e., its volume will be reduced. At the limit of its tensile strength, a material will be stretched to the point where it breaks. At the limit of its shear strength, a material will slide along a plane parallel to the force. With a deck on a bridge, the major concern is compressive strength.

Group 3A

In Uses of Indium

Indium is a soft metal with a low melting point of 157°C. One useful property of indium is its ability to "wet" glass. It spreads out and forms a thin layer on the glass instead of beading up. This property allows indium to be deposited on glass to make mirrors that are as reflective, but more corrosion-resistant, than silvered mirrors.

Low-melting alloys that contain indium are used as solders to join glass to glass. They also can join a metal to a metal at low temperatures. There are products, such as electronics components, that could be damaged if joined at a high temperature. The seals of some fire sprinkler heads are held in place by indium alloys. When heat from a fire melts the alloy, the seal is released, which allows water to pour from the sprinkler head. ■

Al Recycling Aluminum

Lightweight, durable aluminum has many uses. When aluminum parts replace steel parts in vehicles, less fuel is needed to travel the same distance. About 20% of the aluminum produced is used for packaging, including foil and cans.

Aluminum is used in durable consumer goods such as ladders.

In the United States, over 80 billion aluminum soft drink cans are sold each year. Around 50% of these cans are recycled. The energy saved by recycling just one of these cans could be used to operate a television for three hours. The energy needed to recycle aluminum is 5% of the energy needed to obtain new aluminum from ore. An added benefit of recycling is a reduction in solid waste.

Recycled aluminum cans are shredded, crushed, and heated to remove materials other than aluminum. Aluminum pieces about the size of a potato chip are placed in furnaces where they are mixed with new aluminum and melted. The molten aluminum is poured into 7.6-m ingots that have a mass of about 13,600 kg. The ingots pass through rolling mills that reduce their thickness from about 0.5 m to about 0.25 mm. The thin sheets are coiled and shipped to a manufacturer who produces the bodies and tops of the cans. Aluminum from a recycled can is part of a new can within 60 days. ■

Aluminum cans are compressed before being shipped to aluminum companies for recycling.

Al Manufacturing Aluminum

Aluminum is the most abundant metal in Earth's crust (8.3% by mass). It is found in minerals such as bauxite (impure aluminum oxide, Al_2O_3). But for years after its discovery, there wasn't a practical way to extract aluminum from its ores. This rare and expensive metal was used, like gold, mainly for decoration.

A professor at Oberlin College in Ohio challenged his class to find an inexpensive way to produce aluminum. In 1885, Charles Hall responded to the challenge. He set up a laboratory in a woodshed. He knew that other chemists had tried to decompose aluminum oxide by electrolysis. This method was not practical because the melting point of aluminum oxide is quite high (2045°C). Hall found that mixing aluminum oxide with cryolite (Na_3AlF_6) produced a mixture that melted at a much lower temperature of 1012°C.

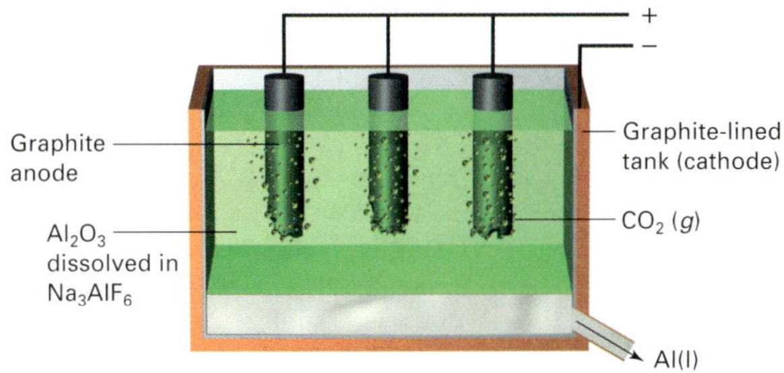

The process Hall invented is often called the Hall-Heroult process. Paul Heroult, a 23-year-old Frenchman, developed the same process almost simultaneously. It is still used today. Pure aluminum oxide is extracted from bauxite through heating to a temperature above 1000°C. The aluminum oxide is dissolved in molten cryolite and contained in a graphite-lined iron tank. The graphite rods used as an anode are consumed during the process.

Anode: $C(s) + 2O^{2-}(l) \longrightarrow CO_2(g) + 4e^-$

Cathode: $3e^- + Al^{3+}(l) \longrightarrow Al(l)$

The products are carbon dioxide and molten aluminum metal. Because the aluminum is more dense than the aluminum oxide-cryolite mixture, it collects at the bottom of the tank and is drawn off periodically. ■

Al Anodized Aluminum

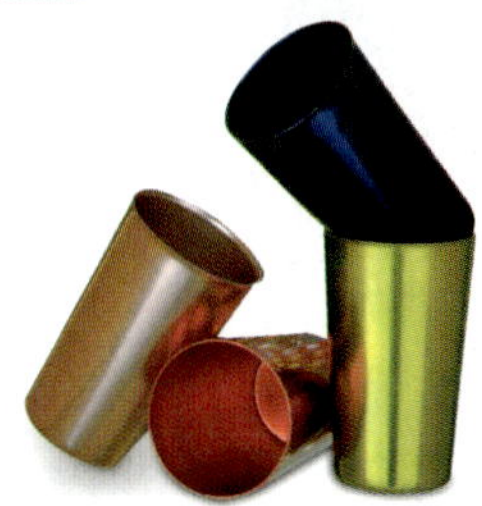

If an aluminum object has a color other than silver, it was anodized before it was dyed. The main goal of anodizing is to protect the aluminum from corrosion by coating it with aluminum oxide. The ability to be dyed is an added benefit.

The aluminum object becomes the anode in an electrolytic cell. The electrolyte is a dilute acid. When an electric current flows through the cell, aluminum oxide forms on the surface of the aluminum.

The oxide layer is thin, hard, and dense. It contains tiny pores, which can absorb an organic dye. If a dyed piece is placed in boiling water, the oxide layer absorbs water and swells, which closes the pores and seals in the dye. Because the oxide layer is so thin, the silver of the aluminum base is visible through the dye and gives the object a metallic sheen. ■

Did You Know...

Indium's name comes from indigo, a bright line in its emission spectrum. Indium will emit a high-pitched sound when it is bent. Tin and gallium also "cry" when bent.

Manufacturing Aluminum

Hall was able to convince a group in Pittsburgh to invest in his process. The investors formed the Pittsburgh Reduction Company, which became the largest producer of aluminum in the United States. The company was later renamed the Aluminum Company of American (Alcoa). Hall bequeathed much of the fortune he made to various universities. Oberlin College received three million dollars in 1914, which is equivalent to more than 50 million dollars today.

Anodized Aluminum

Untreated elemental aluminum cannot absorb enough dye, even if the metal is immersed in a coloring bath, to color the aluminum.

To Do

Aluminum tumblers that were popular in the 1950s are popular again. If you or one of your students has access to these tumblers or to another aluminum object that has been anodized, bring the object to class for other students to observe.

Group 4A

See Roald Hoffmann's article on silicon from the 80th anniversary issue of Chemical & Engineering News for an explanation for why silicon plays so minor a role in biological processes in comparison to carbon.

Physical Properties

Based on their macroscopic properties, it isn't obvious that diamond and graphite have the same chemical composition. In 1772, Lavoisier demonstrated that diamond was carbon by analyzing the products produced when equal masses of diamond and charcoal were burned.

Sources

The orthosilicate ion, SiO_4^{4-}, is found in minerals such as zircon, $ZrSiO_4$. Several minerals contain the disilicate ion, $Si_2O_7^{6-}$, in which an oxygen atom is shared by two silicon atoms. In most silicates, the tetrahedra are linked together in chains or sheets, as in talc, $Mg_3(Si_2O_5)_2(OH)_2$. The repeating unit in a chain is $Si_2O_6^{4-}$, or SiO_3^{2-}. The repeating unit in a sheet is $Si_2O_5^{2-}$. Silicates that contain more than one metal can have a variable composition as long as the total charge on the cations is constant.

To Do

Andalusite and olivine belong to a subclass of silicates called nesosilicates. Pyrophyllite belongs to the subclass phyllosilicates. Students could research the six subclasses of silicates and explain the basis of the classification system. The subclasses are nesosilicates (single unit), sorosilicates (double unit), inosilicates (single and double chains), cyclosilicates (rings), phyllosilicates (sheets), and tectosilicates (frameworks).

Group 4A

6 C Carbon 12.011 — 2 4

Carbon | known since ancient times

Silicon | discovered in 1824 by Jöns Jacob Berzelius

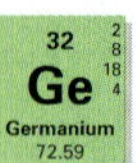

Germanium | discovered in 1886 by Clemens Winkler

Tin | known since ancient times

Lead | known since ancient times

Physical Properties

- Group 4A elements are all solids at room temperature.
- The metallic properties of Group 4A elements increase from carbon to lead.
- Diamond, graphite, and buckminsterfullerene are three allotropes of carbon.

Phase Diagram of Carbon

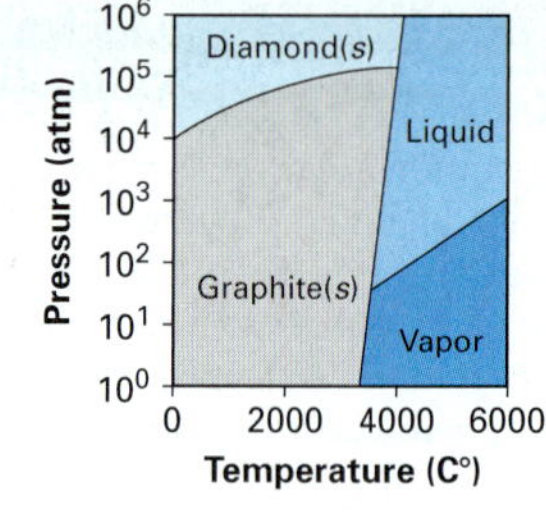

Graphite is more stable than diamond at STP, but the activation energy is too high for diamond to change to graphite at these conditions.

Melting and Boiling Points

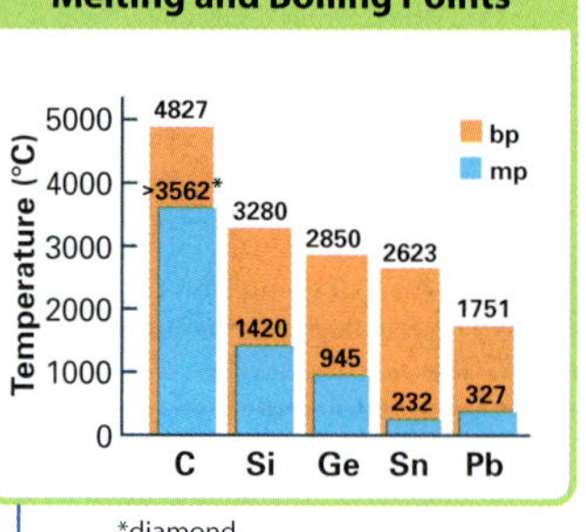

*diamond

Density

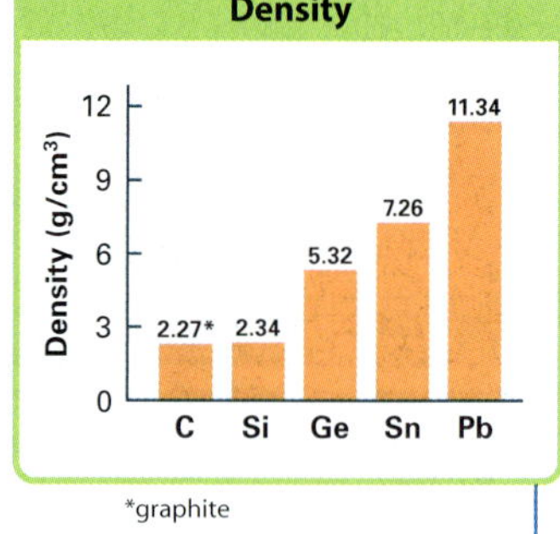

*graphite

Sources

- Carbon is found in nature as an element, in Earth's atmosphere as carbon dioxide, in Earth's crust as carbonate minerals, and in organic compounds produced in cells.
- Silicon can be produced by the reduction of silicon dioxide (silica) with magnesium, carbon, or aluminum. Example:

  $SiO_2(s) + 2Mg(s) \longrightarrow Si(s) + 2MgO(s)$
- Tin is prepared by reduction of the mineral cassiterite, SnO_2.

  $SnO_2(s) + 2C(s) \longrightarrow 2CO(g) + Sn(s)$
- Lead is refined from the mineral galena, PbS. Galena is heated in air to form a mixture of PbO and $PbSO_4$. Lead is produced through further reaction of these compounds with PbS.

Al_2SiO_5 *andalusite*

Fe_2SiO_4 *olivine*

$AlSi_2O_5OH$ *pyrophyllite*

About 90% of the minerals in Earth's crust are silica and silicates. In silicates, each silicon atom is surrounded by three or four oxygen atoms. These units can be linked together in chains, sheets, rings, or crystals.

Atomic Properties

- Group 4A elements have an electron configuration that ends in ns^2np^2.
- For Group 4A elements, the most common oxidation numbers are +4 and +2. For carbon, −4 is also common.
- Silicon and germanium are semiconductors.

First Ionization Energy

Element	C	Si	Ge	Sn	Pb
Energy (kJ/mol)	1086	786	761	708	715

Electronegativity

Element	C	Si	Ge	Sn	Pb
Electronegativity	2.5	1.8	1.8	1.8	1.9

	C	Si	Ge	Sn	Pb
Atomic radius (pm)	77	109	122	139	175
Ionic radius (pm)	15 C^{4+}	41 Si^{4+}	53 Ge^{4+}	71 Sn^{4+}	84 Pb^{4+}

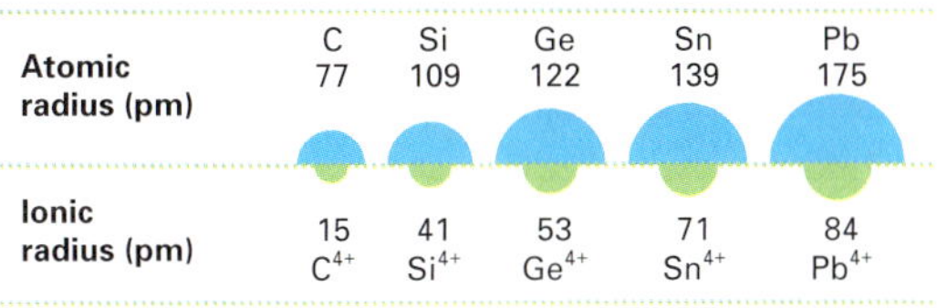

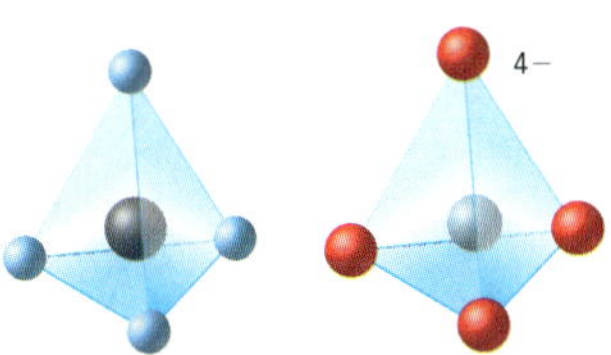

methane molecule *silicate ion*

When carbon and silicon form four covalent bonds, there is often sp^3 hybridization. The result is compounds and ions with tetrahedral structures.

Important Compounds and Reactions

- Group 4A elements are oxidized by halogens. Example:

  $Ge(s) + 2Cl_2(g) \longrightarrow GeCl_4(l)$
- Group 4A elements combine with oxygen to form oxides. Example:

  $Sn(s) + O_2(g) \longrightarrow SnO_2(s)$
- Complete combustion of hydrocarbons yields carbon dioxide and water. Example:

  $CH_4(g) + 2O_2(g) \longrightarrow CO_2(g) + 2H_2O(l)$ $\Delta H = -890$ kJ/mol
- Plants use carbon dioxide to produce carbohydrates and oxygen.
- Aqueous sodium silicate, Na_2SiO_3, is used as an adhesive for paper, as a binder in cement, and to stabilize shale during oil drilling.

  $SiO_2(s) + 2NaOH(aq) \longrightarrow Na_2SiO_3(aq) + H_2O(l)$
- Acetylene is a fuel used for welding. It forms when calcium carbide and water react.

  $CaC_2(s) + 2H_2O(l) \longrightarrow C_2H_2(g) + Ca(OH)_2(aq)$
- Tungsten carbide, WC, is used on the cutting surfaces of drill bits and saw blades.
- Lead(IV) oxide, PbO_2, is used as electrodes in lead acid car batteries.
- Tin(II) fluoride, SnF_2, is used in toothpaste to prevent tooth decay.

Silicon dioxide, SiO_2, is the sand on many beaches and is used to make glass.

Atomic Properties

The graphs for first ionization energy and electronegativity can help you discuss the fact that the first element in a group of representative elements is often the least representative of the group as a whole. The elements in period 2 have only the electrons in the 1*s* orbital to shield the valence electrons from the attraction of the nucleus.

Important Compounds and Reactions

A mixture of sand (SiO_2), lime (CaO), and soda ash (Na_2CO_3) is used to make glass. The addition of small amounts of transition metal oxides can change the color of glass. Adding iron(III) oxide (Fe_2O_3) gives glass the green tint used in some beverage bottles and tinted windows. Blue glass contains CuO or CoO, orange glass contains Cr_2O_3, and red glass contains colloidal gold.

Group 4A **R19**

Group 4A

Green Chemistry

The phrase "green chemistry" was coined in 1992 to define an approach to developing chemical technology. The green chemistry mission statement of the EPA is "to promote innovative chemical technologies that reduce or eliminate the use or generation of hazardous substances in the design, manufacture, and use of chemical products." The EPA program supports research and education through partnerships with industries, academia, other government agencies, and non-government agencies. Los Alamos National Laboratory is one important partner. Los Alamos was instrumental in the development of supercritical carbon dioxide as a recyclable cleaning solvent. Los Alamos is collaborating with industries on projects ranging from fuel cells to a process for producing ethylene that will reduce carbon dioxide emissions and conserve natural gas.

Caffeine is an alkaloid. The amount of caffeine in a coffee bean is typically 1% to 2% by mass. Around 1900, a German chemist, Ludwig Roselius, became interested in extracting caffeine from coffee because he blamed his father's death on the drinking of excessive amounts of coffee. There were a number of solvents that could remove the caffeine from coffee, but most were known to be toxic. By the 1980s, the solvent used for 70 years, methylene chloride, was suspected to be a carcinogen. Ethyl acetate and water are also used to decaffeinate coffee.

Group 4A

C Green Chemistry

The term *green chemistry* was coined in 1992. It describes the effort to design chemical processes that don't use or produce hazardous substances. The goal is to protect the environment and conserve resources. For example, if a catalyst is used to reduce the temperature at which a reaction occurs, the process requires less energy.

Carbon dioxide is at the center of a green chemistry success story. Organic solvents are used to dissolve substances that are insoluble in water. Many of these solvents are toxic. It can be difficult to remove all traces of the toxic solvent from reaction products and safely recycle or dispose of the solvent. Supercritical carbon dioxide can replace some organic solvents.

A gas becomes a supercritical fluid at a temperature and pressure called its critical point. For carbon dioxide, this occurs at 31.1°C and about 100 atmospheres. At its critical point, carbon dioxide is in a hybrid state. It has a high density (like a liquid) but it is easily compressed (like a gas). Many organic compounds dissolve in supercritical carbon dioxide. The solvent is easily separated from a reaction mixture because it evaporates at room temperature and pressure. It is also used to separate substances from mixtures. It can extract caffeine from coffee beans, dry-clean clothes, or clean circuit boards. ■

Decaffeinating Coffee

Caffeine dissolves inside coffee beans soaked in water.

↓ Extraction

Caffeine diffuses into supercritical CO_2.

↓ Absorption

Water droplets leach caffeine from CO_2.

After the caffeine is extracted, the coffee beans are dried and roasted. The aqueous solution of caffeine is sold to soft drink manufacturers.

Si Optical Glass

Glass is a material with the structure of a liquid, but the hardness of a solid. In most solids, the particles are arranged in an orderly lattice. In solid glass, the molecules remain disordered, as in a liquid. The main ingredient in most glass is silica (SiO_2), which is one of the few substances that can cool without crystallizing.

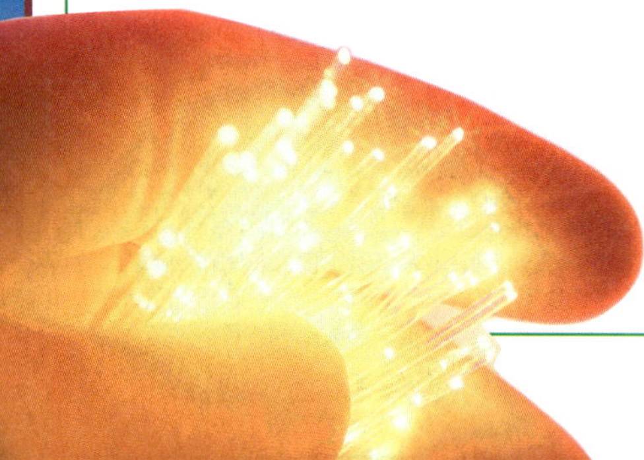

The glass used in eyeglasses, microscopes, and telescopes is called optical glass. It is purer than window glass and transmits more light. Optical glass can be drawn into long fibers that are used like tiny periscopes to view tissues deep within the human body.

In an optical fiber, light travels through a thin glass center called the core. A second glass layer reflects light back into the core. An outer plastic layer protects the fiber from damage.

When the fibers are bundled into cables, they often replace electrical cables in computer networks. They are also used to transmit television signals and phone calls over long distances. ■

Optical Glass

A light signal can travel a great distance through the core of an optical fiber because the cladding doesn't absorb any light. Impurities in the cladding cause the signal to degrade, which is why the glass must be optical glass. Optical fibers are cheaper than copper wires of equivalent length. Because an optical fiber can have a smaller diameter than a copper wire, more fibers can be bundled into a cable with the same diameter.

To Do

Students may want to research what is happening in eyeglass lenses that darken when exposed to light. When lenses containing silver chloride are exposed to bright light, the silver chloride breaks down, releasing opaque silver atoms. These atoms reduce the amount of light that can penetrate the lens. When the amount of available light is reduced, the silver and chlorine atoms re-associate and the glass becomes clear again.

Si Semiconductors

With a cellular phone, you can call your friends from almost any location. You can play a video game, read e-mails, or get the latest news. You may even be able to take and send digital photos. How can such a complex device be small enough to fit in your pocket? Semiconductor technology is responsible.

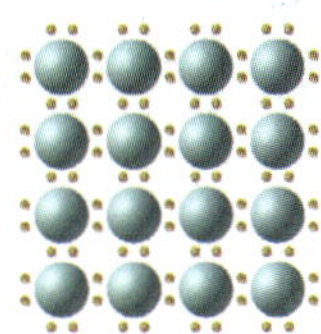
Pure silicon

Silicon is a semiconductor. In its pure form, it conducts an electric current better than most nonmetals but not as well as metals. But its ability to conduct can be changed dramatically by doping, or adding traces of other elements, to the silicon crystal.

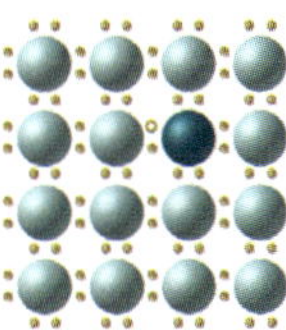
p-type (with boron)

Doping with arsenic produces a donor, or n-type, semiconductor. Each arsenic atom has five valence electrons, compared with four for silicon. So there are extra electrons in the crystal. Doping with boron produces an acceptor, or p-type, semiconductor. Because boron has only three valence electrons, there is a positive "hole" in the crystal for every boron atom. The extra electrons or holes are free to move and conduct an electric current.

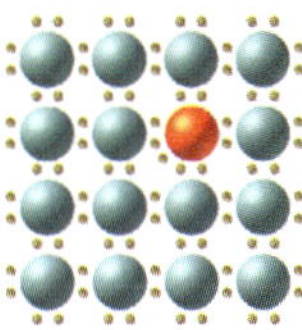
n-type (with arsenic)

Combinations of n-type and p-type semiconductors are used to build tiny electronic components. An integrated circuit containing millions of components can fit on a semiconductor wafer that is smaller than a fingernail! The resulting "chip" can control a computer, portable CD player, calculator, or cellular phone. ■

A cell phone's integrated circuit must process more and more data as features are added to the phone.

C Beyond Buckyballs

Buckminsterfullerene (C_{60}) is one member of a family of fullerenes. These structures are closed-cage spherical or nearly spherical forms of elemental carbon. The cages are networks of 20 to 600 carbon atoms.

Scientists have verified the existence of nesting spheres of fullerenes. C_{60} can be nested inside C_{240}, and this pair can be nested inside C_{540}. These nesting structures are sometimes called bucky-onions because they resemble the layers of an onion. ■

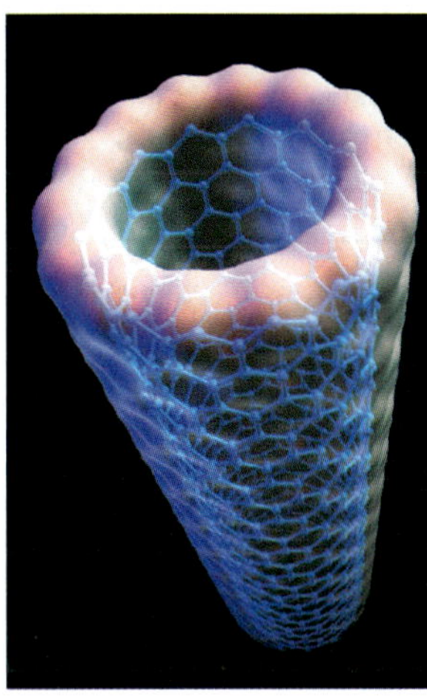
Dr. Sumio Iijima discovered a tubular fullerene, or carbon nanotube, in Japan in 1991.

Did You Know...

One name for diamonds is "ice." A diamond can quickly draw heat from your hand when you touch it. Such a high thermal conductivity is unusual for a substance containing covalent bonds.

Semiconductors

The essential components of a cell phone are a circuit board, antenna, liquid crystal display, keyboard, microphone, speaker, and battery. On the circuit board are analog-to-digital and digital-to-analog chips to translate the outgoing and incoming signals. There is a system processor that performs millions of calculations per second for the manipulation of the signals and a microprocessor that handles the keyboard, display, and all other functions.

Beyond Buckyballs

The intensive properties of a material are degraded by structural defects. For example, if there is a tiny crack in a steel wire, the strength of the wire will be significantly reduced. Because tubular fullerenes are molecules, they don't have the structural defects associated with multi-wall carbon fibers. With molecules, is it also possible to reliably predict chemical changes.

Group 4A

Greenhouse Gases

The Intergovernmental Panel on Climate Change (IPCC) was founded in 1988 to review the evidence and assess the impact of climate change. The IPCC published major reports in 1990, 1995, and 2001 and is planning another major report for 2007. The IPCC concluded that Earth's temperature increased by 0.6°C since the late 1880s and is likely to increase by another 1.9–2.9°C by 2100.

The source for the data in the graph is the Goddard Institute for Space Studies. The 10 warmest years in the 1900s occurred in the last 15 years of the century. 1998 was the warmest year since scientists began recording global mean surface temperatures. During the 1900s, sea level rose 4–8 inches, and there was a decrease in both snow cover in the Northern Hemisphere and floating ice in the Arctic Ocean. Precipitation over land increased worldwide by about 1%.

Carbon dioxide is not the only greenhouse gas. Methane (CH_4), nitrous oxide (N_2O), sulfur hexafluoride (SF_6), and chlorofluorocarbons (CFCs) exist in much smaller quantities but produce a more powerful greenhouse effect than CO_2. For example, PFCs have a global warming potential of up to 7000 times that of CO_2. The atmospheric content of these gases is also rising as a result of human activities.

Group 4A

C Greenhouse Gases

The glass in a greenhouse traps infrared radiation.

There are gases in Earth's atmosphere that are called greenhouse gases because they act like the glass in a greenhouse. Sunlight easily passes through these gases to Earth's surface. Some of the solar energy is reflected off the surface as infrared radiation. This radiation is absorbed by greenhouse gases and radiated back to Earth. By trapping infrared radiation, the greenhouse gases keep Earth's surface about 33°C warmer than it would be otherwise.

Carbon dioxide (CO_2) is the most abundant greenhouse gas. It is released into the air as a product of cellular respiration and removed from the air during photosynthesis. Such interactions normally keep the amount of atmospheric CO_2 in check. But the burning of fossil fuels releases more than 20 billion metric tons of CO_2 every year. Also, as forests are cleared for agriculture, the ability of plants to remove CO_2 from the atmosphere is reduced.

Scientists agree that a rise of only a few degrees in Earth's temperature could cause problems. They disagree on how severe the problems could be. Could climates change so that farmlands become deserts? Could the melting of ice caps cause sea levels to rise until coastal cities are under water? ■

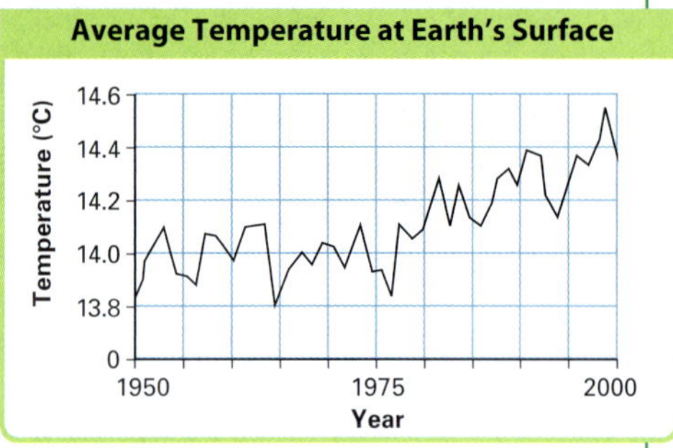

With an increase in greenhouse gases, more infrared radiation is trapped, causing global warming.

Si Composite Materials

Most composites contain two distinctly different materials. The materials can be arranged in layers as when a sheet of plastic is sealed between panes of glass. Or a composite may consist of a matrix in which fibers of a second material are embedded. Often, the matrix is plastic. The fibers can be carbon.

When a hockey stick has a composite shaft, there can be a quicker release of the shot.

Polycrylonitrile (PAN) is used to make some carbon fibers. PAN fibers oxidize when they are heated in air. The chains of aromatic rings that result are then heated in the absence of air. The chains fuse into long thin ribbons of almost pure carbon, which pack together in stacked layers.

Composites reinforced with carbon fibers are stronger than steel, yet light in weight. These composites are used in sports equipment, such as hockey sticks and golf clubs. It is less tiring to swing a tennis racket made from a carbon-fiber composite than one made from wood or metal. Carbon-fiber baseball bats act more like wood bats than do aluminum bats. ■

Composite Materials

Composites are heterogeneous mixtures. Most composites consist of a binder, or matrix, and a reinforcement, which is stronger and stiffer than the matrix. The matrix holds the reinforcement, which can be in the form of particles or fibers, together. In concrete, cement is the matrix and gravel the reinforcement.

To Do

Students who play a string instrument may want to research and report on the properties of bows made from carbon-fiber composites. (If students research composites they are likely to see the terms *carbon fiber* and *graphite fiber* used interchangeably. They are also likely to see *bonded* used to describe how the reinforcement is held in a matrix.)

C Recycling Plastics

At about 2 kg of waste per person per day, the United States leads the world in the production of solid waste. Luckily, the United States is also a leader in recycling. It is important to recycle plastics because they are made from crude oil, a non-renewable resource. Also, some plastics release toxic gases, such as hydrogen cyanide (HCN) and hydrogen chloride (HCl), when they burn in an incinerator. Finally, plastics are used for packaging material because they do not decay when exposed to sunlight, water, or microorganisms. The downside of this resistance to decay is that plastics can remain unchanged in dumps and landfills for decades.

Plastics are usually sorted by type before they are melted and reprocessed. The plastics industry has a code to identify common types of plastics. The numeral 1 is assigned to polyethylene terephthalate (PET), which is used in soft-drink bottles. The numeral 2 refers to high-density polyethylene (HDPE), which is used in milk jugs and shampoo bottles. These are the two types in greatest demand. Carpets and clothing are made from recycled PET fibers. Recycled HDPE is used as a wood substitute for decks and benches. ■

Fleece clothing often contains post-consumer recycled (PCR) plastic.

Si Silicone Polymers

If you have worn hard contact lenses or used shaving cream, you have used a silicone. Silicone polymers have chains in which silicon and oxygen alternate. The properties of silicones depend on the groups bonded to the silicon atoms and the length of the chains.

In silicone rubber and resins, there are cross-links between the chains. These silicones repel water and remain elastic, even at low temperatures. They are used in space suits, as gaskets in airplane windows, and as sealants that are squeezed into place and left to harden. ■

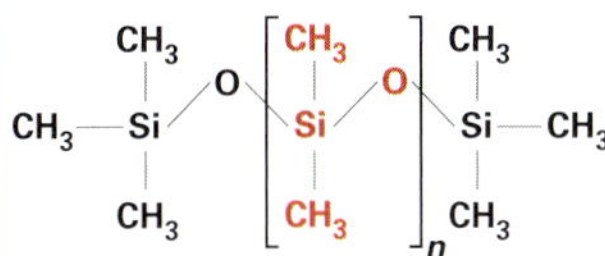

In polydimethylsiloxane, two methyl groups are bonded to each silicon atom in the chain. Polydimethylsiloxane is used as a lubricant in skin and suntan lotions.

C Carbon Monoxide

It is hard to detect colorless, odorless carbon monoxide gas. When it is inhaled, its molecules bind to the hemoglobin in red blood cells. They bind about 200 times more effectively than oxygen molecules do. So less oxygen reaches body tissues. Headaches, dizziness, nausea, and drowsiness are symptoms of low-level carbon monoxide poisoning. Higher levels of carbon monoxide are fatal.

The incomplete combustion of fuel in gas furnaces and space heaters produces carbon monoxide. It also forms in internal combustion engines that are not well maintained. In the United States, cars have catalytic converters, which convert carbon monoxide to carbon dioxide.

Cigarette smoke contains carbon monoxide. The carbon monoxide from one cigarette can remain in a smoker's blood for several hours. Smoking increases the risk of heart attacks because the heart must pump harder to deliver oxygen to cells when the level of oxygen is reduced. ■

Did You Know...

Members of the Scott expedition to the South Pole in 1912 may have died because of **tin**. Their supply of paraffin fuel leaked out through tiny holes in the tin-soldered joints of the storage cans because tin slowly changes to a powder below 13°C.

Recycling Plastics

The numeral 3 is used for polyvinyl chloride, the numeral 4 for low-density polyethylene (LDPE), the numeral 5 for polypropylene, the numeral 6 for polystyrene, and the numeral 7 for all other plastics. As of 2001, the recycling rates for commonly recycled items were 70.1% for corrugated cardboard boxes, 60.2% for newspapers, 58.1% for steel cans, 56.5% for yard trimmings, 49.0% for aluminum beverage cans, 38.6% for tires, 35.6% for plastic soft-drink bottles, 32.0% for magazines, 35.6% for plastic milk and water bottles, and 22.0% for glass containers.

To Do

Students could research recycling programs in your community. They could find out which items are recycled and where these items are collected (curbside or at designated drop-off sites).

Carbon Monoxide

A 2002 report by the Arizona Department of Environmental Quality on air quality in the Greater Phoenix area showed a significant reduction in carbon monoxide between 1990 and 2000 from 13.0 ppm to 5.3 ppm. The report attributed most of the reduction to new emission standards, to a vehicle inspection program, and to the use of oxygenated fuels in the winter.

The Consumer Product Safety Commission recommends that households install a CO_2 detector with an audible alarm near bedrooms and a second monitor near the heating source.

To Do

Students might want to research the relationship between the level of carbon monoxide in the air near highways and emissions from vehicles. In urban areas, the concentration tends to peak during morning and afternoon rush hours. Carbon monoxide tends to accumulate more in cold weather when engines run less efficiently. Other factors to consider are the presence or absence of wind to disperse pollutants or the presence of inversion layers in the atmosphere.

Silicone Polymers

Students will also be familiar with some properties of silicone polymers if they have bounced a superball, stretched Silly Putty, removed a peel off label, placed weather-stripping caulk around a door frame, polished a car, or ingested a remedy for an upset gaseous stomach.

Group 5A

Physical Properties

Cryogen comes from the Greek *kryos*, meaning "icy cold." Foods can be packed, sealed, and sprayed with liquid nitrogen, which absorbs heat as it evaporates. Rapid freezing is also used to preserve tissue samples, such as blood and bone marrow. Surgeons use liquid nitrogen to remove warts and skin cancers.

Sources

Roasting involves heating an ore to a high temperature in a furnace, which causes a reaction of the ore with a gaseous component of the atmosphere in the furnace. The oxidation of sulfide ores is an example of roasting. The products of roasting will be a metal oxide and sulfur dioxide or the elemental metal and sulfur dioxide (depending on the activity of the metal). Roasting of antimony produces the oxide, which can then be reduced with carbon. A flux of sodium sulfate is used to prevent evaporation of molten antimony.

$$2Sb_2S_3 + 9O_2 \rightarrow 2Sb_2O_3 + 6SO_2$$
$$2Sb_2O_3 + 3C \rightarrow 4Sb + 3CO_2$$

Group 5A

7 N Nitrogen 14.007

Nitrogen | discovered in 1772 by Daniel Rutherford

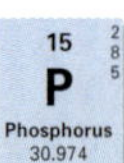

Phosphorus | discovered in 1669 by Hennig Brand

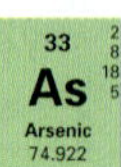

Arsenic | discovered in 1250 by Albertus Magnus

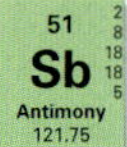

Antimony | discovered in 1600s or earlier

Bismuth | described in 1450 by Basil Valentine; shown to be a distinct element in 1753 by Claude-François Geoffroy

Physical Properties

- Except for nitrogen gas, Group 5A elements are solid at room temperature.
- The metallic properties of Group 5A elements increase from top to bottom within the group. N and P are nonmetals. As and Sb are metalloids. Bi is a metal.
- Liquid nitrogen is a cryogen, a liquid refrigerant that boils below −160°C.

White phosphorus is very reactive and must be stored under water. Red phosphorus is much more stable.

- The 10 allotropes of phosphorus are classified as white, red, or black.

Melting and Boiling Points

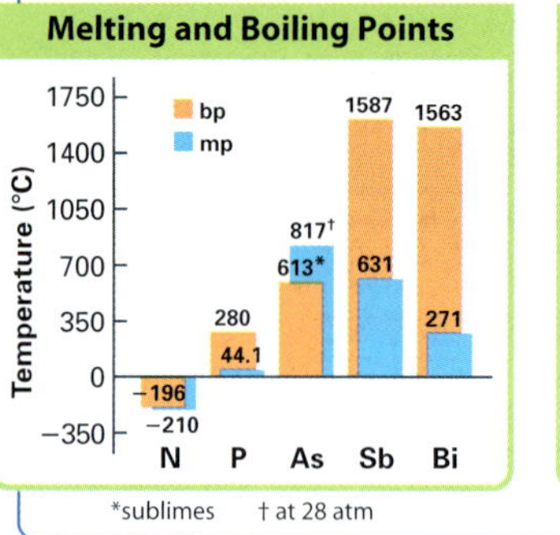

*sublimes † at 28 atm

Density

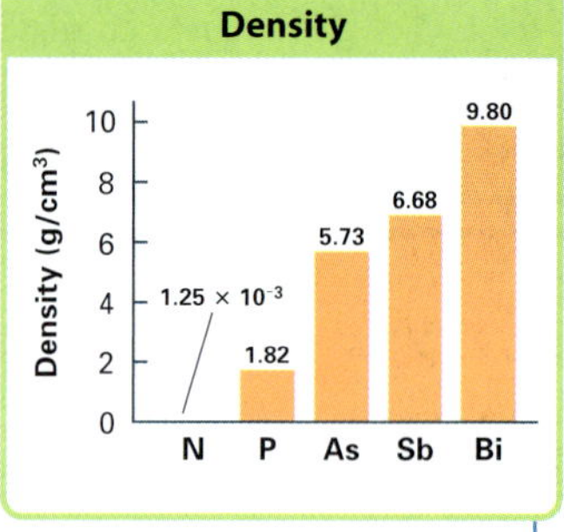

Sources

- Nitrogen is obtained from the fractional distillation of liquefied air.
- Phosphorus is derived from phosphate minerals. Example:
  $$Ca_3(PO_4)_2(s) + 3SiO_2(s) + 5C(s) \longrightarrow 2P(l) + 3CaSiO_3(s) + 5CO(g)$$
- Arsenic is prepared by heating a mixture of $FeAs_2$ and FeS_2 in the absence of air.
- Antimony is prepared by roasting the ore stibnite, Sb_2S_3.

Fractional Distillation of Air

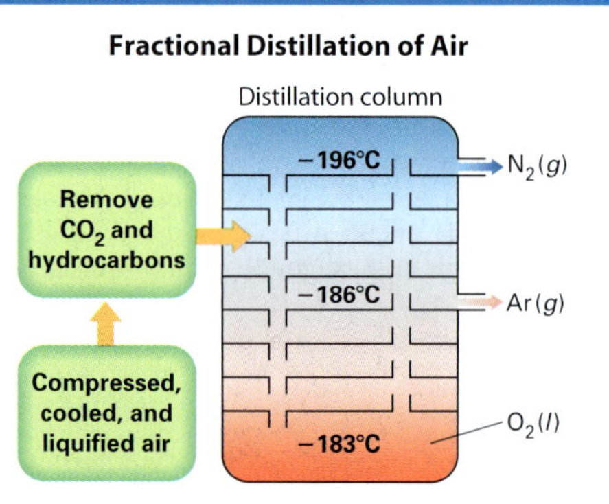

Atomic Properties

- Group 5A elements have an electron configuration that ends in ns^2np^3.
- The most common oxidation numbers for Group 5A elements are +3, +5, and −3.
- Nitrogen has oxidation numbers from −3 to +5 in a variety of stable compounds.
- Elemental nitrogen, N_2, is highly unreactive due to its strong N-to-N triple bond.

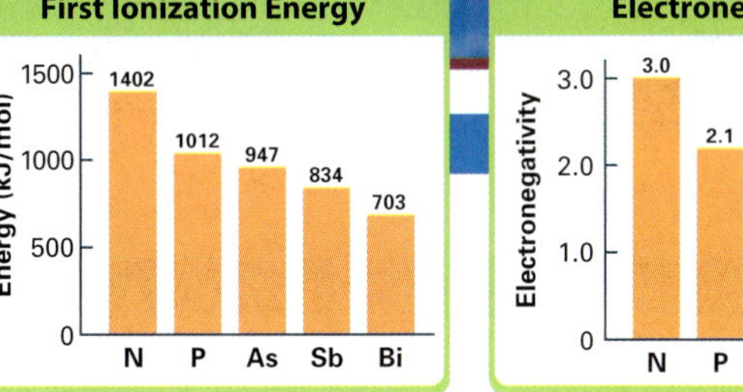

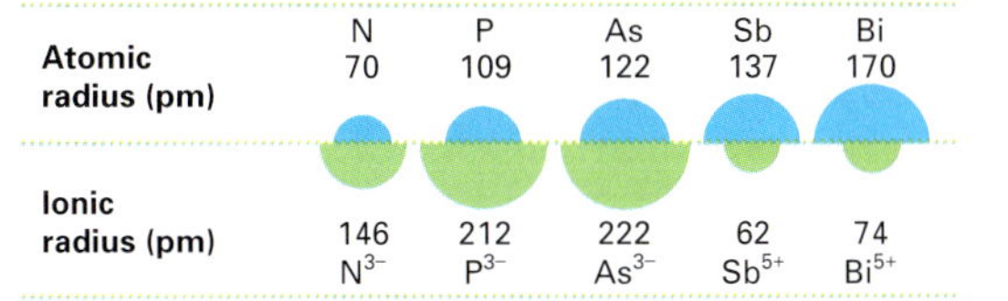

Important Compounds and Reactions

- Nitrous oxide, N_2O, (laughing gas) is an anesthetic. It is made from ammonium nitrate.

  $NH_4NO_3(s) \xrightarrow{\Delta} N_2O(g) + 2H_2O(g)$

- Nitrogen dioxide, NO_2, is an air pollutant produced when fossil fuels burn.
- Lightning causes the nitrogen and oxygen in air to react and form nitric oxide, NO.
- Poisonous hydrazine, N_2H_4, is used as a rocket fuel. It is prepared by the following reaction.

  $2NH_3(aq) + OCl^-(aq) \longrightarrow N_2H_4(aq) + Cl^-(aq) + H_2O(l)$

  The reaction is complex. One intermediate is chloramine, NH_2Cl, which is also poisonous. If you ignore the warning against mixing household ammonia and chlorine bleach, chloramine will be produced.

- Nitric acid, HNO_3, is used in fertilizers and explosives. Nitric acid is made by the Ostwald process.
- Phosphoric acid is used in soft drinks and fertilizers. It is made by a double-replacement reaction.

  $Ca_3(PO_4)_2(s) + 3H_2SO_4(aq) \longrightarrow 3CaSO_4(aq) + 2H_3PO_4(aq)$

  Dentists use phosphoric acid to etch the exposed dentine layer of a drilled tooth to help a filling adhere to the tooth.

- Ammonia is synthesized directly from its elements by the Haber-Bosch process.

  $3H_2(g) + N_2(g) \longrightarrow 2NH_3(g)$ $\Delta H = 46.19$ kJ/mol

- Amino acids such as glycine, H_2NCH_2COOH, are the building blocks of proteins.
- Bismuth subsalicylate, $BiO(C_7H_5O_3)$, is the active ingredient of a pink liquid antacid.
- Arsenic trioxide, As_2O_3, is powerful poison. It is used as a weed killer and insecticide.
- Antimony is added to alloys to increase their hardness.

Ostwald Process

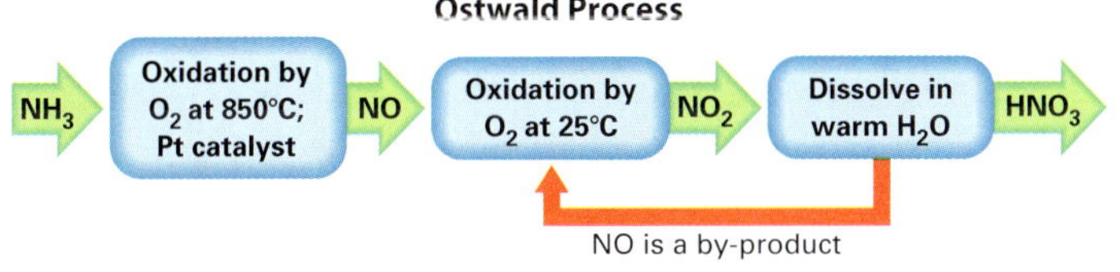

Ammonia is oxidized to nitrogen dioxide in two stages. The nitrogen dioxide forms nitric acid as it passes through warm water.

Atomic Properties

The table compares the bond lengths and bond dissociation energies for single, double, and triple nitrogen-nitrogen bonds.

Bond	Length	Dissociation Energy
single	1.47 Å	145 kJ/mol
double	1.24 Å	456 kJ/mol
triple	1.10 Å	945 kJ/mol

To Do

Students can use the bond lengths and bond dissociation energies to show the relationship between the number of shared electron pairs, bond length, and bond strength.

Important Compounds and Reactions

The approach to healing by traditional physicians during the renaissance was based on the existence of four humors (blood, phlegm, yellow bile, and black bile). Most traditional remedies were derived from plants. Apothecaries and other followers of Paracelsus used inorganic compounds as treatments. For example, a mixture used to treat fever contained powdered nitre (potassium nitrate), carbonate of potash (potassium carbonate), antimonial wine (hydrated potassium antimonyl tartrate), and sweet spirits of nitre (ethyl nitrate).

To Do

Help students write balanced equations for the reactions in the Ostwald Process.

$4NH_3(g) + 5O_2(g) \longrightarrow 4NO(g) + 6H_2O(g)$

$2NO(g) + O_2(g) \longrightarrow 2NO_2(g)$

$3NO_2(g) + H_2O(l) \longrightarrow 2H^+(aq) + 2NO_3^-(aq) + NO(g)$

Group 5A

Ammonia

Haber reacted three-to-one molar mixtures of hydrogen and nitrogen at different conditions of temperature and pressure. The yield of ammonia at equilibrium decreased with an increase in temperature and increased with an increase in pressure. Because the boiling point of ammonia (–33°C) is much higher than the boiling points of nitrogen (–196°C) or hydrogen (–253°C), it is possible to remove the ammonia from the equilibrium mixture and drive the reaction to completion.

The Haber process is sometimes called the Haber-Bosch process. Karl Bosch implemented Haber's process on an industrial scale. It was Bosch who found a catalyst that was cheaper and more available than the osmium and uranium that Haber had used. His team had to develop a furnace that could operate safely at the required temperature and pressure. In 1931, Bosch and Friedrich Bergius were awarded a Nobel Prize for the "invention and development of chemical high pressure methods."

Before the Haber Process, deposits of dried guano in Chile were the main source of fixed nitrogen. (The guano was deposited by fish-eating sea birds.) During World War I, a naval blockade of South America prevented Germany from importing the nitrates it needed to manufacture explosives. It has been argued that World War I would have ended sooner if Haber had not developed his process. Because Haber (as chief of Germany's Chemical Warfare Service) permitted the use of chlorine gas as a weapon, the awarding of his Nobel Prize in 1918 was controversial.

Group 5A

N Ammonia

The nitrogen in air is not in a form that is useful for plants. Natural sources of usable nitrogen in soil aren't sufficient to support current levels of crop production. A German chemist named Fritz Haber came up with a solution. He figured out how to convert atmospheric nitrogen into ammonia. In the Haber process, nitrogen and hydrogen are heated under pressure in the presence of iron.

$$N_2(g) + 3H_2(g) \longrightarrow 2NH_3(g)$$

The ammonia is liquefied. Liquid ammonia, aqueous solutions of ammonia, and ammonium salts are used as fertilizers. Liquid ammonia is also used as a refrigerant. Many cleaning products contain aqueous ammonia, which is a weak base. Ammonia is also used to manufacture explosives. ■

Haber Process

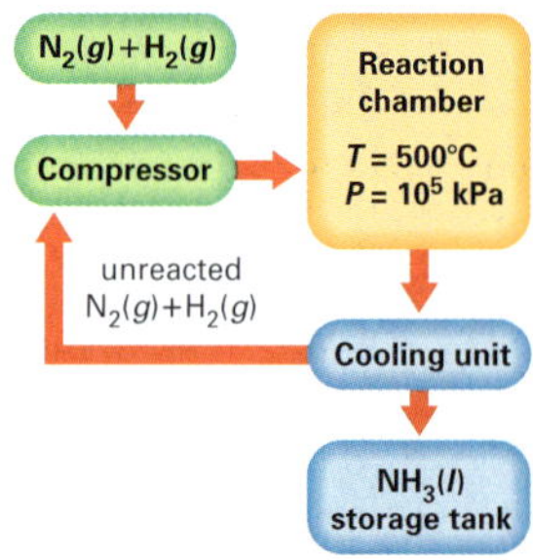

When the reaction mixture is cooled, the ammonia is liquefied and separated from the reactants.

N Explosives

Explosions are extremely rapid exothermic reactions that produce gaseous products. The most forceful explosions occur when the reactants are liquids or solids. Pressure builds up as the reactants are converted to gases, especially if the gases are confined. When the gases expand, the resulting release of pressure causes a shock wave. The kinetic energy of this wave, the wind that follows it, and the heat is the destructive force of the explosion.

An explosion requires fuel and an oxidizer. If the oxidizer and fuel are separate, the reaction is a combustion reaction. This type of explosion could occur if natural gas leaking in a building was ignited by a spark or flame.

$$CH_4 + 2O_2(g) \longrightarrow CO_2(g) + 2H_2O(g) \qquad \Delta H = 890 \text{ kJ/mol}$$

If the fuel is its own oxidizer, a decomposition reaction takes place. Nitroglycerine, $C_3H_5(NO_3)_3$, is a thick, pale, oily liquid. It decomposes to form a mixture of gaseous products.

$$4C_3H_5(NO_3)_3(l) \longrightarrow 6N_2(g) + O_2(g) + 12CO_2(g) + 10H_2O(g) \qquad \Delta H = -1427 \text{ kJ/mol}$$

Self-oxidizing explosives are often unstable. Jarring them may be enough to cause detonation. Alfred Nobel, a Swedish chemist and inventor, found a way to use nitroglycerine with less risk. After his family's nitroglycerine factory exploded in 1864, he moved his experiments to a barge in the middle of a lake. One day he found a cask of nitroglycerine that had leaked. Luckily, the diatomaceous earth in which the cask was packed had absorbed the liquid. Nobel found that the mixture was stable until detonated by a blasting cap. Plus, the mixture was as explosive as the pure liquid. Nobel named his invention dynamite. ■

In his will, Nobel established a fund to provide annual prizes in chemistry, physics, physiology and medicine, literature, and peace.

Explosives

An explosion is highly exothermic when the bonds in the explosive are relatively weak and the bonds in the products are relatively strong. Because the bonds in N_2, CO, and CO_2 are very strong, many explosives are organic molecules with nitro or nitrate groups, which can react to produce N_2, CO, or CO_2.

Nitroglycerin is used to treat angina. Its effect is to cause blood vessels to dilate. The nitroglycerin used in pills, sprays, and patches is very dilute. Nitroglycerin has a short shelf life. It is unstable and must be stored in containers that prevent exposure to light.

N Acid Rain

Normal rainfall has a pH of about 5.6. It is mildly acidic because carbon dioxide in the air dissolves in tiny water droplets and forms carbonic acid (H_2CO_3). In acid rain, the pH is lower due to the emission of nitrogen oxides and sulfur oxides into the atmosphere. Some natural sources, such as volcanoes, emit these oxides. But most come from power plants that burn fossil fuels. In the atmosphere, the oxides form nitric acid (HNO_3) and sulfuric acid (H_2SO_4), which fall to Earth in rain or snow.

Because acid rain leaches nutrients from soil, trees are dying in the Appalachian Mountains. Because aquatic species are highly sensitive to changes in pH, some lakes once full of fish and frogs are nearly lifeless.

Power plants are a main source of sulfur and nitrogen oxides. Chemists developed devices called scrubbers that remove these pollutants from smoke. In one type of scrubber, smoke passes through an aqueous suspension of lime (CaO). The lime reacts with SO_2 to form solid calcium sulfite ($CaSO_3$).

Another type of scrubber is used to remove nitrogen oxides. Ammonia is sprayed onto a surface covered with a mixture of catalysts. As smoke passes over the surface, nitrogen oxides react with ammonia and oxygen, forming nitrogen and water.

Researchers have developed some methods for removing both types of oxides at the same time. For example, the smoke can be passed through an alkaline suspension of yellow phosphorus. The chemical reactions that occur convert almost all the oxides into valuable by-products, including some of the plant nutrients used in fertilizers. ■

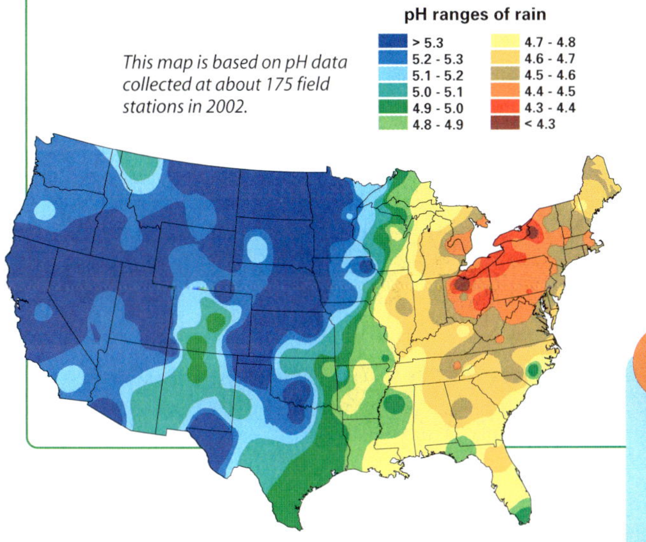

This map is based on pH data collected at about 175 field stations in 2002.

N P Fertilizers

Most fertilizers contain salts of nitrogen, phosphorus, and potassium. These elements are essential to plant growth. On the label, the percents by mass of these nutrients are always listed in the order N-P-K.

Because the salts in a fertilizer can vary, there are rules for reporting the content. Nitrogen is always reported as the percent by mass of elemental nitrogen. Phosphorus and potassium are reported as the percent by mass of phosphorus pentoxide, P_2O_5, and potassium oxide, K_2O.

This system makes it easy to compare the nutrient content of fertilizers. For example, a fertilizer labeled 20-10-10 has twice the mass of nitrogen as a fertilizer labeled 10-10-10. But they both contain the same mass of phosphorus and potassium. The numbers may not add up to 100% because fertilizers can contain ingredients not included in the N-P-K analysis. ■

Did You Know...

If nail polish or lipstick contain **bismuth** oxychloride (BiOCl), they appear lustrous and pearly.

Acid Rain

Acid rain has had a major impact on the ecology of lakes. Most lakes in which organisms live have a pH between 6.5 and 8.5. All of the vertebrates, most of the invertebrates, and many microorganisms cannot survive at a pH below 4.0. Lakes that are most likely to have a serious drop in pH due to acid rain are those with low concentrations of ions such as HCO_3^-, which can act as a buffer.

To Do

Have students use the isopleth map to determine the pH of rainfall in your region as of 2002. (An isopleth is a line connecting points on a graph or map that have equal values for a given variable.) Students can research trends by viewing the isopleth maps produced annually by the National Atmospheric Deposition Program (NADP).

Fertilizers

Mineral plant nutrients can be classified as primary nutrients (N, P, and K), secondary nutrients (Ca, Mg, and S), and micronutrients (B, Cu, Fe, Cl, Mn, Mo, and Zn). Primary and secondary nutrients tend to be less available to plants in soils with a low pH, but micronutrients tend to be more available to plants in soils with a high pH. Most plants do well in soil with a pH between 6.0 and 7.5.

To Do

Invite an expert to talk to the class about why fertilizers with different percent compositions are used with different types of plants or at different stages in plant growth.

Group 6A

Physical Properties

When molten sulfur cools below 119°C, it forms crystals of monoclinic sulfur. This allotrope contains eight-member rings of covalently bonded sulfur atoms (S_8). Below 95.5°C, the sulfur changes to rhombic sulfur, which is also composed of S_8 units. If molten sulfur is poured into cold water, an amorphous form of sulfur results. Within hours, the amorphous sulfur loses its elasticity as it is converted to rhombic crystals, which are more stable. Sulfur can be purified through boiling and condensation. Tiny rhombic crystals in flowerlike patterns form on the wall of the condensation chamber. The powder is known as flowers of sulfur.

Sources

Some of the world's richest deposits of elemental sulfur are on the Gulf Coast in Louisiana and Texas. These deposits are not easily mined because they are buried under several hundred feet of quicksand. Herman Frasch (1851-1914) devised an ingenious method for getting sulfur out of the ground. The method takes advantage of sulfur's low melting point.

Wells are drilled into the sulfur bed. Three concentric tubes (with 2.5-cm, 7.5-cm, and 15-cm diameters) are installed. Compressed air and superheated water are pumped through the inner and outer tubes. The molten sulfur that emerges from the center tube is pumped into large storage vats where it cools and solidifies into huge blocks, which can be 130 m long, 70 m wide, and 30 m high. Dynamite is used to break up the blocks of sulfur for shipping.

Group 6A

Oxygen | discovered in 1772 by Carl Scheele and in 1774 by Joseph Priestly

Sulfur | known since ancient times

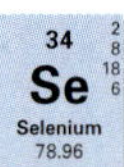

Selenium | discovered in 1817 by Jöns Jacob Berzelius

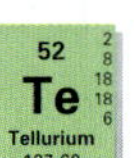

Tellurium | discovered in 1782 by Franz Joseph Müller von Reichenstein

Polonium | discovered in 1898 by Marie Curie

Physical Properties

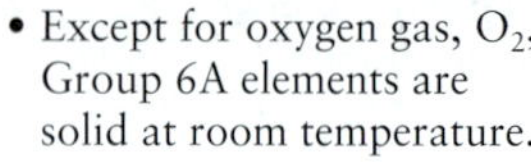

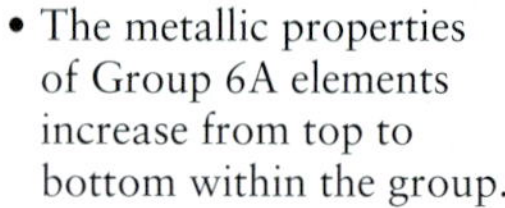

- Except for oxygen gas, O_2, Group 6A elements are solid at room temperature.
- The metallic properties of Group 6A elements increase from top to bottom within the group.
- Polonium is a radioactive metal (half-life = 140 days).

monoclinic *rhombic*

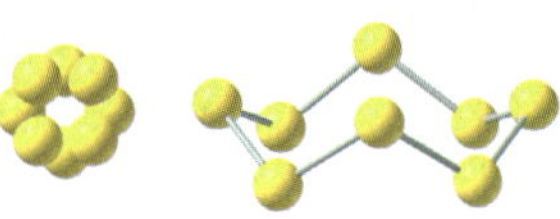

The unit cell in crystalline sulfur is an S_8 molecule.

Melting and Boiling Points

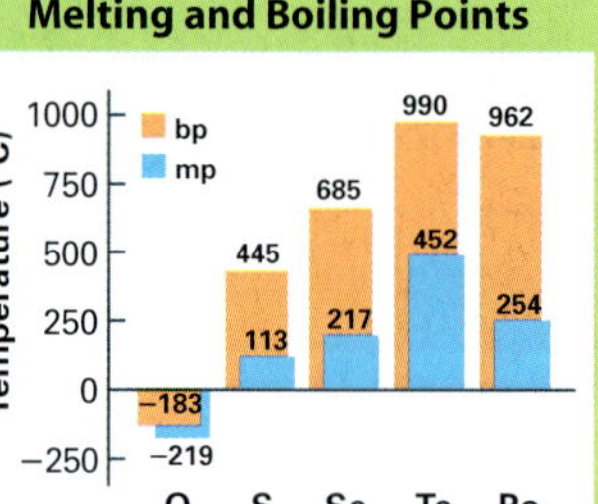

Density

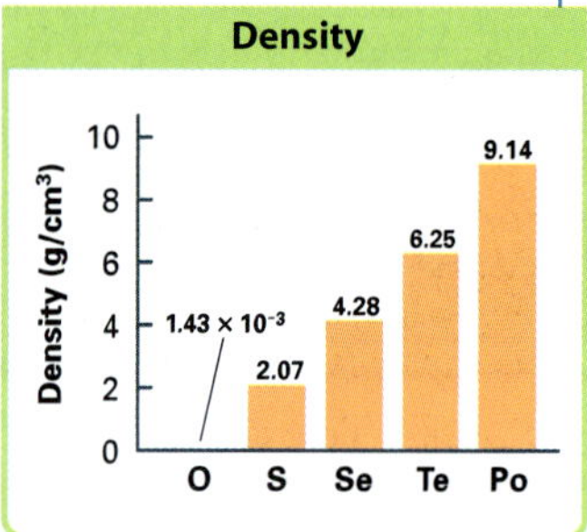

Sources

- Large-scale production of oxygen is by fractional distillation of liquid air. Liquid oxygen is stored and shipped at its boiling point of −183°C in vacuum-walled bottles.
- The Frasch process is used to mine sulfur from underground deposits. A well is drilled into a sulfur bed and a set of concentric tubes installed. Superheated water melts the sulfur. Compressed air forces it to the surface.
- Sulfur is also produced from hydrogen sulfide, H_2S, and sulfur dioxide, SO_2.

$$2H_2S(g) + SO_2(g) \longrightarrow 2H_2O(l) + 3S(s)$$

- Selenium and tellurium are by-products of the processing of sulfide ores for other metals.
- Polonium is formed by the radioactive decay of radium in minerals such as pitchblende.

The Frasch Process

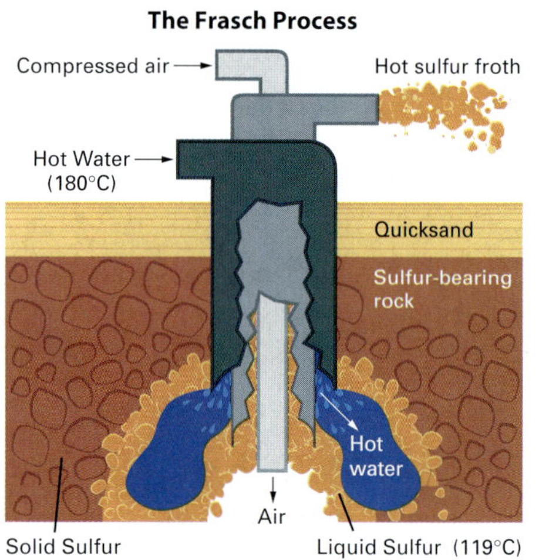

Atomic Properties

- Group 6A elements have an electron configuration that ends in ns^2np^4.
- For Group 6A elements, the most common oxidation numbers are +4, +6, and −2.
- Oxygen is paramagnetic because there are unpaired electrons in O_2 molecules.

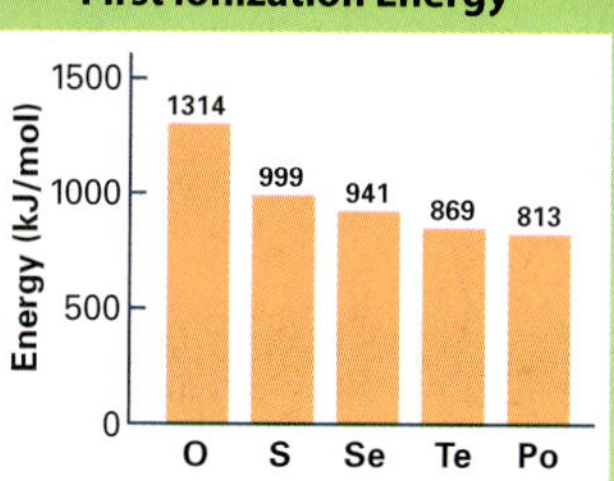

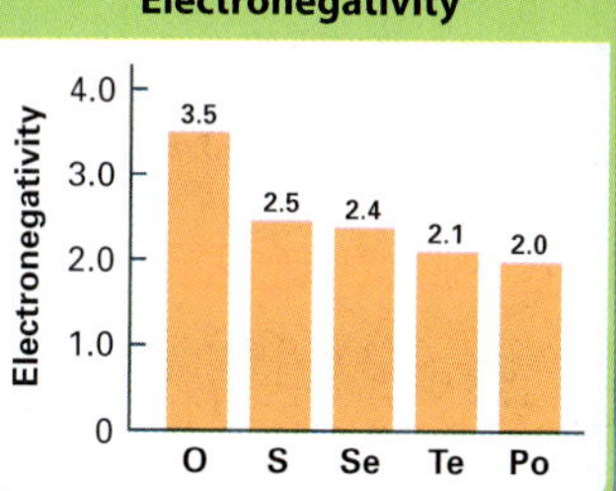

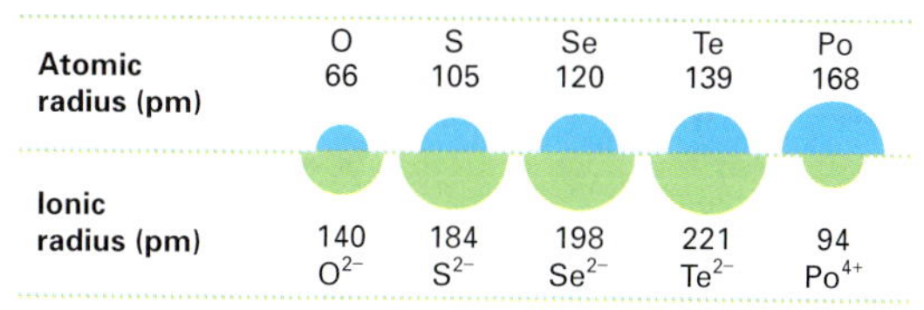

Liquid oxygen is held between the poles of a magnet because of its attraction to the magnet.

Important Compounds and Reactions

- Oxygen reacts with almost all other elements to form oxides. Example:

  $4K(s) + O_2(g) \longrightarrow 2K_2O(s)$ $\quad \Delta H = -363.2$ kJ/mol
- Ozone, O_3, is produced directly from oxygen, O_2, during lightening strikes.

  $3O_2(g) \longrightarrow 2O_3(g)$ $\quad \Delta H = +285$ kJ/mol
- Oxygen is necessary for releasing energy from fuels, such as glucose, in organisms.

  $C_6H_{12}O_6(s) + 6O_2(g) \longrightarrow 6CO_2(g) + 6H_2O(g)$ $\quad \Delta H = -2808$ kJ/mol
- Oxygen is used to produce steel and to oxidize hydrogen in fuel cells.
- Sulfur compounds often have unpleasant odors. Hydrogen sulfide, H_2S, smells like a rotten egg. It forms when metallic sulfides and hydrochloric acid react. Example:

  $FeS(s) + 2HCl(aq) \longrightarrow H_2S(g) + FeCl_2(aq)$

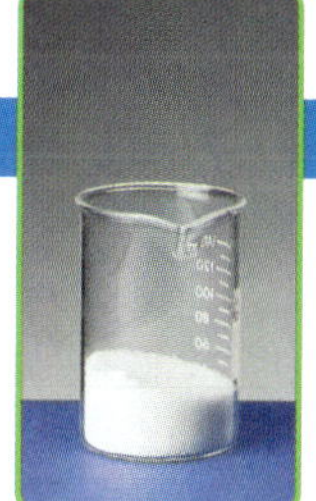

When concentrated sulfuric acid is added to sucrose, water vapor (steam) and carbon are produced. The release of the vapor causes the carbon to expand.

- Concentrated sulfuric acid, H_2SO_4, is a strong dehydrating agent. Example:

  $C_{12}H_{22}O_{11}(s) \xrightarrow{H_2SO_4} 12C(s) + 11H_2O(g)$
- So people will know when there is a natural gas leak, ethyl mercaptan, CH_3CH_2SH, is added to supplies of odorless natural gas.
- Sodium thiosulfate, $Na_2S_2O_3$, also known as *hypo*, is used in the development of film.
- The addition of cadmium selenide, CdSe, gives glass a beautiful ruby color.

Atomic Properties

Students may think that only metals can be affected by a magnetic field. Explain that substances in which all electrons are paired are diamagnetic, i.e., weakly repelled by an external magnetic field. Substances in which there are unpaired electrons are paramagnetic, i.e., attracted to an external magnetic field. The more unpaired electrons, the stronger the force of attraction.

Important Compounds and Reactions

Concentrated sulfuric acid dissolves readily in water. The process is highly exothermic. The ΔH_{soln} for sulfuric acid is −95.28 kJ/mol. The strong affinity of sulfuric acid for water makes sulfuric acid a powerful dehydrating agent, as demonstrated by its effect on sucrose.

Each odorant has a detection threshold, or concentration at which 50% of people can identify the presence of the odor. Ethyl mercaptan's detection threshold is 0.0026 ppm, which is relatively low. Because olfactory acuity, or the ability to detect odors, tends to decrease with age, many elderly people are unable to detect ethyl mercaptan at its concentration in natural gas. Sulfur compounds with unpleasant smells are generally classified as putrid.

Group 6A

Selenium in Food

Keratin is the principal protein in the hoofs and hair of horses, as it is in human hair. Sulfur-containing thiol groups in the amino acid cysteine form the disulfide bonds between adjacent protein chains. Selenium can be substituted for sulfur when present in excess. Chronic selenium poisoning in horses is often called "bob-tail disease" because hair in the mane and tail breaks at the location in the hair shaft where selenium is substituted for sulfur.

Sulfuric Acid

For a long time, researchers were unable to determine how the conversion of sulfur dioxide to sulfur trioxide, which is normally a slow process, occurred so rapidly in air. Research has since shown that dust and other solid particles in the air act as a catalyst for the process.

Group 6A

Se Selenium In Food

Selenium is an antioxidant that protects cell membranes from damage. In the United States, grain is grown on selenium-rich soils. Livestock and people who eat those grains are unlikely to be deficient in selenium.

An essential nutrient may be harmful in large doses. Too much selenium can damage the nervous system. It may also cause anxiety and fatigue.

This milkvetch (Astragalus bisulcatus) accumulates high levels of selenium.

Soils in the Great Plains and Rocky Mountain regions often contain high levels of selenium. If cattle graze on plants that grow in those soils, they may develop chronic selenium poisoning. The symptoms include loss of hair, sore hoofs, lameness, and a lack of energy.

Acute selenium poisoning causes cattle to lose their vision and stumble aimlessly before dying from respiratory failure. Cowboys called this condition the "blind staggers." ■

S Sulfuric Acid

Pure sulfuric acid is a dense, colorless, oily liquid. Concentrated sulfuric acid is 98% H_2SO_4 and 2% H_2O. Dilute sulfuric acid reacts with metals, oxides, hydroxides, or carbonates to form sulfates. The reaction with metals also releases hydrogen gas. Sulfuric acid can be used to produce other acids from their salts. For example, hydrogen chloride can be produced from sulfuric acid and sodium chloride.

$$H_2SO_4(l) + 2NaCl(s) \longrightarrow Na_2SO_4(s) + 2HCl(g)$$

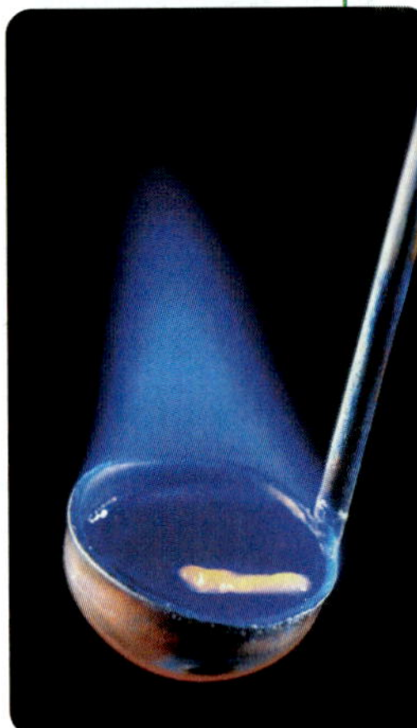

When sulfur burns in air, the product is the irritating gas sulfur dioxide, SO_2.

Sulfuric acid is produced mainly from sulfur dioxide. The process is called the contact process because the key reaction takes place when the reactants are in contact with the surface of the solid catalyst.

(1) Melted sulfur is burned in air.

$$S(l) + O_2(g) \longrightarrow SO_2(g)$$

(2) Sulfur dioxide is oxidized in the presence of a vanadium oxide catalyst, V_2O_5.

$$2SO_2(g) + O_2(g) \xrightarrow{V_2O_5} 2SO_3(g)$$

(3) Sulfur trioxide dissolves in water and forms sulfuric acid.

Much of the sulfuric acid produced in North America is used to make fertilizers. Sulfuric acid is also used in petroleum refining, the production of other chemicals, and for pickling iron and steel. During pickling, oxides are removed from the surface of a metal. ■

This sidegill slug produces sulfuric acid, which discourages predators.

O Ozone

Near Earth's surface, ozone (O_3) is a pollutant. In the stratosphere, ozone is literally a lifesaver. The ozone layer in the stratosphere absorbs 99% of the sun's harmful ultraviolet (UV) radiation. In the 1970s, scientists began to suspect that the ozone layer might be threatened. They based their concerns on laboratory models. In 1985, their suspicions were confirmed when British researchers discovered a "hole" in the ozone layer over Antarctica. In the winter, the amount of ozone was reduced by almost one half.

The scientists traced the thinning of the ozone layer to chlorofluorocarbons (CFCs). These chemicals were used mainly as propellants in aerosol spray cans and as coolants in refrigerators and air conditioners. CFCs are highly stable and inert in the lower atmosphere. Over time, they drift up into the stratosphere, where they are broken down by solar radiation. Through a repeated cycle of reactions (2) and (3), a single chlorine atom can destroy as many as 100,000 molecules of ozone.

(1) $CCl_3F \longrightarrow Cl\cdot + \cdot CCl_2F$

(2) $Cl\cdot + O_3 \longrightarrow ClO\cdot + O_2$

(3) $ClO\cdot + O \longrightarrow Cl\cdot + O_2$

In 1978, the United States banned the use of CFCs in aerosols. By 1990, 40 nations had agreed to eliminate production of CFCs. The ban has had an effect. Concentrations of CFCs have begun to level off in the stratosphere and even decline in the lower atmosphere. ■

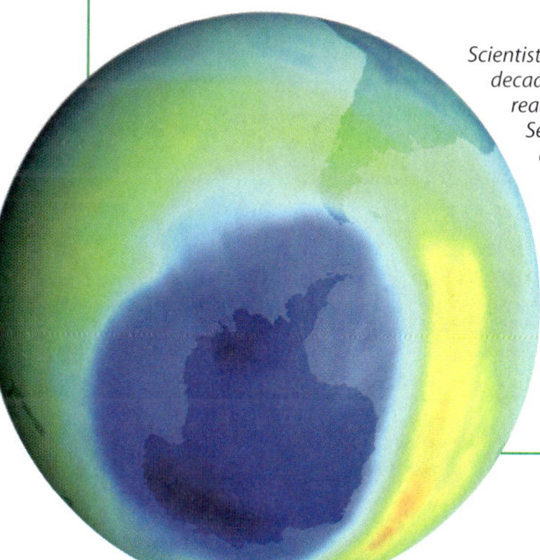

Scientists estimate that it may take decades for the ozone layer to reach pre-1980 levels. On September 9, 2000, the "hole" over Antarctica was record size. The amount of ozone is low in the blue areas.

S O Sulfites

Ancient Egyptians and Romans used sulfur dioxide (SO_2) to preserve food. When SO_2 reacts with water, hydrogen sulfite ions (HSO_3^-) and sulfite ions (SO_3^{2-}) form. The food industry uses the term *sulfite* to describe sulfur-based preservatives. Sulfites are antioxidants that slow down the discoloration or deterioration of food.

Restaurants used to spray fruit and lettuce in salad bars with sulfites to maintain the color of the fruit and keep the lettuce from wilting. Sulfites can cause asthma attacks and headaches. After some people died from severe reactions, the Food and Drug Administration banned the use of sulfites on fresh produce. Sulfites must be listed on the food label if a processed food contains at least 10 parts per million. ■

Did You Know...

Organic compounds containing **sulfur** give onions their taste and smell. When an onion is cut, reactions occur that produce propanethial-S-oxide. When receptors in your eyes are exposed to this irritating gas, they trigger the production of tears.

Group 6A **R31**

Ozone

As sunlight travels through the ozone layer, ozone absorbs light of known wavelengths. The amount of light of these wavelengths that passes through the atmosphere is an indication of the amount of ozone present. In 2003, researchers at the University of Alabama in Huntsville used such data collected by NASA Earth-observing satellites to show that the rate of ozone depletion in the upper stratosphere had slowed since 1997.

When CFCs are broken down by solar radiation, they release chlorine atoms. A chlorine atom can destroy as many as 100,000 molecules of ozone because it is a free radical. A free radical is very reactive because it contains one or more unpaired electrons.

To Do

Encourage interested students to do additional reading on the research that led to the Montreal Protocol, which established an international ban on CFC-based products.

Sulfites

The FDA lists 32 "functional effects" for which additives may be added to food, including antioxidants. Some other examples are anticaking agents; colors and coloring adjuncts; drying agents; curing and pickling agents; emulsifiers; flavor enhancers; humectants; leavening agents; pH control agents; stabilizers and thickeners; sequestrants; synergists; and texturizers.

To Do

Encourage interested students to check the ingredients on food labels and then research the functions of common food additives.

Group 7A: Halogens

Physical Properties

Astatine comes from the Greek *astatos*, meaning "unstable." It was synthesized by bombarding bismuth with alpha particles. Astatine-210 is the longest-lived isotope with a half-life of 8.3 hours. Despite the small quantities available for study, some properties of astatine have been demonstrated, including its ability to dissolve more easily in organic solvents than in water.

Sources

It was difficult for chemists to isolate fluorine because fluorine is such as strong oxidizing agent. The task was further complicated by the toxicity of hydrogen fluoride. Many chemists, including Davy and Gay-Lussac, suffered health problems from their attempts to prepare fluorine. The process Henri Moissan used to prepare fluorine (electrolysis of a solution of potassium fluoride in hydrogen fluoride) is the process used today.

To Do

Invite a mineral collector to class to show off samples of fluorite, which is second only to quartz in its popularity with collectors. You could use the visit to discuss crystal systems.

Group 7A: Halogens

Fluorine | discovered in 1886 by Henri Moissan

Chlorine | discovered in 1774 by Carl Wilhelm Scheele

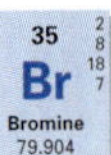

Bromine | discovered in 1826 by Antoine-Jérôme Balard

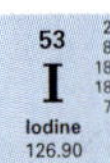

Iodine | discovered in 1811 by Bernard Courtois

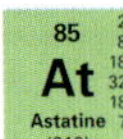

Astatine | discovered in 1940 by Dale R. Corson, K. R. Mackenzie, & Emilio Segrè

Physical Properties

- Halogens are nonmetals. At room temperature, fluorine and chlorine are gases and bromine is a liquid. Iodine and astatine are solids.
- Halogens are very reactive. The reactivity decreases from fluorine to astatine. Halogens do not exist in the elemental form in nature.
- Astatine isotopes are radioactive with short half-lives.

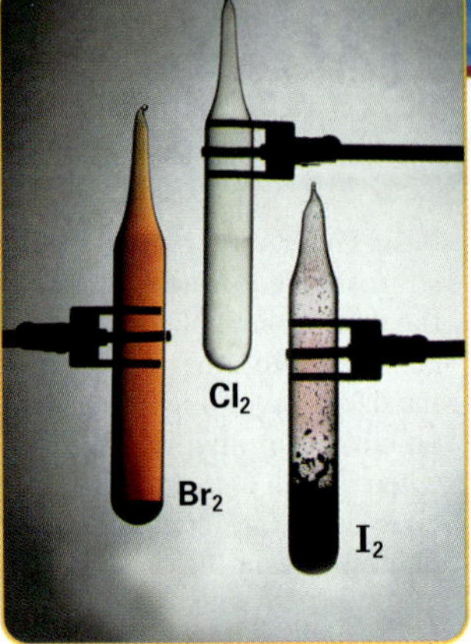

There are vapors in the bromine and iodine tubes because bromine is volatile and iodine sublimes easily at room temperature.

Melting and Boiling Points

Temperature (°C)

	F	Cl	Br	I
bp	−188	−35	59	184
mp	−220	−101	−7	114

Density

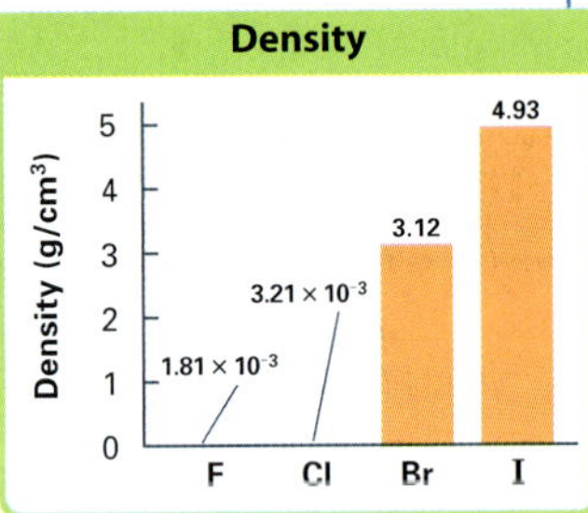

Sources

- Chlorine gas is made commercially by the electrolysis of brine.

$$2NaCl(aq) + 2H_2O(l) \longrightarrow Cl_2(g) + H_2(g) + 2NaOH(aq)$$

- Bromine is obtained from seawater by a displacement reaction with chlorine.

$$2NaBr(aq) + Cl_2(g) \longrightarrow 2NaCl(aq) + Br_2(l)$$

- Iodine is found in brine and in sodium iodate, $NaIO_3$, in deposits of sodium nitrate. Iodine is produced from $NaIO_3$ by this redox reaction.

$$2NaIO_3(aq) + 5NaHSO_3(aq) \longrightarrow I_2(g) + 2Na_2SO_4(aq) + 3NaHSO_4(aq) + H_2O(l)$$

- Fluorine is made by the electrolysis of potassium fluoride, KF, dissolved in liquid hydrogen fluoride, HF.

Fluorite, CaF_2, is the principal mineral of fluorine. The term fluorescent comes from this mineral, which glows in the presence of UV radiation.

Atomic Properties

- Group 7A elements have an electron configuration that ends in ns^2np^5.
- Halogens exist as diatomic molecules.
- Each halogen has the highest electronegativity in its period.
- The most common ionic charge for halogens is 1−. Except for fluorine, halogens also have positive oxidation numbers of +1, +3, +5, and +7.

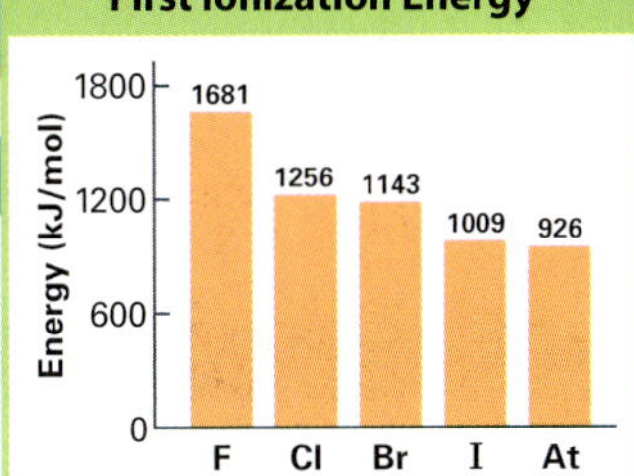

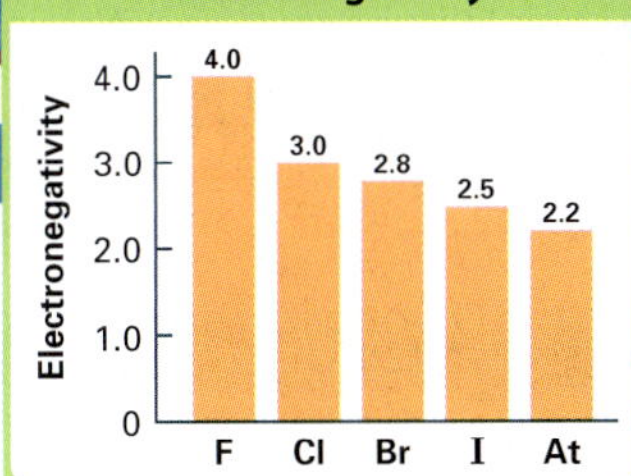

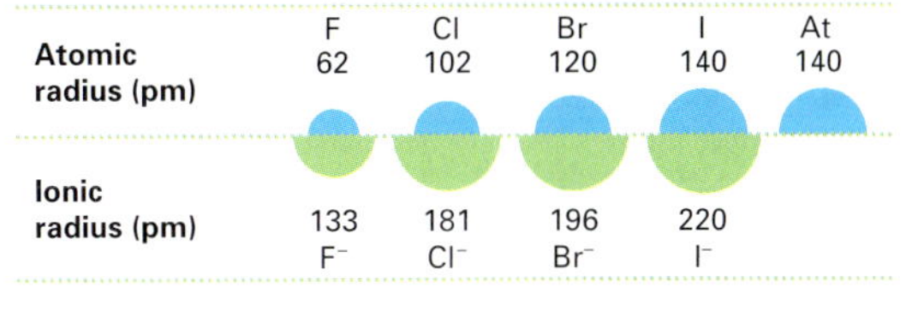

Chlorine forms four anions with oxygen. The oxidation number of chlorine is different in each anion.

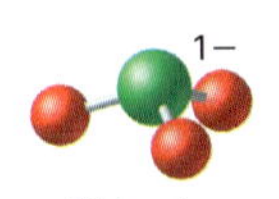
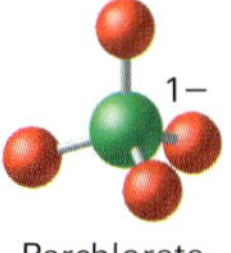

Important Compounds and Reactions

- Halogens form metal halides. Example:

$$2Na(s) + Cl_2(g) \longrightarrow 2NaCl(s) \quad \Delta H = -411 \text{ kJ/mol}$$

- Halogens form hydrogen halides. Example:

$$H_2(g) + Cl_2(g) \longrightarrow 2HCl(g) \quad \Delta H = -92.3 \text{ kJ/mol}$$

- Dry bleach is a mixture of compounds represented by the formula CaCl(ClO). Dry bleach is used to bleach paper and textiles. It also removes stains and disinfects laundry.

$$Ca(OH)_2(aq) + Cl_2(g) \longrightarrow CaCl(ClO)(aq) + H_2O(l)$$

- Small amounts of oxygen are produced in the laboratory by heating potassium chlorate, $KClO_3$. Potassium chlorate is an oxidizing agent in fireworks, matches, and explosives.

$$2KClO_3(s) \longrightarrow 2KCl(s) + 3O_2(g)$$

- Canisters of sodium chlorate are used on submarines to produce oxygen. They are carried on airplanes in case of an emergency.
- Chlorine is used to make the monomer vinyl chloride, $CH_2{=}CHCl$, which reacts to form the polymer polyvinyl chloride (PVC).

Chlorine reacts vigorously with sodium to form solid sodium chloride.

- Hydrofluoric acid (HF) is made from the mineral fluorite and sulfuric acid. Although HF is extremely dangerous, it has many uses, including cleaning metals and frosting glass.

$$CaF_2(s) + H_2SO_4(aq) \longrightarrow 2HF(g) + CaSO_4(s)$$

- Nonstick pans are coated with a polymer of tetrafluoroethene, $F_2C{=}CF_2$.
- Tincture of iodine is a solution of iodine, I_2, and potassium iodide, KI, in alcohol. It is an example of an iodine-based skin disinfectant.

Atomic Properties

The ball-and-stick models of the anions containing chlorine and oxygen could prove useful when students study VSEPR theory in Chapter 8.

To Do

Ask students to calculate the oxidation number of chlorine in the hypochlorite, chlorite, chlorate, and perchlorate ions.

Important Compounds and Reactions

Finding a container to store elemental fluorine was a challenge because fluorine attacks glass and causes most metals to burst into flame. Alloys of copper and nickel are used because when fluorine reacts with the alloy, it forms a protective (passive) fluoride layer on the surface.

Fluorine is used to enrich uranium. Fluorine reacts with uranium to produce the gas uranium hexafluoride (UF_6). Because the rate of diffusion of the gas varies with the isotope, uranium-235 (the fissionable isotope) can be separated from uranium-238.

To Do

Challenge students to explain why HCl, HBr, and HI form strong acids when dissolved in water, but HF forms a weak acid. (The strong hydrogen bonds between the polar HF molecules limit the ionization of hydrogen fluoride in water.) Then challenge students to find out why concentrated hydrofluoric acid, is extremely dangerous. (Hydrofluoric acid causes severe painful burns on contact with body tissues. Because of the small size of its molecules, hydrofluoric acid can penetrate skin and fat quickly. It decalcifies bone and acts as a poison once it is in the blood.)

Group 7A: Halogens

Tooth Decay

Low concentrations of fluoride ions are present in all natural water sources, but the level may not be sufficient for the prevention of tooth decay. In the United States, the optimum concentration for fluoride in water supplies is in the range of 0.7 to 1.2 ppm.

Because fluoride ions are toxic in high concentrations, FDA regulations limit the amount of fluoride in toothpaste. Even so, people should rinse away excess toothpaste. There is a warning on toothpaste labels about the use of toothpaste by children under 6 years of age. Young children need to be supervised until they have learned how to minimize the ingestion of toothpaste.

Swimming Pool Chemistry

Students may wonder why the chlorine that kills bacteria doesn't harm humans who use a swimming pool. The answer lies with the concentration.

To Do

In the early 1900s, health officials in developed countries began to use chlorine to treat drinking water. Students could research the effect that adding chlorine to drinking water had on outbreaks of diseases such as cholera, typhoid fever, and dysentery, which are caused by waterborne pathogens.

Group 7A: Halogens

F Tooth Decay

Your teeth have a hard outer layer called enamel. Enamel is mainly calcium carbonate, $CaCO_3$, and hydroxyapatite, $[Ca_3(PO_4)_2]_3 \cdot Ca(OH)_2$. Lactic acid, $C_3H_6O_3$, is the main cause of tooth decay. It forms when bacteria in saliva feed on sugars present in the sticky plaque on tooth surfaces. An increase in H^+ concentration causes the minerals in tooth enamel to decay faster.

Fluoride ions are added to the water supply in many cities. Most toothpastes contain fluoride ions. The ions replace hydroxide ions in hydroxyapatite to form fluoroapatite, $[Ca_3(PO_4)_2]_3 \cdot CaF_2$. This replacement makes the enamel more resistant.

Fluoride ions alone won't prevent tooth decay. You need to vigorously brush your teeth and floss to keep plaque from building up on tooth enamel. ■

Cl Swimming Pool Chemistry

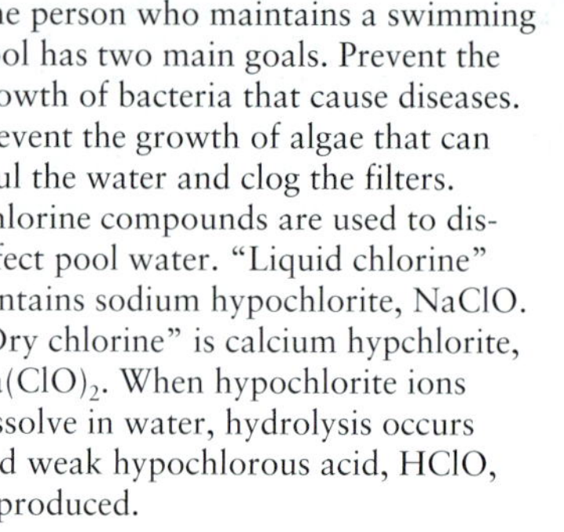

The person who maintains a swimming pool has two main goals. Prevent the growth of bacteria that cause diseases. Prevent the growth of algae that can foul the water and clog the filters. Chlorine compounds are used to disinfect pool water. "Liquid chlorine" contains sodium hypochlorite, NaClO. "Dry chlorine" is calcium hypchlorite, $Ca(ClO)_2$. When hypochlorite ions dissolve in water, hydrolysis occurs and weak hypochlorous acid, HClO, is produced.

$$ClO^-(aq) + H_2O(l) \rightleftharpoons HOCl(aq) + OH^-(aq)$$

The amount of undissociated hypochlorous acid in the pool water depends on the pH. If the pH is too high, the hydrolysis reaction will shift toward the reactants and reduce the concentration of HClO. If the pH is too low, too much acid will form. A high concentration of acid can cause eye irritation, damage plaster, and corrode the metal piping and filters in the pool.

If the pH of the pool water is too high, solid sodium hydrogen sulfate can be used to react with the OH^- ions.

$$NaHSO_4(s) + OH^-(aq) \longrightarrow Na^+(aq) + SO_4^{2-}(aq) + H_2O(l)$$

If the pH is too low, sodium carbonate can be used to neutralize some of the acid.

$$Na_2CO_3(s) + 2H^+(aq) \longrightarrow 2Na^+(aq) + H_2O(l) + CO_2(aq)$$

The use of fluorides in drinking water and toothpastes has caused such a drastic reduction in tooth decay that many young adults have never had a cavity.

R34 Elements Handbook

F Artificial Blood

The most important function of blood is its ability to deliver oxygen from the lungs to cells throughout the body, and carry carbon dioxide from the cells to the lungs. Blood transfusions have been the traditional response to blood loss due to severe injuries or surgery. Blood transfusions have saved millions of lives. But it takes time to determine the recipient's blood type and less common blood types may not be available. Also, blood can carry disease-causing bacteria and viruses.

Some artificial blood contains perfluorocarbons (PFCs). PFCs are organic compounds in which all the hydrogen has been replaced by fluorine. PFCs can dissolve and transport large quantities of oxygen. PFCs are mixed with emulsifiers and salts that mimic the composition of blood, minus the cells.

Large amounts of PFCs can be manufactured and their purity controlled. PFCs can deliver oxygen to areas where blood flow is restricted by narrowed arteries or tumors. In current trials, artificial blood is used to delay the need for blood transfusions during surgery. Because PFCs carry less oxygen than hemoglobin does, the patient must breathe oxygen-rich air.

Because blood is so complex, scientists may never find a substitute to perform all its functions. Therefore, there is still a great need for donations of human blood. ■

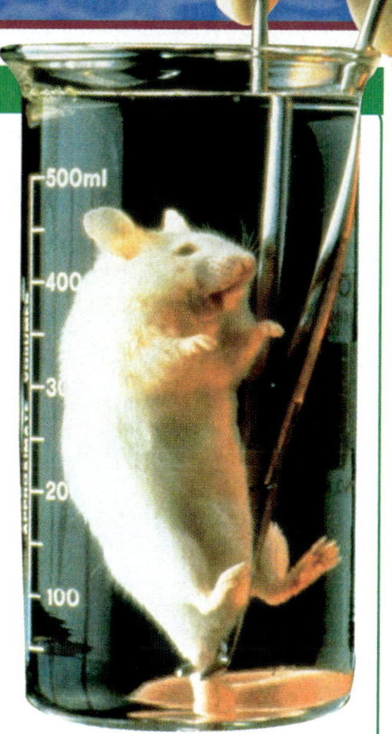

The liquid in the beaker is a perfluorocarbon that is saturated with oxygen. The mouse survived by absorbing oxygen from the liquid in its lungs.

I Iodized Salt

The thyroid gland produces hormones that help to control the body's growth and the energy produced by cells. Trace amounts of iodine are needed to produce thyroid hormones. An adult needs about 150 mg of iodine daily.

Ocean fish are a good source of iodine. When people used to get most of their food from local sources, a person who lived far from the ocean often had an iodine deficiency.

To compensate for the lack of iodine, the thyroid gland might enlarge. A severe deficiency could cause mental retardation. Adding potassium iodide to table salt proved to be a simple solution to this public health problem.

In the United States, iodized salt was first sold in 1924. There is about 400 mg of iodine in a teaspoon of iodized salt.

The use of iodized salt has virtually eliminated the problem of iodine deficiency in the United States. ■

Did You Know...

The purple dye prized by Roman emperors and other rulers contains **bromine**. The dye was extracted from the mollusk *Murex brandaris,* which lives in the Mediterranean Sea.

Artificial Blood

The ideal artificial blood (1) wouldn't require refrigeration, (2) could be used with all blood types, (3) would be free of infectious agents, and (4) would be easy to use. Some artificial blood products are derived from hemoglobin. Others use perfluorocarbons to carry oxygen. PFC-based blood substitutes can be frozen and stored for years. Blood must be refrigerated and used within weeks. PFCs are slow to leave body tissues, and their long-term effects are not known.

Iodized Salt

Recipes that call for large quantities of table salt often specify non-iodized salt. Some cooks claim large amounts of potassium iodate affect the flavor of food. In other cases, the issue is not the iodine, but other additives, e.g., non-caking agents, that the user is trying to avoid.

To Do

Table salt manufacturers in the United States use potassium iodide. But in countries where the climate is hot and humid, potassium iodate may be used instead. Challenge students to find data to demonstrate why potassium iodate might be preferable to potassium iodide in hot, humid climates. (The solubility of potassium iodide is 148 g/100 g H_2O at 25°C; the solubility of potassium iodate is 9.22 g/100g H_2O at 25°C.)

Group 8A: Noble Gases

The names of all the noble gases except radon come from Greek words. Helium comes from *helios,* meaning "sun." Xenon comes from *xenos,* meaning "strange." Krypton comes from *kryptos,* meaning "hidden." Argon comes from *argos,* meaning "inert." Neon comes from *neos,* or "new."

Physical and Chemical Properties

More than 80 compounds of xenon have been produced. Krypton forms the colorless solid krypton(II) fluoride, KrF_2. A compound of argon, argon fluorohydride (HArF), was produced in 2000. But this compound exists only at temperatures below −246°C.

To Do

Show students how to draw an electron dot structure for xenon(II) fluoride (XeF_2). Then have students draw electron dot structures for xenon(IV) fluoride (XeF_4), xenon(IV) oxide (XeO_3), and xenon(VIII) oxide (XeO_4). As a further challenge, students could try drawing the electron dot structures for XeO_2F_2 or $XeOF_4$.

Sources

Although there are measurable amounts of radon in most regions of the United States, the highest levels are recorded in regions where the bedrock is granite. Radon is a product of the natural decay of uranium-238. Radon can seep through cracks in foundations and basements floors. Because it is denser than air, radon tends to collect in the lower levels of a house. Inhaling radon may increase the risk of lung cancer. High levels of radon can be reduced by improved ventilation in basements and other closed areas where radon collects.

Group 8A: Noble Gases

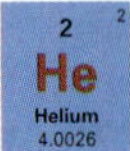

Helium | discovered in 1868 by Pierre Janssen

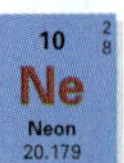

Neon | discovered in 1898 by Sir William Ramsay & Morris Travers

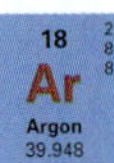

Argon | discovered in 1894 by Lord Rayleigh & Sir William Ramsay

Krypton | discovered in 1898 by Sir William Ramsay & Morris Travers

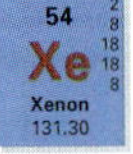

Xenon | discovered in 1898 by Sir William Ramsay & Morris Travers

86 Rn Radon (222)

Radon | discovered in 1900 by Friedrich E. Dorn

Physical and Chemical Properties

- All Group 8A elements are monatomic gases at STP.
- Noble gases are colorless, odorless, and tasteless.
- The first compound of a noble gas, $XePtF_6$, was made in 1962. There are now more than 100 known compounds of fluorine and xenon.

Incandescent light bulbs are filled with argon instead of air to extend the life of the filament.

- In 2000, chemists in Finland made a compound of argon, HArF, that exists only at temperatures below −246°C.

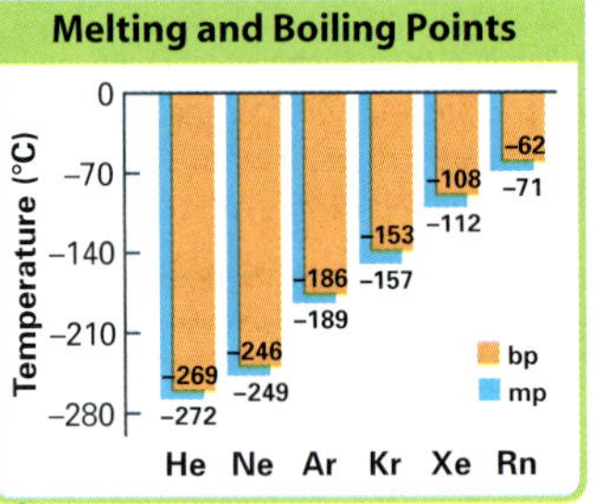

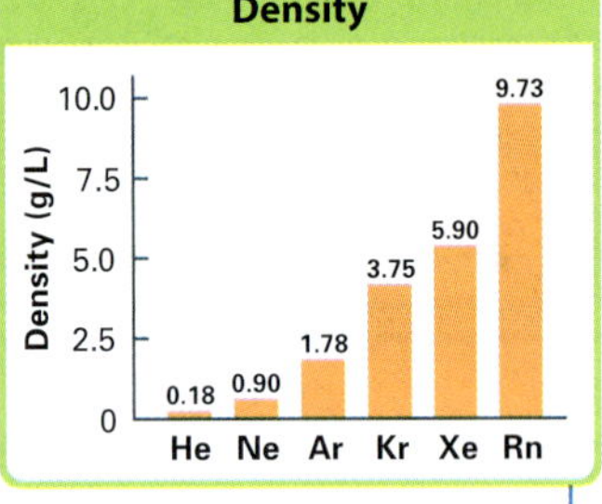

Sources

- Helium is separated from natural gas deposits. Neon, argon, krypton, and xenon are separated from air by fractional distillation.
- Because of its low density, helium is used in weather balloons and airships.
- In addition to "neon" lights, noble gases are used in fluorescent bulbs, strobe lights, and headlights.
- Liquid helium cools the magnets used for magnetic resonance imaging (MRI).

Atomic Properties

- Noble gases have an electron configuration that ends in ns^2np^6. Helium ($1s^2$) is an exception.
- Noble gases rarely form compounds. When they do, the most common oxidation number is +2.
- Each noble gas emits a characteristic color in a gas discharge tube.

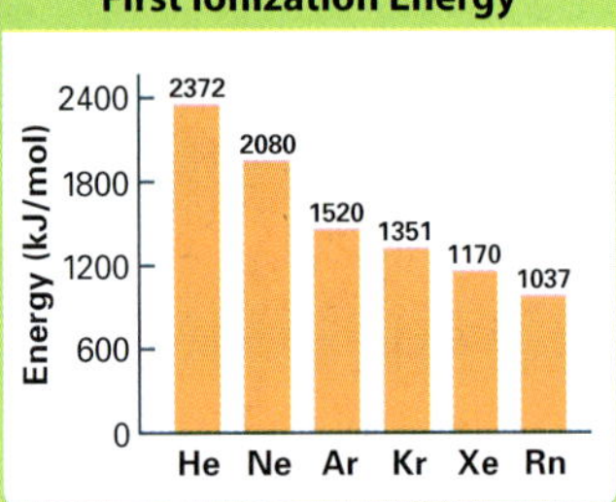

Noble gases have the highest ionization energies because their highest occupied energy levels are filled.

Helium

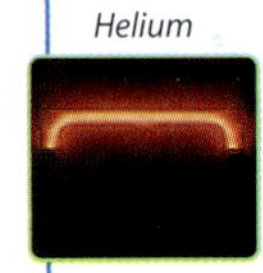

Neon

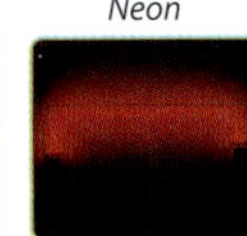

Argon

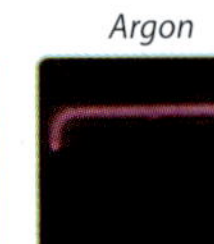

Krypton

Xenon

Atomic Properties

The noble gases that form compounds most readily have larger atoms in which there are more electrons to shield the electrons in the highest occupied energy level from the nuclear charge.

To Do

Encourage students to research the role of krypton-86 in the development of a definition for a meter. Ask them to determine why a definition based on krypton-86 was adopted in 1960 and why it was replaced in 1983.

Ne Neon Lights

By 1855, scientists could produce light by passing an electric current through a gas under low pressure in a sealed glass tube. With the discovery of the noble gases, a new technology emerged. In 1910, George Claude displayed the first neon lamp in Paris, France.

In 1923, a car dealer from Los Angeles bought two signs that spelled out "Packard" for $24,000 (about $250,000 in today's dollars). When he displayed the signs in Los Angeles, people described the light as "liquid fire." By the 1930s, businesses were using neon lights to draw the attention of customers.

Although all noble gases emit visible light when their atoms are excited, neon and argon are the gases most often used in neon lights. Orange-red lights contain only neon. Other colors are produced by adding a bit of mercury to the noble gas. The tube is coated with a material that glows when exposed to UV light emitted by mercury vapor. ■

Ar Taken For a Ride By Argon

Whether riding on a paved city street or on an unpaved mountain trail, a bicyclist is likely to find rough patches. When faced with rough terrain, the cyclist may worry about the tires, but probably not about the bicycle frame. The frame is made of steel or aluminum alloy tubes that are joined together.

The tubes are joined together by Gas Tungsten Arc Welding (GTAW). An electric arc is struck between a tungsten electrode and the parts to be welded. Heat from the arc melts the ends of the tubes and fuses them together. Filler may be placed between the ends of the tubes to increase the strength of the joint or to produce a smoother joint.

During welding in air, there is a danger that the metal tubes or electrode will oxidize. To prevent oxidation, the area around the arc is filled with an inert gas, most often argon. Welding with argon has an added benefit. Because argon is a poor conductor of heat, the arc that forms is narrow. This narrow arc produces a weld that is both neat in appearance and mechanically strong. ■

Xe Xenon-Ion Engines

The signals for a television program may bounce off a communication satellite. The satellite is in orbit above the equator. The position of the satellite may be maintained by a xenon-ion propulsion system.

When electrons strike xenon atoms in a xenon-ion engine, the atoms lose electrons and form positive ions. The ions are accelerated by a charged grid and shot from the engine at about 105 km/h. This action pushes, or thrusts, the satellite in the opposite direction. With multiple engines facing in different directions, a satellite can be moved in any direction.

Although a xenon-ion engine produces a relatively small amount of thrust, it can provide thrust for months or years. This makes xenon-ion engines a good choice for lengthy space missions. In addition, an inert gas poses no hazard for the satellite or the people who handle the propellant tanks. ■

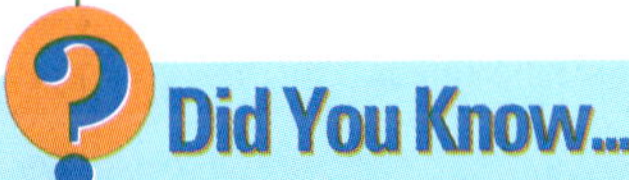

When liquid **helium** is cooled to below 2 K, its viscosity drops to zero. It will escape from an unsealed container by flowing up the sides of the container.

Neon Lights

The development of neon lights can be traced back to an observation made by the French astronomer Jean Picard in 1675. He observed a faint glow in the tube of a mercury barometer when the tube was shaken.

Los Angeles in the 1930s was a low-rise city. Businesses used elaborate, art deco neon signs to attract motorists, who could see the signs from miles away. Many of the vintage signs have been restored. In 1949, Raymond Chandler wrote, "There ought to be a monument to the man who invented neon lights." (A monument to Claude might be a problem given his political choices. He was imprisoned from 1945–1949 for his support of the Vichy government in occupied France.)

Xenon-Ion Engines

Because it completes one orbit in 24 hours, a communication satellite appears to be stationary above Earth. The altitude at which a satellite orbits Earth depends on its task. Stationary weather and communication satellites orbit at an altitude of about 22,300 miles. The region in which they orbit is called the Clarke Belt in honor of the science-fiction writer Arthur C. Clarke. Countries are assigned positions in the Clarke Belt for their communications and weather satellites.

Taken For a Ride By Argon

Welders who use GTAW must be aware of a safety issue. Breathing argon-rich air can deprive the body of oxygen. A person could pass out or even die from asphyxiation. So argon must be used in a well-ventilated room.

Hydrogen

Hydrogen was prepared long before it was recognized as a distinct substance. The Swiss alchemist Theophrastus Paracelsus probably handled hydrogen in the 1500s. He produced a flammable gas when he dissolved a metal in acid. In 1671, Boyle published a paper in which he described a reaction between iron filings and acids that produced "an inflammable gas." Cavendish described hydrogen as "inflammable air from metals." Lavoisier named hydrogen from the Greek *hydro,* meaning "water" and *genes,* meaning to "generate."

Atomic and Physical Properties

Giving unique names to isotopes is limited to the isotopes of hydrogen. Hydrogen is an exception to the rule that there are only minor differences in the physical properties of an element's isotopes. When a neutron is added to an atom with a single proton, the mass is doubled (or tripled when a second neutron is added). These relatively large differences in atomic mass contribute to greater differences in the physical properties of hydrogen isotopes and their compounds. For example, H_2O melts at 0°C and boils at 100°C; D_2O melts at 3.81°C and boils at 101.42°C.

Sources

Much hydrogen is also obtained from petroleum refining. During refining, hydrogen is a by-product of the catalytic break down of large hydrocarbon molecules into smaller molecules. Small quantities of hydrogen are prepared in the laboratory by the reaction of dilute acid with an active metal, such as zinc.

$Zn(s) + 2HCl(aq) \rightarrow H_2(g) + ZnCl_2(aq)$

Hydrogen

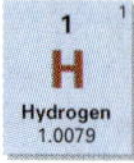

Hydrogen | discovered in 1766 by Henry Cavendish

Sources

- Hydrogen is rarely found on Earth in an uncombined state.
- Electrolysis of water produces the purest hydrogen, but the process requires too much energy to be economical.

$$2H_2O(l) \longrightarrow 2H_2(g) + O_2(g) \quad \Delta H = +572 \text{ kJ}$$

- Hydrogen is produced when methane and steam react at 1100°C over a nickel catalyst.

$$H_2O(g) + CH_4(g) \xrightarrow{Ni} CO(g) + 3H_2(g) \quad \Delta H = +206 \text{ kJ}$$

The products pass over a metal oxide catalyst at 400°C. More hydrogen is produced as carbon monoxide reacts with added steam.

$$CO(g) + H_2O(g) \longrightarrow CO_2(g) + H_2(g) \quad \Delta H = -41 \text{ kJ}$$

Carbon dioxide is removed as the gases flow through a basic solution.

$$CO_2(g) + 2OH^-(aq) \longrightarrow CO_3^{2-}(aq) + H_2O(l)$$

Atomic and Physical Properties

- Hydrogen has an electron configuration of $1s^1$.
- The most common oxidation numbers for hydrogen are +1 and −1.
- Most hydrogen (99.985%) is protium, or hydrogen-1.
- The other stable isotope is deuterium (hydrogen-2), which has the symbol D. Harold Urey discovered heavy hydrogen, D_2, in 1931.
- Tritium (hydrogen-3) was discovered in 1934. Its half-life is 12.3 years.

Properties of Hydrogen

Property	Value
Density at STP	0.09 g/L
Melting point	−259°C
Boiling point	−253°C
Ionization energy	1.312×10^3 kJ/mol
Electronegativity	2.1

Atomic radius (pm) H 30

Ionic radius (pm) 1.2 H^+

Important Compounds and Reactions

- Hydrogen forms covalent binary hydrides with nonmetals. Example:

$$H_2(g) + Cl_2(g) \longrightarrow 2HCl(g)$$

- Two thirds of the hydrogen produced in the U.S. is used to synthesize the covalent binary compound ammonia.
- Hydrogen forms ionic hydrides with alkali metals and alkaline earth metals. These hydrides are powerful reducing agents. Example:

$$Ca(s) + H_2(g) \longrightarrow CaH_2(s)$$

- Hydrogen is used to make methanol, CH_3OH. The reaction takes place at 200–300 atm and 400°C in the presence of a metal oxide catalyst.

$$CO(g) + 2H_2(g) \longrightarrow CH_3OH(g)$$

Methanol is an industrial solvent. It is used to make formaldehyde, CH_2O, which is used to make plastics.

Hydrogen is used to turn liquid oils such as corn oil into solid margarine. Hydrogen is added to carbon–carbon double bonds during hydrogenation.

Important Compounds and Reactions

The oxidation number of hydrogen in covalent binary hydrides is +1. With the exception of water, most low molar mass covalent binary hydrides are gases or liquids with low boiling points. The oxidation number of hydrogen in ionic binary hydrides is −1. These hydrides are white crystalline solids with high melting points. They are formed by direct reaction of alkali metals and alkaline earth metals with hydrogen at about 400°C.

To Do

Ask students to write equations for the dissociation of a covalent binary hydride and an ionic binary hydride in water. Then have them identify the reaction in which hydrogen is acting as a base (in the ionic hydride) and the reaction when hydrogen is acting as an acid. Note that the hydride ion is one of the strongest bases known.

H Heavy Water Reactors

Because 1 in 6000 hydrogen atoms is deuterium, 1 in 20,000,000 water molecules is D_2O. D_2O is called heavy water because it is about 10% heavier than ordinary "light" water. In heavy water nuclear reactors, D_2O is used in place of H_2O as a neutron moderator. Both types of water are good moderators, but D_2O is more efficient than H_2O because D_2O absorbs fewer neutrons. So in a heavy water reactor, the uranium used to make the fuel doesn't have to be enriched. The initial cost of separating the heavy water from light water is offset by the lower cost of the uranium fuel. ■

Lake water passes through a series of extraction towers in which D_2O is separated from H_2O.

H A Hydrogen Economy

Hydrogen is a pollution-free fuel. When it burns in air, the only product is water with trace amounts of nitrogen oxides. No carbon dioxide, carbon monoxide, oxides of sulfur, or unburned hydrocarbons are emitted. In a *hydrogen economy*, hydrogen would replace fossil fuels as the energy source for heating, transportation, and industrial processes. Achieving this goal will require new technology for the production, distribution, and storage of hydrogen.

Large-scale production of hydrogen currently begins with fossil fuels. The process emits CO_2, unless money is spent to trap the CO_2. Some hydrogen is extracted from biomass sources such as wood chips. In the future, green algae may produce hydrogen in the presence of sunlight. Liquid hydrogen may be shipped in vacuum-insulated tanks or through underground pipes.

Cars that run on hydrogen fuel are being built. One challenge is storing the hydrogen. On the space shuttle, it is stored as a liquid. In some vehicles, it is produced from fossil fuels as needed. One promising technology is storage of solid hydrogen in metal hydrides. ■

This is a model of a vehicle in which hydrogen is stored in a metal hydride at low pressure. When the hydride is heated, the hydrogen is released.

H Hydrogen Peroxide

The strips used to whiten teeth usually contain hydrogen peroxide, H_2O_2. It is a powerful oxidizer that doesn't produce toxic gases or unwanted residues. A 3% aqueous solution is safe for use at home. Stronger concentrations are used to treat wastewater, bleach paper, and produce detergents.

Hydrogen peroxide can inhibit the growth of bacteria in water pipes and increase the growth of bacteria that clean up polluted soils. How is this possible? The answer is *selectivity*. Scientists can adjust variables such as pH, temperature, and concentration so that hydrogen peroxide oxidizes one pollutant and not another. ■

Did You Know...

Hydrogen in the center of the sun has a density of about 200 g/mL. The temperature is about 13 million degrees Celsius. Radiation released when hydrogen nuclei fuse takes about a million years to reach the sun's surface.

Heavy Water Reactors

Due to the high neutron absorption factor of D_2O over other moderators, heavy water reactors produce larger numbers of radioisotopes than other types of nuclear reactors. Many of these isotopes are used in medicine and agriculture. Heavy water reactors also permit a higher number of uranium-238 atoms to absorb neutrons, which leads to the production of plutonium-239. If a heavy water reactor ran at full power, it could produce 8–10 kg of plutonium per year, which is enough to build a nuclear weapon. But, a large plutonium separation facility would be needed to produce this much weapons-grade plutonium.

Hydrogen Peroxide

Hydrogen peroxide is a natural metabolite of many organisms, which decompose the H_2O_2 they produce into water and oxygen. Hydrogen peroxide is often used to control odors. It may be applied directly to aqueous wastes containing odorants (such as hydrogen sulfide) or to wet scrubbers that are used to remove pollutants from airstreams. In the case where the odors are the result of biological activity, hydrogen peroxide may be added as a preventive measure to eliminate anoxic (oxygen deficient) conditions, which favor the generation of bad odors.

In 2000, IUPAC published a paper in which *selectivity* was distinguished from *specificity* in relation to analytical tests. A specific test is one that occurs only with the substance of interest. A selective test can occur with multiple substances, but exhibits a degree of preference for the substance of interest. Few tests are specific; many are selective.

A Hydrogen Economy

Metal hydrides are formed by the reaction of hydrogen with inner transition metals and some transition metals. The smaller hydrogen atoms occupy holes in the crystal lattice of the much larger metal atoms. The reaction takes place under pressure. The stored hydrogen is released through a drop in pressure or an increase in temperature.

The nature of the bonding in metal hydrides isn't well understood. They have the general formula MH_x, where x represents the number of hydrogen atoms in the simplest formula. Because of the variable number of holes in the parent alloy, many metallic hydrides are nonstoichiometric. For example, a titanium hydride has the formula $TiH_{1.7}$.

Transition Metals

Physical Properties

The "sea of electrons" model can explain such properties of transition metals as conductivity and malleability. But it is not adequate to explain the periodic trend in melting points, which reflects an increase in the strength of metallic bonding until Group 6B and a subsequent decrease. Chemists explain this trend by applying molecular orbital bonding theory to metallic bonding; i.e., the existence of bonding and antibonding interactions within a metal.

Sources

An ore must contain enough of the desired metal so that the market value of the metal offsets the cost of finding, removing, and refining the ore. For many metals, the ore must contain at least 30% of the metal. However, with high value metals, such as gold, it may be profitable to mine ore containing a much smaller percentage of the metal.

To Do

Students can use sources such as the business sections of newspapers to research which metals are traded on stock exchanges, and to rank the metals from least expensive to most expensive. Ask students to use what they have learned about different metals to propose reasons for the rankings. (Remind students of how the invention of the Hall-Heroult process affected the value of aluminum.)

Transition Metals

Before 1700
- **Au** gold
- **Ag** silver
- **Cu** copper
- **Fe** iron
- **Hg** mercury
- **Zn** zinc

1700–1799
- **Co** cobalt ~1735
- **Pt** platinum 1735
- **Ni** nickel 1751
- **Mn** manganese 1774
- **Mo** molybdenum 1778
- **W** tungsten 1783
- **Ti** titanium 1791
- **Y** yttrium 1794
- **Cr** chromium 1797

1800–1899
- **V** vanadium 1801
- **Nb** niobium 1801
- **Ta** tantalum 1802
- **Pd** palladium 1803
- **Rh** rhodium 1803
- **Os** osmium 1803
- **Ir** iridium 1803
- **Cd** cadmium 1817
- **Zr** zirconium 1824
- **Ru** ruthenium 1844
- **Sc** scandium 1878

After 1900
- **Lu** lutetium 1907
- **Hf** hafnium 1923
- **Re** rhenium 1925

Physical Properties

- Most transition metals are ductile, malleable, and good conductors of heat and electric current.
- For transition metals, density and melting point tend to increase across a series or increase to a peak in Group 6B and then decrease.
- Compounds of transition metals tend to have color.

Except for copper and gold, transition metals, including platinum, have a silvery luster.

Density and Melting Points of Period 4 Transition Metals

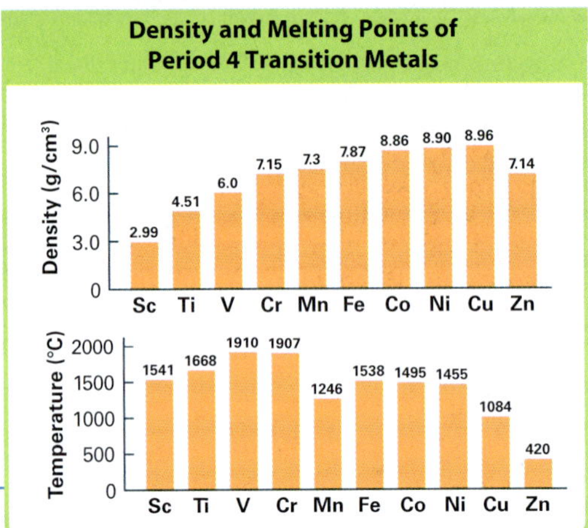

Sources

- Transition metals come from mineral deposits in Earth's crust. Minerals that are used for the commercial production of metals are called ores.
- For centuries, people have developed techniques for separating metals from ores. The ore is concentrated and the metal removed by reduction. Then the metal is refined and purified.

Gold exists as an element in nature. But its ore needs to be concentrated before the gold can be extracted and purified.

Atomic Properties

- Among the transition metals, as atomic number increases, there is an increase in the number of electrons in the second-to-highest occupied energy level.
- In periods 5 and 6, transition metals in the same group have identical or almost identical atomic radii. Consequently, these pairs of elements have very similar chemical properties. They tend to occur together in nature and are difficult to separate.

Trends in Atomic Size for Transition Metals

Atomic radius (pm)

Increasing Atomic Number →

Chemical Properties

- There is great variation in reactivity among transition metals. Scandium and yttrium are similar to Group 1A and 2A metals. They are easily oxidized on exposure to air and react with water to release hydrogen. Platinum and gold are extremely unreactive and resist oxidation.
- In general, transition metals have multiple oxidation states. Compounds in which the elements are in their highest oxidation states are powerful oxidizing agents.
- Most transition metals form compounds with distinctive colors. The color of a transition metal compound or solution can indicate the oxidation state of the metal.

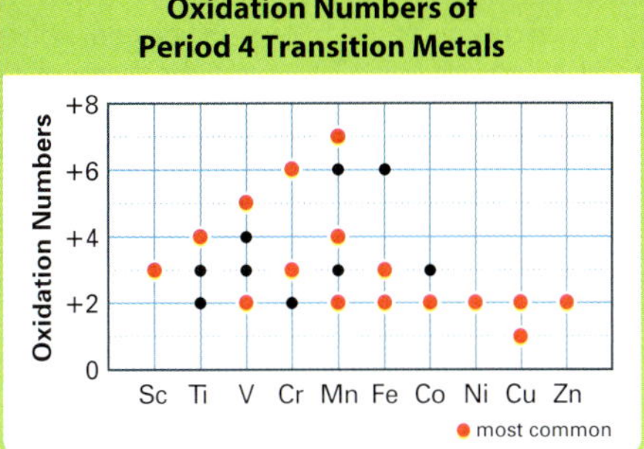

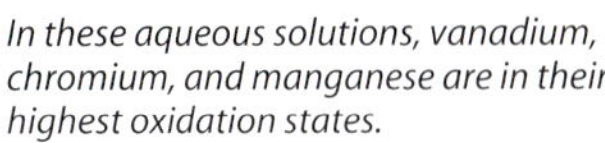

In these aqueous solutions, vanadium, chromium, and manganese are in their highest oxidation states.

The oxidation number of vanadium is +5 in the yellow solution, +4 in the blue solution, +3 in the green solution, and +2 in the purple solution.

Atomic Properties

Metal ions that have a partially filled *d* sublevel are usually colored; metal ions with a completely filled or completely empty *d* sublevel are colorless.

Chemical Properties

The transition metals vary greatly in their reactivity. For example, scandium and yttrium are similar to Group 1A metals in that they are easily oxidized on exposure to air and react with water to produce hydrogen. Yet platinum and gold are extremely unreactive and resist oxidation.

Transition Metals

Copper Alloys

The Bronze Age reigned supreme for many centuries. Because bronze could be cast, it was used for making tools, plowshares, weapons, armor, and decorative objects. About 1200 B.C., there was a disruption of international trade routes and the supply of tin dried up. Metalworkers began to use iron instead of bronze because iron ore was widely available for smelting with charcoal. The substitution of iron for bronze in tools and weapons was the result of iron's abundance rather than any inherent advantages of iron.

Because only 0.13 mm of the Statue of Liberty's copper skin was weathered and oxidized during the statue's first century, the copper skin was one of the few parts of the statue that didn't need to be replaced or rebuilt when the statue was renovated for its centennial. The torch section was the only copper part that required work. It was rebuilt with new copper and patinated before installation to match the rich, green color of the rest of the Statue.

To Do

Encourage students to research the chemistry of patina formation. Have them look at the different compounds that form at various stages and how environmental conditions affect the rate at which a patina forms.

To Do

Beginning in the thirteenth century, some cultures used monumental brasses to commemorate the dead. These large brass plates were engraved with a likeness of the deceased, decorated with heraldic devices and inscriptions, and set on the surface of the tomb. Show students reproductions of rubbings made from plates in the United Kingdom. Compare the rubbings to those made of granite tombstones in New England burial grounds.

Transition Metals

Catalytic Converters

In an internal combustion engine, some hydrocarbon molecules aren't completely oxidized. They form carbon monoxide instead of carbon dioxide. Some hydrocarbon molecules even pass into the exhaust without reacting. At the high temperature and pressure inside an engine's cylinders, some nitrogen from the air reacts with oxygen to produce irritating nitrogen oxides.

A catalytic converter keeps these pollutants from being released into the air. Inside the converter is a porous ceramic cylinder with a honeycomb structure. The ceramic absorbs particles of rhodium and platinum. The metals are catalysts for reactions that occur as exhaust gases pass through channels in the converter. Rhodium helps convert nitrogen oxides to nitrogen. Platinum helps convert hydrocarbons and carbon monoxide to carbon dioxide and water. ■

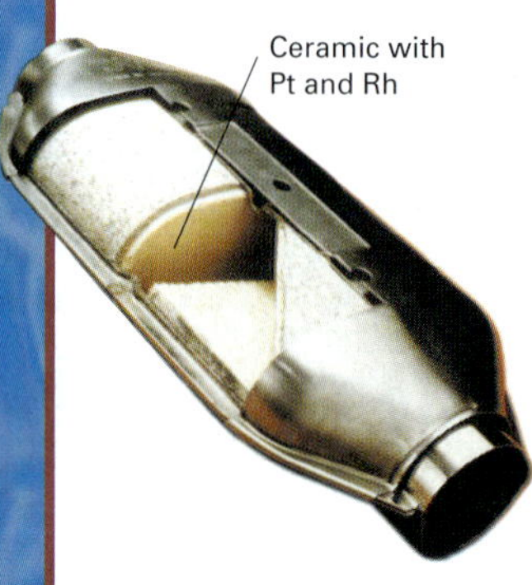

Cu Copper Alloys

When a bronze bell is struck, the clear, loud tone lasts for several seconds.

Copper was one of the first metals to be widely used. It is found uncombined in nature or easily reduced from its ores. The Roman supply of copper came mainly from Cyprus and was known as *aes Cyprium* (metal of Cyprus). This name evolved to *cyprium* and then *cuprium*, which is why copper has the symbol Cu. Pure copper is valued both for its ability to conduct an electric current and its ability to resist corrosion.

Around 3500 BC, people began to add tin to copper to form bronze. This alloy is harder than pure copper and easier to melt for casting. Metalworkers could produce bronze with different properties by varying the amount of tin. Bronze used to make statues might contain as little as 10% tin by mass. Bronze used to make bells would contain 13–25% tin. Most copper coins are bronze with 4% tin and 1% zinc.

Brass is an alloy of copper and zinc. Brass is harder than pure copper and more malleable than bronze. Brass containing at least 65% copper can be worked when it is cold. Brass with 55–65% copper can be worked when it is hot. Before large amounts of gold and silver reached Europe in the 1500s, brass was the metal used for decorative items.

Copper that is exposed to oxygen and water forms a patina of basic copper salts. This thin film protects the underlying metal from further oxidation. The composition of the patina, its color, and the rate at which it forms vary with the climate. The rate is faster near the ocean. ■

Historically, brass was used to make high-quality scientific instruments. This microscope was made in 1765.

Catalytic Converters

Catalytic converters usually have a long life span. However, overheating of the converter can reduce its effectiveness by causing the catalytic particles to merge, reducing the surface area. Note that a single tank of leaded gas can seriously impair the effectiveness of the catalysts. The lead binds to the catalytic particles and inhibits their activity.

To Do

Invite an automobile mechanic to speak to the class about emission testing and the effect of catalytic converters on emissions. Ask the mechanic to bring a new catalytic converter and, if possible, a discarded converter, which has been opened in advance to reveal its contents.

Fe Iron and Steel

Carbon is used to extract iron from its ores. At first, people used charcoal from burnt wood. In 1709, Abraham Darby invented a process that used coke instead of charcoal. Coke is almost pure carbon. It is produced when coal is heated and the impurities removed as gases. With coke, the production of iron became less expensive and more efficient.

Iron ore is reduced to metallic iron in a blast furnace. Ore, coke, and limestone are added at the top of the furnace. Molten iron and slag collect at the bottom. The "pig" iron produced contains 3–5% carbon and smaller amounts of other impurities, which make the iron brittle. Pig iron can't be rolled or welded, but it can be cast. Cast iron is used to make stoves and engine blocks for cars.

Most pig iron is used to make steel. The methods for making steel differ, but they all lower the carbon content to less than 2% and remove other impurities. About 90% of the steel produced is carbon steel, which contains no other metals. Mild steel, which is malleable and ductile, contains less than 0.2% carbon. Medium steel (0.2% to 0.6% carbon) is used for structural components, such as beams and girders. Because high-carbon steel (0.8% to 1.5% carbon) is harder than other carbon steels, it is used to make items such as drill bits and knives.

Transition metals are used to produce alloy steels with a specific set of properties. The most common stainless steel contains about 18% chromium and 8% nickel. ■

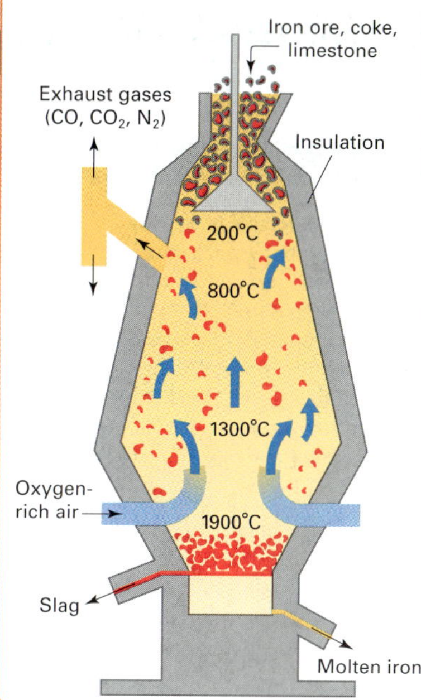

Iron ore typically contains Fe_2O_3, and SiO_2. When heated, limestone produces CaO, which reacts with SiO_2 to form slag. Slag is used to manufacture Portland cement.

In 1779, Abraham Darby III built the world's first cast-iron bridge across the River Severn in England. The bridge is still used by pedestrians.

Did You Know...

Swords made with Damascus steel were highly valued. The source of this quality was the 0.02% **vanadium** in the iron ore that the steel makers used. When they began to use a different source of iron ore, the quality of their steel declined.

Iron and Steel

Various ancient cultures developed techniques for making steel. By 500 B.C., carbon steel was being manufactured in China. There is evidence of ancient forced-draft furnaces on the shores of Lake Victoria, and oral histories indicate that steel was being produced in Africa about 2000 years ago.

In the Bessemer process for making steel, a blast of air is bubbled through molten pig iron in a huge egg-shaped container. The carbon reacts vigorously with the oxygen in the air. With this process, it is not possible to control the final carbon content or the amount of other metals in the alloy.

The open-hearth method uses a shallow lined vessel to produce high quality steel in five to eight hours. The chief advantage of the open-hearth method is that each batch can be analyzed and alloying metals added as needed. When the steel is ready, the furnace is tapped.

In the basic oxygen process, about 300 metric tons of steel can be produced in one hour. The furnace vessel is lined with a basic oxide, such as calcium oxide, and pure oxygen is blown into the molten mass through a narrow tube called a lance at about 10 atmospheres pressure.

Transition Metals

Phytoremediation

Plants used for phytoremediation are called hyperaccumulators because the concentration of a target substance in the plant is greater than its concentration in the soil. One disadvantage of using plants to clean up soil is that the process usually takes longer than the alternatives. One advantage is that the contaminated soil is cleaned up on site. The soil does not have to be dug up and transported to another location for treatment.

Permanent Magnets

Currently, neodymium–iron–boron magnets are the most compact magnetic materials available; that is, they exhibit considerable magnetic strength for their density. These magnets are susceptible to corrosion with the main product of surface corrosion being Nd_2O_3. Because the rate of corrosion increases as the temperature increases, the maximum temperature at which the magnets are useful is 120°C. The magnets are often plated with nickel to prevent corrosion.

Gold

A gold content of 18k and above is rarely used for jewelry because the high gold content makes the gold alloy soft enough to be easily dented. Various metals are added to gold to form alloys. The most common is copper, which preserves the gold color. Metals such as nickel and zinc are used to produce white gold.

Transition Metals

Phytoremediation

Phytoremediation uses plants such as sunflowers, Indian mustard, and dandelions to remove pollutants from contaminated soil and water. The contaminants include organic solvents, pesticides, and toxic metals such as cadmium and chromium.

Plants have a natural ability to absorb nutrients through their roots. Often a plant does not distinguish a toxic metal such as cadmium from a nutrient such as zinc because these metals have similar chemical properties. So cadmium is absorbed and transported to the leaves and stems, where it accumulates. The plants are composted or burned after harvesting. The metal residues are buried in an approved landfill or recovered through smelting. ■

Fe Ni Co Permanent Magnets

Refrigerator magnets contain a barium ferrite, $BaO \cdot 6Fe_2O_3$, or strontium ferrite, $SrO \cdot 6Fe_2O_3$, powder, which is embedded in plastic or rubber. Horseshoe magnets usually contain an alloy of aluminum, nickel, and cobalt.

Iron, nickel, and cobalt are strongly attracted to magnetic fields. When these metals are exposed to a magnetic field, their cations line up in an orderly arrangement. When the field is removed, the ions remain lined up, and the material can act as a magnet. This type of magnetism is called ferromagnetism.

A magnet retains its strength unless it is heated past a point called the Curie temperature. For iron, this temperature is 1043 K. For cobalt, it is 1388 K. For nickel, it is 627 K.

Magnets made from a neodymium, boron, and iron alloy are very powerful. If they are allowed to fly together, they will shatter. They are used to check for counterfeit bills because they can detect tiny magnetic particles placed in the ink of genuine bills. ■

The wrench is strongly attracted to the tiny neodymium-iron-boron cube.

Au Gold

Gold occurs chiefly as small flecks of free metal in veins of quartz. About 5 g of gold is produced from a metric ton (10^6 g) of gold-bearing rock.

Gold can be pounded into sheets so thin that they will transmit light. These sheets, called gold leaf, are used for lettering and decoration in general. Gold is used on the outside surfaces of satellites because it resists corrosion. Its high electrical and thermal conductivity make gold a good choice to plate contacts in microcircuits.

Pure gold is alloyed to make it harder and more durable. Gold alloys are safe to use as fillings for teeth because gold is highly unreactive.

Units called karats (k) are used to describe the purity of gold. Pure gold is 24k or 100% gold. Gold in coins is usually 22k or 92% gold. Gold in rings is often 14k (58%). ■

Micronutrients

Trace amounts of some transition metals are essential for human health.

Iron is found mainly in hemoglobin and myoglobin. Hemoglobin is the protein that transports oxygen in blood. Myoglobin is the protein that stores oxygen in muscle tissue. Vitamin C helps the absorption of iron by promoting the reduction of Fe^{3+} ions to Fe^{2+} ions.

Zinc is a cofactor in many enzymes. It helps protect the immune system. Hormones that control growth and reproduction do not function properly without zinc. A lack of zinc impairs the sense of taste and reduces the appetite.

Copper is a component of enzymes that control the synthesis of melanin, hemoglobin, and phospholipids in the sheath that protects nerves.

Molybdenum affects the absorption of copper. It is also needed for the oxidation of lipids and the metabolism of sulfur and nitrogen.

Chromium assists in the metabolism of glucose and may help to control adult-onset diabetes. A lack of chromium may affect growth.

Manganese is required for the proper function of the nervous system and the thyroid gland. It is needed for glucose metabolism. It helps maintain healthy bones and cartilage.

Cobalt is a component of vitamin B_{12}, which is required for the synthesis of red blood cells. ■

Transition Metal Micronutrients

Element	RDA or AI*	Dietary Sources
Iron	10 mg RDA (M) 20 mg RDA (F)	liver, green vegetables, egg yolk, fish, whole wheat, nuts, oatmeal, molasses, and beans
Zinc	11 mg RDA (M) 8 mg RDA (F)	liver, eggs, meat, milk, whole grains, and shellfish
Copper	900 μg RDA	beans, peas, and shellfish
Molybdenum	45 μg RDA	beans, peas, and whole grains
Chromium	35 μg AI (M) 25 μg AI (F)	meat and whole grains
Manganese	2.3 mg AI (M) 1.8 mg AI (F)	nuts, whole grains, dried fruits, and green leafy vegetables

*Recommended Dietary Allowance or Adequate Intake

Ti Suncreens

You need some exposure to sunlight so your skin cells can make vitamin D, which is needed for healthy bones and teeth. Yet the UV radiation in sunlight can damage skin cells and even lead to skin cancer. The best way to protect your skin is to limit your time in the sun. The next best way is to use a sunscreen.

All the active ingredients in sunscreens protect against UVB light (280–320 nm), which is the primary cause of sunburn. Some protect against UVA light (320–400 nm), which penetrates deeper and causes long-term damage. A sun protective factor rating (SPF) measures only how effective a sunscreen is against UVB, not UVA.

Titanium dioxide, TiO_2, can reflect and scatter UV light. This stable and nonirritating oxide has one drawback. It looks like white paint on the skin. One manufacturer has addressed this problem by decreasing the size of the TiO_2 particles to a diameter of about 21 nm. At this size, TiO_2 appears transparent because its particles are smaller than wavelengths of visible light and light isn't reflected by the particles. ■

Did You Know...

An octopus has blood that is blue, not red, because the compound that transports oxygen in an octopus contains **copper,** not iron. Snails, oysters, and spiders are also bluebloods.

Micronutrients

There are probably more than seven transition metal micronutrients. For example, nickel may affect cell membranes in the liver and vanadium may affect growth. Fe^{2+} ions are absorbed in the stomach and small intestine. Vitamin C helps the absorption by promoting the reduction of Fe^{3+} ions to Fe^{2+} ions.

To Do

Students could research the difference between an RDA and an AI; how these quantities are determined; and why the recommended amounts vary by gender.

Suncreens

Skin responds to UV exposure by producing more melanin, which absorbs some of the damaging radiation. But damage can still occur even if a person has naturally dark skin. An active ingredient in sunscreen absorbs, reflects, or scatters radiation in the UV range of 280–400 nm. Examples of organic active ingredients are avobenzene and oxybenzene, which absorb UV light and dissipate the energy as heat. Inorganic products act by reflecting and scattering UV light. Sunscreens containing both organic and inorganic ingredients can have a very high SPF rating.

To Do

Suggest that students compare the active ingredients and ratings for different sunscreens. Have them check to see if the label claims that the product protects against both UVA and UVB radiation.

Appendix B: Reference Tables

CONTENTS

Table B.1

SI Units and Conversion Factors

Length	**SI Unit: meter (m)**
1 meter	= 1.0936 yards
1 centimeter	= 0.39370 inch
1 inch	= 2.54 centimeters
1 mile	= 5280 feet
	= 1.6093 kilometers

Volume	**SI Unit: cubic meter (m^3)**
1 liter	= 10^{-3} m^3
	= 1.0567 quarts
1 gallon	= 4 quarts
	= 8 pints
	= 3.7854 liters
1 quart	= 32 fluid ounces
	= 0.94635 liter

Temperature	**SI Unit: kelvin (K)**
1 kelvin	= 1 Celsius degree
°C	= (5/9) (°F − 32)
kelvin temperature	= °C + 273.15

Mass	**SI Unit: kilogram (kg)**
1 kilogram	= 1000 grams
	= mass weighing 2.2046 lbs
1 amu	= 1.66057×10^{-27} kilograms

Time	**SI Unit: second (s)**
1 hour	= 60 minutes
1 hour	= 3600 seconds

Energy	**SI Unit: joule (J)**
1 joule	= 1 $kg \cdot m^2/s^2$ (exact)
1 joule	= 0.23901 calorie
1 calorie	= 4.184 joules

Pressure	**SI Unit: Pascal (Pa)**
1 atmosphere	= 101.3 kilopascals
	= 760 mm Hg (torr)
	= 14.70 pounds per square inch

Table B.2

Some Properties of the Elements

Element	Symbol	Atomic number	Atomic mass	Melting point (°C)	Boiling point (°C)	Density (g/cm^3) (gases at STP)	Oxidation numbers
Actinium	Ac	89	(227)	1050	3200	10.07	+3
Aluminum	Al	13	26.98154	660.37	2467	2.6989	+3
Americium	Am	95	243	994	2607	13.67	+3, +4, +5, +6
Antimony	Sb	51	121.75	630.74	1587	6.691	−3, +3, +5
Argon	Ar	18	39.948	−189.2	−185.7	0.0017837	
Arsenic	As	33	74.9216	817	613	5.73	−3, +3, +5
Astatine	At	85	(210)	302	337	—	
Barium	Ba	56	137.33	725	1640	3.5	+2
Berkelium	Bk	97	(247)	986	—	14.78	
Beryllium	Be	4	9.01218	1278	2970	1.848	+2
Bismuth	Bi	83	208.9804	271.3	1560	9.747	+3, +5
Bohrium	Bh	107	(264)	—	—	—	
Boron	B	5	10.81	2075	3675	2.34	+3
Bromine	Br	35	79.904	−7.2	58.78	3.12	−1, +1, +5
Cadmium	Cd	48	112.41	320.9	765	8.65	+2
Calcium	Ca	20	40.08	839	1484	1.55	+2
Californium	Cf	98	(251)	900	—	14	
Carbon	C	6	12.011	3550	4827	2.267	−4, +2, +4
Cerium	Ce	58	140.12	799	3426	6.657	+3, +4
Cesium	Cs	55	132.9054	28.40	669.3	1.873	+1
Chlorine	Cl	17	35.453	−100.98	−34.6	0.003214	−1, +1, +5, +7
Chromium	Cr	24	51.996	1907	2672	7.18	+2, +3, +6
Cobalt	Co	27	58.9332	1495	2870	8.9	+2, +3
Copper	Cu	29	63.546	1083.4	2567	8.96	+1, +2
Curium	Cm	96	(247)	1340	—	13.51	+3
Darmstadtium	Ds	110	(269)	—	—	—	
Dubnium	Db	105	(262)	—	—	—	
Dysprosium	Dy	66	162.50	1412	2562	8.550	+3
Einsteinium	Es	99	(252)	—	—	—	
Erbium	Er	68	167.26	159	2863	9.066	+3
Europium	Eu	63	151.96	822	1597	5.243	+2, +3
Fermium	Fm	100	(257)	—	—	—	
Fluorine	F	9	18.998403	−219.62	−188.54	0.00181	−1
Francium	Fr	87	(223)	27	677	—	+1
Gadolinium	Gd	64	157.25	1313	3266	7.9004	+3
Gallium	Ga	31	69.72	29.78	2204	5.904	+3
Germanium	Ge	32	72.59	937.4	2830	5.323	+2, +4
Gold	Au	79	196.9665	1064.43	2856	19.3	+1, +3
Hafnium	Hf	72	178.49	2227	4602	13.31	+4
Hassium	Hs	108	(265)	—	—	—	
Helium	He	2	4.00260	−272.2	−268.934	0.001785	
Holmium	Ho	67	164.9304	1474	2695	8.795	+3
Hydrogen	H	1	1.00794	−259.14	−252.87	0.00008988	−1, +1
Indium	In	49	114.82	156.61	2080	7.31	+1, +3
Iodine	I	53	126.9045	113.5	184.35	4.93	−1, +1, +5, +7
Iridium	Ir	77	192.22	2410	4130	22.42	+3, +4

Table B.2

Some Properties of the Elements (cont.)							
Element	**Symbol**	**Atomic number**	**Atomic mass**	**Melting point (°C)**	**Boiling point (°C)**	**Density (g/cm^3) (gases at STP)**	**Oxidation numbers**
Iron	Fe	26	55.847	1535	2750	7.874	+2, +3
Krypton	Kr	36	83.80	−156.6	−152.30	0.003733	
Lanthanum	La	57	138.9055	921	3457	6.145	+3
Lawrencium	Lr	103	(262)	—	—	—	+3
Lead	Pb	82	207.2	327.502	1740	11.35	+2, +4
Lithium	Li	3	6.941	180.54	1342	0.534	+1
Lutetium	Lu	71	174.967	1663	3395	9.840	+3
Magnesium	Mg	12	24.305	648.8	1107	1.738	+2
Manganese	Mn	25	54.9380	1244	1962	7.32	+2, +3, +4, +7
Meitnerium	Mt	109	(268)	—	—	—	
Mendelevium	Md	101	257	—	—	—	+2, +3
Mercury	Hg	80	200.59	−38.842	356.58	13.55	+1, +2
Molybdenum	Mo	42	95.94	2617	4612	10.22	+6
Neodymium	Nd	60	144.24	1021	3068	6.90	+3
Neon	Ne	10	20.179	−248.67	−246.048	0.0008999	
Neptunium	Np	93	(237)	640	3902	20.25	+3, +4, +5, +6
Nickel	Ni	28	58.69	1453	2732	8.902	+2, +3
Niobium	Nb	41	92.9064	2468	4742	8.57	+3, +5
Nitrogen	N	7	14.0067	−209.86	−195.8	0.0012506	−3, +3, +5
Nobelium	No	102	(259)	—	—	—	+2, +3
Osmium	Os	76	190.2	3045	5027	22.57	+3, +4
Oxygen	O	8	15.9994	−218.4	−182.962	0.001429	−2
Palladium	Pd	46	106.42	1554	2970	12.02	+2, +4
Phosphorus	P	15	30.97376	44.1	280	1.82	−3, +3, +5
Platinum	Pt	78	195.08	1772	3627	21.45	+2, +4
Plutonium	Pu	94	(244)	641	3232	19.84	+3, +4, +5, +6
Polonium	Po	84	(209)	254	962	9.32	+2, +4
Potassium	K	19	39.0982	63.25	760	0.862	+1
Praseodymium	Pr	59	140.9077	931	3512	6.64	+3
Promethium	Pm	61	(145)	1168	2460	7.22	+3
Protactinium	Pa	91	231.0359	1560	4027	15.37	+4, +5
Radium	Ra	88	(226)	700	1140	5.5	+2
Radon	Rn	86	(222)	−71	−61.8	0.00973	
Rhenium	Re	75	186.207	3180	5627	21.02	+4, +6, +7
Rhodium	Rh	45	102.9055	1966	3727	12.41	+3
Roentgenium	Rg	111	(272)	—	—	—	
Rubidium	Rb	37	85.4678	38.89	686	1.532	+1
Ruthenium	Ru	44	101.07	2310	3900	12.41	+3
Rutherfordium	Rf	104	(261)	—	—	—	
Samarium	Sm	62	150.36	1077	1791	7.520	+2, +3
Scandium	Sc	21	44.9559	1541	2831	2.989	+3
Seaborgium	Sg	106	(263)	—	—	—	
Selenium	Se	34	78.96	217	684.9	4.79	−2, +4, +6
Silicon	Si	14	28.0855	1410	2355	2.33	−4, +2, +4
Silver	Ag	47	107.8682	961.93	2212	10.50	+1

Table B.2

Some Properties of the Elements (cont.)							
Element	**Symbol**	**Atomic number**	**Atomic mass**	**Melting point (°C)**	**Boiling point (°C)**	**Density (g/cm³) (gases at STP)**	**Oxidation numbers**
Sodium	Na	11	22.98977	97.81	882.9	0.971	+1
Strontium	Sr	38	87.62	769	1381	2.63	+2
Sulfur	S	16	32.06	112.8	444.7	2.07	−2, +4, +6
Tantalum	Ta	73	180.9479	2996	5425	16.654	+5
Technetium	Tc	43	(98)	2172	4877	11.50	+4, +6, +7
Tellurium	Te	52	127.60	449.5	989.8	6.24	−2, +4, +6
Terbium	Tb	65	158.9254	1356	3123	8.229	+3
Thallium	Tl	81	204.383	303.5	1457	11.85	+1, +3
Thorium	Th	90	232.0381	1750	4790	11.72	+4
Thulium	Tm	69	168.9342	1545	1947	9.321	+3
Tin	Sn	50	118.69	231.968	2270	7.31	+2, +4
Titanium	Ti	22	47.88	1660	3287	4.54	+2, +3, +4
Tungsten	W	74	183.85	3410	5660	19.3	+6
Ununbium	Uub	112	(277)	—	—	—	
Ununquadium	Uuq	114	—	—	—	—	
Uranium	U	92	238.0289	1132.3	3818	18.95	+3, +4, +5, +6
Vanadium	V	23	50.9415	1890	3380	6.11	+2, +3, +4, +5
Xenon	Xe	54	131.29	−111.9	−107.1	0.005887	
Ytterbium	Yb	70	173.04	819	1194	6.965	+2, +3
Yttrium	Y	39	88.9059	1522	3338	4.469	+3
Zinc	Zn	30	65.38	419.58	907	7.133	+2
Zirconium	Zr	40	91.22	1852	4377	6.506	+4

Table B.3

Physical Constants	
Atomic mass unit	1 amu = 1.6605×10^{-24} g
Avogadro's number	$N = 6.0221 \times 10^{23}$ particles/mol
Gas constant	$R = 8.31$ L·kPa/K·mol
Ideal gas molar volume	$V_m = 22.414$ L/mol
Masses of subatomic particles	
Electron (e^-)	$m_e = 0.0005486$ amu $= 9.1096 \times 10^{-28}$g
Proton (p^+)	$m_p = 1.007277$ amu $= 1.67261 \times 10^{-24}$g
Neutron (n^0)	$m_n = 1.008665$ amu $= 1.67492 \times 10^{-24}$g
Speed of light (in vacuum)	$c = 2.997925 \times 10^8$ m/s

Table B.4

Electron Configurations of the Elements

Elements		Sublevels																		
		1*s*	2*s*	2*p*	3*s*	3*p*	3*d*	4*s*	4*p*	4*d*	4*f*	5*s*	5*p*	5*d*	5*f*	6*s*	6*p*	6*d*	7*s*	7*p*
1	Hydrogen	1																		
2	Helium	2																		
3	Lithium	2	1																	
4	Beryllium	2	2																	
5	Boron	2	2	1																
6	Carbon	2	2	2																
7	Nitrogen	2	2	3																
8	Oxygen	2	2	4																
9	Fluorine	2	2	5																
10	Neon	2	2	6																
11	Sodium	2	2	6	1															
12	Magnesium	2	2	6	2															
13	Aluminum	2	2	6	2	1														
14	Silicon	2	2	6	2	2														
15	Phosphorus	2	2	6	2	3														
16	Sulfur	2	2	6	2	4														
17	Chlorine	2	2	6	2	5														
18	Argon	2	2	6	2	6														
19	Potassium	2	2	6	2	6		1												
20	Calcium	2	2	6	2	6		2												
21	Scandium	2	2	6	2	6	1	2												
22	Titanium	2	2	6	2	6	2	2												
23	Vanadium	2	2	6	2	6	3	2												
24	Chromium	2	2	6	2	6	5	1												
25	Manganese	2	2	6	2	6	5	2												
26	Iron	2	2	6	2	6	6	2												
27	Cobalt	2	2	6	2	6	7	2												
28	Nickel	2	2	6	2	6	8	2												
29	Copper	2	2	6	2	6	10	1												
30	Zinc	2	2	6	2	6	10	2												
31	Gallium	2	2	6	2	6	10	2	1											
32	Germanium	2	2	6	2	6	10	2	2											
33	Arsenic	2	2	6	2	6	10	2	3											
34	Selenium	2	2	6	2	6	10	2	4											
35	Bromine	2	2	6	2	6	10	2	5											
36	Krypton	2	2	6	2	6	10	2	6											
37	Rubidium	2	2	6	2	6	10	2	6			1								
38	Strontium	2	2	6	2	6	10	2	6			2								
39	Yttrium	2	2	6	2	6	10	2	6	1		2								
40	Zirconium	2	2	6	2	6	10	2	6	2		2								
41	Niobium	2	2	6	2	6	10	2	6	4		1								
42	Molybdenum	2	2	6	2	6	10	2	6	5		1								
43	Technetium	2	2	6	2	6	10	2	6	5		2								
44	Ruthenium	2	2	6	2	6	10	2	6	7		1								
45	Rhodium	2	2	6	2	6	10	2	6	8		1								
46	Palladium	2	2	6	2	6	10	2	6	10										

Table B.4

Electron Configurations of the Elements (cont.)

Elements		Sublevels																		
		1*s*	2*s*	2*p*	3*s*	3*p*	3*d*	4*s*	4*p*	4*d*	4*f*	5*s*	5*p*	5*d*	5*f*	6*s*	6*p*	6*d*	7*s*	7*p*
47	Silver	2	2	6	2	6	10	2	6	10		1								
48	Cadmium	2	2	6	2	6	10	2	6	10		2								
49	Indium	2	2	6	2	6	10	2	6	10		2	1							
50	Tin	2	2	6	2	6	10	2	6	10		2	2							
51	Antimony	2	2	6	2	6	10	2	6	10		2	3							
52	Tellurium	2	2	6	2	6	10	2	6	10		2	4							
53	Iodine	2	2	6	2	6	10	2	6	10		2	5							
54	Xenon	2	2	6	2	6	10	2	6	10		2	6							
55	Cesium	2	2	6	2	6	10	2	6	10		2	6			1				
56	Barium	2	2	6	2	6	10	2	6	10		2	6			2				
57	Lanthanum	2	2	6	2	6	10	2	6	10		2	6	1		2				
58	Cerium	2	2	6	2	6	10	2	6	10	1	2	6	1		2				
59	Praseodymium	2	2	6	2	6	10	2	6	10	3	2	6			2				
60	Neodymium	2	2	6	2	6	10	2	6	10	4	2	6			2				
61	Promethium	2	2	6	2	6	10	2	6	10	5	2	6			2				
62	Samarium	2	2	6	2	6	10	2	6	10	6	2	6			2				
63	Europium	2	2	6	2	6	10	2	6	10	7	2	6			2				
64	Gadolinium	2	2	6	2	6	10	2	6	10	7	2	6	1		2				
65	Terbium	2	2	6	2	6	10	2	6	10	9	2	6			2				
66	Dysprosium	2	2	6	2	6	10	2	6	10	10	2	6			2				
67	Holmium	2	2	6	2	6	10	2	6	10	11	2	6			2				
68	Erbium	2	2	6	2	6	10	2	6	10	12	2	6			2				
69	Thulium	2	2	6	2	6	10	2	6	10	13	2	6			2				
70	Ytterbium	2	2	6	2	6	10	2	6	10	14	2	6			2				
71	Lutetium	2	2	6	2	6	10	2	6	10	14	2	6	1		2				
72	Hafnium	2	2	6	2	6	10	2	6	10	14	2	6	2		2				
73	Tantalum	2	2	6	2	6	10	2	6	10	14	2	6	3		2				
74	Tungsten	2	2	6	2	6	10	2	6	10	14	2	6	4		2				
75	Rhenium	2	2	6	2	6	10	2	6	10	14	2	6	5		2				
76	Osmium	2	2	6	2	6	10	2	6	10	14	2	6	6		2				
77	Iridium	2	2	6	2	6	10	2	6	10	14	2	6	7		2				
78	Platinum	2	2	6	2	6	10	2	6	10	14	2	6	9		1				
79	Gold	2	2	6	2	6	10	2	6	10	14	2	6	10		1				
80	Mercury	2	2	6	2	6	10	2	6	10	14	2	6	10		2				
81	Thallium	2	2	6	2	6	10	2	6	10	14	2	6	10		2	1			
82	Lead	2	2	6	2	6	10	2	6	10	14	2	6	10		2	2			
83	Bismuth	2	2	6	2	6	10	2	6	10	14	2	6	10		2	3			
84	Polonium	2	2	6	2	6	10	2	6	10	14	2	6	10		2	4			
85	Astatine	2	2	6	2	6	10	2	6	10	14	2	6	10		2	5			
86	Radon	2	2	6	2	6	10	2	6	10	14	2	6	10		2	6			
87	Francium	2	2	6	2	6	10	2	6	10	14	2	6	10		2	6		1	
88	Radium	2	2	6	2	6	10	2	6	10	14	2	6	10		2	6		2	
89	Actinium	2	2	6	2	6	10	2	6	10	14	2	6	10		2	6	1	2	
90	Thorium	2	2	6	2	6	10	2	6	10	14	2	6	10		2	6	2	2	
91	Protactinium	2	2	6	2	6	10	2	6	10	14	2	6	10	2	2	6	1	2	
92	Uranium	2	2	6	2	6	10	2	6	10	14	2	6	10	3	2	6	1	2	

Table B.4

Electron Configurations of the Elements (cont.)

Elements		Sublevels																		
		1*s*	2*s*	2*p*	3*s*	3*p*	3*d*	4*s*	4*p*	4*d*	4*f*	5*s*	5*p*	5*d*	5*f*	6*s*	6*p*	6*d*	7*s*	7*p*
93	Neptunium	2	2	6	2	6	10	2	6	10	14	2	6	10	4	2	6	1	2	
94	Plutonium	2	2	6	2	6	10	2	6	10	14	2	6	10	6	2	6		2	
95	Americium	2	2	6	2	6	10	2	6	10	14	2	6	10	7	2	6		2	
96	Curium	2	2	6	2	6	10	2	6	10	14	2	6	10	7	2	6	1	2	
97	Berkelium	2	2	6	2	6	10	2	6	10	14	2	6	10	9	2	6		2	
98	Californium	2	2	6	2	6	10	2	6	10	14	2	6	10	10	2	6		2	
99	Einsteinium	2	2	6	2	6	10	2	6	10	14	2	6	10	11	2	6		2	
100	Fermium	2	2	6	2	6	10	2	6	10	14	2	6	10	12	2	6		2	
101	Mendelevium	2	2	6	2	6	10	2	6	10	14	2	6	10	13	2	6		2	
102	Nobelium	2	2	6	2	6	10	2	6	10	14	2	6	10	14	2	6		2	
103	Lawrencium	2	2	6	2	6	10	2	6	10	14	2	6	10	14	2	6	1	2	
104	Rutherfordium	2	2	6	2	6	10	2	6	10	14	2	6	10	14	2	6	2	2	
105	Dubnium	2	2	6	2	6	10	2	6	10	14	2	6	10	14	2	6	3	2	
106	Seaborgium	2	2	6	2	6	10	2	6	10	14	2	6	10	14	2	6	4	2	
107	Bohrium	2	2	6	2	6	10	2	6	10	14	2	6	10	14	2	6	5	2	
108	Hassium	2	2	6	2	6	10	2	6	10	14	2	6	10	14	2	6	6	2	
109	Meitnerium	2	2	6	2	6	10	2	6	10	14	2	6	10	14	2	6	7	2	
110	Darmstadium	2	2	6	2	6	10	2	6	10	14	2	6	10	14	2	6	9	1	
111	Unununium	2	2	6	2	6	10	2	6	10	14	2	6	10	14	2	6	10	1	
112	Ununbium	2	2	6	2	6	10	2	6	10	14	2	6	10	14	2	6	10	2	
114	Ununquadium	2	2	6	2	6	10	2	6	10	14	2	6	10	14	2	6	10	2	2

Table B.5

Common Ions

Positive Ions (Cations)					
1+ charge		2+ charge		3+ charge	
H^+	hydrogen	Mg^{2+}	magnesium	Al^{3+}	aluminum
Na^+	sodium	Fe^{2+}	iron(II)	Fe^{3+}	iron(III)
K^+	potassium	Co^{2+}	cobalt(II)	Co^{3+}	cobalt(III)
NH_4^+	ammonium	Ni^{2+}	nickel(II)	Ni^{3+}	nickel(III)
Li^+	lithium	Ca^{2+}	calcium		
Ag^+	silver	Zn^{2+}	zinc		
Cu^+	copper(I)	Cu^{2+}	copper(II)		
Negative Ions (Anions)					
1− charge		2− charge		3− charge	
F^-	fluoride	O^{2-}	oxide	N^{3-}	nitride
Cl^-	chloride	S^{2-}	sulfide	P^{3-}	phosphide
Br^-	bromide	SO_4^{2-}	sulfate	PO_4^{3-}	phosphate
OH^-	hydroxide	CO_3^{2-}	carbonate		

Table B.6

Other Symbols and Abbreviations					
α	alpha rays	H_f	heat of formation	m	molality
β	beta rays	h	hour	mL	milliliter (*volume*)
γ	gamma rays	h	Planck's constant	mm	millimeter (*length*)
Δ	change in	Hz	hertz (*frequency*)	mol	mole (*amount*)
$\delta+, \delta-$	partial ionic charge	J	joule (*energy*)	mp	melting point
λ	wavelength	K	kelvin (*temperature*)	N	normality
π	pi bond	K_a	acid dissociation constant	n^0	neutron
σ	sigma bond	K_b	base dissociation constant	n	number of moles
ν	frequency	K_b	molal boiling point	n	principal quantum number
amu	atomic mass unit		elevation constant	P	pressure
(*aq*)	aqueous solution	K_{eq}	equilibrium constant	p^+	proton
atm	atmosphere (*pressure*)	K_f	molal freezing point	Pa	pascal (*pressure*)
bp	boiling point		depression constant	R	ideal gas constant
°C	degree Celsius (*temperature*)	K_w	ion product constant	S	entropy
c	speed of light in a vacuum		for water	s	second
cm	centimeter (*length*)	K_{sp}	solubility product constant	(*s*)	solid
E	energy	kcal	kilocalorie (*energy*)	SI	International System
e^-	electron	kg	kilogram (*mass*)		of Units
fp	freezing point	kPa	kilopascal (*pressure*)	STP	standard temperature
G	Gibb's free energy	L	liter (*volume*)		and pressure
g	gram (*mass*)	(*l*)	liquid	T	temperature
(*g*)	gas	M	molarity	$t_{\frac{1}{2}}$	half-life
gfm	gram formula mass	m	meter (length)	V	volume
H	enthalpy	m	mass	v	velocity, speed

Table B.7

Symbols of Common Elements					
Ag	silver	Cu	copper	O	oxygen
Al	aluminum	F	fluorine	P	phosphorus
As	arsenic	Fe	iron	Pb	lead
Au	gold	H	hydrogen	Pt	platinum
Ba	barium	Hg	mercury	S	sulfur
Bi	bismuth	I	iodine	Sb	antimony
Br	bromine	K	potassium	Sn	tin
C	carbon	Mg	magnesium	Sr	strontium
Ca	calcium	Mn	manganese	Ti	titanium
Cl	chlorine	N	nitrogen	U	uranium
Co	cobalt	Na	sodium	W	tungsten
Cr	chromium	Ni	nickel	Zn	zinc

Table B.8

Names and Charges of Polyatomic Ions

1− charge		2− charge		3− charge	
Name	**Formula**	**Name**	**Formula**	**Name**	**Formula**
Chlorate	ClO_3^-	Carbonate	CO_3^{2-}	Phosphate	PO_4^{3-}
Chlorite	ClO_2^-	Chromate	CrO_4^{2-}	Phosphite	PO_3^{3-}
Cyanide	CN^-	Dichromate	$Cr_2O_7^{2-}$		
Dihydrogen phosphate	$H_2PO_4^-$	Oxalate	$C_2O_4^{2-}$	**1+ charge**	
Ethanoate	CH_3COO^-	Peroxide	O_2^{2-}	**Name**	**Formula**
Thiosulfate	$S_2O_3^{2-}$	Silicate	SiO_3^{2-}	Ammonium	NH_4^+
Hydroxide	OH^-	Sulfate	SO_4^{2-}		
Hydrogen carbonate	HCO_3^-	Sulfite	SO_3^{2-}		
Hydrogen sulfate	HSO_4^-				
Hydrogen sulfite	HSO_3^-				
Hypochlorite	ClO^-				
Nitrate	NO_3^-				
Nitrite	NO_2^-				
Perchlorate	ClO_4^-				
Permanganate	MnO_4^-				
Thiocyanate	SCN^-				

Table B.9

Solubilities of Compounds at 25°C and 101.3 kpa

	acetate	bromide	carbonate	chlorate	chloride	hydroxide	iodide	nitrate	oxide	perchlorate	phosphate	sulfate	sulfide
aluminum	S	S	X	S	S	I	S	S	I	S	I	S	d
ammonium	S	S	S	S	S	X	S	S	X	S	S	S	S
barium	S	S	I	S	S	S	S	S	sS	S	I	I	d
calcium	S	S	I	S	S	S	S	S	sS	S	I	sS	I
copper(II)	S	S	X	S	S	I	S	S	I	S	I	S	I
iron(II)	S	S	I	S	S	I	S	S	I	S	I	S	I
iron(III)	S	S	X	S	S	I	S	S	I	S	I	sS	d
lithium	S	S	sS	S	S	S	S	S	S	S	sS	S	S
magnesium	S	S	I	S	S	I	S	S	I	S	I	S	d
potassium	S	S	S	S	S	S	S	S	S	S	S	S	S
silver	sS	I	I	S	I	X	I	S	I	S	I	sS	I
sodium	S	S	S	S	S	S	S	S	S	S	S	S	S
strontium	S	S	I	S	S	S	S	S	S	S	I	I	I
zinc	S	S	I	S	S	I	S	S	I	S	I	S	I

Key:
S = soluble
sS = slightly soluble
I = insoluble
d = decomposes in water
X = no such compound

Table B.10

Rules for Naming Ions

Positive Ions (Cations)

Rule	Example	Name
1. Cations formed from metals have the same name as the element.	Na^{+}	sodium ion
2a. For a cation that can have different charges, indicate the charge by Roman numerals in parentheses.	Fe^{3+} Fe^{2+}	iron(III) ion iron(II) ion
2b. Or, instead of 2a., use a root name with the suffix *-ic* to denote the ion whose positive charge is larger, or the suffix *-ous* to denote the ion whose positive charge is smaller.	Fe^{3+} Fe^{2+}	ferric ion ferrous ion
3. Cations formed from nonmetals end in *-ium.*	NH_4^{+}	ammonium ion

Negative Ions (Anions)

Rule	Example	Name
1. For single-atom ions, replace the end of the element's name with *-ide.*	H^{-}	hydride ion
2. A few simple polyatomic anions have names ending in *-ide.*	OH^{-}	hydroxide ion
3. For polyatomic ions containing oxygen, or *oxyanions,* the most common form of the ion has a name ending in *-ate.*	CO_3^{2-} ClO_3^{-}	carbonate ion chlorate ion
4. Use the suffix *-ite* to indicate one less oxygen than for the suffix *-ate.*	ClO_2^{-}	chlorite ion
5. Use the prefix *hypo-* to indicate one less oxygen atom than for the *-ite* ending.	ClO^{-}	hypochlorite ion
6. Use the prefix *per-* to indicate one more oxygen than for the *-ate* ending.	ClO_4^{-}	perchlorate ion
7. Use the prefix *thio-* to indicate a sulfur atom has replaced an oxygen atom.	SO_4^{2-} $S_2O_3^{2-}$	sulfate ion thiosulfate ion
8. Use *bi-* before the anion name to indicate the presence of a replaceable hydrogen.	CO_3^{2-} HCO_3^{-}	carbonate ion bicarbonate ion

Math Handbook

Scientific Notation

Use Visuals

Point out that 1 micron equals 10^{-6} meter. Ask, **If the lattice has an area of 9.0×10^6 square microns, what is this area in square meters?** *(9.0×10^{-6} m^2)* **In square millimeters?** *(9.0 mm^2)*

Scientific Notation

Scientists use photonic lattices to trap or bend light within extremely tiny spaces. The micrograph above shows a photonic lattice made of silicon rods. Each rod is 1.2 microns, or 1.2×10^{-6} meter, wide.

Very large and very small numbers are often expressed in scientific notation (also known as exponential form). In scientific notation, a number is written as the product of two numbers: a coefficient, and 10 raised to a power. For example, the number 84,000 written in scientific notation is 8.4×10^4. The coefficient in this number is 8.4. In scientific notation, the coefficient is always a number greater than or equal to one and less than ten. The power of ten, or exponent, in this example is 4. The exponent indicates how many times the coefficient 8.4 must be multiplied by 10 to equal the number 84,000.

$$8.4 \times 10^4 = 8.4 \times 10 \times 10 \times 10 \times 10 = 84{,}000$$

exponential form (scientific notation) — standard form

When writing numbers greater than ten in scientific notation, the exponent is equal to the number of places that the decimal point has been moved to the left.

$$6{,}300{,}000 = 6.3 \times 10^6 \qquad 94{,}700 = 9.47 \times 10^4$$

6 places — 4 places

Numbers less than one have a negative exponent when written in scientific notation. For example, the number 0.000 25 written in scientific notation is 2.5×10^{-4}. The negative exponent -4 indicates that the coefficient 2.5 must be divided four times by 10 to equal the number 0.000 25, as shown below.

$$2.5 \times 10^{-4} = \frac{2.5}{10 \times 10 \times 10 \times 10} = 0.000\ 25$$

exponential form (scientific notation) — standard form

When writing numbers less than one in scientific notation, the value of the exponent equals the number of places the decimal has been moved to the right. The sign of the exponent is negative.

$$0.000\ 008 = 8 \times 10^{-6} \qquad 0.00736 = 7.36 \times 10^{-3}$$

6 places — 3 places

If your calculator has an exponent key, you can enter numbers in scientific notation when doing calculations. See the section on using a calculator (pages R62–R65) for more information on calculator operations that involve scientific notation.

Multiplication and Division

To multiply numbers written in scientific notation, multiply the coefficients and add the exponents.

$(3 \times 10^4) \times (2 \times 10^2) = (3 \times 2) \times 10^{4+2} = 6 \times 10^6$
$(2.1 \times 10^3) \times (4.0 \times 10^{-7}) = (2.1 \times 4.0) \times 10^{3+(-7)} = 8.4 \times 10^{-4}$

To divide numbers written in scientific notation, divide the coefficients and subtract the exponent in the denominator from the exponent in the numerator.

$$\frac{3.0 \times 10^5}{6.0 \times 10^2} = \left(\frac{3.0}{6.0}\right) \times 10^{5-2} = 0.5 \times 10^3 = 5.0 \times 10^2$$

Addition and Subtraction

If you want to add or subtract numbers expressed in scientific notation and you are not using a calculator, then the exponents must be the same. For example, suppose you want to calculate the sum of 5.4×10^3 and 8.0×10^2. First, rewrite the second number so that the exponent is a 3.

$$8.0 \times 10^2 = 0.80 \times 10^3$$

Now add the numbers.

$$(5.4 \times 10^3) + (0.80 \times 10^3) = (5.4 + 0.80) \times 10^3 = 6.2 \times 10^3$$

Follow the same rule when you subtract numbers expressed in scientific notation without the aid of a calculator.

$$(3.42 \times 10^{-5}) - (2.5 \times 10^{-6}) = (3.42 \times 10^{-5}) - (0.25 \times 10^{-5})$$
$$= (3.42 - 0.25) \times 10^{-5} = 3.17 \times 10^{-5}$$

SAMPLE PROBLEM MH-1

Using Scientific Notation in Arithmetic Operations

Solve each problem, and express your answer in correct scientific notation.

a. $(8.0 \times 10^{-2}) \times (7.0 \times 10^{-5})$
b. $(7.1 \times 10^{-2}) + (5 \times 10^{-3})$

Solution

Follow the rules described above for multiplying and adding numbers expressed in scientific notation.

a. $(8.0 \times 10^{-2}) \times (7.0 \times 10^{-5}) = (8.0 \times 7.0) \times 10^{-2+(-5)}$
$= 56 \times 10^{-7} = 5.6 \times 10^{-6}$

b. $(7.1 \times 10^{-2}) + (5 \times 10^{-3}) = (7.1 \times 10^{-2}) + (0.5 \times 10^{-2})$
$= (7.1 + 0.5) \times 10^{-2} = 7.6 \times 10^{-2}$

Practice Problems Plus

Solve each problem, and express your answer in scientific notation.

a. $(0.10 \times 10^5) \times (4.9 \times 10^{-2})$ *(4.9×10^2)*
b. $(4.8 \times 10^2) \times (0.15 \times 10^4)$ *(7.2×10^5)*
c. $(5.2 \times 10^4) + (2.7 \times 10^4)$ *(7.9×10^4)*

Math Handbook R57

Answers

1. a. 5×10^5 b. 2.852×10^2 c. 4.2×10^{-8} d. 2×10^{-4} e. 3.006×10^{-2} f. 8.37×10^7
2. a. 0.004 b. 340,000 c. 450 d. 0.000 0059
3. a. 8×10^{12} b. 9.3×10^{-2} c. 1×10^2 d. 8.5×10^{-8}
4. a. 7.3×10^{-2} b. 7.1×10^{-8} c. 6.4×10^{-2}
5. a. 5.7×10^{-2} b. 2.1×10^7 c. 3×10^4 d. 7.2×10^2
6. a. 9.14×10^1 m b. 1.54×10^{-10} m c. 8×10^{-6} m d. 1.496×10^{11} m
7. a. 5.6×10^9 nm b. 6.8×10^{-1} L H_2O c. 2.6×10^{-2} mol NaOH
8. 1.35×10^{-3} m^2
9. 1.04×10^{22} gold atoms
10. 5.79×10^{-8} m^3

Practice Problems Plus

Express each measurement in scientific notation.

a. The radius of Earth: 6,378,000 meters *(6.378×10^6 m)*

b. The height of a two-story house: 6.72 meters *(6.72×10^0 m)*

Practice the Math

1. Express each number in scientific notation
   a. 500,000 b. 285.2
   c. 0.000 000 042 d. 0.0002
   e. 0.030 06 f. 83,700,000
2. Write each number in standard form.
   a. 4×10^{-3} b. 3.4×10^5
   c. 0.045×10^4 d. 5.9×10^{-6}
3. Solve each problem and express your answer in scientific notation.
   a. $(2 \times 10^9) \times (4 \times 10^3)$
   b. $(6.2 \times 10^{-3}) \times (1.5 \times 10^1)$
   c. $(10^{-4}) \times (10^8) \times (10^{-2})$
   d. $(3.4 \times 10^{-3}) \times (2.5 \times 10^{-5})$
4. Solve each problem and express your answer in scientific notation.
   a. $(9.4 \times 10^{-2}) - (2.1 \times 10^{-2})$
   b. $(6.6 \times 10^{-8}) + (5.0 \times 10^{-9})$
   c. $(6.7 \times 10^{-2}) - (3.0 \times 10^{-3})$
5. Solve each problem and express your answer in scientific notation.
   a. $\dfrac{(3.8 \times 10^{-3}) \times (1.2 \times 10^6)}{8 \times 10^4}$
   b. $(1.4 \times 10^2) \times (2 \times 10^8) \times (7.5 \times 10^{-4})$
   c. $\dfrac{6.6 \times 10^6}{(8.8 \times 10^{-2}) \times (2.5 \times 10^3)}$
   d. $\dfrac{(1.2 \times 10^{-3})^2}{(10^{-2})^3 \times (2.0 \times 10^{-3})}$
6. Express each measurement in scientific notation.
   a. The length of a football field: 91.4 m.
   b. The diameter of a carbon atom: 0.000 000 000 154 m.
   c. The diameter of a human hair: 0.000 008 m.
   d. The average distance between the centers of the sun and Earth: 149,600,000,000 m.

Applying Scientific Notation to Chemistry

7. The following expressions are solutions to typical chemistry problems. Calculate the answer for each expression. Make sure to cancel units as you write your solutions.
   a. $5.6 \times 10^3 \text{ mm} \times \left(\dfrac{1 \text{ m}}{10^3 \text{ mm}}\right) \times \left(\dfrac{10^9 \text{ nm}}{1 \text{ m}}\right) = ?$
   b. $6.8 \times 10^4 \text{ cg } H_2O \times \left(\dfrac{1 \text{ g } H_2O}{10^2 \text{ cg } H_2O}\right) \times \left(\dfrac{1 \text{ mL } H_2O}{1 \text{ g } H_2O}\right) \times \left(\dfrac{1 \text{ L } H_2O}{10^3 \text{ mL } H_2O}\right) = ?$
   c. $4.0 \times 10^2 \text{ mL NaOH} \times \left(\dfrac{1 \text{ L NaOH}}{10^3 \text{ mL NaOH}}\right) \times \left(\dfrac{6.5 \times 10^{-2} \text{ mol NaOH}}{1 \text{ L NaOH}}\right) = ?$
8. A cube of aluminum measures 1.50×10^{-2} m on each edge. Use the following expression to calculate the surface area of the cube.

$$\text{Surface area} = 6 \times (1.50 \times 10^{-2} \text{ m})^2 = ?$$

9. A small gold (Au) nugget has a mass of 3.40×10^{-3} kg. Use the following expression to calculate the number of gold atoms contained in the nugget.

$$3.40 \times 10^{-3} \text{ kg Au} \times \left(\frac{10^3 \text{ g Au}}{1 \text{ kg Au}}\right) \times \left(\frac{1 \text{ mol Au}}{197.0 \text{ g Au}}\right) \times \left(\frac{6.02 \times 10^{23} \text{ atoms Au}}{1 \text{ mol Au}}\right) = ?$$

10. The volume of a sphere is given by the formula $V = \frac{4}{3}\pi r^3$, where $\pi = 3.14$ and r = radius. What is the volume of a spherical drop of water with a radius of 2.40×10^{-3} m?

Significant Figures

The significant figures of a measurement are those digits known with certainty plus the rightmost digit that is estimated. Every measurement has a certain number of significant figures. For instance, if you measure the air temperature to be 21°C, this measurement has two significant figures.

Counting Significant Figures

Every nonzero digit in a measurement is significant. For example, the measurement 831 g has three significant figures. The rules for when to count zeros in measurements as significant are as follows:

- Zeros in the middle of a number are always significant. A length that measures 507 m has three significant figures.
- Zeros at the beginning of a number are not significant. The measurement 0.0056 m has two significant figures.
- Zeros at the end of a number are only significant if they follow a decimal point. The measurement 35.00 g has four significant figures; the measurement 2400 g has two significant figures.

Counted values (17 beakers) and the numbers in defined relationships (100 cm = 1 m) are exact numbers and are considered to have an unlimited number of significant figures. Exact numbers never affect the number of significant figures in the results of a calculation.

Each of the measurements listed below has three significant figures. The significant figures are underlined.

456 mL	0.305 g
70.4 mg	0.000457 g
5.64×10^3 km	444,000 ng
1.30×10^{-2} m	0.00406 dm

The mass of this bowl of candy is 298.8 grams. The measurement has four significant figures.

SAMPLE PROBLEM MH-2

Counting Significant Figures in Measurements

How many significant figures are in each measurement?

a. 3300°C	**b.** 110.5 kJ
c. 0.000 0176 g	**d.** 210,000 kcal
e. 5 notebooks	**f.** 0.90 lb

Solution

Follow the rules described above for counting significant figures.

a. 2	**b.** 4
c. 3	**d.** 2
e. unlimited	**f.** 2

Significant Figures

Use Visuals

The mass reading shown on the electronic balance contains four significant figures. Ask, **Which digits are known with certainty?** *(The three digits to the left of the decimal: 2, 9, and 8.)* **Which digit is estimated?** *(The rightmost digit, 8.)*

Practice Problems Plus

How many significant figures are in each measurement?

a. 3.0×10^3 **m** *(2)*

b. **864.2 mL** *(4)*

Math Handbook **R59**

Significant Figures in Calculations

When measurements are used in a calculation, the answer you calculate must be rounded to the correct number of significant figures. How you round your answer depends on the mathematical operation used in the calculation.

- In multiplication and division, the answer can have *no more significant figures* than the least number of significant figures in any measurement in the problem.
- In addition and subtraction, the answer can have *no more decimal places* than the least number of decimal places in any measurement in the problem.

When a calculated answer must be rounded to the appropriate number of significant figures, use the following rules:

- If the first nonsignificant digit is less than 5, drop all nonsignificant digits.
- If the first nonsignificant digit is 5, or greater than 5, increase the last significant digit by one and drop all nonsignificant digits.

The Sample Problem below illustrates how to apply these rules when performing calculations involving measurements.

Practice Problems Plus

What is the area of a rectangular room that is 5.2 meters long and 3.9 meters wide? *(2.0×10^1 m^2)*

SAMPLE PROBLEM MH-3

Rounding Calculated Answers

Solve each problem, and round your answer to the correct number of significant figures.

a. $(5.3\text{ m}) \times (1.54\text{ m})$

b. $23.\,5\text{ m} + 2.1\text{ m} + 7.26\text{ m}$

c. $189.427\text{ g} - 19.00\text{ g}$

d. $\dfrac{0.497\text{ m}^2}{1.50\text{ m}}$

Solution

Follow the rules described above for rounding calculated answers to the appropriate number of significant figures.

a. $(5.3\text{ m}) \times (1.54\text{ m}) = 8.162\text{ m}^2 = 8.2\text{ m}^2$
(5.3 m has two significant figures.)

b. $23.5\text{ m} + 2.1\text{ m} + 7.26\text{ m} = 32.86\text{ m} = 32.9\text{ m}$
(2.1 has one decimal place.)

c. $189.427\text{ g} - 19.00\text{ g} = 170.427\text{ g} = 170.43\text{ g}$
(19.00 has two decimal places.)

d. $\dfrac{0.497\text{ m}^2}{1.50\text{ m}} = 0.3313333\text{ m} = 0.331\text{ m}$
(0.497 and 1.50 each have three significant figures.)

R60 Appendix C

Practice the Math

1. How many significant figures are in each measurement?
   a. 0.723 m
   b. 14.0 g
   c. 123,000 m
   d. 6.00×10^{-2} g
   e. 0.005 12 kg
   f. 1050 cm
2. Round each of the measurements in Question 1 to two significant figures.
3. Multiply or divide the following measurements, and round your answer to the correct number of significant figures.
   a. 3.4 m × 7.8 m
   b. 7.00 cm × 9.8 cm
   c. 1.56 mm × 0.864 mm × 14.00 mm
   d. $6.88\ m^2 \div 2.6$ m
   e. 52.98 g ÷ 1.8 mL
   f. 0.14 kg ÷ 0.0131 L
4. Add or subtract the following measurements, and round your answer to the correct number of significant figures.
   a. 2.34 m + 18.28 m
   b. 828.2 g − 134 g
   c. 0.278 cm + 0.0832 cm + 0.15 cm
   d. 54.2 mg − 12.66 mg
   e. 6.40 ng + 0.450 ng + 1.001 ng
   f. $(5.2 \times 10^{-2}$ dg) + $(1.82 \times 10^{-3}$ dg)

Applying Significant Figures to Chemistry

5. Determine the number of significant figures in each measurement.
   a. The density of mercury (Hg): 13.55 kg/L
   b. The number of milligrams (mg) in one gram (g): 1000 mg = 1 g
   c. The number of protons in an atom of copper (Cu): 29 protons
   d. The mass of a silver (Ag) atom: 1.792×10^{-22} g
   e. The melting point of gallium: 29.8°C
   f. The concentration of a water solution of NaCl: 24% (m/v).
6. The data from Question 5 has been used to set up a solution to each of the following problems. Calculate each answer, and express each answer to the correct number of significant figures. Make sure to cancel units where appropriate.
   a. Calculate the mass of 0.45 L of mercury.

   $$0.45\ \text{L Hg} \times \frac{13.55\ \text{kg Hg}}{1\ \text{L Hg}} = ?$$

   b. How many milligrams are in 6.321 grams?

   $$6.321\ \text{g} \times \frac{1000\ \text{mg}}{1\ \text{g}} = ?$$

   c. How many protons are in 7 copper atoms?

   $$7\ \text{Cu atoms} \times \frac{29\ \text{protons}}{1\ \text{Cu atom}} = ?$$

   d. What is the mass of 35 silver atoms?

   $$35\ \text{Ag atoms} \times \frac{1.792 \times 10^{-22}\ \text{g Ag}}{1\ \text{Ag atom}} = ?$$

   e. What is the melting point of gallium in kelvins (K)?
   K = °C + 273 = 29.8 + 273 = ?

   f. How many grams of NaCl are in 25.8 mL of a 24% (m/v) NaCl solution?

   $$25.8\ \text{mL solution} \times \frac{24\ \text{g NaCl}}{100\ \text{mL solution}} = ?$$

7. Calculate the perimeter ((2 × length) + (2 × width)) and the area (length × width) of a garden plot that measures 32.8 m by 15 m. Round each answer to the correct number of significant figures.

Answers

1. a. 3 **b.** 3 **c.** 3 **d.** 3 **e.** 3 **f.** 3

2. a. 0.72 m **b.** 14 g **c.** 1.2×10^5 m **d.** 6.0×10^{-2} mL **e.** 5.1×10^{-3} kg **f.** 1.1×10^3 cm

3. a. 27 m^2 **b.** 69 cm^2 **c.** 18.9 mm^3 **d.** 2.6 m **e.** 29 g/mL **f.** 11 kg/L

4. a. 20.62 m **b.** 694 g **c.** 0.51 cm **d.** 41.5 mg **e.** 7.85 ng **f.** 5.4×10^{-2} dg

5. a. 4 **b.** unlimited **c.** unlimited **d.** 4 **e.** 3 **f.** 2

6. a. 6.1 kg Hg
b. 6.3211×10^3 mg
c. 203 protons
d. 6.272×10^{-21} g Ag
e. 303 K
f. 6.2 g NaCl

7. area = $4.9 \times 10^2\ m^2$;
perimeter = 96 m

Practice Problems Plus

What is the total mass of an apparatus consisting of a 245.3-g beaker, a 34-g thermometer, and a 18.82-g rubber stopper? *(298 g)*

Using a Calculator

Use Visuals

Point out that the locations of certain function keys on students' calculators may not match those in the calculator pictured. Ask students to locate the **[(]**, **[)]**, and **[EE]** (or **[EXP]**) keys on their calculators. Explain to students how these keys are used; the parentheses keys are used to order arithmetic operations, and the exponential key is used to enter numbers in scientific notation.

Using a Calculator

A scientific calculator has specialized function keys that you can use to manipulate numbers and perform arithmetic operations.

You can readily solve chemistry problems involving numeric calculations with the aid of a hand-held scientific calculator. As you know, a calculator has various keys used for entering numbers and performing operations with those numbers. In the following text, calculator keys are indicated by boldfaced square brackets. As you work out each problem, press the calculator keys that match the operations or numbers within the brackets.

Arithmetic Operations

A scientific calculator has operator keys for basic arithmetic operations such as addition **[+]**, subtraction **[−]**, multiplication **[×]**, and division **[÷]**. To complete an arithmetic operation, you must press the operator key for an equals sign, **[=]**.

A different set of keys is used for entering numbers. For example, to enter the number 27.2 on your calculator, press the numeric entry keys **[2][7][.][2]**.

Try the following arithmetic operations on your calculator. Be sure to clear the calculator using the **[ON/C]**, **[C/CE]**, or **[AC]** key before entering any numbers.

	Numeric Entry Keys	Operator Key	Numeric Entry Keys	Operator Key	Display
a. 94.2 + 71.32	**[9][4][.][2]**	**[+]**	**[7][1][.][3][2]**	**[=]**	*165.52*
b. 301.6 − 58.98	**[3][0][1][.][6]**	**[−]**	**[5][8][.][9][8]**	**[=]**	*242.62*
c. 15.72 × 14.23	**[1][5][.][7][2]**	**[×]**	**[1][4][.][2][3]**	**[=]**	*223.6956*
d. 331.72 ÷ 78.23	**[3][3][1][.][7][2]**	**[÷]**	**[7][8][.][2][3]**	**[=]**	*4.240317014*

If we assume that all the digits in each of the numbers entered are significant figures, then the answers displayed by the calculator have more digits than can be justified as significant. Thus, you must round each calculated answer to the appropriate number of significant figures.

For the addition problem **a**, the displayed value of 165.52 must be rounded to 165.5 (no more than 1 decimal place). For the subtraction problem **b**, the displayed value of 242.62 must be rounded to 242.6 (no more than one decimal place). The answers to multiplication problem **c** and division problem **d** should each be reported to four significant figures (the least number of significant figures in either of the numbers used in each calculation). For **c**, the answer should be reported as 223.7. For **d**, the answer should be reported as 4.240.

For more help with significant figures and rounding, go to page R59.

Exponential Numbers

To enter very large or very small numbers on a scientific calculator, use the exponential key. The exponential key on your calculator may be labeled **[EE]**, **[EXP]**, or **[EEX]**. In the Math Handbook, the exponential key will be represented by **[EE]**.

Suppose you want to enter the exponential number 2.94×10^{24} on your calculator. First enter the coefficient **[2][.][9][4]**. Then press **[EE]** and enter the exponent **[2][4]**. The display reads 2.94^{24}. Notice that the "$\times 10$" part of the exponential form is not entered on (or displayed by) the calculator. Pressing the exponential key indicates to the calculator that the next number entered is a power of 10.

If a number written in exponential form has a negative exponent, use the sign-change key **[+/−]** (not the subtraction key!) to make the sign of the exponent negative. For example, to enter the number 3.08×10^{-12} on your calculator, first enter the coefficient **[3][.][0][8]**. Then press **[EE]** followed by **[+/−]**. Notice how the display changes from 3.08^{00} to 3.08^{-00}. Lastly, enter **[1][2]**, the value of the exponent. The display now reads 3.08^{-12}.

You can practice entering exponential numbers on a calculator by trying the following examples. The keystrokes are listed after each exponential number. The numbers displayed by your calculator should match those in the last column.

	Numeric Entry Keys				Display
a. 2.79×10^{4}	**[2][.][7][9]**	**[EE]**		**[4]**	2.79^{04}
b. 7.43×10^{18}	**[7][.][4][3]**	**[EE]**		**[1][8]**	7.43^{18}
c. 6.89×10^{-22}	**[6][.][8][9]**	**[EE]**	**[+/−]**	**[2][2]**	6.89^{-22}
d. 1.11×10^{-5}	**[1][.][1][1]**	**[EE]**	**[+/−]**	**[5]**	1.11^{-05}

The next two examples show operations involving exponential numbers. For simplification, only the exponential and operator keystrokes are shown. The answer that should appear on your calculator display is written in italics.

e. Given expression: $[2.43 \times 10^{23}] \times 7.30 = 1.77 \times 10^{24}$
Using a calculator: 2.43 **[EE]** 23 **[×]** 7.30 **[=]** 1.7739^{24}

f. Given expression: $9.93 \times 10^{15} \div 2.56 \times 10^{13} = 3.88 \times 10^{2}$
Using a calculator: 9.33 **[EE]** 15 **[÷]** 2.56 **[EE]** 13 **[=]** 3.87890625^{02}

The answer for **e** displayed by the calculator is 1.7739×10^{24}. This number must be rounded to 1.77×10^{24}, according to the rules for significant figures in calculations.

The calculated answer for **f** must be rounded to three significant figures. Note that the final answer on the display of your calculator may be in scientific notation as 3.87890625^{02} or in decimal form as *387.890625*, depending on your calculator settings. If you obtained the latter result, you may be able to set the calculator's display mode so that it automatically converts answers to scientific notation by using the key labeled **[SCI]** or **[MODE]**.

Math Handbook **R63**

Practice Problems Plus

Use a calculator to solve for x. Express the answer in scientific notation and report it to the correct number of significant figures.

$$x = \left(\frac{5.78 \times 10^{32}}{2.43 \times 10^{18}}\right) \times \left(\frac{6.89 \times 10^{3}}{5.41 \times 10^{1}}\right)$$

(3.03×10^{16})

Chain Calculations

Calculations that require more than one arithmetic operation are called chain calculations. Examples of chain calculations are shown below.

a. $2.58 + 7.12 + 9.30 = 19.00$

b. $6.3 \times \left(\frac{7.4 \times 10^{-2}}{1.0 \times 10^{2}}\right) \times \left(\frac{2.79 \times 10^{3}}{1.9 \times 10^{-7}}\right) = 6.8 \times 10^{7}$

c. $2.98 + \left(\frac{8.76 \times 10^{-3}}{6.22 \times 10^{-3}}\right) = 4.39$

To solve problem **a**, enter the following sequence of calculator keystrokes:

2.58 [+] 7.12 [+] 9.30 [=] *19.*

Note that the calculator keystrokes are exactly as they appear in reading the equation from left to right. Also notice that the calculated answer should have two decimal places, but the calculator has dropped these digits in the displayed answer because they are zeros. You must add them in reporting the correct answer, 19.00.

Problem **b** illustrates a kind of problem that comes up with startling regularity in chemistry-related dimensional analysis. The correct calculator keystrokes are as follows.

6.3 [×] 7.4 **[EE]**[+/−] 2 [÷] 1.0 **[EE]** 2 [×] 2.79 **[EE]** 3 [÷] 1.9 **[EE]**[+/−] 7 [=] *6.845778947*^{07}

Note that each operation is performed in the order it appears in the equation. The calculated answer must be rounded to two significant figures.

In problem **c**, however, performing the operations in the order they appear will *not* yield a correct answer. Instead, try completing the division operation within the parentheses first, and then add 2.98 to the result. The calculator keystrokes are as follows.

8.76 **[EE]**[+/−] 3 [÷] 6.22 **[EE]**[+/−] 3 [+] 2.98 [=] *4.388360129*

The calculated answer must be rounded to two decimal places.

Alternatively, you can use the open-parenthesis [(] and closed-parenthesis [)] keys. The [(] and [)] keys allow you to key in equation **c** as written, using the following keystrokes.

2.98 [+][(] 8.76 **[EE]**[+/−] 3 [÷] 6.22 **[EE]**[+/−] 3 [)][=] *4.388360129*

SAMPLE PROBLEM MH-4

Arithmetic Operations with Exponential Numbers

Use a calculator to obtain the answer shown below.

$$\frac{2.88 \times 10^{7}}{5.98 \times 10^{-3}} = 4.82 \times 10^{9}$$

Solution

Use the following keystroke sequence:

2.88 **[EE]** 7 [÷] 5.98 **[EE]**[+/−] 3 [=] *4.816053512*^{09}

The calculated answer must be rounded to three significant figures, yielding the result 4.82×10^{9}.

R64 Appendix C

Practice the Math

Use your calculator to solve for x in each problem. Express your answer in scientific notation and report it to the correct number of significant figures.

1. $x = 5467.4 \div 2.7$
2. $x = 26.54 + 26.8 + 58.33 + 10.00 + 87.3$
3. $x = 3.75 \times 10^{-5} + 7.00 \times 10^{-5}$
4. $x = \left(\frac{6.02 \times 10^{23}}{4.77 \times 10^{17}}\right)$
5. $x = 8.73 + \left(\frac{4.17 \times 10^{-3}}{5.27 \times 10^{-4}}\right)$
6. $x = \left(\frac{3.98 \times 10^{5}}{7.215 \times 10^{13}}\right) \times \left(\frac{6.21 \times 10^{14}}{5.123 \times 10^{12}}\right)$
7. $x = 2.45 \div 0.49$
8. $x = 273.0 - 42.6 + 7.0$
9. $x = \left(\frac{5.44 \times 10^{6}}{6.98 \times 10^{4}}\right) \times 3.89$

Applying Calculator Operations to Chemistry

Use your calculator and the given expression to solve each problem. Make sure to cancel units where appropriate.

10. A gold necklace found in an Egyptian pyramid has a volume of 27.5 cm^3 and a mass of 530.0 g. What is the density of the gold?

$$\text{density} = \frac{\text{mass}}{\text{volume}} = \frac{530.0 \text{ g}}{27.5 \text{ cm}^3} = ?$$

11. A magnesium atom has a diameter of 3.20×10^{-10} meter. What would be the length in meters (m) of 1 mole of magnesium atoms (6.02×10^{23} atoms) laid end-to-end in a straight line?

$$\frac{3.20 \times 10^{-10} \text{ m}}{1 \text{ Mg atom}} \times \frac{6.02 \times 10^{23} \text{ Mg atoms}}{1 \text{ mol Mg}} = ?$$

12. The molecular formula of the sugar glucose is $C_6H_{12}O_6$. The molar masses of the constituent elements of glucose are: carbon, 12.011 g/mol; hydrogen, 1.0079 g/mol; and oxygen, 15.999 g/mol. What is the molar mass of glucose to three decimal places?

$$(6 \times 12.011 \text{ g}) + (12 \times 1.0079 \text{ g}) + (6 \times 15.999 \text{ g}) = ?$$

13. The compound benzene, C_6H_6, has a molar mass of 78.11 g/mol, of which 72.00 g is carbon. What percent of the molar mass of benzene is carbon?

$$\frac{72.00 \text{ g C}}{78.11 \text{ g } C_6H_6} \times 100\% = ?$$

14. A gaseous compound has a density of 1.37 g/L at STP and 1 mol occupies 22.4 L at STP. What is the molar mass of the compound?

$$\frac{1.37 \text{ g}}{1 \text{ L}} \times \frac{22.4 \text{ L}}{1 \text{ mol}} = ?$$

15. How many moles of Fe_2O_3 are in 50.0 g?

$$50.0 \text{ g } Fe_2O_3 \times \frac{1.00 \text{ mol } Fe_2O_3}{159.6 \text{ g } Fe_2O_3} = ?$$

16. The formation of copper(II) sulfide (CuS) from its component elements is given below.

$$Cu(s) + S(s) \longrightarrow CuS(s)$$

How many grams of CuS are produced when 2.73 mol Cu reacts with an excess of sulfur?

$$2.73 \text{ mol Cu} \times \frac{1 \text{ mol CuS}}{1 \text{ mol Cu}} \times \frac{95.6 \text{ g CuS}}{1 \text{ mol CuS}} = ?$$

17. The ideal gas constant, *R*, is calculated from the relationship given below. What is the value of *R* in units of (L•kPa)/(K•mol)?

$$R = \frac{101.3 \text{ kPa} \times 22.4 \text{ L}}{273 \text{ K} \times 1.00 \text{ mol}} = ?$$

18. A student sets up the following relationship to calculate the number of moles/liter of sodium chloride, NaCl, in 100 mL of an aqueous salt solution containing 0.50 g of NaCl. How many moles/liter of sodium chloride does the solution contain?

$$\frac{0.50 \text{ g NaCl}}{100 \text{ mL}} \times \frac{1 \text{ mol NaCl}}{58.5 \text{ g NaCl}} \times \frac{1000 \text{ mL}}{1 \text{ L}} = ?$$

Answers

Each correctly rounded answer is followed by the unrounded calculator display value in parentheses.

1. 2.0×10^3 (2.0249629×10^3)
2. 2.1×10^2 (2.0897×10^2)
3. 1.08×10^{-4} (1.075×10^{-4})
4. 2.07×10^7 (2.0697694×10^7)
5. 1.66×10^1 (1.6642713×10^1)
6. 6.69×10^{-7} (6.686732×10^{-7})
7. 5.00×10^0 (5)
8. 2.374×10^2 (2.374×10^2)
9. 3.03×10^2 (3.0317479×10^2)
10. 19.3 g/cm^3 (1.9272727×10^1)
11. 1.93×10^{14} m/mol Mg atoms (1.9264×10^{14})
12. 180.155 g (1.801548×10^2)
13. 92.18% (9.2177698×10^1)
14. 30.7 g/mol (3.0688×10^1)
15. 3.13×10^{-1} mol Fe_2O_3 (3.132832×10^{-1})
16. 1.30×10^2 g CuS (1.30494×10^2)
17. 8.31 (L•kPa)/(K•mol) (8.3117949)
18. 8.5×10^{-2} mol/L NaCl (8.5470085×10^{-2})

Conversion Problems and Dimensional Analysis

Use Visuals

Ask students if they know their height in inches. Then ask, **How can you determine a conversion factor that allows you to convert inches to centimeters?** *(Acceptable answers include the following: Measure a 1-inch line with a centimeter ruler; based on these equivalent measurements (1 inch = 2.54 cm), write a conversion factor that allows you to convert inches to centimeters (2.54 cm/1 in); then multiply the height in inches by the conversion factor.)*

Conversion Problems and Dimensional Analysis

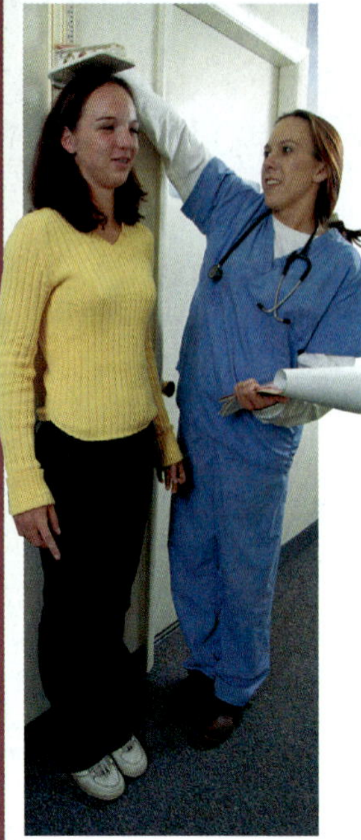

A person's height can be measured in units such as feet, inches, or meters. Conversion factors allow you to convert from one unit to another.

Many problems in both everyday life and in the sciences involve converting measurements. These problems may be simple conversions between the same kinds of measurement. For example:

a. A person is five and one-half feet tall. Express this height in inches.
b. A flask holds 0.575 L of water. How many milliliters of water is this?

In other cases, you may need to convert between different kinds of measurements.

c. How many gallons of gasoline can you buy for \$15.00 if gasoline costs \$1.42/gallon?
d. What is the mass of 254 cm^3 of gold if the density of gold is 19.3 g/cm^3?

More complex conversion problems may require conversions between measurements expressed as ratios of units. Consider the following examples.

e. A car is traveling at 65 miles/hour. What is the speed of the car expressed in feet/second?
f. The density of nitrogen gas is 1.17 g/L. What is the density of nitrogen expressed in micrograms/deciliter (μg/dL)?

Problems **a** through **f** can be solved using a method that is known by a few different names—dimensional analysis, factor label, and unit conversion. These names emphasize the fact that the dimensions, labels, or units of the measurements in a problem—the units in the given measurement(s) as well as the units desired in the answer—can help you write the solution to the problem.

Dimensional analysis makes use of ratios called conversion factors. A conversion factor is a ratio of two quantities equal to one another. For example, to work out problem **a**, you must know the relationship 1 ft = 12 in. The two conversion factors derived from this equality are shown below.

$$\frac{1\ \text{ft}}{12\ \text{in}} = 1\ (\text{unity}) = \frac{12\ \text{in}}{1\ \text{in}}$$

To solve problem **a** by dimensional analysis, you must multiply the given measurement (5.5 ft) by a conversion factor that allows the *feet* units to cancel, leaving the unit *inches*—the unit of the requested answer.

$$5.5\ \cancel{\text{ft}} \times \frac{12\ \text{in}}{1\ \cancel{\text{ft}}} = 66\ \text{in}$$

Carefully study the solutions to the remaining five example problems below. Notice that in each solution, the conversion factors are written so that the unit of the given measurement cancels, leaving the correct unit for each answer. When working conversion problems, the equalities needed to write the conversion factor may be given in the problem. This is true in examples **c** and **d**. In other problems, you need to either know or look up the necessary equalities, as in examples **b**, **e**, and **f**.

b. $0.575\ \cancel{L} \times \frac{10^3\ \text{mL}}{1\ \cancel{L}} = 575\ \text{mL}$

c. $\$15.00 \times \frac{1\ \text{gal}}{\$1.42} = 10.6\ \text{gal}$

d. $254\ \cancel{\text{cm}^3} \times \frac{19.3\ \text{g}}{1\ \cancel{\text{cm}^3}} = 4.90 \times 10^3\ \text{g}$

e. $\frac{65\ \cancel{\text{mi}}}{1\ \cancel{\text{h}}} \times \frac{5280\ \text{ft}}{1\ \cancel{\text{mi}}} \times \frac{1\ \cancel{\text{h}}}{3600\ \text{s}} = 95\ \text{ft/s}$

f. $\frac{1.17\ \cancel{\text{g}}}{1\ \cancel{L}} \times \frac{10^6\ \mu\text{g}}{1\ \cancel{\text{g}}} \times \frac{1\ \cancel{L}}{10\ \text{dL}} = 1.17 \times 10^5\ \mu\text{g/dL}$

Based on the prices advertised below, you can derive conversion factors that relate cost to a certain amount of produce. For example, you can write $1.75/2 Italian artichokes or $0.79/zucchini.

SAMPLE PROBLEM MH-5

Applying Dimensional Analysis

A grocer is selling oranges at "3 for $1." How much would it cost to buy a dozen oranges?

Solution

The following equality is given in the problem.

$$3\ \text{oranges} = \$1$$

You can write two conversion factors based on this relationship.

$$\frac{\$1}{3\ \text{oranges}} \text{ and } \frac{3\ \text{oranges}}{\$1}$$

The given unit is oranges; the desired unit is dollars. Thus, use the conversion factor on the left to convert from oranges to dollars. One dozen equals 12, so you can start the calculation with the measurement 12 oranges.

$$12\ \cancel{\text{oranges}} \times \frac{\$1}{3\ \cancel{\text{oranges}}} = \$4$$

The given unit (oranges) cancels, leaving the desired unit (dollars) in the answer.

Practice Problems Plus

A baker is selling bagels at $3 per dozen. **How much would it cost to buy eight bagels?** *($2)*

Math Handbook **R67**

Answers

1. a. $\frac{1\text{ yd}}{3\text{ ft}}$ **b.** $\frac{365\text{ dy}}{1\text{ yr}}$

c. $\frac{1\text{ rd}}{5.50\text{ yd}}$ **d.** $\frac{24\text{ h}}{1\text{ dy}}$

e. $\frac{1\text{ mi}}{5280\text{ ft}}$ **f.** $\frac{1\text{ min}}{60\text{ s}}$

2. a. $440\text{ yd} \times \frac{3\text{ ft}}{1\text{ yd}} = 1320\text{ ft}$

b. $2\text{ wk} \times \frac{7\text{ dy}}{1\text{ wk}} \times \frac{24\text{ h}}{1\text{ dy}} \times \frac{60\text{ min}}{1\text{ h}} = 20{,}160\text{ min}$

c. $1800\text{ h} \times \frac{1\text{ dy}}{24\text{ h}} = 75\text{ dy}$

d. $660\text{ ft} \times \frac{1\text{ mi}}{5280\text{ ft}} = 0.125\text{ mi}$

e. $100\text{ yd} \times \frac{3\text{ ft}}{1\text{ yd}} \times \frac{12\text{ in}}{1\text{ ft}} = 3600\text{ in}$

f. $1\text{ yr} \times \frac{365\text{ dy}}{1\text{ yr}} \times \frac{24\text{ h}}{1\text{ dy}} = 8760\text{ h}$

g. $12\text{ mi} \times \frac{5280\text{ ft}}{1\text{ mi}} \times \frac{1\text{ yd}}{3\text{ ft}} \times \frac{1\text{ rd}}{5.5\text{ yd}} = 3840\text{ rd}$

h. $7\text{ dy} \times \frac{24\text{ h}}{1\text{ dy}} \times \frac{60\text{ min}}{1\text{ hr}} = 10{,}080\text{ min}$

3. a. $\frac{3.50\text{ mi}}{1\text{ hr}} \times \frac{5280\text{ ft}}{1\text{ mi}} \times \frac{1\text{ yd}}{3\text{ ft}} \times \frac{1\text{ hr}}{60\text{ min}} = 103\ \frac{\text{yd}}{\text{min}}$

b. $\frac{2.5\text{ gal}}{1\text{ min}} \times \frac{60\text{ min}}{1\text{ hr}} \times \frac{24\text{ hr}}{1\text{ dy}} = 3.6 \times 10^3\ \frac{\text{gal}}{\text{dy}}$

c. $\frac{2.6\text{ s gained}}{1\text{ h}} \times \frac{1\text{ min gained}}{60\text{ s gained}} \times \frac{24\text{ hr}}{1\text{ dy}} \times \frac{7\text{ dy}}{1\text{ wk}} = 7.28\ \frac{\text{min gained}}{\text{wk}}$

d. $\frac{115\text{ in}}{1\text{ min}} \times \frac{60\text{ min}}{1\text{ hr}} \times \frac{1\text{ ft}}{12\text{ in}} \times \frac{1\text{ mi}}{5280\text{ ft}} = 0.109\ \frac{\text{mi}}{\text{h}}$

Practice the Math

Use the following equalities for Questions 1–3.

60 s = 1 min	5.50 yd = 1 rod
12 in = 1 ft	7 days = 1 wk
60 min = 1 h	5280 ft = 1 mi
3 ft = 1 yd	365 days = 1 yr
24 h = 1 day	

1. Write the conversion factor need for each unit conversion.
- **a.** feet → yards
- **b.** years → days
- **c.** yards → rods
- **d.** days → hours
- **e.** feet → miles
- **f.** seconds → minutes

2. Solve each problem by dimensional analysis.
- **a.** How many feet long is the 440-yard dash?
- **b.** Calculate the number of minutes in two weeks.
- **c.** Calculate the number of days in 1800 h.
- **d.** How many miles is 660 ft?
- **e.** How many inches long is a 100-yd football field?
- **f.** Calculate the number of hours in one year.
- **g.** How many rods are in 12 miles?
- **h.** Calculate the number of minutes in 7 days.

3. Solve each problem by dimensional analysis.
- **a.** A student walks at a brisk 3.50 mi/h. Calculate the student's speed in yards/minute.
- **b.** Water runs through a hose at the rate of 2.5 gal/min. What is the rate of water flow in units gallons/day?
- **c.** A clock gains 2.60 s each hour (2.6 s gained/h). What is the rate of time gained in minutes/week?
- **d.** A spider travels 115 inches in 1 min (speed = 115 in/min). What is the speed of the spider in miles/hour?

Applying Dimensional Analysis to Chemistry

Use the following metric relationships to work out Questions 4 and 5.

10^3 m = 1 km	10^9 nm = 1 m
10 dm = 1 m	10^{12} pm = 1 m
10^2 cm = 1 m	10^3 cm^3 = 1 L
10^3 mm = 1 m	1 mL = 1 cm^3
10^6 μm = 1 m	1 g H_2O = 1 mL H_2O

4. Perform the following conversions.
- **a.** 45 m to kilometers
- **b.** 4×10^7 nm to meters
- **c.** 8.5 dm to millimeters
- **d.** 8.2×10^{-4} μm to centimeters
- **e.** 0.23 km to decimeters
- **f.** 865 cm^3 to liters
- **g.** 7.28×10^2 pm to micrometers
- **h.** 56 g H_2O to L H_2O

5. Perform the following conversions.
- **a.** 4.5 m/s to millimeters/minute
- **b.** 7.9×10^{-2} km/h to decimeters/minute
- **c.** 77 mL H_2O/s to liters H_2O/hour
- **d.** 3.34×10^4 nm/min to centimeters/second

R68 Appendix C

4. a. 4.5×10^{-2} km
b. 4×10^{-2} m
c. 8.5×10^2 mm
d. 8.2×10^{-8} cm
e. 2.3×10^3 dm
f. 0.865 L
g. 7.28×10^{-4} μm
h. 5.6×10^{-2} L H_2O

5. a. 2.77×10^5 mm/min
b. 13 dm/min
c. 2.8×10^2 L H_2O/h
d. 5.57×10^{-5} cm/s

Algebraic Equations

Many relationships in chemistry can be expressed as simple algebraic equations. However, the equation given is not always in the form that is most useful in figuring out a particular problem. In such a case, you must first solve the equation for the unknown quantity; this is done by rearranging the equation so that the unknown is on one side of the equation, and all the known quantities are on the other side.

Solving Simple Equations

An equation is solved using the laws of equality. The laws of equality are summarized as follows: *If equals are added to, subtracted from, multiplied by, or divided by equals, the results are equal.* In other words, you can perform any of these mathematic operations on an equation and not destroy the equality, *as long as you do the same thing to both sides of the equation.* The laws of equality apply to any legitimate mathematic operation, including squaring, taking square roots, and taking the logarithm.

Consider the following equation, which states the relationship between the Kelvin and Celsius temperature scales.

$$K = {}^{\circ}C + 273$$

Can this equation be used to find the Celsius-temperature equivalent of 400 K? Yes, it can, if the equation is first solved for the unknown quantity, °C.

In the above example, to solve for °C, subtract 273 from both sides of the equation.

$$K = {}^{\circ}C + 273$$
$$K - 273 = {}^{\circ}C + 273 - 273$$
$${}^{\circ}C = K - 273$$

Now you have solved the equation for the unknown quantity. To calculate its value, substitute the known quantity into the solved equation.

$$\begin{aligned} {}^{\circ}C &= K - 273 \\ &= 400 - 273 \\ &= 127^{\circ}C \end{aligned}$$

A technician in a genetic research lab removes frozen cell samples from cryostorage. The cells are stored in liquid nitrogen at −196°C. What is this temperature in kelvins?

Algebraic Equations

Use Visuals

The temperature −196°C is equivalent to 77 K.

Use Visuals

From the photograph, students should observe that –40°F is equivalent to –40°C. The equivalence can be verified by substituting these values into the equation °F = (1.8 × °C) + 32.

$-40°F = (1.8 \times -40°C) + 32$

$-40 = -72 + 32$

$-40 = -40$

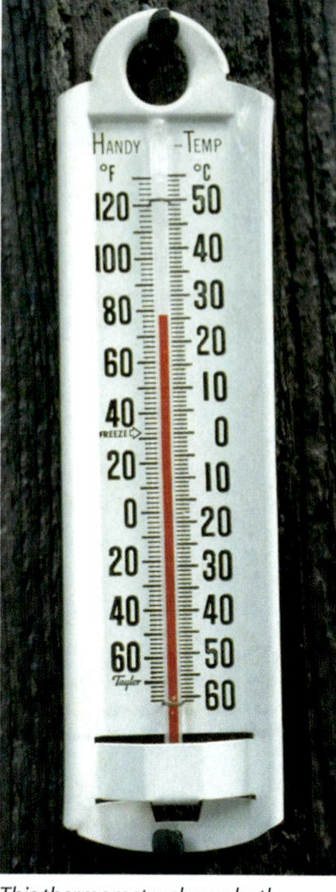

This thermometer shows both the Fahrenheit and Celsius temperature scales. What temperature in Celsius corresponds to −40°F? Verify the conversion by performing a calculation.

Another commonly used temperature scale is the Fahrenheit scale. The relationship between the Fahrenheit and Celsius temperature scales is given by the following equation.

$$°F = (1.8 \times °C) + 32$$

Suppose you want to use this equation to convert 365°F into degrees Celsius. To solve for °C, you must isolate it on one side of the equation. Since the right side of the equation has 32 added to the quantity (1.8 × °C), first subtract 32 from both sides of the equation, and then divide each side by 1.8.

$$°F = (1.8 \times °C) + 32$$
$$°F - 32 = (1.8 \times °C) + 32 - 32$$
$$°F - 32 = (1.8 \times °C)$$
$$\frac{°F - 32}{1.8} = \frac{1.8 \times °C}{1.8}$$
$$\frac{(°F - 32)}{1.8} = °C$$

Now that you have solved the equation for the unknown quantity (°C), you can substitute the known quantity (365°F) into the equation and calculate the answer.

$$°C = \frac{(°F - 32)}{1.8} = \frac{(365 - 32)}{1.8} = \frac{333}{1.8} = 185°C$$

SAMPLE PROBLEM MH-6

Solving Algebraic Equations

The heat (q) absorbed by the water in a calorimeter can be calculated using the following relationship.

$$q = m \times C \times \Delta T$$

In this expression, m is the mass of the water; C is the specific heat of water (4.18 J/g·°C); and ΔT is the change in temperature. If 120 g of water absorb 3500 J of heat, by how much will the temperature of the water increase?

Solution

First solve the equation for the unknown quantity, ΔT.

$$q = m \times C \times \Delta T$$
$$\frac{q}{m \times C} = \frac{m \times C \times \Delta T}{m \times C}$$
$$\frac{q}{m \times C} = \Delta T$$

Now substitute the known values for q, m, and C.

$$\Delta T = \frac{q}{m \times C}$$
$$= \frac{3500 \text{ J}}{(120 \text{ g} \times 4.18 \text{ J/g·°C})} = 7.0°C$$

Practice the Math

1. Solve each equation for z.
   a. $xy + z = 5$
   b. $\frac{z}{a - 4} = t$
   c. $\frac{b}{d} = \frac{2a}{z}$
   d. $\sqrt{z} = 2b$
2. Solve each equation for a. Then calculate a value for a if $b = 4$, $c = 10$, and $d = 2$.
   a. $bd = ac$
   b. $a + b = cd$
   c. $c + b = \frac{a}{d}$
   d. $\frac{bd}{a} = c^2$
3. Solve each equation for h. Then calculate a value for h if $g = 12$, $k = 0.4$, and $m = 1.5$.
   a. $kh = \frac{g}{m}$
   b. $\frac{(g - m)}{h} = k$
   c. $gh - k = m$
   d. $\frac{mk}{(g + h)} = 2$

Applying Algebra to Chemistry

4. Solve for v in the following equation.

$$d = \frac{m}{v}$$

Let d = density, m = mass, and v = volume. What is the volume of 642 g of gold if the density of gold is 19.3 g/cm^3?

5. Solve for n in the following equation.

$$P \times V = n \times R \times T$$

How many moles (n) of helium gas fill a 6.45-L (V) balloon at a pressure (P) of 105 kPa and a temperature (T) of 278 K? (R = 8.31 (L·kPa)/(K·mol))

6. Solve for V_2 in the following equation.

$$\frac{P_1 \times V_1}{T_1} = \frac{P_2 \times V_2}{T_2}$$

A 2.50-L (V_1) sample of nitrogen gas at a temperature (T_1) of 308 K has a pressure (P_1) of 1.15 atm. What is the new volume (V_2) of the gas if the pressure (P_2) is increased to 1.80 atm and the temperature (T_2) decreased to 286 K?

7. Solve for P_{He} in the following equation.

$$P_{total} = P_{Ar} + P_{He} + P_{Kr}$$

A mixture of gases has a total pressure (P_{total}) of 376 kPa. What is the partial pressure of helium (P_{He}), if P_{Ar} = 92 kPa and P_{Kr} = 144 kPa?

8. Solve for T_2 in the following equation.

$$\frac{V_1}{T_1} = \frac{V_2}{T_2}$$

What is the value of T_2 when V_1 = 5.0 L, V_2 = 15 L, and T_1 = 200 K?

Answers

1. a. $z = 5 - xy$
   b. $z = a(t + 4)$
   c. $z = \frac{2ad}{b}$
   d. $z = 4b^2$
2. a. $a = \frac{bd}{c} = 0.8$
   b. $a = cd - b = 16$
   c. $a = d(c + b) = 28$
   d. $a = \frac{bd}{c^2} = 0.08$
3. a. $h = \frac{g}{mk} = 20$
   b. $h = \frac{(g - m)}{k} = 26.25$
   c. $h = \frac{(m + k)}{g} = 0.158$
   d. $h = \frac{(mk - 2g)}{2} = -11.7$
4. $v = \frac{m}{d} = \frac{642\text{ g}}{19.3\text{ g/cm}^3} = 33.3\text{ cm}^3$

   This problem can also be worked by dimensional analysis:

   $642\text{ g} \times \frac{1\text{ cm}^3}{19.3\text{ g}} = 33.3\text{ cm}^3$
5. $n = \frac{P \times V}{R \times T}$

   $= \frac{105\text{ kPa} \times 6.45\text{ L}}{8.31 \frac{\text{L} \times \text{kPa}}{\text{K} \times \text{mol}} \times 278\text{ K}}$

   $= 0.293\text{ mol}$
6. $V_2 = \frac{P_1 \times V_1 \times T_2}{P_2 \times T_1}$

   $= \frac{1.15\text{ atm} \times 2.50\text{ L} \times 286\text{ K}}{1.80\text{ atm} \times 308\text{ K}}$

   $= 1.48\text{ L}$
7. $P_{He} = P_{total} - P_{Ar} - P_{Kr}$

   $= 376\text{ kPa} - 92\text{ kPa} - 144\text{ kPa}$

   $= 140\text{ kPa}$
8. $T_2 = \frac{V_2 \times T_1}{V_1}$

   $= \frac{15\text{ L} \times 200\text{ K}}{5.0\text{ L}}$

   $= 600\text{ K}$

Percents

Use Visuals

Students should be able to make the connection between grade and the slope of a line. Ask, **Which is steeper, a road with a 5% uphill grade or a road with an 8% uphill grade?** (a road with an 8% grade) **What is the grade of a road that has no incline?** (0%)

The grade of a road is the slope expressed as a percent. A downhill 6% grade means that for every 100 units of horizontal distance traveled, there is a corresponding 6-unit descent.

Percents

A percent is a ratio that compares a number to 100. The word *percent* (%) means "parts of 100" or "per 100 parts." Another way to think of a percent is a part of a whole expressed in hundredths. Thus, the number 0.52 ("52 hundredths") can also be expressed as 52%, or "52 per 100 parts."

You should be familiar with percents. For example, test scores are often expressed as percents. Suppose you answer 24 questions correctly on a 30-question exam. The part out of the whole, expressed as a percent, can be calculated as follows.

$$\frac{\text{Part}}{\text{Whole}} = \frac{\text{number of correctly answered questions}}{\text{number of questions asked}}$$
$$= \frac{24}{30}$$
$$= 0.80 = \frac{80}{100} = 80\%$$

Note that 0.80, 80/100, and 80% all express the same value: 80 hundredths (or "80 per 100 parts").

Another way to calculate a percent is to multiply the ratio of the part to the whole by 100%.

$$\text{Percent} = \frac{\text{part}}{\text{whole}} \times 100\%$$

Suppose a high school science club consists of 27 boys and 23 girls. What percent of the club is made up of boys?

$$\text{Percent} = \frac{\text{number of boys}}{\text{number of total members}} \times 100\%$$
$$= \frac{27}{27 + 23} \times 100\%$$
$$= \frac{27}{50} \times 100\%$$
$$= 0.54 \times 100\%$$
$$= 54\%$$

Because a percent represents a relationship between two quantities, it can be used as a conversion factor. For example, suppose a friend tells you that she got a score of 95% on a 40-question exam. How many questions did she answer correctly? A score of 95% means 95 correctly answered questions for every 100 questions asked. By expressing this relationship as a conversion factor, you can calculate the number of correctly answered questions.

$$40 \text{ \sout{questions asked}} \times \left(\frac{95 \text{ correctly answered questions}}{100 \text{ \sout{questions asked}}}\right)$$
$$= 38 \text{ correctly answered questions}$$

SAMPLE PROBLEM MH-7

Using Percents as Conversion Factors

Six students are absent from their class one day. If 80% of the students in the class are present, what is the total class enrollment?

Solution

Because 80% of the class is present, the six absent students make up 20% of the class. If 20% of the students are absent, this means that 20 students are absent for every 100 students enrolled. Use dimensional analysis to calculate the total class enrollment.

$$6 \text{ students absent} \times \left(\frac{100 \text{ students enrolled}}{20 \text{ students absent}}\right) = \frac{600}{20} \text{ students enrolled} = 30 \text{ students enrolled}$$

Practice the Math

1. Write each fraction as a percent.

a. $\frac{3}{4}$ **b.** $\frac{1}{5}$ **c.** $\frac{7}{10}$ **d.** $\frac{5}{8}$

2. Write each decimal as a percent.

a. 0.39 **b.** 0.08 **c.** 4.2 **d.** 0.5

3. Calculate a score for each exam expressed as a percent.

a. Sixteen questions correctly answered on a 25-question exam.

b. Forty-one questions correctly answered on a 50-question exam.

c. Sixty-eight questions correctly answered on an 85-question exam.

4. A quality control inspector found that 7 out of every 200 flashlights produced were defective. What percent of the flashlights were not defective?

5. A student answered 57 questions correctly on an exam and received a score of 95%. How many questions were on the exam?

6. During a flu epidemic, 28% of the students were absent. If 238 students were absent, what is this school's enrollment?

Applying Percents to Chemistry

7. A mining company is abstracting silver from an ore is that 0.014% silver (by mass). How many kilograms of ore must be processed to yield 0.7 kg of silver?

8. A fertilizer is 12.0% (by mass) nitrogen and 5.5% (by mass) phosphorus. How many grams of each element are in 140 g of this fertilizer?

9. A compound is broken down into 34.5 g of element A, 18.2 g of element B, and 2.6 g of element C. What is the percent (by mass) of each element in this compound?

10. What is the percent by mass of sodium chloride (NaCl) in each of the following solutions?

a. 44 g NaCl dissolved in 756 g of water

b. 15 g NaCl dissolved in 485 g of water

c. 135 g NaCl dissolved in 765 g of water

11. The antiseptic hydrogen peroxide is often sold as a 3.0% (by mass) solution, the rest being water. How many grams of hydrogen peroxide are in 250 g of this solution?

12. A nighttime cold medicine is 22% alcohol (by volume). How many mL of alcohol are in a 250 mL bottle of this medicine?

Practice Problems Plus

During the course of a season, a basketball player made 36% of her 3-point field goal attempts, 44% of her 2-point field goal attempts, and 84% of her free throw attempts.

a. How many of her 150 free throw attempts did she make? *(126)*

b. How many of her 75 3-point field goal attempts did she make? *(27)*

c. How many of her 225 2-point field goal attempts did she make? *(99)*

Answers

1. a. 75% **b.** 20% **c.** 70% **d.** 62.5%
2. a. 39% **b.** 8% **c.** 420% **d.** 50%
3. a. 64% **b.** 82% **c.** 80%
4. 96.5%
5. 60 questions
6. 850 students
7. 50 kg ore
8. 16.8 g nitrogen, 7.7 g phosphorus
9. 62.4% A, 32.9% B, 4.7% C
10. a. 5.5% **b.** 3.0 **c.** 15%
11. 7.5 g
12. 55 mL alcohol

Graphing

Use Visuals

Graphs are useful tools for presenting statistics and data. Ask, **Why might a presenter prefer to organize data in a graph rather than in a table?** *(Trends in the data can be spotted more easily when the data is plotted in a graph.)*

Graphing

A line graph can be used to show the relationship between two variables. The manipulated variable is plotted on the x-axis. The responding variable is plotted on the y-axis.

The relationship between two variables in an experiment is often determined by graphing the experimental data. A graph is a "picture" of the data. Once a graph is constructed, additional information can be derived about the variables.

In constructing a graph, you must first label the axes. The manipulated variable (also known as the independent variable) is plotted on the *x*-axis. This is the horizontal axis. The manipulated variable is controlled by the experimenter. When the independent variable is changed, a corresponding change in the responding variable (also known as the dependent variable) is measured. The responding variable is plotted on the *y*-axis. This is the vertical axis. The label on each axis should include the unit of the quantity being graphed.

Before data can be plotted on a graph, each axis must be scaled. Each interval on the scale must represent the same amount. To make it easy to find numbers along the scale, the interval chosen is usually a multiple of 1, 2, 5, or 10. Although each scale can start at zero, this is not always practical.

Data are plotted by putting a point at the intersection of corresponding values of each pair of measurements. Once the data has been plotted, the points are connected by a smooth curve. A smooth curve comes as close as possible to all the plotted points. It may in fact not touch any of them.

Inverse and Direct Proportionalities

Depending on the relationship between two variables, a plotted curve may or may not be a straight line. Two common curves are shown below.

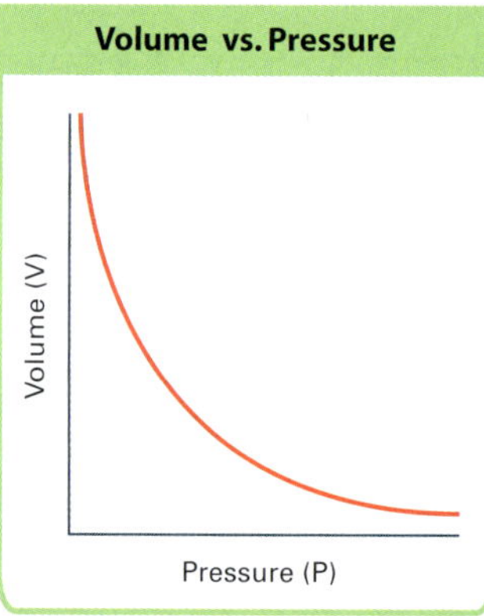

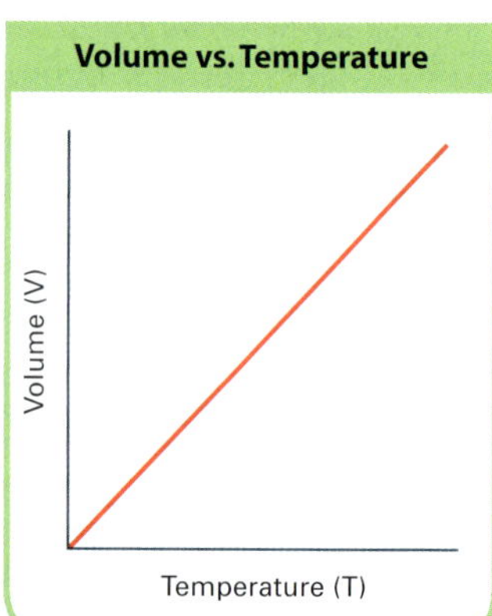

The volume- vs. -pressure curve is typical of an inverse proportionality. As the manipulated variable (P) increases, the responding variable (V) decreases. The product of the two variables at any point on the curve of an inverse proportionality is a constant. Thus, $V \times P$ = constant.

The straight line in the volume- vs. -temperature graph is typical of a direct proportionality. As the manipulated variable (T) increases, there is a corresponding increase in the responding variable (V). A straight line can be represented by the following general equation.

$$y = mx + b$$

The variables y and x are plotted on the vertical and horizontal axes, respectively. The y-intercept, b, is the value of y when x is zero. The slope, m, is the ratio of the change in y (Δy) for a corresponding change in x (Δx).

$$m = \frac{\Delta y}{\Delta x}$$

Plotting and Interpreting Graphs

Consider the following set of data about a bicyclist's trip. Assume that the bicyclist rode at a constant speed.

Distance from home (km)	15	25	35	50	75
Time (h)	1	2	3	4.5	7

Graph these data using time as the manipulated variable, and then use the graph to answer the following questions.

a. How far from home was the bicyclist at the start of the trip?
b. How long did it take for the bicyclist to get 40 km from home?

The plotted points are shown in the graph below. Each point was plotted by finding the value of time on the x-axis, then moving up vertically to the value of the other variable (distance). A smooth curve (in this case a straight line), has been drawn through the points.

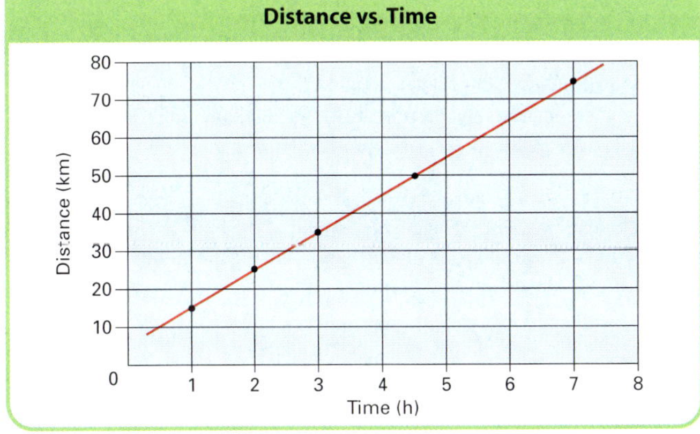

Figure 1

Math Handbook R75

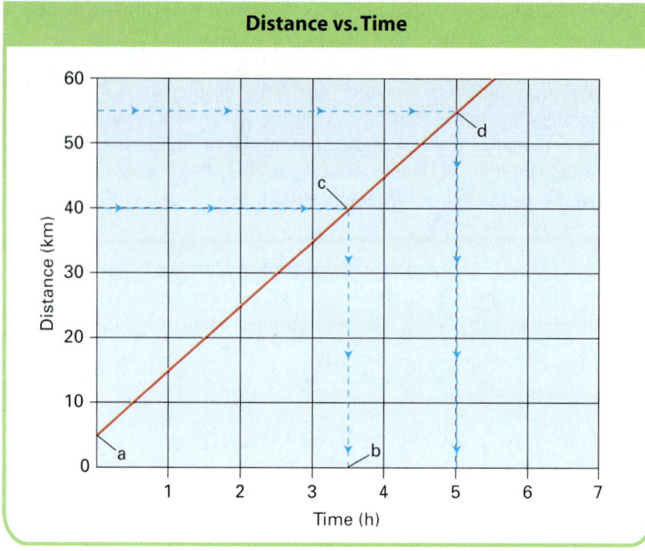

Figure 2

To answer problem **a**, extend the curve so that it intersects the *y*-axis, as in the graph above. The graph shows that the bicyclist started the trip 5 km from home. This is the value of the vertical axis (distance) when the time elapsed is zero (point *a* on the graph).

For problem **b**, find the value 40 km on the *y*-axis of the graph. Move to the right (horizontally) in the graph until you reach the line. Drop down vertically and read the value of time at this point (point *b*). It takes the bicyclist 3.5 h to get 40 km from home.

SAMPLE PROBLEM MH-8

Interpreting Graphs

Use the distance- vs. -time graph above to calculate the bicyclist's average speed in kilometers per hour (km/h).

Solution

Speed is distance/time. The average speed of the bicyclist is the slope of the line in the graph. Calculate the slope using the values for time and distance corresponding to points *c* and *d*.

$$m = \frac{\Delta y}{\Delta x} = \frac{55 \text{ km} - 40 \text{ km}}{5 \text{ h} - 3.5 \text{ h}} = \frac{15 \text{ km}}{1.5 \text{ h}} = 10 \text{ km/h}$$

You can now write an equation for the line.

$$y = mx + b$$

$$\text{Distance} = (10 \text{ km/h})(\text{time}) + 5 \text{ km}$$

R76 Appendix C

Practice the Math

1. A bicyclist wants to ride 100 kilometers. The data below show the time required to ride 100 kilometers at different average speeds.

Time (h)	4	5	8	10	15	20
Avg speed (km/h)	25	20	12.5	10	6.7	5

   a. Graph the data, using average speed as the independent variable.
   b. Is this a direct or inverse proportionality?
   c. What average speed must be maintained to complete the ride in 12 hours?
   d. If a bicyclist's average speed is 18 km/h, how long does it take to ride 100 km?

2. The data below shows how the mass of a baby varies with its age during its first year of life.

Age(days)	40	110	200	270	330
Mass (kg)	4.0	5.4	7.3	8.6	9.9

   a. Graph the data, using age as the independent variable.
   b. Derive an equation in the form of $y = mx + b$. Include units on the values of y and b.
   c. Why would the values of y and b be of interest to both the baby's parents and physician?

Applying Graphs to Chemistry

3. Use the following data to draw a graph that shows the relationship between the Fahrenheit and Celsius temperature scales. Make °F the responding variable. Use the graph to derive an equation relating °F and °C. Then use the graph or the equation to find values for y_1, x_2, and x_3.

Temperature (°F)	50	212	356	−4	y_1	70	400
Temperature (°C)	10	100	180	−20	70	x_2	x_3

4. Different volumes of the same liquid were added to a graduated cylinder sitting on a balance. After each addition of liquid the total volume of liquid and mass of the liquid-filled graduated cylinder was recorded in the table below.

Total volume of liquid (mL)	10	25	45	70	95
Mass of liquid and cylinder (g)	138	159	187	222	257

   a. Graph the data, using volume as the manipulated variable.
   b. What is the y-intercept of the line? Make sure to include the unit.
   c. What does the value of the y-intercept represent?
   d. Calculate the slope of the line, and make sure to include the unit.
   e. What does the slope of the line represent for this liquid?
   f. Write a general equation that represents the line in your graph.

5. A student collected the following data or a fixed volume of gas.

Temperature (°C)	10	20	40	70	100
Pressure (mm Hg)	726	750	800	880	960

   a. Graph the data, using pressure as the responding variable.
   b. Is this a direct or inverse proportionality?
   c. At what temperature is the gas pressure 822 mm Hg?
   d. What is the pressure of the gas at a temperature of 0°C?
   e. How does the pressure of the gas change with a change in temperature?
   f. Write an equation relating the pressure and temperature of the gas.

Answers

1. a.

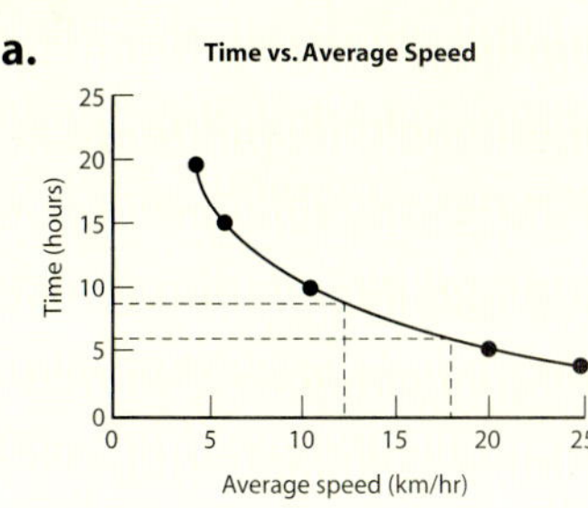

b. inverse **c.** ~8.3 km/hr **d.** ~5.5 h

2. a.

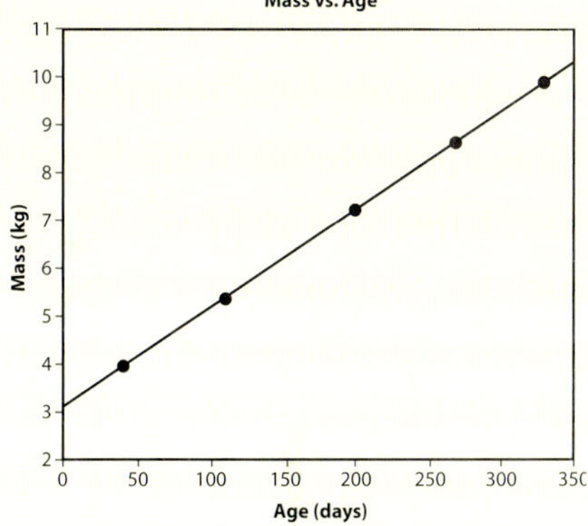

b. mass = (0.02kg/day)(age) + 3.2 kg
c. Measuring the mass of the infant (y) over time helps doctors assess the infant's health. The slope m is the average mass gain per day in the infant's first year of life. The y-intercept b is the infant's mass at birth.

3.

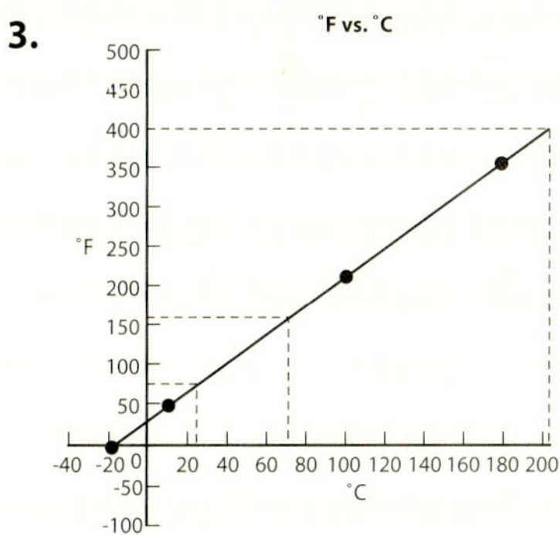

°F = 1.8°C + 32
y_1 = 158°F, x_2 = 21°C, x_3 = 204°C

4. a.

b. 124 g **c.** mass of the empty graduated cylinder **d.** 1.4 g/mL **e.** density
f. mass = (1.4 g/mL)(volume) + 124 g

5. a.

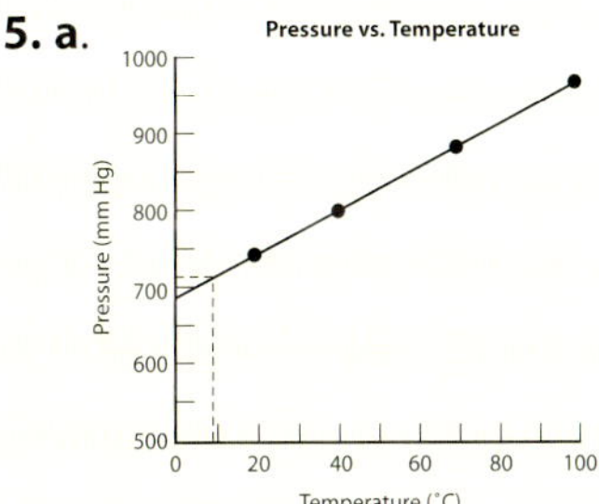

b. direct **c.** ~47°C **d.** ~700 mm Hg
e. m = 2.6 mm Hg/°C. **f.** P = (2.6 mm Hg/°C)T + 700 mm Hg

Logarithms

Use Visuals

The photograph shows a pH meter reading. Explain that the pH scale is a logarithmic scale. Ask, **How can the pH of a solution be expressed as a hydrogen ion concentration?** *(by taking the antilog of the negative of the pH value)*

Logarithms

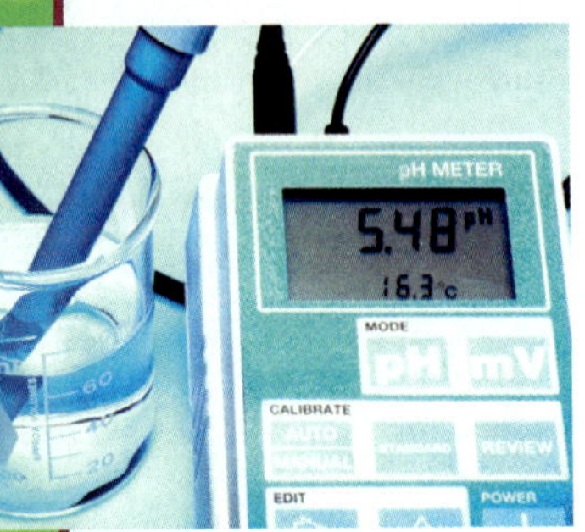

The acidity of a solution is measured with a pH meter. The pH scale is a logarithmic scale based on the relationship $pH = -log[H^+]$.

Certain scales of measurement are based on logarithms, including the pH scale (used to measure acidity) and the decibel scale (used to measure the relative loudness of sounds). Because the pH scale is a logarithmic scale, it allows a vast range of values (from 1.0 to 10^{-14}) to be expressed simply as a number between 0 and 14.

The common logarithm (*log*) of a number is the exponent to which 10 must be raised to produce that number. If $x = 10^y$, then $\log x = y$. Thus, since $0.01 = 10^{-2}$, it follows that $\log 0.01 = -2$. Likewise, $\log 10{,}000 = 4$, because $10{,}000 = 10^4$.

Using a Calculator to Compute Logarithms

You can determine the logarithm of a number by using the **[log]** key on a calculator. (Do not confuse the **[log]** key with the **[ln]** key.) If you are calculating the logarithm of a measured value, then the number of decimal places in the logarithm must be rounded to equal the number of significant figures in the measurement. For example, to calculate the logarithm of 3.45, enter the keystrokes shown below.

log 3.45 = **[3][.][4][5] [log]** = *0.5378191* = 0.538

The number in the display (the logarithm) is *0.5378191*. This value must be rounded to three decimal places so as to equal the number of significant figures in the original number. (3.45 has three significant figures.)

Practice using the logarithm function on your calculator by working out problems **a** and **b**. For simplification, only the **[log]** keystrokes are indicated. Make sure to round your answers to the correct number of decimal places.

a. log 0.0087 = 0.0087 **[log]** = *−2.0604807* = −2.06
(0.0087 has two significant figures, so round to two decimal places.)

b. log 3.11×10^{-5} = 3.11×10^{-5} **[log]** = *−4.5072396* = −4.507
(3.11×10^{-5} has three significant figures, so round to three decimal places.)

Antilogarithms

The reverse process of converting a logarithm into a number is referred to as obtaining the antilogarithm. The antilogarithm of the logarithm of number x is the number x itself.

$$\text{antilog}(\log x) = x$$

If $\log 10^{-7} = -7$, then antilog $(-7) = 10^{-7}$. Likewise, if $\log 10^2 = 2$, then antilog $(2) = 10^2$.

You can calculate the antilog of a number by using the **[10^x]** key on your calculator. To find the antilogarithm of 0.538, enter **[.][5][3][8]** followed by the **[10^x]** key. (Depending on your calculator, you may need to press the **[2nd]** or **[INV]** key first.) The number in the display (the antilogarithm) is *3.45143734*. This value must be reported to three significant figures (as 3.45), because the original number (0.538) was reported to three decimal places.

Practice using the **[10^x] key** on your calculator by working out the antilogarithms of the logarithms you calculated earlier.

a. antilog $(-2.06) = -2.06$ **[10^x]** $=$ *0.0087096* $= 0.0087$
(−2.06 has two decimal places, so round to two significant figures)

b. antilog $(-4.507) = -4.507$ **[10^x]** $=$ *0.000031117* $= 3.11 \times 10^{-5}$
(−4.507 has three decimal places, so round to three significant figures)

SAMPLE PROBLEM MH-9

Logarithms and Antilogarithms

Use your calculator to find the logarithm of 8.10×10^3. Check your answer by computing its antilogarithm.

Solution

log $8.10 \times 10^3 = 8.10 \times 10^3$ **[log]** $=$ *3.9084850* $= 3.908$

antilog $(3.908) = 3.908$ **[10^x]** $= 8090.9590 = 8.09 \times 10^3$

Notice that because you rounded the logarithm to three decimal places, the calculated antilog (8.09×10^3) is not exactly the same as the original number given (8.10×10^3).

Practice the Math

1. Calculate the logarithms of the following numbers.

a. 7.56 **b.** 0.000678
c. 456 **d.** 4.27×10^3
e. 1.485×10^{-6} **f.** 1×10^{-12}

2. Calculate the antilogarithms of the following numbers.

a. 2.56 **b.** 6.111
c. −3.55 **d.** 1.138
e. 0.962 **f.** −0.864

Applying Logarithms to Chemistry

3. The hydrogen ion concentration, $[H^+]$, of an aqueous solution can be expressed as a pH value according to the following relationship.

$$pH = -\log [H^+]$$

Determine pH values for solutions having the following $[H^+]$ values. The symbol *M* represents molarity, a unit of concentration.

a. 1×10^{-8} *M* **b.** 1×10^{-1} *M*
c. 4.8×10^{-7} *M* **d.** 0.0034 *M*

4. The relationship between pH and $[H^+]$ can also be expressed as follows.

$$[H^+] = \text{antilog}(-pH)$$

Determine $[H^+]$ values for aqueous solutions having the following pH values.

a. 14.0 **b.** 7.0
c. 7.86 **d.** 4.05
e. 11.65 **f.** 2.945
g. 3.68 **h.** 12.850

Practice Problems Plus

Use your calculator to determine the logarithms of the following numbers.

a. 2.10×10^{-3} *(−2.678 (three decimal places))*

b. 1.24×10^0 *(0.093 (three decimal places))*

Use your calculator to find the antilogarithms of the following logarithms.

c. −9.0 *(1×10^{-9} (one significant figure))*

d. −2.0054 *(9.876×10^{-3} (four significant figures))*

Answers

1. a. 0.879 (three decimal places)
b. −3.169 (three decimal places)
c. 2.659 (three decimal places)
d. 3.630 (three decimal places)
e. −5.8283 (four decimal places)
f. −12.0 (one decimal places)

2. a. 3.6×10^2 (two significant figures)
b. 1.29×10^6 (three significant figures)
c. 2.8×10^{-4} (two significant figures)
d. 1.37×10^1 (three significant figures)
e. 9.16 (three significant figures)
f. 1.37×10^{-1} (three significant figures)

3. a. 8.0 **b.** 1.0 **c.** 6.32 **d.** 2.47

4. a. 1×10^{-14} **b.** 1×10^{-7}
c. 1.4×10^{-8} **d.** 8.9×10^{-5}
e. 2.2×10^{-12} **f.** 1.14×10^{-3}
g. 2.1×10^{-4} **h.** 1.41×10^{-13}

Safety in the Chemistry Lab

Although the labs in this book were designed to be a safe as possible, students still need to follow appropriate laboratory practice.

Pre-Lab Preparation

- Follow an approved method for the storage of chemicals.
- Lab burners should be connected to outlets with safety tubing, which cannot be easily cut or punctured.
- Use only ground fault circuit interrupters or GFCI protected circuits with electrical equipment.
- At the beginning of the year, explain the exit procedures in case of a fire. Review the procedures periodically during the year.
- Show students where the emergency shutoffs for gas and electricity are located.
- Check to see if students are allergic to any of the materials used in a lab, such as a latex balloon or some substances in air fresheners.
- Avoid having students handle solutions with concentrations greater than 1.0*M*. Dilute concentrated acids yourself by adding them to water.

Proper Laboratory Practices

- Always model good safety practices, such as wearing splash goggles or washing your hands thoroughly at the end of a lab.
- If a reactant is a volatile liquid whose vapor is irritating or a product is an irritating vapor, do the lab in a fume hood or in a well-ventilated room.
- Be aware that some students may be tempted to use pipettes as squirt guns.
- Remind students to clean up any liquids that spill on the floor so that no one will slip and fall.

Appendix D: Safety In the Chemistry Lab

The experiments in this book have been carefully designed to minimize the risk of injury. However, safety is also your responsibility. The following rules are essential for keeping you safe in the laboratory. The rules address pre-lab preparation, proper laboratory practices, and post-lab procedures.

Pre-Lab Preparation

1. Read the entire procedure before you begin. Listen to all of your teacher's instructions. When in doubt about a procedure, ask your teacher.

2. Do only the assigned experiments. Do any experiment only when your teacher is present and has given you permission to work.

3. Know the location and operation of the following safety equipment: fire extinguisher, fire blanket, emergency shower, and eye wash station.

4. Know the location of emergency exits and escape routes. To make it easy to exit quickly, do not block walkways with furniture. Keep your work area orderly and free of personal belongings such as coats and backpacks.

5. Protect your clothing and hair from chemicals and sources of heat. Tie back long hair and roll up loose sleeves when working in the laboratory. Avoid wearing bulky or loose-fitting clothing. Remove dangling jewelry. Wear closed-toe shoes at all times in the laboratory.

Proper Laboratory Practices

6. Even with well-designed and tested laboratory procedures, an accident may occur while you are working in the lab.

Report any accident, no matter how minor, to your teacher.

7. Wear chemical splash goggles at all times when working in the laboratory. These goggles are designed to protect your eyes from injury. While working in the lab, do not rub your eyes, because chemicals are easily transferred from your hands to your eyes.

If, despite these precautions, a chemical gets in your eye, remove any contact lenses and immediately wash your eye with a continuous stream of lukewarm water for at least 15 minutes.

8. To reduce danger, waste, and cleanup, always use the minimal amounts of chemicals specified for an experiment.

9. Never taste any chemical used in the laboratory, including food products that are the subject of an investigation. Treat all items as though they are contaminated with unknown chemicals that may be toxic. Keep all food and drink that is not part of an experiment out of the laboratory. Do not eat, drink, or chew gum in the laboratory.

If you accidentally ingest a substance, notify your teacher immediately.

10. Don't use chipped or cracked glassware. Don't handle broken glass. If glassware breaks, tell your teacher and nearby classmates. Discard broken glass as instructed by your teacher.

If, despite these precautions, you receive a minor cut, allow it to bleed for a short time. Wash the injured area under cold running water and notify your teacher. More serious cuts or puncture wounds require immediate medical attention.

11. Do not handle hot glassware or equipment. You can prevent burns by being aware that hot and cold equipment can look exactly the same.

If you are burned, immediately run cold water over the burned area for several minutes until the pain is reduced. Cooling helps the burn heal. Ask a classmate to notify your teacher.

12. Recognize that the danger of an electrical shock is greater in the presence of water. Keep electrical appliances away from sinks and faucets to minimize the risk of electrical shock. Be careful not to spill water or other liquids in the vicinity of an electrical appliance.

If, despite these precautions, you spill water near an electrical appliance, stand back, notify your teacher, and warn other students in the area.

13. Report any chemical spills immediately to your teacher. Follow your teacher's instructions for cleaning up spills. Warn other students about the identity and location of spilled chemicals.

If, despite these precautions, a corrosive chemical gets on your skin or clothing, notify your teacher. Then wash the affected area with cold running water for several minutes.

Post-Lab Procedures

14. Dispose of chemicals in a way that protects you, your classmates, and the environment. Always follow your teacher's directions for cleanup and disposal. Clean your small-scale reaction surface by draining the contents onto a paper towel. Then wipe the surface with a damp paper towel and dry the surface completely. Dispose of the paper towels in the waste bin.

15. Wash your hands thoroughly with soap and water before leaving the laboratory.

Take appropriate precautions when any of the following safety symbols appears in an experiment.

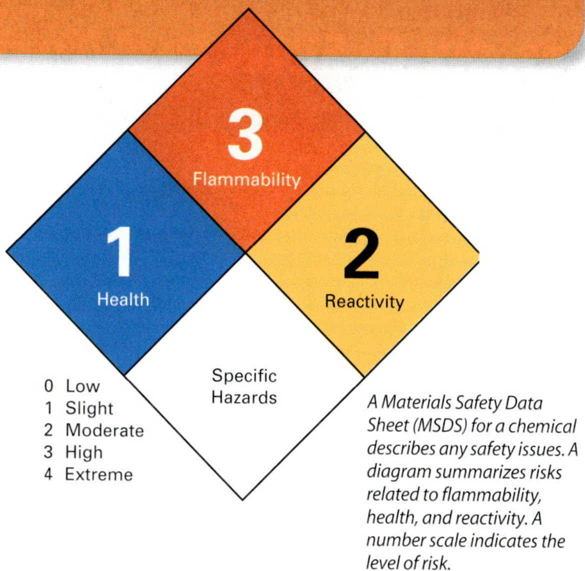

A Materials Safety Data Sheet (MSDS) for a chemical describes any safety issues. A diagram summarizes risks related to flammability, health, and reactivity. A number scale indicates the level of risk.

Safety Symbols

Eye Safety
Wear safety goggles.

Clothing Protection
Wear a lab coat or apron when using corrosive chemicals or chemicals that can stain clothing.

Skin Protection
Wear plastic gloves when using chemicals that can irritate or stain your skin.

Broken Glass
Do not use chipped or cracked glassware. Do not heat the bottom of a test tube.

Open Flame
Tie back hair and loose clothing. Never reach across a lit burner.

Flammable Substance
Do not have a flame near flammable materials.

Corrosive Substance
Wear safety goggles, an apron, and gloves when working with corrosive chemicals.

Poison
Don't chew gum, drink, or eat in the laboratory. Never taste a chemical in the laboratory.

Fume
Avoid inhaling substances that can irritate your respiratory system.

Thermal Burn
Do not touch hot glassware or equipment.

Electrical Equipment
Keep electrical equipment away from water or other liquids.

Sharp Object
To avoid a puncture wound, use scissors or other sharp objects only as intended.

Disposal
Dispose of chemicals only as directed.

Hand Washing
Wash your hands thoroughly with soap and water.

Post-Lab Procedures

- Demonstrate how to use paper towels to clean reaction surfaces to avoid direct contact with chemicals.
- Students should wash their hands thoroughly with soap or detergent before leaving the lab.

Material Safety Data Sheets

Review pertinent information with students from the Material Safety Data Sheet for any hazardous chemical used in a lab. Suppliers general ship a sheet with each reagent. These sheets can also be found at many sites on the Internet. Some of these sites include information on interpreting the sheets plus glossaries of terms related to safety.

Appendix E: Selected Answers

Chapter 1

26. 24 short blocks

27. 24 minutes

28. The order of the morning errands cannot vary, but the order of the afternoon errands can vary, as long as the haircut takes place before 3 pm.

29. Possible answers are to do an errand during his lunch hour or to extend the hours during which he does errands.

35. Chemistry concerns the changes that matter undergoes.

37. Carothers was doing pure chemistry because he did experiments to test the proposal of another chemist. His results led to applied chemistry—large-scale production of nylon.

39. A possible answer is that a firefighter needs to know which chemicals to use to fight different types of fires, and knowledge of chemistry will help a reporter gather information during an interview with a chemist.

41. Insulation acts as a barrier to heat flow. If heat flow is reduced, energy is conserved.

43. gene therapy and production of chemicals such as insulin

45. A pollutant is a material that can be found in air, soil, and water that is harmful to living organisms.

47. by analyzing the light they transmit to Earth

49. the scientific method

51. c

53. Repeat the experiment. If you get the same result, you must propose a new hypothesis.

55. to share knowledge across disciplines and to share resources between industries and universities

57. developing a plan and implementing the plan

59. 54 games (1/3 of 163)

61. 12 days

63. Answers will vary but should demonstrate an understanding that chemistry is the study of matter and the changes it undergoes.

65. You may choose biochemist because biochemistry is the study of processes that take place in organisms. However, you might choose physical chemist because physical chemistry includes the study of energy transfer as matter undergoes a change.

67. Your experiment may be correct, but your hypothesis may be wrong. You should reexamine your hypothesis and repeat the experiment.

69. Answers will vary but should reflect knowledge of the steps in a scientific method including making observations and testing hypotheses.

71. 300 miles

73. A possible answer is that scientists accept hypotheses that are supported by the results of experiments and reject hypotheses that are not supported by experimental results.

75. A person who is educated in the theories and practice of chemistry is more likely to recognize the significance of an accidental discovery and have the means and motivation to develop that accidental discovery into an important scientific contribution.

77. A theory can never be proven. It is a well-tested explanation of a broad set of observations. A theory may need to be changed in the future to explain new observations.

79. Your diagram should show one string that is threaded through both holes A and C. The string at hole B is a separate thread from the string passing through holes A and C.

81. 144,000 eggs

83. **a.** \$1.00 per package
b. number of envelopes in a package

Chapter 2

9. Iron is magnetic; table salt is not. Table salt will dissolve in water; iron will not.

10. By lowering the temperature to below the boiling point of each gas, you could condense each substance and separate the gases.

18. Liquid A is probably a substance. Liquid B is a mixture.

19. The liquid was not an element because a solid was left when the liquid evaporated. A physical process, such as evaporation, cannot be used to break down a compound. Therefore, the liquid was a mixture.

35. An extensive property depends on the amount of matter in a sample; an intensive property depends on the type of matter in a sample. Extensive properties include mass and volume. Intensive properties include color, hardness, melting point, and boiling point.

37. melting point and boiling point

39. a. solid **b.** liquid **c.** gas
d. liquid **e.** gas **f.** liquid

41. The particles in a solid are packed tightly together in an orderly arrangement. The particles in a liquid are in close contact, but not in a rigid or orderly arrangement. The particles in a gas are relatively far apart.

43. Sharpening a pencil is an irreversible physical change. Making ice cubes is a reversible physical change.

45. one; solutions are homogeneous mixtures with uniform composition throughout.

47. The goal of a distillation is to separate the components of a solution. The solution is boiled to produce a vapor, which is then condensed into a liquid. Solids dissolved in the solution are left behind.

49. a. Hydrogen and oxygen are the elements that make up the compound water.
b. Nitrogen and oxygen are both elements present in the mixture air.
c. Sodium and chlorine are both elements in the compound sodium chloride (table salt).
d. Carbon is an element and water is a compound. They are the final products of heating table sugar (sucrose).

51. In the symbol W, the single letter is capitalized. In the two-letter symbol Hg, the first letter is a capital and the second letter is lowercase.

53. When heated, sulfur and iron react and form a new substance, iron sulfide. The composition of the reactants in a chemical change is different from the composition of the products. In a physical change, the chemical composition of a sample doesn't change.

55. chemical property

57. Mass is an extensive property, which depends only on the amount of matter in the sample, not on the composition of the sample.

59. Substances are classified as solids, liquids, or gases according to their state at room temperature, which in this book is 20°C.

61. neon

63. sulfur

65. In both the kitchen and park, you will see mostly mixtures.

67. a. physical **b.** physical **c.** physical
d. physical **e.** chemical

69. In photograph A, bubbles indicate the production of a gas. In photograph B, there is a color change and a precipitate.

71. A gas can be released during a physical change. For example, bubbles form when water boils.

73. A gas expands to fill any space; a gas has no shape or volume without a container. A solid has a definite shape and volume; a solid doesn't need a container to maintain its shape and volume.

75. Gallium will freeze first; mercury will freeze last.

77. Iron rusts when it reacts with oxygen in the air to form an oxide (Fe_2O_3). The mass of the rust is the sum of the mass of the iron and the oxygen that combined with the iron.

79. a. Yes; because the graph is a straight line, the proportion of iron to oxygen is a constant, which is true for a compound.
b. No; a point for the values given wouldn't fall on the line. The mass ratio of iron to oxygen is different.

81. a. mercury and sulfur
b. Sulfur melts at 113°C and boils at 445°C. Between 113°C and 445°C, it exists as a liquid. Mercury melts at −39°C, and boils at 357°C. In between these temperatures, it exists as a liquid.
c. Possibilities include by color, by boiling point, or in alphabetical order.

Chapter 3

1. a. 4 **b.** 4 **c.** 2 **d.** 5

2. a. 3 **b.** 2 **c.** 4 **d.** 4

3. a. 8.71×10^1 m **b.** 4.36×10^8 m
c. 1.55×10^{-2} m **d.** 9.01×10^3 m
e. 1.78×10^{-3} m **f.** 6.30×10^2 m

4. a. 9×10^1 m **b.** 4×10^8 m
c. 2×10^{-2} m **d.** 9×10^3 m
e. 2×10^{-3} m **f.** 6×10^2 m

5. a. 79.2 m **b.** 7.33 m
c. 11.53 m **d.** 17.3 m

6. 23.8 g

7. a. 1.8×10^1 m^2 **b.** 6.75×10^2 m
c. 5.87×10^{-1} min

8. 1.3×10^3 m^3

16. −196°C

17. melting point = 1234 K; boiling point = 2485 K

28. 1.0080×10^4 min

29. 1.44000×10^5 s

30. 67 students

31. 86.4°F

32. a. 44 m **b.** 4.6×10^{-3} g **c.** 10.7 cg

33. a. 1.5×10^{-2} L **b.** 7.38×10^{-3} kg
c. 6.7×10^3 ms **d.** 9.45×10^7 μg

34. 2.27×10^{-8} cm

35. 1.3×10^8 dm

36. 1.93×10^4 kg/m^3

37. 7.0×10^{12} RBC/L

46. density = 2.50 g/cm^3; no

47. 10.5 g/cm^3

48. a. 6.32 cm^3 **b.** 0.342 cm^3

49. See answers for problem 48.

57. Lissa: inaccurate and imprecise; Lamont: accurate and precise; Leigh Anne: inaccurate and precise.

59. **a.** 98.5 L **b.** 0.000763 cg **c.** 57.0 m **d.** 12.2°C

61. **59a.** 9.85×10^1 L **59b.** 7.63×10^{-4} cg **59c.** 5.70×10^1 m **59d.** 1.22×10^1°C
60a. 4.3×10^1 g **60b.** 2.258×10^{-2} L **60c.** 9.20×10^1 kg **60d.** 3.24×10^1 m^3

63. **a.** second **b.** meter **c.** kelvin **d.** kilogram

65. **a.** 2.4 mm **b.** 14.33 cm **c.** 27.50 cm

67. conversion factor

69. The unit of the conversion factor in the denominator must be identical to the unit in the given measurement or the previous conversion factor.

71. **a.** 7.3 μL/s **b.** 78.6 mg/mm^2 **c.** 1.54 g/cm^3

73. **a.** 2.83×10^2 mg **b.** 0.283 g **c.** 2.83×10^{-4} kg **d.** 6.6 g **e.** 6.6×10^2 cg **f.** 6.6×10^{-3} kg **g.** 2.8×10^{-1} mg **h.** 2.8×10^{-2} cg **i.** 2.8×10^{-7} kg

75. Yes; neither mass nor volume changes with location.

77. The carbon dioxide-filled balloon would sink. The neon- and hydrogen-filled balloons would rise, the hydrogen at a much faster rate.

79. e, d, c, f, a, b

81. °F = 1.8°C + 32

83. $\frac{1\text{ g}}{10^2\text{ cg}}, \frac{10^2\text{ cg}}{1\text{ g}}, \frac{1\text{ g}}{10^3\text{ mg}}, \frac{10^3\text{ mg}}{1\text{ g}}, \frac{10^2\text{ cg}}{10^3\text{ mg}}, \frac{10^3\text{ mg}}{10^2\text{ cg}}$

85. 0.69–0.789 g/cm^3

87. 0.804 g/cm^3

89. 0.92 kg/L

91. 8.3 min

93. 5.52 kg/dm^3

95. Yes; the mass of an object is constant, but the weight of an object varies with location.

97. 31.1 m/s

99. Answers will vary. Lakes would freeze solid from the bottom up; aquatic life would be destroyed; possible climate changes.

101.

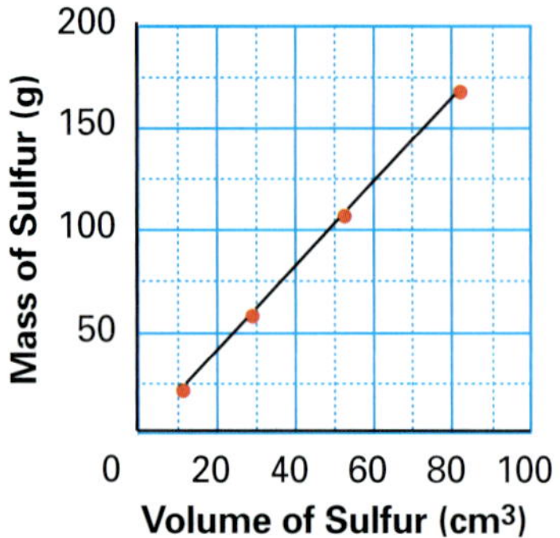

density of sulfur = 2.1 g/cm^3

103. Volume of iron cube = 45.1 cm^3; mass of lead cube = 514 g

105. 1.8×10^3 kg

107. 1.79 mL

Chapter 4

15. **a.** 19 **b.** B **c.** 5 **d.** 5 **e.** 16 **f.** 16 **g.** 23 **h.** 23

16. **a.** 9 protons and 9 electrons
b. 20 protons and 20 electrons
c. 13 protons and 13 electrons

17. **a.** 8 **b.** 16 **c.** 61 **d.** 45 **e.** 125

18. **a.** ${}^{12}_{6}C$ **b.** ${}^{19}_{9}F$ **c.** ${}^{9}_{4}Be$

19. ${}^{16}_{8}O$, ${}^{17}_{8}O$, ${}^{18}_{8}O$

20. Chromium-50 has 26 neutrons, chromium-52 has 28 neutrons, and chromium-53 has 29 neutrons.

21. boron-11

22. Silicon-28 must be by far the most abundant. The other two isotopes must be present in very small amounts.

23. 63.6 amu

24. 79.91 amu

35. Democritus's ideas were not helpful in explaining chemical behavior because they lacked experimental support.

37. The atoms are separated, joined, and rearranged.

39. repel

41. Atoms are neutral: number of protons = number of electrons. Loss of an electron means that the number of protons is greater than the number of electrons, so the remaining particle is positively charged.

43. Rutherford did not expect alpha particles to be deflected over a large angle.

45. protons and neutrons

47. the number of protons in the nucleus

49. The atomic number is the number of protons. The mass number is the sum of the protons and neutrons.

51. mass numbers, atomic masses, number of neutrons, relative abundance

53. which isotopes exist, their masses, and their natural percent abundance

55. The atomic mass is the weighted average of the masses of all the isotopes.

57. Sample answer: The table is set up so that chemical properties recur at regular intervals.

59. The nucleus is very small and very dense compared with the atom.

61. All atoms of the same element are not identical (isotopes). The atom is not the smallest particle of matter.

63. They are the same value.

65. 207 amu

67. Atoms are the smallest particle of an element that retains the properties of that element.

69. $^{14}_{7}$N: 14.003 amu; 99.63%; $^{15}_{7}$N: 15.000 amu; 0.37%; average atomic mass = 14.01 amu

71. Atomic number is the same as the number of protons and electrons; mass number minus atomic number equals number of neutrons.

73. The pattern repeats.

75. Change the metal used as a target and account for differences in deflection patterns.

77. The theory must be modified and then retested.

79. In a chemical change, atoms are neither created nor destroyed; instead, they are rearranged.

81. 92.5%

83. Pure chemistry involves the accumulation of scientific knowledge for its own sake; applied chemistry is accumulating knowledge to attain a specific goal.

85. **a.** element **b.** mixture
c. mixture **d.** mixture

87. 6.38×10^{7} cm^3

Chapter 5

8. **a.** $1s^22s^22p^2$
b. $1s^22s^22p^63s^23p^6$
c. $1s^22s^22p^63s^23p^63d^84s^2$

9. **a.** $1s^22s^22p^1$; one unpaired electron
b. $1s^22s^22p^63s^23p^2$ two unpaired electrons

14. 2.00×10^{-5} m; longer wavelength than red light

15. 6.00×10^{15} s^{-1}; ultraviolet

23. Bohr proposed that electrons traveled in circular paths around the nucleus.

25. An electron is found 90% of the time inside this boundary.

27. 3

29. **a.** 1 **b.** 2 **c.** 3 **d.** 4

31. The Aufbau principle states that electrons occupy the lowest possible energy levels. The Pauli exclusion principle states that an atomic orbital can hold at most two electrons. Hund's rule states that one electron occupies each of a set of orbitals with equal energies before any pairing of electrons occurs.

33. The *p* orbitals in the third quantum level have three electrons.

35. **a.** correct **b.** incorrect
c. incorrect **d.** correct

37. 2*s*, 3*p*, 4*s*, 3*d*

39. **a.** $1s^22s^22p^63s^23p^63d^{10}4s^24p^4$
b. $1s^22s^22p^63s^23p^63d^44s^2$
c. $1s^22s^22p^63s^23p^63d^84s^2$
d. $1s^22s^22p^63s^23p^64s^2$

41. Frequency is the number of wave cycles that pass a given point per unit time. Frequency units are cycles/s or reciprocal seconds or hertz. Wavelength and frequency are inversely related.

43. Classical physics views energy changes as continuous. In the quantum concept, energy changes occur in tiny discrete units called quanta.

45. **a.** v, vi, iv, iii, i, ii
b. It is the reverse.

47. The electron of the hydrogen atom is raised (excited) to a higher energy level.

49. visible spectrum, Balmer series

51. $1s^22s^22p^63s^23p^63d^{10}4s^24p^3$. The first three energy levels are full; the fourth energy level is partially filled.

53. $1s^22s^22p^3$ nitrogen; 3 unpaired electrons

55. **a.** 4.36×10^{-5} cm **b.** visible
c. 6.88×10^{14} s^{-1}

57. **a.** Na, sodium **b.** N, nitrogen
c. Si, silicon **d.** O, oxygen
e. K, potassium **f.** Ti, titanium

59. It is not possible to know both the position and the velocity of a particle at the same time.

61. c.

63. a.

65. An orbit confines the electron to a fixed circular path around the nucleus; an orbital is a region around the nucleus in which electrons are likely to be found.

67. Answers will vary. The model of the atom uses the abstract idea of probability; light is considered a particle and a wave at the same time. Atoms and light cannot be compared to familiar objects or observations because humans cannot experience atoms or photons directly and because matter and energy behave differently at the atomic level than at the level humans can observe directly.

69. **a.** $n = 1$ level **b.** $n = 4$ level
c. $n = 4$ level **d.** $n = 1$ level

71. **a.** potassium, excited state, valence electron has been promoted from 4*s* to 5*p*
b. potassium, ground state, correct electron configuration
c. impossible configuration, 3*p* can hold a maximum of 6 electrons, not 7

73. **a.** **a.** 5.20×10^{12} **b.** 4.40×10^{13}
c. 9.50×10^{13} **d.** 1.70×10^{14}
e. 2.20×10^{14} **f.** 4.70×10^{14}
b.

Energy (x 10^{-19} J): 0, 0.5, 1.0, 1.5, 2.0, 2.5, 3.0
Frequency (x 10^{12} s^{-1}): 0, 100, 200, 300, 400, 500

c. 6.3×10^{-34} J/s
d. The slope is Planck's constant.

75. H: 1312 kJ/mol ($n = 1$); 328 kJ($n = 2$); Li^{2+}: 1.18×10^4 kJ ($n = 1$)

77. a. and b. are heterogeneous; c. is homogeneous.

79. A compound has constant composition; the composition of a mixture can vary.

81. 7.7×10^{-5} μm

83. the piece of lead

85. a. and b. are exact.

87. 8.92 g/cm^3

89. Helium gas is much less dense than the nitrogen gas and oxygen gas in the air.

91. accuracy–how close measured value is to true value; precision–how close a series of measurements are to one another

93. Neon-20 has 10 neutrons in the nucleus, neon-21 has 11 neutrons in the nucleus.

Chapter 6

8. a. $1s^22s^22p^2$
b. $1s^22s^22p^63s^23p^63d^{10}4s^2 4p^6 5s^2$
c. $1s^22s^22p^63s^23p^63d^3 4s^2$

9. a. B, Al, Ga, In, Tl **b.** F, Cl, Br, I, At
c. Ti, Zr, Hf, Rf

25. The close match between the predicted properties and the actual properties of gallium, which was discovered in 1875, helped gain wider acceptance for Mendeleev's periodic table.

27. Yes; both carbon and silicon are in Group 4A and each has four electrons in its highest occupied energy level.

29. Metalloids have properties that are similar to both metals and nonmetals. How a metalloid behaves depends on the conditions.

31. Na, Mg, Cl

33. aluminum

35. a. Ar: $1s^22s^22p^63s^23p^6$ **b.** Si: $1s^22s^22p^63s^23p^2$
c. Mg: $1s^22s^22p^63s^2$

37. The first ionization energy is the energy needed to remove a first electron from an atom. The second ionization energy is the energy needed to remove a second electron.

39. a. Sr, Mg, Be **b.** Cs, Ba, Bi **c.** Na, Al, S

41. The ionic radius of a cation is smaller than the atomic radius of the metal atom.

43. a. F **b.** N **c.** Mg **d.** As

45. a. O **b.** F **c.** O **d.** S

47. a. 1801–1850; 28 elements
b. Mendeleev's periodic table helped scientists predict the existence of undiscovered elements.
c. 75%

49. b. Nitrogen and phosphorus are in the same group (5A).

51. Nonmetals; The trend is for ionization energy to increase from left to right across a period.

53. a. H, Li, Na, K, Rb, Cs, Fr
b. O, S, Se, Te, Po
c. Zn, Cd, Hg, Uub

55. c

57. First ionization energy increases across a period.

59. a. The atomic radius increases from top to bottom within the group.
b. Cations are smaller than their corresponding atoms. The attraction between the nucleus and any remaining electron is greater. There is one fewer occupied energy level.

61. a. As first ionization energy increases, so does electronegativity.
b. Both properties depend on the attraction between the nucleus and electrons. The attraction between the nucleus and the electrons in the highest occupied energy level increases across a period because the nuclear charge increases, but the shielding effect is constant.

63. a. Both electrons in Ca are removed from the same energy level. The second electron removed from a K atom is in a lower energy level.
b. Because Cs has a larger atomic radius than Li, the nuclear charge in a Cs atom has a smaller effect on the electrons in the highest occupied energy level.
c. It is relatively easy to remove all three electrons from an Al atom, but the third electron removed from a Mg atom is in a lower energy level.

65. The ionic radii would decrease from S^{2-} to Sc^{3+}. The number of electrons and the shielding effect do not change, but the number of protons increases from left to right in this series. So the ionic size decreases. The same is true for the series O^{2-} to Mg^{2+}.

67. a.

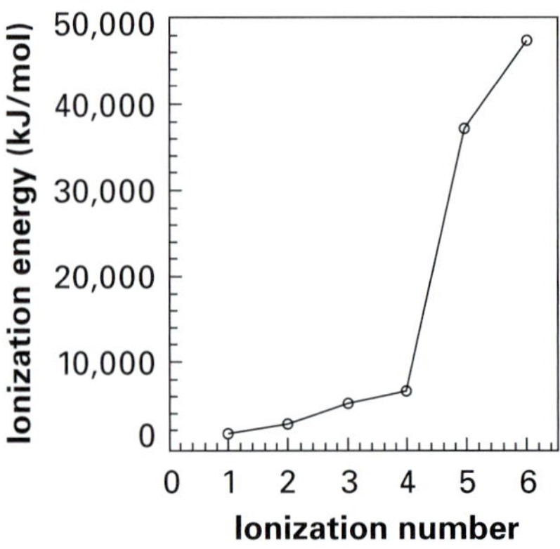

b. The largest increase is between ionization numbers 4 and 5 because carbon easily loses the first four electrons from the second energy level. The fifth electron is removed from the first energy level.

69. Electron affinity should increase (become more negative) from left to right across a period because the nuclear charge increases and the shielding effect is constant.

71. a. physical change **b.** chemical change
c. physical change **d.** chemical change

73. 4

75. The density of the cube is 0.984 g/cm^3. The cube will float on water.

77. 5.2%

79. The density of the olive is 1.05 g/cm^3. The olive will sink in water.

81. The density of sulfur does not vary with mass. The density is constant.

83. 4.54 g/cm^3

85. a. silver, 62 neutrons
b. tin, 50 protons
c. molybdenum, 42 electrons
d. scandium, 21 electrons

Chapter 7

1. a. sulfide ion, S^{2-} **b.** aluminum ion, Al^{3+}

2. a. 2 electrons lost **b.** 3 electrons gained
c. 2 electrons lost

12. a. KI **b.** Al_2O_3

13. $CaCl_2$

31. a. gain of 1 electron **b.** loss of one electron
c. gain of 3 electrons **d.** loss of 2 electrons
e. loss of 1 electron **f.** gain of 1 electron

33. electrons in the highest occupied energy level

35. a. :C̈l̤· **b.** :S̈· **c.** ·Ȧl· **d.** Li·

37. a. Al^{3+} **b.** Li^+ **c.** Ba^{2+}
d. K^+ **e.** Ca^{2+} **f.** Sr^{2+}

39. a. S^{2-} **b.** Na^+ **c.** F^- **d.** P^{3-}

41. a, c, and e

43. The positive charges balance the negative charges.

45. a. K^+, Cl^- **b.** Ba^{2+}, SO_4^{2-}
c. Mg^{2+}, Br^- **d.** Li^+, CO_3^{2-}

47. Ions are free to move in molten $MgCl_2$

49. body-centered cubic: Na, K, Fe, Cr, or W; face-centered cubic: Cu, Ag, Au, Al, or Pb; hexagonal close-packed: Mg, Zn, or Cd

51. The properties of the steel will vary according to its composition. In addition to iron, steel can contain varying amounts of carbon and such metals as chromium, nickel, and molybdenum.

53. a. ·Ċ̣· **b.** ·Be· **c.** :Ö·
d. :F̤̈· **e.** Na· **f.** ·P̣̈·

55. It has lost valence electrons.

57. a. oxygen atom, sulfur atom, oxide ion, sulfide ion
b. sodium ion, potassium ion, sodium atom, potassium atom

59. a. $1s^22s^22p^63s^23p^63d^3$ **b.** $1s^22s^22p^63s^23p^63d^4$
c. $1s^22s^22p^63s^23p^63d^5$

61. a. Br^- **b.** H^- **c.** As^{3-} **d.** Se^{2-}

63. All are $1s^22s^22p^6$. All have the same configuration as neon.

65. a. $1s^22s^22p^63s^23p^6$ **b.** $1s^22s^22p^6$
Each has a noble gas electron configuration.

67. a, c, e, f

69. 12

71. Brass is a mixture of copper and zinc. The properties of a particular sample of brass will vary with the relative proportions of the two metals.

73. By gaining or losing electrons, the atoms of elements achieve a noble gas electron configuration.

75. No. Sodium chloride is composed of equal numbers of sodium ions and chloride ions. The ions are in a 1:1 ratio. Each sodium ion is surrounded by chloride ions and each chloride is surrounded by sodium ions.

77. The spheres are more closely packed in a.; there is less empty space in a., and a rough count shows 25 spheres in a. compared with 22 spheres in b..

79. Both metals and ionic compounds are composed of ions. Both are held together by electrostatic bonds. Metals always conduct electricity, and ionic compounds conduct only when melted or in water solution. Ionic compounds are composed of cations and anions, but metals are composed of cations and free-floating valence electrons. Metals are ductile, but ionic compounds are brittle.

81. Na^+ and Cs^+ differ greatly in size. Na^+ and Cl^- are similar in size to Mn^{2+} and S^{2-}.

83. a. copper and zinc
b. silver and copper
c. copper and tin
d. iron, chromium, nickel, and carbon
e. iron, chromium, nickel, and molybdenum
f. iron, chromium, and carbon

85. an analytical chemist

87. a, b, and d are chemical changes; c is a physical change.

89. a. liquid, vapor **b.** vapor **c.** liquid, vapor **d.** liquid, vapor

91. a

93. 27.0 cm^3

95. 14 amu

97. a. 1 **b.** 3 **c.** 1 **d.** 5

99. chlorine, Cl, $1s^22s^22p^63s^23p^5$

101. a. K, $1s^22s^22p^63s^23p^64s^1$
b. Al, $1s^22s^22p^63s^23p^1$
c. S, $1s^22s^22p^63s^23p^4$
d. Ba, $1s^22s^22p^63s^23p^63d^{10}4s^24p^64d^{10}5s^25p^66s^2$

103. sodium (Na), cesium (Cs), rubidium (Rb), lithium (Li)

Chapter 8

7. a. :Cl:Cl: **b.** :Br:Br: **c.** :I:I:

8. a. H:O:O:H **b.** :Cl:P:Cl: with :Cl: below P

9. [H:O:]$^-$

10. [:F:B:F: with :F: above and below B]$^-$

11. [:O:S:O: with :O: above and below S]$^{2-}$ [:O:C::O with :O: above C]$^{2-}$

12. [H:O:C with :O above C and :O: below C]$^-$

30. a. moderately polar covalent
b. ionic
c. moderately to very polar covalent
d. moderately to very polar covalent
e. ionic
f. nonpolar covalent

31. c and d (tied), b, a

39. ionic

41. Nitrogen and oxygen achieve stability as diatomic molecules; argon exists as individual atoms because it has a stable noble gas electron configuration.

43. a. ionic **b.** ionic **c.** covalent **d.** covalent

45. A double covalent bond has four shared electrons (two bonding pairs); a triple covalent bond has six shared electrons (three bonding pairs).

47. One atom contributes both electrons to a coordinate covalent bond as in CO.

49. [:O:N::O:]$^-$ ↔ [:O::N:O:]$^-$

51. Bond dissociation energy is defined as the energy needed to break one covalent bond.

53. A pi bond is formed by the side-by-side overlap of two half-filled f *p* atomic orbitals to produce a pi molecular orbital. In a pi bond, the bonding electrons are most likely to be found in sausage-shaped regions above and below the bond. See Figure 8.15.

55. The 2*s* and the 2*p* orbitals form two sp^2 hybrid orbitals on the carbon atom. One sp^2 hybrid orbital forms a sigma bond with the carbon atom. Pi bonds between each oxygen atom and the carbon are formed by the unhybridized 2*p* orbitals.

57. The electronegativities of the two atoms will differ by about 0.4 to 2.0.

59. A hydrogen bond is formed by an electrostatic interaction between a hydrogen atom covalently bonded to a very electronegative atom and an unshared electron pair of a nearby atom.

61. More energy is required to separate the molecules.

63. :Cl:S:Cl: with :O: below S

65. a. tetrahedral, 109.5° **b.** trigonal planar, 120° **c.** tetrahedral, 109.5° **d.** bent, 105°

67. a. 109.5° **b.** 120° **c.** 180°

69. a. Phosphorus in PBr_5 has 10 valence electrons.

71. C, O, H, S, N, F, Cl, I, Br: These elements are all nonmetals.

73. a. two covalent bonds to hydrogen
H:C::C:H
b. Fluorine and oxygen have only four electrons.
:F:O:H
c. Halogens form one covalent bond, not three.
:I:Cl:
d. Nitrogens should be forming only three covalent bonds, not four.
H:N:N:H

75. a. bent **b.** tetrahedral **c.** pyramidal

77. H:C:H with H above and below C; H, H, H, H around C with angles 90°, 90°; H, H, H, H around C with angles 109.5°, 109.5°, 109.5°, 109.5°

The first sketch is tetrahedral. The second sketch is a tetrahedron. The bond angles in the first sketch are not all the same, with some being 90°. The bond angles in the second sketch are all 109.5°. The second sketch is correct. (Note: The wedge-shaped lines come out of the page; the dotted lines recede into the page.)

```
       :O:
        ::
79. H:C:O:H
          ··
```

81. H:N̈::N::N̈: ↔ H:N̤:N::N:

83. formation of a gas, a change in color or odor

85. **a.** 2 **b.** 2 **c.** 3 **d.** 3

87. Isotopes have the same number of protons and electrons, but different numbers of neutrons.

89. **a.** 6 **b.** 2 **c.** 5 **d.** 0

91. The *d* sublevel of the third principal energy level contains 5 electrons.

93. The anion is larger than the corresponding neutral atom.

95. **a.** K: $1s^22s^22p^63s^23p^64s^1$
b. Al: $1s^22s^22p^63s^23p^1$
c. S: $1s^22s^22p^63s^23p^4$
d. Ba: $1s^22s^22p^63s^23p^63d^{10}4s^24p^64d^{10}5s^25p^66s^2$

97. **e.** II and III only

99. **b.** cesium

101. **a.** $1s^22s^22p^6$ **b.** $1s^22s^22p^6$
c. $1s^22s^22p^6$ **d.** $1s^22s^22p^63s^23p^6$

Chapter 9

1. **a.** selenide ion, anion
b. barium ion, cation
c. phosphide ion, anion

2. **a.** three electrons lost
b. two electrons gained
c. one electron lost

10. **a.** BaS **b.** Li_2O **c.** Ca_3N_2 **d.** CuI_2

11. **a.** NaI **b.** $SnCl_2$ **c.** K_2S **d.** CaI_2

12. **a.** $(NH_4)_2SO_3$ **b.** $Ca_3(PO_4)_2$

13. **a.** $LiHSO_4$ **b.** $Cr(NO_2)_3$

34. 2:1

43. **a.** 2+ **b.** 2+ **c.** 3+ **d.** 1+

45. cyanide, CN^-, and hydroxide, OH^-

47. zero

49. Determine the charge of the anion, then work backwards to find the charge of the transition metal cation needed to give a net charge of zero for the formula unit.

51. a and b

53. NH_4NO_3 ammonium nitrate; $(NH_4)_2CO_3$ ammonium carbonate; NH_4CN ammonium cyanide; $(NH_4)_3PO_4$ ammonium phosphate; $Sn(NO_3)_4$ tin(IV) nitrate; $Sn(CO_3)_2$ tin(IV) carbonate; $Sn(CN)_4$ tin(IV) cyanide; $Sn_3(PO_4)_4$ tin(IV) phosphate; $Fe(NO_3)_3$ iron(III) nitrate; $Fe_2(CO_3)_3$ iron(III) carbonate; $Fe(CN)_3$ iron(III) cyanide; $FePO_4$ iron(III) phosphate; $Mg(NO_3)_2$ magnesium nitrate; $MgCO_3$ magnesium carbonate; $Mg(CN)_2$ magnesium cyanide; $Mg_3(PO_4)_2$ magnesium phosphate

55. **a.** tri- **b.** mono- **c.** di-
d. hexa- **e.** penta- **f.** tetra-

57. **a.** BCl_3 **b.** dinitrogen pentoxide
c. N_2H_4 **d.** carbon tetrachloride

59. No, to be an acid the compound must produce H^+ ions in water solution.

61. **a.** $Fe(OH)_2$ **b.** lead(II) hydroxide
c. $Cu(OH)_2$ **d.** cobalt(II) hydroxide

63. Whenever two elements form more than one compound, the different masses of one element that combine with the same mass of the other element are in the ratio of small whole numbers.

65. **a.** $KMnO_4$ **b.** $Ca(HCO_3)_2$ **c.** Cl_2O_7
d. Si_3N_4 **e.** NaH_2PO_4 **f.** PBr_5
g. CCl_4

67. **a.** sodium chlorate
b. mercury(I) bromide
c. potassium chromate
d. perchloric acid
e. tin(IV) oxide
f. iron(III) acetate
g. potassium hydrogen sulfate
h. calcium hydroxide
i. barium sulfide

69. **a.** magnesium permanganate
b. beryllium nitrate
c. potassium carbonate
d. dinitrogen tetrahydride
e. lithium hydroxide
f. barium fluoride
g. phosphorus triiodide
h. zinc oxide
i. phosphorous acid

71. binary molecular compound

73. $SnCl_4$

75. **a.** 9.85%
b. nitrogen, oxygen, and chlorine; 54.9 kg
c. 34.7%
d. H_2SO_4, N_2, O_2, NH_3, CaO, H_3PO_4, NaOH, Cl_2, Na_2CO_3, HNO_3

77. on the right hand side

79. The statement is true for the representative metals, but not for the transition metals, which often have multiple charges.

81. **a.** N_2O, dinitrogen monoxide
b. NO_2, nitrogen dioxide
c. NO, nitrogen monoxide
d. N_2O_4, dinitrogen tetroxide

83. **a.** The charges do not balance, CsCl.
b. The charges do not balance, ZnO.
c. Neon does not form compounds.
d. The subscripts are not the lowest whole-number ratio, BaS.

87. Answers will vary but may include: color (physical), solid (physical), magnetic (physical), malleable (physical), conducts electricity (physical), burns (chemical).

89. 5.2 cm

91. 0.538 g/cm^3

93. Both are in the nucleus and have a mass of about 1 amu. A proton is positively charged; a neutron has no charge.

95. **a.** 1 **b.** 6 **c.** 5
d. 2 **e.** 7 **f.** 8

97. **a.** cesium, potassium, sodium, lithium
b. lithium, boron, carbon, fluorine, neon

99. When metallic elements of Group 1A and 2A form ions, they lose all their outer shell electrons. This increases the attraction by the nucleus for the fewer remaining elections and results in ions that are smaller than the corresponding atoms. The electron that a Group 7A element gains in forming an ion enters the outer shell resulting in a decrease in the effective nuclear attraction of the increased number of electrons. The anion is larger than the corresponding atom.

101. **a.** 12 protons and 10 electrons
b. 35 protons and 36 electrons
c. 38 protons and 36 electrons
d. 16 protons and 18 electrons

103. b, d, and f

105. A hydrogen bond is an intermolecular force between a hydrogen atom covalently bonded to a very electronegative atom and an unshared pair of electrons from another electronegative atom.

Chapter 10

1. 5.0 kg

2. 672 seeds

3. 4.65 mol Si

4. 0.360 mol

5. 2.75×10^{24} atoms

6. 7.72 mol NO_2

7. 137.5 g/mol

8. 84.0 g $NaHCO_3$

16. 1.27 g $C_{20}H_{42}$

17. 225 g $Fe(OH)_2$

18. 3.43×10^{-2} mol B

19. 0.987 mol N_2O_3

20. **a.** 7.17×10^{-2} L CO_2 **b.** 82.9 L N_2

21. **a.** 28.0 L He **b.** 7.50 L C_2H_6

22. 80.2 g/mol

23. 3.74 g/L

32. 72.2% Mg, 27.8% N

33. 93.0% Hg, 7.0% O

34. **a.** 80.0% C, 20.0% H
b. 19.2% Na, 0.83% H, 26.7% S, 53.3% O

35. **a.** 82.4% N **b.** 35.0% N

36. **a.** OH **b.** $HgSO_4$

37. C_3H_8N

38. $C_2H_6O_2$

39. **a.** same empirical formula
b. different

47. number, mass, or volume; examples will vary.

49. **a.** 3 **b.** 2 **c.** 9 **d.** 10

51. 1.00 mol C_2H_6

53. **a.** 98.0 g **b.** 76.0 g **c.** 100.1 g
d. 132.1 g **e.** 89.0 g **f.** 159.8 g

55. Answers will vary but should include:
1. Determine the moles of each atom from the formula.
2. Look up the molar mass of each element.
3. Multiply the number of moles of each atom by its molar mass.
4. Sum these products.

57. Answers will vary. For example; if a particle is a 0.1-mm cube, how high would a stack of Avogadro's number of particles be? (6.02×10^{16} km)

59. **a.** 108 g C_5H_{12} **b.** 547 g F_2 **c.** 71.8 g $Ca(CN)_2$
d. 238 g H_2O_2 **e.** 224 g NaOH **f.** 1.88 g Ni

61. **a.** 1.96 g/L **b.** 0.902 g/L **c.** 2.05 g/L

63. **a.** 5.9% H, 94.1% S
b. 22.6% N, 6.5% H, 19.4% C, 51.6% O
c. 41.7% Mg, 54.9% O, 3.4% H
d. 42.1% Na, 18.9% P, 39.0% O

65. **d.** 77.7% Fe in FeO

67. **a.** molecular **b.** molecular **c.** empirical

69. **a.** H_2O_2 **b.** $C_6H_6O_4$

71. **a.** A, $C_2H_4O_2$; D, $C_5H_{10}O_5$; E, $C_6H_{12}O_6$
b. slope = 2.5/1 which is the ratio of the molar mass of the empirical formula to the mass of carbon in the empirical formula: 30/12 = 2.5/1.
c. mass of carbon = 36, molar mass = 90; mass of carbon = 48, molar mass = 120

73. **b.** 0.842 mol C_2H_4

75. **a.** CO **b.** $C_2O_2NH_5$ **c.** Cl_2OC

77. 3.01×10^{13} km

79. $C_3H_6O_3$

81. 2.73×10^{20} F atoms

83. C_2H_6O

85. Sulfur atoms have 16 protons, 16 electrons, and 16 neutrons; carbon has 6 electrons, and 6 neutrons. Therefore, 6.02×10^{23} sulfur atoms will have a greater mass than the same number of carbon atoms.

87. a. $C_9H_{11}O_2N$ **b.** $C_9H_{11}O_2N$

89. 21.9 cm^3

91. a

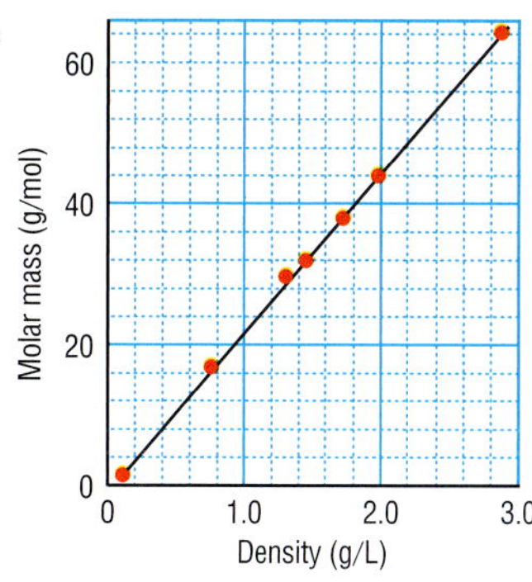

b. 22.4 L/mol
c. 24.6 g/mol
d. 2.5 g/L

93. 6.025×10^{23} formula units/mol

95. a. physical change **b.** chemical change **c.** chemical change **d.** physical change **e.** chemical change **f.** physical change

97. No; the student has ignored the units. The density of sugar is 1.59 g/mL; the density of carbon dioxide is much less, 1.83 g/L (or .00183 g/mL)

99. a. 4.72×10^3 mg **b.** 97 km/h **c.** 4.4×10^{-2} dm

101. a. $1s^22s^22p^5$ **b.** $1s^22s^1$ **c.** $[Kr]5s^1$

103. Cr, Cd, Cu, and Co

105. A molecule is composed of two or more atoms.

107. For single bond a single line connects the atoms (X—X). Atoms are connected by two lines in a double bond (X═X), and three lines in a triple bond (X≡X).

109. Calculate the electronegativity difference between two atoms. If the difference is small (0.0–0.4), the bond is nonpolar covalent. If the difference ≥ 2.0, the bond is most likely ionic. For values between 0.4 and 2.0, the bond is polar covalent.

111. a. iron(III) hydroxide **b.** ammonium iodide **c.** sodium carbonate **d.** carbon tetrachloride

112. a. KNO_3 **b.** CuO **c.** Mg_3N_2 **d.** AgF

Chapter 11

1. When solid sodium is dropped into water, hydrogen gas and aqueous sodium hydroxide are produced.

2. $S(s) + O_2(g) \longrightarrow SO_2(g)$

3. a. $2AgNO_3 + H_2S \longrightarrow Ag_2S + 2HNO_3$
b. $3Zn(OH)_2 + 2H_3PO_4 \longrightarrow Zn_3(PO_4)_2 + 6H_2O$

4. a. $H_2 + S \longrightarrow H_2S$
b. $2FeCl_3 + 3Ca(OH)_2 \longrightarrow 2Fe(OH)_3 + 3CaCl_2$

5. a. $FeCl_3 + 3NaOH \longrightarrow Fe(OH)_3 + 3NaCl$
b. $CS_2 + 3Cl_2 \longrightarrow CCl_4 + S_2Cl_2$

6. $Ca(OH)_2 + H_2SO_4 \longrightarrow CaSO_4 + 2H_2O$

13. $2Be + O_2 \longrightarrow 2BeO$

14. $3Mg + N_2 \longrightarrow Mg_3N_2$

15. $2HI \longrightarrow H_2 + I_2$

16. HBr

17. a. $Fe(s) + Pb(NO_3)_2(aq) \longrightarrow Fe(NO_3)_2(aq) + Pb(s)$
b. $Cl_2(aq) + 2NaI(aq) \longrightarrow 2NaCl(aq) + I_2(aq)$
c. $Ca(s) + 2H_2O(l) \longrightarrow Ca(OH)_2(aq) + H_2(g)$

18. a. $3NaOH(aq) + Fe(NO_3)_3(aq) \longrightarrow Fe(OH)_3(s) + 3NaNO_3(aq)$
b. $3Ba(NO_3)_2(aq) + 2H_3PO_4(aq) \longrightarrow Ba_3(PO_4)_2(s) + 6HNO_3(aq)$

19. a. $3KOH(aq) + H_3PO_4(aq) \longrightarrow K_3PO_4(aq) + 3H_2O(l)$
b. $3H_2SO_4(aq) + 2Al(OH)_3(aq) \longrightarrow Al_2(SO_4)_3(aq) + 6H_2O(l)$

20. a. $2HCOOH + O_2 \longrightarrow 2CO_2 + 2H_2O$
b. $C_7H_{16} + 11O_2 \longrightarrow 7CO_2 + 8H_2O$

21. $C_6H_{12}O_6 + 6O_2 \longrightarrow 6CO_2 + 6H_2O$

28. $H^+(aq) + OH^-(aq) \longrightarrow H_2O(l)$

29. complete ionic equation:
$3Ca^{2+}(aq) + 6OH^-(aq) + 6H^+(aq) + 2PO_4^{3-}(aq) \longrightarrow Ca_3(PO_4)_2(s) + 6H_2O(l)$;
net ionic equation: same as complete ionic equation

37. Dalton said that the atoms of reactants are rearranged to form new substances as products.

39. a. Gaseous ammonia and oxygen react in the presence of a platinum catalyst to produce nitrogen monoxide gas and water vapor.
b. Aqueous solutions of sulfuric acid and barium chloride are mixed to produce a precipitate of barium sulfate and aqueous hydrochloric acid.
c. The gas dinitrogen trioxide reacts with water to produce an aqueous solution of nitrous acid.

41. a. $C + 2F + 2G \longrightarrow CF_2G_2$
b. $F + 3W + S + 2P \longrightarrow FW_3SP_2$

43. a. $2PbO_2 \longrightarrow 2\,PbO + O_2$
b. $2Fe(OH)_3 \longrightarrow Fe_2O_3 + 3H_2O$
c. $(NH_4)_2CO_3 \longrightarrow 2NH_3 + H_2O + CO_2$
d. $2NaCl + H_2SO_4 \longrightarrow Na_2SO_4 + 2HCl$

45. a. $2Mg + O_2 \longrightarrow 2MgO$
b. $4P + 5O_2 \longrightarrow 2P_2O_5$
c. $Ca + S \longrightarrow CaS$

47. a. $2Ag_2O \xrightarrow{\Delta} 4Ag + O_2$
b. $NH_4NO_3 \xrightarrow{\Delta} N_2O + 2H_2O$

49. a. $H_2C_2O_4(aq) + 2KOH(aq) \longrightarrow K_2C_2O_4(aq) + 2H_2O(l)$
b. $CdBr_2(aq) + Na_2S(aq) \longrightarrow CdS(s) + 2NaBr(aq)$

51. a. $C_4H_8 + 6O_2 \longrightarrow 4CO_2 + 4H_2O$
b. $C_3H_6O + 4O_2 \longrightarrow 3CO_2 + 3H_2O$

53. an ion that does not participate in the reaction

55. a. $2Al(s) + 6H^+(aq) \longrightarrow 2Al^{3+}(aq) + 3H_2(g)$
b. $H^+(aq) + OH^-(aq) \longrightarrow H_2O(l)$
c. no reaction

57. a. $Cl_2(g) + 2KI(aq) \longrightarrow I_2(aq) + 2KCl(aq)$
b. $2Fe(s) + 6HCl(aq) \longrightarrow 2FeCl_3(aq) + 3H_2(g)$
c. $P_4O_{10}(s) + 6H_2O(l) \longrightarrow 4H_3PO_4(aq)$

59. a. $Na_2O(s) + H_2O(l) \longrightarrow 2NaOH(aq)$
b. $H_2(g) + Br_2(g) \longrightarrow 2HBr(g)$
c. $Cl_2O_7(l) + H_2O(l) \longrightarrow 2HClO_4(aq)$

61. a. tube A
b. $2Na(s) + 2H_2O(l) \longrightarrow 2NaOH(aq) + H_2(g)$; single-replacement

63. a. $2Al_2O_3 \xrightarrow{\Delta} 4Al + 3O_2$
b. $Sn(OH)_4 \xrightarrow{\Delta} SnO_2 + 2H_2O$
c. $Ag_2CO_3 \xrightarrow{\Delta} Ag_2O + CO_2$

65. a. $CdS(s)$
b. $Na^+(aq)$ and $NO_3^-(aq)$
c. $Cd^{2+}(aq) + S^{2-}(aq) \longrightarrow CdS(s)$

67. a. $2K(s) + 2H_2O(l) \longrightarrow 2KOH(aq) + H_2(g)$
b. $C_2H_5OH(l) + 3O_2(g) \longrightarrow 2CO_2(g) + 3H_2O(g)$
c. $2Bi(NO_3)_3(aq) + 3H_2S(g) \longrightarrow Bi_2S_3(s) + 6HNO_3(aq)$
d. $2Al(s) + 3Br_2(l) \longrightarrow 2AlBr_3(s)$
e. $2HNO_3(aq) + Ba(OH)_2(aq) \longrightarrow Ba(NO_3)_2(aq) + 2H_2O(l)$

69. a. $C_5H_{12} + 8O_2 \longrightarrow 5CO_2 + 6H_2O$;
$C_9H_{20} + 14O_2 \longrightarrow 9CO_2 + 10H_2O$
b. $2C_{12}H_{26} + 37O_2 \longrightarrow 24CO_2 + 26H_2O$;
$C_{17}H_{36} + 26O_2 \longrightarrow 17CO_2 + 18H_2O$
c. $n = CO_2$; $(n + 1) = H_2O$

71. a. (1) combination (2) single-replacement (3) combustion (4) double-replacement
b. (1) $2Al(s) + 3Br_2(l) \longrightarrow 2AlBr_3(s)$;
(2) $Cu(s) + 2AgNO_3(aq) \longrightarrow Cu(NO_3)_2(aq) + 2Ag(s)$;
(3) $C_3H_8(g) + 5O_2(g) \longrightarrow 3CO_2(g) + 4H_2O(g)$;
(4) $Pb(NO_3)_2(aq) + 2KI(aq) \longrightarrow PbI_2(s) + 2KNO_3(aq)$

73. a. water **b.** water vapor in the air
c. physical change

75. 36.6 kg

77. a. $1s^22s^22p^63s^23p^64s^23d^{10}4p^6$
b. $1s^22s^22p^63s^23p^6$
c. $1s^22s^22p^63s^23p^63d^{10}$
d. $1s^22s^22p^63s^23p^63d^{10}$

79. a. incorrect; KBr **b.** correct
c. incorrect; Ca_3N_2 **d.** correct

81. a. 2.41 mol **b.** 6.91×10^{-2} mol
c. 0.934 mol **d.** 7.09 mol

83. $C_8H_{10}O_2N_4$

Chapter 12

1. 288 seats, 864 wheels, 576 pedals

2. Answers will vary but should include the correct number of parts to make the product.

3. 2 molecules H_2 + 1 molecule $O_2 \longrightarrow$ 2 molecules H_2O;
2 mol H_2 + 1 mol $O_2 \longrightarrow$ 2 mol H_2O;
44.8 L H_2 + 22.4 L $O_2 \longrightarrow$ 44.8 L H_2O

4. 2 mol C_2H_2 + 5 mol $O_2 \longrightarrow$ 4 mol CO_2 + 2 mol H_2O;
44.8 L C_2H_2 + 112 L $O_2 \longrightarrow$ 89.6 L CO_2 + 44.8 L H_2O;
212 g reactants $\longrightarrow$ 212 g products

11. a. $\frac{4 \text{ mol Al}}{3 \text{ mol } O_2}, \frac{3 \text{ mol } O_2}{4 \text{ mol Al}}, \frac{4 \text{ mol Al}}{2 \text{ mol } Al_2O_3}, \frac{2 \text{ mol } Al_2O_3}{4 \text{ mol Al}}, \frac{2 \text{ mol } Al_2O_3}{3 \text{ mol } O_2}, \frac{3 \text{ mol } O_2}{2 \text{ mol } Al_2O_3}$
b. 7.4 mol

12. a. 11.1 mol **b.** 0.52 mol

13. 2.03 g C_2H_2

14. 1.36 mol CaC_2

15. 4.82×10^{22} molecules O_2

16. 11.5 g NO_2

17. 1.93 L O_2

18. 0.28 L PH_3

19. 18.6 mL SO_2

20. 1.9 dL CO_2

25. O_2 is the limiting reagent.

26. HCl is the limiting reagent.

27. a. 5.40 mol H_2O is required, so C_2H_4 is the limiting reagent;
b. 5.40 mol H_2O

28. 43.2 g H_2O

29. 59.3 g Fe

30. 17.0 g Ag

31. 83.5%

32. 57.7%

37. a. 2 mol $KClO_3$ decompose to form 2 mol KCl and 3 mol O_2.
b. 4 mol NH_3 react with 6 mol NO to form 5 mol N_2 and 6 mol H_2O.
c. 4 mol K react with 1 mol O_2 to form 2 mol K_2O.

39. Answers will vary but should include the idea of writing a ratio using the coefficients of two substances from a balanced equation as the number of moles of each substance reacting or being formed.

41. a. 11.3 mol CO, 22.5 mol H_2
b. 112 g CO, 16.0 g H_2
c. 11.4 g H_2

43. The coefficients indicate the relative number of moles (or particles) of reactants and products.

45. The amount of the limiting reagent determines the maximum amount of product that can be formed. The excess reagent is only partially consumed in the reaction.

47. **a.** Al **b.** 3.0 mol $AlCl_3$ **c.** 0.8 mol Cl_2

49. **a.** 2.36 g H_3PO_4 **b.** 1.89 g CO_2

51. **a.** 7.0×10^2 L N_2 **b.** no reagent in excess

53. 10.7 kg $CaSO_4$

55. **a.** Initially, the amount of NaCl formed increases as the amount of Na used increases. For this part of the curve sodium is the limiting reagent. Beyond a mass of about 2.5g of Na, the amount of product formed remains constant because chlorine is now the limiting reagent.
b. Chlorine becomes the limiting reagent when the mass of sodium exceeds 2.5 g. This corresponds to a mass of about 3.9 g chlorine.

57. The percent yield is 115%; such a yield could be attributed to experimenter error, or to unreacted starting material, or to outside materials contaminating the product.

59. **a.** 29 frames **b.** 58 wheels
c. 174 pedals **d.** 87 seats

61. 13 days

63. 87.4% $CaCO_3$

65. **a.** 347 g Fe **b.** 239 g CO

67. **a.** 22 e^-, 22 p^+, 25 n^0 **b.** 50 e^-, 50 p^+, 70 n^0
c. 8 e^-, 8 p^+, 10 n^0 **d.** 12 e^-, 12 p^+, 14 n^0

69. **a.** sodium. **b.** arsenic **c.** cesium

71. c and d

73. Yes, an ionic compound with at least one polyatomic ion has covalent bonds.

75. **a.** phosphate ion **b.** aluminum ion
c. selenide ion **d.** ammonium ion

77. **a.** $Al_2(CO_3)_3$ **b.** NO_2 **c.** K_2S
d. $MnCrO_4$ **e.** NaBr

79. 7.38 g Be

81. **a.** 0.473 mol KNO_3 **b.** 9.91×10^{-2} mol SO_2
c. 3.74×10^{-2} mol PCl_3

83. **a.** $Ba(NO_3)_2(aq) + Na_2SO_4(aq) \longrightarrow BaSO_4(s) + 2NaNO_3(aq)$
b. $AlCl_3(aq) + 3AgNO_3(aq) \longrightarrow 3AgCl(s) + Al(NO_3)_3(aq)$
c. $H_2SO_4(aq) + Mg(OH)_2(aq) \longrightarrow MgSO_4(aq) + 2H_2O(l)$

85. **a.** sodium ion and nitrate ion
b. aluminum ion and nitrate ion
c. magnesium ion and sulfate ion

Chapter 13

1. 51.3 kPa, 0.507 atm

2. 33.7 kPa is greater than 0.25 atm

27. a, b, c, e, and f

29. 16.35 atm

31. The Kelvin temperature is directly proportional to the average kinetic energy.

33. STP stands for standard temperature (0°C) and standard pressure (101.3 kPa or 1 atm).

35. The average kinetic energy triples.

37. In both cases, particles with sufficient kinetic energy move from the liquid to the vapor phase. In a closed container, a dynamic equilibrium is set up between the contained liquid and its vapor.

39. More molecules have enough energy to escape the attractions within the liquid.

41. The average kinetic energy increases, which allows more vapor to form above the liquid, which increases the vapor pressure.

43. **a.** about 50 mm Hg
b. about 94°C
c. 760 mm Hg is standard pressure

45. Ionic compounds generally have higher melting points than do molecular solids.

47. The intermolecular attractions between molecules are weaker than the attractions between ions.

49. The temperature remains constant while the liquid boils because the energy that is added is used to vaporize the molecules.

51. **a.** 121°C **b.** chloroform **c.** chloroform
d. The external pressure on ethanol would have to increase; the external pressure on ethanoic acid would have to decrease.

53. Although some molecules are evaporating and an equal number of particles are condensing, the net amounts of vapor and liquid remain constant.

55. decrease; as the attractions become stronger, it is more difficult for molecules to overcome the attractions and vaporize.

57. about 77°C

59. As the temperature drops to −196°C, the average kinetic energy of particles in the air decreases drastically, as does the pressure. So the volume of the balloon, which is a flexible container, decreases. As the balloon warms back to room temperature, the average kinetic energy of the particles increases and the balloon expands to its previous volume.

61. Possible answer: Since the beaker is an open container, the water should boil at 100°C at or close to sea level. Your partner probably misread the thermometer and should recheck the value.

63. The average kinetic energy is the same because the temperature is the same.

65. Because the perspiration absorbs heat as it evaporates, the remaining perspiration and the skin are cooled.

67. The vapor pressure depends only on the kinetic energy of the escaping molecules, which depends on the temperature.

69. The kinetic energy of the molecules in the vapor is the same in both cases; so the vapor pressure is the same.

71. condensation of water vapor on a cold surface

73. a. orthorhombic **b.** rhombohedral **c.** tetragonal **d.** triclinic **e.** cubic

75. no; if (a) = (b) then water vapor will condense at the same rate as the liquid evaporates.

77. inversely

79. a. $1s^2 2s^2 2p^6 3s^2 3p^6$ **b.** $1s^2 2s^2 2p^6 3s^2 3p^6$ **c.** $1s^2$

81. a, c, b

83. a. CO **b.** PBr_3

85. a. 53.7% Fe **b.** 34.6% Al

87. a. 51.2 g Cl_2O_7 **b.** 30.6 mL H_2O

89. $H_2S(aq) + Cd(NO_3)_2(aq) \longrightarrow 2HNO_3(aq) + CdS(s)$

91. a. Mg **b.** Li

93. a. 198 g H_2O **b.** 23 mol **c.** 144 g C

95. 39.4 g FeS (0.448 mol)

Chapter 14

7. 6.48 L

8. 68.3 kPa

9. 3.39 L

10. 8.36 L

11. 2.58 kPa

12. 341 K (68°C)

13. 0.342 L

14. 1.29×10^2 kPa

23. 2.51×10^2 mol He(g)

24. 2.5 g air

31. 93.4 × kPa

32. 3.3 kPa

39. The space between the particles is reduced.

41. The volume decreases. The molecules have less kinetic energy and cause less pressure on the inside of the balloon.

43. The pressure quadruples.

45. $V_1/T_1 = V_2/T_2$

V_1 and V_2 are the initial and final volumes; T_1 and T_2 are the initial and final temperatures.

47. 1.80 L

49. 846 K (573°C)

51. 1.10×10^3 kPa

53. Its particles have no volume, there are no attractions between them, and collisions are elastic. An ideal gas follows the gas laws at all temperatures and pressures.

55. 33.0 L

57. 3.60×10^2 kPa

59. The total pressure of a gaseous mixture is equal to the sum of the individual pressures of each gas.

61. Molecular oxygen

63. 3.08:1

65. Boiling the water fills the can with steam. When the can is plunged upside down into ice water, the steam is trapped and rapidly condenses, reducing gas pressure inside the can. The walls of the can are not strong enough to resist the comparatively high atmospheric pressure, which crushes the can.

67. High temperatures increase the pressure of the contents of the container and may cause it to explode.

69. Temperatures measured on the Kelvin scale are directly proportional to the average kinetic energy of the particles. Celsius temperatures are not.

71. The variables are directly proportional.

73. The particles in a real gas have a finite volume and are attracted to one another.

75. Helium atoms have a smaller molar mass than oxygen and nitrogen molecules and effuse faster through pores in the balloon.

77. 2.0×10^2 g

79. A vacuum contains no matter to allow the transfer of kinetic energy between molecules.

81. Helium gas is composed of small atoms with little attraction for each other.

83. a. 1.63×10^2 kPa **b.** 4.48×10^2 kPa

85. 2 mol KNO_3 for each 1 mol O_2

87. a. 2.0×10^{-3}% **b.** 2.0%

89. K = °C + 273

91. 82 protons, 82 electrons, 124 neutrons

93. a, tungsten

95. b, SO_2

97. 206 g

99. It is the volume occupied by 1 mol of a gas at STP.

101. a. single-replacement **b.** decomposition

103. 60.0% C, 13.3% H, 26.7% O

105. The motion of particles in a gas is constant, random, and rapid.

Chapter 15

6. 36.1% H_2O

7. 49.3% H_2O

23. Surface molecules are attracted to the liquid molecules below but not to the air. Molecules inside the liquid are attracted in all directions.

25. A surfactant is a wetting agent such as a soap or detergent. A surfactant interferes with hydrogen bonding between water molecules and reduces surface tension.

27. The water has low vapor pressure.

29. Bodies of water would freeze from the bottom up. This would kill many forms of aquatic life. Also, ice floats.

31. Solutions are homogeneous mixtures in which a solute is dissolved in a solvent. Aqueous solutions are solutions that have water as the solvent.

33. The polar water molecules attract ions and polar covalent molecules. Nonpolar compounds are unaffected because they have no charges.

35. Solvent molecules surround positively charged and negatively charged ions.

37. Water is polar and gasoline is nonpolar.

39. Its ions are free to move toward an electrode.

41. water in the crystal structure of a substance

43. **a.** tin(IV) chloride pentahydrate
b. iron(II) sulfate heptahydrate
c. barium bromide tetrahydrate
d. iron(III) phosphate tetrahydrate

45. Hygroscopic substances absorb water vapor from the air and create a dry environment in a sealed container.

47. Efflorescence is the loss of water of hydration that occurs when the hydrate has a higher vapor pressure than that of the water vapor in air.

49. solutions, colloids, suspensions

51. The molecules or ions are too small to have reflective surfaces.

53. Brownian motion and repulsion between like charged ions adsorbed on colloidal particles' surfaces

55. the addition of an emulsifier

57. Water molecules at 4°C are tightly packed and have maximum density. Below 4°C the water molecules arrange in a regular network because of the attractions between them. As a result, ice has a lower density than water and floats on water.

59. **a.** Water expands when it freezes to ice.
b. Water is polar and wax is nonpolar, and water has a higher surface tension.
c. A network of hydrogen bonding between water molecules lowers the vapor pressure. Two ethanol molecules can hydrogen bond, but a network is not formed.

61. **a.** gasoline **b.** water
c. gasoline **d.** water

63. **a.** No, both form clear, colorless solutions.
b. dry to examine the crystals, test for electrical conductivity, do a flame test

65. No, nonpolar molecules do not dissolve in polar molecules.

67. **a.** sodium carbonate monohydrate, 14.5% H_2O
b. magnesium sulfate heptahydrate, 51.2% H_2O

69. $CaCl_2 \longrightarrow 2H_2O$

71. Liquid water contains some hydrogen-bonded structures similar to ice, in which the water molecules are spaced fairly wide apart. These structures are disrupted when ethyl alcohol is added because the alcohol competes for hydrogen bonds with water molecules and the water structure collapses. Thus, mixtures of water and ethyl alcohol have less volume than the sum of the volumes of the components. Mixing two liquids could result in a volume greater than the sum of the volumes of the components if the structural ordering in the mixture is greater than in the separated components.

73. Most of the important chemical reactions of life take place in aqueous solutions inside cells.

75. Particles in the atmosphere that are approximately the same size as the wavelengths of visible light cause light from the sun to scatter and split into individual components. Oxygen and nitrogen in the atmosphere scatter violet and blue light due to their small size. Thus the sky appears blue at midday, when the sun is closest to Earth. At sunrise and sunset, the sun is at its greatest distance from Earth. It shines through a layer of atmospheric aerosols consisting of small particles of dust and other pollutant particles mixed with air. The sunlight is reflected and absorbed by these particles. The shorter wavelengths (blue and violet) are scattered away from the path of the sun, but the longer wavelengths (yellow and red) pass through and are the most visible.

77. A surfactant helps to wet the burning material, so less water is needed to put out the fire. Thus less water carries pollutants into the environment.

79. Water enters cracks in pavement and expands when it freezes, creating larger cracks. Continuous freeze-thaw cycles cause pavement to break up and form potholes.

81. **a.** pink **b.** pink **c.** blue
d. 45.4% **e.** water or water vapor

83. **a.** 5 **b.** 2 **c.** 2 **d.** 4

85. $H^+ + H{:}\ddot{\underset{..}{O}}{:}H \rightarrow \underset{\displaystyle H}{H{:}\ddot{\underset{..}{O}}{:}H^+}$

87. **a.** $6CO_2 + 6H_2O \longrightarrow C_6H_{12}O_6 + 6O_2$
b. $2Na + 2H_2O \longrightarrow 2Na^+ + 2OH^- + H_2$

89. 3.60×10^{-2} g H_2O, 2.24×10^{-2} L O_2

91. 636 g C_2H_4O

93. **a.** hydrogen **b.** 0.048 g H_2O
c. oxygen **d.** 0.010 L

95. 1.27 atm

Chapter 16

1. 4.4×10^{-1} g/L
2. 2.6 atm
8. $1.0 \times 10^{-1}M$
9. $2.8M$
10. 1.42×10^{-1} mol NH_4NO_3
11. 5.0×10^{-1} mol $CaCl_2$; 5.6×10^{1} g $CaCl_2$
12. 47.5 mL
13. Use a pipet to transfer 5.0×10^{1} mL of the $1.0M$ solution to a 250-ml volumetric flask. Then add distilled water up to the mark.
14. 5.0% v/v
15. 1.2×10^{1} mL
29. 1.26×10^{1} g NaF
30. $2.85 \times 10^{-1}m$ NaCl
31. $X_{C_2H_5OH} = 0.190$; $X_{H_2O} = 0.810$
32. $X_{CCl_4} = 0.437$; $X_{CHCl_3} = 0.563$
33. 2.06°C
34. 44.1°C
35. 101.37°C
36. 115 g NaCl
43. Random collisions of the solvent molecules with the solute particles provide enough force to overcome gravity.
45. Particles of solute crystallize.
47. 5.6×10^{2} g $AgNO_3$
49. **a.** 1.6×10^{-2} g/L **b.** 4.7×10^{-2} g/L
51. Molarity is the number of moles of solute dissolved in one liter of solution.
    a. $1.3M$ KCl **b.** $3.3 \times 10^{-1}M$ $MgCl_2$
53. **a.** 5.0×10^{-1} mol NaCl, 29 g NaCl
    b. 1.0 mol KNO_3, 1.0×10^{2} g KNO_3
    c. 2.5×10^{-2} mol $CaCl_2$, 2.8 g $CaCl_2$
55. **a.** 16% (v/v) ethanol
    b. 63.6% (v/v) isopropyl alcohol
57. **a.** sea water **b.** $1.5M$ KNO_3 **c.** $0.100M$ $MgCl_2$
59. When vapor pressure is lowered relative to pure solvent, more energy must be supplied to reach the boiling point; therefore the boiling point is increased relative to pure solvent.
61. $1M$ solution: 1 mol of solute in 1 L of solution; $1m$ solution: 1 mol of solute in 1 kg of solvent
63. **a.** 100.26°C **b.** 101.54°C
65. **a.** −1.1°C **b.** −0.74°C **c.** −1.5°C
67. $\Delta T_f = -9.60$°C; $\Delta T_b = +4.74$°C
69. The mole fraction of $NaHCO_3$ is 0.020, and of water is 0.98. The solution is $1.1m$.
71. Add one crystal of KNO_3. If the solution is supersaturated, crystallization will occur. If saturated, the crystal will not dissolve. If unsaturated, the crystal will dissolve.
73. **a.** about 1.14 **b.** about −7.2°C **c.** about −9.5°C
75. $X_{C_2H_5OH} = 0.20$; $X_{H_2O} = 0.80$
77. **a.** 44.2 g KCl **b.** 5.8 g KCl
79. unsaturated
81. **a.** 7.5 g H_2O_2 **b.** $8.8 \times 10^{-1}M$
81. 85.5 g/mol
85. $CaCl_2$ produces three particles upon dissolving; NaCl produces two. Freezing point depression depends on the number of solute particle in the solvent.
87. 6.00 g $Ca(NO_3)_2$ in 30 g of water
89. 1.10×10^{2} mL HNO_3
91. $9.0 \times 10^{-2}M$ Na_2SO_4
93. **a.** 1.98×10^{2} g H_2O
    b. 1.98×10^{5} mg H_2O
    c. 1.98×10^{-1} kg H_2O
95. Rutherford's model contains a nucleus.
97. Calcium permanganate is $Ca(MnO_4)_2$. Four formula units contain 4 Ca atoms, 8 Mn atoms, and 32 O atoms.
99. **a.** 55.8 g Fe, 63.5 g Cu, 201 g Hg, 32.1 g S
    b. Each sample contains 6.02×10^{23} atoms.
    c. 4.48×10^{-1} mol Fe, 3.93×10^{-1} mol Cu, 1.25×10^{-1} mol Hg, 7.80×10^{-1} mol S
101. **a.** combination **b.** decomposition
    c. single-replacement **d.** combustion
    e. single-replacement **f.** double-replacement
103. **a.** $NH_4Cl(s) \longrightarrow NH_4^{+}(aq) + Cl^{-}(aq)$
    b. $Cu(NO_3)_2(s) \longrightarrow Cu^{2+}(aq) + 2NO_3^{-}(aq)$
    c. $HNO_3(aq) \longrightarrow H^{+}(aq) + NO_3^{-}(aq)$
    d. $HC_2H_3O_2(l) \longrightarrow H^{+}(aq) + C_2H_3O_2^{-}(aq)$
    e. $Na_2SO_4(s) \longrightarrow 2Na^{+}(aq) + SO_4^{2-}(aq)$
    f. $HgCl_2(s) \longrightarrow Hg^{2+}(aq) + 2Cl^{-}(aq)$
105. **a.** :Ï· **b.** ·Ṫe· **c.** :Ṡb· **d.** ·Sr·
107. The particles of an ideal gas have neither volume nor inter-particle interactions. Real gases have both.
109. Hydrogen chloride produces hydronium ions (H_3O^{+}) and chloride ions (Cl^{-}) that are stabilized by solvent shells of water in aqueous solution. Nonpolar solvents such as benzene have virtually no interaction with HCl.

Chapter 17

1. Heat flows from the system (paraffin) to the surroundings (air). The process is exothermic.
2. Since the beaker becomes cold, heat is absorbed by the system (chemicals within the beaker) from the surroundings (beaker and surrounding air). The process is endothermic.

3. 2.0 J/(g•°C)

4. 1.8 kJ

12. 1.46 kJ

13. 146 J

14. 6.63 kJ

15. 89.4 kJ

21. 3.34 kJ

22. 1.20 g ice

23. 1.44×10^2 kJ

24. 1.9×10^{-1} kJ

25. -3.01×10^2 kJ

26. 3.42 mol $NH_4NO_3(s)$

32. a. -3.091×10^1 kJ **b.** 1.784×10^2 kJ **c.** -1.130×10^2 kJ

33. CO is a compound and not an element in its standard state.

39. Heat flows from the object at the higher temperature to the object at the lower temperature.

41. the chemical composition of the substance and its mass

43. a. 8.50×10^{-1} Calorie **b.** 1.86×10^3 J **c.** 1.8×10^3 J **d.** 1.1×10^2 cal

45. A negative sign is given to heat flow from the system to the surroundings. A positive sign is given to heat flow to the system from the surroundings.

47. a. exothermic **b.** endothermic **c.** exothermic **d.** endothermic

49. A calorimeter is an instrument used to measure heat changes in physical or chemical processes.

51. bomb calorimeter

53. amount of heat released or absorbed in a chemical change at constant pressure

55. a. -2.10×10^1 kJ **b.** -1.8×10^1 kJ **c.** -5.56×10^2 kJ **d.** 6.5 kJ

57. Hess's law allows the calculation of the enthalpy change for a reaction from the known enthalpy changes for two or more other reactions.

59. 3.02×10^1 kJ

61. The statement is true, since stability implies lower energy. The greater the release of heat, the more stable is the compound relative to its elements (all of which have $\Delta H_f^0 = 0$).

63. 4.00×10^1 g water; 9.60×10^2 g ice

65. 2.44×10^4 cal; 1.02×10^5 J

67. a. -8.902×10^2 kJ **b.** -5.660×10^2 kJ

69. a. -2.21×10^2 kJ **b.** -1.96×10^2 kJ **c.** -9.046×10^2 kJ

71. 2.36×10^1 kJ

73. 2.38×10^2 kJ

75. 6.71×10^1 kJ

77. a. 1.5×10^2 kJ
b. The refrigerator absorbs 1.5×10^2 kJ of heat.
c. assumes the mineral water has the same specific heat as chemically pure water, that no heat is lost by the refrigerator, and the volume of the water is exactly 2 L

79. When a solid reaches its melting point, additional heat must be absorbed to convert it to a liquid. Therefore, fusion of a solid is endothermic. This heat of fusion is released when a liquid freezes, so freezing is exothermic.

81. -1207 kJ

83. a. 3.24×10^1 kcal; 1.36×10^2 kJ
b. 8.13 kg

85. 9.6 g

87. The manipulated variable is the variable you change during an experiment. The responding variable is the variable you observe during an experiment.

89. a. 6.99 **b.** 10.68 **c.** 3.6×10^2 **d.** 4.44

91. 32.2 m

93. a. 2 **b.** 2 **c.** 3 **d.** 1

95. a. K_3N **b.** Al_2S_3 **c.** $Ca(NO_3)_2$ **d.** $CaSO_4$

97. $Ag^+(aq) + Cl^-(aq) \longrightarrow AgCl(s)$

99. 1.18×10^1 g O_2

101. 11.1 L

103. solutions; suspensions

Chapter 18

6. a. reactants favored **b.** reactants favored **c.** products favored **d.** products favored

7. $K_{eq} = 12$

8. $K_{eq} = 8.3 \times 10^{-2}$; one is the inverse of the other.

9. $K_{eq} = 1.6 \times 10^{-3}$

10. $K_{eq} = 0.79$

17. $2 \times 10^{-14}M$

18. $6.7 \times 10^{-5}M$

19. $2 \times 10^{-17}M$

20. $3.2 \times 10^{-6}M$

36. Rate = k[A]; rate is moles per liter per second. [A] is moles per liter. k = rate/[A]. $k = 1/s = s^{-1}$

37. 0.25 mol/L s; 0.125 mol/L s

43. Chemical reactions require collisions between reacting particles with sufficient energy to break and form bonds.

45. A catalyst increases the rate of reactions by providing an alternative reaction mechanism with lower activation energy.

47. Gas molecules and oxygen molecules mix readily but do not have enough energy to react at room temperature. The flame raises the temperature and the energy of collisions, so the reaction rate increases. The heat released by the reaction maintains the high temperature, and the reaction continues spontaneously.

49. The rates are equal.

51. a. $K_{eq} = \frac{[H_2S]^2 \times [CH_4]}{[H_2]^4 \times [CS_2]}$ **b.** $K_{eq} = \frac{[PCl_3] \times [Cl_2]}{[PCl_5]}$

53. a. $[Ni^{2+}][S^{2-}]$ **b.** $[Ba^{2+}][CO_3^{2-}]$

55. c, b, d, a

57. A spontaneous reaction has a negative free energy.

59. unfavorable

61. a. entropy increases **b.** entropy decreases

63. The favorable exothermic change of the condensation process offsets the unfavorable entropy change.

65. a. proportionality constant that relates concentrations of reactants to rate of reaction

b. reaction rate directly proportional to concentration of one reactant

c. expression of rate of reaction in terms of concentrations of reactants

67.

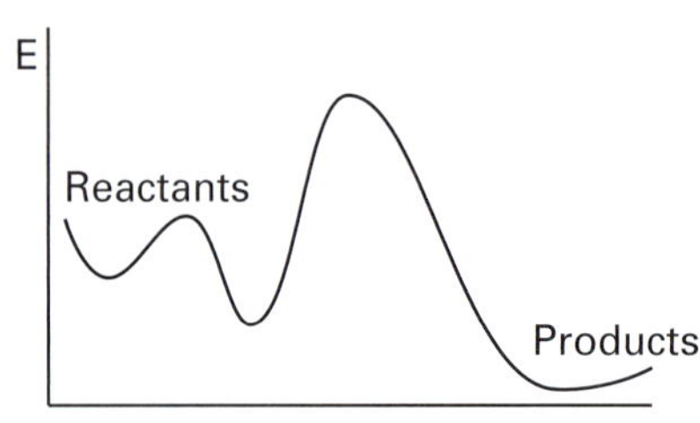

$2NO + O_2 \rightleftharpoons 2NO_2$

69. c

71. The change from Figure a to Figure b is spontaneous, favored by an increase in entropy. The change from Figure b to Figure c will not occur because it would result in a decrease in entropy and a nonspontaneous process.

73. increase in products

75. $K_{eq} = 6.59 \times 10^{-1}$

77. The product of the ion concentrations must be greater than the ion-product constant (K_{sp}).

79.

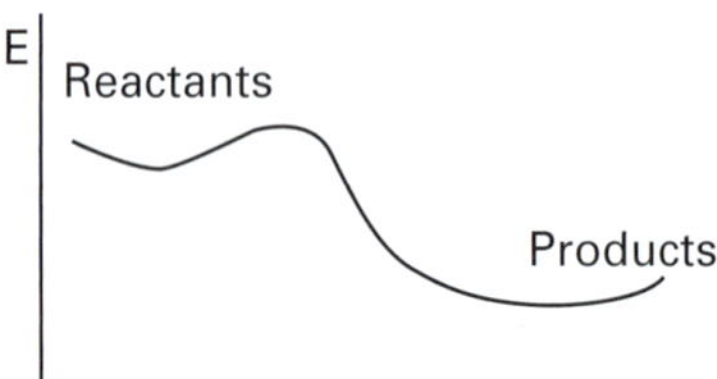

81. a. IO^-

b. No, the I^- is changed in the reaction. A catalyst does not appear in the reaction as a reactant, an intermediate, or a product.

c. two

d. the slow reaction

e. $2H_2O_2 \longrightarrow 2H_2O + O_2$

83. a. increase $[H_2O]$ **b.** decrease $[H_2O]$
c. decrease $[H_2O]$ **d.** no change in $[H_2O]$

85. Possible answers: used a blow dryer, flushed a toilet, mowed the lawn, cooked breakfast, drove a car, and simply breathed

87. first-order in NO_2, first order in NH_4^+, second-order overall

89. a. 3 g **b.** 1.3 g **c.** Rate decreases.

91. Potassium chloride is an ionic compound, not a molecular compound.

93. a. sodium perchlorate, ClO_4^-
b. potassium permanganate, MnO_4^-
c. calcium phosphate, PO_4^{3-}
d. magnesium carbonate, CO_3^{2-}
e. sodium sulfate, SO_4^{2-}
f. potassium dichromate, $Cr_2O_7^{2-}$

95. a. 251 g **b.** 45.9 g
c. 2.99×10^{-22} g **d.** 9.57 g

97. a. $2KClO_3(s) \xrightarrow{\text{heat}} 2KCl(s) + 3O_2(g)$
b. 1.91 g O_2

99. No; the boiling point is the temperature at which the vapor pressure of the liquid equals the atmospheric pressure; it changes if the atmospheric pressure changes.

101. 17.1 L

103. 19.5% H_2O

105. 1.58 mol

107. a. solute, ethanol; solvent, water.
b. below

109. exothermic

Chapter 19

1. a. H^+ is the Lewis acid; H_2O is the Lewis base.
b. $AlCl_3$ is the Lewis acid; Cl^- is the Lewis base.

2. a Lewis base; it has a non-bonding pair of electrons that it can donate.

9. a. basic **b.** basic **c.** acidic **d.** neutral

10. $1.0 \times 10^{-11}M$; basic

11. a. 4.0 **b.** 2.82

12. a. 12.00 **b.** 1.35

13. a. $1.0 \times 10^{-5}M$ **b.** $1.5 \times 10^{-13}M$

14. a. $1.0 \times 10^{-4}M$ **b.** $2.8 \times 10^{-12}M$

15. a. 9.63 **b.** 3.65

16. a. 4.30 **b.** 9.08

22. 1.8×10^{-4}

23. 4.86×10^{-6}

30. 4.68 mol KOH

31. 0.20 mol NaOH

32. 56 mL HCl

33. $0.129M\ H_3PO_4$

38. a. $HPO_4^{2-} + H^+ \longrightarrow H_2PO_4^-$
b. $H_2PO_4^- + OH^- \longrightarrow HPO_4^{2-} + H_2O$

39. $CH_3COO^- + H^+ \longrightarrow CH_3COOH$

45. a. base **b.** acid **c.** base
d. acid **e.** acid **f.** acid

47. a. $2Li + 2H_2O \longrightarrow 2LiOH + H_2$
b. $Ba + 2H_2O \longrightarrow Ba(OH)_2 + H_2$

49. a. HNO_3 with NO_3^-, H_2O with H_3O^+
b. CH_3COOH with CH_3COO^-, H_2O with H_3O^+
c. H_2O with OH^-, NH_3 with NH_4^+
d. H_2O with OH^-, CH_3COO^- with CH_3COOH

51. $H_2O \rightleftharpoons H^+ + OH^-$

53. the negative logarithm of the $[H^+]$

55. a. pH = 2.00, acidic **b.** pH = 12.00, basic
c. pH = 6.00, acidic

57. a. 5.62 **b.** $6.3 \times 10^{-14}M$

59. A strong acid is completely dissociated; *K*a must be large.

61. a. $K_a = \frac{[H^+][F^-]}{[HF]}$ **b.** $K_a = \frac{[H^+][HCO_3^-]}{[H_2CO_3]}$

63. a. $HNO_3 + KOH \longrightarrow KNO_3 + H_2O$
b. $2HCl + Ca(OH)_2 \longrightarrow CaCl_2 + 2H_2O$
c. $H_2SO_4 + 2NaOH \longrightarrow Na_2SO_4 + 2H_2O$

65. a. 1.40*M* **b.** 2.61*M*

67. $HCO_3^-(aq) + H_2O(l) \longrightarrow H_2CO_3(aq) + OH^-(aq)$

69. a. basic **b.** acidic **c.** neutral
d. basic **e.** neutral **f.** acidic

71. yes; acids like acetic acid dissolve well but ionize poorly.

73. 4.6×10^{-4}

75. a. $HClO_2$, chlorous acid
b. H_3PO_4, phosphoric acid
c. H_3O^+, hydronium ion
d. NH_4^+, ammonium ion

77. $H_3PO_4 \rightleftharpoons H^+ + H_2PO_4^-$;
$H_2PO_4^- \rightleftharpoons H^+ + HPO_4^{2-}$;
$HPO_4^{2-} \rightleftharpoons H^+ + PO_4^{3-}$

79. a. HSO_4^- **b.** CN^- **c.** OH^- **d.** NH_3

81. a. $2HCl + Mg(OH)_2 \longrightarrow MgCl_2 + 2H_2O$
b. $2HCl + CaCO_3 \longrightarrow H_2O + CO_2 + CaCl_2$
c. $Al(OH)_3 + 3HCl \longrightarrow AlCl_3 + 3H_2O$

83. a. 8.73
b. phenolphthalein or phenol red

85. b, c, d, a

87. a. You may consider the Arrhenius theory the easiest to understand, and the Lewis theory the best because it is the most general. All three theories provide definitions and describe accepted behavior of a certain group of compounds. The Brønsted-Lowry theory includes a greater number of compounds than the Arrhenius theory because it is more general, and the Lewis theory includes the greatest number of compounds because it is the most general.
b. Each theory has advantages in certain circumstances.

89. The *y*-axis might correspond to $[H^+]$ because HCl is a strong acid.

91. a. false; an indicator determines a range of pH values.
b. false; an Arrhenius base dissociates to give hydroxide ions in aqueous solution. Ammonia does not do this.
c. false; strength is a measure of dissociation or ionization, not concentration.

93. pH = 10.66

95. a. 7.4721, 7.2675, 7.0835, 6.9165, 6.7675, 6.6310

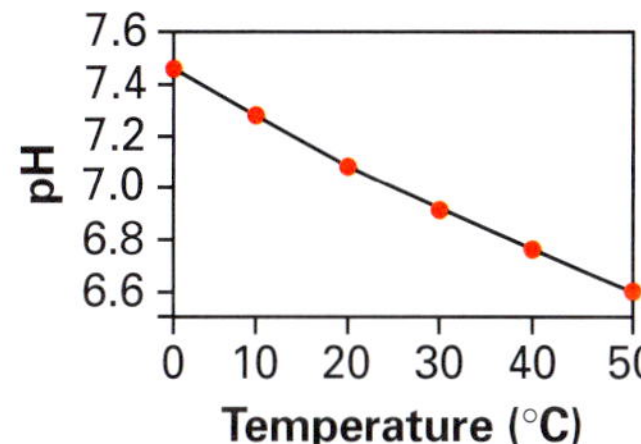

b. 7.37 **c.** 35°C

97. 50.0 mL; the pH = 7 when $[H^+] = [OH^-]$. Because HCl is a strong acid that supplies one hydrogen ion per formula unit and NaOH is a strong base that supplies one hydroxide ion per formula unit, $[H^+] = [OH^-]$ when equal volumes of solutions of the same molarity are combined.

99. $[OH^-] = 4.6 \times 10^{-4}$; pH = 10.66

101. 131 g O_2

103. The total pressure in a mixture of gases is equal to the sum of the partial pressures of each gas in the mixture.

105. suspension

107. hydrogen bond

109. Dissolve 0.272 mol KOH(*s*) in water and add sufficient water to give 400.0 mL of solution.

111. b, c, and d

113. a. 144 J **b.** 1.0×10^3 kJ **c.** 82.9 cal

115. The product of the concentrations of the two ions must be greater than the solubility product.

117. $2.0 \times 10^{-8}M$

119. a. shift right **b.** no change **c.** shift right
d. shift right **e.** shift right

Chapter 20

1. a. Na: oxidized (reducing agent); S: reduced (oxidizing agent)
b. Al: oxidized (reducing agent);O_2: reduced (oxidizing agent)

2. a. oxidation **b.** reduction

9. a. S, +3; O, −2 **b.** Na, +1; O, −1
c. P, +5; O, −2 **d.** N, +5; O, −2

10. a. +5 **b.** 0 **c.** +7 **d.** +1

11. a. H_2 oxidized, O_2 reduced
b. N reduced, O oxidized

12. a. H_2 reducing agent, O_2 oxidizing agent
b. N oxidizing agent, O reducing agent

17. a. redox reaction; Mg is oxidized, Br_2 is reduced.
b. not a redox reaction

18. a. not a redox reaction
b. redox reaction; H_2 is oxidized, Cu is reduced.

19. a. 2, 2, 3 **b.** 2, 2, 2, 1, 2

20. a. 1, 8, 2, 2, 3, 4 **b.** 1, 2, 1, 2, 1

21. $4Zn + NO_3^- + 6H_2O + 7OH^- \longrightarrow 4Zn(OH)_4^{2-} + NH_3$

27. The oxidizing agent is reduced.

29. a. oxidation **b.** oxidation
c. oxidation **d.** oxidation

31. a. H_2 is oxidized; S is reduced.
b. N_2 is reduced; H_2 is oxidized.
c. S is oxidized; O_2 is reduced.
d. H_2 is oxidized; O_2 is reduced.

33. An oxidation number is the charge an atom would have if the electrons in each bond were assigned to the atoms of the more electronegative element.

35. a. +2 **b.** +3 **c.** Na, +1; Cr, +6
d. +5 **e.** +7

37. a. Al is oxidized; Mn is reduced.
b. K is oxidized; H is reduced.
c. Hg is reduced; O is oxidized.
d. P is oxidized; O is reduced.

39. redox: a, b, c, d, e

41. a. $4Al(s) + 3MnO_2(s) \longrightarrow 2Al_2O_3(s) + 3Mn(s)$
b. $2K(s) + 2H_2O(l) \longrightarrow 2KOH(aq) + H_2(g)$
c. $2HgO(s) \longrightarrow 2Hg(l) + O_2(g)$
d. $P_4(s) + 5O_2(g) \longrightarrow P_4O_{10}(s)$

43. a. +4 **b.** +5 **c.** +5
d. +3 **e.** +5 **f.** +3

45. a. Cl oxidized, Mn reduced, Mn oxidizing agent, reducing agent
b. Cu oxidized, N reduced, N oxidizing agent, Cu reducing agent
c. P oxidized, N reduced, N oxidizing agent, P reducing agent
d. Sn oxidized, Bi reduced, Bi oxidizing agent, Sn reducing agent

47. a. $16H^+(aq) + 2Cr_2O_7^{2-}(aq) + C_2H_5OH(aq) \longrightarrow 4Cr^{3+}(aq) + 2CO_2(g) + 11H_2O(l)$
b. oxidizing agent

49. a. oxidized
b. H is the oxidizing agent; Ag is the reducing agent.
c. $2Ag(s) + H_2S(s) \longrightarrow Ag_2S(g) + H_2(g)$

51. a. yes; the oxidation number of bismuth changes from +3 to zero; the oxidation number of carbon changes from zero to +2.
b. no; there is no change in oxidation number of any of the atoms in this reaction.
c. no; there is no change in oxidation number of any of the atoms in this reaction.

53. a. reactant, 0; product, +3
b. reactant, −2; product, −2
c. X
d. H

55. $1s^2\,2s^2\,2p^6\,3s^2\,3p^5$; A chlorine atom can lose its 7 valence electrons or it can gain one electron to complete the third energy level.

57. Double replacement reactions never involve the transfer of electrons; instead they involve the transfer of positive ions in aqueous solution.

59. a. SO_4^{2-} **b.** H_2O_2 **c.** NO_3^-
d. $Cr_2O_7^{2-}$ **e.** H_2O

61. a. $Rb(s) + I_2(s) \longrightarrow RbI_2(s)$; oxidizing agent is I
b. $Ba(s) + 2H_2O(l) \longrightarrow Ba(OH)_2(aq) + H_2(g)$; oxidizing agent is H
c. $2Al(s) + 3FeSO_4(aq) \longrightarrow Al_2(SO_4)_3(aq) + 3Fe(s)$; oxidizing agent is Fe
d. $C_4H_8(g) + 6O_2(g) \longrightarrow 4CO_2(g) + 4H_2O(l)$; oxidizing agent is O
e. $Zn(s) + 2HBr(aq) \longrightarrow ZnBr_2(aq) + H_2(g)$; oxidizing agent is H
f. $Mg(s) + Br_2(l) \longrightarrow MgBr_2(s)$; oxidizing agent is Br

63. MnO_4^-, because the manganese is at its highest oxidation state

65. 104 mL $K_2Cr_2O_7$

67. a. +5 **b.** −3 **c.** +3 **d.** +3
e. +1 **f.** −3 **g.** +2 **h.** +4

69. The nitride ion has the minimum oxidation number of −3, therefore it cannot gain additional electrons and be an oxidizing agent. It can lose electrons, however, and be a reducing agent. The nitrate ion has the maximum oxidation number of +5; therefore, it cannot lose additional electrons and be a reducing agent. It can gain electrons, however, and be an oxidizing agent.

71. a. a. 6.5 b. 4 c. 5 d. 8 e. 5 f. 6 g. 9.5 h. 6 i. 7
b. $C_xH_y + [x + (y/4)]O_2 \longrightarrow xCO_2 + (y/2)H_2O$

73. 1.8×10^2 kPa

75. a, c, and d

77. $Ca(NO_3)_2$; boiling point elevation is a colligative property that depends on the number of particles in solution. $Ca(NO_3)_2$ gives three particles per formula unit; LiF gives two particles per formula unit.

79. solubility: $PbBr_2 = 8.1 \times 10^{-3} M$

81. a. $1.0 \times 10^{-2} M$ **b.** $1.0 \times 10^{-11} M$
c. $1.6 \times 10^{-9} M$

83. a. NH_4^+ and NH_3; H_2O and H_3O^+
b. H_2SO_3 and HSO_3^-; NH_2^- and NH_3
c. HNO_3 and NO_3^-; I^- and HI

85. a. 5.00 **b.** 10.00 **c.** 13.00 **d.** 6.52

Chapter 21

9. nonspontaneous; $E^0_{cell} = -0.02$ V

10. yes; $E^0_{cell} = +0.16$ V

11. $2Al(s) + 3Cu^{2+}(aq) \longrightarrow 2Al^{3+}(aq) + 3Cu(s)$

12. $Cu(s) + 2Ag^+(aq) \longrightarrow Cu^{2+}(aq) + 2Ag(s)$

13. $E^0_{cell} = +2.00$ V

14. $E^0_{cell} = +0.46$ V

27. oxidation: $Al(s) \longrightarrow Al^{3+}(aq) + 3e^-$; reduction: $Cu^{2+}(aq) + 2e^- \longrightarrow Cu(s)$

29. a. Cu **b.** Ca **c.** Mg
d. Sn **e.** Zn **f.** Al

31. The salt bridge allows ions to pass from one half-cell to the other but prevents the solutions from mixing.

33. Water is produced by the redox reaction and sulfuric acid is used up; water has a lower density than sulfuric acid.

35. Fuel cells cannot generate electricity as economically as more conventional forms of electrical generation.

37. It was arbitrarily set at zero.

39. The relative order is the same because both tables rank the elements according to their tendency to undergo oxidation/reduction.

41. a. nonspontaneous; $E^0_{cell} = -0.34$ V
b. nonspontaneous; $E^0_{cell} = -1.24$ V

43. A direct current flows in one direction only.

45. $2H_2O(l) \longrightarrow O_2(g) + 2H_2(g)$

47. Apparatus similar to that on page 687; the small spheres representing molecules of H_2 and Cl_2 will be in 1:1 ratio.

49. Two half-cells are needed because oxidation or reduction cannot occur in isolation. One half-cell gains electrons and one loses them, producing an electric current.

51. Some of the iron dissolves and the nail becomes coated with copper.
Overall reaction: $Fe(s) + CuSO_4(aq) \longrightarrow FeSO_4(aq) + Cu(s)$;
oxidation half-reaction: $Fe \longrightarrow Fe^{2+} + 2e^-$;
reduction half-reaction: $Cu^{2+} + 2e^- \longrightarrow Cu$

53. Lead(II) sulfate and lead dioxide are highly insoluble in sulfuric acid.

55. a. oxidation: $6Cl^-(l) \longrightarrow 3Cl_2(g) + 6e^-$ (anode);
reduction: $2Al(l) + 6e^- \longrightarrow 2Al(l)$ (cathode)
b. overall reaction: $2AlCl_3(l) \longrightarrow 2Al(l) + 3Cl_2(g)$
c. chlorine gas at anode; liquid aluminum at cathode

57. In each type of cell, oxidation occurs at the anode and reduction occurs at the cathode.

59. a. $Zn \longrightarrow Zn^{2+} + 2e^-$ **e.** $Fe \longrightarrow Fe^{2+} + 2e^-$
f. $Na \longrightarrow Na^+ + e^-$

61. a. +0.63 V **e.** +0.21 V **f.** +4.07 V

63. a. possible oxidation reactions at anode:
(i) $2Cl^-(aq) \longrightarrow Cl_2(g) + 2e^-$;
(ii) $2H_2O(l) \longrightarrow O_2(g) + 4H^+(aq) + 4e^-$
b. possible reduction reactions at cathode:
(i) $Na^+(aq) + e^- \longrightarrow Na(s)$;
(ii) $2H_2O(l) + 2e^- \longrightarrow H_2(g) + 2OH^-(aq)$
c. (i) Chloride ions are more readily oxidized to chlorine gas than water molecules are oxidized to water.
d. (ii) Water is reduced to produce hydrogen gas; sodium ions are not reduced to sodium metal because water molecules are more easily reduced than sodium ions.

65. Gold belongs near the bottom, below silver, because it is one of the least active metals.

67. The chemists' definition focuses on the electrons that are produced by oxidation at the anode of a voltaic cell; the dictionary definition is based on an electrolytic cell. The electrodes are defined by the battery terminals to which they are attached.

69. d; the voltage falls steadily.

71. As electrons flow from the anode to the cathode in the external circuit, anions must flow from the cathode compartment to the anode compartment to maintain neutrality in the electrolytes. Anions cannot flow through wire made of copper or any other metal; the cell will not function if the salt bridge is replaced with a metal wire.

73. oxidation: $2Cu(impure) + 2H_2SO_4 \longrightarrow 2Cu^{2+} + 2H_2 + SO_4^{2+}$;
reduction: $2Cu^{2+} + 2SO_4^{2+} + 2H_2O \longrightarrow 2Cu(pure) + 2H_2SO_4 + O_2$;
overall reaction: $2Cu(impure) + 2H_2O \longrightarrow 2Cu(pure) + 2H_2 + O_2$

75. The battery output would not be 12V.

77. 467 mL

79. a. 0.0125g NaCl **b.** 101g KNO_3

81. a. 4.32×10^2 kJ **b.** 2.55×10^5 cal
c. 2.70×10^3 J

83. 267 kJ

85. $[N_2] \times [H_2]^3/[NH_3]^2$

87. a. $[OH^-] = 1 \times 10^{-7} M$ **b.** $[OH^-] = 1 \times 10^{-10} M$
c. $[OH^-] = 1 \times 10^{-6} M$

89. a. +6 **b.** −2 **c.** +4
d. +2 **e.** 0 **f.** +4

91. b (Ca is oxidized, Cl_2 is reduced) and d (Ca is oxidized, H is reduced)

93. **a.** Cr, +6 **b.** I, +5

c. Mn, +7 **d.** Fe, +3

Chapter 22

1.

```
     H   H   H   H   H
     |   |   |   |   |
 H - C - C - C - C - C - H
     |   |   |   |   |
     H   H   H   H   H
```

```
     H   H   H   H   H   H
     |   |   |   |   |   |
 H - C - C - C - C - C - C - H
     |   |   |   |   |   |
     H   H   H   H   H   H
```

2. 10

3. **a.** 3-ethylhexane **b.** 2-methylbutane

4. **a.** 2,3-dimethylpentane **b.** 2-methylpentane

5.

```
   CH3
   |
CH3CHCHCH2CH2CH3
      |
      CH3
```

6.

```
       CH3  CH3  CH3
       |    |    |
CH3 -  CH - CH - C - CH2 - CH2 - CH2 - CH3
                 |
                 CH2
                 |
                 CH3
```

18. Both a and b have an asymmetric carbon, denoted as C*.

```
      *              *
CH3CHCHO       CH3CHOH
    |              |
    Cl             CH3
```

19. **a.** The asymmetric carbon is denoted as C*.

```
            CH3
           *|
CH3CH2 -   C - Br
            |
            F
```

b. no asymmetric carbon

37. pentane: $CH_3CH_2CH_2CH_2CH_3$; hexane: $CH_3CH_2CH_2CH_2\ CH_2CH_3$

39.

```
    H           H   H   H
    |           |   |   |
H - C -     H - C - C - C -
    |           |   |   |
    H           H   H   H
  methyl         propyl
```

```
    H   H
    |   |
H - C - C -
    |   |
    H   H
   ethyl
```

41. The carbon-carbon bonds are nonpolar and the carbon-hydrogen bonds are very weakly polar.

43. $CH_2{=}CHCH_2CH_2CH_3$
1-pentene
$CH_3CH{=}CHCH_2CH_3$
2-pentene

```
       CH3
       |
CH3 -  C = CH - CH3
```

2-methyl-2-butene

45. **a.** Accept any isomer with 5 carbons and 12 hydrogens.
b. Accept any isomer with 7 carbons and 16 hydrogens.

47. No; Only molecules with at least one asymmetric carbon have optical isomers.

49. **a.** benzene ring with CH_2CH_3 on top and CH_2CH_3 on bottom (para)

b.

```
           CH3
           |
     CH3 - CH - CH - CH2 - CH3
                |
             (benzene ring)
```

c. benzene ring with CH_3 at two positions (CH_3– and CH_3)

51. Catalysts are used during cracking to produce more short-chain components, including components that increase the performance of gasoline.

53. The combustion of sulfur in coal produces the air pollutants SO_2 and SO_3.

55. **a.** Ethyne (acetylene) has one triple carbon-carbon bond and two single carbon-hydrogen bonds.
b. All the bonds in propane are single bonds.
c. In methylbenzene, there are hybrid bonds within the ring and single bonds within the substituents and between the substituents and the ring.

57. propane, butane, pentane

59.

```
a.  H H               b.  H H H
    H:C::C:H              H:C:C:C:H
                            H H H

c.  H:C:::C:H         d.  H H
                          H:C:C:H
                          H:C:C:H
                            H H
```

61. The middle structure is most stable due to resonance within the ring.

63. No; the structures are identical; one has been flipped over.

65. The amount of heat per carbon is higher for methane (−890 kJ/mol per mole of carbon burned) than for benzene (−545 kJ/mol per mole of carbon burned). Burning aromatic compounds produces more soot.

67. Because there is no rotation around the double bond, the methyl and ethyl groups can be on the same side of the bond or opposite sides.

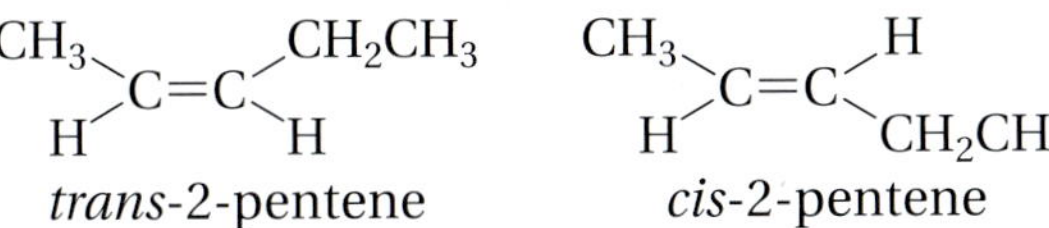

trans-2-pentene *cis*-2-pentene

69. $H_2C{=}C{=}CH_2$

71. a. **b.**

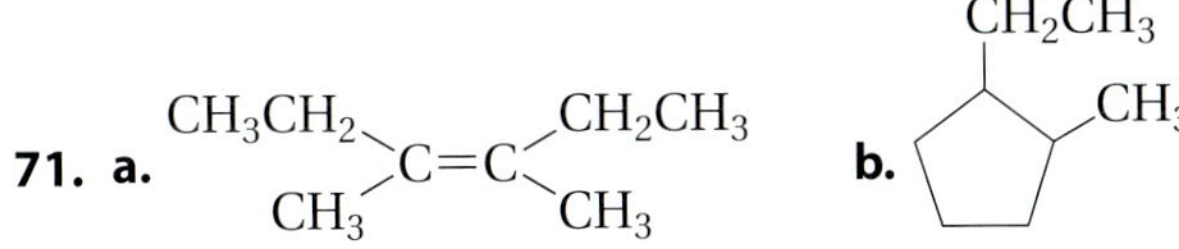

73. Alkanes contain only C—C and C—H single bonds. Alkenes contain at least one C—C double bond. Aromatic hydrocarbons contain a benzene ring or a similar ring. Cycloalkanes contain aliphatic carbon chains linked end-to-end.

75. The graph is not a straight line. The estimated boiling point should be higher than 150°C. The boiling point of undecane is 196°C.

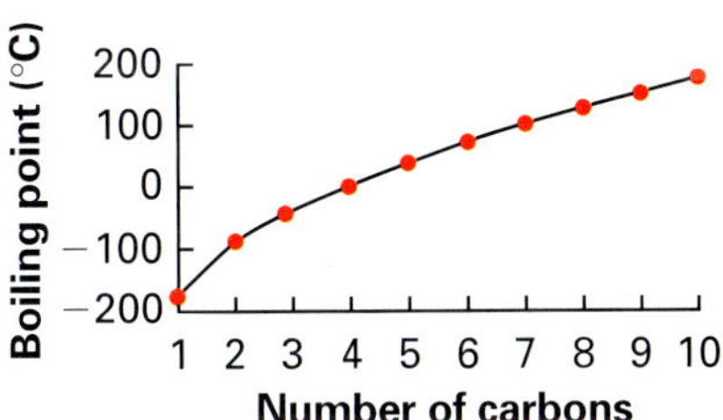

77. a. $C_6 = 5$, $C_7 = 9$, $C_8 = 18$, $C_9 = 35$, $C_{10} = 75$

b. As the number of carbon atoms increases, there is a dramatic increase in the number of ways that the carbon atoms can be arranged in the molecule.

79. a. 13.9 L **b.** 1 L **c.** 20 kPa

81. 1.13 mol KNO_3; 1.14×10^2 g KNO_3

83. 1 cal = 4.184 J; 4.184×10^3 J

85. a. favors reactants **b.** favors products

87. a. 10.00 **b.** 7.59 **c.** 12.00 **d.** 11.70

89. a. H_3PO_4 **b.** CsOH **c.** H_2CO_3 **d.** $Be(OH)_2$

91. a. Ca, +2; C, +4; O, −2 **b.** Cl, 0
c. Li, +1; I, +5; O, −2 **d.** Na, +1; S, +4; O, −2

93. a. +4 **b.** +4 **c.** +3
d. +5 **e.** +5 **f.** +2

95. It is the cell potential when the ion concentrations in the half-cells are 1*M*, the temperature is 25°C, and the pressure of any gases present is 101.3 kPa.

97. The reaction is nonspontaneous.

Chapter 23

27. a. $ClCH_2CCl_2CH_2CH_3$ (Cl above and below the central C)

b. benzene ring with Br at positions 1, 3, 5 (Br, Br, Br)

c. cyclohexane ring with Cl and C substituents (Cl, C)

29.

a. Cl / CH—CH_2—CH_3 / Cl
1, 1-dichloropropane

Cl Cl / CH_2—CH—CH_3
1, 2-dichloropropane

Cl Cl / CH_2—CH_2—CH_2
1, 3-dichloropropane

b. Br / CH_3—CH_2—CH_2—CH_2
1-bromobutane

Br / CH_3—CH_2—CH—CH_3
2-bromobutane

CH_3 Br / CH_3—CH—CH_2
1-bromo-2-methylpropane

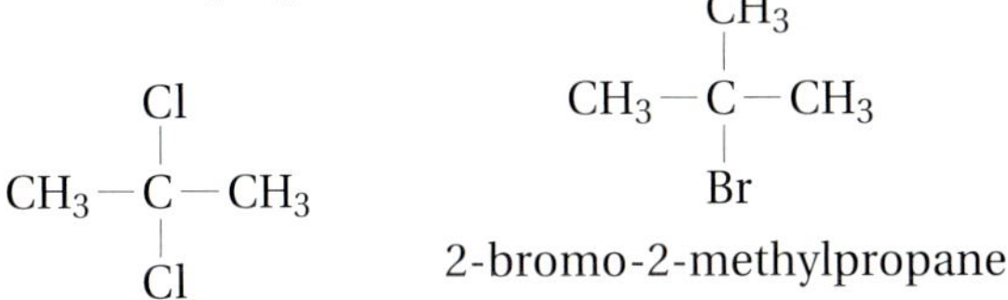

2, 2-dichloropropane

31. a. 2-propanol **b.** 1,2-propanediol

33. a. H Br / CH_2—CH_2
bromoethane

d. H H / CH_2—CH_2
ethane

b. Cl Cl / CH_2—CH_2
1, 2-dichloroethane

e. H Cl / CH_2—CH_2
chloroethane

c. H OH / CH_2—CH_2
ethanol

35. a. propanone or acetone
b. 3-methylbutanal
c. 2-phenylethanal
d. ethanol or acetaldehyde
e. diphenylmethanone or diphenyl ketone or benzophenone
f. 3-hexanone or ethylpropylketone

37. a. —CH_2—CH— / CH_2 / CH_3

b. —CH—CH— / Cl Cl

39. c, a, b

41. Both atoms in a carbon-carbon double bond have the same electronegativity, so the bond is nonpolar. Because oxygen is more electronegative than carbon, a carbon-oxygen bond is very polar.

43. a. phenol **b.** ether **c.** alcohol
d. phenol **e.** alcohol

45. **a.** carboxylic acid, ethanoic acid (acetic acid)
b. ketone carbonyl group, propanone (acetone)
c. ether, diethyl ether (ethyl ether)
d. alcohol, ethanol (ethyl alcohol)

47. **a.** $CH_3CH_2COO^-Na^+$, CH_3CH_2OH

b. $CH_3COO^-K^+$, C_6H_5—OH

c. CH_3CH_2COOH, $CH_3CH(CH_3)CH_2OH$

49. The chemical properties (and toxicity) of organic compounds are determined by the compound as a whole. As a substituent in a molecule, a phenyl group ring does not have the same properties as benzene.

51. The alcohol molecules form hydrogen bonds with one another, resulting in a higher boiling point. They also form hydrogen bonds with water molecules, causing 1-butanol to be more soluble than diethyl ether. (Although diethyl ether is polar, 1-butanol has greater polarity.)

53. The short-chain ethanoic acid has a higher water solubility.

55. $H_2NCH_2(CH_2)_3CH_2NH_2$ cadaverine; $H_2NCH_2(CH_2)_4CH_2NH_2$ putrescine; Both compounds are amines.

57. Cholesterol is an alcohol with a hydroxyl group on a cycloalkane. It has four nonaromatic rings. It has a double bond on one of its rings, as well as a large alkyl group, making it nonpolar.

59. Waving lotion reduces —S—S— bonds to —SH bonds. Hair can be placed in curlers to form the hair in the desired shape. The neutralizing agent is an oxidizing agent that re-forms —S—S— bonds, locking the hair into its curly shape. Similar steps could be used to straighten curly hair.

61. b. 3

63. 2.86 g SO_2

65. Anhydrous $Na_2CO_3(s)$ is the better value; the decahydrate is 63.0% water.

67. $0.117M$ $Ca(NO_3)_2$

69. 71 kJ

71. a, c, d, b

73. Oxidized: H of BH_4 Reduced: H of H_2O
Unaffected: Na, B, O

75. Reduction always occurs at the cathode. In the electrolytic cell, the cathode is the negative electrode.

77. coal

Chapter 24

37. A eukaryotic cell has a cell membrane, a nucleus, and various organelles including mitochondria, lysosomes, an endoplasmic reticulum, and ribosomes. Prokaryotic cells have a cell membrane, but lack organelles.

39. $C_6H_{12}O_6 + 6O_2 \longrightarrow 6CO_2 + 6H_2O$ + energy

41. glucose and fructose

43. Glucose is an aldehyde; fructose is a ketone.

45. **a.** glucose **b.** glucose

47. peptide bond

49. Peptide chains fold into helixes or into sheets in which peptide chains lie side by side.

51. Enzymes catalyze biological reactions.

53. At room temperature, animal fats are solid; plant oils are liquid.

55. alkali metal salt of a fatty acid

57.

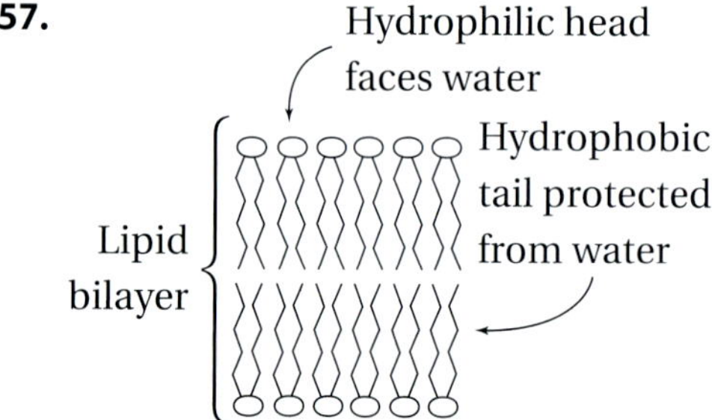

59. protection from water loss and microorganisms in plants; pliability and waterproofing in animals

61. a phosphate group, a 5-carbon sugar unit, and a nitrogen base

63. hydrogen bonding

65. three

67. DNA sequences are unique for each individual.

69. ATP + $H_2O \longrightarrow$ ADP + P_i

71. In catabolism, large biomolecules are broken down and energy is captured through the production of ATP. In anabolism, the products and energy from catabolism are used to make biomolecules.

73. products of catabolism

75. Catabolism of 1 mol glucose yields 2.82×10^3 kJ of energy.

77. The negatively charged hydrophilic phosphate heads interact favorably with water.

79. **a.** triglyceride **b.** glucose **c.** dipeptide

81. G C T A G G T

83. permit passage of nutrients into cells

85. many applications in medicine and agriculture

87. $30.5 \text{ kJ/mol} \times 38 \text{ mol}/2.82 \times 10^3 \text{ kJ/mol} \times 100\% = 41.4\%$

89. The shapes of the cellulose and starch molecules are different; humans lack enzymes necessary to cleave cellulose to glucose monomers.

91. Coenzymes must be present for an enzyme-catalyzed reaction to occur. Many water-soluble vitamins, such as B vitamins, are coenzymes.

93. **a.** iron(II) ion
b. iron(III) ion; methemoglobinemia

95. Photosynthesis provides the carbon compounds that plants and animals need to exist.

97. A carboxylic acid group of one amino acid and amino group of another amino acid undergo condensation polymerization to form an amide (peptide) bond.

99. Blanching the corn destroys most of the enzymes responsible for the conversion of glucose to starch. Freezing the corn slows down the action of any remaining starch-producing enzymes and prevents spoilage.

101. Inside the helix; strands must unwind

103. Bacteria necessary for nitrogen fixation might be killed by sterilization of the soil.

105.

CH$_2$OH
H C—O H
C H C
OH H
HO C—C
H OH
O
OCH$_2$ O
C C
H H HO CH$_2$OH
C—C
OH H

$+ H_2O$

107. Sample answer: GTA-ACA-CCA-GCA-GGG; yes, because GGA, GGG, GGT, and GGC all code for proline.

109.

$$H_2N-\underset{H}{\overset{R}{C}}-\overset{O}{\overset{\|}{C}}-OH + H-\underset{H}{N}-\underset{H}{\overset{R}{C}}-\overset{O}{\overset{\|}{C}}-OH \longrightarrow$$

Amino acid   Amino acid

$$H_2N-\underset{H}{\overset{R}{C}}-\overset{O}{\overset{\|}{C}}-\underset{H}{N}-\underset{H}{\overset{R}{C}}-\overset{O}{\overset{\|}{C}}-OH + H_2O$$

Peptide

An amido group is formed.

111.

$$\begin{array}{l} CH_2OH \\ | \\ CHOH \\ | \\ CH_2OH \end{array} + \begin{array}{l} CH_3(CH_2)_{14}-\overset{O}{\overset{\|}{C}}-O^-Na^+ \\ CH_3(CH_2)_{12}-\overset{O}{\overset{\|}{C}}-O^-Na^+ \\ CH_3(CH_2)_{16}-\overset{O}{\overset{\|}{C}}-O^-Na^+ \end{array}$$

glycerol   salts of fatty acids

113. hydrogen bonds between adjacent parts of folded chains and covalent bonds between side-chain groups of cysteine

115. The particles in real gases have volumes and are attracted to one another.

117. **a.** 0.519°C **b.** 0.209°C **c.** 0.238°C **d.** 0.227°C

119. Besides being composed of highly combustible material, the needles have a large surface area that increases their rate of combustion.

121. **a.** 4.15 **b.** 5.26 **c.** 12.79 **d.** 10.36

123. Oxidation always occurs at the anode. In the voltaic cell the anode is the negative electrode.

125. **a.** Cu **b.** Ni **c.** Ag
d. Fe **e.** Cd **f.** Cu

127. **a.** cyclopentane
b. 2-methyl-2-propanol (*tert*-butyl alcohol)
c. 3-pentanone

129. 1-chlorobutane, 2-chlorobutane, 1-chloro-2-methylpropane, 2-chloro-2-methylpropane

131. **a.** polytetrafluoroethene (Teflon or PTFE)
b. polyethylene
c. polyvinyl chloride (PVC)
d. polystyrene

Chapter 25

7. 0.063 mg Mn-56

8. No; one forth of the sample will remain.

25. Each isotope of an element has the same atomic number, but a different atomic mass. A radioisotope is an isotope that is radioactive.

27. $^{210}_{82}Pb \longrightarrow\ ^{210}_{83}Bi + ^{0}_{-1}e$

29. a. $^{218}_{92}U \longrightarrow\ ^{234}_{90}Th + ^{4}_{2}He$; thorium – 234
b. $^{230}_{90}Th \longrightarrow\ ^{226}_{88}Ra + ^{4}_{2}He$; radium – 226
c. $^{235}_{92}U \longrightarrow\ ^{231}_{90}Th + ^{4}_{2}He$; thorium – 231
d. $^{222}_{86}Rn \longrightarrow\ ^{218}_{84}Po + ^{4}_{2}He$; polonium – 218

31. a. mass number is unchanged; atomic number increases by 1
b. mass number decreases by 4; atomic number decreases by 2
c. mass number and atomic number are both unchanged

33. It undergoes radioactive decay.

35. a. $^{13}_{6}C$ **b.** $^{1}_{1}H$ **c.** $^{16}_{8}O$ **d.** $^{14}_{7}N$

37. so the person is exposed to radioactivity for a limited time

39. Natural radioactivity comes from elements in nature. Artificial radioactivity comes from elements created in nuclear reactors and accelerators.

41. The nuclei of certain isotopes are bombarded with neutrons. The nuclei break into two fragments and release more neutrons. Released neutrons hit other nuclei to start a chain reaction that releases large amounts of energy.

43. Fusion requires extremely high temperatures, making it difficult to start or contain the reaction.

45. The film badge measures radiation exposure; an exposed film badge indicates how much radiation a worker has received.

47. a. $^{30}_{15}P + ^{0}_{-1}e \longrightarrow\ ^{30}_{14}Si$ **b.** $^{13}_{6}C + ^{0}_{1}n \longrightarrow\ ^{14}_{6}C$
c. $^{131}_{53}I \longrightarrow\ ^{131}_{54}Xe + ^{0}_{-1}e$

49. a. $^{32}_{16}S$ **b.** $^{14}_{6}C$ **c.** $^{4}_{2}He$ **d.** $^{141}_{57}La$ **e.** $^{185}_{79}Au$

51. a. about 20% **b.** about 85 g
c. about 83 days **d.** about 25 days

53. a. Curie named radioactivity and discovered several radioactive elements.
b. Becquerel discovered natural radioactivity from uranium ores.
c. Chadwick discovered the neutron.
d. Rutherford artificially transmuted elements.

55. $^{215}_{85}At$

57. 11,460 years old

59. a. $^{231}_{91}Pa \longrightarrow\ ^{227}_{89}Ac + ^{4}_{2}He$
b. $^{241}_{95}Am \longrightarrow\ ^{237}_{93}Np + ^{4}_{2}He$
c. $^{226}_{88}Ra \longrightarrow\ ^{222}_{86}Rn + ^{4}_{2}He$
d. $^{252}_{99}Es \longrightarrow\ ^{248}_{87}Bk + ^{4}_{2}He$

61. 5730 years old

63. The organism would be exposed to less harmful radiation.

65. An alpha particle is much more likely than other kinds of radiation to collide with another particle and be stopped. At the atomic level, the larger the size of a particle, the greater is the chance of it striking another particle. The greater the magnitude of a particle's charge, the more strongly it will be attracted to particles of opposite charge.

67. one neutron

69. uranium; $^{239}_{94}Pu \longrightarrow\ ^{235}_{92}U + ^{4}_{2}He$

71. $^{18}_{8}O$

73. This graph shows the radioactive decay of carbon-14, along with the increase of the nitrogen product.

75. Bismuth-214 remains.

77. 4.2×10^2 cm^3

79. a. 26 protons and 33 neutrons
b. 92 protons and 143 neutrons
c. 24 protons and 28 neturons

81. a. covalent **b.** ionic **c.** covalent **d.** ionic

83. 9.22×10^3 cm^3 H_2; 0.412 mol H_2

85. 6.7 mL

87. a. propanoic acid **b.** propanal
c. 1-propanol **d.** 1-aminopropane
e. 1-chloropropane **f.** ethylmethyl ether

89. 1-propanol and ethanoic acid

91. a. (4) **b.** (3) **c.** (1) **d.** (4) **e.** (1) **f.** (2)
g. (3) **h.** (2) **i.** (4) **j.** (3)

Glossary

A

absolute zero the zero point on the Kelvin temperature scale, equivalent to −273.15°C *(3.2)*

accepted value a quantity used by general agreement of the scientific community *(3.1)*

accuracy the closeness of a measurement to the true value of what is being measured *(3.1)*

acid a compound that produces hydrogen ions in solution; *see also* hydrogen-ion donor, Lewis acid *(9.4)*

acid dissociation constant (K_a) the ratio of the concentration of the dissociated form of an acid to the undissociated form; stronger acids have larger K_a values than weaker acids *(19.3)*

acidic solution any solution in which the hydrogen-ion concentration is greater than the hydroxide-ion concentration *(19.2)*

activated complex an unstable arrangement of atoms that exists momentarily at the peak of the activation-energy barrier; an intermediate or transitional structure formed during the course of a reaction *(18.1)*

activation energy the minimum energy colliding particles must have in order to react *(18.1)*

active site a groove or pocket in an enzyme molecule into which the substrate (reactant molecule) fits; where the substrate is converted to products *(24.3)*

activity series a list of elements in order of decreasing activity; the activity series of halogens is Fl, Cl, Br, I *(11.2)*

actual yield the amount of product that forms when a reaction is carried out in the laboratory *(12.3)*

addition reaction a reaction in which a substance is added at the double bond of an alkene or at the triple bond of an alkyne *(23.2)*

adenosine triphosphate (ATP) a molecule that transmits the energy needed by cells of all living things *(24.6)*

alcohol an organic compound having an —OH (hydroxyl) group; the general structure is R—OH *(23.2)*

aldehyde an organic compound in which the carbon of the carbonyl group is joined to at least one hydrogen; the general formula is RCHO *(23.3)*

aliphatic hydrocarbon any straight-chain or branched-chain alkane, alkene, or alkyne *(22.3)*

alkali metal any metal in Group 1A of the periodic table *(6.2)*

alkaline earth metal any metal in Group 2A of the periodic table *(6.2)*

alkaline solution a basic solution *(19.2)*

alkane a hydrocarbon containing only single covalent bonds; alkanes are saturated hydrocarbons *(22.1)*

alkene a hydrocarbon containing one or more carbon–carbon double bonds; alkenes are unsaturated hydrocarbons *(22.2)*

alkyl group a hydrocarbon substituent; the methyl group (—CH_3) is an alkyl group *(22.1)*

alkyl halide a halocarbon in which one or more halogen atoms are attached to the carbon atoms of an aliphatic chain *(23.1)*

alkyne a hydrocarbon containing a carbon–carbon triple bond; alkynes are unsaturated hydrocarbons *(22.2)*

allotrope one of two or more different molecular forms of an element in the same physical state; oxygen (O_2) and ozone (O_3) are allotropes of the element oxygen *(13.3)*

alloy a mixture composed of two or more elements, at least one of which is a metal *(7.3)*

alpha particle a positively charged particle emitted from certain radioactive nuclei; it consists of two protons and two neutrons and is identical to the nucleus of a helium atom *(25.1)*

amino acid an organic compound having amino (—NH_2) and carboxyl (—COOH) groups in the same molecule; proteins are made from the 20 naturally occurring amino acids *(24.3)*

amorphous solid describes a solid that lacks an ordered internal structure; denotes a random arrangement of atoms *(13.3)*

amphoteric a substance that can act as both an acid and a base *(19.1)*

amplitude the height of a wave's crest *(5.3)*

anabolism synthesis processes in the metabolism of cells; these processes usually require the expenditure of energy *(24.6)*

analytical chemistry the area of chemistry that focuses on the composition of matter *(1.1)*

anion any atom or group of atoms with a negative charge *(6.3)*

anode the electrode at which oxidation occurs *(21.1)*

applied chemistry research that is directed toward a practical goal or application *(1.1)*

aqueous solution water that contains dissolved substances *(15.2)*

aromatic compound an organic compound that contains a benzene ring or other ring in which the bonding is like that of benzene; aromatic compounds are also known as arenes *(22.4)*

aryl halide a halocarbon in which one or more halogens are attached to the carbon atoms of an arene ring *(23.1)*

asymmetric carbon a carbon atom that has four different atoms or groups attached *(22.3)*

atmospheric pressure the pressure exerted by atoms and molecules in the atmosphere surrounding Earth, resulting from collisions of these particles with objects *(13.1)*

atom the smallest particle of an element that retains its identity in a chemical reaction *(4.1)*

atomic emission spectrum the pattern formed when light passes through a prism or diffraction grating to separate it into the different frequencies of light it contains *(5.3)*

atomic mass the weighted average of the masses of the isotopes of an element *(4.3)*

atomic mass unit (amu) a unit of mass equal to one-twelfth the mass of a carbon-12 atom *(4.3)*

atomic number the number of protons in the nucleus of an atom of an element *(4.3)*

atomic orbital a mathematical expression describing the probability of finding an electron at various locations; usually represented by the region of space around the nucleus where there is a high probability of finding an electron *(5.1)*

atomic radius one-half the distance between the nuclei of two atoms of the same element when the atoms are joined *(6.3)*

aufbau principle the rule that electrons occupy the orbitals of lowest energy first *(5.2)*

Avogadro's hypothesis equal volumes of gases at the same temperature and pressure contain equal numbers of particles *(10.2)*

Avogadro's number the number of representative particles contained in one mole of a substance; equal to 6.02×10^{23} particles *(10.1)*

balanced equation a chemical equation in which mass is conserved; each side of the equation has the same number of atoms of each element *(11.1)*

band of stability the location of stable nuclei on a neutron-vs.-proton plot *(25.2)*

barometer an instrument used to measure atmospheric pressure *(13.1)*

base a compound that produces hydroxide ions in solution; *see also* hydrogen-ion acceptor, Lewis base *(9.4)*

base dissociation constant (K_b) the ratio of the concentration of the conjugate acid times the concentration of the hydroxide ion to the concentration of the base *(19.3)*

basic solution any solution in which the hydroxide-ion concentration is greater than the hydrogen-ion concentration *(19.2)*

battery a group of voltaic cells that are connected to one another *(21.1)*

beta particle an electron resulting from the breaking apart of neutrons in an atom *(25.1)*

binary compound a compound composed of two elements; NaCl and Al_2O_3 are binary compounds *(9.2)*

biochemistry the area of chemistry that focuses on processes that take place in organisms *(1.1)*

biotechnology the field that applies science to the production of biological products or processes *(1.2)*

boiling point (bp) the temperature at which the vapor pressure of a liquid is just equal to the external pressure on the liquid *(13.2)*

boiling-point elevation the difference in temperature between the boiling point of a solution and the boiling point of the pure solvent *(16.3)*

bond dissociation energy the energy required to break the bond between two covalently bonded atoms; this value is usually expressed in kJ per mol of substance *(8.2)*

bonding orbital a molecular orbital that can be occupied by two electrons of a covalent bond *(8.3)*

Boyle's law for a given mass of gas at constant temperature, the volume of the gas varies inversely with pressure *(14.2)*

branched-chain alkane an alkane with one or more alkyl groups attached to the parent structure *(22.1)*

Brownian motion the chaotic movement of colloidal particles, caused by collision with particles of the solvent in which they are dispersed *(15.3)*

buffer a solution in which the pH remains relatively constant when small amounts of acid or base are added; a buffer can be either a solution of a weak acid and the salt of a weak acid or a solution of a weak base with the salt of a weak base *(19.5)*

buffer capacity a measure of the amount of acid or base that may be added to a buffer solution before a significant change in pH occurs *(19.5)*

calorie (cal) the quantity of heat needed to raise the temperature of 1 g of pure water 1°C *(3.2)*

calorimeter an insulated device used to measure the absorption or release of heat in chemical or physical processes *(17.2)*

calorimetry the precise measurement of heat flow out of a system for chemical and physical processes *(17.2)*

carbohydrate the name given to monomers and polymers of aldehydes and ketones that have numerous hydroxyl groups; sugars and starches are carbohydrates *(24.2)*

carbonyl group a functional group having a carbon atom and an oxygen atom joined by a double bond; it is found in aldehydes, ketones, esters, and amides *(23.3)*

carboxyl group a functional group consisting of a carbonyl group attached to a hydroxyl group; it is found in carboxylic acids *(23.3)*

carboxylic acid an organic acid containing a carboxyl group; the general formula is RCOOH *(23.3)*

catabolism the reactions in living cells in which substances are broken down and energy is produced *(24.6)*

catalyst a substance that increases the rate of reaction by lowering the activation-energy barrier; the catalyst is not used up in the reaction *(11.1)*

cathode the electrode at which reduction occurs *(21.1)*

cathode ray a stream of electrons produced at the negative electrode (cathode) of a tube containing a gas at low pressure *(4.2)*

cation any atom or group of atoms with a positive charge *(6.3)*

cell potential the difference between the reduction potentials of two half-cells *(21.2)*

Celsius scale the temperature scale on which the freezing point of water is 0°C and the boiling point is 100°C *(3.2)*

Charles's law the volume of a fixed mass of gas is directly proportional to its Kelvin temperature if the pressure is kept constant *(14.2)*

chemical change a change that produces matter with a different composition than the original matter *(2.3)*

chemical equation an expression representing a chemical reaction; the formulas of the reactants (on the left) are connected by an arrow with the formulas for the products (on the right) *(11.1)*

chemical equilibrium a state of balance in which the rates of the forward and reverse reactions are equal; no net change in the amount of reactants and products occurs in the chemical system *(18.2)*

chemical formula an expression that indicates the number and type of atoms present in the smallest representative unit of a substance *(7.2)*

chemical potential energy energy stored in chemical bonds *(17.1)*

chemical property the ability of a substance to undergo a specific chemical change *(2.4)*

chemical reaction a change in which one or more reactants change into one or more products; characterized by the breaking of bonds in reactants and the formation of bonds in products *(2.4)*

chemical symbol a one- or two-letter representation of an element *(2.3)*

chemistry the study of the composition of matter and the changes that matter undergoes *(1.1)*

***cis* configuration** the configuration in which substituent groups are on the same side of a double bond *(22.3)*

coefficient a small whole number that appears in front of a formula in a balanced chemical equation *(11.1)*

coenzyme a small organic molecule or metal ion necessary for an enzyme's biological activity *(24.3)*

colligative property a property of a solution that depends only upon the number of solute particles, and not upon their identities; boiling-point elevation, freezing-point depression, and vapor-pressure lowering are colligative properties *(16.3)*

collision theory atoms, ions, and molecules can react to form products when they collide, provided that the particles have enough kinetic energy *(18.1)*

colloid a mixture whose particles are intermediate in size between those of a suspension and a solution *(15.3)*

combination reaction a chemical change in which two or more substances react to form a single new substance; also called a synthesis reaction *(11.2)*

combined gas law the law that describes the relationship among the pressure, temperature, and volume of an enclosed gas *(14.2)*

combustion reaction a chemical change in which an element or a compound reacts with oxygen, often producing energy in the form of heat and light *(11.2)*

common ion an ion that is common to both salts in a solution; in a solution of silver nitrate and silver chloride, Ag^+ would be a common ion *(18.3)*

common ion effect a decrease in the solubility of an ionic compound caused by the addition of a common ion *(18.3)*

complete ionic equation an equation that shows dissolved ionic compounds as dissociated free ions *(11.3)*

compound a substance that contains two or more elements chemically combined in a fixed proportion *(2.3)*

compressibility a measure of how much the volume of matter decreases under pressure *(14.1)*

concentrated solution a solution containing a large amount of solute *(16.2)*

concentration a measurement of the amount of solute that is dissolved in a given quantity of solvent; usually expressed as mol/L *(16.2)*

condensed structural formula a structural formula that leaves out some bonds and/or atoms; the presence of these atoms or bonds is understood *(22.1)*

conjugate acid the particle formed when a base gains a hydrogen ion; NH_4^+ is the conjugate acid of the base NH_3 *(19.1)*

conjugate acid-base pair two substances that are related by the loss or gain of a single hydrogen ion; ammonia (NH_3) and the ammonium ion (NH_4^+) are a conjugate acid-base pair *(19.1)*

conjugate base the particle that remains when an acid has donated a hydrogen ion; OH^- is the conjugate base of the acid water *(19.1)*

conversion factor a ratio of equivalent measurements used to convert a quantity from one unit to another *(3.3)*

coordinate covalent bond a covalent bond in which one atom contributes both bonding electrons *(8.2)*

coordination number the number of ions of opposite charge that surround each ion in a crystal *(7.2)*

covalent bond a bond formed by the sharing of electrons between atoms *(8.1)*

cracking the controlled process by which hydrocarbons are broken down or rearranged into smaller, more useful molecules *(22.5)*

crystal a solid in which the atoms, ions, or molecules are arranged in an orderly, repeating, three-dimensional pattern called a crystal lattice *(13.3)*

cyclic hydrocarbon an organic compound that contains a hydrocarbon ring *(22.4)*

Dalton's atomic theory the first theory to relate chemical changes to events at the atomic level *(4.1)*

Dalton's law of partial pressures at constant volume and temperature, the total pressure exerted by a mixture of gases is equal to the sum of the partial pressures of the component gases *(14.4)*

decomposition reaction a chemical change in which a single compound is broken down into two or more simpler products *(11.2)*

dehydrogenation reaction a reaction in which hydrogen is lost *(23.3)*

deliquescent describes a substance that removes sufficient water from the air to form a solution; the solution formed has a lower vapor pressure than that of the water in the air *(15.2)*

denatured alcohol ethanol to which a poisonous substance has been added to make it toxic *(23.2)*

density the ratio of the mass of an object to its volume *(3.4)*

dependent variable *see* responding variable

desiccant a hygroscopic substance used as a drying agent *(15.2)*

diatomic molecule a molecule consisting of two atoms *(8.1)*

diffusion the tendency of molecules to move toward areas of lower concentration until the concentration is uniform throughout *(14.4)*

dilute solution a solution that contains a small amount of solute *(16.2)*

dimensional analysis a technique of problem-solving that uses the units that are part of a measurement to help solve the problem *(3.3)*

dipole a molecule that has two poles, or regions, with opposite charges *(8.4)*

dipole interactions intermolecular forces resulting from the attraction of oppositely charged regions of polar molecules *(8.4)*

diprotic acid any acid that contains two ionizable protons (hydrogen ions); sulfuric acid (H_2SO_4) is a diprotic acid *(19.1)*

disaccharide a carbohydrate formed from two monosaccharide units; common table sugar (sucrose) is a disaccharide *(24.2)*

dispersion forces attractions between molecules caused by the electron motion on one molecule affecting the electron motion on the other through electrical forces; these are the weakest interactions between molecules *(8.4)*

displacement reaction *see* single-replacement reaction

distillation a process used to separate dissolved solids from a liquid, which is boiled to produce a vapor that is then condensed into a liquid *(2.2)*

double covalent bond a bond in which two atoms share two pairs of electrons *(8.2)*

double-replacement reaction a chemical change that involves an exchange of positive ions between two compounds *(11.2)*

dry cell a commercial voltaic cell in which the electrolyte is a moist paste; despite their name, the compact, portable batteries used in flashlights are dry cells *(21.1)*

E

effloresce to lose water of hydration; the process occurs when the hydrate has a vapor pressure higher than that of water vapor in the air *(15.2)*

effusion the process that occurs when a gas escapes through a tiny hole in its container *(14.4)*

electrical potential the ability of a voltaic cell to produce an electric current *(21.2)*

electrochemical cell any device that converts chemical energy into electrical energy or electrical energy into chemical energy *(21.1)*

electrochemical process the conversion of chemical energy into electrical energy or electrical energy into chemical energy; all electrochemical processes involve redox reactions *(21.1)*

electrode a conductor in a circuit that carries electrons to or from a substance other than a metal *(21.1)*

electrolysis a process in which electrical energy is used to bring about a chemical change; the electrolysis of water produces hydrogen and oxygen *(21.3)*

electrolyte a compound that conducts an electric current when it is in an aqueous solution or in the molten state; all ionic compounds are electrolytes, but most covalent compounds are not *(15.2)*

electrolytic cell an electrochemical cell used to cause a chemical change through the application of electrical energy *(21.3)*

electromagnetic radiation energy waves that travel in a vacuum at a speed of 2.998×10^8 m/s; includes radio waves, microwaves, infrared waves, visible light, ultraviolet waves, X-rays, and gamma rays *(5.3)*

electron a negatively charged subatomic particle *(4.2)*

electron configuration the arrangement of electrons of an atom in its ground state into various orbitals around the nuclei of atoms *(5.2)*

electron dot structure a notation that depicts valence electrons as dots around the atomic symbol of the element; the symbol represents the inner electrons and atomic nucleus; also called Lewis dot structure *(7.1)*

electronegativity the ability of an atom to attract electrons when the atom is in a compound *(6.3)*

element the simplest form of matter that has a unique set of properties; an element cannot be broken down into simpler substances by chemical means *(2.3)*

elementary reaction a reaction in which reactants are converted to products in a single step *(18.5)*

empirical formula a formula with the lowest whole-number ratio of elements in a compound; the empirical formula of hydrogen peroxide (H_2O_2) is HO *(10.3)*

emulsion the colloidal dispersion of one liquid in another *(15.3)*

endothermic process a process that absorbs heat from the surroundings *(17.1)*

end point the point in a titration at which the indicator changes color *(19.4)*

energy the capacity for doing work or producing heat *(3.2)*

energy level the specific energies an electron in an atom or other system can have *(5.1)*

enthalpy (H) the heat content of a system at constant pressure *(17.2)*

entropy (S) a measure of the disorder of a system; systems tend to go from a state of order (low entropy) to a state of maximum disorder (high entropy) *(18.4)*

enzyme a protein that acts as a biological catalyst *(24.3)*

enzyme–substrate complex the structure formed when a substrate molecule joins an enzyme at its active site *(24.3)*

equilibrium constant (K_{eq}) the ratio of product concentrations to reactant concentrations at equilibrium, with each concentration raised to a power equal to the number of moles of that substance in the balanced chemical equation *(18.2)*

equilibrium position the relative concentrations of reactants and products of a reaction that has reached equilibrium; indicates whether the reactants or products are favored in the reversible reaction *(18.2)*

equivalence point the point in a titration where the number of moles of hydrogen ions equals the number of moles of hydroxide ions *(19.4)*

error the difference between the accepted value and the experimental value *(3.1)*

ester a derivative of a carboxylic acid in which the —OH of the carboxyl group has been replaced by the —OR from an alcohol; the general formula is RCOOR *(23.3)*

ether an organic compound in which oxygen is bonded to two carbon groups; the general formula is R—O—R *(23.2)*

evaporation vaporization that occurs at the surface of a liquid that is not boiling *(13.2)*

excess reagent a reagent present in a quantity that is more than sufficient to react with a limiting reagent; any reactant that remains after the limiting reagent is used up in a chemical reaction *(12.3)*

exothermic process a process that releases heat to its surroundings *(17.1)*

experiment a repeatable procedure that is used to test a hypothesis *(1.3)*

experimental value a quantitative value measured during an experiment *(3.1)*

extensive property a property that depends on the amount of matter in a sample *(2.1)*

fatty acid the name given to continuous-chain carboxylic acids that were first isolated from fats *(23.3)*

fermentation the production of ethanol from sugars by the action of yeast or bacteria *(23.2)*

film badge a radiation detection device that consists of several layers of photographic film covered with black lightproof paper, all encased in a plastic or metal holder *(25.4)*

filtration a process that separates a solid from the liquid in a heterogeneous mixture *(2.2)*

first-order reaction a reaction in which the reaction rate is proportional to the concentration of only one reactant *(18.5)*

fission the splitting of a nucleus into smaller fragments, accompanied by the release of neutrons and a large amount of energy *(25.3)*

formula unit the lowest whole-number ratio of ions in an ionic compound; in magnesium chloride, the ratio of magnesium ions to chloride ions is 1:2 and the formula unit is $MgCl_2$ *(7.2)*

free energy the energy available to do work *(18.4)*

freezing-point depression the difference in temperature between the freezing point of a solution and the freezing point of the pure solvent *(16.3)*

frequency (ν) the number of wave cycles that pass a given point per unit of time; frequency and wavelength are inversely proportional to each other *(5.3)*

fuel cell a voltaic cell that does not need to be recharged; the fuel is oxidized to produce a continuous supply of electrical energy *(21.1)*

functional group a specific arrangement of atoms in an organic compound that is capable of characteristic chemical reactions; the chemistry of an organic compound is determined by its functional groups *(23.1)*

fusion the process of combining nuclei to produce a nucleus of greater mass *(25.3)*

gamma ray a high-energy photon emitted by a radioisotope *(25.1)*

gas a form of matter that takes the shape and volume of its container; a gas has no definite shape or volume *(2.1)*

gas pressure results from the force exerted by a gas per unit surface area of an object; due to collisions of gas particles with the object *(13.1)*

Gay-Lussac's law the pressure of a gas is directly proportional to the Kelvin temperature if the volume is constant *(14.2)*

Geiger counter a device that uses a gas-filled metal tube to detect radiation *(25.4)*

gene a segment of DNA that codes for a single peptide chain *(24.5)*

genetic code the arrangement of bases in the code words in DNA; there are 61 triplet code words used to specify the amino acids in proteins plus 3 triplet code words that signal "stop" after the protein synthesis is complete *(24.5)*

geometric isomers compounds that have atoms in the same order, but differ in the orientation of groups around a double bond *(22.3)*

Gibbs free-energy change (ΔG) the maximum amount of energy that can be coupled to another process to do useful work *(18.4)*

glass transparent fusion product of inorganic materials that have cooled to a rigid state without crystallizing *(13.3)*

Graham's law of effusion the rate of effusion of a gas is inversely proportional to the square root of its molar mass; this relationship is also true for the diffusion of gases *(14.4)*

gram (g) a metric mass unit equal to the mass of 1 cm^3 of water at 4°C *(3.2)*

ground state the lowest possible energy of an atom described by quantum mechanics *(5.3)*

group a vertical column of elements in the periodic table; the constituent elements of a group have similar chemical and physical properties *(4.3)*

half-cell the part of a voltaic cell in which either oxidation or reduction occurs; it consists of a single electrode immersed in a solution of its ions *(21.1)*

half-life ($t_{1/2}$) the time required for one-half of the nuclei of a radioisotope sample to decay to products *(25.2)*

half-reaction an equation showing either the oxidation or the reduction that takes place in a redox reaction *(20.3)*

half-reaction method a method of balancing a redox equation by balancing the oxidation and reduction half-reactions separately before combining them into a balanced redox equation *(20.3)*

halide ion a negative ion formed when a halogen atom gains an electron *(7.1)*

halocarbon any member of a class of organic compounds containing covalently bonded fluorine, chlorine, bromine, or iodine *(23.1)*

halogen a nonmetal in Group 7A of the periodic table *(6.2)*

heat (q) energy that transfers from one object to another because of a temperature difference between the objects *(17.1)*

heat capacity the amount of heat needed to increase the temperature of an object exactly 1°C *(17.1)*

heat of combustion the heat of reaction for the complete burning of one mole of a substance *(17.2)*

heat of reaction the enthalpy change for a chemical equation exactly as it is written *(17.2)*

Heisenberg uncertainty principle it is impossible to know exactly both the velocity and the position of a particle at the same time *(5.3)*

Henry's law at a given temperature the solubility of a gas in a liquid is directly proportional to the pressure of the gas above the liquid *(16.1)*

hertz (Hz) the unit of frequency, equal to one cycle per second *(5.3)*

Hess's law of heat summation if you add two or more thermochemical equations to give a final equation, then you also add the heats of reaction to give the final heat of reaction *(17.4)*

heterogeneous mixture a mixture that is not uniform in composition; components are not evenly distributed throughout the mixture *(2.2)*

homogeneous mixture a mixture that is uniform in composition; components are evenly distributed and not easily distinguished *(2.2)*

homologous series a group of compounds in which there is a constant increment of change in molecular structure from one compound in the series to the next *(22.1)*

Hund's rule electrons occupy orbitals of the same energy in a way that makes the number or electrons with the same spin direction as large as possible *(5.2)*

hybridization the mixing of several atomic orbitals to form the same total number of equivalent hybrid orbitals *(8.3)*

hydrate a compound that has a specific number of water molecules bound to each formula unit *(15.2)*

hydration reaction a reaction in which water is added to an alkene *(23.2)*

hydrocarbon an organic compound that contains only carbon and hydrogen *(22.1)*

hydrogenation reaction a reaction in which hydrogen is added to a carbon–carbon double bond to give an alkane *(23.2)*

hydrogen bonds attractive forces in which a hydrogen covalently bonded to a very electronegative atom is also weakly bonded to an unshared electron pair of another electronegative atom *(8.4)*

hydrogen-ion acceptor a base, according to the Brønsted-Lowry theory; ammonia acts as a base when it accepts hydrogen ions from water *(19.1)*

hydrogen-ion donor an acid, according to the Brønsted-Lowry theory *(19.1)*

hydronium ion (H_3O^+) the positive ion formed when a water molecule gains a hydrogen ion *(19.1)*

hydroxide ion (OH^-) the negative ion formed when a water molecule loses a hydrogen ion *(19.1)*

hydroxyl group the —OH functional group present in alcohols *(23.2)*

hygroscopic a term describing salts and other compounds that remove moisture from the air *(15.2)*

hypothesis a proposed explanation for an observation *(1.3)*

ideal gas constant the constant in the ideal gas law with the symbol R and the value 8.31 (L·kPa)/(K·mol) *(14.3)*

ideal gas law the relationship $PV = nRT$, which describes the behavior of an ideal gas *(14.3)*

immiscible describes liquids that are insoluble in one another; oil and water are immiscible *(16.1)*

independent variable *see* manipulated variable

inhibitor a substance that interferes with the action of a catalyst *(18.1)*

inner transition metal an element in the lanthanide or actinide series; the highest occupied *s* sublevel and nearby *f* sublevel of its atoms generally contain electrons; also called inner transition element *(6.2)*

inorganic chemistry the study of substances that, in general, do not contain carbon *(1.1)*

intensive property a property that depends on the type of matter in a sample, not the amount of matter *(2.1)*

intermediate a product of one of the steps in a reaction mechanism; it becomes a reactant in the next step *(18.5)*

International System of Units (SI) the revised version of the metric system, adopted by international agreement in 1960 *(3.2)*

ion an atom or group of atoms that has a positive or negative charge *(6.3)*

ionic bond the electrostatic attraction that binds oppositely charged ions together *(7.2)*

ionic compound a compound composed of positive and negative ions *(7.2)*

ionization energy the energy required to remove an electron from an atom in its gaseous state *(6.3)*

ionizing radiation radiation with enough energy to knock electrons off some atoms of a bombarded substance to produce ions *(25.4)*

ion-product constant for water (K_w) the product of the concentrations of hydrogen ions and hydroxide ions in water; it is 1×10^{-14} at 25°C *(19.2)*

isomers compounds that have the same molecular formula but different molecular structures *(22.3)*

isotopes atoms of the same element that have the same atomic number but different atomic masses due to a different number of neutrons *(4.3)*

joule (J) the SI unit of energy; 4.184 J equal one calorie *(3.2)*

Kelvin scale the temperature scale in which the freezing point of water is 273 K and the boiling point is 373 K; 0 K is absolute zero *(3.2)*

ketone an organic compound in which the carbon of the carbonyl group is joined to two other carbons; the general formula is RCOR *(23.3)*

kilogram (kg) the mass of 1 L of water at 4°C; it is the base unit of mass in SI *(3.2)*

kinetic energy the energy an object has because of its motion *(13.1)*

kinetic theory a theory explaining the states of matter, based on the concept that all matter consists of tiny particles that are in constant motion *(13.1)*

law of conservation of energy in any chemical or physical process, energy is neither created nor destroyed *(17.1)*

law of conservation of mass in any physical change or chemical reaction, mass is conserved; mass can be neither created nor destroyed *(2.4)*

law of definite proportions in samples of any chemical compound, the masses of the elements are always in the same proportion *(9.5)*

law of disorder it is a natural tendency of systems to move in the direction of maximum chaos or disorder *(18.4)*

law of multiple proportions whenever two elements form more than one compound, the different masses of one element that combine with the same mass of the other element are in the ratio of small whole numbers *(9.5)*

Le Châtelier's principle when a stress is applied to a system in dynamic equilibrium, the system changes in a way that relieves the stress *(18.2)*

Lewis acid any substance that can accept a pair of electrons to form a covalent bond *(19.1)*

Lewis base any substance that can donate a pair of electrons to form a covalent bond *(19.1)*

limiting reagent any reactant that is used up first in a chemical reaction; it determines the amount of product that can be formed in the reaction *(12.3)*

lipid a member of a large class of relatively water-insoluble organic compounds; fats, oils, and waxes are lipids *(24.4)*

liquid a form of matter that flows, has a fixed volume, and an indefinite shape *(2.1)*

liter (L) the volume of a cube measuring 10 centimeters on each edge (1000 cm^3); it is the common unprefixed unit of volume in the metric system *(3.2)*

macroscopic describes the world of objects that are large enough to see with the unaided eye *(1.2)*

manipulated variable the variable that is changed during an experiment; also called independent variable *(1.3)*

mass a measure of the amount of matter that an object contains; the SI base unit of mass is the kilogram *(2.1)*

mass number the total number of protons and neutrons in the nucleus of an atom *(4.3)*

matter anything that has mass and occupies space *(1.1)*

measurement a quantitative description that includes both a number and a unit *(3.1)*

melting point (mp) the temperature at which a substance changes from a solid to a liquid; the melting point of water is 0°C *(13.3)*

metabolism all the chemical reactions carried out by an organism; includes energy-producing (catabolism) reactions and energy-absorbing (anabolism) reactions *(24.6)*

metal one of a class of elements that are good conductors of heat and electric current; metals tend to be ductile, malleable, and shiny *(6.1)*

metallic bond the force of attraction that holds metals together; it consists of the attraction of free-floating valence electrons for positively charged metal ions *(7.3)*

metalloid an element that tends to have properties that are similar to those of metals and nonmetals *(6.1)*

meter (m) the base unit of length in SI *(3.2)*

microscopic describes the world of objects that can be seen only under magnification *(1.2)*

miscible describes liquids that dissolve in one another in all proportions *(16.1)*

mixture a physical blend of two or more substances that are not chemically combined *(2.2)*

molal boiling-point elevation constant (K_b) the change in boiling point for a 1-molal solution of a nonvolatile molecular solute *(16.4)*

molal freezing-point depression constant (K_f) the change in freezing point for a 1-molal solution of a nonvolatile molecular solute *(16.4)*

molality (*m*) the concentration of solute in a solution expressed as the number of moles of solute dissolved in 1 kilogram (1000 g) of solvent *(16.4)*

molar heat of condensation (ΔH_{cond}) the amount of heat released by one mole of a vapor as it condenses to a liquid at a constant temperature *(17.3)*

molar heat of fusion (ΔH_{fus}) the amount of heat absorbed by one mole of a solid substance as it melts to a liquid at a constant temperature *(17.3)*

molar heat of solidification (ΔH_{solid}) the amount of heat lost by one mole of a liquid as it solidifies at a constant temperature *(17.3)*

molar heat of solution (ΔH_{soln}) the enthalpy change caused by the dissolution of one mole of a substance *(17.3)*

molar heat of vaporization (ΔH_{vap}) the amount of heat absorbed by one mole of a liquid as it vaporizes at a constant temperature *(17.3)*

molarity (*M*) the concentration of solute in a solution expressed as the number of moles of solute dissolved in 1 liter of solution *(16.2)*

molar mass a term used to refer to the mass of a mole of any substance *(10.1)*

molar volume the volume occupied by 1 mole of a gas at standard temperature and pressure (STP); 22.4 L *(10.2)*

mole (mol) the amount of a substance that contains 6.02×10^{23} representative particles of that substance *(10.1)*

molecular compound a compound that is composed of molecules *(8.1)*

molecular formula a chemical formula of a molecular compound that shows the kinds and numbers of atoms present in a molecule of a compound *(8.1)*

molecular orbital an orbital that applies to the entire molecule *(8.3)*

molecule a neutral group of atoms joined together by covalent bonds *(8.1)*

mole fraction the ratio of the moles of solute in solution to the total number of moles of both solvent and solute *(16.4)*

mole ratio a conversion factor derived from the coefficients of a balanced chemical equation interpreted in terms of moles *(12.2)*

monatomic ion a single atom with a positive or negative charge resulting from the loss or gain of one or more valence electrons *(9.1)*

monomer a simple molecule that repeatedly combines to form a polymer *(23.4)*

monoprotic acid any acid that contains one ionizable proton (hydrogen ion); nitric acid (HNO_3) is a monoprotic acid *(19.1)*

monosaccharide a carbohydrate consisting of one sugar unit; also called a simple sugar *(24.2)*

net ionic equation an equation for a reaction in solution showing only those particles that are directly involved in the chemical change *(11.3)*

network solid a solid in which all of the atoms are covalently bonded to each other *(8.4)*

neutralization reaction a reaction in which an acid and a base react in an aqueous solution to produce a salt and water *(19.4)*

neutral solution an aqueous solution in which the concentrations of hydrogen and hydroxide ions are equal; it has a pH of 7.0 *(19.2)*

neutron a subatomic particle with no charge and a mass of 1 amu; found in the nucleus of an atom *(4.2)*

neutron absorption a process that decreases the number of slow-moving neutrons in a nuclear reactor; this is accomplished by using control rods made of a material such as cadmium, which absorbs neutrons *(25.3)*

neutron activation analysis a procedure used to detect trace amounts of elements in samples by bombarding the sample with neutrons from a radioactive source *(25.4)*

neutron moderation a process used in nuclear reactors to slow down neutrons so the reactor fuel captures them to continue the chain reaction *(25.3)*

noble gas an element in Group 8A of the periodic table; the *s* and *p* sublevels of the highest occupied energy level are filled *(6.2)*

nonelectrolyte a compound that does not conduct an electric current in aqueous solution or in the molten state *(15.2)*

nonmetal an element that tends to be a poor conductor of heat and electric current; nonmetals generally have properties opposite to those of metals *(6.1)*

nonpolar covalent bond a covalent bond in which the electrons are shared equally by the two atoms *(8.4)*

nonspontaneous reaction a reaction that does not favor the formation of products at the specified conditions *(18.4)*

normal boiling point the boiling point of a liquid at a pressure of 101.3 kPa or 1 atm *(13.2)*

nucleic acid a polymer of ribonucleotides (RNA) or deoxyribonucleotides (DNA) found primarily in cell nuclei; nucleic acids play an important role in the transmission of hereditary characteristics, protein synthesis, and the control of cell activities *(24.5)*

nucleotide one of the monomers that make up DNA and RNA; it consists of a nitrogen-containing base (a purine or pyrimidine), a sugar (ribose or deoxyribose), and a phosphate group *(24.5)*

nucleus the tiny, dense central portion of an atom, composed of protons and neutrons *(4.2)*

observation information obtained through the senses; observation in science often involves a measurement *(1.3)*

octet rule atoms react by gaining or losing electrons so as to acquire the stable electron structure of a noble gas, usually eight valence electrons *(7.1)*

optical isomers molecules that differ from one another in the way that four different groups are arranged around a carbon atom *(22.3)*

organic chemistry the study of compounds containing carbon *(1.1)*

oxidation a process that involves complete or partial loss of electrons or a gain of oxygen; it results in an increase in the oxidation number of an atom *(20.1)*

oxidation number a positive or negative number assigned to an atom to indicate its degree of oxidation or reduction; the oxidation number of an uncombined element is zero *(20.2)*

oxidation-number-change method a method of balancing a redox equation by comparing the increases and decreases in oxidation numbers *(20.3)*

oxidation-reduction reaction a reaction that involves the transfer of electrons between reactants *(20.1)*

oxidizing agent the substance in a redox reaction that accepts electrons; in the reaction, the oxidizing agent is reduced *(20.1)*

partial pressure the contribution each gas in a mixture of gases makes to the total pressure *(14.4)*

pascal (Pa) the SI unit of pressure *(13.1)*

Pauli exclusion principle an atomic orbital may describe at most two electrons, each with opposite spin direction *(5.2)*

peptide an organic compound formed by a combination of amino acids in which the amino group of one acid is united with the carboxyl group of another through an amide bond *(24.3)*

peptide bond the bond between the carbonyl group of one amino acid and the nitrogen of the next amino acid in the peptide chain; the structure is

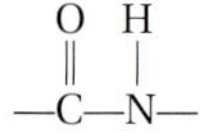

(24.3)

percent composition the percent by mass of each element in a compound *(10.3)*

percent error the percent that a measured value differs from the accepted value *(3.1)*

percent yield the ratio of the actual yield to the theoretical yield for a chemical reaction expressed as a percentage; a measure of the efficiency of a reaction *(12.3)*

period a horizontal row of elements in the periodic table *(4.3)*

periodic law when the elements are arranged in order of increasing atomic number, there is a periodic repetition of their physical and chemical properties *(6.1)*

periodic table an arrangement of elements in which the elements are separated into groups based on a set of repeating properties *(4.3)*

pH a number used to denote the hydrogen-ion concentration, or acidity, of a solution; it is the negative logarithm of the hydrogen-ion concentration of a solution *(19.2)*

phase any part of a sample with uniform composition and properties *(2.2)*

phase diagram a graph showing the conditions at which a substance exists as a solid, liquid, or vapor *(13.4)*

phospholipid a lipid that contains a phosphate group; because phospholipids have hydrophilic heads and hydrophobic tails, they form the lipid bilayers found in cell membranes *(24.4)*

photon a quantum of light; a discrete bundle of electromagnetic energy that interacts with matter similarly to particles *(5.3)*

photon a quantum of light; a discrete bundle of electromagnetic energy that interacts with matter similarly to particles *(5.3)*

photosynthesis the process by which green plants and algae use radiant energy from the sun to synthesize glucose from carbon dioxide and water *(24.1)*

physical change a change during which some properties of a material change, but the composition of the material does not change *(2.1)*

physical chemistry the area of chemistry that deals with the mechanism, the rate, and the energy transfer that occurs when matter undergoes a change *(1.1)*

physical property a quality or condition of a substance that can be observed or measured without changing the substance's composition *(2.1)*

pi bond (π bond) a covalent bond in which the bonding electrons are most likely to be found in sausage-shaped regions above and below the bond axis of the bonded atoms *(8.3)*

polar covalent bond (polar bond) a covalent bond between atoms in which the electrons are shared unequally *(8.4)*

polar molecule a molecule in which one side of the molecule is slightly negative and the opposite side is slightly positive *(8.4)*

pollutant a material found in air, water, or soil that is harmful to humans and other organisms *(1.2)*

polyatomic ion a tightly bound group of atoms that behaves as a unit and has a positive or negative charge *(8.2)*

polymer a very large molecule formed by the covalent bonding of repeating small molecules, known as monomers *(23.4)*

polypeptide a peptide with more than 10 amino acids *(24.3)*

polysaccharide a complex carbohydrate polymer formed by the linkage of many monosaccharide monomers; starch, glycogen, and cellulose are polysaccharides *(24.2)*

positron a particle with the mass of an electron but a positive charge *(25.2)*

precipitate a solid that forms and settles out of a liquid mixture *(2.4)*

precision describes the closeness, or reproducibility, of a set of measurements taken under the same conditions *(3.1)*

product a substance produced in a chemical reaction *(2.4)*

protein any peptide with more than 100 amino acids *(24.3)*

proton a positively charged subatomic particle found in the nucleus of an atom *(4.2)*

pure chemistry the pursuit of chemical knowledge for its own sake *(1.1)*

pure substance *see* substance

quantum the amount of energy needed to move an electron from one energy level to another *(5.1)*

quantum mechanical model the modern description, primarily mathematical, of the behavior of electrons in atoms *(5.1)*

R

radiation the penetrating rays and particles emitted by a radioactive source *(25.1)*

radioactivity the process by which nuclei emit particles and rays *(25.1)*

radioisotope an isotope that has an unstable nucleus and undergoes radioactive decay *(25.1)*

rate describes the speed of change over an interval of time *(18.1)*

rate law an expression relating the rate of a reaction to the concentration of the reactants *(18.5)*

reactant a substance present at the start of a reaction *(2.4)*

reaction mechanism a series of elementary reactions that take place during the course of a complex reaction *(18.5)*

redox reaction another name for an oxidation-reduction reaction *(20.1)*

reducing agent the substance in a redox reaction that donates electrons; in the reaction, the reducing agent is oxidized *(20.1)*

reduction a process that involves a complete or partial gain of electrons or the loss of oxygen; it results in a decrease in the oxidation number of an atom *(20.1)*

reduction potential a measure of the tendency of a given half-reaction to occur as a reduction (gain of electrons) in an electrochemical cell *(21.2)*

representative element an element in an "A" group in the periodic table; as a group these elements display a wide range of physical and chemical properties. In their atoms, the *s* and *p* sublevels in the highest occupied energy level are partially filled *(6.2)*

representative particle the smallest unit into which a substance can be broken down without a change in composition, usually atoms, molecules, or ions *(10.1)*

resonance structure one of the two or more equally valid electron dot structures of a molecule or polyatomic ion *(8.2)*

responding variable the variable that is observed during an experiment; also called dependent variable *(1.3)*

reversible reaction a reaction in which the conversion of reactants into products and the conversion of products into reactants occur simultaneously *(18.2)*

salt bridge a tube containing a strong electrolyte used to separate the half-cells in a voltaic cell; it allows the passage of ions from one half-cell to the other but prevents the solutions from mixing completely *(23.1)*

salt hydrolysis a process in which the cations or anions of a dissociated salt accept hydrogen ions from water or donate hydrogen ions to water *(19.5)*

saponification the hydrolysis of fats or oils by a hot aqueous alkali-metal hydroxide; soaps are made by saponification *(24.4)*

saturated compound an organic compound in which all carbon atoms are joined by single covalent bonds; it contains the maximum number of hydrogen atoms per carbon atom *(22.2)*

saturated solution a solution containing the maximum amount of solute for a given amount of solvent at a constant temperature and pressure; an equilibrium exists between undissolved solute and ions in solution *(16.1)*

scientific law a concise statement that summarizes the results of many observations and experiments *(1.3)*

scientific method a logical, systematic approach to the solution of a scientific problem; steps in the scientific method include making observations, testing hypotheses, and developing theories *(1.3)*

scientific notation an expression of numbers in the form $m \times 10^n$, where m is equal to or greater than 1 and less than 10, and n is an integer *(3.1)*

scintillation counter a device that uses a phosphor-coated surface to detect radiation *(25.4)*

self-ionization a term describing the reaction in which two water molecules react to produce ions *(19.2)*

sigma bond (σ bond) a bond formed when two atomic orbitals combine for form a molecular orbital that is symmetrical around the axis connecting the two atomic nuclei *(8.3)*

significant figures all the digits that can be known precisely in a measurement, plus a last estimated digit *(3.1)*

single covalent bond a bond formed when two atoms share a pair of electrons *(8.2)*

single-replacement reaction a chemical change in which one element replaces a second element in a compound; also called a displacement reaction *(11.2)*

skeleton equation a chemical equation that does not indicate the relative amounts of reactants and products *(11.1)*

solid a form of matter that has a definite shape and volume *(2.1)*

solubility the amount of a substance that dissolves in a given quantity of solvent at specified conditions of temperature and pressure to produce a saturated solution *(16.1)*

solubility product constant (K_{sp}) an equilibrium constant applied to the solubility of electrolytes; it is equal to the product of the concentrations of the ions each raised to a power equal to the coefficient of the ion in the dissociation equation *(18.3)*

solute dissolved particles in a solution *(15.2)*

solution a homogeneous mixture; consists of solutes dissolved in a solvent *(2.2)*

solvation a process that occurs when an ionic solute dissolves; in solution, solvent molecules surround the positive and negative ions. *(15.2)*

solvent the dissolving medium in a solution *(15.2)*

specific heat the amount of heat needed to increase the temperature of 1 g of a substance 1°C; also called specific heat capacity *(17.1)*

specific rate constant a proportionality constant relating the concentrations of reactants to the rate of the reaction *(18.5)*

spectator ion an ion that is not directly involved in a chemical reaction; an ion that does not change oxidation number or composition during a reaction *(11.3)*

spectrum wavelengths of visible light that are separated when a beam of light passes through a prism; range of wavelengths of electromagnetic radiation *(5.3)*

spontaneous reaction a reaction that favors the formation of products at the specified conditions; spontaneity depends on enthalpy and entropy changes *(18.4)*

standard atmosphere (atm) a unit of pressure; it is the pressure required to support 760 mm of mercury in a mercury barometer at 25°C *(13.1)*

standard cell potential (E^0_{cell}) the measured cell potential when the ion concentration in the half-cells are 1.00*M* at 1 atm of pressure and 25°C *(21.2)*

standard heat of formation (ΔH_f^0) the change in enthalpy that accompanies the formation of one mole of a compound from its elements with all substances in their standard states at 25°C *(17.4)*

standard hydrogen electrode an arbitrary reference electrode (half-cell) used with another electrode (half-cell) to measure the standard reduction potential of that cell; the standard reduction potential of the hydrogen electrode is assigned a value of 0.00 V *(21.2)*

standard solution a solution of known concentration used in carrying out a titration *(19.4)*

standard temperature and pressure (STP) the conditions under which the volume of a gas is usually measured; standard temperature is 0°C, and standard pressure is 101.3 kPa, or 1 atmosphere (atm) *(10.2)*

stereoisomers molecules that have atoms in the same order, but which differ in the arrangement of the atoms in space *(22.3)*

stoichiometry that portion of chemistry dealing with numerical relationships in chemical reactions; the calculation of quantities of substances involved in chemical equations *(12.1)*

straight-chain alkane a saturated hydrocarbon that contains any number of carbon atoms arranged one after the other in a chain *(22.1)*

strong acid an acid that is completely (or almost completely) ionized in aqueous solution *(19.3)*

strong base a base that completely dissociates into metal ions and hydroxide ions in aqueous solution *(19.3)*

strong electrolyte a solution in which a large portion of the solute exists as ions *(15.2)*

structural formula a chemical formula that shows the arrangement of atoms in a molecule or a polyatomic ion; each dash between a pair of atoms indicates a pair of shared electrons *(8.2)*

structural isomers compounds that have the same molecular formula, but whose atoms are bonded in a different order *(22.3)*

sublimation the process in which a solid changes to a gas or vapor without passing through the liquid state *(13.4)*

substance matter that has a uniform and definite composition; either an element or a compound; also called pure substance *(2.1)*

substituent an atom or group of atoms that can take the place of a hydrogen atom on a parent hydrocarbon molecule *(22.1)*

substitution reaction a common type of organic reaction; involves the replacement of an atom or group of atoms by another atom or group of atoms *(23.1)*

substrate a molecule on which an enzyme acts *(24.3)*

supersaturated solution a solution that contains more solute than it can theoretically hold at a given temperature; excess solute precipitates if a seed crystal is added *(16.1)*

surface tension an inward force that tends to minimize the surface area of a liquid; it causes the surface to behave as if it were a thin skin *(15.1)*

surfactant any substance that interferes with the hydrogen bonding between water molecules and thereby reduces surface tension; soaps are surfactants. *(15.1)*

surroundings everything in the universe outside of the system *(17.1)*

suspension a mixture from which some of the particles settle out slowly upon standing *(15.3)*

synthesis reaction *see* combination reaction

system a part of the universe on which you focus your attention *(17.1)*

technology the means by which a society provides its members with those things needed and desired *(1.1)*

temperature a measure of the average kinetic energy of particles in matter; temperature determines the direction of heat transfer *(3.2)*

tetrahedral angle a bond angle of 109.5° that results when a central atom forms four bonds directed toward the center of a regular tetrahedron *(8.3)*

theoretical yield the amount of product that could form during a reaction calculated from a balanced chemical equation; it represents the maximum amount of product that could be formed from a given amount of reactant *(12.3)*

theory a well-tested explanation for a broad set of observations *(1.3)*

thermochemical equation a chemical equation that includes the enthalpy change *(17.2)*

thermochemistry the study of energy changes that occur during chemical reactions and changes in state *(17.1)*

titration process used to determine the concentration of a solution (often an acid or base) in which a solution of known concentration (the standard) is added to a measured amount of the solution of unknown concentration until an indicator signals the end point *(19.4)*

***trans* configuration** the configuration in which substituent groups are on the opposite sides of a double bond *(22.3)*

transition metal one of the Group B elements in which the highest occupied *s* sublevel and a nearby *d* sublevel generally contain electrons *(6.2)*

transition state a term sometimes used to refer to the activated complex *(18.1)*

transmutation the conversion of an atom of one element to an atom of another element *(25.2)*

transuranium element any elements in the periodic table with atomic number above 92, the atomic number of uranium *(25.2)*

triglyceride an ester in which all three hydroxyl groups on a glycerol molecule have been replaced by long-chain fatty acids; fats are triglycerides *(24.4)*

triple covalent bond a covalent bond in which three pairs of electrons are shared by two atoms *(8.2)*

triple point the point on a phase diagram that represents the only set of conditions at which all three phases exist in equilibrium with one another *(13.4)*

triprotic acid any acid that contains three ionizable protons (hydrogen ions); phosphoric acid (H_3PO_4) is a triprotic acid *(19.1)*

Tyndall effect scattering of light by particles in a colloid or suspension, which causes a beam of light to become visible *(15.3)*

unit cell the smallest group of particles within a crystal that retains the geometric shape of the crystal *(13.3)*

unsaturated solution a solution that contains less solute than a saturated solution at a given temperature and pressure *(16.1)*

unsaturated compound an organic compound with one or more double or triple carbon–carbon bonds *(22.2)*

unshared pair a pair of valence electrons that is not shared between atoms *(8.2)*

vacuum a space where no particles of matter exist *(13.1)*

valence electron an electron in the highest occupied energy level of an atom *(7.1)*

van der Waals forces the two weakest intermolecular attractions—dispersion interactions and dipole forces *(8.4)*

vapor describes the gaseous state of a substance that is generally a liquid or solid at room temperature *(2.1)*

vaporization the conversion of a liquid to a gas or a vapor *(13.2)*

vapor pressure a measure of the force exerted by a gas above a liquid in a sealed container; a dynamic equilibrium exists between the vapor and the liquid *(13.2)*

voltaic cell an electrochemical cell used to convert chemical energy into electrical energy; the energy is produced by a spontaneous redox reaction *(21.1)*

volume a measure of the space occupied by a sample of matter *(2.1)*

VSEPR theory valence-shell electron-pair repulsion theory; because electron pairs repel, molecules adjust their shapes so that valence electron pairs are as far apart as possible *(8.3)*

water of hydration water molecules that are an integral part of a crystal structure *(15.2)*

wavelength (λ) the distance between adjacent crests of a wave *(5.3)*

wax an ester of a long-chain fatty acid and a long-chain alcohol *(24.4)*

weak acid an acid that is only slightly ionized in aqueous solution *(19.3)*

weak base a base that reacts with water to form the hydroxide ion and the conjugate acid of the base *(19.3)*

weak electrolyte a solution that conducts electricity poorly because only a fraction of the solute exists as ions *(15.2)*

weight a force that measures the pull of gravity on a given mass *(3.2)*

Spanish Glossary

absolute zero/cero absoluto

atmospheric pressure/presión atmosférica presión ejercida por átomos y moléculas de la atmósfera que rodea a la Tierra y que resulta de los choques de dichas partículas con los objetos *(13.1)*

atom/átomo partícula más pequeña de un elemento que conserva su identidad en una reacción química *(4.1)*

atomic emission spectrum/espectro de emisión atómica patrón que se forma cuando la luz atraviesa un prisma o una rejilla de difracción que la separa en las diferentes frecuencias de luz que contiene *(5.3)*

atomic mass/masa atómica promedio ponderado de las masas de los isótopos de un elemento *(4.3)*

atomic mass unit (amu)/unidad de masa atómica (uma) unidad de masa igual a un doceavo de la masa de un átomo de carbono 12 *(4.3)*

atomic number/número atómico número de protones que hay en el núcleo del átomo de un elemento *(4.3)*

atomic orbital/orbital atómico expresión matemática que describe la probabilidad de hallar un electrón en diversos lugares; se suele representar como la región del espacio en torno al núcleo donde hay una probabilidad elevada de hallar un electrón *(5.1)*

atomic radius/radio atómico mitad de la distancia entre los núcleos de dos átomos del mismo elemento cuando dichos átomos están unidos *(6.3)*

aufbau principle/principio de aufbau regla según la cual los electrones primero ocupan los orbitales de energía más baja *(5.2)*

Avogadro's hypothesis/hipótesis de Avogadro volúmenes iguales de gases a la misma temperatura y presión contienen el mismo número de partículas *(10.2)*

Avogadro's number/número de Avogadro número de partículas representativas contenidas en un mol de una sustancia; es igual a 6.02×10^{23} partículas *(10.1)*

B

balanced equation/ecuación balanceada ecuación química en la que se conserva la masa; cada lado de la ecuación tiene el mismo número de átomos de cada elemento *(11.1)*

band of stability/banda de estabilidad región ocupada por los núcleos estables en un diagrama neutrones-protones *(25.2)*

barometer/barómetro instrumento que sirve para medir la presión atmosférica *(13.1)*

base/base compuesto que, en solución, produce iones hidróxido, acepta iones hidrógeno o cede pares de electrones *(9.4)*

base dissociation constant (K_b)/constante de disociación básica (K_b) razón de la concentración del ácido combinado multiplicada por la concentración del ion hidróxido, a la concentración de la base *(19.3)*

basic solution/solución básica cualquier solución en la que la concentración de ion hidróxido es mayor que la concentración de ion hidrógeno *(19.2)*

battery/batería grupo de celdas voltaicas conectadas entre sí *(21.1)*

beta particle/partícula beta electrón que se produce al descomponerse los neutrones de un átomo *(25.1)*

binary compound/compuesto binario compuesto integrado por dos elementos; NaCl y Al_2O_3 son compuestos binarios *(9.2)*

biochemistry/bioquímica rama de la química que se concentra en los procesos que se dan en los organismos *(1.1)*

biotechnology/biotecnología campo que aplica la ciencia a la realización de productos o procesos biológicos *(1.2)*

boiling point (bp)/punto de ebullición (p. eb.) temperatura en la que la presión de vapor de un líquido es apenas igual a la presión externa sobre el líquido *(13.2)*

boiling-point elevation/incremento del punto de ebullición diferencia de temperatura entre el punto de ebullición de una solución y el punto de ebullición del disolvente puro *(16.3)*

bond dissociation energy/energía de disociación de enlaces energía requerida para romper el enlace entre dos átomos unidos de forma covalente; este valor suele expresarse en kJ por mol de sustancia *(8.2)*

bonding orbital/orbital de enlace orbital molecular que puede ser ocupado por los dos electrones de un enlace covalente *(8.3)*

Boyle's law/ley de Boyle para una masa dada de gas a temperatura constante, el volumen del gas varía en proporción inversa con la presión *(14.2)*

branched-chain alkane/alcano de cadena ramificada alcano con uno o más grupos alquilo unidos a la estructura madre *(22.1)*

Brownian motion/movimiento browniano movimiento caótico de partículas coloidales, debido a los choques con las partículas del disolvente en el que están dispersas *(15.3)*

buffer/solución amortiguadora solución cuyo pH permanece relativamente constante si se le añaden pequeñas cantidades de ácido o base; una solución amortiguadora puede ser una solución de un ácido débil y la sal de un ácido débil o una solución de una base débil y la sal de una base débil *(19.5)*

buffer capacity/capacidad amortiguadora medida de la cantidad de ácido o base que se puede añadir a una solución amortiguadora sin que haya un cambio importante del pH *(19.5)*

calorie (cal)/caloría (cal) cantidad de calor necesaria para elevar 1°C la temperatura de 1 g de agua pura *(3.2)*

calorimeter/calorímetro aparato con material aislante que sirve para medir la absorción o desprendimiento de calor durante procesos químicos o físicos *(17.2)*

calorimetry/calorimetría medición precisa del cambio del calor durante procesos químicos y físicos *(17.2)*

carbohydrate/carbohidrato nombre dado a monómeros y polímeros de aldehídos y cetonas que tienen muchos grupos hidroxilo; los azúcares y almidones son carbohidratos *(24.2)*

carbonyl group/grupo carbonilo grupo funcional que consiste en un átomo de carbono y uno de oxígeno unidos por un doble enlace; se le encuentra en aldehídos, cetonas, ésteres y amidas *(23.3)*

carboxyl group/grupo carboxilo grupo funcional que consiste en un grupo carbonilo unido a un grupo hidroxilo; se le encuentra en los ácidos carboxílicos *(23.3)*

carboxylic acid/ácido carboxílico ácido orgánico que contiene un grupo carboxilo; su fórmula general es RCOOH *(23.3)*

catabolism/catabolismo reacción, dentro de las células vivas, por la que diversas sustancias se descomponen y producen energía *(24.6)*

catalyst/catalizador sustancia que aumenta la velocidad de reacción disminuyendo la barrera de energía de activación; el catalizador no se consume en la reacción *(11.1)*

cathode/cátodo electrodo en el que hay reducción *(21.1)*

cathode ray/rayo catódico haz de electrones producido en el electrodo negativo (cátodo) de un tubo que contiene un gas a baja presión *(4.2)*

cation/catión cualquier átomo o grupo de átomos que posee carga positiva *(6.3)*

cell potential/potencial de celda diferencia entre los potenciales de reducción de dos medias celdas *(21.2)*

Celsius scale/escala Celsius escala de temperatura en la que el punto de congelación del agua es 0°C y el punto de ebullición del agua es 100°C *(3.2)*

Charles's law/ley de Charles el volumen de una masa fija de gas es directamente proporcional a su temperatura Kelvin si la presión se mantiene constante *(14.2)*

chemical change/cambio químico cambio que produce materia con una composición diferente que la de la materia original *(2.3)*

chemical equation/ecuación química expresión que representa una reacción química; las fórmulas de los reactantes (a la izquierda) se unen mediante una flecha a las fórmulas de los productos (a la derecha) *(11.1)*

chemical equilibrium/equilibrio químico estado de equilibrio en el que las velocidades de la reacción hacia adelante y la reacción inversa son iguales; no hay un cambio total en la cantidad de reactantes y productos en el sistema químico *(18.2)*

chemical formula/fórmula química expresión que indica el número y tipo de átomos que están presentes en la unidad más pequeña representativa de una sustancia *(7.2)*

chemical potential energy/energía potencial química energía almacenada en los enlaces químicos *(17.1)*

chemical property/propiedad química capacidad de una sustancia para sufrir un cambio químico específico *(2.4)*

chemical reaction/reacción química cambio en el que uno o más reactantes se convierten en uno o más productos; se caracteriza por la ruptura de enlaces en los reactantes y la formación de enlaces en los productos *(2.4)*

chemical symbol/símbolo químico representación de un elemento que emplea una o dos letras *(2.3)*

chemistry/química estudio de la composición de la materia y los cambios que ésta sufre *(1.1)*

cis* configuration/configuración *cis configuración en la cual los grupos sustitutos están del mismo lado de un doble enlace *(22.3)*

coefficient/coeficiente número entero pequeño que aparece antepuesto a una fórmula en una ecuación química balanceada *(11.1)*

coenzyme/coenzima pequeña molécula orgánica o ion metálico que se requiere para que una enzima tenga actividad biológica *(24.3)*

colligative property/propiedad coligativa propiedad de una solución que depende únicamente del número de partículas de soluto, y no de su identidad; el incremento del punto de ebullición, la disminución del punto de congelación y el descenso de la presión de vapor son propiedades coligativas *(16.3)*

collision theory/teoría de choques los átomos, iones y moléculas pueden reaccionar para formar productos cuando chocan, siempre que las partículas tengan suficiente energía cinética *(18.1)*

colloid/coloide mezcla cuyas partículas tienen un tamaño intermedio entre las de una suspensión y una solución *(15.3)*

combination reaction/reacción de combinación cambio químico en el que dos o más sustancias reaccionan para formar una sola sustancia nueva; también llamado reacción de síntesis *(11.2)*

combined gas law/ley combinada de los gases ley que describe las relaciones entre la presión, la temperatura y el volumen de un gas encerrado *(14.2)*

combustion reaction/reacción de combustión cambio químico en el que un elemento o un compuesto reacciona con oxígeno y por lo regular produce energía en forma de luz y calor *(11.2)*

common ion/ion común ion que es común a dos sales disueltas en una solución; en una solución de nitrato de plata y cloruro de plata, Ag^+ sería un ion común *(18.3)*

common ion effect/efecto de ion común disminución en la solubilidad de un compuesto iónico debida a la adición de un ion común *(18.3)*

complete ionic equation/ecuación iónica completa ecuación que muestra los compuestos iónicos disueltos en forma de iones disociados libres *(11.3)*

compound/compuesto sustancia que contiene dos o más elementos combinados químicamente en una proporción fija *(2.3)*

compressibility/compresibilidad medida de cuánto disminuye el volumen de la materia cuando se le aplica presión *(14.1)*

concentrated solution/solución concentrada solución que contiene una gran cantidad de soluto *(16.2)*

concentration/concentración medida de la cantidad de soluto que está disuelto en una cantidad específica de disolvente; suele expresarse en mol/L *(16.2)*

condensed structural formula/fórmula estructural condensada fórmula estructural que no muestra algunos enlaces y/o átomos; la presencia de estos enlaces o átomos se sobreentiende *(22.1)*

conjugate acid/ácido conjugado partícula que se forma cuando una base gana un ion hidrógeno; NH_4^+ es el ácido conjugado de la base NH_3 *(19.1)*

conjugate acid-base pair/par conjugado ácido-base dos sustancias relacionadas entre sí por la pérdida o ganancia de un solo ion hidrógeno; el amoniaco (NH_3) y el ion amonio (NH_4^+) son un par conjugado ácido-base *(19.1)*

conjugate base/base conjugada partícula que queda cuando un ácido transfiere un ion hidrógeno; OH^- es la base conjugada del ácido agua *(19.1)*

conversion factor/factor de conversión razón de medidas equivalentes usadas para convertir una cantidad de una unidad a otra *(3.3)*

coordinate covalent bond/enlace covalente coordinado enlace covalente en el que un átomo aporta dos electrones de enlace *(8.2)*

coordination number/número de coordinación número de iones de carga opuesta que rodean a cada ion en un cristal *(7.2)*

covalent bond/enlace covalente enlace que se forma cuando dos átomos comparten electrones *(8.1)*

cracking/pirólisis proceso controlado por el cual los hidrocarburos se descomponen o reacomodan para obtener moléculas más pequeñas y útiles *(22.5)*

crystal/cristal sólido en el que los átomos, iones o moléculas están dispuestos en un patrón tridimensional ordenado y repetitivo llamado red cristalina *(13.3)*

cyclic hydrocarbon/hidrocarburo cíclico compuesto orgánico que contiene un anillo de hidrocarburo *(22.4)*

Dalton's atomic theory/teoría atómica de Dalton primera teoría en relacionar los cambios químicos con sucesos a nivel atómico *(4.1)*

Dalton's law of partial pressures/teoría de Dalton de las presiones parciales a volumen y temperatura constantes, la presión total ejercida por una mezcla de gases es igual a la suma de las presiones parciales de los gases componentes *(14.4)*

decomposition reaction/reacción de descomposición cambio químico en el que un solo compuesto se descompone en dos o más productos más simples *(11.2)*

dehydrogenation reaction/reacción de deshidrogenación reacción en la que se pierde hidrógeno *(23.3)*

deliquescent/delicuescente término que describe una sustancia que absorbe suficiente humedad del aire como para formar una solución; la solución formada tiene una presión de vapor más baja que la de la humedad del aire *(15.2)*

denatured alcohol/alcohol desnaturalizado etanol al que se ha añadido una sustancia venenosa para hacerlo tóxico *(23.2)*

density/densidad razón de la masa de un objeto a su volumen *(3.4)*

dependent variable/variable dependiente *véase* variable de respuesta

desiccant/desecante sustancia higroscópica empleada como agente secante *(15.2)*

diatomic molecule/molécula diatómica molécula que consta de dos átomos *(8.1)*

diffusion/difusión tendencia de las moléculas a moverse hacia áreas de concentración más baja hasta que la concentración es uniforme en todo el medio *(14.4)*

dilute solution/solución diluida solución que contiene muy poco soluto *(16.2)*

dimensional analysis/análisis dimensional técnica para resolver problemas que se apoya en las unidades de las mediciones para resolver el problema *(3.3)*

dipole/dipolo molécula que tiene dos polos o regiones de carga opuesta *(8.4)*

dipole interaction/interacción dipolar fuerzas intermoleculares que resultan de la atracción de regiones de moléculas polares que tienen cargas opuestas *(8.4)*

diprotic acid/ácido diprótico cualquier ácido que contenga dos protones (iones hidrógeno) ionizables; el ácido sulfúrico (H_2SO_4) es un ácido diprótico *(19.1)*

disaccharide/disacárido carbohidrato formado por dos unidades de monosacárido; el azúcar de mesa común (sacarosa) es un disacárido *(24.2)*

dispersion forces/fuerzas de dispersión atracciones entre moléculas que se dan cuando el movimiento de los electrones de una molécula afecta el movimiento de los electrones de la otra mediante fuerzas eléctricas; se trata de las interacciones más débiles entre moléculas *(8.4)*

displacement reaction/reacción de desplazamiento *véase* reacción de sustitución sencilla

distillation/destilación proceso empleado para separar sólidos disueltos de un líquido, el cual se hace hervir para producir un vapor que después se condensa para convertirlo en líquido *(2.2)*

double covalent bond/enlace covalente doble enlace en el que dos átomos comparten dos pares de electrones *(8.2)*

double-replacement reaction/reacción de sustitución doble cambio químico que implica un intercambio de iones positivos entre dos compuestos *(11.2)*

dry cell/pila seca celda voltaica comercial en la que, a pesar del nombre, el electrolito es una pasta húmeda; las baterías compactas y portátiles que se usan en las linternas son pilas secas *(21.1)*

effloresce/eflorecerse perder agua de hidratación; el proceso se presenta cuando la presión de vapor del hidrato es más alta que la del vapor de agua en el aire *(15.2)*

effusion/efusión proceso en el cual un gas escapa por un agujero diminuto en su recipiente *(14.4)*

electrical potential/potencial eléctrico capacidad de una celda voltaica para producir corriente eléctrica *(21.2)*

electrochemical cell/celda electroquímica cualquier dispositivo que convierte energía química en energía eléctrica o energía eléctrica en energía química *(21.1)*

electrochemical process/proceso electroquímico conversión de energía química en energía eléctrica o energía eléctrica en energía química; en todos los procesos electroquímicos intervienen reacciones redox *(21.1)*

electrode/electrodo en un circuito, un conductor que transporta electrones hacia o desde una sustancia que no es un metal *(21.1)*

electrolysis/electrolisis proceso en el que se usa energía eléctrica para realizar un cambio químico; la electrolisis del agua produce hidrógeno y oxígeno *(21.3)*

electrolyte/electrolito compuesto que conduce una corriente eléctrica cuando está en solución acuosa o derretido; todos los compuestos iónicos son electrolitos, pero muy pocos compuestos covalentes lo son *(15.2)*

electrolytic cell/celda electrolítica celda electroquímica que se usa para efectuar un cambio químico mediante la aplicación de energía eléctrica *(21.3)*

electromagnetic radiation/radiación electromagnética ondas de energía que viajan en el vacío a una velocidad de 2.998×10^8 m/s; incluye las ondas de radio, microondas, ondas infrarrojas, luz visible, ondas ultravioleta, rayos X y rayos gamma *(5.3)*

electron/electrón partícula subátomica con carga negativa *(4.2)*

electron configuration/configuración electrónica distribución de los electrones de un átomo en su estado basal, en diversos orbitales alrededor del núcleo del átomo *(5.2)*

electron dot structure/estructura electrón-punto notación que muestra los electrones de valencia como puntos alrededor del símbolo atómico del elemento; el símbolo representa los electrones internos y el núcleo atómico; también se conoce como estructura de puntos de Lewis *(7.1)*

electronegativity/electronegatividad capacidad de un átomo para atraer electrones cuando el átomo está en un compuesto *(6.3)*

element/elemento forma más simple de materia que posee un conjunto único de propiedades; un elemento no puede descomponerse en sustancias más simples usando métodos químicos *(2.3)*

elementary reaction/reacción básica reacción en la que los reactantes se convierten en productos en un solo paso *(18.5)*

empirical formula/fórmula empírica fórmula que muestra las proporciones de los elementos en un compuesto con los números enteros más pequeños posibles; la fórmula empírica del peróxido de hidrógeno (H_2O_2) es HO *(10.3)*

emulsion/emulsión dispersión coloidal de un líquido en otro *(15.3)*

endothermic process/proceso endotérmico proceso en el que se absorbe calor del entorno *(17.1)*

end point/punto final punto de una valoración química en que el indicador cambia de color *(19.4)*

energy/energía capacidad para efectuar trabajo o producir calor *(3.2)*

energy level/nivel energético las energías específicas que puede tener un electrón en un átomo u otro sistema *(5.1)*

enthalpy (H)/entalpía (H) cantidad de calor en un sistema a presión constante *(17.2)*

entropy (S)/entropía (S) medida del desorden de un sistema; los sistemas tienden a pasar de un estado ordenado (baja entropía) a un estado de máximo desorden (alta entropía) *(18.4)*

enzyme/enzima proteína que actúa como catalizador biológico *(24.3)*

enzyme–substrate complex/complejo enzima-sustrato estructura que se forma cuando una molécula de sustrato se une a una enzima en su sitio activo *(24.3)*

equilibrium constant (K_{eq})/constante de equilibrio (K_{eq}) razón de las concentraciones de los productos a las concentraciones de los reactantes en el equilibrio, con cada concentración elevada a una potencia igual al número de moles de esa sustancia en la ecuación química balanceada *(18.2)*

equilibrium position/posición de equilibrio las concentraciones relativas de reactantes y productos de una reacción que ha alcanzado el equilibrio; indica si se favorecen los reactantes o productos en la reacción reversible *(18.2)*

equivalence point/punto de equivalencia punto de una valoración química en la que el número de moles de iones hidrógeno es igual al número de moles de iones hidróxido *(19.4)*

error/error diferencia entre el valor aceptado y el valor experimental *(3.1)*

ester/éster derivado de un ácido carboxílico en el que el —OH del grupo carboxilo ha sido sustituido por el —OR de un alcohol; la fórmula general es RCOOR *(23.3)*

ether/éter compuesto orgánico en el que el oxígeno está unido a dos grupos carbono; la fórmula general es R—O—R *(23.2)*

evaporation/evaporación vaporización que se da en la superficie de un líquido que no está en ebullición *(13.2)*

excess reagent/reactivo excesivo reactivo que está presente en una cantidad más que suficiente para reaccionar con un reactivo limitante; cualquier reactante que queda después de que se ha usado todo el reactivo limitante en una reacción química *(12.3)*

exothermic process/proceso exotérmico proceso en el que se desprende calor hacia el entorno *(17.1)*

experiment/experimento procedimiento repetido que sirve para probar una hipótesis *(1.3)*

experimental value/valor experimental valor cuantitativo que se mide durante un experimento *(3.1)*

extensive property/propiedad extensiva propiedad que depende de la cantidad de materia en una muestra *(2.1)*

fatty acid/ácido graso nombre que se da a los ácidos carboxílicos de cadena continua que se aislaron originalmente de las grasas *(23.3)*

fermentation/fermentación producción de etanol a partir de azúcares por la acción de levaduras o bacterias *(23.2)*

film badge/placa de película dispositivo para detectar radiación que consta de varias capas de película fotográfica cubiertas con papel negro opaco, todo envuelto en un estuche de plástico o metal *(25.4)*

filtration/filtración proceso para separar un sólido de un líquido en una mezcla heterogénea *(2.2)*

first-order reaction/reacción de primer orden reacción cuya velocidad de reacción es proporcional a la concentración de un solo reactante *(18.5)*

fission/fisión división de un núcleo en fragmentos más pequeños, acompañada por desprendimiento de neutrones y una gran cantidad de energía *(25.3)*

formula unit/unidad de fórmula razón más baja, expresada en números enteros, de los iones de un compuesto iónico; en el cloruro de magnesio, la razón de iones magnesio a iones cloruro es de 1:2, así que la unidad de fórmula es $MgCl_2$ *(7.2)*

free energy/energía libre energía que está disponible para realizar trabajo *(18.4)*

freezing-point depression/disminución del punto de congelación diferencia de temperatura entre el punto de congelación de una solución y el del disolvente puro *(16.3)*

frequency (ν)/frecuencia (ν) número de ciclos de onda que pasan por un punto específico en la unidad de tiempo; la frecuencia y la longitud de onda son inversamente proporcionales *(5.3)*

fuel cell/celda de combustible celda voltaica que no necesita recargarse; el combustible se oxida para producir un suministro continuo de energía eléctrica *(21.1)*

functional group/grupo funcional distribución específica de átomos en un compuesto orgánico que puede participar en reacciones químicas características; la química de un compuesto orgánico está determinada por sus grupos funcionales *(23.1)*

fusion/fusión proceso en el que se combinan núcleos para producir un núcleo con mayor masa *(25.3)*

G

gamma ray/rayo gamma fotón de alta energía emitido por un radioisótopo *(25.1)*

gas/gas estado de la materia que adopta la forma y el volumen del recipiente que la contiene; los gases no tienen forma ni volumen definidos *(2.1)*

gas pressure/presión de gas resultado de fuerza que ejerce un gas por unidad de área total de un objeto; se debe a los choques de las partículas de gas contra el objeto *(13.1)*

Gay-Lussac's law/ley de Gay-Lussac la presión de un gas es directamente proporcional a su temperatura Kelvin si se mantiene constante el volumen *(14.2)*

Geiger counter/contador Geiger dispositivo que usa un tubo de metal lleno de gas para detectar radiación *(25.4)*

gene/gen segmento de ADN que contiene el código para una sola cadena péptida *(24.5)*

genetic code/código genético distribución de bases en las palabras de código del ADN; se usan 61 tripletes de palabras de código para especificar los aminoácidos de las proteínas, además de 3 tripletes de palabras de código que indican "parar" una vez que termina la síntesis de proteínas *(24.5)*

geometric isomers/isómeros geométricos compuestos que tienen átomos en el mismo orden, pero difieren en la orientación de los grupos alrededor de un enlace doble *(22.3)*

Gibbs free-energy change (*ΔG*)/cambio de energía libre de Gibbs (*ΔG*) cantidad máxima de energía que se puede acoplar a otro proceso para efectuar trabajo útil *(18.4)*

glass/vidrio producto transparente que resulta de la fusión de materiales inorgánicos que se han enfriado hasta solidificarse sin cristalizarse *(13.3)*

Graham's laws of effusion/leyes de efusión de Graham la velocidad de efusión de un gas es inversamente proporcional a la raíz cuadrada de su masa molar; esta relación también se cumple en la difusión de gases *(14.4)*

gram (g)/gramo (g) unidad métrica de masa igual a la masa de 1 cm^3 de agua a 4°C *(3.2)*

ground state/estado fundamental energía más baja que puede tener un átomo descrito por la mecánica cuántica *(5.3)*

group/grupo columna vertical de elementos en la tabla periódica; los elementos de un grupo tienen propiedades físicas y químicas similares *(4.3)*

half-cell/semicelda parte de una celda voltaica en la que se lleva a cabo la oxidación o reducción; consta de un solo electrodo sumergido en una solución de sus iones *(21.1)*

half-life ($t_{1/2}$)/semivida ($t_{1/2}$) tiempo que tarda en desintegrarse la mitad de los núcleos de una muestra de un radioisótopo *(25.2)*

half-reaction/semirreacción ecuación que muestra la oxidación o bien la reducción que se da en una reacción redox *(20.3)*

half-reaction method/método de semirreacción método para balancear una ecuación redox equilibrando por separado las semirreacciones de oxidación y reducción antes de combinarlas para obtener una ecuación redox balanceada *(20.3)*

halide ion/ion haluro ion negativo que se forma cuando un átomo de halógeno gana un electrón *(7.1)*

halocarbon/compuesto halocarbonado cualquier miembro de una clase de compuestos orgánicos que contienen flúor, cloro, bromo o yodo unidos mediante enlaces covalentes *(23.1)*

halogen/halógeno no metal del grupo 7A de la tabla periódica *(6.2)*

heat (*q*)/calor (*q*) energía que fluye de un objeto a otro debido a la diferencia de temperaturas entre los objetos *(17.1)*

heat capacity/capacidad calorífica cantidad de calor necesaria para elevar exactamente 1°C la temperatura de un objeto *(17.1)*

heat of combustion/calor de combustión calor de reacción al quemarse totalmente un mol de una sustancia *(17.2)*

heat of reaction/calor de reacción cambio de calor correspondiente a una ecuación química en la forma exacta en que está escrita *(17.2)*

Heisenberg uncertainty principle/principio de incertidumbre de Heisenberg es imposible conocer con exactitud la velocidad y la posición de una partícula al mismo tiempo *(5.3)*

Henry's law/ley de Henry a una temperatura determinada, la solubilidad de un gas en un líquido es directamente proporcional a la presión del gas sobre el líquido *(16.1)*

hertz (Hz)/hertz (Hz) unidad de frecuencia, igual a un ciclo por segundo *(5.3)*

Hess's law of heat summation/ley de Hess de la suma de los calores ley según la cual, si se suman dos o más ecuaciones termoquímicas para obtener una ecuación final, también se suman los calores de reacción para obtener el calor de reacción final *(17.4)*

heterogeneous mixture/mezcla heterogénea mezcla cuya composición no es uniforme; sus componentes no están distribuidos de forma equitativa en toda la mezcla *(2.2)*

homogeneous mixture/mezcla homogénea mezcla cuya composición es uniforme; sus componentes están distribuidos de forma equitativa y no es fácil distinguirlos *(2.2)*

homologous series/serie homóloga grupo de compuestos en el que se observa un incremento constante de cambio en la estructura molecular de un compuesto al siguiente *(22.1)*

Hund's rule/regla de Hund los electrones ocupan orbitales de la misma energía haciendo que el número de electrones cuyo espín tiene la misma dirección sea lo más grande posible *(5.2)*

hybridization/hibridización combinación de varios orbitales atómicos para formar el mismo número total de orbitales híbridos equivalentes *(8.3)*

hydrate/hidrato compuesto que tiene un número específico de moléculas de agua unidas a cada unidad de fórmula *(15.2)*

hydration reaction/reacción de hidratación reacción en la que se añade agua a un alqueno *(23.2)*

hydrocarbon/hidrocarburo compuesto orgánico que contiene sólo carbono e hidrógeno *(22.1)*

hydrogenation reaction/reacción de hidrogenación reacción en la que se añade hidrógeno a un doble enlace carbono–carbono para dar un alcano *(23.2)*

hydrogen bonds/enlaces de hidrógeno fuerzas de atracción en las que un átomo de hidrógeno, unido por un enlace covalente a un átomo muy electronegativo, también se une débilmente a un par no compartido de electrones de otro átomo electronegativo *(8.4)*

hydrogen-ion acceptor/receptor de iones hidrógeno una base, según la teoría de Brønsted-Lowry; el amoniaco actúa como base cuando acepta iones hidrógeno del agua *(19.1)*

hydrogen-ion donor/donador de iones hidrógeno un ácido, según la teoría de Brønsted-Lowry *(19.1)*

hydronium ion (H_3O^+)/ion hidronio (H_3O^+) ion positivo que se forma cuando una molécula de agua gana un ion hidrógeno *(19.1)*

hydroxide ion (OH^-)/ion hidróxido (OH^-) ion negativo que se forma cuando una molécula de agua pierde un ion hidrógeno *(19.1)*

hydroxyl group/grupo hidroxilo grupo funcional —OH presente en los alcoholes *(23.2)*

hygroscopic/higroscópico se llama así a las sales y otros compuestos que absorben humedad del aire *(15.2)*

hypothesis/hipótesis explicación propuesta para una observación *(1.3)*

ideal gas constant/constante del gas ideal constante de la ley del gas ideal; se representa con el símbolo R y tiene un valor de 8.31 (L·kPa)/(K·mol) *(14.3)*

ideal gas law/ley del gas ideal relación $PV = nRT$, que describe el comportamiento del gas ideal *(14.3)*

immiscible/inmiscible se dice de los líquidos que son insolubles uno en el otro; el aceite y el agua son inmiscibles *(16.1)*

independent variable/variable independiente *véase* variable manipulada

inhibitor/inhibidor sustancia que interfiere la acción de un catalizador *(18.1)*

inner transition metal/metal interno de transición elemento de las series de los lantánidos o los actínidos; el subnivel *s* más alto ocupado y el subnivel *f* cercano de sus átomos contienen electrones; también se llama elemento interno de transición *(6.2)*

inorganic chemistry/química inorgánica estudio de sustancias que, en general, no contienen carbono *(1.1)*

intensive property/propiedad intensiva propiedad que depende del tipo de materia de una muestra, no de la cantidad de materia *(2.1)*

intermediate/intermediario producto de uno de los pasos de un mecanismo de reacción; se convierte en reactante en el siguiente paso *(18.5)*

International System of Units (SI)/Sistema Internacional de Unidades (SI) versión modificada del sistema métrico, adoptada por acuerdo internacional en 1960 *(3.2)*

ion/ion átomo o grupo de átomos que tiene carga positiva o negativa *(6.3)*

ionic bond/enlace iónico atracción electrostática que une a iones con carga opuesta *(7.2)*

ionic compound/compuesto iónico compuesto formado por iones positivos y negativos *(7.2)*

ionization energy/energía de ionización energía necesaria para sacar un electrón de un átomo en su estado gaseoso *(6.3)*

ionizing radiation/radiación ionizante radiación que tiene la energía suficiente para desprender electrones de algunos átomos de una sustancia bombardeada, produciendo así iones *(25.4)*

ion-product constant for water (K_w)/constante de producto de iones del agua (K_w) producto de las concentraciones de iones hidrógeno y de iones hidróxido del agua; es 1×10^{-14} a 25°C *(9.2)*

isomers/isómeros compuestos que tienen la misma fórmula molecular, pero diferentes estructuras moleculares *(22.3)*

isotopes/isótopos átomos del mismo elemento que tienen el mismo número atómico pero diferentes masas atómicas porque tienen distinto número de neutrones *(4.3)*

joule (J)/julio (J) unidad de energía en el SI; 4.184 J equivalen a una caloría *(3.2)*

Kelvin scale/escala Kelvin escala de temperatura en la que el punto de congelación del agua es 273 K, y el de ebullición, 373 K; 0 K es el cero absoluto *(3.2)*

ketone/cetona compuesto orgánico en el que el carbono del grupo carbonilo está unido a otros dos carbonos: la fórmula general es RCOR *(23.3)*

kilogram (kg)/kilogramo (kg) masa de 1 L de agua a 4°C; es la unidad base de masa en el SI *(3.2)*

kinetic energy/energía cinética energía que tienen los objetos de acuerdo con su movimiento *(13.1)*

kinetic theory/teoría cinética teoría que explica los estados de la materia basándose en el concepto de que toda la materia está formada por pequeñas partículas que están en constante movimiento *(13.1)*

law of conservation of energy/ley de conservación de la energía ley según la cual en ningún proceso químico o físico se crea ni se destruye energía *(17.1)*

law of conservation of mass/ley de conservación de la masa en cualquier cambio físico o reacción química, la masa se conserva; la masa no puede crearse ni destruirse *(2.4)*

law of definite proportions/ley de las proporciones definidas en muestras de cualquier compuesto químico, las masas de los elementos siempre están en la misma proporción *(9.5)*

law of disorder/ley del desorden tendencia natural de los sistemas a desplazarse en la dirección de máximo caos o desorden *(18.4)*

law of multiple proportions/ley de las proporciones múltiples siempre que dos elementos forman más de un compuesto, las diferentes masas de un elemento que se combinan con la misma masa del otro elemento están en razón de números enteros pequeños *(9.5)*

Le Châtelier's principle/principio de Le Châtelier cuando se aplica una tensión a un sistema que está en equilibrio dinámico, el sistema cambia a modo de aliviar dicha tensión *(18.2)*

Lewis acid/ácido de Lewis cualquier sustancia capaz de aceptar un par de electrones para formar un enlace covalente *(19.1)*

Lewis base/base de Lewis cualquier sustancia capaz de ceder un par de electrones para formar un enlace covalente *(19.1)*

limiting reagent/reactivo limitante cualquier reactante que se haya consumido primero en una reacción química; determina la cantidad de producto que se puede formar en la reacción *(12.3)*

lipid/lípido miembro de una clase amplia de compuestos orgánicos relativamente insolubles en agua; las grasas, aceites y ceras son lípidos *(24.4)*

liquid/líquido forma de materia que fluye; tiene volumen fijo pero forma indefinida *(2.1)*

liter (L)/litro (L) volumen de un cubo cuyas aristas miden 10 centímetros cada una (1000 cm^3); es la unidad común de volumen en el sistema métrico *(3.2)*

M

macroscopic/macroscópico término que describe el mundo de los objetos suficientemente grandes como para verse a simple vista *(1.2)*

manipulated variable/variable manipulada variable que cambia durante un experimento; también se llama variable independiente *(1.3)*

mass/masa medida de la cantidad de materia contenida en un objeto; la unidad base de masa en el SI es el kilogramo *(2.1)*

mass number/número de masa número total de protones y neutrones que contiene el núcleo de un átomo *(4.3)*

matter/materia todo lo que tiene masa y ocupa espacio *(1.1)*

measurement/medición descripción cuantitativa que incluye tantos números como unidades *(3.1)*

melting point (mp)/punto de fusión (p.f.) temperatura a la que una sustancia cambia del estado sólido al líquido; el punto de fusión del agua es 0°C *(13.3)*

metabolism/metabolismo todas las reacciones químicas llevadas a cabo por los organismos; incluyen reacciones que producen energía (catabolismo) y reacciones que consumen energía (anabolismo) *(24.6)*

metal/metal miembro de una clase de elementos que son buenos conductores del calor y la electricidad; los metales suelen ser dúctiles, maleables y brillantes *(6.1)*

metallic bond/enlace metálico fuerza de atracción que mantiene unidos los átomos de un metal; se debe a la atracción entre los electrones de valencia, que flotan libremente, y los iones metálicos de carga positiva *(7.3)*

metalloid/metaloide elemento cuyas propiedades son similares a las de los metales y de los no metales *(6.1)*

meter (m)/metro (m) unidad base de longitud en el SI *(3.2)*

microscopic/microscópico término que describe el mundo de los objetos que sólo pueden verse con instrumentos amplificadores *(1.2)*

miscible/miscible se les llama así a los líquidos que se disuelven uno en el otro en todas las proporciones *(16.1)*

mixture/mezcla incorporación física de dos o más sustancias que no se combinan químicamente *(2.2)*

molal boiling-point elevation constant (K_b)/constante molal de incremento del punto de ebullición (K_b) cambio en el punto de ebullición de una solución 1-molal de un soluto molecular no volátil *(16.4)*

molal freezing-point depression constant (K_f)/constante molal de disminución del punto de congelación (K_f) cambio en el punto de congelación de una solución 1-molal de un soluto molecular no volátil *(16.4)*

molality (m)/molalidad (m) concentración de soluto en una solución expresada como el número de moles de soluto disueltos en 1 kilogramo (1000 g) de disolvente *(16.4)*

molar heat of condensation (ΔH_{cond})/calor molar de condensación (ΔH_{cond}) cantidad de calor que un mol de vapor desprende al condensarse, convirtiéndose en líquido, a temperatura constante *(17.3)*

molar heat of fusion (ΔH_{fus})/calor molar de fusión (ΔH_{fus}) cantidad de calor que un mol de una sustancia sólida absorbe al fundirse, convirtiéndose en líquido, a temperatura constante *(17.3)*

molar heat of solidification (ΔH_{solid})/calor molar de solidificación (ΔH_{solid}) cantidad de calor que un mol de un líquido pierde al solidificarse a temperatura constante *(17.3)*

molar heat of solution (ΔH_{soln})/calor molar de disolución (ΔH_{soln}) cambio de calor debido a la disolución de un mol de una sustancia *(17.3)*

molar heat of vaporization (ΔH_{vap})/calor molar de vaporización (ΔH_{vap}) cantidad de calor absorbida por un mol de un líquido al evaporarse a temperatura constante *(17.3)*

molarity (M)/molaridad (M) concentración de soluto en una solución expresada como el número de moles de soluto disueltos en 1 litro de solución *(16.2)*

molar mass/masa molar término empleado para referirse a la masa de un mol de cualquier sustancia *(10.1)*

molar volume/volumen molar volumen ocupado por 1 mol de un gas a temperatura y presión estándar (TPE); 22.4 L *(10.2)*

mole (mol)/mol cantidad de una sustancia que contiene 6.02×10^{23} partículas representativas de esa sustancia *(10.1)*

molecular compound/compuesto molecular compuesto formado por moléculas *(8.1)*

molecular formula/fórmula molecular fórmula química de un compuesto molecular que indica los tipos y números de átomos presentes en una molécula de un compuesto *(8.1)*

molecular orbital/orbital molecular orbital que abarca toda la molécula *(8.3)*

molecule/molécula grupo neutro de átomos unidos por enlaces covalentes *(8.1)*

mole fraction/fracción molar razón de los moles de soluto en solución al número total de moles de disolvente y soluto *(16.4)*

mole ratio/razón molar factor de conversión derivado de los coeficientes de una ecuación química equilibrada interpretada en términos de moles *(12.2)*

monatomic ion/ion monoatómico un solo átomo con carga positiva o negativa debida a la pérdida o ganancia de uno o más electrones de valencia *(9.1)*

monomer/monómero molécula sencilla que se combina repetidamente para formar un polímero *(23.4)*

monoprotic acid/ácido monoprótico ácido que sólo contiene un protón (ion hidrógeno) ionizable; el ácido nítrico (HNO_3) es un ácido monoprótico *(19.1)*

monosaccharide/monosacárido carbohidrato que consta de una sola unidad de azúcar; también llamado azúcar simple *(24.2)*

N

net ionic equation/ecuación iónica neta ecuación de una reacción en solución que sólo muestra las partículas que intervienen directamente en el cambio químico *(11.3)*

network solid/sólido en cadena sólido en el que todos los átomos están unidos entre sí por enlaces covalentes *(8.4)*

neutralization reaction/reacción de neutralización reacción en la que un ácido y una base reaccionan en solución acuosa para producir una sal y agua *(19.4)*

neutral solution/solución neutral solución acuosa en la que las concentraciones de iones hidrógeno y iones hidróxido son iguales; tiene un pH de 7.0 *(19.2)*

neutron/neutrón partícula subatómica sin carga que tiene una masa de 1 uma; se le encuentra en el núcleo de los átomos *(4.2)*

neutron absorption/absorción de neutrones proceso que reduce el número de neutrones lentos en un reactor nuclear; esto se logra mediante el uso de varillas de control hechas con un material como el cadmio, que absorbe neutrones *(25.3)*

neutron activation analysis/análisis por activación de neutrones procedimiento en el que se bombardea una muestra con neutrones de una fuente radiactiva para detectar cantidades muy pequeñas de elementos en la muestra *(25.4)*

neutron moderation/moderación de neutrones proceso empleado en reactores nucleares para frenar los neutrones de modo que el combustible del reactor los capture para continuar la reacción en cadena *(25.3)*

noble gas/gas noble elemento del grupo 8A de la tabla periódica; los subniveles *s* y *p* del nivel energético ocupado más alto están totalmente llenos *(6.2)*

nonelectrolyte/no electrolito compuesto que no conduce una corriente eléctrica en solución acuosa ni en estado fundido *(15.2)*

nonmetal/no metal elemento que suele ser mal conductor del calor y la electricidad; las propiedades de los no metales generalmente son opuestas a las de los metales *(6.1)*

nonpolar covalent bond/enlace covalente no polar enlace covalente en el que los dos átomos comparten equitativamente los electrones *(8.4)*

nonspontaneous reaction/reacción no espontánea reacción que no favorece la formación de productos en las condiciones especificadas *(18.4)*

normal boiling point/punto normal de ebullición el punto de ebullición de un líquido a una presión de 101.3 kPa o 1 atm *(13.2)*

nuclear force/fuerza nuclear fuerza de atracción que actúa entre todas las partículas nucleares que están extremadamente cerca unas de otras, como los protones y los neutrones en un núcleo.

nucleic acid/ácido nucleico polímero de ribonucleótidos (ARN) o desoxirribonucleótidos (ADN) que se encuentra primordialmente en el núcleo de las células; los ácidos nucleicos desempeñan un papel importante en la transmisión de las características hereditarias, en la síntesis de proteínas y en el control de las actividades celulares *(24.5)*

nucleotide/nucleótido uno de los monómeros que constituyen el ADN y el ARN; consiste en una base nitrogenada (una purina o una pirimidina), un azúcar (ribosa o desoxirribosa) y un grupo fosfato *(24.5)*

nucleus/núcleo la diminuta porción central densa de un átomo; se compone de protones y neutrones *(4.2)*

O

observation/observación información obtenida a través de los sentidos; en la ciencia, la observación suele implicar la medición *(1.3)*

octet rule/regla del octeto los átomos reaccionan ganando o perdiendo electrones a modo de adquirir la estructura electrónica estable de un gas noble, que por lo regular consta de ocho electrones de valencia *(7.1)*

optical isomers/isómeros ópticos moléculas que se diferencian entre ellas por la manera como se distribuyen cuatro grupos distintos en torno a un átomo de carbono *(22.3)*

organic chemistry/química orgánica estudio de los compuestos que contienen carbono *(1.1)*

oxidation/oxidación proceso que implica la pérdida total o parcial de electrones o la ganancia de oxígeno; conduce a un aumento en el número de oxidación de un átomo *(20.1)*

oxidation number/número de oxidación número positivo o negativo que se asigna a un átomo para indicar su grado de oxidación o reducción; el número de oxidación de un elemento no combinado es cero *(20.2)*

oxidation-number-change method/método de cambio del número de oxidación método para balancear una ecuación redox comparando los incrementos y reducciones de los números de oxidación *(20.3)*

oxidation–reduction reaction/reacción de oxidación–reducción reacción en la que hay transferencia de electrones entre los reactantes *(20.1)*

oxidizing agent/agente oxidante en una reacción redox, la sustancia que acepta electrones; en la reacción, el agente oxidante se reduce *(20.1)*

partial pressure/presión parcial contribución de cada gas de una mezcla de gases a la presión total *(14.4)*

pascal (Pa)/pascal (Pa) unidad de presión en el SI *(13.1)*

Pauli exclusion principle/principio de exclusión de Pauli orbital atómico puede describir como máximo a dos electrones, los cuales deben tener espín opuesto *(5.2)*

peptide/péptido compuesto orgánico formado por la combinación de aminoácidos de modo que el grupo amino de un ácido se une al grupo carboxilo de otro creando un enlace amida *(24.3)*

peptide bond/enlace péptido enlace que hay entre el grupo carbonilo de un aminoácido y el nitrógeno del siguiente aminoácido de la cadena péptida; la estructura es

(24.3)

percent composition/composición porcentual porcentaje en masa de cada elemento de un compuesto *(10.3)*

percent error/error porcentual porcentaje en que un valor medido difiere del valor aceptado *(3.1)*

percent yield/rendimiento porcentual razón del rendimiento real al rendimiento teórico de una reacción química, expresando como porcentaje; es una medida de la eficiencia de la reacción *(12.3)*

period/periodo fila horizontal de elementos en la tabla periódica *(4.3)*

periodic law/ley periódica si los elementos se acomodan en orden de menor a mayor número atómico, se observa una repetición periódica de sus propiedades físicas y químicas *(6.1)*

periodic table/tabla periódica distribución de los elementos dividiéndolos en grupos según un conjunto de propiedades repetidas *(4.3)*

pH/pH número empleado para denotar la concentración de ion hidrógeno (acidez) de una solución; es el logaritmo negativo de la concentración de ion hidrógeno en una solución *(19.2)*

phase/fase cualquier parte de una muestra que tiene composición y propiedades uniformes *(2.2)*

phase diagram/diagrama de fases gráfica que muestra las condiciones en las que una sustancia existe como sólido, líquido o vapor *(13.4)*

phospholipid/fosfolípido lípido que contiene un grupo fosfato; como los fosfolípidos tienen una cabeza hidrofílica y una cola hidrofóbica, pueden formar las bicapas lípidas de las membranas celulares *(24.4)*

photon/fotón cuanto de luz; paquete discreto de energía electromagnética que interactúa con la materia de forma similar a como lo hacen las partículas *(5.3)*

photosynthesis/fotosíntesis proceso por el cual las plantas y algas verdes aprovechan la energía radiante del sol para sintetizar glucosa a partir de dióxido de carbono y agua *(24.1)*

physical change/cambio físico cambio durante el cual se alteran algunas propiedades de un material, pero sin que se altere la composición del material *(2.1)*

physical chemistry/fisicoquímica área de la química que se relaciona con el mecanismo, la velocidad y la transferencia de energía que ocurre cuando la materia sufre un cambio *(1.1)*

physical property/propiedad física cualidad o condición de una sustancia que se puede observar o medir sin alterar la composición de la sustancia *(2.1)*

pi bond (π bond)/enlace pi (enlace π) enlace covalente en el que hay una alta probabilidad de encontrar los electrones de enlace en regiones alargadas que están arriba y abajo del eje de enlace de los átomos enlazados *(8.3)*

polar covalent bond (polar bond)/enlace covalente polar (enlace polar) enlace covalente entre átomos que no comparten equitativamente sus electrones *(8.4)*

polar molecule/molécula polar molécula que tiene un lado ligeramente negativo y el otro ligeramente positivo *(8.4)*

pollutant/contaminante material presente en el aire, agua o suelos y que es perjudicial para el ser humano y otros organismos *(1.2)*

polyatomic ion/ion poliatómico grupo fuertemente enlazado de átomos, que se comporta como una unidad y tiene carga positiva o negativa *(8.2)*

polymer/polímero molécula muy grande formada por la unión, mediante enlaces covalentes, de moléculas pequeñas repetidas, llamadas monómeros *(23.4)*

polypeptide/polipéptido péptido que tiene más de 10 aminoácidos *(24.3)*

polysaccharide/polisacárido carbohidrato complejo formado por el encadenamiento de muchos monómeros de monosacárido; el almidón, el glucógeno y la celulosa son polisacáridos *(24.2)*

positron/positrón partícula con la misma masa que un electrón pero con carga positiva *(25.2)*

precipitate/precipitado sólido que se forma a partir de una mezcla líquida y se asienta *(2.4)*

precision/precisión cifra que describe la variabilidad de una serie de mediciones efectuadas en las mismas condiciones *(3.1)*

product/producto sustancia que se obtiene en una reacción química *(2.4)*

protein/proteína cualquier péptido que tiene más de 100 aminoácidos *(24.3)*

proton/protón partícula subatómica con carga positiva que se encuentra en el núcleo de los átomos *(4.2)*

pure chemistry/química pura búsqueda de conocimientos químicos por sí mismos *(1.1)*

pure substance/sustancia pura *véase* sustancia

quantum/cuanto cantidad de energía necesaria para desplazar un electrón de un nivel energético a otro *(5.1)*

quantum mechanical model/modelo según la mecánica cuántica descripción moderna, primordialmente matemática, del comportamiento de los electrones en los átomos *(5.1)*

radiation/radiación rayos y partículas penetrantes emitidos por una fuente radiactiva *(25.1)*

radioactivity/radiactividad proceso por el cual los núcleos emiten partículas y rayos *(25.1)*

radioisotope/radioisótopo isótopo cuyo núcleo es inestable y sufre desintegración radiactiva *(25.1)*

rate/velocidad (de reacción) cifra que describe la velocidad de cambio a lo largo de un intervalo de tiempo *(18.1)*

rate law/ley de velocidad de reacción expresión que relaciona la velocidad de una reacción con la concentración de los reactantes *(18.5)*

reactant/reactante sustancia presente al principio de una reacción *(2.4)*

reaction mechanism/mecanismo de reacción serie de reacciones básicas que se dan durante el curso de una reacción compleja *(18.5)*

redox reaction/reacción redox otro nombre que se da a la reacción de oxidación-reducción *(20.1)*

reducing agent/agente reductor en una reacción redox, la sustancia que cede electrones; en la reacción, el agente reductor se oxida *(20.1)*

reduction/reducción proceso que implica una ganancia total o parcial de electrones o pérdida de oxígeno; provoca una disminución en el número de oxidación de un átomo *(20.1)*

reduction potential/potencial de reducción medida de la tendencia que tiene una semirreacción específica de efectuarse como reducción (con ganancia de electrones) en una celda electroquímica *(21.2)*

representative element/elemento representativo elemento de un grupo "A" de la tabla periódica; en conjunto, estos elementos exhiben una amplia gama de propiedades físicas y químicas. En sus átomos, los subniveles *s* y *p* del nivel energético ocupado más alto están parcialmente llenos *(6.2)*

representative particle/partícula representativa unidad más pequeña en que puede dividirse una sustancia sin que cambie su composición; por lo regular es un átomo, molécula o ion *(10.1)*

resonance structure/estructura de resonancia de las dos o más estructuras electrón-punto igualmente válidas de una molécula o ion poliatómico *(8.2)*

responding variable/variable de respuesta variable que se observa durante un experimento; también llamada variable dependiente *(1.3)*

reversible reaction/reacción reversible reacción en la que se da en forma simultánea la conversión de reactantes en productos y la conversión de productos en reactantes *(18.2)*

S

salt bridge/puente salino tubo que contiene un electrolito fuerte y se usa para separar las semiceldas de una celda voltaica; permite el paso de iones de una semicelda a la otra, pero impide que las soluciones se mezclen totalmente *(23.1)*

salt hydrolysis/hidrólisis de sales proceso por el cual los cationes o aniones de una sal disociada aceptan iones hidrógeno del agua o ceden iones hidrógeno al agua *(19.5)*

saponification/saponificación hidrólisis de grasas o aceites con una solución acuosa caliente de un hidróxido de metal alcalino; los jabones se hacen mediante la saponificación *(24.4)*

saturated compound/compuesto saturado compuesto orgánico en el que todos los átomos de carbono están unidos unos a otros por enlaces covalentes sencillos; contiene el número máximo de átomos de hidrógeno por átomo de carbono *(22.2)*

saturated solution/solución saturada solución que contiene la cantidad máxima de soluto para una cantidad dada de disolvente a temperatura y presión constantes; existe equilibrio entre el soluto no disuelto y los iones en solución. *(16.1)*

scientific law/ley científica expresión concisa que resume los resultados de muchas observaciones y experimentos *(1.3)*

scientific method/método científico enfoque lógico y sistemático para resolver un problema científico; los pasos del método científico incluyen hacer observaciones, probar hipótesis y desarrollar teorías *(1.3)*

scientific notation/notación científica convención por la cual los números se expresan en la forma $m \times 10^n$, donde m es un número mayor o igual que 1 y menor que 10, y n es un entero *(3.1)*

scintillation counter/contador de centelleo dispositivo que utiliza una superficie recubierta con un material fosforescente para detectar radiación *(25.4)*

self-ionization/autoionización reacción en la que dos moléculas de agua reaccionan para producir iones *(19.2)*

sigma bond (σ bond)/enlace sigma (enlace σ) enlace que se forma cuando dos orbitales atómicos se combinan para formar un orbital molecular que es simétrico respecto al eje que conecta a los dos núcleos atómicos *(8.3)*

significant figures/dígitos significativos todos los dígitos de una medición que se pueden conocer con precisión, más un último dígito estimado *(3.1)*

single covalent bond/enlace covalente sencillo enlace que se forma cuando dos átomos comparten un par de electrones *(8.2)*

single-replacement reaction/reacción de sustitución sencilla cambio químico en el que un elemento reemplaza a un segundo elemento en un compuesto; también llamado reacción de desplazamiento *(11.2)*

skeleton equation/ecuación esqueleto ecuación química que no indica las cantidades relativas de los reactantes y productos *(11.1)*

solid/sólido estado de la materia que tiene forma y volumen definidos *(2.1)*

solubility/solubilidad cantidad de una sustancia que se disuelve en una cantidad dada de disolvente en condiciones específicas de temperatura y presión para producir una solución saturada *(16.1)*

solubility product constant (K_{sp})/constante del producto de solubilidad (K_{sp}) constante de equilibrio aplicada a la capacidad disoluble de electrolitos; es igual al producto de las concentraciones de los iones, cada una elevada a una potencia igual al coeficiente que tiene ese ion en la ecuación de disociación *(18.3)*

solute/soluto partículas disueltas en una solución *(15.2)*

solution/solución mezcla homogénea que consiste en solutos disueltos en un disolvente *(2.2)*

solvation/solvatación proceso que tiene lugar cuando se disuelve un soluto iónico; en solución, las moléculas de disolvente rodean a los iones positivos y negativos *(15.2)*

solvent/disolvente medio dispersor en una solución *(15.2)*

specific heat/calor específico cantidad de calor requerida para elevar 1°C la temperatura de 1 g de una sustancia *(17.1)*

specific rate constant/constante específica de velocidad de reacción constante de proporcionalidad que relaciona las concentraciones de los reactantes con la velocidad de la reacción *(18.5)*

spectator ion/ion espectador ion que no interviene directamente en una reacción química; ion que no cambia de número de oxidación ni de composición durante una reacción *(11.3)*

spectrum/espectro longitudes de onda de la luz visible que se separan cuando un haz de luz atraviesa un prisma; gama de longitudes de onda de radiación electromagnética *(5.3)*

spontaneous reaction/reacción espontánea reacción que favorece la formación de productos en las condiciones especificadas; la espontaneidad depende de los cambios de entalpía y de entropía *(18.4)*

standard atmosphere (atm)/atmósfera estándar (atm) unidad de presión; es la presión necesaria para sostener 760 mm de mercurio en un barómetro de mercurio a 25°C *(13.1)*

standard cell potential (E^0_{cel})/potencial estándar de celda (E^0_{cel}) potencial de celda que se mide cuando las concentraciones de los iones en las semiceldas son 1.00M a 1 atm de presión y 25°C *(21.2)*

standard heat of formation (ΔH_f^0)/calor estándar de formación (ΔH_f^0) cambio de entalpía que acompaña a la formación de un mol de un compuesto a partir de sus elementos, estando todas las sustancias en su estado estándar a 25°C *(17.4)*

standard hydrogen electrode/electrodo estándar de hidrógeno electrodo (semicelda) arbitrario de referencia que se usa junto con otro electrodo (semicelda) para medir el potencial estándar de reducción de esa celda; se asigna al potencial estándar de reducción del electrodo de hidrógeno el valor de 0.00 V *(21.2)*

standard solution/solución estándar solución cuya concentración se conoce; se usa para efectuar valoraciones químicas *(19.4)*

standard temperature and pressure (STP)/temperatura y presión estándar (TPE) las condiciones en las que normalmente se mide el volumen de un gas; la temperatura estándar es 0°C y la presión estándar es 101.3 kPa, o sea, 1 atmósfera (atm) *(10.2)*

stereoisomers/estereoisómeros moléculas cuyos átomos están en el mismo orden, pero que difieren en la distribución de los átomos en el espacio *(22.3)*

stoichiometry/estequiometría rama de la química que se ocupa de las relaciones numéricas en las ecuaciones químicas; el cálculo de las cantidades de sustancias presentes en las ecuaciones químicas *(12.1)*

straight-chain alkane/alcano de cadena lineal hidrocarburo saturado que contiene cualquier número de átomos de carbono acomodados uno tras otro en una cadena *(22.1)*

strong acid/ácido fuerte ácido que se ioniza casi totalmente en solución acuosa *(19.3)*

strong base/base fuerte base que se disocia totalmente en una solución acuosa para dar iones metálicos y iones hidróxido *(19.3)*

strong electrolyte/electrolito fuerte solución en la que una porción considerable del soluto existe en forma de iones *(15.2)*

structural formula/fórmula estructural fórmula química que indica la distribución de los átomos en una molécula o ion poliatómico; cada raya entre un par de átomos indica un par de electrones compartidos *(8.2)*

structural isomers/isómeros estructurales compuestos que tienen la misma fórmula molecular, aunque sus átomos están enlazados en un orden distinto *(22.3)*

sublimation/sublimación proceso por el cual un sólido cambia a gas o vapor sin pasar por el estado líquido *(13.4)*

substance/sustancia materia que tiene una composición uniforme y definida; puede ser un elemento o un compuesto; también llamada sustancia pura *(2.1)*

substituent/sustituto átomo o grupo de átomos que puede ocupar el lugar de un átomo de hidrógeno en una molécula precursora de hidrocarburo *(22.1)*

substitution reaction/reacción de sustitución tipo común de reacción orgánica; implica el reemplazo de un átomo o grupo de átomos por otro átomo o grupo de átomos *(23.1)*

substrate/sustrato molécula sobre la que actúa una enzima *(24.3)*

supersaturated solution/solución sobresaturada solución que contiene más soluto del que en teoría puede contener a una temperatura específica; el soluto en exceso se precipita si se añade un cristal que actúa como semilla *(16.1)*

surface tension/tensión superficial fuerza que tiende a reducir al mínimo la superficie total de un líquido y actúa hacia el seno de éste; hace que la superficie se comporte como si fuera una membrana elástica *(15.1)*

surfactant/tensoactivo cualquier sustancia que perturba la formación de enlaces de hidrógeno entre las moléculas de agua y así reduce la tensión superficial; los jabones y detergentes son tensoactivos *(15.1)*

surroundings/entorno todo lo que no forma parte del sistema, es decir, el resto del universo *(17.1)*

suspension/suspensión mezcla de la que se separan lentamente algunas partículas por asentamiento cuando no se agita *(15.3)*

synthesis reaction/reacción de síntesis *véase* reacción de combinación

system/sistema parte del universo en la que centramos nuestra atención *(17.1)*

technology/tecnología los medios por los cuales una sociedad proporciona a sus miembros las cosas que necesitan y desean *(1.1)*

temperature/temperatura medida de la energía cinética promedio de las partículas de la materia; la temperatura determina la dirección de la transferencia de calor *(3.2)*

tetrahedral angle/ángulo tetraédrico ángulo de enlace de 109.5° que se forma cuando un átomo central forma cuatro enlaces dirigidos hacia el centro de un tetraedro regular *(8.3)*

theoretical yield/rendimiento teórico cantidad de producto que podría formarse durante una reacción, calculada a partir de una ecuación química balanceada; representa la cantidad máxima de producto que podría formarse a partir de una cantidad determinada de reactantes *(12.3)*

theory/teoría explicación, probada exhaustivamente, de un conjunto amplio de observaciones *(1.3)*

thermochemical equation/ecuación termoquímica ecuación química que incluye el cambio de calor *(17.2)*

thermochemistry/termoquímica estudio de los cambios de calor que acompañan a las reacciones químicas y a los cambios de estado físico *(17.1)*

titration/valoración química proceso empleado para determinar la concentración de una solución (a menudo un ácido o base) por el cual una solución de concentración conocida (solución estándar) se añade a una cantidad medida de una solución cuya concentración se desconoce, hasta que un indicador marca el punto final *(19.4)*

trans* configuration/configuración *trans configuración en la que los grupos sustitutos están en lados opuestos de un doble enlace *(22.3)*

transition metal/metal de transición uno de los elementos del grupo B en el que el subnivel *s* ocupado más alto y un subnivel *d* cercano generalmente contienen electrones *(6.2)*

transition state/estado de transición término empleado ocasionalmente para referirse al complejo activado *(18.1)*

transmutation/transmutación conversión de un átomo de un elemento en un átomo de otro elemento *(25.2)*

transuranium element/elemento transuránico cualquier elemento de la tabla periódica cuyo número atómico es mayor que 92, el número atómico del uranio *(25.2)*

triglyceride/triglicérido éster en el que los tres grupos hidroxilo de una molécula de glicerol han sido sustituidos por ácidos grasos de cadena larga; las grasas son triglicéridos *(24.4)*

triple covalent bond/enlace covalente triple enlace covalente en el que dos átomos comparten tres pares de electrones *(8.2)*

triple point/punto triple punto de un diagrama de fases que representa el único conjunto de condiciones en el que las tres fases existen en equilibrio *(13.4)*

triprotic acid/ácido triprótico ácido que contiene tres protones (iones hidrógeno) ionizables; el ácido fosfórico (H_3PO_4) es un ácido triprótico *(19.1)*

Tyndall effect/efecto Tyndall dispersión de la luz por las partículas de un coloide o una suspensión, que hace que un haz de luz se vuelva visible *(15.3)*

unit cell/celda unitaria grupo más pequeño de partículas dentro de un cristal que conserva la forma geométrica del cristal *(13.3)*

unsaturated solution/solución insaturada se dice de una solución que contiene menos soluto que una solución saturada a una temperatura y presión específicas *(16.1)*

unsaturated compound/compuesto insaturado compuesto orgánico que tiene uno o más dobles o triples enlaces carbono–carbono *(22.2)*

unshared pair/par no compartido par de electrones de valencia que no es compartido por dos átomos *(8.2)*

vacuum/vacío espacio en el que no existen partículas de materia *(13.1)*

valence electron/electrón de valencia electrón que está en el nivel energético ocupado más alto de un átomo *(7.1)*

van der Waals forces/fuerzas de van der Waals las dos atracciones intermoleculares más débiles —interacciones de dispersión y fuerzas dipolares *(8.4)*

vapor/vapor estado gaseoso de una sustancia que suele ser líquida o sólida a temperatura ambiente *(2.1)*

vaporization/vaporización conversión de un líquido en gas o vapor *(13.2)*

vapor pressure/presión de vapor medida de la fuerza que ejerce un gas sobre un líquido en un contenedor sellado; equilibrio dinámico que existe entre el vapor y el líquido *(13.2)*

voltaic cell/celda voltaica celda electroquímica empleada para convertir energía química en energía eléctrica; la energía se produce por una reacción redox espontánea *(21.1)*

volume/volumen medida del espacio ocupado por una muestra de materia *(2.1)*

VSEPR theory/teoría RPENV teoría de repulsión de pares de electrones del nivel de valencia; como los pares de electrones se repelen, las moléculas ajustan su forma de modo que los pares de electrones de valencia estén lo más alejados posible entre sí *(8.3)*

water of hydration/agua de hidratación moléculas de agua que forman parte integral de una estructura cristalina *(15.2)*

wavelength (λ)/longitud de onda (λ) distancia entre crestas adyacentes de una onda *(5.3)*

wax/cera éster de un ácido graso de cadena larga y un alcohol de cadena larga *(24.4)*

weak acid/ácido débil ácido que se ioniza poco en solución acuosa *(19.3)*

weak base/base débil base que reacciona con agua para formar el ion hidróxido y el ácido combinado de la base *(19.3)*

weak electrolyte/electrolito débil solución que apenas conduce la electricidad porque sólo una fracción del soluto existe en forma de iones *(15.2)*

weight/peso fuerza que mide la atracción de la gravedad sobre una masa específica *(3.2)*

Index

The page on which a term is defined is indicated in **boldface** type.

G

I

J

K

R

T

U

Acknowledgments

Photographs

Table of Contents iiiMR, © Charles O'Rear/CORBIS; iiiBR, © Geoffrey Nilsen Photography/Addison-Wesley; vTR, © James Soots Lawrence/Livermore National Laboratory; vB, © Richard Megna/Fundamental Photographs; viL, © Kevin R. Morris/CORBIS; viR, © Douglas E. Walker/ Masterfile; vii, © Runk/Schoenberger/Grant Heilman Photography; viii, © Ken Karp/Addison-Wesley; ixT, © David Bishop/Phototake; ixB, © Jens J. Jensen/CORBIS; x, © Robert George Young/Masterfile; xii, xiii, © Ken O'Donoghue/Prentice Hall; xiv, © Mitch Hrdlicka/Getty Images; xviiiT, © Lon C Diehl/PhotoEdit; xviiiT(inset), © RAL Studio/Stock Connection/PictureQuest; xviiiM, © Derek Berwin/Getty Images; xviiiB, © Michael & Patricia Fogden/ CORBIS; xixT, © Royalty-Free/PictureQuest; xixBL, © Bob Handelman/Getty Images; xixBR, © George Haling/ Photo Researchers

Chapter 1 6, © Andrew Brookes/CORBIS; 7T, Courtesy of NASA; 7B, © George Ranalli/Photo Researchers; 8C, © Mike Powell/Getty Images; 8TR, © Michelle Joyce/Index Stock Imagery; 8MR, © Tom McCarthy/Rainbow; 8BR, © Michael Abbey/Photo Researchers; 8TL, © Alfred Pasieka/ Photo Researchers; 8BL, © Nicholas Veasey/Getty Images; 9L © Tom Bochsler Photography/ Prentice Hall; 9R, © Stewart Cohen/Index Stock Imagery; 10L, © John Lund/Getty Images; 10R, © Raeanne Rubenstein/IndexStock Imagery; 11, © Marc Pokempner/Getty Images; 12T, © Michael Brohm/Courtesy of the Illinois State Museum; 12B, © Dr. Jeremy Burgess/Photo Researchers; 13T, © James Soots Lawrence/Livermore National Laboratory; 13B, 13B (inset), © Thomas Hovland/Grant Heilman Photography; 14, © Alfred Pasieka/Photo Researchers; 15T, © Andy Caulfield/Getty Images; 15B, © Courtesy of Maggie Sliker; 16, Courtesy of U.S. Environmental Protection Agency; 17ALL, Courtesy of NASA; 18C, © Larry Lefever/Grant Heilman Photography; 18TR, © Dorling Kindersley; 18BR, © Arthur C. Smith III/Grant Heilman Photography; 18BL, © Wally Eberhart/Visuals Unlimited; 19TL, © Gustav Verderber/Visuals Unlimited; 19TR, © Doug Wechsler/Animals Animals; 19M, © Norbert Wu/Minden Pictures; 19BL, © Michael & Patricia Fogden/CORBIS; 19BR, © Jack M. Bostrack/Visuals Unlimited; 20T, © Jack M. Bostrack/ Visuals Unlimited; 20B, © Chassenet Bsip/Photo Researchers; 21BOTH, The Art Archive; 22, © Peter Gregoire/IndexStock Imagery; 23, © Ken O'Donoghue/Prentice Hall; 24T, © AFP/CORBIS; 24B, © J. Nouvok/PhotoEdit; 25, © John Abbott Photography; 26-27, © Ken O'Donoghue/Prentice Hall; 26BR, © Richard Megna/ Fundamental Photographs; 28T © Pat Kepic/Addison-Wesley; 28B, © Amy Etra/PhotoEdit; 30, © Andre Jenny/Alamy Images; 35, © Robert E. Daemmrich/Getty Images

Chapter 2 38, © Wendy Chan/Getty Images; 39T, © BananaStock/PictureQuest; 39B, © Prentice Hall; 40L, © Graham Tim/CORBIS SYGMA; 40R, © Sandro Vannini/CORBIS; 41, © Geoffrey Nilsen Photography/Addison-Wesley; 42, © Lawrence Migdale/Photo Researchers; 43TR, © Dennis Kunkel/Phototake; 43BL&BKGD, © Walter Bibikow/Getty Images; 43BR, © Royalty Free/Getty Images; 44T, © Dorling Kindersley; 44B, © David Young-Wolff/PhotoEdit/PictureQuest; 45, © Ken O'Donoghue/Prentice Hall; 46T, © Cheyenne Rouse/IndexStock Imagery; 46B, © John A. Rizzo/PhotoDisc/PictureQuest; 48T, © Evan Sklar/FoodPix/Getty Images; 48B, © David Lawrence/Addison-Wesley; 49L, © Rich Treptow/Photo Researchers; 49L (inset), © David Lawrence/Addison-Wesley; 49M, © Mike Hewitt/Getty Images; 49M (inset), © Yoav Levy/Phototake; 49R, © Art Stein/Photo Researchers; 49R (inset), © Michael Newman/PhotoEdit; 50L, © E. R. Degginger/Color-Pic; 50ML, © Coco McCoy/Rainbow; 50MR, © PhotoDisc/Getty Images; 50R, © Paul Conklin; 51, © Clive Streeter/Dorling Kindersley; 53T, © Cameron Davidson/Getty Images; 53B, © Tom Bochsler/Prentice Hall; 54L, © Chris Pellant/Dorling Kindersley; 54M, © Saturn Stills/Photo Researchers; 54R, Courtesy of Dr. David Fankhauser; 55, © Richard Megna/Fundamental Photographs; 56, © Ken O'Donoghue/Prentice Hall; 59, © Richard Megna/Fundamental Photographs

Chapter 3 62, © Daryl Benson/Masterfile; 63T, Courtesy of NASA; 63B, © Peter Arnold; 64, © Randy Faris/CORBIS; 65, © Richard Hutchings/Addison-Wesley; 66, © PhotoDisc/Getty Images; 68, © PhotoDisc/Getty Images; 69, © F. Hirdes/Masterfile; 70, © Jeff Greenberg/PhotoEdit; 71, © Andy Sacks/Getty Images; 72, © KenO'Donoghue/Prentice Hall; 73T, © David McLain/Aurora & Quanta Productions; 74, © David Zimmerman/CORBIS; 75L, © Ken Karp/Addison-Wesley; 75M, 75R, © David Lawrence/Addison-Wesley; 76, Courtesy of NASA; 77L, © W. Geiersperger/CORBIS; 77M, ©Eric O'Connell/Getty Images; 77R, © Richard Megna/Fundamental Photographs; 78, © Pierre Arsenault/Masterfile; 79, © Phototake; 80T, © William Whitehurst/CORBIS; 81B, © Fran Heyl Associates; 82, © Tony Freeman/PhotoEdit; 84, © Digital Vision/Getty Images; 85, © Topham/The Image Works; 86, © Yorgos Nikas/Getty Images; 87, © Ken O'Donoghue/Prentice Hall; 88ML, R, © Fox & Fowle Architects; 88M, © Andres Groden Photography, Couresty of Fox & Fowle Architects; 88BL, © Alan Levenson/Getty Images; 88BKGD, © Royalty Free/Getty Images ; 89T, © IFA/eStock Photo/PictureQuest; 90, © Clive Streeter/Dorling Kindersley; 91, © Nancy Richmon/The Image Works; 92T, © PhotoDisc/Getty Images; 92B, © Matthew Donaldson/ImageState/PictureQuest; 93, © TEK Image/Photo Researchers; 94, © Ken O'Donoghue/ Prentice Hall

Chapter 4 100, Courtesy of IBM Research, Almaden Reserch Center; 101T, © V.C.L./Getty Images; 101B, © Archivo Iconografico,S.A./CORBIS; 103, Courtesy of IBM Research, Almaden Research Center; 104T, © G. Schuster/Masterfile; 105, © Richard Megna/Fundamental Photographs; 106, © G.R. "Dick" Roberts Photo Library; 108, © Ken O'Donoghue/Prentice Hall; 109TR, © Volker Steger/Peter Arnold; 109MR, © Hank Morgan/Photo Researchers; 109BR, © K.H. Kjeldsen/Photo Researchers; 109BL, © Andrew Syred/Photo Researchers; 110T, © Royalty-Free/CORBIS; 111T, © Tom Bochsler/Prentice Hall; 111B, © Malcolm Piers/Getty Images; 113T, © John Maher/IndexStock Imagery; 113B, © L. Clarke/CORBIS; 114, © Barry L. Runk/Grant Heilman Photography; 115, © Mike Powell/Getty Images; 116, 117, © Charles D. Winters/Photo Researchers; 119, © James L. Amos/CORBIS; 120, © Ken O'Donoghue/Prentice Hall; 122, © Richard Megna/Fundamental Photographs

Chapter 5 126, © David Parker/Photo Researchers; 129T, © Embry Riddle Aeronautical University; 129B, © Phil Degginger/Color-Pic; 130, © Jim Sugar/CORBIS; 133T, © Joe Bator/CORBIS; 135, © Gary Morrison/Getty Images; 136, © Tony Freeman/PhotoEdit; 137, © Ken O'Donoghue/Prentice Hall; 138T, © Richard Pasley/Stock Boston; 140, © Walter Bibikow/Getty Images; 141, 142, © Richard Megna/Fundamental Photographs; 144, © Oliver Meckes/Photo Researchers; 146, © Mauritius, GMBH/Phototake; 147TL, © PhotoDisc/Getty Images; 147TR, © V.C.L./Getty Images; 147B, © Chris Barry/Phototake; 150, © Richard Megna/Fundamental Photographs; 152, © Carolina Biological Supply Company/Phototake

Chapter 6 154, © James Leynse/CORBIS; 155T, © Glamour International/AGE Fotostock America; 155B, © Carey B. Van Loon; 156T, © Gary J. Shulfer/C. Marvin Lang; 156B, © The Granger Collection, New York; 159BL, © David Muench/CORBIS; 159M, © Ken Davies/Masterfile; 159R, © Dorling Kindersley; 160L, Courtesy of Texas Instruments Incorporated; 160R, © Paul Chesley/Getty Images; 161T, © Susan Van Etten/PhotoEdit; 164, © Woodward Payne/Royalty-Free/PictureQuest; 165T, © PhotoDisc/Getty Images; 165T(inset), © John Durham/Photo Researchers; 165B, © U.S. Geological Survey, Denver; 165B (inset), © Color-Pic/Earth Scenes; 167, © Richard L. Carlton/Visuals Unlimited; 168TR, © Roy Rainford/Robert Harding World Imagery; 168ML, © Geoff Dann/Dorling Kindersley; 168MM, © Harry Taylor/Dorling Kindersley; 168MR, © Mike Dunning/Dorling Kindersley; 168B, © Robert Frerck/Getty Images; 169T, © Vital Pictures/Getty Images; 169T (inset), © Ken O'Donoghue/Prentice Hall; 169BL, Jacob Lawrence/Spike Mafford/Francine Seders Gallery; 169BR, © Jeff Greenberg/PhotoEdit; 170T, © SuperStock; 179, Ken O'Donoghue/Prentice Hall

Chapter 7 186, © Scott Camazine/Photo Researchers; 187T, © Robert Maier/Earth Scene; 187BL, © Steve Dunwell/Getty Images; 187BM, © Kelly Harriger/CORBIS; 187BR, © Geoff Dann/Dorling Kindersley; 189, © PhotoDisc/Getty Images; 190, © Louis B. Wallach/Getty Images; 192, © David Samuel Robbins/CORBIS; 193, © Richard Megna/Fundamental Photographs; 194T, © Erwin C. Nielsen/Visuals Unlimited; 195T, © Ken Karp/Addison-Wesley; 195B, © Bruce Iverson/Photo Researchers; 196, © Kevin R. Morris/CORBIS; 197TL, © Runk/Schoenberger/Grant Heilman Photography; 197TM, © Gim Media/Gemological Institute of America; 197TCL, © Breck P. Kent; 197TCM, © J&L Weber/Peter Arnold; 197TCR, © Jeffrey A. Scovil; 197BCL, © J&L Weber/Peter Arnold, 197BCM, © Breck P. Kent/Earth Scenes; 197BCR, © Grace Davies/Omni-Photo Communications; 197BL, © George Whiteley/Photo Researchers; 197BM, © Tomisch/Photo Researchers; 197BL, © Paul Silverman/Fundamental Photographs; 199, 200, © Ken O'Donoghue/Prentice Hall; 201T, © Lowell Georgia/CORBIS; 202T, © David Muscroft/AGE fotostock, © Paul Silverman/Fundamental Photographs; 202BM, © Digital Vision/PictureQuest; 202BR, © Clive Streeter/Dorling Kindersley; 203, © David Young-Wolff/PhotoEdit; 204BL, © J. Lawrence/ImageState/PictureQuest; 204BM, © Digital Vision/Getty Images; 204TR, © AFP/CORBIS; 205BL, © Nicolas Russell/Getty Images; 205, © EyeWire Collection/Getty Images

Chapter 8 212, © Brett Baunton/Getty Images; 213T, © Ken O'Donoghue/Prentice Hall; 213BL, © Ken Karp/Addison-Wesley; 213BR, © Patti McConville/Getty Images; 214BL, © Robert Frerck/Woodfin Camp; 214BR, © Garry Hunter/Getty Images; 215TL, © David Young-Wolff/PhotoEdit; 217T, © NASA/Photo Researchers; 220, © Tony Freeman/PhotoEdit; 221, © EyeWire Collection/Getty Images; 222, © Davies & Starr/Getty Images; 223, © Michael Newman/PhotoEdit; 225, © G. Biss/Masterfile; 226, © Ken O'Donoghue/Prentice Hall; 227, © Ulf Wallin/Getty Images; 228, © Douglas E. Walker/ Masterfile; 230T, © Andrew Sacks/Getty Images; 237T, © Allan Davey/Masterfile; 237B, © David Young-Wolff/PhotoEdit; 239, © Richard Hutchings/PhotoEdit; 241, © Craig Tuttle/CORBIS; 243, © Leloup Vincent/CORBIS; 245, © Ken O'Donoghue/Prentice Hall

Chapter 9 252, © Brian Pieters/Masterfile; 253T, © John Michael/ImageState; 254, Tom Pantages; 255, © Ken Karp/Addison-Wesley; 256T, © Bill Aron/PhotoEdit; 256B, © Brian Hagiwara/Getty Images; 258L, © Alastair Laidlaw/Getty Images; 258M, Klaus Gludbrandsen Photo Researchers, Inc.; 258R, © Jack Fields/Photo Researchers; 259L, © Lon C Diehl/PhotoEdit; 259 (inset), © RAL Studio/Stock Connection/PictureQuest; 260T, © Brian Leatart/Getty Images; 260B, © Lawrence Migdale/Stock Boston; 261L, © Michael Prince/CORBIS; 261R, © PhotoDisc/Getty Images; 262, Weirton Steel Corporation; 263, © José Manuel Sanchis Calvete/CORBIS; 264L, © Andreas Einsiedel/Dorling Kindersley; 264M, © Tony Freeman/PhotoEdit; 264R, 265, © Richard Megna/Fundamental Photographs; 266, © Yoav Lev/Phototake; 267, © Ken O'Donoghue/Prentice Hall; 268T, © DigitalVision/PictureQuest; 268BL, © Kennan Harvey/Getty Images; 268BR, © Lowell Georgia/CORBIS; 270, © Victor Paris/Getty Images; 271T, © James Robinson/Animals Animals; 271B, © Richard

Megna/Fundamental Photographs; 273L, © Joel Benard/ Masterfile; 273M, © Royalty-Free/CORBIS; 273R, © Rosemary Weller/Getty Images, 274T, © Kathleen Finlay/ Masterfile; 274B, © Ken Karp/Addison-Wesley; 276T, © Jonathan Nourok/PhotoEdit; 276B, © Kelly-Mooney Photography/CORBIS; 277, © Richard Megna/Fundamental Photographs; 279, © Ken O'Donoghue/Prentice Hall

Chapter 10 286, © Francis Dean/The Image Works; 287T, © Peter Gridley/Getty Images; 287B, © Steve Gorton/Dorling Kindersley; 288, © Jon Chomitz/Photo Researchers; 289T, © James Stevenson/Dorling Kindersley; 289B, © Sally Boon/FoodPix/Getty Images; 290T,M, © David Lawrence/Addison-Wesley; 290B, © Richard Hutchins/Addison-Wesley; 291, © Ken Karp/Addison-Wesley; 292, © Lon C Diehl/PhotoEdit; 293, Jose B. Ruiz Nature Picture Library; 294, 295, © Tom Pantages; 296, Courtesy of NASA; 297T, © Evan Sklar/Getty Images; 298, © Jean Miele/ Bettmann/CORBIS; 299, © Tom Schierlitz/Getty Images; 300BL, © Richard Megna/Fundamental Photographs; 300BR, © Tom Pantages; 301, 304, © Ken O'Donoghue/Prentice Hall; 305T, © David Lawrence/Addison-Wesley; 306B, © Digital Vision; 306B (inset), ©Jack Wilburn/EarthScenes; 308, © Ken O'Donoghue/Prentice Hall; 309T, ©Bill Barley/SuperStock; 309B, © Royalty-Free/CORBIS; 311, © Tom Pantages; 313TR, © Dennis O'Clair/Getty Images; 318, © Thomas Brase/Getty Images

Chapter 11 320, © Lewis Kemper/Phototake; 321T, © Sam Shere/Hulton Archive Photos/Getty Images, 321B, © Ken O'Donoghue/Prentice Hall; 322L, © Terry Why/Phototake; 322M, © Michael Dalton/Fundamental Photographs; 322R, © Phil Degginger/Color-Pic; 323, © Ken Karp/Addison-Wesley; 324, © Richard Megna/ Fundamental Photographs; 325, © Jeff Greenberg/Omni-Photo Communications; 326, © Ken O'Donoghue/Prentice Hall; 327, © EAC Images/NASA/Aurora & Quanta Productions; 328, © Peter DeLory/Getty Images; 329, © Digital Vision/PictureQuest; 330T, © Charles O'Rear/CORBIS; 331T, David Lawrence/Addison-Wesley; 331B, © Tom Pantages; 332T, © Ken Karp/Addison-Wesley; 332B, © Reuters NewMedia Inc./CORBIS; 333, David Lawrence/Addison-Wesley; 334, © Runk/Schoenberger/Grant Heilman Photography; 335, 336, © Richard Megna/Fundamental Photographs; 337, © Runk/ Schoenberger/Grant Heilman Photography; 338T, © David Lawrence/Addison-Wesley; 338M, © Ken Karp/Addison-Wesley; 338B, © David Lawrence/Addison-Wesley; 339, © Richard Megna/Fundamental Photographs; 340, © Bill Stormont/CORBIS; 341T, © Patrick Durand/CORBIS SYGMA; 341BL, © David Job/Getty Images; 341BL (inset), © Tony Freeman/PhotoEdit; 341BR, © Prentice Hall; 341BR (inset), ©Ana/The Image Works; 342T, © Demetrio Carrasco/Getty Images; 342B, 343, © Richard Megna/Fundamental Photographs; 345, © Ken O'Donoghue/Prentice Hall; 348, © Ken Karp/Addison-Wesley; 349TL, © Carey B. Van Loon; 349TR, © Stephen Frisch/Stock Boston; 349BL, 349BR, © Richard Megna/Fundamental Photographs

Chapter 12 352, © Mitch Hrdlicka/Getty Images; 353T, © Noriyuki Yoshida/ SuperStock; 353B, © Michael Newman/PhotoEdit; 355, © AP/Wide World Photos; 356, © Bill Ross/CORBIS; 358, © D. Cavagnaro/DRK Photo; 359T, © David Woods/ CORBIS; 359B, © Grant Heilman Photography; 360, © Richard Hutchings/Addison-Wesley; 362, © Courtesy NASA; 364, © Runk/Schoenberger/Grant Heilman Photography; 365, © Steve Craft/Masterfile; 367, © Jon Chomitz/Prentice Hall; 368T, © Patrick Ramsey/ImageState; 368B, © Ken O'Donoghue/Prentice Hall; 370, © Ken Karp/Addison-Wesley; 371, © Peter Christopher/Masterfile; 372, © Ken O'Donoghue/Prentice Hall; 373TL, © Courtesy of The Topps Company, Inc.; 373TR, © Blank Archives/Getty Images; 373B, © Richard Hutchings/Addison-Wesley; 374, © PhotoDisc/Getty Images; 381, © AP/Wide World Photos

Chapter 13 384, © Fukuhara/CORBIS; 385T, © Daniel J Cox/Getty Images; 386TR, © Leonard Arnold; 389, © Wolfgang Ketterle/Courtesy of Ketterle Lab; 390T, © G. Brad Lewis/Getty Images; 390B, © Richard Megna/Fundamental Photographs; 391TL, © Phil Degginger/Color-Pic; 391TR, © Dave King/Dorling Kindersley; 396T, © Scott Camazine/Photo Researchers; 397TL, © Joe McDonald/Earth Scene; 397TM, © Arthur R. Hill/Visuals Unlimited; 397TR, © Mark A. Schneider/Photo Researchers; 397BL, 397BML, © Mark A. Schneider/Visuals Unlimited; 397BMR, © Mark A. Schneider/Photo Researchers; 397BR, © Mark A. Schneider/Visuals Unlimited; 400, © Ken O'Donoghue/Prentice Hall; 401T, © Breck P. Kent/Earth Scenes; 403, © Ken Karp/Addison-Wesley; 404, © John Madere/CORBIS; 405TR, © Mason Morfit/Getty Images; 405ML, © AP/Wide World Photos; 405MR, © PhotoDisc/Getty Images; 405BL, © Peter Dazeley/Getty Images; 405BR, © Thinkstock/PictureQuest; 408, © Richard Megna/Fundamental Photographs

Chapter 14 412, © Federico Cabello/SuperStock; 413T, © Mike Hewitt/Getty Images; 413B, © Romilly Lockyer/Getty Images; 415, © Bill Bachman/Photo Researchers; 418T, © Trevor Collens-Pool/Getty Images; 420, © Ken Karp/Addison-Wesley; 422, © Bonnie Kamin/PhotoEdit; 425, © NASA/Photo Researchers; 426T, © Charles D. Winters/Photo Researchers; 427, © Robert Garvey/CORBIS; 428T, © Ken O'Donoghue/Prentice Hall; 428B, © Richard Megna/Fundamental Photographs; 430B, © Wayne Levin/Getty Images; 431TR, Courtesy of NAUI; 432T, © Davis Barber/PhotoEdit; 433, © Galen Rowell/CORBIS; 434, © Derek Berwin/Getty Images; 435, © Ken Karp/Addison-Wesley; 436, © Joseph Sohm/ChromoSohm/CORBIS; 437, © Ken O'Donoghue/Prentice Hall; 440L, © Robert Brenner/PhotoEdit; 440R, © Andrew Lambert Photography /Photo Researchers

Chapter 15 444, © Brian Sytnyk/Masterfile; 445T, © Donald E. Carrel/ Getty; 445B, © Jeremy Woodhouse/Digital Vision; 447, © Hermann Eisenbeiss/Photo Researchers; 448, © Ken O'Donoghue/Prentice Hall; 449, © Richard Walters/Visuals Unlimited; 450B, © Russ Lappa/Addison-Wesley; 451, © Eunice Harris/Photo Researchers; 452, © David Wasserman/Brand X Pictures/PictureQuest; 453, 454T, © Richard Megna/Fundamental Photographs; 454BL, © Kristen Brochman/ Fundamental Photographs; 455, © Tom Pantages; 457, © Ken Karp/Addison-Wesley; 458, © Ken O'Donoghue/Prentice Hall; 459T, © Paul Poplis/Foodpix; 459B, © Russ Lappa/Addison-Wesley; 460(2nd from B), © Johnathan T. Wright/Bruce Coleman; 460 ALL OTHER, © Ken Karp/Addison-Wesley; 461, © Sanford/Agliolo/CORBIS; 462, © Jon Chomitz/Prentice Hall; 463TR, © Ian O'Leary/Dorling Kindersley; 463ML, © David Wasserman/ Brand X Pictures /PictureQuest; 463MM, © Biophoto Associates/ Photo Researchers; 463MR, © Flip Chalfant/Getty Images; 463B, © Frank Pedrick/The Image Works; 467, © Walter Hodges/CORBIS

Chapter 16 470, © Paul Edmondson/Getty Images; 471T, © St. Petersburg Times/Getty Images; 471B, © Ken Karp/Addison-Wesley; 473L, © Tony Craddock/Photo Researchers; 473R, © Ken O'Donoghue/Prentice Hall; 474, © John Lamb/Getty Images; 475, © Richard Megna/Fundamental Photographs; 479BR, © Hank Morgan/Photo Researchers; 481T, ©Jeff Greenberg/The Image Works; 481, 483T, © Ken Karp/Addison-Wesley; 483B, © Richard Hutchings/Addison-Wesley; 484, © Ken O'Donoghue/Prentice Hall; 485, © Tom Pantages; 487T, © Zig Leszczynski/Animals Animals; 488, © Ken Karp/Addison-Wesley; 489T, © Richard Megna/Fundamental Photographs; 489B, © AP/Wide World Photos; 490, © Tom Pantages; 491T, © Jose Azel/Aurora Photos; 493, © Dan McCoy/ Rainbow; 497, © Ken O'Donoghue/Prentice Hall; 502, © Ken O'Donoghue

Chapter 17 504, © Phil Degginger/Color-Pic; 505T, © Royalty-Free/CORBIS; 505B, © Rich Graves/Getty Images; 506L, © Michael Jones/Alaska Stock Images; 506R, © Pete Saloutos/CORBIS; 507, © Mary Steinbacher/PhotoEdit; 508, © Paul A. Souders/CORBIS; 509L, © AP/Wide World Photos; 509R, © Jon Chomitz/Prentice Hall; 510, © Clive Streeter/Dorling Kindersley; 511T, © Lee Snyder/Photo Researchers; 512TL, © Michael Newman/PhotoEdit; 513, © Tom Pantages; 514, © Alyx Kellington/IndexStock Imagery; 517, © Royalty-Free/CORBIS; 518-519, © Royalty-Free/CORBIS; 518TR, © Richard Cummins/CORBIS; 519BR, © PhotoDisc/Getty Images; 520, © AFP/CORBIS; 521, © PhotoDisc/Getty Images; 522, 525, © Ken O'Donoghue/Prentice Hall; 527T, © Dave King/Dorling Kindersley; 527B, © Horst Klemm/Masterfile; 528L, © Charles D. Winters/Photo Researchers; 528R, © E. R. Degginger/Color-Pic; 531, © 1998 Ben S. Kwiatkowski/Fundamental Photographs; 533, © Jonnie Miles/Getty Images; 538, © Pete Turner/Getty Images

Chapter 18 540, © AP/Wide World Photos; 541T, © Alpine Aperture; 541BL, © Orban Thierry/CORBIS SYGMA; 5410BR, © Dave King/Dorling Kindersley; 542L, © Barry Runk/Grant Heilman Photography; 542M, © Erika Stone/Peter Arnold; 542R, © G. I. Bernard/Earth Scenes; 544, © Ken O'Donoghue/Prentice Hall; 545, © Ken Karp/Addison-Wesley; 546, Science VU/Visuals Unlimited; 548, © Ken O'Donoghue/Prentice Hall; 549T, © PhotoDisc/Getty Images; 551, © Dennis MacDonald/PhotoEdit; 552-553, © Empics/SportsChrome; 555, © Dominic Rouse/Getty Images; 556, © Geoffrey Nilsen/Addison-Wesley; 557, © Thomas Hovland/Grant Heilman Photography; 559, © Yoav Levy/Phototake; 560T, © CNRI/Photo Researchers; 560B, © Ken Karp/Addison-Wesley; 561, © Tony Freeman/PhotoEdit; 562, © Richard Megna/Fundamental Photographs; 563, © Ken Karp/Addison-Wesley; 564, © David Bishop/Phototake; 565, © Michael Dalton/Fundamental Photographs; 566T, © Franz-Marc Frei/CORBIS; 567T, © Annie Griffiths Belt/Getty Images; 567B, © Ken Karp/Addison-Wesley; 568T, © Ron Watts/CORBIS; 569B, © Richard Hutchings/Addison-Wesley; 569R, © Ken O'Donoghue/Prentice Hall; 573, © Leonard Lessin/Peter Arnold; 574, © Ken O'Donoghue/Prentice Hall; 575T, © Robert Laberge/Getty Images; 576L, © Tom Bean/CORBIS; 576-577, © Photo Disc/Getty Images; 579, © Bob Handelman/Getty Images

Chapter 19 586, © Jose Azel/Aurora Photos; 587T, © Virginia Weinland/Photo Researchers; 587BL, © Ken Karp/Addison-Wesley; 258BML, © Richard Megna/Fundamental Photographs, 587BMR, © Russ Lappa/Addison-Wesley, 587BR, © PhotoDisc/Getty Images; 590T, © Leonard Lessin/Peter Arnold; 590B, © Ken Karp/Addison-Wesley; 593, © Scott T. Smith/CORBIS; 594T, © Brian Pieters/Masterfile; 595T, © E.R. Degginger/Color-Pic; 595B, © Ken Karp/Addison-Wesley; 596, © Michael Newman/PhotoEdit; 598, © PhotoDisc/Getty Images; 600, 602, © Richard Megna/Fundamental Photographs; 603TL, © Royalty-Free/CORBIS; 603TR, © Patrick Johns/CORBIS; 603ML, © Robert Mackinlay/Peter Arnold; 603MR, © Phil Schermeister/CORBIS; 603B, © Richard Megna/Fundamental Photographs; 604, © Ken O'Donoghue/Prentice Hall; 605T, © PhotoDisc/Getty Images; 607, © David Young-Wolff/PhotoEdit; 608, © Paul J. Richards/AFP /Getty Images; 611, © Franco Origlia/Getty Images; 612T, © PhotoDisc/Getty Images; 612B, © Garry Black/ Masterfile; 614, © Ken Karp/Addison-Wesley; 615T, © Richard Megna/Fundamental Photographs; 615B, © Ken Karp/Addison-Wesley; 617, © Jon Chomitz/Prentice Hall; 618T, © Andrew Syred/Getty Images; 619T, © Richard Megna/Fundamental Photographys; 619B, 620, © Ken Karp/Addison-Wesley; 622, © Charles D. Winters/ Photo Researchers; 623TL, © Silver Burdett Ginn; 623TR, © Dorling Kindersley; 623 M, 623B, Library of Congress

Chapter 20 630, © Sandi and Bill Langley/IndexStock Imagery; 631T, © Christian Michaels/Getty Images; 631B, © Richard Megna/Fundamental Photographs; 632T, © PhotoDisc/Getty Images; 632BL, 632BR, © E.R. Degginger/Color-Pic; 633L, © Paul Silverman/Fundamental Photographs; 633M, 633R, © Ken Karp/Addison-Wesley; 634, © Dorling Kindersley; 635, Blair Seitz/Photo Researchers; 636, © Richard Hamilton Smith/CORBIS; 637T, © Phil Degginger/Color-Pic; 637BL, © Michael Newman/PhotoEdit; 637BR, © Julie Houck/Stock Boston; 638, © Annette Price - H2O Photography/Alamy Photography; 639T, © Ross M. Horowitz/Getty Images; 640ALL, 642ALL, © Ken Karp/ Addison-Wesley; 643, © Tom Pantages; 644TR, © Ken O'Donoghue/Prentice Hall; 644B, Courtesy of National Association of Underwater Instructors; 645T, The Granger Collection, New York; 645B, © Richard Megna/Fundamental Photographs; 646T, © Adam Jones/Visuals Unlimited; 646B, © E.R. Degginger/Color-Pic; 647, © Charles D. Winters/Photo Researchers; 648, © Photo Researchers; 649, © Ken Karp/Addison-Wesley; 653, © Ken O'Donoghue/Prentice Hall; 654, © George Haling/Photo Researchers; 655, © Richard Megna/Fundamental Photographs; 658, 660, © Ken Karp/Addison-Wesley

Chapter 21 662, © ThinkStock/IndexStock Imagery; 663T, © HBOI/Visuals Unlimited; 663B, © Richard Megna/Fundamental Photographs; 665BOTH, The Granger Collection, New York; 668, © Tim Boyle/Getty Images; 670, © D. Ducros/ Photo Researchers; 671T, © Jeremy Maude/Masterfile; 671B, © Paul Silverman/

Fundamental Photographs; 677, © Prentice Hall; 678T, © Chris Collins/CORBIS; 678BL, © Markku Lahdesmaki/CORBIS; 678BR, © Occidental Chemical Corp.; 680, © Charles D. Winters/Photo Researchers; 682, © Julith Miller/ Dorling Kindersley/ Titus Omega, © Christian Sarramon/CORBIS; 684, 685TR, © Ken O'Donoghue/Prentice Hall

Chapter 22 692, © Berry Bingel/Getty Images; 693T, © Richard Hansen/Photo Researchers; 695TL, © Paul Fusco/Magnum Photos; 695TR, © Leonard Lessin/Peter Arnold; 697, 699, © Ken O'Donoghue/Prentice Hall; 700T, © Davis Barber/PhotoEdit; 700B, © Simon Fraser/Photo Researchers; 701, © Royalty-Free/PictureQuest; 702T, © Gusto/Photo Researchers; 702B, © Alan & Linda Detrich/Photo Researchers; 704T, © Benelux Press/IndexStock Imagery; 706, © Yoav Levy/Phototake; 707, 708, © Ken O'Donoghue/Prentice Hall; 709T, © Barry Runk/Grant Heilman Photography; 711, © Patricia Jordan/Peter Arnold; 712T, Courtesy of NASA; 714BKGD, Earth History Hall, American Museum of Natural History; 714BL, © John Sereafin; 714ML, © E. R. Degginger/Color-Pic; 714MR, © Carolina Biological Supply Company/Phototake; 714BR, © E. R. Degginger/Color-Pic; 715, © Allen Oddie/ PhotoEdit; 716-717, © James Nelson/Getty Images; 716TR, ©Syracuse Newspapers/ Logan Wallace/The Image Works; 717T, © Keith Wood/Getty Images

Chapter 23 724, © Neil Fletcher/Dorling Kindersley; 725T, © AP/Wide World Photos; 725B, © Ken O'Donoghue/Prentice Hall; 727, © John Greim/Photo Researchers; 730T, The Granger Collection, New York; 731, © Ken O'Donoghue/Prentice Hall; 732L, © Michael Keller/ CORBIS; 732M, © Bob Daemmrich/Stock Boston; 732R, © Michael Groen/Addison-Wesley; 733L, © Barry Runk/Grant Heilman Photography; 733R, © Felica Martinez/PhotoEdit, 734, © Richard Megna/Fundamental Photographs; 736, © Hulton-Deutsch Collection/CORBIS; 737, © Ken Karp/Addison-Wesley; 740TL, © Ken O'Donoghue/Prentice Hall; 740TR, © Clayton Sharrard/PhotoEdit; 740B, © Michael Groen/Addison-Wesley; 742L, © Dorling Kindersley; 742M, © Alan & Linda Detrick/Grant Heilman Photography; 742R, © Barry L. Runk/Grant Heilman Photography; 744T, © SuperStock/PictureQuest; 744B, © PhotoDisc/Getty Images; 745, 746, © Richard Megna/Fundamental Photographs; 747T, © William McKellar/ Brand X Pictures/PictureQuest; 747B, © Ken O'Donoghue/Prentice Hall; 748L, Dorling Kindersley; 748R, © Bud Freund/Index Stock Imagery/PictureQuest; 749T, © Stephen Frisch/Stock Boston/PictureQuest; 749BL, © Richard Choy/Peter Arnold; 749MM, © Garry Gay/Getty Images,749MB, © Chris Sorensen/CORBIS; 749BR, © Michel Tcherevkoff/Getty Images, 750BOTH, © Michael English/Custom Medical Stock Photo; 751L, © Paul Souders/Getty Images; 751R, © Amy and Chuck Wiley/Wales/Index Stock Imagery; 752, © Bob Daemmrich/Stock Boston; 753, © Ken O'Donoghue/Prentice Hall; 754-755, © Andrea Pistolesi/Getty Images; 754TR, © Peter Angelo Simon/CORBIS; 754ML, © David McLain/Aurora Photos; 754MR, © Neil Fletcher & Matthew Ward/Dorling Kindersley; 754ML&MR (inset), © Courtesy Louet Sales; 755TL, © Dorling Kindersley; 755TM, © Richard Shiell/Earth Scenes; 755TR, © Howard Rice/Dorling Kindersley; 755TL,TM,TR (inset), © Courtesy Louet Sales

Chapter 24 762, © Laureen March/CORBIS; 763T, © Dr. Dennis Kunkel/Visuals Unlimited; 764, © Dr. George Chapmen/Visuals Unlimited; 765, © Cornelia Doerr/AGE Fotostock America; 766T, © Bill Beatty/Visuals Unlimited; 766M, © Maximilian Stock Ltd./Photo Researchers; 766B, © Brian Hagiwara/FoodPix/Getty Images; 769T, © Tony Garcia/Getty Images; 771BL, © Dennis Kunkel/Phototake; 773, © Dr. Hans Gelderblom/Visuals Unlimited; 774, © Ken O'Donoghue/Prentice Hall; 775T, © Jennifer Cheung/Getty Images; 775B, © John E. Kelly/FoodPix/Getty Images; 776T, © Richard Hutchings/Addison-Wesley; 777TL, © Dennis Kunkel/Phototake; 778T, © Ariel Skelley/CORBIS; 778B, © Phil A. Harrington/Peter Arnold; 780, © Ken O'Donoghue/Prentice Hall; 781, © Tony Freeman/PhotoEdit; 782, © Ken Edward/Photo Researchers; 783, © Dan McCoy/Rainbow; 785TL, © Calgene/Peter Arnold; 785BR, © Reuters/Ho/Archive Photo/Getty Images; 786T, © Alison Wright/ CORBIS; 787, © John Shaw/Photo Researchers; 789L, © Bojan Brecelj/CORBIS; 789M, © Jim Strawer/Grant Heilman Photography; 789R, © Royalty-Free/CORBIS; 790, © Wally Eberhart/Visuals Unlimited; 791TR, © Phototake; 791ML, © Mark Maio/Phototake; 791MR, © Lippincott Williams & Wilkins; 791BL, © Derek Berwin/Getty Images; 791BR, © James Balog/Getty Images; 797, © Leonard Lessin/Peter Arnold

Chapter 25 798, © W. Deuter/Masterfile; 799T, The Granger Collection, New York; 803T, © Steven Cole/Getty Images; 806, Reuters Newmedia Inc/Corbis; 808, CERN/Photo Researchers; 809, © Ken O'Donoghue/Prentice Hall; 810T, © Robert George Young/Masterfile; 811T, © Dr. Jeremy Burgess/Photo Researchers; 812, © Roger Ressmeyer/CORBIS; 816T, © Stephen Marks/Getty Images; 816B, © Royalty-Free/CORBIS; 814BL, © Jonathan Blair/CORBIS; 814-815, © Nicholas Devore III/Photographers/Aspen/PictureQuest; 815TR, © Eduardo Gill/BlackStar Publishing/ PictureQuest; 815BR, © Dave King/Dorling Kindersley; 816T, © Stephen Marks/Getty Images; 816B, © Royalty-Free/CORBIS; 817T, © Mauritius/GMBH/PhotoTake; 817M, © Kevin Schafer/Peter Arnold; 817R, © Yoav Levy/Phototake; 819T, © Ryan McVay/Getty Images; 819B, © CNRI/Phototake

Elements Handbook 826T, © Tom Bochsler/Prentice Hall; 826M2, © Richard Megna/ Fundamental Photographs; 826B2, © Lester V. Bergman/CORBIS; R1, © PhotoDisc/ Getty Images; R2TR, © Rich Treptow/Visuals Unlimited; R2BL, © Phil Silverman/ Fundamental Photographs; R2BR, © Gunter Marx Photography/CORBIS; R4T, R4M, © Corbis Images/ PictureQuest; R4B, © Joe Sohm/ Visions Of America, LLC/ PictureQuest; R5, © Royalty-Free/CORBIS; R6T, © Runk/Schoenberger/Grant Heilman Photography; R7L, © Tom Bochsler/Prentice Hall; M7M2, © Richard Megna/ Fundamental Photographs; R7B2, © Lester V. Bergman/ CORBIS; R8T, © PhotoDisc/ Getty Images; R8B, © Park Street/PhotoEdit; R9T, © Jon Chomitz/Prentice Hall; R9B, © Bruce Forster/Getty Images; R10, © Reg Charity/CORBIS; R11M, © Lester V. Bergman/CORBIS; R11B, © Image Farm/Alamy Images; R12T, © Peter Bowater/Photo Researchers; R12B, © David Greedy/Getty Images; R13, © Jules Cowan/ImageState/ PictureQuest; R14T, © Lester V. Bergman/CORBIS; R14B, © Doug Steley/Alamy Images; R15M, © Phil Silverman/Fundamental Photographs; R15B, © Andrew Syred/ Photo Researchers; R16T, © Tomi/ PhotoLink/PhotoDisc/PictureQuest; R16B, © Jose Fuste Raga/CORBIS; R17, © Ed Degginger/Color-Pic; R18M, © M. Claye/Photo Researchers; R18BL, © Mark A. Schneider/Photo Researchers; R18BR, © Harry Taylor/Dorling Kindersley; R19, © Gabe Palmer/CORBIS; R20, © Elizabeth Simpson/Getty Images; R21BL, © 2003 Jeff J. Daly/Fundamental Photographs; R21MR, © Alfred Pasieka/Photo Researchers; R22T, © Nigel Cattlin/Photo Researchers; R22B, © Stockbyte/PictureQuest; R23T, © Alan Detrick/Photo Researchers; R23M, © PhotoDisc/Getty Images; R24, © Richard Megna/Fundamental Photographs; R26T, © David Harvey/Woodfin Camp; R26B, © Ted Spiegel/CORBIS; R27, Barry L. Runk/Grant Heilman Photography; R28TL, © Prentice Hall, Inc.; R28TR, © Jeffrey A. Scovil; R29TL, © Yoav Levy/Phototake; R29M, © Ken Karp/Addison-Wesley; R30ML, Colin Ladyka; R30MR, © Richard Megna/Fundamental Photographs; R30B, Photo Massimo Boyer/edge-of-reef.com; R31TR, © Dave King/Dorling Kindersley; R31B, Courtesy of NASA; R32T, © Richard Megna/Fundamental Photographs; R32, ©1987 Paul Silverman/Fundamental Photographs; R33, © Richard Megna/Fundamental Photographs; R34T, © Royalty-Free/CORBIS; R34B, © Jose Luis Pelaez/CORBIS; R35T, © Gwen Rosenberg /Alliance Pharmaceutical Corp.; R35B, © Carolina Biological Supply Company/Phototake; R36T, © Dominic Episcopo/ImageState; R36B, © Al Fenn/Time Life Pictures/Getty Images; R37T, © Paul A. Souders/CORBIS; R37B, © Tim McGuire/ CORBIS; R38, ©Micjael Gadomski/Earthscenes; R39T, Courtesy of AECL (Atomic Energy of Canada Limited); R39B, © AFP/Getty Images; R40T, © Harry Taylor/Dorling Kindersley; R40B, © Layne Kennedy/CORBIS; R41BL, © Stephen Frisch; R41BR, © Richard Megna/Fundamental Photographs; R42TR, R42BR, © Burstein Collection/CORBIS; R42BL, © Document General Motors/Reuter/CORBIS; R43, © Robert Estall/CORBIS; R44TR, © Phil Degginger/Color-Pic; R44B, © Gunter Marx Photography/CORBIS; R45, © Microzoa/Getty Images

Math Handbook R56, © SNL/DOE/Photo Researchers; R59, © Richard Megna/ Fundamental Photographs; R62, © Paul Silverman/Fundamental Photographs; R66, © Michelle D. Bridwell/PhotoEdit; R67, © Matthew McVay/Stock Boston; R69, © James King-Holmes/Photo Researchers; R70, © Robert Brenner/PhotoEdit; R72, © Tony Freeman/PhotoEdit; R74, © Syracuse Newspapers /Stephen D. Cannerelli/The Image Works; R78, © Charles D. Winters/Photo Researchers

Illustrations

Academy Artworks 213, 215

Ron Carboni 519

Stephen Durke 141, 300, 386T, 391, 442, 443B, 447, 476, 511, 512, 666, 672, 673, 681, 687, 689BOTH, 713, 818, R6B, R17

GTS Companies xv, 16, 22, 29, 31, 32, 34, 37, 50, 51, 56, 60, 81, 89, 118, 120, 124, 128-129, 133, 138, 143, 145, 153, 157, 162-163, 165, 166, 171, 174, 175, 176, 178, 182, 183, 184, 189, 190, 191, 192, 200, 201, 211, 217, 218, 219, 221, 222, 223, 228, 232, 233, 245, 248, 253, 254, 267, 275, 277, 278, 293, 303, 304, 305, 316, 325, 345, 349, 350, 354, 357, 362, 363, 369, 380, 388, 394, 397, 400, 403, 407, 411, 418, 420, 429, 437, 441, 442, 443, 458, 466, 468, 474, 492, 500, 506, 515, 520, 523, 528, 529, 530, 532, 537, 539, 542, 543, 547, 550, 556, 572, 576, 578, 582, 583, 597, 598, 602, 606, 615, 617, 618, 621, 626, 627, 655, 684, 689, 708, 721, 765, 768, 771, 772, 778, 779, 784, 786, 788, 789, 803, 804, 805, 809, 822, 823, 825, R6T, R7, R10, R11, R12, R14, R15, R18, R19, R20, R22, R23, R24, R25, R26, R28, R29, R32, R33T, R36, R38, R40, R40, R43

J/B Woosley Associates 447

Trevor Johnston 313, 376-377

Rob Kemp 716-717

George Ladas xviii, 259

Richard McMahon 44, 47, 64, 67, 78, 81, 89, 104, 105, 184, 198, 351T, 372, 386B, 393, 394T, 402, 411R, 461, 472, 487, 491, 570, 636, 667, 668, 669, 670

Precision Graphics iii, vii, x, 36B, 41, 43, 61, 99, 102, 107, 113, 115, 123, 125, 128-129, 129B, 130, 131, 132, 139, 153, 161, 170, 172, 175, 176, 185, 195, 198, 202, 208, 209, 214, 215, 216, 217, 218, 219, 221, 222, 223, 228, 229, 230, 231, 233, 234, 235, 236, 237, 238, 239, 240, 241, 243, 248, 251, 257, 274, 283, 284, 285, 295, 309, 311, 319, 325, 326, 327, 331, 332, 333, 335, 336, 351, 357, 364, 369, 383, 386TR, 396B, 398, 399, 401, 414, 415, 416, 417, 418, 420, 422, 430, 431, 433, 439, 446, 449, 451, 453, 469, 478, 479, 483, 487, 488, 503, 543, 549, 554, 560, 568, 573, 585, 588, 591, 592, 594, 613, 618, 629, 631, 632, 633, 661, 663, 680, 694, 695, 698, 702, 703, 704, 705, 708, 709, 710, 720, 723, 726, 732, 735, 737, 741, 743, 758, 759, 761, 763, 771B, 773, 776, 777, 782, 791, 797, 800, 801, 802, 804, 807, 810, 811, 813, 824, 825, R21, R27, R28B, R33M

Prentice Hall 137, 392, 396M, 409, 691, R74, R75, R76, R84, R86, R91

Neil Stewart 242, 644

Bob Young 685